Organic Electronic Spectral Data

Volume XI 1969

Organic Electronic Spectral Data

Volume XI 1969

JOHN P. PHILLIPS, HENRY FEUER
P. M. LAUGHTON & B. S. THYAGARAJAN

EDITORS

CONTRIBUTORS

Dallas Bates
H. Feuer
L. D. Freedman
M. K. Hrenoff
P. M. Laughton
C. M. Martini
F. C. Nachod
J. P. Phillips

AN INTERSCIENCE ® PUBLICATION

JOHN WILEY & SONS
New York • London • Sydney • Toronto

An Interscience ® Publication

Library of Congress Catalog Card Number: 60–16428

ISBN 0–471 68802–9

Printed in the United States of America

10 9 8 7 6 5 4 3 2 1

INTRODUCTION TO THE SERIES

In 1956 a cooperative effort to abstract and publish in formula order all the ultraviolet-visible spectra or organic compounds presented in the journal literature was organized through the enterprise and leadership of M.J. Kamlet and H.E. Ungnade. Organic Electronic Spectral Data was incorporated in 1957 to create a formal structure for the venture, and coverage of the literature from 1946 onward was then carried out by chemists with special interests in spectrophotometry through a page by page search of the major chemical journals. After the first two volumes (covering the literature from 1946 through 1955) were produced, a regular schedule of one volume for each subsequent period of two years was introduced. In 1966 an annual schedule was inaugurated. Eleven volumes have now been published.

Altogether, more than fifty chemists have searched a group of journals totalling more than a hundred titles during the course of this sustained project. Additions and subtractions from both the lists of contributors and of journals have occurred from time to time, and it is estimated that the effort to cover all the literature containing spectra may not be more than 95% successful. However, the total collection is by far the largest ever assembled, amounting to nearly a quarter-million spectra through the volumes so far.

Volume XII is in preparation.

PREFACE

Processing of the data provided by the contributors to Volume XI as to the last several volumes was performed at the University of Louisville.

John P. Phillips
Henry Feuer
P.M. Laughton
B.S. Thyagarajan

ORGANIZATION AND USE OF THE DATA

The data in this volume were abstracted from the journals listed in the reference section at the end. Although a few exceptions were made, the data generally had to satisfy the following requirements: the compound had to be pure enough for satisfactory elemental analysis and for a definite empirical formula; solvent and phase had to be given; and sufficient data to calculate molar absorptivities had to be available. Later it was decided to include spectra even if solvent was not mentioned; experience has shown that the most probable solvent in such circumstances is ethanol.

All entries in the compilation are organized according to the molecular formula index system used by Chemical Abstracts. Most of the compound names have been made to conform with the Chemical Abstracts system of nomenclature.

Solvent or phase appears in the second column of the data lists, often abbreviated according to standard practice; there is a key to less obvious abbreviations on the next page. Anion and cation are used in this column if the spectra are run in relatively basic or acidic conditions respectively but exact specifications can not be ascertained.

The numerical data in the third column present wavelength values in millimicrons (or nanometers) for all maxima, shoulders and inflections, with the logarithms of the corresponding molar absorptivities in parentheses. Shoulders and inflections are marked with a letter s. In spectra with considerable fine structure in the bands a main maximum is listed and labelled with a letter f. Numerical values reported by authors are given to the nearest nanometer for wavelength and nearest 0.01 unit for the logarithm of the molar absorptivity. Spectra that change with time or other common conditions are labelled "anom." or "changing" and temperatures are indicated if unusual.

The reference column contains the code number of the journal, the initial page number of the paper, and in the last two digits, the year (1969). A letter is added for journals with more than one volume or section in a year. The complete list of all articles and authors thereof appears in the References at the end of the volume.

Several journals that were abstracted for previous volumes in this series have been omitted, usually for lack of useful data, and several new ones have been added.

ABBREVIATIONS

s	shoulder or inflection
f	fine structure
n.s.g.	no solvent given in original reference
C_5H_5N	pyridine
$C_6H_{11}Me$	methylcyclohexane
C_6H_{12}	cyclohexane
DMF	dimethylformamide
DMSO	dimethylsulfoxide
THF	tetrahydrofuran

Other solvent abbreviations generally follow the practice of Chemical Abstracts.

Underlined data were estimated from graphs.

JOURNALS ABSTRACTED

Journal	No.
Acta Chem. Scand.	1
Indian J. Chem.	2
Anal. Chem.	3
J. Heterocyclic Chem.	4
Ann. Chem. Liebigs	5
Ann. Chim. (Paris)	6
Ann. chim. (Rome)	7
Appl. Spectroscopy	9
Arch. Biochem. Biophys.	10
Arkiv Kemi	11
Australian J. Chem.	12
Steroids	13
Bol. inst. quim. univ. na. auton. Mex.	16
Boll. sci. fac. chim. ind. Bologna	17
Bull. Chem. Soc. Japan	18
Bull. Acad. Polon. Sci.	19
Bull. soc. chim. Belges	20
Bull. soc. chim. biol.	21
Bull. soc. chim. France	22
Can. J. Chem.	23
Chem. Ber.	24
Chem. and Ind. (London)	25
Chimia	27
Compt. rend.	28
Doklady Akad. Nauk S.S.S.R.	30
Experientia	31
Gazz. chim. ital.	32
Helv. Chim. Acta	33
J. Chem. Eng. Data	34
J. Am. Chem. Soc.	35
J. Pharm. Soc.	36
J. Chem. Phys.	38
J. Chem. Soc., Sect. A	39A
J. Chem. Soc., Sect. B	39B
J. Chem. Soc., Sect. C	39C
Nippon Kagaku Zasshi	40
J. Indian Chem. Soc.	42
J. Org. Chem.	44
J. Pharm. Pharmacol.	45
J. Phys. Chem.	46
J. Polymer Sci.	47
J. prakt. Chem.	48
Monatsh. Chem.	49
Naturwiss.	51
Rec. trav. chim.	54
Ricerca Sci.	55
Roczniki Chem.	56
Science	57
Spectrochim. Acta	59
Trans. Faraday Soc.	60
Ber. Bunsengesellschaft Phys. Chem.	61
Z. phys. Chem. (Frankfurt)	62
Z. physiol. Chem.	63
Z. Naturforsch.	64
Zhur. Obshchei Khim.	65
Proc. Iowa Acad. Sci.	67
Biochemistry	69
Izvest. Akad. Nauk S.S.S.R.	70
Coll. Czech. Chem. Comm.	73
J. Chem. Soc., Sect. D	77
Tetrahedron	78
Revue Romaine Chim.	80
Arch. Pharm.	83
J. Med. Chem.	87
Tetrahedron Letters	88
Angew. Chem.	89
Chem. Pharm. Bull. (Japan)	94
J. Pharm. Soc. Japan	95
Z. Chemie	97
Lloydia	100
J. Organometallic Chem.	101
Phytochemistry	102
Khim. Geterosikl. Soedin.	103
Zhur. Organ. Khim.	104
Khim. Prirodn. Soedin.	105
Die Pharmazie	106
Israel J. Chem.	108
Chim. ther.	111
Spectroscopy Letters	112
Acta Chim. Acad. Sci. Hung.	114
J. Chem. United Arab Republic	115
Org. Preps. and Procedures	117
Synthesis	118
South African Chem. Inst. Journal	119
Pakistan J. Sci. Ind. Research	120
J. Macromol. Sci.	121
Reakt. Sposobnost' Org. Soedin	122
Moscow University Chem. Bulletin	123
Ukrain. Khim. Zhur.	124

Compound	Solvent	$\lambda_{max}(\log \epsilon)$	Ref.
CCl_2OS Thiophosgene, S-oxide	hexane	252(3.07),288(3.30)	88-4461-69
CF_2N_2 Diazirine, difluoro-	gas	185(1.3)	38-0045-69B
CF_3NOS Thionitrous acid, S-(trifluoromethyl) ester	gas	225(4.30),373(2.05), 570f(0.78)	39-1587-69A
$CHBrN_2O_4$ Methane, bromotrinitro-	anion	385(4.19)	122-0413-69
CHF_3 Methane, trifluoro-	gas	114f(3.3)	35-5937-69
CHN_3O_6 Methane, trinitro-	anion	350(4.16)	122-0100-69 +122-0413-69
CH_2F_2 Methane, difluoro-	gas	110f(3.3),119f(3.6), 133(3.3)	35-5937-69
$CH_2N_2O_4$ Methane, dinitro-	anion	363(4.32)	122-0413-69
CH_3ClS Methanesulfenyl chloride	gas	206(2.83),355(1.39)	112-0301-69
CH_3F Methane, fluoro-	gas	109(3.8),119f(3.5), 132(3.6)	35-5937-69
CH_3NO_2 Methane, nitro-	gas	200(3.8),270(1.1)	59-0275-69
CH_4 Methane	gas	119(3.8),129(3.8)	35-5937-69
$CH_4N_2O_2$ Methylamine, N-nitro-, potassium salt	0.005M HCl	233(3.86)	60-0904-69
CH_5N_3 Guanidine, reaction product with biacetyl and 1-naphthol	alkali	525(3.25)	94-0639-69
biacetyl and 8-quinolinol	alkali	520(2.38)	94-0639-69

Compound	Solvent	λ_{max}(log ϵ)	Ref.
$C_2HCl_3O_2$			
Acetic acid, trichloro-, Pb(IV) salt	C_6H_{12}	223(3.87)(end abs.)	33-1495-69
C_2HF_3OS			
Thioacetic acid, trifluoro-	gas	282(--)	44-4173-69
	C_6H_{12}	230(3.74)	44-4173-69
$C_2HN_3O_4$			
Acetonitrile, dinitro-	anion	345(4.19)	122-0100-69
$C_2H_2N_2O_2$			
Acetonitrile, nitro-	acid	285(2.6)	22-0545-69
	base	289(4.02)	22-0545-69
$C_2H_2N_2O_2S$			
1,2,5-Thiadiazole-3,4-diol	H_2O	282(3.83)	117-0255-69
$C_2H_2N_2O_3S$			
1,2,5-Thiadiazolidine-3,4-dione, 1-oxide	EtOH	307(2.2)	117-0255-69
$C_2H_2N_2S$			
1,2,5-Thia(S^{IV})diazole	n.s.g.	255(3.89)	103-0180-69
$C_2H_2N_4O_2$			
s-Triazole, 3-nitro-	pH 1	215(3.78),245(3.70)	12-2251-69
	H_2O	215(3.76),245(3.67)	12-2251-69
	pH 10	289(3.79)	12-2251-69
$C_2H_2O_4$			
Oxalic acid	0.4M HCl	244(1.57)	18-2282-69
C_2H_3BrO			
Acetyl bromide	heptane	250s(1.97)	59-0765-69
C_2H_3ClO			
Acetyl chloride	heptane	240(1.58)	59-0765-69
C_2H_3FO			
Acetyl fluoride	heptane	205(1.57)	59-0765-69
$C_2H_3NO_2$			
Ethylene, nitro-	gas	203(4.0),242(2.9), 295f(1.7)	59-0275-69
$C_2H_3NO_3$			
Glyoxylic acid, 2-oxime, syn-H	1% MeOH	250(4.00)	22-2894-69
monoanion	1% MeOH	under 215 nm	22-2894-69
C_2H_3NS			
Isothiocyanic acid, methyl ester	hexane	192(4),249(3.34)	114-0309-69B
$C_2H_3N_3O_5$			
Acetamide, dinitro-	anion	361(3.97)	122-0100-69
$C_2H_4ClN_3$			
2-Chloroethyl azide	hexane	214(2.78),283(1.53)	114-0309-69B
$C_2H_4Cl_2$			
Ethane, 1,2-dichloro-	MeOH	269(2.47)	73-0819-69

Compound	Solvent	$\lambda_{max}(\log \epsilon)$	Ref.
$C_2H_4N_2$			
Diimide, vinyl-	MeCN	221(3.86),403(1.60)	35-3375-69
$C_2H_4N_2O_4$			
Ethane, 1,1-dinitro-	anion	379(4.25)	122-0413-69
$C_2H_4N_2S$			
1,2,5-Thia(S^{IV})diazole, 3,4-dihydro-	EtOH	257(2.74)	88-4117-69
$C_2H_4O_2$			
Acetic acid	heptane	202(1.72)	59-0765-69
C_2H_5Li			
Lithium, ethyl-	isooctane	210(3.30)	35-4490-69
C_2H_5NO			
Acetamide	heptane	207s(2.00)	59-0765-69
C_2H_5NOS			
Thionitrous acid, S-ethyl ester	gas	343(2.81),550(1.32)	39-1587-69A
$C_2H_5N_3OS$			
Thiobiuret	neutral	255(4.22)	104-0225-69
	basic	223(4.03),260(4.20)	104-0225-69
$C_2H_6ClO_3P$			
Phosphorochloridic acid, dimethyl ester	MeOH	208(4.32),247(3.87)	44-1592-69
$C_2H_6N_2$			
Azomethane, (E)-	gas	340(0.7)	35-1710-69
$C_2H_6N_2O$			
Acetic acid, hydrazide, 1,2-naphthoquinone-4-sulfonic acid product (also other color reactions given)	alkali	475(4.20)	95-0007-69
Glycinamide, copper derivative	H_2O	535(1.67)	18-0168-69
	MeOH	344s(--),530(1.71)	18-0168-69
$C_2H_6N_2O_2$			
Nitramine, N,N-dimethyl-	H_2O	238(3.85)	60-0904-69
$C_2H_6N_4S$			
Urea, 1-amidino-2-thio-	pH 8	238(4.15),266(4.08)	104-0225-69
protonated	n.s.g.	262(4.00)	104-0225-69
$C_2H_8N_4O$			
Triazan-1-carboxamide, 2-methyl-	n.s.g.	218(4.22),284(4.37)	5-0001-69E

Compound	Solvent	$\lambda_{max}(\log \epsilon)$	Ref.
C_3ClF_5O			
Acetone, chloropentafluoro-	gas	305f(1.7)	39-0893-69A
C_3Cl_4O			
Acryloyl chloride, trichloro-	heptane	218(3.7),267(3.8)	5-0001-69H
C_3D_6O			
Acetone-d_6	gas	166(3.18),185(3.04), 188(3.46),191(3.76), 195(3.84)	18-2453-69
C_3HN			
Propiolonitrile	EtOH	210(2.20),219(2.24), 229(2.00),243(0.78)	23-4503-69
$C_3H_2Cl_3N_3S$			
1,3,4-Thiadiazole, 2-amino-5-(trichloromethyl)-, hydrochloride	MeOH	250(3.84)	4-0835-69
$C_3H_2F_3N_3S$			
1,3,4-Thiadiazole, 2-amino-5-(trifluoromethyl)-	MeOH	273(3.79)	4-0835-69
$C_3H_2N_2O_2S$			
Imidazolidine-2-thione, 4,5-dioxo-	EtOH	304(4.02),416(1.60)	18-2323-69
1,2,5-Thia(S^{IV})diazole-3-carboxylic acid	n.s.g.	263(4.01)	103-0180-69
Thiazole, 2-nitro-	EtOH	203(3.59),235(3.61), 308(3.68)	28-1343-69B
Thiazole, 5-nitro-	EtOH	222(3.57),240(--), 299(3.68)	28-1343-69B
$C_3H_2N_2O_3$			
Imidazolidinetrione	EtOH	314(1.60)	18-2323-69
Isoxazole, 4-nitro-	acid	218(3.70),251(3.70)	22-0545-69
$C_3H_3F_2N_3S$			
1,3,4-Thiadiazole, 2-amino-5-(difluoromethyl)-	MeOH	268(3.79)	4-0835-69
$C_3H_3F_3N_4S$			
s-Triazole-5-thiol, 4-amino-3-(trifluoromethyl)-	EtOH	252(4.43)	2-0959-69
$C_3H_3F_3O$			
Acetone, 1,1,1-trifluoro-	heptane	288(1.00)	59-0765-69
$C_3H_3NO_4$			
Malonaldehyde, nitro-, anion	neutral	228(3.90),268(3.93), 328(3.98)	22-0545-69
C_3H_4			
Cyclopropene	gas	171f(3.9)	38-0045-69B
$C_3H_4Cl_3N_3S$			
Chloral, thiosemicarbazone	EtOH	245(4.02)	94-0844-69
$C_3H_4N_2$			
Propene, 3-diazo-	C_6H_{12}	486(1.28)	35-0711-69

Compound	Solvent	λ max (log ε)	Ref.
$C_3H_4N_2O_6$			
Acetic acid, trinitromethyl ester	anion	357(4.12)	122-0100-69
$C_3H_4N_4$			
Propane, 1,3-bis(diazo)-	C_6H_{12}	462(1.43)	35-0706-69
as-Triazine, 3-amino-	EtOH	229(4.19),327(3.45)	22-3675-69
$C_3H_4N_4OS$			
1,3,4-Thiadiazole-2-carboxaldehyde, 5-amino-, oxime	MeOH	301(4.06)	4-0835-69
$C_3H_4N_4O_2$			
s-Triazole, 3-methyl-5-nitro-	pH 1	215(3.74),257(3.62)	12-2251-69
	H_2O	215(3.74),257(3.61)	12-2251-69
	pH 10	217(3.60),303(3.76)	12-2251-69
$C_3H_4O_2$			
Malonaldehyde, sodium derivative	MeOH	266(4.36)	78-4315-69
$C_3H_4O_2S$			
Carbonic acid, thio-, cyclic O,O-ethylene ester	EtOH	235(4.15),304(1.45)	44-3011-69
$C_3H_4O_3S$			
Pyruvic acid, mercapto-	pH 10.6	230(3.68),293(3.83)	28-1844-69A
C_3H_5BrO			
Acetone, bromo-	heptane	301(1.92)	59-0765-69
C_3H_5ClO			
Acetone, chloro-	heptane	289(1.67)	59-0765-69
C_3H_5FO			
Acetone, fluoro-	heptane	279(1.18)	59-0765-69
$C_3H_5NO_2$			
1-Propene, 1-nitro-, (E)-	gas	212(4.1),250s(3.0), 300(1.9)	59-0275-69
1-Propene, 3-nitro-	gas	194(3.8),275(1.6)	59-0275-69
Pyruvaldehyde, 1-oxime, syn-H	1% MeOH	225(4.01)	22-2894-69
anion	1% MeOH	275(4.23)	22-2894-69
$C_3H_5NO_3$			
Pyruvic acid, oxime, anion, syn-Me	1% MeOH	under 215 nm	22-2894-69
dianion	1% MeOH	245(3.91)	22-2894-69
C_3H_5NS			
Isothiocyanic acid, ethyl ester	hexane	250(2.94)	114-0309-69B
$C_3H_5N_3O_4$			
Malonaldehyde, nitro-, dioxime	pH 6	232(3.98),332(4.09)	22-0545-69
	pH 12	275(4.04),315(4.21)	22-0545-69
$C_3H_5N_5$			
Formamidine, N-(s-triazol-4-yl)-	MeOH	227(3.83)	48-0477-69
$C_3H_5OS_2$			
Carbonic acid, dithio-, O-ethyl ester, ion(1-)	n.s.g.	227(4.0),250s(3.3), 302(4.1)	39-1182-69B

Compound	Solvent	λmax(log ε)	Ref.
C_3H_6			
Cyclopropane	gas	159(0.7)	38-0052-69B
$C_3H_6N_2$			
Diimide, isopropenyl-	MeCN	230(3.81),410(1.61)	35-3375-69
$C_3H_6N_2O_5$			
Ethane, 1-methoxy-2,2-dinitro-	anion	365(4.25)	122-0413-69
$C_3H_6N_2S$			
3H-1,2,6-Thia(S^{IV})diazine, 4,5-dihydro-	EtOH	278(3.61)	88-4117-69
1,2,5-Thia(S^{IV})diazole, 3,4-dihydro-3-methyl-	EtOH	257(2.72)	88-4117-69
$C_3H_6N_4$			
1H-1,2,3-Triazole, 5-amino-1-methyl-	pH -0.2	259(3.63)	39-2379-69C
	pH 7.0	238(3.73)	39-2379-69C
$C_3H_6N_6O_6$			
s-Triazine, hexahydro-1,3,5-trinitro-	MeOH	204(4.12),234s(--)	60-0904-69
	EtOH	202(4.12),236s(--)	60-0904-69
	MeCN	235s(--)	60-0904-69
C_3H_6O			
Acetone	gas	166(3.18),187(3.33), 191(3.75),195(3.95)	18-2453-69
	heptane	186(3.31),278(1.18)	59-0765-69
	H_2O	263(1.3)	122-0490-69
	70% H_2SO_4	247(1.3)	122-0490-69
(also other concentrations of H_2SO_4)	76% H_2SO_4	240s(1.3)	122-0490-69
$C_3H_6O_2$			
Acetic acid, methyl ester	heptane	209(1.78)	59-0765-69
Propionic acid, Pb(IV) salt	CH_2Cl_2	240(3.92)	33-1495-69
$C_3H_6O_2S_2$			
1,2-Dithiolane, 1,1-dioxide	EtOH	263(1.62)	44-0036-69
$C_3H_6S_3$			
s-Trithiane	hexane	188(3.69),196s(3.70), 206(3.79),232(2.85), 242(2.96)	59-0407-69
	iso-PrOH	198(3.54),206(3.63), 225(2.82),239(2.98)	59-0407-69
$C_3H_7NO_2$			
Nitrous acid, isopropyl ester	gas	358f(1.7)	35-1085-69
$C_3H_7NS_2$			
Carbamic acid, N,N-dimethyldithio-, dimethylamine salt (also metal complexes)	CH_2Cl_2	255(3.41),289(3.36)	39-1152-69A
$C_3H_7N_3O$			
Acetaldehyde, semicarbazone	EtOH	230(4.08)	94-0844-69
$C_3H_7N_3S$			
Acetaldehyde, thiosemicarbazone	EtOH	241(3.98),269(4.04)	94-0844-69

Compound	Solvent	$\lambda_{max}(\log \epsilon)$	Ref.
$C_3H_8N_2O$			
L-Alaninamide, copper derivative	H_2O	524(1.74)	18-0168-69
	MeOH	510(1.77)	18-0168-69
	EtOH	345s(--),500(1.78)	18-0168-69
$C_3H_8N_2O_2$			
Nitramine, O-ethyl-N-methyl-, trans	H_2O	207(3.78)	39-0397-69C
85% cis	H_2O	215(3.65)	39-0397-69C
2-Propanone, 1-(hydroxyamino)-, oxime, (E)-	n.s.g.	229(3.25)	70-2201-69
$C_3H_8S_2$			
Methane, bis(methylthio)-	hexane	199(3.79),206(3.78), 230s(3.00)	59-0407-69
$C_3H_9Br_2Sb$			
Antimony, dibromotrimethyl-	MeCN	none above 240 nm	101-0335-69A
$C_3H_9Cl_2Sb$			
Antimony, dichlorotrimethyl-	MeCN	none above 240 nm	101-0335-69A
$C_3H_9GeN_3$			
Germane, azidotrimethyl-	hexane	212(2.40),266(1.36)	114-0309-69B
$C_3H_9N_3Si$			
Silane, azidotrimethyl-	hexane	212(2.42),255(1.28)	114-0309-69B
$C_3H_9N_5O$			
2-Tetrazene, 1,1,4-trimethyl-4-nitroso-	EtOH	310(4.11)	44-2997-69
C_3H_9SSb			
Stibine sulfide, trimethyl-	MeCN	267(3.6)	101-0335-69A

Compound	Solvent	$\lambda_{max}(\log \epsilon)$	Ref.
C_4Cl_6O			
3-Buten-2-one, 1,1,1,3,4,4-hexachloro-	MeOH	205(3.71),245(3.87)	44-1592-69
Crotonoyl chloride, pentachloro-	heptane	232(4.0)	5-0001-69H
$C_4F_3IN_2$			
Pyrimidine, 2,4,5-trifluoro-6-iodo-	hexane	222(3.64),243s(3.5), 264(4.01),269s(3.9)	39-1866-69C
C_4F_6			
1,3-Butadiene, hexafluoro-	gas	165(3.83),201(3.49)	112-0369-69
$C_4F_6HgO_2S_2$			
Acetic acid, trifluorothio-, S-mercury salt	C_6H_{12}	246(3.74)	44-4173-69
$C_4F_6O_2S_2$			
Bis(trifluoroacetyl) disulfide	gas	230(5.18)	44-4173-69
$C_4F_6O_3$			
Acetic acid, trifluoro-, anhydride	gas	215(2.11)	44-3438-69
p-Dioxan-2-one, 3,3,5,5,6,6-hexafluoro-	gas	225(1.6)	44-3438-69
C_4HCl_5O			
3-Butenal, 2,2,3,4,4-pentachloro-	heptane	225s(4.12)	5-0001-69H
3-Buten-2-one, 1,1,3,4,4-pentachloro-	MeOH	212(3.90),271(3.49)	44-1592-69
C_4HCl_7O			
Butyraldehyde, heptachloro-	heptane	240s(2.8),280s(1.5)	24-3127-69
$C_4HF_3N_2$			
Pyrimidine, 2,4,5-trifluoro-	hexane	250(3.60)	39-1866-69C
$C_4H_2ClFN_2O$			
2-Pyrimidinone, 4-chloro-5-fluoro-	pH 1	317(3.65)	1-0294-69
	pH 13	311(3.83)	1-0294-69
$C_4H_2Cl_4$			
1,3-Butadiene, 1,2,3,4-tetrachloro-	hexane	211(4.21),235s(3.60)	78-0883-69
$C_4H_2F_2IN_3$			
Pyrimidine, 2,5-difluoro-6-iodo-	EtOH	213s(4.12),223(4.19), 280(3.95)	39-1866-69C
$C_4H_2F_2N_2$			
Pyrimidine, 2,5-difluoro-	hexane	258(3.70)	39-1866-69C
$C_4H_2F_3N_5S$			
s-Triazolo[3,4-b][1,3,4]thiadiazole, 6-amino-3-(trifluoromethyl)-	EtOH	248(4.01)	2-0959-69
$C_4H_2FeO_4$			
Fumaric acid, ferrous salt	nujol	500(--),930(--)	2-0266-69
$C_4H_2N_2O_4S$			
1,2,5-Thia(S^{IV})diazole-3,4-dicarboxylic acid	n.s.g.	266(3.95)	103-0180-69
$C_4H_2N_4O_3$			
[1,2,5]Oxadiazolo[3,4-b]pyrazine, 5,6-dihydroxy-	EtOH	269(4.06)	4-0769-69

Compound	Solvent	λ max (log ε)	Ref.
$C_4H_2O_2S$			
Maleic acid, thioanhydride	hexane	230(3.97),318(2.79)	24-2471-69
$C_4H_3FN_2O$			
2-Pyrimidinone, 5-fluoro-	pH 1	322(3.61)	1-0294-69
	pH 13	312(3.65)	1-0294-69
$C_4H_3N_3O_3$			
4(3H)-Pyrimidinone, 5-nitro-	N HCl	233(3.79),312(3.68)	4-0593-69
	N NaOH	337(3.80)	4-0593-69
$C_4H_3N_5O$			
4H-v-Triazolo[4,5-d]pyridazin-4-one, 1,5-dihydro-	pH 13	211(4.46),230s(4.22), 312(4.02)	4-0093-69
s-Triazolo[2,3-c]triazin-7-ol	EtOH	240(3.58),296(3.80)	22-2492-69
s-Triazolo[4,3-b]triazin-5-ol	EtOH	235(4.07),295(3.88)	22-2492-69
$C_4H_3N_5S$			
8-Azapurine-6-thiol	pH 2.0	226s(3.99),330(4.25)	39-0152-69C
	pH 7.0	226s(2.98),328(4.34)	39-0152-69C
	pH 12.0	221(4.17),323(4.18)	39-0152-69C
4H-v-Triazolo[4,5-d]pyridazine-4-thione, 1,5-dihydro-	pH 13	215(4.59),298(3.59)	4-0093-69
$C_4H_7N_7O_2$			
2H-1,2,3-Triazole, 4-nitro-2-(s-triazol-3-yl)-	pH 1	205(3.98),277(3.97)	12-2251-69
	H_2O	205(3.97),277(3.96)	12-2251-69
	pH 10	302(3.92)	12-2251-69
$C_4H_4BrN_3$			
Pyrimidine, 4-amino-5-bromo-	pH 1.0	254(4.00),276s(3.80)	39-0096-69B
	pH 7.0	237(3.94),281(3.65)	39-0096-69B
$C_4H_4ClNO_2$			
3-Isoxazolone, 5-(chloromethyl)-	MeOH	210(3.95)	33-0720-69B
	MeOH-base	213(3.89),245s(3.35)	33-0720-69B
$C_4H_4N_2OS$			
Uracil, 2-thio-	H_2O	270(4.1)	88-3547-69
Uracil, 4-thio-	H_2O	328(4.2)	88-3547-69
$C_4H_4N_2O_2$			
3-Pyridazinol, 1-oxide	EtOH	256(3.7),320(3.7)	94-0763-69
3(2H)-Pyridazinone, 1-oxide	EtOH	247(3.6),338(3.8)	94-0763-69
Uracil, ketyl radical	pH 5.1	305(3.15)	38-4881-69B
radical anion	pH 11.7	310(3.18)	38-4881-69B
$C_4H_4N_2O_2S$			
Thiazole, 4-methyl-5-nitro-	EtOH	217(3.56),296(3.86)	28-1343-69B
Thiazole, 5-methyl-4-nitro-	EtOH	216(4.00),235(3.60), 281(3.70)	28-1343-69B
$C_4H_4N_2O_3$			
Barbituric acid	pH 6.2	255(3.50)	65-2568-69
	pH 13	258(4.10)	65-2568-69
	base	258(4.31)	39-0802-69B
ketyl radical	pH 5.0	310(3.15),420(2.95)	38-4881-69B
radical anion	pH 12.9	420(2.85)	38-4881-69B
Isobarbituric acid	pH 1	277(3.85)	44-2636-69
	pH 10	240(3.88),305(3.79)	44-2636-69

Compound	Solvent	λ_{max}(log ε)	Ref.
Isobarbituric acid (cont.)	pH 14	240s(3.86),302(3.7)	44-2636-69
ketyl radical	pH 5.0	340(3.30)	38-4881-69B
radical anion	pH 12.2	340(3.26)	38-4881-69B
$C_4H_4N_6$			
1H-v-Triazolo[4,5-d]pyridazine, 4-amino-	H_2O	214(4.37),259(3.94), 275s(3.84),285s(3.59)	4-0093-69
C_4H_4O			
Furan	EtOH	200(4.00),252(3.00)	6-0277-69
$C_4H_4O_3$			
2(5H)-Furanone, 4-hydroxy-	EtOH	249(4.36)	39-0056-69C
$C_4H_5BrN_4O$			
6-Pyrimidinol, 2,4-diamino-5-bromo-	pH 1	275(4.20)	39-0603-69C
	pH 13	270(3.94)	39-0603-69C
C_4H_5BrO			
2,3-Butadien-1-ol, 4-bromo-	heptane	187(4.44)	78-3277-69
$C_4H_5BrO_2$			
2(3H)-Furanone, 3-bromodihydro-	EtOH	225(2.54)	28-2028-69A
$C_4H_5ClN_4$			
Pyridazine, 6-chloro-3-hydrazino-	pH 1	235s(--),302(2.89)	7-0552-69
	pH 7.38	245(3.70),318(2.89)	7-0552-69
Pyrimidine, 2,4-diamino-5-chloro-	pH 2	208(4.40),220s(4.24), 280(3.66)	44-0821-69
	pH 12	233(4.01),294(3.82)	44-0821-69
Pyrimidine, 2,4-diamino-6-chloro-	acid	209(4.15),228(4.03), 298(3.90)	44-0821-69
	neutral	205(4.39),228s(3.98), 282(3.89)	44-0821-69
$C_4H_5ClN_4O$			
Imidazole-4-carboxamide, 5-amino-2-chloro-	pH 1	274(3.80)	18-0750-69
	pH 6	273(3.99)	18-0750-69
	pH 13	284(3.99)	18-0750-69
$C_4H_5Cl_3N_2O_2SSe$			
1,4,2-Thiaselenazine, tetrahydro-3-imino-5-(trichloromethyl)-, 1,1-dioxide	n.s.g.	215(3.31)	30-0355-69A
$C_4H_5NO_5$			
Malonaldehydic acid, nitro-, methyl ester	hexane	222(3.84),254s(3.72)	30-1079-69C
C_4H_5NS			
Isothiocyanic acid, allyl ester	hexane	249(3.10)	114-0309-69B
$C_4H_5NS_3$			
1,3-Thiazane-2,4-dithione	MeOH	317(4.21),344(4.27)	42-0919-69
$C_4H_5N_3$			
as-Triazine, 3-methyl-	MeOH	256(3.56),383(2.58)	88-3147-69
$C_4H_5N_3O$			
Cytosine, ketyl radical	pH 5.5	305(--),320(--)	38-4881-69B

Compound	Solvent	$\lambda_{max}(\log \epsilon)$	Ref.
Cytosine, radical anion	pH 13.3	310(2.90)	38-4881-69B
Isocytosine, ketyl radical	pH 7.5	317(3.18)	38-4881-69B
$C_4H_5N_3OS$			
as-Triazin-5-one, 2,3,4,5-tetrahydro- 6-methyl-3-thioxo-	pH 1	213(3.98),268(4.17)	73-1690-69
	pH 13	223(4.24),257(4.18), 312(3.69)	73-1690-69
	40% EtOH	214(4.03),269(4.28)	73-2306-69
$C_4H_5N_3O_2$			
Uracil, 5-amino-, radical anion	pH 11.7	355(3.20)	38-4881-69B
$C_4H_5N_3O_4$			
Propionitrile, 3,3-dinitro-	anion	373(4.23)	122-0413-69
$C_4H_5N_5$			
1,2,3-Triazole-5-carbonitrile, 4-amino- 3-methyl-	MeOH	225(3.95),251(3.78)	39-2379-69C
$C_4H_5N_5O$			
as-Triazine-5-carboxamide, 6-amino-	pH 7	239(4.07),357(3.61)	44-2102-69
$C_4H_5N_5O_2$			
as-Triazine, 3-formylhydrazino- 5-hydroxy-	EtOH	236(3.76)	22-2492-69
$C_4H_5N_7$			
1H-v-Triazolo[4,5-d]pyridazine, 4-hydrazino-	H_2O	223s(4.26),263(3.93)	4-0093-69
1H-v-Triazolo[4,5-d]pyrimidine, 5,7-diamino-, cation	acid	203(4.26),253(4.09), 270s(3.98)	44-0821-69
$C_4H_5N_7O_3$			
Glyoxal, nitro-, 1-(s-triazol- 3-ylhydrazone), oxime	MeOH	249(3.91),354(4.06), 460(2.64)	12-2251-69
	MeOH-HCl	247(3.93),349(4.04)	12-2251-69
	MeOH-NaOH	295(3.84),453(4.12)	12-2251-69
$C_4H_6MoO_4$			
Acetic acid, Mo(II) salt	EtOH	233(3.54),294(3.88), 440(1.78)	12-1571-69
$C_4H_6N_2$			
2-Butene, 1-diazo-, trans	C_6H_{10}	502(1.38)	35-0711-69
Propene, 3-diazo-2-methyl-	C_6H_{10}	492(1.28)	35-0711-69
$C_4H_6N_2O$			
2-Butanone, 3-diazo-	H_2O	240(3.84),286(4.06)	33-2417-69
	EtOH	245(3.89),286(4.05)	33-2417-69
1,3,4-Oxadiazole, 2-ethyl-	MeOH	270(1.0)	48-0646-69
2-Pyrazolin-5-one, 3-methyl-	MeOH	244(3.67)	22-4159-69
$C_4H_6N_2O_2$			
Imidazole, 1-hydroxy-4-methyl-, 3-oxide	MeOH	232(3.78)	48-0746-69
Uracil, dihydro-, ketyl radical	pH 5.2	275(3.26)	38-4881-69B
$C_4H_6N_2O_2S_2$			
1,4-Dithian, 2,3-bis(hydroxyimino)-	EtOH	213(3.99),262(3.79)	39-2319-69C

Compound	Solvent	$\lambda_{max}(\log \epsilon)$	Ref.
$C_4H_6N_2O_4$			
1-Butene, 4,4-dinitro-	anion	379(4.23)	122-0413-69
$C_4H_6N_4$			
Pyrimidine, 2,4-diamino-	pH 2	266(3.73)	44-0821-69
	pH 12	228(3.96),282(3.84)	44-0821-69
as-Triazine, 3-amino-5-methyl-	EtOH	231(4.18),334(3.42)	22-3675-69
as-Triazine, 3-amino-6-methyl-	EtOH	231(4.18),334(3.42)	22-3675-69
$C_4H_6N_4O$			
Pyruvonitrile, semicarbazone	EtOH	257(4.25)	73-2306-69
as-Triazin-3(2H)-one, 5-amino-6-methyl-	H_2O	253(3.79)	73-2306-69
$C_4H_6N_4OS$			
v-Triazole-4-carbothioic acid, 5-amino-,	pH 5.0	237(3.75),293(4.06)	39-2379-69C
S-methyl ester	pH 10.0	211(4.05),231s(3.67), 255s(3.48),300(4.03)	39-2379-69C
$C_4H_6N_4O_2$			
Uracil, 6-hydrazino-	EtOH	266(4.19)	56-0519-69
$C_4H_6N_4O_3S$			
4-Pyrimidinesulfonic acid, 2,6-diamino-,	acid	270(3.81),310s(3.03)	44-0821-69
sodium salt	neutral	227(4.01),293(3.80)	44-0821-69
$C_4H_6N_4O_8$			
Butane, 1,1,3,3-tetranitro-	anion	365(4.23)	122-0100-69
$C_4H_6N_4S$			
Pyruvonitrile, thiosemicarbazone	40% EtOH	206(3.81),295(4.41)	73-2306-69
as-Triazine-3(2H)-tione, 5-amino-6-methyl-	40% EtOH	195(3.79),211(3.79), 239(4.02),273(4.42)	73-2306-69
$C_4H_6O_2$			
2,3-Butanedione	H_2O	409(0.98)	35-3980-69
$C_4H_7BrN_2O_2$			
Vinylamine, 2-bromo-N,N-dimethyl-2-nitro-	CH_2Cl_2	372(4.25)	33-2641-69
C_4H_7NO			
2-Pyrrolidinone	C_6H_{12}	186(3.78)	46-1642-69
	H_2O	189(3.85)	46-1642-69
$C_4H_7NO_2$			
2,3-Butanedione, oxime, syn-Me	1% MeOH	227(3.94)	22-2894-69
anion	1% MeOH	275(3.85)	22-2894-69
$C_4H_7NO_3$			
Butane, 2,3-epoxy-2-nitro-	MeOH	232(3.78),290(2.90)	77-0369-69
Nitrone, α-carboxy-N-methyl-, methyl ester	MeOH	266(4.02)	24-2346-69
$C_4H_7NO_4$			
Acetic acid, nitro-, ethyl ester, potassium salt	n.s.g.	308(3.2)	78-1929-69
KHF_2 adduct	n.s.g.	304(2.7)	78-1929-69
$C_4H_7N_3$			
s-Triazole, 3,5-dimethyl-	n.s.g.	208(2.63)	39-2251-69C

Compound	Solvent	λ_{max}(log ε)	Ref.
$C_4H_7N_5$			
Acetamidine, N-4H-1,2,4-triazol-4-yl-	MeOH	215(3.60)	48-0477-69
Pyrimidine, 2,4,6-triamino-	acid	214(4.47),272(4.26)	44-0821-69
	neutral	209(4.55),238s(3.60), 268(4.07)	44-0821-69
$C_4H_7N_5O$			
v-Triazole-4-carboxamide, 5-methyl-	pH 5.0	232(3.93),275(3.86)	39-0152-69C
	pH 11.0	277(3.85)	39-0152-69C
$C_4H_7N_7$			
as-Triazine-5-carboxamide, 6-amino-, hydrazone	pH 7	357(3.96)	44-2102-69
C_4H_8ClNOS			
Acetic acid, chlorothio-, S-(2-amino-ethyl) ester	pH 1	234(3.66)	35-2358-69
$C_4H_8N_2$			
Diimide, isopropenylmethyl-	EtOH	232(3.78),386(1.74)	35-3375-69
$C_4H_8N_2O_4$			
Isobutane, 1,1-dinitro-	anion	384(4.04)	122-0413-69
$C_4H_8N_2S$			
1,2,5-Thia(S^{IV})diazole, 3,4-dihydro-3,3-dimethyl-	EtOH	263(2.16)	88-4117-69
$C_4H_8N_4$			
s-Tetrazine, 1,4-dihydro-1,4-dimethyl-	EtOH	237(3.90)	35-2443-69
$C_4H_8N_4O_2$			
2-Tetrazene-1,4-dicarboxaldehyde, 1,4-dimethyl-	EtOH	270(4.39)	44-2997-69
$C_4H_8N_8O_8$			
s-Tetrazocine, 1,3,5,7-tetranitro-octahydro-	MeCN	226(4.26)	60-0904-69
C_4H_8O			
2-Butanone	gas	167(3.18),170(3.43), 172(3.26),188(2.97), 191(3.10),192(3.32), 194(2.97),195(3.27), 197(3.34)	18-2453-69
	C_6H_{12}	279(1.20)	77-1246-69
	MeOH	272(1.23)	77-1246-69
C_4H_8OS			
1,4-Oxathiane	hexane	191s(3.45),197(3.47), 206s(3.28)	59-0407-69
	iso-PrOH	204s(3.18),211s(2.92)	59-0407-69
$C_4H_8O_2$			
Isobutyric acid, Pb(IV) salt	hexane	232(4.18)	33-1495-69
	CH_2Cl_2	235(4.22)	33-1495-69
$C_4H_8S_2$			
m-Dithiane	hexane	190(3.73),193s(3.71), 232(2.62),249(2.68)	59-0407-69

Compound	Solvent	$\lambda_{max}(\log \epsilon)$	Ref.
m-Dithiane (cont.)	iso-PrOH	230(2.65),248(2.67)	59-0407-69
p-Dithiane	hexane	188s(3.49),198(3.83), 228(2.57)	59-0407-69
	iso-PrOH	195(3.60),225(2.48)	59-0407-69
C_4H_9NSSi			
Trimethylsilyl isothiocyanate	hexane	198(4.83),246(3.09)	114-0309-69B
$C_4H_9N_3$			
Butyl azide	hexane	216(2.73),287(1.40)	114-0309-69B
tert-Butyl azide	hexane	216(2.70),288(1.36)	114-0309-69B
$C_4H_9N_3S$			
Propionaldehyde, thiosemicarbazone	EtOH	229(3.88),270(4.38)	94-0844-69
$C_4H_{10}GeO_2$			
Germanecarboxylic acid, trimethyl-	EtOH	250(2.76),254s(2.74)	101-00P5-69A
sodium salt	EtOH	255(3.00),259s(2.9)	101-00P5-69A
$C_4H_{10}N_3O_3P$			
Phosphorazidic acid, diethyl ester	hexane	208(2.50),250s(1.30)	114-0309-69B
$C_4H_{10}N_4$			
s-Tetrazine, 1,2,3,4-tetrahydro-1,4-dimethyl-	EtOH	242(3.76)	35-2443-69
$C_4H_{10}N_4O$			
2-Tetrazene-1-carboxaldehyde, 1,4,4-trimethyl-	EtOH	275(4.27)	44-2997-69
$C_4H_{10}N_4S$			
s-Tetrazine-3(2H)-thione, 6-ethyltetrahydro-	MeOH	251(4.10)	44-0756-69
$C_4H_{10}S_2$			
Ethane, 1,2-bis(methylthio)-	hexane	190(3.75),194s(3.70)	59-0407-69
	iso-PrOH	205(3.52)	59-0407-69
C_4N_4S			
1,2,5-Thia(S^{IV})diazole-3,4-dicarbo-	n.s.g.	272(4.0)	103-0180-69

Compound	Solvent	$\lambda_{max}(\log \epsilon)$	Ref.
C_5BrF_4N			
Pyridine, 2-bromo-3,4,5,6-tetrafluoro-	C_6H_{12}	265(3.56)	39-2559-69C
C_5ClF_4NS			
4-Pyridinesulfenyl chloride, 2,3,5,6-tetrafluoro-	hexane	215(3.76),298(3.45), 385(2.53)	39-1660-69C
$C_5HCl_6N_3OS$			
Acetamide, 2,2,2-trichloro-N-[5-(trichloromethyl)-1,3,4-thiadiazol-2-yl]-	MeOH	282(3.89)	4-0835-69
C_5HF_4NS			
4-Pyridinethiol, 2,3,5,6-tetrafluoro-	hexane	210(3.57),239(4.09), 268(3.49),302s(1.48)	39-1660-69C
$C_5H_2Br_2O_2$			
2H-Pyran-2-one, 3,5-dibromo-	CH_2Cl_2	315(3.91)	44-2239-69
$C_5H_2Cl_2N_2O_3$			
Pyridine, 2,6-dichloro-4-nitro-, 1-oxide	EtOH	330(4.04)	104-1961-69
$C_5H_2N_4O_7$			
Pyridine, 2,4,6-trinitro-, 1-oxide	EtOH	315(3.89)	104-1961-69
$C_5H_3BrN_2O_3$			
Uracil-4-carboxaldehyde, 5-bromo-	pH 7	279(3.91)	63-0457-69
	pH 13	280(3.71)	63-0457-69
$C_5H_3BrN_4O$			
Purine, 6-bromo-, 3-oxide	pH 2	233(4.34),290s(3.96), 303(4.05)	44-2157-69
	pH 8	233(4.46),293s(3.86), 306(3.92)	44-2157-69
$C_5H_3BrO_2$			
2-Furaldehyde, 5-bromo-	30% EtOH	220(3.52),290(4.20)	103-0278-69
2H-Pyran-2-one, 3-bromo-	CCl_4	300(3.76)	44-2239-69
2H-Pyran-2-one, 5-bromo-	CCl_4	307(3.73)	44-2239-69
$C_5H_3Br_2NO$			
4-Pyridinol, 3,5-dibromo-	pH -3.6	221(4.46),255(3.70), 267s(3.54)	12-2581-69
	pH 4	218(4.32),268s(4.00), 273(4.02),283(3.80)	12-2581-69
	pH 11	214(4.47),247(3.77), 278(3.57),284s(3.53)	12-2581-69
$C_5H_3ClN_2S_2$			
Isothiazole-4-carbonitrile, 3-chloro-5-(methylthio)-	MeOH	221(4.12),287(4.04)	88-4265-69
$C_5H_3ClN_4O$			
Purine, 6-chloro-, 3-oxide	pH 1	230(4.31),304(3.90)	18-0750-69
	pH 2	229(4.35),302(3.96)	44-2157-69
	pH 6	231(4.46),308(3.83)	18-0750-69
	pH 8	230(4.44),305(3.89)	44-2157-69
	pH 13	232(4.44),308(3.85)	18-0750-69

Compound	Solvent	λ_{max}(log ε)	Ref.
$C_5H_3ClO_2$			
2-Furaldehyde, 5-chloro-	30% EtOH	225(3.46),288(4.17)	103-0278-69
$C_5H_3Cl_3N_2O$			
4(3H)-Pyrimidinone, 5,6-dichloro-2-(chloromethyl)-	MeOH	242(3.71),289(3.76)	44-2972-69
$C_5H_3Cl_4N_3OS$			
Acetamide, 2,2-dichloro-N-[5-(dichloromethyl)-1,3,4-thiadiazol-2-yl]-	MeOH	266(4.00)	4-0835-69
$C_5H_3FN_4$			
Purine, 6-fluoro-	pH 2	251(3.82)	77-0381-69
	pH 11	260(3.87)	77-0381-69
$C_5H_3F_2IN_2O$			
Pyrimidine, 2,5-difluoro-4-iodo-6-methoxy-	hexane	215(3.80),237(3.65), 265(4.02),272s(3.9)	39-1866-69C
$C_5H_3F_3N_4S$			
7H-s-Triazolo[3,4-b][1,3,4]thiadiazine, 3-(trifluoromethyl)-	EtOH	286(3.31)	2-0959-69
$C_5H_3IN_4O$			
Purine, 6-iodo-, 3-oxide	pH 2	217(4.15),231(4.14), 293s(3.98),311(4.15)	44-2157-69
	pH 8	238(4.30),295s(3.97), 311(4.04),322s(3.93)	44-2157-69
$C_5H_3N_5O$			
Pyrimido[5,4-e]-as-triazin-5(6H)-one	pH 1	232(3.91),264(3.65), 329(3.73)	44-2102-69
$C_5H_3N_5O_2$			
Pyrimido[5,4-e]-as-triazine-5,7(6H,8H)-dione	pH 1	232(4.16),265s(--), 332(3.70)	44-2102-69
	pH 7	236s(--),248(4.06), 264s(3.96),350(3.41), 385(3.46)	44-2102-69
	pH 13	259(4.29),312(3.31), 394(3.59)	44-2102-69
$C_5H_3N_5S$			
Pyrimido[5,4-e]-as-triazine-5(6H)-thione	pH 7	259(4.07),450(3.70)	44-3161-69
$C_5H_3N_7$			
7H-Imidazo[4,5-d]tetrazolo[1,5-b]pyridazine	H_2O	206(4.49),277(3.78)	4-0093-69
C_5H_4BrN			
Pyridine, 2-bromo-	C_6H_{12}	266(3.55)	4-0859-69
	pH 1	272(3.89)	4-0859-69
	MeOH	265(3.26)	4-0859-69
Pyridine, 3-bromo-	C_6H_{12}	269(3.40)	4-0859-69
	pH 1	275(3.67)	4-0859-69
	MeOH	262(3.29)	4-0859-69
Pyridine, 4-bromo-	C_6H_{12}	257(3.21)	4-0859-69
	MeOH	259(3.37)	4-0859-69

Compound	Solvent	$\lambda_{max}(\log \epsilon)$	Ref.
C_5H_4BrNO			
Pyridine, 2-bromo-, N-oxide	C_6H_{12}	282(4.09)	4-0859-69
	pH 1	257(3.28)	4-0859-69
	MeOH	266(3.80)	4-0859-69
Pyridine, 3-bromo-, N-oxide	C_6H_{12}	283(3.77)	4-0859-69
	pH 1	260(3.80)	4-0859-69
	MeOH	270(4.09)	4-0859-69
Pyridine, 4-bromo-, N-oxide	C_6H_{12}	293(4.34)	4-0859-69
	MeOH	275(4.26)	4-0859-69
3-Pyridinol, 2-bromo-	pH 1	225s(--),295(3.81)	1-1704-69
	pH 13	240(3.94),305(3.84)	1-1704-69
$C_5H_4Br_4$			
1,4-Pentadiene, 1,2,4,5-tetrabromo-	EtOH	216(4.27),221(4.26)	78-2823-69
C_5H_4ClN			
Pyridine, 2-chloro-	C_6H_{12}	264(3.54)	4-0859-69
	pH 1	270(3.85)	4-0859-69
	MeOH	263(3.55)	4-0859-69
Pyridine, 3-chloro-	C_6H_{12}	267(3.68)	4-0859-69
	pH 1	270(3.70)	4-0859-69
	MeOH	267(3.16)	4-0859-69
Pyridine, 4-chloro-	C_6H_{12}	258(3.17)	4-0859-69
	pH 1	255(3.69)	4-0859-69
	MeOH	257(3.33)	4-0859-69
C_5H_4ClNO			
Pyridine, 2-chloro-, N-oxide	C_6H_{12}	282(4.01)	4-0859-69
	pH 1	257(3.25)	4-0859-69
	MeOH	265(4.06)	4-0859-69
Pyridine, 3-chloro-, N-oxide	C_6H_{12}	285(3.82)	4-0859-69
	pH 1	258(3.45)	4-0859-69
	MeOH	269(4.11)	4-0859-69
Pyridine, 4-chloro-, N-oxide	C_6H_{12}	288(3.87)	4-0859-69
	pH 1	263(3.83)	4-0859-69
	MeOH	273(4.22)	4-0859-69
$C_5H_4ClN_3O_2$			
Pyridine, 2-amino-4-chloro-3-nitro-	pH 1	232(4.16),329(3.80)	54-1263-69
	pH 13	250s(--)	54-1263-69
Pyridine, 2-amino-4-chloro-5-nitro-	pH 1	308(3.99)	54-1263-69
	pH 13	238s(--)	54-1263-69
$C_5H_4Cl_2O_2$			
2(5H)-Furanone, 3,4-dichloro-5-methoxy-	hexane	226(4.02)	39-0728-69C
$C_5H_4Cl_2O_3$			
Acrylic acid, α,β-dichloro-β-formyl-, methyl ester	hexane	258(3.91)	39-0728-69C
C_5H_4IN			
Pyridine, 3-iodo-	C_6H_{12}	268(3.42)	4-0859-69
	pH 1	280(3.54)	4-0859-69
	MeOH	273(3.17)	4-0859-69
C_5H_4INO			
Pyridine, 3-iodo-, N-oxide	C_6H_{12}	287(4.09)	4-0859-69
	MeOH	269(4.07)	4-0859-69

Compound	Solvent	λ_{max}(log ε)	Ref.
$C_5H_4N_2O_4$			
Imidazole-4,5-dicarboxylic acid	pH 7	253(3.81)	5-0231-69I
Orotic acid, ketyl radical	pH 5.8	328(4.11)	38-4881-69B
$C_5H_4N_4$			
1H-Imidazo[4,5-d]pyridazine	EtOH	203(4.61),216s(4.22), 249s(3.75),280(3.33)	4-0093-69
$C_5H_4N_4O$			
4H-Imidazo[4,5-d]pyridazin-4-one, 1,5-dihydro-	pH 13	212(4.42)	4-0093-69
$C_5H_4N_4OS$			
Acetamide, N-(5-cyano-1,3,4-thiadiazol-2-yl)-	MeOH	280(4.00)	4-0835-69
$C_5H_4N_4O_2$			
Hypoxanthine, 3-oxide	pH -1	212(4.16),275(3.90)	44-2153-69
	pH 1	261(3.81)	18-0750-69
	pH 3	223(4.24),271(3.97)	44-2153-69
	pH 6	280(4.01)	18-0750-69
	pH 8	218(4.29),286(4.08)	44-2153-69
	pH 12	224(4.34),275s(4.00), 285(4.01)	44-2153-69
	pH 13	226(4.40),286(4.02)	18-0750-69
$C_5H_4N_4O_3$			
Uric acid	pH -6.5	228s(3.85),278(3.73), 300(3.85)	44-0978-69
	pH -4.9	229(3.90),283(3.89)	44-0978-69
	pH -3.0	231(3.90),284(4.01)	44-0978-69
	pH 3.0	231(3.92),285(4.08)	44-0978-69
	pH 8.0	207(4.15),236(3.99), 292(4.10)	44-0978-69
	pH 12.3	219(3.40),296(4.13)	44-0978-69
Xanthine, 3-hydroxy-	pH -2	238(3.85),268(3.95)	44-0978-69
	pH 3.0	205(3.38),272(4.00)	44-0978-69
	pH 8.17	218(3.34),257(3.75), 299(3.83)	44-0978-69
	pH 11.4	225(3.45),297(3.93)	44-0978-69
	pH 15	224(3.46),292(3.94)	44-0978-69
$C_5H_4N_4O_4S$			
Purine-6-sulfonic acid, 3-oxide, potassium salt	pH 3	229(4.36),311(4.09)	44-2153-69
	pH 12	230(4.46),316(3.93)	44-2153-69
$C_5H_4N_4S$			
4H-Imidazo[4,5-d]pyridazine-4-thione, 1,5-dihydro-	pH 13	213(4.59)	4-0093-69
$C_5H_4N_6$			
Pyrimido[5,4-e]-as-triazine, 5-amino-	pH 1	217(4.14),243(4.00), 353(4.02),358s(4.01)	44-2102-69
$C_5H_4N_6O$			
Pyrimido[5,4-e]-as-triazine, 5-(hydroxy-amino)-	pH 1	381(3.82)	44-3161-69
$C_5H_4N_6O_2S$			
1,3,4-Thiadiazole, 2-amino-5-(4-nitro-2-imidazolyl)-	MeOH	298(3.90),370(3.99)	4-0835-69

Compound	Solvent	λ_{max}(log ϵ)	Ref.
$C_5H_4O_5$			
4-Cyclopentene-1,3-dione, 2,4,5-trihydroxy-	n.s.g.	230(3.95),292(4.10)	22-4590-69
anion	n.s.g.	250(4.06),335(4.13)	22-4590-69
dianion	n.s.g.	272(4.08),375(4.13)	22-4590-69
$C_5H_4S_2$			
2H-Thiopyran-2-thione	C_6H_{12}	240(4.07),304s(3.77), 312s(3.98),318(4.05), 326(4.04),335(3.85), 433(3.73),587s(1.97)	78-1441-69
	EtOH	240(4.02),315(3.95), 433(3.74)	78-1441-69
C_5H_5Br			
3-Penten-1-yne, 5-bromo-, cis	EtOH	239(4.06)	18-2589-69
3-Penten-1-yne, 5-bromo-, trans	EtOH	238(4.16)	18-2589-69
$C_5H_5BrN_2O$			
4(3H)-Pyrimidinone, 5-bromo-2-methyl-	pH 1	235(3.77),275(3.75)	1-2437-69
	pH 13	235(3.85),275(3.63)	1-2437-69
$C_5H_5BrN_2O_3$			
Uracil, 5-bromo-6-(hydroxymethyl)-	pH 7	278(3.94)	63-0457-69
	pH 13	291(3.78)	63-0457-69
$C_5H_5BrN_2S$			
4(3H)-Pyrimidinethione, 5-bromo-2-methyl-	pH 1	285(3.57),340(3.58)	1-2437-69
	pH 13	280(3.88),315(3.94)	1-2437-69
$C_5H_5ClN_2O$			
4(3H)-Pyrimidinone, 6-chloro-2-methyl-	MeOH	225(3.78),279(3.64)	44-2972-69
C_5H_5ClO			
Cyclobutanone, 2-(chloromethylene)-	MeCN	243(4.17)	35-2375-69
$C_5H_5ClO_2$			
2,4-Pentadienoic acid, 5-chloro-, 2-cis-4-trans	EtOH	255(4.32)	35-1179-69
$C_5H_5Cl_2NO$			
Isoxazole, 3-(dichloromethyl)-5-methyl-	EtOH	223(3.65)	35-4749-69
$C_5H_5Cl_2N_3$			
Pyridine, 2,3-diamino-4,6-dichloro-	pH 1	225(4.07),262(3.79), 329(3.89)	54-1263-69
	pH 13	248(3.92),312(3.86)	54-1263-69
$C_5H_5Cl_3N_2OS$			
Hydrouracil, 2-thio-, 6-(trichloromethyl)-	EtOH	274(4.42)	104-1790-69
$C_5H_5Cl_3N_2OSe$			
4H-1,3-Selenazin-4-one, tetrahydro-2-imino-6-(trichloromethyl)-	EtOH	230(4.00),270(3.41)	104-1790-69
	n.s.g.	230(4.00),270(3.41)	30-0355-69A
$C_5H_5FN_2O_2$			
2-Pyrimidinone, 5-fluoro-4-methoxy-	pH 1	280(3.65)	1-0294-69
	pH 13	290(3.82)	1-0294-69
4-Pyrimidinone, 5-fluoro-2-methoxy-	pH 1	263(3.67)	1-0294-69

Compound	Solvent	λ_{max}(log ε)	Ref.
4-Pyrimidinone, 5-fluoro-2-methoxy-	pH 13	269(3.83)	1-0294-69
C_5H_5N			
2,4-Pentadienenitrile, cis	n.s.g.	242(4.36)	22-1349-69
2,4-Pentadienenitrile, trans	n.s.g.	242(4.42)	22-1349-69
Pyridine	C_6H_{12}	251(3.30)	4-0859-69
	pH 1	255(3.74)	4-0859-69
	MeOH	253(3.62)	4-0859-69
at 5°	alkali	244(3.3),250(3.4), 257(3.5),263(3.3)	12-0721-69
pyridinium ion at 5°	n.s.g.	250(3.7),256(3.7)	12-0721-69
C_5H_5NO			
Pyridine, N-oxide	C_6H_{12}	283(4.11)	4-0859-69
	decalin	283(4.1),314s(3.3), 324s(3.2),338(2.8)	33-0789-69
	pH 1	257(3.57)	4-0859-69
	MeOH	263(4.15)	4-0859-69
2-Pyridinol	benzene	299(3.72)	30-0620-69C
	0.1N HCl	222(3.69),285(3.72)	
	pH 1.0	220(3.68),282(3.76)	
	pH 1.55	224(3.83),290(3.78)	
	pH 3-10	224(3.88),294(3.78)	
	H_2O	224(3.89),294(3.80)	
	0.1N KOH	228(4.06),292(3.84)	
	N KOH	228(4.02),292(3.71)	
	EtOH	228(3.93),299(3.76)	
	EtOH-HCl	282(3.84)	
	EtOH-KOH	232(4.00),295(3.62)	
	$CHCl_3$	230(3.93),297(3.72)	
	dioxan	220(3.95),305(3.75)	
Pyrrole-3-carboxaldehyde	EtOH	204(4.05),247(3.95), 266s(3.86)	78-3879-69
C_5H_5NOS			
2(1H)-Pyridinethione, 3-hydroxy-	pH 1	260(3.77),350(4.02)	1-1704-69
	pH 13	265(3.80),355(3.91)	1-1704-69
$C_5H_5NO_4S$			
4-Pyridinesulfonic acid, 1-oxide	65% H_2SO_4	233(4.20),264(3.60)	104-1945-69
	pH 13	214(4.04),267(4.22)	104-1945-69
$C_5H_5N_3O_2$			
4-Pyridazinecarboxylic acid, 5-amino-	EtOH	272(3.55),301s(3.28)	12-1745-69
Pyridine, 4-nitramino-	pH -0.2	277(4.27)	12-2611-69
	pH 4	224(4.03),266s(3.34), 276s(3.36),328(4.32)	
	pH 9	292(4.00)	
	EtOH-acid	227(4.00),335(4.34)	
	MeCN	227(4.00),339(4.29)	
	DMSO-acid	342(4.35)	
	dioxan	268(3.92),338(3.15)	
$C_5H_5N_5$			
Adenine	4N H_2SO_4	262(4.07)	10-0414-69F
	3N H_2SO_4	262(4.08)	
	2N H_2SO_4	262(4.10)	
	N H_2SO_4	262(4.11)	
	0.5N H_2SO_4	262(4.11)	
	0.1N H_2SO_4	262(4.12)	

Compound	Solvent	$\lambda_{max}(\log \epsilon)$	Ref.
Adenine (cont.)	H_2O	260(4.13)	10-0414-69F
	pH 13	269(4.09)	10-0414-69F
5-Pyrimidinecarbonitrile, 2,4-diamino-	pH 2	220(4.61),238s(4.20), 278(3.69)	44-0821-69
	pH 12	213(4.43),249(4.19), 293(3.98)	44-0821-69
1H-v-Triazolo[4,5-c]pyridine, 1-amino-	EtOH	262(3.68)	39-1758-69C
1H-v-Triazolo[4,5-c]pyridine, 6-amino-	pH 1	226(4.52),260(3.76), 339(3.51)	54-1263-69
	pH 13	220(4.43),267(3.40), 319(3.56)	54-1263-69
3H-v-Triazolo[4,5-b]pyridine, 3-amino-	EtOH	256(3.73),285(3.80)	39-1758-69C
$C_5H_5N_5O$			
Adenine, 3-oxide	pH 0	224s(3.76),277(3.93)	44-2153-69
	pH 5	229(3.95),293(3.85)	44-2153-69
	pH 10	231(4.07),278s(3.76), 290(3.80)	44-2153-69
s-Triazolo[2,3-d]-as-triazin-5-ol, 8-methyl-	EtOH	261(3.92)	22-3670-69
s-Triazolo[3,2-c]-as-triazin-7-ol, 2-methyl-	EtOH	238(3.50),296(3.77)	22-2492-69
s-Triazolo[3,2-c]-as-triazin-7-ol, 6-methyl-	EtOH	242(3.64),297(3.77)	22-2492-69 +22-3670-69
s-Triazolo[3,4-c]-as-triazin-5-ol, 6-methyl-	EtOH	239(3.68),315(3.50)	22-2492-69
	EtOH	238(3.68),315(3.50)	22-3670-69
s-Triazolo[4,3-b]-as-triazin-7-ol, 3-methyl-	EtOH	230(3.99),300(3.62)	22-2492-69
s-Triazolo[4,3-b]-as-triazin-7-ol, 6-methyl-	EtOH	230(3.64),285(3.56)	22-2492-69 +22-3670-69
s-Triazolo[4,3-d]-as-triazin-5-ol, 8-methyl-	EtOH	263(3.83)	22-3670-69
$C_5H_5N_5OS$			
Purine-6,8(1H,9H)-dione, 9-amino-6-thio-, hydrochloride	pH 1	239(4.07),332(4.15)	44-3161-69
4-Pyrimidinol, 2,6-diamino-5-thiocyanato- (hydrate)	pH 1	259(4.18)	39-0603-69C
Thiazolo[4,5-d]pyrimidin-7-ol, 2,5-diamino-	pH 1	223(4.40),255s(3.78), 305(4.11)	39-0603-69C
$C_5H_5N_5O_2$			
Adenine, N-hydroxy-, 3-oxide	pH 2	225(4.00),288(4.17)	87-0717-69
	pH 6.47	232(4.16),302(4.12)	87-0717-69
Guanine, 3-oxide	pH -2	244(4.03),260s(--)	44-0978-69
	pH 1.0	213(4.10),245(3.89), 267(3.98)	44-0978-69
	pH 4.8	217(3.36),270(3.94)	44-0978-69
	pH 8.0	224(3.49),254(3.72), 292(3.82)	44-0978-69
	pH 12-15	226(3.49),283(3.99)	44-0978-69
as-Triazine-5-carboxamide, 6-formamido-	pH 7	244(4.22),315(3.53)	44-2102-69
$C_5H_5N_5O_4$			
Formic acid, 2-(1,6-dihydro-5-nitro-6-oxo-4-pyrimidinyl)hydrazide	pH 7	260(3.77),335(3.79)	44-2102-69
$C_5H_5N_5S$			
8-Azapurine-6-thiol, 9-methyl-	pH 5.0	229(3.97),327(4.30)	39-0152-69C
	pH 10.0	229(4.12),269(3.30), 333(4.19)	39-0152-69C

Compound	Solvent	$\lambda_{max}(\log \epsilon)$	Ref.
Pyrimido[5,4-e]-as-triazine-5(1H)-thione, 2,6-dihydro-	pH 1	238(3.83),256(3.90), 328(3.42),412(3.75)	44-3161-69
1,3,4-Thiadiazole, 2-amino-5-(2-imidazolyl)-, dihydrochloride	MeOH	317(4.08)	4-0835-69
1H-v-Triazolo[4,5-d]pyridazine, 4-(methylthio)-	EtOH	206(4.41),219s(4.24), 275(4.32)	4-0093-69
s-Triazolo[2,3-d]triazine-5-thiol, 8-methyl-	EtOH	237(4.07),299(4.40)	22-3670-69
s-Triazolo[3,2-c]triazine-7-thiol, 6-methyl-	EtOH	236(3.89),270(3.68), 366(4.25)	22-3670-69
s-Triazolo[4,3-b]triazine-7-thiol, 6-methyl-	EtOH	246(3.89),279(3.82), 369(3.98)	22-3670-69
s-Triazolo[4,3-d]triazine-5-thiol, 8-methyl-	EtOH	219(3.83),310(4.21)	22-3670-69
$C_5H_5N_7$			
Pyrimido[5,4-e]-as-triazine, 5-hydrazino-	pH 1	365(3.87)	44-2102-69
$C_5H_6BrN_3$			
Pyrimidine, 4-amino-5-bromo-2-methyl-	EtOH	237(4.03),284(3.81)	88-3553-69
	EtOH-HCl	249(--),273(--)	88-3553-69
$C_5H_6ClN_3$			
Pyridine, 3-chloro-4-hydrazino-	EtOH	257(3.79),291(3.75)	39-1758-69C
Pyridine, 2,3-diamino-4-chloro-	pH 1	258(3.74),319(3.99)	54-1263-69
	pH 13	243(3.76),302(3.81)	54-1263-69
$C_5H_6N_2O$			
Acrylonitrile, 3-formamido-2-methyl-	EtOH	255(4.22)	88-4899-69
Pyrazole-3-carboxaldehyde, 5-methyl-	MeOH	236(3.89)	4-0545-69
Pyrazole-4-carboxaldehyde, 1-methyl-	MeOH	246(4.02)	4-0545-69
Pyrimidine, 5-methyl-, 1-oxide	n.s.g.	280(4.02)	88-4899-69
$C_5H_6N_2OS$			
4(3H)-Pyrimidinethione, 5-hydroxy-2-methyl-	pH 1	245(3.29),335(3.79)	1-2437-69
	pH 13	245(3.70),340(3.78)	1-2437-69
$C_5H_6N_2O_2$			
1-Pyrazoline-3,5-dione, 4,4-dimethyl-	MeCN	620(2.18)	88-4497-69
Pyridazine, 3-methoxy-, 1-oxide	EtOH	260(3.6),319(3.7)	94-0763-69
3-Pyridazinol, 5-methyl-, 1-oxide	EtOH	258(3.6),312(3.6)	94-0763-69
3(2H)-Pyridazinone, 5-methyl-, 1-oxide	EtOH	242(3.5),332(3.7)	94-0763-69
2-Pyrimidinemethanol, 4-hydroxy-	MeOH	223(3.77),268(3.48)	35-7752-69
4(3H)-Pyrimidinone, 5-hydroxy-2-methyl-	pH 1	255(3.36)	1-2437-69
	pH 13	260(3.92),295(3.92)	1-2437-69
Thymine, ketyl radical	pH 5.1	305(3.18),330(3.18)	38-4881-69B
radical anion	pH 11.8	330(3.23)	38-4881-69B
Uracil, 1-methyl-, ketyl radical	pH 5.3	305(3.36)	38-4881-69B
Uracil, 3-methyl-, ketyl radical	pH 6.0	320(3.15)	38-4881-69B
radical anion	pH 11.8	320(3.00)	38-4881-69B
Uracil, 6-methyl-	pH 2	261(3.86)	56-0499-69
	pH 12	280(3.72)	56-0499-69
	5N NaOH	274(3.82)	56-0499-69
ketyl radical	pH 5.3	295(3.11),325(3.08)	38-4881-69B
radical anion	pH 11.8	310(2.95)	38-4881-69B
$C_5H_6N_2O_2S$			
Imidazolidine-2-thione, 1,3-dimethyl-4,5-dioxo-	EtOH	298(4.14),402(1.40)	18-2323-69

Compound	Solvent	λ max (log ε)	Ref.
Thiazole, 2,4-dimethyl-5-nitro-	EtOH	220(3.62),253(3.49), 304(3.54)	28-1343-69B
Thiazole, 2,5-dimethyl-4-nitro-	EtOH	219(4.07),290(3.69)	28-1343-69B
Thiazole, 5-ethyl-4-nitro-	EtOH	217(4.04),235(--), 282(3.71)	28-1343-69B
Uracil, 6-(methylthio)-	pH 1	277(4.20)	39-0791-69C
	pH 11	284(4.15)	39-0791-69C
$C_5H_6N_2O_3$			
1,3,4-Oxadiazole-2-carboxylic acid, ethyl ester	MeOH	243(3.2)	48-0646-69
4H-Pyrazol-4-one, 3,5-dimethyl-, 1,2-dioxide	n.s.g.	280(4.08),358(2.20)	44-0187-69
Uracil, 5-(hydroxymethyl)-	0.05N HCl	261(3.90)	35-7490-69
	0.05N NaOH	286(3.87)	35-7490-69
Uracil, 5-hydroxy-1-methyl-	pH 1	284(3.90)	44-2636-69
	pH 10	240(3.87),310(3.83)	44-2636-69
	pH 14	240s(3.84),304(3.78)	44-2636-69
Uracil, 5-hydroxy-3-methyl-	pH 1	277(3.83)	44-2636-69
	pH 10	242(3.86),304(3.79)	44-2636-69
(changing with time)	pH 14	242(3.89),322(3.90)	44-2636-69
$C_5H_6N_2O_3S$			
Uracil-4-sulfenic acid, 1-methyl-, silver salt	pH 4.6	270(--),368(--)	35-3634-69
	pH 12	260(--),364(--)	35-3634-69
$C_5H_6N_4$			
Imidazole-4-carbonitrile, 5-amino-1-methyl-	pH 1	235(3.98),255(3.92)	39-2198-69C
	H_2O	240(4.06)	39-2198-69C
	pH 13	245(4.06)	39-2198-69C
$C_5H_6N_4O$			
3-Buten-2-one, 4-(2H-tetrazol-2-yl)-	EtOH	250(4.1)	97-0025-69
$C_5H_6N_4O_2$			
Pyridine, 2,4-diamino-3-nitro-	pH 1	235(4.30),327(3.95)	54-1263-69
	pH 13	253s(--)	54-1263-69
Pyrimidine-5-carboxylic acid, 2,4-diamino-	pH 2	220(4.56),233(4.27), 273(3.73)	44-0821-69
	pH 12	215(4.51),230s(4.22), 273(3.73)	44-0821-69
Uracil-5-carboxaldehyde, hydrazone	pH 13	264(4.08),315(4.05)	24-2877-69
$C_5H_6N_6$			
8-Azapurine, 6-amino-9-methyl-	pH 0.2	264(4.03)	39-0152-69C
	pH 7.0	277(4.0)	39-0152-69C
1H-Imidazo[4,5-d]pyridazine, 4-hydrazino-	H_2O	210(4.24),252(3.88)	4-0093-69
C_5H_6O			
2-Cyclopenten-1-one	EtOH	217(4.01),304(1.81)	24-3877-69
	94% H_2SO_4	248(3.96)	23-2263-69
	86.1% H_2SO_4	248(3.94)	
	83.3% H_2SO_4	248(3.94)	
	79.4% H_2SO_4	246(3.91)	
	79.2% H_2SO_4	243(3.89)	
	71.5% H_2SO_4	236(3.86)	
	69.1% H_2SO_4	231(3.87)	
	66.0% H_2SO_4	230(3.89)	
	62.0% H_2SO_4	227(3.90)	

Compound	Solvent	$\lambda_{max}(\log \epsilon)$	Ref.
2-Cyclopenten-1-one (cont.)	57.3% H_2SO_4	225(3.93)	23-2263-69
	51.0% H_2SO_4	223(3.95)	
	46.3% H_2SO_4	223(3.95)	
	38.0% H_2SO_4	222(3.96)	
$C_5H_6O_2$			
Furan, 2-methoxy-	n.s.g.	213(4.62)	88-2767-69
$C_5H_6O_3$			
2(5H)-Furanone, 5-hydroxy-5-methyl-	EtOH	206(4.46)	18-1098-69
$C_5H_6O_4$			
1,3-Dioxepane-5,6-dione	MeCN	275(1.58),386(1.48)	88-2545-69
C_5H_6S			
2-Penten-4-yne-1-thiol	EtOH	229(3.99)	18-2589-69
4H-Thiopyran	pet ether	237s(3.72),278(3.34)	117-0021-69
$C_5H_7BF_2O_2$			
Boron, difluoro(2,4-pentanedionato)-	$CHCl_3$	283(4.23)	39-0526-69A
C_5H_7BrO			
Ether, 4-bromo-2,3-butadienyl methyl	heptane	187(4.47)	78-3277-69
$C_5H_7BrO_2$			
2,4-Pentanedione, 3-bromo-, aluminum	EtOH	304(4.50)	42-0945-69
chelate	$CHCl_3$	305(4.48)	42-0945-69
copper chelate	$CHCl_3$	248(4.19),308(4.42), 670(1.62)	42-0945-69
palladium chelate	$CHCl_3$	265s(--),345(3.94)	42-0945-69
C_5H_7ClO			
Cyclopentanone, 2-chloro-	C_6H_{12}	305(1.56)	77-1246-69
	MeOH	302(1.49)	77-1246-69
$C_5H_7ClO_2$			
2,4-Pentanedione, 3-chloro-, aluminum	EtOH	289(4.45)	42-0945-69
chelate	$CHCl_3$	290(4.42)	42-0945-69
copper chelate	$CHCl_3$	312(4.44),540s(--), 670(1.61)	42-0945-69
palladium chelate	$CHCl_3$	265s(--),345(4.09)	42-0945-69
$C_5H_7Cl_2N_3O_2S$			
Thiosemicarbazide, 1,1-bis(chloro-acetyl)-	MeOH	235(3.69)	4-0835-69
$C_5H_7Cl_3N_2O_4$			
Butane, 1,1,1-trichloro-4,4-dinitro-	anion	380(4.19)	122-0413-69
$C_5H_7Cl_3O_2$			
2-Pentanone, 5,5,5-trichloro-4-hydroxy-	EtOH	219(1.61),275(1.23)	23-2029-69
C_5H_7N			
2-Penten-4-ynylamine	EtOH	226(4.16)	18-2589-69
C_5H_7NO			
Propiolaldehyde, (dimethylamino)-	CH_2Cl_2	282(4.26)	33-2641-69
C_5H_7NOS			
4-Oxazoline-2-thione, 4,5-dimethyl-	2% EtOH	271(4.17)	1-2888-69

Compound	Solvent	λ max(log ε)	Ref.
4-Oxazoline-2-thione, 4,5-dimethyl-, conjugate acid	1% MeOH	258(3.96)	1-2888-69
$C_5H_7NOS_2$			
1,3-Thiazane-2-thione, 3-methyl-4-oxo-	MeOH	267(4.16),310(4.07)	42-0919-69
$C_5H_7NO_2S$			
D-Alanine, methyl ester, isothiocyanato derivative	hexane	248(2.84)	39-1143-69B
L-isomer	hexane	248(2.88)	39-1143-69B
$C_5H_7NO_4$			
2,4-Pentanedione, 3-nitro-, aluminum chelate	EtOH	282(4.50),340(3.25)	42-0945-69
	$CHCl_3$	284(4.51),341(3.23)	42-0945-69
copper chelate	$CHCl_3$	246(4.08),294(4.45), 645(1.77)	42-0945-69
palladium chelate	$CHCl_3$	250(4.32),265s(--), 328(4.05)	42-0945-69
$C_5H_7NS_2$			
4H-1,4-Thiazine-3-thione, 2,3-dihydro-5-methyl-	EtOH	205(3.88),273(3.75), 351(3.50)	4-0247-69
4-Thiazoline-2-thione, 4,5-dimethyl-	2% EtOH	317(4.16)	1-2888-69
conjugate acid	2% EtOH	296(3.99)	1-2888-69
$C_5H_7NS_3$			
2H-1,3-Thiazane-2,4-dithione, dihydro-3-methyl-	MeOH	325(4.09),335(4.06)	42-0919-69
2H-1,3-Thiazane-2,4-dithione, dihydro-5-methyl-	MeOH	317(4.09),345(4.13)	42-0919-69
2H-1,3-Thiazane-2,4-dithione, dihydro-6-methyl-	MeOH	317(4.00),345(4.02)	42-0919-69
$C_5H_7N_3$			
Pyridine, 4-hydrazino-	pH -1	260(4.20)	12-2611-69
	pH 5.6	270(4.22)	12-2611-69
	pH 12	249(4.11)	12-2611-69
as-Triazine, 3,5-dimethyl-	MeOH	259(3.58),368(2.55)	88-3147-69
as-Triazine, 3,6-dimethyl-	MeOH	263(3.64),362(2.60)	88-3147-69
$C_5H_7N_3O$			
4(3H)-Pyrimidinone, 2-amino-3-methyl-	pH 1	256(3.77)	70-0835-69
	pH 13	283(3.95)	70-0835-69
$C_5H_7N_3OS$			
Imidazole-5-thiocarboxylic acid, 4-amino-, S-methyl ester	pH 1.0	243(3.74),300(4.24)	39-2379-69C
	pH 7.0	235(3.60),306(4.25)	39-2379-69C
	pH 13	230s(3.72),318(4.23)	39-2379-69C
as-Triazin-5(2H)-one, 1-methyl-3-(methylthio)-	MeOH	235(4.38)	114-0093-69C
as-Triazin-5(2H)-one, 6-methyl-3-(methylthio)-	EtOH	236(4.30)	114-0181-69C
anion	EtOH	218(4.24),290(3.90)	114-0181-69C
as-Triazin-5(4H)-one, 4-methyl-3-(methylthio)-	MeOH	208(4.11),226s(3.76), 298(3.92)	114-0093-69C
$C_5H_7N_3O_2S$			
as-Triazine-3,5(2H,4H)-dione, 6-(methoxymethyl)-3-thio-	pH 2	214(4.18),268(4.34)	73-1690-69
	pH 13	222(4.34),256(4.27), 306(3.60)	73-1690-69

Compound	Solvent	λ_{max}(log ε)	Ref.
$C_5H_7N_5O$			
Formamide, N-(2,4-diamino-5-pyrimidinyl)-	pH 2	208(4.35),270(3.57)	44-0821-69
	pH 12	222(3.92),287(3.74)	44-0821-69
Pyrimidine-4-carboxamide, 2,6-diamino-	acid	211(4.51),285(3.74)	44-0821-69
	neutral	226(4.09),313(3.70)	44-0821-69
$C_5H_7N_5O_2$			
Formic acid, 2-(5-hydroxy-6-methyl-as-triazin-3-yl)hydrazide	EtOH	236(4.19)	22-2492-69
$C_5H_7N_7$			
8-Azapurine, 6-hydrazino-9-methyl-	pH 1.0	268(4.10)	39-0152-69C
	pH 7.0	211(4.16),285(4.08)	39-0152-69C
C_5H_7P			
Phosphole, 1-methyl-	gas	281(--)	35-3308-69
	isooctane	286(3.89)	35-3308-69
	H_2O	284(4.0)	35-3308-69
	EtOH	285(--)	35-3308-69
C_5H_8BrClO			
2-Pentanone, 3-bromo-5-chloro-	C_6H_{12}	297(2.05)	22-0628-69
C_5H_8BrNO			
Acrolein, 2-bromo-3-(dimethylamino)-	CH_2Cl_2	294(4.39)	33-2641-69
C_5H_8ClNO			
Acrylamide, 3-chloro-N,N-dimethyl-, trans	hexane	216(4.13),246(3.54)	33-2641-69
$C_5H_8N_2O$			
2,3-Diazabicyclo[2.2.1]hept-2-ene, 2-oxide	EtOH	228(3.78)	88-0575-69
2-Pentanone, 3-diazo-	40% dioxan	251(3.74),290(3.90)	33-2417-69
2-Pyrazolin-5-one, 3,4-dimethyl-	MeOH	248(4.00)	22-4159-69
$C_5H_8N_2O_2$			
1-Imidazolol, 2,4-dimethyl-, 3-oxide	MeOH	223(3.73)	48-0746-69
1-Imidazolol, 4,5-dimethyl-, 3-oxide	MeOH	234(3.76)	48-0746-69
Thymine, dihydro-, ketyl radical	pH 5.0	290(3.20)	38-4881-69B
$C_5H_8N_2O_5$			
2-Pentanone, 5,5-dinitro-	EtOH	379(3.95)	44-1470-69
	EtOH-KOH	379(4.24)	44-1470-69
	anion	379(4.23)	122-0413-69
$C_5H_8N_2O_6$			
Butyric acid, 4,4-dinitro-, methyl ester	anion	379(4.23)	122-0413-69
$C_5H_8N_2S$			
4-Imidazoline-2-thione, 4,5-dimethyl-	2% EtOH	261(4.16)	1-2888-69
conjugate acid	2% EtOH	250(3.94)	1-2888-69
$C_5H_8N_2S_2$			
Pseudodithiohydantoin, N,N'-dimethyl-	EtOH	291(4.16)	94-0910-69
$C_5H_8N_4$			
Pyrimidine, 2,4-diamino-5-methyl-	pH 2	271(3.74)	44-0821-69
	pH 12	228(3.99),286(3.84)	44-0821-69
Pyrimidine, 2,4-diamino-6-methyl-	acid	207(4.06),267(3.88)	44-0821-69

Compound	Solvent	λ_{max}(log ε)	Ref.
Pyrimidine, 2,4-diamino-6-methyl-	neutral	228s(3.95),278(3.89)	44-0821-69
$C_5H_8N_4O$			
Imidazole-4-carboxamide, 5-amino-1-methyl-	pH 1	240(3.85),269(3.97)	39-2198-69C
	H_2O	267(4.05)	39-2198-69C
	pH 13	267(4.05)	39-2198-69C
Pyrimidine, 2,4-diamino-5-methoxy-	pH 2	200(4.14),223(4.26), 283(3.70)	44-0821-69
	pH 12	215(4.05),230s(3.97), 296(3.79)	44-0821-69
as-Triazin-5(4H)-one, 3-amino-4,6-dimethyl-	EtOH	222(3.90),299(3.89)	114-0181-69C
$C_5H_8N_4OS$			
1H-1,2,3-Triazole-5-carbothioic acid, 4-amino-1-methyl-, S-methyl ester	EtOH	232s(3.62),256(3.63), 312(3.96)	39-2379-69C
1,2,3-Triazole-5-carbothioic acid, 4-amino-2-methyl-, S-methyl ester	MeOH	247(3.48),307(3.95)	39-2379-69C
1,2,3-Triazole-5-carbothioic acid, 4-amino-3-methyl-, S-methyl ester	EtOH	239(3.77),287(4.13)	39-2379-69C
$C_5H_8N_4O_2$			
Pyrazole-3-carboxamide, 4-amino-5-(hydroxymethyl)-	pH 1	210(3.83)	88-0289-69
	pH 13	220(3.94),250(3.85)	88-0289-69
as-Triazin-5(2H)-one, 3-amino-6-(2-hydroxyethyl)-	EtOH	205(4.43),249(3.78)	114-0181-69C
$C_5H_8N_4O_8$			
Pentane, 1,1,3,3-tetranitro-	anion	367(4.22)	122-0100-69
$C_5H_8N_4S$			
Pyruvonitrile, S-methylisothiosemicarbazone	EtOH	290(4.30)	73-2306-69
as-Triazine, 5-amino-6-methyl-3-(methylthio)-	EtOH	224(4.25),303(3.74)	73-2306-69
s-Triazolo[3,4-b][1,3,4]thiadiazole, 3-ethyl-5,6-dihydro-	EtOH	262(4.04)	2-0959-69
C_5H_8O			
Cyclopentanone	C_6H_{12}	300(1.19)	77-1246-69
	MeOH	286(1.26)	77-1246-69
Ether, 1,3-butadienyl methyl, cis	heptane	242(4.26)	104-1696-69
Ether, 1,3-butadienyl methyl, trans	heptane	234(4.19)	104-1696-69
$C_5H_8O_2$			
2,4-Pentanedione (enol)	$CHCl_3$	272(3.93)	39-0526-69A
anion	$CHCl_3$	295(4.32)	39-0526-69A
aluminum chelate	EtOH	287(4.61)	42-0945-69
	$CHCl_3$	288(4.58)	42-0945-69
copper chelate	C_6H_{12}	201(4.19),245(4.26), 297(4.38)	42-0945-69
	MeOH	241(4.19),294(4.40)	42-0945-69
	$CHCl_3$	246(4.20),296(4.42), 560(1.50),670(1.56)	42-0945-69
palladium chelate	$CHCl_3$	250(4.53),265s(--), 330(4.38),400s(2.47)	42-0945-69

Compound	Solvent	λ_{max}(log ϵ)	Ref.
$C_5H_8O_2S_2$			
o-Dithiane-3-carboxylic acid	EtOH	245s(--),280(2.39)	11-0511-69
C_5H_9NO			
Acrolein, 3-(ethylamino)-	heptane	304(4.16)	77-0861-69
	EtOH	279(4.52)	77-0861-69
	$CHCl_3$	276(4.20),312(3.96)	77-0861-69
3-Penten-2-one, 4-amino-	EtOH	299(4.20)	35-4749-69
2-Piperidone	C_6H_{12}	196(3.73)	46-1642-69
	H_2O	195(3.89)	46-1642-69
$C_5H_9NO_2$			
2,3-Pentanedione, 2-oxime, syn-Me	1% MeOH	255(3.98)	22-2894-69
anion	1% MeOH	293(4.12)	22-2894-69
Valeramide, 4-oxo-	MeCN	182(3.90),215(1.67), 268(1.11)	28-0279-69A
$C_5H_9NO_3$			
2-Pentanone, 5-nitro-	EtOH	235s(2.3)	88-0583-69
	EtOH-NaOH	237(3.9)	88-0583-69
Pyruvic acid, ethyl ester, 2-oxime	1% MeOH	215(3.86)	22-2894-69
anion	1% MeOH	260(4.07)	22-2894-69
C_5H_9NS			
Isothiocyanic acid, butyl ester	hexane	249(3.50)	114-0309-69B
$C_5H_9N_3$			
Pyrazole, 3-(dimethylamino)-	MeOH	240(3.71)	33-2641-69
Pyrimidine, 4-amino-5,6-dihydro-2-methyl-	EtOH	264(3.98)	88-3553-69
	EtOH-HCl	275(--)	88-3553-69
$C_5H_9N_3O$			
2-Pyrazolin-5-one, 3-(dimethylamino)-	MeOH	247(3.88),279(3.76)	33-2641-69
$C_5H_9N_5$			
Propionamidine, N-4H-1,2,4-triazol-4-yl-	MeOH	218(3.62)	48-0477-69
$C_5H_9N_5O$			
Acetamidine, 2-methoxy-N-4H-1,2,4-triazol-4-yl-	MeOH	227(3.68)	48-0477-69
4(3H)-Pyrimidinone, 2,5-diamino-6-(methylamino)-	pH 7.5	216(4.23),289(4.05)	24-4032-69
$C_5H_{10}N_2O_2$			
Acetone, carbomethoxyhydrazone	MeOH	220(4.03)	78-0619-69
$C_5H_{10}N_2O_2S$			
β-Alanine, N-(thiocarbamoyl)-, methyl ester	EtOH	203(4.09),242(4.11)	39-2631-69C
$C_5H_{10}N_2O_4$			
L-Norvaline, 5-nitro-	2N HCl	268(1.46)	33-0388-69
$C_5H_{10}O$			
2-Butanone, 3-methyl-	gas	173(3.54),174(3.58), 175(3.56),192(3.11), 194(3.12),195(3.05), 197(3.15),198(3.03)	18-2453-69
	C_6H_{12}	284(1.30)	77-1246-69
Butyraldehyde, 2-methyl-	C_6H_{12}	295(1.38)	77-1246-69

Compound	Solvent	λ_{max}(log ϵ)	Ref.
2-Pentanone	gas	171(3.22),174(2.78), 188(2.97),192(3.26), 195(3.29)	18-2453-69
3-Pentanone	gas	172(2.95),175(2.92), 192(2.89),194(2.89), 196(2.89),197(2.89)	18-2453-69
$C_5H_{10}O_2$			
Pivalic acid	EtOH	213(1.85)	101-00P5-69A
sodium salt	EtOH	none	101-00P5-69A
Pivalic acid, Pb(IV) salt	hexane	236(4.33)	33-1495-69
	CH_2Cl_2	239(4.24)	33-1495-69
Valeric acid, Pb(IV) salt	C_6H_{12}	230(4.43)	33-1495-69
$C_5H_{10}O_2S_2$			
1,2-Dithiepane, 1,1-dioxide	EtOH	240(1.73)	44-0036-69
$C_5H_{10}S$			
2H-Thiopyran, tetrahydro-	hexane	202(3.43),212s(3.23)	59-0407-69
	iso-PrOH	199(3.12),210s(2.92)	59-0407-69
$C_5H_{11}NS$			
Acetimidic acid, thio-N-ethyl-, S-methyl ester	MeCN	265(4.08)	35-0737-69
$C_5H_{11}NS_2$			
1,2-Dithiepan-4-amine, hydrochloride, (S)-(-)-	H_2O	255(2.60)	87-0620-69
	EtOH	256(2.62)	87-0620-69
1,2-Dithiepan-5-amine, hydrochloride	H_2O	257(2.61)	87-0617-69
	EtOH	258(2.62)	87-0617-69
Dithiocarbamic acid, N,N-diethyl-, diethylamine salt	CH_2Cl_2	259(3.25),290(3.18)	39-1152-69A
$C_5H_{11}NS_3$			
Peroxycarbamic acid, dimethyltrithio-, ethyl ester	EtOH	281(3.86)	34-0119-69
$C_5H_{11}N_3O$			
Isobutyraldehyde, semicarbazone	EtOH	234(4.09)	94-0844-69
$C_5H_{11}N_3S$			
Butyraldehyde, thiosemicarbazone	EtOH	230(3.89),270(4.74)	94-0844-69
Isobutyraldehyde, thiosemicarbazone	EtOH	230(3.85),270(4.28)	94-0844-69
$C_5H_{12}N_2O$			
Methanol, (tert-butylazo)-	C_6H_{12}	340(1.27)	89-0451-69
$C_5H_{12}N_2S$			
Thioformic acid, 2,2-diethylhydrazide	H_2O	255(3.97)	24-2117-69
	2.23N HCl	259(3.96)	24-2117-69
	0.5N NaOH	239(3.96)	24-2117-69
	$CHCl_3$	260(3.98)	24-2117-69
tautomer	H_2O	275(4.11)	24-2117-69
	2.23N HCl	258(3.98)	24-2117-69
	0.5N NaOH	250(3.96)	24-2117-69
	$CHCl_3$	270(4.13)	24-2117-69
Thiourea, 1,1,3,3-tetramethyl-	ether	324(2.41)	18-2323-69
$C_5H_{12}N_4O_3$			
Guanidine, 1-(4-hydroxybutyl)-3-nitro-	MeOH	222(3.80),271(4.26)	78-5155-69

Compound	Solvent	$\lambda_{max}(\log \epsilon)$	Ref.
$C_5H_{12}N_4S$			
s-Tetrazine-3(2H)-thione, 6-ethyl-tetrahydro-6-methyl-	MeOH	249(4.12)	44-0756-69
$C_5H_{12}OSi$			
Silane, acetyltrimethyl-	heptane	195(3.62),371(2.10)	59-0765-69
$C_5H_{12}O_2Si$			
Silanecarboxylic acid, trimethyl-, methyl ester (also fine structure)	C_6H_{12}	245(2.52),250(2.52)	35-2154-69
	EtOH	245(2.50)	35-2154-69
$C_5H_{12}Si$			
Silane, trimethylvinyl-	heptane	179(4.28)	121-0803-69
	MeCN	179(4.19)	121-0803-69
	gas	176s(--),178(--)	121-0803-69
$C_5H_{13}BF_4N_2O$			
1,1-Diethyl-2-methoxydiazenium tetrafluoroborate	MeOH	220(3.57)	24-2093-69
$C_5H_{15}F_3Si_3$			
Trisilane, 1,1,3-trifluoro-1,2,2,3,3-pentamethyl-	isooctane	211(3.70)	35-6613-69
$C_5H_{18}Si_3$			
Trisilane, 1,1,2,2,3-pentamethyl-	isooctane	215(3.84)	35-6613-69

Compound	Solvent	λ_{max}(log ε)	Ref.
$C_6Cl_4O_2$			
p-Benzoquinone, tetrachloro-, barium iodide complex	MeOH	310(2.8),350(3.2)	65-1619-69
barium thiocyanate complex	MeOH	330(3.5),400(3.4)	65-1619-69
cadmium iodide complex	MeOH	320(2.6),400(2.6)	65-1619-69
cadmium thiocyanate complex	MeOH	340(3.0),400(2.9)	65-1619-69
C_6Cl_6O			
2,4-Cyclohexadien-1-one, hexachloro-	C_6H_{12}	265(3.60)	23-1943-69
2,5-Cyclohexadien-1-one, hexachloro-	C_6H_{12}	262(4.20)	23-1943-69
$C_6Cl_6O_2$			
3-Cyclohexene-1,2-dione, hexachloro-	$CHCl_3$	269(3.95)	44-1966-69
hydrate	$CHCl_3$	269(3.93)	44-1966-69
C_6Cl_8O			
2-Cyclohexen-1-one, octachloro-	C_6H_{12}	267(4.02)	23-1943-69
3-Cyclohexen-1-one, octachloro-	C_6H_{12}	236(3.95)	23-1943-69
3,5-Hexadienoyl chloride, 2,2,3,4,5,6,6-heptachloro-	heptane	224(4.2)	5-0001-69H
$C_6Cl_9N_3$			
s-Triazine, 2,4,6-tris(trichloromethyl)-	MeOH	219(3.41),287(2.90)	18-2924-69
C_6F_6			
Benzene, hexafluoro-	EtOH	231(3.13)	25-0695-69
C_6F_7N			
2-Picoline, heptafluoro-	hexane	256(3.37)	39-2559-69C
$C_6F_{10}O$			
Cyclohexanone, decafluoro-	n.s.g.	235(1.95),330f(1.26)	39-0201-69B
C_6HCl_7O			
3-Cyclohexen-1-one, 2,2,3,4,5,6,6-hepta-chloro-	MeOH	215(3.92),230(3.86)	23-1943-69
$C_6H_2Cl_4$			
1,5-Hexadien-3-yne, 1,2,5,6-tetra-chloro-	hexane	214(3.95),235(3.89), 243(3.90),282(4.12), 296(4.14)	78-2223-69
$C_6H_2Cl_6$			
1,3,5-Hexatriene, 1,2,3,4,5,6-hexa-chloro-	hexane	214(4.35),243(3.79)	78-0883-69
$C_6H_2Cl_6O$			
3-Cyclohexen-1-one, 2,2,4,5,6,6-hexa-chloro-	MeOH	215(3.91)	23-1943-69
$C_6H_3BrClNS_2$			
6-Bromo-1,2,3-benzodithiazol-2-ium chloride	H_2SO_4	250(4.05),370(4.11), 422(3.69)	104-0151-69
$C_6H_3BrClN_3$			
7H-Pyrrolo[2,3-d]pyrimidine, 5-bromo-4-chloro-	pH 1	229(4.48),268(3.51), 300(3.45)	4-0207-69
	pH 11	237(4.47),278(3.66)	4-0207-69

Compound	Solvent	λ_{max}(log ϵ)	Ref.
$C_6H_3ClN_2O_2$			
Benzofurazan, 5-chloro-, N-oxide	H_2O	300s(3.39),314(3.56), 329(3.68),356(3.83)	78-4197-69
$C_6H_3ClN_4O_2$			
7H-Pyrrolo[2,3-d]pyrimidine, 4-chloro-5-nitro-	pH 1	334(4.05)	4-0207-69
	pH 11	243s(4.03),328(3.91), 380s(3.83)	4-0207-69
$C_6H_3Cl_2NS_2$			
6-Chloro-1,2,3-benzodithiazol-2-ium chloride	H_2SO_4	242(3.88),365(4.13), 423(3.60)	104-0151-69
$C_6H_3Cl_2N_3$			
1H-Imidazo[4,5-b]pyridine, 5,7-dichloro-	pH 1	244(3.86),282(4.10)	54-1263-69
	pH 13	296(4.09)	54-1263-69
$C_6H_3Cl_3N_2O_2S$			
Barbituric acid, 2-thio-5-(2,2,2-trichloroethylidene)-	EtOH	290(4.10)	103-0830-69
$C_6H_3Cl_6N_3$			
s-Triazine, 2-methyl-4,6-bis(trichloromethyl)-	MeOH	219(3.28),280(2.98)	18-2924-69
$C_6H_3Cl_6N_3S$			
s-Triazine, 2-(methylthio)-4,6-bis-(trichloromethyl)-	MeOH	217(3.52),272(4.18)	18-2931-69
$C_6H_3FN_2O_4$			
Benzene, 1-fluoro-2,4-dinitro-	MeOH	217(4.08),290(3.84)	44-2038-69
$C_6H_3N_3O_7$			
Picric acid	dioxan	249(4.93),338(4.29), 418(2.84)	65-0879-69
azetidine complex	dioxan	240(4.13),346(4.26), 406(4.03)	65-0879-69
diethylamine complex	dioxan	260(4.69),350(4.53), 403(4.00)	65-0879-69
dimethylamine complex	dioxan	240(4.07),344(4.22), 410(3.93)	65-0879-69
morpholine complex	dioxan	240(4.09),344(4.20), 406(3.91)	65-0879-69
piperazine complex	dioxan	256(3.63),346(3.90), 405(3.6)	65-0879-69
piperidine complex	dioxan	240(4.08),344(4.22), 410(3.91)	65-0879-69
$C_6H_4BrNO_2$			
p-Benzoquinone, 2-bromo-, 4-oxime	EtOH	237(3.57),316(4.26)	12-0935-69
	EtOH-NaOH	277(3.81),418(4.50)	12-0935-69
	dioxan	730(0.26)	12-0935-69
$C_6H_4BrN_3O$			
4H-Pyrrolo[2,3-d]pyrimidin-4-one, 5-bromo-3,7-dihydro-	pH 1	267(3.89)	4-0207-69
	pH 11	269(3.89)	4-0207-69
$C_6H_4BrN_3S$			
4H-Pyrrolo[2,3-d]pyrimidine-4-thione, 5-bromo-3,7-dihydro-	pH 1	224(4.22),278(3.88), 327(4.09)	4-0207-69

Compound	Solvent	λ max (log ε)	Ref.
4H-Pyrrolo[2,3-d]pyrimidine-4-thione, 5-bromo-3,7-dihydro- (cont.)	pH 11	228(4.27),300s(3.96), 315(4.01)	4-0207-69
$C_6H_4Br_2N_2OS$			
1-Bromothiazolo[3,2-a]pyrazinium bromide, 2-oxide	n.s.g.	197(4.35),235(4.02), 255s(3.86),276(3.99), 325s(3.96),344s(4.13), 359(4.18)	39-2270-69C
$C_6H_4Br_2N_2O_2$			
Aniline, 2,5-dibromo-4-nitro-	EtOH	217(4.25),268(3.80), 359(3.98)	117-0287-69
$C_6H_4Br_2N_2S$			
1-Bromothiazolo[3,2-c]pyrazinium bromide	n.s.g.	198(4.15),241(4.14), 272s(3.22),330s(4.09), 343(4.14)	39-2270-69C
$C_6H_4Br_2O_2S_2$			
1H,3H-Thieno[3,4-c]thiophene, 4,6-dibromo-, 2,2-dioxide	EtOH	250(3.96),263s(3.91)	44-0333-69
$C_6H_4Br_2S_2$			
1H,3H-Thieno[3,4-c]thiophene, 4,6-dibromo-	EtOH	253(3.97)	44-0333-69
$C_6H_4Br_4S$			
Thiophene, 2,5-dibromo-3,4-bis(bromomethyl)-	EtOH	231(4.29),270s(3.70)	54-0321-69
$C_6H_4ClNO_2$			
p-Benzoquinone, 2-chloro-, 4-oxime	EtOH	234(3.57),313(4.26)	12-0935-69
	EtOH-NaOH	274(3.98),418(4.49)	12-0935-69
	dioxan	731(0.26)	12-0935-69
$C_6H_4ClNS_2$			
1,2,3-Benzodithiazol-2-ium chloride	H_2SO_4	238(3.91),350(4.35), 425(3.29)	104-0151-69
$C_6H_4ClN_3$			
1H-Benzotriazole, 1-chloro-	n.s.g.	252(3.72),257(3.71), 275(3.60)	39-1474-69C
1H-Imidazo[4,5-b]pyridine, 7-chloro-	pH 1	273(3.97),280(4.01)	54-1263-69
	pH 13	287(4.11)	54-1263-69
1H-Imidazo[4,5-c]pyridine, 4-chloro-	pH 1	208(4.47),269(3.83)	54-1263-69
	pH 13	278(3.83)	54-1263-69
$C_6H_4Cl_2I_2S$			
Thiophene, 2,5-dichloro-3,4-bis(iodomethyl)-	EtOH	242(4.33)	44-0333-69
$C_6H_4Cl_2OS$			
1H,3H-Thieno[3,4-c]furan, 4,6-dichloro-	EtOH	247(4.00)	44-0340-69
$C_6H_4FNO_2$			
p-Benzoquinone, 2-fluoro-, 4-oxime	EtOH	227(3.34),309(4.28)	12-0935-69
	EtOH-NaOH	257(3.66),410(4.53)	12-0935-69
	dioxan	727(-0.09)	12-0935-69

Compound	Solvent	λ_{max}(log ϵ)	Ref.
$C_6H_4FN_3O_4$			
Aniline, 2-fluoro-3,5-dinitro-	EtOH	375(3.33)	44-0395-69
$C_6H_4F_3NO$			
Pyridine, 2-(trifluoromethyl)-, 1-oxide	EtOH	271(--)	94-0510-69
	dioxan	279(--)	94-0510-69
Pyridine, 3-(trifluoromethyl)-, 1-oxide	EtOH	272(4.08)	94-0510-69
	dioxan	284(4.11)	94-0510-69
Pyridine, 4-(trifluoromethyl)-, 1-oxide	EtOH	277(4.14)	94-0510-69
	dioxan	287(4.18)	94-0510-69
$C_6H_4F_3N_5O$			
Guanine, 8-(trifluoromethyl)-	pH -0.89	241(3.84),264(4.03)	5-0201-69F
	pH 4.0	254(4.03),270s(3.95)	5-0201-69F
	pH 9.0	255(3.92),279(3.97)	5-0201-69F
	pH 14.0	279(3.96)	5-0201-69F
$C_6H_4F_4N_2$			
Hydrazine, (2,3,4,5-tetrafluorophenyl)-	EtOH	231(3.79),284(3.08)	70-0609-69
$C_6H_4INO_2$			
p-Benzoquinone, 2-iodo-, 4-oxime	EtOH	240(3.64),309(4.18)	12-0935-69
	EtOH-NaOH	278(3.88),416(4.49)	12-0935-69
	dioxan	726(0.11)	12-0935-69
$C_6H_4N_2$			
Isonicotinonitrile	C_6H_{12}	271(3.44)	4-0859-69
	pH 1	276(3.72)	4-0859-69
	MeOH	272(3.46)	4-0849-69
Mucon준nitrile, cis-cis	MeOH	252s(--),259(4.36), 269(4.22)	39-0742-69C
Mucononitrile, cis-trans	MeOH	252s(--),259(4.60), 269(4.22)	39-0742-69C
Nicotinonitrile	isooctane	264(3.32)	4-0859-69
	pH 1	264(3.67)	4-0859-69
	MeOH	265(3.39)	4-0859-69
Picolinonitrile	isooctane	265(3.69)	4-0859-69
	pH 1	266(3.70)	4-0859-69
	MeOH	265(3.55)	4-0859-69
$C_6H_4N_2O$			
Isonicotinonitrile, 1-oxide	C_6H_{12}	300(4.03)	4-0859-69
	pH 1	281(4.25)	4-0859-69
	MeOH	291(4.29)	4-0859-69
Nicotinonitrile, 1-oxide	isooctane	286(4.33)	4-0859-69
	pH 1	264(3.27)	4-0859-69
	MeOH	273(4.07)	4-0859-69
Picolinonitrile, 1-oxide	C_6H_{12}	287(3.96)	4-0859-69
	pH 1	268(3.97)	4-0859-69
	MeOH	275(3.99)	4-0859-69
$C_6H_4N_2OS$			
Furo[2,3-d]pyridazine-4(5H)-thione	H_2O	208(4.46),307(4.24)	22-4004-69
Furo[2,3-d]pyridazine-7(6H)-thione	H_2O	202(4.10),257(4.17), 333(3.89)	22-4004-69
$C_6H_4N_2O_2$			
1,3,4-Oxadiazole, 2-(2-furyl)-	MeOH	265(4.3)	48-0646-69

Compound	Solvent	$\lambda_{max}(\log \epsilon)$	Ref.
$C_6H_4N_2O_4$			
Benzene, m-dinitro-	EtOH	235(4.23)	65-0660-69
$C_6H_4N_2O_5$			
Phenol, 2,4-dinitro-	EtOH	253(4.01),292(3.97)	65-0660-69
$C_6H_4N_2S$			
2,1,3-Benzothia(S^{IV})diazole	n.s.g.	224(4.25),305(4.14), 311(4.14)	103-0180-69
$C_6H_4N_2S_2$			
p-Dithiin-2,3-dicarbonitrile, 5,6-dihydro-	MeOH	257(3.79),356(4.00)	33-2221-69
$C_6H_4N_4O$			
4(3H)-Pteridinone	MeOH	233(3.93),273s(3.51), 314(3.75)	39-2415-69C
Pyrazino[2,3-d]pyridazin-5(6H)-one	pH 13	213(4.27),256(4.21), 312(3.63),342(3.63)	4-0093-69
	EtOH	204(4.54),249(4.03), 307(3.59)	4-0093-69
$C_6H_4N_4O_3$			
4H-Pyrrolo[2,3-d]pyrimidin-4-one, 3,7-dihydro-5-nitro-	pH 1	227(4.05),355(4.13)	4-0207-69
	pH 11	235s(4.00),394(4.05)	4-0207-69
$C_6H_4N_4O_4$			
1,2,4-Oxadiazole, 5-amino-3-(5-nitro-2-furyl)-	EtOH	254(3.89),285(3.97), 317(4.06)	18-3335-69
$C_6H_4N_4O_6$			
Aniline, 2,4,6-trinitro-	MeOH	318(4.08),407(3.90)	35-4155-69
C_6H_4O			
2,4-Hexadiynal	CH_2Cl_2	245(3.40),257(3.78), 272(3.98),288(3.93)	39-0683-69C
$C_6H_4O_2$			
p-Benzoquinone	pH 8	246(4.30),255s(4.23)	39-0207-69B
$C_6H_4O_3S_2$			
p-Dithiin-2,3-dicarboxylic anhydride, 5,6-dihydro-	MeOH	224(3.84),318(3.82)	33-2221-69
$C_6H_5BiBr_2$			
Bismuthine, dibromophenyl-	CH_2Cl_2	341(3.45)	101-0099-69E
$C_6H_5BrN_2$			
Diimide, (o-bromophenyl)-	MeCN	215s(--),265(3.70)	35-2325-69
Diimide, (p-bromophenyl)-	MeCN	217(4.00),272(4.01), 415(1.99)	35-2325-69
$C_6H_5BrN_2OS$			
Thiazolo[3,2-a]pyrazin-4-ium bromide, 2-oxide	n.s.g.	195(4.35),224(4.08), 246s(3.8),268(3.88), 349(4.34)	39-2270-69C
$C_6H_5BrN_2S$			
Thiazolo[3,2-a]pyrazin-4-ium bromide (continued on next page)	n.s.g.	196(4.27),233(3.99), 270s(3.44),315(4.04),	39-2270-69C

Compound	Solvent	λ_{max}(log ϵ)	Ref.
		326(4.06),350s(3.29)	
$C_6H_5BrN_4$			
7H-Pyrrolo[2,3-d]pyrimidine, 4-amino-5-bromo-	pH 1	231(4.31),280(3.94)	4-0207-69
	pH 11	225s(4.09),276(3.94)	4-0207-69
$C_6H_5BrN_4O$			
7H-Pyrrolo[2,3-d]pyrimidine, 5-bromo-4-(hydroxyamino)-	pH 1	237(4.15),282(4.02)	4-0207-69
	pH 11	232(3.89),271(3.99), 280s(3.97)	4-0207-69
$C_6H_5BrO_2$			
Ketone, 3-bromo-2-furyl methyl	n.s.g.	265(4.32)	103-0012-69
$C_6H_5BrO_2S_2$			
Thieno[3,4-b]thiophene, 2-bromo-4,6-dihydro-, 5,5-dioxide	EtOH	241(3.86)	54-0321-69
$C_6H_5Br_2NO$			
Pyridine, 3,5-dibromo-4-methoxy-	pH -0.7	221(4.39),261s(3.58), 267(3.59),274(3.59), 284s(3.44)	12-2581-69
	pH 7	227s(3.89),271(3.24), 278(3.17)	12-2581-69
3-Pyridinol, 2,4-dibromo-6-methyl-	pH 1	290(3.78)	1-1704-69
	pH 13	240(3.80),315(3.86)	1-1704-69
4(1H)-Pyridone, 3,5-dibromo-1-methyl-	pH -3.5	223(4.46),265s(3.70), 273s(3.52)	12-2581-69
	pH 6	221(4.34),278(4.08), 289s(3.59)	12-2581-69
$C_6H_5ClNO_2$			
Benzene, 1-chloro-4-nitro-, half-reduction product radical	pH 1	285(4.25),460(2.95)	64-1336-69
	pH 13	295(4.33),420(2.71)	64-1336-69
$C_6H_5ClN_2O_4S$			
Benzenesulfonamide, 4-chloro-3-nitro-	pH 7.6	224(4.56),263(4.83)	35-4912-69
$C_6H_5ClN_4$			
1H-Pyrazolo[3,4-d]pyrimidine, 4-chloro-1-methyl-	MeOH	258(3.71)	23-1129-69
C_6H_5ClO			
Phenol, p-chloro-	pH 8	280(4.14)	69-1042-69
$C_6H_5Cl_2I$			
Benzene, (dichloroiodo)-	MeOH	262(2.81)	120-0012-69
$C_6H_5Cl_3N_2O$			
4(3H)-Pyrimidinone, 6-methyl-2-(trichloromethyl)-	EtOH	229(3.73),255(3.53)	95-0460-69
$C_6H_5Cl_3OS$			
3-Thiophenemethanol, 2,5-dichloro-4-(chloromethyl)-	EtOH	251(3.77)	44-0340-69
C_6H_5I			
Benzene, iodo-	MeOH	227(4.00),252(2.77), 257(2.76)	120-0012-69

Compound	Solvent	$\lambda_{max}(\log \epsilon)$	Ref.
C_6H_5IO			
Benzene, iodoso-	MeOH	262(3.15)	120-0012-69
$C_6H_5IO_2$			
Benzene, iodoxy-	MeOH	264(3.24)	120-0012-69
C_6H_5NO			
2,5-Cyclohexadien-1-one, 4-imino-	pH 8	254(4.38),260(4.39)	39-0207-69B
$C_6H_5NO_2$			
Benzene, nitro-	C_6H_{12}	252(3.92),278(3.32), 288(3.09),298(2.71)	59-0989-69
	THF	259(3.89)	59-0989-69
p-Benzoquinone, monoxime	EtOH	236(3.42),302(4.23)	12-0935-69
	EtOH-NaOH	267(3.72),406(4.50)	12-0935-69
	dioxan	735(0.90)	12-0935-69
$C_6H_5NO_2$			
Pyrrole-2,4-dicarboxaldehyde	EtOH	235(4.26),290(4.17)	78-3879-69
$C_6H_5NO_2S_2$			
p-Dithiin-2,3-dicarboximide, 5,6-dihydro-	EtOH	249(4.01),403(3.51)	33-2221-69
$C_6H_5NO_3$			
Phenol, p-nitro-	pH 8	400(4.24)	69-1042-69
	EtOH	225(3.85),312(4.02)	65-0660-69
$C_6H_5NO_4$			
Catechol, 3-nitro-	pH 2	215(4.13),238(3.71), 298(3.79),370(3.20)	61-0203-69
	anion	193(4.38),224(4.00), 235(4.05),301(3.79), 393(3.32)	61-0203-69
	0.15N NaOH	201(4.48),250(4.06), 358(3.83),512(4.06)	61-0203-69
Catechol, 4-nitro-	pH 2	209(4.07),242(3.84), 309(3.72),345(3.80)	61-0203-69
	anion	193(4.24),217(3.95), 254(3.93),333(3.65), 385(3.89)	61-0203-69
	0.15N NaOH	205(4.11),273(3.81), 385(3.82),509(4.06)	61-0203-69
Hydroquinone, nitro-	pH 2	215(4.15),247(3.75), 284(3.76),395(3.45)	61-0203-69
	anion	227(4.23),289(3.67), 459(3.60)	61-0203-69
	0.15N NaOH	238(4.23),292(3.89), 534(3.59)	61-0203-69
Resorcinol, 2-nitro-	pH 2	190(4.59),212(4.06), 313(3.68),370(2.98)	61-0203-69
	anion	196(4.40),224(4.00), 327(3.62),442(3.18)	61-0203-69
	0.15N NaOH	207(4.62),235(3.92), 357(3.15),394(3.26)	61-0203-69
Resorcinol, 4-nitro-	pH 2	190(4.52),211(4.03), 229(3.70),347(3.95)	61-0203-69
	anion	199(4.29),216(4.17), 236(3.73),400(4.33)	61-0203-69

Compound	Solvent	λ_{max}(log ε)	Ref.
Resorcinol, 4-nitro- (cont.)	0.15N NaOH	211(4.44),228(4.08), 242(3.85),394(4.31)	61-0203-69
Resorcinol, 5-nitro-	pH 2	190(4.42),219(4.08), 294(3.48),333(3.15)	61-0203-69
	anion	198(4.60),229(4.13), 317(3.45),384(3.06)	61-0203-69
	0.15N NaOH	210(4.49),250(4.05), 335(3.52),442(3.20)	61-0203-69
$C_6H_5N_2O_4$			
Benzene, m-dinitro-, half-reduction product radical	pH 1	280(4.14),390(2.72)	64-1336-69
	pH 13	285(4.23),410(2.78), 650(2.32)	64-1336-69
Benzene, o-dinitro-, half-reduction product radical	pH 1	285(4.17),390(2.96), 550(2.26)	64-1336-69
	pH 13	295(4.23),420(3.03), 650(2.68)	64-1336-69
$C_6H_5N_3O_2$			
Diimide, (p-nitrophenyl)-	MeCN	274(4.09)	35-2325-69
$C_6H_5N_3O_4$			
Aniline, 2,4-dinitro-	MeOH	336(4.16),390s(3.81)	35-4155-69
	EtOH	257(3.99),336(4.17)	65-0660-69
$C_6H_5N_3O_4S$			
3H-Pyrrolo[2,3-d]pyrimidine-5-sulfonic acid, 4,7-dihydro-4-oxo-	pH 1	256(4.02)	4-0207-69
	pH 11	258(4.01)	4-0207-69
$C_6H_5N_5O_2$			
Formamide, N-(1,6-dihydro-6-oxopurin-2-yl)-	pH 1	258(--)	69-0238-69
	pH 7	261(4.18)	69-0238-69
$C_6H_5N_5O_6$			
m-Phenylenediamine, 2,4,6-trinitro-	MeOH	276(4.17),326(4.23), 411(4.17)	35-4155-69
$C_6H_5N_5S$			
Pyrimido[5,4-e]-as-triazine, 5-(methylthio)-	pH 7	238(4.05),262(3.73), 288(3.86)	44-3161-69
$C_6H_5N_7$			
Bis-s-triazolo[4,3-b:4',3'-d]-as-triazine, 6-methyl-	EtOH	259(3.86)	22-3670-69
C_6H_6			
Benzene	MeOH	203(3.87)	87-0146-69
	EtOH	203(3.90),255(2.36)	28-1562-69B
tetracyanoethylene complex	C_6H_{12}	380(3.20)	39-1161-69B
1,3-Hexadien-5-yne, cis	EtOH	253(4.35)	78-2837-69
1,3-Hexadien-5-yne, trans	EtOH	253(4.39),264s(4.43)	78-2837-69
$C_6H_6BBrO_2$			
Benzeneboronic acid, p-bromo-	H_2O	196(4.54),232(4.16), 266(2.60)	39-0392-69A
C_6H_6BrNO			
3-Pyridinol, 2-bromo-6-methyl-	pH 1	235(3.78),300(3.97)	1-1704-69
	pH 13	245(4.05),310(3.77)	1-2065-69

Compound	Solvent	λ_{max}(log ε)	Ref.
$C_6H_6BrN_5$			
Adenine, 8-bromo-2-methyl-	EtOH	288(4.18)	65-2125-69
$C_6H_6Br_2N_2$			
p-Phenylenediamine, 2,5-dibromo-	C_6H_{12}	217(4.32),243s(--), 332(3.62)	117-0287-69
$C_6H_6ClO_4P$			
Phosphoric acid, mono(p-chlorophenyl) ester	pH 8	276(3.88)	69-1042-69
$C_6H_6Cl_2N_2O$			
4(3H)-Pyrimidinone, 2-(dichloromethyl)-6-methyl-	EtOH	230(3.85),280(3.60)	95-0460-69
$C_6H_6Cl_2O_2S$			
3,4-Thiophenedimethanol, 2,5-dichloro-	EtOH	248(3.83)	44-0333-69 +44-0340-69
$C_6H_6Cl_2S$			
Thiophene, 3,4-bis(chloromethyl)-	EtOH	230s(3.72)	44-0333-69
$C_6H_6Cl_3N_3$			
s-Triazine, 2,4-dimethyl-6-(trichloromethyl)-	MeOH	234(3.31),260(2.95)	18-2924-69
$C_6H_6F_3N_5O_2$			
Acetamide, N-(2,4-diamino-1,6-dihydro-6-oxo-5-pyrimidinyl)-2,2,2-trifluoro-	pH 0.0	263(4.32)	5-0201-69F
	pH 6.0	265(4.24)	5-0201-69F
	pH 11.0	262(3.96),282s(3.62)	5-0201-69F
C_6H_6INO			
3-Pyridinol, 2-iodo-6-methyl-	pH 1	225s(--),315(4.10)	1-1704-69
	pH 13	245(4.01),320(3.87)	1-1704-69
$C_6H_6NO_6P$			
Phosphoric acid, mono(p-nitrophenyl) ester	pH 8	308(4.01),326(3.92)	69-1042-69
$C_6H_6N_2$			
2,5-Cyclohexadiene-1,4-diimine	pH 8	257(4.43),266(4.42)	39-0207-69B
Nicotinonitrile, 1,4-dihydro-	EtOH	330(3.75)	88-3101-69
$C_6H_6N_2O$			
Nicotinamide	0.4N HCl	262(3.75)	18-2282-69
$C_6H_6N_2OS_3$			
4-Isothiazolecarbonitrile, 3-(methylsulfinyl)-5-(methylthio)-	MeOH	227(4.12),293(4.02)	88-4265-69
4-Isothiazolecarbonitrile, 5-(methylsulfinyl)-3-(methylthio)-	MeOH	232(4.12),283(3.62), 312(3.58)	88-4265-69
$C_6H_6N_2O_2$			
Aniline, o-nitro-	1%MeOH-NaOH	415(3.65)	39-0922-69B
Aniline, p-nitro-	1%MeOH-NaOH	380(4.11)	39-0922-69B
Isonicotinaldehyde, oxime, 1-oxide, anti	H_2O	218(--),290(4.35)	83-0494-69
anion	H_2O	325(4.27)	83-0494-69
Isonicotinaldehyde, oxime, 1-oxide, syn	H_2O	216(--),290(4.35)	83-0494-69
anion	H_2O	321(4.38)	83-0494-69
Picolinaldehyde, oxime, 1-oxide, anti	H_2O	231(4.30),269(4.12)	83-0494-69

Compound	Solvent	λ_{max}(log ε)	Ref.
Picolinaldehyde, oxime, 1-oxide, anti, anion	H_2O	261(4.07),307(4.00)	83-0494-69
Picolinaldehyde, oxime, 1-oxide, syn	H_2O	238(4.40),275(4.16)	83-0494-69
anion	H_2O	259(4.23),305(4.19)	83-0494-69
$C_6H_6N_2O_2S_2$			
p-Dithiino[2,3-d]pyridazine-5,8-dione, 2,3,6,7-tetrahydro-	EtOH	233(4.21),327(3.94)	33-2236-69
$C_6H_6N_2O_3$			
Barbituric acid, 5-ethylidene-	EtOH	260(3.88)	103-0827-69
Phenol, 3-amino-4-nitro-	1%MeOH-HOAc	397(3.77)	39-0922-69B
5-Pyrimidinecarboxaldehyde, 1,2,3,4-tetrahydro-6-methyl-2,4-dioxo-	H_2O	233(3.97),284(4.10)	24-2877-69
	pH 13	251(3.90),307(4.31)	24-2877-69
$C_6H_6N_2O_3S$			
Acetic acid, thio-, S-ester with 5-thioisobarbituric acid	EtOH	270(3.92)	44-3806-69
$C_6H_6N_2O_3S_2$			
Acetic acid, mercapto-, 4-oxo-2-thiazoline-2-carboximidate	EtOH	314(3.88),365(1.00)	44-2053-69
$C_6H_6N_2O_5S$			
Benzenesulfonamide, 4-hydroxy-3-nitro-	pH 7.6	395(3.62)	35-4912-69
Benzenesulfonic acid, 2-amino-5-nitro-	anion	368(4.15)	22-4569-69
	dianion	460(4.47)	22-4569-69
Benzenesulfonic acid, 4-amino-3-nitro-	anion	405(3.65)	22-4569-69
	dianion	485(3.89)	22-4569-69
$C_6H_6N_2S$			
Thioisonicotinamide	EtOH	290(3.72),306(3.71)	39-0861-69C
$C_6H_6N_2S_3$			
4-Isothiazolecarbonitrile, 3,5-bis-(methylthio)-	MeOH	213(4.06),230(4.09), 284(4.16)	88-4265-69
$C_6H_6N_4$			
Benzotriazole, 1-amino-	EtOH	263(3.76),280s(3.65)	39-0742-69C
Benzotriazole, 2-amino-	EtOH	216(4.12),285(4.08)	39-0742-69C
1H-Imidazo[4,5-c]pyridine, 6-amino-	pH 1	220(4.66),252(3.70), 324(3.62)	54-1263-69
	pH 13	220(4.50),255(3.43), 303(3.58)	54-1263-69C
4H-Pyridazino[6,1-c]-as-triazine	pH 1	274(3.66),325(3.30)	7-0552-69
	pH 13	313(3.72),380s(--)	7-0552-69
$C_6H_6N_4O$			
Furo[2,3-d]pyridazine, 4-hydrazino-	H_2O	198(4.32),257(4.08)	22-4004-69
$C_6H_6N_4OS$			
Purine, 6-(methylthio)-, 3-oxide	pH -1	236(3.91),317(4.32)	44-2153-69
	pH 3	236(3.86),254(3.91), 319(4.28)	44-2153-69
	pH 12	214(4.20),245(4.21), 312(4.23),322s(4.19)	44-2153-69
$C_6H_6N_4O_2$			
Purine, 6-methoxy-, 3-oxide	pH 0	263(3.84)	44-2153-69
	pH 1	268(4.06)	18-0750-69

Compound	Solvent	λmax(log ε)	Ref.
Purine, 6-methoxy-, 3-oxide (cont.)	pH 4	224(4.38),283(4.00)	44-2153-69
	pH 6	226(4.40),285(4.01)	18-0750-69
	pH 9	227(4.45),276(3.96)	44-2153-69
	pH 13	228(4.50),282(3.97)	18-0750-69
1H-Pyrazolo[4,3-d]pyrimidine-3-methanol, 7-hydroxy-	pH 1	222(3.72),277(3.44)	88-0289-69
	0.05N NaOH	228(3.70),291(3.43)	88-0289-69
$C_6H_6N_4O_3$			
1H-Pyrazolo[3,4-d]pyridazine-3-methanol, 4,7-dihydroxy-	pH 1	263(3.74)	88-0289-69
	0.05N NaOH	217(4.34),274(3.79)	88-0289-69
1H-Pyrazolo[4,3-d]pyrimidine-3-methanol, 5,7-dihydroxy-	pH 1	207(4.29),287(3.72)	88-0289-69
	0.05N NaOH	223(4.35),307(3.66)	88-0289-69
$C_6H_6N_4O_3S$			
Purine, 6-(methylsulfonyl)-, 3-oxide	pH 1	234(4.04),300(3.49), 326(3.66)	18-0750-69
	pH 3	232(4.56),299s(4.15), 322(4.36)	44-2153-69
	pH 7	234(4.21),305(3.71), 330(3.81)	18-0750-69
	pH 9	233(4.63),258s(3.95), 302(4.11),332(4.18)	44-2153-69
	pH 13	228(4.28),290(3.85), 336(2.48)	18-0750-69
2H-Pyridazino[4,5-e]-1,2,4-thiadiazin-8(7H)-one, 7-methyl-, 1,1-dioxide	MeOH	227(4.01),292(3.78)	4-0407-69
5H-Pyridazino[4,5-e]-1,2,4-thiadiazin-5-one, 2,6-dihydro-6-methyl-, 1,1-dioxide	MeOH	240(3.78),320(3.88)	4-0407-69
$C_6H_6N_4O_4$			
m-Phenylenediamine, 4,6-dinitro-	MeOH	321(4.47),397(4.01)	35-4155-69
$C_6H_6N_4S$			
1H-Imidazo[4,5-d]pyridazine, 4-(methylthio)-	EtOH	222(4.23),272(3.98)	4-0093-69
9H-Purine-6-thiol, 9-methyl-	pH 1	221(4.08),325(4.34)	73-2114-69
	pH 13	234(4.19),314(4.37)	73-2114-69
$C_6H_6N_8$			
Pyrimido[5,4-e]-as-triazin-5-ylguanidine	pH 7	246(3.16),400(3.88)	44-3161-69
C_6H_6O			
Phenol	C_6H_{12}	272(3.30),278(3.28)	18-1831-69
	H_2O	270(3.16)	28-2217-69A
	pH 2	270(3.17)	30-1113-69A
	pH 8	270(3.23)	69-1042-69
	pH 13	287(3.41)	28-2217-69A
	pH 13	236(3.97),288(3.42)	30-0867-69A
	iso-PrOH	273(3.27),278s(3.19)	30-0867-69A
oxidation product after coupling with 3-methyl-2-benzothiazoline hydrazone	H_2O	520(4.4+)	3-1750-69
C_6H_6OS			
1H,3H-Thieno[3,4-c]furan	EtOH	244(3.82)	44-0340-69
$C_6H_6O_2$			
Catechol	pH 2	194(4.68),212(3.80), 275(3.32)	61-0203-69
	anion	202(4.48),224(3.66), 283(3.60)	61-0203-69

Compound	Solvent	$\lambda_{max}(\log \epsilon)$	Ref.
Catechol (cont.)	0.15N NaOH	203(4.51),235(3.84), 290(3.57)	61-0203-69
2-Cyclohexene-1,4-dione	hexane	368(1.75)	39-0124-69C
2-Furaldehyde, 5-methyl-	30% EtOH	228(--),292(4.23)	103-0278-69
Hydroquinone	pH 2	192(4.54),221(3.72), 287(3.43)	61-0203-69
	anion	194(4.56),231(3.89), 305(3.56)	61-0203-69
	0.15N NaOH	196(4.65),238(3.94), 320(3.52)	61-0203-69
Resorcinol	pH 2	195(4.63),216(3.83), 273(3.28)	61-0203-69
	anion	200(4.51),235(3.87), 286(3.45)	61-0203-69
	0.15N NaOH	211(4.59),240(3.85), 290(3.51)	61-0203-69
$C_6H_6O_2S$			
2-Thiophenecarboxylic acid, 3-methyl-	EtOH	254(4.04)	2-0009-69
$C_6H_6O_2S_2$			
Thieno[3,4-b]thiophene, 4,6-dihydro-, 5,5-dioxide	EtOH	234(3.83)	54-0321-69
1H,3H-Thieno[3,4-c]thiophene, 2,2-dioxide	EtOH	244(3.81)	44-0333-69
$C_6H_6O_3$			
1,2,4-Benzenetriol	pH 2	194(4.57),219(3.81), 288(3.45)	61-0203-69
	anion	202(4.33),228(3.77), 304(3.55)	61-0203-69
	0.15N NaOH	207(4.32),239(3.81), 321(3.55)	61-0203-69
Cyclopentadienone, x-hydroxy-y-(hydroxymethyl)-	EtOH	226(3.46),282(4.04)	94-2370-69
Glutaconic anhydride, β-methyl-	MeOH	222(4.03)	77-0685-69
	MeOH-NaOH	236(3.88),342(4.19)	77-0685-69
2,4-Pentadienoic acid, 5-formyl-	EtOH	267(4.46)	104-0759-69
Phloroglucinol	acid	202(4.69),220(3.90), 267(2.75)	61-0203-69
anion	n.s.g.	203(4.42),224(3.97), 280(3.70)	61-0203-69
dianion	n.s.g.	225(4.27),252(4.31), 348(3.95)	61-0203-69
trianion	n.s.g.	225(4.32),252(4.35), 349(3.88)	61-0203-69
2H-Pyran-3-carboxaldehyde, 4-hydroxy-	MeOH	280(3.71)	69-4172-69
in 10% MeOH	HCl	276(3.75)	69-4172-69
in 10% MeOH	NaOH	284(3.67)	69-4172-69
Pyrogallol	pH 2	198(4.64),217(3.92), 266(2.86)	61-0203-69
	anion	206(4.52),231(3.81), 275(3.16)	61-0203-69
	0,15N NaOH	215(4.60),235(3.97), 281(3.07)	61-0203-69
$C_6H_6O_4$			
2(5H)-Furanone, 3-acetyl-4-hydroxy-	EtOH	211(3.68),232(3.93), 265(4.08)	39-0056-69C

Compound	Solvent	λ_{max}(log ε)	Ref.
$C_6H_6O_6$			
Dehydroascorbic acid	0.4N HCl	302(2.11)	18-2282-69
Isocitric lactone, (-)-	MeOH	214(2.15)	1-0286-69
$C_6H_6O_7$			
Garcinia acid	MeOH	216(2.22)	1-0286-69
Hibiscus acid	MeOH	216(2.20)	1-0286-69
C_6H_6S			
Benzenethiol	hexane	272(2.82),280(2.80), 289(2.62)	39-1305-69A
at 77°K	isopentane-$C_6H_{11}Me$	266(2.78),272(2.83), 278(2.86),280(2.89), 288(2.81)	39-1305-69A
$C_6H_6S_2$			
p-Dithiin, 2-vinyl-	C_6H_{12}	272(3.62),281(3.77), 291(3.70),344(3.52)	97-0184-69
$C_6H_6S_3$			
2-Thiophenecarbodithioic acid, methyl ester	C_6H_{12}	295(3.79),340(4.24), 513(1.96)	48-0045-69
	EtOH	296(3.78),342(4.20), 507(1.98)	48-0045-69
$C_6H_7AsO_3$			
Benzenearsonic acid	H_2O	262(2.90)	25-0381-69
	0.18% H_2SO_4	262(2.97)	
	0.35% H_2SO_4	262(2.97)	
	0.93% H_2SO_4	262(3.00)	
	4.7% H_2SO_4	262(3.03)	
	22.2% H_2SO_4	262(3.09)	
	53.1% H_2SO_4	263(3.11)	
	72% H_2SO_4	264(3.14)	
	93% H_2SO_4	265(3.20)	
	15% $HClO_4$	262(2.95)	
	28% $HClO_4$	263(2.98)	
	59% $HClO_4$	264(3.11)	
$C_6H_7BO_2$			
Benzeneboronic acid	H_2O	190(4.70),218(3.93), 266(2.65)	39-0392-69A
	EtOH	219(4.04),267(2.68)	28-1562-69B
C_6H_7BrOS			
Acetic acid, thio-, S-(4-bromo-2,3-butadienyl) ester	heptane	188(4.48)	78-3277-69
$C_6H_7BrO_2$			
2,3-Butadien-1-ol, 4-bromo-, acetate	heptane	187(4.42)	78-3277-69
$C_6H_7BrO_3S$			
1,2-Oxathiin, 3-bromo-4,6-dimethyl-, 2,2-dioxide	$CHCl_3$	278(3.87)	88-0651-69
1,2-Oxathiin, 5-bromo-4,6-dimethyl-, 2,2-dioxide	$CHCl_3$	285(3.58)	88-0651-69
$C_6H_7ClN_2O$			
4(3H)-Pyrimidinone, 2-(chloromethyl)-6-methyl-	EtOH	230(3.85),277(3.64)	95-0460-69

Compound	Solvent	λ_{max}(log ϵ)	Ref.
$C_6H_7Cl_3O_2$			
2H-Pyran-2-ol, 5,6-dihydro-6-(trichloro-methyl)-	EtOH	225(2.73),310(--)	104-0759-69
$C_6H_7FN_2O_2$			
Pyrimidine, 5-fluoro-2,4-dimethoxy-	EtOH	268(3.95)	1-0294-69
2(1H)-Pyrimidinone, 4-ethoxy-5-fluoro-	pH 1	273(3.68)	1-0294-69
	pH 13	289(3.68)	1-0294-69
C_6H_7N			
Aniline	MeOH	230(3.93)	87-0146-69
2,4-Pentadienenitrile, 3-methyl-, cis	n.s.g.	248(4.28)	22-1349-69
2,4-Pentadienenitrile, 3-methyl-, trans	n.s.g.	248(4.35)	22-1349-69
2,4-Pentadienenitrile, 4-methyl-, cis	n.s.g.	250(4.32)	22-1349-69
2,4-Pentadienenitrile, 4-methyl-, trans	n.s.g.	249(4.39)	22-1349-69
2-Picoline	C_6H_{12}	259(3.40)	4-0859-69
	pH 1	262(3.61)	4-0859-69
	MeOH	262(3.55)	4-0859-69
	MeOH	261(3.53),268(3.42)	73-0819-69
3-Picoline	isooctane	264(3.34)	4-0859-69
	pH 1	263(3.71)	4-0859-69
	MeOH	263(3.42)	4-0859-69
4-Picoline	C_6H_{12}	255(3.20)	4-0859-69
	pH 1	252(3.68)	4-0859-69
	MeOH	255(3.40)	4-0859-69
C_6H_7NO			
Phenol, p-amino-	pH 8	233(3.87),300(3.34)	39-0207-69B
2-Picoline, N-oxide	C_6H_{12}	277(4.08)	4-0859-69
	pH 1	264(3.64)	4-0859-69
	MeOH	259(4.07)	4-0859-69
3-Picoline, N-oxide	isooctane	283(4.4)	4-0859-69
	pH 1	264(3.51)	4-0859-69
	MeOH	264(4.09)	4-0859-69
4-Picoline, N-oxide	C_6H_{12}	286(4.17)	4-0859-69
	pH 1	255(3.47)	4-0859-69
	MeOH	264(4.20)	4-0859-69
3-Pyridinol, 6-methyl-	pH 1	220(3.63),297(3.79)	1-0371-69
	pH 13	240(4.02),306(3.63)	1-0371-69
2(1H)-Pyridone, 1-methyl-	$CHCl_3$	304(3.72)	44-0836-69
Pyrrole, 3-acetyl-	EtOH	202(4.06),243(3.94), 266(3.75)	33-1911-69
Pyrrole-3-carboxaldehyde, 2-methyl-	EtOH	205(4.05),248(3.97), 284(3.82)	78-3879-69
C_6H_7NOS			
2(1H)-Pyridinethione, 3-hydroxy-6-methyl-	pH 1	260(3.78),355(4.10)	1-1704-69
	pH 13	240(3.76),265(3.71), 370(3.82)	1-1704-69
	pH 13	245(3.76),265(3.71), 370(3.80)	1-2065-69
$C_6H_7NOS_2$			
2-Propanone, 1-(2-thiazolylthio)-	MeOH	274(3.87)	4-0397-69
$C_6H_7NO_2$			
Catechol, 3-amino-	cation	194(4.43),214(3.62), 273(3.23)	61-0203-69
	neutral	213(4.34),232(3.83), 276(3.26)	61-0203-69

Compound	Solvent	λ_{max}(log ϵ)	Ref.
Catechol, 3-amino- (cont.)	anion	213(4.43),233(3.96), 279(3.30)	61-0203-69
	dianion	215(4.42),236(3.97), 281(3.33)	61-0203-69
Catechol, 4-amino-	cation	196(4.67),219(3.77), 278(3.46)	61-0203-69
	neutral	200(4.43),227(3.85), 299(3.54)	
	anion	207(4.34),248(3.78), 304(3.54)	
	dianion	211(4.33),266(3.75), 309(3.60)	
Hydroquinone, amino-	cation	190(4.54),220(3.82), 289(3.62)	61-0203-69
	pH 2	193(4.50),228(3.88), 294(3.61)	
	anion	203(4.52),233(3.92), 305(3.74)	
	0.15N NaOH	213(4.38),237(3.99), 322(3.71)	
Resorcinol, 2-amino-	cation	193(4.41),217(3.72), 271(3.20)	61-0203-69
	neutral	205(4.49),235(3.73), 271(3.07)	
	anion	211(4.49),238(3.81), 283(3.31)	
	dianion	220(4.41),242(3.86), 265(3.26)	
Resorcinol, 4-amino-	cation	189(4.40),212(3.78), 276(3.28)	61-0203-69
	neutral	198(4.39),223(3.81), 292(3.48)	
	anion	207(4.42),235(3.79), 302(3.55)	
	dianion	209(4.47),238(3.93), 311(3.59)	
Resorcinol, 5-amino-	cation	197(4.52),219(3.84), 278(3.16)	61-0203-69
	neutral	209(4.41),235(3.80), 280(2.99)	
	anion	220(4.41),239(3.88), 289(3.20)	
	dianion	225(4.36),246(3.92), 296(3.28)	
$C_6H_7NO_2S$			
2,4-Pentanedione, 3-thiocyanato-,	EtOH	280(4.48)	42-0945-69
aluminum chelate	$CHCl_3$	283(4.48)	42-0945-69
copper chelate	$CHCl_3$	245(4.14),294(4.43), 645(1.77)	42-0945-69
palladium chelate	$CHCl_3$	250(4.31),265s(--), 327(4.05)	42-0945-69
$C_6H_7NO_3$			
Nicotinic acid, 1,2,5,6-tetrahydro- 4-hydroxy-2-oxo-	EtOH	231(4.1),238(4.16)	105-0026-69
$C_6H_7N_2O_6S$			
2(?H)-Thiophenone, 3,5-dinitro-, dimethyl acetal, ion(1-)	MeOH	312(3.88),532(4.36)	77-0953-69

Compound	Solvent	$\lambda_{max}(\log \epsilon)$	Ref.
$C_6H_7N_3O$			
3-Buten-2-one, 4-(1H-1,2,3-triazol-1-yl)-	EtOH	265(4.2)	97-0025-69
3-Buten-2-one, 4-(4H-1,2,4-triazol-4-yl)-	EtOH	250(4.1)	97-0025-69
Isonicotinamidoxime	EtOH	276(3.69)	39-0861-69C
Isonicotinic acid hydrazide	pH 4.6	none from 540 to 820 nm	10-0655-69C
$C_6H_7N_3O_2$			
Pyridine, 1,4-dihydro-1-methyl-4-nitrimino-	pH -1	282(4.29)	12-2611-69
	pH 4	225(4.01),334(4.35)	12-2611-69
Pyridine, 4-(N-methylnitramino)-	pH 1	290(4.06)	12-2611-69
	pH 7	263(3.93)	12-2611-69
	dioxan	261(3.88)	12-2611-69
$C_6H_7N_3O_3$			
4-Pyrimidinecarboxylic acid, 2-amino-1,6-dihydro-6-oxo-, methyl ester	EtOH	270(4.23),316(3.96)	44-3705-69
$C_6H_7N_5$			
Adenine, 1-methyl-	pH 1.4	259(4.08),264(4.05)	50-0273-69A
	pH 10.0	272(4.08)	50-0273-69A
	pH 12.3	271(4.16)	50-0273-69A
$C_6H_7N_5O$			
8-Azapurin-6-one, dihydro-1,9-dimethyl-	H_2O	257(3.92)	89-0132-69
Guanine, 8-methyl-	pH 1.6	248(4.11),275(3.92)	5-0201-69F
	pH 7	248(4.01),278(3.93)	24-4032-69
	pH 11.4	247(3.80),275(3.94)	24-4032-69
	pH 15.0	255s(3.84),276(4.02)	5-0201-69F
s-Triazolo[4,3-b]-as-triazin-7-ol, 3,6-dimethyl-	EtOH	230(3.77),289(3.58)	22-2492-69
s-Triazolo[5,1-c]-as-triazin-4-ol, 3,7-dimethyl-	EtOH	241(3.60),296(3.83)	22-2492-69
s-Triazolo[3,4-c]-as-triazin-5(8H)-one, 6,8-dimethyl-	EtOH	243(3.87),329(3.76)	22-2492-69
s-Triazolo[5,1-c]-as-triazin-4(1H)-one, 1,3-dimethyl-	EtOH	243(3.85),300(3.82)	22-2492-69
$C_6H_7N_5O_9$			
Hexanenitrile, 4,4,6,6-tetranitro-	anion	366(4.22)	122-0100-69
$C_6H_7N_5S$			
8-Azapurine, 9-methyl-6-(methylthio)-	aq. EtOH	223(4.00),296(4.18)	39-0152-69C
Pyrimido[5,4-e]-as-triazine, 1,2-dihydro-5-(methylthio)-	pH 7	239(4.02),381(3.72)	44-3161-69
1,3,4-Thiadiazole, 2-amino-5-(1-methyl-2-imidazolyl)-	MeOH	308(4.22)	4-0835-69
s-Triazolo[4,3-b]-as-triazine, 6-methyl-7-(methylthio)-	EtOH	246(3.81),290(3.56), 341(3.72)	22-3670-69
s-Triazolo[4,3-d]-as-triazine, 8-methyl-5-(methylthio)-	EtOH	235(3.82),288(3.87)	22-3670-69
$C_6H_7N_7S$			
2H,5H-1-Thia-2a,4,5,6,8-pentaazaacenaphthylene, 2,2-diamino-, hydrochloride	pH 1	230(4.08),308(3.99)	44-3161-69
$C_6H_7O_3P$			
Phosphonic acid, phenyl-	H_2O	270(2.54)	25-0381-69

Compound	Solvent	$\lambda_{max}(\log \epsilon)$	Ref.
Phosphonic acid, phenyl- (cont.)	0.35% H_2SO_4	270(2.69)	25-0381-69
	0.93% H_2SO_4	270(2.78)	
	4.7% H_2SO_4	270(2.81)	
	22.2% H_2SO_4	270(2.83)	
	34.2% H_2SO_4	270(2.86)	
	70.9% H_2SO_4	272(2.94)	
	93.0% H_2SO_4	273(3.08)	
	24.1% $HClO_4$	270(2.70)	
	35.7% $HClO_4$	270(2.75)	
	65.4% $HClO_4$	271(2.95)	
$C_6H_7O_4P$			
Phosphoric acid, phenyl ester	pH 8	267(2.83)	69-1042-69
C_6H_8BrClO			
Cyclohexanone, 2-bromo-2-chloro-	C_6H_{12}	303(1.97)	22-0628-69
$C_6H_8Br_2$			
1,3-Butadiene, 1,4-dibromo-2,3-dimethyl-	C_6H_{12}	252(4.23)	78-4375-69
C_6H_8ClNO			
Valeronitrile, 2-chloro-2-methyl-3-oxo-	C_6H_{12}	290(1.84)	22-0628-69
$C_6H_8Cl_2$			
1,3-Butadiene, 1,4-dichloro-2,3-dimethyl-	C_6H_{12}	244(4.32)	78-4375-69
$C_6H_8Cl_2N_2$			
2,3-Diazabicyclo[3.2.0]hept-2-ene, 5,6-dichloro-4-methyl-, cis	n.s.g.	326(2.48)	89-0210-69
$C_6H_8I_2$			
1,3-Butadiene, 1,4-diiodo-2,3-dimethyl-	C_6H_{12}	259(4.30)	78-4375-69
1-Cyclobutene, 3,4-diiodo-1,2-dimethyl-, trans	C_6H_{12}	230(4.12)	78-4375-69
$C_6H_8N_2$			
p-Phenylenediamine	pH 8	242(3.98),307(3.30)	39-0207-69B
Pyrrolidin-2-ylideneacetonitrile	n.s.g.	267(4.28)	89-0343-69
$C_6H_8N_2O$			
Ketone, methyl 5-methylpyrazol-3-yl	MeOH	235(3.94)	4-0545-69
Pyrazole-3-carboxaldehyde, 1,5-dimethyl-	MeOH	244(3.97)	4-0545-69
Pyrazole-4-carboxaldehyde, 1,3-dimethyl-	MeOH	248(4.07)	4-0545-69
Pyrazole-4-carboxaldehyde, 1,5-dimethyl-	MeOH	250(4.11)	4-0545-69
Pyrazole-5-carboxaldehyde, 1,3-dimethyl-	MeOH	247(3.80)	4-0545-69
4(3H)-Pyrimidinone, 2,6-dimethyl-	EtOH	225(3.89),270(3.67)	95-0460-69
$C_6H_8N_2O_2$			
Pyridazine, 3-methoxy-5-methyl-, 1-oxide	EtOH	258(3.8),308(3.7)	94-0763-69
Thymine, 1-methyl-, ketyl radical	pH 5.5	310(3.40),340(3.36)	38-4881-69B
radical anion	pH 12.1	310(3.11)	38-4881-69B
Uracil, 1,3-dimethyl-, ketyl radical	pH 5.1	320s(3.28)	38-4881-69B
	pH 12.4	325s(3.56)	38-4881-69B
Uracil, 1,6-dimethyl-	75% dioxan	270(4.02)	95-0266-69
	+ NaOH	267(3.89)	95-0266-69
Uracil, 5,6-dimethyl-	pH 2	268(3.93)	56-0499-69
	pH 12	275(3.76)	56-0499-69
	5N NaOH	279(3.87)	56-0499-69

Compound	Solvent	λ_{max}(log ε)	Ref.
Uracil, 5-ethyl-	pH 2	264(3.90)	56-0499-69
	pH 12	289(3.75)	56-0499-69
$C_6H_8N_2O_2S$			
Thiazole, 2-ethyl-4-methyl-5-nitro-	EtOH	221(3.65),250(3.51), 305(3.54)	28-1343-69B
Thiazole, 2-ethyl-5-methyl-4-nitro-	EtOH	220(4.10),240(--), 241(3.69)	28-1343-69B
Thiazole, 5-ethyl-2-methyl-4-nitro-	EtOH	220(3.98),291(3.58)	28-1343-69B
Thiazole, 5-isopropyl-4-nitro-	EtOH	218(4.03),282(3.68)	28-1343-69B
$C_6H_8N_2O_3$			
Uracil, 5-hydroxy-1,3-dimethyl-	pH 1	283(3.87)	44-2636-69
	pH 10	242(3.84),310(3.83)	44-2636-69
$C_6H_8N_4O$			
Imidazo[1,2-b]-as-triazin-3(5H)-one, 6,7-dihydro-2-methyl-	EtOH	210(4.42),244s(3.86)	114-0181-69C
Imidazo[2,1-c]-as-triazin-4(6H)-one, 7,8-dihydro-3-methyl-, hydrochloride	EtOH	220(4.05),300(3.70)	114-0181-69C
$C_6H_8N_4O_2$			
Pyrimidine-4-carboxylic acid, 2,6-diamino-, methyl ester	neutral	228(4.01),318(3.86)	44-0821-69
	acid	211(4.45),288(3.80)	44-0821-69
$C_6H_8N_4S$			
7H-s-Triazolo[3,4-b][1,3,4]thiadiazine, 3-ethyl-, hydrobromide	EtOH	291(3.09)	2-0959-69
C_6H_8O			
2-Cyclohexen-1-one	C_6H_{12}	334(1.43)	77-1246-69
	MeOH	317(1.53)	77-1246-69
	EtOH	225(4.14)	24-3877-69
	H_2SO_4	231(3.88)	23-2263-69
protonated	H_2SO_4	271(3.97)	23-2263-69
2-Cyclopenten-1-one, 3-methyl-	H_2SO_4	232(4.16)	23-2263-69
protonated	H_2SO_4	254(4.20)	23-2263-69
2-Cyclopropene, 1-acetyl-1-methyl-	hexane	200(3.16),218(3.18), 290(2.08)	88-1235-69
2-Cyclopropene, 1-acetyl-2-methyl-	hexane	206(3.48),288(1.91)	88-1235-69
Furan, 2,3-dimethyl-	hexane	218(3.93)	88-1235-69
Furan, 2,5-dimethyl-	hexane	219(3.88)	78-5357-69
2-Hexen-4-yn-1-ol	MeOH	223(4.02),213(3.95)	35-4824-69
2,4-Pentadienal, 4-methyl-	EtOH	262(4.37)	54-0119-69
C_6H_8OS			
3-Thietanone, 2-isopropenyl-	EtOH	245(2.96),330(2.22)	44-1566-69
2H-Thiopyran-3(6H)-one, 5-methyl-	isooctane	232(3.97),270(2.48), 347(2.99)	44-1566-69
$C_6H_8OS_2$			
2(3H)-Thiophenone, dihydro-3-(thioacetyl)-	EtOH	305(4.06),366(2.57)	78-5703-69
$C_6H_8O_2$			
1,4-Cyclohexanedione	EtOH	281(1.46)	27-0411-69
2-Cyclohexen-1-one, 3-hydroxy-	H_2SO_4	256(4.21)	23-2263-69
protonated	H_2SO_4	273(4.31)	23-2263-69
2-Cyclopenten-1-one, 2-hydroxy-3-methyl-	H_2SO_4	258(4.03)	23-2263-69
protonated	H_2SO_4	292(4.08)	23-2263-69

Compound	Solvent	$\lambda_{max}(\log \epsilon)$	Ref.
2(5H)-Furanone, 4-ethyl-	EtOH	209(4.07)	44-4046-69
2-Hexenal, 4-oxo-	pet ether	220(4.05),225(4.08)	54-0989-69
3-Hexene-2,5-dione	MeOH	228(3.11)	44-0457-69
4H-Pyran-4-one, 2,3-dihydro-6-methyl-	EtOH	264(4.19)	22-1383-69
$C_6H_8O_2S$			
2(3H)-Furanone, dihydro-3-(thioacetyl)-	EtOH	278(3.98),326(3.24)	78-5703-69
3,4-Thiophenedimethanol	EtOH	236(3.72),239(3.72)	44-0333-69
$C_6H_8O_3$			
2,3-Furandione, dihydro-4,4-dimethyl-	C_6H_{12}	219(2.29)	77-1467-69
(after two minutes)	MeOH	215(2.23),368(1.20)	77-1467-69
	$CHCl_3$	382(1.46)	77-1467-69
$C_6H_8O_6$			
Ascorbic acid	0.4N HCl	244(4.00)	18-2282-69
L-Ascorbic acid, 3-pyrophosphate,	pH 1	235(3.99)	94-0381-69
magnesium salt	pH 13	259(4.23)	94-0381-69
$C_6H_8O_6S_2$			
Pyruvic acid, mercapto-, dimer, disodium salt	pH 10.6	230(3.68)	28-1844-69A
C_6H_8S			
Thiophene, 2,5-dimethyl-	hexane	194(3.92),196(3.93), 237(3.86)	78-5357-69
C_6H_8Si			
Silane, phenyl-	C_6H_{12}	265(2.51)	101-0017-69B
$C_6H_9BF_2O_2$			
Boron, difluoro(3-methyl-2,4-pentanedionato)-	$CHCl_3$	304(4.19)	39-0526-69A
$C_6H_9BF_4N_2O$			
3,4-Dihydro-1,3-dimethyl-4-oxopyrimidinium tetrafluoroborate	EtOH	229(3.94),272(3.53)	44-0589-69
$C_6H_9ClN_4O$			
1,2,4-Triazin-5(2H)-one, 3-[(2-chloro-	EtOH	210(4.38),250s(3.8)	114-0181-69C
ethyl)amino]-6-methyl-, hydrochloride	EtOH-NaOH	210(4.42),244s(3.86)	114-0181-69C
C_6H_9ClO			
Cyclohexanone, 2-chloro-	C_6H_{12}	304(1.54)	77-1246-69
$C_6H_9ClO_2$			
2,4-Hexanedione, 5-chloro-	n.s.g.	220(2.1),290(1.8)	65-1954-69
$C_6H_9ClO_3$			
Acetoacetic acid, 2-chloro-, ethyl	isooctane	265(3.77)	22-3694-69
ester	EtOH	262(3.22)	22-3694-69
$C_6H_9Cl_3O$			
2-Pentanone, 2-(trichloromethyl)-	EtOH	285(1.65)	35-2062-69
	EtOH-base	244(3.51)	35-2062-69
$C_6H_9Cl_3O_2$			
3-Hexanone, 6,6,6-trichloro-5-hydroxy-	EtOH	212(1.69),276(1.38)	23-2029-69
2-Pentanone, 5,5,5-trichloro-4-hydroxy-3-methyl-, erythro-	EtOH	214(1.70),276(1.42)	23-2029-69

Compound	Solvent	λ_{max}(log ε)	Ref.
2-Pentanone, 5,5,5-trichloro-4-hydroxy-3-methyl-, threo-	EtOH	210(1.99),278(1.52)	23-2029-69
$C_6H_9IN_2$			
4-Amino-1-methylpyridinium iodide	pH 1-14.2	269(4.27)	12-2595-69
	10N NaOH	275(4.22)	12-2595-69
C_6H_9N			
Pyrrole, 2,5-dimethyl-	hexane	216(3.76)	78-5357-69
C_6H_9NO			
3-Butyn-2-one, 4-(dimethylamino)-	CH Cl	277(4.23)	33-2641-69
C_6H_9NOS			
Oxazole, 4,5-dimethyl-2-(methylthio)-	2% EtOH	251(3.83)	1-2888-69
conjugate acid	2% EtOH	262(4.01)	1-2888-69
4-Oxazoline-2-thione, 3,4,5-trimethyl-	2% EtOH	270(4.16)	1-2888-69
conjugate acid	2% EtOH	257(3.95)	1-2888-69
$C_6H_9NOS_2$			
1,3-Thiazane-2-thione, 3,5-dimethyl-4-oxo-	MeOH	267(4.13),310(4.06)	42-0919-69
1,3-Thiazane-2-thione, 3,6-dimethyl-4-oxo-	MeOH	267(4.11),310(4.07)	42-0919-69
$C_6H_9NO_2$			
1-Butyne, 3,3-dimethyl-1-nitro-	hexane	238(3.9)	89-0273-69
Propiolic acid, (dimethylamino)-, methyl ester	CH_2Cl_2	255(4.07)	33-2641-69
$C_6H_9NO_3S$			
Carbamic acid, 2-isothiocyanatopropyl methyl ester, D-	EtOH	259(4.23)	39-1143-69B
L-isomer	EtOH	259(4.17)	39-1143-69B
C_6H_9NS			
2-Thiophenamine, N,N-dimethyl-	EtOH	278(4.0)	48-0827-69
$C_6H_9NS_2$			
Cyclopentanecarbodithioic acid, 2-imino-	heptane	310(3.39),389(4.10)	44-0730-69
	EtOH	302(4.05),388(4.38)	44-0730-69
Thiazole, 4,5-dimethyl-2-(methylthio)-	2% EtOH	285(3.90)	1-2888-69
conjugate acid	2% EtOH	302(4.02)	1-2888-69
4-Thiazoline-2-thione, 3,4,5-trimethyl-	2% EtOH	312(4.13)	1-2888-69
conjugate acid	2% EtOH	293(3.95)	1-2888-69
$C_6H_9NS_3$			
1,3-Thiazane-2,4-dithione, 3,5-dimethyl-	MeOH	325(4.08),340(4.07)	42-0919-69
1,3-Thiazane-2,4-dithione, 3,6-dimethyl-	MeOH	323(4.10),339(4.07)	42-0919-69
$C_6H_9N_3$			
as-Triazine, 3,5,6-trimethyl-	MeOH	263(3.64),362(2.60)	88-3147-69
s-Triazine, 2,4,6-trimethyl-	MeOH	228(2.43),257(2.59)	18-2924-69
$C_6H_9N_3OS$			
Imidazole-5-thiocarboxylic acid, 4-amino-3-methyl-, S-methyl ester	pH 1.0	247(3.67),302(4.21)	39-2379-69C
	pH 7.0	239(3.59),305(4.22)	39-2379-69C

Compound	Solvent	λ_{max}(log ϵ)	Ref.
as-Triazine-3,5(2H,4H)-dione, 2,4,6-trimethyl-3-thio-	MeOH	223(4.04),266(4.22), 302s(3.60)	114-0093-69C
as-Triazin-5(4H)-one, 4,6-dimethyl-3-(methylthio)-	MeOH	235(4.38)	114-0181-69C
$C_6H_9N_3O_4$			
Hydantoin, 5-(3-nitropropyl)-	50% EtOH	270(1.57)	33-0388-69
$C_6H_9N_5O$			
4H-as-Triazino[4,3-b]-as-triazin-4-one, 6,7,8,9-tetrahydro-3-methyl-	EtOH	226(4.11),311(3.84)	114-0181-69C
$C_6H_9N_5O_2$			
Acetamide, N-(2,4-diamino-1,6-dihydro-6-oxo-5-pyrimidinyl)-	pH 0.0	265(4.30)	5-0201-69F
	pH 6.0	267(4.19)	5-0201-69F
	pH 12.0	263(4.05)	5-0201-69F
1,2,4-Triazin-5-ol, 3-acetylhydrazino-6-methyl-	EtOH	242(3.73)	22-2492-69
1,2,4-Triazin-5-one, 2,5-dihydro-3-formylhydrazino-2,6-dimethyl-	EtOH	252(3.85)	22-2492-69
$C_6H_9N_5O_9$			
Hexanoic acid amide, 4,4,6,6-tetranitro-	anion	367(4.20)	122-0100-69
$C_6H_9O_9P$			
L-Ascorbic acid, 2-phosphate, magnesium salt	pH 1	235(3.94)	94-0387-69
	pH 13	259(4.15)	94-0387-69
C_6H_{10}			
Propene, 2-cyclopropyl-	hexane	196(3.96)	77-0781-69
$C_6H_{10}BrNO$			
3-Buten-2-one, 3-bromo-4-(dimethylamino)-	CH_2Cl_2	305(4.01)	33-2641-69
$C_6H_{10}BrNO_2$			
Acrylic acid, 2-bromo-3-(dimethylamino)-, methyl ester	CH_2Cl_2	293(4.05)	33-2641-69
$C_6H_{10}ClN_5O$			
as-Triazin-5(4H)-one, 3-amino-4-[(2-chloroethyl)amino]-6-methyl-	EtOH	218(3.93),302(3.83)	114-0181-69C
$C_6H_{10}N_2$			
Imidazole, 1,4,5-trimethyl-	C_6H_{12}	213(3.66),234(3.72)	78-3287-69
	EtOH	226(3.67)	78-3287-69
	tert-BuOH	226(3.67)	78-3287-69
Imidazole, 2,4,5-trimethyl-	n.s.g.	233(3.36)	39-2251-69C
Pyrazole, 1,3,5-trimethyl-	EtOH	210(3.58),221(3.67)	78-3287-69
Pyrazole, 3,4,5-trimethyl-	n.s.g.	230(3.30)	39-2251-69C
$C_6H_{10}N_2O$			
2,3-Diazabicyclo[2.2.1]hept-2-ene, 5-methoxy-, endo	C_6H_{12}	342(2.59)	35-6766-69
2-Pentanone, 3-diazo-4-methyl-	40% dioxan	243(3.64),289(3.72)	33-2417-69
2-Pyrazolin-5-one, 3,4,4-trimethyl-	MeOH	240(3.66)	22-4159-69
	EtOH	240(3.86)	22-4159-69
$C_6H_{10}N_2O_2$			
1-Pyrazolol, 3,4,5-trimethyl-, 2-oxide	EtOH	250(λ3.8)	44-0194-69

Compound	Solvent	λ_{max}(log ε)	Ref.
3(2H)-Pyridazinone, 4,5-dihydro-5-(hydroxymethyl)-6-methyl-	MeOH	244(3.85)	5-0217-69D
$C_6H_{10}N_2O_3S$			
Acrylamide, N-ethyl-3-[(nitromethyl)-thio]-, cis	90% EtOH	258(4.08)	12-0765-69
$C_6H_{10}N_2S$			
Imidazole, 4,5-dimethyl-2-(methylthio)-	2% EtOH	254(3.87)	1-2888-69
conjugate acid	2% EtOH	233s(3.69),259(3.95)	1-2888-69
4-Imidazoline-2-thione, 1,4,5-trimethyl-	2% EtOH	209(3.83),259(4.16)	1-2888-69
conjugate acid	2% EtOH	250(3.99)	1-2888-69
$C_6H_{10}N_4$			
Pyrimidine, 4-amino-5-(aminomethyl)-2-methyl-	pH 1.79	241(4.0),250s(3.9)	65-0102-69
	pH 11.07	230(3.9),270(3.6)	65-0102-69
Pyrimidine, 2,4-diamino-5,6-dimethyl-	pH 2	273(3.86)	44-0821-69
	pH 12	228s(3.95),285(3.88)	44-0821-69
$C_6H_{10}N_4O$			
as-Triazin-5(2H)-one, 2,4,6-trimethyl-3-imino-	EtOH	216(3.88),266(3.58), 308s(2.4)	114-0181-69C
$C_6H_{10}N_4O_2$			
as-Triazin-5(2H)-one, 3-[(2-hydroxy-ethyl)amino]-6-methyl-	EtOH	210(4.42),240s(3.86)	114-0181-69C
$C_6H_{10}N_4O_3$			
Uracil, 5-amino-6-[(2-hydroxyethyl)-amino]-	pH -3.6	269(4.30)	24-4032-69
	pH 2	267(4.35)	24-4032-69
	pH 7	285(4.20)	24-4032-69
$C_6H_{10}N_4S$			
5H-s-Triazolo[3,4-b][1,3,4]thiadiazine, 3-ethyl-6,7-dihydro-	EtOH	222(3.88)	2-0959-69
$C_6H_{10}N_6$			
as-Triazine-5-carboxamidine, 6-amino-N-ethyl-	pH 7	240(3.97),347(3.61)	44-2102-69
$C_6H_{10}N_6O$			
Pyrazole-4-carboxamide, 3-(3,3-dimethyl-1-triazeno)-	pH 7	229(3.99),314(4.06)	87-0545-69
$C_6H_{10}O$			
Butadiene, 1-ethoxy-, cis	heptane	245(4.32)	28-1549-69A
	EtOH	245(4.43)	22-2508-69
Butadiene, 1-ethoxy-, trans	EtOH	236(4.30)	22-2508-69
1,3-Butadiene, 1-ethoxy-, cis	C_6H_{12}	244(4.44)	22-2508-69
	heptane	244(4.18)	104-1696-69
1,3-Butadiene, 1-ethoxy-, trans	C_6H_{12}	237(4.35)	22-2508-69
	heptane	236(4.36)	28-1549-69A
	heptane	236(4.13)	104-1696-69
1,3-Butadiene, 2-ethoxy-	EtOH	235(3.98)	22-2508-69
Cyclohexanone	neat	288(1.20)	77-1246-69
	C_6H_{12}	289(1.18)	77-1246-69
	MeOH	281(1.18)	77-1246-69
$C_6H_{10}S$			
6-Thiabicyclo[3.1.1]heptane	EtOH	278(1.15)	44-1233-69

Compound	Solvent	$\lambda_{max}(\log \epsilon)$	Ref.
7-Thiabicyclo[2.2.1]heptane	EtOH	247(1.63)	44-1233-69
$C_6H_{11}NO$			
2H-Azepin-2-one, hexahydro-	C_6H_{12}	194(3.85)	46-1642-69
	H_2O	196(3.89)	46-1642-69
$C_6H_{11}NOS$			
Acrylamide, N-ethyl-3-(methylthio)-, cis	90% EtOH	282(4.18)	12-0765-69
$C_6H_{11}NO_2$			
Acetoacetamide, N,N-dimethyl-	CH_2Cl_2	255(3.47)	33-2641-69
Acrylic acid, 3-(dimethylamino)-, methyl ester	CH_2Cl_2	278(4.28)	33-2641-69
$C_6H_{11}NO_3$			
Malonamic acid, N,N-dimethyl-, methyl ester	CH_2Cl_2	280(3.10)	33-2641-69
$C_6H_{11}NO_3S$			
Alanine, 3-(allylsulfinyl)-	H_2O	none above 210 nm	10-0312-69B
Alanine, 3-(propenylsulfinyl)-, trans	H_2O	230s(3.3)	10-0312-69B
	90% MeOH	230s(3.4)	10-0312-69B
$C_6H_{11}N_3$			
Cyclohexane, azido-	hexane	217(2.64),287(1.42)	114-0309-69B
Pyrazole, 3-(dimethylamino)-5-methyl-	MeOH	237(3.74)	33-2641-69
$C_6H_{11}N_3O_2$			
Pyrimidine, 1,2,3,4-tetrahydro-1,3-dimethyl-5-nitro-	hexane	220(--),365(4.30)	78-1617-69
$C_6H_{11}N_3S$			
Thiosemicarbazide, 4-(cyclopent-2-en-1-yl)-	EtOH	242(4.06)	124-0509-69
$C_6H_{11}N_5$			
Propionamidine, 2-methyl-N-(4H-1,2,4-triazol-4-yl)-	MeOH	227(3.60)	48-0477-69
$C_6H_{11}N_5O_2$			
as-Triazin-5(2H)-one, 3-[2-(2-hydroxyethyl)hydrazino]-6-methyl-	EtOH	211(4.36),244s(3.85)	114-0181-69C
as-Triazin-5(4H)-one, 3-amino-4-[(2-hydroxyethyl)amino]-6-methyl-	EtOH	220(3.88),300(3.78)	114-0181-69C
Urea, [4-(dimethylamino)-2-oxo-3-imidazolin-5-yl]-	pH 7-10	249(4.06)	69-0733-69
C_6H_{12}			
Cyclopropane, isopropyl-, iodine complex	hexane	252(4.16)	35-2279-69
$C_6H_{12}N_2$			
1,4-Diazabicyclo[2.2.2]octane	gas	234(3.45),236f(3.49), 238f(3.52),241f(3.61), 243(3.41),246f(3.72), 252(3.64)	59-0501-69
2-Pyrazoline, 1,4,4-trimethyl-	EtOH	242(3.52)	22-3306-69
3-Pyrazoline, 1,2,3-trimethyl-	EtOH	235(3.43)	22-3316-69
3-Pyrazoline, 1,2,4-trimethyl-	EtOH	240(3.42)	22-3316-69

Compound	Solvent	$\lambda_{max}(\log \epsilon)$	Ref.
$C_6H_{12}N_2O$			
3-Buten-2-one, 4-amino-4-(dimethyl-amino)-	CH_2Cl_2	291(4.41)	33-2641-69
$C_6H_{12}N_2OS_2$			
Carbamic acid, [2-(methylthiocarbamoyl)-ethyl]thio-, O-methyl ester	EtOH	201(3.94),247(4.25), 262s(4.12)	39-2631-69C
$C_6H_{12}N_2O_2$			
Acrylic acid, 3-amino-3-(dimethyl-amino)-, methyl ester	CH_2Cl_2	274(4.46)	33-2641-69
Pyruvic acid, methyl ester, 2-(dimeth-ylhydrazone)	hexane	210s(3.71),310(3.48)	39-2443-69C
	EtOH	213s(3.66),305(3.63)	39-2443-69C
$C_6H_{12}N_2O_2S$			
β-Alanine, N-(methylthiocarbamoyl)-, methyl ester	EtOH	207(4.10),241(4.12)	39-2631-69C
Carbamic acid, [2-(methylcarbamoyl)-ethyl]thio-, O-methyl ester	EtOH	208(4.11),243(4.22)	39-2631-69C
Carbamic acid, [2-(methylthiocarbamoyl)-ethyl]-, methyl ester	EtOH	202(3.62),264(3.92)	39-2631-69C
Hydrouracil, 2-thio-, 4-(dimethyl acetal)	EtOH	203(4.05),247(4.20)	39-2631-69C
$C_6H_{12}N_2S_2$			
1-Cyclopentene-1-dithiocarboxylic acid, 2-amino-, ammonium salt	EtOH	301(3.95),388(4.35)	44-0730-69
$C_6H_{12}N_4O_4$			
2-Tetrazene-1,4-dicarboxylic acid, 1,4-dimethyl-, dimethyl ester	CH_2Cl_2	255(4.15),270(4.27)	27-0405-69
$C_6H_{12}N_4S$			
Carbohydrazide, 1-cyclopentylidene-3-thio-	MeOH	250(4.12)	44-0756-69
$C_6H_{12}O$			
2-Butanone, 3,3-dimethyl-	gas	175(3.76),195(3.04)	18-2453-69
	C_6H_{12}	285(1.40)	35-2154-69
	C_6H_{12}	287(1.32)	77-1246-69
	heptane	185(3.04),287(1.36)	59-0765-69
	MeOH	282(1.38)	77-1246-69
2-Hexanone	gas	171(3.16),175(2.74), 192(3.10),196(3.15)	18-2453-69
	95.3% H_2SO_4	none	104-2046-69
	68.6% H_2SO_4	255(1.5)	104-2046-69
	61.5% H_2SO_4	260(1.5)	104-2046-69
2-Pentanone, 3-methyl-	gas	169(3.13),172(3.16), 178(2.48),188(3.02), 192(3.34),192(3.31), 194(3.25),195(3.32), 196(3.44)	18-2453-69
2-Pentanone, 4-methyl-	gas	174(3.39),191(3.04), 192(3.06),194(3.06), 195(3.09),196(3.00)	18-2453-69
$C_6H_{12}O_2$			
Pivalic acid, methyl ester	C_6H_{12}	212(2.00)	35-2154-69

Compound	Solvent	$\lambda_{max}(\log \epsilon)$	Ref.
$C_6H_{12}O_6$			
Fructose	0.4N HCl	281(0.45)	18-2282-69
$C_6H_{12}S_3$			
s-Trithiane, 2,4,6-trimethyl-, α-	hexane	206(3.82),208s(3.81), 240(2.96)	59-0407-69
	iso-PrOH	200(3.79),220(2.87), 236(2.87)	59-0407-69
β-form	hexane	191s(3.41),205s(3.90), 207(3.91),228(2.94), 239(3.00)	59-0407-69
	iso-PrOH	200(3.90),236(2.86)	59-0407-69
$C_6H_{12}Si$			
Silane, dimethyldivinyl-	gas	176(--)	121-0803-69
	heptane	178(4.45)	121-0803-69
	MeCN	179(4.41)	121-0803-69
$C_6H_{13}BO_2$			
Cyclohexaneboronic acid	EtOH	204(1.30)	28-1562-69B
$C_6H_{13}NO$			
2-Pentanone, 4-amino-4-methyl-, hydrochloride	n.s.g.	270(1.3)	104-1739-69
$C_6H_{13}NO_2S_3$			
Methanesulfonamide, N-(1,2-dithiepan-4-yl)-, S-(-)-	EtOH	259(2.60)	87-0620-69
	$CHCl_3$	261(2.71)	87-0620-69
$C_6H_{13}NO_3S$			
Alanine, 3-(propylsulfinyl)-	H_2O	none above 210 nm	10-0312-69B
$C_6H_{13}N_3O$			
Pivalaldehyde, semicarbazone	EtOH	228(4.08)	94-0844-69
$C_6H_{13}N_3S$			
Isovaleraldehyde, thiosemicarbazone	EtOH	230(3.99),270(4.44)	94-0844-69
Pivalaldehyde, thiosemicarbazone	EtOH	228(3.81),270(4.33)	94-0844-69
Valeraldehyde, thiosemicarbazone	EtOH	230(3.98),271(4.47)	94-0844-69
$C_6H_{14}N_2O$			
Ethanol, 1-(tert-butylazo)-	C_6H_{12}	345(1.28)	89-0451-69
L-Leucinamide, copper derivative	H_2O	515(1.76)	18-0168-69
	MeOH	345s(--),505(1.79)	18-0168-69
Propane, 1-NNO-azoxyethane, 1-methyl-	EtOH	220(3.72),280s(1.62)	88-3897-69
$C_6H_{14}N_4S$			
2-Butanone, 3-methyl-, monohydrazone with thiocarbohydrazide	MeOH	249(4.11)	44-0756-69
Isovaleraldehyde, monohydrazone with thiocarbohydrazide	MeOH	250(4.09)	44-0756-69
Pivalaldehyde, monohydrazone with thiocarbohydrazide	MeOH	249(4.09)	44-0756-69
$C_6H_{14}Si$			
Silane, allyltrimethyl-	gas	189(--)	121-0803-69
	isooctane	192(4.02)	121-0803-69
	EtOH	194(3.96)	121-0803-69

Compound	Solvent	$\lambda_{max}(\log \epsilon)$	Ref.
$C_6H_{14}Sn$			
Stannane, allyltrimethyl-	hexane	200(3.95)	44-3715-69
(corrections)	MeCN	208(4.04)	44-3715-69
$C_6H_{18}Cl_2Si_3$			
Trisilane, 2,2-dichlorohexamethyl-	isooctane	232(3.49)	35-6613-69
	dioxan	233(3.49)	35-6613-69
$C_6H_{18}F_2Si_3$			
Trisilane, 1,3-difluorohexamethyl-	isooctane	217(3.78)	35-6613-69
Trisilane, 2,2-difluorohexamethyl-	isooctane	228(4.06)	35-6613-69
	dioxan	229(4.03)	35-6613-69
$C_6H_{18}N_3OP$			
Phosphoric triamide, hexamethyl-	n.s.g.	240(2.43)	28-2040-69A
cupric chloride complex	n.s.g.	295(3.68)	28-2040-69A
ferric chloride complex	n.s.g.	246(4.11)	28-2040-69A
$C_6H_{19}ClSi_3$			
Trisilane, 2-chloro-1,1,1,3,3,3-hexa-methyl-	isooctane	201(4.15),241(2.75)	35-6613-69
	dioxan	244(--)	35-6613-69
$C_6H_{20}Si_3$			
Silane, bis(trimethylsilyl)-	isooctane	196(4.26)	35-6613-69
$C_6I_2O_6Os_2$			
Osmium, hexacarbonyldi-μ-iododi-(Os-Os)	C_6H_{12}	222(3.92),249(3.86), 337(3.60)	39-0987-69A

Compound	Solvent	λ_{max}(log ϵ)	Ref.
$C_7F_{12}N_2S$			
Δ^3-1,3,4-Thiadiazoline, 2,2-bis(trifluoromethyl)-5-[trifluoro-1-(trifluoromethyl)ethylidene]-	isooctane	327(3.62)	44-3201-69
$C_7F_{12}S$			
2-Pentene, 3,4-epithio-1,1,1,5,5,5-hexafluoro-2,4-bis(trifluoromethyl)-	EtOH	239(4.06)	44-3201-69
C_7HF_5O			
Benzaldehyde, pentafluoro-, [(p-formyltetrafluorophenyl)hydrazone]	EtOH	386(4.5)	65-1607-69
$C_7H_2BrClN_4$			
7H-Pyrrolo[2,3-d]pyrimidine-3-carbonitrile, 6-bromo-4-chloro-	pH 1	225(4.64),285(3.99)	35-2102-69
	pH 11	243(4.33),272(4.22), 307(3.71)	35-2102-69
	EtOH	245(4.40),273(3.91), 312(3.61)	35-2102-69
$C_7H_2Br_2N_4O_8$			
Methane, (3,5-dibromo-2,4-dinitrophenyl)dinitro-	anion	360(4.17)	122-0100-69
$C_7H_2Cl_9N_3$			
s-Triazine, 2-(1,1,2-trichloroethyl)-4,6-bis(trichloromethyl)-	MeOH	223(3.37),285(2.89)	18-2924-69
$C_7H_2F_4O_2$			
Benzaldehyde, 2,3,5,6-tetrafluoro-4-hydroxy-	EtOH	270(3.67),328(4.45)	65-1607-69
$C_7H_2F_4O_3$			
Benzoic acid, 2,3,5,6-tetrafluoro-4-hydroxy-	EtOH	240(3.82),284(3.12)	65-1607-69
$C_7H_3BrN_4O$			
7H-Pyrrolo[2,3-d]pyrimidine-5-carbonitrile, 6-bromo-3,4-dihydro-4-oxo-	pH 1	272(4.21)	35-2102-69
	pH 11	241(4.34),287(4.14)	35-2102-69
	EtOH	245(4.11),277(4.11)	35-2102-69
$C_7H_3BrN_4O_8$			
Methane, (5-bromo-2,4-dinitrophenyl)-dinitro-	anion	368(4.22)	122-0100-69
$C_7H_3Br_2NS$			
Thieno[2,3-b]pyridine, 2,3-dibromo-	EtOH	236(4.49),282(4.05), 288(4.03),299(3.90)	44-0347-69
$C_7H_3ClF_4$			
Toluene, α-chloro-2,3,4,5-tetrafluoro-	C_6H_{12}	265(2.92)	39-0211-69C
$C_7H_3ClN_4$			
7H-Pyrrolo[2,3-d]pyrimidine-5-carbonitrile, 4-chloro-	pH 1	220(4.41),272(3.77)	35-2102-69
	pH 11	243(4.37),272(3.86), 315(3.51)	35-2102-69
	EtOH	220(4.40),246(3.86), 273(3.80)	35-2102-69

Compound	Solvent	λ_{max}(log ε)	Ref.
$C_7H_3ClN_4O_8$			
Methane, (4-chloro-2,6-dinitrophenyl)-dinitro-	anion	370(4.17)	122-0100-69
Methane, (6-chloro-2,4-dinitrophenyl)-dinitro-	anion	366(4.21)	122-0100-69
$C_7H_3Cl_2N_3O_6$			
Methane, (4,5-dichloro-2-nitrophenyl)-dinitro-	anion	369(4.20)	122-0100-69
$C_7H_3Cl_5O$			
2,4-Cyclohexadien-1-one, 2,3,4,5,6-pentachloro-6-methyl-	MeOH	354(3.54)	23-1943-69
2,4-Cyclohexadien-1-one, 2,4,5,6,6-pentachloro-3-methyl-	MeOH	363(3.47)	23-1943-69
2,5-Cyclohexadien-1-one, 2,3,4,4,6-pentachloro-5-methyl-	MeOH	258(4.17)	23-1943-69
$C_7H_3Cl_7O$			
3-Cyclohexen-1-one, 2,2,3,4,5,6,6-heptachloro-5-methyl-	MeOH	231(3.87)	23-1943-69
3-Cyclohexen-1-one, 2,3,4,5,5,6,6-heptachloro-2-methyl-	MeOH	232(3.87)	23-1943-69
$C_7H_3FN_4O_4$			
3H-Benzimidazole, 4-fluoro-5,7-dinitro-	pH 1	275(4.21)	44-0384-69
	pH 7.9	300(4.07)	44-0384-69
	pH 10.7	305(4.14),345(4.04)	44-0384-69
$C_7H_3FN_4O_8$			
Methane, (6-fluoro-2,4-dinitrophenyl)-dinitro-	anion	365(4.24)	122-0100-69
$C_7H_3F_3N_4$			
Pteridine, 4-(trifluoromethyl)-	hexane	231(3.38),285(3.64), 292(3.80),297(3.85), 303(3.89),309(3.79)	39-1751-69C
	pH 7.0	231(3.45),295(3.84), 302(3.92),313(3.86)	39-1751-69C
$C_7H_3N_3O$			
1,3-Oxazepine-2,4-dicarbonitrile	CH_2Cl_2	331(3.38)	78-0295-69
2,6-Pyridinedicarbonitrile, 1-oxide	EtOH	244(4.57),292(3.94), 357(3.13)	78-0295-69
Pyrrole-2-glyoxylonitrile, 5-cyano-	CH_2Cl_2	321(4.24)	78-0295-69
$C_7H_3N_4O_6$			
Cyclohexadienecarbonitrile, 2,4,6-tri-nitro-, ion(1-)	MeOH	428(4.38),540(--)	23-4129-69
	EtOH	435(4.37),544(--)	
	PrOH	435(4.39),542(--)	
	iso-PrOH	436(4.42),546(--)	
	BuOH	435(4.38),540(--)	
	tert-BuOH	437(4.42),549(--)	
	$CHCl_3$	442(4.60),558(4.35)	
$C_7H_3N_5O_{10}$			
Methane, (2,4,6-trinitrophenyl)dinitro-	anion	370(4.17)	122-0100-69
$C_7H_4BrFN_2S$			
Benzothiazole, 2-amino-x-bromo-4-fluoro-	EtOH	228(4.55),264s(4.05), 273(4.28),293s(3.86)	39-0268-69C

Compound	Solvent	λ max (log ε)	Ref.
$C_7H_4BrNO_3$			
Benzaldehyde, 2-bromo-4-nitro-	C_6H_{12}	268(4.01)	42-0148-69
	EtOH	256s(3.88),263(3.89), 271(3.86),299(3.87)	42-0148-69
Benzaldehyde, 4-bromo-2-nitro-	C_6H_{12}	296(4.01)	42-0148-69
	EtOH	224(3.85),252(3.82), 259(3.87),265(3.92), 293s(4.18)	42-0148-69
$C_7H_4BrN_3$			
Pyrido[2,3-d]pyridazine, 3-bromo-	EtOH	225(3.72),254(3.62), 262s(3.57),306(3.67), 316s(3.61)	12-1745-69
$C_7H_4BrN_3O_6$			
Methane, (4-bromo-3-nitrophenyl)-dinitro-	anion	366(4.18)	122-0161-69
Methane, (5-bromo-2-nitrophenyl)-dinitro-	anion	369(4.19)	122-0100-69
$C_7H_4BrN_5$			
7H-Pyrrolo[2,3-d]pyrimidine-5-carbonitrile, 4-amino-6-bromo-	pH 1	230(4.16),280(4.18)	35-2102-69
	pH 11	245(4.21),293(4.07)	35-2102-69
	EtOH	250(3.89),284(4.14)	35-2102-69
$C_7H_4Br_2N_2OS$			
Thiazole, 2-amino-5-bromo-4-(5-bromo-2-furyl)-	EtOH	250(4.39),300(4.17)	103-0372-69
$C_7H_4Br_2O$			
Benzaldehyde, 2,4-dibromo-	C_6H_{12}	232(4.26),265(3.12)	42-0148-69
	EtOH	266(3.92),268(3.72)	42-0148-69
$C_7H_4Br_2O_2$			
Salicylaldehyde, 3,5-dibromo-	C_6H_{12}	230(4.53),257(4.18), 262(4.22),352(3.83)	42-0148-69
	EtOH	228(4.78),251(4.16), 256(4.18),260(4.13), 349(3.77)	42-0148-69
Tropolone, 3,7-dibromo-	EtOH	264(4.4),355(3.38), 373(3.36),391(3.37)	23-2803-69
$C_7H_4ClNO_3$			
Benzaldehyde, 2-chloro-4-nitro-	C_6H_{12}	246(4.11),266(2.87), 295(3.77)	42-0148-69
	EtOH	239(4.18),244(4.18), 268(3.83),315(3.90)	42-0148-69
Benzaldehyde, 4-chloro-2-nitro-	C_6H_{12}	235(3.53),258(3.16)	42-0148-69
	EtOH	230(3.91),255(3.67), 261(3.68)	42-0148-69
$C_7H_4ClN_3$			
Pyrido[2,3-d]pyridazine, 5-chloro-	EtOH	223(4.12),256(3.63), 283s(3.33),288s(3.20), 293s(3.08),302s(2.92)	12-1759-69
Pyrido[2,3-d]pyridazine, 8-chloro-	EtOH	214s(3.91),227(4.13), 243s(3.62),258(3.62), 281s(3.31),287s(3.17), 292s(3.04),300s(2.86)	12-1759-69

Compound	Solvent	λ_{max}(log ε)	Ref.
$C_7H_4ClN_3O$			
Pyrido[2,3-d]pyridazin-5(6H)-one, 8-chloro-	EtOH	219s(3.86),224(3.87), 242(3.84),250s(3.81), 294s(3.65),306(3.67), 316s(3.62)	12-1759-69
	N KOH	224(3.88),252(3.76), 268s(3.63),333(3.51)	12-1759-69
Pyrido[2,3-d]pyridazin-8(7H)-one, 5-chloro-	EtOH	218s(3.75),226(3.82), 239s(3.83),250(3.84), 285s(3.60),290s(3.64), 301(3.66),312s(3.60)	12-1759-69
	N KOH	226(3.86),254(4.03), 325(3.70)	12-1759-69
$C_7H_4ClN_3O_6$			
Methane, (2-chloro-4-nitrophenyl)-dinitro-	anion	366(4.26)	122-0100-69
Methane, (2-chloro-5-nitrophenyl)-dinitro-	anion	365(4.24)	122-0100-69
Methane, (4-chloro-2-nitrophenyl)-dinitro-	anion	370(4.21)	122-0100-69
Methane, (5-chloro-2-nitrophenyl)-dinitro-	anion	369(4.23)	122-0100-69
$C_7H_4Cl_2FeO_3$			
Iron, tricarbonyl(1,4-dichloro-1,3-butadiene)-	C_6H_{12}	280s(3.46)	64-1518-69
$C_7H_4Cl_2N_2O_4$			
Toluene, 2,4-dichloro-α,α-dinitro-	anion	366(4.22)	122-0161-69
$C_7H_4Cl_2S$			
4H-Cyclopenta[c]thiophene, 1,3-dichloro-	EtOH	232(4.08),266(4.02)	1-0703-69
$C_7H_4Cl_4O$			
2,5-Cyclohexadien-1-one, 2,4,4,6-tetra-chloro-3-methyl-	MeOH	253(4.14)	23-1943-69
$C_7H_4Cl_6O$			
3-Cyclohexen-1-one, 2,2,3,4,5,6-hexa-chloro-6-methyl-	MeOH	226(3.92)	23-1943-69
stereoisomer	MeOH	220(3.90)	23-1943-69
3-Cyclohexen-1-one, 2,2,4,5,6,6-hexa-chloro-3-methyl-	MeOH	220(3.91)	23-1943-69
3-Cyclohexen-1-one, 2,3,4,5,6,6-hexa-chloro-2-methyl-	MeOH	228(3.92)	23-1943-69
stereoisomer	MeOH	226(3.89)	23-1943-69
$C_7H_4F_3NO_2$			
p-Benzoquinone, 2-(trifluoromethyl)-, 4-oxime	EtOH	244(3.63),309(4.28)	12-0935-69
	EtOH-NaOH	271(4.08),419(4.48)	12-0935-69
	dioxan	725(0.83)	12-0935-69
$C_7H_4F_3NO_4S$			
Sulfone, p-nitrophenyl trifluoromethyl	benzene	284(3.39)	65-1634-69
	MeOH	242(4.07),282(3.09)	65-1634-69
	EtOH	244(4.09),286(3.32)	65-1634-69
	DMF	288(3.47)	65-1634-69

Compound	Solvent	$\lambda_{max}(\log \epsilon)$	Ref.
$C_7H_4F_3N_5OS$			
Acetic acid, trifluoro-, 2-thiazolo-[5,4-d]pyrimidin-7-ylhydrazide	pH 1	266(4.01)	44-3161-69
	MeOH	262(4.01)	44-3161-69
$C_7H_4F_4$			
Toluene, 2,3,4,5-tetrafluoro-	C_6H_{12}	263(2.84)	39-0211-69C
$C_7H_4F_4O_2S$			
Sulfone, p-fluorophenyl trifluoromethyl	benzene	278(2.35)	65-1634-69
	MeOH	224(4.02),258(2.83), 270(2.70)	65-1634-69
	EtOH	224(4.02),260(2.82), 270(2.66)	65-1634-69
	DMF	314(1.94)	65-1634-69
$C_7H_4K_2N_6O_{10}$			
3,5-Dioxaheptanedinitrile, 2,2,6,6-tetranitro-, dipotassium salt (hydrate)	H_2O	349(4.49)	34-0116-69
$C_7H_4N_2O_2$			
1,4-Cyclohexadiene-1-carbonitrile, 3,6-dioxo-, 3-oxime	EtOH	256(3.81),328(4.15)	12-0935-69
	EtOH-NaOH	250(3.64),305(3.75), 402(4.46)	12-0935-69
	dioxan	729(1.18)	12-0935-69
$C_7H_4N_4O$			
7H-Pyrrolo[2,3-d]pyrimidine-5-carbonitrile, 3,4-dihydro-4-oxo-	pH 1	221(4.17),263(4.09)	35-2102-69
	pH 11	242(4.12),273(4.08)	35-2102-69
	EtOH	263(4.03)	35-2102-69
$C_7H_4N_4O_8$			
Toluene, α,α,2,4-tetranitro-	anion	370(4.23)	122-0100-69
Toluene, α,α,2,6-tetranitro-	anion	372(4.19)	122-0100-69
$C_7H_4N_6O$			
1,2,4-Benzotriazine, 3-azido-, 1-oxide	EtOH	262(4.44),362(3.54)	24-3818-69
$C_7H_4O_3$			
Carbonic acid, cyclic o-phenylene ester (also spectra in other solvents)	C_6H_{12}	253s(3.0),258s(3.2), 262(3.4),268(3.5), 273(3.4)	80-1323-69
	$MeOH-HClO_4$	255s(3.1),261s(3.3), 266(3.4),272(3.3)	80-1323-69
$C_7H_5BF_2O_2$			
Boron, difluoro(tropolonato)-	$CHCl_3$	250(4.11),305(3.96), 355s(3.82),365(3.89)	39-0526-69A
$C_7H_5BrN_2OS$			
Thiazole, 2-amino-4-(5-bromo-2-furyl)-	EtOH	247(4.35),293(4.16)	103-0372-69
$C_7H_5BrN_2O_4$			
Toluene, m-bromo-α,α-dinitro-	anion	369(4.21)	122-0161-69
Toluene, o-bromo-α,α-dinitro-	anion	370(4.21)	122-0100-69
Toluene, p-bromo-α,α-dinitro-	anion	370(4.17)	122-0161-69
$C_7H_5BrN_2S$			
Benzothiazole, 2-amino-4-bromo-	EtOH	232(4.54),257s(3.31), 264(3.77),273(4.04), 298s(3.32)	39-0268-69C

Compound	Solvent	λ_{max}(log ϵ)	Ref.
Benzothiazole, 2-amino-5-bromo-	EtOH	233(4.60),257s(3.51), 264s(3.81),272(4.02), 297(3.56),304(3.62)	39-0268-69C
Benzothiazole, 2-amino-6-bromo-	EtOH	228(4.55),257(3.21), 264s(3.92),273(4.19), 304s(3.45)	39-0268-69C
Benzothiazole, 2-amino-7-bromo-	EtOH	230(4.55),256s(3.48), 265(3.92),273(4.05), 298(3.31)	39-0268-69C
C_7H_5BrO			
Benzaldehyde, p-bromo-	C_6H_{12}	257(4.07),262(4.33)	42-0148-69
	EtOH	250(4.07),255(4.05), 262(4.06),270(4.10)	42-0148-69
Benzoyl bromide	heptane	245(4.22),282(3.18)	59-0765-69
Tropone, 4-bromo-	EtOH	230(4.12),308(3.72)	23-1169-69
$C_7H_5BrO_2$			
Benzaldehyde, 2-bromo-4-hydroxy-	EtOH	238(4.09),284(4.12)	94-0089-69
Salicylaldehyde, 3-bromo-	C_6H_{12}	229(4.21),272(3.81), 362(3.46)	42-0148-69
	EtOH	228(3.93),248(4.03), 356(3.32)	42-0148-69
Salicylaldehyde, 4-bromo-	EtOH	266(4.15),325(3.65)	94-0089-69
Salicylaldehyde, 5-bromo-	C_6H_{12}	215(3.68),230(4.33), 352(3.29)	42-0148-69
	EtOH	230(4.24),256(3.92), 261(3.86),350(3.91)	42-0148-69
Salicylaldehyde, 6-bromo-	EtOH	270(3.90),344(3.50)	94-0089-69
$C_7H_5ClN_2O_2S$			
Sulfone, m-chlorophenyl diazomethyl	C_6H_{12}	251(4.07),277(3.29), 286(3.17)	54-0641-69
	40% dioxan	385(1.84)	54-0641-69
$C_7H_5ClN_2O_4$			
Toluene, m-chloro-α,α-dinitro-	anion	371(4.21)	122-0161-69
Toluene, o-chloro-α,α-dinitro-	anion	369(4.22)	122-0161-69
Toluene, p-chloro-α,α-dinitro-	anion	373(4.17)	122-0161-69
$C_7H_5ClN_2S$			
Benzothiazole, 2-amino-4-chloro-	EtOH	232(4.55),268s(3.89), 271(4.08),300s(3.21)	39-0268-69C
Benzothiazole, 2-amino-5-chloro-	EtOH	232(4.59),257s(3.03), 264s(3.72),272(4.00), 296(3.60),304(3.59)	39-0268-69C
Benzothiazole, 2-amino-6-chloro-	EtOH	227(4.55),258s(3.54), 264s(3.96),273(4.17), 298(3.45)	39-0268-69C
Benzothiazole, 2-amino-7-chloro-	EtOH	228(4.55),258s(3.76), 262s(3.98),272(4.08), 298s(3.26)	39-0268-69C
C_7H_5ClO			
Benzaldehyde, o-chloro-	C_6H_{12}	250(3.88),301(3.76)	42-0148-69
	EtOH	250(3.79),256(3.76), 260s(3.65),302(3.07)	42-0148-69
Benzoyl chloride	heptane	240(4.43),281(3.35)	59-0765-69
Tropone, 3-chloro- (hemihydrate)	EtOH	235(4.34),240(4.40), 248(4.21),305(3.67), 316(3.62)	39-1499-69C

Compound	Solvent	$\lambda_{max}(\log \epsilon)$	Ref.
Tropone, 4-chloro-	EtOH	231(4.37),307(3.94), 316(3.94)	39-1499-69C
$C_7H_5ClO_2$			
Salicylaldehyde, 3-chloro-	C_6H_{12}	227(4.24),261(4.18), 267(3.92),345(3.71)	42-0148-69
	EtOH	225(4.11),250(3.94), 256(4.06),262(4.08), 267(4.01),341(3.57)	42-0148-69
Salicylaldehyde, 5-chloro-	C_6H_{12}	230(4.23),257(3.88), 262(3.76),353(3.64)	42-0148-69
	EtOH	237(3.55),250(3.48), 256(3.46),261(3.42), 290(3.10),348(3.24)	42-0148-69
$C_7H_5ClO_3$			
Salicylic acid, 3-chloro-	MeOH	238(3.77),308(3.63)	39-2418-69C
Salicylic acid, 5-chloro-	MeOH	232(3.88),314(3.54)	39-2418-69C
C_7H_5ClS			
Benzoyl chloride, thio-	C_6H_{12}	224s(3.9),228s(3.9), 243s(3.7),260s(3.6), 272s(3.5),313(4.14), 530(1.82)	48-0045-69
	MeCN	225(3.74),248s(3.5), 317(4.08),518(1.79)	48-0045-69
$C_7H_5Cl_2NS$			
Benzamide, 3,4-dichlorothio-	EtOH	247(4.13),300(3.83)	39-0861-69C
Benzamide, 3,5-dichlorothio-	EtOH	240(4.02),304(3.80)	39-0861-69C
$C_7H_5Cl_6N_3$			
s-Triazine, 2-ethyl-4,6-bis(trichloro-methyl)-	MeOH	221(3.25),278(2.76)	18-2924-69
$C_7H_5Cl_6N_3S$			
s-Triazine, 2-(ethylthio)-4,6-bis-(trichloromethyl)-	MeOH	221(3.46),274(4.13)	18-2931-69
$C_7H_5FN_2O_4$			
Toluene, o-fluoro-α,α-dinitro-	anion	367(4.19)	122-0161-69
Toluene, p-fluoro-α,α-dinitro-	anion	373(4.19)	122-0161-69
$C_7H_5FN_2O_5$			
Anisole, 4-fluoro-2,6-dinitro-, Meisen-heimer complex intermediate	93% dioxan-MeOH	485(4.35)	22-2692-69
$C_7H_5FN_2S$			
Benzothiazole, 2-amino-4-fluoro-	EtOH	223(4.55),262(4.21)	39-0268-69C
Benzothiazole, 2-amino-5-fluoro-	EtOH	222(4.40),268(4.22), 290s(3.99),297s(3.91)	39-0268-69C
Benzothiazole, 2-amino-6-fluoro-	EtOH	223(4.53),264(4.21), 289s(3.86)	39-0268-69C
Benzothiazole, 2-amino-7-fluoro-	EtOH	223(4.51),269(4.24)	39-0268-69C
C_7H_5FO			
Benzoyl fluoride	heptane	229(4.27),277(3.28)	59-0765-69
$C_7H_5F_3N_4O$			
4-Pteridinol, 3,4-dihydro-4-(trifluoro-methyl)-	pH 7.0	226(3.53),264(3.64), 315(3.87)	39-1751-69C

Compound	Solvent	$\lambda_{max}(\log \epsilon)$	Ref.
C_7H_5IOS			
2-Propyn-1-ol, 3-(5-iodo-2-thienyl)-	MeOH	284(4.17)	39-2173-69C
C_7H_5IS			
Thiophene, 2-iodo-5-(1-propynyl)-	MeOH	282(4.15)	39-1813-69C
C_7H_5NOS			
2-Benzothiazolinone	MeOH	213(4.60),243(4.01), 282(3.46),289(3.46)	4-0163-69
$C_7H_5NO_2$			
1,2-Benzisoxazol-3-ol	MeOH	240(3.7),285f(3.5)	24-3775-69
Nicotinic acid, 2-(hydroxymethyl)-, lactone	EtOH	221(3.76),264s(3.63), 270(3.70),276s(3.53)	12-1759-69
Picolinic acid, 3-(hydroxymethyl)-, lactone	EtOH	223(3.87),264s(3.82), 271(3.87),278s(3.68)	12-1759-69
$C_7H_5NO_3$			
Benzaldehyde, 2-nitro-	C_6H_{12}	228(4.15),252(3.87)	42-0148-69
	EtOH	226(3.80),250(3.62), 255(3.63)	42-0148-69
3-Isoxazolol, 5-(2-furyl)-	EtOH	272(4.31)	33-0720-69
$C_7H_5NO_3S$			
Saccharin	0.4N HCl	279(2.89)	18-2282-69
$C_7H_5NO_4$			
1,4-Cyclohexadiene-1-carboxylic acid, 3,6-dioxo-, 3-oxime	EtOH	221(4.04),259(3.91), 334(4.11)	12-0935-69
	EtOH-NaOH	283(3.78),407(4.46)	12-0935-69
	dioxan	724(1.74)	12-0935-69
Salicylaldehyde, 3-nitro-	C_6H_{12}	240(4.24),357(3.50)	42-0148-69
	EtOH	238(4.14),240(4.15), 365(2.67)	42-0148-69
Salicylaldehyde, 5-nitro-	C_6H_{12}	246(4.11),295(3.77)	42-0148-69
	EtOH	239(4.18),244(4.18), 315(4.00)	42-0148-69
$C_7H_5NO_5S$			
3H-1,2-Benzoxathiole, 5-nitro-, 2,2-dioxide	MeCN	276(4.00)	35-4912-69
C_7H_5NS			
Isothiocyanic acid, phenyl ester	hexane	220(4.38),281(4.12)	114-0309-69B
$C_7H_5NS_2$			
2-Benzothiazolinethione	MeOH	233(4.15),238(4.16), 326(4.41)	4-0163-69
$C_7H_5N_2O_2$			
Benzonitrile, m-nitro-, half-reduction radical of nitro group	pH 1	280(4.15),400(2.72)	64-1336-69
	pH 13	290(4.16),400(2.81)	64-1336-69
Benzonitrile, p-nitro-, half-reduction radical of nitro group	pH 1	305(4.28),460(2.65)	64-1336-69
	pH 13	330(4.15),480(3.15)	64-1336-69
$C_7H_5N_3O$			
2,4,6-Cycloheptatrien-1-one, 3-azido-	C_6H_{12}	226(4.33),273(4.19), 299(3.46),312(3.40), 327(3.20)	39-1499-69C

Compound	Solvent	$\lambda_{max}(\log \epsilon)$	Ref.
2,4,6-Cycloheptatrien-1-one, 4-azido-	C_6H_{12}	230(4.35),243(4.28), 252(4.28),261(4.26), 272(4.16),310(4.22), 325(4.22)	39-1499-69C
Nicotinamide, 2-cyano-	EtOH	233(3.82),258s(3.27), 263s(3.22),266s(3.20), 270(3.17),279(3.03)	12-1745-69
Pyrido[2,3-d]pyridazine, 6-oxide	EtOH	213s(3.80),232(4.02), 249s(4.20),255(4.22), 296(4.03),307s(3.89), 355s(3.10)	12-1745-69
Pyrido[2,3-d]pyridazine, 7-oxide	EtOH	218s(3.78),233s(3.86), 248s(3.92),253(3.92), 294(3.96),300s(3.90), 304s(3.76),355s(3.00)	12-1745-69
Pyrido[2,3-d]pyridazin-5(6H)-one	EtOH	212s(3.75),223(3.78), 236(3.73),243s(3.70), 275s(3.52),282s(3.57), 292(3.58),303s(3.56), 307s(3.53)	12-1745-69
Pyrido[2,3-d]pyridazin-8(7H)-one	EtOH	213s(3.73),226(3.82), 238s(3.82),243(3.83), 275s(3.56),293(3.62), 304s(3.58),308s(3.55)	12-1745-69
v-Triazolo[1,5-a]pyridine-3-carbox-aldehyde	EtOH	258(3.59),290s(--), 313(4.13)	88-1549-69
$C_7H_5N_3O_2$			
Benzimidazole, 5-nitro-	MeOH	300(4.00)	22-3002-69
anion	MeOH	350(4.02)	22-3002-69
1H-Indazole, 4-nitro-	MeOH	335(3.83)	22-3002-69
anion	MeOH	420(3.69)	22-3002-69
1H-Indazole, 5-nitro-	MeOH	305(3.86)	22-3002-69
anion	MeOH	390(3.86)	22-3002-69
1H-Indazole, 6-nitro-	MeOH	250(4.14)	22-3002-69
anion	MeOH	280(4.19)	22-3002-69
1H-Indazole, 7-nitro-	MeOH	355(3.90)	22-3002-69
anion	MeOH	435(3.80)	22-3002-69
$C_7H_5N_3O_4$			
1,2,4-Oxadiazole, 5-methyl-3-(5-nitro-2-furyl)-	EtOH	304(3.97)	18-3335-69
$C_7H_5N_3O_4S$			
Sulfone, diazomethyl m-nitrophenyl	C_6H_{12}	248(4.23)	54-0641-69
	40% dioxan	385(1.97)	54-0641-69
$C_7H_5N_3O_6$			
Toluene, α,α,2-trinitro-	anion	371(4.20)	122-0100-69
Toluene, α,α,3-trinitro-	anion	370(4.18)	122-0161-69
Toluene, α,α,4-trinitro-	anion	369(4.25)	122-0161-69
$C_7H_5N_5O_8$			
Toluene, 4-amino-α,α,3,5-tetranitro-	anion	367(4.18)	122-0161-69
C_7H_6BrIOS			
1-Propanone, 2-bromo-1-(5-iodo-2-thienyl)-	MeOH	316(4.10)	39-1813-69C

Compound	Solvent	$\lambda_{max}(\log \epsilon)$	Ref.
C_7H_6BrNO			
2,4,6-Cycloheptatrien-1-one, 3-amino-2-bromo-	MeOH	216(4.06),264(4.33), 273(4.30),313(3.71)	94-2548-69
$C_7H_6BrNO_2$			
p-Benzoquinone, 2-bromo-6-methyl-, 4-oxime	EtOH	230(3.52),316(4.30)	12-0935-69
	EtOH-NaOH	272(3.96),418(4.51)	12-0935-69
	dioxan	725(-1.1)	12-0935-69
Toluene, α-bromo-p-nitro-	hexane	205(3.87),263(3.94)	39-1178-69B
$C_7H_6BrNO_3$			
m-Cresol, α-bromo-6-nitro-	hexane	215(4.26),279(4.15), 348(3.81)	39-1178-69B
$C_7H_6Br_4O$			
2-Cyclopenten-1-one, 2,3,5,5-tetra-bromo-4,4-dimethyl-	MeOH	269(3.93)	20-0017-69
$C_7H_6ClNOS_2$			
6-Methoxy-1,2,3-benzodithiazol-2-ium chloride	H_2SO_4	260(3.87),375(3.90), 470(4.03)	104-0151-69
$C_7H_6ClNO_2$			
p-Benzoquinone, 2-chloro-6-methyl-, 4-oxime	EtOH	230(3.49),315(4.28)	12-0935-69
	EtOH-NaOH	271(4.08),419(4.48)	12-0935-69
	dioxan	725(-1)	12-0935-69
Toluene, α-chloro-p-nitro-	hexane	211(4.08),258(4.20)	39-1178-69B
C_7H_6ClNS			
Benzamide, m-chlorothio-	EtOH	237(4.01),298(3.84)	39-0861-69C
Benzamide, p-chlorothio-	EtOH	248(4.09),304(3.84)	39-0861-69C
$C_7H_6Cl_2N_2O$			
Benzamidoxime, 2,4-dichloro-	EtOH	251(4.37)	39-0861-69C
Benzamidoxime, 3,4-dichloro-	EtOH	249(4.08),276(3.92)	39-0861-69C
Benzamidoxime, 3,5-dichloro-	EtOH	280(3.82)	39-0861-69C
C_7H_6FNS			
Benzamide, p-fluorothio-	EtOH	240(3.98),300(3.85)	39-0861-69C
$C_7H_6F_3N_5O_3$			
Guanine, 1-methyl-8-(trifluoromethyl)-	pH 0.0	248s(3.81),269(3.97)	5-0201-69F
	pH 4.0	257(4.05),270s(3.97)	5-0201-69F
	pH 10.0	258(3.89),278(3.95)	5-0201-69F
Guanine, 7-methyl-8-(trifluoromethyl)-	pH 0.0	245s(3.72),270(3.93)	5-0201-69F
	pH 6.0	245(3.72),292(3.76)	5-0201-69F
	pH 12.0	240s(3.72),292(3.76)	5-0201-69F
Guanine, 9-methyl-8-(trifluoromethyl)-	pH -1.9	256(4.01)	5-0201-69F
	pH 7.0	258(4.20),270s(4.11)	5-0201-69F
	pH 12.0	260s(3.78),277(3.98)	5-0201-69F
$C_7H_6INO_2$			
Toluene, α-iodo-p-nitro-	hexane	210(4.06),259(4.12)	39-1178-69B
$C_7H_6NO_3$			
Phenol, m-nitro-, half-reduction radical of nitro group	pH 1	285(4.12),400(2.80)	64-1336-69
	base	290(4.14),400(2.75)	64-1336-69
Phenol, p-nitro-, half-reduction radical of nitro group	pH 1	330(4.29),500(2.85)	64-1336-69
	base	360(4.25),560(3.44)	64-1336-69

Compound	Solvent	$\lambda_{max}(\log \epsilon)$	Ref.
$C_7H_6NO_4$			
Benzoic acid, m-nitro-, half-reduction	pH 1	280(4.08),400(2.70)	64-1336-69
radical of nitro group	pH 13	285(4.20),400(2.70)	64-1336-69
$C_7H_6N_2$			
Benzimidazole	MeOH	244(4.11),250(4.11), 273(4.15),289(4.15)	22-1926-69
	MeOH	279(3.77)	22-3002-69
anion	MeOH	284(3.84)	22-3002-69
1H-Indazole	MeOH	285(3.65)	22-3002-69
anion	MeOH	300(--)	22-3002-69
Picolinonitrile, 6-methyl-	isooctane	271(3.64)	4-0859-69
	pH 1	274(3.89)	4-0859-69
	MeOH	271(3.64)	4-0859-69
Toluene, α-diazo-	pentane	274(4.44),490(1.46)	47-3313-69
$C_7H_6N_2O$			
Picolinonitrile, 6-methyl-, 1-oxide	isooctane	279(3.91)	4-0859-69
	pH 1	264(3.92)	4-0859-69
	MeOH	271(3.91)	4-0859-69
$C_7H_6N_2O_2$			
Benzoxazolin-2-one, 3-amino-	EtOH	229(3.97),275(3.70), 280s(3.62)	39-0772-69C
Furo[2,3-d]pyridazine, 4-methoxy-	H_2O	200(4.17),243(3.78), 269(3.73)	22-4004-69
Furo[2,3-d]pyridazine, 7-methoxy-	H_2O	208(4.45),239(3.79)	22-4004-69
Pyrrole-2-carboxylic acid, 5-cyano-, methyl ester	EtOH	266(4.58),274s(4.50)	78-0295-69
$C_7H_6N_2O_3$			
Benzaldehyde, o-nitro-, oxime, trans-	H_2O	233(4.2),264s(3.8)	80-0225-69
Ph/OH	pH 13	260(4.2),345(3.5)	80-0225-69
Benzofurazan, 5-methoxy-, N-oxide	H_2O	295s(3.61),306(3.79), 320(3.79),367(3.61)	78-4197-69
1,4-Cyclohexadiene-1-carboxamide,	EtOH	257(3.72),320(4.20)	12-0935-69
3,6-dioxo-, 3-oxime	EtOH-NaOH	255(3.52),293(3.76), 401(4.48)	12-0935-69
	dioxan	722(1.43)	12-0935-69
$C_7H_6N_2O_7$			
4H-Pyrazole-3,5-dicarboxylic acid, 4-oxo-, dimethyl ester, 1,2-dioxide	n.s.g.	233(4.20),285(3.84), 362(2.23)	44-0187-69
$C_7H_6N_2S$			
2-Benzimidazolinethione	MeOH	246(4.19),297s(4.39), 304(4.49)	97-0152-69
1,2-Benzisothiazole, 3-amino-	MeOH	226(4.14),245(3.99), 281(3.34),319(3.71)	24-1961-69
	90% MeOH-HCl	231(4.35),255s(3.96), 323(3.85),333s(3.79)	24-1961-69
Benzothiazole, 2-amino-	EtOH	224(4.50),265(4.13), 295s(3.51)	2-0694-69
	EtOH	224(4.55),277(4.28), 293s(3.86)	39-0268-69C
$C_7H_6N_4$			
2,3-Diazabicyclo[3.2.0]hept-2-ene-1,5-dicarbonitrile, cis	ether	322(2.46)	27-0037-69
1H-1,2-Diazepine-3,6-dicarbonitrile, 4,5-dihydro-	ether	239(3.40),328(3.75)	27-0037-69

Compound	Solvent	λ_{max}(log ε)	Ref.
$C_7H_6N_4O$			
4(1H)-Pteridinone, 1-methyl-	MeOH	232(4.05),328(3.88)	39-2415-69C
4(3H)-Pteridinone, 3-methyl-	MeOH	237(4.03),277s(3.57), 315(3.73)	39-2415-69C
$C_7H_6N_4OS$			
Purine-8(9H)-thione, 9(7)-acetyl-	MeOH	323(4.10)	4-0023-69
$C_7H_6N_4O_4$			
Xanthine, 3-acetoxy-	pH 5	267(4.15)	88-0785-69
$C_7H_6N_4O_7$			
Aniline, N-methoxy-2,4,6-trinitro-	$C_2H_4Cl_2$	262s(3.78),332(4.02), 413s(3.48)	80-0941-69
$C_7H_6N_6O_3$			
6-Pteridinecarboxaldehyde, 2-amino-3,4-dihydro-4-oxo-, 6-oxime 8-oxide	pH 13	270s(4.20),302(4.34), 410(3.86)	5-0100-69F
$C_7H_6N_6O_{10}$			
3,5-Dioxaheptanedinitrile, 2,2,6,6-tetranitro-, dipotassium salt (hydrate)	H_2O	349(4.49)	34-0116-69
$C_7H_6N_6S$			
Acetonitrile, [(1,2-dihydropyrimido-[5,4-e]-as-triazin-5-yl)thio]-	pH 7	258(4.12),356(3.54)	44-3161-69
C_7H_6O			
2,4,6-Cycloheptatrien-1-one	H_2O	225(4.33),228(4.34), 232(4.34),239(4.10), 304(3.90),313(3.92)	35-0219-69
C_7H_6OS			
Benzoic acid, thio-	n.s.g.	238(4.0),286s(3.8), 415(1.7)	39-1182-69B
anion	n.s.g.	232s(4.0),278(4.0)	39-1182-69B
$C_7H_6O_2$			
Benzaldehyde, p-hydroxy-	EtOH	221(4.14),284(4.24)	42-0148-69
Benzoic acid	heptane	232(4.11),274(3.02)	59-0765-69
	EtOH	225(3.99),272(2.93)	28-1562-69B
Pb(IV) salt	C_6H_{12}	233(4.31)	33-1495-69
2-Furanacrolein	30% EtOH	236(3.37),322(4.26)	103-0278-69
Salicylaldehyde	hexane	253(4.0),258(4.1), 398(3.6)	18-0960-69
	C_6H_{12}	229(3.44),255(3.64), 262(3.63),339(3.20)	42-0148-69
	EtOH	228(3.57),250(3.54), 255(3.61),327(3.57)	42-0148-69
$C_7H_6O_3$			
6,7-Dioxabicyclo[3.2.2]nona-3,8-dien-2-one	isooctane	214(3.84),248s(3.00), 290s(2.45)	88-3295-69
C_7H_6S			
4H-Cyclopenta[c]thiophene	EtOH	252(4.01)	1-0703-69
$C_7H_6S_2$			
Benzoic acid, dithio-	C_6H_{12}	244s(3.9),298(4.00), 538(1.85)	48-0045-69

Compound	Solvent	λ max (log ε)	Ref.
Benzoic acid, dithio- (cont.)	EtOH	300(--),518(--)	48-0045-69
	MeCN	302(3.99),522(1.85)	48-0045-69
	n.s.g.	294s(4.0),345(4.0), 505(2.0)	39-1182-69B
anion	n.s.g.	294(3.9),358(3.9), 495(2.1)	39-1182-69B
$C_7H_7BO_4$			
p-Boronobenzoic acid	EtOH	232(4.15),280(3.02)	28-1562-69B
$C_7H_7BrN_2OS$			
6-Methylthiazolo[3,2-a]pyrazin-4-ium bromide, 7-oxide	n.s.g.	198(4.35),226(4.16), 234s(4.10),250s(3.87), 267(3.87),275s(3.81), 339s(4.19),349(4.2), 359s(3.81),362s(3.62)	39-2270-69C
8-Methylthiazolo[3,2-a]pyrazin-4-ium bromide, 7-oxide	n.s.g.	198(4.32),220(4.08), 249(3.93),266s(3.85), 331s(4.15),347(4.32)	39-2270-69C
$C_7H_7BrN_2S$			
8-Methylthiazolo[3,2-a]pyrazin-4-ium bromide	n.s.g.	195(4.24),233(4.06), 270(3.6),315(4.03), 326(4.03)	39-2270-69C
Thiourea, N-(m-bromophenyl)-	EtOH	238s(4.16),276(4.16)	39-0268-69C
Thiourea, N-(o-bromophenyl)-	EtOH	276(3.98)	39-0268-69C
Thiourea, N-(p-bromophenyl)-	EtOH	245(4.23),277(4.36)	39-0268-69C
$C_7H_7BrO_3$			
1,2,4-Cyclopentanetrione, 5-bromo-3,3-dimethyl-	MeOH-HCl	285(4.14)	20-0277-69
	MeOH-NaOH	227(4.01),325(4.10)	20-0277-69
$C_7H_7Br_2NO_2$			
2-Hexenimide, 4,5-dibromo-N-methyl-	MeOH	215(4.02),245(3.81)	44-1695-69
$C_7H_7Br_3O$			
2-Cyclopenten-1-one, 2,3,5-tribromo-4,4-dimethyl-	MeOH	263(3.87)	20-0017-69
$C_7H_7ClNO_6P$			
o-Cresol, α-chloro-4-nitro-, dihydrogen phosphate	H_2O	293(3.93)	35-4532-69
$C_7H_7ClN_2$			
1H-Pyrrolo[3,2-c]pyridine, 6-chloro-2,3-dihydro-	EtOH	260(3.96)	88-1909-69
	EtOH	260(3.96)	103-0410-69
$C_7H_7ClN_2O$			
Aniline, p-chloro-N-methyl-N-nitroso-	MeCN	274(3.99),375(3.88)	24-2093-69
Benzamidoxime, m-chloro-	EtOH	269(3.69)	39-0861-69C
Benzamidoxime, p-chloro-	EtOH	225(4.08),267(3.79)	39-0861-69C
Pyrrolo[1,2-a]pyrimidin-4(6H)-one, 2-chloro-7,8-dihydro-	EtOH	224(3.66),275(3.61)	22-3133-69
$C_7H_7ClN_2S$			
Thiourea, N-(m-chlorophenyl)-	EtOH	240(4.09),276(4.18)	39-0268-69C
Thiourea, N-(o-chlorophenyl)-	EtOH	240(4.01),272(3.99)	39-0268-69C
Thiourea, N-(p-chlorophenyl)-	EtOH	242(4.13),277(4.18)	39-0268-69C

Compound	Solvent	$\lambda_{max}(\log \epsilon)$	Ref.
$C_7H_7Cl_3O_3S_2$			
3-Thiophenemethanol, 2,5-dichloro-4-(chloromethyl)-, methanesulfonate	EtOH	257(3.76)	44-0340-69
$C_7H_7FN_2O$			
Aniline, p-fluoro-N-methyl-N-nitroso-	MeCN	270(3.85),377(2.30)	24-2093-69
Benzamidoxime, p-fluoro-	EtOH	258(3.71)	39-0861-69C
$C_7H_7FN_2S$			
Thiourea, N-(m-fluorophenyl)-	EtOH	269(4.17)	39-0268-69C
Thiourea, N-(o-fluorophenyl)-	EtOH	236(4.05),268(4.04)	39-0268-69C
Thiourea, N-(p-fluorophenyl)-	EtOH	238(4.00),269(4.09)	39-0268-69C
$C_7H_7IN_2O$			
Benzamidoxime, p-iodo-	EtOH	241(4.10),268(3.91)	39-0861-69C
C_7H_7IOS			
1-Propanone, 1-(5-iodo-2-thienyl)-	MeOH	217(3.65),270s(3.90), 298(3.15)	39-1813-69C
$C_7H_7IO_2S$			
1-Propanone, 2-hydroxy-1-(5-iodo-2-thienyl)-	MeOH	218(3.70),272s(3.85), 303(4.16)	39-1813-69C
C_7H_7IS			
Thiophene, 2-iodo-5-propenyl-	MeOH	218(3.82),290(4.00)	39-1813-69C
C_7H_7NO			
Anthranilaldehyde	C_6H_{12}	255(3.75),261s(3.59), 352(3.67)	22-3523-69
Benzaldoxime, m. 132° (anti)	EtOH	244(4.1),290(2.8)	28-0078-69A
Benzaldoxime, m. 35° (syn)	EtOH	205(4.3),210s(4.2), 251(4.2),279s(3.3), 292(3.0)	28-0078-69A
Benzamide	heptane	224(4.00),275(2.75)	59-0765-69
Pyridine, 2-(vinyloxy)-	hexane	224(4.25),276(4.10)	30-0620-69C
2(1H)-Pyridone, N-vinyl-	EtOH	322(3.85)	30-0620-69C
Tropone, 3-amino-	MeOH	263(4.49),274(4.43), 303(3.81),313(3.79)	94-2548-69
	EtOH	263(4.36),275(4.31), 303(3.72),313(3.69)	39-1499-69C
Tropone, 4-amino-	EtOH	227(4.23),260(3.76), 362(4.30)	39-1499-69C
$(C_7H_7NO)_n$			
Pyridine, 2-vinyl-, 1-oxide, polymer	H_2O	215(3.89),261(3.58)	39-0054-69B
	0.01M silicic acid	214(3.90),259(3.58)	39-0054-69B
Pyridine, 4-vinyl-, 1-oxide, polymer	H_2O	208(4.05),261(4.02)	39-0054-69B
	0.01M silicic acid	208(4.05),260(4.02)	39-0054-69B
C_7H_7NOS			
2,3-Dihydro-8-hydroxythiazolo[3,2-a]-pyridinium hydroxide, inner salt	pH 1	230(3.68),330(3.91)	1-1704-69
	pH 13	245(3.84),350(3.98)	1-1704-69
$C_7H_7NO_2$			
Anthranilic acid	MeOH	245(3.88),335(3.66)	2-1051-69
(also rare earth chelate spectra)	EtOH	245(4.04),335(3.87)	2-1051-69
Benzohydroxamic acid	EtOH	280(4.30)	44-0984-69

Compound	Solvent	$\lambda_{max}(\log \epsilon)$	Ref.
Benzohydroxamic acid, sodium salt	EtOH	305(4.30)	44-0984-69
Benzoic acid, p-amino-	EtOH	220(3.99),284(4.14)	9-0254-69
p-Benzoquinone, 2-methyl-, 4-oxime	EtOH	228(3.40),305(4.28)	12-0935-69
	EtOH-NaOH	263(3.66),407(4.48)	12-0935-69
	dioxan	730(0.0)	12-0935-69
2,4-Hexadienimide, N-methyl-	MeOH	220(4.18),287(3.56)	44-1695-69
Maleimide, 2-methyl-2-vinyl-	MeOH	227(4.18),317(3.36)	25-0107-69
Pyrrole-3-carboxaldehyde, 4-acetyl-	EtOH	211(4.28),227s(3.80), 267(3.97),282s(3.90)	33-1911-69
$C_7H_7NO_2S$			
p-Benzoquinone, 2-(methylthio)-, 4-oxime	EtOH	238(3.85),317(4.15), 390(3.60)	12-0935-69
	EtOH-NaOH	274(3.81),425(4.30)	12-0935-69
	dioxan	725(-1)	12-0935-69
$C_7H_7NO_3$			
Anisole, p-nitro-	H_2O	317(4.01)	88-4205-69
	MeOH	226(3.90),306(4.06)	65-1634-69
	MeCN	310(4.04)	88-4205-69
p-Benzoquinone, 2-methoxy-, 4-oxime	EtOH	321(3.23)	12-0935-69
	EtOH-NaOH	258(3.65),402(4.48)	12-0935-69
	dioxan	725(-0.5)	12-0935-69
Pyrrole-2-carboxylic acid, 3-acetyl-	EtOH	214(4.16),259(3.87), 307(3.76)	33-1911-69
$C_7H_7NO_4S$			
4,5-Thiazoledicarboxylic acid, dimethyl ester	MeOH	252(3.89)	32-0029-69
C_7H_7NS			
Benzamide, thio-	C_6H_{12}	240(--),295(--), 305(--),418(--)	48-0045-69
	EtOH	241(3.98),293(3.85), 370s(2.4)	48-0045-69
	MeCN	240(4.00),294(3.86)	48-0045-69
$C_7H_7N_3O_2S$			
Thiazolo[4,5-d]pyrimidine-5,7(4H,6H)-dione, 4,6-dimethyl-	MeOH	299(3.77)	44-3285-69
$C_7H_7N_3O_3$			
Benzamidoxime, p-nitro-	MeOH	232(3.92),257(3.93), 328(3.71)	12-0161-69
	MeOH-NaOH	265(4.01),440(3.97)	12-0161-69
$C_7H_7N_3O_4$			
Aziridine, 1-(5-nitro-2-furoyl)-, oxime	EtOH	250(3.90),348(4.06)	18-0556-69
4H-1,2,4-Oxadiazine, 5,6-dihydro-3-(5-nitro-2-furyl)-	EtOH	240(3.83),304(3.88), 348(3.86)	18-0556-69
3,5-Pyridinedicarboxylic acid, 2,6-diamino-	pH 2	275(4.38),333(4.14)	39-0133-69C
	pH 12	223(4.39),277(4.20), 325(4.20)	39-0133-69C
$C_7H_7N_3O_5$			
Aniline, N-methoxy-2,4-dinitro-	$C_2H_4Cl_2$	262(4.03),337(4.18), 395s(3.65)	80-0941-69
Aniline, N-methoxy-2,6-dinitro-	$C_2H_4Cl_2$	268(3.66),387(3.49)	80-0941-69

Compound	Solvent	$\lambda_{max}(\log \epsilon)$	Ref.
$C_7H_7N_5O$			
2-Furamidine, N-4H-1,2,4-triazol-4-yl-	MeOH	258(4.11)	48-0477-69
9H-Purine-9-acetaldehyde, 6-amino-, hydrochloride hydrate	pH 1	259(4.14)	88-2285-69
	pH 12	261(4.15)	88-2285-69
$C_7H_7N_5OS$			
Acetamide, N-(5-imidazol-2-yl-1,3,4-thiadiazol-2-yl)-	MeOH	304(4.22)	4-0835-69
$C_7H_7N_5O_2$			
Acetamide, N-(1,6-dihydro-6-oxo-9H-purin-9-yl)-	pH 13	256(4.07)	44-2102-69
Glycine-^{15}N, N-purin-6-yl-	pH 1	274(4.20)	44-3498-69
	pH 5.8	268(4.20)	44-3498-69
	pH 13	274(4.26)	44-3498-69
Guanine, N^2-acetyl-	pH 1	260(--)	69-0238-69
	pH 7	260(4.15),275s(--)	69-0238-69
	pH 11	265(--)	69-0238-69
	pH 14	268(--)	69-0238-69
3H-Purine-3-acetic acid, 6-amino-	pH 1	275(4.25)	4-0955-69
	pH 10	274(4.11)	4-0955-69
9H-Purine-9-acetic acid, 6-amino-	pH 1	257(4.15)	4-0955-69
	pH 10	260(4.16)	4-0955-69
Pyrimido[5,4-e]-as-triazine-5,7(6H,8H)-dione, 6,8-dimethyl- (fervenulin)	EtOH	239(4.23),275(3.20), 340(3.64)	44-2102-69
7H-Pyrrolo[2,3-d]pyrimidine, 4-(methyl-amino)-5-nitro-	pH 1	350(4.14)	4-0207-69
	pH 11	247(4.05),415(4.11)	4-0207-69
$C_7H_7N_5O_3$			
9H-Imidazo[1,2-a]purin-9-one, 3,5,6,7-tetrahydro-6,7-dihydroxy-	pH 5	246(4.02),279(3.80)	69-0238-69
$C_7H_7N_5O_6$			
m-Phenylenediamine, N-methyl-2,4,6-tri-nitro-	MeOH	292(4.26),333(4.27), 409(4.15)	35-4155-69
C_7H_8			
1,3,5-Cycloheptatriene	EtOH	261(3.57)	18-2013-69
1,3-Heptadien-5-yne, cis	EtOH	258(4.34)	78-2837-69
1,3-Heptadien-5-yne, trans	EtOH	258(4.50),268s(4.39)	78-2837-69
Toluene	C_6H_{12}	262(2.39)	101-0017-69B
	EtOH	206(3.86),263(2.43)	28-1562-69B
tetracyanoethylene complex	C_6H_{12}	403(3.29)	39-1161-69B
$C_7H_8BF_4N_3$			
2-Methyl-s-triazolo[4,3-a]pyridinium tetrafluoroborate	n.s.g.	218f(3.6),290f(3.7)	24-3159-69
$C_7H_8BaN_5O_5P$			
6-Purinol, 9-(2-hydroxyethyl)-2-amino-, 2'-phosphate, barium salt	pH 1	255(3.79),280s(--)	94-1268-69
	pH 13	255s(--),269(3.98)	94-1268-69
C_7H_8BrNO			
2-Picoline, 6-bromo-5-methoxy-	pH 1	230(3.74),300(3.91)	1-2065-69
	pH 13	225(3.87),285(3.82)	1-2065-69
$C_7H_8BrN_3O_3$			
as-Triazine-3,5(2H,4H)-dione, 6-bromo-2-(tetrahydro-2-furyl)-	pH 3-4	277(3.84)	103-0283-69

Compound	Solvent	λ_{max}(log ε)	Ref.
$C_7H_8BrN_5$			
Adenine, 8-bromo-N,N-dimethyl-	EtOH	281(4.24)	65-2125-69
$C_7H_8Br_2N_2OS$			
7-Bromo-2,3-dihydro-8-hydroxy-5-methyl-	pH 1	250(3.59),345(3.73)	1-2437-69
thiazolo[3,2-c]pyrimidin-4-ium	pH 13	260(3.61),370(3.61)	1-2437-69
bromide			
$C_7H_8Br_2N_2S$			
8-Bromo-2,3-dihydro-5-methylthiazolo-[3,2-c]pyrimidin-4-ium bromide	pH 1	265(3.5),325(--)	1-2437-69
$C_7H_8Br_2O_2$			
1,3-Cyclopentanedione, 2,5-dibromo-	MeOH-HCl	262(4.14)	20-0017-69
4,4-dimethyl-	MeOH-NaOH	282(4.25)	20-0017-69
$C_7H_8Br_3NO_2$			
Hexanimide, 2,2,5-tribromo-N-methyl-	MeOH	252(2.80)	44-1695-69
C_7H_8ClN			
Pyridine, 3-chloro-2,6-dimethyl-	n.s.g.	274(3.61)	39-2249-69C
$C_7H_8ClNO_3$			
Acetoacetic acid, 4-chloro-2-cyano-, ethyl ester	EtOH	266(3.30)	104-1713-69
$C_7H_8ClN_3OS$			
3H,6H-[1,3]Thiazino[3,2-b]-as-triazin-3-one, 8-chloro-7,8-dihydro-2-methyl-	EtOH	240(4.36)	114-0093-69C
$C_7H_8ClN_3O_2$			
Formimidic acid, N-(4-chloro-1,6-dihydro-6-oxo-5-pyrimidinyl)-, ethyl ester	EtOH	259(3.74),300(3.90)	44-2102-69
$C_7H_8ClN_5$			
1-Methyl-4-(1H-tetrazol-5-yl)pyridinium chloride	MeOH	277(4.13)	87-0944-69
$C_7H_8ClN_5OS$			
Acetamide, 2-chloro-N-(3-ethyl-s-triazolo[3,4-b][1,3,4]thiadiazol-6-yl)-	EtOH	252(3.97)	2-0959-69
$C_7H_8Cl_2O$			
Cyclohexanone, 2-(dichloromethylene)-	EtOH	248(3.73)	35-2062-69
$C_7H_8F_3N_5O_2$			
Acetamide, N-[2-amino-1,6-dihydro-	pH -1.0	269(4.34)	5-0201-69F
4-(methylamino)-6-oxo-5-pyrimi-	pH 7.0	268(4.23)	5-0201-69F
dinyl]-2,2,2-trifluoro-	pH 12.0	266(3.94),276s(3.86)	5-0201-69F
Acetamide, N-(2,4-diamino-1,6-dihydro-	pH 0.0	263(4.23)	5-0201-69F
1-methyl-6-oxo-5-pyrimidinyl)-	pH 7.0	265(4.11)	5-0201-69F
2,2,2-trifluoro-	pH 13.0	286(4.08)	5-0201-69F
Acetamide, N-(2,4-diamino-1,6-dihydro-	pH 0.0	262(4.22)	5-0201-69F
6-oxo-5-pyrimidinyl)-2,2,2-tri-	pH 7.0	265(4.13)	5-0201-69F
fluoro-N-methyl-	pH 13.0	260(3.93)	5-0201-69F
$C_7H_8NO_2$			
Toluene, m-nitro-, half-reduction	pH 1	270(4.05),440(2.81)	64-1336-69
radical	pH 13	285(4.21),400(2.60)	64-1336-69

Compound	Solvent	λ_{max}(log ε)	Ref.
Toluene, o-nitro-, half-reduction	pH 1	295(4.18),440(2.87)	64-1336-69
radical	pH 13	295(4.38),400(3.00)	64-1336-69
Toluene, p-nitro-, half-reduction	pH 1	280(4.08),440(2.82)	64-1336-69
radical	pH 13	285(4.24),440(2.60)	64-1336-69
$C_7H_8NO_3$			
Anisole, m-nitro-, half-reduction	pH 1	280(4.03),480(2.78)	64-1336-69
radical	pH 13	285(4.19),460(2.43)	64-1336-69
Anisole, p-nitro-, half-reduction	pH 1	290(4.10),480(3.20)	64-1336-69
radical	pH 13	295(4.26),460(2.78)	64-1336-69
$C_7H_8N_2$			
1H-Pyrrolo[3,2-c]pyridine, 2,3-dihydro-	EtOH	260(3.96)	88-1909-69
	EtOH	260(3.96)	103-0410-69
hydrochloride	EtOH	282(4.04)	103-0410-69
$C_7H_8N_2O$			
Aniline, N-methyl-N-nitroso-	MeCN	268(3.79),370(3.28)	24-2093-69
3-Buten-2-one, 4-imidazol-1-yl-	EtOH	270(4.2)	97-0025-69
3-Buten-2-one, 4-pyrazol-1-yl-	EtOH	286(4.4)	97-0025-69
Diimide, (p-methoxyphenyl)-	MeCN	228(3.95),287s(--), 298(4.01)	35-2325-69
$C_7H_8N_2OS$			
2,3-Dihydro-8-hydroxy-5-methylthiazolo-[3,2-c]pyrimidin-4-ium hydroxide,	pH 1	240(3.66),335(3.79)	1-2437-69
inner salt	pH 13	265(3.78),360(3.84)	1-2437-69
$C_7H_8N_2O_2$			
Aniline, N-methyl-o-nitro-	1%MeOH-NaOH	447(3.78)	39-0922-69B
Aniline, N-methyl-p-nitro-	1%MeOH-NaOH	407(4.25)	39-0922-69B
5H-Oxazolo[3,2-a]pyrimidin-5-one,	pH 1.3	217(4.35),265(4.35)	70-0835-69
2,3-dihydro-7-methyl-	pH 5.5	212(4.54),263(4.58)	70-0835-69
	pH 10.35	221(4.15),267(4.26)	70-0835-69
Pyrrolo[1,2-a]pyrimidin-4(6H)-one,	5N HCl	246(3.87)	22-3133-69
7,8-dihydro-2-hydroxy-	H_2O	255(4.03)	22-3133-69
	N NaOH	255(3.90)	22-3133-69
	EtOH	256(3.77)	22-3133-69
$C_7H_8N_2O_3$			
Barbituric acid, 5-propylidene-	EtOH	255(3.91)	103-0827-69
Phenol, 3-(methylamino)-4-nitro-	1%MeOH-HOAc	427(3.87)	39-0922-69B
4-Picoline, 2-methoxy-5-nitro-	EtOH	289(3.83)	35-2338-69
1(2H)-Pyrimidineacetaldehyde, 3,4-di-	pH 1	270(3.99)	88-2285-69
hydro-5-methyl-2,4-dioxo- (hydrate)	pH 12	267(3.83)	88-2285-69
$C_7H_8N_2S$			
Thiourea, phenyl-	EtOH	242s(4.02),270(4.24)	39-0268-69C
$C_7H_8N_2S_3$			
4-Isothiazolecarbonitrile, 3-(ethyl-thio)-5-(methylthio)-	MeOH	213(4.06),230(4.08), 284(4.16)	88-4265-69
4-Isothiazolecarbonitrile, 5-(ethyl-thio)-3-(methylthio)-	MeOH	215(4.08),230(4.10), 286(4.13)	88-4265-69
$C_7H_8N_3NaO_6$			
Pyridine, 3,5-dinitro-4-methoxy-, sodium methoxide adduct	MeOH	308(3.69),455(4.26)	35-6742-69

Compound	Solvent	λ_{max}(log ϵ)	Ref.
$C_7H_8N_4$			
Benzotriazole, 1-amino-5-methyl-	EtOH	213(3.93),259(3.72), 291(3.60)	39-0742-69C
Benzotriazole, 1-amino-7-methyl-	EtOH	214(4.03),268(3.72), 283(3.61)	39-0742-69C
$C_7H_8N_4OS$			
9H-Purine, 9-methyl-2-(methylsulfinyl)-	pH 7.0	273(3.93)	39-2620-69C
9H-Purine, 9-methyl-6-(methylsulfinyl)-	pH 7.0	277(3.95)	39-2620-69C
9H-Purine, 9-methyl-8-(methylsulfinyl)-	pH 7.0	274(4.11)	39-2620-69C
$C_7H_8N_4O_2$			
Purine, 6-ethoxy-, 3-oxide	pH 1	268(4.10)	18-0750-69
	pH 6	226(4.40),285(4.03)	18-0750-69
	pH 13	228(4.50),282(3.99)	18-0750-69
2-Pyrimidineacetamide, 4-amino-5-formyl-	MeOH	249(3.92),310(3.78)	39-0133-69C
$C_7H_8N_4O_2S$			
9H-Purine, 9-methyl-2-(methylsulfonyl)-	pH 7.0	266(3.91)	39-2620-69C
9H-Purine, 9-methyl-6-(methylsulfonyl)-	pH 7.0	280(3.93)	39-2620-69C
9H-Purine, 9-methyl-8-(methylsulfonyl)-	pH 7.0	271(4.10)	39-2620-69C
$C_7H_8N_4O_3$			
1,2,3,6-Tetrahydro-1-hydroxy-7,9-dimethyl-2,6-dioxopurinium hydroxide, inner salt	pH 0	262(4.19)	44-0978-69
	pH 4	247(3.59),283(3.68)	44-0978-69
	pH 10	204(4.04),238(3.96), 291(3.61)	44-0978-69
1,2,3,6-Tetrahydro-3-hydroxy-7,9-dimethyl-2,6-dioxopurinium hydroxide, inner salt	pH 2	245(3.73),270(3.83)	44-0978-69
	pH 6.5	218(4.26),276(3.82), 313(3.72)	44-0978-69
	pH 11	227(4.30),285(3.64), 308(3.58)	44-0978-69
Uric acid, 7,9-dimethyl-	pH -7.0	215(4.11),307(3.86)	44-0978-69
	pH -5.0	235(3.94),289(3.98)	44-0978-69
	pH -2.7	235(3.87),288(4.06)	44-0978-69
	pH 2.0	236(3.90),285(4.08)	44-0978-69
	pH 10.0	242(3.99),294(4.06)	44-0978-69
$C_7H_8N_4O_3S$			
1,3,4-Thiadiazole-2-carboxamide, 5-acetamido-N-acetyl-	MeOH	284(4.04)	4-0835-69
$C_7H_8N_4S$			
3H-Purine, 3-methyl-6-(methylthio)-	pH 7.0	237(3.99),311(4.22)	39-2620-69C
3H-Purine, 3-methyl-8-(methylthio)-	pH 5	318(4.32)	39-2620-69C
9H-Purine, 9-methyl-2-(methylthio)-	pH 7.0	236(4.5+),257(4.26), 261(4.25),302(4.26)	39-2620-69C
9H-Purine, 9-methyl-6-(methylthio)-	pH 7.0	220(4.08),242(4.25), 286(4.24)	39-2620-69C
9H-Purine, 9-methyl-8-(methylthio)-	pH 7.0	221(4.13),289(4.25)	39-2620-69C
$C_7H_8N_6$			
Pyrimido[5,4-e]-as-triazine, 5-(ethylamino)-	pH 1	224(4.10),249s(3.65), 363(4.03)	44-2102-69
C_7H_8O			
Anisole	EtOH	218(3.90),271(3.28)	28-1562-69B
Benzyl alcohol	MeOH	225(2.21),258(2.29), 264(2.17),268(1.98)	73-0819-69
Bicyclo[3.1.0]hex-3-en-2-one, 4-methyl-	hexane	203s(--),246(3.54)	22-2095-69

Compound	Solvent	λ max (log ε)	Ref.
Bicyclo[3.1.0]hex-3-en-2-one, 4-methyl- (cont.)	EtOH	208(3.74),258(3.54)	22-2095-69
m-Cresol	pH 1	271(3.16)	30-1113-69A
	pH 13	238(3.92),288(3.42)	30-1113-69A
	MeOH	274(3.24)	30-1113-69A
	MeOH-Na	240(3.95),291(3.44)	30-1113-69A
o-Cresol	pH 2	270(3.20)	30-1113-69A
	pH 13	237(3.96),289(3.51)	30-1113-69A
	MeOH	274(3.29)	30-1113-69A
	MeOH-Na	239(3.98),290(3.51)	30-1113-69A
oxidized coupling product with 3-methyl-2-benzothiazoline hydrazone	H_2O	500(4.38+)	3-1750-69
p-Cresol	C_6H_{12}	286(3.32)	18-1831-69
	pH 2	277(3.24)	30-1113-69A
	pH 8	277(3.24)	69-1042-69
	pH 13	237(3.94),295(3.41)	30-1113-69A
	MeOH	280(3.29)	30-1113-69A
	MeOH-Na	240(3.98),298(3.40)	30-1113-69A
oxidized coupling product with 3-methyl-2-benzothiazoline hydrazone	H_2O	550(3.95+)	3-1750-69
2,6-Cycloheptadien-1-one	EtOH	235(4.03),266(3.40), 339(1.52)	35-0215-69
C_7H_8OS			
Benzenethiol, p-methoxy-	EtOH	239(4.00),286(3.11), 291s(3.10)	117-0043-69
4H-Pyran-4-thione, 2,6-dimethyl-	EtOH	254(3.93),341(4.32)	35-4749-69
4H-Pyran-4-thione, 3,5-dimethyl-	EtOH	251(3.82),342(4.28)	44-0589-69
Sulfoxide, methyl phenyl	10M $HClO_4$	230(3.9),270(3.1)	35-6703-69
	4M $HClO_4$	226(3.7)	35-6703-69
	H_2O	229s(3.6)	35-6703-69
4H-Thiopyran-4-one, 3,5-dimethyl-	EtOH	227(3.89),299(4.25), 305(4.25)	44-0589-69
$C_7H_8O_2$			
Catechol, 3-methyl-	pH 2	196(4.65),215(3.88), 273(3.17)	61-0203-69
	anion	205(4.53),227(3.70), 280(3.51)	61-0203-69
	0.15N NaOH	205(4.55),237(3.77), 286(3.52)	61-0203-69
Catechol, 4-methyl-	pH 2	195(4.68),215(3.83), 277(3.38)	61-0203-69
	anion	203(4.55),227(3.78), 284(3.57)	61-0203-69
	0.15N NaOH	204(4.50),235(3.86), 293(3.62)	61-0203-69
2-Cyclohexene-1,4-dione, 5-methyl-	hexane	366(1.73)	39-0124-69C
Hydroquinone, methyl-	pH 2	192(4.61),213(3.81), 286(3.46)	61-0203-69
	anion	196(4.61),232(3.91), 304(3.59)	61-0203-69
	0.15N NaOH	203(4.59),240(3.96), 317(3.57)	61-0203-69
Resorcinol, 2-methyl-	pH 2	197(4.70),216(3.90), 271(2.95)	61-0203-69
	anion	203(4.57),237(3.87), 285(3.30)	61-0203-69
	0.15N NaOH	215(4.61),239(3.93), 288(3.30)	61-0203-69

Compound	Solvent	$\lambda_{max}(\log \epsilon)$	Ref.
Resorcinol, 4-methyl-	pH 2	194(4.69),216(3.88), 278(3.36)	61-0203-69
	anion	201(4.45),229(3.83), 290(3.53)	61-0203-69
	0.15N NaOH	209(4.56),236(3.88), 294(3.55)	61-0203-69
Resorcinol, 5-methyl-	pH 2	200(4.66),217(3.89), 272(3.14)	61-0203-69
	anion	206(4.50),235(3.84), 287(3.33)	61-0203-69
	0.15N NaOH	216(4.52),238(3.86), 292(3.41)	61-0203-69
$C_7H_8O_2S$			
Acetic acid, (2-penten-4-ynylthio)-	EtOH	229(4.09)	18-2589-69
$C_7H_8O_3$			
1,2,4-Cyclopentanetrione, 3,3-dimethyl-	MeOH-HCl	265(4.14)	20-0277-69
	MeOH-NaOH	225(4.10),309(4.16)	20-0277-69
1-Cyclopentene-1-carboxylic acid, 2-methyl-5-oxo-	EtOH	232(4.02)	35-4739-69
2,4-Pentadienal, 5-acetoxy-, di-trans	EtOH	278(4.57),363(3.67)	78-4315-69
2,4-Pentadienoic acid, 5-formyl-3-methyl-	EtOH	271(4.34)	104-0759-69
Versicolin	pH 1	285(3.57)	88-4871-69
	pH 13	288(3.66)	88-4871-69
	EtOH	222(4.4),390(2.6), 520(2.2)	88-4871-69
$C_7H_8O_3S$			
p-Toluenesulfonic acid	99.9% HOAc	273(3.52)	44-3638-69
	+$HClO_4$	273(3.52)	44-3638-69
	+KOAc	272(3.23)	44-3638-69
	HOAc-Ac_2O	273(3.54)	44-3638-69
	+$HClO_4$	273(3.52)	44-3638-69
	+KOAc	272(3.20)	44-3638-69
$C_7H_8O_4$			
1-Cyclopentene-1-acetic acid, 2-hydroxy-5-oxo-	MeOH	248(4.19)	39-1625-69C
2H-Pyran-2-one, 3-acetyl-5,6-dihydro-4-hydroxy-	pH 13	248(4.03),267(3.99)	22-4091-69
	EtOH	217(3.94),273(4.03)	22-4091-69
$C_7H_8O_5$			
4H-Pyran-2-carboxylic acid, 5,6-dihydro-3-hydroxy-4-oxo-, methyl ester	EtOH	325(3.92)	88-2383-69
C_7H_8S			
p-Toluenethiol (also spectra in other solvents)	hexane	277(2.84),286(2.85), 295(2.69)	39-1305-69A
at 77°K	isopentane-$C_6H_{11}Me$	268(2.86),271(2.86), 277(2.92),280(2.91), 286(2.98),295(2.88)	39-1305-69A
$C_7H_8S_2$			
2H-Thiopyran-2-thione, 3,4-dimethyl-	C_6H_{12}	241(4.07),313(3.92), 437(3.84)	78-1441-69
	EtOH	241(4.10),307(3.88), 437(3.90)	78-1441-69

Compound	Solvent	$\lambda_{max}(\log \epsilon)$	Ref.
2H-Thiopyran-2-thione, 3,5-dimethyl-	C_6H_{12}	247(4.09),317(3.97), 438(3.81)	78-1441-69
	EtOH	247(4.11),311(3.93), 437(3.86)	78-1441-69
4H-Thiopyran-4-thione, 3,5-dimethyl-	EtOH	262(3.57),383(4.40)	44-0589-69
$C_7H_9AsO_3$			
α-Toluenearsonic acid	H_2O	260(2.40)	25-0381-69
	22% H_2SO_4	260(2.43)	25-0381-69
	34.6% H_2SO_4	260(2.45)	25-0381-69
	51.4% H_2SO_4	260(2.43)	25-0381-69
$C_7H_9BO_2$			
o-Tolueneboronic acid	EtOH	210(3.92),270(2.72)	28-1562-69B
p-Tolueneboronic acid	EtOH	206(3.90),263(2.59)	28-1562-69B
$C_7H_9BO_3$			
Benzeneboronic acid, o-methoxy-	EtOH	225(3.91),280(3.41)	28-1562-69B
Benzeneboronic acid, p-methoxy-	EtOH	236(4.10),272(3.08)	28-1562-69B
C_7H_9BrINO			
6-Bromo-5-hydroxy-1-methyl-2-picolinium iodide	pH 1	225(4.30),310(4.09)	1-2065-69
	pH 13	340(4.08)	1-2065-69
$C_7H_9BrO_2$			
1,3-Cyclopentanedione, 5-bromo-4,4-dimethyl-	MeOH-HCl	247(4.20)	20-0017-69
	MeOH-NaOH	270(4.33)	20-0017-69
$C_7H_9Br_2NO_2$			
Hexanimide, 2,5-dibromo-N-methyl-, trans	MeOH	245(2.76)	44-1695-69
$C_7H_9ClN_2$			
Pyridazine, 4-chloro-3,5,6-trimethyl-	n.s.g.	227(3.41),271(3.22), 306(2.48)	39-2251-69C
Pyrimidine, 2-chloro-4,5,6-trimethyl-	n.s.g.	226(3.35),261(3.50)	39-2251-69C
Pyrimidine, 5-chloro-2,4,6-trimethyl-	n.s.g.	217(3.82),261(3.52)	39-2251-69C
$C_7H_9ClN_2O$			
Pyrimidine, 4-chloro-5-ethoxy-2-methyl-	pH 1	290(3.68)	1-2437-69
	pH 13	285(3.66)	1-2437-69
4(3H)-Pyrimidinone, 6-chloro-2-ethyl-5-methyl-	MeOH	230(3.78),279(3.83)	44-2972-69
$C_7H_9ClO_2$			
2,4-Pentadienoic acid, 5-chloro-, ethyl ester, 2,4-di-trans	EtOH	258(4.49)	78-4315-69
$C_7H_9Cl_2N$			
2H-Pyrrole, 2-(dichloromethyl)-2,5-dimethyl-	EtOH	230(3.23)	39-2249-69C
$C_7H_9Cl_3O_2$			
Acetic acid, trichloro-, 1-ethylpropenyl ester	C_6H_{12}	220(2.94)	78-1679-69
C_7H_9N			
2,4-Lutidine	C_6H_{12}	260(3.31)	4-0859-69
	pH 1	260(3.77)	
	MeOH	260(3.44)	

Compound	Solvent	λ_{max}(log ϵ)	Ref.
2,5-Lutidine	isooctane	269(3.47)	4-0859-69
	pH 1	270(3.83)	
	MeOH	269(3.55)	
2,6-Lutidine	C_6H_{12}	267(3.52)	4-0859-69
	pH 1	270(3.93)	
	MeOH	267(3.64)	
3,4-Lutidine	C_6H_{12}	260(3.29)	4-0859-69
	pH 1	258(3.69)	
	MeOH	259(3.39)	
Pyridine, 2-ethyl-	C_6H_{12}	258(3.42)	4-0859-69
	pH 1	263(3.85)	
	MeOH	261(3.56)	
Pyridine, 4-ethyl-	isooctane	255(3.19)	4-0859-69
	pH 1	254(3.67)	
	MeOH	256(3.31)	
C_7H_9NO			
Hydroxylamine, N-benzyl-	EtOH	241(2.1),247(2.2), 251(2.2),256(2.3), 262(2.2)	28-0078-69A
2,4-Lutidine, N-oxide	isooctane	280(4.4)	4-0859-69
	pH 1	263(3.56)	
	MeOH	261(4.11)	
2,5-Lutidine, N-oxide	isooctane	278(4.05)	4-0859-69
	pH 1	271(3.67)	
	MeOH	259(4.04)	
2,6-Lutidine, N-oxide	C_6H_{12}	274(4.04)	4-0859-69
	pH 1	268(3.78)	
	MeOH	258(4.01)	
3,4-Lutidine, N-oxide	isooctane	284(4.04)	4-0859-69
	pH 1	261(3.49)	
	MeOH	264(4.16)	
Pyridine, 2-ethyl-, 1-oxide	C_6H_{12}	279(4.36)	4-0859-69
	pH 1	265(3.67)	4-0859-69
	H_2O	211(4.38),255(4.10)	39-0054-69B
	MeOH	261(4.04)	4-0859-69
Pyridine, 4-ethyl-, 1-oxide	isooctane	286(4.18)	4-0859-69
	pH 1	257(3.49)	4-0859-69
	H_2O	207(4.28),258(4.18)	39-0054-69B
	MeOH	265(4.18)	4-0859-69
3-Pyridinol, 2,6-dimethyl-	MeOH	221(3.95),288(3.84)	22-0948-69
Pyrrole, 2-acetyl-5-methyl-	ether	290(4.27)	22-0948-69
Pyrrole, 3-acetyl-2-methyl-	EtOH	206(4.33),244(3.91), 279(3.78)	78-3879-69
Pyrrole, 4-acetyl-2-methyl-	EtOH	205(4.14),245(4.01), 280(3.72)	33-1911-69
C_7H_9NOS			
2(1H)-Pyridinethione, 3-hydroxy-1,6-dimethyl-	pH 1	260(3.68),350(4.10)	1-2065-69
	pH 13	245(3.68),270(3.65), 370(3.98)	1-2065-69
3-Pyridinol, 6-methyl-2-(methylthio)-	pH 1	240(3.58),325(3.95)	1-2065-69
	pH 13	250(3.86),325(3.96)	1-2065-69
2-Pyrrolethiocarboxylic acid, S-ethyl ester	EtOH	239(3.72),299(4.27)	78-3879-69
2-Thiophenecarboxaldehyde, 5-(dimethylamino)-	EtOH	370(4.5)	48-0827-69
$C_7H_9NOS_2$			
2-Butanone, 4-(2-thiazolylthio)-	MeOH	276(3.88)	4-0397-69

Compound	Solvent	λ_{max}(log ϵ)	Ref.
2-Butanone, 4-(2-thioxo-4-thiazolin-3-yl)-	MeOH	317(4.15)	4-0397-69
1,3-Thiazane-2-thione, 3-allyl-4-oxo-	MeOH	267(4.10),313(3.97)	42-0919-69
$C_7H_9NO_2$			
2-Furaldehyde, 5-(dimethylamino)-	n.s.g.	364(4.52)	103-0434-69
Ketone, 1-(hydroxymethyl)-3-pyrrolyl methyl	EtOH	195(4.16),205(4.18), 246(4.13),270s(3.91)	33-1911-69
2-Propanone, 1-(5-methyl-3-isoxazolyl)-	EtOH	218(3.76)	35-4749-69
4H-Pyran-4-one, 2,6-dimethyl-, oxime	EtOH	261(4.18)	35-4749-69
Pyrrole, 3-(3-hydroxypropionyl)-	EtOH	203(4.10),245(3.98), 268s(3.80)	33-1911-69
2-Pyrrolemethanol, 4-acetyl-	EtOH	210(4.20),213s(4.16), 245(3.93),274(3.74)	78-3879-69
$C_7H_9NO_2S$			
1H-Azepine, 1-(methylsulfonyl)-	EtOH	205(4.24),307(2.88)	44-2688-69
$C_7H_9NO_3$			
Acetamide, N-(2-hydroxy-5-oxo-1-cyclopenten-1-yl)-	MeOH	245(4.08)	88-0141-69
1-Cyclopentene-1-acetic acid, 2-amino-5-oxo-	MeOH	247(4.20)	39-1625-69C
4-Isoxazolecarboxylic acid, 3-methyl-, ethyl ester	MeOH	211(3.75),271(2.88)	4-0783-69
$C_7H_9NO_4$			
3-Pyrroline-3-carboxylic acid, 4-hydroxy-2-oxo-, ethyl ester	EtOH	246(3.39)	104-1713-69
$C_7H_9N_3$			
Triazene, 3-methyl-1-phenyl-	MeOH	280(4.10)	65-0303-69
mercury complex	benzene	305(3.82),398(4.18)	65-0303-69
	DMF	300(3.76),390(4.10)	65-0303-69
$C_7H_9N_3O$			
Pyrido[2,3-d]pyridazin-5(1H)-one, 2,3,4,6-tetrahydro-	EtOH	214s(3.87),233(4.26), 283(3.51),302(3.63), 320(3.40)	12-1759-69
Pyrido[2,3-d]pyridazin-8(2H)-one, 1,3,4,7-tetrahydro-	EtOH	213s(3.49),228(3.58), 234s(3.57),270s(3.36), 280s(3.51),295s(3.64), 306(3.65),318s(3.48)	12-1759-69
$C_7H_9N_3OS$			
3H,6H-[1,3]Thiazino[3,2-b]-as-triazin-3-one, 7,8-dihydro-2-methyl-	EtOH	238(4.36)	114-0093-69C
$C_7H_9N_3O_2$			
Glutaconimide, N-methyl-4-oxo-, 4-(methylhydrazone)	MeOH	270(3.95),374(4.32)	44-2011-69
5-Pyrimidinecarboxaldehyde, 4-amino-2-(2-hydroxyethyl)-	MeOH	249(3.93),308(3.75)	39-0133-69C
$C_7H_9N_3O_2S$			
3H,6H-[1,3]Thiazino[3,2-b]-as-triazin-3-one, 7,8-dihydro-8-hydroxy-2-methyl-	EtOH	238(4.38)	114-0093-69C
$C_7H_9N_3O_3$			
Cytosine, N(4)-carbethoxy-	pH 1	233(4.00),294(3.95)	39-1860-69C

Compound	Solvent	$\lambda_{max}(\log \epsilon)$	Ref.
Cytosine, 5-carbethoxy-	pH 1	227(3.89)	39-1860-69C
2-Pyrimidineacetic acid, 4-amino-5-(hydroxymethyl)-	pH 2	248(4.07)	39-0133-69C
	0.2M NH_3	235(3.95),274(3.69)	39-0133-69C
5-Pyrimidinecarboxylic acid, 4-amino-2-(2-hydroxyethyl)-	pH 1	243(4.10),280s(3.60)	39-0133-69C
	pH 13	240(4.05),288(3.75)	39-0133-69C
as-Triazine-3,5(2H,4H)-dione, 2-(tetrahydro-2-furyl)-	pH 3-4	262(3.79)	103-0283-69
	pH 7-8	261(3.77)	103-0283-69
	pH 11-12	255(3.81)	103-0283-69
$C_7H_9N_5O$			
Guanine, 1,8-dimethyl-	pH 2.0	251(4.10),274(3.94)	5-0201-69F
	pH 7.0	250(4.02),276(3.95)	5-0201-69F
	pH 14.0	259s(3.87),278(3.96)	5-0201-69F
Guanine, 7,8-dimethyl-	pH 0.0	249(4.10),277(3.92)	5-0201-69F
	pH 6.0	250(3.81),283(3.93)	5-0201-69F
	pH 12.0	235s(3.83),280(3.93)	5-0201-69F
Guanine, 8,9-dimethyl-	pH 1.0	252(4.14),277(3.93), 287s(3.85)	5-0201-69F
	pH 5.0	252(4.13),273(3.96)	5-0201-69F
	pH 7	251(4.12),278s(3.95)	24-4032-69
	pH 12	252s(4.03),268(4.06)	5-0201-69F
	pH 12	252s(4.03),268(4.06)	24-4032-69
s-Triazolo[3,4-c]-as-triazin-5(8H)-one, 3,6,8-trimethyl-	EtOH	245(3.92),333(3.84)	22-2492-69
s-Triazolo[5,1-c]-as-triazin-4(1H)-one, 1,3,7-trimethyl-	EtOH	246(3.69),304(3.92)	22-2492-69
$C_7H_9N_5O_2$			
Guanine, 9-(2-hydroxyethyl)-	pH 1	255(3.99),280(3.79)	94-1268-69
	pH 13	255s(--),269(3.98)	94-1268-69
$C_7H_9O_3P$			
Phosphonic acid, benzyl-	H_2O	259(2.30)	25-0381-69
	13.4% H_2SO_4	258(2.26)	25-0381-69
	22.2% H_2SO_4	258(2.28)	25-0381-69
$C_7H_9O_4P$			
p-Cresyl phosphate	pH 8	274(2.96)	69-1042-69
C_7H_{10}			
1,3-Cycloheptadiene	EtOH	246(3.88)	18-2013-69
1,3-Cyclohexadiene, 1-methyl-	heptane	263(2.64)	78-4933-69
Cyclopentene, 3-ethylidene-	C_6H_{12}	240(4.17)	25-0619-69
Cyclopentene, 4-methyl-3-methylene-	C_6H_{12}	234(4.18)	25-0619-69
1,2,4-Hexatriene, 5-methyl-	heptane	231(4.41)	78-4933-69
1,3,5-Hexatriene, 2-methyl-, cis	heptane	249(4.46),259(4.54), 269(4.45)	78-4933-69
$C_7H_{10}ClN_3O_2$			
Uracil, 5-chloro-1,3-dimethyl-6-(methylamino)-	MeOH	243(3.85),278(4.17)	44-3285-69
$C_7H_{10}Cl_2N_2$			
4H-Pyrazole, 4-(dichloromethyl)-3,4,5-trimethyl-	n.s.g.	227(3.44)	39-2251-69C
$C_7H_{10}N_2O$			
Ketone, 1,5-dimethylpyrazol-3-yl methyl	MeOH	242(3.99)	4-0545-69
Pyrazole-4-carboxaldehyde, 1,3,5-trimethyl-	MeOH	252(4.12)	4-0545-69

Compound	Solvent	$\lambda_{max}(\log \epsilon)$	Ref.
Pyrrolo[1,2-a]pyrimidin-2(6H)-one, 3,4,7,8-tetrahydro-, hydrochloride	EtOH	233(3.93)	22-3139-69
Pyrrolo[1,2-a]pyrimidin-4(3H)-one, 2,6,7,8-tetrahydro-	EtOH	232(3.80)	22-3139-69
$C_7H_{10}N_2OS$			
4(3H)-Pyrimidinethione, 5-ethoxy-2-methyl-	pH 1	245(3.29),335(4.01)	1-2437-69
	pH 13	265(3.57),320(3.26)	1-2437-69
$C_7H_{10}N_2O_2$			
Acrylic acid, 3-amino-3-cyano-2-methyl-, ethyl ester	EtOH	305(4.07)	39-2186-69C
Acrylic acid, 2-cyano-3-(dimethylamino)-, methyl ester	isooctane	240(3.90),297(4.32)	35-6689-69
Benzimidazole, 4,5,6,7-tetrahydro-1-hydroxy-, 3-oxide	MeOH	235(3.76)	48-0746-69
Nicotinamide, 1,4,5,6-tetrahydro-2-methyl-6-oxo-	EtOH	271(3.7)	94-2411-69
4H-1,3-Oxazin-4-one, 2,3-dihydro-3,6-dimethyl-2-(methylimino)-	EtOH	214(4.05),255(3.38)	32-1000-69
Pyrano[3,4-c]pyrazol-7(2H)-one, 3,3a,4,5-tetrahydro-5-methyl-	MeCN	294(3.87)	78-2041-69
Pyrano[3,4-c]pyrazol-7(3H)-one, 3a,4,5,7a-tetrahydro-5-methyl-	MeCN	322(2.43)	78-2041-69
Pyrazole-3-carboxylic acid, 1-methyl-, ethyl ester	MeOH	223(3.95)	4-0545-69
Pyrazole-4-carboxylic acid, 1-methyl-, ethyl ester	MeOH	224(4.04)	4-0545-69
Pyrazole-5-carboxylic acid, 1-methyl-, ethyl ester	MeOH	227(4.02)	4-0545-69
3H-Pyrazole-4-carboxylic acid, 3,3-dimethyl-, methyl ester	MeOH	236(3.51),349(2.13)	88-0015-69
3H-Pyrazole-5-carboxylic acid, 3,3-dimethyl-, methyl ester	hexane	232(3.54),270s(2.86), 367(2.08)	39-1065-69C
	EtOH	235(3.49),353(2.13)	39-1065-69C
1-Pyrazoline-3,5-dione, 4,4-diethyl-	MeCN	629(2.15)	88-4497-69
	CH_2Cl_2	621(2.14),644(2.19), 680(2.04),724(1.57)	44-3181-69
4(3H)-Pyrimidinone, 5-ethoxy-2-methyl-	pH 1	255(3.93)	1-2437-69
	pH 13	235(3.82),275(3.87)	1-2437-69
2(1H)-Pyrimidinone, 4-methoxy-5,6-dimethyl-	pH 7	205(4.14),278(3.87)	73-2316-69
Thymine, 2-O-ethyl-	EtOH	271(3.84)	78-5989-69
Uracil, 5-isopropyl-	pH 2	265(3.85)	56-0499-69
	pH 12	288(3.69)	56-0499-69
Uracil, 5-propyl-	pH 2	263(3.90)	56-0499-69
	pH 12	286(3.79)	56-0499-69
Uracil, 3,5,6-trimethyl-	75% dioxan	267(3.89)	95-0266-69
	+NaOH	292(4.01)	95-0266-69
$C_7H_{10}N_2O_2S$			
Imidazolidine-2-thione, 4,5-dioxo-1,3-diethyl-	EtOH	301(4.16),405(1.54)	18-2323-69
Thiazole, 2,5-diethyl-4-nitro-	EtOH	220(4.10),240(--), 292(3.56)	28-1343-69B
Thiazole, 2-isopropyl-4-methyl-5-nitro-	EtOH	220(3.18),260(--), 300(3.36)	28-1343-69B
$C_7H_{10}N_2O_4$			
Barbituric acid, 5-(2-methoxyethyl)-	pH 6.2	266(2.88)	65-2568-69

Compound	Solvent	$\lambda_{max}(\log \epsilon)$	Ref.
Barbituric acid, 5-(2-methoxyethyl)- (cont.)	pH 13	268(2.82)	65-2568-69
$C_7H_{10}N_2O_6$			
Malonamic acid, 2-(nitroacetyl)-, ethyl ester (3-keto form)	EtOH	251(4.04),274(4.13)	104-1713-69
2-en-3-ol form	EtOH	257(4.04),269(4.05)	104-1713-69
$C_7H_{10}N_2S_2$			
1H-1,2,4-Benzothiadiazine-3-thiol, 5,6,7,8-tetrahydro-	EtOH	326(4.15)	95-0689-69
2-Benzothiazolinethione, 3-amino-4,5,6,7-tetrahydro-	$C_4H_8O_2$	283(3.92)	95-0689-69
$C_7H_{10}N_4O$			
3H-Pyrimido[1,2-c]-as-triazin-3-one, 5,6,7,8-tetrahydro-2-methyl-	EtOH	212(4.42),255(3.82)	114-0181-69C
4H-Pyrimido[2,1-c]-as-triazin-4-one, 6,7,8,9-tetrahydro-3-methyl-, hydrochloride	EtOH	231(4.05),316(3.85)	114-0181-69C
$C_7H_{10}N_4O_2$			
Nicotinic acid, 2,6-diamino-3-(aminomethyl)- (hydrate)	pH 2	268(4.10),335(4.32)	39-0133-69C
	pH 12	263(4.04),321(4.11)	39-0133-69C
2-Pyrimidineacetamide, 4-amino-5-(hydroxymethyl)-	MeOH	236(3.65),273(3.95)	39-0133-69C
5-Pyrimidinecarboxylic acid, 2,4-diamino-, ethyl ester	pH 2	240s(4.20),278(3.76)	44-0821-69
	pH 12	215(4.38),252(4.19), 293(4.10)	44-0821-69
$C_7H_{10}N_4O_2S$			
as-Triazin-2(5H)-acetamide, 6-methyl-3-(methylthio)-5-oxo-	EtOH	238(4.11)	114-0181-69C
$C_7H_{10}N_4O_3$			
1(2H)-Pyrimidinepropionic acid, α,4-diamino-2-oxo-, DL-	pH 1	279(4.08)	88-2285-69
	pH 12	275(3.93)	88-2285-69
$C_7H_{10}N_4O_4$			
Formamide, N-methyl-N-[1,2,3,4-tetrahydro-1-hydroxy-6-(methylamino)-2,4-dioxo-5-pyrimidinyl]-	pH 3	270(3.97)	44-0978-69
	pH 9	231(4.35),286(4.01)	44-0978-69
$C_7H_{10}N_4S$			
7H-s-Triazolo[3,4-b][1,3,4]thiadiazine, 3-ethyl-6-methyl-	EtOH	280(4.42)	2-0959-69
$C_7H_{10}N_5O_5P$			
6-Purinol, 9-(2-hydroxyethyl)-2-amino-, 2'-phosphate, barium salt	pH 1	255(3.79),280s(--)	94-1268-69
	pH 13	255s(--),269(3.98)	94-1268-69
$C_7H_{10}N_6O_{12}$			
Propionamide, 2,2'-(methylenedioxy)bis-[3,3-dinitro-	dil. KOH	356(4.48)	34-0116-69
$C_7H_{10}O$			
Bicyclo[3.1.0]hexan-2-one, 5-methyl-	EtOH	194(3.57),280(1.75)	44-2512-69
2-Cyclohepten-1-one	EtOH	227(4.01),317(1.70)	24-3877-69
2-Cyclohexen-1-one, 3-methyl-	H_2SO_4	241(4.12)	23-2263-69
protonated	H_2SO_4	277(4.14)	23-2263-69

Compound	Solvent	λmax(log ε)	Ref.
2-Cyclopenten-1-one, 2,3-dimethyl-	H_2SO_4	240(4.04)	23-2263-69
	EtOH	234(4.17),300(1.81)	39-0449-69B
protonated	H_2SO_4	266(4.13)	23-2263-69
2-Cyclopenten-1-one, 4,4-dimethyl-	EtOH	219(4.11),315(1.60)	39-0449-69B
2-Hepten-4-yn-1-ol	MeOH	227(4.17),235(4.10)	35-4824-69
1-Heptyn-3-one	EtOH	207(3.79),320(1.90)	33-2465-69
Norbornanone	C_6H_{12}	294(1.36)	77-1246-69
1,2,3-Pentatriene, 1-methoxy-4-methyl-	EtOH	238(4.00),265(4.33)	54-0119-69
1,2,4-Pentatriene, 1-methoxy-4-methyl-	EtOH	227(4.34)	54-0119-69
$C_7H_{10}OS_2$			
2H-Thiopyran-2-one, tetrahydro-3-(thio-acetyl)-	EtOH	304(4.02),365(3.02)	78-5703-69
$C_7H_{10}O_2$			
1-Cyclohexene-1-carboxylic acid, anion	iso-PrOH	285s(4.0)	35-5792-69
1-Cyclopropene-1-carboxylic acid,	C_6H_{12}	225(3.64)	88-0015-69
3,3-dimethyl-, methyl ester	MeOH	329(2.30)	88-2659-69
Furan, 2-ethoxy-4-methyl-	EtOH	221(3.72)	35-6432-69
Furan, 2-ethoxy-5-methyl-	EtOH	225(3.63)	35-6432-69
3-Oxabicyclo[4.1.0]heptan-2-one, 4-methyl-	EtOH	210(2.38)(end abs.)	78-2041-69
2,4-Pentadienoic acid, ethyl ester	MeOH	247(4.39)	78-4187-69
2H-Pyran-2-one, 5,6-dihydro-4,6-di-methyl-	EtOH	212(3.90)	78-2041-69
4H-Pyran-4-one, 2,3-dihydro-3,6-di-methyl-	EtOH	262(4.08)	22-1383-69
Sorbic acid, 4-methyl-	n.s.g.	260(4.29)	5-0191-69J
$C_7H_{10}O_2S$			
2(3H)-Furanone, dihydro-5-methyl-3-(thioacetyl)-	EtOH	281(4.05),348(3.61)	78-5703-69
2(3H)-Furanone, dihydro-3-(thio-propionyl)-	EtOH	279(4.03),332(2.90)	78-5703-69
$C_7H_{10}O_3$			
δ-Caprolactone, α-(hydroxymethylene)-	n.s.g.	252(4.04)	102-1797-69
Dehydroolivomal	EtOH	263(3.81)	105-0089-69
4-Pentenoic acid, 3-oxo-, ethyl ester	EtOH	224(3.94),274(3.93)	22-4091-69
	base	225(3.81),308(4.15)	22-4091-69
$C_7H_{10}O_4$			
Crotonic acid, β-(carbomethoxymethyl)-	MeOH	213(4.07)	77-0685-69
	MeOH-NaOH	204(4.06)	77-0685-69
Hexanoic acid, 3,5-dioxo-. methyl ester	MeOH	276(3.88)	77-0685-69
	MeOH-NaOMe	296(4.29)	77-0685-69
$C_7H_{10}O_4S_2$			
Methanetricarboxylic acid, C,C-dithio-, trimethyl ester	MeOH	315(3.76),334(3.90)	78-4649-69
$C_7H_{10}S_3$			
Carbonic acid, trithio-, cyclic ethyl-methylmethyleneethylene ester	EtOH	308(3.62),327(4.15)	94-2442-69
3,4,5-Trithiatricyclo[5.2.1.0^{2,6}]decane, exo-	C_6H_{12}	284(3.58)	35-5415-69
$C_7H_{10}Si$			
Silane, methylphenyl-	C_6H_{12}	264(2.45)	101-0017-69B

Compound	Solvent	λ_{max}(log ε)	Ref.
$C_7H_{11}Cl_3O_2$			
3-Hexanone, 6,6,6-trichloro-5-hydroxy-4-methyl-, erythro-	EtOH	213(1.71),280(1.49)	23-2029-69
threo-	EtOH	213(1.77),279(1.53)	23-2029-69
$C_7H_{11}NO$			
Isoxazole, 3,5-diethyl-	EtOH	215(3.82)	22-2820-69
$C_7H_{11}NO_2$			
2(3H)-Furanone, dihydro-3-[1-(methylamino)ethylidene]-	hexane	295(4.22)	104-0998-69
Hexanimide, N-methyl-	MeOH	235(3.20)	44-1695-69
3-Isoxazoleethanol, α,5-dimethyl-	EtOH	216(3.66)	35-4749-69
$C_7H_{11}NO_3S$			
Carbonic acid, 2-isothiocyanatobutyl methyl ester, D-	EtOH	259(4.20)	39-1143-69B
L-isomer	EtOH	259(4.21)	39-1143-69B
$C_7H_{11}NO_4$			
Malonamic acid, 2-acetyl-, ethyl ester	EtOH	228(2.70),259(3.13)	104-1713-69
$C_7H_{11}NO_5$			
D-glycero-Hex-2-enopyranosid-4-ulose, methyl 2-amino-2-deoxy-, α-	n.s.g.	314(4.24)	51-0328-69
Nitrone, α-carboxy-α-(carboxymethyl)-N-methyl-, dimethyl ester	MeOH	273(4.10)	24-2346-69
$C_7H_{11}NS_2$			
2H-1,4-Thiazine, 3-(ethylthio)-5-methyl-	EtOH	212(3.71),251(3.83), 336(3.61)	4-0247-69
$C_7H_{11}N_3O$			
1,3-Diazaspiro[4.4]non-1-en-4-one, 2-amino-	H_2O	226(3.86)	18-3314-69
$C_7H_{11}N_3O_2$			
4H-Cyclopentapyrimidin-4-one, 2-amino-3,4a,5,6,7,7a-hexahydro-7a-hydroxy-	H_2O	227(4.00)	18-3314-69
2-Pyrimidineethanol, 4-amino-5-(hydroxymethyl)-	MeOH	235(4.02),271(3.72)	39-0133-69C
$C_7H_{11}N_3O_3$			
4-Isoxazolecarboxylic acid, 5-hydrazino-3-methyl-, ethyl ester	MeOH	252(4.13)	4-0783-69
Uramil, 1,3,7-trimethyl-	pH 1	255(4.08)	104-0547-69
$C_7H_{11}N_5O_2$			
Acetic acid, (4H-1,2,4-triazol-4-ylamidino)-, ethyl ester	MeOH	232(3.33)	48-0477-69
Propionamide, N-(2,4-diamino-1,6-dihydro-6-oxo-5-pyrimidinyl)-	pH 0.0	266(4.23)	5-0201-69F
	pH 7.0	268(4.13)	5-0201-69F
	pH 12.0	263(4.00)	5-0201-69F
C_7H_{12}			
Bicyclo[4.1.0]heptane, iodine complex	hexane	255(4.17)	35-2279-69
Cyclohexane, methylene-	isooctane	191(4.07),198(3.96)	121-0803-69
	EtOH	192(4.03)	121-0803-69

Compound	Solvent	$\lambda_{max}(\log \epsilon)$	Ref.
$C_7H_{12}ClNO_4S$			
Trimethyl-2-thienylammonium perchlorate	EtOH	227(4.1)	48-0827-69
$C_7H_{12}N_2$			
2,3-Diazabicyclo[2.2.1]hept-5-ene, 2,3-dimethyl-	C_6H_{12}	266(2.83)	78-0657-69
	H_2O	230s(--)	78-0657-69
	EtOH	242(2.76)	78-0657-69
	dioxan	263(2.78)	78-0657-69
$C_7H_{12}N_2O$			
Imidazole, 2-ethyl-4,5-dimethyl-, 1-oxide	MeOH	225(3.80)	48-0746-69
Propiolamide, 3-(dimethylamino)-N,N-dimethyl-	CH_2Cl_2	252(4.05)	33-2641-69
2-Pyrazoline-1-carboxaldehyde, 5-ethyl-4-methyl-	MeOH	235(4.32)	104-0736-69
2-Pyrazoline-1-carboxaldehyde, 3,5,5-trimethyl-	MeOH	235(4.14)	104-0736-69
$C_7H_{12}N_2O_2$			
Imidazole, 2-ethyl-3-hydroxy-4,5-dimethyl-, 1-oxide	MeOH	235(3.63)	48-0746-69
2-Pyrazoline-3-carboxylic acid, 5,5-dimethyl-, methyl ester	hexane	276(4.05)	39-0787-69C
3,5-Pyrazolidinedione, 4,4-diethyl-	H_2O	253(3.56)	44-3181-69
$C_7H_{12}N_2O_4$			
1,3,4-Oxadiazol-2-ine-4-carboxylic acid, 2-methoxy-5,5-dimethyl-, methyl ester	C_6H_{12}	214(3.76)	78-0619-69
$C_7H_{12}N_2O_6$			
1,3-Dioxolane, 2-(3,3-dinitropropyl)-2-methyl-	EtOH	381(3.66)	44-1470-69
$C_7H_{12}N_2S$			
Imidazole, 1,4,5-trimethyl-2-(methylthio)-	2% EtOH	257(3.82)	1-2888-69
conjugate acid	2% EtOH	259(3.89)	1-2888-69
4-Imidazoline-2-thione, 1,3,4,5-tetramethyl-	2% EtOH	257(4.11)	1-2888-69
conjugate acid	2% EtOH	251(3.88)	1-2888-69
$C_7H_{12}N_4O_2$			
as-Triazin-5(2H)-one, 3-[(3-hydroxypropyl)amino]-6-methyl-	EtOH	211(4.41),240s(3.86)	114-0181-69C
$C_7H_{12}N_4S_3$			
1,3-Thiazane-2,4-dithione, 3-ethyl-, 4-(thiosemicarbazone)	EtOH	250(4.05),310s(--)	103-0049-69
$C_7H_{12}O$			
Butadiene, 1-ethoxy-2-methyl-, cis	EtOH	244(4.19)	22-2508-69
Butadiene, 1-ethoxy-2-methyl-, trans	EtOH	240(4.27)	22-2508-69
Butadiene, 1-ethoxy-3-methyl-, cis	heptane	243(4.23)	104-1696-69
Butadiene, 1-ethoxy-3-methyl-, trans	heptane	240(4.30)	104-1696-69
	EtOH	243(4.35)	22-2508-69
1,3-Butadiene, 2-ethoxy-3-methyl-	EtOH	232(3.94)	22-2508-69
Cyclohexanone, 2-methyl-	C_6H_{12}	288(1.26)	77-1246-69
	MeOH	284(1.28)	77-1246-69

Compound	Solvent	$\lambda_{max}(\log \epsilon)$	Ref.
1,3-Pentadiene, 3-ethoxy-, cis	EtOH	246(3.94)	22-2508-69
1,3-Pentadiene, 3-ethoxy-, trans	EtOH	230(4.11)	22-2508-69
$C_7H_{12}OS$			
Crotonic acid, 2,3-dimethyl-thio-, S-methyl ester	hexane	240(3.79)	54-0465-69
3-Penten-2-one, 4-methyl-3-(methylthio)-	hexane	237(3.53),274(3.20)	54-0465-69
$C_7H_{12}O_2$			
Cyclohexanecarboxylic acid, Pb(IV) salt	C_6H_{12}	243(4.38)	33-1495-69
$C_7H_{12}Si$			
Silane, methyltrivinyl-	gas	176(--)	121-0803-69
	heptane	177(4.53)	121-0803-69
	MeCN	179(4.50)	121-0803-69
$C_7H_{13}N$			
Quinuclidine	gas	215f(2.29),223(2.1), 226(2.14),229(2.1)	59-0501-69
$C_7H_{13}NO$			
2(1H)-Azocinone, hexahydro-	C_6H_{12}	195(3.83)	46-1642-69
	H_2O	197(3.90)	46-1642-69
Cyclohexane, 1-methyl-1-nitroso- (dimer)	EtOH	286(1.69)	39-1073-69C
$C_7H_{13}NO_2$			
Acrolein, 3-(dimethylamino)-3-ethoxy-	CH_2Cl_2	289(4.34)	33-2641-69
$C_7H_{13}N_3O_3$			
3-Imidazolin-5-one, 2,2-dimethoxy-1-methyl-4-(methylamino)-	EtOH	218(4.45),252(3.79)	78-0557-69
$C_7H_{13}N_5O_2$			
Urea, 1-[4-(dimethylamino)-2-oxo-3-imidazolin-5-yl]-3-methyl-	pH 7-10	249(4.04)	69-0733-69
C_7H_{14}			
1-Pentene, 4,4-dimethyl-	gas	179(--)	121-0803-69
	heptane	181(4.06)	121-0803-69
	MeCN	182(3.98)	121-0803-69
$C_7H_{14}N_2$			
2-Pyrazoline, 1,3,5,5-tetramethyl-	EtOH	244(3.62)	22-3306-69
3-Pyrazoline, 1,2,3,4-tetramethyl-	EtOH	237(3.49)	22-3316-69
$C_7H_{14}N_2O$			
3-Penten-2-ol, 2-methyl-4-(methylazo)-, cis	n.s.g.	389(2.07)	89-0459-69
3-Penten-2-ol, 2-methyl-4-(methylazo)-, trans	n.s.g.	395(1.91)	89-0459-69
$C_7H_{14}N_2O_2$			
Methanol, (tert-butylazo)-, acetate	C_6H_{12}	359(1.21)	89-0451-69
$C_7H_{14}N_2O_2S$			
Hydrouracil, 1-methyl-2-thio-, 4-(dimethyl acetal)	EtOH	209(4.12),254(4.41)	39-2631-69C

Compound	Solvent	λ_{max}(log ϵ)	Ref.
$C_7H_{14}N_2O_4$			
4-Isoxazoline, 3,5-diethyl-5-(dihydroxy-amino)-, 1-oxide, sodium salt	EtOH	275(4.21)	44-0984-69
$C_7H_{14}N_2S_2$			
2H-1,3,5-Thiadiazine-2-thione, 3,5-di-ethyltetrahydro-	MeOH	250(3.92),293(3.93)	73-2952-69
$C_7H_{14}N_4O_2S$			
2-Hydroxyethylammonium 6-methyl-3-(meth-ylthio)-1,2,4-triazin-5-olate	EtOH	220(4.18),230(4.17), 286(3.59)	114-0181-69C
$C_7H_{14}O$			
2-Heptanone	gas	170(3.18),176(2.85), 193(3.00),196(3.04)	18-2453-69
4-Heptanone	gas	167(3.06),175(2.90), 192(2.78),196(2.81)	18-2453-69
3-Pentanone, 2,4-dimethyl-	gas	167(3.38),167(3.40), 168(3.36),169(3.35), 179(3.35),183(2.98), 195(2.00)	18-2453-69
$C_7H_{14}O_2$			
Heptanoic acid, Pb(IV) salt	C_6H_{12}	230(4.40)	33-1495-69
$C_7H_{14}Si$			
Silane, allyldimethylvinyl-	gas	178(--),191(--)	121-0803-69
	heptane	177(4.31),194(4.18)	121-0803-69
	MeCN	178(4.26),194(4.14)	121-0803-69
$C_7H_{15}B_{10}Cs$			
Cesium [7,10^2]hemiousenide	H_2O	217(4.60),262(4.04), 439(4.13)	35-0323-69
	12N HCl	217(--),262(--), 427(--)	35-0323-69
	96% H_2SO_4	215(4.60),261(4.03), 373(3.81)(changing)	35-0323-69
	MeCN	220(4.22),264(3.92), 512(4.16)	35-0323-69
	acetone	536(4.19)	35-0323-69
$C_7H_{15}NS_2$			
Carbamic acid, N,N-dipropyldithio- (also spectra of metal salts)	CH_2Cl_2	263(3.50),290(3.18)	39-1152-69A
$C_7H_{15}NS_3$			
Peroxycarbamic acid, diethyltrithio-, ethyl ester	EtOH	281(3.94)	34-0119-69
Peroxycarbamic acid, N,N-dimethyl-trithio-, S-(butylthio) ester	EtOH	278(3.92)	34-0119-69
Peroxycarbamic acid, N,N-dimethyl-trithio-, S-(tert-butylthio) ester	EtOH	240(3.88),281(3.83)	34-0119-69
$C_7H_{15}N_3O$			
Hexanal, semicarbazone	EtOH	230(4.07)	94-0844-69
$C_7H_{15}N_3S$			
Butyraldehyde, 2-ethyl-, thiosemicarb-azone	EtOH	230(3.83),270(4.32)	94-0844-69

Compound	Solvent	$\lambda_{max}(\log \epsilon)$	Ref.
$C_7H_{16}B_9Ni$			
Nickel(1+), π-cyclopentadienyl(undeca-hydro-7,8-dicarbaundecaborato(2-)]-, ion	$CHCl_3$	327(4.35),434(3.42)	35-0758-69
$C_7H_{16}N_2$			
1-Propene-1,1-diamine, N,N,N',N'-tetra-methyl-	C_6H_{12}	207(3.81)	39-0016-69C
$C_7H_{16}N_4S$			
Carbohydrazide, 3-thio-1-(1,2,2-tri-methylpropylidene)-	MeOH	267(4.37)	44-0756-69
$C_7H_{17}B_{12}Cs$			
Cesium, [7,12]hemiousenide	H_2O	215(4.53),262(4.11), 349(4.01)	35-0323-69
	96% H_2SO_4	213(4.59),258(4.15), 332(4.00)(changing)	35-0323-69
	MeCN	216(4.60),267(4.04), 384(4.07)	35-0323-69
$C_7H_{21}BrSi_3$			
Trisilane, 1-bromoheptamethyl-	isooctane	217(3.87)	35-6613-69
Trisilane, 2-bromoheptamethyl-	isooctane	217(3.89),240s(2.57)	35-6613-69
	MeCN	216(3.92),240s(2.58)	35-6613-69
	dioxan	217(3.92),240s(2.57)	35-6613-69
$C_7H_{21}ClSi_3$			
Trisilane, 1-chloroheptamethyl-	isooctane	217(3.95)	35-6613-69
Trisilane, 2-chloroheptamethyl-	isooctane	216(3.90),233s(2.98)	35-6613-69
	MeCN	216(3.86),233s(2.95)	35-6613-69
	dioxan	216(3.94),233s(3.00)	35-6613-69
$C_7H_{21}FSi_3$			
Trisilane, 1-fluoroheptamethyl-	isooctane	217(3.84)	35-6613-69
Trisilane, 2-fluoroheptamethyl-	isooctane	216(3.89),240(3.36)	35-6613-69
	MeCN	216(3.86),241(3.33)	35-6613-69
	dioxan	216(3.96),240(3.38)	35-6613-69
$C_7H_{21}ISi_3$			
Trisilane, 2-iodoheptamethyl-	isooctane	224(3.81)	35-6613-69
$C_7H_{22}OSi_3$			
Trisilane, 2-methoxy-1,1,1,3,3,3-hexa-methyl-	isooctane	205(4.10),245(3.09)	35-6613-69
	MeOH	204(4.12),244(3.10)	35-6613-69
1-Trisilanol, heptamethyl-	isooctane	219(3.78)	35-6613-69
sodium derivative	isooctane	200(3.85)(end abs.)	35-6613-69
2-Trisilanol, heptamethyl-	isooctane	221(3.67),242(3.19)	35-6613-69
	MeCN	222(3.68)	35-6613-69
	dioxan	223(3.77)	35-6613-69
$C_7H_{22}SSi_3$			
1-Trisilanethiol, heptamethyl-	isooctane	219(3.86)	35-6613-69
2-Trisilanethiol, heptamethyl-	isooctane	198(3.93)(end abs.)	35-6613-69
$C_7H_{22}Si_3$			
Trisilane, 1,1,1,2,2,3,3-heptamethyl-	hexane	218(3.91)	77-0004-69
Trisilane, heptamethyl-	isooctane	202(4.13)	35-6613-69

Compound	Solvent	λ_{max}(log ϵ)	Ref.
$C_7H_{23}NSi_3$			
Trisilane, 1-aminoheptamethyl-	isooctane	213s(3.81)	35-6613-69
	MeCN	213s(--)	35-6613-69
	dioxan	213s(3.81)	35-6613-69
Trisilane, 2-aminoheptamethyl-	isooctane	205(3.96)(end abs.)	35-6613-69

Compound	Solvent	$\lambda_{max}(\log \epsilon)$	Ref.
$C_8Cl_2F_4N_4S$			
s-Triazine, 2,4-dichloro-6-[(2,3,5,6-tetrafluoro-4-pyridyl)thio]-	hexane	214(4.13),241(4.08), 263(4.08)	39-1660-69C
C_8Cl_8			
Cyclooctatetraene, octachloro-	n.s.g.	237(4.47),275s(--)	88-2137-69
Tricyclo[4.2.0.0^{2,5}]octa-3,7-diene, 1,2,3,4,5,6,7,8-octachloro-	n.s.g.	239(3.72)	89-0759-69
$C_8F_6N_4$			
4,4'-Bipyrimidine, 2,2',5,5',6,6'-hexafluoro-	hexane	210(3.74),275(3.97)	39-1866-69C
C_8F_{10}			
o-Xylene, decafluoro-	C_6H_{12}	272(3.19)	39-0211-69C
$C_8F_{14}MoO_4$			
Butyric acid, heptafluoro-, Mo(II) salt	EtOH	228(2.90),340(3.48), 435(2.00)	12-1571-69
$C_8HN_3O_2S_2$			
1,4-Dithiin-2,3-dicarboximide, 5,6-dicyano-	THF	244(4.11),343(3.88)	88-4273-69
$C_8H_2Cl_6$			
1,3,7-Octatrien-5-yne, 1,2,3,4,7,8-hexachloro-	hexane	217(4.25),292(4.23), 305(4.23)	78-2223-69
$C_8H_2Cl_8$			
1,3,5,7-Octatetraene, 1,2,3,4,5,6,7,8-octachloro-	hexane	214(4.43),248s(3.95)	78-0883-69
$C_8H_2F_4$			
Acetylene, (2,3,4,5-tetrafluorophenyl)-	C_6H_{12}	275(3.16),279(3.17), 284(3.13)	39-0211-69C
$C_8H_2F_4O_3$			
Terephthalaldehydic acid, tetrafluoro-	EtOH	264(3.36)	65-1607-69
$C_8H_2F_5N$			
Acetonitrile, (pentafluorophenyl)-	EtOH	262(3.00)	65-2071-69
$C_8H_3F_4N_3O_2$			
Benzimidazole, 7-fluoro-4-nitro-2-(trifluoromethyl)-	pH 1	316(3.99)	44-0384-69
	pH 6.3	252(3.90)	44-0384-69
	pH 11	335(3.94)	44-0384-69
$C_8H_3F_7$			
o-Xylene, α,α,α,3,4,5,6-heptafluoro-	C_6H_{12}	270(3.15)	39-0211-69C
$C_8H_4BF_5O_2S$			
Boron, difluoro[4,4,4-trifluoro-1-(2-thienyl)-1,3-butanedionato]-	$CHCl_3$	325(4.27),365(4.37)	39-0526-69A
$C_8H_4Cl_2N_2OS$			
1,3,4-Oxadiazole-2-thiol, 5-(2,4-di-	n.s.g.	211(4.54),240(4.26), 310(4.17)	2-0583-69
$C_8H_4F_6O_4S_2$			
Benzene, p-bis[(trifluoromethyl)sulfonyl]-	benzene	284(3.46)	65-1634-69

Compound	Solvent	λmax(log ε)	Ref.
Benzene, p-bis[(trifluoromethyl)sulfonyl]- (cont.)	MeOH	224(4.12),281(3.43), 288(3.35)	65-1634-69
	EtOH	224(4.14),282(3.44), 290(3.34)	65-1634-69
	DMF	286(3.45)	65-1634-69
$C_8H_4N_2O_4$			
4H-1,2-Benzoxazin-4-one, 3-nitro-	MeOH	242(3.90),257s(3.83), 322(3.76)	89-0979-69
$C_8H_4OS_4$			
Dithieno[3,4-b:3',4'-e]-p-dithiin, 4-oxide	EtOH	244(4.19)	89-0598-69
$C_8H_4O_2S_4$			
Dithieno[3,4-b:3',4'-e]-p-dithiin, 4,8-dioxide	EtOH	250s(3.93)	89-0598-69
$C_8H_4S_4$			
Dithieno[3,4-b:3',4'-e]-p-dithiin	EtOH	270(3.96)	89-0598-69
$C_8H_5BrClNO_4$			
3-Buten-2-one, 1-bromo-3-chloro-4-(5-nitro-2-furyl)-	EtOH	231(3.85),346(4.28)	103-0414-69
$C_8H_5BrIN_3$			
Styrene, α-azido-β-bromo-β-iodo-	MeOH	206(4.12),264(3.81)	35-6122-69
$C_8H_5BrN_2O_6$			
Benzoic acid, 4-bromo-3,5-dinitro-, methyl ester	EtOH	225(4.52),258s(4.38)	4-0533-69
C_8H_5BrS			
Benzothiophene, 3-bromo-	MeOH	230(4.32),261(3.68), 282(3.38),291(3.50), 300(3.65)	73-2959-69
$C_8H_5ClN_2O$			
Furazan, 3-chloro-4-phenyl-	EtOH	251(3.80)	39-2794-69C
$C_8H_5Cl_2HgNO_3$			
Anthranil, 5,6-(methylenedioxy)-, mercuric chloride complex	EtOH	310(3.69)	78-5365-69
$C_8H_5Cl_2N_3O$			
1,2,3-Benzotriazin-4(3H)-one, 6,8-dichloro-3-methyl-	EtOH	226(4.39),298(4.03)	4-0809-69
$C_8H_5FN_4O_4$			
Benzimidazole, 7-fluoro-4,6-dinitro-2-methyl-	pH 1	276(4.16)	44-0384-69
	pH 7.9	282(4.13)	44-0384-69
	pH 11.7	310(4.16)	44-0384-69
$C_8H_5F_3N_2O_2S$			
Sulfone, diazomethyl α,α,α-trifluoro-m-tolyl	C_6H_{12}	251(3.97)	54-0641-69
	40% dioxan	383(1.84)	54-0641-69
$C_8H_5F_3N_4$			
Pteridine, 7-methyl-4-(trifluoromethyl)-	pH 7.0	296(3.96),302(3.95), 314(3.91)	39-1751-69C

Compound	Solvent	λ_{max}(log ε)	Ref.
Pteridine, 7-methyl-4-(trifluoromethyl)- (cont.)	hexane	231(3.40),294(3.81), 298(3.86),304(3.88), 309(3.80)	39-1751-69C
$C_8H_5F_3O$			
Acetophenone, 2,2,2-trifluoro-	heptane	251(3.74),330(1.62)	59-0765-69
$C_8H_5F_3O_2$			
Acetophenone, 2,2,2-trifluoro-2'-hydroxy-	hexane	263(4.1),425(3.6)	18-0960-69
$C_8H_5NO_2S$			
Benzothiophene, 3-nitro-	MeOH	216(4.56),240(4.09), 340(3.90)	73-2959-69
$C_8H_5NO_3$			
Phthalic anhydride, 3-amino-	EtOH	203(4.14),249(4.22), 388(3.67)	16-0241-69
$C_8H_5NO_4$			
7-Benzofuranol, 6-nitro-	MeOH	205(4.29),227(4.07), 310(3.94)	32-1177-69
$C_8H_5NO_6$			
Isophthalaldehydic acid, 2-hydroxy-5-nitro-	pH 1	226(--),319(--)	22-0817-69
	pH 13	237(--),385s(--)	22-0817-69
	EtOH	225(4.19),308(3.99)	22-0817-69
$C_8H_5N_3$			
Malononitrile, (pyrrol-2-ylmethylene)-	MeOH	375(4.51)	22-0948-69
Pyridinium dicyanomethylide	MeCN	393(4.33)	22-0948-69
$C_8H_5N_3O$			
1,3-Oxazepine-2,4-dicarbonitrile, 6-methyl-	CH_2Cl_2	334(3.37)	78-0295-69
2,6-Pyridinedicarbonitrile, 4-methyl-, 1-oxide	EtOH	244(4.53),298(4.02), 365(3.20)	78-0295-69
$C_8H_5N_3O_3$			
Acetophenone, 2-diazo-4'-nitro-	EtOH	241s(4.09),245s(4.09), 264(4.14),307(4.10)	12-0577-69
4,5-Isoxazoledione, 3-(4-pyridyl)-, 4-oxime	EtOH	258(4.24),335s(3.61)	36-0460-69
1,3,4-Oxadiazole, 2-(p-nitrophenyl)-	MeOH	280(4.3)	48-0646-69
$C_8H_5N_3O_4$			
m-Tolunitrile, α,α-dinitro-	anion	368(4.19)	122-0161-69
p-Tolunitrile, α,α-dinitro-	anion	368(4.18)	122-0161-69
$C_8H_5N_3O_5$			
p-Anisonitrile, 3,5-dinitro-, Meisenheimer complex intermediate	93% DMSO-MeOH	460(4.32)	22-2692-69
final complex	93% DMSO-MeOH	540(4.39)	22-2692-69
$C_8H_5N_5$			
Pyrido[2,3-d]-s-triazolo[4,3-b]pyridazine	EtOH	230s(4.23),237(4.24), 248s(4.03),267(3.69), 280s(3.53),285s(3.40), 311(3.29)	12-1759-69

Compound	Solvent	$\lambda_{max}(\log \epsilon)$	Ref.
Pyrido[3,2-d]-s-triazolo[4,3-b]pyridazine	EtOH	212s(3.66),234(4.06), 246s(3.88),265s(3.56), 271s(3.49),276s(3.43), 281s(3.35),308(3.07), 320s(3.02)	12-1759-69
s-Triazolo[3,4-c][1,2,4]benzotriazine	EtOH	216(3.98),320(3.58)	24-3818-69
$C_8H_5N_5O$			
s-Triazolo[3,4-c][1,2,4]benzotriazine, 5-oxide	EtOH	226(4.39),332(3.98)	24-3818-69
$C_8H_5N_5OS$			
s-Triazolo[3,4-c][1,2,4]benzotriazine-1-thiol, 5-oxide	EtOH	225(4.45),307(4.29)	24-3818-69
$C_8H_5N_5S$			
s-Triazolo[3,4-c][1,2,4]benzotriazine-1-thiol	EtOH	214(4.27),300(4.09)	24-3818-69
C_8H_6BrN			
Indole, 5-bromo-	MeOH	280(3.73)	22-3002-69
anion	MeOH	298(--)	22-3002-69
$C_8H_6BrNO_2$			
Styrene, p-bromo-β-nitro-	n.s.g.	314(4.34)	67-0127-69
$C_8H_6BrNO_3$			
Acetophenone, 2-bromo-2-nitro-	acid	257(4.00)	22-0534-69
	base	365(3.88)	22-0534-69
$C_8H_6BrN_3O_7$			
Anisole, 2-bromo-4-(dinitromethyl)-6-nitro-	anion	366(4.17)	122-0161-69
$C_8H_6Br_2N_2O_3$			
Acetanilide, 2',5'-dibromo-4'-nitro-	C_6H_{12}	223(4.29),246(4.14), 254s(--),287(3.91)	117-0287-69
$C_8H_6ClN_3O$			
1,2,3-Benzotriazin-4(3H)-one, 6-chloro-3-methyl-	EtOH	290(3.87)	4-0809-69
1,2,3-Benzotriazin-4(3H)-one, 7-chloro-3-methyl-	EtOH	236(4.43),290(3.74)	4-0809-69
$C_8H_6Cl_2N_4$			
Quinazoline, 2,4-diamino-5,8-dichloro-	acid	228s(4.41),231(4.42), 244(4.47),263s(4.00), 330(3.66),338s(3.56)	44-0821-69
	neutral	238(4.47),245s(4.44), 270s(3.86),279(3.94), 288(3.91),350(3.70)	44-0821-69
$C_8H_6Cl_2O_2$			
p-Benzoquinone, 2,6-dichloro-3,5-dimethyl-	MeOH	276(4.23)	64-0547-69
$C_8H_6Cl_4$			
Bicyclo[2.2.0]hexa-2,5-diene, 1,2,3,4-tetrachloro-5,6-dimethyl-	n.s.g.	240(3.43)	89-0759-69

Compound	Solvent	$\lambda_{max}(\log \epsilon)$	Ref.
$C_8H_6Cl_6O_4$			
Acetic acid, trichloro-, cyclobutylidene ester	EtOH	220(2.81)(end abs.)	78-1679-69
$C_8H_6FNO_2$			
Styrene, p-fluoro-β-nitro-	n.s.g.	313(4.21)	67-0127-69
$C_8H_6F_3NO$			
Benzaldoxime, o-(trifluoromethyl)-, syn	EtOH	250(4.14)	36-0490-69
$C_8H_6F_3N_3O_2S$			
Thiazolo[4,5-d]pyrimidine-5,7(4H,6H)-dione, 4,6-dimethyl-2-(trifluoromethyl)-	MeOH	317(3.68)	44-3285-69
$C_8H_6F_3N_5O_2$			
Acetamide, N-[1,6-dihydro-6-oxo-8-(trifluoromethyl)purin-2-yl]-	pH 2.0	263(4.26)	5-0201-69F
	pH 8.0	223(4.27),270(4.21)	5-0201-69F
	pH 13.0	272(4.11)	5-0201-69F
$C_8H_6F_4$			
o-Xylene, 3,4,5,6-tetrafluoro-	C_6H_{12}	260(--),264(2.66)	39-0211-69C
$C_8H_6INO_2$			
Styrene, p-iodo-β-nitro-	n.s.g.	323(4.24)	67-0127-69
$C_8H_6N_2O$			
1,6-Naphthyridine, 1-oxide	EtOH	217(4.33),319(3.43), 334(3.51)	95-1260-69
1,6-Naphthyridine, 6-oxide	EtOH	214(4.28),249(4.03), 303(3.60),349(3.65)	95-1260-69
1,6-Naphthyridin-2(1H)-one	EtOH	229(4.47),273(3.54), 309(3.74)	95-1260-69
1,3,4-Oxadiazole, 2-phenyl-	MeOH	247(4.3)	48-0646-69
$C_8H_6N_2OS$			
1,3,4-Oxadiazole-2-thiol, 5-phenyl-	n.s.g.	225(4.34),310(4.25)	2-0583-69
$C_8H_6N_2O_2$			
Acetonitrile, 2-nitro-2-phenyl-	anion	327(4.13)	122-0100-69
Benzonitrile, 2-methoxy-5-nitroso-	dioxan	254(4.04),332(4.15), 730(1.70)	12-0935-69
3-Furazanol, 4-phenyl-	EtOH	258(3.96)	39-2794-69C
Indole, 3-nitro-	MeOH	347(3.99)	22-3002-69
anion	MeOH	395(4.34)	22-3002-69
Indole, 5-nitro-	MeOH	322(3.92)	22-3002-69
anion	MeOH	405(--)	22-3002-69
1,6-Naphthyridin-2(1H)-one, 6-oxide	EtOH	222(4.46),305(3.97), 317(3.92)	95-1260-69
1,2,4-Oxadiazol-3-ol, 5-phenyl-	EtOH	256(4.10)	39-2794-69C
1,4-Phthalazinedione, 2,3-dihydro-, sodium salt	DMSO	364(3.55)	44-2462-69
	pH 13	310(--)	44-2462-69
	90% DMSO	342(--)	44-2462-69
	+KOH	342(--)	44-2462-69
Sydnone, 3-phenyl-	MeOH	310(3.70)	42-0182-69
	EtOH	311(3.79)	
	iso-PrOH	313(3.69)	
	CHCl	315(3.68)	
	+20% C_6H_{12}	315(3.68)	
	+40% C_6H_{12}	315(3.68)	

Compound	Solvent	λ_{max}(log ε)	Ref.
Sydnone, 3-phenyl- (cont.)	+60% C_6H_{12}	316(3.67)	42-0182-69
	+80% C_6H_{12}	317(3.63)	
	+90% C_6H_{12}	320(3.65)	
	CCl_4	323(3.69)	
	MeCN	313(3.67)	
$C_8H_6N_2O_3$			
1,4-Cyclohexadiene-1-carbonitrile, 5-methoxy-3,6-dioxo-, 3-oxime	EtOH	249(3.75),329(4.15)	12-0935-69
	EtOH-NaOH	230s(4.00),256(3.67), 307(3.77),412(4.43)	12-0935-69
	dioxan	725(0.23)	12-0935-69
$C_8H_6N_2O_4$			
Benzonitrile, 2-hydroxy-3-methoxy-5-nitro-	EtOH	256(3.98),330(3.89)	12-0935-69
	EtOH-NaOH	286(3.86),335(3.88), 408(4.20)	12-0935-69
Isoxazole, 5-methyl-3-(5-nitro-2-furyl)-	EtOH	320(4.16)	18-0258-69
Styrene, p,β-dinitro-	n.s.g.	301(4.32)	67-0127-69
$C_8H_6N_2O_6$			
Acetic acid, (2,4-dinitrophenyl)-	pH 10	253(4.15)	35-2467-69
Phenol, 2,4-dinitrophenyl-, acetate	EtOH	237(4.18)	65-0660-69
$C_8H_6N_2Se$			
1,2,3-Selenadiazole, 4-phenyl-	EtOH	245(4.15),316(3.38)	88-5105-69
$C_8H_6N_4$			
Benzene, m-bis(diazomethyl)-	CH_2Cl_2	279(4.81),486(1.80)	47-3313-69
Benzene, p-bis(diazomethyl)-	hexane	313(4.66),505(2.05)	47-3313-69
3,3'-Bipyridazine	EtOH	234(3.60),261(3.26)	88-2359-69
Malononitrile, [(1-methylpyrazol-4-yl)-methylene]-	MeOH	317(3.34)	4-0545-69
$C_8H_6N_4O_2$			
Pyrazino[2,3-d]pyridazine, 5-acetoxy-	EtOH	204(4.47),250(4.19), 276(4.05)	4-0093-69
$C_8H_6N_4O_3$			
1,2,4-Oxadiazole, 5-amino-3-(m-nitro-phenyl)-	EtOH	253(4.01)	18-3335-69
1,2,4-Oxadiazole, 5-amino-3-(p-nitro-phenyl)-	EtOH	261(4.16),298(3.90)	18-3335-69
$C_8H_6N_4O_9$			
Anisole, 2-(dinitromethyl)-4,6-dinitro-	anion	364(4.22)	122-0100-69
Anisole, 5-(dinitromethyl)-2,4-dinitro-	anion	366(4.23)	122-0100-69
C_8H_6OS			
Thiepino[4,5-c]furan	MeOH	225(4.20),233(4.24), 240(4.43),249(4.38), 275(3.05),345(2.67), 364(2.71),415(2.15)	88-4361-69
$C_8H_6O_2S$			
Thiepino[4,5-c]furan, 6-oxide	MeOH	225(4.18),270(3.60)	88-4361-69
$C_8H_6O_2S_2$			
[3,3'-Bithiophene]-2,2'(5H,5'H)-dione	EtOH	241(4.08)	118-0170-69

Compound	Solvent	λ_{max}(log ε)	Ref.
$C_8H_6O_3$			
Glyoxylic acid, phenyl-	hexane	278(3.77)	18-2614-69
	EtOH	253(4.08)	
	ether	254(4.06)	
	CH_2Cl_2	279(3.95)	
	$CHCl_3$	282(3.84)	
	MeCN	253(4.01)	
$C_8H_6O_3S$			
Benzo[b]thiophen-3(2H)-one, 1,1-dioxide	dioxan	245(4.00),292(2.92), 320(1.53)	59-0393-69
Thiepino[4,5-c]furan, 6,6-dioxide	MeOH	229(4.41),270(3.44)	88-4361-69
$C_8H_6O_4$			
Benzo[1,2-d:3,4-d']bis[1,3]dioxole	EtOH	286(2.30)	5-0099-69E
Salicylic acid, 3-formyl-	pH 1	223(--),333(--)	22-0817-69
	pH 13	223(--),392(--)	22-0817-69
	EtOH	221(3.98),334(3.72)	22-0817-69
C_8H_6S			
Benzo[b]thiophene	MeOH	226(4.72),256(3.77), 280(3.28),288(3.42), 297(3.60)	73-2959-69
$C_8H_7BF_2O_3$			
Boron, difluoro(hydrogen salicylato)-, methyl ester	$CHCl_3$	247(3.61),308(3.65)	39-0526-69A
$C_8H_7BF_4O_2$			
Carboxycycloheptatrienylium tetrafluoroborate	EtOH	272(3.76)	35-6391-69
$C_8H_7BrN_2O_3$			
Acetophenone, 2'-bromo-4'-nitro-, oxime	H_2O	270(3.93)	80-0481-69
	pH 13	260(3.86),320(3.71)	80-0481-69
$C_8H_7BrN_2O_5$			
Anisole, 2-bromo-4-(dinitromethyl)-	anion	370(4.15)	122-0161-69
C_8H_7BrO			
Acetophenone, 2-bromo-	heptane	249(4.07),331(2.12)	59-0765-69
$C_8H_7BrO_2$			
Acetophenone, 3'-bromo-2'-hydroxy-	C_6H_{12}	226(3.36),253(3.19), 335(2.80)	42-0148-69
	EtOH	225(3.99),250(4.02), 261(3.93),335(2.18)	42-0148-69
Acetophenone, 3'-bromo-4'-hydroxy-	EtOH	229(4.23),262(3.65)	42-0148-69
Acetophenone, 5'-bromo-2'-hydroxy-	C_6H_{12}	230(3.92),252(3.60), 352(3.31)	42-0148-69
	EtOH	230(3.92),250(3.77), 255(3.72),260(3.57), 350(3.48)	42-0148-69
Benzaldehyde, 4-bromo-2-methoxy-	EtOH	265(4.20),320(3.84)	94-0089-69
Benzaldehyde, 6-bromo-2-methoxy-	EtOH	265(3.74),325(3.52)	94-0089-69
$C_8H_7BrO_3$			
Benzaldehyde, 5-bromo-4-hydroxy-3-methoxy-	EtOH	238(4.18),298(4.02)	42-0148-69

Compound	Solvent	$\lambda_{max}(\log \epsilon)$	Ref.
$C_8H_7BrO_4$			
p-Benzoquinone, 5-bromo-2,3-dimethoxy-	EtOH	208(4.01),289(3.85)	39-1353-69C
$C_8H_7BrS_2$			
Benzoic acid, p-bromodithio-, methyl ester	C_6H_{12}	231s(3.8),257s(3.4), 306(4.32),506(2.14)	48-0045-69
	EtOH	230s(3.9),305(4.27), 502(2.18)	48-0045-69
$C_8H_7Br_2NO$			
Acetanilide, 2',5'-dibromo-	C_6H_{12}	218(4.51),243(4.13), 285(3.14),293(3.14)	117-0287-69
$C_8H_7Br_2NO_2S$			
4-Thia-1-azabicyclo[3.2.0]heptane-3,7-dione, 6,6-dibromo-2-isopropylidene-	MeOH	215(3.72),272(4.08)	39-2123-69C
$C_8H_7ClN_2O_2$			
Ethylene, 1-(p-chloroanilino)-2-nitro-	dioxan	245(4.33),370(4.41)	78-1617-69
$C_8H_7ClN_2O_3$			
Acetophenone, 2'-chloro-4'-nitro-, oxime	H_2O	270(3.94)	80-0481-69
	pH 13	260(3.80),323(3.71)	80-0481-69
$C_8H_7ClN_4O_2$			
9H-Purine-9-carboxylic acid, 6-chloro-, ethyl ester	C_6H_{12}	248(3.92)	44-0973-69
C_8H_7ClO			
Acetophenone, 2-chloro-	heptane	245(4.05),325(1.85)	59-0765-69
C_8H_7ClOS			
Benzoic acid, m-chlorothio-, O-methyl ester	C_6H_{12}	262s(--),283(4.07), 319s(--),420(2.04)	18-3556-69
Benzoic acid, p-chlorothio-, O-methyl ester	C_6H_{12}	248(3.88),292(4.11), 422(2.08)	18-3556-69
$C_8H_7ClO_2$			
Acetic acid, p-chlorophenyl ester	C_6H_{12}	269(2.70)	18-1831-69
Acetophenone, 3'-chloro-2'-hydroxy-	C_6H_{12}	225(3.10),260(2.96), 335(2.61)	42-0148-69
	EtOH	229(3.71),250(3.77), 256(3.71),262(3.27), 267(3.66),340(2.55)	42-0148-69
Acetophenone, 3'-chloro-4'-hydroxy-	EtOH	228(4.07),276(4.07)	42-0148-69
Acetophenone, 3'-chloro-5'-hydroxy-	C_6H_{12}	228(4.00),252(3.66), 354(3.21)	42-0148-69
	EtOH	229(3.81),250(3.70), 255(3.68),260(3.50), 349(3.48)	42-0148-69
Acetophenone, 5'-chloro-2'-hydroxy-	C_6H_{12}	254(3.63),343(3.37)	18-1831-69
$C_8H_7ClO_3$			
p-Benzoquinone, 2-chloro-5-hydroxy-3,6-dimethyl-	EtOH	283(4.26),430(2.70)	39-1245-69C
$C_8H_7ClS_2$			
Benzoic acid, o-chlorodithio-, methyl ester	C_6H_{12}	239s(3.9),297(3.95), 324s(3.7),501(1.88)	48-0045-69
	EtOH	239s(3.9),297(3.91), 497(1.85)	48-0045-69

Compound	Solvent	$\lambda_{max}(\log \epsilon)$	Ref.
Benzoic acid, p-chlorodithio-, methyl ester	C_6H_{12}	228s(3.8),257s(3.5), 304(4.20),506(2.16)	48-0045-69
	EtOH	229s(3.72),304(4.14), 498(2.14)	48-0045-69
$C_8H_7Cl_6N_3$			
s-Triazine, 2-isopropyl-4,6-bis(trichloromethyl)-	MeOH	214(3.51),265(2.76)	18-2924-69
s-Triazine, 2-propyl-4,6-bis(trichloromethyl)-	MeOH	220(3.26),278(2.82)	18-2924-69
$C_8H_7Cl_6N_3S$			
s-Triazine, 2-(isopropylthio)-4,6-bis-(trichloromethyl)-	MeOH	215(3.56),274(4.14)	18-2931-69
s-Triazine, 2-(propylthio)-4,6-bis-(trichloromethyl)-	MeOH	215(3.54),274(4.12)	18-2931-69
C_8H_7FO			
Acetophenone, 2-fluoro-	heptane	243(4.04),316(1.70)	59-0765-69
$C_8H_7FS_2$			
Benzoic acid, p-fluorodithio-, methyl ester	C_6H_{12}	226s(3.9),254s(3.4), 298(4.20),327s(3.9), 502(2.06)	48-0045-69
	EtOH	224s(3.9),300(4.23), 496(2.07)	48-0045-69
$C_8H_7F_3N_4O$			
4-Pteridinol, 3,4-dihydro-7-methyl-4-(trifluoromethyl)-	pH 7.0	227(3.50),267(3.53), 316(3.97)	39-1751-69C
$C_8H_7F_3O_3S$			
Sulfone, p-methoxyphenyl trifluoromethyl	MeOH	250(4.25)	65-1634-69
$C_8H_7IN_2O_3$			
Acetophenone, 2'-iodo-4'-nitro-, oxime	H_2O	270(3.92)	80-0481-69
	pH 13	262(3.88),320(3.62)	80-0481-69
$C_8H_7IS_2$			
Benzoic acid, p-iododithio-, methyl ester	C_6H_{12}	235s(3.9),255s(3.6), 316(4.37),506(2.25)	48-0045-69
	EtOH	230s(4.0),315(4.32), 497(2.28)	48-0045-69
C_8H_7N			
Indole	EtOH	271(3.64),278(3.63), 287(3.54)	11-0159-69B
C_8H_7NO			
1,2-Benzisoxazole, 3-methyl-	EtOH	237(3.92),243(3.86), 282(3.49),292s(--)	4-0279-69
2,1-Benzisoxazole, 3-methyl-	EtOH	245-280f(--),316(3.76)	35-4749-69
Benzoxazole, 2-methyl-	EtOH	232(3.89),263(3.40), 270(3.59),277(3.64)	4-0279-69
	EtOH	222(3.94),266s(3.58), 271(3.70),279(3.68)	65-0392-69
C_8H_7NOS			
Benzothiazole, 2-methyl-, 3-oxide	heptane	235(4.1),260(3.8), 263(4.0),330(3.7)	94-1598-69

Compound	Solvent	λ_{max}(log ϵ)	Ref.
Benzothiazole, 2-methyl-, 3-oxide (cont.)	H_2O	237(4.2),268s(3.7), 279(3.8),299(3.7)	94-1598-69
	EtOH	247(4.1),270s(3.6), 280(3.7),305(3.8)	94-1598-69
2-Benzothiazolinone, 3-methyl-	MeOH	215(4.68),245(3.76), 283(3.49),289(3.49)	4-0163-69
2H-1,4-Benzoxazine-3(4H)-thione	EtOH	253(3.97),278(3.56), 323(4.07)	78-0517-69
$C_8H_7NO_2$			
2H-1,4-Benzoxazin-3(4H)-one	EtOH	256(3.66),283(3.42)	78-0517-69
Glyoxal, phenyl-, 1-oxime, syn-H	1% MeOH	265(4.03)	22-2894-69
anion	1% MeOH	295(4.17)	22-2894-69
$C_8H_7NO_2S$			
Terephthalamic acid, 4-thio-	EtOH	251(4.16),301(3.84)	39-0861-69C
$C_8H_7NO_3$			
Acetophenone, 2-nitro-	hexane	245(4.00),330(3.85)	22-0534-69
	pH 0	250(4.13)	22-0534-69
	$CHCl_3$	250(4.11),345(3.48)	22-0534-69
p-Benzoquinone, 2-acetyl-, 4-oxime	EtOH	236(3.96),266(3.87), 324(4.11)	12-0935-69
	EtOH-NaOH	234(4.04),407s(4.48)	12-0935-69
	dioxan	720(1.73)	12-0935-69
2H-1,4-Benzoxazin-3(4H)-one, 2-hydroxy-	EtOH	249(3.96),280(3.65)	25-0328-69
Glyoxylic acid, phenyl-, 2-oxime, anti, anion	1% MeOH	246(3.99)	22-2894-69
dianion	1% MeOH	265(3.92)	22-2894-69
syn, anion	1% MeOH	235(3.72)	22-2894-69
syn, dianion	1% MeOH	254(3.83)	22-2894-69
o-Tolualdehyde, 3-nitro-	EtOH	205(4.16),237(4.24)	39-1935-69C
$C_8H_7NO_3S$			
Benzoic acid, p-nitrothio-, O-methyl ester	C_6H_{12}	299(4.05),433(2.25)	18-3556-69
2-Carboxy-2,3-dihydro-8-hydroxythiazolo-[3,2-a]pyridinium hydroxide, inner salt	pH 1	230(3.76),330(3.96)	1-1704-69
	pH 13	245(3.79),350(4.00)	1-1704-69
Δ^4-1,3,4-Oxathiazoline, 4-phenyl-, 3,3-dioxide	n.s.g.	250(4.25),290(3.10)	39-0652-69C
2-Thiopheneacrolein, α-methyl-5-nitro-	EtOH	249(4.01),362(4.36)	94-1757-69
$C_8H_7NO_4$			
Acetic acid, p-nitrophenyl ester	EtOH	268(3.99)	65-0660-69
Acetic acid, (m-nitrophenyl)-	pH 10	273(3.89)	35-2467-69
Acetic acid, (o-nitrophenyl)-	pH 10	268(3.74)	35-2467-69
Acetic acid, (p-nitrophenyl)-	pH 10	285(3.98)	35-2467-69
1,4-Cyclohexadiene-1-carboxylic acid, 3,6-dioxo-, methyl ester, 3-oxime	EtOH	259(3.88),333(4.04)	12-0935-69
	EtOH-NaOH	261(3.42),295(3.62), 317(3.60),403(4.46)	12-0935-69
	dioxan	732(1.70)	12-0935-69
C_8H_7NS			
Benzothiazole, 2-methyl-	EtOH	220s(4.47),254(3.95), 288(3.20),295(3.10)	65-0392-69
2-Indolinethione	EtOH	228(4.11),293(4.03), 317(4.22)	94-0550-69

Compound	Solvent	λ_{max}(log ε)	Ref.
C_8H_7NSSe			
2-Benzothiazolinone, 3-methyl-2-seleno-	EtOH	249(4.04),283s(3.30), 348(4.46)	65-0941-69
$C_8H_7NS_2$			
Benzothiazole, 2-(methylthio)-	MeOH	225(4.35),243(3.96), 278(4.13),289(4.07), 300(3.96)	4-0163-69
2-Benzothiazolinethione, 3-methyl-	MeOH	208(4.18),230(4.15), 240(4.15),257(3.74), 325(4.43)	4-0163-69
$C_8H_7N_3$			
Glyoxylonitrile, phenylhydrazone	EtOH	232(4.04),286s(3.83), 294(3.92),328(4.30)	12-1915-69
Pyridine, 4-(imidazol-2-yl)-	MeOH	290(4.19)	87-0944-69
Quinazoline, 4-amino-	EtOH	277s(3.78),284(3.81), 302s(3.66),312(3.76), 324(3.64)	39-1282-69C
Triazole, 1-phenyl-	EtOH	244(3.89)	70-0131-69
$C_8H_7N_3O$			
1,2,4-Oxadiazole, 5-amino-3-phenyl-	EtOH	254(3.53)	18-3335-69
Pyridine, 4-(5-methyl-1,2,4-oxadiazol-3-yl)-	MeOH	226(4.01),273(3.40)	87-0381-69
$C_8H_7N_3OS$			
1,2-Benzisothiazole, 3-(methylnitrosamino)-	MeOH	230(4.28),319(4.08)	24-1961-69
$C_8H_7N_3O_2$			
1,4-Phthalazinedione, 5-amino-2,3-dihydro-	THF	290(3.9),360(3.9)	24-3241-69
$C_8H_7N_3O_3$			
Benzimidazole, 3-methyl-6-nitro-, 1-oxide	H_2O	202(4.30),267(4.18), 295s(3.74)	39-2127-69C
	0.5N HCl	239(4.23),286(3.89)	39-2127-69C
	pH 10	267(4.17),300s(3.74)	39-2127-69C
	0.5N NaOH	399(4.08)	39-2127-69C
2H-1,4-Benzoxazine-2,3(4H)-dione, dioxime	EtOH	217(4.47),253(4.15), 298(3.98)	39-2319-69C
$C_8H_7N_3O_4$			
Indoline, 5,6-dinitro-	MeOH	235(4.42),262(4.98), 390(4.18)	103-0196-69
$C_8H_7N_3O_7$			
Anisole, 2-(dinitromethyl)-4-nitro-	anion	365(4.21)	122-0161-69
Anisole, 4-(dinitromethyl)-2-nitro-	anion	369(4.19)	122-0161-69
$C_8H_7N_5O$			
3H-Imidazo[2,1-i]purin-8(7H)-one, 3-methyl-	pH 1	282(4.04)	44-3492-69
	pH 7	274(4.08),303(4.16)	44-3492-69
	pH 13	303(4.08)	44-3492-69
C_8H_8			
Bicyclo[3.2.0]hepta-1,4,6-triene, 2-methyl-	n.s.g.	204(4.16),277(3.32)	35-5694-69
Bicyclo[4.2.0]octa-1,3,5-triene	n.s.g.	273(3.49),277(3.48)	77-0680-69

Compound	Solvent	λ_{max}(log ϵ)	Ref.
1-Octene-3,5-diyne	dioxan	228(3.64),238(3.77), 252(3.89),266(4.07), 281(3.98)	48-0153-69
2-Octene-4,6-diyne, cis-trans mixture	dioxan	227(2.98),239(3.34), 252(3.71),266(3.89), 282(3.76)	48-0153-69
Semibullvalene	EtOH	230s(3.38)	35-3316-69
$C_8H_8BrN_3$			
Benzimidazole, 2-amino-5-bromo-1-methyl-	n.s.g.	293(4.02)	103-0529-69
Benzimidazole, 2-amino-6-bromo-1-methyl-	n.s.g.	257(3.95),295(4.01)	103-0529-69
$C_8H_8ClNO_2$			
3H-Azepine-3-carboxaldehyde, 6-chloro-2-methoxy-	MeOH	264(3.72)	78-5205-69
$C_8H_8ClNO_3S$			
7-Oxabicyclo[4.2.1]nona-2,4-diene-$\Delta^{8,N}$-sulfamoyl chloride	EtOH	265(3.60)	88-5325-69
$C_8H_8Cl_2O$			
2,5-Cyclohexadien-1-one, 4-(dichloromethyl)-4-methyl-	H_2O	235(4.17)	44-0224-69
	90.5% H_2SO_4	262(4.12),294(3.56)	44-0224-69
$C_8H_8Cl_2O_2$			
2-Cyclopenten-1-one, 3,5-dichloro-4-hydroxy-2-propenyl-	MeOH	219(4.22),272(3.91)	39-2187-69C
$C_8H_8Cl_2O_4$			
2-Cyclopentene-1-carboxylic acid, 3,5-dichloro-1-hydroxy-2-methyl-4-oxo-, methyl ester	EtOH	243(4.07)	88-3613-69
$C_8H_8N_2$			
5-Azaindole, 3-methyl-	EtOH	222(4.53),275(3.39), 283s(3.28)	23-2061-69
	EtOH-HCl	228(4.57),284(3.34), 312s(3.04)	23-2061-69
Benzimidazole, 2-methyl-	MeOH	280(3.85)	22-3002-69
anion	MeOH	281(--)	22-3002-69
2H-Cyclopenta[d]pyridazine, 2-methyl-	ether	248(4.47),253(4.46), 268(4.19),307s(4.09), 312(3.58),317(3.56), 324(3.52),395(2.91)	35-0924-69
9,10-Diazapentacyclo[$4.4.0.0^{2,5}.0^{3,10}.0^{4,7}$]dec-9-ene	MeOH	388(2.28)	24-3304-69
$C_8H_8N_2O$			
Phthalimidine, N-amino-	pH 1	232(4.10)	7-0451-69
	pH 7.38	232s(3.46)	7-0451-69
$C_8H_8N_2O_2$			
Ethylenimine, N-(p-nitrophenyl)-	DMSO	321(3.07)	39-0544-69B
Glyoxylamide, 2-phenyl-, 2-oxime, anti-Ph	1% MeOH	252(4.05)	22-2894-69
anion	1% MeOH	275(4.03)	22-2894-69
syn-Ph	1% MeOH	240(3.72)	22-2894-69
anion	1% MeOH	267(3.94)	22-2894-69
Pyrrole-2-carboxylic acid, 5-cyano-, ethyl ester	EtOH	268(4.50),275s(4.38)	78-0295-69

Compound	Solvent	$\lambda_{max}(\log \epsilon)$	Ref.
Pyrrole-2-carboxylic acid, 5-cyano-3-methyl-, methyl ester	EtOH	270(4.36)	78-0295-69
$C_8H_8N_2O_2S$			
Sulfone, diazomethyl m-tolyl	C_6H_{12}	250(4.09),281(3.12)	54-0641-69
	40% dioxan	390(1.79)	54-0641-69
Δ^2-1,2,4-Thiadiazoline, 3-phenyl-, 1,1-dioxide	n.s.g.	240(4.22),269(3.64)	39-0652-69C
Thiourea, N-benzoyl-N'-hydroxy-	EtOH	243(4.17)	39-2794-69C
$C_8H_8N_2O_3$			
Acetophenone, 2-nitro-, oxime	pH 0	250(4.02)	22-0534-69
	pH 12.5	260(4.06),305(4.12)	22-0534-69
Acetophenone, o-nitro-, oxime	H_2O	263(3.8)	80-0225-69
	pH 13	250(4.0)	80-0225-69
o-Anisamide, 5-nitroso-	EtOH	251(3.95),330(4.04)	12-0935-69
	dioxan	727(1.68)	12-0935-69
Benzamidoxime, p-carboxy-	EtOH	232(4.12),286(3.79)	39-0861-69C
2-Isoxazolin-5-one, 3-methyl-4-(3-methyl-3-isoxazolin-5-ylidene)-	EtOH	250(3.8),290(4.2)	39-0245-69C
	pH 2.06	250(4.2),282(4.5)	39-0245-69C
$C_8H_8N_2O_3S$			
Anisole, m-[(diazomethyl)sulfonyl]-	C_6H_{12}	249(4.02),288(3.50)	54-0641-69
	40% dioxan	390(1.87)	54-0641-69
3-Carboxy-2,3-dihydro-8-hydroxy-5-methylthiazolo[3,2-c]pyrimidin-4-ium hydroxide, inner salt	pH 1	250(3.79),340(3.95)	1-2437-69
	pH 13	265(3.93),365(3.93)	1-2437-69
$C_8H_8N_2O_4$			
m-Xylene, α,α-dinitro-	anion	371(4.18)	122-0161-69
o-Xylene, α,α-dinitro-	anion	371(4.21)	122-0161-69
p-Xylene, α,α-dinitro-	anion	370(4.17)	122-0161-69
$C_8H_8N_2O_5$			
Anisole, m-(dinitromethyl)-	anion	370(4.18)	122-0161-69
Anisole, o-(dinitromethyl)-	anion	369(4.15)	122-0161-69
Anisole, p-(dinitromethyl)-	anion	371(4.15)	122-0161-69
$C_8H_8N_2S$			
Benzimidazole, 2-(methylthio)-	MeOH	248(3.88),255s(3.83), 283(4.18),291(4.19)	97-0152-69
	EtOH	248(3.91),284(4.15), 291(4.17)	48-0997-69
2-Benzimidazolinethione, 1-methyl-	MeOH	245(4.50),298s(4.65), 306(4.75)	97-0152-69
	EtOH	220(4.27),244(4.25), 307(4.48)	48-0997-69
1,2-Benzisothiazole, 3-(methylamino)-	MeOH	228(4.14),248(3.83), 287(3.45),323(3.71)	24-1961-69
hydrochloride	pH 1	232(4.32),246s(3.93), 329(3.84),333s(3.80)	24-1961-69
1,2-Benzisothiazoline, 3-imino-2-methyl-, hydrochloride	pH 1	231(4.36),318(3.84), 328(3.82)	24-1961-69
Benzothiazole, 2-(methylamino)-	EtOH	225(4.51),268(4.13), 298s(3.46)	2-0964-69
2-Benzothiazoline, 2-imino-3-methyl-	EtOH	223(4.53),264(4.08), 290(3.84),300(3.83)	2-0964-69
$C_8H_8N_3NaO_8$			
Anisole, 2,4,6-trinitro-, NaOMe adduct	MeOH	414(4.42),487(4.26)	35-6742-69

Compound	Solvent	λ_{max}(log ϵ)	Ref.
$C_8H_8N_4$			
Quinazoline, 2,4-diamino-	acid	226(4.58),231s(4.55), 248s(4.09),267s(3.74), 315(3.67),325s(3.58)	44-0821-69
	neutral	230(4.65),265(3.91), 273(3.86),330(3.62)	44-0821-69
$C_8H_8N_4O$			
Acetamide, N-1H-imidazo[4,5-c]pyridin-6-yl-	pH 13	222(4.32),269(3.74)	54-1263-69
4(3H)-Pteridinone, 6,7-dimethyl-	EtOH	232(4.42),273(4.11), 315(4.28)	39-2415-69C
4H-Pyrazolo[3,4-d]pyrrolo[1,2-a]pyrimidin-4-one, 1,6,7,8-tetrahydro-	MeOH	209(4.24),252(3.87)	24-2739-69
$C_8H_8N_4OS$			
Purine-8(9H)-thione, 9(7)-propionyl-	MeOH	323(4.13)	4-0023-69
$C_8H_8N_4O_2$			
2,3-Quinoxalinedione, 1,4-dihydro-, dioxime	EtOH	232(4.57),265(4.15), 321(4.16)	39-2319-69C
$C_8H_8N_4O_2S$			
9H-Purine-9-carboxylic acid, 1,6-dihydro-6-thioxo-, ethyl ester	MeOH	322(4.28)	44-0973-69
9H-Purine-9-carboxylic acid, 7,8-dihydro-8-thioxo-, ethyl ester	dioxan	319(4.38)	4-0023-69
$C_8H_8N_4O_2Se$			
9H-Purine-9-carboxylic acid, 1,6-dihydro-6-selenoxo-	MeOH	354(4.20)	44-0973-69
$C_8H_8N_4O_7$			
Aniline, N-ethoxy-2,4,6-trinitro-	$C_2H_4Cl_2$	337(4.10)	80-0941-69
$C_8H_8N_4S$			
4(3H)-Pteridinethione, 6,7-dimethyl-	pH 5	260(4.06),385(4.04)	39-0114-69C
$C_8H_8N_6$			
3,3'-Bipyridazine, 6,6'-diamino-	EtOH	218(3.77)	88-2359-69
Isonicotinamidine, N-4H-1,2,4-triazol-4-yl-	MeOH	224(3.73),266(3.75)	48-0477-69
Nicotinamidine, N-4H-1,2,4-triazol-4-yl-	MeOH	227(3.83),260(3.75)	48-0477-69
Picolinamidine, N-4H-1,2,4-triazol-4-yl-	MeOH	228(3.87),271(--)	48-0477-69
C_8H_8O			
Acetophenone	heptane	235(4.16),320(1.68)	59-0765-69
	EP	319(1.7)	18-0010-69
	EP at 77°K	309(1.6),321(1.6)	18-0010-69
2,3-Homotropone	EtOH	293(3.65),340(3.15)	35-0215-69
4,5-Homotropone (bicyclo[5.1.0]octa-2,5-dien-4-one)	C_6H_{12}	248(3.89),275s(3.65), 360(1.52),390s(1.29), 410s(0.80)	35-0215-69
	EtOH	264(3.88),290(3.66)	35-0215-69
2,4-Octadiynal	ether	245(3.48),257(3.74), 272(3.93),288(3.85)	39-0683-69C
1-Octene-4,6-diyn-3-ol	ether	228(2.60),241(2.60), 255(2.48)	39-0685-69C

Compound	Solvent	λmax(log ε)	Ref.
2-Octene-4,6-diyn-1-ol, trans	MeOH	230(3.30),239(3.77), 252(4.11),265(4.25), 281(4.17)	39-0685-69C
Oxirane, 2-phenyl-, (R)-(+)-	EtOH	210(4.0),265f(2.1)	88-2717-69
Oxonin	pentane	251(3.32)	35-7769-69
C_8H_8OS			
Benzoic acid, thio-, O-methyl ester	C_6H_{12}	247(3.77),287(4.07), 417(2.08)	18-3556-69
	C_6H_{12}	245(3.84),287(4.02), 418(2.04)	48-0045-69
	EtOH	244(3.92),286(4.05), 411(2.09)	48-0045-69
	MeCN	245(3.95),285(4.11), 407(2.06)	48-0045-69
$C_8H_8O_2$			
Acetic acid, phenyl ester	C_6H_{12}	259(2.46)	18-1831-69
Acetophenone, 2'-hydroxy-	C_6H_{12}	249(3.97),325(3.57)	18-1831-69
	EtOH	212(4.4),250(4.0), 325(3.6)	28-0730-69A
Acetophenone, 4'-hydroxy-	C_6H_{12}	259(4.09)	18-1831-69
Benzoic acid, methyl ester	heptane	227(4.09),276(2.93)	59-0765-69
	EtOH	228(4.05),272(2.96)	65-0660-69
o-Benzoquinone, 4,5-dimethyl-	MeOH	400(3.01)	64-0547-69
p-Benzoquinone, 2,6-dimethyl-	MeOH	252(4.27)	64-0547-69
Dispiro[2.0.2.2]octane-7,8-dione	C_6H_{12}	439(1.08),470(1.65)	88-3545-69
7-Oxabicyclo[4.2.1]nona-2,4-dien-8-one	EtOH	265(3.69)	88-5325-69
Tricyclo[5.1.0.0^{3,5}]octane-2,6-dione, anti	EtOH	271(1.75)	27-0411-69
syn	EtOH	280(1.67)	27-0411-69
$C_8H_8O_3$			
Acetophenone, 2',5'-dihydroxy-	MeOH	255(3.7),360(3.6)	2-0040-69
Acetophenone, 2',6'-dihydroxy-	EtOH	222(4.0),269(4.0), 343(3.4)	28-0730-69A
Salicylic acid, methyl ester	$CHCl_3$	236(3.95),306(3.62)	39-0526-69A
$C_8H_8O_4$			
Benzoic acid, 2,3-dihydroxy-, methyl ester	pH 2	207(4.43),246(3.92), 315(3.45)	61-0203-69
anion	buffer	227(4.35),253(3.74), 338(3.57)	61-0203-69
	0.15N NaOH	227(4.35),253(3.78), 320(3.64)	61-0203-69
Benzoic acid, 2,4-dihydroxy-, methyl ester	pH 2	208(4.53),222(4.13), 257(4.13),294(3.75)	61-0203-69
anion	buffer	207(4.28),235(4.00), 281(4.11),302(4.28)	61-0203-69
	0.15N NaOH	228(4.36),243(4.18), 282(4.11),316(3.98)	61-0203-69
Benzoic acid, 2,5-dihydroxy-, methyl ester	pH 2	196(4.21),212(4.44), 237(3.86),328(3.63)	61-0203-69
anion	buffer	212(4.19),222(4.27), 249(3.76),360(3.58)	61-0203-69
	0.15N NaOH	220(4.33),260(3.66), 379(3.61)	61-0203-69
Benzoic acid, 2,6-dihydroxy-, methyl ester	pH 2	208(4.30),214(4.34), 250(3.94),318(3.48)	61-0203-69

Compound	Solvent	λ_{max}(log ϵ)	Ref.
Benzoic acid, 2,6-dihydroxy-, methyl ester, anion	buffer	192(4.48),229(4.26), 254(3.71),343(3.55)	61-0203-69
	0.15N NaOH	209(4.49),238(4.04), 300(3.54)	61-0203-69
Benzoic acid, 3,4-dihydroxy-, methyl ester	pH 2	205(4.39),217(4.33), 259(4.03),294(3.74)	61-0203-69
anion	buffer	205(4.23),235(4.12), 286(3.94),312(4.19)	61-0203-69
	0.15N NaOH	213(4.30),250(4.16), 295(3.85),339(4.01)	61-0203-69
Benzoic acid, 3,5-dihydroxy-, methyl ester	pH 2	198(4.44),206(4.50), 249(3.82),307(3.42)	61-0203-69
anion	buffer	202(4.30),227(4.34), 263(3.58),331(3.43)	61-0203-69
	0.15N NaOH	219(4.38),229(4.27), 283(3.49),338(3.43)	61-0203-69
Phloroacetophenone ($CO^{14}Me$)	EtOH	227(4.11),286(4.24)	94-2054-69
4H-Pyran-4-one, 3-acetyl-6-hydroxy-2-methyl-	EtOH	225(4.00),261(3.91)	44-2480-69
$C_8H_8O_5$			
4H-Pyran-2-carboxylic acid, 5-hydroxy-4-oxo-, ethyl ester	MeOH	230(4.3),302(3.5)	83-0628-69
$C_8H_8S_2$			
Benzoic acid, dithio-, methyl ester	C_6H_{12}	246s(3.5),296(4.12), 329s(3.8),504(2.11)	48-0045-69
	EtOH	244s(3.4),298(4.16), 330s(3.8),498(2.07)	48-0045-69
	MeCN	227s(3.8),298(4.11), 330s(3.84),495(2.10)	48-0045-69
$C_8H_9ClN_2O$			
Cyclopentylideneacetamide, N-chloro-α-cyano-	EtOH	236(4.06)	22-1724-69
$C_8H_9ClN_2O_3S$			
Acetanilide, 4'-chloro-2'-sulfamyl-	MeOH	252(4.10),295(3.38)	44-1780-69
$C_8H_9Cl_3N_2$			
Pyrazole, 3,4,5-trimethyl-1-(trichlorovinyl)-	n.s.g.	232(3.8)	39-2251-69C
$C_8H_9Cl_3O_2$			
Acetic acid, trichloro-, 1-cyclohexen-1-yl ester	C_6H_{12}	225(3.08)	78-1679-69
$C_8H_9IN_4O$			
1,4-Dihydro-1,3-dimethyl-4-oxopteridinium iodide	EtOH	222(4.23),249(4.15), 310s(3.28),355(3.76)	39-2415-69C
C_8H_9N			
1,3-Cyclohexadiene-1-carbonitrile, 4-methyl-	EtOH	293(3.98)	12-2037-69
Indoline	EtOH	210(4.06),240(3.97), 291(3.53)	11-0159-69B
Pyridine, 2-cyclopropyl-	EtOH	269(3.60)	44-3191-69

Compound	Solvent	$\lambda_{max}(\log \epsilon)$	Ref.
C_8H_9NO			
Acetophenone, 2'-amino-	C_6H_{12}	225(4.40),232(4.32), 252(3.78),260s(3.60), 350(3.66)	22-3523-69
	EtOH	227(4.40),256(3.80), 262s(3.69),364(3.70)	22-3523-69
Acetophenone, 3'-amino-	EtOH	230(4.33),256s(3.82), 334(3.27)	22-3523-69
Pyridine, 4-propionyl-	EtOH	221(3.82),272(3.51)	39-2134-69C
$(C_8H_9NO)_n$			
Poly(3-methyl-2-vinylpyridine N-oxide)	H_2O	218(3.46),265(3.11)	39-0054-69B
	0.01M silicic acid	218(3.48),265(3.11)	39-0054-69B
Poly(3-methyl-4-vinylpyridine N-oxide)	H_2O	213(4.12),264(4.07)	39-0054-69B
	0.01M silicic acid	212(4.14),262(4.07)	39-0054-69B
Poly(6-methyl-2-vinylpyridine N-oxide)	H_2O	214(3.71),260(3.28)	39-0054-69B
	0.01N silicic acid	214(3.65),260(3.20)	39-0054-69B
Poly(6-methyl-3-vinylpyridine N-oxide)	H_2O	212(3.97),257(3.64)	39-0054-69B
	0.01M silicic acid	212(3.97),257(3.64)	39-0054-69B
C_8H_9NOS			
2,3-Dihydro-8-hydroxy-5-methylthiazolo-[3,2-a]pyridinium hydroxide, inner salt	pH 1	230(3.70),245s(3.69), 340(3.98)	1-0371-69 +1-2437-69
	pH 1	240(3.70),340(3.98)	1-1704-69
	pH 13	245(3.89),260s(3.81), 360(4.02)	1-0371-69 +1-2437-69
	pH 13	245(3.89),360(4.02)	1-1704-69
$C_8H_9NO_2$			
Acetophenone, 2'-hydroxy-, oxime	EtOH	214(4.4),254(3.9), 262s(3.8),303(3.5)	28-0730-69A
p-Anisaldehyde, oxime, α-form	n.s.g.	212s(4.3),267(4.3)	23-0051-69
p-Anisaldehyde, oxime, β-form	n.s.g.	262(4.3)	23-0051-69
Anthranilic acid, methyl ester	EtOH	247(3.70),337(3.70)	22-3523-69
Anthranilic acid, N-methyl-	EtOH	252(3.70),353(3.70)	22-3523-69
3H-Azepine-3-carboxaldehyde, 2-methoxy-	MeOH	258(3.71)	78-5205-69
1H-Azepine-1-carboxylic acid, methyl ester	hexane	211(4.36),238s(3.50), 330(2.76)	44-2866-69
	EtOH	208(4.34),242s(3.44), 318(2.83)	44-2866-69
2H-Azepin-2-one, 1,3-dihydro-3-acetyl-	MeOH	255(3.66)	78-5205-69
Benzene, 1-ethyl-2-nitro-	H_2O	270(3.6)	80-0225-69
Benzene, 1-ethyl-4-nitro-	H_2O	282(3.9)	80-0225-69
2,1-Benzisoxazol-4(5H)-one, 6,7-dihydro-3-methyl-	EtOH	242(3.97)	70-2680-69
p-Benzoquinone, 2,6-dimethyl-, 4-oxime	EtOH	228s(3.40),308(4.32)	12-0935-69
	EtOH-NaOH	261(3.65),400(4.43)	12-0935-69
	dioxan	725(-1.3)	12-0935-69
p-Benzoquinone, 2-ethyl-, 4-oxime	EtOH	232(3.52),305(4.26)	12-0935-69
	EtOH-NaOH	263(3.72),403(4.46)	12-0935-69
	dioxan	726(-0.19)	12-0935-69
Furo[2,3-b]pyridin-6-ol, 2,3-dihydro-4-methyl-	EtOH	305(3.98),343(3.78)	39-1678-69C
	dioxan	301(4.08),350(2.72)	39-1678-69C
Glycine, 2-phenyl-, hydrochloride, (R)-(-)-	EtOH	252(2.20),258(2.33), 262(2.26),264s(2.25), 268(2.11)	60-2611-69

Compound	Solvent	$\lambda_{max}(\log \epsilon)$	Ref.
Pyridine, 5-acetyl-3-hydroxy-2-methyl-	pH 13	265(3.70),345(3.74)	78-3527-69
2-Pyrrolidinone, 5-(2-furyl)-	EtOH	215(4.34)	117-0271-69
Salicylidenimine, 5-(hydroxymethyl)-, copper complex	$CHCl_3$	273(4.42),360(3.97), 550(1.97)	65-0401-69
$C_8H_9NO_2S$			
Pyrrole-2-carbothioic acid, 4-formyl-, S-ethyl ester	EtOH	228(4.28),261(3.90), 270(3.92),298(4.21)	78-3879-69
$C_8H_9NO_3$			
Acetophenone, 2',6'-dihydroxy-, oxime	EtOH	206(4.3),247(3.8), 279(3.3)	28-0730-69A
p-Benzoquinone, 2-(methoxymethyl)-, 4-oxime	EtOH	229(3.45),306(4.28)	12-0935-69
	EtOH-NaOH	268(3.70),405(4.49)	12-0935-69
	dioxan	725(0.0)	12-0935-69
p-Benzoquinone, 2-methoxy-6-methyl-, 4-oxime	EtOH	322(4.20)	12-0935-69
	EtOH-NaOH	257(3.64),398(4.43)	12-0935-69
	dioxan	725(-1.4)	12-0935-69
Benzyl alcohol, 2-methyl-3-nitro-	EtOH	209(4.1),258(3.55)	39-1935-69C
1-(1-Carboxyethyl)-3-hydroxypyridinium hydroxide, inner salt	N HCl	295(3.48)	1-2488-69
	N NaOH	250(3.88),325(3.72)	1-2488-69
2-Pyrrolecarboxylic acid, 3-acetyl-5-methyl-	EtOH	222(4.24),261(3.96), 323(3.82)	33-1911-69
$C_8H_9NO_3S$			
Acetic acid, [(3-hydroxy-6-methyl-2-pyridyl)thio]-	pH 1	240(3.57),325(3.94)	1-2065-69
	pH 13	250(3.86),325(3.95)	1-2065-69
$C_8H_9NO_3S_2$			
p-Dithiin-2,3-dicarboximide, 5,6-dihydro-N-(2-hydroxyethyl)-	MeOH	261(4.03),411(3.40)	33-2221-69
4-Thiazolecarboxylic acid, 2-[(3-oxobutyl)thio]-	MeOH	282(3.81)	4-0397-69
$C_8H_9NO_4$			
Benzene, 1,4-dimethoxy-2-nitro-	EtOH	219(4.23),242s(--), 256(3.41),268s(--)	117-0287-69
$C_8H_9NO_5$			
4H-Pyran-2-carboxylic acid, 6-amino-5-hydroxy-4-oxo-, ethyl ester	MeOH	298(3.9),415(3.9)	83-0628-69
	$CHCl_3$	270(3.8),355(3.8)	83-0628-69
C_8H_9NS			
Benzamide, N-methyl-thio-	C_6H_{12}	236(4.00),288(3.81), 296s(3.7),400(2.37)	48-0045-69
	EtOH	239(4.04),285(3.86), 363(2.32)	48-0045-69
	MeCN	237(4.05),285(3.81), 376(2.46)	48-0045-69
$C_8H_9N_3$			
Benzimidazole, 2-(aminomethyl)-, dihydrochloride	EtOH	275(4.09),281(4.05)	94-2381-69
Glutacononitrile, 4-(1-aminoethylidene)-3-methyl-	EtOH	220(3.84),256(3.66), 327(4.00)	18-2319-69
Nicotinonitrile, 6-amino-2,4-dimethyl-	EtOH	272(3.86)	18-2319-69
7H-Pyrrolo[2,3-d]pyrimidine, 5,6-dimethyl-	EtOH	232(4.39),275(3.28), 308(3.26)	23-2061-69
	EtOH-HCl	235(4.38),293(3.30), 323s(3.15)	23-2061-69

Compound	Solvent	$\lambda_{max}(\log \epsilon)$	Ref.
7H-Pyrrolo[2,3-d]pyrimidine, 5,6-dimethyl- (cont.)	concd. HCl	241(4.38),296(3.50), 326s(3.22)	23-2061-69
$C_8H_9N_3O_4$			
4-Pyrimidinecarboxylic acid, 2-acetamido-1,6-dihydro-6-oxo-, methyl ester	EtOH	273(4.15),322(4.08)	44-3705-69
as-Triazine-3,6-dicarboxylic acid, 5-methyl-, dimethyl ester	MeOH	264(3.65),355(2.63)	4-0497-69
$C_8H_9N_3O_5$			
Aniline, N-ethoxy-2,4-dinitro-	$C_2H_4Cl_2$	262(3.98),340(4.15), 413s(3.64)	80-0941-69
Aniline, N-ethoxy-2,6-dinitro-	$C_2H_4Cl_2$	269s(3.81),403(3.81)	80-0941-69
$C_8H_9N_5O$			
Acetamide, N-(9-methyl-9H-purin-6-yl)-	pH 1	278(4.19)	44-3492-69
	pH 6.6	274(4.20)	44-3492-69
	pH 13	288(4.06)	44-3492-69
4H-Pyrazolo[3,4-d]pyrrolo[1,2-a]pyrimidin-4-one, 3-amino-1,6,7,8-tetrahydro-	MeOH	226(4.25),265(3.58)	24-2739-69
$C_8H_9N_5O_2$			
Acetamide, N-(1,6-dihydro-8-methyl-6-oxopurin-2-yl)-	pH 0.0	259(4.30)	5-0201-69F
	pH 6.0	260(4.23)	5-0201-69F
	pH 11.0	265(4.13)	5-0201-69F
Glycine, N-(9-methyl-9H-purin-6-yl)-	pH 1	267(4.20)	44-3492-69
	pH 5.6	268(4.25)	44-3492-69
	pH 13	268(4.27)	44-3492-69
Isoxanthopterin, 6,8-dimethyl-	pH 5	216(4.53),286s(4.01), 293(4.05),338(4.15)	24-4032-69
	pH 11	211(4.33),255s(4.05), 283(3.64),289(3.61), 350(4.17)	24-4032-69
9H-Purine-1(6H)-acetic acid, 6-imino-9-methyl-	pH 1	261(4.20)	44-3492-69
	pH 7	261(4.22)	44-3492-69
	pH 11.5	260(4.22),267s(4.18)	44-3492-69
7H-Pyrrolo[2,3-d]pyrimidine, 4-(dimethylamino)-5-nitro-	pH 1	357(4.18)	4-0207-69
	pH 11	255(4.01),430(4.05)	4-0207-69
Sarcosine, N-methyl-N-(purin-6-yl)-	pH 1	280(4.23)	44-3498-69
	pH 5.8	277(4.26)	44-3498-69
	pH 13	281(4.31)	44-3498-69
$C_8H_9N_5O_6$			
m-Phenylenediamine, N,N-dimethyl-2,4,6-trinitro-	MeOH	298(4.08),351(4.14), 405(4.04)	35-4155-69
m-Phenylenediamine, N,N'-dimethyl-2,4,6-trinitro-	MeOH	307(4.32),347(4.23), 413(4.10)	35-4155-69
m-Phenylenediamine, N-ethyl-2,4,6-trinitro-	MeOH	293(4.26),334(4.28), 410(4.13)	35-4155-69
C_8H_{10}			
Benzene, ethyl-	C_6H_{12}	262(2.32)	101-0017-69B
tetracyanoethylene complex	C_6H_{12}	405(3.30)	39-1161-69B
$C_8H_{10}BrN_3O_3$			
as-Triazine-3,5(2H,4H)-dione, 6-bromo-2-(tetrahydro-2H-pyran-2-yl)-	pH 3-4	275(3.86)	103-0283-69
	pH 7-8	270(3.81)	103-0283-69
$C_8H_{10}ClNO$			
2,3-Dihydro-2-hydroxy-1H-indolizinium chloride	EtOH	263(3.9),270(3.8)	24-1309-69

Compound	Solvent	$\lambda_{max}(\log \epsilon)$	Ref.
$C_8H_{10}ClNO_2$			
Glutaconimide, 2-(2-chloroethyl)-3-methyl-	EtOH	312(3.7)	39-1678-69C
$C_8H_{10}Cl_2N_4S$			
s-Triazine, 2,4-dichloro-6-[2-(dimethylamino)-2-(methylthio)vinyl]-	C_6H_{12}	277(3.61),372(4.04)	5-0073-69E
$C_8H_{10}Cl_2O_2$			
Acetic acid, trichloro-, 1-cyclohexen-1-yl ester	EtOH	200s(3.46),220(3.14)	35-2062-69
$C_8H_{10}Cl_2O_6S_3$			
3,4-Thiophenedimethanol, 2,5-dichloro-, dimethanesulfonate	EtOH	256(3.76)	44-0340-69
$C_8H_{10}NO_3$			
Benzene, 1-ethoxy-2-nitro-, half-reduction radical of nitro group	pH 1	290(4.15),420(2.92)	64-1336-69
	pH 13	295(4.36),400(2.95)	64-1336-69
Benzene, 1-ethoxy-4-nitro-, half-reduction radical of nitro group	pH 1	290(4.14),480(3.23)	64-1336-69
	pH 13	290(4.30),460(2.89)	64-1336-69
$C_8H_{10}NO_7P$			
Isonicotinaldehyde, 3-hydroxy-2,5-bis-(hydroxymethyl)-, 5-(dihydrogen phosphate)	pH 1	295(3.96)	10-0001-69D
	pH 7	338s(3.5),387(3.70)	10-0001-69D
	pH 13	388(3.85)	10-0001-69D
$C_8H_{10}N_2$			
Benzaldehyde, methylhydrazone	EtOH	218(3.98),284(4.20)	39-1703-69C
	EtOH-HCl	250(4.13),282s(3.20)	39-1703-69C
7,8-Diazatricyclo[4.2.2.0^{2,5}]deca-3,7-diene	MeOH	378(1.89)	24-3304-69
Quinazoline, 5,6,7,8-tetrahydro-	pH 7.5	255(3.58)	39-1635-69C
$C_8H_{10}N_2O$			
Acetamide, N-(3-pyridylmethyl)-	EtOH	254(--),262(3.60),269(--)	27-0155-69
Acetic acid, 1-phenylhydrazide	EtOH	238(3.7)	28-1703-69A
Acetophenone, 2'-hydroxy-, hydrazone	EtOH	216(4.3),266(4.0),307(3.8)	28-0730-69A
3-Buten-2-one, 4-(3-methylpyrazol-1-yl)-	EtOH	295(4.5)	97-0025-69
Pyrrolo[1,2-a]pyrimidin-2(6H)-one, 7,8-dihydro-4-methyl-	EtOH	239(4.12)	22-3139-69
Pyrrolo[1,2-a]pyrimidin-4(6H)-one, 7,8-dihydro-2-methyl-	EtOH	226(3.87),267(3.72)	22-3139-69
o-Toluidine, N-methyl-N-nitroso-	C_6H_{12}	212(4.04),251(3.79)	35-3383-69
p-Toluidine, N-methyl-N-nitroso-	MeCN	275(3.89),440(3.23)	24-2093-69
$C_8H_{10}N_2OS$			
Thiourea, (p-methoxyphenyl)-	EtOH	270(4.13)	39-0268-69C
$C_8H_{10}N_2O_2$			
Acetophenone, 2',6'-dihydroxy-, hydrazone	EtOH	220(4.2),272(4.0)	28-0730-69A
Acetophenone, 2-(hydroxyamino)-, oxime, anti	n.s.g.	234(3.85)	70-2201-69
Aniline, N,N-dimethyl-p-nitro-	1%MeOH-NaOH	422(4.30)	39-0922-69B
Aniline, N-ethyl-o-nitro-	1%MeOH-NaOH	447(3.78)	39-0922-69B
Aniline, N-ethyl-p-nitro-	1%MeOH-NaOH	410(4.25)	39-0922-69B
p-Anisidine, N-methyl-N-nitroso-	MeCN	273(3.94),470(3.25)	24-2093-69

Compound	Solvent	$\lambda_{max}(\log \epsilon)$	Ref.
1H-1,2-Diazepine-1-carboxylic acid, ethyl ester	hexane	220(3.96),373(2.37)	77-0432-69
	MeOH	217(4.05),362(2.45)	22-0948-69 +22-2175-69
2-Pyridinecarbamic acid, ethyl ester	EtOH	229(4.24),287(3.82)	44-3672-69
4H,6H-Pyrimido[2,1-b][1,3]oxazin-4-one, 7,8-dihydro-2-methyl-	EtOH	258(3.4)	32-1000-69
Pyrrolidin-2-ylideneacetic acid, α-cyano-, methyl ester	n.s.g.	267(4.33)	89-0343-69
Pyrrolo[1,2-a]pyrimidin-4(6H)-one, 7,8-dihydro-2-hydroxy-3-methyl-	EtOH	263(3.89)	22-3133-69
Pyrrolo[1,2-a]pyrimidin-4(6H)-one, 7,8-dihydro-2-methoxy-	5N HCl	243(3.85)	22-3133-69
	H_2O	254(3.67)	22-3133-69
	EtOH	230s(--),264(3.59)	22-3133-69
4,6,7,8-Tetrahydro-2-hydroxy-1-methyl-4-oxopyrrolo[1,2-a]pyrimidinium hydroxide, inner salt	5N HCl	245(3.75)	22-3133-69
	H_2O	257(3.92)	22-3133-69
	EtOH	253(3.87)	22-3133-69
$C_8H_{10}N_2O_2S$			
Barbituric acid, 5-isobutylidene-2-thio-	EtOH	280(4.23)	103-0830-69
$C_8H_{10}N_2O_2S_2$			
p-Dithiin-2,3-dicarboximide, N-(dimethylamino)-5,6-dihydro-	EtOH	259(3.97),416(3.46)	33-2236-69
p-Dithiino[2,3-d]pyridazine-5,8-dione, 2,3,6,7-tetrahydro-6,7-dimethyl-	MeOH	223(4.14),338(4.02)	33-2236-69
$C_8H_{10}N_2O_3$			
Barbituric acid, 5-butylidene-	EtOH	260(4.09)	103-0827-69
Barbituric acid, 5-isobutylidene-	EtOH	240(3.95)	103-0827-69
2H-Indazole-4-carboxylic acid, 3,3a,4,5,6,7-hexahydro-3-oxo-	EtOH	202(3.57),215s(3.37), 247(3.75)	39-2783-69C
2H-Indazole-7-carboxylic acid, 3,3a,4,5,6,7-hexahydro-3-oxo-	EtOH	203(3.64),231s(3.58), 252(3.66)	39-2783-69C
5H-Oxazolo[3,2-a]pyrimidin-5-one, 2,3-dihydro-2-(hydroxymethyl)-7-methyl-	pH 1.35	217(4.46),266(4.45)	70-0835-69
	pH 6.05	215(4.47),267(4.49)	70-0835-69
	pH 10.43	221(4.46),267(4.48)	70-0835-69
Phenol, 3-(dimethylamino)-4-nitro-	1%MeOH-HOAc	430(3.45)	39-0922-69B
Phenol, 3-(ethylamino)-4-nitro-	1%MeOH-HOAc	430(3.85)	39-0922-69B
Uracil, 1-(tetrahydro-2-furyl)-	pH 2	262(4.05)	103-0130-69
	pH 7	262(3.99)	103-0130-69
	pH 12	259(3.86)	103-0130-69
$C_8H_{10}N_2O_4$			
Isobarbituric acid, 1,3-dimethyl-, acetate	EtOH	208(4.14),271(4.13)	35-5264-69
Pyrazole-3,4-dicarboxylic acid, 1-methyl-, dimethyl ester	MeOH	198(4.35),224(4.06)	77-0066-69
Uracil, 1-acetyl-5-(methoxymethyl)-	EtOH	211(3.89),260(3.97)	73-1690-69
$C_8H_{10}N_2O_5$			
N-Acetyl-3-deoxy-2,3-didehydro-O-carbamoylpolyoxaminolactone	MeOH	243(3.94)	35-7490-69
$C_8H_{10}N_2S$			
4H-Cyclopenta[d]pyrimidine-4-thione, 1,5,6,7-tetrahydro-2-methyl-	EtOH	290(3.97),338(4.03)	4-0037-69
Formic acid, thio-, 2-methyl-2-phenylhydrazide, (E)-	$CHCl_3$	275(3.56)	89-0989-69
(Z)-	$CHCl_3$	252(3.91)	89-0989-69

Compound	Solvent	λ_{max}(log ε)	Ref.
$C_8H_{10}N_2S_3$			
4-Isothiazolecarbonitrile, 3,5-bis-(ethylthio)-	MeOH	216(4.07),231(4.09), 286(4.14)	88-4265-69
$C_8H_{10}N_4OS$			
Caffeine, 2-thio-	$CHCl_3$	289(4.15)	24-3000-69
$C_8H_{10}N_4O_2$			
Caffeine	0.4N HCl	270(3.90)	18-2282-69
Crotonaldehyde, (2,6-dihydroxypyrimidin-4-yl)hydrazone	EtOH	289(4.19)	56-0519-69
1,3,4-Oxadiazole, 2,2'-tetramethylene-bis-	MeOH	252(2.0)	48-0646-69
$C_8H_{10}N_4O_3$			
Xanthine, 9-(2-hydroxyethyl)-8-methyl-	pH 0.0	236(3.90),262(4.04), 270s(3.94)	24-4032-69
	pH 4.2	236(3.97),243s(3.93), 267(4.01)	24-4032-69
	pH 9	246(4.05),280(4.00)	24-4032-69
$C_8H_{10}N_4O_4$			
Pyrazinecarboxylic acid, 3-amino-6-formyl-, ethyl ester, oxime, 4-oxide	EtOH	277(4.08),295s(3.75), 390(3.50)	5-0100-69F
$C_8H_{10}N_4S$			
Carbohydrazide, 1-benzylidene-3-thio-	MeOH	313(4.44)	44-0756-69
Purine, 8,9-dimethyl-2-(methylthio)-	pH 7.0	242(4.35),311(4.34)	39-2620-69C
$C_8H_{10}N_6O$			
Acetamide, 2-amino-N-(9-methyl-9H-purin-6-yl)-, trihydrobromide	pH 1	273s(--),282s(--)	44-3492-69
	pH 6.4	273(4.11)	44-3492-69
	pH 11.5	283(--)	44-3492-69
$C_8H_{10}N_6O_2$			
9H-Purine-9-propionic acid, α,6-di-amino-, DL-	pH 1	259(4.14)	78-5983-69
	pH 1	257(4.17)	88-2285-69
	pH 7	261(4.15)	78-5983-69
	pH 12	262(4.18)	88-2285-69
	pH 13	261(4.18)	78-5983-69
$C_8H_{10}N_8O_2S$			
4-Pyrimidinol, 5,5'-thiobis[2,6-di-amino-	pH 1	258(4.41)	39-0603-69C
	pH 13	266(4.55)	39-0603-69C
$C_8H_{10}O$			
Benzyl alcohol, α-methyl-, (-)-(R)-(-)-	MeOH	252(2.35),257(2.37)	19-0475-69
	EtOH	248(2.17),252(2.24), 258(2.28),260s(2.16), 264(2.15),268(1.96)	60-2611-69
Bicyclo[3.1.0]hex-3-en-2-one, 1,4-di-methyl-	hexane	210(3.73),251(3.41)	22-2095-69
	EtOH	215(3.77),262(3.43)	22-2095-69
Bicyclo[3.1.0]hex-3-en-2-one, 4,5-di-methyl-	hexane	205(3.83),249(3.51)	22-2095-69
	EtOH	211(3.83),261(3.46)	22-2095-69
Bicyclo[5.1.0]oct-2-en-4-one	isooctane	237(3.89),322(1.70), 337s(1.60)	88-0995-69
2,4-Cyclooctadien-1-one	EtOH	278(3.76)	39-0160-69C
2-Norbornene-5-carboxaldehyde, endo-	isooctane	298(2.42)	44-2346-69
2-Norbornene-5-carboxaldehyde, exo-	isooctane	301(2.63)	44-2346-69

Compound	Solvent	$\lambda_{max}(\log \epsilon)$	Ref.
9-Oxabicyclo[4.2.1]nona-2,4-diene	EtOH	254(3.67)	39-0160-69C
1(3aH)-Pentalenone, 4,5,6,6a-tetrahydro-	isooctane	303(1.85)	88-0995-69
Tricyclo[3.3.0.0^{4,6}]octan-3-one	EtOH	215(2.18),292(1.46)	35-3020-69
2,3-Xylenol	pH 2	271(3.09)	30-1113-69A
	pH 13	239(3.91),289(3.47)	
	MeOH	273(3.20)	
	MeOH-Na	240(3.91),291(3.48)	
2,4-Xylenol	pH 2	277(3.28)	30-1113-69A
	pH 13	238(3.93),296(3.50)	
	MeOH	281(3.33)	
	MeOH-Na	240(3.91),298(3.50)	
2,5-Xylenol	pH 2	274(3.26)	30-1113-69A
	pH 13	239(3.90),291(3.54)	
	MeOH	277(3.32)	
	MeOH-Na	241(3.90),294(3.55)	
2,6-Xylenol	pH 2	270(3.08)	
	pH 13	239(3.93),288(3.56)	
	MeOH	273(3.18)	
	MeOH-Na	242(3.94),292(3.58)	
in 1% MeOH	acid	268(3.18)	28-2217-69A
in 1% MeOH	NaOH	287(3.56)	28-2217-69A
3,4-Xylenol	pH 2	277(3.26)	30-1113-69A
	pH 13	237(3.95),295(3.45)	
	MeOH	280(3.30)	
	MeOH-Na	241(3.95),296(3.45)	
3,5-Xylenol	pH 2	272(3.09)	30-1113-69A
	pH 13	240(3.90),290(3.41)	
	MeOH	275(3.15)	
	MeOH-Na	243(3.89),293(3.42)	
$C_8H_{10}OS$			
2-Thiabicyclo[6.1.0]non-6-en-3-one	EtOH	229(3.95),283(3.32)	35-4000-69
9-Thiabicyclo[3.3.1]non-6-en-2-one	EtOH	240(2.46),295(2.44)	35-4000-69
$C_8H_{10}OS_2$			
Carbamic acid, dithio-, S-2-penten-4-ynyl ester	EtOH	228(4.33),283(4.11)	18-2589-69
$C_8H_{10}O_2$			
2-Cyclohexene-1,4-dione, 2,5-dimethyl-	hexane	357(1.76)	39-0124-69C
2-Cyclohexene-1,4-dione, 5,6-dimethyl-, trans	hexane	365(1.76)	39-0124-69C
2-Cyclohexen-1-one, 3-acetyl-	EtOH	247(4.22)	70-1664-69
1,2-Ethanediol, 1-phenyl-, (S)-(-)-	EtOH	236(1.60),242(1.81), 247(1.99),252(2.15), 258(2.26),264(2.13), 268s(1.85)	60-2611-69
$C_8H_{10}O_2S$			
2-Thiophenecarboxylic acid, 3-methyl-, ethyl ester	EtOH	255(4.09)	2-0009-69
$C_8H_{10}O_3$			
4-Cyclopentene-1,3-dione, 4-methoxy-2,2-dimethyl-	MeOH	260(4.04)	20-0277-69
7-Oxabicyclo[4.1.0]heptan-2-one, 1-acetyl-	EtOH	297(1.51)	70-1664-69
7-Oxabicyclo[4.1.0]heptan-2-one, 6-acetyl-	EtOH	298(1.90)	70-1664-69
4H-Pyran-4-one, 3-acetyl-5,6-dihydro-2-methyl-	EtOH	238(3.72),271(4.03)	22-1383-69

Compound	Solvent	λ max(log ε)	Ref.
$C_8H_{10}O_3S$			
p-Toluenesulfonic acid, methyl ester	99.9% HOAc	273(2.70)	44-3638-69
	HOAc-Ac_2O	273(2.71)	44-3638-69
$C_8H_{10}O_4$			
Cyclobutene-1,2-dicarboxylic acid, dimethyl ester	C_6H_{12}	228(4.08)	80-1191-69
Malonic acid, ethylidene-, cyclic	MeOH	261(4.15),296(3.11)	88-4983-69
isopropylidene ester	MeOH-NaOMe	262(4.23)	88-4983-69
2H-Pyran-2-one, 3-acetyl-5,6-dihydro-	pH 13	250(4.16),270(4.13)	22-4091-69
4-hydroxy-5-methyl-	EtOH	218(3.90),273(4.05)	22-4091-69
2H-Pyran-2-one, 3-acetyl-5,6-dihydro-	pH 13	247(4.15),268(4.19)	22-4091-69
4-hydroxy-6-methyl-	EtOH	216(3.95),271(4.00)	22-4091-69
$C_8H_{10}O_4S$			
2-Thiopheneacetic acid, 4-carboxy-2,5-dihydro-, 4-methyl ester	EtOH	245s(3.19),265s(2.98)	35-7780-69
$C_8H_{10}O_4S_2$			
Cyclopropanecarboxylic acid, 1,1'-dithiodi-	C_6H_{12}	254(2.68)	11-0277-69A
$C_8H_{10}O_5$			
2-Heptenedioic acid, 2-methyl-4-oxo-	MeOH	235(4.20)	40-0096-69
$C_8H_{10}S$			
Benzo[c]thiophene, 4,5,6,7-tetrahydro-	EtOH	243(3.78)	2-0009-69
$C_8H_{10}S_3$			
Cyclohepta-1,2-dithiole-3(4H)-thione, 5,6,7,8-tetrahydro-	EtOH	229(4.04),244(3.81), 279(3.86),312(3.73), 413(4.00)	44-0730-69
$C_8H_{11}Cl_3O_2$			
Cyclohexanone, 2-(2,2,2-trichloro-1-hydroxyethyl)-, erythro-	EtOH	214(1.69),283(1.32)	23-2029-69
$C_8H_{11}I_2NO$			
1-Ethyl-5-hydroxy-6-iodo-2-picolinium	pH 1	220(4.27),320(4.08)	1-2065-69
iodide	pH 13	225(4.39),345(4.03)	1-2065-69
$C_8H_{11}N$			
Pyridine, 3-ethyl-4-methyl-	MeOH	259(3.42),267(3.32)	106-0196-69
Pyridine, 3-ethyl-6-methyl-	isooctane	269(3.47)	4-0859-69
	pH 1	270(3.85)	
	MeOH	269(3.55)	
Pyridine, 4-isopropyl-	C_6H_{12}	254(3.23)	4-0859-69
	pH 1	282(3.66)	
	MeOH	255(3.31)	
Pyridine, 4-propyl-	C_6H_{12}	255(3.21)	4-0859-69
	pH 1	252(3.66)	
	MeOH	255(3.31)	
	EtOH	214(3.20),257(3.35)	39-2134-69C
Pyridine, 2,4,6-trimethyl-	C_6H_{12}	265(3.46)	4-0859-69
	pH 1	267(3.90)	
	MeOH	264(3.58)	
$C_8H_{11}NO$			
Pyridine, 2-ethyl-3-methyl-, 1-oxide	H_2O	213(4.43),256(4.00)	39-0054-69B
Pyridine, 2-ethyl-6-methyl-, 1-oxide	H_2O	212(4.44),253(3.99)	39-0054-69B

Compound	Solvent	$\lambda_{max}(\log \epsilon)$	Ref.
Pyridine, 3-ethyl-6-methyl-, 1-oxide	isooctane	277(4.05)	4-0859-69
	pH 1	272(3.66)	4-0859-69
	H_2O	213(4.43),253(4.05)	39-0054-69B
	MeOH	260(4.04)	4-0859-69
Pyridine, 4-ethyl-3-methyl-, 1-oxide	H_2O	210(4.37),258(4.18)	39-0054-69B
Pyridine, 4-isopropyl-, 1-oxide	C_6H_{12}	286(4.11)	4-0859-69
	pH 1	250(3.43)	
	MeOH	266(4.12)	
Pyridine, 4-propyl-, 1-oxide	C_6H_{12}	286(4.20)	4-0859-69
	pH 1	253(3.57)	
	MeOH	265(4.18)	
Pyridine, 2,4,6-trimethyl-, 1-oxide	C_6H_{12}	274(4.09)	4-0859-69
	pH 1	268(3.77)	
	MeOH	259(4.10)	
2-Pyridinemethanol, 4,6-dimethyl-	ether	265(3.52)	39-2246-69C
2-Pyridinemethanol, α-ethyl-	EtOH	257(3.30),262(3.33), 269(3.24)	23-4393-69
$C_8H_{11}NOS$			
3-Hydroxy-6-methyl-2-(methylthio)pyridinium hydroxide, inner salt	pH 1	325(3.86)	1-2065-69
	pH 13	255(3.86),355(3.90)	1-2065-69
2(1H)-Pyridinethione, 1-ethyl-3-hydroxy-6-methyl-	pH 1	260(3.70),355(4.11)	1-2065-69
	pH 13	250(3.74),270(3.81), 375(4.14)	1-2065-69
3-Pyridinol, 2-(ethylthio)-6-methyl-	pH 1	240(3.55),325(3.90)	1-2065-69
	pH 13	250(3.85),325(3.94)	1-2065-69
$C_8H_{11}NOS_2$			
2-Butanone, 4-[(4-methyl-2-thiazolyl)-thio]-	MeOH	281s(4.11)	4-0397-69
2-Butanone, 4-(4-methyl-2-thioxo-4-thiazolin-3-yl)-	MeOH	321(4.13)	4-0397-69
2H-1,3-Thiazine-2,4(3H)-dione, 3-allyl-dihydro-5-methyl-2-thio-	MeOH	267(3.99),312(3.96)	42-0919-69
2H-1,3-Thiazine-2,4(3H)-dione, 3-allyl-dihydro-6-methyl-2-thio-	MeOH	267(4.10),313(4.06)	42-0919-69
$C_8H_{11}NO_2$			
Acetamide, N-(6-oxo-1-cyclohexen-1-yl)-	MeOH	212(3.99),267(3.75)	104-1193-69
2,6-Pyridinediol, 3-ethyl-4-methyl-	EtOH	314(3.63)	39-1678-69C
Retronecine, 3,8-didehydro-	n.s.g.	222(3.84)	39-1155-69C
$C_8H_{11}NO_2S$			
3-Pyridinol, 2-[(2-hydroxyethyl)thio]-6-methyl-	pH 1	240(3.53),330(3.94)	1-1704-69
	pH 13	245(3.75),325(3.88)	1-1704-69
$C_8H_{11}NO_3$			
1-Cyclopentene-1-acetic acid, 2-amino-5-oxo-, methyl ester	MeOH	267(4.66)	39-1625-69C
1-Propanone, 3-hydroxy-2-(hydroxymethyl)-1-pyrrol-3-yl-	EtOH	203(4.12),247(4.00), 270s(3.84)	33-1911-69
$C_8H_{11}NO_4$			
Isonicotinic acid, 1,2,3,6-tetrahydro-5-hydroxy-6-methyl-2-oxo-, methyl ester	90% EtOH	250(3.91)	94-0467-69
$C_8H_{11}NO_4S$			
Malonic acid, thiazolidin-2-ylidene-, dimethyl ester	MeOH	227(4.13),283(4.37)	78-4649-69

Compound	Solvent	λ_{max}(log ε)	Ref.
$C_8H_{11}NO_5$			
4-Isoxazoline-3,4-dicarboxylic acid, 2-methyl-, dimethyl ester	MeOH	262(3.74)	24-2346-69
$C_8H_{11}NS$			
4H-Thieno[3,4-c]pyrrole, 5-ethyl-5,6-dihydro-	EtOH	233s(3.77),243(3.79)	44-0333-69
$C_8H_{11}N_3$			
Triazene, 3-methyl-1-p-tolyl-	MeOH	280(4.15)	65-0303-69
$C_8H_{11}N_3O_2S$			
7,8-Dihydro-6-hydroxy-2,8-dimethyl-3H,6H-[1,3]thiazino[3,2-b]-as-triazin-3-one	EtOH	239(4.38)	114-0093-69C
	dioxan	238(4.38)	114-0093-69C
$C_8H_{11}N_3O_3$			
as-Triazine-3,5(2H,4H)-dione, 6-methyl-2-(tetrahydro-2-furyl)-	pH 3-4	264(3.81)	103-0283-69
	pH 7-8	263(3.82)	103-0283-69
	pH 11-12	251(3.86)	103-0283-69
as-Triazine-3,5(2H,4H)-dione, 2-(tetrahydro-2H-pyran-2-yl)-	pH 3-4	258(3.84)	103-0283-69
	pH 7-8	256(3.81)	103-0283-69
	pH 11-12	245(3.86)	103-0283-69
$C_8H_{11}N_3O_4$			
1(2H)-Pyrimidinepropionic acid, α-amino-3,4-dihydro-5-methyl-2,4-dioxo-, DL-	pH 1	269(3.99)	88-2285-69
	pH 12	271(3.87)	88-2285-69
1(2H)-Pyrimidinebutyric acid, α-amino-3,4-dihydro-2,4-dioxo-	pH 1	265(3.98)	78-5989-69
	pH 7	265(4.01)	78-5989-69
	pH 13	264(3.86)	78-5989-69
$C_8H_{11}N_3O_5S$			
Pentitol, 2,5-anhydro-1-C-(2,3,4,5-tetrahydro-5-oxo-3-thioxo-as-triazin-6-yl)-, D-allo-	H_2O	214(4.04),268(4.30)	73-1673-69
D-altro-	H_2O	215(4.05),267(4.33)	73-1673-69
as-Triazine-3,5(2H,4H)-dione, 6-β-D-ribofuranosyl-3-thio-	pH 1	214(4.05),268(4.31)	73-1690-69
	pH 13	258(4.17),313(3.75)	73-1690-69
$C_8H_{11}N_3O_6$			
Pentitol, 2,5-anhydro-1-C-(2,3,4,5-tetrahydro-3,5-dioxo-as-triazin-6-yl)-, D-allo-	H_2O	263(3.84)	73-1673-69
D-altro-	H_2O	263(3.84)	73-1673-69
as-Triazine-3,5(2H,4H)-dione, 2-β-D-lyxofuranosyl-	H_2O	263(3.67)	73-0618-69
as-Triazine-3,5(2H,4H)-dione, 6-β-D-ribofuranosyl-	H_2O	263(3.84)	73-1690-69
$C_8H_{11}N_5$			
Adenine, 9-ethyl-1-methyl-	pH 1	262(4.16)	95-0591-69
	H_2O	262(4.11)	95-0591-69
	pH 10	261(4.09),267s(4.04)	95-0591-69
	pH 13	267(4.15)	95-0591-69
Adenine, 9-propyl-	pH 7	262(4.18)	44-3240-69
	MeOH-HOAc	262(4.17)	44-3240-69
$C_8H_{11}N_5O_3$			
Morpholine, 4-(6-amino-5-nitro-4-pyrimidinyl)-	EtOH	360(4.03)	65-2125-69

Compound	Solvent	$\lambda_{max}(\log \epsilon)$	Ref.
C_8H_{12}			
Cyclobutane, 1,2-diethylidene-	n.s.g.	251(4.18)	25-0269-69
Cyclobutene, 1-ethyl-2-vinyl-	n.s.g.	243(4.28)	25-0269-69
Cyclohexane, 1,4-dimethylene-	isooctane	198(4.32),203(4.27), 209(4.01)	121-0803-69
	EtOH	199(4.29),202(4.23), 209(3.98)	121-0803-69
Ethylene, 1,1-dicyclopropyl-	hexane	202(3.94)	77-0781-69
Ethylene, 1,2-dicyclopropyl-, cis	hexane	204(4.22)	77-0781-69
Ethylene, 1,2-dicyclopropyl-, trans	hexane	205(4.16)	77-0781-69
1,2,5-Hexatriene, 3,5-dimethyl-	isooctane	214(3.27),223(2.89), 254(2.11)	35-3299-69
$C_8H_{12}N_2$			
7,8-Diazatricyclo[4.2.2.0^{2,5}]dec-7-ene	MeOH	378(1.94)	24-3304-69
$C_8H_{12}N_2O$			
Pyrazole, 4-acetyl-1,3,5-trimethyl-	MeOH	248(4.07)	4-0545-69
Pyrrolo[1,2-a]pyrimidin-2(6H)-one, 3,4,7,8-tetrahydro-3-methyl-	EtOH	262(4.02)	22-3139-69
Pyrrolo[1,2-a]pyrimidin-2(6H)-one, 3,4,7,8-tetrahydro-4-methyl-, hydrochloride	EtOH	233(3.91)	22-3139-69
Pyrrolo[1,2-a]pyrimidin-4(3H)-one, 2,6,7,8-tetrahydro-2-methyl-	EtOH	232(3.78)	22-3139-69
$C_8H_{12}N_2OS$			
1,3-Diazaspiro[4.5]decane-2,4-dione, 4-thio-	H_2O	229(3.57),273(4.03)	23-1117-69
	pH 13	287(4.07)	23-1117-69
	96% H_2SO_4	240(4.02),269(3.90)	23-1117-69
	EtOH	228(3.68),277(4.17), 375(1.41)	23-1117-69
$C_8H_{12}N_2O_2$			
Benzimidazole, 4,5,6,7-tetrahydro-1-hydroxy-2-methyl-, 3-oxide	MeOH	227(3.79)	48-0746-69
Crotonic acid, 3-(dimethylamino)-2-cyano-, methyl ester	isooctane	301(4.26)	35-6689-69
Nicotinamide, 1,4,5,6-tetrahydro-2,4-dimethyl-6-oxo-	EtOH	272(3.7)	94-2411-69
Nicotinamide, 1,4,5,6-tetrahydro-2,5-dimethyl-6-oxo-	EtOH	271(3.3)	94-2411-69
p-Phenylenediamine, 2,5-dimethoxy-	EtOH	241(3.87),282(3.79), 310(3.71),352s(--)	117-0287-69
2,5-Piperazinedione, 3-isobutylidene-	KBr	220(4.22),240(4.15)	18-0191-69
3H-Pyrazole-4-carboxylic acid, 3,3,5-trimethyl-, methyl ester	hexane	257(3.67),381(2.16)	39-1065-69C
	EtOH	262(3.66),369(2.15)	39-1065-69C
3H-Pyrazole-5-carboxylic acid, 3,3,4-trimethyl-, methyl ester	hexane	245(3.63),360(2.18)	39-1065-69C
	EtOH	252(3.64),345(2.26)	39-1065-69C
2-Pyrazoline-3-carboxylic acid, 5,5-dimethyl-4-methylene-, methyl ester	EtOH	223(3.95),330(4.00)	39-2443-69C
Pyrimidine, 5-ethoxy-4-methoxy-2-methyl-	pH 1	240(3.79),270(3.80)	1-2437-69
	pH 13	230(3.85),270(3.66)	1-2437-69
2(1H)-Pyrimidinone, 4-isopropoxy-6-methyl-	H_2O	268(4.22)	73-2316-69
4(3H)-Pyrimidinone, 5-ethoxy-2,3-dimethyl-	pH 1	255(3.76)	1-2437-69
	pH 13	240(3.71),275(3.74)	1-2437-69
Uracil, 5-butyl-	pH 2	266(3.92)	56-0499-69
	pH 12	292(3.79)	56-0499-69
Uracil, 5-tert-butyl-	pH 2	265(3.93)	56-0499-69

Compound	Solvent	$\lambda_{max}(\log \epsilon)$	Ref.
Uracil, 5-tert-butyl- (cont.)	pH 12	289(3.86)	56-0499-69
$C_8H_{12}N_2O_2S$			
Thiazole, 2-tert-butyl-4-methyl-5-nitro-	EtOH	227(4.62),250(3.48), 306(3.95)	28-1343-69B
Thiazole, 2-tert-butyl-5-methyl-4-nitro-	EtOH	220(4.11),240(--), 291(3.69)	28-1343-69B
Thiazole, 5-ethyl-2-isopropyl-4-nitro-	EtOH	220(4.00),240(--), 292(3.56)	28-1343-69B
3-Thiophenecarboxylic acid, 4-hydrazino-2-methyl-, ethyl ester	MeOH	226(4.01)	103-0424-69
Uracil, 5-(butylthio)-	pH 13	292(3.84)	44-3806-69
$C_8H_{12}N_2O_3$			
Hydrouracil, 1-(tetrahydro-2-furyl)-	pH 12	230(3.82)	103-0130-69
Veronal	base	241(4.00)	39-0802-69B
$C_8H_{12}N_2O_4$			
Barbituric acid, 5-(1-methoxyethyl)-	pH 6.2	218(3.78)	65-2568-69
5-methyl-	pH 13	240(3.69)	65-2568-69
Barbituric acid, 5-(2-methoxyethyl)-	pH 6.2	267(2.97)	65-2568-69
5-methyl-	pH 13	241(3.60)	65-2568-69
$C_8H_{12}N_2O_5$			
Pyrazine, 2-(D-arabino-tetrahydroxybutyl)-, 4-N-oxide	H_2O	217(4.04),265(4.08)	44-3842-69
$C_8H_{12}N_2O_6S_3$			
Metanilamide, 4,6-bis(methylsulfonyl)-	MeOH	226(4.44),271(4.25), 325(3.70)	44-1780-69
$C_8H_{12}N_2S$			
2-Pyrrolidinecarbonitrile, 2,3,3-trimethyl-5-thio-	EtOH	269(4.2)	89-0343-69
$C_8H_{12}N_2S_2$			
1,3-Diazaspiro[4.5]decane-2,4-dithione	H_2O	209(3.97),295(4.45)	23-1117-69
	pH 9.11	318(4.51)	23-1117-69
	71% H_2SO_4	245(4.09),290(4.23)	23-1117-69
	EtOH	227(3.74),297(4.49), 393(1.66)	23-1117-69
$C_8H_{12}N_4O_2$			
Butyraldehyde, (2,6-dihydroxy-4-pyrimidinyl)hydrazone	EtOH	286(4.13)	56-0519-69
$C_8H_{12}N_4O_3$			
1(2H)-Pyrimidinebutyric acid, α,4-diamino-2-oxo-	pH 1	281(4.14)	78-5989-69
	pH 7	273(3.98)	78-5989-69
	pH 13	274(3.93)	78-5989-69
4-Pyrimidinol, 6-(butylamino)-5-nitro-	pH 1	231(4.24),291(3.59), 345(3.91)	73-2114-69
	pH 13	222(4.16),347(3.99)	73-2114-69
$C_8H_{12}N_6O$			
Ethanol, 2-[(6-amino-2-methylpurin-8-yl)amino]-	pH 1	288(4.14)	65-2125-69
1-Propanol, 3-[(6-aminopurin-8-yl)-amino]-	pH 1	284(4.14)	65-2125-69
	EtOH	280(4.15)	65-2125-69

Compound	Solvent	$\lambda_{max}(\log \epsilon)$	Ref.
$C_8H_{12}O$			
2-Cyclohexen-1-one, 3,5-dimethyl-	H_2SO_4	243(4.08)	23-2263-69
protonated	H_2SO_4	278(4.12)	23-2263-69
2-Cycloocten-1-one	EtOH	230(3.89),310(1.90)	24-3877-69
2,4-Hexadienal, 2,5-dimethyl-	EtOH	291(4.76)	77-1096-69
1,3,5-Hexatriene, 1-methoxy-4-methyl-	EtOH	282(4.56)	54-0119-69
2-Hexen-4-yne, 2-ethoxy-	EtOH	236(4.14)	78-2837-69
Ketone, 1-cyclohexen-1-yl methyl	MeOH	232(4.10)	78-0223-69
Ketone, methyl 2-methyl-1-cyclopenten-1-yl	MeOH	253(3.97)	39-0608-69C
4-Pentenal, 5-cyclopropyl-	EtOH	208(3.83)	28-1808-69A
2H-Pyran, 3,4-dihydro-6-propenyl-	EtOH	242(4.00)	22-2508-69
Spiro[2.3]hexan-4-one, 1,1-dimethyl-	isooctane	212(3.62),248(2.21), 255(2.25),272(2.28), 275(2.27)	22-3981-69
$C_8H_{12}OS$			
Acetic acid, thio-, S-1-hexynyl ester	H_2O	257(3.54)	54-0307-69
$C_8H_{12}O_2$			
2-Cyclohexen-1-one, 3-ethoxy-	H_2SO_4	257(4.25)	23-2263-69
protonated	H_2SO_4	273(4.32)	23-2263-69
1-Cyclopropene-1-carboxylic acid, 2,3,3-trimethyl-, methyl ester	hexane	230(3.84)	39-1065-69C
2,3,4-Hexatrien-1-ol, 2-methoxy-5-methyl-	EtOH	267(4.35)	54-0119-69
4H-Pyran-4-one, 2,3-dihydro-2,5,6-trimethyl-	EtOH	275(4.08)	22-4091-69
4H-Pyran-4-one, 2,3-dihydro-3,5,6-trimethyl-	EtOH	274(4.01)	22-4091-69
$C_8H_{12}O_2S$			
2(3H)-Furanone, dihydro-3-(2-methylthiopropionyl)-	EtOH	285(3.85),339(2.79)	78-5703-69
2(3H)-Furanone, dihydro-3-(thiobutyryl)-	EtOH	281(3.97),332(2.90)	78-5703-69
9-Oxabicyclo[3.3.1]non-2-en-1-ol, 6-mercapto-	EtOH	none	33-0725-69
9-Thiabicyclo[3.3.1]nonan-2-one, 6-hydroxy-	C_6H_{12}	242(2.36),305(2.03)	33-0725-69
	H_2O	245(2.47),299(2.45)	33-0725-69
	MeOH	246(2.42),302(2.28)	33-0725-69
	EtOH	247(2.43),303(2.31)	33-0725-69
	MeCN	246(2.44),301(2.21)	33-0725-69
$C_8H_{12}O_3$			
Cyclopropanecarboxylic acid, β-oxo-, ethyl ester	isooctane	254(3.61)	22-3694-69
	EtOH	260(2.86)	22-3694-69
2(5H)-Furanone, 5-tert-butyl-5-hydroxy-	EtOH	204(4.27)	18-1098-69
3-Furoic acid, 4,5-dihydro-5,5-dimethyl-, methyl ester	MeOH	254(4.04)	44-2782-69
4-Hexenoic acid, 3-oxo-, ethyl ester	EtOH	225(3.96),280(3.86)	22-4091-69
	base	228(4.01),306(4.23)	22-4091-69
4-Pentenoic acid, 4-methyl-3-oxo-, ethyl ester	EtOH	224(3.87),274(3.56)	22-4091-69
	base	227(3.82),289(3.98)	22-4091-69
$C_8H_{12}O_4$			
Acetoacetic acid, 2-acetyl-, ethyl ester	C_6H_{12}	275(4.00)	22-3281-69
	MeOH	275(3.97)	22-3281-69
	EtOH	276(3.98)	22-3281-69
	ether	275(3.99)	22-3281-69
	MeCN	275(4.03)	22-3281-69

Compound	Solvent	$\lambda_{max}(\log \epsilon)$	Ref.
Acetoacetic acid, 2-acetyl-, ethyl ester (cont.)	$CHCl_3$	277(4.07)	22-3281-69
1,1-Cyclobutanediol, diacetate	EtOH	220(2.20)	78-1699-69
m-Dioxan, 5,5-diacetyl-	EtOH	302(1.77)	39-0879-69C
2,4-Hexanedione, 6-acetoxy-	EtOH	275(3.96)	22-1383-69
	EtOH	275(3.96)	22-4091-69
Hexanoic acid, 3,5-dioxo-, ethyl ester	EtOH	276(3.96)	22-0231-69
Malonic acid, isopropylidene-, dimethyl ester	EtOH	218(4.03)	22-0313-69
Malonic acid, propylidene, dimethyl ester	EtOH	210(3.89)	22-0313-69
2H-Pyran-2-carboxylic acid, 3,6-dihydro-6-hydroxy-, ethyl ester	EtOH	220(2.94)	104-0759-69
3H-Pyran-3-one, 2,4,5,6-tetrahydro-4,5-dihydroxy-, acetonide	n.s.g.	300(1.88)	77-0090-69
$C_8H_{12}Si$			
Silane, dimethylphenyl-	C_6H_{12}	264(2.37)	101-0017-69B
Silane, tetravinyl-	gas	175(--)	121-0803-69
	heptane	177(4.59)	121-0803-69
	MeCN	179(4.56)	121-0803-69
$C_8H_{13}BF_2O_2$			
Boron, (5,5-dimethyl-2,4-hexanedionato)difluoro-	$CHCl_3$	289(4.26)	39-0526-69A
$C_8H_{13}BrN_2$			
1-Pyrazoline, 4-(bromomethylene)-3,3,5,5-tetramethyl-	hexane	327(2.39)	39-2443-69C
	EtOH	325(2.29)	39-2443-69C
$C_8H_{13}Cl_3O_2$			
4-Heptanone, 1,1,1-trichloro-2-hydroxy-6-methyl-	EtOH	213(1.67),239(1.52), 280(1.43)	23-2029-69
$C_8H_{13}NO$			
2-Cyclohexen-1-one, 3-amino-6,6-dimethyl-	MeOH	280(4.56)	104-1165-69
Isoxazole, 3,5-diethyl-4-methyl-	EtOH	224(3.66)	22-2820-69
2(1H)-Pyridone, 5,6-dihydro-4,6,6-trimethyl-	H_2O	213(4.19),235s(3.63)	23-2781-69
$C_8H_{13}NO_2$			
2(3H)-Furanone, 3-[1-(ethylamino)-ethylidene]dihydro-	hexane	296(4.18)	104-0998-69
4(1H)-Pyridone, 6-ethoxy-2,3-dihydro-1-methyl-	EtOH	298(4.26)	94-2314-69
Retronecine	EtOH	214(3.34)	73-1832-69
$C_8H_{13}NO_2S$			
Acrylamide, 3-(acetonylthio)-N-ethyl-, (Z)-	90% EtOH	274(4.15)	12-0765-69
Butyric acid, 2-isothiocyanato-3-methyl-, ethyl ester, L-	hexane	250(3.12)	39-1143-69B
$C_8H_{13}NO_3$			
3-Piperidinecarboxylic acid, N-methyl-4-oxo-, methyl ester	hexane	252(--)	103-0062-69
	MeOH	252(3.86)	
	EtOH	252(3.88)	
	iso-BuOH	252(3.90)	
hydrochloride	H_2O	245(3.52)	103-0062-69

Compound	Solvent	λ max(log ε)	Ref.
3-Piperidinecarboxylic acid, N-methyl-4-oxo-, methyl ester, hydrochloride, (cont.)	MeOH	245(4.00)(changing)	103-0062-69
	EtOH	245(3.96)(changing)	103-0062-69
	iso-BuOH	245(3.99)(changing)	103-0062-69
$C_8H_{13}NO_5$			
Nitrone, α-carboxy-α-(carboxymethyl)-N-ethyl-, dimethyl ester	MeOH	273(4.11)	24-2336-69
$C_8H_{13}N_2O_6P$			
Isopyridoxamine, 4'-phosphate (hydrate)	pH 1	291(3.94)	69-5181-69
	pH 7	258(3.56),285(3.55), 324(3.66)	69-5181-69
	pH 13	243(3.88),309(3.91)	69-5181-69
2,5-Pyridinedimethanol, 4-(aminomethyl)-3-hydroxy-, 5-(dihydrogen phosphate)	pH 1	295(3.98)	10-0001-69D
	pH 7	327(3.94)	10-0001-69D
	pH 13	312(3.89)	10-0001-69D
$C_8H_{13}N_3O_2$			
4(3H)-Pyrimidinone, 2-amino-3-(2-hydroxypropyl)-6-methyl-	pH 1.4	215(4.61),254(4.56)	70-0835-69
	pH 6.0	226(4.87),284(4.90)	70-0835-69
	pH 10.5	227(4.50),282(4.52)	70-0835-69
$C_8H_{13}N_3O_3$			
4(3H)-Pyrimidinone, 2-amino-3-(2,3-dihydroxypropyl)-6-methyl-	pH 2.3	225(4.35),259(4.22)	70-0835-69
	pH 6.8	228(4.29),284(4.38)	70-0835-69
	pH 11.7	228(4.22),283(4.36)	70-0835-69
$C_8H_{13}N_3O_6S$			
as-Triazine-3,5(2H,4H)-dione, 6-(D-allo-1,2,3,4,5-pentahydroxypentyl)-3-thio- (same spectrum for altro-, gluco- and galacto- isomers)	H_2O	215(4.02),267(4.30)	73-1673-69
$C_8H_{13}N_3O_7$			
as-Triazine-3,5(2H,4H)-dione, 6-(D-allo-1,2,3,4,5-pentahydroxypentyl)-	H_2O	263(3.84)	73-1673-69
as-Triazine-3,5(2H,4H)-dione, 6-(D-altro-1,2,3,4,5-pentahydroxypentyl)-	H_2O	263(3.80)	73-1673-69
as-Triazine-3,5(2H,4H)-dione, 6-(D-galacto-1,2,3,4,5-pentahydroxypentyl)-	H_2O	263(3.81)	73-1673-69
as-Triazine-3,5(2H,4H)-dione, 6-(D-gluco-1,2,3,4,5-pentahydroxypentyl)-	H_2O	263(3.80)	73-1673-69
$C_8H_{13}N_5O$			
4-Pteridinol, 2-amino-5,6,7,8-tetrahydro-5,6-dimethyl-	pH 1	216(4.26),263(4.18)	10-0436-69E
	pH 7.0	218(4.30),270(3.88), 285(3.90)	10-0436-69E
	pH 13	280(3.94)	10-0436-69E
4-Pteridinol, 2-amino-5,6,7,8-tetrahydro-5,7-dimethyl-	pH 1	216(4.31),263(4.18)	10-0436-69E
	pH 7.0	218(4.33),287(4.03)	10-0436-69E
	pH 13	279(3.95)	10-0436-69E
$C_8H_{13}N_7$			
1H-v-Triazolo[4,5-d]pyridazine, 4-[[2-2-(dimethylamino)ethyl]amino]-, hydrochloride	EtOH	204(4.59),229s(4.25), 268(3.84),318s(3.36)	4-0093-69
$C_8H_{14}BrN$			
2,3-Butadienylamine, 4-bromo-N,N-diethyl-	heptane	187(4.53)	78-3277-69

Compound	Solvent	$\lambda_{max}(\log \epsilon)$	Ref.
2,3-Butadienylamine, 4-bromo-N-isobutyl-	heptane	186(4.43)	78-3277-69
$C_8H_{14}BrNO$			
1,2,3,6-Tetrahydro-1,1,5-trimethyl-3-oxopyridinium bromide	EtOH	230(4.08)	70-2627-69
$C_8H_{14}ClNO_2$			
Nipecotic acid, N-chloro-, ethyl ester, (R)-(-)-	dioxan	273(2.35)	24-2864-69
$C_8H_{14}ClNO_5$			
N-(2-Formylvinyl)-N-methylpyrrolidinium perchlorate	H_2O	204(3.89)	24-2609-69
$C_8H_{14}N_2$			
2,3-Diazabicyclo[2.2.1]hept-5-ene, 1,2,3-trimethyl-	C_6H_{12}	266(2.87)	78-0657-69
Pyrazole, 1-ethyl-3,4,5-trimethyl-	n.s.g.	230(3.68)	39-2251-69C
$C_8H_{14}N_2O$			
Pyrazoline, 1-acetyl-3-ethyl-4-methyl-	MeOH	240(4.21)	123-0041-69A
Pyrazoline, 1-acetyl-5-ethyl-4-methyl-	MeOH	240(4.06)	123-0041-69A
$C_8H_{14}N_2O_2$			
1H-1,2-Diazepine-1-carboxylic acid, 4,5,6,7-tetrahydro-, ethyl ester	MeOH	260(3.37)	22-2175-69
1-Pyrazoline-3-carboxylic acid, 3,5,5-trimethyl-, methyl ester	hexane	327(2.23)	39-0787-69C
	EtOH	324(2.18)	39-0787-69C
1-Pyrazoline-4-carboxylic acid, 3,3,5-trimethyl-, methyl ester	hexane	329(2.43)	39-0787-69C
	EtOH	325(2.40)	39-0787-69C
2-Pyrazoline-3-carboxylic acid, 4,5,5-trimethyl-, methyl ester	hexane	277(3.91)	39-0787-69C
$C_8H_{14}N_2O_3$			
Nipecotic acid, N-nitroso-, ethyl ester, (R)-(-)-	dioxan	237(3.94),360(2.03)	24-2864-69
5-Pyrimidinecarboxaldehyde, 1,4,5,6-tetrahydro-2-methyl-4-oxo-, 5-(dimethyl acetal), hydrochloride	MeOH	228(3.82)	88-3553-69
	MeOH-NaOH	210(--)	88-3553-69
$C_8H_{14}N_2O_4$			
Oxalacetic acid, dimethyl ester, 2-(dimethylhydrazone)	MeOH	291(3.80)	44-2248-69
$C_8H_{14}N_2O_5$			
Norvaline, N-acetyl-5-hydroxy-, methyl ester, nitrite	n.s.g.	none	33-0388-69
Norvaline, N-acetyl-5-nitro-, methyl ester	EtOH	278(1.36)	33-0388-69
$C_8H_{14}N_4OS$			
Acetimidic acid, N-(3-ethyl-5-mercapto-4H-1,2,4-triazol-4-yl)-, ethyl ester	EtOH	258(4.37)	2-0959-69
$C_8H_{14}N_4S$			
4-Pyrimidinethiol, 5-amino-6-(butylamino)-	pH 1	247(4.08),309(4.07)	73-2114-69
	pH 13	232(4.22),244(4.24), 310(4.15)	73-2114-69

Compound	Solvent	λmax(log ε)	Ref.
$C_8H_{14}O$			
1,3-Butadiene, 1-ethoxy-2,3-dimethyl-, trans	EtOH	247(4.29)	22-2508-69
Ketone, methyl 1,2,2-trimethylcyclopropyl	isooctane	277(1.79)	22-3981-69
1,3-Pentadiene, 1-ethoxy-4-methyl-	EtOH	243(4.30)	22-2508-69
1,3-Pentadiene, 4-ethoxy-3-methyl-, cis	EtOH	244(4.13)	22-2508-69
trans	EtOH	247(4.27)	22-2508-69
$C_8H_{14}O_2S$			
Acetic acid, thio-, S-ester with 1-mercapto-2-hexanone	H_2O	233(3.54)	54-0307-69
1,3-Oxathiol-2-ol, 5-butyl-2-methyl-	H_2O	210s(3.85)	54-0307-69
$C_8H_{14}O_3S$			
2-Thiophenecarboxylic acid, tetrahydro-3-hydroxy-3-methyl-, ethyl ester	EtOH	247(2.84)	2-0009-69
$C_8H_{14}S$			
9-Thiabicyclo[3.3.1]nonane	EtOH	235s(2.22)	44-1233-69
9-Thiabicyclo[4.2.1]nonane	EtOH	239s(1.94)	44-1233-69
$C_8H_{14}S_2$			
Cyclohexanecarbodithioic acid, methyl ester	C_6H_{12}	302(4.00),456(1.24)	48-0045-69
	EtOH	302(4.08),450(1.24)	48-0045-69
$C_8H_{15}NO$			
2H-Azonin-2-one, octahydro-	C_6H_{12}	194(3.78)	46-1642-69
	H_2O	197(3.89)	46-1642-69
Cyclohexane, 1-ethyl-1-nitroso- (dimer)	EtOH	300(2.02),699(1.20)	39-1073-69C
Nitrone, N-(1-ethylcyclopentyl)-	EtOH	241(3.82)	39-1073-69C
$C_8H_{15}NO_2$			
2H-Azonin-2-one, octahydro-6-hydroxy-	H_2O	197(3.89)	39-1358-69C
Cyclohexane, 1-ethyl-1-nitro-	EtOH	272(2.82)	39-1073-69C
Formic acid, (N-tert-butylformimidoyl)-, ethyl ester	C_6H_{12}	240(2.04),294(1.91)	54-0005-69
$C_8H_{15}N_3O_2$			
5-Pyrimidinecarboxaldehyde, 4-amino-5,6-dihydro-2-methyl-, dimethyl acetal, hydrochloride	MeOH	278(4.06)	88-3553-69
	MeOH	260(--)	88-3553-69
$C_8H_{15}N_3O_3$			
Imidazole-2,5-dione, 4-(ethylamino)-1-methyl-, 2-(dimethyl acetal)	EtOH	218(4.42),253(3.78)	78-0557-69
Imidazole-2,5-dione, 1-methyl-4-(methylamino)-, 2-(ethyl methyl acetal)	EtOH	218(4.44),252(3.79)	78-0557-69
$C_8H_{15}N_3O_7$			
Streptozotocin	EtOH	229(3.80),380(1.88), 394(2.01),412(1.92)	33-2555-69
$C_8H_{15}N_5O$			
s-Triazine, 2,4-bis(ethylamino)-6-methoxy-	pH 1	211s(4.36),236s(4.13)	59-1167-69
	pH 10.0	217(4.55)	59-1167-69
s-Triazine, 2-(isopropylamino)-4-methoxy-6-(methylamino)-	pH 1.0	211s(4.33),236s(4.13)	59-1167-69
	pH 10.0	217(4.53)	59-1167-69

Compound	Solvent	λ_{max}(log ε)	Ref.
$C_8H_{15}N_5O_2$			
Ethanol, 2,2'-[(6-methyl-as-triazine-3,5-diyl)diimino]di-	EtOH	212(4.36),234(4.12), 309(3.75)	114-0181-69C
C_8H_{16}			
Pentane, 1-cyclopropyl-, iodine complex	hexane	253(4.18)	35-2279-69
$C_8H_{16}ClNO_5$			
Diethyl(2-formylvinyl)methylammonium perchlorate	H_2O	204(3.81)	24-2609-69
$C_8H_{16}NO_6P$			
Isopyridoxal 4'-phosphate	pH 1	293(3.86)	69-5181-69
	pH 7	221(4.17),313(3.73)	69-5181-69
	pH 13	239(4.28),356(3.64)	69-5181-69
$C_8H_{16}N_2$			
1,3-Butadiene-1,1-diamine, N,N,N',N'-tetramethyl-	C_6H_{12}	234(3.84),288(4.27)	39-0016-69C
$C_8H_{16}N_2O$			
Formaldehyde, (1,1-dimethyl-3-oxobutyl)-methylhydrazone	EtOH	245(3.61)	22-3306-69
Formaldehyde, (2,2-dimethyl-3-oxobutyl)-methylhydrazone	EtOH	245(3.83)	22-3306-69
$C_8H_{16}N_2O_2$			
Ethanol, 1-(tert-butylazo)-, acetate	C_6H_{12}	360(1.21)	89-0451-69
$C_8H_{16}N_4O_2$			
Acetamide, N,N'-azobis[N-ethyl-	EtOH	285(4.38)	44-2997-69
$C_8H_{16}Si$			
Silane, diallyldimethyl-	gas	190(--)	121-0803-69
	isooctane	195(4.28)	121-0803-69
	EtOH	196(4.27)	121-0803-69
$C_8H_{16}Sn$			
Stannane, diallyldimethyl-	hexane	208(4.30)	44-3715-69
	MeCN	212(4.26)	44-3715-69
$C_8H_{17}NO$			
4-Piperidinemethanol, α-ethyl-	EtOH	225(2.51),268(2.15)	39-2134-69C
$C_8H_{17}N_3S$			
Heptanal, thiosemicarbazone	EtOH	230(3.93),271(4.40)	94-0844-69
$C_8H_{17}O_4P$			
Phosphonic acid, (2-ethoxyvinyl)-, diethyl ester	EtOH	206(4.17)	39-0460-69C
$C_8H_{18}N_2$			
Azoethane, 1,1,1',1'-tetramethyl-	gas	356(0.90)	35-1710-69
$C_8H_{18}N_2O$			
1-Butanol, 1-(tert-butylazo)-	C_6H_{12}	343(1.29)	89-0451-69
1-Propanol, 1-(tert-butylazo)-2-methyl-	C_6H_{12}	346(1.32)	89-0451-69
$C_8H_{18}N_4S$			
s-Tetrazine-3(2H)-thione, 6-hexyltetra-hydro-	MeOH	250(4.10)	44-0756-69

Compound	Solvent	$\lambda_{max}(\log \epsilon)$	Ref.
$C_8H_{19}ClN_2Pt$			
Platinum, (2-methyl-3-dimethylamino-propenyl)-, chloride, dimethyl-amine adduct	$CHCl_3$	245(4.14),295s(3.82), 350(3.70)	101-0309-69A
$C_8H_{19}NOSi$			
Silanecarboxamide, N,N-diethyl-1,1,1-trimethyl-	C_6H_{12}	218(3.72),264(2.43)	77-0462-69
$C_8H_{24}O_2Si_3$			
Trisilane, 2,2-dimethoxyhexamethyl-	isooctane	234(3.87)	35-6613-69
	dioxan	235(3.91)	35-6613-69
$C_8H_{24}Si_3$			
Trisilane, octamethyl-	isooctane	216(3.88)	35-6613-69
$C_8H_{25}NSi_3$			
Trisilane, 1-(methylamino)heptamethyl-	isooctane	200(4.04)(end abs.)	35-6613-69
	dioxan	210(3.91)(end abs.)	35-6613-69
$C_8I_2O_8Os_2$			
Osmium, octacarbonyldiiododi(Os-Os)-	C_6H_{12}	232(4.00),337(3.74)	39-0987-69A

Compound	Solvent	λ_{max}(log ε)	Ref.
$C_9Cl_4N_4$			
1,2-Diaza[4.4]nona-1,3,6,8-tetraene-3,4-dicarbonitrile, 6,7,8,9-tetrachloro-	$CHCl_3$	244(4.13),258(4.03),263(4.02),315(4.13)	64-0536-69
$C_9F_{10}N_2$			
2-Pyridinebutyronitrile, α,α,β,β,γ,γ,3,4,5,6-decafluoro-	C_6H_{12}	258(3.52)	39-2559-69C
$C_9H_2Cl_2N_2O_2$			
1-Oxaspiro[2.5]octa-4,7-diene-4,5-dicarbonitrile, 7,8-dichloro-6-oxo-	MeOH	283(3.8)	44-1203-69
$C_9H_2F_4O_2$			
Chromone, 5,6,7,8-tetrafluoro-	EtOH	201(3.93),221(4.15),298(3.63)	70-0812-69
$C_9H_3F_5O$			
Cinnamaldehyde, 2,3,4,5,6-pentafluoro-	EtOH	270(4.27)	65-1347-69
$C_9H_3N_3O_2S_2$			
p-Dithiin-2,3-dicarboximide, 5,6-dicyano-N-methyl-	THF	242(4.13),336(3.72)	88-4273-69
$C_9H_4Br_3NO$			
9H-Pyrrolo[1,2-a]azepin-9-one, 1,2,3-tribromo-	EtOH	208(4.05),244s(--),252s(--),279(4.40)	39-1028-69C
	100% H_2SO_4	239(4.22),269(4.27),297(4.59),378(3.87),477(3.20)	39-1028-69C
$C_9H_4F_5NO_2$			
Styrene, 2,3,4,5,6-pentafluoro-α-methyl-β-nitro-	EtOH	250(3.39)	65-1766-69
$C_9H_4N_2O_8$			
2-Benzofurancarboxylic acid, 7-hydroxy-4,6-dinitro-	MeOH	213(4.19),288(4.31)	32-1177-69
$C_9H_5BrN_4$			
s-Triazolo[3,4-a]phthalazine, 3-bromo-	MeOH	243(4.68),250(4.61),273s(3.84)	44-3221-69
$C_9H_5ClN_4$			
s-Triazolo[3,4-a]phthalazine, 3-chloro-	MeOH	209(4.33),234s(4.58),241(4.70),249(4.63),274s(3.83)	44-3221-69
$C_9H_5F_4NO$			
Indole-2-methanol, 4,5,6,7-tetrafluoro-	EtOH	206(4.56),260(3.82)	65-1615-69
$C_9H_5F_4NO_2$			
Cinnamic acid, 4-amino-2,3,5,6-tetrafluoro-, trans	EtOH	230(3.94),325(4.25)	44-0534-69
C_9H_5N			
Propiolonitrile, phenyl-	EtOH	248(4.18),260(4.33),274(4.20)	22-0210-69

Compound	Solvent	λ max (log ε)	Ref.
$C_9H_5NO_3$			
5,8-Isoquinolinedione, 7-hydroxy-	MeCN	245(4.05),250(4.04), 295(3.64)	4-0697-69
$C_9H_5NO_6$			
2-Benzofurancarboxylic acid, 7-hydroxy-6-nitro-	MeOH	232(4.35),273(4.17)	32-1177-69
C_9H_5NS			
Benzo[b]thiophene-3-carbonitrile	MeOH	209(4.58),263(3.84), 272(3.88),292(3.78), 301(3.76)	73-2959-69
$C_9H_5N_3$			
Malononitrile, (phenylimino)-	C_6H_{12}	235s(3.75),281(3.72), 367(3.69)	44-2146-69
$C_9H_5N_3O_4S$			
Isothiazole, 4-(2,4-dinitrophenyl)-	EtOH	215(4.16),241(4.14), 290(4.01)	4-0841-69
$C_9H_5N_5$			
Ethenetricarbonitrile, (1-methylpyrazol-4-yl)-	MeOH	354(4.28)	4-0545-69
$C_9H_5N_5O_2$			
Imidazo[5,1-c][1,2,4]benzotriazine, 7-nitro-	5N H_2SO_4	233(4.15),277(4.33), 337(3.77)	103-0683-69
	MeOH	238(4.18),287(4.31), 363(3.80)	103-0683-69
$C_9H_6BrClO_4S_2$			
4-(p-Bromophenyl)-1,3-dithiol-1-ium perchlorate	50% H_2SO_4	247(4.25),272s(4.13), 345(3.61)	94-1924-69
C_9H_6BrNO			
Carbostyril, 3-bromo-	EtOH	233(4.49),280(3.93), 336(4.07)	1-0159-69
Carbostyril, 4-bromo-	EtOH	231(4.48),273(3.74), 335(3.70)	1-0159-69
Carbostyril, 6-bromo-	EtOH	239(4.56),277(3.56), 338(3.61)	1-0159-69
C_9H_6BrNS			
Isothiazole, 4-(p-bromophenyl)-	EtOH	206(4.35),248(4.18), 270(4.21)	4-0841-69
$C_9H_6BrN_5O$			
Acetamide, N-(6-bromo-5-cyano-7H-pyrrolo[2,3-d]pyrimidin-4-yl)-	pH 1	297(3.95)	35-2102-69
	pH 11	248(4.15),282(3.85), 315(3.72)	35-2102-69
	EtOH	250(4.11),282(3.85), 312(3.65)	35-2102-69
C_9H_6ClN			
Atroponitrile, β-chloro-, cis	EtOH	220s(--),262(4.01)	22-0217-69
Atroponitrile, β-chloro-, trans	EtOH	221(4.06),268(4.06)	22-0217-69
C_9H_6ClNO			
Carbostyril, 4-chloro-	EtOH	234(4.45),272(3.81), 335(3.78)	1-0159-69

Compound	Solvent	λmax(log ε)	Ref.
Carbostyril, 6-chloro-	EtOH	234(4.50),275(3.47), 336(3.52)	1-0159-69
Carbostyril, 7-chloro-	EtOH	231(4.52),280(3.69), 327(3.92)	1-0159-69
Isocarbostyril, 3-chloro-	pH 4	228(4.33),247(3.58), 274(4.00),282(3.98), 319s(3.63),328(3.70), 340s(3.54)	24-3666-69
	pH 13	222(4.23),238s(4.09), 245s(4.05),297(3.99), 330(3.65)	24-3666-69
	pH -4.9	230(4.69),260(3.66), 268(3.70),279(3.68), 332(3.74),343(3.69)	24-3666-69
8-Quinolinol, 5-chloro-	pH 1.1	226s(3.98),261(4.66), 305(3.34),380(3.41)	23-0105-69
	pH 8.4	246(4.56),264s(3.70), 383(3.60),329(3.56)	23-0105-69
	pH 14.0	258(4.51),346(3.56), 383(3.60)	23-0105-69
$C_9H_6ClNO_3$			
Indole-2,3-dione, 5-chloro-6-methoxy-	EtOH	220(4.22),267(4.41), 273(4.40),317(3.87), 418(2.04)	18-3016-69
$C_9H_6ClNO_4S_2$			
Thiazolo[2,3-b]benzothiazolium perchlorate	EtOH	217(4.50),260(4.60), 299(3.83),307(4.06)	95-0469-69
$C_9H_6ClNO_6S_2$			
4-(p-Nitrophenyl)-1,3-dithiol-1-ium perchlorate	50% H_2SO_4	233s(4.04),254(4.02), 285(4.09),323s(3.93)	94-1924-69
$C_9H_6ClN_3O_3S$			
Thiazole, 2-amino-5-[1-chloro-2-(5-nitro-2-furyl)vinyl]-	EtOH	251(4.08),305(4.02), 390(4.28)	103-0414-69
	dioxan	249(4.04),302(3.97), 390(4.27)	103-0414-69
$C_9H_6Cl_2N_2OS$			
Acetamide, 2-chloro-N-(4-chloro-2-benzothiazolyl)-	EtOH	218(4.38),285(4.30), 304(4.26)	23-4483-69
Acetamide, 2-chloro-N-(6-chloro-2-benzothiazolyl)-	EtOH	210(4.39),280(4.30)	23-4483-69
1,3,4-Oxadiazole, 2-(2,4-dichlorophenyl)-5-(methylthio)-	n.s.g.	215(4.35)	2-0583-69
1,3,4-Oxadiazoline-2-thione, 3-(hydroxymethyl)-5-(2,4-dichlorophenyl)-	n.s.g.	215(4.58),238(4.25), 311(4.15)	2-0583-69
$C_9H_6Cl_2N_4$			
s-Triazine, 2-anilino-4,6-dichloro-	n.s.g.	263(4.20)	42-0595-69
$C_9H_6Cl_2O_4S_2$			
4-(p-Chlorophenyl)-1,3-dithiol-1-ium perchlorate	50% H_2SO_4	245(4.27),270s(4.08), 344(3.60)	94-1924-69
C_9H_6FNO			
Carbostyril, 6-fluoro-	EtOH	229(4.37),264(3.83), 336(3.73)	1-0159-69

Compound	Solvent	$\lambda_{max}(\log \epsilon)$	Ref.
$C_9H_6F_5NO$			
Acetamide, N-(2,3,4,5,6-pentafluoro-benzyl)-	EtOH	262(2.74)	65-1607-69
$C_9H_6F_5NO_4$			
1,3-Propanediol, 2-nitro-1-(pentafluoro-phenyl)-	EtOH	204(4.01),263(2.70)	65-1615-69
$C_9H_6I_2O_3$			
2,4-Cyclohexadienepropionic acid, 1-hydroxy-3,5-diiodo-4-oxo-, lactone	EtOH	265(3.55),295(3.49)	69-1844-69
$C_9H_6N_2$			
Indole-5-carbonitrile	MeOH	276(3.65)	22-3002-69
anion	MeOH	298(--)	22-3002-69
2,4-Norcaradiene-7,7-dicarbonitrile	MeOH	273(3.48)	22-0948-69
$C_9H_6N_2O$			
1-Indanone, 2-diazo-	40% dioxan	259(5.15),320(4.78)	33-2417-69
$C_9H_6N_2O_2$			
4-Cinnolinecarboxylic acid	EtOH	232(4.55),308(3.58), 330(3.54),390s(2.42)	95-1167-69
Styrene, p-cyano-β-nitro-	n.s.g.	299(4.32)	67-0127-69
$C_9H_6N_2O_2S$			
Imidazolidine-2-thione, 4,5-dioxo-1-phenyl-	EtOH	302(4.12),422(1.69)	18-2323-69
Isothiazole, 4-(p-nitrophenyl)-	EtOH	204(4.34),219(4.08), 307(4.22)	4-0841-69
$C_9H_6N_2O_3$			
1,3,4(2H)-Isoquinolinetrione, 3-oxime	H_2O	267(3.9),307(3.4)	95-0418-69
	pH 13	270(4.0),320(3.7)	95-0418-69
Ketone, diazomethyl 3,4-methylenedioxy-phenyl	EtOH	233(4.23),315(4.23)	12-1011-69
$C_9H_6N_2O_3S$			
Barbituric acid, 5-furfurylidene-2-thio-	EtOH	395(4.32)	103-0830-69
$C_9H_6N_2O_4$			
Barbituric acid, 5-furfurylidene-	EtOH	255(3.58),370(3.92)	103-0827-69
$C_9H_6N_2S$			
Thiazolo[3,2-a]benzimidazole	EtOH	250(3.78),286(3.84), 293(3.94),306(3.75)	4-0797-69
	EtOH	215(4.48),242(4.18), 247(4.16),275(3.99), 283s(3.90)	39-1334-69C
$C_9H_6N_2Se$			
Indole, 5-selenocyanato-	EtOH	230(4.39),280(3.69)	11-0159-69B
$C_9H_6N_4$			
Benzonitrile, o-(s-triazol-3-yl)-	MeOH	210(4.56),230(4.15), 263(3.86)	44-3221-69
Imidazo[5,1-c][1,2,4]benzotriazine	octane	233(4.01),262(3.88), 362(3.48)	103-0683-69
	5N H_2SO_4	220(4.20),244(4.22), 265s(--),325(3.83)	103-0683-69

Compound	Solvent	λ_{max}(log ϵ)	Ref.
Imidazo[5,1-c][1,2,4]benzotriazine (cont.)	H_2O	222(4.26),240(4.12), 262(4.02),385(3.73)	103-0683-69
	0.05N NaOH	221(4.27),242(4.10), 258(4.01),388(3.72)	103-0683-69
	MeOH	244(4.09),261(4.02), 378(3.67)	103-0683-69
s-Triazolo[3,4-a]phthalazine	MeOH	237(4.53),267s(3.60)	44-3221-69
1H-v-Triazolo[4,5-b]quinoline	EtOH	237(4.55),322(4.00), 351(3.67)	39-1758-69C
1H-v-Triazolo[4,5-c]quinoline	EtOH	217(4.43),239(4.52), 281(3.74),306(3.40), 320(3.33)	39-1758-69C
$C_9H_6N_4O$			
3H-as-Triazino[5,6-b]indol-3-one, 2,5-dihydro-	EtOH	260(4.4),330(3.7), 380(2.8)	65-2339-69
s-Triazolo[3,4-a]phthalazin-3-ol	MeOH	205(4.54),254s(4.32), 264(4.45)	44-3221-69
$C_9H_6N_4O_3S$			
5H-as-Triazino[5,6-b]indole-3-sulfonic acid	EtOH	260(4.3),310(3.8)	65-2111-69
$C_9H_6N_4S$			
5H-as-Triazino[5,6-b]indol-3-thiol	pH 13	260(4.5),290(4.5), 370(4.0)	65-0078-69
	EtOH	260(4.1),300(4.6)	65-0078-69
	EtOH	260(4.2),310(4.7)	65-0640-69
s-Triazolo[3,4-a]phthalazine-3-thiol	MeOH	220(4.62),257(4.67), 284(4.34)	44-3221-69
C_9H_6O			
Propiolaldehyde, phenyl-	EtOH	238(4.23),249(4.22), 278(3.56)	22-0210-69
	ether	234(3.93),264s(--), 275(4.22),285s(--)	39-0683-69C
C_9H_6OS			
Benzothiophene-3-carboxaldehyde	MeOH	218(4.58),240(4.05), 282(3.86),299(3.96)	73-2959-69
$C_9H_6OS_2$			
4-Hydroxy-2-phenyl-1,3-dithiol-1-ium	MeOH	203(4.50),220(4.24), 265(3.96),285(3.89), 325(3.46)	77-0569-69
$C_9H_6O_2$			
Coumarin	EtOH	274(3.69),313(3.98)	39-0016-69C
$C_9H_6O_2S$			
Benzo[c]thiophene-2-carboxylic acid	EtOH	237(4.27),286s(3.41), 299(3.59),313(3.66), 352(3.81)	2-0009-69C
$C_9H_6O_3$			
4,7-Benzofurandione, 3-methyl-	EtOH	217(4.14),248(4.20), 292(4.19),394(3.31)	18-3318-69
Chromone, 2-hydroxy-	MeOH	262(3.91),279(3.99),	64-0225-69
Coumarin, 5-hydroxy-	EtOH	300(4.2)	39-0029-69C
Coumarin, 7-hydroxy-	EtOH	325(4.1)	39-0029-69C

Compound	Solvent	λ_{max}(log ε)	Ref.
Coumarin, 7-hydroxy- (cont.)	EtOH	325(4.19)	39-0016-69C
$C_9H_6O_5$			
2H,5H-Pyrano[4,3-b]pyran-2,5-dione, 4-hydroxy-7-methyl-	EtOH	271(4.17),330(3.93)	44-2480-69
$C_9H_7BrN_2O$			
3-Pyrazolol, 1-(p-bromophenyl)-	EtOH	289(4.44)	22-1683-69
	EtOH	281(4.32)	103-0527-69
$C_9H_7BrN_2O_2S$			
Acetamide, N-[4-(5-bromo-2-furyl)-2-thiazolyl]-	EtOH	250(4.30),276(4.36)	103-0372-69
$C_9H_7BrO_2$			
Benzofuran, 5-bromo-6-methoxy-	EtOH	244(3.94),252(3.92), 292(3.75),302(3.68)	44-2311-69
C_9H_7BrS			
Benzo[b]thiophene, 2-(bromomethyl)-	EtOH	206(4.40),229(4.40), 273(4.18),300(3.70)	89-0456-69
$C_9H_7Br_3O_2$			
1,3-Cyclopentadiene, 2,3-diacetyl-1,4,5-tribromo-	EtOH	258(4.24),347(4.21), 400(4.00)	39-2464-69C
Propiophenone, 2',3',5'-tribromo-4'-hydroxy-	C_6H_{12}	289(3.34),295(3.34)	22-2079-69
Propiophenone, 2',4',6'-tribromo-3'-hydroxy-	C_6H_{12}	292(3.45),298(3.53)	22-2079-69
	EtOH	299(3.52)	22-2079-69
Propiophenone, 3',4',5'-tribromo-2'-hydroxy-	C_6H_{12}	271(4.08),279(4.10), 351(3.69)	22-2079-69
	EtOH	270(4.05),345(3.65)	22-2079-69
$C_9H_7ClN_2$			
Pyrazole, 3(5)-(p-chlorophenyl)-	3N HCl	259(4.33)	35-0711-69
$C_9H_7ClN_2O$			
Pyrazol-3-ol, 5-(p-chlorophenyl)-	EtOH	254(4.21)	39-0836-69C
$C_9H_7ClN_2OS$			
Acetamide, N-2-benzothiazolyl-2-chloro-	EtOH	207(4.32),277(4.21)	23-4483-69
$C_9H_7ClN_2S_2$			
4H-1,3,4-Thiadiazine-2-thiol, 5-(p-chlorophenyl)-	EtOH	235(4.05),270(3.64), 352(4.30)	95-0689-69
4-Thiazoline-2-thione, 3-amino-4-(p-chlorophenyl)-	EtOH	238(4.22),314(4.21)	95-0689-69
$C_9H_7ClN_4O_4$			
Acrolein, 3-chloro-, 2,4-dinitrophenylhydrazone, trans	$CHCl_3$	370(4.47)	54-0989-69
$C_9H_7ClO_4S_2$			
4-Phenyl-1,3-dithiol-1-ium perchlorate	50% H_2SO_4	241(4.23),265s(3.97), 338(3.51)	94-1924-69
$C_9H_7ClO_5S_2$			
4-(p-Hydroxyphenyl)-1,3-dithiol-1-ium perchlorate	50% H_2SO_4	245(4.08),282(4.12), 376(3.50)	94-1924-69

Compound	Solvent	$\lambda_{max}(\log \epsilon)$	Ref.
$C_9H_7Cl_2N_3O$			
1,2,3-Benzotriazin-4(3H)-one, 6,8-dichloro-3-ethyl-	EtOH	226(4.33),299(3.99)	4-0809-69
$C_9H_7FN_4O_4$			
Benzimidazole, 4-fluoro-1,2-dimethyl-5,6-dinitro-	pH 7.9	238(4.13),330(3.66)	44-0384-69
Benzimidazole, 4-fluoro-1,2-dimethyl-5,7-dinitro-	pH 7.9	280(4.12)	44-0384-69
$C_9H_7FO_2$			
Cinnamic acid, o-fluoro-, trans	EtOH	214(3.92),268(4.12)	44-0534-69
$C_9H_7F_3N_4$			
Pteridine, 6,7-dimethyl-4-(trifluoromethyl)-	hexane	287(3.70),293(3.84), 297(3.88),304(3.97), 316(3.77)	39-1751-69C
	pH 5.0	232(3.70),300(3.95), 306(4.06),318(4.01)	39-1751-69C
$C_9H_7F_4NO$			
Benzaldehyde, 4-(dimethylamino)-2,3,5,6-tetrafluoro-	EtOH	258(3.66),334(4.18)	65-1607-69
C_9H_7N			
Heptafulvene, 8-cyano-	isooctane	237(4.00),253s(3.96), 264s(3.67),326(4.30), 337(4.31),353s(4.01), 412s(2.71),444s(2.67), 478s(2.43),518s(2.30), 562s(1.67)	77-0352-69
C_9H_7NO			
Atroponitrile, β-hydroxy-	EtOH	207s(--),264(4.06)	22-0198-69
sodium salt	EtOH	221(3.77),250(3.72), 286(3.97)	22-0198-69
Carbostyril	EtOH	229(4.54),268(3.86), 329(3.81)	1-0159-69
Indole-3-carboxaldehyde	MeOH	295(4.13)	22-3002-69
anion	MeOH	325(4.31)	22-3002-69
Isoquinoline, N-oxide	H_2O	297(3.96)	78-5761-69
3-Isoquinolinol	100% EtOH	344(3.32),405(3.29)	39-1729-69C
	95% EtOH	347(3.24),405(3.41)	39-1729-69C
	ether	341(3.55)	39-1729-69C
	1:1 EtOH:ether	343(3.45),410(2.72)	39-1729-69C
Oxazole, 2-phenyl-	MeOH	263(4.21)	39-0270-69B
Oxazole, 4-phenyl-	MeOH	243(4.24)	39-0270-69B
Oxazole, 5-phenyl-	MeOH	261(4.30),267(4.29)	39-0270-69B
C_9H_7NOS			
4-Oxazoline-2-thione, 4-phenyl-	2% EtOH	271(4.33)	1-2888-69
conjugate acid	2% EtOH	249(4.16)	1-2888-69
Thieno[2,3-b]pyridine, 2-acetyl-	EtOH	241(4.15),292(4.24)	44-0347-69
	EtOH-HCl	248(4.25),294(4.16)	44-0347-69
Thieno[2,3-b]pyridine, 5-acetyl-	EtOH	247(4.61),275(3.77)	44-0347-69
Thieno[3,2-b]pyridine, 6-acetyl-	EtOH	246(4.37),292(4.04)	44-0347-69
	EtOH-HCl	252(4.40),307(3.78), 325(3.71)	44-0347-69

Compound	Solvent	λmax(log ε)	Ref.
$C_9H_7NO_2$			
2H-Cyclohepta[b]furan-6-ol, 2-imino-	EtOH	232(4.23),285(3.62), 301(3.69),338(3.95), 414(3.90)	89-0673-69
Indene, 1-nitro-, potassium salt	H_2O	240(3.9),250s(3.8), 290(3.8),301(3.7), 359(4.0)	35-0366-69
(also other solvents)	EtOH	378(--)	35-0366-69
	tert-BuOH	376(--)	35-0366-69
	acetone	414(--)	35-0366-69
$C_9H_7NO_3$			
1,2-Benzisoxazole-3-acetic acid	EtOH	238(3.78),244(3.67), 283(3.40),293s(--)	4-0279-69
1,2-Benzoxazolin-3-one, 2-acetyl-	MeOH	240(4.2),300f(3.7)	24-3775-69
Benzoxazole-2-acetic acid	EtOH	234(3.92),265(3.41), 272(3.58),279(3.59)	4-0279-69
Phthalimidine-3-carboxylic acid	H_2O	274(2.87),281(2.80)	95-0418-69
	pH 13	none	95-0418-69
$C_9H_7NO_4$			
1,2-Benz-isoxazoline-2-carboxylic acid, 3-oxo-, methyl ester	MeOH	240(4.1),290f(3.7)	24-3775-69
Benzofuran, 7-methoxy-5-nitro-	MeOH	204(4.28),237(4.40), 295(3.74)	32-1177-69
Carbonic acid, 1,2-benzisoxazol-3-yl methyl ester	MeOH	240(4.0),280f(3.5)	24-3775-69
3(2H)-Isoxazolone, N-acetyl-5-(2-furyl)-	EtOH	299(4.39)	33-0720-69
$C_9H_7NO_6$			
Homophthalic acid, 4-nitro-	pH 10	285(3.94)	35-2467-69
Isophthalaldehydic acid, 2-hydroxy-5-nitro-, methyl ester	EtOH	228(4.21),303(3.98)	22-0817-69
	pH 1	227(--),237(--)	22-0817-69
	pH 13	237(--),385s(--)	22-0817-69
C_9H_7NS			
Thiophene, 2-(2-pyridyl)-	2N HCl	268(3.83),334(4.18)	44-3175-69
	EtOH	262(3.89),303(4.12)	44-3175-69
Thiophene, 2-(3-pyridyl)-	2N HCl	242(3.92),295(4.03)	44-3175-69
	EtOH	262s(3.93),287(4.12)	44-3175-69
Thiophene, 2-(4-pyridyl)-	2N HCl	239(3.72),275(3.73), 335(4.31)	44-3175-69
	EtOH	226(3.72),266(3.92), 295(4.20)	44-3175-69
Thiophene, 3-(2-pyridyl)-	2N HCl	253(3.86),312(4.07)	44-3175-69
	EtOH	228(4.05),255(4.00), 285(3.93)	44-3175-69
Thiophene, 3-(3-pyridyl)-	2N HCl	221(4.17),228s(4.11), 265(4.12),305s(3.66)	44-3175-69
	EtOH	227(4.11),259(4.14)	44-3175-69
Thiophene, 3-(4-pyridyl)-	2N HCl	245(3.58),270s(3.90), 307(4.23)	44-3175-69
	EtOH	228(4.05),268(4.22)	44-3175-69
$C_9H_7NS_2$			
4-Thiazoline-2-thione, 4-phenyl-	2% EtOH	316(4.20)	1-2888-69
conjugate acid	2% EtOH	247(4.07),302(3.84)	1-2888-69
4-Thiazoline-2-thione, 5-phenyl-	2% EtOH	334(4.37)	1-2888-69
conjugate acid	2% EtOH	315(4.17)	1-2888-69

Compound	Solvent	λ_{max}(log ϵ)	Ref.
$C_9H_7N_3$			
as-Triazine, 3-phenyl-	MeOH	254(4.26),370(2.59)	88-3147-69
$C_9H_7N_3O$			
4-Cinnolinecarboxamide	EtOH	230(4.56),296(3.85), 326(3.80),380s(2.79)	95-1167-69
$C_9H_7N_3O_2$			
Pyrazole, 3(5)-(m-nitrophenyl)-	3N HCl	250(4.39)	35-0711-69
$C_9H_7N_3O_2S$			
as-Triazine-3,5(2H,4H)-dione, 6-[2-(2-furyl)vinyl]-3-thio-	EtOH	253(3.79),310(4.18), 365(4.30)	103-0427-69
$C_9H_7N_3O_3$			
5-Isoxazolone, 3-anilino-4-oximino-	EtOH	238(4.39),314(4.03), 375(3.70)	4-0317-69
1,2,4-Oxadiazole, 5-methyl-3-(m-nitrophenyl)-	EtOH	224(4.33)	18-3335-69
1,2,4-Oxadiazole, 5-methyl-3-(p-nitrophenyl)-	EtOH	276(4.02)	18-3335-69
4(3H)-Quinazolinone, 2-methyl-6-nitro-	EtOH	255s(4.02),320(4.01)	18-3198-69
4(3H)-Quinazolinone, 2-methyl-7-nitro-	EtOH	255(4.34),340(3.47)	18-3198-69
as-Triazine-3,5(2H,4H)-dione, 6-[2-(2-furyl)vinyl]-	EtOH	275(4.51),325(4.05)	103-0427-69
$C_9H_7N_3O_3S$			
Thiazole, 2-amino-5-[2-(5-nitro-2-furyl)vinyl]-	EtOH	247(4.12),300(4.09), 405(4.28)	103-0414-69
$C_9H_7N_3O_4$			
5-Hydroxy-3-(3-nitro-p-tolyl)-1,2,3-oxadiazolium hydroxide, inner salt	EtOH	230(4.34),240(4.26), 311(3.74)	88-0579-69
$C_9H_7N_3O_5$			
5-Hydroxy-3-(4-methoxy-3-nitrophenyl)-1,2,3-oxadiazolium hydroxide, inner salt	EtOH	251(4.07),267(4.05), 315(3.94)	88-0579-69
2-Propanone, 1-[3-(5-nitro-2-furyl)-1,2,4-oxadiazol-5-yl]-	EtOH	310(4.11)	18-3008-69
$C_9H_7N_5$			
Pyrido[2,3-d]-s-triazolo[4,3-b]pyridazine, 3-methyl-	EtOH	214s(3.82),218s(3.83), 227s(3.96),234s(4.04), 241(4.05),275s(3.45), 281s(3.43),292s(3.39), 308s(3.24),320s(3.13)	12-1759-69
Pyrido[3,2-d]-s-triazolo[4,3-b]pyridazine, 3-methyl-	EtOH	238(4.08),248s(3.92), 272(3.51),284s(3.38), 318(3.04)	12-1759-69
s-Triazolo[3,4-a]phthalazine, 3-amino-	MeOH	209(4.53),258(4.41), 266(4.43)	44-3221-69
1H-v-Triazolo[4,5-c]quinoline, 1-amino-	EtOH	217(4.36),238(4.58), 307(3.30),322(3.26)	39-1758-69C
3H-v-Triazolo[4,5-b]quinoline, 3-amino-	EtOH	237(4.57),325(4.00), 360(3.62)	39-1758-69C
C_9H_8			
Indene	n.s.g.	248(3.98)	39-0944-69C

Compound	Solvent	λmax(log ε)	Ref.
$C_9H_8BrNO_2$			
1-Indolinecarboxaldehyde, 5-bromo-2-hydroxy-	EtOH	260(4.18),288(3.58),296(3.54)	1-0159-69
Styrene, p-bromo-β-methyl-β-nitro-	n.s.g.	308(4.23)	67-0127-69
$C_9H_8BrNO_4$			
3-Buten-2-one, 1-bromo-3-methyl-4-(5-nitro-2-furyl)-	EtOH	226(3.91),348(4.25)	103-0414-69
$C_9H_8BrNS_2$			
2,3-Dihydrothiazolo[2,3-b]benzothiazolium bromide	MeOH	224(3.96),252(3.75),258(3.86),311(4.19)	4-0163-69
$C_9H_8Br_2O_2$			
Bicyclo[3.3.1]non-3-ene-2,6-dione, 3,7-dibromo-	MeOH	261(3.48)	88-5135-69
Propiophenone, 3',5'-dibromo-2'-hydroxy-	C_6H_{12}	260(3.81),266(3.81),252(3.64)	22-2079-69
	EtOH	257(3.81),344(3.61)	22-2079-69
Propiophenone, 3',5'-dibromo-4'-hydroxy-	C_6H_{12}	265(4.03),287(3.23),295(2.90)	22-2079-69
C_9H_8ClNOS			
2-Benzothiazolinone, 3-(2-chloroethyl)-	MeOH	214(4.64),244(3.74),283(3.45),289(3.45)	4-0163-69
$C_9H_8ClNO_2$			
1-Indolinecarboxaldehyde, 3-chloro-2-hydroxy-	EtOH	251(3.94),288(3.07),295(3.01)	1-0159-69
1-Indolinecarboxaldehyde, 6-chloro-2-hydroxy-	EtOH	250(3.97),288(3.37),295(3.41)	1-0159-69
Styrene, p-chloro-β-methyl-β-nitro-	n.s.g.	310(4.15)	67-0127-69
$C_9H_8ClNS_2$			
2,3-Dihydrothiazolo[2,3-b]benzothiazolium chloride	MeOH	225(3.92),252(3.74),258(3.85),311(4.16)	4-0163-69
$C_9H_8ClN_3O$			
1,2,3-Benzotriazin-4(3H)-one, 6-chloro-3-ethyl-	EtOH	291(3.98)	4-0809-69
1,2,3-Benzotriazin-4(3H)-one, 7-chloro-3-ethyl-	EtOH	238(4.41),290(3.72)	4-0809-69
$C_9H_8ClN_3OS$			
1,2-Benzisothiazole, 3-[(2-chloroethyl)nitrosamino]-	MeOH	229(4.29),319(4.09)	24-1961-69
$C_9H_8ClN_5$			
Benzamidine, p-chloro-N-4H-1,2,4-triazol-4-yl-	MeOH	236(4.00)	48-0477-69
$C_9H_8Cl_4O$			
Bicyclo[5.1.0]oct-4-en-3-one, 1,7,8,8-tetrachloro-5-methyl-, cis	$C_6H_{11}Me$	232(4.11)	44-3645-69
$C_9H_8FNO_2$			
1-Indolinecarboxaldehyde, 5-fluoro-2-hydroxy-	EtOH	250(3.90),287(3.33),294(3.28)	1-0159-69
Styrene, p-fluoro-β-methyl-β-nitro-	n.s.g.	308(4.04)	67-0127-69

Compound	Solvent	$\lambda_{max}(\log \epsilon)$	Ref.
$C_9H_8FN_3O_2$			
Benzimidazole, 4-fluoro-1,2-dimethyl-5-nitro-	EtOH	300(3.95)	44-0384-69
$C_9H_8F_3NO$			
Acetophenone, 2'-(trifluoromethyl)-, oxime	EtOH	263(3.17),270(3.15)	36-0490-69
$C_9H_8F_3N_5O_2$			
Acetamide, N-[1,6-dihydro-1-methyl-6-	pH 3.0	261(4.31)	5-0201-69F
oxo-8-(trifluoromethyl)purin-2-yl]-	pH 8.0	264(4.04)	5-0201-69F
Acetamide, N-[1,6-dihydro-7-methyl-6-	pH 5.0	221(4.28),268(4.15)	5-0201-69F
oxo-8-(trifluoromethyl)purin-2-yl]-	pH 11.0	228(4.33),276(3.96)	5-0201-69F
Acetamide, N-[1,6-dihydro-9-methyl-6-	pH 5.0	267(4.25)	5-0201-69F
oxo-8-(trifluoromethyl)purin-2-yl]-	pH 12.0	277(4.08)	5-0201-69F
$C_9H_8F_5NO_2$			
1,3-Propanediol, 2-amino-1-(pentafluorophenyl)-	EtOH	201(3.74),262(2.59)	65-1615-69
$C_9H_8INO_2$			
Styrene, p-iodo-β-methyl-β-nitro-	n.s.g.	318(4.19)	67-0127-69
$C_9H_8N_2$			
Atroponitrile, β-amino-	EtOH	223(3.77),248(3.70), 287(4.13)	22-0210-69
hydrochloride	EtOH	263(3.98)	22-0210-69
Imidazole, 4-phenyl-	EtOH	258(4.11)	23-1123-69
anion	H_2O	277(4.05)	23-1123-69
cation	H_2O	248(4.25)	23-1123-69
Pyrazole, 3(5)-phenyl-	3N HCl	248(4.15)	35-0711-69
$C_9H_8N_2O$			
1,3,4-Oxadiazole, 2-benzyl-	MeOH	226(4.2)	48-0646-69
2-Pyrazolin-5-one, 3-phenyl-	EtOH	202(4.28),250(4.26)	22-4159-69
3-Pyrazolin-5-one, 3-phenyl-	EtOH	252(4.17)	39-0836-69C
Quinazoline, 4-methoxy-	n.s.g.	225(4.4),262(3.7), 298(3.5),309(3.5)	106-0035-69
$C_9H_8N_2OS$			
2-Benzimidazolinethione, 1-acetyl-	MeOH	315(3.69)	4-0023-69
4H-1,3-Benzothiazin-4-one, 2,3-dihydro-2-imino-3-methyl-	n.s.g.	230(4.5),325(3.3)	106-0100-69
4H-1,3-Benzothiazin-4-one, 2,3-dihydro-2-(methylimino)-	n.s.g.	230(4.4),250s(4.3), 275s(4.0),315(3.5)	106-0100-69
Benzo[b]thiophene-2,3-dione, 3-(methylhydrazone)	MeOH	219(4.51),258(4.01), 282(3.66),290(3.80), 346(3.94),372(3.95)	88-4995-69
1,2,4-Oxadiazole, 3-(methylthio)-5-phenyl-	EtOH	247(3.36)	39-2794-69C
1,2,4-Oxadiazole, 5-(methylthio)-3-phenyl-	EtOH	236(3.12)	39-2794-69C
1,3,4-Oxadiazole, 2-(methylthio)-5-phenyl-	n.s.g.	225(3.70),274(4.24)	2-0583-69
Δ^2-1,3,4-Oxadiazoline-5-thione, 2-benzyl-	n.s.g.	257(2.81)	2-0583-69
Thiazolo[3,2-a]benzimidazol-3-ol, 2,3-dihydro-	EtOH	250(3.96),282(4.02), 291(4.06)	4-0797-69

Compound	Solvent	$\lambda_{max}(\log \epsilon)$	Ref.
$C_9H_8N_2O_2$			
Acetonitrile, (3-nitro-o-tolyl)-	EtOH	208(4.14),253(3.58)	39-1935-69C
1H-1,4-Benzodiazepine-2,5-dione, 3,4-dihydro-	EtOH	215(4.62),272(4.10), 293(3.54)	44-1359-69
Furazan, 3-methoxy-4-phenyl-	EtOH	256(4.05)	39-2794-69C
Imidazole, 1-hydroxy-4-phenyl-, 3-oxide	MeOH	233(4.26)	48-0746-69
Isosydnone, 4-methyl-5-phenyl-	EtOH	223(3.84),296(4.04)	39-1185-69B
Isosydnone, 5-methyl-4-phenyl-	EtOH	217(3.96),262(3.59)	39-1185-69B
1,2,4-Oxadiazole, 3-methoxy-5-phenyl-	EtOH	255(4.19)	39-2794-69C
1,3,4-Oxadiazole, 2-(p-methoxyphenyl)-	MeOH	270(4.5)	48-0646-69
1,4-Phthalazinedione, 2,3-dihydro-	pH 13	313(--)	44-2462-69
2-methyl-, sodium salt	DMSO	368(3.56)	44-2462-69
	90% DMSO	345(--)	44-2462-69
	+KOH	345(--)	44-2462-69
Phthalimide, 3-amino-N-methyl-	EtOH	389(3.72)	104-1231-69
Phthalimide, 4-amino-N-methyl-	EtOH	375(3.68)	104-1231-69
Phthalimidine, 3-carbamoyl-	H_2O	273(3.14),281(3.09)	95-0418-69
	pH 13	none	95-0418-69
2-Pyridineacetic acid, α-cyano-, methyl methyl ester	EtOH	223(4.10),294(4.32), 371(4.03)	95-0203-69
$C_9H_8N_2O_2S$			
Δ^2-1,3,4-Oxadiazoline-5-thione, 4-(hydroxymethyl)-2-phenyl-	n.s.g.	224(4.24),310(4.16)	2-0583-69
$C_9H_8N_2O_2S_2$			
Benzo-1,4-dithiane, 2,3-bis(hydroxy-imino)-6-methyl-	EtOH	225(4.18),245(4.26), 277s(3.79)	39-2319-69C
$C_9H_8N_2O_3$			
o-Veratronitrile, 5-nitroso-	EtOH	264(3.94),320(3.76), 357(3.83)	12-0935-69
	dioxan	740(1.63)	12-0935-69
$C_9H_8N_2O_3S$			
Acetic acid, (furo[2,3-d]pyridazin-7-ylthio)-, methyl ester	H_2O	196(4.21),238(4.18)	22-4004-69
$C_9H_8N_2O_4$			
Styrene, β-methyl-p,β-dinitro-	n.s.g.	301(4.22)	67-0127-69
$C_9H_8N_2S$			
4-Imidazoline-2-thione, 4(5)-phenyl-	2% EtOH	216(4.13),285(4.27)	1-2888-69
conjugate acid	2% EtOH	264(4.24)	1-2888-69
Isothiazole, 4-(p-aminophenyl)-	EtOH	226(4.07),266(4.15), 287(4.09)	4-0841-69
Thione, 2,2'-dipyrryl-	10N HCl	257(3.31),363(4.07), 464(4.65)	12-0239-69
	$CHCl_3$	303(3.47),406(4.56)	12-0239-69
$C_9H_8N_2S_2$			
Cyclopenta[c]thiopyran-4-carbonitrile, 3-amino-1,5,6,7-tetrahydro-1-thioxo-	EtOH	245(4.35),325(4.07), 457(4.33)	78-1441-69
	C_6H_{12}	240(--),299(--), 345s(--),447(--)	78-1441-69
4-Imidazoline-2-thione, 5-mercapto-4-phenyl-	EtOH	277(4.21),345s(--)	23-1123-69
anion	H_2O	280(4.15),315s(4.06)	23-1123-69
cation	H_2O	262(3.97)	23-1123-69

Compound	Solvent	$\lambda_{max}(\log \epsilon)$	Ref.
$C_9H_8N_2Se$			
Indoline, 5-selenocyanato-	EtOH	207(4.20),271(3.94)	11-0159-69B
$C_9H_8N_4$			
as-Triazine, 3-amino-5-phenyl-	EtOH	225(4.18),274(4.06), 340(3.79)	22-3675-69
as-Triazine, 3-amino-6-phenyl-	EtOH	210(4.06),267(4.24), 350(3.63)	22-3675-69
s-Triazolo[3,4-a]phthalazine, 5,6-dihydro-	MeOH	250(4.14),286s(3.15)	44-3221-69
$C_9H_8N_4O$			
4-Cinnolinecarboxamidoxime	EtOH	229(4.56),297(3.46), 332(3.56)	95-1167-69
Glyoxylonitrile, phenyl-, semicarbazone	EtOH	230(4.51),313(4.38)	73-2306-69
as-Triazin-3(2H)-one, 5-amino-6-phenyl-	EtOH	212(4.23),234(4.05), 276(3.71)	73-2306-69
$C_9H_8N_4OS$			
Indole-2,3-dione, 3-(thiosemicarbazone)	EtOH	250(4.0),270(3.9), 360(4.2)	65-0070-69
	EtOH-NaOH	260(4.0),290(4.0), 370(4.2),450(4.1)	65-0070-69
$C_9H_8N_4O_2$			
Indole-2,3-dione, 3-semicarbazone	EtOH	270(4.0),310(4.2), 370(3.4)	65-2345-69
	EtOH-HCl	280(3.7),390(4.1)	65-2345-69
as-Triazine-3,5(2H,4H)-dione, 6-(o-aminophenyl)-	EtOH	260(4.2),330(4.2), 390s(2.0)	65-2345-69
	EtOH-HCl	300(3.8)	65-2345-69
$C_9H_8N_4O_3$			
Pyrimidine, 4-amino-6-methyl-2-(5-nitro-2-furyl)-	EtOH	327(4.12)	56-0519-69
$C_9H_8N_4O_6$			
Barbituric acid, 5,5'-methylenedi-	EtOH	255(4.00)	103-0827-69
$C_9H_8N_4O_8$			
Benzene, 1-(dinitromethyl)-2,4-dimethyl-3,5-dinitro-	anion	365(4.22)	122-0100-69
$C_9H_8N_6$			
s-Triazolo[3,4-a]phthalazine, 3-hydrazino-	MeOH	213(4.43),264(4.26)	44-3221-69
s-Triazolo[3,4-a]phthalazine, 6-hydrazino-	MeOH	223s(4.48),229(4.53), 255(3.85)	44-3221-69
C_9H_8O			
1-Indanone	dioxan	246(4.07),282(3.41), 292(3.49),322(1.56)	59-0393-69
C_9H_8OS			
Benzo[b]thiophene, 3-methoxy-	MeOH	240(4.35),292(3.66), 302(3.75)	73-2959-69
Thiochroman-3-one	isooctane	254(3.84),357(2.18)	44-1566-69
$C_9H_8OS_2$			
1,3-Dithiole, 2-hydroxy-4-phenyl-	EtOH	233(4.06),309(3.92)	94-1931-69

Compound	Solvent	λ_{max}(log ϵ)	Ref.
$C_9H_8O_2$			
Cinnamic acid, anion	80% iso-PrOH	217f(4.3),263(4.3), 298s(3.7)	35-5792-69
$C_9H_8O_2S$			
Benzo[b]thiophen-2(3H)-one, 5-methoxy-	EtOH	216(4.05),235(4.05), 271(3.76),298(2.97)	44-1566-69
$C_9H_8O_2S_2$			
Benzo[b]thiophene-3-sulfinic acid, methyl ester	MeOH	217(4.56),260(3.90), 268(3.88),290(3.76), 299(3.80)	73-2959-69
$C_9H_8O_3$			
Acetophenone, 3',4'-(methylenedioxy)-	hexane	224(4.30),227(4.30), 265s(3.90),270(3.91), 299(3.85)	59-0313-69
$C_9H_8O_3S$			
2(3H)-Furanone, dihydro-3-(2-thiofuroyl)-	EtOH	318(4.14)	78-5703-69
$C_9H_8O_4$			
Benzoic acid, 3-formyl-2-hydroxy-, methyl ester	pH 1	224(--),332(--)	22-0817-69
	EtOH	223(4.03),333(3.85)	22-0817-69
	pH 13	228(--),395(--)	22-0817-69
2H,5H-Pyrano[4,3-b]pyran-2,5-dione, 3,4-dihydro-7-methyl-	EtOH	288(3.78)	4-0917-69
$C_9H_8O_5$			
7-Oxabicyclo[2.2.1]hepta-2,5-diene-2,3-dicarboxylic acid, 1-methyl-	H_2O	229(3.78),293(3.15)	33-0584-69
Phthalic, 3-hydroxy-6-methyl-	H_2O	242s(3.56),308(3.61)	33-0584-69
Phthalide, 4,5,7-trihydroxy-6-methyl-	MeOH	217(4.13),268(3.84), 300(3.33)	88-4675-69
$C_9H_8O_6$			
7-Oxabicyclo[2.2.1]heptadiene-2,3-dicarboxylic acid, 1-methoxy-	isooctane	213(3.58),273s(3.10)	33-0584-69
C_9H_8S			
Benzothiophene, 3-methyl-	MeOH	229(4.40),259(3.66), 290(3.41),298(3.44)	73-2959-69
$C_9H_9BrN_2$			
2-Pyrazoline, 3-bromo-1-phenyl-	EtOH	246(3.95),289(4.27)	22-1683-69
$C_9H_9BrN_2O$			
3-Pyrazolidinone, 1-(p-bromophenyl)-	EtOH	225(4.13)	103-0527-69
$C_9H_9BrO_2$			
1,3-Cyclopentadiene, 2,3-diacetyl-5-bromo-	EtOH	257(4.39),329(4.00), 404(3.91)	39-2464-69C
Propiophenone, 2'-bromo-4'-hydroxy-	C_6H_{12}	257(3.89)	22-2079-69
	EtOH	271(3.96)	22-2079-69
Propiophenone, 3'-bromo-2'-hydroxy-	C_6H_{12}	259(3.95),263(3.93), 338(3.64)	22-2079-69
	EtOH	259(3.95),332(3.61)	22-2079-69
Propiophenone, 3'-bromo-4'-hydroxy-	C_6H_{12}	262(4.15),282(3.43), 293(3.26)	22-2079-69

Compound	Solvent	$\lambda_{max}(\log \epsilon)$	Ref.
Propiophenone, 3'-bromo-4'-hydroxy- (cont.)	EtOH	272(4.12)	22-2079-69
Propiophenone, 4'-bromo-2'-hydroxy-	C_6H_{12}	262(4.20),269(4.19), 325(3.73)	22-2079-69
	EtOH	260(4.19),319(3.70)	22-2079-69
Propiophenone, 5'-bromo-2'-hydroxy-	C_6H_{12}	243(3.87),249(3.88), 256(3.83),344(3.58)	22-2079-69
	EtOH	248(3.88),338(3.54)	22-2079-69
C_9H_9Cl			
Styrene, p-chloro-α-methyl-	pet ether	244(3.99)	88-4933-69
$C_9H_9ClN_2O_3$			
Acetanilide, 4-chloro-3-methoxyisoni- troso-	EtOH	218s(4.25),290s(3.98), 300(4.00)	18-3016-69
2-Propen-1-ol, 3-(p-chloroanilino)- 2-nitro-	dioxan	245(4.77),385(4.38)	78-1617-69
$C_9H_9ClN_2S$			
1,2-Benzisothiazole, 3-[(2-chloroethyl)- amino]-	MeOH	227(4.21),248(3.88), 286(4.23),322(3.74)	24-1961-69
hydrochloride	MeOH-HCl	232(4.24),247s(3.89), 324(3.81),334s(3.74)	24-1961-69
1,2-Benzisothiazoline, 2-(2-chloroeth- yl)-3-imino-	MeOH-HCl	232(4.41),321(3.81), 332(3.80)	24-1961-69
C_9H_9ClOS			
Benzoic acid, m-chloro-thio-, O-ethyl ester	C_6H_{12}	259(3.76),285(4.00), 293s(--),424(--)	18-3556-69
Benzoic acid, p-chloro-thio-, O-ethyl ester	C_6H_{12}	255(3.88),293(4.14), 422(2.15)	18-3556-69
$C_9H_9ClO_3S$			
Acetophenone, 2-chloro-2-(methylsulfon- yl)-	ether	256(4.14)	24-4017-69
$C_9H_9ClO_4S_2$			
4,6-Dimethylthieno[2,3-b]thiapyrylium perchlorate	CH_2Cl_2	231(4.19),264(4.53), 333(3.79),369s(3.57)	88-0239-69
5,7-Dimethylthieno[3,2-b]thiapyrylium perchlorate	CH_2Cl_2	230(4.38),262(4.50), 329(3.88),381(3.75)	88-0239-69
$C_9H_9Cl_6N_3$			
s-Triazine, 2-sec-butyl-4,6-bis(tri- chloromethyl)-	MeOH	222(3.28),276(2.78)	18-2924-69
s-Triazine, 2-tert-butyl-4,6-bis(tri- chloromethyl)-	MeOH	224(3.35),283(2.79)	18-2924-69
s-Triazine, 2-isobutyl-4,6-bis(tri- chloromethyl)-	MeOH	216(3.58),281(2.86)	18-2924-69
$C_9H_9Cl_6N_3S$			
s-Triazine, 2-(butylthio)-4,6-bis(tri- chloromethyl)-	MeOH	221(3.46),275(4.14)	18-2931-69
s-Triazine, 2-(isobutylthio)-4,6-bis- (trichloromethyl)-	MeOH	220(3.47),274(4.15)	18-2931-69
$C_9H_9F_3N_4O$			
4-Pteridinol, 3,4-dihydro-6,7-dimethyl- 4-(trifluoromethyl)-	pH 5.0	227(3.45),272(3.68), 318(3.98)	39-1751-69C

Compound	Solvent	λ_{max}(log ε)	Ref.
$C_9H_9F_3O_3S$			
Sulfone, p-ethoxyphenyl trifluoromethyl	EtOH	252(4.33)	65-1634-69
$C_9H_9IN_2O_4S_2$			
Δ^2-1,2,4-Thiadiazoline, 4-[(iodomethyl)sulfonyl]-3-phenyl-, 1,1-dioxide	n.s.g.	241(4.23),267(3.68)	39-0652-69C
C_9H_9N			
Skatole	EtOH	222(4.35),275(3.70), 281(3.73),290(3.66)	11-0031-69B
C_9H_9NO			
Atropaldehyde, β-amino-	EtOH	277(4.28)	22-0205-69
7-Azabicyclo[4.2.2]deca-2,4,9-trien-8-one	MeCN	266(3.49)	35-4714-69
C_9H_9NOS			
Benzamide, N-acetyl-thio-	EtOH	208(3.91),218s(3.87), 246s(3.84),268(3.96), 298(3.94),476(2.19)	48-0045-69
2H-1,4-Benzoxazine, 3-(methylthio)-	EtOH	238(4.10),284(4.02), 293(4.03),309(3.90)	78-0517-69
2H-1,4-Benzoxazine-3(4H)-thione, 2-methyl-	EtOH	252(4.08),327(4.25)	78-0517-69
2H-1,4-Benzoxazine-3(4H)-thione, 4-methyl-	EtOH	256(4.00),323(4.02)	78-0517-69
$C_9H_9NOS_2$			
Benzothiazole, 2-(2-hydroxyethylthio)-	MeOH	225(4.31),244(3.93), 280(4.11),290(4.06), 301(3.96)	4-0163-69
2-Benzothiazolinethione, 3-(2-hydroxyethyl)-	MeOH	208(4.24),230(4.16), 239(4.12),325(4.42)	4-0163-69
Carbamic acid, dithio-N-benzoyl-, methyl ester	EtOH	240(3.87),270(4.34), 311(4.04)	39-2794-69C
$C_9H_9NO_2$			
Acetamide, N-(7-oxo-1,3,5-cycloheptatrien-1-yl)-	MeOH	218(4.16),262(4.60), 268(4.57),307(3.92)	94-2548-69
Benzene, 1-cyclopropyl-3-nitro-	hexane	260(3.91)	35-3558-69
	EtOH	270(3.84)	35-3558-69
Benzene, 1-cyclopropyl-4-nitro-	C_6H_{12}	283(4.05)	35-3558-69
	EtOH	297(4.03)	35-3558-69
2H-1,4-Benzoxazin-3(4H)-one, 2-methyl-	EtOH	255(3.86),282(3.74)	78-0517-69
2H-1,4-Benzoxazin-3(4H)-one, 4-methyl-	EtOH	256(3.79),283(3.63)	78-0517-69
Benzoylacetaldehyde, monooxime	MeOH	243(3.56)	77-1062-69
Cinnamic acid, p-amino-, trans	EtOH	235(3.96),340(4.43)	44-0534-69
Indan, β-nitro-	C_6H_{12}	273(3.94)	35-3558-69
	EtOH	282(3.90)	35-3558-69
1-Indolinecarboxaldehyde, 2-hydroxy-	EtOH	248(4.13),278(3.61), 287(3.60)	1-0159-69
Malonaldehyde, phenyl-, monooxime	EtOH	240-270(c.2.3)	44-3451-69
1,2-Propanedione, 1-phenyl-, 1-oxime	1% MeOH	235s(3.51)	22-2894-69
anion	1% MeOH	284(4.08)	22-2894-69
5H-2-Pyrindin-5-one, 6,7-dihydro-4-hydroxy-3-methyl-	pH 1	326(3.87)	70-2127-69
	pH 7	382(3.85)	70-2127-69
	pH 10	305(2.98)	70-2127-69
	pH 13	380(3.87)	70-2127-69
	pH 13	380(3.87)	78-3527-69
	dioxan	330(3.73)	70-2127-69

Compound	Solvent	$\lambda_{max}(\log \epsilon)$	Ref.
Styrene, β-methyl-β-nitro-	n.s.g.	307(4.09)	67-0127-69
$C_9H_9NO_2S$			
2-Benzothiazolinone, 3-(2-hydroxyethyl)-	MeOH	215(4.64),246(3.76), 283(3.45),290(3.45)	4-0163-69
$C_9H_9NO_3$			
Acetophenone, 2'-methoxy-5'-nitroso-	MeOH	230(3.89),261(3.94), 334(4.11)	12-0935-69
	dioxan	732(1.67)	12-0935-69
Benzene, (1,2-epoxy-2-nitropropyl)-	MeOH	230(3.65),290(2.65)	77-0369-69
p-Benzoquinone, 2-acetyl-6-methyl-,	EtOH	245(3.91),320(4.18)	12-0935-69
4-oxime	EtOH-NaOH	240(3.94),409(4.49)	12-0935-69
	dioxan	725(1.66)	12-0935-69
Glyoxylic acid, phenyl-, methyl ester, 2-oxime	1% MeOH	245(3.62)	22-2894-69
anion	1% MeOH	269(4.02)	22-2894-69
$C_9H_9NO_3S$			
2-Carboxy-2,3-dihydro-8-hydroxy-5-methylthiazolo[3,2-a]pyridinium hydroxide, inner salt	pH 1	240(3.83),340(4.00)	1-1704-69
	pH 13	245(3.99),360(4.02)	1-1704-69
3-Carboxy-2,3-dihydro-8-hydroxy-5-methylthiazolo[3,2-a]pyridinium hydroxide, inner salt	pH 1	240(3.85),340(4.01)	1-0371-69
	pH 13	245(4.01),360(4.05)	1-0371-69 +1-1704-69
$C_9H_9NO_4$			
Acetic acid, (3-nitro-o-tolyl)-	EtOH	206(4.1),258(3.6)	39-1935-69C
Acetic acid, (6-nitro-o-tolyl)-	EtOH	207(4.13),263(3.63)	39-1935-69C
o-Anisic acid, 5-nitroso-, methyl ester	EtOH	254(3.96),334(4.15)	12-0935-69
	dioxan	730(1.68)	12-0935-69
$C_9H_9NO_4S$			
3-Carboxy-2,3-dihydro-8-hydroxy-5-methylthiazolo[3,2-a]pyridinium hydroxide, inner salt, S-oxide	pH 1	235(3.82),312(3.87)	1-0371-69
	pH 13	230(4.17),250s(3.91), 350(3.83)	1-0371-69
$C_9H_9NO_5$			
Terephthalamic acid, 3,5-dihydroxy-, methyl ester	n.s.g.	224(4.13),260(3.97), 342(3.71)	39-2805-69C
$C_9H_9NO_6S$			
2,4,5-Thiazoletricarboxylic acid, trimethyl ester	MeOH	276(3.91)	32-0029-69
C_9H_9NS			
7-Azabicyclo[4.2.2]deca-2,4,9-triene-8-thione	MeCN	216(3.91),278(4.11)	35-5296-69
Benzothiazoline, 3-methyl-2-methylene- (solvent has 4% tetramethylguanidine)	MeCN	294(4.06)	88-2709-69
Indole, 2-(methylthio)-	EtOH	281(4.08),289(4.10), 297(4.05)	94-0550-69
2-Indolinethione, 1-methyl-	EtOH	232(4.15),295s(4.05), 315(4.22)	94-0550-69
2-Indolinethione, 3-methyl-	EtOH	230(4.15),293s(4.01), 318(4.21)	94-0550-69
$C_9H_9N_3$			
Quinazoline, 4-amino-6-methyl-	EtOH	278(3.65),285(3.80), 306s(3.56),317(3.73), 330(3.61)	39-1282-69C

Compound	Solvent	λ max(log ε)	Ref.
$C_9H_9N_3O$			
Isoxazole, 5-hydrazino-3-phenyl-	MeOH	237(4.29),277(3.98)	4-0783-69
$C_9H_9N_3OS$			
1,2-Benzisothiazole, 3-(ethylnitros-amino)-	MeOH	230(4.25),319(4.04)	24-1961-69
$C_9H_9N_3O_2$			
1-(Carboxyamino)-2-cyanopyridinium hydroxide, inner salt, ethyl ester	MeOH	224(4.04),246(3.89), 360(3.63)	22-2175-69
$C_9H_9N_3O_2S$			
2-Benzimidazolinethione, 1,3-dimethyl-5-nitro-	EtOH	246(4.17),282(4.43), 368(4.10)	65-0941-69
$C_9H_9N_3O_2Se$			
Benzimidazoline, 1,3-dimethyl-5-nitro-2-seleno-	EtOH	222(4.22),237(4.22), 292(4.30),380(4.10)	65-0941-69
$C_9H_9N_3O_3$			
2-Benzimidazolinone, 1,3-dimethyl-5-nitro-	EtOH	230(4.31),254(4.22), 343(4.03)	65-0941-69
2H-1,4-Benzoxazine-2,3(4H)-dione, 4-methyl-, dioxime	EtOH	216(4.44),259(4.04), 312(3.79)	39-2319-C
$C_9H_9N_3O_5$			
Aziridine, 1-(5-nitro-2-furoyl)-, O-acetyl oxime	EtOH	250(4.01),326(4.22)	18-0556-69
$C_9H_9N_3O_6$			
Benzene, 2,4-dimethyl-1-(dinitromethyl)-5-nitro-	anion	367(4.18)	122-0100-69
$C_9H_9N_5$			
Benzamidine, N-4H-1,2,4-triazol-4-yl-	MeOH	229(4.03)	48-0477-69
$C_9H_9N_5O$			
4H-1,2,4-Triazole, 4-(benzylnitros-amino)-	MeOH	241(3.82)	48-0009-69
$C_9H_9N_5O_2$			
4-Pteridinol, 6-acetyl-2-amino-7-methyl-	pH 1	269(3.99),321(4.14), 380s(3.08)	18-2662-69
	pH 13	273(4.25),310(3.85), 366(4.16)	18-2662-69
C_9H_{10}			
1,3,5,7-Cyclononatetraene, cis	hexane	205(3.7+),248s(3.4)	88-4491-69
Indene, 3a,7a-dihydro-, trans	n.s.g.	260(3.54)	77-0833-69
Spiro[2.4]hepta-4,6-diene, 1-vinyl-	hexane	234(3.88),271s(3.24)	24-1789-69
Spiro[2.6]nona-4,6,8-triene	EtOH	262(3.43)	88-3957-69
Spiro[4.4]nona-1,3,7-triene	hexane	254(3.35)	24-1789-69
$C_9H_{10}BrNO_2$			
3-Pyridineacetic acid, 2-bromo-6-methyl-, methyl ester	pH 1	300(3.72)	1-2065-69
	pH 13	225(3.84),285(3.83)	1-2065-69
$C_9H_{10}BrN_3$			
Benzimidazole, 2-amino-5-bromo-1-ethyl-	n.s.g.	293(3.98)	103-0529-69
Benzimidazole, 2-amino-6-bromo-1-ethyl-	n.s.g.	257(3.92),296(3.97)	103-0529-69

Compound	Solvent	$\lambda_{max}(\log \epsilon)$	Ref.
$C_9H_{10}BrN_5O$			
Purine, 8-bromo-6-morpholino-	EtOH	286(4.26)	65-2125-69
$C_9H_{10}ClN_3$			
4-(Inidazol-2-yl)-1-methylpyridinium chloride	MeOH	345(4.33)	87-0944-69
$C_9H_{10}ClN_3O$			
1-Methyl-4-(5-methyl-1,2,4-oxadiazol-3-yl)pyridinium chloride	MeOH	253(4.03)	87-0381-69
$C_9H_{10}ClN_5O_2$			
9H-Purine-9-propionic acid, 6-amino-α-chloro-, methyl ester	EtOH	260(4.18)	78-5983-69
$C_9H_{10}Cl_2O_2S$			
2-Thiophenecarboxylic acid, 4,5-bis-(chloromethyl)-3-methyl-, methyl ester	EtOH	218(4.15),270(4.13)	44-0333-69
$C_9H_{10}Cl_6O_4$			
Acetic acid, trichloro-, 1-ethylpropylidene ester	EtOH	220(2.83)(end abs.)	78-1679-69
$C_9H_{10}IN_3O$			
1-Methyl-4-(5-methyl-1,3,4-oxadiazol-2-yl)pyridinium iodide	MeOH	217(4.25),278(4.28)	87-0944-69
$C_9H_{10}IN_3S$			
1-Methyl-4-(5-methyl-1,3,4-thiadiazol-2-yl)pyridinium iodide	MeOH	218(4.26),287(4.22)	87-0944-69
$C_9H_{10}NO_4$			
Benzoic acid, m-nitro-, ethyl ester,	pH 1	280(4.15),400(2.79)	64-1336-69
half-reduction radical of nitro group	pH 13	285(4.18),390(2.79)	64-1336-69
Benzoic acid, p-nitro-, ethyl ester,	pH 1	310(4.42),460(2.79)	64-1336-69
half-reduction radical of nitro group	base	330(4.32),500(3.26)	64-1336-69
$C_9H_{10}NO_5P$			
Phosphonic acid, [2-(4-formyl-5-hydroxy-6-methyl-3-pyridyl)vinyl]-	pH 1	227(4.18),258(3.92), 313(3.88)	87-0058-69
	pH 7.5	216(4.14),244(4.12), 307(3.80),380(2.95)	87-0058-69
	pH 13	225(4.31),296(3.70), 402(3.55)	87-0058-69
$C_9H_{10}N_2$			
2-Pyrazoline, 1-phenyl-	EtOH	240(3.76),280(4.11)	39-1703-69C
2-Pyrazoline, 3-phenyl-	EtOH	220(3.91),283(3.92)	39-1703-69C
1H-Pyrrolo[3,2-c]pyridine, 2,3-dimethyl-	EtOH	226(4.54),278(3.67), 286s(3.54)	23-2061-69
	EtOH-HCl	232(4.62),290(3.53), 314s(3.17)	23-2061-69
$C_9H_{10}N_2O$			
2-Benzimidazolinone, 1,3-dimethyl-	EtOH	232(3.79),283(3.87)	65-0941-69
3-Buten-2-one, 4-amino-4-(3-pyridyl)-	EtOH	225(3.90),248(3.78), 325(4.15)	35-4749-69
1H-Indazol-5-ol, 4,7-dimethyl-	EtOH	215(4.30),255(3.83), 314(3.74)	103-0254-69

Compound	Solvent	λ max(log ε)	Ref.
3-Pyrazolidinone, 1-phenyl-	N HCl	257(2.78)	65-1835-69
	H_2O	240(3.90)	65-1835-69
1H-Pyrrolo[3,2-c]pyridine, 1-acetyl-2,3-dihydro-	EtOH	257(4.08)	103-0410-69
1H-Pyrrolo[3,2-b]pyridine, 5-ethoxy-	EtOH	302(4.07)	88-1909-69
2(1H)-Quinoxalinone, 3,4-dihydro-3-methyl-	EtOH	228(4.32),265s(3.55),305(3.67)	106-0308-69
$C_9H_{10}N_2O_2$			
Azetidine, N-(p-nitrophenyl)-	DMSO	384(3.29)	39-0544-69B
Carbazic acid, 3-benzylidene-, methyl ester	MeOH	211(4.23),216(4.25),220s(4.11),277(4.34),296s(4.00)	78-0619-69
Carbostyril, 3-amino-3,4-dihydro-1-hydroxy-, hydrochloride	pH 2	257(4.11)	4-0937-69
	pH 7	258(4.15)	4-0937-69
	pH 10	267s(3.77),290(3.86)	4-0937-69
Carbostyril, 3-amino-3,4-dihydro-6-hydroxy-, hydrochloride	pH 2	258(4.66),290s(4.12)	4-0937-69
	pH 7	258(4.68),291s(4.15)	4-0937-69
	pH 10	273(4.68)	4-0937-69
Furo[2,3-d]pyrrolo[1,2-a]pyrimidin-4(2H)-one, 3,6,7,8-tetrahydro-	MeOH	209(4.27),250(3.57),267(3.53)	24-2739-69
5H-2-Pyrindin-5-one, 6,7-dihydro-4-hydroxy-3-methyl-, oxime	pH 1	293(3.92),305s(--),315s(--)	70-2127-69
	pH 7	241(4.21),303(3.75),359(3.91)	70-2127-69
	pH 13	239(4.13),290(3.79),354(3.78)	70-2127-69
$C_9H_{10}N_2O_2S$			
Barbituric acid, 5-cyclopentylidene-2-thio-	EtOH	295(4.08)	103-0830-69
$C_9H_{10}N_2O_3$			
4,5'-Biisoxazole, 5-methoxy-3,3'-dimethyl-	EtOH	257(4.2)	39-0245-69C
1,4-Cyclohexadiene-1-carboxamide, N,N-dimethyl-3,6-dioxo-, 3-oxime	EtOH	311(4.20)	12-0935-69
	EtOH-NaOH	282(3.78),427(4.49)	12-0935-69
	dioxan	725(1.04)	12-0935-69
5(4H)-Isoxazolinone, 3-methyl-4-(2,3-dimethyl-5-isoxazolinylidene)-	EtOH	253(3.8),293(4.1)	39-0245-69C
$C_9H_{10}N_2O_3S$			
3-Carboxy-2,3-dihydro-8-hydroxy-5-methylthiazolo[3,2-c]pyrimidin-4-ium hydroxide, inner salt, methyl ester	pH 1	250(3.51),340(3.73)	1-2437-69
	pH 13	265(3.67),365(3.81)	1-2437-69
$C_9H_{10}N_2O_4$			
Benzene, 2,4-dimethyl-1-(dinitromethyl)-	anion	370(4.21)	122-0161-69
Benzohydroximic acid, p-nitro-, ethyl ester	MeOH	222(4.00),304(3.97)	12-0161-69
	MeOH-NaOH	247(3.97),363(3.93)	12-0161-69
$C_9H_{10}N_2O_4S_2$			
6H-p-Dithiino[2,3-c]pyrrole-6-carbamic acid, 2,3,5,7-tetrahydro-5,7-dioxo-, ethyl ester	MeOH	259(3.89),409(3.56)	33-2236-69
Δ^2-1,2,4-Thiadiazoline, 4-(methylsulfonyl)-3-phenyl-, 1,1-dioxide	n.s.g.	241(4.18),268(3.70)	39-0652-69C
$C_9H_{10}N_2O_5$			
Uracil, 1-(3,5-anhydro-β-D-xylofuranosyl)-	H_2O	261(3.99)	94-0775-69
	pH 13	262(3.89)	

Compound	Solvent	λ_{max}(log ε)	Ref.
$C_9H_{10}N_2O_7$			
4H-Pyrazole-3,5-dicarboxylic acid, 4-oxo-, diethyl ester, 1,2-dioxide	n.s.g.	229(4.16),285(3.74), 368(2.87)	44-0187-69
$C_9H_{10}N_2S$			
Benzimidazole, 1-methyl-2-(methylthio)-	MeOH	252(3.72),258(3.71), 284(4.02),292(4.03)	97-0152-69
	EtOH	252(3.89),259(3.88), 285(4.15),292(4.16)	48-0997-69
2-Benzimidazolinethione, 1,3-dimethyl-	MeOH	244(4.29),298s(4.33), 308(4.48)	97-0152-69
	EtOH	227(4.28),243(4.28), 308(4.46)	48-0997-69
	EtOH	228(4.28),245(4.27), 310(4.46)	65-0941-69
	Bu_2O	232(4.25),243(4.15), 319(4.44)	48-0997-69
1,2-Benzisothiazole, 2-ethyl-3-imino-, hydrochloride	MeOH-HCl	231(4.30),318(3.72), 328(3.70)	24-1961-69
1,2-Benzisothiazole, 3-(ethylamino)-	MeOH	228(4.12),250(3.75), 289(3.40),325(3.81)	24-1961-69
hydrochloride	MeOH-NaOH	233(4.34),248s(3.95), 324(3.86),334s(3.81)	24-1961-69
Benzothiazole, 2-(dimethylamino)-	EtOH	226(4.49),276(4.21), 298s(3.64)	2-0964-69
2-Benzothiazolinimine, N,N'-dimethyl-	EtOH	222(4.55),264(4.01), 304(3.77)	2-0964-69
$C_9H_{10}N_2S_2$			
Benzothiazole, 2-[(2-aminoethyl)thio]-, hydrochloride	MeOH	223(4.31),243(3.94), 276(4.09),289(4.00), 300(3.88)	4-0163-69
2H-Thiopyran-5-carbonitrile, 6-amino-	C_6H_{12}	241(--),332(--), 342s(--),456(--)	78-1441-69
	EtOH	244(4.31),325(4.06), 464(4.23)	78-1441-69
$C_9H_{10}N_2Se$			
2-Benzimidazolinone, 1,3-dimethyl-2-seleno-	EtOH	231(4.18),252(4.12), 318(4.43)	65-0941-69
$C_9H_{10}N_2Te$			
2-Benzimidazolinone, 1,3-dimethyl-2-telluro-	EtOH	233(4.15),257(3.96), 336(4.22),366(3.92)	65-0941-69
$C_9H_{10}N_4O$			
4(1H)-Pteridinone, 1,6,7-trimethyl-	EtOH	235(4.11),330(3.96)	39-2415-69C
4(3H)-Pteridinone, 3,6,7-trimethyl-	EtOH	239(4.10),280s(3.72), 315(3.86)	39-2415-69C
$C_9H_{10}N_4O_2$			
Nicotinic acid, 2,6-diamino-5-cyano-, ethyl ester	MeOH	228(4.58),276(4.25), 328(4.26)	39-0133-69C
$C_9H_{10}N_4O_3$			
Propionic hydrazide, α,β-dioximino-, β-phenyl-, syn	EtOH	248(4.13)	55-0859-69
1H-Pyrazolo[4,3-d]pyrimidine-5,7-diol, 3-(tetrahydro-2-furyl)-	pH 1	208(4.36),288(3.82)	88-0289-69
	0.05N NaOH	222(4.46),302(3.65)	88-0289-69

Compound	Solvent	λmax(log ε)	Ref.
$C_9H_{10}N_4O_4$			
2,4,7(1H,3H,8H)-Pteridinetrione, 8-(2-hydroxyethyl)-6-methyl-	pH -4.14	256(3.78),273(3.83), 354(4.00),363s(3.99)	24-4032-69
	pH 1	282(4.08),327(4.09), 345s(3.87)	24-4032-69
	pH 7	212(4.48),256s(3.67), 289(4.04),346(4.13)	24-4032-69
	pH 14.7	261(4.08),280s(3.70), 362(4.16)	24-4032-69
$C_9H_{10}N_4S$			
Pteridine, 6,7-dimethyl-4-(methylthio)-	pH 5	287s(--),346(3.96)	39-0114-69C
4(1H)-Pteridinethione, 1,6,7-trimethyl-	pH 5	271(4.09),409(4.11)	39-0114-69C
$C_9H_{10}N_6$			
9H-Purine-9-butyronitrile, 6-amino-	EtOH	262(4.15)	44-3240-69
	EtOH-HCl	260(4.16)	44-3240-69
	EtOH-NaOH	260(4.16)	44-3240-69
$C_9H_{10}N_6O_4$			
Semicarbazide, 1-[syn-α,β-dioximino-β-(4-pyridyl)]propionyl-	EtOH	258(4.24),275s(4.18)	55-0859-69
$C_9H_{10}N_6O_{12}$			
Butyronitrile, 2,2'-(methylenedioxy)-bis[4-hydroxy-3,5-dinitro-	Na_2CO_3	349(4.49)	34-0116-69
$C_9H_{10}O$			
Cumene, α,β-epoxy-, (R)-(-)-	n.s.g.	215(4.0),260f(2.2)	88-2717-69
(1S,2R)-(+)-	n.s.g.	212(4.0),255f(2.2)	88-2717-69
4H-Cyclopenta[c]furan, 1,3-dimethyl-	EtOH	218(4.00),242(3.68)	88-1299-69
$C_9H_{10}OS$			
Benzoic acid, thio-, O-ethyl ester	C_6H_{12}	248(3.85),288(4.09), 420(2.09)	18-3556-69
Benzo[c]thiophen-4(5H)-one, 6,7-di-hydro-1-methyl-	EtOH	215(4.07),265(4.06), 300(3.19)	22-0991-69
m-Toluic acid, thio-, O-methyl ester	C_6H_{12}	249(3.78),259(3.78), 288(4.15),417(2.11)	18-3556-69
p-Toluic acid, thio-, O-methyl ester	C_6H_{12}	247(3.74),294(4.13), 414(2.13)	18-3556-69
$C_9H_{10}OS_2$			
p-Anisic acid, dithio-, methyl ester	C_6H_{12}	240(3.93),246(3.95), 329(4.34),502(2.37)	48-0045-69
	EtOH	241s(3.9),246(3.90), 333(4.33),492(2.33)	48-0045-69
$C_9H_{10}O_2$			
Acetic acid, p-tolyl ester	C_6H_{12}	266(2.73)	18-1831-69
Acetophenone, 2'-hydroxy-5'-methyl-	C_6H_{12}	258(3.87),340(3.49)	18-1831-69
Acetophenone, 2'-methoxy-	EtOH	212(4.2),246(3.8), 307(3.5)	28-0730-69A
Acetophenone, 3'-methoxy-	hexane	214(4.39),216(4.38), 245(3.91),301(3.43)	59-0313-69
Acetophenone, 4'-methoxy-	hexane	215(4.02),219(4.02), 264(4.25)	59-0313-69
4-Cyclopentene-1,3-dione, 4-allyl-5-methyl-	EtOH	242(4.10)	88-0373-69
p-Toluic acid, methyl ester	heptane	235(4.20)	44-0077-69

Compound	Solvent	λ_{max}(log ε)	Ref.
p-Toluic acid, methyl ester (cont.)	MeOH	238(4.18)	44-0077-69
$C_9H_{10}O_2S$			
Acetophenone, 2-(methylsulfinyl)-	EtOH	247(4.10)	108-0057-69
m-Anisic acid, thio-, O-methyl ester	C_6H_{12}	287(4.03),320s(--), 418(2.09)	18-3556-69
p-Anisic acid, thio-, O-methyl ester	C_6H_{12}	258(3.73),311(4.34), 412(2.29)	18-3556-69
Benzo[c]thiophene-2-carboxylic acid, tetrahydro-	EtOH	261(4.14)	2-0009-69
$C_9H_{10}O_3$			
Acetophenone, 2'-hydroxy-4'-(hydroxymethyl)-	ether	213(4.17),256(4.14), 324(3.28)	24-0864-69
p-Anisic acid, methyl ester	EtOH	256(4.26),300s(2.0)	65-0660-69
2-Furanacrylic acid, α-methyl-, methyl ester	MeOH	300(4.40)	40-0096-69
$C_9H_{10}O_3S$			
Acetophenone, 2'-(methylsulfonyl)-	EtOH	222(3.8),268(3.10), 274(3.12),300s(1.9)	59-0393-69
$C_9H_{10}O_4$			
Acetophenone, 2',6'-dihydroxy-4'-methoxy-	EtOH	227(4.14),286(4.26), 326(3.61)	95-0372-69
2-Furanacrylic acid, 5-methoxy-, methyl ester	MeOH	335(4.42)	40-0096-69
2-Furanacrylic acid, 5-methoxy-α-methyl-	H_2O	325(4.45)	40-0096-69
4,5,7-Phthalantriol, 6-methyl-, (methanol adduct)	MeOH	273(3.56),280(3.51), 310(3.19)	88-4675-69
Phloroacetophenone, methyl-, ($CO^{14}Me$)	EtOH	223(4.02),292(4.20), 330(3.44)	94-2054-69
$C_9H_{10}O_6$			
4H-Pyran-2-carboxylic acid, 5,6-dihydro-3-hydroxy-4-oxo-, methyl ester, acetate	EtOH	285(3.91)	88-2383-69
$C_9H_{10}O_7$			
3-Cyclohexene-1,2,3-tricarboxylic acid, 6-hydroxy-	n.s.g.	257(2.91)	104-2164-69
$C_9H_{10}S_2$			
Benzoic acid, dithio-, ethyl ester	C_6H_{12}	247s(3.6),299(4.11), 330s(3.9),508(2.07)	48-0045-69
	EtOH	246s(3.5),299(4.16), 333s(3.88),501(2.07)	48-0045-69
	MeCN	244s(3.5),299(4.17), 332s(3.91),497(2.10)	48-0045-69
m-Toluic acid, dithio-, methyl ester	C_6H_{12}	232s(3.8),243s(3.7), 304(4.03),498(1.97)	48-0045-69
	EtOH	231(3.77),243(3.69), 308(4.03),489(1.99)	48-0045-69
o-Toluic acid, dithio-, methyl ester	C_6H_{12}	245(3.63),310(4.02), 494(2.00)	48-0045-69
	EtOH	245(3.60),310(4.03), 481(1.96)	48-0045-69
p-Toluic acid, dithio-, methyl ester	C_6H_{12}	230s(3.8),245s(3.5), 307(4.20),501(2.08)	48-0045-69

Compound	Solvent	$\lambda_{max}(\log \epsilon)$	Ref.
p-Toluic acid, dithio-, methyl ester (cont.)	EtOH	230s(3.8),237s(3.8), 245s(3.6),310(4.24), 492(2.16)	48-0045-69
$C_9H_{10}S_2Se$			
Benzoic acid, p-(methylselenyl)-dithio-, methyl ester	C_6H_{12}	252(3.82),309(4.13), 357s(3.8),506(2.34)	48-0045-69
	EtOH	250(--),309(--), 500(--)	48-0045-69
$C_9H_{10}S_3$			
Benzoic acid, p-(methylthio)-dithio-, methyl ester	C_6H_{12}	220(4.01),248(3.84), 257(3.84),349(4.38), 503(2.39)	48-0045-69
	EtOH	252(3.81),355(4.30), 497(2.46)	48-0045-69
$C_9H_{11}BaN_2O_8P$			
3-Pyridazone, 2-β-D-ribofuranosyl-, 2'(3')phosphate, barium salt	pH 2	286(3.53)	73-0089-69
$C_9H_{11}Br$			
1,3,6-Cyclononatriene, 2-bromo-	C_6H_{12}	250s(3.20)	39-1808-69C
$C_9H_{11}BrN_2O_3$			
Uracil, 1-(tetrahydropyran-2-yl)-5-bromo-	pH 3-4	278(4.09)	103-0283-69
	pH 7-8	276(4.07)	103-0283-69
	pH 11-12	272(3.96)	103-0283-69
$C_9H_{11}BrN_2O_5$			
Uracil, 1-(5-bromo-5-deoxy-β-D-xylofuranosyl)-	H_2O	262(4.00)	94-0775-69
	pH 13	263(3.92)	94-0775-69
Uridine, 5'-bromo-5'-deoxy-	pH 5	207(4.00),261(4.03)	44-1627-69
$C_9H_{11}BrO_2$			
Bicyclo[3.3.1]nonane-2,6-dione, 3-bromo-	MeOH	278(1.91)	88-5135-69
$C_9H_{11}Br_2NO_4S$			
Penicillanic acid, 6,6-dibromo-, methyl ester, 1α-oxide	MeOH	214(3.75)	39-2123-69C
1β-oxide	MeOH	216(3.72)	39-2123-69C
$C_9H_{11}ClN_2O_3$			
Uracil, 1-(tetrahydropyran-2-yl)-5-chloro-	pH 3-4	273(4.01)	103-0283-69
	pH 7-8	273(4.00)	103-0283-69
	pH 11-12	273(3.86)	103-0283-69
$C_9H_{11}ClN_2O_5$			
Uracil, 1-(5-chloro-5-deoxy-β-D-xylofuranosyl)-	H_2O	263(3.98)	94-0775-69
	pH 13	263(3.86)	94-0775-69
Uridine, 3'-chloro-3'-deoxy-	MeOH	258(3.87)	44-1627-69
Uridine, 5'-chloro-5'-deoxy-	pH 5	207(4.00),261(4.01)	44-1627-69
$C_9H_{11}ClN_4$			
1-Methyl-4-(5-methyl-s-triazol-3-yl)-pyridinium chloride	MeOH	283(4.13)	87-0944-69
$C_9H_{11}ClN_4O_2$			
Pyrimidine, 4-chloro-6-(cyclopentylamino)-5-nitro-	pH 1	234(4.31),287(3.70), 348(3.91)	73-2114-69

Compound	Solvent	λ_{max}(log ε)	Ref.
Pyrimidine, 4-chloro-6-(cyclopentyl-amino)-5-nitro- (cont.)	pH 13	224(4.18),343(4.02)	73-2114-69
$C_9H_{11}ClOS$			
Sulfoxide, p-chlorophenethyl methyl	n.s.g.	268(2.40)	39-0481-69B
$C_9H_{11}ClO_4S$			
(Ethylthio)tropylium perchlorate	MeCN	260(4.17),382(4.25)	64-1353-69
$C_9H_{11}FN_2O_3$			
Uracil, 5-fluoro-1-(tetrahydropyran-2-yl)-	pH 3-4	267(3.94)	103-0283-69
	pH 7-8	265(3.78)	103-0283-69
	pH 11-12	268(3.79)	103-0283-69
$C_9H_{11}IN_2O_3$			
Uracil, 5-iodo-1-(tetrahydropyran-2-yl)-	pH 3-4	285(3.94)	103-0283-69
	pH 7-8	285(3.84)	103-0283-69
	pH 11-12	281(3.77)	103-0283-69
$C_9H_{11}IN_2O_5$			
Uracil, 1-(5-deoxy-5-iodo-β-D-xylo-furanosyl)-	H_2O	261(4.03)	94-0775-69
	pH 13	260(3.95)	94-0775-69
$C_9H_{11}NO$			
Acetophenone, 2'-(methylamino)-	n.s.g.	230(4.3),260(3.8), 380(3.7)	24-0342-69
1-Butanone, 1-(4-pyridyl)-	EtOH	222(3.78),272(3.47)	39-2134-69C
Oxazirane, 2-ethyl-3-phenyl-	n.s.g.	231(4.1)	23-0051-69
Tropone, 3-(dimethylamino)-	EtOH	277(4.36),283(4.36), 310(4.04),324(3.96)	39-1499-69C
$(C_9H_{11}NO)_n$			
Pyridine, 5-ethyl-2-vinyl-, 1-oxide, polymer	H_2O	217(3.90),263(3.59)	39-0054-69B
	0.01M silicic acid	216(3.92),260(3.62)	39-0054-69B
$C_9H_{11}NOS$			
Ethanol, 2-[(p-mercaptobenzylidene)-amino]-	EtOH	210(4.0),290(4.1), 350(3.5),450(3.7)	30-0605-69A
	dioxan	210(4.0),290(4.3)	30-0605-69A
$C_9H_{11}NO_2$			
Acetophenone, 2'-methoxy-, oxime	EtOH	232(3.9),280(3.4)	28-0730-69A
Anthranilic acid, N-methyl-, methyl ester	EtOH	252(3.70),353(3.70)	22-3523-69
2-Azaadamantane-4,8-dione	dioxan	316(2.13)	88-5135-69
5-Azaspiro[2.5]oct-7-ene-4,6-dione, 5,8-dimethyl-	EtOH	255(3.86)	39-1678-69C
3H-Azepine, 3-acetyl-2-methoxy-	MeOH	259(3.71)	78-5205-69
1H-Azepine-1-carboxylic acid, 2-methyl-, methyl ester	hexane	212(4.34),302(3.01)	44-2866-69
	EtOH	210(4.38),291(3.21)	44-2866-69
1H-Azepine-1-carboxylic acid, 3-methyl-, methyl ester	hexane	212(4.37),238s(3.48), 321(2.81)	44-2866-69
	EtOH	212(4.38),238s(3.38), 309(2.87)	44-2866-69
1H-Azepine-1-carboxylic acid, 4-methyl-, methyl ester	hexane	211(4.36),239s(3.51), 323(2.83)	44-2866-69
	EtOH	207(4.40),241s(3.67), 309(2.99)	44-2866-69
p-Benzoquinone, 2-isopropyl-, 4-oxime	EtOH	228(3.43),305(4.23)	12-0935-69

Compound	Solvent	λmax(log ε)	Ref.
p-Benzoquinone, 2-isopropyl-, 4-oxime (cont.)	EtOH-NaOH	263(3.70),403(4.49)	12-0935-69
	dioxan	730(-0.25)	12-0935-69
2,4,6-Cycloheptatriene-1-carbamic acid, methyl ester	EtOH	253(3.64)	35-6391-69
Ketone, 3-amino-5,6-dihydro-4H-cyclopenta[b]furan-2-yl methyl	EtOH	215(3.38),274(3.88), 320(4.14)	28-0536-69A
Pyridine, 3,5-diacetyl-1,2-dihydro-	n.s.g.	217(4.08),281(4.21), 386(3.77)	77-1348-69
Pyrrole-2,5-dicarboxaldehyde, 3-ethyl-4-methyl-	MeOH	243(4.03),316(4.31)	63-1291-69
$C_9H_{11}NO_2S$			
Pyrrole-2-carbothioic acid, 4-acetyl-, S-ethyl ester	EtOH	228(4.36),263(3.91), 270(4.14),298(4.27)	78-3879-69
Pyrrole-2-carbothioic acid, 5-acetyl-, S-ethyl ester	EtOH	200(3.84),230s(4.01), 236(4.08),315(4.50)	78-3879-69
2-Thiophenecarboxaldehyde, 3-morpholino-	n.s.g.	285(3.8),370(3.8)	48-0827-69
$C_9H_{11}NO_3$			
Anthranilic acid, x-methoxy-, methyl ester	n.s.g.	222(4.34),255(3.88), 355(3.85)	119-0035S-69
2H-1,2-Benzoxazine-3,5(4H,6H)-dione, 7,8-dihydro-4-methyl-	EtOH	218(3.52),269(4.09)	78-2393-69
Benzylidenimine, 2-hydroxy-5-(hydroxymethyl)-3-methoxy-, copper complex	$CHCl_3$	285(4.41),373(3.84), 680(1.80)	65-0401-69
2-Furaldehyde, 5-morpholino-	n.s.g.	356(4.18)	103-0434-69
2-(3-Hydroxy-6-methylpyridinium)propionic acid, inner salt, L-(+)-	acid	300(3.79)	1-2475-69
	pH 1	220s(3.72),295(3.80)	1-0371-69
	pH 13	246(3.91),325(3.75)	1-0371-69
	NaOH	332(3.75)	1-2475-69
4H-Pyran-4-one, 2,6-dimethyl-, O-acetyloxime	EtOH	214(3.88),266(4.23)	35-4749-69
2-Pyridineacetic acid, 1-acetyl-1,2-dihydro-	MeOH	211(3.72),298(3.81)	39-2509-69C
Pyridoxal, 3-O-methyl-	pH 1	285(3.93)	69-5181-69
	pH 7	275(3.69)	69-5181-69
	pH 13	273(3.65)	69-5181-69
$C_9H_{11}NO_3S$			
6-[(Carboxymethyl)thio]-5-hydroxy-1-methyl-2-picolinium hydroxide, inner salt	pH 1	325(3.97)	1-2065-69
	pH 13	255(3.89),355(3.97)	1-2065-69
2-(3-Hydroxy-6-methylpyridinium)-3-mercaptopropionic acid, inner salt	aq. HCl	302(3.78)	1-2475-69
	aq. NaOH	333(3.75)	1-2475-69
$C_9H_{11}NO_4$			
3,5-Pyridinedicarboxylic acid, 1,2-dihydro-, dimethyl ester	EtOH	213(3.99),281(4.03), 386(3.64)	73-0427-69
3,5-Pyridinedicarboxylic acid, 1,4-dihydro-, dimethyl ester	EtOH	213(4.69),242s(3.61), 374(3.95)	73-0427-69
$C_9H_{11}NO_5$			
Acetamide, N-[2,5-dihydro-5-(hydroxymethyl)-2-oxo-3-furyl]-, acetate	MeOH	243(3.92)	35-7490-69
$C_9H_{11}N_3$			
Benzimidazole, 2-(1-aminoethyl)-, dihydrochloride	EtOH	275(4.10),281(4.08)	94-2381-69

Compound	Solvent	$\lambda_{max}(\log \epsilon)$	Ref.
$C_9H_{11}N_3O$			
2-Imidazolidinone, 1-amino-3-phenyl-	MeOH	246(4.27)	44-0372-69
2H-Imidazo[4,5-b]pyridin-2-one, 1,3-dihydro-3-isopropyl-	EtOH	229(3.54),292(4.09)	4-0735-69
$C_9H_{11}N_3OS$			
Acetophenone, 2'-hydroxy-, thiosemicarbazone	EtOH	231(4.2),289(4.2), 330s(3.9)	28-0730-69A
$C_9H_{11}N_3O_2$			
Acetophenone, 2'-hydroxy-, semicarbazone	EtOH	218(4.3),271(4.2), 280s(4.1),312(3.8)	28-0730-69A
$C_9H_{11}N_3O_2S$			
Acetophenone, 2',6'-dihydroxy-, thiosemicarbazone	EtOH	228s(4.2),272(4.3)	28-0730-69A
$C_9H_{11}N_3O_3$			
Acetophenone, 2',6'-dihydroxy-, semicarbazone	EtOH	221(4.3),252s(3.7), 280(3.4)	28-0730-69A
$C_9H_{11}N_3O_3S$			
3H,6H-[1,3]Thiazino[3,2-b]-as-triazin-3-one, 7,8-dihydro-8-hydroxy-2-methyl-, acetate	EtOH	238(4.44)	114-0093-69C
7H-Thiazolo[3,2-b]-as-triazine-3-acetic acid, 2,3-dihydro-6-methyl-7-oxo-, methyl ester	EtOH	231(4.12)	114-0093-69C
$C_9H_{11}N_3S$			
Malononitrile, [3-(dimethylamino)-3-(methylthio)allylidene]-	C_6H_{12}	381(4.59)	5-0073-69E
Thiazolidine, 3-amino-2-(phenylimino)-	n.s.g.	250(4.00)	44-0372-69
$C_9H_{11}N_5O_2$			
Acetamide, N-(1,6-dihydro-1,8-dimethyl-6-oxopurin-2-yl)-	pH 0.0	254(4.06)	5-0201-69F
	pH 7.0	258(4.02)	5-0201-69F
	pH 11.0	263(4.07)	5-0201-69F
Acetamide, N-(1,6-dihydro-7,8-dimethyl-6-oxopurin-2-yl)-	pH 1.0	260(4.26)	5-0201-69F
	pH 5.0	222(4.41),265(4.22)	5-0201-69F
	pH 12.0	224(4.42),268(4.09)	5-0201-69F
Acetamide, N-(1,6-dihydro-8,9-dimethyl-6-oxopurin-2-yl)-	pH 1.0	262(4.23)	5-0201-69F
	pH 5.0	262(4.22),277s(4.05)	5-0201-69F
	pH 12.0	262(4.11)	5-0201-69F
6-Pteridinemethanol, 2-amino-4-hydroxy-α,7-dimethyl-	pH 1	216(4.34),253(4.02), 322(4.01)	18-2662-69
	pH 13	253(4.38),358(3.95)	18-2662-69
3H-Purine-3-acetic acid, 6-amino-, ethyl ester	pH 1	274(4.37)	4-0955-69
	pH 7	275(4.17)	4-0955-69
	pH 10	273(4.13)	4-0955-69
9H-Purine-9-acetic acid, 6-amino-, ethyl ester	pH 1	257(4.16)	4-0955-69
	pH 7	260(4.17)	4-0955-69
	pH 10	259(4.16)	4-0955-69
$C_9H_{11}N_5O_3$			
Biopterin	pH 1	248(4.06),323(3.93)	39-0928-69C
	pH 13	255(4.39),365(3.89)	39-0928-69C
9H-Purine-9-butyric acid, α-amino-1,6-dihydro-6-oxo-	pH 1	250(4.02)	78-5971-69
	pH 7.0	250(4.04)	78-5971-69
	pH 13	255(4.07)	78-5971-69

Compound	Solvent	λ_{max}(log ϵ)	Ref.
$C_9H_{11}N_5O_4$			
Lentinacin	N HCl	260(4.15)	31-1237-69
	H_2O	262(4.16)	31-1237-69
	N NaOH	262(4.16)	31-1237-69
Lentysine	0.5N HCl	260(4.15)	88-4729-69
	H_2O	261(4.16)	88-4729-69
	0.5N NaOH	261(4.16)	88-4729-69
4(3H)-Pteridinone, 2-amino-6-(D-erythro-1,2,3-trihydroxypropyl)-	pH 1	249(4.06),323(3.90)	39-0928-69C
	pH 13	256(4.38),365(3.88)	39-0928-69C
L-isomer	pH 1	248(4.06),322(3.90)	39-0928-69C
	pH 13	255(4.39),366(3.89)	39-0928-69C
9H-Purine-9-butyric acid, α-amino-1,2,3,6-tetrahydro-2,6-dioxo-	pH 1	234(3.88),262(3.98)	78-5971-69
	pH 7	247(4.00),277(3.98)	78-5971-69
	pH 13	247(3.99),279(4.00)	78-5971-69
$C_9H_{11}N_5O_5$			
Uracil, 1-(5-azido-5-deoxy-β-D-xylofuranosyl)-	H_2O	261(4.05)	94-0798-69
	pH 13	261(3.95)	94-0798-69
$C_9H_{11}N_5O_6$			
m-Phenylenediamine, N-isopropyl-2,4,6-trinitro-	MeOH	297(4.27),335(4.28), 408(4.08)	35-4155-69
C_9H_{12}			
Benzene, propyl-, tetracyanoethylene complex	C_6H_{12}	405(3.32)	39-1161-69B
Bicyclo[4.3.0]nona-1,6-diene	heptane	245(4.11)	88-3347-69
Bicyclo[5.1.0]oct-2-ene, 4-methylene-	isooctane	237(4.23)	88-0999-69
Cumene	C_6H_{12}	261(2.26)	101-0017-69B
Cyclopentene, 2,3-divinyl-	EtOH	234(3.99)	39-1808-69C
Indan, 3a,7a-dihydro-, cis	heptane	260(3.62)	18-2033-69
Indan, tetrahydro-	heptane	254(3.59)	88-3347-69
$C_9H_{12}BaN_5O_5P$			
6-Purinol, 2-amino-9-(4-hydroxybutyl)-, 4'-phosphate, barium salt	pH 1	254(4.03),281(3.81)	94-1268-69
	pH 13	270(4.01)	94-1268-69
$C_9H_{12}BrFN_2O_4$			
Hydrouracil, 5-bromo-5-fluoro-6-methoxy-1-(tetrahydro-2-furyl)-	pH 12	243(3.78)	103-0130-69
$C_9H_{12}BrN_5$			
Adenine, 8-bromo-N,N-diethyl-	EtOH	283(4.26)	65-2125-69
$C_9H_{12}Br_2$			
1,5-Cyclononadiene, 1,9-dibromo-	EtOH	236(3.41)	39-1808-69C
$C_9H_{12}ClN$			
Pyridine, 3-chloro-2,4,5,6-tetramethyl-	n.s.g.	221(3.67),274(3.59)	39-2249-69C
$C_9H_{12}ClNO$			
1,2,3,4-Tetrahydro-3-hydroxyquinolizinium chloride	EtOH	265(3.8),276(3.7)	24-1309-69
$C_9H_{12}Cl_2O_3$			
2-Cyclopentene-1-methanol, 2-allyl-3,5-dichloro-1,4-dihydroxy-	MeOH	245(4.31)	39-2187-69C
$C_9H_{12}NO_5P$			
Phosphonic acid, [2-(4-formyl-5-hydroxy-6-methyl-3-pyridyl)ethyl]-	pH 1	258(3.24),295(3.78), 340(3.30)	87-0058-69

Compound	Solvent	$\lambda_{max}(\log \epsilon)$	Ref.
Phosphonic acid, [2-(4-formyl-5-hydroxy-6-methyl-3-pyridyl)ethyl]- (cont.)	pH 7.5	220(4.00),327(3.39), 380(3.54)	87-0058-69
	pH 13	232(4.07),268(3.44), 301(3.22),391(3.72)	87-0058-69
Phosphonic acid, [2-[5-hydroxy-4-(hydroxymethyl)-6-methyl-3-pyridyl]-vinyl]-	pH 1	226(4.27),303(3.90)	87-0058-69
	pH 7.5	238(4.33),334(3.80)	87-0058-69
phosphate borate buffer	pH 7.5	238(4.33),310(3.86), 340(3.18)	87-0058-69
	pH 13	233(4.32),325(3.81)	87-0058-69
$C_9H_{12}NO_7P$			
Phosphoric acid, monoethyl mono(α-hydroxy-4-nitro-o-tolyl) ester (as ammonium salt)	H_2O	292(3.93)	35-4532-69
$C_9H_{12}N_2$			
Acetone, phenylhydrazone	EtOH	270(4.23)	39-1703-69C
	EtOH-HCl	273(3.48)	39-1703-69C
Acetophenone, methylhydrazone	MeOH	218(3.90),278(3.98)	39-1703-69C
Benzaldehyde, dimethylhydrazone	EtOH	222(3.94),295(4.24)	39-1703-69C
Benzaldehyde, ethylhydrazone	hexane	291(4.02),366(1.55)	104-0257-69
Ethaneazomethane, 1'-phenyl-	hexane	355(1.48)	104-0257-69
Indazoline, 1,2-dimethyl-	N HCl	231(3.79),280(3.19)	104-0547-69
	EtOH	258(3.76),292s(3.34)	104-0547-69
Phthalazine, 1,2,3,4-tetrahydro-2-methyl-, hydrochloride	EtOH	225(4.17),244(3.77), 253(3.77),287(3.84), 313(3.48)	44-2715-69
$C_9H_{12}N_2O$			
Aniline, N-isopropyl-N-nitroso-	C_6H_{12}	209(3.98),226(3.52), 255(3.72)	35-3383-69
5H-2-Pyrindin-4-ol, 5-amino-6,7-dihydro-3-methyl-	pH 1	292(3.87)	70-2127-69
	pH 7	253(3.62),319(3.92)	70-2127-69
	pH 13	246(3.83),301(3.94)	70-2127-69
2,6-Xylidine, N-methyl-N-nitroso-	C_6H_{12}	212(4.04),239s(3.79), 275s(2.52)	35-3383-69
$C_9H_{12}N_2O_2$			
Aniline, N-isopropyl-o-nitro-	1%MeOH-NaOH	450(3.78)	39-0922-69B
Aniline, N-isopropyl-p-nitro-	1%MeOH-NaOH	412(4.19)	39-0922-69B
1H-1,2-Diazepine-1-carboxylic acid, 3-methyl-, ethyl ester	EtOH	220(3.99),325(2.63)	77-0432-69
1H-1,2-Diazepine-1-carboxylic acid, 5-methyl-, ethyl ester	hexane	220(3.87),368(2.43)	77-0432-69
$\Delta^{2,\alpha}$-Pyrrolidineacetic acid, α-cyano-1-methyl-, methyl ester	isooctane	284(4.32)	35-6689-69
3-Pyrrolidine-3-carbonitrile, 1-butyl-4-hydroxy-5-oxo-	H_2O	241(4.02)	70-0694-69
	aq. KOH	234(3.86),289(3.99)	70-0694-69
	MeOH	240(4.07)	70-0694-69
	EtOH	242(4.12)	70-0694-69
	BuOH	242(4.11)	70-0694-69
	dioxan	242(4.13)	70-0694-69
6H-Pyrrolo[1,2-a]pyrimidin-4(6H)-one, 3-ethyl-7,8-dihydro-2-hydroxy-	EtOH	264(3.87)	22-3133-69
$C_9H_{12}N_2O_3$			
Barbituric acid, 5-isopentylidene-	EtOH	253(3.99)	103-0827-69
Phenol, 3-(isopropylamino)-4-nitro-	1%MeOH-HOAc	432(3.85)	39-0922-69B

Compound	Solvent	λ max (log ε)	Ref.
2H-Pyrano[2,3-d]pyrimidine-2,4(3H)-dione, 1,5,6,7-tetrahydro-3,7-dimethyl-	75% dioxan	262(3.98)	95-0266-69
	+NaOH	277(4.06)	95-0266-69
Uracil, 1-(tetrahydro-2H-pyran-2-yl)-	pH 3-4	256(3.83)	103-0283-69
	pH 7-8	256(3.84)	103-0283-69
	pH 11-12	255(3.70)	103-0283-69
$C_9H_{12}N_2O_4$			
2,3-Diazabicyclo[2.2.1]hept-5-ene-2,3-dicarboxylic acid, dimethyl ester	C_6H_{12}	230s(--)	78-0657-69
2,3-Diazabicyclo[3.2.0]hept-2-ene-1,5-dicarboxylic acid, dimethyl ester	ether	322(2.38)	27-0037-69
1H-1,2-Diazepine-3,6-dicarboxylic acid, 4,5-dihydro-, dimethyl ester	MeOH	250(3.61),344(3.95)	27-0037-69
4,5-Pyrazoleninedicarboxylic acid, 3,3-dimethyl-, dimethyl ester	MeOH	244(3.65),358(2.23)	88-0015-69
$C_9H_{12}N_2O_5$			
Uracil, 1-(5-deoxy-β-D-xylofuranosyl)-	H_2O	263(3.87)	94-0775-69
	pH 13	263(3.72)	94-0775-69
$C_9H_{12}N_2O_6$			
4,6(1H,5H)-Pyrimidinedione, 1-β-D-ribofuranosyl-	pH 6.8	256(3.86)	69-1344-69
Uracil, arabinoside	pH 2	263(4.04)	21-1511-69
	pH 7	263(4.03)	21-1511-69
	pH 12	263(3.89)	21-1511-69
Uracil, 3-β-D-ribofuranosyl-	pH 1	262(3.89)	39-0791-69C
	pH 11	292(4.03)	39-0791-69C
$C_9H_{12}N_2O_6S$			
Uracil, 6-(β-D-ribofuranosylthio)-	pH 1	275(4.06)	87-0653-69
	pH 11	226s(4.11),287(4.09)	87-0653-69
$C_9H_{12}N_2O_7$			
Barbituric acid, 1-β-D-ribofuranosyl-	pH 0	214s(3.87),260(2.48)	44-1390-69
	pH 1	258(2.70-)	39-0791-69C
	pH 4	260(4.32)	39-0791-69C
	pH 7	260(4.35)	44-1390-69
	pH 11	260(4.32)	39-0791-69C
	pH 14	265(4.16)	39-0791-69C
	pH 14	265(4.18)	44-1390-69
$C_9H_{12}N_2S$			
4H-Cyclopenta[d]pyrimidine-4-thione, 1,5,6,7-tetrahydro-1,2-dimethyl-	EtOH	239(3.84),339(4.38)	4-0037-69
$C_9H_{12}N_3O_7P$			
ara-Cytidine, 2',5'-cyclic phosphate	pH 2.0	210(4.01),279(4.15)	44-0244-69
Cytidine, O^2,2'-anhydro-, 3'-phosphate	pH 1-7	231(3.98),263(4.03)	35-5409-69
$C_9H_{12}N_3O_8P$			
ara-Cytidine, 3'-phosphate	pH 1	212(3.99),279(4.14)	88-3481-69
	H_2O	211(4.02),275(4.03)	88-3481-69
$C_9H_{12}N_4$			
Indole-3-carbonitrile, 1,2-diamino-4,5,6,7-tetrahydro-	MeOH	253(3.73)	48-0388-69
$C_9H_{12}N_4O_3$			
Semicarbazide, 1,1-dimethyl-4-(p-nitrophenyl)-	MeOH	205(4.18),221(4.07), 328(4.18)	5-0001-69E

Compound	Solvent	$\lambda_{max}(\log \epsilon)$	Ref.
$C_9H_{12}N_4O_7$			
Uridine, 6-amino-5-nitroso-	pH 7	213(4.07),223(4.17)	4-0995-69
$C_9H_{12}N_4S$			
Carbohydrazide, 1-phenethylidene-3-thio-	MeOH	252(4.11)	44-0756-69
9H-Purine-6-thiol, 9-butyl-	pH 1	220(4.05),325(4.33)	73-2114-69
	pH 13	234(4.16),314(4.36)	73-2114-69
$C_9H_{12}N_6$			
Pyrimido[5,4-e]-as-triazine, 5-(diethyl-amino)-	pH 1	233(4.06),373(4.06)	44-2102-69
$C_9H_{12}N_6O_2$			
9H-Purine-9-butyric acid, α,6-diamino-	pH 1	259(4.16)	78-5971-69
	pH 7	262(4.14)	78-5971-69
	pH 13	262(4.14)	78-5971-69
$C_9H_{12}N_6O_3$			
9H-Purine-9-butyric acid, α,2-diamino-1,6-dihydro-6-oxo-	pH 1	254(4.08),277(3.92)	78-5971-69
	pH 13	257(4.00),269(4.03)	78-5971-69
$C_9H_{12}O$			
Cycloheptanone, 2-ethynyl-	isooctane	290(1.81),296s(1.74), 308s(1.57)	77-0674-69
1,3-Cyclohexadiene, 1-acetyl-4-methyl-	EtOH	285(4.03)	12-2037-69
2,3-Cyclononadien-1-one	isooctane	204(4.17),220s(3.89), 287(2.58)	77-0674-69
2,4-Cyclopentadien-1-one, 3-tert-butyl-	isooctane	200(4.71),380(1.9+)	35-6785-69
	MeOH	210(--)	35-6785-69
Dispiro[2.0.2.3]nonan-8-one	C_6H_{12}	279s(1.40),288(1.52), 298(1.56)	88-0979-69
Phenol, m-isopropyl-	pH 2	270(3.27)	30-1113-69A
	pH 13	238(3.92),287(3.47)	30-1113-69A
	MeOH	273(3.33)	30-1113-69A
	MeOH-Na	240(3.95),290(3.49)	30-1113-69A
Phenol, o-propyl-	pH 2	270(3.30)	30-1113-69A
	pH 13	237(3.96),289(3.57)	30-1113-69A
	MeOH	273(3.33)	30-1113-69A
	MeOH-Na	239(3.97),290(3.58)	30-1113-69A
Phenol, p-propyl-	pH 2	276(3.26)	30-1113-69A
	pH 13	237(3.94),294(3.43)	30-1113-69A
	MeOH	279(3.31)	30-1113-69A
	MeOH-Na	240(3.97),297(3.44)	30-1113-69A
Phenol, 3,4,5-trimethyl-	pH 2	277(3.23)	30-1113-69A
	pH 13	238(3.92),293(3.46)	30-1113-69A
	MeOH	280(3.27)	30-1113-69A
	MeOH-Na	241(3.93),297(3.48)	30-1113-69A
Tricyclo[3.3.1^{1,2},0]nonan-3-one	EtOH	280(2.04)	78-5287-69
$C_9H_{12}O_2$			
2-Cyclohexene-1,4-dione, 2,5,6-trimeth-yl-	hexane	358(1.72)	39-0124-69C
4H-Pyran-4-one, 2,6-diethyl-	EtOH	248(4.16)	44-4046-69
$C_9H_{12}O_3$			
3-Cyclohexene-1-carboxylic acid, 2-oxo-, ethyl ester	EtOH	225(3.95)	2-0876-69
4H-Pyran-4-one, 5-acetyl-2,3-dihydro-3,6-dimethyl-	EtOH	232(3.92),269(4.10)	22-1383-69

Compound	Solvent	$\lambda_{max}(\log \epsilon)$	Ref.
$C_9H_{12}O_4$			
1-Cyclopropene-1,2-dicarboxylic acid,	C_6H_{12}	243(3.71)	88-0015-69
3,3-dimethyl-, dimethyl ester	MeOH	320(2.61)	88-2659-69
Hex-2-enoic acid, 2,3,6-trideoxy-3-methyl-, γ-lactone, acetate	n.s.g.	211(4.08)	77-1149-69
Hex-2-enoic acid, 2,3,6-trideoxy-3-methyl-, δ-lactone, acetate	n.s.g.	211(4.12)	77-1149-69
Malonic acid, propylidene-, cyclic	MeOH	226(4.20)	88-4983-69
isopropylidene ester	MeONa	262(4.24)	88-4983-69
2H-Pyran-5-carboxylic acid, 3,4-dihydro-6-methyl-4-oxo-, ethyl ester	EtOH	266(4.08)	22-1383-69
2H-Pyran-5-carboxylic acid, 3,4-dihydro-2,2,6-trimethyl-4-oxo-	EtOH	211(4.02),275(4.03)	22-4091-69
2H-Pyran-2-one, 3-acetyl-5,6-dihydro-	pH 13	250(4.12),266(4.13)	22-4091-69
4-hydroxy-5,6-dimethyl-	EtOH	219(3.86),273(3.91)	22-4091-69
2H-Pyran-2-one, 3-acetyl-5,6-dihydro-	pH 13	249(4.14),268(4.13)	22-4091-69
4-hydroxy-6,6-dimethyl-	EtOH	218(3.75),273(3.91)	22-4091-69
2H-Pyran-2-one, 5,6-dihydro-4-hydroxy-	pH 13	250(4.17),264(4.11)	22-4091-69
6-methyl-3-propionyl-	EtOH	220(3.96),272(4.07)	22-4091-69
$C_9H_{12}O_5$			
2-Heptenedioic acid, 2-methyl-4-oxo-, 7-methyl ester	MeOH	233(3.93)	40-0096-69
2-Heptenedioic acid, 4-oxo-, dimethyl ester	MeOH	219(4.08)	40-0096-69
α-D-glycero-Hex-4-enofuranos-3-ulose, 5-deoxy-1,2-O-isopropylidene-	MeOH	276(3.75)	24-4199-69
Octanoic acid, 3,5,7-trioxo-, methyl ester	EtOH	265(3.98),318s(3.62)	35-0517-69
4H-Pyran-2-carboxylic acid, 5,6-dihydro-4-hydroxy-, methyl ester, acetate	EtOH	240(3.87)	88-2383-69
$C_9H_{12}S_2$			
2H-Thiopyran-2-thione, 3,5-diethyl-	C_6H_{12}	248(4.11),321(3.98), 437(3.83)	78-1441-69
	EtOH	248(4.13),314(3.98), 438(3.89)	78-1441-69
$C_9H_{12}S_3$			
Carbonic acid, trithio-, cyclic ester with 1-mercapto-α-methylenecyclohexanemethanethiol	EtOH	328(4.24)	94-2442-69
$C_9H_{13}BF_4S_3$			
3-[2-(Ethylthio)propenyl]-5-methyl-1,2-dithiol-1-ium tetrafluoroborate	$CHCl_3$	244(3.99),294(3.82), 446(4.48)	24-1580-69
$C_9H_{13}Br$			
1,3-Cyclononadiene, 2-bromo-	EtOH	260s(2.92)	39-1803-69C
$C_9H_{13}BrN_2OS$			
8-Ethoxy-2,3-dihydro-5-methylthiazolo-	pH 1	240(3.64),335(3.83)	1-2437-69
[3,2-c]pyrimidin-4-ium bromide	pH 13	235(3.60),285(3.71)	1-2437-69
$C_9H_{13}BrN_2O_4$			
Orotaldehyde, 5-bromo-, 4-(diethyl	pH 7	280(3.93)	63-0457-69
acetal)	pH 13	298(3.86)	63-0457-69
$C_9H_{13}BrO$			
2-Cyclopenten-1-one, 4-bromo-2-tert-butyl-	isooctane	219(4.32),334(1.55)	35-6785-69

Compound	Solvent	λ_{max}(log ε)	Ref.
2-Cyclopenten-1-one, 4-bromo-2-tert-butyl- (cont.)	MeOH	222(4.03),323(1.59)	35-6785-69
5(4H)-Indanone, 6α-bromo-3a,6,7,7a-tetrahydro-, axial	EtOH	310(1.97)	22-4321-69
conformer	EtOH	293(1.57)	22-4321-69
$C_9H_{13}BrO_2$			
2-Cyclopenten-1-one, 3-bromo-5-hydroxy-4,4,5,5-tetramethyl-	EtOH	267(4.03)	39-1346-69C
	NaOH	303(3.96)	39-1346-69C
$C_9H_{13}BrO_3$			
Cyclohexanecarboxylic acid, 1-bromo-2-oxo-, ethyl ester	EtOH	220(2.95)	2-0876-69
$C_9H_{13}ClN_2O$			
4(3H)-Pyrimidinone, 6-chloro-5-ethyl-2-propyl-	MeOH	232(3.76),282(3.84)	44-2972-69
$C_9H_{13}ClO$			
5(4H)-Indanone, 6α-chloro-3,6,7,7a-tetrahydro-, axial	EtOH	303(1.69)	22-4321-69
conformer	EtOH	285(1.46)	22-4321-69
$C_9H_{13}Cl_2N$			
2H-Pyrrole, 2-(dichloromethyl)-2,3,4,5-tetramethyl-	n.s.g.	284(3.34)	39-2249-69C
$C_9H_{13}IO_2S$			
2-Thiophenecarboxaldehyde, 5-iodo-, diethyl acetal	MeOH	253(4.04)	39-1813-69C
$C_9H_{13}N$			
1H-Azepine, 1,2,7-trimethyl-	isooctane	215(4.19),257(3.22)	44-2866-69
Pyridine, 4-butyl-	EtOH	213(3.33),256(3.37)	39-2134-69C
Pyridine, 3-ethyl-2,6-dimethyl-	MeOH	270(3.69)	106-0196-69
$C_9H_{13}NO$			
Isocarbostyril, 3,4,5,6,7,8-hexahydro-	EtOH	216(4.08)	94-1578-69
3(2H)-Isoquinolone, 1,5,6,7,8,8a-hexahydro-	EtOH	218(4.16)	94-1578-69
Isoxazole, 3-cyclopropyl-5-ethyl-4-methyl-	EtOH	231(2.90)	22-2820-69
Pyridine, 2,5-diethyl-, 1-oxide	H_2O	214(4.38),254(3.96)	39-0054-69B
Pyridine, 2-(ethoxymethyl)-5-methyl-	n.s.g.	214(3.56),266(3.46)	39-2255-69C
5H-Pyrrolo[1,2-a]azepin-9-ol, 6,7,8,9-tetrahydro-	EtOH	220(3.83)	39-1028-69C
$C_9H_{13}NOS_2$			
2-Butanone, 4-[(4,5-dimethyl-2-thiazolyl)thio]-	MeOH	287(3.90)	4-0397-69
2-Butanone, 4-(4,5-dimethyl-2-thioxo-4-thiazolin-3-yl)-	MeOH	325(4.18)	4-0397-69
$C_9H_{13}NO_2$			
2-Butanone, 1-(5-ethyl-3-isoxazolyl)-	EtOH	217(3.82)	35-4749-69
2-Furaldehyde, 5-(diethylamino)-	n.s.g.	365(4.55)	103-0434-69
2(3H)-Furanone, 3-[1-(allylamino)ethylidene]dihydro-	hexane	295(4.18)	104-0998-69
4H-Pyran-4-one, 2,6-diethyl-, oxime	EtOH	265(4.09)	35-4749-69

Compound	Solvent	$\lambda_{max}(\log \epsilon)$	Ref.
$C_9H_{13}NO_2S$			
2H-1,4-Thiazine-6-carboxylic acid, 3,4-dihydro-3-isopropenyl-, methyl ester, (3R)-	EtOH	314(4.05)	77-1368-69
$C_9H_{13}NO_2Si$			
Silane, trimethyl(p-nitrophenyl)-	hexane	260(4.02)	78-4825-69
	MeOH	270(4.03)	78-4825-69
$C_9H_{13}NO_3$			
2,4-Pentadienoic acid, 5-(methylcarbamoyl)-, ethyl ester, cis-cis	EtOH	267(4.39)	44-1695-69
$C_9H_{13}NO_4$			
Maleic acid, (isopropylideneamino)-, dimethyl ester	MeOH	282(3.94)	44-2248-69
Pyrrole, 1-α-D-ribofuranosyl-	H_2O	210(3.81)	18-3539-69
Pyrrole, 1-β-D-ribofuranosyl-	H_2O	212(3.83)	18-3539-69
Pyrrole, 1-β-D-ribopyranosyl-	H_2O	209(3.91)	18-3539-69
$C_9H_{13}NO_4S$			
Δ^2,α-Thiazolidinemalonic acid, 3-methyl-, dimethyl ester	MeOH	249(3.83),295(4.18)	78-4649-69
$C_9H_{13}NO_5S$			
D-Serine, methyl propyl dicarbonate, isothiocyanato derivative	EtOH	256(4.21)	39-1143-69B
L-isomer	EtOH	256(4.21)	39-1143-69B
D-Threonine, methyl ethyl dicarbonate, isothiocyanato derivative	EtOH	256(4.36)	39-1143-69B
L-isomer	EtOH	256(4.23)	39-1143-69B
$C_9H_{13}NS_2$			
Cyclopenta[d][1,3]thiazine-4(2H)-thione, 1,5,6,7-tetrahydro-2,2-dimethyl-	EtOH	336(3.67),406(4.26)	44-0730-69
$C_9H_{13}N_3$			
1-Pyrazoline-4-carbonitrile, 5-isopropylidene-3,3-dimethyl-	n.s.g.	252(4.02),353(2.46)	39-2443-69C
2-Pyrazoline-4-carbonitrile, 5,5-dimethyl-3-isopropenyl-	n.s.g.	267(4.09)	39-2443-69C
$C_9H_{13}N_3O$			
5-Norbornene-2-carboxaldehyde, semicarbazone, endo	EtOH	233(4.34)	44-2346-69
exo	EtOH	234(4.20)	44-2346-69
$C_9H_{13}N_3OS$			
Benzo[b]thiophene-3-carboxylic acid, 2-amino-4,5,6,7-tetrahydro-, hydrazide	MeOH	229(4.51),266s(3.81), 304(3.84)	48-0402-69
$C_9H_{13}N_3O_3$			
as-Triazine-3,5(2H,4H)-dione, 6-methyl-2-(tetrahydro-2H-pyran-2-yl)-	pH 3-4	260(3.84)	103-0283-69
	pH 7-8	260(3.81)	103-0283-69
	pH 11-12	250(3.87)	103-0283-69
$C_9H_{13}N_3O_3S$			
as-Triazine-2(5H)-acetic acid, 6-methyl-3-(methylthio)-5-oxo-, ethyl ester	EtOH	238(4.34)	114-0181-69C

Compound	Solvent	λ_{max}(log ε)	Ref.
$C_9H_{13}N_3O_4$			
1(2H)-Pyrimidinebutyric acid, α-amino-3,4-dihydro-5-methyl-2,4-dioxo-	pH 1	270(3.97)	78-5989-69
	pH 7	270(4.00)	78-5989-69
	pH 13	269(3.93)	78-5989-69
4(3H)-Pyrimidinone, 6-amino-3-(2-deoxy-α-D-ribofuranosyl)-	pH 1	258(3.95)	44-0431-69
	pH 4	258(3.79)	44-0431-69
	pH 11	258(3.79)	44-0431-69
	pH 14	260(3.79)	44-0431-69
4(3H)-Pyrimidinone, 6-amino-3-(2-deoxy-β-D-ribofuranosyl)-	pH 1	258(3.94)	44-0431-69
	pH 4	258(3.79)	44-0431-69
	pH 11	258(3.79)	44-0431-69
	pH 14	260(3.79)	44-0431-69
$C_9H_{13}N_3O_5$			
4(3H)-Pyrimidinone, 6-amino-3-β-D-ribofuranosyl-	pH 1	257(3.86)	44-0431-69
	pH 4	257(3.78)	44-0431-69
	pH 11	257(3.81)	44-0431-69
	pH 14	257(3.80)	44-0431-69
$C_9H_{13}N_3O_5S$			
4(3H)-Pyrimidinone, 6-amino-2-(β-D-ribofuranosylthio)-	pH 1	269(4.17)	39-0791-69C
	pH 4	262(3.86)	39-0791-69C
	pH 11	262(3.86)	39-0791-69C
	pH 14	262(3.86)	39-0791-69C
$C_9H_{13}N_3O_6$			
Uracil, 6-amino-3-β-D-ribofuranosyl-	pH 1	267(4.34)	39-0791-69C
	pH 4	268(4.28)	39-0791-69C
	pH 11	272(4.19)	39-0791-69C
	pH 14	272(4.19)	39-0791-69C
Uridine, 5-amino-	pH 7	272(4.39)	4-0995-69
$C_9H_{13}N_5O_2$			
Guanine, 9-(4-hydroxybutyl)-	pH 1	254(3.99),280(3.81)	94-1268-69
	pH 13	255s(--),270(3.99)	94-1268-69
Pyrimidine, 4-amino-5-nitro-6-piperidino-	EtOH	360(4.06)	65-2125-69
$C_9H_{13}N_5O_6$			
L-erythro-Pentulose, 1-[(2-amino-1,6-dihydro-5-nitro-6-oxo-4-pyrimidinyl)-amino]-1,5-dideoxy-	pH 1	335(4.15)	39-0928-69C
$C_9H_{13}N_5O_7$			
L-erythro-Pentulose, 1-[(2-amino-1,6-dihydro-5-nitro-6-oxo-4-pyrimidinyl)-amino]-1-deoxy-	pH 1	334(4.14)	39-0928-69C
$C_9H_{13}O_7P$			
1,2-Oxaphosphol-3-ine-3-carboxylic acid, 5-acetyl-2,4-dimethoxy-5-methyl-, 2-oxide	MeCN	236(3.97)	35-2293-69
C_9H_{14}			
1,3,5-Heptatriene, 2,4-dimethyl-, trans-trans (all absorbancies in this section are approximate)	isooctane	212(3.85),249s(4.34), 259(4.44),268(4.43), 269s(4.26)	35-3299-69
1,3,5-Heptatriene, 4,6-dimethyl-, cis	isooctane	214(4.06),247(4.16)	35-3299-69
1,4,5-Heptatriene, 2,4-dimethyl-	isooctane	214(3.29),223(2.89), 254(2.11)	35-3299-69

Compound	Solvent	$\lambda_{max}(\log \epsilon)$	Ref.
1,4,5-Heptatriene, 2,6-dimethyl-	isooctane	211(3.33),238(2.98)	35-3299-69
$C_9H_{14}ClNO_5$			
1-tert-Butoxypyridinium perchlorate	n.s.g.	258(3.63)	78-4291-69
$C_9H_{14}INOS$			
6-(Ethylthio)-5-hydroxy-1-methyl-2-picolinium iodide	pH 1	325(3.85)	1-2065-69
	pH 13	255(3.87),355(3.90)	1-2065-69
$C_9H_{14}NO_5P$			
Phosphonic acid, [2-[5-hydroxy-4-(hydroxymethyl)-6-methyl-3-pyridyl]ethyl]- (in phosphate-borate buffer)	pH 1	292(3.97)	87-0058-69
	pH 7.5	254(3.52),324(3.93)	87-0058-69
	pH 7.5	297(3.88)	87-0058-69
	pH 13	242(3.85),305(3.92)	87-0058-69
$C_9H_{14}N_2$			
Phenethylamine, o-amino-β-methyl-, dihydrochloride	MeOH-NaOH	235(3.85),287(3.31)	4-0491-69
$C_9H_{14}N_2OS$			
1,3-Diazaspiro[4.5]decane-2,4-dione, 1-methyl-4-thio-	H_2O	237(3.82),286(4.10)	23-1117-69
	pH 10.80	240(3.97),283(3.67)	23-1117-69
	EtOH	235(3.76),286(4.03), 389(1.48)	23-1117-69
	96% H_2SO_4	240(3.97),283(3.67)	23-1117-69
$C_9H_{14}N_2O_2$			
Benzimidazole, 2-ethyl-4,5,6,7-tetrahydro-1-hydroxy-, 3-oxide	MeOH	232(3.73)	48-0746-69
2-Cyclopenten-1-one, 2-morpholino-, oxime	hexane	264(3.7)	48-0162-69
Pyridoxamine, 3-O-methyl-, dihydrochloride	pH 1	282(3.88)	69-5181-69
	pH 7	275(3.69)	69-5181-69
	pH 13	270(3.65)	69-5181-69
Uracil, 5-ethyl-6-propyl-	pH 2	268(4.08)	56-0499-69
	pH 12	270(3.88)	56-0499-69
	5N NaOH	280(3.89)	56-0499-69
$C_9H_{14}N_2O_2S$			
Ethanol, 2-[(5-ethoxy-2-methyl-4-pyrimidinyl)thio]-	pH 1	240(3.56),315(3.83)	1-2437-69
	pH 13	235(3.64),275(3.56)	1-2437-69
2(1H)-Pyrimidinone, tetrahydro-1-[3-(methylthio)crotonoyl]-, cis	EtOH	307(4.2)	32-1000-69
trans	EtOH	293(4.15)	32-1000-69
Thiazole, 4-tert-butyl-2-ethyl-5-nitro-	EtOH	220(3.68),250(3.52), 314(3.82)	28-1343-69B
$C_9H_{14}N_2O_3$			
Barbituric acid, 5-(1-methylbutyl)-, (S)-(+)-	MeOH	208s(3.93),268(4.08)	44-2676-69
	50% MeOH-HCl	209s(3.96)	44-2676-69
2-Pyrazolin-5-one, 3-acetoxy-4,4-diethyl-	EtOH	219(3.95),275(3.61)	44-3181-69
$C_9H_{14}N_2O_4$			
Barbituric acid, 5-ethyl-5-(1-methoxyethyl)-	pH 6.2	218(3.85)	65-2568-69
	pH 13	240(3.76)	65-2568-69
Barbituric acid, 5-ethyl-5-(2-methoxyethyl)-	pH 6.2	267(3.41)	65-2568-69
	pH 13	241(3.78)	65-2568-69
1-Pyrazoline-3,3-dicarboxylic acid, 4-ethyl-, dimethyl ester	$CHCl_3$	325(2.34)	22-0313-69

Compound	Solvent	λ_{max}(log ϵ)	Ref.
$C_9H_{14}N_2O_5$			
Malonamic acid, 2-(N-acetylglycyl)-, ethyl ester	EtOH	232(3.95),264(4.18)	104-1713-69
$C_9H_{14}N_2O_5S$			
1(6H)-Pyrimidineethanesulfonic acid, 5-ethoxy-2-methyl-6-oxo-	pH 1	255(3.96)	1-2437-69
	pH 13	240(3.77),275(3.81)	1-2437-69
$C_9H_{14}N_2S_2$			
1,3-Diazaspiro[4.5]decane-2,4-dithione, 1-methyl-	H_2O	206(4.08),297(4.39)	23-1117-69
	pH 9.11	316(4.38)	23-1117-69
	EtOH	226(3.90),300(4.46), 404(1.56)	23-1117-69
	71% H_2SO_4	245(4.16),299(4.20)	23-1117-69
$C_9H_{14}N_4O_2$			
Nicotinic acid, 2,6-diamino-5-(aminomethyl)-, ethyl ester	pH 2	267(4.30),333(4.16)	39-0133-69C
Valeraldehyde, (2,6-dihydroxypyrimidinyl)hydrazone	EtOH	288(4.12)	56-0519-69
$C_9H_{14}N_4O_3$			
1H-1,2,3-Triazole-1-carboxylic acid, 4-acetyl-5-(dimethylamino)-	MeOH	218(4.05),248(3.79), 326(3.57)	33-2641-69
$C_9H_{14}N_4O_4$			
Cytosine, 1-(3-amino-3-deoxy-β-D-arabinofuranosyl)-, (sulfate)	H_2O	228s(3.96),272(4.03)	94-1188-69
1H-1,2,3-Triazole-1,4-dicarboxylic acid, 5-(dimethylamino)-, 1-ethyl methyl ester	MeOH	213(4.05),234(3.94), 320(3.64)	33-2641-69
$C_9H_{14}N_4O_5$			
6-Azacytidine, 5-methyl-	pH 6.2	260(3.91)	73-1104-69
$C_9H_{14}N_4S$			
4-Pyrimidinethiol, 5-amino-6-(cyclopentylamino)-	pH 1	251(4.21),313(4.22)	73-2114-69
	pH 13	232(4.28),244(4.30), 311(4.32)	73-2114-69
$C_9H_{14}N_6$			
1H-Imidazo[4,5-d]pyridazine, 4-[[2-(dimethylamino)ethyl]amino]-, dihydrochloride	EtOH	213(4.36),258(3.93)	4-0093-69
$C_9H_{14}N_6O$			
Ethanol, 2-[[6-(dimethylamino)purin-8-yl]amino]-	pH 1	306(4.29)	65-2125-69
	EtOH	292(4.29)	65-2125-69
$C_9H_{14}N_6O_{14}$			
Butyramide, 2,2'-(methylenedioxy)bis-[4-hydroxy-3,3-dinitro-	aq. NaOH	356(4.48)	34-0116-69
$C_9H_{14}O$			
Bicyclo[6.1.0]nonan-2-one, cis	hexane	293(1.56)	88-0059-69
	MeCN	199(3.46),289(1.61)	88-0059-69
Bicyclo[6.1.0]nonan-2-one, trans	hexane	286(1.71)	88-0059-69
	MeCN	192(3.69),282(1.83)	88-0059-69
Bicyclo[6.1.0]nonan-3-one, cis	hexane	298(1.53)	88-0059-69
	MeCN	294(1.63)	88-0059-69

Compound	Solvent	λ max(log ε)	Ref.
Bicyclo[6.1.0]nonan-3-one, trans	hexane	296(1.45)	88-0059-69
	MeCN	293(1.58)	88-0059-69
Bicyclo[6.1.0]nonan-4-one, cis	hexane	288(1.15)	88-0059-69
	MeCN	286(1.15)	88-0059-69
Bicyclo[6.1.0]nonan-4-one, trans	hexane	289(1.15)	88-0059-69
	MeCN	289(1.15)	88-0059-69
Cyclohexene, 1-acetyl-2-methyl-	MeOH	248(3.81)	78-0223-69
2-Cyclononen-1-one	EtOH	231(3.91),318(1.93)	24-3877-69
Cyclopentene, 1-acetyl-2-ethyl-	MeOH	253(3.97)	39-0608-69C
	EtOH	256(3.72)	22-2415-69
2-Cyclopenten-1-one, 2,3,4,5-tetramethyl-, cis	MeOH	240(4.08)	35-6404-69
trans	MeOH	240(4.08)	35-6404-69
2-Cyclopenten-1-one, 4,4,5,5-tetramethyl-	EtOH	221(3.90)	39-0449-69B
5(4H)-Indanone, 3a,6,7,7a-tetrahydro-, trans	EtOH	284(1.32)	22-4321-69
4H-Pyran, 2,4,4,6-tetramethyl-	MeOH	225(3.08)	44-3169-69
$C_9H_{14}O_2$			
5-Nonenal, 8-oxo-	EtOH	224(3.94),284(2.82)	104-1722-69
4H-Pyran-4-one, 2,3-dihydro-2,2,5,6-tetramethyl-	EtOH	280(4.09)	22-4091-69
4H-Pyran-4-one, 2,3-dihydro-2,3,5,6-tetramethyl-	EtOH	275(4.07)	22-4091-69
$C_9H_{14}O_3$			
Cyclobutanepropionic acid, β-oxo-, ethyl ester	isooctane	248(3.85)	22-3694-69
	EtOH	250(3.27)	22-3694-69
2-Cyclononen-1-one, 2,3-dihydroxy-	DMSO-acid	264(4.35)	24-0388-69
	DMSO-base	292(4.33)	24-0388-69
Cyclopropanepropionic acid, α-methyl-β-oxo-, ethyl ester	isooctane	262(2.63)	22-3694-69
	EtOH	265(2.48)	22-3694-69
4-Hexenoic acid, 4-methyl-3-oxo-, ethyl ester	EtOH	232(4.09),276(3.45)	22-4091-69
	base	238(4.06),283(3.91)	22-4091-69
4-Hexenoic acid, 5-methyl-3-oxo-, ethyl ester	EtOH	240(4.05),288(3.51)	22-4091-69
	base	244(3.94),287(4.09)	22-4091-69
$C_9H_{14}O_4$			
1-Cyclobutene-1-carboxylic acid, 2,4-diethyl-3,4-dihydroxy-	EtOH	214(3.89)	44-4046-69
1,1-Cyclopropanedicarboxylic acid, 2,2-dimethyl-, dimethyl ester	C_6H_{12}	214(2.68)	22-0313-69
1,1-Cyclopropanedicarboxylic acid, 2-ethyl-, dimethyl ester	C_6H_{12}	214(2.87)	22-0313-69
2,4-Heptanedione, 6-acetoxy-	EtOH	276(4.06)	22-4091-69
2,4-Hexanedione, 6-acetoxy-5-methyl-	EtOH	275(4.03)	22-1383-69
	EtOH	275(4.04)	22-4091-69
Malonic acid, butylidene-, dimethyl ester	EtOH	212(3.85)	22-0313-69
Malonic acid, sec-butylidene-, dimethyl ester	EtOH	219(3.91)	22-0313-69
$C_9H_{14}S$			
Thiophene, 4-tert-butyl-2-methyl-	EtOH	235(3.80)	22-0991-69
Thiophene, 5-tert-butyl-2-methyl-	EtOH	237(3.91)	22-0991-69
$C_9H_{14}Si$			
Silane, trimethylphenyl-	C_6H_{12}	266(2.35)	101-0017-69B

Compound	Solvent	$\lambda_{max}(\log \epsilon)$	Ref.
$C_9H_{15}BrN_2O$			
[5-(Dimethylamino)furfurylidene]dimethylammonium bromide	n.s.g.	396(4.70)	103-0434-69
$C_9H_{15}ClN_2O$			
[5-(Dimethylamino)furfurylidene]dimethylammonium chloride	n.s.g.	396(4.74)	103-0434-69
$C_9H_{15}IN_2O$			
[5-(Dimethylamino)furfurylidene]dimethylammonium iodide	n.s.g.	398(4.73)	103-0434-69
$C_9H_{15}NO$			
2H-Azepin-2-one, 1,5,6,7-tetrahydro-4,6,6-trimethyl-	H_2O	220(4.11)	23-2781-69
$C_9H_{15}NO_2$			
2(3H)-Furanone, dihydro-3-[1-(isopropylamino)ethylidene]-	hexane	296(4.06)	104-0998-69
2(3H)-Furanone, dihydro-3-[1-(propylamino)ethylidene]-	hexane	296(4.19)	104-0998-69
2-Nopinol, nitrite, cis-(-)-	C_6H_{12}	327(1.45),337(1.58), 347(1.73),358(1.84), 371(1.83)	44-3392-69
Piperidin-2-ylideneacetic acid, ethyl ester	EtOH	292(4.27)	94-2306-69
$C_9H_{15}NO_2S$			
Valeric acid, 2-isothiocyanato-4-methyl-, ethyl ester, D-	hexane	251(4.28)	39-1143-69B
L-	hexane	251(4.28)	39-1143-69B
$C_9H_{15}NO_3$			
Cyclohexane, (1,2-epoxy-2-nitropropyl)-	MeOH	none	77-0369-69
Nipecotic acid, 1-ethyl-4-oxo-, methyl ester	MeOH	252(3.86)	103-0062-69
	EtOH	252(3.86)	103-0062-69
	isobutanol	252(3.89)	103-0062-69
hydrochloride	H_2O	245(3.59)	103-0062-69
	MeOH	245(3.87)(changing)	103-0062-69
	EtOH	245(3.91)(changing)	103-0062-69
	isobutanol	245(3.96)(changing)	103-0062-69
$C_9H_{15}NO_3S$			
Acetic acid, [[2-(ethylcarbamoyl)vinyl]thio]-, ethyl ester, cis	90% EtOH	276(4.17)	12-0765-69
$C_9H_{15}NO_4$			
Malonic acid, [1-(dimethylamino)ethylidene]-, dimethyl ester	isooctane	236(3.72),297(4.14)	35-6683-69
$C_9H_{15}NO_4S$			
Malonic acid, [(dimethylamino)(methylthio)methylene]-, dimethyl ester	MeOH	247(4.22),338(3.85)	78-4649-69
$C_9H_{15}N_2O_5P$			
N-Methylpyridoxamine 5'-phosphate	pH 1	295(3.91)	69-5181-69
	pH 7	252(3.56),331(3.96)	69-5181-69
	pH 13	252(3.64),327(3.97)	69-5181-69

Compound	Solvent	λ_{max}(log ε)	Ref.
$C_9H_{15}N_2O_7P$			
Phosphoric acid, monoethyl mono(α-hydroxy-4-nitro-o-tolyl) ester, ammonium salt	H_2O	292(3.93)	35-4532-69
$C_9H_{15}N_2O_9PS$			
4-Thiouridine, 2'(3')-phosphate	pH 1,5,6	245(3.60),331(4.31)	94-0181-69
	pH 13	315(4.26)	94-0181-69
$C_9H_{15}N_3O_7$			
as-Triazine-3,5(2H,4H)-dione, 2-methyl-6-(D-altro-1,2,3,4,5-pentahydroxypentyl)-	H_2O	263(3.77)	73-1673-69
6-D-gluco-	H_2O	263(3.77)	73-1673-69
$C_9H_{15}N_5O_2$			
6-Pteridinemethanol, 2-amino-5,6,7,8-tetrahydro-4-hydroxy- ,7-dimethyl-	pH 1	218s(4.08),266(4.26)	18-2662-69
	pH 13	255s(3.80),288(3.93)	18-2662-69
C_9H_{16}			
Bicyclo[6.1.0]nonane, iodine complex	hexane	257(4.18)	35-2279-69
1,6-Heptadiene, 4,4-dimethyl-	gas	180(--)	121-0803-69
	heptane	183(4.34)	121-0803-69
	MeCN	183(4.30)	121-0803-69
$C_9H_{16}ClN$			
Quinoline, N-chlorodecahydro-, (8aS)-trans	dioxan	274(2.63)	78-0737-69
$C_9H_{16}ClNO_5$			
N-(2-Formylvinyl)-N-methylpiperidinium perchlorate	H_2O	205(3.89)	24-2609-69
$C_9H_{16}ClN_5$			
s-Triazine, 2-chloro-4,6-bis(isopropylamino)-	pH 0.5	220(4.37),276(3.72)	59-1167-69
	pH 10.0	221(4.58),262(3.56)	59-1167-69
$C_9H_{16}NO_8P$			
3-O-Methylpyridoxal 5'-phosphate	pH 1	279(3.85)	69-5181-69
	pH 7	313(3.43)	69-5181-69
	pH 13	313(3.28)	69-5181-69
$C_9H_{16}N_2$			
Indazoline, 4,5,6,7-tetrahydro-1,2-dimethyl-	EtOH	237(3.67)	22-3316-69
$C_9H_{16}N_2O$			
2-Pyrazoline-1-carboxaldehyde, 4-ethyl-5-propyl-	MeOH	249(3.90)	104-0736-69
2-Pyrazoline-1-carboxaldehyde, 5-isopropyl-4,4-dimethyl-	MeOH	242(4.05)	104-0736-69
Pyrazoline, 1-propionyl-3,5,5-trimethyl-	MeOH	239(4.1)	123-0041-69A
Quinoline, decahydro-N-nitroso-, (8aS)-trans	dioxan	235(3.89),364(1.89)	78-0737-69
$C_9H_{16}N_2O_2$			
1-Pyrazoline-3-carboxylic acid, 3,4,5,5-tetramethyl-, methyl ester, cis	hexane	328(2.23)	39-0787-69C
	EtOH	327(2.18)	39-0787-69C
trans	hexane	328(2.24)	39-0787-69C
	EtOH	327(2.17)	39-0787-69C

Compound	Solvent	$\lambda_{max}(\log \epsilon)$	Ref.
1-Pyrazoline-4-carboxylic acid, 3,3,5,5-tetramethyl-, methyl ester	hexane	327(2.48)	39-0787-69C
$C_9H_{16}N_4$			
Pyrimidine, 2,4-diamino-5-isobutyl-6-methyl-	pH 2	276(3.86)	44-0821-69
	pH 12	230(3.98),287(3.89)	44-0821-69
$C_9H_{16}N_4O$			
4-Imidazolecarboxamide, 1-isobutyl-5-amino-2-methyl-	pH 1	267(4.03)	39-2198-69C
	H_2O	271(4.07)	39-2198-69C
	pH 13	271(4.07)	39-2198-69C
$C_9H_{16}O$			
Cyclononanone	C_6H_{12}	293(1.15)	77-1246-69
	MeOH	285(1.23)	77-1246-69
Cyclooctanone, 2-methylene-	MeOH	230(4.0)	83-0387-69
$C_9H_{16}O_3S$			
2-Thiophenecarboxylic acid, tetrahydro-3-hydroxy-3,4-dimethyl-, ethyl ester	EtOH	260(3.57)	2-0009-69
$C_9H_{17}NO$			
2(1H)-Azecinone, octahydro-	C_6H_{12}	190(3.77)	46-1642-69
	H_2O	193(3.85)	46-1642-69
Nitrone, N-(1-ethylcyclohexyl)-	EtOH	241(3.81)	39-1073-69C
$C_9H_{17}NOS$			
Acrylamide, 3-(butylthio)-N-ethyl-, cis	90% EtOH	282(4.18)	12-0765-69
$C_9H_{17}N_2O_6P$			
3-O-Methylpyridoxamine 5'-phosphate	pH 1	282(3.90)	69-5181-69
	pH 7	275(3.73)	69-5181-69
	pH 13	270(3.69)	69-5181-69
$C_9H_{17}N_3O_3$			
Imidazole-2,5-dione, 1-methyl-4-(methylamino)-, 2-(diethyl acetal)	EtOH	218(4.45),252(3.79)	78-0557-69
$C_9H_{17}N_5O$			
s-Triazine, 2-(ethylamino)-4-(isopropylamino)-6-methoxy-	pH 1.0	217s(4.35),235s(4.15)	59-1167-69
	pH 10.0	217(4.56)	59-1167-69
s-Triazin-2-ol, 4,6-bis(isopropylamino)-	pH 1.0	205s(4.30),242(4.42)	59-1167-69
	pH 8.0	218(4.38),236s(4.33)	59-1167-69
	pH 11.5	211s(4.53)	59-1167-69
$C_9H_{18}ClNO_5$			
(2-Formylvinyl)triethylammonium perchlorate	H_2O	204(3.87)	24-2609-69
$C_9H_{18}N_2$			
1,3-Butadiene-1,1-diamine, N,N,N',N',3-pentamethyl-	C_6H_{12}	235(3.66),290(4.13)	39-0016-69C
$C_9H_{18}N_2O$			
Pyrazolidine, 2-acetyl-5-ethyl-1,4-dimethyl-	MeOH	255(3.71)	104-0736-69
Pyrazolidine, 2-acetyl-1,3,5,5-tetramethyl-	MeOH	255(4.00)	104-0736-69

Compound	Solvent	$\lambda_{max}(\log \epsilon)$	Ref.
$C_9H_{19}ClN_2O_4$			
3-Ethyl-1,2,5,5-tetramethyl-2-pyrazolinium perchlorate	EtOH	277(3.40)	22-3316-69
$C_9H_{19}NO$			
Nitrone, N-(1,1,3,3-tetramethylbutyl)-	EtOH	241(3.83)	39-1073-69C
4-Piperidinemethanol, α-propyl-	EtOH	222(2.32),261(2.05)	39-2134-69C
$C_9H_{19}NO_2$			
Pivalamide, N,N-diethyl-	C_6H_{12}	201(3.96)	77-0462-69
$C_9H_{19}NS_3$			
Peroxycarbamic acid, diethyltrithio-, butyl ester	EtOH	281(3.91)	34-0119-69
Peroxycarbamic acid, diethyltrithio-, tert-butyl ester	EtOH	243(3.88),284(3.88)	34-0119-69
Peroxycarbamic acid, dipropyltrithio-, ethyl ester	EtOH	282(3.91)	34-0119-69
$C_9H_{19}N_3S$			
Octanal, thiosemicarbazone	EtOH	230(3.98),271(4.37)	94-0844-69
$C_9H_{20}Sn$			
Stannane, allyltriethyl-	MeCN	203(4.11)	44-3715-69
$C_9H_{24}O_2Si_3$			
Trisilane, 1-acetoxyheptamethyl-	isooctane	216(3.86)	35-6613-69
Trisilane, 2-acetoxyheptamethyl-	isooctane	212(3.90),233(3.48)	35-6613-69
	dioxan	212(3.91),232(3.47)	35-6613-69
$C_9H_{25}NOSi_3$			
Trisilane, 1-acetamidoheptamethyl-	isooctane	216(3.95)	35-6613-69
	dioxan	216(4.00)	35-6613-69
Trisilane, 2-acetamidoheptamethyl-	isooctane	200(4.09)(end abs.)	35-6613-69
	dioxan	208(3.97)(end abs.)	35-6613-69
$C_9H_{26}OSi_3$			
Trisilane, 1-ethoxyheptamethyl-	isooctane	220(3.84)	35-6613-69
	EtOH	219(3.85)	35-6613-69
Trisilane, 2-ethoxyheptamethyl-	isooctane	225(3.79)	35-6613-69
	EtOH	224(3.76)	35-6613-69
$C_9H_{27}FSi_4$			
Trisilane, 2-fluoro-1,1,1,3,3,3-hexamethyl-2-(trimethylsilyl)-	isooctane	250(3.30),260s(3.30)	35-6613-69
	MeCN	250(3.29),260s(3.26)	35-6613-69
$C_9H_{28}SSi_4$			
Trisilane, 1,1,1,3,3,3-hexamethyl-2-[(trimethylsilyl)thio]-	isooctane	228s(3.41),250s(2.79)	35-6613-69
$C_9H_{28}Si_4$			
1H-Tetrasilane, nonamethyl-	hexane	236(4.10)	77-0004-69

Compound	Solvent	$\lambda_{max}(\log \epsilon)$	Ref.
$C_{10}F_8N_2OS$			
Pyridine, 4,4'-sulfinylbis[2,3,5,6-tetrafluoro-	hexane	215(3.94),255(3.54), 290(3.86)	39-1660-69C
$C_{10}F_8N_2O_2S$			
Pyridine, 4,4'-sulfonylbis[2,3,5,6-tetrafluoro-	hexane	213(4.02),298(3.92)	39-1660-69C
$C_{10}F_8N_2S$			
Pyridine, 4,4'-thiobis[2,3,5,6-tetrafluoro-	hexane	210(3.79),245(3.87), 273(4.13)	39-1660-69C
$C_{10}F_8N_2S_2$			
Pyridine, 4,4'-dithiobis[2,3,5,6-tetra-	hexane	206(4.07),228(4.14), 293(3.87)	39-1660-69C
$C_{10}H_2F_4O_3$			
4H-1-Benzopyran-3-carboxaldehyde, 5,6,7,8-tetrafluoro-4-oxo-	EtOH	203(4.17),250(3.85), 298(3.93)	70-0812-69
$C_{10}H_4Br_2N_2$			
Heptafulvalene, 2,4-dibromo-8,8-dicyano-	MeOH	257(4.21),281(4.19), 394(4.35),414s(4.22)	25-0075-69
$C_{10}H_4Cl_6N_4O_4$			
Benzeneazo-1'-[1]butene, 2',3',3',4',-4',4'-hexachloro-2,4-dinitro-	heptane	330(4.5),450(2.7)	24-3135-69
$C_{10}H_4Cl_8N_2$			
Benzeneazo-1'-[1]butene, 2,2',3',3',-4,4',4',4'-octachloro-	heptane	250(3.8),350(4.4), 470(2.6)	24-3135-69
$C_{10}H_4I_2N_2$			
Heptafulvalene, 2,4-diiodo-8,8-dicyano-	MeOH	289(4.11),303(4.09), 404(4.38),422s(4.34)	25-0075-69
$C_{10}H_4N_6O_6$			
Alloxazine, 7,9-dinitro-	EtOH	240(4.3),280(4.3), 380(4.1)	65-1630-69
$C_{10}H_4O_2S_2$			
Benzo[1,2-b:4,3-b']dithiophene-4,5-dione	EtOH	257(4.18),363(3.95)	54-1244-69
Benzo[1,2-c:3,4-c']dithiophene-4,5-dione	EtOH	226(4.21),254s(4.15), 264(4.20),272(4.20), 319(3.58)	54-1244-69
Benzo[2,1-b:3,4-b']dithiophene-4,5-dione	EtOH	270(4.16),280s(4.11), 319(3.83)	54-1244-69
$C_{10}H_5BrN_2$			
Heptafulvalene, 2-bromo-8,8-dicyano-	MeOH	240(4.06),275(4.16), 389(4.35),416s(4.22)	25-0075-69
Heptafulvalene, 3-bromo-8,8-dicyano-	MeOH	240(4.01),261(4.04), 393(4.40),412s(4.28)	25-0075-69
$C_{10}H_5ClHgO_2$			
Mercury, chloro[2-oxo-6-(1,3-pentadiynyl)-2H-pyran-5-yl]-	MeOH	251(3.81),265(3.81), 341(4.19)	1-1461-69

Compound	Solvent	λ_{max}(log ε)	Ref.
$C_{10}H_5ClN_2$			
Heptafulvalene, 3-chloro-8,8-dicyano-	MeOH	235(4.12),260(4.14), 395(4.39),406s(4.31)	25-0075-69
$C_{10}H_5IN_2$			
Heptafulvalene, 3-iodo-8,8-dicyano-	MeOH	252(4.02),310(3.76), 400(4.39),416s(4.32)	25-0075-69
$C_{10}H_5N_3O_2S_2$			
p-Dithiin-2,3-dicarboximide, 5,6-dicyano-N-ethyl-	THF	246(4.08),345(3.74)	88-4273-69
$C_{10}H_5N_5O_4$			
Alloxazine, 7-nitro-	EtOH	240(4.1),280(4.3), 370(4.0)	65-1630-69
Alloxazine, 9-nitro-	EtOH	250(4.5),320(3.7), 380(4.0)	65-1630-69
$C_{10}H_6$			
1,4,6,9-Decatetrayne	isooctane	225(2.69),237(2.68), 251(2.38)	78-2823-69
$C_{10}H_6BrNO_2$			
5,8-Isoquinolinedione, 7-bromo-3-methyl-	MeCN	248(4.07),263(4.05), 309(3.77)	4-0697-69
$C_{10}H_6ClN_5$			
3H-v-Triazolo[4,5-d]pyrimidine, 7-chloro-3-phenyl-	MeOH	236(4.51)	23-1129-69
$C_{10}H_6Cl_2N_4$			
s-Triazolo[3,4-a]phthalazine, 3-(dichloromethyl)-	MeOH	209(4.43),234s(4.68), 248(4.76),272(4.02)	44-3221-69
$C_{10}H_6Cl_4N_2$			
Pyrazole, 4,5-dichloro-3-(dichloromethyl)-1-phenyl-	MeOH	247(4.06)	44-1592-69
$C_{10}H_6Cl_6N_2$			
Benzeneazo-1'-[1]butene, 2',3',3',4',-4',4'-hexachloro-	heptane	240(3.9),330(4.4), 460(2.6)	24-3135-69
$C_{10}H_6CrO_4$			
Benzaldehyde, $Cr(CO)_3$ complex	H_2O	213(4.48),267(3.77), 320(3.95),423(3.53)	35-5871-69
	94% H_2SO_4	210(4.33),295(4.17), 333(3.67),453(3.59)	35-5871-69
$C_{10}H_6F_6$			
Bicyclo[2.2.2]octa-2,5,7-triene, 2,3-bis(trifluoromethyl)-	C_6H_{12}	220s(2.44),262(2.16)	44-1271-69
1,3,5,7-Cyclooctatetraene, 1,2-bis(trifluoromethyl)-	C_6H_{12}	265(2.85)	44-1271-69
$C_{10}H_6N_2$			
Heptafulvalene, 8,8-dicyano-	MeOH	225(4.01),255(4.08), 384(4.36)	18-2386-69
	MeOH	225(4.01),255(4.08), 384(4.36)	25-0075-69

Compound	Solvent	λ_{max}(log ε)	Ref.
$C_{10}H_6N_2O_2S_2$			
6-Benzothiazolol, 2-(4-hydroxy-2-thiazolyl)-	pH 4	372(4.19)	88-4683-69
	pH 7	382(4.06)	
	pH 9	420(4.07)	
	pH 13	427(4.33)	
	MeOH	371(4.28)	
	MeOH-HCl	371(4.26)	
	MeOH-KOH	425(4.28)	
	DMSO	377(4.30)	
	+tert-BuOK	490(4.01),570(3.68)	
$C_{10}H_6N_2O_3S_2$			
2-Thiazolin-4-one, 5-hydroxy-2-(6-hydroxybenzothiazol-2-yl)-	MeOH	286(3.91),296(3.91), 305(3.85),388(3.79)	88-4683-69
	MeOH-HCl	282(3.94),296(3.92), 305(3.85),382(3.96)	88-4683-69
	MeOH-KOH	322(4.00),437(3.94)	88-4683-69
$C_{10}H_6N_2O_8$			
2-Benzofurancarboxylic acid, 7-methoxy-4,6-dinitro-	MeOH	213(4.19),281(4.20), 325(4.13)	32-1177-69
$C_{10}H_6N_4O$			
s-Triazolo[3,4-a]phthalazine-3-carboxaldehyde	MeOH	206(4.30),233s(4.60), 239(4.60),246(4.60), 263s(3.88),272s(3.85)	44-3221-69
$C_{10}H_6N_4O_2$			
Alloxazine	EtOH	250(4.6),330(3.8), 380(3.8)	65-1630-69
1,3,4-Oxadiazole, 2,2'-p-phenylenebis-	MeOH	282(4.45)	48-0646-69
$C_{10}H_6N_6$			
Bis-s-triazolo[3,4-a:4',3'-c]phthalazine	MeOH	220(4.77),227(4.88), 235(4.82),260(4.25)	44-3221-69
$C_{10}H_6N_8O_2S_2$			
Purine, 6,6'-dithiodi-, 3,3'-dioxide	pH 6	238(4.42),323(4.31)	44-2153-69
$C_{10}H_6O$			
1-Decene-4,6,8-triyn-3-one	ether	238(4.20),247(4.19), 258s(--),279(3.40), 297(3.65),316(3.74), 338(3.60)	39-1096-69C
$C_{10}H_6O_2$			
Benzo[1,2-b:5,4-b']difuran	EtOH	252s(3.72),260(3.89), 269(3.93),282(3.52), 287(3.68),293(3.83), 299(3.86),305(3.93)	4-0191-69
4,6,8-Decatriyn-3-one, 1,2-epoxy- (relative ε measured)	ether	225(--),234(--), 241(--),279(--), 295(--),315(--), 336(--)	39-1096-69C
1,4-Naphthoquinone	dioxan	245(4.22),249s(4.19), 330(3.43),413(1.67)	88-1169-69
$C_{10}H_6O_8$			
1,2,4,5-Benzenetetracarboxylic acid	pH 0.30	291(3.4)	40-1048-69
	pH 2.79	294(3.2)	40-1048-69

Compound	Solvent	$\lambda_{max}(\log \epsilon)$	Ref.
1,2,4,5-Benzenetetracarboxylic acid (cont.)	pH 4.18	295(3.2)	40-1048-69
	pH 5.01	294s(3.2)	40-1048-69
	pH 7.45	291s(3.0)	40-1048-69
	pH 13.70	291s(3.0)	40-1048-69
$C_{10}H_6O_{10}$			
[Bi-3-cyclopenten-1-yl]-2,2',5,5'-tetrone, 1,1',3,3',4,4'-hexahydroxy-	n.s.g.	242(3.91),293(3.98)	22-4590-69
anion	n.s.g.	251(4.06),335(4.02)	22-4590-69
dianion	n.s.g.	273(4.04),368(4.06)	22-4590-69
$C_{10}H_7BrN_2O_2$			
Hydantoin, 5-(m-bromobenzylidene)-	neutral	234(4.04),319(4.29)	122-0455-69
	anion	242(4.00),273s(--), 328(4.27)	122-0455-69
Hydantoin, 5-(p-bromobenzylidene)-	neutral	226(4.08),326(4.44)	122-0455-69
	anion	232(3.91),273s(--), 330(4.39)	122-0455-69
$C_{10}H_7ClHgO_2$			
Mercury, chloro[2-oxo-6-(3-penten-1-ynyl)-2H-pyran-5-yl]-	MeOH	255(3.93),341(4.18)	1-1059-69
$C_{10}H_7ClN_2O_2$			
Hydantoin, 5-(m-chlorobenzylidene)-	neutral	232(4.11),318(4.34)	122-0455-69
	anion	238(3.62),273s(--), 326(4.18)	122-0455-69
Hydantoin, 5-(p-chlorobenzylidene)-	neutral	226(4.04),326(4.41)	122-0455-69
	anion	232(3.90),273s(--), 330(4.37)	122-0455-69
$C_{10}H_7ClN_2O_2S$			
Pyrimidine, 2-[(p-chlorophenyl)sulfonyl]-	EtOH	240(3.19)	39-2720-69C
$C_{10}H_7ClN_2S$			
Pyrimidine, 2-[(p-chlorophenyl)thio]-	pH 5	224(4.08),243(4.16)	39-2720-69C
$C_{10}H_7ClN_4$			
2-Amino-4-chloro-1H-naphtho[1,8-de]-v-triazinium hydroxide, inner salt	EtOH	236(4.50),270(3.85), 350(4.02)	39-0756-69C
s-Triazolo[3,4-a]phthalazine, 3-(chloromethyl)-	MeOH	234(4.48),249(4.50), 274s(3.76)	44-3221-69
$C_{10}H_7ClO_3$			
Isocoumarin, 4-chloro-3-methoxy-	EtOH	233(4.28),280(3.89), 350(3.47)	2-1114-69
$C_{10}H_7Cl_2N_3O$			
1,2,3-Benzotriazin-4(3H)-one, 3-allyl-6,8-dichloro-	EtOH	227(4.34),298(3.95)	4-0809-69
$C_{10}H_7IN_2O_2$			
Hydantoin, 5-(p-iodobenzylidene)-	neutral	232(4.09),328(4.40)	122-0455-69
	anion	238(4.12),273s(--), 334(4.44)	122-0455-69
$C_{10}H_7N$			
Pyrrolo[3,2,1-hi]indole	n.s.g.	255s(4.09),262(4.20), 271(4.22),305(3.87)	4-0415-69

Compound	Solvent	λ_{max}(log ϵ)	Ref.
$C_{10}H_7NO_2$			
5,8-Isoquinolinedione, 3-methyl-	MeCN	237(4.34),254(4.17), 340(3.60)	4-0697-69
Naphthalene, 1-nitro-	H_2O	342(3.60)	88-4205-69
	MeCN	334(3.60)	88-4205-69
1,2-Naphthoquinone, 4-amino-	pH 4.0	230(5.03),270(4.98), 295s(4.92)	78-5807-69
1,4-Naphthoquinon-4-imine, 2-hydroxy-	pH 0.3	248(4.99),255(4.99), 296(4.92)	78-5807-69
$C_{10}H_7NO_2S_2$			
Thiophene, 4-nitro-2-(phenylthio)-	EtOH	250(4.1),260s(4.0)	32-0535-69
$C_{10}H_7NO_3$			
1H-1-Benzazepine-2,5-dione, 4-hydroxy-	EtOH	237(4.39),277(4.20)	88-1243-69
5,8-Isoquinolinedione, 7-methoxy-	MeCN	240(4.03),255(3.97), 307(3.63)	4-0697-69
$C_{10}H_7NO_4S_2$			
Thiophene, 4-nitro-2-(phenylsulfonyl)-	EtOH	250(4.3),270(3.9)	32-0535-69
$C_{10}H_7NO_6$			
2-Benzofurancarboxylic acid, 7-methoxy-5-nitro-	MeOH	207(4.30),262(4.47)	32-1177-69
$C_{10}H_7N_3O$			
Isoxazolo[3,4-h]quinoline, 1-amino-	MeOH	242(4.55),270s(3.81), 323(3.86),360(3.76)	94-0140-69
3H-Pyridazino[3,4-b]indol-3-one, 2,9-dihydro-	EtOH	238(4.41),260s(4.48), 265(4.52),300(4.00), 430(3.55)	95-0058-69
$C_{10}H_7N_3O_2S$			
Pyrimidine, 2-[(p-nitrophenyl)thio]-	pH 5	245(4.14),319(3.87)	39-2720-69C
$C_{10}H_7N_3O_3S$			
Pyrimidine, 2-[(p-nitrophenyl)sulfinyl]-	EtOH	254(4.11)	39-2720-69C
$C_{10}H_7N_3O_4$			
Hydantoin, 5-(m-nitrobenzylidene)-	neutral	230(4.13),314(4.29)	122-0455-69
	anion	272(4.17),322(4.29)	122-0455-69
$C_{10}H_7N_5O$			
s-Triazolo[4,3-b]-as-triazin-7-ol, 6-phenyl-	EtOH	230(4.04),299(3.92)	22-2492-69
s-Triazolo[5,1-c]-as-triazin-4-ol, 6-phenyl-	EtOH	267(4.18),322(4.22)	22-2492-69
$C_{10}H_7N_5O_2$			
s-Triazolo[3,4-a]phthalazine, 6-methyl-8-nitro-	MeOH	224(4.40),240(4.35)	44-3221-69
$C_{10}H_8ClNO$			
Bicyclo[3.2.2]nona-3,6-diene-8β-carbonitrile, 1-chloro-2-oxo-	MeOH	226(3.81),332(1.93)	88-0443-69
Bicyclo[3.2.2]nona-3,6-diene-8α-carbonitrile, 3-chloro-2-oxo-	MeOH	237(3.66),262(3.49), 332(1.97)	88-0443-69
Bicyclo[3.2.2]nona-3,6-diene-8β-carbonitrile, 3-chloro-2-oxo-	MeOH	237(3.69),262s(3.47), 335(1.90)	88-0443-69

Compound	Solvent	λmax(log ε)	Ref.
Bicyclo[3.2.2]nona-3,6-diene-9α-carbonitrile, 3-chloro-2-oxo-	MeOH	234(3.72),260s(3.44), 330(1.84)	88-0443-69
Isocarbostyril, 3-chloro-2-methyl-	pH -6.1	231(4.71),254(3.63), 263(3.68),273(3.69), 284(3.63),333(3.76), 346(3.73)	24-3666-69
	pH 4	228(4.35),248(3.90), 280(4.00),288(4.00), 320s(3.64),330(3.71), 341s(3.54)	24-3666-69
Isoquinoline, 3-chloro-1-methoxy-	pH -4.06	232(4.71),260(3.70), 269(3.68),280(3.62), 330(3.73),343(3.68)	24-3666-69
	pH 4	221(4.53),255s(3.51), 264(3.75),273(3.90), 284(3.89),318(3.56), 328(3.51)	24-3666-69
$C_{10}H_8ClNO_2$			
Indole-3-carboxaldehyde, 2-chloro-5-methoxy-	EtOH	252(4.24),280(4.04), 306(4.0)	2-0662-69
Isocarbostyril, 3-chloro-6-methoxy- (in 10% MeOH)	pH 4	249(4.71),258s(4.21), 279(3.82),290(3.77), 316(3.54),326(3.48)	24-3666-69
Isocarbostyril, 3-chloro-7-methoxy- (in 10% MeOH)	pH -4.9	236(4.62),268s(3.85), 274(3.91),284(3.89), 350(3.62),360(3.62)	24-3666-69
	pH 4	231(4.49),247s(3.82), 268s(3.39),276(4.09), 286(4.05),330s(3.53), 345(3.65),357(3.57)	24-3666-69
	pH 12	250(4.04),292(3.98), 330s(3.52),344(3.64), 356(3.54)	24-3666-69
$C_{10}H_8ClNO_3$			
Indole-2,3-dione, 5-chloro-6-methoxy-1-methyl-	EtOH	221(4.24),270(4.44), 276(4.42),319(3.84), 430(2.90)	18-3016-69
$C_{10}H_8ClNO_4S_2$			
2-Methylthiazolo[2,3-b]benzothiazolium perchlorate	EtOH	219(4.25),265(3.57), 299(3.62),312(3.73)	95-0469-69
3-Methylthiazolo[2,3-b]benzothiazolium perchlorate	EtOH	214(4.93),260(3.90), 312(4.06)	95-0469-69
2H-[1,3]Thiazino[2,3-b]benzothiazolium perchlorate	EtOH	226(4.30),245s(3.96), 282(4.02),291(3.98), 301(3.90),328(3.47)	95-0469-69
$C_{10}H_8ClN_3O$			
1,2,3-Benzotriazin-4(3H)-one, 3-allyl-6-chloro-	EtOH	291(3.96)	4-0809-69
1,2,3-Benzotriazin-4(3H)-one, 3-allyl-7-chloro-	EtOH	244(4.49),289(3.79)	4-0809-69
$C_{10}H_8Cl_2$			
1,3-Butadiene, 1-chloro-4-(p-chlorophenyl)-	EtOH	221(4.2),291(4.7)	104-0678-69

Compound	Solvent	λ_{max}(log ε)	Ref.
$C_{10}H_8Cl_2O_6$			
Terephthalic acid, 2,5-dichloro-3,6-dihydroxy-, dimethyl ester	hexane	264(3.80),275s(--), 364(3.41),400s(--)	35-6102-69
	EtOH	313(3.55)	35-6102-69
	CCl_4	278s(--),361(3.65), 400s(--)	35-6102-69
	$CHCl_3$	261(3.88),276s(--), 356(3.69),400s(--)	35-6102-69
	THF	313(3.57),350(3.45)	35-6102-69
	DMSO	313(3.75)	35-6102-69
$C_{10}H_8CrO_4$			
Benzyl alcohol, $Cr(CO)_3$ complex	EtOH	217(4.41),254(3.81), 316(3.99)	35-5870-69
	41% H_2SO_4	216(4.43),253(3.78), 313(3.99)	35-5870-69
	86% H_2SO_4	201(4.44),278(4.06), 348(3.18),514(2.56)	35-5870-69
$C_{10}H_8D_2O_2$			
Cinnamic-α,β-d_2 acid, methyl ester	n.s.g.	274(4.06)	39-1542-69C
$C_{10}H_8N_2$			
2,2'-Bipyridine, sodium complex	THF	268(3.94),386(4.47), 423s(3.80),532(3.79), 554(3.81),750(3.06), 833(3.19),995(3.13)	18-2264-69
Indole-2-carbonitrile, 3-methyl-	EtOH	287(4.21),300s(3.97), 312s(3.69)	35-6199-69
$C_{10}H_8N_2O$			
1,3,4-Oxadiazole, 2-styryl-	MeOH	290(4.4)	48-0646-69
$C_{10}H_8N_2O_2$			
Hydantoin, 5-benzylidene-	neutral	226(4.00),318(4.36)	122-0455-69
	anion	276(3.83),324(4.29)	122-0455-69
4H-Pyrazol-4-one, 3-methyl-5-phenyl-, 1-oxide	EtOH	238(3.86),323(3.83) 415(2.90)	44-0187-69
Styrene, p-cyano-β-methyl-β-nitro-	n.s.g.	294(4.22)	67-0127-69
$C_{10}H_8N_2O_2S$			
Imidazolidine-2-thione, 1-benzyl-4,5-dioxo-	EtOH	302(4.11),415(1.48)	18-2323-69
Isothiazole, 3-anthranoyloxy-	EtOH	224(4.44),248(4.20), 351(3.82)	12-2497-69
$C_{10}H_8N_2O_3$			
Pyrazol-3-ol, 5-[3,4-(methylenedioxy)-phenyl]-	EtOH	266(4.31),300(4.12)	39-0836-69C
4H-Pyrazol-4-one, 3-methyl-5-phenyl-, 1,2-dioxide	n.s.g.	256(4.32),303(4.77), 418(2.86)	44-0187-69
$C_{10}H_8N_2O_3S$			
Barbituric acid, 5-(4-methylfurfurylidene)-2-thio-	EtOH	260(3.82),405(4.47)	103-0830-69
$C_{10}H_8N_2O_4$			
Barbituric acid, 5-(5-methylfurfurylidene)-	EtOH	256(3.87),380(4.39)	103-0827-69
Benzene, p-bis(2-nitrovinyl)-	EtOH	230(4.03),345(4.55)	23-4076-69

Compound	Solvent	λmax(log ε)	Ref.
3-Pyrrolin-2-one, 3-hydroxy-4-nitro-5-phenyl-	EtOH	358(4.05)	44-3279-69
$C_{10}H_8N_2O_4S_2$			
[2,2'-Bithiazole]-4,4'-diol, diacetate	EtOH	345(4.40)	44-2053-69
$C_{10}H_8N_2O_5$			
1H-Pyrrolo[2,3-c]pyridine-3-acetic acid, 2-carboxy-5,6-dihydro-5-oxo-	EtOH	292(3.91),300(3.85)	35-2338-69
$C_{10}H_8N_2S$			
Thiazolo[3,2-a]benzimidazole, 2-methyl-	EtOH	244(3.89),250(3.95), 277(4.09)	4-0797-69
Thiazolo[3,2-a]benzimidazole, 3-methyl-	EtOH	240(4.17),247(4.46), 280(4.91)	4-0797-69
$C_{10}H_8N_2S_2$			
Pyridine, 2,2'-dithiodi-	C_6H_{12}	240(4.2),285(4.0)	5-0035-69G
Pyridine, 3,3'-dithiodi-	C_6H_{12}	235(4.2),280s(3.7)	5-0035-69G
Pyridine, 4,4'-dithiodi-	C_6H_{12}	243(4.2)	5-0035-69G
$C_{10}H_8N_4$			
Benzonitrile, o-(5-methyl-s-triazol-3-yl)-	MeOH	223(4.82),255s(4.25), 290(3.84)	44-3221-69
Imidazo[5,1-c][1,2,4]benzotriazine, 1-methyl-	5N H_2SO_4	217(4.24),245(4.16), 270s(--),336(3.76)	103-0683-69
	0.05N NaOH	235(4.10),264(3.97), 393(3.73)	103-0683-69
	MeOH	233(4.13),264(4.02), 285(3.74)	103-0683-69
1H-Naphtho[1,8-de]-v-triazine, 1-amino-	EtOH	233(4.42),339(4.01)	39-0756-69C
1H-Naphtho[1,8-de]-v-triazinium, 2-amino-, hydroxide, inner salt	EtOH	234(4.50),272(3.85), 350(4.06)	39-0756-69C
1-Naphthylamine, 8-azido-	EtOH	232(4.20),253(4.19), 346(3.94)	39-0756-69C
2-Naphthylamine, 3-azido-	EtOH	257(4.46),355(3.57)	39-0756-69C
Pyridine, 4,4'-azodi-	iso-PrOH	283(4.20),453(2.32)	88-2733-69
s-Triazolo[3,4-a]phthalazine, 3-methyl-	MeOH	231(4.53),239(4.44)	44-3221-69
	EtOH	240(4.63),248(4.6), 268(3.8),275(3.7), 310(2.8)	63-0085-69
s-Triazolo[3,4-a]phthalazine, 6-methyl-	MeOH	227(4.53)	44-3221-69
$C_{10}H_8N_4O$			
Imidazo[5,1-c][1,2,4]benzotriazine, 7-methoxy-	5N H_2SO_4	240(--),263(4.42), 330(3.83),387(3.54)	103-0683-69
	MeOH	231(4.35),252(4.28), 343(3.68),400(3.61)	103-0683-69
3H-as-Triazino[5,6-b]indol-3-one, 2,5-dihydro-2-methyl-	EtOH	260(4.4),330(3.7), 380(3.0)	65-2339-69
	EtOH-NaOH	290(4.5),340(4.0), 440(3.0)	65-2339-69
3H-as-Triazino[5,6-b]indol-3-one, 2,5-dihydro-5-methyl-	EtOH	260(4.4),330(3.7), 380s(2.7)	65-2339-69
	EtOH-NaOH	280(4.4),350(4.0)	65-2339-69
1H-1,2,3-Triazole-5-carboxaldehyde, 1-(benzylideneamino)-	EtOH	210(3.98),230(3.96), 273(4.19)	39-1416-69C
s-Triazolo[3,4-a]phthalazin-3(2H)-one, 2-methyl-	MeOH	208(4.59),255(4.44), 265(4.57)	44-3221-69

Compound	Solvent	$\lambda_{max}(\log \epsilon)$	Ref.
$C_{10}H_8N_4O_2$			
Benzamide, N-(4-formyl-1H-1,2,3-triazol-1-yl)-	EtOH	206(3.91),233(4.14)	39-1416-69C
$C_{10}H_8N_4O_2S$			
5H-as-Triazino[5,6-b]indole, 3-(methylsulfonyl)-	EtOH	260(4.5),320(3.9), 360s(3.2)	65-2111-69
$C_{10}H_8N_4S$			
3-Mercapto-1-methyl-s-triazolo[3,4-a]phthalazinium hydroxide, inner salt	MeOH	210(4.46),243s(4.16), 255(4.10),265(4.13), 280s(3.70),290s(3.16)	44-3221-69
3H-as-Triazino[5,6-b]indole, 3-(methylthio)-	EtOH	260(4.5),350(4.0)	65-0640-69
5H-as-Triazino[5,6-b]indole-3-thiol,	EtOH	260(4.1),300(4.6)	65-0078-69
5-methyl-	EtOH	260(4.2),310(4.7)	65-0640-69
3H-as-Triazino[5,6-b]indole-3-thione, 2,5-dihydro-2-methyl-	EtOH	260(4.1),300(4.3), 370(3.3)	65-0640-69
s-Triazolo[3,4-a]phthalazine, 3-(methylthio)-	MeOH	214(4.63),257(4.53), 266(4.53)	44-3221-69
$C_{10}H_8N_8$			
Bis-s-triazolo[3,4-a:4',3'-c]phthalazine, 3,6-diamino-	MeOH	205(4.26),267(4.51)	44-3221-69
$C_{10}H_8O$			
3-Butyn-2-one, 4-phenyl-	EtOH	272(4.16)	22-0210-69
1-Decene-4,6,8-triyn-3-ol	ether	212(5.24),240(2.30), 257(2.40),273(2.54), 290(2.60),310(2.40)	39-1096-69C
Furan, 2-phenyl-	EtOH	279(4.29)	6-0277-69
1-Naphthol	pH 8	295(3.65),322(3.43)	69-1042-69
2-Naphthol	pH 8	274(3.65),328(3.23)	69-1042-69
$C_{10}H_8OS$			
Benzothiophene, 3-acetyl-	MeOH	218(4.60),237(4.08), 282(3.88),301(3.96)	73-2959-69
Benzo[b]thiophene, 5-acetyl-	EtOH	241(4.67),248s(4.61), 278(4.01),313(3.40)	22-0607-69
Furan, 2-[2-(2-thienyl)vinyl]-	MeOH	240s(3.48),332(4.48)	117-0209-69
$C_{10}H_8O_2$			
Chromone, 2-methyl-	MeOH	223(4.41),245(4.07), 296(3.95)	64-0225-69
Coumarin, 3-methyl-	EtOH	275(4.18),307(3.97)	39-0016-69C
Coumarin, 4-methyl-	EtOH	213(4.28),274(3.95), 315(3.70)	18-2319-69
2,8-Decadiene-4,6-diynal, 10-hydroxy-	ether	247(4.40),261(4.35), 297(4.14),316(4.27), 338(4.21)	24-1679-69
4,6,8-Decatriyn-3-one, 1-hydroxy-	ether	222(4.88),230(4.92), 261(3.30),276(3.48), 293(3.70),312(3.81), 334(3.65)	39-1096-69C
Furan, 2,2'-vinylenedi-, cis	C_6H_{12}	307s(4.29),322(4.45), 339(4.37)	44-0073-69
Furan, 2,2'-vinylenedi-, trans	C_6H_{12}	307s(4.46),321(4.62), 339(4.57)	44-0073-69
2H-Pyran-2-one, 6-(3-penten-1-ynyl)-, cis	MeOH	247(3.91),254(3.91), 333(4.22)	1-1059-69

Compound	Solvent	$\lambda_{max}(\log \epsilon)$	Ref.
$C_{10}H_8O_2S$			
Benzothiophene-3-carboxylic acid, methyl ester	MeOH	223(4.53),277(3.77), 294(3.74),302(3.71)	73-2959-69
Benzo[c]thiophene-2-carboxylic acid, methyl ester	EtOH	244(4.32),287s(3.51), 300(3.71),314(3.82), 356(3.88)	2-0009-69
Benzothiophene-3-ol, acetate	MeOH	226(4.45),260(3.67), 289(3.54),298(3.62)	73-2959-69
1,2-Propanedione, 1-[5-(1-propynyl)-2-thienyl]-	MeOH	250(3.83),295(3.75), 347(4.35)	39-1813-69C
2H-Pyran-2-one, 6-(5-methyl-2-thienyl)-	hexane	260(3.78),360(4.19)	1-1461-69
$C_{10}H_8O_2S_2$			
4-Hydroxy-2-(p-methoxyphenyl)-1,3-dithiol-1-ium hydroxide, inner salt	MeOH	200(4.36),230(4.21), 255(3.83),330(3.83)	77-0569-69
$C_{10}H_8O_3$			
Acrolein, 3-hydroxy-, benzoate, (E)-	MeOH	247(4.46)	78-4315-69
Benzo[b]furan, 5,6-(methylenedioxy)-3-methyl-	EtOH	249(3.92),256(3.83), 303(3.90),308(3.90)	18-1971-69
Chromone, 3-hydroxy-2-methyl-	MeOH	233(4.31),281(3.71), 319(3.93)	64-0225-69
Chromone, 5-hydroxy-2-methyl-	MeOH	226(4.30),252(4.08), 325(3.62)	64-0225-69
Chromone, 6-hydroxy-2-methyl-	MeOH	226(4.35),328(3.79)	64-0225-69
Chromone, 7-hydroxy-2-methyl-	MeOH	241(4.23),248(4.30), 294(4.06)	64-0225-69
Chromone, 8-hydroxy-2-methyl-	MeOH	231(4.28),252(4.18), 291(3.95)	64-0225-69
Cinnamaldehyde, 2-formyl-3-hydroxy-	EtOH	248(4.23),291(4.07), 361(3.77)	88-3977-69
Coumarin, 7-hydroxy-3-methyl-	EtOH	325(4.5)	39-1749-69C
Coumarin, 7-hydroxy-4-methyl-	EtOH	222(4.15),254(3.36), 326(4.11)	18-2319-69
1-Indanone, 4,5-(methylenedioxy)-	EtOH	285(3.95),300s(3.86)	23-2501-69
1(2H)-Naphthalenone, r-2,3-epoxy-3,4-dihydro-c-4-hydroxy-	EtOH	256(3.97),292(3.16)	39-2053-69C
t-4-hydroxy-	EtOH	255(3.89),290(3.09)	39-2053-69C
Phthalide, 3-acetyl-	EtOH	230(4.10),273(2.95)	12-0577-69
$C_{10}H_8O_3S_2$			
Glycolic acid, di-2-thienyl-	1:1 CH_2Cl_2: $ClSO_3H$	380(4.3),405(4.3), 518(4.52)	88-4313-69
$C_{10}H_8O_4$			
Chromone, 3,5-dihydroxy-2-methyl-	MeOH	243(4.40),345(3.89)	64-0225-69
Chromone, 5,7-dihydroxy-2-methyl-	MeOH	227(4.21),248(4.29), 255(4.28),294(3.89)	64-0225-69
Coumarin, 6-hydroxy-7-methoxy-	EtOH	231(4.25),255(3.80), 259s(3.78),297(3.81), 349(4.06)	39-0526-69C
	EtOH-NaOH	254(4.33),278s(3.93), 314(3.89),403(3.91)	39-0526-69C
1(2H)-Naphthalenone, 2α,3α-epoxy-3,4-dihydro-4α,8-dihydroxy-	EtOH	268(3.92),343(3.58)	39-2053-69C
1(2H)-Naphthalenone, 2β,3β-epoxy-3,4-dihydro-4β,5-dihydroxy-	EtOH	265(3.86),325(3.45)	39-2053-69C
Scopoletin (7-hydroxy-6-methoxycoumarin)	EtOH	230(4.18),255(3.71), 261s(3.68),299(3.76), 347(4.08)	39-0526-69C

Compound	Solvent	λ_{max}(log ε)	Ref.
Scopoletin (cont.)	EtOH-NaOH	242(4.05),278s(3.65), 294s(3.51),400(4.37)	39-0526-69C
$C_{10}H_8O_5$			
Chromone, 3,5,7-trihydroxy-2-methyl-	MeOH	249(4.57),295(3.99), 332(4.03)	64-0225-69
$C_{10}H_8S$			
Thiophene, 2,5-di-1-propynyl-	MeOH	292(4.33),300(4.33), 309(4.37)	39-1813-69C
$C_{10}H_8S_2$			
Thiophene, 2,2'-vinylenedi-	MeOH	260s(3.30),325s(4.38), 339(4.45)	117-0209-69
Thiophene, 2,2'-vinylenedi-, cis	MeCN	250(4.00),322(3.90)	44-0073-69
Thiophene, 2,2'-vinylenedi-, trans	MeCN	263(3.72),325s(4.34), 338(4.41),357s(4.21)	44-0073-69
$C_{10}H_9BF_2O_2$			
Boron, difluoro(1-phenyl-1,3-butane-dionato)-	$CHCl_3$	265(3.69),330(4.44), 342s(4.29)	39-0526-69A
$C_{10}H_9Br$			
Indene, 1-(bromomethyl)-	ether	256(3.87),288(2.65), 294(2.23)	44-0528-69
$C_{10}H_9BrN_4$			
Pyrimidine, 2,4-diamino-5-(o-bromo-phenyl)-	pH 2	210(4.53),268(3.68)	44-0821-69
	pH 12	252s(3.81),290(3.85)	44-0821-69
$C_{10}H_9BrO$			
1-Indanone, 2-bromo-6-methyl-	EtOH	255(4.15),308(3.53)	2-0215-69
$C_{10}H_9BrOS_2$			
1,3-Dithiole, 4-(p-bromophenyl)-2-methoxy-	EtOH	242(4.12),304(4.08)	94-1931-69
$C_{10}H_9BrO_3$			
1-Indanone, 4-bromo-7-hydroxy-6-methoxy-	MeOH	226(4.23),262(3.79), 343(3.45)	78-5475-69
	MeOH-KOH	241(4.27),267s(--), 384(3.68)	78-5475-69
$C_{10}H_9Br_3O_2$			
Propiophenone, 2',3',5'-tribromo-4'-methoxy-	C_6H_{12}	225(4.45),290(2.81)	22-2079-69
Propiophenone, 2',4',6'-tribromo-3'-methoxy-	C_6H_{12}	284(3.04),291(3.08)	22-2079-69
	EtOH	284(3.04),291(3.08)	22-2079-69
Propiophenone, 3',4',5'-tribromo-2'-methoxy-	C_6H_{12}	225(4.55),250(3.92), 304(3.20)	22-2079-69
	EtOH	224(4.51),251(3.90), 305(3.23)	22-2079-69
$C_{10}H_9Cl$			
1,3-Butadiene, 1-chloro-4-phenyl-	EtOH	225(4.0),280(4.5)	104-0678-69
$C_{10}H_9ClHgO_2$			
Mercury, chloro[2-oxo-6-(1-pentynyl)-2H-pyran-5-yl]-	MeOH	232(4.01),323(4.02)	1-1059-69

Compound	Solvent	$\lambda_{max}(\log \epsilon)$	Ref.
$C_{10}H_9ClN_2OS$			
Acetamide, 2-chloro-N-(3-methyl-2-benzothiazolinylidene)-	EtOH	215(4.31),313(4.35)	23-4483-69
Propionamide, N-2-benzothiazolyl-2-chloro-	EtOH	210(4.39),278(4.27)	23-4483-69
$C_{10}H_9ClN_2O_2S$			
Acetamide, 2-chloro-N-(6-methoxy-2-benzothiazolyl)-	EtOH	217(4.36),308(4.28)	23-4483-69
$C_{10}H_9ClN_4$			
Pyrimidine, 2,4-diamino-5-(m-chlorophenyl)-	pH 2	213(4.59),235s(4.22),272s(3.83)	44-0821-69
	pH 12	208(4.53),262(4.03),292(4.00)	44-0821-69
Pyrimidine, 2,4-diamino-5-(p-chloro-	pH 2	210(4.53),240s(4.25),278s(3.83)	44-0821-69
	pH 12	262(4.11),290(4.03)	44-0821-69
$C_{10}H_9ClN_4O$			
4-Pyrimidinol, 2,6-diamino-5-(p-chlorophenyl)-	pH 2	270(4.26)	44-0821-69
	pH 12	273(4.16)	44-0821-69
$C_{10}H_9ClN_4O_5$			
Hydantoin, 5-(2-chloroethyl)-1-[(5-nitrofurfurylidene)amino]-	H_2O	265(4.04),368(4.24)	44-3213-69
$C_{10}H_9ClO$			
Crotonophenone, 2-chloro-	EtOH	250(4.08)	44-1474-69
$C_{10}H_9ClOS_2$			
1,3-Dithiole, 4-(p-chlorophenyl)-2-methoxy-	EtOH	236(4.11),312(4.05)	94-1931-69
$C_{10}H_9ClO_3$			
1(2H)-Naphthalenone, 2α-chloro-3,4-dihydro-3β,4β-dihydroxy-	EtOH	254(4.00),294(3.17)	39-2053-69C
$C_{10}H_9ClO_4S_2$			
4-p-Tolyl-1,3-dithiol-1-ium perchlorate	50% H_2SO_4	245(4.19),270(4.09),357(3.53)	94-1924-69
$C_{10}H_9ClO_5S_2$			
4-(p-Methoxyphenyl)-1,3-dithiol-1-ium perchlorate	50% H_2SO_4	247(4.03),283(4.08),377(3.47)	94-1924-69
$C_{10}H_9F_3N_2O_4$			
Uridine, 2',3'-didehydro-2',3'-dideoxy-5-(trifluoromethyl)-	pH 1	260(3.91)	87-0543-69
	pH 12	259(3.74)	87-0543-69
$C_{10}H_9F_5O$			
Benzyl alcohol, α-ethyl-2,3,4,5,6-pentafluoro-α-methyl-	EtOH	262(2.71)	65-1766-69
$C_{10}H_9FeNO_5S$			
Iron, tricarbonyl[1-(methylsulfonyl)-1H-azepine]-	EtOH	232(4.16)	44-2866-69
$C_{10}H_9IN_4$			
2-Methyl-s-triazolo[3,4-a]phthalazinium iodide	MeOH	220(4.60),266(4.07)	44-3221-69

Compound	Solvent	$\lambda_{max}(\log \epsilon)$	Ref.
$C_{10}H_9N$			
Isoquinoline, 3-methyl-	MeOH	218(4.89),258(3.57), 268(3.56),313(3.45), 322(3.46),327(3.50)	106-0196-69
$C_{10}H_9NO$			
Atroponitrile, β-methoxy-, cis	EtOH	214(4.15),269(4.10)	22-0198-69
Atroponitrile, β-methoxy-, trans	EtOH	210(4.17),266(4.20)	22-0198-69
Bicyclo[3.2.2]nona-3,6-diene-8α-carbonitrile, 2-oxo-	MeOH	226(3.89),338(2.05)	88-0443-69
Bicyclo[3.2.2]nona-3,6-diene-9α-carbonitrile, 2-oxo-	MeOH	224(3.95),340(1.95)	88-0443-69
Carbostyril, 3-methyl-	EtOH	220(4.25),269(3.91), 324(3.91)	1-0159-69
Carbostyril, 4-methyl-	EtOH	230(4.58),268(3.82), 327(3.84)	1-0159-69
Carbostyril, 5-methyl-	EtOH	234(4.51),281(3.97), 333(3.76)	1-0159-69
Carbostyril, 6-methyl-	EtOH	233(4.61),271(3.88), 337(3.81)	1-0159-69
Carbostyril, 7-methyl-	EtOH	232(4.61),276(3.95), 329(3.99)	1-0159-69
Carbostyril, 8-methyl-	EtOH	233(4.64),276(4.05), 332(3.88)	1-0159-69
Crotononitrile, 3-hydroxy-2-phenyl-	EtOH	266(4.12)	22-0198-69
sodium salt	EtOH	256(3.64),286(3.66)	22-0198-69
Indole, 3-acetyl-	MeOH	295(4.10)	22-3002-69
anion	MeOH	329(4.30)	22-3002-69
Indole-3-carboxaldehyde, 2-methyl-	MeOH	303(4.12)	22-3002-69
anion	MeOH	328(4.30)	22-3002-69
Isoquinoline, 3-methoxy-	EtOH	339(3.56)	39-1729-69C
3(2H)-Isoquinolone, 1-methyl-	100% EtOH	345(3.30),412(3.51)	39-1729-69C
	95% EtOH	350(3.20),410(3.63)	39-1729-69C
	ether	344(3.66)	39-1729-69C
	1:1 EtOH: ether	345(3.50),414(3.25)	39-1729-69C
3(2H)-Isoquinolone, 2-methyl-	EtOH	410(3.65)	39-1729-69C
Isoxazole, 3-methyl-4-phenyl-	MeOH	228(4.0)	4-0783-69
Isoxazole, 4-methyl-3-phenyl-	MeOH	236(4.03)	4-0783-69
4(1H)-Quinolone, 2-methyl-	MeOH	215(4.27),236(4.37), 322(4.02),335(3.99)	78-0255-69
	0.1N HCl	234(4.50),305(3.83)	78-0255-69
	0.1N $HClO_4$	235(4.37),299(3.94)	78-0255-69
	pH 13	244(4.35),319(3.91)	78-0255-69
$C_{10}H_9NOS$			
Oxazole, 2-(methylthio)-4-phenyl-	2% EtOH	251(4.17)	1-2888-69
conjugate acid	2% EtOH	258(4.19)	1-2888-69
Oxazole, 2-(methylthio)-5-phenyl-	2% EtOH	248s(3.85),289(4.33)	1-2888-69
conjugate acid	2% EtOH	287(4.30)	1-2888-69
4-Oxazoline-2-thione, 3-methyl-4-phenyl-	2% EtOH	266(4.34)	1-2888-69
conjugate acid	2% EtOH	250(4.06)	1-2888-69
4-Oxazoline-2-thione, 3-methyl-5-phenyl-	2% EtOH	218(4.13),300(4.38)	1-2888-69
conjugate acid	2% EtOH	282(4.24)	1-2888-69
$C_{10}H_9NOS_2$			
2-Propanone, 1-(2-benzothiazolylthio)-	EtOH	225(4.37),244(3.99), 277(4.12),289(4.04), 300(3.93)	95-0469-69

Compound	Solvent	λmax(log ε)	Ref.
2-Propen-1-ol, 3-(2-benzothiazolylthio)-, hydrochloride	EtOH	226(3.90),246(3.58), 281(3.74),291(3.68), 302(3.62)	95-0469-69
$C_{10}H_9NO_2$			
2H-Azirine-2-carboxylic acid, 3-phenyl-, methyl ester	EtOH	204(4.22),243(3.96), 288s(2.69)	88-2049-69
Carbostyril, 3-methoxy-	EtOH	273(3.84),316(4.06), 329(3.94)	94-2083-69
Carbostyril, 4-methoxy-	EtOH	264(3.88),274(3.87), 313(3.81)	94-2083-69
Carbostyril, 6-methoxy-	EtOH	234(4.43),268(3.73), 349(3.50)	1-0159-69
Carbostyril, 8-methoxy-	EtOH	235(4.01),256(4.04), 273(3.56),336(3.15)	1-0159-69
Cycloprop[a]indene, 1,1a,6,6a-tetrahydro-3-nitro-	C_6H_{12}	275(3.89)	35-3558-69
	EtOH	282(3.84)	35-3558-69
Cycloprop[a]indene, 1,1a,6,6a-tetrahydro-4-nitro-	C_6H_{12}	287(3.98)	35-3558-69
	EtOH	301(3.94)	35-3558-69
Isoxazole, 5-methoxy-3-phenyl-	EtOH	204(4.27),238(3.89)	88-2049-69
Quinoline, 3-methoxy-, 1-oxide	EtOH	297(3.67),309(3.73), 345(3.76),359(3.80)	94-2083-69
$C_{10}H_9NO_2S$			
Quinoline, 2-(methylsulfonyl)-	EtOH	236(4.72),293(3.53), 310(3.48),320(3.36)	95-0074-69
$C_{10}H_9NO_3$			
Noroxyhydrastinine	MeOH	223(4.31),261(3.58), 304(3.67)	95-0074-69
$C_{10}H_9NO_3S_2$			
1,3-Dithiole, 2-methoxy-4-(p-nitrophenyl)-	EtOH	258(4.02),373(4.11)	94-1931-69
$C_{10}H_9NO_4$			
1,3-Butanedione, 2-nitro-1-phenyl-, Al chelate	$CHCl_3$	260(4.14),311(4.18), 345s(--)	42-0945-69
Pd chelate	$CHCl_3$	260(4.62),284(4.70), 348(4.36)	42-0945-69
1,4-Cyclohexadiene-1-acrylic acid, 3,6-dioxo-, methyl ester, 3-oxime	EtOH	233(4.18),300(4.23) 427(4.41)	12-0935-69
	dioxan	728(0.46)	12-0935-69
2,4-Pentadienal, 2-methyl-5-(5-nitro-2-furyl)-	EtOH	285(4.20),384(4.46)	94-0306-69
2,4-Pentadienal, 4-methyl-5-(5-nitro-2-furyl)-	EtOH	275(4.09),380(4.44)	94-0306-69
$C_{10}H_9NS_2$			
Thiazole, 2-(methylthio)-4-phenyl-	2% EtOH	238(4.26),268(4.09)	1-2888-69
conjugate acid	2% EtOH	240(4.09),261(4.01), 303(4.01)	1-2888-69
Thiazole, 2-(methylthio)-5-phenyl-	2% EtOH	309(4.36)	1-2888-69
conjugate acid	2% EtOH	252(3.68),318(4.34)	1-2888-69
4-Thiazoline-2-thione, 3-methyl-4-phenyl-	2% EtOH	309(4.19)	1-2888-69
conjugate acid	2% EtOH	234(3.95),290(3.94)	1-2888-69
4-Thiazoline-2-thione, 3-methyl-5-phenyl-	2% EtOH	331(4.31)	1-2888-69
conjugate acid	2% EtOH	244(3.76),314(4.15)	1-2888-69

Compound	Solvent	$\lambda_{max}(\log \epsilon)$	Ref.
$C_{10}H_9N_3$			
6H-Pyrrolo[2,3-g]quinoxaline, 7,8-dihydro-	MeOH	265(4.37),410(3.92)	103-0199-69
as-Triazine, 5-methyl-3-phenyl-	MeOH	255(4.26),368(2.62)	88-3147-69
as-Triazine, 3-p-tolyl-	MeOH	265(4.34),385(2.59)	88-3147-69
$C_{10}H_9N_3O$			
1,2,3-Benzotriazin-4(3H)-one, 3-allyl-	EtOH	284(3.83)	4-0809-69
$C_{10}H_9N_3O_3S$			
Thiazole, 2-amino-5-[1-methyl-2-(5-nitro-2-furyl)vinyl]-	EtOH	251(4.10),297(4.00), 405(4.24)	103-0414-69
$C_{10}H_9N_3O_4$			
2-Benzimidazolinone, 3-acetyl-1-methyl-5-nitro-	n.s.g.	199(4.14),222(4.08), 280(4.09)	39-0070-69C
$C_{10}H_9N_3O_5$			
Indoline, 1-acetyl-5,6-dinitro-	MeOH	253(4.25),335(4.04)	103-0196-69
$C_{10}H_9N_5O_2$			
Formic acid, 2-(5-hydroxy-6-phenyl-as-triazin-3-yl)hydrazide	EtOH	286(4.10)	22-2492-69
Picolinaldehyde, (2,6-dihydroxy-4-pyrimidinyl)hydrazone	EtOH	342(4.13)	56-0519-69
$C_{10}H_9N_5O_7$			
2(1H)-Quinoxalinone, 3-glycinyl-3,4-dihydro-5,7-dinitro-	EtOH	266(4.15),374(4.03)	44-0395-69
$C_{10}H_9N_5S$			
10H-Pyrimido[5,4-b][1,4]benzothiazine, 2,4-diamino-	acid	225s(4.26),266(4.46), 290s(3.83),370(3.29)	44-0821-69
	neutral	255(4.47),293(3.82), 330s(3.30)	44-0821-69
$C_{10}H_9N_7O$			
Benzamide, o-(3-amino-7H-s-triazolo-[4,3-b]-s-triazol-6-yl)-	MeOH	243(4.38)	44-3221-69
$C_{10}H_9O_4P$			
1-Naphthyl phosphate	pH 8	286(3.73)	69-1042-69
2-Naphthyl phosphate	pH 8	274(3.66)	69-1042-69
$C_{10}H_{10}BrNO$			
Acrolein, 2-bromo-3-(N-methylanilino)-	CH_2Cl_2	310(4.43)	33-2641-69
$C_{10}H_{10}BrNO_2$			
Styrene, p-bromo-β-ethyl-β-nitro-	n.s.g.	313(4.11)	67-0127-69
$C_{10}H_{10}Br_2N_2O_2$			
Acetamide, N,N-(2,5-dibromo-p-phenylene)bis-	dioxan	269(4.27),305s(--)	117-0287-69
$C_{10}H_{10}Br_2O_2$			
Propiophenone, 3',5'-dibromo-2'-methoxy-	C_6H_{12}	300(3.20)	22-2079-69
	EtOH	300(3.18)	22-2079-69
Propiophenone, 3',5'-dibromo-4'-methoxy-	C_6H_{12}	257(3.99),291(2.85), 303(2.70)	22-2079-69
	EtOH	257(3.96)	22-2079-69

Compound	Solvent	λ max (log ε)	Ref.
$C_{10}H_{10}ClNO_2$			
Crotonohydroxamic acid, N-(p-chloro-phenyl)-	EtOH	216(4.29),280(3.91)	34-0278-69
Styrene, p-chloro-β-ethyl-β-nitro-	n.s.g.	312(4.09)	67-0127-69
$C_{10}H_{10}ClN_3O$			
1,3,4-Oxadiazol-3-ine, 2-[(p-chloro-phenyl)imino]-5,5-dimethyl-	EtOH	227(4.02),325(3.82)	44-3230-69
$C_{10}H_{10}ClN_5$			
as-Triazine, 3,5-diamino-6-(p-chloro-benzyl)-	acid	207(4.54),219s(4.43), 253(3.86)	44-0821-69
	neutral	223(4.31),303(3.79)	44-0821-69
$C_{10}H_{10}Cl_2O_4$			
Cryptosporiopsin	EtOH	292(4.36)	23-2087-69
2-Cyclopentene-1-carboxylic acid, 3,5-dichloro-1-hydroxy-4-oxo-2-propenyl-, methyl ester	MeOH	289(4.35)	39-2187-69C
1S,5S-(E)-isomer	MeOH	289(4.35)	35-0157-69
$C_{10}H_{10}Cl_6O_4$			
Acetic acid, trichloro-, cyclohexylidene ester	EtOH	220(2.90)(end abs.)	78-1679-69
$C_{10}H_{10}FNO_2$			
Styrene, p-fluoro-β-ethyl-β-nitro-	n.s.g.	309(4.01)	67-0127-69
$C_{10}H_{10}F_4N_2S$			
Piperidine, 1-[(2,3,5,6-tetrafluoro-4-pyridyl)thio]-	hexane	214(3.75),253s(3.59), 276(3.67)	39-1660-69C
$C_{10}H_{10}F_6$			
Bicyclo[2.2.2]oct-2-ene, 2,3-bis(tri-fluoromethyl)-	MeOH	218(4.30),225(4.04) (not maxima)	44-1271-69
$C_{10}H_{10}INO_2$			
Styrene, p-iodo-β-ethyl-β-nitro-	n.s.g.	320(4.18)	67-0127-69
$C_{10}H_{10}N_2$			
Crotononitrile, 3-amino-2-phenyl-	EtOH	251(3.97),287(4.07)	22-0210-69
hydrochloride	EtOH	266(4.16)	22-0210-69
Pyrazole, 3-p-tolyl-	3N HCl	260(4.28)	35-0711-69
$C_{10}H_{10}N_2O$			
4-Imidazolin-2-one, 1-methyl-4-phenyl-	pH 7	213s(4.08),281(4.17)	44-1133-69
2-Pyrazolin-5-one, 3-benzyl-	EtOH	245(3.48)	22-4159-69
2-Pyrazolin-5-one, 4-methyl-3-phenyl-	EtOH	205(4.32),250(4.17)	22-4159-69
3-Pyrazolin-5-one, 3-methyl-1-phenyl-	MeOH	243(4.2)	94-1467-69
3-Pyrazolin-5-one, 3-methyl-2-phenyl-	MeOH	260(4.1)	94-1467-69
Pyrazol-3-ol, 4-methyl-1-phenyl-	EtOH	279(4.25)	103-0527-69
Urea, 1-inden-3-yl-	EtOH	212(3.72),268(2.99), 275(3.04)	36-0047-69
$C_{10}H_{10}N_2OS$			
Benzimidazole, 1-acetyl-2-(methylthio)-	EtOH	243(4.35),286(4.08), 295(4.06)	48-0997-69
1,3,4-Oxadiazole, 2-benzyl-5-(methyl-thio)-	n.s.g.	220(4.07)	2-0583-69
1,2,4-Oxadiazole, 3-(ethylthio)-5-phenyl-	EtOH	247(4.35)	39-2794-69C

Compound	Solvent	$\lambda_{max}(\log \epsilon)$	Ref.
Quinoxaline, 2-(ethylsulfinyl)-	pH 6.0	239(4.55),325(3.90)	39-0333-69B
Thiazolo[3,2-a]benzimidazol-3-ol, 2,3-dihydro-2-methyl-	EtOH	250(3.98),284(4.04), 291(4.08)	4-0797-69
Thiazolo[3,2-a]benzimidazol-3-ol, 2,3-dihydro-3-methyl-	EtOH	249(3.99),284(4.12), 291(4.15)	4-0797-69
$C_{10}H_{10}N_2O_2$			
2-Benzimidazolinone, 1-acetyl-3-methyl-	n.s.g.	222(3.84),287(3.72)	39-0070-69C
1H-1,4-Benzodiazepine-2,5-dione, 3,4-dihydro-4-methyl-	EtOH	215(4.51),291(3.34)	44-1359-69
Furazan, 3-ethoxy-4-phenyl-	EtOH	256(4.04)	39-2794-69C
Imidazole, 1-hydroxy-2-methyl-4-phenyl-, 3-oxide	MeOH	230(3.93),260(3.97)	48-0746-69
Imidazole, 1-hydroxy-4-methyl-2-phenyl-, 3-oxide	MeOH	275(4.06)	48-0746-69
Isosydnone, 5-benzyl-4-methyl-	EtOH	253(3.84)	39-1185-69B
Isosydnone, 4-methyl-5-p-tolyl-	EtOH	227(3.85),298(4.13)	39-1185-69B
1,2,4-Oxadiazole, 3-ethoxy-5-phenyl-	EtOH	255(4.17)	39-2794-69C
1,3,4-Oxadiazole, 2-[(benzyloxy)methyl]-	MeOH	235(2.9)	48-0646-69
Phthalimide, N-methyl-3-(methylamino)- (two shorter wavelength maxima)	EtOH	410(3.71)	104-1231-69
Phthalimide, N-methyl-4-(methylamino)-	EtOH	391(3.71)	104-1231-69
2-Pyrazolin-5-one, 3-(p-methoxyphenyl)-	EtOH	263(4.43)	22-4159-69
$C_{10}H_{10}N_2O_2S$			
1-Benzimidazolinecarboxylic acid, 2-thioxo-, ethyl ester	dioxan	313(3.98)	4-0023-69
1,2,4-Oxadiazole, 3-(ethylsulfinyl)-5-phenyl-	EtOH	254(4.30)	39-2794-69C
1,3,4-Oxadiazole-2-thione, 5-benzyl-3-(hydroxymethyl)-	n.s.g.	253(2.95)	2-0583-69
Quinoxaline, 2-(ethylsulfonyl)-	pH 6.0	241(4.65),321(3.85)	39-0333-69B
$C_{10}H_{10}N_2O_2S_4$			
$\Delta^{5,5'}$-Birhodanine, 3,3'-diethyl-	heptane	241(--),271(--), 279(--),405s(--), 428(4.44),454s(--)	51-0515-69
$C_{10}H_{10}N_2O_3$			
Furoxan, 3-(p-methoxyphenyl)-4-methyl-	n.s.g.	243(4.26),291(3.99)	39-1901-69C
Furoxan, 4-(p-methoxyphenyl)-3-methyl-	n.s.g.	238(4.18),272(4.11)	39-1901-69C
$C_{10}H_{10}N_2O_3S$			
1,2,4-Oxadiazole, 3-(ethylsulfonyl)-5-phenyl-	EtOH	256(4.32)	39-2794-69C
$C_{10}H_{10}N_2O_4$			
Styrene, β-ethyl-p,β-dinitro-	n.s.g.	300(4.22)	67-0127-69
$C_{10}H_{10}N_2O_4S$			
Δ^2-1,2,4-Thiadiazoline-4-carboxylic acid, 3-phenyl-, methyl ester, 1,1-dioxide	n.s.g.	253(4.10)	39-0652-69C
$C_{10}H_{10}N_2O_5$			
1H-Pyrrolo[2,3-c]pyridine-3-acetic acid, 2-carboxy-4,5,6,7-tetrahydro-5-oxo-	EtOH	270(4.20)	35-2338-69
$C_{10}H_{10}N_2S$			
Imidazole, 2-(methylthio)-4-phenyl-	2% EtOH	269(4.29)	1-2888-69

Compound	Solvent	λmax(log ε)	Ref.
Imidazole, 2-(methylthio)-4-phenyl-, conjugate acid	2% EtOH	270(4.34)	1-2888-69
4-Imidazoline-2-thione, 1-methyl-4-phenyl-	2% EtOH	218(4.14),288(4.24)	1-2888-69
conjugate acid	2% EtOH	263(4.24)	1-2888-69
4-Imidazoline-2-thione, 1-methyl-5-phenyl-	2% EtOH	266(4.22)	1-2888-69
conjugate acid	2% EtOH	250(4.14)	1-2888-69
Quinoxaline, 2-(ethylthio)-	pH 6.0	242(4.16),265(4.14), 358(3.95)	39-0333-69B
$C_{10}H_{10}N_2S_2$			
1H-2-Benzothiopyran-4-carbonitrile, 3-amino-5,6,7,8-tetrahydro-1-thioxo-	C_6H_{12}	241(--),328(--), 340s(--),446(--)	78-1441-69
	EtOH	245(4.35),322(4.10), 459(4.22)	78-1441-69
2-Pyridineacetonitrile, α-carbonyl-, dimethyl mercaptole	EtOH	232(3.94),284(3.80), 331(4.07)	95-0203-69
4-Thiazoline-2-thione, 3-anilino-4-methyl-	EtOH	232(4.09),322(4.12)	95-0689-69
$C_{10}H_{10}N_4$			
3,3'-Bipyridazine, 5,5'-dimethyl-	EtOH	239(3.35),263(3.05)	88-2359-69
3,3'-Bipyridazine, 6,6'-dimethyl-	EtOH	241(3.80),267(3.51)	88-2359-69
Malononitrile, [(1,3,5-trimethylpyrazol-4-yl)methylene]-	heptane	332(4.30)	4-0545-69
	MeOH	331(4.33)	4-0545-69
2-Pyridineacetonitrile, α-(2-imidazolidinylidene)-	EtOH	252(4.16),288(4.39), 327(4.06)	95-0203-69
Pyrimidine, 2,4-diamino-5-phenyl-	pH 2	208(4.53),235s(4.18), 275s(3.71)	44-0821-69
	pH 12	257(3.99),292(3.87)	44-0821-69
Pyrimidine, 4,5-diamino-2-phenyl-	pH 3.8	248(4.38),300(4.10)	39-1408-69C
	pH 7.9	222(4.14),296(4.07)	39-1408-69C
Pyrimidine, 4,5-diamino-6-phenyl-	pH 3.6	216s(4.18),307(4.04)	39-1883-69C
	pH 7.6	230(4.18),302(3.88)	39-1883-69C
4H-1,2,4-Triazole, 4-(benzylideneamino)-3-methyl-	MeOH	281(4.31)	48-0897-69
s-Triazolo[3,4-a]phthalazine, 5,6-dihydro-3-methyl-	MeOH	240(4.32),260(4.00), 291(3.24)	44-3221-69
$C_{10}H_{10}N_4O$			
2-Pyrazolin-4-one, 5-imino-3-methyl-1-phenyl-, 4-oxime	EtOH	225(4.08),307(4.16), 350s(3.63)	4-0317-69
Pyrimidine, 2,4-diamino-6-phenoxy-	acid	223(4.15),279(4.10)	44-0821-69
	neutral	238(3.93),268(3.97)	44-0821-69
$C_{10}H_{10}N_4OS$			
Indole-2,3-dione, 3-(3-methyl-3-thioisosemicarbazone)	EtOH	250(4.0),300(4.1), 370(3.5)	65-0070-69
Indole-2,3-dione, 3-(2-methyl-3-thiosemicarbazone)	EtOH	250(4.1),270s(3.7), 350(4.2)	65-0070-69
Indole-2,3-dione, 1-methyl-, 3-(thiosemicarbazone)	EtOH	240(4.0),270(4.0), 360(4.3)	65-0070-69
$C_{10}H_{10}N_4O_2$			
3,3'-Bipyridazine, 6,6'-dimethoxy-	EtOH	239(3.42),263(3.12)	88-2359-69
Indole-2,3-dione, 1-methyl-, 3-semicarbazone, anti	EtOH	270(4.2),310(4.2), 380s(3.2)	65-2345-69
syn	EtOH	280(4.0),370(4.2), 390s(3.0)	65-2345-69

Compound	Solvent	$\lambda_{max}(\log \epsilon)$	Ref.
Indole-2,3-dione, 1-methyl-, 3-semicarbazone, syn (cont.)	EtOH-HCl	290(3.8),390(4.2)	65-2345-69
Pyrazole-4,5-dione, 3-methyl-1-phenyl-, dioxime	EtOH	258(4.17),398(3.52)	4-0317-69
unstable isomer	EtOH	220(3.86),257(4.13), 397(3.54)	4-0317-69
$C_{10}H_{10}N_4O_2S_2$			
2(1H)-Pyrimidinone, 4,4'-dithiobis-[1-methyl-	pH 5.6	257(4.07),308(4.26)	35-3634-69
$C_{10}H_{10}N_4O_6$			
Hydantoin, 5-(2-hydroxyethyl)-1-[(5-nitrofurfurylidene)amino]-	H_2O	265(4.03),365(4.23)	44-3213-69
$C_{10}H_{10}O$			
Benzaldehyde, o-isopropenyl-	C_6H_{12}	224(4.09),239(4.05), 245(4.02),288(3.20)	12-1457-69
3-Buten-2-one, 4-phenyl-	EtOH	286(4.44)	22-3719-69
2,8-Decadiene-4,6-diyn-1-ol, cis-cis	ether	236(4.46),246(4.39), 260(3.91),275(4.14), 292(4.31),312(4.23)	24-1679-69
Inden-1-ol, 3-methyl-	EtOH	219(4.37),226(4.32), 262(3.77)	12-1457-69
$C_{10}H_{10}OS$			
Benzo[b]thiophene-2-methanol, 5-methyl-	EtOH	206(4.48),233(4.57), 265(4.11),295(3.60), 305(3.60)	89-0456-69
1-Propanone, 1-[5-(1-propynyl)-2-thienyl]-	MeOH	283s(3.79),310(4.06)	39-1813-69C
$C_{10}H_{10}OS_2$			
1,3-Dithiole, 2-methoxy-4-phenyl-	EtOH	231(4.08),305(3.94)	94-1931-69
$C_{10}H_{10}O_2$			
2-Benzofuranol, 2,3-dihydro-6-methyl-3-methylene-	n.s.g.	225(4.07),255(4.08), 264(4.00),317(3.91), 328(3.86)	88-4703-69
2(3H)-Benzofuranone, 5-ethyl-	MeOH	228s(3.81),270s(3.14), 278(3.25),284(3.23)	12-0977-69
1,3-Butanedione, 1-phenyl-	$CHCl_3$	310(4.14)	39-0526-69A
Al chelate	$CHCl_3$	250(3.95),320(4.30)	42-0945-69
Cu chelate	MeOH	250(3.85),320(4.32), 640(1.70),680s(1.67)	42-0945-69
Pd chelate	$CHCl_3$	260(4.70),285(4.69), 352(4.49)	42-0945-69
2,8-Decadiene-4,6-diyne-1,10-diol, cis-cis	ether	236(4.46),246(4.39), 261(3.92),276(4.16), 293(4.33),313(4.25)	24-1679-69
4,6,8-Decatriyne-1,3-diol	ether	210(5.12),240s(--), 253(2.65),269(2.70), 287(2.70),306(2.54)	39-1096-69C
(R)-	ether	210(5.15)	39-1096-69C
2(3H)-Furanone, 5-(4-hexen-2-ynylidene)-dihydro-, cis	ether	224(4.61),253s(--), 266(4.22),279(4.33), 293(4.24)	39-1096-69C
trans	ether	223(4.44),249s(--), 262(4.04),276(4.22), 292(4.11)	39-1096-69C

Compound	Solvent	λ max(log ε)	Ref.
2H-Pyran-2-one, 6-(1-pentynyl)-	MeOH	219(3.86),225(3.98), 318(4.05)	1-1059-69
$C_{10}H_{10}O_2S$			
Isothiochroman-4-one, 7-methoxy-	EtOH	229(4.04),285(4.08)	44-1566-69
$C_{10}H_{10}O_2S_2$			
Terephthalic acid, 1,1-dithio-, dimethyl ester	C_6H_{12}	243s(3.6),295(4.16), 335s(3.7),511(2.17)	48-0045-69
	EtOH	241s(3.7),290(4.20), 337s(3.8),508(2.23)	48-0045-69
$C_{10}H_{10}O_3$			
2-Benzofuranol, 2,3-dihydro-, acetate	EtOH	275(3.31),282(3.27)	12-1923-69
Ketone, 1,4-benzodioxan-6-yl methyl	hexane	224(4.30),229(4.33), 263(4.05),269(4.06), 297(3.66)	59-0313-69
1,2-Naphthalenediol, 2,3-epoxy-1,2,3,4-tetrahydro-, c-1,c-4	EtOH	252(2.31),262(2.26)	39-2053-69C
c-1,t-4	EtOH	260(2.37)	39-2053-69C
$C_{10}H_{10}O_3S$			
2H-Thiet-3-ol, 2-methyl-4-phenyl-, 1,1-dioxide	MeOH	264(4.19)	88-0647-69
	N KOH	280(4.26)	88-0647-69
$C_{10}H_{10}O_4$			
Acetophenone, 2-methoxy-3',4'-(methylenedioxy)-	EtOH	229(4.19),275(3.80), 310(3.86)	12-1011-69
1,4-Benzodioxan-6-ol, acetate	EtOH	283(3.51)	104-0468-69
1,4-Benzodioxan-6-ol, 7-acetyl-	EtOH	239(4.14),277(4.12), 348(3.82)	104-0468-69
Cinnamic acid, 3-hydroxy-4-methoxy-	EtOH	220(4.08),243(4.03), 295(4.12),324(4.17)	78-5475-69
Cyclopenta[c]pyran-4-carboxylic acid, 1,5,6,7-tetrahydro-1-oxo-, methyl ester	MeOH	265(3.90),291(3.66)	18-1776-69
1,4,5-Naphthalenetriol, 2β,3β-epoxy-1α,2,3,4α-tetrahydro-	EtOH	277(3.35),282s(3.34)	39-2053-69C
$C_{10}H_{10}O_5$			
7-Oxabicyclo[2.2.1]heptadiene-2,3-dicarboxylic acid, 1,4-dimethyl-	H_2O	228(3.73),289(3.15)	33-0584-69
$C_{10}H_{10}O_6$			
Glucopyranose oxidation product	MeOH	252(3.03)	23-0511-69
Phthalic acid, 4,5-dimethoxy-	EtOH	223(4.43),266(3.97)	39-1921-69C
Terephthalic acid, 2,5-dihydroxy-, dimethyl ester	EtOH	252(4.20)	35-6102-69
	$CHCl_3$	255(4.19)	35-6102-69
	DMSO	365(3.64)	35-6102-69
$C_{10}H_{10}S$			
Benzo[b]thiophene, 2,3-dimethyl-	n.s.g.	232(4.47),267(3.88), 291(3.43),300(3.33)	22-0601-69
Thiophene, 2-propenyl-5-(1-propynyl)-	MeOH	281(3.86),314(3.86)	39-1813-69C
$C_{10}H_{11}BaN_4O_7P$			
Hypoxanthine, 9-β-D-xylofuranosyl-, 3',5'-cyclic phosphate, barium salt	pH 1	250(4.04)	44-1547-69
	H_2O	249(4.05)	44-1547-69
	pH 13	254(4.09)	44-1547-69

Compound	Solvent	λ_{max}(log ϵ)	Ref.
$C_{10}H_{11}BaN_4O_8P$			
Hypoxanthine, 9-β-D-xylofuranosyl-,	pH 1	249(4.02)	44-1547-69
	H_2O	249(4.06)	44-1547-69
	pH 13	254(4.07)	44-1547-69
$C_{10}H_{11}Br$			
1-Butene, 4-bromo-1-phenyl-, trans	ether	250(4.29),283(3.33), 292(3.21)	44-0528-69
$C_{10}H_{11}BrN_2S_2$			
2H-1,3,5-Thiadiazine-2-thione, 3-(p-bromophenyl)-tetrahydro-5-methyl-	MeOH	250(4.05),292(3.98)	73-2952-69
$C_{10}H_{11}BrO_2$			
Propiophenone, 2'-bromo-4'-methoxy-	C_6H_{12}	258(3.94)	22-2079-69
	EtOH	266(3.94)	22-2079-69
Propiophenone, 3'-bromo-2'-methoxy-	C_6H_{12}	239(3.73),289(3.08)	22-2079-69
	EtOH	240(3.71),291(3.08)	22-2079-69
Propiophenone, 3'-bromo-4'-methoxy-	C_6H_{12}	266(4.17),285(3.57), 296(3.38)	22-2079-69
	EtOH	270(4.15)	22-2079-69
Propiophenone, 3'-bromo-6'-methoxy-	C_6H_{12}	240(3.96),310(3.47)	22-2079-69
	EtOH	244(3.89),316(3.48)	22-2079-69
Propiophenone, 4'-bromo-2'-methoxy-	C_6H_{12}	252(4.04),257(3.95), 297(3.67)	22-2079-69
	EtOH	254(4.06),302(3.66)	22-2079-69
$C_{10}H_{11}Br_3O$			
2,5-Cyclohexadien-1-one, 2,4,6-tribromo-4-tert-butyl-	hexane	273(4.11)	55-0074-69
$C_{10}H_{11}Br_3O_2$			
Benzyl alcohol, 2,3,5-tribromo-α-ethyl-4-methoxy-	C_6H_{12}	281(2.91),289(2.92)	22-2079-69
	EtOH	281(2.92),289(2.94)	22-2079-69
Benzyl alcohol, 2,4,6-tribromo-α-ethyl-3-methoxy-	C_6H_{12}	285(3.00),292(3.02)	22-2079-69
	EtOH	284(3.00),292(3.02)	22-2079-69
Benzyl alcohol, 3,4,5-tribromo-α-ethyl-2-methoxy-	C_6H_{12}	230(4.20),254(3.15), 283(2.90),292(2.95)	22-2079-69
	EtOH	227(4.21),254(3.11), 282(2.90),291(2.95)	22-2079-69
$C_{10}H_{11}ClN_4O_3$			
1-Propene-1,3-diamine, N^1-(p-chlorophenyl)-N^3-methyl-2-nitro-N^3-nitroso-	dioxan	245(4.31),380(4.33)	78-1617-69
$C_{10}H_{11}ClN_4O_5$			
Inosine, 2-chloro-	pH 1	254(4.01)	94-2581-69
	pH 13	258(4.06)	94-2581-69
$C_{10}H_{11}ClO_2$			
2,4,6-Cycloheptatrien-1-one, 3-chloro-2-hydroxy-4-propyl-	EtOH	245(4.43),328(3.96), 360(3.80),378(3.72)	12-1011-69
	KOH	252(4.43),343(4.18), 408(4.09)	12-1011-69
$C_{10}H_{11}ClO_3$			
1,2,4-Naphthalenetriol, 3α-chloro-1α,2α,3,4α-tetrahydro-	EtOH	261(2.35)	39-2053-69C

Compound	Solvent	$\lambda_{max}(\log \epsilon)$	Ref.
$C_{10}H_{11}ClO_3S$			
Propiophenone, 2-chloro-2-(methylsulfonyl)-	ether	256(3.98)	24-4017-69
$C_{10}H_{11}ClO_4$			
Cryptosporiopsin (cf. $C_{10}H_{10}Cl_2O_4$)	EtOH	284(4.34)	23-3700-69
2-Cyclopentene-1-carboxylic acid, 3-chloro-1-hydroxy-4-oxo-2-propenyl-, methyl ester, (1S,5R)-, trans	MeOH	277(4.34)	35-0157-69
(1S,5S)-	MeOH	277(4.36)	35-0157-69
2-Cyclopentene-1-carboxylic acid, 3-chloro-5-hydroxy-4-oxo-2-propenyl-, methyl ester	EtOH	285(4.34)	23-2087-69
$C_{10}H_{11}Cl_6N_3$			
s-Triazine, 2-pentyl-4,6-bis(trichloromethyl)-	MeOH	216(3.54),280(2.81)	18-2924-69
$C_{10}H_{11}Cl_6N_3S$			
s-Triazine, 2-(pentylthio)-4,6-bis-(trichloromethyl)-	MeOH	221(3.46),275(4.14)	18-2931-69
$C_{10}H_{11}FN_4O_4$			
9H-Purine, 6-fluoro-9-β-D-ribofuranosyl-	pH 2	248(3.82)	77-0381-69
	pH 14	254(4.13)	77-0381-69
$C_{10}H_{11}FO$			
2,5-Cyclohexadien-1-one, 4-allyl-2-fluoro-4-methyl-	EtOH	239(4.13)	33-1354-69
2,5-Cyclohexadien-1-one, 4-allyl-3-fluoro-4-methyl-	EtOH	230(4.16)	33-1354-69
$C_{10}H_{11}F_3N_2O_6$			
Uridine, 5-(trifluoromethyl)-	pH 2	263(4.02)	87-0543-69
	pH 11	262(3.85)	87-0543-69
$C_{10}H_{11}FeN$			
Iron, π-cyclopentadienyl(2-methyl-π-pyrrolyl)-	C_6H_{12}	325(2.0),440(1.7)	89-0135-69
$C_{10}H_{11}N$			
Isoquinoline, 7,8-dihydro-8-methyl-	MeOH	260(4.00)	106-0196-69
$C_{10}H_{11}NO$			
7-Azabicyclo[4.2.2]deca-2,4,7,9-tetraene, 8-methoxy-	MeCN	271(3.30)	35-4714-69
4-Azatricyclo[3.3.2.0^{2,8}]deca-3,6,9-triene, 3-methoxy-	MeCN	271(3.30)	35-5296-69
3-Buten-2-one, 4-amino-3-phenyl-	EtOH	300(4.02)	22-0205-69
Hydrocinnamonitrile, p-methoxy-	EtOH	277(3.33)	88-0073-69
$C_{10}H_{11}NOS$			
4,1-Benzothiazepin-2(3H)-one, 1,5-dihydro-	iso-PrOH	236(3.81)	44-0365-69
2H-1,4-Benzoxazine, 2-methyl-3-(methylthio)-	EtOH	238(4.15),284(4.05),293(4.10),308(3.88)	78-0517-69
2H-1,4-Benzoxazine-3-thione, 2,4-dimethyl-	EtOH	252(4.10),325(4.20)	78-0517-69

Compound	Solvent	$\lambda_{max}(\log \epsilon)$	Ref.
$C_{10}H_{11}NOS_2$			
Carbamic acid, N-benzoyl-dithio-, ethyl ester	EtOH	271(4.36)	39-2794-69C
Imidocarbonic acid, benzoyldithio-, dimethyl ester	EtOH	246(4.09)	39-2794-69C
$C_{10}H_{11}NO_2$			
7-Azabicyclo[4.2.2]deca-2,4,7-trien-9-one, 8-methoxy-	hexane	264(3.54),273(3.53)	35-4714-69
7-Azabicyclo[4.2.2]deca-3,7,9-trien-2-one, 8-methoxy-	hexane	221(3.73)	35-4714-69
4-Azatricyclo[3.3.2.0^{2,8}]deca-3,9-dien-7-one, 3-methoxy-, perchlorate	hexane	273s(2.10),280s(2.04)	35-5296-69
2H-1,4-Benzoxazin-3(4H)-one, 2,4-dimethyl-	EtOH	255(3.90),282(3.78)	78-0517-69
1,3-Butanedione, 1-phenyl-, oxime	EtOH	254(2.98)	77-1062-69
Carbanilic acid, o-vinyl-, methyl ester	n.s.g.	221(4.31),247(4.15), 295(3.04)	23-2751-69
Crotonohydroxamic acid, N-phenyl-	EtOH	216(4.29),278(3.94)	34-0278-69
1-Indolinecarboxaldehyde, 2-hydroxy-5-methyl-	EtOH	253(4.16),286(3.58), 293(3.58)	1-0159-69
1-Indolinecarboxaldehyde, 2-hydroxy-7-methyl-	EtOH	249(4.05),281(3.45), 302(3.28)	1-0159-69
Naphthalene, 1,2,3,4-tetrahydro-6-nitro-	C_6H_{12}	272(3.97)	35-3558-69
	EtOH	281(3.95)	35-3558-69
Pedicularine	n.s.g.	272(2.89)	105-0383-69
Propionamide, 3-benzoyl-	MeCN	198(4.45),240(4.11), 278(2.92),311s(1.40)	28-0279-69A
Styrene, β-ethyl-β-nitro-	n.s.g.	306(3.99)	67-0127-69
$C_{10}H_{11}NO_2S$			
4,1-Benzothiazepin-2(1H)-one, 3,5-dihydro-3-methyl-, 4-oxide, cis	iso-PrOH	226(4.33),269(3.43)	44-0365-69
trans	iso-PrOH	265(3.48)	44-0365-69
Imidocarbonic acid, benzoylthio-, O-ethyl ester	EtOH	272(4.20)	39-2794-69C
$C_{10}H_{11}NO_3$			
Acetophenone, 2'-methoxy-3'-methyl-5'-nitroso-	EtOH	235(4.00),294(3.78), 321(3.91)	12-0935-69
	dioxan	742(1.64)	12-0935-69
2H-1,4-Benzoxazine-6,7-dione, 3,4-dihydro-5,8-dimethyl-	EtOH	302(4.11),560(3.07)	39-1325-69C
Glyoxylic acid, phenyl-, ethyl ester, 2-oxime	1% MeOH	245(3.62)	22-2894-69
anion	1% MeOH	269(4.02)	22-2894-69
1-Indolinecarboxaldehyde, 2-hydroxy-	EtOH	257(4.12),294(3.58), 303(3.51)	1-0159-69
$C_{10}H_{11}NO_4$			
Benzoic acid, 2,3,4-trimethyl-5-nitro-	EtOH	224(3.91),269(3.32)	78-1047-69
3,5-Pyridinedicarboxylic acid, 2-methyl-, dimethyl ester	EtOH	207(4.40),229(4.11), 271(3.51)	73-0427-69
$C_{10}H_{11}NO_4S$			
Propionic acid, 2-[(o-nitrobenzyl)thio]-	iso-PrOH	250(3.72)	44-0365-69
$C_{10}H_{11}NS$			
7-Azabicyclo[4.2.2]deca-2,4,7,9-tetraene, 8-(methylthio)-	MeCN	214(3.86),282(3.23)	35-5296-69

Compound	Solvent	$\lambda_{max}(\log \epsilon)$	Ref.
4-Azatricyclo[3.3.2.0^{2,8}]deca-3,6,9-triene, 3-(methylthio)-	MeCN	235(3.86)	35-5296-69
Benzothiazoline, 3-ethyl-2-methylene-	MeCN	294(4.05)	88-2709-69
Indole, 1-methyl-2-(methylthio)-	EtOH	222(4.51),287(4.10), 293(4.09)	94-0550-69
2-Indolinethione, 1,3-dimethyl-	EtOH	230(4.17),295s(4.05), 315(4.23)	94-0550-69
2-Thiazoline, 4-methyl-2-phenyl-	MeOH	244(4.3),278(3.4)	39-1120-69C
Thieno[2,3-b]pyridine, 4,5,6-trimethyl-	EtOH	234(4.47),272(3.75), 285(3.62),291(3.62), 297(3.49),302(3.57)	44-0347-69
	EtOH-HCl	242(4.49),306(3.87)	44-0347-69
$C_{10}H_{11}NSe$			
Benzoselenazoline, 3-ethyl-2-methylene-	MeCN	295(3.92)	88-2709-69
$C_{10}H_{11}N_3$			
6,8-Diazacarbazole, 1,2,3,4-tetrahydro-	EtOH	233(4.50),275(3.38), 305(3.35)	23-2061-69
	EtOH-HCl	237(4.47),293(3.44), 325s(3.03)	23-2061-69
	concd HCl	242(4.50),300(3.73), 325s(3.50)	23-2061-69
$C_{10}H_{11}N_3O$			
Isocyanic acid, 1-methyl-1-(phenylazo)-ethyl ester	C_6H_{12}	266(3.98),388(2.08)	49-1479-69
Isoxazole, 5-hydrazino-3-methyl-4-phenyl-	MeOH	257(4.12)	4-0783-69
Isoxazole, 5-hydrazino-4-methyl-3-phenyl-	MeOH	230(4.14),285(4.08)	4-0783-69
1,3,4-Oxadiazol-3-ine, 2,2-dimethyl-5-(phenylimino)-	EtOH	221(3.90),322(3.82)	44-3230-69
2-Pentenedinitrile, 4-(morpholinomethylene)-	EtOH	318(4.52)	39-1086-69C
2-Propen-1-ol, 2-methyl-3-v-triazolo-[1,5-a]pyridin-3-yl-	EtOH	224(4.23),262(4.10), 293(3.97),320s(--)	88-1549-69
geometric isomer	EtOH	224(4.22),252s(4.07), 259(4.08),269(3.99), 293(3.92)	88-1549-69
$C_{10}H_{11}N_3OS$			
4H-Cyclopenta[4,5]thieno[2,3-d]pyrimidin-4-one, 3-amino-3,5,6,7-tetrahydro-2-methyl-	MeOH	217(4.03),271(3.73), 313(3.96),326s(3.85)	48-0402-69
$C_{10}H_{11}N_3O_2$			
1,4-Phthalazinedione, 5-(dimethylamino)-2,3-dihydro-	THF	270(3.4),310(3.7)	24-3241-69
Pyrimido[1,2-a]azepine-3-carbonitrile, 4,6,7,8,9,10-hexahydro-2-hydroxy-4-oxo-	MeOH	216(4.43),247(3.46), 287(3.54)	24-2739-69
Spiro[oxazolidine-2,1'(4'H)-phthalazin]-4'-one, 2',3'-dihydro-	pH 1	214(4.5),233s(4.0), 300(3.7)	114-0151-69B
	pH 13	233s(3.9),322(3.7)	114-0151-69B
$C_{10}H_{11}N_3O_2S$			
Anthranilic acid, N-(5,6-dihydro-4H-1,3,4-thiadiazin-2-yl)-	EtOH	222(4.77),311(3.69)	44-0372-69

Compound	Solvent	λ_{max}(log ε)	Ref.
$C_{10}H_{11}N_3O_4S$			
Thiazolo[4,5-d]pyrimidine-2-carboxylic acid, 4,5,6,7-tetrahydro-4,6-dimethyl-5,7-dioxo-, ethyl ester	MeOH	226(4.30),252(3.71), 342(3.81)	44-3285-69
$C_{10}H_{11}N_3S$			
Malononitrile, [[1-methyl-2-(methylthio)-2-pyrrolin-3-yl]methylene]-	C_6H_{12}	411(4.49)	5-0073-69E
$C_{10}H_{11}N_4Na_2O_9P$			
Inosine, 1-oxide, 5'-phosphate, disodium salt	pH 1	252(3.92)	94-1128-69
	pH 7	228(4.65),254(3.74)	94-1128-69
	pH 13	258(3.81),295(3.60)	94-1128-69
$C_{10}H_{11}N_5$			
Acetamidine, 2-phenyl-N-4H-1,2,4-triazol-4-yl-	MeOH	218(3.82)	48-0477-69
p-Toluamidine, N-4H-1,2,4-triazol-4-yl-	MeOH	237(4.11)	48-0477-69
$C_{10}H_{11}N_5OS$			
as-Triazine-2,4(3H,5H)-dipropionitrile, 6-methyl-5-oxo-3-thioxo-	EtOH	222(4.13),270(4.21), 300(3.69)	114-0093-69C
$C_{10}H_{11}N_5O_3$			
Acetamide, N-(7-acetyl-1,6-dihydro-8-methyl-6-oxopurin-2-yl)-	MeOH	257(4.21),292(3.97)	5-0201-69F
Acetamide, N,N'-(2-methyloxazolo[5,4-d]-pyrimidine-5,7-diyl)bis-	pH -0.89	267(4.32),301(4.09)	5-0201-69F
	pH 5.0	234(4.50),256(4.19), 284(4.13)	5-0201-69F
Isoxanthopterin, 6-acetonyl-8-methyl-	pH -3.32	222s(4.24),287(3.96), 320(3.87),398(4.01), 419(4.12),446(4.10)	24-4032-69
	pH 5	218(4.52),287s(3.94), 292(3.97),345(4.14), 358s(4.09)	24-4032-69
	pH 12	259(3.99),276s(3.77), 364(4.10),402(3.81)	24-4032-69
$C_{10}H_{11}N_5O_4$			
Hydrocinnamic acid, α,β-dioxo-, semicarbazide, dioxime	EtOH	245(4.19)	55-0859-69
$C_{10}H_{12}$			
Bicyclo[3.2.2]nona-2,6,8-triene, 3-methyl-	hexane	220(3.53)(end abs.)	35-3998-69
2-Butene, 2-phenyl-, (E)-	n.s.g.	243(4.01)	28-1654-69B
2-Butene, 2-phenyl-, (Z)-	n.s.g.	235(3.91)	28-1654-69B
1,3,5-Cycloheptatriene, 1-isopropenyl-	hexane	222(4.32),227s(4.31), 296(3.86)	35-3998-69
1,3,5-Cycloheptatriene, 2-isopropenyl-	hexane	229(4.28)	35-3998-69
1,3,5-Cycloheptatriene, 3-isopropenyl-	hexane	225(4.03),277(3.98)	35-3998-69
1,3,5-Cycloheptatriene, 7-isopropenyl-	hexane	258(3.50)	35-3998-69
$C_{10}H_{12}BaNO_8P$			
2(1H)-Pyridone, 1-β-D-ribofuranosyl-, 2'(3')-phosphate, barium salt	pH 2	300(3.73)	73-0089-69
$C_{10}H_{12}BrNO_4$			
Tyrosine, 3-bromo-5-methoxy-, hydrochloride	n.s.g.	280s(3.36),284(3.37)	39-0559-69C

Compound	Solvent	λ max (log ε)	Ref.
$C_{10}H_{12}BrN_5$			
Purine, 8-bromo-6-piperidino-	EtOH	286(4.28)	65-2125-69
$C_{10}H_{12}Br_2O_2$			
Benzyl alcohol, 3,5-dibromo-α-ethyl-2-methoxy-	C_6H_{12}	280(2.95),287(2.95)	22-2079-69
	EtOH	277(2.96),285(2.95)	22-2079-69
Benzyl alcohol, 3,5-dibromo-α-ethyl-4-methoxy-	C_6H_{12}	224(3.98),277(2.85),284(2.87)	22-2079-69
	EtOH	224(3.98),276(2.85),283(2.86)	22-2079-69
2-Cyclohexene-1,4-dione, 5,6-dibromo-2-tert-butyl-	MeOH	269(4.07),370(2.16)	39-2164-69C
$C_{10}H_{12}CaIN_2O_7P$			
Thymidine, 3'-deoxy-3'-iodo-, 5'-phosphate, calcium salt	H_2O	267(3.99)	69-4889-69
$C_{10}H_{12}ClNO_3S_3$			
Metanilyl chloride, N-acetyl-4,6-bis-(methylthio)-	MeOH	266(4.37)	44-1780-69
$C_{10}H_{12}ClNO_7S_3$			
Metanilyl chloride, N-acetyl-4,6-bis-(methylsulfonyl)-	MeOH	225(4.44),263(4.16),303(3.82)	44-1780-69
$C_{10}H_{12}Cl_2N_2O_2$			
p-Benzoquinone, 2,5-bis[(2-chloroethyl)-amino]-	EtOH	338(4.47),490(2.51)	39-1325-69C
$C_{10}H_{12}Cl_2N_4S$			
s-Triazine, 2,4-dichloro-6-[1,4,5,6-tetrahydro-1-methyl-2-(methylthio)-3-pyridyl]-	C_6H_{12}	308s(3.64),389(4.61)	5-0073-69E
$C_{10}H_{12}Cl_2O$			
2,5-Cyclohexadien-1-one, 4-(dichloro-methyl)-3,4,5-trimethyl-	H_2O	242(4.24)	44-0224-69
	90.5% H_2SO_4	263(4.22),320(3.84)	44-0224-69
$C_{10}H_{12}Cl_2O_3$			
3-Cyclohexene-1-acetic acid, 2-(di-chloromethyl)-2-methyl-5-oxo-	MeOH	221(4.04)	35-2299-69
$C_{10}H_{12}Cl_2O_4$			
2-Cyclopentene-1-carboxylic acid, 2-allyl-3,5-dichloro-1,4-dihydroxy-, dimethyl ester	MeOH	247(4.35),256s(4.20)	39-2187-69C
2-Cyclopentene-1-carboxylic acid, 1,3-dichloro-5-hydroxy-4-oxo-2-propyl-, methyl ester	EtOH	244(3.92)	23-2087-69
$C_{10}H_{12}FN_5O_3$			
Adenine, 9-(2-deoxy-2-fluoro-α-D-arabinofuranosyl)-	pH 1	257(4.17)	44-2632-69
	H_2O	259(4.18)	44-2632-69
β-isomer	pH 1	258(4.16)	44-2632-69
	H_2O	260(4.16)	44-2632-69
Adenine, 9-(2-deoxy-α-D-ribofuranosyl)-2-fluoro-	pH 1	263(4.11),267s(4.10)	87-0498-69
	pH 7,13	261(4.17),268s(4.07)	87-0498-69
Adenosine, 2'-deoxy-2-fluoro-	pH 1	262(4.12),267s(4.08)	87-0498-69
	pH 7,13	261(4.16),267s(4.06)	87-0498-69

Compound	Solvent	λmax(log ε)	Ref.
$C_{10}H_{12}FN_5O_4$			
Adenine, 9-α-D-arabinofuranosyl-	pH 1	262(4.13)	87-0498-69
2-fluoro-	pH 7,13	262(4.17)	87-0498-69
Adenine, 9-β-D-arabinofuranosyl-	pH 1	262(4.12)	87-0498-69
2-fluoro-	pH 7	261(4.17)	87-0498-69
	pH 13	262(4.18)	87-0498-69
Adenine, 2-fluoro-9-β-D-xylofuranosyl-	pH 1	262(4.12),267s(--)	44-1396-69
	pH 7	262(4.17)	44-1396-69
$C_{10}H_{12}I_2N_2O_2$			
p-Benzoquinone, 2,5-bis[(2-iodoethyl)-amino]-	EtOH	342(4.48),495(2.49)	39-1325-69C
$C_{10}H_{12}N_2$			
Phthalazine, 5,8-dihydro-1,4-dimethyl-	EtOH	254(3.41)	77-0795-69
2-Pyrazoline, 3-methyl-1-phenyl-	EtOH	244s(3.74),276(4.08)	39-1703-69C
	EtOH-HCl	282(2.95)	39-1703-69C
3H-Pyrrolo[1,2-b][1,2]diazepine, 2,5-dimethyl-	EtOH	245(4.1),273(4.0)	24-3268-69
1H-Pyrrolo[3,2-c]pyridine, 2-ethyl-3-methyl-	EtOH	228(4.57),278(3.70), 286s(3.65)	23-2061-69
	EtOH-HCl	232(4.65),289(3.65), 316s(3.28)	23-2061-69
3H-Pyrrolo[3,2-c]pyridine, 2,3,3-tri-	EtOH	244(3.56),299s(2.78)	23-2061-69
methyl-	EtOH-HCl	278(3.85),308s(2.94)	23-2061-69
	concd. HCl	256(3.84),313s(2.43)	23-2061-69
$C_{10}H_{12}N_2O$			
3-Pyrazolidinone, 4-methyl-1-phenyl-	N HCl	260(2.78)	65-1835-69
	H_2O	240(3.91)	65-1835-69
3-Pyrazolidinone, 4-methyl-2-phenyl-	N HCl	235(3.79)	65-1835-69
	H_2O	255(3.88)	65-1835-69
3-Pyrazolidinone, 5-methyl-1-phenyl-	N HCl	260(2.78)	65-1835-69
	H_2O	235(3.90)	65-1835-69
3-Pyrazolidinone, 5-methyl-2-phenyl-	N HCl	235(3.80)	65-1835-69
	H_2O	255(3.89)	65-1835-69
$C_{10}H_{12}N_2O_2$			
Furo[2,3-d]pyrrolo[1,2-a]pyrimidin-4(2H)-one, 3,6,7,8-tetrahydro-2-methyl-	MeOH	209(4.31),250(3.62), 267(3.54)	24-2739-69
Pyrrolidine, N-(p-nitrophenyl)-	DMSO	392(3.34)	39-0544-69B
$C_{10}H_{12}N_2O_3$			
p-Acetophenetidide, 2'-nitroso-	MeOH	250(4.2),298(3.8), 310(3.9),430(3.6), 750(1.7)	83-0043-69
o-Anisamide, N,N-dimethyl-5-nitroso-	EtOH	246(3.94),339(4.15)	12-0935-69
	dioxan	730(1.67)	12-0935-69
2-Isoxazolin-5-one, 3-methyl-4-(N-ethyl-3-methyl-3-isoxazolin-5-ylidene)-	EtOH	250(3.63),295(4.14)	39-0245-69C
1H-Pyrrolo[2,3-c]pyridine-2-carboxylic acid, 4,5,6,7-tetrahydro-5-oxo-, ethyl ester	EtOH	276(4.27)	35-2338-69
$C_{10}H_{12}N_2O_3S$			
3-Carboxy-8-ethoxy-2,3-dihydro-5-methyl-	pH 1	245(3.50),335(3.59)	1-2437-69
thiazolo[3,2-c]pyrimidin-4-ium hydroxide, inner salt	pH 13	235(3.56),285(3.60)	1-2437-69

Compound	Solvent	$\lambda_{max}(\log \epsilon)$	Ref.
Thymine, (S)2,3'-anhydro-1-(2,3-dideoxy-β-D-threo-pentofuranosyl)-2-thio-	N HCl	232(4.10),278(3.95)	44-1020-69
	pH 6.34	237(4.46),270s(--)	44-1020-69
$C_{10}H_{12}N_2O_3S_2$			
Thymine, 1-(2,3,5-trideoxy-3,5-epidithio-β-D-threo-pentofuranosyl)-	pH 0-6.34	267(3.99)	44-1020-69
	pH 12	267(3.88)	44-1020-69
$C_{10}H_{12}N_2O_4$			
o-Anisamide, N,N-dimethyl-5-nitro-	EtOH	300(4.02)	12-0935-69
$C_{10}H_{12}N_2O_5$			
Acetanilide, 2',5'-dimethoxy-4'-nitro-	EtOH	226s(4.12),258(3.91), 296(3.77),369(3.90)	117-0287-69
Uracil, 1-(3,5-anhydro-β-D-xylofuranosyl)-3-methyl-	H_2O	262(3.95)	94-0785-69
	pH 13	260(3.94)	94-0785-69
$C_{10}H_{12}N_2O_6$			
3,6-Pyridazinedicarboxylic acid, 4-(2-hydroxyethoxy)-, dimethyl ester	MeOH	275(3.49)	4-0497-69
$C_{10}H_{12}N_2O_7$			
Uracil, 2'-deoxy-β-D-arabinohexopyranuronosyl-	H_2O	260(3.72)	88-1961-69
$C_{10}H_{12}N_2O_7S$			
Uracil, 2,2'-anhydro-1-(5-O-mesyl-β-D-arabinofuranosyl)-	MeOH	226(4.03),247(3.95)	94-1188-69
$C_{10}H_{12}N_4$			
Quinazoline, 2,4-diamino-8-ethyl-	acid	210s(4.17),228(4.52), 236(4.54),250s(4.14), 260s(3.99),270s(3.71), 318(3.65),327s(3.60)	44-0821-69
	neutral	233(4.59),263s(3.84), 270(3.90),280(3.86), 335(3.64)	44-0821-69
$C_{10}H_{12}N_4O$			
2(1H)-Naphthalenone, 1,3-bis(diazo)-3,4,4a,5,6,7,8,8a-octahydro-, cis	EtOH	323(4.40)	44-1480-69
trans	EtOH	324(4.38)	44-1480-69
3-Pyrazolin-5-one, 1-(4,6-dimethyl-2-pyrimidinyl)-3-methyl-	MeOH	250(4.3)	94-1467-69
3-Pyrazolin-5-one, 2-(4,6-dimethyl-2-pyrimidinyl)-3-methyl-	MeOH	278(4.3)	94-1467-69
4H-Pyrazolo[3',4':4,5]pyrimido[1,2-a]azepin-4-one, 1,6,7,8,9,10-hexahydro-	MeOH	211(4.38),252(3.98)	24-2739-69
$C_{10}H_{12}N_4O_2$			
3-Pyrazolin-5-one, 1-(4-methoxy-6-methyl-2-pyrimidinyl)-3-methyl-	pH 7.0	250(4.1)	94-1485-69
	$CHCl_3$	262(4.3)	94-1485-69
Pyrazol-3-ol, 1-(4-methoxy-6-methyl-2-pyrimidinyl)-5-methyl-	pH 7.0	280(4.3)	94-1485-69
	$CHCl_3$	285(4.4)	94-1485-69
2,3-Quinoxalinedione, 1,4-dihydro-1,4-dimethyl-, dioxime	EtOH	233(4.49),276(3.81), 329(3.89)	39-2319-69C
$C_{10}H_{12}N_4O_3$			
4,5-Pyrazolidinedione, 3-hydroxy-3-methyl-1-phenyl-, dioxime	EtOH	239(4.22),309s(3.51)	4-0317-69

Compound	Solvent	$\lambda_{max}(\log \epsilon)$	Ref.
$C_{10}H_{12}N_4O_4S$			
9H-Purine-6-thiol, 9-α-D-arabinopyrano-syl-	pH 1	322(4.38)	44-0416-69
	pH 7	317(4.37)	44-0416-69
	pH 13	310(4.36)	44-0416-69
9H-Purine-6-thiol, 9-β-D-arabinopyrano-syl-	pH 1	321(4.37)	44-0416-69
	pH 7	317(4.18)	44-0416-69
	pH 13	309(4.37)	44-0416-69
9H-Purine-6-thiol, 9-β-D-xylopyranosyl-	pH 1	322(4.40)	44-0416-69
	pH 7	318(4.38)	44-0416-69
	pH 13	310(4.42)	44-0416-69
Purine-6(1H)-thione, 7-β-D-ribofurano-syl-	pH 1	221(3.95),331(4.20)	44-2646-69
	pH 7	221(3.95),328(4.19)	44-2646-69
	pH 13	230(4.03),318(4.18)	44-2646-69
$C_{10}H_{12}N_4O_5$			
Hypoxanthine, 9-α-D-arabinopyranosyl-	pH 1	247(4.04)	44-0416-69
	pH 7	247(4.05)	44-0416-69
	pH 13	253(4.08)	44-0416-69
Hypoxanthine, 9-β-D-arabinopyranosyl-	pH 1	248(4.11)	44-0416-69
	pH 7	248(4.10)	44-0416-69
	pH 13	253(4.14)	44-0416-69
Hypoxanthine, 7-α-D-ribofuranosyl-	pH 1	226(3.60),252(3.97)	44-2646-69
	pH 7	229(3.57),257(3.92)	44-2646-69
	pH 13	230(3.60),263(3.95)	44-2646-69
Hypoxanthine, 7-β-D-ribofuranosyl-	pH 1	226(3.60),252(3.96)	44-2646-69
	pH 7	229(3.61),256(3.93)	44-2646-69
	pH 13	229(3.62),263(3.97)	44-2646-69
Hypoxanthine, 9-β-D-xylopyranosyl-	pH 1	247(4.07)	44-0416-69
	pH 7	247(4.09)	44-0416-69
	pH 13	253(4.11)	44-0416-69
$C_{10}H_{12}N_4O_6$			
Xanthine, 3-β-D-ribofuranosyl-	pH 5	267(4.07)	4-0995-69
$C_{10}H_{12}N_4S$			
9H-Purine-6-thiol, 9-cyclopentyl-	pH 1	221(4.08),325(4.34)	73-2114-69
	pH 13	234(4.17),314(4.36)	73-2114-69
$C_{10}H_{12}O$			
2,5-Cyclohexadien-1-one, 4-allyl-4-methyl-	EtOH	239(4.14)	33-1354-69
2H-Cycloprop[c]inden-2-one, 1,1a,4,5,6,7-hexahydro-	hexane	210(3.81),247(3.48)	22-2095-69
	EtOH	216(3.81),260(3.43)	22-2095-69
Isobutyrophenone	EP	327(1.8)	18-0010-69
	EP at 77°K	325(1.8)	18-0010-69
2-Octene-4,6-diyne, 2-ethoxy-	EtOH	222(4.46),251s(3.88), 263(4.04),276(4.17), 292(4.06)	78-2823-69
Propiophenone, 2'-methyl-	2:1C_6H_{11}Me: isopentane	238(3.95),283(3.09), 318s(1.8)	59-0393-69
Trispiro[2.0.2.0.2.1]decan-10-one	C_6H_{12}	210(4.05),284(1.75), 296s(1.48)	88-3545-69
4,7-Methanoinden-1-ol, 3a,4,7,7a-tetra-hydro-	isooctane	183(4.09)	39-0710-69B
$C_{10}H_{12}OS$			
m-Toluic acid, thio-, O-ethyl ester	C_6H_{12}	251(3.79),259(3.79), 288(4.11),419(2.12)	18-3556-69
p-Toluic acid, thio-, O-ethyl ester	C_6H_{12}	252(3.83),295(4.17), 419(2.16)	18-3556-69

Compound	Solvent	λ max (log ε)	Ref.
$C_{10}H_{12}OSe$			
1-Benzoselenepin-5-ol, 2,3,4,5-tetrahydro-	MeOH	236s(3.55),276(3.66)	73-3801-69
$C_{10}H_{12}O_2$			
Acetophenone, 3'-ethyl-4'-hydroxy-	EtOH	226(4.13),283(4.16)	22-0781-69
1,4-Cyclohexadiene-1-carboxaldehyde, 4,6,6-trimethyl-3-oxo-	ether	244(3.93),286(3.53)	24-2211-69
3-Cyclohexene-1-acetaldehyde, 4-methyl-α-methylene-5-oxo-	EtOH	219(4.15),234(3.98)	44-0857-69
o-Eugenol	EtOH	276(3.35)	12-1011-69
	KOH	243(3.89),292(3.65)	12-1011-69
Isobutyric acid, α-phenyl-, Pb(IV) salt	hexane	230(4.39)	33-1495-69
Isobutyrophenone, 2'-hydroxy-	hexane	250(4.0),256(4.0), 385(3.6)	18-0960-69
2H-Pyran-2-one, 4-methyl-6-(2-methylpropenyl)-	n.s.g.	330(3.57)	114-0397-69A
$C_{10}H_{12}O_2S$			
Benzoic acid, m-methoxy-thio-, O-ethyl ester	C_6H_{12}	288(4.04),318s(--), 420(2.08)	18-3556-69
Benzoic acid, p-methoxy-thio-, O-ethyl ester	C_6H_{12}	261(3.68),312(4.25), 415(2.32)	18-3556-69
Benzo[c]thiophene-2-carboxylic acid, 4,5,6,7-tetrahydro-	EtOH	263(4.0)	2-0009-69
$C_{10}H_{12}O_3$			
Acetophenone, 3',4'-dimethoxy-	hexane	225(4.29),229s(4.24), 267(4.09),271(4.08), 296(3.83),305(3.76)	59-0313-69
1,4-Benzodioxan-6-ol, 7-ethyl-	EtOH	294(3.73)	104-0468-69
Benzoic acid, 3-ethyl-4-hydroxy-, methyl ester	EtOH	262(4.15)	22-0781-69
Butyric acid, 2-(p-hydroxyphenyl)-, L-	H_2O	197(4.23),221(3.80), 275(3.20),280s(--)	54-0805-69
	N KOH	199(4.39),238(3.96), 293(3.42)	54-0805-69
	MeOH	211(4.54),225(--), 277(3.20),283s(--)	54-0805-69
	dioxan	224(3.89),278(3.32), 285s(--)	54-0805-69
	HOAc	277(3.18),283s(--)	54-0805-69
	DMF	280(3.28),287s(--)	54-0805-69
1,2,4-Naphthalenetriol, 1,2,3,4-tetrahydro-	EtOH	264(2.23)	39-2053-69C
Propiophenone, 4'-hydroxy-3'-methoxy-	EtOH	275(4.00),302(3.93)	44-0585-69
2H-Pyran-2-one, 6-(2-oxopentyl)-	hexane	219(3.62),299(3.76)	1-1059-69
$C_{10}H_{12}O_3S$			
Acetic acid, [(m-methoxybenzyl)thio]-	EtOH	276(3.35),283(3.30)	44-1566-69
Benzene, [1,2-epoxy-2-(methylsulfonyl)-propyl]-	ether	216(4.02),254(2.25), 260(2.36),266(2.28)	24-4017-69
$C_{10}H_{12}O_4$			
Acetophenone, 2'-hydroxy-4',6'-dimethoxy-	EtOH	224(4.16),288(4.29), 325(3.63)	95-0372-69
2-Furanacrylic acid, 5-methoxy-α-methyl-, methyl ester	MeOH	330(4.49)	40-0096-69
	MeOH	332(4.54)	40-0716-69
Phthalic acid, 1,2-dihydro-4,5-dimethyl-	n.s.g.	232(3.40),260(3.55), 300(3.10)	104-2164-69

Compound	Solvent	λ_{max}(log ε)	Ref.
4H-Pyran-3-carboxylic acid, 2,6-dimethyl-4-oxo-, ethyl ester	EtOH	265(4.10)	22-0231-69
$C_{10}H_{12}S_2$			
Benzoic acid, dithio-, isopropyl ester	C_6H_{12}	248s(3.7),296(4.24), 330s(3.99),509(2.07)	48-0045-69
	EtOH	245s(3.6),299(4.20), 334s(3.9),503(2.06)	48-0045-69
	MeCN	245s(3.5),298(4.18), 334s(3.9),500(2.11)	48-0045-69
$C_{10}H_{13}$			
Isopropyltropylium (chloroplatinate)	EtOH	271(--),362s(--)	18-3277-69
$C_{10}H_{13}BrO_2$			
Anisole, 2-bromo-4-(1-hydroxypropyl)-	C_6H_{12}	224(3.98),274(3.32), 281(3.43),288(3.42)	22-2079-69
	EtOH	222(3.97),282(3.38), 289(3.34)	22-2079-69
Anisole, 2-bromo-6-(1-hydroxypropyl)-	C_6H_{12}	258(3.13),270(3.04), 277(2.87)	22-2079-69
	EtOH	258(3.12),269(3.04), 276(2.83)	22-2079-69
Anisole, 3-bromo-4-(1-hydroxypropyl)-	C_6H_{12}	279(3.36),287(3.33)	22-2079-69
	EtOH	279(3.34),287(3.30)	22-2079-69
Anisole, 3-bromo-6-(1-hydroxypropyl)-	C_6H_{12}	226(3.96),277(3.41), 284(3.40)	22-2079-69
	EtOH	226(3.96),277(3.43), 284(3.41)	22-2079-69
Anisole, 4-bromo-2-(1-hydroxypropyl)-	C_6H_{12}	229(4.03),282(3.28), 289(3.26)	22-2079-69
	EtOH	229(4.00),282(3.30), 288(3.27)	22-2079-69
$C_{10}H_{13}BrO_3$			
Bicyclo[3.3.1]nonane-2,6-dione, 3-bromo-7-methoxy-, m. 91°	MeOH	302(1.08)	88-5135-69
isomer	MeOH	300(1.34)	88-5135-69
$C_{10}H_{13}BrO_3S$			
1,2-Benzoxathiin, 3-bromo-5,6,7,8-tetrahydro-4,7-dimethyl-, 2,2-dioxide	$CHCl_3$	284(3.94)	88-0651-69
$C_{10}H_{13}ClN_4O_2$			
Pyrimidine, 4-chloro-6-(cyclohexylamino)-5-nitro-	pH 1	233(4.30),287(3.74), 345(3.92)	73-2114-69
	pH 13	224(4.21),339(4.01)	73-2114-69
$C_{10}H_{13}Cl_2N$			
3-Chloro-1,2,3,4-tetrahydro-6-methylquinolizinium chloride	EtOH	272(4.0),280(3.9)	24-1309-69
$C_{10}H_{13}IN_2O_5$			
Uracil, 1-(5-deoxy-5-iodo-β-D-xylofuranosyl)-3-methyl-	H_2O	262(4.03)	94-0785-69
	pH 13	261(4.01)	94-0785-69
Uridine, 2'-deoxy-2'-iodo-3-methyl-	H_2O	260(3.93)	94-0785-69
	pH 13	264(4.00)	94-0785-69

Compound	Solvent	$\lambda_{max}(\log \epsilon)$	Ref.
$C_{10}H_{13}IN_4O$			
1,4-Dihydro-1,3,6,7-tetramethyl-4-oxo-pteridinium iodide	EtOH	223(4.34),250s(3.92), 310(3.79),325s(3.67), 356s(3.18)	39-2415-69C
$C_{10}H_{13}N$			
2-Picoline, 5-(2-butenyl)-, cis	MeOH	268(3.63),275(3.52)	106-0196-69
2-Picoline, 5-(2-butenyl)-, trans	MeOH	268(3.62),275(3.49)	106-0196-69
Pyridine, 5-ethyl-2-propenyl-, cis	MeOH	239(4.21),281(3.85)	106-0196-69
Pyridine, 5-ethyl-2-propenyl-, trans	MeOH	244(4.27),287(3.89)	106-0196-69
Styrylamine, N,N-dimethyl-	C_6H_{12}	226(3.93),295(4.33)	22-0903-69
$C_{10}H_{13}NO$			
7-Azabicyclo[4.2.2]deca-2,4,7,9-tetraene, 8-methoxy-9-methyl-	MeCN	269(3.43)	35-4714-69
Nitrone, N-mesityl-	EtOH	241(3.87)	39-1073-69C
1-Pentanone, 1-(4-pyridyl)-	EtOH	221(3.53),270(3.47)	39-2134-69C
2H-Quinolizin-2-one, 6,7,8,9-tetrahydro-4-methyl-	MeOH	261(4.12)	49-0136-69
4(1H)-Quinolone, 5,6,7,8-tetrahydro-2-methyl-	pH 1	213(4.14),242(3.90), 261(3.77)	78-0255-69
	pH 13	224(4.09),252(4.03), 261s(3.88)	78-0255-69
	MeOH	217(4.22),223(4.24), 267(4.20)	78-0255-69
$C_{10}H_{13}NO_2$			
2-Azaadamantane-4,8-dione, 2-methyl-	dioxan	321(2.08)	88-5135-69
1H-Azepine-1-carboxylic acid, 2,7-dimethyl-, methyl ester	hexane	215(4.29),230s(3.58), 285(3.32)	44-2866-69
	EtOH	208(4.32),276(3.40)	44-2866-69
1H-Azepine-1-carboxylic acid, 3,6-dimethyl-, methyl ester	hexane	215(4.36),242s(3.43), 316(2.72)	44-2866-69
	EtOH	215(4.36),244s(3.35), 301(2.83)	44-2866-69
1H-Azepine-1-carboxylic acid, 4,5-dimethyl-, methyl ester	hexane	247(3.74),313(3.05)	44-2866-69
	EtOH	208(4.35),251(3.70), 306(3.07)	44-2866-69
Benzoic acid, p-(dimethylamino)-, methyl ester	heptane	298(4.45)	44-0077-69
	MeOH	310(4.46)	44-0077-69
p-Benzoquinone tert-butylimine, N-oxide	EtOH	384(4.33)	39-1459-69C
p-Benzoquinone, 2-tert-butyl-, 4-oxime	EtOH	230(3.48),303(4.30)	12-0935-69
	EtOH-NaOH	262(3.72),397(4.46)	12-0935-69
	dioxan	729(-0.52)	12-0935-69
2-Furaldehyde, 5-piperidino-	n.s.g.	365(4.51)	103-0434-69
Ketone, 3-amino-4,5,6,7-tetrahydro-2-benzofuranyl methyl	EtOH	216(3.52),275(3.97), 320(4.22)	28-0536-69A
Oxazirane, 2-ethyl-3-(p-methoxyphenyl)-	n.s.g.	231(4.14)	23-0051-69
Pyridine, 1,4-diacetyl-1,4-dihydro-4-methyl-	MeOH	254(4.23)	88-4829-69
Tyramine, N-acetyl-	EtOH	226(4.21),280(3.40), 286(3.33)	78-0937-69
$C_{10}H_{13}NO_2S_2$			
2-Butanone, 4-[(5-acetyl-4-methyl-2-thiazolyl)thio]-	MeOH	319(4.20)	4-0397-69
2-Butanone, 4-(5-acetyl-4-methyl-2-thioxo-4-thiazolin-3-yl)-	MeOH	355(4.26)	4-0397-69

Compound	Solvent	λ max (log ε)	Ref.
$C_{10}H_{13}NO_3$			
Acetanilide, 2',5'-dimethoxy-	EtOH	245(4.01),299(3.79)	117-0287-69
1-(1-Carboxy-2-methylpropyl)-3-hydroxy-pyridinium hydroxide, inner salt	N HCl	295(3.72)	1-2488-69
	N NaOH	250(3.92),325(3.67)	1-2488-69
Diacetamide, N-(6-oxo-1-cyclohexen-1-yl)-	MeOH	209(4.14),316(1.65)	104-1193-69
$C_{10}H_{13}NO_3S$			
Acetic acid, [[6-methyl-2-(methylthio)-3-pyridyl]oxy]-	pH 1	245(3.41),335(3.85)	1-2065-69
	pH 13	245(3.75),305(3.76)	1-2065-69
$C_{10}H_{13}NO_3S_2$			
4-Thiazolecarboxylic acid, 2-[(3-oxo-butyl)thio]-, ethyl ester	MeOH	280(3.79)	4-0397-69
$C_{10}H_{13}NO_4$			
2-Azabicyclo[3.1.0]hex-3-ene-2,6-dicarb-oxylic acid, 6-ethyl methyl ester	EtOH	244(2.29)	77-1359-69
p-Benzoquinone, 2-hydroxy-5-[(2-hydroxy-ethyl)amino]-3,6-dimethyl-	EtOH	314(4.26),530(3.15)	39-1325-69C
1-Cyclopentene-1-acetic acid, 2-acet-amido-5-oxo-, methyl ester	MeOH	275(4.37)	39-1625-69C
3,5-Pyridinedicarboxylic acid, 1,2-di-hydro-2-methyl-, dimethyl ester	EtOH	210(4.20),282(4.27), 379(3.77)	73-0427-69
3,5-Pyridinedicarboxylic acid, 1,4-di-hydro-4-methyl-, dimethyl ester	EtOH	208(5.27),242s(3.79), 360(3.96)	73-0427-69
$C_{10}H_{13}NO_7$			
4-Isoxazoline-3,4,5-tricarboxylic acid, 2-methyl-, trimethyl ester	MeOH	265(3.28)	24-2346-69
$C_{10}H_{13}NS_2$			
Benzoic acid, p-(dimethylamino)-dithio-, methyl ester	C_6H_{12}	225s(4.0),229(3.99), 258(3.90),265s(3.8), 320s(3.8),331(3.86), 394(4.54),495(2.83)	48-0045-69
	EtOH	226(3.93),260s(3.8), 332(3.83),412(4.47), 480s(3.4)	48-0045-69
$C_{10}H_{13}N_3O$			
Acetone, 2-phenylsemicarbazone	MeOH	242(4.28)	5-0226-69D
Indoline, 1-acetyl-5,6-diamino-	MeOH	272(4.11),321(4.04)	103-0196-69
Pyrrole-2-carboxamide, 5-cyano-N,N-di-ethyl-	EtOH	268(4.30)	78-0295-69
1,2,4-Triazolid-3-one, 5,5-dimethyl-2-phenyl-	MeOH	253(4.26)	5-0226-69D
$C_{10}H_{13}N_3OS$			
Acetophenone, 2'-methoxy-, thiosemicarb-azone	EtOH	203(4.3),279(4.3)	28-0730-69A
$C_{10}H_{13}N_3O_2$			
Acetophenone, 2'-methoxy-, semicarba-zone	EtOH	253(4.1),285s(3.7)	28-0730-69A
Piperazine, N-(p-nitrophenyl)-	DMSO	383(3.25)	39-0544-69B
Pyrido[2,3-d]pyrimidin-7(8H)-one, 5,6-dihydro-2-methyl-6-(methoxymethyl)-	EtOH	240(3.92),272(4.19)	88-1825-69
	EtOH-HCl	277(--)	88-1825-69
	EtOH-NaOH	299(--)	88-1825-69

Compound	Solvent	$\lambda_{max}(\log \epsilon)$	Ref.
$C_{10}H_{13}N_3O_2S$			
5H-Cyclopenta[b]thiophene-3-carboxylic acid, 2-acetamido-5,6-dihydro-, hydrazide	MeOH	224(4.26),254(4.03), 313(4.03)	48-0402-69
Hydrazine, 1-acetyl-2-[(2-amino-5,6-dihydro-4H-cyclopenta[b]thien-2-yl)-carbonyl]-	MeOH	229(4.58),266s(3.84), 315(3.90)	48-0402-69
$C_{10}H_{13}N_3O_3$			
Acrylic acid, 2-cyano-3-hydroxy-3-(2-pyrrolidinylideneamino)-	MeOH	219(4.27),284(4.24)	24-2739-69
Urea, 1-[2-(benzyloxy)ethyl]-1-nitroso-	EtOH	211(4.00),235(3.74)	88-0289-69
$C_{10}H_{13}N_3O_3S$			
3H,6H-[1,3]Thiazino[3,2-b]-as-triazin-3-one, 7,8-dihydro-6-hydroxy-2,8-dimethyl-, acetate	dioxan	238(4.34)	114-0093-69C
$C_{10}H_{13}N_3O_4$			
as-Triazine-5,6-dicarboxylic acid, 3-methyl-, diethyl ester	MeOH	263(3.82),390(2.54)	88-3147-69
$C_{10}H_{13}N_3O_5$			
Acetamide, N-[1-(N-acetylglycyl)-2,4-dioxo-3-pyrrolidinyl]-	EtOH	212(4.18),277(4.27)	24-2153-69
$C_{10}H_{13}N_3O_5S$			
Benzenesulfonamide, 2-acetyl-N,N-dimethyl-5-nitro-, 2-oxime	H_2O	250(3.97)	80-0481-69
	pH 13	250(3.98),333(3.36)	80-0481-69
$C_{10}H_{13}N_3O_7$			
Allofuranuronic acid, 5-amino-1,5-dideoxy-1-(3,4-dihydro-2,4-dioxo-1(2H)-pyrimidinyl)-, β-D-	0.05N HCl	258(3.98)	35-7490-69
	0.05N NaOH	262(3.86)	35-7490-69
$C_{10}H_{13}N_3S_2$			
Malononitrile, [3-(dimethylamino)-1,3-bis(methylthio)allylidene]-	C_6H_{12}	420(4.35)	5-0073-69E
$C_{10}H_{13}N_4O_7PS$			
9H-Purine-6(1H)-thione, 9-β-D-arabinofuranosyl-, 5'-phosphate	pH 7.0	226(3.99),319(4.38)	23-1095-69
$C_{10}H_{13}N_4O_9P$			
Inosine, 1-oxide, 5'-phosphate, (disodium salt)	pH 1	252(3.92)	94-1128-69
	pH 7	228(4.65),254(3.74)	94-1128-69
	pH 13	258(3.81),295(3.60)	94-1128-69
$C_{10}H_{13}N_5O$			
Formamide, N-(6-methyl-8-propyl-s-triazolo[4,3-a]pyrazin-3-yl)-	H_2O	221(4.32),238s(4.28), 307s(3.55)	39-1593-69C
4H-Pyrazolo[3',4':4,5]pyrimido[1,2-a]-azepin-4-one, 3-amino-1,6,7,8,9,10-hexahydro-	MeOH	226(4.34),271(3.74)	24-2739-69
$C_{10}H_{13}N_5O_2$			
Propionamide, N-(8-ethyl-1,6-dihydro-6-oxopurin-2-yl)-	pH 0.0	260(4.31)	5-0201-69F
	pH 6.0	261(4.26)	5-0201-69F
	pH 12	268(4.15)	5-0201-69F
9H-Purine-9-carboxylic acid, 6-(dimethylamino)-, ethyl ester	MeOH	272(4.26)	44-0973-69

Compound	Solvent	λmax(log ε)	Ref.
$C_{10}H_{13}N_5O_3$			
Propionamide, N-(1,6-dihydro-6-oxopurin-2-yl)-2-ethoxy-	pH 1	260(4.28)	69-0238-69
	pH 7	259(4.14),280s(--)	69-0238-69
	pH 10.8	267(--)	69-0238-69
$C_{10}H_{13}N_5O_4$			
Acetamide, N,N',N"-(1,6-dihydro-6-oxo-2,4,5-pyrimidinetriyl)tris-	pH -2.7	253(3.96),281(4.05)	5-0201-69F
	pH 6.0	226(4.30),287(3.90)	5-0201-69F
	pH 12.0	250s(3.87),281(3.82)	5-0201-69F
Adenine, 9-α-D-arabinopyranosyl-	pH 1	256(4.17)	44-0092-69
	pH 7	258(4.17)	44-0092-69
	pH 13	258(4.17)	44-0092-69
Adenine, 9-β-D-arabinopyranosyl-	pH 1	256(4.15)	44-0092-69
	pH 2	257(4.07)	21-1511-69
	pH 7	258(4.15)	21-1511-69
	pH 7	258(4.16)	44-0092-69
	pH 12	259(4.15)	21-1511-69
	pH 13	258(4.17)	44-0092-69
Adenine, arabinopyranosyl-	pH 1	272(4.10)	44-0092-69
	pH 7	267(3.94)	44-0092-69
	pH 13	267(3.95)	44-0092-69
Adenine, 8-β-D-ribofuranosyl-	H_2O	211(4.35),265(4.19)	73-0247-69
	pH 12	217(4.36),272(4.20)	73-0247-69
Adenine, 9-α-D-xylopyranosyl-	pH 1	256(4.18)	44-0092-69
	pH 7	258(4.20)	44-0092-69
	pH 13	258(4.18)	44-0092-69
Adenine, 9-β-D-xylopyranosyl-	pH 1	256(4.21)	44-0092-69
	pH 7	258(4.20)	44-0092-69
	pH 13	258(4.22)	44-0092-69
Adenine, xylopyranosyl-	pH 1	273(4.14)	44-0092-69
	pH 7	270(3.95)	44-0092-69
	pH 13	270(3.96)	44-0092-69
Guanine, 9-(2-deoxy-α-D-erythro-pentofuranosyl)-	pH 1	274(3.89)	44-2160-69
	pH 11	263(4.00)	44-2160-69
	MeOH	253(4.08),268(3.95)	44-2160-69
Guanosine, 2'-deoxy-	pH 1	255(4.08),272s(3.93)	44-2160-69
	pH 11	262(4.08)	44-2160-69
	MeOH	253(4.16),267(4.03)	44-2160-69
$C_{10}H_{13}N_5O_4S$			
Adenine, 8-(β-D-ribofuranosylthio)-	pH 1	281(4.33)	87-0653-69
	pH 11	225s(4.30),282(4.33)	87-0653-69
$C_{10}H_{13}N_5O_5$			
Adenine, 9-β-D-arabinofuranosyl-N-hydroxy-	pH 1	265(4.17)	4-0405-69
	pH 6.7	267(4.02)	4-0405-69
Guanine, 9-β-D-xylopyranosyl-	pH 1	256(4.08)	44-0416-69
	pH 7	252(4.12)	44-0416-69
	pH 13	262(4.03)	44-0416-69
Inosine, 1-amino-	pH 1	250(4.00)	44-1025-69
	pH 11	251(4.00)	44-1025-69
	MeOH	251(3.98),266s(--)	44-1025-69
Uracil, 1-(5-azido-5-deoxy-β-D-xylofuranosyl)-3-methyl-	H_2O	261(3.93)	94-0798-69
	pH 13	265(3.94)	94-0798-69
$C_{10}H_{13}N_5O_6$			
m-Phenylenediamine, N-tert-butyl-2,4,6-trinitro-	MeOH	311(4.25),335s(4.23),407(4.00)	35-4155-69
m-Phenylenediamine, N,N-diethyl-2,4,6-trinitro-	MeOH	302(4.03),358(4.12),400(4.01)	35-4155-69

Compound	Solvent	$\lambda_{max}(\log \epsilon)$	Ref.
m-Phenylenediamine, N,N'-diethyl-2,4,6-trinitro-	MeOH	308(4.34),348(4.25), 413(4.08)	35-4155-69
m-Phenylenediamine, N,N,N',N'-tetramethyl-2,4,6-trinitro-	MeOH	345(4.11)	35-4155-69
$C_{10}H_{14}$			
Benzene, butyl-, tetracyanoethylene complex	C_6H_{12}	402(3.13)	39-1161-69B
Benzene, tert-butyl-	C_6H_{12}	258(2.30)	101-0017-69B
1,3-Cyclohexadiene, 1-(2-methylpropenyl)-	EtOH	296(4.05)	33-1249-69
1,3-Cyclohexadiene, 2-(2-methylpropenyl)-	EtOH	224(4.21)	33-1249-69
Cyclohexene, 3-(2-methylallylidene)-	EtOH	273(4.03)	33-1249-69
Dispiro[2.4.2.0]dec-5-ene	C_6H_{12}	242s(2.51)	108-0479-69
o-Mentha-1(7),5,8-triene	EtOH	234(4.20)	33-1249-69
$C_{10}H_{14}AsBr$			
Arsine, (p-bromophenyl)diethyl-	dioxan	227(4.1),252(3.8)	70-1381-69
$C_{10}H_{14}AsBrO$			
Arsine oxide, (p-bromophenyl)diethyl-	pH 2	235(4.1),270(2.8)	70-1381-69
	H_2O	232(4.2),265(2.8)	70-1381-69
	dioxan	230(4.2),270(2.8)	70-1381-69
$C_{10}H_{14}AsCl$			
Arsine, (p-chlorophenyl)diethyl-	dioxan	225(4.3),250(3.9)	70-1381-69
$C_{10}H_{14}AsClO$			
Arsine oxide, (p-chlorophenyl)diethyl-	pH 2	230(4.2),265(3.0)	70-1381-69
	H_2O	227(4.2),265(2.8)	70-1381-69
	dioxan	225(4.2),263(2.8)	70-1381-69
$C_{10}H_{14}AsNO_2$			
Arsine, diethyl(p-nitrophenyl)-	dioxan	265(4.0),325s(3.8)	70-1381-69
$C_{10}H_{14}AsNO_3$			
Arsine oxide, diethyl(p-nitrophenyl)-	pH 2	260(3.9),340s(2.6)	70-1381-69
	H_2O	262(4.0),340s(2.6)	70-1381-69
	dioxan	260(4.2),340s(2.9)	70-1381-69
$C_{10}H_{14}BrFN_2O_4$			
Hydrouracil, 5-bromo-6-ethoxy-5-fluoro-1-(tetrahydro-2-furyl)-	pH 12	243(3.79)	103-0130-69
$C_{10}H_{14}Br_2N_2$			
p-Phenylenediamine, 2,5-dibromo-N,N,N',N'-tetramethyl-	EtOH	223(4.33),269(3.94), 310s(--)	117-0287-69
$C_{10}H_{14}Br_2O$			
2-Cyclopenten-1-one, 2-bromo-5-(bromomethyl)-5-isopropyl-3-methyl-	MeOH	247(4.04),309(2.00)	44-0136-69
$C_{10}H_{14}ClNO$			
1,2,3,4-Tetrahydro-3-hydroxy-6-methylquinolizinium chloride	EtOH	272(3.9),280(3.8)	24-1309-69
$C_{10}H_{14}ClN_5NaO_8P$			
Inosine, 2-chloro-, 5'-phosphate, sodium salt	pH 1	253(4.06)	94-2581-69
	pH 6	255(4.06)	94-2581-69

Compound	Solvent	λ_{max}(log ϵ)	Ref.
$C_{10}H_{14}Cl_2$			
Cyclobutane, 1,2-dichloro-3,4-diisopropylidene-	n.s.g.	273(4.13)	35-6038-69
$C_{10}H_{14}Cl_2N_4S$			
s-Triazine, 2,4-dichloro-6-[2-(dimethylamino)-1-ethyl-2-(methylthio)vinyl]-	C_6H_{12}	229(3.94),300(3.64), 401(4.47)	5-0073-69E
$C_{10}H_{14}GeO$			
Germane, benzoyltrimethyl-	heptane	248(4.05),418(2.15)	59-0765-69
$C_{10}H_{14}IN_2O_7P$			
Thymidine, 3'-deoxy-3'-iodo-, 5'-phosphate (calcium salt)	H_2O	267(3.99)	69-4889-69
$C_{10}H_{14}NO_8P$			
2(1H)-Pyridone, 1-β-D-ribofuranosyl-, 2'(3')-phosphate (barium salt)	pH 2	300(3.73)	73-0089-69
$C_{10}H_{14}N_2$			
Acetone, methylphenylhydrazone	EtOH	249(4.04),283(3.53)	39-1703-69C
	EtOH-HCl	229(3.82),270(2.85)	39-1703-69C
Acetophenone, dimethylhydrazone	MeOH	234(4.00),308(3.34)	39-1703-69C
	EtOH	228(4.12),308(3.32)	39-1703-69C
	EtOH-HCl	247(4.07),280s(3.15)	39-1703-69C
Acetophenone, ethylhydrazone	hexane	266(4.03),361(1.99)	104-0257-69
Azoethane, 1-phenyl-	hexane	356(1.58)	104-0257-69
Benzaldehyde, propylhydrazone	hexane	289(4.05),369(1.50)	104-0257-69
5,7-Methano-1H-indazole, 4,5,6,7-tetrahydro-6,6-dimethyl-	C_6H_{12}	223(3.79),300s(1.00)	28-1315-69B
Nicotine	EtOH	262(3.46)	23-4393-69
Propane-1-azomethane, 1'-phenyl-	hexane	358(1.51)	104-0257-69
$C_{10}H_{14}N_2O$			
Aniline, N-tert-butyl-N-nitroso-	C_6H_{12}	209(3.86),228(3.66), 251s(3.57),276s(2.62)	35-3383-69
o-Toluidine, N-isopropyl-N-nitroso-	C_6H_{12}	211(4.04),228s(3.08), 244s(3.72),274s(--)	35-3383-69
Urea, 3-(2,4,6-cycloheptatrien-1-yl)-1,1-dimethyl-	EtOH	258(3.62)	35-6391-69
$C_{10}H_{14}N_2OS$			
4-Cyclopenta[d]pyrimidinethione, 1,5,6,7-tetrahydro-1-(2-hydroxyethyl)-2-methyl-	DMF	343(4.40)	4-0037-69
$C_{10}H_{14}N_2O_2$			
Aniline, N,N-diethyl-4-nitro-	1%MeOH-NaOH	430(4.36)	39-0922-69B
Aniline, 3,4,5,6-tetramethyl-2-nitro-	MeOH	209(4.31),232(4.02), 273(3.31),404(3.00)	44-0758-69
p-Benzoquinone, 2-amino-5-(butylamino)-	$CHCl_3$	337(4.48),500(2.65)	18-2695-69
1H,5H-Cyclopenta[d]pyrazolo[1,2-a]pyridazine-5,9(5aH)-dione, hexahydro-	$C_2H_3F_3O$	202(3.65),235(2.69)	35-1256-69
$C_{10}H_{14}N_2O_2S_2$			
p-Dithiin-2,3-dicarboximide, N-[2-(dimethylamino)ethyl]-5,6-dihydro-	MeOH	260(3.98),412(3.38)	33-2221-69
$C_{10}H_{14}N_2O_3$			
Acetanilide, 4'-amino-2',5'-dimethoxy-	C_6H_{12}	261(4.15),313(3.94)	117-0287-69

Compound	Solvent	$\lambda_{max}(\log \epsilon)$	Ref.
2H-Indazole-4-carboxylic acid, 3,3a,4,5,6,7-hexahydro-3-oxo-, ethyl ester	EtOH	203(3.64),224s(3.52), 250(3.66)	39-2783-69C
2H-Indazole-7-carboxylic acid, 3,3a,4,5,6,7-hexahydro-3-oxo-, ethyl ester	EtOH	203(3.67),229(3.61), 252(3.69)	39-2783-69C
Thymine, 1-(tetrahydro-2H-pyran-2-yl)-	pH 3-4	267(3.91)	103-0283-69
	pH 7-8	267(3.88)	103-0283-69
	pH 11-12	267(3.78)	103-0283-69
$C_{10}H_{14}N_2O_3S$			
Acetic acid, cyano[[2-(ethylcarbamoyl)-vinyl]thio]-, ethyl ester, (Z)-	90% EtOH	267(4.08)	12-0765-69
Alanine, 3-[[2-hydroxy-2-(3-pyridyl)-ethyl]thio]-, dihydrochloride	pH 1	262(3.66)	63-0473-69
$C_{10}H_{14}N_2O_3S_3$			
Acetanilide, 2',4'-bis(methylthio)-5'-sulfamoyl-	MeOH	262(4.44)	44-1780-69
$C_{10}H_{14}N_2O_4$			
Glutaric acid, 2-diazo-3,3-dimethyl-4-methylene-, dimethyl ester	C_6H_{12}	262(3.65),407(2.37)	88-2659-69
Pyridazine-1,2-dicarboxylic acid, 1,2-dihydro-, diethyl ester	EtOH	298(3.50)	77-0677-69
Pyridazine-3,5-dicarboxylic acid, 1,4-dihydro-4,4-dimethyl-, dimethyl ester	MeOH	241(3.46),360(3.49)	88-2659-69
Pyridazine-3,5-dicarboxylic acid, 1,4-dihydro-4,6-dimethyl-, dimethyl ester	EtOH	251(3.80),350(3.76)	27-0037-69
$C_{10}H_{14}N_2O_4S$			
Thymine, 1-(2,3-dideoxy-3-thio-β-D-threo-pentofuranosyl)-	pH 0-6.34	267(3.99)	44-1020-69
	pH 9.47	267(4.00)	44-1020-69
	pH 12	267(3.92)	44-1020-69
$C_{10}H_{14}N_2O_5$			
Uracil, 1-(5-deoxy-β-D-xylofuranosyl)-5-methyl-	H_2O	262(3.86)	94-0785-69
	pH 13	264(3.89)	94-0785-69
$C_{10}H_{14}N_2O_5S$			
4(3H)-Pyrimidinone, 6-(methylthio)-3-β-D-ribofuranosyl-	pH 1	239(4.28),270(3.99)	44-0431-69
	pH 4	238(4.28),269(4.00)	44-0431-69
	pH 11	239(4.54),265s(4.14)	44-0431-69
Uridine, 2'-deoxy-5'-(methylthio)-	pH 1	232(3.83),282(3.71)	44-3806-69
	pH 13	282(3.66)	44-3806-69
$C_{10}H_{14}N_2O_6$			
D-Allitol, 2,5-anhydro-1-deoxy-1-(3,4-dihydro-2,4-dioxo-1(2H)-pyrimidinyl)-	H_2O	207(4.03),265(4.11)	73-1684-69
4(3H)-Pyrimidinone, 6-methoxy-3-β-D-ribofuranosyl-	pH 1	262(3.83)	44-0431-69
	pH 4	262(3.83)	44-0431-69
	pH 11	262(3.83)	44-0431-69
	pH 14	253(3.77)	44-0431-69
Thymine, arabinoside	pH 2	270(4.00)	21-1511-69
	pH 7	268(3.93)	21-1511-69
	pH 12	268(3.90)	21-1511-69
Uracil, 1-β-D-arabinofuranosyl-3-methyl-	H_2O	262(3.97)	94-0785-69
	pH 13	261(3.97)	94-0785-69
Uracil, 1-β-D-xylofuranosyl-3-methyl-	H_2O	262(4.11)	94-0785-69
	pH 13	261(4.10)	94-0785-69

Compound	Solvent	$\lambda_{max}(\log \epsilon)$	Ref.
Uridine, 6-methyl-	pH 7	207(3.91),261(4.06)	73-2316-69
	pH 11	262(3.98)	73-2316-69
$C_{10}H_{14}N_2O_6S$			
Uracil, 6-(methylthio)-3-β-D-ribofuran-osyl-	pH 1	281(4.20)	39-0791-69C
	pH 11	247(4.15),292(4.13)	39-0791-69C
Uridine, 2'-deoxy-5-(methylsulfinyl)-	EtOH	273(3.91)	44-3806-69
$C_{10}H_{14}N_2O_7S$			
Uracil, 6-(β-D-glucopyranosylthio)-	pH 1	276(4.04)	87-0653-69
	pH 11	287(4.07)	87-0653-69
$C_{10}H_{14}N_2S_2$			
Propionitrile, 3-[(4-tert-butyl-2-thia-zolyl)thio]-	MeOH	276(3.84)	4-0397-69
$C_{10}H_{14}N_4$			
Imidazole-4-carbonitrile, 5-amino-1-cyclohexyl-	pH 1	235(3.95),256(3.89)	39-2198-69C
	H_2O	245(4.00)	39-2198-69C
	pH 13	245(4.04)	39-2198-69C
$C_{10}H_{14}N_4O_2$			
Imidazo[1,2-b]-as-triazine-3,6(5H,7H)-dione, 5-butyl-2-methyl-	EtOH	215(4.34),235(4.13)	114-0181-69C
	dioxan	214(4.28),268s(3.84)	114-0181-69C
$C_{10}H_{14}N_4O_3$			
4-Pyrimidinol, 6-(cyclohexylamino)-5-nitro-	pH 1	233(4.26),291(3.64), 350(3.93)	73-2114-69
	pH 13	350(3.99)	73-2114-69
$C_{10}H_{14}N_4O_4$			
m-Phenylenediamine, N-tert-butyl-4,6-dinitro-	EtOH	311(4.25),335s(4.23), 407(4.10)	35-4155-69
m-Phenylenediamine, N,N'-diethyl-4,6-dinitro-	MeOH	333(4.47),417(4.08)	35-4155-69
$C_{10}H_{14}N_4O_5$			
1H-1,3,5,7-Tetrazonine-2,4,6,8,9-(3H,5H,7H)-pentone, 1-ethyl-3,5,7-trimethyl-	EtOH	210(4.57)(end abs.)	78-0549-69
1H-1,3,5,7-Tetrazonine-2,4,6,8,9-(3H,5H,7H)-pentone, 3-ethyl-1,5,7-trimethyl-	EtOH	210(4.58)(end abs.)	78-0549-69
$C_{10}H_{14}N_6O_2$			
Adenosine, 5'-amino-2',5'-dideoxy-	pH 1	256(4.09)	87-0658-69
	pH 11	259(4.11)	87-0658-69
$C_{10}H_{14}N_6O_3$			
9H-Purine, 2,6-diamino-9-(2-deoxy-α-D-erythro-pentofuranosyl)-	pH 1	252(4.03),291(3.99)	87-0498-69
	pH 7,13	256(3.96),279(4.01)	87-0498-69
$C_{10}H_{14}N_6O_4$			
Guanosine, 1-amino-2'-deoxy-	pH 1	257(4.07)	44-1025-69
(also shoulders)	pH 11	254(4.15)	44-1025-69
	MeOH	256(4.20)	44-1025-69
9H-Purine, 2,6-diamino-9-α-D-arabino-furanosyl-	pH 1	253(4.05),292(4.00)	87-0498-69
9H-Purine, 2,6-diamino-9-β-D-xylo-furanosyl-	pH 1	252(4.05),290(4.00)	44-1396-69
	pH 7	255(3.98),278(4.02)	44-1396-69

Compound	Solvent	λ max(log ε)	Ref.
9H-Purine, 2,6-diamino-9-β-D-xylo-furanosyl- (cont.)	pH 13	255(3.95),278(4.01)	44-1396-69
9H-Purine, 6-hydrazino-9-β-D-ribo-furanosyl-	pH 1	261(4.20)	94-2373-69
	pH 7	265(4.17)	94-2373-69
$C_{10}H_{14}N_6O_5$			
Adenosine, 7-amino-8-oxo-	pH 1	269(4.09)	44-1025-69
	pH 7	273(4.13)	44-1025-69
	pH 11	273(4.13)	44-1025-69
Cytosine, 1-(4-azido-4-deoxy-β-D-gluco-pyranosyl-	pH 1	276(4.10)	94-0416-69
	pH 6.8	198(4.32),235(3.91), 267(3.93)	94-0416-69
	pH 14	268(3.99)	94-0416-69
Guanosine, 1-amino-	pH 1	257(4.06)	44-1025-69
	pH 11	255(4.16)	44-1025-69
	MeOH	256(4.21)	44-1025-69
$C_{10}H_{14}N_6O_5S$			
Adenosine, 2'-deoxy-, 5'-sulfamate	pH 1	258(4.19)	35-3391-69
	pH 11	259(4.23)	35-3391-69
$C_{10}H_{14}N_6O_6$			
Guanosine, 1-amino-8-oxo-	pH 1	248(4.08),294(4.02)	44-1025-69
	pH 11	259(4.09),300(3.98)	44-1025-69
	MeOH	252(4.13),293(4.01)	44-1025-69
$C_{10}H_{14}N_6O_6S$			
Adenosine, 5'-sulfamate	pH 1	257(4.17)	35-3391-69
	pH 11	259(4.19)	35-3391-69
$C_{10}H_{14}O$			
Bicyclo[3.1.0]hex-3-en-2-one, 1-isopro-pyl-4-methyl-	hexane	211(3.72),253(3.48)	22-2095-69
	EtOH	218(3.77),264(3.51)	22-2095-69
Bicyclo[3.1.0]hex-3-en-2-one, 5-isopro-pyl-4-methyl-	hexane	206(3.83),247(3.43)	22-2095-69
	EtOH	214(3.86),259(3.49)	22-2095-69
Bicyclo[3.2.0]hept-2-en-6-one, 2,7,7-trimethyl-, (-)-	$CHCl_3$	301(1.54)	35-0779-69
4-Caren-3-one, (+)-	isooctane	262(3.79)	56-0943-69
d-Carvone	H_2SO_4	293(3.91)	23-2263-69
2-Cyclopenten-1-one, 5-isopropyl-3-methyl-4-methylene-	MeOH	325(2.00),267(4.15)	78-3161-69
Lyratal	EtOH	232(4.21)	78-3217-69
2-Norcarene-3-carboxaldehyde, 7,7-di-methyl-	EtOH	263(4.07)	35-6473-69
2-Norpinanone, 6,6-dimethyl-3-methyl-ene-, (+)-	C_6H_{12}	225(3.95),332(1.48)	28-1315-69B
2-Norpinene-3-carboxaldehyde, 6,6-di-methyl-, (-)-	C_6H_{12}	243(4.08),328(1.65)	28-1315-69B
2(1H)-Pentalenone, 4,5,6,6a-tetrahydro-1,4-dimethyl-	n.s.g.	240(4.33)	40-0507-69
4-Pentenal, 5-(1-cyclopenten-1-yl)-	EtOH	247(4.33)	28-1535-69A
Phenol, 2-butyl-	pH 2	271(3.30)	30-1113-69A
	pH 13	238(3.95),290(3.58)	30-1113-69A
	MeOH	274(3.33)	30-1113-69A
	MeOH-Na	240(3.98),291(3.58)	30-1113-69A
Phenol, 2-tert-butyl-	pH 2	270(3.23)	30-1113-69A
	pH 13	241(3.82),292(3.37)	30-0867-69A
	pH 13	240(3.92),290(3.53)	30-1113-69A
	MeOH	274(3.35)	30-1113-69A
	MeOH-Na	244(3.95),291(3.56)	30-1113-69A

Compound	Solvent	λ_{max}(log ε)	Ref.
Phenol, 2-tert-butyl- (cont.)	iso-PrOH	223(3.62),274(3.27), 278(3.22)	30-0867-69A
Phenol, 3-tert-butyl-	pH 2	270(3.18)	30-1113-69A
	pH 13	237(3.95),288(3.37)	30-0867-69A
	pH 13	236(3.92),288(3.47)	30-1113-69A
	MeOH	274(3.26)	30-1113-69A
	MeOH-Na	240(3.95),291(3.48)	30-1113-69A
	iso-PrOH	274(3.28),280s(3.22)	30-0867-69A
Phenol, 4-tert-butyl-	pH 2	274(3.24)	30-1113-69A
	pH 8	275(3.20)	69-1042-69
	pH 13	237(4.08),292(3.40)	30-0867-69A
	pH 13	238(4.08),292(3.43)	30-1113-69A
	MeOH	277(3.27)	30-1113-69A
	MeOH-Na	239(4.08),294(3.40)	30-1113-69A
	iso-PrOH	224(3.89),278(3.28), 285s(3.15)	30-0867-69A
Phenol, 2-isopropyl-3-methyl-	pH 2	272(3.16)	30-1113-69A
	pH 13	239(3.91),292(3.54)	30-1113-69A
	MeOH	274(3.25)	30-1113-69A
	MeOH-Na	240(3.91),293(3.56)	30-1113-69A
Tricyclo[4.3.1^{1,2}.0]decan-3-one	EtOH	276(2.08)	78-5287-69
$C_{10}H_{14}OS$			
Sulfoxide, methyl p-methylphenethyl	n.s.g.	273(2.53),274(2.53)	39-0481-69B
Sulfoxide, methyl 3-phenylpropyl	EtOH	220(3.94)	39-0581-69B
$C_{10}H_{14}OSi$			
Silane, benzoyltrimethyl-	hexane	200(4.29),250(4.05), 282(3.06),423f(2.08)	35-0355-69
	heptane	251(4.07),423(2.00)	59-0765-69
$C_{10}H_{14}O_2$			
Bicyclo[4.3.1]decane-7,10-dione	MeOH	286(2.02)	39-0592-69C
2-Cyclohexen-1-one, 5-hydroxy-5-isopropenyl-2-methyl-	EtOH	236(3.89)	44-0857-69
2-Cyclohexene-1,4-dione, 6-tert-butyl-	hexane	369(1.83)	39-0124-69C
2-Cyclopenten-1-one, 2-(2-butenyl)-4-hydroxy-3-methyl-, cis	EtOH	230(4.06)	39-1016-69C
4,6,8-Decatrienoic acid	EtOH	258(4.59),267(4.72), 278(4.61)	44-1147-69
1,2-Naphthalenedione, octahydro-	EtOH	277(4.08)	94-1578-69
1,8-Naphthalenedione, octahydro-	EtOH	299(4.03)	94-1572-69
2-Norpinanone, 3-(hydroxymethylene)-6,6-dimethyl-, (+)-	C_6H_{12}	275(3.74)	28-1315-69B
Spiro[norpinane-3,2'-oxirane]-2-one, 6,6-dimethyl-	EtOH	294(1.30)	28-1315-69B
$C_{10}H_{14}O_2S$			
Anisole, p-[2-(methylsulfinyl)ethyl]-	n.s.g.	277(3.15)	39-0481-69B
$C_{10}H_{14}O_3$			
Benzyl alcohol, α-[(2-hydroxyethoxy)-methyl]-	EtOH	206(3.73),255(1.80)	22-2048-69
1-Cycloheptene-1-carboxylic acid, 2-methyl-7-oxo-, methyl ester	EtOH	210s(3.6),237(3.94)	94-0629-69
1-Cyclopentene-1-carboxylic acid, 3-oxo-5-propyl-, methyl ester	EtOH	237(4.00)	23-2087-69
1-Cyclopentene-1-propionic acid, β-oxo-, ethyl ester	EtOH	237(3.88),281(3.67)	33-1996-69

Compound	Solvent	$\lambda_{max}(\log \epsilon)$	Ref.
$C_{10}H_{14}O_3S$			
9-Oxabicyclo[3.3.1]non-2-en-1-ol, 6-mercapto-, S-acetate	EtOH	234(3.60)	33-0725-69
$C_{10}H_{14}O_4$			
Cyclopentanecarboxylic acid, 4-acetyl-2-methyl-3-oxo-, methyl ester	pH 13	283(3.88),307(4.36)	39-1845-69C
2,4-Dioxabicyclo[3.2.0]hept-6-ene-6-carboxylic acid, 3,3,5,7-tetramethyl-	EtOH	214(3.93)	35-4739-69
2-Hexenedioic acid, 5-ethylidene-, dimethyl ester, trans	EtOH	212(4.30)	12-2351-69
2-Hexenedioic acid, 5-vinyl-, dimethyl ester, trans	EtOH	210(4.15)	12-2351-69
Malonic acid, butylidene-, cyclic isopropylidene ester	MeOH	230(4.23)	88-4983-69
	MeONa	262(4.18)	88-4983-69
2,6-Octadienedioic acid, dimethyl ester, trans-trans	EtOH	214(4.40)	12-2351-69
2H-Pyran-5-carboxylic acid, 3,4-dihydro-3,6-dimethyl-4-oxo-, ethyl ester	EtOH	264(4.01)	22-1383-69
2H-Pyran-2-one, 5,6-dihydro-4-hydroxy-6,6-dimethyl-3-propionyl-	pH 13	254(4.20),265(4.08)	22-4091-69
	EtOH	221(3.94),273(4.04)	22-4091-69
Succinic acid, 2,3-divinyl-, dimethyl ester	EtOH	217s(2.98),227(3.04), 252s(2.36)	12-2351-69
$C_{10}H_{14}O_4S_2$			
p-Dithiin-2,3-dicarboxylic acid, 5,6-dihydro-, diethyl ester	C_6H_{12}	235(3.80),322(3.84)	33-2221-69
	EtOH	238(3.76),328(3.86)	33-2221-69
$C_{10}H_{14}O_5$			
2-Heptenedioic acid, 2-methyl-4-oxo-, dimethyl ester	MeOH	235(3.91)	40-0096-69
$C_{10}H_{15}As$			
Arsine, diethylphenyl-	dioxan	207(4.3),245(3.2)	70-1381-69
$C_{10}H_{15}AsO$			
Arsine oxide, diethylphenyl-	pH 2	215(4.0),262(3.0)	70-1381-69
	H_2O	214(3.9),260(2.9)	70-1381-69
	dioxan	210(4.1),265(2.8)	70-1381-69
$C_{10}H_{15}BrN_2O_4$			
Hydrouracil, 5-bromo-6-methoxy-5-methyl-1-(tetrahydro-2-furyl)-	pH 12	246(3.74)	103-0130-69
$C_{10}H_{15}BrO$			
Camphor, α-bromo-, (+)-	MeOH	306(1.99)	18-2593-69
2-Cyclopenten-1-one, 2-bromo-5-isopropyl-3,5-dimethyl-	MeOH	244(4.08),307(2.00)	44-0136-69
2-Cyclopenten-1-one, 5-(bromomethyl)-5-isopropyl-3-methyl-, (R)-(-)-	MeOH	230(4.11),310(1.90)	44-0136-69
2-Decalone, 1-bromo-, axial, trans	EtOH	307(2.09)	22-3199-69
equatorial form	EtOH	280(1.52)	22-3199-69
$C_{10}H_{15}BrO_4$			
3-Cyclopentene-1,2-dione, 3-bromo-4-methoxy-5,5-dimethyl-, 1-(dimethyl acetal)	MeOH	266(4.08)	20-0277-69
$C_{10}H_{15}BrO_4S_3$			
Thiophene, 5-bromo-2,3-bis[(ethylsulfonyl)methyl]-	EtOH	253(4.01)	54-0321-69

Compound	Solvent	λ_{max}(log ϵ)	Ref.
$C_{10}H_{15}ClN_2O_2$			
3-Carbamoyl-1-(propoxymethyl)pyridinium chloride	EtOH	266(3.55)	39-0192-69B
$C_{10}H_{15}ClN_4O_2$			
Cyclopentanemethanol, 3-[(5-amino-6-chloro-4-pyrimidinyl)amino]-5-hydroxy-	pH 1	308(4.11)	88-2231-69
	pH 7,13	264(3.95),292(3.98)	88-2231-69
Pyrimidine, 4-chloro-6-(hexylamino)-5-nitro-	pH 1	225(4.25),287(3.68), 342(3.88)	73-2114-69
	pH 13	223(4.05),347(3.94)	73-2114-69
$C_{10}H_{15}ClN_4O_3$			
1,2-Cyclopentanediol, 3-[(5-amino-6-chloro-4-pyrimidinyl)amino]-5-(hydroxymethyl)-	pH 1	207(4.0),220s(--), 285s(--),307(4.12)	35-3075-69
	pH 7	207(4.29),263(3.95), 293(3.98)	35-3075-69
	pH 13	263(3.94),292(3.97)	35-3075-69
$C_{10}H_{15}ClN_5O_8P$			
Inosine, 2-chloro-, 5'-phosphate, (sodium salt)	pH 1	253(4.06)	94-2581-69
	pH 6	255(4.06)	94-2581-69
	pH 13	258(4.13)	94-2581-69
$C_{10}H_{15}ClO$			
2-Decalone, 1-chloro-, trans, axial	EtOH	303(1.62)	22-3199-69
equatorial form	EtOH	280(1.40)	22-3199-69
$C_{10}H_{15}FO$			
2-Decalone, 1-fluoro-, trans, axial	EtOH	300(1.04)	22-3199-69
equatorial form	EtOH	283(1.36)	22-3199-69
$C_{10}H_{15}I$			
2-Bornene, 2-iodo-	EtOH	248(2.02)	77-0174-69
Norbornane, 1-iodo-3,3-dimethyl-2-methylene-	EtOH	253(2.77)	77-0174-69
$C_{10}H_{15}N$			
Cyclopropaneacrylonitrile, α,1,2,2-tetramethyl-	n.s.g.	220(3.11)	39-2443-69C
2-Picoline, 5-butyl-	MeOH	268(3.19),275(3.07)	106-0196-69
Pyridine, 5-ethyl-2-propyl-	MeOH	268(3.63),275(3.50)	106-0196-69
Pyridine, 4-pentyl-	EtOH	212(3.41),256(3.25)	39-2134-69C
$C_{10}H_{15}NO$			
1H-2-Benzazepin-1-one, 2,3,4,5,6,7,8,9-octahydro-	EtOH	222(4.12)	94-1578-69
2H-3-Benzazepin-2-one, 3,4,5,5a,6,7,8,9-octahydro-	EtOH	219(4.25)	94-1578-69
Isoxazole, 3-cyclobutyl-5-ethyl-4-methyl-	EtOH	226(3.72)	22-2820-69
$C_{10}H_{15}NOS_2$			
2-Propanone, 1-[(4-tert-butyl-2-thiazolyl)thio]-	MeOH	275(3.84)	4-0397-69
$C_{10}H_{15}NO_2$			
Pyrrole-3-carboxylic acid, 2,4,5-trimethyl-, ethyl ester	MeOH	232(3.97),271(3.67)	35-3931-69

Compound	Solvent	$\lambda_{max}(\log \epsilon)$	Ref.
5H-Pyrrolizine, 6,7-dihydro-1-(methoxymethyl)-7-methoxy-	n.s.g.	220(3.88),275(2.08)	39-1155-69C
$C_{10}H_{15}NO_2Si$			
Silane, trimethyl(p-nitrobenzyl)-	hexane	284(4.04)	78-4825-69
	MeOH	282(4.01)	78-4825-69
$C_{10}H_{15}NO_3$			
Phenol, 4-(2-aminoethyl)-2,6-dimethoxy-	EtOH	230s(3.81),272(3.11)	117-0171-69
$C_{10}H_{15}NO_3S$			
Acrylamide, 3-[(1-acetylacetonyl)thio]-N-ethyl-, (Z)-	90% EtOH	258(4.20)	12-0765-69
$C_{10}H_{15}NO_4$			
2-Azetine-2,3-dicarboxylic acid, 1-ethyl-4-methyl-, d-methyl ester	MeOH	244(3.98)	24-2336-69
$\Delta^{2,\alpha}$-Pyrrolidinemalonic acid, 1-methyl-, dimethyl ester	isooctane	233(3.71),281(4.25)	35-6683-69
$C_{10}H_{15}NO_4S$			
Glutaric acid, 2-isothiocyanato-, diethyl ester, L-	hexane	257(3.22)	39-1143-69B
$C_{10}H_{15}NO_5$			
Pyrrole, 1-β-D-glucopyranosyl-	H_2O	210(3.84)	18-3539-69
$C_{10}H_{15}NO_6$			
Malonamic acid, 2-malonyl-, diethyl ester	EtOH	231(2.85),266(4.17)	104-1713-69
$C_{10}H_{15}N_3$			
1-Pyrazoline-4-carbonitrile, 5-isopropylidene-3,3,4-trimethyl-	n.s.g.	255(3.93),357(2.42)	39-2443-69C
Triazene, 3-butyl-1-phenyl-, mercury	benzene	288(4.18),353(4.07)	65-0303-69
complex	DMF	285(4.20),335(4.18)	65-0303-69
Triazene, 3-tert-butyl-1-phenyl-	MeOH	298(3.75),360(4.08)	65-0303-69
mercury complex	benzene	303(4.11),393(4.35)	65-0303-69
	DMF	295(4.15),360(4.40)	65-0303-69
$C_{10}H_{15}N_3O$			
Tricyclo[3.3.1^{1,2}.0]nonan-3-one, semicarbazone	EtOH	232(4.06)	78-5287-69
$C_{10}H_{15}N_3O_4$			
Glutaric acid, 2-diazo-3,3-dimethyl-4-(methylimino)-, dimethyl ester	C_6H_{12}	216(3.67),259(4.04), 405(1.70)	88-2659-69
$C_{10}H_{15}N_3O_5$			
D-Allitol, 1-(4-amino-2-oxo-1(2H)-pyrimidinyl)-2,5-anhydro-1-deoxy-	H_2O	272(3.99)	73-1684-69
Uracil, 1-(5-amino-5-deoxy-β-D-xylo-	H_2O	261(3.91)	94-0798-69
furanosyl)-3-methyl-	pH 13	264(3.92)	94-0798-69
$C_{10}H_{15}N_3O_6$			
Uracil, 6-amino-1-methyl-3-β-D-ribofuranosyl-	pH 1	270(4.32)	39-0791-69C
	pH 4	270(4.32)	39-0791-69C
	pH 11	270(4.32)	39-0791-69C
	pH 14	270(4.32)	39-0791-69C

Compound	Solvent	$\lambda_{max}(\log \epsilon)$	Ref.
$C_{10}H_{15}N_3O_{10}S_2$			
as-Triazine-3,5(2H,4H)-dione, 2-β-D-ribofuranosyl-, 2',3'-dimethanesulfonate	50% EtOH	257(3.81)	73-0618-69
$C_{10}H_{15}N_5O_2$			
Pyrazole-4-carboxylic acid, 3-amino-5-(2-pyrrolidinylideneamino)-, ethyl ester	MeOH	220(4.20),239(4.32), 254(4.18)	24-2739-69
$C_{10}H_{15}N_5O_5$			
Pyrimidine, 4,6-diamino-5-(2,5-anhydro-D-allonoyl)amino-	H_2O	219(4.64),260(3.74)	73-0247-69
$C_{10}H_{15}N_7O_4$			
Purine, 2-amino-6-hydrazino-9-β-D-ribofuranosyl-	pH 7	259(3.98),282(4.10)	94-2373-69
$C_{10}H_{15}N_7O_6$			
Guanosine, 1,7-diamino-8-oxo-	pH 1	248(4.09),296(4.02)	44-1025-69
	pH 7	249(4.09),296(4.02)	44-1025-69
	pH 11	253(4.07),297(4.00)	44-1025-69
$C_{10}H_{15}O_2P$			
Phenol, p-(diethylphosphinyl)-	pH 4.2	233(4.15),270(2.98), 277(2.84)	65-0373-69
$C_{10}H_{15}O_4P$			
p-tert-Butylphenyl phosphate	pH 8	272(2.90)	69-1042-69
$C_{10}H_{15}O_7P$			
Phosphonic acid, (5-acetyl-2,5-dihydro-4-methoxy-5-methyl-2-oxo-3-furyl)-, dimethyl ester	MeCN	212(3.71),242(4.03)	35-2293-69
$C_{10}H_{16}$			
Cyclohexene, 1,5,5-trimethyl-3-methylene-	EtOH	236(4.01)	77-1440-69
Dispiro[2.4.2.0]decane	C_6H_{12}	242s(1.38)	108-0479-69
1,4,5-Heptatriene, 2,4,6-trimethyl-	isooctane	215(3.40)	35-3299-69
β-Myrcene	n.s.g.	225(4.30)	2-0450-69
1,3,5-Octatriene, 3,7-dimethyl-, cis-cis	EtOH	257s(4.15),266(4.17), 276s(4.11)	35-6444-69
cis-trans	EtOH	259(4.45),268(4.56), 278(4.45)	35-6444-69
trans-trans	EtOH	259(4.53),268(4.66), 279(4.55)	35-6444-69
1,5,7-Octatriene, 2,6-dimethyl-	EtOH	230(4.46)	2-1111-69
$C_{10}H_{16}Cl_2N_2S$			
3,4-Thiophenebis(methylamine), 2,5-dichloro-N,N'-diethyl-	EtOH	247(3.81)	44-0333-69
$C_{10}H_{16}N_2$			
4-Picoline, 3-[1-(ethylamino)ethyl]-	MeOH	259(3.33)	106-0196-69
$C_{10}H_{16}N_2O$			
2-Cyclohexen-1-one, 2-pyrrolidino-, oxime	hexane	219(4.05),290(3.58)	48-0162-69

Compound	Solvent	$\lambda_{max}(\log \epsilon)$	Ref.
$C_{10}H_{16}N_2O_2$			
Acetamide, N-[(1-acetyl-1,4,5,6-tetra-hydro-3-pyridyl)methyl]-	EtOH	238(4.68)	27-0155-69
2-Cyclohexen-1-one, 2-morpholino-, oxime	hexane	222(4.0),274(3.51)	48-0162-69
Nicotinamide, 1,4-dihydro-1-(propoxy-methyl)-	EtOH	338(3.71)	39-0192-69B
Nicotinamide, 1,6-dihydro-1-(propoxy-methyl)-	EtOH	262(3.68),343(3.71)	39-0192-69B
1-Pyrazoline-3-carboxylic acid, 3,5,5-trimethyl-4-methylene-, ethyl ester	hexane	328(2.28)	39-2443-69C
	EtOH	326(2.22)	39-2443-69C
$C_{10}H_{16}N_2O_2S$			
4H-4a,7-Methano-2H-3,1,2-benzothiadia-zine, 5,6,7,8-tetrahydro-9,9-dimeth-yl-, 3,3-dioxide	EtOH	212(3.72),229s(3.54)	39-0120-69C
$C_{10}H_{16}N_2O_4$			
Barbituric acid, 5-isopropyl-5-(1-meth-oxyethyl)-	pH 6.2	210s(4.02)	65-2568-69
	pH 13	242(3.90)	65-2568-69
Barbituric acid, 5-isopropyl-5-(2-meth-oxyethyl)-	pH 6.2	270(3.57)	65-2568-69
	pH 13	270(4.18)	65-2568-69
Barbituric acid, 5-(1-methoxyethyl)-5-propyl-	pH 6.2	210s(4.04)	65-2568-69
	pH 13	244(3.89)	65-2568-69
Barbituric acid, 5-(2-methoxyethyl)-5-propyl-	pH 6.2	270(3.14)	65-2568-69
	pH 13	244(3.83)	65-2568-69
1-Pyrazoline-3,3-dicarboxylic acid, 4-propyl-, dimethyl ester	$CHCl_3$	325(2.23)	22-0313-69
2-Pyrazoline-3,5-dicarboxylic acid, 5-methyl-, diethyl ester	hexane	270(3.80)	39-2443-69C
$C_{10}H_{16}N_2O_6S_2$			
Methanesulfonamide, N,N-(2,5-dimethoxy-p-phenylene)bis-	EtOH	210(4.49),238(4.08), 300(3.84)	117-0287-69
$C_{10}H_{16}N_2O_6S_3$			
Metanilamide, N^3,N^3-dimethyl-4,6-bis-(methylsulfonyl)-	MeOH	220(4.32),268(4.27), 327(3.71)	44-1780-69
$C_{10}H_{16}N_4$			
Imidazole-4-carbonitrile, 5-amino-1-iso-pentyl-2-methyl-	pH 1	253(3.98)	39-2198-69C
	pH 13	249(3.98)	39-2198-69C
	H_2O	247(3.95)	39-2198-69C
$C_{10}H_{16}N_4O$			
4-Imidazolecarboxamide, 5-amino-1-cyclo-hexyl-	pH 1	240(3.76),269(3.86)	39-2198-69C
	H_2O	267(3.95)	39-2198-69C
	pH 13	267(3.95)	39-2198-69C
$C_{10}H_{16}N_4O_3$			
4-Pyrimidinol, 6-(hexylamino)-5-nitro-	pH 1	226(4.28),290(3.58), 342(3.91)	73-2114-69
	pH 13	224(4.06),348(3.99)	73-2114-69
$C_{10}H_{16}N_4O_4$			
Cytosine, 1-(3-amino-3-deoxy-β-D-ara-binofuranosyl)-N-methyl-, sulfate	H_2O	234s(3.90),273(4.07)	94-1188-69
$C_{10}H_{16}N_4O_5$			
Cytosine, 1-(4-amino-4-deoxy-β-D-gluco-pyranosyl)-	pH 2	275(4.07)	94-0416-69

Compound	Solvent	$\lambda_{max}(\log \epsilon)$	Ref.
Cytosine, 1-(4-amino-4-deoxy-β-D-glucopyranosyl)- (cont.)	pH 6.8	198(4.33),235(3.90), 267(3.91)	94-0416-69
	pH 13	368(3.93)	94-0416-69
$C_{10}H_{16}N_4S$			
4-Pyrimidinethiol, 5-amino-6-(cyclohexylamino)-	pH 1	250(4.25),314(4.21)	73-2114-69
	pH 13	232(4.27),243(4.29), 311(4.22)	73-2114-69
$C_{10}H_{16}N_6$			
1H-Imidazo[4,5-d]pyridazine, 4-[[3-(dimethylamino)propyl]amino]-, dihydrochloride	EtOH	212(4.64),258(4.32)	4-0093-69
$C_{10}H_{16}N_6O$			
1-Propanol, 3-[[6-(dimethylamino)purin-8-yl]amino]-	pH 1	306(4.17)	65-2125-69
	EtOH	294(4.18)	65-2125-69
$C_{10}H_{16}O$			
Bicyclo[2.1.0]pentane, 5-acetyl-1,4,5-trimethyl-	isooctane	300(1.90)	77-1103-69
2-Buten-1-one, 1-(2,2-dimethylcyclopropyl)-3-methyl-	isooctane	235(4.37),330(1.87)	22-3981-69
D-Camphor, (+)-	C_6H_{12}	282s(1.32),294(1.40), 313s(1.15),322(1.38)	60-2611-69
	EtOH	289(1.43)	60-2611-69
2-Cyclodecen-1-one	EtOH	227(3.54),300(1.81)	24-3877-69
1-Cyclohexene-, 1-acetyl-3,3-dimethyl-	n.s.g.	232(4.97),305(2.62)	97-0063-69
1-Cyclohexene, 1-acetyl-2-ethyl-	MeOH	248(3.76)	78-0223-69
	EtOH	249(3.75)	23-2403-69
2-Cyclohexen-1-one, 3-isopropyl-5-methyl-	EtOH	234(4.05),310(1.6)	77-0726-69
2-Cyclohexen-1-one, 4-isopropyl-3-methyl-	EtOH	238(4.19)	39-2634-69C
Cyclopentene, 1-acetyl-2-propyl-	MeOH	256(3.94)	39-0608-69C
Cyclopentene, 2-methyl-1-propionyl-	MeOH	253(3.92)	39-0608-69C
2-Cyclopenten-1-one, 3-ethyl-2-propyl-	EtOH	236(4.20)	104-1993-69
2-Cyclopenten-1-one, 2-isopropyl-3,4-dimethyl-, (±)-	MeOH	236(4.15),306(1.58)	78-3161-69
2-Cyclopenten-1-one, 5-isopropyl-3,4-dimethyl-	MeOH	228(4.15),308(1.76)	78-3161-69
2-Cyclopenten-1-one, 5-isopropyl-3,5-dimethyl-, (S)-(-)-	MeOH	228(4.15),310(1.92)	44-0136-69
Cyclopropanecarboxaldehyde, 2,2-dimethyl-3-(2-methylpropenyl)-	EtOH	198(4.15),229s(3.65), 283s(2.63)	44-2301-69
Furan, 2,3-dihydro-3,3-dimethyl-2-(2-methylpropenyl)-	EtOH	200(4.10),220(3.55) (end absorptions)	44-2301-69
Lyratol	EtOH	211(2.53)	78-3217-69
2(1H)-Naphthalenone, octahydro-, trans	EtOH	280(1.30)	22-3199-69
2-Norpinanone, 3,6,6-trimethyl-	C H	287(1.40)	28-1315-69B
Spiro[2.3]hexan-5-one, 1,1,4,4-tetramethyl-	hexane	198(3.14),243(--)	28-0346-69B
$C_{10}H_{16}O_2$			
Cyclobutanol, 1,2,2-trimethyl-3-methylene-, acetate	isooctane	214(2.48),227s(2.20), 265(1.23)	35-3299-69
Cyclobutanol, 1,2,2-trimethyl-4-methylene, acetate	isooctane	212(2.56),227s(2.20), 257(1.78)	35-3299-69
Cyclobutanol, 1,3,3-trimethyl-2-methylene-, acetate	isooctane	215(2.73),226s(2.38), 265(1.90)	35-3299-69

Compound	Solvent	$\lambda_{max}(\log \epsilon)$	Ref.
1,3-Dioxolane, 2-(3,5-hexadienyl)-2-methyl-, cis	EtOH	228(4.31)	39-1016-69C
4,5-Heptadien-2-ol, 2-methyl-, acetate	isooctane	210(2.87),267(2.18), 273(2.11)	35-3299-69
3,4,6-Heptatrien-2-ol, 3-methoxy-2,6-dimethyl-	EtOH	226(4.47)	54-0119-69
4,5-Hexadien-2-ol, 2,4-dimethyl-, acetate	isooctane	213(2.67),227s(2.42)	35-3299-69
$C_{10}H_{16}O_3$			
1,2-Cyclodecanedione, 3-hydroxy-, enol	DMSO-acid	268(4.15)	24-0388-69
	DMSO-base	311(4.25)	24-0388-69
Cyclopentanepropionic acid, β-oxo-, ethyl ester	isooctane	247(3.72)	22-3694-69
	EtOH	248(3.19)	22-3694-69
2-Octenoic acid, 8-formyl-, ethyl ester	EtOH	217(4.07)	104-1722-69
$C_{10}H_{16}O_4$			
1,1-Cyclopropanedicarboxylic acid, 2-propyl-, dimethyl ester	C_6H_{12}	214(3.75)	22-0313-69
2,4-Heptanedione, 6-acetoxy-6-methyl-	EtOH	275(4.07)	22-4091-69
Malonic acid, (1-methylbutylidene)-, dimethyl ester	EtOH	221(3.97)	22-0313-69
Malonic acid, pentylidene-, dimethyl ester	EtOH	212(4.00)	22-0313-69
$C_{10}H_{16}O_4S_3$			
Thiophene, 2,3-bis[(ethylsulfonyl)-methyl]-	EtOH	243(4.03)	54-0321-69
$C_{10}H_{17}BrO$			
Cyclohexanone, 2-bromo-4-tert-butyl-, cis	EtOH	283(1.34)	22-2013-69
trans	EtOH	310(2.02)	22-2013-69
Cyclohexanone, 2-bromo-5-tert-butyl-, axial, cis	EtOH	308(2.02)	22-2021-69
equatorial, trans	EtOH	288(1.58)	22-2021-69
$C_{10}H_{17}ClO$			
Cyclohexanone, 4-tert-butyl-2-chloro-, cis	EtOH	280(1.30)	22-2013-69
trans	EtOH	303(1.68)	22-2013-69
Cyclohexanone, 5-tert-butyl-2-chloro-, axial	EtOH	302(1.65)	22-2021-69
equatorial	EtOH	282(1.45)	22-2021-69
m-Menthan-5-one, 8-chloro-, cis-(+)-	EtOH	285(1.25)	77-0726-69
$C_{10}H_{17}FO$			
Cyclohexanone, 4-tert-butyl-2-fluoro-, cis	EtOH	285(1.67)	22-2013-69
trans	EtOH	302(1.28)	22-2013-69
Cyclohexanone, 5-tert-butyl-2-fluoro-, axial	EtOH	297(1.38)	22-2021-69
equatorial	EtOH	284(1.74)	22-2021-69
$C_{10}H_{17}I$			
Norbornane, 1-iodo-2,3,3-trimethyl-, endo	EtOH	253(2.84)	77-0174-69
$C_{10}H_{17}NO$			
2-Cyclohexen-1-one, 3-(diethylamino)-	EtOH	300(4.53)	77-0861-69

Compound	Solvent	λ_{max}(log ϵ)	Ref.
$C_{10}H_{17}NO_2$			
Crotonic acid, 3-(N-tert-butylformimid-oyl)-, methyl ester	C_6H_{12}	240(4.35)	54-0005-69
2(3H)-Furanone, 3-[1-(butylamino)ethyl-idene]dihydro-	hexane	295(4.13)	104-0998-69
$C_{10}H_{17}NO_3$			
Nipecotic acid, 4-oxo-1-propyl-, methyl ester	hexane	252(4.00)	103-0062-69
	MeOH	252(3.88)	103-0062-69
	EtOH	252(3.92)	103-0062-69
	iso-BuOH	252(3.96)	103-0062-69
hydrochloride	H_2O	245(3.63)	103-0062-69
	MeOH	245(3.95)(changing)	103-0062-69
	EtOH	245(3.93)(changing)	103-0062-69
	iso-BuOH	245(3.93)(changing)	103-0062-69
$C_{10}H_{17}NO_4$			
Formic acid, [N-(2-hydroxy-1,1-dimethyl-ethyl)formimidoyl]-, ethyl ester, acetate	n.s.g.	240s(--),290(2.49)	54-0005-69
$C_{10}H_{17}NO_5$			
Maleic acid, (diethylamino)(hydroxy)-, dimethyl ester	MeOH	264(4.18)	24-2336-69
Maleic acid, [(diethylamino)oxy]-, dimethyl ester	MeOH	230(3.92)	24-2336-69
$C_{10}H_{17}N_3O$			
v-Triazolo[1,5-a]pyridine-3-propanol, 4,5,6,7-tetrahydro-β-methyl-	EtOH	225(3.40)	88-1549-69
$C_{10}H_{17}N_3O_3$			
1(2H)-Pyrimidineacetaldehyde, 4-amino-2-oxo-, 1-(diethyl acetal)	pH 1	283(4.13)	88-2285-69
	pH 12	273(3.94)	88-2285-69
$C_{10}H_{17}N_3O_3S$			
as-Triazine-3,5(2H,4H)-dione, 2-(3-eth-oxy-3-hydroxy-1-methylpropyl)-6-meth-yl-3-thio-	EtOH	220(4.12),270(4.32), 305s(3.64)	114-0093-69C
$C_{10}H_{17}N_3O_6S$			
as-Triazin-5(2H)-one, 2-methyl-3-(methylthio)-6-(D-allo-1,2,3,4,5-pentahydroxypentyl)-	H_2O	237(4.30)	73-1673-69
6-D-altro-	H_2O	237(4.30)	73-1673-69
6-D-gluco-	H_2O	237(4.30)	73-1673-69
$C_{10}H_{18}ClN_5$			
s-Triazine, 2-chloro-4-(diethylamino)-6-(isopropylamino)-	pH 0.5	228(4.35),282s(3.58)	59-1167-69
	pH 10.0	228(4.54),271s(3.58)	59-1167-69
$C_{10}H_{18}N_2$			
2,3-Diazabicyclo[2.2.1]hept-5-ene, 1-isopropyl-2,3-dimethyl-	C_6H_{12}	265(2.85)	78-0657-69
2,3-Diazabicyclo[2.2.1]hept-5-ene, 5-isopropyl-2,3-dimethyl-	C_6H_{12}	253(2.93)	78-0657-69
2,3-Diazabicyclo[2.2.1]hept-5-ene, 1,2,3,7,7-pentamethyl-	C_6H_{12}	281(2.79)	78-0657-69
	EtOH	276(2.79)	78-0657-69
	dioxan	277(2.78)	78-0657-69

Compound	Solvent	$\lambda_{max}(\log \epsilon)$	Ref.
$C_{10}H_{18}N_2O$			
2H-Pyrrol-3-ol, 5-(diethylamino)-2,4-dimethyl-	MeOH	228(4.00),289(4.29)	78-5721-69
$C_{10}H_{18}N_2O_2$			
1-Pyrazoline-3-carboxylic acid, 4-tert-butyl-5-methyl-, methyl ester	EtOH	324(2.37)	39-0787-69C
$C_{10}H_{18}N_2O_3$			
2-Butanone, 3-(1-hexenyl-ONN-azoxy)-1-hydroxy-	n.s.g.	238(3.95)	35-2808-69
$C_{10}H_{18}N_4O_2$			
Uracil, 5,6-bis(dimethylamino)-1,3-dimethyl-	H_2O	297(3.97)	104-2004-69
$C_{10}H_{18}N_4S$			
4-Pyrimidinethiol, 5-amino-6-(hexylamino)-	pH 1	247(4.19),312(4.16)	73-2114-69
	pH 13	232(4.34),310(4.25)	73-2114-69
$C_{10}H_{18}O$			
Cyclohexanone, 3-tert-butyl-	EtOH	284(1.30)	22-2021-69
Cyclohexanone, 4-tert-butyl-	C_6H_{12}	290(1.26)	77-1246-69
	EtOH	283(1.26)	22-2013-69
Cyclopentanone, 2-isopropyl-3,4-dimethyl-	MeOH	292(1.40)	78-3161-69
2-Hexenal, 4-ethyl-2,5-dimethyl- (tetrahydrolyratal)	EtOH	232(4.14)	78-3217-69
p-Menthan-3-one	EtOH	290(1.37)	30-0338-69F
$C_{10}H_{18}OS$			
Cyclohexanone, 2-(tert-butylthio)-	EtOH	245(2.63),304(2.47)	44-1566-69
$C_{10}H_{18}OS_2$			
Morpholinium 2-amino-1-cyclopentene-1-dithiocarboxylate	EtOH	303(3.40),393(3.94)	44-0730-69
$C_{10}H_{18}O_3$			
2-Hexenal, 4-oxo-, 1-(diethyl acetal)-, (E)-	pet ether	212(4.04)	54-0989-69
$C_{10}H_{18}Si$			
Silane, triallylmethyl-	gas	190(--)	121-0803-69
	isooctane	196(4.46)	121-0803-69
	EtOH	197(4.44)	121-0803-69
$C_{10}H_{18}Sn$			
Stannane, triallylmethyl- (correction)	hexane	211(4.40)	44-3715-69
	MeCN	217(4.40)	44-3715-69
$C_{10}H_{19}NO$			
Azacycloundecan-2-one	C_6H_{12}	190(3.80)	46-1642-69
	H_2O	192(3.88)	46-1642-69
2-Aziridinone, 1,3-di-tert-butyl-	hexane	251(2.03)	35-1176-69
	MeOH	234(1.98)	35-1176-69
	EtOH	240(2.19)	35-1176-69
$C_{10}H_{19}NO_2$			
Azacycloundecan-2-one, 7-hydroxy-	H_2O	192(3.87)	39-1358-69C

Compound	Solvent	$\lambda_{max}(\log \epsilon)$	Ref.
$C_{10}H_{19}N_5O$			
s-Triazine, 2,4-bis(isopropylamino)-	pH 1.0	211s(4.36),235s(4.16)	59-1167-69
6-methoxy-	pH 10.0	217(4.57)	59-1167-69
s-Triazine, 2-(diethylamino)-4-(ethyl-	pH 1.0	216(4.31),243(4.27)	59-1167-69
amino)-6-methoxy-	pH 10.0	222(4.53)	59-1167-69
s-Triazin-2-ol, 4-(diethylamino)-	pH 1.0	250(4.41)	59-1167-69
6-(isopropylamino)-	pH 8.0	220(4.34),238(4.35)	59-1167-69
	pH 11.5	222s(4.49)	59-1167-69
$C_{10}H_{19}N_5O_2$			
1-Propanol, 3,3'-[(6-methyl-as-tria-zine-3,5-diyl)diimino]di-	EtOH	212(4.38),230s(4.09), 314(3.64)	114-0181-69C
$C_{10}H_{19}N_5S$			
s-Triazine, 2,4-bis(isopropylamino)-	pH 1.0	222(4.42),255s(4.21)	59-1167-69
6-(methylthio)-	pH 10.0	222(4.60),264s(3.68)	59-1167-69
$C_{10}H_{19}O_5P$			
Crotonic acid, 3-methyl-4-phosphono-, P,P-diethyl methyl ester	EtOH	221(4.08)	22-3252-69
$C_{10}H_{20}N_2O_2$			
1-Butanol, 2-(tert-butylazo)-, acetate	C_6H_{12}	363(1.24)	89-0451-69
1-Propanol, 1-(tert-butylazo)-2-methyl-, acetate	C_6H_{12}	360(1.29)	89-0451-69
$C_{10}H_{20}N_2O_6$			
Malonamic acid, 2-N-carbobenzoxysarco-syl-, ethyl ester	EtOH	264(4.41)	104-1713-69
$C_{10}H_{20}Sn$			
Stannane, diallyldiethyl-	hexane	208(4.29)	44-3715-69
(correction)	MeCN	212(4.30)	44-3715-69
$C_{10}H_{21}NO$			
4-Piperidinemethanol, α-butyl-	EtOH	225(2.65),266(2.38)	39-2134-69C
$C_{10}H_{21}N_3S$			
Nonanal, thiosemicarbazone	EtOH	230(4.01),271(4.45)	94-0844-69
$C_{10}H_{30}OSi_4$			
Trisilane, heptamethyl-2-(trimethyl-	isooctane	224(3.75),236(3.51)	35-6613-69
siloxy)-	MeCN	223(3.69),236(3.50)	35-6613-69

Compound	Solvent	λ_{max}(log ε)	Ref.
$C_{11}F_9N$			
Pyridine, tetrafluoro-2-(pentafluoro-phenyl)-	C_6H_{12}	265(3.96)	39-2559-69C
$C_{11}H_2Cl_9N_3S$			
s-Triazine, 2,4-bis(trichloromethyl)-6-[(2,4,5-trichlorophenyl)thio]-	MeOH	213(4.37),266(3.85)	18-2931-69
$C_{11}H_3Cl_7N_4O_2S$			
s-Triazine, 2-[(4-chloro-3-nitrophenyl)-thio]-4,6-bis(trichloromethyl)-	MeOH	263(4.04)	18-2931-69
$C_{11}H_3Cl_8N_3$			
s-Triazine, 2-(2,4-dichlorophenyl)-4,6-bis(trichloromethyl)-	MeOH	216(4.36),283(3.42)	18-2924-69
$C_{11}H_3Cl_8N_3S$			
s-Triazine, 2-[(2,4-dichlorophenyl)-thio]-4,6-bis(trichloromethyl)-	MeOH	213(4.32),267(4.12)	18-2931-69
s-Triazine, 2-[(3,4-dichlorophenyl)-thio]-4,6-bis(trichloromethyl)-	MeOH	212(4.38),271(3.96)	18-2931-69
$C_{11}H_4BrCl_6N_3S$			
s-Triazine, 2-[(m-bromophenyl)thio]-4,6-bis(trichloromethyl)-	MeOH	215(4.24),270(4.03)	18-2931-69
$C_{11}H_4Cl_6N_4O_2S$			
s-Triazine, 2-[(p-nitrophenyl)thio]-4,6-bis(trichloromethyl)-	MeOH	288(4.21)	18-2931-69
$C_{11}H_4Cl_7N_3S$			
s-Triazine, 2-[(m-chlorophenyl)thio]-4,6-bis(trichloromethyl)-	MeOH	213(4.25),270(4.04)	18-2931-69
s-Triazine, 2-[(o-chlorophenyl)thio]-4,6-bis(trichloromethyl)-	MeOH	210(4.20),270(4.01)	18-2931-69
s-Triazine, 2-[(p-chlorophenyl)thio]-4,6-bis(trichloromethyl)-	MeOH	210(4.28),268(4.04)	18-2931-69
$C_{11}H_5ClN_2O_4$			
Benzoxazole, 5-chloro-2-(5-nitro-2-fur-yl)-	EtOH	233(3.98),269(4.03),352(4.36)	18-3335-69
$C_{11}H_5Cl_6N_3$			
s-Triazine, 2-phenyl-4,6-bis(trichloro-methyl)-	MeOH	282(4.37)	18-2924-69
$C_{11}H_5Cl_6N_3S$			
s-Triazine, 2-(phenylthio)-4,6-bis-(trichloromethyl)-	MeOH	271(3.99)	18-2931-69
$C_{11}H_5F_4NOS$			
Pyridine, 2,3,5,6-tetrafluoro-4-(phen-ylsulfinyl)-	hexane	210(3.35),238(3.42),281(2.97)	39-1660-69C
$C_{11}H_5F_4NO_2S$			
Pyridine, 2,3,5,6-tetrafluoro-4-(phen-ylsulfonyl)-	hexane	213(3.80),237(4.06),272s(3.52),278(3.62),290(3.60)	39-1660-69C

Compound	Solvent	λ_{max}(log ϵ)	Ref.
$C_{11}H_5F_4NS$			
Pyridine, 2,3,5,6-tetrafluoro-4-(phen-	hexane	211(4.17),248(3.85), 275(3.93)	39-1660-69C
$C_{11}H_6ClN_5O_2$			
1H-Pyrazolo[3,4-d]pyrimidine, 4-chloro-1-(p-nitrophenyl)-	MeOH	224(4.33),263(3.93), 310(4.23)	23-1129-69
$C_{11}H_6Cl_4N_2O_4$			
1,2-Diazaspiro[4.4]nona-1,3,6,8-tetraene-3,4-dicarboxylic acid, 6,7,8,9-tetrachloro-, dimethyl ester	$CHCl_3$	245(3.78),295(3.56)	64-0536-69
$C_{11}H_6F_4$			
Naphthalene, 1,2,3,4-tetrafluoro-5-methyl-	EtOH	272(3.72),280(3.78), 290(3.65),306(3.07), 320(3.06)	70-0131-69
Naphthalene, 1,2,3,4-tetrafluoro-6-methyl-	EtOH	276(3.66),306(2.79), 304(2.67),320(2.72)	70-0131-69
$C_{11}H_6F_4O$			
Naphthalene, 1,2,3,4-tetrafluoro-5-methoxy-	EtOH	292(3.82),310(3.67), 324(3.60)	70-0131-69
$C_{11}H_6N_2O_3S$			
Benzothiazole, 2-(5-nitro-2-furyl)-	EtOH	280(3.85),364(4.38)	18-3335-69
$C_{11}H_6N_2O_4$			
Benzoxazole, 2-(5-nitro-2-furyl)-	EtOH	230(4.10),265(4.15), 349(4.44)	18-3335-69
$C_{11}H_6N_4$			
1-Phthalazinemalononitrile	EtOH	216(4.64),282(3.96), 291(4.09),357(4.17)	95-0959-69
$C_{11}H_6N_4O$			
4H-Quinolizine-1,3-dicarbonitrile, 2-amino-4-oxo-	EtOH	249(4.75),295(4.25), 386(4.09)	95-0203-69
$C_{11}H_6N_4O_7$			
Barbituric acid, 5-(2,4-dinitrobenzylidene)-	EtOH	255(4.57)	103-0827-69
1-(Hydroxypicryl)pyridinium hydroxide, inner salt	MeOH	350(5.0)	56-1843-69
Pyridine, 2-(picryloxy)-	MeOH	220s(4.20),300s(3.5)	56-1653-69
2(1H)-Pyridone, 1-picryl-	MeOH	220(4.4),298(3.8), 350s(3.2)	56-1653-69
$C_{11}H_6O_4$			
Xanthotoxol	EtOH	251(4.24),268(4.25), 308(4.08)	88-5223-69
$C_{11}H_7BrN_2O_2S$			
Barbituric acid, 5-(p-bromobenzylidene)-2-thio-	EtOH	265(4.32),310s(--)	103-0830-69
$C_{11}H_7BrN_4$			
1H-Pyrazolo[3,4-d]pyrimidine, 4-bromo-1-phenyl-	MeOH	244(4.50)	23-1129-69

Compound	Solvent	λ max(log ε)	Ref.
$C_{11}H_7BrO$			
5H-Benzocyclohepten-5-one, 6-bromo-	EtOH	226(4.2),315(4.1), 325(4.1)	23-1169-69
	EtOH	231(4.50),263(4.02), 312(3.90),325(3.90)	42-0855-69
	EtOH	210(4.07),239(4.37), 262s(--),270s(--), 332(3.94),360(3.85)	39-2656-69C
7H-Benzocyclohepten-7-one, 6-bromo-	EtOH	237(4.4),276(4.6), 341(3.6),358(3.3)	23-1169-69
	EtOH	238(4.43),275(4.47), 340(3.57)	42-0855-69
$C_{11}H_7BrO_4$			
Bicyclo[3.2.2]nona-2,8-diene-6,7-dicarboxylic anhydride, 3-bromo-4-oxo-, endo-	MeOH	245(3.55),275(3.47), 337(2.13)	88-0443-69
$C_{11}H_7ClN_2O_2S$			
Barbituric acid, 5-(p-chlorobenzylidene)-2-thio-	EtOH	263(4.24),290(4.27)	103-0830-69
$C_{11}H_7ClN_2O_3$			
Barbituric acid, 5-(p-chlorobenzylidene)-	EtOH	255(4.33),374(3.25)	103-0827-69
$C_{11}H_7ClN_4$			
1H-Pyrazolo[3,4-d]pyrimidine, 4-chloro-1-phenyl-	MeOH	243(4.51)	23-1129-69
$C_{11}H_7ClN_4O$			
1H-Benzotriazole, 1-(6-chloro-2-pyridyl)-, 2-oxide	n.s.g.	262(4.07),284(4.13), 305(4.09)	20-0553-69
1H-Benzotriazole, 1-(6-chloro-2-pyridyl)-, 3-oxide	n.s.g.	223(4.26),256(4.21), 334(4.03),346(4.15), 361(4.04)	20-0553-69
$C_{11}H_7ClO_4$			
Bicyclo[3.2.2]nona-2,8-diene-6,7-dicarboxylic anhydride, 3-chloro-4-oxo-, endo-	MeOH	234(3.81),260s(3.45), 330(2.07)	88-0443-69
Cyclopenta[c]pyran-4-carboxylic acid, 7-(chloromethylene)-1,7-dihydro-1-oxo-, methyl ester	pentane	234(3.70),273(3.56), 314(4.01),327(3.99), 343(3.71)	35-6470-69
$C_{11}H_7IN_4O_4S$			
2-Thiophenecarboxaldehyde, 5-iodo-, 2,4-dinitrophenylhydrazone	MeOH	256(4.16),311(3.80),	39-1813-69C
$C_{11}H_7NO$			
3H-Pyrrolo[1,2-a]indol-3-one	EtOH	355(4.03)	77-0991-69
$C_{11}H_7NO_3$			
5H-Benzocyclohepten-5-one, 3-nitro-	EtOH	219(4.31),253(4.3), 346(4.02)	39-2656-69C
$C_{11}H_7NO_4SSe$			
1,3-Propanedione, 1-(5-nitroselenophene-2-yl)-3-(2-thienyl)-	H_2O	260(4.17),307(4.02), 405(5.56)	123-0074-69C

Compound	Solvent	λ_{max}(log ϵ)	Ref.
$C_{11}H_7NO_4Se_2$			
1,3-Propanedione, 1-(5-nitroselenophene-2-yl)-3-selenophene-2-yl-	H_2O	265(4.15),310(4.02), 410(5.62)	123-0074-69C
$C_{11}H_7NS$			
Thiopyrano[4,3-b]indole	EtOH	222(4.23),272s(4.53), 279(4.58),296s(4.20), 303s(4.16),351s(3.44), 402(3.39)	64-0024-69
$C_{11}H_7N_3O_2$			
1,2-Propanedione, 3-diazo-1-indol-3-yl-	EtOH	217(4.15),272(4.18), 278(4.13),338(4.02)	97-0269-69
$C_{11}H_7N_3O_4S$			
Barbituric acid, 5-(p-nitrobenzylidene)-2-thio-	EtOH	272(4.20),340(4.12)	103-0830-69
$C_{11}H_7N_3O_5$			
Barbituric acid, 5-(m-nitrobenzylidene)-	EtOH	260(4.25)	103-0827-69
Barbituric acid, 5-(p-nitrobenzylidene)-	EtOH	260(4.23)	103-0827-69
$C_{11}H_8ClNO$			
5H-Benzocyclohepten-5-one, 7-amino-6-chloro-	EtOH	248(4.39),293(4.21), 375(4.09),416(3.78)	39-1499-69C
Furo[3,2-c]quinoline, 7-chloro-2,3-dihydro-	EtOH	233(4.76),292s(3.76), 303(3.84),315(3.80), 329(3.71)	2-0952-69
Furo[3,2-c]quinoline, 8-chloro-2,3-dihydro-	EtOH	230s(4.68),235(4.70), 289s(3.74),298(3.80), 315s(3.60),322s(3.36)	2-0952-69
$C_{11}H_8ClNO_2$			
8-Quinolinol, 7-acetyl-5-chloro-	pH 1.2	230s(4.10),271(4.49), 364(3.50)	23-0105-69
	pH 8.5	230s(4.10),269(4.41), 288s(3.87),366(3.66)	23-0105-69
	pH 14.0	285(4.39),373(3.95)	23-0105-69
$C_{11}H_8ClN_3O$			
Pyrazolo[1,5-a]quinazolin-5(4H)-one, 7-chloro-2-methyl-	EtOH-HCl	203(4.46),222s(4.33), 227(4.35),264(4.44)	4-0947-69
	EtOH	203(4.47),222s(4.13), 228(4.36),264(4.44)	4-0947-69
	EtOH-KOH	225(4.46),239(3.96), 266(4.43),291s(3.79), 328(3.63)	4-0947-69
$C_{11}H_8ClN_3OS$			
Δ^2-1,3,4-Oxadiazoline-4-propionitrile, 2-(o-chlorophenyl)-5-thioxo-	EtOH	253(4.20),295(4.26)	2-0583-69
$C_{11}H_8ClN_3O_4S$			
Acetamide, N-[5-[1-chloro-2-(5-nitro-2-furyl)vinyl]-2-thiazolyl]-	dioxan	246(4.03),280(4.04), 381(4.20)	103-0414-69
	HOAc	280(4.09),370(4.24)	103-0414-69
$C_{11}H_8Cl_2F_5NO_3$			
Acetamide, 2,2-dichloro-N-[2,3,4,5,6-pentafluoro-β-hydroxy-α-(hydroxymethyl)phenethyl]-	EtOH	206(3.95),260(2.74)	65-1615-69

Compound	Solvent	$\lambda_{max}(\log \epsilon)$	Ref.
$C_{11}H_8Cl_2N_4$			
9H-Indeno[2,1-d]pyrimidine, 2,4-diamino-6,7-dichloro-	pH 1	276(4.50),297s(4.24)	4-0613-69
	EtOH	282(4.47),303(4.40)	4-0613-69
$C_{11}H_8Cl_2O$			
1H-Cyclobut[a]inden-1-one, 2,2-dichloro-2,2a,7,7a-tetrahydro-, cis	EtOH	205(4.22),213(3.93)	44-2792-69
$C_{11}H_8F_6$			
Bicyclo[2.2.2]octa-2,5,7-triene, 5-methyl-2,3-bis(trifluoromethyl)-	hexane	232(2.54),268(2.23)	44-1271-69
$C_{11}H_8N_2$			
2,4,6-Cycloheptatriene-$\Delta^{1,\alpha}$-succinonitrile	EtOH	247(3.98),345(4.23)	35-6391-69
Pyrido[1,2-a]benzimidazole	EtOH	240(4.45),245(4.46), 258(3.99),266(3.92)	39-1334-69C
Spiro[2.6]nona-4,6,8-triene-1,2-dicarbonitrile, cis	EtOH	262(3.43)	35-6391-69
trans	EtOH	262(3.44)	35-6391-69
$C_{11}H_8N_2O_2S$			
Barbituric acid, 5-benzylidene-2-thio-	EtOH	250(4.18),280(4.24)	103-0830-69
$C_{11}H_8N_2O_3$			
Barbituric acid, 5-benzylidene-	EtOH	255(4.02),330(2.88)	103-0827-69
$C_{11}H_8N_2O_4$			
Barbituric acid, 5-[3-(2-furyl)allylidene]-	EtOH	245(3.47),368(4.43)	103-0827-69
Barbituric acid, 5-(p-hydroxybenzylidene)-	EtOH	260(4.39),336(2.78), 395(3.11)	103-0827-69
$C_{11}H_8N_4O$			
1H-Benzotriazole, 1-(2-pyridyl)-, N-oxide	n.s.g.	242(4.22),258(4.22), 293(3.85)	20-0553-69
1H-Benzotriazole, 1-(2-pyridyl)-, 2-oxide	n.s.g.	211(4.28),277(4.16), 300(4.07)	20-0553-69
1H-Benzotriazole, 1-(2-pyridyl)-, 3-oxide	n.s.g.	223(4.26),251(4.21), 333s(4.11),345(4.17), 358(4.01)	20-0553-69
$C_{11}H_8N_4O_2S$			
Acetic acid, (5H-as-triazino[5,6-b]indol-3-ylthio)-	EtOH	270(4.5),350(4.0)	65-2111-69
$C_{11}H_8O$			
7H-Cyclobut[a]inden-7-one, 2a,7a-dihydro-	EtOH	215(4.17),248(3.98), 295(3.37)	39-2656-69C
3,5-Heptadiynal, 2-(2-butynylidene)-, (E)-	ether	242(3.75),255(3.91), 317(4.09)	39-0683-69C
(Z)-	ether	227(3.65),239(3.79), 251(3.79),260s(--), 268(3.67),319(4.19)	39-0683-69C
1-Naphthaldehyde	EtOH	203(4.6),242(4.2), 248s(4.2),314(3.8), 328(3.8)	28-0248-69B
2-Naphthaldehyde	EtOH	224(4.3),242s(4.5), 248(4.6),271s(3.8), 282(4.0),292(3.9),	28-0248-69B

(see next page)

Compound	Solvent	λ max (log ε)	Ref.
2-Naphthaldehyde (cont.)		332s(3.2),343(3.3)	
$C_{11}H_8OSSe$			
2-Propen-1-one, 1-selenophene-2-yl-3-(2-thienyl)-	hexane	240(3.50),287(3.96), 338(4.44)	103-0055-69
	EtOH	232(3.75),289(3.96), 357(4.36)	103-0055-69
2-Propen-1-one, 3-selenophene-2-yl-1-(2-thienyl)-	hexane	308(4.12),343(4.32)	103-0055-69
	EtOH	314(4.05),365(4.30)	103-0055-69
$C_{11}H_8OSe_2$			
2-Propen-1-one, 1,3-diselenophene-2-yl-	hexane	228(3.72),305s(--), 346(4.38)	103-0055-69
	EtOH	237(3.84),310s(--), 365(4.33)	103-0055-69
$C_{11}H_8O_2$			
Benzaldehyde, o-2-furyl-	EtOH	327(3.71)	6-0277-69
1-Naphthaldehyde, 4-hydroxy-	C_6H_{12}	213(4.17),239(4.58), 310s(4.04),320(4.10), 335(4.02)	19-0253-69
	aq. MeOH	212(4.20),236(4.38), 242(4.38),337(4.14)	19-0253-69
1,4-Naphthoquinone, 5-methyl-	EtOH	204(4.29),248(4.33), 348(3.51)	22-3612-69
$C_{11}H_8O_2S$			
2-Thiophenecarboxylic acid, 3-phenyl-	EtOH	270(4.03)	2-0009-69
$C_{11}H_8O_2SSe$			
1,3-Propanedione, 1-selenophene-3-yl-3-(2-thienyl)-	H_2O	259(4.00),350(5.64)	123-0074-69C
$C_{11}H_8O_2Se$			
2-Propen-1-one, 1-(2-furyl)-3-selenophene-2-yl-	hexane	235s(--),307(4.18), 344(4.34)	103-0055-69
	EtOH	235s(--),316(4.11), 363(4.33)	103-0055-69
2-Propen-1-one, 3-(2-furyl)-1-selenophene-2-yl-	hexane	238(3.74),290s(--), 336(4.50),349(4.44)	103-0055-69
	EtOH	241(3.75),295s(--), 355(4.43)	103-0055-69
$C_{11}H_8O_3$			
Naphthaldehyde, 2,3-dihydroxy-	EtOH	223(4.88)	25-1738-69
1,4-Naphthoquinone, 2-hydroxy-8-methyl-	EtOH	220(4.14),251(4.06), 271(4.27),348(4.48)	22-3612-69
1,2-Naphthoquinone, 4-methoxy-	pH 7	250(5.08),275(4.71)	78-5807-69
1,2-Naphthoquinone, 8-methoxy-	EtOH	243(4.29),418(3.84)	39-2059-69C
1,4-Naphthoquinone, 2-methoxy-	pH 7.0	240(4.89),248(4.77)	78-5807-69
$C_{11}H_8O_3Se$			
1,3-Propanedione, 1-(2-furyl)-3-selenophene-3-yl-	H_2O	277(4.95),363(4.49)	123-0074-69C
$C_{11}H_8O_4$			
Benzo[b]furan, 2-acetyl-5,6-(methylenedioxy)-	EtOH	267(3.76),292(4.00), 340(4.27)	18-1971-69
Bicyclo[3.2.2]nona-3,6-diene-8,9-dicarboxylic anhydride, 2-oxo-	MeOH	225(3.80),337(2.14)	88-0443-69

Compound	Solvent	λ_{max}(log ε)	Ref.
1,4-Naphthoquinone, 2,3-epoxy-2,3-dihydro-5-methoxy-	EtOH	233(4.08),345(3.61)	39-2059-69C
$C_{11}H_8O_5$			
Cyclopenta[c]pyran-4-carboxylic acid, 1,7-dihydro-7-(hydroxymethylene)-1-oxo-, methyl ester	EtOH	357(3.86)	35-6470-69
	EtOH-NaOH	304(3.97),378(4.38)	35-6470-69
$C_{11}H_8O_6$			
2H-1-Benzopyran-4-carboxylic acid, 3,6-dihydroxy-2-oxo-, methyl ester	EtOH	297(3.99),332(3.86)	48-0786-69 +48-0800-69
$C_{11}H_8S_2$			
2H-Thiopyrano[3,2-b][1]benzothiophene	EtOH	235(4.22),254(4.07), 266(3.97),300(3.70), 311(3.78),322(3.61), 360(3.52)	22-0601-69
$C_{11}H_9BrN_2$			
4-Cyano-2-methylisoquinolinium bromide	MeOH	240(4.19),307(4.03)	22-2045-69
$C_{11}H_9BrO$			
5H-Benzocyclohepten-5-one, 6-bromo-8,9-dihydro-	EtOH	209(3.86),261(3.81), 280s(--)	39-2656-69C
5H-Benzocyclohepten-5-one, 8-bromo-8,9-dihydro-	EtOH	217(3.93),236(3.94)	39-2656-69C
$C_{11}H_9ClN_2O$			
Cinnamamide, N-chloro-α-cyano-β-methyl-, cis	EtOH	272(4.02)	22-1724-69
trans	EtOH	271(4.03)	22-1724-69
$C_{11}H_9ClN_2O_4$			
4-Cyano-2-methylisoquinolinium perchlorate	EtOH	205(4.24),230(4.20), 307(4.04)	22-2045-69
$C_{11}H_9ClN_4$			
9H-Indeno[2,1-d]pyrimidine, 2,4-diamino-6-chloro-	pH 1	218(4.58),272(4.44), 298s(4.04)	4-0613-69
	pH 10	277(4.45)	4-0613-69
	EtOH	238s(4.10),277(4.44), 297(4.31)	4-0613-69
9H-Indeno[2,1-d]pyrimidine, 2,4-diamino-8-chloro-	pH 1	267(4.40)	4-0613-69
	pH 10	277(4.41),293s(4.27)	4-0613-69
	EtOH	277(4.40),295s(4.29)	4-0613-69
$C_{11}H_9ClN_4O$			
Ketone, p-chlorophenyl 2,4-diamino-5-pyrimidinyl	pH 2	228(4.46),260(4.26), 285(4.14)	44-0821-69
	pH 12	217(4.28),258(4.14), 313(4.29)	44-0821-69
$C_{11}H_9ClN_4O_2$			
Benzaldehyde, p-chloro-, (2,6-dihydroxy-4-pyrimidinyl)hydrazone	EtOH	330(4.39)	56-0519-69
$C_{11}H_9ClO$			
7H-Cyclobut[a]inden-1-one, 2-chloro-	EtOH	207(4.15),216(3.88)	44-2792-69

Compound	Solvent	λ_{max}(log ε)	Ref.
$C_{11}H_9ClOS$			
1-Benzothiepin-3-one, 4-chloro-2,3-dihydro-2-methyl-	EtOH	247(4.27),264(4.10), 372(3.69)	44-0056-69
1-Thiochromone, 3-chloro-2-ethyl-	EtOH	216(3.99),227(4.03), 250(4.36),256(4.35), 281(3.49),291(3.62), 344(4.04)	44-0056-69
$C_{11}H_9ClO_2$			
2-Cyclopropene-1-carboxylic acid, 2-(m-chlorophenyl)-3-methyl-	EtOH	216(4.20),222s(4.11), 264(4.18)	44-0505-69
2-Cyclopropene-1-carboxylic acid, 2-(p-chlorophenyl)-3-methyl-	EtOH	218s(4.37),228s(4.34), 268(4.36)	44-0505-69
$C_{11}H_9ClO_3$			
Isocoumarin, 4-chloro-3-ethoxy-	EtOH	233(4.39),280(4.04), 350(3.59)	2-1114-69
$C_{11}H_9ClO_4$			
Chromone, 8-chloro-5,7-dihydroxy-2,6-dimethyl-	MeOH	225(4.5),260(4.5), 297(4.1),327(3.9)	64-0750-69
$C_{11}H_9ClO_4S_4$			
(2-Thienyl)[3-(2-thienylthio)allylidene]sulfonium perchlorate	CH_2Cl_2	259s(4.12),295(3.66), 411(3.59)	88-0239-69
$C_{11}H_9Cl_2N$			
Isoquinoline, 1-chloro-3-(chloromethyl)-4-methyl-	EtOH	243(4.07),282(3.13), 290(3.17)	44-3664-69
$C_{11}H_9Cl_3N_2O$			
4(3H)-Pyrimidinone, 5,6-dihydro-2-phenyl-6-(trichloromethyl)-	EtOH	232(3.97)	104-1797-69
$C_{11}H_9FO_2$			
2-Cyclopropene-1-carboxylic acid, 2-(p-fluorophenyl)-3-methyl-	EtOH	215s(4.10),260(4.21)	44-0505-69
$C_{11}H_9F_4NO_2$			
Benzaldehyde, 2,3,5,6-tetrafluoro-4-morpholino-	EtOH	260(3.87),328(3.97)	65-1607-69
$C_{11}H_9FeNO_5$			
Iron, (1H-azepine-1-carboxylic acid)-tricarbonyl-, methyl ester	EtOH	250(4.25),297s(3.73)	44-2866-69
$C_{11}H_9N$			
Cyclobuta[b]quinoline, 1,2-dihydro-	H_2O	304(3.76),310(3.74), 317(3.87)	44-4131-69
Pyridine, 2-phenyl-	2N HCl	242(3.94),295(4.14)	44-3175-69
	EtOH	244(4.05),277(3.95)	44-3175-69
Pyridine, 3-phenyl-	2N HCl	232(4.13),255(4.05), 285s(3.77)	44-3175-69
	EtOH	246(4.14),275s(3.79)	44-3175-69
Pyridine, 4-phenyl-	2N HCl	288(4.23)	44-3175-69
	EtOH	256(4.20)	44-3175-69
$C_{11}H_9NO$			
Cyclobuta[b]quinoline, 1,2-dihydro-, 3-oxide	MeOH	238(4.82),314(4.04), 327(3.99)	44-4131-69

Compound	Solvent	$\lambda_{max}(\log \epsilon)$	Ref.
Furo[3,2-c]quinoline, 2,3-dihydro-	EtOH	228(4.58),237s(4.56), 296(3.62),307(3.54), 320(3.43)	2-0952-69
1-Naphthaldehyde, oxime	EtOH	213s(4.3),228(4.7), 245s(4.1),292s(3.9), 306(4.0),318s(3.8)	28-0248-69B
2-Naphthaldehyde, oxime	EtOH	204(4.0),231s(4.5), 239(4.6),248(4.6), 251s(4.5),273(4.1), 285(4.2),294(4.1), 316(3.1),324(2.8), 334s(2.6),342(2.8)	28-0248-69B
$C_{11}H_9NOS$			
Acetophenone, 4'-(4-isothiazolyl)-	EtOH	210(4.21),285(4.22)	4-0841-69
$C_{11}H_9NOSe$			
2-Propen-1-one, 1-pyrrol-2-yl-3-selenophene-2-yl-	EtOH	245(3.80),364(4.40)	103-0055-69
2-Propen-1-one, 3-pyrrol-2-yl-1-selenophene-2-yl-	EtOH	288(3.91),406(4.38)	103-0055-69
$C_{11}H_9NO_2$			
Cinchoninic acid, methyl ester	EtOH	238(4.30),315(3.66)	44-4131-69
$C_{11}H_9NO_3$			
Indolin-2-ylideneacetic acid, 3-oxo-, methyl ester	MeOH	243(4.20),278(3.08), 290(2.95),352(4.01)	23-3545-69
Naphthalene, 1-methoxy-4-nitro-	H_2O	378(3.92)	88-4205-69
	MeCN	366(3.91)	88-4205-69
1,2-Propanedione, 3-hydroxy-1-(indol-3-yl)-	EtOH	216(4.12),247(3.95), 264(3.89),313(3.92)	97-0269-69
Quinaldic acid, 8-hydroxy-, methyl ester	EtOH	258(4.43),309(3.04), 364(3.06)	64-0038-69
$C_{11}H_9NO_4$			
Coumarin, 4-acetyl-3-amino-6-hydroxy-	EtOH	214(4.42),262(4.16), 295(3.94),373(3.97)	48-0786-69
$C_{11}H_9NO_5$			
2H-1-Benzopyran-4-carboxylic acid, 3-amino-6-hydroxy-2-oxo-, methyl ester	EtOH	214(4.56),249(4.03), 280s(3.57),287(3.61), 359(3.97)	48-0786-69
$C_{11}H_9NS$			
Pyridine, 2-[2-(2-thienyl)vinyl]-	C_6H_{12}	288(4.01),302(3.98), 336(4.19),352s(4.07)	117-0209-69
	MeOH	228(3.88),272(3.92), 320(3.94)	117-0209-69
	CCl_4	292(4.02),305(4.01), 339(4.18),354s(4.03)	117-0209-69
Pyridine, 4-[2-(2-thienyl)vinyl]-	MeOH	235s(3.81),330(4.46)	117-0209-69
8H-Thieno[2,3-b]indole, 2-methyl-	EtOH	234(5.29),272(4.04), 285s(3.94)	95-0058-69
$C_{11}H_9N_3$			
2H-Pyrrolo[3,4-b]quinoxaline, 2-methyl-	MeOH	258(4.53),344(3.88), 352(3.96),356(3.90), 504(3.24)	88-1581-69

Compound	Solvent	λ_{max}(log ϵ)	Ref.
2H-Pyrrolo[3,4-b]quinoxaline, 2-methyl- (cont.)	MeOH-HCl	250(4.26),265(4.36), 362(4.05),372(4.03), 578(3.18)	88-1581-69
$C_{11}H_9N_3O$			
Pyrazolo[1,5-a]quinazolin-5(4H)-one, 2-methyl-	EtOH	226(4.39),231(4.42), 258(4.35)	4-0947-69
	EtOH-HCl	225(4.38),231(4.42), 258(4.35)	4-0947-69
	EtOH-KOH	224(4.45),259(4.35), 282s(3.87),313(3.70), 321s(3.70)	4-0947-69
3H-Pyridazino[3,4-b]indol-3-one, 2,9-dihydro-9-methyl-	EtOH	242(4.14),264s(4.24), 268(4.31),301(3.63), 442(3.25)	95-0058-69
2H-Pyrrolo[3,4-b]quinoxaline, 2-methyl-, 4-oxide	MeOH	261(4.78),348(3.97), 506(3.63)	88-1581-69
	MeOH-HCl	263(4.54),362(4.08), 584(3.38)	88-1581-69
$C_{11}H_9N_3OS$			
1,3,4-Oxadiazole-2-thione, 3-(2-cyanoethyl)-5-phenyl-	EtOH	254(4.15),259(4.15), 298(4.38)	2-0583-69
$C_{11}H_9N_3O_2$			
2-Pyridinol, 3-[(p-hydroxyphenyl)azo]-	EtOH	395(4.3)	119-0037-69
$C_{11}H_9N_3O_2S$			
as-Triazine-3,5(2H,4H)-dione, 6-[4-(2-furyl)-1,3-butadienyl]-3-thio-	EtOH-DMF	277(4.00),397(4.31)	103-0427-69
$C_{11}H_9N_3O_3$			
2-Propen-1-one, 1-(5-nitro-2-pyrrolyl)-3-pyrrol-2-yl-	EtOH	260(3.79),330(4.03), 421(4.49)	103-0383-69
	dioxan	268(3.59),328(3.97), 424(4.39)	103-0383-69
	concd. H_2SO_4	255(3.76),347(3.83), 523(4.78)	103-0383-69
as-Triazine-3,5(2H,4H)-dione, 6-[4-(2-furyl)-1,3-butadienyl]-	EtOH-DMF	355(4.44)	103-0427-69
$C_{11}H_9N_3O_4$			
2-Isoxazolin-5-one, 3-p-acetamidophenyl-4-isonitroso-	EtOH	258(4.30)	55-0859-69
1,2,4-Oxadiazole, 3-(m-nitrophenyl)-5-(2-oxopropyl)-	EtOH	262(4.00)	18-3008-69
1,2,4-Oxadiazole, 3-(p-nitrophenyl)-5-(2-oxopropyl)-	EtOH	274(4.15)	18-3008-69
Resorcinol, 4-[(2,6-dihydroxy-3-pyridyl)azo]-	EtOH	477(4.5)	119-0037-69
$C_{11}H_9N_3O_4S$			
as-Triazine-6-propionic acid, β-furfurylidene-2,3,4,5-tetrahydro-5-oxo-3-thioxo-	EtOH	220(3.81),260(3.65), 299(3.90),370(4.07)	103-0427-69
$C_{11}H_9N_3O_5$			
as-Triazine-6-propionic acid, β-furfurylidene-2,3,4,5-tetrahydro-3,5-dioxo-	EtOH	221(3.85),317(4.29)	103-0427-69

Compound	Solvent	$\lambda_{max}(\log \epsilon)$	Ref.
$C_{11}H_9N_3O_5S$			
Sulfanilic acid, N-(5-nitro-2-pyridyl)-	anion	372(4.25)	22-4569-69
	dianion	466(4.38)	22-4569-69
$C_{11}H_9N_5O$			
3H-v-Triazolo[4,5-d]pyrimidin-7-ol, 3-benzyl-	pH 5.0	255(4.03)	39-0152-69C
	pH 11	275(4.06)	39-0152-69C
s-Triazolo[5,1-c]-as-triazin-4-ol, 3-methyl-7-phenyl-	EtOH	242(4.21),310(3.85)	22-2492-69
$C_{11}H_9N_5O_2S$			
s-Triazolo[3,4-b][1,3,4]benzothiadiazepine, 3-ethyl-8-nitro-	EtOH	237(3.24),278s(3.08)	2-0959-69
$C_{11}H_9N_5O_4$			
Benzaldehyde, m-nitro-, (2,6-dihydroxy-4-pyrimidinyl)hydrazone	EtOH	322(4.29)	56-0519-69
Benzaldehyde, p-nitro-, (2,6-dihydroxy-4-pyrimidinyl)hydrazone	EtOH	359(4.23)	56-0519-69
$C_{11}H_9N_5S$			
[1,2,3]Thiadiazolo[5,4-d]pyrimidine, 7-(benzylamino)-	EtOH	248(4.02),268(4.73), 325(3.86)	39-0152-69C
1H-v-Triazolo[4,5-d]pyridazine, 4-(benzylthio)-	EtOH	208(4.56),285(3.87)	4-0093-69
3H-v-Triazolo[4,5-d]pyrimidine-7-thiol, 3-benzyl-	pH 13	231(4.12),335(4.24)	39-0152-69C
	aq. MeOH	232(3.97),332(4.27)	39-0152-69C
$C_{11}H_{10}$			
Bicyclo[6.3.0]undeca-1(8),2,4,6,9-pentaene	C_6H_{12}	253s(3.69),281s(3.26)	24-1789-69
Bicyclo[6.3.0]undeca-1,4,6,8,10-pentaene	C_6H_{12}	267(4.16),325s(3.26), 425(2.96)	24-1789-69
Spiro[2,4-cyclopentadiene-1,1'-[2,4]-norcaradiene]	hexane	228(3.61),234(3.61), 267(3.84)	24-1789-69
$C_{11}H_{10}BF_4N_3$			
2-Methyl-s-triazolo[4,3-a]isoquinolinium tetrafluoroborate	n.s.g.	235f(4.6),270f(3.8), 310f(3.2)	24-3159-69
2-Methyl-s-triazolo[4,3-a]quinolinium tetrafluoroborate	n.s.g.	210f(4.1),240(4.4), 290f(4.0),300(3.4), 320(3.3)	24-3159-69
$C_{11}H_{10}BrN_3O_2$			
Cinnamic acid, α-azido-o-bromo-, ethyl ester	C_6H_{12}	231(4.05),308(4.28)	49-1599-69
Cinnamic acid, α-azido-p-bromo-, ethyl ester	C_6H_{12}	229(4.09),318(4.31)	49-1599-69
$C_{11}H_{10}Br_2N_2O$			
2-Indolinone, 5,7-dibromo-3-[(dimethylamino)methylene]-	EtOH	235(4.25),282(4.23), 327(4.04),355(4.17)	2-0662-69
	5% NaOH	275(--),315(--)	2-0662-69
$C_{11}H_{10}ClN$			
Pyrrole, 2-(m-chlorophenyl)-4-methyl-	EtOH	305(4.27)	94-0582-69
$C_{11}H_{10}ClNO_2$			
2(3H)-Furanone, 3-[(m-chloroanilino)-methylene]-dihydro-	EtOH	226(6.26),295(4.8), 320(5.00)	2-0952-69

Compound	Solvent	λ_{max}(log ε)	Ref.
2(3H)-Furanone, 3-[(p-chloroanilino)-methylene]dihydro-	EtOH	219(4.00),304(4.42), 320(4.49)	2-0952-69
Isocarbostyril, 3-chloro-6-methoxy-2-methyl- (in 10% MeOH)	pH 4	250(4.68),260s(4.26), 284(3.80),295(3.79), 320(3.58)	24-3666-69
Isocarbostyril, 3-chloro-7-methoxy-2-methyl- (in 10% MeOH)	pH -6.1	239(4.66),270s(3.87), 277(3.94),287(3.90), 354(3.68),365(3.68)	24-3666-69
	pH 4	226(4.52),247s(3.93), 271s(3.93),281(4.14), 291(4.11),330s(3.58), 344(3.69),359(3.57)	24-3666-69
Isoquinoline, 3-chloro-1,6-dimethoxy- (in 10% MeOH)	pH -2.7	244(4.74),250(4.84), 289(3.72),299(3.71), 325(3.64)	24-3666-69
	pH 4	240(4.66),249s(4.52), 270(3.59),280(3.60), 302s(3.38),320(3.24)	24-3666-69
Isoquinoline, 3-chloro-1,7-dimethoxy- (in 10% MeOH)	pH -3.8	240(4.70),270s(3.85), 275(3.88),284(3.92), 354(3.67),364(3.69)	24-3666-69
	pH 4	223(4.64),240s(4.24), 261s(3.89),270(4.07), 280(4.10),334(3.58), 345(3.56)	24-3666-69
$C_{11}H_{10}ClNO_4S_2$			
2-Ethylthiazolo[2,3-b]benzothiazolium perchlorate	EtOH	225(4.65),258(4.64), 289(4.40),301(4.28)	95-0469-69
$C_{11}H_{10}ClN_3O_2$			
Cinnamic acid, α-azido-o-chloro-, ethyl ester	C_6H_{12}	227(4.03),307(4.05)	49-1599-69
Cinnamic acid, α-azido-p-chloro-, ethyl ester	C_6H_{12}	229(3.98),317(4.31)	49-1599-69
$C_{11}H_{10}ClN_5O_2S$			
4H-1,2,4-Triazole-3-thiol, 4-[(2-chloro-5-nitrobenzylidene)amino]-5-ethyl-	EtOH	250(4.43),350(3.62)	2-0959-69
$C_{11}H_{10}Cl_2$			
4a,8a-Methanonaphthalene, 9,9-dichloro-1,4-dihydro-	n.s.g.	278(3.20)	88-4475-69
$C_{11}H_{10}Cl_2N_2O$			
2-Indolinone, 4,7-dichloro-3-[(dimethyl-amino)methylene]-	EtOH	270(4.25),278(4.28), 347(4.12)	2-0662-69
	5% NaOH	267(--),338(--)	2-0662-69
$C_{11}H_{10}Cl_2N_4$			
Pyrimidine, 2,4-diamino-5-(3,4-dichloro-phenyl)-6-methyl-	pH 2	201(4.63),208(4.63), 271(3.92)	44-0821-69
	pH 12	285(4.01)	44-0821-69
$C_{11}H_{10}Cl_6O_4$			
Acetic acid, trichloro-, 2-norbornylidene ester	EtOH	230s(2.57)	78-1679-69
$C_{11}H_{10}FN_3O_2$			
Cinnamic acid, α-azido-p-fluoro-, ethyl ester	C_6H_{12}	225(4.02),311(4.32)	49-1599-69

Compound	Solvent	$\lambda_{max}(\log \epsilon)$	Ref.
$C_{11}H_{10}FeO_4$			
Iron, tricarbonyl(4,4-dimethyl-2,5-cyclohexadien-1-one)-	isooctane	219(4.29)	101-0342-69A
$C_{11}H_{10}N_2$			
1H-Benz[g]indazole, 4,5-dihydro-	EtOH	223s(3.95),250(3.80), 260(3.47),268(3.97), 285(3.68),294(3.57)	4-0771-69
$C_{11}H_{10}N_2O$			
2-Propanone, 1-(1-phthalazinyl)-	EtOH	217(4.47),252(3.61), 260(3.59),283(4.01), 291(4.32),360s(4.04), 382(4.24),400(4.11)	95-0959-69
$C_{11}H_{10}N_2OS$			
4-Pyrazolecarboxaldehyde, 5-mercapto-3-methyl-1-phenyl-	C_6H_{12}	270(4.1)	104-1634-69
	MeOH	232(4.1),265(4.1), 330(3.6)	104-1634-69
	EtOH	232(4.3),269(4.3), 330(3.7)	104-1634-69
	pH 12	228(3.9),265(3.8), 330(3.5)	104-1634-69
Pyrimidine, 2-[(p-methoxyphenyl)thio]-	pH 5	248(4.23)	39-2720-69C
Pyrimidine, 2-(p-tolylsulfinyl)-	EtOH	233(4.20)	39-2720-69C
$C_{11}H_{10}N_2O_2$			
Hydantoin, 5-(p-methylbenzylidene)-	neutral	228(4.03),329(4.42)	122-0455-69
	anion	233(4.00),331(4.37)	122-0455-69
2-Indolinone, 3-(2-oxopropylidene)-, 3-oxime	n.s.g.	255(4.20),322(4.04)	77-0125-69
5,8-Isoquinolinedione, 3-methyl-7-(methylamino)-	MeCN	235(4.24),276(4.21), 322(3.70)	4-0697-69
4H-Pyrazol-4-one, 5-ethyl-3-phenyl-, 1-oxide	EtOH	240(3.88),325(3.80), 418(2.93)	44-0187-69
Styrene, p-cyano-β-ethyl-β-nitro-	n.s.g.	292(4.18)	67-0127-69
$C_{11}H_{10}N_2O_2S$			
1H-1,2-Diazepine, 1-(phenylsulfonyl)-	MeOH	214(4.20),342(2.45)	22-2175-69
Pyrimidine, 2-[(p-methoxyphenyl)sulfinyl]-	EtOH	245(4.21)	39-2720-69C
Pyrimidine, 2-(p-tolylsulfonyl)-	EtOH	238(3.09)	39-2720-69C
Uracil, 5-(benzylthio)-	pH 1	258s(--)	44-3806-69
	pH 13	294(3.85)	44-3806-69
$C_{11}H_{10}N_2O_3$			
Hydantoin, 5-(p-methoxybenzylidene)-	neutral	228(3.99),337(4.13)	122-0455-69
	anion	234(3.99),336(4.01)	122-0455-69
2-Indolinone, 1-methyl-3-(2-nitrovinyl)-cis	EtOH	255(4.16),453(4.08)	7-0712-69
trans	EtOH	255(4.11),453(3.72)	7-0712-69
6(5H)-Isoquinolone, 5,8-dimethyl-5-nitro-	EtOH	275(3.83),284(3.83), 312(3.97)	12-2489-69
4H-Pyrazol-4-one, 3-ethyl-5-phenyl-, 1,2-dioxide	n.s.g.	254(4.31),301(3.74), 420(2.80)	44-0187-69
$C_{11}H_{10}N_2O_3S$			
Pyrimidine, 2-[(p-methoxyphenyl)sulfonyl]-	EtOH	256(3.04)	39-2720-69C

Compound	Solvent	λmax(log ε)	Ref.
$C_{11}H_{10}N_2O_4$			
3-Pyrrolin-2-one, 3-methoxy-4-nitro-5-phenyl-	EtOH	257(3.85)	44-3279-69
$C_{11}H_{10}N_2O_5$			
1H-Pyrrolo[2,3-c]pyridine-3-acetic acid, 2-carboxy-5-methoxy-	EtOH	282(4.09),291(4.14)	35-2338-69
1H-Pyrrolo[2,3-c]pyridine-3-propionic acid, 2-carboxy-5,6-dihydro-5-oxo-	EtOH	292(3.62),302(3.64)	35-2338-69
$C_{11}H_{10}N_2S$			
Pyrimidine, 2-(p-tolylthio)-	pH 5	250(4.11)	39-2720-69C
$C_{11}H_{10}N_4$			
9H-Indeno[2,1-d]pyrimidine, 2,4-diamino-	pH 1	269(4.34),286s(4.13)	4-0613-69
	pH 10	274(4.34),291(4.31)	4-0613-69
	EtOH	274(4.34),291(4.31)	4-0613-69
Picolinaldehyde, 2-pyridylhydrazone,	benzene	335(4.34)	39-0819-69A
(E)-	EtOH	343(4.49)	39-0819-69A
(Z)-	benzene	364(4.38)	39-0819-69A
	EtOH	355(4.34)	39-0819-69A
s-Triazolo[3,4-a]phthalazine, 3,6-dimethyl-	MeOH	240(4.69),247(4.60)	44-3221-69
$C_{11}H_{10}N_4O$			
Ketone, 2,4-diamino-5-pyrimidinyl phenyl	pH 2	228(4.45),255(4.24), 285(4.06)	44-0821-69
	pH 12	219(4.22),245(4.09), 265(4.05),312(4.24)	44-0821-69
as-Triazino[2,3-a]indole-10-carbonitrile, 1,2,6,7,8,9-hexahydro-2-oxo-	$CHCl_3$	263(4.34),270(4.35), 309(3.40),399(2.93)	48-0388-69
5H-as-Triazino[5,6-b]indole, 3-methoxy-5-methyl-	EtOH	260(4.4),320(4.0)	65-2339-69
3H-as-Triazino[5,6-b]indol-3-one, 2,5-dihydro-2,5-dimethyl-	EtOH	260(4.4),340(3.9)	65-2339-69
3H-as-Triazino[5,6-b]indol-3-one, 2,5-dihydro-4,5-dimethyl-	EtOH	290(3.8),400(3.1)	65-2339-69
$C_{11}H_{10}N_4O_2$			
Benzaldehyde, (2,6-dihydroxy-4-pyrimidinyl)hydrazone	EtOH	323(4.28)	56-0519-69
4H-Pyrido[1,2-a]pyrimidine-3-carbonitrile, 2-[(2-hydroxyethyl)amino]-4-oxo-	EtOH	255(4.58),345(4.00), 359s(3.84)	95-0203-69
$C_{11}H_{10}N_4S$			
5H-as-Triazino[5,6-b]indole, 5-methyl-2-(methylthio)-	EtOH	270(4.5),350(4.0)	65-2111-69
3H-as-Triazino[5,6-b]indole-3-thione, 2,5-dihydro-2,5-dimethyl-	EtOH	260(4.2),300(4.6), 370(3.4)	65-0640-69
s-Triazolo[3,4-b][1,3,4]thiadiazole, 3-ethyl-6-phenyl-	EtOH	215(4.08),270(4.32)	2-0959-69
$C_{11}H_{10}O$			
7H-Cyclobut[a]inden-7-one, 1,2,2a,7a-tetrahydro-	EtOH	213(4.02),246(4.03), 292(3.39)	39-2656-69C
1,3,5,7-Cyclooctatetraene-1-acrolein	EtOH	265(4.34)	39-0978-69C
2-Cyclopenten-1-one, 2-phenyl-	n.s.g.	223(4.42),259(4.07)	88-0909-69

Compound	Solvent	λ_{max}(log ϵ)	Ref.
$C_{11}H_{10}OS$			
6H-Cyclohepta[c]thiophen-6-one, 1,3-dimethyl-	EtOH	228(4.14),276(4.63), 331(4.01)	104-0559-69
	42% $HClO_4$	241(4.03),253(4.05), 304(4.76),351(4.09), 365(4.24),517(3.28)	104-0559-69
$C_{11}H_{10}OS_2$			
2H-Thiopyrano[3,2-b][1]benzothiophen-4-ol, 3,4-dihydro-	EtOH	220(4.38),233(4.19), 254(4.12),290(3.60), 303(3.77),314(3.86)	22-0601-69
$C_{11}H_{10}O_2$			
Coumarin, 4,6-dimethyl-	EtOH	218(4.38),276(4.08), 324(3.78)	18-2319-69
Coumarin, 4,7-dimethyl-	EtOH	222(4.28),282(4.04), 319(4.00)	18-2319-69
2-Cyclopropene-1-carboxylic acid, 2-methyl-3-phenyl-	EtOH	214s(4.23),260(4.25)	44-0505-69
4,6,8-Decatriyn-3-one, 1-methoxy- (relative ε given instead of log)	ether	222(34.5),229(39), 260(1.0),275(1.0), 292(1.3),311(1.6), 333(1.1)	39-1096-69C
2(3H)-Furanone, 3-benzylidenedihydro-	EtOH	218(4.16),223(4.10), 282(4.40)	44-3792-69
1-Indeneacetic acid, endo-	n.s.g.	253(3.99)	44-4182-69
2-Indeneacetic acid	EtOH	259(4.14)	88-2331-69
1,2,4-Methenonaphthalene-5,8-diol	EtOH	295(3.53)	35-7763-69
2,4-Pentadiynoic acid, 5-(1-cyclohexen-1-yl)-	EtOH	218(3.78),250(3.11), 268(3.19),271(3.20), 284(3.21),304(3.08)	70-2117-69
Propiolic acid, ethyl ester	EtOH	257(4.14)	22-0210-69
$C_{11}H_{10}O_2S$			
2(3H)-Furanone, dihydro-3-(thiobenzoyl)-	EtOH	298(3.96)	78-5703-69
Hydrocoumarin, 3-(thioacetyl)-	EtOH	281(3.96),352(3.48)	78-5703-69
Thiochroman-4-one, 3-(hydroxymethylene)-2-methyl-	EtOH	250(4.23),268s(3.82), 323(3.89)	44-0056-69
$C_{11}H_{10}O_3$			
Acrylic acid, 3-benzoyl-, methyl ester, cis	EtOH	251(4.07),280s(3.58), 342(2.18)	18-1353-69
Benzo[b]furan, 5,6-(methylenedioxy)-2,3-dimethyl-	EtOH	251(3.96),307(3.95), 311(3.95)	18-1971-69
Chromone, 5-methoxy-2-methyl-	MeOH	224(4.36),253(4.05), 313(3.76)	64-0225-69
Coumarin, 7-methoxy-4-methyl-	EtOH	222(4.20),252(3.30), 324(4.15)	18-2319-69
Isocoumarin, 7-methoxy-3-methyl-	MeOH	230(4.53),269(4.03), 348(3.60)	42-0935-69
Umbelliferone, 6-ethyl-	n.s.g.	336(4.18)	42-0275-69
$C_{11}H_{10}O_4$			
Chromone, 3-hydroxy-5-methoxy-2-methyl-	MeOH	241(4.42),334(3.89)	64-0225-69
Chromone, 5-hydroxy-3-methoxy-2-methyl-	MeOH	241(4.38),334(3.77)	64-0225-69
Chromone, 5-hydroxy-7-methoxy-2-methyl-	MeOH	229(4.33),248(4.40), 255(4.41),289(3.98)	64-0225-69
2-Naphthoic acid, 1,2,3,4-tetrahydro-	EtOH	224(4.24),254(3.95), 327(3.48)	2-0561-69

Compound	Solvent	λ_{max}(log ϵ)	Ref.
$C_{11}H_{10}O_5$			
Chromone, 5,7-dihydroxy-3-methoxy-2-methyl-	MeOH	250(4.36),296(3.88), 322(3.78)	64-0225-69
$C_{11}H_{10}O_7$			
Phthalonic acid	EtOH	224(4.37),262(3.83), 298(3.61)	39-1921-69C
$C_{11}H_{10}S$			
Thiophene, 2-benzyl-	MeOH	235(3.89)	111-0228-69
1H-Thiopyran, 1-phenyl-	EtOH	202(4.01),246(3.64)	35-1206-69
$C_{11}H_{10}S_2$			
2H-Thiopyrano[3,2-b][1]benzothiophene, 3,4-dihydro-	EtOH	215(4.41),237(4.16), 253(4.03),290(3.64), 299(3.74),310(3.79)	22-0601-69
$C_{11}H_{11}BF_2O_2$			
Boron, difluoro(2-methyl-1-phenyl-1,3-butanedionato)-	$CHCl_3$	260(3.65),334(4.36)	39-0526-69A
$C_{11}H_{11}BO_4$			
Boron, (2,4-pentanedionato)[pyrocatecholato(2-)]-	CH_2Cl_2	213(3.32),262(4.20)	39-0173-69A
$C_{11}H_{11}BrClN_3O_4$			
7H-Pyrrolo[2,3-d]pyrimidine, 5-bromo-4-chloro-7-β-D-ribofuranosyl-	pH 1	231(4.82),270(3.96), 296(3.92)	4-0215-69
	pH 11	231(4.80),270(3.97), 296(3.92)	4-0215-69
$C_{11}H_{11}BrN_2O$			
4-Carbamoyl-2-methylisoquinolinium bromide	MeOH	235(4.59),325(4.67)	22-2045-69
$C_{11}H_{11}BrN_2O_2S_2$			
2H-1,3,5-Thiadiazine-3(4H)-carboxylic acid, 5-(p-bromophenyl)dihydro-6-thioxo-, methyl ester	MeOH	253(4.08),295(3.98)	73-2952-69
triethylamine salt	MeOH	250(4.15),293(3.98)	73-2952-69
$C_{11}H_{11}BrO_2$			
Benzofuran, 5-bromo-2-(1-hydroxy-1-methylethyl)-	EtOH	251(4.08),278s(3.45), 285(3.59),293(3.58)	12-1923-69
$C_{11}H_{11}Cl$			
1,3-Butadiene, 1-chloro-4-p-tolyl-	EtOH	226(4.1),285(4.4)	104-0678-69
$C_{11}H_{11}ClIN_3O_4$			
7H-Pyrrolo[2,3-d]pyrimidine, 4-chloro-5-iodo-7-β-D-ribofuranosyl-	pH 1	234(4.39),269(3.46), 305(3.46)	4-0215-69
	pH 11	234(4.39),269(3.46), 305(3.46)	4-0215-69
$C_{11}H_{11}ClN_2OS$			
Acetamide, 2-chloro-N-(5,6-dimethyl-2-benzothiazolyl)-	EtOH	210(4.38),284(4.28), 305(4.26)	23-4483-69
Propionamide, 2-chloro-N-(3-methyl-2-benzothiazolinylidene)-	EtOH	215(4.32),314(4.36)	23-4483-69

Compound	Solvent	$\lambda_{max}(\log \epsilon)$	Ref.
$C_{11}H_{11}ClN_2O_2S$			
Acetamide, 2-chloro-N-(6-ethoxy-2-benzothiazolyl)-	EtOH	216(4.44),308(4.34)	23-4483-69
$C_{11}H_{11}ClN_2O_5$			
Isoquinoline-4-carboxamide, 2-methoperchlorate	MeOH	215(4.42),240(3.47), 305(3.69)	22-2045-69
$C_{11}H_{11}ClN_4$			
Pyrimidine, 2,4-diamino-5-(p-chlorobenzyl)-	pH 2	223(4.47),270(3.74)	44-0821-69
	pH 12	220(4.33),235s(4.15), 287(3.86)	44-0821-69
Pyrimidine, 2,4-diamino-5-(p-chlorophenyl)-6-methyl-	pH 2	208(4.54),222s(4.41), 270(3.90)	44-0821-69
	pH 12	250s(3.93),285(3.99)	44-0821-69
$C_{11}H_{11}ClO_4$			
5H-Benzofurancarboxaldehyde, 7-chloro-2,3-dihydro-4-hydroxy-2-methoxy-6-methyl-	MeOH	266(4.21),293(4.12), 335s(3.60)	78-1323-69
Isocoumarin, 5-chloro-3,4-dihydro-8-hydroxy-6-methoxy-3-methyl-	MeOH	219(4.36),265(4.07), 312(3.76)	39-2187-69C
$C_{11}H_{11}ClO_4S$			
4,6-Dimethyl-1-benzothiopyrylium perchlorate	HOAc-$HClO_4$	263(4.55),345(3.84), 390(3.57)	2-0017-69
$C_{11}H_{11}ClO_5S$			
6-Hydroxy-1,3-dimethylcyclohepta[c]-thiolium perchlorate	42% $HClO_4$	241(4.03),253(4.05), 304(4.76),351(4.09), 365(4.24),517(3.28)	104-0947-69
6-Methoxy-4-methyl-1-benzothiopyrylium perchlorate	HOAc-$HClO_4$	276(4.20),353(3.87), 407(3.50)	2-0017-69
7-Methoxy-4-methyl-1-benzothiopyrylium perchlorate	HOAc-$HClO_4$	271(4.53),339(3.70), 413(3.82)	2-0017-69
8-Methoxy-4-methyl-1-benzothiopyrylium perchlorate	HOAc-$HClO_4$	281(4.36),332(3.32), 344(3.35),450(3.28)	2-0017-69
$C_{11}H_{11}Cl_2N_3O$			
1H-Benzotriazole, 5,6-dichloro-1-(tetrahydro-2H-pyran-2-yl)-	EtOH	265(3.76),271(3.75), 297(3.62)	4-0005-69
1,2,3-Benzotriazin-4(3H)-one, 3-butyl-6,8-dichloro-	EtOH	226(4.38),299(4.03)	4-0809-69
$C_{11}H_{11}Cl_2N_3O_4$			
7H-Pyrrolo[2,3-d]pyrimidine, 4,5-dichloro-7-β-D-ribofuranosyl-	pH 1	230(4.44),270(3.53), 296(3.53)	4-0215-69
	pH 11	230(4.42),270(3.53), 296(3.53)	4-0215-69
$C_{11}H_{11}F_5O_2$			
Benzaldehyde, pentafluoro-, diethyl acetal	EtOH	262(2.85)	65-1607-69
$C_{11}H_{11}Fe$			
Ferrocenylmethylium cation	58% H_2SO_4	256(3.95)	35-0509-69
$C_{11}H_{11}I_2NO_4$			
Glycine, N-(4-hydroxy-3,5-diiodohydrocinnamoyl)-	EtOH	287(3.51),294s(3.47)	69-1844-69

Compound	Solvent	λmax(log ε)	Ref.
$C_{11}H_{11}KN_2O_4$			
Glutaconic acid, 2,4-dicyano-, diethyl ester, potassium salt	EtOH	222(4.22),355(4.76)	88-3279-69
$C_{11}H_{11}N$			
Indole, N-allyl-	EtOH	220(4.53),280(3.78), 292(3.65)	39-0595-69C
$C_{11}H_{11}NO$			
Atroponitrile, β-ethoxy-	EtOH	212(4.16),268(4.18)	22-0198-69
4-Azatetracyclo[5.3.2.0^{2,10}.0^{3,6}]dodeca-8,11-dien-5-one	EtOH	247s(2.81)	35-3970-69
Carbostyril, 3-ethyl-	EtOH	229(4.45),269(3.87), 324(3.87)	1-0159-69
Crotononitrile, 3-methoxy-2-phenyl-, cis	EtOH	209(4.09),265(4.13)	22-0198-69
trans	EtOH	210(4.18),268(4.21)	22-0198-69
Indole, 3-acetyl-2-methyl-	MeOH	301(4.07)	22-3002-69
anion	MeOH	330(4.29)	22-3002-69
Isoquinoline, 3-methoxy-1-methyl-	EtOH	340(3.62)	39-1729-69C
3(2H)-Isoquinolone, 1,2-dimethyl-	EtOH	416(3.73)	39-1729-69C
	ether	444(3.64)	39-1729-69C
Quinaldine, 4-methoxy-	pH 1	234(4.54),308(3.91)	78-0255-69
	MeOH	231(4.47),281(3.82)	78-0255-69
4(1H)-Quinolone, 1,2-dimethyl-	pH 1	237(4.53),309(4.00)	78-0255-69
	MeOH	215(4.30),239(4.40), 327(4.14),341(4.15)	78-0255-69
$C_{11}H_{11}NOS$			
4-Thiazolin-2-one, 3-benzyl-4-methyl-	EtOH	242(3.69)	10-0214-69F
$C_{11}H_{11}NOS_2$			
2-Butanone, 4-(2-benzothiazolylthio)-	EtOH	231(4.21),242(4.18), 261(3.86),326(4.41)	95-0469-69
$C_{11}H_{11}NO_2$			
2H-Azirine-2-carboxylic acid, 3-phenyl-, ethyl ester	EtOH	204(4.24),243(4.02), 288s(2.91)	88-2049-69
Bicyclo[3.2.2]nona-3,6-diene-8α-carbonitrile, 1-methoxy-2-oxo-	MeOH	225(3.87),345(2.03)	88-0443-69
Bicyclo[3.2.2]nona-3,6-diene-9α-carbonitrile, 3-methoxy-2-oxo-	MeOH	231(3.51),276(3.58) 325s(2.13)	88-0443-69
9β-isomer	MeOH	231(3.43),276(3.48)	88-0443-69
Carbostyril, 6-methoxy-4-methyl-	EtOH	235(4.51),278(3.62), 347(3.67)	1-0159-69
1H-Cycloprop[a]naphthalene, 1a,2,3,7b-tetrahydro-4-nitro-	C_6H_{12}	290(4.00)	35-3558-69
	EtOH	306(3.99)	35-3558-69
2(3H)-Furanone, 3-(anilinomethylene)-dihydro-	EtOH	292s(4.10),312(4.33)	2-0952-69
Isoxazole, 5-ethoxy-3-phenyl-	EtOH	204(4.14),238(3.81)	88-2049-69
Spiro[cyclopropane-1,1'-indan], 5'-nitro-	C_6H_{12}	293(3.97)	35-3558-69
	EtOH	313(3.96)	35-3558-69
Spiro[cyclopropane-1,1'-indan], 6'-nitro-	C_6H_{12}	275(3.85)	35-3558-69
	EtOH	282(3.80)	35-3558-69
$C_{11}H_{11}NO_2S_2$			
2-Benzothiazolinethione, 3-(2-acetoxyethyl)-	MeOH	229(4.13),240(4.10), 325(4.38)	4-0163-69

Compound	Solvent	λ_{max}(log ε)	Ref.
$C_{11}H_{11}NO_3$			
1H-1-Benzazepine-7-carboxylic acid, 2,3,4,5-tetrahydro-5-oxo-	MeOH	242(4.12),288(4.16), 345(2.46)	44-1070-69
Phthalic anhydride, 3-(isopropylamino)-	EtOH	206(4.19),254(4.13), 408(3.78)	16-0241-69
$C_{11}H_{11}NO_3S$			
2-Benzothiazolinone, 3-(2-acetoxy-ethyl)-	MeOH	215(4.65),244(3.74), 282(3.46),289(3.45)	4-0163-69
Carbonic acid, β-isothiocyanophenethyl methyl ester, D-	EtOH	259(4.24)	39-1143-69B
2-Cyclopenten-1-one, 3,5-dimethyl-2-(5-nitro-2-thienyl)-	EtOH	230(3.89),360(4.24)	94-1757-69
2,4-Pentadienal, 2,4-dimethyl-5-(5-nitro-2-thienyl)-	EtOH	222(3.89),280(4.00), 310(3.93),394(4.39)	94-1757-69
$C_{11}H_{11}NO_4$			
Cinnamic acid, 2-methoxy-5-nitroso-,	EtOH	284(4.34),328(4.08)	12-0935-69
methyl ester	dioxan	728(1.71)	12-0935-69
2-Cyclopenten-1-one, 3,5-dimethyl-2-(5-nitro-2-furyl)-	EtOH	227(3.15),271(3.73), 353(4.14)	94-0306-69
	EtOH	228(4.15),272(3.73), 353(4.14)	94-1757-69
3-Indolepropionic acid, α,β-dihydroxy-, sodium salt	H_2O	219(4.48),273(3.74), 280(3.76),288(3.69)	97-0269-69
2,4-Pentadienal, 2,4-dimethyl-5-(5-nitro-2-furyl)-	EtOH	221(4.08),280(4.00), 380(4.32)	94-1757-69
$C_{11}H_{11}NO_4S$			
2,4-Pentanedione, 3-(p-nitrophenylthio)-, Al chelate	CH_2Cl_2	292(4.54),340(4.56)	88-2255-69
Co(III) chelate	CH_2Cl_2	255(4.53),341(4.69)	88-2255-69
Cr chelate	CH_2Cl_2	340(4.72)	88-2255-69
$C_{11}H_{11}NO_4S_3$			
4-Methyl-2H-1,3-thiazino[2,3-b]benzothiazolium bisulfate	EtOH	247(3.77),333(4.08)	95-0469-69
$C_{11}H_{11}NO_5$			
Cyclopentanone, 2,3-epoxy-3,5-dimethyl-2-(5-nitro-2-furyl)-	EtOH	316(4.01)	94-0306-69
$C_{11}H_{11}NO_6$			
2,4,5-Pyridinetricarboxylic acid, trimethyl ester	MeOH	278(3.41)	32-0431-69
$C_{11}H_{11}NS$			
2H-1,4-Benzothiazine, 2-ethylidene-3-methyl-	MeOH	360(3.56)	124-1278-69
perchlorate	MeOH	440(3.40)	124-1278-69
$C_{11}H_{11}N_3$			
Pyridine, 4,4'-(aminomethylene)di-	EtOH	265(3.46),278(3.35)	88-3101-69
as-Triazine, 5,6-dimethyl-3-phenyl-	MeOH	255(4.40),365(2.62)	88-3147-69
as-Triazine, 5-methyl-3-p-tolyl-	MeOH	264(4.31),360(2.68)	88-3147-69
$C_{11}H_{11}N_3O$			
Quinoxaline, 2-acetyl-3-methyl-, oxime, anti	EtOH	213(4.42),241(4.42), 329(4.00)	39-0600-69C
syn	EtOH	214(4.04),237(4.32), 319(3.95)	39-0600-69C

Compound	Solvent	$\lambda_{max}(\log \epsilon)$	Ref.
$C_{11}H_{11}N_3OS$			
3(2H)-Pyridazinone, 5-amino-4-(benzylthio)-	EtOH	207(4.56),216(4.44), 227(4.35),311(3.63)	4-0093-69
$C_{11}H_{11}N_3O_2$			
Cinnamic acid, α-azido-, ethyl ester	C_6H_{12}	225(4.02),310(4.31)	49-1599-69
Pyrido[2,3-b]pyrazin-2(1H)-one, 3-acetonyl-1-methyl-	pH -0.6	211(4.40),230s(4.00), 234(4.01),246(3.79), 289(3.84),309s(3.51), 383(4.43),400s(4.34), 435s(3.03)	24-4032-69
	pH 5	212(4.40),251s(3.77), 293(3.81),375s(4.26), 393(4.38),413(4.23)	24-4032-69
Pyrido[2,3-b]pyrazin-3(4H)-one, 2-acetonyl-4-methyl-	pH -1.4	212(4.29),238s(3.84), 246s(3.78),269s(3.78), 277(3.80),319(3.94), 330s(3.89),394(4.33), 412(4.29)	24-4032-69
$C_{11}H_{11}N_3O_3$			
Pyrido[2,3-d]pyrimidine-6-carboxylic acid, 7,8-dihydro-2-methyl-7-oxo-, ethyl ester	EtOH	265(3.53),274(3.49), 324(3.97)	88-1825-69
	EtOH-HCl	265(--),275(--), 320(--)	88-1825-69
	EtOH-NaOH	256(--),272(--), 353(--)	88-1825-69
$C_{11}H_{11}N_3O_6$			
Aziridine, 1-(5-nitro-2-furoyl)-, O-acetoacetyloxime	EtOH	245(4.04),320(4.15)	18-3008-69
$C_{11}H_{11}N_4S$			
1-Methyl-3-(methylthio)-s-triazolo-[3,4-a]phthalazinium (methosulfate)	MeOH	210(4.57),243(4.46), 255(3.39),266s(4.42), 280s(4.06),290s(3.93)	44-3221-69
$C_{11}H_{11}N_5O$			
Acetic acid, [(3-phenyl-3H-1,2,4-triazol-1-yl)methylene]hydrazide	MeOH	290(4.40)	48-0646-69
$C_{11}H_{11}N_5O_2$			
Benzoic acid, 2-(5-hydroxy-6-methyl-as-triazin-3-yl)hydrazide	EtOH	230(4.11)	22-2492-69
$C_{11}H_{11}N_5O_7$			
2(1H)-Quinoxalinone, 3-alaninyl-3,4-dihydro-5,7-dinitro-	EtOH	264(4.13),374(3.99)	44-0395-69
$C_{11}H_{11}N_5S$			
7H-s-Triazolo[3,4-b][1,3,4]thiadiazine, 3-ethyl-7-(2-pyridyl)-	EtOH	286(4.16)	2-0959-69
$C_{11}H_{12}$			
Indene, 4,6-dimethyl-	C_6H_{12}	205(4.36),213(4.39), 218(4.41),226(4.37), 232(4.19),262(4.29), 291(2.85),298(2.58), 303(2.55)	22-1981-69

Compound	Solvent	$\lambda_{max}(\log \epsilon)$	Ref.
Indene, 4,7-dimethyl-	C_6H_{12}	212(4.44),218(4.44), 224(4.38),232(4.24), 258(4.04),285(3.86), 290(3.00),296(2.76), 301(2.99)	22-1981-69
Indene, 5,6-dimethyl-	C_6H_{12}	212(4.66),258(4.01), 261(4.01),288(3.99), 390(2.79)	22-1981-69
$C_{11}H_{12}BrN_3O_4S$			
4H-Pyrrolo[2,3-d]pyrimidine-4-thione, 5-bromo-5,7-dihydro-7-β-D-ribofuranosyl-	pH 1	277(3.90),327(4.27)	4-0215-69
	pH 11	229(4.27),317(4.24)	4-0215-69
$C_{11}H_{12}BrN_3O_5$			
4H-Pyrrolo[2,3-d]pyrimidin-4-one, 5-bromo-5,7-dihydro-7-β-D-ribofuranosyl-	pH 1	265(3.95)	4-0215-69
	pH 11	271(4.00)	4-0215-69
$C_{11}H_{12}Br_3ClO$			
Ether, 3-bromo-2,2-bis(bromomethyl)-propyl o-chlorophenyl	hexane	218(3.99),273(3.31), 280(3.26)	56-1641-69
$C_{11}H_{12}ClNO_6$			
7,8-Dimethoxyisoquinolinium perchlorate	EtOH	236(4.33),252(4.35), 290s(3.49),360(3.28)	78-1881-69
$C_{11}H_{12}ClN_3O$			
1,2,3-Benzotriazin-4(3H)-one, 3-butyl-6-chloro-	EtOH	292(3.95)	4-0809-69
1,2,3-Benzotriazin-4(3H)-one, 3-butyl-7-chloro-	EtOH	238(4.42),291(3.74)	4-0809-69
$C_{11}H_{12}ClN_3O_2$			
Pyrimidine, 1-(p-chlorophenyl)-1,2,3,4-tetrahydro-3-methyl-5-nitro-	dioxan	250(4.32),385(4.42)	78-1617-69
$C_{11}H_{12}FN_3O_6$			
Carbanilic acid, 2-fluoro-3,5-dinitro-, tert-butyl ester	EtOH	330(3.35)	44-0395-69
$C_{11}H_{12}F_3N_5O$			
Acetamide, 2,2,2-trifluoro-N-(6-methyl-8-propyl-s-triazolo[4,3-a]pyrazin-3-yl-	EtOAc	230s(4.01),252(4.06), 287(4.06),338(3.47)	39-1593-69C
$C_{11}H_{12}F_4O_2$			
Benzaldehyde, 2,3,5,6-tetrafluoro-, diethyl acetal	EtOH	268(3.25)	65-1607-69
$C_{11}H_{12}N$			
2,4-Dimethylquinolizinium (picrate)	EtOH	226(4.5),295(3.9), 319(4.2),328(4.1), 336(4.3)	24-1309-69
$C_{11}H_{12}N_2$			
Benzeneazo-1'-cyclopentene	THF	312(4.37)	24-3647-69
Isoquinoline, 1-(dimethylamino)-	EtOH	235(3.90),302(3.78), 332(3.70)	95-0510-69
1-Naphthaldehyde, hydrazone	EtOH	230(4.5),313(4.1)	28-0248-69B

Compound	Solvent	λ max (log ε)	Ref.
2-Naphthaldehyde, hydrazone	EtOH	212(4.2),240(4.4), 264(4.4),292(4.3), 300(4.3)	28-0248-69B
Pyrazole, 1,5-dimethyl-4-phenyl-	MeOH	241(4.20)	44-3639-69
$C_{11}H_{12}N_2O$			
Imidazole, 2,4-dimethyl-5-phenyl-, 3-oxide	MeOH	264(4.15)	48-0746-69
Imidazole, 4,5-dimethyl-2-phenyl-, 3-oxide	MeOH	292(4.18)	48-0746-69
2-Pyrazolin-5-one, 3-benzyl-4-methyl-	EtOH	250(3.68)	22-4159-69
2-Pyrazolin-5-one, 4,4-dimethyl-3-phenyl-	EtOH	210(4.06),285(4.19)	22-4159-69
3-Pyrazolin-5-one, 1,3-dimethyl-2-phenyl-	MeOH	264(4.0)	94-1467-69
3-Pyrazolin-5-one, 2,3-dimethyl-1-phenyl-	MeOH	241(3.9),265(3.9)	94-1467-69
2H-2-Pyrindine-4-carbonitrile, 1-ethyl-3,5,6,7-tetrahydro-3-hydroxy-	EtOH	220(4.27),223(4.29), 240(3.80),340(4.07)	44-3670-69
$C_{11}H_{12}N_2OS$			
Acetamide, N-(1,2-benzisothiazol-3-yl)-N-ethyl-	MeOH	224(4.41),308(3.80), 312(3.80)	24-1961-69
$C_{11}H_{12}N_2OS_2$			
2H-Thiopyran-2-thione, 6-acetylamino-5-cyano-4-ethyl-3-methyl-	C_6H_{12}	245(--),288(--), 352(--),460(--)	78-1441-69
	EtOH	248(4.43),302s(3.53), 347(4.17),458(3.82)	78-1441-69
$C_{11}H_{12}N_2O_2$			
1H-1,4-Benzodiazepine-2,5-dione, 3,5-dihydro-1,4-dimethyl-	EtOH	288(3.29)	44-1359-69
2-Benzoxazolinone, 3-(2,3-dimethyl-1-aziridinyl)-, cis	EtOH	236(3.96),277(3.73)	39-0772-69C
trans	EtOH	237(3.94),278(3.71)	39-0772-69C
Imidazole, 2-ethyl-1-hydroxy-4-phenyl-, 3-oxide	MeOH	229(4.21)	48-0746-69
Imidazole, 1-hydroxy-2,5-dimethyl-4-phenyl-, 3-oxide	MeOH	228(4.16)	48-0746-69
Imidazole, 1-hydroxy-4,5-dimethyl-2-phenyl-, 3-oxide	MeOH	283(4.16)	48-0746-69
Phthalimide, 3-(dimethylamino)-N-methyl-	EtOH	415(3.64)	104-1231-69
Phthalimide, 4-(dimethylamino)-N-methyl-	EtOH	397(3.78)	104-1231-69
Pyrazole, 1-hydroxy-4,5-dimethyl-3-phenyl-, 2-oxide	EtOH	230(4.1),288(4.0)	44-0194-69
2-Pyrazolin-5-one, 3-(p-methoxyphenyl)-4-methyl-	EtOH	224(4.07),258(4.01)	22-4159-69
3(2H)-Pyridazinone, 4,5-dihydro-5-(hydroxymethyl)-6-phenyl-	MeOH	215(4.01),288(4.23)	5-0217-69D
$C_{11}H_{12}N_2O_2S$			
Quinoxaline, 2-(isopropylsulfonyl)-	pH 6.0	242(4.64),321(3.87)	39-0333-69B
$C_{11}H_{12}N_2O_3$			
1H-Pyrrolo[2,3-c]pyridine-2-carboxylic acid, 5-methoxy-, ethyl ester	EtOH	278(4.18),287(4.23), 344(3.56)	35-2338-69
Pyruvanilide, 2'-(methylcarbamoyl)-	MeOH	241(3.98),302(3.77)	44-3035-69

Compound	Solvent	λ_{max}(log ε)	Ref.
$C_{11}H_{12}N_2O_4$			
Glutaconic acid, 2,4-dicyano-, diethyl ester (potassium salt)	EtOH	222(4.22),355(4.76)	88-3279-69
1,1-Hydrazinedicarboxylic acid, 2-benzylidene-, dimethyl ester	MeOH	252(4.15),276s(3.98)	78-0619-69
$C_{11}H_{12}N_2O_4S$			
Δ^2-1,2,4-Thiadiazoline-4-carboxylic acid, 3-phenyl-, ethyl ester, 1,1-dioxide	n.s.g.	253(4.11)	39-0652-69C
$C_{11}H_{12}N_2O_5$			
1H-Pyrrolo[2,3-c]pyridine-3-propionic acid, 2-carboxy-4,5,6,7-tetrahydro-5-oxo-	EtOH	276(4.10)	35-2338-69
$C_{11}H_{12}N_2O_6$			
4-Pyridinepyruvic acid, 2-methoxy-5-nitro-, ethyl ester	EtOH	287(4.00)	35-2338-69
$C_{11}H_{12}N_2S$			
Imidazole, 1-methyl-2-(methylthio)-4-phenyl-	2% EtOH	265(4.17)	1-2888-69
conjugate acid	2% EtOH	260(4.16)	1-2888-69
Imidazole, 1-methyl-2-(methylthio)-5-phenyl-	2% EtOH	266(4.24)	1-2888-69
conjugate acid	2% EtOH	270(4.22)	1-2888-69
4-Imidazoline-2-thione, 1,3-dimethyl-4-phenyl-	2% EtOH	261(4.28)	1-2888-69
conjugate acid	2% EtOH	252(4.13)	1-2888-69
Quinoxaline, 2-(isopropylthio)-	pH 6.0	242(4.19),266(4.10), 358(3.94)	39-0333-69B
Thione, bis(1-methylpyrrol-2-yl)-	CHCl	320(3.36),393(4.36)	12-0239-69
	10N HCl	257(4.50),380(4.06), 472(4.60)	12-0239-69
$C_{11}H_{12}N_4$			
Pyrimidine, 2,4-diamino-5-benzyl-	pH 2	206(4.50),218s(4.41), 273(3.72)	44-0821-69
	pH 12	230s(4.11),288(3.86)	44-0821-69
Pyrimidine, 2,4-diamino-6-methyl-5-phenyl-	pH 2	208(4.51),225s(4.30), 273(3.86)	44-0821-69
	pH 12	240s(3.94),285(3.95)	44-0821-69
Pyrimidine, 4,5-diamino-2-methyl-6-phenyl-	pH 4.5	218s(4.18),305(3.99)	39-1883-69C
	pH 8.6	229(4.16),302(3.86)	39-1883-69C
Pyrimidine, 4,5-diamino-6-methyl-2-phenyl-	pH 3.5	245(4.37),298(4.10)	39-1408-69C
	pH 7.9	219s(4.35),291(4.10)	39-1408-69C
s-Triazolo[3,4-a]phthalazine, 5,6-dihydro-3,6-dimethyl-	MeOH	205(4.38),254(4.22), 286(3.22)	44-3221-69
$C_{11}H_{12}N_4OS$			
Indole-2,3-dione, 1-methyl-, 3-(3-methyl-3-thioisosemicarbazone)	EtOH	250(4.4),300(3.7), 370(3.9)	65-0070-69
Indole-2,3-dione, 1-methyl-, 3-(2-methyl-3-thiosemicarbazone)	EtOH	250(4.1),270s(3.9), 350(4.3)	65-0070-69
$C_{11}H_{12}N_4O_2$			
Formamidine, N,N-dimethyl-N'-(1,2,3,4-tetrahydro-1,4-dioxophthalazin-5-yl)-, hydrochloride	H_2O	228(4.53),283(4.17), 318(4.17)	18-2090-69

Compound	Solvent	λ_{max}(log ε)	Ref.
Uracil, 6-amino-5-(benzylamino)-	pH -7.0	267(4.16)	24-4032-69
	pH 2.0	260(4.23)	24-4032-69
	pH 6.4	276(4.11)	24-4032-69
$C_{11}H_{12}N_4O_4S_2$			
Barbituric acid, 5,5'-isopropylidene-bis[2-thio-	EtOH	280(4.43)	103-0830-69
$C_{11}H_{12}N_4O_5$			
4,7(3H,8H)-Pteridinedione, 6-acetonyl-2-hydroxy-8-(2-hydroxyethyl)-	pH 1	279(4.03),335(4.11)	24-4032-69
	pH 5.5	260(3.73),288(3.98), 354(4.15)	24-4032-69
$C_{11}H_{12}N_4O_6$			
Barbituric acid, 5,5'-isopropylidenedi-	EtOH	255(4.30)	103-0827-69
$C_{11}H_{12}N_6O$			
Benzamide, N-[(4H-1,2,4-triazol-4-yl-amidino)methyl]-	MeOH	233(4.00)	48-0477-69
$C_{11}H_{12}N_6S_2$			
5H-Imidazo[4,5-d]thiazole-5-thione, 2-(4-amino-2-methyl-5-pyrimidinyl)-4,6-dihydro-4,6-dimethyl-	EtOH	225(4.11),283(4.15), 400(4.31)	94-0910-69
$C_{11}H_{12}O$			
Benzofuran, 2-isopropyl-	EtOH	245(4.0),269(3.3), 275(3.4),281(3.5)	12-1923-69
3-Butenal, 2-methyl-4-phenyl-	EtOH	240(4.01),278s(2.77)	22-3719-69
1(2H)-Naphthalenone, 3,4-dihydro-5-methyl-	EtOH	210(4.26),252(4.08), 296(3.29)	22-0985-69
4-Pentenal, 2-phenyl-	C_6H_{12}	294(2.23)	22-0903-69
4-Pentenophenone	EtOH	243(4.00)	35-0456-69
2-Propen-1-ol, 3-(1,3,5,7-cyclooctatetraen-1-yl)-	EtOH	235(4.30),293(3.76)	39-0978-69C
$C_{11}H_{12}OS$			
Thiochroman-4-one, 2-ethyl-	EtOH	241(4.39),262(3.88), 346(3.47)	44-0056-69
$C_{11}H_{12}OS_2$			
1,3-Dithiole, 2-methoxy-4-p-tolyl-	EtOH	232(4.02),303(3.91)	94-1931-69
$C_{11}H_{12}O_2$			
2(3H)-Benzofuranone, 5-ethyl-3-methyl-	MeOH	228s(3.67),269s(3.06), 276(3.17),284(3.13)	12-0977-69
Cinnamic acid, ethyl ester, trans	EtOH	216(4.20),222(4.13), 276(4.35)	35-3517-69
1,4-Methanonaphthalene-5,8-dione, 1,4,4a,6,7,8a-hexahydro-, endo-cis	EtOH	228(4.11),385(1.79)	39-0105-69C
2,7(1H,4aH)-Naphthalenedione, 8,8a-dihydro-4a-methyl-	MeOH	217(4.22),228s(4.19)	35-2299-69
$C_{11}H_{12}O_2S$			
2,4-Pentanedione, 3-(phenylthio)-, Al chelate	CH_2Cl_2	252(4.53),291(4.43)	88-2255-69
Co(III) chelate	CH_2Cl_2	253(4.77)	88-2255-69
Cr chelate	CH_2Cl_2	255(4.53),341(4.69)	88-2255-69

Compound	Solvent	λ_{max}(log ε)	Ref.
$C_{11}H_{12}O_2S_2$			
1,3-Dithiole, 2-ethoxy-4-(p-hydroxyphenyl)-, perchlorate	EtOH	234(3.95),289(4.07)	94-1931-69
1,3-Dithiole, 2-methoxy-4-(p-methoxyphenyl)-	EtOH	237(3.98),290(4.07)	94-1931-69
$C_{11}H_{12}O_3$			
Acetophenone, 3',4'-(trimethylenedioxy)-	hexane	220(4.30),227(4.23), 264(4.07),269(4.06), 292(3.56)	59-0313-69
Benzaldehyde, 3-allyl-4-hydroxy-5-methoxy-	EtOH	233(4.23),296(4.07)	12-1011-69
	KOH	256(4.03),359(4.50)	12-1011-69
Benzaldehyde, 2-allyloxy-3-methoxy-	EtOH	217(4.28),261(3.92), 322(3.42)	12-1011-69
Benzene, 1-methoxy-4,5-(methylenedioxy)-2-propenyl-, trans	MeOH	260(4.13),322(3.95)	12-1803-69
Bicyclo[4.2.0]octa-3,7-diene-2,5-dione, 1-methoxy-7,8-dimethyl-	EtOH	223(4.11),283s(2.48), 387(2.19)	44-0520-69
2(3H)-Furanone, 3-(α-hydroxybenzyl)-	EtOH	280(2.73)	44-3792-69
2,4-Pentadiynoic acid, 5-(1-hydroxycyclohexyl)-	EtOH	222(3.08),236(3.20), 246(3.52),260(3.60), 276(3.38)	70-2117-69
Salicylaldehyde, 5-allyl-3-methoxy-	EtOH	224(4.30),268(4.00), 352(3.50)	12-1011-69
	KOH	243(4.31),280(3.78), 400(3.79)	12-1011-69
Spiro[2H-1,5-benzodioxepin-3(4H),3'-oxetane]	MeOH	273(3.13)	25-1271-69
$C_{11}H_{12}O_4$			
2H-1-Benzo[b]pyran-2,5(6H)-dione, 7,8-dihydro-4-hydroxy-7,7-dimethyl-	EtOH	232(4.27),270(4.13)	18-3233-69
Chromanone, 5,7-dihydroxy-2,2-dimethyl-	EtOH	290(4.23),325s(--)	39-1540-69C
	EtOH-base	240s(--),324(--)	39-1540-69C
Chromanone, 5,7-dihydroxy-6,8-dimethyl-	EtOH	215(4.11),297(4.08), 343(3.30)	95-0851-69
p-Dioxan-2-one, 6-(p-methoxyphenyl)-	EtOH	225(4.03),272(3.17), 278(3.09)	22-2048-69
2,5-Norbornadiene-2,3-dicarboxylic acid, dimethyl ester	MeCN	238(3.6)	5-0052-69E
$C_{11}H_{12}O_4S$			
Benzoic acid, o-[(carboxymethyl)thio]-, o-ethyl ester	EtOH	222(4.31),257(3.88), 314(3.47)	44-1566-69
$C_{11}H_{12}O_5$			
Bicyclo[3.1.0]hex-2-ene-6-carboxaldehyde, 1,2-diacetoxy-	EtOH	224(3.60)	89-0276-69
2-Norbornene-2,3-diol, 5,6-epoxy-, diacetate	EtOH	254(3.58)	89-0276-69
Phthalic acid, 3-hydroxy-6-methyl-, dimethyl ester	EtOH	208(4.40),238s(3.71), 317(3.64)	33-0584-69
Phthalic acid, 4-hydroxy-5-methyl-, dimethyl ester	EtOH	218(4.37),262(3.98)	33-0584-69
Succinic acid, (p-hydroxybenzyl)-	EtOH	225(3.93),279(3.24)	2-0561-69
$C_{11}H_{12}O_6$			
1H-2-Benzopyran-4-carboxylic acid, 5,6,7,8-tetrahydro-6,7-dihydroxy-1-oxo-, methyl ester	pH 12	251(3.95)	35-6470-69
	EtOH	253(3.95),285(3.67)	35-6470-69

Compound	Solvent	λ_{max}(log ϵ)	Ref.
Phthalic acid, 3-hydroxy-6-methoxy-, dimethyl ester	EtOH	218(3.90),240s(3.57), 334(3.54)	33-0584-69
$C_{11}H_{12}O_7$			
2-Furoic acid, 2-acetonyl-3-acetyl-2,5-dihydro-4-hydroxy-5-oxo-, methyl ester	100% EtOH	274(4.01),320s(3.48)	1-0597-69
	85% EtOH	320(4.09)	1-0597-69
$C_{11}H_{12}S_2$			
Benzo[b]thiophene, 2-ethyl-3-(methylthio)-	EtOH	228(4.37),266(3.83), 291(3.59),300(3.67)	22-0601-69
$C_{11}H_{13}BrN_2O$			
2-Pyrazoline, 1-(p-bromophenyl)-3-ethoxy-	EtOH	280(4.07)	22-1683-69
$C_{11}H_{13}BrN_2S_2$			
2H-1,3,5-Thiadiazine-2-thione, 3-(p-bromophenyl)-5-ethyltetrahydro-	MeOH	245(4.13),290(3.93)	73-2952-69
$C_{11}H_{13}BrN_4O_4$			
7H-Pyrrolo[2,3-d]pyrimidine, 4-amino-5-bromo-7-β-D-ribofuranosyl-	pH 1	234(4.32),282(3.94)	4-0215-69
	pH 11	228(4.10),278(3.97)	4-0215-69
$C_{11}H_{13}BrN_4O_5$			
7H-Pyrrolo[2,3-d]pyrimidine, 5-bromo-4-(hydroxyamino)-7-β-D-ribofuranosyl-	pH 1	239(4.20),280(4.02)	4-0215-69
	pH 11	276(3.90)	4-0215-69
$C_{11}H_{13}BrO_2$			
2,6-Adamantanedione, 4-(bromomethyl)-, (1R)-	MeOH	290(1.58)	78-5601-69
Benzofuran, 5-bromo-2,3-dihydro-2-(1-hydroxy-1-methylethyl)-	EtOH	292(3.41),300(3.33)	12-1923-69
Tropolone, 7-bromo-4-tert-butyl-	MeOH	256(4.30),331(3.48), 367(3.48),379(3.50)	18-3277-69
$C_{11}H_{13}BrS$			
Dimethyl(3-phenyl-2-propynyl)sulfonium bromide	EtOH	246(4.23),255s(4.12)	94-0966-69
$C_{11}H_{13}ClN_4O_4$			
7H-Pyrrolo[2,3-d]pyrimidine, 4-amino-5-chloro-7-β-D-ribofuranosyl-	pH 1	232(4.34),279(3.92)	4-0215-69
	pH 11	224s(4.02),277(3.96)	4-0215-69
$C_{11}H_{13}ClO$			
Benzofuran, 2-(1-chloro-1-methylethyl)-2,3-dihydro-	EtOH	280(3.54),287(3.50)	12-1923-69
Phenol, m-(4-chloro-3-methyl-3-butenyl)-	EtOH	274(3.24),282(3.19)	35-2299-69
$C_{11}H_{13}Cl_2NOS$			
2-Propen-1-one, 3-[bis(2-chloroethyl)-amino]-1-(2-thienyl)-	EtOH	214(3.40),263(3.74), 289(3.65),347(4.32)	24-3139-69
$C_{11}H_{13}Cl_6N_3S$			
s-Triazine, 2-(hexylthio)-4,6-bis(trichloromethyl)-	MeOH	218(3.41),275(4.15)	18-2931-69
$C_{11}H_{13}IN_4O_4$			
7H-Pyrrolo[2,3-d]pyrimidine, 4-amino-5-iodo-7-β-D-ribofuranosyl-	pH 1	237(4.22),285(3.90)	4-0215-69
	pH 11	280(3.97)	4-0215-69

Compound	Solvent	λ_{max}(log ε)	Ref.
$C_{11}H_{13}IN_4O_5$			
7H-Pyrrolo[2,3-d]pyrimidine, 5-iodo-4-(hydroxyamino)-7-β-D-ribofuranosyl-	pH 1	244(4.10),285(3.92)	4-0215-69
	pH 11	283(3.92)	4-0215-69
$C_{11}H_{13}N$			
Indole, 1,3,7-trimethyl-	hexane	226(4.59),281(3.84)	23-0785-69
3H-Indole, 2,3,3-trimethyl-	EtOH	222s(4.04),256(3.76)	39-2703-69C
$C_{11}H_{13}NO$			
Acetamide, N-methyl-N-styryl-, cis	C_6H_{12}	219(3.98),271(4.11)	24-2987-69
	MeOH	260(4.10)	78-5745-69
Acetamide, N-methyl-N-styryl-, trans	MeOH	220(--),284(4.41)	78-5745-69
Anthranil, 3,5,6,7-tetramethyl-	EtOH	268(3.11),279(3.31), 294(3.44),321(3.75)	78-1047-69
7-Azabicyclo[4.2.2]deca-2,4,7,9-tetraene, 8-methoxy-2-methyl-	MeCN	273(3.64)	35-4714-69
7-Azabicyclo[4.2.2]deca-2,4,7,9-tetraene, 8-methoxy-9-methyl-	MeCN	269(3.43)	35-4714-69
3-Buten-2-one, 4-(methylamino)-3-phenyl-, cis	C_6H_{12}	216(3.74),264(3.86), 315(4.05)	24-2987-69
Cinnamamide, N,N-dimethyl-	EtOH	280(4.30)	39-0016-69C
4a(2H)-Naphthalenecarbonitrile, 3,4,5,6,7,8-hexahydro-2-oxo-	EtOH	228(4.24)	28-0864-69A
Propionitrile, 3-(4-methoxy-3-methylphenyl)-	EtOH	276(3.39)	88-0073-69
$C_{11}H_{13}NOS_2$			
Hippuric acid, 1,1-dithio-, ethyl ester	MeOH	309(4.04)	5-0237-69I
	$CHCl_3$	445(1.43)	5-0237-69I
$C_{11}H_{13}NO_2$			
7-Azabicyclo[4.2.2]deca-2,4,7,9-tetraene, 2,8-dimethoxy-	hexane	274(3.78)	35-4714-69
7-Azabicyclo[4.2.2]deca-2,4,7,9-tetraene, 8,9-dimethoxy-	hexane	268(3.48)	35-4714-69
4-Azabicyclo[5.2.0]nona-2,5,8-triene-4-carboxylic acid, ethyl ester	n.s.g.	228(4.09)	88-1619-69
10-Azatricyclo[4.3.1.0^{1,6}]deca-2,4-diene-10-carboxylic acid, methyl ester	hexane	246(3.47),255(3.46)	44-2866-69
	EtOH	242(3.38),258(3.45)	44-2866-69
1H-Azonine-1-carboxylic acid, ethyl ester	hexane	220(3.95),270s(3.34)	88-5239-69
	MeOH	265(3.53)	77-1204-69
1H-1-Benzazepine-7-carboxylic acid, 2,3,4,5-tetrahydro-	MeOH	230(3.30),287(4.14)	44-1070-69
1H-1-Benzazepine-8-carboxylic acid, 2,3,4,5-tetrahydro-	MeOH	218(4.41),294(3.33)	44-2235-69
5H-Benzocycloheptene, 6,7,8,9-tetrahydro-2-nitro-	C_6H_{12}	273(3.96)	35-3558-69
	EtOH	282(3.92)	35-3558-69
Crotonohydroxamic acid, O-benzyl-, trans	MeOH	208(4.4),220(4.13)	12-0161-69
	MeOH-NaOH	219(3.84),254(3.91)	12-0161-69
Crotonohydroxamic acid, N-p-tolyl-	EtOH	216(4.29),280(3.92)	34-0278-69
Cyclopent[b]azepine-1(6H)-carboxylic acid, 7,8-dihydro-, methyl ester	hexane	217(4.32),253s(3.07), 324(3.03)	44-2879-69
	EtOH	217(4.28),249s(2.96), 312(3.10)	44-2879-69
Formanilide, 2'-oxobutyl-	EtOH	250(4.07),279(3.53), 288(3.48)	1-0159-69
Levulinanilide	MeCN	200(4.45),242(4.20)	28-0279-69A
Propionanilide, 4'-(vinyloxy)-	n.s.g.	250(4.24)	30-0108-69A
stannic chloride complex	n.s.g.	250(4.12),585(2.62)	30-0108-69A

Compound	Solvent	λ_{max}(log ε)	Ref.
2-Pyrrolidinone, 5-hydroxy-1-methyl-5-phenyl-	MeCN	188(4.77),257f(2.30)	28-0279-69A
Styrene, β-ethyl-p-methyl-β-nitro-	n.s.g.	325(3.93)	67-0127-69
$C_{11}H_{13}NO_3$			
5H-2-Pyrindine-4-carboxylic acid, 6,7-dihydro-6 -hydroxy-7 -methyl-, methyl ester	EtOH	271(3.39),278s(3.31)	12-1283-69
Styrene, β-ethyl-p-methoxy-β-nitro-	n.s.g.	346(4.19)	67-0127-69
Thalifoline	MeOH	224(4.41),261(3.87), 302(3.77)	78-0469-69
	MeOH-KOH	238(4.40),270(3.73), 330(3.65)	78-0469-69
$C_{11}H_{13}NO_4$			
Acetic acid, (3-nitro-o-tolyl)-, ethyl ester	EtOH	208(4.1),255(3.52)	39-1935-69C
Benzonitrile, 2,3,4,6-tetramethoxy-	EtOH	255(3.92),281(3.37)	39-0887-69
Pyrrole-2,5-dicarboxaldehyde, 3-methyl-4-(2-carbomethoxyethyl)-	MeOH	244(4.00),315(4.27)	63-1291-69
$C_{11}H_{13}NO_6$			
3-Carboxy-1-D-ribofuranosylpyridinium hydroxide, inner salt	MeOH	266(3.62)	39-0918-69C
Pyrrole-2,3,4-tricarboxylic acid, 1-methyl-, trimethyl ester	MeOH	262(3.90)	24-2346-69
Pyrrole-2,3,5-tricarboxylic acid, 1-methyl-, trimethyl ester	MeOH	263(3.93)	24-2346-69
$C_{11}H_{13}NS$			
Indole, 1,3-dimethyl-2-(methylthio)-	EtOH	227(4.53),287(4.04), 295(4.03)	94-0550-69
2-Indolinethione, 1,3,3-trimethyl-	EtOH	230(4.15),295s(4.11), 312(4.28)	94-0550-69
$C_{11}H_{13}NS_2$			
1,3-Dithiole-2-amine, N,N-dimethyl-4-phenyl-	EtOH	317(--)	94-1924-69
$C_{11}H_{13}N_3$			
1H-1,5-Benzodiazepine, 7-amino-2,4-dimethyl-, hydrochloride	EtOH	230(4.11),273(4.01), 308(3.92),360(3.48), 560(2.75)	104-0158-69
3H-1,5-Benzodiazepine, 7-amino-2,4-dimethyl-	EtOH	232(4.29),274(3.97), 300(3.89)	104-0158-69
$C_{11}H_{13}N_3O$			
1H-Benzotriazole, 1-(tetrahydro-2H-pyran-2-yl)-	EtOH	255(3.83),260(3.80), 280(3.64)	4-0005-69
1,3,4-Oxadiazol-3-ine, 5-(benzylimino)-2,2-dimethyl-	EtOH	260(3.61),329(2.49)	44-3230-69
1,3,4-Oxadiazol-3-ine, 2,2-dimethyl-5-(p-tolylimino)-	EtOH	226(4.00),332(3.89)	44-3230-69
2-Pyrazolin-5-one, 3-(dimethylamino)-1-phenyl-	MeOH	254(4.19),280(3.80)	33-2641-69
1,2,4-Triazolin-3-one, 1-ethyl-5-methyl-2-phenyl-	$CHCl_3$	263(3.94)	89-0456-69
$C_{11}H_{13}N_3O_2$			
3-Indolineacetic acid, 1-methyl-2-oxo-, hydrazide	EtOH	277(4.35),318(4.23)	95-0058-69

Compound	Solvent	λ_{max}(log ϵ)	Ref.
1,3,4-Oxadiazol-3-ine, 5-[(p-methoxy-phenyl)imino]-2,2-dimethyl-	EtOH	231(4.16),357(4.06)	44-3230-69
$C_{11}H_{13}N_3O_3$			
3-Penten-2-one, 4-(2-amino-5-nitroanil-ino)-, (hydrate)	EtOH	314(4.32),360(4.23)	104-0158-69
Pyrido[2,3-d]pyrimidine-6-carboxylic acid, 5,6,7,8-tetrahydro-2-methyl-7-oxo-, ethyl ester	EtOH	240(3.91),271(4.15)	88-1825-69
	EtOH-HCl	277(--)	88-1825-69
	EtOH-NaOH	298(--)	88-1825-69
$C_{11}H_{13}N_3O_9$			
Uracil-5-carboxylic acid, 1-(5-amino-5-deoxy-β-D-allofuranuronosyl)-	0.05N HCl	220(4.02),275(4.07)	35-7490-69
	0.05N NaOH	270(3.85)	35-7490-69
$C_{11}H_{13}N_3S_2$			
Malononitrile, [[1-methyl-2-(methyl-thio)-2-pyrrolin-3-yl](methylthio)-methylene]-	C_6H_{12}	456(4.26)	5-0073-69E
$C_{11}H_{13}N_4Na_2O_9P$			
Inosine, 2-methoxy-, 5'-phosphate, disodium salt	pH 1	254(3.96)	94-2581-69
	pH 6	248(4.00),264s(--)	94-2581-69
	pH 13	261(4.07)	94-2581-69
$C_{11}H_{13}N_5O$			
Pyrimidine, 2,4,5-triamino-6-(benzyl-oxy)-	pH 1	224s(3.95),277(4.10)	44-2160-69
	pH 11	243(3.94),283(3.90)	44-2160-69
	EtOH	245(3.86),285(3.95)	44-2160-69
4(3H)-Pyrimidinone, 2,6-diamino-5-(benzylamino)-	pH -1	262(4.19)	24-4032-69
	pH 3.2	267(4.11)	24-4032-69
	pH 8	246(3.83),280(4.05)	24-4032-69
	pH 13	216(4.14),250s(3.83), 253(3.84),273(3.97)	24-4032-69
4H-1,2,4-Triazole, 4-(benzylnitros-amino)-3,5-dimethyl-	MeOH	241(3.60)	48-0009-69
1H-1,2,3-Triazole-4-carboxamide, 1-benzyl-5-(methylamino)-	aq. EtOH	244(3.99),265(3.92)	39-0152-69C
$C_{11}H_{13}N_5O_4S$			
Carbonic acid, thio-, O-ethyl ester, S-ester with ethyl 2-amino-9H-purine-9-carboxylate	pH 1	323(3.63)	44-0973-69
	MeOH	336(3.87)	44-0973-69
$C_{11}H_{13}N_5S$			
Pyrimidine, 5-amino-4-(benzylthio)-6-hydrazino-	pH 1	277(3.74),328(3.95)	44-3161-69
$C_{11}H_{14}$			
Bicyclo[6.1.0]nona-2,4,6-triene, 9,9-dimethyl-	hexane	248(3.57)	35-1239-69
Indene, 3a,4-dihydro-1,1-dimethyl-	hexane	266(3.67)	35-1239-69
Indene, 3a,7a-dihydro-1,1-dimethyl-, trans	hexane	260(3.52)	35-1239-69
Indene, 7,7a-dihydro-1,1-dimethyl-	hexane	302(3.99)	35-1239-69
$C_{11}H_{14}BrNO_4$			
4-Oxazoleacrylic acid, α-bromo-5-eth-oxy-2-methyl-, ethyl ester	EtOH	229(3.78),312(4.31)	94-2424-69

Compound	Solvent	$\lambda_{max}(\log \epsilon)$	Ref.
$C_{11}H_{14}BrO_4P$			
Phosphonic acid, (p-bromobenzoyl)-, diethyl ester	C_6H_{12}	271(4.27),385(1.98)	18-0821-69
$C_{11}H_{14}Br_2O$			
2(3H)-Naphthalenone, 4a-(dibromomethyl)-4,4a,5,6,7,8-hexahydro-	MeOH	239(4.11)	44-0496-69
$C_{11}H_{14}ClO_4P$			
Phosphonic acid, (p-chlorobenzoyl)-, diethyl ester	C_6H_{12}	268(4.10),385(1.97)	18-0821-69
$C_{11}H_{14}Cl_2N_4S$			
1H-Azepine, 3-(4,6-dichloro-s-triazin-2-yl)-4,5,6,7-tetrahydro-1-methyl-2-(methylthio)-	C_6H_{12}	298(3.64),403(4.49)	5-0073-69E
$C_{11}H_{14}Cl_2O$			
2(3H)-Naphthalenone, 4a-(dichloromethyl)-4,4a,5,6,7,8-hexahydro-	MeOH	237(4.11)	44-0496-69
$C_{11}H_{14}Cl_2O_2$			
2,5-Cyclohexadien-1-one, 4-(dichloromethyl)-2-methoxy-4-propyl-	EtOH	239(3.97),289(3.42)	12-1011-69
$C_{11}H_{14}F_4N_2O_2$			
Benzaldehyde, 2,3,5,6-tetrafluoro-4-hydrazino-, diethyl acetal	EtOH	248(4.06)	65-1607-69
$C_{11}H_{14}N$			
N-Mesitylacetonitrilium ion	95% H_2SO_4	258(4.15)	77-0645-69
$C_{11}H_{14}NO_4P$			
Phosphonic acid, 2-benzoxazolyl-, diethyl ester	EtOH	242(3.86),270s(3.51), 283(3.64),290(3.59)	65-0392-69
$C_{11}H_{14}N_2$			
Isoindole, 2-(aminoethyl)-1-methyl-	EtOH	227(4.35),269(3.35), 280(3.30),341(3.34), 404(2.65),428(2.78), 443(2.88)	44-1720-69
Isoindole, 2-(3-aminopropyl)-	EtOH	224(4.50),266(3.20), 270(3.18),277(3.24), 289(3.14),327(3.62), 340s(3.51)	44-1720-69
3-Pyrazoline, 1,2-dimethyl-3-phenyl-	EtOH	229(4.01),281(3.61)	22-3316-69
Pyridazine, 1,4,5,6-tetrahydro-1-methyl-3-phenyl-	EtOH	224(3.87),292(4.00)	39-1703-69C
Pyridine, 3-(5,5-dimethyl-1-pyrrolin-2-yl)-	EtOH	235(4.05),265s(3.58)	36-0860-69
	EtOH-HCl	250(4.08),275s(3.83)	36-0860-69
$C_{11}H_{14}N_2O$			
3-Pyrazolidinone, 5,5-dimethyl-1-phenyl-	N HCl	260(2.78)	65-1835-69
	H_2O	240(3.90)	65-1835-69
2-Pyrazoline, 3-ethoxy-1-phenyl-	EtOH	272(4.15)	22-1683-69
Pyridine, 3-(5,5-dimethyl-1-pyrrolin-2-yl)-, N-oxide	EtOH	236(3.69),300(4.05)	36-0860-69
	EtOH-HCl	248(3.89),306(3.96)	36-0860-69
$C_{11}H_{14}N_2O_2$			
Piperidine, N-(p-nitrophenyl)-	DMSO	389(3.28)	39-0544-69B

Compound	Solvent	λmax(log ε)	Ref.
Unnamed pyrrole acyl derivative	MeOH	290(4.00)	88-5135-69
$C_{11}H_{14}N_2O_2S$			
1H-Cyclopenta[d]pyrimidine-1-propionic acid, 4,5,6,7-tetrahydro-2-methyl-4-thioxo-	EtOH	239(3.87),338(4.35)	4-0037-69
$C_{11}H_{14}N_2O_3S$			
p-Toluenesulfonic acid, (1-methylacetonyl)hydrazide, sodium salt	pH 11	225(4.2),245s(3.9), 287(4.09)	44-1751-69
$C_{11}H_{14}N_2O_5$			
1H-1,2-Diazepine-3,6-dicarboxylic acid, 1-acetyl-4,5-dihydro-, dimethyl ester	EtOH	270(4.06),303(3.69)	27-0037-69
$C_{11}H_{14}N_2O_7$			
Uracil, 1-β-D-arabinofuranosyl-, 3'-acetate	MeOH	263(4.10)	94-1188-69
$C_{11}H_{14}N_2S$			
1,2-Benzisothiazole, 3-(diethylamino)-	MeOH	227(4.21),246(3.76), 336(3.67)	24-1961-69
	90% MeOH-HCl	236(4.20),253(3.78), 335(3.81),347s(3.73)	24-1961-69
Pyrimidine, 1,4,5,6-tetrahydro-1-methyl-2-[2-(2-thienyl)vinyl]-, (tartrate)	H_2O	312(4.27)	87-1066-69
$C_{11}H_{14}N_2S_2$			
2H-1,3,5-Thiadiazine-2-thione, 5-ethyldihydro-3-phenyl-	MeOH	248(3.93),295(3.96)	73-2952-69
$C_{11}H_{14}N_4O$			
3-Pyrazolin-5-one, 1-(4,6-dimethyl-2-pyrimidinyl)-2,3-dimethyl-	MeOH	241(4.2),281(3.9)	94-1467-69
3-Pyrazolin-5-one, 2-(4,6-dimethyl-2-pyrimidinyl)-1,3-dimethyl-	MeOH	286(4.3)	94-1467-69
$C_{11}H_{14}N_4O_2$			
Pyrazole, 3-methoxy-1-(4-methoxy-6-methyl-2-pyrimidinyl)-5-methyl-	pH 7.0	270(4.3)	94-1485-69
	$CHCl_3$	284(4.4)	94-1485-69
Pyrazole, 5-methoxy-1-(4-methoxy-6-methyl-2-pyrimidinyl)-3-methyl-	pH 7.0	250(4.3)	94-1485-69
	$CHCl_3$	259(4.3)	94-1485-69
3-Pyrazolin-5-one, 1-(4-methoxy-6-methyl-2-pyrimidinyl)-2,3-dimethyl-	pH 7.0	243(4.2),280s(3.9)	94-1485-69
	$CHCl_3$	245(4.1),290(4.0)	94-1485-69
3-Pyrazolin-5-one, 2-(4-methoxy-6-methyl-2-pyrimidinyl)-1,3-dimethyl-	pH 7.0	280(4.3)	94-1485-69
	$CHCl_3$	293(4.2)	94-1485-69
$C_{11}H_{14}N_4O_2S$			
Valeric acid, 5-[(9-methyl-9H-purin-6-yl)thio]-	pH 1	222(4.08),293(4.26)	73-2114-69
	pH 13	226(4.07),287(4.29), 295(4.30)	73-2114-69
$C_{11}H_{14}N_4O_3S$			
9H-Purine-6(1H)-thione, 9-[2α,3α-dihydroxy-4β-(hydroxymethyl)cyclopentyl]-	pH 1	224(3.98),323(4.34)	35-3075-69
	pH 7	225(4.02),321(4.36)	35-3075-69
	pH 13	232(4.16),311(4.36)	35-3075-69
$C_{11}H_{14}N_4O_4$			
Hypoxanthine, 9-[2α,3α-dihydroxy-4β-(hydroxymethyl)cyclopentyl]-	pH 1	250(4.06)	35-3075-69
	pH 7	249(4.08)	35-3075-69

Compound	Solvent	$\lambda_{max}(\log \epsilon)$	Ref.
Hypoxanthine, 9-[2α,3α-dihydroxy-4β-(hydroxymethyl)cyclopentyl]- (cont.)	pH 13	254(4.11)	35-3075-69
$C_{11}H_{14}N_4O_4S$			
6-Purinethiol, 9-β-L-fucopyranosyl-	pH 1	224(3.98),322(4.40)	4-0949-69
	pH 7	226(4.02),318(4.38)	4-0949-69
	pH 13	232(4.18),310(4.37)	4-0949-69
$C_{11}H_{14}N_4O_5$			
Hypoxanthine, 9-β-L-fucopyranosyl-	pH 1	248(4.08)	4-0949-69
	pH 7	248(4.08)	4-0949-69
	pH 13	252(4.11)	4-0949-69
$C_{11}H_{14}N_4O_6S$			
Hypoxanthine, 8-(β-D-glucopyranosylthio)-	pH 1	269(4.26)	87-0653-69
	pH 11	274(4.31)	87-0653-69
$C_{11}H_{14}N_4S$			
9H-Purine-6-thiol, 9-cyclohexyl-	pH 1	220(4.08),325(4.29)	73-2114-69
	pH 13	234(4.12),314(4.34)	73-2114-69
$C_{11}H_{14}O$			
2-Butene, 2-(p-methoxyphenyl)-, (E)-	n.s.g.	252(4.19)	28-1654-69B
(Z)-	n.s.g.	243(4.04)	28-1654-69B
Cyclopentaneacetaldehyde, 2-(2-butynylidene)-	EtOH	240(4.07)	28-1808-69A
2-Cyclopenten-1-one, 3-methyl-2-(2,4-pentadienyl)-, cis	EtOH	228(4.50)	39-1024-69C
Furan, 2-(cyclohexylidenemethyl)-	MeOH	269(4.24)	44-0212-69
3H-2,8a-Methanonaphthalen-3-one, 1,2,5,6,7,8-hexahydro-	EtOH	258(3.50)	78-5267-69
7H-1,4a-Methanonaphthalen-7-one, 1,2,3,4,5,6-hexahydro-	EtOH	244(4.12)	78-5281-69
Phenol, o-(dimethylallyl)-	EtOH	275(3.35)	12-1923-69
Pivalophenone	hexane	198(4.40),238(3.94), 274(2.82),320(2.01)	35-0355-69
	heptane	240(3.94),320(2.00)	59-0765-69
Propionaldehyde, 2,2-dimethyl-3-phenyl-	C_6H_{12}	299(1.52)	22-0903-69
Tricyclo[5.3.0^{1,7}.0^{2,7}]dec-8-en-10-one, 2-methyl-	EtOH	237(3.72),333(1.54)	33-0971-69
10-Undecene-6,8-diyn-5-ol	n.s.g.	215(4.34),228(3.41), 240(3.71),253(3.92), 267(4.18),283(4.02)	83-0100-69
$C_{11}H_{14}O_2$			
Acetophenone, 3'-ethyl-4'-methoxy-	EtOH	233(4.07),275(4.11)	22-0781-69
p-Benzoquinone, 2-tert-butyl-5-methyl-	MeOH	253(4.28)	64-0547-69
2-Cyclopenten-1-one, 4-hydroxy-3-methyl-2-(2,3-pentadienyl)-	EtOH	229(3.96)	39-1016-69C
1,5(6H)-Indandione, 7a-ethyl-7,7a-dihydro-, dl-	EtOH	240(4.03)	44-0107-69
1,6(2H,7H)-Naphthalenedione, 3,4,8,8a-tetrahydro-8a-methyl-	H_2SO_4	250(4.07)	23-2263-69
protonated	H_2SO_4	279(4.05)	23-2263-69
Tropolone, 4-tert-butyl-	MeOH	240(4.23),325(3.52), 349(3.45)	18-3277-69
Tropolone, 5-tert-butyl-	MeOH	235(4.48),245s(4.40), 250s(4.27),323(4.04), 351(3.87),367(3.78)	18-3277-69

Compound	Solvent	$\lambda_{max}(\log \epsilon)$	Ref.
$C_{11}H_{14}O_3$			
Acetophenone, 4',5'-dimethoxy-2'-methyl-	EtOH	229(4.33),272(4.03), 306(3.8)	39-1921-69C
2H-1-Benzopyran-2,5(3H)-dione, 5,6,7,8-tetrahydro-7,7-dimethyl-	EtOH	255(4.04)	4-0917-69
1,3-Cyclohexanedione, 2-(2-oxocyclopentyl)-	EtOH	270(4.11)	104-1196-69
	5N NaOH	290(4.34)	104-1196-69
1,3-Cyclopentanedione, 2-methyl-2-(3-oxo-4-pentenyl)-	EtOH	211(3.95)	35-2806-69
p-Dioxane, 2-(p-methoxyphenyl)-	EtOH	223(2.99),272(3.16), 278(3.09)	22-2048-69
Isovalerophenone, 2',5'-dihydroxy-	MeOH	255(3.7),360(3.5)	2-0040-69
2H-Naphtho[1,8-bc]furan-2,6(2aβH)-dione, 3,4,5,5a,7,8,8a,8b-octahydro-	MeOH	291(1.34)	39-1632-69C
Salicylaldehyde, 3-methoxy-5-propyl-	EtOH	224(4.20),268(3.92), 347(3.48)	12-1011-69
	KOH	242(4.24),283(3.70), 402(3.75)	12-1011-69
$C_{11}H_{14}O_3S_2$			
2-Thiophanecarboxylic acid, 3-hydroxy-3-(2-thienyl)-, ethyl ester	EtOH	258(3.93),285s(3.82)	2-0009-69
$C_{11}H_{14}O_4$			
3-Cyclohexen-1-ylideneacetic acid, 6-carboxy-, dimethyl ester	EtOH	214(4.05)	35-6470-69
Propiophenone, 4'-hydroxy-3',5'-dimethoxy-	EtOH	299(4.05)	44-0585-69
$C_{11}H_{14}O_5$			
2,5-Methano-4H,5H-pyrano[2,3-d]-m-dioxin-8-carboxylic acid, 4a,8a-dihydro-4-methyl-, methyl ester	EtOH	233(3.98)	88-2725-69
$C_{11}H_{14}O_6$			
2-Oxabicyclo[2.2.2]oct-5-ene-6-carboxylic acid, 3,8-dioxo-, methyl ester, 8-(dimethyl acetal)	$CHCl_3$	241(3.70)	24-2835-69
$C_{11}H_{14}O_7$			
L-erythro-Hex-4-enodialdo-1,5-pyranoside, methyl 4-deoxy-, diacetate, β-	MeOH	250(3.56)	23-0511-69
$C_{11}H_{14}O_8$			
L-threo-Hex-2-enonic acid, 2,6-anhydro-, methyl ester, 3,4-diacetate	EtOH	246(4.17)	88-2383-69
$C_{11}H_{14}S_2$			
Benzoic acid, dithio-, tert-butyl ester	C_6H_{12}	228s(4.0),296(4.18), 329s(3.9),526(2.01)	48-0045-69
	EtOH	227(3.91),298(4.08), 332s(3.8),520(1.97)	48-0045-69
	MeCN	231(3.87),296(4.08), 330s(3.8),515(1.88)	48-0045-69
$C_{11}H_{15}BrO$			
2-Adamantanone, 4-(bromomethyl)-	EtOH	292(1.34)	78-5601-69
$C_{11}H_{15}BrO_4$			
m-Dioxane-5-methanol, 2-(5-bromo-2-furyl)-5-ethyl-	30% EtOH	220(4.07)	103-0278-69

Compound	Solvent	$\lambda_{max}(\log \epsilon)$	Ref.
$C_{11}H_{15}ClN_2O_5$			
3-Carbamoyl-1-D-ribofuranosylpyridinium chloride	MeOH	265(3.73)	39-0199-69C
$C_{11}H_{15}ClO_4$			
m-Dioxane-5-methanol, 2-(5-chloro-2-furyl)-5-ethyl-	30% EtOH	224(4.04)	103-0278-69
$C_{11}H_{15}Cl_2N_5O$			
Hypoxanthine, 9-[2-[bis(2-chloroethyl)-amino]ethyl]-	H_2O	250(4.05)	87-0540-69
$C_{11}H_{15}F_5Sn$			
Stannane, triallyl(pentafluoroethyl)-(correction)	hexane	193(4.36)	44-3715-69
	MeCN	203(4.38)	44-3715-69
$C_{11}H_{15}IN_2O_4$			
Uridine, 2',3'-dideoxy-3'-iodo-3,5-dimethyl-	H_2O	267(3.96)	94-0785-69
	pH 13	267(3.97)	94-0785-69
$C_{11}H_{15}N$			
Benzylamine, N,N-dimethyl-o-vinyl-	EtOH	242(4.39),285(3.11), 310(2.93)	33-2004-69
tert-Butylamine, N-benzylidene-	EtOH	244(4.20)	35-2653-69
Styrylamine, N,N,β-trimethyl-	C_6H_{12}	211(3.93),291(3.93)	22-0903-69
$C_{11}H_{15}NO$			
Acetophenone, 2'-amino-3',4',5'-trimethyl-	EtOH	270(3.78),376(3.42)	78-1047-69
3H-Cyclohepta[c]pyridin-3-one, 2,5,6,7,8,9-hexahydro-1-methyl-	EtOH	201(4.19),237s(3.79), 244(3.72),305(3.88)	44-3670-69
2-Decalone, 1(e)-cyano-, trans	EtOH	282(1.38)	22-3199-69
2-Decalone, 3(e)-cyano-, trans	EtOH	280(1.40)	22-3199-69
Quinaldine, 5,6,7,8-tetrahydro-4-methoxy-	pH 1	214(3.75),242(3.98), 263(3.86)	78-0255-69
	MeOH	228(3.83),261(3.37), 269s(3.30)	78-0255-69
4(1H)-Quinolone, 5,6,7,8-tetrahydro-1,2-dimethyl-	pH 1	213(4.24),246(3.90), 262(3.66)	78-0255-69
	MeOH	218(4.22),224(4.20), 272(4.17)	78-0255-69
$C_{11}H_{15}NO_2$			
1H-Azepine-1-carboxylic acid, tert-butyl ester	isooctane	206(4.36),238s(3.55), 330(2.72)	44-2866-69
	EtOH	209(4.35),240s(3.51), 317(2.80)	44-2866-69
Benzaldehyde, p-(tert-butylhydroxy-amino)-	EtOH	315(4.01)	39-1459-69C
Benzoic acid, dimethylaminoethyl ester	dioxan	273(3.01),280(2.96)	65-1861-69
2,5-Cyclohexadien-1-one, 4-(tert-butyl-imino)-, N-oxide	EtOH	377(4.36)	39-1459-69C
p-Toluic acid, α-(dimethylamino)-, methyl ester	heptane	233(4.21)	44-0077-69
	MeOH	234(4.19)	44-0077-69
Valerohydroxamic acid, N-phenyl-	EtOH	247(4.02)	42-0831-69
$C_{11}H_{15}NO_2S$			
4H-Thieno[2,3-c]pyrrole-2-carboxylic acid, 5-ethyl-5,6-dihydro-3-methyl-, methyl ester	EtOH	258(4.04),286(3.95)	44-0333-69

Compound	Solvent	$\lambda_{max}(\log \epsilon)$	Ref.
$C_{11}H_{15}NO_2S_3$			
Benzenesulfonamide, N-(1,2-dithiepan-4-yl)-, (S)-(-)-	EtOH	259(2.91),265(2.98), 272(2.90)	87-0620-69
$C_{11}H_{15}NO_3$			
1-(1-Carboxy-2-methylpropyl)-5-hydroxy-2-picolinium hydroxide, inner salt, L-(+)-	aq. HCl	302(3.83)	1-2475-69
	aq. NaOH	334(3.76)	1-2475-69
3-Furanacetamide, α-acetyl-N,N,2-trimethyl-	EtOH	275s(2.4)	33-1030-69
$C_{11}H_{15}NO_3S_2$			
5-Thiazolecarboxylic acid, 4-methyl-2-[(3-oxobutyl)thio]-, ethyl ester	MeOH	304(4.16)	4-0397-69
4-Thiazoline-5-carboxylic acid, 4-methyl-3-(3-oxobutyl)-2-thioxo-, ethyl ester	MeOH	341(4.29)	4-0397-69
$C_{11}H_{15}NO_4$			
3,5-Pyridinedicarboxylic acid, 1,2-dihydro-, diethyl ester	EtOH	210(4.55),283(4.27), 392(3.73)	73-0427-69
3,5-Pyridinedicarboxylic acid, 1,4-dihydro-, diethyl ester	EtOH	210(4.51),245s(3.68), 374(3.83)	73-0427-69
$C_{11}H_{15}NO_4S_2$			
Malonic acid, [(2-thioxo-4-thiazolin-3-yl)methyl]-, diethyl ester	MeOH	318(4.12)	4-0397-69
$C_{11}H_{15}NO_6$			
m-Dioxane-5-methanol, 5-ethyl-2-(5-nitro-2-furyl)-	30% EtOH	224(3.40),310(3.78)	103-0278-69
$C_{11}H_{15}NO_7$			
4-Isoxazoline-3,4-dicarboxylic acid, 3-(carboxymethyl)-2-methyl-, trimethyl ester	MeOH	280(3.64)	24-2346-69
$C_{11}H_{15}NS_2$			
Spiro[cyclopentane-1,2'(4'H)-cyclopenta[d][1,3]thiazine]-4'-thione, 1',5',6',7'-tetrahydro-	EtOH	336(3.89),407(4.36)	44-0730-69
$C_{11}H_{15}N_3$			
Benzimidazole, 2-(1-amino-2-methylpropyl)-, dihydrochloride	EtOH	274(3.86),281(3.85)	94-2381-69
$C_{11}H_{15}N_3O_2$			
Carbamic acid, [1-methyl-1-(phenylazo)-ethyl]-, methyl ester	EtOH	265(3.99),398(2.10)	49-1479-69
$C_{11}H_{15}N_3O_7$			
Cytidine, 5'-(O-carboxymethyl)-	pH 7	270(3.95)	28-1160-69A
$C_{11}H_{15}N_3O_8$			
Polyoxin C	0.05N HCl	262(3.97)	35-7490-69
	0.05N NaOH	264(3.87)	35-7490-69
$C_{11}H_{15}N_4O_9P$			
Inosine, 2-methoxy-, 5'-phosphate, (disodium salt)	pH 1	254(3.96)	94-2581-69
	pH 6	248(4.00),264s(--)	94-2581-69

Compound	Solvent	λ_{max}(log ε)	Ref.
$C_{11}H_{15}N_5$			
1H-Pyrazolo[3,4-d]pyrimidine, 1-methyl-4-piperidino-	MeOH	293(4.22)	23-1129-69
$C_{11}H_{15}N_5O$			
Acetamide, N-(5-methyl-8-propyl-s-triazolo[4,3-a]pyrazin-3-yl)-	EtOAc	220(--),298(3.76)	39-1593-69C
Acetamide, N-(6-methyl-8-propyl-s-triazolo[4,3-a]pyrazin-3-yl)-	BuOH	223(4.19),302(3.52)	39-1593-69C
$C_{11}H_{15}N_5OS_5$			
2H-1,3-Thiazine-2,4(3H)-dione, 3,3'-ethylenebis[dihydro-2-thio-, 4-(thiosemicarbazone)	EtOH	255(4.23),310s(--)	103-0049-69
$C_{11}H_{15}N_5O_2$			
Cyclopentanemethanol, 3-(6-amino-9H-purin-9-yl)-4-hydroxy-	pH 1	259(4.15)	88-2231-69
	pH 7,13	262(4.16)	88-2231-69
$C_{11}H_{15}N_5O_3$			
1,2-Cyclopentanediol, 3-(6-amino-9H-purin-9-yl)-5-(hydroxymethyl)-	pH 1	212(4.31),258(4.16)	35-3075-69
	pH 7	260(4.17)	35-3075-69
	pH 13	260(4.18)	35-3075-69
$C_{11}H_{15}N_5O_3S$			
Adenosine, 5'-deoxy-5'-(methylthio)-	pH 7.0	260(4.19)	10-0414-69F
$C_{11}H_{15}N_5O_4$			
Adenine, 9-β-L-fucopyranosyl-	pH 1	256(4.17)	4-0949-69
	pH 7	258(4.17)	4-0949-69
	pH 13	258(4.17)	4-0949-69
L-Allitol, 6-(6-amino-9-purinyl)-1,5-anhydro-6-deoxy-	pH 7	261(4.15)	87-0175-69
Guanine, 9-(2-deoxy-α-D-erythro-pentofuranosyl)-N-methyl-	pH 1	257(4.06),280s(3.80)	44-2160-69
	pH 11	257(4.01),269s(3.95)	44-2160-69
	MeOH	254(4.12),273s(3.92)	44-2160-69
Guanosine, 2'-deoxy-N-methyl-	pH 1	258(4.13),281s(3.87)	44-2160-69
	pH 11	258(4.05),270s(4.01)	44-2160-69
	MeOH	254(4.16),273(3.96)	44-2160-69
9H-Imidazo[1,2-a]purin-9-one, 6-(1-ethoxyethyl)-3,5,6,7-tetrahydro-6,7-dihydroxy-	pH 7	246(4.00),279(3.79)	69-0238-69
$C_{11}H_{15}N_5O_5$			
Adenosine, 2-methyl-, 1-oxide	pH 1	260(4.12)	94-1128-69
	pH 7	233(4.58),264(3.91)	94-1128-69
	pH 13	233(4.49),268(3.94)	94-1128-69
$C_{11}H_{15}N_5O_5S$			
Adenine, 8-(β-D-glucopyranosylthio)-	pH 1	278(4.31)	87-0653-69
	pH 11	226(4.26),281(4.30)	87-0653-69
$C_{11}H_{15}N_5O_6$			
Erythronic acid, 4-[(6-amino-5-nitro-4-pyrimidinyl)amino]-4-deoxy-2,3-O-isopropylidene-	EtOH	230s(4.25),339(4.00)	88-4729-69
$C_{11}H_{15}N_5S$			
Adenine, N-(3-methyl-2-butenyl)-2-(methylthio)-	EtOH	242(4.40),279(4.20)	69-3071-69
	EtOH-acid	253(4.34),292(4.20)	69-3071-69

Compound	Solvent	λ max (log ε)	Ref.
Adenine, N-(3-methyl-2-butenyl)-2-(methylthio)- (cont.)	EtOH-base	287(4.17)	69-3071-69
$C_{11}H_{15}O_4P$			
Phosphonic acid, benzoyl-, diethyl ester	C_6H_{12}	258(4.05),379(1.92)	18-0821-69
$C_{11}H_{16}$			
Benzene, pentyl-, tetracyanoethylene complex	C_6H_{12}	404(3.21)	39-1161-69B
Ethylene, tricyclopropyl-	hexane	213(4.09)	77-0781-69
4H-Indene, 3a,5,6,7-tetrahydro-3aα,7α-dimethyl-	MeOH	256(3.59)	35-3664-69
Nopadiene	hexane	240(4.32)	44-0742-69
2-Norcarene, 7,7-dimethyl-3-vinyl-	EtOH	245(4.22)	35-6473-69
2-Norpinene, 4-ethylidene-6,6-dimethyl-	hexane	245(4.20),251(4.20)	44-0742-69
$C_{11}H_{16}ClNO_4$			
5,6,7,8-Tetrahydro-1,4-dimethylquinolinium perchlorate	EtOH	271(3.83)	78-4161-69
$C_{11}H_{16}Cl_2N_4S$			
s-Triazine, 2,4-dichloro-6-[2-(diethylamino)-1-methyl-2-(methylthio)vinyl]-	C_6H_{12}	302(3.41),408(4.17)	5-0073-69E
s-Triazine, 2,4-dichloro-6-[2-(dimethylamino)-2-(methylthio)-1-propylvinyl]-	C_6H_{12}	232(3.89),301(3.61), 403(4.45)	5-0073-69E
$C_{11}H_{16}Cl_2N_6$			
Adenine, 9-[2-[bis(2-chloroethyl)amino]-ethyl]-, dihydrochloride	EtOH	260(4.18)	87-0540-69
$C_{11}H_{16}Cl_2N_6O$			
Guanine, 9-[2-[bis(2-chloroethyl)amino]-ethyl]-, hydrochloride	H_2O	253(4.04),269(3.91)	87-0540-69
$C_{11}H_{16}NO$			
1,2,3,4-Tetrahydro-3-hydroxy-6,8-dimethylquinolizinium (picrate)	EtOH	268(4.0),279(3.9)	24-1309-69
$C_{11}H_{16}N_2$			
Acetophenone, propylhydrazone	hexane	264(4.04),357(2.05)	104-0257-69
Propane-1-azoethane, 1'-phenyl-	hexane	353(1.42)	104-0257-69
Pyridazine, 1,4,5,6-tetrahydro-1-methyl-3-phenyl-	EtOH	224(3.87),292(4.00)	39-1703-69C
Pyrrolidine, 2,2-dimethyl-5-(3-pyridyl)-	EtOH	258s(3.42),260(3.50), 268s(3.35)	36-0860-69
	EtOH-HCl	258(3.69),266s(3.61)	36-0860-69
$C_{11}H_{16}N_2O$			
1-Pyrrolidinol, 2,2-dimethyl-5-(3-pyridyl)-	EtOH	257s(3.38),262(3.43), 268(3.29)	36-0860-69
	EtOH-HCl	260(3.67)	36-0860-69
2,6-Xylidine, N-isopropyl-N-nitroso-	C_6H_{12}	213(4.08),242s(3.70), 275s(2.41)	35-3383-69
$C_{11}H_{16}N_2O_2$			
2(1H)-Naphthalenone, 1-diazooctahydro-8a-hydroxy-4a-methyl-	MeOH	262s(3.59),291(3.78)	88-5049-69
1H-Pyrazolo[1,2-b]phthalazine-5,10-dione, octahydro-	$C_2H_3F_3O$	202(3.42),248(3.51)	35-1256-69

Compound	Solvent	λ_{max}(log ε)	Ref.
3,4-Xylidine, 6-nitro-N-propyl-	EtOH	237(4.38),294(3.79), 443(3.80)	44-3240-69
	EtOH-HCl	237(4.39),294(3.82), 443(3.81)	44-3240-69
	EtOH-NaOH	235s(4.43),294(3.85), 443(3.82)	44-3240-69
$C_{11}H_{16}N_2O_2S_2$			
2,4-Pentadienoic acid, 2-cyano-5-(dimethylamino)-3,5-bis(methylthio)-, methyl ester	C_6H_{12}	421(4.25)	5-0073-69E
$C_{11}H_{16}N_2O_4$			
2,3-Diazabicyclo[3.2.0]hept-2-ene-6,7-dicarboxylic acid, 4,4-dimethyl-, dimethyl ester	n.s.g.	328(2.34)	89-0210-69
4H-Pyrazole-3,5-diol, 4,4-diethyl-, diacetate	EtOH	217(3.98)	44-3181-69
$C_{11}H_{16}N_2O_5$			
Uridine, 2'-deoxy-5-ethyl-, α-	pH 2	268(3.99)	87-0533-69
	pH 12	268(3.87)	87-0533-69
β-form	pH 2	268(3.98)	87-0533-69
	pH 12	268(3.86)	87-0533-69
$C_{11}H_{16}N_2O_5S$			
Uridine, 2'-deoxy-5-(ethylthio)-	EtOH	233(3.82),288(3.71)	44-3806-69
$C_{11}H_{16}N_2O_6$			
Uridine, 3,6-dimethyl-	pH 7	206(3.91),261(3.96)	73-2316-69
Uridine, 5,6-dimethyl-	pH 7	206(4.12),267(4.04)	73-2316-69
	pH 14	270(3.89)	73-2316-69
$C_{11}H_{16}N_2O_7$			
Uridine, 5'-O-(2-hydroxyethyl)-	MeOH	262(3.98)	28-1160-69A
$C_{11}H_{16}N_2O_8S$			
Uridine, 3-methyl-, 3'-methanesulfonate	H_2O	260(3.93)	94-0785-69
	pH 13	262(3.93)	94-0785-69
$C_{11}H_{16}N_2O_{10}S_2$			
Uracil, 1-β-D-arabinofuranosyl-, 2',5'-dimethanesulfonate	MeOH	260(4.06)	94-1188-69
$C_{11}H_{16}N_4OS_2$			
4-Morpholinecarbodithioic acid, (4-amino-2-methyl-5-pyrimidinyl)methyl ester	EtOH	248(4.17),282(4.18)	94-2299-69
$C_{11}H_{16}N_4S$			
9H-Purine-6-thiol, 9-hexyl-	pH 1	220(4.10),325(4.33)	73-2114-69
	pH 13	234(4.17),314(4.36)	73-2114-69
$C_{11}H_{16}N_5O_6P$			
Aristeromycin, 6'-phosphate	pH 1	261(4.16)	44-1547-69
	H_2O	262(4.17)	44-1547-69
	pH 13	263(4.16)	44-1547-69
$C_{11}H_{16}N_6O_2$			
Ethanol, 2-[(6-morpholinopurin-8-yl)amino]-	pH 1	316(4.35)	65-2125-69
	EtOH	290(4.30)	65-2125-69

Compound	Solvent	λmax(log ε)	Ref.
$C_{11}H_{16}N_6O_4S$			
Adenosine, 2',5'-dideoxy-5'-N-methane-sulfonamido-	pH 1	256(4.18)	87-0658-69
	pH 11	258(4.20)	87-0658-69
$C_{11}H_{16}O$			
2-Adamantanone, 4-methyl-	EtOH	288(1.36)	78-5601-69
Bicyclo[5.1.0]oct-5-en-2-one, 1,4,4-trimethyl-, cis	EtOH	207(3.57),272(2.42)	88-0995-69
o-Cresol, 4-butyl-	pH 2	277(3.31)	30-1113-69A
	pH 13	238(3.93),295(3.53)	
	MeOH	281(3.36)	
	MeOH-Na	240(3.92),297(3.52)	
o-Cresol, 4-tert-butyl-	pH 2	274(3.28)	30-1113-69A
	pH 13	238(3.93),293(3.52)	
	MeOH	278(3.35)	
	MeOH-Na	240(3.94),295(3.52)	
o-Cresol, 6-butyl-	pH 2	270(3.15)	30-1113-69A
	pH 13	239(3.93),289(3.63)	
	MeOH	273(3.24)	
	MeOH-Na	242(3.94),293(3.64)	
p-Cresol, 2-tert-butyl-	pH 2	277(3.34)	30-1113-69A
	pH 13	239(3.93),297(3.57)	
	MeOH	281(3.40)	
	MeOH-Na	240(3.94),298(3.58)	
2-Cyclopenten-1-one, 3-(1-methylcyclopentyl)-	EtOH	232(4.13)	33-0971-69
Inden-5(4H)-one, 2,6,7,7a-tetrahydro-4,4-dimethyl-	isooctane	289(1.95),298(1.40), 308(1.93),319(1.65)	44-0450-69
7H-1,4a-Methanonaphthalen-7-one, octahydro-	EtOH	282(1.49)	78-5281-69
2(1H)-Naphthalenone, 4a,5,6,7,8,8a-hexahydro-4a-methyl-	MeOH	229(3.96)	23-4299-69
2(3H)-Naphthalenone, 4,4a,5,6,7,8-hexahydro-1-methyl-	H_2SO_4	258(4.04)	23-2263-69
protonated	H_2SO_4	297(4.05)	23-2263-69
2(3H)-Naphthalenone, 4,4a,5,6,7,8-hexahydro-3-methyl-	H_2SO_4	249(4.04)	23-2263-69
protonated	H_2SO_4	287(4.05)	23-2263-69
2(3H)-Naphthalenone, 4,4a,5,6,7,8-hexahydro-4a-methyl-	EtOH	239(4.25),309(1.85)	23-2263-69
1(2H)-Pentalenone, 3,3a,6,6a-tetrahydro-3,3,6a-trimethyl-	EtOH	286(1.60)	88-0995-69
4-Pentenal, 5-(1-cyclohexen-1-yl)-	EtOH	235(4.37)	28-1535-69A
4-Pentenal, 5-(1-cyclopenten-1-yl)-3-methyl-	EtOH	242(4.19)	28-1535-69A
Phenol, 4-(diethylmethyl)-	pH 2	274(3.25)	30-1113-69A
	pH 13	237(4.03),292(3.45)	
	MeOH	278(3.26)	
	MeOH-Na	238(4.07),293(3.41)	
Phenol, m-isopentyl-	EtOH	274(3.21),280(3.17)	35-2299-69
Spirohexan-4-one, 5-isopropylidene-1,1-dimethyl-	isooctane	240(4.24),246(4.25), 374(1.57)	22-3981-69
Tricyclo[$4.4.1^{1,3}.0$]undecan-4-one	EtOH	273(1.62)	78-5267-69
Tricyclo[$5.4.0^{1,4}.0^{1,7}$]undecan-5-one	EtOH	295(1.65)	78-5281-69
$C_{11}H_{16}O_2$			
1-Adamantanecarboxylic acid, Pb(IV) salt	C_6H_{12}	262(4.21)	33-1495-69
1-Cyclopentene-1-carboxylic acid, 5-isopropenyl-2-methyl-, methyl ester	n.s.g.	230(3.90)	78-3767-69

Compound	Solvent	λ_{max}(log ϵ)	Ref.
2-Cyclopenten-1-one, 4-hydroxy-3-methyl-2-(2-pentenyl)-, cis	EtOH	230(4.05)	39-1016-69C
1,4-Dioxaspiro[4.4]nona-6,8-diene, 7-tert-butyl-	isooctane	196(3.35),280(3.04)	35-6785-69
	MeOH	203(3.14),280(3.01)	35-6785-69
Indene-2-carboxylic acid, 3a,4,5,6,7,7a-hexahydro-, methyl ester	EtOH	227(4.00)	44-1480-69
Indene-3-carboxylic acid, 3a,4,5,6,7,7a-hexahydro-, methyl ester	n.s.g.	227(4.15)	44-1480-69
isomer	EtOH	227(4.04)	44-1480-69
2,3-Naphthalenedione, octahydro-1-methyl-	EtOH	273(4.00)	94-1578-69
2(1H)-Naphthalenone, 4a,5,6,7,8,8a-hexahydro-4a-(hydroxymethyl)-, trans	EtOH	228(4.08)	78-5281-69
4H-Pyran, 3-acetyl-2,4,4,6-tetramethyl-	MeOH	215(3.21),284(2.93)	44-3169-69
$C_{11}H_{16}O_3$			
2(4H)-Benzofuranone, 5,6,7,7a-tetrahydro-6-hydroxy-4,4,7a-trimethyl-	EtOH	213(4.06)	44-3858-69
1-Cyclooctene-1-carboxylic acid, 2-methyl-8-oxo-, methyl ester	MeOH	205(3.95),207(3.92) 225(3.83)	94-0629-69
8,10-Undecadiene-2,5-dione, 3-hydroxy-, cis, (±)	EtOH	228(4.27)	39-1016-69C
$C_{11}H_{16}O_4$			
m-Dioxane-5-methanol, 5-ethyl-2-(2-furyl)-	30% EtOH	214(4.03)	103-0278-69
1α-Naphthoic acid, decahydro-5α-hydroxy-8-oxo-, cis	MeOH	288(1.34)	39-1632-69C
trans	MeOH	282(1.49)	39-1632-69C
$C_{11}H_{17}As$			
Arsine, diethyl-p-tolyl-	dioxan	223(4.1),244(3.6)	70-1381-69
$C_{11}H_{17}AsO$			
Arsine oxide, diethyl-p-tolyl-	pH 2	227(4.1),265(2.6)	70-1381-69
	H_2O	225(4.1),265(2.7)	70-1381-69
	dioxan	223(4.1),263(2.4)	70-1381-69
$C_{11}H_{17}AsO_2$			
Arsine oxide, diethyl(p-methoxyphenyl)-	pH 2	240(4.2),270(3.2)	70-1381-69
	H_2O	235(4.2),270(3.0)	70-1381-69
	dioxan	235(4.2),275(3.2)	70-1381-69
$C_{11}H_{17}BrN_4$			
5-Bromo-3-penten-1-yne hexamethylenetetramine adduct	EtOH	219(4.08),228(4.19), 237(4.08)	18-2589-69
$C_{11}H_{17}ClN_2O$			
4(3H)-Pyrimidinone, 2-butyl-6-chloro-5-propyl-	MeOH	231(3.82),281(3.86)	44-2972-69
$C_{11}H_{17}ClN_2O_2$			
3-Carbamoyl-1-(propoxymethyl)-4-picolinium chloride	EtOH	263(3.53)	39-0192-69B
$C_{11}H_{17}ClN_4O_8P$			
Uridine, 5'-[hydrogen (2-chloroethyl)-phosphoramidate]	pH 7	262(4.00)	65-0668-69
	pH 12	261(3.89)	65-0668-69

Compound	Solvent	λ max(log ε)	Ref.
$C_{11}H_{17}IN_2$			
2-Ethyl-1,2-dimethylindazolinium iodide	n.s.g.	222(4.22),282(3.23)	104-2004-69
$C_{11}H_{17}N$			
m-Toluidine, N-tert-butyl-	EtOH	249(3.86)	39-1459-69C
$C_{11}H_{17}NO$			
Isoxazole, 3-cyclopentyl-5-ethyl-4-methyl-	EtOH	225(3.78)	22-2820-69
$C_{11}H_{17}NOS$			
3-Pyridinol, 6-methyl-2-(pentylthio)-	pH 1	240(3.50),325(3.90)	1-2065-69
	pH 13	250(3.86),325(3.94)	1-2065-69
$C_{11}H_{17}NOS_2$			
2-Butanone, 4-[(4-tert-butyl-2-thiazolyl)thio]-	MeOH	278(3.83)	4-0397-69
$C_{11}H_{17}NO_2$			
Pyrrole-3-carboxylic acid, 5-ethyl-2,4-dimethyl-, ethyl ester	MeOH	270(3.66)	35-3931-69
Pyrrole-3-carboxylic acid, 1,2,4,5-tetramethyl-, ethyl ester	MeOH	240(3.97),269(3.64)	35-3931-69
2H-Quinolizin-2-one, 4-ethoxy-1,6,7,8,9,9a-hexahydro-	EtOH	298(4.16)	94-2314-69
$C_{11}H_{17}NO_2S$			
Acrylamide, N-ethyl-3-[(2-oxocyclohexyl)thio]-, (Z)-	90% EtOH	277(4.12)	12-0765-69
$C_{11}H_{17}NO_3$			
2,4-Pentadienoic acid, 2-acetyl-5-(dimethylamino)-, ethyl ester	EtOH	241(3.81),265(3.86), 395(4.77)	70-0363-69
$C_{11}H_{17}NO_3S_2$			
2-Pyridinol, 1-acetyl-1,2,3,6-tetrahydro-3, 6-bis(methylthio)-, acetate	MeOH	200(4.11)	44-0660-69
$C_{11}H_{17}NO_4$			
3,5-Pyridinedicarboxylic acid, 1,2,3,4-tetrahydro-, diethyl ester	n.s.g.	281(4.31)	77-1348-69
3-Pyrrolidine-3-carboxylic acid, 1-butyl-4-hydroxy-5-oxo-, ethyl ester	H_2O	247(4.04),300(3.26)	70-0694-69
	MeOH	246(4.04),300(3.45)	70-0694-69
	EtOH	247(4.13)	70-0694-69
	BuOH	248(4.10)	70-0694-69
	MeCN	246(4.07)	70-0694-69
	dioxan	247(4.13)	70-0694-69
	aq. KOH	239(3.76),302(4.11)	70-0694-69
3-Pyrroline-3-carboxylic acid, 1-tert-butyl-4-hydroxy-2-oxo-, ethyl ester	EtOH	251(3.39)	104-1713-69
$C_{11}H_{17}NO_4S$			
Acetoacetic acid, 2-[[2-(ethylcarbamoyl)vinyl]thio]-, ethyl ester, (Z)	90% EtOH	267(4.17)	12-0765-69
$C_{11}H_{17}NO_5$			
Maleic acid, (piperidinooxy)-, dimethyl ester	MeOH	233(4.06)	24-2336-69

Compound	Solvent	λ max (log ε)	Ref.
$C_{11}H_{17}NS_2$			
Cyclopenta[d][1,3]thiazine-4(2H)-thione, 2,2-diethyl-1,5,6,7-tetrahydro-	EtOH	338(3.76),411(4.19)	44-0730-69
Cyclopenta[d][1,3]thiazine-4(2H)-thione, 1,5,6,7-tetrahydro-2-isopropyl-2-methyl-	EtOH	338(3.79),409(4.31)	44-0730-69
$C_{11}H_{17}N_3O$			
Tricyclo[4.3.1^{1,2}.0]decan-3-one, semicarbazone	EtOH	232(4.14)	78-5287-69
$C_{11}H_{17}N_3OS_3$			
Spiro[furo[2,3-d]thiazole-2(3H),4'-imidazolidine]-2',5'-dithione, tetrahydro-1',3,3',3a-tetramethyl-	EtOH	321(4.22)	94-0910-69
$C_{11}H_{17}N_3O_2$			
Piperidine, 1-(1,4-dihydro-1-methyl-3-nitro-4-pyridyl)-	$CHCl_3$	400(4.13)	24-2163-69
$C_{11}H_{17}N_3O_3$			
4-Pyrimidinecarbamic acid, 1,2-dihydro-2-oxo-, hexyl ester	2N HCl	227(3.56),298(3.51)	39-1860-69C
$C_{11}H_{17}N_3O_5$			
Cytidine, 5,6-dimethyl-	pH 7	216(4.05),232s(3.95), 281(3.96)	73-2316-69
3-Pyrroline-1,3-dicarboxylic acid, 4-ureido-, diethyl ester	EtOH	273(4.36)	36-1038-69
$C_{11}H_{17}N_3O_{12}S_3$			
as-Triazine-3,5(2H,4H)-dione, 2-β-D-ribofuranosyl-, 2',3',5'-trimethanesulfonate	50% EtOH	255(3.76)	73-0618-69
$C_{11}H_{17}N_5$			
Adenine, 9-isopentyl-8-methyl-	pH 1	218(4.05),264(4.11)	39-2198-69C
	H_2O	215(4.07),263(4.11)	39-2198-69C
	pH 13	221(3.97),263(4.15)	39-2198-69C
$C_{11}H_{17}N_5O_2$			
9H-Purine-9-acetaldehyde, 6-amino-, diethyl acetal	pH 1	259(4.15)	88-2285-69
	pH 12	261(4.15)	88-2285-69
$C_{11}H_{17}N_5O_4$			
Erythronic acid, 4-deoxy-4-[(5,6-diamino-4-pyrimidinyl)amino]-2,3-O-isopropylidene-, D-	H_2O	218(4.43),280(4.03)	88-4729-69
$C_{11}H_{18}$			
1,6-Heptadiene, 4-allyl-4-methyl-	gas	181(--)	121-0803-69
	heptane	184(4.49)	121-0803-69
	MeCN	184(4.30)	121-0803-69
$C_{11}H_{18}ClNO_4$			
2,3,5,6,7,8-Hexahydro-1,4-dimethylquinolinium perchlorate	$CHCl_3$	304(3.67)	78-4161-69
Malonamic acid, N-tert-butyl-2-(chloroacetyl)-, ethyl ester	EtOH	252(2.17)	104-1713-69

Compound	Solvent	λmax(log ε)	Ref.
$C_{11}H_{18}CrGe_2O_5S$			
Chromium, pentacarbonyl(hexamethyldigermthiane)-	n.s.g.	420(2.95)	89-0271-69
$C_{11}H_{18}CrO_5SSn_2$			
Chromium, pentacarbonyl(hexamethyldistannthiane)-	n.s.g.	435(2.91)	89-0271-69
$C_{11}H_{18}N_2O$			
2-Cyclohexen-1-one, 2-piperidino-, oxime	hexane	225(4.1),280(3.60)	48-0162-69
$C_{11}H_{18}N_2O_2$			
Nicotinamide, 1,4-dihydro-4-methyl-1-(propoxymethyl)-	EtOH	326(3.78)	39-0192-69B
Nicotinamide, 1,6-dihydro-4-methyl-1-(propoxymethyl)-	EtOH	261(3.86),339(3.65)	39-0192-69B
1-Pyrazoline-3-carboxylic acid, 4-ethylidene-3,5,5-trimethyl-	hexane	327(2.24)	39-2443-69C
	EtOH	325(2.18)	39-2443-69C
1-Pyrazoline-4-carboxylic acid, 5-isopropylidene-3,3-dimethyl-	hexane	255(4.00),350(2.48)	39-2443-69C
	EtOH	259(3.98),344(3.51)	39-2443-69C
Uracil, 6-butyl-5-propyl-	pH 2	268(3.88)	56-0499-69
	pH 12	270(3.80)	56-0499-69
	5N NaOH	279(3.81)	56-0499-69
$C_{11}H_{18}N_2O_2S$			
Thiazole, 2,4-di-tert-butyl-5-nitro-	EtOH	220(3.65),245(--), 314(3.83)	28-1343-69B
$C_{11}H_{18}N_2O_3$			
Barbituric acid, 5-ethyl-5-(1-methylbutyl)-, S-(-)-	MeOH	210(3.98)	44-2676-69
	pH 1.4	212(3.87)	44-2676-69
	pH 12.4	240(4.00)	44-2676-69
$C_{11}H_{18}N_2O_4$			
Barbituric acid, 5-butyl-5-(2-methoxyethyl)-	pH 6.2	218(3.80)	65-2568-69
	pH 13	242(3.69)	65-2568-69
Barbituric acid, 5-ethyl-5-(3-hydroxy-1-methylbutyl)-	0.05N NaOH	255(3.82)	87-0180-69
Barbituric acid, 5-ethyl-5-(1-propoxyethyl)-	pH 6.2	266(3.86)	65-2568-69
	pH 13	244(3.77)	65-2568-69
Barbituric acid, 5-(1-isobutoxyethyl)-5-methyl-	pH 6.2	266(3.77)	65-2568-69
	pH 13	242(3.96)	65-2568-69
1(2H)-Pyrimidineacetaldehyde, 3,4-dihydro-5-methyl-2,4-dioxo-, 1-(diethyl acetal)	pH 1	271(3.94)	88-2285-69
	pH 12	269(3.81)	88-2285-69
$C_{11}H_{18}N_4O_2$			
Carbamic acid, dimethyl-, 2-(dimethylamino)-5,6-dimethyl-4-pyrimidinyl ester	MeOH	209(3.73),246(4.20), 310(3.47)	25-1018-69
1,2,7,8-Tetraazaspiro[4.4]nona-1,6-diene-6-carboxylic acid, 3,3,9,9-tetramethyl-, methyl ester	benzene	348(3.31)	39-2443-69C
	EtOH	223(3.53),300(3.96), 348(3.42)	39-2443-69C
$C_{11}H_{18}N_4O_4$			
Cytosine, 1-(3-amino-3-deoxy-β-D-arabino-N,N-dimethyl-, hydrochloride	H_2O	220s(4.00),280(4.14)	94-1188-69

Compound	Solvent	$\lambda_{max}(\log \epsilon)$	Ref.
$C_{11}H_{18}N_6$			
Adenine, 9-(6-aminohexyl)-	EtOH	262(4.15)	44-3240-69
	EtOH-HCl	259(4.16)	44-3240-69
	EtOH-NaOH	262(4.16)	44-3240-69
$C_{11}H_{18}N_6O$			
Ethanol, 2-[[6-(diethylamino)purin-8-yl]amino]-	pH 1	301(4.31)	65-2125-69
	EtOH	292(4.32)	65-2125-69
$C_{11}H_{18}O$			
Cyclohexene, 1-acetyl-2-propyl-	MeOH	248(3.71)	78-0223-69
Cyclohexene, 2-methyl-1-propionyl-	MeOH	247(3.70)	78-0223-69
2-Cyclohexen-1-one, 2-pentyl-	n.s.g.	237(3.92)	78-6025-69
Cyclopentene, 1-acetyl-2-butyl-	MeOH	256(3.92)	39-0608-69C
2-Cyclopenten-1-one, 2-butyl-3-ethyl-	EtOH	236(4.18)	104-1993-69
2-Cyclopenten-1-one, 2-methyl-3-pentyl-	EtOH	236(4.23)	104-1993-69
2-Cyclopenten-1-one, 3-methyl-2-pentyl-	EtOH	236(4.16)	104-1993-69
2-Cycloundecen-1-one, cis	EtOH	230(3.79),310(1.73)	24-3877-69
2-Cycloundecen-1-one, trans	EtOH	227(4.08),314(1.81)	24-3877-69
2-Decalone, 1(a)-methyl-, trans	EtOH	291(1.49)	22-3199-69
1(e)-	EtOH	280(1.45)	22-3199-69
2-Decalone, 3(a)-methyl-, trans	EtOH	290(1.56)	22-3199-69
3(e)-	EtOH	284(1.46)	22-3199-69
Furan, 2,3-dihydro-3,3,5-trimethyl-2-(1-methylpropenyl)-	EtOH	210(3.87),220(3.66) (end absorptions)	44-2301-69
Ketone, bis(2,2-dimethylcyclopropyl)	isooctane	213(3.90),291(2.19)	22-3981-69
Ketone, 2,2-dimethyl-3-(2-methylpropenyl)cyclopropyl methyl	EtOH	210(3.99),233s(3.75), 280(2.31)	44-2301-69
Ketone, 2-ethyl-1-cyclopentenyl isopropyl	MeOH	253(3.89)	39-0608-69C
3,7-Nonandien-2-one, 4,8-dimethyl- (isomeric mixture)	EtOH	206(3.70),240(4.11)	16-0226-69
trans	EtOH	238(3.92),315(1.91)	16-0226-69
3-Penten-1-ol, 5-cyclohexylidene-	MeOH	243(3.91)	44-0212-69
Spirohexan-4-one, 1-ethyl-1,5,5-trimethyl-	hexane	205(3.81),303(1.71)	28-0346-69B
Spirohexan-5-one, 1-ethyl-1,4,4-trimethyl-	hexane	199(3.08),240(--), 295(1.51)	28-0346-69B
$C_{11}H_{18}O_2$			
Cyclohexanone, 3-acetyl-2,2,4-trimethyl-	EtOH	207(2.98),282(2.89)	16-0226-69
2,4-Cyclopentadien-1-one, 3-tert-butyl-, dimethyl acetal	isooctane	195(3.34),270(3.09)	35-6785-69
	MeOH	203(3.11),271(3.01)	35-6785-69
2,4-Decadienoic acid, methyl ester, (E,E)-	n.s.g.	261(4.44)	39-2477-69C
2-Decalone, 1(a)-methoxy-, trans	EtOH	306(1.66)	22-3199-69
1(e)-	EtOH	287(1.52)	22-3199-69
2-Decalone, 3(a)-methoxy-, trans	EtOH	304(1.54)	22-3199-69
3(e)-	EtOH	287(1.38)	22-3199-69
4,5-Heptadien-2-ol, 2,4-dimethyl-, acetate	isooctane	216(3.21),229(3.04), 285(1.36)	35-3299-69
4,5-Heptadien-2-ol, 2,6-dimethyl-, acetate	isooctane	210(3.04),240s(2.64), 270(3.18),285(2.00)	35-3299-69
$C_{11}H_{18}O_3$			
Cyclohexanepropionic acid, β-oxo-, ethyl ester	isooctane	246(3.82)	22-3694-69
	EtOH	246(3.33)	22-3694-69
Cyclopentaneacetic acid, α-acetyl-, ethyl ester	isooctane	290(1.75)	22-3694-69
	EtOH	285(1.81)	22-3694-69

Compound	Solvent	λ max (log ε)	Ref.
1,2-Cycloundecanedione, 3-hydroxy-, (enol)	DMSO-acid	267(4.09)	24-0388-69
	DMSO-base	331(4.04)	24-0388-69
2(5H)-Furanone, 5-hexyl-4-hydroxy-5-methyl-	EtOH	249(4.30)	39-0056-69C
Umbellulone, dihydro-3-hydroxy-4-methoxy-, (3S,4R)-(-)-	MeOH	282(1.53)	78-3161-69
$C_{11}H_{18}O_4$			
Malonic acid, butylidene-, diethyl ester	EtOH	210(3.99)	22-0313-69
$C_{11}H_{18}O_8$			
Tuliposide A	H_2O	207s(4.23)	24-2057-69
$C_{11}H_{19}BF_2O_2$			
Boron, difluoro(2,2,6,6-tetramethyl-3,5-heptanedionato)-	$CHCl_3$	293(4.35)	39-0526-69A
$C_{11}H_{19}ClN_2O_4$			
1-[3-(1-Pyrrolidinyl)allylidene]pyrrolidinium perchlorate	EtOH	320(4.70)	24-2609-69
$C_{11}H_{19}ClN_2O_6$			
4-(3-Morpholinoallylidene)morpholinium perchlorate	EtOH	319(4.79)	24-2609-69
$C_{11}H_{19}ClN_4O_6$			
7-Ethyl-1,2,3,6,8,9-hexahydro-1,3,7,9-tetramethyl-2,6-dioxopurinium perchlorate	H_2O	277(4.19)	104-2004-69
$C_{11}H_{19}NO_2$			
2(3H)-Furanone, dihydro-3-[1-(pentylamino)ethylidene]-	hexane	296(4.14)	104-0998-69
$C_{11}H_{19}NO_3$			
3-Piperidinecarboxylic acid, 1-butyl-4-oxo-, methyl ester	MeOH	252(3.73)	103-0062-69
	EtOH	252(3.76)	103-0062-69
	iso-BuOH	252(3.78)	103-0062-69
hydrochloride	H_2O	245(3.59)	103-0062-69
	MeOH	245(3.98)(changing)	103-0062-69
	EtOH	245(4.00)(changing)	103-0062-69
	iso-BuOH	245(3.98)(changing)	103-0062-69
$C_{11}H_{19}NO_4$			
Malonamic acid, 2-acetyl-N-tert-butyl-, ethyl ester	EtOH	225(2.86),262(2.93)	104-1713-69
$C_{11}H_{19}NS_2$			
1(2H)-Quinolinecarbodithioic acid, octahydro-, methyl ester, trans (8aS)	dioxan	252(3.92),278(4.03)	78-0737-69
$C_{11}H_{19}N_3O_2$			
2-Butanone, 3-(2-ethyl-4,5-dimethylimidazol-1-yl)-, oxime, N-oxide	MeOH	222(3.77)	48-0746-69
$C_{11}H_{19}N_5OS$			
5H-s-Triazolo[3,4-b][1,3,4]thiadiazine-5-carboxamide, N-butyl-3-ethyl-6,7-dihydro-	EtOH	223(3.98),242s(3.71)	2-0959-69

Compound	Solvent	λ_{max}(log ϵ)	Ref.
$C_{11}H_{19}N_7$			
1H-v-Triazolo[4,5-d]pyridazine, 4-[[3-(diethylamino)propyl]amino]-	EtOH	202(4.52),225s(4.30), 278(3.93)	4-0093-69
$C_{11}H_{20}N_2O$			
2H-Pyrrol-3-ol, 5-(diethylamino)-2,2,4-trimethyl-	MeOH	226(4.00),287(4.35)	78-5721-69
$C_{11}H_{20}N_2O_2$			
Pyrazolidine, 1,2-diisobutyryl-	$C_2H_3F_3O$	202(3.49),230(2.65)	35-1256-69
1-Pyrazoline-4-carboxylic acid, 5-tert-butyl-3,3-dimethyl-, methyl ester	EtOH	326(2.41)	39-0787-69C
$C_{11}H_{20}N_2O_3$			
Piperidine-4-acetic acid, N-nitroso-2,2-dimethyl-, ethyl ester	dioxan	233(3.84),370(1.92)	78-0737-69
$C_{11}H_{20}N_2S$			
Thiazole, 2-(dimethylamino)-5-isopropyl-4-propyl-	n.s.g.	268(4.10)	5-0121-69A
$C_{11}H_{20}N_4O_2$			
Uracil, 6-(dimethylamino)-5-(ethylmethylamino)-1,3-dimethyl-	EtOH	292(3.97)	104-2004-69
$C_{11}H_{20}O$			
3-Cyclohexene-1-methanol, α,α,2,4-tetramethyl-	EtOH	206(3.22)	16-0226-69
$C_{11}H_{20}O_2$			
1-Cyclohexanol, 3-acetyl-2,2,4-trimethyl-	EtOH	209(2.72)	16-0226-69
3-Nonen-2-one, 8-hydroxy-4,8-dimethyl-	EtOH	238(3.98)	16-0226-69
$C_{11}H_{20}O_3$			
Butyric acid, 2-isopropyl-3-oxo-, ethyl ester	EtOH	285(1.88)	78-5443-69
Crotonic acid, α-ethyl-β-isopropoxy-, ethyl ester	EtOH	250(3.95)	78-5443-69
$C_{11}H_{20}Sn$			
Stannane, triallylethyl-	hexane	210(4.44)	44-3715-69
	MeCN	217(4.47)	44-3715-69
$C_{11}H_{21}N_5O$			
s-Triazine, 2-(diethylamino)-4-(isopropylamino)-6-methoxy-	pH 1.0	217(4.31),243(4.27)	59-1167-69
	pH 10.0	223(4.52)	59-1167-69
$C_{11}H_{21}N_5S$			
s-Triazine, 2-(diethylamino)-4-(isopropylamino)-6-(methylthio)-	pH 1.0	234(4.39),261s(4.19)	59-1167-69
	pH 10.0	230(4.56),271s(3.58)	59-1167-69
$C_{11}H_{23}NS_3$			
Peroxycarbamic acid, dibutyltrithio-, ethyl ester	EtOH	282(3.91)	34-0119-69
Peroxycarbamic acid, dipropyltrithio-, butyl ester	EtOH	282(3.87)	34-0119-69
Peroxycarbamic acid, dipropyltrithio-, tert-butyl ester	EtOH	243(3.86),284(3.87)	34-0119-69

Compound	Solvent	$\lambda_{max}(\log \epsilon)$	Ref.
$C_{11}H_{24}O_2Si$			
3-Heptanone, 5-methyl-5-(trimethylsil-oxy)-	EtOH	239(1.83),284(1.50)	44-2324-69
$C_{11}H_{30}OSi_3$			
Trisilane, 1-butoxyheptamethyl-	isooctane	220(3.84)	35-6613-69
$C_{11}H_{31}NSi_3$			
Trisilane, 1-(tert-butylamino)hepta-methyl-	isooctane	210s(3.94),238s(3.47)	35-6613-69
	MeCN	210s(3.89),238(3.45)	35-6613-69
	dioxan	238(3.53)	35-6613-69
Trisilane, 2-(tert-butylamino)hepta-methyl-	isooctane	213s(3.85),243s(3.35)	35-6613-69
	dioxan	240s(3.39)	35-6613-69
$C_{11}H_{34}Si_5$			
Pentasilane, 1H-undecamethyl-	hexane	251(4.28)	77-0004-69

Compound	Solvent	$\lambda_{max}(\log \epsilon)$	Ref.
$C_{12}BrF_9$			
Biphenyl, 2-bromononafluoro-	EtOH	230(3.94),269(3.31)	25-0695-69
$C_{12}Br_2F_8$			
Biphenyl, 2,2'-dibromooctafluoro-	EtOH	218(4.25),273(3.46)	25-0695-69
$C_{12}Br_3F_7$			
Biphenyl, 2,2',3-tribromoheptafluoro-	EtOH	222(4.31),272(3.22)	25-0695-69
Biphenyl, 2,2',6-tribromoheptafluoro-	EtOH	220(4.28),273(3.27)	25-0695-69
$C_{12}Cl_8O_4$			
Spiro[1,3-benzodioxole-2,1'-[5]cyclohex-ene]-3',4'-dione, 2,2',4,5,5',6,6',7-octachloro-	EtOH	234(4.25),295(3.40), 302(3.45)	44-1966-69
$C_{12}F_9I$			
Biphenyl, nonafluoro-2-iodo-	EtOH	225(4.01),270(3.38)	25-0695-69
$C_{12}F_{10}$			
Biphenyl, decafluoro-	EtOH	231(4.13),268(3.38)	25-0695-69
$C_{12}F_{18}$			
Benzene, hexakis(trifluoromethyl)-	n.s.g.	212(4.01),283(2.15)	77-0202-69
Bicyclo[2.2.0]hexadiene, hexakis(tri-fluoromethyl)-	isooctane	203(3.22)	35-3373-69
	gas	203(--)	35-3373-69
Bicyclo[2.2.0]hexa-2,5-diene, 1,2,3,4,5,6-hexakis(trifluoromethyl)-	n.s.g.	211(2.79)	77-0202-69
Tricyclo[3.1.0.0^{2,6}]hex-3-ene, 1,2,3,4,5,6-hexakis(trifluoromethyl)-	hexane	221(3.26)	77-0202-69
	isooctane	221(3.35)	35-3373-69
$C_{12}HF_9$			
Biphenyl, 2,2',3,3',4,4',5,5',6-nona-fluoro-	EtOH	230(4.03),267(3.34)	25-0695-69
$C_{12}H_2Cl_{12}$			
1,3,5,7,9,11-Dodecahexaene, 1,2,3,4,5,-6,7,8,9,10,11,12-dodecachloro-	hexane	214(4.58),252(4.16)	78-0883-69
$C_{12}H_2F_8$			
Biphenyl, 2,2',3,3',4,4',5,5'-octa-fluoro-	EtOH	234(4.09),268(3.45)	25-0695-69
$C_{12}H_4Cl_2N_4O_8$			
Biphenyl, 4,4'-dichloro-2,2',6,6'-tetra-nitro-	EtOH	207(4.79),240(4.56)	4-0533-69
$C_{12}H_4Cl_6N_2$			
Azobenzene, 2,2',4,4',5,5'-hexachloro-	MeOH	251(4.16),348(4.12)	64-0997-69
Azobenzene, 2,2',4,4',6,6'-hexachloro-	MeOH	258(4.14),293(4.10)	64-0997-69
$C_{12}H_4F_5NO_2$			
Biphenyl, 2,3,4,5,6-pentafluoro-2'-nitro-	EtOH	273(4.37)	39-2747-69C
$C_{12}H_4N_2O_2$			
2-Indanylidenemalononitrile, 1,3-dioxo-	CH_2Cl_2	261(4.44),271(4.46), 350(3.82),470(1.40)	39-0725-69B
$C_{12}H_5Cl_4NO$			
Pyridine, 4-benzoyl-2,3,5,6-tetrachloro-	C_6H_{12}	233(4.20),260(4.20), 293(3.83)	117-0187-69

Compound	Solvent	$\lambda_{max}(\log \epsilon)$	Ref.
$C_{12}H_5F_4N$			
Carbazole, 1,2,3,4-tetrafluoro-	EtOH	228(4.68),280(4.04), 316(3.53),328(3.54)	70-0609-69
$C_{12}H_5N_5O_9$			
Phenoxazine, 1,3,7,9-tetranitro-	EtOH-HCl	268(4.30),297s(--), 339(4.44),380s(--), 480(4.02)	73-3732-69
	EtOH-KOH	255(3.98),309(3.96), 475s(--),618(4.35)	73-3732-69
$C_{12}H_6Br_2F_4$			
Benzo[3,4]tricyclo[3.2.1.0^{2,7}]oct-3-ene, 6,8-dibromotetrafluoro-, trans	EtOH	270(2.91)	65-2332-69
$C_{12}H_6Cl_4N_2$			
Azobenzene, 2,2',3,3'-tetrachloro-	MeOH	241(4.16),328(4.17)	64-0997-69
Azobenzene, 2,2',4,4'-tetrachloro-	MeOH	246(4.20),345(4.27)	64-0997-69
Azobenzene, 2,2',5,5'-tetrachloro-	MeOH	242(.18),313(4.08)	64-0997-69
Azobenzene, 2,2',6,6'-tetrachloro-	MeOH	245(4.03),292(4.02)	64-0997-69
Azobenzene, 3,3',4,4'-tetrachloro-	MeOH	241(4.29),232(4.39)	64-0997-69
Azobenzene, 3,3',5,5'-tetrachloro-	MeOH	236(4.29),318(4.29)	64-0997-69
$C_{12}H_6Cl_6N_4O_2S$			
s-Triazine, 2-[(p-nitrobenzyl)thio]-4,6-bis(trichloromethyl)-	MeOH	295(4.31)	18-2931-69
$C_{12}H_6Cl_6N_4O_3S$			
s-Triazine, 2-[(2-methoxy-4-nitrophenyl)thio]-4,6-bis(trichloromethyl)-	MeOH	286(4.00)	18-2931-69
$C_{12}H_6Cl_7N_3S$			
s-Triazine, 2-[(p-chlorobenzyl)thio]-4,6-bis(trichloromethyl)-	MeOH	232(3.56),279(4.18)	18-2931-69
s-Triazine, 2-[(4-chloro-m-tolyl)thio]-4,6-bis(trichloromethyl)-	MeOH	270(4.03)	18-2931-69
$C_{12}H_6F_4$			
Biphenyl, 2,3,4,5-tetrafluoro-	C_6H_{12}	241(4.14)	39-0211-69C
$C_{12}H_6N_2O_2S$			
Benzo[b]thiophene-2,3-dicarbonitrile, 4,7-dihydro-5,6-dimethyl-4,7-dioxo-	$CHCl_3$	258(4.64),276(4.69), 334s(3.90),378(3.85)	24-2378-69
$C_{12}H_6N_2O_2S_2$			
1,4-Benzodithiin-2,3-dicarbonitrile, 5,8-dihydro-6,7-dimethyl-5,8-dioxo-	$CHCl_3$	245(4.50),328(4.08), 384s(3.90)	24-2378-69
$C_{12}H_6N_4O_7$			
Phenoxazine, 1,3,7-trinitro-	EtOH-HCl	238(4.24),278(4.36), 355(3.67),465(4.24)	73-3732-69
	EtOH-KOH	258(4.17),313(4.19), 420(3.59),617(4.48)	73-3732-69
Phenoxazine, 1,3,9-trinitro-	EtOH-HCl	239(4.23),273(4.33), 478(4.09)	73-3732-69
	EtOH-KOH	221(4.52),267(4.15), 300s(--),577(4.26)	73-3732-69
$C_{12}H_6O_4$			
2H,8H-Benzo[1,2-b:3,4-b']dipyran-2,8-dione	MeCN	278s(4.3),292(4.4)	39-0029-69C

Compound	Solvent	$\lambda_{max}(\log \epsilon)$	Ref.
2H,8H-Benzo[1,2-b:5,4-b']dipyran-2,8-dione	MeCN	274(5.1),338(4.8), 355(4.7)	39-0029-69C
2,3-Biphenylenedione, 6,7-dihydroxy-	EtOH at pH 0 and 2	264(4.56),288s(4.31), 442(4.46)	39-2579-69C
	EtOH at pH 10 and 12	274s(4.45),285(4.74), 298(4.83),353(4.05), 372(4.15),492(4.80)	39-2579-69C
$C_{12}H_6O_5$			
2H,5H-Pyrano[3,2-c][1]benzopyran-2,5-dione, 4-hydroxy-	EtOH	261(4.19),339(4.03)	18-3233-69
$C_{12}H_6O_5S_2$			
1,4-Benzodithiin-2,3-dicarboxylic anhydride, 5,8-dihydro-6,7-dimethyl-5,8-dioxo-	THF	248(3.22),324(2.81), 433(2.58)	24-2378-69
$C_{12}H_7BrF_4$			
Tetrafluorobenzo[2,3]bicyclo[2.2.2]octa-2,7-diene, 6-bromo-	EtOH	264(2.69)	65-2326-69
Tetrafluorobenzo[3,4]bicyclo[3.2.1]octa-3,6-diene, 8-bromo-	EtOH	262(2.57)	65-2326-69
$C_{12}H_7ClFO_2PS$			
Phenothiaphosphine, 2-chloro-8-fluoro-10-hydroxy-, 10-oxide	EtOH	219(4.34),256s(3.81), 269(4.13),288s(3.70), 316(3.50)	78-3919-69
$C_{12}H_7Cl_6N_3O$			
s-Triazine, 2-(p-methoxyphenyl)-4,6-bis(trichloromethyl)-	MeOH	240(3.90),328(4.35)	18-2924-69
$C_{12}H_7Cl_6N_3OS$			
s-Triazine, 2-[(o-methoxyphenyl)thio]-4,6-bis(trichloromethyl)-	MeOH	272(4.04)	18-2931-69
s-Triazine, 2-[(p-methoxyphenyl)thio]-4,6-bis(trichloromethyl)-	MeOH	271(4.04)	18-2931-69
$C_{12}H_7Cl_6N_3S$			
s-Triazine, 2-(benzylthio)-4,6-bis-(trichloromethyl)-	MeOH	231(3.38),277(4.14)	18-2931-69
s-Triazine, 2-(m-tolylthio)-4,6-bis-(trichloromethyl)-	MeOH	270(3.99)	18-2931-69
s-Triazine, 2-(o-tolylthio)-4,6-bis-(trichloromethyl)-	MeOH	270(3.99)	18-2931-69
s-Triazine, 2-(p-tolylthio)-4,6-bis-(trichloromethyl)-	MeOH	269(4.01)	18-2931-69
$C_{12}H_7NO$			
1H-Benz[de]isoquinolin-1-one, hydrochloride	H_2SO_4	234(4.28),310(3.35), 419(4.11),440(4.11)	4-0681-69
$C_{12}H_7NO_2$			
3H-Phenoxazin-3-one	EtOH	445(4.04)	73-0221-69
$C_{12}H_7NO_3$			
Naphth[2,1-d]oxazole-4,5-dione, 2-methyl-	MeOH	207(4.33),257(4.50), 264(4.51),33?(3.23), 417(3.26)	77-0923-69
3H-Phenoxazin-3-one, 2-hydroxy-	EtOH	404(4.17)	73-0221-69

Compound	Solvent	$\lambda_{max}(\log \epsilon)$	Ref.
3H-Phenoxazin-3-one, 2-hydroxy-, anion	EtOH	434(4.08)	73-0221-69
3H-Phenoxazin-3-one, 7-hydroxy-	EtOH	476(4.00)	73-0221-69
anion	EtOH	573(4.90)	73-0221-69
$C_{12}H_7NO_4$			
3H-Phenoxazin-3-one, 7-hydroxy-, 10-oxide	EtOH	517(4.28)	73-0221-69
anion	EtOH	615(4.60)	73-0221-69
$C_{12}H_7NO_5$			
3H-Phenoxazin-3-one, 1,7,9-trihydroxy-	EtOH	468(4.28)	73-0221-69
trianion	EtOH	528(4.41)	73-0221-69
$C_{12}H_7N_3OS$			
4H-Quinolizine-1,3-dicarbonitrile, 2-(methylthio)-4-oxo-	EtOH	269(4.29),288(4.09), 313(4.09),357(3.81), 416(4.31)	95-0203-69
$C_{12}H_7N_3O_5$			
Phenoxazine, 1,3-dinitro-	EtOH-HCl	253(4.18),283(4.18), 450(4.05)	73-3732-69
	EtOH-KOH	231(4.43),296(4.09), 495(4.27),520(4.25), 595(4.06)	73-3732-69
Phenoxazine, 3,7-dinitro-	EtOH-HCl	285(4.29),380(3.76), 465(4.23)	73-3732-69
	EtOH-KOH	222(4.36),262(3.88), 330(4.12),460(3.87), 680(4.59)	73-3732-69
$C_{12}H_8$			
Acenaphthylene	dioxan	264(3.42),275(3.38), 312s(3.94),324(4.04), 340(3.64),412(2.28)	24-3599-69
Naphthalene, 1-ethynyl-	MeOH	235(4.18),285(3.84), 296(3.96),307(3.79)	39-2173-69C
$C_{12}H_8BrClOS$			
Phenol, p-[(3-bromo-2-chlorophenyl)-thio]-	EtOH	222(4.42),252(4.15)	104-1903-69
Phenol, p-[(4-bromo-2-chlorophenyl)-thio]-	EtOH	230(4.18),256(4.22)	104-1903-69
$C_{12}H_8BrClO_2S$			
Phenol, p-[(3-bromo-2-chlorophenyl)-sulfinyl]-	EtOH	252(4.25)	104-1903-69
Phenol, p-[(4-bromo-2-chlorophenyl)-sulfinyl]-	EtOH	256(4.37)	104-1903-69
$C_{12}H_8BrClO_3S$			
Phenol, p-[(3-bromo-2-chlorophenyl)-sulfonyl]-	EtOH	236(4.28),264(4.16)	104-1903-69
Phenol, p-[(4-bromo-2-chlorophenyl)-sulfonyl]-	EtOH	236(4.16),266(4.19)	104-1903-69
$C_{12}H_8BrN_3O_2S$			
Benzimidazole, 2-(5-bromo-2-thienyl)-1-methyl-5-nitro-	EtOH	218s(4.12),246(4.22), 295(4.40),330s(4.27)	73-0572-69

Compound	Solvent	λ_{max}(log ε)	Ref.
$C_{12}H_8BrN_3O_3$			
Benzimidazole, 2-(5-bromo-2-furyl)-1-methyl-5-nitro-	EtOH	248s(3.95),289(4.46), 335(4.24)	73-0572-69
$C_{12}H_8Br_2N_2$			
Azobenzene, 4,4'-dibromo-	EtOH	326(4.39)	35-2325-69
$C_{12}H_8Br_2O_4S_2$			
Disulfone, bis(p-bromophenyl)-	dioxan	267(4.50)	35-5510-69
$C_{12}H_8ClIOS$			
Phenol, p-[(2-chloro-3-iodophenyl)-thio]-	EtOH	230(4.45)	104-1903-69
Phenol, p-[(2-chloro-4-iodophenyl)-thio]-	EtOH	228(4.27),264(4.36)	104-1903-69
$C_{12}H_8ClIO_2S$			
Phenol, p-[(2-chloro-3-iodophenyl)-sulfinyl]-	EtOH	244(4.43)	104-1903-69
Phenol, p-[(2-chloro-4-iodophenyl)-sulfinyl]-	EtOH	260(4.40)	104-1903-69
$C_{12}H_8ClIO_3S$			
Phenol, p-[(2-chloro-3-iodophenyl)-sulfonyl]-	EtOH	246(4.41)	104-1903-69
Phenol, p-[(2-chloro-4-iodophenyl)-sulfonyl]-	EtOH	272(4.38)	104-1903-69
$C_{12}H_8ClNO$			
1H-Benz[de]isoquinolin-1-one, hydro-chloride	H_2SO_4	234(4.28),310(3.35), 419(4.11),440(4.11)	4-0681-69
$C_{12}H_8ClNO_2S_2$			
p-Dithiin-2,3-dicarboximide, N-(o-chlo-rophenyl)-5,6-dihydro-	MeOH	238(3.94),260(3.95), 411(3.48)	33-2221-69
p-Dithiin-2,3-dicarboximide, N-(p-chlo-rophenyl)-5,6-dihydro-	MeOH	219(4.05),244(4.26), 416(3.40)	33-2221-69
$C_{12}H_8ClNO_3S$			
Phenol, p-[(2-chloro-3-nitrophenyl)-thio]-	EtOH	236(4.31),260(4.26)	104-1903-69
Phenol, p-[(2-chloro-4-nitrophenyl)-thio]-	EtOH	234(4.29),342(4.08)	104-1903-69
$C_{12}H_8ClNO_4S$			
Phenol, p-[(2-chloro-3-nitrophenyl)-sulfinyl]-	EtOH	254(4.51)	104-1903-69
Phenol, p-[(2-chloro-4-nitrophenyl)-sulfinyl]-	EtOH	248(4.20),278(4.07)	104-1903-69
$C_{12}H_8ClNO_5S$			
Phenol, p-[(2-chloro-3-nitrophenyl)-sulfonyl]-	EtOH	266(4.31)	104-1903-69
Phenol, p-[(2-chloro-4-nitrophenyl)-sulfonyl]-	EtOH	244(4.26),298(3.92)	104-1903-69
$C_{12}H_8ClNS_2$			
6-Phenyl-1,2,3-benzodithiazol-2-ium chloride	H_2SO_4	278(4.29),318(3.52), 435(3.92)	104-0151-69

Compound	Solvent	λ_{max}(log ε)	Ref.
$C_{12}H_8ClN_3O_2S$			
Sulfur diimide, (m-chlorophenyl)-(p-nitrophenyl)-	benzene	425(4.12)	104-0459-69
Sulfur diimide, (p-chlorophenyl)-(p-nitrophenyl)-	benzene	430(4.16)	104-0459-69
$C_{12}H_8ClN_3O_3$			
Benzimidazole, 2-(5-chloro-2-furyl)-1-methyl-5-nitro-	EtOH	243s(3.88),291(4.43), 333(4.16)	73-0572-69
Phenol, 4-chloro-2-[(m-nitrophenyl)azo]-	HCl	392(3.90)	73-3740-69
(in 50% EtOH)	NaOH	494(4.05)	73-3740-69
Phenol, 4-chloro-2-[(p-nitrophenyl)azo]-	HCl	404(3.98)	73-3740-69
(in 50% EtOH)	NaOH	530(4.16)	73-3740-69
$C_{12}H_8ClN_5O_2$			
4H-Pyridazino[6,1-c]-as-triazine, 7-chloro-3-(m-nitrophenyl)-,	pH 1	227(4.25),295(4.13), 350s(--)	7-0552-69
hydrochloride	pH 13	238(4.33),343(4.04), 410s(--)	7-0552-69
$C_{12}H_8Cl_2N_2$			
Azobenzene, 2,2'-dichloro-	MeOH	237(4.10),327(4.09)	64-0997-69
Azobenzene, 3,3'-dichloro-	MeOH	233(4.15),317(4.16)	64-0997-69
Azobenzene, 4,4'-dichloro-	MeOH	234(4.08),327(4.11)	64-0997-69
$C_{12}H_8Cl_2N_2O$			
Phenol, 4-chloro-2-[(m-chlorophenyl)-	HCl	388(3.91)	73-3740-69
azo]- (in 50% EtOH)	NaOH	482(4.05)	73-3740-69
Phenol, 4-chloro-2-[(p-chlorophenyl)-	HCl	388(3.99)	73-3740-69
azo]- (in 50% EtOH)	NaOH	480(4.09)	73-3740-69
$C_{12}H_8Cl_2N_2O_2$			
Phthalimide, N-(2,3-dichloro-5-azabicyclo[2.1.0]pent-5-yl)-	MeCN	231(3.4),295(2.15), 305(2.11)	4-0987-69
$C_{12}H_8Cl_2N_2S$			
Sulfur diimide, bis(m-chlorophenyl)-	hexane	413(4.3)	104-0459-69
Sulfur diimide, bis(p-chlorophenyl)-	benzene	432(4.09)	104-0459-69
$C_{12}H_8Cl_2OS$			
Phenol, p-[(2,3-dichlorophenyl)thio]-	EtOH	250(4.18)	104-1903-69
Phenol, p-[(2,4-dichlorophenyl)thio]-	EtOH	232(4.15),254(4.20)	104-1903-69
$C_{12}H_8Cl_2O_2S$			
Phenol, p-[(2,3-dichlorophenyl)sulfinyl]-	EtOH	252(4.19)	104-1903-69
Phenol, p-[(2,4-dichlorophenyl)sulfinyl]-	EtOH	252(4.19)	104-1903-69
$C_{12}H_8Cl_2O_3S$			
Phenol, p-[(2,3-dichlorophenyl)sulfonyl]-	EtOH	232(4.19),266(4.13)	104-1903-69
Phenol, p-[(2,4-dichlorophenyl)sulfonyl]-	EtOH	230(4.16),266(4.20)	104-1903-69
$C_{12}H_8Cl_2O_4S_2$			
Disulfone, bis(p-chlorophenyl)	dioxan	263(4.52)	35-5510-69
$C_{12}H_8Cl_2S$			
Thiophene, 2-(3,4-dichlorostyryl)-	MeOH	240(3.92),248s(3.81), 322(4.47)	117-0209-69

Compound	Solvent	$\lambda_{max}(\log \epsilon)$	Ref.
$C_{12}H_8Cl_4O$			
4H-Pyran, 2,6-dimethyl-4-(2,3,4,5-tetrachloro-2,4-cyclopentadien-1-ylidene)-	CH_2Cl_2	267(3.94),425(4.59)	83-0886-69
$C_{12}H_8Cl_4O_2S_2$			
4H,6H,10H,12H-Dithieno[3,4-c:3',4'-h]-[1,6]dioxecin, 1,3,7,9-tetrachloro-	CH_2Cl_2	246(4.16)	44-0340-69
$C_{12}H_8Cl_5NO_2$			
Indole-3-carboxylic acid, 4,5,6,7-tetrachloro-2-(chloromethyl)-, ethyl ester	MeOH	237(4.87),302(4.13)	24-2684-69
$C_{12}H_8FN_3O_2S$			
Sulfur diimide, (p-fluorophenyl)-(p-nitrophenyl)-	benzene	426(4.24)	104-0459-69
$C_{12}H_8F_4$			
1,2,4-Methenonaphthalene, 5,6,7,8-tetrafluoro-1,2,3,4-tetrahydro-	hexane	210(3.79),262(2.48)	88-3273-69
Naphthalene, 1,2,3,4-tetrafluoro-5,7-dimethyl-	EtOH	278(3.74),308(3.04),322(2.98)	70-0131-69
Naphthalene, 1,2,3,4-tetrafluoro-6,7-dimethyl-	EtOH	270(3.61),278(3.63)	70-0131-69
$C_{12}H_8F_6N_2O_3$			
Pyridine, 1,4-dihydro-4-[4-methyl-5-oxo-2-(trifluoromethyl)-3-oxazolin-2-yl]-1-(trifluoroacetyl)-	MeCN	264(4.01)	24-1129-69
$C_{12}H_8IN_3O_3$			
Benzimidazole, 2-(5-iodo-2-furyl)-1-methyl-5-nitro-	EtOH	250s(4.01),293(4.42),339(4.26)	73-0572-69
$C_{12}H_8N_2$			
Malononitrile, cinnamylidene-	EtOH	235s(--),241(3.87),248(3.87),350(4.52)	23-4076-69
$C_{12}H_8N_2O_2$			
Benzo[c]cinnoline, 5,6-dioxide	EtOH	214(4.16),221(4.12),240(4.48),250s(3.52),258(4.58),269s(4.34),291(4.12),304(4.12),345(3.06)	4-0523-69
Benzo[c]cinnoline-3,8-diol	pH 1	206(4.32),222s(4.23),272(4.75),350(3.97),490(3.40)	4-0523-69
	pH 13	230s(4.46),255s(4.24),280(4.92),300s(4.39),497(3.51)	4-0523-69
	EtOH	202(4.49),219(4.11),250(4.82),277s(4.16),295s(3.92),317s(3.68),420(3.51)	4-0523-69
1H-Naphth[2,3-d]imidazole-4,9-dione, 2-methyl-	pH 1	245(4.49),249s(--),276(4.22)	4-0909-69
	pH 7	247(4.61),283(4.20)	4-0909-69
	pH 13	261(4.57)	4-0909-69
	MeOH	244(4.58),277(4.18),330(3.46)	4-0909-69
3H-Phenoxazin-3-one, 2-amino-	EtOH	464(4.30)	73-0221-69

Compound	Solvent	$\lambda_{max}(\log \epsilon)$	Ref.
3H-Phenoxazin-3-one, 7-amino-	EtOH	520(4.09),545(4.18)	73-0221-69
$C_{12}H_8N_2O_2S_2$			
1,4-Benzodithiin-2,3-dicarbonitrile, 5,8-dihydroxy-6,7-dimethyl-	MeOH	215(4.68),259(4.45), 310s(4.11),402(4.08)	24-2378-69
$C_{12}H_8N_2O_3$			
Benzo[c]cinnoline-3,8-diol, 5-oxide	pH 13	213s(4.45),243s(4.34), 286(4.78),323(4.16), 356s(3.73),521(3.57)	4-0523-69
	EtOH	235s(4.34),252s(4.65), 260(4.66),301s(4.16), 327(3.95),357s(3.64), 430(3.64)	4-0523-69
Phenoxazine, 1-nitro-	EtOH-HCl	232(4.68),261(3.80), 297(3.88),488(3.75)	73-3732-69
Phenoxazine, 3-nitro-	EtOH-HCl	226(4.41),280(4.09), 455(4.06)	73-3732-69
	EtOH-KOH	232(4.30),277(3.91), 325(3.89),570(4.24), 605(4.26),655s(--)	73-3732-69
3H-Phenoxazin-3-one, 7-amino-1-hydroxy-	EtOH	430(4.45)	73-0221-69
anion	EtOH	480(4.36)	73-0221-69
$C_{12}H_8N_2O_4$			
Benzo[c]cinnoline-3,8-diol, 5,6-dioxide	pH 13	214(4.66),256s(4.67), 275s(4.80),297(4.77), 358(3.77),529(3.59)	4-0523-69
	EtOH	221s(4.33),243s(4.38), 255s(4.53),267(4.64), 286(4.48),329(3.98), 348(3.93),431(3.59)	4-0523-69
Benzoxazole, 5-methyl-2-(5-nitro-2-furyl)-	EtOH	230(4.05),269(4.11), 356(4.40)	18-3335-69
2(1H)-Pyridone, 1,1'-oxalyldi-	$CHCl_3$	330(3.57)	44-0836-69
$C_{12}H_8N_2O_4S$			
Sulfide, 2,4-dinitrophenyl phenyl	MeOH	215(4.29),242s(--), 265s(--),328(4.00)	44-2038-69
Thiophene, 2-(2,4-dinitrostyryl)-	MeOH	385(4.34)	117-0209-69
$C_{12}H_8N_2O_4S_2$			
p-Dithiin-2,3-dicarboximide, 5,6-dihydro-N-(m-nitrophenyl)-	MeOH	249(4.36),415(3.43)	33-2221-69
p-Dithiin-2,3-dicarboximide, 5,6-dihydro-N-(p-nitrophenyl)-	MeOH	272(4.25),413(3.55)	33-2221-69
$C_{12}H_8N_2O_6$			
4,4'-Biphenyldiol, 2,2'-dinitro-	EtOH	218(4.45),235s(4.38), 273s(3.85),340s(3.64)	4-0523-69
$C_{12}H_8N_2S_2$			
2H-Thiopyran-5-carbonitrile, 6-amino-4-phenyl-2-thioxo-	EtOH	248(4.24),299(4.25), 325s(4.19),469(4.27)	78-1441-69
$C_{12}H_8N_4$			
Pteridine, 2-phenyl-	hexane	253(4.36),265(4.27), 278(4.26),325(4.01), 339(4.15),355(4.06)	39-1408-69C
	pH 2.0	253(4.42),311(4.14)	39-1408-69C

Compound	Solvent	$\lambda_{max}(\log \epsilon)$	Ref.
Pteridine, 2-phenyl- (cont.)	pH 6.0	247(4.31),264s(4.20), 335(4.05)	39-1408-69C
Pteridine, 4-phenyl-	hexane	231(4.19),277s(3.63), 288s(3.81),296(3.91), 326(4.10)	39-1883-69C
	pH 1.9	226(4.20),325(4.05)	39-1883-69C
	pH 6.1	228(4.12),301s(3.89), 323(4.02)	39-1883-69C
$C_{12}H_8N_4O$			
4-Pteridinol, 2-phenyl-	pH 5.0	243(4.14),285(4.15), 302s(4.10)	39-1408-69C
	pH 10.0	226(4.05),268(4.41), 341(3.93)	39-1408-69C
$C_{12}H_8N_4O_4S$			
Sulfur diimide, bis(m-nitrophenyl)-	benzene	415(4.33)	104-0459-69
Sulfur diimide, bis(p-nitrophenyl)-	benzene	431(4.20)	104-0459-69
Sulfur diimide, (m-nitrophenyl)-(p-nitrophenyl)-	benzene	423(4.23)	104-0459-69
$C_{12}H_8N_4O_7$			
1-(Hydroxypicryl)-3-methylpyridinium hydroxide, inner salt	MeOH	362(4.1)	56-1219-69
1-(Hydroxypicryl)-4-methylpyridinium hydroxide, inner salt	MeOH	230(4.3),360(4.1)	56-1219-69
$C_{12}H_8N_4O_9S$			
Benzenesulfonic acid, o-(2,4,6-trinitro-anilino)-, anion	base	382(4.19)	22-4569-69
dianion	base	440(4.34)	22-4569-69
Metanilic acid, N-picryl-, anion	base	372(4.14)	22-4569-69
dianion	base	435(4.33)	22-4569-69
$C_{12}H_8N_6O_8$			
Benzidine, 2,2',6,6'-tetranitro-	EtOH	238(4.96),305s(--)	4-0533-69
$C_{12}H_8O$			
Naphtho[2,3-b]furan	EtOH	236(4.95),296(4.06), 310(4.18),320(4.07), 336(3.95)	2-0536-69
$C_{12}H_8O_3$			
Bicyclo[4.2.2]deca-2,4,7,9-tetraene-3,4-dicarboxylic anhydride	MeOH	255(3.95),310s(3.14)	88-1125-69
Furo[3,2-g]coumarin, 6-methyl-	EtOH	246(4.38),292(4.08), 325(3.95)	39-1749-69C
Spiro[1,3-benzodioxole-2,1'-cyclohexa-dien]-4'-one	C_6H_{12}	217(4.30),235s(3.73), 270s(3.60),274(3.66), 279(3.67),285(3.54)	39-1982-69C
Tricyclo[3.3.2.0^{2,8}]deca-3,6,9-triene-3,4-dicarboxylic anhydride	MeOH	248s(3.92),255(3.97), 264s(3.86),306s(3.26)	108-0435-69
$C_{12}H_8O_4$			
Isobergapten	EtOH	223(4.2),250(4.2), 308(3.9)	2-0115-69
Xanthotoxin	EtOH	218(4.44),249(4.44), 299(4.08)	88-5223-69

Compound	Solvent	$\lambda_{max}(\log \epsilon)$	Ref.
$C_{12}H_8O_5$			
7H-Furo[3,2-g][1]benzopyran-7-one, 4-hydroxy-9-methoxy-	EtOH	224(4.22),241s(3.94), 248s(3.87),268s(4.00), 275(4.10),297s(3.80), 316(3.87),328s(3.81)	36-0675-69
	EtOH-NaOH	232(4.20),292(4.17), 324(3.73),332s(3.70), 337s(3.67)	36-0675-69
$C_{12}H_9BF_2O_2$			
Naphtho[1,2-e]-1,3,2-dioxaborin, 2,2-difluoro-4-methyl-	MeCN	205(4.37),218(4.70), 242s(--),282(3.78), 350(4.00),395s(--)	4-0029-69
Naphtho[2,1-e]-1,3,2-dioxaborin, 2,2-difluoro-4-methyl-	MeCN	215(4.55),264(4.75), 298(4.16),310(4.22), 410(3.62)	4-0029-69
$C_{12}H_9BO_2$			
2-Biphenyleneboronic acid	EtOH	247(4.64),254(4.85), 342(3.59),346(3.59), 360(3.76)	39-2789-69C
$C_{12}H_9BrN_2OS$			
6-Phenylthiazolo[3,2-a]pyrazin-4-ium bromide, 7-oxide	n.s.g.	199(4.38),254(4.14), 340s(4.16),350(4.19)	39-2270-69C
$C_{12}H_9BrN_2S$			
6-Phenylthiazolo[3,2-a]pyrazin-4-ium bromide	n.s.g.	200(4.39),218s(4.14), 250s(4.2),269(4.29), 350(4.1)	39-2270-69C
8-Phenylthiazolo[3,2-a]pyrazin-4-ium bromide	n.s.g.	198(4.41),245(4.05), 279(3.97),335(3.99)	39-2270-69C
$C_{12}H_9BrN_4S$			
7H-Thiazolo[3,2-c]pyrimidine, 5-amino-3-(p-bromophenyl)-7-imino-, sulfate	pH 1	230(4.49),328(4.25)	87-0227-69
$C_{12}H_9BrOS$			
Sulfoxide, p-bromophenyl phenyl	H_2O	243(4.28)	18-1964-69
	90% H_2SO_4	268(4.07)	18-1964-69
$C_{12}H_9BrO_5$			
Bicyclo[3.2.2]nona-3,6-diene-8β,9β-dicarboxylic anhydride, 7-bromo-1-methoxy-2-oxo-	MeOH	233(3.81),270s(2.5), 345(2.14)	88-0443-69
isomer	MeOH	240(3.81),320(1.81)	88-0443-69
$C_{12}H_9Br_2N_3$			
Triazene, 1,3-bis(p-bromophenyl)-	MeOH	237(3.86),300(3.67), 370(4.22)	65-0059-69
copper complex	benzene	355(4.61),660(2.98)	65-0059-69
mercury complex	benzene	307(4.54),400(4.81)	65-0059-69
$C_{12}H_9ClN_2O$			
Phenol, 4-chloro-2-(phenylazo)-	HCl	384(3.87)	73-3740-69
(in 50% EtOH)	NaOH	470(3.92)	73-3740-69
$C_{12}H_9ClN_2O_2S_2$			
Sulfur diimide, (p-chlorophenyl)(phenylsulfonyl)-	benzene	393(4.16)	104-0459-69

Compound	Solvent	$\lambda_{max}(\log \epsilon)$	Ref.
$C_{12}H_9ClN_2O_5S$			
Benzenesulfonic acid, 2-(p-chloroanil-ino)-5-nitro-, anion	base	386(4.28)	22-4569-69
dianion	base	478(4.40)	22-4569-69
8-Phenylthiazolo[3,2-a]pyrazin-4-ium perchlorate, 7-oxide	n.s.g.	198(4.39),225s(4.12), 249s(4.02),274(4.16), 352(4.15)	39-2270-69C
$C_{12}H_9ClN_4$			
1H-Pyrazolo[3,4-d]pyrimidine, 4-chloro-6-methyl-1-phenyl-	MeOH	245(4.50)	23-1129-69
1H-Pyrazolo[3,4-d]pyrimidine, 4-chloro-1-p-tolyl-	MeOH	246(4.47)	23-1129-69
4H-Pyridazino[6,1-c]-as-triazine, 7-chloro-3-phenyl-	pH 1	271(3.89),307(3.95), 370s(--)	7-0552-69
	pH 13	257(3.95),350(3.95), 420s(--)	7-0552-69
$C_{12}H_9ClOS$			
Phenol, p-[(o-chlorophenyl)thio]-	EtOH	234(4.14),248(4.19)	104-1903-69
Sulfoxide, m-chlorophenyl phenyl	H_2O	232(4.10)	18-1964-69
	90.2% H_2SO_4	245(3.94)	18-1964-69
Sulfoxide, p-chlorophenyl phenyl	H_2O	240(4.23)	18-1964-69
	91.3% H_2SO_4	262(4.07)	18-1964-69
$C_{12}H_9ClOS_2$			
Benzenesulfinic acid, p-chloro-thio-, S-phenyl ester	60% EtOH	286(4.04)	18-2899-69
Benzenesulfinic acid, thio-, S-(p-chlorophenyl) ester	60% EtOH	286(3.96)	18-2899-69
$C_{12}H_9ClO_2S$			
Phenol, p-[(o-chlorophenyl)sulfinyl]-	EtOH	250(4.25)	104-1903-69
$C_{12}H_9ClO_3S$			
Phenol, p-[(o-chlorophenyl)sulfonyl]-	EtOH	260(4.18)	104-1903-69
$C_{12}H_9Cl_2N_3$			
Aniline, 2-chloro-4-[(o-chlorophenyl)-azo]-	n.s.g.	259(3.99),396(4.41)	40-0207-69
$C_{12}H_9Cl_4NO_2$			
Indole-3-carboxylic acid, 4,5,6,7-tetrachloro-2-methyl-, ethyl ester	MeOH	229(4.71),290(4.08)	24-2685-69
$C_{12}H_9FN_4O$			
Purine, 6-(benzyloxy)-2-fluoro-	pH 1	256(4.13)	44-2160-69
	pH 11	263(4.11)	44-2160-69
	MeOH	239s(3.88),256(4.10), 261(4.06),271(3.55)	44-2160-69
$C_{12}H_9F_3O_3$			
3-Benzofurancarboxylic acid, 4,5,7-trifluoro-2-methyl-, ethyl ester	EtOH	250(3.91)	103-0574-69
$C_{12}H_9F_4N$			
Carbazole, 1,2,3,4-tetrafluoro-5,6,7,8-tetrahydro-	EtOH	268(3.79)	70-0609-69

Compound	Solvent	λ_{max}(log ε)	Ref.
$C_{12}H_9F_5O_3$			
Crotonic acid, 3-(pentafluorophenoxy)-, ethyl ester	EtOH	238(4.16)	103-0574-69
$C_{12}H_9IOS$			
Sulfoxide, p-iodophenyl phenyl	H_2O	240(4.40)	18-1964-69
	90% H_2SO_4	263(4.19)	18-1964-69
$C_{12}H_9N$			
Carbazole	$CHCl_3$	292(4.23)	18-0210-69
$C_{12}H_9NO$			
2-Carbazolol	EtOH	210(4.41),238(4.66), 260(4.34),304(4.15)	39-1518-69C
Phenoxazine	EtOH	315(3.91)	73-0221-69
$C_{12}H_9NOSe$			
2-Propen-1-one, 1-(2-pyridyl)-3-selenophene-2-yl-	hexane	247(3.98),272(3.90), 310s(--),353(4.22)	103-0055-69
	EtOH	240s(--),277(3.85), 310s(--),362(4.27)	103-0055-69
2-Propen-1-one, 1-(3-pyridyl)-3-selenophene-2-yl-	hexane	249(3.98),275s(--), 310s(--),350(4.19)	103-0055-69
	EtOH	240s(--),275(3.75), 310s(--),360(4.27)	103-0055-69
$C_{12}H_9NO_2$			
Bicyclo[4.2.2]deca-2,4,7,9-tetraene-3,4-dicarboximide	MeOH	254(3.86),295s(3.14)	88-1125-69
	MeOH	254(3.86),295s(3.17)	108-0435-69
Dicyclopropa[de,ij]naphthalene-4c,4d-dicarboximide, 2a,2b,4a,4b-tetrahydro-	MeOH	220(3.81)(end abs.)	88-1125-69
Naphthalene, 2-(2-nitrovinyl)-	EtOH	325(3.90)	87-0157-69
Pyrrolo[1,2-b]isoquinoline-1,5-dione, 2,3-dihydro-	EtOH	250(3.98),334(4.08)	78-2275-69
$C_{12}H_9NO_2S$			
Sulfide, m-nitrophenyl phenyl	EtOH	250(4.3),275s(4.1), 340(3.0)	32-0535-69
$C_{12}H_9NO_2S_2$			
p-Dithiin-2,3-dicarboximide, 5,6-dihydro-N-phenyl-	MeOH	234(4.14),265(3.91), 416(3.38)	33-2221-69
$C_{12}H_9NO_2Se$			
1,3-Propanedione, 1-(2-pyridyl)-3-selenophene-2-yl-	H_2O	280(4.73),365(4.66)	123-0074-69C
1,3-Propanedione, 1-(2-pyridyl)-3-selenophene-3-yl-	H_2O	275(4.17),355(5.66)	123-0074-69C
1,3-Propanedione, 1-(4-pyridyl)-3-selenophene-2-yl-	H_2O	270(4.21),362(3.56)	123-0074-69C
1,3-Propanedione, 1-(4-pyridyl)-3-selenophene-3-yl-	H_2O	270(5.98),365(5.60)	123-0074-69C
$C_{12}H_9NO_3$			
Pyrrolo[1,2-b]isoquinoline-1,5-dione, 2,3-dihydro-10-hydroxy-	EtOH	258(3.52),365(3.45)	78-2275-69
$C_{12}H_9NO_3S_2$			
p-Dithiin-2,3-dicarboximide, 5,6-dihydro-N-(m-hydroxyphenyl)-	MeOH	234(4.12),273(4.04), 416(3.39)	33-2221-69

Compound	Solvent	λ_{max}(log ε)	Ref.
p-Dithiin-2,3-dicarboximide, 5,6-dihydro-N-(o-hydroxyphenyl)-	MeOH	270(4.07),410(3.46)	33-2221-69
p-Dithiin-2,3-dicarboximide, 5,6-dihydro-N-(p-hydroxyphenyl)-	MeOH	233(4.27),263(3.91), 317(3.19),417(3.32)	33-2221-69
$C_{12}H_9NO_4$			
2'-Acetonaphthone, 1'-hydroxy-4'-nitro-	C_6H_{12}	207(4.33),250(4.48), 278(4.10),289(4.02), 340s(3.97),356(4.02), 374s(3.90)	19-0253-69
	MeOH	210(4.29),247(4.49), 280(3.97),357(4.01)	19-0253-69
	aq. MeOH	209(4.29),247(4.52), 284(3.97),361(4.00), 380s(3.93)	19-0253-69
Isoquinoline-4-carboxylic acid, 6,7-(methylenedioxy)-, hydrochloride	EtOH	246(4.62)	78-0101-69
Isoquinoline-4-carboxylic acid, 7,8-(methylenedioxy)-, hydrochloride	EtOH	240(4.37),302(3.34), 379(3.39)	78-0101-69
$C_{12}H_9NO_4S$			
Sulfone, m-nitrophenyl phenyl	EtOH	240(4.2),260s(3.7)	32-0535-69
$C_{12}H_9NO_5$			
Indolin-2-ylideneacetic acid, 7-carboxy-3-oxo-, 2-methyl ester	MeOH	219(4.38),242(4.24), 354(4.00),373(4.07)	23-3545-69
$C_{12}H_9NS$			
Phenothiazine	EtOH	254(4.63),319(3.64)	9-0249-69
$C_{12}H_9N_3O$			
Phenoxazin-3-imine, 7-amino-	EtOH	498(4.40)	73-0221-69
hydrochloride	EtOH	583(4.86)	73-0221-69
4-Pyridazinecarbonitrile, 2,3-dihydro-2-methyl-3-oxo-4-phenyl-	$CHCl_3$	312(4.10),344s(3.80)	48-0388-69
$C_{12}H_9N_3O_2$			
Isoxazolo[3,4-d]pyrimidin-3(1H)-one, 1-methyl-6-phenyl-	n.s.g.	284(4.20)	25-0458-69
1,2,4-Triazine, 5,6-di-2-furyl-3-methyl-	MeOH	238(3.75),270(3.95), 313(4.14)	88-3147-69
$C_{12}H_9N_3O_2S$			
Benzimidazole, 1-methyl-5-nitro-2-(2-thienyl)-	EtOH	217s(3.88),247s(4.02), 271s(4.26),288(4.33), 325s(4.14)	73-0572-69
Sulfur diimide, (p-nitrophenyl)phenyl-	benzene	422(4.15)	104-0459-69
$C_{12}H_9N_3O_3$			
Benzimidazole, 2-(2-furyl)-1-methyl-5-nitro-	EtOH	238s(3.94),281(4.49), 330(4.16)	73-0572-69
Pyridazine, 3-(m-nitrostyryl)-, 1-oxide	EtOH	274(4.36),350(3.83)	95-0132-69
Pyridazine, 3-(p-nitrostyryl)-, 1-oxide	EtOH	284(4.17),316(4.14)	95-0132-69
Pyridazine, 4-(m-nitrostyryl)-, 1-oxide	EtOH	246(4.08),320(4.36), 349(4.37)	95-0132-69
Pyridazine, 4-(p-nitrostyryl)-, 1-oxide	EtOH	366(4.54)	95-0132-69
Pyridazine, 5-(m-nitrostyryl)-, 1-oxide	EtOH	270(4.45),302(4.22)	95-0132-69
Pyridazine, 5-(p-nitrostyryl)-, 1-oxide	EtOH	271(4.33)	95-0132-69
Pyridazine, 6-(m-nitrostyryl)-, 1-oxide	EtOH	273(4.36)	95-0132-69

Compound	Solvent	$\lambda_{max}(\log \epsilon)$	Ref.
Pyridazine, 6-(p-nitrostyryl)-, 1-oxide	EtOH	239(4.08),291(4.17), 335(4.21)	95-0132-69
4H-Quinolizine-1-carboxylic acid, 2-amino-3-cyano-4-oxo-, methyl ester	EtOH	253(4.61),295(4.16), 385(4.08)	95-0203-69
$C_{12}H_9N_3O_4$			
Benzoic acid, m-[(5-nitro-2-pyridyl)-amino]-, anion	base	380(4.29)	22-4569-69
dianion	base	465(4.44)	22-4569-69
Benzoic acid, o-[(5-nitro-2-pyridyl)-amino]-, anion	base	386(4.26)	22-4569-69
dianion	base	466(4.34)	22-4569-69
Benzoic acid, p-[(5-nitro-2-pyridyl)-amino]-, anion	base	382(4.35)	22-4569-69
dianion	base	470(4.46)	22-4569-69
Diphenylamine, 4,4'-dinitro-	acid	240(4.07),400(4.28)	73-3732-69
anion	base	240(4.12),286(4.12), 440(4.00),600(4.79)	73-3732-69
Pyridine, 2-(2,4-dinitrobenzyl)-	EtOH	246(4.56)	22-4425-69
Pyridine, 4-(2,4-dinitrobenzyl)-	EtOH	241(4.57)	22-4425-69
$C_{12}H_9N_3O_4S_2$			
p-Dithiin-2,3-dicarboximide, 5,6-dihydro-N-(o-nitroanilino)-	MeOH	221(4.43),266(4.14), 383(3.88)	33-2236-69
p-Dithiin-2,3-dicarboximide, 5,6-dihydro-N-(p-nitroanilino)-	MeOH	334(4.37)	33-2236-69
Sulfur diimide, (m-nitrophenyl)-(phenylsulfonyl)-	benzene	371(4.03)	104-0459-69
Sulfur diimide, (p-nitrophenyl)-(phenylsulfonyl)-	benzene	376(4.02)	104-0459-69
$C_{12}H_9N_3O_7S$			
Metanilic acid, N-(2,6-dinitrophenyl)-, anion	base	365(4.24)	22-4569-69
dianion	base	420(4.26)	22-4569-69
Sulfanilic acid, N-(2,4-dinitrophenyl)-, anion	base	365(4.25)	22-4569-69
dianion	base	420(4.26)	22-4569-69
$C_{12}H_9N_5$			
Pyrrole, 1-(isopropylideneamino)-2-(tricyanovinyl)-	$CHCl_3$	434(4.4)	24-3268-69
1H-v-Triazolo[4,5-c]pyridine, 1-(benzylideneamino)-	EtOH	261(4.18),318(4.28)	39-1758-69C
hydrochloride, hydrate	EtOH	261(4.28),318(4.37)	39-1758-69C
3H-v-Triazolo[4,5-b]pyridine, 3-(benzylideneamino)-	EtOH	279(4.42),316(4.21)	39-1758-69C
$C_{12}H_9N_5O_2S$			
Purine, 8-[(p-nitrobenzyl)thio]-	MeOH	289(4.31)	4-0023-69
$C_{12}H_9N_5O_4$			
Triazene, 1,3-bis(p-nitrophenyl)-	MeOH	237(4.27),395(4.40)	65-0059-69
mercury complex	benzene	335(4.62),430(4.85)	65-0059-69
$C_{12}H_9N_5S$			
Pyrimido[5,4-e]-as-triazine, 5-(benzylthio)- (unstable)	pH 7	240(4.13),391(3.92)	44-3161-69

Compound	Solvent	λ max (log ε)	Ref.
$C_{12}H_9N_7O_7$			
Benzotriazole, 1-amino-, picrate	EtOH	263(3.76),280s(3.65)	39-0742-69C
$C_{12}H_{10}$			
Benzocyclooctene	EtOH	234(4.33),275(3.96)	35-4734-69
Biphenyl	C_6H_{12}	248(4.3)	64-0524-69
	EtOH	247(4.26)	6-0277-69
1,4-Ethenonaphthalene, 1,4-dihydro-	heptane	175(4.52),188(4.29), 191(4.30),206(4.45), 231s(3.08),263(2.70), 270(2.84),277(2.89)	78-2417-69
4a,8a-Ethenonaphthalene, 4a,8a-dihydro-	isooctane	247(3.40),290(3.32)	35-3973-69
Indene, 3-(2-propynyl)-	n.s.g.	252(3.92)	39-0944-69C
$C_{12}H_{10}BF_4N_3$			
2-Phenyl-s-triazolo[4,3-a]pyridinium tetrafluoroborate	n.s.g.	240(4.3),380s(4.0)	24-3159-69
$C_{12}H_{10}B_2O_4$			
2,6(or 7)-Biphenylenediboronic acid	EtOH	254(4.79),260(4.90), 364(3.78),366s(3.66)	39-2789-69C
$C_{12}H_{10}BiBr$			
Bismuthine, bromodiphenyl-	benzene	325(3.36)	101-0099-69E
	CH_2Cl_2	325(3.36)	101-0099-69E
$C_{12}H_{10}BiCl$			
Bismuthine, chlorodiphenyl-	benzene	316(3.34)	101-0099-69E
	CH_2Cl_2	316(3.34)	101-0099-69E
$C_{12}H_{10}BrI$			
Diphenyliodonium bromide	MeOH	225(4.09),264(3.47), 271(3.37)	120-0012-69
$C_{12}H_{10}BrNO_2S$			
1H-Azepine, 1-[(p-bromophenyl)sulfonyl]-	EtOH	233(4.12),268s(3.58)	44-2866-69
$C_{12}H_{10}BrNO_4$			
2,3-Indolizinedicarboxylic acid, 1-bromo-, dimethyl ester	MeOH	221(2.24),249s(2.33), 257(2.33),335s(1.16), 349(1.23)	39-1143-69C
$C_{12}H_{10}BrN_3O$			
Benzimidazole, 5-amino-2-(5-bromo-2-furyl)-1-methyl-	EtOH	210(4.35),241(4.04), 294(4.20),338(4.20)	73-0572-69
$C_{12}H_{10}BrN_3S$			
Benzimidazole, 5-amino-2-(5-bromo-2-thienyl)-1-methyl-	EtOH	218(4.51),252s(3.95), 294s(3.96),344(4.18)	73-0572-69
$C_{12}H_{10}ClNO_4$			
Isoquinoline-4-carboxylic acid, 6,7-(methylenedioxy)-, hydrochloride	EtOH	246(4.62)	78-0101-69
Isoquinoline-4-carboxylic acid, 7,8-(methylenedioxy)-, hydrochloride	EtOH	240(4.37),302(3.34), 379(3.39)	78-0101-69
$C_{12}H_{10}ClN_3$			
Aniline, p-[(o-chlorophenyl)azo]-	n.s.g.	259(3.99),402(4.40)	40-0207-69
Aniline, 2-chloro-4-(phenylazo)-	n.s.g.	263(3.99),384(4.42)	40-0207-69

Compound	Solvent	λmax(log ε)	Ref.
$C_{12}H_{10}ClN_3O$			
Benzimidazole, 5-amino-2-(5-chloro-2-furyl)-1-methyl-	EtOH	211(4.39),241(4.07), 295(4.14),338(4.18)	73-0572-69
5H-Pyrrolo[3,4-d]pyrimidin-4-ol, 6-(p-chlorophenyl)-6,7-dihydro-	10% NaOH	205s(3.32),257(4.28)	4-0507-69
$C_{12}H_{10}ClN_3O_3S$			
Thiazole, 2-(allylamino)-5-[1-chloro-2-(5-nitrofuryl)vinyl]-	EtOH	255(4.18),308(4.02), 396(4.20)	103-0414-69
$C_{12}H_{10}ClN_5$			
1H-v-Triazolo[4,5-c]pyridine, 1-(benzylideneamino)-, hydrochloride, hydrate	EtOH	261(4.28),318(4.37)	39-1758-69C
$C_{12}H_{10}Cl_4O_3S_2$			
3-Thiophenemethanol, 4,4'-(oxydimethylene)bis[2,5-dichloro-	EtOH	248(4.09)	44-0340-69
$C_{12}H_{10}F_6$			
Bicyclo[2.2.2]octa-2,5,7-triene, 2,3-dimethyl-5,6-bis(trifluoromethyl)-	C_6H_{12}	230s(2.54),275(1.96)	44-1271-69
Bicyclo[2.2.2]octa-2,5,7-triene, 5,7-dimethyl-2,3-bis(trifluoromethyl)-	hexane	226(2.52),272(2.04)	44-1271-69
$C_{12}H_{10}INO_2$			
1,4-Naphthoquinone, 2-[(2-iodoethyl)-amino]-	EtOH	218(4.23),232s(4.18), 269(4.40),330(3.42), 447(3.57)	39-1325-69C
$C_{12}H_{10}INS$			
5-Methylthiopyrano[4,3-b]indolium iodide	EtOH	212(4.27),240s(3.80), 270s(4.13),278(4.15), 300s(3.77),332s(3.41), 370s(3.25)	64-0024-69
$C_{12}H_{10}IN_3O$			
Benzimidazole, 5-amino-2-(5-iodo-2-furyl)-1-methyl-	EtOH	213(4.40),238s(3.99), 292(4.22),340(4.24)	73-0572-69
$C_{12}H_{10}I_2$			
Diphenyliodonium iodide	MeOH	222(4.49),258(3.22), 264(3.29),270(3.21)	120-0012-69
$C_{12}H_{10}NO$			
Nitroxide, diphenyl-	$CHCl_3$	314(4.24)	18-0210-69
$C_{12}H_{10}N_2$			
Azobenzene	MeOH	231(4.15),317(4.26)	64-0997-69
	$CHCl_3$	319(4.27)	18-0210-69
	20% EtOH	322(4.31),428(3.00)	54-0562-69
	+ H_2SO_4	236(3.60),420(4.43)	54-0562-69
Heptafulvalene, 8,8-dicyano-1,6-dimethyl-	MeOH	245(4.04),390(3.96)	18-2386-69
Pyrido[1,2-a]benzimidazole, 4-methyl-	n.s.g.	240(4.54),245(4.53), 254(4.11),266(4.00)	39-1334-69C
$C_{12}H_{10}N_2O$			
Azoxybenzene	$CHCl_3$	326(4.17)	18-0210-69
1H-1,2-Diazepine, 1-benzoyl-	MeOH	224(4.14)	22-2175-69
Diphenylamine, N-nitroso-	$CHCl_3$	292(3.88)	18-0210-69

Compound	Solvent	λmax(log ε)	Ref.
Diphenylamine, 4-nitroso-	$CHCl_3$	414(4.42)	18-0210-69
Phenol, o-(phenylazo)-	$CHCl_3$	325(4.26)	18-0210-69
Phenol, p-(phenylazo)-	100% H_2SO_4	426(4.52)	23-4011-69
	95% H_2SO_4	460(4.58)	23-4011-69
	neutral	348(4.40)	23-4011-69
	anion	402(4.33),440(4.34)	23-4011-69
in 5% dioxan	95% H_2SO_4	460(4.60)	23-3631-69
in 5% dioxan	neutral	348(4.34)	23-3631-69
$C_{12}H_{10}N_2O_2$			
Diphenylamine, 2-nitro-	acid	228(4.19),262(4.29), 425(3.98),540s(3.45)	73-3732-69
	$CHCl_3$	263(4.16)	18-0210-69
Diphenylamine, 4-nitro-	acid	228(3.88),258(4.08), 400(4.29),490s(3.45)	73-3732-69
	anion	258(3.97),300(3.59), 440s(4.23),490(4.31)	73-3732-69
	$CHCl_3$	380(4.28)	18-0210-69
Furazan, 3-(p-methoxyphenyl)-4-methyl-	n.s.g.	267(4.03)	39-1901-69C
4-Isoquinolinecarbonitrile, 6,8-dimethoxy-	EtOH	217(4.26),254(4.51)	95-1492-69
5,8-Isoquinolinedione, 7-(1-aziridinyl)-3-methyl-	MeCN	232(4.21),262(4.13), 325(3.87)	4-0697-69
$C_{12}H_{10}N_2O_2S_2$			
p-Dithiin-2,3-dicarboximide, N-(m-aminophenyl)-5,6-dihydro-	MeOH	224(4.30),269(3.91), 415(3.35)	33-2221-69
p-Dithiin-2,3-dicarboximide, N-(o-aminophenyl)-5,6-dihydro-	MeOH	236(4.20),267(3.94), 410(3.41)	33-2221-69
p-Dithiin-2,3-dicarboximide, N-(p-aminophenyl)-5,6-dihydro-	MeOH	250(4.38),353(3.24), 413(3.23)	33-2221-69
p-Dithiin-2,3-dicarboximide, N-anilino-5,6-dihydro-	MeOH	230(4.15),260(3.92), 412(3.41)	33-2236-69
Sulfur diimide, phenyl(phenylsulfonyl)-	benzene	384(3.83)	104-0459-69
$C_{12}H_{10}N_2O_3$			
Acetic acid, 2-(1,4-dihydro-1,4-dioxo-2-naphthyl)hydrazide	alkali	540(3.82)	95-0007-69
Acetic acid, 2-(3,4-dihydro-3,4-dioxo-1-naphthyl)hydrazide	alkali	475(4.20)	95-0007-69
$C_{12}H_{10}N_2O_3S$			
Barbituric acid, 5-[3-(2-furyl)-2-methylallylidene]-2-thio-	EtOH	250(4.20),400(4.52)	103-0830-69
$C_{12}H_{10}N_2O_3S_2$			
Terephthalamic acid, N-(4-methyl-2-oxo-4-thiazolin-3-yl)-4-thio-	MeOH	251(4.27),291(3.90), 411(2.52)	89-0758-69
$C_{12}H_{10}N_2O_4S$			
Benzenesulfonic acid, p-[(p-hydroxyphenyl)azo]-	98% H_2SO_4	463(4.67)	23-4011-69
	89% H_2SO_4	465(4.66)	23-4011-69
diprotonated ion	n.s.g.	420(4.59)	23-4011-69
neutral form	n.s.g.	350(4.38)	23-4011-69
anion	n.s.g.	448(4.41)	23-4011-69
$C_{12}H_{10}N_2O_5$			
4H-Pyrazole-3-carboxylic acid, 4-oxo-5-phenyl-, ethyl ester, 1,2-dioxide	n.s.g.	254(4.28),290(3.68), 424(3.08)	44-0187-69

Compound	Solvent	$\lambda_{max}(\log \epsilon)$	Ref.
$C_{12}H_{10}N_2O_5S$			
Diphenylamine-2-sulfonic acid, 4-nitro-, anion	base	390(4.24)	22-4569-69
dianion	base	480(4.42)	22-4569-69
Diphenylamine-2-sulfonic acid, 4'-nitro-, anion	base	395(4.17)	22-4569-69
dianion	base	475(4.24)	22-4569-69
Diphenylamine-4-sulfonic acid, 4'-nitro-, anion	base	400(4.34)	22-4569-69
dianion	base	490(4.47)	22-4569-69
$C_{12}H_{10}N_2O_7S_2$			
Azobenzene-3,4'-disulfonic acid, 4-hydroxy-	H_2SO_4	452(4.65)	23-4011-69
dication	H_2SO_4	436(4.57)	23-4011-69
neutral	H_2SO_4	348(4.40)	23-4011-69
anion	H_2SO_4	440(4.43)	23-4011-69
$C_{12}H_{10}N_2S$			
Sulfur diimide, diphenyl-	benzene	420(4.07)	104-0459-69
	EtOH	418(--)	104-0459-69
	acetone	418(--)	104-0459-69
	$CHCl_3$	420(--)	104-0459-69
1H-Thieno[2,3-c]pyrazole, 3-methyl-1-phenyl-	EtOH	232(4.30),285(4.23)	104-1460-69
$C_{12}H_{10}N_4$			
Pyrazine, 2-amino-5-(3-indolyl)-	MeOH	228(4.42),274(4.34), 369(3.83)	95-1646-69
	MeOH-HCl	224(4.46),308(4.40), 420(3.70)	95-1646-69
4H-Pyridazino[3,2-c]-as-triazine, 3-phenyl-, hydrobromide	pH 1	260(3.87),300(3.98)	7-0552-69
	pH 13	243(4.05),347(3.96)	7-0552-69
$C_{12}H_{10}N_4O$			
1H-Pyrazolo[3,4-b]pyrazin-6-ol, 1-methyl-3-phenyl-	EtOH	241(4.18),321(3.96)	32-0463-69
1H-Pyrazolo[3,4-b]pyrazin-6-ol, 3-methyl-1-phenyl-	EtOH	257(4.42),321(3.98)	32-0463-69
$C_{12}H_{10}N_4O_2$			
Benzo[c]cinnoline, 3,8-diamino-, 5,6-dioxide	EtOH	222(3.08),255(3.13), 305(4.56),320s(3.33), 496(3.42)	4-0523-69
Pyrazole-5-carbonitrile, 3-ethyl-1-(p-nitrophenyl)-	MeOH	226(4.13),247s(3.65), 303(4.17)	39-2587-69C
$C_{12}H_{10}N_4O_2S$			
Acetic acid, [(5-methyl-5H-as-triazino-[5,6-b]indol-3-yl)thio]-	EtOH	270(4.5),350(4.0)	65-2111-69
$C_{12}H_{10}N_4O_2S_2$			
5H-Imidazo[4,5-d]thiazole-5-thione, 4,6-dihydro-4,6-dimethyl-2-(p-nitrophenyl)-	EtOH	283(4.17),442(4.26)	94-0910-69
$C_{12}H_{10}N_4O_4$			
Hydrazine, 1,2-bis(p-nitrophenyl)-	MeCN	194(4.54),223(4.14), 368(4.43)	35-2325-69

Compound	Solvent	λ_{max}(log ε)	Ref.
$C_{12}H_{10}N_4O_6$			
2,4-Pentadienoic acid, 5-formyl-, 5-(2,4-dinitrophenylhydrazone)	EtOH	390(4.51)	104-0759-69
$C_{12}H_{10}N_4O_9$			
β-Alanine, N-(5-nitro-2-furoyl)-, 5-nitro-2-furyl ester, N-oxime	EtOH	234(4.01),294(4.22), 360(3.79)	18-0556-69
$C_{12}H_{10}N_4S$			
1H-Imidazo[4,5-d]pyridazine, 4-(benzylthio)-	EtOH	218(4.31),273(4.03)	4-0093-69
$C_{12}H_{10}O$			
1'-Acetonaphthone	C_6H_{12}	212(4.49),222s(4.45), 239(4.43),246(4.42), 282(3.69),293(3.75), 299(3.70),341s(2.79)	19-0253-69
	MeOH	213(4.53),240(4.36), 280s(3.69),293(3.77), 320s(3.55)	19-0253-69
	aq. MeOH	212(4.56),243(4.29), 295(3.78),320s(3.60)	19-0253-69
2'-Acetonaphthone	hexane	237(4.42),246(4.48), 280(3.95),291(3.85), 324(3.15),340(3.21)	35-0355-69
	C_6H_{12}	230(4.70),247(4.79), 270s(3.82),281(3.96), 288(3.85),292(3.86), 310s(2.89),326(3.14), 341(3.22)	19-0253-69
	MeOH	208(4.29),240s(4.67), 246(4.71),282(4.04), 332(3.25),340(3.25)	19-0253-69
	aq. MeOH	208(4.33),230s(4.38), 250(4.65),284(3.94), 344(3.19)	19-0253-69
5H-Benzocycloheptene-5-carboxaldehyde	EtOH	280(3.77)	35-4734-69
Bicyclo[3.1.0]hex-3-en-2-one, 4-phenyl-	hexane	219(4.15),289(4.23)	22-2095-69
	EtOH	222(4.13),301(4.17)	22-2095-69
Furan, 2-styryl-, cis	MeOH	303(4.15)	44-0212-69
Furan, 2-styryl-, trans	MeOH	318(4.51)	44-0212-69
$C_{12}H_{10}OS$			
Phenol, p-(phenylthio)-	EtOH	234(4.07),248(4.14)	104-1913-69
Phenyl sulfoxide	H_2O	233(4.15)	18-1964-69
	91% H_2SO_4	255(3.91)	18-1964-69
$C_{12}H_{10}OS_2$			
Benzenesulfinic acid, thio-, S-phenyl ester	60% EtOH	282(3.90)	18-2899-69
$C_{12}H_{10}O_2$			
1'-Acetonaphthone, 2'-hydroxy-	C_6H_{12}	226(4.60),238s(4.28), 253s(4.02),313(3.85), 358(3.74)	19-0253-69
	MeOH	224(4.71),278(3.51), 290(3.56),301(3.57), 336(3.56)	19-0253-69

Compound	Solvent	λ_{max}(log ε)	Ref.
1'-Acetonaphthone, 2'-hydroxy- (cont.)	aq. MeOH	223(4.75),290(3.53), 308(3.55),318(3.54), 332(3.56)	19-0253-69
2'-Acetonaphthone, 1'-hydroxy-	C_6H_{12}	218(4.39),256(4.49), 266(4.48),285(3.74), 296(3.69),308s(3.32), 366(3.78)	19-0253-69
	MeOH	214(4.45),255(4.51), 263(4.49),284(3.77), 294(3.77),306s(3.40), 366(3.76)	19-0253-69
	aq. MeOH	214(4.43),254(4.51), 262(4.51),284(3.81), 294(3.85),306s(3.57), 368(3.75)	19-0253-69
2'-Acetonaphthone, 3'-hydroxy-	C_6H_{12}	218(4.37),252(4.63), 260s(4.50),282s(3.78), 292(3.97),304(4.01), 390(3.34)	19-0253-69
	MeOH	216(4.37),249(4.62), 292(3.92),302(3.95), 384(3.27)	19-0253-69
	aq. MeOH	214(4.37),250(4.61), 292s(3.96),302(4.00), 382(3.24)	19-0253-69
2'-Acetonaphthone, 6'-hydroxy-	C_6H_{12}	215(4.23),240(4.63), 248(4.60),257(4.61), 302(4.01)	19-0253-69
	MeOH	214(4.30),242(4.50), 260(4.37),317(4.05)	19-0253-69
	aq. MeOH	213(4.30),242(4.49), 258(4.40),316(4.07), 350s(3.56)	19-0253-69
Acetophenone, o-(2-furyl)-	EtOH	231s(4.27),240(4.31), 274(4.24),308(3.81)	6-0277-69
Benzo[1,2-b:5,4-b']difuran, 3,5-dimethyl-	EtOH	256s(3.41),267(3.85), 276(3.89),283(3.46), 289(3.74),295(3.87), 301(3.87),308(3.92)	4-0191-69
Naphthalene, 6-methyl-2,3-(methylenedioxy)-	EtOH	227(4.76),263(3.72), 275(3.71),286(3.53), 301(3.14),306(3.27), 314(3.53),322(3.50), 328(3.77)	12-1721-69
4H-Pyran-4-one, 2-methyl-6-phenyl-	EtOH	272(4.31)	94-2126-69
$C_{12}H_{10}O_2S$			
Phenol, p-(phenylsulfonyl)-	EtOH	246(4.24)	104-1903-69
$C_{12}H_{10}O_3$			
Benzo[b]furan, 2-isopropenyl-5,6-(methylenedioxy)-	EtOH	221(4.16),276(4.07), 320(4.26),332(4.22)	18-1971-69
2,3-Benzo-6-oxabicyclo[3.2.1]octane-4,7-dione, 1-methyl-	EtOH	259(3.92),293(3.28)	36-0894-69
2(5H)-Furanone, 5-(p-methoxybenzylidene)-	EtOH	241(4.04),360(4.46)	77-1479-69
2,4-Pentadienal, 5-hydroxy-, benzoate, trans-trans	MeOH	238(3.93),284(4.53), 362(3.52)	78-4315-69
2H-Pyran-2-one, 6-(p-methoxyphenyl)-	EtOH	257(3.96),352(4.28)	77-1479-69

Compound	Solvent	$\lambda_{max}(\log \epsilon)$	Ref.
$C_{12}H_{10}O_3S$			
6H-Cyclohepta[c]thiophene-5-carboxylic acid, 1,3-dimethyl-6-oxo-	EtOH	288(4.46),344(3.77)	104-0559-69
Phenol, p-(phenylsulfonyl)-	EtOH	256(4.14)	104-1903-69
$C_{12}H_{10}O_4$			
Bicyclo[3.2.2]nona-3,6-diene-8α,9α-dicarboxylic anhydride, 3-methyl-2-oxo-	MeOH	235(3.70),260s(3.2), 328(1.99)	88-0403-69
1(2H)-Naphthalenone, 2,3-epoxy-3,4-dihydro-4-hydroxy-, acetate, cis	EtOH	254(3.96),292(3.11)	39-2053-69C
trans	EtOH	254(3.88),289(3.10)	39-2053-69C
$C_{12}H_{10}O_4S$			
Sulfone, 2,5-dihydroxyphenyl phenyl	H_2O	320(3.69)	78-2715-69
$C_{12}H_{10}O_5$			
Cyclopenta[c]pyran-4-carboxylic acid, 1,7-dihydro-7-(methoxymethylene)-1-oxo-, methyl ester	EtOH	347(4.24),357(4.25)	35-6470-69
Isocoumarin-4-carboxylic acid, 7-methoxy-3-methyl-	MeOH	230(4.46),270(4.06), 348(3.71)	42-0935-69
$C_{12}H_{10}O_6$			
Succinic acid, piperonylidene-	EtOH	232(4.16),287(4.07), 315(4.11)	39-2470-69C
$C_{12}H_{10}O_7$			
2H,5H-Pyrano[4,3-b]pyran-8-carboxylic acid, 4-hydroxy-7-methyl-2,5-dioxo-, ethyl ester	EtOH	268(4.05),329(3.74)	44-2480-69
$C_{12}H_{10}S$			
Phenyl sulfide	MeOH	249(4.06)	87-0146-69
Thiophene, 2-styryl-	MeOH	229(4.05),268s(3.74), 323(4.47)	117-0209-69
$C_{12}H_{10}S_2$			
Benzo[4,5]thieno[3,2-b]thiepin, 2,3-dihydro-	EtOH	222(4.33),238s(4.24), 260(4.12),275s(3.97), 303(3.96),312(3.98), 325(3.71)	22-0601-69
1-Naphthoic acid, dithio-, methyl ester	C_6H_{12}	221(4.78),274(3.73), 306(4.02),494(2.11)	48-0045-69
	EtOH	308(4.06),488(2.05)	48-0045-69
2-Naphthoic acid, dithio-, methyl ester	C_6H_{12}	273(4.49),318(4.44), 506(2.30)	48-0045-69
	EtOH	271(4.37),319(4.35), 495(2.33)	48-0045-69
Thiophene, 2,2'-(1,3-butadienylene)di-	n.s.g.	267(3.88),342(4.62), 357(4.76),375(4.63)	22-2076-69
2H-Thiopyran-2-thione, 3-methyl-4-phenyl-	C_6H_{12}	244(4.48),301(3.95), 329(4.02),456(4.03)	78-1441-69
	EtOH	244(4.49),318(4.06), 457(4.07)	78-1441-69
$C_{12}H_{11}BrCuP$			
Diphenylphosphinecopper(I) bromide	MeOH	224(4.22),235s(3.18), 253(2.90),259(2.98), 265(3.03),272(2.97)	39-0133-69A

Compound	Solvent	$\lambda_{max}(\log \epsilon)$	Ref.
$C_{12}H_{11}BrN_2O_2$			
1-Indolinecarboxaldehyde, 5-bromo-3-[(dimethylamino)methylene]-2-oxo-	EtOH	243(4.51),247(4.15), 280(4.14),300(4.14), 312(4.14),360(4.08)	2-0662-69
$C_{12}H_{11}BrN_4OS$			
Acetophenone, 4'-bromo-2-[(2,6-diamino-4-pyrimidinyl)thio]-	neutral	220s(4.4),263(4.3), 285s(4.2)	87-0227-69
cation	n.s.g.	263(4.3),292s(4.2)	87-0227-69
$C_{12}H_{11}BrO_5$			
Bicyclo[3.2.2]nona-3,6-diene-8α,9α-dicarboxylic anhydride, 3-bromo-1-methoxy-2-oxo-	MeOH	245(3.50),275(3.45), 340(2.15)	88-0443-69
$C_{12}H_{11}ClCuP$			
Diphenylphosphinecopper(I) chloride	MeOH	224(4.18),235s(3.18), 253(2.91),259(2.99), 265(3.03),272(2.99)	39-0133-69A
$C_{12}H_{11}ClINO$			
Furo[3,2-c]quinoline, 7-chloro-2,3-dihydro-, methiodide	EtOH	243(5.89),312(4.75), 328(4.93)	2-0952-69
Furo[3,2-c]quinoline, 8-chloro-2,3-dihydro-, methiodide	EtOH	241(5.78),316(4.85), 328(4.93)	2-0952-69
$C_{12}H_{11}ClN_2O$			
Cinnamamide, N-chloro-α-cyano-β-ethyl-, cis	EtOH	268(3.99)	22-1724-69
trans	EtOH	269(4.00)	22-1724-69
Crotonamide, N-chloro-2-cyano-3-methyl-4-phenyl-, trans	EtOH	220(4.08)	22-1724-69
$C_{12}H_{11}ClN_2O_2$			
1-Indolinecarboxaldehyde, 4-chloro-3-[(dimethylamino)methylene]-2-oxo-	EtOH	232(4.37),280(4.18), 317(3.86),358(4.17)	2-0662-69
$C_{12}H_{11}ClN_2O_3$			
1-(5-Nitrosalicyl)pyridinium chloride	H_2O	260(3.85),267s(--), 316(3.93)	35-4532-69
$C_{12}H_{11}ClN_4$			
5H-Pyrrolo[3,4-d]pyrimidine, 4-amino-6-(p-chlorophenyl)-6,7-dihydro-	10% HCl	252(4.30),299(3.45)	4-0507-69
$C_{12}H_{11}ClN_4O$			
5H-Pyrrolo[3,4-d]pyrimidin-4-ol, 2-amino-6-(p-chlorophenyl)-6,7-dihydro-	10% NaOH	255(4.27),310s(3.26)	4-0507-69
$C_{12}H_{11}ClO$			
4,8-Ethenoazulen-5(3H)-one, 4-chloro-3a,4,8,8a-tetrahydro-	MeOH	231(3.59)	88-0775-69
isomer	MeOH	225(3.69),255(3.11)	88-0775-69
4,8-Ethenoazulen-5(3H)-one, 6-chloro-3a,4,8,8a-tetrahydro-	MeOH	237(3.54),271(3.48)	88-0775-69
Tricyclo[4.4.1.1^{2,5}]dodeca-3,7,9-trien-11-one, 1-chloro-	MeOH	252(3.49),261(3.61), 270(3.60)	88-0775-69

Compound	Solvent	λ_{max}(log ε)	Ref.
$C_{12}H_{11}Cl_3N_2O_2$			
4(3H)-Pyrimidinone, 5,6-dihydro-2-(p-methoxyphenyl)-6-(trichloromethyl)-	EtOH	263(3.19)	104-1797-69
$C_{12}H_{11}CuIP$			
Diphenylphosphinecopper(I) iodide (also other complexes)	MeOH	222(4.19),235s(3.18), 253(2.90),259(2.98), 265(3.02),272(2.98)	39-0133-69A
$C_{12}H_{11}F$			
Naphthalene, 2-fluoro-1,3-dimethyl-	pentane	224(4.32),278(3.78), 306(3.04),320(3.04)	88-3867-69
$C_{12}H_{11}F_3N_2O_3$			
Pyridine, 1-acetyl-1,4-dihydro-4-[4-methyl-5-oxo-2-(trifluoromethyl)-3-oxazolin-2-yl]-	MeCN	244(4.06)	24-1129-69
$C_{12}H_{11}F_3N_2O_4$			
1(4H)-Pyridinecarboxylic acid, 4-[4-methyl-5-oxo-2-(trifluoromethyl)-3-oxazolin-2-yl]-, methyl ester	MeCN	226(4.23)	24-1129-69
$C_{12}H_{11}N$			
Diphenylamine	C_6H_{12}	236s(3.65),282(4.22), 292s(4.16)	22-3523-69
	MeOH	281(4.28)	87-0146-69
	EtOH	240s(3.67),286(4.34)	22-3523-69
	$CHCl_3$	290(4.23)	18-0210-69
Pyridine, 2-benzyl-	C_6H_{12}	260(3.55)	4-0859-69
	pH 1	265(3.88)	4-0859-69
	MeOH	262(3.63)	4-0859-69
Pyridine, 4-benzyl-	C_6H_{12}	255(3.26)	4-0859-69
	pH 1	253(3.76)	4-0859-69
	MeOH	256(3.39)	4-0859-69
$C_{12}H_{11}NO$			
Carbazol-4(1H)-one, 2,3-dihydro-	EtOH	242(4.26),265(4.16), 295(4.10)	94-1290-69
Furo[3,2-c]quinoline, 2,3-dihydro-7(9)-methyl-	EtOH	234(4.72),282s(3.14), 322(3.78)	2-0952-69
Furo[3,2-c]quinoline, 2,3-dihydro-8-methyl-	EtOH	233(4.8),238s(4.78), 257s(3.54),257s(3.54), 320(3.78)	2-0952-69
Pyridine, 2-benzyl-, N-oxide	C_6H_{12}	278(4.05)	4-0859-69
	pH 1	266(3.73)	4-0859-69
	MeOH	262(4.20)	4-0859-69
Pyridine, 4-benzyl-, N-oxide	C_6H_{12}	287(4.24)	4-0859-69
	pH 1	250(3.71)	4-0859-69
	MeOH	267(4.18)	4-0859-69
4(1H)-Pyridone, 2-methyl-6-phenyl-	EtOH	240(4.39),262(4.18)	18-2389-69
$C_{12}H_{11}NOS$			
Nitrone, N-benzyl-α-2-thienyl-	EtOH	205(4.1),314(4.3)	28-0078-69A
$C_{12}H_{11}NOS_2$			
2-Propanone, 1-[(4-phenyl-2-thiazolyl)-thio]-	MeOH	271(4.12)	4-0397-69

Compound	Solvent	λ_{max}(log ϵ)	Ref.
$C_{12}H_{11}NO_2$			
2-Furaldehyde, 5-(N-methylanilino)-	n.s.g.	365(4.45)	103-0434-69
Furo[3,2-c]quinoline, 2,3-dihydro-8-methoxy-	EtOH	230s(4.5),243(4.64), 250s(4.44),300s(3.56), 311(3.68),337(3.75)	2-0952-69
Naphthalene, 1,2-dimethyl-3(6)-nitro-	EtOH	270(4.33),320(3.91)	39-0873-69B
Naphthalene, 1,2-dimethyl-4-nitro-	EtOH	250(3.92)	39-0873-69B
Naphthalene, 1,2-dimethyl-5-nitro-	EtOH	252s(3.81),350(3.32)	39-0873-69B
Naphthalene, 1,3-dimethyl-2(6)-nitro-	EtOH	270(4.36),320(3.91)	39-0873-69B
Naphthalene, 1,3-dimethyl-5-nitro-	EtOH	250s(3.94),300(3.55)	39-0873-69B
Naphthalene, 1,4-dimethyl-5-nitro-	EtOH	260(3.75)	39-0873-69B
Naphthalene, 1,5-dimethyl-2-nitro-	EtOH	260(4.32),268(4.33)	39-0873-69B
Naphthalene, 1,5-dimethyl-4-nitro-	EtOH	250s(3.83),340(3.61)	39-0873-69B
Naphthalene, 1,8-dimethyl-2-nitro-	EtOH	259(4.00)	39-0873-69B
Naphthalene, 2,3-dimethyl-1-nitro-	EtOH	270(3.67)	39-0873-69B
Naphthalene, 2,3-dimethyl-6-nitro-	EtOH	260(4.35),268(4.36), 315(4.04)	39-0873-69B
Naphthalene, 2,4-dimethyl-1-nitro-	EtOH	250(3.92)	39-0873-69B
Naphthalene, 2,6-dimethyl-1-nitro-	EtOH	264(3.65)	39-0873-69B
Naphthalene, 2,6-dimethyl-3-nitro-	EtOH	258(4.16),262(4.12)	39-0873-69B
Naphthalene, 2,7-dimethyl-1-nitro-	EtOH	268(3.64)	39-0873-69B
Naphthalene, 3,6-dimethyl-1-nitro-	EtOH	250s(3.83),344(3.61)	39-0873-69B
Naphthalene, 3,6-dimethyl-2-nitro-	EtOH	262(4.21),268(4.21), 316(3.77)	39-0873-69B
Naphthalene, 3,7-dimethyl-1-nitro-	EtOH	248s(3.87),350(3.57)	39-0873-69B
Naphthalene, 4,5-dimethyl-1-nitro-	EtOH	250(3.63)	39-0873-69B
Naphthalene, 6,7-dimethyl-1-nitro-	EtOH	245s(3.94),340(3.65)	39-0873-69B
1,4-Naphthoquinone, 2-(ethylamino)-	EtOH	232(4.20),271(4.37), 330(3.42),454(3.56)	39-1325-69C
Nitrone, N-benzyl-α-2-furyl-	EtOH	204(4.1),308(4.4)	28-0078-69A
3-Pyrrolin-2-one, 5-(p-methoxybenzylidene)-, m. 146-8°	90% EtOH	247(4.04),363(4.40)	77-1479-69
	+HCl	247(4.04),363(4.40)	77-1479-69
	+NaOH	247(4.04),363(4.43)	77-1479-69
m. 151-2°	90% EtOH	241(4.04),355(4.54)	77-1479-69
	+HCl	240(3.96),355(4.58)	77-1479-69
	+NaOH	241(4.11),355(4.62)	77-1479-69
3-Quinolinecarboxylic acid, 4-methyl-, methyl ester	EtOH	215(4.35),235(4.71), 284(3.74)	77-0463-69
$C_{12}H_{11}NO_2S$			
4H-Furo[3,4-d][1,3]thiazin-7(2H)-one, 1,5-dihydro-2-phenyl-	EtOH	269(3.63)	44-1582-69
Ketone, 2-methoxy-3H-azepin-3-yl 2-thienyl	MeOH	261(4.16)	78-5205-69
$C_{12}H_{11}NO_3$			
Bicyclo[3.2.2]nona-3,6-diene-8α-carbonitrile, 1-acetoxy-2-oxo-	MeOH	224(3.86),332(2.09)	88-0443-69
8β-	MeOH	225(3.84),332(1.79)	88-0443-69
Ketone, 2-furyl 2-methoxy-3H-azepin-3-yl	MeOH	272(4.25)	78-5205-69
2-Oxazolepropionic acid, 5-phenyl-	acid	263(4.29)	4-0707-69
	neutral	265(4.33),271(4.32)	4-0707-69
	base	266(4.32),272(4.32)	4-0707-69
$C_{12}H_{11}NO_3S$			
3-Quinolinecarboxylic acid, 1,4-dihydro-4-hydroxy-2-(methylthio)-, methyl ester	MeOH	228(4.22),264(4.39), 320s(3.90),332s(3.18)	78-4649-69

Compound	Solvent	$\lambda_{max}(\log \epsilon)$	Ref.
$C_{12}H_{11}NO_4$			
2H-1,4-Benzoxazine-3-acetic acid, 2-oxo-, ethyl ester	$CHCl_3$	375(4.23)	49-1274-69
Chromone, 2-(carbethoxyamino)-	EtOH-piperidine	231(4.32),284(4.10)	103-0019-69
	$CHCl_3$	273(4.12),285s(4.00)	103-0019-69
2,4,6-Heptatrienal, 2-methyl-7-(5-nitro-2-furyl)-	dioxan	230(3.82),309(4.38), 402(4.65)	94-0306-69
2,4,6-Heptatrienal, 4-methyl-7-(5-nitro-2-furyl)-	dioxan	233(3.92),311(4.31), 404(4.64)	94-0306-69
2,4,6-Heptatrienal, 6-methyl-7-(5-nitro-2-furyl)-	dioxan	235(3.99),312(4.32), 399(4.67)	94-0306-69
Isoindole-1,3-dicarboxylic acid, dimethyl ester	EtOH	223(4.27),248(4.58), 345(4.35),362(4.37)	32-1115-69
$C_{12}H_{11}N_3$			
Aniline, p-(phenylazo)-	n.s.g.	252(3.96),387(4.41)	40-0207-69
Triazene, 1,3-diphenyl-	MeOH	237(4.18),295(3.94), 360(4.29)	65-0059-69
copper complex	benzene	335(4.33),615(2.69)	65-0059-69
mercury complex	benzene	305(4.18),390(4.40)	65-0059-69
$C_{12}H_{11}N_3O$			
Benzimidazole, 5-amino-2-(2-furyl)-1-methyl-	EtOH	210(4.32),240(4.10), 289(4.07),334(4.12)	73-0572-69
2,5-Cyclohexadien-1-imine, 3-amino-4-(p-hydroxyphenyl)imino-	pH 10.0	633(4.46)	39-0823-69B
Indolino[5,6]pyrazine, 1-acetyl-	MeOH	261(4.47),360(4.08)	103-0199-69
5H-Pyrrolo[3,4-d]pyrimidin-4-ol, 6,7-dihydro-6-phenyl-	10% NaOH	238(4.29),298s(3.32)	4-0507-69
$C_{12}H_{11}N_3O_3$			
1H-Pyrrolo[2,3-c]pyridine-2-carboxylic acid, 3-(cyanomethyl)-5-methoxy-, methyl ester	EtOH	282(4.16),290(4.23)	35-2338-69
$C_{12}H_{11}N_3O_3S$			
Thiazole, 2-(allylamino)-5-[2-(5-nitro-2-furyl)vinyl]-	EtOH	253(4.19),304(4.05), 410(4.21)	103-0414-69
$C_{12}H_{11}N_3O_4$			
2-Propanone, 1-[4-methyl-3-(p-nitrophenyl)-Δ^2-1,2,4-oxadiazolin-5-ylidene]-	EtOH	246(4.08),294(4.56)	18-3008-69
$C_{12}H_{11}N_3O_4S$			
Acetamide, N-[5-[1-methyl-2-(5-nitro-2-furyl)vinyl]-2-thiazolyl]-	EtOH	245(4.10),280(4.16), 395(4.28)	103-0414-69
$C_{12}H_{11}N_3S$			
Benzimidazole, 5-amino-1-methyl-2-(2-thienyl)-	EtOH	215(4.53),236s(4.15), 297s(4.00),355(4.14)	73-0572-69
$C_{12}H_{11}N_3S_2$			
5H-Imidazo[4,5-d]thiazole-5-thione, 4,6-dihydro-4,6-dimethyl-2-phenyl-	EtOH	249(4.07),280(4.15), 375(4.32)	94-0910-69
$C_{12}H_{11}N_3S_3$			
3a,6a-Epithio-4H-imidazo[4,5-d]thiazole-5(6H)-thione, 4,6-dimethyl-2-phenyl-	EtOH	274(3.99)	88-2875-69

Compound	Solvent	λ max (log ε)	Ref.
$C_{12}H_{11}N_5O$			
Purine, 2-amino-6-(benzyloxy)-	pH 1	286(4.08)	44-2160-69
	pH 11	282(3.97)	44-2160-69
	MeOH	241(3.88),282(3.96)	44-2160-69
3H-v-Triazolo[4,5-d]pyrimidine, 3-benzyl-7-methoxy-	aq. EtOH	252(4.05)	39-0152-69C
$C_{12}H_{11}N_5O_2$			
Acetamide, N-(6,7-dihydro-5-phenyl-[1,2,5]oxadiazolo[3,4-d]pyrimidin-7-yl)-	EtOH	242(4.08),292(3.95)	35-5181-69
$C_{12}H_{11}N_5O_8$			
Alanine, N-[(5,7-dinitro-2-oxo-1-benzimidazolinyl)acetyl]-	pH 10.07	310(4.00),380s(3.70)	44-0395-69
	EtOH	275(3.94),350(3.84)	44-0395-69
$C_{12}H_{11}N_5S$			
Pyrimido[5,4-e]-as-triazine, 5-(benzylthio)-1,2-dihydro-	pH 1	356(3.76)	44-3161-69
$C_{12}H_{12}$			
Benzene, 2,5-cyclohexadien-1-yl-	C_6H_{12}	257(2.6),263(2.7), 270(2.6)	64-0524-69
Benzocyclooctene, 4a,10a-dihydro-, trans	C_6H_{12}	252(3.78)	89-0069-69
Benzocyclooctene, 5,6-dihydro-	hexane	264(3.81)	39-0474-69C
Benzonornornene, 2-methylene-, (+)-	isooctane	196(4.53),216s(3.96), 224s(3.89),228s(3.73), 254s(2.61),261(2.88), 267(3.07),274(3.09)	35-0645-69
Biphenylene, 4a,4b,8a,8b-tetrahydro-	C_6H_{12}	266(3.61)	89-0069-69
$C_{12}H_{12}BrN_5O_4$			
7H-Pyrrolo[2,3-d]pyrimidine-5-carbonitrile, 4-amino-6-bromo-7-β-D-ribofuranosyl-	pH 1	231(4.21),281(4.24)	35-2102-69
	pH 11	283(4.25)	35-2102-69
	EtOH	284(4.26)	35-2102-69
$C_{12}H_{12}ClNO_2$			
Sorbohydroxamic acid, N-(p-chlorophenyl)-	EtOH	270(4.40)	34-0278-69
$C_{12}H_{12}ClNO_4S_2$			
2-Ethyl-3-methylthiazolo[2,3-b]benzothiazolium perchlorate	EtOH	217(4.35),262(3.83), 315(3.96)	95-0469-69
3-Ethyl-2-methylthiazolo[2,3-b]benzothiazolium perchlorate	EtOH	219(4.72),263(4.18), 315(4.34)	95-0469-69
$C_{12}H_{12}Cl_2N_4$			
Pyrimidine, 2,4-diamino-5-(3,4-dichlorophenyl)-6-ethyl-	pH 2	201(4.65),208(4.65), 272(3.92)	44-0821-69
	pH 12	285(4.00)	44-0821-69
$C_{12}H_{12}Cl_2O_2$			
2-Indancarboxylic acid, 1-(dichloromethyl)-, methyl ester, trans	EtOH	232(4.32),236(4.29), 264(4.30),280(4.02)	44-2792-69
$C_{12}H_{12}F_4N_2$			
Cyclohexanone, (2,3,4,5-tetrafluorophenyl)hydrazone	EtOH	260(4.20),294(3.95)	70-0609-69

Compound	Solvent	λ_{max}(log ϵ)	Ref.
$C_{12}H_{12}INO$			
2,3-Dihydro-5-methylfuro[3,2-c]quinolinium iodide	EtOH	230(4.75),294(3.69), 305(3.69),320(3.66)	2-0952-69
$C_{12}H_{12}Mg_3O_{18}P_2$			
L-Ascorbic acid, 3-phosphate, magnesium derivative	pH 1	237(3.99)	94-0381-69
	pH 13	261(4.21)	94-0381-69
$C_{12}H_{12}N_2$			
2,2'-Bipyridine, 4,4'-dimethyl-, sodium derivative	THF	282(4.3),394(4.0), 445s(3.8),546(3.5), 585(2.7),805(2.7), 900(2.6)	18-2264-69
2,2'-Bipyridine, 5,5'-dimethyl-, sodium derivative	THF	282(4.08),395(3.94), 435s(3.25),560(3.36), 590(3.35),811s(2.90), 916(3.07),1060(3.06)	18-2264-69
$C_{12}H_{12}N_2O$			
1(4H)-Naphthalenone, 2-(methylamino)-4-(methylimino)-	EtOH	238(4.41),270(4.18), 280s(4.14),337(3.67), 425(3.68)	39-1799-69C
2-Pyrrolidinone, 5-indol-3-yl-	EtOH	220(4.55),280(3.79), 287s(3.70)	117-0271-69
$C_{12}H_{12}N_2O_2$			
Acrylic acid, 2-cyano-3-(N-methylanilino)-, methyl ester	n.s.g.	230(3.78),304(4.39)	35-6689-69
Carbamic acid, (α-cyano-β-methylstyryl)-, methyl ester, cis	EtOH	263(4.03)	22-1724-69
trans	EtOH	260(4.05)	22-1724-69
Cinnamamide, α-cyano-p-methoxy-β-methyl-, cis	EtOH	224(4.11),300(4.08)	22-1724-69
trans	EtOH	222(4.08),304(4.04)	22-1724-69
5,8-Isoquinolinedione, 7-(ethylamino)-3-methyl-	MeCN	240(4.24),277(4.23), 320(3.72)	4-0697-69
Phthalimide, N-(2,3-dimethyl-1-aziridinyl)-, cis	EtOH	230(4.46),270(3.92)	27-0405-69
trans	EtOH	236(4.40),270s(3.77)	27-0405-69
1H-Pyrrolo[1,2-a]indol-2(3H)-one, 7-methoxy-, oxime	EtOH	203(4.49),220(4.49), 276(3.90)	88-0101-69
3-Quinolinecarboxaldehyde, 4-(dimethylamino)-1,2-dihydro-2-oxo-	dioxan	238(4.51),287(3.73), 359(3.99)	33-2641-69
2(1H)-Quinoxalinone, 3-acetonyl-1-methyl-	pH -3.0	216(4.42),254(3.90), 259(3.89),370s(3.94), 392(4.19),412(4.40), 438(4.37)	24-4032-69
	pH 4.43	223(4.45),227s(4.43), 255(3.82),272(3.87), 279(3.85),382s(4.21), 401(4.35),424(4.25)	24-4032-69
$C_{12}H_{12}N_2O_2S$			
1H-1,2-Diazepine, 1-(p-tolylsulfonyl)-	MeOH	223(4.22),350(2.46)	22-2175-69
$C_{12}H_{12}N_2O_2S_2$			
3-Pyridinol, 2,2'-dithiobis[6-methyl-	pH 1	310(3.94),345(3.95)	1-1704-69
	pH 13	240(4.12),335(3.96)	1-1704-69

Compound	Solvent	λ_{max}(log ϵ)	Ref.
$C_{12}H_{12}N_2O_3$			
2,6-Adamantanedione, 4-(diazoacetyl)-	MeOH	250(3.95),275(3.87)	78-5601-69
2-Quinoxalineacetic acid, 3,4-dihydro-3-oxo-, ethyl ester	$CHCl_3$	390(4.34)	49-1274-69
Uracil, 3-benzyl-5-hydroxy-1-methyl-	pH 1	286(3.86)	44-2636-69
	pH 10	243(3.85),312(3.82)	44-2636-69
$C_{12}H_{12}N_2O_3S$			
Acetic acid, [[o-(diazoacetyl)phenyl]-thio]-, ethyl ester	EtOH	237(4.23),290(4.04), 326s(3.51)	44-1566-69
$C_{12}H_{12}N_2O_4$			
Benzene, p-bis(2-nitropropenyl)-	EtOH	226(4.06),335(4.40)	23-4076-69
Carbamic acid, [(3-hydroxy-5-isoxazol-yl)methyl]-, benzyl ester	pH 12	210(4.09),235(3.46), 257(2.69),262(2.51), 267(2.32)	33-0720-69
	EtOH	204(4.21),235s(3.28), 236(2.69),262(2.51), 267(2.32)	33-0720-69
Hydantoin, 5-(2,4-dimethoxybenzylidene)-	neutral	232(3.64),344(4.28)	122-0455-69
	anion	245(3.83),340(4.28)	122-0455-69
$C_{12}H_{12}N_2O_4S_2$			
[2,2'-Bithiazole]-4,4'-diol, 5,5'-di-methyl-, diacetate	EtOH	346(4.40)	44-2053-69
$C_{12}H_{12}N_2O_5$			
1H-Pyrrolo[2,3-c]pyridine-3-propionic acid, 2-carboxy-5-methoxy-	EtOH	283(4.11),292(4.20), 350(3.64)	35-2338-69
$C_{12}H_{12}N_4$			
9H-Indeno[2,1-d]pyrimidine, 2,4-diamino-6(7)-methyl-	pH 1	272(4.34),285s(4.13)	4-0613-69
	EtOH	275(4.36),298(4.29)	4-0613-69
p-Phenylenediamine, N-(2-amino-4-imino-2,5-cyclohexadien-1-ylidene)-	pH 8	552(4.13)	39-0827-69B
5H-Pyrrolo[3,4-d]pyrimidine, 4-amino-6,7-dihydro-6-phenyl-	10% HCl	247(4.26),300s(3.28)	4-0507-69
$C_{12}H_{12}N_4O$			
5H-Pyrrolo[3,4-d]pyrimidin-4-ol, 2-ami-no-6,7-dihydro-6-phenyl-	10% NaOH	242(4.35),270s(4.08), 305s(3.38)	4-0507-69
$C_{12}H_{12}N_4O_3$			
p-Anisaldehyde, (2,6-dihydroxy-4-pyrim-idinyl)hydrazone	EtOH	331(4.27)	56-0519-69
$C_{12}H_{12}N_4O_4$			
2-Cyclohexen-1-one, 2,4-dinitrophenyl-hydrazone	MeOH	254(4.25),382(4.46)	24-3647-69
$C_{12}H_{12}N_4S$			
7H-s-Triazolo[3,4-b][1,3,4]thiadiazine, 3-ethyl-7-phenyl-	EtOH	284(4.18)	2-0959-69
s-Triazolo[3,4-b][1,3,4]thiadiazole, 3-ethyl-6-p-tolyl-	EtOH	212s(4.31),278(4.34)	2-0959-69
$C_{12}H_{12}N_6O_7$			
Glycine, N-[N-(5,7-dinitro-4-benzimid-azolyl)-L-alanyl]-	10%EtOH-HCl	228(3.90),295(3.89), 362(4.05),415(3.98)	44-0384-69

Compound	Solvent	$\lambda_{max}(\log \epsilon)$	Ref.
$C_{12}H_{12}N_8O$			
Imidazole-4-carbonitrile, 5,5'-azoxybis-[1,2-dimethyl-	EtOH	208(3.75),260s(--), 406(3.74)	4-0053-69
Imidazole-5-carbonitrile, 4,4'-azoxybis-[1,2-dimethyl-	EtOH	214(3.80),262s(--), 411(3.69)	4-0053-69
$C_{12}H_{12}N_8O_4$			
Formamide, N-[9-(5-azido-2,5-dideoxy-β-D-erythropentofuranosyl)-9H-purin-6-yl]-, formate	MeOH	272(4.30)	87-0658-69
$C_{12}H_{12}O$			
2-Buten-1-one, 1-(1,3,5,7-cycloocta-tetraen-1-yl)-	EtOH	244(4.11)	39-0978-69C
5,8-Ethenocycloocta[c]furan, 1,3,5,8-tetrahydro-	MeOH	265(3.58),272(3.51), 286s(3.34)	88-1125-69
1-Naphthalenemethanol, α-methyl-, (-)	MeOH	224(4.30),281(3.84)	19-0475-69
2-Naphthalenemethanol, α-methyl-, (-)	MeOH	224(4.90),274(3.58)	19-0475-69
5-Oxatricyclo[7.2.2.0^{3,7}]trideca-2,7,10,12-tetraene	MeOH	265(3.53),272(3.52), 286s(3.34)	108-0435-69
2,4-Pentadienal, 2-methyl-5-phenyl-	CH_2Cl_2	237(3.95),245s(3.87), 323(4.66)	24-0623-69
2-Propanone, 1-(1-indanylidene)-	n.s.g.	294(4.07),317(4.15)	39-0944-69C
$C_{12}H_{12}OS$			
Benzo[b]thiophene, 6-acetyl-2,3-dimeth-yl-	EtOH	220(4.19),230(4.06), 255(4.14),296(4.05), 333(3.81)	22-0607-69
6H-Cyclohepta[c]thiophen-6-one, 1,3,5-trimethyl-	EtOH	230(4.07),277(4.58), 335(3.94)	104-0559-69
	$HOAc-HClO_4$	257(3.92),307(4.61), 355(3.98),370(4.06), 525(3.20)	104-0559-69
$C_{12}H_{12}OS_2$			
Thiepino[3,2-b][1]benzothiophene-5-ol, 2,3,4,5-tetrahydro-	EtOH	232(4.42),278(3.82), 294(3.77),303(3.81)	22-0601-69
$C_{12}H_{12}O_2$			
2-Cyclopropene-1-carboxylic acid, 2-methyl-3-m-tolyl-	EtOH	219s(4.40),227s(4.32), 261(4.28)	44-0505-69
2,8-Decadiene-4,6-diyn-1-ol, acetate	ether	236(4.46),246(4.38), 260(3.90),275(4.14), 292(4.31),312(4.22)	24-1679-69
4a,8a-Ethenonaphthalene-9,10-dione, 1,4,5,8-tetrahydro-	n.s.g.	538(1.86)	88-1529-69
Indene-2-acetic acid, α-methyl-	EtOH	260(4.16)	88-2331-69
2H-Indeno[1,2-b]furan-2-one, 3,3a,4,8b-tetrahydro-3-methyl-	EtOH	260(2.84),266(3.03), 273(3.06)	88-2331-69
Ketone, 3-oxabicyclo[3.1.0]hex-6-yl phenyl, cis	EtOH	244(4.05),278(2.90)	88-2775-69
2,7-Methano-1H-cyclopropa[b]naphthalene-3,6-diol, 1a,2,7,7a-tetrahydro-	EtOH	293(3.36)	35-7763-69
2,4-Pentadienoic acid, 5-methyl-5-phen-yl-, 2-cis-4-trans	EtOH	228(4.06),305(4.26)	22-4447-69
$C_{12}H_{12}O_3$			
Benzene, 1,3,5-triacetyl-	EtOH	226(4.75),246s(3.56), 292(2.74),321s(2.20)	1-0751-69

Compound	Solvent	$\lambda_{max}(\log \epsilon)$	Ref.
Benzo[b]furan, 2-isopropyl-5,6-(methylenedioxy)-	EtOH	247(4.04),251(4.03), 255(4.04),305(3.97), 309(3.97)	18-1971-69
2-Cyclopropene-1-carboxylic acid, 2-(p-methoxyphenyl)-3-methyl-	EtOH	228s(4.45),269(4.43)	44-0505-69
3-Furanol, 2-(2,4-hexadiynylidene)-tetrahydro-, acetate, (R)-trans (only relative absorbancies)	ether	251s(--),264(1.0), 277(1.4),293(1.1)	39-1096-69C
1,3,5-Hexanetrione, 1-phenyl-	EtOH	247(3.75),345(4.16)	94-2126-69
1-Indanylideneacetic acid, 5-methoxy-, exo	MeOH	320(4.28)	44-4182-69
Indene-3-acetic acid, 6-methoxy-, endo	MeOH	263(4.09)	44-4182-69
1,4-Methanonaphthalene-5,8-dione, 1,4,4a,8a-tetrahydro-6-methoxy-, cis-endo	EtOH	268(4.01)	39-0105-69C
1-Naphthol, 2,4-dimethoxy-	MeOH	242(4.44),313(3.61)	44-2788-69
$C_{12}H_{12}O_3S$			
Thiochroman-2-carboxylic acid, 3-oxo-, ethyl ester	EtOH	229(4.03),261(3.95), 299(3.42)	44-1566-69
$C_{12}H_{12}O_4$			
Benzo[b]furan, 2-(1-hydroxyisopropyl)-5,6-(methylenedioxy)-	EtOH	247(4.03),251(4.03), 255(4.03),306s(3.98), 309(3.98),331s(3.10)	18-1971-69
Chromone, 5-hydroxy-7-methoxy-2,6-dimethyl- (eugenitin)	MeOH	211(4.24),230(4.20), 251(4.16),257(4.16), 292(3.93),317s(3.57)	39-0704-69C
	MeOH-NaOH	241(4.03),251(4.05), 268(4.14),363(3.44)	39-0704-69C
Spiro[1,3-dioxolan-2,5'(6'H)-indene]-4,6'-dione, 4'-methyl-	EtOH	355(3.49)	23-0515-69
$C_{12}H_{12}O_5$			
Bicyclo[3.2.2]nona-3,6-diene-8α,9α-dicarboxylic acid, 3-methyl-2-oxo-	MeOH	236(3.78),260s(3.3), 327(2.24)	88-0443-69
Chromone, 5-hydroxy-3,7-dimethoxy-2-methyl-	MeOH	248(4.30),292(3.84), 322(3.69)	64-0225-69
$C_{12}H_{12}S$			
Sulfide, 2-penten-4-ynyl p-tolyl	EtOH	227(4.32)	18-2589-69
$C_{12}H_{12}S_2$			
Thiepino[3,2-b][1]benzothiophene, 2,3,4,5-tetrahydro-	EtOH	233(4.46),278(3.75), 296(3.68),305(3.71)	22-0601-69
$C_{12}H_{12}Si$			
Silane, diphenyl-	C_6H_{12}	265(3.98)	101-0017-69B
$C_{12}H_{13}BF_2O_2$			
Boron, difluoro(4-methyl-1-phenyl-1,3-pentanedionato)-	$CHCl_3$	265(3.69),333(4.50), 346s(4.32)	39-0526-69A
Boron, difluoro(p-phenyl-1,3-hexanedionato)-	$CHCl_3$	265(3.69),330(4.51), 346s(4.35)	39-0526-69A
$C_{12}H_{13}BF_4OS$			
6-Methoxy-1,3-dimethylcyclohepta[c]-thiolium tetrafluoroborate	$C_2H_4Cl_2$	264(4.00),312(4.67), 359(4.03),374(4.32), 535(3.26)	104-0947-69

Compound	Solvent	$\lambda_{max}(\log \epsilon)$	Ref.
$C_{12}H_{13}Br$			
Indene, 3-(3-bromopropyl)-	EtOH	223(3.95),252(3.95), 279(3.09),289(2.95)	39-0345-69C
$C_{12}H_{13}BrN_2O_2S_2$			
2H-1,3,5-Thiadiazine-3(4H)-carboxylic acid, 5-(p-bromobenzyl)dihydro-6-thioxo-, methyl ester	MeOH	240(4.14),290(4.10)	73-2952-69
$C_{12}H_{13}BrN_4S$			
4H-1,2,4-Triazole-3-thiol, 4-[[-(bromomethyl)benzylidene]amino]-5-ethyl-	EtOH	282(4.15)	2-0959-69
$C_{12}H_{13}ClN_2$			
2,3-Dihydro-7-methyl-1H-imidazo[1,2-a]-quinolinium chloride	EtOH	243(4.59),261(4.19), 295(4.08),354(4.17)	95-0767-69
2,3-Dihydro-8-methyl-1H-imidazo[1,2-a]-quinolinium chloride	EtOH	243(4.62),261(4.22), 295(4.12),354(4.21)	95-0767-69
$C_{12}H_{13}ClN_4$			
Pyrimidine, 2,4-diamino-5-(p-chlorobenzyl)-6-methyl-	pH 2	275(3.87)	44-0821-69
	pH 12	218(4.33),238s(4.07), 285(3.91)	44-0821-69
Pyrimidine, 2,4-diamino-5-(p-chlorophenyl)-6-ethyl-	pH 2	223s(4.36),272(3.86)	44-0821-69
	pH 12	220s(4.30),248s(3.87), 285(3.94)	44-0821-69
$C_{12}H_{13}ClO_4S_2$			
4-Phenyl-2-propyl-1,3-dithiol-1-ium perchlorate	50% H_2SO_4	240(4.21),271s(3.96), 340(3.71)	94-1931-69
$C_{12}H_{13}ClO_5S$			
6-Hydroxy-1,3,5-trimethylcyclohepta[c]-thiolium perchlorate	HOAc-$HClO_4$	257(3.92),303(4.60), 355(3.98),370(4.06), 525(3.20)	104-0947-69
$C_{12}H_{13}Cl_2N_3O_2$			
2H-Pyran-2-methanol, 6-(5,6-dichloro-1H-benzotriazol-1-yl)tetrahydro-, cis	EtOH	264(3.79),271(3.77), 296(3.63)	4-0005-69
trans	EtOH	266(3.79),272(3.79), 296(3.56)	4-0005-69
$C_{12}H_{13}Fe$			
α-Ferrocenylethyl cation	58% H_2SO_4	258(4.04),375(2.62)	35-0509-69
$C_{12}H_{13}IN_2O$			
1-Methyl-2-phenacylpyrazolium iodide	EtOH	221(4.38),245(4.28), 282(3.34),330(2.60)	44-4134-69
$C_{12}H_{13}N$			
Lepidine, 6-ethyl-	MeOH	227(4.71),280(3.68), 304(3.56),317(3.56)	106-0196-69
3,3-Spirocyclopentylindolenine	EtOH	215(4.13),220(4.18), 226(4.12),267(3.96)	78-4843-69
	EtOH-HCl	227(4.13),276(3.81)	78-4843-69
$C_{12}H_{13}NO$			
Atroponitrile, β-propoxy-	EtOH	212(4.13),268(4.15)	22-0198-69
Crotononitrile, 3-ethoxy-2-phenyl-, cis	EtOH	208(4.03),264(4.07)	22-0198-69
trans	EtOH	211(4.14),269(4.22)	22-0198-69

Compound	Solvent	λ_{max}(log ϵ)	Ref.
3,6-Ethenocyclohepta[b]pyrrole, 3,3a,6,8a-tetrahydro-2-methoxy-	EtOH	245s(2.22)	35-3970-69
Indole-2-acetaldehyde, α,α-dimethyl-	EtOH	222(4.31),282(3.90), 290(3.79)	39-0595-69C
Indole-3-acetaldehyde, α,α-dimethyl-	EtOH	220(4.52),280(3.80), 289(3.71)	39-0595-69C
Isoquinoline, 6-methoxy-5,8-dimethyl-	EtOH	240(4.81),309(3.80)	12-2489-69
Ketone, 3,3-dimethyl-3H-indol-2-yl methyl	EtOH	235(3.85),240(3.87), 303(4.00)	78-2757-69
Pyrrole, 4-(p-methoxyphenyl)-2-methyl-	EtOH	273(4.02)	94-0582-69
2-Pyrrolidinone, N-styryl-, cis	MeOH	263(4.20)	78-5745-69
2-Pyrrolidinone, N-styryl-, trans	MeOH	220(--),286(4.40)	78-5745-69
$C_{12}H_{13}NOS_2$			
2-Pentanone, 3-(2-benzothiazolylthio)-	EtOH	226(4.33),245(3.99), 278(4.12),290(4.05), 301(3.96)	95-0469-69
3-Pentanone, 2-(2-benzothiazolylthio)-	EtOH	224(4.12),245(3.77), 280(3.73),290(3.73), 301(3.67)	95-0469-69
$C_{12}H_{13}NO_2$			
4-Azatricyclo[7.2.1.0^{2,8}]dodeca-6,10-diene-3,5-dione, 4-methyl-, endo	MeOH	220(3.81)	44-1695-69
exo	MeOH	220(3.84)	44-1695-69
2H-1-Benzazepin-2-one, 7-acetyl-1,3,4,5-tetrahydro-	MeOH	218(4.16),275(4.22)	44-1070-69
Benzo[a]cyclopropa[c]cycloheptene, 1,1a,2,3,4,8b-hexahydro-6-nitro-	C_6H_{12}	273(3.97)	35-3558-69
	EtOH	280(3.94)	35-3558-69
Benzo[a]cyclopropa[c]cycloheptene, 1,1a,2,3,4,8b-hexahydro-7-nitro-	C_6H_{12}	279(3.99)	35-3558-69
	EtOH	285(3.96)	35-3558-69
2-Cyclohexen-1-one, 4-(N-hydroxyanilino)-	EtOH	237(4.1),282(3.1)	12-2493-69
2(3H)-Furanone, dihydro-3-(m-toluidinomethylene)-	EtOH	236(5.50),295(3.78), 320(4.66)	2-0952-69
2(3H)-Furanone, dihydro-3-(p-toluidinomethylene)-	EtOH	223(4.00),298s(4.25), 320(4.41)	2-0952-69
3-Morpholinone, N-styryl-, cis	MeOH	259(4.17)	78-5745-69
3-Morpholinone, N-styryl-, trans	MeOH	220(--),285(4.33)	78-5745-69
2,4-Pyrrolidinedione, 3,3-dimethyl-1-phenyl-	EtOH	207(4.22),250(3.92)	39-0595-69C
Pyrrolo[2,1-b]oxazol-5(6H)-one, tetrahydro-7a-phenyl-	EtOH	251(2.70),259(2.74), 267(2.70)	44-0165-69
Sorbohydroxamic acid, N-phenyl-	EtOH	266(4.39)	34-0278-69
Spiro[cyclopropane-1,1'(2'H)-naphthalene], 3',4'-dihydro-6'-nitro-	C_6H_{12}	292(3.99)	35-3558-69
	EtOH	306(3.98)	35-3558-69
$C_{12}H_{13}NO_3$			
Acrylic acid, 2-maino-3-benzoyl-, ethyl ester	EtOH	230(3.80),259(3.84), 346(4.31)	48-0786-69
	$HClO_4$	247(3.57),257s(3.56), 373(4.24),455s(2.68)	48-0786-69
Carbostyril, 3-ethyl-4-hydroxy-7-methoxy-	EtOH	217(4.65),235s(4.02), 245(3.91),270s(3.73), 277(3.77),309(4.09), 321(4.10)	12-0447-69
2(3H)-Furanone, 3-(p-methoxyanilinomethylene)dihydro-	EtOH	214(4.01),306(4.46), 324(4.47)	2-0952-69
Phthalimide, N-(4-hydroxybutyl)-	MeOH	222(4.40),233s(4.09), 242(3.95),297(3.20)	78-5155-69

Compound	Solvent	$\lambda_{max}(\log \epsilon)$	Ref.
2,4-Quinolinedimethanol, α^4-methyl-8-hydroxy-, hydrochloride	H_2O	243(4.57)	35-4934-69
	acid	256(--)	35-4934-69
	base	258(--)	35-4934-69
$C_{12}H_{13}NO_3S$			
4,1-Benzothiazepin-2(3H)-one, 3-acetoxy-1,5-dihydro-3-methyl-	iso-PrOH	207(4.46),243(3.76)	44-0365-69
Butyronitrile, 4-benzoyl-4-(methylsulfonyl)-	EtOH	255(4.12)	108-0057-69
Carbonic acid, 2-isothiocyanato-3-phenylpropyl methyl ester, D-	EtOH	259(4.20)	39-1143-69B
L-isomer	EtOH	259(4.22)	39-1143-69B
$C_{12}H_{13}NO_4$			
2-Cyclopenten-1-one, 3-ethyl-5-methyl-2-(5-nitro-2-furyl)-	EtOH	227(4.18),272(3.81),352(4.18)	94-1757-69
2-Cyclopenten-1-one, 5-ethyl-3-methyl-2-(5-nitro-2-furyl)-	EtOH	226(4.15),270(3.76),350(4.14)	94-1757-69
1(2H)-Isoquinolone, 5-hydroxy-6,7-dimethoxy-2-methyl-	EtOH	247(4.56),270(3.75),281(3.78),293(3.81),322(3.51)	88-1951-69
2,4-Pentadienal, 2-ethyl-4-methyl-5-(5-nitro-2-furyl)-	EtOH	220(3.94),285(4.02),383(4.40)	94-1757-69
2,4-Pentadienal, 4-ethyl-2-methyl-5-(5-nitro-2-furyl)-	EtOH	222(4.06),285(4.00),383(4.37)	94-1757-69
$C_{12}H_{13}NO_6$			
1H-Azepine-1,4,5-tricarboxylic acid, trimethyl ester	EtOH	207(4.24),346(3.15)	23-2391-69
o-Xylene-α,α-diol, 3-nitro-, diacetate	EtOH	209(4.1),243(3.55)	39-1935-69C
$C_{12}H_{13}NS$			
2H-1,4-Benzothiazine, 3-methyl-2-propylidene-	MeOH	360(3.46)	124-1278-69
	$MeOH-HClO_4$	440(3.40)	124-1278-69
$C_{12}H_{13}NS_4$			
1,3-Dithiole, 2-(dimethylaminothiocarbonylthio)-4-phenyl-	EtOH	234(4.12),315(3.91)	94-1931-69
$C_{12}H_{13}N_3$			
6H-Pyrrolo[2,3-g]quinoxaline, 7,8-dihydro-2,3-dimethyl-	MeOH	220(4.64),262(4.46),397(3.99)	103-0199-69
as-Triazine, 5,6-dimethyl-3-p-tolyl-	MeOH	264(4.32),362(2.61)	88-3147-69
$C_{12}H_{13}N_3O$			
2-Pyrazolin-5-one, 3-methyl-4-[(methylamino)methylene]-1-phenyl-	heptane	225(3.99),258(4.35),298(4.40),345(3.53)	104-1249-69
	EtOH	232(3.99),258(4.24),292(4.36)	104-1249-69
Pyrrolo[2,3-f]benzimidazole, 5-acetyl-1,5,6,7-tetrahydro-2-methyl-	MeOH	232(4.42),260(3.90),270(3.87),302(4.12),312(4.14)	103-0196-69
$C_{12}H_{13}N_3O_2$			
Benzeneazo-1'-cyclohexene, 4-nitro-	THF	330(4.40)	24-3467-69
Cinnamic acid, α-azido-m-methyl-, ethyl ester	C_6H_{12}	228(4.02),313(4.34)	49-1599-69
Cinnamic acid, α-azido-o-methyl-, ethyl ester	C_6H_{12}	227(4.00),307(4.23)	49-1599-69
Cinnamic acid, α-azido-p-methyl-, ethyl ester	C_6H_{12}	229(3.98),316(4.38)	49-1599-69

Compound	Solvent	$\lambda_{max}(\log \epsilon)$	Ref.
2-Cyclohexen-1-one, (o-nitrophenyl)hydrazone	MeOH	280(4.23),308(4.35), 456(3.89)	24-3647-69
Hydantoin, 5-[(p-dimethylamino)benzylidene]-	neutral	237(3.87),390(4.37)	122-0455-69
anion	n.s.g.	248(4.17),368(4.39)	122-0455-69
$C_{12}H_{13}N_3O_3$			
Cinnamic acid, α-azido-m-methoxy-, ethyl ester	C_6H_{12}	232(4.07),326(4.15)	49-1599-69
Cinnamic acid, α-azido-o-methoxy-, ethyl ester	C_6H_{12}	282(3.57)	49-1599-69
Cinnamic acid, α-azido-p-methoxy-, ethyl ester	C_6H_{12}	227(4.13),311(4.31)	49-1599-69
Hydantoin, 1-(benzylideneamino)-5-(2-hydroxyethyl)-	H_2O	288(4.32)	44-3213-69
4(3H)-Quinazolinone, 3-butyl-6-nitro-	EtOH	212(4.44),224(4.41), 320(4.11)	2-1166-69
$C_{12}H_{13}N_3O_4$			
Acetamide, N-(1-acetyl-5-nitro-6-indolinyl)-	MeOH	236(4.63),267(4.52), 332(3.97),348(3.99)	103-0196-69
Acetamide, N-(1-acetyl-6-nitro-5-indolinyl)-	MeOH	262(4.36),375(3.35)	103-0196-69
2-Benzimidazolemethanol, 1-ethyl-5-nitro-, acetate	n.s.g.	242(4.52),303(4.12)	39-0070-69C
$C_{12}H_{13}N_3O_4S_2$			
4,6-Dimethyl-2-phenyl-4H-imidazo[4,5-d]thiazolium bisulfate	EtOH	227(3.93),240(3.88), 308(4.37)	94-0910-69
$C_{12}H_{13}N_5$			
5H-Pyrrolo[3,4-d]pyrimidine, 2,4-diamino-6,7-dihydro-6-phenyl-	10% HCl	230s(--),272(--)	4-0507-69
$C_{12}H_{13}N_5O$			
2-Cyclohexen-1-one, 3-[(purin-6-ylamino)methyl]-	pH 1	213(4.24),229s(4.16), 278(4.21)	44-3814-69
	pH 13	273(4.12)	44-3814-69
	EtOH	212(4.34),233(4.15), 268(4.25)	44-3814-69
Propionic acid, [(3-phenyl-1H-1,2,4-triazol-1-yl)methylene]hydrazide	MeOH	290(4.40)	48-0646-69
Purino[1,6-a]quinazolin-5(1H)-one, 5a,6,7,8,9,9a-hexahydro-	pH 1	220(4.47),286(4.20)	44-3814-69
	pH 13	243(4.26),316(4.40)	44-3814-69
	EtOH	230(4.28),310(4.27), 320s(4.20)	44-3814-69
$C_{12}H_{13}N_5O_4$			
7H-Pyrrolo[2,3-d]pyrimidine-5-carbonitrile, 4-amino-7-β-D-ribofuranosyl-	pH 1	232(4.20),272(4.09)	35-2102-69
	pH 11	227(4.01),277(4.16), 287s(3.97)	35-2102-69
	EtOH	231(3.97),278(4.18), 288s(3.99)	35-2102-69
$C_{12}H_{14}$			
Benzene, (1-cyclohexen-1-yl)-	C_6H_{12}	245(4.0)	64-0524-69
Benzene, (2-cyclohexen-1-yl)-	C_6H_{12}	248(2.4),253(2.5), 255(2.5),260(2.5), 264(2.2),268(2.3)	64-0524-69
Benzene, (3-cyclohexen-1-yl)-	C_6H_{12}	260(2.5)	64-0524-69

Compound	Solvent	λ_{max}(log ϵ)	Ref.
Benzene, o-dipropenyl-, (E,Z)-	n.s.g.	225(4.18),255(4.04)	28-0986-69A
Benzene, o-dipropenyl-, (Z,Z)-	n.s.g.	220(3.93),250(3.76)	28-0986-69A
5H-Benzocycloheptene, 6,7-dihydro-6-methyl-	n.s.g.	210(4.46),253(4.17)	28-0986-69A
4a,8a-Ethenonaphthalene, 1,2,3,4-tetrahydro-	isooctane	268(3.38),276s(3.32)	35-3973-69
	isooctane	268(3.38),276s(3.32)	77-0680-69
$C_{12}H_{14}BF_4N_5O$			
6a,7,8,9,10a,11,12-Octahydro-10-oxo-3H-quinazolino[2,1-i]purin-6-ium tetrafluoroborate	pH 1	212(4.24),264(4.13)	44-3814-69
	pH 13	274(4.23),283(4.16)	44-3814-69
	EtOH	212(4.27),268(4.23)	44-3814-69
7-Oxo-4-aza-2-azoniabicyclo[4.2.2]dec-2-eno[3,2-i]purine tetrafluoroborate	pH 1	211(4.22),265(4.14)	44-3814-69
	pH 13	276(4.24),284s(4.10)	44-3814-69
	EtOH	212(4.23),267(4.26)	44-3814-69
7-Oxo-3-aza-1-azoniaspiro[4.5]dec-1-eno[2,1-i]purine tetrafluoroborate	pH 1	214(4.36),263(4.09)	44-3814-69
	pH 13	272(4.11)(changing)	44-3814-69
	EtOH	214(4.31),232s(3.87), 264(4.05)	44-3814-69
$C_{12}H_{14}BrN$			
3H-Indole, 3-(2-bromoethyl)-2,3-dimethyl-	MeOH	260(3.66)	23-3647-69
$C_{12}H_{14}BrNO_2$			
Crotonic acid, 3-(benzylamino)-4-bromo-, methyl ester	EtOH	305(4.15)	78-3277-69
$C_{12}H_{14}BrN_3O_4S$			
7H-Pyrrolo[2,3-d]pyrimidine, 5-bromo-4-(methylthio)-7-β-D-ribofuranosyl-	pH 1	227(4.29),269(4.08), 295(3.96)	4-0215-69
	pH 11	227(4.37),257(3.82), 305(4.01)	4-0215-69
$C_{12}H_{14}BrN_5O_5$			
7H-Pyrrolo[2,3-d]pyrimidine-5-carboxamide, 4-amino-6-bromo-7-β-D-ribofuranosyl-	pH 1	280(4.20)	35-2102-69
	pH 11	283(4.18)	35-2102-69
	EtOH	285(4.19)	35-2102-69
$C_{12}H_{14}ClNO_5$			
4-Methoxy-1,2-dimethylquinolinium perchlorate	MeOH	229(4.54),285(3.73)	78-0255-69
$C_{12}H_{14}ClN_5O$			
6a,7,8,9,10,10a,11,12-Octahydro-8-oxo-3H-quinazolino[2,1-i]purin-6-ium chloride	pH 1	212(4.31),263(4.14)	44-3814-69
	pH 13	275(4.23),283s(4.11)	44-3814-69
	EtOH	212(4.27),264(4.11)	44-3814-69
6a,7,8,9,10,10a,11,12-Octahydro-11-oxo-3H-quinazolino[2,1-i]purin-6-ium chloride	pH 1	220(4.47),286(4.20)	44-3814-69
	pH 13	242(4.25),316(4.39)	44-3814-69
	EtOH	222(4.29),288s(4.08), 296(4.11),309(4.29), 320s(4.01)	44-3814-69
$C_{12}H_{14}F_3NO_2S$			
Piperidine, 1-[p-[(trifluoromethyl)-sulfonyl]phenyl]-	benzene	312(4.46)	65-1634-69
	MeOH	228(3.82),312(4.51)	65-1634-69
	DMF	316(4.53)	65-1634-69
$C_{12}H_{14}FeO_3$			
Iron, tricarbonyl(1-isopropyl-2,3-dimethyl-1,3-cyclobutadiene)-	hexane	206(4.35),290s(--)	78-1089-69

Compound	Solvent	$\lambda_{max}(\log \epsilon)$	Ref.
$C_{12}H_{14}IN$			
1-Ethylquinaldinium iodide	n.s.g.	320(4.01)	22-1284-69
$C_{12}H_{14}INO$			
4-Methoxy-1,2-dimethylquinolinium iodide	MeOH	236(4.60),307(3.95)	78-0255-69
$C_{12}H_{14}I_3N_3O_2$			
Isophthalamide, 5-amino-2,4,6-triiodo-N,N,N',N'-tetramethyl-, cis	EtOH	234(4.53),267s(4.00), 321(3.73)	88-3879-69
trans	EtOH	235(4.53),267s(4.00), 321(3.70)	88-3879-69
$C_{12}H_{14}N_2$			
Benzeneazo-1'-cyclohexene	THF	304(4.32)	24-3467-69
$C_{12}H_{14}N_2O$			
6H-Azepino[1,2-a]benzimidazol-6-ol, 7,8,9,10-tetrahydro-	n.s.g.	216(3.94),254(3.98), 269(3.86),276(3.94), 283(3.89)	39-0070-69C
Crotonaldehyde, α-acetylphenylhydrazone	EtOH	256s(4.4),263(4.4), 271s(4.4)	28-1703-69A
5H-Cyclohepta[c]pyridine-4-carbonitrile, 6,7,8,9-tetrahydro-3-hydroxy-1-methyl-	EtOH	218(4.21),222(4.21), 244s(3.85),252(3.73), 341(4.11)	44-3670-69
Ethanol, 2-[(7-methyl-2-quinolyl)-amino]-	EtOH	246(4.63),335(4.11), 338(4.06)	95-0767-69
Imidazole, 2-ethyl-4-methyl-5-phenyl-, 3-oxide	MeOH	263(4.06)	48-0746-69
1H-Imidazo[1,2-a]quinolin-3a(1H)-ol, 2,3-dihydro-8-methyl-	EtOH	242(4.41),261(3.91), 296(3.79),354(3.88)	95-0767-69
2-Pyrazoline, 3-phenyl-1-propionyl-	MeOH	290(4.26)	123-0041-69A
2-Pyrazolin-5-one, 3-benzyl-4,4-dimethyl-	EtOH	240(3.83)	22-4159-69
2H-2-Pyrindine-4-carbonitrile, 3,5,6,7-tetrahydro-1-isopropyl-3-oxo-	EtOH	218(4.27),223(4.28), 240(3.81),340(4.09)	44-3670-69
2H-2-Pyrindine-4-carbonitrile, 3,5,6,7-tetrahydro-3-oxo-1-propyl-	EtOH	218(4.26),223(4.28), 241(3.77),341(4.08)	44-3670-69
5H-Pyrrolo[1,2-a]imidazol-5-one, hexahydro-6a-phenyl-	EtOH	212(3.88),246(2.04), 251(2.19),257(2.29), 263(2.17)	44-0165-69
Pyrrol-3-ol, 5-(dimethylamino)-4-phenyl-	MeOH	222(4.08),287(4.30)	78-5721-69
	MeOH-HCl	241(4.21)	78-5721-69
4(3H)-Quinazolinone, 3-butyl-	EtOH	208(4.37),226(4.42), 267(3.43),278(3.44), 302(3.57),315(3.50)	2-1166-69
$C_{12}H_{14}N_2OS$			
Quinoxaline, 2-(tert-butylsulfinyl)-	pH 6.0	240(4.58),327(3.91)	39-0333-69B
$C_{12}H_{14}N_2O_2$			
2(3H)-Furanone, 4-acetyl-4,5-dihydro-, 4-(phenylhydrazone)	MeOH	205(4.30),272(4.35), 295s(3.90)	5-0217-69D
Imidazole, 2-ethyl-1-hydroxy-5-methyl-4-phenyl-, 3-oxide	MeOH	229(4.13)	48-0746-69
2-Pyrazolin-5-one, 3-(p-methoxyphenyl)-4,4-dimethyl-	EtOH	218(4.08),295(4.21)	22-4159-69
3(2H)-Pyridazinone, 4,5-dihydro-5-(hydroxymethyl)-6-methyl-2-phenyl-	MeOH	251s(3.75),280(4.13)	5-0217-69D

Compound	Solvent	$\lambda_{max}(\log \epsilon)$	Ref.
$C_{12}H_{14}N_2O_2S$			
Quinoxaline, 2-(tert-butylsulfonyl)-		242(4.64),321(3.86)	39-0333-69B
$C_{12}H_{14}N_2O_2S_2$			
2H-1,3,5-Thiadiazine-3(4H)-acetic acid, 5-benzyldihydro-6-thioxo-, triethyl amine salt	MeOH	250(4.03),290(4.15)	73-2952-69
$C_{12}H_{14}N_2O_4$			
Benzimidazole, 2-β-D-ribofuranosyl-, hydrochloride	H_2O	213(4.07),245(3.72), 272(3.89),278(3.89)	73-0247-69
hydrogen oxalate	H_2O	212(4.22),246(3.79), 275(3.90),282(3.88)	73-0247-69
$C_{12}H_{14}N_2O_6$			
Uridine, 5',6-anhydro-2',3'-O-isopropylidene-6-hydroxy-	pH 1	261(4.12)	44-1390-69
	pH 7	261(4.12)	44-1390-69
	pH 12	262(3.97)	44-1390-69
$C_{12}H_{14}N_2S$			
Pyrrolo[1,2-a]pyrimidine, 2,3,4,6,7,8-hexahydro-8-(2-thenylidene)-	H_2O	318(4.38)	87-1066-69
Quinoxaline, 2-(tert-butylthio)-	pH 6.0	243(4.43),266s(3.81), 333(3.79),350s(3.75)	39-0333-69B
$C_{12}H_{14}N_4$			
Pyrimidine, 2,4-diamino-5-benzyl-6-methyl-	pH 2	275(3.83)	44-0821-69
	pH 12	230s(4.05),285(3.88)	44-0821-69
$C_{12}H_{14}N_4OS$			
Acetophenone, 2-[(4-amino-5-ethyl-4H-1,2,4-triazol-3-yl)thio]-	EtOH	245(4.12)	2-0959-69
Indole-2,3-dione, 1-methyl-, 3-(4,4-dimethylthiosemicarbazone)	EtOH	240(4.1),270(4.0), 350(4.2)	65-0070-69
$C_{12}H_{14}N_4O_2S_2$			
2H-1,3-Thiazine-2,4(3H)-dione, 3-ethyldihydro-2-thio-, 4-[(p-nitrophenyl)-hydrazone]	EtOH	239(3.89),303(3.92), 405(4.41)	103-0049-69
$C_{12}H_{14}N_4O_3$			
4(3H)-Quinazolinone, 6-amino-3-butyl-5-nitro-	EtOH	225(4.36),293(4.18)	2-1166-69
$C_{12}H_{14}N_4O_4$			
Cyclopentanone, 3-methyl-, 2,4-dinitrophenylhydrazone	EtOH	225(4.22),260s(4.02), 360(4.32)	35-4739-69
$C_{12}H_{14}N_4O_4S_2$			
Barbituric acid, 5,5'-butylidenebis-[2-thio-	EtOH	287(4.11)	103-0830-69
Barbituric acid, 5,5'-sec-butylidene-bis[2-thio-	EtOH	285(4.36)	103-0830-69
1-Methyl-3-(methylthio)-s-triazolo-[3,4-a]phthalazinium methosulfate	MeOH	210(4.57),243(4.46), 255(3.39),266s(4.42), 280s(4.06),290s(3.93)	44-3221-69
2-Methyl-3-(methylthio)-s-triazolo-[3,4-a]phthalazinium methosulfate	MeOH	220(4.73),268(4.43)	44-3221-69

Compound	Solvent	λ max(log ε)	Ref.
$C_{12}H_{14}N_4O_6$			
Barbituric acid, 5,5'-sec-butylidenedi-	EtOH	255(4.35)	103-0827-69
7H-Pyrrolo[2,3-d]pyrimidine-5-carboxylic acid, 4-amino-7-β-D-ribofuranosyl-	pH 1	227(4.12),274(4.07)	35-2102-69
	pH 11	277(4.14)	35-2102-69
	EtOH	231(3.88),279(4.13)	35-2102-69
$C_{12}H_{14}O$			
4-Pentenal, 2-methyl-2-phenyl-	C_6H_{12}	299(2.11)	22-0903-69
$C_{12}H_{14}OS$			
2H-1-Benzothiopyran, 4-ethoxy-2-methyl-	EtOH	223(4.24),254(4.12), 318(3.19)	44-0056-69
$C_{12}H_{14}O_2$			
2-Coumaranone, 5-tert-butyl-	MeOH	227s(3.71),264s(3.02), 276(3.13),284(3.11)	12-0977-69
2-Decene-4,6-diyn-1-ol, acetate	ether	254(4.12),267(4.26), 284(4.18)	24-1682-69
1(2H)-Naphthalenone, 3,4-dihydro-7-methoxy-5-methyl-	EtOH	218(4.30),257(3.92), 321(3.51)	18-3011-69
$C_{12}H_{14}O_2S$			
2,4-Pentanedione, 3-(p-tolylthio)-, aluminum chelate	CH_2Cl_2	254(4.56),290(4.49)	88-2255-69
Co(III) chelate	CH_2Cl_2	255(4.83)	88-2255-69
Cr(III) chelate	CH_2Cl_2	252(4.51),335(3.94)	88-2255-69
$C_{12}H_{14}O_3$			
Acetophenone, 5'-allyl-2'-hydroxy-3'-methoxy-	EtOH	224(4.34),265(3.97), 345(3.53)	12-1011-69
	KOH	241(4.37),383(3.70)	12-1011-69
Acetophenone, 2'-(allyloxy)-3'-methoxy-	EtOH	251(3.76),305(3.33)	12-1011-69
Benzo[b]furan, 2,3-dihydro-2-isopropyl-5,6-(methylenedioxy)-	EtOH	240(3.51),313(3.82)	18-1971-69
Bicyclo[4.2.2]dec-2-ene-3,4-dicarboxylic anhydride	MeOH	229(3.99)	108-0435-69
Bicyclo[4.2.2]dec-3-ene-3,4-dicarboxylic anhydride	MeOH	254(3.73)	108-0435-69
Bicyclo[4.2.2]dec-7-ene-7,8-dicarboxylic anhydride	MeOH	259(3.51)	108-0435-69
1,4-Methano-5,8-naphthalenedione, 1,2,3,4,4a,8a-hexahydro-6-methoxy-, endo-cis	C_6H_{12}	264(4.08),340(--), 355(1.89)	39-0105-69C
	EtOH	268(4.08),330(2.03)	39-0105-69C
2,4-Pentadiynoic acid, 5-(1-hydroxy-cyclohexyl)-, methyl ester	EtOH	222(3.15),235(3.34), 249(3.51),262(3.71), 276(3.58)	70-2117-69
2,4-Pentadiynoic acid, 5-(1-hydroxy-2-methylcyclohexyl)-	EtOH	224(3.11),236(3.36), 248(3.61),262(3.72), 276(3.58)	70-2117-69
$C_{12}H_{14}O_4$			
1,4-Benzodioxan-6-ol, butyrate	EtOH	230(3.61),284(3.48)	104-0468-69
1,4-Benzodioxan-6-ol, 7-butyryl-	EtOH	238(4.10),279(4.09), 346(4.08)	104-0468-69
Dillapiole	EtOH	235s(3.74),286(3.37)	12-1531-69
Propionic acid, 3-benzoyl-2-ethoxy-	EtOH	244(3.97),273s(3.53), 345(2.30)	18-1353-69
Propionic acid, 3-benzoyl-2-methoxy-, methyl ester	EtOH	244(4.04),280(3.04), 323(1.82)	18-1353-69

Compound	Solvent	λ_{max}(log ε)	Ref.
Propionic acid, 3-(o-carboxyphenyl)-, dimethyl ester	EtOH	208(3.77),231(3.91), 278(3.13)	39-2656-69C
$C_{12}H_{14}O_5$			
7-Oxabicyclo[2.2.1]hepta-2,5-diene-2,3-dicarboxylic acid, 1,4,5,6-tetramethyl-	H_2O	234(3.76),307s(2.67)	33-0584-69
4,5-Oxepinedicarboxylic acid, 2,7-dimethyl-, dimethyl ester	isooctane	265(3.26),325(3.34)	33-0584-69
Phthalic acid, 3-hydroxy-4,6-dimethyl-, dimethyl ester	H_2O	242s(3.64),311(3.62)	33-0584-69
Phthalic acid, 4-hydroxy-3,5-dimethyl-, dimethyl ester	EtOH	264(4.03)	33-0584-69
$C_{12}H_{14}O_5S$			
Butyric acid, 4-benzoyl-4-(methylsulfonyl)-	EtOH	253(4.12)	108-0057-69
$C_{12}H_{14}O_6$			
m-Hemipinic acid, dimethyl ester	EtOH	223(4.44),266(3.97)	39-1921-69C
$C_{12}H_{14}O_7$			
2-Furoic acid, 2-acetonyl-3-acetyl-2,5-dihydro-4-methoxy-5-oxo-, methyl ester	EtOH	277(4.01)	1-0597-69
$C_{12}H_{15}BrN_2O_5$			
Uridine, 5'-bromo-5'-deoxy-2',3'-O-isopropylidene-	pH 5	258(4.07)	44-1627-69
$C_{12}H_{15}BrN_2S_2$			
2H-1,3,5-Thiadiazine-2-thione, 3-(p-bromophenyl)tetrahydro-5-isopropyl-	MeOH	248(4.15),290(3.66)	73-2952-69
$C_{12}H_{15}BrN_4O_4$			
7H-Pyrrolo[2,3-d]pyrimidine, 5-bromo-4-(methylamino)-7-β-D-ribofuranosyl-	pH 1	237(4.27),278(4.05)	4-0215-69
	pH 11	231(3.81),281(4.06)	4-0215-69
$C_{12}H_{15}ClN_2O_5$			
Uridine, 5'-chloro-5'-deoxy-2',3'-O-isopropylidene-	pH 5	261(4.06)	44-1627-69
$C_{12}H_{15}ClN_4OS$			
4-Thiazolin-2-one, 3-[(4-amino-2-methyl-5-pyrimidinyl)methyl]-5-(2-chloroethyl)-4-methyl-	EtOH	236(4.17),276s(--)	88-3279-69
$C_{12}H_{15}ClO_3$			
1,4-Benzodioxan, 6-(2-chloroethoxy)-7-ethyl-	EtOH	293(3.65)	104-0468-69
$C_{12}H_{15}Cl_2N_3$			
Benzimidazole, 5-[bis(2-chloroethyl)-amino]-2-methyl-	EtOH	228(4.40),314(3.86)	104-0158-69
hydrochloride hydrate	EtOH	228(4.41),260(3.92), 318(3.83)	104-0158-69
$C_{12}H_{15}F_7Sn$			
Stannane, triallyl(heptafluoropropyl)-(corrections)	hexane	193(4.42)	44-3715-69
	MeCN	202(4.40)	44-3715-69

Compound	Solvent	$\lambda_{max}(\log \epsilon)$	Ref.
$C_{12}H_{15}N$			
3H-Indole, 3-ethyl-2,3-dimethyl-	MeOH	255(3.82)	23-3647-69
perchlorate	EtOH	257(3.93)	78-0095-69
	EtOH-HCl	231(4.00),237(3.98), 280(3.90)	78-0095-69
2H-1,4-Methanoquinoline, 3,4-dihydro-4,9-dimethyl-	MeOH	257(2.77),263(2.90), 270(2.89)	23-3647-69
	MeOH-HCl	253(2.51),259(2.56), 265(2.54)	23-3647-69
1-Naphthaleneethylamine, 3,4-dihydro-	n.s.g.	256(3.93)	39-0217-69C
Pyridine, 1-benzyl-1,2,5,6-tetrahydro-	EtOH	251(2.48),257(2.53), 263(2.52)	44-3672-69
Quinoline, 3-ethyl-5,6-dihydro-5-methyl-	MeOH	258(3.87),295(3.98)	106-0196-69
Quinoline, 7-ethyl-5,6-dihydro-4-methyl-	MeOH	273(4.07)	106-0196-69
$C_{12}H_{15}NO$			
1H-1-Benzazepine, 1-acetyl-2,3,4,5-tetrahydro-	MeOH	226(3.84),265(2.65)	44-2235-69
7H-1-Benzo[b]azonin-7-one, 1,2,3,4,5,6-hexahydro-	n.s.g.	230(4.2),260s(3.5), 340(3.3)	24-0342-69
Cyclohexanone, 4-anilino- (only one wavelength cited)	EtOH	248(4.13),?(3.28)	12-2493-69
Indole-2-ethanol, β,β-dimethyl-	EtOH	222(4.48),280(3.85), 290(3.77)	39-0595-69C
Indole-3-ethanol, β,β-dimethyl-	EtOH	222(4.56),282(3.79), 290(3.72)	39-0595-69C
3-Isoxazoline, 2,4,5-trimethyl-3-phenyl-	n.s.g.	221(4.0),290(3.5)	94-2201-69
4-Isoxazoline, 3-ethyl-2-methyl-5-phenyl-	EtOH	226(3.92),275(3.80)	88-4875-69
4-Isoxazoline, 2,3,4-trimethyl-5-phenyl-	EtOH	224(3.95),275(3.79)	88-4875-69
Mesitylene, 2-(2-nitrosopropenyl)-	n.s.g.	732(1.66)	48-0260-69
$C_{12}H_{15}NO_2$			
Anisole, 3,5-dimethyl-2-(2-nitrosopropenyl)-	n.s.g.	732(1.75)	48-0260-69
11-Azabicyclo[4.4.1]undeca-1,3,5-triene-	hexane	213(4.28),258(3.43)	44-2866-69
11-carboxylic acid, methyl ester	EtOH	209(4.31),252(3.50)	44-2866-69
5-Azatricyclo[7.2.2.0^{3,7}]tridec-2-ene, 4,6-dioxo-	MeOH	227(3.19)	108-0435-69
5-Azatricyclo[7.2.2.0^{3,7}]tridec-3-ene, 4,6-dioxo-	MeOH	220(3.54),227s(3.52)	108-0435-69
2H-1-Benzazepin-2-one, 7-(1-hydroxyethyl)-1,3,4,5-tetrahydro-	MeOH	240(4.25),280s(3.35)	44-1070-69
3H,10H-Benzo[ij]quinolizine-3,10-dione, 1,2,5,6,7,7a,8,9-octahydro-	n.s.g.	298(4.1)	23-0433-69
1,3-Butanedione, 1-[p-(dimethylamino)-phenyl]-	n.s.g.	240(3.91),247(3.90), 354(4.60)	44-2817-69
Propionamide, 3-benzoyl-N-ethyl-	MeCN	198(4.51),240(4.12), 278(2.95),311(1.43)	28-0279-69A
2-Pyrrolidinone, 5-hydroxy-1-methyl-5-p-tolyl-	MeCN	193(4.76),263f(2.49)	28-0279-69A
$C_{12}H_{15}NO_2S_2$			
Propionaldehyde, 2-(2-benzothiazolylthio)-, dimethyl acetal	EtOH	225(4.33),282(4.06), 291(3.97),301(3.92)	95-0469-69
2-Pyridineacetic acid, α-carbonyl-, ethyl ester, α-(dimethyl mercaptole)	EtOH	251(3.90),297(3.91), 343(4.35)	95-0203-69
$C_{12}H_{15}NO_3$			
Propionamide, 3-p-anisoyl-N-methyl- (plus a shoulder)	MeCN	195(4.39),268(4.20)	28-0279-69A

Compound	Solvent	$\lambda_{max}(\log \epsilon)$	Ref.
$C_{12}H_{15}NO_4$			
Isoindole-1,3-dicarboxylic acid, 4,5,6,7-tetrahydro-, dimethyl ester	EtOH	221(4.26),284(4.36)	32-1115-69
Isonicotinic acid, 3,5-diacetyl-1,4-dihydro-2,6-dimethyl-	pH 1	252(4.20),376(3.85)	103-0762-69
	pH 4.2	253(4.13),378(3.84)	103-0762-69
	pH 13	260(4.09),383(3.84)	103-0762-69
ammonium salt	EtOH-HCl	254(4.15),285(4.13), 370(3.79)	103-0762-69
	EtOH	260(4.08),295(4.19), 380(3.80)	103-0762-69
	EtOH-NaOH	260(3.99),295(4.10), 380(3.71)	103-0762-69
3,5-Pyridinedicarboxylic acid, 2-methyl-, diethyl ester	EtOH	206(4.54),227(4.13), 272(3.59)	73-0427-69
$C_{12}H_{15}NO_6$			
3-Carboxy-1-D-ribofuranosyl-4-picolinium hydroxide, inner salt	MeOH	273s(3.41)	39-0918-69C
3-Pyrroline-2,2-dicarboxylic acid, 1-acetyl-5-oxo-, diethyl ester	EtOH	229(3.83)	94-2417-69
$C_{12}H_{15}NO_8$			
Glucopyranoside, o-nitrophenyl-, β-D-	H_2O	260(3.58),320(3.32)	18-1052-69
Glucopyranoside, p-nitrophenyl-, α-D-	H_2O	222(3.78),306(4.00)	18-1052-69
Glucopyranoside, p-nitrophenyl-, β-D-	H_2O	222(3.78),304(3.95)	18-1052-69
$C_{12}H_{15}N_3O$			
2-Imidazolidinone, 1-(isopropylideneamino)-3-phenyl-	MeOH	248(3.65)	44-0372-69
4(3H)-Quinazolinone, 6-amino-3-butyl-	EtOH	228(4.40),300(3.98), 362(4.05)	2-1166-69
$C_{12}H_{15}N_3O_2$			
Acetamide, N-(1-acetyl-5-amino-6-indolinyl)-	MeOH	270(4.26),312(3.79)	103-0196-69
Acetamide, N-(1-acetyl-6-amino-5-indolinyl)-	MeOH	262(4.15),315(3.91)	103-0196-69
2H-Pyran-2-methanol, 6-(1H-benzotriazol-1-yl)tetrahydro-, trans	EtOH	256(3.80),262(3.78), 280(3.59)	4-0005-69
$C_{12}H_{15}N_3O_2S$			
3-Butenoic acid, 2-(dicyanomethylene)-4-(dimethylamino)-4-(methylthio)-, ethyl ester	C_6H_{12}	398(4.18)	5-0073-69E
$C_{12}H_{15}N_3O_3$			
4-Isoxazolecarboxylic acid, 5-(3,5-dimethyl-1-pyrazolyl)-3-methyl-	MeOH	260(3.89)	4-0783-69
$C_{12}H_{15}N_5O_5$			
7H-Pyrrolo[2,3-d]pyrimidine-5-carboxamide, 4-amino-7-β-D-ribofuranosyl-	pH 1	228(3.98),273(4.11)	35-2102-69
	pH 11	227(4.15),277(4.16)	35-2102-69
	EtOH	229(3.91),278(4.18)	35-2102-69
$C_{12}H_{15}O_{14}P$			
L-Ascorbic acid, 3,3'-(hydrogen phosphate), calcium salt	pH 1	235(4.26)	94-0387-69
	pH 13	260(4.45)	94-0387-69

Compound	Solvent	$\lambda_{max}(\log \epsilon)$	Ref.
$C_{12}H_{16}$			
5H-Benzocycloheptene, 6,7,8,9-tetra-hydro-6-methyl-	n.s.g.	210(4.15),263(2.51), 271(2.60)	28-0986-69A
1,2-Benzocyclooct-1-ene	hexane	252(--),257(2.96), 265(--),272(2.91)	39-0474-69C
Tetraspiro[2.0.2.0.2.0.2.0]dodecane	C_6H_{12}	271(1.53),281s(1.40)	88-3545-69
$C_{12}H_{16}BrNO_2$			
Benzylidenimine, 3-bromo-N-butyl-6-hy-droxy-5-methoxy-, copper complex	$CHCl_3$	280(4.38),385(3.92)	65-0401-69
$C_{12}H_{16}BrN_5$			
4-[(α-Aminobenzylidene)amino]-1-propyl-1H-1,2,4-triazolium bromide	MeOH	233(4.34)	48-0477-69
$C_{12}H_{16}ClNO_2$			
1,2,3,4-Tetrahydro-3-hydroxy-6-methyl-quinolizinium chloride, acetate	EtOH	272(3.9),280(3.7)	24-1309-69
$C_{12}H_{16}ClNO_5$			
Benzyl(2-formylvinyl)dimethylammonium perchlorate	H_2O	205s(4.18)	24-2609-69
$C_{12}H_{16}Cl_2N_2O_2$			
p-Benzoquinone, 2,5-bis[(2-chloroethyl)-amino]-3,6-dimethyl-	EtOH	347(4.13),520(2.16)	39-1325-69C
$C_{12}H_{16}NO_3PS$			
Phosphonic acid, (2-benzothiazolylmeth-yl)-, diethyl ester	EtOH	232s(4.33),254(4.10), 288(3.59),299(3.57)	65-0392-69
$C_{12}H_{16}NO_4P$			
Phosphonic acid, (2-benzoxazolylmeth-yl)-, diethyl ester	EtOH	247(3.94),271s(3.56), 285(3.74),292(3.68)	65-0392-69
$C_{12}H_{16}N_2$			
Isoindole, 2-(3-aminopropyl)-1-methyl-	EtOH	227(4.64),269(3.30), 280(3.20),342(3.60)	44-1720-69
3-Pyrazoline, 1,2,3-trimethyl-4-phenyl-	EtOH	221s(3.83),301(3.88)	22-3316-69
3-Pyrazoline, 1,2,3-trimethyl-5-phenyl-	EtOH	245(3.58)	22-3316-69
3-Pyrazoline, 1,2,4-trimethyl-3-phenyl-	EtOH	222(4.15),281(3.75)	22-3316-69
3-Pyrazoline, 1,2,5-trimethyl-3-phenyl-	EtOH	235(4.03),282(3.45)	22-3316-69
$C_{12}H_{16}N_2O$			
Acetamide, N-(1,2,3,4-tetrahydro-2-meth-yl-4-isoquinolyl)-	MeOH	226(2.98),264(2.53), 272(2.38)	22-2045-69
Ketone, 1-adamantyl diazomethyl	EtOH	250(3.87),340(2.51)	18-1617-69
$C_{12}H_{16}N_2O_2$			
1H-Azepine, hexahydro-1-(p-nitrophenyl)-	DMSO	395(3.35)	39-0544-69B
Furo[2',3':4,5]pyrimido[1,2-a]azepin-4(2H)-one, 3,6,7,8,9,10-hexahydro-2-methyl-	MeOH	212(4.38),245s(3.47), 278(3.61)	24-2739-69
Glycine, N-benzoyl-, propylamide	H_2O	194(4.57),228(4.06)	35-1822-69
	24% H_2SO_4	196(4.54),230(4.04)	
	40% H_2SO_4	195(4.50),232(4.04)	
	57% H_2SO_4	195(4.50),237(4.05)	
	77% H_2SO_4	195(4.39),252(4.13)	
	96% H_2SO_4	195(4.33),258(4.18)	
	99.7% H_2SO_4	195(4.29),258(4.19)	

Compound	Solvent	λ_{max}(log ε)	Ref.
Glycine, N-benzoyl-, propylamide (cont.) (also other solvent mixtures)	53% $MeSO_3H$	195(4.48),232(4.05)	35-1822-69
	85% $MeSO_3H$	196(4.41),242(4.06)	
	95% $MeSO_3H$	197(4.33),249(4.12)	
	99% $MeSO_3H$	197(4.30),252(4.15)	
5,8-Methano-1H-pyrazolo[1,2-a]pyridazine-1,3(2H)-dione, 2,2-diethyl-5,8-dihydro-	C_6H_{12}	269(3.48)	44-3181-69
	EtOH	254(3.62)	44-3181-69
$C_{12}H_{16}N_2O_2S_2$			
2-Pyrroline-3-acrylic acid, α-cyano-1-methyl-β,2-bis(methylthio)-, methyl ester	C_6H_{12}	458(4.02)	5-0073-69E
$C_{12}H_{16}N_2O_7$			
Isobarbituric acid, 1-(2,3-O-isopropylidene-β-D-ribofuranosyl)-	pH 1	280(3.94)	44-2636-69
	pH 10	247(3.81),306(3.87)	44-2636-69
	pH 14	303(3.85)	44-2636-69
$C_{12}H_{16}N_2S$			
Pyrimidine, 1,4,5,6-tetrahydro-1-methyl-2-[1-methyl-2-(2-thienyl)vinyl]-	H_2O	287(4.24)	87-1066-69
Pyrimidine, 1,4,5,6-tetrahydro-1-methyl-2-[2-(3-methyl-2-thienyl)vinyl]-	H_2O	318(4.26)	87-1066-69
Pyrimidine, 1,4,5,6-tetrahydro-1-methyl-2-[2-(2-thienyl)propenyl]-	H_2O	292(4.10)	87-1066-69
$C_{12}H_{16}N_2S_2$			
2H-1,3,5-Thiadiazine-2-thione, 5-benzyl-3-ethyltetrahydro-	MeOH	248(3.96),285(4.24)	73-2952-69
$C_{12}H_{16}N_2S_4$			
Cyclopent-1-ylidenemethanethiol, α,α-dithiobis[2-imino-	EtOH	308(4.01),398(4.64)	44-0730-69
$C_{12}H_{16}N_4O$			
4a(2H)-Naphthalenecarbonitrile, 3,4,5,6,7,8-hexahydro-2-oxo-, semicarbazone	EtOH	270(4.47)	28-0864-69A
$C_{12}H_{16}N_4O_2$			
2-Pyrazolin-5-one, 1-(4-methoxy-6-methyl-2-pyrimidinyl)-3,4,4-trimethyl-	pH 7.0	235(4.1),255(4.0)	94-1485-69
	$CHCl_3$	248(4.1)	94-1485-69
$C_{12}H_{16}N_4O_3S$			
1,3-Cyclopentanediol, 3-(hydroxymethyl)-5-[6-(methylthio)-9H-purin-9-yl]-	pH 1	221(4.06),290s(--), 295(4.23),303s(--)	35-3075-69
	pH 7	221(4.07),286(4.27), 293(4.26)	35-3075-69
	pH 13	221(4.07),286(4.27), 293(4.26)	35-3075-69
$C_{12}H_{16}O$			
Bicyclo[4.2.0]octa-1,5-dien-3-one, 2,4,4,5-tetramethyl-	MeOH	320(3.68)	88-4841-69
Cyclohexaneacetaldehyde, 2-(2-butynylidene)-	EtOH	234(4.11)	28-1808-69A
Cyclohexanone, 3-(1,5-cyclohexadien-1-yl)-	EtOH	263(3.72)	28-1170-69A
Cyclohexanone, 3-(cyclohex-2-enylidene)-	EtOH	258(3.57)	28-1170-69A

Compound	Solvent	$\lambda_{max}(\log \epsilon)$	Ref.
2-Cyclohexen-1-one, 3-(1-cyclohexen-1-yl)-	EtOH	283(4.32)	28-1170-69A
2-Decalone, 1(a)-ethynyl-, trans	EtOH	292(1.94)	22-3199-69
2-Decalone, 3(a)-ethynyl-, trans	EtOH	289(2.03)	22-3199-69
2-Octene-4,6-diyne, 2-tert-butoxy-	pentane	221(4.46),252s(3.90), 263(4.05),275(4.20), 291(4.11)	78-2823-69
$C_{12}H_{16}O_2$			
Acetophenone, 4'-(1-ethoxyethyl)-	hexane	248(4.18),320(2.83)	23-0687-69
p-Benzoquinone, 2,6-diisopropyl-	MeOH	253(4.23)	64-0547-69
2-Butene, 2-(2,4-dimethoxyphenyl)-, (E)-	n.s.g.	237(3.97)	28-1654-69B
(Z)-	n.s.g.	230(3.92)	28-1654-69B
2-Buten-1-ol, 1-(6-hydroxy-m-tolyl)-3-methyl-	EtOH	264(3.55),273s(--), 315(3.40)	32-0616-69
3-Buten-2-ol, 4-(6-hydroxy-m-tolyl)-2-methyl-	EtOH	251(4.02),310(3.63)	32-0612-69
2,6,8,10-Dodecatetraenoic acid	EtOH	260(4.56),268(4.68), 279(4.58)	44-1147-69
4a,8a-Ethanonaphthalene-9,10-dione, octahydro-	n.s.g.	461(1.86)	88-1529-69
1,3(2H,4H)-Naphthalenedione, 4a,5,6,7-tetrahydro-4a,5-dimethyl-	EtOH	297(4.17)	24-2697-69
	EtOH-NaOH	312(4.26)	24-2697-69
1,6(2H,7H)-Naphthalenedione, 3,4,8,8a-tetrahydro-2,8a-dimethyl-	H_2SO_4	259(4.08)	23-2263-69
protonated	H_2SO_4	299(4.09)	23-2263-69
Thymoquinol, dimethyl ether	C_6H_{12}	290(3.58),292(3.58)	12-0495-69
$C_{12}H_{16}O_3$			
Acetophenone, 2'-hydroxy-3'-methoxy-5'-propyl-	EtOH	264(3.98),350(3.49)	12-1011-69
	KOH	383(3.84)	12-1011-69
Acetophenone, 5'-hydroxy-4'-methoxy-2'-propyl-	EtOH	232(4.30),273(3.82), 305(3.69)	12-1011-69
	KOH	250(4.40),286(3.72), 355(3.63)	12-1011-69
1,4-Benzodioxan-6-ol, 7-butyl-	EtOH	295(3.73)	104-0468-69
2H-1-Benzopyran-2,5(3H)-dione, 4,6,7,8-tetrahydro-4,7,7-trimethyl-	EtOH	254(4.08)	4-0917-69
Bicyclo[4.2.1]non-6-ene-1-carboxylic acid, 9-oxo-, ethyl ester	n.s.g.	241(3.59),310(2.32)	77-0832-69
Butyrophenone, 4'-ethoxy-2'-hydroxy-	n.s.g.	280(4.2)	104-1609-69
Calythrone	EtOH	240(4.33),265(4.28)	39-1845-69C
Ethanol, 1-[2-(allyloxy)-3-methoxyphenyl]-	EtOH	272(3.25),277(3.25)	12-1011-69
2-Furaldehyde, 5-(2-methyl-5-furylvinylene)-	EtOH	277(3.92),282(3.92), 375(4.41)	12-1951-69
Tricyclo[2.2.1.0^{2,6}]heptane-3-propionic acid, 2,3-dimethyl-7-oxo-	MeOH	207(3.53)	88-3169-69
$C_{12}H_{16}O_4$			
Amitenone (see also $C_{53}H_{72}O_8$)	EtOH	286(4.47)	39-2398-69C
1-Cyclohexene-1-glyoxylic acid, 2,4,4-trimethyl-6-oxo-, methyl ester	pH 2.80	242(4.00),320(3.58)	5-0091-69D
	pH 9.15	246(3.78),410(4.28)	5-0091-69D
1,2,4-Cyclopentanetrione, 5-isovaleryl-3,3-dimethyl-	MeOH-HCl	284(3.99)	20-0277-69
	MeOH-NaOH	253(4.06),323(3.99)	20-0277-69
Ether, ethyl ethynyl, polymerization intermediate	C_6H_{12}	228(3.97),275s(3.76), 282s(3.75),298s(3.69)	34-0125-69
	MeOH	232(3.97)	34-0125-69
	EtOH	236(3.99),278s(3.80), 284s(3.79),292s(3.76)	34-0125-69

Compound	Solvent	$\lambda_{max}(\log \epsilon)$	Ref.
Propiophenone, 2',3,4'-trimethoxy-	EtOH	228(4.15),267(4.09), 303(3.91)	23-1529-69
2H-Pyran-2-one, 4-hydroxy-6-methyl-3-(4-methylvaleryl)-	EtOH	225(4.01),310(4.10)	88-2279-69
$C_{12}H_{16}O_4S$			
Acetic acid, thio-, S-(5-hydroxy-9-oxabicyclo[3.3.1]non-6-en-2-yl)ester, acetate	EtOH	232(3.64)	33-0725-69
$C_{12}H_{16}O_5$			
Cyclopenta[c]pyran-4-carboxylic acid, 1,4a,5,6,7,7a-hexahydro-1α-methoxy-7α-methyl-6-oxo-, methyl ester	EtOH	235(4.06)	39-0721-69C
$C_{12}H_{16}S_2$			
Benzoic acid, 4-isopropyl-2-methyldithio-, methyl ester	C_6H_{12}	247(3.63),311(4.06), 493(2.06)	48-0045-69
	EtOH	245(3.67),312(4.06), 484(2.00)	48-0045-69
$C_{12}H_{17}BrO_2$			
2(3H)-Naphthalenone, 4a-(bromomethoxymethyl)-4,4a,5,6,7,8-hexahydro-	MeOH	225(3.56)	44-0496-69
$C_{12}H_{17}BrO_3$			
1,3-Cyclopentanedione, 5-bromo-2-isovaleryl-4,4-dimethyl-	MeOH-HCl	228(4.02),278(4.02)	20-0017-69
	MeOH-NaOH	255(4.27),272s(4.21)	20-0017-69
$C_{12}H_{17}BrSi$			
Silane, (α-bromocinnamyl)trimethyl- (graph shows differences)	heptane	208(4.37),217s(4.28), 270(4.44),291(4.13), 299(3.87)	28-1878-69A
$C_{12}H_{17}ClNO_6P$			
Phosphoric acid, mono(α-chloro-4-nitro-o-tolyl) monopentyl ester	MeOH	290(3.94)	35-4532-69
$C_{12}H_{17}ClN_2O_4$			
2-Allyl-1,2-dimethylindazolinium perchlorate	n.s.g.	236(3.81),282(3.25)	104-2004-69
$C_{12}H_{17}ClN_2O_5$			
3-Carbamoyl-4-methyl-1-D-ribofuranosylpyridinium chloride	MeOH	264(3.63)	39-0199-69C
$C_{12}H_{17}IN_2$			
1,1,2-Trimethyl-3-phenyl-3-pyrazolinium iodide	EtOH	218(4.26),248(3.80)	22-3316-69
$C_{12}H_{17}N$			
Bornane-$\Delta^{2,\alpha}$-acetonitrile	EtOH	223(3.98)	7-0335-69
Pyridine, 2-(1,3,5-heptatrienyl-3,4,5,6-tetrahydro- (nigrifactin)	MeOH	354(4.56)	88-2535-69
Quinoline, 3-ethyl-5,6,7,8-tetrahydro-7-methyl-	MeOH	273(3.75)	106-0196-69
$C_{12}H_{17}NO_2$			
Benzoic acid, (dimethylaminopropyl) ester	dioxan	273(3.21),280(3.14)	65-1861-69

Compound	Solvent	λmax(log ε)	Ref.
Benzoic acid, p-[2-(dimethylamino)-ethyl]-, methyl ester	heptane	236(4.17)	44-0077-69
	MeOH	237(4.19)	44-0077-69
p-Benzoquinone, 2,6-diisopropyl-, 4-oxime	EtOH	229(3.45),304(4.28)	12-0935-69
	EtOH-NaOH	256(3.85),382(4.59)	12-0935-69
	dioxan	725(-1.5)	12-0935-69
Hexanohydroxamic acid, N-phenyl-	EtOH	247(4.01)	42-0831-69
Salicylidenimine, N-butyl-5-(hydroxy-methyl)-, copper complex	$CHCl_3$	305(4.01),370(4.07), 588(2.20)	65-0401-69
Valerohydroxamic acid, N-p-tolyl-	EtOH	248(4.06)	42-0831-69
$C_{12}H_{17}NO_3$			
Propionamide, 3-hydroxy-3-(o-hydroxy-phenyl)-N,N,2-trimethyl-	EtOH	273(4.31),279(4.24)	39-0016-69C
$C_{12}H_{17}NO_4$			
1H-Azepine-3,6-dicarboxylic acid, 4,5-dihydro-2,7-dimethyl-, dimethyl ester	hexane	226(4.14),321(4.12)	27-0037-69
3,5-Pyridinedicarboxylic acid, 1,2-di-hydro-2-methyl-, diethyl ester	EtOH	210(4.52),281(3.94), 371(3.49)	73-0427-69
3,5-Pyridinedicarboxylic acid, 1,4-di-hydro-4-methyl-, diethyl ester	EtOH	212(5.15),244s(3.73), 360(3.94)	73-0427-69
Pyrrole, 1-(2-C,2'-O-isopropylidene-α-D-ribofuranosyl)-	H_2O	217(3.86)	18-3539-69
Pyrrole, 1-(2',3'-O-isopropylidene-α-D-ribofuranosyl)-	EtOH	214(3.81)	18-3539-69
β-isomer	EtOH	215(3.83)	18-3539-69
Pyrrole, 1-(2',3'-O-isopropylidene-β-D-ribopyranosyl)-	EtOH	215(3.85)	18-3539-69
$C_{12}H_{17}NS_2$			
Spiro[cyclohexane-1,2'(4'H)-cyclopenta-[d][1,3]thiazine]-4'-thione, 1',5',6',7'-tetrahydro-	EtOH	334(3.91),406(4.41)	44-0730-69
$C_{12}H_{17}N_3$			
Benzimidazole, 2-(1-amino-2-methyl-butyl)-, dihydrochloride	EtOH	274(3.84),281(3.82)	94-2381-69
Benzimidazole, 2-(1-amino-3-methyl-butyl)-, dihydrochloride	EtOH	274(3.85),281(3.82)	94-2381-69
$C_{12}H_{17}N_3O_2$			
Cyclopentanone, 2-(5,6-dihydro-2-methyl-1(4H)-cyclopentimidazolyl)-, oxime, 3-oxide	MeOH	223(3.83)	48-0746-69
Ethanol, 2,2'-[(2-methyl-5-benzimidazol-yl)imino]di-	EtOH	227(4.43),319(3.81)	104-0158-69
$C_{12}H_{17}N_3O_3$			
Acrylic acid, 2-cyano-3-[(hexahydro-2H-azepin-2-ylidene)amino]-3-hydroxy-, ethyl ester	MeOH	222(4.29),292(4.24)	24-2739-69
$C_{12}H_{17}N_3O_5S$			
Benzenesulfonamide, 2-acetyl-N,N-dieth-yl-5-nitro-, oxime	H_2O	248(3.99)	80-0481-69
	pH 13	248(4.00),332(3.42)	80-0481-69
$C_{12}H_{17}N_4OS$			
Thiamine, diphosphate	pH 1.79	241(4.3),250s(4.2)	65-0102-69

Compound	Solvent	λ_{max}(log ε)	Ref.
$C_{12}H_{17}N_5O$			
Propionamide, N-(6-methyl-8-propyl-s-triazolo[4,3-a]pyrazin-3-yl)-	EtOH	225(4.25),305(3.59)	39-1593-69C
$C_{12}H_{17}N_5O_2$			
s-Triazolo[4,3-a]pyrazine-3-carbamic acid, 6-methyl-8-propyl-, ethyl ester	H_2O	222(4.40),303(3.63)	39-1593-69C
$C_{12}H_{17}N_5O_4$			
Adenosine, 3'-deoxy-3'-(2-hydroxyethyl)-	pH 1	208(4.23),257(4.07)	44-1029-69
	pH 7	208(4.23),260(4.06)	44-1029-69
	pH 13	211(4.11),260(4.05)	44-1029-69
$C_{12}H_{17}N_5O_5$			
Adenosine, 2-ethyl-, 1-oxide	pH 1	260(4.09)	94-1128-69
	pH 7	233(4.65),264(3.94)	94-1128-69
Adenosine, 5'-O-(2-hydroxyethyl)-	n.s.g.	258(4.16)	28-1160-69A
D-Erythronic acid, 4-[(6-amino-5-formamido-4-pyrimidinyl)amino]-4-deoxy-2,3-O-isopropylidene-	H_2O	222(4.55),265(3.85)	88-4729-69
$C_{12}H_{17}N_5O_6$			
Guanosine, 5'-O-(2-hydroxyethyl)-	pH 7	252(4.08)	28-1160-69A
m-Phenylenediamine, N,N'-diisopropyl-2,4,6-trinitro-	MeOH	313(4.35),348(4.26), 408(4.02)	35-4155-69
$C_{12}H_{17}O_5P$			
Phosphonic acid, p-anisoyl-, diethyl ester	C_6H_{12}	294(4.01),373(2.18)	18-0821-69
$C_{12}H_{18}$			
Bicyclo[3.1.0]hex-2-ene, 1,2,3,5,6-pentamethyl-4-methylene-, endo	EtOH	245(3.68)	39-1299-69A
Bicyclo[3.1.0]hex-2-ene, 1,3,5,6,6-pentamethyl-4-methylene-	MeOH	246(3.91)	88-4929-69
Bicyclo[3.1.0]hex-2-ene, 2,3,5,6,6-pentamethyl-4-methylene-	MeOH	259(4.01)	88-4929-69
Bicyclo[5.1.0]oct-2-ene, 4,4,7-trimethyl-6-methylene-	EtOH	no maxima	88-0999-69
1-Butene, 4-(1,2,3-trimethyl-4-methylene-2-cyclobuten-1-yl)-	hexane	234(4.07)	24-0275-69
1,3,6,9-Decatetraene, 5,8-dimethyl-	hexane	192(4.15),228(4.33)	39-0227-69C
1,4,6,9-Decatetraene, 3,8-dimethyl-	hexane	196(3.99),231(4.25)	39-0227-69C
as-Indacene, 1,2,3,4,5,5a,6,7,8,8a-decahydro-, cis	EtOH	206(3.73)	22-4501-69
1,5,7,9-Undecatetraene, 3-methyl-	hexane	249s(4.10),258(4.50), 267(4.59),278(4.45)	39-0227-69C
1,3,6,10-Undecatetraene, 9-methyl-	hexane	192(4.08),227(4.37)	39-0227-69C
$C_{12}H_{18}Cl_2N_2O_2$			
Hydroquinone, 2,5-dichloro-3,6-bis-[(dimethylamino)methyl]-	EtOH	312(3.69)	39-1245-69C
$C_{12}H_{18}Cl_2O$			
2-Cyclobuten-1-one, 2,3-di-tert-butyl-4,4-dichloro-	EtOH	227(3.87),307(2.12)	35-4766-69
$C_{12}H_{18}Cl_2O_6Pd_2$			
Palladium, bis(1-carboxy-2-hydroxy-π-allyl)di-μ-chlorodi-, diethyl ester	EtOH	250(4.2),310(3.7), 365(3.3)	18-0443-69

Compound	Solvent	$\lambda_{max}(\log \epsilon)$	Ref.
$C_{12}H_{18}INO$			
5,6,7,8-Tetrahydro-4-methoxy-1,2-dimeth-ylquinolinium iodide	MeOH	215(4.08),226(4.12), 243s(3.99),263(3.71)	78-0255-69
$C_{12}H_{18}NO$			
Nitroxide, tert-butyl 3,5-xylyl	n.s.g.	296(3.95),451(2.87)	39-1459-69C
$C_{12}H_{18}N_2$			
3,4-Diazabicyclo[4.2.0]octa-2,4,7-tri-ene, 1,2,5,6,7,8-hexamethyl-	ether	231(3.40),243(3.39)	24-1928-69C
Pyrrolidine, 1,2,2-trimethyl-5-(3-pyri-dyl)-	EtOH	257(3.43),262(3.46), 267s(3.34)	36-0860-69
	EtOH-HCl	260(3.69)	36-0860-69
$C_{12}H_{18}N_2NaO_7PS$			
Thymidine, 5'-O-ester with S-ethyl phosphorothioate, sodium salt	H_2O	266(3.96)	35-1522-69
$C_{12}H_{18}N_2OS$			
1-Cyclopentene-1-carboxamide, N-acetyl-2-(1-pyrrolidinyl)thio-	$CHCl_3$	300(3.99),406(4.17)	4-0037-69
$C_{12}H_{18}N_2O_2$			
Acrolein, 3,3'-(1,2-cyclohexylenedi-imino)di-, hydrate	n.s.g.	240(2.74),284(3.14)	39-2044-69C
5,8-Methano-1H-pyrazolo[1,2-a]pyrid-azine-1,3(2H)-dione, 2,2-diethyl-tetrahydro-	C_6H_{12}	277(3.33)	44-3181-69
	EtOH	266(3.38)	44-3181-69
1H-Pyrazolo[1,2-a]pyridazine-1,3(2H)-dione, 2,2-diethyl-5,8-dihydro-6-methyl-	C_6H_{12}	256(3.34)	44-3181-69
	EtOH	256(3.45)	44-3181-69
$C_{12}H_{18}N_2O_3$			
Phenol, 5-tert-butyl-4-(dimethylamino)-2-nitro-	MeOH	222(3.78),286(3.77), 368(3.44)	39-2164-69C
$C_{12}H_{18}N_2O_3S_3$			
Acetanilide, 5'-(dimethylsulfamoyl)-2',4'-bis(methylthio)-	MeOH	264(4.45),281s(--), 316s(--)	44-1780-69
$C_{12}H_{18}N_2O_4$			
p-Benzoquinone, 2,5-bis[(2-hydroxyeth-yl)amino]-3,6-dimethyl-	EtOH	222(4.06),350(4.11), 535(2.70)	39-1325-69C
1-Pyrazoline-3,5-dicarboxylic acid, 3,5-dimethyl-4-methylene-, diethyl ester	n.s.g.	327(2.21)	39-2443-69C
$C_{12}H_{18}N_2O_7$			
Uracil, 3-D-glucopyranosyl-1,6-dimethyl-	H_2O	211(3.91),272(4.05)	73-2316-69
$C_{12}H_{18}N_2O_7S_3$			
Acetanilide, 5'-(dimethylsulfamoyl)-2',4'-bis(methylsulfonyl)-	MeOH	224(4.44),262(4.24), 304(3.87)	44-1780-69
$C_{12}H_{18}N_4O_3$			
[4,5'-Bipyrimidine]-2,2',4'(1H,1'H,3'H)-trione, 3,4,5,6-tetrahydro-1,1',3,3'-tetramethyl-	H_2O	204(4.27),271(3.95)	35-5264-69
	pH 12	271(3.95)	35-5264-69

Compound	Solvent	λ_{max}(log ε)	Ref.
$C_{12}H_{18}N_4O_6$			
Cytosine, 1-(4-acetamido-4-deoxy-β-D-glucopyranosyl)-	pH 2	276(4.08)	94-0416-69
	pH 6.8	198(4.35),235(3.90), 267(3.91)	94-0416-69
	pH 13	268(3.92)	94-0416-69
$C_{12}H_{18}N_4S_2$			
1-Piperidinecarbodithioic acid, (4-amino-2-methyl-5-pyrimidinyl)-methyl ester	EtOH	250(4.19),280(4.26)	94-2299-69
$C_{12}H_{18}N_6O$			
Ethanol, 2-[(6-piperidinopurin-8-yl)-amino]-	pH 1	310(4.38)	65-2125-69
	EtOH	293(4.29)	65-2125-69
$C_{12}H_{18}N_8S_6$			
2H-1,3-Thiazine-2,4(3H)-dione, 3,3'-ethylenebis[dihydro-2-thio-, 4,4'-bis(thiosemicarbazone)	EtOH	247(4.39),310s(--)	103-0049-69
$C_{12}H_{18}O$			
6H-Benzocyclohepten-6-one, 1,2,3,4,-7,8,9,9a-octahydro-9a-methyl-	EtOH	240(3.90)	35-3676-69
Bicyclo[2.1.0]pentene, 5-acetyl-1,2,3,4,5-pentamethyl-	EtOH	227s(3.26),290(1.42)	89-0880-69
Cyclohexaneacetaldehyde, 2-(2-butenylidene)-	EtOH	236(4.38)	28-1535-69A
2-Cyclohexen-1-one, 4-methyl-3-(3-pentenyl)-	EtOH	237(4.04)	35-2371-69
Dispiro[2.1.2.1]octan-4-one, 1,1,6,6-tetramethyl-	isooctane	284(1.98),291(2.00), 298(1.98)	22-3981-69
2(1H)-Naphthalenone, 3,4,4a,5,6,7-hexahydro-1,1-dimethyl-	isooctane	289(1.90),297(1.93), 306(1.88),317(1.60)	44-0450-69
2(1H)-Naphthalenone, 4a,5,6,7,8,8a-hexahydro-4a,5-dimethyl-	MeOH	230(3.97)	23-4307-69
2(3H)-Naphthalenone, 4,4a,5,6,7,8-hexahydro-4a,5-dimethyl-	MeOH	240(4.08)	23-4307-69
2(3H)-Naphthalenone, 4,4a,5,6,7,8-hexahydro-5,5-dimethyl-	EtOH	240(4.00)	35-2371-69
2-Norcaranone, 1-methyl-4-(1-methylcyclopropyl)-	isooctane	278(1.62)	22-3981-69
13-Oxabicyclo[10.1.0]trideca-4,8-diene	hexane	225(3.91)	78-5357-69
Phenol, 2,6-diisopropyl-	pH 2	269(3.18)	30-1113-69A
	pH 13	239(3.93),291(3.66)	30-1113-69A
	MeOH	272(3.26)	30-1113-69A
	MeOH-Na	242(3.94),294(3.67)	30-1113-69A
in 1% MeOH	acid	268(3.28)	28-2217-69A
in 1% MeOH	base	290(3.63)	28-2217-69A
Phenol, 2,4-dipropyl-	pH 2	277(3.35)	30-1113-69A
	pH 13	238(3.92),296(3.57)	30-1113-69A
	MeOH	281(3.41)	30-1113-69A
	MeOH-Na	240(3.94),298(3.57)	30-1113-69A
Phenol, 2,6-dipropyl-	pH 2	270(3.19)	30-1113-69A
	pH 13	239(3.95),291(3.66)	30-1113-69A
	MeOH	273(3.24)	30-1113-69A
	MeOH-Na	242(3.94),294(3.68)	30-1113-69A
Phenol, o-hexyl-	pH 2	272(3.35)	30-1113-69A
	pH 13	239(3.95),289(3.63)	30-1113-69A
	MeOH	275(3.38)	30-1113-69A
	MeOH-Na	241(3.97),290(3.63)	30-1113-69A

Compound	Solvent	λmax(log ε)	Ref.
Spiro[bornane-2,1'-cyclopropan]-3-one	EtOH	212(3.17),271(1.95), 279(1.96),284(1.93)	22-3981-69
2,4-Xylenol, 6-tert-butyl-	pH 2	276(3.23)	30-1113-69A
	pH 13	239(3.92),297(3.62)	30-1113-69A
	MeOH	280(3.30)	30-1113-69A
	MeOH-Na	242(3.94),301(3.62)	30-1113-69A
2,5-Xylenol, 4-butyl-	pH 2	278(3.37)	30-1113-69A
	pH 13	239(3.93),296(3.59)	30-1113-69A
	MeOH	281(3.41)	30-1113-69A
	MeOH-Na	243(3.92),299(3.40)	30-1113-69A
$C_{12}H_{18}OS$			
Sulfoxide, 2-benzylbutyl methyl	EtOH	223(3.85)	39-0581-69B
$C_{12}H_{18}OSi$			
Silane, dimethyl[(α-methylstyryl)oxy]-, (E)-	EtOH	254(4.19),286s(3.17)	44-2324-69
(Z)-	EtOH	258(4.28)	44-2324-69
$C_{12}H_{18}O_2$			
Benzyl alcohol, o-(1-ethoxyethyl)-α-methyl-	hexane	206(3.93),258(2.28)	23-0687-69
diastereoisomer	hexane	206(3.92),258(2.30)	23-0687-69
4,6,8-Decatrienoic acid, ethyl ester	EtOH	259(4.58),267(4.69), 277(4.60)	44-1147-69
1,3(2H,4H)-Naphthalenedione, hexahydro-4a,5-dimethyl-	EtOH	258(3.80)	24-2697-69
	EtOH-NaOH	285(4.34)	24-2697-69
$C_{12}H_{18}O_3$			
1-Cyclononene-1-carboxylic acid, 2-methyl-9-oxo-, methyl ester	EtOH	206s(3.85),209(3.88), 223(3.92)	94-0629-69
4,8-Nonadienoic acid, 9-formyl-4,8-dimethyl-	C_6H_{12}	235(3.96)	77-0086-69
2,6-Octadienoic acid, 3,7-dimethyl-5-oxo-, ethyl ester	H_2O	243(4.12)	114-0397-69A
4H-Pyran-3-carboxylic acid, 2,4,4,6-tetramethyl-, ethyl ester	MeOH	208(3.45),270(3.34)	44-3169-69
$C_{12}H_{18}O_4$			
1,3-Cyclopentanedione, 5-hydroxy-2-isovaleryl-4,4-dimethyl-	MeOH-HCl	226(4.04),269(3.98)	20-0277-69
	MeOH-NaOH	252(4.30),268s(4.21)	20-0277-69
2,6-Decadienedioic acid, 3,7-dimethyl-, trans-trans	pentane	225(4.16)	77-0086-69
m-Dioxane-5-methanol, 5-ethyl-2-(5-methyl-2-furyl)-	30% EtOH	220(3.97)	103-0278-69
2,5-Hexanedione, 3,4-diacetyl-3,4-dimethyl-	ether	280(2.25)	24-1928-69
$C_{12}H_{18}O_5$			
2H-Pyran-2-ol, 3,5,5-triacetyl-2-methyl-	EtOH	285(2.78)	39-0879-69C
$C_{12}H_{18}O_6$			
3-Butene-1,2,3-tricarboxylic acid, 1,2-diethyl 3-methyl ester	hexane	213(3.76)	18-2732-69
$C_{12}H_{18}O_6S_3$			
2-Thiophenecarboxylic acid, 4,5-bis-[(ethylsulfonyl)methyl]-, methyl ester	EtOH	262(4.12),276s(4.04)	54-0321-69

Compound	Solvent	λ_{max}(log ϵ)	Ref.
$C_{12}H_{18}O_7$			
2,4-Pentanedione, 3-[(hydroxymethoxy)-methyl]-3-(hydroxymethyl)-, diacetate	EtOH	302(2.03)	39-0879-69C
$C_{12}H_{18}S$			
13-Thiabicyclo[10.1.0]trideca-4,8-diene	hexane	199(3.90),250(3.79)	78-5357-69
$C_{12}H_{18}Si$			
Silane, cinnamyltrimethyl-	heptane	223(3.6),251s(4.0), 259(4.1),288s(3.4), 297s(3.0)	28-1878-69A
$C_{12}H_{19}BrN_4O_2$			
7-Allyl-1,2,3,6,8,9-hexahydro-1,3,7,9-tetramethylpurinium bromide	H_2O	246s(3.79)	104-2004-69
$C_{12}H_{19}BrO_5$			
Glutaric acid, 2-acetyl-2-bromo-3-methyl-, diethyl ester	EtOH	289(2.04)	35-4739-69
$C_{12}H_{19}Cl_2N_3$			
Benzimidazole, 2-(1-amino-2-methylbutyl)-, dihydrochloride	EtOH	274(3.84),281(3.82)	94-2381-69
Benzimidazole, 2-(1-amino-3-methylbutyl)-, dihydrochloride	EtOH	274(3.85),281(3.82)	94-2381-69
$C_{12}H_{19}IN_2$			
4-Amino-2,3,5,5a,6,7,8,9-octahydro-1H-cyclopenta[c]quinolizinium iodide	EtOH	348(4.31)	44-0698-69
$C_{12}H_{19}N$			
13-Azabicyclo[10.1.0]trideca-4,8-diene	hexane	223(3.86)	78-5357-69
Pyridine, 2-heptyl-	EtOH	257(3.46),263(3.51), 269(3.37)	88-2535-69
$C_{12}H_{19}NO$			
Isoxazole, 3-cyclohexyl-5-ethyl-4-methyl-	EtOH	224(4.09)	22-2820-69
Phenol, 3-tert-butyl-4-(dimethylamino)-	MeOH	230(3.52),277(3.10)	39-2164-69C
$C_{12}H_{19}NO_4$			
Crotonic acid, 3-[N-(2-hydroxy-1,1-dimethylethyl)formimidoyl]-, methyl ester, acetate	C_6H_{12}	240(4.27)	54-0005-69
$C_{12}H_{19}NO_5S$			
Malonic acid, 2-[[2-(ethylcarbamoyl)-vinyl]thio]-, diethyl ester	90% EtOH	272(4.15)	12-0765-69
$C_{12}H_{19}NO_6$			
Maleic acid, (diethylamino)hydroxy-, dimethyl ester, acetate	MeOH	293(4.07)	24-2336-69
$C_{12}H_{19}N_2O_7PS$			
Thymidine, 5'-O-ester with S-ethyl phosphorothioate, sodium salt	H_2O	266(3.96)	35-1522-69
$C_{12}H_{19}N_3O$			
Bicyclo[3.2.1]octan-2-one, 8-isopropylidene-, semicarbazone	isooctane	288s(2.56),296(2.62), 305(2.59),317(2.35)	44-0450-69

Compound	Solvent	λ_{max}(log ε)	Ref.
2H-1,4a-Methanonaphthalen-2-one, octahydro-, semicarbazone	EtOH	225(4.17)	78-5267-69
Pyrazole, 4-acetyl-1-(1-methyl-4-piperidyl)-5-methyl-, hydrochloride	EtOH	246(4.04)	36-0432-69
Tricyclo[4.4.1^{1,3},0]undecan-4-one, semicarbazone	EtOH	225(4.13)	78-5267-69
$C_{12}H_{19}N_3O_2$			
Cyclopentanone, 2-(2-ethyl-4,5-dimethylimidazol-1-yl)-, oxime, 3-oxide	MeOH	220(3.80)	48-0746-69
Isonicotinamidine, N-(formylmethyl)-, diethyl acetal	MeOH	265(3.59)	87-0944-69
$C_{12}H_{19}N_4O_7P_2S$			
Thiamine, O-diphosphate	pH 1.79	241(4.3),250s(4.2)	65-0102-69
	pH 11.07	230(4.1),270(4.0)	65-0102-69
$C_{12}H_{19}N_5$			
Adenine, N,8-dimethyl-9-isopentyl-, picrate	pH 1	215(4.43),264(4.33)	39-2198-69C
	H_2O	214(3.40),265(4.32)	39-2198-69C
	pH 13	221(4.35),265(4.31)	39-2198-69C
$C_{12}H_{20}AsNO$			
Arsine oxide, [p-(dimethylamino)-phenyl]diethyl-	pH 2	293(4.0),300s(3.0)	70-1381-69
	H_2O	277(4.3),320s(3.2)	70-1381-69
	dioxan	275(4.4),310(3.4)	70-1381-69
$C_{12}H_{20}NOP$			
Phosphine oxide, [(p-dimethylamino)-phenyl]diethyl-	pH 5.90	276(4.3)	65-0321-69
$C_{12}H_{20}N_2$			
3,4-Diazabicyclo[4.2.0]octa-2,4-diene, 1,2,5,6,7,8-hexamethyl-	ether	236s(--),243(3.50)	24-1928-69
$C_{12}H_{20}N_2O$			
Phenol, 2-amino-5-tert-butyl-4-(dimethylamino)-	MeOH	231(3.83),293(3.51)	39-2164-69C
$C_{12}H_{20}N_2O_2$			
Acetic acid, 2-(5,5-dimethyl-3-oxo-1-cyclohexen-1-yl)-1,2-dimethylhydrazide	MeOH	287(4.41)	27-0036-69
1-Pyrazoline-4-carboxylic acid, 5-isopropylidene-3,3,4-trimethyl-, ethyl ester	hexane	258(4.00),355(2.44)	39-2443-69
$C_{12}H_{20}N_2O_4$			
1-Pyrazoline-3,3-dicarboxylic acid, 4-propyl-, diethyl ester	$CHCl_3$	326(2.32)	22-0313-69
$C_{12}H_{20}N_4O_2$			
2H-Azepin-2-one, 1,1'-azobis[hexahydro-	EtOH	283(4.23)	44-2997-69
$C_{12}H_{20}N_4O_4$			
Ornithine, N-(octahydro-7a-hydroxy-2H-cyclopentapyrimidin-2-ylidene)-, L-	H_2O	217(4.23)	18-3314-69

Compound	Solvent	λ_{max}(log ϵ)	Ref.
$C_{12}H_{20}N_6$			
1H-Imidazo[4,5-d]pyridazine, 4-[[3-(diethylamino)propyl]amino]-, dihydrochloride	EtOH	208(4.48),260(3.88)	4-0093-69
$C_{12}H_{20}N_6O$			
1-Propanol, 3-[[6-(diethylamino)purin-8-yl]amino]-	pH 1	302(4.28)	65-2125-69
	EtOH	292(4.38)	65-2125-69
$C_{12}H_{20}O$			
Benzofuran, 2,3,3a,4,5,6-hexahydro-2,2,3aα,4α-tetramethyl-	C_6H_{12}	200(3.94)	23-0831-69
Bicyclo[2.1.0]pentane, 5-acetyl-1,2,3,4,5-pentamethyl-	EtOH	220(3.74)(end abs.)	89-0880-69
2-Cyclododecen-1-one, cis	EtOH	228(3.72),320(1.67)	24-3877-69
2-Cyclododecen-1-one, trans	EtOH	230(4.05),321(2.24)	24-3877-69
Cyclohexene, 1-acetyl-2-butyl-	MeOH	248(3.61)	78-0223-69
Cyclohexene, 1-isobutyryl-2-methyl-	MeOH	247(3.60)	78-0223-69
Cyclopentene, 1-acetyl-2-pentyl-	MeOH	257(3.92)	39-0608-69C
2-Cyclopenten-1-one, 3-ethyl-2-pentyl-	EtOH	236(4.16)	104-1993-69
Cycloundecanone, 2-methylene-	n.s.g.	227(3.79),332(1.65)	97-0303-69
$C_{12}H_{20}O_2$			
1,2-Cyclobutanedione, 3,4-di-tert-butyl-	n.s.g.	536(1.81)	88-1529-69
4,5-Heptadien-2-ol, 2,4,6-trimethyl-, acetate	isooctane	215(3.29)	35-3299-69
$C_{12}H_{20}O_3$			
1,2-Cyclododecanedione, 3-hydroxy-, (enol)	DMSO-acid	267(4.57)	24-0388-69
	DMSO-base	336(4.27)	24-0388-69
Cyclohexaneacetic acid, α-acetyl-, ethyl ester	isooctane	290(1.85)	22-3694-69
	EtOH	290(1.91)	22-3694-69
2(5H)-Furanone, 3,5-di-tert-butyl-5-hydroxy-	EtOH	210(3.99)	88-1673-69
$C_{12}H_{20}O_4$			
1,1-Cyclopropanedicarboxylic acid, 2-propyl-, diethyl ester	C_6H_{12}	212(2.88)	22-0313-69
Malonic acid, (1-methylbutylidene)-, diethyl ester	EtOH	224(4.02)	22-0313-69
Malonic acid, pentylidene-, diethyl ester	EtOH	213(4.01)	22-0313-69
$C_{12}H_{20}O_6S_4$			
α-D-Mannopyranose, 1,2-bis(O-ethyl dithiocarbonate)	EtOH	274(4.26)	94-2571-69
$C_{12}H_{20}Si$			
Silane, tetraallyl-	isooctane	196(4.57)	121-0803-69
	EtOH	197(4.55)	121-0803-69
$C_{12}H_{20}Sn$			
Stannane, tetraallyl- (correction)	hexane	217(4.50)	44-3715-69
	MeCN	220(4.50)	44-3715-69
$C_{12}H_{21}NO$			
2-Aziridinone, 1-tert-butyl-3-(1-methylcyclopentyl)-	hexane	246(1.77)	35-1176-69
	EtOH	233(2.17)	35-1176-69

Compound	Solvent	λ_{max}(log ε)	Ref.
$C_{12}H_{21}NO_2$			
2(3H)-Furanone, 3-[1-(hexylamino)ethylidene]dihydro-	hexane	296(4.23)	104-0998-69
2,4-Piperidinedione, 1-methyl-5,5-dipropyl-	pH 1	275(2.98)	7-0658-69
	pH 13	283(4.13)	7-0658-69
	MeOH	265(--)	7-0658-69
$C_{12}H_{21}NO_4$			
Malonamic acid, N-tert-butyl-2-(1-methoxyethylidene)-, ethyl ester	EtOH	224(3.33),254(3.14)	104-1713-69
$C_{12}H_{21}N_3$			
1H-1,4-Diazepine, 2,3-dihydro-5,7-dimethyl-6-piperidino-, perchlorate	H_2O	340(3.86)	39-1449-69C
$C_{12}H_{22}N_2O$			
2H-Pyrrol-3-ol, 5-(diethylamino)-4-methyl-2-propyl-	MeOH	228(4.00),289(4.29)	78-5721-69
$C_{12}H_{22}O_{11}$			
Sucrose	0.4N HCl	258(0.11)	18-2282-69
$C_{12}H_{22}Si$			
Silane, triallylpropyl-	isooctane	196(4.44)	121-0803-69
	EtOH	297(4.42)	121-0803-69
$C_{12}H_{22}Sn$			
Stannane, triallylpropyl- (corrections)	hexane	208(4.47)	44-3715-69
	MeCN	215(4.46)	44-3715-69
$C_{12}H_{23}NO$			
Azacyclotridecan-2-one	C_6H_{12}	188(3.82)	46-1642-69
	H_2O	191(3.89)	46-1642-69
$C_{12}H_{23}N_5O$			
s-Triazine, 2,4-bis(diethylamino)-6-methoxy-	pH 1.0	225(4.32),250(4.29)	59-1167-69
	pH 10.0	230(4.53)	59-1167-69
$C_{12}H_{24}Cl_2N_4NiO_8$			
1,4,8,11-Tetraazacyclotetradeca-4,11-diene, 5,12-dimethyl-, nickel perchlorate complex	MeOH	208(4.26),227s(4.10), 290(3.67),448(1.90)	19-0013-69
$C_{12}H_{24}CoNS_4$			
Tetraethylammonium bis(dimercaptoethylene)cobaltate	DMSO	280(3.94),325(3.84), 354s(3.78),440(3.34), 515(3.15),630s(3.05), 731(3.10)	24-0603-69
$C_{12}H_{24}NO_3P$			
Phosphonic acid, [2-(cyclohexylamino)-vinyl]-, diethyl ester	EtOH	242(4.20)	39-0460-69C
$C_{12}H_{24}N_4$			
1H-Azepine, 1,1'-azobis[hexahydro-	EtOH	289(4.02)	44-2997-69
1,4,8,11-Tetraazacyclotetradeca-4,11-diene, 5,12-dimethyl-, diperchlorate	MeOH	198(2.37),208s(2.00), 232(2.44)	19-0013-69
$C_{12}H_{24}N_4O_2$			
Propionamide, N,N'-azobis[N-propyl-	EtOH	287(4.38)	44-2997-69

Compound	Solvent	λ_{max}(log ε)	Ref.
$C_{12}H_{24}Si$			
Silane, diallyldipropyl-	isooctane	196(4.27)	121-0803-69
	EtOH	196(4.25)	121-0803-69
$C_{12}H_{24}Sn$			
Stannane, diallyldipropyl-	hexane	208(4.28)	44-3715-69
(corrections)	MeCN	211(4.26)	44-3715-69
$C_{12}H_{26}Si$			
Silane, allyltripropyl-	isooctane	195(4.01)	121-0803-69
	EtOH	195(3.99)	121-0803-69
$C_{12}H_{26}Sn$			
Stannane, allyltripropyl-	MeCN	205(4.04)	44-3715-69
$C_{12}H_{28}Cl_2N_4NiO_8$			
1,4,8,11-Tetraazacyclotetradecane, 5,12-dimethyl-, nickel diperchlorate derivative, isomer A	MeOH	211s(4.04),227(4.09), 461(1.79)	19-0013-69
isomer B	MeOH	210s(4.02),227(4.10), 456(1.83)	19-0013-69
$C_{12}O_{12}Ru_3$			
Ruthenium, dodecacarbonyltri-, triangulo-	$C_6H_{11}Me$	390(3.9)	101-0289-69A

Compound	Solvent	λmax(log ε)	Ref.
$C_{13}F_{14}N_2$			
Pyridine, 2,2'-(hexafluorotrimethylene)-bis[tetrafluoro-	C_6H_{12}	259(3.81)	39-2559-69C
$C_{13}H_2F_9NO$			
Benzophenone, 4-aminononafluoro-	EtOH	240(4.02),324(4.33)	65-1766-69
$C_{13}H_6BrClN_2O_2$			
Acridine, 2-bromo-9-chloro-6-nitro-	EtOH	256(4.58),290(4.42), 356(3.81),375(3.92)	103-0707-69
$C_{13}H_7Br_3O_2$			
2,4,6-Cycloheptatrien-1-one, 3,5,7-tribromo-2-hydroxy-4-phenyl-	n.s.g.	273(4.23),356(3.91), 440(3.82)	40-0085-69
$C_{13}H_7ClN_2O_3$			
Benzoxazole, 5-chloro-2-(m-nitrophenyl)-	EtOH	263(4.33),307(4.31)	18-3335-69
Benzoxazole, 5-chloro-2-(p-nitrophenyl)-	EtOH	232(3.91),330(4.23)	18-3335-69
$C_{13}H_7ClOSe$			
Selenoxanthen-9-one, 2-chloro-	MeOH	216(4.12),221(4.11), 256(4.53),262(4.52), 277s(3.72),306(3.56), 401(3.82)	73-3792-69
$C_{13}H_7Cl_2N_3O_6$			
Benzoic acid, 4-(2,4-dichloroanilino)-3,5-dinitro-, anion	base	405(3.77)	22-4569-69
dianion	base	555(3.78)	22-4569-69
$C_{13}H_7N$			
1-Naphthalenepropiolonitrile	EtOH	222(4.59),241(4.30), 315(4.10),333(4.10)	22-0210-69
2-Naphthalenepropiolonitrile	EtOH	217(4.58),247(4.57), 259(4.78),294(4.20), 305(4.30)	22-0210-69
$C_{13}H_8BrNO_2S_2$			
Isothiocyanic acid, p-[(p-bromophenyl)-sulfonyl]phenyl ester	EtOH	219(4.36),245(4.32), 292(4.45),297(4.49)	73-4005-69
$C_{13}H_8BrNS_2$			
Isothiocyanic acid, p-[(p-bromophenyl)-thio]phenyl ester	EtOH	220(4.64),268(4.32), 294s(--),302(4.44)	73-4005-69
$C_{13}H_8BrN_3O_2$			
Acridine, 9-amino-2-bromo-6-nitro-	EtOH	234(4.37),254(4.39), 285(4.61),320(4.18), 366s(3.45),385(3.59), 452(3.67)	103-0707-69
$C_{13}H_8BrN_3O_2S$			
Resorcinol, 4-[(6-bromo-2-benzothiazol-yl)azo]- (at 0.1 ionic strength)	pH 2.50	470(4.44)	123-0027-69A
	pH 6.00	500(4.45)	123-0027-69A
	pH 12.0	520(4.65)	123-0027-69A
$C_{13}H_8BrN_3S$			
Isothiocyanic acid, p-[(p-bromophenyl)-azo]phenyl ester	heptane	239(3.63),353(3.99), 445(2.69)	73-3912-69

Compound	Solvent	$\lambda_{max}(\log \epsilon)$	Ref.
$C_{13}H_8Br_2Cl_4O$			
Tetracyclo[5.4.2.0^{2,6}.0^{8,11}]trideca-3,12-diene, 9,10-dibromo-2,3,4,6-tetrachloro-5-oxo-	EtOH	251(3.93)	39-2710-69C
$C_{13}H_8Br_2O_2$			
2,4,6-Cycloheptatrien-1-one, 3,7-dibromo-2-hydroxy-5-phenyl-	n.s.g.	270(4.53),360(4.30), 430(3.98)	40-0085-69
$C_{13}H_8ClNO_2S_2$			
Isothiocyanic acid, p-[(p-chlorophenyl)-sulfonyl]phenyl ester	EtOH	228(4.40),242(4.29), 303(4.56)	73-4005-69
$C_{13}H_8ClNS_2$			
Isothiocyanic acid, p-[(p-chlorophenyl)-thio]phenyl ester	EtOH	218(4.66),269(4.26), 294s(--),302(4.44)	73-4005-69
$C_{13}H_8ClN_3O_6$			
Benzoic acid, 4-(p-chloroanilino)-3,5-dinitro-, anion	base	418(3.81)	22-4569-69
dianion	base	570(3.76)	22-4569-69
$C_{13}H_8ClN_3S$			
Isothiocyanic acid, p-[(p-chlorophenyl)-azo]-phenyl ester	heptane	238(3.80),350(4.07), 445(2.74)	73-3912-69
$C_{13}H_8Cl_2O$			
Benzophenone, 3,3'-dichloro-	n.s.g.	252(4.54)	78-4919-69
Benzophenone, 4,4'-dichloro-	n.s.g.	260(4.32)	78-4919-69
$C_{13}H_8F_3NO_2S_2$			
p-Dithiin-2,3-dicarboximide, 5,6-dihydro-N-(α,α,α-trifluoro-m-tolyl)-	MeOH	247(4.14),415(3.43)	33-2221-69
$C_{13}H_8F_3N_3O_2S$			
Sulfur diimide, (p-nitrophenyl)(α,α,α-trifluoro-p-tolyl)-	benzene	424(4.22)	104-0459-69
$C_{13}H_8F_3N_3O_4S_2$			
Sulfur diimide, (p-nitrophenyl)[p-(trifluoromethyl)sulfonyl]phenyl]-	benzene	422(4.14)	104-0459-69
$C_{13}H_8N_2$			
1,2-Naphthalenedicarbonitrile, 4-methyl-	MeCN	219(4.46),246(4.84), 307(3.78),318(3.84), 332(3.61),348(3.70)	44-1923-69
$C_{13}H_8N_2O$			
1,2-Naphthalenedicarbonitrile, 7-methoxy-	MeCN	222(4.59),257(4.73), 286s(3.56),295(3.64), 306(3.54),373(3.66)	44-1923-69
$C_{13}H_8N_2O_2S$			
Benzothiazole, 2-(m-nitrophenyl)-	EtOH	271(4.04),294(4.06)	18-3335-69
Benzothiazole, 2-(p-nitrophenyl)-	EtOH	305(3.51),342(4.12)	18-3335-69
$C_{13}H_8N_2O_2S_2$			
Isothiocyanic acid, p-[(p-nitrophenyl)-thio]phenyl ester	EtOH	226(4.62),261(4.12), 289(4.34),297s(--), 332(4.27)	73-4005-69

Compound	Solvent	λ max (log ε)	Ref.
$C_{13}H_8N_2O_2S_3$			
p-Dithiin-2,3-dicarboximide, N-2-benzothiazolyl-5,6-dihydro-	MeCN	218(4.38),274(4.45), 419(3.41)	33-2221-69
$C_{13}H_8N_2O_3$			
1,2-Benzisoxazole, 3-(m-nitrophenyl)-	EtOH	224(4.29),285(3.81)	18-0826-69
Benzoxazole, 2-(m-nitrophenyl)-	EtOH	263(4.31),295(4.36)	18-3335-69
Benzoxazole, 2-(p-nitrophenyl)-	EtOH	230(4.15),326(4.37)	18-3335-69
$C_{13}H_8N_2O_4S_2$			
Isothiocyanic acid, p-[(p-nitrophenyl)-sulfonyl]phenyl ester	EtOH	224(4.41),290(4.46), 304(4.54)	73-4005-69
$C_{13}H_8N_2O_5$			
Benzophenone, 3,3'-dinitro-	n.s.g.	232(4.34)	78-4919-69
Benzophenone, 4,4'-dinitro-	n.s.g.	266(4.29)	78-4919-69
$C_{13}H_8N_2O_6$			
Benzoic acid, 2,4-dinitrophenyl ester	EtOH	237(4.48)	65-0660-69
2,4,6-Cycloheptatrien-1-one, 2-hydroxy-3,7-dinitro-5-phenyl-	n.s.g.	250(3.95),330(3.44), 431(3.34)	40-0085-69
2,4,6-Cycloheptatrien-1-one, 2-hydroxy-5,7-dinitro-6-phenyl-	n.s.g.	225(4.43),410(4.21)	40-0085-69
$C_{13}H_8N_2S$			
Benzimidazo[2,1-b]benzothiazole	EtOH	222(4.30),242(4.47), 263s(4.00),283(3.93), 291(3.97),300(3.85)	39-1334-69C
Pyrazolo[4,3-c]dibenzothiophene	EtOH	249(4.70),295(4.3), 319(4.16),333(4.17)	4-0771-69
$C_{13}H_8N_4O_2S$			
Isothiocyanic acid, p-[(p-nitrophenyl)-azo]phenyl ester	heptane	282(3.49),361(4.17), 445(2.62)	73-3912-69
$C_{13}H_8N_4O_4$			
Imidazo[1,2-a]pyridine, 3-nitro-2-(p-nitrophenyl)-	EtOH-2%DMF	277(4.40),361(4.07)	103-0093-69
$C_{13}H_8N_4O_4S$			
Benzothiazole, 6-nitro-2-(o-nitroanilino)-	EtOH	226(4.39),250(4.25), 376(4.21)	17-0095-69
$C_{13}H_8N_4O_8$			
Anthranilic acid, 3,5-dinitro-, anion	base	390(4.32)	22-4569-69
dianion	base	520(4.42)	22-4569-69
Anthranilic acid, N-picryl-, anion	base	397(4.23)	22-4569-69
dianion	base	437(4.32)	22-4569-69
Benzoic acid, m-(2,4,6-trinitroanilino)-, anion	base	375(4.06)	22-4569-69
dianion	base	440(4.24)	22-4569-69
$C_{13}H_8O$			
1-Naphthalenepropiolaldehyde	MeOH	230(4.26),246(4.14), 322(4.01),339(3.99)	39-2173-69C
$C_{13}H_8OS_2$			
1-Thiacoumarin, 4-(2-thienyl)-	EtOH	240(3.8),310(4.1), 360s(3.51)	2-0315-69

Compound	Solvent	$\lambda_{max}(\log \epsilon)$	Ref.
$C_{13}H_8OSe$			
Selenoxanthen-9-one	MeOH	219(4.06),255(4.55), 284s(3.75),296s(3.50)	73-3792-69
$C_{13}H_8O_2$			
5,6-Benzocoumarin	EtOH	231(4.8),250(4.1), 317(4.0)	39-2618-69C
7,8-Benzocoumarin (naphtho[1,2-b]pyran-2-one)	EtOH	219(4.6),265(4.4), 275(4.6),308(3.9), 322(4.0)	39-2618-69C
$C_{13}H_8O_3$			
7,8-Benzocoumarin, 6-hydroxy-	EtOH	226(4.5),278(4.4), 288(4.4),317(3.8), 390(3.7)	78-4207-69
Xanthen-9-one, 2-hydroxy-	EtOH	248(4.52),299(3.65), 303s(3.62),365(3.86)	39-0281-69C
Xanthen-9-one, 4-hydroxy-	EtOH	235s(4.37),251(4.53), 280(3.75),291(3.68), 350s(3.68)	39-0281-69C
$C_{13}H_8O_4$			
Xanthen-9-one, 1,5-dihydroxy-	MeOH	235s(4.34),248(4.53), 315(3.81),370(3.55)	39-2421-69C
Xanthen-9-one, 3,5-dihydroxy-	EtOH	213(4.40),235(4.59), 245s(4.51),270(4.17), 307(4.02),332s(3.86)	39-0281-69C
$C_{13}H_8O_4S$			
Resorcinol-captan fusion product	MeOH	228(4.24),320(3.90), 353(3.84),423(4.48)	88-5307-69
$C_{13}H_8O_5$			
2H,5H-Pyrano[3,2-c]benzo[b]pyran-2,5-dione, 4-hydroxy-7-methyl-	EtOH	264(4.23),348(4.05)	18-3233-69
2H,5H-Pyrano[3,2-c]benzo[b]pyran-2,5-dione, 4-hydroxy-9-methyl-	EtOH	261(4.16),351(4.02)	18-3233-69
2H,5H-Pyrano[3,2-c]benzo[b]pyran-2,5-dione, 4-hydroxy-10-methyl-	EtOH	266(4.15),339(4.10)	18-3233-69
Xanthen-9-one, 1,3,5-trihydroxy-	EtOH	220s(4.20),247(4.53), 313(4.21),360s(3.62)	39-0281-69C
Xanthen-9-one, 1,3,7-trihydroxy-	EtOH	220s(4.22),238(4.47), 260(4.55),310(4.18), 373(3.82)	39-0281-69C
Xanthen-9-one, 1,5,6-trihydroxy-	MeOH	251(4.58),315(3.81), 332(4.18)	39-2201-69C
Xanthen-9-one, 1,6,7-trihydroxy-	MeOH	250(4.34),268s(3.96), 291(3.93),313s(3.84), 360(4.00)	39-2201-69C
$C_{13}H_8O_6$			
Norswertianin	EtOH	239(4.42),267(4.50), 332(4.10),392s(3.70)	94-0155-69
	EtOH-NaOAc	270(4.40),360(4.30)	94-0155-69
$C_{13}H_9BrN_2O$			
Benzimidazole, 2-[2-(5-bromo-2-furyl)-vinyl]-	MeOH	348(4.51)	103-0246-69

Compound	Solvent	$\lambda_{max}(\log \epsilon)$	Ref.
$C_{13}H_9BrN_2O_4$			
Anthranilic acid, N-(p-bromophenyl)-4-nitro-	EtOH	238(4.11),295(4.44),418(3.57)	103-0707-69
$C_{13}H_9BrO_2$			
Benzoic acid, 2-bromo-5-phenyl-	n.s.g.	257(4.36)	40-0085-69
2,4,6-Cycloheptatrien-1-one, 2-hydroxy-7-bromo-4-phenyl-	n.s.g.	278(4.51),342(4.01),400(4.00)	40-0085-69
$C_{13}H_9BrO_4$			
1-Naphthol, 2-bromo-3,4-(methylenedioxy)-, acetate	EtOH	244(4.66),281(3.40),299(3.54),312(3.58),350(3.56)	77-0167-69
$C_{13}H_9ClN_2O$			
Benzimidazole, 2-[2-(5-chloro-2-furyl)-vinyl]-	MeOH	344(5.03)	103-0246-69
$C_{13}H_9ClN_4O_2$			
9H-Purine-9-carboxylic acid, 6-chloro-, benzyl ester	MeCN	245(3.85)	44-0973-69
$C_{13}H_9ClO_2S$			
Fluorene-9-sulfonyl chloride	isooctane	235(4.38),273(4.00)	44-2901-69
$C_{13}H_9ClO_3S$			
Benzoic acid, 2-chloro-3-[(p-hydroxyphenyl)thio]-	EtOH	230(4.53)	104-1903-69
Benzoic acid, 3-chloro-4-[(p-hydroxyphenyl)thio]-	EtOH	232(4.23),280(4.23)	104-1903-69
$C_{13}H_9ClO_4S$			
Benzoic acid, 2-chloro-3-[(p-hydroxyphenyl)sulfinyl]-	EtOH	238(4.28)	104-1903-69
Benzoic acid, 3-chloro-4-[(p-hydroxyphenyl)sulfinyl]-	EtOH	258(4.28)	104-1903-69
$C_{13}H_9ClO_4S_2$			
4-(2-Thienyl)-1-benzothiopyrylium perchlorate	HOAc-1% $HClO_4$	262(4.05),340(3.55),380(3.67),455(4.17)	2-0191-69
$C_{13}H_9ClO_5S$			
Benzoic acid, 2-chloro-3-[(p-hydroxyphenyl)sulfonyl]-	EtOH	240(4.30)	104-1903-69
Benzoic acid, 3-chloro-4-[(p-hydroxyphenyl)sulfonyl]-	EtOH	232(4.22),274(4.05)	104-1903-69
$C_{13}H_9Cl_6N_3OS$			
s-Triazine, 2-[(o-ethoxyphenyl)thio]-4,6-bis(trichloromethyl)-	MeOH	213(4.28),282(3.96)	18-2931-69
s-Triazine, 2-[(p-ethoxyphenyl)thio]-4,6-bis(trichloromethyl)-	MeOH	212(4.18),283(4.07)	18-2931-69
$C_{13}H_9FN_2$			
Benzimidazole, 2-(o-fluorophenyl)-	MeOH	231(4.16),238(4.13),293(4.28),307(4.30),327(3.97)	4-0605-69
Indole, 5-fluoro-2-(3-pyridyl)-	EtOH	227(4.27),240s(4.12),255(3.95),316(4.42)	22-4154-69

Compound	Solvent	λ_{max}(log ϵ)	Ref.
$C_{13}H_9F_3N_2O_2S_2$			
Sulfur diimide, (phenylsulfonyl)(α,α,α-trifluoro-p-tolyl)-	benzene	372(4.21)	104-0459-69
$C_{13}H_9F_3N_2O_4S_3$			
Sulfur diimide, (phenylsulfonyl)[p-[(trifluoromethyl)sulfonyl]phenyl]-	benzene	372(4.03)	104-0459-69
$C_{13}H_9F_5O_3$			
Cinnamic acid, α-acetyl-2,3,4,5,6-penta-fluoro-, ethyl ester	EtOH	266(3.88)	65-1347-69
$C_{13}H_9IN_2$			
Benzimidazole, 2-(o-iodophenyl)-	MeOH	229(4.26),276(4.06), 282(4.11)	4-0605-69
$C_{13}H_9N$			
Pyridine, 2-(phenylethynyl)-	MeOH	266s(4.03),271(4.17), 303(5.40)	39-1143-69C
$C_{13}H_9NO$			
9-Acridinol	EtOH	214(4.39),250s(4.76), 254(4.77),260s(4.60), 270s(4.27),294(3.31), 306(3.04),362s(3.65), 379(3.93),398(3.97)	39-0096-69B
	10% EtOH	213(4.30),235s(4.20), 254(4.82),290s(3.13), 370s(3.72),385(3.92), 404(3.87)	39-0096-69B
Fluorene, 4-nitroso-	EtOH	240s(4.14),256(4.30), 301(3.68),326(3.79), 400(3.74)	39-0345-69C
1-Naphthalenemalonaldehydonitrile	EtOH	223(4.85),282(3.85), 292(3.81)	22-0198-69
2-Naphthalenemalonaldehydonitrile	EtOH	217(4.68),223(4.60), 254(4.42),263(4.41), 291(4.10),303(4.06)	22-0198-69
sodium salt	EtOH	220(4.61),261(4.18), 274(4.08),309(3.98)	22-0198-69
$C_{13}H_9NOS$			
8-Quinolinol, 2-(2-thienyl)-, nickel or cobalt chelate	50% dioxan	300(4.3),340s(3.7), 400s(3.0)	23-4655-69
$C_{13}H_9NOS_2$			
Isothiocyanic acid, p-(phenylsulfinyl)-phenyl ester	EtOH	234(4.40),287(4.41), 294(4.42)	73-4005-69
$C_{13}H_9NO_2$			
1,8-Azulenedicarboximide, N-methyl-	MeOH	214(4.30),255(4.42), 265(4.43),308s(4.03), 326s(3.96),335s(3.92), 405(3.76),593(2.94), 640(2.92),705(2.55)	108-0435-69
Carbazole-1-carboxaldehyde, 2-hydroxy-	EtOH	228(4.58),257(3.94), 285s(4.18),295(4.25), 377(3.82)	39-1518-69C

Compound	Solvent	$\lambda_{max}(\log \epsilon)$	Ref.
Carbazole-3-carboxaldehyde, 2-hydroxy-	EtOH	234(4.42),247(4.21), 278(4.54),298(4.56), 345(4.07)	39-1518-69C
Fluorene, 1-nitro-	EtOH	260(4.43),270s(4.30), 288s(3.81),333(3.51)	39-0345-69C
Fluoren, 9-nitro-, potassium salt (also other solvents not listed)	H_2O	250s(4.2),266(3.9), 300(3.7),357(4.0)	35-0366-69
	EtOH	373(--)	35-0366-69
	tert-BuOH	371(--)	35-0366-69
	ether	381s(--),396(--)	35-0366-69
	DMSO	396s(--),419(--)	35-0366-69
$C_{13}H_9NO_2S$			
Dibenzothiophene, 2-methyl-4-nitro-	n.s.g.	233(4.54),267(4.15), 292(3.71),302(3.88), 382(3.71)	88-4483-69
$C_{13}H_9NO_2S_2$			
Isothiocyanic acid, p-(phenylsulfonyl)-phenyl ester	EtOH	213(4.48),238(4.43), 289s(--),296(4.45)	73-4005-69
$C_{13}H_9NO_3$			
Carbazole-1-carboxylic acid, 2-hydroxy-	EtOH	235(4.72),240(4.68), 275(4.78),311(4.57), 331(4.56)	2-1065-69
Carbazole-3-carboxylic acid, 2-hydroxy-	EtOH	237(4.67),243(4.74), 273(4.75),311(4.0), 320(4.0),335(3.91)	2-1065-69
$C_{13}H_9NO_4$			
Benzoic acid, p-nitrophenyl ester	EtOH	237(4.16),267(4.16)	65-0660-69
2,4,6-Cycloheptatrien-1-one, 2-hydroxy-3-nitro-4-phenyl-	n.s.g.	243(4.4),328(3.87), 383(3.88)	40-0085-69
2,4,6-Cycloheptatrien-1-one, 2-hydroxy-7-nitro-4-phenyl-	n.s.g.	240(4.21),283(4.36), 395(3.88)	40-0085-69
$C_{13}H_9NO_4Se$			
1,3-Propanedione, 1-(5-nitroselenophene-2-yl)-3-phenyl-	H_2O	272(4.95),305(4.99), 370(4.66)	123-0074-69C
$C_{13}H_9NS_2$			
Isothiocyanic acid, p-(phenylthio)-phenyl ester	EtOH	258(4.07),292(4.29), 303(4.30)	73-4005-69
$C_{13}H_9N_3O$			
Pyridine, 2-(5-phenyl-1,3,4-oxadiazol-2-yl)-	H_2SO_4	241(3.74),271(4.09), 314(4.42)	4-0965-69
$C_{13}H_9N_3O_2$			
Imidazo[1,2-a]pyridine, 2-(p-nitrophenyl)-	EtOH-2% DMF	272(4.20),340(4.24)	103-0093-69
$C_{13}H_9N_3O_2S$			
Benzothiazole, 2-(o-nitroanilino)-	EtOH	226(4.38),278s(4.25), 292s(4.24),302s(4.21), 404(3.79)	17-0095-69
Resorcinol, 4-(2-benzothiazolylazo)- (all at 0.1 ionic strength)	pH 1.0	460(4.28)	123-0027-69A
	pH 4.00	440(4.28)	123-0027-69A
	pH 7.50	500(4.34)	123-0027-69A
	pH 12.0	520(4.46)	123-0027-69A

Compound	Solvent	$\lambda_{max}(\log \epsilon)$	Ref.
$C_{13}H_9N_3O_3$			
Benzoxazole, 2-(o-nitroanilino)-	EtOH	253(4.33),281(4.33), 289(4.32),388(3.83)	17-0095-69
$C_{13}H_9N_3O_4S$			
1H-Thieno[2,3-c]pyrazole-5-carboxylic acid, 3-methyl-1-(p-nitrophenyl)-	EtOH	222(4.05),245(4.07), 272(3.92),355(4.11)	104-1460-69
$C_{13}H_9N_3O_5$			
Acrylophenone, 4'-nitro-3-(5-nitropyr-rol-2-yl)-	EtOH	270(4.21),396(4.45)	103-0383-69
	dioxan	273(4.20),392(4.40)	103-0383-69
2-Propen-1-one, 3-(p-nitrophenyl)-1-(4-nitro-2-pyrrolyl)-	EtOH	322(4.37)	103-0383-69
	dioxan	323(4.35)	103-0383-69
	H_2SO_4	327(4.06),424(4.58)	103-0383-69
2-Propen-1-one, 3-(p-nitrophenyl)-1-(5-nitro-2-pyrrolyl)-	EtOH	245(4.07),277(4.12), 333(4.08)	103-0383-69
	dioxan	242(4.09),275(4.04), 333(4.16)	103-0383-69
	H_2SO_4	317(3.73),345(3.75), 443(4.56)	103-0383-69
$C_{13}H_9N_3O_6$			
Benzoic acid, m-(2,4-dinitroanilino)-, anion	base	366(4.27)	22-4569-69
dianion	base	420(4.27)	22-4569-69
Benzoic acid, p-(2,4-dinitroanilino)-, anion	base	370(4.30)	22-4569-69
dianion	base	420(4.32)	22-4569-69
$C_{13}H_9N_3S$			
Isothiocyanic acid, p-(phenylazo)phenyl ester	heptane	242(4.23),340(4.48), 448(3.13)	73-3912-69
$C_{13}H_{10}BrClO_4$			
(o-Bromophenyl)cycloheptatrienylium perchlorate	MeCN	269(3.94),343(3.75)	49-0001-69
$C_{13}H_{10}BrN$			
Aniline, N-benzylidene-p-bromo-	80% MeOH-acid	335(4.31)	78-0057-69
$C_{13}H_{10}BrNO$			
Nitrone, α-(p-bromophenyl)-N-phenyl-	C_6H_{12}	329(4.37)	18-3306-69
	EtOH	323(4.39)	18-3306-69
$C_{13}H_{10}BrN_3O_2S$			
Benzenesulfenyl bromide, o-[(4-nitro-o-tolyl)azo]-	EtOH	335(5.73),431(4.32)	39-1828-69C
	$CHCl_3$	332(5.22)	39-1828-69C
$C_{13}H_{10}Br_2ClN$			
Aniline, N-(p-chlorobenzylidene)-, bromine adduct	MeCN	251(4.48),283s(3.89)	25-0135-69
$C_{13}H_{10}Br_2N_2O_2$			
Aniline, N-benzylidene-p-nitro-, bromine adduct	MeCN	246(4.41),279(4.32), 356(4.15)	25-0135-69
$C_{13}H_{10}ClN$			
Aniline, N-benzylidene-p-chloro-	80% MeOH-acid	330(4.28)	78-0057-69

Compound	Solvent	$\lambda_{max}(\log \epsilon)$	Ref.
Pyridine, 2-(m-chlorostyryl)-, trans	H_2O	308(4.42)	23-2355-69
hydrochloride	H_2O	329(4.45)	23-2355-69
Pyridine, 2-(p-chlorostyryl)-, trans	H_2O	312(4.49)	23-2355-69
hydrochloride	H_2O	339(4.51)	23-2355-69
Pyridine, 3-(m-chlorostyryl)-, trans	50% MeOH	300(4.47)	23-2355-69
hydrochloride	n.s.g.	332(4.47)	23-2355-69
Pyridine, 3-(p-chlorostyryl)-, trans	50% MeOH	311(4.51)	23-2355-69
hydrochloride	n.s.g.	343(4.53)	23-2355-69
$C_{13}H_{10}ClNO$			
2-Methyl-1-oxo-1H-benz[de]isoquinolin-ium chloride	H_2SO_4	232(4.36),313(3.54), 417(4.25),441(4.29)	4-0681-69
Nitrone, α-(p-chlorophenyl)-N-phenyl-	C_6H_{12}	327(4.35)	18-3306-69
	EtOH	321(4.37)	18-3306-69
$C_{13}H_{10}ClNO_2$			
2H-Azepin-2-one, 3-benzoyl-5-chloro-1,3-dihydro-	MeOH	247(4.26)	78-5205-69
HI salt	MeOH	252(4.50)	78-5205-69
$C_{13}H_{10}ClNO_3$			
2-Furanacrylohydroxamic acid, N-p-chlorophenyl-	EtOH	315(4.43)	34-0278-69
$C_{13}H_{10}ClNO_6$			
(o-Nitrophenyl)cycloheptatrienylium perchlorate	MeCN	255s(4.14),325(3.78)	49-0001-69
(p-Nitrophenyl)cycloheptatrienylium perchlorate	MeCN	267(4.41),338(4.27)	49-0001-69
$C_{13}H_{10}ClNS$			
Benzothiazoline, 2-(o-chlorophenyl)-, nickel compound	n.s.g.	460(3.46)	97-0385-69
$C_{13}H_{10}ClN_3O_2S$			
Benzenesulfenyl chloride, o-[(4-nitro-o-tolyl)azo]-	EtOH	335(5.32),434(4.83)	39-1828-69C
	$CHCl_3$	330(5.26)	39-1828-69C
	CCl_4	338(5.29)	39-1828-69C
$C_{13}H_{10}ClN_3O_6S$			
Benzenesulfenyl perchlorate, o-[(4-nitro-o-tolyl)azo]-	EtOH	335(5.30),434(4.82)	39-1828-69C
$C_{13}H_{10}Cl_2$			
Methane, bis(m-chlorophenyl)-	n.s.g.	269(2.74)	78-4919-69
Methane, bis(p-chlorophenyl)-	n.s.g.	269(2.78)	78-4919-69
$C_{13}H_{10}Cl_2O_4$			
(o-Chlorophenyl)cycloheptatrienylium perchlorate	MeCN	267(3.96),346(3.81)	49-0001-69
$C_{13}H_{10}Cl_5NO_2$			
Indole-3-carboxylic acid, 4,5,6,7-tetrachloro-2-(chloromethyl)-1-methyl-	MeOH	233(4.77),298(4.17)	24-2684-69
$C_{13}H_{10}FN$			
Aniline, N-benzylidene-p-fluoro-	80% MeOH-acid	328(4.29)	78-0057-69

Compound	Solvent	$\lambda_{max}(\log \epsilon)$	Ref.
$C_{13}H_{10}IN$			
Aniline, N-benzylidene-p-iodo-	80% MeOH-acid	340(4.34)	78-0057-69
$C_{13}H_{10}IN_3O_2S$			
Benzenesulfenyl iodide, o-[(4-nitro-o-tolyl)azo]-	EtOH	335(5.38),429(4.84)	39-1828-69C
	$CHCl_3$	335(5.39)	39-1828-69C
$C_{13}H_{10}N_2$			
Acridine, 9-amino-	EtOH	260(4.8),310(3.2),320(3.1),400(4.0)	65-1156-69
2H-Cyclopenta[d]pyridazine, 2-phenyl-	ether	248(4.35),258s(4.30),287(4.45),318s(3.87),408(3.45)	35-0924-69
Indole, 2-(3-pyridyl)-	EtOH	227(4.29),240s(4.16),255(4.03),314(4.38)	22-4154-69
Methane, diazodiphenyl-	dioxan	285(4.32),525(2.00)	47-3313-69
1-Naphthaleneacetonitrile, α-(aminomethylene)-	EtOH	221(4.76),250(4.18)	22-0210-69
2-Naphthaleneacetonitrile, α-(aminomethylene)-	EtOH	221(4.67),263(4.30),275(4.30),340(4.24)	22-0210-69
$C_{13}H_{10}N_2O$			
Benzimidazole, 2-[2-(2-furyl)vinyl]-	MeOH	342(4.37)	103-0246-69
Benzimidazole, 2-phenoxy-	MeCN	276(3.91),282(3.90)	77-1110-69
4-Isoxazolecarbonitrile, 3-methyl-5-styryl-	MeOH	225s(--),229(4.03),236(3.99),318(4.58)	32-0753-69
$C_{13}H_{10}N_2OS$			
1H-Thieno[2,3-b]pyrazole-5-carboxaldehyde, 3-methyl-1-phenyl-	EtOH	228(4.18),270(4.32),325(4.25)	104-1460-69
$C_{13}H_{10}N_2O_2$			
Glyoxal, phenyl-2-pyridyl-, 1-oxime	n.s.g.	251(4.3)	36-0857-69
Glyoxal, phenyl-3-pyridyl-, 1-oxime	n.s.g.	251(4.3)	36-0857-69
Glyoxal, phenyl-4-pyridyl-, 1-oxime	n.s.g.	251(4.3)	36-0857-69
3H-Phenoxazin-3-one, 7-amino-1-methyl-	EtOH	534(4.48),551(4.48)	73-0221-69
$C_{13}H_{10}N_2O_2$			
Pyridine, 2-(p-nitrostyryl)-, trans	H_2O	344(4.42)	23-2355-69
hydrochloride	H_2O	341(4.54)	23-2355-69
Pyridine, 3-(p-nitrostyryl)-, trans	50% MeOH	341(4.56)	23-2355-69
hydrochloride	n.s.g.	333(4.43)	23-2355-69
$C_{13}H_{10}N_2O_2S$			
Barbituric acid, 5-cinnamylidene-2-thio-	EtOH	260(4.17),405(4.58)	103-0830-69
Benzothiazoline, 2-(o-nitrophenyl)-, cadmium derivative	n.s.g.	440(3.34)	97-0385-69
nickel derivative	n.s.g.	460(3.56)	97-0385-69
zinc derivative	n.s.g.	440(3.11)	97-0385-69
Benzothiazoline, 2-(p-nitrophenyl)-, mercury derivative	n.s.g.	424(3.82)	97-0385-69
nickel derivative	n.s.g.	530(3.65)	97-0385-69
1H-Thieno[2,3-c]pyrazole-5-carboxylic acid, 3-methyl-1-phenyl-	EtOH	260(4.24),288(4.23)	104-1460-69
$C_{13}H_{10}N_2O_3$			
Acrylophenone, 3-(5-nitro-2-pyrrolyl)-	EtOH	288(3.98),390(4.42)	103-0383-69
	dioxan	285(3.96),386(4.45)	103-0383-69

Compound	Solvent	λ_{max}(log ϵ)	Ref.
Acrylophenone, 3-(5-nitro-2-pyrrolyl)- (cont.)	H_2SO_4	260(3.77),330(3.96), 493(4.65)	103-0383-69
	20% EtOH-KOH	268(4.06),398(4.11), 450(4.18)	103-0383-69
Barbituric acid, 5-cinnamylidene-	EtOH	250(4.08),375(4.60)	103-0827-69
Nitrone, α-(p-nitrophenyl)-N-phenyl-	C_6H_{12}	363(4.25)	18-3306-69
	EtOH	356(4.29)	18-3306-69
Phenoxazine, 1-methyl-3-nitro-	EtOH-HCl	228(4.45),278(4.06), 455(4.05)	73-3732-69
	EtOH-KOH	234(4.25),280(3.93), 325(3.83),558(4.25), 595s(--),655s(--)	73-3732-69
Phenoxazine, 9-methyl-3-nitro-	EtOH-HCl	230(4.46),283(4.07), 455(4.06)	73-3732-69
	EtOH-KOH	234(4.27),280(3.90), 320(3.83),565(4.22), 595s(--),650s(--)	73-3732-69
2-Propen-1-one, 1-(4-nitro-2-pyrrolyl)-	EtOH	230(4.10),325(4.44)	103-0383-69
3-phenyl-	dioxan	255(4.02),323(4.45)	103-0383-69
	H_2SO_4	255(4.04),335(4.03), 463(4.76)	103-0383-69
2-Propen-1-one, 1-(5-nitro-2-pyrrolyl)-	EtOH	232(3.99),360(4.38)	103-0383-69
3-phenyl-	dioxan	354(4.50)	103-0383-69
	H_2SO_4	255(3.80),337(3.79), 480(4.73)	103-0383-69
	20% EtOH-KOH	270(3.97),312(4.11), 410(4.40)	103-0383-69
4H-Quinolizine-1-carboxylic acid, 3-cyano-4-oxo-, ethyl ester	EtOH	267(4.22),346(3.80), 402(4.26)	95-0203-69
$C_{13}H_{10}N_2O_3S$			
4H-Quinolizine-1-carboxylic acid, 3-cyano-2-(methylthio)-4-oxo-, methyl ester	EtOH	272(4.17),315(4.01), 420(4.47)	95-0203-69
$C_{13}H_{10}N_2O_4$			
Anthranilic acid, N-(p-nitrophenyl)-, anion	base	410(4.32)	22-4569-69
dianion	base	480(4.44)	22-4569-69
Benzoic acid, p-(p-nitroanilino)-, anion	base	405(4.36)	22-4569-69
dianion	base	490(4.47)	22-4569-69
Methane, bis(m-nitrophenyl)-	n.s.g.	264(3.90)	78-4919-69
Methane, bis(p-nitrophenyl)-	n.s.g.	278(3.91)	78-4919-69
$C_{13}H_{10}N_2S$			
Pyrazolo[4,3-c]dibenzothiophene, 4,5-dihydro-	EtOH	233(4.04),248(3.75), 257(3.91),266(3.97), 299s(4.07),309(4.15), 324(4.07)	4-0771-69
$C_{13}H_{10}N_2S_2$			
1,2,3,4-Dithiadiazolidine, 4,5-diphenyl-, meso-ionic didehydro derivative	MeOH	397(3.72)	42-0388-69
$C_{13}H_{10}N_4$			
1H-Benzotriazole, 1-(benzylideneamino)-	EtOH	274(4.29),332(4.27)	39-1758-69C
Pteridine, 2-methyl-4-phenyl-	hexane	230(4.22),288s(3.81), 297(3.90),328(4.10)	39-1883-69C

Compound	Solvent	λ_{max}(log ϵ)	Ref.
Pteridine, 2-methyl-4-phenyl- (cont.)	pH 2.0	228(4.18),286s(3.69), 325(4.00)	39-1883-69C
	pH 6.3	226(4.11),302s(3.87), 325(4.03)	39-1883-69C
Pteridine, 4-methyl-2-phenyl-	hexane	254(4.43),275s(4.26), 279(4.26),323(4.01), 337(4.14),352(4.06)	39-1408-69C
	pH 0.7	251(4.39),313(4.12)	39-1408-69C
	pH 5.4	247(4.32),266s(4.16), 333(4.01)	39-1408-69C
Pteridine, 7-methyl-2-phenyl-	hexane	252(4.39),273(4.21), 281s(4.16),323s(4.06), 334(4.21),349(4.15)	39-1408-69C
	pH -0.3	253(4.22),290s(4.04), 311(4.19)	39-1408-69C
	pH 6.0	249(4.33),333(4.12)	39-1408-69C
Pteridine, 7-methyl-4-phenyl-	hexane	231(4.24),289s(3.94), 301s(4.04),321(4.17)	39-1883-69C
	pH 0	227(4.20),325(4.03)	39-1883-69C
	pH 4.2	227(4.17),304s(4.04), 319(4.11)	39-1883-69C
Pyridine, 2-(5-phenyl-s-triazol-3-yl)-	H_2SO_4	264(4.16),295(4.26)	4-0965-69
$C_{13}H_{10}N_4O$			
4-Pteridinol, 7-methyl-2-phenyl-	pH 5.0	246(4.21),290(4.12), 303s(4.11)	39-1408-69C
	pH 10.0	227(4.00),266(4.41), 341(3.99)	39-1408-69C
$C_{13}H_{10}N_4O_2$			
Benzimidazole, 2-(o-nitroanilino)-	95% EtOH	230s(4.27),241s(4.29), 282(4.32),402(3.58)	17-0095-69
Benzimidazole, 2-(p-nitroanilino)-	95% EtOH	232s(4.21),242(4.23), 282(3.94),382(4.30)	17-0095-69
4H-Quinolizine-1,3-dicarbonitrile, 2-[(2-hydroxyethyl)amino]-4-oxo-	EtOH	258(4.62),303(4.02), 392(4.01)	95-0203-69
$C_{13}H_{10}N_4O_5$			
1,3,4-Oxadiazole, 2-[1-(2-furyl)-5-(methylamino)-2-(5-nitro-2-furyl)-vinyl]-, cis	n.s.g.	310(4.04),433(4.24)	40-0713-69
trans	n.s.g.	318(4.04),423(4.28)	40-0713-69
$C_{13}H_{10}O$			
Benzophenone	n.s.g.	252(4.38)	78-4919-69
5,10-Methanobenzocycloocten-11-one, 5,10-dihydro-	EtOH	220(4.01)(end abs.)	88-1761-69
Naphtho[2,3-b]furan, 2-methyl-	MeOH	243(4.85),297(3.93), 307(4.04),321(3.90)	2-0536-69
2-Propyn-1-ol, 3-(1-naphthyl)-	MeOH	234(4.37),286(3.99), 298(4.12),307(3.96), 311(3.98)	39-2173-69C
$C_{13}H_{10}OS$			
Acrylophenone, 3-(2-thienyl)-	MeOH	338(4.35)	73-2771-69
Benzoic acid, thio-, O-phenyl ester	C_6H_{12}	219(4.07),237(3.92), 288(4.03),441(1.96)	48-0045-69
	EtOH	237(3.93),289(4.01), 435(1.99)	48-0045-69
	MeCN	235(3.93),290(4.02), 430(2.05)	48-0045-69

Compound	Solvent	λmax(log ε)	Ref.
2-Propen-1-one, 3-phenyl-1-(2-thienyl)-	MeOH	322(4.49)	73-2771-69
3-Thiaphenanthrene, 1,2,3,4-tetrahydro-1-oxo-	EtOH	212(4.48),251(4.60), 289(3.91),345(3.33)	44-1566-69
$C_{13}H_{10}O_2$			
Acrylophenone, 3-(2-furyl)-	MeOH	342(4.45)	73-2771-69
Biphenyl, 3,4-(methylenedioxy)-	EtOH	265(4.08),288(3.93)	78-4315-69
2-Propen-1-one, 1-(2-furyl)-3-phenyl-	MeOH	320(4.48)	73-2771-69
$C_{13}H_{10}O_2S$			
Thiophene, 2-[(3,4-methylenedioxy)styryl]-	n.s.g.	255s(3.60),313s(4.30), 338(4.50)	117-0209-69
$C_{13}H_{10}O_3$			
Benzophenone, 2,3'-dihydroxy-	EtOH	224(4.21),263(4.09), 333(3.76)	39-0281-69C
Benzophenone, 2,4-dihydroxy-	EtOH	244(4.01),290(4.13), 326(4.01)	39-0034-69C
$C_{13}H_{10}O_4$			
Benzophenone, 2,3',6-trihydroxy-	EtOH	230(4.15),260(3.97), 287(3.80),312(3.67)	39-0281-69C
2H-Pyran-2-one, 4-hydroxy-6-phenacyl-	EtOH	245(4.16),284(3.96)	78-2687-69
Pyrocanesin	EtOH	249(4.6),289(4.19), 301(4.30),346(3.75)	12-1933-69
$C_{13}H_{10}O_5$			
Benzophenone, 2,3',4,6-tetrahydroxy-	EtOH	220s(4.33),261(3.73), 306(4.20)	39-0281-69C
$C_{13}H_{10}O_5S$			
6H-Cyclohepta[c]thiophene-5,7-dicarboxylic acid, 1,3-dimethyl-6-oxo-	EtOH	237(4.14),289(4.46), 348(3.78)	104-0559-69
$C_{13}H_{10}S_2$			
Benzoic acid, dithio-, phenyl ester	C_6H_{12}	234s(4.1),290s(3.7), 521(1.49)	48-0045-69
	EtOH	231(--),298(--), 519(--)	48-0045-69
	MeCN	231(4.2),296(3.70), 511(1.30)	48-0045-69
2H-1-Benzothiopyran, 4-(2-thienyl)-	EtOH	235(3.86),260s(3.81), 300(4.1),340s(3.4)	2-0315-69
$C_{13}H_{11}$			
Diphenyl methyl cation	acid	304(3.45),442(4.78)	59-0447-69
$C_{13}H_{11}BrN_2$			
Benzamidine, N-(p-bromophenyl)-	50% EtOH	234(4.20)	49-1307-69
$C_{13}H_{11}BrN_2OS$			
8-Methyl-6-phenylthiazolo[3,2-a]pyrazin-4-ium bromide, 7-oxide	n.s.g.	202(4.42),251(4.2), 350(4.22)	39-2270-69C
$C_{13}H_{11}Br_2N$			
Aniline, N-benzylidene-, bromine adduct	MeCN	243(4.47),288(3.71)	25-0135-69
$C_{13}H_{11}ClINO_2$			
2H-Azepin-2-one, 3-benzoyl-5-chloro-1,3-dihydro-, HI salt	MeOH	252(4.50)	78-5205-69

Compound	Solvent	λ_{max}(log ε)	Ref.
$C_{13}H_{11}ClN_2$			
Benzamidine, N-(p-chlorophenyl)-	50% EtOH	256(4.24)	49-1307-69
$C_{13}H_{11}ClN_2O$			
p-Cresol, 2-[(p-chlorophenyl)azo]-	50% EtOH	332(4.31),388(3.95)	73-1087-69
	50% EtOH-54% $HClO_4$	305s(--),403(4.35), 496(4.14)	73-1087-69
Phenol, 4-chloro-2-(m-tolylazo)- (in 50% EtOH)	HCl	384(3.96)	73-3740-69
	NaOH	470(3.99)	73-3740-69
Phenol, 4-chloro-2-(p-tolylazo)- (in 50% EtOH)	HCl	385(4.02)	73-3740-69
	NaOH	470(4.01)	73-3740-69
Pyrrolo[1,2-a]pyrimidin-4(6H)-one, 2-chloro-7,8-dihydro-3-phenyl-	EtOH	233s(--),287(3.95)	22-3133-69
$C_{13}H_{11}ClN_2O_2$			
Phenol, 4-chloro-2-[(m-methoxyphenyl)-azo]- (in 50% EtOH)	HCl	389(4.02)	73-3740-69
	NaOH	474(4.00)	73-3740-69
Phenol, 4-chloro-2-[(p-methoxyphenyl)-azo]- (in 50% EtOH)	HCl	390(4.19)	73-3740-69
	NaOH	490(4.09)	73-3740-69
$C_{13}H_{11}ClN_2O_4S$			
3-Methyl-2-(2-pyridyl)benzothiazolium perchlorate	MeCN	251(3.98),314(4.23)	24-0568-69
3-Methyl-2-(3-pyridyl)benzothiazolium perchlorate	MeCN	249(4.03),300(4.16)	24-0568-69
3-Methyl-2-(4-pyridyl)benzothiazolium perchlorate	MeCN	246(4.12),296(4.12)	24-0568-69
$C_{13}H_{11}ClO_3$			
1-Naphthol, 2-chloro-4-methoxy-, acetate	MeOH	235(4.63),295(3.96)	44-2788-69
$C_{13}H_{11}Cl_2N$			
Pyridine, 2-(m-chlorostyryl)-, trans, hydrochloride (see also $C_{13}H_{10}ClN$)	H_2O	329(4.45)	23-2355-69
$C_{13}H_{11}Cl_3NO_2S_2$			
4H-Thieno[3,4-c]pyrrole, 1,3-dichloro-5,6-dihydro-5-(p-tolylsulfonyl)-	EtOH	233(4.27)	44-0333-69
$C_{13}H_{11}Cl_3N_2O_3$			
4-Carboxamido-2-methylisoquinolinium trichloroacetate	MeOH	240(4.42),285(3.47), 342(3.69)	22-2045-69
$C_{13}H_{11}Cl_4N$			
Pyridine, 1,4-dihydro-1,2,6-trimethyl-4-(2,3,4,5-tetrachloro-2,4-cyclopentadien-1-ylidene)-	CH_2Cl_2	259(4.10),432(4.58)	83-0886-69
$C_{13}H_{11}Cl_4NO_2$			
3-Indolecarboxylic acid, 4,5,6,7-tetrachloro-1,2-dimethyl-, ethyl ester	MeOH	233(4.55),298(3.96)	24-2685-69
$C_{13}H_{11}IN_2$			
Benzamidine, N-(m-iodophenyl)-	50% EtOH	232(4.27),262(--)	49-1307-69
Benzamidine, N-(p-iodophenyl)-	50% EtOH	220s(--),236(4.26)	49-1307-69
$C_{13}H_{11}IN_2O$			
p-Cresol, 2-[(o-iodophenyl)azo]-	50% EtOH	340(4.33),390s(--), 430s(--)	73-1087-69

Compound	Solvent	λmax(log ε)	Ref.
p-Cresol, 2-[(o-iodophenyl)azo]- (cont.)	50% EtOH- 52% $HClO_4$	335s(--),422(4.30), 505(4.22)	73-1087-69
$C_{13}H_{11}N$			
Aniline, N-benzylidene-	80% MeOH- acid	328(4.34)	78-0057-69
Pyridine, 2-styryl-, cis	hexane	220(4.2),285(4.0)	62-0029-69A
	H_2O	288(3.98)	23-2355-69
hydrochloride	H_2O	317(3.99)	23-2355-69
Pyridine, 2-styryl-, trans	hexane	226(4.1),285s(4.3), 315(4.5)	62-0029-69A
	H_2O	309(4.42),334(4.45)	23-2355-69
hydrochloride	H_2O	334(4.45)	23-2355-69
Pyridine, 3-styryl-, trans	50% MeOH	308(4.49)	23-2355-69
hydrochloride	n.s.g.	341(4.50)	23-2355-69
$C_{13}H_{11}NO$			
4,5-Benzo-7-azabicyclo[4.2.2]deca-2,4,9-trien-8-one	MeCN	269(3.99)	35-4714-69
2,3-Benzo-9-azatricyclo[5.3.0.0^{6,10}]-deca-2,4-dien-8-one	MeCN	265(3.81)	35-5296-69
Benzophenone, 2-amino-	EtOH	235(4.33),260s(3.99), 378(3.78)	22-3523-69
Carbazol-2-ol, 1-methyl-	EtOH	216(4.41),238(4.59), 252(4.44),256s(4.42), 300(4.13),315s(3.87), 328s(3.41)	35-5872-69
Carbazol-2-ol, 3-methyl-	EtOH	236(4.64),258(4.20), 305(4.19),332(3.62)	77-1120-69
	n.s.g.	236(4.54),254(4.29), 260(4.31),320(4.15)	25-1662-69
Ketone, benzyl 4-pyridyl	EtOH	219(3.79),260(3.52)	39-2134-69C
Nitrone, N,α-diphenyl-	C_6H_{12}	320(4.27)	18-3306-69
	EtOH	315(4.32)	18-3306-69
$C_{13}H_{11}NOS$			
Benzothiazoline, 2-(o-hydroxyphenyl)-, 1:1 cobalt derivative trihydrate	n.s.g.	418(4.23)	97-0385-69
1:2 mercury derivative	n.s.g.	342(3.88)	97-0385-69
1:1 nickel derivative	n.s.g.	412(4.47)	97-0385-69
1:2 tin derivative	n.s.g.	412(4.36)	97-0385-69
1:1 zinc derivative	n.s.g.	516(3.72)	97-0385-69
$C_{13}H_{11}NO_2$			
Acetamide, N-(5-oxo-5H-benzocyclohepten-3-yl)-	EtOH	207(4.04),242(4.42), 330(3.99),375(3.72)	39-2656-69C
Anthranilic acid, N-phenyl-	C_6H_{12}	198(4.47),203(4.40), 240s(3.99),286(4.11), 359(3.86)	22-3523-69
	EtOH	202(4.49),220s(4.22), 235s(3.94),292(4.22), 335(3.74)	22-3523-69
2H-Azepin-2-one, 3-benzoyl-1,3-dihydro-	MeOH	246(4.24)	78-5205-69
4H-Benzo[de]quinoline-8,9-dione, 5,6-dihydro-6-methyl-	MeOH	472(3.53)	83-0487-69
	acetone	448(--)	83-0487-69
Bicyclo[4.2.2]deca-2,4,7,9-tetraene-3,4-dicarboximide, N-methyl-	MeOH	248(4.12),293s(3.41)	88-1125-69
	MeOH	248(4.11),293s(3.42)	108-0435-69
Cinchoninic acid, 2,3-dimethyl-	n.s.g.	214(4.88),230(4.72), 286(4.55),316(4.51)	39-0539-69B

Compound	Solvent	λ_{max}(log ε)	Ref.
Cinchoninic acid, 2-ethyl-	n.s.g.	209(4.86),231(4.82), 294(4.66),315(4.61)	39-0539-69B
Cyclobuta[b]quinoline-8-carboxylic acid, 1,2-dihydro-, methyl ester	EtOH	246(4.24),251(4.22), 325(3.86)	44-4131-69
2,4,6-Cycloheptatrien-1-one, 3-amino-2-hydroxy-4-phenyl-	n.s.g.	274(4.85),344(3.91), 408(3.91)	40-0085-69
2,4,6-Cycloheptatrien-1-one, 5-amino-2-hydroxy-4-phenyl-	n.s.g.	338(4.40),345(4.14), 420s(3.87)	40-0085-69
2,4,6-Cycloheptatrien-1-one, 7-amino-2-hydroxy-4-phenyl-	n.s.g.	266(4.51),355(4.23), 416(4.03)	40-0085-69
2,4-Pentadienoic acid, 2-cyano-5-phenyl-, methyl ester	EtOH	341(4.49)	23-4076-69
9H-Pyrrolo[1,2-a]indole-2-carboxaldehyde, 7-methoxy-	EtOH	242(3.95),289(4.32), 311(4.31)	88-0101-69
$C_{13}H_{11}NO_3$			
2-(1-Carboxy-2-hydroxyvinyl)isoquinolinium hydroxide, inner salt	EtOH	225(4.64),254(4.28), 340(3.60),395(3.40)	24-0915-69
	EtOH-H_2SO_4	232(4.65),269(3.50), 276(3.52),341(3.64)	24-0915-69
2-Furohydroxamic acid, N-(p-vinylphenyl)-	EtOH	315(4.44)	34-0278-69
4-Isoxazolecarboxylic acid, 3-methyl-5-styryl-	MeOH	225s(--),229(4.08), 237(4.03),318(4.50)	32-0753-69
$C_{13}H_{11}NO_3S_2$			
p-Dithiin-2,3-dicarboximide, 5,6-dihydro-N-(m-methoxyphenyl)-	MeOH	215(4.29),237(4.11), 273(4.03),415(3.39)	33-2221-69
p-Dithiin-2,3-dicarboximide, 5,6-dihydro-N-(o-methoxyphenyl)-	MeOH	270(4.02),410(3.47)	33-2221-69
p-Dithiin-2,3-dicarboximide, 5,6-dihydro-N-(p-methoxyphenyl)-	MeOH	234(4.29),263(3.90), 314(3.33),417(3.32)	33-2221-69
$C_{13}H_{11}NO_4S$			
Acetic acid, thio-, S-ester with 4-(mercaptomethylene)-2-(phenoxymethyl)-2-oxazolin-5-one	EtOH	320(3.92),365(4.05)	88-3381-69
$C_{13}H_{11}NO_5$			
2-Cyclopenten-1-one, 4-hydroxy-2-methyl-, p-nitrobenzoate	EtOH	218(4.25),260(4.20)	95-0750-69
$C_{13}H_{11}NS$			
Benzanilide, thio-	C_6H_{12}	241(4.20),252s(4.1), 320(3.95),438(2.42)	48-0045-69
	EtOH	234(4.16),256s(4.1), 316(3.91),399s(2.6)	48-0045-69
	MeCN	237(4.12),255s(4.1), 316(3.85),412(2.53)	48-0045-69
Benzothiazoline, 2-phenyl-, copper derivative	n.s.g.	420(3.57)	97-0385-69
nickel derivative	n.s.g.	460(3.58)	97-0385-69
zinc derivative	n.s.g.	454(3.48)	97-0385-69
$C_{13}H_{11}N_3$			
Acridine, 3,6-diamino-, sulfate	H_2O	260(4.5),444(4.3)	10-0205-69D
	EtOH	260f(4.5),455(4.5)	10-0205-69D
$C_{13}H_{11}N_3O$			
Acrylophenone, 3-(2-pyrimidinylamino)-, cis	$CHCl_3$	348(3.70)	49-1993-69
	DMSO	360(3.99)	49-1993-69

Compound	Solvent	λmax(log ε)	Ref.
Acrylophenone, 3-(2-pyrimidinylamino)-, trans	$CHCl_3$	248(2.90)	49-1993-69
	DMSO	260(3.20)	49-1993-69
$C_{13}H_{11}N_3O_2$			
Azobenzene, 2-hydroxy-5-methyl-3'-nitro-	50% EtOH	323(4.30),396(3.85)	73-1087-69
	50% EtOH-61% $HClO_4$	388(4.38),502(3.97)	73-1087-69
Azobenzene, 2-hydroxy-5-methyl-4'-nitro-	50% EtOH	340(4.34),414(3.97)	73-1087-69
	50% EtOH-64% $HClO_4$	315s(--),398(4.43), 513(4.04)	73-1087-69
Benzamidine, N-(m-nitrophenyl)-	50% EtOH	224s(--),260(4.23)	49-1307-69
Benzamidine, N-(p-nitrophenyl)-	50% EtOH	230(--),324(3.91)	49-1307-69
Glyoxal, phenyl-2-pyridyl-, dioxime, syn	n.s.g.	247-252(4.28)	36-0857-69
1-Phthalazineacetic acid, α-cyano-, ethyl ester	EtOH	212(4.64),282s(3.96), 291(4.17),368(4.25)	95-0959-69
$C_{13}H_{11}N_3O_2S$			
Sulfur diimide, (p-nitrophenyl)-(p-tolyl)-	benzene	437(4.29)	104-0459-69
Thiazolo[4,5-d]pyrimidine-5,7(4H,6H)-dione, 4,6-dimethyl-2-phenyl-	MeOH	248(4.26),270(4.14), 342(4.18)	44-3285-69
$C_{13}H_{11}N_3O_3$			
Benzimidazole, 1-methyl-2-(5-methyl-2-furyl)-5-nitro-	EtOH	249s(3.97),275s(4.31), 293(4.44),340(4.15)	73-0572-69
4H-Quinolizine-1-carboxylic acid, 2-amino-3-cyano-4-oxo-, ethyl ester	EtOH	252(4.62),294(4.20), 384(4.08)	95-0203-69
$C_{13}H_{11}N_3O_3S$			
2,3-Dihydro-8-hydroxy-5-methyl-7-(p-nitrophenyl)thiazolo[3,2-c]pyrimidin-4-ium hydroxide, inner salt	pH 1	260(3.95),365(3.96), 365(4.07)[sic]	1-2437-69
	pH 13	270(3.83),430(3.86)	1-2437-69
Sulfur diimide, (p-methoxyphenyl)-(p-nitrophenyl)-	benzene	445(4.39)	104-0459-69
$C_{13}H_{11}N_3O_4$			
2-Picoline, 4-(2,4-dinitrobenzyl)-	EtOH	245(4.53)	22-4425-69
2-Picoline, 6-(2,4-dinitrobenzyl)-	EtOH	246(4.52)	22-4425-69
4-Picoline, 2-(2,4-dinitrobenzyl)-	EtOH	251(4.53)	22-4425-69
as-Triazine-3,6-dicarboxylic acid, 5-phenyl-, dimethyl ester	MeOH	295(3.98),360(2.64)	4-0497-69
$C_{13}H_{11}N_5$			
2-Naphthamidine, N-4H-1,2,4-triazol-4-yl-	MeOH	240(4.20)	48-0477-69
$C_{13}H_{11}OP$			
5H-Dibenzophsophole, 5-methyl-, 5-oxide	EtOH	229(4.38),235(4.50), 243(4.52),278(3.91), 289(3.85),323(3.27)	39-0252-69C
$C_{13}H_{11}O_2P$			
Phenoxaphosphine, 10-methyl-, 10-oxide	EtOH	225(4.25),242(4.09), 277(3.51),289(3.66), 297(3.71)	39-0252-69C
$C_{13}H_{12}$			
Biphenyl, 2-methyl-	EtOH	237(4.00)	6-0277-69
Methane, diphenyl-	C_6H_{12}	262(2.63)	101-0017-69B

Compound	Solvent	$\lambda_{max}(\log \epsilon)$	Ref.
Methane, diphenyl- (cont.)	n.s.g.	262(2.61)	78-4919-69
Naphthalene, 1-cyclopropyl-	EtOH	225(4.86),283(3.82)	35-3558-69
Naphthalene, 2-cyclopropyl-	EtOH	228(4.90),277(3.72)	35-3558-69
Spiro[bicyclo[4.2.1]nona-2,4,7-triene-9,1'-[2,4]cyclopentadiene]	hexane	226(3.71),237(3.72), 260s(3.64)	24-1789-69
Spiro[bicyclo[6.1.0]nona-2,4,6-triene-9,1'-[2,4]cyclopentadiene]	hexane	229(4.20),258s(3.85)	24-1789-69
$C_{13}H_{12}AsN$			
Phenarsazine, 5,10-dihydro-5-methyl-	EtOH	280(4.08),310(3.98)	101-0117-69E
$C_{13}H_{12}BF_4N_3$			
3-Methyl-2-phenyl-s-triazolo[4,3-a]-pyridinium tetrafluoroborate	n.s.g.	235(4.2),280f(3.8)	24-3159-69
$C_{13}H_{12}BrNO_2S$			
1H-Azepine, 1-[(p-bromophenyl)sulfonyl]-2-methyl-	EtOH	234(4.11),269s(3.55)	44-2866-69
$C_{13}H_{12}BrN_3$			
Triazene, 3-(p-bromophenyl)-1-p-tolyl-	MeOH	237(3.83),300(3.99), 360(4.33)	65-0059-69
mercury complex	benzene	307(4.54),400(4.79)	65-0059-69
$C_{13}H_{12}ClN$			
Pyridine, 2-styryl-, cis, hydrochloride (see also $C_{13}H_{11}N$)	H_2O	317(3.99)	23-2355-69
$C_{13}H_{12}ClNO$			
Acridine, 9-chloro-1,2,3,4-tetrahydro-, 10-oxide	EtOH	235s(4.56),245(4.71), 330(4.01),344(3.98)	95-1305-69
$C_{13}H_{12}ClNO_2$			
Pyrrole-2-carboxylic acid, 5-(m-chlorophenyl)-, ethyl ester	EtOH	308(4.50)	94-0582-69
$C_{13}H_{12}ClN_3$			
Aniline, p-[(p-chlorophenyl)azo]-N-methyl-	EtOH	266(3.92),411(4.48)	18-3565-69
1H-Imidazo[4,5-g]quinoline, 8-chloro-1,2,6-trimethyl-	EtOH	248(4.76),332(4.00)	94-2455-69
3H-Imidazo[4,5-f]quinoline, 9-chloro-2,3,7-trimethyl-	EtOH	256s(4.60),260(4.64), 320(3.48)	94-2455-69
Triazene, 1-(p-chlorophenyl)-3-methyl-3-phenyl-	EtOH	236(4.24),295(3.99), 346(4.33)	18-3565-69
$C_{13}H_{12}ClN_3O_4$			
1-Amino-2-phenyl-1H-imidazo[1,2-a]pyridin-4-ium perchlorate	EtOH	205(4.49),235(4.35), 293(4.12)	44-2129-69
1,4-Dihydro-3-phenylpyrido[2,1-c]-as-triazinium perchlorate	EtOH	202(4.36),223s(4.08), 243s(3.96),293(4.02), 360(3.99)	44-2129-69
$C_{13}H_{12}ClN_5O_7$			
Pyrimidine, 5-chloro-2,4,6-trimethyl-, picrate	n.s.g.	217(3.82),261(3.52)	39-2251-69C
$C_{13}H_{12}INO_2$			
Indole-3-acrylic acid, α-iodo-, ethyl ester	EtOH	354(4.31)	104-0678-69

Compound	Solvent	$\lambda_{max}(\log \epsilon)$	Ref.
$C_{13}H_{12}NOPS$			
Formamide, 1-(diphenylphosphinyl)thio-	EtOH	295(3.76),406(1.58)	18-2975-69
$C_{13}H_{12}NPS$			
Formamide, 1-(diphenylphosphino)thio-	EtOH	292(3.94),376(2.13)	18-2975-69
$C_{13}H_{12}NPS_2$			
Formamide, 1-(diphenylphosphinothioyl)-thio-	EtOH	290(3.77),412(1.61)	18-2975-69
$C_{13}H_{12}N_2$			
Azobenzene, 2-methyl-, trans	20% EtOH	229(4.11),328(4.27), 435(2.98)	54-0562-69
	+ H_2SO_4	239(3.81),435(4.41)	54-0562-69
Azobenzene, 4-methyl-, trans	20% EtOH	232(4.07),330(4.33), 420(3.15)	54-0562-69
	+ H_2SO_4	240(3.88),430(4.42)	54-0562-69
Benzaldehyde, phenylhydrazone	EtOH	237(4.15),302(4.04), 344(4.38)	39-1703-69C
	EtOH-HCl	232(4.02),292s(3.15), 364(1.74)	39-1703-69C
Benzamidine, N-phenyl-	50% EtOH	220s(--),236(4.16)	49-1307-69
Benzophenone, hydrazone	EtOH	272(4.07)	39-1703-69C
Pyridine, 2-(p-aminostyryl)-, cis	H_2O	313(4.11)	23-2355-69
hydrochloride	H_2O	378(4.02)	23-2355-69
Pyridine, 2-(p-aminostyryl)-, trans	H_2O	337(4.48)	23-2355-69
hydrochloride	H_2O	385(4.51)	23-2355-69
Pyridine, 3-(p-aminostyryl)-, trans	50% MeOH	344(4.44)	23-2355-69
hydrochloride	50% MeOH	414(4.44)	23-2355-69
$C_{13}H_{12}N_2O$			
Acetophenone, 2-(2-pyridyl)-, oxime, syn (same spectrum for 3- and 4-pyridyl isomers)	n.s.g.	251(4.04)	36-0857-69
p-Cresol, 2-(phenylazo)-	50% EtOH	327(4.29),385(3.90)	73-1087-69
	50% EtOH-48% $HClO_4$	300s(--),394(4.33), 486(4.07)	73-1087-69
Nitrone, N-benzyl-α-3-pyridyl-	EtOH	238(3.7),303(4.1)	28-0078-69A
Nitrone, N-benzyl-α-4-pyridyl-	EtOH	222(3.8),300(4.2)	28-0078-69A
Pseudourea, 1,2-diphenyl-	iso-PrOH	225(4.20)	77-1110-69
Pyrrolo[1,2-a]pyrimidin-2(6H)-one, 7,8-dihydro-4-phenyl-	EtOH	238(4.27)	22-3139-69
Pyrrolo[1,2-a]pyrimidin-4(6H)-one, 7,8-dihydro-2-phenyl-	EtOH	238(4.31),286(3.89)	22-3139-69
$C_{13}H_{12}N_2OS$			
Acetic acid, phenyl-2-thenylidenehydrazide	EtOH	263(3.9),274(3.9), 303(4.3),320s(4.1)	28-1703-69A
Acrylophenone, 4'-methyl-3-(2-thiazolylamino)-, cis	$CHCl_3$	371(3.87)	49-1993-69
	DMSO	366(4.08)	49-1993-69
trans	$CHCl_3$	268(3.10)	49-1993-69
	DMSO	262(3.37)	49-1993-69
$C_{13}H_{12}N_2O_2$			
Acetic acid, furfurylidenephenylhydrazide	EtOH	203(4.2),294(4.4), 306s(4.4)	28-1703-69A
4-Isoquinolinecarbonitrile, 6,8-dimethoxy-3-methyl-	EtOH	216(4.45),255(4.53)	95-1492-69
5,8-Isoquinolinedione, 7-(cyclopropylamino)-3-methyl-	MeCN	236(4.13),258(4.01), 275(4.01),324(3.57)	4-0697-69

Compound	Solvent	$\lambda_{max}(\log \epsilon)$	Ref.
Pyridazine, 3-(p-methoxystyryl)-, 1-oxide	EtOH	240(3.90),332(4.11)	95-0132-69
Pyridazine, 4-(p-methoxyphenyl)-, 1-oxide	EtOH	229(4.02),276(4.05), 326(3.85)	95-0132-69
Pyridazine, 5-(p-methoxystyryl)-, 1-oxide	EtOH	245(4.48),270(4.46), 340(4.29)	95-0132-69
Pyridazine, 6-(p-methoxystyryl)-, 1-oxide	EtOH	241(4.17),348(4.42)	95-0132-69
Pyrrolo[1,2-a]pyrimidin-4(6H)-one,	N NaOH	263(4.08)	22-3133-69
7,8-dihydro-2-hydroxy-3-phenyl-	EtOH	268(3.69)	22-3133-69
	5N HCl	258(3.94)	22-3133-69
sodium salt	EtOH	267(3.93),292(3.80)	22-3133-69
$C_{13}H_{12}N_2O_2S$			
2H-1,2,4-Benzothiadiazine, 3,4-dihydro-2-phenyl-, 1,1-dioxide	EtOH	210(4.55),252(4.06), 314(3.48)	7-0590-69
$C_{13}H_{12}N_2O_2S_2$			
Sulfur diimide, (phenylsulfonyl)-(p-tolyl)-	benzene	397(4.31)	104-0459-69
p-Dithiin-2,3-dicarboximide, 5,6-dihydro-N-(N-methylanilino)-	MeOH	235(4.35),409(3.52)	33-2236-69
$C_{13}H_{12}N_2O_3$			
Acetic acid, 2-(1,4-dihydro-1,4-dioxo-2-naphthyl)hydrazide	alkali	615(3.66)	95-0007-69
Acridine, 1,2,3,4-tetrahydro-9-nitro-, 10-oxide	EtOH	243(4.58),330(3.96)	95-1305-69
5,8-Isoquinolinedione, 7-morpholino-	MeCN	251(4.21),284(4.05), 324(3.54)	4-0697-69
$C_{13}H_{12}N_2O_3S$			
Formanilide, 4'-sulfanilyl-	MeOH	256(4.35),294(4.42)	87-0357-69
$C_{13}H_{12}N_2O_3S_2$			
2-Butanone, 4-[[4-(p-nitrophenyl)-2-thiazolyl]thio]-	MeOH	340(4.11)	4-0397-69
p-Dithiin-2,3-dicarboximide, N-p-anisidino-5,6-dihydro-	MeOH	233(4.24),260(3.97), 410(3.43)	33-2236-69
Sulfur diimide, (p-methoxyphenyl)-(phenylsulfonyl)-	benzene	423(4.11)	104-0459-69
$C_{13}H_{12}N_2O_4S$			
Azobenzene-3'-sulfonic acid, 2-hydroxy-5-methyl-	50% EtOH	326(4.34),394(3.82)	73-1087-69
	+58% $HClO_4$	393(4.30),500(3.95)	73-1087-69
$C_{13}H_{12}N_2O_5$			
Pyrazole-4,5-dicarboxylic acid, 3-(benzyloxy)methyl-	EtOH	211(4.04)	88-0289-69
$C_{13}H_{12}N_2S$			
4H-Cyclopenta[d]pyrimidine-4-thione, 1,5,6,7-tetrahydro-2-phenyl-	EtOH	243(4.29),308(4.09), 354(3.74)	4-0037-69
$C_{13}H_{12}N_4$			
Benzimidazo[2,1-c][1,2,4]benzotriazine, 8,9,10,11-tetrahydro-	5N H_2SO_4	218(4.35),249(4.19), 289(3.45),380(4.10)	103-0683-69
	0.05N NaOH	225(4.48),260(3.94), 395(4.12)	103-0683-69
	MeOH	225(4.64),262(3.96), 385(4.23)	103-0683-69

Compound	Solvent	$\lambda_{max}(\log \epsilon)$	Ref.
Formazan, 1,5-diphenyl-	MeOH	420(4.44)	24-1379-69
	$MeOH-HClO_4$	519(4.47)	24-1379-69
	DMF	427(4.42)	24-1379-69
	$DMF-HClO_4$	541(4.75)	24-1379-69
$C_{13}H_{12}N_4O$			
Hydrazine, 1-benzoyl-2-picolinimidoyl-	H_2SO_4	267(4.35)	4-0965-69
1H-Pyrazolo[3,4-b]pyrazin-6-ol, 1,5-dimethyl-3-phenyl-	EtOH	245(4.11),321(3.94)	32-0463-69
1H-Pyrazolo[3,4-b]pyrazin-6-ol, 3,5-dimethyl-1-phenyl-	EtOH	258(4.30),319(3.95)	32-0463-69
$C_{13}H_{12}N_4O_2$			
Cinnamaldehyde, (2,6-dihydroxy-4-pyrimidinyl)hydrazone	EtOH	345(4.58)	56-0519-69
Theophylline, 9-phenyl-	EtOH	265(4.33)	78-0541-69
Xanthine, 7-benzyl-8-methyl-	pH -1.2	231(3.86),263(3.98), 274s(3.82)	24-4032-69
	pH 5	271(4.04)	24-4032-69
	pH 12	211(4.48),216s(4.44), 289(3.97)	24-4032-69
$C_{13}H_{12}N_4O_3$			
1H-Pyrazolo[3,4-d]pyridazine-4,7-diol, 3-[(benzyloxy)methyl]-	pH 13	277(3.78)	88-0289-69
1H-Pyrazolo[4,3-d]pyrimidine-5,7-diol, 3-[(benzyloxy)methyl]-	pH 1	208(4.36),284(3.83)	88-0289-69
	0.05N NaOH	222(4.39),302(3.65)	88-0289-69
$C_{13}H_{12}N_4O_6$			
1-Cyclopentene-1-carboxylic acid, 2-methyl-5-oxo-, 5-(2,4-dinitrophenylhydrazone)	EtOH	218(4.26),252(4.20), 376(4.44)	35-4739-69
$C_{13}H_{12}O$			
Bicyclo[3.1.0]hex-3-en-2-one, 5-methyl-4-phenyl-	EtOH	221(4.07),244s(3.85), 297(4.09)	22-2095-69
$C_{13}H_{12}OS$			
Sulfoxide, phenyl m-tolyl	H_2O	234(4.13)	18-1964-69
	91% H_2SO_4	253(3.88)	18-1964-69
Sulfoxide, phenyl p-tolyl	H_2O	237(4.20)	18-1964-69
	91% H_2SO_4	264(4.02)	18-1964-69
$C_{13}H_{12}OS_2$			
Benzenesulfinic acid, thio-, S-p-tolyl ester	60% EtOH	285(3.97)	18-2899-69
Ketone, 3,4-dihydro-2H-thiopyrano-[3,2-b][1]benzothien-7-yl methyl	EtOH	225(4.39),267(4.34), 302(3.81),335(3.79)	22-0607-69
p-Toluenesulfinic acid, thio-, S-phenyl ester	60% EtOH	287(3.96)	18-2899-69
$C_{13}H_{12}O_2$			
1'-Acetonaphthone, 2'-methoxy-	C_6H_{12}	226(4.72),284(3.58), 292(3.60),324(3.37), 336(3.45)	19-0253-69
	MeOH	225(4.74),280(3.59), 291(3.59),320s(3.41), 334(3.49)	19-0253-69
	aq. MeOH	224(4.76),280(3.56), 292(3.56),333(3.47)	19-0253-69

Compound	Solvent	λ_{max}(log ϵ)	Ref.
1'-Acetonaphthone, 4'-methoxy-	C_6H_{12}	214(4.11),238(4.59), 316(4.05)	19-0253-69
	MeOH	212(4.19),236(4.51), 321(4.05)	19-0253-69
	aq. MeOH	210(4.21),234(4.46), 324(4.05)	19-0253-69
2'-Acetonaphthone, 1'-methoxy-	C_6H_{12}	214(4.35),234(4.52), 242(4.61),250(4.60), 276s(3.76),284(3.84), 294s(3.70),330(3.32), 342s(3.23)	19-0253-69
	MeOH	214(4.41),245(4.56), 250s(4.55),276s(3.78), 285(3.89),296s(3.77), 334(3.34)	19-0253-69
	aq. MeOH	213(4.44),247(4.54), 252s(4.53),287(3.87), 298s(3.78),340(3.36)	19-0253-69
2'-Acetonaphthone, 3'-methoxy-	C_6H_{12}	242(4.67),279(3.73), 346(3.17)	19-0253-69
	MeOH	218(4.39),244(4.59), 282(3.77),349(3.15)	19-0253-69
	aq. MeOH	218(4.43),246(4.56), 283(3.81),354(3.14)	19-0253-69
2'-Acetonaphthone, 6'-methoxy-	C_6H_{12}	214(4.21),240(4.62), 249(4.56),259(4.57), 302(4.10)	19-0253-69
	MeOH	213(4.31),241(4.54), 246s(4.51),257(4.41), 309(4.09)	19-0253-69
	aq. MeOH	213(4.31),242(4.49), 258(4.39),311(4.10)	19-0253-69
Acetophenone, 2'-(2-furyl)-5'-methyl-	EtOH	240(4.36),274(4.28), 313(3.87)	6-0277-69
8H-Benzo[b]cyclobuta[c]pyran-8-one, 2a,8a-dihydro-1,2-dimethyl-	MeOH	216(4.27),254(3.89), 322(3.48)	35-4494-69
Benzo[1,2-b:5,4-b']difuran, 2,6,8-trimethyl-	EtOH	251s(3.81),260(4.01), 269(3.98),283s(3.43), 289(3.64),294(3.84), 300(3.85),306(3.98)	4-0191-69
Benzo[1,2-b:5,4-b']difuran, 3,5,8-trimethyl-	EtOH	256s(3.74),266(3.97), 275(4.02),287(3.69), 293(3.81),299(3.78), 306(3.86)	4-0191-69
4H-Pyran-4-one, 2-ethyl-6-phenyl-	EtOH	272(4.36)	94-2126-69
$C_{13}H_{12}O_2S$			
Acetic acid, [(2-methyl-1-naphthyl)-thio]-	EtOH	225(4.89),284(3.91)	44-1566-69
Anisole, p-(phenylsulfinyl)-	H_2O	245(4.21)	18-1964-69
	91% H_2SO_4	281(4.05)	18-1964-69
2-Furanacrolein, 5-[5-(methylthio)-4-penten-2-ynyl]-, trans	ether	278(4.24),320(4.53)	24-4209-69
2-Thiophenecarboxylic acid, 3-phenyl-, ethyl ester	EtOH	275(4.55)	2-0009-69
$C_{13}H_{12}O_2S_2$			
Benzenesulfinic acid, p-methoxy-thio-, S-phenyl ester	60% EtOH	295(4.00)	18-2899-69

Compound	Solvent	λ_{max}(log ε)	Ref.
Benzenesulfinic acid, thio-, S-(p-methoxyphenyl) ester	60% EtOH	293(4.05)	18-2899-69
$C_{13}H_{12}O_3$			
2,3-Benzo-6-oxabicyclo[3.3.1]nonane-4,7-dione, 1-methyl-	EtOH	253(4.02),290(3.20)	36-0894-69
$C_{13}H_{12}O_3S$			
Thiochroman-4-one, 3-(acetoxymethylene)-2-methyl-	EtOH	250(4.23),268s(3.82), 323(3.89)	44-0056-69
$C_{13}H_{12}O_4$			
2'-Acetonaphthone, 1',6',8'-trihydroxy-3'-methyl-	EtOH	235(4.24),273(4.28), 398(3.75)	88-0471-69
$C_{13}H_{12}O_5$			
Bicyclo[3.2.2]nona-2,6,8-triene-6,7-dicarboxylic acid, 4-oxo-, dimethyl ester	C_6H_{12}	218(4.07)	88-3943-69
Indene-3a,7a-dicarboxylic acid, 1-oxo-, dimethyl ester, cis	C_6H_{12}	216(4.08),256(3.50)	88-3943-69
2-Naphthoic acid, 6-hydroxy-4,5-dimethoxy-	n.s.g.	242(4.52),247(4.55), 252(4.52),258(4.47), 263(4.39),319(3.96), 332(3.94),348(3.89)	102-0789-69
$C_{13}H_{12}O_6$			
7H-Furo[3,2-g][1]benzopyran-7-one, 2,3-dihydro-5-hydroxy-6,9-dimethoxy-	EtOH	244s(3.95),295s(3.98), 319(4.23)	31-0354B-69
Spiro[furan-2(5H),1'-phthalan]-3',5-dione, 3,4-dihydro-5',7'-dimethoxy-	MeOH	252(3.48),308(3.54)	22-2365-69
$C_{13}H_{13}BrN_2O$			
1-Benzyl-3-carbamoylpyridinium bromide	EtOH	264(3.63)	39-0192-69B
$C_{13}H_{13}ClN_2$			
Pyridine, 2-(p-aminostyryl)-, cis, hydrochloride (see also $C_{13}H_{12}N_2$)	H_2O	378(4.02)	23-2355-69
$C_{13}H_{13}ClO_4S$			
1,2,3,4-Tetrahydro-10-thiaanthracenium perchlorate	HOAc-1% $HClO_4$	260(4.14),355(3.65), 396(3.66)	2-0737-69
1,2,3,4-Tetrahydro-6-thiaphenanthrenium perchlorate	HOAc-1% $HClO_4$	263(3.41),345(3.72), 405(4.62)	2-0637-69
$C_{13}H_{13}ClO_4S_2$			
2,4-Dimethylthionaphtheno[3,2-b]thiapyrylium perchlorate	CH_2Cl_2	261(4.07),301(3.98), 382(4.09),435s(3.46)	88-0239-69
$C_{13}H_{13}ClO_5$			
Bicyclo[3.2.2]nona-3,6-diene-8α,9α-dicarboxylic acid, 3-chloro-2-oxo-	MeOH	235(3.68),260s(3.46), 333(2.00)	88-0443-69
1,2,3,4-Tetrahydro-6-oxaphenanthrenium perchlorate	$HClO_4$	240(4.33),262(3.45), 324(3.99)	78-1939-69
$C_{13}H_{13}Cl_2NO_4$			
4-Penten-1-ol, 4-(3,4-dichlorophenyl)-4-nitro-, acetate	EtOH	301(3.93)	87-0157-69

Compound	Solvent	$\lambda_{max}(\log \epsilon)$	Ref.
$C_{13}H_{13}Cl_2N_3O_2S$			
Δ^3-1,3,4-Oxadiazoline-5-thione, 2-(2,4-dichlorophenyl)-4-(morpholinomethyl)-	n.s.g.	205(4.59),238(4.31), 321(4.14)	2-0583-69
$C_{13}H_{13}Cl_4NO$			
Acrylophenone, 3-[bis(2-chloroethyl)-amino]-3',4'-dichloro-	EtOH	217(4.14),253(4.05), 346(4.32)	24-3139-69
$C_{13}H_{13}N$			
Pyridine, 4-phenethyl-	EtOH	214(3.09),252(3.62)	39-2134-69C
o-Toluidine, N-phenyl-	EtOH	235s(3.81),283(4.18)	22-2355-69
p-Toluidine, N-phenyl-	EtOH	238s(3.76),287(4.32)	22-2355-69
$C_{13}H_{13}NO$			
o-Anisidine, N-phenyl-	EtOH	230s(3.92),278(4.19), 295s(4.06)	22-2355-69
p-Anisidine, N-phenyl-	EtOH	241(3.77),285(4.26)	22-2355-69
1-Carbazolone, 1,2,3,4-tetrahydro-9-methyl-	MeOH	239(4.27),310(4.30)	24-1198-69
Cyclohept[b]indol-6(5H)-one, 7,8,9,10-tetrahydro-	EtOH	244(4.34),264(4.08), 293(3.67),301(3.67)	94-1290-69
Cyclohept[b]indol-10(5H)-one, 6,7,8,9-tetrahydro-	EtOH	245(4.19),268(4.07), 302(4.09)	94-1290-69
Cyclopenta[b]pyrrol-2(1H)-one, 3,3aβ,4,6aβ-tetrahydro-3α-phenyl-	EtOH	242(2.00),248(2.06), 253(2.19),259(2.30), 265(2.18),268(1.98)	44-2058-69
2-Pyridineethanol, β-phenyl-	MeOH	261(3.64),268(3.52)	73-0819-69
$C_{13}H_{13}NOS$			
3-Pyridinol, 6-methyl-2-(p-tolylthio)-	pH 1	240(3.91),305(3.80), 330(3.75)	1-2065-69
	pH 13	245(4.10),325(3.86)	1-2065-69
$C_{13}H_{13}NOS_2$			
2-Butanone, 4-[(4-phenyl-2-thiazolyl)-thio]-	MeOH	274(4.11)	4-0397-69
2-Butanone, 4-(4-phenyl-2-thioxo-4-thiazolin-3-yl)-	MeOH	323(4.29)	4-0397-69
$C_{13}H_{13}NO_2$			
2-Carbazolol, 5,6,7,8-tetrahydro-3-methyl-8-oxo-	n.s.g.	234(4.07),284(3.82)	25-1662-69
Cyclopenta[b]pyrrol-2(1H)-one, 3,3aβ,4,6aβ-tetrahydro-1-hydroxy-3α-phenyl-	EtOH	248s(2.18),253(2.24), 259(2.30),265(2.18), 268s(1.86)	44-2058-69
2-Isoxazolin-5-one, 4-allyl-4-methyl-3-phenyl-	EtOH	257(4.07)	88-0543-69
3-Isoxazolin-5-one, N-allyl-4-methyl-3-phenyl-	EtOH	248(4.00),283(3.98)	88-0543-69
Pierardine	EtOH	227(4.09),273(3.29), 281(3.29)	1-2177-69
2(1H)-Pyridone, 3-ethyl-4-hydroxy-1-phenyl-	MeOH	293(3.79)	88-0287-69
Pyrrole-2-carboxylic acid, 5-phenyl-, ethyl ester	EtOH	308(4.39)	94-0582-69
1H-Pyrrolo[1,2-a]indole, 2-acetyl-2,3-dihydro-7-hydroxy-	EtOH	218(4.56),277(4.01)	88-0101-69
Quinoline-4-carboxylic acid, 2,3-dimethyl-, methyl ester	EtOH	230s(4.44),235(4.46), 306(3.63),319(3.69)	44-4131-69

Compound	Solvent	λ_{max}(log ϵ)	Ref.
Spiro[cyclopentane-1,3'(2'H)-quinoline]-2',4'(1'H)-dione	EtOH	233(4.66),257(3.84), 340(3.57)	94-1290-69
$C_{13}H_{13}NO_2S$			
Benzothiophene-3-carboxaldehyde, 2-morpholino-	EtOH	227(4.5),260(4.3), 300(3.8),355(3.9)	48-0827-69
$C_{13}H_{13}NO_2S_2$			
4H-Thieno[3,4-c]pyrrole, 5,6-dihydro-5-(p-tolylsulfonyl)-	EtOH	232(4.27)	44-0333-69
$C_{13}H_{13}NO_3$			
Carbostyril, 4-hydroxy-3-isobutyryl-	MeOH	220(4.25),237(4.44), 306(4.04)	44-2183-69
	MeOH-acid	220(4.25),237(4.44), 306(4.04)	44-2183-69
2-(1-Carboxy-2-hydroxyvinyl)-3,4-dihydroisoquinolinium hydroxide, inner salt, methyl ester	$CHCl_3$	240(3.98),247(4.12), 253(4.24),259(4.30), 264(4.29),324(3.57), 440(3.87)	24-0904-69
Indole-2-carboxylic acid, 3-(allyloxy)-, methyl ester	EtOH	231(4.34),298(4.23)	39-0595-69C
2-Indolinecarboxylic acid, 2-allyl-3-oxo-, methyl ester	EtOH	232(4.39),256s(3.77), 394(3.56)	39-0595-69C
Phthalic anhydride, 3-(cyclopentylamino)-	EtOH	206(4.17),255(4.06), 403(3.57)	16-0241-69
$C_{13}H_{13}NO_4$			
Acetamide, N-acetoxy-N-(2-benzoylvinyl)-	MeOH	297(4.24)	77-1062-69
1H-Benzazepine-7-carboxylic acid, 1-acetyl-2,3,4,5-tetrahydro-5-oxo-	MeOH	231(4.15),261(3.07), 305(3.07)	44-1070-69
2-Cyclopenten-1-one, 3,5-dimethyl-2-[2-(5-nitro-2-furyl)vinyl]-	EtOH	220(3.89),265(4.25), 383(4.26)	94-1757-69
2,4,6-Heptatrienal, 2,6-dimethyl-7-(5-nitro-2-furyl)-	dioxan	238(3.91),316(4.34), 403(4.69)	94-0306-69
Isoindole-1,3-dicarboxylic acid, 2-methyl-, dimethyl ester	EtOH	224(4.29),252(4.54), 348(4.32),366(4.33)	32-1115-69
$C_{13}H_{13}NO_5$			
2H-1-Benzopyran-4-carboxylic acid, 3-(ethylamino)-6-hydroxy-2-oxo-, ethyl ester	EtOH	290(3.90),299s(3.89), 335(3.79)	48-0786-69
$C_{13}H_{13}NO_6$			
Fumaric acid, (o-carboxyanilino)-, 1,4-dimethyl ester	MeOH	218(4.25),238(4.11), 339(4.21)	23-3545-69
Propionic acid, 3-[(6-carbethoxy-2,4-dimethoxybenzoyl)-, lactol, oxime	MeOH	230(4.35),268s(3.92)	22-2365-69
$C_{13}H_{13}NO_7S$			
2-Oxazolepropionic acid, 5-(4-methoxy-3-sulfophenyl)-, hemihydrate	acid	215(4.31),280(4.35)	4-0707-69
	neutral	216(4.29),274(4.39)	4-0707-69
	base	274(4.39)	4-0707-69
$C_{13}H_{13}NS$			
1H-Phenothiazine, 2,3-dihydro-4-methyl-	MeOH	405(3.45)	124-1278-69
	$MeOH-HClO_4$	458(3.57)	124-1278-69
$C_{13}H_{13}NS_2$			
Aniline, N-[1-methyl-2-(5-methyl-3H-1,2-dithiol-3-ylidene)ethylidene]-	$CHCl_3$	250(4.51),420(4.14)	24-1580-69

Compound	Solvent	λ_{max}(log ε)	Ref.
$C_{13}H_{13}N_3$			
3H-Imidazo[4,5-f]quinoline, 2,3,7-tri-methyl-	EtOH	255s(4.56),259(4.59), 310(3.60)	94-2455-69
Triazene, 3-benzyl-1-phenyl-	MeOH	280(4.20)	65-0303-69
mercury complex	benzene	293(4.19),350(4.14)	65-0303-69
	DMF	290(4.23),335(4.15)	65-0303-69
Triazene, 3-phenyl-1-p-tolyl-	MeOH	235(4.25),295(3.96), 360(4.30)	65-0059-69
copper complex	benzene	340(4.13),620(2.67)	65-0059-69
mercury complex	benzene	305(4.29),397(4.53)	65-0059-69
$C_{13}H_{13}N_3O$			
Aniline, p-[(o-methoxyphenyl)azo]-	n.s.g.	251(3.95),394(4.40)	40-0207-69
o-Anisidine, 4-(phenylazo)-	n.s.g.	255(4.01),401(4.37)	40-0207-69
Benzimidazole, 5-amino-1-methyl-2-(5-methyl-2-furyl)-	EtOH	212(4.43),240s(4.07), 293(4.14),335(4.25)	73-0572-69
2,5-Cyclohexadien-1-one, 4-[(2,4-diami-nophenyl)imino]-2-methyl-	pH 10.0	646(4.43)	39-0823-69B
2,5-Cyclohexadien-1-one, 4-[(2,4-diami-nophenyl)imino]-3-methyl-	pH 10.0	652(4.43)	39-0823-69B
2,5-Cyclohexadien-1-one, 4-[(2,4-diami-no-m-tolyl)imino]-	pH 10.0	628(4.44)	39-0823-69B
2,5-Cyclohexadien-1-one, 4-[(4,6-diami-no-m-tolyl)imino]-	pH 10.0	630(4.36)	39-0823-69B
Triazene, 1-(p-methoxyphenyl)-3-phenyl-	MeOH	237(4.00),295(3.98), 365(4.28)	65-0059-69
mercury complex	benzene	307(4.36),405(4.51)	65-0059-69
$C_{13}H_{13}N_3O_2S$			
Barbituric acid, 5-[p-(dimethylamino)-benzylidene]-2-thio-	EtOH	245(4.06),285(4.10), 342(4.18)	103-0830-69
$C_{13}H_{13}N_3O_3$			
L-Alanine, N-methyl-N-(2-quinoxalinyl-carbonyl)-	EtOH	207(4.24),240(4.40), 313(3.77),324(3.83)	87-0141-69
Barbituric acid, 5-[p-(dimethylamino)-benzylidene]-	EtOH	255(4.0),340(4.39)	103-0827-69
$C_{13}H_{13}N_3O_3S$			
Thiazole, 2-(allylamino)-5-[1-methyl-2-(5-nitro-2-furyl)vinyl]-	EtOH	255(4.21),299(4.04), 410(4.26)	103-0414-69
$C_{13}H_{13}N_3O_4S$			
Acrylic acid, 2-cyano-3-(N-methyl-p-ni-troanilino)-3-(methylthio)-, methyl ester	MeOH	318(4.23),377(4.21)	78-4649-69
Propionamide, N-[5-[1-methyl-2-(5-nitro-2-furyl)vinyl]-2-thiazolyl]-	EtOH	246(4.06),281(4.14), 395(4.24)	103-0414-69
$C_{13}H_{13}N_5$			
Adenine, 9-benzyl-1-methyl-	pH 1	260(4.05)	95-0591-69
	H_2O	260(4.06)	95-0591-69
	pH 13	268(4.16)	95-0591-69
$C_{13}H_{13}N_5O$			
Guanine, 7-benzyl-8-methyl-	pH -3.6	239(4.09),257s(4.01)	24-4032-69
	pH 1	250(4.09),278(3.91)	24-4032-69
	pH 7	217(4.37),247(3.84), 284(3.93)	24-4032-69
	pH 12	211(4.48),280(3.89)	24-4032-69

Compound	Solvent	λ_{max}(log ε)	Ref.
$C_{13}H_{13}N_5O_7$			
2(1H)-Quinoxalinone, 3,4-dihydro-5,7-dinitro-3-prolinyl-	EtOH	276(4.04),392(3.98)	44-0395-69
$C_{13}H_{13}N_7O_5S$			
Propionaldehyde, 3-[(2,5-dihydro-6-methyl-5-oxo-as-triazin-3-yl)thio]-, 1-(2,4-dinitrophenylhydrazone)	EtOH	235(4.47),359(4.32)	114-0093-69C
$C_{13}H_{14}$			
5,8-Methano-5H-benzocycloheptene, 8,9-dihydro-10-methyl-, endo	heptane	193(3.59),259(2.67), 266(2.81),274(2.85)	78-0335-69
$C_{13}H_{14}BF_4NOS_2$			
4-(4-Phenyl-1,3-dithiol-2-ylidene)-morpholinium tetrafluoroborate	EtOH	325(4.09)	94-1924-69
$C_{13}H_{14}BrCl_2NO$			
Acrylophenone, 3-[bis(2-chloroethyl)-qmino]-4'-bromo-	EtOH	220(3.86),255(4.12), 341(4.39)	24-3139-69
$C_{13}H_{14}BrNO_4$			
4-Penten-1-ol, 5-(p-bromophenyl)-4-nitro-, acetate	EtOH	310(3.97)	87-0157-69
$C_{13}H_{14}BrN_3OS$			
Salicylaldehyde, (6-bromo-3a,5,6,6a-tetrahydro-4H-cyclopentathiazol-2-yl)hydrazone, hydrobromide	EtOH	237(3.6),335(4.1)	124-0509-69
$C_{13}H_{14}BrN_3S$			
Benzaldehyde, (6-bromo-3a,5,6,6a-tetrahydro-4H-cyclopentathiazol-2-yl)-hydrazone, hydrobromide	EtOH	220(4.2),325(5.0)	124-0509-69
$C_{13}H_{14}BrN_5OS$			
s-Triazolo[3,4-b][1,3,4]thiadiazole, 3-ethyl-6-(α-phenyl-β-bromoethylidene)amino-	EtOH	251(4.46)	2-0959-69
$C_{13}H_{14}ClNO_2$			
Morpholine, 4-(p-chlorocinnamoyl)-	EtOH	220(4.18),280(4.49)	39-0016-69C
$C_{13}H_{14}ClNO_2S$			
Crotonic acid, 3-(p-chlorobenzamido)-thio-, ethyl ester	MeCN	345(4.18)	24-0269-69
Crotonic acid, 3-(p-chlorothiobenzamido)-, ethyl ester	MeCN	460(2.48)	24-0269-69
$C_{13}H_{14}ClNO_4$			
4-Penten-1-ol, 5-(p-chlorophenyl)-4-nitro-, acetate	EtOH	305(4.19)	87-0157-69
$C_{13}H_{14}ClNO_4S_2$			
2-Isopropyl-3-methylthiazolo[2,3-b]-benzothiazolium perchlorate	EtOH	218(4.62),262(4.09), 315(4.23)	95-0469-69
$C_{13}H_{14}Cl_2FNO$			
Acrylophenone, 3-[bis(2-chloroethyl)-amino]-4'-fluoro-	EtOH	212(3.94),247(3.92), 335(4.36)	24-3139-69

Compound	Solvent	λ_{max}(log ε)	Ref.
$C_{13}H_{14}Cl_2INO$			
Acrylophenone, 3-[bis(2-chloroethyl)-amino]-4'-iodo-	EtOH	220(3.86),267(3.99), 342(4.38)	24-3139-69
$C_{13}H_{14}Cl_2N_2O_3$			
Acrylophenone, 3-[bis(2-chloroethyl)-amino]-3'-nitro-	EtOH	235(4.23),343(4.28)	24-3139-69
Acrylophenone, 3-[bis(2-chloroethyl)-amino]-4'-nitro-	EtOH	206(3.89),247(4.18), 365(4.11)	24-3139-69
$C_{13}H_{14}Cl_2OS$			
Benzo[b]cyclopropa[d]thiopyran, 1,1-dichloro-7b-ethoxy-1,1a,2,7b-tetrahydro-2-methyl-	EtOH	211(4.07),225(4.10), 261(3.75),291(3.15)	44-0056-69
$C_{13}H_{14}Cl_3NO$			
Acrylophenone, 3-[bis(2-chloroethyl)-amino]-4'-chloro-	EtOH	213(3.94),253(4.04), 340(4.34)	24-3139-69
$C_{13}H_{14}FNO_4$			
4-Penten-1-ol, 5-(p-fluorophenyl)-4-nitro-, acetate	EtOH	306(4.07)	87-0157-69
$C_{13}H_{14}INO$			
2,3-Dihydro-5,7(9)-dimethylfuro[3,2-c]-quinolinium iodide	EtOH	241(5.94),309(4.72), 332(4.91)	2-0952-69
2,3-Dihydro-5,8-dimethylfuro[3,2-c]-quinolinium iodide	EtOH	241(5.90),312(4.83), 328(4.92)	2-0952-69
$C_{13}H_{14}INO_2$			
2,3-Dihydro-8-methoxy-5-methylfuro-[3,2-c]quinolinium iodide	EtOH	246(5.41),315(4.74), 335(4.75)	2-0952-69
$C_{13}H_{14}N_2O$			
1H-Indazol-3-ol, 4,5,6,7-tetrahydro-1-phenyl-	MeOH	268(4.19)	88-1661-69
2H-Indazol-3-ol, 4,5,6,7-tetrahydro-2-phenyl-	MeOH	248(4.15),272(3.99)	88-1661-69
3H-Indazol-3-one, 2,3a,4,5,6,7-hexahydro-2-phenyl-	EtOH	248(4.14)	22-4159-69
Nicotinamide, 1-benzyl-1,4-dihydro-	EtOH	352(3.80)	39-0192-69B
Nicotinamide, 1-benzyl-1,6-dihydro-	EtOH	267(3.79),358(3.75)	39-0192-69B
3-Pentanone, 2-(1-phthalazinyl)-	EtOH	219(4.78),266s(3.69), 275(3.67),307(3.13)	95-0959-69
Pyrrolo[1,2-a]pyrimidin-2(6H)-one, 3,4,7,8-tetrahydro-4-phenyl-	EtOH	264(4.03)	22-3139-69
Pyrrolo[1,2-a]pyrimidin-4(3H)-one, 2,6,7,8-tetrahydro-2-phenyl-	EtOH	234(3.80)	22-3139-69
3-Quinuclidinone, 2-(4-pyridylmethyl-ene)-	EtOH	277(4.27)	22-1251-69
$C_{13}H_{14}N_2OS$			
2-Pyrazolin-5-one, 4-[(ethylthio)meth-ylene]-3-methyl-1-phenyl-	hexane	250(4.27),320(4.43), 407(3.11)	104-1249-69
	EtOH	250(4.36),327(4.45)	104-1249-69
$C_{13}H_{14}N_2O_2$			
Carbamic acid, (α-cyano-β-ethylstyryl)-, methyl ester, trans	EtOH	256(4.03)	22-1724-69

Compound	Solvent	$\lambda_{max}(\log \epsilon)$	Ref.
Carbamic acid, (1-cyano-2-methyl-3-phenylpropenyl)-, methyl ester, trans	EtOH	233(4.12)	22-1724-69
Carbostyril, 3-acetyl-4-(dimethylamino)-	dioxan	233(4.52),265(3.82), 335(3.95)	33-2641-69
Crotonic acid, 2-cyano-3-(N-methylanilino)-, methyl ester	isooctane	306(4.26)	35-6689-69
Cyclopenta[b]pyrrol-2(1H)-one, 3,3aβ,4,5,6,6aβ-hexahydro-1-nitroso-3α-phenyl-	EtOH	251(3.78),430(1.7), 451(1.7)	44-2058-69
Indone, 3-acetyl-2-methoxy-, 1-(methylhydrazone)	C_6H_{12}	242(4.37),340(4.22), 420s(3.36)	88-1697-69
Isoquinoline-4-carboxamide, 1,2-dihydro-2,N,N-trimethyl-1-oxo-	MeOH	242(3.98),252(3.90), 285?(3.91),292(3.89), 322(3.70),335(3.56)	22-2045-69
5,8-Isoquinolinedione, 3-methyl-7-(propylamino)-	MeCN	236(4.35),277(4.34), 321(3.78)	4-0697-69
3H-Pyrazole-4-carboxylic acid, 3,3-dimethyl-5-phenyl-, methyl ester	MeOH	234(4.02),303(3.84)	88-0015-69
3H-Pyrazole-5-carboxylic acid, 3,3-dimethyl-4-phenyl-, methyl ester	MeOH	293(3.67)	88-0015-69
Pyrrolo[2,3-b]indole, 1-acetyl-1,2,3,4-tetrahydro-5-methoxy-	n.s.g.	223(4.22),318(4.32)	87-0636-69
$C_{13}H_{14}N_2O_2S$			
Acrylic acid, 2-cyano-3-(N-methylanilino)-3-(methylthio)-, methyl ester	MeOH	328(4.18)	78-4649-69
$C_{13}H_{14}N_2O_2S_2$			
1,3,5-Thiadiazine-2-thione, 3,5-difurfuryltetrahydro-	MeOH	248(3.97),290(4.16)	73-2952-69
$C_{13}H_{14}N_2O_3$			
Cyclopenta[b]pyrrol-2(1H)-one, 3,3aβ,4,5,6,6aβ-hexahydro-3α-(p-nitrophenyl)-	EtOH	273(4.03)	44-2058-69
1H-Indazol-5-ol, 1-acetyl-4,7-dimethyl-, acetate	EtOH	232(4.39),255s(--), 308(3.91)	103-0254-69
2H-Indazol-5-ol, 2-acetyl-4,7-dimethyl-, acetate	EtOH	230(4.35),293(3.98), 336(3.61)	103-0254-69
2,4-Pentadienamide, N,N-dimethyl-5-(p-nitrophenyl)-, trans-trans	EtOH	342(4.51)	39-0016-69C
3-Quinolinecarboxylic acid, 4-(dimethylamino)-1,2-dihydro-2-oxo-, methyl ester	dioxan	233(4.60),263(3.79), 322(3.91),329(3.92)	33-2641-69
Uracil, 3-benzyl-5-methoxy-1-methyl-	pH 1-14	283(3.85)	44-2636-69
$C_{13}H_{14}N_2O_3S$			
4-Pyridinesulfinic acid, 1-benzyl-3-carbamoyl-1,4-dihydro-, Na salt	N NaOH	258(4.04),367(3.56) (anom.)	22-1299-69
$C_{13}H_{14}N_2O_4$			
Morpholine, 4-(o-nitrocinnamoyl)-	EtOH	245(4.43)	39-0016-69C
Morpholine, 4-(p-nitrocinnamoyl)-	EtOH	309(4.38)	39-0016-69C
$C_{13}H_{14}N_2O_5$			
4(3H)-Quinazolinone, 3-β-D-ribofuranosyl-	pH 1	232(4.44),274(3.85), 292s(3.72)	4-0089-69
	pH 11	226(4.46),263(3.83), 272(3.76),301(3.53), 312(3.43)	4-0089-69

Compound	Solvent	λ_{max}(log ε)	Ref.
4(3H)-Quinazolinone, 3-β-D-ribofuranosyl- (cont.)	MeOH	225(4.49),264(3.89), 273(3.85),301(3.57), 312(3.48)	4-0089-69
$C_{13}H_{14}N_2O_5S$			
2,4(1H,3H)-Quinazolinedione, 1-(5-thio-β-D-5-deoxyribofuranosyl)-	pH 1	305(3.58)	4-0089-69
	pH 11	306(3.62)	4-0089-69
	MeOH	305(3.58)	4-0089-69
$C_{13}H_{14}N_4$			
9H-Indeno[2,1-d]pyrimidine, 2,4-diamino-6,7-dimethyl-	pH 1	274(4.28),290s(4.10), 295s(4.09)	4-0613-69
	pH 10	277(4.30),297(4.23)	4-0613-69
	EtOH	277(4.30),297(4.24)	4-0613-69
p-Phenylenediamine, N-(2-amino-4-imino-3-methyl-2,5-cyclohexadien-1-ylidene)-	pH 8	552(4.09)	39-0827-69B
p-Phenylenediamine, N-(2-amino-4-imino-5-methyl-2,5-cyclohexadien-1-ylidene)-	pH 8	549(4.09)	39-0827-69B
Toluene-2,5-diamine, N^5-(2-amino-4-imino-2,5-cyclohexadien-1-ylidene)-	pH 8	572(4.06)	39-0827-69B
$C_{13}H_{14}N_4O_2$			
9H-Indeno[2,1-d]pyrimidine, 2,4-diamino-6,7-dimethoxy-	pH 1	278(4.24),302(4.23)	4-0613-69
	pH 10	278(4.44),292s(4.28), 304s(4.27)	4-0613-69
	EtOH	279(4.30),306(4.31)	4-0613-69
$C_{13}H_{14}N_4O_3$			
2-Propanone, 1-(5-methyl-3-isoxazolyl)-, p-nitrophenylhydrazone	EtOH	218s(4.16),250(4.11), 384(4.45)	35-4749-69
$C_{13}H_{14}N_4O_5$			
Inosine, 8,5'-anhydro-2',3'-O-isopropylidene-	pH 1	250(4.04)	2-0001-69
	H_2O	251(4.07)	2-0001-69
	pH 13	253(4.07)	2-0001-69
Uric acid, dihydro-4,5-dimethoxy-9-phenyl-	EtOH	218(4.56),248s(4.27)	78-0541-69
$C_{13}H_{14}N_4O_6$			
Barbituric acid, 5,5'-cyclopentylidenedi-	EtOH	255(4.55)	103-0827-69
2(3H)-Furanone, 5-acetyldihydro-4-methyl-, 2,4-dinitrophenylhydrazone, cis	EtOH	226(4.18),250(4.08), 353(4.32)	35-4739-69
trans	EtOH	225(4.10),252(3.98), 353(4.23)	35-4739-69
$C_{13}H_{14}N_4O_7$			
Barbituric acid, 5,5'-(1-methyl-3-oxobutylidene)di-	EtOH	260(4.48)	103-0827-69
$C_{13}H_{14}O$			
3,4-Benzobicyclo[3.2.1]octen-6-one, 8-methyl-	heptane	264(2.57),269(2.64), 276(2.65),289(2.61), 298(2.73),309(2.72), 320(2.45)	78-0335-69
3,4-Benzobicyclo[3.2.1]octen-7-one, 8-methyl-, endo	heptane	260(2.51),267(2.69), 274(2.73),314(1.29)	78-0335-69
5,10-Methanobenzocycloocten-11-one, 5,6,7,8,9,10-hexahydro-	EtOH	264(3.48),270(3.62), 276(3.64)	88-1761-69
Sorbophenone, 5-methyl-1-phenyl-	EtOH	219(3.98),315(4.40)	22-3638-69

Compound	Solvent	$\lambda_{max}(\log \epsilon)$	Ref.
$C_{13}H_{14}OS_2$			
Ketone, 2-ethyl-3-(methylthio)benzo[b]-thien-5-yl methyl	EtOH	223(4.32),250(4.27), 295(4.12),330(3.82)	22-0607-69
$C_{13}H_{14}O_2$			
Cyclohexanone, 2-benzoyl-	EtOH	248(4.10),320s(--)	94-1572-69
1-Cyclopropene-1-carboxylic acid, 3,3-dimethyl-2-phenyl-, methyl ester	C_6H_{12}	221(3.56),227(3.46), 291(3.83)	88-0015-69
1-Indanylideneacetic acid, ethyl ester, exo	MeOH	285(4.21)	44-4182-69
1-Indeneacetic acid, ethyl ester, endo	MeOH	253(3.99)	44-4182-69
5-Indenecarboxylic acid, 2,7-dimethyl-, methyl ester	EtOH	234(4.58),256s(3.92), 273s(3.67),309(2.99)	24-2502-69
5-Indenecarboxylic acid, 3,7-dimethyl-, methyl ester	EtOH	233(4.58),255s(3.85), 267s(3.47),305(2.88)	24-2493-69
Naphthalene, 2,3-dimethoxy-6-methyl-	EtOH	233(4.87),266(3.66), 275(3.59),297(3.08), 304(3.11),312(3.40), 317(3.30),325(3.61)	12-1721-69
2-Naphthaleneacetic acid, 3,4-dihydro-α-methyl-	EtOH	266(4.04)	88-2331-69
Naphtho[1,2-b]furan-2(3H)-one, 3aα,4,5,9bβ-tetrahydro-3-methyl-	EtOH	265(2.73),274(2.81)	88-2331-69
2,4-Pentadienoic acid, 5-phenyl-, ethyl ester	EtOH	228s(3.91),233(4.00), 240s(3.87),310(4.58)	23-4076-69
$C_{13}H_{14}O_2S$			
1-Benzothiepin-3(2H)-one, 5-ethoxy-2,3-dihydro-2-methyl-	EtOH	220(4.33),246(3.94), 266(3.79),296(3.03)	44-0056-69
2H-1-Benzothiopyran-3-carboxaldehyde, 4-ethoxy-2-methyl-	EtOH	212(4.03),251(4.18), 264(4.12),307(4.04), 370(3.20)	44-0056-69
$C_{13}H_{14}O_3$			
3-Chromene-6-acetic acid, 2,2-dimethyl-	EtOH	265(3.48),312(3.37)	12-1923-69
2(3H)-Furanone, dihydro-3-(p-methoxy-α-methylbenzylidene)-, cis	EtOH	220(4.17),302(3.85)	44-3792-69
trans	EtOH	220(4.12),289(4.17)	44-3792-69
1,3,5-Heptanetrione, 1-phenyl-	EtOH	247(3.80),345(4.21)	94-2126-69
Indan-3-acetic acid, 3,6-dimethyl-1-oxo-	EtOH	248(4.02),302(3.17)	2-0215-69
Indan-3-carboxylic acid, 3,6-dimethyl-1-oxo-, methyl ester	EtOH	250(4.04),300(3.4)	2-0215-69
Ketone, 5-hydroxy-2,2-dimethyl-2H-1-benzopyran-6-yl methyl	MeOH	265(4.49),303(4.02)	2-1072-69
$C_{13}H_{14}O_4$			
Acetoacetic acid, 2-benzoyl-, ethyl ester	C_6H_{12}	245(4.17),284(4.06)	22-3281-69
	MeOH	246(4.13),286(3.86)	22-3281-69
	ether	244(4.10),284(3.95)	22-3281-69
	MeCN	245(4.04),284(3.83)	22-3281-69
	$CHCl_3$	250(4.07),285(3.88)	22-3281-69
	CCl_4	284(4.02)	22-3281-69
Chroman-6-acetic acid, 2,2-dimethyl-4-oxo-	EtOH	253(3.75),330(3.39)	12-1923-69
2,4,6-Cycloheptatriene-$\Delta^{1,\alpha}$-succinic acid, dimethyl ester	EtOH	245(4.00),340(4.11)	35-6391-69
Spiro[cyclopropane-1,7'-[2,5]norbornadiene]	MeCN	240(3.4),275s(3.0)	5-0052-69E

Compound	Solvent	$\lambda_{max}(\log \epsilon)$	Ref.
$C_{13}H_{14}O_5$			
Bicyclo[3.2.2]nona-3,6-diene-8,9-dicarboxylic acid, 2-oxo-, dimethyl ester, endo	MeOH	227(3.90),355(2.30)	88-0443-69
Chromone, 6-(methoxymethyl)-5-hydroxy-7-methoxy-2-methyl-	MeOH	210(4.21),230(4.21), 248(4.21),257(4.22), 291(3.77),313(3.55)	39-0704-69C
	MeOH-NaOH	245s(4.04),265(4.20), 356(3.53)	39-0704-69C
$C_{13}H_{14}O_7$			
Bicyclo[3.2.0]hept-6-ene-6,7-dicarboxylic acid, 1-hydroxy-4-oxo-, methyl ester, acetate	n.s.g.	204(3.91),255(3.48)	44-0794-69
1-Phthalanpropionic acid, 1-hydroxy-5,7-dimethoxy-3-oxo-	MeOH	244(4.57),302(3.48)	22-2365-69
Phthalonic acid, 4,5-dimethoxy-, (dimethyl ester)	EtOH	224(4.37),262(3.83), 298(3.61)	39-1921-69C
$C_{13}H_{15}BF_2O_2$			
Boron, (4,4-dimethyl-1-phenyl-1,3-pentanedionato)difluoro-	$CHCl_3$	265(3.69),332(4.49), 346s(4.33)	39-0526-69A
$C_{13}H_{15}BF_4OS$			
6-Methoxy-1,3,5-trimethylcyclohepta[c]-thiolium tetrafluoroborate	$C_2H_4Cl_2$	267(4.05),316(4.72), 362(4.10),378(4.32), 550(3.35)	104-0947-69
$C_{13}H_{15}BrCl_3NO_5$			
(p-Bromo-γ-hydroxycinnamylidene)bis(2-chloroethyl)ammonium perchlorate	EtOH	277(4.01),349(4.21)	24-3139-69
$C_{13}H_{15}BrN_4O_5$			
Inosine, 8-bromo-2',3'-O-isopropylidene-	pH 1	254(4.23)	2-0001-69
	H_2O	254(4.21)	2-0001-69
	pH 13	259(4.17)	2-0001-69
$C_{13}H_{15}BrO$			
Anisole, 4-bromo-2-(1-cyclopenten-1-yl)-5-methyl-	MeOH	260(4.09),302(3.68), 313(3.63)	78-3509-69
$C_{13}H_{15}BrO_2$			
Cyclopentanone, 2-(5-bromo-2-methoxy-p-tolyl)-	MeOH	285(3.30)	78-3509-69
$C_{13}H_{15}BrS$			
1,3,5,7-Tetramethylcyclohepta[c]thiolium bromide	heptane	241(4.20),268s(4.08)	104-2014-69
	MeCN	244(4.16),280(4.43), 323(4.26),378(3.42), 682(3.38)	104-2014-69
	$C_2H_4Cl_2$	283(4.28),328(4.06), 378s(3.36),692(2.80)	104-2014-69
$C_{13}H_{15}ClN_2$			
Quinoline, 2-[(3-chloropropyl)amino]-7-methyl-, hydrochloride	EtOH	243(5.40),256(5.11), 335(4.94),346(4.83)	95-0759-69
$C_{13}H_{15}ClN_4$			
Pyrimidine, 2,4-diamino-5-(p-chlorobenzyl)-6-ethyl-	pH 2	223(4.45),276(3.89)	44-0821-69
	pH 12	218(4.33),238s(4.06), 286(3.91)	44-0821-69

Compound	Solvent	$\lambda_{max}(\log \epsilon)$	Ref.
Pyrimidine, 2,4-diamino-5-(p-chloro-phenyl)-6-propyl-	pH 2	208(4.55),223s(4.41), 273(3.91)	44-0821-69
	pH 12	220s(4.33),248s(3.91), 286(3.98)	44-0821-69
$C_{13}H_{15}ClN_4O_5$			
Inosine, 2-chloro-2',3'-O-isopropyli-dene-	pH 1	254(4.09)	94-2581-69
	pH 13	258(4.12)	94-2581-69
$C_{13}H_{15}ClO$			
2-Hepten-1-one, 2-chloro-1-phenyl-	EtOH	251(4.11)	44-1474-69
1-Penten-3-one, 1-(p-chlorophenyl)-4,4-dimethyl-	iso-PrOH	294(4.33)	104-0826-69
$C_{13}H_{15}ClO_4$			
1H-2-Benzopyran-4-carboxylic acid, 1-chloro-4a,5,6,8a-tetrahydro-3,8a-dimethyl-6-oxo-, methyl ester	MeOH	223(4.23),240s(4.00)	35-2299-69
$C_{13}H_{15}ClO_4S$			
1-tert-Butyl-2-benzothiopyrylium perchlorate	MeOH	205(4.41),240(3.90), 290(3.57),302(3.67), 320(3.72)	44-3202-69
$C_{13}H_{15}ClS$			
1,3,5,7-Tetramethylcyclohepta[c]thiol-ium chloride	heptane	237(4.22),306s(3.67)	104-2014-69
	MeCN	242(4.20),279(4.34), 322(4.08),368s(3.47), 690(3.20)	104-2014-69
	$C_2H_4Cl_2$	280s(4.00),325s(3.68), 700(2.55)	104-2014-69
$C_{13}H_{15}Cl_2NO$			
Acrylophenone, 3-[bis(2-chloroethyl)-amino]-	EtOH	207(3.81),245(3.92), 336(4.34)	24-3139-69
$C_{13}H_{15}Cl_2NO_5$			
Gluosamine, 2,4-dichlorobenzylidene-	neutral	258(3.62)	94-0770-69
	alkaline	392(3.64)	94-0770-69
$C_{13}H_{15}FeNO_4S$			
Alanine, N-(ferrocenylsulfonyl)-	EtOH	432(2.13)	49-1552-69
$C_{13}H_{15}I_3S$			
1,3,5,7-Tetramethylcyclohepta[c]thiol-ium triiodide	MeCN	282(4.18),285(4.82), 356(4.44),690(3.39)	104-2014-69
	$C_2H_4Cl_2$	288(4.74),358(4.38), 698(3.33)	104-2014-69
$C_{13}H_{15}N$			
1H-Carbazole, 2,3,4,4a-tetrahydro-4a-methyl-, perchlorate	EtOH	231(3.87),237(3.84), 277(3.89)	78-2757-69
Indole, 2-(1,1-dimethylallyl)-	EtOH	225(4.30),280(3.86), 291(3.77)	39-0595-69C
2,8-Methanobenzo[4,5]cyclohept[1,2-b]-azirine, 1,1a,2,7,8,8a-hexahydro-9-methyl-, endo	heptane	260(2.56),266(2.74), 274(2.79)	78-0335-69
exo	heptane	260(2.72),266(2.69), 273(2.74)	78-0335-69
Spiro[cyclohexane-1,3'-[3H]indole	EtOH	225(3.72),248(3.45)	78-4843-69

Compound	Solvent	λ_{max}(log ϵ)	Ref.
Spiro[cyclohexane-1,3'-[3H]indole	EtOH-HCl	233(3.60),285(3.26)	78-4843-69
$C_{13}H_{15}NO$			
Acridine, 1,2,3,4,7,8-hexahydro-, N-oxide	EtOH	250(3.51),290(3.26)	103-0226-69
Atroponitrile, β-butoxy-	EtOH	212(4.18),269(4.20)	22-0198-69
Carbazole-9-carboxaldehyde, 1,2,3,4,4a,-9a-hexahydro-	MeOH	227(3.47),273(3.55), 287(3.64)	20-0271-69
Crotononitrile, 2-phenyl-3-propoxy-, cis	EtOH	209(4.07),265(4.15)	22-0198-69
trans	EtOH	210(4.15),269(4.20)	22-0198-69
Cyclopenta[b]pyrrol-2(1H)-one, 3,3aβ,4,5,6,6aβ-hexahydro-3α-phenyl-	EtOH	243(1.88),248(2.08), 253(2.21),259(2.30), 265(2.18),268s(1.88)	44-2058-69
3β-phenyl-	EtOH	242(2.00),248(2.09), 252(2.22),259(2.31), 262s(2.20),265(2.19), 268(2.02)	44-2058-69
4-Isoxazoline, 3-allyl-2-methyl-5-phenyl-	EtOH	223(4.12),276(3.91)	88-4875-69
2,4-Pentadienamide, N,N-dimethyl-5-phenyl-, (E,E)-	EtOH	225(4.00),307(4.43)	39-0016-69C
(E,Z)-	EtOH	227(4.06),306(4.40)	39-0016-69C
2-Piperidone, 1-styryl-, cis	MeOH	259(4.12)	78-5745-69
2-Piperidone, 1-styryl-, trans	MeOH	222(--),286(4.37)	78-5745-69
$C_{13}H_{15}NOS$			
5H-Thiazolo[3,2-a]pyridin-5-one, hexahydro-8a-phenyl-	EtOH	249(2.81),261(2.82), 267(2.73)	44-0165-69
$C_{13}H_{15}NOS_2$			
Morpholine, 4-(4-phenyl-1,3-dithiol-2-yl)-	EtOH	317(3.93)	94-1924-69
2-Pentanone, 3-(2-benzothiazolylthio)-4-methyl-	EtOH	225(4.31),245(3.93), 280(4.04),291(3.98), 301(3.90)	95-0469-69
$C_{13}H_{15}NO_2$			
6H-Azepino[1,2-a]indol-6-one, 7,8,9,10,-10a,11-hexahydro-10a-hydroxy-	EtOH	252(4.11),281(3.62), 289(3.56)	94-1290-69
2H-Azirine-2-carboxylic acid, 3-phenyl-, butyl ester	EtOH	203(4.22),243(3.96), 288s(2.82)	88-2049-69
Cyclopenta[b]pyrrol-2(1H)-one, 3,3aβ,4,5,6,6aβ-hexahydro-1-hydroxy-3α-phenyl-	EtOH	247s(2.18),253(2.23), 259(2.29),265(2.17), 268s(1.88)	44-2058-69
2(3H)-Furanone, 3-[1-(benzylamino)ethylidene]dihydro-	hexane	294(4.19)	104-0998-69
Isoxazole, 5-butoxy-3-phenyl-	EtOH	204(4.09),238(3.79)	88-2049-69
Morpholine, 4-cinnamoyl-, trans	EtOH	282(4.36)	39-0016-69C
2-Oxa-3-azabicyclo[2.2.2]oct-5-ene, 1-methoxy-N-phenyl-	EtOH	245(3.85)	12-2493-69
5H-Oxazolo[3,2-a]pyridin-5-one, hexahydro-8a-phenyl-	EtOH	251(2.89),257(2.85), 263(2.82)	44-0165-69
2H-Pyrrolo[2,1-b][1,3]oxazin-6(7H)-one, tetrahydro-8a-phenyl-	EtOH	252(2.68),258(2.72), 264(2.69)	44-0165-69
Sorbohydroxamic acid, N-m-tolyl-	EtOH	265(4.38)	34-0278-69
Sorbohydroxamic acid, N-p-tolyl-	EtOH	265(4.39)	34-0278-69
Spiro[cyclopentane-1,3')2'H)-quinolin]-2'-one, 1',4'-dihydro-4'-hydroxy-	EtOH	253(4.08),281(3.41), 291(3.26)	94-1290-69

Compound	Solvent	$\lambda_{max}(\log \epsilon)$	Ref.
$C_{13}H_{15}NO_2S$			
Acrylamide, N-ethyl-3-(phenacylthio)-, (Z)-	90% EtOH	248(4.23),272(4.23)	12-0765-69
Crotonic acid, 3-benzamido-thio-, O-ethyl ester	MeCN	344(4.17)	24-0269-69
Indoline-3-acetic acid, 1-methyl-2-thioxo-, ethyl ester	EtOH	250s(3.83),281(3.72), 346(3.86)	95-0058-69
$C_{13}H_{15}NO_2S_2$			
4-Morpholinecarbodithioic acid, ester with 2-mercaptoacetophenone	EtOH	251(--),282(--)	94-1924-69
$C_{13}H_{15}NO_3$			
1H-1-Benzazepine-7-carboxylic acid, 1-acetyl-2,3,4,5-tetrahydro-	MeOH	213(3.88),246(3.59)	44-1070-69
1H-1-Benzazepine-7-carboxylic acid, 9-acetyl-2,3,4,5-tetrahydro-	MeOH	230s(4.19),245(4.33), 302(4.18),373(3.69)	44-1070-69
1H-1-Benzazepine-8-carboxylic acid, 1-acetyl-2,3,4,5-tetrahydro-	MeOH	223(4.27),276(2.98)	44-2235-69
Carbostyril, 3-ethyl-4,7-dimethoxy-	EtOH and EtOH-HCl	217(4.71),226s(4.54), 254(3.81),278(3.79), 310s(4.00),321(4.20), 335(4.15)	12-0447-69
3-Furoic acid, 2-amino-4,5-dihydro-5-phenyl-, ethyl ester	MeOH	208(4.22),271(4.34)	24-2739-69
Morpholine, 4-(m-hydroxycinnamoyl)-	EtOH	215(4.18),234(4.12), 280(4.33)	39-0016-69C
Morpholine, 4-(p-hydroxycinnamoyl)-	EtOH	228(4.02),314(4.31)	39-0016-69C
Phthalic anhydride, 3-[(1,2-dimethylpropyl)amino]-	EtOH	207(4.24),256(4.32), 355(3.86),405(3.63)	16-0241-69
$C_{13}H_{15}NO_4$			
1H-1-Benzazepine-7-carboxylic acid, 1-acetyl-2,3,4,5-tetrahydro-	MeOH	244(3.99)	44-1070-69
Isocarbostyril, 5,6,7-trimethoxy-2-methyl-	EtOH	247(4.62),270(3.60), 281(3.68),293(3.74), 332(3.48)	88-1951-69
Malonic acid, [(N-methylanilino)methylene]-, dimethyl ester	isooctane	221(4.00),298(4.43)	35-6683-69
4-Penten-1-ol, 4-nitro-5-phenyl-, acetate	EtOH	305(4.02)	87-0157-69
$C_{13}H_{15}NO_4S$			
Malonic acid, [anilino(methylthio)-methylene]-, dimethyl ester	MeOH	316(4.16)	78-4649-69
$C_{13}H_{15}NO_5S_3$			
N-(4-Phenyl-1,3-dithiol-2-ylidene)-morpholinium bisulfate	EtOH	324(4.09)	94-1924-69
$C_{13}H_{15}NS$			
2H-1,4-Benzothiazine, 2-isobutylidene-3-methyl-	MeOH	370(3.54)	124-1278-69
	$MeOH-HClO_4$	440(3.56)	124-1278-69
$C_{13}H_{15}N_3O$			
2-Pyrazolin-5-one, 4-[(dimethylamino)-methylene]-3-methyl-1-phenyl-	heptane	222(3.85),262(4.24), 300(4.36),355(3.43)	104-1249-69
	EtOH	228(4.05),292(4.36)	104-1249-69

Compound	Solvent	$\lambda_{max}(\log \epsilon)$	Ref.
$C_{13}H_{15}N_3OS$			
Salicylaldehyde, 4-(2-cyclopenten-1-yl)-3-thiosemicarbazone	EtOH	242(4.0),338(4.5)	124-0509-69
$C_{13}H_{15}N_3O_2S$			
Δ^2-1,3,4-Oxadiazoline-5-thione, 4-(morpholinomethyl)-2-phenyl-	n.s.g.	222(4.20),309(4.12)	2-0583-69
2-Pyrroline-3-acetic acid, α-(dicyanomethylene)-1-methyl-2-(methylthio)-, ethyl ester	C_6H_{12}	430(4.03)	5-0073-69E
$C_{13}H_{15}N_3O_4$			
Cinnamic acid, α-azido-2,4-dimethoxy-, ethyl ester	C_6H_{12}	240(4.07),305(4.10), 342(4.40)	49-1599-69
Cinnamic acid, α-azido-2,6-dimethoxy-, ethyl ester	C_6H_{12}	299(4.09)	49-1599-69
Cinnamic acid, α-azido-3,4-dimethoxy-, ethyl ester	C_6H_{12}	237(3.89),338(4.21)	49-1599-69
1H-1,2,3-Triazole, 4-phenyl-1-β-D-ribofuranosyl-	EtOH	274(4.14),284(2.70)	4-0639-69
1H-1,2,3-Triazole, 5-phenyl-1-β-D-ribofuranosyl-	EtOH	235(4.04)	4-0639-69
$C_{13}H_{15}N_3O_4S$			
Pyridine, 3-(2,4-dinitrophenyl)-1,4,5,6-tetrahydro-1-methyl-2-(methylthio)-	$CHCl_3$	438(3.60)	5-0073-69E
Pyrrolidine, 1-[α-(methylthio)-2,4-dinitrostyryl]-	$CHCl_3$	495(4.00)	5-0073-69E
$C_{13}H_{15}N_3S$			
Benzaldehyde, 4-(2-cyclopenten-1-yl)-3-thiosemicarbazone	EtOH	220(4.1),322(4.4)	124-0509-69
$C_{13}H_{15}N_5O$			
Adenine, N-[(2-methoxy-1,4-cyclohexadien-1-yl)methyl]-	pH 1	272(4.15)	44-3814-69
	pH 13	276(4.28),285s(4.14)	44-3814-69
	EtOH	210(4.33),270(4.27)	44-3814-69
Adenine, N-[(2-methoxy-2,5-cyclohexadien-1-yl)methyl]-	pH 13	276(4.25),285s(4.12)	44-3814-69
	EtOH	210(4.33),270(4.26)	44-3814-69
Adenine, N-[(4-methoxy-1,4-cyclohexadien-1-yl)methyl]-	pH 1	209(4.09),277(4.24)	44-3814-69
	pH 13	276(4.26),284s(4.12)	44-3814-69
	EtOH	211(4.29),268(4.26)	44-3814-69
Adenine, N-[(5-methoxy-1,4-cyclohexadien-1-yl)methyl]-	pH 1	278(4.24)	44-3814-69
	pH 13	276(4.27),284s(4.14)	44-3814-69
	EtOH	210(4.34),269(4.26)	44-3814-69
Butyric acid, [(3-phenyl-1H-1,2,4-triazol-1-yl)methylene]hydrazide	MeOH	290(4.36)	48-0646-69
2-Cyclohexen-1-one, 3-methyl-4-[(purin-6-ylamino)methyl]-, hydrochloride	pH 1	281(4.28)	44-3814-69
	pH 13	274(4.20)	44-3814-69
	EtOH	213(4.35),232(4.09), 270(4.29)	44-3814-69
Purino[1,6-a]quinazoline, 4,5,5a,6,9,9a-hexahydro-9a-methoxy-	pH 1	265(4.15)	44-3814-69
	pH 13	273(4.18),281(4.16)	44-3814-69
	EtOH	231(4.29),278(4.15)	44-3814-69
$C_{13}H_{15}N_5OS$			
5H-s-Triazolo[3,4-b][1,3,4]thiadiazine-5-carboxanilide, 3-ethyl-6,7-dihydro-	EtOH	240(4.27)	2-0959-69

Compound	Solvent	λmax(log ε)	Ref.
$C_{13}H_{15}N_5O_2$			
Benzaldehyde, p-(dimethylamino)-, (2,6-dihydroxy-4-pyrimidinyl)hydrazone	EtOH	351(4.62)	56-0519-69
$C_{13}H_{15}N_5O_3$			
Pyrazole-4-carboxylic acid, 3-[(benzyl-oxy)methyl]-5-carbamoyl-, hydrazide	EtOH	210(4.10)	88-0289-69
$C_{13}H_{15}N_5O_5$			
Guanosine, 8,5'-anhydro-2',3'-O-isopro-pylidene-	pH 1	250(4.17),280(3.99)	2-0001-69
	H_2O	250(4.16),282(3.99)	2-0001-69
	pH 13	252(4.12),268(4.09)	2-0001-69
$C_{13}H_{15}N_5O_6$			
Adenosine, 3'-acetate 5'-formate	H_2O	260(4.17)	78-4057-69
$C_{13}H_{15}N_5O_7$			
2(1H)-Quinoxalinone, 3,4-dihydro-5,7-dinitro-3-valinyl-	EtOH	268(4.14),374(4.01)	44-0395-69
$C_{13}H_{15}O_4P$			
Phosphonic acid, benzoyl-, diallyl ester	C_6H_{12}	265(4.31),378(2.18)	18-0821-69
$C_{13}H_{16}ClNO_4$			
Dimethyl(5-phenyl-2,4-pentadienylidene)-ammonium perchlorate	MeCN	255(3.97),388(4.76)	24-0623-69
$C_{13}H_{16}ClNO_5$			
2-tert-Butyl-4-phenylisoxazolium perchlorate	CH_2Cl_2	235(4.24)	44-3451-69
Glucosamine, p-chlorobenzylidene-	neutral	255(3.84)	94-0770-69
	alkaline	375(3.58)	94-0770-69
$C_{13}H_{16}ClN_5O$			
4,5,5a,6,7,8,9,9a-Octahydro-9a-methyl-8-oxo-1H-purino[1,6-a]quinazolin-10-ium chloride	pH 1	212(4.25),265(4.14)	44-3814-69
	pH 13	275(4.24),284s(4.13)	44-3814-69
	EtOH	212(4.27),265(4.14)	44-3814-69
$C_{13}H_{16}Cl_2N_2$			
Quinoline, 2-[(3-chloropropyl)amino]-7-methyl-, hydrochloride	EtOH	243(5.40),256(5.11), 335(4.94),346(4.83)	95-0759-69
$C_{13}H_{16}Cl_2O_4$			
Bicyclo[3.3.1]nona-2,6-diene-2-carbox-ylic acid, 8-(dichloromethyl)-3,5-dihydroxy-8-methyl-, methyl ester	MeOH	255(4.86)	35-2299-69
2-Oxabicyclo[3.3.1]non-3-ene-4-carbox-ylic acid, 9-(dichloromethyl)-3,9-dimethyl-7-oxo-, methyl ester	MeOH	244(3.75)	35-2299-69
$C_{13}H_{16}Cl_2O_5$			
3-Cyclohexene-1-malonic acid, 2-(di-chloromethyl)-2-methyl-5-oxo-, dimethyl ester	MeOH	222(4.03)	35-2299-69
$C_{13}H_{16}N_2$			
Toluene-4-azo-1'-cyclohexene	THF	314(4.36)	24-3647-69
$C_{13}H_{16}N_2O$			
Anisole, p-(1-cyclohexen-1-ylazo)-	THF	332(4.33)	24-3647-69

Compound	Solvent	$\lambda_{max}(\log \epsilon)$	Ref.
Azocino[4,5,6-cd]indole, 1,3,4,5,6,7-hexahydro-8-methoxy-	n.s.g.	279(2.78),289(2.70)	87-0636-69
Cycloocta[c]pyridine-4-carbonitrile, 5,6,7,8,9,10-hexahydro-3-hydroxy-1-methyl-	EtOH	217(4.25),222(4.24), 242s(3.82),249(3.67), 342(4.09)	44-3670-69
Cyclopenta[b]pyrrol-2(1H)-one, 3α-(p-aminophenyl)-3,3aβ,4,5,6,6aβ-hexahydro-	EtOH	241(4.02),290(2.9)	44-2058-69
1H-1,4-Diazepine, 2,3-dihydro-6-methoxy-5-methyl-7-phenyl-, perchlorate	MeOH	265(3.82),367(4.21)	39-1449-69C
Imidazo[1,2-a]pyridin-5(1H)-one, hexahydro-8a-phenyl-	EtOH	247(3.02),251(3.18), 257(3.21),267(2.76)	44-0165-69
Isoquinoline-4-carboxamide, 1,2-dihydro-N,N,2-trimethyl-	MeOH	237(4.02),276(3.86), 337(3.95)	22-2045-69
2-Pyrrolin-4-one, 2-(dimethylamino)-1-methyl-3-phenyl-	MeOH	243(3.98),263(3.98), 302(4.10)	78-5721-69
	MeOH-HCl	250(4.15)	78-5721-69
2H-Pyrrol-3-ol, 5-(dimethylamino)-2-methyl-4-phenyl-	MeOH	222(4.08),287(4.25)	78-5721-69
	MeOH-HCl	240(4.23)	78-5721-69
Urea, 1-inden-3-yl-3-propyl-	EtOH	210(4.07),268(3.03), 275(3.06)	36-0047-69
$C_{13}H_{16}N_2O_2$			
Phthalimide, 3-(diethylamino)-N-methyl- (other maxima present but not given)	EtOH	421(3.65)	104-1231-69
Phthalimide, 4-(diethylamino)-N-methyl-	EtOH	402(3.81)	104-1231-69
$C_{13}H_{16}N_2O_2S_2$			
Pyridine, 1,2,3,4-tetrahydro-1-methyl-6-(methylthio)-5-[(o-nitrophenyl)-thio]-	MeOH	390(3.58)	5-0073-69E
$C_{13}H_{16}N_2O_5$			
5aH-[1,3]Dioxolo[3',4']furo[3',2':4,5]-pyrrolo[1,2-a]pyrimidin-9(11H)-one, 3a ,4 -dihydro-4-(hydroxymethyl)-2,2-dimethyl-	MeOH	239(4.06)	35-7752-69
$C_{13}H_{16}N_2O_7$			
Glucosamine, p-nitrobenzylidene-	neutral	285(4.12)	94-0770-69
	alkaline	485(4.33)	94-0770-69
$C_{13}H_{16}N_2S$			
2,2'-Dipyrrylthione, 3,4,3',4'-tetramethyl-	$CHCl_3$	332(3.37),414(4.34)	12-0239-69
	10N HCl	263(3.54),357(4.03), 482(4.61)	12-0239-69
$C_{13}H_{16}N_4O_2$			
Formic acid, [N-(2-amino-3-cyano-4,5,6,7-tetrahydroindol-1-yl)-formimidoyl]-, ethyl ester	MeOH	261(4.20),297(4.19), 390s(2.84)	48-0388-69
$C_{13}H_{16}N_4O_3$			
Histidine, 1-[2-hydroxy-2-(3-pyridyl)-ethyl]-, diformate	H_2O	262(3.66)	63-0473-69
Pyrimidine, 2,4-diamino-5-(3,4,5-trimethoxyphenyl)-	pH 2	208(4.56),240s(4.09)	44-0821-69
	pH 12	261(3.97),288(3.87)	44-0821-69
$C_{13}H_{16}N_4O_4S_2$			
Barbituric acid, 5,5'-isopentylidenebis-[2-thio-	EtOH	285(4.23)	103-0830-69

Compound	Solvent	λmax(log ε)	Ref.
$C_{13}H_{16}N_4O_5$			
Hypoxanthine, 1-(2,3-O-isopropylidene-β-D-ribofuranosyl)-	pH 1	248(3.97)	44-2646-69
	pH 7	251(3.95)	44-2646-69
	pH 13	261(3.98)	44-2646-69
Hypoxanthine, 9-propenyl-1-β-D-ribofuranosyl-	pH 1	223(4.33),253s(--)	44-2646-69
	pH 7	226(4.42),270s(--)	44-2646-69
	pH 13	226(4.41),270s(--)	44-2646-69
$C_{13}H_{16}N_4O_5S$			
Imidazole, 2-[2,4-dihydroxy-1-α-D-(2-deoxyribofuranosyl)pyrimidin-5-yl)methylthio]-	pH 7	260(4.23)	63-0809-69
β-form	pH 7	258(4.25)	63-0809-69
$C_{13}H_{16}N_4O_6$			
Barbituric acid, 5,5'-(1-ethylpropylidene)di-	EtOH	255(4.39)	103-0827-69
Inosine, 2',3'-O-isopropylidene-, 1-N-oxide	pH 1	252(3.86)	94-1128-69
	pH 7	228(4.64),252(3.87)	94-1128-69
	pH 13	257(3.84),295(3.63)	94-1128-69
$C_{13}H_{16}N_4O_6S$			
Uridine, 5-[(imidazol-2-ylthio)methyl]-	pH 7	260(4.25)	63-0710-69
$C_{13}H_{16}N_8O_4$			
Guanosine, 5'-azido-5'-deoxy-2',3'-O-isopropylidene-	pH 1	256(4.11)	87-0658-69
	pH 11	258(4.11)	87-0658-69
	MeOH	253(4.18)	87-0658-69
$C_{13}H_{16}O$			
3,4-Benzobicyclo[3.2.1]octen-6-ol, 8-methyl-, exo-endo	EtOH	260(2.59),266(2.77), 273(2.83)	78-0335-69
endo	EtOH	259(2.56),266(2.69), 273(2.72)	78-0335-69
3,4-Benzobicyclo[3.2.1]octen-7-ol, 8-methyl-, endo-endo	EtOH	261(2.62),267(2.82), 274(2.89)	78-0335-69
exo-endo	EtOH	260(2.60),267(2.79), 274(2.85)	78-0335-69
1-Hexen-3-one, 4-methyl-1-phenyl-	EtOH	289(4.29)	44-0444-69
1-Penten-3-one, 4,4-dimethyl-1-phenyl-	iso-PrOH	291(4.29)	104-0826-69
Phenol, m-(2-norbornyl)-	pH 13	239(3.90),290(3.41)	30-0867-69A
	iso-PrOH	275(3.32),281(3.28)	30-0867-69A
Phenol, o-(2-norbornyl)-	pH 13	240(3.96),291(3.63)	30-0867-69A
	iso-PrOH	275(3.42),281(3.37)	30-0867-69A
Phenol, p-(2-norbornyl)-	pH 13	240(4.05),297(3.31)	30-0867-69A
	iso-PrOH	280(3.10),287(3.28)	30-0867-69A
Tetraspiro[2.0.2.0.2.0.2.1]tridecan-13-one	C_6H_{12}	210(3.49),245s(2.64)	88-0979-69
$C_{13}H_{16}O_2$			
7(6H)-Benzocyclooctenone, 5,8,9,10-tetrahydro-4-methoxy-	EtOH	279(3.13)	78-2661-69
3,5-Cycloheptadien-1-one, 2-(2-oxo-cyclohexyl)-	EtOH	235(3.73),285(2.74)	77-0737-69
3-Hexenophenone, 5-hydroxy-5-methyl-	EtOH	241(4.10)	22-3638-69
as-Indacene-4-carboxylic acid, 1,2,3,3a,5a,6,7,8-octahydro-	EtOH	207(4.05)	22-4068-69
1,4-Methanonaphthalene-5,8-dione, 1,2,3,4,4a,8a-hexahydro-6,7-di-methyl-, endo-cis	EtOH	252(4.20),362(1.77)	39-0105-69C

Compound	Solvent	λ_{max}(log ε)	Ref.
1(2H)-Naphthalenone, 3,4-dihydro-6-methoxy-4,4-dimethyl-	EtOH	278(4.24)	36-0340-69
1(2H)-Naphthalenone, 3,4-dihydro-7-methoxy-4,4-dimethyl-	EtOH	225(3.92),316(3.45)	36-0340-69
$C_{13}H_{16}O_3$			
Acetophenone, 2',4'-dihydroxy-3'-(3-methyl-2-butenyl)-	MeOH	285(4.25)	2-1072-69
Acetophenone, 2'-hydroxy-4'-[(3-methyl-2-butenyl)oxy]-	MeOH	229(4.32),276(4.53), 315(4.21)	2-1072-69
2,4-Pentadiynoic acid, 5-(1-hydroxy-2-methylcyclohexyl)-, methyl ester	EtOH	224(3.15),236(3.38), 248(3.63),262(3.76), 278(3.60)	70-2117-69
$C_{13}H_{16}O_3S$			
2-Thiophenecarboxylic acid, tetrahydro-3-hydroxy-3-phenyl-, ethyl ester	EtOH	243(4.02)	2-0009-69
$C_{13}H_{16}O_4$			
4-Indancarboxylic acid, 5,6,7,7a-tetrahydro-7a-methyl-1,5-dioxo-, ethyl ester	EtOH	285(4.01)	104-1298-69
Ketone, 5,7-dihydroxy-2,2-dimethyl-6-chromanyl methyl	EtOH	294(4.29),332s(3.46)	2-0873-69
	EtOH-NaOH	303(4.26),380(3.67)	2-0873-69
Ketone, 2-ethyl-2,3-dihydro-4,6-dihydroxy-2-methyl-7-benzofuranyl methyl	EtOH	288(4.29),340(3.51)	2-0873-69
	EtOH-NaOH	324(4.48)	2-0873-69
$C_{13}H_{16}O_5$			
6-Chromancarboxylic acid, 7-hydroxy-5-methoxy-2,2-dimethyl-	EtOH	266(4.10),305(3.59)	39-0365-69C
$C_{13}H_{16}O_6$			
2,3-Furandicarboxylic acid, 4-acetyl-5-methyl-, diethyl ester	EtOH	274(4.09)	88-2053-69
$C_{13}H_{16}O_9$			
1-threo-Hex-2-enonic acid, 2,6-anhydro-1-deoxy-, methyl ester, triacetate	EtOH	243(3.88)	88-2383-69
$C_{13}H_{17}BrN_4O_4$			
7H-Pyrrolo[2,3-d]pyrimidine, 5-bromo-4-(dimethylamino)-7-β-D-ribofuranosyl-	pH 1	249(4.14),291(4.07)	4-0215-69
	pH 11	293(4.14)	4-0215-69
$C_{13}H_{17}Br_2NO_5S$			
Carbonic acid, thio-, O-ethyl ester, S-ester with ethyl 3,3-dibromo-α-isopropylidene-2-mercapto-4-oxo-1-azetidineacetate	MeOH	225(4.00)	39-2123-69C
$C_{13}H_{17}ClN_2O_4S_2$			
Bis[5-(dimethylamino)-2-thienyl]methylium perchlorate	CH_2Cl_2	309(4.23),353(3.40), 376(3.33),544s(4.69), 591(5.11)	48-0827-69
$C_{13}H_{17}FN_2NaO_8PS$			
Uridine, 2'-deoxy-5-fluoro-, 3'-acetate 5'-O-ester with S-ethylphosphorothioate, sodium salt	H_2O	265(3.92)	35-1522-69

Compound	Solvent	λ_{max}(log ϵ)	Ref.
$C_{13}H_{17}F_4NO_2$			
Benzaldehyde, 4-(dimethylamino)-2,3,5,6-tetrafluoro-, diethyl acetal	EtOH	268(4.04),322(2.76)	65-1607-69
$C_{13}H_{17}IN_2$			
4-Amino-2-benzyl-2,3-dihydro-1-methyl-pyridinium iodide	EtOH	345(4.40)	44-4154-69
$C_{13}H_{17}N$			
1-Naphthaleneethylamine, 3,4-dihydro-N-methyl-, hydrochloride	n.s.g.	260(3.94)	39-0217-69C
Spiro[cyclopentane-1,3'(2'H)-quinoline], 1',4'-dihydro-	EtOH	250(3.76),302(3.18)	94-1290-69
$C_{13}H_{17}NO$			
Benzene, 1,2,4,5-tetramethyl-3-(2-nitro-sopropenyl)-	n.s.g.	736(1.73)	48-0260-69
2-Naphthonitrile, 1,2,3,4,4a,5,6,7-octahydro-4a,8-dimethyl-7-oxo-	n.s.g.	245(4.12)	44-1949-69
Quinoline, 1,2-dihydro-1-methoxy-1,3,4-trimethyl-	EtOH	233(4.48),325(3.54)	2-0419-69
$C_{13}H_{17}NOS$			
4-Thiazolin-2-one, 4-(1-adamantyl)-	EtOH	217(3.82)	18-1617-69
$C_{13}H_{17}NOS_2$			
Alanine, N-acetyl-3-phenyldithio-, ethyl ester	MeOH	310(3.98)	5-0237-69I
	$CHCl_3$	455(1.58)	5-0237-69I
2-Benzothiazolinethione, 3-(2-butoxy-ethyl)-	MeOH	229(4.15),240(4.10), 278(3.36),326(4.41)	4-0163-69
$C_{13}H_{17}NO_2$			
Acetamide, 2-(o-acetonylphenyl)-N,N-dimethyl-	EtOH	253(2.45),260(2.45), 266(2.37),282(2.14)	33-1030-69
Acetoacetamide, N,N-dimethyl-2-o-tolyl-	EtOH	265(2.62),271(2.58), 284s(2.28)	33-1030-69
	EtOH-NaOEt	292(3.33)	33-1030-69
4,5-Acridinediol, 1,2,3,4,5,6,7,8-octa-hydro-	EtOH	276(4.33)	103-0226-69
Bicyclo[4.2.2]dec-2-ene-3,4-dicarbox-imide, N-methyl-	MeOH	229(3.44)	108-0435-69
Bicyclo[4.2.2]dec-3-ene-3,4-dicarbox-imide, N-methyl-	MeOH	225(3.37),237s(3.20)	108-0435-69
Cycloheptazepine-N-carboxylic acid, ethyl ester	hexane	214(4.29),299(3.13)	44-2879-69
	EtOH	212(4.29),290(3.18)	44-2879-69
$C_{13}H_{17}NO_2S_2$			
Acetaldehyde, (2-benzothiazolylthio)-, diethyl acetal	EtOH	225(4.30),282(4.05), 291(4.01),301(3.98)	95-0469-69
$C_{13}H_{17}NO_3$			
4-Acridinol, 1,2,3,4,5,6,7,8-octahydro-, N-oxide	EtOH	265(3.48)	103-0226-69
Carbamic acid, (o-hydroxystyryl)-, tert-butyl ester	EtOH	273(4.27),282(4.19), 305(4.04),313(4.05)	44-2667-69
$C_{13}H_{17}NO_4$			
3-Furanglyoxylamide, 2-formyl-N,N-di-propyl-	pH 8	303(4.44)	35-7187-69

Compound	Solvent	λ_{max}(log ε)	Ref.
Isoindole-1,3-dicarboxylic acid, 4,5,6,7-tetrahydro-2-methyl-, dimethyl ester	EtOH	221(4.17),287(4.33)	32-1115-69
Phthalic acid, 3-(isopropylamino)-, dimethyl ester	EtOH	217(4.23),261(3.74), 361(3.54)	16-0241-69
2,5-Pyridinedicarboxylic acid, 3,6-dimethyl-, diethyl ester	MeOH	227(3.91),283(3.66)	77-0140-69
$C_{13}H_{17}NO_5$			
D-Allitol, 2,5-anhydro-1-deoxy-1-(salicylideneamino)-	H_2O	216(4.24),257(4.05), 328(3.50),375(3.41)	73-1684-69
Glucosamine, benzylidene-	neutral	248(3.54)	94-0770-69
	alkaline	362(2.84)	94-0770-69
$C_{13}H_{17}NO_6$			
Glucosamine, o-hydroxybenzylidene-	neutral	256(4.09),317(3.59)	94-0770-69
	alkaline	382(3.76)	94-0770-69
$C_{13}H_{17}NO_7$			
Pyruvic acid, (2,3,4,6-tetramethoxyphenyl)-, 2-oxime	EtOH	282(3.33)	39-0887-69C
$C_{13}H_{17}NO_8$			
2-Pyrroline-2,3,4,5-tetracarboxylic acid, 1-methyl-, tetramethyl ester	MeOH	302(4.01)	24-2346-69
$C_{13}H_{17}NO_9$			
1-Butene-1,2,3,4-tetracarboxylic acid, 1-(methylamino)-4-oxo-, tetramethyl ester	MeOH	296(4.14)	24-2346-69
4-Isoxazoline-3,4,5-tricarboxylic acid, 3-(carboxymethyl)-2-methyl-, tetramethyl ester	MeOH	284(3.38)	24-2346-69
$C_{13}H_{17}N_3$			
Phthalazine, 1-[3-(dimethylamino)-propyl]-	EtOH	216(4.8),264(3.7)	33-1376-69
$C_{13}H_{17}N_3O$			
1H-Benzotriazole, 5,6-dimethyl-1-(tetrahydro-2H-pyran-2-yl)-	EtOH	261(3.86),268(3.84), 287(3.67)	4-0005-69
Formanilide, 1-(cyclohexylazo)-	MeOH	206(4.37),228(4.08), 290(3.69)	5-0001-69E
Morpholine, 4-(1-phenyl-2-pyrazolin-3-yl)-	EtOH	287(4.16)	22-1683-69
$C_{13}H_{17}N_3OS$			
4H-Cyclopenta[b]thiophene-3-carboxylic acid, 2-amino-5,6-dihydro-, cyclopentylidenehydrazide	MeOH	228(4.21),254(4.24), 327(3.90)	48-0402-69
$C_{13}H_{17}N_3O_3$			
Cyclohexanone, 2-methoxy-, (p-nitrophenylhydrazone), anti	MeOH	252(4.07),391(4.40)	24-3647-69
syn	MeOH	251(4.07),388(4.37)	24-3647-69
$C_{13}H_{17}N_3O_5$			
Hydroperoxide, 2-methoxy-1-[(p-nitrophenyl)azo]cyclohexyl-	MeOH	283(4.17)	24-3647-69

Compound	Solvent	λ max(log ε)	Ref.
$C_{13}H_{17}N_3O_7S$			
Homocysteine, S-uridyl-	H_2O	260(4.21)	65-0434-69
$C_{13}H_{17}N_5O_3$			
Cyclopentanemethanol, 3-(6-amino-9H-purin-9-yl)-4-hydroxy-, 4-acetate	pH 1	258(4.15)	88-2231-69
	pH 7, 13	262(4.17)	88-2231-69
$C_{13}H_{18}Br_2O_3$			
2,4-Cycloheptadien-1-one, 2,6-dibromo-5-tert-butyl-7-ethoxy-7-hydroxy-	EtOH	240(3.83),348(3.57)	18-3277-69
$C_{13}H_{18}ClN$			
1-Naphthaleneethylamine, 3,4-dihydro-N-methyl-, hydrochloride	n.s.g.	260(3.94)	39-0217-69C
$C_{13}H_{18}NOSb$			
Antimony, tetramethyl(8-quinolinolato)-	benzene	325s(3.2),338(3.3), 380(3.3)	101-0071-69B
	EtOH	320(3.2),370s(2.8)	101-0071-69B
	$CHCl_3$	320(3.2),338(3.2), 380(3.1)	101-0071-69B
$C_{13}H_{18}N_2$			
3-Pyrazoline, 4-ethyl-1,2-dimethyl-3-phenyl-	EtOH	222(3.96),282(3.54)	22-3316-69
$C_{13}H_{18}N_2O_2$			
Acetamide, N-(1,1-dimethyl-3-nicotinoylpropyl)-	EtOH	229(3.99),264s(3.55), 266(3.57),273s(3.48)	36-0860-69
	EtOH-HCl	225(3.67),262s(3.65), 265(3.67),270s(3.60)	36-0860-69
5,8-Ethano-1H-pyrazolo[1,2-a]pyridazine-1,3(2H)-dione, 2,2-diethyl-5,8-dihydro-	C_6H_{12}	268(3.59)	44-3181-69
	EtOH	256(3.65)	44-3181-69
Heptamethylenimine, N-(p-nitrophenyl)-	DMSO	390(3.36)	39-0544-69B
Nicotine, 1'-demethyl-1'-hydroxy-5',5'-dimethyl-, acetate	EtOH	256(3.72),262(3.44), 268(3.41),299(3.04)	36-0860-69
	EtOH-HCl	259(3.72)	36-0860-69
$C_{13}H_{18}N_2O_3$			
L-Leucinamide, N-3-(2-furyl)acryloyl-	pH 8	305(4.43)	35-7187-69
2H-Pyrano[2,3-d]pyrimidine-2,4(3H)-dione, 3-cyclohexyl-1,5,6,7-tetrahydro-	75% dioxan	262(4.02)	95-0266-69
	+NaOH	277(4.05)	95-0266-69
$C_{13}H_{18}N_2O_4S$			
Thymine, 1-(2-deoxy-3,5-S,O-isopropylidene-3-thio-β-D-threo-pentofuranosyl)-	pH 6.34	212(3.97),268(4.00)	44-1020-69
	pH 12	267(3.90)	44-1020-69
$C_{13}H_{18}N_2O_5$			
4(1H)-Pyrimidinone, 1-(2,3-O-isopropylidene-β-D-ribofuranosyl)-2-methyl-	MeOH	242(4.20)	35-7752-69
$C_{13}H_{18}N_2O_6$			
D-Allitol, 2,5-anhydro-1-deoxy-1-(3,4-dihydro-2,4-dioxo-1(2H)-pyrimidinyl)-3,4-O-isopropylidene-	EtOH	211(3.73),266(3.92)	73-1684-69
$C_{13}H_{18}N_2S$			
Thiazole, 4-(1-adamantyl)-2-amino-	EtOH	253(3.76)	18-1617-69

Compound	Solvent	$\lambda_{max}(\log \epsilon)$	Ref.
$C_{13}H_{18}N_4O_6$			
Theophylline, 7-(2-deoxy-β-D-arabino-hexopyranosyl)-	n.s.g.	274(3.94)	22-3927-69
Theophylline, 7-(3-deoxy-β-D-xylo-hexopyranosyl)-	n.s.g.	274(3.89)	22-3927-69
$C_{13}H_{18}N_6O_3$			
Adenosine, 5'-amino-5'-deoxy-2',3'-O-isopropylidene-	pH 1	256(4.20)	87-0658-69
	pH 11	259(4.20)	87-0658-69
$C_{13}H_{18}N_6O_4$			
Adenosine, 5'-(carboxyamino)-2',5'-dideoxy-, ethyl ester	pH 1	256(4.17)	87-0658-69
	pH 11	259(4.17)	87-0658-69
$C_{13}H_{18}N_6O_5S$			
Adenosine, 5'-S-(2-amino-2-carboxyethyl)-5'-thio-	H_2O	260(4.15)	65-0434-69
$C_{13}H_{18}O$			
Inden-2(4H)-one, 5,6,7,7a-tetrahydro-1-isopropylidene-7aα-methyl-	MeOH	255(4.24)	23-4299-69
$C_{13}H_{18}OS$			
Benzo[c]thiophen-4(5H)-one, 3-tert-butyl-6,7-dihydro-1-methyl-	EtOH	219(4.01),269(4.05), 305(3.40)	22-0991-69
$C_{13}H_{18}OS_2$			
2-Bornanone, 3-(1,3-dithiolan-2-ylidene)-	dioxan	270(3.60),322(4.29), 345s(2.88)	28-0186-69A
$C_{13}H_{18}O_2$			
3-Buten-2-one, 4-(2,4,4,6-tetramethyl-4H-pyran-3-yl)-	MeOH	216(4.52),336(4.70)	44-3169-69
as-Indacene-4-carboxylic acid, 1,2,3,-3a,4,5,5a,6,7,8-decahydro-, cis-cis	EtOH	209(3.91)	22-4068-69
as-Indacene-4-carboxylic acid, 1,2,3,-4,5a,6,7,8,8a-decahydro-, cis-cis	EtOH	209(3.54)	22-4068-69
cis-trans	EtOH	207(3.62)	22-4068-69
1(4H)-Naphthalenone, 4a,5,6,7-tetrahydro-3-methoxy-4a,5-dimethyl-	EtOH	274(4.34)	24-2697-69
2(1H)-Naphthalenone, 6,7,8,8a-tetrahydro-4-methoxy-8,8a-dimethyl-	EtOH	283(4.27)	24-2697-69
Propiophenone, 2'-hydroxy-3'-isopropyl-2-methyl-	hexane	256(4.0),263s(4.0), 405(3.6)	18-0960-69
$C_{13}H_{18}O_2S$			
2-Bornanone, 3-(1,3-oxathiolan-2-ylidene)-, (+)-	dioxan	290(4.22),345s(2.40)	28-0186-69A
$C_{13}H_{18}O_3$			
2-Adamantanone, 4-(hydroxymethyl)-, acetate	EtOH	286(1.42)	78-5601-69
1,3-Cyclohexanedione, 5,5-dimethyl-2-(2-oxocyclopentyl)-	EtOH	291(4.00)	104-1196-69
	N NaOH	290(4.32)	104-1196-69
2-Cyclohexen-1-one, 3-ethoxy-2-(2-oxocyclopentyl)-	EtOH	266(4.15)	104-1196-69
2-Cyclohexen-1-one, 3-hydroxy-5,5-dimethyl-2-(3-oxocyclopentyl)-	EtOH	272(4.04)	104-1566-69
	2N KOH	292(4.43)	104-1566-69
4a(2H)-Naphthalenecarboxylic acid, 1,3,4,7,8,8a-hexahydro-7-oxo-, ethyl ester, trans	EtOH	222(3.98)	78-5281-69

Compound	Solvent	λ_{max}(log ε)	Ref.
$C_{13}H_{18}O_4$			
Acetophenone, 2',4',6'-trihydroxy-3-(3-methylbutyl)-	EtOH	294(4.26),345s(3.47)	2-0873-69
	EtOH-NaOH	328(4.39)	2-0873-69
Benzaldehyde, 2-(allyloxy)-3-methoxy-, dimethyl acetal	EtOH	276(3.36)	12-1011-69
m-Dioxane-5-methanol, 5-ethyl-2-[2-(2-furyl)vinyl]-	30% EtOH	268(4.22)	103-0278-69
1α-Naphthoic acid, 1,2,3,4,5,8-hexahydro-2α-hydroxy-7-methoxy-1β-methyl-	EtOH	206(3.69)	44-0126-69
Propiophenone, 3-ethoxy-2',4'-dimethoxy-	EtOH	228(4.15),267(4.09), 302(3.91)	23-1529-69
$C_{13}H_{18}O_5$			
Butyric acid, 4-[2-(3-carboxyethyl)-3-oxo-1-cyclohexen-1-yl]-	EtOH	244(4.17)	78-2661-69
1-Cyclopentene-1-heptanoic acid, 2-formyl-3-hydroxy-5-oxo-	MeOH	228(4.00)	88-1615-69
2,4-Dioxabicyclo[3.2.0]hept-6-ene-6-carboxylic acid, 5,7-dimethyl-3-(1-methyl-3-oxobutyl)-	EtOH	213(4.00)	35-4739-69
2,6-Heptanedione, 3-acetyl-3-(hydroxymethyl)-5-methylene-, acetate	EtOH	220(3.90),298(2.01)	39-0879-69C
2,4,10-Trioxatricyclo[4.4.0.0^{3,8}]decane, 6,8-diacetyl-1,3-dimethyl-	EtOH	288(1.57)	39-0879-69C
$C_{13}H_{18}O_6$			
Acetophenone, 2-ethoxy-6'-hydroxy-2',3',4'-trimethoxy-	EtOH	235(3.75),283(4.07), 334(3.68)	44-1460-69
2-Oxabicyclo[2.2.2]oct-5-ene-6-carboxylic acid, 3,8-dioxo-, methyl ester, 8-(diethyl acetal)	$CHCl_3$	240(3.89)	24-2835-69
$C_{13}H_{18}S$			
Thioxanthene, 1,2,3,4,5,6,7,8-octahydro-	hexane	234(3.82)	103-0172-69
$C_{13}H_{19}BrN_2O_3$			
4-(5-Morpholinofurfurylidene)morpholinium bromide	n.s.g.	393(4.42)	103-0434-69
$C_{13}H_{19}F_3N_2O_2$			
2-Pyrrolin-4-one, 2-(diethylamino)-3,5,5-trimethyl-1-(trifluoroacetyl)-	MeOH	266(4.06),305(4.07)	78-5721-69
$C_{13}H_{19}IN_2$			
1,1,2,3-Tetramethyl-4-phenyl-3-pyrazolinium iodide	EtOH	221(4.26),270(4.06)	22-3316-69
1,1,2,4-Tetramethyl-3-phenyl-3-pyrazolinium iodide	EtOH	221(4.42),252(3.91)	22-3316-69
$C_{13}H_{19}N$			
$\Delta^{1,\alpha}$-Cyclohexaneacetonitrile, 2-methyl-2-(2-methylpropenyl)-	MeOH	219(4.04)	23-4299-69
Piperidine, 4-benzyl-N-methyl-	EtOH	260(3.80)	33-2004-69
$C_{13}H_{19}NO$			
Cyclopenta[b]pyrrol-2(1H)-one, 3,3a,4,6a-tetrahydro-3,3a,5,6,6a-pentamethyl-4-methylene-	EtOH	237(4.12)	35-6107-69
4a(2H)-Naphthalenecarbonitrile, octahydro-4,8a-dimethyl-3-oxo-, cis	n.s.g.	286(1.30)	44-1942-69

Compound	Solvent	λ_{max}(log ε)	Ref.
4a(2H)-Naphthalenecarbonitrile, octa-hydro-4,8a-dimethyl-3-oxo-, trans	n.s.g.	287(1.26)	44-1942-69
Piperidine, 4-(1-hydroxyphenethyl)-	EtOH	220(3.11),261(2.79)	39-2134-69C
$C_{13}H_{19}NO_2$			
Benzoic acid, 2-(diethylamino)ethyl ester	dioxan	257(2.94),267(2.98), 273(3.03),280(2.93)	65-1861-69
Benzoic acid, p-[3-(dimethylamino)-propyl]-, methyl ester	heptane	237(4.19)	44-0077-69
	MeOH	238(4.22)	44-0077-69
Hexanohydroxamic acid, N-p-tolyl-	EtOH	252(4.03)	42-0831-69
Hydroquinone, 2-methyl-5-(piperidino-methyl)-	EtOH	298(3.61)	39-1245-69C
$C_{13}H_{19}NO_3$			
Benzylidenimine, N-butyl-2-hydroxy-5-(hydroxymethyl)-3-methoxy-, copper complex	$CHCl_3$	280(4.48),376(3.88), 635(2.18)	65-0401-69
Codonopsinine	EtOH	228(4.14),277(3.22), 284s(3.14)	105-0530-69
$C_{13}H_{19}NO_5$			
Benzoic acid, 4-hydroxy-3,5-dimethoxy-, 4-aminobutyl ester (leonuramine)	MeOH	222(4.32),279(4.04)	78-5155-69
	MeOH-NaOH	239(4.15),327(4.30)	78-5155-69
$C_{13}H_{19}NO_6$			
2-Propene-1,1,2-tricarboxylic acid, 3-(tert-butylimino)-, trimethyl ester	ether	247(4.07)	24-1656-69
$C_{13}H_{19}NS_2$			
Spirocycloheptane-1,2'(4'H)-cyclopenta-[d][1,3]thiazine-4'-thione, 1',5',6',7'-tetrahydro-	EtOH	338(3.80),408(4.32)	44-0730-69
$C_{13}H_{19}N_3$			
Phthalazine, 1-[3-(dimethylamino)pro-pyl]-1,2-dihydro-	EtOH	219(4.30),305(3.45)	33-1376-69
$C_{13}H_{19}N_3O$			
Cyclohexanone, 2-(4,5,6,7-tetrahydro-benzimidazolyl)-, oxime	MeOH	224(3.75)	48-0746-69
$C_{13}H_{19}N_3O_2$			
Cyclopentanone, 2-(2-ethyl-5,6-dihydro-1(4H)-cyclopentimidazolyl)-, oxime, 3-oxide	MeOH	226(3.85)	48-0746-69
$C_{13}H_{19}N_3O_4$			
Pyrido[2,3-d]pyrimidine-6-carboxalde-hyde, 5,6,7,8-tetrahydro-6-(methoxy-methyl)-2-methyl-7-oxo-, 6-(dimethyl acetal)	EtOH	242(3.75),275(4.03)	88-1825-69
	EtOH-HCl	280(--)	88-1825-69
	EtOH-NaOH	280(--),301(--)	88-1825-69
$C_{13}H_{19}N_4O_2S$			
Thiamine, (hydroxymethyl)-, bromide, hydrobromide	EtOH	239(4.10),265(4.19)	94-0343-69
$C_{13}H_{19}N_5O$			
Butyramide, N-(6-methyl-8-propyl-3-s-triazolo[4,3-a]pyrazin-3-yl)-	BuOH	224(4.28),304(3.59)	39-1593-69C

Compound	Solvent	λ_{max}(log ε)	Ref.
$C_{13}H_{19}N_5O_5$			
Adenine, 9-[3-deoxy-3-(2-hydroxyethyl)-β-D-allofuranosyl]-	pH 1	208(4.28),257(4.12)	44-1029-69
	H_2O	208(4.28),260(4.14)	44-1029-69
	pH 13	211(4.14),260(4.13)	44-1029-69
$C_{13}H_{19}N_9$			
1H-1,2,4-Triazole, 1,1',1"-methylidynetris[3,5-dimethyl-	n.s.g.	217(3.37)	39-2251-69C
$C_{13}H_{19}O_4P$			
Phosphonic acid, benzoyl-, diisopropyl ester	C_6H_{12}	258(4.11),379(1.91)	18-0821-69
$C_{13}H_{20}$			
Bicyclo[4.2.0]octa-2,7-diene, 1,3,6,7,8-pentamethyl-	hexane	240(2.47)(end abs.)	24-0275-69
1-Butene, 2-methyl-4-(1,2,3-trimethyl-4-methylene-2-cyclobuten-1-yl)-	hexane	234(4.05)	24-0275-69
1,4-Cyclohexadiene, 1,2,3,3,4,5-hexamethyl-6-methylene-	isooctane	254(4.29)	88-4929-69
1-Pentene, 5-(1,2,3-trimethyl-4-methylene-2-cyclobuten-1-yl)-	hexane	235(4.14)	24-0275-69
$C_{13}H_{20}BF_4NS$			
[1-(Ethylthio)butylidene]methylphenylammonium tetrafluoroborate	MeCN	232(3.70)	35-0737-69
$C_{13}H_{20}Br_2N_4O_2S$			
Thiamine, (hydroxymethyl)-, HBr salt	EtOH	239(4.10),265(4.19)	94-0343-69
$C_{13}H_{20}ClNO_4$			
2,3,5,5a,6,7,8,9-Octahydro-4-methyl-1H-cyclopenta[c]quinolizinium perchlorate	$CHCl_3$	306(3.88)	78-4161-69
$C_{13}H_{20}ClNO_5$			
2,3,5,5a,6,7,8,9-Octahydro-4-methoxy-1H-cyclopenta[c]quinolizinium perchlorate	EtOH	334(4.07)	44-0698-69
$C_{13}H_{20}Cl_2N_4O_2S$			
Thiamine, (hydroxymethyl)-, HCl salt	EtOH	239(4.12),265(4.20)	94-0343-69
$C_{13}H_{20}INO$			
2,3,5,5a,6,7,8,9-Octahydro-4-methoxy-1H-cyclopenta[c]quinolizinium iodide	EtOH	334(4.10)	44-0698-69
$C_{13}H_{20}N_2O_2$			
Benzoic acid, p-amino-, 2-(diethylamino)ethyl ester	EtOH	221(4.08),295(4.36)	9-0254-69
5,8-Ethano-1H-pyrazolo[1,2-a]pyridazine-1,3(2H)-dione, 2,2-diethyltetrahydro-	C_6H_{12}	270(3.44)	44-3181-69
	EtOH	264(3.56)	44-3181-69
1H-Pyrazolo[1,2-a]pyridazine-1,3(2H)-dione, 2,2-diethyl-5,8-dihydro-6,7-dimethyl-	C_6H_{12}	256(3.31)	44-3181-69
	EtOH	257(3.35)	44-3181-69
$C_{13}H_{20}N_2O_3$			
2,4-Piperidinedione, 3-carbamoyl-5-cyclohexyl-1-methyl-	pH 1	232(4.11),275(3.91)	7-0658-69
	pH 13	242(4.09),278(4.03)	7-0658-69
	MeOH	238(--),272(--)	7-0658-69

Compound	Solvent	λ_{max}(log ε)	Ref.
$C_{13}H_{20}N_2O_3S_2$			
Acetone, 1,1-bis[(N-ethyl-3-acrylamido)-mercapto]-, cis	90% EtOH	282(4.47)	12-0765-69
Acetone, 1,3-bis[(N-ethyl-3-acrylamido)-mercapto]-, cis	90% EtOH	277(4.43)	12-0765-69
$C_{13}H_{20}N_2O_5S$			
Uridine, 5-(butylthio)-2'-deoxy-	EtOH	235(3.74),292(3.69)	44-3806-69
$C_{13}H_{20}N_4O_3$			
Malonaldehydonitrile, [(4-amino-2-methyl-5-pyrimidinyl)methyl](methoxymethyl)-, dimethyl acetal	EtOH	235(3.84),278(3.53)	88-1825-69
Pyrido[2,3-d]pyrimidine-6-carboxaldehyde, 7-amino-5,6-dihydro-6-(methoxymethyl)-2-methyl-, dimethyl acetal	EtOH	221(4.12),301(4.15)	88-1825-69
	EtOH-HCl	290(--),307(--)	88-1825-69
$C_{13}H_{20}N_6O$			
1-Propanol, 3-[(6-piperidinopurin-8-yl)-amino]-	pH 1	310(4.31)	65-2125-69
	EtOH	296(4.31)	65-2125-69
$C_{13}H_{20}O$			
6H-Benzocyclohepten-6-one, 1,2,3,4,-7,8,9,9a-octahydro-7,9a-dimethyl-	EtOH	243(3.88)	35-3676-69
4H-1-Benzopyran, 5,6,7,8-tetrahydro-2,4,4,7-tetramethyl-	C_6H_{12}	221(3.66),230(3.44),275(2.20),286(2.15),303(1.56),318(1.32)	44-1465-69
	MeOH	211(3.30),229s(3.14)	44-0380-69
Camphor, isopropylidene-, (+)-	MeOH	248(3.96),332(1.85)	18-2593-69
Cyclopentadienone, 2,4-di-tert-butyl-	isooctane	210(3.86),390(2.43)	35-6785-69
	EtOH	217(3.87),395(2.36)	35-6785-69
as-Indacene-4-methanol, 1,2,3,3a,4,5,-5a,6,7,8-decahydro-, cis-cis	EtOH	209(3.93)	22-4068-69
cis-trans	EtOH	209(3.94)	22-4068-69
as-Indacene-4-methanol, 1,2,3,4,5,5a,-6,7,8,8a-decahydro-, cis-cis	EtOH	207(3.85)	22-4068-69
cis-trans	EtOH	207(3.85)	22-4068-69
2(1H)-Naphthalenone, 3,4,4a,5,6,7-hexahydro-1,1,4a-trimethyl-	isooctane	228(1.54),296s(1.47),305s(1.48)	44-0450-69
2(1H)-Naphthalenone, 4a,5,6,7,8,8a-hexahydro-4a,8,8-trimethyl-, cis	MeOH	235(3.89)	77-1396-69
trans	MeOH	228(3.92)	77-1396-69
$C_{13}H_{20}OS$			
Sulfoxide, 2-benzyl-3-methylbutyl methyl	EtOH	221(3.86)	39-0581-69B
$C_{13}H_{20}OS_2$			
2-Bornanone, 3-carbonyl-, 3-(dimethyl mercaptole)	dioxan	323(4.11),375s(2.80)	28-0186-69A
$C_{13}H_{20}O_2$			
Camphor oxide, isopropylidene-, (+)-	MeOH	310(1.70)	18-2593-69
Camphor oxide, isopropylidene-, (-)-	MeOH	310(1.69)	18-2593-69
1-Oxaspiro[4.5]dec-6-en-8-one, 2,6,10,10-tetramethyl-, cis	EtOH	234(3.95)	88-1995-69
trans	EtOH	234(3.99)	88-1995-69
Theaspirone	EtOH	234(4.20)	88-1995-69

Compound	Solvent	$\lambda_{max}(\log \epsilon)$	Ref.
$C_{13}H_{20}O_3$			
3-Buten-2-one, 4-(2,4-dihydroxy-2,6,6-trimethylcyclohexylidene)-	EtOH	233(4.12)	77-0085-69
1-Cyclodecene-1-carboxylic acid, 2-methyl-10-oxo-, methyl ester	EtOH	205(3.95),208(3.94), 221(3.88)	94-0629-69
Vomifoliol	EtOH	237(4.05)	88-1173-69
$C_{13}H_{20}O_4$			
Cyclopentanecarboxylic acid, 4-isovaleryl-2-methyl-3-oxo-, methyl ester	pH 13	286(3.95),309(4.38)	39-1845-69C
Malonic acid, (1-methyl-4-penten-1-ylidene)-, diethyl ester	EtOH	222(4.0)	34-0112-69
$C_{13}H_{20}O_6$			
3,7-Dioxabicyclo[3.3.1]nonane-2,6-diol, 1,5-diacetyl-2,6-dimethyl-, (dihydrate)	EtOH	290(1.69)	39-0879-69C
$C_{13}H_{21}N$			
11-Azabicyclo[4.4.1]undec-1-ene-, 11-methyl-3-vinyl-	isooctane	248(3.61)	35-3616-69
2,4-Cyclohexadiene-1-methylamine, N,N,1,3,5-pentamethyl-6-methylene-	EtOH	314(3.8)	44-2393-69
$C_{13}H_{21}NO_2$			
Alloanodendrine	MeOH	212(2.81)(end abs.)	88-4065-69
Anodendrine	MeOH	212(2.81)(end abs.)	88-4065-69
$C_{13}H_{21}NO_3$			
2,4-Pentadienoic acid, 2-acetyl-5-(diethylamino)-, ethyl ester	EtOH	240(3.84),265(3.90), 396(4.76)	70-0363-69
$C_{13}H_{21}N_3O$			
Bicyclo[3.3.1]nonan-2-one, 9-isopropylidene-, semicarbazone	isooctane	293(2.11),302(2.11), 311s(2.00)	44-0450-69
3H-2,8a-Methanonaphthalen-3-one, octahydro-2-methyl-, semicarbazone	EtOH	228(4.17)	78-5267-69
$C_{13}H_{21}N_3OS_3$			
Spiro[furo[2,3-d]thiazole-2(3H),4'-imidazolidine]-2',5'-dithione, 1',3'-diethyltetrahydro-3,3a-dimethyl-	EtOH	323(4.16)	94-0910-69
$C_{13}H_{22}N_2O_2$			
3-Penten-2-one, 4,4'-(trimethylenediimino)di-	MeOH	318(3.85)	39-2044-69C
1H-Pyrazolo[1,2-a]pyridazine-1,3(2H)-dione, 2,2-diethyltetrahydro-6,7-dimethyl-, cis-endo	C_6H_{12}	256(3.40)	44-3181-69
	EtOH	258(3.41)	44-3181-69
Uracil, 5-butyl-6-pentyl-	pH 2	269(3.82)	56-0499-69
	pH 12	271(3.82)	56-0499-69
$C_{13}H_{22}N_2O_3$			
2,4-Piperidinedione, 3-carbamoyl-1-methyl-5,5-dipropyl-	pH 1	237(4.13),274(3.93)	7-0658-69
	pH 13	245(4.12),278(4.06)	7-0658-69
	MeOH	237(--),270(--)	7-0658-69
3,5-Pyrazolidinedione, 4,4-diethyl-1-(2-ethylbutyryl)-	EtOH	260(3.54)	44-3181-69

Compound	Solvent	λ_{max}(log ε)	Ref.
$C_{13}H_{22}N_2SSi_2$			
2-Benzimidazolinethione, 1,3-bis(trimethylsilyl)-	Bu_2O	226(4.33),263(3.73), 316(4.46)	48-0997-69
$C_{13}H_{22}N_4O_2$			
1,2,7,8-Tetraazaspiro[4.4]nona-1,7-diene-6-carboxylic acid, 3,3,9,9-tetramethyl-, ethyl ester	EtOH	327(2.60)	39-2443-69C
$C_{13}H_{22}O$			
Cyclododecanone, 2-methylene-	n.s.g.	227(3.63),322(1.57)	97-0303-69
Cyclohexene, 1-acetyl-2-pentyl-	MeOH	248(3.58)	78-0223-69
2-Cyclopenten-1-one, 3-ethyl-2-hexyl-	EtOH	234(4.18)	104-1993-69
$C_{13}H_{22}OS$			
Cyclohexanone, 6-[(butylthio)methylene]-2,3-dimethyl-	MeOH	309(4.19)	23-0831-69
$C_{13}H_{22}OSi_2$			
Silane, trimethyl[p-(trimethylsilyl)-benzoyl]-	hexane	258(4.27),285s(3.40), 427(2.04)	35-0355-69
$C_{13}H_{22}O_2$			
2-Cyclopenten-1-one, 2,4-di-tert-butyl-4-hydroxy-	EtOH	224(3.92)	88-1673-69
1,2-Cyclotridecanedione (may have an impurity)	hexane	267(3.22),446(1.3)	23-3266-69
$C_{13}H_{22}O_3$			
2-Cyclohexen-1-one, 4-hydroxy-4-(3-hydroxybutyl)-3,5,5-trimethyl-	EtOH	243(4.00)	88-1803-69
$C_{13}H_{22}S$			
Thiophene, 3,5-di-tert-butyl-2-methyl-	EtOH	235(3.98)	22-0991-69
$C_{13}H_{23}ClN_2O_4$			
1-(3-Piperidinoallylidene)piperidinium perchlorate	EtOH	316(4.77)	24-2609-69
$C_{13}H_{23}IN_2O$			
[5-(Diethylamino)furfurylidene]diethylammonium iodide	n.s.g.	400(4.32)	103-0434-69
$C_{13}H_{23}NO$			
Aziridinone, 1-tert-butyl-3-(1-methylcyclohexyl)-	hexane	247(2.16)	35-1176-69
Nitrone, N-(1-cyclohexylcyclohexyl)-	EtOH	242(3.81)	39-1073-69C
$C_{13}H_{24}Si$			
Silane, trimethyl(2,6,6-trimethyl-2-hepten-4-ynyl)-	n.s.g.	240(4.36)	101-0291-69C
$C_{13}H_{24}Sn$			
Tin, triallylbutyl-	hexane	209(4.48)	44-3715-69
	MeCN	214(4.44)	44-3715-69
$C_{13}H_{26}NO_3P$			
Phosphonic acid, [2-(cyclohexylamino)-1-methylvinyl]-, diethyl ester	EtOH	244(4.30)	39-0460-69C

Compound	Solvent	λ_{max}(log ε)	Ref.
$C_{13}H_{26}N_2O_2$			
Crotonic acid, 3,4-bis(butylamino)-, methyl ester	EtOH	278(4.25)	78-3277-69
$C_{13}H_{27}NS_3$			
Peroxycarbamic acid, dibutyl-trithio-, butyl ester	EtOH	282(3.86)	34-0119-69
Peroxycarbamic acid, dibutyl-trithio-, tert-butyl ester	EtOH	246(3.85),284(3.87)	34-0119-69
$C_{13}H_{27}N_3$			
Diaziridine, 1,2-di-tert-butyl-3-(tert-butylimino)-	hexane	226s(2.95)	89-0448-69

Compound	Solvent	λ_{max}(log ϵ)	Ref.
$C_{14}HF_{10}NO_2$			
Ethylene, 2-nitro-1,1-bis(pentafluoro-phenyl)-	EtOH	202(4.22),211(4.21), 266(3.74)	65-1766-69
$C_{14}H_2F_{10}$			
Stilbene, decafluoro-, cis	EtOH	201(4.13),210(4.06), 260(4.05)	70-0134-69
Stilbene, decafluoro-, trans	EtOH	200(4.04),213(3.96), 283(4.36),292(4.35), 304(4.07)	70-0134-69
$C_{14}H_2F_{10}O$			
Acetophenone, 2',3',4',5',6'-penta-fluoro-2-(pentafluorophenyl)-	EtOH	196(4.45)	65-1774-69
$C_{14}H_4F_8$			
Stilbene, 2,2',3,3',4,4',5,5'-octa-fluoro-, trans	C_6H_{12}	293(4.27)	39-0211-69C
$C_{14}H_4N_2O_2S_2$			
Naphtho[2,3-b]-p-dithiin-2,3-dicarbo-nitrile, 5,10-dihydro-5,10-dioxo-	morpholine	562(3.69)	83-0075-69
	piperidine	554(3.65)	83-0075-69
$C_{14}H_6Br_2O_2$			
Pyracene-1,2-dione, 5,6-dibromo-	EtOH	225(4.52),231(4.53), 243s(4.27),324(3.89), 332(3.89)	35-0918-69
$C_{14}H_6Cl_4$			
Phenanthrene, 1,2,3,4-tetrachloro-	MeOH	256(4.54),267(4.74), 282(4.08),305(4.06), 317(4.11),350(2.78), 359(2.73)	39-1684-69C
$C_{14}H_6Cl_4O_5$			
Xanthen-9-one, 2,4,5,7-tetrachloro-1,3,6-trihydroxy-8-methyl- (thio-phanic acid)	MeOH	248(4.60),320(4.13), 360(4.22)	64-0750-69
$C_{14}H_6F_4$			
Phenanthrene, 1,2,3,4-tetrafluoro-	MeOH	236(4.66),244(4.83), 257(4.36),273(4.19), 283(4.20),295(4.21), 330(2.94)	39-1684-69C
$C_{14}H_6N_2O_4$			
Anthra[1,2-c][1,2,5]oxadiazole-6,11-di-one, 4-hydroxy-	EtOH	435(3.62)	104-1642-69
$C_{14}H_6N_4$			
Malononitrile, (p-phenylenedimethyli-dyne)di-	EtOH	204(4.05),209s(--), 226(3.95),335s(--), 349(4.56),365s(--)	23-4076-69
$C_{14}H_6N_4O$			
Acenaphtho[1,2-b]oxadiazolo[3,4-e]pyra-zine	dioxan	233(4.77),294(4.54), 304(4.61),370(3.15)	4-0769-69

Compound	Solvent	$\lambda_{max}(\log \epsilon)$	Ref.
$C_{14}H_6O_2$			
Pyracyloquinone	EtOH	230(4.47),247(4.19), 307(4.25),314(4.23), 346(3.85)	35-0918-69
$C_{14}H_7F_4N$			
Indole, 4,5,6,7-tetrafluoro-2-phenyl-	EtOH	236(4.26),296(4.28)	70-0609-69
$C_{14}H_8Br_2O_4$			
2,3-Biphenylenedione, 1,4-dibromo-6,7-dimethoxy-	EtOH	225(4.06),292(4.42), 322s(4.03),433(4.45)	39-0646-69C
$C_{14}H_8Br_8O_4$			
4H,8H-Cyclobuta[1,2-b:3,4-b']dipyran-4,8-dione, 3,4a,7,8a-tetrabromo-2,6-bis(dibromomethyl)-4a,4b,8a,8b-tetrahydro-4b,8b-dimethyl-	$CHCl_3$	306(4.08)	44-4052-69
$C_{14}H_8Cl_2N_2$			
3H-Indole, 5-chloro-3-(chloroimino)-2-phenyl-	EtOH	269(4.55),442(3.49)	22-2008-69
$C_{14}H_8Cl_2N_2O_2$			
Isosydnone, 4,5-bis(p-chlorophenyl)-	EtOH	229(4.19),320(3.77)	39-1185-69B
$C_{14}H_8Cl_2O$			
Anthrone, 1,8-dichloro-	n.s.g.	273(4.17)	44-3093-69
$C_{14}H_8Cl_2O_5$			
Xanthen-9-one, 2,7-dichloro-1,3,6-trihydroxy-8-methyl-	EtOH	247(4.38),267s(4.07), 322s(3.86),359(4.11)	11-0455-69A
at higher concentration	EtOH	321(4.08),357(4.04)	11-0455-69A
$C_{14}H_8Cl_4$			
Phenanthrene, 1,2,3,4-tetrachloro-9,10-dihydro-	MeOH	237(4.25),270(4.28), 313(3.52)	39-1684-69C
$C_{14}H_8F_2MoO_4$			
Benzoic acid, p-fluoro-, Mo(II) salt	EtOH	229(3.95+),420(3.95)	12-1571-69
$C_{14}H_8F_4$			
Phenanthrene, 1,2,3,4-tetrafluoro-9,10-dihydro-	MeOH	264(4.22),290(3.35)	39-1684-69C
$C_{14}H_8F_6O_4S_3$			
Sulfur diimide, bis(α,α,α-trifluoro-p-tolylsulfonyl)-	benzene	424(3.67)	104-0459-69
$C_{14}H_8N_2O$			
6H-Azepino[1,2-a]indole-11-carbonitrile, 6-oxo-	CH_2Cl_2	256(4.04),285(4.18), 312(4.11),415(3.59)	94-1294-69
Cyclobuta[4,5]furo[2,3-b]quinoline-9-carbonitrile, 2a,9b-dihydro-	CH_2Cl_2	260(4.24),304(3.68), 315(3.71),345(3.88), 359(3.89)	94-1294-69
2,7-Epoxy-1-benzazecine-8-carbonitrile	CH_2Cl_2	276(4.64),375(3.89)	94-1294-69
Oxepino[2,3-b]quinoline-6-carbonitrile	CH_2Cl_2	258(4.32),345(3.97), 367(3.95)	94-1294-69
$C_{14}H_8N_2O_2S_3$			
Isothiocyanic acid, sulfonyldi-p-phenylene ester	EtOH	288(4.55),305(4.68)	73-4005-69

Compound	Solvent	λ_{max}(log ε)	Ref.
$C_{14}H_8N_2O_6$			
Phenanthrene-9,10-diol, 2,7-dinitro-	dioxan	296(4.59),340s(4.05), 454(3.13)	24-2384-69
$C_{14}H_8N_2S_3$			
Isothiocyanic acid, thiodi-p-phenylene ester	EtOH	219(4.78),292(4.60), 300(4.57)	73-4005-69
$C_{14}H_8N_4S_2$			
Isothiocyanic acid, azodi-p-phenylene ester	heptane	240(3.82),364(4.15), 440(2.93)	73-3912-69
$C_{14}H_8N_6$			
1,3-Butadiene-1,1,3-tricarbonitrile, 2-amino-4-(1-benzimidazolyl)-	EtOH	202(4.5),240(3.75), 261(3.72),267(3.84), 273(3.89),278(3.50), 344(4.51)	44-2983-69
$C_{14}H_8N_6O_{10}$			
4,4'-Biphenyldicarboximide, 2,2',6,6'-tetranitro-	EtOH	217(4.75)	4-0533-69
$C_{14}H_8O_2$			
Anthracene peroxide	dioxan	270(2.9),278(2.8)	18-1377-69
Anthraquinone	EtOH	253(4.68),263s(4.25), 272(4.18),326(3.70), 384(2.15)	114-0109-69A
	dioxan	252(4.66),263s(4.27), 270s(4.13),324(3.67), 386(2.00)	114-0109-69A
	$CHCl_3$	254(4.71),262s(4.32), 274(4.23),327(3.71), 384(2.10)	114-0109-69A
	CCl_4	263(4.28),272s(4.26), 323(3.72),394(2.00), 411(1.92),439(1.67)	114-0109-69A
	pyridine	327(3.70),384(2.08)	114-0109-69A
Cyclopent[fg]acenaphthylene-1,2-dione, 5,6-dihydro-	EtOH	213(3.63),238(4.72), 246(4.68),318s(3.69), 332(3.81),354(3.83), 364s(3.75)	35-0918-69
$C_{14}H_8O_2S$			
1,3-Indandione, 2-(2-thienylmethylene)-	C_6H_{12}	234(4.28),241(4.23), 253(4.16),264(4.19), 305(3.78),378(4.49), 437s(3.10),500(2.38)	108-0099-69
	MeCN	236(4.32),246(4.24), 262(4.22),304(3.89), 382(4.52),500(2.50)	108-0099-69
$C_{14}H_8O_3$			
Anthraquinone, 1-hydroxy-	EtOH	219(4.34),253(4.39), 272s(4.18),331(3.39), 390s(3.72),406(3.77), 423(3.71)	114-0109-69A
	dioxan	255(4.40),267s(4.42), 329(3.44),392s(3.76), 405(3.79),420s(3.72)	114-0109-69A

Compound	Solvent	λ_{max}(log ε)	Ref.
Anthraquinone, 1-hydroxy- (cont.)	$CHCl_3$	254(4.37),258s(4.35), 271(4.28),278s(4.24), 334(3.40),392s(3.72), 408(3.77),423(3.70)	114-0109-69A
	pyridine	337(3.40),394s(3.72), 408(3.75),424s(3.68)	114-0109-69A
$C_{14}H_8O_4$			
Anthraquinone, 1,2-dihydroxy-	EtOH	234s(4.27),249(4.50), 260s(4.38),278s(4.19), 327s(3.53),438(3.75)	114-0109-69A
	dioxan	233(4.26),247(4.49), 278(4.29),328(3.52), 426(3.72)	114-0109-69A
	$CHCl_3$	249(4.46),254(4.45), 258(4.40),277(4.23), 331(3.52),414s(3.71), 425(3.72)	114-0109-69A
	pyridine	329(3.58),440(3.75)	114-0109-69A
$C_{14}H_8O_5$			
Xanthen-9-one, 4-hydroxy-2,3-(methylene-dioxy)-	EtOH	245(4.49),286(3.78), 319(4.07)	24-2414-69
$C_{14}H_8S$			
Acenaphtho[1,2-b]thiophene	C_6H_{12}	239(4.63),279(4.03), 290(4.10),316(3.84), 321(3.91),328(4.06), 335(4.00),342(4.13), 350(3.97),357(4.11), 395(3.06)	48-0614-69
Acenaphtho[1,2-c]thiophene	C_6H_{12}	273(3.35),282(3.28), 342(3.76),357(3.51)	97-0022-69
Acenaphtho[5,6-cd]thiopyran	benzene	515(2.95),590(2.91), 635(2.78)	77-1214-69
$C_{14}H_9Br$			
Phenanthrene, 4-bromo-	MeOH	256(4.78)	78-4339-69
$C_{14}H_9BrF_4O_2$			
5,8-Methano-5H-benzocyclohepten-9-ol, 10-bromo-1,2,3,4-tetrafluoro-8,9-dihydro-, acetate	EtOH	268(2.79)	65-2332-69
1,3-Methano-1H-cyclopropa[a]naphthalen-2-ol, 8-bromo-4,5,6,7-tetrafluoro-1a,2,3,7b-tetrahydro-, acetate	EtOH	268(2.85)	65-2332-69
isomer	EtOH	266(2.93)	65-2332-69
$C_{14}H_9BrO_4$			
2,3-Biphenylenedione, 1-bromo-6,7-di-methoxy-	EtOH	272(4.46),285(4.34), 320s(3.93)	39-0646-69C
$C_{14}H_9Br_7O_4$			
4H,8H-Cyclobuta[1,2-b:3,4-b']dipyran-4,8-dione, 3,4aβ,7-tribromo-2,6-bis-(dibromomethyl)-4a,4b,8aβ,8b-tetra-hydro-4bβ,8bβ-dimethyl-	CHCl	302(4.20)	44-4052-69
$C_{14}H_9Cl$			
Phenanthrene, 4-chloro-	MeOH	254(4.79)	78-4339-69

Compound	Solvent	$\lambda_{max}(\log \epsilon)$	Ref.
$C_{14}H_9ClF_4O_2$			
5,8-Methano-5H-benzocyclohepten-9-ol, 10-chloro-1,2,3,4-tetrafluoro-8,9-dihydro-, acetate	EtOH	268(2.87)	65-2326-69
5,9-Methano-5H-benzocyclohepten-6-ol, 10-chloro-1,2,3,4-tetrafluoro-6,9-dihydro-, acetate	EtOH	266(2.70)	65-2326-69
$C_{14}H_9ClN_2$			
Phthalazine, 1-(p-chlorophenyl)-	EtOH	219(4.8),281(4.0)	33-1376-69
$C_{14}H_9ClN_2O_2$			
Acridine, 1-chloro-2-methyl-6-nitro-	EtOH	242(4.88),299(4.69), 357(4.02),373(4.13)	103-0707-69
Isosydnone, 4-(p-chlorophenyl)-5-phenyl-	EtOH	223(4.13),315(3.87)	39-1185-69B
Isosydnone, 5-(p-chlorophenyl)-4-phenyl-	EtOH	230(4.16),320(3.89)	39-1185-69B
$C_{14}H_9ClOSe$			
Dibenzo[b,f]selenepin-10(11H)-one, 8-chloro-	MeOH	221s(4.25),234(4.16), 278(3.92),343(3.61)	73-3801-69
$C_{14}H_9ClO_6$			
2-(2-Hydroxy-1-cycloheptatrienylium)-benzoic acid lactone, perchlorate	MeCN	258s(4.14),302(4.03), 334(4.00)	49-0001-69
$C_{14}H_9ClSe$			
Dibenzo[b,f]selenepin, 2-chloro-	MeOH	219(4.56),270(4.34), 290(3.78)	73-3801-69
$C_{14}H_9Cl_5$			
Ethane, 1,1,1-trichloro-2,2-bis(p-chlorophenyl)-	hexane	240(3.93),270(2.93)	57-1083-69B
$C_{14}H_9F$			
Anthracene, 9-fluoro-	$CHCl_3$	258(5.17),325(3.43), 348(3.70),366(3.88), 386(3.79)	22-3538-69
$C_{14}H_9I$			
Phenanthrene, 4-iodo-	MeOH	260(4.51)	78-4339-69
$C_{14}H_9N$			
Fluoren-9-yl isocyanide	dioxan	224(4.34),231(4.18), 267(4.28),279(4.15), 292(3.44),302(3.16)	89-0772-69
Spiro[2H-azirine-2,9'-fluorene]	dioxan	234(4.31),260(4.19), 270(4.43),284(4.10)	89-0772-69
$C_{14}H_9NO_2$			
Anthraquinone, monooxime	$CHCl_3$	280(4.18),332(3.61)	22-3538-69
Anthrone, 4-hydroxy-10-imino-	EtOH	221(4.23),250(4.34), 276(4.15),320(3.65), 409(3.38),519(3.75)	114-0109-69A
	dioxan	226(4.28),239(4.29), 252(4.28),268(4.24), 319(3.52),398(3.59), 501(2.86),535(2.92), 569(2.82)	114-0109-69A
	pyridine	395(3.56),530(3.34), 561(3.23)	114-0109-69A

Compound	Solvent	λmax(log ε)	Ref.
$C_{14}H_9NO_3$			
Anthrone, 3,4-dihydroxy-10-imino-	EtOH	236(4.23),260(4.48), 290s(4.02),406(3.40), 514(3.79),544(3.87), 575(3.74)	114-0109-69A
	dioxan	251(4.39),264(4.46), 283(3.99),406(3.40), 519(3.71),544(3.79), 577(3.63)	114-0109-69A
1,3-Indandione, 2-hydroxy-2-(2-pyridyl)-	MeOH	233(4.23),267(3.87), 343(4.24)	44-4150-69
4-Phenanthridinecarboxylic acid, 5,6-dihydro-6-oxo-	EtOH	332(2.91),347(2.90)	28-0781-69B
7-Phenanthridinecarboxylic acid, 5,6-dihydro-6-oxo-	EtOH	326(2.87),341(2.80)	28-0781-69B
$C_{14}H_9N_3$			
6H-Indolo[2,3-b]quinoxaline	H_2SO_4	272(4.79),427(4.68), 441(4.68)	47-1803-69
$C_{14}H_9N_3O_2$			
3,5-Pyridinedicarbonitrile, 1,2-dihydro-6-hydroxy-1-methyl-2-oxo-4-phenyl-	n.s.g.	259(4.41),337(4.36)	24-4147-69
$C_{14}H_9N_3O_3$			
Furo[2,3-d]pyrimidine, 4-amino-5,6-di-2-furyl-	EtOH	233(4.10),281(3.99), 334(4.30)	95-1434-69
$C_{14}H_{10}BF_2NO_2$			
Boron, difluoro[1-phenyl-3-(2-pyridyl)-1,3-propanedionato]-	$CHCl_3$	368(4.54),385(4.49)	39-0526-69A
$C_{14}H_{10}BF_4N_3S$			
2-Phenyl-s-triazolo[3,4-b]benzothiazolium tetrafluoroborate	MeCN	226(4.50),270(4.18)	24-3159-69
$C_{14}H_{10}BrNO_2$			
Pyrrolo[1,2-a]quinoline-3-carboxylic acid, 7-bromo-, methyl ester	MeOH	211s(4.40),223(4.54), 232s(4.44),250s(4.37), 257(4.48),273(4.30), 279s(4.23),343s(4.10), 357(4.23),375(4.09)	39-2311-69C
$C_{14}H_{10}BrN_3O_3$			
Glyoxal, p-bromophenyl-, o-nitrophenylhydrazone	$CHCl_3$	408(4.15),423(4.21)	28-0642-69B
$C_{14}H_{10}Br_2O_2S$			
1,4-Butanedione, 2,3-dibromo-1-phenyl-4-(2-thienyl)-	EtOH	264(4.58),295(4.42)	104-0511-69
$C_{14}H_{10}Br_2O_3$			
1,4-Butanedione, 2,3-dibromo-1-(2-furyl)-4-phenyl-	EtOH	265(4.34),288(4.43)	104-0511-69
$C_{14}H_{10}Br_6O_4$			
4H,8H-Cyclobuta[1,2-b:3,4-b']dipyran-4,8-dione, 3,7-dibromo-2,6-bis(dibromomethyl)-4aβ,4b,8aβ,8b-tetrahydro-4bβ,8bβ-dimethyl-	$CHCl_3$	300(4.23)	44-4052-69

Compound	Solvent	λ_{max}(log ε)	Ref.
$C_{14}H_{10}ClNO_2$			
Ethylene, 2-chloro-1-(p-nitrophenyl)-1-phenyl-, cis	C_6H_{12}	252(4.32),297(3.92)	39-0932-69B
	DMF	305s(3.79)	39-0932-69B
Ethylene, 2-chloro-1-(p-nitrophenyl)-1-phenyl-, trans	C_6H_{12}	229(4.21),251(4.16), 304(4.16)	39-0932-69B
	DMF	315(4.09)	39-0932-69B
$C_{14}H_{10}ClN_3O_3$			
Glyoxal, p-chlorophenyl-, o-nitrophenylhydrazone	$CHCl_3$	405(4.20),420(4.23)	28-0642-69A
$C_{14}H_{10}FN_3O_3$			
Glyoxal, p-fluorophenyl-, o-nitrophenylhydrazone	$CHCl_3$	400(4.18),410(4.19)	28-0642-69B
$C_{14}H_{10}F_4N_2$			
Acetophenone, (2,3,4,5-tetrafluorophenyl)hydrazone	EtOH	228(4.22),290(4.10), 320(4.27)	70-0609-69
$C_{14}H_{10}MoO_4$			
Benzoic acid, Mo(II) salt	EtOH	226(3.95+),422(3.95)	12-1571-69
$C_{14}H_{10}N_2$			
Dibenzo[c,g][1,2]diazocine	n.s.g.	235(4.3)	44-3237-69
$C_{14}H_{10}N_2O$			
Anthrone, 10-imino-, oxime	$CHCl_3$	279(4.19)	22-3538-69
4-Cinnolinol, 3-phenyl-	n.s.g.	260(4.3),315(4.1), 355(4.1)	39-0796-69C
Ketone, indol-2-yl 4-pyridyl	EtOH	220(4.31),265(3.89), 329(4.24)	39-2738-69C
Nitrone, α-(p-cyanophenyl)-N-phenyl-	C_6H_{12}	337(4.44)	18-3306-69
	EtOH	331(4.43)	18-3306-69
1,2,4-Oxadiazole, 3,5-diphenyl-	EtOH	258(4.55)	18-3335-69
4(3H)-Quinazolinone, 2-phenyl-	EtOH	233(4.42),291(4.18)	78-0783-69
$C_{14}H_{10}N_2OS$			
2-Benzimidazolinethione, 1-benzoyl-	MeOH	245(4.27),297s(4.27), 305(4.32)	97-0152-69
	dioxan	306(4.10)	4-0023-69
$C_{14}H_{10}N_2O_2$			
1,4,2,5-Dioxadiazine, 3,6-diphenyl-	EtOH	249(4.54)	32-0165-69
Isosydnone, 4,5-diphenyl-	EtOH	226(4.05),312(3.92)	39-1185-69B
1,2,4-Oxadiazole, 3,5-diphenyl-, 4-oxide, boron trifluoride adduct	EtOH	243(4.42),322(4.05)	32-0165-69
5-Oxazolol, 2-phenyl-4-(4-pyridyl)-	acetone	428(4.70),450(4.65)	24-1129-69
(in acetone)	HCl	418(4.66),440(4.67)	24-1129-69
(in acetone)	NaOH	420(4.58)	24-1129-69
$C_{14}H_{10}N_2O_3$			
Benzoxazole, 5-methyl-2-(m-nitrophenyl)-	EtOH	264(4.11),304(4.17)	18-3335-69
Benzoxazole, 5-methyl-2-(p-nitrophenyl)-	EtOH	232(4.18),334(4.39)	18-3335-69
$C_{14}H_{10}N_2O_4$			
Benzoic acid, 4,4'-azodi-	pH 13	233(4.08),332(4.41), 435(3.10)	78-4241-69
Stilbene, 4,4'-dinitro-, trans	DMF	368(4.57)	39-0554-69B

Compound	Solvent	λ_{max}(log ϵ)	Ref.
$C_{14}H_{10}N_2O_5$			
Benzoic acid, 4,4'-azoxydi-	pH 13	262(4.01),334(4.25)	78-4241-69
$C_{14}H_{10}N_2O_6$			
Acetic acid, bis(p-nitrophenyl)-	pH 10	285(4.30)	35-2467-69
3-Biphenylcarboxylic acid, 4,6-dinitro-, methyl ester	n.s.g.	295s(3.95)	40-0085-69
$C_{14}H_{10}N_2O_7$			
p-Anisic acid, 2,4-dinitrophenyl ester	EtOH	262(4.53)	65-0660-69
Phenol, p-methoxy-, 2,4-dinitrobenzoate	EtOH	300s(3.48),330s(3.14)	65-0660-69
$C_{14}H_{10}N_4$			
5,5'-Bibenzimidazole	EtOH	228(4.68),292(4.22)	22-1926-69
Cinnolino[5,4,3-cde]cinnoline, 2,7-dimethyl-	EtOH	224(4.80),243(4.85), 255s(4.49),263(4.44), 342(4.38),353(4.33), 391(3.82),411(4.06)	4-0533-69
$C_{14}H_{10}N_4O$			
7H-Pyrrolo[2,3-d]pyrimidine-6-carbonitrile, 5-hydroxy-7-methyl-2-phenyl-	EtOH	235(4.11),278(4.53)	4-0819-69
$C_{14}H_{10}N_4O_2$			
Cinnolino[5,4,3-cde]cinnoline, 2,7-dimethyl-, 4,9-dioxide	EtOH	228(4.67),250s(4.46), 267(4.32),277(4.33), 289s(4.28),310(3.93), 324(3.91),368(3.94), 411(3.97),435(3.99)	4-0533-69
$C_{14}H_{10}N_4O_2S$			
Thiocyanic acid, o-[(4-nitro-o-tolyl)-azo]phenyl ester	EtOH	333(5.37),456(--)	39-1828-69C
	$CHCl_3$	333(5.43)	39-1828-69C
$C_{14}H_{10}N_4O_2S_2$			
Benzenesulfenic acid, o-[(4-nitro-o-tolyl)azo]-, anhydride with thiocyanic acid	EtOH	334(5.47),434(4.84)	39-1828-69C
	$CHCl_3$	325(5.43)	39-1828-69C
$C_{14}H_{10}N_4O_3$			
Furo[2,3-d]pyrimidine, 3-amino-3,4-dihydro-5,6-di-2-furyl-4-imino-	EtOH	229(4.16),341(4.32)	95-1434-69
Furo[2,3-d]pyrimidine, 5,6-di-2-furyl-4-hydrazino-	EtOH	235(4.25),292(4.06), 341(4.42)	95-1434-69
$C_{14}H_{10}N_4O_4$			
Benzo[c]cinnoline, 2,9-dimethyl-1,10-dinitro-	EtOH	232(4.65),284(4.34), 319s(4.00)	4-0533-69
$C_{14}H_{10}N_4O_8$			
p,p'-Bitolyl, 2,2',6,6'-tetranitro-	EtOH	247(4.59)	4-0533-69
$C_{14}H_{10}N_4O_{10}$			
Biphenyl, 4,4'-dimethoxy-2,2',6,6'-tetranitro-	EtOH	208(4.85),245s(4.43), 340(3.72)	4-0533-69
$C_{14}H_{10}N_8O_{10}$			
4,4'-Biphenyldicarboxylic acid, 2,2',6,6'-tetranitro-, dihydrazide	EtOH	208(4.69),228(4.66), 239(4.66)	4-0533-69

Compound	Solvent	$\lambda_{max}(\log \epsilon)$	Ref.
$C_{14}H_{10}O$			
1,4-Epoxyanthracene, 1,4-dihydro-	n.s.g.	236(4.48),260(3.89), 268(3.78),276(3.53)	39-0765-69C
1,3-Tridecadiene-5,7,9,11-tetrayne, 3-methoxy-, cis	ether	283(4.70),302(4.84), 322(3.99),344(4.21), 367(4.38),396(4.24)	24-3765-69
$C_{14}H_{10}OSe$			
Dibenzo[b,f]selenepin-10(11H)-one	MeOH	237(4.18),246(4.15), 277s(3.82),284s(3.70), 339(3.64)	73-3801-69
$C_{14}H_{10}O_2$			
5,6-Benzocoumarin, 3-methyl-	EtOH	232(4.7),317(4.0), 345(4.1)	39-2618-69C
7,8-Benzocoumarin, 3-methyl-	EtOH	222(4.5),267(4.5), 277(4.6),310(3.9), 323(4.0)	39-2618-69C
2-Biphenylenol, acetate	EtOH	242(4.52),249(4.54), 341(3.76),361(3.88)	39-0742-69C
Phthalide, 3-phenyl-	EtOH	226s(4.06),237s(3.76), 274(3.25),281(3.25)	12-0577-69
$C_{14}H_{10}O_2S$			
2-Butene-1,4-dione, 1-phenyl-4-(2-thienyl)-	EtOH	228(4.15),265(4.27), 290(4.24)	104-0511-69
$C_{14}H_{10}O_3$			
2-Butene-1,4-dione, 1-(2-furyl)-4-phenyl-	EtOH	252(4.15),308(4.20)	104-0511-69
Naphtho[1,2-b]furan-4,5-dione, 3,6-dimethyl-	EtOH	220(4.30),252s(4.40), 265(4.36),270(4.36), 455(3.36)	39-1184-69C
Naphtho[1,2-b]furan-4,5-dione, 3,9-dimethyl-	EtOH	220(4.39),255s(4.29), 267(4.44),274(4.45), 355(3.32),475(3.38)	39-1184-69C
Naphtho[2,3-b]furan-4,9-dione, 3,5-dimethyl- (maturinone)	EtOH	210(4.01),251(4.39), 266(4.00),287s(3.74), 355(3.58)	39-1184-69C
	EtOH	251(4.40),267s(4.00), 294(3.67),352(3.67)	88-1929-69
isomer	EtOH	251(4.34),267s(3.91), 294(3.56),357(3.55)	88-1929-69
Naphtho[2,3-b]furan-4,9-dione, 3,8-dimethyl-	EtOH	206(4.26),250(4.47), 265s(4.06),285s(3.80), 358(3.68)	39-1184-69C
	EtOH	250(4.30),265(4.18), 290(4.00)	22-3612-69
2H-Naphtho[1,2-b]pyran-2-one, 6-methoxy-	EtOH	223(4.6),274(4.4), 285(4.5),318(3.7), 327s(3.5),380(3.7)	78-4207-69
Xanthen-9-one, 1-methoxy-	MeOH	236(4.60),278(3.91), 346(3.87)	39-2421-69C
Xanthen-9-one, 2-methoxy-	EtOH	242(4.35),249s(4.38), 297(3.50),302(3.48), 357(3.71)	39-0281-69C
$C_{14}H_{10}O_4$			
Benzoic acid, 2-piperonyl-	EtOH	262(3.80),295(3.87)	78-4315-69
Benzoic acid, 3-piperonyl-	EtOH	269(4.03),295(4.01)	78-4315-69

Compound	Solvent	λ_{max}(log ϵ)	Ref.
2,3-Biphenylenedione, 6,7-dihydroxy-1,8-dimethyl-	EtOH	269(4.60),288(4.56), 366(3.92),433(4.44), 517(3.70)	39-2579-69C
2,3-Biphenylenedione, 6,7-dimethoxy-	EtOH	263(4.47),277s(4.34), 306s(3.92),420(4.38)	39-0646-69C +39-2579-69C
4a,8a-[2]Butenonaphthalene-1,4,5,8-tetrone	MeOH	215s(4.03),226s(3.90), 285(3.32),354(2.46)	78-5115-69
Cyperaquinone	EtOH	259(4.46),347(3.50), 473(3.63)	88-4669-69
Isophthalic acid, 5-phenyl-	EtOH	231(4.50),254(4.18), 299(3.12)	44-4134-69
Xanthen-9-one, 1,3-dihydroxy-2-methyl-	MeOH	236(4.54),311(4.22)	78-0275-69
Xanthen-9-one, 1-hydroxy-5-methoxy-	MeOH	235s(4.51),247(4.65), 310(4.02),370(3.83)	39-2421-69C
Xanthen-9-one, 5-hydroxy-1-methoxy-	MeOH	236s(4.49),245(4.57), 304(3.93),351(3.66)	39-2421-69C
$C_{14}H_{10}O_5$			
Benzo[1,2-b:5,4-b']difuran-4,8-dione, 2-[1-(hydroxymethyl)vinyl]-5-methyl-	EtOH	262(4.62),347(3.49), 473(3.62)	88-4669-69
Xanthen-9-one, 1,7-dihydroxy-3-methoxy- (gentisin)	MeOH	258(4.35),308(3.91), 370(3.61)	39-2201-69C
Xanthen-9-one, 1,3,8-trihydroxy-2-methyl-	MeOH	248(4.45),329(4.28)	78-0275-69
$C_{14}H_{10}O_6$			
Bellidifolin	EtOH	255(4.22),279(4.10), 335(3.90),400s(3.70)	94-0155-69
Isobellidifolin	EtOH	253(4.3),277(4.2), 336(4.1),380s(3.7)	94-0155-69
	EtOH-NaOAc	247(4.20),274(4.15), 362(4.20)	94-0155-69
Swertianin	EtOH	242(4.40),269(4.48), 314s(4.10),330(4.12), 400s(3.70)	94-0155-69
Xanthen-9-one, 1,3,8-trihydroxy-7-methoxy-	EtOH	238(4.45),262(4.50), 334(4.20),380s(3.67)	94-0155-69
	EtOH-NaOAc	263s(4.41),270(4.43), 361(4.40)	94-0155-69
Xanthen-9-one, 1,4,7-trihydroxy-3-methoxy-	MeOH	263(4.44),314(4.00), 380(3.71)	39-2201-69C
$C_{14}H_{10}S_2$			
Dibenzo[e,g][1,4]dithiocin	EtOH	273(3.69)	97-0184-69
Thiophene, 2,2'-p-phenylenedi-	n.s.g.	235(3.95),257(--), 322(4.48)	22-2076-69
$C_{14}H_{10}S_3$			
Benzoic acid, dithio-, anhydrosulfide	C_6H_{12}	228s(4.2),290(4.34), 539(2.34)	48-0045-69
	EtOH	230s(4.1),243(4.32), 302(4.23),524(2.22)	48-0045-69
	MeCN	228s(4.1),307(4.36), 522(2.28)	48-0045-69
$C_{14}H_{10}Se$			
Dibenzo[b,f]selenepin	MeOH	226(4.44),268(4.35), 295(3.77)	73-3801-69

Compound	Solvent	$\lambda_{max}(\log \epsilon)$	Ref.
$C_{14}H_{11}BrN_2O$			
Benzimidazole, 2-[2-(5-bromo-2-furyl)-vinyl]-1-methyl-	MeOH	340(4.23)	103-0246-69
$C_{14}H_{11}Br_2N_3$			
Pyrido[1,2-a]pyrido[1',2':3,4]imidazo-[2,1-c]pyrazine-5,8-diinium dibromide	H_2O	243(4.34),250s(4.32), 268(4.24),274(4.24), 296(4.18),308(4.3), 355s(3.75),370(3.9), 416(4.19)	39-1987-69C
$C_{14}H_{11}ClN_2O$			
Benzimidazole, 2-[2-(5-chloro-2-furyl)-vinyl]-1-methyl-	MeOH	336(4.16)	103-0246-69
$C_{14}H_{11}ClN_2O_2$			
Aniline, p-chloro-N-(α-methyl-p-nitro-benzylidene)-	EtOH	274(4.28),339(3.51)	56-0749-69
4(1H)-Quinazolinone, 2-(o-chlorophenyl)-2,3-dihydro-3-hydroxy-	MeOH	230(4.5),250s(3.9), 350(3.4)	24-3735-69
$C_{14}H_{11}ClN_2O_3$			
Benzoic acid, p-[(5-chloro-2-hydroxy-phenyl)azo]-, methyl ester (in 50% EtOH)	HCl	396(3.97)	73-3740-69
	NaOH	500(4.10)	73-3740-69
$C_{14}H_{11}ClOSe$			
Dibenzo[b,f]selenepin-10-ol, 8-chloro-10,11-dihydro-	MeOH	257(3.72),280(3.80)	73-3801-69
$C_{14}H_{11}ClO_4S$			
4-Methyl-1-thiaphenanthrenium perchlorate	$HOAc-HClO_4$	253(4.36),290(4.01), 309(4.02),380s(3.45), 432(3.86)	2-0017-69
isomer	$HOAc-HClO_4$	249(4.32),267(3.91), 300s(4.19),309(4.29), 383(3.78),420s(3.57)	2-0017-69
$C_{14}H_{11}Cl_2N$			
Aniline, p-chloro-N-(p-chloro-α-methyl-benzylidene)-	EtOH	254(4.36),315(3.51)	56-0749-69
$C_{14}H_{11}Cl_2NO$			
Nitrone, N-benzyl-α-(2,6-dichlorophen-yl)-	EtOH	265(3.8)	28-0078-69A
$C_{14}H_{11}Cl_2NO_3$			
Acetohydroxamic acid, 2-(2,4-dichloro-phenoxy)-N-phenyl-	EtOH	232(4.23),244(--)	42-0831-69
$C_{14}H_{11}Cl_3O_8S_2$			
6-Chloro-3,9-dimethylbenzo[1,2-b:3,4-b]-bisthiopyrylium diperchlorate	$HOAc-HClO_4$	251(3.00),303(4.12), 380(3.59),470(4.09)	2-0948-69
$C_{14}H_{11}F$			
Ethylene, 2-fluoro-1,1-diphenyl-	hexane	219(4.27),244(4.17)	39-1100-69B
$C_{14}H_{11}F_4N$			
4a(8bH)-Biphenylenamine, 5,6,7,8-tetra-fluoro-N,N-dimethyl-	n.s.g.	263(3.85)	77-0810-69

Compound	Solvent	$\lambda_{max}(\log \epsilon)$	Ref.
$C_{14}H_{11}F_5O_4$			
Malonic acid, (pentafluorobenzylidene)-, diethyl ester	EtOH	262(3.55)	65-1347-69
$C_{14}H_{11}N$			
Benzo[h]isoquinoline, 10-methyl-	hexane	247(4.6),290(3.9), 300(3.9),330f(3.0)	4-0465-69
Benzo[f]quinoline, 5-methyl-	hexane	240f(4.6),267(4.4), 330f(3.3)	4-0465-69
Indole, 1-phenyl-	EtOH	256(4.30),290(3.92)	39-1537-69C
1-Phenanthrylamine	EtOH	216(4.59),226s(4.53), 242(4.47),249(4.56), 313(3.89),362(3.51)	24-2384-69
	50% EtOH-HCl	211(4.55),219(4.33), 246(4.71),253(4.80), 275(4.10),285(4.04), 296(4.16),317(2.52), 325(2.48),332(2.57), 340(2.42),348(2.56)	24-2384-69
$C_{14}H_{11}NO$			
Acrylophenone, 3-(2-pyridyl)-	EtOH	270s(4.13),305(4.29)	95-0375-69
	EtOH-HCl	253(3.94),305(4.33)	95-0375-69
Acrylophenone, 3-(3-pyridyl)-	EtOH	300(4.37)	95-0375-69
	EtOH-HCl	283(4.36)	95-0375-69
Acrylophenone, 3-(4-pyridyl)-	EtOH	283(4.43)	95-0375-69
	EtOH-HCl	284(4.45)	95-0375-69
10H-Azepino[1,2-a]indol-10-one, 11-methyl-	EtOH	208(4.14),226(4.28), 292(4.55),329(3.99)	39-1028-69C
	H_2SO_4	233(4.04),272(4.06), 327s(--),336(4.64), 550s(--),611(3.04)	39-1028-69C
Carbazole, 1-acetyl-	EtOH	223(4.68),253(4.11), 289(4.26),320s(3.68), 378(3.94)	22-3523-69
Carbazole, 2-acetyl-	EtOH	253(4.56),274s(3.91), 319(4.36),370s(3.60)	22-3523-69
Carbazole, 3-acetyl-	EtOH	233(4.49),244s(4.31), 273(4.50),289(4.51), 330(4.04)	22-3523-69
Formamide, N-fluoren-9-yl-	dioxan	227(4.28),236(4.12), 270(4.21),293(3.66), 305(3.67)	89-0772-69
$C_{14}H_{11}NO_2$			
Benzil, monooxime, anti-phenyl	1% MeOH	256(4.36)	22-2894-69
anion	1% MeOH	280s(4.18)	22-2894-69
Benzil, monooxime, syn-phenyl	1% MeOH	256(4.09)	22-2894-69
anion	1% MeOH	301(4.03)	22-2894-69
Benz[b]indolizine-10-carboxylic acid, methyl ester	EtOH	223(4.31),247(4.59), 254(4.73),278(4.16), 288(4.22),310(3.82), 322(3.74)	12-0997-69
Benz[e]indolizine-3-carboxylic acid, methyl ester	MeOH	223(4.53),229s(4.41), 245s(4.20),252(4.31), 263(4.20),271(4.22), 279s(4.04),323s(3.92), 339s(4.13),351(4.26), 369(4.11)	39-2311-69C

Compound	Solvent	λ_{max}(log ϵ)	Ref.
$C_{14}H_{11}NO_2S_2$			
Sulfone, p-isothiocyanatophenyl p-tolyl	EtOH	212(4.58),238(4.37), 287(4.43),296(4.45)	73-4005-69
$C_{14}H_{11}NO_3$			
Acrylophenone, 2',4'-dihydroxy-3-(2-pyridyl)-	EtOH	265(4.02),275(4.02), 313(4.28),350(4.24)	95-0375-69
	EtOH-HCl	250(3.97),265s(3.98), 304(4.26),371(4.12)	95-0375-69
	EtOH-NaOH	270s(4.09),292(4.28), 408(4.40)	95-0375-69
Acrylophenone, 2',4'-dihydroxy-3-(3-pyridyl)-	EtOH	270s(4.03),308(4.27), 347(4.22)	95-0375-69
	EtOH	270s(4.19),287(4.26), 355(4.12)	95-0375-69
	EtOH-NaOH	277(4.23),293(4.26), 413(4.39)	95-0375-69
Acrylophenone, 2',4'-dihydroxy-3-(4-pyridyl)-	EtOH	270s(4.25),282(4.27), 350(4.14)	95-0375-69
	EtOH-HCl	285(4.37),376(4.03)	95-0375-69
	EtOH-NaOH	278(4.30),414(4.29)	95-0375-69
Carbazole-1-carboxylic acid, 2-hydroxy-, methyl ester	EtOH	228(4.64),278(4.34), 286(4.36),360(3.95)	39-1518-69C
Carbazole-3-carboxylic acid, 1-methoxy- (mukoeic acid)	n.s.g.	235(4.51),270(4.58), 320(3.92)	25-0549-69
3H-Phenoxazin-3-one, 7-ethoxy-	EtOH	560(4.00)	73-0221-69
Phthalimide, N-(6-oxo-1-cyclohexen-1-yl)-	MeOH	218(4.60),292(3.26)	104-1193-69
$C_{14}H_{11}NO_4$			
Bicyclo[4.2.0]octa-1,3,5-triene-$\Delta^{7,\alpha}$-acetic acid, 8-(cyanomethylene)-3,4-dimethoxy-	EtOH	261(4.39),283s(4.28), 344s(3.78),360(3.74), 384(3.85),407s(3.79)	39-0646-69C
2,3-Biphenylenedione, 6,7-dimethoxy-, monooxime	EtOH	268s(4.76),273(4.77), 287s(4.68),325(4.30), 334s(4.28),429(4.49)	39-0646-69C
$C_{14}H_{11}NO_5$			
p-Anisic acid, p-nitrophenyl ester	EtOH	277(4.46)	65-0660-69
Phenol, p-methoxy-, p-nitrobenzoate	EtOH	259(4.20),310s(3.45), 330s(3.0)	65-0660-69
$C_{14}H_{11}NS$			
2H-1,4-Benzothiazine, 3-phenyl-	n.s.g.	261(4.4),300(3.9), 320(3.9)	4-0635-69
$C_{14}H_{11}NS_2$			
Isothiocyanic acid, p-(p-tolylthio)-phenyl ester	EtOH	221(4.55),263(4.07), 295s(--),303(4.35)	73-4005-69
$C_{14}H_{11}N_3OS$			
Isothiocyanic acid, p-[(p-methoxyphenyl)azo]phenyl ester	heptane	238(3.89),361(3.48), 440(3.20)	73-3912-69
$C_{14}H_{11}N_3O_2$			
Acridine, 9-amino-2-methyl-6-nitro-	EtOH	246(4.45),283(4.46), 320(4.23),364s(3.46), 384(3.59),450(3.60)	103-0707-69

Compound	Solvent	$\lambda_{max}(\log \epsilon)$	Ref.
Acridine, 9-amino-5-methyl-3-nitro-	EtOH	244(4.51),282(4.51), 321(4.20),362s(3.54), 382(3.64),455(3.63)	103-0707-69
Aeruginosin A (2-amino-6-carboxy-10-methylphenazinium hydroxide inner salt)	0.5N HCl	235(4.49),285(4.64), 380(4.05),538(4.13)	39-2514-69C
	0.1N HCl	231(4.51),281(4.65), 383(4.04),515(4.08)	39-2514-69C
	pH 3.87	235(4.49),283(4.64), 396(4.05),515(4.13)	39-2514-69C
	pH 6.95	235(4.50),282(4.65), 396(4.07),515(4.15)	39-2514-69C
	pH 9.91	235(4.51),283(4.65), 395(4.05),515(4.14)	39-2514-69C
	0.5N NaOH	235(4.41),275(4.55), 375(3.85),525(4.04)	39-2514-69C
$C_{14}H_{11}N_3O_2S$			
Benzimidazole, 2-(p-nitrobenzylthio)-	dioxan	283(4.28),291(4.26)	4-0023-69
$C_{14}H_{11}N_3O_3$			
Benzamide, o-[(m-nitrobenzylidene)-amino]-	MeOH	220(4.4),250(4.4), 320(3.7)	24-3735-69
Benzamide, o-[(o-nitrobenzylidene)-amino]-	MeOH	210(4.3),230(4.3), 340s(3.7)	24-3735-69
Benzamide, o-[(p-nitrobenzylidene)-amino]-	MeOH	210(4.5),250(4.1), 290(4.2),350s(3.9)	24-3735-69
Glyoxal, phenyl-, o-nitrophenylhydrazone	$CHCl_3$	400(4.15),418(4.22)	28-0642-69B
2-Pyridinepropionanilide, α,β-dioxo-, α-oxime, anti	EtOH	243(3.98),268(3.90)	36-0460-69
syn	EtOH	222(4.22),273(4.10)	36-0460-69
3-Pyridinepropionanilide, α,β-dioxo-, α-oxime, anti	EtOH	245(4.31)	36-0460-69
syn	EtOH	223(4.19),267(3.99)	36-0460-69
4-Pyridinepropionanilide, α,β-dioxo-, α-oxime, anti	EtOH	243(4.21)	36-0460-69
syn	EtOH	224(4.25),251s(4.02)	36-0460-69
4(1H)-Quinazolinone, 2,3-dihydro-2-(m-nitrophenyl)-	MeOH	225(4.6),260s(4.1), 340(3.4)	24-3735-69
4(1H)-Quinazolinone, 2,3-dihydro-2-(o-nitrophenyl)-	MeOH	225(4.6),250s(4.0), 340(3.5)	24-3735-69
4(1H)-Quinazolinone, 2,3-dihydro-2-(p-nitrophenyl)-	MeOH	225(4.6),260(4.2), 340(3.4)	24-3735-69
$C_{14}H_{11}N_3O_3S$			
Benzothiazole, 6-methoxy-2-(p-nitro-anilino)-	EtOH	229(4.46),290(4.30), 312s(4.21),416(3.87)	17-0095-69
Benzothiazole, 6-methoxy-2-(p-nitro-anilino)-	EtOH	226(4.35),272(4.15), 312s(3.65),384(4.35)	17-0095-69
$C_{14}H_{11}N_3O_4$			
Aniline, N-(α-methyl-p-nitrobenzylidene)-m-nitro-	EtOH	274(4.32),332(3.62)	56-0749-69
4(1H)-Quinazolinone, 2,3-dihydro-3-hydroxy-2-(m-nitrophenyl)-	MeOH	225(4.5),350(3.5)	24-3735-69
4(1H)-Quinazolinone, 2,3-dihydro-3-hydroxy-2-(o-nitrophenyl)-	MeOH	225(4.5),350(3.5)	24-3735-69
4(1H)-Quinazolinone, 2,3-dihydro-3-hydroxy-2-(p-nitrophenyl)-	MeOH	225(4.5),260(4.1), 340(3.5)	24-3735-69

Compound	Solvent	$\lambda_{max}(\log \epsilon)$	Ref.
$C_{14}H_{11}N_3O_5$			
Aziridine, 1-(5-nitro-2-furoyl)-, O-benzoyloxime	EtOH	232(4.17),254(4.08), 325(4.18)	18-0556-69
2-Furoic acid, 5-(1-methyl-5-nitro-2-benzimidazolyl)-	EtOH	242(3.84),295(4.39), 335(4.28)	73-0572-69
$C_{14}H_{11}N_3O_5S$			
Aeruginosin B (7-amino-1-carboxy-5-methyl-3-sulfophenazinium hydroxide S-inner salt)	0.5N HCl	240(4.52),290(4.58), 390(4.05),522(4.14)	39-2517-69C
	0.1N HCl	231(4.51),281(4.65), 383(4.04),515(4.08)	39-2517-69C
	pH 3.87	240(4.54),290(4.60), 390(4.05),522(4.14)	39-2517-69C
	pH 9.91	240(4.54),288(4.59), 392(4.05),522(4.14)	39-2517-69C
	0.5N NaOH	240(4.55),295(4.60), 378(4.05),542(4.11)	39-2517-69C
$C_{14}H_{11}N_3O_6$			
p-Anisaldehyde, 2',4'-dinitro-	EtOH	251(4.22),316(4.29)	65-0660-69
p-Benzanisidide, 2,4-dinitro-	EtOH	250(4.42),305s(3.49), 330s(3.28)	65-0660-69
$C_{14}H_{11}N_3S$			
Isothiocyanic acid, p-(p-tolylazo)-phenyl ester	heptane	237(4.29),349(4.54), 444(3.19)	73-3912-69
$C_{14}H_{11}N_5O_3S$			
9H-Purine, 9-acetyl-6-[(p-nitrobenzyl)-thio]-	MeCN	286(4.49)	44-0973-69
Purine, 9(7)-acetyl-8-[(p-nitrobenzyl)-thio]-	dioxan	287(4.39)	4-0023-69
$C_{14}H_{12}$			
Cyclobuta[4,5]benzo[1,2]cyclooctene, 1,2-dihydro-	C_6H_{12}	212(4.20),236(4.24), 290(3.33)	35-4734-69
Ethylene, 1,1-diphenyl-	C_6H_{12}	231(4.20),251(4.04)	59-0487-69
Stilbene, α,β-deuteriated	n.s.g.	218(4.23),222(4.24), 280(4.44),289(4.41), 298(4.46),308(4.20)	39-1542-69C
$C_{14}H_{12}BF_4N_3$			
5,6-Dihydropyrido[1',2':4,3]-s-triazolo-[1,5-a]quinolin-13-ium tetrafluoroborate	n.s.g.	210(4.4),250f(4.3), 300(3.9),310(3.6)	24-3159-69
$C_{14}H_{12}BrNO_2$			
Benz[b]indolizine-10-carboxylic acid, 3-bromo-6,7-dihydro-, methyl ester	EtOH	227(4.45),256(4.43), 320(4.28),330s(4.25)	12-0997-69
Benz[b]indolizine-10-carboxylic acid, 8-bromo-6,7-dihydro-, methyl ester	EtOH	222(4.37),245(4.26), 326(4.26)	12-0997-69
5H-Pyrano[3,2-c]quinolin-5-one, 4-bromo-2,6-dihydro-2,2-dimethyl-	EtOH	235(4.84),341(4.50), 355(4.62),372(4.48)	32-0711-69
$C_{14}H_{12}BrN_3$			
Benzimidazole, 5-amino-1-(p-bromophenyl)-2-methyl-	EtOH	210(4.78),250s(4.22), 303(3.73)	73-0572-69
Benzimidazole, 5-amino-2-(p-bromophenyl)-1-methyl-	EtOH	209(4.66),240s(4.41), 272(4.19),325(4.01)	73-0572-69

Compound	Solvent	λmax(log ε)	Ref.
$C_{14}H_{12}Br_4O_2$			
Benzene, 1,2,3,4-tetrabromo-5,6-[(2,5-dimethyl-2,4-hexadienylidene)dioxy]-	ether	248(4.69),295(3.53), 306(3.53)	77-1096-69
$C_{14}H_{12}Br_4O_4$			
4H,8H-Cyclobuta[1,2-b:3,4-b']dipyran-4,8-dione, 3,7-dibromo-2,6-bis(bromomethyl)-4aβ,4b,8aβ,8b-tetrahydro-4bβ,8bβ-dimethyl-	$CHCl_3$	298(4.25)	44-4052-69
$C_{14}H_{12}ClN$			
Aniline, N-(p-chloro-α-methylbenzylidene)-	EtOH	254(4.34),314(3.34)	56-0749-69
Aniline, p-chloro-N-(α-methylbenzylidene)-	EtOH	246(4.32),312(3.40)	56-0749-69
$C_{14}H_{12}ClNO$			
Nitrone, N-benzyl-α-(m-chlorophenyl)-	EtOH	204(4.3),231(3.9), 294(4.3)	28-0078-69A
Nitrone, N-benzyl-α-(o-chlorophenyl)-	EtOH	231(3.9),299(4.2)	28-0078-69A
Nitrone, N-benzyl-α-(p-chlorophenyl)-	EtOH	225(3.9),233(3.9), 301(4.3)	28-0078-69A
$C_{14}H_{12}ClNO_2$			
3H-Azepine, 3-benzoyl-5-chloro-2-methoxy-	MeOH	247(4.30)	78-5205-69
3H-Azepine, 3-benzoyl-6-chloro-2-methoxy-	MeOH	247(4.28)	78-5155-69
Benzamide, N-(benzyloxy)-o-chloro-	MeOH	208(4.4),259(2.43)	12-0161-69
	MeOH-NaOH	259(3.16)	12-0161-69
Benzamide, N-(benzyloxy)-p-chloro-	MeOH	208(4.4),237(4.15)	12-0161-69
	MeOH-NaOH	224(4.06),274(3.80)	12-0161-69
$C_{14}H_{12}ClN_3$			
Benzimidazole, 5-amino-1-(p-chlorophenyl)-2-methyl-	EtOH	212(4.59),250s(4.12), 305(3.61)	73-0572-69
Benzimidazole, 5-amino-2-(p-chlorophenyl)-1-methyl-	EtOH	206(4.61),235(4.37), 274(4.20),325(3.97)	73-0572-69
$C_{14}H_{12}Cl_2NOPS$			
Formamide, 1-[bis(p-chlorophenyl)phosphinyl]-N-methyl-thio-	EtOH	290(3.86),394(1.90)	18-2975-69
$C_{14}H_{12}Cl_2N_2$			
p,p'-Azotoluene, 3,3'-dichloro-	MeOH	238(4.29),332(4.40)	64-0997-69
Benzaldehyde, 2,6-dichloro-, methylphenylhydrazone	EtOH	207(4.4),252(4.0), 300s(3.9),338(4.2)	28-0137-69B
$C_{14}H_{12}INO_2$			
Benzamide, N-(benzyloxy)-o-iodo-	MeOH	208(4.4),227s(4.01), 254s(3.19)	12-0161-69
	MeOH-NaOH	225s(4.07),253s(3.22)	12-0161-69
Benzamide, N-(benzyloxy)-p-iodo-	MeOH	208(4.4),251(4.27)	12-0161-69
	MeOH-NaOH	237(4.16),277(3.90)	12-0161-69
$C_{14}H_{12}IN_3$			
Benzimidazole, 5-amino-1-(p-iodophenyl)-	EtOH	208(4.86),252s(4.30), 300(3.83)	73-0572-69
Benzimidazole, 5-amino-2-(p-iodophenyl)-	EtOH	211(4.50),242(4.39), 275(4.25),325(4.12)	73-0572-69

Compound	Solvent	λ_{max}(log ϵ)	Ref.
$C_{14}H_{12}N_2$			
Acridine, 9-(methylamino)-	pH 7.0	220(4.37),256s(4.68), 266(4.75),310s(3.18), 325s(2.94),370s(3.49), 389(3.84),408(4.02), 431(3.93)	39-0096-69B
	pH 12.2	223(4.37),263(4.71), 399(3.86),417(3.90)	39-0096-69B
Dibenzo[c,g][1,2]diazocine, 5,6-dihydro-	n.s.g.	246(4.5),300(3.8), 345(3.7)	44-3237-69
Indole, 5-methyl-2-(3-pyridyl)-	EtOH	227(4.39),242s(4.19), 257(4.04),320(4.45)	22-4154-69
$C_{14}H_{12}N_2O$			
Benzaldehyde, azine, monooxide	EtOH	224(3.91),275(4.09), 340(4.41)	44-0155-69
Benzimidazole, 2-[2-(2-furyl)vinyl]-1-methyl-	MeOH	342(4.43)	103-0246-69
Oxepino[2,3-b]quinoline-6-carbonitrile, 2,3,4,5-tetrahydro-	CH_2Cl_2	250(4.51),314(3.81), 326(3.88),338(3.92)	94-1294-69
2(1H)-Quinoxalinone, 3,4-dihydro-3-phenyl-	EtOH	228(4.48),260s(3.66), 305(3.74)	106-0384-69
$C_{14}H_{12}N_2OS$			
2-Benzothiazolinone, 3-(anilinomethyl)-	MeOH	213(4.65),242(4.29), 282(3.62),288(3.64)	4-0163-69
$C_{14}H_{12}N_2OS_2$			
6-(2-Benzothiazolylthio)-5-hydroxy-1-methyl-2-picolinium hydroxide, inner salt	EtOH	280(4.11),365(3.95)	1-2065-69
$C_{14}H_{12}N_2O_2$			
Aniline, N-(α-methylbenzylidene)-m-nitro-	EtOH	250(4.40),312(3.53)	56-0749-69
Aniline, N-(α-methylbenzylidene)-p-nitro-	EtOH	276(4.18),340s(3.40)	56-0749-69
Benzo[c]cinnoline, 2,9-dimethoxy-	EtOH	235s(4.35),244(4.48), 255(4.51),275(4.27), 330(3.93),350(3.89)	4-0523-69
Benzo[c]cinnoline, 3,8-dimethoxy-	pH 1	205(4.39),220s(4.28), 272(4.66),352(4.07), 475(3.52)	4-0523-69
	EtOH	220s(4.12),259(4.87), 290s(4.00),311(3.83), 400(3.51)	4-0523-69
Benzo[c]cinnoline, 2,9-dimethyl-, 5,6-dioxide	EtOH	207s(2.74),211(3.62), 217(3.75),223(3.49), 242(4.46),251s(4.51), 258(4.61),267s(4.39), 293(4.18),304(4.21), 347(4.11),379s(3.45), 398s(2.98)	4-0523-69
Benzo[c]cinnoline, 3,8-dimethyl-, 5,6-dioxide	EtOH	211(3.86),216(3.93), 224(3.90),242s(4.41), 256(4.60),262(4.64), 275s(4.52),293s(4.17), 307s(4.01),347(4.04), 385s(3.43),407s(3.24)	4-0523-69

Compound	Solvent	$\lambda_{max}(\log \epsilon)$	Ref.
Glyoxime, N,N-diphenyl-	EtOH	245(3.98),285(3.81), 300s(3.79),387(4.21)	39-1073-69C
Phthalimide, N-(7-azabicyclo[4.1.0]-hept-2-en-7-yl)-	EtOH	239(4.45),293(3.12)	27-0405-69
Salicylaldehyde, azine	MeCN	237s(4.0),246s(3.9), 291(4.2),294s(4.2), 352(4.2)	39-0742-69B
$C_{14}H_{12}N_2O_2S$			
1,2-Benzisothiazoline, 2-methyl-3-(phenylimino)-, 1,1-dioxide	EtOH	238(4.5),335(3.7)	44-1786-69
3H-Phenothiazin-3-one, 2-hydroxy-7-(dimethylamino)-	EtOH	275(5.32),565(5.22)	5-0106-69G
$C_{14}H_{12}N_2O_3$			
Azobenzene-4'-carboxylic acid, 2-hydroxy-5-methyl-	50% EtOH	331(4.35),398(4.20)	73-1087-69
	50% EtOH-$HClO_4$	305s(--),397(4.39), 505(4.05)	73-1087-69
Barbituric acid, 5-cinnamylidene-1-methyl-	EtOH	265(3.95)	103-0827-69
Benzo[c]cinnoline, 2,9-dimethoxy-, 5-oxide	EtOH	242(4.25),246(4.54), 259(4.50),287s(4.18), 296s(4.19),307(4.27), 354(4.12),362(4.12), 381(4.01)	4-0523-69
Benzo[c]cinnoline, 3,8-dimethoxy-, 5-oxide	EtOH	253s(4.47),260(4.51), 300s(3.95),334(3.69), 410(3.51),430(3.51)	4-0523-69
Nitrone, N-benzyl-α-(m-nitrophenyl)-	EtOH	204(4.3),274(4.2), 293(4.2),333s(3.2)	28-0078-69A
Nitrone, N-benzyl-α-(o-nitrophenyl)-	EtOH	262(4.0),331(3.6)	28-0078-69A
Nitrone, N-benzyl-α-(p-nitrophenyl)-	EtOH	248(4.0),343(4.2)	28-0078-69A
Nitrone, N-(o-nitrobenzyl)-α-phenyl-	EtOH	222(4.0),295(4.2)	28-0078-69A
Nitrone, N-(p-nitrobenzyl)-α-phenyl-	EtOH	218s(4.0),233s(3.9), 293(4.3)	28-0078-69A
Nitrone, α-(4-nitro-o-tolyl)-N-phenyl-	EtOH	358(4.23)	18-3306-69
Phenoxazine, 1,7-dimethyl-3-nitro-	EtOH-HCl	228(4.47),278(4.07), 465(4.05)	73-3732-69
	EtOH-KOH	236s(--),283(3.92), 340(3.85),545(4.29)	73-3732-69
Phenoxazine, 1,9-dimethyl-3-nitro-	EtOH-HCl	230(4.49),278(4.05), 445(4.01)	73-3732-69
	EtOH-KOH	242(4.21),288(3.93), 339(3.86),515(4.32), 660s(--)	73-3732-69
$C_{14}H_{12}N_2O_3S$			
4H-Quinolizine-1-carboxylic acid, 3-cyano-2-(methylthio)-4-oxo-, ethyl ester	EtOH	272(4.09),316(3.91), 422(4.39)	95-0203-69
$C_{14}H_{12}N_2O_3S_2$			
Acetanilide, 4'-(2,3,5,7-tetrahydro-5,7-dioxo-6H-p-dithiino[2,3-c]-pyrrol-6-yl)-	MeOH	255(4.43),416(3.32)	33-2221-69
$C_{14}H_{12}N_2O_4$			
Acrylophenone, 4'-methoxy-3-(5-nitro-2-pyrrolyl)-	EtOH	323(4.14),392(4.48)	103-0383-69
	dioxan	320(4.11),386(4.48)	103-0383-69
	H_2SO_4	365(4.08),495(4.63)	103-0383-69

Compound	Solvent	λ_{max}(log ε)	Ref.
p-Anisanilide, 4'-nitro-	EtOH	253(4.25),300s(3.85), 327(3.78)	65-0660-69
Anthranilic acid, 4-nitro-N-o-tolyl-	EtOH	234(4.11),278(4.32), 420(3.47)	103-0707-69
Anthranilic acid, 4-nitro-N-p-tolyl-	EtOH	237(3.98),282(4.28), 425(3.38)	103-0707-69
Benzamide, N-(benzyloxy)-p-nitro-	MeOH	263(4.06)	12-0161-69
	MeOH-NaOH	246(3.95),348(3.64)	12-0161-69
Benz[cd]indazole-1,2-dicarboxylic acid, 1,2-dihydro-, dimethyl ester	n.s.g.	237(4.60),325(4.06), 335(4.05)	39-0760-69C
Benzo[c]cinnoline, 2,9-dimethoxy-, 5,6-dioxide	EtOH	212(4.25),268(4.62), 315(4.30),358(4.14)	4-0523-69
Benzo[c]cinnoline, 3,8-dimethoxy-, 5,6-dioxide	EtOH	224s(4.94),268(4.62), 282s(4.34),349(3.93), 410(3.51),430(3.51)	4-0523-69
Indazole C from altersolanol A	EtOH	263(4.33),406(3.58)	23-0767-69
Indazole D from altersolanol A	EtOH	220(4.29),255(4.26), 270(4.37),391(3.63)	23-0767-69
2-Propen-1-one, 3-(p-methoxyphenyl)-1-(4-nitro-2-pyrrolyl)-	EtOH	245(4.18),355(4.44)	103-0383-69
	dioxan	243(4.18),350(4.45)	103-0383-69
	H_2SO_4	265(4.07),345(4.05), 480(4.88)	103-0383-69
2-Propen-1-one, 3-(p-methoxyphenyl)-1-(5-nitro-2-pyrrolyl)-	EtOH	247(3.97),377(4.44)	103-0383-69
	dioxan	245(4.02),372(4.54)	103-0383-69
	H_2SO_4	270(3.87),350(3.90), 505(4.74)	103-0383-69
$C_{14}H_{12}N_2O_4S_2$			
Δ^2-1,2,4-Thiadiazoline, 3-phenyl-4-(phenylsulfonyl)-, 1,1-dioxide	n.s.g.	222(4.08),252(4.08)	39-0652-69C
$C_{14}H_{12}N_2O_6$			
Biphenyl, 5,5'-dimethoxy-2,2'-dinitro-	EtOH	213(4.36),264s(4.08), 296s(2.75),344(3.65)	4-0523-69
$C_{14}H_{12}N_2S$			
Lepidine, 2-(2-methyl-4-thiazolyl)-	n.s.g.	340(3.99)	97-0186-69
$C_{14}H_{12}N_2S_2$			
2-Benzothiazolinethione, 3-(anilino-methyl)-	MeOH	240(4.43),325(4.42)	4-0163-69
1,2,3,4-Dithiadiazolidine, 5-benzyl-4-phenyl- ,meso-ionic didehydro deriv.	MeOH	365(3.51)	42-0388-69
$C_{14}H_{12}N_4$			
Azobenzene, 4-cyano-4'-(methylamino)-	EtOH	271(4.09),440(4.52)	18-3565-69
Benzonitrile, p-(3-methyl-3-phenyl-1-triazeno)-	EtOH	241(4.28),358(4.42)	18-3565-69
Pteridine, 2,7-dimethyl-4-phenyl-	hexane	230(4.22),290s(3.89), 299(3.96),324(4.13)	39-1883-69C
Pteridine, 4,7-dimethyl-2-phenyl-	hexane	255(4.42),275(4.17), 321s(4.01),333(4.16), 348(4.10)	39-1408-69C
	pH -1.0	252(4.37),311(4.14)	39-1408-69C
	pH 3.9	248(4.39),268s(4.18), 331(4.15)	39-1408-69C
Pteridine, 6,7-dimethyl-2-phenyl-	hexane	251(4.42),257s(4.40), 271(4.32),277s(4.30), 302s(3.76),322s(4.06), 334(4.22)	39-1408-69C

Compound	Solvent	λ_{max}(log ϵ)	Ref.
Pteridine, 6,7-dimethyl-2-phenyl- (cont.)	pH -0.3	249(4.11),289s(4.03), 313(4.25)	39-1408-69C
	pH 5.0	248(4.34),334(4.27)	39-1408-69C
Pteridine, 6,7-dimethyl-4-phenyl-	hexane	230(4.31),286s(3.89), 298(3.99),323(4.22)	39-1883-69C
s-Tetrazine, 1,4-dihydro-1,4-diphenyl-	EtOH	295(4.43)	35-2443-69
$C_{14}H_{12}N_4O$			
4-Pteridinol, 6,7-dimethyl-2-phenyl-	pH 5.0	244(4.22),291(4.18), 307s(4.17)	39-1408-69C
	pH 10.0	226(4.05),264(4.44), 340(4.03)	39-1408-69C
Pyrazole, 4-acetyl-1-(2-quinoxalinyl)-5-methyl-	EtOH	234(4.33),264(4.51), 330(4.14)	36-0432-69
$C_{14}H_{12}N_4O_2$			
1H-Pyrazolo[3,4-b]pyrazin-6-ol, 1-methyl-3-phenyl-, acetate	EtOH	244(4.32),288(3.95), 335(3.83)	32-0463-69
1H-Pyrazolo[3,4-b]pyrazin-6-ol, 3-methyl-1-phenyl-, acetate	EtOH	258(4.44),324s(3.68)	32-0463-69
$C_{14}H_{12}N_4O_3$			
3-Pyridinepropionanilide, α,β-dioxo-, α,β-dioxime, syn	EtOH	252(4.45),275s(4.38)	36-0460-69
4-Pyridinepropionanilide, α,β-dioxo-, α,β-dioxime, syn	EtOH	258(4.50)	36-0460-69
$C_{14}H_{12}N_4O_5$			
Aziridine, 1-(5-nitro-2-furoyl)-, O-(phenylcarbamoyl)oxime	EtOH	237(4.34),330(4.19)	18-0556-69
1,3,4-Oxadiazole, 2-(ethylamino)-5-[1-(2-furyl)-2-(5-nitro-2-furyl)vinyl]-, cis	n.s.g.	310(4.09),432(4.27)	40-0713-69
trans	n.s.g.	318(4.06),422(4.28)	40-0713-69
$C_{14}H_{12}N_4O_8$			
β-Alanine, N-(5-nitro-2-furoyl)-, p-nitrophenyl ester, oxime	EtOH	254(4.28),295(3.98), 360(3.77)	18-0556-69
Tropone imine, 3-methoxy-, picrate	MeOH	214(4.39),253(4.59), 260(4.51),304(4.03), 316(4.14),350(4.22)	94-2548-69
$C_{14}H_{12}N_4S$			
9-Acridanone, thiosemicarbazone	EtOH-HCl	270(4.7),390(3.8), 400(4.0),440(3.9)	65-1156-69
	EtOH	240(4.7),280(4.1), 300(4.0),420(4.2)	65-1156-69
1,3,4-Thiadiazole, 2,5-dianilino-	EtOH	231(4.0),261s(3.9), 313(4.4)	28-0730-69A
$C_{14}H_{12}N_6O_4$			
Hydrazine, N-[α,β-dioximino-β-4-pyridyl)propionyl]-N'-isonicotinoyl-, syn	$HOCH_2CH_2OH$	253(4.12)	55-0859-69
$C_{14}H_{12}O$			
Acetophenone, o-phenyl-	EtOH	228(4.25),244(4.02), 286(3.51)	6-0277-69
2H-Cyclopropa[2,3]cyclopenta[1,2-a]-naphthalen-2-one, 2a,3,4,5-tetrahydro-	EtOH	229(4.06),234(4.02), 255(3.85),310(4.13)	22-2095-69

Compound	Solvent	λ max (log ε)	Ref.
Dibenzofuran, 1,4-dimethyl-	EtOH	224(4.71),247(4.07), 255(4.28),279(4.36), 294(3.80),306(3.73)	4-0379-69
Fluoren-7-ol, 1-methyl-	EtOH	274(4.2),283(4.2), 308(3.9),314(3.7)	42-0415-69
Ketone, benzyl phenyl	EP	322(2.1),336(2.1)	18-0010-69
	EP at 201° K	312s(2.0),323(2.1), 337(2.1),350(1.9), 367(1.5)	18-0010-69
	ET at 77°K	310(1.8),320(1.9), 333(1.9),349(1.7), 365(1.3)	18-0010-69
$C_{14}H_{12}OSe$			
Dibenzo[b,f]selenepin-10-ol, 10,11-dihydro-	MeOH	219(4.19),256(3.71), 279(3.73),286(3.74)	73-3801-69
$C_{14}H_{12}O_2S$			
Sulfone, methyl 1-methyl-6-phenyl-1-hexene-3,5-diynyl	ether	250(4.45),263(4.43), 278(4.10),294(4.22), 312(4.31),332(4.24)	24-4017-69
$C_{14}H_{12}O_3$			
Benzophenone, 2-hydroxy-3'-methoxy-	EtOH	225(4.21),263(4.06), 332(3.68)	39-0281-69C
Benzophenone, 4-hydroxy-2-methoxy-	$CHCl_3$	280(3.96),309(3.90)	39-0034-69C
Cyclopent[a]indene-3,8-dione, 3a,8a-dihydro-3a-hydroxy-1,2-dimethyl-	EtOH	225(4.14),245(4.10), 297(2.87),325s(2.52), 339(2.62),354(2.69), 370s(2.47)	44-0520-69
2-Furaldehyde, 5-[2-(1-propenyl)fur-5-ylvinylene]-	EtOH	300(3.86),325(3.81), 428(4.14)	12-1951-69
1,3-Indandione, 2-(2-oxocyclopentyl)-	pH 13	254(4.50),264(4.50), 290(3.11),304(3.20), 316(3.28),330(3.27), 435(3.32)	104-1196-69
	EtOH	224(4.48),254(4.28), 264(4.23),292(3.04), 302(3.07),316(3.02), 330(2.98),435(3.04)	104-1196-69
Naphtho[1,2-b]furan-4,5-dione, 2,3-dihydro-3,6-dimethyl-	EtOH	218(4.26),260(4.42), 269s(4.35),355(3.34), 440(3.36)	39-1184-69C
Naphtho[1,2-b]furan-4,5-dione, 2,3-dihydro-3,9-dimethyl-	EtOH	220(4.45),258s(4.49), 267(4.57),275s(4.52), 293s(4.03),360(3.49), 465(3.62)	39-1184-69C
Naphtho[2,3-b]furan-4,9-dione, 2,3-dihydro-3,8-dimethyl-	EtOH	210(4.12),248s(4.15), 253(4.21),290(3.99), 358(3.44)	39-1184-69C
Psoralen, 4,4',8-trimethyl-	MeOH	220(3.87),248(4.63), 298(4.28)	42-1014-69
$C_{14}H_{12}O_4$			
Benzo[1,2-b:5,4-b']difuran-4,8-dione, 2,3-dihydro-2-isopropenyl-5-methyl-	EtOH	275(4.28),334(3.93), 463(2.75)	88-4669-69
Coumarin, 4-hydroxy-3-(3-oxocyclopentyl)-	2N NaOH	310(4.13)	104-1566-69
	EtOH	310(4.09)	104-1566-69
Coumarin, 4-[(2-oxocyclopentyl)oxy]-	EtOH	268(4.03),278(4.02), 304(3.85)	104-1196-69

Compound	Solvent	λ_{max}(log ε)	Ref.
Cyclopenta[c]pyran-4-carboxylic acid, 7-(2-butenylidene)-1,7-dihydro-1-oxo-, methyl ester	EtOH	272(3.85),365(4.53)	35-6470-69
1,2-Naphthalenedicarboxylic acid, dimethyl ester	EtOH	238(4.76),283(3.82), 294(3.72)	12-1721-69
1-Naphthoic acid, 6,7-(methylenedioxy)-3-methyl-, methyl ester	EtOH	235(4.76),294(3.70), 332(3.67)	12-1721-69
1,4-Naphthoquinone, 2-p-dioxan-2-yl-	dioxan	246(4.24),252(4.25), 300(3.53),323(3.51), 408(1.86)	88-1169-69
Pyrocanescin, methyl-	EtOH	249(4.59),287(4.19), 300(4.32),346(3.58)	12-1933-69
3,3',4,5'-Stilbenetetrol	EtOH	221(4.35),305(4.20), 326(4.24)	39-1109-69C
$C_{14}H_{12}O_4S$			
2,4,6-Cycloheptatrien-1-one, 3-hydroxy-, p-toluenesulfonate	MeOH	227(4.33),293(3.73)	94-2548-69
3,4-Thiophenedicarboxylic acid, 2-phenyl-, dimethyl ester	MeOH	207(4.28),230(4.28), 275(3.91)	77-0569-69
$C_{14}H_{12}O_5$			
Benzophenone, 2,4,6-trihydroxy-3'-methoxy-	EtOH	220(4.36),261(3.67), 306(4.22)	39-0281-69C
Naphtho[2,3-c]furan-1(3H)-one, 6-hydroxy-5,7-dimethoxy-	EtOH	261(4.67),322(3.97)	78-3223-69
$C_{14}H_{12}O_6$			
Coumarin, 3-acetoxy-4-(acetoxymethyl)-	EtOH	279(4.10),315s(--)	87-0531-69
Isohalfordin	EtOH	240(4.33),245s(4.33), 259s(4.19),299(4.11)	31-0354B-69
$C_{14}H_{13}$			
1,1-Diphenylethylium cation	acid	314(4.08),427(4.59)	59-0447-69
o-Methyl-α-phenylbenzylium cation	acid	306(3.53),446(4.69)	59-0447-69
$C_{14}H_{13}BrO_4$			
6-Oxabicyclo[3.2.1]octan-7-one, 4-bromo-8-[3,4-(methylenedioxy)phenyl]-	EtOH	236(3.67),286(3.65)	78-4315-69
$C_{14}H_{13}Br_2N_3$			
6,7-Dihydropyrido[1,2-a]pyrido[1',2':-3,4]imidazo[2,1-c]pyrazine-5,8-diium dibromide	H_2O	248(3.99),253(3.98), 315(3.55),330s(3.41), 410(4.34)	39-1987-69C
$C_{14}H_{13}Br_2N_3O$			
6,7-Dihydro-6-hydroxypyrido[1,2-a]pyrido[1',2':3,4]imidazo[2,1-c]pyrazine-5,8-diium dibromide	5N HCl	248(4.02),255s(3.98), 315(3.57),330s(3.45), 415(3.32)	39-1987-69C
	H_2O	316(4.05)	39-1987-69C
$C_{14}H_{13}Br_3$			
Naphthalene, 1,2,3-tris(bromomethyl)-4-methyl-	EtOH	235(4.66),247(4.57), 295(3.85)	94-1591-69
$C_{14}H_{13}ClN_4O_2$			
Aniline, p-[(4-chloro-3-nitrophenyl)-azo]-N,N-dimethyl-	benzene	325(3.51),350(3.38)	73-2092-69

Compound	Solvent	λ_{max}(log ε)	Ref.
$C_{14}H_{13}ClO_3$			
1,4-Naphthoquinone, 3-(2-chloro-1-methylethyl)-2-hydroxy-5-methyl-	EtOH	209(4.16),251(4.26), 280(4.12),352(3.53)	39-1184-69C
$C_{14}H_{13}ClO_4$			
1-Methyl-2-phenylcycloheptatrienylium perchlorate	MeCN	277(3.80),352(3.59)	49-0001-69
1-Methyl-3-phenylcycloheptatrienylium perchlorate	MeCN	275(4.30),359(4.13)	49-0001-69
o-Tolylcycloheptatrienylium perchlorate	MeCN	268(3.88),366(3.85)	49-0001-69
$C_{14}H_{13}Cl_2NO_2$			
Pyrrole-2-carboxylic acid, 3-(2,4-dichlorophenyl)-5-methyl-, ethyl ester	EtOH	245(4.14),282(4.13)	94-0582-69
Pyrrole-2-carboxylic acid, 5-(2,4-dichlorophenyl)-3-methyl-, ethyl ester	EtOH	240(4.06),305(4.37)	94-0582-69
Pyrrole-2-carboxylic acid, 3-(3,4-dichlorophenyl)-5-methyl-, ethyl ester	EtOH	283(4.20)	94-0582-69
Pyrrole-2-carboxylic acid, 5-(3,4-dichlorophenyl)-3-methyl-, ethyl ester	EtOH	300(4.36)	94-0582-69
Pyrrole-3-carboxylic acid, 4-(3,4-dichlorophenyl)-2-methyl-, ethyl ester	EtOH	278(3.98)	94-0582-69
$C_{14}H_{13}IO_4$			
6-Oxabicyclo[3.2.1]octan-7-one, 4-iodo-8-[3,4-(methylenedioxy)phenyl]-, cis-trans	EtOH	237(3.68),286(3.68)	78-4315-69
$C_{14}H_{13}N$			
Aniline, N-(α-methylbenzylidene)-	EtOH	244(4.28),310(3.32)	56-0749-69
Pyridine, 2-(m-methylstyryl)-, trans	H_2O	309(4.41)	23-2355-69
hydrochloride	H_2O	337(4.43)	23-2355-69
Pyridine, 2-(p-methylstyryl)-, cis	H_2O	292(4.00)	23-2355-69
hydrochloride	H_2O	329(3.95)	23-2355-69
Pyridine, 2-(p-methylstyryl)-, trans	H_2O	313(4.47)	23-2355-69
hydrochloride	H_2O	343(4.49)	23-2355-69
Pyridine, 2-(α-methylstyryl)-, cis	hexane	288(4.0)	4-0465-69
Pyridine, 2-(α-methylstyryl)-, trans	hexane	290(4.3)	4-0465-69
Pyridine, 3-(m-methylstyryl)-, trans	50% MeOH	308(4.47)	23-2355-69
hydrochloride	50% MeOH	342(4.49)	23-2355-69
Pyridine, 3-(p-methylstyryl)-, trans	50% MeOH	313(4.52)	23-2355-69
hydrochloride	50% MeOH	353(4.53)	23-2355-69
Pyridine, 4-(α-methylstyryl)-, cis	hexane	240(4.2),260(4.0)	4-0465-69
Pyridine, 4-(α-methylstyryl)-, trans	hexane	275(4.3)	4-0465-69
$C_{14}H_{13}NO$			
Acetophenone, 2'-anilino-	C_6H_{12}	228(4.34),252(4.06), 288(4.06),380(3.88)	22-3523-69
	EtOH	228(4.25),250(4.08), 288(4.00),389(3.81)	22-3523-69
Acetophenone, 3'-anilino-	EtOH	239(4.16),259(4.13), 292(4.25),375(3.14)	22-3523-69
Acetophenone, 4'-anilino-	C_6H_{12}	227(4.01),321(4.41)	22-3523-69
	EtOH	241(4.06),300s(3.82), 345(4.49)	22-3523-69
4,5-Benzo-7-azabicyclo[4.2.2]deca-2,4,7,9-tetraene, 8-methoxy-	hexane	274(3.93)	35-4714-69
6,7-Benzo-3-azatricyclo[3.3.2.0^{2,8}]-deca-3,6,9-triene, 4-methoxy-	hexane	270(2.90),281(2.71)	35-5296-69

Compound	Solvent	$\lambda_{max}(\log \epsilon)$	Ref.
Carbazole, 2-methoxy-1-methyl-	EtOH	240(4.58),253(4.45), 258(4.39),301(4.16), 319(3.74),330(3.38)	77-1120-69
Carbazole, 2-methoxy-3-methyl-	EtOH	236(4.73),250(4.30), 257(4.26),302(4.25), 331(3.65)	77-1120-69
Fluorene-4a(2H)-carbonitrile, 1,3,4,9a-tetrahydro-9-oxo-	EtOH	244(4.09),283(3.24), 288(3.24)	44-1899-69
Indol-4(5H)-one, 6,7-dihydro-3-phenyl-	EtOH	218(4.34),242s(3.93), 263(4.02),287(3.82)	78-4005-69
Nitrone, N-benzyl-α-phenyl-	EtOH	226s(3.9),294(4.3), 303s(4.1)	28-0078-69A
Nitrone, N-phenyl-α-p-tolyl-	C_6H_{12}	327(4.32)	18-3306-69
	EtOH	319(4.37)	18-3306-69
1-Propanone, 3-phenyl-1-(4-pyridyl)-	EtOH	216(3.85),260(3.32)	39-2134-69C
Pyridine, 2-(m-methoxystyryl)-, trans	H_2O	308(4.38)	23-2355-69
hydrochloride	H_2O	333(4.38)	23-2355-69
Pyridine, 2-(p-methoxystyryl)-, cis	H_2O	295(4.02)	23-2355-69
hydrochloride	H_2O	346(3.96)	23-2355-69
Pyridine, 2-(p-methoxystyryl)-, trans	H_2O	322(4.48)	23-2355-69
hydrochloride	H_2O	358(4.49)	23-2355-69
Pyridine, 3-(m-methoxystyryl)-, trans	50% MeOH	300(4.41)	23-2355-69
hydrochloride	50% MeOH	338(4.39)	23-2355-69
Pyridine, 3-(p-methoxystyryl)-, trans	50% EtOH	324(4.50)	23-2355-69
hydrochloride	50% EtOH	372(4.53)	23-2355-69
$C_{14}H_{13}NOS$			
Benzenethiol, o-[N-(p-methoxyphenyl)-formimidoyl]-	EtOH	240(4.3),340(4.0), 510(3.5)	30-0605-69A
	dioxan	240(4.3),340(4.0), 510(3.5)	30-0605-69A
Benzothiazoline, 2-(p-methoxyphenyl)-, cobalt derivative	n.s.g.	430(3.61),500s(3.38)	97-0385-69
nickel derivative	n.s.g.	458(3.70)	97-0385-69
zinc derivative	n.s.g.	444(3.90)	97-0385-69
$C_{14}H_{13}NO_2$			
Anthranilic acid, N-o-tolyl-	C_6H_{12}	195(4.47),205(4.39), 223(4.45),240s(4.00), 280(3.95),359(3.87)	22-3523-69
	EtOH	205(4.48),220(4.34), 289(4.10),337(3.76)	22-3523-69
3H-Azepine, 3-benzoyl-2-methoxy-	MeOH	246(4.21)	78-5205-69
Benzamide, N-(benzyloxy)-	MeOH	208(4.4),217(4.12), 224(4.09)	12-0161-69
	MeOH-NaOH	267(3.65)	12-0161-69
Furo[2,3-b]quinolin-4(9H)-one, 3-isopropyl-	MeOH	238(4.42),249(4.36), 257(4.39),282(3.40), 294(3.40),320s(3.80), 332(3.92),340s(3.84)	44-2183-69
	MeOH-acid	241(4.68),257s(4.10), 286(3.33),299s(3.57), 318s(3.78),334(3.88), 342s(3.80)	44-2183-69
	MeOH-base	232(4.29),255s(4.52), 262(4.59),307s(3.53), 320s(3.70),331s(3.82), 342(3.93),355(3.85)	44-2183-69
Nitrone, N-benzyl-α-(m-hydroxyphenyl)-	EtOH	222s(4.1),290(4.2)	28-0078-69A
Nitrone, N-benzyl-α-(o-hydroxyphenyl)-	EtOH	228s(4.0),286(4.0), 335(3.7)	28-0078-69A

Compound	Solvent	$\lambda_{max}(\log \epsilon)$	Ref.
Nitrone, N-benzyl-α-(p-hydroxyphenyl)-	EtOH	229(3.9),310(4.4)	28-0078-69A
Nitrone, α-(p-methoxyphenyl)-N-phenyl-	C_6H_{12}	334(4.31)	18-3306-69
	EtOH	331(4.38)	18-3306-69
5H-Pyrano[3,2-c]quinolin-5-one, 2,6-dihydro-2,2-dimethyl-	EtOH	222(4.46),234(4.36), 334(3.99),350(4.08), 365(3.92)	32-0711-69
9H-Pyrrolo[1,2-a]indole, 2-acetyl-7-methoxy-	EtOH	241(3.90),289(4.30), 311(4.29)	88-0101-69
$C_{14}H_{13}NO_2S_2$			
2H-Thiopyran-5-carboxylic acid, 6-amino-4-phenyl-2-thioxo-, ethyl ester	EtOH	227(4.24),303(4.16), 330s(4.07),468(4.23)	78-1441-69
$C_{14}H_{13}NO_3$			
Acenaphtho[1,2-c]pyrrole-2-carboxylic acid, N-methyl-, anhydride with acetic acid	$CHCl_3$	258(4.30),315(3.42), 329(3.57),344(3.89), 386(4.02)	97-0022-69
Anthranilic acid, N-p-methoxyphenyl-	EtOH	202(4.50),210s(4.47), 225s(4.23),290(4.15), 340(3.74)	22-3523-69
	EtOH	288(3.97),333(3.54)	42-0103-69
1-(α-Carboxyphenethyl)-3-hydroxypyridinium hydroxide, inner salt	N HCl	295(3.61)	1-2488-69
	N NaOH	250(3.85),325(3.60)	1-2488-69
Furanacrylohydroxamic acid, N-m-tolyl-	EtOH	315(4.42)	34-0278-69
Furanacrylohydroxamic acid, N-p-tolyl-	EtOH	315(4.43)	34-0278-69
Indoline-2,4(5H)-dione, 6,7-dihydro-1-hydroxy-3-phenyl-	EtOH	268(4.03)	78-4005-69
1-Isoquinolineacrylic acid, α-hydroxy-, ethyl ester, (Z)-	$CHCl_3$	231(4.21),234(4.15), 303(4.00),418(4.36), 439(4.40)	24-0915-69
$C_{14}H_{13}NO_4$			
4-Oxazoleacrylic acid, 5-ethoxy-2-phenyl-, trans	EtOH	223(4.26),312(4.39)	94-2424-69
$C_{14}H_{13}NO_4S_2$			
p-Dithiin-2,3-dicarboximide, 5,6-dihydro-N-(2,5-dimethoxyphenyl)-	MeOH	221(4.22),265(3.97), 410(3.43)	33-2221-69
$C_{14}H_{13}N_3$			
Benzimidazole, 5-amino-1-methyl-2-phenyl-	EtOH	211(4.57),229s(4.38), 270(4.00),321(3.94)	73-0572-69
Benzimidazole, 5-amino-2-methyl-1-phenyl-	EtOH	210(4.51),252s(4.00), 305(3.61)	73-0572-69
Lepidine, 2-(2-pyrrolyl)-	n.s.g.	360(4.12)	97-0186-69
$C_{14}H_{13}N_3O$			
Benzaldehyde, 4-phenylsemicarbazone	MeOH	232(4.34),292(4.44)	5-0001-69E
Isonicotinic acid, (α-methylbenzylidene)hydrazide	n.s.g.	286(4.16)	40-0073-69
Nicotinic acid, (α-methylbenzylidene)-hydrazide	n.s.g.	292(4.20)	40-0073-69
Picolinic acid, (α-methylbenzylidene)-hydrazide	n.s.g.	222(4.31),303(4.35)	40-0073-69
$C_{14}H_{13}N_3OS$			
6-(2-Benzimidazolylthio)-5-hydroxy-1-methyl-2-picolinium hydroxide, inner salt	pH 1	280(4.09),325(4.11)	1-2065-69
	pH 13	275(4.17),355(4.03)	1-2065-69

Compound	Solvent	λ_{max}(log ϵ)	Ref.
$C_{14}H_{13}N_3O_2$			
Benzaldehyde, m-nitro-, methylphenylhydrazone	EtOH	203(4.3),226(4.3), 232s(4.2),250s(4.1), 289(3.9),338(4.4)	28-0137-69B
Benzaldehyde, o-nitro-, methylphenylhydrazone	EtOH	224(4.2),247(4.1), 290(3.9),300(3.9), 333(4.2),400(3.9)	28-0137-69B
Benzaldehyde, p-nitro-, methylphenylhydrazone	EtOH	264(4.1),415(4.4)	28-0137-69B
2-Pyridinepropionanilide, β-oxo-, β-oxime, anti	EtOH	243(4.32)	36-0460-69
3-Pyridinepropionanilide, β-oxo-, β-oxime, anti	EtOH	243(4.38)	36-0460-69
4-Pyridinepropionanilide, β-oxo-, β-oxime, anti	EtOH	243(4.35)	36-0460-69
$C_{14}H_{13}N_3O_3$			
Benzaldehyde, α-methoxy-, p-nitrophenylhydrazone [sic]	dioxan	225(4.09),263s(3.80), 291(3.88),321s(3.66), 398(4.48)	39-2587-69C
Benzoic acid, p-(5,7-dihydro-4-hydroxy-6H-pyrrolo[3,4-d]pyrimidin-4-yl)-, methyl ester	10% NaOH	291(4.29),295s(4.29)	4-0507-69
2-Furoic acid, 5-(5-amino-1-methyl-2-benzimidazolyl)-, methyl ester	EtOH	208(4.38),245(3.95), 314s(4.05),355(4.26)	73-0572-69
$C_{14}H_{13}N_3O_3S$			
Sulfur diimide, (p-ethoxyphenyl)-(p-nitrophenyl)-	benzene	460(4.29)	104-0459-69
$C_{14}H_{13}N_3O_4$			
Pyridine, 2-(2,4-dinitrobenzyl)-4-ethyl-	EtOH	249(4.54)	22-4425-69
$C_{14}H_{13}N_3S$			
Lepidine, 2-(2-amino-4-methyl-5-thiazolyl)-	n.s.g.	366(4.12)	97-0186-69
$C_{14}H_{13}OP$			
Dibenzo[b,d]phosphorin, 5,6-dihydro-5-methyl-, 5-oxide	EtOH	215(4.46),228s(4.08), 267(4.08),281s(3.86)	39-0252-69C
$C_{14}H_{14}$			
Bicyclohepta-1,3,5-trien-1-yl	EtOH	262(3.60),340(4)	88-0663-69
Bicyclohepta-2,4,6-trien-1-yl	pentane	252(3.83)	77-1159-69
Cyclobuta[4,5]benzo[1,2]cyclooctene, 1,2,3,10-tetrahydro-	EtOH	209(4.42),258s(3.08)	35-4734-69
1,3,5-Cycloheptatriene, 1-(2,4,6-cycloheptatrienyl)-	EtOH	262(3.79)	88-0663-69
$C_{14}H_{14}BrI$			
Di-p-tolyliodonium bromide	MeOH	235(4.43),260(3.68), 263(3.62),275(3.39)	120-0012-69
$C_{14}H_{14}BrNO_2$			
Pyrido[1,2-a]indole-10-carboxylic acid, 8-bromo-6,7,8,9-tetrahydro-, methyl ester	EtOH	216(4.38),233(4.26), 288(4.05)	12-0997-69
Pyrrole-2-carboxylic acid, 3-(m-bromophenyl)-5-methyl-, ethyl ester	EtOH	288(4.14)	94-0582-69

Compound	Solvent	λ_{max}(log ε)	Ref.
$C_{14}H_{14}BrNO_3$			
Pyrido[1,2-a]indole-10-carboxylic acid, 8-bromo-9-hydroxy-6,7,8,9-tetrahydro-, methyl ester	EtOH	215(4.48),230(4.37), 292(4.12)	12-0997-69
$C_{14}H_{14}BrN_3O$			
2-[2-(2-Oxo-1(2H)-pyridyl)ethyl]imidazo-[1,5-a]pyridinium bromide	H_2O	227(3.93),263s(3.77), 274(3.95),282(4.03), 296(3.98),301s(3.96)	39-1987-69C
$C_{14}H_{14}ClN$			
6H-Cyclohepta[b]quinoline, 11-chloro-7,8,9,10-tetrahydro-	EtOH	230(4.88),235s(4.76), 277(3.87),293s(3.80), 305(3.79),310s(3.60), 319(3.89)	88-0557-69
$C_{14}H_{14}ClNO_2$			
Pyrido[1,2-a]indole-10-carboxylic acid, 8-chloro-6,7,8,9-tetrahydro-, methyl ester	EtOH	215(4.48),229(4.34), 251s(3.95),288(4.11), 296s(3.99)	12-0997-69
Pyrrole-2-carboxylic acid, 3-(o-chlorophenyl)-5-methyl-, ethyl ester	EtOH	281(4.19)	94-0582-69
Pyrrole-2-carboxylic acid, 3-(p-chlorophenyl)-5-methyl-, ethyl ester	EtOH	283(4.20)	94-0582-69
Pyrrole-3-carboxylic acid, 2-(m-chlorophenyl)-4-methyl-, ethyl ester	EtOH	300(4.0)	94-0582-69
Pyrrole-x-carboxylic acid, y-(3-chlorophenyl)-z-methyl-, ethyl ester	EtOH	232(4.20),310(4.48)	94-0582-69
$C_{14}H_{14}ClNO_4$			
(N-Methylanilino)tropylium perchlorate	MeCN	240(4.33),340(4.23)	64-1353-69
$C_{14}H_{14}ClN_3$			
Aniline, p-[(p-chlorophenyl)azo]-N,N-dimethyl-	EtOH	272(4.05),412(4.48)	18-3565-69
1H-Imidazo[4,5-g]quinoline, 8-chloro-1-ethyl-2,6-dimethyl-	EtOH	248(4.76),332(4.00)	94-2455-69
3H-Imidazo[4,5-f]quinoline, 9-chloro-3-ethyl-2,7-dimethyl-	EtOH	256s(4.60),260(4.64), 320(3.57)	94-2455-69
$C_{14}H_{14}ClN_3O_4$			
8,9,10,11-Tetrahydropyrido[1',2':1,5]-s-triazolo[3,4-a]isoquinolin-12-ium perchlorate	MeCN	227s(4.45),233(4.61), 239(4.67),246(4.68), 263(3.99),272(3.99), 283(3.87),294(3.15), 306(3.50),319(3.20)	24-3159-69
$C_{14}H_{14}Cl_2N_4$			
2-Tetrazene, 1,4-bis(m-chlorophenyl)-1,4-dimethyl-	$(MeOCH_2)_2$	248(4.11),310s(4.23), 352(4.60)	35-6452-69
2-Tetrazene, 1,4-bis(p-chlorophenyl)-1,4-dimethyl-	$CHCl_3$	252(3.91),318s(4.23), 357(4.52)	35-6452-69
	$(MeOCH_2)_2$	254(4.23),315s(4.20), 354(3.49)	35-6452-69
$C_{14}H_{14}CrGeO_5S$			
Chromium, pentacarbonyl[trimethyl(phenylthio)germane]-	n.s.g.	413(2.89)	89-0271-69

Compound	Solvent	$\lambda_{max}(\log \epsilon)$	Ref.
$C_{14}H_{14}CrO_5SSn$			
Chromium, pentacarbonyl[trimethyl(phenylthio)stannane]-	n.s.g.	417(3.08)	89-0271-69
$C_{14}H_{14}FeO$			
3-Buten-2-one, 4-ferrocenyl-	EtOH	210(4.20),258(3.99), 305(4.16),369(3.33), 486(3.24)	18-3273-69
Ferrocenecarboxaldehyde, 1',2-trimethylene-	EtOH	338(3.06),448(2.87)	49-1540-69
Ferrocenecarboxaldehyde, 1',3-trimethylene-	EtOH	338(3.29),458(3.06)	49-1540-69
1-Propanone, 1,3-(2-methyl-1,1'-ferrocenediyl)-	EtOH	336(2.94),448(2.46)	49-1540-69
isomer	EtOH	334(2.40),446(2.48)	49-1540-69
1-Propanone, 1,3-(3-methyl-1,1'-ferrocenediyl)-	EtOH	332(3.02),444(2.51)	49-1540-69
isomer	EtOH	334(2.94),446(2.50)	49-1540-69
$C_{14}H_{14}FeO_2$			
Ferrocenecarboxylic acid, 1',2-trimethylene-	EtOH	297(3.00),338(2.55), 428(2.49)	49-1540-69
Ferrocenecarboxylic acid, 1',3-trimethylene-	EtOH	304(3.20),340(2.70), 442(2.66)	49-1540-69
$C_{14}H_{14}IN_3$			
1-Methyl-4-(5-methyl-3-pyrazolyl)quinolinium iodide	pH 13	392(4.15)	87-1124-69
	MeOH	243(4.49),352(4.10)	87-1124-69
$C_{14}H_{14}I_2$			
Di-p-tolyliodonium iodide	MeOH	220(4.37),235(4.24), 263(3.70),275(3.42), 286(3.41)	120-0012-69
$C_{14}H_{14}NOPS$			
Formamide, 1-(diphenylphosphinyl)-N-methyl-thio-	EtOH	291(3.84),395(1.92)	18-2975-69
$C_{14}H_{14}NPS_2$			
Formamide, 1-(diphenylphosphinothioyl)-N-methyl-thio-	EtOH	280(3.88),400(1.83)	18-2975-69
$C_{14}H_{14}N_2$			
Acetophenone, phenylhydrazone	EtOH	232(4.11),302s(4.14), 330(4.30)	39-1703-69C
	EtOH-HCl	230(4.08),276s(3.59), 346(3.16)	39-1703-69C
Azobenzene, 2,2'-dimethyl-, trans	20% EtOH	234(4.02),335(4.22), 440(2.88)	54-0562-69
	+ H_2SO_4	239(3.73),440(4.39)	54-0562-69
Azobenzene, 2,6-dimethyl-, trans	20% EtOH	312(4.04),435(2.93)	54-0562-69
Azobenzene, 4,4'-dimethyl-, trans	20% EtOH	237(4.18),340(4.43), 420(3.34)	54-0562-69
	+ H_2SO_4	245(3.97),450(4.54)	54-0562-69
Benzaldehyde, methylphenylhydrazone	EtOH	202(4.4),236(4.2), 300s(4.0),335(4.4)	28-0137-69B
	EtOH	236(4.20),304(4.10), 334(4.42)	39-1703-69C
	EtOH-HCl	286(3.30),342(2.49)	39-1703-69C
Benzamidine, N-p-tolyl-	50% EtOH	270(4.06)	49-1307-69

Compound	Solvent	$\lambda_{max}(\log \epsilon)$	Ref.
Dibenzo[c,g][1,2]diazocine, 5,6,11,12-tetrahydro-	EtOH	222(4.50),257(3.71), 305(3.60)	39-0882-69C
	EtOH-HCl	249(3.97),294(3.27)	39-0882-69C
	n.s.g.	250(4.2),280(3.6)	44-3237-69
3-Picoline, 6,6'-vinylenedi-, trans	n.s.g.	271(4.16),319(4.41)	39-2255-69C
$C_{14}H_{14}N_2O$			
3H-Azepine, 2-anilino-3-acetyl-	MeOH	305(4.18)	78-5205-69
Cinchoninonitrile, 3-butyl-1,2-dihydro-2-oxo-	CH_2Cl_2	233(4.36),294(3.87), 342s(3.84),357(3.87)	94-1294-69
p-Cresol, 2-(p-tolylazo)-	50% EtOH	333(4.31),390s(--)	73-1087-69
	50% EtOH-46% $HClO_4$	305s(--),412(4.31), 488(4.22)	73-1087-69
Cyclohexanone, 2-(1-phthalazinyl)-	EtOH	218(4.79),257s(3.67), 266(3.73),274(3.71), 306(3.10)	95-0959-69
4H-Indazol-4-one, 2,5,6,7-tetrahydro-3-methyl-2-phenyl-	EtOH	260(4.61)	70-2680-69
$C_{14}H_{14}N_2O_2$			
2,1-Benzisoxazole, 3,4,6-trimethyl-5-(5-methylisoxazol-3-yl)-	EtOH	210(4.62),260-90f(3), 318(3.79)	35-4749-69
p-Cresol, 2-[(o-methoxyphenyl)azo]-	50% EtOH	323(4.15),382(4.12), 396(4.11)	73-1087-69
	50% EtOH-42% $HClO_4$	307(3.55),386(4.17), 445s(--),508(4.28)	73-1087-69
p-Cresol, 2-[(p-methoxyphenyl)azo]-	50% EtOH	351(4.28),387(4.17), 425s(--)	73-1087-69
	50% EtOH-44% $HClO_4$	307(3.61),433(4.27), 508(4.39)	73-1087-69
4-Isoquinolinecarbonitrile, 3-ethyl-6,8-dimethoxy-	EtOH	218(4.43),256(4.54)	95-1492-69
5,8-Isoquinolinedione, 3-methyl-7-(1-pyrrolidinyl)-	MeCN	245(4.06),250(4.05), 295(3.64)	4-0697-69
$\Delta^{2,\alpha}$-Pyrrolidineacetic acid, α-cyano-1-phenyl-, methyl ester	isooctane	237s(3.58),292(4.20)	35-6689-69
Pyrrolo[1,2-a]pyrimidin-4(6H)-one, 7,8-dihydro-2-methoxy-3-phenyl-	5N HCl	258(3.99)	22-3133-69
	H_2O	262s(--),277(3.93)	22-3133-69
	EtOH	258(3.76),290(3.96)	22-3133-69
4,6,7,8-Tetrahydro-2-hydroxy-1-methyl-4-oxo-3-phenylpyrrolo[1,2-a]pyrimidinium hydroxide, inner salt	5N HCl	262(3.89)	22-3133-69
	H_2O	264(3.95),290s(--)	22-3133-69
	EtOH	268(3.91),315(3.74)	22-3133-69
$C_{14}H_{14}N_2O_2S$			
2H-1,2,4-Benzothiadiazine, 3,4-dihydro-4-methyl-2-phenyl-, 1,1-dioxide	EtOH	211(4.49),258(4.10), 323(3.56)	7-0590-69
1-Indolineacrylic acid, α-cyano-β-(methylthio)-, methyl ester	MeOH	240(3.88),305(3.96), 364(4.24)	78-4649-69
Sulfur diimide, bis(p-methoxyphenyl)-	benzene	458(4.02)	104-0459-69
p-Toluenesulfonic acid, 2,4,6-cycloheptatrien-1-ylidenehydrazide	EtOH	245(3.90),315(4.12)	35-6391-69
hydrochloride	EtOH	229(4.12),315(4.20)	35-6391-69
$C_{14}H_{14}N_2O_2S_2$			
p-Dithiin-2,3-dicarboximide, N-[p-(dimethylamino)phenyl]-5,6-dihydro-	MeOH	264(4.50),374(3.38)	33-2221-69
$C_{14}H_{14}N_2O_3$			
2H-Indazole-4-carboxylic acid, 3,3a,4,-5,6,7-hexahydro-3-oxo-2-phenyl-	EtOH	204(4.27),247(4.11), 273s(3.98)	39-2783-69C

Compound	Solvent	$\lambda_{max}(\log \epsilon)$	Ref.
2H-Indazole-7-carboxylic acid, 3,3a,4,-5,6,7-hexahydro-3-oxo-2-phenyl-	EtOH	204(4.25),248(4.14), 280s(3.92)	39-2783-69C
Isobutyric acid, (3-hydroxy-4-oxo-1(4H)-naphthylidene)hydrazide (in 30% MeOH)	pH 1.2	224(4.30),257(3.87), 305(4.03),394(4.23)	44-2750-69
	pH 12.2	294(3.98),450(4.04),	44-2750-69
5,8-Isoquinolinedione, 3-methyl-7-morpholino-	MeCN	243(4.10),275(4.01), 325(3.64)	4-0697-69
1-Phthalazineacetic acid, α-acetyl-, ethyl ester	EtOH	220(4.75),255(4.06), 384(3.10)	95-0959-69
2H-Pyrano[2,3-d]pyrimidine-2,4(3H)-dione, 1,5,6,7-tetrahydro-7-methyl-3-phenyl-	75% dioxan	264(4.03)	95-0266-69
	+ NaOH	280(4.06)	95-0266-69
Pyridazine, 3-(3,4-dimethoxystyryl)-, 1-oxide	EtOH	261(4.22),368(4.27)	95-0132-69
Pyridazine, 4-(3,4-dimethoxystyryl)-, 1-oxide	EtOH	267(4.20),359(4.10)	95-0132-69
Pyridazine, 6-(3,4-dimethoxystyryl)-, 1-oxide	EtOH	256(3.81),354(3.88)	95-0132-69
$C_{14}H_{14}N_2O_3S_2$			
Sulfur diimide, (p-ethoxyphenyl)-(phenylsulfonyl)-	benzene	434(4.32)	104-0459-69
$C_{14}H_{14}N_2O_4$			
1,6-Indolinedicarboxaldehyde, 3-[(dimethylamino)methylene]-5-methoxy-2-oxo-	EtOH	234(4.10),271(3.87), 320(3.94),380(4.38)	2-0662-69
$C_{14}H_{14}N_2S$			
4H-Cyclopenta[d]pyrimidine-4-thione, 1,5,6,7-tetrahydro-1-methyl-2-phenyl-	EtOH	249(3.91),342(4.24)	4-0037-69
Sulfur diimide, di-p-tolyl-	benzene	434(4.11)	104-0459-69
$C_{14}H_{14}N_4O$			
1H-Pyrazolo[3,4-b]pyrazin-6-ol, 5-ethyl-1-methyl-3-phenyl-	EtOH	245(4.19),330(4.10)	32-0463-69
1H-Pyrazolo[3,4-b]pyrazin-6-ol, 5-ethyl-3-methyl-1-phenyl-	EtOH	258(4.36),317(3.98)	32-0463-69
4H-Pyrido[1,2-a]pyrimidine-3-carbonitrile, 4-oxo-2-piperidino-	EtOH	271(4.63),349(3.96)	95-0203-69
2-Tetrazene-1-carboxaldehyde, 4-methyl-1,4-diphenyl-	EtOH	314(4.00)	44-2997-69
$C_{14}H_{14}N_4O_2S$			
Sulfur diimide, (p-dimethylaminophenyl)-(p-nitrophenyl)-	benzene	531(4.40)	104-0459-69
$C_{14}H_{14}N_4O_3$			
as-Triazino[4,5-a]benzimidazole-4-carboxylic acid, 2-ethyl-1,2-dihydro-1-oxo-, ethyl ester	n.s.g.	326(3.93)	88-1203-69
N-ind-Et isomer	n.s.g.	359(4.07)	88-1203-69
$C_{14}H_{14}N_6O_4$			
2-Tetrazene, 1,4-dimethyl-1,4-bis(p-nitrophenyl)-	$CHCl_3$	263(3.99),293(3.74), 435(4.73)	35-6452-69
$C_{14}H_{14}O$			
Fluoren-2(1H)-one, 3,4-dihydro-8-methyl-	EtOH	254(4.0),262(3.9)	42-0415-69

Compound	Solvent	$\lambda_{max}(\log \epsilon)$	Ref.
1-Phenanthrol, 1,2,3,4-tetrahydro-, (+)-	MeOH	228(4.96),272(3.79), 281(3.82)	19-0475-69
(-)-	MeOH	228(4.93),271(3.72), 281(3.79)	19-0475-69
4-Phenanthrol, 1,2,3,4-tetrahydro-, (-)-	MeOH	228(4.95),278(3.72), 289(2.58),322(2.68)	19-0475-69
$C_{14}H_{14}OS$			
Dibenzothiophene, 2-acetyl-5,6,7,8-tetrahydro-	EtOH	220(4.32),230s(4.17), 256(4.26),298(4.18), 333(3.93)	22-0607-69
$C_{14}H_{14}OS_2$			
Ketone, methyl 2,3,4,5-tetrahydrothiepino[3,2-b][1]benzothien-9-yl	EtOH	228(4.04),262(4.22), 300(3.97),338(3.81)	22-0607-69
$C_{14}H_{14}O_2$			
Acetophenone, 2'-(2-furyl)-3',5'-dimethyl-	EtOH	239(4.26),265s(4.02), 295(3.63)	6-0277-69
Benzo[b]cyclopenta[3,4]cyclobuta[1,2-e]pyran-9(1H)-one, 2,3,3a,3b,9a,9b-hexahydro-	MeOH	216(4.31),255(3.88), 330(3.45)	35-4494-69
4-Chromanone, 2-(2-cyclopenten-1-yl)-	MeOH	214(4.34),252(3.86), 321(3.45)	35-4494-69
2,4-Pentanedione, 3-cinnamylidene-	EtOH	207(4.30),225s(--), 326(4.38)	23-4076-69
$C_{14}H_{14}O_2S$			
Phenetole, p-(phenylsulfonyl)-	H_2O	246(4.22)	18-1964-69
	90% H_2SO_4	283(4.11)	18-1964-69
2-Thiophenecarboxylic acid, 3-methyl-5-phenyl-, ethyl ester	EtOH	225(3.98),312(4.30)	2-0009-69
$C_{14}H_{14}O_3$			
2-Biphenylcarboxylic acid, 2'-hydroxy-5'-methoxy-, lactone	EtOH	275(4.47),334(4.18)	12-1721-69
Fluorene-4a(2H)-carboxylic acid, 1,3,4,9a-tetrahydro-9-oxo-	EtOH	242(4.04),290(3.32)	44-1899-69
Naphtho[2,3-b]furan-4,9-dione, 5,6,7,8-tetrahydro-3,8-dimethyl-	MeOH	216(4.12),260(4.08), 266(4.06),302(3.53)	23-2465-69
1-Naphthoic acid, 7-methoxy-3-methyl-, methyl ester	EtOH	222(4.76),233(4.75), 293(3.75),347(3.85)	12-1721-69
9H-Xanthen-9-one, 1,2,3,4-tetrahydro-7-methoxy-	EtOH	235(4.38),279(3.83), 322(3.80)	12-1721-69
Xanthyletin, dihydro-	EtOH	223(4.2),249(3.6), 260(3.6),334(4.2)	39-0033-69C
$C_{14}H_{14}O_4$			
2'-Acetonaphthone, 1',8'-dihydroxy-6'-methoxy-3'-methyl-	EtOH	232(4.45),267(4.43), 315(3.69),383(3.75)	94-0454-69
Cyclopentanecarboxylic acid, 4-benzoyl-2-methyl-3-oxo-	EtOH	246(3.91),320(4.10)	39-1845-69C
Cyperaquinone, tetrahydro-	EtOH	264(3.66),323(4.36), 480(2.50)	88-4669-69C
Lomatin, (+)-	EtOH	247(3.42),256(3.36), 327(4.09)	33-1165-69
1-Naphthol, 2,4-dimethoxy-, acetate	MeOH	253(4.60),301(3.67)	44-2788-69
2,4-Pentadienoic acid, 5-hydroxy-, ethyl ester, benzoate, di-trans	MeOH	242s(4.20),275(4.59)	78-4315-69
2-cis-4-trans	MeOH	235(4.25),273(4.39)	78-4315-69

Compound	Solvent	$\lambda_{max}(\log \epsilon)$	Ref.
Prenyletin	EtOH	231(4.26),256(3.81), 260s(3.80),298(3.86), 350(4.12)	39-0526-69C
	EtOH-NaOH	254(4.41),279s(4.05), 315(3.97),403(3.94)	39-0526-69C
$C_{14}H_{14}O_5$			
Cyclopenta[c]pyran-4-carboxylic acid, 1,7-dihydro-7-(isopropoxymethylene)-1-oxo-, methyl ester	EtOH	344(4.32),355(4.33)	35-6470-69
D-glycero-Hex-3-enopyranosidulose, methyl 3,4-dideoxy-, benzoate	n.s.g.	219(4.09),277(4.17)	35-5392-69
7H-Pyrano[2,3-g]-1,4-benzodioxin-7-one, 2,3-dihydro-2-hydroxy-2-isopropyl-	EtOH	229(4.18),252(3.75), 259(3.75),298(3.98), 345(4.19)	39-0526-69C
	EtOH-NaOH	250(4.35),281(3.94), 300(3.84),402(3.80)	39-0526-69C
Yellow compound, m. 166-167°	$CHCl_3$	273(3.95)	22-3609-69
$C_{14}H_{14}O_5S_2$			
Acrylic acid, 3-(benzoylmercapto)-2-carboxy-3-(methylthio)-, dimethyl ester	MeOH	249(4.26),317(4.01)	78-4649-69
$C_{14}H_{14}O_6$			
Isohalfordin, dihydro-	EtOH	320(4.21)	31-0354B-69
$C_{14}H_{14}O_6S_2$			
Anisole, 4,4'-disulfonyldi-	dioxan	286(4.32)	35-5510-69
$C_{14}H_{15}AsIN$			
10,10-Dimethylphenarsazinium iodide	EtOH	270(4.20),325(4.00)	101-0117-69E
$C_{14}H_{15}BBrF_4NS_2$			
1-[4-(p-Bromophenyl)-1,3-dithiol-2-ylidene]piperidinium tetrafluoroborate	EtOH	243(4.20),322(4.17)	94-1924-69
$C_{14}H_{15}BF_4N_2O_2S_2$			
1-[4-(p-Nitrophenyl)-1,3-dithiol-2-ylidene]piperidinium tetrafluoroborate	EtOH	330(4.31)	94-1924-69
$C_{14}H_{15}BrN_2O$			
1-Benzyl-3-carbamoyl-4-picolinium bromide	EtOH	263(3.60)	39-0192-69B
$C_{14}H_{15}BrO$			
Fluorene, 6-bromo-1,2,3,4-tetrahydro-7-methoxy-	EtOH	217(4.45),269(4.13), 280s(4.03),306(3.44), 316s(3.29)	44-2209-69
$C_{14}H_{15}BrO_2$			
3H-Cyclopenta[b]benzofuran-3-one, 7-bromo-1,2,3a,8b-tetrahydro-3a,6,8b-trimethyl-	MeOH	294(3.52)	78-3509-69
$C_{14}H_{15}BrO_6$			
Bicyclo[3.2.2]nona-3,6-diene-8α,9α-dicarboxylic acid, 3-bromo-1-methoxy-2-oxo-, dimethyl ester	MeOH	248(3.50),275(3.47), 347(1.80)	88-0443-69

Compound	Solvent	$\lambda_{max}(\log \epsilon)$	Ref.
$C_{14}H_{15}BrO_8$			
1,2,4-Benzenetriol , 3-bromo-5,6-dimethoxy-, triacetate	EtOH	216(4.00),270(2.69)	39-1353-69C
$C_{14}H_{15}ClN_2O_2S$			
Benzenesulfonanilide, 4-chloro-4'-(dimethylamino)-	EtOH-1% $CHCl_3$	230(4.26),262(4.26), 300s(5.64)	44-2083-69
p-Phenylenediamine, 2-[(p-chlorophenyl)-sulfonyl]-N^4,N^4-dimethyl-	EtOH-1% $CHCl_3$	241(4.45),265s(4.14)	44-2083-69
$C_{14}H_{15}Cl_2NO$			
1-Indanone, 2-[[bis(2-chloroethyl)-amino]methylene]-	EtOH	210(4.01),254(3.92), 297(3.48),350(4.40)	24-3139-69
$C_{14}H_{15}Cl_2N_3O_3$			
2H-Pyran-2-methanol, 6-(5,6-dichloro-1H-benzotriazol-1-yl)tetrahydro-, acetate, cis	EtOH	264(3.78),271(3.76),	4-0005-69
trans	EtOH	264(3.79),278(3.78), 296(3.62)	4-0005-69
$C_{14}H_{15}Cl_3N_2O_4$			
4(3H)-Pyrimidinone, 5,6-dihydro-6-(trichloromethyl)-2-(3,4,5-trimethoxyphenyl)-	EtOH	269(4.15)	104-1797-69
$C_{14}H_{15}FeNO_5$			
Iron, (1H-azepine-1-carboxylic acid)-(tricarbonyl)-, tert-butyl ester	EtOH	250(4.27),298s(3.77)	44-2866-69
$C_{14}H_{15}N$			
6H-Cyclohepta[b]quinoline, 7,8,9,10-tetrahydro-	EtOH	227(4.66),232(4.69), 235(4.71),269(3.68), 284s(3.64),290(3.66), 296(3.62),303(3.79), 309(3.66),316(3.95)	88-0557-69
Diphenylamine, 2,4-dimethyl-	EtOH	242(3.86),282(4.17)	22-2355-69
Diphenylamine, 2,6-dimethyl-	EtOH	243(4.05),275(3.89)	22-2355-69
Fulvene, 6-(dimethylamino)-6-phenyl-	EtOH	240(3.64),340(4.18)	39-0111-69B
Lepidine, 3-(2-methylpropenyl)-	EtOH	228(4.34),278(3.56), 318(3.22)	77-0463-69
Pyridine, 4-(3-phenylpropyl)-	EtOH	221(3.51),257(3.51), 290s(--)	39-2134-69C
$C_{14}H_{15}NO$			
6H-Azepino[1,2-a]indol-6-one, 7,8,9,10-tetrahydro-11-methyl-	EtOH	246(4.10),268(3.88), 298(3.54),304(3.54)	94-1290-69
10H-Azepino[1,2-a]indol-10-one, 6,7,8,9-tetrahydro-7-methyl-	EtOH	214(4.16),242(4.22), 313(4.26)	39-1028-69C
6H-Cyclopent[g]isoquinolin-9-ol, 7,8-dihydro-7,7-dimethyl-	pH 13	252(4.61),325(3.72)	44-0240-69
	MeOH	238(4.76),330(3.69)	44-0240-69
	MeOH-HCl	253(4.71),378(3.72)	44-0240-69
4(5H)-Indolinone, 6,7-dihydro-3-phenyl-	6N HCl	293(4.25)	78-4005-69
	EtOH	308(4.33)	78-4005-69
3-Quinuclidinone, 2-benzylidene-	EtOH	297(4.55)	22-1251-69
Spiro[cyclopentane-1,3'(3H)-indole], 2-acetyl-	EtOH	237(3.99),243(4.01), 305(3.98)	78-2757-69
Spiro[cyclopentane-1,3'(2'H)-quinolin]-2'-one, 1',4'-dihydro-4'-methylene-	EtOH	235(4.38),267(3.91), 310(3.43)	94-1290-69

Compound	Solvent	$\lambda_{max}(\log \epsilon)$	Ref.
$C_{14}H_{15}NOS_4$			
4-Morpholinecarbodithioic acid, anhydro-sulfide with cyclic phenylvinylene trithioorthoformate	EtOH	234(4.15),316(3.94)	94-1931-69
$C_{14}H_{15}NO_2$			
Carbazol-1(2H)-one, 3,4-dihydro-6-methoxy-3-methyl-	EtOH	232(4.23),315(4.39)	102-0773-69
Cyclopenta[b]pyrrol-2(1H)-one, 3,3aβ,4,8aβ-tetrahydro-1-methoxy-3α-phenyl-	EtOH	248s(2.20),253(2.25), 259(2.30),265(2.19), 268s(1.84)	44-2058-69
2-Isoxazolin-5-one, 4-(2-butenyl)-4-methyl-3-phenyl-	EtOH	256(4.06)	88-0543-69
3-Isoxazolin-5-one, N-(2-butenyl)-4-methyl-3-phenyl-	EtOH	247(3.98),282(3.96)	88-0543-69
2-Isoxazolin-5-one, 4-methyl-4-(1-methylallyl)-3-phenyl-	EtOH	254(4.03)	88-0543-69
2-Isoxazolin-5-one, 4-methyl-4-(2-methylallyl)-3-phenyl-	EtOH	259(4.10)	88-0543-69
3-Isoxazolin-5-one, 4-methyl-N-(2-methylallyl)-3-phenyl-	EtOH	246(3.98),286(3.98)	88-0543-69
Pyrido[1,2-a]indole-10-carboxylic acid, 6,7,8,9-tetrahydro-, methyl ester	EtOH	216(4.49),232(4.34), 251s(4.01),289(4.13), 298s(4.01)	12-0997-69
Pyrrole-2-carboxylic acid, 5-methyl-3-phenyl-, ethyl ester	EtOH	293(4.22)	94-0582-69
Pyrrole-3-carboxylic acid, 1-methyl-2-phenyl-, ethyl ester	MeOH	203(4.26),220(4.07), 273(3.79)	77-0066-69
1H-Pyrrolo[1,2-a]indole, 2-acetyl-2,3-dihydro-7-hydroxy-6-methyl-	EtOH	213(4.51),276(3.92)	88-0101-69
1H-Pyrrolo[1,2-a]indole, 2-acetyl-2,3-dihydro-7-methoxy-	EtOH	220(4.52),277(3.92)	88-0101-69
$C_{14}H_{15}NO_2S_2$			
2-Butanone, 4-[[4-(p-methoxyphenyl)-2-thiazolyl)thio]-	MeOH	270(4.31)	4-0397-69
$C_{14}H_{15}NO_3$			
5H-1-Benzazepin-5-one, 1,7-diacetyl-1,2,3,4-tetrahydro-	MeOH	231(4.27),270(3.91)	44-1070-69
Carbostyril, 3-isobutyryl-4-methoxy-	MeOH and MeOH-acid	228(4.40),273(3.70), 278(2.71),328(3.63)	44-2183-69
	MeOH-base	236(4.41),273s(3.69), 338(3.50)	44-2183-69
3-Indolepentanoic acid, 5-oxo-, methyl ester	EtOH	243(4.08),264(4.09), 294(4.16)	78-4843-69
2(1H)-Isoquinolineacrylic acid, 1-methoxy-, methyl ester, trans	MeOH	209(4.32),240(4.15), 302s(4.16),334(4.50)	39-2311-69C
	MeOH-acid	216(4.28),252(4.58), 305(4.15),357(3.75)	39-2311-69C
Phthalic anhydride, 3-(cyclohexylamino)-	EtOH	206(4.17),256(4.07), 405(3.63)	16-0241-69
Pyrido[1,2-a]indole-10-carboxylic acid, 6,7,8,9-tetrahydro-9-hydroxy-, methyl ester	EtOH	215(4.68),234(4.41), 292(4.23)	12-0997-69
Pyrrole-2-carboxylic acid, 5-(m-methoxyphenyl)-, ethyl ester	EtOH	310(4.44)	94-0582-69

Compound	Solvent	$\lambda_{max}(\log \epsilon)$	Ref.
$C_{14}H_{15}NO_4$			
3-Buten-2-one, 4-hydroxy-4-phenyl-, O-acetyloxime, acetate	EtOH	282(4.13)	77-1062-69
Carbostyril, 4-hydroxy-3-isobutyryl-8-methoxy-	MeOH and MeOH-acid	248(4.30),308(4.12), 316(4.13)	44-2183-69
	MeOH-base	242(4.49),267s(4.03), 310(3.96)	44-2183-69
2,4,6-Heptatrienal, 7-(5-nitro-2-furyl)-2,4,6-trimethyl-, 2,4-cis-6-trans	EtOH	242(4.05),313(4.17), 400(4.46)	94-1757-69
Isonicotinic acid, 2-benzyl-1,2,5,6-tetrahydro-3-hydroxy-6-oxo-, methyl ester	90% EtOH	251(3.92)	94-0467-69
3-Pyrroline-3-carboxylic acid, 4-hydroxy-1-methyl-5-oxo-2-phenyl-, ethyl ester	EtOH	242(4.40),272(4.04)	44-3187-69
	EtOH-NaOH	229(4.32),307(4.38)	44-3187-69
$C_{14}H_{15}NO_5$			
Indole-3-glyceric acid, 1-acetyl-, methyl ester	EtOH	240(4.32),292(3.88), 300(3.92)	97-0269-69
$C_{14}H_{15}NO_6$			
4-Penten-1-ol, 5-[3,4-(methylenedioxy)-phenyl]-4-nitro-, acetate	EtOH	359(4.20)	87-0157-69
$C_{14}H_{15}NS$			
4H-Cyclohepta[c]thiophene-4-carbonitrile, 1,3,5,7-tetramethyl-	heptane	250(3.76),302(3.91)	104-1115-69
4H-Cyclohepta[c]thiophene-6-carbonitrile, 1,3,5,7-tetramethyl-	heptane	289(3.74),315(3.73)	104-1115-69
6H-Cyclohepta[c]thiophene-4-carbonitrile, 1,3,5,7-tetramethyl-	heptane	234(4.30),252(4.30)	104-1115-69
6H-Cyclohepta[c]thiophene-6-carbonitrile, 1,3,5,7-tetramethyl-	heptane	242(4.57),282(3.77)	104-1115-69
1H-Phenothiazine, 2,3-dihydro-2,4-dimethyl-	MeOH	385(3.57)	124-1278-69
	MeOH-$HClO_4$	458(3.66)	124-1278-69
1H-Phenothiazine, 2,3-dihydro-3,3-dimethyl-	MeOH	390(3.50)	124-1278-69
	MeOH-$HClO_4$	458(3.50)	124-1278-69
$C_{14}H_{15}N_2O_6P$			
1-(5-Nitrosalicyl)pyridinium hydroxide, ethyl hydrogen phosphate, inner salt	H_2O	261(3.91),268(3.90), 293(0.92)	35-4532-69
$C_{14}H_{15}N_3$			
Aniline, N-methyl-p-(m-tolylazo)-	EtOH	253(3.99),402(4.45)	18-3565-69
Aniline, N-methyl-p-(p-tolylazo)-	EtOH	253(3.99),401(4.46)	18-3565-69
3H-Imidazo[4,5-f]quinoline, 3-ethyl-2,7-dimethyl-	EtOH	255s(4.60),259(4.62), 310(3.62)	94-2455-69
3H-Imidazo[4,5-f]quinoline, 2,3,7,9-tetramethyl-	EtOH	253s(4.59),257(4.61), 310(3.65)	94-2455-69
Triazene, 1,3-di-p-tolyl-	MeOH	237(3.87),295(3.99), 360(4.34)	65-0059-69
copper complex	benzene	345(4.52),670(2.96)	65-0059-69
mercury complex	benzene	305(4.22),397(4.48)	65-0059-69
Triazene, 3-(α-methylbenzyl)-1-phenyl-, mercury complex	benzene	293(4.26),350(4.14)	65-0303-69
	DMF	285(4.24),340(4.17)	65-0303-69
Triazene, 3-methyl-3-phenyl-1-m-tolyl-	EtOH	236(4.23),343(4.27)	18-3565-69
Triazene, 3-methyl-3-phenyl-1-p-tolyl-	EtOH	237(4.14),296(3.96), 345(4.30)	18-3565-69

Compound	Solvent	λ_{max}(log ϵ)	Ref.
$C_{14}H_{15}N_3O$			
Aniline, p-[(p-methoxyphenyl)azo]-N-methyl-	EtOH	251(4.01),401(4.48)	18-3565-69
2,5-Cyclohexadien-1-one, 4-[(2,4-diamino-m-tolyl)imino]-2-methyl-	pH 10.0	648(4.41)	39-0823-69B
2,5-Cyclohexadien-1-one, 4-[(2,4-diamino-m-tolyl)imino]-3-methyl-	pH 10.0	652(4.43)	39-0823-69B
2,5-Cyclohexadien-1-one, 4-[(4,6-diamino-m-tolyl)imino]-2-methyl-	pH 10.0	655(4.43)	39-0823-69B
2,5-Cyclohexadien-1-one, 4-[(4,6-diamino-m-tolyl)imino]-3-methyl-	pH 10.0	655(4.37)	39-0823-69B
2,5-Cyclohexadien-1-one, 4-[(2,4-diamino-3,5-xylyl)imino]-	pH 10.0	627(4.35)	39-0823-69B
Nicotinamide, 1,2-dihydro-2-imino-1,5-dimethyl-6-phenyl-	EtOH	251(3.99),341(4.03)	94-2209-69
6H-Pyrrolo[2,3-g]quinoxaline, 6-acetyl-7,8-dihydro-2,3-dimethyl-	MeOH	260(4.39),357(4.05)	103-0199-69
Triazene, 1-(p-methoxyphenyl)-3-methyl-3-phenyl-	EtOH	236(4.17),296(4.05), 308(4.02),351(4.28)	18-3565-69
$C_{14}H_{15}N_3O_2$			
o-Anisidine, 5-[(o-methoxyphenyl)azo]-	n.s.g.	260(4.00),407(4.35)	40-0207-69
Triazene, 1,3-bis(p-methoxyphenyl)-	MeOH	237(3.95),302(4.13), 370(4.27)	65-0059-69
mercury complex	benzene	310(4.26),400(4.28)	65-0059-69
Tryptophan, alanyl-, anhydride, trans	EtOH	221(4.47),281(3.75), 290(3.70)	39-1003-69C
$C_{14}H_{15}N_3O_2S_2$			
Sulfur diimide, [p-(dimethylamino)-phenyl](phenylsulfonyl)-	benzene	512(4.62)	104-0459-69
$C_{14}H_{15}N_3O_3$			
3-Pyrrolin-2-one, 4-nitro-5-phenyl-3-pyrrolidino-	EtOH	242(4.00),402(4.19)	44-3279-69
$C_{14}H_{15}N_3O_4$			
6H-Azepino[1,2-a]benzimidazol-6-ol, 7,8,9,10-tetrahydro-3-nitro-, acetate	n.s.g.	240(4.58),307(4.13)	39-0070-69C
Pyrazole-4-carboxylic acid, 3-[(benzyloxy)methyl]-5-carbamoyl-, methyl ester	EtOH	213(4.16)	88-0289-69
$C_{14}H_{15}N_3O_4S_2$			
5-Thia-1-azabicyclo[4.2.0]oct-2-ene-2-carboxylic acid, 1-[2-amino-2-(2-thienyl)acetamido]-3-methyl-8-oxo-	H_2O	234(4.10),260(3.86)	87-0310-69
$C_{14}H_{15}N_5O_3$			
Acetamide, N-[6,7-dihydro-7-(1-hydroxyethyl)-5-phenyl[1,2,5]oxadiazolo-[3,4-d]pyrimidin-7-yl]-	EtOH	239(4.07),288(3.94)	35-5181-69
$C_{14}H_{15}N_5O_4S$			
Adenosine, N-2-thienyl-	pH 1.0	270(4.27)	87-1056-69
	pH 7.0	211(4.30),272(4.27)	87-1056-69
	pH 12.0	272(4.28)	87-1056-69
$C_{14}H_{15}N_7O_2$			
Adenine, 9-[3-(2-nitroanilino)propyl]-	EtOH	235(4.33),260(4.26), 425(3.79)	44-3240-69

Compound	Solvent	λ_{max}(log ε)	Ref.
Adenine, 9-[3-(2-nitroanilino)propyl]- (cont.)	EtOH-HCl	235(4.35),258(4.26), 422(3.77)	44-3240-69
	EtOH-NaOH	235s(--),258(4.27), 425(3.51)	44-3240-69
$C_{14}H_{15}N_7O_5S$			
as-Triazine-2(3H)-propionaldehyde, 4,5-dihydro-β,6-dimethyl-5-oxo-3-thioxo-, α-(2,4-dinitrophenylhydrazone)	EtOH	221(4.38),270(4.40), 354(4.32)	114-0093-69C
$C_{14}H_{16}$			
Naphthalene, 1,2,3,4-tetramethyl-	EtOH	234(4.89),284s(3.74), 293(3.79)	94-1591-69
Tricyclo[4.4.0.0^{2,5}]deca-3,6,8,10-tetraene, 2,3,4,5-tetramethyl-	hexane	264(2.78),270(2.93), 277(2.88)	88-4753-69
$C_{14}H_{16}BF_4NS_2$			
N-(4-Phenyl-1,3-dithiol-2-ylidene)-piperidinium tetrafluoroborate	EtOH	237(4.12),323(4.08)	94-1924-69
$C_{14}H_{16}BrN$			
Carbazole, 9-(2-bromoethyl)-1,2,3,4-tetrahydro-	MeOH	230(4.41),286(3.83), 293s(--)	23-3647-69
1H-Carbazole, 4a-(2-bromoethyl)-2,3,4,4a-tetrahydro-	MeOH	257(3.90)	23-3647-69
$C_{14}H_{16}BrNOS_2$			
1-Piperidinecarbodithioic acid, ester with 4'-bromo-2-mercaptoacetophenone	EtOH	259(4.43)	94-1924-69
$C_{14}H_{16}BrNS_2$			
Piperidine, 1-[4-(p-bromophenyl)-1,3-dithiol-2-yl]-	EtOH	245(4.19),326(4.05)	94-1924-69
$C_{14}H_{16}ClNOS_2$			
1-Piperidinecarbodithioic acid, ester with 4'-chloro-2-mercaptoacetophenone	EtOH	245(4.42)	94-1924-69
$C_{14}H_{16}ClNO_3S_3$			
N-[4-(p-Chlorophenyl)-1,3-dithiol-2-ylidene]piperidinium bisulfate	EtOH	231(3.91),240(3.91), 308(3.83),322(3.83)	94-1924-69
$C_{14}H_{16}ClNO_4S_2$			
2-Butyl-3-methylthiazolo[2,3-b]benzothiazolium perchlorate	EtOH	218(4.60),262(4.05), 315(4.20)	95-0469-69
3-Ethyl-2-propylthiazolo[2,3-b]benzothiazolium perchlorate	EtOH	218(4.49),262(3.93), 316(4.09)	95-0469-69
$C_{14}H_{16}Cl_2O_6$			
Crotepoxide, dideacetyl-, dichlorohydrin	MeOH	274(3.00),281(2.92)	44-3898-69
$C_{14}H_{16}FN_5O_5$			
Adenosine, 2'-deoxy-2-fluoro-, 3',5'-diacetate	pH 1	262(4.13),267s(--)	87-0498-69
	pH 7, 13	261(4.17),267s(--)	87-0498-69
$C_{14}H_{16}Fe$			
Ferrocene, 1-methyl-1',2-trimethylene-	EtOH	312(2.06),436(2.23)	49-1540-69
Ferrocene, 1-methyl-1',3-trimethylene-	EtOH	314(2.00),436(2.26)	49-1540-69

Compound	Solvent	λ_{max}(log ε)	Ref.
$C_{14}H_{16}N_2$			
3,5-Hexadienenitrile, 2-(dimethylamino)-6-phenyl-	EtOH	228(4.05),290(4.40)	70-0873-69
$C_{14}H_{16}N_2O$			
3,4-Diazabicyclo[4.1.0]hept-4-en-2-one, 3,7,7-trimethyl-5-phenyl-	EtOH	282(4.26)	88-1063-69
3H-1,2-Diazepin-3-one, 2,4-dihydro-2,4,4-trimethyl-7-phenyl-	EtOH	228(4.19),250(4.03), 315(2.66)	88-1063-69
1H-Indazol-5-ol, 1-benzyl-4,5,6,7-tetrahydro-	MeOH	234(3.82),258(3.51)	88-1661-69
3H-Indazol-3-one, 2,3a,4,5,6,7-hexahydro-3a-methyl-2-phenyl-	EtOH	243(4.34)	22-4159-69
4-Morpholineacetonitrile, α-styryl-	EtOH	253(4.12)	70-0873-69
Nicotinamide, 1-benzyl-1,4-dihydro-4-methyl-	EtOH	339(3.81)	39-0192-69B
Nicotinamide, 1-benzyl-1,6-dihydro-4-methyl-	EtOH	266(3.92),350(3.75)	39-0192-69B
3-Picoline, 6,6'-(oxydimethylene)di-	n.s.g.	213(4.12),262(3.93)	39-2255-69C
3(2H)-Pyridazinone, 4-isopropyl-2-methyl-6-phenyl-	EtOH	246(4.40)	88-1063-69
$C_{14}H_{16}N_2O_2$			
5,8-Isoquinolinedione, 7-(butylamino)-3-methyl-	MeCN	234(4.33),276(4.29), 320(3.74)	4-0697-69
4-Morpholinecarboxamide, N-inden-3-yl-	EtOH	210(4.21),268(3.03), 275(3.02)	36-0047-69
$\Delta^{2(1H),N}$-Pyridinecarbamic acid. 3,4-dihydro-1-phenyl-, ethyl ester	EtOH	245(3.98),284(3.95)	44-3672-69
$C_{14}H_{16}N_2O_2S$			
Benzenesulfonanilide, 4'-(dimethylamino)-	EtOH-1% $CHCl_3$	262(4.23),300s(3.59)	44-2083-69
2,6-Lutidine, 4,4'-thiodi-, 1,1'-dioxide	EtOH	221(4.57),287(4.45), 315(4.27)	35-4749-69
p-Phenylenediamine, N ,N -dimethyl-2-(phenylsulfonyl)-	EtOH-1% $CHCl_3$	233(4.39),265(4.10)	44-2083-69
$\Delta^{4(1H),\alpha}$-Quinolineacetic acid, 1-methyl-α-(methylthio)-, ethyl ester	EtOH	250(4.28),272(4.18), 363(4.42)	88-1451-69
$C_{14}H_{16}N_2O_3S_2$			
1-Piperidinecarbodithioic acid, ester with 2-mercapto-4'-nitroacetophenone	EtOH	268(4.39)	94-1924-69
$C_{14}H_{16}N_2O_4S$			
2,6-Lutidine, 4,4'-sulfonyldi-, 1,1'-dioxide	EtOH	230(4.43),312(4.47)	35-4749-69
$C_{14}H_{16}N_2O_6S$			
Malonic acid, [(N-methyl-p-nitroanilino)(methylthio)methylene]-, dimethyl ester	MeOH	301(4.18),372(4.32)	78-4649-69
$C_{14}H_{16}N_2O_6S_3$			
N-[4-(p-Nitrophenyl)-1,3-dithiol-2-ylidene]piperidinium bisulfate	EtOH	320(4.23)	94-1924-69
$C_{14}H_{16}N_4$			
p-Phenylenediamine, N-(2-amino-4-imino-2,5-cyclohexadien-1-ylidene)-2,5-dimethyl-	pH 8	600(4.00)	39-0827-69B

Compound	Solvent	λ_{max}(log ε)	Ref.
p-Phenylenediamine, N[4]-(2-amino-4-imino-2,5-cyclohexadien-1-ylidene)-2,6-dimethyl-	pH 8	580(4.02)	39-0827-69B
p-Phenylenediamine, N-(2-amino-4-imino-3,5-dimethyl-2,5-cyclohexadien-1-ylidene)-	pH 8	547(4.08)	39-0827-69B
2-Tetrazene, 1,4-dimethyl-1,4-diphenyl-	$CHCl_3$	247(3.92),310s(4.08), 347(4.45)	35-6452-69
	$(MeOCH_2)_2$	251(4.18),305s(4.11), 347(4.49)	35-6452-69
Toluene-2,5-diamine, N[5]-(2-amino-4-imino-3-methyl-2,5-cyclohexadien-1-ylidene)-	pH 5	567(4.09)	39-0827-69B
Toluene-2,5-diamine, N[5]-(2-amino-4-imino-5-methyl-2,5-cyclohexadien-1-ylidene)-	pH 8	567(4.06)	39-0827-69B
$C_{14}H_{16}N_4O$			
1-Azaspiro[3.5]nonan-2-one, 3-azido-1-phenyl-	EtOH	255(4.13)	78-4421-69
1H-Imidazo[4,5-f]quinazolin-9-one, 8-butyl-1,8-dihydro-2-methyl-	EtOH	213(4.19),247(4.46), 272(3.71),282(3.71), 294(3.78),322(3.99), 335(3.95)	2-1166-69
$C_{14}H_{16}N_4O_3$			
2-Pyrazolin-3-acetic acid, 5-imino-α,α-dimethyl-4-oxo-1-phenyl-, methyl ester, 4-oxime	EtOH	225(3.88),310(3.85), 350s(3.45)	4-0317-69
$C_{14}H_{16}N_4O_4$			
Acetamide, N-(3-butyl-3,4-dihydro-5-nitro-4-oxo-6-quinazolinyl)-	EtOH	208(4.43),228(4.46), 285(4.16),327(3.45)	2-1166-69
Carbonic acid, monoanhydride with [(2-amino-3-cyano-4,5,6,7-tetrahydroindol-1-yl)imino]acetic acid, ethyl ester	EtOH	243(4.33),260s(4.31), 269(4.37),311s(3.23), 387(3.84)	48-0388-69
2-Cyclohexen-1-one, 5,5-dimethyl-, 2,4-dinitrophenylhydrazone	EtOH	255(4.17)	95-0506-69
Malonic acid, [(5-methyl-1H-benzotriazol-1-yl)imino]-, diethyl ester	EtOH	220(3.94),266(4.14), 341(3.89)	39-0742-69C
Malonic acid, [(7-methyl-1H-benzotriazol-1-yl)imino]-, diethyl ester	EtOH	222(3.94),256(4.08), 343(3.94)	39-0742-69C
2-Propanone, 1-cyclopent-1-en-1-yl-, 2,4-dinitrophenylhydrazone	EtOH	360(4.36)	28-0075-69A
2-Propanone, 1-cyclopentylidene-, 2,4-dinitrophenylhydrazone	EtOH	374(4.37)	28-0075-69A
$C_{14}H_{16}N_4O_5$			
2-Cyclohexen-1-one, 4-(hydroxymethyl)-4-methyl-, 2,4-dinitrophenylhydrazone	EtOH	229(4.04)	78-5287-69
Uric acid, dihydro-4-hydroxy-5-methoxy-1,3-dimethyl-9-phenyl-	EtOH	232(4.44),271(3.23)	78-0541-69
$C_{14}H_{16}N_4O_6$			
Barbituric acid, 5,5'-cyclohexylidenedi-	EtOH	263(4.32)	103-0827-69
$C_{14}H_{16}N_4O_6S$			
9H-Purine-6(1H)-thione, 9-β-D-arabinofuranosyl-, 2',3'-diacetate	EtOH	225s(3.88),323(4.36)	23-1095-69

Compound	Solvent	$\lambda_{max}(\log \epsilon)$	Ref.
$C_{14}H_{16}N_6O_7$			
Glycine, N-[(N-1,2-dimethyl-5,7-dinitro-4-benzimidazolyl)-L-alanyl]-	10% EtOH	230(3.96),250(3.97), 375(3.92),425s(3.68)	44-0384-69
Morpholine, 4-(4-methyl-1-picryl-2-pyrazolin-3-yl)-	EtOH	248(4.04),440(4.29)	22-1683-69
$C_{14}H_{16}O$			
2-Cyclopenten-1-one, 4-isopropyl-2-phenyl-	n.s.g.	223(4.45),263(4.11)	88-0909-69
Dibenzofuran, 1,2,3,4-tetrahydro-6,9-dimethyl-	EtOH	257(4.13)	4-0379-69
Fluorene, 1,2,3,4-tetrahydro-7-methoxy-	EtOH	266(4.21),274s(4.13)	44-2209-69
1H-Fluorene, 2,3,4,4aα-tetrahydro-7-methoxy-	EtOH	226(4.41),264(3.86), 293(3.51),303(3.48)	44-2209-69
Fluoren-9-ol, 1,2,3,4-tetrahydro-9-methyl-	EtOH	272(3.83),276s(3.85)	44-1899-69
Fluoren-9-one, 1,2,3,4,4a,9a-hexahydro-4a-methyl-	EtOH	245(4.03),290(3.35)	44-1899-69
3,6-Heptadien-2-one, 5-methyl-7-phenyl-	EtOH	246s(3.97),279s(3.36)	22-3719-69
2,4-Hexadien-1-one, 2,5-dimethyl-1-phenyl-	EtOH	234(4.06),302(4.28)	22-3638-69
2H-Pyran, 2,2,4-trimethyl-6-phenyl-	EtOH	224(3.90),242s(--), 325(3.88)	22-3638-69
$C_{14}H_{16}OS$			
6-Thiaphenanthrene, 1,2,3,4,5,6-hexahydro-8-methoxy-	EtOH	256(4.14),330(3.05)	78-1939-69
$C_{14}H_{16}OSi$			
Silane, trimethyl-2-naphthoyl-	hexane	210(4.44),244(4.63), 252(4.75),283(4.05), 283(4.05),292(4.02), 330(3.18),334(3.20), 426(2.01)	35-0355-69
$C_{14}H_{16}O_2$			
3H-Benz[e]indene-3,7(5H)-dione, 1,2,3a,4,8,9-hexahydro-3a-methyl-	EtOH	296(4.24)	35-2806-69
Naphthalene, 4-ethoxy-2-methoxy-1-methyl-	EtOH	212(4.46),239(4.69), 293(3.74),303(3.76), 330(3.52)	44-0056-69
1,2-Naphthoquinone, 3,4-dihydro-3,3,6,8-tetramethyl-	EtOH	265s(3.82),292(3.87), 410s(1.64)	5-0094-69D
2,8-Nonadiene-4,6-diyn-1-ol, 9-(tetrahydro-2H-pyran-2-yl)-, trans-trans	ether	233(4.39),238(4.40), 248(4.32),262(3.90), 277(4.15),294(4.30), 313(4.23)	39-0830-69C
2-Propenone, 3-(m-methoxyphenyl)-1-methyl-	MeOH	221(4.05),240(3.99), 290(4.19)	35-5862-69
2H-Pyran, tetrahydro-2-[(3-phenylpropynyl)oxy]-	MeOH	242(4.04),251(3.96)	39-2173-69C
$C_{14}H_{16}O_3$			
2'-Acetonaphthone, 3',4',4'a,7-tetrahydro-1'-hydroxy-4'a,8'-dimethyl-7'-oxo-	EtOH	261(3.98),346(4.04)	39-1088-69C +39-1619-69C
Fluorene-9α-carboxylic acid, 1,2,3,4,4aα,9aα-hexahydro-7-hydroxy-	EtOH	222(3.85),283(3.48)	44-2209-69
3-Indanacetic acid, 3,6-dimethyl-1-oxo-, methyl ester	EtOH	248(4.11),301(3.48)	2-0215-69

Compound	Solvent	λ_{max}(log ε)	Ref.
Indene-3-acetic acid, 6-methoxy-, ethyl ester, endo	MeOH	263(--)	44-4182-69
exo	MeOH	321(4.41)	44-4182-69
1,2,3-Naphthalenetrimethanol, 4-methyl-	EtOH	229s(4.84),235(5.01), 279s(3.79),289(3.85), 299s(3.73)	94-1591-69
Naphtho[2,1-b]furan-2(1H)-one, 3a,4,5,-9b-tetrahydro-8-methoxy-9-methyl-	EtOH	279(3.30)	88-4559-69
trans	EtOH	280(3.30)	88-4559-69
$C_{14}H_{16}O_4$			
4H,8H-Cyclobuta[1,2-b:3,4-b']dipyran-4,8-dione, 4a,4b,8a,8b-tetrahydro-2,4b,6,8b-tetramethyl-	EtOH	266(4.34)	44-4052-69
Isoevodionol	EtOH	272(4.81),294s(4.38), 306s(4.31),350(3.74)	88-4673-69
1-Naphthaleneacetic acid, 1,2,3,4-tetrahydro-7-methoxy-8-methyl-2-oxo-	EtOH	230(4.30),285(3.40)	88-4559-69
2,5-Norbornadiene-2,3-dicarboxylic acid, 7-isopropylidene-, dimethyl ester	benzene	304(2.93)	33-0956-69
	heptane	301(2.95)	33-0956-69
	MeOH	305(2.92)	33-0956-69
	ether	300(2.90)	33-0956-69
	MeCN	301(2.93)	33-0956-69
	MeCN	300(3.0)	5-0052-69E
	CCl_4	306(2.94)	33-0956-69
	EtBr	303(2.92)	33-0956-69
Obliquin, tetrahydro-	EtOH	221(4.11),238s(3.71), 291(3.59)	39-0526-69C
Pyriculol	EtOH	233(4.19),280(3.78), 358(3.45)	88-3977-69
Remirol	EtOH	216(4.22),239(3.98), 294(4.26)	88-4673-69
$C_{14}H_{16}O_5$			
Chromone, 6-(ethoxymethyl)-5-hydroxy-7-methoxy-2-methyl-	MeOH	209(4.26),232(4.26), 248(4.25),256(4.26), 292(3.82),312s(3.60)	39-0704-69C
	MeOH-NaOH	245s(4.05),266(4.20), 357(3.54)	39-0704-69C
Succinic acid, 2-(2-benzoylethyl)-3-methyl-	EtOH	242(4.07)	39-1845-69C
$C_{14}H_{16}O_6$			
Phthalic anhydride, 4-isopropyl-3,5,6-trimethoxy-	EtOH	235(4.45),287(3.61), 341(3.72)	33-1685-69
Trimesic acid, 2-methyl-	EtOH	215(4.62),235(4.14), 285(2.70),298(2.60)	44-4134-69
$C_{14}H_{17}BF_4OS$			
6-Methoxy-1,3,5,7-tetramethylcyclohepta-[c]thiolium tetrafluoroborate	$C_2H_4Cl_2$	274(4.16),325(4.54), 373(3.92),390(4.03), 600(3.25)	104-0947-69
$C_{14}H_{17}BF_4O_2S$			
5,6-Dimethoxy-1,3,7-trimethylcyclohepta-[c]thiolium tetrafluoroborate	$C_2H_4Cl_2$	277(4.14),289(4.10), 329(4.50),411(4.17), 600(3.28)	104-0947-69
$C_{14}H_{17}BrO_2$			
Cyclopentanone, 2-(5-bromo-2-methoxy-p-tolyl)-2-methyl-	MeOH	283(3.34),289(3.33)	78-3509-69

Compound	Solvent	$\lambda_{max}(\log \epsilon)$	Ref.
$C_{14}H_{17}BrO_3$			
1-Adamantanecrotonolactone, α-bromo-γ-hydroxy-	50% MeOH-HCl	224(3.82)	73-3343-69
$C_{14}H_{17}ClN_2$			
Quinoline, 2-[(3-chloropropyl)amino]-4,6-dimethyl-, hydrochloride	EtOH	246(4.35),254(4.28),336(3.81)	95-0759-69
1,2,3,4-Tetrahydro-6,8-dimethylpyrimido-[1,2-a]quinolin-11-ium chloride	EtOH	242(4.53),247(4.47),256(4.41),282(3.94),340(3.97)	95-0759-69
$C_{14}H_{17}ClN_2O$			
Indole-2-carboxamide, N-tert-butyl-6-chloro-3-methyl-	EtOH	228(4.5),304(4.3)	24-0678-69
$C_{14}H_{17}ClN_4$			
Pyrimidine, 2,4-diamino-5-(p-chlorobenzyl)-6-propyl-	pH 2	223(4.45),276(3.91)	44-0821-69
	pH 12	219(4.33),239s(4.04),286(3.92)	44-0821-69
Pyrimidine, 2,4-diamino-5-(p-chlorophenyl)-6-isobutyl-	pH 2	208(4.54),220s(4.40),273(3.89)	44-0821-69
	pH 12	220s(4.32),245s(3.91),287(3.95)	44-0821-69
$C_{14}H_{17}ClO_3$			
1-Adamantanecrotonolactone, α-chloro-γ-hydroxy-	50% MeOH-HCl	217(3.78)	73-3343-69
$C_{14}H_{17}Cl_2NO$			
Acrylophenone, 3-[bis(2-chloroethyl)-amino]-4'-methyl-	EtOH	214(3.87),260(3.90),337(4.38)	24-3139-69
$C_{14}H_{17}Cl_2NO_2$			
Acrylophenone, 3-[bis(2-chloroethyl)-amino]-4'-methoxy-	EtOH	223(3.95),281(3.89),337(4.43)	24-3139-69
$C_{14}H_{17}N$			
1H-Carbazole, 4a-ethyl-2,3,4,4a-tetrahydro-	MeOH	256(3.81)	23-3647-69
	MeOH-HCl	231(3.82),276(3.74)	23-3647-69
perchlorate	EtOH	233(3.87),239(3.85),279(3.79)	78-0095-69
	EtOH-NaOH	259(3.80)	78-0095-69
2H-4a,9-Ethanocarbazole, 1,3,4,9a-tetrahydro-	MeOH	257(2.70),266(2.75),273(2.75)	23-3647-69
	MeOH-HCl	256(2.30),261(2.34),268(2.32)	23-3647-69
2H-Pyrrole, 2,5-dimethyl-2-(α-methylbenzyl)-	MeOH	210(3.96)	88-2215-69
Quinoline, 6,8-diethyl-4-methyl-	MeOH	293(3.73),305(3.63),319(3.46)	106-0196-69
$C_{14}H_{17}NO$			
2H-Azepin-2-one, hexahydro-N-styryl-, cis	MeOH	267(4.26)	78-5745-69
trans	MeOH	221(--),288(4.33)	78-5745-69
Carbazole, 1,2,3,4-tetrahydro-6-methoxy-3-methyl-	EtOH	222(4.32),278(3.84)	102-0773-69
1H-Carbazole, 2,3,4,4a-tetrahydro-5-methoxy-4a-methyl-	MeOH	247(3.62),278(3.73)	95-1061-69

Compound	Solvent	$\lambda_{max}(\log \epsilon)$	Ref.
1H-Carbazole, 2,3,4,4a-tetrahydro-6-methoxy-4a-methyl-	MeOH	274(3.80)	95-1061-69
1H-Carbazole, 2,3,4,4a-tetrahydro-7-methoxy-4a-methyl-	MeOH	262(3.59),295(3.35)	95-1061-69
1H-Carbazole, 2,3,4,4a-tetrahydro-8-methoxy-4a-methyl-	MeOH	256(3.78),300(3.59)	95-1061-69
Crotononitrile, 3-butoxy-2-phenyl-	EtOH	210(4.15),268(4.20)	22-0198-69
2-Cyclohexen-1-one, 3-(phenethylamino)-	C_6H_{12}	268(4.55)	39-0299-69B
	N HCl	283(4.36)	39-0299-69B
	H_2O	293(4.52)	39-0299-69B
	EtOH	289(4.51)	39-0299-69B
5H-Dibenz[c,e]azepin-5-one, 6,7,7aα,8,-9,10,11,11aβ-octahydro-	EtOH	246(3.80),271(3.69)	4-0131-69
6H-Dibenzo[b,d]azepin-6-one, 5,7,7aα,8,-9,10,11,11aβ-octahydro-	EtOH	240(4.01)	4-0131-69
2H-2β,4aβ-Methanonaphthalene-5-carbonitrile, 1,3,4,5β,6,7-hexahydro-1,1-dimethyl-7-oxo-	n.s.g.	240(4.12)	77-1335-69
2,4-Pentadienamide, N,N,3-trimethyl-5-phenyl-, 2-cis-4-trans	EtOH	227(4.15),301(4.47)	39-0016-69C
9(1H)-Phenanthrone, 2,3,4,4a,10,10a-hexahydro-, oxime, 4a,10a-trans	EtOH	253(4.05)	4-0131-69
$C_{14}H_{17}NOS_2$			
2-Heptanone, 3-(2-benzothiazolylthio)-	EtOH	225(4.36),245(3.98), 280(4.00),291(3.96), 301(3.90)	95-0469-69
3-Heptanone, 4-(2-benzothiazolylthio)-	EtOH	224(4.32),245(3.91), 280(3.97),291(3.92), 301(3.85)	95-0469-69
Phenol, p-(2-piperidino-1,3-dithiol-5-yl)-	EtOH	243(3.77),300(3.71)	94-1924-69
1-Piperidinecarbodithioic acid, ester with 2-mercaptoacetophenone	EtOH	248(4.24),281(4.02)	94-1924-69
$C_{14}H_{17}NO_2$			
6H-Azepino[1,2-a]indol-6-one, 7,8,9,10,10a,11-hexahydro-10a-hydroxy-11-methyl-	EtOH	252(4.11),280(3.62), 288(3.53)	94-1290-69
1H-1-Benzazepine, 1,7-diacetyl-2,3,4,5-tetrahydro-	MeOH	262(4.06)	44-1070-69
1H-1-Benzazepine, 1,8-diacetyl-2,3,4,5-tetrahydro-	MeOH	234(4.16),248(4.18), 284s(3.10)	44-2235-69
Indole-2-acetic acid, α,α-dimethyl-, ethyl ester	EtOH	220(4.49),284(3.89), 290(3.80)	39-0595-69C
3-Indolevaleric acid, methyl ester	EtOH	226(4.36),276(3.88), 282(3.89),291(3.80)	78-4843-69
2H,6H-Pyrido[2,1-b][1,3]oxazin-6-one, hexahydro-9a-phenyl-	EtOH	252(2.82),258(2.84), 264(2.82)	44-0165-69
Spiro[cyclopentane-1,2'-indolin]-3'-one, 5'-methoxy-2-methyl-	EtOH	229(4.3),262(3.95)	2-0135-69
Spiro[cyclopentane-1,3'(2'H)-quinolin]-2'-one, 1',4'-dihydro-4'-methoxy-	EtOH	253(4.15),282(3.49), 291(3.39)	94-1290-69
$C_{14}H_{17}NO_2S$			
Crotonic acid, thio-3-p-toluamido-, O-ethyl ester	MeCN	346(4.26)	24-0269-69

Compound	Solvent	λ_{max}(log ε)	Ref.
$C_{14}H_{17}NO_2S_2$			
1-Piperidinecarbodithioic acid, ester with 4'-hydroxy-2-mercaptoacetophenone	EtOH	220(3.88),282(4.04)	94-1924-69
$C_{14}H_{17}NO_3$			
2H-1-Benzazepin-2-one, 7-acetyl-1,3,4,5-tetrahydro-, ketal	MeOH	249(3.90),280s(2.52)	44-1070-69
3-Furoic acid, 2-amino-4,5-dihydro-5,5-dimethyl-4-phenyl-, methyl ester	EtOH	210(3.86),269(4.06), 273(4.06)	87-0339-69
3-Indolineacetic acid, 1-acetyl-2,3-dimethyl-	MeOH	251(4.17),278s(--), 286(3.70)	23-3647-69
Morpholine, 4-(p-methoxycinnamoyl)-	EtOH	226(4.08),310(4.41)	39-0016-69C
$C_{14}H_{17}NO_4$			
Malonic acid, [1-(N-methylanilino)ethylidene]-, dimethyl ester	isooctane	235(4.15),305(4.33)	35-6683-69
4-Penten-1-ol, 4-nitro-5-p-tolyl-, acetate	EtOH	318(3.94)	87-0157-69
$C_{14}H_{17}NO_4S$			
Malonic acid, [(N-methylanilino)(methylthio)methylene]-, dimethyl ester	MeOH	238(3.99),294(3.88), 339(3.95)	78-4649-69
$C_{14}H_{17}NO_4S_2$			
N-(4-Phenyl-1,3-dithiol-2-ylidene)piperidinium bisulfate	EtOH	322(4.11)	94-1924-69
$C_{14}H_{17}NO_5$			
4-Penten-1-ol, 5-(p-methoxyphenyl)-4-nitro-, acetate	EtOH	340(4.19)	87-0157-69
$C_{14}H_{17}NO_5S_3$			
N-[4-(p-Hydroxyphenyl)-1,3-dithiol-2-ylidene]-piperidinium bisulfate	EtOH	265(3.81),290(3.81), 336(3.70)	94-1924-69
$C_{14}H_{17}NO_9$			
2H-1,4-Benzoxazin-3(4H)-one, 2-(β-D-glucopyranosyloxy)-7-hydroxy-	H_2O	262(3.96),282s(--)	88-5001-69
$C_{14}H_{17}NS$			
2H-1,4-Benzothiazine, 2-isopentylidene-3-methyl-	MeOH	370(3.45)	124-1278-69
	MeOH-$HClO_4$	440(3.46)	124-1278-69
$C_{14}H_{17}NS_2$			
Piperidine, 1-(4-phenyl-1,3-dithiol-2-yl)-	EtOH	235(4.13),319(3.89)	94-1924-69
$C_{14}H_{17}N_3OS$			
1-Benzimidazolecarboxamide, N-cyclohexyl-2-thioxo-	dioxan	313(4.34)	4-0023-69
$C_{14}H_{17}N_3O_2$			
Acetamide, N-(3-butyl-3,4-dihydro-4-oxo-6-quinazolinyl)-	EtOH	230(4.55),290(4.26), 324(3.59),337(3.39)	2-1166-69
Cinnamic acid, α-azido-p-isopropyl-, ethyl ester	C_6H_{12}	228(4.07),317(4.22)	49-1599-69
Cinnamic acid, α-azido-2,4,6-trimethyl-, ethyl ester	C_6H_{12}	243(3.82),288(3.99)	49-1599-69
Hydantoin, 5-(p-diethylaminobenzylidene)-	neutral	236(3.81),403(4.46)	122-0455-69

Compound	Solvent	$\lambda_{max}(\log \epsilon)$	Ref.
Hydantoin, 5-(p-diethylaminobenzylidene)-, anion	n.s.g.	248(4.13),375(4.43)	122-0455-69
Uracil, 5-(benzylmethylamino)-1,3-dimethyl-	EtOH	306(3.71)	104-2004-69
$C_{14}H_{17}N_3O_2S$			
Δ^2-1,3,4-Oxadiazoline-2-thione, 2-benzyl-4-(morpholinomethyl)-	n.s.g.	258(4.22)	2-0583-69
$C_{14}H_{17}N_3O_3$			
Acetamide, N,N'-(1-acetyl-5,6-indolinediyl)bis-	MeOH	268(4.39)	103-0196-69
$C_{14}H_{17}N_3O_4$			
Glycine, N-benzoyl-2-(N-carbamoylformimidoyl)-, ethyl ester	acid or neutral	266(4.34)	88-2539-69
	aq. NaOH	308(4.38)	88-2539-69
$C_{14}H_{17}N_3O_4S$			
1H-Azepine, 3-(2,4-dinitrophenyl)-4,5,6,7-tetrahydro-1-methyl-2-(methylthio)-	$CHCl_3$	460(3.66)	5-0073-69E
$C_{14}H_{17}N_3O_5$			
1H-1,2,3-Triazole, 1-β-D-galactopyranosyl-4-phenyl-	EtOH	243(4.22),274s(3.00)	4-0639-69
1H-1,2,3-Triazole, 1-β-D-galactopyranosyl-5-phenyl-	EtOH	242(4.02)	4-0639-69
1H-1,2,3-Triazole, 1-β-D-glucopyranosyl-4-phenyl-	EtOH	241(4.25)	4-0639-69
1H-1,2,3-Triazole, 1-β-D-glucopyranosyl-5-phenyl-	EtOH	241(4.08)	4-0639-69
$C_{14}H_{17}N_3O_6$			
Mesoxalic acid, diethyl ester, 2-[(2-nitro-p-tolyl)hydrazone]	EtOH	219(411),278(3.95), 330(4.10),394(4.06)	39-0742-69C
Mesoxalic acid, diethyl ester, 2-[(6-nitro-o-tolyl)hydrazone]	EtOH	220(4.05),315(4.05), 360(3.97)	39-0742-69C
$C_{14}H_{17}N_3O_9$			
as-Triazine-3,5(2H,4H)-dione, 2-β-D-lyxofuranosyl-, 2',3',5'-triacetate	EtOH	263(3.88)	73-0618-69
$C_{14}H_{17}N_5O$			
Adenine, [(4-methoxy-2-methyl-1,4-cyclohexadien-1-yl)methyl]-	pH 1	209(4.15),281(4.28)	44-3814-69
	pH 13	277(4.28),285s(4.15)	44-3814-69
	EtOH	211(4.34),270(4.29)	44-3814-69
$C_{14}H_{17}N_5O_5$			
Morpholine, 4-[1-(2,4-dinitrophenyl)-5-methyl-2-pyrazolin-3-yl]-	EtOH	268(4.14),451(4.22)	22-1683-69
$C_{14}H_{17}N_5O_7$			
2(1H)-Quinoxalinone, 3-isoleucinyl-3,4-dihydro-5,7-dinitro-	EtOH	265(4.12),374(4.00)	44-0395-69
$C_{14}H_{17}O_5P$			
1,2-Oxaphosphol-4-ene-4-carboxylic acid, 2-methoxy-5-methyl-3-phenyl-, ethyl ester, 2-oxide	ether	233(3.82),285(3.10)	70-1012-69

Compound	Solvent	λ_{max}(log ε)	Ref.
$C_{14}H_{18}$			
Azulene, 6-ethyl-1,6-dihydro-3,8-dimethyl-	n.s.g.	233(4.21),307(3.69)	73-2288-69
Benzo[3,4]cyclobuta[1,2]cyclooctene, 4b,5,6,7,8,9,10,10a-octahydro-, cis	isooctane	259(3.16),266(3.32), 272(3.29)	88-1089-69
trans	isooctane	259(3.11),266(3.27), 272(3.24)	88-1089-69
Ethylene, 1-cyclohexyl-2-phenyl-, trans	MeOH	252(4.29)	48-0091-69
$C_{14}H_{18}BNO_4$			
Boron, (N,N-diethylacetoacetamidato)-[pyrocatecholato(2-)]-	CH_2Cl_2	233(3.65),274(4.30)	39-0173-69A
$C_{14}H_{18}ClNO_4$			
1H-Carbazole, 4a-ethyl-2,3,4,4a-tetrahydro-, perchlorate	EtOH	233(3.87),239(3.85), 279(3.79)	78-0095-69
	EtOH-NaOH	259(3.80)	78-0095-69
$C_{14}H_{18}ClNO_5$			
2-tert-Butyl-4-methyl-5-phenylisoxazolium perchlorate	CH_2Cl_2	296(4.32)	44-3451-69
$C_{14}H_{18}Cl_2N_2$			
Quinoline, 2-[(3-chloropropyl)amino]-4,6-dimethyl-, hydrochloride	EtOH	246(4.35),254(4.28), 336(3.81)	95-0759-69
$C_{14}H_{18}N_2$			
2-Cyclopropene-$\Delta^{1,\alpha}$-malononitrile, 2,3-di-tert-butyl-	MeCN	245(4.33)	35-4766-69
Indene-$\Delta^{1,\alpha}$-methanediamine, N,N,N',N'-tetramethyl-	hexane	255(3.96),279(4.08), 375(4.36)	89-0593-69
$C_{14}H_{18}N_2O$			
3-Buten-2-one, 4-amino-4-(2,4-dimethyl-6-propenyl-3-pyridyl)-	EtOH	246(4.09),303(4.43)	35-4749-69
Imidazo[1,2-a]pyridin-5(1H)-one, hexahydro-8a-methyl-8-phenyl-	EtOH	212(3.97),251(2.26),	44-0165-69
Indole-2-carboxamide, N-tert-butyl-3-methyl-	EtOH	223(4.4),293(4.2)	24-0678-69
2,8a(1H)-Naphthalenedicarbonitrile, 2,3,4,4a,5,6,7,8-octahydro-4a,8-dimethyl-7-oxo-	n.s.g.	288(1.28)	44-1949-69
4-Quinolone, 2-(diethylamino)-3-methyl-	MeOH	230(4.32),260(4.20), 322(4.15)	78-5721-69
$C_{14}H_{18}N_2OS$			
Uracil, 6-(1-adamantyl)-2-thio- (all spectra in 50% EtOH)	0.1N HCl	275(4.29)	73-2278-69
	pH 8.2	276(4.25)	73-2278-69
	0.1N NaOH	238(4.10),258(4.10), 314(3.84)	73-2278-69
$C_{14}H_{18}N_2O_2$			
Uracil, 6-(1-adamantyl)- (all spectra in 50% EtOH)	0.1N HCl	264(4.04)	73-2278-69
	pH 8.2	263(4.04)	73-2278-69
	0.1N NaOH	228(3.85),264(3.83)	73-2278-69
$C_{14}H_{18}N_2O_3$			
Nitrone, α-cyclohexyl-N-(p-nitrobenzyl)-	EtOH	244(4.1),324s(2.6)	28-0078-69A

Compound	Solvent	$\lambda_{max}(\log \epsilon)$	Ref.
$C_{14}H_{18}N_2O_4$			
4H,8H-Cyclobuta[1,2-b:3,4-b']dipyran-4,8-dione, 4a,4b,8a,8b-tetrahydro-2,4b,6,8b-tetramethyl-, dioxime	EtOH	262(4.47)	35-4749-69
1,3-Cyclobutanediol, 1,3-dimethyl-2,4-bis(5-methyl-3-isoxazolyl)-	EtOH	218(4.10)	35-4749-69
3,9-Dioxapentacyclo[6.4.0.0^{2,7}.0^{4,11}.0^{5,10}]dodecane-6,12-dione, 2,4,8,10-tetramethyl-, dioxime	EtOH	232(4.24)	35-4749-69
$C_{14}H_{18}N_2S_2$			
Uracil, 6-(1-adamantyl)-2,4-dithio- (all spectra in 50% EtOH)	0.1N HCl	286(4.38),361(4.12)	73-2278-69
	pH 8.2	258(4.18),285(4.30), 365(4.14)	73-2278-69
	0.1N NaOH	272(4.33),322(3.87), 364(3.76)	73-2278-69
$C_{14}H_{18}N_4O_2$			
Acetamide, N-(5-amino-3-butyl-3,4-dihydro-4-oxo-6-quinazolinyl)-	EtOH	226(4.15),250(3.97), 292(3.31),355(3.46)	2-1166-69
$C_{14}H_{18}N_4O_3$			
Pyrimidine, 2,4-diamino-5-(3,4,5-trimethoxybenzyl)-	pH 2	269(3.78)	44-0821-69
	pH 12	228(4.32),287(3.86)	44-0821-69
$C_{14}H_{18}N_4O_4$			
Tubercidin, N^6-allyl-	pH 1.0	231(4.32),275(4.14)	87-1056-69
	pH 7.0	210(4.42),275(4.15)	87-1056-69
	pH 12.0	275(4.16)	87-1056-69
$C_{14}H_{18}N_4O_6$			
Inosine, 2',3'-O-isopropylidene-2-methoxy-	pH 1	254(4.03)	94-2581-69
	pH 13	261(4.01)	94-2581-69
$C_{14}H_{18}N_4O_6S$			
Uridine, 5-[(4-methylimidazol-2-yl)-thio]methyl]-	pH 7	257(4.35)	63-0710-69
$C_{14}H_{18}N_4O_7$			
Hexanoic acid, 4-hydroxy-3-methyl-5-oxo-, methyl ester, 5-(2,4-dinitrophenylhydrazone)	EtOH	249(4.00),359(4.30)	35-4739-69
$C_{14}H_{18}N_6O_5$			
9H-Purine, 2,6-diamino-9-(2-deoxy-β-D-erythro-pentofuranosyl)-, 3',5'-diacetate	pH 1	252(4.07),292(4.00)	87-0498-69
	pH 7	256(3.97),279(4.00)	87-0498-69
	pH 13	256(3.97),279(4.00)	87-0498-69
$C_{14}H_{18}O$			
2α,9aα-Cyclo-7α,8aα-cyclo-4aβ,10aβ-perhydro-9-anthrone	EtOH	275(1.84)	44-3085-69
Cyclopent[a]inden-2-ol, 3,5-dimethyl-1,2,3,3a,8,8a-hexahydro-	EtOH	266(3.12),274(3.23), 280(3.28)	2-0215-69
1(2H)-Naphthalenone, 3,4-dihydro-8-isopropyl-5-methyl-	EtOH	215(4.39),256(4.07), 303(3.37)	22-2110-69
1(2H)-Naphthalenone, 3,4-dihydro-3,3,6,8-tetramethyl-	EtOH	260(4.11),300(3.28)	5-0094-69D
1-Penten-3-one, 4,4-dimethyl-1-p-tolyl-	iso-PrOH	301(4.32)	104-0826-69

Compound	Solvent	$\lambda_{max}(\log \epsilon)$	Ref.
Tricyclo[5.2.1.0^{2,6}]deca-4,8-dien-3-one, 5-tert-butyl-, endo	isooctane	193(3.88),225(4.02), 318(1.56),331(1.64), 345(1.56)	35-6785-69
	MeOH	201(3.73),233(4.07), 309(1.81)	35-6785-69
$C_{14}H_{18}OS$			
4H-Cyclohepta[c]thiophene, 4-methoxy-1,3,5,7-tetramethyl-	heptane	244s(3.89),306(3.90)	104-1115-69
6H-Cyclohepta[c]thiophene, 6-methoxy-1,3,5,7-tetramethyl-	heptane	241(4.47),283(3.66)	104-1115-69
6-Thiaphenanthrene, 1,2,3,4,4a,5,6,10b-octahydro-8-methoxy-	EtOH	221(4.45),257(3.96), 291(3.47),299(3.40)	78-1939-69
$C_{14}H_{18}O_2$			
1H-Benz[e]indene-3,7-diol, 2,3β,3a,4,-5,9bβ-hexahydro-3aα-methyl-	EtOH	283(3.30)	22-1920-69
Fluoren-9β-ol, 1,2,3,4,4aα,9aα-hexahydro-7-methoxy-	EtOH	218(4.29),247(4.05), 301(3.35)	44-2209-69
6-Hepten-2-one, 4-hydroxy-5-methyl-7-phenyl-	EtOH	239s(3.45),278s(3.18) (changing)	22-3719-69
3-Hexenophenone, 5-hydroxy-2,5-dimethyl-	EtOH	241(4.08)	22-3638-69
3-Hexenophenone, 5-hydroxy-4,5-dimethyl-	EtOH	241(4.11)	22-3638-69
as-Indacene-4-carboxylic acid, 1,2,3,3a,5a,6,7,8-octahydro-, methyl ester	EtOH	215(4.48),251(3.98), 298(3.42)	22-4068-69
2(1H)-Naphthalenone, 7β-ethynyl-4a,5,6,7,8,8aβ-hexahydro-1β-hydroxy-1,4aβ-dimethyl-	MeOH	228(3.87)	88-4689-69
8-Nonene-4,6-diyn-1-ol, 9-(tetrahydro-pyran-2-yl)-	ether	239(3.78),252(4.11), 266(4.26),282(4.18)	39-0830-69C
1-Penten-3-one, 1-(p-methoxyphenyl)-4,4-dimethyl-	iso-PrOH	324(4.35)	104-0826-69
Phenol, 2-methoxy-4-(2-norbornyl)-	pH 13	243(4.00),296(3.63)	30-0867-69A
	iso-PrOH	229(3.84),281(3.49), 289s(3.41)	30-0867-69A
Phenol, 2-methoxy-5-(2-norbornyl)-	pH 13	243(3.91),293(3.73)	30-0867-69A
	iso-PrOH	282(3.55),290s(3.39)	30-0867-69A
Phenol, 2-methoxy-6-(2-norbornyl)-	pH 13	245(3.83),290(3.61)	30-0867-69A
	iso-PrOH	275(3.28),279(3.27)	30-0867-69A
$C_{14}H_{18}O_3$			
2(1H)-Naphthalenone, 1-ethoxy-3,4-dihydro-7-methoxy-8-methyl-	EtOH	221(4.08),279(4.33)	88-4559-69
$C_{14}H_{18}O_4$			
Acetophenone, 5'-acetoxy-4'-methoxy-2'-propyl-	EtOH	223(4.28),269(4.07)	12-1011-69
4-Indancarboxylic acid, 7a-ethyl-5,6,7,7a-tetrahydro-1,5-dioxo-	EtOH	285(3.84)	104-1298-69
1-Naphthaleneacetic acid, 1,2,3,4-tetrahydro-2-hydroxy-7-methoxy-8-methyl-	EtOH	279(3.30)	88-4559-69
Propionic acid, 3-benzoyl-2-ethoxy-, ethyl ester	EtOH	243(4.18)	18-1353-69
$C_{14}H_{18}O_4S$			
Acetic acid, (o-phenylenethio)di-, diethyl ester	EtOH	250s(3.54)	44-1566-69

Compound	Solvent	$\lambda_{max}(\log \epsilon)$	Ref.
$C_{14}H_{18}O_5$			
1,3-Cyclohexadiene-1,2-dicarboxylic acid, 3,4,5,5-tetramethyl-6-oxo-, dimethyl ester	EtOH	221s(3.83),336(3.79)	33-0584-69
Hydrocinnamic acid, 2-(3-carboxypropyl)-4-methoxy-	EtOH	277(3.20)	88-0073-69
$C_{14}H_{18}O_7$			
2,3,4-Furantricarboxylic acid, 5-methyl-, triethyl ester	EtOH	256(4.09)	88-2053-69
$C_{14}H_{19}ClN_2O_4S$			
2-[2-(Dimethylamino)propyl]-3-ethylbenzothiazolium perchlorate	EtOH	384(2.76)	65-1732-69
$C_{14}H_{19}ClN_2O_5$			
6-Ethoxy-2,3-dihydro-5-methyl-7-phenyl-1H-1,4-diazepinium perchlorate	MeOH	260(3.86),367(4.16)	39-1449-69C
$C_{14}H_{19}ClO_3$			
1,4-Benzodioxan, 6-butyl-7-(2-chloroethoxy)-	EtOH	294(3.66)	104-0468-69
$C_{14}H_{19}Cl_6N_3$			
s-Triazine, 2-nonyl-4,6-bis(trichloromethyl)-	MeOH	216(3.37),262(2.65)	18-2924-69
$C_{14}H_{19}N$			
Indole, 2-isopentyl-6-methyl-	C_6H_{12}	223(4.59),270(3.91), 284(3.73),289(3.75), 295(3.78)	1-2578-69
1-Naphthaleneethylamine, 3,4-dihydro-N,N-dimethyl-, hydrochloride	n.s.g.	261(3.92)	39-0217-69C
1-Naphthalenepropylamine, 3,4-dihydro-N-methyl-	n.s.g.	260(3.91)	39-0217-69C
Piperidine, 1-(β-methylstyryl)-	C_6H_{12}	219(4.04),290(3.91)	22-0903-69
$C_{14}H_{19}NO$			
1H-1-Benzazepine, 1-acetyl-7-ethyl-2,3,4,5-tetrahydro-	MeOH	226s(3.99),265(2.60), 274(2.46)	44-1070-69
1H-1-Benzazepine, 1-acetyl-8-ethyl-2,3,4,5-tetrahydro-	MeOH	208s(4.40),266(2.66)	44-2235-69
2H-1-Benzopyran-4-propylamine, N,N-dimethyl-	n.s.g.	258(3.78)	39-0217-69C
1-Hepten-3-one, 1-(N-methylanilino)-	EtOH	226(3.92),321(4.47)	33-2465-69
Nitrone, N-benzyl-α-cyclohexyl-	C_6H_{12}	253(3.9)	28-0078-69A
	EtOH	239(3.9)	28-0078-69A
$C_{14}H_{19}NOS$			
Indole, 2-[1,1-dimethyl-2-[(methylthio)-methoxy]ethyl]-	EtOH	223(4.41),280(3.83), 290(3.76)	39-0595-69C
$C_{14}H_{19}NO_2$			
Bicyclo[3.2.2]nona-3,6-dien-2-one, 8-methyl-9-morpholino-	EtOH	215s(3.81),291(3.39)	77-0737-69
	EtOH-HCl	224(3.82),258s(3.10)	77-0737-69
Indoline, 1-acetyl-3-(2-hydroxyethyl)-2,3-dimethyl-	MeOH	252(3.87),276s(--), 286s(--)	23-3647-69
Phenol, m-1-oxa-4-azaspiro[4.5]dec-2-yl-	EtOH	277(3.43)	94-2353-69
Spiro[cyclohexane-1,1'(2'H)-isoquinoline]-4',6'-diol, 3',4'-dihydro-	EtOH	282(3.34)	94-2353-69

Compound	Solvent	λ max(log ε)	Ref.
$C_{14}H_{19}NO_2S$			
2H-Thiet-3-amine, N,N-diethyl-2-methyl-4-phenyl-, 1,1-dioxide	EtOH	200(4.24),244(4.05)	44-1136-69
$C_{14}H_{19}NO_2S_2$			
Propionaldehyde, 3-(2-benzothiazolylthio)-, diethyl acetal	EtOH	226(4.30),245s(3.95), 282(4.02),291(3.98), 301(3.90),328(3.47)	95-0469-69
$C_{14}H_{19}NO_4$			
2,3-Pyridinedicarboxylic acid, 5-ethyl-6-methyl-, diethyl ester	MeOH	231(4.03),272(3.70)	77-0140-69
2,5-Pyridinedicarboxylic acid, 3-ethyl-6-methyl-, diethyl ester	MeOH	225(4.04),282(3.74)	77-0140-69
$C_{14}H_{19}NO_6$			
Glucosamine, p-methoxybenzylidene-	neutral	265(3.64)	94-0770-69
	alkaline	366(3.00)	94-0770-69
3,4,5-Pyridinetricarboxylic acid, 1,4-dihydro-2,6-dimethyl-, 3,5-diethyl ester	pH 1	231(4.29),348(3.79)	103-0762-69
	H_2O	232(4.23),350(3.81)	103-0762-69
	pH 13	234(4.21),352(3.79)	103-0762-69
	EtOH	230(4.24),342(3.80)	103-0762-69
	EtOH-HCl	230(4.30),342(3.84)	103-0762-69
	EtOH-NaOH	234(4.22),351(3.77)	103-0762-69
$C_{14}H_{19}NS$			
Thiazole, 4-(1-adamantyl)-2-methyl-	EtOH	242(3.61)	18-1617-69
$C_{14}H_{19}N_3$			
Aniline, p-(1-cyclohexen-1-ylazo)-N,N-dimethyl-	THF	263(4.05),385(4.44)	24-3467-69
$C_{14}H_{19}N_3O$			
Cytosine, 6-(1-adamantyl)- (all spectra in 50% EtOH)	0.1N HCl	281(4.15)	73-2278-69
	pH 8	271(4.00)	73-2278-69
	0.1N NaOH	225(3.96),272(3.95)	73-2278-69
4-Pyrimidinol, 6-(1-adamantyl)-2-amino-(all spectra in 50% EtOH)	0.1N HCl	258(3.96)	73-2278-69
	pH 8.2	265(3.75),285(3.68)	73-2278-69
	0.1N NaOH	230(3.98),275(3.90)	73-2278-69
$C_{14}H_{19}N_3O_2$			
2H-Pyran-2-methanol, 6-(5,6-dimethyl-1H-benzotriazol-1-yl)tetrahydro-, cis	EtOH	260(3.88),287(3.68)	4-0005-69
$C_{14}H_{19}N_3O_2S$			
Valeric acid, 2-(dicyanomethylene)-3-[(dimethylamino)(methylthio)-methylene]-, ethyl ester	C_6H_{12}	471(3.85)	5-0073-69E
$C_{14}H_{19}N_3O_4$			
Mesoxalic acid, diethyl ester, 2-[(2-amino-p-tolyl)hydrazone]	EtOH	215(4.16),237(4.02), 280(3.83),340s(--), 400(4.11)	39-0742-69C
Mesoxalic acid, diethyl ester, 2-[(6-amino-o-tolyl)hydrazone]	EtOH	215(4.28),232(4.10), 277(3.84),340(3.89), 410(4.05)	39-0742-69C
$C_{14}H_{19}N_3S$			
2-Thiocytosine, 6-(1-adamantyl)-(all spectra in 50% EtOH)	0.1N HCl	228(4.24),281(4.49)	73-2278-69
	pH 8.2	245(4.34),272(4.26)	73-2278-69

Compound	Solvent	λ_{max}(log ϵ)	Ref.
2-Thiocytosine, 6-(1-adamantyl)- (cont.)	0.1N NaOH	226(4.25),265(4.19), 294s(3.87)	73-2278-69
$C_{14}H_{19}N_5O_3$			
4H-Cyclopenta-1,3-dioxole-4-methanol, 6-(6-amino-9H-purin-9-yl)tetrahydro-2,2-dimethyl-	pH 1	259(4.17)	35-3075-69
	pH 7	262(4.17)	35-3075-69
	pH 13	261(4.17)	35-3075-69
$C_{14}H_{19}N_5O_5$			
Adenosine, 2',3'-O-isopropylidene-N-methoxy-	pH 1	268(4.19)	94-1128-69
	pH 7	269(4.16)	94-1128-69
	pH 13	284(4.05)	94-1128-69
Adenosine, 2',3'-O-isopropylidene-2-methyl-, 1'-oxide	pH 1	260(4.07)	94-1128-69
	pH 7	233(4.61),264(3.90)	94-1128-69
	pH 13	232(4.51),269(3.94)	94-1128-69
$C_{14}H_{20}$			
Ethylene, tetracyclopropyl-	hexane	219(4.08)	77-0781-69
Tetralin, 2,2,5,7-tetramethyl-	EtOH	269(2.75),278(2.66)	5-0094-69D
$C_{14}H_{20}BrClO$			
2,5-Cyclohexadien-1-one, 4-bromo-2,6-di-tert-butyl-4-chloro-	hexane	247(4.16)	55-0074-69
$C_{14}H_{20}BrNO_3$			
2,4-Cyclohexadien-1-one, 6-bromo-2,6-di-tert-butyl-4-nitro-	C_6H_{12}	332(3.38)	64-0547-69
$C_{14}H_{20}Br_2O$			
2,5-Cyclohexadien-1-one, 2,4-dibromo-4,6-di-tert-butyl-	hexane	262(4.09)	55-0074-69
2,5-Cyclohexadien-1-one, 4,4-dibromo-2,6-di-tert-butyl-	hexane	250(4.15)	55-0074-69
	MeOH	251(4.05)	64-0547-69
$C_{14}H_{20}ClN$			
1-Naphthaleneethylamine, 3,4-dihydro-N,N-dimethyl-, hydrochloride	n.s.g.	261(3.92)	39-0217-69C
$C_{14}H_{20}ClNO_4$			
1,3-Diethyl-2,3-dimethyl-3H-indolium perchlorate	EtOH	239(3.74),280(3.97)	78-0095-69
	EtOH-NaOH	280(4.41)	78-0095-69
1,2,3,4,7,8,9,10-Octahydro-6-methyl-benzo[c]quinolizinium perchlorate	$CHCl_3$	280(3.86)	78-4161-69
$C_{14}H_{20}N_2$			
Indene-$\Delta^{1,\alpha}$-methanediamine, 3a,7a-dihydro-N,N,N',N'-tetramethyl-	hexane	260(4.06),270(4.12), 302(4.17)	89-0593-69
$C_{14}H_{20}N_2NaO_8PS$			
Thymidine, 3'-acetate, 5'-O-ester with S-ethyl phosphorothioate, sodium salt	H_2O	265(3.97)	35-1522-69
$C_{14}H_{20}N_2O$			
3-Buten-2-one, 4-amino-4-(2,4-dimethyl-6-propyl-3-pyridyl)-	EtOH	275s(3.92),303(4.30)	35-4749-69
3-Buten-2-one, 4-anilino-4-(tert-butyl-amino)-	EtOH	232(3.94),248s(--), 302(4.31)	44-3451-69
$C_{14}H_{20}N_2O_2$			
Octamethylenimine, N-(p-nitrophenyl)-	DMSO	393(3.35)	39-0544-69B

Compound	Solvent	λ_{max}(log ϵ)	Ref.
Pyrrole-2-carboxylic acid, 4-(2-cyano-ethyl)-3,5-dimethyl-, tert-butyl ester	$CHCl_3$	278(4.25)	39-0176-69C
$C_{14}H_{20}N_2O_2S_2$			
Propenylamine, N,N-diethyl-1-(methyl-thio)-2-[(o-nitrophenyl)thio]-	MeOH	386(3.62)	5-0073-69E
$C_{14}H_{20}N_2O_3$			
Acetanilide, 2',4'-diisopropyl-5'-nitro-	C_6H_{12}	235(4.24)	78-1423-69
Acetanilide, 2',5'-diisopropyl-4'-nitro-	C_6H_{12}	287(3.69)	78-1423-69
Acetanilide, 4',5'-diisopropyl-2'-nitro-	C_6H_{12}	290(3.83),364(3.61)	78-1423-69
Acetimidic acid, N-(2,3,4,5-tetramethyl-6-nitrophenyl)-, ethyl ester	MeOH	208(4.42)	44-0758-69
2-Furanacrylamide, N-(1-carbamoylhexyl)-	pH 8	304(4.41)	35-7187-69
2H-Pyrano[2,3-d]pyrimidine-2,4(3H)-di-one, 3-cyclohexyl-1,5,6,7-tetra-hydro-7-methyl-	75% dioxan	262(4.01)	95-0266-69
	+NaOH	277(4.11)	95-0266-69
$C_{14}H_{20}N_2O_4$			
Benzene, 1,4-di-tert-butyl-2,3-dinitro-	EtOH	280(2.89),329(2.58)	54-0386-69
Benzene, 1,4-di-tert-butyl-2,5-dinitro-	EtOH	281(3.16),330(2.74)	54-0386-69
Benzene, 1,4-di-tert-butyl-2,6-dinitro-	EtOH	336(2.70)	54-0386-69
Cinnamamide, N-(4-amino-2-hydroxybutyl)-4-hydroxy-3-methoxy-	acid or neutral	293(3.85),316(3.92)	88-2807-69
	base	309(--),363(--)	88-2807-69
2H-Pyrano[2,3-d]pyrimidine-2,4(3H)-di-one, 1,5,6,7-tetrahydro-3-(4-hydroxy-cyclohexyl)-7-methyl-	75% dioxan	262(4.03)	95-0266-69
	+NaOH	277(4.07)	95-0266-69
1H,5H-Pyrazolo[1,2-a]pyrazole-1,3,5,7-(2H,6H)-tetrone, 2,2,6,6-tetraethyl-	EtOH	233(4.05),260(3.04), 270(3.00)	44-3181-69
	dioxan	233(3.22),260(3.19), 272(3.15)	88-4497-69
$C_{14}H_{20}N_2O_5$			
Bicarbamic acid, (3,6,6-trimethyl-2-methylene-5-oxo-3-cyclohexen-1-yl)-, dimethyl ester	MeOH	272(4.11)	88-4781-69
$C_{14}H_{20}N_4$			
Pyrimidine, 4-(1-adamantyl)-2,6-diamino-(all spectra in 50% EtOH)	0.1N HCl	269(3.99)	73-2278-69
	pH 8.2	227s(4.09),276(3.92)	73-2278-69
	0.1N NaOH	229(4.04),280(3.93)	73-2278-69
$C_{14}H_{20}N_4O_2S$			
Valeric acid, 5-[(9-bromo-9H-purin-6-yl)thio]-	pH 1	222(4.15),295(4.30)	73-2114-69
	pH 13	227(3.95),287(4.26), 294(4.26)	73-2114-69
$C_{14}H_{20}N_4O_2S_3$			
Thiamine, S-dithiomethoxycarbonyl-	EtOH	231(4.08),270(3.93), 312(4.10)	94-2299-69
Thiamine, methylxanthogenate	EtOH	229(3.96),271(3.48), 303(3.75)	94-2299-69
	EtOH-HCl	259(--)	94-2299-69
$C_{14}H_{20}N_4O_7$			
Benzoic acid, 4-hydroxy-3,5-dimethoxy-, ester with 1-(4-hydroxybutyl)-3-ni-troguanidine (nitroleonurine)	MeOH	222(4.39),273(4.39)	78-5155-69
	MeOH-NaOH	240(4.25),271(4.24), 328(4.28)	78-5155-69

Compound	Solvent	λ_{max}(log ε)	Ref.
$C_{14}H_{20}N_6O_4S$			
Uracil, 5,5'-thiobis[1,3-dimethyl-6-(methylamino)-	MeOH	286(4.43)	44-3285-69
$C_{14}H_{20}N_6O_5S$			
Homocysteine, S-adenosyl-	H_2O	260(4.18)	65-0434-69
$C_{14}H_{20}O$			
Aristolone, 4-demethyl-, (+)-	MeOH	235(4.06)	23-4299-69
1(2H)-Biphenylenone, 3,4,4b,5,6,7,8,8a-octahydro-3,3-dimethyl-	EtOH	249(4.00)	44-0794-69
3-Buten-2-one, 4-(3,3-dimethyl-2-methylene-1-norbornyl)-	n.s.g.	230(4.00)	77-1335-69
7α,8aα-Cycloanthran-2β-ol, 1,2,3,4,4aβ,-5,6,8,10,10aβ-decahydro-	C_6H_{12}	218(4.09)	44-3085-69
5-Epiaristolone, 4-demethyl-, (+)-	MeOH	232(4.07)	23-4299-69
2(3H)-Naphthalenone, 4,4a,5,6,7,8-hexahydro-7-isopropenyl-1-methyl-	MeOH	249(4.16)	23-0137-69
2(3H)-Naphthalenone, 4,4a,5,6-tetrahydro-6,7-methano-4a,9,9-trimethyl-	$CHCl_3$	247(4.15)	32-0219-69
2(3H)-Naphthalenone, 4,4a,5,6-tetrahydro-7,8-methano-1,4a,9-trimethyl-	EtOH	275(4.28)	88-3871-69
Norcedrenone	EtOH	230(4.03)	88-0227-69
$C_{14}H_{20}OS$			
2-Bornanone, 3-(m-dithian-2-ylidene)-, (+)-	dioxan	270(3.57),322(4.27), 370s(2.94)	28-0186-69A
$C_{14}H_{20}O_2$			
2'-Acetonaphthone, 1',5',6',7',8',8'a-hexahydro-3'-hydroxy-8',8'a-dimethyl-	EtOH	252(3.95),337(3.83)	77-0893-69
o-Benzoquinone, 3,5-di-tert-butyl-	MeOH	256s(3.50),402(3.28), 550(1.77)	64-0547-69
p-Benzoquinone, 2,5-di-tert-butyl-	MeOH	253(4.25)	64-0547-69
p-Benzoquinone, 2,6-di-tert-butyl-	MeOH	254(4.25),318(2.67), 446(1.57)	64-0547-69
Cyclohexanone, 3-(5-isopropyl-2-furyl)-6-methyl-	EtOH	229(4.30)	44-0857-69
as-Indacene-4-carboxylic acid, 1,2,3,-4a,5,5a,6,7,8,8a-decahydro-, methyl ester, cis-cis	EtOH	206(3.76)	22-4068-69
cis-trans	EtOH	206(3.78)	22-4068-69
as-Indacene-4-carboxylic acid, 1,2,3,-3a,4,5,5b,6,7,8-decahydro-, methyl ester, cis-cis	EtOH	209(3.91)	22-4068-69
cis-trans	EtOH	208(3.89)	22-4068-69
7αH-12-Noreudesm-4-ene-3,11-dione	MeOH	247(4.22)	44-0732-69
Tricyclo[5.2.1.0^{2,6}]decane-3,8-dione, 5,5,10,10-tetramethyl-, endo	EtOH	300(1.76)	39-0449-69B
$C_{14}H_{20}O_3$			
4H-1-Benzopyran-3-carboxylic acid, 5,6,7,8-tetrahydro-2,4,4,7-tetramethyl-	EtOH	209(3.83),270(3.41)	44-1465-69
Confertifolin, 6-oxo-	isooctane	211(3.96)	78-3895-69
2-Cyclohexen-1-one, 3-methoxy-5,5-dimethyl-2-(3-oxocyclopentyl)-	EtOH	268(4.17)	104-1566-69
2,6-Methano-7H-1-benzoxocin-7-one, 2,3,4,5,6,8,9,10-octahydro-2-hydroxy-9,9-dimethyl-	EtOH	266(4.13)	104-1566-69

Compound	Solvent	λ_{max}(log ε)	Ref.
2-Naphthol, 1-ethoxy-1,2,3,4-tetrahydro-7-methoxy-8-methyl-	EtOH	278(3.29)	88-4559-69
$C_{14}H_{20}O_4$			
Chroman-3,5-diol, 2,2-dimethyl-6-(3-hydroxypropyl)-	EtOH	279(3.29)	33-1165-69
$C_{14}H_{20}O_4S$			
α-D-Lyxopyranoside, ethyl 4-S=benzyl-4-thio-	MeOH	261s(3.53),267s(3.34)	44-2643-69
β-L-Xylopyranoside, ethyl 3-S-benzyl-3-thio-	MeOH	260(3.49),265(3.32)	44-2643-69
$C_{14}H_{20}O_5$			
1-Cyclohexene-1-butyric acid, 2-(2-carboxyethyl)-6-methyl-3-oxo-	EtOH	243(4.15)	78-2661-69
4H-Pyran-3,5-dicarboxylic acid, 2,4,6-trimethyl-, diethyl ester	MeOH	214(4.04),221(4.04), 285 (3.46)	44-3169-69
$C_{14}H_{20}O_6$			
Acetophenone, 2,2'-diethoxy-6'-hydroxy-3',4'-dimethoxy-	EtOH	235(3.93),283(4.06), 335(3.53)	44-1460-69
1,4-Cycloheptanedicarboxylic acid, 6-methyl-2,3-dioxo-, diethyl ester	EtOH	305(4.18)	12-2025-69
5-Epiloganin aglucone, 5-O-acetyl-1-O-methyl-	EtOH	236(4.04)	39-0721-69C
α-D-galacto-Oct-5-enos-7-ulose, 6,8-dideoxy-1,2:3,4-di-O-isopropylidene-, cis	EtOH	248(4.20)	23-0081-69
2-Oxabicyclo[2.2.2]oct-5-ene-6-carboxylic acid, 3,8-dioxo-, ethyl ester, 8-(diethyl acetal)	$CHCl_3$	242(3.71)	24-2835-69
$C_{14}H_{20}O_7$			
α-D-erythro-Hex-3-enofuranose, 3-O-acetyl-1,2:5,6-di-O-isopropylidene-	MeOH	219(3.8)	24-1071-69
$C_{14}H_{20}S$			
Thioxanthene, octahydro-9-methyl-	hexane	238(3.79)	103-0172-69
$C_{14}H_{21}BrO$			
2,5-Cyclohexadien-1-one, 4-bromo-2,6-di-tert-butyl-	hexane	250(4.11)	55-0074-69
$C_{14}H_{21}ClN_4$			
1-Cyclobutene-1,3-dicarbonitrile, 4-chloro-2,4-bis(diethylamino)-	EtOH	212(4.38),256(4.52), 336(4.00)	88-3639-69
$C_{14}H_{21}ClO$			
2,5-Cyclohexadien-1-one, 2,6-di-tert-butyl-4-chloro-	hexane	239(4.11)	55-0074-69
$C_{14}H_{21}Cl_2N_3O_6$			
Uridine, 5-[[bis(2-chloroethyl)amino]-methyl]-, hydrochloride	H_2O	267(4.02)	73-1696-69
$C_{14}H_{21}N$			
$\Delta^{1,\alpha}$-Cyclohexaneacetonitrile, 2,3-dimethyl-2-(2-methylpropenyl)-	MeOH	221(4.09)	23-0831-69

Compound	Solvent	λ_{max}(log ϵ)	Ref.
Indoline, 2-isopentyl-6-methyl-	C_6H_{12}	214(4.29),248(3.91), 302(3.54)	1-2578-69
$C_{14}H_{21}NO$			
Benzene, 1,4-di-tert-butyl-2-nitroso-	hexane	790(1.54)	18-3611-69
Benzene, 2,4-di-tert-butyl-1-nitroso-	hexane	790(1.74)	18-3611-69
Cyclopenta[b]pyrrole, 3,3a,4,6a-tetra-hydro-2-methoxy-3,3a,5,6,6a-penta-methyl-4-methylene-	EtOH	240(4.15)	35-6107-69
isomer	EtOH	240(4.20)	35-6107-69
4-Piperidinemethanol, α-phenethyl-	EtOH	219(3.38),257(3.18)	39-1234-69C
Spiro[cyclohexane-1,2'[2H]-indol]-3'(1'H)-one, 4',5',6',7'-tetra-hydro-1'-methyl-	EtOH	320(4.11)	78-4161-69
$C_{14}H_{21}NO_2$			
Benzoic acid, diethylaminopropyl ester	dioxan	267(2.92),273(2.98), 280(2.89)	65-1861-69
Benzoic acid, p-[4-(dimethylamino)-butyl]-, methyl ester	heptane	237(4.21)	44-0077-69
	MeOH	238(4.22)	44-0077-69
Octanohydroxamic acid, N-phenyl-	EtOH	247(3.99)	42-0831-69
$C_{14}H_{21}NO_3$			
2,4-Pentadienoic acid, 2-acetyl-5-piper-idino-, ethyl ester	EtOH	240(3.82),270(3.85), 394(4.64)	70-0363-69
$C_{14}H_{21}NO_4$			
Codonopsine	EtOH	226(4.1),278(3.26)	105-0024-69
3,5-Pyridinedicarboxylic acid, 1,4-dihy-dro-1,2,6-trimethyl-, diethyl ester	EtOH	232(4.21),258(4.16), 345(3.86)	27-0037-69
3,5-Pyridinedicarboxylic acid, 1,4-dihy-dro-2,4,6-trimethyl-, diethyl ester	EtOH	349(3.91)	27-0037-69
$C_{14}H_{21}NO_7$			
α-D-galacto-Oct-6-enopyranose, 6,7,8-trideoxy-1,2:3,4-di-O-isopropyli-dene-7-nitro-, cis	EtOH	245(4.08)	23-0075-69
trans	EtOH	232(3.67)	23-0075-69
$C_{14}H_{21}N_3NaO_8P$			
Thymidine, 5'-(hydrogen morpholinophos-phonate), sodium salt	H_2O	268(3.96)	35-1522-69
$C_{14}H_{21}N_3O$			
Cyclohexanone, 2-(4,5,6,7-tetrahydro-2-methyl-1-benzimidazolyl)-, oxime	MeOH	227(3.82)	48-0746-69
$C_{14}H_{21}N_3O_2$			
Cyclohexanone, 2-(4,5,6,7-tetrahydro-2-methyl-1-benzimidazolyl)-, oxime, 3-oxide	MeOH	221(3.74)	48-0746-69
$C_{14}H_{21}N_3O_2S$			
Cyclohexanone, 2-[4,5,6,7-tetrahydro-2-(methylthio)-1-benzimidazolyl]-, oxime, 3-oxide	MeOH	226(3.70),262(3.88)	48-0746-69
$C_{14}H_{21}N_3O_4$			
Acetamide, N,N'-[4-[bis(2-hydroxyethyl)-amino]-o-phenylene]bis-	EtOH	273(4.20)	104-0158-69

Compound	Solvent	$\lambda_{max}(\log \epsilon)$	Ref.
$C_{14}H_{21}N_3O_5$			
Benzoic acid, 4-hydroxy-3,5-dimethoxy-, ester with (4-hydroxybutyl)guanidine, hydrochloride (leonurine)	MeOH	222(4.30),278(4.03)	78-5155-69
	MeOH-NaOH	239(4.15),327(4.29)	78-5155-69
$C_{14}H_{21}N_5O_5$			
Adenosine, N-(2-ethoxyethyl)-	pH 1.0	263(4.27)	87-1056-69
	pH 7.0	211(4.29),267(4.25)	87-1056-69
	pH 12.0	267(4.26)	87-1056-69
$C_{14}H_{21}N_5O_6$			
m-Phenylenediamine, N,N'-di-tert-butyl-2,4,6-trinitro-	MeOH	323(4.21),350s(4.13), 410s(3.82)	35-4155-69
m-Phenylenediamine, N,N,N',N'-tetra-ethyl-2,4,6-trinitro-	MeOH	352(4.02)	35-4155-69
$C_{14}H_{22}$			
Cyclobutene, 3-(5-hexenyl)-1,2,3-tri-methyl-4-methylene-	hexane	334(4.13)	24-0275-69
Cyclobutene, 3-(4-methyl-4-pentenyl)-1,2,3-trimethyl-4-methylene-	hexane	235(4.14)	24-0275-69
$C_{14}H_{22}BrP$			
Allyl(tert-butyl)methylphenylphosphon-ium bromide	MeOH	220(4.05),256s(2.83), 260(2.83),266(2.97), 273(2.89)	24-3546-69
$C_{14}H_{22}Br_2N_2$			
p-Phenylenediamine, 2,5-dibromo-N,N,N',N'-tetraethyl-	C_6H_{12}	226(4.45),275(4.06)	117-0287-69
$C_{14}H_{22}ClNO_4$			
4-tert-Butyl-5,6,7,8-tetrahydro-1-meth-ylquinolinium perchlorate	EtOH	276(3.84)	78-4161-69
1,2,3,4,4a,5,6,7,8,9,10-Decahydro-6-methylbenzo[c]quinolizinium perchlorate	$CHCl_3$	310(3.77)	78-4161-69
$C_{14}H_{22}INO$			
4-Ethoxy-2,3,5,5a,6,7,8,9-octahydro-1H-cyclopenta[c]quinolizinium iodide	EtOH	334(4.04)	44-0698-69
$C_{14}H_{22}NO$			
Nitroxide, N-tert-butyl-N-(p-tert-butyl-phenyl)-	EtOH	294(4.02),450(2.62)	39-1459-69C
$C_{14}H_{22}N_2O_2$			
Benzoic acid, p-amino-, 3-(dimethyl-amino)-1,2-dimethylpropyl ester	EtOH	220(4.02),294(4.35)	9-0254-69
p-Benzoquinone-1-imine, 3-(tert-butyl-amino)-N-tert-butyl-, N-oxide	EtOH	394(4.26)	35-3724-69
Nitroxide, m-phenylenebis[tert-butyl-	EtOH	289(4.28)	35-3724-69
$C_{14}H_{22}N_2O_5$			
1-Pyrazolidinepropionic acid, α,α,4,4-tetraethyl-β,3,5-trioxo-	EtOH	255(3.32)	44-3181-69
$C_{14}H_{22}N_4$			
1-Cyclobutene-1,3-dicarbonitrile,2,4-bis(diethylamino)-, hydrochloride	EtOH	258(4.60),310(3.89)	88-3639-69

Compound	Solvent	λ_{max}(log ε)	Ref.
$C_{14}H_{22}N_4O_4$			
m-Phenylenediamine, N,N,N',N'-tetra-ethyl-4,6-dinitro-	MeOH	333(4.47),417(4.08)	35-4155-69
Piperidine, 1,1'-(2,3-dinitro-1,3-buta-dienylene)di-	n.s.g.	360(4.5)	77-0549-69
$C_{14}H_{22}O$			
Bicyclo[6.2.0]dec-9-en-3-one, 1,8,9,10-tetramethyl-	MeOH	291(2.93)	24-0275-69
3H-Cyclopenta[1,3]cyclopropa[1,2]benzen-3-one, octahydro-4-isopropyl-7-methyl-	MeOH	206(3.79)	88-1251-69
2H-Cyclopropa[a]naphthalen-2-one,	MeOH	213(3.72)	23-4299-69
decahydro-1,1,7a-trimethyl-	MeOH	210(3.68)	23-4299-69
2,6,10-Dodecatrienal, 7,11-dimethyl-, 6-cis-2-trans	n.s.g.	218(4.30)	35-3281-69
2(3H)-Naphthalenone, 4,4a,5,6,7,8-hexa-hydro-7-isopropyl-5-methyl-	MeOH	240(4)	88-3729-69
6-Octenal, 2-ethylidene-3,7-dimethyl-3-vinyl-	n.s.g.	234(3.93)	35-3281-69
Phenol, 2,4-di-tert-butyl-	pH 2	274(3.30)	30-1113-69A
	pH 12	240(4.01),293(3.53)	30-1113-69A
	MeOH	278(3.56)	30-1113-69A
	MeOH-Na	245(4.03),296(3.53)	30-1113-69A
Phenol, 2,6-di-tert-butyl-	MeOH	272(3.21)	30-1113-69A
	MeOH-Na	251(3.96),298(3.72)	30-1113-69A
$C_{14}H_{22}OS$			
2-Norpinanone, 3-[(butylthio)methylene]-6,6-dimethyl-, (-)-	C_6H_{12}	293(4.32)	28-1315-69B
Sulfoxide, 2-benzyl-3,3-dimethylbutyl methyl	EtOH	221(3.82)	39-0581-69B
$C_{14}H_{22}O_2$			
5,9(1H,6H)-Benzocyclooctenedione, octahydro-7,7-dimethyl-	EtOH	305(1.64)	44-0794-69
4H-1-Benzopyran-3-methanol, 5,6,7,8-	MeOH	240s(3.32)	44-0380-69
tetrahydro-2,4,4,7-tetramethyl-	EtOH	235(3.42),285(1.30), 300(1.00)	44-1465-69
Bicyclo[5.2.0]non-8-en-3-ol, 1,7,8,9-tetramethyl-, formate	MeOH	291(2.36)	24-0275-69
β-Ionone, 3-methoxy-	EtOH	220(3.72),293(3.87)	22-0232-69
Naphthalene, decahydro-1,3-dimethoxy-4a,5-dimethyl-	EtOH	250(4.10)	24-2697-69
$C_{14}H_{22}O_4$			
Bicyclo[2.2.2]oct-2-ene-2-carboxylic acid, 5,?-diethoxy-, methyl ester	$CHCl_3$	233(3.84)	24-2835-69
Cyclopentaneheptanoic acid, 2,5-dioxo-, ethyl ester	EtOH	250(4.22),270s(3.76)	94-0408-69
$C_{14}H_{22}Si_2$			
2,4-Disila-5,7-dodecadiene-3,9-diyne, 2,2,11,11-tetramethyl-, cis-cis	EtOH	292(4.47),304(4.60), 320(4.57)	35-7520-69
cis-trans	EtOH	291(4.51),303(4.66), 319(4.61)	35-7520-69
trans-trans	EtOH	290(4.57),302(4.76), 318(4.80)	35-7520-69
$C_{14}H_{23}BO_2$			
Benzeneboronic acid, dibutyl ester	hexane	192(4.82),220(4.01), 268(2.52)	39-0392-69A

Compound	Solvent	$\lambda_{max}(\log \epsilon)$	Ref.
$C_{14}H_{23}N$			
2,4-Cyclohexadiene-1-methylamine, N,N,1,2,3,5-hexamethyl-6-methylene-	EtOH	319(4.00)	44-2393-69
$C_{14}H_{23}N_3O$			
Bicyclo[3.3.1]nonan-2-one, 9-isopropylidene-5-methyl-, semicarbazone	isooctane	286(2.16),300s(2.10)	44-0450-69
4,9-Cyclodecadiene-7,7,10-trimethyl-3-oxo-, semicarbazone	EtOH	232(4.34)	78-5275-69
$C_{14}H_{24}$			
2,4,9-Undecatriene, 2,6,10-trimethyl-	EtOH	238(4.40)	44-3789-69
$C_{14}H_{24}BrP$			
Butylmethylphenylpropylphosphonium bromide	MeOH	219(4.00),255s(2.61), 259(2.80),265(2.94), 272(2.86)	24-3546-69
tert-Butylmethylphenylpropylphosphonium bromide	MeOH	220(4.03),254s(2.65), 260(2.85),266(2.98), 273(2.91)	24-3546-69
$C_{14}H_{24}ClNO_4$			
4-tert-Butyl-2,3,5,6,7,8-hexahydro-1-methylquinolinium perchlorate	$CHCl_3$	306(3.73)	78-4161-69
$C_{14}H_{24}ClN_4O_8P$			
Uridine, 5'-[hydrogen[2-(2-chloroethyl)-methylamino]ethyl]phosphoramidate]	pH 7	262(4.00)	65-0668-69
	pH 12	262(3.84)	65-0668-69
$C_{14}H_{24}N_2O$			
Phenol, 5-tert-butyl-2,4-bis(dimethylamino)-	MeOH	232(3.84),288(3.50)	39-2164-69C
$C_{14}H_{24}N_2O_2$			
5-tert-Butyl-4-(dimethylamino)-N,N-dimethyl-1,2-quinonimonium hydroxide	H_2O	212(3.78),266(3.67), 310s(3.19),514(3.02)	39-2164-69C
m-Phenylenediamine, N,N'-di-tert-butyl-N,N'-dihydroxy-	EtOH	243(3.95)	35-3724-69
$C_{14}H_{24}N_2O_6$			
Hydrazine, N,N'-bis(diethylmalonoyl)-	EtOH	233(3.69)	44-3181-69
$C_{14}H_{24}O$			
2-Cyclododecen-1-one, 4-ethyl-	EtOH	231(4.05)	35-1264-69
2-Cyclododecen-1-one, 12-ethyl-	EtOH	230(4.01)	35-1264-69
Cyclotridecanone, 2-methylene-	n.s.g.	227(3.79),323(1.56)	97-0303-69
$C_{14}H_{24}OSi$			
Silane, [(3,4,4a,5,6,7-hexahydro-4a-methyl-2-naphthyl)oxy]trimethyl-	EtOH	241(4.32),283s(2.37)	44-2324-69
$C_{14}H_{24}O_2$			
1,2-Cyclotetradecanedione	hexane	278(1.53),443(1.37)	23-3266-69
$C_{14}H_{25}N_4O_8P$			
Uridine, 5'-[hydrogen[3-(dimethylamino)-propyl]phosphoramidate]	pH 7	262(4.00)	65-0668-69
	pH 12	260(3.88)	65-0668-69

Compound	Solvent	$\lambda_{max}(\log \epsilon)$	Ref.
$C_{14}H_{25}N_5O_2$			
as-Triazine-2(5H)-acetamide, N-butyl-3-(butylamino)-6-methyl-5-oxo-	EtOH	212(4.43),248s(3.94)	114-0181-69C
$C_{14}H_{26}N_2$			
Piperidine, 1,1'-(ethylvinylene)di-	C_6H_{12}	208(4.00),230(3.98)	22-3883-69
$C_{14}H_{26}O_4$			
Undecanoic acid, monoester with 1,3-dihydroxy-2-propanone	dioxan	224(1.60),275(1.30)	23-1249-69
$C_{14}H_{27}N_7O_9$			
L-Alanine, N(ω)-nitro-L-arginyl-L-seryl-, acetate	HOAc	269(3.81)	39-0401-69C
$C_{14}H_{28}Sn$			
Stannane, diallyldibutyl- (correction)	hexane	208(4.30)	44-3715-69
	MeCN	212(4.31)	44-3715-69
$C_{14}H_{42}OSi_6$			
Ether, bis(heptamethyltrisilan-1-yl)	isooctane	220(4.15)	35-6613-69
Ether, bis(heptamethyltrisilan-2-yl)	isooctane	226(3.94),254(3.27)	35-6613-69
	EtOH	225(3.97),254(3.28)	35-6613-69
$C_{14}H_{43}NSi_6$			
Bis(heptamethyltrisilan-1-yl)amine	isooctane	218s(4.12)	35-6613-69

Compound	Solvent	λ_{max}(log ε)	Ref.
$C_{15}HF_9O_2$			
2(3H)-Benzofuranone, 4,5,6,7-tetra-fluoro-3-(pentafluorobenzylidene)-	EtOH	201(4.28),281(3.92)	65-1774-69
Coumarin, 5,6,7,8-tetrafluoro-3-(penta-fluorophenyl)-	EtOH	201(4.39),284(4.20), 400(2.81)	65-1774-69
$C_{15}HF_{10}N$			
Acrylonitrile, 2,3-bis(pentafluorophen-yl)-, cis	EtOH	200(4.11),273(4.18)	65-2071-69
trans	EtOH	200(4.14),272(4.16)	65-2071-69
$C_{15}H_2F_{10}O_2$			
Acrylic acid, 2,3-bis(pentafluorophen-yl)-	EtOH	226(4.12),250(4.14)	65-1774-69
$C_{15}H_5N_3O_3$			
Anthra[1,2-c][1,2,5]oxadiazole-4-carbo-nitrile, 6,11-dihydro-6,11-dioxo-	EtOH	370s(3.64)	104-1642-69
$C_{15}H_6N_4O_{10}$			
Coumarin, 3-(2,4-dinitrophenyl)-6,8-di-nitro-	EtOH	250(4.00)	2-0049-69
$C_{15}H_7F_5O$			
Chalcone, 2,3,4,5,6-pentafluoro-	EtOH	288(4.47)	65-1347-69
$C_{15}H_7N_3O_8$			
Coumarin, 6,8-dinitro-3-(p-nitrophenyl)-	EtOH	256-274(3.96)	2-0049-69
Coumarin, 3-(2,4-dinitrophenyl)-	EtOH	268(4.81)	2-0049-69
6-nitro-	EtOH-NaOEt	327(4.63)	2-0049-69
$C_{15}H_7N_3O_9$			
Coumarin, 7-hydroxy-6,8-dinitro-	EtOH	272(4.39)	2-0049-69
3-(p-nitrophenyl)-	EtOH-NaOAc	335(4.10)	2-0049-69
$C_{15}H_8$			
1,4,6,9,11,14-Pentadecahexayne	isooctane	227(2.88),238(2.86), 252(2.66)	78-2823-69
$C_{15}H_8ClN_3O_2$			
Benzimidazo[2,1-a]isoquinoline, 1-chlo-ro-10-nitro-	EtOH	209(4.27),280(4.35)	24-1529-69
Benzimidazo[2,1-a]isoquinoline, 4-chlo-ro-10-nitro-	EtOH	212(4.28),283(4.38)	24-1529-69
$C_{15}H_8Cl_2O$			
Anthrone, 10-(dichloromethylene)-	C_6H_{12}	234(4.39),252(4.29), 278(4.18),330(3.78), 340(3.75)	28-0340-69A
Cyclopropenone, bis(p-chlorophenyl)-	CH_2Cl_2	227(4.29),236s(--), 243(4.25),300s(--), 310(4.53)	24-0319-69
$C_{15}H_8Cl_4$			
Phenanthrene, 1,2,3,4-tetrachloro-9-methyl-	MeOH	259(4.51),267(4.63), 290(4.00),305(3.92), 317(3.90),350(2.58), 360(2.54)	39-1684-69C

Compound	Solvent	$\lambda_{max}(\log \epsilon)$	Ref.
$C_{15}H_8F_4$			
Phenanthrene, 1,2,3,4-tetrafluoro-6-methyl-	MeOH	240(4.7),259(4.35) 276(3.88),287(4.02), 299(4.12),319(2.77), 334(3.03),349(3.07)	39-1684-69C
Phenanthrene, 1,2,3,4-tetrafluoro-7-methyl-	MeOH	244(4.78),249(4.87), 259(4.36),275(4.04), 285(3.98),318(2.53), 332(2.62),348(2.63)	39-1684-69C
Phenanthrene, 1,2,3,4-tetrafluoro-8-methyl-	MeOH	240(4.67),247(4.79), 262(4.23),280(3.91), 291(4.07),303(4.13), 330(2.69),350(2.57)	39-1684-69C
Phenanthrene, 1,2,3,4-tetrafluoro-9-methyl-	MeOH	240(4.65),246(4.76), 259(4.32),274(4.12), 285(4.13),297(4.11), 348(2.90)	39-1684-69C
Phenanthrene, 1,2,3,4-tetrafluoro-10-methyl-	MeOH	240(4.69),246(4.81), 260(4.26),276(3.96), 287(4.02),299(4.07), 336(3.08),351(3.14)	39-1684-69C
$C_{15}H_8N_2O_2$			
Isoindolo[1,2-b]quinazoline-10,12-dione	EtOH	275(4.01),303(3.89), 315(3.76)	44-2123-69
$C_{15}H_8N_2O_3$			
9H-Naphtho[1',2':5,6]pyrano[2,3-d]pyrimidine-9,11(10H)-dione	EtOH	258(4.14),420(2.94), 440(2.94)	103-0827-69
$C_{15}H_8N_2O_4$			
Anthra[1,2-c][1,2,5]oxadiazole-6,11-dipne, 4-methoxy-	EtOH	417(3.73)	104-1642-69
$C_{15}H_8N_2O_6$			
Coumarin, 6,8-dinitro-3-phenyl-	EtOH	256(3.89),335(3.89)	2-0049-69
Coumarin, 3-(2,4-dinitrophenyl)-	EtOH	286(4.03),335(3.99)	2-0049-69
Coumarin, 6-nitro-3-(p-nitrophenyl)-	EtOH	280(4.26),320(4.24)	2-0049-69
	EtOH-NaOAc	232(3.89),287(4.11), 317(4.12),415(4.17)	2-0049-69
Coumarin, 8-nitro-3-(p-nitrophenyl)-	EtOH	315(4.31)	2-0049-69
	EtOH-NaOAc	235(4.05),315(4.15), 455(4.09)	2-0049-69
	EtOH-NaOEt	286(4.04),455(4.09)	2-0049-69
$C_{15}H_8N_2O_7$			
Coumarin, 7-hydroxy-3-(2,4-dinitrophenyl)-	EtOH	272(3.44)	2-0049-69
Coumarin, 7-hydroxy-6-nitro-3-(p-nitrophenyl)-	EtOH	283(4.15)	2-0049-69
	EtOH-NaOAc	335(4.28)	2-0049-69
Coumarin, 7-hydroxy-8-nitro-3-(p-nitrophenyl)-	EtOH	332(3.73)	2-0049-69
$C_{15}H_8N_4O_3$			
Furo[3,2-e]-s-triazolo[1,5-c]pyrimidine, 8,9-di-2-furyl-	EtOH	238(4.19),252s(4.09), 298(4.28),346(4.34)	95-1434-69
Furo[3,2-e]-s-triazolo[4,3-c]pyrimidine, 8,9-di-2-furyl-	EtOH	243(4.07),252(4.08), 310(4.19),344(4.24)	95-1434-69

Compound	Solvent	λ max(log ε)	Ref.
$C_{15}H_8N_4O_7$			
2-(Hydroxypicryl)isoquinolinium hydroxide, inner salt	MeOH	233(4.7),295s(4.0), 345(4.3)	56-1219-69
$C_{15}H_8O$			
6-Tetradecene-2,4,8,10,12-pentayne, 7-formyl-, (E)-	ether	232(4.99),282(4.19), 300(4.37),354(4.59), 375(4.54)	39-0683-69C
$C_{15}H_9BrClNO_4S_2$			
3-(p-Bromophenyl)thiazolo[2,3-b]benzothiazolium perchlorate	EtOH	221(4.30),255(3.98), 314(3.96)	95-0469-69
$C_{15}H_9BrO_2$			
Benzofuran, 2-(p-bromobenzoyl)-	n.s.g.	318(4.41)	32-1273-69
Isocoumarin, 4-bromo-3-phenyl-	EtOH	236(4.22),254s(4.10), 259s(4.00),288(3.99), 330s(3.60)	12-0577-69
$C_{15}H_9BrO_3$			
Phthalide, 3-benzoyl-3-bromo-	EtOH	237s(3.90),257s(3.81), 290s(3.42)	12-0577-69
$C_{15}H_9ClN_2O$			
4H-Indolo[3,2,1-de][1,5]naphthyridin-4-one, 5-chloro-6-methyl-	MeOH	215(4.63),262(4.46), 288(4.43),323(3.85), 390(4.05)	44-4199-69
$C_{15}H_9ClO_2$			
Benzofuran, 2-(p-chlorobenzoyl)-	n.s.g.	318(4.39)	32-1273-69
Flavone, 3-chloro-	EtOH	249(4.78),305(4.42)	88-3589-69
Flavone, 6-chloro-	EtOH	255(4.23),298(4.22)	23-0105-69
$C_{15}H_9Cl_2NO_4S_2$			
3-(p-Chlorophenyl)thiazolo[2,3-b]benzothiazolium perchlorate	EtOH	221(3.56),255(3.05), 313(3.12)	95-0469-69
$C_{15}H_9Cl_2N_3O_3$			
2-Indolinone, 4,7-dichloro-3-[(p-nitroanilino)methylene]-	EtOH	252(4.17),265(4.15), 283(4.13),393(4.65)	2-0667-69
$C_{15}H_9Cl_3N_2O$			
2-Indolinone, 4,7-dichloro-3-[(p-chloroanilino)methylene]-	EtOH	274(4.39),308(3.97), 367(4.47)	2-0667-69
$C_{15}H_9Cl_3O_5$			
Xanthen-9-one, 2,4,7-trichloro-1,6-dihydroxy-3-methoxy-8-methyl-	MeOH	246(4.53),314(4.18), 356(4.01)	64-0756-69
	MeOH-NaOH	240(4.54),288(4.06), 326(4.13),375(4.25)	64-0756-69
$C_{15}H_9FO_2$			
Benzofuran, 2-(p-fluorobenzoyl)-	n.s.g.	316(4.39)	32-1273-69
$C_{15}H_9F_3OS$			
Dibenzo[b,f]thiepin-10(11H)-one, 7-(trifluoromethyl)-	MeOH	222(4.27),227(4.25), 240s(4.18),263s(3.92), 336(3.60)	73-3936-69
Dibenzo[b,f]thiepin-10(11H)-one, 8-(trifluoromethyl)-	MeOH	228(4.24),242(4.22), 267s(3.90),327(3.61)	73-3936-69

Compound	Solvent	λ_{max}(log ε)	Ref.
$C_{15}H_9F_3S$			
Dibenzo[b,f]thiepin, 2-(trifluorometh-yl)-	MeOH	213(4.43),267(4.21), 282(3.82)	73-3936-69
$C_{15}H_9NO$			
Benzofuro[2,3-g]isoquinoline	EtOH	220(4.68),244(4.50), 267(4.84),273s(4.77), 324(4.24),338(4.18)	4-0875-69
$C_{15}H_9NO_2$			
1,4-Isoquinolinedione, 3-phenyl-	CH_2Cl_2	316(4.01),500(2.30)	33-1810-69
$C_{15}H_9NO_4$			
Coumarin, 3-(p-nitrophenyl)-	EtOH	335(3.8)	2-0049-69
	EtOH-NaOEt	335(3.46)	2-0049-69
	EtOH-NaOAc	335(3.80)	2-0049-69
Coumarin, 6-nitro-3-phenyl-	EtOH	280(4.01),320(3.91)	2-0049-69
	EtOH-NaOEt	240(3.71),350(3.61), 415(3.50)	2-0049-69
	EtOH-NaOAc	280(4.02),320(3.92)	2-0049-69
Coumarin, 8-nitro-3-phenyl-	EtOH	310(3.59)	2-0049-69
	EtOH-NaOEt	326(3.47)	2-0049-69
	EtOH-NaOAc	310(3.65)	2-0049-69
$C_{15}H_9NO_5$			
Coumarin, 7-hydroxy-3-(p-nitrophenyl)-	EtOH	269(3.60)	2-0049-69
Coumarin, 7-hydroxy-6-nitro-3-phenyl-	EtOH	313(4.13)	2-0049-69
Coumarin, 8-hydroxy-5-nitro-3-phenyl-	EtOH	290(3.67)	2-0049-69
	EtOH-NaOAc	355(3.84)	2-0049-69
Coumarin, 8-hydroxy-6-nitro-3-phenyl-	EtOH	312-325(3.10)	2-0049-69
Coumarin, 8-hydroxy-7-nitro-3-phenyl-	EtOH	290(3.64)	2-0049-69
	EtOH-NaOAc	344(3.72)	
$C_{15}H_9NS$			
[1]Benzothieno[2,3-g]isoquinoline	EtOH	231(4.61),238(4.51), 268s(4.61),276(4.82), 284(4.90)	4-0875-69
	n.s.g.	231(4.62),238(4.51), 268s(4.62),276(4.82), 284(4.90)	77-0598-69
$C_{15}H_9N_3O_2$			
Benzimidazo[2,1-a]isoquinoline, 10-ni-tro-	EtOH	211(3.49),268(4.49), 290(4.46)	24-1529-69
$C_{15}H_9N_3O_4$			
4(3H)-Quinazolinone, 2-benzoyl-6-nitro-	EtOH	258(4.41),325s(3.86), 370(4.06)	18-3198-69
4(3H)-Quinazolinone, 2-benzoyl-7-nitro-	EtOH	258(4.43),320s(3.68), 340s(--)	18-3198-69
$C_{15}H_{10}BrNOS_2$			
Acetophenone, 2-(2-benzothiazolylthio)-4'-bromo-	EtOH	221(4.45),263(4.43), 289(4.18),300(4.03)	95-0469-69
$C_{15}H_{10}BrNO_4$			
Norevoxanthine, 4-bromo-	EtOH	209(4.02),218(4.02), 260s(4.19),282(4.32), 411(3.48)	12-1493-69

Compound	Solvent	$\lambda_{max}(\log \epsilon)$	Ref.
$C_{15}H_{10}Br_2O_2$			
2-Biphenylenecarboxylic acid, 6,7-dibromo-, ethyl ester	EtOH	267(4.59),354(3.62), 370(3.76),394(3.49)	39-2789-69C
$C_{15}H_{10}ClNOS$			
3-(p-Chlorophenyl)-4-hydroxy-2-phenylthiazolium hydroxide, inner salt	MeOH	200(4.22),220s(3.87), 242(3.66),385(3.46)	77-1128-69
$C_{15}H_{10}ClNOS_2$			
Acetophenone, 2-(2-benzothiazolylthio)-4'-chloro-	EtOH	218(4.54),262(3.99), 289(4.18),301(4.13)	95-0469-69
$C_{15}H_{10}ClNO_2$			
Flavone, 8-amino-6-chloro-	EtOH	278(4.49)	23-0105-69
$C_{15}H_{10}ClNO_4S_2$			
2-Phenylthiazolo[2,3-b]benzothiazolium perchlorate	EtOH	235(4.73),262(4.63), 321(4.51)	95-0469-69
$C_{15}H_{10}ClN_3$			
Benzimidazo[2,1-a]isoquinoline, 10-amino-1-chloro-	EtOH	199(4.22),222(4.27), 268(4.25),276(4.26), 302(4.27),352(3.90)	24-1529-69
Benzimidazo[2,1-a]isoquinoline, 10-amino-4-chloro-	EtOH	222(4.24),240(4.22), 268(4.23),276(4.22), 303(4.24)	24-1529-69
Quinoline, 2-[(p-chlorophenyl)azo]-	MeCN	236(4.27),254s(--), 291(4.11),335(4.39), 451(2.65)	24-3176-69
$C_{15}H_{10}ClN_3O_3S$			
Thiazole, 2-anilino-5-[1-chloro-2-(5-nitro-2-furyl)vinyl]-	EtOH	294(4.47),400(4.24)	103-0414-69
$C_{15}H_{10}Cl_2N_2O$			
2-Indolinone, 3-(anilinomethylene)-4,7-dichloro-	EtOH	278(4.24),362(4.45)	2-0667-69
$C_{15}H_{10}Cl_2O$			
Anthrone, 1,8-dichloro-10-methyl-	EtOH	275(4.15)	44-3093-69
$C_{15}H_{10}Cl_2O_5$			
Xanthen-9-one, 2,4-dichloro-1,3-dihydroxy-6-methoxy-8-methyl-	EtOH	215(4.39),247(4.64), 273s(4.27),310(4.38), 350(3.82)	11-0445-69A
	MeOH	245(4.5),310(4.1), 342(4.0)	64-0750-69
$C_{15}H_{10}Cl_4$			
Phenanthrene, 1,2,3,4-tetrachloro-9,10-dihydro-9-methyl-	MeOH	237(4.29),278(4.29), 310(3.38)	39-1684-69C
Propene, 2-phenyl-3-(2,3,4,5-tetrachlorophenyl)-	MeOH	235(4.00),279(3.42), 290(3.27)	39-1684-69C
$C_{15}H_{10}D_2O$			
Chalcone-α,β-d_2	n.s.g.	237(4.07),253(4.08), 301(4.03)	39-1542-69C

Compound	Solvent	$\lambda_{max}(\log \epsilon)$	Ref.
$C_{15}H_{10}F_4$			
Phenanthrene, 1,2,3,4-tetrafluoro-9,10-dihydro-6-methyl-	MeOH	265(4.22),298(3.50)	39-1684-69C
Phenanthrene, 1,2,3,4-tetrafluoro-9,10-dihydro-8-methyl-	MeOH	266(3.98),300(2.85)	39-1684-69C
Phenanthrene, 1,2,3,4-tetrafluoro-9,10-dihydro-9-methyl-	MeOH	267(4.26),293(3.35)	39-1684-69C
Propene, 2-phenyl-3-(2,3,4,5-tetrafluorophenyl)-	MeOH	244(4.00),284(2.80), 297(2.49)	39-1684-69C
$C_{15}H_{10}FeO_3$			
Iron, (benzocyclooctene)tricarbonyl-	ether	238(4.38),270(4.31), 305s(4.09)	35-4734-69
$C_{15}H_{10}KNO_2$			
Fluorene, 9-(2-aci-nitroethylidene)-, potassium salt	MeOH	208(4.23),244(4.55), 400(4.44)	24-1707-69
$C_{15}H_{10}N_2$			
Benzimidazo[2,1-a]isoquinoline	EtOH	221(4.39),232(4.40), 248(4.16),258(4.30), 269(4.34),279(4.59), 316(3.95),331(3.84), 348(3.72)	24-1529-69
1H-Phenanthro[9,10-d]imidazole	EtOH	255(4.92),301(3.89)	77-0200-69
$C_{15}H_{10}N_2O$			
14H-Indolo[3,2,1-de][1,5]naphthyridin-4-one (norisotuboflavine)	MeOH	225s(4.28),260(4.45), 282(4.42),321(3.78), 385(4.05)	44-4199-69
$C_{15}H_{10}N_2O_2$			
4H-Pyrazol-4-one, 3,5-diphenyl-, 1-oxide	EtOH	250(4.28),303(3.76), 333(3.78),483(3.14)	44-0187-69
$C_{15}H_{10}N_2O_2S$			
Imidazolidine-2-thione, 4,5-dioxo-1,3-diphenyl-	EtOH	305(4.19),432(1.64)	18-2323-69
$C_{15}H_{10}N_2O_3$			
Benzamide, o-phthalimido-	EtOH	218(4.55),282(3.71)	44-2123-69
4H-Pyrazol-4-one, 3,5-diphenyl-, 1,2-dioxide	n.s.g.	256(4.59),330(3.62), 452(3.23)	44-0187-69
$C_{15}H_{10}N_2S$			
Thiazolo[3,2-a]benzimidazole, 2-phenyl-	EtOH	218(4.19),277(4.14), 311(4.11)	4-0797-69
Thiazolo[3,2-a]benzimidazole, 3-phenyl-	EtOH	235(4.30),251(4.13), 270(4.15),285s(4.05)	4-0797-69
$C_{15}H_{10}N_4$			
s-Triazolo[3,4-a]phthalazine, 6-phenyl-	MeOH	237(4.48)	44-3221-69
$C_{15}H_{10}N_4O$			
2-Propanone, 1,3-bis(diazo)-1,3-diphenyl-	hexane	264(4.37),324(3.85)	35-7534-69
	MeOH	264(4.32),328(4.00)	35-7534-69
4H-Pyrazolo[3,4-c][1,2,5]oxadiazole, 4,6-diphenyl-	EtOH	227(4.22),268(4.43), 383(3.94)	4-0317-69

Compound	Solvent	λ_{max}(log ϵ)	Ref.
$C_{15}H_{10}N_4O_2$			
Quinoline, 2-[(p-nitrophenyl)azo]-	MeCN	215s(--),283(--), 290(4.24),338(4.41), 463(2.85)	24-3176-69
$C_{15}H_{10}N_4O_3$			
Benzonitrile, p-glyoxyloyl-, o-nitro-phenylhydrazone	$CHCl_3$	405(4.17),420(4.19)	28-0642-69B
$C_{15}H_{10}O$			
Anthrone, 10-methylene-	EtOH	234(4.59),262(4.18), 276(4.20),352(3.76)	28-0340-69A
$C_{15}H_{10}OS$			
Benzothiophene, 3-benzoyl-	MeOH	220(4.51),250(4.40), 308(4.02)	73-2959-69
Benzo[b]thiophen-3(2H)-one, 2-benzylidene-	MeOH	253(4.1),278(4.1), 313(4.2),434(3.9)	83-0401-69
1-Thiocoumarin, 4-phenyl-	EtOH	242s(3.8),298(3.5), 350s(3.28)	2-0315-69
$C_{15}H_{10}O_2$			
Benzofuran, 2-benzoyl-	n.s.g.	316(4.36)	32-1273-69
Coumarin, 3-phenyl-	EtOH	325(--)	2-0049-69
Isocoumarin, 3-phenyl-	EtOH	234(4.34),246s(4.16), 254s(4.06),286s(4.24), 297(4.36),309(4.47), 337(4.01)	12-0577-69
1,4-Methanoanthracene-9,10-dione, cage drivative	iso-PrOH	218(3.44),282(3.40), 385(1.48)	39-0616-69B
$C_{15}H_{10}O_2S$			
Benzo[b]thiophen-3(2H)-one, 2-(p-hydroxybenzylidene)-	MeOH	259(4.2),272s(4.0), 346(4.2),439(4.3)	83-0401-69
$C_{15}H_{10}O_3$			
Anthraquinone, 1-methoxy-	EtOH	220(4.43),254(4.41), 268s(4.20),311(3.34), 387(3.77)	114-0109-69A
	dioxan	255(4.42),266s(4.21), 330(3.36),376(3.75)	114-0109-69A
	$CHCl_3$	256(4.42),271s(4.21), 328(3.34),383(3.76)	114-0109-69A
	CCl_4	256(4.40),268s(4.18), 328(3.03),375(3.75)	114-0109-69A
	pyridine	329(3.36),380(3.76)	114-0109-69A
3(2H)-Benzofuranone, 2-benzylidene-7-hydroxy-	MeOH	234(4.1),267s(3.7), 312s(4.2),320(4.3), 393(4.1)	83-0401-69
Coumaranone, 2-benzoyl-	EtOH	241(3.99),257(3.98), 328(4.08),349(4.15)	88-3589-69
Phthalide, 3-benzoyl-	EtOH	230s(4.19),236(4.22), 246(4.14),274(3.55), 282(3.55)	12-0577-69
$C_{15}H_{10}O_3S$			
Benzo[b]thiophen-3(2H)-one, 2-(3,4-dihydroxybenzylidene)-	MeOH	260(4.3),271s(4.2), 359(4.1),445(4.3)	83-0401-69
Dibenzo[b,f]thiepin-3-carboxylic acid, 10,11-dihydro-11-oxo-	MeOH	227(4.23),247(4.33), 348(3.53)	73-3936-69

Compound	Solvent	λ_{max}(log ε)	Ref.
$C_{15}H_{10}O_4$			
Anthraquinone, 1,3-dihydroxy-2-methyl- (rubiadin)	EtOH-acid	242s(4.35),246(4.40), 282(4.46),335(3.36), 412(3.84)	102-0501-69
	EtOH-base	270(4.41),315(4.48), 512(4.02)	102-0501-69
Anthraquinone, 1,6(or 7)-dihydroxy-3-methyl-	MeOH	268(4.19),412(3.57)	102-0315-69
Anthraquinone, 1-hydroxy-2-methoxy-	EtOH	231s(4.28),248(4.51), 246s(3.45),278s(4.16), 325s(3.50),409s(3.74), 429(3.81),439s(3.78)	114-0109-69A
	dioxan	234s(4.31),248(4.52), 255s(4.44),275s(4.17), 321(3.51),425(3.80)	114-0109-69A
	$CHCl_3$	249(4.50),255s(4.48), 261s(4.41),278(4.20), 327(3.50),409s(3.76), 427(3.82),440s(3.78)	114-0109-69A
	pyridine	326(3.52),411s(3.73), 428(3.79),440s(3.76)	114-0109-69A
Aurone, 6,7-dihydroxy-	MeOH	317(4.1),373(4.1)	2-0543-69
3(2H)-Benzofuranone, 7-hydroxy-3-(p-hydroxyphenyl)-	MeOH	237s(--),251s(--), 260(3.8),310s(3.9), 363s(4.1),413(4.4)	83-0401-69
Coumarin, 5,7-dihydroxy-4-phenyl-	EtOH	260(4.07),340(4.01)	44-3784-69
Flavanol, 7-hydroxy-	EtOH	255(4.26),345(4.33)	104-1803-69
$C_{15}H_{10}O_4S$			
Benzo[b]thiophen-3(2H)-one, 2-(3,4,5-trihydroxybenzylidene)-	MeOH	259(4.3),276s(4.2), 365(4.0),450(4.3)	83-0401-69
$C_{15}H_{10}O_5$			
3(2H)-Benzofuranone, 2-benzylidene-4,6,7-trihydroxy-	MeOH	255(3.9),310(4.0)	2-0543-69
3(2H)-Benzofuranone, 2-(3,4-dihydroxybenzylidene)-7-hydroxy-	MeOH	278(3.9),286(3.8), 316(3.8),420(4.3)	83-0401-69
Flavone, 2',6,6'-trihydroxy-	EtOH	226(4.38),317(3.92)	2-0746-69
Flavonol, 2',7-dihydroxy-	EtOH	253(4.16),319(4.11)	104-1803-69
	EtOH-NaOAc	269(--),388(--)	104-1803-69
Xanthen-9-one, 4-methoxy-2,3-(methylenedioxy)-	EtOH	245(4.53),279(3.72), 310(4.05)	24-2414-69
$C_{15}H_{10}O_6$			
Anthraquinone, 1,5,6,8-tetrahydroxy-3-methyl-	EtOH	237(4.50),259(4.33), 305(4.04),494(4.16), 517(4.03),529(4.05), 568(3.31)	1-0144-69
Flavone, 2',5,6,6'-tetrahydroxy-	EtOH	226(4.31),269(4.25)	2-0746-69
	EtOH-$AlCl_3$	277(4.17),296(4.16)	2-0746-69
	EtOH-NaOAc-H_3BO_3	229(4.21),267(4.29)	2-0746-69
Flavone, 2',5,6',8-tetrahydroxy-	MeOH	255(4.32),261(4.31), 348(3.51)	2-0746-69
$C_{15}H_{10}O_7$			
Flavone, 2',3',5,6,7-pentahydroxy-	MeOH	276(4.49),325(4.37)	2-0110-69
	MeOH-$AlCl_3$	300(4.4),390(4.39)	2-0110-69
Flavone, 2',3',5,7,8-pentahydroxy-	EtOH	281(4.32)	2-0110-69
	EtOH-$AlCl_3$	288(4.29),306(4.21)	2-0110-69

Compound	Solvent	λmax(log ε)	Ref.
Flavone, 2',3',5,7,8-pentahydroxy- (cont.)	EtOH-NaOAc	290(4.22)	2-0110-69
	EtOH-NaOAc-H_3BO_3	288(4.40)	2-0110-69
$C_{15}H_{11}BF_2O_2$			
Boron, (1,3-diphenyl-1,3-propanedionato)difluoro-	$CHCl_3$	270(3.69),365(4.62), 378s(4.57)	39-0526-69A
$C_{15}H_{11}Br$			
Anthracene, 9-bromo-10-methyl-	EtOH	252s(4.89),260(5.09), 343(3.53),360(3.80), 378(3.98),399(3.93)	78-3485-69
$C_{15}H_{11}BrO$			
2-Propen-1-one, 1-(m-bromophenyl)-3-phenyl-	MeOH	310(4.33)	73-2771-69
2-Propen-1-one, 1-(p-bromophenyl)-3-phenyl-	MeOH	310(4.34)	73-2771-69
$C_{15}H_{11}BrO_3$			
3-Benzofuranol, 2-(p-bromobenzoyl)-2,3-dihydro-	n.s.g.	263(4.32)	32-1273-69
isomer, m. 223°	n.s.g.	263(4.30)	32-1273-69
$C_{15}H_{11}Br_3N_4O_5$			
Propiophenone, 3',4',5'-tribromo-2'-hydroxy-, 2,4-dinitrophenylhydrazone	EtOH	380(4.38)	22-2079-69
$C_{15}H_{11}Cl$			
Anthracene, 9-chloro-10-methyl-	EtOH	253s(4.86),260(5.17), 342(3.36),359(3.71), 378(3.94),400(3.92)	78-3485-69
Cyclopropene, 3-chloro-1,3-diphenyl-	MeCN	288(4.37),301(4.47), 316(4.37)	44-2728-69
$C_{15}H_{11}ClN_2O$			
2H-1,4-Benzodiazepin-2-one, 7-chloro-1,3-dihydro-5-phenyl-	pH 1.5	240(4.5),280(4.0), 360(3.5)	65-0443-69
	pH 13.5	230(4.5),330(3.7)	65-0443-69
3H-Indole, 3-(chloroimino)-5-methoxy-2-phenyl-	EtOH	281(4.55),495(3.36)	22-2008-69
2-Indolinone, 3-[(p-chloroanilino)methylene]-	EtOH	241(4.38),245(4.42), 350(4.74)	2-0667-69
Pyrazol-3-ol, 5-(m-chlorophenyl)-1-phenyl-	EtOH	240(4.28)	39-0836-69C
Pyrazol-3-ol, 5-(p-chlorophenyl)-1-phenyl-	EtOH	244(4.26)	39-0836-69C
Quinazoline, 6-chloro-4-methoxy-2-phenyl-	EtOH	212(4.54),254(4.51), 292(4.35)	22-2008-69
$C_{15}H_{11}ClO$			
2-Propen-1-one, 1-(m-chlorophenyl)-3-phenyl-	MeOH	312(4.25)	73-2771-69
2-Propen-1-one, 1-(p-chlorophenyl)-3-phenyl-	MeOH	310(4.42)	73-2771-69
2-Propen-1-one, 3-(m-chlorophenyl)-1-phenyl-	MeOH	299(4.34)	73-2771-69
2-Propen-1-one, 3-(p-chlorophenyl)-1-phenyl-	MeOH	312(4.46)	73-2771-69

Compound	Solvent	λ_{max}(log ε)	Ref.
$C_{15}H_{11}ClO_2$			
Flavanone, 6-chloro-	EtOH	220(4.45),246s(3.9), 284s(3.4),327(3.62)	23-0105-69
$C_{15}H_{11}ClO_3$			
3-Benzofuranol, 2-(p-chlorobenzoyl)-2,3-dihydro-, d. 173°	n.s.g.	260(4.28)	32-1273-69
isomer, d. 216°	n.s.g.	260(4.26)	32-1273-69
$C_{15}H_{11}ClO_4S$			
1-Phenyl-2-benzothiopyrylium perchlorate	MeOH	235(3.94),242(3.90), 267(3.66),293(3.81), 308(3.80)	44-3202-69
2-Phenyl-1-benzothiopyrylium perchlorate	HOAc-1% $HClO_4$	265(4.52),300s(3.77), 396(4.30)	2-0191-69
4-Phenyl-1-benzothiopyrylium perchlorate	HOAc-1% $HClO_4$	262(4.54),344(3.94), 404(4.04)	2-0191-69
$C_{15}H_{11}ClO_4S_2$			
2,4-Diphenyl-1,3-dithiol-1-ium perchlorate	50% H_2SO_4	243(4.38),277s(4.05), 310(3.95),394(4.35)	94-1931-69
4,5-Diphenyl-1,3-dithiol-1-ium perchlorate	50% H_2SO_4	239(4.26),349(3.44)	94-1924-69
$C_{15}H_{11}Cl_2NO$			
Acridan, 1-(dichloroacetyl)-	MeOH	256s(4.10),295s(3.95), 435(3.73)	22-3523-69
$C_{15}H_{11}FN_2O_2$			
Indole-3-acetic acid, 5-fluoro-2-(3-pyridyl)-	EtOH	221(4.32),310(4.28)	22-4154-69
$C_{15}H_{11}FO$			
2-Propen-1-one, 1-(p-fluorophenyl)-3-phenyl-	MeOH	308(4.42)	73-2771-69
2-Propen-1-one, 3-(p-fluorophenyl)-1-phenyl-	MeOH	310(4.39)	73-2771-69
$C_{15}H_{11}FO_3$			
3-Benzofuranol, 2,3-dihydro-2-(p-fluorobenzoyl)-, d. 159°	n.s.g.	250(4.13),281(3.65)	32-1273-69
isomer, d. 211-214°	n.s.g.	250(4.11),281(3.65)	32-1273-69
$C_{15}H_{11}N$			
7H-Acenaphtho[1,2-b]pyrrole, 7-methyl-	C_6H_{12}	234(4.61),242(4.75), 296(3.99),305(3.90), 315(3.82),323(3.82), 331(3.80),336(3.93), 349(3.81),353(3.96), 430(3.05)	48-0614-69
Cyclopenta[ij]pyrido[2,1,6-de]quinolizine, 3-methyl-	EtOH	246(4.55),254s(4.48), 282s(4.06),341s(4.34), 353(4.49),425(3.72), 444(4.08),537s(2.40), 571(2.54),620(2.52), 681(2.06)	39-0239-69C
8,16-Imino[2.2]metacyclophane-1,9-diene	C_6H_{12}	230s(4.45),250s(4.15), 304(4.10)	35-1672-69
Indeno[2,1-b]indole, 5,6-dihydro-	EtOH	226(4.39),235s(4.38), 240(4.42),278(4.36), 292s(4.25)	44-2988-69

Compound	Solvent	λ max (log ε)	Ref.
$C_{15}H_{11}NO$			
Benz[d][1,3]oxazepine, 4-phenyl-	C_6H_{12}	258(4.32),311(3.88)	1-2149-69
Carbostyril, 4-phenyl-	n.s.g.	228(4.52),280(3.85), 331(3.76)	35-6083-69
Indole-2-carboxaldehyde, 3-phenyl-	EtOH	227(4.14),248(4.29), 318(4.24),350s(3.86)	1-2149-69
Indole-3-carboxaldehyde, 2-phenyl-	EtOH	257(4.46),314(4.16)	1-2149-69
3-Isoquinolinol, 1-phenyl-	EtOH	350(3.78),420(3.34)	39-1729-69C
	ether	350(3.91)	39-1729-69C
	1:1 EtOH: ether	350(3.85),420(3.01)	39-1729-69C
Oxazole, 2,4-diphenyl-	MeOH	232(4.24),276(4.24)	39-0270-69B
Oxazole, 2,5-diphenyl-	MeOH	244(4.08),314(4.34)	39-0270-69B
Quinoline, 4-phenyl-, N-oxide	EtOH	233(4.40),334(3.90), 343s(3.89)	1-2149-69
$C_{15}H_{11}NOS$			
4-Hydroxy-2,3-diphenylthiazolium hydroxide, inner salt	MeOH	202(4.72),240(4.27), 350(2.62)	77-1128-69
$C_{15}H_{11}NO_2$			
Anthrone, 10-imino-4-methoxy-	EtOH	213(4.42),254(4.46), 270s(4.16),324(3.11), 373(4.00)	114-0109-69A
	dioxan	254(4.45),268s(4.17), 323(3.59),367(3.58)	114-0109-69A
	$CHCl_3$	237(4.18),256(4.48), 267s(4.21),325(3.54), 362(3.59)	114-0109-69A
	pyridine	320(3.54),372(3.58)	114-0109-69A
1-Cyclopenin	EtOH	211(4.58),290(3.31)	44-1359-69
Fluorene, 9-(2-nitroethylidene)-	$CHCl_3$	252(4.44),260(4.56), 300(3.88)	24-1707-69
Viridicatin	n.s.g.	220(4.63),285(3.96), 307s(3.96),317(4.05), 329(3.92)	44-1359-69
$C_{15}H_{11}NO_2S_2$			
Spiro[1,3-dithiolane-2,9'-fluorene], 3'-nitro-	EtOH	261(4.24),285s(4.06), 300s(3.34)	39-0345-69C
$C_{15}H_{11}NO_3$			
Anthrone, 4-hydroxy-10-imino-3-methoxy-	EtOH	259(4.59),290s(3.88), 391(3.18),507(3.92), 535(4.01),565(3.87)	114-0109-69A
	dioxan	248(4.47),258s(4.39), 282(4.01),420(3.52), 514(3.59),544(3.68), 579(3.52)	114-0109-69A
	$CHCl_3$	251(4.47),262s(4.45), 421(3.49),510(3.68), 536(3.78),575(3.61)	114-0109-69A
	pyridine	400(3.34),510(3.81), 540(3.92),575(3.78)	114-0109-69A
Cinnamaldehyde, o-nitro-α-phenyl-, trans	EtOH	222(4.27),268(4.07)	44-2285-69
Phthalic anhydride, 3-(benzylamino)-	EtOH	212(4.18),224(4.22), 257(4.10),396(3.92)	16-0241-69
2-Propen-1-one, 1-(m-nitrophenyl)-3-phenyl-	MeOH	315(4.34)	73-2771-69

Compound	Solvent	λ_{max}(log ε)	Ref.
2-Propen-1-one, 1-(p-nitrophenyl)-3-phenyl-	MeOH	318(4.30)	73-2771-69
2-Propen-1-one, 3-(m-nitrophenyl)-1-phenyl-	MeOH	290(4.39)	73-2771-69
$C_{15}H_{11}NO_3S$			
Spiro[benzo[b]thiophene-3(2H),1'-isoindolin]-3'-one, 1,1-dioxide	EtOH	267(3.27),275(3.23)	39-1149-69C
$C_{15}H_{11}NO_4$			
Carbazole-1-carboxylic acid, 2-acetoxy-	EtOH	222(3.66),240(3.28), 280(3.27),305(2.94), 355(2.83)	39-1518-69C
Norevoxanthine	EtOH	204(4.27),218(4.18), 241(4.35),263s(4.52), 280(4.66),330s(3.67), 406(3.87)	39-2327-69C
	EtOH-NaOH	234(4.32),250(4.29), 283(4.56),324(3.86), 422(3.65)	39-2327-69C
Phthalimide, 3-carboxy-6-hydroxy-N-phenyl-1,2,3,6-tetrahydro-, lactone cis-trans-cis	EtOH	217(3.99)	78-4315-69
$C_{15}H_{11}NO_5$			
Azulene-1,2-dicarboxylic acid, 3-cyano-, dimethyl ester	EtOH	233(4.51),266(4.29), 292(4.59),302(4.71), 337(3.81),528(2.90)	77-0352-69
$C_{15}H_{11}N_3$			
Benzimidazo[2,1-a]isoquinoline, 10-amino-	EtOH	221(4.35),240(4.33), 262(4.37),297(4.31), 332(4.03),346(4.04)	24-1529-69
as-Triazine, 3,5-diphenyl-	MeOH	266(4.41),311(4.02), 380(2.60)	88-3147-69
$C_{15}H_{11}N_3O$			
Formic acid, cyano-, benzylidenephenylhydrazide	C_6H_{12}	289(4.33)	44-2146-69
$C_{15}H_{11}N_3O_2$			
Furazan, 3-anilino-4-benzoyl-	EtOH	254(4.48)	4-0317-69
Pyrazole, 1-(p-nitrophenyl)-3-phenyl-	MeOH	228(4.12),250(4.07), 337(4.38)	39-2587-69C
Resorcinol, 4-(8-quinolylazo)-(aqueous solutions at 0.1 ionic	pH 1.00	420(4.01)	123-0027-69A
	pH 12.0	450(4.20)	123-0027-69A
	H_2SO_4	490(4.18)	123-0027-69A
	concd. HCl	420(4.00)	123-0027-69A
$C_{15}H_{11}N_3O_2S$			
as-Triazine-3,5(2H,4H)-dione, 6-[2-(2-furyl)vinyl]-4-phenyl-3-thio-	EtOH	241(3.48),309(3.94), 389(4.13)	103-0427-69
$C_{15}H_{11}N_3O_3$			
2H-1,4-Benzodiazepin-2-one, 1,3-dihydro-	pH 1.5	270(4.3),340(4.2)	65-0443-69
	pH 13.5	230(4.3),250s(3.7), 370(4.0)	65-0443-69
	MeOH	201(4.41),258(4.26), 309(4.02)	88-3201-69

Compound	Solvent	$\lambda_{max}(\log \epsilon)$	Ref.
2-Indolinone, 3-[(p-nitroanilino)methylene]-	EtOH	265(4.19),405(4.50)	2-0667-69
4(3H)-Quinazolinone, 2-benzyl-6-nitro-	EtOH	260s(3.90),325(4.01)	18-3198-69
4(3H)-Quinazolinone, 2-benzyl-7-nitro-	EtOH	253(4.44),350(3.47)	18-3198-69
$C_{15}H_{11}N_3O_3S$			
Thiazole, 2-anilino-5-[2-(5-nitro-2-furyl)vinyl]-	EtOH	294(4.32),416(4.18)	103-0414-69
$C_{15}H_{11}N_3O_5$			
2-Propanone, 1-[3-(5-nitro-2-furyl)-4-phenyl-Δ^2-1,2,4-oxadiazolin-5-ylidene]-	EtOH	304(4.54)	18-3008-69
$C_{15}H_{12}$			
Anthracene, 9-methyl-	EtOH	256(4.8),386(3.2)	9-0249-69
	EtOH	253s(4.96),258(5.21),333(3.39),349(3.72),367(3.94),387(3.93)	78-3485-69
Fluorene, 9-ethylidene-	pentane	229(4.66),246(4.51),256(4.69),270(4.14),279(4.16),296(4.04),311(4.02)	88-3867-69
$C_{15}H_{12}BrNO_2$			
Benzanilide, 4-bromo-4'-(vinyloxy)-	n.s.g.	276(4.55)	30-0108-69A
stannic chloride complex	n.s.g.	276(4.49),585(1.55)	30-0108-69A
Stilbene, α-(bromomethyl)-2'-nitro-, trans	EtOH	239(4.09),263(4.03)	44-2285-69
$C_{15}H_{12}BrN_3O_4$			
Aniline, N-[2-(p-bromophenyl)cyclopropyl]-2,4-dinitro-, trans	EtOH	348(3.24)	39-1135-69B
Benzoic acid, p-[(2-bromo-4-nitrophenyl)azo]-, ethyl ester	EtOH	333(4.41)	5-0130-69E
$C_{15}H_{12}Br_2N_2$			
2-Pyrazoline, 3-bromo-1-(p-bromophenyl)-5-phenyl-	EtOH	253(4.09),294(4.25)	22-1683-69
$C_{15}H_{12}Br_2N_2O_2$			
1,3-Cyclopentadiene, 1,2-diacetyl-5-[(2,4-dibromophenyl)azo]-	EtOH	248s(4.23),267(4.33),472(4.35)	39-2464-69C
$C_{15}H_{12}Br_2N_4O_5$			
Propiophenone, 3',5'-dibromo-2'-hydroxy-, 2,4-dinitrophenylhydrazone	EtOH	378(4.37)	22-2079-69
Propiophenone, 3',5'-dibromo-4'-hydroxy-, 2,4-dinitrophenylhydrazone	EtOH	383(4.43)	22-2079-69
$C_{15}H_{12}Br_2O_6$			
6-Oxabicyclo[3.2.1]octane-2-carboxylic acid, 4-bromo-8-[2-bromo-4,5-(methylenedioxy)phenyl]-7-oxo-	EtOH	242(3.74),293(3.71)	78-4315-69
$C_{15}H_{12}ClNO_2$			
Cinnamohydroxamic acid, N-(p-chlorophenyl)-	EtOH	295(4.38)	34-0278-69
Flavanone, 8-amino-6-chloro-	EtOH	245(4.22),280s(3.76),350(3.57)	23-0105-69

Compound	Solvent	λmax(log ε)	Ref.
$C_{15}H_{12}ClNO_4S_2$			
2-Thiopheneacrylic acid, α-[(p-chloro-phenyl)thio]-5-nitro-, ethyl ester	EtOH	256(4.2),353s(4.0), 403(4.1)	95-1446-69
geometric isomer	EtOH	256(4.1),403(4.1)	95-1446-69
$C_{15}H_{12}ClN_3O$			
5H-Imidazo[2',1':5,1]pyrrolo[3,4-b]pyr-idin-5-one, 1,2,3,9b-tetrahydro-9b-(p-chlorophenyl)-	EtOH	225(4.20),268(3.70), 273(3.70)	44-0165-69
$C_{15}H_{12}ClN_3O_4$			
Aniline, N-[2-(p-chlorophenyl)cyclopro-pyl]-2,4-dinitro-, trans	EtOH	346(3.23)	39-1135-69B
$C_{15}H_{12}ClN_5$			
s-Triazine, 2-chloro-4,6-dianilino-	n.s.g.	268(3.53)	42-0595-69
$C_{15}H_{12}ClN_5O_3S$			
s-Triazine, 2-chloro-4-(m-sulfoanilino)-6-anilino-	n.s.g.	272(3.34)	42-0595-69
s-Triazine, 2-chloro-4-(p-sulfoanilino)-6-anilino-	n.s.g.	282(2.64)	42-0595-69
$C_{15}H_{12}Cl_2N_2$			
Malondianil, m,m'-dichloro-	EtOH	359(4.61)	24-2609-69
Malondianil, p,p'-dichloro-	EtOH	365(4.66)	24-2609-69
hydrochloride	MeOH	247(4.18),283s(--), 293(3.79),303(3.76), 391(4.80)	59-1075-69
$C_{15}H_{12}Cl_2N_2O$			
Acetic acid, (2,6-dichlorobenzylidene)-phenylhydrazide	EtOH	206(4.4),219(4.3), 278(4.2)	28-1703-69A
$C_{15}H_{12}FN_3O$			
Indole-3-acetamide, 5-fluoro-2-(3-pyri-dyl)-	EtOH	223(4.36),310(4.28)	22-4154-69
$C_{15}H_{12}FN_3O_4$			
Aniline ,N-[2-(p-fluorophenyl)cyclopro-pyl]-2,4-dinitro-, trans	EtOH	348(3.23)	39-1135-69B
$C_{15}H_{12}I_2N_2$			
Malondianil, p,p'-diiodo-, hydrochlor-ide	MeOH	251(4.10),287(3.64), 295(3.68),304s(--), 385(4.48)	59-1075-69
$C_{15}H_{12}N_2$			
Imidazole, 4,5-diphenyl-	EtOH	285(4.06)	77-0200-69
3H-Pyrazole, 3,3-diphenyl-	MeOH	245(3.47),354(2.05)	24-1865-69
Quinoline, 2-(p-aminophenyl)-	EtOH	223(4.56),279(4.22), 306(4.18),353(4.20)	78-0837-69
Quinoline, 4-(p-aminophenyl)-	EtOH	230(4.55),256s(--), 305s(--),317(3.95), 337(3.99)	78-0837-69
$C_{15}H_{12}N_2O$			
Acrylophenone, 3-(phenylazo)-	benzene	320(4.5),470(2.6)	7-0315-69
Cinnoline, 4-methoxy-3-phenyl-	EtOH	250(4.6),290(3.9), 335(3.6)	39-0796-69C

Compound	Solvent	λ max (log ε)	Ref.
4-Cinnolinone, 1-methyl-3-phenyl-	EtOH	268(4.2),316(4.1), 365(4.2)	39-0796-69C
4-Hydroxy-2-methyl-3-phenylcinnolinium hydroxide, inner salt	EtOH	260(4.0),360(4.1), 380(4.1)	39-0796-69C
Indole, 2-isonicotinoyl-1-methyl-	EtOH	223(4.47),265(3.72), 324(4.31)	39-2738-69C
2-Indolinone, 3-(anilinomethylene)-	EtOH	241(4.14),277(4.19), 362(4.48)	2-0667-69
4H-Indolo[3,2,1-de][1,5]naphthyridin-4-one, 5,6-dihydro-6-methyl-	MeOH	250(4.12),268(3.95), 288(4.10),312(3.92), 372(3.72)	44-4199-69
3-Pyrazolin-5-one, 2,3-diphenyl-	EtOH	238(4.24)	39-0836-69C
4(1H)-Quinazolinone, 1-methyl-2-phenyl-	EtOH	240(4.41),280s(3.80), 308(4.00),318s(3.95)	78-0783-69
4(3H)-Quinazolinone, 3-methyl-2-phenyl-	EtOH	228(4.46),278(4.02), 306s(3.80),318s(3.60)	78-0783-69
3(4H)-Quinoxalinone, 2-benzyl-	$CHCl_3$	340(3.82)	49-1274-69
$C_{15}H_{12}N_2OS$			
2-Benzimidazolinethione, 1-benzoyl-3-methyl-	MeOH	243(4.30),297s(4.26), 306(4.32)	97-0152-69
1,2,4-Oxadiazole, 3-(benzylthio)-5-phenyl-	EtOH	248(4.42)	39-2794-69C
1,2,4-Oxadiazole, 5-(benzylthio)-3-phenyl-	EtOH	236(3.00)	39-2794-69C
Pseudothiohydantoin, 5,5-diphenyl-	EtOH	225(4.40),255(3.83)	54-0905-69
Thiazolo[3,2-a]benzimidazol-3-ol, 2,3-dihydro-2-phenyl-	EtOH	250(4.08),284(4.07), 291(4.10)	4-0797-69
$C_{15}H_{12}N_2O_2$			
Benzohydroxamic acid, O-benzyl-p-cyano-	MeOH	208(4.4),235(4.24)	12-0161-69
	MeOH-NaOH	230(4.21),299(3.85)	12-0161-69
1-Imidazolol, 2,4-diphenyl-, 3-oxide	MeOH	257(3.95)	48-0746-69
Indole-3-acetic acid, 2-(3-pyridyl)-	EtOH	219(4.56),309(4.39)	22-4154-69
Isosydnone, 4-phenyl-5-p-tolyl-	EtOH	231(4.22),317(4.17)	39-1185-69B
Isosydnone, 5-phenyl-4-p-tolyl-	EtOH	282(4.13),315(4.01)	39-1185-69B
Naphth[2',3':4,5]imidazo[1,2-a]pyridine-6,11-dione, 1 2 3 4-tetrahydro-	pH 1	244(4.50),250(4.55), 279(4.20)	4-0909-69
	pH 7	248(4.66),282(4.19)	4-0909-69
	pH 13	249(4.65),283(4.18)	4-0909-69
	MeOH	247(4.64),275(4.20), 280(4.18),332(3.48)	4-0909-69
2(1H)-Quinazolinone, 4-(p-methoxyphenyl)-	EtOH	228(4.46),277(3.72), 322(3.8)	42-0103-69
4(1H)-Quinazolinone, 1-(p-methoxyphenyl)-	EtOH	235(4.47),280(3.88), 304(4.08),314(4.00)	42-0103-69
4(3H)-Quinazolinone, 2-(p-methoxyphenyl)-	EtOH	297(4.19)	42-0103-69
4(3H)-Quinazolinone, 3-(p-methoxyphenyl)-	EtOH	230(4.72),266(4.2), 303(3.75),312(3.69)	42-0103-69
$C_{15}H_{12}N_2O_3$			
2-Pyrazoline-3-carboxylic acid, 1-(2-naphthyl)-5-oxo-	EtOH	215(4.57),250(4.47), 280(4.17)	22-4159-69
$C_{15}H_{12}N_2O_3S$			
Δ^2-1,2,4-Thiadiazoline, 4-benzoyl-3-phenyl-, 1,1-dioxide	n.s.g.	225(3.08),251(4.24)	39-0652-69C

Compound	Solvent	λ_{max}(log ϵ)	Ref.
$C_{15}H_{12}N_2O_4$			
4H-3,1-Benzoxazin-4-one, 1,2-dihydro-2-hydroxy-N-methyl-2-(o-nitrosophenyl)-	EtOH	283(3.80),310(3.70), 348s(3.15)	88-0375-69
4H-3,1-Benzoxazin-4-one, 1,2-dihydro-N-methyl-2-(o-nitrophenyl)-	EtOH	347(3.50)	88-0375-69
$C_{15}H_{12}N_2O_4S$			
Phthalimide, N-[1,2,5,7-tetrahydro-7-oxo-4H-furo[3,4-d][1,3]thiazin-2-yl)methyl]-	EtOH	272(3.65)	44-1582-69
$C_{15}H_{12}N_2O_6$			
3-Biphenylcarboxylic acid, 4,6-dinitro-, ethyl ester	n.s.g.	266(4.15),295s(3.85)	40-0085-69
$C_{15}H_{12}N_2S_2$			
1,2,3,4-Dithiadiazolidine, 4-phenyl-5-styryl-, meso-ionic didehydro derivative	MeOH	445(4.04)	42-0388-69
$C_{15}H_{12}N_4O$			
2-Pyrazolin-4-one, 5-imino-1,3-diphenyl-, oxime	EtOH	246(4.28),322(4.14), 367s(3.58)	4-0317-69
7H-Pyrrolo[2,3-d]pyrimidine-6-carbonitrile, 7-ethyl-5-hydroxy-2-phenyl-	EtOH	236(4.17),279(4.55)	4-0819-69
7H-Pyrrolo[2,3-d]pyrimidine-6-carbonitrile, 5-methoxy-7-methyl-2-phenyl-	EtOH	228(4.18),270(4.53), 308(4.15)	4-0819-69
$C_{15}H_{12}N_4O_5S_2$			
Barbituric acid, 5,5'-salicylidenebis-[2-thio-	EtOH	260(4.17),310s(--)	103-0830-69
$C_{15}H_{12}N_4O_7$			
Barbituric acid, 5,5'-(m-hydroxybenzylidene)di-	EtOH	260(4.55),320(2.46)	103-0827-69
Barbituric acid, 5,5'-salicylidenedi-	EtOH	260(4.20)	103-0827-69
$C_{15}H_{12}N_4S$			
9-Acridinecarboxaldehyde, thiosemicarbazone	EtOH	370(4.1),410(4.1)	65-1151-69
	EtOH-HCl	260(4.7),360(4.1), 490(4.0)	65-1151-69
$C_{15}H_{12}N_6O_2$			
2,3-Diazabicyclo[3.2.2]nona-3,6-diene-2-carboxylic acid, 8,8,9,9-tetracyano-4-methyl-, ethyl ester	EtOH	248(3.65)	77-0432-69
2,3-Diazabicyclo[3.2.2]nona-3,6-diene-2-carboxylic acid, 8,8,9,9-tetracyano-6-methyl-, ethyl ester	EtOH	238(3.63)	77-0432-69
$C_{15}H_{12}N_6O_6$			
Hydrazine, N-(α,β-dioximino-β-p-nitrophenyl)propionyl-N'-isonicotinoyl-	EtOH	280(4.14)	55-0859-69
$C_{15}H_{12}O$			
Benzyl alcohol, o-(phenylethynyl)-	EtOH	278(4.33),283(4.48), 292(4.32),302(4.37)	44-0874-69
Chalcone	MeOH	308(4.36)	73-2771-69
	iso-PrOH	227(4.01),308(4.35)	104-0826-69

Compound	Solvent	$\lambda_{max}(\log \epsilon)$	Ref.
Ketone, 1,3,5,7-cyclooctatetraen-1-yl phenyl	EtOH	250(4.09)	39-0978-69C
$C_{15}H_{12}O_2$			
Chalcone, 2'-hydroxy-	MeOH	221(4.07),314(4.3)	78-5415-69
Fluorene-9-carboxylic acid, 9-methyl-	MeOH	256(4.24),265(4.30), 277s(4.08),288(3.75), 299(3.73)	44-2799-69
1,4-Methanoanthracene-9,10-dione, 1,4,4a,9a-tetrahydro-	iso-PrOH	225(4.48),253(4.00), 297(2.38),307(2.48), 340(1.90)	39-0616-69B
1,4-bridge derivative	EPA at 77°K	225(4.46),253(4.02), 297(2.57),307(2.48), 340(1.90)	39-0616-69B
6,12-6H,12H-Methanodibenzo[b,f][1,5]-dioxocin	isooctane	277(3.63),286(3.53)	44-1907-69
(+)-	MeOH	277(3.58),286(3.49)	44-1907-69
1,3-Propanedione, 1,3-diphenyl-	$CHCl_3$	245(3.65),342(4.38)	39-0526-69A
aluminum chelate	$CHCl_3$	258(4.16),348(4.65)	42-0945-69
copper chelate	dioxan	260(3.28),350(4.58), 650(1.88),685s(1.86)	42-0945-69
palladium chelate	$CHCl_3$	265(4.80),300(4.74), 372(4.50),405(4.30)	42-0945-69
$C_{15}H_{12}O_2S$			
2H-Thiete, 2,4-diphenyl-, 1,1-dioxide	EtOH	257(4.24)	44-1136-69
$C_{15}H_{12}O_3$			
3-Benzofuranol, 2-benzoyl-2,3-dihydro-	n.s.g.	248(4.13),280(3.63)	32-1273-69
isomer	n.s.g.	248(4.16),280(3.64)	32-1273-69
Bicyclo[4.2.0]octa-3,7-diene-2,5-dione, 7-methoxy-8-phenyl-	EtOH	214(4.18),233(4.04), 252(4.13),260(3.94)	88-0737-69
Chalcone, 2',3'-dihydroxy-	EtOH	232(4.32),322(4.63)	18-0560-69
Chalcone, 2',4-dihydroxy-	MeOH	241(4.08),368(4.45)	78-5415-69
Chalcone, 2',4'-dihydroxy-	EtOH	220(4.00),263(3.69), 323(4.29),349(4.29)	95-0375-69
	EtOH-HCl	221(3.99),262(3.70), 322(4.29),348(4.29)	95-0375-69
	EtOH-NaOH	280(4.17),302(4.12), 395(4.39)	95-0375-69
Chalcone, 2',5'-dihydroxy-	EtOH	229(4.12),316(4.37), 408(3.69)	18-0560-69
Cinnamic acid, m-hydroxyphenyl ester	EtOH	284(4.42)	18-0560-69
Cinnamic acid, o-hydroxyphenyl ester	EtOH	283(4.49)	18-0560-69
Cinnamic acid, p-hydroxyphenyl ester	EtOH	283(4.49)	18-0560-69
Dibenzo[b,e]oxepin-11(6H)-one, 3-methoxy-	EtOH	213(4.24),249(3.91), 292(4.07),321(3.99)	39-1873-69C
Flavanone, 4'-hydroxy-, (±)-	MeOH	251(4.00),319(3.54)	78-5415-69
Fluorene-6-carboxylic acid, 7-hydroxy-1-methyl-	EtOH	282(4.3),325(4.5)	42-0415-69
Fluorenone, 2,3-dimethoxy-	MeOH	213(4.19),270(4.5), 323(3.4),428(2.6)	2-0557-69
Fluorenone, 3,4-dimethoxy-	MeOH	213(4.2),253(4.36), 300(3.63),350(3.8)	2-0557-69
Ketone, phenyl o-carboxybenzyl	EtOH	240(4.18),279(3.26)	12-0577-69
α-Lapachone, dehydro-	C_6H_{12}	233s(3.97),251s(4.14), 258(4.16),273s(4.16), 282(4.21),292s(4.10), 330(3.38),340s(3.27), 418(3.21),426(3.21), 434(3.21),457s(3.05)	44-0120-69

Compound	Solvent	$\lambda_{max}(\log \epsilon)$	Ref.
β-Lapachone, dehydro-	C_6H_{12}	226(4.30),244(4.09), 285(4.29),297(4.25), 330s(4.29),465(3.28), 530s(2.92)	44-0120-69
2-Oxatricyclo[6.2.0.0^{1,4}]deca-5,9-dien-7-one, 4-hydroxy-10-phenyl-	EtOH	226(4.00),292(3.24)	88-0737-69
Spiro[cyclohepta-2,4,6-triene-1,7'-[5]-norbornene]-2',3'-dicarboxylic anhydride	MeOH	258(3.58)	24-1789-69
$C_{15}H_{12}O_4$			
2H-Naphtho[2,3-b]pyran-5,10-dione, 9-hydroxy-2,2-diphenyl- (α-caropteron)	isooctane	240(4.19),280(4.08), 426(3.62)	33-0808-69
1,3-Propanedione, 2-hydroxy-1-(o-hydroxyphenyl)-3-phenyl-	EtOH	253(4.00),321(3.50)	18-3345-69
Xanthen-9-one, 1,5-dimethoxy-	MeOH	235s(4.54),350(3.75)	39-2421-69C
Xanthen-9-one, 1-hydroxy-3-methoxy-2-methyl-	MeOH	240(4.50),306(4.24)	78-0275-69
Xanthen-9-one, 1-hydroxy-3-methoxy-4-methyl-	MeOH	233(4.44),258(4.39), 308(4.10)	78-0275-69
$C_{15}H_{12}O_4S$			
Acetic acid, [(o-carboxyphenylthio)-phenyl]-	MeOH	234s(4.03),264(4.12), 287(4.18)	73-3936-69
Xanthene, 3,6-dimethoxy-9-sulfinyl-	EtOH	420(4.54)	23-2898-69
$C_{15}H_{12}O_5$			
Acetic acid, (4-benzoyl-3-hydroxyphenoxy)-	EtOH	244(3.95),287(4.13), 326(3.93)	39-0037-69C
1H-Naphtho[2,3-c]pyran-5,10-dione, 7,9-dihydroxy-1,3-dimethyl-	$CHCl_3$	285(4.06),362(3.61), 523(3.62)	39-0631-69C
Xanthen-9-one, 1,3-dihydroxy-7-methoxy-	MeOH	231(4.58),261(4.49), 312(4.21)	78-0275-69
Xanthen-9-one, 1,8-dihydroxy-3-methoxy-2-methyl-	MeOH	249(4.07),325(3.84)	78-0275-69
Xanthen-9-one, 1,8-dihydroxy-3-methoxy-4-methyl-	MeOH	230(4.15),250(4.19), 337(3.98)	78-0275-69
Xanthen-9-one, 5-hydroxy-1,3-dimethoxy-	EtOH	247(4.62),302(4.26)	2-0463-69
Xanthen-9-one, 8-hydroxy-1,2-dimethoxy-	MeOH	237(4.31),269(4.36), 289(3.61),319(3.53), 386(3.51)	39-1567-69C
$C_{15}H_{12}O_6$			
Bellidifolin, methyl-	EtOH	254(4.40),278(4.20), 335(4.00),390s(3.80)	95-0410-69
Swertianin, methyl-	EtOH	240(4.42),265(4.48), 312s(4.12),330(4.14), 386s(3.68)	94-0155-69
Xanthen-9-one, 1,3-dihydroxy-4,5-dimethoxy- (also spectra in other solvents)	EtOH	243(4.49),260(4.36), 290s(4.26),318(4.14), 366(3.61)	78-1947-69
Xanthen-9-one, 1,3-dihydroxy-4,7-dimethoxy-	EtOH	234(4.41),266(4.42), 316(3.96),376(3.79)	78-1961-69
$C_{15}H_{12}O_7$			
1H-Naphtho[2,3-c]pyran-1,6,9-trione, 3,4-dihydro-7,10-dihydroxy-8-methoxy-3-methyl-	EtOH	218(4.17),230s(4.10), 272(3.98),385(3.60)	23-1561-69

Compound	Solvent	$\lambda_{max}(\log \epsilon)$	Ref.
$C_{15}H_{12}S$			
1H-2-Benzothiopyran, 1-phenyl-	MeOH	207(4.55),245(4.07), 322(3.70)	44-3202-69
2H-1-Benzothiopyran, 4-phenyl-	EtOH	235(3.4),325(3.7)	2-0315-69
6H-Cyclopenta[5,6]naphtho[2,1-b]thiophene, 7,8-dihydro-	n.s.g.	247(4.62),256(4.54), 297(4.06),307(4.00), 324(3.44),340(3.39)	28-0194-69A
$C_{15}H_{12}S_3$			
1,3-Dithiole, 4-phenyl-2-(phenylthio)-	EtOH	221(4.25),320(3.85)	94-1931-69
$C_{15}H_{13}$			
1-Phenylindanylium cation	acid	302(4.11),412(4.57)	59-0447-69
$C_{15}H_{13}BF_4O_2$			
(o-Carboxyphenyl)cycloheptatrienylium tetrafluoroborate, methyl ester	MeCN	264s(3.95),341(3.88)	49-0001-69
$C_{15}H_{13}BrN_2$			
2-Pyrazoline, 3-bromo-1,5-diphenyl-	EtOH	244(4.03),289(4.17)	22-1683-69
$C_{15}H_{13}BrN_2O_2$			
1,3-Cyclopentadiene, 1,2-diacetyl-5-[(p-bromophenyl)azo]-	EtOH	262(4.37),488(4.35)	39-2464-69C
$C_{15}H_{13}BrN_2S_2$			
2H-1,3,5-Thiadiazine-2-thione, 3-(p-bromophenyl)tetrahydro-5-phenyl-	MeOH	245(4.32),293(4.07)	73-2952-69
$C_{15}H_{13}BrN_4O_5$			
Propiophenone, 3'-bromo-2'-hydroxy-, 2,4-dinitrophenylhydrazone	EtOH	380(4.39)	22-2079-69
Propiophenone, 3'-bromo-4'-hydroxy-, 2,4-dinitrophenylhydrazone	EtOH	388(4.44)	22-2079-69
Propiophenone, 4'-bromo-2'-hydroxy-, 2,4-dinitrophenylhydrazone	EtOH	376(4.43)	22-2079-69
Propiophenone, 5'-bromo-2'-hydroxy-, 2,4-dinitrophenylhydrazone	EtOH	378(4.39)	22-2079-69
$C_{15}H_{13}BrO_3$			
Benzophenone, 4-(2-bromoethoxy)-2-hydroxy-	EtOH	240(4.08),285(4.22), 325(4.00)	39-0034-69C
$C_{15}H_{13}Br_2N$			
Aniline, cinnamylidene-, compound with bromine	MeCN	300(4.65)	25-0135-69
$C_{15}H_{13}ClN_2O$			
Acetic acid, (m-chlorobenzylidene)phenylhydrazide	EtOH	203(4.3),225(4.1), 231s(4.0),290(4.2), 313(3.9)	28-1703-69A
Acetic acid, (o-chlorobenzylidene)phenylhydrazide	EtOH	203(4.4),218s(4.2), 224(4.2),231(4.1), 279(4.3),290(4.3)	28-1703-69A
Acetic acid, (p-chlorobenzylidene)phenylhydrazide	EtOH	218(4.2),226s(4.1), 286(4.4),305s(4.3)	28-1703-69A
$C_{15}H_{13}ClN_2O_2$			
1H-Naphth[2,3-d]imidazole-4,9-dione, 2-(4-chlorobutyl)-	pH 1	246(4.51),251(4.61), 278(4.21)	4-0909-69

Compound	Solvent	λ_{max}(log ϵ)	Ref.
1H-Naphth[2,3-d]imidazole-4,9-dione, 2-(4-chlorobutyl)- (cont.)	pH 7	247(4.64),281(4.16)	4-0909-69
	pH 13	262(4.59)	4-0909-69
	MeOH	246(4.62),275(4.16), 279(4.18),332(3.46)	4-0909-69
2(1H)-Quinazolinone, 6-chloro-3,4-dihydro-4-hydroxy-3-methyl-4-phenyl-	n.s.g.	255(4.20),300(3.36)	40-0917-69
$C_{15}H_{13}ClN_2O_5$			
4-Imino-3,4-dihydro-2,3-diphenyl-1,3-oxazolium perchlorate	MeCN	298(4.13)	18-2310-69
$C_{15}H_{13}ClO$			
Anisole, p-(p-chlorostyryl)-	DMF	309(4.47),326(4.52)	33-2521-69
$C_{15}H_{13}ClO_3$			
Benzophenone, 2'-(chloromethyl)-2-hydroxy-4-methoxy-	C_6H_{12}	224(4.04),240(3.95), 287(4.18),328(3.93)	39-1873-69C
$C_{15}H_{13}ClO_3S$			
Benzoic acid, 2-chloro-3-[(p-hydroxyphenyl)thio]-, ethyl ester	EtOH	232(4.54)	104-1903-69
Benzoic acid, 3-chloro-4-[(p-hydroxyphenyl)thio]-, ethyl ester	EtOH	234(4.28),293(4.28)	104-1903-69
$C_{15}H_{13}ClO_4S$			
Benzoic acid, 2-chloro-3-[(p-hydroxyphenyl)sulfinyl]-, ethyl ester	EtOH	242(4.43)	104-1903-69
Benzoic acid, 3-chloro-4-[(p-hydroxyphenyl)sulfinyl]-, ethyl ester	EtOH	262(4.19)	104-1903-69
$C_{15}H_{13}ClO_5S$			
Benzoic acid, 2-chloro-3-[(p-hydroxyphenyl)sulfonyl]-, ethyl ester	EtOH	246(4.41)	104-1903-69
Benzoic acid, 3-chloro-4-[(p-hydroxyphenyl)sulfonyl]-, ethyl ester	EtOH	234(4.28),280(4.11)	104-1903-69
$C_{15}H_{13}ClO_6$			
(p-Carboxyphenyl)cycloheptatrienylium perchlorate, methyl ester	MeCN	268(4.24),353(4.27)	49-0001-69
$C_{15}H_{13}Cl_2NO_3$			
Acetohydroxamic acid, 2-(2,4-dichlorophenoxy)-N-p-tolyl-	EtOH	230(4.17),248(--)	42-0831-69
Valeramide, 5-chloro-N-(3-chloro-1,4-dihydro-1,4-dioxo-2-naphthyl)-	MeOH	247(4.24),252(4.31), 286(3.98),337(3.52)	4-0909-69
$C_{15}H_{13}Cl_2N_3$			
Benzimidazo[2,1-a]isoquinoline, 10-amino-, dihydrochloride	EtOH	222(4.25),240(4.23), 260(4.24),267(4.23), 297(4.21),331(4.00), 346(3.96)	24-1529-69
$C_{15}H_{13}Cl_2N_3O_2$			
1-Propene, 1,3-bis(p-chloroanilino)-2-nitro-, hydrochloride	dioxan	240(4.25),380(4.36)	78-1617-69
$C_{15}H_{13}Cl_3N_2O_4$			
Malondianil, m,m'-dichloro-, perchlorate	EtOH	362(4.64)	24-2609-69
Malondianil, p,p'-dichloro-, perchlorate	EtOH	369(4.60)	24-2609-69

Compound	Solvent	λ_{max}(log ϵ)	Ref.
$C_{15}H_{13}IO_6$			
6-Oxabicyclo[3.2.1]octane-2-carboxylic acid, 4-iodo-8-[3,4-(methylenedioxy)-phenyl]-7-oxo-	EtOH	238(3.67),287(3.67)	78-4315-69
$C_{15}H_{13}NO$			
Acetamide, N-(diphenylmethylene)-	EtOH	254(4.28),280(3.78), 350(2.08)	88-0927-69
Acrolein, 2-amino-3,5-diphenyl-	EtOH	232(4.09),254s(3.99), 307s(3.94),340(3.98)	77-0228-69
Formimidic acid, N-fluoren-9-yl-, methyl ester	dioxan	225(4.35),233(4.16), 270(4.24),293(3.72), 305(3.74)	89-0772-69
Isocyanic acid, 1,2-diphenyl-, ethyl ester	isooctane	248(2.51),252(2.62), 258(2.72),263(2.64), 267(2.48)	30-0559-69F
1-Isoindolone, 2,3-dihydro-3-methyl-3-phenyl-	EtOH	272s(3.04),281(2.94)	39-1149-69C
Ketone, 4-acridanyl methyl	MeOH	230s(4.08),250(4.04), 297(4.00),400(3.85)	22-3523-69
1-Propanone, 1-carbazol-1-yl-	EtOH	224(4.67),254(4.02), 289(4.25),320s(3.71), 379(3.95)	22-3523-69
1-Propanone, 1-carbazol-2-yl-	EtOH	253(4.58),274s(3.93), 318(4.35),370s(3.57)	22-3523-69
1-Propanone, 1-carbazol-3-yl-	EtOH	234(4.49),244s(4.34), 273(4.50),288(4.53), 330(4.02)	22-3523-69
2-Propen-1-one, 1-(p-aminophenyl)-3-phenyl-	MeOH	367(4.30)	73-2771-69
2-Propen-1-one, 3-(p-aminophenyl)-1-phenyl-	MeOH	397(4.39)	73-2771-69
$C_{15}H_{13}NOS$			
Benzo[b]thiophen-4(5H)-one, 5-(anilino-methylene)-6,7-dihydro-	EtOH	259(4.29),381(4.45)	39-2750-69C
$C_{15}H_{13}NOS_2$			
Carbamic acid, N-benzoyldithio-, benzyl ester	EtOH	273(4.36)	39-2794-69C
$C_{15}H_{13}NO_2$			
Acetohydroxamic acid, N-fluoren-1-yl-	EtOH	257(4.30),262(4.30), 273s(4.11),291(3.69), 295s(3.66),303(3.80)	39-0345-69C
Acetohydroxamic acid, N-fluoren-3-yl-	EtOH	252(4.46),302(3.80), 313(3.71)	39-0345-69C
Acetohydroxamic acid, N-fluoren-4-yl-	EtOH	251s(4.07),263s(4.10), 266(4.11),290(3.85), 301(3.78)	39-0345-69C
Benzanilide, 2'-acetyl-	EtOH	244(4.04),282s(3.05)	1-2149-69
Benzanilide, 4'-(vinyloxy)-	n.s.g.	276(4.16)	30-0108-69A
stannic chloride complex	n.s.g.	276(4.11),585(1.47)	30-0108-69A
Indoline-N-carboxaldehyde, 2-hydroxy-3-phenyl-	EtOH	248(4.09),279(3.50), 286(3.45)	1-2149-69
Pyrrolo[2,1-a]isoquinoline-1-carboxylic acid, 5-methyl-, methyl ester	MeOH	218(4.49),243(4.13), 266(4.48),275(4.69), 313s(3.91),323(4.00), 337(4.00),353(0.85)	39-2311-69C

Compound	Solvent	λ_{max}(log ε)	Ref.
Pyrrolo[1,2-a]quinoline-3-carboxylic acid, 7-methyl-, methyl ester	MeOH	209s(4.37),223(4.62), 232s(4.40),247s(4.27), 254(4.37),266(4.27), 272(4.29),280s(4.16), 329s(4.04),341s(4.16), 354(4.30),371(4.18)	39-2311-69C
Pyrrolo[1,2-a]quinoline-3-carboxylic acid, 9-methyl-, methyl ester	MeOH	209s(4.27),225(4.58), 247s(4.24),255(4.30), 266(4.29),283s(3.97), 335s(4.18),347(4.30), 364(4.19)	39-2311-69C
Stilbene, α-methyl-2'-nitro-, cis	EtOH	240(4.03),256(4.00)	44-2285-69
$C_{15}H_{13}NO_2S$			
8-Thia-4H-pyrido[3,2,1-de]phenanthridine, 5,6-dihydro-, 8,8-dioxide	EtOH	232(4.40),270(4.13), 314(3.71)	2-0848-69
$C_{15}H_{13}NO_3$			
Anisole, p-(p-nitrostyryl)-, trans	MeOH	374(4.45)	39-0554-69B
	DMF	384(4.44)	39-0554-69B
Anthranilic acid, N-(p-acetylphenyl)-	isooctane	224(4.04),244s(3.61), 319(3.91),356(3.87)	22-3523-69
	EtOH	202(4.44),218s(4.22), 232s(4.18),330s(4.17), 358(4.46)	22-3523-69
Carbazole-1-carboxylic acid, 2-methoxy-, methyl ester	EtOH	225(4.9),284(4.36), 320(3.96),350(3.97)	39-1518-69C
Nitrone, α-(p-carboxyphenyl)-N-phenyl-, methyl ester	C_6H_{12}	340(4.42)	18-3306-69
	EtOH	331(4.44)	18-3306-69
2-Propen-1-ol, 3-(o-nitrophenyl)-2-phenyl-, trans	EtOH	240(4.03),255(3.97)	44-2285-69
Pyrido[1,2-a]indole-10-carboxylic acid, 3-hydroxy-, ethyl ester	EtOH	248(4.60),255(4.58), 288s(4.03),297(4.10), 316(3.94),329(3.94), 397(3.82),417(3.83), 440s(3.58)	12-1525-69
$C_{15}H_{13}NO_4$			
Evoxanthine	EtOH	215s(4.14),240s(4.11), 275(4.58),303s(3.57), 320s(3.41),387s(3.83), 400(3.86)	39-2327-69C
	EtOH-HCl	240(4.26),267(4.45), 283(4.55),346s(3.83), 361(4.03),403(3.73), 430(3.52)	39-2327-69C
$C_{15}H_{13}NO_5$			
2-(1,2-Dicarboxy-2-hydroxyvinyl)isoquinolinium hydroxide, inner salt, dimethyl ester	$CHCl_3$	231(4.44),234(4.40), 246(4.10),252(4.14), 259(4.25),265(4.28), 300(3.68),342(3.56), 410(3.63)	24-0915-69
	10% EtOH-H_2SO_4	233(4.68),270s(3.54), 277(3.58),336s(3.67), 342(3.68)	24-0915-69
$C_{15}H_{13}NO_6$			
4H-Pyran-2-carboxylic acid, 5-hydroxy-4-oxo-, ethyl ester, carbanilate	MeOH	235(4.7),265s(3.8), 360(3.6)	83-0628-69

Compound	Solvent	λmax(log ε)	Ref.
$C_{15}H_{13}NS$			
4H-1,4-Benzothiazine, 4-methyl-3-phenyl-	n.s.g.	251(4.1),300(3.8)	4-0635-69
2-Thiazoline, 2,4-diphenyl-	MeOH	246(4.3),279s(3.5)	39-1120-69C
hydrochloride	MeOH	246(4.2),279s(3.4)	39-1120-69C
$C_{15}H_{13}N_3O$			
Benzamide, N-(2-benzimidazolylmethyl)-	EtOH	275(3.94),282(3.94)	94-2381-69
Indole-3-acetamide, 2-(3-pyridyl)-	EtOH	219(4.53),309(4.39)	22-4154-69
$C_{15}H_{13}N_3O_3$			
Acetic acid, (m-nitrobenzylidene)phenylhydrazide	EtOH	279(4.4),333s(3.2)	28-1703-69A
Acetic acid, (o-nitrobenzylidene)phenylhydrazide	EtOH	202(4.4),262(4.2), 286s(4.1),316s(3.8)	28-1703-69A
Glyoxal, p-tolyl-, 2-(o-nitrophenyl)-hydrazone	$CHCl_3$	402(4.19),418(4.27)	28-0642-69B
7H-Pyrrolo[2,3-d]pyrimidine-6-carboxylic acid, 5-hydroxy-2-phenyl-, ethyl ester	EtOH	239(4.13),270(4.53)	4-0819-69
$C_{15}H_{13}N_3O_4$			
Aniline, 2,4-dinitro-N-(2-phenylcyclopropyl)-, trans	EtOH	348(3.23)	39-1135-69B
Glyoxal, p-methoxyphenyl-, 2-(o-nitrophenyl)hydrazone	$CHCl_3$	424(4.31)	28-0642-69B
$C_{15}H_{13}N_5O_3$			
Isoxanthopterin, 6-acetonyl-8-phenyl-	pH -3.3	276s(4.01),283(4.03), 326(3.95),376s(3.64), 398(3.83),423(--), 449(3.79)	24-4032-69
	pH 5	285s(3.98),293(4.03), 345(4.16)	24-4032-69
	pH 12	259(4.12),290s(3.66), 360(4.17)	24-4032-69
$C_{15}H_{13}N_5O_4$			
Hydrazine, N-(α,β-dioximino-β-phenyl)-propionyl-N'-isonicotinoyl-, syn	EtOH	248(4.20)	55-0859-69
$C_{15}H_{13}N_5O_4S$			
Carbonic acid, thio-, O-ethyl S-[9-(p-nitrobenzyl)-9H-purin-6-yl] ester	MeOH	277(3.99)	44-0973-69
Purine-9(7)-carboxylic acid, 8-[(p-nitrobenzyl)thio]-, ethyl ester	dioxan	296(4.38)	4-0023-69
9H-Purine-9-carboxylic acid, 6-[(p-nitrobenzyl)thio]-, ethyl ester	MeOH	289(4.48)	44-0973-69
$C_{15}H_{13}N_5O_4Se$			
9H-Purine-9-carboxylic acid, 6-[(p-nitrobenzyl)selenyl]-, ethyl ester	MeOH	299(4.29)	44-0973-69
$C_{15}H_{13}N_5O_7$			
Barbituric acid, 5,5'-(5-aminosalicylidene)di-	EtOH	245(4.14),325(4.26)	103-0827-69
$C_{15}H_{14}$			
Fluorene, 1,6-dimethyl-	EtOH	212(4.6),269(4.3), 274(4.2),280(3.5), 290(3.7),302(3.8)	42-0415-69

Compound	Solvent	λ_{max}(log ε)	Ref.
Fluorene, 1,6-dimethyl- (cont.)	EtOH	270(4.4),300(4.3)	42-0415-69
Fluorene, 1,7-dimethyl-	EtOH	270(4.5),297(3.9), 304(3.9)	42-0415-69
Phenanthrene, 4,4a-dihydro-4a-methyl-	C_6H_{12}	247(4.17),252(4.27), 261(4.21),345(3.95), 358(4.04),376(3.88)	35-0902-69
Propene, 1,1-diphenyl-	C_6H_{12}	225s(4.1),249(4.16)	59-0487-69
Propene, 2,3-diphenyl-	isooctane	242(4.01)	35-6707-69
Stilbene, α-methyl-, cis	isooctane	263(4.03)	35-6707-69
Stilbene, α-methyl-, trans	isooctane	272(4.31)	35-6707-69
$C_{15}H_{14}BF_2NO_2$			
Naphtho[1,2-e]-1,3,2-dioxaborin, 2,2-difluoro-4-(β-dimethylaminovinyl)-	MeCN	245(4.38),273(3.81), 315(3.81),340(3.72), 412(4.49),428(4.44)	4-0029-69
Naphtho[2,1-e]-1,3,2-dioxaborin, 2,2-difluoro-4-(β-dimethylaminovinyl)-	MeCN	245s(3.53),272(4.85), 338(4.26),355(4.35), 398(4.55),416(4.57)	4-0375-69
$C_{15}H_{14}BrIN_2O$			
2-[2-(5-Bromo-2-furyl)vinyl]-1,3-dimethylbenzimidazolium iodide	MeOH	310(3.98)	103-0246-69
$C_{15}H_{14}ClIN_2O$			
2-[2-(5-Chloro-2-furyl)vinyl]-1,3-dimethylbenzimidazolium iodide	MeOH	316(3.99)	103-0246-69
$C_{15}H_{14}ClN$			
Aniline, p-chloro-N-(p,α-dimethylbenzylidene)-	EtOH	256(4.32),306(3.50)	56-0749-69
p-Toluidine, N-(p-chloro-α-methylbenzylidene)-	EtOH	254(4.30),318(3.42)	56-0749-69
$C_{15}H_{14}ClNO$			
Aniline, p-chloro-N-(p-methoxy-α-methylbenzylidene)-	EtOH	273(4.32),309(3.73)	56-0749-69
p-Anisidine, N-(p-chloro-α-methylbenzylidene)-	EtOH	254(4.23),331(3.48)	56-0749-69
$C_{15}H_{14}ClNO_4$			
Acetanilide, 4'-chloro-N-(4-formyl-2-hydroxy-1,3-butadienyl)-, acetate	EtOH	324(4.40)	78-2035-69
Acetanilide, 4'-chloro-N-(4-formyl-4-hydroxy-1,3-butadienyl)-, acetate	EtOH	320(4.43)	78-2035-69
1-Indolinyltropylium perchlorate	MeCN	245(4.18),398(4.29)	64-1353-69
$C_{15}H_{14}ClNS$			
2-Thiazoline, 2,4-diphenyl-, hydrochloride	MeOH	246(4.2),279s(3.4)	39-1120-69C
$C_{15}H_{14}ClN_3O_2$			
Acetanilide, 5'-[(p-chlorophenyl)azo]-2'-hydroxy-N-methyl-	EtOH	355(4.40)	22-4390-69
	EtOH-KOH	422(4.48)	22-4390-69
$C_{15}H_{14}ClN_3O_5$			
1-Acetamido-2-phenyl-1H-imidazo[1,2-a]pyridin-4-ium perchlorate	EtOH	204(4.47),234(4.37), 295(4.01)	44-2129-69

Compound	Solvent	$\lambda_{max}(\log \epsilon)$	Ref.
$C_{15}H_{14}ClN_5O_5$			
1-[2-[3-[(5-Nitrofurfurylidene)amino]-2,5-dioxo-4-imidazolidinyl]ethyl]-pyridinium chloride	H_2O	250(4.13),365(4.22)	44-3213-69
$C_{15}H_{14}Cl_3N_3O_2$			
1-Propene, 1,3-bis(p-chloroanilino)-2-nitro-, hydrochloride	dioxan	240(4.25),380(4.36)	78-1617-69
$C_{15}H_{14}Fe$			
Ferrocene, 1-ethynyl-1',2-trimethylene-	EtOH	325(2.32),433(2.42)	49-1540-69
Ferrocene, 1-ethynyl-1',3-trimethylene-	EtOH	328(2.53),443(2.66)	49-1540-69
$C_{15}H_{14}IN_3O_3$			
Propionanilide, 3-(2-pyridyl)-3-oxo-2-(hydroxyimino)-, methiodide, syn	EtOH	221(4.30),242(4.25)	36-0460-69
Propionanilide, 3-(3-pyridyl)-3-oxo-2-(hydroxyimino)-, methiodide, anti	EtOH	218(4.26),285(4.05)	36-0460-69
syn	EtOH	227(4.42),290(4.09)	36-0460-69
Propionanilide, 3-(4-pyridyl)-3-oxo-2-(hydroxyimino)-, methiodide, anti	EtOH	225(4.38),238(4.39)	36-0460-69
syn	EtOH	220(4.43),242s(--)	36-0460-69
$C_{15}H_{14}N_2$			
1,1'-Bipyrrole, 2-methyl-5-phenyl-	EtOH	283(4.3)	24-3268-69
Dibenzo[c,g][1,2]diazocine, 5,6-dihydro-5-methyl-	n.s.g.	252(4.8),300(3.9), 350(3.8)	44-3237-69
Malondianil	EtOH	365(4.60)	24-2609-69
Malondianil, hydrochloride	MeOH	244(4.07),277(3.40), 285(3.38),295s(--), 382(4.69)	59-1075-69
5,12-Methanodibenzo[b,f][1,4]diazocine, 6,11-dihydro-	EtOH	282(3.31)	39-0882-69C
	EtOH-HCl	247(3.84)	39-0882-69C
1H-Pyrazino[3,2,1-jk]carbazole, 2,3-di-hydro-3-methyl-	MeOH	230(4.59),255(4.59), 297(3.97),350(3.73)	103-0568-69
2-Pyrazoline, 1,3-diphenyl-	EtOH	243(4.12),303(3.88), 356(4.29)	39-1703-69C
3H-Pyrrolo[1,2-b][1,2]diazepine, 2-meth-yl-5-phenyl-	EtOH	238(4.3)	24-3268-69
3H-Pyrrolo[1,2-b][1,2]diazepine, 5-meth-yl-2-phenyl-	EtOH	268(4.3)	24-3268-69
$C_{15}H_{14}N_2O$			
Acetanilide, 4'-[2-(2-pyridyl)vinyl]-, (E)-	H_2O	323(4.56)	23-2355-69
hydrochloride	H_2O	353(4.54)	23-2355-69
Acetic acid, benzylidenephenylhydrazide	EtOH	204(4.3),216(4.2), 272(4.3),291s(4.3), 298s(4.2)	28-1703-69A
Acrylophenone, 3-amino-3-anilino-	EtOH	237(4.06),333(4.38)	44-3451-69
Benzimidazole, 2-(o-ethoxyphenyl)-	MeOH	219(4.42),251s(4.27), 294s(3.57)	4-0605-69
Cinnamic acid, 2-phenylhydrazide	N HCl	285(4.93)	65-1835-69
	H_2O	280(4.94)	65-1835-69
3-Pyrazolidinone, 1,5-diphenyl-	N HCl	260(3.10)	65-1835-69
	H_2O	240(3.91)	65-1835-69
1-Pyrazoline, 3,5-diphenyl-, 1-oxide	EtOH	225s(3.85)	88-0575-69
2-Pyrazolin-5-one, 4,4-dimethyl-3-(2-naphthyl)-	EtOH	224(4.57),245(4.30), 266(4.46),276(4.46), 305(4.46)	22-4159-69

Compound	Solvent	λ_{max}(log ε)	Ref.
2(1H)-Quinoxalinone, 3,4-dihydro-1-methyl-3-phenyl-	EtOH	228(4.79),255s(4.36), 308(3.69)	106-0384-69
$C_{15}H_{14}N_2O_2$			
Acetic acid, (m-hydroxybenzylidene)-phenylhydrazide	EtOH	281(4.3),292s(4.3), 308s(3.9)	28-1703-69A
Acetic acid, (p-hydroxybenzylidene)-phenylhydrazide	EtOH	222(4.2),290(4.3), 305(4.3)	28-1703-69A
Acetic acid, phenylsalicylidenehydrazide	EtOH	246s(4.1),278(4.3), 288(4.3),322(4.1)	28-1703-69A
Aniline, N-(p,α-dimethylbenzylidene)-m-nitro-	EtOH	256(4.42),310s(3.60)	56-0749-69
Benzoic acid, (α-methylsalicylidene)-hydrazide	EtOH	231(4.3),280(4.2), 292s(4.2),324(4.1)	28-0730-69A
9H-Benzo[g]pyrrolo[1,2-b]indazol-9-one, 5,6,8,11-tetrahydro-3-methoxy-	EtOH	274(4.06),337(4.18)	88-3693-69
Benzyl alcohol, α-(phenylazo)-, acetate	dioxan	275(4.13),403(2.32)	39-2587-69C +77-0179-69
1,3-Cyclopentadiene, 1,2-diacetyl-5-(phenylazo)-	EtOH	258(4.39),274s(4.20), 488(4.33)	39-2464-69C
Furo[2,3-d]pyrrolo[1,2-a]pyrimidin-4(2H)-one, 3,6,7,8-tetrahydro-2-phenyl-	MeOH	208(4.20),271(3.43)	24-2739-69
Methane, diazobis(p-methoxyphenyl)-	dioxan	290(4.62),525(2.29)	47-3313-69
3H-Phenoxazin-3-one, 7-(dimethylamino)-1-methyl-	EtOH	560(4.59)	73-0221-69
Pyridine, 5-ethyl-2-(p-nitrostyryl)-, trans	H_2O	350(4.38)	23-2355-69
hydrochloride	H_2O	344(4.57)	23-2355-69
p-Toluidine, N-(α-methyl-p-nitrobenzylidene)-	EtOH	274(4.20),352(3.46)	56-0749-69
$C_{15}H_{14}N_2O_2S_2$			
Isothiocyanic acid, p-[[p-(dimethylamino)phenyl]sulfonyl]phenyl ester	EtOH	222(4.48),292(4.51), 305(4.60)	73-4005-69
$C_{15}H_{14}N_2O_3$			
Aniline, N-(p-methoxy-α-methylbenzylidene)-m-nitro-	EtOH	274(4.38),324s(3.62)	56-0749-69
p-Anisidine, N-(α-methyl-p-nitrobenzylidene)-	EtOH	274(4.26),368(3.56)	56-0749-69
Benzoic acid, (2,6-dihydroxy-α-methylbenzylidene)hydrazide	EtOH	221(4.3),301(4.3)	28-0730-69A
2-Isoxazolin-5-one, 3-methyl-4-(2-benzyl-3-methyl-3-isoxazolin-5-ylidene)-	EtOH	249(3.58),298(4.07)	39-0245-69C
Pyrrolo[1,2-a]pyrimidin-4(6H)-one, 2-acetoxy-7,8-dihydro-3-phenyl-	EtOH	235s(--),286(4.00)	22-3133-69
$C_{15}H_{14}N_2O_4$			
Azobenzene-3-carboxylic acid, 4-hydroxy-4'-methoxy-, methyl ester	EtOH	222s(4.26),274(3.86), 350(3.94),400(4.36)	12-0935-69
	EtOH-KOH	254s(3.95),387(4.40), 425(4.23)	12-0935-69
Benzaldehyde, 2,4,6-trihydroxy-, N-acetylphenylhydrazone	EtOH	240(4.0),317(4.4)	28-1703-69A
$C_{15}H_{14}N_2O_6$			
4H-Pyran-2-carboxylic acid, 6-amino-5-hydroxy-4-oxo-, ethyl ester, carbanilate	MeOH	240(4.5),295(4.0), 410(4.0)	83-0628-69

Compound	Solvent	$\lambda_{max}(\log \epsilon)$	Ref.
$C_{15}H_{14}N_2S$			
Lepidine, 2-(2,4-dimethyl-5-thiazolyl)-	n.s.g.	345(4.27)	97-0186-69
$C_{15}H_{14}N_2S_2$			
Benzothiazole, 2-[(2-anilinoethyl)thio]-	MeOH	226(4.40),246(4.32), 283(4.16),291(4.15), 301(4.08)	4-0163-69
1,2,3,4-Dithiadiazolidine, 5-phenethyl-4-phenyl-, meso-ionic didehydro derivative	MeOH	362(3.56)	42-0388-69
Isothiocyanic acid, p-[[p-(dimethylamino)phenyl]thio]phenyl ester	EtOH	255(4.25),284s(--), 290(4.58),305(4.57)	73-4005-69
2H-1,3,5-Thiadiazine-2-thione, tetrahydro-3,5-diphenyl-	MeOH	245(4.27),293(4.07)	73-2952-69
$C_{15}H_{14}N_4$			
Pteridine, 2,6,7-trimethyl-4-phenyl-	hexane	228(4.27),286s(3.83), 295(3.90),328(4.13)	39-1883-69C
Pteridine, 4,6,7-trimethyl-2-phenyl-	hexane	252(4.45),274(4.34), 321s(4.02),333(4.17), 348(4.11)	39-1408-69C
	pH 2.9	247(4.38),265s(4.25), 332(4.14)	39-1408-69C
$C_{15}H_{14}N_4O$			
Acetic acid, 2-(o-2-benzimidazolylphenyl)hydrazide	MeOH	240s(4.15),247s(4.09), 291(4.25),298(4.18), 314(4.06),328(3.98)	4-0605-69
$C_{15}H_{14}N_4O_2$			
7H-Pyrrolo[2,3-d]pyrimidine-6-carboxamide, 5-methoxy-7-methyl-2-phenyl-	EtOH	231(4.15),274(4.57), 309(4.05)	4-0819-69
$C_{15}H_{14}N_4O_4$			
Acetanilide, 2'-hydroxy-N-methyl-5'-[(p-nitrophenyl)azo]-	EtOH	374(4.39)	22-4390-69
	EtOH-KOH	494(4.56)	22-4390-69
$C_{15}H_{14}N_4S$			
9-Acridanone, 3-methyl-3-thioisosemicarbazone	EtOH	240(4.6),250(4.5), 300(3.9),420(4.0)	65-1156-69
	EtOH-HCl	260(4.8),340(3.6), 400(3.9),410(3.9), 450(3.8)	65-1156-69
9-Acridanone, 2-methyl-3-thiosemicarbazone	EtOH	250(4.6),420(4.0)	65-1156-69
	EtOH-HCl	260(4.8),330(3.6), 400(3.9),420(4.1), 450(4.0)	65-1156-69
9-Acridanone, 10-methyl-, thiosemicarbazone	EtOH	240(4.7),280(4.1), 400(4.2)	65-1156-69
	EtOH-HCl	270(4.7),430(4.0), 450(3.9)	65-1156-69
Isothiocyanic acid, p-[[p-(dimethylamino)phenyl]azo]phenyl ester	heptane	308(4.15),420(4.59)	73-3912-69
$C_{15}H_{14}O$			
Acetophenone, 5'-methyl-2'-phenyl-	EtOH	216(4.27),230(4.29), 250(4.09),292(3.52)	6-0277-69
Acetophenone, 2'-p-tolyl-	EtOH	218(4.08),232(4.20), 250(4.00),293(3.49)	6-0277-69

Compound	Solvent	$\lambda_{max}(\log \epsilon)$	Ref.
Anisole, p-(1,3,5,7-cyclooctatetraen-1-yl)-	hexane	245(4.23)	35-4714-69
Anisole, p-styryl-	DMF	307(4.47),321(4.49)	33-2521-69
Benzophenone, m,m'-dimethyl-	n.s.g.	254(4.21)	78-4919-69
Benzophenone, p,p'-dimethyl-	n.s.g.	258(4.22)	78-4919-69
3,5,7-Cyclooctatrien-1-ol, 2-benzylidene-	EtOH	260(3.84),300(3.89)	39-0978-69C
Propionaldehyde, 2,3-diphenyl-	C_6H_{12}	291(2.30)	22-0903-69
Stilbene, α-methoxy-, trans	MeOH	220(4.09),292(4.32)	44-1746-69
$C_{15}H_{14}O_2$			
Dispiro[5.0.5.3]pentadeca-1,4,8,11-tetraene-3,10-dione	EtOH	234(4.28),258(4.20)	77-0996-69
4α-Flavanol, (-)	MeOH	277(3.35),284(3.31)	78-5415-69
2-Propanone, 1-hydroxy-1,3-diphenyl-	EtOH	253(2.61),259(2.67), 264(2.63),293(2.56)	44-1430-69
Resorcinol, 4-cinnamyl-	n.s.g.	253(4.38),278(4.06), 285s(4.02)	78-1407-69
Spiro[2,5-cyclohexadiene-1,1'-(2'H)-naphthalen]-4-one, 3',4'-dihydro-6'-hydroxy-	EtOH	237(4.48),280(3.41)	35-2800-69
Tetracyclo[3.2.0.0^{2,7}.0^{4,6}]heptane-1-carboxylic acid, 5-phenyl-, methyl ester	MeCN	290(3.83)	33-0956-69
m-Toluic acid, m-tolyl ester	C_6H_{12}	233(4.18),268s(--), 278s(--),289(2.83)	5-0216-69F
m-Toluic acid, o-tolyl ester	C_6H_{12}	233(4.18),269(2.99), 279(2.92),289(2.85)	5-0216-69F
m-Toluic acid, p-tolyl ester	C_6H_{12}	233(4.21),272(3.47), 289(3.19)	5-0216-69F
o-Toluic acid, m-tolyl ester	C_6H_{12}	202s(--),212(4.35), 232(4.19),270s(--), 279s(--),289s(--)	5-0216-69F
o-Toluic acid, o-tolyl ester	C_6H_{12}	232(4.16),269(3.02), 280(2.99),288s(--)	5-0216-69F
o-Toluic acid, p-tolyl ester	C_6H_{12}	232(4.18),278(3.17), 288s(--)	5-0216-69F
p-Toluic acid, m-tolyl ester	C_6H_{12}	201(--),239(4.29), 270s(--),282s(--)	5-0216-69F
p-Toluic acid, o-tolyl ester	C_6H_{12}	239(4.29),269(3.37), 283(2.99)	5-0216-69F
p-Toluic acid, p-tolyl ester	C_6H_{12}	239(4.30),272s(--), 283s(--)	5-0216-69F
$C_{15}H_{14}O_2S$			
Thietane, 2,4-diphenyl-, 1,1-dioxide, cis	EtOH	226(4.31)	44-1136-69
$C_{15}H_{14}O_3$			
Benzophenone, 2,3'-dimethoxy-	EtOH	205s(4.41),218(4.51), 256(3.98),306(3.62)	39-0281-69C
Benzophenone, 4,4'-dimethoxy-	n.s.g.	290(4.25)	78-4919-69
Benzophenone, 2-hydroxy-4-methoxy-6-methyl-	EtOH	213(4.16),250(4.16), 284(3.6),350(2.99)	39-1721-69C
Benzophenone, 2-hydroxy-6-methoxy-4-methyl-	EtOH	214(4.14),227s(4.01), 249(4.07),289(3.73), 350(2.9)	39-1721-69C
Benzophenone, 4-hydroxy-2-methoxy-6-methyl-	EtOH	213(4.18),248(4.19), 284(3.50),315(3.29)	39-1721-69C

Compound	Solvent	$\lambda_{max}(\log \epsilon)$	Ref.
Cyclobuta[b]naphthalene-3,8-dione, 2a,8a-dihydro-2a-methoxy-1,2-dimethyl-	EtOH	227(4.37),301(3.18), 350(2.37)	44-0520-69
2,4-Cyclohexadien-1-one, 6-hydroxy-2,6-dimethyl-, benzoate	n.s.g.	231(4.26),275(3.48), 283(3.56),304(3.67)	77-0550-69
4α,4'-Flavandiol, (-)	MeOH	225(4.25),275(3.52), 281s(3.47)	78-5415-69
Frutescine	ether	282(3.48)	24-3298-69
Ketone, p-hydroxyphenethyl o-hydroxyphenyl	MeOH	252(4.04),278(3.37), 324(3.62)	78-5415-69
	MeOH-$AlCl_3$	273(--),312s(--), 382(--)	78-5415-69
Phenalen-1-one, 2,3-dihydro-6,8-dimethoxy-	EtOH	266(4.66),318(4.00)	88-0073-69
Psoralen, 4'-ethyl-4,8-dimethyl-	MeOH	220(3.80),249(4.50), 299(4.19)	42-1014-69
$C_{15}H_{14}O_3S$			
Benzene, [1,2-epoxy-2-(p-tolylsulfonyl)-ethyl]-	ether	229(4.41),255(2.94), 262(3.02),265(2.98), 267(2.97),273(2.86)	24-4017-69
$C_{15}H_{14}O_4$			
o-Anisic acid, 6-hydroxy-, benzyl ester	ether	208(3.59),255(4.02), 317(4.43)	24-1682-69
Benzophenone, 2-hydroxy-4-(2-hydroxyethoxy)-	EtOH	240(4.06),290(4.23), 320(4.04)	39-0034-69C
α-Caryopteron, dihydro-	ether	244(4.19),280(4.08), 391(3.59),406(3.59)	33-0808-69
β-Caryopteron, dihydro-	ether	252(4.30),278(3.83), 409(3.56)	33-0808-69
2,6-Methano-2H,7H-oxocino[3,2-c][1]benzopyran-7-one, 2-hydroxy-3,4,5,6-tetrahydro-	EtOH	284(3.99),306(4.06)	104-1566-69
$C_{15}H_{14}O_5$			
1,2-Naphthalenedicarboxylic acid, 6-methoxy-, dimethyl ester	EtOH	238(4.59),253(4.62), 307(3.96)	12-1721-69
$C_{15}H_{14}O_5S$			
3,4-Thiophenedicarboxylic acid, 2-(p-methoxyphenyl)-, dimethyl ester	MeOH	205(4.47),240(4.41), 275(4.38)	77-0569-69
$C_{15}H_{14}S_2$			
1-Thiocoumarin, 5,6,7,8-tetrahydro-3-phenyl-2-thio-	C_6H_{12}	233(4.36),324(3.85), 449(3.83)	78-1441-69
	EtOH	231(4.30),319(3.84), 452(3.87)	78-1441-69
$C_{15}H_{15}$			
α,2-Dimethyl-α-phenylbenzylium cation	acid	320(4.01),431(4.42)	59-0447-69
o-Ethyl-α-phenylbenzylium cation	acid	307(3.62),447(4.68)	59-0447-69
$C_{15}H_{15}BrO_5$			
2H,8H-Benzo[1,2-b:3,4-b']dipyran-2-one, 3-bromo-9,10-dihydro-9-hydroxy-10-methoxy-8,8-dimethyl-	EtOH	252(3.75),263(3.72), 338(4.30)	33-1165-69

Compound	Solvent	λ_{max}(log ϵ)	Ref.
$C_{15}H_{15}ClN_2O_2$			
Quinaldine, 5-chloro-4-hydroxy-3-(morpholinomethyl)-	EtOH	245(4.36),258s(4.25), 325(3.96),338s(3.90)	2-1186-69
$C_{15}H_{15}ClN_2O_3$			
Valeramide, N-(3-amino-1,4-dihydro-1,4-dioxo-2-naphthyl)-5-chloro-	MeOH	268(4.33),327(3.34), 452(3.38)	4-0909-69
$C_{15}H_{15}ClN_2O_4$			
Malondianil, perchlorate	EtOH	378(4.64)	24-2609-69
$C_{15}H_{15}Cl_2N_3O_9$			
Pyrido[1,2-a]pyrido[1',2':3,4]imidazo-[2,1-c]pyrazine-5,8-diium, 6,7-dihydro-6-methoxy-, perchlorate	H_2O	249(4.00),255(3.99), 315(3.56),330s(3.42), 419(4.34)	39-1987-69C
$C_{15}H_{15}FN_2O$			
2H-3,7-Methanoazonino[5,4-b]indol-2-one, 11-fluoro-1,4,5,6,7,8-hexahydro-	EtOH	223(4.43),281(3.94), 288s(3.97),297(3.89)	22-4154-69
$C_{15}H_{15}IN_2O$			
2-[2-(2-Furyl)vinyl]-1,3-dimethylbenzimidazolium iodide	MeOH	314(4.09)	103-0246-69
$C_{15}H_{15}IN_4O_3$			
Propionanilide, α,β-bis(hydroxyimino)-β-(2-pyridyl)-, methiodide, syn	EtOH	220(4.27),290(3.78)	36-0460-69
Propionanilide, α,β-bis(hydroxyimino)-β-(3-pyridyl)-, methiodide, syn	EtOH	220(4.64),262(4.39), 285s(4.24)	36-0460-69
Propionanilide, α,β-bis(hydroxyimino)-β-(4-pyridyl)-, methiodide, syn	EtOH	222(4.53),282(4.58)	36-0460-69
$C_{15}H_{15}KN_2O_6S_2$			
2,5-Cyclohexadien-1-one, 4-[[4-(dimethylamino)-2-mercaptophenyl]imino]-2-hydroxy-5-methoxy-, S-(hydrogen sulfate), potassium salt	EtOH	545(3.68)	5-0106-69G
$C_{15}H_{15}Li$			
Propane, 1,1-diphenyl-, 1-lithium deriv.	THF	497(4.49)	35-2456-69
$C_{15}H_{15}N$			
Acridan, 9,9-dimethyl-	hexane	280(4.19)	78-1125-69
Acridan, 9-ethyl-	hexane	279(4.20)	78-1125-69
Aniline, N-(p,α-dimethylbenzylidene)-	EtOH	254(4.26),308(3.36)	56-0749-69
Pyridine, 5-ethyl-3-styryl-, trans	H_2O	312(4.47)	23-2355-69
hydrochloride	H_2O	334(4.50)	23-2355-69
o-Tolualdehyde, N-benzylimine	EtOH	250(4.26)	35-2653-69
p-Toluidine, N-(α-methylbenzylidene)-	EtOH	244(4.28),316(3.40)	56-0749-69
$C_{15}H_{15}NO$			
Acetophenone, 4'-anilino-2'-methyl-	heptane	212(4.08),239(3.98), 317(4.42)	22-3523-69
	EtOH	204(4.51),244(4.04), 296s(3.85),337(4.46)	22-3523-69
Acetophenone, 2'-m-toluidino-	EtOH	228(4.32),248(4.14), 288(4.05),387(3.87)	22-3523-69
Acetophenone, 2'-o-toluidino-	C_6H_{12}	208(4.32),228(4.38), 244s(4.08),283(3.92), 296s(3.80),379(3.85)	22-3523-69

Compound	Solvent	λ max (log ε)	Ref.
Acetophenone, 2'-o-toluidino- (cont.)	EtOH	229(4.34),245s(4.09), 283(3.89),385(3.86)	22-3523-69
Acetophenone, 2'-p-toluidino-	EtOH	205(4.32),228(4.27), 248(4.10),286(4.05), 384(3.82)	22-3523-69
Acetophenone, 4'-o-toluidino-	C_6H_{12}	202(4.40),226(4.03), 309(4.35)	22-3523-69
	EtOH	203(4.45),235(3.92), 295s(3.82),334(4.38)	22-3523-69
Acetophenone, 4'-p-toluidino-	EtOH	244(3.97),295s(3.67), 348(4.44)	22-3523-69
Aniline, N-(p-methoxy-α-methylbenzylidene)-	EtOH	270(4.32),310s(3.60)	56-0749-69
p-Anisidine, N-(α-methylbenzylidene)-	EtOH	242(4.32),326(3.45)	56-0749-69
Fulvene, 6-hydroxy-6-methyl-2-(1-N-phenyliminoethyl)-	EtOH	266(4.40),345(4.07), 415(4.19)	39-2464-69C
Nitrone, N-phenyl-α-(2,4-xylyl)-	C_6H_{12}	329(4.27)	18-3306-69
	EtOH	325(4.32)	18-3306-69
Propiophenone, 2'-anilino-	EtOH	229(4.21),251(4.04), 289(4.12),389(3.73)	22-3523-69
Propiophenone, 3'-anilino-	EtOH	239(4.22),256(4.21), 292(4.27),365(3.28)	22-3523-69
Propiophenone, 4'-anilino-	EtOH	242(3.91),300s(3.74), 345(4.42)	22-3523-69
$C_{15}H_{15}NOS$			
2-Propen-1-one, 3-[p-(dimethylamino)-phenyl]-1-(2-thienyl)-	EtOH	275(3.93),425(4.09)	2-0311-69
Quinoline, 1,2-dihydro-7-methoxy-1-methyl-4-(2-thienyl)-	EtOH	236(4.38),342(3.55)	2-0419-69
$C_{15}H_{15}NO_2$			
Acetophenone, 2'-o-anisidino-	C_6H_{12}	210(4.35),230(4.34), 244s(4.21),283(3.97), 305(4.00),384(3.88)	22-3523-69
	EtOH	212(4.34),229(4.30), 248(4.21),283(3.96), 305(3.95),395(3.83)	22-3523-69
Acetophenone, 2'-p-anisidino-	C_6H_{12}	214(4.31),229(4.32), 248(4.11),283(4.02), 381(3.85)	22-3523-69
	EtOH	228(4.38),246(4.25), 282(4.12),385(3.95)	22-3523-69
Acetophenone, 4'-p-anisidino-	EtOH	243(3.98),290s(3.66), 347(4.40)	22-3523-69
Anthranilic acid, N-2,6-xylyl-	C_6H_{12}	196(4.62),223(4.51), 257(3.91),354(3.86)	22-3523-69
	EtOH	203(4.53),215(4.58), 252(3.91),273(3.85), 336(3.76)	22-3523-69
Benzyl alcohol, α-[(benzylideneamino)-methyl]-m-hydroxy-	EtOH	249(4.38)	94-2353-69
Furo[2,3-b]quinolin-4(9H)-one, 3-isopropyl-9-methyl-	MeOH	240(4.23),252(4.20), 260(4.23),285(3.23), 295(3.23),328s(3.67), 339(3.79),353s(3.71)	44-2183-69
	MeOH-acid	243(4.57),257s(3.88), 305(3.47),318(3.69), 333s(3.74),343(3.82), 355s(3.77)	44-2183-69

Compound	Solvent	λmax(log ε)	Ref.
Furo[2,3-b]quinolin-4(9H)-one, 3-isopropyl-9-methyl- (cont.)	MeOH-base	240(4.23),252(4.20), 260(4.23),285(3.23), 295(3.23),328s(3.67), 339(3.79),353s(3.74)	44-2183-69
Isoquinoline-4,6-diol, 1,2,3,4-tetrahydro-1-phenyl-	EtOH	285(3.37)	94-2353-69
1-Naphthalenepropionitrile, 4,6-dimethoxy-	EtOH	248(4.50),285(3.79)	88-0073-69
Nitrone, α-(4-methoxy-o-tolyl)-N-phenyl-	EtOH	338(4.36)	18-3306-69
Pyridine, 3,5-diacetyl-1,4-dihydro-4-phenyl-	EtOH	230(4.14),258s(3.84), 381(3.93)	73-3336-69
p-Toluamide, N-(benzyloxy)-	MeOH	208(4.4),234(4.20)	12-0161-69
	MeOH-NaOH	227(4.20),263(3.91)	12-0161-69
$C_{15}H_{15}NO_2S$			
3-Thiophenecarboxaldehyde, 3-morpholino-5-phenyl-	EtOH	255(4.3),302(4.1), 360s(3.7)	48-0827-69
$C_{15}H_{15}NO_2S_2$			
2H-Thiopyran-5-carboxylic acid, 6-amino-4-methyl-3-phenyl-2-thioxo-, ethyl ester	EtOH	232(4.35),326(4.08), 461(4.30)	78-1441-69
$C_{15}H_{15}NO_3$			
Acetohydroxamic acid, 2-phenoxy-N-p-tolyl-	EtOH	250(4.20)	42-0831-69
	EtOH	232(3.94),240(--)	42-0831-69
p-Anisamide, N-(benzyloxy)-	MeOH	208(4.4),254(4.23)	12-0161-69
	MeOH-NaOH	235(4.08),254(4.05)	12-0161-69
1-(α-Carboxyphenethyl)-5-hydroxy-2-picolinium hydroxide, inner salt, L-(-)-	aq. HCl	302(3.80)	1-2475-69
	aq. NaOH	333(3.75)	1-2475-69
Nicotinic acid, 1,4-dihydro-1,2,5-trimethyl-4-oxo-6-phenyl-	EtOH	220(4.62),261(4.02)	94-2209-69
Pyrido[1,2-a]indole-10-carboxylic acid, 1,2,3,4-tetrahydro-4-oxo-, ethyl ester	EtOH	235(4.53),250s(4.36), 258(4.48),270(4.49), 276(4.60),310s(4.25), 325(4.33),340(4.38), 350(4.37)	12-1525-69
$C_{15}H_{15}NO_3S$			
6,11-Epoxy-3H-naphtho[2,3-d]azepine, 5a,6,11,11a-tetrahydro-3-(methylsulfonyl)-, endo	EtOH	235(3.49)	44-2888-69
$C_{15}H_{15}NO_4$			
Acetanilide, N-(4-formyl-2-hydroxy-1,3-butadienyl)-, acetate	EtOH	325(4.34)	78-2035-69
Acetanilide, N-(4-formyl-4-hydroxy-1,3-butadienyl)-, acetate	EtOH	320(4.42)	78-2035-69
$C_{15}H_{15}N_3$			
Benzimidazole, 5-amino-1-methyl-2-p-tolyl-, dihydrochloride	EtOH	207(4.62),229s(4.39), 257s(4.08),320(3.89)	73-0572-69
Benzimidazole, 5-amino-2-methyl-1-p-tolyl-	EtOH	212(4.58),252s(4.09), 305(3.72)	73-0572-69
Benzimidazole, 2-(1-amino-2-phenylethyl)-, dihydrochloride	EtOH	275(4.13),282(4.11)	94-2381-69
$C_{15}H_{15}N_3O$			
Acetophenone, 4'-[[p-(methylamino)phenyl]azo]-	EtOH	279(4.07),438(4.51)	18-3565-69

Compound	Solvent	λ max (log ε)	Ref.
Acetophenone, 4'-(3-methyl-3-phenyl-1-triazeno)-	EtOH	241(4.21),365(4.45)	18-3565-69
Benzimidazole, 5-amino-1-(p-methoxyphenyl)-2-methyl-	EtOH	209(4.78),252s(4.22), 306(3.73)	73-0572-69
Benzimidazole, 5-amino-2-(p-methoxy henyl)-1-methyl-	EtOH	211(4.64),241(4.30), 270(4.21),320(4.43)	73-0572-69
Isoxazole, 5-(3,5-dimethyl-1-pyrazolyl)-4-methyl-3-phenyl-	MeOH	234(4.17),240s(4.1)	4-0783-69
Nicotinic acid, 6-methyl-, (α-methylbenzylidene)hydrazide	n.s.g.	219(4.32),292(4.36)	40-0073-69
$C_{15}H_{15}N_3O_2$			
Isonicotinic acid, (m-methoxy-α-methylbenzylidene)hydrazide	n.s.g.	215(4.53),290(4.21)	40-0073-69
Isonicotinic acid, (o-methoxy-α-methylbenzylidene)hydrazide	n.s.g.	275(4.08)	40-0073-69
Isonicotinic acid, (p-methoxy-α-methylbenzylidene)hydrazide	n.s.g.	218(4.31),310(4.27)	40-0073-69
Nicotinic acid, (m-methoxy-α-methylbenzylidene)hydrazide	n.s.g.	214(4.52),294(4.26)	40-0073-69
Nicotinic acid, (o-methoxy-α-methylbenzylidene)hydrazide	n.s.g.	272(4.11)	40-0073-69
Nicotinic acid, (p-methoxy-α-methylbenzylidene)hydrazide	n.s.g.	220(4.33),305(4.34)	40-0073-69
Picolinic acid, (m-methoxy-α-methylbenzylidene)hydrazide	n.s.g.	219(4.37),306(4.27)	40-0073-69
Picolinic acid, (o-methoxy-α-methylbenzylidene)hydrazide	n.s.g.	280(4.13)	40-0073-69
Picolinic acid, (p-methoxy-α-methylbenzylidene)hydrazide	n.s.g.	225(4.33),317(4.36)	40-0073-69
$C_{15}H_{15}N_3O_2S$			
Benzenesulfonanilide, 4-cyano-4'-(dimethylamino)-	EtOH-1% $CHCl_3$	240(4.31),261(4.32), 305s(3.53)	44-2083-69
Benzonitrile, p-[[2-amino-5-(dimethylamino)phenyl]sulfonyl]-	EtOH-1% $CHCl_3$	248(4.39),275s(4.10)	44-2083-69
$C_{15}H_{15}N_3O_4$			
Benzaldehyde, α-[(p-nitrophenyl)azo]-, dimethyl acetal	dioxan	286(4.24),442(2.20)	39-2587-69C
Pyridine, 2-(2,4-dinitrobenzyl)-4-isopropyl-	EtOH	243(4.54)	22-4425-69
Pyridine, 2-(2,4-dinitrobenzyl)-4-propyl-	EtOH	247(4.54)	22-4425-69
4H-Quinolizine-1-carboxylic acid, 3-cyano-2-[(2-hydroxyethyl)amino]-4-oxo-	EtOH	264(4.65),302(4.19), 394(4.14)	95-0203-69
as-Triazine-5,6-dicarboxylic acid, 3-phenyl-, diethyl ester	MeOH	275(2.02),394(2.56)	88-3147-69
$C_{15}H_{15}N_3O_7$			
as-Triazine-3,5(2H,4H)-dione, 2-lyxofuranosyl-, 5'-benzoate	EtOH	230(4.21),266(3.85)	73-0618-69
as-Triazine-3,5(2H,4H)-dione, 2-ribofuranosyl-, 5'-benzoate	EtOH	230(4.18),267(3.83)	73-0618-69
$C_{15}H_{15}N_5O_2$			
Formazan, 3,5-dimethyl-5-(p-nitrophenyl)-1-phenyl-	dioxan	230(4.18),298(4.08), 358(4.03),435(4.36)	24-3082-69

Compound	Solvent	λmax(log ε)	Ref.
$C_{15}H_{15}O_3PS$			
Phenothiaphosphine, 2,8,10-trimethyl-, 5,5,10-trioxide	EtOH	229(4.31),254s(3.52), 268s(3.36),277(3.46), 285(3.55)	78-3919-69
$C_{15}H_{16}$			
Methane, di-m-tolyl-	n.s.g.	263(2.83)	78-4919-69
Methane, di-p-tolyl-	n.s.g.	266(2.82)	78-4919-69
Propane, 1,1-diphenyl-, 1-lithium deriv.	THF	497(4.49)	35-2456-69
$C_{15}H_{16}BrNO_3$			
Pyrido[1,2-a]indole-10-carboxylic acid, 8-bromo-6,7,8,9-tetrahydro-9-methoxy-	EtOH	216(4.49),231(4.39), 292(4.11)	12-0997-69
$C_{15}H_{16}ClN$			
2,9-Metheno-9H-benzazacycloundecine, 14-chloro-3,4,5,6,7,8-hexahydro-	EtOH	241(4.47),293(3.61), 320s(3.26),336(3.22)	88-0557-69
$C_{15}H_{16}ClNO_4$			
5H-1-Pyrindinium, 6,7-dihydro-1-methyl-2-phenyl-, perchlorate	EtOH	295(4.10)	94-2209-69
$C_{15}H_{16}ClNO_5$			
2-Phenyl-6-(1-pyrrolidinyl)pyrylium perchlorate	MeOH	252(4.15),368(4.16)	5-0073-69E
5,6,7,8-Tetrahydro-8-methyl-2-phenyl-pyrano[2,3-b]pyridin-1-ium perchlorate	MeOH	255(4.16),362(4.19)	5-0073-69E
$C_{15}H_{16}F_3N_5O_3$			
6a,7,10,10a,11,12-Hexahydro-6a-methoxy-3H-quinazolino[2,1-i]purin-6-ium trifluoroacetate	pH 1	264(4.12)	44-3814-69
	pH 13	273(4.17),281(4.16)	44-3814-69
	EtOH	228(3.76),266(4.08)	44-3814-69
$C_{15}H_{16}Fe$			
Ferrocene, 1,1'-trimethylene-2-vinyl-	EtOH	335(2.65),430(2.48)	49-1540-69
Ferrocene, 1,1'-trimethylene-3-vinyl-	EtOH	443(2.56)	49-1540-69
$C_{15}H_{16}FeO$			
3-Buten-2-one, 4-ferrocenyl-3-methyl-	EtOH	218(3.81),256(3.70), 308(3.76),357(3.11), 479(2.94)	18-3273-69
Ketone, methyl 1',2-trimethyleneferrocenyl	EtOH	334(3.22),445(2.95)	49-1540-69
	EtOH	226(4.18),265(3.80), 337(3.00),440(2.73)	78-5245-69
Ketone, methyl 1,3'-trimethyleneferrocenyl	EtOH	338(3.20),450(2.91)	49-1540-69
	EtOH	230(4.15),267(3.84), 338(3.25),456(2.94)	78-5245-69
3-Pentanone, 1,5-(1,1'-ferrocenediyl)-	n.s.g.	308s(2.81),450(2.10)	78-0861-69
1-Penten-3-one, 1-ferrocenyl-	EtOH	211(4.42),258(3.98), 305(4.39),371(3.29), 486(3.20)	18-3273-69
$C_{15}H_{16}FeO_2$			
Ferroceneacrylic acid, ethyl ester	n.s.g.	250(4.3),293(4.3), 356(3.2),460(2.9)	65-1008-69
Ferrocenecarboxylic acid, 1',2-trimethylene-, methyl ester	EtOH	296(2.96),324(2.57), 428(2.70)	49-1540-69
Ferrocenecarboxylic acid, 1',3-trimethylene-, methyl ester	EtOH	303(3.08),336(2.76), 442(2.65)	49-1540-69

Compound	Solvent	λ_{max}(log ε)	Ref.
$C_{15}H_{16}IN_3O_2$			
Propionanilide, β-(hydroxyimino)-β-(2-pyridyl)-, methiodide, anti	EtOH	222(4.37)	36-0460-69
Propionanilide, β-(hydroxyimino)-β-(3-pyridyl)-, methiodide, anti	EtOH	225(4.37),235(4.38)	36-0460-69
Propionanilide, β-(hydroxyimino)-β-(4-pyridyl)-, methiodide, anti	EtOH	222(4.33),237s(4.28), 285(4.07)	36-0460-69
$C_{15}H_{16}N_2$			
Acetophenone, methylphenylhydrazone	EtOH	250(4.32),288(3.42), 347(3.59)	39-1703-69C
	EtOH-HCl	244(4.16),278(3.30), 358(2.30)	39-1703-69C
Benzophenone, dimethylhydrazone	EtOH	239(4.12),320(3.58)	39-1703-69C
Dibenzo[b,f][1,4]diazocine, 5,6,11,12-tetrahydro-5-methyl-	EtOH	229(4.46),269(3.83), 308(3.66)	39-0882-69C
2-Propanone, 1-phenyl-, phenylhydrazone	EtOH	207(4.3),276(4.4)	39-0446-69B
	EtOH-H_2SO_4	206(4.3),226(4.4), 295(4.0)	39-0446-69B
Pyridine, 2-[p-(dimethylamino)styryl]-, cis	H_2O	323(4.04)	23-2355-69
hydrochloride	H_2O	410(3.90)	23-2355-69
Pyridine, 2-[p-(dimethylamino)styryl]-, trans	H_2O	348(4.42)	23-2355-69
hydrochloride	H_2O	425(4.25)	23-2355-69
Pyridine, 3-[p-(dimethylamino)styryl]-, trans	50% MeOH	369(4.44)	23-2355-69
hydrochloride	50% MeOH	462(4.53)	23-2355-69
$C_{15}H_{16}N_2O$			
o-Anisaldehyde, methylphenylhydrazone	EtOH	204(4.4),238(4.1), 252s(4.0),306(4.0), 345(4.4)	28-0137-69B
Dasycarpidone, deethyl- (spectrum unchanged in acid)	EtOH	237(4.02),314(4.21)	39-2738-69C
Ketone, indol-2-yl 1,2,5,6-tetrahydro-1-methyl-4-pyridyl	EtOH	241(4.00),319(4.23)	39-2738-69C
2H-3,7-Methanoazonino[5,4-b]indol-2-one, 1,4,5,6,7,8-hexahydro-	EtOH	223(4.56),284(3.90), 291(3.86)	22-4154-69
1,5-Methano[1,3]diazocino[1,8-a]indol-6(1H)-one, 2,3,4,5-tetrahydro-2-methyl-	EtOH	237(4.03),315(4.17)	39-2738-69C
	EtOH-HCl	233(3.94),310(4.16)	39-2738-69C
Nitrone, α-[p-(dimethylamino)phenyl]-N-phenyl-	C_6H_{12}	372(4.38)	18-3306-69
	EtOH	385(4.51)	18-3306-69
$C_{15}H_{16}N_2OS$			
4H-Cyclopenta[d]pyrimidine-4-thione, 1,5,6,7-tetrahydro-1-(2-hydroxyethyl)-2-phenyl-	EtOH	245(3.99),342(4.39)	4-0037-69
$C_{15}H_{16}N_2OS_2$			
2H-1,3,5-Thiadiazine-2-thione, 5-benzyl-3-furfuryltetrahydro-	MeOH	248(3.96),290(4.11)	73-2952-69
$C_{15}H_{16}N_2O_2$			
Benzaldehyde, α-(phenylazo)-, dimethyl acetal	dioxan	276(3.98),424(2.08)	39-2587-69C
5,8-Isoquinolinedione, 3-methyl-7-piperidino-	MeCN	245(4.16),279(4.07)	4-0697-69

Compound	Solvent	$\lambda_{max}(\log \epsilon)$	Ref.
2,5-Piperazinedione, 3-benzylidene-6-isobutylidene-	DMF	321(4.4)	18-0191-69
$\Delta^{2,\alpha}$-Pyrrolidineacetic acid, 1-benzyl-α-cyano-, methyl ester	isooctane	286(4.38)	35-6689-69
$C_{15}H_{16}N_2O_2S$			
2H-1,2,4-Benzothiadiazine, 3-ethyl-3,4-dihydro-2-phenyl-, 1,1-dioxide	EtOH	210(4.54),252(4.08), 314(3.53)	7-0590-69
$C_{15}H_{16}N_2O_3S$			
Acetamide, N-(3-methyl-8-oxo-5-thia-1-azabicyclo[4.2.0]oct-2-en-7-yl)-2-phenoxy-	n.s.g.	256(3.96)	35-1401-69
$C_{15}H_{16}N_2O_4$			
4-Penten-1-ol, 5-indol-3-yl-4-nitro-, acetate	EtOH	396(4.23)	87-0157-69
1-Phthalazinemalonic acid, diethyl ester	EtOH	210(4.76),255s(3.60), 267(3.68),275s(3.65), 288s(3.48),307(3.03), 372(2.83)	95-0959-69
$C_{15}H_{16}N_2O_5S_2$			
2-Thiopheneacetamide, N-[3-(hydroxymethyl)-8-oxo-5-thia-1-azabicyclo[4.2.0]-oct-2-en-7-yl]-, acetate, 5-oxide	MeOH	240(4.18)	88-3993-69
$C_{15}H_{16}N_2S$			
Benzothiazoline, 2-[p-(dimethylamino)-phenyl]-, cobalt derivative	n.s.g.	410(4.66),450s(4.38)	97-0385-69
copper derivative	n.s.g.	370(4.85)	97-0385-69
nickel derivative	n.s.g.	384(4.46),480s(4.12)	97-0385-69
zinc derivative	n.s.g.	390(4.72),436(4.50)	97-0385-69
$C_{15}H_{16}N_4$			
Formazan, 5-methyl-1-phenyl-5-p-tolyl-	dioxan	259(4.21),295s(3.90), 412(4.48)	24-3082-69
Formazan, 5-methyl-5-phenyl-1-p-tolyl-	dioxan	259(4.20),295(3.85), 408(4.52)	24-3082-69
$C_{15}H_{16}N_4O_2$			
Isoalloxazine, 7,8-dimethyl-10-propyl-	pH 7	266(4.58),368(4.14), 445(4.22)	44-3240-69
	MeOH-HOAc	267(4.54),350(3.92), 443(4.09)	44-3240-69
$C_{15}H_{16}N_4O_4$			
Cyclopropa[c]pentalen-2(3H)-one, hexahydro-, 2,4-dinitrophenylhydrazone	$CHCl_3$	370(4.43)	78-5287-69
1,3-Propanediamine, N,N-bis(o-nitrophenyl)-	EtOH	232(4.63),280(3.96), 425(4.09)	44-3240-69
	EtOH-HCl	232(4.66),280(4.06), 425(4.20)	44-3240-69
	EtOH-NaOH	235s(4.67),280(4.03), 425(4.10)	44-3240-69
$C_{15}H_{16}N_4O_5$			
Allethrolone, 2,4-dinitrophenylhydrazone	EtOH	256(4.21),287(3.94), 380(4.42)	78-1117-69

Compound	Solvent	λ_{max}(log ε)	Ref.
$C_{15}H_{16}N_4O_8$			
Pyridine, 2-ethoxymethyl-5-methyl-, picrate	n.s.g.	214(3.56),266(3.46)	39-2255-69C
$C_{15}H_{16}N_6$			
3H-v-Triazolo[4,5-d]pyrimidine, 3-phenyl-7-piperidino-	MeOH	246(4.35),309(4.28)	23-1129-69
$C_{15}H_{16}N_6S_2$			
5H-Imidazo[4,5-d]thiazole-5-thione, 4,6-diallyl-2-(4-amino-2-methylpyrimidin 5-yl)-4,6-dihydro-	EtOH	225(4.09),287(4.08), 399(4.25)	94-0910-69
$C_{15}H_{16}O$			
3,5-Nonadiynal, 2-(2-hexynylidene)-	ether	230(3.81),242(3.88), 254(3.90),263s(3.88), 273(3.81),325(4.32)	39-0683-69C
$C_{15}H_{16}OS$			
8H-Cyclopenta[5,6]naphtho[2,1-b]thiophene, 5,5a,6,7,8a,8b,9,10-octahydro-5-oxo-, m. 126°	n.s.g.	307(4.27)	28-0194-69A
isomer m. 155°	n.s.g.	310(4.28)	28-0194-69A
isomer m. 169°	n.s.g.	304(4.29)	28-0194-69A
$C_{15}H_{16}O_2$			
Cyclopent[a]indene-2-carboxylic acid, 1,3a,8,8a-tetrahydro-3,5-dimethyl-	EtOH	274(3.33),280(3.42)	2-0215-69
1,6-Etheno-2,5-methano-7H-cyclobut[e]-inden-7-one, 1,2,2a,4a,5,6,7a,7b-octahydro-7a-methoxy-	MeOH	308(2.60)	88-0691-69
Lindenone	EtOH	224(4.21),272(3.89), 303(3.80)	39-2786-69C
Methane, bis(p-methoxyphenyl)-	n.s.g.	279(3.24)	78-4919-69
Naphtho[2,3-b]furan-9-ol, 5,6-dihydro-3,4,5-trimethyl-	EtOH	228(4.18),228s(4.05), 285(4.25),315(3.76), 330(3.60)[sic]	16-0092-69
Pentacyclo[7.5.0.0^{2,7}.0^{5,13}.0^{6,12}]tetradeca-3,10-dien-8-one, 7-methoxy-	MeOH	310(2.25)	88-0691-69
2-Phenanthrenecarboxylic acid, 1,2,3,4,9,10-hexahydro-	EtOH	267(4.01)	77-1253-69
4-Phenanthrol, 1,2,3,4-tetrahydro-7-methoxy-, (-)-	MeOH	232(4.85),282(3.56), 322(3.20),336(3.40)	19-0475-69
(+)-	MeOH	232(4.94),282(3.64), 322(3.38),336(3.48)	19-0475-69
Spiro[1,3-dioxolane-2,9'-fluorene], 1',2',3',4'-tetrahydro-	EtOH	219(4.49),225(4.45), 277(3.68)	44-1899-69
$C_{15}H_{16}O_3$			
Acetophenone, 4'-[(2-hydroxy-6-oxo-1-cyclohexen-1-yl)methyl]-	EtOH	205(4.59),260(4.73)	44-2192-69
Cacalone	MeOH	207(3.75),249(3.89), 321(3.72)	23-2465-69
1-Cyclohexene-1-carboxylic acid, 6-(2,4-hexadiynylidene)-2-methoxy-, methyl ester, cis	ether	239s(3.95),248(3.98), 314(4.43)	24-3298-69
trans	ether	241(3.99),314(4.53)	24-3298-69
isomeric mixture	ether	224(4.14),235(4.14), 260s(3.90),293s(4.39), 304(4.24)	24-3298-69

Compound	Solvent	$\lambda_{max}(\log \epsilon)$	Ref.
2-Cyclohexene-1-carboxylic acid, 6-(2,4-hexadiynylidene)-2-methoxy-, methyl ester	ether	211(4.55),217(4.26), 245(3.85),258(4.05), 272(4.36),289(4.29)	24-3298-69
2-Cyclohexen-1-one, 2-benzyl-3-hydroxy-, acetate	EtOH	207(3.98),217s(3.94), 238(3.87)	44-2192-69
Fluorene-4a(2H)-acetic acid, 1,3,4,9a-tetrahydro-9-oxo-	EtOH	245(4.06),291(3.36)	44-1899-69
Fluorene-2-carboxylic acid, 1,2,3,4-tetrahydro-7-methoxy-	EtOH	266(4.31)	77-1253-69
Fluorene-9-carboxylic acid, 1,2,3,4-tetrahydro-7-methoxy-	EtOH	268(4.16),276s(4.12)	44-2209-69
1H-Fluorene-8-carboxylic acid, 2,3,4,4a-tetrahydro-7-methoxy-	EtOH	232(4.34),264(4.06), 318(3.62)	44-2209-69
4aH-Fluorene-4a-carboxylic acid, 1,2,3,4-tetrahydro-7-methoxy-	EtOH	230(4.52),265(3.78), 294(3.41),305(3.37)	44-2209-69
Neolinderalactone	EtOH	212(4.16)	39-2783-69C
2,4-Pentadienoic acid, 2-acetyl-5-phenyl-, ethyl ester	EtOH	229s(--),235(3.91), 331(4.49)	23-4076-69
Santonene	MeOH	290(4.30)	39-1088-69C
$C_{15}H_{16}O_4$			
2'-Acetonaphthone, 1'-hydroxy-6',8'-dimethoxy-3'-methyl-	EtOH	233(4.53),260(4.30), 338(3.75)	94-0454-69
2'-Acetonaphthone, 8'-hydroxy-1',6'-dimethoxy-3'-methyl-	EtOH	231(4.51),242(4.51), 337(3.59)	94-0454-69
Cyclopentanecarboxylic acid, 4-benzoyl-2-methyl-3-oxo-, methyl ester	EtOH	245(3.82),320(3.98)	39-1845-69C
Fluorene-1-carboxylic acid, 4b,5,6,7,-8,8aβ-hexahydro-2-methoxy-9-oxo-	EtOH	218(4.37),248(3.81), 323(3.61)	44-2209-69
Indene-3-propionic acid, 2-(carboxymethyl)-7-methyl-	EtOH	262(4.2)	42-0415-69
Isabelin	EtOH	211(4.28)	78-4767-69
Isoisabelin	EtOH	207(4.13)	78-4767-69
Isopeucenin	MeOH	229(4.42),252(4.46), 299(4.19)	2-1072-69
1-Naphthalenepropionic acid, 4,6-dimethoxy-	EtOH	250(4.46),285(3.69)	88-0073-69
1-Naphthoic acid, 6,7-dimethoxy-3-methyl-, methyl ester	EtOH	230(4.61),315(3.68), 338(3.77)	12-1721-69
Perezinone, oxo-	MeOH	204(3.47),225(3.48), 268(2.87),325(3.69), 343(4.05),361(4.14)	23-2465-69
Peucenin	MeOH	231(4.33),259(4.26), 299(4.00)	2-1072-69
Santonene, 4α-hydroxy-	MeOH	219(3.98),289(4.32)	39-1088-69C
	EtOH	219(3.98),288(4.32)	39-1619-69C
Santonene, 4β-hydroxy-	EtOH	219(3.94),287(4.34), 332s(3.13)	39-1619-69C
Santonene, 6β-hydroxy-	EtOH	241(4.11)	39-1619-69C
$C_{15}H_{16}O_5$			
6-Benzofuranacrylic acid, 7-acetyl-2,4,5,6-tetrahydro-3,6-dimethyl-2-oxo-	EtOH	216(3.82),303(4.31)	39-1619-69C
2H-Furo[2,3-h]-1-benzopyran-2-one, 8,9-dihydro-6-hydroxy-5-methoxy-8,9,9-trimethyl- (nieshoutol)	EtOH	230(4.18),251(3.64), 258(3.55),340(4.08)	88-4031-69
	EtOH-NaOH	255(4.26),340(3.98), 400(3.79)	88-4031-69
Illudalic acid	H_2O	247(4.46),270s(4.08), 332(2.88)	44-0240-69
Pentalenolactone	MeOH	219(3.94)	88-2737-69

Compound	Solvent	λ_{max}(log ϵ)	Ref.
Vernolepin	MeOH	207(4.31)(end abs.)	44-3903-69
Vernomenin	MeOH	210(4.30)(end abs.)	44-3903-69
$C_{15}H_{16}O_6$			
Elephantol	MeOH	209(4.26)	44-3867-69
Malonic acid, diester with 2-hydroxy-2-cyclohexen-1-one	EtOH	230(4.27),304(2.57)	18-3233-69
$C_{15}H_{16}O_7$			
4,5,7-Phthalantriol, 6-methyl-, triacetate	MeOH	263(2.86),270(2.83)	88-4675-69
$C_{15}H_{16}S$			
Thiophene, 2-(p-isopropylstyryl)-	MeOH	231(4.06),240s(3.95), 328(4.52)	117-0209-69
$C_{15}H_{17}BrO_2$			
2-Cyclopenten-1-one, 4-(5-bromo-2-methoxy-p-tolyl)-4,5-dimethyl-	MeOH	284(3.41),289(3.40)	78-3509-69
$C_{15}H_{17}ClN_2$			
Pyridine, 2-[p-(dimethylamino)styryl]-, cis, hydrochloride (see $C_{15}H_{16}N_2$)	H_2O	410(3.90)	23-2355-69
trans, hydrochloride	H_2O	425(4.25)	23-2355-69
$C_{15}H_{17}ClN_2O$			
4-Quinolinol, 5-chloro-2-methyl-3-(1-pyrrolidinylmethyl)-	EtOH	242(4.34),300(3.73), 328(3.96),342(3.86)	2-1186-69
$C_{15}H_{17}ClN_2O_5$			
4,9-Dihydro-2-(3-oxobutyl)-3H-pyrido-[3,4-b]indolium perchlorate	H_2O	203(4.34),247(4.09), 357(4.25)	24-3959-69
$C_{15}H_{17}Cl_2N_3$			
Benzimidazole, 5-amino-1-methyl-2-p-tolyl-, dihydrochloride (see $C_{15}H_{15}N_3$)	EtOH	207(4.62),229s(4.39), 257s(4.08),320(3.89)	73-0572-69
$C_{15}H_{17}FeNO_4S$			
Proline, N-(ferrocenylsulfonyl)-	EtOH	434(2.21)	49-1552-69
$C_{15}H_{17}N$			
Diphenylamine, 2,4,6-trimethyl-	EtOH	245(4.07),270(3.93)	22-2355-69
$C_{15}H_{17}NO$			
Carbazol-4(1H)-one, 2,3-dihydro-2,2,9-trimethyl-	MeOH	214(4.43),245(4.25), 266(4.06),302(4.10)	27-0036-69
Fulvene, 6-(dimethylamino)-6-(p-methoxyphenyl)-	EtOH	241(4.00),323(4.23)	39-0111-69B
2,4,6-Heptatrienamide, N,N-dimethyl-7-phenyl-, 2,4-di-trans	EtOH	243(4.05),340(4.79)	39-0016-69C
1-Naphthaleneacetamide, N,N,2-trimethyl-	EtOH	228(4.92),275(3.73), 284(3.76),292(3.63), 307(2.90),313s(2.68), 321(2.82)	33-1030-69
2-Naphthalenepropionamide, N,N-dimethyl-	EtOH	226(5.05),267(3.67), 276(3.70),285(3.53), 304(2.64),318(2.52)	33-1030-69
Phenol, p-(4-methyl-6-propyl-3-pyridyl)-	EtOH	254(3.50),278s(--)	77-0709-69
	EtOH-base	240(4.43),278(3.47)	77-0709-69

Compound	Solvent	λ_{max}(log ε)	Ref.
$C_{15}H_{17}NO_2$			
Carbostyril, 3-(1,2-dimethylallyl)-4-hydroxy-1-methyl-	EtOH	229(4.51),242s(4.06), 275(3.67),286(3.70), 320(3.81),334(3.63)	2-0678-69
Carbostyril, 4-methoxy-3-(3-methyl-2-butenyl)- (atanine)	EtOH	224(4.57),228(4.56), 270(3.77),278(3.82), 311(3.76),321(3.90), 338(3.71)	2-0678-69
Carbostyril, 1-methyl-4-[(3-methyl-2-butenyl)oxy]- (ravenine)	EtOH	224(4.69),229(4.72), 268(3.85),278(3.80), 316(3.79),330(3.67)	2-0678-69
Furo[2,3-b]quinoline, 2,3-dihydro-2-isopropyl-4-methoxy-	MeOH	229(4.56),232s(4.44), 252(3.74),262(3.75), 272(3.78),283(3.70), 309(3.48),323(3.51)	44-2183-69
	MeOH-acid	216(4.44),234(4.47), 239s(4.45),293(3.93), 304(3.88),317(3.76)	44-2183-69
	MeOH-base	229(4.56),232s(4.44), 252(3.74),262(3.75), 272(3.78),283(3.70), 309(3.48),323(3.51)	44-2183-69
2-Isoxazolin-5-one, 4-methyl-4-(3-methyl-2-butenyl)-3-phenyl-	EtOH	256(4.06)	88-0543-69
3-Isoxazolin-5-one, 4-methyl-2-(3-methyl-2-butenyl)-3-phenyl-	EtOH	245(3.99),284(3.97)	88-0543-69
Morpholine, 4-(5-phenyl-2,4-pentadienoyl)-, 2,4-di-trans	EtOH	232(3.96),312(4.59)	39-0016-69C
Pyrido[1,2-a]indole-10-carboxylic acid, 1,2,3,4-tetrahydro-, ethyl ester	EtOH	234(4.59),255(4.07), 261(4.12),271(4.12), 303s(4.07),315(4.18), 340(4.08),352(4.09)	12-1525-69
Spectabiline	EtOH	230(4.30),249s(4.15), 308(4.04),317(4.02)	88-4789-69
$C_{15}H_{17}NO_3$			
Carbostyril, 4-methoxy-3-(3-methyl-2-oxobutyl)-	MeOH and MeOH-acid	231(4.28),245s(3.80), 264(3.59),271(3.70), 279(3.65),314(3.44), 324(3.61),336(3.55)	44-2183-69
Cyclopenta[b]pyrrol-2(1H)-one, 1-acetoxy-3,3aβ,4,5,6,6aβ-hexahydro-3α-phenyl-	EtOH	247(2.17),252(2.24), 258(2.31),264(2.17)	44-2058-69
Furo[2,3-b]quinolin-3-ol, 2,3-dihydro-3-isopropyl-4-methoxy-	MeOH	228(4.52),239s(4.45), 264(3.67),274(3.73), 285(3.67),305s(3.41), 314(3.63),328(3.67)	44-2183-69
	MeOH-acid	238(4.49),293(3.87), 314s(3.77)	44-2183-69
	MeOH-base	228(4.52),239s(4.45), 264(3.67),274(3.73), 285(3.67),305s(3.41), 314(3.63),328(3.67)	44-2183-69
3-Isoquinuclidone, 6-anisoyl-	EtOH	280(1.87)	22-0781-69
Piperidine, 1-[3,4-(methylenedioxy)-cinnamoyl]-	EtOH	228(4.26),282(4.13), 292(4.12),318(4.18)	12-1531-69
Pyrrole-2-carboxylic acid, 3-(p-methoxyphenyl)-5-methyl-, ethyl ester	EtOH	282(4.21)	94-0582-69

Compound	Solvent	$\lambda_{max}(\log \epsilon)$	Ref.
$C_{15}H_{17}NO_3S$			
Spiro[benzo[b]thiophen-3(2H),1'-isoindolin]-3'-one, 3a,4,5,6,7,7a-hexahydro-, 1,1-dioxide	EtOH	223s(4.08),273(3.11), 280(3.05)	39-1149-69C
$C_{15}H_{17}NO_4$			
6H-Azepino[1,2-a]indole-11-carboxylic acid, 7,8,9,10,10a,11-hexahydro-10a-hydroxy-6-oxo-, methyl ester	EtOH	252(4.09),279(3.61), 287(3.55)	94-1290-69
Carbostyril, 3-isobutyryl-4,8-dimethoxy-	MeOH and MeOH-acid	225(4.10),253(4.22), 287(3.69)	44-2183-69
	MeOH-base	253(4.36),275s(3.84)	44-2183-69
Isonicotinic acid, 1-benzyl-1,2,3,6-tetrahydro-5-hydroxy-6-methyl-2-oxo-, methyl ester	90% EtOH	250(3.90)	94-0467-69
2,4-Pentadienoic acid, 2,4-dimethyl-5-(p-nitrophenyl)-, ethyl ester	EtOH	332(4.26)	88-0355-69
$\Delta^{2,\alpha}$-Pyrrolidinemalonic acid, 1-phenyl-, dimethyl ester	isooctane	240(3.90),297(4.10)	35-6683-69
$C_{15}H_{17}NO_8S$			
8aH-Thiazolo[3,2-a]pyridine-5,6,7,8-tetracarboxylic acid, 2,3-dihydro-, tetramethyl ester	EtOH	236(3.98),282(3.92), 392(3.52)	28-0870-69A
$C_{15}H_{17}NS_4$			
1-Piperidinecarbodithioic acid, anhydrosulfide with cyclic phenylvinylene trithioorthoformate	EtOH	234(4.12),317(3.90)	94-1931-69
$C_{15}H_{17}N_3$			
3H-Imidazo[4,5-f]quinoline, 3-ethyl-2,7,9-trimethyl-	EtOH	253s(4.61),257(4.63), 310(3.67)	94-2455-69
$C_{15}H_{17}N_3O$			
2,5-Cyclohexadien-1-one, 4-[(2,4-diamino-3,5-xylyl)imino]-2-methyl-	pH 10.0	657(4.35)	39-0823-69B
2,5-Cyclohexadien-1-one, 4-[(2,4-diamino-3,5-xylyl)imino]-3-methyl-	pH 10.0	658(4.21)	39-0823-69B
Pyrazole-3-carbonitrile, 5-(1-adamantylcarbonyl)-	EtOH	237(3.94)	18-1617-69
$C_{15}H_{17}N_3O_3$			
2-Cyclopenten-1-one, 2-allyl-4-hydroxy-3-methyl-, o-nitrophenylhydrazone	EtOH	267(4.24),306(4.38), 455(3.91)	78-1117-69
2-Cyclopenten-1-one, 2-allyl-4-hydroxy-3-methyl-, p-nitrophenylhydrazone	EtOH	285(3.96),318(3.61), 408(4.44)	78-1117-69
$C_{15}H_{17}N_3O_4$			
Glutaric acid, 2-diazo-3,3-dimethyl-4-(phenylimino)-, dimethyl ester	C_6H_{12}	258(4.03),305(3.22), 425(1.38)	88-2659-69
$C_{15}H_{17}N_5O_3$			
Acetamide, N-[6,7-dihydro-7-(1-hydroxy-1-methylethyl)-5-phenyl[1,2,5]oxadiazolo[3,4-d]pyrimidin-7-yl]-	EtOH	239(4.11),292(3.96)	35-5181-69

Compound	Solvent	λ_{max}(log ϵ)	Ref.
$C_{15}H_{17}N_5O_4$			
7H-Pyrrolo[2,3-d]pyrimidine-5-carbonitrile, 4-amino-7-(2,3-O-isopropylidene-β-D-ribofuranosyl)-	pH 1	233(4.25),271(4.20)	35-2102-69
	pH 11	227(4.05),276(4.20), 286s(4.02)	35-2102-69
	EtOH	230(4.01),278(4.22), 288s(4.04)	35-2102-69
$C_{15}H_{17}N_5O_5$			
2-Butanone, 1-(5-ethyl-3-isoxazolyl)-, 2,4-dinitrophenylhydrazone	EtOH	220(4.13),267(3.85), 356(4.36)	35-4749-69
$C_{15}H_{17}N_7O_2$			
Adenine, 9-[2-(6-nitro-3,5-xylidino)-ethyl]-	EtOH	238(4.36),260(4.27), 290s(3.76),430(3.76)	44-3240-69
	EtOH-HCl	237(4.40),258(4.30), 290s(3.83),430(3.76)	44-3240-69
	EtOH-NaOH	235s(4.43),260(4.29), 295s(3.81),430(3.76)	44-3240-69
$C_{15}H_{17}O_2P$			
Phosphine oxide, 2-biphenylyl(methoxymethyl)methyl-	EtOH	213(4.33),240s(3.82), 277(3.38)	39-0252-69C
$C_{15}H_{17}O_6P$			
Phosphinic acid, (5-acetyl-2,5-dihydro-4-methoxy-5-methyl-2-oxo-3-furyl)-phenyl-, methyl ester	MeCN	216(4.16),244(4.04)	35-2293-69
$C_{15}H_{18}$			
Fluorene, 1,2,3,4-tetrahydro-1,6-dimethyl-	EtOH	267(4.2)	42-0415-69
S-Guaiazulene, trinitrobenzene adduct	n.s.g.	604(2.64),630(2.58), 658(2.55),740(2.08)	78-4751-69
$C_{15}H_{18}BrClO_5$			
6-Bromo-2-hexyl-1-benzopyrylium perchlorate	HOAc-$HClO_4$	224(4.43),263(3.72), 315(3.30)	5-0162-69B
$C_{15}H_{18}BrNO_2$			
Fluorene-9α-carboxamide, 6-bromo-1,2,3,4,4aα,9a -hexahydro-7-methoxy-	EtOH	288(3.66),294(3.62)	44-2209-69
$C_{15}H_{18}ClNO_4S_2$			
2-sec-Butyl-3-ethylthiazolo[2,3-b]benzothiazolium perchlorate	EtOH	219(4.50),263(3.95), 289(3.70),315(4.13)	95-0469-69
3-Methyl-2-pentylthiazolo[2,3-b]benzothiazolium perchlorate	EtOH	217(4.65),262(4.12), 315(4.02)	95-0469-69
$C_{15}H_{18}INO$			
7,8-Dihydro-9-hydroxy-2,7,7-trimethyl-6H-cyclopent[g]isoquinolinium iodide	MeOH	271(4.57),343(3.68), 465(3.68)	44-0240-69
	MeOH-HCl	217(4.61),256(4.78), 320(3.58),380(3.80)	44-0240-69
$C_{15}H_{18}N_2$			
3,7-Methanoazonino[5,4-b]indole, 1,2,4,5,6,7-hexahydro-	EtOH	227(4.48),284(3.80), 291(3.78)	22-4154-69
1-Piperidineacetonitrile, α-phenylethylidene-	EtOH	269(3.90)	70-0873-69
1-Piperidineacetonitrile, α-styryl-	EtOH	251(4.22)	70-0873-69

Compound	Solvent	$\lambda_{max}(\log \epsilon)$	Ref.
$C_{15}H_{18}N_2O$			
Indole-2-methanol, α-(1,2,3,6-tetrahydro-1-methyl-4-pyridyl)-	EtOH	220(4.46),273(3.90), 282(3.90),290(3.76)	39-2738-69C
Ketone, 2-indolyl 1-methyl-4-piperidyl	EtOH	225(4.02),307(4.31)	39-2738-69C
1,5-Methano[1,3]diazocino[1,8-a]indol-6-ol, 1,2,3,4,5,6-hexahydro-2-methyl-	EtOH	225(4.9),278(3.9), 284(3.9),293(3.8)	39-2738-69C
	EtOH-HCl	220(--),268(--), 279(--),290(--)	39-2738-69C
$C_{15}H_{18}N_2OS$			
2-Pyrazolin-5-one, 4-[(butylthio)methylene]-3-methyl-1-phenyl-	hexane	250(4.37),320(4.36), 405(2.90)	104-1249-69
	EtOH	250(4.32),328(4.39)	104-1249-69
Pyrrolo[2,1-b]thiazol-7(7aH)-one, 5-(dimethylamino)-2,3-dihydro-7a-methyl-6-phenyl-	MeOH	243(4.11),260(4.11), 316(4.03)	78-5721-69
$C_{15}H_{18}N_2O_2$			
2-Benzoxazolinone, 3-[(2,5-dimethylhexa-2,4-dienyl)amino]-	EtOH	236(4.17),244(4.16), 275(3.42),280s(3.33)	39-0778-69C
Isoindolo[2,1-a]quinazoline-5,11-dione	EtOH	230(4.42),245(4.27), 255(4.29),265(4.19), 283(3.96),310(3.64)	44-2123-69
Isoquinoline-4-carboxamide, 1,2-dihydro-2-methyl-1-oxo-N,N-diethyl-	MeOH	242(3.94),252(3.89), 283(3.92),290(3.96), 320(3.70),355(3.54)	22-2045-69
5,8-Isoquinolinedione, 3-methyl-7-(pentylamino)-	MeCN	236(4.33),277(4.30), 320(3.77)	4-0697-69
1,4-Naphthoquinone, 2-[[3-(dimethylamino)propyl]amino]-	EtOH	325(3.28),445(3.58)	39-1799-69C
Pyridazino[1,2-c]pyridazin-1(2H)-one, 6,7,8,9-tetrahydro-4-(p-methoxyphenyl)-	n.s.g.	246(4.26)	44-2720-69
2H-Pyrido[4,3-b]indole-2-carboxylic acid, 1,3,4,9b-tetrahydro-9b-methyl-	MeOH	220s(4.2),255s(3.8)	33-0629-69
$C_{15}H_{18}N_2O_2S$			
Benzamide, N-[[2-[(2-hydroxyethyl)-amino]-1-cyclopenten-1-yl]thiocarbonyl]-	$CHCl_3$	288(4.15),405(4.32)	4-0037-69
p-Phenylenediamine, N ,N -dimethyl-2-(p-tolylsulfonyl)-	EtOH-1% $CHCl_3$	238(4.40),265s(4.08)	44-2083-69
2(1H)-Pyrimidinone, 1-[3-(benzylthio)-crotonoyl]-tetrahydro-, cis	EtOH	304(4.16)	32-1000-69
trans	EtOH	294(4.23)	32-1000-69
p-Toluenesulfonanilide, 4'-(dimethyl-amino)-	EtOH-1% $CHCl_3$	225(4.18),262(4.24), 300s(3.57)	44-2083-69
$C_{15}H_{18}N_2O_6$			
Malonamic acid, 2-(N-carboxyglycyl)-, N-benzyl ethyl ester	EtOH	232(3.84),264(4.08)	104-1713-69
$C_{15}H_{18}N_2O_9$			
Uracil, 5-β-D-ribofuranosyl-, 2',3',5'-triacetate	EtOH	262(3.90)	73-1690-69
$C_{15}H_{18}N_4$			
p-Phenylenediamine, N-(2-amino-4-imino-3-methyl-2,5-cyclohexadien-1-ylidene)-2,5-dimethyl-	pH 8	585(4.04)	39-0827-69B

Compound	Solvent	λ_{max}(log ϵ)	Ref.
p-Phenylenediamine, N-(2-amino-4-imino-5-methyl-2,5-cyclohexadien-1-ylidene)-2,5-dimethyl-	pH 8	586(4.00)	39-0827-69B
p-Phenylenediamine, N^4-(2-amino-4-imino-3-methyl-2,5-cyclohexadien-1-ylidene)-2,6-dimethyl-	pH 8	573(4.11)	39-0827-69B
p-Phenylenediamine, N^4-(2-amino-4-imino-5-methyl-2,5-cyclohexadien-1-ylidene)-2,6-dimethyl-	pH 8	565(4.02)	39-0827-69B
$C_{15}H_{18}N_4O$			
Phenol, 5-(diethylamino)-2-(2-pyridylazo)-, U(IV) complex	pH 8.2	564(4.88)	3-1652-69
$C_{15}H_{18}N_4O_3$			
Benzamide, N-methyl-N-[1,2,3,4-tetrahydro-1,3-dimethyl-6-(methylamino)-2,4-dioxo-5-pyrimidinyl]-	H_2O	277(4.17)	104-0547-69
$C_{15}H_{18}N_4O_4$			
Cyclohexane-$\Delta^{1,\alpha}$-acetaldehyde, α-methyl-, 2,4-dinitrophenylhydrazone	EtOH	222(4.20),259(4.23), 295(4.04),389(4.50)	39-0460-69C
Pyrazole-4-carbamic acid, 3-[(benzyloxy)methyl]-5-carbamoyl-, ethyl ester	EtOH	210(4.36),251(3.53)	88-0289-69
	0.05N NaOH	218(4.19),248(3.94)	88-0289-69
$C_{15}H_{18}N_4O_5$			
Uric acid, dihydro-4,5-dimethoxy-1,3-dimethyl-9-phenyl-	EtOH	220(4.47),275(4.18)	78-0541-69
$C_{15}H_{18}N_4O_6$			
Barbituric acid, 5,5'-(4-methylcyclohexylidene)di-	EtOH	255(4.46)	103-0827-69
3(2H)-Furanone, 4-ethyldihydro-5-propionyl-, mono(2,4-dinitrophenylhydrazone), cis	EtOH	216(4.15),252(4.03), 353(4.30)	35-4739-69
trans	EtOH	227(3.99),252s(3.85), 352(4.18)	35-4739-69
$C_{15}H_{18}N_4S_2$			
Carbamic acid, benzylmethyldithio-, (4-amino-2-methyl-5-pyrimidinyl)methyl ester	EtOH	247(4.28),279(4.25)	94-2299-69
$C_{15}H_{18}O$			
2-Cyclohexen-1-one, 3-methyl-2-(p-methylbenzyl)-	EtOH	241(4.0)	42-0415-69
Fluorene, 1,2,3,4,4aα,9aα-hexahydro-7-methoxy-9-methylene-	EtOH	213(4.38),252(4.04), 306(3.73),314(3.68)	44-2209-69
$C_{15}H_{18}O_2$			
5H-Benzo[b]cyclobuta[e]pyran-8-one, 1,2,2a,8a-tetrahydro-1,1,2,2-tetramethyl-	MeOH	217(4.31),257(3.89), 331(3.45)	35-4494-69
Cacalol	MeOH	218(4.28),254(3.88), 262(3.84),285(3.20), 296(3.20),325(3.11)	23-2465-69
4-Chromanone, 2-(2,3-dimethyl-2-butenyl)-	MeOH	214(4.34),251(3.86), 320(3.45)	35-4494-69
4-Chromanone, 2-(1,1,2-trimethyl-2-propenyl)-	MeOH	214(4.29),251(3.81), 320(3.36)	35-4494-69

Compound	Solvent	$\lambda_{max}(\log \epsilon)$	Ref.
o-Cresol, 5-(5-isobutyl-3-furyl)-	EtOH	210(4.62),255(3.91), 292(3.56)	44-0857-69
Crotonic acid, 2-methyl-, 2,6,8-decatrien-4-ynyl ester	ether	294(4.56),310(4.52)	24-1682-69
Crotonic acid, 2-methyl-, 2-decene-4,6-diynyl ester	ether	254(4.12),267(4.27), 284(4.19)	24-1682-69
2-Cyclohexen-1-one, 2-(2,5-dimethylbenzyl)-3-hydroxy-	EtOH	204(4.28),263(4.18)	44-2192-69
Fluorene-9-carboxaldehyde, 1,2,3,4,4aα,9aα-hexahydro-7-methoxy-	EtOH	281(3.45),287s(3.40)	44-2209-69
Furanodienone	EtOH	241(3.98),269(3.86)	77-0662-69
Isofuranodienone	EtOH	223(4.17),248(3.95)	77-0662-69
$C_{15}H_{18}O_3$			
5-Azuleneacetic acid, 1,5,6,7-tetrahydro-α,3,8-trimethyl-1-oxo-	EtOH	237(4.39),245(4.45), 254(4.47),320(3.88)	78-2117-69
Azuleno[4,5-b]furan-2,7-dione, 3,3a,4,5,6,6a,9a,9bα-octahydro-3β,6aα-dimethyl-6-methylene-	EtOH	205(4.03)	23-2849-69
1H-Benz[e]inden-3,5,7(2H)-trione, 3a,4,8,9,9aα,9bβ-hexahydro-3aα,6-dimethyl-	MeOH	261(3.99)	44-0112-69
1H-Cyclopropa[a]naphthalene-1-carboxylic acid, 1a,2,3,7b-tetrahydro-6-methoxy-1,7-dimethyl-	EtOH	280(3.31)	88-4559-69
Epipulchellin, anhydrodehydro-	EtOH	213(4.07)	88-2073-69
Eudesma-3,5,7(11)-trien-13-oic acid, 3,6-dihydroxy-, 13→6-lactone	MeOH	374(4.05)	39-1088-69C
Fluorene-9α-carboxylic acid, 1,2,3,4,-4aα,9aα-hexahydro-7-methoxy-	EtOH	218s(3.94),282(3.45), 287(3.41)	44-2209-69
9β-isomer	EtOH	218(3.80),227(3.83), 281(3.39),288(3.33)	44-2209-69
Furoeremophil-10(1)-ene, 6-hydroxy-8-oxo-	EtOH	302(4.22)	73-1739-69
Naphtho[2,1-b]furan-2(1H)-one, 3a,4,5,9b-tetrahydro-8-methoxy-1,9-dimethyl-	EtOH	220(4.00),280(3.30)	88-4559-69
Perezinone	MeOH	208(4.23),323(4.42), 330(4.38),354(3.75)	23-2465-69
2H-Pyran-2,4(3H)-dione, 5,6-dihydro-3,3,5,5-tetramethyl-6-phenyl-	EtOH	259(2.47)	39-2799-69C
Pyrolumisantonin	EtOH	219(4.06)	18-2736-69
Warburgiadione	EtOH	292(4.33)	78-2865-69
$C_{15}H_{18}O_4$			
Bahia I	EtOH	207(4.02)	23-2849-69
Cumambrin B, dehydro-	EtOH	244(4.04)	102-0305-69
Eleganin	EtOH	203(4.26)	105-0056-69
Euryopsol, chromic acid oxidation product diketone	EtOH	270(3.53)	78-5227-69
Fluorene-1-carboxylic acid, 4bβ,5,6,7,-8,8aβ-hexahydro-9α-hydroxy-2-methoxy-	EtOH	291(3.46)	44-2209-69
Fluorene-3-carboxylic acid, 4bβ,5,6,7,-8,8aβ-hexahydro-9α-hydroxy-2-methoxy-	EtOH	213(4.45),244(3.90), 304(3.62)	44-2209-69
Grosshemin	n.s.g.	210(4.04)	105-0158-69
1-Indanpropionic acid, 2-(carboxymethyl)-4-methyl-	EtOH	265(2.84),273(3.1), 279(3.19)	2-0215-69
Isabelin, dihydro-	EtOH	209(3.97)	78-4767-69
1-Naphthaleneacetic acid, 1,2,3,4-tetrahydro-7-methoxy-α,8-dimethyl-2-oxo-	EtOH	286(3.45)	88-4559-69

Compound	Solvent	$\lambda_{max}(\log \epsilon)$	Ref.
Santonene, 1,2-dihydro-4α-hydroxy-	MeOH	285(4.27),358(1.79)	39-1088-69C
	EtOH	285(4.29)	39-1619-69C
Santonene, 1,2-dihydro-4β-hydroxy-	MeOH	284(4.30)	39-1088-69C
	EtOH	284(4.30)	39-1619-69C
Santonene, 1,2-dihydro-6α-hydroxy-	EtOH	243(4.16)	39-1619-69C
Santonenic acid	MeOH	250(4.1),310(2.2)	39-1088-69C
β-Santonenic acid	MeOH	250(4.12)	39-1088-69C
Succinic acid, benzylidene-, diethyl ester	EtOH	210(3.90),265(4.26)	1-1151-69
$C_{15}H_{18}O_5$			
Canin	EtOH	208(3.95)	102-1515-69
1,2-Naphthalenedicarboxylic acid, 5,6,7,8-tetrahydro-3-methoxy-, dimethyl ester	n.s.g.	213(4.36),304(3.63)	12-2255-69
Oxiraneacetic acid, 2-carboxy-3-phenyl-, diethyl ester	EtOH	219(3.90),254(2.70), 260(2.74),265(2.68)	1-1151-69
$C_{15}H_{18}O_6$			
Elephantol, dihydro-	MeOH	211(3.98)	44-3867-69
Elephantolide, dihydro-	MeOH	211(3.94)	44-3867-69
$C_{15}H_{18}O_7$			
Benzoic acid, 2-(3-carboxypropionyl)-3,5-dimethoxy-, dimethyl ester	MeOH	252(3.57),306(3.55)	22-2365-69
$C_{15}H_{19}BrN_2S_2$			
2H-1,3,5-Thiadiazine-2-thione, 3-(p-bromophenyl)-5-cyclohexyltetrahydro-	MeOH	248(4.14),290(3.72)	73-2952-69
$C_{15}H_{19}BrO$			
Anisole, 4-bromo-2-(1,2-dimethyl-2-cyclopenten-1-yl)-5-methyl-	MeOH	283(3.37),289(3.36)	78-3509-69
Aplysin	EtOH	232(3.91),293(3.64)	18-0843-69
Aplysin, (±)	MeOH	235(3.86),295(3.59)	78-3509-69
Aplysin, isomer	MeOH	236(3.68),294(3.57)	78-3509-69
Laurenisol	EtOH	278(3.42),285(3.42)	88-1343-69
2,5-Methano-1-benzoxepin, 2-(bromomethyl)-2,3,4,5-tetrahydro-5,8,10-trimethyl-	EtOH	278(3.53),287(3.54)	88-1343-69
$C_{15}H_{19}BrO_2$			
Cyclopentanone, 2-(5-bromo-2-methoxy-p-tolyl)-2,5-dimethyl-	MeOH	283(3.36)	78-3509-69
Cyclopentanone, 2-(5-bromo-2-methoxy-p-tolyl)-3,3-dimethyl-	MeOH	285(3.43),292(3.42)	78-3509-69
Cyclopentanone, 3-(5-bromo-2-methoxy-p-tolyl)-2,3-dimethyl-	MeOH	282(3.38),289(3.39)	78-3509-69
$C_{15}H_{19}BrO_4$			
Aplysinol	EtOH	233(4.01),292(3.73)	18-0843-69
$C_{15}H_{19}ClN_2O$			
4-Quinolinol, 5-chloro-3-[(diethylamino)methyl]-2-methyl-	EtOH	245(4.21),295(4.01), 325(3.97),340(3.84)	2-1186-69
$C_{15}H_{19}ClN_2O_4S$			
2-[4-(Dimethylamino)-1,3-butadienyl]-3-ethylbenzothiazolium perchlorate	EtOH	481(5.06)	65-1732-69

Compound	Solvent	$\lambda_{max}(\log \epsilon)$	Ref.
5-(Dimethylamino)-α-[p-(dimethylamino)-phenyl-2-thenylium perchlorate	CH_2Cl_2	293(4.20),330(3.72), 346(3.75),369(3.65), 533s(4.62),574(4.92)	48-0827-69
$C_{15}H_{19}ClO$			
1,3-Heptadiene, 2-chloro-1-ethoxy-1-phenyl-	EtOH	285(4.11)	44-1474-69
$C_{15}H_{19}Cl_2NO$			
Acrylophenone, 3-[bis(2-chloroethyl)-amino]-2',4'-dimethyl-	EtOH	213(3.90),262(3.40), 302(4.32)	24-3139-69
Acrylophenone, 3-[bis(2-chloroethyl)-amino]-2',5'-dimethyl-	EtOH	213(3.90),262(3.54), 301(4.30)	24-3139-69
Acrylophenone, 3-[bis(2-chloroethyl)-amino]-3',4'-dimethyl-	EtOH	213(3.97),264(3.86), 336(4.36)	24-3139-69
$C_{15}H_{19}Cl_2NO_3$			
Acrylophenone, 3-[bis(2-chloroethyl)-amino]-3',4'-dimethoxy-	EtOH	235(4.06),280(3.74), 343(4.45)	24-3139-69
$C_{15}H_{19}FeNO_4S$			
Alanine, N-(ferrocenesulfonyl)-, ethyl ester	EtOH	431(2.17)	49-1552-69
$C_{15}H_{19}NO$			
2-Cyclohexen-1-one, 3-(methylphenethyl-amino)-	C_6H_{12}	282(4.41)	39-0299-69B
	N HCl	289(4.35)	39-0299-69B
	H_2O	302(4.50)	39-0299-69B
	EtOH	300(4.54)	39-0299-69B
Ipalbidine	EtOH	236(4.00),278(3.24)	77-0709-69
	EtOH-base	248(4.39),295s(--)	77-0709-69
Phenanthridine, 5,6,7,8,9,10-hexahydro-3-methoxy-5-methyl-	EtOH	233(4.56),328(3.61)	2-0419-69
Phenethylamine, N-(3-methoxy-2-cyclohex-en-1-ylidene)-	C_6H_{12}	242(4.37)	39-0299-69B
	N HCl	280(4.33)	39-0299-69B
	H_2O	280(4.38)	39-0299-69B
	EtOH	250(4.20)	39-0299-69B
$C_{15}H_{19}NOS_2$			
1,3-Dithiole, 4-(p-methoxyphenyl)-2-piperidino-	EtOH	244(4.11),307(4.01)	94-1924-69
3-Heptanone, 4-(2-benzothiazolylthio)-5-methyl-	EtOH	225(4.30),244(3.94), 280(4.58),291(3.93), 301(3.81)	95-0469-69
2-Octanone, 3-(2-benzothiazolylthio)-	EtOH	221(4.10),243(3.99), 280(4.03),293(3.97), 302(3.90)	95-0469-69
1-Piperidinecarbodithioic acid, ester with 2-mercapto-4'-methylacetophenone	EtOH	257(4.31),282s(4.07)	94-1924-69
$C_{15}H_{19}NO_2$			
10a,10b-Ethano-3H,8H-benzo[ij]quinolizine-3,8-dione, 1,2,5,6,7,7a-hexahydro-12-methyl-	n.s.g.	230(4)	23-0433-69
Fluorene-9α-carboxamide, 1,2,3,4,4a,9a-hexahydro-7-methoxy-	EtOH	218(3.92),282(3.46), 287(3.41)	44-2209-69
2,4-Pentadienamide, 5-(p-methoxyphenyl)-N,N,3-trimethyl-, 2-cis-4-trans	EtOH	232(4.00),317(4.34)	39-0016-69C
Spiro[cyclopentane-1,3'(2'H)-quinolin]-2'-one, 1',4'-dihydro-4'-methoxy-4'-methyl-	EtOH	254(4.08),284(3.38), 294(3.26)	94-1290-69

Compound	Solvent	λ_{max}(log ϵ)	Ref.
Spiro[cyclopentane-1,3'(2'H)-quinolin]-2'-one, 4'-ethoxy-1',4'-dihydro-	EtOH	253(4.14),280(3.43), 290(3.30)	94-1290-69
$C_{15}H_{19}NO_2S_2$			
1-Piperidinecarbodithioic acid, ester with 2-mercapto-4'-methoxyacetophenone	EtOH	219(4.29),279(4.43)	94-1924-69
$C_{15}H_{19}NO_3$			
Carbazole, 1,2,3,4-tetrahydro-5,6,7-trimethoxy-	MeOH	231(4.39),277(3.77), 295s(3.64)	95-1061-69
Carbazole, 1,2,3,4-tetrahydro-6,7,8-trimethoxy-	MeOH	225(4.50),279(3.97), 290s(3.96)	95-1061-69
Carbostyril, 3-(2-hydroxy-3-methylbutyl)-4-methoxy- (spectrum unchanged by acid and base; partial data listed here)	MeOH	231(4.39),245s(3.98), 272(3.82),280(3.77), 311s(3.63),324(3.76), 336s(3.63)	44-2183-69
	MeOH-acid	231(4.39),245s(3.98), 272(3.82),280(3.77)	44-2183-69
	MeOH-base	231(4.39),245s(3.98), 272(3.82),280(3.77)	44-2183-69
$C_{15}H_{19}NO_4S_3$			
N-(4-p-Tolyl-1,3-dithiol-2-ylidene)-piperidinium bisulfate	EtOH	225(4.14),326(4.08)	94-1924-69
$C_{15}H_{19}NO_5S_3$			
N-(4-p-Methoxyphenyl-1,3-dithiol-2-ylidene)piperidinium bisulfate	EtOH	261(4.09),292(4.07), 333(4.00)	94-1924-69
$C_{15}H_{19}NO_6$			
4-Penten-1-ol, 5-(3,4-dimethoxyphenyl)-4-nitro-, acetate	EtOH	360(4.13)	87-0157-69
$C_{15}H_{19}NS_2$			
1,3-Dithiole, 2-piperidino-4-p-tolyl-	EtOH	245(4.14),326(4.08)	94-1924-69
$C_{15}H_{19}N_3O$			
Fluoren-2(1H)-one, 3,4,4a,9a-tetrahydro-8-methyl-, semicarbazone	EtOH	250(4.0)	42-0415-69
$C_{15}H_{19}N_3O_2$			
2-Butanone, 3-(4,5-dimethyl-2-phenylimidazol-1-yl)-, oxime, N-oxide	MeOH	283(4.13)	48-0746-69
$C_{15}H_{19}N_3O_9$			
Uracil, 6-amino-3-β-D-ribofuranosyl-, 2',3',5'-triacetate	pH 1	267(4.36)	39-0791-69C
	pH 11	272(4.25)	39-0791-69C
$C_{15}H_{19}N_5O$			
4H-Pyrido[1,2-a]pyrimidine-3-carbonitrile, 2-[[2-(diethylamino)ethyl]-amino]-4-oxo-	EtOH	230s(4.13),256(4.55), 345(3.99),360s(3.83)	95-0203-69
$C_{15}H_{19}N_5O_2$			
Benzaldehyde, p-(diethylamino)-, (2,6-dihydroxy-4-pyrimidinyl)hydrazone	EtOH	335(4.59)	56-0519-69

Compound	Solvent	λ_{max}(log ε)	Ref.
$C_{15}H_{20}$			
Cyclodecane, 1,2,4,6,9-pentakis(methylene)-	n.s.g.	230(4.09)	88-0297-69
1H-Cyclopentacyclooctene, 4-butyl-4,5-dihydro-	dioxan	233(3.74),302(3.80)	24-1789-69
Laurine	EtOH	253(2.45),259(2.45), 265(2.45),274(2.38)	78-0459-69
Pentaspiro[2.0.2.0.2.0.2.0.2.0]pentadecane	C_6H_{12}	250s(1.85),257(1.84), 263(1.82),269(1.74), 276(1.60)	88-0979-69
$C_{15}H_{20}BrNO$			
1,5-Cyclohexadiene-1-carbonitrile, 3-bromo-3,5-di-tert-butyl-4-oxo-	C_6H_{12}	329(3.35)	64-0547-69
	MeOH	328(3.41)	64-0547-69
$C_{15}H_{20}ClNO_4$			
4a-Ethyl-2,3,4,4a-tetrahydro-9-methyl-1H-carbazolium perchlorate	EtOH	232(3.80),239(3.78), 276(3.85)	78-0095-69
	EtOH-NaOH	259(3.92),277(4.02)	78-0095-69
$C_{15}H_{20}ClN_5OS$			
Formamidine, N'-[5-[[5-(2-chloroethyl)-4-methyl-2-oxo-4-thiazolin-3-yl]-methyl]-2-methyl-4-pyrimidinyl]-N,N-dimethyl-	EtOH	223(4.05),265(4.15), 314(4.35)	88-3279-69
$C_{15}H_{20}Cl_2NRh$			
Rhodium, dichloro(pentamethyl-π-cyclopentadienyl)(pyridine)-	H_2O	206(3.89),228(4.18), 376(3.36)	39-1299-69A
$C_{15}H_{20}N_2$			
Carbazole, 9-[(dimethylamino)methyl]-1,2,3,4-tetrahydro-	EtOH	231(4.46),285(3.80), 293(3.75)	39-2738-69C
$C_{15}H_{20}N_2O$			
Acetic acid, (cyclohexylmethylene)phenylhydrazide	EtOH	233(4.1),303(4.2)	28-1703-69A
Indole-2-methanol, α-(1-methyl-4-piperidyl)-	EtOH	220(4.45),273(3.91), 280(3.92),289(3.79)	39-2738-69C
4-Isoquinolinecarboxamide, N,N-diethyl-1,2-dihydro-2-methyl-	MeOH	237(3.86),280(3.69), 336(3.96)	22-2045-69
2H-Pyrrol-3-ol, 5-(dimethylamino)-4-phenyl-2-propyl-	MeOH	223(4.13),287(4.29)	78-5721-69
	MeOH-HCl	242(4.26)	78-5721-69
$C_{15}H_{20}N_2OS$			
4-Pyrimidinol, 6-(1-adamantyl)-2-(methylthio)- (all spectra in 50% EtOH)	0.1N HCl	238(4.03),282(3.89)	73-2278-69
	pH 8.2	237(4.04),285(3.88)	73-2278-69
	0.1N NaOH	244(4.05),278(3.87)	73-2278-69
$C_{15}H_{20}N_2O_2$			
Indole-2-carboxamide, N-tert-butyl-4-methoxy-3-methyl-	EtOH	290(4.2),235(4.5)	24-0678-69
Morpholine, 4-[p-(dimethylamino)cinnamoyl]-	EtOH	231(4.28),316(4.50)	39-0016-69C
Morpholine, 4-(2,4-dimethyl-3-phenyl-3-isoxazolin-5-yl)-	n.s.g.	221(4.1),285(3.5)	94-2201-69
$C_{15}H_{20}N_2O_4$			
4-Penten-1-ol, 5-[p-(dimethylamino)phenyl]-4-nitro-, acetate	EtOH	416(4.36)	87-0157-69

Compound	Solvent	$\lambda_{max}(\log \epsilon)$	Ref.
$C_{15}H_{20}N_2S$			
Thiazole, 2-anilino-5-isopropyl-4-propyl-	n.s.g.	295(4.22)	5-0121-69A
Thione, bis(3,4,5-trimethylpyrrol-2-yl)	$CHCl_3$	332(3.37),439(4.50)	12-0239-69
	10N HCl	269(3.58),370(3.79), 497(4.98)	12-0239-69
mercury complex	$CHCl_3$	404(4.10),494(4.43)	12-0239-69
$C_{15}H_{20}N_2S_2$			
2H-1,3,5-Thiadiazine-2-thione, 3-cyclohexyltetrahydro-5-phenyl-	MeOH	243(3.95),293(3.79)	73-2952-69
$C_{15}H_{20}N_4O_2$			
Uracil, 5-(benzylmethylamino)-1,3-dimethyl-6-(methylamino)-	EtOH	285(4.06)	104-2004-69
$C_{15}H_{20}N_4O_2S$			
Valeric acid, 5-[(9-cyclopentyl-9H-purin-6-yl)thio]-	pH 1	295(4.28)	73-2114-69
	pH 13	228(3.93),287(4.30), 294(4.30)	73-2114-69
$C_{15}H_{20}N_4O_6S$			
Uridine, 5-[[(4,5-dimethylimidazol-2-yl)thio]methyl]-	pH 7	260(4.44)	63-0710-69
$C_{15}H_{20}N_4S$			
Benzaldehyde, p-(dimethylamino)-, 4-(2-cyclopenten-1-yl)-3-thiosemicarbazone	EtOH	242(4.0),365(4.6)	124-0509-69
$C_{15}H_{20}O$			
Cyclohexanone, 3-methyl-2-(p-methylbenzyl)-	EtOH	266(2.3),275(2.8)	42-0415-69
1(2H)-Naphthalenone, 7-tert-butyl-3,4-dihydro-5-methyl-	EtOH	213(4.46),258(4.10), 306(3.34)	22-0985-69
1(2H)-Naphthalenone, 8-tert-butyl-3,4-dihydro-5-methyl-	EtOH	215(4.27),255s(3.87), 298(3.27)	22-0985-69
Nuciferal	EtOH	222(4.28),231(4.28), 264(2.97),267(2.90), 273(2.85)	44-1122-69
2,4-Pentadienal, 3-methyl-5-(2,6,6-trimethyl-1,3-cyclohexadien-1-yl)-, cis	EtOH	272(4.03),367(4.09)	22-3252-69
trans	EtOH	273(4.05),376(4.18)	22-3252-69
$C_{15}H_{20}O_2$			
Bemadienolide	n.s.g.	273(3.42)	78-3903-69
Brachylaenalone A	EtOH	244(3.95)	77-0630-69
β-Cyclocostunolide	n.s.g.	205(4.18)(end abs.)	88-2917-69
2-Cyclohexen-1-one, 5-(2-hydroxy-5-methyl-1-methylene-3-hexynyl)-2-methyl-	EtOH	237(4.00)	44-0857-69
2-Cyclohexen-1-one, 5-[1-(hydroxymethyl)-5-methyl-1-hexen-3-ynyl]-2-methyl-	EtOH	234(4.32)	44-0857-69
2-Cyclohexen-1-one, 5-(5-isobutyl-3-furyl)-2-methyl- (bilobanone)	EtOH	224(4.13)	44-0857-69
Eremophila-1(10),7-diene-2,9-dione	MeOH	274(4.19)	94-1324-69
Eremophila-1(10),7(11)-diene-2,9-dione	MeOH	248(4.00)	94-1324-69
Fluorene-9α-methanol, 1,2,3,4,4aα,9aα-hexahydro-7-methoxy-	EtOH	219(3.85),227(3.86), 281(3.47),287s(3.40)	44-2209-69
9β-isomer	EtOH	228(3.89),282(3.48), 288(3.42)	44-2209-69
Furoeremophilane, 9-oxo-	EtOH	278(4.10)	73-1739-69

Compound	Solvent	λ_{max}(log ε)	Ref.
Furoeremophilane, 9-oxo-, cis	EtOH	282(4.19)	73-0336-69
Furoeremophilane, 9-oxo-, trans	EtOH	278(4.10)	73-0336-69
4,6-Heptadien-3-one, 2-methyl-6-(4-methyl-5-oxo-3-cyclohexen-1-yl)-	EtOH	243(4.18)	44-0857-69
Inunolide	n.s.g.	210(4.16)	2-0310-69
Isopetasone	EtOH	244(4.09),281(3.85)	78-2865-69
1(2H)-Naphthalenone, 3,4-dihydro-6-isopropyl-7-methoxy-5-methyl-	EtOH	224(4.30),269(4.08), 323(3.60)	18-3011-69
1(2H)-Naphthalenone, 3,4-dihydro-8-isopropyl-7-methoxy-5-methyl-	EtOH	220(4.28),259(3.78), 325(3.38)	18-3011-69
Pinguisone	n.s.g.	218(3.87),292(1.68)	73-0582-69
$C_{15}H_{20}O_3$			
Crotonic acid, ester with 3-hydroxy-4,6,6-trimethyl-1,4-cyclohexadiene-1-carboxaldehyde	ether	219(4.38)	24-2211-69
4αH-Eudesma-5,7(11)-dien-13-oic acid, 3β,6-dihydroxy-, 13→6-lactone	MeOH	291(4.27)	39-1088-69C
4βH-isomer	MeOH	289(4.46)	39-1088-69C
Fragrolide	isooctane	211(3.96)	78-3903-69
2(3H)-Furanone, 4-(3,7-dimethyl-5-oxo-2,6-octadienyl)dihydro-	n.s.g.	203(4.16),232(4.15)	88-2417-69
10αH-Furoeremophilane, 6α-hydroxy-9-oxo-	EtOH	282(4.06)	73-2792-69
Guai-1-en-12-oic acid, 8-hydroxy-3-oxo-, lactone	EtOH	237(4.20)	16-0066-69
2H-2β,4aβ-Methanonaphthalene-5-carboxylic acid, 1,3,4,5β,6,7-hexahydro-1,1-dimethyl-7-oxo-, methyl ester	n.s.g.	241(4.19)	77-1335-69
Naphtho[1,2-c]furan-3(1H)-one, 5aα,6,7,8,9,9a-hexahydro-1-hydroxy-6,6,9aβ-trimethyl-	n.s.g.	284(3.55)	78-3903-69
Spiro[benzofuran-2(4H),2'(5'H)-furan]-5'-one, 5,6,7,7a-tetrahydro-3',4,4a,7a-tetramethyl-	EtOH	210(4.24)	22-0232-69
$C_{15}H_{20}O_4$			
Abscisin II, (±)-	H_2O	254(4.30)	28-0865-69B
3-Buten-2-one, 3-[(3,3-diacetyl-3,4-dihydro-6-methyl-2H-pyran-5-yl)methyl]-	EtOH	217(3.77),257(3.09)	39-0879-69C
Cumambrin B	EtOH	197(4.13)	102-0305-69
Epipulchellin, dehydro-	EtOH	209(4.05)	88-2073-69
4βH-Eudesma-5,7(11)-dien-13-oic acid, 3α,4β,6-trihydroxy-, 13→6 lactone	MeOH	288(4.27)	39-1088-69C
Guai-1(10)-en-12-oic acid, 2,6-dihydroxy-3-oxo-, 12→6 lactone	EtOH	263(4.09)	78-2099-69
	aq. KOH	304(3.95)	78-2099-69
Isabelin, tetrahydro-	EtOH	208(3.47)	78-4767-69
Isophoto-α-santonic lactone	EtOH	240(4.10)	16-0082-69
1-Naphthaleneacetic acid, 1,2,3,4-tetrahydro-2-hydroxy-7-methoxy-α,8-dimethyl-	EtOH	279(3.31)	88-4559-69
Naphtho[1,2-c]furan-3,5-dione, 1,5a,6,7,8,9,9a,9b-octahydro-9b-hydroxy-6,6,9a-trimethyl-	MeOH	235(4.04)	78-3895-69
Neohelenalin, dihydrohydroxy-	EtOH	237(4.26),304(1.90)	16-0066-69
Pleniradin	EtOH	210(4.16)(end abs.)	102-1753-69
Plenolin	EtOH	225(3.97)	102-2371-69
$C_{15}H_{20}O_5$			
Epilaccishellolic acid	EtOH	224(3.68)	78-3855-69
Jalaric acid	EtOH	220(3.79)	78-3841-69

Compound	Solvent	λ_{max}(log ϵ)	Ref.
Laccishellolic acid	EtOH	223(3.62)	78-3855-69
Malic acid, 2-benzyl-, diethyl ester	EtOH	210(3.85),242(2.00), 248(2.11),252(2.23), 258(2.36),264(2.23)	1-1151-69
Phaseic acid	MeOH	258(4.16)	78-5893-69
Vernolepin, tetrahydro-	MeOH	214(3.72)	44-3903-69
$C_{15}H_{20}O_7$			
Nivalenol	MeOH	218(3.80)	88-2823-69
$C_{15}H_{20}O_9$			
Pleoside	EtOH	224(4.27),285(4.33), 324(3.73)	95-0372-69
$C_{15}H_{21}BrO$			
4,7-Methanoazulen-3(3aH)-one, 2-bromo-4,5,6,7β,8,8aβ-hexahydro-3aα,4β,8,8-tetramethyl-	n.s.g.	260(3.72)	88-3761-69
$C_{15}H_{21}BrO_2$			
1,5-Cyclohexadiene-1-carboxaldehyde, 3-bromo-3,5-di-tert-butyl-4-oxo-	C_6H_{12}	324(3.28)	64-0547-69
$C_{15}H_{21}BrO_3$			
1,5-Cyclohexadiene-1-carboxylic acid, 3-bromo-3,5-di-tert-butyl-4-oxo-	C_6H_{12}	329(3.41)	64-0547-69
$C_{15}H_{21}N$			
Indene-5-carbonitrile, 1,6,7,7a-tetrahydro-3-isobutyl-1-methyl-	EtOH	284(4.21)	2-0643-69
Indole, 3-isohexyl-6-methyl-	C_6H_{12}	225(4.53),275(3.77), 283(3.71),288(3.67), 295(3.61)	1-2578-69
Indole, 2-isopentyl-3,6-dimethyl-	C_6H_{12}	228(4.53),277(3.78), 285(3.71),291(3.62), 298(3.53)	1-2578-69
1-Naphthalenepropylamine, 3,4-dihydro-N,N-dimethyl-, hydrochloride	n.s.g.	260(3.91)	39-0217-69C
$C_{15}H_{21}NO$			
2H-1-Benzopyran-4-propylamine, N,N,3-trimethyl-	n.s.g.	259(3.86)	39-0217-69C
Cinnamamide, N,N,α-triethyl-	ether	247(4.16)	5-0064-69I
Ipalbidine, dihydro-	EtOH	223(3.92),277(2.83), 283(2.76)	77-0709-69
	EtOH-base	240(3.88),286(2.83), 296(2.80)	77-0709-69
1-Naphthaleneethylamine, 3,4-dihydro-7-methoxy-N,N-dimethyl-	n.s.g.	261(3.80)	39-0217-69C
hydrochloride	n.s.g.	260(3.82)	39-0217-69C
$C_{15}H_{21}NO_2$			
2H-1-Benzopyran-4-propylamine, 6-methoxy-N,N-dimethyl-	n.s.g.	254(3.72)	39-0217-69C
2H-1-Benzopyran-4-propylamine, 7-methoxy-N,N-dimethyl-, maleate	n.s.g.	266(3.94)	39-0217-69C
Benzylamine, p-methoxy-N-(3-methoxy-2,4-pentadienyl)-N-methyl-	EtOH	210(4.06),228(4.22), 247(4.11),281(3.31)	44-4158-69

Compound	Solvent	λ_{max}(log ε)	Ref.
$C_{15}H_{21}NO_2S_2$			
Butyraldehyde, 2-(2-benzothiazolylthio)-, diethyl acetal	EtOH	226(4.29),283(4.05), 291(4.01),301(3.93)	95-0469-69
$C_{15}H_{21}NO_4$			
4,5-Oxazolidinedione, 2-(1-adamantyl)-2-ethoxy-	EtOH	222(3.73)	78-2909-69
$C_{15}H_{21}N_3O$			
5-Pyrroline-2-carbonitrile, 2,3,3-trimethyl-5-(3,3-dimethyl-5-oxopyrrolidin-2-ylidene)methyl-	EtOH	292(4.3)	89-0343-69
$C_{15}H_{21}N_3OS$			
5-Pyrroline-2-carbonitrile, 2,3,3-trimethyl-5-[(3,3-dimethyl-5-oxo-2-pyrrolidinylidene)methyl]thio-	EtOH	259(4.1)	89-0343-69
$C_{15}H_{21}N_3O_2$			
Cyclohexanone, 2-(4,5,6,7-tetrahydro-2-vinyl-1-benzimidazolyl)-, oxime, 3-oxide	MeOH	287(4.13)	48-0746-69
1,2,4-Oxadiazolin-5-one, 2-tert-butyl-3-(isopropylimino)-4-phenyl-	isooctane	238(4.05)	44-2269-69
2-Pyrazoline-5-carboxamide, 3-(1-adamantylcarbonyl)-	EtOH	209(4.22),234(3.89)	18-1617-69
$C_{15}H_{21}N_3O_2S$			
Hexanoic acid, 2-(dicyanomethylene)-3-[(dimethylamino)(methylthio)methylene]-, ethyl ester	C_6H_{12}	474(4.06)	5-0073-69E
$C_{15}H_{21}N_5O_2S$			
Formamidine, N'-[5-[[5-(2-hydroxyethyl)-4-methyl-2-oxo-4-thiazolin-3-yl]-methyl]-2-methyl-4-pyrimidinyl]-N,N-dimethyl-	EtOH	222(4.09),265(4.17), 314(4.38)	88-3279-69
$C_{15}H_{21}N_5O_5$			
Adenosine, 2-ethyl-2',3'-O-isopropylidene-, 1-oxide	pH 1	260(4.06)	94-1128-69
	pH 7	233(4.50),264(3.88)	94-1128-69
	pH 13	232(4.51),268(3.89)	
$C_{15}H_{21}O_6P$			
Hydrocinnamic acid, α-acetyl-β-phosphono-, C-ethyl dimethyl ester	H_2O	262f(2.76)	70-1012-69
	MeOH	285f(2.87)	70-1012-69
	ether	280f(3.03)	70-1012-69
Phosphoric acid, diethyl ester, ester with 2-hydroxy-2-methyl-1-phenyl-1,3-butanedione	n.s.g.	250(4.0),290(3.0), 330s(2.0)	65-1954-69
$C_{15}H_{22}$			
Cyclopropa[d]naphthalene, 1,1a,2,4a,5,-6,7,8-octahydro-4a,8,8-trimethyl-2-methylene-	MeOH	235(4.38)	88-3091-69
Indan, 3-isobutyl-1,5-dimethyl-	EtOH	262(3.16),268(3.21), 278(3.16)	2-0643-69
Naphthalene, 1,2,3,4-tetrahydro-7-isopropyl-1,1-dimethyl-	EtOH	269(2.81),277(2.87)	54-0313-69
	EtOH	267(2.78),277(2.80)	88-2323-69

Compound	Solvent	$\lambda_{max}(\log \epsilon)$	Ref.
$C_{15}H_{22}ClN$			
1-Naphthalenepropylamine, 3,4-dihydro-N,N-dimethyl-, hydrochloride	n.s.g.	260(3.91)	39-0217-69C
$C_{15}H_{22}ClNO$			
1-Naphthalenethylamine, 3,4-dihydro-7-methoxy-N,N-dimethyl-, hydrochloride	n.s.g.	260(3.82)	39-0217-69C
$C_{15}H_{22}ClNO_4$			
2,3,4,7,8,9,10,11-Octahydro-6-methyl-1H-cyclohepta[c]quinolizinium perchlorate	$CHCl_3$	282(3.94)	78-4161-69
$C_{15}H_{22}N_2O$			
Darvasine	n.s.g.	244(4.3)	105-0115-69
Indole-2-ethanol, 3-[(dimethylamino)-methyl]-β,β-dimethyl-	EtOH	222(4.40),282(3.84), 290(3.78)	39-1003-69C
$C_{15}H_{22}N_2O_2$			
5,10-Etheno-1H-pyrazolo[1,2-a][1,2]di-azocine-1,3(2H)-dione, 2,2-diethyl-5,6,7,8,9,10-hexahydro-	C_6H_{12}	261(3.58)	44-3181-69
	MeOH	257(3.45)	44-3181-69
$C_{15}H_{22}N_2O_5$			
Glucosamine, p-(dimethylamino)benzyli-dene-	neutral	328(4.10)	94-0770-69
	alkaline	327(4.09)	94-0770-69
$C_{15}H_{22}N_2O_6S$			
Dimethyl-1-sulfoxonium [1-(2,3-O-iso-propylidene-β-D-ribofuranosyl)-1,4-dihydro-4-oxo-2-pyrimidinyl]methylide	MeOH	236(4.30),278(4.38)	35-7752-69
$C_{15}H_{22}N_2O_{10}$			
Uracil, 6-methyl-1,3-di-β-D-ribofurano-syl-	pH 7	208(3.96),265(4.07)	73-2316-69
$C_{15}H_{22}N_4O_2S_3$			
Thiamine, ethylxanthogenate	EtOH	230(4.29),270(3.93), 304(4.22)	94-2299-69
$C_{15}H_{22}N_6O_5S$			
Adenosine, 5'-[(3-amino-3-carboxypro-pyl)methylsulfonio]-5'-deoxy-, hydroxide, inner salt	N HCl	257(4.17)	10-0414-69F
	pH 7.0	260(4.19)	10-0414-69F
	H_2O	260(4.18)	65-0434-69
$C_{15}H_{22}O$			
Aristolone, (+)-	MeOH	236(4.01)	23-0831-69
Bicyclo[5.3.1]undec-2-en-4-one, 2,6,6-trimethyl-8-methylene-	EtOH	239(3.92)	31-1139-69
Chiloscyphone	EtOH	220(3.95)	88-1599-69
4H-Cyclopropa[b]naphthalen-4-one, 1,1a,2,5,6,6a,7,7a-octahydro-1,1,3,6a-tetramethyl-	$CHCl_3$	253(4.24)	32-0219-69
6,7-Epiaristolone	MeOH	233(4.00)	23-0831-69
Eremophila-1,11-dien-9-one	MeOH	285(2.18)	88-0307-69
1,7-Ethanoazulen-10-one, decahydro-8,8-dimethyl-4-methylene-	EtOH	275(1.64)	78-4895-69
5H-2,4aβ-Ethanonaphthalen-5-one, 1,2α,6,7,8,8a-hexahydro-3,8a,8a-trimethyl-	n.s.g.	242(4.19)	88-3761-69

Compound	Solvent	$\lambda_{max}(\log \epsilon)$	Ref.
2(1H)-Naphthalenone, 4a,5,6,8a-tetrahydro-1-isopropyl-4,7-dimethyl-	hexane	226(3.99)	88-1601-69
antipode	hexane	237(4.05)	88-1601-69
2(1H)-Naphthalenone, 4a,5,6,8a-tetrahydro-5-isopropyl-3,8-dimethyl-	hexane	231(3.95)	88-1601-69
antipode	hexane	240(3.96)	88-1601-69
2,4-Pentadien-1-ol, 3-methyl-5-(2,6,6-trimethyl-1,3-cyclohexadien-1-yl)-	EtOH	238(4.02),245(3.98), 307(4.12)	22-3252-69
isomer	EtOH	239(3.95),312(4.05)	22-3252-69
α-Vetivone	C_6H_{12}	233(4.20)	94-1324-69
$C_{15}H_{22}OS_2$			
2-Bornanone, 3-(1,3-dithiepan-2-ylidene)-, (+)-	dioxan	275(3.65),328(4.32), 370s(2.90)	28-0186-69A
$C_{15}H_{22}O_2$			
Cinnamolide	MeOH	224(3.94)	78-3895-69
1,4-Cyclopentadiene, 1,2-dipivaloyl-	EtOH	253(4.17),334(3.96), 397(3.98)	39-2464-69C
Indene-5-carboxylic acid, 2,6,7,7a-tetrahydro-3-isobutyl-1-methyl-	EtOH	282(4.30)	2-0643-69
Ishwarone	EtOH	211(2.44),288(1.48)	88-0133-69
Isopetasol	EtOH	246(4.11),280(3.89)	78-2865-69
3-Norcaren-2-one, 3-isopropyl-6-methyl-7-(3-oxobutyl)-	MeOH	225(3.96),315(2.15)	88-3357-69
α-Rotunol	n.s.g.	235(4.03)	88-2741-69
β-Rotunol	n.s.g.	232(4.11)	88-2741-69
Zerumbone oxide	n.s.g.	232(3.97),250(3.94), 328(2.36)	25-0047-69
$C_{15}H_{22}O_3$			
Confertifolin, 6β-hydroxy-	MeOH	218(4.05)	78-3895-69
2-Cyclohexene-1,4-dione, 5-[1-(4-hydroxy-4-methylpentyl)vinyl]-2-methyl-	MeOH	240(3.86)	88-3185-69
2-Cyclohexen-1-one, 3-ethoxy-5,5-dimethyl-2-(2-oxocyclopentyl)-	EtOH	267(4.12),340(3.02)	104-1196-69
2,6-Methano-7H-1-benzoxocin-7-one, 2,3,4,5,6,8,9,10-octahydro-2-methoxy-9,9-dimethyl-	EtOH	264(3.93)	104-1566-69
2(3H)-Naphthalenone, 4,4a,5,6,7,8-hexahydro-7β-(1-hydroxyethyl)-1,4aβ-dimethyl-, formate	EtOH	247(4.26)	88-3871-69
Nardosinone	EtOH	249(3.67),315(1.58)	24-2691-69
Valdiviolide	MeOH	221(4.01)	78-3903-69
$C_{15}H_{22}O_4$			
Apo-9'-fucoxanthinone(C_{15})	EtOH	232(4.09)	39-0429-69C
Epilaccilaksholic acid	EtOH	218(3.80)	78-3855-69
Epipulchellin	EtOH	208(3.99)	88-2075-69
Euryopsol	EtOH	220(3.83)	78-5227-69
Inunolide, dihydro-, diepoxide	n.s.g.	212(2.39)	2-0310-69
Isabelin, hexahydro-	EtOH	209(2.56)	78-4767-69
Naphtho[1,2-c]furan-3(1H)-one, 5,5aα,6,7,8,9,9a,9b-octahydro-5β,9bα-dihydroxy-6,6,9aβ-trimethyl-	MeOH	208(3.91)	78-3895-69
Propiophenone, 2',4',6'-trihydroxy-3-isopentyl-2-methyl-	EtOH	292(4.24)	39-0938-69C

Compound	Solvent	$\lambda_{max}(\log \epsilon)$	Ref.
$C_{15}H_{22}O_5$			
1-Cyclopentene-1-heptanoic acid, 2-carboxy-5-oxo-, dimethyl ester	MeOH	246(3.98)	88-4639-69
$C_{15}H_{22}O_6$			
1-Cyclopentene-1-heptanoic acid, 2-carboxy-3-hydroxy-5-oxo-, dimethyl ester	MeOH	237(4.07)	88-4639-69
2,4-Hexadiene-1,1,2-tricarboxylic acid, 3,4,5-trimethyl-, trimethyl ester	isooctane	243s(3.40)	33-0584-69
$C_{15}H_{22}O_8$			
Laksholic acid	EtOH	230(3.82)	78-3841-69
2H-Pyran-5-carboxylic acid, 3,4-diformylmethyl-2-oxo-, methyl ester, bis-(dimethylacetal)	EtOH	254(3.78),292(3.76)	35-6470-69
$C_{15}H_{23}BrO$			
2,5-Cyclohexadien-1-one, 4-bromo-2,6-di-tert-butyl-4-methyl-	hexane	251(4.13)	55-0074-69
$C_{15}H_{23}Cl$			
Phenanthrene, 9-chloro-1,2,3,4,5,6,7,8,-8a,9,10,10a-dodecahydro-10-methyl-	EtOH	209(4.06)	22-0962-69
$C_{15}H_{23}ClO$			
2,5-Cyclohexadien-1-one, 2,6-di-tert-butyl-4-chloro-4-methyl-	hexane	240(4.12)	55-0074-69
$C_{15}H_{23}IN_2O$			
1-(5-Piperidinofurfurylidene)piperidinium iodide	n.s.g.	405(4.38)	103-0434-69
$C_{15}H_{23}N$			
Indoline, 3-isohexyl-6-methyl-	C_6H_{12}	211(4.28),251(3.77), 300(3.48)	1-2578-69
Indoline, 2-isopentyl-3,6-dimethyl-	C_6H_{12}	211(4.33),248(3.84), 300(3.53)	1-2578-69
$C_{15}H_{23}NO$			
Toluene, 3,5-di-tert-butyl-2-nitroso-	hexane	810(1.59)	18-3611-69
$C_{15}H_{23}NO_2$			
Bicyclo[5.3.1]undec-7(11)-en-8-one, 11-morpholino-	EtOH	355(4.21)	39-0592-69C
Hydroquinone, 2-[(cyclohexylmethylamino)methyl]-5-methyl-	EtOH	297(3.63)	39-1245-69C
Hydroquinone, 2-[(cyclohexylmethylamino)methyl]-6-methyl-	EtOH	298(3.56)	39-1245-69C
Hydroquinone, 2-[(hexahydro-1(2H)-azocinyl)methyl]-5-methyl-	EtOH	299(3.25)	39-1245-69C
Hydroquinone, trimethyl(piperidinomethyl)-	EtOH	294(3.55)	39-1245-69C
	ether	297(3.56)	39-1245-69C
Octanohydroxamic acid, N-p-tolyl-	EtOH	248(4.04)	42-0831-69
$C_{15}H_{23}NO_3$			
2,5-Cyclohexadien-1-one, 2,6-di-tert-butyl-4-methyl-4-nitro-	MeOH	234(4.02)	64-0547-69

Compound	Solvent	λ_{max}(log ϵ)	Ref.
$C_{15}H_{23}NS_2$			
Spiro[cycloheptane-1,2'(4'H)-cyclohepta-[d][1,3]thiazine]-4'-thione, 1',5',6',7',8',9'-hexahydro-	EtOH	335(3.67),409(4.32)	44-0730-69
$C_{15}H_{23}N_3O_2$			
Cyclohexanone, 2-(2-ethyl-4,5,6,7-tetra-hydro-1-benzimidazolyl)-, oxime, 3-oxide	MeOH	223(3.87)	48-0746-69
$C_{15}H_{23}N_5O_4$			
Adenosine, N-pentyl-	pH 1.0	263(4.27)	87-1056-69
	pH 7.0	210(4.26),268(4.22)	87-1056-69
	pH 12.0	268(4.24)	87-1056-69
$C_{15}H_{23}O_4P$			
Phosphonic acid, benzoyl-, dibutyl ester	C_6H_{12}	259(4.11),380(2.01)	18-0821-69
Phosphonic acid, (p-tert-butylbenzoyl)-, diethyl ester	C_6H_{12}	271(4.06),375(2.16)	18-0821-69
$C_{15}H_{24}$			
Allofarnesene	EtOH	280(4.08)	44-3789-69
C_{10}-cis	EtOH	266(4.45),277(4.56), 288(4.47)	18-3615-69
C_{10}-trans	EtOH	270(4.43),281(4.55), 292(4.43)	18-3615-69
β-Cubebene	MeOH	209(3.99)	88-1251-69
1,5,7-Cyclodecatriene, 8-isopropyl-1,5-dimethyl-	hexane	255(4.04)	88-1799-69
α-Farnesene, cis	EtOH	238(4.05)	18-3615-69
α-Farnesene, trans-trans	EtOH	232(4.37)	44-3789-69
β-Farnesene, trans	heptane	224(4.15)	44-3789-69
Iso-α-gurjunene A	EtOH	255(4.17)	78-1785-69
Iso-α-gurjunene B	EtOH	250(4.23)	78-1785-69
$C_{15}H_{24}ClNO_4$			
2,3,4,4a,5,7,8,9,10,11-Octahydro-6-meth-yl-1H-cyclohepta[c]quinolizinium perchlorate	$CHCl_3$	304(3.78)	78-4161-69
4',5',6',7'-Tetrahydro-1',3'-dimethyl-spiro[cyclohexane-1,2'-[2H]indolium] perchlorate	EtOH	282(3.85)	78-4161-69
$C_{15}H_{24}N_2O_2$			
Benzoic acid, p-(butylamino)-, 2-(di-methylamino)ethyl ester	EtOH	225(4.08),298(4.56)	9-0254-69
Nuttalline	EtOH	208(3.75)	77-0660-69
$C_{15}H_{24}N_2O_4$			
2,3-Diazabicyclo[4.1.0]oct-4-ene-2,3-dicarboxylic acid, 1,4,5,6,7,7-hexa-methyl-, dimethyl ester	EtOH	218(3.97),260s(2.82), 294s(2.44)	88-4785-69
$C_{15}H_{24}O$			
Aristolone, 9,10-dihydro-, (+)-	MeOH	213(3.63)	23-4307-69
p-Cresol, 2,6-di-tert-butyl-	MeOH	279(3.26)	30-1113-69A
	MeOH-Na	254(3.91),307(3.69)	30-1113-69A
	50% MeOH-acid	275(3.22)	28-2217-69A

Compound	Solvent	$\lambda_{max}(\log \epsilon)$	Ref.
2,6-Cyclodecadien-1-one, 10-isopropyl-4,7-dimethyl-	MeOH	208(3.57),241(3.74)	88-4295-69
2,7,9,11-Dodecatetraen-6-ol, 2,6,10-trimethyl-, (E,E)-	n.s.g.	251s(--),260(--), 270(4.63),281(--)	12-2403-69
1,7-Ethanoazulen-10-one, decahydro-4,8,8-trimethyl-	EtOH	280(1.42)	78-4895-69
Isofuranogermacrene, tetrahydro-, (+)-	dioxan	222(3.89)	39-1491-69C
2(4aH)-Naphthalenone, 7-isopropyl-1,4a-dimethyl-3,4,5,6,7,8-hexahydro-	EtOH	251(4.20)	78-1037-69
isomer	EtOH	252(4.28)	78-1037-69
Phenol, 2,4,6-tripropyl-	pH 2	277(3.28)	30-1113-69A
	pH 13	238(3.93),297(3.64)	30-1113-69A
	MeOH	281(3.33)	30-1113-69A
	MeOH-Na	242(3.93),301(3.65)	30-1113-69A
$C_{15}H_{24}OS_2$			
2-Bornanone, 3-carbonyl-, 3-(diethyl mercaptole), (+)-	dioxan	325(4.08),375s(2.80)	28-0186-69A
$C_{15}H_{24}O_2$			
Bicyclo[6.2.0]dec-9-en-3-ol, 1,8,9,10-tetramethyl-, formate	hexane	278(2.88)	24-0275-69
1,3-Cyclohexadiene-1-propanol, α,2,6,6-tetramethyl-, acetate	n.s.g.	242(3.62),250(3.64), 263(3.66)	88-1803-69
2,5-Cyclohexadien-1-one, 2,6-di-tert-butyl-4-hydroxy-4-methyl-	MeOH	235(4.01)	64-0547-69
Hydroxyketone from tricyclohumuludiol oxidation	hexane	203(3.45)	18-2405-69
5(6H)-Indanone, 1β-tert-butoxy-7aβ-ethyl-7,7a-dihydro-	EtOH	241(4.11)	44-0107-69
$C_{15}H_{24}O_3$			
Allotorilolone	EtOH	291(1.48)	78-4751-69
Cryptomerone	MeOH	234(3.73)	88-3185-69
2-Cyclododecene-1-carboxylic acid, 2-methyl-12-oxo-, methyl ester	EtOH	204(3.87),207s(3.8), 245f(3.39)	94-0629-69
1-Epiallotorilolone	EtOH	290(1.49)	78-4751-69
Nardosinonediol	EtOH	250(3.81),320(1.77)	24-2691-69
Sesquiterpene from Fungus Lavicis	EtOH	205(3.28),225(1.4)	83-0965-69
Torilolone	EtOH	241(4.0)	78-4751-69
$C_{15}H_{24}O_4$			
Aspertetronin A, de-O-methylhexahydro-	EtOH	211(3.70),233(3.73), 268(4.01)	39-0056-69C
1,3-Cyclopentanedione, 4-hydroxy-5-isopentyl-2-isovaleryl-, cis	MeOH-acid	226(4.06),266(3.98)	20-0017-69
	MeOH-base	254(4.31),266s(4.31)	20-0017-69
trans	MeOH-acid	226(4.06),266(3.98)	20-0017-69
	MeOH-base	254(4.31),266s(4.24)	20-0017-69
3-Cyclopentene-1-carboxylic acid, 2,4-di-tert-butyl-2-hydroxy-5-oxo-, methyl ester	EtOH	228(3.91)	88-1673-69
1-Cyclopentene-1-heptanoic acid, 2-methoxy-5-oxo-, ethyl ester	EtOH	252(4.26)	94-0408-69
$C_{15}H_{25}N$			
Bicyclo[3.1.0]hex-2-ene-6-methylamine, N,N,1,2,3,5,6-heptamethyl-4-methylene-, perchlorate	EtOH	257(3.88)	35-6107-69
Cyclohexylamine, N-(2-cyclohexylidenepropylidene)-	EtOH	247(4.38)	39-0460-69C

Compound	Solvent	$\lambda_{max}(\log \epsilon)$	Ref.
4-Norcarene-2-methylamine, N,N,1,2,4,5-hexamethyl-3-methylene-	EtOH	257(4.2)	44-2393-69
4-Norcarene-2-methylamine, N,N,1,2,4,6-hexamethyl-3-methylene-	EtOH	257(4.0)	44-2393-69
$C_{15}H_{25}NO_3S_3$			
Capronic acid, ε-aminodithio-, ethyl ester, p-toluenesulfonate	MeOH	306(3.88)	5-0237-69I
	$CHCl_3$	445(1.36)	5-0237-69I
$C_{15}H_{25}NO_4$			
Supinine	EtOH	212(3.27)	73-1832-69
$C_{15}H_{25}N_2O_6P$			
Triethyl(5-nitrosalicyl)ammonium hydroxide, ethyl hydrogen phosphate, inner salt	H_2O	288(3.95)	35-4532-69
$C_{15}H_{26}ClN_4O_8P$			
Uridine, 5'-[hydrogen[3-[(2-chloroethyl)methylamino]propyl]phosphoramidate]	pH 2	262(4.00)	65-0668-69
	pH 12	225(4.19),262(3.92)	65-0668-69
$C_{15}H_{26}N_2O$			
Sparteine, 4α-hydroxy-, (-)-	EtOH	207(3.70)	77-0660-69
$C_{15}H_{26}N_2O_2$			
3-Penten-2-one, 4,4'-[(1,3-dimethyltrimethylene)dimino]di-	MeOH	318(3.42)	39-2044-69C
$C_{15}H_{26}O_2$			
1,2-Cyclopentadecanedione	hexane	276(1.58),442(1.31)	23-3266-69
Isocalamendiol	MeOH	203(3.2)	88-3729-69
$C_{15}H_{26}O_3$			
Nardosinonediol, dihydro-	EtOH	290(3.76)	24-2691-69
$C_{15}H_{27}ClN_2O_4$			
1-[3-(Hexahydro-1H-azepin-1-yl)allylidene]hexahydro-1H-azepinium perchlorate	EtOH	317(4.73)	24-2609-69
$C_{15}H_{28}N_2$			
11-Azabicyclo[4.4.1]undec-1-ene, 5-[2-(dimethylamino)ethyl]-11-methyl-	isooctane	248(3.55)	35-3616-69
$C_{15}H_{28}N_2O_2$			
1-Heptene, 1,2-dimorpholino-	C_6H_{12}	200(3.99),225(3.94)	22-3883-69
1-Pentene, 4,4-dimethyl-1,2-dimorpholino-	C_6H_{12}	231(3.86)	22-3883-69
$C_{15}H_{28}O$			
Cyclohexanone, 2-ethyl-3,5-diisopropyl-2-methyl-	MeOH	290(1.54)	88-0869-69
$C_{15}H_{32}Sn$			
Stannane, allyltributyl-	MeCN	200(4.04)	44-3715-69
$C_{15}H_{45}NSi_6$			
Methylamine, N,N-bis(heptamethyltrisilan-1-yl)-	isooctane	223s(4.22)	35-6613-69

Compound	Solvent	$\lambda_{max}(\log \epsilon)$	Ref.
$C_{16}HClF_{10}O$			
Furan, 3-chloro-2,4-bis(pentafluorophen-yl)-	EtOH	214(4.23),261(4.14)	4-0253-69
$C_{16}H_2Cl_{16}$			
1,3,5,7,9,11,13,15-Hexadecaoctaene, 1-16-hexadecachloro-	hexane	215(4.71),255(4.32)	78-0883-69
$C_{16}H_4N_6O_8$			
Fluorene-$\Delta^{9,\alpha}$-malononitrile, 2,4,5,7-tetranitro-, radical ion(1-),			
1,3-dimethylbenzimidazolium complex	acetone	380(3.95),515(4.03)	28-0557-69A
2,3-dimethylbenzothiazolium complex	acetone	380(3.97),515(4.05)	28-0557-69A
3-ethylbenzothiazolium complex	acetone	380(3.92),515(4.03)	28-0557-69A
3-ethyl-1-methylbenzimidazolium complex	acetone	380(3.94),515(4.02)	28-0557-69A
1-ethylquinolinium complex	acetone	380(3.95),515(4.04)	28-0557-69A
1-methylquinolinium complex	acetone	380(3.95),515(4.02)	28-0557-69A
$C_{16}H_5N_5O_6$			
Fluorene-$\Delta^{9,\alpha}$-malononitrile, 2,4,7-tri-nitro-, radical ion(1-),			
cyclohexylethylammonium complex	acetone	380(3.86),515(4.05)	28-0557-69A
1,3-dimethylbenzimidazolium complex	acetone	380(3.81),515(4.02)	28-0557-69A
3-ethylbenzothiazolium complex and other heterocyclic cations	acetone	380(3.76),515(3.99)	28-0557-69A
$C_{16}H_6Cl_8O_7$			
Spiro[1,3-benzodioxole-2,1'-[2]cyclohex-en]-4'-one, 2',3',4,5,6,6',6',7-octa-chloro-5,5'-dihydroxy-, diacetate	$CHCl_3$	257(4.23),292(3.50), 301(3.48)	44-1966-69
$C_{16}H_6F_5NO_2$			
2-Oxazolin-5-one, 4-(pentafluorobenzyli-dene)-2-phenyl-	ether	237(4.18),254(4.17), 310(4.46)	44-0534-69
$C_{16}H_6F_6$			
Fluoranthene, 6b,7,8,9,10,10a-hexa-fluoro-6b,10a-dihydro-	n.s.g.	222(4.74),289(3.83)	39-0760-69C
$C_{16}H_6F_{10}O_2$			
2,5-Cyclohexadien-1-one, 2,3,4,5,6-pen-tafluoro-4-[2-(pentafluorophenoxy)-3-butenyl]-	heptane	222(3.52)	70-2324-69
$C_{16}H_8BrClO_3$			
1,3-Indandione, 2-(3-bromo-5-chlorosali-cylidene)-	dioxan	240(4.82),330(4.25), 387(4.11),545(2.51)	65-1377-69
1,3-Indandione, 2-(5-bromo-3-chlorosali-cylidene)-	dioxan	245(4.43),332(4.34), 380(4.19),545(2.20)	65-1377-69
$C_{16}H_8BrNO_5$			
1,3-Indandione, 2-(5-bromo-3-nitrosali-cylidene)-	dioxan	245(4.34),325(4.27), 360s(4.00),500(2.85)	65-1377-69
$C_{16}H_8Br_2O_3$			
1,3-Indandione, 2-(3,5-dibromo-4-hydr-oxybenzylidene)-	dioxan	257(4.51),370(4.54), 480(3.33)	65-1373-69
1,3-Indandione, 2-(3,5-dibromo-salicylidene)-	dioxan	240(4.37),333(4.32), 387(4.16),550(2.86)	65-1377-69

Compound	Solvent	λ_{max}(log ε)	Ref.
$C_{16}H_8Br_4$			
Cyclobuta[1]phenanthrene, 1,1,2,2-tetra-bromo-1,2-dihydro-	n.s.g.	256(4.81),275(4.69), 324(4.47)	88-0457-69
$C_{16}H_8ClNO_5$			
1,3-Indandione, 2-(5-chloro-3-nitro-salicylidene)-	dioxan	263(4.40),330(4.11), 395(4.11),560(3.25)	65-1377-69
$C_{16}H_8Cl_2O_2$			
1,3-Indandione, 2-(2,6-dichlorobenzyli-dene)-	C_6H_{12}	240(4.45),250(4.46), 259(4.28),320(3.98), 370s(2.87),389s(2.59)	108-0099-69
	MeCN	240(4.45),250(4.48), 260(4.35),321(4.02), 365s(3.04),391s(2.62)	108-0099-69
$C_{16}H_8Cl_2O_3$			
1,3-Indandione, 2-(3,5-dichloro-4-hydr-oxybenzylidene)-	dioxan	258(4.43),365(4.48), 480(3.20)	65-1373-69
1,3-Indandione, 2-(3,5-dichlorosalicyli-dene)-	dioxan	240(4.43),330(4.41), 390(4.24),540(1.97)	65-1377-69
$C_{16}H_8F_8$			
[2,2]Paracyclophane, octafluoro-	n.s.g.	222(4.08),242s(3.62), 291(3.21)	35-1862-69
$C_{16}H_8N_2O_4$			
Isocoumarin, 4-cyano-3-(p-nitrophenyl)-	EtOH	222s(4.39),233s(4.30), 250s(4.20),303s(4.09), 320s(4.20),335(4.22)	12-0577-69
$C_{16}H_8N_4O$			
Dibenzo[f,h][1,2,5]oxadiazolo[3,4-b]-quinoxaline	dioxan	238(4.78),265(4.31), 272(4.29),289(4.12), 299(4.28),377(4.12)	4-0769-69
$C_{16}H_8N_4O_4$			
Phthalimide, N,N-azodi-	CH_2Cl_2	251(4.45),268(4.57), 337(3.40)	27-0405-69
$C_{16}H_8O_2$			
Cyclohepta[def]fluorene-4,8-dione	EtOH	231(4.41),252s(4.47), 260(4.59),312(3.81)	39-1427-69C
$C_{16}H_8O_4$			
Oxindigo, cis	C_6H_{12}	396(4.12)	77-0133-69
Oxindigo, trans	C_6H_{12}	413(4.14)	77-0133-69
$C_{16}H_9BrO_3$			
1,3-Indandione, 2-(3-bromo-4-hydroxy-benzylidene)-	dioxan	255(4.46),375(4.55), 480(2.92)	65-1373-69
1,3-Indandione, 2-(5-bromosalicylidene)-	dioxan	243(4.27),327(4.09), 395(4.05)	65-1377-69
$C_{16}H_9Cl$			
Pyrene, 2-chloro-	EtOH	237(4.98),246(5.27), 254(4.47),265(4.65), 277(4.89),295(3.93), 307(4.29),320(4.69), 336(4.92)	24-2301-69

Compound	Solvent	λ_{max}(log ε)	Ref.
$C_{16}H_9ClK_2N_2O_3$			
1H-1,4-Benzodiazepine-3-carboxylic acid, 7-chloro-2-hydroxy-5-phenyl-, dipotassium salt	H_2O	231(4.53),311(3.39)	111-0239-69
$C_{16}H_9ClO_2$			
1,3-Indandione, 2-(m-chlorobenzylidene)-	C_6H_{12}	225(4.23),243(4.39), 253(4.32),264(4.33), 336(4.55),352(4.43), 412s(2.66)	108-0099-69
	MeCN	223(4.15),241(4.30), 263(4.24),336(4.46), 400s(2.79)	108-0099-69
1,3-Indandione, 2-(o-chlorobenzylidene)-	C_6H_{12}	241(4.32),262(4.23), 334(4.40),408s(2.66)	108-0099-69
	MeCN	241(4.32),262(4.28), 335(4.53),400s(2.86)	108-0099-69
1,3-Indandione, 2-(p-chlorobenzylidene)-	C_6H_{12}	239(4.42),245(4.45), 253(4.35),263(4.36), 332(4.54),346(4.68), 362(4.49),405s(2.96)	108-0099-69
	MeCN	243(4.39),263(4.29), 343(4.60),363(4.40), 410s(3.00)	108-0099-69
	dioxan	240(4.49),345(4.66)	65-1373-69
$C_{16}H_9ClO_3$			
1,3-Indandione, 2-(5-chlorosalicylidene)-	dioxan	263(4.27),325(4.24), 394(4.24)	65-1377-69
$C_{16}H_9Cl_3O_2$			
Spiro[oxetane-2,9'(10'H)-phenanthren]-10'-one, 3,3,4-trichloro-	dioxan	336(3.41)	24-3747-69
$C_{16}H_9FO_2$			
1,3-Indandione, 2-(p-fluorobenzylidene)-	C_6H_{12}	237(4.35),242(4.39), 250(4.27),262(4.27), 344(4.54),403s(2.79)	108-0099-69
	MeCN	242(4.34),260(4.23), 340(4.53),405s(2.97)	108-0099-69
$C_{16}H_9IO_2$			
1,3-Indandione, 2-(p-iodobenzylidene)-	C_6H_{12}	254(--),264(--), 342(--),357(--), 375(--),410s(--)	108-0099-69
	MeCN	246(4.33),263(4.28), 353(4.55),420s(3.08)	108-0099-69
$C_{16}H_9NO_2$			
5H-Benzo[a]phenoxazin-5-one	EtOH	430(4.11)	73-0221-69
9H-Benzo[a]phenoxazin-9-one	EtOH	485(4.27)	73-0221-69
$C_{16}H_9NO_3$			
9H-Benzo[a]phenoxazin-9-one, 5-hydroxy-	EtOH	461(3.99)	73-0221-69
anion	EtOH	578(4.30)	73-0221-69
$C_{16}H_9NO_4$			
1,3-Indandione, 2-(o-nitrobenzylidene)-	C_6H_{12}	259(4.55),333(4.20), 372(3.13),405s(2.48), 426s(2.40)	108-0099-69

Compound	Solvent	$\lambda_{max}(\log \epsilon)$	Ref.
1,3-Indandione, 2-(o-nitrobenzylidene)-	MeCN	259(4.53),323(4.14), 370(3.26),400s(2.67), 420s(2.51)	108-0099-69
1,3-Indandione, 2-(p-nitrobenzylidene)-	C_6H_{12}	237(--),245(--), 260(--),325(--), 337(--),353(--), 420s(--)	108-0099-69
	MeCN	243(4.28),255(4.22), 264(4.25),338(4.54), 420s(2.76)	108-0099-69
$C_{16}H_9N_3O_2$			
Cinnamonitrile, α-cyano-β-(p-nitrophenyl)-	EtOH	280(4.24),315(4.17)	2-0412-69
$C_{16}H_9N_3O_8$			
Coumarin, 3-(2,4-dinitrophenyl)-8-methoxy-6-nitro-	EtOH	340(--)	2-0049-69
Coumarin, 3-(2,4-dinitrophenyl)-8-methoxy-7-nitro-	EtOH	320(--)	2-0049-69
$C_{16}H_{10}$			
Naphtho[2,1,8-cde]azulene	EtOH	241(4.7),278(4.2), 345(3.7),399(3.3), 416(3.4),445(3.3), 570(2.9)	5-0061-69C
$C_{16}H_{10}BrNS$			
5H-Thieno[2,3-b]indole, 2-(p-bromophenyl)-	EtOH	240(4.71),287(4.24), 293(4.25),345(4.57)	95-0058-69
$C_{16}H_{10}Br_2$			
Butatriene, 1,4-dibromo-1,4-diphenyl-	CH_2Cl_2	251(4.11),305(3.72), 357(4.32),385(4.56), 402(4.68)	88-0435-69
$C_{16}H_{10}ClKN_2O_3$			
1H-1,4-Benzodiazepine-3-carboxylic acid, 7-chloro-2-hydroxy-5-phenyl-, potassium salt	H_2O	229(4.52),310(3.33)	111-0239-69
$C_{16}H_{10}Cl_2O_2$			
Phenanthro[9,10-b]-p-dioxin, 2,2-dichloro-2,3-dihydro-	benzene	357(3.03)	24-3747-69
Spiro[oxetane-2,9'(10'H)-phenanthren]-10'-one, 3,3-dichloro-	dioxan	333(3.41)	24-3747-69
Spiro[oxetane-2,9'(10'H)-phenanthren]-10'-one, 3,4-dichloro-	dioxan	333(3.44)	24-3747-69
isomer	dioxan	333(3.43)	24-3747-69
$C_{16}H_{10}Cl_2O_4$			
Flavone, 3',5'-dichloro-7-hydroxy-4'-methoxy-	EtOH	232(4.40),260s(4.10), 312(4.26)	111-0298-69
	EtOH-NaOH	270(4.47),375(4.18)	111-0298-69
$C_{16}H_{10}Cl_2O_5$			
Flavone, 3',5'-dichloro-5,7-dihydroxy-4'-methoxy-	EtOH	274(4.37),325(4.03)	111-0298-69
	EtOH-NaOH	278(4.42),372(4.01)	111-0298-69

Compound	Solvent	λ_{max}(log ϵ)	Ref.
$C_{16}H_{10}Cl_4O$			
1,4,7-Metheno-5H-dibenzo[a,c]cyclohepten-5-one, 1,2,3,4-tetrachloro-1,4,4a,6,7,11b-hexahydro-	C_6H_{12}	261(2.51),268(2.63), 276(2.64)	88-1185-69
$C_{16}H_{10}Cl_8$			
Benzo[g]biphenylene, 1,2,3,4,7,8,9,10-octachloro-4a,5,6,6a,6b,10a,10b,10c-octahydro-	EtOH	287(3.65),296(3.75), 309(3.63)	39-2710-69C
$C_{16}H_{10}F_4$			
Phenanthrene, 1,2,3,4-tetrafluoro-6,8-dimethyl-	MeOH	240(4.63),249(4.78), 263(4.32),272(4.21), 295(4.05),307(4.10), 338(3.02),352(2.95)	39-1684-69C
$C_{16}H_{10}N_2$			
Benzobicyclo[2.2.2]octatriene, dicyanoacetylene adduct	EtOH	254(3.99),295(3.65)	78-2417-69
Malononitrile, (diphenylmethylene)-	C_6H_{12}	228(4.02),275s(4.02), 315(4.20)	44-2146-69
$C_{16}H_{10}N_2O$			
9H-Benzo[a]phenoxazine, 9-imino-	EtOH	535(3.99)	73-0221-69
cation	EtOH	542(4.11)	73-0221-69
2,2-Oxiranedicarbonitrile, 3,3-diphenyl-	MeCN	226(4.08),256(2.88), 258(2.88),262(2.93), 269(2.84)	44-2146-69
$C_{16}H_{10}N_2OS_2$			
2H-1,4-Benzothiazin-3(4H)-one, 2-(2-benzothiazolylmethylene)-	EtOH	234s(4.43),286(4.14), 397s(4.24),410(4.26)	32-0323-69
	EtOH-0.1N HCl	231s(4.44),288(4.06), 297s(4.22),410(4.26)	32-0323-69
	EtOH-2N HCl	230s(4.37),292(3.85), 422s(4.15),444(4.18)	32-0323-69
12H-p-Dithiino[2',3':3,4]pyrrolo[1,2-a]perimidin-12-one, 9,10-dihydro-	MeCN	223(4.77),252(4.14), 284(4.23),335(3.70), 338(3.70),351(3.71), 405(4.12),496(3.68)	33-2221-69
$C_{16}H_{10}N_2O_2$			
5H-Benzo[a]phenoxazin-5-one, 9-amino-	EtOH	498s(--)	73-0221-69
9H-Benzo[a]phenoxazin-9-one, 5-amino-	EtOH	573(4.65),595(4.79), 615(4.92)	73-0221-69
$C_{16}H_{10}N_2O_3$			
Isoindigo, 1-hydroxy-	dioxan	476(3.55)	24-3691-69
$C_{16}H_{10}N_2O_4$			
Anthra[1,2-a][1,2,5]oxadiazole-6,11-dione, 4-ethoxy-	EtOH	420(3.67)	104-1642-69
Isoindigo, 1,1'-dihydroxy-	dioxan	480(3.43)	24-3691-69
1(4H)-Naphthalenone, 2-hydroxy-4-(m-nitroanil)-	pH 0.00	242(5.17),248(5.17), 315(4.68)	78-5807-69
1,2-Naphthoquinone, 4-(m-nitroanilino)-	pH 9.0	237(5.21),280(4.92)	78-5807-69
$C_{16}H_{10}N_2O_7$			
Coumarin, 3-(2,4-dinitrophenyl)-8-methoxy-	EtOH	252(4.22)	2-0049-69
	EtOH-NaOAc	311(4.11)	2-0049-69

Compound	Solvent	$\lambda_{max}(\log \epsilon)$	Ref.
Coumarin, 7-methoxy-6-nitro-3-(p-nitro-phenyl)-	EtOH	270(3.64)	2-0049-69
	EtOH-NaOAc	345(4.82)	2-0049-69
Coumarin, 8-methoxy-6-nitro-3-(p-nitro-phenyl)-	EtOH	310(4.33)	2-0049-69
Coumarin, 8-methoxy-7-nitro-3-(p-nitro-phenyl)-	EtOH	325(4.16)	2-0049-69
$C_{16}H_{10}N_2S_2$			
Thieno[3,4-c][1,2,5]thiadiazole-5-S^{IV}, 4,6-diphenyl-	CH_2Cl_2	275(4.22),312(4.33), 330(4.30),558(3.94)	35-6891-69
$C_{16}H_{10}N_4O$			
[1,2,5]Oxadiazolo[3,4-b]pyrazine, 5,6-diphenyl-	EtOH	250s(3.88),354(3.93)	4-0769-69
$C_{16}H_{10}N_4O_2$			
2,2'-Bi-1,3,4-oxadiazole, 5,5'-diphenyl-	EtOH	207(4.00),225(3.78), 298(4.19)	39-1416-69C
Pyrazole-5-carbonitrile, 1-(p-nitrophen-yl)-3-phenyl-	MeOH	228(4.34),237(4.34), 285s(3.75),323(4.25)	39-2587-69C
$C_{16}H_{10}N_4O_2S_2$			
1,2,4-Oxadiazole, 3,3'-dithiobis-[5-phenyl-	EtOH	252(4.57)	39-2794-69C
$C_{16}H_{10}N_4O_3$			
Furo[3,2-e]-s-triazolo[1,5-c]pyrimidine, 8,9-di-2-furyl-2-methyl-	EtOH	238(4.28),251s(4.19), 297(4.30),343(4.40)	95-1434-69
Furo[3,2-e]-s-triazolo[4,3-c]pyrimidine, 8,9-di-2-furyl-3-methyl-	EtOH	246(4.19),256(4.21), 313(4.31),344(4.36)	95-1434-69
$C_{16}H_{10}N_4O_4$			
Cinnolino[5,4,3-cde]cinnoline-2,7-di-carboxylic acid, dimethyl ester	EtOH	262(4.70),284s(4.32), 323s(3.94),334(4.06), 344(4.06),376(3.25), 395(3.15)	4-0533-69
$C_{16}H_{10}N_4O_{12}$			
4,4'-Biphenyldicarboxylic acid, 2,2',6,6'-tetranitro-, dimethyl ester	EtOH	220(4.53)	4-0533-69
$C_{16}H_{10}O$			
Cyclohepta[def]fluoren-8(4H)-one	EtOH	213(4.34),245s(4.32), 253(4.40),267(4.42), 291(4.15),307(4.19), 357(3.61)	39-1427-69C
8,16-Oxido[2,2]metacyclophane-1,9-diene	C_6H_{12}	239(4.20),302(4.21)	35-1665-69
$C_{16}H_{10}O_2$			
Benzoditropone	MeOH	240(4.17),246(4.20), 296(5.07),320s(4.33), 337(4.21),350s(3.79), 376(3.65),398(3.58)	64-0464-69
	H_2SO_4	326(5.12)	64-0464-69
2-Butyne-1,4-dione, 1,4-diphenyl-	EtOH	239(4.38),245(4.45), 282(4.02),291(4.06), 333(3.96)	44-0874-69
Cyclohepta[def]fluorene-4,8-dione, 9,10-dihydro-	EtOH	229(4.66),261s(4.39), 270(4.47),279(4.42), 313(3.88)	39-1427-69C

Compound	Solvent	λ_{max}(log ϵ)	Ref.
1,3-Indandione, 2-benzylidene-	C_6H_{12}	239(4.27),251(4.22), 261(4.22),343(4.48), 405s(2.56)	108-0099-69
	MeCN	239(4.30),263(4.24), 342(4.53),405s(2.91)	108-0099-69
	dioxan	240(4.29),340(4.53)	65-1373-69
Unidentified phenol	EtOH	215(4.22),237(4.55), 257(4.49),264(4.65), 285(5.04),288(4.00), 296(4.05),310(3.72), 317s(3.53),325(3.61), 332(3.34),340(3.72)	39-1982-69C
$C_{16}H_{10}O_3$			
1,3-Indandione, 2-(p-hydroxybenzylidene)-	dioxan	250(4.41),380(4.66)	65-1373-69
1,3-Indandione, 2-salicylidene-	dioxan	260(4.24),333(4.21), 387(4.31)	65-1377-69
Spiro[1,3-benzodioxole-2,1'(2'H)-naphthalen]-2'-one	C_6H_{12}	275s(3.72),280(3.78), 285(3.81),291(3.79), 322(3.88),350s(3.28)	39-1982-69C
Spiro[1,3-benzodioxole-2,1'(4'H)-naphthalen]-4'-one	C_6H_{12}	265(3.83),274(3.88), 278(3.89),284(3.88), 290(3.79),310s(2.95)	39-1982-69C
Spiro[2,5-cyclohexadien-1,2'-naphtho-[2,3-d][1,3]dioxol]-4-one	C_6H_{12}	256(3.87),266(3.88), 276(3.85),287(3.69), 299(3.36),305(3.45), 312(3.60),318(3.58), 326(3.70)	39-1982-69C
$C_{16}H_{10}O_6$			
2H,5H-Pyrano[4,3-b]pyran-2,5-dione, 8-acetyl-4-hydroxy-7-phenyl-	EtOH	245(4.20),278(4.20), 357(4.16)	44-2480-69
2H,5H-Pyrano[4,3-b]pyran-2,5-dione, 8-benzoyl-4-hydroxy-7-methyl-	EtOH	255(4.09),315(3.77)	44-2480-69
$C_{16}H_{10}O_7$			
Xanthen-9-one, 1-methoxy-2,3:6,7-bis-(methylenedioxy)-	EtOH	248(4.51),259s(--), 282s(--),321(4.31), 350s(--)	78-4415-69
$C_{16}H_{10}S$			
Benzo[b]naphtho[2,3-d]thiophene	EtOH	226(4.31),244(4.62), 263(4.56),275(4.59), 283(4.49),315(3.78), 330(3.78),355(3.5), 377(3.47)	4-0885-69
$C_{16}H_{11}BF_4$			
4H-Cyclohepta[def]fluorenylium tetrafluoroborate	20N H_2SO_4	234(4.08),245(4.12), 281(4.48),300(4.29), 323(4.22),481(3.85)	39-1427-69C
$C_{16}H_{11}Br$			
Naphthalene, 1-(o-bromophenyl)-	MeOH	262(3.68),272(3.86), 281(3.94),287s(3.85), 291(3.84),313(2.87)	35-6049-69

Compound	Solvent	λmax(log ε)	Ref.
$C_{16}H_{11}BrN_4O_8$			
2-Pyrazoline, 3-(bromoacetyl)-4-(5-nitro-2-furyl)-1-[3-(5-nitro-2-furyl)-acryloyl]-	$C_2H_4Cl_2$	238(4.17),316(4.48), 360(4.43)	94-2105-69
$C_{16}H_{11}BrO_2$			
2-Butene-1,4-dione, 1-(p-bromophenyl)-4-phenyl-	EtOH	228(4.24),232(4.25), 270(4.44)	104-0511-69
$C_{16}H_{11}Br_2ClO_2$			
1,4-Butanedione, 2,3-dibromo-1-(p-chlorophenyl)-4-phenyl-	EtOH	268(4.32)	104-0511-69
$C_{16}H_{11}Br_2NO_4$			
1,4-Butanedione, 2,3-dibromo-1-(p-nitrophenyl)-4-phenyl-	EtOH	230(4.12),265(4.53)	104-0511-69
$C_{16}H_{11}Br_3O_2$			
1,4-Butanedione, 2,3-dibromo-1-(p-brcmophenyl)-4-phenyl-	EtOH	269(4.45)	104-0511-69
$C_{16}H_{11}Cl$			
Naphthalene, 1-(o-chlorophenyl)-	MeOH	262s(3.65),272(3.84), 281(3.93),287s(3.85), 291(3.83),313(2.81)	35-6049-69
Naphthalene, 1-(p-chlorophenyl)-	MeOH	263s(3.84),275s(3.99), 281(4.00)	35-6049-69
$C_{16}H_{11}ClN_2O_3$			
1H-1,4-Benzodiazepine-3-carboxylic acid, 7-chloro-2-hydroxy-5-phenyl-, monopotassium salt	H_2O	229(4.52),310(3.33)	111-0239-69
dipotassium salt	H_2O	231(4.53),311(3.39)	111-0239-69
$C_{16}H_{11}ClO$			
Naphthalene, 1-(o-chlorophenoxy)-	MeOH	278s(3.82),281(3.85), 287(3.82),292s(3.80), 306s(3.51),312s(3.30), 320(3.14)	88-0625-69
$C_{16}H_{11}ClO_2$			
2-Butene-1,4-dione, 1-(p-chlorophenyl)-4-phenyl-	EtOH	232(4.18),269(4.34)	104-0511-69
$C_{16}H_{11}ClO_4$			
4H-Cyclohepta[def]fluorenylium perchlorate	20N H_2SO_4	234(4.06),246(4.09), 281(4.47),300(4.28), 323(4.19),481(3.82)	39-1427-69C
$C_{16}H_{11}I$			
Naphthalene, 1-iodo-8-phenyl-	MeOH	289s(3.90),299(3.97), 307s(3.96),313s(3.86)	35-6049-69
Naphthalene, 1-(o-iodophenyl)-	MeOH	262s(3.68),272(3.85), 282(3.93),288(3.83), 292(3.82),314(2.82)	35-6049-69
$C_{16}H_{11}N$			
7H-Benzo[c]carbazole	dioxan	264(4.75),286(4.06), 326(4.18),344(3.74), 362(3.68)	24-2728-69

Compound	Solvent	$\lambda_{max}(\log \epsilon)$	Ref.
11H-Benzo[a]carbazole	dioxan	251(4.68),280(4.65), 300(4.30),306(4.32), 320(3.80),337(3.78), 354(3.83)	24-2728-69
2H-Dibenz[e,g]isoindole	EtOH	210(4.34),230(4.38), 250(4.41),260(4.41), 276(4.24),288(4.06), 301(4.04),314(4.02), 352(3.25),385(2.94)	35-5192-69
$C_{16}H_{11}NO$			
7H-Benzo[c]carbazol-4-ol	dioxan	250(4.68),290(4.32), 310(4.03),320(4.06), 333(4.12),348(3.78), 364(3.77)	24-2728-69
11H-Benzo[a]carbazol-3-ol	dioxan	259(4.58),285(4.66), 300(4.30),310(4.28), 325(3.80),342(3.67), 359(3.72)	24-2728-69
11H-Benzo[a]carbazol-4-ol	dioxan	255(4.70),275(4.55), 300(4.32),310(4.35), 325(3.80),343(3.95), 360(4.13)	24-2728-69
Benzonitrile, p-(2-benzoylvinyl)-	MeOH	302(4.56)	73-2771-69
12H-Benzo[b]phenoxazine	n.s.g.	237(4.52),262(4.49), 315(4.00),360(3.93)	44-1691-69
5H-Dibenzo[a,d]cycloheptene-5-carbonitrile, 10,11-dihydro-10,11-epoxy-	EtOH	268(2.91)	23-2827-69
5H-Dibenzo[a,d]cycloheptene-5-carbonitrile, 10,11-dihydro-10-oxo-	EtOH	240(3.91),290(3.72)	23-2827-69
Phenanthro[9,10-d]oxazole, 2-methyl-	EtOH	255(4.71),302(3.93)	77-0200-69
$C_{16}H_{11}NOSe$			
2-Propen-1-one, 1-(2-quinolyl)-3-selenophene-2-yl-	hexane	252(4.29),292(4.15), 355(4.30)	103-0055-69
	EtOH	255(4.23),305(4.15), 366(4.30)	103-0055-69
2-Propen-1-one, 3-(2-quinolyl)-1-selenophene-2-yl-	hexane	235(4.29),280(4.34), 334(4.37)	103-0055-69
	EtOH	232(3.75),282(3.96), 337(4.44)	103-0055-69
$C_{16}H_{11}NO_2$			
8H-Acenaphtho[1,2-c]pyrrole-7-carboxylic acid, 8-methyl-	$CHCl_3$	349(3.40),400(3.82), 506(3.15),560(3.12)	97-0022-69
9-Anthronitrile, 10-hydroxy-2-methoxy-	EtOH	263(4.65),274s(4.52), 284s(4.34),329(3.16), 344(3.45),363(3.61), 399(3.75),418(3.81), 441(3.62)	39-1873-69C
Indolo[1,7-ab][1]benzazepine-1,2-dione, 6,7-dihydro-	EtOH	249(4.35),284s(3.64), 430(2.65)	35-1672-69
Isocarbostyril, 4-benzoyl-	C_6H_{12}	240(4.28),310s(4.00), 321(4.08),335s(3.93)	24-2987-69
1(4H)-Naphthalenone, 2-hydroxy-4-(phen-	pH 0.3	252(5.02),258(5.02), 295(4.82),350(4.71)	78-5807-69
1,2-Naphthoquinone, 4-anilino-	pH 1.2	244(5.07),276(4.83)	78-5807-69
	pH 7.5	241(5.08),272(4.84)	78-5807-69
	pH 11.0	240(5.11),278(4.85)	78-5807-69
Phenanthrene, 9-(2-nitrovinyl)-	EtOH	360(4.07)	87-0157-69

Compound	Solvent	λ_{max}(log ε)	Ref.
$C_{16}H_{11}NO_3$			
10,5-(Iminomethano)-5H-dibenzo[a,d]-cycloheptene-11,13(10H)-dione, 10 -hydroxy-	EtOH	263(3.88)	23-2827-69
$C_{16}H_{11}NO_3S$			
Ketone, 2-furyl 2-(2-furyl-2-benzothiazolinyl)	MeOH	220(4.4),290(4.2), 380(2.9)	24-0728-69
	MeOH-0.01N NaOH	220(4.4),290(4.2), 330(3.8),380(3.2)	24-0728-69
	MeOH-0.1N NaOH	220(4.3),280(4.3), 330(4.1),380(3.6)	24-0728-69
	MeOH-N NaOH	220(4.2),280(4.3), 330(4.2),380(3.7)	24-0728-69
$C_{16}H_{11}NO_4$			
2-Butene-1,4-dione, 1-(p-nitrophenyl)-4-phenyl-	EtOH	275(4.47)	104-0511-69
$C_{16}H_{11}NO_5$			
Coumarin, 7-methoxy-6-nitro-3-phenyl-	EtOH	269(3.51)	2-0049-69
	EtOH-NaOAc	335(3.69)	2-0049-69
Coumarin, 7-methoxy-8-nitro-3-phenyl-	EtOH	269(3.67)	2-0049-69
	EtOH-NaOAc	347(4.14)	2-0049-69
Coumarin, 8-methoxy-3-(p-nitrophenyl)-	EtOH	245(4.12)	2-0049-69
Coumarin, 8-methoxy-5-nitro-3-phenyl-	EtOH	290(4.05)	2-0049-69
	EtOH-NaOAc	344(4.07)	2-0049-69
Coumarin, 8-methoxy-6-nitro-3-phenyl-	EtOH	246(4.14)	2-0049-69
	EtOH-NaOAc	308(4.37)	2-0049-69
Coumarin, 8-methoxy-7-nitro-3-phenyl-	EtOH	345(4.62)[sic]	2-0049-69
$C_{16}H_{11}NO_5S$			
Spiro[benzo[b]thiophene-3(2H),1'-isoindoline]-2-carboxylic acid, 3'-oxo-, 1,1-dioxide	EtOH	260s(3.30),269(3.31), 275(3.28)	39-1149-69C
Spiro[benzo[b]thiophene-3(2H),1'-phthalan]-2-carboxamide, 3'-oxo-, 1,1-dioxide	EtOH	220s(4.11),270s(3.48), 276(3.52),284(3.50), 314(3.45)	39-1149-69C
$C_{16}H_{11}NO_6$			
2-Anthramide, 1,4,9,10-tetrahydro-3,8-dihydroxy-5-methyl-1,4,9-trioxo-	n.s.g.	242(4.12),260(4.09), 386(3.53)	39-2805-69C
$C_{16}H_{11}NS$			
7H-Benzo[c]phenothiazine	dioxan	274(4.56),335(3.54), 400s(3.08)	24-2728-69
[1]Benzothieno[2,3-g]isoquinoline, 5-methyl-	EtOH	229(4.54),240(4.45), 268(4.61),278(4.81), 285(4.85)	4-0875-69
[1]Benzothieno[3,2-g]isoquinoline, 11-methyl-	EtOH	230(4.40),244(4.48), 267(4.66),269(4.84)	4-0875-69
	n.s.g.	230(4.40),244(4.48), 267(4.66),269(4.84)	77-0598-69
Phenanthro[9,10-d]thiazole, 2-methyl-	EtOH	254(4.69),304(3.90)	77-0200-69
8H-Thieno[2,3-b]indole, 2-phenyl-	EtOH	239(4.62),287s(4.13), 293(4.15),336(4.36)	95-0058-69
$C_{16}H_{11}N_3$			
1-Phthalazineacetonitrile, α-phenyl- (continued on next page)	EtOH	219(4.70),260(3.69), 267(3.71),276(3.72),	95-0959-69

Compound	Solvent	$\lambda_{max}(\log \epsilon)$	Ref.
		286s(3.65),380(3.30)	95-0959-69
Pyrazole-4-carbonitrile, 1,3-diphenyl-	MeOH	229s(4.21),270(4.32)	39-2587-69C
Pyrazole-5-carbonitrile, 1,3-diphenyl-	MeOH	223(4.38),275(4.01), 291(3.91)	39-2587-69C
$C_{16}H_{11}N_3O_3$			
Isonicolinic acid, 2-(1,4-dihydro-1,4-dioxo-2-naphthyl)hydrazide	alkali	594(4.07)	95-0007-69
Isonicotinic acid, 2-(3,4-dihydro-3,4-dioxo-1-naphthyl)hydrazide	alkali	493(4.11)	95-0007-69
4(3H)-Quinazolinone, 6-nitro-2-styryl-	EtOH	255s(4.12),310s(4.14), 360s(4.2)	18-3198-69
4(3H)-Quinazolinone, 7-nitro-2-styryl-	EtOH	210(4.7)(end abs.)	18-3198-69
$C_{16}H_{11}N_3O_4$			
Pyrazole-5-carboxylic acid, 1-(p-nitrophenyl)-3-phenyl-	MeOH	229s(4.37),242s(4.35), 284s(3.85),322(4.15)	39-2587-69C
$C_{16}H_{11}N_3O_5$			
1H-1,4-Benzodiazepine-3-carboxylic acid, 2-hydroxy-7-nitro-5-phenyl-, monopotassium salt	H_2O	218(4.38),261(4.23), 308(4.00)	111-0239-69
dipotassium acid	H_2O	218(4.37),261(4.24), 302(4.01)	111-0239-69
$C_{16}H_{12}$			
Butatriene, 1,4-diphenyl-	CH_2Cl_2	250(4.27),257(4.08), 319(3.97),364(4.67), 382(4.75)	88-0435-69
Cyclohepta[def]fluorene, 4,8-dihydro-	EtOH	224(4.44),239(4.42), 250(4.40),255s(4.33), 301(3.72)	39-1427-69C
Fluoranthene, 6b,10a-dihydro-	n.s.g.	225(4.64)	39-0760-69C
Naphthalene, 1,5-di-1-propynyl-	EtOH	212(4.45),236(4.81), 307(4.12),319(4.28), 337(4.19)	35-0371-69
Naphthalene, 1,8-di-1-propynyl-	EtOH	213(4.31),238(4.79), 262(3.98),310(4.07), 324(4.20),339(4.20)	35-0371-69
$C_{16}H_{12}BrN_3OS_2$			
Rhodanine, 3-benzyl-5-[(o-bromophenyl)-azo]-	MeOH	243(4.02),269(3.78), 315(3.82),427(4.29)	2-0129-69
Rhodanine, 3-benzyl-5-[(p-bromophenyl)-azo]-	$CHCl_3$	424(4.49)	2-0129-69
$C_{16}H_{12}Br_2O_2$			
1,4-Butanedione, 2,3-dibromo-1,4-diphenyl-	EtOH	262(4.44)	104-0511-69
$C_{16}H_{12}ClNO_4S_2$			
2-Methyl-3-phenylthiazolo[2,3-b]benzothiazolium perchlorate	EtOH	218(4.52),263(3.97), 288(3.75),315(4.16)	95-0469-69
4-Phenyl-2H-[1,3]thiazino[2,3-b]benzothiazolium perchlorate	EtOH	254(4.17),281s(3.77), 341(3.62)	95-0469-69
3-p-Tolylthiazolo[2,3-b]benzothiazolium perchlorate	EtOH	216(4.62),252(4.04), 312(4.17)	95-0469-69

Compound	Solvent	$\lambda_{max}(\log \epsilon)$	Ref.
$C_{16}H_{12}ClN_3OS_2$			
Rhodanine, 3-benzyl-5-[(o-chlorophenyl)-azo]-	MeOH	242(4.04),268(3.82),295(3.89),422(4.15)	2-0129-69
	$CHCl_3$	446(4.56)	2-0129-69
Rhodanine, 3-benzyl-5-[(p-chlorophenyl)-azo]-	$CHCl_3$	424(4.49)	2-0129-69
$C_{16}H_{12}Cl_2$			
Cyclohepta[def]fluorene, 4,8-dichloro-4,8,9,10-tetrahydro-	EtOH	219(4.61),238(4.31),280(4.06),287(3.99)	39-1427-69C
$C_{16}H_{12}Cl_2N_2O_2$			
2-Indolinone, 4,7-dichloro-3-(p-methoxy-anilinomethylene)-	EtOH	278(4.11),310(3.42),365(4.85)	2-0667-69
$C_{16}H_{12}Cl_2O$			
Anthrone, 1,8-dichloro-10,10-dimethyl-	EtOH	277(4.15)	44-3093-69
$C_{16}H_{12}F_4$			
[2,2]Paracyclophane, 4,5,7,8-tetra-fluoro-	n.s.g.	221(4.13),288(2.72),297(2.75)	35-1862-69
Phenanthrene, 1,2,3,4-tetrafluoro-9,10-dihydro-6,8-dimethyl-	MeOH	265(4.21),305(3.48)	39-1684-69C
$C_{16}H_{12}F_4O_4$			
1,3-Methano-1H-cyclopropa[a]naphthalene-2,8-dione, 4,5,6,7-tetrafluoro-1a,2,3,7b-tetrahydro-, diacetate	EtOH	266(2.81)	65-2332-69
$C_{16}H_{12}IN_3OS_2$			
Rhodanine, 3-benzyl-5-[(o-iodophenyl)-azo]-	MeOH	243(3.98),268(3.78),320(3.79),432(4.22)	2-0129-69
	$CHCl_3$	450(4.52)	2-0129-69
Rhodanine, 3-benzyl-5-[(p-iodophenyl)-azo]-	$CHCl_3$	422(4.50)	2-0129-69
$C_{16}H_{12}N_2$			
Fluorene-$\Delta^{9,\alpha}$-malononitrile, 1,2,3,4-tetrahydro-	EtOH	251(4.29),290(4.24),353(4.03)	44-1899-69
6H-Pyrido[4,3-b]carbazole, 5-methyl-	EtOH	224(4.46),238(4.33),273(4.72),284(4.83),293(4.85),310(3.69),328(3.77),373(3.61),390(3.57)	12-0185-69
$C_{16}H_{12}N_2O$			
Acetophenone, 2-(1-phthalazinyl)-	EtOH	219(4.48),263(3.89),276(3.85),297(4.06),371s(4.24),398(4.50),420(4.47)	95-0959-69
6H-Azepino[1,2-a]indole-11-carbonitrile, 2,9-dimethyl-6-oxo-	CH_2Cl_2	263(4.16),293(4.37),316(4.36),415(3.76)	94-1294-69
Cyclobuta[4,5]furo[2,3-b]quinoline-9-carbonitrile, 2a,9b-dihydro-1,7-dimethyl-	CH_2Cl_2	264(4.34),306(3.82),318(3.83),356(4.01),371(4.05)	94-1294-69
2,7-Epoxy-1-benzazecine-8-carbonitrile, 5,10-dimethyl-	CH_2Cl_2	283(4.63),385(3.91)	94-1294-69
Naphthalene, 1-(p-hydroxyphenylazo)-	EtOH	386(4.53)	124-0179-69
Naphthalene, 2-(p-hydroxyphenylazo)-	EtOH	362(4.49)	124-0179-69

Compound	Solvent	λ_{max}(log ϵ)	Ref.
Oxepino[2,3-b]quinoline-6-carbonitrile, 4,8-dimethyl-	CH_2Cl_2	236(4.56),264(4.34), 354(4.06),385(4.12)	94-1294-69
Pyrazole-4-carboxaldehyde, 1,3-diphenyl-	n.s.g.	248(4.42)	28-1063-69B
6H-Pyrido[4,3-b]carbazole, 5-methyl-, 2-oxide	EtOH	234(4.30),253(4.31), 293(4.62),303(4.63), 314(4.82),347(3.77)	4-0389-69
$C_{16}H_{12}N_2O_2$			
Isoquinoline, 1-(o-nitrobenzyl)-	MeOH	261(3.96),268(3.95), 278s(3.84),307(3.63), 321(3.62)	44-3786-69
5,8-Isoquinolinedione, 7-anilino-3-methyl-	MeCN	268(4.45),328(3.86)	4-0697-69
1,4-Naphthoquinone, 2-(2-phenylhydrazino)-	MeOH	227(4.26),298(4.11), 453(4.01),525(4.09)	44-2750-69
$C_{16}H_{12}N_2O_2S$			
Imidazolidine-2-thione, 1-benzyl-4,5-dioxo-3-phenyl-	EtOH	302(4.15),421(1.58)	18-2323-69
$C_{16}H_{12}N_2O_3$			
Pyridine, 1-acetyl-1,4-dihydro-4-(5-oxo-2-phenyl-2-oxazolin-4-ylidene)-	acetone	428(4.55),455(4.68), 483(4.54)	24-1129-69
$C_{16}H_{12}N_2O_4$			
Benzo[c]cinnoline-3,8-dicarboxylic acid, dimethyl ester	EtOH	205s(3.15),209s(4.20), 217(4.28),264(4.78), 280(4.55),297s(4.29), 315(4.03)	4-0523-69
1,4-Isoquinolinedione, 2,3-dihydro-3-(nitromethyl)-3-phenyl-	EtOH	218(4.65)	33-1810-69
1(4H)-Pyridinecarboxylic acid, 4-(5-oxo-2-phenyl-2-oxazolin-4-ylidene)-, methyl ester	acetone	420(4.54),445(4.66), 472(4.50)	24-1129-69
$C_{16}H_{12}N_2O_5$			
Benzo[c]cinnoline-3,8-dicarboxylic acid, dimethyl ester, 5-oxide	EtOH	204(3.38),218s(4.29), 228s(4.22),261(4.62), 272(4.63),282s(4.55), 294(4.54),309s(4.19), 314s(4.03),345(3.89), 373(3.53),391(3.30)	4-0523-69
$C_{16}H_{12}N_2O_8$			
4,4'-Biphenyldiol, 2,2'-dinitro-, diacetate	EtOH	219(4.52),261s(4.06), 305s(3.60)	4-0523-69
$C_{16}H_{12}N_2Se_2$			
5,5'-Diindolyl diselenide	EtOH	265(4.26)	11-0159-69B
$C_{16}H_{12}N_4$			
Benzimidazole, 2,2'-vinylenebis-, cis	EtOH	241(4.59),273(4.71), 279(4.79),340(2.88), 360(3.03),378(2.34)	87-0818-69
trans	EtOH	225s(4.21),245(4.05), 275(4.09),280(4.12), 345(4.51),360(4.58), 380(4.43)	87-0818-69
s-Triazolo[3,4-a]phthalazine, 3-methyl-6-phenyl-	MeOH	244(4.52),250(4.50)	44-3221-69

Compound	Solvent	$\lambda_{max}(\log \epsilon)$	Ref.
$C_{16}H_{12}N_4O$			
4H-Pyrido[1,2-a]pyrimidine-3-carbonitrile, 2-(benzylamino)-4-oxo-	EtOH	256(4.61),346(3.97), 359s(3.76)	95-0203-69
$C_{16}H_{12}N_4O_2$			
Acetophenone, 2-(5-benzoyl-1H-tetrazol-1-yl)-	EtOH	252(4.29),272s(4.19)	23-3997-69
Acetophenone, 2-(5-benzoyl-2H-tetrazol-2-yl)-	EtOH	250(4.39)	23-3997-69
Pyrazoline-5-carbonitrile, 1-(p-nitrophenyl)-3-phenyl-	MeOH	225(4.23),263s(3.80), 308(3.82),385(4.46)	39-2587-69C
7H-Pyrrolo[2,3-d]pyrimidine-6-carbonitrile, 5-hydroxy-7-methyl-2-phenyl-, 5-p-toluenesulfonate	EtOH	262(4.47),315(4.27)	4-0819-69
$C_{16}H_{12}N_4O_2S$			
Thiourea, 1-(m-nitrophenyl)-3-(8-quinolyl)-	dioxan	307(4.11),323(4.20), 338(4.05),357(4.23)	34-0506-69
Thiourea, 1-(p-nitrophenyl)-3-(8-quinolyl)-	dioxan	271(4.25),323(4.20), 342(4.23)	34-0506-69
$C_{16}H_{12}N_4O_4$			
Spiro[oxaziridine-3,4'-[2]pyrazolin]-5'-one, 3'-methyl-2-(p-nitrophenyl)-1'-phenyl-	MeOH	238(4.11),268(4.20)	44-1687-69
$C_{16}H_{12}N_4O_5S$			
1,2-Propanedione, 1-[5-(1-propynyl)-2-thienyl]-, 2-(2,4-dinitrophenylhydrazone)	MeOH	253s(4.12),393(4.27), 399(4.33)	39-1813-69C
$C_{16}H_{12}N_4S$			
5H-s-Triazolo[3,4-b][1,3,4]thiadiazine, 3,6-diphenyl-	EtOH	253(4.15),291(4.19)	95-0689-69
$C_{16}H_{12}N_6O_4$			
Benzoic acid, m-azido-, ethylene ester	dioxan	251(4.39),297(3.71)	73-3811-69
Benzoic acid, p-azido-, ethylene ester	dioxan	271(4.72)	73-3811-69
$C_{16}H_{12}N_6O_{10}$			
Acetamide, N,N'-(2,2',6,6'-tetranitro-4,4'-biphenylylene)bis-	EtOH	246(4.85),295s(4.22), 342(3.74)	4-0533-69
$C_{16}H_{12}O$			
Cyclohepta[def]fluoren-8(4H)-one, 9,10-dihydro-	EtOH	218(4.42),243(4.26), 248(4.25),271(4.05), 280(4.02),334(3.84)	39-1427-69C
9-Fluorenylideneacetaldehyde, α-methyl-	EtOH	260(4.21),270(4.35)	77-0773-69
Furan, 2-(4-biphenylyl)-	EtOH	301(4.60)	6-0277-69
Indone, 2-methyl-3-phenyl-	hexane	240(4.66),273(3.95), 315(3.20),324(3.04), 390(3.00)	54-0465-69
Naphthalene, 1-phenoxy-	MeOH	277(3.83),283(3.84), 289(3.81),307s(3.44), 312s(3.22),321(3.05)	88-0625-69
2-Naphthol, 1-phenyl-	EtOH	231(4.79),267s(--), 278(3.82),290s(--), 333(3.58)	24-3405-69

Compound	Solvent	λ_{max}(log ε)	Ref.
$C_{16}H_{12}O_2$			
Benzaldehyde, o-[(α-hydroxy-o-tolyl)-ethynyl]-	EtOH	254(4.40),278(4.23), 284(4.23),293(4.26), 328(3.86)	44-0874-69
2-Butene-1,4-dione, 1,4-diphenyl-	EtOH	232(4.31),260(4.42)	104-0511-69
3-Phenanthrol, acetate	EtOH	212(4.54),245(4.75), 251(4.83),267s(4.21), 275(4.20),281(4.06), 293(4.11),318(--), 332(--),348(--)	24-2384-69
$C_{16}H_{12}O_2S$			
1-Thiocoumarin, 4-(p-methoxyphenyl)-	EtOH	235(3.91),300(3.82), 340s(3.56)	2-0315-69
$C_{16}H_{12}O_3$			
Acenaphthene-$\Delta^{1,\alpha}$-acetic acid, 2-oxo-, ethyl ester	EtOH	225(4.73),232(4.83), 274(4.22),313(4.0)	18-0181-69
geometric isomer	EtOH	220(4.46),260(3.90)	18-0181-69
Flavone, 7-methoxy-	n.s.g.	243(4.44),305(4.30)	2-0940-69
$C_{16}H_{12}O_4$			
1,2-Acenaphthylenedicarboxylic acid,	EtOH	229(4.32),334(4.08)	39-0769-69C
dimethyl ester	n.s.g.	233(4.63),253(4.33), 337(4.45),363(4.26)	39-0760-69C
Anthraquinone, 1-hydroxy-3-methoxy-2-methyl-	EtOH-acid	240s(4.31),245(4.34), 276(4.45),334(3.42), 410(3.79)	102-0501-69
	EtOH-base	261(4.50),284s(4.14), 315s(3.61),395(3.35), 510(3.76)	102-0501-69
Anthraquinone, 1-hydroxy-3-methoxy-6-methyl-	EtOH	226(4.12),252(4.25), 261(4.23),279(4.23), 409(3.71)	23-0767-69
Anthraquinone, 3-hydroxy-1-methoxy-2-methyl- (rubiadin 1-methyl ether)	EtOH-acid	238(4.26),244s(4.23), 282(4.55),358(3.48)	102-0501-69
	EtOH-base	247(4.42),314(4.40), 480(3.72)	102-0501-69
Isoflavone, 7-hydroxy-4'-methoxy-	MeOH	249(4.40),298s(4.05)	100-0108-69
	MeOH-base	258(--),340(--)	100-0108-69
Xanthen-9-one, 2-allyl-1,3-dihydroxy-	MeOH	238(4.55),311(4.23)	2-1182-69
Xanthen-9-one, 4-allyl-1,3-dihydroxy-	MeOH	235(4.52),258(4.42), 311(4.17)	2-1182-69
Xanthen-9-one, 3-(allyloxy)-1-hydroxy-	MeOH	234(4.51),305(4.23), 350(3.75)	2-1182-69
$C_{16}H_{12}O_4S$			
1,4,6,9-Dibenzothiophenetetrone, 2,3,7,8-tetramethyl-	THF	268(4.58),382(3.95)	24-1739-69
$C_{16}H_{12}O_4S_2$			
1,4,6,9-Thianthrenetetrone, 2,3,7,8-tetramethyl-	THF	265(4.37),521(3.72)	24-1739-69
$C_{16}H_{12}O_5$			
Anthracene-9,10-dicarboxylic acid, 9,10-dihydro-9-hydroxy-	EtOH	253(2.67),262(2.61)	23-2827-69
Anthraquinone, 1,7-dihydroxy-3-methoxy-6-methyl- (5 λ,6 ε)	EtOH	225(4.23),251s(4.00), 284(4.03),305(4.45), 379(4.03),?(3.74)	23-0767-69

Compound	Solvent	$\lambda_{max}(\log \epsilon)$	Ref.
3(2H)-Benzofuranone, 4,6-dihydroxy-2-(p-methoxybenzylidene)-	MeOH	400(4.2)	2-0543-69
Flavone, 5,7-dihydroxy-4'-methoxy-(acacetin)	EtOH	269(4.30),298(4.15), 330(4.27)	105-0397-69
	EtOH-AlCl$_3$	280(4.24),302(4.26), 335(4.27),380(4.01)	105-0397-69
Isoflavone, 3',7-dihydroxy-4'-methoxy-	MeOH	249(4.25),257s(4.23), 292(4.03)	2-0118-69
Pterocarpan, 3-hydroxy-8,9-(methylenedioxy)-	EtOH	281(3.89),287(3.93), 309(4.11)	39-1109-69C
Xanthen-9-one, 5-acetoxy-1-methoxy-	MeOH	237(4.59),278(3.91), 346(3.88)	39-2421-69C
$C_{16}H_{12}O_5S_2$			
1,4,6,9-Thianthrenetetrone, 2,3,7,8-tetramethyl-, 5-oxide	THF	240(4.26),270(4.43), 310s(--),427(3.78)	24-1939-69
$C_{16}H_{12}O_6$			
Anthraquinone, 1,3,5-trihydroxy-7-methoxy-2-methyl-	EtOH	227(4.39),256(4.19), 275(4.17),300s(3.91), 428(3.91)	23-0767-69
Anthraquinone, 1,5,7-trihydroxy-6-methoxy-3-methyl-	EtOH	256(4.14),280(4.45), 305(3.98),421(4.02)	24-4104-69
	EtOH-NaOH	267(4.04),305(4.45), 469(4.16)	24-4104-69
Anthraquinone, 1,5,8-trihydroxy-6-methoxy-3-methyl-	EtOH	235(4.40),258(4.55), 306(4.10),463(4.05), 485(4.16),490(4.25), 511(4.06),525(4.06)	1-0144-69
Anthraquinone, 3,4,5-trihydroxy-7-methoxy-2-methyl-	EtOH	229(4.36),280(4.43), 312(4.00),425(4.03)	23-0767-69
3(2H)-Benzofuranone, 4,6,7-trihydroxy-2-(p-methoxybenzylidene)-	MeOH	255(4.1),340(4.1)	2-0543-69
Dermoglaucine	EtOH	248(3.94),254(3.97), 282(4.29),435(3.86), 580(3.11)	24-4104-69
	EtOH-NaOH	271(4.19),293(4.06), 515(3.83)	24-4104-69
Flavone, 3,4',5-trihydroxy-7-methoxy-	EtOH	268(3.75),368(4.01)	105-0397-69
	EtOH-AlCl$_3$	270(4.06),303s(3.64), 350(3.76),422(4.07)	105-0397-69
Flavone, 3,5,7-trihydroxy-4'-methoxy-(kaempferide)	EtOH	269(4.38),367(4.28)	105-0397-69
	EtOH-AlCl$_3$	278(4.42),302(4.17), 350(4.21),420(4.29)	105-0397-69
Isoflavone, 2',4',5,7-tetrahydroxy-6-methyl-	MeOH	266(4.38)	2-0061-69
	MeOH-NaOAc	277(4.35)	2-0061-69
Naphtho[1,2-c]furan-4,5-dicarboxylic acid, 1,3-dihydro-3-oxo-, dimethyl ester	EtOH	243(4.79),288(3.73), 299s(3.70),320s(3.35), 334(3.25)	94-1591-69
$C_{16}H_{12}O_7$			
Eupafolin	EtOH	253(4.01),272(4.00), 342(4.24)	78-1603-69
	EtOH-NaOAc	276(4.17),380(4.09)	78-1603-69
	EtOH-NaOAc-AlCl$_3$	276(4.12),432(4.37)	78-1603-69
Luteolin, 6-methoxy-	EtOH	255(4.21),273(4.21), 350(4.42)	31-0355-69
	EtOH-NaOAc	276(4.34),366(4.28)	31-0355-69

Compound	Solvent	$\lambda_{max}(\log \epsilon)$	Ref.
$C_{16}H_{13}BF_2O_2$			
Boron, difluoro(2-methyl-1,3-diphenyl-1,3-propanedionato)-	$CHCl_3$	260(3.65),360(4.48)	39-0526-69A
$C_{16}H_{13}BrO_3$			
Benzoic acid, o-(α-bromophenacyl)-, methyl ester	EtOH	244(4.12),255s(4.03), 287s(3.39),315s(2.95)	12-0577-69
$C_{16}H_{13}Br_3N_4O_5$			
Propiophenone, 2',3',5'-tribromo-4'-methoxy-, 2,4-dinitrophenylhydrazone	EtOH	357(4.35)	22-2079-69
Propiophenone, 3',4',5'-tribromo-2'-methoxy-, 2,4-dinitrophenylhydrazone	EtOH	362(4.43)	22-2079-69
$C_{16}H_{13}Cl$			
Anthracene, 9-chloro-10-ethyl-	EtOH	253s(4.90),258(5.19), 341(3.51),358(3.81), 377(4.02),398(4.00)	78-3485-69
Indene, 2-(α-chlorobenzyl)-	hexane	267(4.14)	88-1717-69
$C_{16}H_{13}ClN_2O$			
1,5-Benzodiazocin-2(1H)-one, 8-chloro-3,4-dihydro-6-phenyl-	EtOH	240(4.30)	44-0183-69
Carbostyril, 3-(aminomethyl)-6-chloro-4-phenyl-	MeOH	238(4.64)	36-0830-69
Isoindole-1-carboxamide, 5-chloro-N-methyl-3-phenyl-	iso-PrOH	261(4.67),296(3.91), 370(4.29)	44-0649-69
$C_{16}H_{13}ClN_2S$			
Isoindole-1-carboxamide, 5-chloro-N-methyl-3-phenyl-	iso-PrOH	224(4.34),275(4.32), 320s(3.95),333s(3.86), 417(4.23)	44-0649-69
$C_{16}H_{13}ClO_4S$			
4-Methyl-2-phenyl-1-benzothiopyrylium perchlorate	HOAc-1% $HClO_4$	264(3.43),296s(3.99), 390(3.39)	2-0191-69
$C_{16}H_{13}ClO_5S$			
4-(p-Methoxyphenyl)-1-benzothiopyrylium perchlorate	HOAc-1% $HClO_4$	262(3.26),335(3.63), 457(3.48)	2-0315-69
$C_{16}H_{13}Cl_2NO_4S$			
2-(β-Chlorostyryl)-3-methylbenzothiazolium perchlorate	MeCN	245s(3.83),254s(3.74), 366(4.48)	24-0568-69
$C_{16}H_{13}Cl_2N_3O_2$			
Pyrimidine, 1,3-bis(p-chlorophenyl)-1,2,3,4-tetrahydro-5-nitro-	dioxan	250(4.39),385(4.34)	78-1617-69
$C_{16}H_{13}N$			
Acridine, 9-propenyl-, cis	EtOH	253(4.2),360(3.0)	40-1263-69
Acridine, 9-propenyl-, trans	EtOH	254(4.1),360(3.0)	40-1263-69
2,3-Benzocarbazole, 1,4-dihydro-	EtOH	225(4.52),272(4.26), 282(4.19),292(4.03)	78-0227-69
1H-Dibenz[e,g]isoindole, 2,3-dihydro-	EtOH	210(4.63),222(4.49), 248(4.78),255(4.85), 270(4.05),277(4.00), 288(3.95),300(3.98), 321(2.92),327(2.86), 336(2.91),344(2.80), 353(2.91)	35-5192-69

Compound	Solvent	$\lambda_{max}(\log \epsilon)$	Ref.
Indole, 3-styryl-	n.s.g.	234(4.34),330(4.42)	103-0190-69
$C_{16}H_{13}NO$			
Acetamide, N-1-phenanthryl-	EtOH	212(4.60),254(4.71), 285(4.01),297(4.10), 335(2.61),350(2.40)	24-2384-69
Acetoacetonitrile, 2,4-diphenyl-	EtOH	268(--),272(4.22)	39-2132-69C
	EtOH-NaOH	253(3.98),305(4.29)	39-2132-69C
5H-Dibenzo[a,d]cycloheptene-5-carbonitrile, 10,11-dihydro-10-hydroxy-, anti	EtOH	262(2.71)	23-2827-69
syn	EtOH	263(2.68)	23-2827-69
10,5-(Epoxymethano)-5H-benzo[a,d]cycloheptene-13-imine, 10,11-dihydro-	EtOH	265(2.76)	23-2827-69
Oxazole, 2-methyl-4,5-diphenyl-	EtOH	288(4.19)	77-0200-69
4H,8H-Pyrido[3,2,1-de]phenanthridin-8-one, 5,6-dihydro-	EtOH	235(4.63),263(4.31), 326(3.88),340(3.84)	2-0848-69
2H-Pyrrol-2-ol, 2,5-diphenyl-	ether	252(3.90)	22-1667-69
$C_{16}H_{13}NOS_2$			
Acetophenone, 2-(2-benzothiazolylthio)-4'-methyl-	EtOH	211(4.46),259(4.40), 289(4.20),301(4.07)	95-0469-69
Propiophenone, 2-(2-benzothiazolylthio)-	EtOH	225(4.53),246(4.34), 278(4.23),290(4.19), 301(4.11)	95-0469-69
Propiophenone, 3-(2-benzothiazolylthio)-	EtOH	243(4.40),327(4.40)	95-0469-69
$C_{16}H_{13}NO_2$			
Acetamide, N-(2-hydroxy-1-phenanthryl)-	EtOH	257(4.82),267s(4.61), 291s(4.12),324s(2.86), 340(3.12),357(3.15)	24-2384-69
	50% EtOH-0.1N NaOH	268(4.72),278s(4.54), 301(4.13),316(3.92), 360s(3.28),374(3.30)	24-2384-69
Hydroperoxide, 2,5-diphenyl-2H-pyrrol-2-yl-	ether	255(4.08)	22-1667-69
10,5-(Iminomethano)-5H-dibenzo[a,d]-cyclohepten-13-one, 10,11-dihydro-10-hydroxy-	EtOH	260(2.95),272(2.84)	23-2827-69
Indole, 3-(p-methoxybenzoyl)-	EtOH	221(4.21),272(4.25), 314(4.18)	78-0227-69
Oxazolo[2,3-a]isoindol-5(9bH)-one, 2,3-dihydro-9b-phenyl-	EtOH	225s(4.28)	44-0165-69
$C_{16}H_{13}NO_2S_2$			
2-Benzothiazolinethione, 3-(2-hydroxyethyl)-, benzoate	MeOH	229(4.41),282(3.49), 326(4.41)	4-0163-69
$C_{16}H_{13}NO_3$			
Dibenzofuran, 1,4-dimethyl-2-(2-nitrovinyl)-	EtOH	227(4.48),253(4.20), 277(4.35),296(4.06), 350(4.21)	4-0379-69
10,5-(Iminomethano)-5H-dibenzo[a,d]-cyclohepten-13-one, 10,11-dihydro-10,11-dihydroxy-	EtOH	263(2.62)	23-2827-69
1,4-Isoquinolinedione, 2,3-dihydro-3-methoxy-3-phenyl-	EtOH	222(4.58)	33-1810-69

Compound	Solvent	$\lambda_{max}(\log \epsilon)$	Ref.
$C_{16}H_{13}NO_3S$			
2-Benzothiazolinone, 3-(2-hydroxyethyl)-, benzoate	MeOH	215(4.66),282(3.53), 289(3.45)	4-0163-69
$C_{16}H_{13}NO_4$			
Phthalic anhydride, 3-(p-methoxybenzylamino)-	EtOH	207(4.24),256(4.32), 354(3.88)	16-0241-69
$C_{16}H_{13}NO_5$			
1,3-Dioxolo[4,5-b]acridine-10,11(5H,-11aH)-dione, 11a-methoxy-5-methyl-	EtOH	244(4.09),280s(4.25), 287(4.29),315(4.03)	12-1493-69
Nortecleanthine	EtOH	208(4.36),222(4.23), 240(4.28),269(4.65), 290(4.64),318s(3.86), 413(3.90)	39-2327-69C
	EtOH-NaOH	232s(4.28),260(4.42), 289(4.47),327(4.01), 422(3.68)	39-2327-69C
$C_{16}H_{13}NO_6$			
Anthraquinone, 4-amino-1,3,5-trihydroxy-2-methoxy-7-methyl-	EtOH	247(4.21),254(4.29), 261(4.38),263(4.41), 266(4.39),270(4.35), 304(3.75),445(3.60), 467(3.71),490(3.69), 519(4.13),559(4.16)	24-4104-69
	EtOH-NaOH	248(4.18),257(4.13), 2644.15),267(4.18), 278(4.24),290(4.23), 525(4.09),556(4.13)	24-4104-69
$C_{16}H_{13}NS$			
Thiazole, 2-methyl-4,5-diphenyl-	EtOH	288(4.04)	77-0200-69
$C_{16}H_{13}N_3$			
Isoquinoline, 1-[(p-tolyl)azo]-	MeCN	244s(4.16),308(4.06), 338s(4.02),347s(4.00), 447s(3.00)	24-3176-69
Quinoline, 2-[(p-tolyl)azo]-	MeCN	238(4.35),295s(--), 337(3.44),451(2.83)	24-3176-69
as-Triazine, 3-methyl-5,6-diphenyl-	MeOH	258(4.14),304(3.95), 372(2.60)	88-3147-69
$C_{16}H_{13}N_3OS_2$			
Rhodanine, 3-benzyl-5-(phenylazo)-	$CHCl_3$	416(4.48)	2-0129-69
$C_{16}H_{13}N_3O_3$			
3-Pyrrolin-2-one, 3-anilino-4-nitro-5-phenyl-	EtOH	234(3.97),385(3.97)	44-3279-69
4(3H)-Quinazolinone, 2-benzyl-3-methyl-6-nitro-	EtOH	260s(4.00),335(4.14)	18-3198-69
4(3H)-Quinazolinone, 2-benzyl-3-methyl-7-nitro-	EtOH	248(4.38),326(3.56)	18-3198-69
$C_{16}H_{13}N_3O_3S$			
Benzimidazole, 1-acetyl-2-[(p-nitrobenzyl)thio]-	MeOH	287(4.25)	4-0023-69
Thiazole, 2-anilino-5-[1-methyl-2-(5-(5-nitro-2-furyl)vinyl]-	EtOH	295(4.46),413(4.27)	103-0414-69

Compound	Solvent	$\lambda_{max}(\log \epsilon)$	Ref.
$C_{16}H_{13}N_3O_4$			
[1,3]Dioxolo[4,5-g]-1,2,3-benzotriazin-4(3H)-one, 3-(p-hydroxyphenethyl)-	EtOH	243(4.24),248(4.27), 254(4.26),260(4.16), 318(3.47)	44-2667-69
$C_{16}H_{13}N_3S$			
Thiourea, 1-phenyl-3-(8-quinolyl)-	EtOH	244(4.53),270(4.36), 332(3.96)	34-0506-69
$C_{16}H_{13}N_5O_4S$			
Urea, [[p-(8-hydroxy-5-quinolyl)azo]-phenyl]sulfonyl]-	benzene	398(4.20)	123-0041-69C
	MeOH	400(4.28)	
	EtOH	400(4.36)	
	BuOH	402(4.21)	
	pentanol	403(4.22)	
	$CHCl_3$	398(4.21)	
	$C_2H_4Cl_2$	395(4.24)	
	dioxan	398(4.40)	
	acetone	396(4.39)	
	DMF	408(4.09),515(4.43)	
	80% DMF	498(4.43)	
	60% DMF	490(4.41)	
	40% DMF	470(4.34)	
	20% DMF	455(4.28)	
Urea, [[p-(8-hydroxy-7-quinolyl)azo]-phenyl]sulfonyl]-	benzene	398(4.40)	123-0041-69C
	MeOH	400(4.44)	
	EtOH	400(4.49)	
	BuOH	395(4.41)	
	pentanol	395(4.42)	
	$CHCl_3$	398(4.41)	
	$C_2H_4Cl_2$	395(4.40)	
	dioxan	398(4.53)	
	acetone	396(4.52)	
	DMF	508(4.53)	
	80% DMF	510(4.49)	
	60% DMF	510(4.47)	
	40% DMF	430(4.23),505(4.21)	
(also other solvent mixtures)	20% DMF	410(4.29)	
$C_{16}H_{14}$			
Anthracene, 9-ethyl-	EtOH	249s(5.00),255(5.27), 332(3.46),347(3.74), 365(3.93),385(3.91)	78-3485-69
Fluorene, 9-isopropylidene-	EtOH at 80°K	234(4.62),250(4.3), 260(4.58),284f(4.15), 323(4.10),345s(2.9)	88-2157-69
Indan, 1-benzylidene-	n.s.g.	205(4.40),228(4.13), 284(4.33),296(4.36), 305s(4.36),316(4.48), 330(4.35)	39-0944-69C
Indene, 3-benzyl-	n.s.g.	250(4.02)	39-0944-69C
[2,2]Metacyclophane, 5,8,13,16-tetra-hydro-	EtOH	none	33-1624-69
$C_{16}H_{14}BF_2NO$			
Boron, difluoro[3-(phenylimino)butyro-phenonato]-	$CHCl_3$	255(3.61),345(4.45)	39-0526-69A

Compound	Solvent	λ_{max}(log ε)	Ref.
$C_{16}H_{14}BF_4NS$			
3-Methyl-2-styrylbenzothiazolium tetrafluoroborate	MeCN	235s(4.03),245s(3.89), 371(4.51)	24-0568-69
$C_{16}H_{14}Br_2N_4O_5$			
Propiophenone, 3',5'-dibromo-2'-methoxy-, 2,4-dinitrophenylhydrazone	EtOH	361(4.39)	22-2079-69
Propiophenone, 3',5'-dibromo-4'-methoxy-, 2,4-dinitrophenylhydrazone	EtOH	374(4.46)	22-2079-69
$C_{16}H_{14}ClNO_3$			
Benzamide, N-acetyl-N-(benzyloxy)-o-chloro-	MeOH	208(--),211(4.30)	12-0161-69
	MeOH-NaOH	263(3.46)	12-0161-69
Benzamide, N-acetyl-N-(benzyloxy)-p-chloro-	MeOH	208(--),244(4.09)	12-0161-69
	MeOH-NaOH	228(4.08),271(3.70)	12-0161-69
$C_{16}H_{14}ClNO_5$			
4-Methyl-2,5-diphenylisoxazolium perchlorate	CH_2Cl_2	332(4.20)	44-1468-69
$C_{16}H_{14}Cl_2N_2$			
Acetophenone, 4'-chloro-, azine	C_6H_{12}	274(4.48)	39-0836-69C
$C_{16}H_{14}Cl_2N_4O_7$			
2-Cyclopentene-1-carboxylic acid, 3,5-dichloro-1-hydroxy-4-oxo-2-propenyl-, methyl ester, 4-(2,4-dinitrophenylhydrazone)	MeOH	269(4.14),303(4.02), 385(4.56)	35-0157-69
$C_{16}H_{14}Cl_2O$			
1,4,7-Metheno-5H-dibenzo[a,c]cyclohepten-5-one, 1,4-dichloro-1,2,3,4,4a,6,-7,11b-octahydro-	C_6H_{12}	268(2.65),275(2.67)	88-1185-69
$C_{16}H_{14}Cl_2O_5$			
Benzophenone, 3,5-dichloro-2,2',6-trihydroxy-4'-methoxy-4,6'-dimethyl-	EtOH	221(4.28),288(4.15), 350(3.81)	39-1721-69C
2,6-Cresotic acid, 4-methoxy-, 2,4-dichloro-5-hydroxy-m-tolyl ester	EtOH	230(3.77),282(3.96), 308(3.50)	39-1721-69C
$C_{16}H_{14}FeO_2$			
1,5-Pentanedione, 1,5-(1,1'-ferrocenediyl)-2-methylene-	n.s.g.	228(4.31),254s(3.98), 273s(3.78),349(3.28), 473(2.62)	78-0861-69
$C_{16}H_{14}INO_3$			
Benzamide, N-acetyl-N-(benzyloxy)-p-iodo-	MeOH	208(--),258(4.15)	12-0161-69
	MeOH-NaOH	244(4.10),253s(4.06)	12-0161-69
$C_{16}H_{14}N_2$			
Quinaldine, 4-(p-anilino)-	EtOH	233(4.55),256s(--), 305s(--),319(4.04), 331(4.18)	78-0847-69
$C_{16}H_{14}N_2O$			
2H-1,4-Benzodiazepin-2-one, 1,3-dihydro-7-methyl-5-phenyl-	pH 1.5	230(4.4),290(4.1), 370(3.6)	65-0443-69
	pH 13.5	240(4.4),340(3.5)	65-0443-69
Fluorene-4a(2H)-malononitrile, 1,3,4,9a-tetrahydro-9-oxo-	EtOH	203(3.26),243(4.11), 289(3.26)	44-1899-69

Compound	Solvent	$\lambda_{max}(\log \epsilon)$	Ref.
5H-Imidazo[2,1-a]isoindol-5-one, 1,2,3,9b-tetrahydro-9b-phenyl-	EtOH	225(4.21)	44-0165-69
2-Indolinone, 3-[(N-methylanilino)methylene]-	EtOH	228(4.22),275(3.98), 358(4.14)	2-0667-69
Isoindole-1-carboxamide, N-methyl-3-phenyl-	iso-PrOH	221(4.16),230(4.18), 258(4.52),295(3.77), 369(4.35)	44-0649-69
1H-Isoindole-1-carboxamide, N-methyl-3-phenyl-	iso-PrOH	253(4.14)	44-0649-69
Ketone, 3-ethyl-4-pyridyl indol-2-yl	EtOH	227(4.33),322(4.26)	39-2738-69C
$C_{16}H_{14}N_2OS$			
Benzamide, N-(1,2-benzisothiazol-3-yl)-N-ethyl-	MeOH	222(4.35),301(3.73), 306(3.73),311(3.72)	24-1961-69
$C_{16}H_{14}N_2O_2$			
p-Benzophenetedide, 2'-cyano-	EtOH	214(4.52),233(4.33), 258(3.95),307(3.84)	22-2008-69
Carbamic acid, [1-cyano-2-(1-naphthyl)-propenyl]-, methyl ester, trans	EtOH	282(3.89)	22-1724-69
Carbamic acid, [1-cyano-2-(2-naphthyl)-propenyl]-, methyl ester, trans	EtOH	263(4.37),290(3.99)	22-1724-69
Indole-3-acetic acid, 5-methyl-2-(3-pyridyl)-	EtOH	219(4.48),312(4.30)	22-4154-69
2-Indolinone, 3-(anilinomethylene)-5-methoxy-	EtOH	238(3.22),283(3.30), 353(3.39)	2-0667-69
2-Indolinone, 3-(p-methoxyanilinomethylene)-	EtOH	278(4.16),367(4.48)	2-0667-69
1,4-Isoquinolinedione, 1,2,3,4-tetrahydro-3-(methylamino)-3-phenyl-	EtOH	220(4.57),296s(3.34), 350(2.70)	33-1810-69
Isosydnone, 4,5-di-p-tolyl-	EtOH	232(4.15),316(4.10)	39-1185-69B
Pyrazole, 1-hydroxy-4-methyl-3,5-diphenyl-, 2-oxide	EtOH	262(4.4),285(4.3)	44-0194-69
$C_{16}H_{14}N_2O_2S$			
Imidazole, 5-(benzylsulfonyl)-4-phenyl-	EtOH	252(3.99)	23-1123-69
anion	H_2O	260(3.89)	23-1123-69
cation	H_2O	243(3.97)	23-1123-69
Quinoline, 8-p-toluenesulfonamido-, cupric complex	$CHCl_3$	370(3.89)	3-0529-69
$C_{16}H_{14}N_2O_5$			
Acetic acid, anhydride with N-(benzyloxy)-p-nitrobenzimidic acid	MeOH	300(4.20)	12-0161-69
	MeOH-NaOH	245(4.07),350(3.79)	12-0161-69
Benzamide, N-acetyl-N-(benzyloxy)-p-nitro-	MeOH	263(4.07)	12-0161-69
	MeOH-NaOH	258(4.07),350(3.46)	12-0161-69
Biphenyleno[2,3-c][1,2,3]oxadiazole, 4,5,6,7-tetramethoxy-	EtOH	270(4.64),277(4.66), 291s(4.45),389(4.03)	39-0646-69C
$C_{16}H_{14}N_2O_6$			
6(3aH)-Indolinone, 4,5-dihydro-1-(6-nitropiperonoyl)-	EtOH	246(4.1),289(4.3), 340s(3.8)	44-2667-69
Piperonylamide, N-(o-hydroxyphenethyl)-6-nitro-	EtOH	225(4.41),268(3.69), 342(3.66)	44-2667-69
Piperonylamide, N-(p-hydroxyphenethyl)-6-nitro-	EtOH	225(4.47),250(4.68), 340(3.70)	44-2667-69
Uracil, 2,2'-anhydro-1-(5-O-benzoyl-β-D-arabinofuranosyl)-	MeOH	229(4.42)	94-1188-69

Compound	Solvent	$\lambda_{max}(\log \epsilon)$	Ref.
$C_{16}H_{14}N_2S$			
Imidazole, 5-(benzylthio)-4-phenyl-	EtOH	250(3.96),277(3.98)	23-1123-69
anion	H_2O	290(3.95)	23-1123-69
cation	H_2O	245(4.05)	23-1123-69
$C_{16}H_{14}N_2S_2$			
4-Imidazoline-2-thione, 5-(benzylthio)-4-phenyl-	EtOH	277(4.23),313s(--)	23-1123-69
anion	H_2O	260(4.08),302s(3.94)	23-1123-69
cation	H_2O	260(4.24)	23-1123-69
$C_{16}H_{14}N_4$			
5,5'-Bibenzimidazole, 2,2'-dimethyl-	EtOH	228(4.78),296(4.30)	22-1926-69
$C_{16}H_{14}N_4OS$			
Benzoic acid, 2-(5-phenyl-4H-1,3,4-thiadiazin-2-yl)hydrazide	EtOH	225(4.27),314(4.11)	95-0689-69
$C_{16}H_{14}N_4O_3$			
4H-Quinolizine-1-carboxylic acid, 3-cyano-2-[(2-cyanoethyl)amino]-4-oxo-, ethyl ester	EtOH	262(4.72),301(4.26), 391(4.15)	95-0203-69
$C_{16}H_{14}N_4O_4$			
Acetamide, N,N'-benzo[c]cinnoline-3,8-diyl)bis-, 5,6-dioxide	EtOH	280s(3.66),310(3.47)	4-0523-69
$C_{16}H_{14}N_4O_4S$			
1-Propanone, 1-[5-(1-propynyl)-2-thienyl]-, 2,4-dinitrophenylhydrazone	MeOH	320(3.88),404(4.45)	39-1813-69C
$C_{16}H_{14}N_4O_6$			
Barbituric acid, 5,5'-(α-methylbenzylidene)di-	EtOH	277(4.46)	103-0827-69
$C_{16}H_{14}N_4S$			
9-Acridinecarboxaldehyde, 2-methyl-3-thiosemicarbazone	EtOH-HCl	260(4.7),360(4.1), 460(4.0)	65-1151-69
	EtOH	370(4.1),400(4.0)	65-1151-69
9-Acridinecarboxaldehyde, S-methylthiosemicarbazone	EtOH-HCl	260(4.7),360(4.1), 450(3.7)	65-1151-69
	EtOH	370(4.1),410(4.1)	65-1151-69
$C_{16}H_{14}N_6O_2$			
Phthalimide, N-[3-(6-amino-9H-purin-9-yl)propyl]-	EtOH	232(4.27),234(4.21), 262(4.20),295s(3.53)	44-3240-69
	EtOH-HCl	232(4.30),243(4.22), 260(4.24),295s(3.53)	44-3240-69
	EtOH-NaOH	260(4.24),295s(3.25)	44-3240-69
$C_{16}H_{14}O$			
3-Butenal, 2,4-diphenyl-	EtOH	243s(3.93),281s(3.68)	22-3719-69
Cyclohepta[def]fluoren-8-ol, 4,8,9,10-tetrahydro-	EtOH	214(4.56),225s(4.21), 262s(4.20),268(4.26), 289s(4.15),301(3.65)	39-1427-69C
1-Indanol, 2-benzylidene-, trans	hexane	258(4.30),270s(4.10)	88-1717-69
8,16-Oxido[2,2]metacyclophane	C_6H_{12}	236(3.66),263s(2.77)	35-1665-69
2-Propen-1-one, 1-phenyl-3-p-tolyl-	MeOH	320(4.42)	73-2771-69
	iso-PrOH	232(4.02),322(4.38)	104-0826-69
2-Propen-1-one, 3-phenyl-1-p-tolyl-	MeOH	308(4.39)	73-2771-69

Compound	Solvent	λ_{max}(log ε)	Ref.
2-Propen-1-one, 3-phenyl-1-p-tolyl-	iso-PrOH	312(4.41)	104-0826-69
$C_{16}H_{14}OS$			
2H-1-Benzothiopyran, 4-(p-methoxy-phenyl)-	EtOH	235(3.8),300(3.71)	2-0315-69
$C_{16}H_{14}OS_2$			
1,3-Dithiole, 2-methoxy-4,5-diphenyl-	EtOH	294(3.46),327(3.46)	94-1931-69
$C_{16}H_{14}O_2$			
Anthrone, 10-ethyl-10-hydroxy-	EtOH	216(4.13),270(4.20)	39-2266-69C
Benz[e]-as-indacene-7,10-dione, 1,2,3,4,5,6-hexahydro-	C_6H_{12} or EtOH	216(4.58),253(4.33), 262(4.21),374(3.67), 455s(2.04)	39-0061-69B
Benzyl alcohol, 2,2'-ethynylenedi-	EtOH	285(4.39),294(4.26), 304(4.34)	44-0874-69
1H-Cyclopropa[a]phenanthrene-1-carbox-ylic acid, tetrahydro-	n.s.g.	275(4),310(2.3)	39-0068-69C
o-Dioxin, 3,6-dihydro-3,6-diphenyl-, cis	EtOH	252(2.63),258(2.68), 264(2.62),268(2.45)	22-1664-69
2-Propen-1-one, 1-(m-methoxyphenyl)-3-phenyl-	MeOH	308(4.40)	73-2771-69
2-Propen-1-one, 1-(p-methoxyphenyl)-3-phenyl-	MeOH	319(4.51)	73-2771-69
	iso-PrOH	227(4.12),317(4.43)	104-0826-69
2-Propen-1-one, 3-(p-methoxyphenyl)-1-phenyl-	MeOH	340(4.42)	73-2771-69
	iso-PrOH	246(4.05),342(4.37)	104-0826-69
$C_{16}H_{14}O_3$			
Benz[e]-as-indacene-7,10-dione, 1,2,3,4,5,6-hexahydro-8-hydroxy-	C_6H_{12} or EtOH	217(4.41),254f(4.08), 290f(4.08),378f(3.66)	39-0061-69B
p-Benzoquinone, 2-cinnamyl-5-methoxy-	EtOH	257(4.45),368s(3.24)	88-2863-69
Bicyclo[4.2.0]octa-3,7-diene-2,5-dione, 1-methoxy-7-methyl-8-phenyl-	EtOH	232(4.30),258(4.18), 311(2.74),385(2.46)	44-0520-69
Dibenzofuran-2-carboxylic acid, 1,4-di-methyl-, methyl ester	EtOH	219(4.66),240(4.64), 283(4.02),297(3.61), 307(3.22)	4-0379-69
Dibenzofuran-3-carboxylic acid, 1,4-di-methyl-, methyl ester	EtOH	225(4.59),269(4.15), 292(4.43)	4-0379-69
p-Dioxan-2-one, 6,6-diphenyl-	EtOH	212(4.27),256(2.66)	22-2048-69
1,3-Dioxolan-4-one, 2-benzhydryl-	EtOH	205(4.20),257(2.61)	22-2048-69
Fluorene-6-carboxylic acid, 7-hydroxy-1-methyl-, methyl ester	EtOH	248(4.5),272(4.3), 288(4.2)	42-0415-69
$C_{16}H_{14}O_4$			
1,2-Acenaphthene-1,2-dicarboxylic acid, dimethyl ester, trans	EtOH	226(4.93),267(3.58), 277(3.81),287(3.89), 291(3.76),298(3.71), 304(3.25),315(2.82), 319(2.63)	44-2418-69
2-Biphenylcarboxylic acid, 3',4'-(meth-ylenedioxy)-, ethyl ester	EtOH	263(3.80),295(3.92)	78-4315-69
3-Biphenylcarboxylic acid, 3',4'-(meth-ylenedioxy)-, ethyl ester	EtOH	270(3.99),295(3.99)	78-4315-69
2,3-Biphenylenedione, 6,7-dimethoxy-1,8-dimethyl-	EtOH	263(4.46),287(4.50), 320s(3.94),415(4.32)	39-2579-69C
Chalcone, 2',6'-dihydroxy-4'-methoxy-	n.s.g.	214(4.31),342(4.35)	25-1779-69
1,4-Ethenonaphthalene-2,3-dicarboxylic acid, 1,4-dihydro-, dimethyl ester	EtOH	204(4.41),208(4.40), 238(4.23),274(3.33), 283(3.30),318(2.68), 334(2.65)	44-2418-69

Compound	Solvent	λ_{max}(log ϵ)	Ref.
Flavanone, 5-hydroxy-7-methoxy-, (-)-	EtOH	289(4.36),320(3.61)	105-0397-69
Isophthalic acid, 5-phenyl-, dimethyl ester	EtOH	231(4.49),257(4.14), 301(3.40)	44-4134-69
1-Naphthalenefumaric acid, dimethyl ester	EtOH	223(4.85),264s(3.64), 272(3.73),281(3.76), 288s(3.67)	44-2418-69
2H-Naphtho[1,2-b]pyran-6-propionic acid, 3,4-dihydro-2-oxo-	dioxan	230(4.6),286(3.8), 296(3.9),325(3.4)	39-2618-69C
Phthalide, 3-(2,4-dimethoxyphenyl)-	EtOH	215(4.14),233(4.23), 278(3.64),283(3.64)	39-1873-69C
Tricyclo[4.2.2.0^{2,5}]deca-3,9-diene-7,8-dicarboxylic anhydride, 3-crotonoyl-	EtOH	256(4.17)	39-0978-69C
Xanthen-9-one, 1,3-dimethoxy-2-methyl-	MeOH	238(4.59),277(4.03), 300(4.20)	78-0275-69
Xanthen-9-one, 1-hydroxy-6-methoxy-3,8-dimethyl-	EtOH	213(3.91),237(4.22), 251(4.10),267(3.93), 301(4.01),345(3.46)	39-1721-69C
$C_{16}H_{14}O_4S_2$			
1,4-Thianthrenedione, 6,9-dihydroxy-2,3,7,8-tetramethyl-	THF	243(4.45),287(4.28), 330(4.23),438(3.78)	24-1739-69
$C_{16}H_{14}O_5$			
Acetic acid, (4-benzoyl-3-hydroxyphenoxy)-, methyl ester	EtOH	240(3.91),285(4.08), 325(3.82)	39-0037-69C
Flavanone, 4',5,7-trihydroxy-6-methyl-(poriol)	MeOH	293(4.22)	78-0283-69
	EtOH	298(4.42),336s(3.66)	12-0483-69
	EtOH-NaOEt	336(4.64)	12-0483-69
	EtOH-NaOAc	301(4.30),336(4.29)	12-0483-69
	EtOH-$AlCl_3$	315(4.50),392(3.58)	12-0483-69
Isoflavanone, 4',5-dihydroxy-7-methoxy-, (±)-	EtOH	289(3.44),335s(4.21)	78-1013-69
Rubrofusarin, 6-O-methyl-	EtOH	225(4.44),254s(4.43), 275(4.69),322(3.51), 395(3.81)	94-0458-69
Xanthen-9-one, 1-hydroxy-3,7-dimethoxy-2-methyl-	MeOH	235(4.50),307(4.21)	78-0275-69
Xanthen-9-one, 1-hydroxy-3,7-dimethoxy-4-methyl-	MeOH	232(4.28),263(4.34), 310(3.87)	78-0275-69
Xanthen-9-one, 1,6,7-trimethoxy-	MeOH	239s(4.19),248(4.39), 283(3.89),303(3.49), 353(3.94)	39-2201-69C
$C_{16}H_{14}O_6$			
Aflatoxin B_3	MeOH	229(4.00),253(3.86), 262(3.88),326(3.97)	78-1497-69
	MeOH	229(4.00),253(3.86), 262(3.88),326(3.97)	119-0107S-69
Bellidifolin, di-O-methyl-	EtOH	254(4.4),278(4.2), 300s(3.8),335(4.0)	94-0155-69
2,3-Biphenylenedione, 1,6,7,8-tetramethoxy-	EtOH	264(4.47),303(4.58), 417(4.24)	39-2579-69C
Naphtho[2,3-d]-1,3-dioxole-5,6-dicarboxylic acid, 7-methyl-, dimethyl ester	EtOH	245(4.67),296(3.76), 342(3.69)	12-1721-69
Naphtho[1,2-c]furan-4,5-dicarboxylic acid, 1,3,4,5-tetrahydro-3-oxo-, dimethyl ester	EtOH	223(4.21),229(4.23), 287(4.19)	94-1591-69
Swertianin, di-O-methyl-	EtOH	240(4.56),260(4.64), 315(4.16),380s(3.68)	94-0155-69

Compound	Solvent	$\lambda_{max}(\log \epsilon)$	Ref.
Xanthen-9-one, 1,3-dihydroxy-6,7-dimethoxy-2-methyl-	MeOH	241(4.33),259(4.38), 318(4.16)	78-0275-69
Xanthen-9-one, 1-hydroxy-2,3,7-trimethoxy-	EtOH	238(4.44),262(4.46), 300(4.07),320s(4.03), 363(3.71)	78-1947-69
Xanthen-9-one, 1-hydroxy-3,4,5-trimethoxy-	EtOH	246(4.47),256(4.43), 318(4.12),371(3.65)	78-1961-69
$C_{16}H_{14}S$			
6H-Cyclopenta[5,6]naphtho[2,1-b]thiophene, 7,8-dihydro-5-methyl-	n.s.g.	258(4.72),300(4.14), 312(4.08),322(3.59), 338(3.21)	28-0194-69A
Styryl sulfide, (E,Z)-	pentane	229(4.38),320(4.56)	44-0896-69
Sulfide, benzyl 3-phenyl-2-propynyl	EtOH	246(4.26),250(4.21)	94-0966-69
Thiophene, 2,3-dihydro-3,4-diphenyl-	pentane	226(4.21),306(4.18)	44-0896-69
$C_{16}H_{15}$			
3-Methyl-1-phenylindanylium cation	acid	304(4.08),415(4.54)	59-0447-69
$C_{16}H_{15}BrN_2O_4S$			
5-Thia-1-azabicyclo[4.2.0]oct-2-ene-2-carboxylic acid, 7-[2-(m-bromophenyl)-acetamido]-3-methyl-8-oxo-, potassium salt	H_2O	260(3.87)	87-0310-69
$C_{16}H_{15}BrN_2O_6$			
Piperonylamide, N-[2-(1-bromo-4-oxo-2-cyclohexen-1-yl)ethyl]-6-nitro-	EtOH	228(3.8),243(3.7), 285(3.3),342(3.3)	44-2667-69
$C_{16}H_{15}BrN_2S_2$			
2H-1,3,5-Thiadiazine-2-thione, 5-benzyl-3-(p-bromophenyl)tetrahydro-	MeOH	252(4.15),292(4.08)	73-2952-69
$C_{16}H_{15}BrN_4O_5$			
Propiophenone, 2'-bromo-4'-methoxy-, 2,4-dinitrophenylhydrazone	EtOH	363(4.37)	22-2079-69
Propiophenone, 3'-bromo-2'-methoxy-, 2,4-dinitrophenylhydrazone	EtOH	364(4.40)	22-2079-69
Propiophenone, 3'-bromo-4'-methoxy-, 2,4-dinitrophenylhydrazone	EtOH	384(4.43)	22-2079-69
Propiophenone, 4'-bromo-2'-methoxy-, 2,4-dinitrophenylhydrazone	EtOH	368(4.40)	22-2079-69
Propiophenone, 5'-bromo-2'-methoxy-, 2,4-dinitrophenylhydrazone	EtOH	366(4.39)	22-2079-69
$C_{16}H_{15}Br_2NO$			
p-Anisidine, N-cinnamylidene-, compound with bromine	MeCN	293(4.55),348(4.42)	25-0135-69
$C_{16}H_{15}ClN_2O$			
6H-Indolo[2,3-a]quinolizin-5-ium, 7,12-dihydro-3-(hydroxymethyl)-, chloride	MeOH	254(3.88),314(4.15), 390(4.16)	44-3545-69
$C_{16}H_{15}ClN_2OS$			
2(1H)-Quinazolinethione, 6-chloro-3-ethyl-3,4-dihydro-4-hydroxy-4-phenyl-	n.s.g.	262(3.98),290(4.44)	40-0917-69

Compound	Solvent	λ_{max}(log ε)	Ref.
$C_{16}H_{15}ClN_2O_2$			
2(1H)-Quinazolinone, 6-chloro-3-ethyl-3,4-dihydro-4-hydroxy-4-phenyl-	n.s.g.	254(4.14),304(3.15)	40-0917-69
$C_{16}H_{15}ClN_2O_4S$			
5-Thia-1-azabicyclo[4.2.0]oct-2-ene-2-carboxylic acid, 7-[2-(p-chlorophenyl)acetamido]-3-methyl-8-oxo-, potassium salt	H_2O	262(3.87)	87-0310-69
$C_{16}H_{15}ClN_2O_4S_2$			
5-Thia-1-azabicyclo[4.2.0]oct-2-ene-2-carboxylic acid, 7-[2-[(m-chlorophenyl)thio]-acetamido]-3-methyl-8-oxo-, potassium salt	H_2O	255(4.13)	87-0310-69
$C_{16}H_{15}F$			
Ethylene, 2-fluoro-1,1-di-p-tolyl-	MeOH	249s(3.99)	39-1100-69B
$C_{16}H_{15}Li$			
1-Butene, 4,4-diphenyl-, 4-lithium derivative	THF	486(4.48)	35-2456-69
$C_{16}H_{15}N$			
8,16-Imino[2,2]metacyclophane	C_6H_{12}	248(3.77)	35-1672-69
Indole, 3-phenethyl-	n.s.g.	224(4.55),282(3.76), 290(3.69)	103-0190-69
9,10-(Methaniminomethano)anthracene, 9,10-dihydro-	EtOH	256(3.03),267(3.22), 275(3.06)	33-2197-69
$C_{16}H_{15}NO$			
Acetanilide, N-styryl-, cis	MeOH	254(4.14)	78-5745-69
Acetanilide, N-styryl-, trans	MeOH	222(--),285(4.41)	78-5745-69
Acrylophenone, 3-(methylamino)-2-phenyl-	C_6H_{12}	218(3.92),240(3.95), 268(3.91),350(4.01)	24-2987-69
7-Azabicyclo[4.2.2]deca-2,4,7,9-tetraene, 8-methoxy-2-phenyl-	hexane	299(4.00)	35-4714-69
7-Azabicyclo[4.2.2]deca-2,4,7,9-tetraene, 8-methoxy-9-phenyl-	hexane	247(4.14)	35-4714-69
Benzyl alcohol, o-(indol-3-ylmethyl)-	EtOH	223(4.56),275(3.76), 283(3.78),292(3.72)	78-0227-69
10,5-(Iminomethano)-5H-dibenzo[a,d]-cyclohepten-10(11H)-ol	MeCN	273(2.83),290(2.35)	23-2827-69
	MeCN-HCl	271s(2.92),293(2.89)	23-2827-69
2-I dolinone, 7-methyl-1-o-tolyl-	MeOH	250(3.91)	24-3486-69
Nitrone, N-benzyl-α-styryl-	EtOH	237(4.0),332(4.4)	28-0078-69A
Phthalimidine, 2,3-dimethyl-3-phenyl-	EtOH	246(3.78),272s(3.41), 281(3.26)	39-1149-69C
Spiro[indan-2,3'-indolin]-1-ol	EtOH	244(3.76),272(3.34), 295(3.43)	78-0227-69
	EtOH-H_2SO_4	263(3.04),267(3.04), 272(2.98)	78-0227-69
$C_{16}H_{15}NOS$			
Benzo[b]thiophene-5-carboxaldehyde, 6,7-dihydro-4-(N-methylanilino)-	EtOH	247(4.40),336(4.06), 418(3.59)	39-2750-69C
Benzo[b]thiophen-4(5H)-one, 6,7-dihydro-5-[(N-methylanilino)methylene]-	EtOH	254(4.42),374(4.19)	39-2750-69C

Compound	Solvent	λ_{max}(log ε)	Ref.
$C_{16}H_{15}NO_2$			
Acetamide, N-(1,2,3,4-tetrahydro-3-oxo-4-phenanthryl)-	EtOH	224(4.82),264(3.64), 273(3.77),280(3.80), 283(3.80),290(3.68), 306(2.93),313(2.74), 320(2.84)	24-2384-69
Acetophenone, 2-phenyl-, O-acetyloxime	EtOH	246(4.12)	44-1430-69
Benzo[g]isoquinoline-5,10-dione, 6,9-dihydro-3,7,8-trimethyl-	MeCN	249(4.47),271(4.76), 278(4.73),317(4.13)	4-0697-69
Cinnamamide, N-(benzyloxy)-	MeOH	208(4.4),216(4.09), 222(3.98),277(4.15)	12-0161-69
	MeOH-NaOH	265(3.92),296(3.94)	12-0161-69
Cinnamohydroxamic acid, N-m-tolyl-	EtOH	292(4.33)	34-0278-69
Cinnamohydroxamic acid, N-o-tolyl-	EtOH	217(4.21),282(4.37)	34-0278-69
5H-Dibenzo[b,f]azepine-4-carboxylic acid, 10,11-dihydro-, methyl ester	EtOH	223s(4.24),293(4.23), 367(3.93)	35-1672-69
Propionanilide, 3-benzoyl-	MeCN	200(4.76),243(4.48), 311(1.42)	28-0279-69A
p-Toluanilide, 4'-(vinyloxy)-	n.s.g.	250(4.34)	30-0108-69A
stannic chloride complex	n.s.g.	250(4.27),585(1.67)	30-0108-69A
$C_{16}H_{15}NO_3$			
Acetic acid, anhydride with N-(benzyloxy)benzimidic acid	MeOH	208(--),256(3.97)	12-0161-69
	MeOH-NaOH	265(3.72)	12-0161-69
p-Anisanilide, 4'-(vinyloxy)-	n.s.g.	276(4.44)	30-0108-69A
stannic chloride complex	n.s.g.	276(4.44),585(1.00)	30-0108-69A
Anthranilic acid, N-(4-acetyl-m-tolyl)-	isooctane	226(4.22),240s(3.93), 316(4.12),357(4.02)	22-3523-69
	EtOH	205(4.52),234(4.09), 324s(4.14),356(4.49)	22-3523-69
Benzamide, N-acetyl-N-(benzyloxy)-	MeOH	208(--),234(3.78)	12-0161-69
	MeOH-NaOH	264(3.41)	12-0161-69
6,11-Epoxy-3H-naphth[2,3-d]azepine-3-carboxylic acid, 5a,6,11,11a-tetrahydro-, methyl ester	EtOH	230(4.28)	44-2888-69
Phthalimidine, 3-hydroxy-2-(2-hydroxyethyl)-3-phenyl-	EtOH	252(3.50),264(3.40), 285s(2.89)	44-0165-69
Pyrido[1,2-a]indole-10-carboxylic acid, 3-hydroxy-1-methyl-, ethyl ester	EtOH	253(4.58),259(4.62), 290s(4.21),300(4.23), 316(4.21),330(4.20)	12-1525-69
$C_{16}H_{15}NO_3S_3$			
2,3-Dihydrothiazolo[2,3-b]benzothiazolium p-toluenesulfonate	MeOH	223(4.31),252(3.76), 258(3.87),311(4.18)	4-0163-69
$C_{16}H_{15}NSe$			
Selenide, benzyl 3-indolylmethyl	EtOH	223(4.59),273(3.74), 281(3.75),290(3.67)	11-0031-69B
$C_{16}H_{15}N_3$			
Anthranilonitrile, N-[α-(dimethylamino)-benzylidene]-	EtOH	215(4.47),259(3.85), 307(3.68)	22-2008-69
$C_{16}H_{15}N_3O$			
Benzamide, N-[1-(2-benzimidazolyl)-ethyl]-	EtOH	275(3.93),282(3.92)	94-2381-69
Indole-3-acetamide, 5-methyl-2-(3-pyridyl)-	EtOH	218(4.55),306(4.36)	22-4154-69

Compound	Solvent	$\lambda_{max}(\log \epsilon)$	Ref.
$C_{16}H_{15}N_3O_2$			
Benzoic acid, p-(5-amino-1-methyl-2-benzimidazolyl)-, methyl ester	EtOH	212(4.54),234s(4.38), 296(4.00),316s(3.81)	73-0572-69
$C_{16}H_{15}N_3O_2S_2$			
2H-1,3,5-Thiadiazine-2-thione, 3-benzyl-tetrahydro-5-(p-nitrophenyl)-	MeOH	230(4.14),290(4.14)	73-2952-69
$C_{16}H_{15}N_3O_4$			
Aniline, N-(2-p-tolylcyclopropyl)-2,4-dinitro-, trans	EtOH	348(3.23)	39-1135-69B
$C_{16}H_{15}N_3O_5$			
Aniline, N-[2-(p-methoxyphenyl)cyclopropyl]-2,4-dinitro-, trans	EtOH	348(3.22)	39-1135-69B
$C_{16}H_{15}N_3O_6S$			
5-Thia-1-azabicyclo[4.2.0]oct-2-ene-2-carboxylic acid, 3-methyl-7-[2-(m-nitrophenyl)acetamido]-8-oxo-, potassium salt	H_2O	264(4.12)	87-0310-69
5-Thia-1-azabicyclo[4.2.0]oct-2-ene-2-carboxylic acid, 3-methyl-7-[2-(p-nitrophenyl)acetamido]-8-oxo-, potassium salt	H_2O	257(4.24)	87-0310-69
$C_{16}H_{15}N_5O_3$			
4,6-Pteridinedione, 7-acetonyl-2-amino-5-benzyl-3,5-dihydro-	pH -1.2	235(4.09),243s(4.05), 264(3.89),313s(3.82), 324(3.83),380s(4.11), 394(4.13),415s(4.01), 442s(3.66)	24-4032-69
	pH 6	240(4.25),312(4.05), 406(4.10)	24-4032-69
	pH 12	236(4.29),262s(3.86), 296s(3.88),307(3.89), 330s(3.79),424(4.05), 450s(3.97)	24-4032-69
$C_{16}H_{16}$			
1-Butene, 4,4-diphenyl-, 4-lithium derivative	THF	486(4.48)	35-2456-69
Ethylene, 1-(p-ethylphenyl)-1-phenyl-	hexane	205(4.47),232(4.30), 250(4.11)	59-0447-69
Isobutene, 1,1-diphenyl-	C_6H_{12}	229(4.09),245(4.14)	59-0487-69
[2,2]Paracyclophane	n.s.g.	225(4.38),244s(3.52), 286(2.41),302s(2.19)	35-1862-69
tetracyanoethylene complex	CH_2Cl_2	521(--)	35-3553-69
1-Propene, 1-phenyl-1-o-tolyl-, cis	C_6H_{12}	250(4.15)	59-0487-69
trans	C_6H_{12}	233(4.17)	59-0487-69
1-Propene, 1-phenyl-1-p-tolyl-, cis	C_6H_{12}	242(4.13)	59-0487-69
trans	C_6H_{12}	226(4.26),252(4.24)	59-0487-69
Stilbene, 2,5-dimethyl-, trans	EtOH	224(4.30),297(4.32)	35-7166-69
$C_{16}H_{16}BrNO$			
2,5-Cyclohexadien-1-one, 4-[(4-bromo-2,6-xylyl)imino]-2,6-dimethyl-	EtOH	235(3.90),275(4.50), 470(3.04)	78-2291-69

Compound	Solvent	$\lambda_{max}(\log \epsilon)$	Ref.
$C_{16}H_{16}BrN_3O_4S$			
5-Thia-1-azabicyclo[4.2.0]oct-2-ene-2-carboxylic acid, 7-[2-amino-2-(m-bromophenyl)acetamido]-3-methyl-8-oxo-	H_2O	261(3.85)	87-0310-69
$C_{16}H_{16}BrN_5$			
4-[(α-Aminobenzylidene)imino]-1-benzyl-4H-1,2,4-triazolium bromide	MeOH	235(4.26)	48-0477-69
$C_{16}H_{16}ClNO$			
2,5-Cyclohexadien-1-one, 4-[4-chloro-2,6-xylyl)imino]-2,6-dimethyl-	EtOH	275(4.49),475(3.03)	78-2291-69
$C_{16}H_{16}ClNO_3$			
Hexanamide, N-(3-chloro-1,4-dihydro-1,4-dioxo-2-naphthyl)-	MeOH	247(4.27),252(4.31), 286(3.96),336(3.52)	4-0909-69
$C_{16}H_{16}ClN_3O_4S$			
5-Thia-1-azabicyclo[4.2.0]oct-2-ene-2-carboxylic acid, 7-[2-amino-2-(m-chlorophenyl)acetamido]-3-methyl-8-oxo-	H_2O	261(3.82)	87-0310-69
5-Thia-1-azabicyclo[4.2.0]oct-2-ene-2-carboxylic acid, 7-[2-amino-2-(p-chlorophenyl)acetamido]-3-methyl-8-oxo-	H_2O	260(3.84)	87-0310-69
$C_{16}H_{16}Cl_2N_4OS$			
s-Triazine, 2,4-dichloro-6-[2-(methylthio)-2-morpholino-1-phenylvinyl]-	C_6H_{12}	235(3.92),290(4.17), 419(4.29)	5-0073-69E
$C_{16}H_{16}FN_3$			
Indole, 3-[(dimethylamino)methyl]-5-fluoro-2-(3-pyridyl)-	EtOH	220(4.29),310(4.14)	22-4154-69
$C_{16}H_{16}FN_3O_4S$			
5-Thia-1-azabicyclo[4.2.0]oct-2-ene-2-carboxylic acid, 7-[2-amino-2-(m-fluorophenyl)acetamido]-3-methyl-8-oxo-	H_2O	263(3.90)	87-0310-69
$C_{16}H_{16}Fe$			
[3]Ferrocenophane, α,β-(Δ^2-trimethylene)-	EtOH	273(3.29),328(2.75), 448(2.49)	49-1540-69
$C_{16}H_{16}FeO$			
1-Pentanone, 1,5-(1,1'-ferrocenediyl)-2-methylene-	n.s.g.	235(4.08),283(3.72), 356(3.17),464(2.92)	78-0861-69
1-Propanone, 1,3-(2,2'-trimethylene-1,1'-ferrocenediyl)-	EtOH	264(3.25),329(2.83), 426(2.40)	49-1540-69
1-Propanone, 1,3-(3,3'-trimethylene-1,1'-ferrocenediyl)-	EtOH	328(3.03),468(2.79)	49-1540-69
$C_{16}H_{16}FeO_2$			
1,3-Pentanedione, 1,5-(1,1'-ferrocenediyl)-2-methyl-	n.s.g.	226(4.39),256(3.98), 341(3.27),465(2.56)	78-0861-69
$C_{16}H_{16}INO$			
2,5-Cyclohexadien-1-one, 4-[(4-iodo-2,6-xylyl)imino]-2,6-dimethyl-	EtOH	240(4.02),276(4.45), 474(3.01)	78-2291-69

Compound	Solvent	λ_{max}(log ϵ)	Ref.
$C_{16}H_{16}NOPS$			
Formamide, N-allyl-1-(diphenylphosphinyl)thio-	EtOH	292(3.90),401(1.92)	18-2975-69
$C_{16}H_{16}N_2$			
Acetophenone, azine	C_6H_{12}	271(4.42),296(4.41)	39-0836-69C
Benzimidazole, 2,6-dimethyl-1-p-tolyl-	EtOH	270(4.30)	95-1566-69
Cinnamaldehyde, methylphenylhydrazone	EtOH	259(4.0),357(4.5)	28-0137-69B
Cinnamaldehyde, o-tolylhydrazone	EtOH	253(4.0),291s(3.9), 364(4.5)	28-0137-69B
Dibenzo[c,g][1,2]diazocine, 5,6-dihydro-5,6-dimethyl-	n.s.g.	260(4.9),305(3.8), 315(3.8)	44-3237-69
Phthalazino[2,3-b]phthalazine, 5,7,12,14-tetrahydro-, hydrobromide	EtOH	252(2.76),258(2.84), 266(2.95),273(2.99)	44-2715-69
Pyridazine, 1,4,5,6-tetrahydro-1,3-diphenyl-	EtOH	243(4.11),310s(3.52), 340(4.27)	39-1703-69C
Pyridazine, 2,3,4,5-tetrahydro-3,6-diphenyl-	MeOH	285(4.14)	23-4041-69
Pyridazine, 3,4,5,6-tetrahydro-3,6-diphenyl-	n.s.g.	285(2.45),385(2.56)	23-4041-69
$C_{16}H_{16}N_2O$			
Acetic acid, (m-methylbenzylidene)phenylhydrazide	EtOH	220(4.3),286(4.4)	28-1703-69A
Acetic acid, (o-methylbenzylidene)phenylhydrazide	EtOH	221s(4.2),280(4.3)	28-1703-69A
Acetic acid, (p-methylbenzylidene)phenylhydrazide	EtOH	220(4.2),285(4.4), 288s(4.4),300s(4.3)	28-1703-69A
6H-Azepino[1,2-a]indole-11-carbonitrile, 7,8,9,10-tetrahydro-2,9-dimethyl-6-oxo-	CH_2Cl_2	259(4.16),284(4.05), 299(4.04),306(4.02)	94-1294-69
Carbostyril, 3,4-dihydro-3-(methylamino)-4-phenyl-	n.s.g.	253(4.17)	35-6083-69
Cinnamic acid, 2-methyl-2-phenylhydrazide	H_2O	237(4.15),280(4.40)	65-1835-69
$C_{16}H_{16}N_2O_2$			
Acetic acid, (m-methoxybenzylidene)phenylhydrazide	EtOH	203(4.4),217s(4.3), 279(4.3),318s(3.8)	28-1703-69A
Acetic acid, (o-methoxybenzylidene)phenylhydrazide	EtOH	203(4.4),227s(4.1), 278(4.2),286(4.1), 316(4.1)	28-1703-69A
Acetic acid, (p-methoxybenzylidene)phenylhydrazide	EtOH	205(4.3),221(4.2), 288(4.4),304(4.4)	28-1703-69A
Acetophenone, 2'-hydroxy-, azine	EtOH	220(4.4),240s(4.0), 293(4.3),364(4.2)	28-0730-69A
Benzamidine, N-(p-ethoxycarbonylphenyl)-	50% EtOH	284(4.46)	49-1307-69
Benzimidazole, 6-methoxy-1-(p-methoxyphenyl)-2-methyl-	EtOH	270(4.30)	95-1566-69
1H-Naphth[2,3-d]imidazole-4,9-dione, 2-pentyl-	pH 1	245(4.50),250(4.62), 278(4.21)	4-0909-69
	pH 7	248(4.64),283(4.16)	4-0909-69
	pH 13	264(4.61)	4-0909-69
	MeOH	246(4.63),277(4.16), 332(3.48)	4-0909-69
$C_{16}H_{16}N_2O_3$			
Acetophenone, 2'-methyl-5'-[(p-methoxyphenyl)azo]-	EtOH	244(4.23),353(4.28), 439s(3.34)	12-0935-69

Compound	Solvent	$\lambda_{max}(\log \epsilon)$	Ref.
Alaninamide, N-[3-(2-furyl)acryloyl]-α-phenyl-	pH 8	305(4.41)	35-7187-69
1-Methylpyrazolium 2-[α-(2-carboxy-vinyl)phenacylide]	EtOH	220(4.16),228(4.15), 274(3.72),344(4.46)	44-4134-69
$C_{16}H_{16}N_2O_4$			
Acetophenone, 2',6'-dihydroxy-, azine	EtOH	223(4.4),333(4.4)	28-0730-69A
4,4'-Biphenyldicarboxylic acid, 2,2'-diamino-, dimethyl ester	EtOH	232(4.69),276(3.90), 333(3.95)	4-0523-69
$C_{16}H_{16}N_2O_6$			
2,3-Biphenylenedione, 1,6,7,8-tetrameth-oxy-, dioxime	EtOH	233s(4.16),258s(4.46), 287(4.70),362(4.03), 379s(3.99),418(3.85)	39-0646-69C
Piperonylamide, 6-nitro-N-[2-(4-oxo-1-cyclohexen-1-yl)ethyl]-	EtOH	242(4.0),343(3.6)	44-2667-69
$C_{16}H_{16}N_2O_7$			
Uracil, 1-β-D-xylofuranosyl-, 2'-ben-zoate	EtOH	230(4.22),260(4.08)	94-0798-69
$C_{16}H_{16}N_2S_2$			
2H-1,3,5-Thiadiazine-2-thione, 3-benzyl-tetrahydro-5-phenyl-	MeOH	250(4.01),295(4.06)	73-2952-69
$C_{16}H_{16}N_2Se_2$			
Diselenide, 5,5'-diindolinyl	EtOH	206(4.48),247(4.12), 309(3.99)	11-0159-69B
$C_{16}H_{16}N_3O_8P$			
Benzamide, N-(1-β-D-arabinofuranosyl-1,2-dihydro-2-oxo-4-pyrimidinyl)-, cyclic hydrogen phosphate, compound with N,N'-dicyclohexyl-4-morpholino-carboxamidine	n.s.g.	258(4.26),302(4.03)	44-0244-69
$C_{16}H_{16}N_4$			
Stilbene, α,α'-bis(methylazo)-	MeOH	235(4.27),302(3.58), 394(2.97)	44-4118-69
$C_{16}H_{16}N_4OS$			
1,3,4-Thiadiazolidine-3-ethanol, 2,5-bis(phenylimino)-	EtOH	232(4.21),257(4.30)	44-0372-69
$C_{16}H_{16}N_4O_2$			
1H-Pyrazolo[3,4-b]pyrazin-6-ol, 1,5-di-methyl-3-phenyl-, propionate	EtOH	245(4.28),288(3.94), 336(3.82)	32-0463-69
1H-Pyrazolo[3,4-b]pyrazin-6-ol, 3,5-di-methyl-1-phenyl-, propionate	EtOH	258(4.45),323s(3.65)	32-0463-69
5-Pyrimidinecarboxylic acid, 4-[(cyano-methyl)methylamino]-2-phenyl-	EtOH	258(4.47)	4-0819-69
Urea, 1-(2-oxo-3-phenyl-1-imidazolidin-yl)-3-phenyl-	MeOH	242(3.46),270s(--), 280s(--)	44-0372-69
$C_{16}H_{16}N_4O_5$			
Acetophenone, 3'-ethyl-4'-hydroxy-, 2,4-dinitrophenylhydrazone	$CHCl_3$	394(4.38)	22-0781-69
1,3,4-Oxadiazole, 2-[1-(2-furyl)-2-(5-nitro-2-furyl)vinyl]-5-(isobutyl-amino)-, cis	n.s.g.	308(4.11),428(4.31)	40-0713-69

Compound	Solvent	λ_{max}(log ε)	Ref.
1,3,4-Oxadiazole, 2-[1-(2-furyl)-2-(5-nitro-2-furyl)vinyl]-5-(isobutylamino)-, trans	n.s.g.	316(4.10),418(4.28)	40-0713-69
$C_{16}H_{16}N_4S$			
9-Acridanone, 4,4-dimethyl-3-thiosemicarbazone	EtOH-HCl	250(4.3),330(3.5), 400(3.9),420(4.0), 450(4.0),490(3.0)	65-1156-69
	EtOH	250(5.0),310(3.8), 490(4.5),510(4.5)	65-1156-69
$C_{16}H_{16}N_4S_2$			
Urea, 1-phenyl-3-[2-(phenylimino)-3-thiazolidinyl]-2-thio-	EtOH	250(4.43)	44-0372-69
$C_{16}H_{16}N_6O_2$			
1H-Pyrazolo[3,4-d]pyrimidine, 1-(p-nitrophenyl)-4-piperidino-	MeOH	281(4.36),327(4.30)	23-1129-69
$C_{16}H_{16}N_6O_3$			
Adenine, N^6-(N-benzyloxycarbonylglycyl)-9-methyl-	pH 1	276(4.15)	44-3492-69
	pH 5.6	274(4.19)	44-3492-69
	pH 13	290(4.04)	44-3492-69
Adenine, N^6-(N-benzyloxycarbonylsarcosyl)-	pH 1	276(4.13),283(4.13)	44-3498-69
	pH 5.8	280(4.10)	44-3498-69
	pH 13	279(4.12)	44-3498-69
$C_{16}H_{16}O$			
Acetophenone, 3',5'-dimethyl-2'-phenyl-	EtOH	224(4.15),228(4.15), 245(3.86),293(3.27)	6-0277-69
Acetophenone, 5'-methyl-2'-p-tolyl-	EtOH	223(4.18),233(4.27), 253(4.04),298(3.42)	6-0277-69
1H-2,10a-Ethanophenanthren-12-one, 2,3,9,10-tetrahydro-	EtOH	265(4.17)	77-1253-69
1H-3,4bβ-Methanobenzo[1,3]cyclopropa-[1,2-a]naphthalen-4(4aβH)-one, 2,3β,5,6-tetrahydro-	EtOH	238(4.1)	77-1253-69
2,4,6,8-Nonatetraenal, 3-methyl-9-phenyl-	$CHCl_3$	281(3.82),386(4.78)	88-4049-69
1-Propene, 1-(p-methoxyphenyl)-3-phenyl-	EtOH	260(5.03),265s(5.02), 270(4.98),295s(4.16), 303(3.92)	1-2403-69
1-Propene, 2-(p-methoxyphenyl)-1-phenyl-	MeCN	287(4.37)	33-1010-69
1-Propene, 2-(p-methoxyphenyl)-3-phenyl-	EtOH	257(4.08)	33-1010-69
1-Propene, 3-(p-methoxyphenyl)-1-phenyl-	EtOH	254(4.26),276s(3.71), 284(3.55),292(3.07)	1-2403-69
Propionaldehyde, 2-methyl-2,3-diphenyl-	C_6H_{12}	291(2.24)	22-0903-69
$C_{16}H_{16}OS_2$			
Dimethylsulfonium α-(phenylthio)phenacylide	EtOH	247(4.07),282(3.84)	88-3179-69
$C_{16}H_{16}O_2$			
Anthracene, 9,10-dihydro-2,6-dimethoxy-	hexane	207(4.47),230(4.20), 282(3.44),289(3.44)	23-4489-69
p-Dioxane, 2,2-diphenyl-	EtOH	202(4.25),256(2.56)	22-2048-69
Dispiro[5.0.5.4]hexadeca-1,4,8,11-tetraene-3,10-dione	EtOH	240(4.34)	77-0996-69

Compound	Solvent	λ_{max}(log ε)	Ref.
5H-3,4bα-Methano-1H-benzo[1,3]cycloprop-[1,2-a]inden-4(4aβH)-one, 2,3β-dihydro-7-methoxy-	EtOH	248(4.01)	77-1253-69
Naphtho[1,8-bc]pyran, 2,3-dihydro-9-methoxy-2-methyl-4-vinyl-	EtOH	301(3.78),356(3.48)	49-0163-69
1,4-Naphthoquinone, 2-methyl-3-(1-pentenyl)-, cis	EtOH	249(4.33),286(3.57),	35-0731-69
trans	EtOH	252(4.36),292(3.87), 333(3.44)	35-0731-69
Stilbene, 4,4'-dimethoxy-	DMF	309(4.46),328(4.48)	33-2521-69
Stilbene, 4,4'-dimethoxy-, trans	MeOH	322(4.53)	39-0554-69B
	DMF	322(4.49)	39-0554-69B
$C_{16}H_{16}O_3$			
Benzophenone, 2,6-dimethoxy-4-methyl-	EtOH	213(4.21),227(3.99), 248(4.13),282(3.52)	39-1721-69C
2,4-Cyclohexadien-1-one, 6-hydroxy-2,4,6-trimethyl-, benzoate	n.s.g.	231(4.25),275(3.38), 283(3.40),312(3.51)	77-0550-69
Dibenz[b,f]oxepine-4,6-dimethanol, 10,11-dihydro-	EtOH	236(3.84),268s(3.23)	35-1665-69
Fluorene-3-carboxylic acid, 1,2,3,4-tetrahydro-8-methyl-2-oxo-, methyl ester	EtOH	249(4.3)	42-0415-69
1,4-Naphthoquinone, 2-(3-hydroxy-3-methyl-1-butenyl)-3-methyl-	EtOH	253(4.36),292(3.89), 335(3.52)	35-0731-69
Phthalan, 1-(2,4-dimethoxyphenyl)-	EtOH	215(4.11),233(4.01), 267(3.34),273(3.47), 279(3.46),285(3.39)	39-1873-69C
$C_{16}H_{16}O_4$			
Acetophenone, 2'-hydroxy-4',5'-dimethoxy-2-phenyl-	EtOH	239(4.21),277(4.12), 344(4.03)	39-1787-69C
Benzophenone, 2'-(hydroxymethyl)-2,4-dimethoxy-	EtOH	213(4.20),235(4.04), 278(3.91),309(3.83)	39-1873-69C
Benzophenone, 2,3',6-trimethoxy-	EtOH	226(4.22),255(3.89), 312(3.39)	39-0281-69C
2,4,6-Heptatrienoic acid, 7-hydroxy-, ethyl ester, benzoate, all trans	MeOH	235(4.07),308(4.72)	78-4315-69
2H-Naphtho[1,2-b]pyran-5,6-dione, 3,4-dihydro-7-methoxy-2,2-dimethyl-	ether	227(4.32),251(4.25), 386(3.69)	33-0808-69
Santonene, acetyl-	MeOH	268(4.6),381(4.3)	39-1088-69C
Warburgin	EtOH	370(4.30)	78-2865-69
$C_{16}H_{16}O_4S$			
2,5-Thiophenedicarboxylic acid, 3-phenyl-, diethyl ester	EtOH	240(4.19),280(4.12)	2-0009-69
$C_{16}H_{16}O_5$			
Benzophenone, 2-hydroxy-3',4,6-trimethoxy-	EtOH	209(4.32),221s(4.05), 254(3.61),310(3.27)	78-1507-69
Benzophenone, 2,2',4-trihydroxy-4'-methoxy-6,6'-dimethyl-	EtOH	213(3.3),224s(4.24), 292(4.16),345(3.84)	39-1721-69C
Benzophenone, 2,2',6-trihydroxy-4'-methoxy-4,6'-dimethyl-	EtOH	213(4.29),227(4.19), 287(4.10),350(3.55)	39-1721-69C
Benzophenone, 2,4,4'-trihydroxy-2-methoxy-6,6'-dimethyl-	EtOH	215(4.16),225(4.18), 236s(4.06),295(4.16), 342(3.90)	39-1721-69C
Benzophenone, 2,4',6-trihydroxy-2'-methoxy-4,6'-dimethyl-	EtOH	216(4.12),227(4.13), 286(4.11),349(3.59)	39-1721-69C

Compound	Solvent	λ_{max}(log ϵ)	Ref.
2,6-Cresotic acid, 4-methoxy-, 5-hydroxy-m-tolyl ester	EtOH	220(4.41),267(4.25), 302(3.8)	39-1721-69C
Eudesma-1,5,7(11)-trien-12-oic acid, 4,6-dihydroxy-3-oxo-, γ-lactone, formate	MeOH	288(4.47)	39-1088-69C
Fluorene-1,9-dicarboxylic acid, 5,6,7,8-tetrahydro-2-methoxy-	EtOH	273(4.13),333(3.35)	44-2209-69
7H-Furo[3,2-g][1]benzopyran-7-one, 4-(2-hydroxy-3-methylbutoxy)-	n.s.g.	222(4.19),250(4.24), 258(4.09),266(4.16), 310(4.04)	105-0001-69
1,2-Naphthalenedicarboxylic acid, 7-methoxy-3-methyl-, dimethyl ester	EtOH	240(4.66),279(3.64), 348(3.55)	12-1721-69
$C_{16}H_{16}O_6$			
Altersolanol B	EtOH	217(4.52),265(4.14), 285(3.97),421(3.58)	23-0767-69
Coumarin, 6-acetoxy-7-[(3-methyl-2-oxobutyl)oxy]-	EtOH	222(3.95),250s(3.28), 292(3.58),328(3.83)	39-0526-69C
2H-Furo[2,3-h]-1-benzopyran-2-one, 8,9-dihydro-5,6-dihydroxy-8,9,9-trimethyl-, 6-acetate	EtOH	226(4.03),252(3.67), 260(3.70),335(3.95)	88-4031-69
1,2-Naphthalenedicarboxylic acid, 5,7-dimethoxy-, dimethyl ester	EtOH	257(4.69),282(3.66), 290(3.66)	12-1721-69
Spiro[furan-2(3H),2'(1'H)-naphthalene]-4-acetic acid, 3',4,4',5-tetrahydro-6'-methoxy-1',5-dioxo-	EtOH	277(4.17)	2-1169-69
$C_{16}H_{16}O_7$			
Isocoumarin, 6(or 8)-hydroxy-8(or 6)-methoxy-3-methyl-7-(tetrahydro-4-methoxy-5-oxo-2-furyl)-	EtOH	246(4.71),258(4.15), 280(3.92),290s(--), 329(3.76)	12-1933-69
$C_{16}H_{16}O_8$			
Altersolanol A	EtOH	219(4.57),240(3.96), 268(4.15),285s(3.84), 422(3.65)	23-0767-69
$C_{16}H_{16}O_{10}$			
Benzenepentol, pentaacetate	MeOH	268(2.78)	34-0118-69
$C_{16}H_{17}$			
1,1-Diphenylbutylium cation	acid	321(4.11),431(4.60)	59-0447-69
o-Ethyl-α-methyl-α-phenylbenzylium cation	acid	323(4.04),434(4.43)	59-0447-69
p-Ethyl-α-methyl-α-phenylbenzylium cation	acid	318(3.99),444(4.70)	59-0447-69
$C_{16}H_{17}BrN_2$			
Phthalazino[2,3-b]phthalazine, 5,7,12,14-tetrahydro-, hydrobromide	EtOH	252(2.76),258(2.84), 266(2.95),273(2.99)	44-2715-69
$C_{16}H_{17}BrO_3$			
2-Cyclopenten-1-one, 3-bromo-2-hydroxy-4,4,5,5-tetramethyl-, benzoate	EtOH	239(4.0)	39-1346-69C
$C_{16}H_{17}ClFeO$			
1-Pentanone, 4-(chloromethyl)-1,5-(1,1'-ferrocenediyl)-	n.s.g.	229(4.10),277(3.66), 344(3.13),456(2.69)	78-0861-69

Compound	Solvent	$\lambda_{max}(\log \epsilon)$	Ref.
$C_{16}H_{17}ClN_2O$			
Phenol, 4-tert-butyl-2-[(p-chlorophenyl)azo]-	50% EtOH	334(4.81),390(4.44)	73-1087-69
	+ $HClO_4$	307(4.04),407(4.82), 496(4.63)	73-1087-69
$C_{16}H_{17}ClN_4O_2$			
Aniline, p-[(4-chloro-3-nitrophenyl)-azo]-N,N-diethyl-	benzene	290(3.96),430(4.44)	73-2092-69
$C_{16}H_{17}Cl_4NO_4$			
6,9-Methano-3H-3-benzazepine-3-carboxylic acid, 6,7,8,9-tetrachloro-5a,6,9,9a-tetrahydro-10-oxo-, ethyl ester, 10-(dimethyl acetal)	n.s.g.	226(4.08)	88-1619-69
$C_{16}H_{17}FN_4O_7$			
9H-Purine, 6-fluoro-9-β-D-ribofuranosyl-, 2',3',5'-triacetate	EtOH	243(3.81)	44-1396-69
$C_{16}H_{17}N$			
Acridan, 9-ethyl-9-methyl-	hexane	278(4.30)	78-1125-69
Pyridine, 5-ethyl-2-(p-methylstyryl)-, trans	H_2O	316(4.46)	23-2355-69
hydrochloride	H_2O	343(4.50)	23-2355-69
p-Toluidine, N-(p,α-dimethylbenzylidene)-	EtOH	256(4.23),312(3.45)	56-0749-69
$C_{16}H_{17}NO$			
Acetophenone, 4'-anilino-2',5'-dimethyl-	heptane	214(4.24),242(4.06), 318(4.36)	22-3523-69
	EtOH	203(4.48),244(4.08), 295s(3.00),335(4.39)	22-3523-69
Acetophenone, 4'-anilino-2',6'-dimethyl-	C_6H_{12}	240s(3.82),294(4.17)	22-3523-69
	EtOH	204(4.51),240s(3.87), 295(4.16),328s(3.97)	22-3523-69
Acetophenone, 4'-p-toluidino-3'-methyl-	C_6H_{12}	244(4.28),300s(3.99), 316(4.07)	22-3523-69
	EtOH	205(4.59),249(4.30), 300s(3.81),338(3.96)	22-3523-69
Acetophenone, 2'-(2,5-xylidino)-	C_6H_{12}	214(4.36),229(4.40), 244s(4.11),282(3.88), 296s(3.72),372(3.86)	22-3523-69
	EtOH	229(4.38),245s(4.13), 260s(4.02),282(3.84), 300s(3.64),385(3.85)	22-3523-69
p-Anisidine, N-(p,α-dimethylbenzylidene)-	EtOH	254(4.28),328(3.54)	56-0749-69
o-Cresol, α-(1-methyl-2-isoindolinyl)-	EtOH	266(3.44),272(3.56)	44-1720-69
4-Hexen-2-ynophenone, 5-(1-pyrrolidinyl)-	EtOH	263(4.21),408(4.57)	94-2126-69
Propiophenone, 4'-p-toluidino-	EtOH	242(3.95),300s(3.80), 346(4.44)	22-3523-69
Pyridine, 5-ethyl-2-(p-methoxystyryl)-, trans	H_2O	324(4.50)	23-2355-69
hydrochloride	H_2O	358(4.52)	23-2355-69
p-Toluidine, N-(p-methoxy-α-methylbenzylidene)-	EtOH	270(4.34),306s(3.63)	56-0749-69
$C_{16}H_{17}NO_2$			
p-Anisidine, N-(p-methoxy-α-methylbenzylidene)-	EtOH	272(4.26),316(3.65)	56-0749-69

Compound	Solvent	λ_{max}(log ε)	Ref.
Anthranilic acid, N-mesityl-	hexane	222(4.50),258(3.96), 268s(3.83),353(3.78)	22-3523-69
	MeOH	212(4.49),256(3.94), 274s(3.87),336(3.71)	22-3523-69
4-Hexen-2-ynophenone, 5-morpholino-	EtOH	262(4.25),395(4.48)	94-2126-69
3H-Naphth[2,3-d]azepine-3-carboxylic acid, 1,2,4,5-tetrahydro-, methyl ester	EtOH	261(3.65),269(3.75), 279(3.75),289(3.53)	44-2888-69
Propiophenone, 2'-o-anisidino-	EtOH	212(4.34),230(4.22), 246s(4.08),284(4.00), 307(3.98),370s(3.58), 395(3.68)	22-3523-69
Propiophenone, 4'-p-anisidino-	EtOH	243(3.96),293s(3.70), 343(4.38)	22-3523-69
$C_{16}H_{17}NO_2S$			
3H-3-Benzazepine, 6,7,8,9-tetrahydro-3-(phenylsulfonyl)-	EtOH	205(4.30),274(3.51), 350(2.54)	44-2866-69
$C_{16}H_{17}NO_3$			
Aponorscopolamine, hydrochloride	EtOH	253(3.55)	5-0152-69I
6H-Cyclopent[g]isoquinoline-5-carboxylic acid, 7,8-dihydro-9-methoxy-7,7-dimethyl- (illudinine)	MeOH	233(4.65),300(3.67), 332(3.69)	44-0240-69
	MeOH-HCl	228s(4.34),251(4.61), 295(3.56),360(3.72)	44-0240-69
	pH 13	238(4.72),298(3.85), 332(3.65)	44-0240-69
Furo[2,3-b]quinolin-4(9H)-one, 3-isopropyl-8-methoxy-9-methyl-	MeOH	235(4.40),241s(4.45), 247(4.53),262s(3.86), 295s(3.66),303(3.73), 327s(3.80),339(3.98), 354(3.93)	44-2183-69
	MeOH-acid	248(4.59),303(3.60), 328s(3.71),340(3.88), 354(3.88)	44-2183-69
	MeOH-base	235(4.40),241s(4.45), 247(4.53),262s(3.86), 295s(3.66),303(3.73), 327s(3.80),339(3.98), 354(3.93)	44-2183-69
1-Isoquinolineacetic acid, α-(ethoxymethylene)-, ethyl ester	EtOH	281s(3.64),310(3.53), 322(3.60)	24-0915-69
	EtOH-H_2SO_4	229(4.70),279(3.51), 288(3.47),341(3.80)	24-0915-69
Nicotinic acid, 1,2-dihydro-1,5-dimethyl-2-oxo-6-phenyl-, ethyl ester	EtOH	243(3.86),349(4.08)	94-2209-69
$C_{16}H_{17}NO_3S$			
1-[2-(Benzylthio)-1-carboxyethyl]-5-hydroxy-2-picolinium hydroxide, inner salt	aq. HCl	304(3.77)	1-2475-69
	aq. NaOH	335(3.74)	1-2475-69
$C_{16}H_{17}NO_4$			
p-Acetotoluidide, N-(4-formyl-2-hydroxy-1,3-butadienyl)-, acetate	EtOH	320(4.40)	78-2035-69
p-Acetotoluidide, N-(4-formyl-4-hydroxy-1,3-butadienyl)-, acetate	EtOH	326(4.39)	78-2035-69
Nicotinic acid, 1,2-dihydro-4-hydroxy-1,5-dimethyl-2-oxo-6-phenyl-, ethyl ester	EtOH	229(4.23),326(4.11)	94-2209-69

Compound	Solvent	$\lambda_{max}(\log \epsilon)$	Ref.
4-Oxazoleacrylic acid, 5-ethoxy-2-phenyl-, ethyl ester, trans	EtOH	222(4.30),316(4.43)	94-2424-69
$C_{16}H_{17}NO_5S$			
L-Tyrosine, N-(p-tolylsulfonyl)-	H_2O	198(4.57),225(4.32), 274f(3.32)	54-0805-69
	N KOH	200(4.62),232(4.33), 293(3.43)	54-0805-69
	MeOH	211(4.53),225(--), 274f(3.19)	54-0805-69
	dioxan	227(4.27),278f(3.22)	54-0805-69
	HOAc	274f(3.26)	54-0805-69
	DMF	279f(3.25)	54-0805-69
$C_{16}H_{17}N_2O_7P$			
Thymine, 1-(2-deoxy-β-D-threo-pentofuranosyl)-, cyclic 3',5'-(phenyl phosphate)	MeOH	266(3.96)	69-4889-69
isomer	MeOH	265(3.99)	69-4889-69
$C_{16}H_{17}N_3$			
Indole, 3-(dimethylaminomethyl)-2-(3-pyridyl)-	EtOH	219(4.29),312(4.14)	22-4154-69
$C_{16}H_{17}N_3O$			
Benzimidazole, 5-amino-2-methyl-1-(p-ethoxyphenyl)-	EtOH	209(4.68),254s(4.12), 306(3.69)	73-0572-69
$C_{16}H_{17}N_3O_2$			
Acetanilide, 2'-hydroxy-N-methyl-5'-(p-tolylazo)-	EtOH	352(4.35)	22-4390-69
	EtOH-KOH	407(4.40)	22-4390-69
Isonicotinic acid, (p-methoxy-α-methylphenethylidene)hydrazide	n.s.g.	225s(4.18),271(4.10)	40-0073-69
Nicotinic acid, (p-methoxy-α-methylphenethylidene)hydrazide	n.s.g.	225(4.25),267(4.12)	40-0073-69
Nicotinic acid, 6-methyl-, (m-methoxy-α-methylbenzylidene)hydrazide	n.s.g.	216(4.41),294(4.21)	40-0073-69
Nicotinic acid, 6-methyl-, (o-methoxy-α-methylbenzylidene)hydrazide	n.s.g.	276(4.17)	40-0073-69
Nicotinic acid, 6-methyl-, (p-methoxy-α-methylbenzylidene)hydrazide	n.s.g.	222(4.32),305(4.34)	40-0073-69
Picolinic acid, (p-methoxy-α-methylphenethylidene)hydrazide	n.s.g.	225(4.33),275(4.19)	40-0073-69
$C_{16}H_{17}N_3O_3$			
Acetanilide, 2'-hydroxy-5'-[(p-methoxyphenyl)azo]-N-methyl-	EtOH	360(4.40)	22-4390-69
	EtOH-KOH	405(4.42)	22-4390-69
Isonicotinic acid, (α-methylveratrylidene)hydrazide	n.s.g.	216(4.41),318(4.26)	40-0073-69
Nicotinic acid, (α-methylveratrylidene)-hydrazide	n.s.g.	217(4.37),316(4.27)	40-0073-69
Phenol, 4-tert-butyl-2-[(m-nitrophenyl)-azo]-	50% EtOH	325(4.78),397(4.32)	73-1087-69
	+59% $HClO_4$	391(4.85),498(4.43)	73-1087-69
Phenol, 4-tert-butyl-2-[(p-nitrophenyl)-azo]-	50% EtOH	338(4.87),416(4.55)	73-1087-69
	+64% $HClO_4$	400(4.92),415(4.54)	73-1087-69
Picolinic acid, (α-methylveratrylidene)-hydrazide	n.s.g.	218(4.35),326(4.34)	40-0073-69
$C_{16}H_{17}N_3O_4$			
Pyridine, 4-tert-butyl-2-(2,4-dinitrobenzyl)-	EtOH	243(4.53)	22-4425-69

Compound	Solvent	$\lambda_{max}(\log \epsilon)$	Ref.
as-Triazine-5,6-dicarboxylic acid, 3-p-tolyl-, diethyl ester	MeOH	290(4.34),390(2.60)	88-3147-69
$C_{16}H_{17}N_3O_4S$			
5-Thia-1-azabicyclo[4.2.0]oct-2-ene-2-carboxylic acid, 7-(2-amino-2-phenylacetamido)-3-methyl-8-oxo-	H_2O	260(3.89)	87-0310-69
$C_{16}H_{17}N_3O_5S$			
5-Thia-1-azabicyclo[4.2.0]oct-2-ene-2-carboxylic acid, 7-[2-amino-2-(m-hydroxyphenyl)acetamido]-3-methyl-8-oxo-	H_2O	264(3.85)	87-0310-69
$C_{16}H_{17}N_3S$			
Benzimidazole, 5-amino-2-methyl-1-(p-ethylthiophenyl)-	EtOH	213(4.61),264(4.30), 308s(3.68)	73-0572-69
$C_{16}H_{17}N_4O_{12}P$			
Thymidine, 3'-(2,4-dinitrophenyl)phosphate, ammonium salt	pH 7	265(4.30),295s(3.87)	69-3067-69
Thymidine, 5'-(2,4-dinitrophenyl)phosphate, ammonium salt	pH 7	263(4.27),295s(3.87)	69-3067-69
$C_{16}H_{17}N_5$			
1H-Pyrazolo[3,4-d]pyrimidine, 1-phenyl-4-piperidino-	MeOH	243(4.35),300(4.28)	23-1129-69
$C_{16}H_{17}N_5O$			
Benzamide, N-(6-methyl-8-propyl-s-triazolo[4,3-a]pyrazin-3-yl)-	BuOH	224(4.23),336s(4.18)	39-1593-69C
Formic acid, (α-aminobenzylidene)hydrazide, benzylformylhydrazone	MeOH	230(4.04),286(4.11)	48-0477-69
$C_{16}H_{17}N_5O_3S_4$			
2H-1,3-Thiazine-2,4(3H)-dione, 3,3'-ethylenebis[dihydro-2-thio-, 4-[(p-nitrophenyl)hydrazone]	EtOH	265(4.11),316(4.23), 404(4.33)	103-0049-69
$C_{16}H_{17}N_5O_4$			
Adenosine, N-phenyl-	pH 1.0	273(4.26)	87-1056-69
	pH 7.0	288(4.28)	87-1056-69
	pH 12.0	288(4.28)	87-1056-69
$C_{16}H_{18}$			
Biphenyl, 2,2'-diethyl-	EtOH	230s(3.78)	6-0277-69
$C_{16}H_{18}BrNO_3$			
Pyrido[1,2-a]indole-10-carboxylic acid, 8-bromo-9-ethoxy-6,7,8,9-tetrahydro-, methyl ester	EtOH	215(4.67),231(4.51), 292(4.24)	12-0997-69
$C_{16}H_{18}ClN$			
Pyridine, 5-ethyl-2-(p-methylstyryl)-, trans, hydrochloride	H_2O	343(4.50)	23-2355-69
$C_{16}H_{18}ClNO$			
Pyridine, 5-ethyl-2-(p-methoxystyryl)-, trans, hydrochloride	H_2O	358(4.52)	23-2355-69

Compound	Solvent	λ_{max}(log ϵ)	Ref.
$C_{16}H_{18}ClNO_3$			
Aponorscopolamine, hydrochloride	EtOH	253(3.55)	5-0152-I
$C_{16}H_{18}ClNO_4$			
6,7-Dihydro-1,3-dimethyl-2-phenyl-5H-1-pyrindinium perchlorate	EtOH	295(4.10)	94-2209-69
5,6,7,8-Tetrahydro-1-methyl-4-phenylquinolinium perchlorate	EtOH	284(4.19)	78-4161-69
$C_{16}H_{18}ClNO_5$			
6,7,8,9-Tetrahydro-9-methyl-2-phenyl-5H-pyrano[2,3-b]azepin-1-ium perchlorate	MeOH	259(4.17),369(4.19)	5-0073-69E
$C_{16}H_{18}Cl_4N_2S_2$			
4H,10H-Dithieno[3,4-c:3',4'-h][1,6]diazecine, 1,3,7,9-tetrachloro-5,11-diethyl-5,6,11,12-tetrahydro-	EtOH	245(4.18)	44-0333-69
$C_{16}H_{18}FN_5O_7$			
Adenine, 2-fluoro-9-β-D-xylofuranosyl-, 2',3',5'-triacetate	pH 1	261(4.06),268s(--)	44-1396-69
	pH 7	261(4.09),268s(--)	44-1396-69
	pH 13	261(4.09),268s(--)	44-1396-69
$C_{16}H_{18}F_6N_6O_4S$			
Uracil, 5,5'-thiobis[1,3-dimethyl-6-[(2,2,2-trifluoroethyl)amino]-	MeOH	280(4.32)	44-3285-69
$C_{16}H_{18}Fe$			
Ferrocene, 1,1'-(2-methylenepentamethylene)-	n.s.g.	327(1.93),448(2.04)	78-0861-69
$C_{16}H_{18}FeO$			
Ketone, methyl 1',2-tetramethyleneferrocenyl	EtOH	228(4.11),268(3.72), 339(3.03),455(2.62)	78-5245-69
Ketone, methyl 1',3-tetramethyleneferrocenyl	EtOH	233(4.16),274(3.75), 338(3.13),454(2.79)	78-5245-69
$C_{16}H_{18}FeO_2$			
Ferroceneacrylic acid, propyl ester	n.s.g.	249(4.1),291(4.2), 356(3.2),460(2.9)	65-1008-69
$C_{16}H_{18}NO$			
Nitroxide, 4-biphenylyl tert-butyl	n.s.g.	322(4.26),512(2.71)	39-1459-69C
$C_{16}H_{18}N_2$			
Cycloclavine	MeOH	224(4.55),275(3.86), 283(3.89),295(3.79)	78-5879-69
3H-[1,2]Diazepino[1,7-a]indole, 2,5,7,10-tetramethyl-	EtOH	246(5.3),272(4.1), 311(4.1)	24-3268-69
Dibenzo[b,f][1,4]diazocine, 5,6,11,12-tetrahydro-5,12-dimethyl-	EtOH	234(4.47),278(3.90), 312(3.76)	39-0882-69C
$C_{16}H_{18}N_2O$			
3,4-Diazabicyclo[4.2.0]octa-4,7-dien-2-one, 1,6,7,8-tetramethyl-5-phenyl-	MeOH	275(3.96)	24-1928-69
Ketone, 1-methylindol-2-yl 1,2,5,6-tetrahydro-1-methyl-4-pyridyl	EtOH	241(4.07),315(4.26)	39-2738-69C
2H-3,7-Methanoazonino[5,4-b]indol-2-one, 1,4,5,6,7,8-hexahydro-11-methyl-	EtOH	224(4.53),279s(3.87), 286(3.88),297(3.78)	22-4154-69

Compound	Solvent	λ_{max}(log ϵ)	Ref.
Nitrone, α-[4-(dimethylamino)-o-tolyl]-N-phenyl-	EtOH	392(4.54)	18-3306-69
Phenol, 4-tert-butyl-2-(phenylazo)-	50% EtOH	327(4.28),388(3.91)	73-1087-69
	50% EtOH-$HClO_4$	303s(--),395(4.32), 484(4.08)	73-1087-69
$C_{16}H_{18}N_2O_2$			
Azecino[4,5,6-cd]indole-11-carboxylic acid, 2,6,7,10,11,12-hexahydro-8-methyl- (clavicipitic acid)	EtOH	225(4.58),288(3.81)	88-1857-69
$C_{16}H_{18}N_2O_2S$			
2H-1,2,4-Benzothiadiazine, 3,4-dihydro-2-phenyl-3-propyl-, 1,1-dioxide	EtOH	210(4.55),252(4.07), 314(3.48)	7-0590-69
2H-1,2,4-Benzothiadiazine, 3,4-dihydro-2-phenyl-4-propyl-, 1,1-dioxide	EtOH	211(4.48),258(4.14), 323(3.56)	7-0590-69
Sulfur diimide, bis(p-ethoxyphenyl)-	benzene	457(4.21)	104-0459-69
$C_{16}H_{18}N_2O_3$			
Hexanamide, N-(3-amino-1,4-dihydro-1,4-dioxo-2-naphthyl)-	MeOH	268(4.33),323(3.32), 452(3.38)	4-0909-69
2H-Indazole-4-carboxylic acid, 3,3a,4,5,6,7-hexahydro-3-oxo-2-phenyl-, ethyl ester	EtOH	204(4.27),247(4.11), 273s(3.98)	39-2783-69C
2H-Indazole-7-carboxylic acid, 3,3a,4,5,6,7-hexahydro-3-oxo-2-phenyl-, ethyl ester	EtOH	204(4.26),248(4.14), 280s(3.84)	39-2783-69C
$C_{16}H_{18}N_2O_4S$			
Azobenzene-4'-sulfonic acid, 5-tert-butyl-2-hydroxy-	50% EtOH	326(4.28),388(3.86)	73-1087-69
	50% EtOH-$HClO_4$	300s(--),391(4.35), 493(4.00)	73-1087-69
Benzylpenicillenic acid	EtOH	322(4.37)	87-0483-69
$C_{16}H_{18}N_2O_5$			
4(1H)-Quinazolinone, 1-(2,3-O-isopropylidene-β-D-ribofuranosyl)-	pH 1	235(4.38),292(3.74), 302(3.76)	4-0089-69
	pH 11	232(4.16),302(3.90), 313(3.83)	4-0089-69
	MeOH	229(4.22),264(3.63), 273(3.65),302(3.95), 313(3.89)	4-0089-69
$C_{16}H_{18}N_2O_5S$			
2,4(1H,3H)-Quinazolinedione, 1-(2,3-O-isopropylidene-β-D-ribofuranosyl)-2-thio-	pH 1	282(4.08)	4-0089-69
	pH 11	223s(4.40),277(4.05), 310(3.74)	4-0089-69
	MeOH	288(4.06)	4-0089-69
2,4(1H,3H)-Quinazolinedione, 1-(2,3-O-isopropylidene-5-thio-β-D-ribofuranosyl)-	pH 1	306(3.53)	4-0089-69
	pH 11	306(3.58)	4-0089-69
	MeOH	306(3.53)	4-0089-69
$C_{16}H_{18}N_2O_8S$			
Uridine, 5'-p-toluenesulfonate	EtOH	260(4.16)	65-0434-69
$C_{16}H_{18}N_2S$			
Sulfur diimide, di-3,4-xylyl-	hexane	375(3.95),445(4.05)	48-0621-69

Compound	Solvent	λ_{max}(log ϵ)	Ref.
$C_{16}H_{18}N_4$			
Benzimidazole, 5-amino-2-[p-(dimethyl-amino)phenyl]-1-methyl-	EtOH	205(4.52),218s(4.47), 296s(4.26),330(4.43)	73-0572-69
Benzo[c]cinnoline, 3,8-bis(dimethyl-amino)-	C_6H_{12}	211(4.38),280(4.84), 305(4.41),320s(4.23), 450(3.42)	4-0523-69
	EtOH	213(4.26),290(4.94), 336s(4.31),498(3.50)	4-0523-69
$C_{16}H_{18}N_4O$			
Benzo[c]cinnoline, 3,8-bis(dimethyl-amino)-, 5-oxide	C_6H_{12}	204(4.20),245s(4.19), 265(4.35),307(4.54), 322s(4.48),343(4.26), 370s(3.26),460(3.46), 495s(3.26)	4-0523-69
	EtOH	303s(4.14),313s(4.24), 336(4.69),352s(4.41), 519(3.46)	4-0523-69
$C_{16}H_{18}N_4O_2$			
Aniline, N,N-diethyl-p-[(p-nitrophen-yl)azo]-	n.s.g.	280(4.08),485(4.53)	88-1303-69
	hexane	455(--)	88-1303-69
	DMSO	510(--)	88-1303-69
$C_{16}H_{18}N_4O_4$			
4,4'-Biphenyldicarboxylic acid, 2,2',6,6'-tetramino-, dimethyl ester	EtOH	222(4.75),241(4.59), 281(4.18),338(3.87)	4-0533-69
2-Cyclopententen-1-one, 2-isopropenyl-4,4-dimethyl-, 2,4-dinitrophenyl-hydrazone	THF	378(4.40),450s(3.08)	78-2145-69
Tricyclo[4.3.1^{1,2}.0]decan-3-one, 2,4-dinitrophenylhydrazone	$CHCl_3$	373(4.43)	78-5287-69
$C_{16}H_{18}N_4O_5$			
2-Cyclopenten-1-one, 2-allyl-4-methoxy-3-methyl-, 2,4-dinitrophenylhydrazone	EtOH	256(4.20),287(3.92), 380(4.40)	78-1117-69
$C_{16}H_{18}N_4O_6$			
1-Cyclopentene-1-propionic acid, β-oxo-, ethyl ester, β-(2,4-dinitrophenyl-hydrazone)	EtOH	377(4.30)	33-1996-69
$C_{16}H_{18}N_4O_7S$			
9H-Purine-6-thiol, 9-α-D-arabinopyrano-syl-, 2',3',4'-triacetate	pH 1	322(4.39)	44-0416-69
	pH 7	317(4.34)	44-0416-69
	pH 13	312(4.36)	44-0416-69
9H-Purine-6-thiol, 9-β-D-arabinopyrano-syl-, 2',3',4'-triacetate	pH 1	322(4.39)	44-0416-69
	pH 7	318(4.34)	44-0416-69
	pH 13	310(4.36)	44-0416-69
9H-Purine-6-thiol, 9-β-D-xylopyranosyl-, 2',3',4'-triacetate	pH 1	322(4.40)	44-0416-69
	pH 7	318(4.33)	44-0416-69
	pH 13	311(4.36)	44-0416-69
$C_{16}H_{18}N_4O_8$			
2-Heptenedioic acid, 2-methyl-4-oxo-, dimethyl ester, 4-(2,4-dinitro-phenylhydrazone)	$CHCl_3$	375(4.51)	40-0096-69
Hypoxanthine, 9-α-D-arabinopyranosyl-, 2',3',4'-triacetate	pH 1	247(4.10)	44-0416-69
	pH 7	247(4.10)	44-0416-69
	pH 13	252(4.15)	44-0416-69

Compound	Solvent	λ_{max}(log ε)	Ref.
Hypoxanthine, 9-β-D-xylopyranosyl-, 2',3',4'-triacetate	pH 1	243s(--),247(4.15)	44-0416-69
	pH 7	243s(--),247(4.15)	44-0416-69
	pH 13	253(4.17)	44-0416-69
$C_{16}H_{18}O$			
1H-2,10aα-Ethanophenanthren-12-one, 2β,3,4,4aβ,9,10-hexahydro-	EtOH	266(2.74),274(2.73)	77-1253-69
2,4,6,8-Nonatetraen-1-ol, 3-methyl-9-phenyl-	EtOH	328(--),342(--), 362(--)	88-4049-69
$C_{16}H_{18}O_2$			
Acetophenone, 3',5'-diethyl-2'-(2-furyl)-	EtOH	238(4.20),263s(3.93), 297(3.59)	6-0277-69
18,19-Bisnorpodocarpa-4,8,11,13-tetraen-3-one, 12-methoxy-	n.s.g.	227(4.13),278(3.41), 286(3.37)	12-1711-69
4bα-Gibba-1,3,4a(10a)-trien-8-one, 2-methoxy-	EtOH	230(3.75)	77-1253-69
3-Oxa-A-norestra-1,5(10),9(11)-trien-17-one	n.s.g.	208(4.12),237(4.09)	44-1151-69
1-Phenanthrenecarboxylic acid, 1,2,3,4,9,10-hexahydro-1-methyl-	EtOH	265(4.09)	44-3739-69
$C_{16}H_{18}O_3$			
Benzhydrol, α-[(2-hydroxyethoxy)methyl]-	EtOH	201(4.26),256(2.62)	22-2048-69
Benzyl alcohol, o-hydroxy-α-(p-methoxyphenethyl)-	EtOH	265(4.31),295s(3.38), 305s(3.59)	32-0612-69
2-Cyclohexen-1-one, 2-benzyl-3-hydroxy-, propionate	EtOH	207(4.09),230s(3.92), 243(3.96)	44-2192-69
3-Dibenzofurancarboxylic acid, 6,7,8,9-tetrahydro-1,4-dimethyl-, methyl ester	EtOH	216(4.29),283(4.26)	4-0379-69
Eudesma-1,3,5,7(11)-tetraen-12-oic acid, 6-hydroxy-3-methoxy-, γ-lactone	MeOH	273(3.77),419(4.39)	39-1088-69C
1H-Fluorene-8-carboxylic acid, 2,3,4,4a-tetrahydro-7-methoxy-, methyl ester	EtOH	235(4.33),262(4.05), 316(3.63)	44-2209-69
2,8-Nonadiene-4,6-diyn-1-ol, 9-(tetrahydro-2H-pyran-2-yl)-, acetate, di-trans	ether	238(4.45),248(4.36), 262(3.85),278(4.15), 295(4.32),314(4.23)	39-0830-69C
3-Oxabicyclo[3.2.0]hept-6-en-2-one, 4-hydroxy-1,5,6,7-tetramethyl-4-phenyl-	MeOH	236(3.30)	24-1928-69
Propiophenone, 4'-[(2-hydroxy-6-oxo-1-cyclohexen-1-yl)methyl]-	EtOH	205(4.28),260(4.40)	44-2192-69
$C_{16}H_{18}O_4$			
2'-Acetonaphthone, 1',6',8'-trimethoxy-3'-methyl-	EtOH	237(4.62),303(3.71), 334s(3.62)	94-0454-69
p-Benzenediacrylic acid, diethyl ester	EtOH	202(4.36),223(4.18), 228(4.20),234s(--), 319(4.63),331(--)	23-4076-69
Benzophenone, 2-(hydroxymethyl)-2',4'-dimethoxy-	EtOH	213(4.20),235(4.04), 278(3.91),309(3.83)	39-1873-69C
2,3-Butanediol, 2,3-bis(p-hydroxyphenyl)-	EtOH	275(3.57),282(3.55)	12-0761-69
3-Cyclohexene-1-carboxylic acid, 2-[3,4-(methylenedioxy)phenyl]-, ethyl ester, cis	EtOH	237(3.68),286(3.62)	78-4315-69
Eudesma-1,5,7(11)-trien-12-oic acid, 6-hydroxy-4α-methoxy-3-oxo-, γ-lactone	MeOH	290(4.39)	39-1088-69C
4β- form	MeOH	287(4.55)	39-1088-69C

Compound	Solvent	λmax(log ε)	Ref.
Fluorene-8aβ(4bβH)-acetic acid, 5,6,7,8-tetrahydro-2-methoxy-9-oxo-	EtOH	220(4.43),249(3.91), 320(3.56)	44-2209-69
Fluorene-1-carboxylic acid, 4bβ,5,6,7,-8,8aβ-hexahydro-2-methoxy-9-oxo-, methyl ester	EtOH	220(4.37),244(3.82), 323(3.63)	44-2209-69
Malonic acid, cinnamylidene-, diethyl ester	EtOH	229s(--),235(3.94), 241s(--),321(4.54)	23-4076-69
Methanol, (2-hydroxymethylphenyl)(2,4-dimethoxyphenyl)-	EtOH	215(4.15),234(3.98), 273(3.38),279(3.47), 284(3.42)	39-1873-69C
1-Naphthol, 2,4-diethoxy-, acetate	MeOH	236(4.55),301(3.62)	44-2788-69
Peucenin, 7-O-methyl-	MeOH	232(4.29),259(4.19), 293(3.91)	2-1072-69
$C_{16}H_{18}O_4S$			
2,5-Thiophenedicarboxylic acid, 2,5-dihydro-3-phenyl-, diethyl ester	EtOH	240(4.24)	2-0009-69
$C_{16}H_{18}O_5$			
8-Benzocyclodecenecarboxylic acid, 5,6,7,8,9,10,11,12-octahydro-2-methoxy-4,10-dioxo-	EtOH	271(3.89)	2-1169-69
Benzo[1,2-b:3,4-b']dipyran-4,10-dione, 2,3,8,9-tetrahydro-5-hydroxy-2,2,8,8-tetramethyl-	EtOH	268(4.63)	39-1540-69C
	EtOH-base	283(--),317(--)	39-1540-69C
Fluorene-1,9α-dicarboxylic acid, 4bβ,5,6,7,8,8aβ-hexahydro-2-methoxy-	EtOH	300(3.53)	44-2209-69
Fluorene-1,9β-dicarboxylic acid, 4bβ,5,6,7,8,8aβ-hexahydro-2-methoxy-	EtOH	299(3.52)	44-2209-69
Fluorene-3,9β-dicarboxylic acid, 4bβ,5,6,7,8,8aβ-hexahydro-2-methoxy-	EtOH	212(4.48),243(3.93), 304(3.65)	44-2209-69
$C_{16}H_{18}O_6$			
6-Isochromanacetic acid, 4aβ,5,6β,7,-8,8a-hexahydro-5β-hydroxy-α,4-dimethylene-3,7-dioxo-8aβ-vinyl-, methyl ester	MeOH	211(4.07)	44-3903-69
Succinic acid, piperonylidene-, diethyl ester	EtOH	235(4.12),280(4.00), 314(4.06)	39-2470-69C
Succinic acid, [(1,2,3,4-tetrahydro-6-methoxy-1-oxo-2-naphthyl)methyl]-	EtOH	274(4.20)	2-1169-69
Tetracyclo[3.3.2.0^{2,4}.0^{6,8}]dec-9-ene-3,9,10-tricarboxylic acid, trimethyl ester	EtOH	239s(3.49)	89-0883-69
Tricycloprop[cd,f,hi]indene-2,2d,2e-tricarboxylic acid, octahydro-, trimethyl ester	EtOH	230(3.78)	24- 164-69
$C_{16}H_{18}O_9$			
1,3-Cyclohexadiene-1,2,3,4-tetracarboxylic acid, 5,5-dimethyl-6-oxo-	EtOH	233s(3.61),306(3.74), 375(3.08)	33-0584-69
7-Oxabicyclo[2.2.1]hepta-2,5-diene-2,3,5,6-tetracarboxylic acid, 1,4-dimethyl-, tetramethyl ester	isooctane	223(4.09),283s(2.89)	33-0584-69
Oxepine-3,4,5,6-tetracarboxylic acid, 2,7-dimethyl-, tetramethyl ester	EtOH	213(4.24),293(3.42)	33-0584-69
$C_{16}H_{19}$			
(4-tert-Butyl-1,3-cyclopentadien-1-yl)-cycloheptatrienylium cation	CH_2Cl_2	230(--),283(--), 500(--)	89-0881-69

Compound	Solvent	$\lambda_{max}(\log \epsilon)$	Ref.
$C_{16}H_{19}BrN_2O_2S$			
2(1H)-Pyrimidinone, 1-[[3-(p-bromobenzyl)thio]crotonoyl]-3-methyl-, cis	EtOH	307(4.23)	32-1000-69
trans	EtOH	294(4.25)	32-1000-69
$C_{16}H_{19}ClN_2O$			
4-Quinolinol, 5-chloro-2-methyl-3-(piperidinomethyl)-	EtOH	240(4.39),300(3.83), 325(3.98),337(3.89)	2-1186-69
$C_{16}H_{19}ClN_2O_2S$			
Benzenesulfonanilide, 4-chloro-4'-(diethylamino)-	EtOH-1% $CHCl_3$	230(4.25),268(4.32), 300s(3.69)	44-2083-69
p-Phenylenediamine, 2-[(p-chlorophenyl)-sulfonyl]-N^4,N^4-diethyl-	EtOH-1% $CHCl_3$	242(4.45),270(4.20)	44-2083-69
$C_{16}H_{19}ClN_2O_4$			
2-Benzyl-1,2-dimethylindazolinium perchlorate	n.s.g.	240(3.89),270(3.15), 290(3.32)	104-2004-69
1,1,2-Trimethyl-3-phenylindazolinium perchlorate	n.s.g.	260(2.90),266(2.80)	104-2004-69
$C_{16}H_{19}Cl_2NO$			
5H-Benzocyclohepten-5-one, 6-[[bis(2-chloroethyl)amino]methylene]-	EtOH	210(3.93),250(3.81), 345(4.20)	24-3139-69
$C_{16}H_{19}NO$			
Sorbophenone, 5-(1-pyrrolidinyl)-	EtOH	264(3.95),447(4.74)	94-2126-69
$C_{16}H_{19}NO_2$			
Carbostyril, 4-methoxy-1-methyl-3-(3-methyl-2-butenyl)-	ether	270(3.84),280(3.83), 317(3.73),329(3.86), 344(3.71)	24-1774-69
Eleocarpine, (±)-	EtOH	253(3.94),320(3.49)	12-0775-69
4-Hexene-1,3-dione, 1-hexyl-5-(1-pyrrolidinyl)-	EtOH	243(3.85),301(3.90), 390(4.50)	94-2126-69
Isoeleacarpine, (±)-	EtOH	258(3.97),326(3.49)	12-0775-69
2(3H)-Naphthalenone, 4,4a,5,6,7,8-hexahydro-4a-(N-hydroxyanilino)-	EtOH	240(3.62)	12-2493-69
$C_{16}H_{19}NO_3$			
Coumarin, 4-[(cyclohexylamino)methyl]-3-hydroxy-	EtOH	328(4.18)	87-0531-69
2,4-Pentadienamide, 5-(p-acetoxyphenyl)-N,N,3-trimethyl-, 2-cis-4-trans	EtOH	230(4.21),302(4.47)	39-0016-69C
$C_{16}H_{19}NO_4$			
Carbostyril, 3-isobutyryl-4,8-dimethoxy-1-methyl- (spectrum unchanged by acid or base)	MeOH	232(4.55),258(4.41), 291(3.93)	44-2183-69
2-Cyclopenten-1-ol, 2-butyl-, p-nitrobenzoate	EtOH	260(4.24)	95-0506-69
Norscopolamine, (-)-	EtOH	252(1.8),258(2.07), 264(1.9)	5-0152-69I
4,5-Oxazolidinedione, 2-(1-adamantyl)-2-(2-propynoxy)-	EtOH	220s(3.79)	78-2909-69
Pyrrole-2-carboxylic acid, 3-(3,4-dimethoxyphenyl)-5-methyl-, ethyl ester	EtOH	287(4.26)	94-0582-69
$\Delta^{2,\alpha}$-Pyrrolidinemalonic acid, 1-benzyl-, dimethyl ester	isooctane	282(4.29)	35-6683-69

Compound	Solvent	$\lambda_{max}(\log \epsilon)$	Ref.
Spiro[cyclopentane-3'-isoquinoline]-4'-carboxylic acid, 2-hydroxy-3,4-dihydro-4-methoxy-, methyl ester	EtOH	256(4.07),286(3.56), 295(3.46)	94-1290-69
$C_{16}H_{19}NO_4S$			
Acetic acid, benzoyl[[2-(ethylcarbamoyl)vinyl]thio]-, ethyl ester, (Z)-	90% EtOH	254(4.27),275s(4.22)	12-0765-69
$C_{16}H_{19}NO_4S_2$			
3-Thiophenecarboxylic acid, 4,4'-iminobis[2-methyl-, diethyl ester	MeOH	248(4.36),279(4.02), 345(3.61)	103-0424-69
$C_{16}H_{19}NO_8S$			
Thiazolo[3,2-a]azepine-5,6,7,8-tetracarboxylic acid, 2,3,5,6-tetrahydro-, tetramethyl ester	EtOH	212(3.90),267(3.63), 382(4.31)	28-0615-69B
	EtOH-$HClO_4$	208(3.71),272(3.99), 332(3.61)	28-0615-69B
8aH-Thiazolo[3,2-a]pyridine-5,6,7,8-tetracarboxylic acid, 2,3-dihydro-8a-methyl-, tetramethyl ester	EtOH	219(4.04),280(4.10), 370(3.78)	28-0870-69A
$C_{16}H_{19}N_3$			
Quinoxaline, 1,2,3,4-tetrahydro-2-(p-toluidinomethyl)-	EtOH	220(4.46),252(4.19), 310(3.71)	35-5270-69
$C_{16}H_{19}N_3O$			
3,5-Nonadiynal, 2-(2-hexynylidene)-, semicarbazone	ether	231(4.18),259s(--), 276(4.24),317(4.55), 329(4.55)	39-0683-69C
$C_{16}H_{19}N_3O_2$			
Benzo[h]pyrrolo[1,2-c]quinazolin-8(9H)-one, 6-amino-4bβ,10,10aβ,10bβ,11,12-hexahydro-2-methoxy-	EtOH	225(4.00),276(3.29), 283(3.28)	88-3693-69
Cyclopentanone, 2-(2,4-dimethyl-5-phenylimidazol-1-yl)-, oxime, 3-oxide	MeOH	261(4.18)	48-0746-69
$C_{16}H_{19}N_3O_3$			
2-Cyclopenten-1-one, 2-allyl-4-methoxy-3-methyl-, o-nitrophenylhydrazone	EtOH	270(4.13),307(4.22), 455(3.76)	78-1117-69
2-Cyclopenten-1-one, 2-allyl-4-methoxy-3-methyl-, p-nitrophenylhydrazone	EtOH	286(3.94),318(3.62), 408(4.45)	78-1117-69
3-Pyrrolin-2-one, 3-(cyclohexylamino)-4-nitro-5-phenyl-	EtOH	372(4.21)	44-3279-69
$C_{16}H_{19}N_3S$			
10H-Pyrido[3,2-b][1,4]benzothiazine, 10-[3-(dimethylamino)propyl]-	EtOH	207(4.13),252(4.25), 319(3.53)	9-0249-69
$C_{16}H_{19}N_5O_3$			
Acetamide, N-[6,7-dihydro-7-(2-hydroxy-2-methylpropyl)-5-phenyl[1,2,5]oxadiazolo[3,4-d]pyrimidin-7-yl]-	EtOH	239(4.12),290(3.98)	35-5181-69
$C_{16}H_{19}N_5O_7$			
Adenine, 9-α-D-arabinopyranosyl-, 2',3',4'-triacetate	pH 1	255(4.18)	44-0416-69
	pH 7	258(4.18)	44-0416-69
	pH 13	258(4.18)	44-0416-69

Compound	Solvent	λ_{max}(log ε)	Ref.
$C_{16}H_{19}N_5O_8$			
9H-Purin-2(3H)-one, 6-amino-9-β-D-xylo-furanosyl-, 2',3',5'-triacetate	pH 1	235(3.90),280(4.00)	44-1396-69
	pH 7	248(3.98),292(3.92)	44-1396-69
	pH 13	253(3.89),283(3.89)	44-1396-69
$C_{16}H_{19}N_7$			
Guanidine, [3-(3-amino-6-indol-3-yl)-pyrazinyl)propyl]-, (etioluciferin)	MeOH	225(4.4),275(4.2), 370(3.8)	95-1646-69
$C_{16}H_{19}N_7O_2$			
Adenine, 9-[3-(6-nitro-3,4-xylidino)-propyl]-	EtOH	238(4.43),260(4.29), 293s(3.81),435(3.80)	44-3240-69
	EtOH-HCl	238(4.43),258(4.16), 293s(3.83),435(3.78)	44-3240-69
	EtOH-NaOH	235s(4.48),259(4.32), 290s(3.87),438(3.84)	44-3240-69
$C_{16}H_{20}BrNO$			
Carbazole, 9-acetyl-4a-(2-bromoethyl)-1,2,3,4,4a,9a-hexahydro-	MeOH	252(4.12),286(3.43)	23-3647-69
$C_{16}H_{20}BrP$			
Benzylethylmethylphenylphosphonium bromide	MeOH	217s(4.17),255s(2.81), 260(2.97),266(3.07), 273(2.93)	24-3546-69
$C_{16}H_{20}ClNO_4$			
2,3,5,6,7,8-Hexahydro-1-methyl-4-phenyl-quinolinium perchlorate	$CHCl_3$	332(4.07)	78-4161-69
$C_{16}H_{20}ClNO_5$			
2-(Diethylamino)-3-methyl-6-phenylpyryl-ium perchlorate	MeOH	256(4.17),368(4.08)	5-0073-69E
$C_{16}H_{20}N_2$			
Ethylenediamine, N,N'-dimethyl-1,2-di-phenyl-	EtOH	247(2.63)	35-2653-69
2H-3,7-Methanoazonino[5,4-b]indole, 1,4,5,6,7,8-hexahydro-11-methyl-	EtOH	225(4.48),276s(3.88), 285(3.48),296(3.80)	22-4154-69
$C_{16}H_{20}N_2O$			
Benz[cd]indole-2-carboxamide, N-tert-butyl-1,3,4,5-tetrahydro-	EtOH	232(4.4),294(4.3)	24-0678-69
Indole-2-methanol, 1-methyl-α-(1,2,3,6-tetrahydro-1-methyl-4-pyridyl)-	EtOH	223(4.61),277(3.95), 284(3.95),290(3.84)	39-2738-69C
1(4H)-Naphthalenone, 2-(propylamino)-	EtOH	338(3.66),425(3.68)	39-1799-69C
Pyrrocolin-3-one, 1-(dimethylamino)-3,4,5,6,7,8-hexahydro-2-phenyl-	MeOH	249(4.17),308(4.11)	78-5721-69
	MeOH-HCl	250(4.15)	78-5721-69
$C_{16}H_{20}N_2O_2$			
2,3-Butanediol, 2,3-bis(o-aminophenyl)-	pH 1	256s(3.20)	1-3567-69
	pH 14	301(3.96)	1-3567-69
isomer	pH 1	258(2.60)	1-3567-69
	pH 13	299(3.66)	1-3567-69
2,3-Diazabicyclo[3.1.0]hex-2-ene-1-carb-oxylic acid, 4,4,6,6-tetramethyl-, methyl ester	MeOH	331(2.51)	88-2659-69
2,3-Diazabicyclo[3.1.0]hex-2-ene-5-carb-oxylic acid, 4,4,6,6-tetramethyl-, methyl ester	MeOH	332(2.45)	88-2659-69

Compound	Solvent	λmax(log ε)	Ref.
Isoquinaldonitrile, 3-allyl-1,2,3,4-tetrahydro-6,7-dimethoxy-2-methyl-	EtOH	235(3.95),286(3.62), 315(3.40),372(3.32)	78-0101-69
2,6-Methano-2,5,6,7-tetrahydro-1(1H)-benzazonine, 3-acetyl-4-amino-7-hydroxy-7-methyl-	pH 1	298(3.48)	1-3567-69
	pH 13	317(4.03)	1-3567-69
isomer	pH 1	308(3.95)	1-3567-69
	pH 13	318(4.07)	1-3567-69
$C_{16}H_{20}N_2O_2S$			
Benzenesulfonanilide, 4'-(diethylamino)-	EtOH-1% $CHCl_3$	268(4.26),300s(3.62)	44-2083-69
p-Phenylenediamine, N^4,N^4-diethyl-2-(phenylsulfonyl)-	EtOH-1% $CHCl_3$	234(4.31),270(4.14)	44-2083-69
p-Phenylenediamine, N^4,N^4-dimethyl-2-(3,4-xylylsulfonyl)-	EtOH-1% $CHCl_3$	241(4.41),265s(4.11)	44-2083-69
3,4-Xylenesulfonanilide, 4'-(dimethylamino)-	EtOH-1% $CHCl_3$	228(4.13),262(4.23), 300s(--)	44-2083-69
$C_{16}H_{20}N_2O_3S$			
Acetic acid, [[4-(1-adamantyl)-6-hydroxy-2-pyrimidinyl]thio]-	0.1N HCl	234(3.98),281(3.95)	73-2278-69
(all spectra in 50% EtOH)	pH 8.2	240(4.03),287(3.90)	73-2278-69
	0.1N NaOH	243(4.03),278(3.94)	73-2278-69
$C_{16}H_{20}N_4$			
Azobenzene, 4-amino-4'-(diethylamino)-	iso-PrOH	415(4.42),444(4.40)	88-1303-69
2-Tetrazene, 1,4-dimethyl-1,4-di-p-tolyl-	$CHCl_3$	248(4.04),317s(4.23), 350(4.46)	35-6452-69
	$(MeOCH_2)_2$	253(4.23),315s(4.20), 348(4.49)	35-6452-69
$C_{16}H_{20}N_4O_2$			
2-Tetrazene, 1,4-bis(p-methoxyphenyl)-1,4-dimethyl-	$(MeOCH_2)_2$	247(4.04),319s(4.32), 350(4.46)	35-6452-69
$C_{16}H_{20}N_4O_3$			
Benzamide, N-[6-(dimethylamino)-1,2,3,4-tetrahydro-1,3-dimethyl-2,4-dioxo-5-pyrimidinyl]-N-methyl-	H_2O	295(4.18)	104-0547-69
$C_{16}H_{20}N_4O_5$			
1H-1,2,3-Triazole, 1-(2-acetamido-2-deoxy-β-D-glucopyranosyl)-4-phenyl-	EtOH	245(4.28),275s(2.94), 286s(2.32)	4-0639-69
1H-1,2,3-Triazole, 1-(2-acetamido-2-deoxy-β-D-glucopyranosyl)-5-phenyl-	EtOH	243(4.15)	4-0639-69
Uric acid, dihydro-4,5-dimethoxy-1,3,7-trimethyl-9-phenyl-	EtOH	221(4.55),275(3.38)	78-0541-69
$C_{16}H_{20}N_4O_6$			
Cyclohexanecarboxylic acid, 3-ethyl-4-oxo-, methyl ester, 2,4-dinitrophenylhydrazone	$CHCl_3$	365(4.32)	22-0781-69
$C_{16}H_{20}N_4S$			
Sulfur diimide, bis[p-(dimethylamino)-phenyl]-	benzene	532(4.50)	104-0459-69
$C_{16}H_{20}N_6$			
Adenine, 9-[3-(3,4-xylidino)propyl]-	EtOH-HCl	243(4.13),258(4.21), 298s(3.26)	44-3240-69
	EtOH	254(4.33),298s(3.26)	44-3240-69

Compound	Solvent	λ_{max}(log ϵ)	Ref.
Adenine, 9-[3-(3,4-xylidino)propyl]-	EtOH-NaOH	252(4.41),295s(3.52)	44-3240-69
$C_{16}H_{20}N_6O_2$			
Adenine, 9-[3-(3,5-dimethoxyanilino)-propyl]-	EtOH-HCl	258(4.21)	44-3240-69
	EtOH	255(4.34)	44-3240-69
	EtOH-NaOH	254(4.38)	44-3240-69
$C_{16}H_{20}O$			
Phenanthrene, 1,2,3,4a,10a-hexahydro-6-methoxy-1-methyl-, (-)-, isomer A	EtOH	273(4.17)	94-1158-69
isomer B	EtOH	272(4.14)	94-1158-69
isomer C	EtOH	274(4.14)	94-1158-69
$C_{16}H_{20}O_2$			
Cacalol, methyl ether	MeOH	218(4.52),256(4.13), 261(4.11),282(3.32), 292(3.20)	23-2465-69
6H-Dibenzo[b,d]pyran-1-ol, 6a,7,10,10a-tetrahydro-6,6,9-trimethyl-	EtOH	227s(4.04),276(3.17), 282(3.17)	33-1102-69
Fluorene, 1,2,3,4,4a,9a-hexahydro-7-methoxy-9-(methoxymethylene)-	EtOH	214(4.30),267(4.11), 274(4.05),309(3.92)	44-2209-69
Guaia-1(5),3,6,9-tetraen-12-oic acid, methyl ester	n.s.g.	236(4.31),308(3.61)	73-2288-69
Guaia-1(10),2,4,6-tetraen-12-oic acid, methyl ester	n.s.g.	253(4.23),284(3.91), 423(3.14)	73-2288-69
20-Nordeisopropyldehydroabietic acid	EtOH	266(2.5),274(2.5)	44-3739-69
Phenanthrene-1α-carboxylic acid, 1,2,3,4,9,10,11,12α-octahydro-1β-methyl-, cis	EtOH	266(2.58),273(2.58)	44-3739-69
Phenanthrene-1β-carboxylic acid, 1,2,3,4,9,10,11,12α-octahydro-1α-methyl-, cis	EtOH	266(2.71),273(2.77)	44-3739-69
9(1H)-Phenanthrone, 2,3,4,4a,10,10a-hexahydro-6-methoxy-1-methyl-, (-)-, isomer A	EtOH	278(4.20)	94-1158-69
isomer B	EtOH	277(4.21)	94-1158-69
isomer C	EtOH	277(4.20)	94-1158-69
9(1H)-Phenanthrone, 2,3,4,4a,10,10a-hexahydro-6-methoxy-4a-methyl-, (-)-	EtOH	280(4.18)	94-1158-69
Resorcinol, 2-p-mentha-1,8-dien-3-yl-, trans-(-)-	EtOH	275(3.1),281(3.08)	33-1102-69
Resorcinol, 4-p-mentha-1,8-dien-3-yl-, trans-(-)-	EtOH	226(4.0),281(3.5)	33-1102-69
$C_{16}H_{20}O_3$			
4H-1-Benzopyran-3-carboxylic acid, 2,4,4,7-tetramethyl-, ethyl ester	MeOH	215(3.68),258(3.20)	44-0380-69
Cyclopentanone, 5-hydroxy-2,2,3,3-tetramethyl-, benzoate	EtOH	231(4.16)	39-1346-69C
Eudesma-3,5,7(11)-trien-12-oic acid, 6-hydroxy-3-methoxy-, γ-lactone	MeOH	238(3.60),278(3.59), 294(3.58),365(4.5)	39-1088-69C
Fluorene-9α-carboxylic acid, 1,2,3,4,-4aα,9aα-hexahydro-7-methoxy-	EtOH	218(3.91),282(3), 287s(3.43)	44-2209-69
Fluorene-9β-carboxylic acid, 1,2,3,4,-4aα,9aα-hexahydro-7-methoxy-	EtOH	227(3.91),282(3.45), 288(3.39)	44-2209-69
$C_{16}H_{20}O_4$			
Aspertetronin A	EtOH	230(4.38),240s(4.35), 300(4.04)	39-0056-69C

Compound	Solvent	$\lambda_{max}(\log \epsilon)$	Ref.
Benzoic acid, 2,5-dimethyl-3-(2-oxo-cyclohexyloxy)-, methyl ester	EtOH	298(3.36)	4-0379-69
Fluorene-1-carboxylic acid, 4bβ,5,6,7-8,8aβ-hexahydro-9α-hydroxy-2-meth-oxy-, methyl ester	EtOH	292(3.50)	44-2209-69
Furocyclodec-9-ene-6-carboxylic acid, 3,10-dimethyl-4-oxo-, methyl ester	EtOH	211(4.12),275(3.54)	39-2786-69C
Santonenic acid, methyl ester	MeOH	250(4.1)	39-1088-69C
β-Santonenic acid, methyl ester	MeOH	250(4.23)	39-1088-69C
Warburgin, tetrahydro-	EtOH	225(3.42)	78-2865-69
$C_{16}H_{20}O_4S$			
Naphtho[1,2-c]thiophene-1-carboxylic acid, 3-ethoxy-1,3,3a,4,5,9b-hexa-hydro-9b-hydroxy-, methyl ester	EtOH	228(4.03),268(3.31)	2-0009-69
$C_{16}H_{20}O_5S$			
2,5-Thiophenedicarboxylic acid, tetra-hydro-3-hydroxy-3-phenyl-, diethyl ester	EtOH	245(4.13)	2-0009-69
$C_{16}H_{20}O_6$			
6-Isochromanacetic acid, 4aβ,5,6β,7,-8,8a-hexahydro-5β,7β-dihydroxy-α,4-dimethylene-3-oxo-8aβ-vinyl-, methyl ester (vernolepin methanol adduct)	MeOH	208(4.31)(end abs.)	44-3903-69
Succinic acid, piperonyl-, diethyl ester	EtOH	237(3.56),287(3.57)	39-2470-69C
$C_{16}H_{20}O_{10}$			
Asperuloside, deacetyl-	EtOH	238(3.78)	94-1942-69
$C_{16}H_{21}BrO$			
Anisole, 4-bromo-5-methyl-2-(1,2,3-trimethyl-2-cyclopenten-1-yl)-	MeOH	282(3.37),288(3.37)	78-3509-69
	EtOH	282(3.53),288(3.51)	18-3342-69
Anisole, 4-bromo-5-methyl-2-(4,4,5-trimethyl-1-cyclopenten-1-yl)-	EtOH	284(3.60),290(3.61)	18-3342-69
$C_{16}H_{21}BrO_3$			
1-Adamantanecrotonic acid, α-bromo-γ-oxo-, ethyl ester, trans	50% MeOH	243(3.85)	73-3343-69
$C_{16}H_{21}ClN_4O_6$			
7-Benzyl-1,2,3,6,8,9-hexahydro-1,3,7,9-tetramethyl-2,6-dioxopurinium perchlorate	H_2O	280(4.15)	104-2004-69
$C_{16}H_{21}ClO_3$			
1-Adamantanecrotonic acid, α-chloro-γ-oxo-, ethyl ester, trans	50% MeOH	236(3.95)	73-3343-69
$C_{16}H_{21}ClO_5$			
2-Hexyl-6-methyl-1-benzopyrylium perchlorate	HOAc-HClO	219(4.37),255(3.91), 270(3.78),320(3.38), 347(3.38)	5-0162-69B
$C_{16}H_{21}ClO_7$			
1-Butanol, 4-chloro-, carbethoxysyring-ate	MeOH	213(4.47),257(4.00), 299(3.38)	78-5155-69

Compound	Solvent	$\lambda_{max}(\log \epsilon)$	Ref.
$C_{16}H_{21}Cl_2NO$			
Acrylophenone, 3-[bis(2-chloroethyl)]-2',4',5'-trimethyl-	EtOH	215(3.90),300(4.25)	24-3139-69
Acrylophenone, 3-[bis(2-chloroethyl)]-2',4',6'-trimethyl-	EtOH	207(3.93),215(3.90),297(4.45)	24-3139-69
$C_{16}H_{21}N$			
Piperidine, 4-(3,4-dihydro-1-naphthyl)-1-methyl-	n.s.g.	259(3.93)	39-0217-69C
$C_{16}H_{21}NO$			
2,5-Cyclohexadien-1-ylideneacetonitrile, 3,5-di-tert-butyl-4-oxo-	hexane	300(4.43),312(4.54)	70-0580-69
$C_{16}H_{21}NO_2$			
Alolycopine	MeOH	237(3.70)	23-2457-69
Carbazole-4a(2H)-ethanol, 9-acetyl-1,3,4,9a-tetrahydro-	MeOH	254(4.17),280(3.62),288(3.53)	23-3647-69
Eleocarpiline	EtOH	221(3.70),241s(3.66),323(3.86)	12-0793-69
Isoeleocarpiline	EtOH	224(3.67),280s(3.54),323(3.89)	12-0793-69
1-Naphthaleneacetamide, α-acetyl-5,6,7,8-tetrahydro-N,N-dimethyl-	EtOH	266(3.93),276(2.89),294(2.60)	33-1030-69
	EtOH-NaOEt	275(3.20),287(3.20)	33-1030-69
1-Naphthaleneacetamide, 8-acetyl-5,6,7,8-tetrahydro-N,N-dimethyl-	EtOH	265(3.00)	33-1030-69
Spiro[cyclopentane-1,3'(2'H)-quinolin]-2'-one, 4'-ethoxy-1',4'-dihydro-4'-methyl-	EtOH	254(4.12),283(3.40),292(3.23)	94-1290-69
$C_{16}H_{21}NO_3$			
3-Indolineethanol, 1-acetyl-2,3-dimethyl-, acetate	MeOH	250(4.09),280s(--),286(3.81)	23-3647-69
Isoeleocarpicine, (+)-	EtOH	248(3.50),294(3.32)	12-0775-69
2-Pentenamide, N-isobutyl-5-[3,4-(methylenedioxy)phenyl]-	EtOH	232(4.10),287(3.63)	12-1531-69
3-Pentenamide, N-isobutyl-5-[3,4-(methylenedioxy)phenyl]-	EtOH	234(3.71),287(3.60)	12-1531-69
$C_{16}H_{21}NO_6$			
Monocrotaline, didehydro-	n.s.g.	224(3.64)	39-1155-69C
$C_{16}H_{21}N_3O_2$			
5,8-Isoquinolinedione, 7-[[2-(diethylamino)ethyl]amino]-3-methyl-	MeCN	235(4.23),277(4.19),322(3.67)	4-0697-69
1,4-Phthalazinedione, 5-[2-(diisopropylamino)vinyl]-2,3-dihydro-	THF	300s(4.0),370(4.2)	24-3241-69
$C_{16}H_{21}N_3O_2S$			
Acetic acid, [[4-(1-adamantyl)-6-amino-2-pyrimidinyl]thio]- (all spectra in 50% EtOH)	0.1N HCl	242(4.46),266s(4.10)	73-2278-69
	pH 8.2	222(4.30),247(4.08),284(3.82)	73-2278-69
	0.1N NaOH	227(4.21),248(4.06),284(3.81)	73-2278-69
$C_{16}H_{21}N_3O_3$			
2H-Pyran-2-methanol, 6-(5,6-dimethyl-1H-benzotriazol-1-yl)tetrahydro-, acetate, cis	EtOH	261(3.87),267(3.84),290(3.65)	4-0005-69

Compound	Solvent	λ_{max}(log ε)	Ref.
2H-Pyran-2-methanol, 6-(5,6-dimethyl-1H-benzotriazol-1-yl)tetrahydro-, acetate, trans	EtOH	263(3.90),287(3.74)	4-0005-69
$C_{16}H_{21}N_5O_9$			
Carbanilic acid, 2-[[[(1-carboxyethyl)-carbamoyl]methyl]amino]-3,5-dinitro-, mono-tert-butyl ester, L-	EtOH	354(3.99)	44-0395-69
$C_{16}H_{22}$			
1,3-Hexadien-5-yne, 3-methyl-1-(2,6,6-trimethyl-1-cyclohexen-1-yl)-	EtOH	287(4.30)	22-3247-69
$C_{16}H_{22}BNO_4$			
Boron, (N,N-diisopropylacetoacetamid-ato)[pyrocatecholato(2-)]-	CH_2Cl_2	235(3.71),276(4.34)	39-0173-69A
$C_{16}H_{22}Br_2O_6$			
Unidentified acetate from L. Okamurai Yamuda	EtOH	290(2.26)	18-0843-69
$C_{16}H_{22}ClNO_4$			
4a,9-Diethyl-2,3,4,4a-tetrahydro-1H-carbazolium perchlorate	EtOH	233(3.80),240(3.80), 279(3.85)	78-0095-69
	EtOH-NaOH	275(3.89),281(4.02)	78-0095-69
$C_{16}H_{22}N_2$			
Indole, 2-(1,1-dimethylallyl)-3-[(di-methylamino)methyl]-	EtOH	223(4.55),282(3.98), 290(3.92)	39-1003-69C
$C_{16}H_{22}N_2O$			
2H-Pyrrol-3-ol, 5-(dimethylamino)-2-isobutyl-4-phenyl-	MeOH	225(4.08),288(4.24)	78-5721-69
$C_{16}H_{22}N_2O_2$			
Costunolide, pyrazoline derivative	EtOH	350(2.15)	36-0877-69
1,4-Naphthalenediol, 2,3-bis[(dimethyl-amino)methyl]-	EtOH	251(4.33),337(3.54)	39-1245-69C
$C_{16}H_{22}N_2O_3$			
3-Buten-2-one, 4-amino-4-[6-(2-hydroxy-propyl)-2,4-dimethyl-3-pyridyl]-, acetate	EtOH	275s(3.90),304(4.31)	35-4749-69
$C_{16}H_{22}N_2O_3SSi$			
Nicotinamide, 1,4-dihydro-N-(p-tolyl-sulfonyl)-1-(trimethylsilyl)-	EtOH	224(4.47),355(4.10)	44-3672-69
$C_{16}H_{22}N_2O_8S_2$			
Ethylene, 1,2-bis(4-pyridyl)-N,N'-di-methyl-, dimethosulfate	n.s.g.	318(4.58)	39-1643-69C
$C_{16}H_{22}N_4O_2$			
Uracil, 5-(benzylmethylamino)-6-(di-methylamino)-1,3-dimethyl-	EtOH	293(3.95)	104-2004-69
$C_{16}H_{22}N_4O_2S$			
Valeric acid, 5-[(9-cyclohexyl-9H-purin-6-yl)thio]-	pH 1	223(4.09),295(4.31)	73-2114-69
	pH 13	226(4.09),291(4.36)	73-2114-69

Compound	Solvent	λmax(log ε)	Ref.
$C_{16}H_{22}N_4O_2S_3$			
Thiamine, S-dithioallyloxycarbonyl-	EtOH	236s(4.09),268(3.92), 313(4.09)	94-2299-69
$C_{16}H_{22}N_4O_4$			
3-Hepten-2-one, 4-propyl-, 2,4-dinitrophenylhydrazone	EtOH	379(4.37)	28-0075-69A
4-Hepten-2-one, 4-propyl-, 2,4-dinitrophenylhydrazone	EtOH	230(4.31),275(4.20)	28-0075-69A
Tubercidin, N^6-isopentenyl-	pH 1.0	230(4.31),275(4.20)	87-1056-69
	pH 7.0	275(4.21)	87-1056-69
	pH 12.0	275(4.21)	87-1056-69
$C_{16}H_{22}N_4O_6$			
Barbituric acid, 5,5'-(1-methylheptylidene)di-	EtOH	255(4.21)	103-0827-69
$C_{16}H_{22}N_6OS_3$			
Spiro[furo[2,3-d]thiazole-2(3H),4'-imidazolidine]-2',5'-dithione, 3-[(4-amino-2-methyl-5-pyrimidinyl)methyl]-Tetrahydro-1',3',3a-trimethyl-	EtOH	325(4.18)	94-0910-69
$C_{16}H_{22}O$			
Phenol, m-isobornyl-	pH 13	238(3.88),292(3.45)	30-0867-69A
	iso-PrOH	276(3.35),282(3.32)	30-0867-69A
Phenol, o-isobornyl-	pH 13	243(3.91),294(3.62)	30-0867-69A
	iso-PrOH	277(3.46),284(3.40)	30-0867-69A
Phenol, p-isobornyl-	pH 13	242(4.24),294(3.57)	30-0867-69A
	iso-PrOH	229(3.96),280(3.23), 286s(3.34)	30-0867-69A
Phenol, m-(5,5,6-trimethyl-2-norbornyl)-	pH 13	239(3.95),289(3.49)	30-0867-69A
	iso-PrOH	275(3.37),281(3.29)	30-0867-69A
Phenol, o-(5,5,6-trimethyl-2-norbornyl)-	pH 13	239(3.86),295(3.50)	30-0867-69A
	iso-PrOH	275(3.43),283s(3.31)	30-0867-69A
Phenol, p-(5,5,6-trimethyl-2-norbornyl)-	pH 13	241(4.09),295(3.38)	30-0867-69A
	iso-PrOH	227(3.96),280(3.26), 287s(3.14)	30-0867-69A
$C_{16}H_{22}O_2$			
2,5-Cyclohexadien-1-one, 2,6-di-tert-butyl-4-ethynyl-4-hydroxy-	MeOH	235(3.98)	64-0547-69
2,5-Cyclohexadien-1-ylideneacetic acid, 3,5-di-tert-butyl-4-oxo-	hexane	302(4.41),315(4.44)	70-0580-69
4-Hexen-3-one, 2-(3-hydroxymesityl)-4-methyl-	MeOH	223(4.19),285(3.30)	77-0162-69
	MeOH-base	295(3.56)	77-0162-69
$C_{16}H_{22}O_2S_2$			
3,3'-Bithiophene, 2,2'-di-tert-butoxy-	EtOH	269(4.09)	118-0170-69
$C_{16}H_{22}O_3$			
2H,8H-Benzo[1,2-b:3,4-b']dipyran-5-ol, 3,4,9,10-tetrahydro-2,2,8,8-tetramethyl-	EtOH	270(2.86)	39-1540-69C
	EtOH-base	282(--)	39-1540-69C
10αH-Furoeremophilane, 6α-methoxy-9-oxo-	EtOH	284(4.11)	73-2792-69
cis anomer	EtOH	285(3.95)	73-2792-69
$C_{16}H_{22}O_4$			
Hydrinda-3(3a),4-diene-5-carboxylic acid, 1-methyl-3-isobutyl-, methyl ester	EtOH	296(4.30)	2-0643-69

Compound	Solvent	λ max (log ε)	Ref.
Naphtho[2,1-b]furan-6-carboxylic acid, 2,3aα,4,5,5aα,6,7,8,9,9a-decahydro-6β,9aβ-dimethyl-2-oxo-	EtOH	215(4.18)	44-3477-69
$C_{16}H_{22}O_5$			
Aspertetronin B	EtOH	235(4.04),265s(3.83)	39-0056-69C
$C_{16}H_{22}O_7$			
3,6-Cyclohexadiene-1,3-dicarboxylic acid, 2-(carboxymethyl)-4-hydroxy-, triethyl ester	n.s.g.	263(4.10),300(3.87), 323(3.88)	35-7342-69
$C_{16}H_{22}O_9S_4$			
α-D-Mannopyranose, 1,2-dithio-, 3,4,6-triacetate 1,2-bis(O-methyl dithiocarbonate)	EtOH	275(4.26)	94-2571-69
$C_{16}H_{22}O_{11}$			
Cyclopenta[c]pyran-4-carboxylic acid, 1α-(β-D-glucopyranosyloxy)-1,4aα,5,7aα-tetrahydro-5β-hydroxy-7-(hydroxymethyl)-	MeOH	235(4.05)	94-1942-69
Scandoside	EtOH	235(4.16)	94-1942-69
$C_{16}H_{23}BrO_2$			
Cyclopentanol, 2-(5-bromo-2-methoxy-p-tolyl)-1,2,5-trimethyl-	MeOH	282(3.31),288(3.31)	78-3509-69
$C_{16}H_{23}BrO_3$			
1,5-Cyclohexadiene-1-carboxylic acid, 3-bromo-3,5-di-tert-butyl-4-oxo-, methyl ester	C_6H_{12}	330(3.36)	64-0547-69
$C_{16}H_{23}IN_2$			
Trimethyl[(1,2,3,4-tetrahydrocarbazol-9-yl)methyl]ammonium iodide	EtOH	233(4.33),270(3.94)	39-2738-69C
$C_{16}H_{23}N$			
Naphthalene, 1,2,3,4-tetrahydro-2,2-dimethyl-1-[2-(dimethylamino)ethylidene]-, hydrochloride	n.s.g.	245(4.14)	39-0217-69C
Naphthalene, 1,2,3,4-tetrahydro-2-methyl-1-[3-(dimethylamino)propylidene]-	n.s.g.	254(3.85)	39-0217-69C
$C_{16}H_{23}NO$			
2-Azabicyclo[9.2.2]pentadeca-11,13,14-trien-10-one, 13,14-dimethyl-, diradical open chain intermediate	MeOH	236(3.79),306(3.72)	88-2281-69
1-Naphthalenepropylamine, 3,4-dihydro-5-methoxy-N,N-dimethyl-, hydrochloride	n.s.g.	265(3.95)	39-0217-69C
1-Naphthalenepropylamine, 3,4-dihydro-6-methoxy-N,N-dimethyl-, hydrochloride	n.s.g.	270(4.11)	39-0217-69C
1-Naphthalenepropylamine, 3,4-dihydro-7-methoxy-N,N-dimethyl-, hydrochloride	n.s.g.	259(3.84)	39-0217-69C

Compound	Solvent	$\lambda_{max}(\log \epsilon)$	Ref.
$C_{16}H_{23}NO_2$			
2-Cyclopropene-$\Delta^{1,\alpha}$-acetic acid, 2,3-di-tert-butyl-, ester with hydracrylonitrile	MeCN	255(4.37)	35-4766-69
Eleocarpine, 13,14,15,16-tetrahydro-, (+)-	EtOH	273(3.95)	12-0793-69
Isoeleocarpine, 13,14,15,16-tetrahydro-, (-)-	EtOH	275(3.95)	12-0793-69
Spiro[cyclohexane-1,4'(1'H)-isoquinolin]-4-ol, 2',3'-dihydro-6'-methoxy-2'-methyl-	EtOH	279(3.29),285s(3.22)	94-1564-69
hydrochloride	EtOH	222(3.89),278(3.34), 285s(3.27)	94-1564-69
$C_{16}H_{23}NO_6$			
Monocrotaline	EtOH	217(3.32)	73-1832-69
$C_{16}H_{23}N_3O_2$			
1,4,2-Dioxazolidine, 2-tert-butyl-3-(tert-butylimino)-5-(phenylimino)-	isooctane	243(4.08),276s(3.23), 284s(2.97)	44-2269-69
1,2,4-Oxadiazolin-5-one, 2-tert-butyl-3-(tert-butylimino)-4-phenyl-	isooctane	238(4.05)	44-2269-69
$C_{16}H_{23}N_5O_4S$			
Adenosine, N-(3-methyl-2-butenyl)-2-(methylthio)-	EtOH-acid	246(4.27),286(4.21)	69-3071-69
	EtOH	244(4.40),283(4.26)	69-3071-69
	EtOH-base	243(4.39),283(4.26)	69-3071-69
$C_{16}H_{23}N_5O_{12}$			
β-D-Allofuranuronic acid, 5-(2-amino-2-deoxy-L-xylonamido)-1,5-dideoxy-1-(3,4-dihydro-2,4-dioxo-1(2H)-pyrimidinyl)-, 5-carbamate (polyoxin L)	0.05N HCl	259(3.96)	35-7490-69
	0.05N NaOH	262(3.85)	35-7490-69
$C_{16}H_{23}O_6P$			
1,2-Oxaphosphol-4-ene-4-carboxylic acid, 2,3-dihydro-2,2,2-trimethoxy-5-methyl-3-phenyl-, ethyl ester	isooctane	253(4.17),285(3.02)	70-1012-69
$C_{16}H_{24}ClN$			
Naphthalene, 1,2,3,4-tetrahydro-2,2-dimethyl-1-[2-(dimethylamino)ethylidene]-, hydrochloride	n.s.g.	245(4.14)	39-0217-69C
$C_{16}H_{24}ClNO_4$			
1,2,3,4,7,8,9,10,11,12-Decahydro-6-methylcycloocta[c]quinolizinium perchlorate	$CHCl_3$	281(3.96)	78-4161-69
$C_{16}H_{24}N_2O$			
1,6-Diazecine, 1,2,3,4,5,8,9,10-octahydro-7-(p-methoxyphenyl)-1-methyl-	n.s.g.	226(3.96),276(3.26), 283(3.16)	44-2720-69
$C_{16}H_{24}N_2O_5$			
2,3-Diazabicyclo[2.2.2]oct-5-ene-2,3-dicarboxylic acid, 1,4,5,6,7,7-hexamethyl-8-oxo-, dimethyl ester	EtOH	210(3.79),300(2.63)	88-4785-69
2,3-Diazabicyclo[4.2.0]oct-4-ene-2,3-dicarboxylic acid, 1,4,5,6,8,8-hexamethyl-7-oxo-, dimethyl ester	C_6H_{12}	210(3.81),311(2.20)	88-4785-69

Compound	Solvent	$\lambda_{max}(\log \epsilon)$	Ref.
$C_{16}H_{24}N_4O_2S$			
Valeric acid, 5-[(9-hexyl-9H-purin-6-yl)thio]-	pH 1	225(4.12),287(4.27), 295(4.29)	73-2114-69
	pH 13	228(4.07),287(4.32), 295(4.32)	73-2114-69
$C_{16}H_{24}O$			
Benzo[1,2:3,4]dicyclohepten-6(1H)-one, 2,3,4,5,5a,7,7a,8,9,10,11,12-dodecahydro-	EtOH	211(3.87)	22-4493-69
Benzo[1,2:3,4]dicyclohepten-6(1H)-one, 2,3,4,5,7,7a,8,9,10,11,12,12a-dodecahydro-	EtOH	253(3.96)	22-4493-69
2,5-Cyclohexadien-1-one, 2,6-di-tert-butyl-4-ethylidene-	hexane	300(4.29)	70-0580-69
$C_{16}H_{24}O_2$			
2,5-Cyclohexadien-1-one, 2,6-di-tert-butyl-4-(methoxymethylene)-	hexane	325(4.12)	70-0580-69
$C_{16}H_{24}O_3$			
4H-1-Benzopyran-3-carboxylic acid, 5,6,7,8-tetrahydro-2,4,4,7-tetramethyl-, ethyl ester	EtOH	206(3.77),272(3.40)	44-1465-69
2βH-Cedr-8-en-15-oic acid, 10β-hydroxy-, methyl ester	EtOH	232(3.82)	78-3855-69
2-Cyclohexen-1-one, 3-methoxy-5,5-dimethyl-2-[3(or 4)-oxocycloheptyl]-	EtOH	266(4.36)	104-1566-69
Octanophenone, 4'-ethoxy-2'-hydroxy-	n.s.g.	275(4.1)	104-1609-69
2,4-Pentadienoic acid, 5-(1-hydroxy-2,2-dimethyl-6-methylenecyclohexyl)-3-methyl-, methyl ester	EtOH	267(4.25)	22-0232-69
isomer	EtOH	268(4.16)	22-0232-69
2,4-Pentadienoic acid, 5-(2-hydroxy-2,6,6-trimethylcyclohexylidene)-3-methyl-, methyl ester	EtOH	269(4.23)	22-0232-69
isomer	EtOH	270(4.23)	22-0232-69
$C_{16}H_{24}O_4$			
1,2-Benzodioxin-3-acrylic acid, 3,5,6,7,8,8a-hexahydro-β,5,5,8a-tetramethyl-, methyl ester	EtOH	214(4.15)	22-0232-69
isomer	EtOH	222(4.12)	22-0232-69
2-Benzofuranacrylic acid, 2,4,5,6,7,7a-hexahydro-2α-hydroxy-β,4,4,7a-tetramethyl-, methyl ester	EtOH	222(4.17)	22-0232-69
2-Oxabicyclo[3.3.1]non-3-ene-4-carboxylic acid, 1-hydroxy-7-isopropenyl-3,9-dimethyl-, ethyl ester, (-)-	hexane	243(4.06)	78-3139-69
	EtOH	251(4.09)	78-3139-69
Photosantonic acid lactone, O-methyl-	EtOH	238(3.99)	33-1237-69
$C_{16}H_{24}O_4S_2$			
2H-Thiopyran-2-propionic acid, tetrahydro-3-hydroxy-, ester with 6-hydroxy-9-thiabicyclo[3.3.1]nonan-2-one	n.s.g.	244s(2.38),303(2.15)	33-0967-69
$C_{16}H_{24}O_9S_2$			
α-D-Glucopyranoside, methyl 6-thio-, 2,3,4-triacetate 6-(O-ethyl dithiocarbonate)	n.s.g.	221(3.92),276(4.06), 350(1.74)	24-0494-69

Compound	Solvent	$\lambda_{max}(\log \epsilon)$	Ref.
β-D-Glucopyranoside, methyl 6-thio-, 2,3,4-triacetate 6-(O-ethyl dithiocarbonate)	n.s.g.	221(3.88),277(4.03), 351(1.69)	24-0494-69
$C_{16}H_{25}N$			
Pentylamine, 3-benzylidene-N,N,4,4-tetramethyl-, trans, hydrochloride	H_2O	238(3.93)	78-0641-69
$C_{16}H_{25}NO$			
2-Aziridinone, 1-(1-adamantyl)-3-tert-butyl-	hexane	252(2.16)	77-0049-69
	EtOH	241(2.33)	77-0049-69
2-Aziridinone, 3-(1-adamantyl)-1-tert-butyl-	hexane	252(2.13)	77-0049-69
	EtOH	242(2.27)	77-0049-69
Lycopodine	octene	217(3.38)	23-0449-69
	MeOH	217(3.38)	23-0449-69
$C_{16}H_{25}NO_2$			
Alkaloid L.20	MeOH	220(3.40)	23-0449-69
Bicyclo[6.3.1]dodec-8(12)-en-9-one, 12-morpholino-	EtOH	346(4.22)	39-0592-69C
Decanohydroxamic acid, N-phenyl-	EtOH	247(4.01)	42-0831-69
Flabelliformine	MeOH	228s(2.70)	23-0449-69
Hydroquinone, 5-[(cyclohexylmethylamino)methyl]-2,3-dimethyl-	EtOH	295(3.35)	39-1245-69C
Lycodoline	MeOH	221(3.18)	23-0449-69
$C_{16}H_{25}NO_3$			
2,4,6-Heptatrienoic acid, 2-acetyl-5-(diethylamino)-6-methyl-, ethyl ester	EtOH	263(3.81),405(4.70)	70-0363-69
$C_{16}H_{25}NO_5$			
Heliotrine, didehydro-	n.s.g.	221(3.83),297(3.10)	39-1155-69C
$C_{16}H_{25}NO_7S_3$			
β-D-Mannopyranoside, methyl 1,2-dithio-, 3,4,6-triacetate 2-(dimethyldithiocarbamate)	EtOH	240(3.92),277(3.93)	94-2571-69
$C_{16}H_{25}N_3$			
Quinazoline, 2-[3-(dimethylamino)propyl]-1,2-dihydro-4-isopropyl-, hydrochloride	EtOH	229(4.35),260s(3.7), 367(3.25)	33-2351-69
$C_{16}H_{25}N_3O_2$			
Cycloheptanone, 2-(5,6,7,8-tetrahydro-2-methyl-1(4H)-cycloheptimidazolyl)-, oxime, 3-oxide	MeOH	220(3.83)	48-0746-69
$C_{16}H_{25}N_5O_4$			
Adenosine, N-hexyl-	pH 1.0	263(4.31)	87-1056-69
	pH 7.0	210(4.30),268(4.22)	87-1056-69
	pH 12.0	268(4.26)	87-1056-69
$C_{16}H_{26}$			
Benzo[1,2:3,4]dicycloheptene, 1,2,3,4,-5,5a,6,7,8,9,10,11,12,12b-tetradecahydro-	EtOH	209(4.01)	22-4501-69
Benzo[1,2:3,4]dicycloheptene, 1,2,3,4,-5,5a,6,7,7a,8,9,10,11,12-tetradecahydro-	EtOH	212(3.76)	22-4501-69

Compound	Solvent	$\lambda_{max}(\log \epsilon)$	Ref.
Phenanthrene, 1,2,3,4,5,6,7,8,8a,9,10,-10a-dodecahydro-9,10-dimethyl-	EtOH	211(4.02)	22-0962-69
$C_{16}H_{26}ClNO_4$			
1,2,3,4,4a,5,7,8,9,10,11,12-Dodecahydro-6-methylcycloocta[c]quinolizinium perchlorate	$CHCl_3$	311(3.74)	78-4161-69
$C_{16}H_{26}N_4O$			
1-Cyclobutene-1,3-dicarbonitrile, 2,4-bis(diethylamino)-4-ethoxy-	EtOH	217(4.47),256(4.60), 318(4.13)	88-3639-69
$C_{16}H_{26}N_8S_6$			
2H-1,3-Thiazine-2,4(3H)-dione, 3,3'-hexamethylenebis[dihydro-2-thio-, 4,4'-bis(thiosemicarbazone)	EtOH	250(4.30),310s(--)	103-0049-69
$C_{16}H_{26}O$			
Benzo[1,2:3,4]dicyclohepten-6-ol, 1,2,3,4,5,5a,6,7,7a,8,9,10,11,12-tetradecahydro-	EtOH	209(3.95)	22-4493-69
4-Octen-1-one, 1-(1-cyclopenten-1-yl)-5-propyl-	EtOH	242(3.92)	33-1996-69
9-Phenanthrenemethanol, 1,2,3,4,5,6,7,-8,8a,9,10,10a-dodecahydro-10-methyl-, cis-anti-trans	EtOH	211(4.05)	22-0962-69
9-Tridecen-5-ynal, 10-propyl-	EtOH	360(4.28)	33-1996-69
$C_{16}H_{26}OS_2$			
Spiro[1,3-dithiolane-2,2'(3'H)-naphthalene]-7'-methanol, 4',4'a,5',6',-7',8'-hexahydro-α,α,4'a-trimethyl-	hexane	244(2.90)	32-0231-69
$C_{16}H_{26}O_2$			
2(3H)-Naphthalenone, 5β-tert-butoxy-4aβ-ethyl-4,4a,5,6,7,8-hexahydro-	EtOH	246(4.08)	44-0107-69
9,10-Phenanthrenedimethanol, 1,2,3,4,5,6,7,8,8a,9,10,10a-dodecahydro-	EtOH	209(3.95)	22-0962-69
$C_{16}H_{26}O_3$			
Decarboxyportentol	n.s.g.	242(3.74)	77-0162-69
$C_{16}H_{26}O_4$			
Aspertetronin A, hexahydro-	EtOH	215(3.88),264(4.00)	39-0056-69C
2-Oxabicyclo[3.3.1]non-3-ene-4-carboxylic acid, 1-hydroxy-7-isopropyl-3,9-dimethyl-, ethyl ester	EtOH	251(4.07)	78-3139-69
$C_{16}H_{27}NO$			
Lycopodine, dihydro-	MeOH	none	23-0449-69
$C_{16}H_{27}NO_5$			
Heliotrine	EtOH	213(3.29)	73-1832-69
$C_{16}H_{28}N_2O_2$			
p-Phenylenediamine, N,N,N',N'-tetraethyl-2,5-dimethoxy-	C_6H_{12}	215(3.96),265(3.96), 306(3.78)	117-0287-69

Compound	Solvent	λ_{max}(log ϵ)	Ref.
$C_{16}H_{29}Br_2N_5O_4$			
N-[2-(2,4-Dinitroanilino)ethyl]-N,N,N',N',N'-pentamethyl-N,N'-trimethylenediammonium dibromide (also spectra of mixtures)	pH 6.50	350(4.22)	35-5136-69
	EtOH	338(4.22)	35-5136-69
$C_{16}H_{30}Br_2N_4O_2$			
N,N,N,N',N'-Pentamethyl-N'-[2-(o-nitro-anilino)ethyl]-N,N'-trimethylene-diammonium dibromide (also spectra of mixtures)	pH 6.5	422(4.71)	35-5136-69
	EtOH	415(3.74)	35-5136-69
N,N,N,N',N'-Pentamethyl-N'-[2-(p-nitro-anilino)ethyl]-N,N'-trimethylene-diammonium dibromide (also spectra of mixtures)	pH 6.5	383(4.21)	35-5136-69
	EtOH	366(4.22)	35-5136-69
$C_{16}H_{30}N_2O$			
Morpholine, 4-(2-piperidino-1-heptenyl)-	C_6H_{12}	200(3.99),225(3.97)	22-3883-69
$C_{16}H_{31}IN_2$			
Trimethyl[2-(11-methyl-11-azabicyclo-[4.4.1]undec-5-en-2-yl)ethyl]ammonium iodide	EtOH	220(4.17),250s(--)	35-3616-69
$C_{16}H_{32}O_2$			
Palmitic acid, Pb(IV) salt	hexane	230(4.33)	33-1495-69
$C_{16}H_{32}Si_2$			
2,6-Disilahept-3-ene, 3-(3,3-dimethyl-1-butynyl)-2,2,4,6,6-pentamethyl-	n.s.g.	255(3.99)	101-0291-69C

Compound	Solvent	$\lambda_{max}(\log \epsilon)$	Ref.
$C_{17}H_6Cl_8O$			
4,9-Etheno-1H-cyclopenta[b]biphenylen-1-one, 2,3,3a,5,6,7,8,9a-octachloro-3a,4,4a,4b,8a,8b,9,9a-octahydro-	$CHCl_3$	257(3.92),287(3.65), 297(3.74),312(3.66)	39-2710-69C
$C_{17}H_8N_4$			
Spiro[fluoren-9,3'-[3H]pyrazole]-4',5'-dicarbonitrile	$CHCl_3$	255(4.32),270(4.38), 283(4.23),296(3.36), 312(2.91)	64-0536-69
$C_{17}H_9ClN_2O_2$			
1H-Naphth[2,3-d]imidazole-4,9-dione, 2-(p-chlorophenyl)-	pH 13	292(4.72)	4-0909-69
	MeOH	285(4.64),294(4.63), 390(3.23)	4-0909-69
$C_{17}H_9Cl_2NO_2$			
Benzamide, p-chloro-N-(3-chloro-4-hydroxy-1-oxo-2(1H)-naphthylidene)-	MeOH	236(4.39),287(3.96)	4-0909-69
$C_{17}H_9Cl_3O_3$			
Benzoic acid, 2,4-dichloro-, ester with 4-chloro-2-hydroxy-3-phenyl-2-cyclobuten-1-one	MeCN	293(4.35)	88-1443-69
$C_{17}H_9NO_3$			
Naphtho[2,1-d]isoxazole-4,5-dione, 3-phenyl-	EtOH	258(4.55),384(3.40)	32-0565-69
$C_{17}H_{10}BrClO_3$			
1,3-Indandione, 2-(3-bromo-5-chloro-2-methoxybenzylidene)-	dioxan	246(4.34),332(4.36)	65-1377-69
1,3-Indandione, 2-(5-bromo-3-chloro-2-methoxybenzylidene)-	dioxan	245(4.33),333(4.36)	65-1377-69
$C_{17}H_{10}Br_2OS$			
4H-Thiopyran-4-one, 3,5-dibromo-2,6-diphenyl-	$CHCl_3$	273(4.34),317(4.17)	18-3005-69
$C_{17}H_{10}Br_2O_3$			
1,3-Indandione, 2-(3,5-dibromo-2-methoxybenzylidene)-	dioxan	247(4.36),335(4.40)	65-1377-69
$C_{17}H_{10}ClN$			
Benzo[b]phenanthridine, 5-chloro-	EtOH	220(4.85),249(4.99), 257(5.03),265(4.94), 276(4.91),287(5.05), 299(4.98),326(3.98), 341(4.05)	44-3664-69
$C_{17}H_{10}Cl_2O_3$			
1,3-Indandione, 2-(3,5-dichloro-2-methoxybenzylidene)-	dioxan	245(4.30),332(4.37)	65-1377-69
1,3-Indandione, 2-(3,5-dichloro-4-methoxybenzylidene)-	dioxan	250(4.42),345(4.51)	65-1373-69
$C_{17}H_{10}O$			
2,4,6,12,14-Cycloheptadecapentaene-8,10,16-triyn-1-one	ether	293(4.79),304(4.87), 463(3.00),500s(2.91), 540s(2.56)	35-7518-69

Compound	Solvent	λ_{max}(log ϵ)	Ref.
$C_{17}H_{10}O_3$			
6H-Cyclopenta[a]phenanthrene-6,7,17-trione, 15,16-dihydro-	EtOH	275(4.03),331(3.31), 406(3.21)	39-2484-69C
$C_{17}H_{10}O_4$			
Benzo[i]xanthone, 1,3-dihydroxy-	$CHCl_3$	265(4.69)	49-1368-69
$C_{17}H_{11}BrOS$			
4H-Thiopyran-4-one, 3-bromo-2,6-diphenyl-	$CHCl_3$	273(4.37),308(4.23)	18-3005-69
$C_{17}H_{11}BrO_3$			
2-Cyclobuten-1-one, 4-bromo-2-hydroxy-3-phenyl-, benzoate	MeCN	297(4.36)	88-1443-69
1,3-Indandione, 2-(3-bromo-4-methoxybenzylidene)-	dioxan	255(4.59),380(4.63)	65-1373-69
1,3-Indandione, 2-(5-bromo-2-methoxybenzylidene)-	dioxan	240(4.28),330(4.14), 390(4.13)	65-1377-69
$C_{17}H_{11}ClN_2O$			
3-Quinolineacetonitrile, 6-chloro-1,2-dihydro-2-oxo-4-phenyl-	EtOH	234s(--),239(4.61), 279(3.77),342(3.82), 355s(--)	4-0599-69
$C_{17}H_{11}ClN_2O_3$			
Benzamide, N-(3-amino-1,4-dihydro-1,4-dioxo-2-naphthyl)-p-chloro-	MeOH	267(4.39)	4-0909-69
$C_{17}H_{11}ClOS$			
4H-Thiopyran-4-one, 3-chloro-2,6-diphenyl-	EtOH	268(4.28),305(4.18)	39-0315-69C
$C_{17}H_{11}ClO_3$			
2-Cyclobuten-1-one, 4-chloro-2-hydroxy-3-phenyl-, benzoate	MeCN	294(4.40)	88-1443-69
1,3-Indandione, 2-(5-chloro-2-methoxybenzylidene)-	dioxan	243(4.26),323(4.16), 392(4.17)	65-1377-69
$C_{17}H_{11}ClO_4S$			
1,2-Benzo-6-thiaphenanthrenium perchlorate	MeCN-1% $HClO_4$	246s(4.18),270(4.08), 296s(4.21),312(4.33), 402(3.06),490(3.44)	2-0637-69
9,10-Benzo-6-thiaphenanthrenium perchlorate	MeCN-1% $HClO_4$	244(4.38),272(4.30), 302s(4.47),312(4.54), 455(4.46),498s(3.22)	2-0637-69
6-Thiachrysenium perchlorate	MeCN-1% $HClO_4$	254s(4.60),261(4.67), 305(3.92),320s(3.83), 337s(3.73),451(3.68)	2-0637-69
$C_{17}H_{11}ClO_7$			
2-Anthroic acid, 5-chloro-9,10-dihydro-1,6-dihydroxy-8-methoxy-3-methyl-9,10-dioxo-	EtOH	230(4.52),253(4.23), 290(4.28),436(4.00)	24-4104-69
	EtOH-NaOH	242(4.53),314(4.37), 515(4.06)	24-4104-69
$C_{17}H_{11}ClO_8$			
2-Anthroic acid, 5-chloro-9,10-dihydro-1,4,6-trihydroxy-8-methoxy-3-methyl-9,10-dioxo-	EtOH	255s(4.14),283(4.25), 486(4.09)	24-4104-69
	EtOH-NaOH	246(4.44),307(4.31), 320s(4.28),537(4.16)	24-4104-69

Compound	Solvent	λ_{max}(log ϵ)	Ref.
$C_{17}H_{11}IO_3$			
2-Cyclobuten-1-one, 2-hydroxy-4-iodo-3-phenyl-, benzoate	MeCN	303(4.23)	88-1443-69
$C_{17}H_{11}N$			
Dibenzo[a,e]cyclooctene-5-carbonitrile	EtOH	237(4.45)	22-0217-69
$C_{17}H_{11}NO$			
Benzo[b]phenanthridin-5(6H)-one	EtOH	224(4.47),252(4.68), 260(4.76),269(4.75), 358(3.28),375(2.89)	44-3664-69
2-Penten-4-ynenitrile, 3-hydroxy-2,5-diphenyl-	EtOH	210(4.03),229(4.09), 344(4.23)	39-0915-69C
	NaOH	228(4.18),250s(4.11), 350(4.19)	39-0915-69C
$C_{17}H_{11}NO_2$			
3H-Dibenz[f,ij]isoquinoline-2,7-dione, 3-methyl-	EtOH	240(4.18),265(4.08), 340(3.95)	78-5365-69
Pyrrolo[1,2-f]phenanthridine-1-carboxylic acid	MeOH	209(4.44),233(4.57), 242(4.50),253s(4.35), 264(4.31),275(4.30), 287(4.26),306(4.13), 331(4.00),340(3.97)	39-2311-69C
$C_{17}H_{11}NO_3$			
Spiro[1,4,2-dioxazole-5,1'(2'H)-naphthalen]-2'-one, 3-phenyl-	EtOH	238(4.44),268(3.94), 330(3.80)	32-0565-69
Spiro[1,4,2-dioxazole-5,2'(1'H)-naphthalen]-1'-one, 3-phenyl-	EtOH	238(4.75),268(4.03), 350(3.13)	32-0565-69
$C_{17}H_{11}NO_4$			
1,2-Naphthoquinone, 4-(m-carboxyanilino)-	pH 0.80	250(5.22),255(5.22), 264(5.21)	78-5807-69
$C_{17}H_{11}N_3$			
α,α'-Bi-o-tolunitrile, α-cyano-	EtOH	226(4.28),260s(2.93), 267s(3.18),274(3.36), 283(3.41)	12-0577-69
$C_{17}H_{11}N_3O_2$			
2,8-Anthrydinediol, 3-phenyl-	NaOH	225(2.9),240(2.9), 290(1.9),385(2.6)	32-0677-69
$C_{17}H_{11}N_3O_5$			
Benzoic acid, p-nitro-, 2-(1,4-dihydro-1,4-dioxo-2-naphthyl)hydrazide	alkali	602(4.13)	95-0007-69
Benzoic acid, p-nitro-, 2-(3,4-dihydro-3,4-dioxo-1-naphthyl)hydrazide	alkali	520(4.22)	95-0007-69
$C_{17}H_{11}N_5O_8S$			
Thiocyanic acid, [[4-(5-nitro-2-furyl)-1-[3-(5-nitro-2-furyl)acryloyl]-2-pyrazolin-3-yl]carbonyl]methyl ester	$C_2H_4Cl_2$	240(4.20),318(4.51), 360s(4.40)	94-2105-69
$C_{17}H_{12}$			
Cyclohepta[def]fluorene, 4,8-dihydro-8-methylene-	EtOH	258(4.13),269s(4.01), 315(3.56)	39-1427-69C

Compound	Solvent	λmax(log ε)	Ref.
Indene, 2-(phenylethynyl)-	EtOH	232(3.98),238(3.95), 251(--),302(--), 308s(--),315(4.55), 321(4.42),337(4.35)	87-0513-69
Pyrene, 1-methyl-	EtOH	232(4.80),242(5.06), 255(4.08),265(4.47), 276(4.81),299(3.68), 314(4.10),326(4.50), 342(4.68)	24-2301-69
Pyrene, 2-methyl-	EtOH	234(4.79),243(5.12), 254(4.22),264(4.48), 275(4.74),295(3.71), 307(4.13),320(4.55), 337(4.81)	24-2301-69
$C_{17}H_{12}BrNO_3$			
4H-Quinolizine-3-carboxylic acid, 1-bromo-4-oxo-2-phenyl-, methyl ester	MeOH	206(4.32),236(4.32), 266(4.35),414(4.41)	39-1143-69C
$C_{17}H_{12}BrNS$			
8H-Thieno[2,3-b]indole, 2-(p-bromophenyl)-8-methyl-	EtOH	242(4.71),288s(4.22), 294(4.24),346(4.56)	95-0058-69
$C_{17}H_{12}Br_2ClNOS$			
Spiro[2H-1-benzopyran-2,2'-benzothiazoline], 6,8-dibromo-7-chloro-3,3'-dimethyl-	$(MeOCH_2)_2$	240(4.46),279(4.07), 290(4.08),324(3.56)	78-3251-69
$C_{17}H_{12}ClNOS$			
4H-Thiopyran-4-one, 3-chloro-2,6-diphenyl-, oxime	EtOH	248(4.27),312(4.15)	39-0315-69C
$C_{17}H_{12}ClNO_3$			
Oxazole, 2-(p-chlorophenyl)-4-phenyl-5-acetoxy-	EtOH	240(4.31),299(4.28)	88-1557-69
$C_{17}H_{12}Cl_2O_6$			
Flavone, 3',5'-dichloro-5,7-dihydroxy-3,4'-dimethoxy-	EtOH	273(4.35),358(3.98)	111-0298-69
	EtOH-NaOH	279(4.39),375(3.98)	111-0298-69
$C_{17}H_{12}Cl_4O_5$			
Xanthen-9-one, 2,4,5,7-tetrachloro-1,3,6-trimethoxy-8-methyl-	MeOH	250(5.2),290(4.6), 345(4.2)	64-0750-69
$C_{17}H_{12}N_2$			
2-Quinolineacetonitrile, α-phenyl-	EtOH	235(4.67),276(3.73), 302(3.74),317(3.79), 424(3.08)	95-0074-69
$C_{17}H_{12}N_2O$			
2-Indolinone, 3-(indol-2-ylmethylene)-	MeOH	410(4.39)	24-1347-69
2-Indolinone, 3-(indol-3-ylmethylene)-	MeOH	410(4.21)	24-1347-69
4-Isoquinolinecarbonitrile, 2-benzyl-1,2-dihydro-1-oxo-	EtOH	255(3.94),297(4.16), 322(3.73),335(3.51)	22-2045-69
$C_{17}H_{12}N_2O_2$			
Carbamic acid, (cyanofluoren-9-ylidenemethyl)-	EtOH	256(4.54),265(4.71), 287(3.94),330(4.26)	22-1724-69

Compound	Solvent	$\lambda_{max}(\log \epsilon)$	Ref.
$C_{17}H_{12}N_2O_3$			
Benzoic acid, 2-(1,4-dihydro-1,4-dioxo-2-naphthyl)hydrazide	pH 1	435(3.48)	44-2750-69
	pH 13	555(4.06)	44-2750-69
	alkali	603(4.05)	95-0007-69
Benzoic acid, 2-(3,4-dihydro-3,4-dioxo-1-naphthyl)hydrazide	alkali	504(4.13)	95-0007-69
Indigo, 1'-hydroxy-1-methyl-	dioxan	480(3.49)	24-3691-69
$C_{17}H_{12}N_2O_4$			
Benzoic acid, p-hydroxy-, 2-(1,4-dihydro-1,4-dioxo-2-naphthyl)hydrazide	alkali	548(4.01)	95-0007-69
Benzoic acid, p-hydroxy-, 2-(3,4-dihydro-3,4-dioxo-1-naphthyl)hydrazide	alkali	464(4.26)	95-0007-69
Carbostyril, 4-methyl-3-(o-nitrobenzoyl)-	EtOH	231(4.51),284(3.95), 351(3.75)	18-2952-69
$C_{17}H_{12}N_2S_2$			
Imidazole, 4-phenyl-2,5-di-2-thienyl-	EtOH	238(4.23),320(4.42)	95-0783-69
Pyrazole, 4-phenyl-3,5-di-2-thienyl-	EtOH	255s(4.3),284(4.51)	95-0783-69
$C_{17}H_{12}N_4O$			
1H-Pyrazolo[3,4-b]pyrazin-6-ol, 1,3-diphenyl-	EtOH	250(4.24),269(4.26), 325s(4.04)	32-0463-69
$C_{17}H_{12}N_6O$			
1,4-Pentadien-3-one, 1,5-bis(p-azidophenyl)-	dioxan	357(4.65)	73-3811-69
$C_{17}H_{12}O$			
Cinnamaldehyde, α-(phenylethynyl)-, (Z)-	ether	259(3.95),322(4.08)	39-0683-69C
isomer	ether	258(4.04),329(4.08)	39-0683-69C
2,4,6,12,14-Cycloheptadecapentaene-8,10,16-triyn-1-ol, (E,E,Z,Z,E)-	ether	253(4.18),282(4.68), 291(4.78),408(3.61), 433(3.46)	35-7518-69
Pyrene, 2-methoxy-	EtOH	251(4.92),270(4.35), 281(4.37),307(4.05), 321(4.47),337(4.71)	24-2301-69
$C_{17}H_{12}OS$			
4H-1-Benzothiopyran-$\Delta^{4,\alpha}$-acetaldehyde, 2-phenyl-	MeCN	255(4.43),299(4.13), 401(4.29)	44-2736-69
4H-Pyran-4-thione, 2,6-diphenyl-	EtOH	259(4.21),326(4.27), 377(4.26)	39-0315-69C
Thione, 1-naphthyl phenyl, S-oxide, cis	$CHCl_3$	293(3.97),333(4.00)	88-4513-69
trans	$CHCl_3$	323(4.09)	88-4513-69
4H-Thiopyran-4-one, 2,6-diphenyl-	benzene	278(4.35)	18-3005-69
	EtOH	265(4.28),304(4.33)	18-3005-69
	EtOH	265(4.29),302(4.33)	39-0315-69C
	ether	276(4.35)	18-3005-69
	CH_2Cl_2	266(4.43),300(4.20)	18-3005-69
$C_{17}H_{12}OS_2$			
Acetophenone, 2-(4-phenyl-1,3-dithiol-2-ylidene)-	$CHCl_3$	238(4.28),335(3.59), 418(4.49)	24-1580-69
2-Propen-1-one, 2-phenyl-1,3-di-2-thienyl-	EtOH	224(4.07),306(4.53)	95-0783-69
$C_{17}H_{12}O_2$			
4H-1-Benzopyran-$\Delta^{4,\alpha}$-acetaldehyde, 2-phenyl-	MeCN	244(4.34),295(4.20), 374(4.33)	44-2736-69

Compound	Solvent	λ_{max}(log ε)	Ref.
Cyclopenta[def]cyclopropa[l]phenanthrene-8-carboxylic acid, 4,7b,8,8a-tetrahydro-, exo	EtOH	217(4.56),230(4.56), 236s(4.47),278(4.15)	39-1427-69C
17H-Cyclopenta[a]phenanthren-17-one, 15,16-dihydro-3-hydroxy-	EtOH	270s(4.94),278(5.02), 292s(4.30),324(4.08), 366(3.28),400(2.91)	39-2484-69C
17H-Cyclopenta[a]phenanthren-17-one, 15,16-dihydro-6-hydroxy-	EtOH	274(4.79),285s(4.54), 386(3.45)	39-2484-69C
17H-Cyclopenta[a]phenanthren-17-one, 15,16-dihydro-16-hydroxy-	EtOH	265(4.95),284(4.49), 296(4.39),335(3.15), 351(3.32),369(3.33)	39-2484-69C
1,3-Indadione, 2-(o-methylbenzylidene)-	C_6H_{12}	243(4.31),262(4.30), 342(4.37),354(4.40), 371(4.18),405s(2.95), 440s(2.57)	108-0099-69
	MeCN	244(4.39),260(4.30), 343(4.36),420s(2.95)	108-0099-69
1,3-Indandione, 2-(p-methylbenzylidene)-	C_6H_{12}	243(4.36),261(4.26), 342(4.48),353(4.60), 370(4.45),405s(3.06), 435s(2.65)	108-0099-69
	MeCN	245(4.36),263(4.24), 349(4.57),420s(3.06)	108-0099-69
Ketone, 1-hydroxy-2-naphthyl phenyl	C_6H_{12}	220(4.45),262(4.49), 299(4.02),310s(3.61), 381(3.83)	19-0253-69
	MeOH	204s(4.32),220(4.45), 262(4.48),298(4.01), 310s(3.81),379(3.84), 384s(3.84)	19-0253-69
	aq. MeOH	219(4.44),265(4.50), 300(4.05),310(3.97), 388(3.86)	19-0253-69
Ketone, 2-hydroxy-1-naphthyl phenyl	C_6H_{12}	228(4.75),247(4.04), 259(4.09),321(3.72), 370(3.77)	19-0253-69
	MeOH	208(4.48),226(4.83), 250(4.24),276s(3.83), 288(3.73),320s(3.42), 334(3.49),360s(3.12)	19-0253-69
	aq. MeOH	226(4.85),256(4.23), 288(3.80),320s(3.43), 330(3.48),360s(3.17)	19-0253-69
Ketone, 4-hydroxy-1-naphthyl phenyl	C_6H_{12}	212(4.60),235(4.56), 250s(4.31),320(3.97), 330s(3.89)	19-0253-69
	MeOH	210(4.63),236(4.52), 254s(4.26),339(3.98)	19-0253-69
	aq. MeOH	210(4.66),234(4.51), 252(4.30),342(4.02), 348(4.02)	19-0253-69
4H-Pyran-4-one, 2,6-diphenyl-	EtOH	257(4.36),283(4.41)	39-0315-69C
$C_{17}H_{12}O_3$			
1,3-Indandione, 2-(m-methoxybenzylidene)-	C_6H_{12}	242(4.30),254(4.29), 262(4.33),342(4.35), 370(4.23),425s(2.70)	108-0099-69
	MeCN	247(4.32),259(4.32), 336(4.44),375(4.13)	108-0099-69

Compound	Solvent	λ_{max}(log ε)	Ref.
1,3-Indandione, 2-(o-methoxybenzylidene)-	MeCN	244(4.32),260(4.29), 333(4.18),388(4.30)	108-0099-69
	dioxan	245(4.40),333(4.25), 378(4.41)	65-1377-69
1,3-Indandione, 2-(p-methoxybenzylidene)-	C_6H_{12}	238(4.30),248(4.39), 255(4.36),378(4.58), 428s(3.30)	108-0099-69
	MeCN	238(4.35),248(4.45), 254(4.43),383(4.60)	108-0099-69
	dioxan	242(4.53),380(4.72)	65-1373-69
Spiro[1,3-benzodioxole-2,1'(4'H)-naphthalen]-4'-one, 5-methyl-	C_6H_{12}	265(3.89),279(3.92), 284(3.92),290(3.89), 296(3.78)	39-1982-69C
$C_{17}H_{12}O_5$			
Anhydropisatin	EtOH	215(4.49),232s(4.22), 243s(4.13),251(4.09), 294(3.82),340(4.54), 359(4.57)	18-1408-69
6H-Benzofuro[3,2-c][1]benzopyran-6-one, 3,9-dimethoxy-	EtOH	244(4.23),303(3.81), 342(4.16)	39-1109-69C
$C_{17}H_{12}O_6$			
2H-Naphtho[2,3-b]pyran-8,9-dicarboxylic acid, 2-oxo-, methyl ester	EtOH	233(4.52),264(4.50), 272s(4.43),314(4.32), 327(4.40)	12-1721-69
3,3'-Spirobiphthalide, 4,6-dimethoxy-	MeOH	252(3.65),308(3.63)	22-2370-69
$C_{17}H_{12}O_7$			
2-Anthroic acid, 9,10-dihydro-1,6-dihydroxy-8-methoxy-3-methyl-9,10-dioxo-	EtOH	247(4.31),286(3.36), 432(3.95)	24-4104-69
	EtOH-NaOH	234s(4.56),302s(4.30), 314(4.36),516(3.98)	24-4104-69
$C_{17}H_{12}O_8$			
Aflatoxin GM_1	MeOH	235(4.33),262(4.21), 358(4.08)	119-0107S-69
Dermorubin	EtOH	233(4.53),279(4.30), 266(4.40),276s(4.47), 292(4.50),524s(4.14), 546(4.18),575s(4.09)	24-4104-69
1H,2H-Furo[3',2':4,5]furo[2,3-h]pyrano-[3,4-c][1]benzopyran-1,12-dione, 3,4,7a,10a-tetrahydro-10a-hydroxy-5-methoxy-	MeOH	235(4.33),262(4.21), 358(4.08)	78-1497-69
$C_{17}H_{12}S$			
Benzo[b]naphtho[2,3-d]thiophene, 6-methyl-	EtOH	226(4.37),243(4.54), 265(4.56),275(4.78), 283(4.13),315(3.69), 330(3.65),353(3.33), 376(3.44)	4-0885-69
Benzo[b]naphtho[2,3-d]thiophene, 7-methyl-	EtOH	227(4.38),246(4.54), 265(4.56),275(4.92), 283(4.34),314(3.66), 330(3.53),353(3.25), 370(3.37)	4-0885-69

Compound	Solvent	λ_{max}(log ϵ)	Ref.
Benzo[b]naphtho[2,3-d]thiophene, 8-methyl-	EtOH	228(4.37),245(4.62), 265(4.59),275(4.9), 282(4.46),318(3.7), 332(3.68),347(3.44), 367(3.25)	4-0885-69
1,2-Benzo-6-thiaphenanthrene, 5,6-dihydro-	EtOH	225(4.46),239(4.53), 275(2.74),313(3.87), 330s(3.79)	2-0637-69
6-Thiachrysene, 5,6-dihydro-	EtOH	225(4.33),264s(4.25), 272s(4.31),278(4.34), 325(3.45),335s(3.43), 355s(3.31)	2-0637-69
$C_{17}H_{12}S_2$			
2H-Thiopyran-2-thione, 3,6-diphenyl-	C_6H_{12}	251(4.48),336(3.93), 472(3.86)	78-1441-69
	EtOH	251(4.44),330(3.99), 472(3.91)	78-1441-69
2H-Thiopyran-2-thione, 4,6-diphenyl-	C_6H_{12}	245(4.35),302(4.46), 336s(3.96),474(3.88)	78-1441-69
	EtOH	245(4.16),308(4.17), 477(3.64)	48-0061-69
	EtOH	243(4.27),311(4.49), 474(3.87)	78-1441-69
2H-Thiopyran-2-thione, 5,6-diphenyl-	C_6H_{12}	245(4.36),340(4.17), 454(3.86)	78-1441-69
	EtOH	235(4.40),338(4.13), 454(3.94)	78-1441-69
4H-Thiopyran-4-thione, 2,6-diphenyl-	EtOH	245(4.20),310(4.17), 405(3.93)	39-0315-69C
$C_{17}H_{13}BF_4$			
8-Methyl-4H-cyclohepta[def]fluorenylium tetrafluoroborate	20N H_2SO_4	245(4.10),287(4.48), 291s(4.32),325(4.20), 474(3.86)	39-1427-69C
$C_{17}H_{13}BrN_2O_3S$			
Spiro[2H-1-benzopyran-2,2'-benzothiazoline], 6-bromo-3,3'-dimethyl-8-nitro-	$(MeOCH_2)_2$	248s(--),273(4.39), 302s(--),322s(--)	78-3251-69
$C_{17}H_{13}BrN_2O_5S$			
2-Isoxazolin-5-one, 3-(p-acetamidophenyl)-4-(p-bromophenylsulfonyl)-	$HOCH_2CH_2OH$	229(4.29),260(4.24)	55-0859-69
$C_{17}H_{13}BrO_6$			
Anthraquinone, 1-bromo-2,4-dihydroxy-5,7-dimethoxy-3-methyl-	EtOH	227(4.50),278(4.41), 306(4.10),425(3.96)	39-2763-69C
$C_{17}H_{13}Br_2NOS$			
Spiro[2H-1-benzopyran-2,2'-benzothiazoline], 6,8-dibromo-3,3'-dimethyl-	$(MeOCH_2)_2$	237(4.70),272(4.01), 283(3.97),303(3.82), 330s(--)	78-3251-69
$C_{17}H_{13}Cl$			
Anthracene, 9-allyl-10-chloro-	EtOH	252s(4.91),260(5.15), 342(3.51),358(3.84), 377(4.04),399(4.01)	78-3485-69
Naphthalene, 2-chloro-1-methyl-3-phenyl-	pentane	219(4.54),233(4.73), 276(3.84),286(3.82)	88-3867-69

Compound	Solvent	λ max (log ε)	Ref.
$C_{17}H_{13}ClN_2O_2$			
1H-1,4-Benzodiazepine-2,5-dione, 3-benzylidene-1-chloro-3,4-dihydro-4-methyl-, trans	n.s.g.	213(4.60),243(4.21), 283(4.12)	44-1359-69
3-Quinolineacetamide, 6-chloro-1,2-dihydro-2-oxo-4-phenyl-	EtOH	234s(--),239(4.61), 279(3.78),338(3.85), 352s(--)	4-0599-69
$C_{17}H_{13}ClN_2O_3S$			
Spiro[2H-1-benzopyran-2,2'-benzothiazoline], 6'-chloro-3,3'-dimethyl-6-nitro-	$(MeOCH_2)_2$	229(4.60),266(4.42), 312(4.06),340s(--)	78-3251-69
$C_{17}H_{13}ClN_2S$			
4H-Thiopyran-4-one, 3-chloro-2,6-diphenyl-, hydrazone	EtOH	240(4.37),331(4.20)	39-0315-69C
$C_{17}H_{13}ClO$			
Naphthalene, 1-(o-chlorophenyl)-5-methoxy-	MeOH	290s(3.89),298(3.94), 308s(3.86),314s(3.79), 323(3.66)	35-6049-69
$C_{17}H_{13}ClO_4S$			
3,4-Dihydro-1,2-benzo-6-thiaphenanthrenium perchlorate	HOAc-1% $HClO_4$	258s(3.98),275(4.25), 367(3.77),438(3.30)	2-0637-69
5,6-Dihydro-12-thiabenz[a]anthracenium perchlorate	HOAc-1% $HClO_4$	255s(4.04),269(4.23), 315s(4.13),345s(3.62), 431(4.33)	2-0737-69
$C_{17}H_{13}Cl_3O_5$			
Xanthen-9-one, 2,4,7-trichloro-1,3,6-trimethoxy-8-methyl-	MeOH	246(4.60),302(4.23), 340(3.79)	64-0756-69
$C_{17}H_{13}F$			
Naphthalene, 2-fluoro-1-methyl-3-phenyl-	pentane	212(4.59),245(3.70), 284(3.99)	88-3867-69
$C_{17}H_{13}I_2NOS$			
Spiro[2H-1-benzopyran-2,2'-benzothiazoline], 6,8-diiodo-3,3'-dimethyl-	$(MeOCH_2)_2$	246(4.88),275(4.22), 285(4.19),304s(--), 326s(--)	78-3251-69
$C_{17}H_{13}I_2NO_5$			
1-Oxaspiro[4.5]deca-6,9-diene-3-carbamic acid, 7,9-diiodo-2,8-dioxo-, benzyl ester	EtOH	262(3.62),295(3.56)	69-1844-69
$C_{17}H_{13}N$			
Aniline, N-(1-naphthylmethylene)-	EtOH	222(4.5),249s(4.2), 330(4.1)	28-0248-69B
Aniline, N-(2-naphthylmethylene)-	EtOH	212(4.3),256(4.5), 284(4.2),294(4.2), 333s(4.1)	28-0248-69B
8,16-Imino[2,2]metacyclophane, N-methyl-	C_6H_{12}	228s(4.50),253s(4.18), 304(4.10)	35-1672-69
1-Pyrenamine, N-methyl-	C_6H_{12}	231(4.53),243(4.62), 277s(4.23),292(4.26), 320s(3.63),367(4.24), 382(4.22),386(4.22), 404s(4.25),407(4.26)	35-1672-69

Compound	Solvent	λ max (log ε)	Ref.
$C_{17}H_{13}NO$			
Acetophenone, 2-(2-quinolyl)-	EtOH	225(4.45),265(4.00), 302(4.00),325(4.05), 427(4.44),452(4.37)	95-0074-69
Benzofuro[2,3-g]isoquinoline, 5,11-dimethyl-	EtOH	264(4.80),269(4.82), 273(4.80),283(4.14), 333(3.96),349(3.92)	4-0379-69
	pH 1	229s(--),283(4.79), 332(3.67)	4-0379-69
Benzofuro[3,2-g]isoquinoline, 5,11-dimethyl-	EtOH	221(4.45),264(4.90), 293s(--),305(4.05), 318(4.16),356(3.89)	4-0379-69
	pH 1	225(4.38),276(4.70), 333(4.13)	4-0379-69
5H-Dibenzo[a,d]cycloheptene-5-carbonitrile, 10-methoxy-	EtOH	288(4.10)	23-2827-69
4(1H)-Pyridone, 2,6-diphenyl-	EtOH	249(4.51),280(4.18)	18-2389-69
$C_{17}H_{13}NOS$			
8H-Thieno[2,3-b]indole, 2-(p-methoxyphenyl)-	EtOH	245(4.79),287(4.41), 294(4.44),332(4.60)	95-0058-69
2H-Thiopyran-2-one, 4,6-diphenyl-, oxime	EtOH	268(4.35),402(3.53)	48-0061-69
4H-Thiopyran-4-one, 2,6-diphenyl-, oxime	EtOH	247(4.36),322(4.14)	39-0315-69C
$C_{17}H_{13}NO_2$			
Indole, 5-acetyl-N-benzoyl-	EtOH	236(4.45),260(4.44), 312s(3.67)	2-0848-69
1(4H)-Naphthalenone, 2-methoxy-4-(phenylimino)-	pH 7.0	246(4.74),251(4.74), 291(4.93),335(4.40)	78-5807-69
1,2-Naphthoquinone, 4-(N-methylanilino)-	pH 7.0	247(5.02),278(4.83), 323(4.57)	78-5807-69
1,2-Naphthoquinone, 4-m-toluidino-	pH 4.0	240(5.03),274(4.79)	78-5807-69
	pH 11.0	238(5.10),278(4.85)	78-5807-69
4H-Pyran-4-one, 2,6-diphenyl-, oxime	EtOH	248(4.38),272(4.35), 308(4.14)	39-0315-69C
Spiro[indan-2,4'(1'H)-isoquinoline]-1',3'(2'H)-dione	EtOH	243(4.05),267(3.45), 275(3.46),285(3.35), 292(3.33)	44-3664-69
$C_{17}H_{13}NO_2S$			
5-Acetyl-4-hydroxy-2,3-diphenylthiazolium hydroxide, inner salt	MeOH	200(4.25),237s(3.98), 263(4.10),401(4.04)	77-1128-69
[1]Benzothieno[2,3-g]isoquinoline, 5,11-dimethyl-, 6,6-dioxide	EtOH	260(4.66),291(4.05), 303(4.18),315(4.18)	4-0389-69
$C_{17}H_{13}NO_3$			
Maleamic acid, N-fluoren-1-yl-	EtOH	246s(4.18),257s(4.31), 264(4.36),275s(4.23), 283s(4.05),299s(3.92)	39-0345-69C
1,2-Naphthoquinone, 4-m-anisidino-	pH 9.0	240(5.06),281(4.81)	78-5807-69
$C_{17}H_{13}NO_4$			
1,4-Isoquinolinedione, 3-acetoxy-2,3-dihydro-3-phenyl-	EtOH	220(4.55),297s(3.28)	33-1810-69
$C_{17}H_{13}NO_5$			
Carbazole-1-carboxylic acid, 9-acetyl-2-hydroxy-, acetate	EtOH	225(4.5),235(4.5), 270(4.2)	39-1518-69C

Compound	Solvent	$\lambda_{max}(\log \epsilon)$	Ref.
$C_{17}H_{13}NO_5S$			
Spiro[benzo[b]thiophene-3(2H),1'-isoindoline]-2-carboxylic acid, 3'-oxo-, methyl ester, 1,1-dioxide	EtOH	202(4.86),267(3.36), 275(3.36),317(3.08)	39-1149-69C
$C_{17}H_{13}NO_6$			
2-Anthramide, 9,10-dihydro-1,3-dihydroxy-8-methoxy-5-methyl-9,10-dioxo-	n.s.g.	228(4.26),285(3.88), 448(3.42)	39-2805-69C
Benzoic acid, 3,5-dimethoxy-2-(1-oxo-1H-2,3-benzoxazin-4-yl)-	MeOH	228(4.31),294(3.74), 304(3.73)	22-2370-69
$C_{17}H_{13}NS$			
[1]Benzothieno[2,3-g]isoquinoline, 1,5-dimethyl- (thiaolivacine)	EtOH	230(4.49),239(4.47), 257(4.43),270s(4.62), 280(4.83),286(4.84)	4-0875-69 +77-0598-69
[1]Benzothieno[3,2-g]isoquinoline, 5,11-dimethyl-	EtOH	243(4.47),267(4.71), 276(4.94),302(3.85), 312(3.82)	4-0379-69
8H-Thieno[2,3-b]indole, 8-methyl-2-phenyl-	EtOH	241(4.33),288s(3.81), 294(3.84),338(4.07)	95-0058-69
$C_{17}H_{13}N_3$			
2,4,6-Cycloheptatriene-$\Delta^{1,\alpha}$-malononitrile, 4-p-toluidino-	MeOH	244(4.10),340(3.99), 476(4.57)	25-0075-69
$C_{17}H_{13}N_3O$			
Anthranilonitrile, N-(α-cyanobenzylidene)-5-ethoxy-	EtOH	214(4.46),278(4.13), 391(3.92)	22-2008-69
$C_{17}H_{13}N_3O_3$			
Isonicotinic acid, 2-(1,4-dihydro-3-methyl-1,4-dioxo-2-naphthyl)hydrazide	alkali	624(3.99)	95-0007-69
Oxazole, 4-methyl-5-[(p-nitrobenzylidene)amino]-2-phenyl-	n.s.g.	396(4.46)	22-4108-69
$C_{17}H_{13}N_3O_4$			
1H-Azirino[1,2-a]cyclopropa[c]quinoline-1,7-dicarboxylic acid, 1,7-dicyano-6bα,7,7aα,7bβ-tetrahydro-, dimethyl ester	n.s.g.	250s(3.82)	88-5337-69
Benzimidazo[2,1-a]isoquinoline, 2,3-dimethoxy-10-nitro-	EtOH	227(4.17),262(3.37), 304(4.35)	24-1529-69
2-Propanone, 1-[3-(p-nitrophenyl)-4-phenyl-Δ^2-1,2,4-oxadiazolin-4-ylidene]-	EtOH	250(4.08),292(4.57)	18-3008-69
$C_{17}H_{13}N_5O_2$			
as-Triazino[2,3-a]indole-10-carbonitrile, 6,7,8,9-tetrahydro-2-(p-nitrophenyl)-	CH_2Cl_2	263(4.13),283s(4.07), 339(4.26),461(4.11), 477s(4.09),514s(3.70)	48-0388-69
$C_{17}H_{13}N_5O_4$			
Quinoline, 1,2-dihydro-1-methyl-6-nitro-2-[[(p-nitrophenyl)azo]methylene]-	EtOH-acid	472(1.72)	124-0370-69
$C_{17}H_{14}$			
Anthracene, 9-allyl-	EtOH	252s(4.93),255(4.99), 333(3.41),347(3.62), 365(3.76),385(3.74)	78-3485-69

Compound	Solvent	λ_{max}(log ε)	Ref.
9,10-Cyclopentenophenanthrene	EtOH	213(4.17),220(4.06), 239(4.20),245(4.27), 254(4.36),270(3.88), 280(3.73),289(3.68), 300(3.76),339(2.87), 354(2.79)	44-0763-69
Phenanthrene, 9-propenyl-, trans	pentane	239(4.58),255(4.42), 266(4.42),314(4.30), 328(4.49),342(4.45)	88-3867-69
$C_{17}H_{14}BrNO_4$			
Benzoic acid, 2-(3-bromo-3-phenylpyruv-amido)-, methyl ester	EtOH	218(4.45),250(4.06), 308(3.75)	35-6083-69
Pyrrolo[1,2-a]quinoline-3-acetic acid, 7-bromo-2-carboxy-, dimethyl ester	MeOH	210(4.50),232(4.44), 251(4.61),261s(4.57), 269s(4.53),278s(4.30), 290(3.99),346s(3.93), 359(3.99),373s(3.85)	39-2311-69C
$C_{17}H_{14}BrNO_5$			
1,3-Dioxolo[4,5-c]acridin-6(11H)-one, 8-bromo-4,5-dimethoxy-11-methyl- (7-bromomelicopine)	EtOH	216(4.34),254(4.40), 280(4.71),310(4.12), 420(3.73)	12-1503-69
$C_{17}H_{14}BrN_3OS_2$			
Rhodanine, 3-benzyl-5-[(2-bromo-p-tol-yl)azo]-	$CHCl_3$	450(4.53)	2-0129-69
$C_{17}H_{14}Br_2O_2$			
1,4-Butanedione, 2,3-dibromo-1-phenyl-4-p-tolyl-	EtOH	267(4.45)	104-0511-69
$C_{17}H_{14}Br_2O_3$			
1,4-Butanedione, 2,3-dibromo-1-(p-meth-oxyphenyl)-4-phenyl-	EtOH	265(4.08),296(4.24)	104-0511-69
$C_{17}H_{14}ClIN_4O_2$			
2-Formyl-1-methyl-6-nitroquinolinium iodide, (p-chlorophenyl)hydrazone	EtOH	275(3.63),313(3.3), 492(1.6)	124-0370-69
$C_{17}H_{14}ClN$			
Pyrrole, 3-(p-chlorophenyl)-1-methyl-4-phenyl-	EtOH	247(4.34),271s(--)	88-4875-69
$C_{17}H_{14}ClNO_2$			
1,3-Pentadiene, 1-(p-chlorophenyl)-4-(p-nitrophenyl)-	EtOH	296(4.16),379(4.36)	104-0111-69
$C_{17}H_{14}Cl_2$			
Anthracene, 10-(dichloromethylene)-9,10-dihydro-9,9-dimethyl-	C_6H_{12}	268(4.23)	44-3093-69
1,3-Pentadiene, 1,4-bis(p-chlorophenyl)-	EtOH	325(4.45)	104-0111-69
$C_{17}H_{14}Cl_2N_2O$			
1(2H)-Phthalazinone, 4-(p-chlorophenyl)-2-(3-chloropropyl)-	EtOH	245(4.24),295(4.02)	44-2715-69
$C_{17}H_{14}FNO_3S_2$			
Spiro[2H-1-benzopyran-2,2'-benzothiazo-line]-6-sulfonyl fluoride, 3,3'-di-methyl-	$(MeOCH_2)_2$	300(4.00)	78-3251-69

Compound	Solvent	λ_{max}(log ε)	Ref.
$C_{17}H_{14}FeOS$			
2-Propen-1-one, 1-ferrocenyl-3-(2-thien-yl)-	EtOH	235(4.01),336(4.35)	73-2235-69
2-Propen-1-one, 3-ferrocenyl-1-(2-thien-yl)-	EtOH	275(4.00),330(4.35)	73-2235-69
$C_{17}H_{14}FeO_2$			
2-Propen-1-one, 1-ferrocenyl-3-(2-fur-yl)-	heptane	318(4.89),380(3.94)	73-2235-69
	EtOH	235(4.03),340(4.38)	73-2235-69
2-Propen-1-one, 3-ferrocenyl-1-(2-fur-yl)-	heptane	315(4.73)	73-2235-69
	EtOH	330(4.31)	73-2235-69
$C_{17}H_{14}N_2$			
4-Isoquinolinecarbonitrile, 2-benzyl-1,2-dihydro-	EtOH	220(4.21),245(4.19), 345(4.06)	22-2045-69
1-Naphthaldehyde, phenylhydrazone	EtOH	211(4.6),231(4.4), 260(3.9),364(4.3)	28-0248-69B
2-Naphthaldehyde, phenylhydrazone	EtOH	204(4.4),232(4.6), 248s(4.3),282(4.1), 291(4.0),354(4.5)	28-0248-69B
7H-Pyrido[2,3-c]carbazole, 3,5-dimethyl-	EtOH	227(4.49),246(4.38), 275(4.52),333(3.87)	12-0185-69
$C_{17}H_{14}N_2O$			
2-Furaldehyde, diphenylhydrazone	EtOH	238(4.1),277s(3.9), 342(4.4)	28-0137-69B
Oxazole, 5-(benzylideneamino)-4-methyl-2-phenyl-	n.s.g.	286s(--),368(4.52)	22-4108-69
Propiophenone, α-(1-phthalazinyl)-	EtOH	218(4.72),247(4.15), 275s(3.76),306(3.13)	95-0959-69
Pyrazole, 1-benzoyl-3-methyl-4-phenyl-	MeOH	233(4.23),283(4.04)	44-3639-69
$C_{17}H_{14}N_2OS$			
2-Pyrazolin-5-one, 3-methyl-1-phenyl-4-[(phenylthio)methylene]-	hexane	250(4.38),324(4.40)	104-1249-69
	EtOH	250(4.57),330(4.41)	104-1249-69
$C_{17}H_{14}N_2O_2$			
1H-1,4-Benzodiazepine-2,5-dione, 3-benz-ylidene-3,4-dihydro-4-methyl-, cis	EtOH	215(4.64),245(4.26), 276(4.19)	44-1359-69
trans	EtOH	213(4.57),239(4.20), 255(4.15),286(4.09)	44-1359-69
Carbamic acid, (1-cyano-2,2-diphenyl-vinyl)-, methyl ester	EtOH	293(4.11)	22-1724-69
Carbostyril, 3-anthraniloyl-4-methyl-	EtOH	230(4.62),270(4.08), 335(3.87),350(3.88), 380(3.89)	18-2952-69
5,8-Isoquinolinedione, 7-(benzylamino)-3-methyl-	MeCN	233(4.46),275(4.34), 323(3.79)	4-0697-69
2-Pyrazolin-5-one, 4-benzoyl-3-methyl-1-phenyl-	H_2O	255(4.13),260(4.08)	123-0074-69C
	EtOH	270(4.74)	123-0074-69C
	iso-AmOH	275(4.24)	123-0074-69C
	$CHCl_3$	275(4.24)	123-0074-69C
4(3H)-Quinazolinone, 2-(α,β-epoxyphen-ethyl)-3-methyl-	EtOH	228(4.51),275(4.01), 303(3.67),315(3.56)	35-6083-69
$C_{17}H_{14}N_2O_3$			
Cyclopenine, (±)-	EtOH	211(4.58),290(3.34)	44-1359-69
6H-Indolo[3,2,1-de][1,5]naphthyridine-5-carboxylic acid, 4-hydroxy-6-meth-yl-, methyl ester	MeOH	245(4.39),285(4.08), 322(3.80),375(3.96), 383(3.91)	44-4199-69

Compound	Solvent	λ_{max}(log ε)	Ref.
Isocyclopenine, (±)-	EtOH	212(4.34),290(4.53)	44-1359-69
$C_{17}H_{14}N_2O_3S$			
Spiro[2H-1-benzopyran-2,2'-benzothiazoline], 3,3'-dimethyl-6-nitro-	$(MeOCH_2)_2$	253s(--),269(4.45), 303(4.14),336s(--)	78-3251-69
Spiro[2H-1-benzopyran-2,2'-benzothiazoline], 3,3'-dimethyl-8-nitro-	$(MeOCH_2)_2$	249s(--),291s(--), 338(3.55)	78-3251-69
5H-Thiazolo[3,2-a]pyridine-5-carboxylic acid, 6-cyano-2,3-dihydro-5-oxo-7-phenyl-, ethyl ester	EtOH	211(4.2),220s(4.2), 270(4.1),278(4.1), 345s(4.2),357(4.2)	24-0522-69
	0.09N NaOH	222(4.2),265(4.1), 280(4.1),340s(4.2), 350(4.2)	24-0522-69
$C_{17}H_{14}N_2O_4$			
Cyclopenol	n.s.g.	285(3.57)	44-1359-69
$C_{17}H_{14}N_2O_5$			
Cinnamic acid, α-(N-methyl-o-nitrobenzamido)-	n.s.g.	264(4.30)	44-1359-69
$C_{17}H_{14}N_2O_5S$			
2-Isoxazolin-5-one, 3-(p-acetamidophenyl)-4-phenylsulfonyl-	$HOCH_2CH_2OH$	275(4.45)	55-0859-69
$C_{17}H_{14}N_2S$			
4(1H)-Pyridinethione, 1-methyl-2,5-diphenyl-	$CHCl_3$	351(4.23)	4-0037-69
2H-Thiopyran-2-one, 4,6-diphenyl-, hydrazone	EtOH	272(4.37),405(3.63)	48-0061-69
$C_{17}H_{14}N_4$			
as-Triazino[2,3-a]indole-10-carbonitrile, 6,7,8,9-tetrahydro-2-phenyl-	CH_2Cl_2	250s(4.03),292(4.46), 337s(3.62),441(3.86), 463s(3.81),499(3.35)	48-0388-69
$C_{17}H_{14}N_4O_3$			
Formanilide, 4'-(3'-methyl-5'-oxo-1'-phenylspiro[oxaziridine-3,4'-[2]-pyrazolin]-2-yl)-	MeOH	241(4.31),280(3.93)	44-1687-69
$C_{17}H_{14}N_4O_4$			
1,3,5,7-Cyclooctatetraene-1-acrolein, 2,4-dinitrophenylhydrazone	$CHCl_3$	395(4.64)	39-0978-69C
$C_{17}H_{14}N_6O_2$			
Benzoic acid, [(1-benzamido-1H-1,2,3-triazol-4-yl)methylene]hydrazide	EtOH	207(4.14),232(4.20), 290(4.29)	39-1416-69C
Benzoic acid, [(1-benzamido-1H-1,2,3-triazol-5-yl)methylene]hydrazide	EtOH	210(4.24),228(4.17), 278(4.43)	39-1416-69C
$C_{17}H_{14}O$			
5,13-Cycloheptadecadiene-2,8,10,16-tetrayn-1-ol, (E,E)-	EtOH	227(2.53),240(2.54), 253(2.30)	35-0760-69
Cyclopropenone, di-p-tolyl-	CH_2Cl_2	227(4.33),234s(--), 240(4.24),297s(--), 308(4.51)	24-0319-69
2,4-Pentadienal, 2,5-diphenyl-	CH_2Cl_2	241(4.22),272(4.02), 352(4.61)	24-0623-69

Compound	Solvent	λ_{max}(log ϵ)	Ref.
2,4-Pentadienophenone, 5-phenyl-	EtOH	235(3.92),266(3.95), 344(4.57)	23-4076-69
$C_{17}H_{14}OS_2$			
2-Propanone, 1-(methylthio)-1-(thio-xanthen-9-ylidene)-	hexane	339s(3.82)	54-0465-69
$C_{17}H_{14}O_2$			
2-Butene-1,4-dione, 1-phenyl-4-p-tolyl-	EtOH	228(4.18),233(4.18), 275(4.28)	104-0511-69
2(3H)-Furanone, 3-(diphenylmethylene)-dihydro-	EtOH	284(4.07)	44-3792-69
14β-Gona-1,3,5(10),6,8-pentaene-11,17-dione, (±)-	MeOH	248(4.32),314(3.86)	19-0145-69
4H-Pyran-4-one, 2,3-dihydro-5,6-diphen-yl-	EtOH	248(3.86)	22-4151-69
$C_{17}H_{14}O_2S$			
2-Propanone, 1-(methylthio)-1-(xanthen-9-ylidene)-	hexane	336(4.00)	54-0465-69
$C_{17}H_{14}O_2S_2$			
Acetic acid, [(4-phenyl-2H-1-benzothio-pyran-2-yl)thio]-	EtOH	235(4.25),251(4.33), 325s(3.08)	2-0315-69
$C_{17}H_{14}O_3$			
2H-1-Benzopyran-3-ol, 4-phenyl-, acetate	MeOH	269(3.94)	2-0196-69
2-Butene-1,4-dione, 1-(p-methoxyphen-yl)-4-phenyl-	EtOH	250(4.20)	104-0511-69
Cyclopropenone, bis(p-methoxyphenyl)-	CH_2Cl_2	233s(--),256(4.11), 296(4.41),315(4.47), 342(4.20)	24-0319-69
9-Phenanthrenecarboxylic acid, 10-hydr-oxy-, ethyl ester	EtOH	246(4.63),255(4.36), 261(4.33),270(4.12), 292(3.85),314(3.77), 325(3.70),326(3.85), 360(3.70)	44-1845-69
Spiro[bicyclo[4.2.1]nona-2,4,7-triene-9,7'[5]norbornene]-2',3'-dicarboxylic anhydride	dioxan	263(3.59)	24-1789-69
Spiro[bicyclo[6.1.0]nona-2,4,6-triene-9,7'[5]norbornene]-2',3'-dicarboxylic anhydride	dioxan	246(3.91)	24-1789-69
$C_{17}H_{14}O_4$			
Anthraquinone, 1,3-dimethoxy-2-methyl-	EtOH-acid	239(4.18),245(4.17), 276(4.55),350(3.63)	102-0501-69
3,6(2H,5H)-Benzofurandione, 2-benzyli-dene-4-hydroxy-5,5-dimethyl-	MeOH	390(4.3)	2-0540-69
3(2H)-Benzofuranone, 2-benzylidene-6,7-dimethoxy-	MeOH	242(4.2),340(4.5)	2-0543-69
2H-1-Benzopyran-2-carboxylic acid, 3-hydroxy-4-phenyl-, methyl ester	MeOH	292(3.80)	2-0196-69
Cyclopenta[a]phenanthren-17-one, 6,7,15,16-tetrahydro-6,7-dihydroxy-	EtOH	245(4.02),313(4.43)	39-2484-69C
Flavone, 4',7-dimethoxy-	n.s.g.	233(4.41),310(4.31)	2-0940-69
Flavone, 5,7-dimethoxy-	n.s.g.	261(4.55),308(4.00)	2-0940-69
2(3H)-Furanone, 3-[bis(p-hydroxyphenyl)-methylene]dihydro-	EtOH	230(4.21),310(4.18)	44-3792-69

Compound	Solvent	λ_{max}(log ε)	Ref.
2H-Naphtho[1,2-b]pyran-6-propionic acid, 2-oxo-, methyl ester	MeOH	268(4.4),279(4.4), 296(3.6),309(3.8)	39-2618-69C
$C_{17}H_{14}O_5$			
Anthraquinone, 1-hydroxy-3,7-dimethoxy-6-methyl- (5 λ,4 ε)	EtOH	225(4.35),281(4.56), 308(4.12),377(3.83), 410s(?)	23-0767-69
Flavone, 5-hydroxy-4',7-dimethoxy-	EtOH	270(4.47),330(4.53)	105-0397-69
	EtOH-AlCl$_3$	278(4.41),300(4.40), 337(4.47),380(4.20)	105-0397-69
Flavone, 6-hydroxy-2',6'-dimethoxy-	EtOH	323(4.88)	2-0746-69
Phenanthro[3,4-d]-1,3-dioxol-9-ol, 4,8-dimethoxy-	CHCl$_3$	308(3.98),320(3.86), 339(3.34),355(3.66), 373(3.75)	88-0067-69
Pterocarpan, 3-methoxy-8,9-(methylenedioxy)-	EtOH	281(3.91),287(3.97), 311(4.17)	39-1109-69C
Pterocarpin, (±)-	EtOH	281(3.60),287(3.66), 311(3.89)	18-1408-69
$C_{17}H_{14}O_6$			
Anthraquinone, 1,3-dihydroxy-6,8-dimethoxy-2-methyl-	EtOH	222(4.53),251(4.12), 284(4.53),374(3.68), 445(3.93)	39-2763-69C
Anthraquinone, 1,5-dihydroxy-2,3-dimethoxy-7-methyl-	EtOH	278(4.45),304(3.97), 420(4.01)	24-4104-69
	EtOH-NaOH	244(4.55),259(4.31), 286(4.20),302(4.12), 494(4.11)	24-4104-69
Anthraquinone, 1,8-dihydroxy-2,3-dimethoxy-6-methyl-	EtOH	277(4.21),305(3.78), 415(3.76),435(3.84), 447(3.79),525(2.60), 560(2.30)	24-4104-69
	EtOH-NaOH	242(4.21),262(4.16), 317(3.78),520(3.79)	24-4104-69
Benzo[b]furan, 3-(hydroxymethyl)-2-(2-hydroxy-4-methoxyphenyl)-5,6-(methylenedioxy)-	EtOH	271(4.11),322(4.36)	18-1408-69
Flavone, 3,5-dihydroxy-4',7-dimethoxy-	EtOH	269(4.46),320s(4.24), 369(4.48)	105-0397-69
	EtOH-AlCl$_3$	270(4.48),303(4.01), 350(4.19),425(4.49)	105-0397-69
Flavone, 5,7-dihydroxy-3,4'-dimethoxy-	EtOH	270(4.35),351(4.25)	105-0397-69
	EtOH-AlCl$_3$	279(4.34),303s(4.18), 343(4.28),401(4.11)	105-0397-69
Gnaphalium	EtOH	245(4.27),283(4.61), 363(3.99)	88-0431-69
Isoflavone, 2',4',5-trihydroxy-7-methoxy-6-methyl-	MeOH	265(4.34)	2-0061-69
	MeOH-NaOAc	265(4.33)	2-0061-69
Pectolinarigenin	EtOH	277(4.30),332(4.40)	78-1603-69
$C_{17}H_{14}O_7$			
Aflatoxin B$_{2a}$	MeOH	228(4.24),256(4.01), 363(4.31)	119-0107S-69
2-Epiolivinolide	EtOH	226(4.24),282(4.64), 326(3.74),420(4.22)	105-0494-69
Flavone, 3',4',6-trihydroxy-5,7-dimethoxy-	EtOH	243s(4.26),277(4.20), 336(4.43)	40-1270-69
Flavone, 3,4',7-trihydroxy-5,8-dimethoxy-	EtOH	270(4.25),294(3.98), 369(4.24)	18-2380-69

Compound	Solvent	λmax(log ε)	Ref.
Flavone, 3,4',7-trihydroxy-5,8-dimethoxy- (cont.)	EtOH-NaOH	297(4.22),316(4.24), 340(4.23)	18-2380-69
	EtOH-NaOAc	280(4.33),381(4.15)	18-2380-69
	EtOH-AlCl3	267(4.35),341(3.79), 428(4.28)	18-2380-69
Flavone, 4',5,7-trihydroxy-3,6-dimethoxy-	EtOH	271(4.16),340(4.22)	31-0349-69
Flavone, 4',5,7-trihydroxy-3,8-dimethoxy-	EtOH	274(4.36),310s(4.16), 326(4.19),360(4.15)	18-2380-69 +31-0349-69
	EtOH-NaOH	284(4.42),334(4.15), 412(4.42)	18-2380-69
	EtOH-NaOAc	284(4.42),304s(4.23), 371(4.06)	18-2380-69
	EtOH-AlCl3	284(4.30),313(4.19), 352(4.25),417(4.02)	18-2380-69
Flavone, 4',5,7-trihydroxy-3',8-dimethoxy-	EtOH	245(4.18),253(4.19), 276(4.27),343(4.28)	18-2327-69
Olivinic acid, anhydro-	EtOH	225(4.27),258(3.98), 308(4.18),362(4.01)	105-0257-69
Xanthen-9-one, 1,2,3-trimethoxy-6,7-(methylenedioxy)-	EtOH	248(4.57),273s(4.00), 314(4.28),339(4.07), 353s(4.03)	78-5295-69
$C_{17}H_{14}O_8$			
Aflatoxin G_{2a}	MeOH	223(4.27),242(4.00), 262(3.94),365(4.26)	119-0107S-69
Flavone, 3',4',5,7-tetrahydroxy-3,6-dimethoxy- (axillarin)	EtOH	259(4.25),295(3.91), 358(4.32)	18-1649-69
	EtOH-NaOEt	273(4.32),410(4.35)	18-1649-69
	EtOH-NaOAc	272(4.27),300(3.95), 367(4.24)	18-1649-69
	EtOH-AlCl3	278(--),410(--)	18-1649-69
Flavone, 3',4',5,7-tetrahydroxy-3,8-dimethoxy-	EtOH	262(4.34),275(4.31), 298(3.98),368(4.27)	18-1649-69
	EtOH-NaOAc	266s(4.27),277(4.34), 374(4.17)	18-1649-69
	EtOH-AlCl3	283(--),380s(--), 440(--)	18-1649-69
$C_{17}H_{15}BrN_2O_3$			
1H-1,4-Benzodiazepine-2,5-dione, 3-(α-bromobenzyl)-3,4-dihydro-3-hydroxy-4-methyl-	n.s.g.	217(4.46),295(3.43)	35-6083-69
Pyruvanilide, 3-bromo-2'-(methylcarbamoyl)-3-phenyl-	n.s.g.	250(4.09),298(3.65)	35-6083-69
$C_{17}H_{15}BrO_2$			
4,7-Methanoinden-1α-ol, 3aα,4β,7β,7aα-tetrahydro-, p-bromobenzoate	isooctane	198(4.60),243(4.35), 270(3.03),273s(--), 282(2.77)	39-0710-69B
$C_{17}H_{15}BrO_7$			
1,4-Naphthalenediol, 2-bromo-3-(hydroxymethoxy)-, triacetate	EtOH	232(4.87),285(3.74), 297(3.60),326(2.65)	77-0167-69
$C_{17}H_{15}Cl$			
Anthracene, 9-chloro-10-isopropyl-	EtOH	253s(4.84),259(4.11), 341(3.44),358(3.71), 376(3.96),397(3.94)	78-3485-69

Compound	Solvent	λ_{max}(log ε)	Ref.
Anthracene, 9-chloro-10-propyl-	EtOH	253s(4.90),260(5.22), 342(3.53),358(3.84), 377(4.04),399(4.02)	78-3485-69
1,3-Pentadiene, 4-(p-chlorophenyl)-1-phenyl-	EtOH	325(4.47)	104-0111-69
$C_{17}H_{15}ClFeO_3$			
4-Penten-1-one, 5-chloro-1,5-(1,1'-ferrocenediyl)-3-hydroxy-, acetate	n.s.g.	228s(4.19),253s(3.87), 284(3.59),335(3.13), 464(2.53)	78-0861-69
$C_{17}H_{15}ClIN_3$			
2-Formyl-1-methylquinolinium iodide, (p-chlorophenyl)hydrazone	EtOH	262(3.25),315(2.79), 462(1.4)	124-0370-69
$C_{17}H_{15}ClN_2$			
1H-Isoindole, 5-chloro-1-[(dimethylamino)methylene]-3-phenyl-	iso-PrOH	253(4.49),320s(3.80), 400(4.59)	44-0649-69
	20% iso-PrOH-HCl	255(4.37),273(4.28), 290s(4.21),413(4.59)	44-0649-69
$C_{17}H_{15}ClN_2O$			
1,5-Benzodiazocin-2(1H)-one, 8-chloro-3,4-dihydro-1-methyl-6-phenyl-	MeOH	243(4.24)	36-0830-69
	EtOH	245(4.23)	44-0183-69
1H-Isoindole-1-carboxamide, 5-chloro-N,1-dimethyl-3-phenyl-	iso-PrOH	252(4.14)	44-0649-69
Isoindole-1-carboxamide, 5-chloro-N,2-dimethyl-3-phenyl-	iso-PrOH	226(4.39),262(4.51), 297(3.88),358(4.20)	44-0649-69
$C_{17}H_{15}ClN_2OS$			
2(1H)-Quinazolinethione, 3-allyl-6-chloro-3,4-dihydro-4-hydroxy-4-phenyl-	n.s.g.	261(3.97),291(4.44)	40-0917-69
$C_{17}H_{15}ClN_2O_2$			
2(1H)-Quinazolinone, 3-allyl-6-chloro-3,4-dihydro-4-hydroxy-4-phenyl-	n.s.g.	255(4.22),298(3.36)	40-0917-69
$C_{17}H_{15}ClN_2O_3$			
4-(3-Anthranilyl)-6,7-dimethoxyisoquinolinium chloride	EtOH	247(4.39),312(4.10)	78-5365-69
Benzamide, 2-(3-chloro-3-phenylpyruvamido)-N-methyl-	n.s.g.	216(4.33),250(4.12), 298(3.57)	35-6083-69
$C_{17}H_{15}ClN_2O_6$			
4-Acetamido-2,3-diphenyloxazolium perchlorate	MeCN	244(3.96),308(3.96)	18-2310-69
2-Methyl-4-(p-nitrobenzyl)isoquinolinium perchlorate	EtOH	232(4.23),270(3.62), 340(3.37)	78-0101-69
$C_{17}H_{15}ClO_4S$			
1,2,3,4-Tetrahydro-9,10-benzo-6-thiaphenanthrenium perchlorate	HOAc-1% $HClO_4$	254(4.59),301s(4.30), 309(4.37),440(4.02)	2-0637-69 +2-0737-69
8,9,10,11-Tetrahydro-12-thiabenz[a]anthracenium perchlorate	HOAc-1% $HClO_4$	250(4.23),302(4.11), 395(3.66)	2-0737-69
1,2,3,4-Tetrahydro-6-thiachrysenium perchlorate	HOAc-1% $HClO_4$	252(4.06),298(4.14), 390(3.64)	2-0637-69
$C_{17}H_{15}ClO_5S$			
Spiro[benzo[b]thiophene-2(3H),1'-[2,5]-cyclohexadiene]-3,4'-dione, 7-chloro-2',4,6-trimethoxy-6'-methyl-	MeOH	235(4.64),306(4.27), 348(3.66)	44-1463-69

Compound	Solvent	$\lambda_{max}(\log \epsilon)$	Ref.
$C_{17}H_{15}Cl_2N$			
Isoquinoline, 4-(2,6-dichlorobenzyl)-1,2-dihydro-2-methyl-	MeOH	223(4.21),240(4.10),330(3.94)	28-1550-69B
$C_{17}H_{15}FN_2O$			
3-Quinolinecarboxamide, 1-(p-fluorobenzyl)-1,4-dihydro-	MeOH	234(3.95),340(3.76)	88-3117-69
$C_{17}H_{15}FN_2O_2$			
Indole-3-acetic acid, 5-fluoro-2-(3-pyridyl)-, ethyl ester	EtOH	221(4.38),309(4.31)	22-4154-69
$C_{17}H_{15}F_3N_2O_4S$			
5-Thia-1-azabicyclo[4.2.0]oct-2-ene-2-carboxylic acid, 3-methyl-8-oxo-7-[2-(α,α,α-trifluoro-m-tolyl)acetamido]-, potassium salt	H_2O	260(3.92)	87-0310-69
$C_{17}H_{15}Fe$			
α-Ferrocenylbenzylium cation	58% H_2SO_4	227(4.03),252(4.00),280(3.97),329(4.08),405(3.50)	35-0509-69
$C_{17}H_{15}IN_4O_2$			
2-Formyl-1-methyl-6-nitroquinolinium iodide, phenylhydrazone	EtOH	274(3.38),314(2.97),490(1.62)	124-0370-69
$C_{17}H_{15}I_2NO_5$			
Tyrosine, N-carboxy-3,5-diiodo-, N-benzyl ester	EtOH	287(3.50),294(3.45)	69-1844-69
$C_{17}H_{15}N$			
Acridine, 9-(1-butenyl)-, cis	EtOH	253(4.1),359(3.0)	40-1263-69
Acridine, 9-(1-butenyl)-, trans	EtOH	254(4.0),360(3.0)	40-1263-69
[2,2]Paracyclophane, 4-cyano-, tetracyanoethylene complex	CH_2Cl_2	475(3.30)	35-3553-69
Pyrrole, 1-methyl-3,4-diphenyl-	EtOH	244(4.34),265s(--)	88-4875-69
$C_{17}H_{15}NO$			
Benzofuro[3,2-g]isoquinoline, 3,4-dihydro-5,11-dimethyl-	EtOH	229(3.90),270(3.92),279(4.04),308(4.46)	4-0379-69
Cyclohepta[c]pyrrol-6(2H)-one, 1,3-dimethyl-2-phenyl-	EtOH	293(4.74),328(4.06),343(3.96),386(3.40)	104-0559-69
	$HOAc-HClO_4$	304(4.94),340(4.09),538(3.21)	104-0559-69
4H,8H-Pyrido[3,2,1-de]phenanthridin-8-one, 5,6-dihydro-6-methyl-	EtOH	235(4.68),263(4.38),327(3.91),342(3.87)	2-0848-69
$C_{17}H_{15}NO_2$			
1,3-Butadiene, 2-methyl-4-(p-nitrophenyl)-1-phenyl-	EtOH	282(4.16),368(4.35)	104-0111-69
5H-Dibenzo[a,d]cycloheptene-5-carboxamide, 10-methoxy-	EtOH	286(4.11)	23-2827-69
Indole-2-carboxylic acid, 1-phenyl-, ethyl ester	EtOH	217(4.41),293(4.29)	44-2988-69
2H-[1,3]Oxazino[2,3-a]isoindol-6(10bH)-one, 3,4-dihydro-10b-phenyl-	EtOH	219(4.29),255s(3.59)	44-0165-69
1,3-Pentadiene, 4-(p-nitrophenyl)-1-phenyl-	EtOH	290(4.11),376(4.13)	104-0111-69

Compound	Solvent	λ_{max}(log ε)	Ref.
2,4-Pentadienohydroxamic acid, N,5-diphenyl-	EtOH	323(4.61)	34-0278-69
Phthalimidine, 2-acetyl-5-methyl-3-phenyl-	EtOH	236(4.06),282(3.26), 287s(3.24)	39-1149-69C
4H,8H-Pyrido[3,2,1-de]phenanthridin-8-one, 5,6-dihydro-2-methoxy-	EtOH	236(4.68),268(4.30), 344(3.95),356s(3.87)	2-0848-69
$C_{17}H_{15}NO_3$			
Acetonitrile, [o-(2,4-dimethoxybenzoyl)-phenyl]-	EtOH	215(4.15),241(4.11), 281(3.95),314(3.89)	39-1873-69C
Hydroxylamine, N,O-diacetyl-N-fluoren-1-yl-	EtOH	250s(4.06),254s(4.16), 259s(4.19),263(4.22), 273s(4.06),294(3.76), 303(3.83)	39-0345-69C
Indoline, 5-acetyl-N-salicyloyl-	EtOH	220s(4.25),296s(4.25), 316(4.39)	2-0848-69
$C_{17}H_{15}NO_4$			
Benz[e]indolizine-3-acetic acid, 2-carboxy-, dimethyl ester	MeOH	209(4.41),223(4.43), 239s(4.48),247(4.56), 257s(4.49),264s(4.43), 271s(4.29),282(4.03), 339s(3.92),348(3.95), 364s(3.77)	39-2311-69C
Benz[g]indolizine-1-acetic acid, 2-carboxy-, dimethyl ester	MeOH	216(4.39),257s(4.68), 264(3.83),298(3.67), 311(3.70),324(3.70), 340s(3.43)	39-2311-69C
$C_{17}H_{15}NO_5$			
4H-1-Benzopyran-$\Delta^{4,\alpha}$-acetic acid, 2-carboxy- -cyano-, diethyl ester	EtOH	225(4.40),258(3.90), 282(3.96),383(4.25)	103-0316-69
Melicopidine	EtOH and EtOH-NaOH	207(4.38),220s(4.31), 257s(4.44),277(4.71), 300s(3.96),403(3.94)	39-2327-69C
	EtOH-HCl	207(4.32),220s(4.24), 258s(4.42),277(4.67), 293s(4.31),373(3.81), 406(3.85)	39-2327-69C
Melicopine (spectrum unchanged in acid or base)	EtOH	216(4.37),252s(4.49), 271(4.74),302(4.19), 409(3.89)	39-2327-69C
Tecleanthine	EtOH and EtOH-NaOH	220s(4.29),236(4.23), 269(4.74),281s(4.68), 306s(3.83),401(4.05)	39-2327-69C
	EtOH-HCl	220s(4.30),240s(4.33), 270(4.64),284(4.67), 293(4.76),366(4.02), 406(3.93),453s(3.57)	39-2327-69C
$C_{17}H_{15}NO_6$			
Anthraquinone, 4-amino-1,5-dihydroxy-2,3-dimethoxy-7-methyl-	EtOH	241(4.45),246(4.53), 251(4.59),257(4.65), 262(4.59),269(4.33), 310(3.81),495(3.96), 525(4.19),564(4.17)	24-4104-69
	EtOH-NaOH	241(4.55),246(4.59), 252(4.62),256(4.62), 262(4.52),269(4.27), 498(3.70),525(3.99),	24-4104-69

continued on next page

Compound	Solvent	λ_{max}(log ϵ)	Ref.
		555(4.24),594(4.27)	24-4104-69
1,3-Dioxolo[4,5-b]acridine-10,11(5H,-11aH)-dione, 4,11a-dimethoxy-5-methyl-	EtOH	217(4.28),244s(4.17), 291(4.26),324(4.15)	12-1477-69
$C_{17}H_{15}NO_7$			
Terephthalamic acid, 3,5-dihydroxy-2-(5-methoxy-o-anisoyl)-	n.s.g.	225(4.32),259(4.19), 340(3.89)	39-2805-69C
$C_{17}H_{15}N_3OS_2$			
Rhodanine, 3-benzyl-5-(o-tolylazo)-	MeOH	240(3.98),264(3.75), 295(4.23),425(4.23)	2-0129-69
	$CHCl_3$	458(4.59)	2-0129-69
Rhodanine, 3-benzyl-5-(p-tolylazo)-	$CHCl_3$	428(4.45)	2-0129-69
$C_{17}H_{15}N_3O_2$			
Benzimidazo[2,1-a]isoquinoline, 10-amino-2,3-dimethoxy-	EtOH	203(4.20),263(4.34), 296(4.24),322(4.10), 336(4.17),351(4.15)	24-1529-69
$C_{17}H_{15}N_3O_2S$			
Rhodanine, 3-benzyl-5-[(o-methoxyphenyl)azo]-	MeOH	241(4.00),264(3.84), 297(2.92),444(4.28)	2-0129-69
	$CHCl_3$	467(4.52)	2-0129-69
Rhodanine, 3-benzyl-5-[(p-methoxyphenyl)azo]-	$CHCl_3$	441(4.42)	2-0129-69
$C_{17}H_{15}N_3O_4S$			
1-Benzimidazolecarboxylic acid, 2-[(p-nitrobenzyl)thio]-, ethyl ester	MeOH	275(4.26),282s(4.25), 293s(4.19)	4-0023-69
5-Thia-1-azabicyclo[4.2.0]oct-2-ene-2-carboxylic acid, 7-[2-(p-cyanophenyl)-acetamido]-3-methyl-8-oxo-, K salt	H_2O	233(4.35),262(3.90)	87-0310-69
$C_{17}H_{15}N_3O_4S_2$			
5-Thia-1-azabicyclo[4.2.0]oct-2-ene-2-carboxylic acid, 7-[2-[(p-cyanophenyl)thio]acetamido]-3-methyl-, K salt	H_2O	219(4.19),272(4.28)	87-0310-69
$C_{17}H_{15}N_3O_6$			
Toluene-α,α-diol, α-[(p-nitrophenyl)-azo]-, diacetate	dioxan	283(4.23),409(2.67)	39-1571-69C
$C_{17}H_{15}N_5O_3$			
Indole-3-carbonitrile, 2-amino-4,5,6,7-tetrahydro-1-[(p-nitrophenacylidene)-amino]-	$CHCl_3$	266(4.35),327(4.26), 440s(2.82)	48-0388-69
$C_{17}H_{16}$			
Anthracene, 9-isopropyl-	EtOH	249s(4.97),255(5.25), 333(3.52),348(3.81), 366(3.98),385(3.95)	78-3485-69
Anthracene, 9-propyl-	EtOH	249s(4.96),255(5.26), 332(3.47),347(3.78), 365(3.98),386(3.97)	78-3485-69
Cyclohepta[def]fluorene, 4,8,9,10-tetrahydro-8-methyl-	EtOH	213(4.52),226s(4.17), 258s(4.15),268(4.29), 279(4.12),299(4.29)	39-1427-69C

Compound	Solvent	λ max(log ε)	Ref.
$C_{17}H_{16}BF_2NO$			
Boron, [3-(benzylimino)butyrophenonato]-difluoro-	$CHCl_3$	255(3.53),337(4.42)	39-0526-69A
$C_{17}H_{16}BF_4NS_2$			
Benzylmethyl(4-phenyl-1,3-dithiol-2-ylidene)ammonium tetrafluoroborate	EtOH	322(4.09)	94-1924-69
$C_{17}H_{16}BrNO_4$			
Narwedine, 9-bromo-8-oxo-, (±)-, enone	EtOH	224(4.06),265s(3.28), 313(2.14)	39-2602-69C
$C_{17}H_{16}BrNO_5$			
1,3-Dioxolo[4,5-b]acridin-10(5H)-one, 11-bromo-11,11a-dihydro-11,11a-dimethoxy-5-methyl-	EtOH	212(4.04),229(4.06), 274s(4.23),282(4.26), 321(3.95)	12-1493-69
$C_{17}H_{16}Br_2Cl_2O_2$			
Propane, 2,2-bis(bromomethyl)-1,3-bis-(o-chlorophenoxy)-	hexane	216(4.38),273(3.63), 280(3.55)	56-1641-69
$C_{17}H_{16}Br_2N_2S_2$			
2H-1,3,5-Thiadiazine-2-thione, 3,5-bis-(p-bromobenzyl)tetrahydro-	MeOH	230(3.77),295(3.15)	73-2952-69
$C_{17}H_{16}Br_5NO_5$			
1,3-Dioxolo[4,5-b]acridinium, 11-bromo-11,11a-dihydro-10-hydroxy-11,11a-dimethoxy-5-methyl-, tribromide, hypobromite	EtOH	212(4.40),249(4.31), 285(4.21),344(4.30)	12-1493-69
$C_{17}H_{16}ClNO$			
4-Isoxazoline, 3-(p-chlorobenzyl)-2-methyl-5-phenyl-	EtOH	223(4.34),276(3.97)	88-4875-69
7-Methoxy-1-methyl-4-phenylquinolinium chloride	EtOH	250(4.14),278(3.77), 320(3.98),360s(3.83)	2-0422-69
$C_{17}H_{16}ClNO_3$			
Cyclohexanecarboxamide, N-(3-chloro-1,4-dihydro-1,4-dioxo-2-naphthyl)-	MeOH	247(4.30),253(4.34), 287(3.98)	4-0909-69
$C_{17}H_{16}ClNO_5$			
2-Benzyl-4-methyl-5-phenylisoxazolium perchlorate	CH_2Cl_2	300(4.28)	44-1468-69
6-Hydroxy-1,3-dimethyl-2-phenylcyclohepta[c]pyrrolium perchlorate	$HOAc-HClO_4$	304(4.93),340(4.09), 538(3.21)	104-0947-69
2-(α-Methylbenzyl)-4-phenylisooxazolium perchlorate	CH_2Cl_2	234(4.28)	44-3451-69
$C_{17}H_{16}ClN_3O_2$			
Pyrimidine, 3-benzyl-1-(p-chlorophenyl)-1,2,3,4-tetrahydro-5-nitro-	dioxan	250(4.18),385(4.38)	78-1617-69
$C_{17}H_{16}F_4N_4O_7$			
9H-Purine, 6-fluoro-9-β-D-ribofuranosyl-2-(trifluoromethyl)-, 2',3',5'-triacetate	pH 1	248(3.86)	44-1396-69
	pH 7	248(3.88)	44-1396-69
	pH 13	253(4.05)	44-1396-69
	EtOH	248(3.88)	44-1396-69

Compound	Solvent	$\lambda_{max}(\log \epsilon)$	Ref.
$C_{17}H_{16}IN$			
5,6-Dihydro-8-methyl-4H-pyrido[3,2,1-de]phenanthridinium iodide	EtOH	206(4.46),236s(4.32), 408(3.74)	2-0848-69
$C_{17}H_{16}N_2$			
1H-1,4-Diazepine, 2,3-dihydro-5,7-di-phenyl-	MeOH	265(4.20),358(4.38)	39-1081-69C
Quinoline, 4-[p-(dimethylamino)phenyl]-	EtOH	228(4.55),263(4.22), 313s(--),358(4.08)	78-0837-69
$C_{17}H_{16}N_2O$			
Acetic acid, cinnamylidenephenylhydra-zide	EtOH	220s(4.0),229(4.0), 237s(3.9),300s(4.6), 312(4.6),319s(4.5)	28-1703-69A
6,7-Diazabicyclo[3.2.0]hept-6-ene, 1,5-diphenyl-, 6-oxide	n.s.g.	210(4.25)	35-2818-69
Isoindole-1-carboxamide, N,2-dimethyl-3-phenyl-	iso-PrOH	222(4.37),260(4.41), 298(3.70),306(3.70), 356(4.28)	44-0649-69
Pyrimido[2,1-a]isoindol-6(2H)-one, 1,3,4,10b-tetrahydro-10b-phenyl-	EtOH	220(4.30),253(3.54), 259(3.54)	44-0165-69
4(1H)-Quinolone, 2-(dimethylamino)-3-phenyl-	EtOH	238(4.40),258(4.36), 320(4.15)	78-5721-69
3-Quinuclidinone, 2-(4-quinolylmethyl-ene)-	EtOH	251(4.26),324(4.05)	22-1251-69
$C_{17}H_{16}N_2O_2$			
Benzimidic acid, N-(2-cyano-4-ethoxy-phenyl)-, methyl ester	EtOH	215(4.53),327(3.74)	22-2008-69
Indole-3-acetic acid, 2-(3-pyridyl)-, ethyl ester	EtOH	219(4.51),307(4.32)	22-4154-69
1H-Naphtho[2,3-d]imidazole-4,9-dione, 2-cyclohexyl-	pH 1	245(4.53),251(4.65), 279(4.19)	4-0909-69
	pH 7	243(4.64),283(4.13)	4-0909-69
	pH 13	264(4.61)	
	MeOH	246(4.63),277(4.13), 333(3.46)	4-0909-69
3(2H)-Pyridazinone, 4,5-dihydro-5-(hydroxymethyl)-2,6-diphenyl-	MeOH	206(4.28),218s(4.14), 249(4.02),300(4.06)	5-0217-69D
$C_{17}H_{16}N_2O_3$			
1H-Pyrazolo[2,3-c]pyridine-2-carboxylic acid, 5-(benzyloxy)-, ethyl ester	EtOH	278(4.11),287(4.18)	35-2338-69
Pyruvanilide, 2'-(methylcarbamoyl)-3-phenyl-	EtOH	249(4.11),302(3.90)	44-3035-69
$C_{17}H_{16}N_2O_3S$			
5H-Thiazolo[3,2-a]pyridine-8-carboxylic acid, 6-cyano-2,3,6,7-tetrahydro-5-oxo-7-phenyl-, ethyl ester	EtOH	245(4.0),305(4.0)	24-0522-69
	0.09N NaOH	240(4.2),290(4.1), 360(3.5)	24-0522-69
$C_{17}H_{16}N_2O_4$			
1(4H)-Pyridinecarboxylic acid, 4-(4-methyl-5-oxo-2-phenyl-2-oxazolin-4-yl)-, methyl ester	MeCN	234(4.39)	24-1129-69
Toluene-α,α-diol, α-(phenylazo)-, diacetate	dioxan	280(4.13),401(2.48)	39-1571-69C
$C_{17}H_{16}N_2O_4S$			
Penicillene, anhydro(1-phenoxyethyl)-	$CHCl_3$	269(4.00),319(4.18)	88-3385-69

Compound	Solvent	λ_{max}(log ε)	Ref.
$C_{17}H_{16}N_2O_6$			
4-Pyridinepyruvic acid, 2-(benzyloxy)-5-nitro-, ethyl ester	EtOH	288(4.07)	35-2338-69
$C_{17}H_{16}N_2S_3$			
2,4-Thiazolidinedithione, 3-ethyl-5-[2-(1-methyl-2(1H)-quinolylidene)ethylidene]-	BuOH	250(4.29),295(4.03), 340(4.01)	65-2116-69
$C_{17}H_{16}N_4O$			
Indole-3-carbonitrile, 2-amino-4,5,6,7-tetrahydro-1-(phenacylideneamino)-	CH_2Cl_2	266(4.15),311(4.18), 434s(2.96)	48-0388-69
2-Pyrazolin-5-one, 3-methyl-4-[[p-(methylamino)phenyl]imino]-1-phenyl-	MeOH	434(4.19),511(4.51)	44-1687-69
7H-Pyrrolo[2,3-d]pyrimidine-6-carbonitrile, 7-butyl-5-hydroxy-2-phenyl-	EtOH	233(4.15),277(4.54)	4-0819-69
$C_{17}H_{16}N_4O_2$			
1-Oxa-2,5,6-triazaspiro[2.4]hept-6-en-4-one, 7-methyl-2-[p-(methylamino)-phenyl]-5-phenyl-	MeOH	255(4.34),287(3.94)	44-1687-69
$C_{17}H_{16}N_4O_4$			
3-Butenal ,2-methyl-4-phenyl-, 2,4-dinitrophenylhydrazone	$CHCl_3$	380(4.58)	22-3719-69
$C_{17}H_{16}N_4S$			
9-Acridinecarboxaldehyde, 4,4-dimethyl-3-thiosemicarbazone	EtOH	260(4.5),300(4.3), 350s(3.7),410(3.5)	65-1151-69
	EtOH-HCl	260(4.8),360(4.1), 460(4.0)	65-1151-69
$C_{17}H_{16}O$			
Cyclohepta[def]fluoren-8-ol, 4,8,9,10-tetrahydro-8-methyl-	EtOH	214(4.56),225s(4.39), 257(4.20),268(4.22), 279(4.16),298(3.79)	39-1427-69C
1H-Cyclopenta[l]phenanthren-1-one, 2,3,4,5,6,7-hexahydro-	EtOH	222(4.44),242(3.22), 249(4.20),314(3.91), 320(3.80)	44-0763-69
Ketone, phenyl 2-phenylcyclobutyl, cis	EtOH	247(4.03),318(1.89)	35-0456-69
trans	EtOH	245(4.17),318(2.06)	35-0456-69
$C_{17}H_{16}OS_2$			
1,3-Dithiole, 2-ethoxy-4,5-diphenyl-	EtOH	290(3.80),328(3.79)	94-1931-69
$C_{17}H_{16}O_2$			
Gona-1,3,5(10),6,8-pentaen-11-one, 17-hydroxy-, (±)-	MeOH	243(4.35),310(3.88)	19-0145-69
Gona-1,3,5(10),6,8-pentaen-17-one, 11-hydroxy-	MeOH	229(4.91),278(3.78)	19-0145-69
isomer	MeOH	230(4.96),278(3.84)	19-0145-69
14β-Gona-1,3,5(10),8-tetraene-11,17-dione, (+)-	MeOH	240(4.13),291(3.77)	19-0469-69
(-)-	MeOH	240(4.12),291(3.73)	19-0469-69
1,4-Methanonaphthalene-5,8-dione, 1,2,3,4,4a,8a-hexahydro-6-phenyl-, cis-endo	EtOH	225(4.16),303(3.96)	39-0105-69C

Compound	Solvent	$\lambda_{max}(\log \epsilon)$	Ref.
$C_{17}H_{16}O_3$			
2(3H)-Furanone, dihydro-3-(hydroxydiphenylmethyl)-	EtOH	253(2.86),259(2.88)	44-3792-69
2-Propanone, 1-acetoxy-1,3-diphenyl-	EtOH	253(2.80),258(2.83), 264(2.79),289(2.68)	44-1430-69
$C_{17}H_{16}O_4$			
Chalcone, 4-hydroxy-2',4'-dimethoxy-	n.s.g.	238(4.14),350(4.45)	7-0624-69
Eleutherinol, dimethyl ether	EtOH	238(4.60),270(4.57), 340s(3.97),353(4.07)	94-0454-69
3-Phenanthrol, 2,5,6-trimethoxy-	$CHCl_3$	293(4.03),306(3.86), 328(3.36),345(3.58), 362(3.70)	88-0067-69
Pterocarpan, 3,9-dimethoxy-	EtOH	281s(3.79),286(3.85)	39-1109-69C
2-Stilbenecarboxylic acid, 3,5-dimethoxy-	MeOH	225(4.26),233(4.21), 243(4.16),300(4.45)	44-3192-69
$C_{17}H_{16}O_5$			
Acetic acid, (4-benzoyl-3-hydroxyphenoxy)-, ethyl ester	EtOH	240(4.05),285(4.20), 325(3.97)	39-0037-69C
2'-Acetonaphthone, 1,8-diacetoxy-3-methyl-	EtOH	225(4.82),286(3.80), 324(3.13)	94-0454-69
2,3-Biphenylenedione, 4,5,6-trimethoxy-1,8-dimethyl-	EtOH	253(3.97),308(4.60), 414(4.09)	39-0646-69C
Chalcone, 4,4',6'-trihydroxy-2'-methoxy-3'-methyl-	MeOH	360(4.30)	78-0283-69
Flavanone, 4',7-dihydroxy-5-methoxy-6-methyl-	MeOH	281(4.27)	78-0283-69
Flavanone, 5-hydroxy-4',7-dimethoxy-, (-)-	EtOH	216(4.55),220(4.33), 320s(3.64)	105-0397-69
Flavanone, 4',5,7-trihydroxy-6,8-dimethyl-	EtOH	216(4.39),227(4.38), 297(4.24),350(3.51)	95-0851-69
7H-Furo[3,2-g][1]benzopyran-7-one, 9-methoxy-4-[(3-methyl-2-butenyl)oxy]-	EtOH	223(4.36),242(4.11), 249(4.12),270(4.20), 313(4.02)	36-0675-69
4H-Naphtho[2,3-b]pyran-4-one, 5,6,8-trimethoxy-2-methyl-	EtOH	226(4.47),252s(4.46), 271(4.67),326(3.53), 342(3.62),373(3.78)	94-0458-69
Pterocarpan, 3-hydroxy-8,9-dimethoxy-	EtOH	288(3.85),298(3.85)	18-0233-69
Pterocarpin, dihydro-, (±)-	EtOH	287(3.79),298(3.78)	18-1408-69
Xanthen-9-one, 1,3,7-trimethoxy-2-methyl-	MeOH	238(4.50),255(4.51), 282(4.61),310(4.12)	78-0275-69
$C_{17}H_{16}O_6$			
Acetophenone, 2'-hydroxy-4',5'-dimethoxy-2-[3,4-(methylenedioxy)phenyl]-	EtOH	239(4.25),281(4.14), 345(3.95)	39-1787-69C
Ketone, 2-hydroxy-4,6-dimethoxyphenyl piperonyl	EtOH	225s(4.19),291(4.32)	39-0365-69C
Ougenin	MeOH	289(4.16)	2-0061-69
Philenopteran	EtOH	280(3.59),286(3.54)	39-0887-69C
Xanthen-9-one, 1-hydroxy-3,6,7-trimethoxy-2-methyl-	MeOH	242(4.42),318(4.23)	78-0275-69
Xanthen-9-one, 1-hydroxy-3,6,7-trimethoxy-4-methyl-	MeOH	258(4.45),310(4.11)	78-0275-69
Xanthen-9-one, 1,2,3,5-tetramethoxy-	EtOH	243s(4.53),249(4.60), 268s(4.22),283(4.14), 300s(4.00),343(3.80)	78-1947-69
Xanthen-9-one, 1,3,4,7-tetramethoxy-	MeOH	258(4.60),298s(3.98), 308(4.00),357(3.88)	39-2201-69C

Compound	Solvent	λ_{max}(log ε)	Ref.
Xanthen-9-one, 1,3,4,8-tetramethoxy-	EtOH	236(4.56),242s(4.49), 313(4.18),340s(3.74)	78-1947-69
$C_{17}H_{16}O_7$			
Benzophenone, 2',4,4'-trihydroxy-2-methoxy-6-carbomethoxy-6'-methyl-	EtOH	232(4.07),295(4.08), 333(4.01)	39-1721-69C
1H-Naphtho[2,3-b]pyran-1,6,9-trione, 3,4-dihydro-7,8,10-trimethoxy-3-methyl-	EtOH	214(4.38),267(4.22), 354(3.62)	23-1561-69
Xanthen-9-one, 1-hydroxy-2,3,4,5-tetramethoxy-	EtOH	243(4.40),260(4.43), 275(4.24),312(4.03), 380(3.57)	78-1947-69
Xanthen-9-one, 7-hydroxy-1,2,3,4-tetramethoxy-	EtOH	243(4.43),261(4.55), 288(3.92),310s(--), 370(3.75)	78-4415-69
$C_{17}H_{17}BF_4N_2$			
1,3-Dimethyl-2-styrylbenzimidazolium tetrafluoroborate	MeCN	228s(4.08),265(4.04), 319(4.44)	24-0568-69
$C_{17}H_{17}BrN_2O_2$			
Piperidine, 1-(3'-bromo-3-nitro-4-biphenylyl)-	MeOH	418(3.15)	17-0145-69
Piperidine, 1-(4'-bromo-3-nitro-4-biphenylyl)-	MeOH	420(3.14)	17-0145-69
$C_{17}H_{17}BrN_4O_2$			
Piperidine, 1-[4-[(p-bromophenyl)azo]-2-nitrophenyl]-	benzene	400(4.09)	73-2092-69
$C_{17}H_{17}ClFN_5O_3$			
Adenine, 9-(5-O-benzyl-2-deoxy-2-fluoro-α-D-arabinofuranosyl)-2-chloro-	EtOH	264(4.18)	44-2632-69
β-form	EtOH	265(4.16)	44-2632-69
$C_{17}H_{17}ClN_2$			
Isoindole, 1-(p-chlorophenyl)-2-(3-aminopropyl)-, hydrochloride	EtOH	222(4.48),274(3.59), 280(3.59),349(3.96)	44-1720-69
$C_{17}H_{17}ClN_2OS$			
2(1H)-Quinazolinethione, 6-chloro-3,4-dihydro-4-hydroxy-3-isopropyl-4-phenyl-	n.s.g.	261(4.02),292(4.35)	40-0917-69
2(1H)-Quinazolinethione, 6-chloro-3,4-dihydro-4-hydroxy-4-phenyl-3-propyl-	n.s.g.	256(3.85),290(4.44)	40-0917-69
$C_{17}H_{17}ClN_2O_2$			
Glycine, N-(2-amino-5-chloro-α-phenylbenzylidene)-, ethyl ester, anti	EtOH	234(4.42),370(3.66)	111-0239-69
syn	EtOH	249(4.40)	111-0239-69
Piperidine, 1-(3'-chloro-3-nitro-4-biphenylyl)-	MeOH	420(3.16)	17-0145-69
Piperidine, 1-(4'-chloro-3-nitro-4-biphenylyl)-	MeOH	408(3.16)	17-0145-69
$C_{17}H_{17}ClN_2O_3$			
Benzamide, 2-(3-chloro-2-hydroxy-3-phenylpropionamido)-N-methyl-	n.s.g.	217(4.45),254(4.23), 295(3.54)	35-6083-69

Compound	Solvent	$\lambda_{max}(\log \epsilon)$	Ref.
$C_{17}H_{17}ClN_4O_2$			
Piperidine, 1-[4-[(p-chlorophenyl)azo]-2-nitrophenyl]-	benzene	395(4.22)	73-2092-69
$C_{17}H_{17}ClO_4$			
2H-1-Benzopyran-2-carboxylic acid, 4-chloro-2-(2-oxocyclohexyl)-, methyl ester	heptane	224(4.45),270(3.66), 319(3.48)	103-0321-69
isomer	heptane	222(4.54),268(3.58), 312(3.48)	103-0321-69
$C_{17}H_{17}ClO_5S$			
Benzophenone, 3-chloro-4'-hydroxy-2-mercapto-2',4,6-trimethoxy-6'-methyl-	MeOH	210(4.61),240s(4.33), 300(3.97)	44-1463-69
Spiro[benzo[b]thiophene-2(3H),1'-[2]-cyclohexene]-3,4'-dione, 5-chloro-2',4,6-trimethoxy-	MeOH	230(4.34),250s(4.64), 258(4.70),287(4.22), 360(3.58)	44-3484-69
Spiro[benzo[b]thiophene-2(3H),1'-[2]-cyclohexene]-3,4'-dione, 7-chloro-2',4,6-trimethoxy-	MeOH	231(4.52),248(4.49), 306(4.11),345(3.60)	44-3484-69
$C_{17}H_{17}Cl_2NO$			
1'-Acrylonaphthone, 3-[bis(2-chloroethyl)amino]-	EtOH	221(4.60),304(4.25)	24-3139-69
2'-Acrylonaphthone, 3-[bis(2-chloroethyl)amino]-	EtOH	215(4.51),245(4.32), 251(4.38),283(4.00), 293(3.48),340(4.36)	24-3139-69
$C_{17}H_{17}FN_4O_2$			
Piperidine, 1-[4-[(m-fluorophenyl)azo]-2-nitrophenyl]-	benzene	400(4.23)	73-2092-69
Piperidine, 1-[4-[(p-fluorophenyl)azo]-2-nitrophenyl]-	benzene	395(4.22)	73-2092-69
$C_{17}H_{17}IN_2O_2$			
Piperidine, 1-(3'-iodo-3-nitro-4-biphenylyl)-	MeOH	418(3.15)	17-0145-69
Piperidine, 1-(4'-iodo-3-nitro-4-biphenylyl)-	MeOH	418(3.14)	17-0145-69
$C_{17}H_{17}IO_6$			
6-Oxabicyclo[3.2.1]octane-2-carboxylic acid, 4-iodo-8-[3,4-(methylenedioxy)-phenyl]-7-oxo-, ethyl ester	EtOH	237(3.69),286(3.70)	78-4315-69
$C_{17}H_{17}Li$			
1-Pentene, 5,5-diphenyl-, 5-lithium derivative	THF	495(4.48)	35-2456-69
$C_{17}H_{17}N$			
8,16-Imino[2,2]metacyclophane, N-methyl-	C_6H_{12}	248(3.82)	35-1672-69
$C_{17}H_{17}NO$			
5H-Dibenzo[a,d]cycloheptene-5-methylamine, 10-methoxy-, hydrochloride	EtOH	228(4.12),292(4.10)	23-2827-69
4-Isoxazoline, 3-benzyl-2-methyl-5-phenyl-	EtOH	222(4.17),279(3.90)	88-4875-69
1-Naphthaleneacetonitrile, α-(butoxymethylene)-	EtOH	222(4.78),292(3.88)	22-0198-69

Compound	Solvent	λ_{max}(log ϵ)	Ref.
2-Naphthaleneacetonitrile, α-(butoxy-methylene)-, cis	EtOH	218(4.70),257(4.48),266(4.50),293(4.22),305(4.19)	22-0198-69
trans	EtOH	218(4.68),257(4.48),266(4.48),293(4.34),305(4.26)	22-0198-68
2-Propen-1-one, 1-[p-(dimethylamino)-phenyl]-3-phenyl-	MeOH	390(4.41)	73-2771-69
2-Propen-1-one, 3-[p-(dimethylamino)-phenyl]-1-phenyl-	MeOH	420(4.47)	73-2771-69
	EtOH	247(4.21)	2-0311-69
Quinoline, 1,2-dihydro-7-methoxy-N-methyl-4-phenyl-	EtOH	232(4.47),327(3.49)	2-0419-69
$C_{17}H_{17}NOS$			
Indole, 3-benzyl-2-(ethylsulfinyl)-	EtOH	228(4.48),294(4.18)	35-4598-69
3H-Indol-3-ol, 3-benzyl-2-(ethylthio)-	EtOH	229(4.22),286(3.84),296(3.88),308(3.89)	35-4598-69
$C_{17}H_{17}NOS_2$			
Carbamic acid, benzylmethyldithio-, ester with 2-mercaptoacetophenone	EtOH	249(4.39),279(4.09)	94-1924-69
$C_{17}H_{17}NO_2$			
7-Azabicyclo[4.2.2]deca-2,4,7,9-tetraene, 8-methoxy-2-(p-methoxyphenyl)-	hexane	307(4.18)	35-4714-69
7-Azabicyclo[4.2.2]deca-2,4,7,9-tetraene, 8-methoxy-9-(p-methoxyphenyl)-	hexane	256(4.26)	35-4714-69
2-Indolinol, 1-benzoyl-3,3-dimethyl-	EtOH	258(4.15),284s(3.99)	2-0848-69
$C_{17}H_{17}NO_2S_2$			
Acetaldehyde, (2-benzothiazolylthio)-phenyl-, dimethyl acetal	EtOH	225(4.25),283(4.36),291(4.35),301(4.29)	95-0469-69
$C_{17}H_{17}NO_3$			
Acrophylline	C_6H_{12}	222(4.19),249(4.31),254(4.32),260s(4.34),264(4.53),280s(3.60),323s(3.52)	12-0447-69
	EtOH	227(4.29),253s(4.66),258s(4.69),262(4.78),290s(3.73),323(4.05)	12-0447-69
Anthranilic acid, N-(4-acetyl-2,5-xylyl)-	EtOH	204(4.57),232(4.16),244s(4.10),325s(4.12),358(4.43)	22-3523-69
Anthranilic acid, N-(4-acetyl-2,6-xylyl)-	EtOH	210(4.61),254(4.25),304s(3.76),330s(3.96),346(4.04)	22-3523-69
Anthranilic acid, N-(4-acetyl-3,5-xylyl)-	EtOH	208(4.45),296(4.06),347(4.03)	22-3523-69
6,11-Epoxy-3H-naphtho[2,3-d]azepine-3-carboxylic acid, 5a,6,11,11a-tetrahydro-, ethyl ester, endo	EtOH	232(4.18)	44-2888-69
Hydroxylamine, N-acetyl-O-benzyl-N-p-toluoyl-	MeOH	208(--),253(4.10)	12-0161-69
	MeOH-NaOH	228s(4.00),260s(3.76)	12-0161-69
Phthalimidine, 3-hydroxy-2-(3-hydroxypropyl)-3-phenyl-	EtOH	253(3.39),265(3.45)	44-0165-69
2H-Pyran-2,4(3H)-dione, 6-phenyl-3-[1-(1-pyrrolidinyl)ethylidene]-	MeOH	220(4.18),354(4.19)	44-2527-69

Compound	Solvent	λ_{max}(log ε)	Ref.
$C_{17}H_{17}NO_4$			
Acetic acid, anhydride with N-(benzyl-oxy)-p-anisimidic acid	MeOH	208(--),216(4.28), 277(4.09)	12-0161-69
	MeOH-NaOH	241(4.03),256(4.09)	12-0161-69
5H-Dibenzo[a,d]cycloheptene-2,3,7,8-tetrol, 5-[(methylamino)methyl]-, hydrochloride	EtOH	215(4.51),235(4.49), 312(4.20)	45-0786-69
4-Penten-1-ol, 5-(2-naphthyl)-4-nitro-, acetate	EtOH	324(4.09)	87-0157-69
$C_{17}H_{17}NO_5$			
Haemanthidine, 11-oxo-	EtOH	251(3.59),295(3.66), 314s(3.44),326s(3.30)	35-0150-69
$C_{17}H_{17}NO_6$			
1-Naphthalenemalonic acid, 8-nitro-, diethyl ester	EtOH	217(4.63),242(4.06), 335(3.50)	39-0756-69C
Pyrrole-2,3,4-tricarboxylic acid, 1-benzyl-, trimethyl ester	EtOH	218(4.53),263(4.02)	94-2461-69
$C_{17}H_{17}NS_2$			
1,3-Dithiol-2-amine, N-benzyl-N-methyl-4-phenyl-	EtOH	234(4.18),318(3.92)	94-1924-69
$C_{17}H_{17}NSe$			
Selenoxanthene-$\Delta^{9,\gamma}$-propylamine, N-meth-yl-	MeOH	231(4.44),268(4.07), 316(3.49)	73-3792-69
$C_{17}H_{17}N_3O$			
2H-Imidazo[4,5-b]pyridin-2-one, 1,3-di-hydro-3-isopropyl-1-(1-phenylvinyl)-	EtOH	241(4.15),294(4.04)	4-0735-69
$C_{17}H_{17}N_3OS$			
Malononitrile, [3-(methylthio)-3-morpho-lino-2-phenylallylidene]-	C_6H_{12}	398(4.37)	5-0073-69E
$C_{17}H_{17}N_3O_2$			
Isolongistrobine, anhydro-	EtOH	257(4.44)	78-2767-69
	EtOH-HCl	244(4.37)	78-2767-69
Longistrobine, anhydro-	EtOH	255(4.34)	78-2767-69
	EtOH-HCl	245(4.36)	78-2767-69
$C_{17}H_{17}N_3O_3$			
Isolongistrobine, didehydro-	EtOH	258(4.24)	78-2767-69
	EtOH-HCl	235(4.16)	78-2767-69
Longistrobine, didehydro-	EtOH	255(4.10)	78-2767-69
	EtOH-HCl	236(4.16)	78-2767-69
$C_{17}H_{17}N_3O_4$			
Piperidine, 1-(3,3'-dinitro-4-biphenyl-yl)-	MeOH	406(3.18)	17-0145-69
Piperidine, 1-(3,4'-dinitro-4-biphenyl-yl)-	MeOH	354(4.18)	17-0145-69
$C_{17}H_{17}N_3O_4S_3$			
Ketene, (dimethylamino)(methylthio)-, bis(o-nitrophenyl) mercaptole	MeOH	390(3.92)	5-0073-69E

Compound	Solvent	λ_{max}(log ϵ)	Ref.
$C_{17}H_{17}N_5O_4$			
4a(2H)-Naphthalenecarbonitrile, 3,4,5,6,7,8-hexahydro-2-oxo-, 2,4-dinitrophenylhydrazone	EtOH	375(4.42)	28-0864-69A
Piperidine, 1-[2-nitro-4-(p-nitrophenyl)azo]phenyl]-	benzene	433(4.29)	73-2092-69
$C_{17}H_{18}$			
1H-Cyclopenta[1]phenanthrene, 2,3,4,5,6,7-hexahydro-	EtOH	216(4.41),237(4.75), 256(3.32),270(3.52), 282(3.68),293(3.72), 322(2.97)	44-0763-69
Isobutene, 1-phenyl-1-o-tolyl-	C_6H_{12}	246(4.12)	59-0487-69
1-Pentene, 5,5-diphenyl-, 5-lithium	THF	495(4.48)	35-2456-69
Propene, 1,1-di-o-tolyl-	C_6H_{12}	236(4.15)	59-0487-69
Propene, 1,1-di-p-tolyl-	C_6H_{12}	230(4.22),238(4.22), 252(4.21)	59-0487-69
Propene, 1-phenyl-1-(2,4-xylyl)-, cis	C_6H_{12}	240(4.16),251(4.15)	59-0487-69
trans	C_6H_{12}	235(4.22)	59-0487-69
Propene, 1-phenyl-1-(2,6-xylyl)-, cis	C_6H_{12}	252(4.16)	59-0487-69
trans	C_6H_{12}	237(4.11)	59-0487-69
Propene, 1-o-tolyl-1-p-tolyl-, cis	C_6H_{12}	254(4.18)	59-0487-69
trans	C_6H_{12}	235(4.19)	59-0487-69
$C_{17}H_{18}ClNO_4$			
(4-Cinnamylidene-2,5-cyclohexadien-1-ylidene)dimethylammonium perchlorate	HOAc-1% $HClO_4$	246(2.23),545(1.95), 580(1.87)	2-0311-69
5H-Dibenzo[a,d]cycloheptene-2,3,7,8-tetrol, 5-[(methylamino)methyl]-, hydrochloride	EtOH	215(4.51),235(4.49), 312(4.20)	45-0786-69
Pyrrole-2,4-dicarboxylic acid, 3-(m-chlorophenyl)-5-methyl-, diethyl ester	EtOH	277(4.17)	94-0582-69
$C_{17}H_{18}ClN_3$			
Benzylidenimine, 4-chloro-N-[2-(3-pyridyl)-1-piperidyl]-	n.s.g.	300(4.23)	65-2723-69
$C_{17}H_{18}ClN_5O_4$			
1-(2-Oxocyclohexyl)pyridinium chloride, 2,4-dinitrophenylhydrazone	MeOH	261(4.13),355(4.29)	24-3647-69
$C_{17}H_{18}Cl_2N_2$			
Isoindole, 1-(p-chlorophenyl)-2-(3-aminopropyl)-, hydrochloride	EtOH	222(4.48),274(3.59), 280(3.59),349(3.96)	44-1720-69
$C_{17}H_{18}FN_5O_5S$			
Adenine, 9-(2-deoxy-2-fluoro-α-D-arabinofuranosyl)-, 5'-p-toluenesulfonate	EtOH	262(4.14)	44-2632-69
β-form	EtOH	259(4.16)	44-2632-69
3,5'-Cycloadenine, 9-(2-deoxy-2-fluoro-β-D-arabinofuranosyl)-, p-toluenesulfonate	H_2O	274(4.09)	44-2632-69
$C_{17}H_{18}FeO_2$			
Ferrocenecarboxylic acid, 1',2:3',4-bis(trimethylene)-	EtOH	290(3.12),325(2.92), 405(2.64),440s(2.61)	49-1540-69

Compound	Solvent	λmax(log ε)	Ref.
$C_{17}H_{18}NO_3P$			
Phosphonic acid, 9-acridinyl-, diethyl ester	MeOH	210(4.18),255(4.20), 369(4.06)	44-1420-69
$C_{17}H_{18}N_2$			
Benzo[g]quinoline, 4-(butylamino)-	EtOH	243(4.73),253(4.67), 304(3.52),390(3.96)	103-0223-69
Dibenzo[c,h][1,6]diazecine, 5,6,7,12-tetrahydro-6-methyl-	EtOH	250(3.99)	44-2715-69
Isoindole, 2-[2-(methylamino)ethyl]-1-phenyl-	EtOH	222(4.48),272(3.72), 283(3.72),320s(3.72), 345(3.88)	44-1720-69
Pyridine, 2-[4-[p-(dimethylamino)phenyl]-1,3-butadienyl]-, 1-cis-3-trans	EtOH	235(4.02),279(4.08), 382(4.48)	78-0837-69
m-Toluidine, N,N'-1-propen-1-yl-3-ylidenedi-, hydrochloride	MeOH	248(4.14),299s(--), 382(4.67)	59-1075-69
o-Toluidine, N,N'-1-propen-1-yl-3-ylidenedi-, hydrochloride	MeOH	241(4.16),380(4.59)	59-1075-69
p-Toluidine, N,N'-1-propen-1-yl-3-ylidenedi-	EtOH	373(4.59)	24-2609-69
hydrochloride	MeOH	247(4.06),281s(--), 291(3.57),300(3.54), 394(4.63)	59-1075-69
perchlorate	EtOH	386(4.65)	24-2609-69
$C_{17}H_{18}N_2O$			
1-Isoindolinecarboxamide, N,2-dimethyl-3-phenyl-	iso-PrOH	252(2.84),258(2.90), 264(2.94),271(2.87)	44-0649-69
$C_{17}H_{18}N_2O_2$			
o-Anisidine, N,N'-1-propen-1-yl-3-ylidenedi-, hydrochloride	MeOH	219s(--),243(4.20), 257(--),298(3.77), 396(4.44)	59-1075-69
p-Anisidine, N,N'-1-propen-1-yl-3-ylidenedi-	EtOH	394(4.56)	24-2609-69
perchlorate	EtOH	398(4.60)	24-2609-69
Piperidine, 1-(3-nitro-4-biphenylyl)-	MeOH	420(3.10)	17-0145-69
$C_{17}H_{18}N_2O_3$			
Acetophenone, 2'-methoxy-5'-[(p-methoxyphenyl)azo]-3'-methyl-	EtOH	241(4.23),352(4.28), 427s(3.28)	12-0935-69
Benzoic acid, p-[(5-tert-butyl-2-hydroxyphenyl)azo]-	50% EtOH	333(4.36),396(3.98)	73-1087-69
	50% EtOH-$HClO_4$	310s(--),399(4.39), 503(4.07)	73-1087-69
Cyclohexanecarboxamide, N-(3-amino-1,4-dihydro-1,4-dioxo-2-naphthyl)-	MeOH	268(4.30),325(3.34), 452(3.36)	4-0909-69
Cyclohexanecarboxylic acid, 2-(2-hydroxy-4-oxo-1(4H)-naphthylidene)hydrazide	pH 1.3	262(4.05),269(4.10), 304(4.29),370(4.21)	44-2750-69
	pH 12.2	266(4.10),311(4.22), 355(4.21)	44-2750-69
Cyclohexanecarboxylic acid, 2-(3-hydroxy-4-oxo-1(4H)-naphthylidene)hydrazide	pH 1.5	395(4.30)	44-2750-69
	pH 12.1	450(4.10)	44-2750-69
Lactanilide, 2'-(methylcarbamoyl)-3-phenyl-	n.s.g.	212(4.45),253(4.22), 295(3.57)	35-6083-69
$C_{17}H_{18}N_2O_4S$			
Acetamide, N-[1-(4-isopropylidene-5-oxo-2-oxazolin-2-yl)-2-(methylthio)-vinyl]-2-phenoxy-	EtOH	269(3.89),275(4.12), 335(4.48)	88-3381-69
	EtOH-base	285(4.18)	88-3381-69

Compound	Solvent	$\lambda_{max}(\log \epsilon)$	Ref.
Penicillin, anhydro-α-phenoxyethyl-	$CHCl_3$	269(4.08)	88-3385-69
$C_{17}H_{18}N_2O_4S_2$			
5-Thia-1-azabicyclo[4.2.0]oct-2-ene-2-carboxylic acid, 3-methyl-7-[2-[p-(methylthio)phenyl]acetamido]-8-oxo-, potassium salt	H_2O	256(4.27)	87-0310-69
$C_{17}H_{18}N_2O_5$			
1H-Pyrrolo[3,4-b]quinoline-3-acetic acid, 2-carboxy-2,3-dihydro-5-methoxy-, 2-ethyl ester	MeOH	253(4.55),321(3.74)	44-1364-69
1H-Pyrrolo[3,4-b]quinoline-3-acetic acid, 2-carboxy-2,3-dihydro-6-methoxy-, 2-ethyl ester	MeOH	217(4.62),242s(4.34), 330(3.80)	44-1364-69
1H-Pyrrolo[3,4-b]quinoline-3-acetic acid, 2-carboxy-2,3-dihydro-8-methoxy-, 2-ethyl ester	MeOH	232(4.50),321(3.66)	44-1364-69
$C_{17}H_{18}N_2O_5S$			
4-Isothiazoline-2-acetic acid, α-isopropenyl-3-oxo-4-(2-phenoxyacetamido)-, methyl ester	EtOH	296(3.00)	35-1401-69
4-Isothiazoline-2-acetic acid, α-isopropylidene-3-oxo-4-(2-phenoxyacetamido)-, methyl ester	EtOH	294(4.09)	35-1401-69
2H-1,3-Thiazine-2-carboxylic acid, 3,4-dihydro-2-isopropenyl-4-oxo-5-(2-phenoxyacetamido)-, methyl ester	EtOH	323(3.77)	35-1401-69
$C_{17}H_{18}N_2S_2$			
2H-1,3,5-Thiadiazine-2-thione, 5-benzyltetrahydro-3-p-tolyl-	MeOH	250(3.99),295(4.02)	73-2952-69
2H-1,3,5-Thiadiazine-2-thione, 3,5-dibenzyltetrahydro-	MeOH	250(3.85),287(4.04)	73-2952-69
2H-1,3,5-Thiadiazine-2-thione, tetrahydro-3,5-di-p-tolyl-	MeOH	248(4.31),293(4.10)	73-2952-69
$C_{17}H_{18}N_3O_2$			
2-(2,4-Diaminophenyl)-6,7-dimethoxyisoquinolinium chloride, hydrochloride	EtOH	215(4.45),256(4.52), 306(4.06)	24-1529-69
$C_{17}H_{18}N_4O_2$			
Piperidine, 1-[2-nitro-4-(phenylazo)-phenyl]-	benzene	400(4.23)	73-2092-69
$C_{17}H_{18}N_4O_4$			
3H-2,8a-Methanonaphthalen-3-one, 1,2,5,6,7,8-hexahydro-, 2,4-dinitrophenylhydrazone	$CHCl_3$	378(4.31)	78-5267-69
$C_{17}H_{18}O$			
Acetophenone, 3',5'-dimethyl-2'-o-tolyl-	EtOH	218(4.29),226(4.20), 245(4.10),290(3.55)	6-0277-69
Acetophenone, 3',5'-dimethyl-2'-p-tolyl-	EtOH	222(4.26),228(4.25), 245(4.07),294(3.50)	6-0277-69
5,7,9,15-Heptadecatetraene-11,13-diyn-4-one, all trans	ether	270(4.01),283(4.28), 357(4.66)	24-3293-69

Compound	Solvent	λ_{max}(log ε)	Ref.
$C_{17}H_{18}O_2$			
4(5H)-Benzofuranone, 6,7-dihydro-2,6,6-trimethyl-3-phenyl-	EtOH	280(3.64)	78-2393-69
Gona-1,3,5(10),6,8-pentaene-11,17-diol, (±)-	MeOH	229(5.12),279(3.86)	19-0145-69
isomer	MeOH	229(5.03),278(3.79)	19-0145-69
14β-Gona-1,3,5(10),8-tetraen-11-one, 17α-hydroxy-, (-)-	MeOH	239(4.21),292(3.80)	19-0469-69
(+)-	MeOH	240(4.28),291(3.92)	19-0469-69
2,4,6,8-Nonatetraenoic acid, 3-methyl-9-phenyl-, methyl ester, all trans	EtOH	272(4.19),345(4.72), 362(4.82),377(4.76)	88-4049-69
$C_{17}H_{18}O_2S$			
5,8-Seco-B-dinorestra-1,3,5(10),8-tetra-en-17-one, 5,8-epithio-3-methoxy-	MeCN	238(4.26),259(4.05), 269(4.07),301(4.43), 314(4.37)	88-4495-69
isomer	MeCN	234(4.44),269(3.98), 277(3.98)	88-4495-69
$C_{17}H_{18}O_3$			
2-Propanone, 1-(7,8-dimethoxy-2-vinyl-1-naphthyl)-	EtOH	297(3.86),346(3.29)	49-0163-69
Propionic acid, 3-methoxy-2,3-diphenyl-, methyl ester, erythro	MeOH	252(3.85)	35-7534-69
threo	MeOH	252(3.85)	35-7534-69
$C_{17}H_{18}O_4$			
Cacalone, acetate	MeOH	206(3.80),258(3.95), 317(3.83)	23-2465-69
Dalbergione, dihydro-4,4'-dimethoxy-, (±)-	n.s.g.	264(4.23),350(3.04)	78-1407-69
$C_{17}H_{18}O_5$			
Acetophenone, 2'-hydroxy-4',5'-dimeth-oxy-2-(o-methoxyphenyl)-	EtOH	238(4.17),278(4.08), 345(3.96)	39-1787-69C
Acetophenone, 2-(3-hydroxy-4-methoxy-phenyl)-3',4'-dimethoxy-	EtOH	229(4.46),278(4.24), 320(4.10)	44-1062-69
4-Chromanone, 5,7-dihydroxy-6,8-dimeth-yl-2-(4-oxo-1-cyclohexen-1-yl)-	EtOH	215(4.50),298(4.25), 344(3.53)	95-0851-69
4-Chromone, 5,7-dihydroxy-6,8-dimethyl-2-(4-oxocyclohexyl)-	EtOH	230(4.22),264(4.16), 302(3.90),332(3.54)	95-0851-69
Eudesma-1,4,7(11)-trien-12-oic acid, 6,6-dihydroxy-3-oxo-, 12→6β-lactone, acetate	MeOH	241(4.18)	39-1088-69C
Isosequirin	EtOH	224(4.24),287(3.76)	39-1921-69C
Melanoxin	EtOH	207(4.53),233s(4.08), 291(3.73),305(3.80)	78-4409-69
Propiophenone, 4'-hydroxy-3'-methoxy-2-(o-methoxyphenoxy)-	EtOH	280(4.80),308(4.73)	44-0585-69
Santonene, 4α-acetoxy-	MeOH	219(4.05),290(4.29)	39-1088-69C
	EtOH	219(4.05),290(4.29)	39-1619-69C
Santonene, 4β-acetoxy-	MeOH	223(4.04),288(4.32)	39-1088-69C
	EtOH	223(4.04),288(4.32)	39-1619-69C
Santonene, 6α-acetoxy-	EtOH	241(4.18)	39-1619-69C
Santonene, 6β-acetoxy-	EtOH	242(4.21)	39-1619-69C
Santonene, 11-acetoxy-	EtOH	246(3.95),316(4.01)	39-1619-69C
Santonene, 11ξ-acetoxy-	MeOH	246(3.95),316(4.01)	39-1088-69C
Sequirin B, (2R,4S,5S)-	EtOH	224(4.18),280(3.51)	39-1921-69C

Compound	Solvent	λ_{max}(log ε)	Ref.
$C_{17}H_{18}O_5S$			
6H-Cyclohepta[c]thiophene-5,7-dicarboxylic acid, 1,3-dimethyl-6-oxo-, diethyl ester	EtOH	234(4.20),288(4.75)	104-0559-69
$C_{17}H_{18}O_6$			
Acetophenone, 2',4'-dihydroxy-2-(3,4-dimethoxyphenyl)-5'-methoxy-	EtOH	208(4.35),233(4.20), 282(4.17),344(3.93)	78-3887-69
p-Benzoquinone, 2-[2-hydroxy-5-(1-hydroxypropyl)-3-methoxyphenyl]-6-methoxy-	EtOH	258(4.08),286(3.64), 328(3.56)	44-0580-69
4-Chromanone, 5,7-dihydroxy-2-(1-hydroxy-4-oxo-2-cyclohexen-1-yl)-6,8-dimethyl- (protofarrerol)	EtOH	214(4.46),296(4.24),	95-0851-69
Isocladrastin	EtOH	207(4.48),218(4.46), 263(4.37),326(4.00)	78-3887-69
1,2-Naphthalenedicarboxylic acid, 6,7-dimethoxy-3-methyl-, dimethyl ester	EtOH	248(4.65),302(3.73), 341(3.67)	12-1721-69
Propiophenone, 3,4'-dihydroxy-3'-methoxy-2-(o-methoxyphenoxy)-	EtOH	280(5.12),310(5.08)	44-0585-69
Vernolepin, acetate	MeOH	208(4.26)(end abs.)	44-3903-69
Vernomenin, acetate	MeOH	210(4.24)	44-3903-69
$C_{17}H_{18}O_7$			
Canescin a,dimethyl-	EtOH	248(4.86),270(4.31), 279(4.35),290(4.25), 325(4.06)	12-1933-69
Elephantol, acetate	MeOH	209(4.34)	44-3867-69
$C_{17}H_{19}ClN_2O$			
3H-Azepine, 2-(butylamino)-3-benzoyl-5-chloro-	MeOH	248(4.10),297(3.75)	78-5205-69
3H-Azepine, 2-(diethylamino)-3-benzoyl-5-chloro-	MeOH	237(4.29),310(4.00)	78-5205-69
$C_{17}H_{19}ClN_2O_4$			
1-Butyl-2-phenyl-1H-imidazo[1,2-a]pyridin-4-ium perchlorate	EtOH	204(4.51),228(4.32),	44-2129-69
Methyl[3-(N-methylanilino)allylidene]-phenylammonium perchlorate	EtOH	348(4.60)	24-2609-69
$C_{17}H_{19}ClN_2S$			
Phenothiazine, 2-chloro-10-[3-(dimethylamino)propyl]-	EtOH	258(4.46),310(3.59)	9-0249-69
$C_{17}H_{19}ClN_4O_2$			
1-(2-Oxocyclohexyl)pyridinium chloride, (o-nitrophenyl)hydrazone	MeOH	264(4.29),424(3.43)	24-3647-69
1-(2-Oxocyclohexyl)pyridinium chloride, (p-nitrophenyl)hydrazone	MeOH	256(4.03),377(4.31)	24-3647-69
$C_{17}H_{19}ClN_4O_7$			
Purine, 6-chloro-9-(3-deoxy-xylohexopyranosyl)-, 2',4',6'-triacetate	n.s.g.	265(3.93)	22-3927-69
$C_{17}H_{19}N$			
Acridan, 9-butyl-	hexane	276(4.27)	78-1125-69
Acridan, 9-sec-butyl-	hexane	280(4.23)	78-1125-69
Acridan, 9-tert-butyl-	hexane	275(4.18)	78-1125-69
Acridan, 9,9-diethyl-	hexane	281(4.28)	78-1125-69

Compound	Solvent	λmax(log ε)	Ref.
$C_{17}H_{19}NO$			
Acetophenone, 2',6'-dimethyl-4'-o-toluidino-	C_6H_{12}	234s(4.00),288(4.17)	22-3523-69
	EtOH	208(4.51),238s(3.93), 294(4.06),330(3.92)	22-3523-69
Acetophenone, 3',5'-dimethyl-4'-o-toluidino-	C_6H_{12}	242(4.23),292(3.95), 314(4.02)	22-3523-69
	EtOH	206(4.66),246(4.25), 298s(3.76),332(4.04)	22-3523-69
Acetophenone, 2'-(2,4,6-trimethylanilino)-	C_6H_{12}	202(4.47),229(4.46), 248s(3.95),265(3.92), 374(3.87)	22-3523-69
	EtOH	199(4.69),230(4.41), 245s(4.03),265(3.96), 340s(3.34),382(3.87)	22-3523-69
Acetophenone, 4'-(2,4,6-trimethylanilino)-	C_6H_{12}	203(4.43),230s(4.00), 306(4.44)	22-3523-69
	EtOH	202(4.58),236(3.91), 295s(3.95),330(4.49)	22-3523-69
Benzofuro[2,3-g]isoquinoline, 3,4,7,8-9,10-hexahydro-5,11-dimethyl-	EtOH	249(4.67)	4-0379-69
4-Hepten-2-ynophenone, 5-(1-pyrrolidinyl)-	EtOH	263(4.24),408(4.60)	94-2126-69
Nitrone, N-benzyl-α-mesityl-	EtOH	234s(3.9),270(4.0)	28-0078-69A
$C_{17}H_{19}NO_2$			
Acetophenone, 4'-p-anisidino-2',6'-dimethyl-	heptane	214(4.03),240s(3.87), 294(4.03)	22-3523-69
	EtOH	204(4.54),240s(3.92), 295(4.08),330s(3.93)	22-3523-69
Benzamide, N-(benzyloxy)-2,4,6-trimethyl-	MeOH	261(2.73)	12-0161-69
	MeOH-NaOH	264(3.00)	12-0161-69
Benzyl alcohol, α-[(dimethylamino)methyl]-, benzoate	dioxan	257(3.07),264(3.05), 273(3.04),281(2.93)	65-1861-69
5H-Dibenz[b,f]azepine-4,6-dimethanol, 10,11-dihydro-N-methyl-	EtOH	214(4.33),250(3.70)	35-1672-69
1-Naphthaleneacetamide, 2-acetonyl-N,N-dimethyl-	EtOH	229(4.86),275(3.80), 285(3.85),295(3.69)	33-1030-69
1,4-Naphthoquinone, 2-(cyclohexylmethylamino)-	EtOH	237(4.17),273(4.30), 331s(3.36),465(3.65)	39-1245-69C
$C_{17}H_{19}NO_3$			
Crotsparinine	n.s.g.	228(4.28),285(3.10)	31-0354-69
6H-Cyclopent[g]isoquinoline-5-carboxylic acid, 7,8-dihydro-9-methoxy-7,7-dimethyl-, methyl ester	MeOH	231(4.65),298(3.41), 332(3.41)	44-0240-69
	MeOH-HCl	252(4.57),360(3.40)	44-0240-69
Furo[2,3-b]quinolin-4(9H)-one, 3-isobutyl-8-methoxy-9-methyl-	MeOH and MeOH-base	235(4.40),241s(4.45), 247(4.53),262s(3.86), 295s(3.66),303(3.73), 327s(3.80),339(3.98), 354(3.93)	44-2183-69
	MeOH-acid	248(4.59),303(3.60), 328s(3.71),340(3.88), 354(3.88)	44-2183-69
Nicotinic acid, 1,4-dihydro-1,2,5-trimethyl-4-oxo-6-phenyl-, ethyl ester	EtOH	216(4.36),271(4.20)	94-2209-69
$C_{17}H_{19}NO_4$			
Acrophylline, hydration product	EtOH	227(4.30),253s(4.66), 257s(4.69),262(4.79), 290(3.74),323(4.05)	12-0447-69

Compound	Solvent	λmax(log ε)	Ref.
Alkaloid from Galanthus Caucasicus	n.s.g.	210(2.72),228(4.32), 269(3.91)	105-0281-69
Narwedine	EtOH	225(4.22),265(3.55), 295(3.08)	39-2602-69C
Nitrone, N-benzyl-α-(2,4,6-trimethoxy-phenyl)-	EtOH	203(4.5),235(4.0), 298(4.1)	28-0078-69A
$C_{17}H_{19}NO_5$			
6a-Epi-N-demethyl-3-epimacronine	EtOH	228(4.40),268(3.76), 310(3.77)	35-0150-69
Pyrrole-2,5-dicarboxylic acid, 3-meth-oxy-4-phenyl-, diethyl ester	EtOH	283(4.24)	94-0582-69
$C_{17}H_{19}NS$			
Azuleno[1,2-b]thiophen-4-amine, N,N,3,5,9-pentamethyl-	C_6H_{12}	240(3.68),289(4.30), 310s(4.42),318(4.49), 330s(4.36),365(3.62), 382(3.65),398(3.43), 552s(2.57),559(2.66), 630(2.61),690(2.28), 710(2.20)	18-1404-69
$C_{17}H_{19}N_3$			
Indole, 3-[(dimethylamino)methyl]-5-methyl-2-(3-pyridyl)-	EtOH	217(4.26),312(4.04)	22-4154-69
$C_{17}H_{19}N_3O_2S$			
Benzenesulfonanilide, 4-cyano-4'-(di-ethylamino)-	EtOH-1% $CHCl_3$	240(4.29),268(4.41), 300s(3.62)	44-2083-69
Benzonitrile, p-[[2-amino-5-(diethyl-amino)phenyl]sulfonyl]-	EtOH-1% $CHCl_3$	248(4.38),275s(4.21)	44-2083-69
$C_{17}H_{19}N_3O_3$			
Acetanilide, 5'-[(p-ethoxyphenyl)azo]-2'-hydroxy-N-methyl-	EtOH	364(4.27)	22-4390-69
	EtOH-KOH	415(4.35)	22-4390-69
o-Anisamide, 5-[(p-methoxyphenyl)azo]-N,N-dimethyl-	EtOH	243(4.18),355(4.32), 435s(3.36)	12-0935-69
Isolongistrobine	EtOH	253(4.22)	78-2767-69
	EtOH-HCl	235(4.16)	78-2767-69
Longistrobine	EtOH	257(4.05)	78-2767-69
	EtOH-HCl	235(4.17)	78-2767-69
Nicotinic acid, 6-methyl-, (α-methyl-veratrylidene)hydrazide	n.s.g.	218(4.38),316(4.31)	40-0073-69
$C_{17}H_{19}N_3O_4$			
Isonicotinic acid, (3,4,5-trimethoxy-α-methylbenzylidene)hydrazide	n.s.g.	219(4.49),310(4.23)	40-0073-69
Nicotinic acid, (3,4,5-trimethoxy-α-methylbenzylidene)hydrazide	n.s.g.	219(4.47),307(4.28)	40-0073-69
Picolinic acid, (3,4,5-trimethoxy-α-methylbenzylidene)hydrazide	n.s.g.	220(4.44),316(4.31)	40-0073-69
$C_{17}H_{19}N_3O_5S$			
5-Thia-1-azabicyclo[4.2.0]oct-2-ene-2-carboxylic acid, 7-[2-amino-2-(p-methoxyphenyl)acetamido]-3-methyl-8-oxo-	H_2O	263(3.91)	87-0310-69

Compound	Solvent	$\lambda_{max}(\log \epsilon)$	Ref.
$C_{17}H_{19}N_3O_{11}S_2$			
as-Triazine-3,5(2H,4H)-dione, 2-β-D-ribofuranosyl-, 5'-benzoate, 2',3'-dimethanesulfonate	EtOH	232(4.23),261(3.85)	73-0618-69
$C_{17}H_{19}N_5$			
1H-Pyrazolo[3,4-d]pyrimidine, 6-methyl-1-phenyl-4-piperidino-	MeOH	298(4.25)	23-1129-69
Pyrazolo[3,4-d]pyrimidine, 4-piperidino-1-p-tolyl-	MeOH	244(4.49),301(4.31)	23-1129-69
$C_{17}H_{19}N_5O$			
4H-Quinolizine-1,3-dicarbonitrile, 2-[[2-(diethylamino)ethyl]amino]-4-oxo-	EtOH	258(4.61),303(4.18), 394(4.02)	95-0203-69
$C_{17}H_{19}N_5O_4$			
Adenosine, N-benzyl-	pH 1.0	264(4.30)	87-1056-69
	pH 7.0	268(4.29)	87-1056-69
	pH 12.0	268(4.29)	87-1056-69
9H-Purine, 2-amino-6-(benzyloxy)-9-(2-deoxy-α-D-erythro-pentofuranosyl)-	pH 1	287(4.10)	44-2160-69
	pH 11	249(4.00),280(4.08)	44-2160-69
	MeOH	249(4.04),282(4.10)	44-2160-69
$C_{17}H_{19}N_5O_6S$			
Adenosine, 5'-p-toluenesulfonate	MeOH	260(4.11)	78-0477-69
$C_{17}H_{20}BrP$			
Allylbenzylmethylphenylphsophonium bromide	MeOH	216s(4.20),255s(2.83), 260(2.99),266(3.08), 273(2.95)	24-3546-69
$C_{17}H_{20}ClN$			
2,11-Metheno-11H-1-benzazacyclotridecine, 16-chloro-3,4,5,6,7,8,9,10-octahydro-	EtOH	237(4.65),239s(4.63), 288(3.79),298(3.78), 312(3.62),327(3.61)	88-0557-69
$C_{17}H_{20}ClNO_4$			
5,6,7,8-Tetrahydro-1,3-dimethyl-2-phenylquinolinium perchlorate	EtOH	292(4.08)	94-2209-69
$C_{17}H_{20}FeO$			
Ketone, methyl 1',2-pentamethyleneferrocenyl	EtOH	230(4.18),269(3.83), 340(3.01),464(2.62)	78-5245-69
Ketone, methyl 1',3-pentamethyleneferrocenyl	EtOH	235(4.16),274s(3.78), 340(3.06),460(2.71)	78-5245-69
$C_{17}H_{20}FeO_2$			
Ferroceneacrylic acid, butyl ester	n.s.g.	247(4.3),291(4.3), 356(3.2),460(2.9)	65-1008-69
$C_{17}H_{20}NPS$			
Phenothiaphosphine-10-propylamine, N,N-dimethyl-	MeOH	263(3.68)	87-0146-69
$C_{17}H_{20}N_2$			
Benzimidazole, 2-(1-adamantyl)-	EtOH	243(3.85),275(3.97), 281(4.02)	18-1617-69
Benzophenone, tert-butylhydrazone	EtOH	222s(4.10),288(4.11)	39-1703-69C
Isoquinoline, 1-(o-aminobenzyl)-1,2,3,4-tetrahydro-2-methyl-	EtOH	261(3.80),287(3.28)	44-3784-69

Compound	Solvent	λ_{max}(log ϵ)	Ref.
$C_{17}H_{20}N_2O$			
Azobenzene, 5-tert-butyl-2-hydroxy-4'-methyl-	50% EtOH	334(4.30),386(4.01)	73-1087-69
	50% EtOH-$HClO_4$	305(3.54),407(4.31),491(4.21)	73-1087-69
Ketone, 3-ethyl-1,2,5,6-tetrahydro-1-methyl-4-pyridyl indol-2-yl	EtOH	236(4.10),315(4.28)	39-2738-69C
1,5-Methano[1,3]diazocino[1,8-a]indol-6(1H)-one, 13-ethyl-2,3,4,5-tetrahydro-2-methyl-	EtOH	242(4.09),322(4.20)	39-2738-69C
	EtOH-HCl	235(3.95),315(4.22)	39-2738-69C
$C_{17}H_{20}N_2O_2$			
Azobenzene, 5-tert-butyl-2-hydroxy-2'-methoxy-	50% EtOH	323(4.15),382(4.12),396(4.11)	73-1087-69
	50% EtOH-$HClO_4$	307(3.55),386(4.17),445s(--),508(4.28)	73-1087-69
Azobenzene, 5-tert-butyl-2-hydroxy-4'-methoxy-	50% EtOH	352(4.33),390s(--),425s(--)	73-1087-69
	50% EtOH-$HClO_4$	308(3.66),436(4.30),480s(--),508(4.42)	73-1087-69
$C_{17}H_{20}N_2O_2S$			
Benzamide, N-[(2-morpholino-1-cyclopenten-1-yl)thiocarbonyl]-	$CHCl_3$	308(4.06),436(4.00)	4-0037-69
1,2,4-Benzothiadiazine, 4-butyl-3,4-dihydro-2-phenyl-, 1,1-dioxide	EtOH	211(4.47),258(4.14),323(3.55)	7-0590-69
$C_{17}H_{20}N_2O_3$			
Benzaldehyde, 2,4,6-trimethoxy-, methylphenylhydrazone	EtOH	214s(4.3),246(4.2),335(4.4)	28-0139-69B
$C_{17}H_{20}N_2O_6$			
Uridine, 3-benzyl-6-methyl-	pH 7	210(4.22),263(4.10)	73-2316-69
$C_{17}H_{20}N_2O_7$			
Antimycin A_1, dehexyldeisovaleryloxy-	MeOH	226(4.62),320(3.93)	18-0854-69
$C_{17}H_{20}N_2S$			
5H,10H-Dipyrrolo[1,2-c:2',1'-f]pyrimidine-10-thione, 1,2,3,7,8,9-hexamethyl-5-methylene-	$CHCl_3$	305(4.30),424s(3.87),450(3.89)	12-0239-69
Phergan	EtOH	254(4.63),304(3.79)	9-0249-69
Promazine	MeOH	254(4.48)	87-0146-69
$C_{17}H_{20}N_4O$			
11H-Pyrimido[5,4-g]-1,5-benzodiazepin-11-one, 5,10-dihydro-10-butyl-2,4-dimethyl-	EtOH	212(4.76),243(4.63),275(4.57),303(4.43),368(4.50)	2-1166-69
$C_{17}H_{20}N_4O_2$			
6H-Pyrrolo[1,2-a]pyrimidin-4-one, 2-hydroxy-7,8-dihydro-3-phenyl-, 2-aminopyrroline salt	EtOH	267(3.93),292(3.80)	22-3133-69
$C_{17}H_{20}N_4O_3S_3$			
Spiro[furo[2,3-d]thiazole-2(3H),4'-imidazolidine]-2',5'-dithione, tetrahydro-1',3',3a-trimethyl-3-(p-nitrobenzyl)-	EtOH	321(4.18)	94-0910-69

Compound	Solvent	λ_{max}(log ε)	Ref.
$C_{17}H_{20}N_4O_4$			
2H-1,4a-Methanonaphthalen-2-one, octahydro-, 2,4-dinitrophenylhydrazone	EtOH	225(4.17)	78-5267-69
3H-2,8a-Methanonaphthalen-3-one, octahydro-, 2,4-dinitrophenylhydrazone	EtOH	225(4.13)	78-5267-69
7H-1,4a-Methanonaphthalen-7-one, octahydro-, 2,4-dinitrophenylhydrazone	$CHCl_3$	368(4.38)	78-5281-69
$C_{17}H_{20}N_4O_5$			
5(6H)-Indanone, 7aβ-ethyl-7,7a-dihydro-1β-hydroxy-, 2,4-dinitrophenylhydrazone	EtOH	256(4.22),286(4.44)	44-0107-69
2(1H)-Naphthalenone, 4a,5,6,7,8,8a-hexahydro-4a-(hydroxymethyl)-, 2,4-dinitrophenylhydrazone, trans	$CHCl_3$	380(4.47)	78-5281-69
$C_{17}H_{20}N_4O_7S$			
9H-Purine-6-thiol, 9-β-L-fucopyranosyl-, 2',3',4'-triacetate	pH 1	321(4.40)	4-0949-69
	pH 7	317(4.35)	4-0949-69
	pH 13	311(4.38)	4-0949-69
$C_{17}H_{20}N_4O_8$			
Hypoxanthine, 9-β-L-fucopyranosyl-, 2',3',4'-triacetate	pH 1	246(4.13)	4-0949-69
	pH 7	245(4.13)	4-0949-69
	pH 13	252(4.14)	4-0949-69
$C_{17}H_{20}N_6O_6$			
Inosine, 1,2-diamino-8-(benzyloxy)-	pH 1	248(3.89),294(3.81)	44-1025-69
	pH 11	251(3.99),283(3.75)	44-1025-69
	MeOH	253(3.75),284(3.75)	44-1025-69
$C_{17}H_{20}O$			
2-Bornanone, 3-benzylidene-, (Z)-	MeOH	206(4.25),228(3.99), 310(4.25)	56-0943-69
4,8,10,14,16-Heptadecapentaen-6-yn-3-one	EtOH	328(4.49)	24-1037-69
5,7,9,15-Heptadecatetraene-11,13-diyn-4-ol	ether	259(4.31),267(4.39), 275(4.35),320(4.58), 337(4.69),358(4.60)	24-3293-69
Pentalen-2-one, 3-benzylidene-perhydro-1,4-dimethyl-	n.s.g.	303(4.11)	40-0507-69
$C_{17}H_{20}O_2$			
14β-Gona-1,3,5(10),8-tetraene-11ξ,17β-diol	MeOH	228(4.48),264(4.21)	19-0469-69
5,7,9-Heptadecatriene-11,13-diyn-4-one, 17-hydroxy-, all trans	ether	216(4.22),342(4.73), 360(4.64)	24-3293-69
1-Phenanthrenecarboxylic acid, 1,2,3,4,9,10-hexahydro-1-methyl-, methyl ester	EtOH	264(4.09)	44-3739-69
$C_{17}H_{20}O_2S$			
5,8-Seco-B-dinorestra-1,3,5(10),8-tetraen-17β-ol, 5,8-epithio-3-methoxy-, cis	EtOH	233(4.46),268(4.00), 273(4.00)	88-4495-69
trans	EtOH	235(4.54),269(4.03), 274(4.03)	88-4495-69

Compound	Solvent	λ_{max}(log ϵ)	Ref.
$C_{17}H_{20}O_3$			
Cacalol, acetate	MeOH	215(4.49),255(4.13), 280(3.27),291(3.20)	23-2465-69
2-Cyclobutene-1-carboxylic acid, 4-benzoyl-1,2,3,4-tetramethyl-, methyl ester	MeOH	242(4.04)	24-1928-69
2-Cyclohexen-1-one, 2-(2,5-dimethylbenzyl)-3-hydroxy-, acetate	EtOH	203(4.20),218s(4.05), 238(3.96)	44-2192-69
1-Naphthaleneethanol, 7,8-dimethoxy-α-methyl-2-vinyl-	EtOH	297(3.87),346(3.30)	49-0163-69
1-Phenanthrenecarboxylic acid, 1,2,3,4,4a,9,10,10a-octahydro-1-methyl-9-oxo-, methyl ester	EtOH	248(4.08),294(3.27)	44-3739-69
isomer	EtOH	250(4.14),293(3.38)	44-3739-69
isomer	EtOH	250(4.11),290(3.33)	44-3739-69
$C_{17}H_{20}O_4$			
$\Delta^{10(1)}$-Furoeremophilene, 6-acetoxy-9-oxo-	EtOH	251(3.65),303(4.24)	73-1739-69
Indene-3-propionic acid, 2-(carboxymethyl)-7-methyl-, dimethyl ester	EtOH	258(4.1)	42-0415-69
Santonene, 3-O-acetyldihydro-	MeOH	219(3.63),330(4.46)	39-1088-69C
Spiro[cyclohexane-1,1'-indan]-3'-one, 4-hydroxy-6'-methoxy-, acetate	EtOH	225(4.26),271(4.17), 287(4.06),295s(4.04)	94-1564-69
$C_{17}H_{20}O_5$			
8-Benzocyclodecenecarboxylic acid, 5,6,7,8,9,10,11,12-octahydro-2-methoxy-5,10-dioxo-, methyl ester	EtOH	271(3.89)	2-1169-69
Santonene, 4α-acetoxy-1,2-dihydro-	MeOH	290(4.26)	39-1088-69C +39-1619-69C
Santonene, 6α-acetoxy-1,2-dihydro-	EtOH	247(4.23)	39-1619-69C
$C_{17}H_{20}O_6$			
Protofarrelol, dihydro-	EtOH	212(4.32),297(4.21), 348(3.48)	95-0851-69
Tetraneurin A	n.s.g.	210(3.99)	78-0805-69
$C_{17}H_{21}BrClP$			
(o-Chlorobenzyl)methylphenylpropylphosphonium bromide	MeOH	216s(4.16),255(2.81), 262(2.99),267(3.10), 273(3.01)	24-3546-69
(p-Chlorobenzyl)methylphenylpropylphosphonium bromide	MeOH	223(4.24),254s(2.81), 260(2.95),266(3.05), 273(2.96)	24-3546-69
$C_{17}H_{21}BrN_2O$			
1-Butyl-2,3-dihydro-2-hydroxy-2-phenyl-1H-imidazo[1,2-a]pyridin-4-ium bromide	EtOH	203(4.46),241(4.19), 333(3.70)	44-2129-69
$C_{17}H_{21}BrO_2$			
Laurenisol, acetate	EtOH	267(3.10),275(3.08)	88-1343-69
$C_{17}H_{21}BrO_3$			
Aplysinol, acetate	EtOH	233(3.71),292(3.42)	18-0843-69
$C_{17}H_{21}BrO_5$			
Gaillardin, bromo-	MeOH	211(4.19)	88-0973-69

Compound	Solvent	λ max (log ε)	Ref.
$C_{17}H_{21}ClN_2O_4$			
Michler's Hydrol Blue, perchlorate	CH_2Cl_2	302(4.39),368(3.96), 391(4.01),612(5.20)	48-0827-69
$C_{17}H_{21}ClN_2O_5S$			
α-[p-(Dimethylamino)phenyl]-3-morpholino-2-thenylium perchlorate	CH_2Cl_2	254s(3.64),306(4.06), 337(3.56),355(3.63), 377(3.63),591(4.92)	48-0827-69
α-[p-(Dimethylamino)phenyl]-5-morpholino-2-thenylium perchlorate	CH_2Cl_2	253s(3.77),305(4.08), 335(3.59),353(3.67), 376(3.62),588(4.91)	48-0827-69
$C_{17}H_{21}ClN_2O_6S_2$			
Bis(3-morpholino-2-thienyl)methylium perchlorate	CH_2Cl_2	249s(3.72),276s(3.52), 340(3.72),393(3.89), 447(3.87),586(4.45)	48-0827-69
Bis(5-morpholino-2-thienyl)methylium perchlorate	CH_2Cl_2	315(4.26),358(3.54), 379(3.44),599(5.16)	48-0827-69
(3-Morpholino-2-thienyl)(5-morpholino-2-thienyl)methylium perchlorate	CH_2Cl_2	257(3.88),298(3.63), 312s(3.59),331s(3.39), 357(3.36),426(3.82), 540s(4.51),580(4.67)	48-0827-69
$C_{17}H_{21}FN_2O_2$			
Indole-3-acetic acid, 5-fluoro-2-(3-piperidyl)-, ethyl ester	EtOH	222(4.36),280s(3.91), 287(3.93),296(3.78)	22-4154-69
$C_{17}H_{21}FeNO_4S$			
Proline, 1-(ferrocenylsulfonyl)-, ethyl ester	EtOH	433(2.16)	49-1552-69
$C_{17}H_{21}N$			
2,11-Metheno-11H-benzazacyclotridecine, 3,4,5,6,7,8,9,10-octahydro-	EtOH	229(4.62),278(3.65), 292s(3.61),304(3.49), 317(3.48)	88-0557-69
$C_{17}H_{21}NO_3$			
Acrophylline, tetrahydro-	EtOH	222(4.35),249(4.45), 300s(4.03),306(4.07)	12-0447-69
	0.2M HCl	225(4.45),246(4.46), 302s(3.86),321(4.09)	12-0447-69
2-Epigalanthamine, (±)-	EtOH	234(3.80),290(3.28)	39-2602-69C
Galanthamine, (±)-	EtOH	234(3.78),290(3.72)	39-2602-69C
Piperidine, 1-[5-[3,4-(methylenedioxy)-phenyl]-2-pentenoyl]-	EtOH	232(4.07),285(3.64)	12-1531-69
Piperidine, 1-[5-[3,4-(methylenedioxy)-phenyl]-3-pentenoyl]-	EtOH	232(3.75),287(3.60)	12-1531-69
$C_{17}H_{21}NO_4$			
Spiro[cyclohexane-1,4'(1'H)-isoquinolin]-1'-one, 2',3'-dihydro-4-hydroxy-6'-methoxy-, acetate	EtOH	263(4.14)	94-1564-69
$C_{17}H_{21}NO_5$			
Propionic acid, 2,3-dihydroxy-3-phenyl-, scopine ester, hydrochloride	EtOH	251(1.95),257(2.10), 263(2.00)	5-0152-69I

Compound	Solvent	λ_{max}(log ϵ)	Ref.
$C_{17}H_{21}NO_8$			
3,4-Furandicarboxylic acid, 5-(tert-butylimino)-2-(carboxyethynyl)-2,5-dihydro-2-methoxy-, trimethyl ester	ether	256(4.06)	24-1656-69
$C_{17}H_{21}NO_8S$			
Thiazolo[3,2-a]azepine-5,6,7,8-tetracarboxylic acid, 2,3,5,6-tetrahydro-9-methyl-, tetramethyl ester	EtOH	218(4.06),270(3.80), 397(4.33)	28-0615-69B
	EtOH-$HClO_4$	214(3.84),290(4.08)	28-0615-69B
8aH-Thiazolo[3,2-a]pyridine-5,6,7,8-tetracarboxylic acid, 2,3-dihydro-3,3-dimethyl-, tetramethyl ester	EtOH	225(4.09),275(4.14), 376(3.75)	28-0870-69A
8aH-Thiazolo[3,2-a]pyridine-5,6,7,8-tetracarboxylic acid, 3-ethyl-2,3-dihydro-, tetramethyl ester	EtOH	225(4.09),276(4.14), 375(3.77)	28-0870-69A
8aH-Thiazolo[3,2-a]pyridine-5,6,7,8-tetracarboxylic acid, 8a-ethyl-2,3-dihydro-, tetramethyl ester	EtOH	221(4.06),280(4.08), 370(3.78)	28-0870-69A
$C_{17}H_{21}N_2$			
Michler's Hydrol Blue (cation)	98% HOAc	608(5.17)	39-1068-69B
$C_{17}H_{21}N_2O_6P$			
1-(5-Nitrosalicyl)pyridinium hydroxide, pentyl hydrogen phosphate, inner salt	H_2O	261(3.95),268(3.95), 292(3.96)	35-4532-69
$C_{17}H_{21}N_3$			
Eleocarpidine	EtOH	226(4.60),283(3.90), 290s(3.83)	12-0801-69
$C_{17}H_{21}N_3OS_3$			
Spiro[furo[2,3-d]thiazole-2(3H),4'-imidazolidine]-2',5'-dithione, 3-benzyltetrahydro-1',3',3a-trimethyl-	EtOH	322(4.19)	94-0910-69
$C_{17}H_{21}N_3O_2$			
Cyclohexanone, 2-(4,5-dimethyl-2-phenylimidazol-1-yl)-, oxime, 3-oxide	MeOH	287(4.18)	48-0746-69
$C_{17}H_{22}BrP$			
Benzylisopropylmethylphenylphosphonium bromide	MeOH	216s(4.19),255s(2.79), 260(2.95),266(3.03), 273(2.90)	24-3546-69
Benzylmethylphenylpropylphosphonium bromide	MeOH	217s(4.16),255s(2.87), 260(3.02),266(3.10), 273(2.99)	24-3546-69
$C_{17}H_{22}NP$			
Propylamine, 3-(diphenylphosphino)-N,N-dimethyl-	MeOH	248(3.95)	87-0146-69
$C_{17}H_{22}N_2$			
1,3-Propanediamine, N',N'-dimethyl-1,1-diphenyl-	MeOH	248(3.92)	87-0146-69
$C_{17}H_{22}N_2O$			
Indole-2-methanol, α-(3-ethyl-1,2,5,6-tetrahydro-1-methyl-4-pyridyl)-	EtOH	221(4.47),274(3.85), 282(3.85),291(3.76)	39-2738-69C

Compound	Solvent	λ_{max}(log ε)	Ref.
1,5-Methano[1,3]diazocino[1,8-a]indol-6-ol, 13-ethyl-1,2,3,4,5,6-hexahydro-2-methyl-	EtOH	228(4.31),277(3.89), 283(3.90),292(3.80)	39-2738-69C
	EtOH-HCl	223(4.18),267(3.99), 279(3.83),290(3.53)	39-2738-69C
$C_{17}H_{22}N_2O_2$			
Indole-3-acetic acid, 2-(3-piperidyl)-, ethyl ester	EtOH	228(4.54),282(3.97), 290(3.90)	22-4154-69
$C_{17}H_{22}N_2O_2S$			
p-Phenylenediamine, N^4,N^4-diethyl-2-(p-tolylsulfonyl)-	EtOH-1% $CHCl_3$	239(4.41),270(4.20)	44-2083-69
p-Toluenesulfonanilide, 4'-(diethylamino)-	EtOH-1% $CHCl_3$	226(4.19),268(4.30), 300s(3.68)	44-2083-69
$C_{17}H_{22}N_2O_3$			
2H-Pyrrole-2-propionic acid, 5-(dimethylamino)-3-hydroxy-4-phenyl-, ethyl ester	MeOH	224(4.18),287(4.32)	78-5721-69
$C_{17}H_{22}N_2O_5$			
Cinnamamide, N-(3-acetamidopropyl)-4-hydroxy-3-methoxy-, acetate, trans	n.s.g.	276(4.31),299s(4.04)	88-2807-69
$C_{17}H_{22}N_4O_6$			
Bicarbamic acid, (2-benzimidazolinylidenecarboxymethyl)-, triethyl ester	n.s.g.	323(4.16)	88-1203-69
$C_{17}H_{22}O$			
Cyclopent[a]inden-8(1H)-one, 2,3,3a,8a-tetrahydro-2-isopropyl-3,5-dimethyl-	EtOH	256(4.17),290(3.43)	2-0215-69
4,8,10,14,16-Heptadecapentaen-6-yn-3-ol, all trans	ether	294(4.51),310(4.55)	24-1037-69
2,7,9,13-Heptadecatetraen-11-ynal, all trans	ether	294(4.56),310(4.52)	24-1037-69
1,7,9,13-Heptadecatetraen-6-yn-3-one, all trans	ether	328(4.49)	24-1037-69
Nervogenic acid	MeOH	258(4.08)	78-2723-69
	MeOH-base	291(4.23)	78-2723-69
Phenanthrene, 1-ethyl-1,2,3,4,4a,10a-hexahydro-6-methoxy-, (-)-, A	EtOH	273(4.17)	94-1158-69
isomer B	EtOH	273(4.16)	94-1158-69
$C_{17}H_{22}O_2$			
6H-Dibenzo[b,d]pyran-1-ol, 6a,7,8,10a-tetrahydro-3,6,6,9-tetramethyl-	EtOH	230s(3.98),276(3.15), 283(3.17)	33-1102-69
6H-Dibenzo[b,d]pyran-1-ol, 6a,7,10,10a-tetrahydro-3,6,6,9-tetramethyl-	EtOH	230s(4.02),275(3.18), 282(3.21)	33-1102-69
20-Nordeisopropyldehydroabietic acid, methyl ester	EtOH	266(2.74),274(2.67)	44-3739-69
1α-Phenanthrenecarboxylic acid, 1,2,3,4,4a,9,10,10a-octahydro-1β-methyl-, methyl ester	EtOH	266(2.75),273(2.73)	44-3739-69
1β-Phenanthrenecarboxylic acid, 1,2,3,4,4a,9,10,10a-octahydro-1α-methyl-, methyl ester	EtOH	266(2.62),274(2.69)	44-3739-69
9(1H)-Phenanthrone, 1-ethyl-2,3,4,4a,-10,10a-hexahydro-6-methoxy-, (-), A	EtOH	278(4.21)	94-1158-69
isomer B	EtOH	227(4.14),277(4.22)	94-1158-69

Compound	Solvent	$\lambda_{max}(\log \epsilon)$	Ref.
$C_{17}H_{22}O_3$			
6-Chromancarboxylic acid, 2,2-dimethyl-8-(3-methyl-2-butenyl)-	MeOH	265(4.11)	78-2723-69
	MeOH-base	255(4.06)	78-2723-69
1H-Cyclopropa[a]naphthalene-1-carboxylic acid, 1a,2,3,7b-tetrahydro-6-methoxy-1,7-dimethyl-, ethyl ester	EtOH	280(3.33)	88-4559-69
$\Delta^{10(1)}$-Furoeremophilan-9-one, 6-ethoxy-	EtOH	303(4.16)	73-1739-69
1,4-Phenanthrenedione, 4a,4b,5,6,7,8,-10,10a-octahydro-3-methoxy-8,8-dimethyl-	EtOH	273(3.93)	18-3318-69
β-Resorcylaldehyde, 3-(3,7-dimethyl-2,6-octadienyl)-	MeOH	231(3.95),290(4.13)	32-0308-69
β-Resorcylaldehyde, 5-(3,7-dimethyl-2,6-octadienyl)-	MeOH	238(4.01),285(4.05), 324(3.74)	32-0308-69
$C_{17}H_{22}O_4$			
2-Naphthoic acid, 1,2,3,4-tetrahydro-8-isopropyl-7-methoxy-5-methyl-1-oxo-, methyl ester	EtOH	262(3.85),328(3.46)	18-3011-69
$C_{17}H_{22}O_5$			
Artecalin	n.s.g.	210(4.0)	102-1297-69
Baileyin, acetate	EtOH	210(3.92)	102-2371-69
Crotonic acid, 2-(hydroxymethyl)-, ester with 3-hydroxy-4,6,6-trimethyl-1,4-cyclohexadiene-1-carboxaldehyde, acetate, cis	ether	219(4.35)	24-2211-69
β-Cyclopyrethrosin	MeOH	210(3.94)(end abs.)	23-1139-69
Naphtho[1,2-c]furan-1,4-dione, 3,5,5a,6,7,8,9,9a-octahydro-5β-hydroxy-6,6,9a-trimethyl-, acetate	EtOH	250(4.00)	78-2887-69
	EtOH-HCl	224(3.70),250(3.40)	78-2887-69
	EtOH-NaOH	245(3.93)(changing)	78-2887-69
Naphtho[1,2-c]furan-1,5(3H,4H)-dione, 5a,6,7,8,9,9a-hexahydro-4-hydroxy-6,6,9a-trimethyl-, acetate	EtOH	210(4.04)	78-2887-69
	EtOH-HCl	223(4.04),252(3.60), 347(3.30)	78-2887-69
	EtOH-NaOH	243(4.11),455(3.18)	78-2887-69
Pleniradin, acetate	EtOH	209(4.15)(end abs.)	102-1753-69
Pulchellin E	n.s.g.	210(4.15)	102-0661-69
15,16,17-Trinortiglia-1,6-diene-3,13-dione, 4,9,20-trihydroxy-	MeOH	195(3.96),242(3.63), 330(1.85)	5-0158-69H
$C_{17}H_{22}O_6$			
6-Isochromanacetic acid, 4a,5,6,7,8,8a-hexahydro-5β,7β-dihydroxy-α,4-dimethylene-3-oxo-8aβ-vinyl-, ethyl ester (vernolepin ethanol adduct)	MeOH	208(4.18)	44-3903-69
$C_{17}H_{22}O_{10}$			
2,4-Hexadiene-1,1,2,3,4-pentacarboxylic acid, 5-methyl-, pentamethyl ester	EtOH	230(4.04)	33-0584-69
$C_{17}H_{23}ClO_4$			
1,2-Di-tert-butyl-3-phenylcyclopropenium perchlorate	MeCN	264(4.29)	35-4766-69
Naphtho[1,2-c]furan-3(1H)-one, 4α-chloro-4,5,5a,6,7,8,9,9a-octahydro-5β-hydroxy-6,6,9aβ-trimethyl-, acetate	n.s.g.	216(4.12)	78-3903-69
$C_{17}H_{23}Cl_2NO$			
Acrylophenone, 3-[bis(2-chloroethyl)-amino]-2',3',5',6'-tetramethyl-, (E)-	EtOH	208(4.00),215(3.97), 297(4.43)	24-3139-69

Compound	Solvent	$\lambda_{max}(\log \epsilon)$	Ref.
$C_{17}H_{23}NO$			
9-Aza-D-homogona-5(10),13(14)-dien-4-one	MeOH	318(4.54)	19-0079-69
$C_{17}H_{23}NO_2$			
Morpholine, 4-[2-(3,4-dihydro-7-methoxy-1-naphthyl)ethyl]-, hydrochloride	n.s.g.	259(3.82)	39-0217-69C
$C_{17}H_{23}NO_3$			
Acrophylline, hexahydro-	EtOH and EtOH-HCl	226(4.76),243s(4.18), 252(4.13),286(3.91), 316(4.18),329(4.19)	12-0447-69
Carbostyril, 3-ethyl-4-(isopentyloxy)-7-methoxy-	EtOH and EtOH-HCl	218(4.76),222s(4.59), 254(3.89),277(3.86), 309s(4.04),321(4.24), 335(4.20)	12-0447-69
Littorine	EtOH	243(2.02),248(2.10), 253(2.22),259(2.30), 264(2.20),268(2.03)	12-0221-69
$C_{17}H_{23}NO_8$			
3,4-Furandicarboxylic acid, 2-(tert-butylamino)-5-(2-carboxy-1-methoxy-vinyl)-, trimethyl ester	MeOH	224(4.18),320(4.21), 395(4.12)	24-1656-69
$C_{17}H_{23}N_3O_2$			
5,8-Isoquinolinedione, 7-[[3-(diethyl-amino)propyl]amino]-3-methyl-	MeCN	238(4.26),278(4.21), 320(3.68)	4-0697-69
$C_{17}H_{23}N_3O_2S$			
Acetamide, N-(1,4,5,6,7,8-hexahydro-4-oxospiro[[1]benzothieno[2,3-d]pyrim-idin-2(3H),1'-cyclohexan]-3-yl)-	MeOH	232(4.17),261s(3.68), 319(3.38)	48-0402-69
$C_{17}H_{23}N_5O_{13}$			
Polyoxin E	0.05N HCl	276(4.00)	35-7490-69
	0.05N NaOH	271(3.81)	35-7490-69
$C_{17}H_{23}N_5O_{14}$			
Polyoxin D	0.05N HCl	276(4.05)	35-7490-69
	0.05N NaOH	271(3.85)	35-7490-69
$C_{17}H_{24}$			
Cyclopent[a]indene, 1,2,3,3a,8,8a-hexa-hydro-2-isopropyl-3,5-dimethyl-	EtOH	259(3.54),266(3.60), 270(3.60),274(3.59), 280(3.53)	2-0215-69
$C_{17}H_{24}N_2O$			
Cyclododeca[c]pyridine-4-carbonitrile, 5,6,7,8,9,10,11,12,13,14-decahydro-3-hydroxy-1-methyl-	EtOH	217(4.29),221(4.29), 239s(3.83),249(3.59), 341(4.10)	44-3670-69
$C_{17}H_{24}N_2O_8$			
2,4-Hexadiene-1,2,3,4-tetracarboxylic acid, 5-methyl-1-oxo- , tetramethyl ester, mono(dimethylhydrazone)	MeOH	282(3.89),349(3.51)	44-2248-69
2,3,4,5-Pyridinetetracarboxylic acid, 1-(dimethylamino)-1,6-dihydro-6,6-dimethyl-, tetramethyl ester	MeOH	271(4.04),354(3.76)	44-2248-69

Compound	Solvent	$\lambda_{max}(\log \epsilon)$	Ref.
$C_{17}H_{24}O$			
1'-Acetonaphthone, 5',6',7',8'-tetra-hydro-3'-isopropyl-5',5'-dimethyl-	EtOH	254(3.85),297(2.90)	54-0313-69
2'-Acetonaphthone, 5',6',7',8'-tetra-hydro-3'-isopropyl-5',5'-dimethyl-	EtOH	256(4.02),293(3.21)	54-0313-69
Cyclopent[a]indene, 2-hydroxy-2-isopro-pyl-3,5-dimethyl-1,2,3,3a,8,8a-hexa-hydro-	EtOH	266(3.16),274(3.28), 280(3.31)	2-0215-69
1,9-Heptadecadiene-4,6-diyn-3-ol, cis	ether	230(2.60),242(2.60), 256(2.30)	39-0685-69C
1,7,9,13-Heptadecatetraen-11-yn-3-ol	ether	294(4.56),310(4.52)	24-1037-69
2,7,9,13-Heptadecatetraen-11-yn-1-ol	ether	294(4.57),310(4.52)	24-1037-69
4,8,10,16-Heptadecatetraen-6-yn-3-ol, all trans	ether	294(4.51),310(4.55)	24-1037-69
$C_{17}H_{24}O_2$			
Benzo[1,2:3,4]dicycloheptane-6-carbox-ylic acid, 1,2,3,4,5,5a,7,8,9,10,11,-12-dodecahydro-	EtOH	212(4.00)	22-4493-69
A,B-Dinorandrost-3(5)-en-2-one, 17β-hydroxy-	EtOH	235(4.13)	44-2297-69
1,9-Heptadecadiene-4,6-diyne-3,8-diol, cis	ether	232(2.60),244(2.60), 258(2.30)	39-0685-69C
Phenol, 2-isobornyl-6-methoxy-	pH 13	248(3.88),293(3.72)	30-0867-69A
	iso-PrOH	224(3.91),277(3.38), 282(3.37)	30-0867-69A
Phenol, 4-isocamphyl-2-methoxy-	pH 13	242(4.07),296(3.68)	30-0867-69A
	iso-PrOH	225(3.86),282(3.45), 287s(3.38)	30-0867-69A
Phenol, 5-isocamphyl-2-methoxy-	pH 13	242(3.85),293(3.67)	30-0867-69A
	iso-PrOH	225(3.82),282(3.50), 290s(3.38)	30-0867-69A
Phenol, 6-isocamphyl-2-methoxy-	pH 13	246(3.88),291(3.68)	30-0867-69A
	iso-PrOH	275(3.34),279(3.33)	30-0867-69A
$C_{17}H_{24}O_2S_2$			
10α-Eremophila-7,11-dien-9-one, 8,12-epoxy-2α-hydroxy-, cyclic ethylene mercaptole	EtOH	239(3.93)	78-2865-69
$C_{17}H_{24}O_3$			
10βH-Furoeremophilan-9-one, 6α-ethoxy-	EtOH	222(3.60),287(4.09)	73-2792-69
Shogaol	C_6H_{12}	227(3.91),282(3.42), 287(3.37)	12-1033-69
Unnamed lactone	n.s.g.	231(4.10)	78-2233-69
$C_{17}H_{24}O_4$			
Bemarivolide	isooctane	217(3.96)	78-3903-69
Confertifolin, 6β-acetoxy-	MeOH	216(4.15)	78-3895-69
3-Cyclohexen-1-ol, 4-(3-hydroxy-1-but-yryl)-3,3,5-trimethyl-, diacetate	EtOH	229(4.05)	77-0085-69
$C_{17}H_{24}O_4S$			
α-D-Lyxopyranoside, ethyl 4-S-benzyl-2,3-O-isopropylidene-4-thio-	MeOH	260(3.49),265(3.34)	44-2643-69
$C_{17}H_{24}O_5$			
Cinnamodial (ugandensidial)	isooctane	219(4.07)	78-3895-69
	C_6H_{12}	219(4.15)	78-2887-69
Epilaccishellolic acid, dimethyl ester	EtOH	226(3.75)	78-3855-69

Compound	Solvent	λ max (log ε)	Ref.
Ugandensolide	EtOH	214(4.11)	78-2887-69
$C_{17}H_{24}O_7$			
1-Cyclopentene-1-heptanoic acid, 2-carboxy-3-hydroxy-5-oxo-, dimethyl ester, acetate	MeOH	238(4.11)	88-4639-69
$C_{17}H_{24}O_{10}$			
Secologanin	EtOH	235(3.96)	39-1187-69C
$C_{17}H_{24}O_{11}$			
Scandoside, methyl ester	EtOH	239(4.04)	94-1942-69
Theviridoside	n.s.g.	234(3.9)	33-0478-69
$C_{17}H_{25}Cl_6N_3S$			
s-Triazine, 2-(dodecylthio)-4,6-bis-(trichloromethyl)-	MeOH	214(3.43),274(4.13)	18-2931-69
$C_{17}H_{25}N$			
Naphthalene-$\Delta^{1(2H)}$,γ-propylamine, 3,4-dihydro-N,N,2,2-tetramethyl-, hydrochloride	n.s.g.	245(4.09)	39-0217-69C
$C_{17}H_{25}NO_2$			
Undecenohydroxamic acid, N-phenyl-	EtOH	247(4.11)	42-0831-69
$C_{17}H_{25}NO_8S_3$			
α-D-Mannopyranose, 1,2-dithio-, 2,3,4,6-tetraacetate 1-(dimethyldithiocarbamate)	EtOH	278(4.01)	94-2571-69
$C_{17}H_{25}N_3$			
Cyclohexanone, 2-piperidino-, phenylhydrazone	MeOH	280(4.24)	24-3647-69
$C_{17}H_{25}N_5O_{12}$			
Polyoxin G	0.05N HCl	262(3.92)	35-7490-69
	0.05N NaOH	264(3.82)	35-7490-69
Polyoxin J	0.05N HCl	264(3.91)	35-7490-69
	0.05N NaOH	267(3.81)	35-7490-69
$C_{17}H_{25}N_5O_{13}$			
Polyoxin B	0.05N HCl	262(3.94)	35-7490-69
	0.05N NaOH	264(3.82)	35-7490-69
$C_{17}H_{26}ClNO_4$			
6-tert-Butyl-1,2,3,4,7,8,9,10-octahydrobenzo[c]quinolizinium perchlorate	EtOH	284(3.79)	78-4161-69
$C_{17}H_{26}N_2$			
Benzimidazole, 4,5,7-triisopropyl-2-methyl-	EtOH	250(3.90),274(3.63), 283(3.59)	78-1423-69
$C_{17}H_{26}N_2O_2$			
5,8-Ethano-1H-pyrazolo[1,2-a]pyridazine-1,3(2H)-dione, 2,2-diethyl-5,8-dihydro-10-isopropyl-6-methyl-	C_6H_{12}	256(3.58)	44-3181-69
	EtOH	247(3.62)	44-3181-69
$C_{17}H_{26}N_2O_3$			
Acetanilide, 2',4',5'-triisopropyl-3'-nitro-	C_6H_{12}	242(3.96)	78-1423-69

Compound	Solvent	λ_{max}(log ε)	Ref.
$C_{17}H_{26}N_4O_2S_3$			
Thiamine, isobutylxanthogenate	EtOH	230(4.24),272(3.89), 304(4.16)	94-2299-69
$C_{17}H_{26}O$			
2,5-Cyclohexadien-1-one, 2,6-di-tert-butyl-4-propylidene-	hexane	300(4.21)	70-0580-69
$C_{17}H_{26}O_2$			
Benzo[1,2:3,4]dicycloheptene-6-carboxylic acid, 1,2,3,4,5,5a,6,7,7a,8,9,-10,11,12-tetradecahydro-, cis-cis	EtOH	208(3.91)	22-4493-69
cis-trans	EtOH	213(3.86)	22-4493-69
2,5-Cyclohexadien-1-one, 2,6-di-tert-butyl-4-(ethoxymethylene)-	hexane	325(4.18)	70-0580-69
2-Heptene, 6-methoxy-2-(3-methoxy-p-tolyl)-6-methyl-	MeOH	249(3.92),280(3.42)	88-3185-69
2-Norcarene-3-carboxylic acid, 7-methyl-7-(4-methyl-3-pentenyl)-, ethyl ester	EtOH	252(4.02)	88-1837-69
1-Oxaspiro[2.5]octa-4,7-dien-6-one, 5,7-di-tert-butyl-2,2-dimethyl-	n.s.g.	261(4.16)	70-1732-69
9-Phenanthrenecarboxylic acid, 1,2,3,4,5,6,7,8,8a,9,10,10a-dodecahydro-10-methyl-, methyl ester, cis-anti-trans	EtOH	209(4.06)	22-0962-69
2(3H)-Phenanthrone, 8aβ-ethyl-4,4a,4bα,5,6,7,8,8a,9,10-decahydro-8-hydroxy-1-methyl-	EtOH	250(4.23)	44-0107-69
$C_{17}H_{26}O_3$			
Butyrophenone, 2'-hydroxy-4'-(heptyloxy)-	n.s.g.	280(3.84)	104-1609-69
2,4-Cyclohexadien-1-one, 6-acetonyl-2,4-di-tert-butyl-6-hydroxy-	C_6H_{12}	312(3.53)	64-0547-69
Hydronervogenic acid	MeOH	258(4.06)	78-2723-69
	MeOH-base	291(4.23)	78-2723-69
β-Ionylideneacetic acid, 3-methoxy-, methyl ester	EtOH	255(3.88),305(3.95)	22-0232-69
2,4-Pentadienoic acid, 5-(2-methoxy-2,6,6-trimethylcyclohexylidene)-3-methyl-, methyl ester	EtOH	271(4.32)	22-0232-69
$C_{17}H_{26}O_4$			
2-Norcaranone, 4-hydroxy-3-isopropyl-6-methyl-7-(3-oxobutyl)-, acetate	MeOH	207(3.88)	88-3357-69
2,4-Pentadienoic acid, 5-(1-hydroxy-4-methoxy-2,2-dimethyl-6-methylene-cyclohexyl)-3-methyl-, methyl ester	EtOH	262(4.21)	22-0232-69
isomer	EtOH	262(4.24)	22-0232-69
2,4-Pentadienoic acid, 5-(2-hydroxy-4-methoxy-2,6,6-trimethylcyclohexylidene)-3-methyl-, methyl ester	EtOH	268(4.25)	22-0232-69
isomer	EtOH	268(4.22)	22-0232-69
Photosantonic acid, lactone, O-ethyl-	EtOH	240(4.00)	33-1237-69
$C_{17}H_{26}O_5$			
1-Cyclohexenebutyric acid, 2-(2-carboxyethyl)-3-oxo-, diethyl ester	EtOH	244(4.12)	78-2661-69
β-Ionylideneacetic acid, 3-methoxy-, methyl ester, peroxide	EtOH	218(4.19)	22-0232-69

Compound	Solvent	$\lambda_{max}(\log \epsilon)$	Ref.
β-Ionylideneacetic acid, 3-methoxy-, methyl ester, peroxide, isomer	EtOH	218(4.18)	22-0232-69
$C_{17}H_{26}O_{11}$			
Ipolamiide	MeOH	229(4.03)	32-1150-69
$C_{17}H_{26}O_{12}$			
Lamiide	MeOH	229(4.02)	32-1150-69
$C_{17}H_{26}Si_2$			
2,6-Disilahepta-3,4-diene, 2,2,6,6-tetramethyl-3-phenyl-5-vinyl-	n.s.g.	246(4.30)	101-0291-69C
2,6-Disilahept-3-ene, 2,2,6,6-tetramethyl-3-(phenylethynyl)-	n.s.g.	219(4.28),286(4.25)	101-0291-69C
$C_{17}H_{27}NO$			
Acetanilide, 2',4',5'-triisopropyl-	hexane	240(3.72)	78-1423-69
2,5-Cyclohexadien-1-one, 2,6-di-tert-butyl-4-[(dimethylamino)methylene]-	hexane	380(4.39)	70-0580-69
$C_{17}H_{27}NO_2$			
Bicyclo[7.3.1]tridec-9(13)-en-10-one, 13-morpholino-	EtOH	342(4.29)	39-0592-69C
Decanohydroxamic acid, N-m-tolyl-	EtOH	248(4.01)	42-0831-69
Decanohydroxamic acid, N-p-tolyl-	EtOH	248(4.02)	42-0831-69
$C_{17}H_{27}NO_5S$			
2,6-Pyridinediol, 1-acetyl-3-(tert-butylthio)-1,2,3,6-tetrahydro-3,5-dimethyl-, diacetate	hexane	198(4.38)	44-0660-69
	MeOH	204(4.33)	44-0660-69
$C_{17}H_{27}NO_8$			
3,4-Furandicarboxylic acid, 5-(tert-butylamino)-2-(2-carboxyethyl)-2,5-dihydro-2-methoxy-, trimethyl ester	MeOH	242(3.95)	24-1656-69
$C_{17}H_{28}ClNO_4$			
6-tert-Butyl-1,2,3,4,4a,5,7,8,9,10-decahydrobenzo[c]quinolizinium perchlorate	EtOH	305(3.68)	78-4161-69
$C_{17}H_{28}N_2O_3$			
Benzoic acid, 4-amino-3-butoxy-, 2-(diethylamino)ethyl ester	EtOH	207(4.25),231(4.11), 308(4.19)	9-0254-69
$C_{17}H_{28}OS$			
Cyclohexanone, 6-[(butylthio)methylene]-2,3-dimethyl-2-(2-methylallyl)-	MeOH	281(4.17)	23-0831-69
$C_{17}H_{28}OS_2$			
2-Bornanone, 3-carbonyl-, 3-(dipropyl mercaptole)	dioxan	324(4.05),380s(2.63)	28-0186-69A
$C_{17}H_{28}O_2$			
Benzo[1,2:3,4]dicycloheptene-6-methanol, 1,2,3,4,5,5a,6,7,7a,8,9,10,11,12-tetradecahydro-6-hydroxy-	EtOH	211(3.94)	22-4493-69

Compound	Solvent	λ_{max}(log ε)	Ref.
$C_{17}H_{28}O_3$			
2-Cyclotetradecene-1-carboxylic acid, 2-methyl-14-oxo-, methyl ester	EtOH	204(3.83),207s(3.8), 248f(3.52)	94-0629-69
$C_{17}H_{28}O_3S$			
9-Phenanthrenemethanol, 1,2,3,4,5,6,7,-8,8a,9,10,10a-dodecahydro-9-methyl-, methanesulfonate	EtOH	210(4.09)	22-0962-69
$C_{17}H_{28}O_4$			
2-Butanone, 4-(3,5-dihydroxy-4-isopropyl-1-methyl-7-norcaryl)-, 3-acetate	MeOH	275(1.85)	88-3357-69
$C_{17}H_{29}ClN_2O$			
4(3H)-Pyrimidinone, 6-chloro-2-heptyl-5-hexyl-	EtOH	231(3.80),281(3.87)	44-2972-69
$C_{17}H_{29}NO_3S_2$			
2-Pyridinol, 1-acetyl-3,6-bis(tert-butylthio)-1,2,3,6-tetrahydro-, acetate	hexane	194(4.28)	44-0660-69
	MeOH	204(4.28)	44-0660-69
$C_{17}H_{31}Br_2N_5O_4$			
N-[3-(2,4-Dinitroanilino)propyl-N,N,N',N',N'-pentamethyl-N,N'-trimethylenediammonium dibromide (also spectra in biological media)	pH 6.5	356(4.21)	35-5136-69
	H_2O	356(4.21)	35-5136-69
	EtOH	343(4.21)	35-5136-69
	MeCN	348(4.20)	35-5136-69
	HCOOH	350(4.21)	35-5136-69
	$HCONH_2$	358(4.24)	35-5136-69
$C_{17}H_{32}Br_2N_4O_2$			
N,N,N,N',N'-Pentamethyl-N'-[3-(o-nitroanilino)propyl]-N,N'-trimethylenediammonium dibromide (also spectra in biological media)	pH 6.5	437(3.74)	35-5136-69
	EtOH	427(3.77)	35-5136-69
N,N,N,N',N'-Pentamethyl-N'-[3-(p-nitroanilino)propyl]-N,N'-trimethylenediammonium dibromide	pH 6.5	397(4.28)	35-5136-69
	EtOH	379(4.26)	35-5136-69
N,N,N,N',N'-Pentamethyl-N'-[2-(4-nitro-m-toluidino)ethyl]-N,N'-trimethylenediammonium dibromide	pH 6.5	381(4.13)	35-5136-69
	EtOH	366(4.13)	35-5136-69
N,N,N,N',N'-Pentamethyl-N'-[2-(4-nitro-o-toluidino)ethyl]-N,N'-trimethylenediammonium dibromide (also spectra in biological media)	pH 6.5	387(4.16)	35-5136-69
	EtOH	371(4.17)	35-5136-69
$C_{17}H_{32}N_2$			
Piperidine, 1,1'-(pentylvinylene)di-	C_6H_{12}	210(3.94),228(3.94)	22-3883-69

Compound	Solvent	λ_{max}(log ε)	Ref.
$C_{18}F_{28}$			
Bicyclo[4.2.0]octa-1,3,5-triene, 7,8-difluoro-2,3,4,5-tetrakis(pentafluoroethyl)-7,8-bis(trichloromethyl)-	n.s.g.	278(2.88)	77-0203-69
1,4-Cyclohexadiene, 1,2,4,5-tetrakis-(pentafluoroethyl)-3,6-bis(tetrafluoroethylidene)-	n.s.g.	266(3.91)	77-0203-69
$C_{18}F_{30}$			
Prismane, hexakis(pentafluoroethyl)-	n.s.g.	210(2.46)	77-0202-69
$C_{18}H_8Cl_8O_2$			
4,9-Etheno-5,8-methano-1H-cyclopenta-[b]biphenylene-1,12-dione, 2,3,3a,5,6,7,8,9a-octahydro-	EtOH	253(3.92)	39-2710-69C
$C_{18}H_8F_4$			
Triphenylene, 1,2,3,4-tetrafluoro-	MeOH	245(4.68),252(4.91), 259(4.97),276(4.20), 286(4.15),317(2.88), 332(3.02),347(2.91)	39-1684-69C
$C_{18}H_8Fe_2O_9$			
Iron, hexacarbonyl[μ-4a,8a-naphthalene-dicarboxylic anhydride)]di-	MeOH	220s(4.52),303(3.89)	78-5115-69
$C_{18}H_9ClN_2O_4$			
3H-Benz[e]indole-2,5-dione, 4-chloro-1-(p-nitrophenyl)-	MeOH	261(4.24),327(4.27), 370(4.16)	77-0925-69
$C_{18}H_9Fe_2NO_8$			
Iron, hexacarbonyl[μ-(4a,8a-naphthalene-dicarboximide)]di-	MeOH	220(4.65),243s(4.48), 303(4.01)	78-5115-69
$C_{18}H_{10}Br_2$			
Triphenylene, 2,11-dibromo-	n.s.g.	253(3.89),263(4.12), 277s(3.30),287s(3.21), 310s(2.47),319s(1.94), 326(1.75),335(1.71), 341(1.56),350(1.43)	18-0766-69
$C_{18}H_{10}ClNO_2$			
4H-Pyrano[3,2-h]quinolin-4-one, 6-chloro-2-phenyl-	EtOH	241(4.32),282(4.56)	23-0105-69
$C_{18}H_{10}ClNO_3$			
4H-Pyrano[3,2-h]quinolin-4-one, 6-chloro-3-hydroxy-2-phenyl-	EtOH	235(4.43),280(4.4), 337s(4.1)	23-0105-69
$C_{18}H_{10}Cl_2N_2$			
3,3'-Biquinoline, 4,4'-dichloro-	MeOH	235(4.89),321(3.88)	94-2389-69
$C_{18}H_{10}Cl_2N_2O_2$			
Benzo[g]phthalazine-1,4-dione, 7-chloro-5-(p-chlorophenyl)-2,3-dihydro-	EtOH	251(4.92)	39-0838-69C
Benzo[g]phthalazine-1,4-dione, 9-chloro-5-(o-chlorophenyl)-2,3-dihydro-	EtOH	251(4.83),344(3.80)	39-0838-69C

Compound	Solvent	λ_{max}(log ε)	Ref.
$C_{18}H_{10}F_4$			
Chrysene, 1,2,3,4-tetrafluoro-11,12-dihydro-	MeOH	245(4.40),253(4.69), 263(4.83),297(4.01), 309(3.97),323(3.47), 340(3.14)	39-1684-69C
$C_{18}H_{10}Fe_2O_6$			
Iron, (μ-benzocyclooctene)hexacarbonyldi-	ether	225(4.42),260s(4.20), 320s(3.64)	35-4734-69
$C_{18}H_{10}N_2$			
1,2-Naphthalenedicarbonitrile, 4-phenyl-	MeCN	215(4.49),250(4.80), 322(3.96),353(3.92)	44-1923-69
$C_{18}H_{10}N_4$			
1,4-Ethenobiphenylene-2,2,3,3-tetracarbonitrile, 1,4,4a,8b-tetrahydro-	EtOH	211(3.99),261(3.18), 267(3.32),273(3.30)	35-4734-69
Pyridazino[4,3-c:5,6-c']diquinoline	$CHCl_3$	295(4.70),322(4.19), 368(4.05)	94-2389-69
$C_{18}H_{10}N_4O$			
4-Oxazoline-2,2,5-tricarbonitrile, 3,4-diphenyl-	MeCN	252(4.33),332(3.81)	44-2146-69
Pyridazino[4,3-c:5,6-c']diquinoline, monooxide	$CHCl_3$	262(4.67),294(4.60), 384(4.30)	94-2389-69
$C_{18}H_{10}N_4O_3$			
Pyridazino[4,3-c:5,6-c']diquinoline, 5,8,13-trioxide	$CHCl_3$	270(4.83),342(4.39), 410(4.45)	94-2389-69
$C_{18}H_{10}O_3$			
Naphthalene-2,3-dicarboxylic anhydride, 1-phenyl-	HOAc	261(4.44),347(3.33), 360(3.44)	39-0838-69C
$C_{18}H_{11}Br$			
Triphenylene, 2-bromo-	n.s.g.	252(5.18),261(5.41), 275(4.56),287(4.57), 308(3.80),316s(3.10), 324(2.96),331(2.91), 338(2.82),347(2.48)	18-0766-69
$C_{18}H_{11}BrN_2OS$			
Phenothiazone, 2-anilino-1-bromo-	MeOH	262(4.46),296(4.40), 474(4.24)	80-1617-69
Phenothiazone, 2-anilino-4-bromo-	MeOH	275(--),295(--), 476(--)	80-1617-69
Phenothiazone, 2-anilino-7-bromo-	MeOH	276(4.67),292(4.56), 476(4.44)	80-1617-69
$C_{18}H_{11}BrO_3$			
1-Naphthoic acid, 8-(p-bromobenzoyl)-	M H_2SO_4	305(4.3),327(4.2)	35-0477-69
	dioxan	297(4.6),308(4.6), 326(4.5)	35-0477-69
$C_{18}H_{11}Cl$			
Triphenylene, 2-chloro-	n.s.g.	251(4.56),259(4.79), 275(3.38),286(3.83), 305(3.02),315s(2.46), 323(2.32),331(2.30), 338(2.19),347(1.97)	18-0766-69

Compound	Solvent	λ max (log ε)	Ref.
$C_{18}H_{11}ClN_2OS$			
Phenothiazone, 2-anilino-7-chloro-	MeOH	276(4.68),290(4.56), 476(4.43)	80-1617-69
$C_{18}H_{11}ClN_2O_2$			
1H-Naphth[2,3-d]imidazole-4,9-dione, 2-(p-chlorobenzyl)-	pH 1	247s(--),251(4.63), 275(4.16)	4-0909-69
	pH 7	248(4.65),277(4.16)	4-0909-69
	pH 13	263(4.63)	4-0909-69
	MeOH	244(4.64),277(4.16), 329(3.48)	4-0909-69
$C_{18}H_{11}Cl_2NO_3$			
Acetamide, N-(3-chloro-1,4-dihydro-1,4-dioxo-2-naphthyl)-2-(p-chlorophenyl)-	MeOH	247(4.31),252(4.32), 286(3.98),336(3.40)	4-0909-69
$C_{18}H_{11}F$			
Benz[a]anthracene, 7-fluoro-	$CHCl_3$	257(4.15),270(4.29), 279(4.51),290(4.56), 302(3.89),319(3.96), 328(3.87),342(3.66), 361(3.30),368(3.21), 380(2.36),388(2.54)	78-3501-69
Triphenylene, 2-fluoro-	n.s.g.	248(4.85),257(5.08), 271s(4.10),283(4.08), 303(3.32),314(2.83), 321(2.67),329(2.87), 335(2.60),344(2.84)	18-0766-69
$C_{18}H_{11}NO_2$			
1,3-Butadiene-1-carboxylic acid, 3-hydroxy-2,4-diphenyl-1-cyano-, lactone	CH_2Cl_2	245(3.89),250s(--), 377(4.42)	24-0319-69
$C_{18}H_{11}NO_4$			
Lanuginosine	EtOH	246(4.54),271(4.44), 315(3.89)	25-1056-69
	EtOH-HCl	258(4.57),283(4.47), 315(3.89)	25-1076-69
$C_{18}H_{11}N_5$			
Acetonitrile, di-1-phthalazinyl-	EtOH	220(4.72),290(4.02), 419(4.31),438(4.28)	95-0959-69
$C_{18}H_{12}$			
Naphthacene	benzene	476(3.98)	44-1734-69
$C_{18}H_{12}B_3Br_3O_3$			
Boroxine, tris(p-bromophenyl)-	hexane	205(4.91),254(4.94), 271(3.49)	39-0392-69A
$C_{18}H_{12}Br_2O_3$			
1,4-Butanedione, 1-(2-benzofuranyl)-2,3-dibromo-4-phenyl-	EtOH	230(4.13),260(4.20), 318(4.45)	104-0511-69
$C_{18}H_{12}ClNO_2$			
4H-Pyrano[3,2-h]quinolin-4-one, 6-chloro-2,3-dihydro-2-phenyl-	EtOH	224s(4.19),268(4.43), 314(3.93)	23-0105-69
$C_{18}H_{12}Cl_2N_2$			
Quinoline, 4,7-dichloro-3-(3,4-dihydro-1-isoquinolyl)-	EtOH	236(4.71),312(3.63), 328(3.61)	2-1010-69

Compound	Solvent	λ max(log ε)	Ref.
$C_{18}H_{12}FI$			
Naphthalene, 1-fluoro-2-(o-iodostyryl)-, trans	EtOH	217(4.57),252(4.30), 273(4.30),308(4.30), 345s(4.00),365(3.48)	78-3501-69
$C_{18}H_{12}F_4$			
Triphenylene, 1,2,3,4-tetrafluoro-	MeOH	240(4.72),249(4.53), 262(4.22),277(3.89), 289(3.90),322(2.96), 337(3.06),351(3.05)	39-1684-69C
$C_{18}H_{12}NOP$			
5,10-o-Benzenophenophosphazine, P-oxide	EtOH	221(3.98),238s(3.67)	89-0987-69
$C_{18}H_{12}NP$			
5,10-o-Benzenophenophosphazine	EtOH	217(4.31),245s(3.35)	89-0987-69
	$CHCl_3$	248(3.57),279s(2.94)	89-0987-69
$C_{18}H_{12}N_2$			
Quinoline, 3-(1-isoquinolyl)-	EtOH	222s(4.63),242s(4.35), 275(4.07),325(3.98)	2-1010-69
$C_{18}H_{12}N_2O$			
1,3-Butadiene-1-carboxylic acid, 3-amino-2,4-diphenyl-1-cyano-, lactam	CH_2Cl_2	245(4.01),252s(--), 372(4.41)	24-0319-69
$C_{18}H_{12}N_2OS$			
Phenothiazone, 2-anilino-	MeOH	272(4.60),289(4.53), 476(4.40)	80-1617-69
$C_{18}H_{12}N_2O_2$			
Benzo[g]phthalazine-1,4-dione, 2,3-dihydro-5-phenyl-	EtOH	247(4.76),308s(3.71), 345(3.85)	39-0838-69C
[2,2'-Biquinoline]-4,4'-diol, cuprous complex	N NaOH	517(3.77)	3-0344-69
	2N NaOH	521(3.78)	3-0344-69
	3N NaOH	523(3.80)	3-0344-69
	4N NaOH	525(3.83)	3-0344-69
	5N NaOH	527(3.85)	3-0344-69
	6N NaOH	528(3.85)	3-0344-69
	7N NaOH	528(3.86)	3-0344-69
	8N NaOH	528(3.85)	3-0344-69
	iso-AmOH	525(3.84)	3-0344-69
2,3-Naphthalenedicarboximide, N-amino-1-phenyl-	EtOH	261(4.75)	39-0838-69C
$C_{18}H_{12}N_2O_4$			
Isoindigo, 1-acetoxy-	dioxan	472(3.63)	24-3691-69
$C_{18}H_{12}N_4O$			
1,1,2,2-Dibenzofurantetracarbonitrile, 3,9b-dihydro-4,9b-dimethyl-	90% EtOH	233(3.99),284(3.36)	88-2901-69
4(3H)-Pteridinone, 6,7-diphenyl-	EtOH	224(4.30),261(4.26), 353(3.90)	39-2415-69C
$C_{18}H_{12}O_2$			
p-Benzoquinone, 2,6-diphenyl-	MeOH	227(4.40),333(3.72)	64-0547-69
Dibenzo[a,e]cyclodecene-5,8-diol, 6,7,13,14-tetradehydro-5,8-dihydro-	EtOH	285(4.33)	44-0874-69
Dibenzo[a,e]cyclodecene-5,8-dione, 13,14-didehydro-6,7-dihydro-	EtOH	244(4.40),283(3.93), 291(4.03),326(3.99)	44-0874-69

Compound	Solvent	λ_{max}(log ε)	Ref.
1,3-Indandione, 2-cinnamylidene-	EtOH	254(4.33),270(4.17), 384(4.65)	23-4076-69
	MeCN	252(4.34),269(4.16), 380(4.62)	108-0099-69
	C_6H_{12}	240(--),247(--), 254(--),267(--), 371(--),387(--), 420(--),435s(--)	108-0099-69
$C_{18}H_{12}O_3$			
Benzo[i]xanthone, 1-hydroxy-3-methyl-	$CHCl_3$	265(4.61)	49-1368-69
2-Butene-1,4-dione, 1-(2-benzofuranyl)-4-phenyl-	EtOH	248(4.33),334(4.39)	104-0511-69
Cyclohepta[def]fluoren-8(4H)-one, 4-acetoxy-	EtOH	215(4.32),254(4.37), 267(4.40),294(4.08)	39-1427-69C
6H-Cyclopenta[a]phenanthrene-6,7,17-trione, 15,16-dihydro-11-methyl-	EtOH	263(4.15),345(3.44), 410s(2.94)	39-2484-69C
1-Naphthoic acid, 8-benzoyl-	M H_2SO_4	305(4.2),327(4.1)	35-0477-69
	dioxan	297(4.1),308(4.1), 326(4.0)	35-0477-69
Phenanthro[3,2-b]furan-7,11-dione, 4,8-dimethyl- (isotanshinone I)	EtOH	212(4.39),234(4.55), 283s(4.29),293(4.40), 347(3.70),455(3.35)	18-3318-69
	EtOH	234(4.58),293(4.45), 346(3.78),450(3.28)	88-0301-69
Tanshinone I	EtOH	245(4.50),330(3.60), 420(3.67)	88-0301-69
$C_{18}H_{12}O_6$			
Atromentin	EtOH	275(4.47),349(3.48)	39-2398-69C
$C_{18}H_{12}O_7$			
6H-[1,3]Dioxolo[5,6]benzofuro[3,2-c]-[1]benzopyran-6-one, 3,4-dimethoxy-	$CHCl_3$	268(3.95),284(3.93), 299(3.91),312(3.98), 348(4.47),362s(4.42)	31-0122-69
Versicolorin C	MeOH	222(4.45),265(4.22), 290(4.36),312(4.05), 450(3.84)	78-1497-69
$C_{18}H_{12}O_9$			
Variegatic acid	H_2O	258(4.26),376(3.95)	64-0941-69
	EtOH	261(4.20),400(3.96)	64-0941-69
$C_{18}H_{13}AsClN$			
Phenarsazine, 10-(o-chlorophenyl)-5,10-dihydro-	EtOH	284(4.08),311(3.91)	101-0117-69E
$C_{18}H_{13}BrO_5$			
Osajaxanthone, 4'-bromo-	MeOH	230(4.14),289(4.58), 381(3.59)	78-2787-69
$C_{18}H_{13}Br_2ClO_7$			
Flavone, 5',6-dibromo-3'-chloro-2',5-dihydroxy-3,7,8-trimethoxy-	EtOH	269(4.20),350s(3.83)	39-2418-69C
$C_{18}H_{13}ClN_2$			
Quinoline, 4-chloro-3-(3,4-dihydro-1-isoquinolyl)-	EtOH	228(4.61),318s(3.47)	2-1010-69

Compound	Solvent	λ_{max}(log ε)	Ref.
$C_{18}H_{13}ClN_2O$			
1H-Imidazo[5,1-a]isoindol-1-one, 7-chloro-2,3-dihydro-2-methyl-3-methylene-	iso-PrOH	223(4.37),270(4.35), 333(3.80),413(3.79)	44-0649-69
4-Quinolinol, 7-chloro-3-(3,4-dihydro-1-isoquinolyl)-	EtOH	245(4.30),280(4.34), 295(4.26),398(4.03)	2-1010-69
$C_{18}H_{13}ClN_2O_3$			
Acetamide, N-(3-amino-1,4-dihydro-1,4-dioxo-2-naphthyl)-2-(p-chlorophenyl)-	MeOH	268(4.36)	4-0909-69
4-[1,3]Dioxolo[4,5-f]-2,1-benzisoxazol-3-yl-2-methylisoquinolinium chloride	EtOH	240(4.03)	78-5365-69
$C_{18}H_{13}ClN_4O_2$			
Azobenzene, 4'-anilino-4-chloro-3-nitro-	benzene	350(3.61),430(4.46)	73-2092-69
$C_{18}H_{13}F$			
Naphthalene, 1-fluoro-2-styryl-, (E)-	EtOH	226(4.48),240(4.23), 252(4.15),274(4.38), 283(4.42),319(4.51), 335s(4.11),360(3.53)	78-3501-69
$C_{18}H_{13}N$			
Carbazole, N-phenyl-, tetranitromethane complex	C_6H_{12}	410(2)	77-0364-69
$C_{18}H_{13}NO$			
2-Penten-4-ynenitrile, 3-methoxy-2,5-diphenyl-	EtOH	210(4.07),225(4.21), 337(4.49)	39-0915-69C
$C_{18}H_{13}NO_2$			
1,4-Naphthoquinone, 2-(1-aziridinyl)-3-phenyl-	EtOH	261(4.28),366s(3.56), 420(3.34)	39-1325-69C
2-Penten-4-ynenitrile, 3-hydroxy-2-(p-methoxyphenyl)-5-phenyl-	EtOH	209(4.25),233(4.29), 348(4.47)	39-0915-69C
	NaOH	238(4.35),265s(4.35), 354(--)	39-0915-69C
Pyrrolo[1,2-f]phenanthridine-1-carboxylic acid, methyl ester	MeOH	209(4.43),233(4.57), 242(4.64),252s(4.43), 275(4.40),282s(4.36), 290(4.23),304(4.26), 320s(4.11),333(4.24), 348(4.19)	39-2311-69C
$C_{18}H_{13}NO_3$			
4-Isoxazolecarboxylic acid, 3-phenyl-5-styryl-	MeOH	225s(--),230(4.33), 236(4.26),318(4.54)	32-0753-69
$C_{18}H_{13}NO_4$			
Indolo[1,7-ab][1]benzazepine-11-carboxylic acid, 1,2,6,7-tetrahydro-1,2-dioxo-, methyl ester	EtOH	247(4.26),284s(3.80)	35-1672-69
	$CHCl_3$	440(2.40)	35-1672-69
$C_{18}H_{13}N_3$			
Atroponitrile, β,β'-iminodi-	EtOH	230(4.18),357(4.59)	22-0198-69
Quinoline, 2,2'-iminodi-	EtOH	213(4.80),267(4.66), 365(4.47),399(3.17), 419(3.05)	95-0074-69

Compound	Solvent	λmax(log ε)	Ref.
$C_{18}H_{13}N_3O_2$			
2,6-Anthyridinediol, 8-methyl-3-phenyl-	NaOH	225(3.4),240(3.2), 290(2.1),385(2.9)	32-0677-69
$C_{18}H_{13}N_3O_3S_2$			
Butenedinitrile, [(1,4-dihydro-3-morpholino-1,4-dioxo-2-naphthyl)thio]-mercapto-, compound with morpholine	morpholine	552(3.91)	83-0075-69
$C_{18}H_{13}N_3O_5$			
Benzoic acid, p-nitro-, 2-(1,4-dihydro-1,4-dioxo-2-naphthyl)hydrazide	alkali	645(4.01)	95-0007-69
2,6-Pyridinediol, 3-[(2,4-dihydroxyphenyl)azo]-, 4-benzoate	EtOH	448(4.5)	119-0037-69
$C_{18}H_{13}N_5$			
Imidazole, 2,4,5-tris(2-pyridyl)-	EtOH	320(4.71)	88-3101-69
Imidazole, 2,4,5-tris(4-pyridyl)-	EtOH	315(4.74)	88-3101-69
$C_{18}H_{14}$			
Fluoranthene, 7,10-dimethyl-	n.s.g.	239(4.56),261(4.19), 280(4.05),286(3.85), 291(3.95),322(3.77), 352(3.88),370(3.85)	39-0760-69C
Naphthalene, 1-styryl-	DMF	329(4.34)	33-2521-69
Naphthalene, 2-styryl-	DMF	274(4.48),284(4.51), 320(4.61),334(4.51)	33-2521-69
$C_{18}H_{14}AsN$			
Phenarsazine, 5,10-dihydro-10-phenyl-	EtOH	280(4.22),310(4.07), 338(3.83)	101-0117-69E
$C_{18}H_{14}BrCl$			
Butatriene, 1-bromo-4-chloro-1,4-di-p-tolyl-, trans	CH_2Cl_2	255(4.07),264(4.03), 295(3.64),390(4.64), 409(4.75)	88-0435-69
$C_{18}H_{14}Br_2$			
Butatriene, 1,4-dibromo-1,4-di-p-tolyl-, cis	CH_2Cl_2	244(4.25),259(4.31), 266(4.33),305(3.63), 394(4.50),412(4.57)	88-0435-69
trans	CH_2Cl_2	258(4.05),309(3.64), 393(4.59),412(4.72)	88-0435-69
$C_{18}H_{14}Br_2N_2$			
Cinnamaldehyde, p-bromo-, azine	$CHCl_3$	355(4.81),369(4.80)	22-1367-69
$C_{18}H_{14}Br_2O$			
Bicyclo[3.1.0]hexan-3-one, 2,4-dibromo-6,6-diphenyl-	C_6H_{12}	222(4.23),267(3.40), 273(3.27),329s(2.41), 340(2.45),353(2.39)	35-0434-69
$C_{18}H_{14}ClN$			
Aniline, p-chloro-N-[1-(2-naphthyl)-ethylidene]-	EtOH	247(4.72),284(4.15), 324(3.69)	56-0749-69
$C_{18}H_{14}ClNO_6$			
3H-Phenoxazine-1,6-dicarboxylic acid, 2-chloro-4,9-dimethyl-3-oxo-, dimethyl ester	$CHCl_3$	372(4.13),479(3.97)	24-3205-69

Compound	Solvent	λ_{max}(log ε)	Ref.
$C_{18}H_{14}Cl_2$			
Butatriene, 1,4-dichloro-1,4-di-p-tolyl-	CH_2Cl_2	255(4.15),265(4.13), 289(3.95),337(4.05), 387(4.67),407(4.76)	88-0435-69
$C_{18}H_{14}Cl_4$			
Anthracene, 1,8-dichloro-9-(dichloromethylene)-10-ethyl-9,10-dihydro-10-methyl-	C_6H_{12}	259(4.18)	44-3093-69
$C_{18}H_{14}F_4$			
Triphenylene, 5,6,7,8-tetrafluoro-1,2,3,4,4a,12b-hexahydro-	MeOH	268(4.20),295(3.39)	39-1684-69C
$C_{18}H_{14}N_2$			
Indolo[2,3-a]carbazole, 5,6,11,12-tetrahydro-	CH_2Cl_2	257(4.39),347(4.33), 362(4.52),381(4.50)	24-1198-69
$C_{18}H_{14}N_2O$			
1H-Benz[g]indazol-5-ol, 3-methyl-4-phenyl-	EtOH	224(4.54),256(4.50), 270s(--),315(3.92)	103-0254-69
Benzoic acid, (1-naphthylmethylene)-hydrazide	EtOH	229(4.5),265s(4.0), 334(4.3)	28-0248-69B
Benzoic acid, (2-naphthylmethylene)-hydrazide	EtOH	235(4.4),261s(4.2), 271(4.3),279(4.4), 313(4.5),321s(4.5), 349s(3.8)	28-0248-69B
2-Indolinone, 3-(indol-2-ylmethylene)-1-methyl-	MeOH	405(4.53)	24-1347-69
4-Quinolinol, 3-(3,4-dihydro-1-isoquinolyl)-	EtOH	235(4.33),272(4.22), 295(4.13),332(3.82), 400(3.95)	2-1010-69
$C_{18}H_{14}N_2O_2$			
Aniline, N-[1-(2-naphthyl)ethylidene]-m-nitro-	EtOH	253(4.68),282(4.28), 322(3.74)	56-0749-69
5H-Benzo[a]phenoxazin-5-one, 9-(dimethylamino)-	EtOH	543(4.36)	73-0221-69
1,3-Cyclohexanedione, 2-diazo-4,5-diphenyl-	EtOH	230(4.18),259(3.96)	23-2853-69
4-Isoquinolinecarbonitrile, 6,8-dimethoxy-3-phenyl-	EtOH	278(4.67),326(3.71)	95-1492-69
1-Naphthylamine, N-(α-methyl-p-nitrobenzylidene)-	EtOH	288(4.30),370s(3.32)	56-0749-69
2-Naphthylamine, N-(α-methyl-p-nitrobenzylidene)-	EtOH	230(4.67),276(4.38), 360(3.46)	56-0749-69
7H-Pyrido[2,3-c]carbazole-1-carboxylic acid, 3,5-dimethyl-	EtOH-NaOH	226(4.38),249(4.42), 283(4.43),339(3.93)	12-0185-69
$C_{18}H_{14}N_2O_2S_2$			
p-Dithiin-2,3-dicarboximide, N-(p-anilinophenyl)-5,6-dihydro-	MeOH	289(4.52),377(3.32)	33-2221-69
p-Dithiin-2,3-dicarboximide, N-(diphenylamino)-5,6-dihydro-	MeOH	232(4.26),296(4.24), 405(3.47)	33-2236-69
$C_{18}H_{14}N_2O_3$			
Acetic acid, phenyl-, 2-(1,4-dihydro-1,4-dioxo-2-naphthyl)hydrazide	alkali	540(3.87)	95-0007-69
Acetic acid, phenyl-, 2-(3,4-dihydro-3,4-dioxo-1-naphthyl)hydrazide	alkali	455(4.11)	95-0007-69

Compound	Solvent	λ_{max}(log ϵ)	Ref.
Benzoic acid, 2-(1,4-dihydro-3-methyl-1,4-dioxo-2-naphthyl)hydrazide	alkali	645(3.97)	95-0007-69
[1,3]Dioxolo[4,5-f]-2,1-benzisoxazole, 3-(1,2-dihydro-2-methyl-4-isoquinolyl)-	EtOH	240(4.26)	78-5365-69
$C_{18}H_{14}N_2O_4$			
Anthra[1,2-c][1,2,5]oxadiazole-6,11-dione, 4-butoxy-	EtOH	420(3.69)	104-1642-69
Benzoic acid, p-hydroxy-, 2-(1,4-dihydro-3-methyl-1,4-dioxo-2-naphthyl)-hydrazide	alkali	650(3.86)	95-0007-69
Isoindigo, 1,1'-dimethoxy-	dioxan	472(3.54)	24-3691-69
$C_{18}H_{14}N_2O_6$			
Glyoxylic acid, (azodi-p-phenylene)di-, dimethyl ester	$CHCl_3$	341(4.55),474(2.98)	78-4241-69
$C_{18}H_{14}N_2O_7$			
Glyoxylic acid, (azoxydi-p-phenylene)-di-, dimethyl ester	$CHCl_3$	276(4.05),351(4.26)	78-4241-69
$C_{18}H_{14}N_3OP$			
Phosphine oxide, diphenyl-v-triazolo-[1,5-a]pyridin-3-yl-	MeOH	268s(3.86),273(4.03), 281(4.03)	24-2216-69
$C_{18}H_{14}N_4$			
5H,11H-Benzotriazolo[2,1-a]benzotriazole, 5-phenyl-	n.s.g.	215(4.26),225s(--), 285(4.33),370(3.95)	39-0752-69C
$C_{18}H_{14}N_4O$			
1H-Pyrazolo[3,4-b]pyrazin-6-ol, 5-methyl-1,3-diphenyl-	EtOH	244(4.24),270(4.35), 316s(4.13)	32-0463-69C
$C_{18}H_{14}N_6$			
Pyrimido[5,4-e]-as-triazine, 5-[(diphenylmethyl)amino]-	pH 1	253s(--),369(4.02)	44-3161-69
$C_{18}H_{14}O$			
1,3,5,9,11-Cycloheptadecapentaene-7,13,15-triyne, 6-methoxy-	ether	254(4.58),289(4.70), 298(4.75),402(3.72), 455s(3.45),485s(3.26), 525s(2.72)	35-7518-69
1,3,5,11,13-Cycloheptadecapentaene-7,9,15-triyne, 17-methoxy-	ether	253(4.16),281(4.67), 290(4.77),407(3.61), 433(3.46)	35-7518-69
15H-Cyclopenta[a]phenanthren-15-one, 16,17-dihydro-11-methyl-	n.s.g.	217(4.57),253(4.63), 284(4.15),323(4.14), 363(3.41)	39-2484-69C
Phenanthro[1,2-b]furan, 1,2-dimethyl-	C_6H_{12}	200(4.37),234(4.22), 263(4.75),272(4.94), 300(4.35),322(3.08), 331(2.90),338(3.27), 346(2.82),354(3.27)	33-1461-69
$C_{18}H_{14}O_2$			
9H-Benzo[a]xanthene-8-carboxaldehyde, 10,11-dihydro-	MeCN	257(4.34),267(4.32), 278(4.31),327(4.02), 342(4.12),357(4.03), 404(4.15),419(4.22), 438(4.11)	44-2736-69

Compound	Solvent	λ_{max}(log ϵ)	Ref.
17H-Cyclopenta[a]phenanthren-17-one, 15,16-dihydro-6-hydroxy-11-methyl-	EtOH	270(4.70),289s(4.58), 378(3.32),390(3.35)	39-2484-69C
17H-Cyclopenta[a]phenanthren-17-one, 15,16-dihydro-16-hydroxy-11-methyl-	EtOH	264(4.88),288(4.52), 301(4.35),345(3.14), 361(3.36),379(3.39)	39-2484-69C
17H-Cyclopenta[a]phenanthren-17-one, 15,16-dihydro-6-methoxy-	EtOH	270(4.81),287(4.48), 302(4.28),361(3.37), 378(3.42)	39-2484-69C
Naphthalene, 1-benzoyl-2-methoxy-	C_6H_{12}	229(4.84),246s(4.31), 269(3.79),280(3.83), 290s(3.74),310s(3.32), 322(3.41),336(3.47), 384s(1.71)	19-0253-69
	MeOH	206(4.51),228(4.84), 249(4.25),278(3.84), 290(3.76),320s(3.41), 334(3.48),360s(2.87)	19-0253-69
	aq. MeOH	207(4.54),228(4.84), 257(4.22),280s(3.91), 290s(3.79),320s(3.39), 330(3.44),350s(3.09)	19-0253-69
Naphthalene, 1-benzoyl-4-methoxy-	C_6H_{12}	212(4.61),237(4.56), 250s(4.33),310s(3.95), 321(4.01),336s(3.88), 360s(3.71)	19-0253-69
	MeOH	211(4.62),235(4.51), 250s(4.30),325(3.98)	19-0253-69
	aq. MeOH	210(4.60),234(4.45), 255(4.22),340(3.94)	19-0253-69
Phenanthro[1,2-b]furan-1-methanol, 2-methyl-	C_6H_{12}	198(4.40),222(4.26), 229(4.25),236(4.25), 261(4.81),269(4.99), 298(4.42),325(3.06), 333(2.95),341(3.24), 349(2.91),358(3.25)	33-1461-69
2-Propanone, 1-(2-phenyl-4H-1-benzopyran-4-ylidene)-	MeCN	243(4.40),296(4.18), 380(4.26)	44-2736-69
2,2'-Spirobi[2H-1-benzopyran], 3-methyl-	dioxan	257(4.26),266(4.17), 298(3.68),307(3.64)	5-0162-69B
$C_{18}H_{14}O_2S_2$			
Thiophene, 2,2'-p-phenylenebis[5-acetyl-	n.s.g.	245(3.18),292(3.00), 368(3.65)	22-2076-69
$C_{18}H_{14}O_3$			
1,3-Indandione, 2-acetonyl-2-phenyl-	EtOH	227(4.66),250s(--)	103-0251-69
5,11-Naphthacenedione, 5a,6,11a,12-tetrahydro-3-hydroxy-	EtOH	253(4.47),323(3.68)	2-0561-69
Phenanthro[1,2-b]furan-10,11-dione, 1,2-dihydro-1,6-dimethyl-	EtOH	216(4.30),242(4.40), 292(4.20),335(3.48), 414(3.61)	18-3318-69
$C_{18}H_{14}O_4$			
2-Naphthoic acid, 3-benzylidene-1,2,3,4-tetrahydro-6-hydroxy-4-oxo-	EtOH	228(4.40),305(4.17)	2-0561-69
$C_{18}H_{14}O_5$			
2-Naphthoic acid, 1,2,3,4-tetrahydro-6-hydroxy-3-(p-hydroxybenzylidene)-4-oxo-	EtOH	228(4.34),342(4.24)	2-0561-69

Compound	Solvent	λ_{max}(log ε)	Ref.
Osajaxanthone	MeOH	235(4.28),285(4.61), 339(3.87),380(3.68)	78-2787-69
$C_{18}H_{14}O_6$			
2,3-Biphenylenedione, 6,7-diacetoxy-1,8-dimethyl-	EtOH	217(4.43),247s(4.30), 309s(4.11)	39-2579-69C
Derrustone	EtOH	258(4.41),290(4.13)	39-0365-69C
Isoflavone, 6,7-dimethoxy-3',4'-(methylenedioxy)-	$CHCl_3$	266(4.30),295(4.15), 318(4.04)	39-1787-69C
Pterocarpan, 3-acetoxy-8,9-(methylenedioxy)-	EtOH	285(3.79),311(4.14)	39-1109-69C
Pterocarpan, 3,4-dimethoxy-8,9-(methylenedioxy)-6a,11a-dehydro-	EtOH	271(4.08),321(4.32), 356(3.87)	31-0122-69
$C_{18}H_{14}O_7$			
Coumarin, 4-hydroxy-5,7-dimethoxy-3-[3,4-(methylenedioxy)phenyl]-	EtOH	213(4.57),237s(4.10), 283s(3.85),330(4.26)	39-0365-69C
2H-Naphtho[2,3-b]pyran-8,9-dicarboxylic acid, 5-methoxy-2-oxo-, dimethyl ester	EtOH	235(4.52),270(4.58), 278(4.54),320(4.31), 333(4.35)	12-1721-69
$C_{18}H_{14}S$			
Benzo[b]naphtho[2,3-d]thiophene, 6,7-dimethyl-	EtOH	228(4.37),248(4.53), 267(4.62),275(4.88), 283(4.34),314(3.65), 328(3.62),353(3.34), 370(3.35)	4-0885-69
Benzo[b]naphtho[2,3-d]thiophene, 6,8-dimethyl-	EtOH	231(4.37),243(4.51), 267(4.62),276(4.90), 283(4.37),317(3.65), 334(3.72),350(3.49), 368(3.38)	4-0885-69
Benzo[b]naphtho[2,3-d]thiophene, 7,8-dimethyl-	EtOH	228(4.37),247(4.56), 267(4.62),278(5.0), 283(4.44),321(3.69), 335(3.68),352(3.49), 373(3.4)	4-0885-69
$C_{18}H_{15}As$			
Arsine, triphenyl-	n.s.g.	247(3.7),259(3.6), 264(3.4),272(3.2)	120-0195-69
$C_{18}H_{15}AuBrP$			
Gold, bromo(triphenylphosphine)-	dioxan	269(--),275(3.31)	39-0276-69B
$C_{18}H_{15}AuClP$			
Gold, chloro(triphenylphosphine)-	dioxan	269(--),275(3.27)	39-0276-69B
$C_{18}H_{15}AuIP$			
Gold, iodo(triphenylphosphine)-	dioxan	269(--),275(3.54)	39-0276-69B
$C_{18}H_{15}B_3O_3$			
Boroxin, triphenyl-	hexane	193(5.05),236(4.88), 273(3.40)	39-0392-69A
$C_{18}H_{15}Br$			
Pyrene, 4-bromo-10b,10c-dihydro-10b,10c-dimethyl-, trans	C_6H_{12}	342(4.95),356(4.39), 382(4.68),431s(3.67), 454(3.80),470(3.83), 531(1.93),539(1.88),	35-0902-69

(continued on next page)

Compound	Solvent	λmax(log ε)	Ref.
		582s(2.07),589s(2.13), 601(2.24),614(2.39), 630s(2.46),638s(2.51), 644(2.69)	35-0902-69
$C_{18}H_{15}BrO_4$			
Acetic acid, bromodibenzoyl-, ethyl ester	MeOH	225(4.40)	22-3281-69
$C_{18}H_{15}ClN_2O$			
3-Quinolinecarboxamide, 4-chloro-N-phenethyl-	EtOH	230(4.1),295(3.74)	2-1010-69
$C_{18}H_{15}ClN_2O_2$			
Acetamide, N-[(6-chloro-1,2-dihydro-2-oxo-4-phenyl-2-quinolyl)methyl]-	MeOH	239(4.66)	36-0830-69
Isoindole-1-carboxamide, N-acetyl-5-chloro-N-methyl-3-phenyl-	iso-PrOH	264(4.37),392(4.44)	44-0649-69
Propionanilide, 2'-benzoyl-4'-chloro-2-cyano-2-methyl-	MeOH	238(4.30)	36-0830-69
3-Quinolinecarboxamide, 7-chloro-4-hydroxy-N-phenethyl-	EtOH	225(4.48),252s(4.39), 262(4.50),302s(4.01), 315(4.06),322s(3.98)	2-1010-69
$C_{18}H_{15}ClN_2O_3$			
1H-1,4-Benzodiazepine-3-carboxylic acid, 7-chloro-2,3-dihydro-2-oxo-5-phenyl-	EtOH	230(4.55),323(3.39)	111-0239-69
$C_{18}H_{15}ClO_7$			
Flavone, 3'-chloro-2',5-dihydroxy-3,7,8-trimethoxy-	EtOH	266(4.44),305s(3.85), 350(3.85)	39-2418-69C
	EtOH-$AlCl_3$	279(4.45),323s(3.85), 416(3.79)	39-2418-69C
	EtOH-$Zr(NO_3)_4$	267s(--),277(--), 320s(--),410(--)	39-2418-69C
$C_{18}H_{15}ISn$			
Stannane, iodotriphenyl-	EtOH	none above 240 nm	101-0307-69C
$C_{18}H_{15}N$			
Aniline, N-[1-(2-naphthyl)ethylidene]-	EtOH	246(4.70),282(4.15), 320s(3.60)	56-0749-69
Benzo[a]phenanthridine, 5,6-dihydro-6-methyl-	EtOH	262(4.48),308(3.47), 388(3.47)	2-0419-69
Benzylamine, N-(2-naphthylmethylene)-	EtOH	247(4.7),252(4.7), 273(4.0),283(4.2), 293(4.1),307s(3.4), 325(3.1),339(3.1)	28-0248-69B
1-Naphthylamine, N-(α-methylbenzylidene)-	EtOH	231(4.48),291(3.91), 336s(3.41)	56-0749-69
2-Naphthylamine, N-(α-methylbenzylidene)-	EtOH	236(4.78),328(3.54)	56-0749-69
$C_{18}H_{15}NO$			
5H-Dibenzo[a,d]cycloheptene-5-carbonitrile, 10-ethoxy-	EtOH	287(4.09)	23-2827-69
Nitrone, N-benzyl-α-1-naphthyl-	EtOH	203(4.7),238(4.3), 267(3.9),333(4.2), 348(4.2)	28-0078-69A

Compound	Solvent	λ_{max}(log ε)	Ref.
Nitrone, N-benzyl-α-2-naphthyl-	EtOH	214(4.5),235(4.1), 244(4.2),266(4.4), 318(4.3),334(4.3), 358(3.0)	28-0078-69A
2(1H)-Pyridone, 1-methyl-4,6-diphenyl-	EtOH	244(4.30),325(3.85)	48-0061-69
$C_{18}H_{15}NOS$			
5H-Thieno[2,3-b]indole, 2-(p-methoxyphenyl)-8-methyl-	EtOH	247(4.64),290(4.27), 294(4.28),333(4.44)	95-0058-69
$C_{18}H_{15}NO_2$			
Cyclopenta[ij]pyrido[2,1,6-de]quinolizine-4-carboxylic acid, 9-methyl-, ethyl ester	EtOH	235s(4.31),255(4.44), 306(4.31),335(4.23), 384(4.11),414(3.97), 437(4.40),518s(2.41), 552(2.56),598(2.54), 652(2.20)	39-0239-69C
5H-Dibenzo[a,d]cycloheptene-5-carbonitrile, 10-[(2-hydroxyethyl)oxy]-	EtOH	288(4.06)	23-2827-69
1,3-Indandione, 2-[p-(dimethylamino)-benzylidene]-	C_6H_{12}	238(4.24),248(4.36), 265(4.20),434(4.42), 461(4.89)	108-0099-69
	MeCN	238(4.29),250(4.37), 271(4.15),484(4.83)	108-0099-69
	dioxan	252(4.62),472(5.00)	65-1373-69
1-Naphthaleneacetohydroxamic acid, N-phenyl-	EtOH	250(4.09)	42-0831-69
$C_{18}H_{15}NO_2S$			
Thieno[2,3-b]indole, 5-methoxy-2-(p-methoxyphenyl)-	EtOH	240(4.56),308(4.44), 336(4.42)	95-0058-69
$C_{18}H_{15}NO_3$			
Aporphine-8,11-dione, 6a,7-didehydro-10-methoxy-	MeOH	555(3.52)	83-0487-69
	C_6H_{12}	510(--)	83-0487-69
10,5β-(Iminomethano)-5H-dibenzo[a,d]-cycloheptene-11,13(10H)-dione, 10α-methoxy-12-methyl-	EtOH	268(3.83)	23-2827-69
2,6(1H,5H)-Pyridinedione, 5-(p-methoxyphenyl)-4-phenyl-	EtOH	215s(4.15),225(4.24), 286(4.29)	39-0915-69C
	NaOH	228(4.21),290(4.27)	39-0915-69C
$C_{18}H_{15}NO_4$			
10,5β-(Iminomethano)-5H-dibenzo[a,d]-cyclohepten-13-one, 11-acetoxy-10,11-dihydro-10-hydroxy-	EtOH	256(2.78),262(2.78)	23-2827-69
Indole-2-carboxylic acid, 3-benzoyl-5-hydroxy-, ethyl ester	EtOH	213s(4.46),246s(4.17), 281(4.16),333(3.95)	48-0786-69
4H-Pyrido[3,2,1-de]phenanthridine, 5,6-dihydro-2-methoxy-10,11-(methylenedioxy)-	EtOH	244(4.65),284(4.31), 344(3.84),358(3.80)	2-0848-69
$C_{18}H_{15}NO_5$			
Acrylophenone, 2',4'-dihydroxy-3-(2-pyridyl)-, diacetate	EtOH	270s(4.06),306(4.34)	95-0375-69
Acrylophenone, 2',4'-dihydroxy-3-(3-pyridyl)-, diacetate	EtOH	299(4.34)	95-0375-69

Compound	Solvent	λ_{max}(log ε)	Ref.
$C_{18}H_{15}NO_7$			
Phenoxazine-1,6-dicarboxylic acid, 2-hydroxy-4,9-dimethyl-6-oxo-, dimethyl ester	$CHCl_3$	404(3.99)	24-3205-69
$C_{18}H_{15}NS_2$			
4H-Thiopyran-4-thione, 3-(methylamino)-2,6-diphenyl-	EtOH	309(4.41),455(3.65)	39-0315-69C
$C_{18}H_{15}N_3$			
s-Triazolo[4,3-a]pyridine, 2,3-dihydro-2,3-diphenyl-	MeCN	214(4.30),298(4.18), 449(3.39)	24-3176-69
$C_{18}H_{15}N_3O$			
Cinnamaldehyde, α-(phenylethynyl)-, semicarbazone, (Z?)-	ether	285(4.04),318(4.11)	39-0683-69C
Ketone, 1,8-dihydro-8-methyl-1-phenylpyrazolo[3,4-b]indol-3-yl methyl	EtOH	222(4.59),270(4.23), 317(4.03)	32-0588-69
1-Naphthaldehyde, 4-phenylsemicarbazone	EtOH	235(4.6),261s(4.0), 326(4.3),350s(4.0)	28-0248-69B
2-Naphthaldehyde, 4-phenylsemicarbazone	EtOH	210(4.3),238(4.5), 244s(4.5),264(4.3), 278(4.4),292s(4.4), 305(4.5),321s(4.5), 345s(3.5),350s(3.4)	28-0248-69B
$C_{18}H_{15}N_3O_5$			
1H-1,4-Benzodiazepine-3-carboxylic acid, 2,3-dihydro-7-nitro-2-oxo-5-phenyl-, ethyl ester	EtOH	219(4.39),262(4.22), 309(4.02)	111-0239-69
$C_{18}H_{15}N_3O_6$			
1,4-Phenazinedicarboxylic acid, 7-nitro-, diethyl ester	EtOH	234(4.53),366(4.18)	33-0322-69
$C_{18}H_{15}N_3S$			
1-Naphthaldehyde, 4-phenyl-3-thiosemicarbazone	EtOH	206(4.7),236(4.5), 270(4.0),279(4.0), 348(4.4),361s(4.4)	28-0248-69B
2-Naphthaldehyde, 4-phenyl-3-thiosemicarbazone	EtOH	205(4.5),232(4.4), 250s(4.3),277s(4.2), 287(4.3),330(4.6), 341s(4.6)	28-0248-69B
$C_{18}H_{15}P$			
Phosphine, triphenyl-	MeOH	264(4.08)	87-0146-69
	methyl methacrylate	292(3.86)	47-0265-69
$C_{18}H_{16}$			
Benzo[ghi]fluoranthene, 1,2,9,10,10a,10b-hexahydro-	n.s.g.	273(4.20),292(3.98), 306(3.91)	39-2489-69C
Bicyclo[3.1.0]hex-2-ene, 6,6-diphenyl-	C_6H_{12}	227(4.11),260(3.19), 268(3.06),276(2.85)	35-0434-69
9,10-Cyclopentenophenanthrene, 1'-methyl-	EtOH	212(4.47),222(4.33), 246(4.63),254(4.72), 270(4.23),278(4.06), 289(4.01),302(4.08), 338(3.05),354(2.06)	44-0763-69

Compound	Solvent	λ max (log ε)	Ref.
2H-Dibenz[cd,h]azulene, 6,7-dihydro-2-methyl-	EtOH	244(4.33),250s(4.31), 288(3.67),296(3.66), 306(3.62)	87-0444-69
Pyrene, dihydro-13,15-dimethyl-, trans	hexane	251(4.36),260(4.35), 272(4.22),293(3.62), 306(3.72),320(3.63), 368(3.72),382(3.74), 402s(3.55)	35-0902-69
$C_{18}H_{16}BrCl$			
Butadiene, 2-bromo-3-chloro-1,4-di-p-tolyl-	CH_2Cl_2	316(4.49)	88-0435-69
$C_{18}H_{16}BrNO_3$			
1,4-Benzoxazepine, 5-(p-bromobenzyl)-8,9-dimethoxy-	MeOH	216(4.57),256(4.09), 344(4.23)	95-1048-69
$C_{18}H_{16}Br_2N_4$			
2-Cyclohexen-1-one, 3-[(p-bromophenyl)-azo]-, (p-bromophenyl)hydrazone	$CHCl_3$	312(4.17),443(4.59)	24-1379-69
$C_{18}H_{16}Br_2O_2$			
1,4-Butanedione, 2,3-dibromo-1-phenyl-4-(2,4-xylyl)-	EtOH	266(4.29)	104-0511-69
$C_{18}H_{16}Br_3ClO_4$			
Ethylene, 1-bromo-2,2-bis(2-bromo-4,5-dimethoxyphenyl)-1-chloro-	EtOH	251s(4.12),290(3.91)	7-0922-69
$C_{18}H_{16}ClFeNO_4S$			
2-Ferrocenyl-3-methylbenzothiazolium perchlorate	MeCN	223s(4.23),250(3.99), 280(3.89),329(4.17), 386(3.75),524(3.43)	24-0568-69
$C_{18}H_{16}ClNO_2$			
Carbostyril, 6-chloro-3-ethoxy-1-methyl-4-phenyl-	MeOH	241(4.56),302(4.18)	36-0830-69
$C_{18}H_{16}ClNO_3$			
1,4-Benzoxazepine, 5-(m-chlorobenzyl)-8,9-dimethoxy-	MeOH	215(4.50),259(4.12), 346(4.15)	95-1048-69
1,4-Benzoxazepine, 5-(p-chlorobenzyl)-8,9-dimethoxy-	MeOH	215(4.49),255(4.02), 345(4.15)	95-1048-69
$C_{18}H_{16}ClN_3$			
Isoindole, 5-chloro-3-phenyl-1-(1,4,5,6-tetrahydro-2-pyrimidinyl)-	iso-PrOH	218(4.31),276(4.31), 323(4.09),401(4.55)	44-0649-69
$C_{18}H_{16}Cl_2N_4$			
2-Cyclohexen-1-one, 3-[(p-chlorophenyl)-azo]-, (p-chlorophenyl)hydrazone	$CHCl_3$	312(4.19),441(4.59)	24-1379-69
$C_{18}H_{16}INO$			
4(1H)-Pyridone, 2,3-dihydro-5-iodo-6-methyl-1,2-diphenyl-	EtOH	302(3.51),352(4.07)	18-2690-69
$C_{18}H_{16}I_2N_4$			
2-Cyclohexen-1-one, 3-[(p-iodophenyl)-azo]-, (p-iodophenyl)hydrazone	$CHCl_3$	320(4.20),450(4.63)	24-1379-69

Compound	Solvent	λ_{max}(log ε)	Ref.
$C_{18}H_{16}NOP$			
Phosphine oxide, 1H-azepin-1-yldiphenyl-	EtOH	238(3.99),261(3.33), 267(3.34),273(3.24), 310s(2.77)	44-2866-69
$C_{18}H_{16}N_2$			
Cinnamaldehyde, azine	$CHCl_3$	349(4.79),362(4.78)	22-1367-69
1-Naphthaldehyde, methylphenylhydrazone	EtOH	210(4.6),231(4.4), 283(4.0),363(4.3)	28-0248-69B
2-Naphthaldehyde, methylphenylhydrazone	EtOH	205(4.4),235(4.6), 254(4.2),280(4.0), 291(4.0),345(4.5)	28-0248-69B
6H-Pyrido[4,3-b]carbazole, 3,5,11-trimethyl-	EtOH	240(4.39),247(4.37), 276(4.71),287(4.86), 295(4.92),335(3.63)	12-0185-69
$C_{18}H_{16}N_2O$			
Acetophenone, 2-(5-methyl-4-phenylpyrazol-1-yl)-	MeOH	248(4.43)	44-3639-69
Ketone, methyl 4-methyl-1,3-diphenylpyrazol-5-yl	n.s.g.	243(4.30),288(3.82)	28-1063-69B
3-Penten-2-one, 1-diazo-4-methyl-1,3-diphenyl-	C_6H_{12}	206(4.51),261(4.48), 443(1.88)	88-2659-69
2H-Pyrrolo[2,3-b]quinolin-2-one, 1,3,3a,4-tetrahydro-1-methyl-4-phenyl-	EtOH	275(4.1)	33-1929-69
$C_{18}H_{16}N_2OS$			
2-Pyrazolin-5-one, 4-[(benzylthio)methylene]-3-methyl-1-phenyl-	hexane	250(4.22),320(4.32), 410(3.28)	104-1249-69
	EtOH	255(4.23),325(4.13)	104-1249-69
4(1H)-Pyrimidinethione, 1-(2-hydroxyethyl)-2,5-diphenyl-	EtOH	243(4.20),348(4.26)	4-0037-69
$C_{18}H_{16}N_2OS_4$			
2,3-Dihydrothiazolo[2,3-b]benzothiazolium 2-(2-oxobenzothiazolin-3-yl)ethyl sulfide	MeOH	215(4.84),247(3.97), 256(3.89),285(3.80), 291(3.88),309(3.93)	4-0163-69
$C_{18}H_{16}N_2O_2$			
Ethanol, 2-[(3,6-diphenyl-4-pyridazinyl)oxy]-	MeOH	237(4.33),273(4.38)	4-0497-69
Pyrazole-5-carboxylic acid, 4-methyl-1,3-diphenyl-, methyl ester	n.s.g.	230(4.34),278(3.99)	28-1063-69B
3-Quinolinecarboxamide, 4-hydroxy-N-phenethyl-	EtOH	248(4.36),255(4.34), 303s(4.09),312(4.10), 322s(3.98)	2-1010-69
$C_{18}H_{16}N_2O_2S_2$			
Benzo[b]thiophen-4(5H)-one, 5,5'-(hydrazodimethylidyne)bis[6,7-dihydro-	EtOH	255(4.41),371(4.41)	39-2750-69C
$C_{18}H_{16}N_2O_2S_4$			
2-Benzothiazolinone, 3,3'-(dithiodiethylene)bis-	MeOH	215(4.91),245(4.05), 283(3.75),290(3.75)	4-0163-69
$C_{18}H_{16}N_2O_3$			
1H-1,4-Benzodiazepine-3-carboxylic acid, 2,3-dihydro-2-oxo-5-phenyl-, ethyl ester	EtOH	228(4.48)	111-0239-69

Compound	Solvent	$\lambda_{max}(\log \epsilon)$	Ref.
Carbostyril, 3-anthraniloyl-1-methoxy-4-methyl-	EtOH	230(4.68),268(4.14), 336(3.98),384(4.00)	18-2952-69
Cyclopenin, methyl-, dl-	n.s.g.	212(4.49),285(3.30)	35-6083-69
Maleimide, 2-(benzylamino)-N-(p-methoxyphenyl)-	EtOH	233(4.40),280(3.91), 365(3.32)	94-0980-69
$C_{18}H_{16}N_2O_3S$			
Spiro[2H-1-benzopyran-2,2'-benzothiazoline], 3'-ethyl-3-methyl-6-nitro-	$(MeOCH_2)_2$	245s(--),269(4.39), 303(4.05),333s(--)	78-3251-69
Spiro[2H-1-benzopyran-2,2'-benzothiazoline], 3,3',4'-trimethyl-6-nitro-	$(MeOCH_2)_2$	266(4.38),303(4.00), 333s(--)	78-3251-69
Spiro[2H-1-benzopyran-2,2'-benzothiazoline], 3,3',6-trimethyl-8-nitro-	$(MeOCH_2)_2$	253s(--),292s(--), 348(3.52)	78-3251-69
Spiro[2H-1-benzopyran-2,2'-benzothiazoline], 3,3',6'-trimethyl-6-nitro-	$(MeOCH_2)_2$	268(4.36),307(4.05), 336s(--)	78-3251-69
Spiro[2H-1-benzopyran-2,2'-benzothiazoline], 3,3',8-trimethyl-6-nitro-	$(MeOCH_2)_2$	250(4.38),272(4.32), 302s(--),335s(--)	78-3251-69
$C_{18}H_{16}N_2O_4$			
Acetophenone, 3',4'-(methylenedioxy)-, azine	C_6H_{12}	275s(4.21),325(4.58)	39-0836-69C
Anthranilic acid, N-(α-acetamidocinnamoyl)-	n.s.g.	295(4.31),310(4.33)	35-6083-69
p-Benzenediacrylic acid, α,α'-dicyano-, diethyl ester	EtOH	202(4.15),227(4.05), 347(4.62),362s(--)	23-4076-69
1,4-Phenazinedicarboxylic acid, diethyl ester	EtOH	246(4.86),365(4.22)	33-0322-69
$C_{18}H_{16}N_2O_4S$			
Spiro[2H-1-benzopyran-2,2'-benzothiazoline], 6'-methoxy-3,3'-dimethyl-6-nitro-	$(MeOCH_2)_2$	226(4.53),268(4.33), 318(4.05),350s(--)	78-3251-69
Spiro[2H-1-benzopyran-2,2'-benzothiazoline], 8-methoxy-3,3'-dimethyl-6-nitro-	$(MeOCH_2)_2$	217(4.53),253(4.29), 293s(4.01),342(3.93)	22-3329-69
$C_{18}H_{16}N_2O_5$			
Benzo[c]cinnoline-3,8-dicarboxylic acid, diethyl ester, 5-oxide	EtOH	204s(3.38),218s(4.27), 264(4.61),272(4.60), 280s(4.53),294(4.49), 310(4.11),340s(3.85), 372(4.43),391(3.15)	4-0523-69
1(4H)-Pyridinecarboxylic acid, 4-(5-hydroxy-2-phenyl-4-oxazolyl)-, methyl ester, acetate	MeCN	275(4.31)	24-1129-69
$C_{18}H_{16}N_2O_5S$			
5-Thia-1-azabicyclo[4.2.0]oct-2-ene-2-carboxylic acid, 7-[2-(2-benzofuranyl)acetamido]-3-methyl-8-oxo-, potassium salt	H_2O	247(4.30),274(3.90), 282(3.79)	87-0310-69
$C_{18}H_{16}N_2O_6$			
Benzo[c]cinnoline-3,8-dicarboxylic acid, diethyl ester, 5,6-dioxide	EtOH	218(3.92),240(4.36), 265s(4.45),276(4.59), 292(4.67),314s(4.14), 374(3.90)	4-0523-69
3H-1,6-Phenoxazinedicarboxylic acid, 2-amino-4,9-dimethyl-3-oxo-, dimethyl ester	$CHCl_3$	435(4.50)	24-3205-69

Compound	Solvent	$\lambda_{max}(\log \epsilon)$	Ref.
3H-4,6-Phenoxazinedicarboxylic acid, 2-amino-1,9-dimethyl-3-oxo-, dimethyl ester	$CHCl_3$	420(4.40),438(4.41)	24-3205-69
$C_{18}H_{16}N_2O_8$			
[2,2'-Bipyridine]-3,3',6,6'-tetrol, tetraacetate	$CHCl_3$	281(4.06)	83-0264-69
$C_{18}H_{16}N_2O_{10}S$			
2,4,5-Pyridinetricarboxylic acid, 6-(4,5-dicarboxy-2-thiazolyl)-, pentamethyl ester	MeOH	296(4.21),317(4.09)	32-0431-69
$C_{18}H_{16}N_2S_2$			
[3,3'-Biindoline]-2,2'-dithione, 1,1'-dimethyl-	CCl_4	293s(4.18),299(4.30), 330(4.41)	94-0550-69
$C_{18}H_{16}N_2S_5$			
2,3-Dihydrothiazolo[2,3-b]benzothiazolium 2-(2-benzothiazolinethione-3-yl)ethyl sulfide	MeOH	225(4.51),257(4.00), 322(4.44)	4-0163-69
$C_{18}H_{16}N_2S_6$			
2-Benzothiazolinethione, 3,3'-(dithiodiethylene)bis-	MeOH	229(4.44),325(4.62)	4-0163-69
$C_{18}H_{16}N_2Se$			
Indole, 3,3'-(selenodimethylene)di-	EtOH	224(4.95),281(4.23), 290(4.17)	11-0031-69B
$C_{18}H_{16}N_2Se_2$			
Indole, 3,3'-(diselenodimethylene)di-	EtOH	220(4.54),282(4.11)	11-0031-69B
$C_{18}H_{16}N_4$			
Benzimidazole, 2,2'-vinylenebis[5-methyl-, cis	EtOH	223s(4.47),255s(4.55), 263(4.61),349(4.61), 368(4.70),388(4.61), 418(4.07)	87-0818-69
trans	EtOH	225s(3.45),273(3.92), 350(3.52),368(3.67), 388(3.52)	87-0818-69
$C_{18}H_{16}N_4O_2$			
Formanilide, N-methyl-4'-[(3-methyl-5-oxo-1-phenyl-2-pyrazolin-4-ylidene)-amino]-	MeOH	246(4.43),357(3.75), 493(3.08)	44-1687-69
$C_{18}H_{16}N_4O_3$			
Formanilide, N-methyl-4'-(3'-methyl-5'-oxo-1'-phenylspiro[oxaziridine-3,4'-[2]pyrazolin]-2-yl)-	MeOH	246(4.37),280(4.01)	44-1687-69
$C_{18}H_{16}N_6O_3$			
Propionitrile, 3,3'-[oxybis(p-phenylenenitrosimino)]di-	EtOH	221(4.33),285(4.22)	103-0623-69
$C_{18}H_{16}N_6O_4$			
Benzoic acid, m-azido-, tetramethylene ester	dioxan	250(4.43),297(3.72)	73-3811-69
Benzoic acid, p-azido-, tetramethylene ester	dioxan	272(4.74)	73-3811-69

Compound	Solvent	$\lambda_{max}(\log \epsilon)$	Ref.
$C_{18}H_{16}O$			
Bicyclo[3.1.0]hexan-2-one, 6,6-diphenyl-	C_6H_{12}	224(4.18),254(2.84), 261(2.89),268(2.89), 275(2.70),282(1.87), 289(1.85),309(1.69), 320s(1.30)	35-0434-69
Bicyclo[3.1.0]hexan-3-one, 6,6-diphenyl-	C_6H_{12}	223(4.16),255(2.72), 260(2.81),267(2.84), 274(2.66),290(1.30), 300(1.30),311(1.18), 323(1.78)	35-0434-69
2-Naphthol, 3,4-dimethyl-1-phenyl-	EtOH	235(4.82),285(3.84), 296(3.83),320(3.40), 334(3.49)	24-3428-69
$C_{18}H_{16}O_2$			
9-Anthraceneacetic acid, ethyl ester	EtOH	256(5.27),317(4.07), 347(3.78),365(3.97)	33-2197-69
9-Anthroic acid, 10-methyl-, ethyl ester	EtOH	257(5.24),352(3.79), 370(3.97),391(3.94)	33-2197-69
2-Butene-1,4-dione, 1-phenyl-4-(2,4-xylyl)-	EtOH	250(4.30)	104-0511-69
Dibenzo[a,e]cyclodecene-5,8-dione, 6,7,13,14-tetrahydro-	EtOH	242(3.99),278(3.25)	44-0874-69
4-Oxatricyclo[4.1.0.0^{2,7}]heptan-2-ol, 1,7-diphenyl-	MeOH	218(4.09),274(4.09)	35-4612-69
$C_{18}H_{16}O_2S$			
Thiepin, 2,7-dihydro-4,5-diphenyl-, 1,1-dioxide	EtOH	245(4.40)	77-1159-69
$C_{18}H_{16}O_3$			
17H-Cyclopenta[a]phenanthren-17-one, 6,7,15,16-tetrahydro-6,7-dihydroxy-11-methyl-	EtOH	232s(4.13),311(4.22)	39-2484-69C
$C_{18}H_{16}O_4$			
Acetic acid, dibenzoyl-, ethyl ester	C_6H_{12}	246(4.31),284(4.15)	22-3281-69
	MeOH	248(4.39),282(3.82)	
	EtOH	248(4.37),282(3.90)	
	ether	248(4.33),285(4.03)	
	MeCN	247(4.41),282(3.69)	
	$CHCl_3$	252(4.34),285(4.03)	
	CCl_4	285(4.13)	
5,10-Ethenocyclobuta[b]naphthalene-4,7-diol, 2a,3,8,8a-tetrahydro-, diacetate	EtOH	213(4.08),272(3.49), 282(3.34)	35-4734-69
Itaconic acid, diphenyl-	EtOH	270(4.06)	22-1344-69
2-Naphthoic acid, 3-benzyl-1,2,3,4-tetrahydro-6-hydroxy-4-oxo-	EtOH	226(4.23),253(3.93), 323(3.38)	2-0561-69
$C_{18}H_{16}O_4S$			
Benzo[b]thiophen-3(2H)-one, 2-(3,4,5-trimethoxybenzylidene)-	MeOH	258(4.2),277(4.2), 348(4.1),439(4.2)	83-0401-69
$C_{18}H_{16}O_5$			
3,6(2H,5H)-Benzofurandione, 4-hydroxy-2-(p-methoxybenzylidene)-5,5-dimethyl-	MeOH	382(4.5)	2-0540-69
3(2H)-Benzofuranone, 2-benzylidene-4,6,7-trimethoxy-	MeOH	272(4.4),310(4.4), 390(4.2)	2-0543-69

Compound	Solvent	λ_{max}(log ϵ)	Ref.
3(2H)-Benzofuranone, 6,7-dimethoxy-2-(p-methoxybenzylidene)-	MeOH	249(--),390(4.5)	2-0543-69
3(2H)-Benzofuranone, 2-(3,4,5-trimethoxybenzylidene)-	MeOH	265(3.9),348s(3.8), 393(4.1)	83-0401-69
Dibenz[b,f]oxepine-4,6-dicarboxylic acid, 10,11-dihydro-, dimethyl ester	EtOH	282(3.56)	35-1665-69
Flavanone, 5-acetoxy-7-methoxy-	EtOH	273(4.13),304(3.75)	105-0397-69
Flavone, 3',4',7-trimethoxy-	n.s.g.	238(4.52),313(4.31)	2-0940-69
Flavone, 4',5,7-trimethoxy-	n.s.g.	263(4.47),322(4.23)	2-0940-69
Isoflavone, 3',4',7-trimethoxy-	EtOH	262(4.28),288(4.09)	2-0118-69
Isojaxanthone, dihydro-	MeOH	239(4.41),25 (4.46), 310(4.11),369(3.80)	78-2787-69
2-Naphthoic acid, 1,2,3,4-tetrahydro-6-hydroxy-3-(p-hydroxybenzyl)-4-oxo-	EtOH	224(4.36),253(3.97), 323(3.45)	2-0561-69
Osajaxanthone, dihydro-	MeOH	234(4.34),262(4.39), 317(4.08),381(3.64)	78-2787-69
Phenanthrene, 3,4-(methylenedioxy)-2,7,8-trimethoxy-	$CHCl_3$	306(4.15),317(3.98), 342(3.37),357(3.64), 375(3.70)	88-0067-69
Pterocarpan, 6a,11a-dehydro-3,8,9-trimethoxy-	EtOH	250s(4.11),274(3.95), 337(4.39),355(4.32)	18-2395-69
Succinic acid, [p-(benzyloxy)benzylidene]-	EtOH	286(4.35)	2-0561-69
Xanthen-9-one, 1,3,7-trihydroxy-2-(3-methyl-2-butenyl)-	MeOH	241(3.53),263(3.52), 314(3.23),337(2.81)	39-0486-69C
$C_{18}H_{16}O_6$			
Anthraquinone, 1-hydroxy-2,6,8-trimethoxy-3-methyl-	EtOH	226(4.54),274(4.49), 300s(4.07),418(4.04)	23-0767-6
3(2H)-Benzofuranone, 4-hydroxy-2-(3,4,5-trimethoxybenzylidene)-	MeOH	264(4.0),322(4.0), 405(4.4)	83-0401-69
3(2H)-Benzofuranone, 5-hydroxy-2-(3,4,5-trimethoxybenzylidene)-	MeOH	269(4.2),361(4.3), 405(4.2)	83-0401-69
3(2H)-Benzofuranone, 6-hydroxy-2-(3,4,5-trimethoxybenzylidene)-	MeOH	258(4.1),368(4.5)	83-0401-69
3(2H)-Benzofuranone, 7-hydroxy-2-(3,4,5-trimethoxybenzylidene)-	MeOH	244(3.8),270(3.9), 353s(4.0),404(4.3)	83-0401-69
2,3-Biphenylenediol, 6,7-dimethoxy-, diacetate	EtOH	255(4.91),360(4.16),	39-0646-69C
Cladrastin	EtOH	206(4.50),220(4.42), 262(4.37),320(4.07)	78-3887-69
Flavone, 5-hydroxy-2',6,6'-trimethoxy-	MeOH	230(4.36),265(4.27), 338(3.57),345(3.58)	2-0746-69
Flavone, 5-hydroxy-2',6',8-trimethoxy-	MeOH	229(4.47),348(3.89), 352s(3.88)	2-0746-69
Flavone, 5-hydroxy-3,4',7-trimethoxy-	EtOH	269(4.30),349(4.25)	105-0397-69
Flavone, 5-hydroxy-3,7,8-trimethoxy-	MeOH	273(4.38),358(3.79)	88-0431-69
Isoflavanone, 6,7-dimethoxy-3',4'-(methylenedioxy)-	$CHCl_3$	246(4.19),278(4.18), 338(3.96)	39-1787-69C
Pterocarpin, 4-methoxy-, (±)-	EtOH	311(3.90)	31-0122-69
$C_{18}H_{16}O_7$			
Benzo[b]furan-3-methanol, 2-(3,4-dimethoxy-2-hydroxyphenyl)-5,6-(methylenedioxy)-	EtOH	271(4.16),320(4.38)	31-0122-69
3(2H)-Benzofuranone, 4,6-dihydroxy-2-(3,4,5-trimethoxybenzylidene)-	MeOH	257(4.0),355s(4.3), 388(4.4)	83-0401-69
3(2H)-Benzofuranone, 5,6-dihydroxy-2-(3,4,5-trimethoxybenzylidene)-	MeOH	270(4.1),367(4.5)	83-0401-69

Compound	Solvent	λmax(log ε)	Ref.
3(2H)-Benzofuranone, 6,7-dihydroxy-2-(3,4,5-trimethoxybenzylidene)-	MeOH	249(4.1),361s(4.3), 393(4.4)	83-0401-69
Benzoic acid, 3,5-dimethoxy-2,2'-carbonyldi-, monomethyl ester	MeOH	240s(4.03),296(3.57)	22-2370-69
Eupatilin	EtOH	243(4.25),277(4.23), 340(4.42)	78-1603-69
	EtOH-NaOEt	238(4.46),278(4.45), 312(4.16),376(4.24)	78-1603-69
	EtOH-NaOAc	237(4.38),278(4.38), 322(4.17),364(4.18)	78-1603-69
	EtOH-$AlCl_3$	259(4.19),290(4.26), 363(4.32)	78-1603-69
Eupatorin	EtOH	243(4.24),254(4.29), 274(4.30),342(4.44)	78-1603-69
	EtOH-NaOEt	271(4.12),372(3.82)	78-1603-69
	EtOH-NaOAc	243(4.30),253(4.28), 276(4.28),342(4.40)	78-1603-69
	EtOH-$AlCl_3$	238(4.21),262(4.19), 290(4.25),362(4.37)	78-1603-69
Flavone, 3',5-dihydroxy-4',7,8-trimethoxy-	EtOH	257(4.29),274(4.29), 292s(4.20),350(4.58)	24-0112-69
	EtOH-$AlCl_3$	274(4.27),283(4.36), 303(4.29),351(4.33), 406(4.14)	24-0112-69
Flavone, 3',6-dihydroxy-4',5,7-trimethoxy-	EtOH	243(4.28),277(4.21), 334(4.42)	40-1270-69
Flavone, 4',6-dihydroxy-3',5,7-trimethoxy-	EtOH	242s(4.27),278(4.18), 336(4.47)	40-1270-69
	EtOH-NaOAc	244s(4.27),277(4.13), 335(4.31),398(4.06)	40-1270-69
Gris-2',5'-diene-3,4'-dione, 2',6-dimethoxy-6'-carbomethoxy-4-methyl-	EtOH	214(4.37),230s(4.25), 283(4.20),315(4.07)	39-1721-69C
Olivinic acid, anhydro-, methyl ester	EtOH	230(4.38),257(4.06), 273(3.95),314(4.18), 365(4.00)	105-0257-69
$C_{18}H_{16}O_8$			
Eupatin	EtOH	258(4.34),273(4.19), 366(4.37)	44-1460-69
Flavone, 3,4',5-trihydroxy-3',6,7-trimethoxy-	EtOH	259(4.34),273s(4.17), 370(4.36)	18-1398-69
	EtOH-NaOH	326(4.12)	18-1398-69
	EtOH-NaOAc	260(4.33),271s(4.26), 378(4.27),402s(3.82)	18-1398-69
	EtOH-$AlCl_3$	270(4.37),304s(3.75), 390s(4.23),428(4.38)	18-1398-69
	EtOH-NaOAc-H_3BO_3	259(4.35),273s(4.20), 373(4.34)	18-1398-69
Flavone, 3',4',5-trihydroxy-3,6,7-trimethoxy-	EtOH	260(4.32),356(4.37)	18-1649-69
	EtOH-$AlCl_3$	279(--),415(--)	18-1649-69
Flavone, 3',5,7-trihydroxy-3,4',6-trimethoxy- (centaureidin)	EtOH	257(4.29),271(4.22), 351(4.35)	18-2701-69
	EtOH-NaOAc	274(4.31),364(4.23)	18-2701-69
	EtOH-$AlCl_3$	266(4.25),279(4.22), 369(4.29)	18-2701-69
Flavone, 3',5,7-trihydroxy-3,4',8-trimethoxy-	EtOH	260(4.44),276(4.47), 360(4.33)	18-2701-69
Flavonoid K	EtOH	258(4.34),273(4.19), 366(4.37)	78-1603-69

Compound	Solvent	$\lambda_{max}(\log \epsilon)$	Ref.
$C_{18}H_{16}Si$			
Silane, triphenyl-	C_6H_{12}	265(3.11)	101-0017-69B
$C_{18}H_{17}BrO_6$			
1,2,3-Naphthalenetricarboxylic acid, 4-(bromomethyl)-3-ethyl-, 1,2-dimethyl ester	EtOH	246(4.78),297s(3.81), 304(3.82)	94-1591-69
$C_{18}H_{17}Cl$			
Anthracene, 9-(1-chloroethyl)-10-ethyl-	hexane	253s(4.81),261(5.03), 341(3.39),358(3.73), 373(3.91),397(3.89)	39-2266-69C
1,3-Pentadiene, 4-(p-chlorophenyl)-1-p-tolyl-	EtOH	320(4.43)	104-0111-69
$C_{18}H_{17}ClN_2O$			
1,5-Benzodiazocin-2(1H)-one, 8-chloro-3,4-dihydro-3,3-dimethyl-6-phenyl-, hydrochloride	MeOH	240(4.41)	36-0830-69
3H-Indole, 5-butoxy-3-(chlorimino)-2-phenyl-	EtOH	283(4.50),500(3.27)	22-2008-69
$C_{18}H_{17}ClN_2O_6$			
1,4-Dihydro-2,3-dimethyl-4-(m-nitrobenzylidene)isoquinolinium perchlorate	EtOH	245(4.59),265(3.69), 350(3.49)	78-0101-69
1,4-Dihydro-2,3-dimethyl-4-(p-nitrobenzylidene)isoquinolinium perchlorate	EtOH	232(4.49),350(3.95)	78-0101-69
$C_{18}H_{17}ClN_4$			
2-Cyclohexen-1-one, 3-[(p-chlorophenyl)-azo]-, phenylhydrazone	$CHCl_3$	309(4.14),438(4.56)	24-1379-69
$C_{18}H_{17}ClO_8$			
4',5,7-Trimethoxyflavylium perchlorate	aq. HCl	277(4.32),322(3.77), 469(4.57)	78-2367-69
	EtOH-HCl	278(4.31),325(3.79), 477(4.63)	78-2367-69
$C_{18}H_{17}FN_2O$			
Urea, 1-(p-fluorophenethyl)-3-inden-3-yl-	EtOH	210(4.10),268(3.29), 275(3.27)	36-0047-69
$C_{18}H_{17}IN_4$			
2-Cyclohexen-1-one, 3-[(p-iodophenyl)-azo]-, phenylhydrazone	$CHCl_3$	312(4.16),440(4.57)	24-1379-69
$C_{18}H_{17}IN_4O_2$			
2-Formyl-1-methyl-6-nitroquinolinium iodide, o-tolylhydrazone	EtOH	267(3.29),314(2.88), 492(1.69)	124-0370-69
2-Formyl-1-methyl-6-nitroquinolinium iodide, p-tolylhydrazone	EtOH	280(3.57),315(3.16), 504(1.69)	124-0370-69
$C_{18}H_{17}IN_4O_3$			
2-Formyl-1-methyl-6-nitroquinolinium iodide, (o-methoxyphenyl)hydrazone	EtOH	274(3.69),320(3.01), 512(1.73)	124-0370-69
2-Formyl-1-methyl-6-nitroquinolinium iodide, (p-methoxyphenyl)hydrazone	EtOH	275(3.68),325(3.06), 526(1.50)	124-0370-69
$C_{18}H_{17}N$			
Acridine, 9-(3-methyl-1-butenyl)-, cis	EtOH	253(4.2),359(3.0)	40-1263-69

Compound	Solvent	$\lambda_{max}(\log \epsilon)$	Ref.
Acridine, 9-(3-methyl-1-butenyl)-, trans	EtOH	254(4.0),361(3.0)	40-1263-69
Acridine, 9-(1-pentenyl)-, cis	EtOH	253(4.1),359(3.0)	40-1263-69
Acridine, 9-(1-pentenyl)-, trans	EtOH	254(4.1),361(3.0)	40-1263-69
2-Naphthylamine, N,N-dimethyl-3-phenyl-	MeOH	211(4.60),237(4.48), 263(4.64),345(3.34)	89-0675-69
Phenalene, 1-[3-(dimethylamino)-2-propenylidene]-	hexane	272(4.32),474(4.60)	24-2301-69
	MeCN	278(4.53),496(4.23)	24-2301-69
Pyrrole, 1,2-dimethyl-3,4-diphenyl-	EtOH	244(4.31),266s(--)	88-4875-69
Pyrrole, 3,4-dimethyl-2,5-diphenyl-	EtOH	208(4.11),321(4.34)	39-2249-69C
Pyrrole, 1-methyl-3-phenyl-4-p-tolyl-	EtOH	244(4.35),264s(--)	88-4875-69
$C_{18}H_{17}NO$			
Cyclohepta[c]pyrrol-6(2H)-one, 1,3,5-trimethyl-2-phenyl-	EtOH	293(4.74),330(4.02), 387(3.40)	104-0559-69
	HOAc-$HClO_4$	306(4.95),344(4.06), 554(3.21)	104-0559-69
Girinimbin	n.s.g.	237(4.76),278(4.53), 315(3.88),328(4.00), 340(4.03),355(3.98)	2-0307-69
1,4-Hexadien-3-one, 5-anilino-1-phenyl-	EtOH	232(4.07),297(4.10), 383(4.46)	18-3596-69
8,16-Imino[2,2]metacyclophane, N-acetyl-	EtOH	264(3.86),280s(2.90)	35-1672-69
Isoxazole, 5-mesityl-3-phenyl-	EtOH	230s(4.27),254(4.28), 295(4.16),312(4.13)	88-3329-69
4(1H)-Pyridone, 2,3-dihydro-5-methyl-2,6-diphenyl-	EtOH	243(3.65),347(3.98)	18-3596-69
4(1H)-Pyridone, 2,3-dihydro-6-methyl-1,2-diphenyl-	EtOH	240(3.34),327(4.02)	18-2690-69 +18-3596-69
	EtOH-HCl	242(3.22),325(4.02)	18-1357-69
	EtOH-HCl	327(4.23)	18-2690-69
	EtOH-HBr	327(4.15)	18-2690-69
4(1H)-Pyridone, 2,3-dihydro-6-methyl-2,5-diphenyl-	EtOH	343(4.15)	18-3596-69
3-Quinuclidinone, 2-(1-naphthylmethylene)-	EtOH	252(4.24),343(4.17)	22-1251-69
$C_{18}H_{17}NO_2$			
1-Cyclohexene-1-glyoxylamide, N-1-naphthyl-	EtOH	222(5.0),245s(--), 285(3.9)	24-1876-69
Isoquinoline, 4-benzyl-7,8-dimethoxy-, hydrochloride	EtOH	236(4.13),255(3.92), 280s(3.22)	78-0101-69
[1,3]Oxazepino[2,3-a]isoindol-7(11bH)-one, 2,3,4,5-tetrahydro-11b-phenyl-	EtOH	216(4.32),251(3.60)	44-0165-69
1,3-Pentadiene, 4-(p-nitrophenyl)-1-p-tolyl-	EtOH	298(4.23),382(4.43)	104-0111-69
Roemerine	EtOH	234(4.16),264s(--), 273(4.28),285s(--), 293s(--),318(3.56)	95-1691-69
$C_{18}H_{17}NO_3$			
1,4-Benzoxazepine, 5-benzyl-8,9-dimethoxy-	MeOH	220(4.26),254(3.91), 344(4.03)	95-1048-69
Isoquinoline, 1-(2,4-dimethoxyphenyl)-3-methoxy-	EtOH	221(4.53),279(3.73), 343(3.84)	39-1873-69C
	EtOH-HCl	229(3.84),268(3.89), 383(--)	39-1873-69C
Mecambroline, dl-	EtOH	265(4.02),275(4.04), 310(3.95)	2-0746-69

Compound	Solvent	λ_{max}(log ϵ)	Ref.
6-Phenanthridinelactic acid, ethyl ester	$CHCl_3$	252(4.60),272s(4.07), 290s(3.70),330(3.30), 345(3.23)	24-0915-69
4H,8H-Pyrido[3,2,1-de]phenanthridine, 5,6-dihydro-2-methoxy-10,11-(methylenedioxy)-	EtOH	210(4.53),262(4.48), 280(4.34),368(4.05)	2-0848-69
Steporphine	EtOH	238(4.25),273(4.25), 293(3.87),312(3.55)	88-3287-69
Ushinsunine	EtOH	217(4.2),274(4.1), 323(3.4)	95-1313-69
hydrobromide	EtOH	272(4.44),317(3.90)	95-1313-69
$C_{18}H_{17}NO_4$			
Benzo[e]indolizine-1-acetic acid, 2-carboxy-5-methyl-, dimethyl ester	MeOH	219(4.42),258s(4.66), 266(4.83),276s(4.48), 300s(3.70),312(3.76), 325(3.73),342(3.56), 358s(3.26)	39-2311-69C
Benzo[e]indolizine-3-acetic acid, 2-carboxy-7-methyl-, dimethyl ester	MeOH	213(4.49),231(4.47), 241s(4.46),251(4.68), 260s(4.58),269s(4.51), 277s(4.40),288(4.12), 341s(4.00),352(4.05), 366s(3.88)	39-2311-69C
Benzo[e]indolizine-3-acetic acid, 2-carboxy-9-methyl-, dimethyl ester	MeOH	212(4.48),234(4.53), 254(4.66),262s(4.63), 283s(4.01),339(4.04), 345s(4.03),366(3.88)	39-2311-69C
Dibenz[b,f]azepine-4,6-dicarboxylic acid, 10,11-dihydro-, dimethyl ester	EtOH	224(4.39),250s(4.00), 291(4.06),363(4.03)	35-1672-69
Isoquinoline, 3,4-dihydro-1-[3,4-(methylenedioxy)phenyl]-6,7-dimethoxy-	EtOH	233(4.50),305(4.12)	1-0244-69
Launobine	EtOH	224(4.49),270(4.19), 307(3.94)	95-0737-69
$C_{18}H_{17}NO_5$			
Anthranilic acid, N-[2-(α-hydroxybenzyl)glycidoyl]-, methyl ester	EtOH	224(4.38),255(4.10), 306(3.71)	35-6083-69
Benzoic acid, p-(acetylsalicylamino)-, ethyl ester	MeOH	283(4.33)	18-2044-69
$C_{18}H_{17}NO_7$			
Terephthalamic acid, 3,5-dihydroxy-2-(5-methyl-o-anisoyl)-, methyl ester	n.s.g.	224(4.36),260(4.19), 342(3.92)	39-2805-69C
$C_{18}H_{17}N_3$			
Aniline, N,N-dimethyl-p-(1-naphthylazo)-	EtOH	432(4.40)	124-0179-69
Anthranilonitrile, N-(α-1-pyrrolidinylbenzylidene)-	EtOH	213(4.51),259(3.89), 309(3.74)	22-2008-69
Carbazol-1(2H)-one, 3,4-dihydro-, phenylhydrazone	MeOH	353(4.53)	24-1198-69
$C_{18}H_{17}N_3O$			
Ketone, methyl 1,3a,8,8a-tetrahydro-8-methyl-1-phenylpyrazolo[3,4-b]indol-3-yl	EtOH	248(4.35),304s(3.82), 358(4.19)	32-0588-69
1,2-Propanedione, 1-(1-methylindol-3-yl)-, 1-(phenylhydrazone)	EtOH	220(4.34),289(3.87), 360(4.12)	32-0588-69

Compound	Solvent	λmax(log ε)	Ref.
$C_{18}H_{17}N_3O_2$			
Indole-3-glyoxylic acid, ethyl ester, α-(phenylhydrazone)	EtOH	230(4.51),290(3.99), 390(4.23)	32-0588-69
$C_{18}H_{17}N_3O_3$			
Indole-3-propionamide, α-[3-(2-furyl)-acrylamido]-, L-	pH 8	305(4.44)	35-7187-69
$C_{18}H_{17}N_3O_5$			
Pyrrolo[2,3-b]indole-2-carboxylic acid, 1,2,3,3a,8,8a-hexahydro-3a-(5-nitro-salicyl)-	pH 4	320(3.99)	88-0857-69
	pH 8.9	417(4.31)	88-0857-69
$C_{18}H_{17}N_3O_6$			
1,4-Phenazinedicarboxylic acid, 5,10-dihydro-7-nitro-, diethyl ester	EtOH	235(4.48),281(4.34), 365(3.64),534(4.06)	33-0322-69
$C_{18}H_{18}$			
Anthracene, 9-tert-butyl-	EtOH	250s(4.43),256(5.12), 338(3.19),355(3.31), 372(3.46),390(3.37)	78-3485-69
1H-Cyclopenta[l]phenanthrene, 8,9,10,11-tetrahydro-3-methyl-	EtOH	202(4.54),220(4.57), 241(4.84),247(4.83), 314(4.12),326(4.02), 331(4.01)	44-0763-69
9,10-Ethanoanthracene, 9,10-dihydro-11,12-dimethyl-, trans	EtOH	266(3.24),273(3.31)	78-1661-69
$C_{18}H_{18}BF_4NO$			
6-Methoxy-1,3-dimethyl-N-phenylcyclo-hepta[c]pyrrolium tetrafluoroborate	$C_2H_4Cl_2$	309(4.90),345(4.15), 357(4.05),558(3.26)	104-0947-69
$C_{18}H_{18}BrClO_4$			
Ethylene, 1-bromo-1-chloro-2,2-bis-(3,4-dimethoxyphenyl)-	EtOH	270s(4.16),288(4.20)	7-0922-69
$C_{18}H_{18}BrClO_{12}$			
1,3-Cyclohexadiene-1,2,3,4,5,6-hexa-carboxylic acid, 5-bromo-6-chloro-, hexamethyl ester	C_6H_{12}	297(3.67)	80-1191-69
$C_{18}H_{18}BrNO_3$			
Ushinsunine, hydrobromide	EtOH	272(4.44),317(3.90)	95-1313-69
$C_{18}H_{18}BrNO_7$			
1,3-Dioxolo[4,5-b]acridine-4,10,11(5H)-trione, 3a-bromo-3a,11a-dihydro-11a-methoxy-5-methyl-, 4-(dimethyl acetal)	EtOH	224(4.51),262s(3.99), 329(4.12)	12-1477-69
$C_{18}H_{18}Br_2O_4$			
Ethylene, 1,1-dibromo-2,2-bis(3,4-di-methoxyphenyl)-	EtOH	234s(4.21),289(4.05)	7-0922-69
$C_{18}H_{18}ClN$			
Aniline, p-chloro-N-[1-(1,2,3,4-tetra-hydro-2-naphthyl)ethylidene]-	EtOH	259(4.28),310s(3.53)	56-0749-69
$C_{18}H_{18}ClNO_2$			
Isoquinoline, 4-benzyl-7,8-dimethoxy-, hydrochloride	EtOH	236(4.13),255(3.92), 280s(3.22)	78-0101-69

Compound	Solvent	$\lambda_{max}(\log \epsilon)$	Ref.
$C_{18}H_{18}ClNO_4$			
4-Benzylidene-1,4-dihydro-2,3-dimethylisoquinolinium perchlorate	aq. EtOH	232(4.49),350(3.95)	78-0101-69
$C_{18}H_{18}ClNO_4S$			
4'-(Dimethylamino)-4-methylthioflavylium perchlorate	HOAc-1% $HClO_4$	264(3.85),288s(3.75), 375(3.65),580(2.88)	2-0191-69
$C_{18}H_{18}ClNO_5$			
6-Hydroxy-1,3,5-trimethyl-2-phenylcyclohepta[c]pyrrolium perchlorate	9:1 HOAc-$HClO_4$	308(4.85),347(4.11), 555(3.27)	104-0947-69
$C_{18}H_{18}ClNS$			
Thioxanthene, 2-chloro-9-[3-(dimethylamino)propylidene]-	EtOH	230(4.60),270(4.17), 327(3.55)	9-0249-69
$C_{18}H_{18}ClNSe$			
Selenoxanthene, 2-chloro-9-[3-(dimethylamino)propylidene]-, cis	heptane	231(4.43),273(4.07), 324(3.44)	73-3792-69
trans	heptane	231(4.42),246s(4.18), 273(4.06),323(3.44)	73-3792-69
$C_{18}H_{18}Cl_2O_{12}$			
1,3-Cyclohexadiene-1,2,3,4,5,6-hexacarboxylic acid, 5,6-dichloro-, hexamethyl ester	C_6H_{12}	300(3.50)	80-1191-69
$C_{18}H_{18}FeO_6$			
Iron, tricarbonyl(6α-hydroxy-3-oxoeudesma-1,4-dien-12-oic acid γ-lactone)	CH_2Cl_2	231(4.16)	101-0342-69A
$C_{18}H_{18}IN_3$			
2-Formyl-1-methylquinolinium iodide, o-tolylhydrazone	EtOH	266(3.43),319(3.06), 470(0.76)	124-0370-69
2-Formyl-1-methylquinolinium iodide, p-tolylhydrazone	EtOH	273(3.42),320(3.00), 478(0.88)	124-0370-69
$C_{18}H_{18}IN_3O$			
2-Formyl-1-methylquinolinium iodide, (o-methoxyphenyl)hydrazone	EtOH	263(3.38),327(3.04), 482(0.78)	124-0370-69
2-Formyl-1-methylquinolinium iodide, (p-methoxyphenyl)hydrazone	EtOH	276(3.31),318(3.16), 496(0.6)	124-0370-69
$C_{18}H_{18}N_2$			
1H-1,4-Diazepine, 2,3-dihydro-1-methyl-5,7-diphenyl-, perchlorate	n.s.g.	265(4.10),356(4.36)	39-1081-69C
3H-Indole, 3-(tert-butylimino)-2-phenyl-	EtOH	258(4.47),393(3.49)	78-1467-69
Pyrimido[1,6-a]indole, 1,2,3,4-tetrahydro-2-methyl-5-phenyl-	MeOH	228(4.5),282(4.2)	33-0629-69
$C_{18}H_{18}N_2O$			
1H-1,4-Diazepine, 2,3-dihydro-6-methoxy-5,7-diphenyl-, perchlorate	EtOH or MeOH	265(3.99),377(4.23)	39-1449-69C
7H-[1,3]Diazepino[2,1-a]isoindol-7-one, 1,2,3,4,5,11b-hexahydro-11b-phenyl-	EtOH	259s(3.47)	44-0165-69
Indole, 2-benzoyl-3-[(dimethylamino)-methyl]-	EtOH	248(3.99),324(4.00)	39-2738-69C
4-Isoquinolinecarboxamide, 1,2-dihydro-2-phenethyl-	MeOH	220(4.14),245(3.97), 280(3.95),350(3.96)	22-2045-69

Compound	Solvent	λ_{max}(log ϵ)	Ref.
2-Pyrazolin-5-one, 3-benzyl-4,4-dimethyl-1-phenyl-	EtOH	242(4.26)	22-4159-69
2-Pyrrolin-4-one, 2-(dimethylamino)-1,3-diphenyl-	MeOH	261(4.16),306(4.26)	78-5721-69
3-Pyrrolin-2-one, 1-benzyl-4-(benzylamino)-	EtOH	275(4.15)	78-3277-69
2H-Pyrrolo[2,3-b]quinolin-2-one, 1,3,3a,4,9,9a-hexahydro-1-methyl-4-phenyl-	EtOH	242(4.0),294(3.4)	33-1929-69
$C_{18}H_{18}N_2O_2$			
Carbostyril, 3,4-dihydro-3-methylacetamido-4-phenyl-	EtOH	252(4.17)	35-6083-69
Indole-3-acetic acid, 5-methyl-2-(3-pyridyl)-, ethyl ester	EtOH	220(4.38),313(4.23)	22-4154-69
$C_{18}H_{18}N_2O_4$			
2-Aziridinecarboxylic acid, 1-benzyl-3-[(p-methoxyphenyl)carbamoyl]-, cis	EtOH	258(4.15)	94-0980-69
Cinnamic acid, 2-methoxy-5-(p-methoxyphenylazo)-, methyl ester	EtOH	249(4.23),280(4.32), 354(4.34),450s(3.34)	12-0935-69
1,4-Phenazinedicarboxylic acid, 5,10-dihydro-, diethyl ester	EtOH	239(4.78),282(4.06), 472(3.91),501(4.03), 538(3.88)	33-0322-69
$C_{18}H_{18}N_2O_4S$			
Sulfur diimide, bis(p-carboxyphenyl)-, diethyl ester	hexane	375(4.1),425(4.2)	48-0621-69
$C_{18}H_{18}N_2O_5$			
Azoxybenzene, 3,3'-diacetyl-4,4'-dimethoxy-	EtOH	245(4.34),346(4.38)	12-0935-69
2-Propanone, 1-(p-methoxyphenyl)-3-(p-nitrophenyl)-, O-acetyloxime	MeCN	215(4.19),273(4.07)	44-1430-69
isomer	MeCN	274(4.04)	44-1430-69
$C_{18}H_{18}N_2O_5S$			
Acetic acid, thio-, S-ester with N-[1-(4-isopropylidene-5-oxo-2-oxazolin-2-yl)-2-mercaptovinyl]-2-phenoxyacetamide	EtOH	230(4.46)	88-3381-69
geometric isomer	EtOH	232(4.31)	88-3381-69
$C_{18}H_{18}N_2S_3$			
2,4-Thiazolidinedithione, 3-ethyl-5-(2-ethyl-4(1H)-quinolylidene)ethylidene]-	BuOH	240(4.25),310(4.05)	65-2116-69
$C_{18}H_{18}N_3Na_2O_9P$			
Cytidine, 3'-O-acetyl-N^4-benzoyldeoxy-, 5'-phosphate, disodium salt	pH 1	257(4.14),315(4.32)	35-1522-69
	H_2O	257(4.34),303(4.09)	35-1522-69
$C_{18}H_{18}N_4$			
Benzylideneimine, 4-cyano-N-[2-(3-pyridyl)-1-piperidyl]-	n.s.g.	328(4.42)	65-2723-69
5,5'-Bibenzimidazole, 2,2'-diethyl-	EtOH	228(4.75),298(4.32)	22-1926-69
5,5'-Bibenzimidazole, 1,1',2,2'-tetramethyl-	MeOH	230(4.73),296(4.29)	22-1926-69
2-Cyclohexen-1-one, 3-(phenylazo)-, diphenylhydrazone	MeOH	427(4.57)	24-1379-69
	MeOH-acid	623(4.78+)	24-1379-69
	$CHCl_3$	303(4.12),434(4.52)	24-1379-69

Compound	Solvent	λ_{max}(log ϵ)	Ref.
2-Cyclohexen-1-one, 3-(phenylazo)-, diphenylhydrazone (cont.)	DMF	456(4.59)	24-1379-69
	DMF-acid	663(5.04)	24-1379-69
$C_{18}H_{18}N_4O$			
Ether, bis[4-(N-β-cyanoethylamino)phenyl]	EtOH	257(4.43),306(3.64)	103-0623-69
2-Pyrazolin-5-one, 4-[[p-(dimethylamino)phenyl]imino]-3-methyl-1-phenyl-	MeOH	440(4.09),525(4.48)	44-1687-69
$C_{18}H_{18}N_4O_2$			
4-[2-(4-N,N-Dimethylaminophenyl)oxaziridine-3-spiro]-3-methyl-1-phenyl-2-pyrazolin-5-one	MeOH	261(4.36),287(3.96)	44-1687-69
$C_{18}H_{18}N_4O_3$			
Indole, 1-acetyl-3-[2-nitro-1-(2-phenylhydrazino)ethyl]-	EtOH	239(4.38),290(3.87), 359(3.63)	103-0602-69
3-Pyrazolidinone, 1,1'-(oxydi-p-phenylene)di-	EtOH	257(4.40)	103-0623-69
$C_{18}H_{18}N_4O_4S$			
Cyclohexanone, 2-[(2,4-dinitrophenyl)thio]-, phenylhydrazone	MeOH	278(4.26)	24-3647-69
$C_{18}H_{18}N_4O_7$			
β-Alanine, N,N-(oxydi-p-phenylene)bis-[N-nitroso-	EtOH	223(4.32),285(4.23)	103-0623-69
$C_{18}H_{18}O$			
Bicyclo[3.1.0]hexan-2-ol, 6,6-diphenyl-	C_6H_{12}	223(4.17),248s(2.55), 255(2.67),261(2.79), 268(2.83),274(2.66)	35-0434-69
[2,2]Paracyclophane, 4-acetyl-, tetracyanoethylene complex	CH_2Cl_2	496(3.16)	35-3553-69
$C_{18}H_{18}O_2$			
2,3-Anthracenedione, 1,4-dihydro-1,1,4,4-tetramethyl-	C_6H_{12}	225(4.94),275(3.68), 287(3.62),307(3.13), 323(2.75)	23-4313-69
1,3,5,7,9,11,13,15-Cyclohexadecaoctaene-1-carboxylic acid, methyl ester	ether	295(4.57),450(2.96)	88-4575-69
2-Epiolivinoic acid, methyl ester	EtOH	258(4.74),302(3.86), 358(3.49)	105-0494-69
B-Norestra-1,3,5(10),8,14-pentaen-17-one, 3-methoxy-	EtOH	308(4.44)	88-1207-69
$C_{18}H_{18}O_3$			
2,3-Naphthalenediacetic acid anhydride, α,α,α',α'-tetramethyl-	C_6H_{12}	228(4.9),275f(3.7)	23-4313-69
2-Stilbenol, 3,4'-dimethoxy-6-vinyl-, trans	EtOH	209(4.43),269(4.38), 305(4.32)	78-5475-69
	N KOH	254(4.34),272(4.34), 365(3.97)	78-5475-69
$C_{18}H_{18}O_4$			
4-Cyclopentene-1,3-dione, 4,4'-(1,5-hexadienylene)bis[5-methyl-	EtOH	297(4.41)	88-0373-69
3-Flaven-4'-ol, 2,8-dimethoxy-3-methyl-	EtOH	264(4.11),274(4.08)	78-2367-69
2,4-Pentanedione, 3,3'-(p-phenylenedimethylidyne)di-	EtOH	207(4.30),225s(--), 326(4.53)	23-4076-69

Compound	Solvent	λ_{max}(log ϵ)	Ref.
2-Stilbenecarboxylic acid, 3,5-dimethoxy-, methyl ester	MeOH	218(4.27),233(4.21), 243(4.17),299(4.45)	44-3192-69
$C_{18}H_{18}O_5$			
Chalcone, 2-hydroxy-4,4',6-trimethoxy-, trans	EtOH	255s(--),301s(--), 374(4.49)	78-2367-69
1,2-Dibenzofurandicarboxylic acid, 3,9b-dihydro-4,9b-dimethyl-, dimethyl ester	90% EtOH	230s(4.06),280(3.50), 285s(3.49)	88-2901-69
Flavanone, 5-hydroxy-4',7-dimethoxy-6-methyl-	MeOH	291(4.35)	78-0283-69
Isoflavan, 2',7-dimethoxy-4',5'-(methylenedioxy)-	EtOH	291(3.86),301(3.82)	18-1408-69
Pterocarpan, 3,8,9-trimethoxy-	EtOH	288(3.83),300(3.83)	18-0233-69
Ta V from Tamus communis (a phenanthrene derivative)	$CHCl_3$	295(4.03),309(4.02), 330(2.81),347(3.15), 364(3.23)	88-0067-69
$C_{18}H_{18}O_6$			
Acetophenone, 2',4',5'-trimethoxy-2-[3,4-(methylenedioxy)phenyl]-	EtOH	236(4.29),273(4.04), 329(3.94)	39-1787-69C
Benzo[b]furan, 5,6-dimethoxy-2-(2-hydroxy-4-methoxyphenyl)-3-(hydroxymethyl)-	EtOH	271(4.25),314(4.40)	18-2395-69
Benzoic acid, 2-anisoyl-4,5-dimethoxy-, methyl ester	EtOH	222(4.43),284(4.31)	39-1921-69C
Fluoren-9-one, 2,3,4,5,6-pentamethoxy-	MeOH	280(4.47),320(3.78), 358(3.75)	2-0557-69
Fluoren-9-one, 2,3,4,6,7-pentamethoxy-	MeOH	215(4.11),280(4.76)	2-0557-69
2,3-Furandicarboxylic acid, 4-acetyl-5-phenyl-, diethyl ester	EtOH	218(4.08),296(4.24)	88-2053-69
2,3-Furandicarboxylic acid, 4-benzoyl-5-methyl-, diethyl ester	EtOH	218(4.04),264(4.02), 294(4.24)	88-2053-69
2'-Isoflavanol, 7,8-dimethoxy-4',5'-(methylenedioxy)-	EtOH	286s(--),302(4.15)	39-1109-69C
1,2,3-Naphthalenetricarboxylic acid, 4-methyl-, 3-ethyl dimethyl ester	EtOH	241(4.72),278s(3.69), 288(3.79),297s(3.75), 334(3.15)	94-1591-69
Philenopteran, 9-O-methyl-	EtOH	280(3.51),286(3.45)	39-0887-69C
Torachrysone, diacetate	EtOH	233(4.65),295(3.70), 333(3.36)	94-0454-69
Xanthen-9-one, 1,3,6,7-tetramethoxy-	MeOH	251(4.55),311(4.26)	78-0275-69
$C_{18}H_{18}O_7$			
Acetophenone, 2'-hydroxy-4',5'-dimethoxy-2-[2-methoxy-4,5-(methylenedioxy)phenyl]-	EtOH	236(4.31),280(4.12), 344(3.93)	39-1787-69C
Acetophenone, 2'-hydroxy-4',5'-dimethoxy-2-[3-methoxy-4,5-(methylenedioxy)phenyl]-	EtOH	238(4.26),279(4.09), 345(3.95)	39-1787-69C
2H-Furo[2,3-h]-1-benzopyran-2-one, 8,9-dihydro-9-hydroxy-8-(1-hydroxy-1-methylethyl)-, diacetate	ether	318(4.15)	24-1673-69
Smirniovin	n.s.g.	220(4.33),246(3.57), 257(3.50),300s(3.99), 323(4.22)	103-0506-69
Xanthen-9-one, 1,2,3,4,7-pentamethoxy-	EtOH	240(4.27),260(4.38), 287(3.95),307s(--), 365(3.83)	78-4415-69

Compound	Solvent	λ_{max}(log ε)	Ref.
Xanthen-9-one, 1,2,3,4,8-pentamethoxy-	EtOH	240s(4.54),248(4.56), 290s(4.02),305(4.11), 350(3.79)	78-1947-69
$C_{18}H_{18}O_8$			
Crotepoxide	MeOH	274(3.02),281(2.93)	44-3898-69
2-Cyclohexen-1-one, 4,5,6-trihydroxy-4-(hydroxymethyl)-, 5,6-diacetate, α-benzoate	MeOH	227(4.28),273(3.34), 280(2.94)	44-3898-69
Xanthen-9-one, 1-hydroxy-2,3,4,6,7-pentamethoxy-	EtOH	242(4.46),263(4.47), 318(4.25),365(3.71)	78-4415-69
	EtOH-NaOH	281(--),421(--)	78-4415-69
$C_{18}H_{19}BrN_2O$			
Dibenzo[b,f][1,4]diazocine, 5-(3-bromopropionyl)-5,6,11,12-tetrahydro-12-methyl-	EtOH	261(4.03),305(3.52)	39-0882-69C
$C_{18}H_{19}BrN_2O_8$			
Ethane, 2-bromo-1,1-bis(4,5-dimethoxy-2-nitrophenyl)-	EtOH	243(4.35),317s(3.90), 337(3.94)	7-0922-69
$C_{18}H_{19}BrO_4$			
Biphenylene, 2-bromo-3,4,5,6-tetramethoxy-1,8-dimethyl-	EtOH	275(5.06),331(3.49), 348(3.52),366(3.35)	39-0646-69C
$C_{18}H_{19}ClN_2$			
1H-2,6-Benzodiazinone, 7-(p-chlorophenyl)-2-methyl-2,3,4,5-tetrahydro-, dihydrochloride	EtOH	253(4.20)	44-2715-69
$C_{18}H_{19}ClN_2O$			
Benzamide, o-[(o-chlorobenzylidene)-amino]-N,N-diethyl-	MeOH	220(4.5),260(4.1), 340s(3.7)	24-3735-69
1(2H)-Quinolinecarboxamide, 3,4-dihydro-N,2-dimethyl-4-phenyl-	n.s.g.	258(4.19)	40-1047-69
$C_{18}H_{19}ClN_2OS$			
2(1H)-Quinazolinethione, 3-sec-butyl-6-chloro-3,4-dihydro-4-hydroxy-4-phenyl-	n.s.g.	263(4.02),292(4.43)	40-0917-69
$C_{18}H_{19}ClN_2O_6$			
3-Carbamoyl-1-D-ribofuranosylpyridinium chloride, 5'-benzoate	MeOH	268(3.74)	39-0199-69C
$C_{18}H_{19}ClO_8$			
Crotepoxide chlorohydrin (7-oxabicyclo-[4.1.0]heptane-2,3,4-triol, 5-chloro-1-(hydroxymethyl)-, 2,3-diacetate, 1-benzoate)	MeOH	274(3.01),281(2.92)	44-3898-69
$C_{18}H_{19}ClO_{12}$			
1,3-Cyclohexadiene-1,2,3,4,5,6-hexacarboxylic acid, 5-chloro-, hexamethyl ester	C_6H_{12}	300(3.50)	80-1191-69
$C_{18}H_{19}IO_8$			
7-Oxabicyclo[4.1.0]heptane-2,3,4-triol, 1-(hydroxymethyl)-5-iodo-, 2,3-diacetate 1-benzoate	MeOH	274(3.15),281(3.05)	44-3898-69

Compound	Solvent	$\lambda_{max}(\log \epsilon)$	Ref.
$C_{18}H_{19}N$			
Aniline, N-[1-(1,2,3,4-tetrahydro-2-naphthyl)ethylidene]-	EtOH	258(4.23),312(3.43)	56-0749-69
9,10-Benzophenanthridine, 1,2,3,4,5,6-hexahydro-6-methyl-	EtOH	220(4.21),260(4.52)	2-0419-69
tert-Butylamine, N-(diphenylvinylidene)-	C_6H_{12}	276(4.25)	88-5093-69
$C_{18}H_{19}NO$			
4-Isoxazoline, 3-benzyl-2,4-dimethyl-5-phenyl-	EtOH	222(4.31),279(3.88)	88-4875-69
Isoxazoline, 5-mesityl-3-phenyl-	EtOH	267(4.20)	88-3329-69
4-Isoxazoline, 2-methyl-3-(p-methylbenzyl)-5-phenyl-	EtOH	223(4.16),276(3.88)	88-4875-69
Quinoline, 1,2-dihydro-7-methoxy-1,3-dimethyl-4-phenyl-	EtOH	233(4.48),330(3.57)	2-0419-69
$C_{18}H_{19}NO_2$			
Caaverine, N-methyl-, (±)-	MeOH	273(4.07),311(3.56)	39-0502-69C
$C_{18}H_{19}NO_3$			
Cinnamolaurine	EtOH	287(3.72)	88-5055-69
Phthalimidine, 3-hydroxy-2-(4-hydroxybutyl)-3-phenyl-	EtOH	252(3.65),263(3.51)	44-0165-69
Spiro[2,4-cyclohexadiene-1,7'-(1'H)-cyclopent[ij]isoquinolin]-6-one, 2',3',8',8'a-tetrahydro-6'-hydroxy-5'-methoxy-1'-methyl-	MeOH	290(3.84),312s(3.65)	39-0502-69C
Tuduranine, hydrochloride	EtOH	267(4.12),276(4.13), 301(3.82)	2-0945-69
	EtOH	267(4.13),276(4.14), 304(3.86)	95-1691-69
	EtOH-NaOH	256(4.22),280(4.00), 307(3.46)	2-0945-69
$C_{18}H_{19}NO_3S$			
Alanine, 3-[(α-phenacylbenzyl)thio]-, hydrochloride	n.s.g.	210(3.98),244(3.90)	7-0624-69
Indole-3-propanol, p-toluenesulfonate	EtOH	224(4.66),274(3.79), 282(3.80),291(3.74)	78-4843-69
$C_{18}H_{19}NO_4$			
Flavinine	EtOH	238(4.12),285(3.91)	39-1681-69C
Laurolitsine	EtOH	283(4.09),304(4.07)	95-0737-69
Norsinoacutine	EtOH	241(4.19),275s(3.83)	100-0001-69
$C_{18}H_{19}NO_6$			
1,2,9-Acridantrione, 3,4-dimethoxy-10-methyl-, 2-(dimethyl acetal)	EtOH	224(4.23),241s(4.15), 291(4.30)	12-1477-69
1,3-Dioxolo[4,5-b]acridine-10,11-(5H,11aH)-dione, 11a-methoxy-3-methyl-, 11-(dimethyl acetal)	EtOH	212(3.97),226(3.93), 236s(3.89),275(4.32), 283(4.36),319(3.93), 348(3.77)	12-1493-69
Pyrrole-3-carboxylic acid, 4,5-dihydroxy-1-methyl-2-phenyl-, ethyl ester, diacetate	EtOH	220s(4.21),273(3.85)	44-3187-69
$C_{18}H_{19}NO_7$			
Indole-3-glyceric acid, 1-acetyl-, methyl ester, diacetate	EtOH	240(4.40),292(3.97), 300(4.00)	97-0269-69

Compound	Solvent	λ_{max}(log ε)	Ref.
$C_{18}H_{19}NS$			
Dibenzo[b,e]thiepin, 11-[3-(methyl-amino)propylidene]-6,11-dihydro-, cis	heptane	230(4.29),260(3.89), 305(3.26)	73-1963-69
trans	heptane	231(3.47),262(3.96), 305(3.35)	73-1963-69
$C_{18}H_{19}NSe$			
Selenoxanthene, 9-[3-(dimethylamino)-propylidene]-	MeOH	231(4.42),268(4.03), 316(3.45)	73-3792-69
$C_{18}H_{19}N_3$			
Benzo[g]quinoline, 4-(4-methyl-1-pipera-zinyl)-	EtOH	254(4.88),325(3.43), 360(3.73),375(3.81)	103-0223-69
$C_{18}H_{19}N_3NaO_{11}P$			
Thymidine, 3'-acetate 5'-(p-nitrophenyl) hydrogen phosphate, sodium salt	pH 2	270(4.20)	35-1522-69
$C_{18}H_{19}N_3O$			
Benzamide, N-[1-(2-benzimidazolyl)-2-methylpropyl]-	EtOH	275(3.99),282(4.03)	94-2381-69
$C_{18}H_{19}N_3O_2$			
Indole, 3-(tert-butylamino)-2-(p-nitro-phenyl)-	EtOH	229(4.33),275(3.98), 355s(3.96),420(4.10)	78-1467-69
Indole, 3-[2-(6-nitro-3,4-xylidino)-ethyl]-	EtOH	224(4.63),237s(4.46), 283(4.06),291(4.06), 445(3.81)	44-3240-69
	EtOH-HCl	225(3.32),237s(4.43), 283(4.06),291(4.08), 455(3.81)	44-3240-69
	EtOH-NaOH	237s(4.52),291(4.12), 445(3.83)	44-3240-69
Pyrimidine, 1,3-dibenzyl-1,2,3,4-tetra-hydro-5-nitro-	hexane	220(--),365(4.36)	78-1617-69
$C_{18}H_{19}N_3O_2S_2$			
Butenedinitrile, [(5,6-dimethyl-3-piper-idino-p-benzoquinon-2-yl)thio]-(methylthio)-	$CHCl_3$	244(4.23),329(4.29), 518(3.74)	24-2378-69
$C_{18}H_{19}N_3O_3$			
Benzamide, N,N-diethyl-o-[(m-nitrobenz-ylidene)amino]-	MeOH	210(4.3),250(4.3), 340s(3.7)	24-3735-69
Benzamide, N,N-diethyl-o-[(o-nitrobenz-ylidene)amino]-	MeOH	210(4.5),290(4.2), 350s(4.0)	24-3735-69
Benzamide, N,N-diethyl-o-[(p-nitrobenz-ylidene)amino]-	MeOH	210(4.3),230(4.2), 260(4.1),330(3.7)	24-3735-69
$C_{18}H_{19}N_3O_6$			
1,4-Phenazinedicarboxylic acid, 2,3,5,10-tetrahydro-7-nitro-, diethyl ester	EtOH	254(4.21),307(4.00), 391(4.29),425(4.28), 451(4.22)	33-0322-69
$C_{18}H_{19}N_3O_7$			
as-Triazine-3,5(2H,4H)-dione, 2-(2,3-O-isopropylidene-β-D-lyxofuranosyl)-, 5'-benzoate	EtOH	231(4.16),267(3.84)	73-0618-69

Compound	Solvent	$\lambda_{max}(\log \epsilon)$	Ref.
$C_{18}H_{20}$			
Biphenyl, 2-cyclohexyl-	EtOH	204(4.63),232(4.04)	22-3232-69
Cyclohexane, 1,2-diphenyl-, cis	EtOH	201(4.38),210(3.94), 249(2.51),254(2.62), 259(2.68),266s(--), 265(2.58),269(2.47)	22-3232-69
Cyclohexane, 1,2-diphenyl-, trans	EtOH	201(4.32),250(2.60), 255(2.71),259(2.78), 262(2.76),265(2.70), 269(2.67)	22-3232-69
1H-Cyclopenta[l]phenanthrene, 2,3,4,5,6,7-hexahydro-1-methyl-	EtOH	218(4.57),237(4.91), 326(3.90),329(3.93), 355(3.34)	44-0763-69
Isobutene, 1,1-di-p-tolyl-	C_6H_{12}	225(4.18),246(4.08)	59-0487-69
[2,2]Paracyclophane, 4-ethyl-, tetracyanoethylene complex	CH_2Cl_2	540(3.21)	35-3553-69
[2,4]Paracyclophane	hexane	216(4.17),271(2.63), 282(2.43)	35-3517-69
[3,3]Paracyclophane, tetracyanoethylene complex	CH_2Cl_2	599(3.26)	35-3553-69
1-Pentene, 4-methyl-1,1-diphenyl-	EtOH	251(4.12)	35-6362-69
1-Propene, 1-mesityl-1-phenyl-, cis	C_6H_{12}	252(4.15)	59-0487-69
1-Propene, 1-mesityl-1-phenyl-, trans	C_6H_{12}	238s(4.2)	59-0487-69
Stilbene, α-tert-butyl-, cis	EtOH	228(3.90),255(4.14), 264s(4.03)	35-0388-69
trans	EtOH	233s(4.14)	35-0388-69
$C_{18}H_{20}ClNO_3$			
Tuduranine, hydrochloride (see also ($C_{18}H_{19}NO_3$))	EtOH	267(4.12),276(4.13), 301(3.82)	2-0945-69
$C_{18}H_{20}ClN_3O_6S_3$			
Sporidesmin E (etherate)	MeOH	217(4.52),252(4.22), 295(3.50)	39-1665-69C
$C_{18}H_{20}Fe$			
Ferrocene, 1,1':3,3'-bis(trimethylene)-4-vinyl-	EtOH	320(2.87),410(2.49), 430s(2.48)	49-1540-69
$C_{18}H_{20}FeO$			
Ketone, 1',2:3',4-bis(trimethylene)-ferrocenyl methyl	EtOH	334(3.29),428(3.02), 450s(2.98)	49-1540-69
$C_{18}H_{20}N_2$			
1,2-Diazocine, 3,4,5,6,7,8-hexahydro-3,8-diphenyl-, cis	$CHCl_3$	381(2.05)	35-3226-69
trans	$CHCl_3$	381(2.05)	35-3226-69
Indole, 3-(tert-butylamino)-2-phenyl-	EtOH	206(4.39),241(4.37), 309(4.15)	78-1467-69
Indole, 2-[2-(dimethylamino)ethyl]-1-phenyl-, (hydrogen oxalate)	EtOH	263(4.09),279(--), 289(--)	39-1537-69C
Isoindole, 2-[2-(ethylamino)ethyl]-1-phenyl-	EtOH	274(3.72),285(3.70), 345(3.90)	44-1720-69
Quinoxaline, 2-(1-adamantyl)-	EtOH	235(3.47),309(3.82), 319(3.87)	18-1617-69
$C_{18}H_{20}N_2O$			
Acetic acid, phenyl(2,4,6-trimethylbenzylidene)hydrazide	EtOH	220s(4.2),224(4.2), 231s(4.1),289(4.3)	28-1703-69A

Compound	Solvent	$\lambda_{max}(\log \epsilon)$	Ref.
9-Acridancarboxamide, N,N,9,10-tetra-methyl-	MeOH	288(4.23)	36-0335-69
1,12-(3-Hydroxytrimethylene)indolo[2,3-a]quinolizine, 2,3,4,6,7,12-hexahydro-	MeOH	232(4.34),315(4.26)	44-0330-69
Quinoline, 8-amino-1-benzoyl-1,2,3,4-tetrahydro-2,6-dimethyl-, hydro-chloride	EtOH	208(4.46),242(4.35), 298(4.15)	2-0848-69
$C_{18}H_{20}N_2O_2$			
Acetophenone, p-methoxy-, azine	C_6H_{12}	315(4.39)	39-0836-69C
Azobenzene, 4'-acetyl-5-tert-butyl-2-hydroxy-	50% EtOH	336(4.36),402(3.98)	73-1087-69
	50% EtOH-$HClO_4$	315s(--),402(4.40), 506(4.09)	73-1087-69
Benzoic acid, 2,4,6-trimethyl-, α-meth-ylsalicylidenehydrazide	EtOH	217(4.4),274(4.2), 286s(4.2),319(3.9)	28-0730-69A
Benzophenonimine, N-(4-aminobutyl)-2-carboxy-	EtOH	240(4.09)	44-0165-69
$C_{18}H_{20}N_2O_3$			
Benzoic acid, 2,4,6-trimethyl-, (2,6-di-hydroxy-α-methylbenzylidene)hydrazide	EtOH	220s(4.4),290(4.3)	28-0730-69A
$C_{18}H_{20}N_2O_4$			
Acetic acid, phenyl(2,4,6-trimethoxy-benzylidene)hydrazide	EtOH	203(4.4),235(4.2), 292s(4.4),306(4.4)	28-1703-69A
1,4-Phenazinedicarboxylic acid, 2,3,5,10-tetrahydro-, diethyl ester	EtOH	267(4.65),301(4.35), 315(4.39),377(4.25), 422(4.23),450(4.02)	33-0322-69
1H-Pyrrolo[3,4-b]quinoline-3-acetic acid, 2-carboxy-2,3-dihydro-, diethyl ester	EtOH	230(4.54),293(3.60), 300(3.60),307(3.75), 313(3.69),321(3.91)	44-3853-69
$C_{18}H_{20}N_2O_4S$			
Piperidine, 1-[3'-(methylsulfonyl)-3-nitro-4-biphenylyl]-	MeOH	420(3.16)	17-0145-69
Piperidine, 1-[4'-(methylsulfonyl)-3-nitro-4-biphenylyl]-	MeOH	414(3.20)	17-0145-69
$C_{18}H_{20}N_2O_5$			
β-Alanine, N,N'-(oxydi-p-phenylene)di-	EtOH	257(4.39),307(3.58)	103-0623-69
$C_{18}H_{20}N_2O_8$			
3,3'-Biphenyldipropionic acid, α,α'-diamino-5,5',6,6'-tetrahydroxy-	H_2O	287(3.67)	18-1752-69
Fumaric acid, (m-phenylenediimino)di-, tetramethyl ester	MeOH	225(4.26),243(4.14), 326(4.49)	23-3545-69
Fumaric acid, (p-phenylenediimino)di-, tetramethyl ester	MeOH	250(4.07),348(4.41)	23-3545-69
$C_{18}H_{20}N_4O_2$			
Butyric acid, 5-ethyl-1-methyl-3-phenyl-1H-pyrazolo[3,4-b]pyrazin-6-yl ester	EtOH	246(4.29),288(3.97), 336(3.83)	32-0463-69
Butyric acid, 5-ethyl-3-methyl-1-phenyl-1H-pyrazolo[3,4-b]pyrazin-6-yl ester	EtOH	258(4.50),321s(3.70)	32-0463-69
Piperidine, 1-[2-nitro-4-(p-tolylazo)-phenyl]-	benzene	400(4.38)	73-2092-69
$C_{18}H_{20}O$			
Acetophenone, 3',5'-diethyl-2'-phenyl-	EtOH	221(4.24),226(4.23), 245(3.94),292(3.30)	6-0277-69

Compound	Solvent	λ_{max}(log ϵ)	Ref.
Anisole, p-(p-isopropylstyryl)-	DMF	309(4.48),324(4.51)	33-2521-69
$C_{18}H_{20}O_2$			
9,10-Anthracenediol, 9,10-diethyl-9,10-dihydro-, cis	EtOH	216(4.20),259(2.71),	39-2266-69C
Estra-1,3,5(10)-triene-6,17-dione	EtOH	252(4.23),294(3.34)	94-1725-69
[2,2]Metacyclophane, 5,13-dimethoxy-	EtOH	210(4.63),292(3.39)	44-1960-69
$C_{18}H_{20}O_3$			
Furan, tetrahydro-2,2-bis(p-methoxyphenyl)-	MeOH	232(4.32),275(3.59), 282(3.50)	44-3792-69
$C_{18}H_{20}O_4$			
Biphenylene, 1,2,7,8-tetramethoxy-4,5-dimethyl-	EtOH	262s(4.54),271(4.73), 287(4.06),309(3.20), 325(3.34),341(3.36), 359(3.12)	39-0646-69C
6H,15H-Dibenzo[b,i][1,4,8,11]tetraoxacyclotetradecin, 7,8,16,17-tetrahydro-	MeOH	277(3.69)	25-0171-69
2,3-Naphthalenediacetic acid, α,α,α',α'-tetramethyl-	C_6H_{12}	230(4.87),274(3.57)	23-4313-69
$C_{18}H_{20}O_5$			
Dibenzofuran, 1,3,7,9-tetramethoxy-2,8-dimethyl-	EtOH	236(4.60),272(4.21), 284(4.36),292(4.29)	39-2403-69C
Fluorene-1,9-dicarboxylic acid, 5,6,7,8-tetrahydro-2-methoxy-, dimethyl ester	EtOH	273(4.15)	44-2209-69
4aH-Fluorene-4a,8-dicarboxylic acid, 1,2,3,4-tetrahydro-7-methoxy-, dimethyl ester	EtOH	235(4.46),317(3.59)	44-2209-69
2'-Isoflavanol, 4',7,8-trimethoxy-	EtOH	279(3.97)	39-1109-69C
$C_{18}H_{20}O_6$			
Acetic acid, (3,4-dimethoxyphenyl)-, 3,4-dimethoxyphenyl ester	EtOH	230(4.14),279(3.78)	39-1787-69C
Acetophenone, 2-(3,4-dimethoxyphenyl)-2'-hydroxy-4',5'-dimethyl-	EtOH	234(4.30),279(4.14), 344(3.97)	39-1787-69C
4-Chromanone, 5-hydroxy-2-(1-hydroxy-4-oxo-2-cyclohexen-1-yl)-7-methoxy-6,8-dimethyl-	EtOH	213(4.59),286(4.21), 359(3.61)	95-0851-69
4H,9H-Furo[2',3',4':4,5]naphtho[2,1-c]-pyran-4,9-dione, 1,2,3,3a,5aβ,7,10b,-10cβ-octahydro-7β-hydroxy-3aβ,10bα-dimethyl-	MeOH	257(4.15)	35-2134-69
Lonchocarpan	EtOH	284(3.66)	39-0887-69C
6-Oxabicyclo[3.2.1]oct-2-ene-4,7-dione, 2-(1,2-epoxy-3,6-dioxoheptyl)-8-isopropenyl-3-methyl-	MeOH	228(3.68),268(3.84)	78-4835-69
	MeOH-KOH	247(4.00)	78-4835-69
$C_{18}H_{20}O_7$			
Acetophenone, 2',4'-dihydroxy-2-(2,3,4,6-tetramethoxyphenyl)-	EtOH	228(4.25),278(4.24), 313(3.96)	39-0887-69C
3-Butenoic acid, 4-(6,8-dimethoxy-3-methyl-1-oxo-1H-2-benzopyran-7-yl)-2-methoxy-, methyl ester	EtOH	259(4.72),303(4.30), 310(4.32),346s(--)	12-1933-69
Eudesma-3,5,7(11)-trien-12-oic acid, 1α,2β,3,6-tetrahydroxy-, γ-lactone, 3-acetate 2-formate	MeOH	222(3.58),324(4.53)	39-1088-69C

Compound	Solvent	λ_{max}(log ε)	Ref.
$C_{18}H_{20}O_8$			
Cyclohexanone, 2,3,4-trihydroxy-4-(hydroxymethyl)-, 2,3-diacetate, α-benzoate	MeOH	229(4.05),274(2.98), 281(2.89)	44-3898-69
5-Cyclohexene-1,2,3,4-tetrol, 1-(hydroxymethyl)-, 2,3-diacetate, α-benzoate	MeOH	274(2.99),281(2.90)	44-3898-69
$C_{18}H_{21}BrO_4$			
Ethane, 2-bromo-1,1-bis(3,4-dimethoxyphenyl)-	EtOH	232(4.13),281(3.76)	7-0922-69
$C_{18}H_{21}ClN_2O$			
Quinoline, 8-amino-1-benzoyl-1,2,3,4-tetrahydro-2,6-dimethyl-, hydrochloride	EtOH	208(4.46),242(4.35), 298(4.15)	2-0848-69
$C_{18}H_{21}Cl_3N_2$			
1H-2,6-Benzodiazonine, 7-(p-chlorophenyl)-2-methyl-2,3,4,5-tetrahydro-, dihydrochloride	EtOH	253(4.20)	44-2715-69
$C_{18}H_{21}F_3N_2O_2$			
2-Pyrrolin-4-one, 2-(diethylamino)-3,5-dimethyl-5-phenyl-1-(trifluoroacetyl)-	MeOH	274(4.03),311(4.00)	78-5721-69
$C_{18}H_{21}Li$			
Hexane, 1,1-diphenyl-, 1-lithium deriv.	THF	495(4.45)	35-2456-69
$C_{18}H_{21}NO$			
Acetophenone, 2',6'-dimethyl-4'-(2,5-xylidino)-	C_6H_{12}	236s(4.06),292(4.16)	22-3523-69
	EtOH	208(4.53),235s(4.00), 295(4.04),330s(3.93)	22-3523-69
4-Piperidinol, 6-methyl-1,2-diphenyl-	EtOH	253(3.10),259(3.01), 265(2.89)	18-2690-69
$C_{18}H_{21}NOSe$			
Selenoxanthen-9-ol, 9-[3-(dimethylamino)propyl]-	MeOH	227s(3.96),253(4.00), 275(3.88)	73-3792-69
$C_{18}H_{21}NO_2$			
Benzyl alcohol, α-[2-(dimethylamino)-ethyl]-, benzoate	dioxan	258(3.04),264(3.03), 274(3.02),281(2.93)	65-1861-69
Hydrocinnamic acid, β-anilino-α,α-dimethyl-, methyl ester	EtOH	247(4.16),296(3.29)	77-0936-69
$C_{18}H_{21}NO_3$			
Hydroxylamine, O-acetyl-N-(3,4,5,6,7,8-hexahydro-2-oxo-4a(2H)-naphthyl)-N-phenyl-	EtOH	235(4.28)	12-2493-69
1,4-Naphthoquinone, 3-hydroxy-2-[(cyclohexylmethylamino)methyl]-	EtOH	223(4.21),231(4.21), 270(4.39),320s(3.31), 440(3.42)	39-1245-69C
$C_{18}H_{21}NO_5$			
1,4-Cyclohexanedicarboxylic acid, 2-oxo-3-(phenylimino)-, diethyl ester	EtOH	252(4.16),320(3.74), 385(3.87)	33-0322-69
6a-Epipretazettine	EtOH	242(3.78),291(3.65)	35-0150-69
Estra-1,3,5(10)-trien-17-one, 3,9α-dihydroxy-11β-nitro-	EtOH	279(3.29),285(3.23)	31-1018-69

Compound	Solvent	λ_{max}(log ϵ)	Ref.
Precriwelline, N-demethyl-O-methyl-	EtOH	243(3.84),292(3.67)	35-0150-69
Pretazettine, N-demethyl-O-methyl-	EtOH	242(3.75),291(3.61)	35-0150-69
$C_{18}H_{21}NO_5S$			
1,4-Cyclohexanedicarboxylic acid, 2-[(o-mercaptophenyl)imino]-3-oxo-, diethyl ester	EtOH	219(4.11),269(4.04), 343(4.29)	33-0322-69
$C_{18}H_{21}NO_6$			
1,4-Cyclohexanedicarboxylic acid, 2-[(o-hydroxyphenyl)imino]-3-oxo-, diethyl ester	EtOH	235(3.86),338(4.41)	33-0322-69
$C_{18}H_{21}NO_7$			
3a,4-Secolycoran-4-oic acid, 2,3-dihydroxy-5-oxo-, ethyl ester	EtOH	290(3.72)	44-2667-69
$C_{18}H_{21}N_3$			
Benzylidenimine, α-methyl-N-[2-(3-pyridyl)-1-piperidyl]-	n.s.g.	246(3.94),315(3.20)	65-2723-69
$C_{18}H_{21}N_3O$			
Benzylidenimine, 4-methoxy-N-[2-(3-pyridyl)-1-piperidyl]-	n.s.g.	288(4.20)	65-2723-69
$C_{18}H_{21}N_3OS$			
Acetophenone, 2'-hydroxy-, 4-mesityl-3-thiosemicarbazone	EtOH	230s(4.3),285(4.3), 324s(4.1)	28-0730-69A
$C_{18}H_{21}N_3O_2$			
Acetophenone, 2'-hydroxy-, 4-mesityl-semicarbazone	EtOH	214s(4.5),272(4.3), 283s(4.2),312(3.9)	28-0730-69A
Benzoic acid, p-[(1,2,3,4-tetrahydro-2-quinoxalinyl)methyl]amino]-, ethyl ester	EtOH	221(4.49),308(4.40)	35-5270-69
2-Norbornanone, 3-(4,5-dimethyl-2-phenylimidazol-1-yl)-, oxime, N-oxide	MeOH	281(4.11)	48-0746-69
$C_{18}H_{21}N_3O_2S$			
Acetophenone, 2',6'-dihydroxy-, 4-mesityl-3-thiosemicarbazone	EtOH	240s(4.3),274(4.4)	28-0730-69A
$C_{18}H_{21}N_3O_4$			
Nicotinic acid, 6-methyl-, (3,4,5-trimethoxy-α-methylbenzylidene)hydrazide	n.s.g.	220(4.32),307(4.30)	40-0073-69
$C_{18}H_{21}N_3S_2$			
2H-1,3,5-Thiadiazine-2-thione, 3-benzyl-5-[p-(dimethylamino)phenyl]tetrahydro-	MeOH	265(4.56),284(4.32)	73-2952-69
$C_{18}H_{21}N_5O_5$			
Adenosine, N-(2-phenoxyethyl)-	pH 1.0	263(4.32)	87-1056-69
	pH 7.0	268(4.30)	87-1056-69
	pH 12.0	268(4.32)	87-1056-69
$C_{18}H_{21}N_5O_9$			
Acetamide, N-(6-hydroxy-9-β-D-xylopyranosyl-9H-purin-2-yl)-, triacetate	pH 1	262(3.94),343(4.36),	44-0416-69
	pH 7	265s(--),342(4.38)	44-0416-69
	pH 13	251(4.17),272(3.85)	44-0416-69

Compound	Solvent	$\lambda_{max}(\log \epsilon)$	Ref.
$C_{18}H_{21}NaO_5S$			
Estrone, sulfate, sodium salt	EtOH	270(2.88),276(2.87)	13-0067-69B
$C_{18}H_{22}$			
Biphenyl, p-hexyl-	MeOH	260(4.43)	97-0342-69
Hexane, 1,1-diphenyl-, 1-lithium deriv.	THF	495(4.45)	35-2456-69
$C_{18}H_{22}ClNO_4$			
1H-Cyclopenta[c]quinolizinium perchlorate, 2,3,5,5a,6,7,8,9-octahydro-4-phenyl-	$CHCl_3$	348(4.22)	78-4161-69
$C_{18}H_{22}Cl_2N_8O_3$			
Imidazole-4-carbonitrile, 5,5'-azoxybis[1-[2-(2-chloroethoxy)ethyl]-2-methyl-	EtOH	212(3.79),265s(--), 362(3.70)	4-0053-69
Imidazole-5-carbonitrile, 4,4'-azoxybis[1-[2-(2-chloroethoxy)ethyl]-2-methyl-	EtOH	216(3.80),268s(--), 388(3.70)	4-0053-69
$C_{18}H_{22}HgO_2$			
Mercury, (acetato)(2-phenyl-2-bornen-3-yl)-	n.s.g.	265(3.89)	88-3521-69
$C_{18}H_{22}NO_3P$			
Phosphonic acid, (10-methyl-9-acridanyl)-, diethyl ester	MeOH	210(4.72),287(4.16)	44-1420-69
$C_{18}H_{22}N_2$			
Benzophenone, tert-butylmethylhydrazone	EtOH	250s(4.06),330(3.61)	39-1703-69C
$C_{18}H_{22}N_2O_3$			
1H,10H-3a,9b-Diazabenzo[a]naphth[2,1,8-cde]azulen-10-one, 2,3,4,5,5a,11,12,-12a,12b,12c-decahydro-5a,12c-dihydroxy-	MeOH	212(3.99),257(4.04)	44-0330-69
Tryptophan, N-acetyl-2-(1,1-dimethylallyl)-	EtOH	226(4.52),284(3.92), 291(3.88)	39-1003-69C
$C_{18}H_{22}N_2O_3S_2$			
Acrylamide, 3,3'-(phenacylidenedithio)-bis[N-ethyl-, (Z,Z)-	90% EtOH	255s(4.39),278(4.49)	12-0765-69
$C_{18}H_{22}N_2O_4$			
Azepine-N-carboxylic acid, 3-methyl-, methyl ester, dimer	EtOH	239(4.17)	35-3616-69
Azepine-N-carboxylic acid, 4-methyl-, methyl ester, dimer	EtOH	238(4.15)	35-3616-69
4H-1-Benzopyran-2-carbamic acid, 4-oxo-3-(piperidinomethyl)-, ethyl ester	$CHCl_3$	237(4.38),290(4.18)	103-0019-69
9,14-Diazatricyclo[6.3.2.1^{2,7}]tetradeca-3,5,10,12-tetraene-9,14-dicarboxylic acid, diethyl ester	EtOH	213(4.14),241(3.97)	35-3616-69
$C_{18}H_{22}N_2O_5$			
3,4-Xylidine, 6-nitro-N-(3,4,5-trimethoxybenzyl)-	EtOH	237(4.54),292(3.90), 435(3.86)	44-3240-69
	EtOH-HCl	235(4.58),290(3.92), 435(3.90)	44-3240-69
	EtOH-NaOH	235s(4.61),295(3.32), 435(3.91)	44-3240-69

Compound	Solvent	$\lambda_{max}(\log \epsilon)$	Ref.
$C_{18}H_{22}N_2O_5S$			
Crotonic acid, 3-methyl-2-[3-(methylthio)-2-(2-phenoxyacetamido)a rylamido]-, methyl ester	EtOH	285(4.18)	88-3381-69
$C_{18}H_{22}N_2O_{11}$			
Uracil, 1-β-D-glucopyranosyl-, 2',3',4',6'-tetraacetate	50% EtOH	264(3.97)	39-0203-69C
Uracil, 3-β-D-glucopyranosyl-, 2',3',4',6'-tetraacetate	50% EtOH	260(3.97)	39-0203-69C
$C_{18}H_{22}N_2O_{11}S$			
Uracil, 6-(β-D-glucopyranosylthio)-, 2',3',4',6'-tetraacetate	pH 1	275(4.03)	87-0653-69
	pH 11	229(4.06),288(4.07)	87-0653-69
$C_{18}H_{22}N_4$			
Benzimidazole, 5-amino-2-methyl-1-[p-(diethylamino)phenyl]-	EtOH	212(4.70),274(4.46), 306s(3.96)	73-0572-69
$C_{18}H_{22}N_4O_2$			
Benzoic acid, p-(1,4-dimethyl-4-p-tolyl-2-tetrazeno)-, ethyl ester	$(MeOCH_2)_2$	305s(3.90),367(4.63)	35-6452-69
$C_{18}H_{22}N_4O_4$			
Ethylenediamine, N,N'-bis(6-nitro-3,4-xylyl)-	EtOH	237(4.56),292(4.05), 435(4.00)	44-3240-69
	EtOH-HCl	238(4.57),293(4.07), 435(4.01)	44-3240-69
	EtOH-NaOH	235s(4.61),293(4.09), 440(4.01)	44-3240-69
3H-2,8a-Methanonaphthalen-3-one, octahydro-2-methyl-, 2,4-dinitrophenylhydrazone	$CHCl_3$	372(4.40)	73-0572-69
$C_{18}H_{22}N_4O_4S_2$			
Biacetyl, di-p-toluenesulfonate, disodium salt	n.s.g.	228(4.3),310(4.4)	44-1746-69
$C_{18}H_{22}O$			
1,4-Ethanonaphthalen-9-one, 1,4-dihydro-1,2,3,4,10,10-hexamethyl-	EtOH	263s(3.16),270(3.14), 278(3.06)	88-0379-69 +35-2183-69
1,4-Ethenonaphthalen-2(1H)-one, 3,4-dihydro-1,3,3,4,9,10-hexamethyl-	EtOH	247s(3.47),292(3.11), 301(3.09),313(2.90)	35-2183-69
$C_{18}H_{22}O_2$			
Estra-4,9(11)-diene-3,17-dione, L-	MeOH-HCl	243(4.15),313(3.45)	44-3530-69
Estra-1,3,5(10)-trien-6-one, 3-hydroxy-	EtOH	203(4.20),222(4.29), 255(3.93),327(3.46)	39-1234-69C
B-Norandrosta-1,4-diene-3,17-dione	n.s.g.	246(4.20)	73-0681-69
$C_{18}H_{22}O_3$			
2-Cyclohexene-1-carboxylic acid, 2-methyl-3-(p-methylbenzyl)-4-oxo-, ethyl ester	EtOH	241(4.0)	42-0415-69
2-Cyclohexen-1-one, 2-(2,5-dimethylbenzyl)-3-hydroxy-, propionate	EtOH	203(4.23),218s(4.11), 239(4.03)	44-2192-69
Estra-1,3,5(10),9(11)-tetraene-3,6,17β-triol	MeOH	266(4.19)	44-0116-69
Estr-4-ene-3,11,17-trione, dl-	EtOH	239(4.16)	44-3530-69

Compound	Solvent	λ_{max}(log ϵ)	Ref.
Propiophenone, 4'-[(2-hydroxy-6-oxo-1-cyclohexen-1-yl)methyl]-2',5'-dimethyl-	EtOH	213(4.35),261(4.34)	44-2192-69
$C_{18}H_{22}O_4$			
2,3-Butanediol, 2,3-bis(o-methoxyphenyl)-	EtOH	274(3.63),279(3.56)	12-0761-69
3,4-Hexanediol, 3,4-bis(p-hydroxyphenyl)-	EtOH	276(3.59),284(3.56)	12-0761-69
$C_{18}H_{22}O_5$			
Fluorene-1,9-dicarboxylic acid, 4b,5,6,7,8,8a-hexahydro-2-methoxy-, dimethyl ester	EtOH	302(3.57)	44-2209-69
2-Naphthaleneglyoxylic acid, 1,2,3,4-tetrahydro-8-isopropyl-7-methoxy-5-methyl-1-oxo-, methyl ester	EtOH	330(4.08)	18-3011-69
$C_{18}H_{22}O_6$			
Succinic acid, [(1,2,3,4-tetrahydro-6-methoxy-1-oxo-2-naphthyl)-, dimethyl ester	EtOH	275(4.21)	2-1169-69
$C_{18}H_{22}O_7$			
Phenalene-1,3-dicarboxylic acid, dodecahydro-9b-methyl-2,5,8-trioxo-, dimethyl ester	EtOH	253(3.96)	35-2299-69
$C_{18}H_{22}O_{10}$			
Quinic acid, 3-O-sinapoyl-	n.s.g.	327(4.29)	102-0203-69
$C_{18}H_{22}O_{11}$			
Asperuloside	MeOH	236(3.80)	94-1942-69
$C_{18}H_{23}KO_3S$			
17β-Estradiol, 17-sulfate, K salt	EtOH	281(3.30)	13-0067-69B
$C_{18}H_{23}N$			
13-Azabicyclo[8.2.1]trideca-10,12-diene, 13-phenyl-	hexane	241(3.08)	78-5357-69
$C_{18}H_{23}NO_3$			
Alolycopine, acetate	MeOH	237(3.78)	23-2457-69
4a(2H)-Carbazoleethanol, 9-acetyl-	MeOH	252(4.08),285(3.39)	23-3647-69
Futoamide	EtOH	261(4.18),269s(--), 304(3.83)	94-1225-69
$\Delta^{4,\alpha}$-Piperidineacetic acid, 1-benzoyl-3-ethyl-, ethyl ester	EtOH	230(4.23)	44-3273-69
$C_{18}H_{23}NO_4$			
2-Pentanone, 4-methyl-1-(2,4,8-trimethoxy-3-quinolyl)-	MeOH and MeOH-base	246(4.66),247s(3.73), 280(3.73),290(3.65), 315(3.35),328(3.28)	44-2183-69
	MeOH-acid	247(4.58),257s(4.29), 283(3.70),315(3.62), 333(3.46)	44-2183-69
Spiro[cyclohexane-1,4'(1'H)-isoquinolin]-1'-one, 2',3'-dihydro-4-hydroxy-6'-methoxy-2'-methyl-, acetate	EtOH	264(4.07)	94-1564-69

Compound	Solvent	$\lambda_{max}(\log \epsilon)$	Ref.
$C_{18}H_{23}NO_5$			
Pretazettine, dihydro-	EtOH	238(3.61),291(3.63)	35-0150-69
Ptelefoline	MeOH	218(4.56),236(4.53), 257(4.45),284(4.03), 294(3.99),350(3.73)	88-3803-69
$C_{18}H_{23}NO_6$			
Jacozine	EtOH	218(3.34)	73-1832-69
Retrorsine, didehydro-	n.s.g.	220(3.89)	39-1155-69C
$C_{18}H_{23}NO_9$			
Pyrrole, 1-β-D-glucopyranosyl-, 2',3',4',6'-tetraacetate	EtOH	211(3.91)	18-3539-69
$C_{18}H_{23}N_3O_{10}S$			
Cytosine, 6-(β-D-glucopyranosylthio)-, 2',3',4',6'-tetraacetate	pH 1	290(4.18)	87-0653-69
	pH 11	287(3.98)	87-0653-69
$C_{18}H_{23}OP$			
Phosphine oxide, dimesityl-	C_6H_{12}	239(4.33),276(3.30), 285(3.34)	65-1544-69
$C_{18}H_{23}O_2P$			
Phosphinic acid, dimesityl-	C_6H_{12}	236(4.30),277(3.28), 285(3.33)	65-1544-69
$C_{18}H_{24}$			
1-Cyclohexene, 1-cyclohexyl-2-phenyl-	EtOH	201(4.28),235(3.70)	22-3232-69
1-Cyclohexene, 1-cyclohexyl-6-phenyl-	EtOH	201(4.20),248(2.48), 254(2.51),259(2.54), 262(2.54),265(2.44), 269(2.44)	22-3232-69
1-Cyclohexene, 6-cyclohexyl-1-phenyl-	EtOH	201(4.40),240(4.08)	22-3232-69
$C_{18}H_{24}BrP$			
Benzylbutylmethylphenylphosphonium bromide	MeOH	217s(4.12),254s(2.81), 260(2.96),266(3.05), 273(2.91)	24-3546-69
Benzyl-tert-butylmethylphenylphosphonium bromide	MeOH	216s(4.17),225s(2.80), 260(2.97),266(3.06), 273(2.93)	24-3546-69
β-Phenethylmethylphenylpropylphosphonium bromide	MeOH	216s(4.09),254(2.74), 259(2.88),265(2.95), 272(2.84)	24-3546-69
$C_{18}H_{24}N_2$			
Indolo[2,3-a]quinolizine, 1,2,3,4,6,7,-12,12b-octahydro-1-propyl-	MeOH	227(4.45),283(3.81)	44-0330-69
$C_{18}H_{24}N_2O$			
1(2H)-Naphthalenone, 2-(butylamino)-4-(butylimino)-	EtOH	238(4.40),272(4.12), 280s(4.09),335(3.67), 425(3.67)	39-1799-69C
$C_{18}H_{24}N_2O_2$			
Indole-3-acetic acid, 5-methyl-2-(3-piperidyl)-, ethyl ester	EtOH	223(4.50),278(3.89), 286(3.88),296(3.74)	22-4154-69

Compound	Solvent	$\lambda_{max}(\log \epsilon)$	Ref.
$C_{18}H_{24}N_2O_2S$			
p-Phenylenediamine, N⁴,N⁴-diethyl-2-(3,4-xylylsulfonyl)-	EtOH-1% $CHCl_3$	241(4.32),270(4.21)	44-2083-69
3,4-Xylenesulfonanilide, 4'-(diethylamino)-	EtOH-1% $CHCl_3$	228(4.16),268(4.32), 300s(3.69)	44-2083-69
$C_{18}H_{24}N_2O_4$			
Tulipinolide, pyrazoline derivative	EtOH	351(2.15)	36-0877-69
$C_{18}H_{24}N_4O_4$			
Δ^2-1,2,4-Triazolin-5-one, 1,4-dicyclohexyl-3-(5-nitro-2-furyl)-	EtOH	295(3.95),350s(3.53)	18-0258-69
$C_{18}H_{24}N_5O_{10}P$			
Adenosine, 2'-O-methoxytetrahydropyranyl-3'-O-acetyl-, 5'-phosphate, ammonium salt	H_2O	260(4.16)	78-4057-69
$C_{18}H_{24}O$			
3,5,7-Octatrien-2-one, 6-methyl-8-(2,6,6-trimethyl-1,3-cyclohexadien-1-yl)-, 9-cis	EtOH	297(4.19),376(4.32)	22-3242-69
all trans	EtOH	300(4.02),386(4.41)	22-3242-69
$C_{18}H_{24}O_2$			
Estra-1(10),5-dien-17-one, 3β-hydroxy-	MeOH	238(4.16)	73-0458-69
Estra-4,9(10)-dien-3-one, 17α-hydroxy-	EtOH	304(4.30)	44-3022-69
Estra-4,9(10)-dien-3-one, 17β-hydroxy-	EtOH	304(4.31)	44-1447-69
9β-Estr-4-ene-3,17-dione	EtOH-base	239(3.43),243(4.21)	44-3022-69
2,4,6-Heptatrienoic acid, 5-methyl-7-(2,6,6-trimethyl-1,3-cyclohexadien-1-yl)-, methyl ester, (E,E,E)-	EtOH	290(4.16),368(4.40)	22-3252-69
(E,E,Z)-	EtOH	292(4.26),357(4.32)	22-3252-69
4,7-Methanoindene-1,8-dione, 2,4-di-tert-butyl-3a,4,7,7a-tetrahydro-, endo	isooctane	226(3.82),322(1.79), 336(1.88),351(1.82)	35-6785-69
4,7-Methanoindene-1,8-dione, 3,6-di-tert-butyl-3a,4,7,7a-tetrahydro-, endo	isooctane	197(4.14),223(4.00), 316(1.67),329(1.77), 343(1.71)	35-6785-69
	MeOH	202(3.94),231(4.04), 312(1.89)	35-6785-69
$C_{18}H_{24}O_2S$			
Benzo[b]thiophen-4(5H)-one, 6,7-dihydro-5-(2-isobutyrylcyclohexyl)-	EtOH	220(4.06),252(4.02), 280(3.37)	22-2110-69
$C_{18}H_{24}O_3$			
Acetophenone, 2',4'-dihydroxy-3',5'-bis(3-methyl-2-butenyl)-	MeOH	285(4.33),328(3.92)	2-1072-69
2-Pentanone, 5,5-bis(3,5-dimethyl-2-furyl)-4-methyl-	EtOH	225(4.15)	35-4739-69
Xanthene-4,5-dione, 2,2,7,7,9-pentamethyl-	EtOH	232(4.18)	23-1981-69
$C_{18}H_{24}O_4$			
4H,8H-Cyclobuta[1,2-b:3,4-b']dipyran-4,8-dione, 2,4b,6,8b-tetraethyl-4a,4b,8a,8b-tetrahydro-	EtOH	265(4.36)	44-4046-69

Compound	Solvent	$\lambda_{max}(\log \epsilon)$	Ref.
3,9-Dioxapentacyclo[6.4.0.0^{2,7}.0^{4,11}.-0^{5,10}]dodecane-6,12-dione, 2,4,8,10-tetraethyl-	EtOH	236(3.83)	44-4046-69
2,4,6-Heptatrienoic acid, 7-(1-hydroxy-2,6,6-trimethyl-4-oxo-2-cyclohexen-1-yl)-5-methyl-, methyl ester	EtOH	313(4.51)	22-3252-69
2,4,6-Heptatrienoic acid, 7-(6,7,7-trimethyl-2,3-dioxabicyclo[2.2.2]oct-5-en-1-yl)-, methyl ester, endo	EtOH	309(4.51)	22-3252-69
exo	EtOH	275(4.45)	22-3252-69
Sorbic acid, 5-(2,4,5,7a-tetrahydro-2-hydroxy-4,4,7a-trimethyl-2-benzofuranyl)-, methyl ester	EtOH	278(4.34)	22-3252-69
Spiro[chroman-2,1'-cyclohexane]-2',5,6'(6H)-trione, 7,8-dihydro-4',4',7,7-tetramethyl-	EtOH	260(4.15)	1-2989-69
$C_{18}H_{24}O_8$			
Malaxinic acid	MeOH	256(4.2),286(3.3)	78-2723-69
	MeOH-base	243(4.1),272(3.4), 286(3.3)	78-2723-69
$C_{18}H_{24}O_{11}S$			
6-Epipaederosidic acid	EtOH	233(3.91)	94-1949-69
Paederosidic acid	MeOH	233(3.98)	94-1942-69
	EtOH	234(4.04)	94-1949-69
$C_{18}H_{24}O_{12}$			
Asperulosidic acid	EtOH	234(3.95)	94-1949-69
Scandoside, 10-acetyl-	EtOH	235(3.97)	94-1949-69
$C_{18}H_{25}ClN_2O$			
Indole-2-carboxamide, N,1-di-tert-butyl-6-chloro-3-methyl-	EtOH	233(4.5),295(3.95)	24-0678-69
$C_{18}H_{25}FeNO_4S$			
Isoleucine, N-(ferrocenesulfonyl)-, ethyl ester	EtOH	431(2.22)	49-1552-69
$C_{18}H_{25}NO$			
Piperidine, 1-[2-(3,4-dihydro-7-methoxy-1-naphthyl)ethyl]-, hydrochloride	n.s.g.	259(3.81)	39-0217-69C
$C_{18}H_{25}NO_2$			
2,4-Decadienamide, N-(p-hydroxyphenethyl)-, (E,E)-	n.s.g.	229s(4.15),260(4.54)	39-2477-69C
$C_{18}H_{25}NO_6$			
Jacobine	EtOH	219(3.34)	73-1832-69
$C_{18}H_{25}N_3O_2$			
1,4-Phthalazinedione, 5-[2-(di-sec-butylamino)vinyl]-2,3-dihydro-	THF	270s(4.0),300(3.8), 370(4.2)	24-3241-69
$C_{18}H_{25}N_5O_7$			
Adenosine, 2'-O-(tetrahydromethoxy-2H-pyran-4-yl)-, 3'-acetate	EtOH	260(4.18)	78-4057-69

Compound	Solvent	λ_{max}(log ϵ)	Ref.
$C_{18}H_{26}$			
Benzene, o-dicyclohexyl-	EtOH	200(4.26),209(4.08), 256(2.41),262(2.78), 264(2.78),271(2.74)	23-3232-69
Bicyclohexyl, 2-phenyl-, cis	EtOH	206(4.00),249(2.45), 253(2.48),259(2.51), 262s(--),265(2.39), 269(2.31)	22-3232-69
Bicyclohexyl, 2-phenyl-, trans	EtOH	206(3.93),210s(--), 249(2.38),254(2.46), 260(2.52),262(2.52), 265(2.42),269(2.44)	22-3232-69
$C_{18}H_{26}ClNO_6$			
Jaconine	EtOH	219(3.28)	73-1832-69
$C_{18}H_{26}N_2O$			
Indole-2-carboxamide, N,1-di-tert-butyl-3-methyl-	EtOH	227(4.4),286(3.9)	24-0678-69
$C_{18}H_{26}N_2O_2$			
Indole-2-carboxamide, N,1-di-tert-butyl-6-hydroxy-3-methyl-	EtOH	229(4.3),315(4.0)	24-0678-69
1-Phenanthrylamine, 1,2,3,4,4a,9,10,10a-octahydro-6-methoxy-N,1α,4aβ-trimethyl-N-nitroso-	EtOH	279(3.51),287(3.46), 352(1.86)	23-0497-69
$C_{18}H_{26}N_2O_4$			
Compound F	EtOH	263(4.53)	35-4749-69
Compound G	EtOH	264(4.53)	35-4749-69
benzene solvate	EtOH	265(4.44)	35-4749-69
$C_{18}H_{26}N_6OS_3$			
Spiro[furo[2,3-d]thiazole-2(3H),4'-imidazolidine]-2',5'-dithione, 3-(4-amino-2-methylpyrimidin-5-yl)methyl-1',3'-diethyltetrahydro-3a-methyl-	EtOH	325(4.16)	94-0910-69
$C_{18}H_{26}O$			
3,5,7-Octatrien-2-one, 6-methyl-8-(2,6,6-trimethyl-1-cyclohexen-1-yl)-, 9-cis	EtOH	286(4.06),343(4.34)	22-3242-69
all trans	EtOH	348(4.45)	22-3242-69
$C_{18}H_{26}O_2$			
Benzo[1,2:3,4]dicycloheptene-6-carboxylic acid, 1,2,3,4,5,5a,7a,8,9,10,11,12-dodecahydro-, methyl ester	EtOH	209(4.24)	22-4493-69
Estra-1(10),5-diene-3β,17β-diol	MeOH	240(4.18)	73-0458-69
Estr-4-en-3-one, 17β-hydroxy-, dl-	EtOH	241(4.22)	44-3530-69
8α-Estr-4-en-3-one, 17β-hydroxy-	EtOH	243(4.20)	22-1920-69
8α,10α-Estr-4-en-3-one, 17β-hydroxy-	EtOH	242(4.20)	22-1920-69
9β-Estr-4-en-3-one, 17α-hydroxy-	EtOH	244(4.19)	44-3022-69
	EtOH-base	240(3.69)	44-3022-69
10α-Estr-4-en-3-one, 17α-hydroxy-	EtOH	243(4.18)	44-3022-69
10α-Estr-4-en-3-one, 17β-hydroxy-	EtOH	243(4.19)	44-1447-69
A-Norestr-3(5)-en-2-one, 17β-hydroxy-1β-methyl-	EtOH	235(4.25)	78-1367-69
A-Nor-5α-estr-1(10)-en-2-one, 17β-hydroxy-1-methyl-	EtOH	243(4.25)	78-1367-69

Compound	Solvent	λ_{max}(log ε)	Ref.
$C_{18}H_{26}O_3$			
7H-Benz[e]inden-7-one, 3aα-ethyl-1,2,3,3a,4,5,8,9,9aα,9bβ-decahydro-1α-hydroxy-, acetate, dl-	EtOH	248(4.20)	44-0107-69
Estr-4-en-3-one, 6β,17β-dihydroxy-	EtOH	238(4.13)	44-3530-69
	EtOH-H_2SO_4	297(--),377(--), 397(--),459(--)	44-3530-69
	EtOH-NaOH	242(3.58),266(3.26), 313(3.18),367(3.27)	44-3530-69
Estr-4-en-3-one, 11β,17β-dihydroxy-	EtOH	242(4.17)	44-3530-69
Podocarp-8(14)-en-16-oic acid, 12-oxo-, methyl ester	MeOH	282(2.00)	23-3661-69
$C_{18}H_{26}O_4$			
1,3-Cyclohexanedione, 2,2'-ethylenebis-[5,5-dimethyl-	EtOH-acid	261(4.39)	1-2989-69
	EtOH-base	288(4.60)	1-2989-69
2-Cyclohexen-1-one, 3,3'-(ethylenedioxy)bis[5,5-dimethyl-	EtOH	251(4.53)	1-2989-69
3,9-Dioxapentacyclo[6.4.0.$0^{2,7}$.$0^{4,11}$.$0^{5,10}$]dodecan-6-one, 2,4,8,10-tetraethyl-12-hydroxy-	EtOH	225(3.48)	44-4046-69
Estr-4-en-3-one, 10β,11β,17β-trihydroxy-, d-	EtOH	237(4.11)	44-3530-69
Phthalic acid, butyl 2-ethyl butyl ester	EtOH	226(3.89),278(3.13), 283(3.11)	31-0907-69
$C_{18}H_{26}O_5$			
2-Oxabicyclo[3.3.1]non-3-ene-4-carboxylic acid, 1-hydroxy-7-isopropenyl-3,9-dimethyl-, ethyl ester, acetate	EtOH	251(4.01)	78-3139-69
2,3-Seco-B-norandrost-4-ene-2,3-dioic acid, 17β-hydroxy-	EtOH	223(4.10)	44-2297-69
$C_{18}H_{26}O_6$			
Tolypolide F_3	EtOH	206(3.14)	88-0091-69
$C_{18}H_{26}O_6S$			
[6-[Carboxy(carboxycarbonyl)methylene]-2-hydroxy-4,4-dimethyl-1-cyclohexen-1-yl]dimethylsulfonium hydroxide, inner salt, diethyl ester	EtOH	222(4.07),274(3.93), 392(4.17)	88-2053-69
$C_{18}H_{26}O_7$			
D-1-Glucooctulose, 3,6-anhydro-2-deoxy-1-(p-tolyl)-5,7,8-tri-O-methyl-	MeOH	273(4.23)	65-0119-69
$C_{18}H_{26}O_{10}$			
α-D-erythro-Hex-4-enofuranose, 5-deoxy-1,2-O-isopropylidene-2-C-[tetrahydro-6-hydroxy-5-(2-hydroxyethylidene)-2,2-dimethylfuro[2,3-d]-1,3-dioxol-6-yl]-, (E,E)-	MeOH	205(3.90),283(3.78)	24-4199-69
$C_{18}H_{27}Ir$			
Iridium, (1,5-cyclooctadiene)(pentamethyl-π-cyclopentadienyl)-	hexane	214(4.34)	101-0491-69A
	EtOH	214(4.34)	101-0491-69A
$C_{18}H_{27}N$			
Naphthalene, 3,4-dihydro-2-ethyl-2-methyl-1-[3-(dimethylamino)propylidene]-	n.s.g.	246(4.04)	39-0217-69C

Compound	Solvent	$\lambda_{max}(\log \epsilon)$	Ref.
$C_{18}H_{27}NO$			
13H-1-Benzazacyclopentadecin-13-one, 1,2,3,4,5,6,7,8,9,10,11,12-dodecahydro-	n.s.g.	230(4.2),260s(3.5), 325(3.2)	24-0342-69
Naphthalene, 3,4-dihydro-6-methoxy-2,2-dimethyl-1-[3-(dimethylamino)propylidene]-	n.s.g.	249(3.74)	39-0217-69C
hydrochloride	n.s.g.	255(4.21)	39-0217-69C
Naphthalene, 3,4-dihydro-7-methoxy-2,2-dimethyl-1-[3-(dimethylamino)propylidene]-	n.s.g.	245(3.99)	39-0217-69C
$C_{18}H_{27}NO_2$			
Undecenohydroxamic acid, N-p-tolyl-	EtOH	252(4.04)	42-0831-69
$C_{18}H_{27}NO_3$			
Futoamide, tetrahydro-	EtOH	234(3.59),288(3.58)	94-1225-69
$C_{18}H_{27}NO_4$			
3H-Cyclopenta[f]quinoline-$\Delta^{2,\alpha}$-acetic acid, 1,2,4,4a,5,6,6a,7,8,9,9a,9b-dodecahydro-7β-hydroxy-6aβ-methyl-, ethyl ester, formate	EtOH	291(4.37)	22-0117-69
$C_{18}H_{27}NO_6$			
Alkaloid M_1 from Senecio mikanioides	EtOH	218(3.87)	100-0503-69
Alkaloid P_1 from Senecio petasitis	EtOH	219(3.13)	100-0503-69
$C_{18}H_{27}NO_7$			
Jacoline	EtOH	219(3.33)	73-1832-69
$C_{18}H_{28}N_2O_4$			
9,14-Diazatricyclo[6.3.2.1^{2,7}]tetradec-10-ene-9,14-dicarboxylic acid, diethyl ester	EtOH	226(4.18)	35-3616-69
$C_{18}H_{28}N_4$			
Phthalazine, 1,4-bis[3-(dimethylamino)-propyl]-, dihydrochloride	EtOH	219(4.7),265(3.7)	33-1376-69
$C_{18}H_{28}O$			
Benzo[1,2:3,4]dicycloheptene-6-carboxaldehyde, 1,2,3,4,5,5a,6,7,7a,8,9,10,-11,12-tetradecahydro-6-methyl-, cis-trans	EtOH	210(4.29)	22-4493-69
Spiro[2.5]octa-4,7-dien-6-one, 5,7-di-tert-butyl-1,2-dimethyl-	C_6H_{12}	352(1.58)	35-1580-69
isomer	C_6H_{12}	352(1.59)	35-1580-69
$C_{18}H_{28}O_2$			
Abiet-7?-en-18-oic acid	n.s.g.	204(3.81)	44-1550-69
Abiet-8(14)-en-18-oic acid	n.s.g.	207(3.94)	44-1550-69
13β-Abiet-8(14)-en-18-oic acid	n.s.g.	201(4.03)	44-1550-69
Abiet-13(15)-en-18-oic acid	n.s.g.	197(4.09)	44-1550-69
Benzo[1,2:3,4]dicycloheptene-6-carboxylic acid, 1,2,3,4,5,5a,6,7,7a,8,9,-10,11,12-tetradecahydro-, methyl ester, cis-cis	EtOH	209(3.92)	22-4493-69
cis-trans	EtOH	213(3.88)	22-4493-69

Compound	Solvent	λ_{max}(log ϵ)	Ref.
Benzo[1,2:3,4]dicycloheptene-6-carboxylic acid, 1,2,3,4,5,5a,6,7,7a,8,9,-10,11,12-tetradecahydro-6-methyl-, cis-trans	EtOH	210(3.99)	22-4493-69
A-Nor-5α-estr-9-ene-2α,17β-diol, 1β-methyl-	MeCN	220(3.80)	78-1367-69
A-Norestr-5(10)-ene-2α,17β-diol, 1β-methyl-	MeCN	193s(3.85)	78-1367-69
$C_{18}H_{28}O_3$			
Decanophenone, 4'-ethoxy-2'-hydroxy-	n.s.g.	275(3.8)	104-1609-69
Octanophenone, 4'-butoxy-2'-hydroxy-	n.s.g.	275(4.08)	104-1609-69
$C_{18}H_{28}O_4$			
Gingeryl methyl ether	EtOH	282(3.42)	12-1033-69
$C_{18}H_{28}O_5$			
1-Cyclohexene-1-butyric acid, 2-(2-carboxyethyl)-6-methyl-3-oxo-, diethyl ester	EtOH	248(4.28)	78-2661-69
2-Oxabicyclo[3.3.1]non-3-ene-4-carboxylic acid, 1-hydroxy-7-isopropyl-3,9-dimethyl-, ethyl ester, acetate	EtOH	251(4.08)	78-3139-69
$C_{18}H_{28}O_{11}$			
Cyclopenta[c]pyran-4-carboxylic acid, 1-(β-D-glucopyranosyloxy)-1,3,5,6,-7,7a-hexahydro-7-hydroxy-3-methoxy-7-methyl-, methyl ester	MeOH	222(4.08)	32-1150-69
$C_{18}H_{28}S_2$			
Spiro[benzo[1,2:3,4]dicycloheptene-6(1H),2'-[1,3]dithiolane], 2,3,4,-5,7,7a,8,9,10,11,12,12a-dodecahydro-	EtOH	211(4.10)	22-4501-69
$C_{18}H_{28}Si_2$			
2,6-Disilahepta-3,4-diene, 3-isopropenyl-2,2,6,6-tetramethyl-5-phenyl-	n.s.g.	246(4.45)	101-0291-69C
2,6-Disilahepta-3,4-diene, 2,2,6,6-tetramethyl-3-phenyl-5-propenyl-	n.s.g.	247(4.45)	101-0291-69C
$C_{18}H_{29}BrO$			
2,5-Cyclohexadien-1-one, 4-bromo-2,4,6-tri-tert-butyl-	hexane	251(4.13)	55-0074-69
	MeOH	251(3.97)	64-0547-69
$C_{18}H_{29}ClO$			
2,5-Cyclohexadien-1-one, 2,4,6-tri-tert-butyl-4-chloro-	hexane	241(4.13)	55-0074-69
	MeOH	243(4.11)	64-0547-69
	$CHCl_3$	245(4.11)	64-0547-69
$C_{18}H_{29}NO$			
Benzene, 1,3,5-tri-tert-butyl-2-nitroso-	hexane	765(1.75)	18-3611-69
$C_{18}H_{29}NO_3$			
2,5-Cyclohexadien-1-one, 2,4,6-tri-tert-butyl-4-nitro-	MeOH	241(4.02)	64-0547-69
$C_{18}H_{30}$			
Benzo[1,2:3,4]dicycloheptene, 1,2,3,4,-5,5a,6,7,7a,8,9,10,1,12-tetradecahydro-6,6-dimethyl-	EtOH	210(3.85)	22-4493-69

Compound	Solvent	λ max (log ε)	Ref.
$C_{18}H_{30}ClNO_4$			
Spiro[cyclohexane-1,2'[2H]indolium], 3'-tert-butyl-4',5',6',7'-tetrahydro-1'-methyl-, perchlorate	EtOH	284(3.83)	78-4161-69
$C_{18}H_{30}O$			
Benzo[1,2:3,4]dicycloheptene-6-methanol, 1,2,3,4,5,5a,6,7,7a,8,9,10,11,12-tetradecahydro-6β-methyl-	EtOH	210(3.98)	22-4493-69
Phenol, 2,4,6-tri-tert-butyl-	MeOH	274(3.20)	30-1113-69A
	MeOH-Na	252(4.01),302(3.64)	30-1113-69A
$C_{18}H_{30}O_2$			
2-Cyclobuten-1-one, 2,4-di-tert-butyl-4-(3,3-dimethylbut-1-enyl)-3-hydroxy-	MeOH	254(3.85)	77-1003-69
2,5-Cyclohexadien-1-one, 2,4,6-tri-tert-butyl-4-hydroxy-	MeOH	239(3.96)	64-0547-69
$C_{18}H_{30}O_3$			
2-Cyclopentadecene-1-carboxylic acid, 2-methyl-15-oxo-, methyl ester	EtOH	204(3.91),207s(3.85), 248f(3.40)	94-0629-69
$C_{18}H_{30}O_3S$			
Cyclohexanepropionic acid, 3-[(butylthio)methylene]-1,6-dimethyl-2-oxo-, ethyl ester	MeOH	311(4.17)	23-4307-69
$C_{18}H_{31}Cl$			
Bicyclo[2.2.0]hex-5-ene, 1,2,5-tri-tert-butyl-2-chloro-	C_6H_{12}	258(4.24)	24-3996-69
$C_{18}H_{31}NO_3S_2$			
2-Pyridinol, 1-acetyl-3,6-bis(tert-butylthio)-1,2,3,6-tetrahydro-4-methyl-, acetate	hexane	195(4.39)	44-0660-69
	MeOH	205(4.39)	44-0660-69
$C_{18}H_{31}N_3O$			
Phenol, 2,6-di-tert-butyl-4-(3,3-diethyl-1-triazeno)-	n.s.g.	234(4.07),290(4.13), 325(4.13)	70-2460-69
$C_{18}H_{33}Br_2Sb$			
Antimony, dibromotricyclohexyl-	MeCN	none above 240 nm	101-0335-69A
$C_{18}H_{33}NO_2$			
2(3H)-Furanone, 3-[1-(dodecylamino)-ethylidene]dihydro-	hexane	295(4.19)	104-0998-69
$C_{18}H_{33}SSb$			
Stibine sulfide, tricyclohexyl-	hexane	282(3.7)	101-0335-69A
	MeCN	274(3.8)	101-0335-69A
$C_{18}H_{34}Br_2N_4O_2$			
N,N,N,N',N'-Pentamethyl-N'-[3-(4-nitro-m-toluidino)propyl]-N,N'-trimethylenediammonium dibromide	pH 6.5	393(4.17)	35-5136-69
	EtOH	373(4.18)	35-5136-69
N,N,N,N',N'-Pentamethyl-N'-[3-(4-nitro-o-toluidino)propyl]-N,N'-trimethylenediammonium dibromide (also spectra of these compounds in biological media)	pH 6.5	400(4.18)	35-5136-69
	EtOH	381(4.22)	35-5136-69

Compound	Solvent	$\lambda_{max}(\log \epsilon)$	Ref.
$C_{18}H_{36}Si_2$			
2,6-Disilahepta-3,4-diene, 3-hexyl-5-isopropenyl-2,2,6,6-tetramethyl-	n.s.g.	233(3.81)	101-0291-69C

Compound	Solvent	$\lambda_{max}(\log \epsilon)$	Ref.
$C_{19}H_{10}Br_2O$			
Biphenylene, 2-benzoyl-6,7-dibromo-	EtOH	249s(4.47),264(4.75), 274s(4.67),355(3.76), 372(3.94)	39-2789-69C
$C_{19}H_{10}O_3$			
1H-Naphtho[2,1,8-mna]xanthen-1-one, 5-hydroxy-	n.s.g.	230(4.1),280(3.46), 317(3.36),329(3.29), 362(3.22),379(3.15), 523(3.6)	88-4325-69
$C_{19}H_{11}BrO$			
Biphenylene, 2-benzoyl-6-bromo-	EtOH	221(4.16),263(4.69), 276s(4.58),332(3.43), 348(3.70),366(3.87)	39-2789-69C
Biphenylene, 2-benzoyl-7-bromo-	EtOH	244s(4.36),260(4.59), 275s(4.41),337s(3.40), 351(3.69),368(3.89)	39-2789-69C
$C_{19}H_{11}BrO_2$			
2-Biphenylenol, 6-bromo-, benzoate	EtOH	250s(4.91),256(5.06), 340s(3.55),348(3.90), 365(4.06)	39-2789-69C
$C_{19}H_{11}Br_2NO$			
Biphenylene, 2-benzoyl-6,7-dibromo-, oxime	EtOH	257s(4.84),261(4.89), 356(4.08),370(4.15)	39-2789-69C
2-Biphenylenecarboxanilide, 6,7-dibromo-	EtOH	267(4.74),358(4.15), 378(4.19)	39-2789-69C
$C_{19}H_{11}Br_3O_2$			
1,3-Cyclopentadiene, 2,3-dibenzoyl-1,4,5-tribromo-	EtOH	244(4.27),344(4.04)	39-2464-69C
$C_{19}H_{11}Fe_2NO_8$			
Iron, hexacarbonyl[μ-(N-methyl-4a,8a-naphthalenedicarboximide)]di-	MeOH	222s(4.51),303(3.04)	78-5115-69
$C_{19}H_{11}NO_3$			
Biphenylene, 2-(p-nitrobenzoyl)-	EtOH	248(4.47),273(4.57), 348(3.64),364(3.77), 384(3.71)	39-2789-69C
$C_{19}H_{11}N_5O_2$			
Pyrrolo[1,2-b]-as-triazine-8-carbonitrile, 2-(p-nitrophenyl)-7-phenyl-	CH_2Cl_2	237(4.47),269(4.43), 327(4.15),389s(3.96), 446(4.25),490s(3.65)	48-0388-69
$C_{19}H_{12}BrNO$			
Isoxazole, 3-(p-bromophenyl)-5-(1-naphthyl)-	EtOH	228(4.62),286(4.17), 323(4.19),342(4.16)	88-3329-69
$C_{19}H_{12}Br_2O_2$			
Cyclopenta[1,2-b:1,5-b']bis[1]benzopyran, 3,10-dibromo-6,7-dihydro-	dioxan	218(4.67),258(4.23), 312(3.70)	5-0162-69B
$C_{19}H_{12}ClNO_3$			
4H-Pyrano[3,2-h]quinolin-4-one, 6-chloro-2-(p-methoxyphenyl)-	EtOH	240(4.30),305(4.2)	23-0105-69

Compound	Solvent	$\lambda_{max}(\log \epsilon)$	Ref.
$C_{19}H_{12}ClNO_4$			
4H-Pyrano[3,2-h]quinolin-4-one, 6-chloro-3-hydroxy-2-(p-methoxyphenyl)-	EtOH	236(4.50),284(4.21), 355(3.63)	23-0105-69
$C_{19}H_{12}N_2O$			
Benzo[6,7]indolizino[1,2-b]quinolin-11(13H)-one	EtOH	283(4.34),363(4.49), 376(4.52)	78-2275-69
$C_{19}H_{12}N_4O_4$			
5H-Cyclohepta[b]quinoxaline, 8,10-dinitro-7-phenyl-	n.s.g.	250(3.86),260(3.87), 315(3.52),360(3.30), 476(3.59)	40-0085-69
4-Pyridazinecarbonitrile, 2,3-dihydro-2-(p-nitrophenacyl)-3-oxo-5-phenyl-	DMF	299(4.10),317s(3.70), 365s(3.48)	48-0388-69
	DMF-KOH	630(3.40)	48-0388-69
$C_{19}H_{12}O$			
Biphenylene, 2-benzoyl-	EtOH	260(4.74),355(3.82), 369(3.98)	39-2789-69C
$C_{19}H_{12}OS$			
3H-Naphtho[2,1-b]thiopyran-3-one, 1-phenyl-	EtOH	235(3.87),335(3.9)	2-0315-69
$C_{19}H_{12}O_2$			
2-Biphenylenecarboxylic acid, phenyl ester	EtOH	219(4.48),258(4.60), 346(3.41),356(3.57)	39-2789-69C
5,8-Methanocyclopenta[cd]fluoranthene-1,2-dione, 4b,5,8,8a-tetrahydro-	EtOH	232(4.60),244(4.35), 327(4.00)	35-0918-69
$C_{19}H_{12}O_3$			
4H-Naphtho[1,2-b]pyran-4-one, 3-hydroxy-2-phenyl-	$CHCl_3$	282(4.40),338(4.29)	115-0017-69
$C_{19}H_{12}O_4$			
Benzo[h]xanthone, 3-acetoxy-	$CHCl_3$	250(4.65)	49-1368-69
$C_{19}H_{12}O_5$			
Compound, d. 260-280°	EtOH	408(4.33)	1-2583-69
$C_{19}H_{13}BF_2O_2$			
Naphtho[2,1-e]-1,3,2-dioxaborin, 2,2-difluoro-4-styryl-	MeCN	253s(4.31),269(4.32), 278(4.43),378(4.81), 465(4.48)	4-0029-69
$C_{19}H_{13}BrO_2$			
1,3-Cyclopentadiene, 2,3-dibenzoyl-5-bromo-	EtOH	273(4.20),285(4.21), 346(3.95),432(4.08)	39-2464-69C
$C_{19}H_{13}BrO_3$			
1H,3H-Naphtho[1,8-cd]pyran-1-one, 3-(p-bromophenyl)-3-methoxy-	dioxan	297(4.3),308(4.3), 326(4.2)	35-0477-69
$C_{19}H_{13}ClN_2O_2$			
Carbazole-1-carboxanilide, 4'-chloro-2-hydroxy-	EtOH	230(4.63),292(4.60), 325(4.05),360(4.10)	39-1518-69C
Carbazole-3-carboxanilide, 4'-chloro-2-hydroxy-	dioxan	289(4.68),410(4.54), 445(4.61)	2-1065-69
2,4,6-Cycloheptatrien-1-one, 5-[(p-chlorophenyl)azo]-2-hydroxy-4-phenyl-	n.s.g.	231(4.03),247(4.02), 300(3.78),410(4.04), 460(4.03)	40-0085-69

Compound	Solvent	$\lambda_{max}(\log \epsilon)$	Ref.
$C_{19}H_{13}ClO_4S$			
3-Phenylnaphtho[2,1-b]thiopyrylium perchlorate	HOAc-1% $HClO_4$	250(4.00),307(4.06), 420(3.43)	2-0191-69
$C_{19}H_{13}ClO_4S_2$			
4-(2-Thienyl)thioflavylium perchlorate	HOAc-1% $HClO_4$	265(3.69),310s(3.44), 400(3.48),470(3.55)	2-0191-69
$C_{19}H_{13}F$			
Benz[a]anthracene, 12-fluoro-7-methyl-	EtOH	264(4.38),274(4.51), 284(4.72),294(4.71), 323(3.72),343(3.70), 362(3.70),365(3.70), 376(3.43),382(3.53), 396(3.20)	78-3501-69
$C_{19}H_{13}N$			
Benzo[f]quinoline, 3-phenyl-	dioxan	262(4.16),282(4.20), 283(4.09),312(3.83), 329(3.63),347(3.85), 361(3.87)	24-1202-69
$C_{19}H_{13}NO$			
Benzamide, N-2-biphenylenyl-	EtOH	230(4.33),241(4.34), 261(4.50),351(3.87), 371(3.97)	39-2789-69C
Biphenylene, 2-benzoyl-, oxime	EtOH	256(4.31),263s(4.29), 275s(4.20),365(3.43)	39-2789-69C
2-Biphenylenecarboxanilide	EtOH	238s(4.15),247(4.48), 256(4.63),329(3.61), 347(3.83),362(3.96)	39-2789-69C
Carbazole, 2-benzoyl-	EtOH	257(4.54),283(4.04), 326(4.32)	22-3523-69
Carbazole, 3-benzoyl-	EtOH	234(4.54),244s(4.46), 260(4.17),283(4.42), 297(4.40),335(4.08), 350(4.02)	22-3523-69
Isoxazole, 5-(1-naphthyl)-3-phenyl-	EtOH	230(4.33),321(4.29), 339(4.23)	88-3329-69
$C_{19}H_{13}N_3O_2$			
Benzo[g]quinoline, 4-(p-nitroanilino)-	EtOH	257(4.47),360(3.83), 408(4.12)	103-0223-69
2-Naphthol, 1-(4-quinolylazo)-, N-oxide	MeOH	217(4.79),268(4.18), 515(4.39)	94-2181-69
$C_{19}H_{13}N_3O_4$			
2,4,6-Cycloheptatrien-1-one, 2-hydroxy-5-[(p-nitrophenyl)azo]-4-phenyl-	n.s.g.	230(4.33),255s(4.26), 408(4.24),476(4.44)	40-0085-69
$C_{19}H_{13}N_5O_3S$			
Benzoic acid, thio-, S-[9-(p-nitrobenzyl)-9H-purin-6-yl] ester	MeCN	275(4.39)	44-0973-69
9H-Purine, 9-benzoyl-6-[(p-nitrobenzyl)-thio]-	MeCN	245(4.66),285(4.79)	44-0973-69
$C_{19}H_{13}N_5O_8$			
Pyridine, 2,4-bis(2,4-dinitrobenzyl)-	EtOH	244(4.59)	22-4425-69

Compound	Solvent	λ max(log ε)	Ref.
$C_{19}H_{14}BrN$			
Benzo[f]quinoline, 3-(p-bromophenyl)-5,6-dihydro-	dioxan	233(3.94),295(4.29), 306(4.32),327(4.42)	24-1202-69
$C_{19}H_{14}BrNO_4$			
Pyrrole-2-carboxylic acid, 5-benzoyl-1-(p-bromophenyl)-4-hydroxy-, methyl ester	MeOH	258(4.09),304(3.97), 338(3.88)	78-0527-69
$C_{19}H_{14}ClN$			
Benz[c]acridine, 7-chloro-5,6-dimethyl-	isooctane	224(4.57),240(4.47), 278s(4.65),287(4.71), 299(4.71),333(3.66), 348(3.79),362(3.85), 378(3.86),397(3.79)	4-0361-69
	MeOH	224(4.57),238(4.47), 279s(4.67),285(4.71), 298(4.70),334s(3.58), 349(3.72),364(3.76), 379(3.78),398(3.72)	4-0361-69
Benzo[f]quinoline, 3-(p-chlorophenyl)-5,6-dihydro-	dioxan	287(4.31),295(4.33), 326(4.42)	24-1202-69
$C_{19}H_{14}ClNO_3$			
4H-Pyrano[3,2-h]quinolin-4-one, 6-chloro-2,3-dihydro-2-(p-methoxyphenyl)-	EtOH	226(4.45),268(4.51), 360(3.95)	23-0105-69
$C_{19}H_{14}ClNO_4$			
Pyrrole-2-carboxylic acid, 5-benzoyl-1-(p-chlorophenyl)-4-hydroxy-, methyl ester	MeOH	258(4.11),300(3.96), 336(3.89)	78-0527-69
$C_{19}H_{14}Cl_2N_2$			
Benzaldehyde, 2,6-dichloro-, diphenylhydrazone	EtOH	240(4.3),307(4.0), 349(4.4)	28-0137-69B
$C_{19}H_{14}Cl_2O_2S_2$			
4H-Thiopyran-4-thione, 3,5-dichloro-2,6-bis(p-methoxyphenyl)-	EtOH	262(4.23),394(4.41)	39-0315-69C
$C_{19}H_{14}Cl_2O_3S$			
4H-Thiopyran-4-one, 3,5-dichloro-2,6-bis(p-methoxyphenyl)-	EtOH	226(4.45),280(4.41), 317(4.39)	39-0315-69C
$C_{19}H_{14}FI$			
Naphthalene, 4-fluoro-2-(o-iodostyryl)-1-methyl-, (E)-	EtOH	217(4.56),230(4.51), 275(4.28),314(4.23), 356s(3.70)	78-3501-69
$C_{19}H_{14}N_2$			
Benzo[g]quinoline, 4-anilino-	EtOH	253(4.76),320(3.60), 396(4.05)	103-0223-69
Quinoline, 4-phenyl-2-pyrrol-2-yl-	n.s.g.	364(4.18)	97-0186-69
$C_{19}H_{14}N_2O$			
Benzo[g]quinoline, 4-(p-hydroxyanilino)-	EtOH	248(4.80),403(4.04)	103-0223-69
$C_{19}H_{14}N_2O_2$			
2,4,6-Cycloheptatrien-1-one, 2-hydroxy-4-phenyl-5-(phenylazo)-	n.s.g.	230(3.58),253(3.54), 295s(3.37),412(3.64), 445(3.57)	40-0085-69

Compound	Solvent	$\lambda_{max}(\log \epsilon)$	Ref.
$C_{19}H_{14}N_2O_2S$			
1,2-Benzisothiazoline, 2-phenyl-3-(phenylimino)-, 1,1-dioxide	EtOH	238s(4.6),338(3.7)	44-1786-69
Quinoline, 2,2'-(methylenesulfonyl)di-	EtOH	239(4.92),290(3.90), 305(3.92),318(3.91)	95-0074-69
$C_{19}H_{14}N_2O_5$			
[1,3]Dioxolo[4,5-f]-2,1-benzisoxazole, 3-(7,8-dimethoxy-4-isoquinolyl)-	EtOH	246(4.60)	78-5365-69
$C_{19}H_{14}N_2S$			
Lepidine, 2-(2-phenyl-4-thiazolyl)-	n.s.g.	340(4.07)	97-0186-69
Quinoline, 2-(2-methyl-4-thiazolyl)-4-phenyl-	n.s.g.	344(3.84)	97-0186-69
$C_{19}H_{14}N_4O$			
4(1H)-Pteridinone, 1-methyl-6,7-diphenyl-	EtOH	227(4.48),258s(4.15), 281(4.13),361(4.16)	39-2415-69C
4(3H)-Pteridinone, 3-methyl-6,7-diphenyl-	EtOH	224(4.58),264(4.21), 353(4.05)	39-2415-69C
$C_{19}H_{14}N_4O_2$			
1H-Pyrazolo[3,4-b]pyrazin-6-ol, 1,3-diphenyl-	EtOH	252(4.52),268(4.54)	32-0463-69
$C_{19}H_{14}N_4O_4$			
7,8-Benzobicyclo[4.2.1]nona-2,4,7-trien-9-one, 2,4-dinitrophenylhydrazone	EtOH	362(4.03)	88-1761-69
$C_{19}H_{14}O$			
Triphenylene, 2-methoxy-	n.s.g.	254(4.86),262(5.04), 277s(4.20),288(4.16), 320(3.16),327(2.94), 335(3.26),343(2.84), 351(3.25)	18-0766-69
$C_{19}H_{14}O_2$			
Cyclopenta[1,2-b:1,5-b']bis[1]benzopyran, 6,7-dihydro-	dioxan	257(4.24),270(4.19), 300(3.81)	5-0162-69B
15H-Cyclopenta[a]phenanthren-17-ol, acetate	n.s.g.	222(4.48),269(4.91), 273(4.92),291(4.28), 312(3.89),329(2.91), 345(2.83),363(2.66)	39-2484-69C
Phenanthro[1,2-b]furan, 1-acetyl-2-methyl-	C_6H_{12}	223(4.37),229(4.41), 259(4.83),267(5.00), 284(4.38),296(4.36), 303(3.98),322(3.08), 331(2.90),338(3.27), 346(2.82),354(3.27)	33-1461-69
2H-Pyran-$\Delta^{2,\alpha}$-acetaldehyde, 4,6-diphenyl-	MeCN	224(4.13),286(4.52), 327(4.00),445(3.99)	44-2736-69
4H-Pyran-$\Delta^{4,\alpha}$-acetaldehyde, 2,6-diphenyl-	MeCN	241(4.07),272(4.10), 298(4.26),370(4.41)	44-2736-69
$C_{19}H_{14}O_3$			
17H-Cyclopenta[a]phenanthren-17-one, 15,16-dihydro-16-hydroxy-, acetate	EtOH	266(4.92),284(4.50), 297(4.39),352(3.31), 370(3.32)	39-2484-69C

Compound	Solvent	λmax(log ε)	Ref.
Cyclopent[a]indene-3,8-dione, 3a,8a-dihydro-3a-hydroxy-1-methyl-2-phenyl-	EtOH	227(4.44),278(3.74), 324(2.83),340(2.91), 355(2.19)	44-0520-69
1-Naphthoic acid, 8-benzoyl-, methyl ester	M H_2SO_4	290(4.3)	35-0477-69
	dioxan	290(4.5)	35-0477-69
1-Naphthoic acid, 8-p-toluoyl-	M H_2SO_4	305(4.2),327(4.0)	35-0477-69
	dioxan	297(4.4),308(4.4), 326(4.3)	35-0477-69
1H,3H-Naphtho[1,8-cd]pyran-1-one, 3-methoxy-3-phenyl-	M H_2SO_4	305(4.2),327(4.2)	35-0477-69
	dioxan	297(4.2),308(4.3), 326(4.2)	35-0477-69
Psoralen, 4,8-dimethyl-4'-phenyl-	MeOH	222(3.89),250(4.43), 299(4.18)	42-1014-69
Psoralen, 4',8-dimethyl-4-phenyl-	MeOH	225(3.89),253(4.60), 307(3.75)	42-1014-69
$C_{19}H_{14}O_6$			
5H-Furo[3,2-g][1]benzopyran-5-one, 6-(2-hydroxy-4,5-dimethoxyphenyl)-	EtOH	239(4.56),260s(4.36), 304(4.15)	18-1693-69
Unknown compound from xylerythin and sodium hydroxide	dioxan	249(4.09),265s(4.03), 346(4.06)	1-2583-69
$C_{19}H_{14}S$			
3H-Naphtho[2,1-b]thiopyran, 1-phenyl-	EtOH	223(3.73),257(3.62)	2-0315-69
$C_{19}H_{15}AuP$			
Gold, cyano(triphenylphosphine)-	dioxan	275(3.54)	39-0276-69B
$C_{19}H_{15}BrFeO$			
Acrylophenone, 3'-bromo-3-ferrocenyl-	MeOH	330(4.21)	73-2771-69
Acrylophenone, 4'-bromo-3-ferrocenyl-	MeOH	330(4.30)	73-2771-69
	n.s.g.	271(4.2),318(4.2), 385(3.4),475(3.3)	65-1008-69
$C_{19}H_{15}BrO_2$			
Cyclopent[a]inden-8(3H)-one, 2-(bromomethyl)-3a,8a-dihydro-3a-hydroxy-8a-phenyl-	EtOH	249(4.06),291(2.95)	104-2090-69
$C_{19}H_{15}ClFeO$			
Acrylophenone, 3'-chloro-3-ferrocenyl-	MeOH	330(4.22)	73-2771-69
Acrylophenone, 4'-chloro-3-ferrocenyl-	MeOH	330(4.29)	73-2771-69
	n.s.g.	273(4.2),320(4.2), 387(3.4),480(3.4)	65-1008-69
2-Propen-1-one, 3-(m-chlorophenyl)-1-ferrocenyl-	MeOH	299(4.32)	73-2771-69
2-Propen-1-one, 3-(p-chlorophenyl)-1-ferrocenyl-	MeOH	310(4.44)	73-2771-69
$C_{19}H_{15}ClN_2$			
Benzaldehyde, m-chloro-, diphenylhydrazone	EtOH	239(4.2),249s(4.1), 303(3.9),349(4.3)	28-0137-69B
Benzaldehyde, o-chloro-, diphenylhydrazone	EtOH	239(4.2),248s(4.2), 303(4.0),349(4.3)	28-0137-69B
Benzaldehyde, p-chloro-, diphenylhydrazone	EtOH	204(4.6),239(4.1), 300(4.0),344(4.2)	28-0137-69B
$C_{19}H_{15}ClN_2O$			
3H-Azepine, 2-anilino-3-benzoyl-5-chloro-	MeOH	247(4.29),311(4.13)	78-5205-69

Compound	Solvent	λ_{max}(log ε)	Ref.
$C_{19}H_{15}Cl_2NO_3S$			
4H-Thiopyran-4-one, 3,5-dichloro-2,6-bis(p-methoxyphenyl)-, oxime	EtOH	284(4.34)	39-0315-69C
$C_{19}H_{15}Cl_4N$			
Pyridine, 1-benzyl-1,4-dihydro-2,6-dimethyl-4-(2,3,4,5-tetrachloro-cyclopentadien-1-ylidene)-	CH_2Cl_2	258(4.24),436(4.64)	83-0886-69
$C_{19}H_{15}F$			
Naphthalene, 1-fluoro-2-(p-methylstyryl)-, trans	EtOH	229(4.45),242(4.26), 253(4.15),275(4.34), 286(4.38),323(4.50), 339(4.23),361(3.56)	78-3501-69
$C_{19}H_{15}FFeO$			
Acrylophenone, 3-ferrocenyl-4'-fluoro-	MeOH	330(4.24)	73-2771-69
2-Propen-1-one, 1-ferrocenyl-3-(p-fluorophenyl)-	MeOH	308(4.34)	73-2771-69
$C_{19}H_{15}F_5N_2O_2$			
Cinnamic acid, α-acetyl-2,3,4,5,6-pentafluoro-, ethyl ester, α-(phenylhydrazone)	EtOH	256(4.16),360(4.37)	65-1347-69
$C_{19}H_{15}FeNO_3$			
Acrylophenone, 3-ferrocenyl-3'-nitro-	MeOH	335(4.19)	73-2771-69
Acrylophenone, 3-ferrocenyl-4'-nitro-	MeOH	335(4.20)	73-2771-69
2-Propen-1-one, 1-ferrocenyl-3-(p-nitrophenyl)-	MeOH	318(4.40)	73-2771-69
$C_{19}H_{15}N$			
Benzo[f]quinoline, 5,6-dihydro-3-phenyl-	dioxan	223(3.94),299s(4.24), 332(4.42)	24-1202-69
Pyridine, 2-(1,2-diphenylvinyl)-, cis	hexane	235(4.3),295(4.3)	4-0465-69
trans	hexane	235(4.2),245s(4.1), 280s(4.2),315(4.3)	4-0465-69
Pyridine, 4-(1,2-diphenylvinyl)-, cis	hexane	237(4.2),300(4.2)	4-0465-69
trans	hexane	240(4.3),300(4.3)	4-0465-69
$C_{19}H_{15}NO$			
Benz[c]acridan-7-one, 5,6-dimethyl-	MeOH	207(4.56),213(4.54), 233(4.48),259(4.48), 275s(4.51),284(4.66), 298s(4.08),311(4.04), 324(3.85),343(3.96), 373s(3.68),389(3.86)	4-0361-69
Benzophenone, 2-anilino-	EtOH	233(4.06),255(4.24), 283s(4.02),405(3.80)	22-3523-69
Benzophenone, 3-anilino-	EtOH	259(4.26),282(4.28)	22-3523-69
Benzophenone, 4-anilino-	EtOH	248(4.11),266s(3.92), 367(4.37)	22-3523-69
$C_{19}H_{15}NO_2$			
8,16-Imino[2,2]metacyclophane-1,9-diene, N-carbethoxy-	EtOH	227s(4.45),247s(4.23), 298(4.13)	35-1672-69
Isoxazole, 4-benzoyl-3-methyl-5-styryl-	MeOH	229(4.12),237(4.13), 256(4.19),326(4.41)	32-0753-69
2-Penten-4-ynenitrile, 3-methoxy-2-(p-methoxyphenyl)-5-phenyl-	EtOH	208(4.12),230(4.22), 251(4.47)	39-0915-69C

Compound	Solvent	λ_{max}(log ε)	Ref.
$C_{19}H_{15}NO_4$			
Cyclopenta[ij]pyrido[2,1,6-de]quinolizine-3,4-dicarboxylic acid, 9-methyl-, dimethyl ester	EtOH	232(4.32),268(4.44), 304(4.29),334(4.15), 378(4.21),441(4.11), 615(2.7)	39-0239-69C
Indole-3-acrylic acid, α-hydroxy-, methyl ester, benzoate	EtOH	336(4.25)	104-0730-69
Pyrrole-2-carboxylic acid, 5-benzoyl-4-hydroxy-1-phenyl-, methyl ester	MeOH	260(4.11),305(3.99)	78-0527-69
$C_{19}H_{15}N_3O_2$			
Benzaldehyde, m-nitro-, diphenylhydrazone	EtOH	203(4.6),231(4.4), 275(4.0),344(4.4)	28-0137-69B
Benzaldehyde, o-nitro-, diphenylhydrazone	EtOH	226(4.3),288(4.0), 336(4.1),391(3.9)	28-0137-69B
Benzaldehyde, p-nitro-, diphenylhydrazone	EtOH	238s(4.0),264(4.2), 415(4.3)	28-0137-69B
Benzophenone, 4-nitro-, phenylhydrazone	EtOH	256(4.41),332(4.12)	2-0412-69
Carbazol-2-ol, 1-[(p-methoxyphenyl)-azo]-	EtOH	230(4.57),244(4.54), 292(4.53),316(4.36), 330(4.36),420(4.57)	2-1065-69
Carbazol-2-ol, 3-[(p-methoxyphenyl)-azo]-	EtOH	224(4.57),241(4.54), 296(4.11),316(4.18), 332(4.0),432(4.32)	2-1065-69
$C_{19}H_{15}N_3O_3$			
4H-Quinolizine-1-carboxylic acid, 2-(benzylamino)-3-cyano-4-oxo-	EtOH	266(4.70),303(4.27), 392(4.15)	95-0203-69
$C_{19}H_{15}N_3S$			
Quinoline, 2-(2-amino-4-methyl-5-thiazolyl)-4-phenyl-	n.s.g.	372(4.30)	97-0186-69
$C_{19}H_{15}OP$			
Dibenzo[b,d]phosphorin, 5,6-dihydro-5-phenyl-, 5-oxide	EtOH	213(4.54),271(4.06), 286(3.84)	39-0252-69C
$C_{19}H_{16}$			
Methane, triphenyl-	C_6H_{12}	262(2.89)	101-0017-69B
$C_{19}H_{16}AsClIN$			
10-(o-Chlorophenyl)-10-methylphenarsazinium iodide	EtOH	275(4.19),331(4.00)	101-0117-69E
$C_{19}H_{16}AsN$			
5H-Dibenz[b,e][1,4]azarsepine, 10,11-dihydro-5-phenyl-	EtOH	240(4.09),320(3.52)	101-0117-69E
$C_{19}H_{16}ClN$			
Benz[c]acridine, 7-chloro-5,6-dihydro-5,5-dimethyl-	isooctane	228(4.43),260(4.57), 268(4.66),290s(4.02), 302(4.04),315(4.01), 329(4.06),344(4.12)	4-0361-69
	MeOH	229(4.43),261s(4.52), 268(4.62),303(4.00), 317(4.00),331(4.07), 346(4.12)	4-0361-69
3,4-Lutidine, 5-chloro-2,6-diphenyl-	n.s.g.	214(4.29),232(4.20), 287(3.82)	39-2249-69C

Compound	Solvent	λ_{max}(log ϵ)	Ref.
$C_{19}H_{16}ClNO_2$			
Indole-2-carboxylic acid, 5-chloro-1-(cyclopropylmethyl)-3-phenyl-	MeOH	242(4.50),303(4.08)	94-1263-69
$C_{19}H_{16}Cl_2N_2O_2S$			
4H-Thiopyran-4-one, 3,5-dichloro-2,6-bis(p-methoxyphenyl)-, hydrazone	EtOH	270(4.27),327(4.13)	39-0315-69C
$C_{19}H_{16}FeO$			
Acrylophenone, 3-ferrocenyl-	MeOH	325(4.23)	73-2771-69
	EtOH	266(4.13),326(4.20)	73-2235-69
	n.s.g.	265(4.2),315(4.3),385(3.6),485(3.5)	65-1008-69
2-Propen-1-one, 1-ferrocenyl-3-phenyl-	MeOH	306(4.38)	73-2771-69
	EtOH	306(4.38),390(3.39)	73-2235-69
$C_{19}H_{16}FeO_2S$			
2-Propen-1-one, 1-(1'-acetylferrocenyl)-3-(2-thienyl)-	EtOH	284(4.19),345(4.41)	73-2235-69
2-Propen-1-one, 3-(1'-acetylferrocenyl)-1-(2-thienyl)-	EtOH	276(4.19),322(4.35)	73-2235-69
2-Propen-1-one, 1-(5-acetyl-2-thienyl)-3-ferrocenyl-	EtOH	234(4.09),340(4.33)	73-2235-69
2-Propen-1-one, 3-(5-acetyl-2-thienyl)-1-ferrocenyl-	EtOH	243(4.03),348(4.59)	73-2235-69
$C_{19}H_{16}FeO_3$			
2-Propen-1-one, 1-(1'-acetylferrocenyl)-3-(2-furyl)-	heptane	323(4.77)	73-2235-69
	EtOH	346(4.46)	73-2235-69
2-Propen-1-one, 3-(1'-acetylferrocenyl)-1-(2-furyl)-	heptane	260(4.34),315(4.78)	73-2235-69
	EtOH	286(4.38),343(4.35)	73-2235-69
2-Propen-1-one, 3-(2-acetylferrocenyl)-1-(2-furyl)-	heptane	273(4.51),305(4.58)	73-2235-69
	EtOH	282(4.26),315(4.20)	73-2235-69
2-Propen-1-one, 1-(5-acetyl-2-furyl)-3-ferrocenyl-	heptane	320(4.65)	73-2235-69
	EtOH	302(4.12),343(4.19)	73-2235-69
2-Propen-1-one, 3-(5-acetyl-2-furyl)-1-ferrocenyl-	heptane	329(5.16)	73-2235-69
	EtOH	237(4.10),343(4.40)	73-2235-69
$C_{19}H_{16}NOPS$			
Formanilide, 1-(diphenylphosphinyl)thio-	EtOH	330(4.11),443(2.08)	18-2975-69
$C_{19}H_{16}NPS_2$			
Formanilide, 1-(diphenylphosphinothioyl)thio-	EtOH	333(4.05),447(4.00)	18-2975-69
$C_{19}H_{16}N_2$			
Benzaldehyde, diphenylhydrazone	EtOH	236(4.3),300s(4.0),341(4.3)	28-0137-69B
Benzophenone, phenylhydrazone	C_6H_{12}	240(4.21),297(4.05),350(4.30)	39-1703-69C
	EtOH	237(4.27),298(4.06),340(4.31)	39-1703-69C
	EtOH-HCl	228(4.08),272(4.15),384(3.49)	39-1703-69C
Indole, 3-methyl-2-[(2-methyl-3H-indol-3-ylidene)methyl]-, perchlorate	MeOH	535(3.45)	24-1347-69
$C_{19}H_{16}N_2O$			
Acetic acid, (1-naphthylmethylene)phenylhydrazide	EtOH	230(4.3),260s(4.0),312s(--),324(--),356(3.3)	28-1703-69A

Compound	Solvent	λ_{max}(log ε)	Ref.
Acetic acid, (2-naphthylmethylene)phenylhydrazide	EtOH	227s(4.3),235(4.4), 244(4.4),262(4.4), 270(4.5),286s(4.3), 298(4.5),309(4.5), 324s(3.8),342(3.4)	28-1703-69A
3-Indolinone, 2-allyl-2-indol-3-yl-	EtOH	221(4.69),258(4.04), 280(3.88),288(3.78), 400(3.58)	39-0595-69C
2-Indolinone, 3-[(1,3-dimethylindol-2-yl)methylene]-	MeOH	415(4.09)	24-1347-69
2-Indolinone, 1-methyl-3-[(1-methylindol-3-yl)methylene]-	MeOH	410(4.28)	24-1347-69
1-Naphthol, 2-(5-phenyl-2-pyrazolin-3-yl)-	$CHCl_3$	271(4.48),282(4.55), 358(3.96)	115-0001-69
4H-Naphtho[1,2-b]pyran-4-one, 2,3-dihydro-2-phenyl-, hydrazone	$CHCl_3$	295(4.39),412(4.18)	115-0001-69
Quinoline, 3-(3,4-dihydro-1-isoquinolyl)-4-methoxy-	EtOH	225(4.61)	2-1010-69
Salicylaldehyde, diphenylhydrazone	EtOH	203(4.6),236(4.2), 299(4.0),344(4.3)	28-0137-69B
$C_{19}H_{16}N_2O_2$			
7H-Pyrido[2,3-c]carbazole-1-carboxylic acid, 3,5-dimethyl-, methyl ester	EtOH	225(4.52),249(4.47), 299(4.45),362(3.95)	12-0185-69
$C_{19}H_{16}N_2O_2S$			
2H-1,2,4-Benzothiadiazine, 3,4-dihydro-2,3-diphenyl-, 1,1-dioxide	EtOH	208(4.64),254(4.11), 312(3.66)	7-0590-69
$C_{19}H_{16}N_2O_3$			
Acetanilide, 2'-(1,2-dihydro-4-methyl-2-oxo-3-quinolyl)carbonyl]-	EtOH	230(4.68),270(4.28), 330(4.08)	18-2952-69
Acetic acid, phenyl-, 2-(1,4-dihydro-3-methyl-1,4-dioxo-2-naphthyl)hydrazide	alkali	622(3.70)	95-0007-69
Isoquinaldonitrile, 1,2-dihydro-2-(2,3-dimethoxybenzoyl)-	EtOH	228(4.43),290(4.19)	78-1881-69
Isoquinaldonitrile, 1,2-dihydro-2-(3,4-dimethoxybenzoyl)-	EtOH	228(4.41),296(4.18)	78-1881-69
$C_{19}H_{16}N_2O_3S$			
Spiro[2H-1-benzopyran-2,2'-benzothiazoline], 3'-allyl-3-methyl-6-nitro-	$(MeOCH_2)_2$	253s(--),267(4.36), 300(4.03),335s(--)	78-3251-69
$C_{19}H_{16}N_2O_4$			
2,4-Biphenyldicarboxylic acid, 6-pyrazol-1-yl-, dimethyl ester	EtOH	225(4.35),296(3.39)	44-4134-69
$C_{19}H_{16}N_2S$			
4H-Cyclopenta[d]pyrimidine-4-thione, 1,5,6,7-tetrahydro-1,2-diphenyl-	EtOH	253(4.02),345(4.28)	4-0037-69
$C_{19}H_{16}N_4$			
Benzaldehyde, [3-(benzylideneamino)-2-pyridyl]hydrazone	EtOH	272(4.00),368(4.28)	39-1758-69C
$C_{19}H_{16}N_4O$			
1H-Pyrazolo[3,4-b]pyrazin-6-ol, 5-ethyl-1,3-diphenyl-	EtOH	245(4.25),271(4.35), 317s(4.16)	32-0463-69

Compound	Solvent	λ_{max}(log ε)	Ref.
$C_{19}H_{16}N_4O_2$			
Propane, 1,3-bis(3-oxoindolinylidene-amino)-	MeOH	204(4.64),246(4.59), 251(4.56),292(3.89), 391(3.35)	39-2044-69C
$C_{19}H_{16}N_4O_4$			
Pyrazole-4,5-dione, 1,3-diphenyl-, bis-(O-acetyloxime)	EtOH	226(4.11),267(4.27), 279s(4.12),387(3.74)	4-0317-69
$C_{19}H_{16}O$			
2,4,6-Heptatrienal, 2,7-diphenyl-, (Z,E,E)-	CH_2Cl_2	247(4.26),255s(4.23), 370(5.02)	24-0623-69
Naphthalene, 1-(p-methoxystyryl)-	DMF	339(4.39)	33-2521-69
Naphthalene, 2-(p-methoxystyryl)-	DMF	288(4.41),331(4.62)	33-2521-69
Phenanthro[1,2-b]furan, 1-ethyl-2-methyl-	C_6H_{12}	201(4.38),235(4.26), 263(4.79),272(4.98), 300(4.40),328(3.13), 335(2.97),343(3.30), 351(2.90),360(3.32)	33-1461-69
4-Pyrenecarboxaldehyde, 10b,10c-dihydro-10b,10c-dimethyl-, trans	C_6H_{12}	369(4.83),394(4.34), 429(3.86),447(3.85), 470(3.57),559(2.42), 613(2.69),670(3.19), 679(3.33)	35-0902-69
$C_{19}H_{16}O_2$			
Cyclopenta[def]cyclopropa[l]phenanthrene-8-carboxylic acid, 4,7b,8,8a-tetrahydro-, ethyl ester, endo	EtOH	216(4.58),229(4.57), 236s(4.49),278(4.15)	39-1427-69C
exo	EtOH	216(4.59),229(4.57), 236s(4.49),278(4.15)	39-1427-69C
17H-Cyclopenta[a]phenanthren-17-one, 15,16-dihydro-6-methoxy-11-methyl-	EtOH	265(4.76),289s(4.51), 306s(4.27),368(3.37), 384(3.41)	39-2484-69C
1,3-Indandione, 2-(p-isopropylbenzylidene)-	C_6H_{12}	243(--),260(--), 340(--),354(--), 370(--),400s(--), 430s(--)	108-0099-69
	MeCN	244(4.36),261(4.24), 352(4.54),420s(3.13)	108-0099-69
1,3-Indandione, 2-(2-methallyl)-2-phenyl-	EtOH	226(3.56),336(3.30)	104-2090-69
4-Pyrenecarboxylic acid, 10b,10c-dihydro-10b,10c-dimethyl-, trans	EtOH	343(4.67),384(4.61), 460(3.75),472(3.75), 538(2.37),593(2.52), 640s(2.93),651(3.19)	35-0902-69
$C_{19}H_{16}O_2S$			
2-Thiophenecarboxylic acid, 3,4-diphenyl-, ethyl ester	EtOH	237(4.35),270s(3.91)	2-0009-69
2-Thiophenecarboxylic acid, 3,5-diphenyl-, ethyl ester	EtOH	230(4.14),256(4.22), 313(4.33)	2-0009-69
$C_{19}H_{16}O_4$			
Benzofuran, 5-allyl-7-methoxy-2-[3,4-(methylenedioxy)phenyl]-	EtOH	220(4.58),302(4.47), 317(4.55),330(4.43)	12-1329-69
5,11-Naphthacenedione, 5a,6,11a,12-tetrahydro-3-hydroxy-9-methoxy-	EtOH	224(4.64),253(4.34), 322(3.81)	2-0561-69
Xanthen-9-one, 2-allyl-3-(allyloxy)-1-hydroxy-	MeOH	240(4.52),306(4.25), 350(3.72)	2-1182-69

Compound	Solvent	$\lambda_{max}(\log \epsilon)$	Ref.
Xanthen-9-one, 4-allyl-3-(allyloxy)-1-hydroxy-	MeOH	235(4.51),258(4.48), 310(4.19)	2-1182-69
Xanthen-9-one, 2,4-diallyl-1,3-dihydroxy-	MeOH	236(4.49),260(4.37), 313(4.16)	2-1182-69
$C_{19}H_{16}O_5$			
Furano[2',3':3,4]pterocarpan, 8,9-dimethoxy-	EtOH	248(4.17),256(4.13), 293(3.90)	18-1693-69
	$CHCl_3$	251(4.12),256(4.12), 293(3.89),300s(3.86)	18-1693-69
Furano[3',2':2,3]pterocarpan, 8,9-dimethoxy-	EtOH	247(4.24),255(4.12), 299(4.12)	18-1693-69
	$CHCl_3$	250(4.05),257(4.06), 302(4.10)	18-1693-69
1,3-Indandione, 2-(3,4,5-trimethoxybenzylidene)-	C_6H_{12}	250(--),262(--), 420(--)	108-0099-69
	MeCN	250(4.35),263(4.38), 389(4.47)	108-0099-69
2-Naphthoic acid, 1,2,3,4-tetrahydro-6-hydroxy-3-(p-methoxybenzylidene)-4-oxo-	EtOH	229(4.32),334(4.21)	2-0561-69
$C_{19}H_{16}O_6$			
Flavone, 5-acetoxy-4',7-dimethoxy-	EtOH	248(4.25),257(4.26), 321(4.53)	105-0397-69
5H-Furo[3,2-g][1]benzopyran-5-one, 2,3-dihydro-6-(2-hydroxy-4,5-dimethoxyphenyl)-	EtOH	264(3.92),306(4.10)	18-1693-69
$C_{19}H_{16}O_7$			
Anthraquinone, 3-acetoxy-1-hydroxy-6,8-dimethoxy-2-methyl-	EtOH	226(4.63),249s(4.17), 272(4.34),423(4.03), 434s(4.02)	39-2763-69C
Derrusnin	EtOH	242s(4.12),287s(3.85), 335(4.28)	39-0365-69C
Isoflavone, 2',6,7-trimethoxy-4',5'-(methylenedioxy)-	$CHCl_3$	257(4.22),312(4.29)	39-1787-69C
Isoflavone, 3',6,7-trimethoxy-4',5'-(methylenedioxy)-	$CHCl_3$	272(4.34),319(4.05)	39-1787-69C
Pterocarpan, 3-acetoxy-4-methoxy-8,9-(methylenedioxy)-	EtOH	307(3.82)	39-1109-69C
$C_{19}H_{16}O_9$			
1H-Naphtho[2,3-c]pyran-1,6,9-trione, 3,4-dihydro-7,10-dihydroxy-8-methoxy-3-methyl-, diacetate	EtOH	212(4.15),260(4.11), 284s(3.73),344(3.33)	23-1561-69
$C_{19}H_{16}S$			
Benzo[b]naphtho[2,3-d]thiophene, 9,10-dihydro-6,7,8-trimethyl-	EtOH	230(4.40),243(4.60), 265(4.60),276(5.03), 283(4.38),318(3.71), 333(3.67),350(3.31), 367(3.25)	4-0885-69
$C_{19}H_{17}AsIN$			
10-Methyl-10-phenylphenarsazinium iodide	EtOH	278(4.26),308(4.09), 330(3.90)	101-0117-69E
$C_{19}H_{17}BrO$			
Cyclopentanone, 2-benzyl-5-(p-bromobenzylidene)-, trans	EtOH	305(4.47)	39-1868-69C

Compound	Solvent	$\lambda_{max}(\log \epsilon)$	Ref.
$C_{19}H_{17}ClN_2O$			
2H-1,4-Benzodiazepin-2-one, 7-chloro-1-(cyclopropylmethyl)-1,3-dihydro-5-phenyl-	MeOH	229(4.48),314(3.28)	94-1263-69
Indole-2-carboxamide, 5-chloro-1-(cyclopropylmethyl)-3-phenyl-	MeOH	239(4.46),301(4.01)	94-1263-69
$C_{19}H_{17}ClN_2O_2$			
Propionanilide, 2'-benzoyl-4'-chloro-2-cyano-N,2-dimethyl-	MeOH	253(4.17)	36-0830-69
$C_{19}H_{17}ClO_7$			
Flavone, 3'-chloro-5-hydroxy-2',3,7,8-tetramethoxy-	EtOH	266(4.46),305s(3.74), 353(3.80)	39-2418-69C
	EtOH-$AlCl_3$	280(4.44),320s(3.74), 418(3.73)	39-2418-69C
$C_{19}H_{17}Cl_2FN_4O_4$			
9H-Purine, 9-(5-O-benzyl-2-deoxy-2-fluoro-α-D-arabinofuranosyl)-2,6-dichloro-, 3'-acetate	EtOH	253s(3.72),274(3.95)	44-2632-69
β-form	EtOH	254s(3.68),275(3.89)	44-2632-69
$C_{19}H_{17}Cl_2N$			
2H-Pyrrole, 2-(dichloromethyl)-3,4-dimethyl-2,5-diphenyl-	n.s.g.	209(4.15),248(3.98)	39-2249-69C
$C_{19}H_{17}Cl_2NO_5$			
[1-Chloro-2-(2-phenyl-4H-1-benzopyran-4-ylidene)ethylidene]dimethylammonium perchlorate	MeCN	242(4.28),268s(3.90), 340(4.09),445(4.57), 461(4.57)	4-0803-69
$C_{19}H_{17}N$			
1-Naphthylamine, N-(p,α-dimethylbenzylidene)-	EtOH	232(4.53),254(4.23), 287(3.97),330s(3.50)	56-0749-69
2-Naphthylamine, N-(p,α-dimethylbenzylidene)-	EtOH	226(4.63),254(4.49), 317(3.62)	56-0749-69
p-Toluidine, N-[1-(2-naphthyl)ethylidene]-	EtOH	250(4.70),282(4.18), 320(3.65)	56-0749-69
$C_{19}H_{17}NO$			
p-Anisidine, N-[1-(2-naphthyl)ethylidene]-	EtOH	250(4.68),283(4.18), 329(3.71)	56-0749-69
Benz[c]acridan-7-one, 5,6-dihydro-5,5-dimethyl-	MeOH	265(4.48),273(4.52), 305s(3.87),317(4.05), 337s(3.92),350(4.05), 365(3.95)	4-0361-69
Carbazol-1(2H)-one, 9-benzyl-3,4-dihydro-	MeOH	240(4.23),310(4.28)	24-1198-69
1-Naphthylamine, N-(p-methoxy-α-methylbenzylidene)-	EtOH	273(4.23),290(4.18), 338s(3.54)	56-0749-69
2-Naphthylamine, N-(p-methoxy-α-methylbenzylidene)-	EtOH	228(4.73),272(4.49), 314s(3.81)	56-0749-69
4(1H)-Pyridone, 2-phenethyl-6-phenyl-	EtOH	241(4.57),261(4.40)	18-2389-69
$C_{19}H_{17}NOS$			
2-Propen-1-one, 1-benzo[b]thien-3-yl-3-[p-(dimethylamino)phenyl]-	EtOH	255(4.41),412(4.12)	2-0311-69

Compound	Solvent	λ_{max}(log ϵ)	Ref.
$C_{19}H_{17}NO_2$			
4H-1-Benzopyran-$\Delta^{4,\alpha}$-acetamide, N,N-dimethyl-2-phenyl-	MeCN	240(4.45),283(4.28), 360(4.29)	4-0803-69
Phenanthrene, 9-(2-nitro-1-pentenyl)-	EtOH	338(3.72)	87-0157-69
Spiro[azetidine-2,2'-indan]-1',4-dione, 1,3-dimethyl-3-phenyl-	EtOH	250(4.12),293(3.26)	28-2034-69A
$C_{19}H_{17}NO_2S$			
2-Pyrrolidinethione, 3-(α-hydroxybenzylidene)-1-methyl-, benzoate	MeOH	231(4.57),283(4.21)	5-0073-69E
$C_{19}H_{17}NO_3$			
Glutaconimide, 4-(p-methoxyphenyl)-N-methyl-3-phenyl-	EtOH	212(4.12),227(4.23), 289(4.27)	39-0915-69C
$C_{19}H_{17}NO_4$			
Isoindole-1,3-dicarboxylic acid, 2-benzyl-, dimethyl ester	EtOH	223(4.26),251(4.58), 350(4.25),366(4.26)	32-1115-69
$C_{19}H_{17}NO_5$			
Spiro[7H-indeno[4,5-d]-1,3-dioxole-7,1'(2'H)-isoquinolin]-6(8H)-one, 3',4'-dihydro-6'-hydroxy-7'-methoxy-	EtOH	236(4.46),290(4.08), 315s(3.85)	23-2501-69
$C_{19}H_{17}NO_7$			
3H-Phenoxazine-1,6-dicarboxylic acid, 2-methoxy-4,9-dimethyl-3-oxo-, dimethyl ester	$CHCl_3$	390(4.02),450(3.67)	24-3205-69
$C_{19}H_{17}NS$			
1H-Phenothiazine, 2,3-dihydro-4-methyl-2-phenyl-	MeOH	385(3.57)	124-1278-69
	MeOH-$HClO_4$	458(3.59)	124-1278-69
$C_{19}H_{17}N_3O_2$			
Pyrazolo[3,4-b]indole-3-carboxylic acid, 1,8-dihydro-8-methyl-1-phenyl-, ethyl ester	EtOH	221(4.59),265(4.34), 302(4.12)	32-0588-69
$C_{19}H_{17}N_9O_2$			
Isoalloxazine, 10-[2-(6-amino-9H-purin-9-yl)ethyl]-7,8-dimethyl-	propylene glycol	268(4.56),355(3.83), 445(4.01)	44-3240-69
$C_{19}H_{17}O_2P$			
Phosphine oxide, methyl(o-phenoxyphenyl)phenyl-	EtOH	223(4.24),280(3.59)	39-0252-69C
$C_{19}H_{18}$			
Cyclohexene, 6-methylene-3,3-diphenyl-	EtOH	232(4.49),238(4.50), 263s(3.24),270s(3.06)	35-5307-69
Pyrene, 10b,10c-dihydro-4,10b,10c-trimethyl-, trans	hexane	339(4.94),353s(4.35), 379(4.62),429s(3.59), 453s(3.74),468(3.79), 538(1.90),598(2.19), 613(2.30),629(2.35), 643(2.55)	35-0902-69
$C_{19}H_{18}AsI$			
Methyltriphenylarsonium iodide	n.s.g.	228(4.3),252(3.2), 258(3.4),264(3.5), 270(3.4)	120-0195-69

Compound	Solvent	$\lambda_{max}(\log \epsilon)$	Ref.
$C_{19}H_{18}BrNO_4$			
1,4-Benzoxazepine, 7-bromo-8,9-dimethoxy-5-(p-methoxybenzyl)-	MeOH	220(4.49),252s(3.99), 280(3.85),355(3.99)	95-1048-69
$C_{19}H_{18}Br_2N_4$			
2-Cyclohexen-1-one, 3-[(p-bromophenyl)-azo]-2-methyl-, (p-bromophenyl)hydrazone	$CHCl_3$	320(4.14),443(4.65)	24-1379-69
$C_{19}H_{18}ClN$			
4-Cinnamyl-2-methylisoquinolinium chloride	EtOH	208(4.29),230(4.17), 305(3.98)	78-0101-69
$C_{19}H_{18}ClNO_2$			
Isoquinoline, 1-(p-chlorostyryl)-3,4-dihydro-6,7-dimethoxy-, hydrochloride	EtOH-HCl	221(4.35),254(4.79), 313(3.97),329s(3.88)	39-0094-69C
$C_{19}H_{18}ClNO_4S$			
Dimethyl[2-(2-phenyl-4H-1-benzothiopyran-4-ylidene)ethylidene]ammonium perchlorate	MeCN	244(4.36),264(4.21), 284(3.92),326(4.12), 464(4.55)	44-2736-69
$C_{19}H_{18}ClNO_5$			
Dimethyl[2-(2-phenyl-4H-1-benzopyran-4-ylidene)ethylidene]ammonium perchlorate	MeCN	241(4.29),268(3.92), 320(4.15),428(4.60), 450(4.50)	4-0803-69 +44-2736-69
$C_{19}H_{18}Cl_2N_4$			
2-Cyclohexen-1-one, 3-[(p-chlorophenyl)-azo]-2-methyl-, (p-chlorophenyl)-hydrazone	$CHCl_3$	319(4.17),442(4.59)	24-1379-69
$C_{19}H_{18}Fe$			
Ferrocene, cinnamyl-, (E)-	EtOH	251(4.33)	101-0361-69C
$C_{19}H_{18}I_2N_4$			
2-Cyclohexen-1-one, 3-[(p-iodophenyl)-azo]-2-methyl-, (p-iodophenyl)-hydrazone	$CHCl_3$	324(4.19),450(4.67)	24-1379-69
$C_{19}H_{18}N_2$			
Benz[c]acridine, 7-amino-5,6-dihydro-5,5-dimethyl-	MeOH	222s(4.28),270(4.57), 276s(4.56),314s(3.97), 325(4.00),346s(3.85), 362(3.64)	4-0361-69
Diphenylamine, 4-(2-aminobenzyl)-	EtOH	230(4.12),289(4.41)	78-0847-69
6H-Pyrido[4,3-b]carbazole, 3,5,9,11-tetramethyl-	EtOH	246(4.34),281s(4.64), 291(4.79),299(4.84), 336(3.71)	12-0185-69
6H-Pyrido[4,3-b]carbazole, 5,7,10,11-tetramethyl-	EtOH	244(4.46),285(4.79), 293(4.95),318s(3.78), 337s(3.63)	12-0185-69
$C_{19}H_{18}N_2OS$			
Benzamide, N-[(2-anilino-1-cyclopenten-1-yl)thiocarbonyl]-	$CHCl_3$	303(4.07),425(4.18)	4-0037-69
$C_{19}H_{18}N_2O_2$			
Carbamic acid, (2-benzyl-1-cyano-3-phenylpropenyl)-, methyl ester	EtOH	233(4.12)	22-1724-69

Compound	Solvent	λ_{max}(log ε)	Ref.
Hydantoin, 5-butylidene-1,3-diphenyl-	EtOH	221(4.40)	94-2436-69
trans isomer	EtOH	221(4.35)	94-2436-69
Pyrazole-3-propionic acid, 4,5-diphenyl-, methyl ester	EtOH	220(4.08),250(3.88)	44-0670-69
3H-Pyrazole-3-propionic acid, 3,5-diphenyl-, methyl ester	EtOH	273(4.07),280s(3.25)	44-0670-69
$C_{19}H_{18}N_2O_3$			
1,5-Benzazepine, 2,3-dihydro-2,2-dimethyl-4-(p-nitrostyryl)-	MeOH	300(4.60)	124-1178-69
$C_{19}H_{18}N_2O_3S$			
Spiro[2H-1-benzopyran-2,2'-benzothiazoline], 3'-isopropyl-3-methyl-6-nitro-	$(MeOCH_2)_2$	254s(--),268(4.34), 303(4.01),334s(--)	78-3251-69
Spiro[2H-1-benzopyran-2,2'-benzothiazoline], 3-methyl-6-nitro-3'-propyl-	$(MeOCH_2)_2$	253s(--),269(4.33), 303(4.00),338s(--)	78-3251-69
$C_{19}H_{18}N_2O_4$			
Anthranilic acid, N-(α-acetamidocinnamoyl)-, methyl ester	EtOH	292(4.36),318(4.32)	35-6083-69
Cinnamic acid, α-(o-acetamido-N-methylbenzamido)-	n.s.g.	245(4.21),280s(4.13)	44-1359-69
Cyclopenol, N,O-dimethyl-	EtOH	284(3.52)	44-1359-69
4,8-Diazatricyclo[5.3.2.0^{2,6}]dodeca-9,11-diene-8-carboxylic acid, 3,5-dioxo-4-phenyl-, ethyl ester	EtOH	215(4.20),248(3.89)	44-2888-69
$C_{19}H_{18}N_2O_4S$			
Spiro[2H-1-benzopyran-2,2'-benzothiazoline], 3-ethyl-8-methoxy-3'-methyl-6-nitro-	EtOH	216(4.54),251(4.27), 285s(3.97),344(3.87)	22-3329-69
$C_{19}H_{18}N_2O_5$			
Pyruvanilide, 3-hydroxy-2'-(methylcarbamoyl)-3-phenyl-, acetate	EtOH	252(4.05),300(3.62)	35-6083-69
$C_{19}H_{18}N_2S$			
4(1H)-Pyrimidinethione, 2,5-diphenyl-1-propyl-	EtOH	240(4.17),348(4.23)	4-0037-69
$C_{19}H_{18}N_4O_4$			
7,8-Benzobicyclo[4.2.1]non-7-en-9-one, 2,4-dinitrophenylhydrazone	EtOH	362(4.30)	88-1761-69
$C_{19}H_{18}N_6O_4$			
2-Cyclohexen-1-one, 2-methyl-3-[(p-nitrophenyl)azo]-, (p-nitrophenyl)-hydrazone	$CHCl_3$	467(4.69)	24-1379-69
$C_{19}H_{18}O$			
Cyclopentanone, 2-benzyl-5-benzylidene-, trans	EtOH	296(4.45)	39-1868-69C
2-Cyclopenten-1-one, 2,5-dibenzyl-	EtOH	226s(4.04)	39-1868-69C
$C_{19}H_{18}OS$			
2-Propanone, 1-(10,11-dihydro-5H-dibenzo[a,d]cyclohepten-5-ylidene)-1-(methylthio)-	hexane	282(3.66),310s(3.52)	54-0465-69

Compound	Solvent	λmax(log ε)	Ref.
$C_{19}H_{18}O_2$			
Cyclopropanecarboxylic acid, 2-phenyl-2-styryl-, methyl ester, cis	EtOH	262(4.33),287s(3.69), 296(3.36)	44-0670-69
trans	EtOH	261(4.44),286s(3.85), 296(3.61)	44-0670-69
$C_{19}H_{18}O_2S$			
2-Thiophenecarboxylic acid, 2,5-dihydro-3,5-diphenyl-, ethyl ester	EtOH	252(4.25)	2-0009-69
$C_{19}H_{18}O_3$			
Acetophenone, 2-(3,3-dimethyl-1-phthalanylidene)-4'-methyl-	EtOH	235(4.12),340(4.54)	104-0543-69
Isotanshinone II	EtOH	227(4.02),253(4.33), 256(4.34),303(3.59), 361(3.63)	88-0301-69
Phenanthro[3,2-b]furan-7,11-dione, 1,2,3,4-tetrahydro-4,4,8-trimethyl-	EtOH	252s(4.80),256(4.81), 270s(4.22),305(4.02), 360(4.08)	18-3318-69
1-Phthalanol, 1-[(p-methoxyphenyl)ethynyl]-3,3-dimethyl-	EtOH	255(4.34)	104-0543-69
Tanshinone II	EtOH	224(4.34),252(4.30), 269(4.44),352(3.22), 460(3.43)	88-0301-69
$C_{19}H_{18}O_4$			
Cinnamic acid, α-benzoyl-β-methoxy-, ethyl ester	MeOH	253(4.36)	22-3281-69
2(3H)-Furanone, 3-[bis(p-methoxyphenyl)-methylene]dihydro-	EtOH	231(4.29),309(4.21)	44-3792-69
$C_{19}H_{18}O_5$			
3,6(2H,5H)-Benzofurandione, 4-methoxy-2-(p-methoxybenzylidene)-5,5-dimethyl-	MeOH	308(3.65),375(4.2)	2-0540-69
3(2H)-Benzofuranone, 4,6-dimethoxy-2-(p-methoxybenzylidene)-5-methyl-	MeOH	255(4.2),360(4.5)	2-0540-69
Furano[3',2':2,3]pterocarpan, 4',5'-dihydro-8,9-dimethoxy-	EtOH	294(4.19),298(4.19)	18-1693-69
2-Naphthoic acid, 1,2,3,4-tetrahydro-6-hydroxy-3-(p-methoxybenzyl)-4-oxo-	EtOH	224(4.48),253(3.98), 323(3.47)	2-0561-69
Osajaxanthone, dihydro-, monomethyl ether	MeOH	235(4.50),262(4.53), 316(4.21),377(3.79)	78-2787-69
Xanthen-9-one, 1,3-dihydroxy-7-methoxy-2-(3-methyl-2-butenyl)-	MeOH	238(4.39),262(4.36), 314(4.06),370(3.65)	78-2787-69
$C_{19}H_{18}O_6$			
Anthraquinone, 1,2,6,8-tetramethoxy-3-methyl-	EtOH	223(4.34),276(4.36), 367(3.71)	23-0767-69
3H-Benz[e]indene-1-carboxylic acid, 4,8,9,9a-tetrahydro-5-hydroxy-2-isopropyl-6,9aβ-dimethyl-3,4,8-trioxo-	EtOH	216(4.28),325(3.95), 490(3.32)	33-1685-69
3(2H)-Benzofuranone, 4,6,7-trimethoxy-2-(p-methoxybenzylidene)-	MeOH	255(4.2),335(4.3), 405(4.5)	2-0543-69
Cladrastin, O-methyl-	EtOH	208(4.39),220(4.39), 263(4.33),317(3.97)	78-3887-69
Flavanone, 5-acetoxy-4',7-dimethoxy-	EtOH	224(4.17),275(4.35), 305s(3.93)	105-0397-69
Flavone, 2',5,6,6'-tetramethoxy-	MeOH	229(4.48),324(3.84)	2-0746-69
Flavone, 2',5,6',8-tetramethoxy-	EtOH	221(4.47),259(4.21), 334(3.74)	2-0746-69

Compound	Solvent	λ_{max}(log ϵ)	Ref.
Flavone, 3,4',5,7-tetramethoxy-	EtOH	266(4.38),337(4.40)	105-0397-69
Flavone, 3',4',5,7-tetramethoxy-	n.s.g.	268(4.39),330(4.25)	2-0940-69
Flavone, 3',4',5',7-tetramethoxy-	n.s.g.	209(4.80),311(4.30)	2-0940-69
Flavone, 4',5,6,7-tetramethoxy-	EtOH	265(4.23),370(4.50)	105-0156-69
Isoflavone, 3',4',6,7-tetramethoxy-	$CHCl_3$	266(4.40),319(4.08)	39-1787-69C
1,4,6,10-Phenanthrenetetrone, 4b,5-dihydro-3,9-dihydroxy-2-isopropyl-4b ,8-dimethyl-	EtOH	257(4.26),320(3.87), 410(3.47)	33-1685-69
1-Phthalanpropionic acid, 5,7-dimethoxy-3-oxo-1-phenyl-	MeOH	248(3.56),308(3.57)	22-2365-69
Pterocarpan, 3-acetoxy-4,9-dimethoxy-	EtOH	283(3.64)	39-1109-69C
Pterocarpan, 3-acetoxy-8,9-dimethoxy-	EtOH	288(3.76),300(3.83)	18-0233-69
Pterocarpin, dihydro-, acetate, (±)-	EtOH	284(3.85)	18-1408-69
α,2-Stilbenedicarboxylic acid, 3,5-dimethoxy-, α-methyl ester	MeOH	279(4.31)	44-3192-69
$C_{19}H_{18}O_7$			
2'-Acetonaphthone, 1',6',8'-trihydroxy-3'-methyl-, triacetate	EtOH	229(4.69),289(3.78)	88-0471-69
Benzoic acid, 3,5-dimethoxy-2,2'-carbonyldi-, dimethyl ester	MeOH	244(4.31),312(3.94)	22-2370-69
Cyclopenta[c]furo[3',2':4,5]furo[2,3-h]-[1]benzopyran-1,11-dione, 8-ethoxy-2,3,6a,8,9,9a-hexahydro-4-methoxy-	MeOH	226(4.17),266(4.08), 364(4.10)	25-0983-69
1,2-Dibenzofurandicarboxylic acid, 3,7-dimethoxy-9-methyl-, dimethyl ester	EtOH	229(4.33),255(4.23), 298s(3.85),305(3.95), 324(3.90)	88-4139-69
Flavone, 3'-hydroxy-2',5,7,8-tetramethoxy-	EtOH	268(4.22),325s(3.82)	2-0110-69
	EtOH-NaOEt	254(4.24),330(3.93)	2-0110-69
Flavone, 3'-hydroxy-4',5,6,7-tetramethoxy-	EtOH	243(4.29),263s(4.16), 331(4.32)	24-0112-69
Flavone, 5-hydroxy-2',3',6,7-tetramethoxy-	EtOH	270(4.12),312(3.96)	2-0110-69
	EtOH-$AlCl_3$	283(4.13),330(3.96)	2-0110-69
Flavone, 5-hydroxy-3',4',6,7-tetramethoxy-	EtOH	242(4.31),275(4.29), 339(4.44)	78-1603-69
Flavone, 6-hydroxy-3',4,4',6-tetramethoxy-	EtOH	241(4.30),278(4.33), 330(4.54)	40-1270-69
$C_{19}H_{18}O_8$			
Chrysosplenol B	EtOH	258(4.33),272(4.29), 351(4.35)	95-0702-69
Chrysosplenol E	EtOH	259(4.57),302(3.99), 354(4.33)	95-0129-69
	EtOH-$AlCl_3$	267(4.54),316(3.92), 367(4.26)	95-0129-69
Eupatoretin	EtOH	255(4.31),356(4.37)	44-1460-69
Flavone, 3,4'-dihydroxy-3',5,6,7-tetramethoxy-	EtOH	237s(4.24),255(4.33), 364(4.37)	18-1398-69
Flavone, 4',5-dihydroxy-3,3',6,7-tetramethoxy-	EtOH	258(4.28),272s(4.23), 353(4.37)	18-1398-69
	EtOH-NaOH	270(4.29),410(4.42)	18-1398-69
	EtOH-NaOAc	259(4.27),272s(4.23), 353(4.33),420(3.77)	18-1398-69
	EtOH-$AlCl_3$	269(4.25),282(4.27), 380(4.36)	18-1398-69
Flavonoid L	EtOH	255(4.31),356(4.37)	78-1603-69
1H,12H-Furo[3',2':4,5]furo[2,3-h]pyrano-[3,4-c][1]benzopyran-1,12-dione, 3,4,7a,9,10,10a-hexahydro-9-ethoxy-5-methoxy-	MeOH	223(4.23),244(4.11), 266(4.07),366(4.29)	25-0983-69

Compound	Solvent	λ_{max}(log ε)	Ref.
$C_{19}H_{18}O_9$			
Carbonic acid, methyl ester, 5-ester with methyl 5-hydroxy-2-(4-methyl-γ-resorcyloyl)-m-anisate	EtOH	221(4.30),301(4.10), 347(3.77)	39-1721-69C
Isocoumarin, 6,8-dihydroxy-3-methyl-7-(tetrahydro-4-methoxy-5-oxo-2-furyl)-, diacetate	EtOH	238(4.58),272(4.11), 282(4.81),329(3.69)	12-1933-69
$C_{19}H_{18}O_{11}$			
Mangiferin	EtOH	241(4.43),258(4.57), 316(4.20),366(4.17)	95-0410-69
Norswertianolin	EtOH	252(4.45),275(4.12), 332(4.15)	95-1276-69
	EtOH-NaOAc	248(4.39),275(4.24), 357(4.26)	95-1276-69
	EtOH-$AlCl_3$	254(4.48),267(4.47), 282(4.45),324(4.19), 362(4.36)	95-1276-69
$C_{19}H_{18}S$			
2H-Thiopyran, 3,5-dimethyl-2,6-diphenyl-	n.s.g.	217(4.4),275(3.5), 313(2.7)	104-1660-69
$C_{19}H_{19}BF_4OS$			
6-Methoxy-1,3,7-trimethyl-5-phenylcyclohepta[c]thiolium tetrafluoroborate	$C_2H_4Cl_2$	277(4.17),325(4.60), 397(3.90),595(3.34)	104-0947-69
$C_{19}H_{19}BrN_2O$			
1-Benzyl-2-methylimidazolinium 3-(p-bromophenacylide)	n.s.g.	245(4.28),312(4.18)	103-0638-69
$C_{19}H_{19}ClN_2$			
Indole, 2-(aminomethyl)-5-chloro-1-(cyclopropylmethyl)-3-phenyl-, hydrochloride	MeOH	232(4.56),238(4.58), 267(4.00)	94-1263-69
$C_{19}H_{19}ClN_2O$			
1,5-Benzodiazocin-2(1H)-one, 8-chloro-3,4-dihydro-1,3,3-trimethyl-6-phenyl-	MeOH	240(4.24)	36-0830-69
$C_{19}H_{19}ClN_4O_4$			
2-Heptenophenone, 2-chloro-, 2,4-dinitrophenylhydrazone	EtOH	374(4.46)	44-1474-69
$C_{19}H_{19}Cl_2NO$			
Acrylophenone, 3-[bis(2-chloroethyl)-amino]-4'-phenyl-	EtOH	207(4.17),217(4.06), 285(4.13),344(4.36)	24-3139-69
$C_{19}H_{19}Cl_2NO_2$			
Isoquinoline, 1-(p-chlorostyryl)-3,4-dihydro-6,7-dimethoxy-, hydrochloride	EtOH-HCl	221(4.35),254(4.79), 313(3.97),329s(3.88)	39-0094-69C
$C_{19}H_{19}IN_2O$			
Methyl[5-(N-methylanilino)furfurylidene]phenylammonium iodide	n.s.g.	400(3.64)	103-0434-69
$C_{19}H_{19}NOS$			
Formamide, N-(3-dibenzo[b,e]thiepin-11(6H)-ylidenepropyl)-N-methyl-	heptane	231(4.34),262(3.9), 306(3.35)	73-1963-69

Compound	Solvent	λ max (log ε)	Ref.
$C_{19}H_{19}NO_2$			
6H-Azepino[1,2-a]indol-6-one, 7,8,9,10,10a,11-hexahydro-10a-hydroxy-11-phenyl-	EtOH	255(4.20),281(3.59), 290(3.50)	94-1290-69
Spiro[naphthalene-2(1H),2'-oxiran]-1-one, 3'-(o-aminophenyl)-1,4-dihydro-4,4-dimethyl-	MeOH	247(4.23),259s(4.18),	4-0361-69
hydrochloride	MeOH	246(4.22),259s(4.16), 301(3.67)	4-0361-69
$C_{19}H_{19}NO_3$			
1H-1-Benzazepine-4-carboxylic acid, 2,3,4α,5-tetrahydro-1-methyl-2-oxo-5α-phenyl-, methyl ester	EtOH	242(4.0)	33-1929-69
Isoquinoline, 1-(2,4-dimethoxyphenyl)-3-ethoxy-	EtOH	226(4.60),282(3.76), 348(3.85)	39-1873-69C
	EtOH-HCl	230(3.89),268(3.97), 383(--)	39-1873-69C
$C_{19}H_{19}NO_4$			
Alkaloid J	EtOH	232(4.49),277(3.66), 313(3.70)	12-2219-69
Alkaloid K	EtOH	241(4.25),285(3.77)	12-2219-69
Amurine, (+)-	MeOH	240(4.24),280(3.95)	39-0801-69C
Domesticine, dl-	EtOH	221(4.56),283(4.01), 310(4.17)	2-0841-69
Phanostenine	EtOH	221(4.41),232s(--), 274s(--),282(3.67), 309(4.10),316s(--)	95-1691-69
Spiro[dibenz[cd,f]indole-1(10bH),2'-[1,3]dioxolane]-10b-carboxaldehyde, 2,3,3a,4,5,10c-hexahydro-3a-methyl-4-oxo-	n.s.g.	228(4.38),292(4.30)	88-1071-69
$C_{19}H_{19}NO_5$			
Luteoreticulin	EtOH	225(4.11),256s(3.78), 368(4.16)	88-0355-69
$C_{19}H_{19}NO_6$			
Hemanthidine, 6-acetyl-11-oxo-	EtOH	251(3.59),296(3.64)	35-0150-69
$C_{19}H_{19}NO_7$			
3-Carboxy-1-D-ribofuranosyl-4-picolinium hydroxide, inner salt, 5'-benzoate	MeOH	268(3.74)	39-0918-69C
$C_{19}H_{19}NO_8$			
9H-Pyrrolo[1,2-a]azepine-5,6,7,8-tetracarboxylic acid, 9-vinyl-, tetramethyl ester	MeOH	220(4.00),285(4.04), 330(3.60),412(3.70)	39-2316-69C
	$MeOH-HClO_4$	230(4.17),250s(4.04), 313(3.73)	39-2316-69C
$C_{19}H_{19}N_3O_2$			
Indole-3-glyoxylic acid, 1-methyl-, ethyl ester, α-(phenylhydrazone)	EtOH	232(4.52),299(3.99), 394(4.26)	32-0588-69
Pyrazolo[3,4-b]indole-3-carboxylic acid, 1,3a,8,8a-tetrahydro-8-methyl-1-phenyl-, ethyl ester	EtOH	248(4.52),303s(4.13), 340(4.30)	32-0588-69

Compound	Solvent	λ_{max}(log ϵ)	Ref.
$C_{19}H_{19}N_3O_3$			
Cinnamanilide, α-acetamido-2'-(methylcarbamoyl)-	EtOH	292(4.22),310s(4.20)	35-6083-69
$C_{19}H_{20}BF_2NO_2$			
Naphtho[1,2-e]-1,3,2-dioxaborin, 2,2-difluoro-4-[4-(diethylamino)butadien-1-yl]-	MeCN	240(4.13),315(3.81), 345(3.81),484(4.74), 512(4.76)	4-0029-69
$C_{19}H_{20}BF_4NO$			
6-Methoxy-1,3,5-trimethyl-N-phenylcyclohepta[c]pyrrolium tetrafluoroborate	$C_2H_4Cl_2$	311(4.90),349(4.16), 575(3.28)	104-0947-69
$C_{19}H_{20}BrNO$			
1-Indanone, 3-bromo-2-[α-(isopropylamino)benzyl]-, hydrobromide	$CHCl_3$	253(4.21),293(3.48)	44-0596-69
$C_{19}H_{20}Br_2O_7$			
Propiophenone, 2,3-dibromo-2'-hydroxy-3-(3-hydroxy-4-methoxyphenyl)-3',4',6'-trimethoxy-	CH_2Cl_2	314(4.36)	24-0112-69
$C_{19}H_{20}ClN$			
1,3,5,7-Tetramethyl-2-phenylcyclohepta[c]pyrrolium chloride	$C_2H_4Cl_2$	281(4.47),365(3.88), 712(3.25)	104-2014-69
$C_{19}H_{20}ClNO_2$			
Isoquinoline, 1-(o-chlorostyryl)-1,2,3,4-tetrahydro-6,7-dimethoxy-, hydrochloride	EtOH-HCl	204(4.75),218s(4.35), 261(4.44),287s(3.96), 298s(3.57)	39-0094-69C
Spiro[naphthalene-2(1H),2'-oxiran]-1-one, 3'-(o-aminophenyl)-1,4-dihydro-4,4-dimethyl-, hydrochloride	MeOH	246(4.22),259s(4.16), 301(3.67)	4-0361-69
$C_{19}H_{20}ClN_3$			
Phthalazine, 1-(p-chlorophenyl)-4-[3-(dimethylamino)propyl]-	EtOH	219(4.7),282(4.1)	33-1376-69
$C_{19}H_{20}Cl_2N_2$			
Indole, 2-(aminomethyl)-5-chloro-1-(cyclopropylmethyl)-3-phenyl-, hydrochloride	MeOH	232(4.56),238(4.58), 267(4.00)	94-1263-69
$C_{19}H_{20}INO_3$			
Michepressine, dl-	EtOH	266(4.07),276(4.13), 310(3.85)	2-0746-69
$C_{19}H_{20}INO_4$			
3,4-Dihydro-6,7-dimethoxy-2-methyl-1-[3,4-(methylenedioxy)phenyl]isoquinolinium iodide	EtOH	251(4.23),315(4.01), 367(4.00)	1-0244-69
$C_{19}H_{20}N_2$			
Benzo[g]quinoline, 4-(hexahydro-1H-azepin-1-yl)-, (salicylate)	EtOH	248(4.71),307(3.86), 415(3.95)	103-0223-69
Pyrazino[2,1-a]isoindole, 1,2,3,4-tetrahydro-1,1-dimethyl-6-phenyl-	dioxan	335(3.44),365(3.59)	44-0249-69

Compound	Solvent	$\lambda_{max}(\log \epsilon)$	Ref.
$C_{19}H_{20}N_2O$			
1,5-Benzoxazepine, 4-[p-(dimethylamino)-styryl]-2,3-dihydro-, hydrochloride	MeOH	520(5.18)	124-1178-69
1-Benzyl-2-methylimidazolinium 3-phenacylide	n.s.g.	235(4.13),306(4.15)	103-0638-69
3-Epimeloscine	EtOH	211(4.34),253(4.01), 279s(3.44),287s(3.18)	33-1886-69
Indole-2-carboxanilide, N-tert-butyl-3-phenyl-	EtOH	299(4.2)	24-0678-69
2H-Pyrrol-3-ol, 2-benzyl-5-(dimethyl-amino)-4-phenyl-	MeOH	225(4.08),290(4.19)	78-5721-69
$C_{19}H_{20}N_2OS$			
2-Thiohydantoin, 5-butyl-1,3-diphenyl-	EtOH	241(4.20),280(4.08)	94-2436-69
$C_{19}H_{20}N_2O_2$			
Hydantoin, 5-butyl-1,3-diphenyl-	EtOH	239(4.26)	94-2436-69
$C_{19}H_{20}N_2O_3$			
Acetophenone, 4'-(3-nitro-4-piperidino-phenyl)-	MeOH	414(3.21)	17-0145-69
$C_{19}H_{20}N_2O_3S_2$			
Morpholine, 4-[α-(methylthio)-β-[(o-nitrophenyl)thio]styryl]-	MeOH	400(3.60)	5-0073-69E
$C_{19}H_{20}N_2O_4$			
1,4-Phenazinedicarboxylic acid, 5,10-dihydro-7-methyl-, diethyl ester	EtOH	250(4.82),295(4.00), 478(3.89),508(4.01), 547(3.86)	33-0322-69
$C_{19}H_{20}N_2O_5$			
Glyceranilide, 2'-(methylcarbamoyl)-3-phenyl-, 3-acetate	EtOH	252(4.04),300(3.62)	35-6083-69
$C_{19}H_{20}N_2O_7$			
Uracil, 1-(3,5-O-isopropylidene-β-D-xylofuranosyl)-, 2'-benzoate	EtOH	230(4.31),260(4.17)	94-0798-69
$C_{19}H_{20}N_4$			
Anthranilonitrile, N-[α-(4-methyl-1-piperazinyl)benzylidene]-	EtOH	215(4.48),309(3.74)	22-2008-69
Propionitrile, 3,3'-[methylenebis-(p-phenyleneimino)]di-	EtOH	256(4.49),298(3.64)	103-0623-69
Pyrimidine, 2,4-bis(benzylamino)-6-methyl-	EtOH	244(4.49)	70-2601-69
$C_{19}H_{20}N_4O_3$			
Indole, 1-acetyl-3-[2-nitro-1-(2-phenyl-hydrazino)propyl]-	EtOH	239(4.42),290(3.99), 344(3.7)	103-0602-69
$C_{19}H_{20}O$			
Anthracene, 9-ethyl-10-(1-methoxyethyl)-	hexane	253s(4.94),260(5.19), 323s(3.04),338(3.49), 354(3.84),373(4.06), 394(4.06)	39-2266-69C
$C_{19}H_{20}O_2$			
Anthrone, 10-hydroxy-10-neopentyl-	EtOH	217(4.21),237s(3.81), 276(4.17)	39-2266-69C

Compound	Solvent	λ_{max}(log ϵ)	Ref.
$C_{19}H_{20}O_3$			
Cryptotanshinone, dl-	EtOH	219(4.19),264(4.37), 272(4.30),290s(3.78), 358(3.36),450(3.37)	18-3318-69
Isocryptotanshinone	EtOH	251(3.97),258(4.10), 299(3.79),365(3.30)	88-0301-69
1-Oxaspiro[5.5]undecane-2,4-dione, 3-cinnamylidene-	MeOH	249(4.30),375(4.99)	83-0075-69
Phenanthrene, 4-acetyl-9-ethyl-9,10-	EtOH	312(4.44),372(3.57)	2-0873-69
dihydro-1,3-dihydroxy-9-methyl-	EtOH-NaOH	320(4.48),330(4.46), 434(3.73)	2-0873-69
$C_{19}H_{20}O_3S$			
2-Thiophenecarboxylic acid, tetrahydro-3-hydroxy-3,4-diphenyl-, ethyl ester	EtOH	247(4.31)	2-0009-69
$C_{19}H_{20}O_4$			
2H,8H-Benzo[1,2-b:5,4-b']dipyran-2-one, 5-hydroxy-8,8-dimethyl-10-(3-methyl-2-butenyl)- (trachy phyllin)	EtOH	223(4.27),278(4.37), 339(4.11)	12-2175-69
7-Oxaestra-1,3,5(10),8-tetraen-6-one, 3-acetoxy-	EtOH	201(4.24),221(4.17), 227s(4.11),272s(3.98), 280(3.99)	39-1234-69C
$C_{19}H_{20}O_5$			
Chalcone, 6'-hydroxy-2',4,4'-trimethoxy-3'-methyl-	MeOH	361(4.52)	78-0283-69
Chalcone, 2,4,4',6-tetramethoxy-	EtOH	227(4.31),330(4.45)	105-0397-69
Chalcone, 2,4,4',6-tetramethoxy-, trans	n.s.g.	253(3.98),371(4.47)	78-2367-69
Crotonic acid, 3-methyl-, ester with 8,9-dihydro-8-(1-hydroxy-1-methylethyl)-2H-furo[2,3-h]-1-benzopyran-2-one (libanorin)	ether	248(3.68),259(3.69), 323(4.25)	24-1673-69
	n.s.g.	237(4.21),250(3.65), 261(3.66)	105-0189-69
Decursin	EtOH	220(4.50),258(3.59), 330(4.26)	95-0549-69
Flavanone, 4',5,7-trimethoxy-6-methyl-	MeOH	278(4.14)	78-0283-69
Melanoxin, O-dimethyldehydro-	MeOH	286(4.18),324(4.49)	78-4409-69
$C_{19}H_{20}O_6$			
Altersolanol B, acetonide	EtOH	217(4.53),266(4.11), 287(3.90),425(3.62)	23-0767-69
6H-Benzofuro[3,2-c][1]benzopyran, 6a,11a-dihydro-3,7,9,10-tetramethoxy-	EtOH	279(3.62),285(3.59)	39-0887-69C
1,2-Dibenzofurandicarboxylic acid, 3,9b-dihydro-7-methoxy-4,9b-dimethyl-, dimethyl ester	90% EtOH	222(4.02),286(3.50), 291s(3.44)	88-2901-69
2,3-Dibenzofurandione, 6-acetyl-4,4a-dihydro-7,9-dimethoxy-4,4,4a-trimethyl-	EtOH	256(4.23),390(4.39)	54-0851-69
Isoflavan, 2',6,7-trimethoxy-4',5'-(methylenedioxy)-	EtOH	235(4.08),299(4.16)	39-1787-69C
$C_{19}H_{20}O_7$			
Chalcone, 2',3-dihydroxy-3',4,4',6'-tetramethoxy-	EtOH	262(4.04),374(4.55)	24-0112-69
1,2-Dibenzofurandicarboxylic acid, 3,4-dihydro-3,7-dimethoxy-9-methyl-, dimethyl ester	EtOH	246(4.34),256(4.35), 288(3.91),295s(3.91), 353(3.68)	88-4139-69

Compound	Solvent	$\lambda_{max}(\log \epsilon)$	Ref.
Elephantol, methacrylate	MeOH	209(4.33)	44-3867-69
Elephantopin	MeOH	210(4.43)	44-3867-69
$C_{19}H_{20}O_8$			
Acetic acid, [[α-(2-hydroxy-p-anisoyl)-4,5-dimethoxy-o-tolyl]oxy]-	EtOH	230(4.27),277(4.24),316(3.97)	18-0199-69
Altersolanol A, acetonide	EtOH	218(4.58),268(4.11),285s(3.75),429(3.68)	23-0777-69
Lepraric acid, methyl ester	MeOH	209(4.36),232(4.35),247(4.27),257(4.29),292(3.76),313s(3.54)	39-0704-69C
	MeOH-NaOH	238(4.07),266(4.18),358(3.54)	39-0704-69C
Spiro[furan-2(5H),7'-[2,5]nethano[7H]oxireno[3,4]cyclopent[1,2-d]oxepin]-3',5,6'(2'H.5'H)-trione, 4-acetyl-1'a,1'b,3,4,4'a,7'a-hexahydro-1'b-hydroxy-8'-isopropenyl-6'a-methyl-	MeOH	262(3.17)	78-4835-69
	MeOH-KOH	283(4.32)	78-4835-69
Xanthen-9-one, 1,2,3,4,6,7-hexamethoxy-	EtOH	247s(--),257(4.52),280(4.19),312(4.25),344(3.91)	78-4415-69
$C_{19}H_{21}Br_2NO$			
1-Indanone, 3-bromo-2-[α-(isopropylamino)benzyl]-, hydrobromide	$CHCl_3$	253(4.21),293(3.48)	104-0947-69
$C_{19}H_{21}Br_4NO_7$			
1,3-Dioxolo[4,5-b]acridinium, 11,11a-dihydro-10-hydroxy-4,11a-dimethoxy-5-methyl-11-oxo-, tribromide, hypobromite, dimethyl acetal	EtOH	216(4.34),248(4.37),290(3.74),365(4.06)	12-1477-69
$C_{19}H_{21}ClNO_2$			
2-Benzyl-1-(chloromethyl)-3,4-dihydro-6,7-dimethoxyisoquinolinium (chloride)	EtOH	255(4.10),327(3.97),393(3.99)	39-0064-69C
$C_{19}H_{21}ClN_2O$			
1,5-Benzoxazepine, 4-[p-(dimethylamino)-styryl]-2,3-dihydro-, hydrochloride	MeOH	520(5.18)	124-1178-69
1(2H)-Quinolinecarboxamide, 6-chloro-3,4-dihydro-N,2,3-trimethyl-4-phenyl-	n.s.g.	256(4.09)	40-1047-69
1(2H)-Quinolinecarboxamide, 6-chloro-N-ethyl-3,4-dihydro-2-methyl-4-phenyl-	n.s.g.	259(4.20)	40-1047-69
$C_{19}H_{21}ClN_2O_2$			
Wieland-Gumlich aldehyde, 10-chloro-	EtOH	254(4.00),310(3.48)	33-1564-69
$C_{19}H_{21}ClO_4S$			
1-tert-Butyl-2-phenyl-1H-2-benzothiopyranium perchlorate	MeOH	232(5.34),275(3.70),297(3.87)	44-3202-69
$C_{19}H_{21}Cl_2NO_2$			
Isoquinoline, 1-(o-chlorostyryl)-1,2,3,4-tetrahydro-6,7-dimethoxy-, hydrochloride	EtOH-HCl	204(4.75),218s(4.35),261(4.44),287s(3.96),298s(3.57)	39-0094-69C
$C_{19}H_{21}N$			
Indene, 3-[2-(dimethylamino)ethyl]-2-phenyl-, hydrochloride	EtOH	227(4.13),288(4.28)	87-0513-69

Compound	Solvent	$\lambda_{max}(\log \epsilon)$	Ref.
p-Toluidine, N-[1-(1,2,3,4-tetrahydro-2-naphthyl)ethylidene]-	EtOH	258(4.26),312s(3.53)	56-0749-69
$C_{19}H_{21}NO$			
p-Anisidine, N-[1-(1,2,3,4-tetrahydro-2-naphthyl)ethylidene]-	EtOH	260(4.26),320(3.57)	56-0749-69
Azetidine, 3-benzoyl-1-isopropyl-2-phenyl-, cis	isooctane	240(4.02)	44-0310-69
trans	isooctane	241(4.19)	44-0310-69
$C_{19}H_{21}NO_2$			
Koenimbin, dihydro-	EtOH	243(4.21),258(4.07), 312(4.03),332(3.62), 346(3.22)	31-0790-69
Nuciferine	EtOH	230(4.15),274(4.23), 304s(--),312(3.57)	95-1691-69
$C_{19}H_{21}NO_3$			
Alkaloid G	EtOH	228(4.22),289(3.63)	12-2219-69
Benzamide, N-acetyl-N-(benzyloxy)-	MeOH	208(--),216(4.30)	12-0161-69
2,4,6-trimethyl-	MeOH-NaOH	210(4.38),263s(3.13)	12-0161-69
Morphinan-7-one, 5,6,8,14-tetradehydro-6,8-dimethoxy-17-methyl-	EtOH	208(4.27),263(4.05)	88-1771-69
Thebaine, (±)-	EtOH	225(4.16),285(3.85)	39-2030-69C
$C_{19}H_{21}NO_3S$			
4,5-Oxazolidinedione, 2'-(1-adamantyl)-2-(phenylthio)-	EtOH	224s(4.06)	78-2909-69
$C_{19}H_{21}NO_4$			
Bracteolin	MeOH	218(4.56),268s(4.02), 278(4.12),304(4.15)	106-0635-69
Cryptostyline I	EtOH	235s(4.10),287(3.89)	1-0244-69
Hernovine, 10-O-methyl-, hydrochloride	EtOH	220(4.55),273(4.11), 304(3.69)	100-0001-69
Isoboldine, (±)-	MeOH	280(4.09),305(4.11)	78-3667-69
Isoindole-1,3-dicarboxylic acid, 2-benzyl-4,5,6,7-tetrahydro-, dimethyl ester	EtOH	221(4.24),287(4.31)	32-1115-69
Isosalutaridine	MeOH	235(4.08),283(3.81)	39-2034-69C
Orientalinone, (-)-	MeOH	231s(4.30),242s(4.14), 284(3.77)	106-0635-69
4,5-Oxazolidinedione, 2-(1-adamantyl)-2-phenoxy-	EtOH	220(3.73)	78-2909-69
Pallidine	MeOH	235(4.08),283(3.81)	77-1301-69
Salutaridine, (±)-	MeOH	236(4.23),279(3.76)	39-2030-69C
	MeOH	240(4.27),277(3.77)	106-0635-69
$C_{19}H_{21}NO_6$			
1,2,9-Acridantrione, 3,4-dimethoxy-10-methyl-, 2-(methyl ethyl acetal)	EtOH	226(3.97),245s(3.89), 292(4.06)	12-1477-69
1,3-Dioxolo[4,5-b]acridine-10,11(5H,11aH)-dione, 11a-methoxy-5-methyl-, 11-(ethyl methyl acetal)	EtOH	212(4.09),226(4.00), 235s(3.97),274(4.41), 278(4.45),318(4.01), 348(3.85)	12-1493-69
$C_{19}H_{21}NO_7$			
1,3-Dioxolo[4,5-b]acridine-10,11(5H,11aH)-dione, 4,11a-dimethoxy-5-methyl-, 11-(dimethyl acetal)	EtOH	214(4.10),287(4.12), 325(4.03)	12-1477-69

Compound	Solvent	$\lambda_{max}(\log \epsilon)$	Ref.
$C_{19}H_{21}NO_8$			
5H-Pyrrolo[1,2-a]azepine-5,6,7,8-tetracarboxylic acid, 3,9-dimethyl-, tetramethyl ester	MeOH	215(4.30),263(4.18), 409(3.95)	39-2316-69
$C_{19}H_{21}NS$			
Dibenzo[b,e]thiepin, 11-[3-(dimethylamino)propylidene]-6,11-dihydro-, cis	heptane	230(4.32),261(3.89), 302(3.28)	73-1963-69
trans	heptane	232(4.38),262(3.95), 303(3.37)	73-1963-69
$C_{19}H_{21}N_3$			
Phthalazine, 1-[3-(dimethylamino)propyl]-4-phenyl-, dihydrochloride	EtOH	220(4.7)	33-1376-69
$C_{19}H_{21}N_3NaO_9P$			
Benzamide, N-[1-(2-deoxy- -D-erythro-pentofuranosyl)-1,2-dihydro-2-oxo-4-pyrimidinyl]-, monoacetate mono-(methyl hydrogen phosphate), Na salt	pH 1	257(4.14),315(4.30)	35-1522-69
	H_2O	258(4.33),302(4.07)	35-1522-69
$C_{19}H_{21}N_3O$			
Benzamide, N-[1-(2-benzimidazolyl)-2-methylbutyl]-	EtOH	275(4.02),282(4.06)	94-2381-69
Benzamide, N-[1-(2-benzimidazolyl)-3-methylbutyl]-	EtOH	275(4.01),282(4.03)	94-2381-69
$C_{19}H_{21}N_3O_2$			
Indole, 3-[3-(6-nitro-3,4-xylidino)-propyl]-	EtOH and EtOH-HCl	224(4.66),237s(4.44), 284(4.00),292(4.08), 443(3.84)	44-3240-69
	EtOH-NaOH	237s(4.20),289(3.84), 295(4.10),445(3.87)	44-3240-69
$C_{19}H_{21}N_5O_3$			
Acetanilide, 4'-[(3-nitro-4-piperidino-phenyl)azo]-	benzene	440(4.74)	73-2092-69
$C_{19}H_{22}$			
1-Hexene, 1,1-diphenyl-4-methyl-	EtOH	250(4.13)	35-6362-69
[3,4]Paracyclophane, tetracyanoethylene complex	CH_2Cl_2	538(3.28)	35-3553-69
1-Pentene, 1,1-diphenyl-4,4-dimethyl-	EtOH	249(4.14)	35-6362-69
$C_{19}H_{22}Br_2O_2$			
Propane, 2,2-bis(bromomethyl)-1,3-bis-(o-tolyloxy)-	hexane	220(4.16),271(3.48), 277(3.44)	56-1641-69
$C_{19}H_{22}ClNO_3$			
7,8-Dehydrometathebainone methochloride	EtOH	304(3.94)	77-0092-69
	CF_3COOH-2% H_2SO_4	407(4.07)	77-0092-69
$C_{19}H_{22}ClNO_6$			
2-Benzyl-3,4-dihydro-6,7-dimethoxy-1-methylisoquinolinium perchlorate	EtOH	249(4.26),312(4.03), 366(4.05)	39-0094-69C
$C_{19}H_{22}ClN_3O_6S$			
Sporidesmin F (^{35}Cl)	MeOH	216(4.46),250(4.14), 298(3.30)	39-1564-69C

Compound	Solvent	λ_{max}(log ε)	Ref.
$C_{19}H_{22}F_2O_2$			
Estr-4-ene-3,17-dione, 6α,7α-(difluoro-methylene)-	EtOH	247(4.21)	78-1219-69
Estr-4-ene-3,17-dione, 6β,7β-(difluoro-methylene)-	EtOH	248(4.25)	78-1219-69
$C_{19}H_{22}N_2$			
Indole, 1-[α-[2-(dimethylamino)ethyl]-benzyl]-	EtOH	273(3.82),281(3.81), 285s(--),292(3.67)	39-1537-69C
Malondianil, p,p'-diethyl-, hydrochlor-ide	MeOH	248(4.10),272s(--), 291(3.59),300s(--), 393(4.68)	59-1075-69
$C_{19}H_{22}N_2O$			
Acrylophenone, 3-anilino-3-(tert-butyl-amino)-	EtOH	237(4.15),334(4.34)	44-3451-69
5H-Benz[i]indolo[2,3-a]quinolizin-13(12H)-one, 1,2,3,6,11,14,15a-octahydro-	MeOH	226(4.51),283(3.90)	44-0330-69
7,13b-Methano-13bH-indolo[3,2-e][2]-benzazocin-2(1H)-one, 3,4,4a,5,6,7,-8,13-octahydro-6-methyl-, cis	EtOH	225(4.56),283(3.92), 290(3.88)	44-3165-69
trans	EtOH	225(4.56),282(3.90), 290(3.85)	44-3165-69
$C_{19}H_{22}N_2O_2$			
Ajmalicine, 16-de(methoxycarbonyl)-16,17-dihydro-17-oxo-	MeOH	223(4.58),283(3.83), 290(3.78)	24-3558-69
Benzoic acid, 2,4,6-trimethyl-, (o-meth-oxy-α-methylbenzylidene)hydrazide	EtOH	212s(4.4),265(4.1)	28-0730-69A
19-Epiajmalicine, 16-de(methoxycarbon-yl)-16,17-dihydro-17-oxo-	MeOH	224(4.57),281(3.82), 289(3.74)	24-3558-69
19-Epi-3-isoajmalicine, 16- e(methoxy-carbonyl)-16,17-dihydro-17-oxo-	MeOH	224(4.54),281(3.88), 289(3.81)	24-3558-69
6(2H)-Isoquinolone, 1,3,4,7,8,8a-hexa-hydro-7-hydroxy-2-(2-indol-3-yleth-yl)-, (acetate)	n.s.g.	222(4.56),282(3.70)	24-0310-69
Quebrachamine, 3,10-dioxo-	EtOH	245(4.10),312(4.25)	35-2342-69
Wieland-Gumlich aldehyde	EtOH	244(3.83),298(3.43)	33-1564-69
$C_{19}H_{22}N_2O_4$			
1,4-Phenazinedicarboxylic acid, 2,3,5,10-tetrahydro-7-methyl-, diethyl ester	EtOH	206(4.26),269(4.62), 303(3.87),316(3.81), 403(4.24),430(4.22), 459(4.03)	33-0322-69
$C_{19}H_{22}N_2O_5S$			
2-(p-Methoxystyryl)-1-methylbenzimid-azole methosulfate	MeOH	325(3.99)	22-1926-69
$C_{19}H_{22}N_2Se$			
Piperazine, 1-(10,11-dihydrodibenzo-[b,f]selenepin-10-yl)-4-methyl-	MeOH	238s(3.98),253s(3.81), 278(3.75)	73-3801-69
$C_{19}H_{22}N_4O_6$			
4a(2H)-Naphthalenecarboxylic acid, 1,3,4,7,8,8a-hexahydro-7-oxo-, ethyl ester, 2,4-dinitrophenyl-hydrazone, trans	$CHCl_3$	374(4.41)	78-5281-69

Compound	Solvent	$\lambda_{max}(\log \epsilon)$	Ref.
$C_{19}H_{22}N_4O_{10}S$			
Hypoxanthine, 8-(β-D-glucopyranosyl-thio)-, 2',3',4',6'-tetraacetate	pH 1	267(4.21)	87-0653-69
	pH 11	225s(4.08),275(4.25)	87-0653-69
$C_{19}H_{22}O_3$			
Coumarin, 6-(3,7-dimethyl-2,6-octadien-yl)-7-hydroxy- (ostrutine)	MeOH	225(4.09),250(3.55), 259(3.48),295(3.64), 338(4.01)	32-0308-69
Coumarin, 8-(3,7-dimethyl-2,6-octadien-yl)-7-hydroxy-	MeOH	250(3.58),260(3.58), 329(4.01)	32-0308-69
[2,2]Metacyclophane-8-methanol, 5,13-dimethoxy-	EtOH	247(4.01),292(3.44)	44-1956-69
Phenanthro[3,2-b]furan-7,11-dione, 1,2,3,4,6,6a,11a,11b-octahydro-4,4,8-trimethyl-	EtOH	214(4.05),244(3.71), 303(3.90)	18-3318-69
$C_{19}H_{22}O_4$			
2H,6H-Benzo[1,2-b:5,4-b']dipyran-2-one, 7,8-dihydro-5-hydroxy-8,8-dimethyl-10-(3-methyl-2-butenyl)-	EtOH	229s(4.13),255s(3.90), 263(3.98),335(4.20)	12-2175-69
	EtOH-KOH	240s(4.26),277(4.19), 338(4.08),415(3.86)	12-2175-69
Estra-1,3,5(10)-triene-6,11-dione, 9β-hydroxy-3-methoxy-	EtOH	225(4.29),320(3.37)	39-1234-69C
7α-Gibba-3,4a(4b)-diene-10β-carboxylic acid, 1,7-dimethyl-2,8-dioxo-, methyl ester	EtOH	302(4.14)	78-1293-69
$C_{19}H_{22}O_5$			
Laxifloran, dimethyl ether	EtOH	280(3.55),288(3.44)	39-0887-69C
Propiophenone, 3-(2,4-dimethoxyphenyl)-2',4'-dimethoxy-	EtOH	226(4.34),269(4.16), 300(3.93)	23-1529-69
$C_{19}H_{22}O_6$			
Acetophenone, 2-(3,4-dimethoxyphenyl)-2',4',5'-trimethoxy-	EtOH	232(4.40),272(4.12), 328(4.01)	39-1787-69C
	EtOH	207(4.30),231(4.26), 270(4.02),325(3.84)	78-3887-69
Coleon B, dihydro-	EtOH	267(4.15),283(4.02), 329(3.81),389(3.85)	33-1685-69
Indene-3-succinic acid, 2-(carboxymeth-yl)-7-methyl-, trimethyl ester	EtOH	272(4.1)	42-0415-69
4,7-Methanocyclobuta[b]naphthalene-1,2-dicarboxylic acid, 2a,3,3a,4,5,6,7,-7a,8,8a-decahydro-2a,8a-dimethyl-3,8-dioxo-, dimethyl ester	EtOH	220(3.62)	39-0105-69C
1,4,10(4bH)-Phenanthrenetrione, 5,6,7,8-tetrahydro-3,6,9-trihydroxy-2-isopro-pyl-4b,8-dimethyl-	EtOH	246(4.25),275(3.92), 423(3.26)	33-1685-69
Santonene, 2β-acetoxy-3-acetyl-1,2-di-hydro-	EtOH	224(3.73),324(4.53)	39-1617-69C
Vernodalin	MeOH	210(4.30)(end abs.)	44-3908-69
$C_{19}H_{22}O_7$			
Acetophenone, 2'-hydroxy-4',5'-dimeth-oxy-2-(2,4,5-trimethoxyphenyl)-	EtOH	237(4.25),279(4.05), 344(3.86)	39-1787-69C
Acetophenone, 2'-hydroxy-4'-methoxy-2-(2,3,4,6-tetramethoxyphenyl)-	EtOH	228(4.27),274(4.28), 314(3.93)	39-0887-69C
Diosbulbic acid	EtOH	210(3.72)	2-0452-69

Compound	Solvent	λ_{max}(log ϵ)	Ref.
$C_{19}H_{22}O_8$			
Vernolide, hydroxy-	n.s.g.	210(4.42)	28-0082-69A
$C_{19}H_{22}O_9$			
β-D-Glucopyranoside, 7-acetyl-3,8-dihydroxy-6-methyl-1-naphthyl-	EtOH	236(4.47),268(4.25), 332(3.72),345(3.72)	88-0471-69
$C_{19}H_{23}Cl_2N_3$			
Phthalazine, 1-[3-(dimethylamino)propyl]-4-phenyl-, dihydrochloride	EtOH	220(4.7)	33-1376-69
$C_{19}H_{23}NO$			
Acetophenone, 2',6'-dimethyl-4'-(2,4,6-trimethylanilino)-	C_6H_{12}	236s(4.03),286(4.09)	22-3523-69
	EtOH	202(4.60),240(3.99), 306(3.97),328s(3.92)	22-3523-69
$C_{19}H_{23}NO_2$			
Benzyl alcohol, α-[(diethylamino)methyl]-, benzoate	dioxan	272(3.19),281(3.09)	65-1861-69
Benzyl alcohol, α-[(dimethylamino)methyl]-α-ethyl-, benzoate	dioxan	273(3.14),280(3.04)	65-1861-69
Benzyl alcohol, α-[3-(dimethylamino)-propyl]-, benzoate	dioxan	257(3.01),265(3.00), 273(3.01),281(2.93)	65-1861-69
A-Homoestra-1(10),2,4a-triene-4,17-dione, 4-oxime	EtOH	233(4.29),303(4.09)	33-0121-69
$C_{19}H_{23}NO_2S$			
Thietane, 3-(diethylamino)-2,4-diphenyl-, 1,1-dioxide	EtOH	226(4.31)	44-1136-69
$C_{19}H_{23}NO_3$			
Alkaloid A from S. multiflora R.Br	EtOH	236(3.63),289(3.59)	12-2219-69
Alkaloid E	EtOH	237(3.59),290(3.58)	12-2219-69
3-Epischelhammericine	EtOH	239(3.70),293(3.66)	28-0639-69B
Schelhammericine	EtOH	236(3.68),288(3.60)	12-2219-69
$C_{19}H_{23}NO_4$			
Alkaloid H	EtOH	238(3.70),290(3.63)	12-2219-69
3-Epischelhammerine	EtOH	238(3.74),294(3.70)	28-0639-69B
5-Isoquinolinol, 1,2,3,4-tetrahydro-1-(p-hydroxybenzyl)-6,7-dimethoxy-2-methyl-, L-	EtOH	279(3.44)	44-3884-69
6-Isoquinolinol, 1,2,3,4-tetrahydro-1-(p-hydroxybenzyl)-5,7-dimethoxy-2-methyl-, dl-	EtOH	281(3.50)	44-3884-69
$C_{19}H_{23}NO_5$			
Estra-1,3,5(10)-trien-17-one, 9α-hydroxy-3-methoxy-11β-nitro-	EtOH	278(3.27),285(3.21)	31-1018-69
$C_{19}H_{23}NO_7$			
2-Propene-1,1,3-tricarboxylic acid, 1-benzamido-, triethyl ester	EtOH	225(4.19)	94-2417-69
$C_{19}H_{23}NO_8$			
3,4-Furandicarboxylic acid, 2-(carboxyethynyl)-5-(cyclohexylimino)-2,5-dihydro-2-methoxy-, trimethyl ester	ether	258(4.06)	24-1656-69

Compound	Solvent	λ max(log ε)	Ref.
$C_{19}H_{23}N_3OS$			
Acetophenone, 2'-methoxy-, 4-mesityl-3-thiosemicarbazone	EtOH	280(4.3)	28-0730-69A
4H-Cyclopentapyrimidine-4-thione, 1,5,6,7-tetrahydro-1-(morpholinoethyl)-2-phenyl-	EtOH	246(4.01),343(4.39)	4-0037-69
$C_{19}H_{23}N_3O_2$			
Acetophenone, 2'-methoxy-, 4-mesityl-semicarbazone	EtOH	279(4.2),291s(3.8)	28-0730-69A
Tryptophan, L-alanyl-2-(1,1-dimethylallyl)-, anhydride, (+)-	EtOH	225(4.51),283(3.89), 291(3.85)	39-1003-69C
(-)-	EtOH	225(4.51),283(3.95), 291(3.89)	39-1003-69C
$C_{19}H_{23}N_5O_2$			
Piperidine, 1-[4-[p-(dimethylamino)-phenyl]azo]-2-nitrophenyl-	benzene	325(3.56),350(3.62)	73-2092-69
$C_{19}H_{23}N_5O_9S$			
Adenine, 8-(β-D-glucopyranosylthio)-, 2',3',4',6'-tetraacetate	pH 1	277(4.28)	87-0653-69
	pH 11	228(4.28),281(4.29)	87-0653-69
$C_{19}H_{24}$			
Biphenyl, p-heptyl-	MeOH	260(4.42)	97-0342-69
$C_{19}H_{24}ClN$			
2,13-Metheno-13H-1-benzazacyclopentadecine, 18-chloro-3,4,5,6,7,8,9,10,-11,12-decahydro-	EtOH	235(4.64),285(3.62), 296(3.62),309(3.60), 323(3.62)	88-0557-69
$C_{19}H_{24}ClNO_4$			
1,2,3,4,4a,5,7,8,9,10-Decahydrobenzo-[c]quinolizinium perchlorate	$CHCl_3$	338(3.97)	78-4161-69
$C_{19}H_{24}ClNO_7$			
Corine, perchlorate	50% MeOH	280(3.81)	102-1559-69
$C_{19}H_{24}F_2O_2$			
Estr-4-en-3-one, 6α,7α-(difluoromethylene)-17β-hydroxy-	EtOH	245(4.21)	78-1219-69
$C_{19}H_{24}INO_3$			
Corine, iodide	pH 13	298(3.96)	102-1559-69
	EtOH	283(3.76)	102-1559-69
Tiliacine, iodide	pH 13	295(3.88)	102-1559-69
	EtOH	280(3.68)	102-1559-69
$C_{19}H_{24}NOPS$			
Phenothiaphosphine, 2,8-dimethyl-10-[3-(dimethylamino)propyl]-, 10-oxide, hydrochloride	EtOH	223(4.28),267(4.03), 294(3.68),305s(3.64)	78-3919-69
$C_{19}H_{24}N_2$			
Imipramine	MeOH	253(3.94)	87-0146-69
17,13b-Methano-13bH-indolo[3,2-e][2]-benzazocine, 1,2,3,4,4a,5,6,7,8,13-decahydro-6-methyl-, cis	EtOH	227(4.57),282(3.88), 290(3.83)	44-3165-69
trans	EtOH	227(4.56),282(3.87), 289(3.81)	44-3165-69

Compound	Solvent	$\lambda_{max}(\log \epsilon)$	Ref.
Quebrachamine, 3,4-didehydro-	EtOH	226(4.63),279(3.90), 288(3.83)	35-2342-69
$C_{19}H_{24}N_2OS$			
Phenothiazine, 10-[3-(dimethylamino)-2-methylpropyl]-2-methoxy- (nozinan)	EtOH	241(3.85),253(3.85), 308(3.07)	9-0249-69
$C_{19}H_{24}N_2O_2$			
Quebrachamine, 3-hydroxy-10-oxo-	MeOH	223(4.54),283(3.85), 292(3.78)	35-2342-69
$C_{19}H_{24}N_2O_3S$			
Tiliacine, thiocyanate	EtOH	280(3.74)	102-1559-69
$C_{19}H_{24}N_2O_5$			
3,4-Xylidine, 6-nitro-N-(3,4,5-trimethoxyphenethyl)-	EtOH	237(4.45),292(3.75), 440(3.73)	44-3240-69
	EtOH-HCl	237(4.45),295(3.76), 443(3.73)	44-3240-69
	EtOH-NaOH	293(3.82),443(3.76)	44-3240-69
$C_{19}H_{24}N_2O_7$			
Malonic acid, [(2-carboxy-5-methoxy-1H-pyrrolo[2,3-c]pyridin-3-yl)methyl]-, triethyl ester, hydrochloride	EtOH	284(4.27),294(4.31), 350(3.66)	35-2338-69
$C_{19}H_{24}N_2O_{11}$			
Thymine, N(3)-β-D-glucopyranosyl-, 2',3',4',6'-tetraacetate	pH 13	265(--)	39-0203-69C
	EtOH	262(4.01)	39-0203-69C
$C_{19}H_{24}N_4$			
Malondianil, p,p'-bis(dimethylamino)-	EtOH	449(4.56)	24-2609-69
triperchlorate	EtOH	451(4.48)	24-2609-69
$C_{19}H_{24}N_4O_3$			
4H-Quinolizine-1-carboxylic acid, 3-cyano-2-[[2-(diethylamino)ethyl]amino]-4-oxo-	EtOH	265(4.44),303(3.98), 395(3.89)	95-0203-69
$C_{19}H_{24}N_4O_4$			
4,9-Cyclodecadien-3-one, 7,7,10-trimethyl-, 2,4-dinitrophenylhydrazone	$CHCl_3$	374(4.36)	78-5275-69
1,3-Propanediamine, N,N'-bis(6-nitro-3,4-xylyl)-	EtOH	237(4.65),292(4.06), 435(4.07)	44-3240-69
	EtOH-HCl	237(4.64),293(4.08), 440(4.06)	44-3240-69
	EtOH-NaOH	235(4.68),293(4.10), 440(4.03)	44-3240-69
$C_{19}H_{24}N_4O_9$			
Theophylline, 7-(2-deoxy-β-D-arabino-hexopyranosyl)-, 3',4',6'-triacetate	n.s.g.	274(3.97)	22-3927-69
$C_{19}H_{24}O$			
Estra-1,3,5(10)-trien-6-one, 2-methyl-	EtOH	206(4.21),255(4.33), 292(3.58)	39-1234-69C
A-Homoestra-2,4,5(10)-trien-17-one	EtOH	268(3.60)	33-0121-69
$C_{19}H_{24}O_2$			
Estra-1,3,5(10)-trien-6-one, 3-methoxy-	EtOH	224(4.20),257(3.74), 326(3.29)	39-1234-69C

Compound	Solvent	$\lambda_{max}(\log \epsilon)$	Ref.
Estra-1,3,5(10)-trien-17-one, 3-hydroxy-6β-methyl-	EtOH	283(3.36)	33-0453-69
B-Homoestra-4,6-diene-3,17-dione	EtOH	289(4.31)	33-0453-69
B-Homoestra-1,3,5(10)-trien-17-one, 3-hydroxy-	EtOH	226(3.85),278(3.29)	31-0571-69
$C_{19}H_{24}O_3$			
9β-Androsta-4,8(14)-diene-3,17-dione, 11α-hydroxy-	n.s.g.	245(4.20)	23-1989-69
Androst-4-ene-3,17-dione, 9α,11α-epoxy-	n.s.g.	244(4.20)	23-1989-69
5β,19-Cycloandrostane-3,6,17-trione	MeOH	205(3.63)	44-3837-69
Estra-1,3,5(10),9(11)-tetraene-3,6ξ,17β-triol, 6-methyl-	MeOH	266(4.15)	44-0116-69
Estra-1,3,5(10)-trien-11-one, 9β-hydroxy-3-methoxy-	EtOH	225(3.83),277(3.16), 284(3.12)	39-1234-69C
9β-Estra-1,3,5(10)-trien-17-one, 3,11α-dihydroxy-9-methyl-	n.s.g.	281(3.29)	23-1989-69
	base	298(3.50)	23-1989-69
12α-Etiojerv-4-ene-3,11,17-trione	MeOH	234(4.26)	78-3145-69
Gon-4-ene-3,12,17-trione, 13β-ethyl-, L-	MeOH	238(4.24)	44-3530-69
	0.04N NaOH	238(4.23)	44-3530-69
$C_{19}H_{24}O_4$			
Adenostylone	EtOH	243(3.69),302(4.17)	73-1739-69
Furoeremophil-10(1)-ene, 6α-isobutyryl-9-oxo-	EtOH	251(3.54),303(4.10)	73-2792-69
Isoadenostylone	EtOH	243(3.61),284(4.16)	73-1739-69
4H-Phenanthro[10,1-bc]furan-4-one, 1,2,3,3a,5a,6,10b,10c-octahydro-2,6-dihydroxy-3,7,10,10c-tetramethyl-(eurycomol)	EtOH	224(4.08),270(2.68), 279(2.57),308(1.55)	77-0821-69
Propane, 1,3-bis(2,4-dimethoxyphenyl)-	EtOH	226(4.24),279(3.80)	23-1529-69
9,11-Secoestra-1,3,5(10)-trien-11-oic acid, 3-methoxy-9-oxo-, hydrate	EtOH	203(4.43),224(4.24), 274(4.35)	39-1234-69C
$C_{19}H_{24}O_5$			
2-Oxaandrosta-4,6-diene-1β-carboxylic acid, 17β-hydroxy-3-oxo-	EtOH	274(4.29)	19-0275-69
$C_{19}H_{24}O_6$			
1H-Benz[c]indene-5,7-dione, 3α-glycoloyl-2,3,3a,4,8,9,9a,9b-octahydro-3-hydroxy-3aα,6-dimethyl-, 3α-acetate	MeOH	261(4.04)	44-0112-69
Erioflorin	EtOH	210(4.31)(end abs.)	102-2381-69
Eurycomalactone	EtOH	241(3.85)	77-0821-69
9(1H)-Phenanthrone, 2,3,4,4a-tetrahydro-3,5,6,8,10-pentahydroxy-7-isopropyl-1β,4aβ-dimethyl-	EtOH	262(3.99),285(3.93), 333(3.68),390(3.86)	33-1685-69
Radiatin	EtOH	215(4.11)	102-1753-69
15,16,71-Trinortiglia-1,6-diene-3,13-dione, 4,9-dihydroxy-20-acetoxy-	MeOH	196(4.04),240(3.71), 320(2.09)	5-0158-69H
$C_{19}H_{24}O_6S_4$			
1,2-Dithio-2-(ethoxydithiocarbonyl)-4,6-O-benzylidene-α-D-mannopyranosyl ethyl xanthate	EtOH	275(4.26)	94-2571-69
$C_{19}H_{24}O_7$			
Elephantol, dihydro-, isobutyrate	MeOH	211(4.01)	44-3867-69
Elephantopin, tetrahydro-	MeOH	210(4.02)	44-3867-69
Vernomygdin	MeOH	210(4.30)(end abs.)	44-3908-69

Compound	Solvent	λ_{max}(log ε)	Ref.
$C_{19}H_{25}ClNOPS$			
Phenothiaphosphine, 2,8-dimethyl-10-[3-(dimethylamino)propyl]-, 10-oxide, hydrochloride	EtOH	223(4.28),267(4.03), 294(3.68),305s(3.64)	78-3919-69
$C_{19}H_{25}ClN_2O_7$			
Malonic acid, [(2-carboxy-5-methoxy-1H-pyrrolo[2,3-c]pyridin-3-yl)methyl]-, triethyl ester, hydrochloride	EtOH	284(4.27),294(4.31), 350(3.66)	35-2338-69
$C_{19}H_{25}N$			
13-Azabicyclo[8.2.1]trideca-10,12-diene, 13-o-tolyl-	hexane	237(4.07)	78-5357-69
13-Azabicyclo[8.2.1]trideca-10,12-diene, 13-p-tolyl-	hexane	241(4.16)	78-5357-69
Quinoline, 2,4-(decamethylene)-	EtOH	229(4.65),232s(4.61), 279(3.69),289s(3.67), 302(3.60),316(3.64)	88-0557-69
$C_{19}H_{25}NO_2$			
10-Aza-19-nor-17α-pregn-4-en-20-yn-3-one, 17-hydroxy-	EtOH	321(4.32)	22-0117-69
Podocarpa-8,11,13-trien-7-one, 13-acetyl-15-amino-, hydrochloride	MeOH	251(4.17)	5-0193-68H
$C_{19}H_{25}NO_3$			
Alkaloid B	EtOH	235s(3.90),283(3.57), 289s(3.52)	12-2219-69
Erythroculinol	EtOH	280(3.40),284(3.41)	88-0153-69
2,4-Pentadienoic acid, 2-acetyl-5-(diethylamino)-5-phenyl-, ethyl ester	EtOH	251(3.88),410(4.77)	70-0363-69
2-Propanone, 1-[2-[2-(dimethylamino)-ethyl]-7,8-dimethoxy-1-naphthyl]-, perchlorate	H_2O	289(3.76),301(3.72), 338(3.36)	49-0163-69
$C_{19}H_{25}NO_4$			
10-Azaestra-2,4-dien-1-one, 17β-hydroxy-3-methoxy-, formate	EtOH	214(4.49),235s(--), 291(3.75)	22-0117-69
Bucharidine	n.s.g.	228(4.43),274(3.79), 282(3.76),314(3.77), 328(3.66)	105-0380-69
Naphthalene, 1,2,3,4-tetrahydro-2,2-dimethyl-1-[2-(methylamino)ethylidene]-, maleate	n.s.g.	243(4.17)	39-0217-69C
$C_{19}H_{25}NO_6$			
2H-1-Benzopyran, 7-methoxy-4-[3-(dimethylamino)propyl]-, maleate	n.s.g.	266(3.94)	39-0217-69C
$C_{19}H_{25}N_7O_2$			
Adenine, 9-[6-(6-nitro-3,4-xylidino)-hexyl]-	EtOH	238(4.45),257(4.29), 293s(3.81),490(3.81)	44-3240-69
	EtOH-HCl	238(4.45),256(4.30), 295s(3.81),440(3.81)	44-3240-69
	EtOH-NaOH	235s(4.49),257(4.31), 293s(3.85),440(3.81)	44-3240-69
$C_{19}H_{26}$			
Estra-1,3,5(10)-triene, 2-methyl-	C_6H_{12}	213(3.82),272(3.00)	39-1234-69C

Compound	Solvent	$\lambda_{max}(\log \epsilon)$	Ref.
$C_{19}H_{26}ClNO$			
Androsta-3,5-dien-17-one, 3-chloro-, oxime	EtOH	236s(4.34),243(4.38), 251s(4.23)	94-1749-69
Podocarpa-8,11,13-trien-7-one, 13-acetyl-15-amino-, hydrochloride	MeOH	251(4.17)	5-0193-69H
$C_{19}H_{26}ClNO_5$			
Acutuminine	n.s.g.	246(4.34),272(4.02)	88-1933-69
$C_{19}H_{26}ClNO_6$			
[2-(2,3-Dihydro-9-methoxy-2-methyl-naphtho[-,8-bc]pyran-4-yl)ethyl]-trimethylammonium perchlorate	H_2O	303(3.71),331(3.45)	49-0163-69
$C_{19}H_{26}ClNO_7$			
2-Propanone, 1-[2-[2-(dimethylamino)-ethyl]-7,8-dimethoxy-1-naphthyl]-, perchlorate	H_2O	289(3.76),301(3.72), 338(3.36)	49-0163-69
$C_{19}H_{26}N_2O$			
Antirhine, dihydro-, (hydrate)	EtOH	226(4.53),276s(3.85), 284(3.87),291(3.82)	78-5319-69
Corynantheol, dihydro-	EtOH	226(4.57),275s(3.85), 283(3.87),291(3.80)	78-5329-69
$C_{19}H_{26}N_2O_2$			
Corynantheol, dihydro-10-hydroxy-	EtOH	226(4.44),280(3.96), 296s(3.86)	78-5329-69
Hunterburnine, dihydro-	EtOH	223(4.40),281(3.94), 296s(3.85)	78-5319-69
Quinamine, dihydro-	MeOH	242(3.93),300(3.43)	102-0645-69
$C_{19}H_{26}N_2O_3$			
Pyrrole-2-carboxaldehyde, 5-[[3,4-di-ethyl-5-(hydroxymethyl)pyrrol-2-yl]carbonyl]-3,4-diethyl-	EtOH	271(4.15),309(4.15), 351(4.18)	39-0564-69C
$C_{19}H_{26}N_4O_4$			
2(1H)-Naphthalenone, octahydro-6,6,8a-trimethyl-, 2,4-dinitrophenylhydra-zone	$CHCl_3$	374(4.36)	78-5275-69
$C_{19}H_{26}N_4O_5$			
2(1H)-Naphthalenone, octahydro-5-hydr-oxy-4a,7,7-trimethyl-, 2,4-dinitro-phenylhydrazone, cis	EtOH	368(4.13)	78-5275-69
$C_{19}H_{26}N_4O_6$			
Bicarbamic acid, [carboxy(1-ethyl-2-benzimidazolyl)methyl]-, triethyl ester	MeOH	258(3.89),270(3.85), 277(3.91),285(3.84)	88-1203-69
$C_{19}H_{26}O$			
Estra-1,3,5(10)-trien-3-ol, 2-methyl-	EtOH	204(4.16),278(3.49)	39-1234-69C
B(9α)-Homoestra-1,3,5(10)-trien-17-ol	EtOH	269(2.66),274s(2.57), 278(2.60)	77-0347-69
$C_{19}H_{26}O_2$			
Androsta-1,4-dien-3-one, 17β-hydroxy-	EtOH	244(4.19)	94-1206-69
Androsta-3,5-dien-17-one, 19-hydroxy-	MeOH	234(4.27)	44-3837-69

Compound	Solvent	λ_{max}(log ε)	Ref.
Androsta-4,14-dien-3-one, 17β-hydroxy-	EtOH	240(4.27)	94-1782-69
9β-Androst-4-ene-3,17-dione	EtOH	244(4.16)	73-2459-69
14β-Androst-4-ene-3,17-dione	EtOH	240(4.21)	94-1782-69
Cannabidivarin	MeOH	212s(4.53),230(4.04), 272(3.9),281(3.85)	88-0145-69
5,19-Cyclo-5β-androst-3-en-17-one, 6β-hydroxy-	MeOH	205(3.75)	44-3837-69
3α,5-Cyclo-D-homo-18-nor-5α-androst-13(17a)-en-17-one, 6β-hydroxy-	EtOH	240(4.24)	22-1673-69
B-Homoestr-4-ene-3,17-dione	EtOH	245(4.18)	33-0453-69
18-Nor-14ξ-androsta-5,13(14)-dien-16-one, 3β-hydroxy-17-methyl-	EtOH	239(4.18)	70-0826-69
$C_{19}H_{26}O_3$			
Androsta-1,4-dien-3-one, 2,17β-dihydroxy-	EtOH	254(4.16)	94-1206-69
9β,10α-Androst-4-ene-3,17-dione, 9-hydroxy-	EtOH	241(4.19)	33-1157-69
A,B-Dinorandrost-3(5)-en-2-one, 17β-hydroxy-, acetate	EtOH	235(4.13)	44-2297-69
Heptadeca-1,9-diene-4,6-diyne-3,8-diol, 3-acetate, cis	ether	233(2.60),245(2.60), 259(2.30)	39-0685-69C
B-Norandrost-4-en-3-one, 17β-hydroxy-2-(hydroxymethylene)-	EtOH	250(4.05),307(3.83)	44-2297-69
2-Oxaandrosta-4,6-dien-3-one, 17β-hydroxy-17-methyl-	EtOH	272(4.30)	78-4257-69
Rubrosterone	EtOH	240(4.07)	78-1241-69
$C_{19}H_{26}O_4$			
Adenostylone, dihydro-	EtOH	244(3.60),278(4.10)	73-1739-69
2H,8H-Benzo[1,2-b:5,4-b']dipyran-2-one, 3,4,7,8-tetrahydro-5-hydroxy-10-isopentyl-8,8-dimethyl-	EtOH	263(3.23),279(3.22), 285(3.26),335(2.99)	12-2175-69
Isoadenostylone, dihydro-	EtOH	242(3.54),282(4.08)	73-1739-69
2-Oxaandrosta-4,6-dien-3-one, 1ξ,17β-dihydroxy-17α-methyl-	EtOH	274(4.34)	78-4257-69
$C_{19}H_{26}O_4S$			
2(1H)-Naphthalenone, octahydro-4a-(hydroxymethyl)-3-methyl-, p-toluenesulfonate, trans	EtOH	226(4.19)	78-5267-69
$C_{19}H_{26}O_5$			
7,9a-Methano-9aH-cyclopenta[b]heptalene-4,11(1H,10H)-dione, 2,3,4a,5,6,7,8,9-octahydro-2,8,12-trihydroxy-1,1,8-trimethyl-	EtOH	265(3.90)	94-2036-69
$C_{19}H_{26}O_6$			
Erioflorin, tetrahydro-, oxidation product	EtOH	246(3.7)	102-2381-69
$C_{19}H_{26}O_7$			
2,4-Pentanedione, 3-[(3,3-diacetyl-3,4-dihydro-6-methyl-2H-pyran-5-yl)methyl]-3-(hydroxymethyl)-, acetate	EtOH	253s(3.00),285s(2.64)	39-0879-69C
$C_{19}H_{26}O_{11}S$			
Paederosidic acid, methyl ester	EtOH	236(4.02)	94-1949-69

Compound	Solvent	$\lambda_{max}(\log \epsilon)$	Ref.
$C_{19}H_{26}O_{12}$			
Daphylloside	EtOH	236(3.95)	94-1949-69
$C_{19}H_{27}N$			
Piperidine, 1-[2-(3,4-dihydro-2,2-dimethyl-1(2H)-naphthylidene)ethyl]-, hydrochloride	n.s.g.	245(4.07)	39-0217-69C
$C_{19}H_{27}NO_3$			
10-Azaestr-4-en-3-one, 17β-acetoxy-	EtOH	322(4.27)	22-0117-69
$C_{19}H_{27}NO_4$			
Androst-4-en-3-one, 17β-hydroxy-, nitrate	C_6H_{12}	230(4.26)	78-0761-69
Riddelline, cis	EtOH	214(3.97)	73-1832-69
$C_{19}H_{27}NO_5$			
Othosenine	EtOH	214s(3.46)	73-1832-69
$C_{19}H_{28}N_2O_2$			
Indole-2-carboxamide, N,1-di-tert-butyl-4-methoxy-3-methyl-	EtOH	227(4.5),278(3.9), 298(3.9)	24-0678-69
$C_{19}H_{28}N_4O_2$			
Azetidine, 3,4-bis(tert-butylimino)-2,2-dimethyl-1-(p-nitrophenyl)-	EtOH	216(4.09),243(3.89), 292(3.45),309(3.40), 385(4.38)	78-1467-69
$C_{19}H_{28}N_4O_4$			
Cyclodecan-3-one, 7,7,10-trimethyl-, 2,4-dinitrophenylhydrazone	$CHCl_3$	374(4.36)	78-5275-69
$C_{19}H_{28}N_6$			
Methane, tris(3,4,5-trimethylpyrazol-1-yl)-	n.s.g.	237(3.69)	39-2251-69C
$C_{19}H_{28}O$			
Androst-4-en-6-one	isooctane	232(3.89)	94-2586-69
Androst-5-en-4-one	isooctane	233(3.84)	94-2586-69
5α-Androst-14-en-17-one	hexane	214(3.20)	88-0483-69
5α,14β-Androst-15-en-17-one	hexane	226(3.81)	88-0483-69
5β,14α-Androst-15-en-17-one	EtOH	228(3.73)	39-2767-69C
3α,5-Cycloandrostan-6-one	hexane	201(3.66)	22-1673-69
A-Norandrosta-2,5-dien-17β-ol, 3-methyl-	n.s.g.	238(4.3),247(4.3), 256(4.3)	22-0189-69
$C_{19}H_{28}O_2$			
5α-Androst-15-en-17-one, 14β-hydroxy-	n.s.g.	213(3.64)	88-0483-69
9β-Estr-4-en-3-one, 17β-hydroxy-17α-methyl-	EtOH	243(4.19)	44-3022-69
	EtOH-base	235(3.39)	44-3022-69
9β,10α-Estr-4-en-3-one, 17β-hydroxy-17α-methyl-	EtOH	242(4.21)	44-3022-69
10α-Estr-4-en-3-one, 17β-hydroxy-17α-methyl-	EtOH	242(4.18)	44-3022-69
Gon-4-en-3-one, 13β-ethyl-17β-hydroxy-	EtOH	241(4.20)	44-3530-69
18-Nor-5β-androst-12-en-11-one, 3α-hydroxy-2-methyl-	EtOH	248(4.08)	23-3489-69
Pipataline	EtOH	262(4.10),269(4.09), 307(3.79)	18-0569-69

Compound	Solvent	$\lambda_{max}(\log \epsilon)$	Ref.
$C_{19}H_{28}O_3$			
Androst-4-en-3-one, 4,17β-dihydroxy-	EtOH	277(4.12)	94-1206-69
5α,14β-Androst-15-en-17-one, 3β,16-dihydroxy-	EtOH	260(3.88)	39-0554-69C
	EtOH-NaOH	298(3.78)	39-0554-69C
9β,10α-Androst-4-en-3-one, 8,17β-dihydroxy-	EtOH	238(4.22)	33-1157-69
9β,10α-Androst-4-en-3-one, 9,17β-dihydroxy-	EtOH	241(4.20)	33-1157-69
5α-Androst-15-en-17-one, 14β-hydroperoxy-	n.s.g.	216(3.83)	88-0483-69
5β-Androst-15-en-17-one, 14β-hydroperoxy-	EtOH	212(3.84)	39-2767-69C
Gon-4-en-3-one, 13β-ethyl-12α,17β-dihydroxy-	EtOH	241(4.24)	44-3530-69
Phenanthren-2(3H)-one, 8aβ-ethyl-8-hydroxy-4,4a,4b,5,6,7,8,8a,9,10-decahydro-1-methyl-, acetate	EtOH	250(4.21)	44-0107-69
$C_{19}H_{28}O_4S$			
2-Bornaneethanol, 2-hydroxy-, p-toluenesulfonate	EtOH	225(4.08),255s(2.63), 257(2.64),262(2.74), 265s(2.69),267(2.71), 273(2.66)	22-2372-69
$C_{19}H_{28}O_5$			
7,9a-Methano-9aH-cyclopenta[b]heptalen-4(1H)-one, 2,3,4a,5,6,7,8,9,10,11-decahydro-2,8,11,12-tetrahydroxy-1,1,8-trimethyl-	EtOH	257(3.95)	94-2036-69
$C_{19}H_{28}O_7$			
D-1-Glucooctulose, 3,6-anhydro-2-deoxy-1-(p-methoxyphenyl)-4,5,7,8-tetra-O-methyl-	MeOH	273(4.23)	65-0119-69
$C_{19}H_{29}N$			
Androsta-3,5-dien-17-amine	EtOH	228(4.24),235(4.28), 243s(4.08)	94-1749-69
$C_{19}H_{29}NO_2$			
2,5-Cyclohexadien-1-one, 2,6-di-tert-butyl-4-(morpholinomethylene)-	hexane	380(4.49)	70-0580-69
$C_{19}H_{29}NO_3$			
7a-Aza-B-homoandrost-5-en-7-one, 3β,17β-dihydroxy-	EtOH	221(4.22)	2-1084-69
2,4-Pentadienoic acid, 2-acetyl-5-(1-cyclohexen-1-yl)-5-(diethylamino)-, ethyl ester	EtOH	243(4.05),265(4.04), 405(4.78)	70-0363-69
$C_{19}H_{29}NO_4$			
5α-Androstan-3-one, 17β-hydroxy-, nitrate	C_6H_{12}	186(3.90)(end abs.)	78-0761-69
Androsterone, nitrate	C_6H_{12}	186(3.83)(end abs.)	78-0761-69
$C_{19}H_{29}NO_8$			
3,4-Furandicarboxylic acid, 2-(2-carboxyethyl)-5-(cyclohexylamino)-2,5-dihydro-2-methoxy-, trimethyl ester	ether	242(4.00)	24-1656-69

Compound	Solvent	$\lambda_{max}(\log \epsilon)$	Ref.
$C_{19}H_{30}ClNO$			
17a-Aza-D-homoandrost-5-en-3β-ol, N-chloro-	dioxan	277(2.59)	78-0737-69
$C_{19}H_{30}NO_3P$			
Phosphonic acid, [1-benzyl-2-cyclohexylamino)vinyl]-, diethyl ester	EtOH	248(4.26)	39-0460-69C
$C_{19}H_{30}N_2O_2$			
17a-Aza-D-homoandrost-5-en-3β-ol, N-nitroso-	dioxan	235(3.95),368(1.95)	78-0737-69
2-Cyclohexen-1-one, 3,3'-(trimethylenediimino)bis[5,5-dimethyl-	MeOH	290(4.27)	39-2044-69C
$C_{19}H_{30}O_2$			
Benzo[1,2:3,4]dicycloheptene-6-carboxylic acid, 1,2,3,4,5,5a,6,7,7a,8,9,-10,11,12-tetradecahydro-6-methyl-, methyl ester	EtOH	207(4.12)	22-4493-69
7H-Benzo[e]inden-7-one, 3α-tert-butoxy-1,2,3,3a,4,5,8,9,9a,9b-decahydro-3aα,6-dimethyl-	EtOH	249(4.20)	44-1457-69
$C_{19}H_{30}O_3$			
Butyrophenone, 2'-hydroxy-4'-(nonyloxy)-	n.s.g.	275(4.05)	104-1609-69
$C_{19}H_{30}O_4S$			
Cyclohexaneethanol, 4-tert-butyl-1-hydroxy-, p-toluenesulfonate	EtOH	225(4.09),255s(2.65), 257(2.66),262(2.76), 265s(2.71),267(2.72), 273(2.68)	22-2372-69
$C_{19}H_{30}O_6$			
7,9a-Methano-9aH-cyclopenta[b]heptalen-4(1H)-one, dodecahydro-2,8,11,11a,12-pentahydroxy-1,1,8-trimethyl-	EtOH	286(1.48)	94-2036-69
$C_{19}H_{31}NO_2$			
Androstane, 6α-nitro-	EtOH	282(1.88)	22-1632-69
Bicyclo[9.3.1]pentadec-11(15)-en-12-one, 15-morpholino-	EtOH	339(4.30)	39-0592-69C
perchlorate	EtOH	334(4.21)	39-0592-69C
$C_{19}H_{31}NO_3$			
Androstan-3β-ol, 6α-nitro-	EtOH	282(1.74)	22-1632-69
$C_{19}H_{31}NO_4$			
5α-Androstane-3β,17β-diol, 17-nitrate	C_6H_{12}	188(3.74)	78-0761-69
$C_{19}H_{32}OS_2$			
Camphor, α-bis(butylthio)methylene-	dioxan	325(4.10),375s(2.65)	28-0186-69A
$C_{19}H_{32}O_2$			
2,5-Cyclohexadien-1-one, 2,4,6-tri-tert-butyl-4-methoxy-	MeOH	239(3.98)	64-0547-69
$C_{19}H_{32}O_4$			
2,6-Nonadienoic acid, 3,7-dimethyl-9-(2,2,5,5-tetramethyl-1,3-dioxolan-4-yl)-, methyl ester, cis-cis	EtOH	214(4.05)	12-1737-69

Compound	Solvent	$\lambda_{max}(\log \epsilon)$	Ref.
2,6-Nonadienoic acid, 3,7-dimethyl-9-(2,2,5,5-tetramethyl-1,3-dioxolan-4-yl)-, methyl ester, 2-cis-6-trans	EtOH	214(4.07)	12-1737-69
6-cis-2-trans	EtOH	219(4.14)	12-1737-69
trans-trans	EtOH	219(4.12)	12-1737-69
$C_{19}H_{36}N_2O$			
1,3,4-Oxadiazole, 2-heptadecyl-	MeOH	252(2.3)	48-0646-69
$C_{19}H_{42}BrIrOP_2$			
Iridium, bromocarbonylbis(triisopropyl-phosphine)-	toluene	340(3.37),383(3.48), 436(2.79)	64-0770-69
$C_{19}H_{42}ClIrOP_2$			
Iridium, carbonylchlorobis(triisopropyl-phosphine)-	toluene	336(3.38),379(3.51), 432(2.79)	64-0770-69

Compound	Solvent	$\lambda_{max}(\log \epsilon)$	Ref.
$C_{20}H_8O_4S_2$ Dibenzo[b,i]thianthrene-5,7,12,14-tetrone	morpholine	576(3.79)	83-0075-69
	piperidine	588(3.28)	83-0075-69
$C_{20}H_{10}F_4$ Phenanthrene, 1,2,3,4-tetrafluoro-9-phenyl-	MeOH	240(4.45),248(4.56), 266(4.10),290(3.94), 300(3.99),335(2.96), 350(2.88)	39-1684-69C
$C_{20}H_{10}HgO_4$ Mercury, bis[2-oxo-6-(1,3-pentadiynyl)-2H-pyran-5-yl]-	MeOH	252(4.09),267(4.18), 341(4.49)	1-1461-69
$C_{20}H_{10}N_2$ 2,5-Cyclopentadiene-1,2-dicarbonitrile, 4-phenalen-1-ylidene-	MeCN	270(4.08),287(4.14), 300s(4.10),348s(3.67), 365(3.71),394s(3.65), 587(4.37)	89-0882-69
	C_6H_{12}	555(--)	89-0882-69
	benzene	569(--)	89-0882-69
	EtOH	585(--)	89-0882-69
	CH_2Cl_2	585(--)	89-0882-69
	$CHCl_3$	579(--)	89-0882-69
	CCl_4	560(--)	89-0882-69
	MeCN	587(--)	89-0882-69
	CF_3COOH	596(--)	89-0882-69
protonated form	80% H_2SO_4	226(4.44),314(3.91), 327(3.92),401(4.21), 458(4.21),528(3.90)	89-0882-69
$C_{20}H_{10}O_7$ [1,1'-Binaphthalene]-3,4,5,8'-tetrone, 6',7'-epoxy-6',7'-dihydro-4',5'-dihydroxy- (dehydromycochrysone)	EtOH	243(4.47),365(3.78), 438(3.75)	39-2059-69C
$C_{20}H_{11}BrO_6$ Furo[3',4':6,7]naphtho[1,2-d]-1,3-dioxol-7(9H)-one, bromo-10-[3,4-(methylenedioxy)phenyl]-	EtOH	223(4.57),269(4.67), 290(3.71),360(3.93)	39-0693-69C
$C_{20}H_{11}NO$ 2-Biphenylenecarbonitrile, 6-benzoyl-	EtOH	261(4.64),280(4.51), 344s(3.59),353(3.83), 371(3.97)	39-2789-69C
$C_{20}H_{11}NO_8$ Furo[3',4':6,7]naphtho[1,2-d]-1,3-dioxol-7(9H)-one, 10-[3,4-(methylenedioxy)phenyl]nitro-	EtOH	220(4.34),265(4.46), 411(3.99)	39-0693-69C
$C_{20}H_{12}$ Azulene, 8(4)-ethynyl-1-(p-ethynylphenyl)-	n.s.g.	746(1.73)	88-3003-69
Azuleno[5,6,7-cd]phenalene	EtOH	247(4.38),309(5.06), 316(5.11),360(3.85), 380(3.72),384(3.71), 397(3.90),402(3.96), 424(4.50),441(4.65), 449(4.97),613(2.85),	24-2301-69

(continued on next page)

Compound	Solvent	λ_{max}(log ϵ)	Ref.
Azuleno[5,6,7-cd]phenalene (cont.)		631(2.93),641(2.90), 665(3.15),674(3.08), 700(3.11),714(2.70), 726(2.78),739(3.26)	
Benzo[j]fluoranthene	EtOH	226(4.58),242(4.60), 308(4.38),318(4.48), 332(4.04),349(3.56), 365(3.82),376(3.73), 383(4.00)	35-0371-69
Benzo[k]fluoranthene	EtOH	215(4.56),240(4.75), 269(4.36),282(4.42), 296(4.67),308(4.75), 361(3.86),380(4.12)	35-0371-69
$C_{20}H_{12}Cl_2$			
Anthracene, 9-chloro-10-(p-chloro-phenyl)-	EtOH	250s(4.94),258(5.14), 340(3.54),356(3.86), 374(4.07),395(4.05)	78-3485-69
$C_{20}H_{12}Cl_2N_2O_3$			
Acetamide, N-[5-chloro-1-(o-chlorophen-yl)-2,3-naphthalenedicarboximido]-	HOAc	264(4.53)	39-0838-69C
Acetamide, N-[7-chloro-1-(p-chlorophen-yl)-2,3-naphthalenedicarboximido]-	EtOH	264(4.85),350(3.61), 364(3.70)	39-0838-69C
$C_{20}H_{12}Cl_4O$			
Phenalene, 3-ethoxy-1-(2,3,4,5-tetra-chloro-2,4-cyclopentadien-1-ylidene)-	pentane	244s(3.71),305(3.80), 387(3.52),408(3.66), 587(3.95)	88-1921-69
	MeCN	245s(3.79),313(3.84), 370s(3.57),409(3.71), 606(3.97)	88-1921-69
	H_2SO_4	368(4.07),417(4.12), 460s(4.01)	88-1921-69
$C_{20}H_{12}F_4$			
Phenanthrene, 1,2,3,4-tetrafluoro-9,10-dihydro-10-phenyl-	MeOH	266(4.19),293(3.34)	39-1684-69C
$C_{20}H_{12}N_2S_2$			
Malononitrile, [α-(4-phenyl-1,3-dithiol-2-ylidene)methyl]benzylidene]-	$CHCl_3$	251(4.22),308(3.80), 492(4.67)	24-1580-69
Malononitrile, [α-[(5-phenyl-3H-1,2-di-thiol-3-ylidene)methyl]benzylidene]-	$CHCl_3$	330(4.19),534(4.52)	24-1580-69
$C_{20}H_{12}N_2S_2Se$			
Benzothiazoline, 2,2'-(2,5-selenophene-diyldimethylidyne)bis-	ether	307(4.29),319(4.51), 432(4.34)	124-0824-69
sulfate	HOAc	339(4.34),491(4.67), 525(4.74)	124-0824-69
$C_{20}H_{12}N_6O_2$			
Pyridine, 2,2'-[m-phenylenebis(1,3,4-oxadiazole-5,2-diyl)]di-	H_2SO_4	234(4.17),264(4.33), 301(4.71)	4-0965-69
$C_{20}H_{12}N_6O_4$			
Benzoic acid, m-azido-, m-phenylene ester	dioxan	234(4.99),299(4.20)	73-3811-69
Benzoic acid, m-azido-, p-phenylene ester	dioxan	232(4.68),298(3.76)	73-3811-69

Compound	Solvent	$\lambda_{max}(\log \epsilon)$	Ref.
Benzoic acid, p-azido-, m-phenylene ester	dioxan	278(4.79)	73-3811-69
Benzoic acid, p-azido-, p-phenylene ester	dioxan	278(4.77)	73-3811-69
$C_{20}H_{12}O$			
3H-Benzo[cd]pyren-3-one, 5-methyl-	EtOH	400f(4.3)	104-1628-69
	EtOH-HCl	460(4.5)	104-1628-69
$C_{20}H_{12}O_2$			
1,3-Indandione, 2-(1-naphthylmethylene)-	C_6H_{12}	222(4.72),263(4.39), 395(4.29),420(4.15)	108-0099-69
	MeCN	221(4.77),263(4.39), 395(4.20),425(3.98)	108-0099-69
1,3-Indandione, 2-(2-naphthylmethylene)-	C_6H_{12}	220(4.48),262(4.39), 280(4.19),290(4.24), 360(4.47),381(4.30), 400(4.29),440s(3.04)	108-0099-69
	MeCN	220(4.52),261(4.39), 287(4.23),354(4.45), 400(4.22)	108-0099-69
$C_{20}H_{12}O_3$			
2-Biphenylenecarboxylic acid, 6-benzoyl-	EtOH	262(4.67),281(4.49), 345s(3.62),353(3.89), 372(4.00)	39-2789-69C
Furano[2,3-c]benzo[h]xanthone, 2-methyl-	$CHCl_3$	248(4.47)	49-1368-69
$C_{20}H_{12}O_6$			
Furo[3',4':6,7]naphtho[1,2-d]-1,3-diox-ol-7(9H)-one, 10-[3,4-(methylenedi-oxy)phenyl]- (helioxanthin)	EtOH	220(4.48),267(4.66), 290(3.70),354(3.88)	39-0693-69C
	EtOH-base	252(4.67),293(3.96), 320(3.94),350(3.81)	39-0693-69C
$C_{20}H_{13}Br$			
Anthracene, 9-bromo-10-phenyl-	EtOH	260(5.01),342(3.52), 358(3.86),377(4.08), 399(4.05)	78-3485-69
$C_{20}H_{13}BrN_2O_2$			
Benzo[3,4]cyclobuta[1,2-b]phenazine, 6-bromo-8,9-dimethoxy-	EtOH	244(4.38),285s(4.72), 291(4.74),321(4.37), 356(3.84),437(4.46), 460(4.52)	39-0646-69C
$C_{20}H_{13}Cl$			
Anthracene, 9-chloro-10-phenyl-	EtOH	260(5.06),341(3.54), 357(3.89),377(4.01), 398(4.10)	78-3485-69
Anthracene, 9-(p-chlorophenyl)-	EtOH	247s(4.93),254(5.13), 331(3.57),346(3.84), 364(4.02),383(4.00)	78-3485-69
$C_{20}H_{13}Cl_2N_3O$			
Δ^3-1,3,4-Oxadiazoline, 2,2-bis(p-chloro-phenyl)-5-(phenylimino)-	hexane	226(4.54),329(3.85), 357s(--)	44-3230-69
$C_{20}H_{13}N$			
Cyclopenta[ij]pyrido[2,1,6-de]quinoli-zine, 3-phenyl- (continued)	EtOH	255(4.55),277(4.61), 339s(4.13),363(4.32),	39-0239-69C

Compound	Solvent	λ_{max}(log ε)	Ref.
Cyclopenta[ij]pyrido[2,1,6-de]quinolizine, 3-phenyl- (cont.)	EtOH	436(3.68),449(4.02), 559s(2.40),599(2.56), 650(2.59),719(2.27)	
$C_{20}H_{13}NOS$			
Isoindolo[1,2-b]benzothiazol-11(4bH)-one, 4b-phenyl-	EtOH	284s(3.25)	44-0165-69
$C_{20}H_{13}NO_2$			
Isoindolo[1,2-b]benzoxazol-11(4bH)-one, 4b-phenyl-	EtOH	279(4.39)	44-0165-69
$C_{20}H_{13}N_3O_2$			
Phenanthrene, 1-[(p-nitrophenyl)azo]-	EtOH	216(4.57),244(4.56), 290(4.31),394(4.08)	24-2384-69
$C_{20}H_{13}N_3S$			
Benzimidazo[1,2-c]quinazoline-6(5H)-thione, 5-phenyl-	MeOH	239(4.60),297(4.32), 310(4.31),323(4.14), 340(4.07)	4-0605-69
$C_{20}H_{14}$			
6,7-Acechrysene	C_6H_{12}	263(5.03),272(5.32), 302(4.20),315(4.30), 329(4.30)	44-1176-69
Anthracene, 9-phenyl-	EtOH	250s(4.90),256(5.09), 331(3.44),347(3.77), 365(3.97),385(3.94)	78-3485-69
$C_{20}H_{14}Br_2N_2$			
p-Phenylenediamine, N,N'-dibenzylidene-2,5-dibromo-	EtOH	207(4.55),264(4.59), 355(4.26)	117-0287-69
$C_{20}H_{14}Br_2O_2$			
1,4-Butanedione, 2,3-dibromo-1-(2-naphthyl)-4-phenyl-	EtOH	260(4.74),285(4.34)	104-0511-69
$C_{20}H_{14}Cl_2N_2O$			
1,3-Butadiene-1,1-dicarbonitrile, 2,4-bis(p-chlorophenyl)-3-ethoxy-	CH_2Cl_2	267(4.21),333(4.19), 400s(--)	24-0319-69
$C_{20}H_{14}Cl_2N_2OS$			
2(1H)-Quinazolinethione, 6-chloro-3-(p-chlorophenyl)-3,4-dihydro-4-hydroxy-4-phenyl-	n.s.g.	255(4.13),295(4.45)	40-0917-69
$C_{20}H_{14}FeO_2$			
1,3-Indandione, 2-(ferrocenylmethylene)-	C_6H_{12}	223(4.36),246(4.37), 352(4.40),410s(3.48), 560(3.72)	108-0555-69
	MeCN	228(4.53),245(4.38), 360(4.36),435s(3.58), 575(3.72)	108-0555-69
$C_{20}H_{14}HgO_4$			
Mercury, bis[2-oxo-6-(3-penten-1-ynyl)-2H-pyran-5-yl]-	MeOH	260(4.26),341(4.48)	1-1059-69

Compound	Solvent	λ_{max}(log ε)	Ref.
$C_{20}H_{14}N_2$			
1-Cyclopropene, 1,2-di-p-tolyl-3-(dicyanomethylene)-	CH_2Cl_2	246(4.19),262s(--), 275s(--),284(4.50), 293(4.51),302(4.47), 361(4.20)	24-0319-69
9H-Dibenzo[c,e]benzimidazo[1,2-a]azepine	EtOH	230(4.43),260(4.30), 300(4.28)	78-3789-69
1,2-Naphthalenedicarbonitrile, 4-(α-methylbenzyl)-	MeCN	219(4.51),242(4.80), 305s(3.82),315(3.89), 330(3.67),348(3.75)	44-1923-69
$C_{20}H_{14}N_2O$			
Benzimidazole, 1-benzoyl-2-phenyl-	EtOH	271(4.53)	88-4957-69
3H-Indole, 2-phenyl-3-(phenylimino)-, 1-oxide	benzene	425(3.6)	7-0315-69
11H-Isoindolo[2,1-a]benzimidazol-11-one, 4b,5-dihydro-4b-phenyl-	EtOH	250s(4.04),305s(3.47)	44-0165-69
4(1H)-Quinazolinone, 1,2-diphenyl-	EtOH	236(4.40),306(4.05), 316s(3.98)	78-0783-69
4(3H)-Quinazolinone, 2,3-diphenyl-	EtOH	229(4.52),279(4.20)	78-0783-69
$C_{20}H_{14}N_2O_2$			
Benzo[3,4]cyclobuta[1,2-b]phenazine, 8,9-dimethoxy-	EtOH	220(4.32),244(4.58), 282(4.82),290(4.85), 320(4.36),355(3.84), 443(4.56),514(4.60)	39-0646-69C
1-Cyclopropene, 1,2-bis(p-methoxyphenyl)-3-(dicyanomethylene)-	CH_2Cl_2	226(4.21),238s(--), 247(4.08),297(4.60), 314(4.51),372(4.39)	24-0319-69
3H-Indole, 2-phenyl-3-(phenylimino)-, N,1-dioxide	benzene	388(4.29),500(3.5)	7-0315-69
$C_{20}H_{14}N_2O_3$			
Acetamide, N-(1-phenyl-2,3-naphthalenedicarboximido)-	EtOH	261(4.72),295s(3.97), 350(3.59),363(3.66)	39-0838-69C
$C_{20}H_{14}N_2O_5$			
[$\Delta^{3,3'}$-Biindoline]-2,2'-dione, 1-acetoxy-1'-acetyl-	dioxan	456(3.83)	24-3691-69
$C_{20}H_{14}N_2O_6$			
[$\Delta^{3,3'}$-Biindoline]-2,2'-dione, 1,1'-diacetoxy-	dioxan	459(3.68)	24-3691-69
$C_{20}H_{14}N_2S$			
Sulfur diimide, di-1-naphthyl-	hexane	285(4.1),460(4.1)	48-0621-69
Sulfur diimide, di-2-naphthyl-	hexane	295(4.15),370(4.05), 455(4.2)	48-0621-69
$C_{20}H_{14}N_4$			
Benzene, m-bis(diazobenzyl)-	dioxan	290(4.65),525(2.31)	47-3313-69
Benzene, p-bis(diazobenzyl)-	dioxan	300(4.45),330(4.39), 535(2.52)	47-3313-69
$C_{20}H_{14}N_6O_6$			
2-Pyrazoline, 3-(diazoacetyl)-1-(p-nitrocinnamoyl)-4-(p-nitrophenyl)-	$C_2H_4Cl_2$	310(4.45),336(4.36)	94-2105-69

Compound	Solvent	λ_{max}(log ε)	Ref.
$C_{20}H_{14}N_8$			
Pyridine, 2,2'-[m-phenylenebis(s-triazole-5,3-diyl)di-	H_2SO_4	248(4.41),289(4.65)	4-0965-69
$C_{20}H_{14}O$			
Dibenzo[5,6:7,8]cyclodeca[1,2-c]furan	EtOH	243(4.42),285s(3.42)	12-1449-69
minor conformer	EtOH	241(4.23),287s(3.53)	12-1449-69
$C_{20}H_{14}OS$			
1-Azulenyl sulfoxide	CHCl	282(4.78),298s(4.40), 336(4.14),373(4.20), 557(2.91)	44-2022-69
Benzo[b]furan, 2-[p-[2-(2-thienyl)vinyl]phenyl]-	DMF	366(4.78)	33-1282-69
$C_{20}H_{14}O_2$			
Benzyl alcohol, 2,2'-ethynylenebis[α-ethynyl-	EtOH	286(4.43),304(4.34)	44-0874-69
5,12,6,11-[1,2,3,4]Butanetetrayldibenzo-[a,e]cyclooctene-13,16-dione, 5,6,11,12-tetrahydro-	MeOH	266(2.64),274(2.63), 292(2.54),310s(2.32)	88-0621-69
2-Butene-1,4-dione, 1-(2-naphthyl)-4-phenyl-	EtOH	227(4.60),260(4.45)	104-0511-69
1-Fluorenol, benzoate	MeOH	218s(--),235(4.11), 253(4.15),258(4.18), 263(4.38),285(3.66), 297(3.67)	39-0345-69C
$C_{20}H_{14}O_2S$			
1-Azulenyl sulfone	$CHCl_3$	276(4.67),296(4.57), 306(4.53),366(4.09), 381(4.10),541(2.95), 625(2.45)	44-2022-69
$C_{20}H_{14}O_3$			
3,5-Cyclohexadiene-1,2-dicarboxylic anhydride, 3,6-diphenyl-, cis	EtOH	235s(4.08),340(4.44)	22-3662-69
$C_{20}H_{14}O_4$			
Benzophenone, 2,4-dihydroxy-, 4-benzoate	$CHCl_3$	275(4.09),330(3.68)	39-0034-69C
Isophthalic acid, 4,5-diphenyl-	EtOH	231(4.38),296(3.46)	44-4134-69
4H-Naphtho[1,2-b]pyran-4-one, 3-hydroxy-2-(p-methoxyphenyl)-	$CHCl_3$	283(4.75),332?(4.65)	115-0017-69
$C_{20}H_{14}O_5$			
2,3-Naphthalenecarboxylic anhydride, 1-(p-methoxyphenyl)-7-methoxy-	HOAc	269(4.63),329(3.93)	39-0838-69C
$C_{20}H_{14}O_6$			
Isoelliptone, dehydro-	EtOH	238(4.51),273(4.39), 307(4.30)	31-0789-69
Mycochrysone	EtOH	236(4.49),303(3.75), 438(3.80)	39-2059-69C
$C_{20}H_{14}S$			
Anthracene, 2-[2-(2-thienyl)vinyl]-	DMF	323(4.77),338(4.85), 377(4.40),397(4.38)	33-2521-69

Compound	Solvent	$\lambda_{max}(\log \epsilon)$	Ref.
1-Azulenyl sulfone	C_6H_{12}	238(4.56),278(4.77), 293s(4.66),337(3.93), 371(4.07),598(2.73), 622(2.73),680s(2.52)	44-2022-69
$C_{20}H_{14}S_2$			
Benzo[b]thiophene, 2-[p-[2-(2-thienyl)-vinyl]phenyl]-	DMF	367(4.72)	33-1282-69
$C_{20}H_{15}BF_2O_3$			
Naphtho[1,2-e]-1,3,2-dioxaborin, 2,2-difluoro-4-(p-methoxystyryl)-	MeCN	224(4.64),374(4.09), 470(4.32)	4-0029-69
Naphtho[2,1-e]-1,3,2-dioxaborin, 2,2-difluoro-4-(p-methoxystyryl)-	MeCN	265(4.28),327(3.90), 408(4.47),478(4.55)	4-0029-69
$C_{20}H_{15}BrN_2O_2$			
2,4,6-Cycloheptatrien-1-one, 3-bromo-2-hydroxy-6-phenyl-5-(p-tolylazo)-	n.s.g.	265(4.25),307(4.23), 414(4.34),460s(4.20)	40-0085-69
$C_{20}H_{15}ClN_2OS$			
2(1H)-Quinazolinethione, 6-chloro-3,4-dihydro-4-hydroxy-3,4-diphenyl-	n.s.g.	256(4.10),292(4.47)	40-0917-69
$C_{20}H_{15}ClN_2O_2$			
2(1H)-Quinazolinone, 6-chloro-3,4-dihydro-4-hydroxy-3,4-diphenyl-	n.s.g.	254(4.35),296(3.43)	40-0917-69
Urea, N-(2-benzoyl-4-chlorophenyl)-N'-phenyl-	n.s.g.	252(4.65),280s(--), 360(3.71)	40-0917-69
$C_{20}H_{15}ClO_4S$			
4-Methyl-2-phenylnaphtho[1,2-b]thiopyrylium perchlorate	HOAc-1% $HClO_4$	258(4.4),310(4.55), 450(4.84)	2-0191-69
1-Methyl-3-phenylnaphtho[2,1-b]thiopyrylium perchlorate	HOAc-1% $HClO_4$	325(4.25),400(4.1), 450s(4.01)	2-0191-69
$C_{20}H_{15}FeNO$			
2-Propen-1-one, 1-(p-cyanophenyl)-3-ferrocenyl-	MeOH	330(4.20)	73-2771-69
2-Propen-1-one, 3-(p-cyanophenyl)-1-ferrocenyl-	MeOH	303(4.41)	73-2771-69
$C_{20}H_{15}N$			
Benzo[f]quinoline, 3-p-tolyl-	dioxan	233(4.29),262(4.61), 282(4.64),290(4.58), 312(4.31),330(3.78), 346(3.96),361(3.99)	24-1202-69
Indole, 1,3-diphenyl-	EtOH	221(4.19),255(4.32), 305(4.19)	44-2988-69
$C_{20}H_{15}NO$			
Benzamide, N-(diphenylmethylene)-	EtOH	254(4.36),280(4.08), 350(1.87)	88-0927-69
2-Indolinone, 1,3-diphenyl-	EtOH	257(4.28),290(3.91)	44-2988-69
$C_{20}H_{15}NOS$			
Benzothiazoline, 2-benzoyl-2-phenyl-	MeOH	220(4.5),250(4.2), 320(3.7),380(2.8)	24-0728-69
	MeOH-0.01N NaOH	220(4.4),250(4.2), 320(3.7),370(2.8)	24-0728-69

Compound	Solvent	$\lambda_{max}(\log \epsilon)$	Ref.
Benzothiazoline, 2-benzoyl-2-phenyl- (cont.)	MeOH-0.1N NaOH	220(4.4),260(4.2), 320s(3.8),370(3.1)	24-0728-69
	MeOH-NaOH	220(4.3),260(4.3), 320s(3.8),350(3.3)	24-0728-69
$C_{20}H_{15}NO_2$			
5H-Benzo[c]furo[4,3,2-mn]acridin-5-one, 6a,7-dihydro-7,7-dimethyl-	MeOH	212(4.74),226s(4.52), 270(4.58),334s(4.09), 346(4.21),360(4.10)	4-0361-69
$C_{20}H_{15}NO_4$			
4H-Quinolizine-3-acrylic acid, 1-benzoyl-4-oxo-, methyl ester, trans	MeOH	208(4.52),240s(4.18), 284(4.35),325s(3.75), 380(4.09),428(4.40)	39-1143-69C
$C_{20}H_{15}N_3$			
Pyrazole, 4-(2-pyridyl)-3,5-diphenyl-	EtOH	248(4.4)	95-0783-69
Pyrazole, 4-(3-pyridyl)-3,5-diphenyl-	EtOH	246(4.4)	95-0783-69
$C_{20}H_{15}N_3O$			
Δ^3-1,3,4-Oxadiazoline, 2,2-diphenyl-5-(phenylimino)-	hexane	220(4.14),326(3.79), 357s(--)	44-3230-69
$C_{20}H_{15}N_3O_3S$			
Thiourea, 1-(4-benzoylphenyl)-3-(m-nitrophenyl)-	dioxan	304(4.17)	34-0506-69
Thiourea, 1-(4-benzoylphenyl)-3-(p-nitrophenyl)-	dioxan	320(4.43)	34-0506-69
$C_{20}H_{15}N_3S$			
Thiourea, 1-(1-naphthyl)-3-(8-quinolyl)-	dioxan	313(4.16),323(4.15), 355(4.17)	34-0506-69
Thiourea, 1-(2-naphthyl)-3-(8-quinolyl)-	dioxan	243(4.61),263(4.56), 332(4.05)	34-0506-69
$C_{20}H_{15}N_5$			
as-Triazine, 5,6-di-2-pyridyl-3-p-tolyl-	MeOH	289(4.49),380(2.62)	88-3147-69
$C_{20}H_{15}N_5O$			
1-Phthalazinecarbonitrile, 4-acetonyl-3-(1-phthalazinyl)-	EtOH	250s(4.02),373(4.26)	95-0959-69
$C_{20}H_{15}N_5O_8$			
2-Picoline, 4,6-bis(2,4-dinitrobenzyl)-	EtOH	244(4.42)	22-4425-69
4-Picoline, 2,6-bis(2,4-dinitrobenzyl)-	EtOH	244(4.42)	22-4425-69
$C_{20}H_{16}$			
Azulene, 8(4)-vinyl-1-(p-vinylphenyl)-	n.s.g.	694(2.29)	88-3003-69
Benz[a]pyrene, 7,8,9,10-tetrahydro-	EtOH	238(4.77),247(5.04), 257(4.23),269(4.51), 278(4.74),303(3.72), 315(4.15),329(4.56), 345(4.74)	24-2301-69
Fluorene, 2-o-tolyl-	C_6H_{12}	272(4.49),287s(4.45), 304(4.24)	33-1091-69
Stilbene, 4-phenyl-	DMF	328(4.66)	33-2521-69
$C_{20}H_{16}BrNO_4$			
Pyrrole-2,3-dicarboxylic acid, 1-(p-bromophenyl)-4-phenyl-, dimethyl ester	MeOH	243(4.51),286s(3.97)	78-0527-69

Compound	Solvent	λ_{max}(log ε)	Ref.
$C_{20}H_{16}ClNO$			
Benz[c]acridine-7-carbonyl chloride, 5,6-dihydro-5,5-dimethyl-	MeOH	214(4.64),225s(4.42), 260s(4.43),268(4.54), 302(3.85),318(3.87), 333(4.03),347(4.10)	4-0361-69
$C_{20}H_{16}ClN_3O_5$			
Furo[2,3-d]pyrimidine-2,4(1H,3H)-dione, 5-(p-chlorophenyl)-5,6-dihydro-6,6-dimethyl-3-(p-nitrophenyl)-	EtOH	209(4.43),218(4.43), 267(4.33)	87-0339-69
$C_{20}H_{16}Cl_2N_2O_3$			
Furo[2,3-d]pyrimidine-2,4(1H,3H)-dione, 3,5-bis(p-chlorophenyl)-5,6-dihydro-6,6-dimethyl-	EtOH	209(4.45),223(4.43), 268(3.90)	87-0339-69
$C_{20}H_{16}F_8O_2$			
2,4-Cyclohexadien-1-one, 4-[2,3-dimethyl-2-[(2,3,5,6-tetrafluoro-p-tolyl)-oxy]-3-butenyl]-2,3,5,6-tetrafluoro-4-methyl-	n.s.g.	244(4.06)	70-2324-69
$C_{20}H_{16}Fe_2$			
1,1'-Biferrocenylene	C_6H_{12}	218(--)	35-1258-69
	$CHCl_3$	360s(2.75),466(2.41)	35-1258-69
$C\ _0H_{16}INO_4$			
Pyrrole-2,3-dicarboxylic acid, 1-(p-iodophenyl)-4-phenyl-, dimethyl ester	MeOH	246(4.51),284s(3.97)	78-0527-69
$C_{20}H_{16}N_2$			
Benzeneazoethene, 1',2'-diphenyl-	THF	348(4.52)	24-3647-69
Isoindole, 2-(o-aminophenyl)-1-phenyl-	EtOH	215(4.53),284(3.87), 304(3.87),320s(3.81), 350(3.81)	44-1720-69
1,2-Naphthalenedicarbonitrile, 3,4-dihydro-4-(α-methylbenzyl)-, erythro	MeCN	242(4.09),308(4.04)	44-1923-69
threo	MeCN	243(4.07),311(3.99)	44-1923-69
3H-Pyrrolo[1,2-b][1,2]diazepine, 2,5-diphenyl-	EtOH	240(4.4),281(4.3)	24-3268-69
$C_{20}H_{16}N_2O$			
Benzo[g]quinoline, 4-p-anisidino-	EtOH	248(4.83),398(4.06)	103-0223-69
1,3-Butadiene-1,1-dicarbonitrile, 3-ethoxy-2,4-diphenyl-	CH_2Cl_2	258(4.09),318(4.02), 380s(--)	24-0319-69
$C_{20}H_{16}N_2OS$			
Thiourea, 1-(4-benzoylphenyl)-3-phenyl-	EtOH	315(4.46)	34-0506-69
$C_{20}H_{16}N_2O_2$			
Benzo[g]phthalazine-1,4-dione, 2,3-dihydro-7-methyl-5-p-tolyl-	EtOH	251(4.82),320(3.75), 338(3.82)	39-0838-69C
Benzyl alcohol, α-(phenylazo)-, benzoate	dioxan	275(4.09),405(2.32)	39-2587-69C
2,2'-Biquinoline, 4,4'-dimethoxy-, Cu(I) chelate	iso-AmOH	538(3.85)	3-0344-69
2,4,6-Cycloheptatrien-1-one, 2-hydroxy-4-phenyl-5-(p-tolylazo)-	n.s.g.	225(4.32),255(4.27), 300(4.15),400(4.38)	40-0085-69

Compound	Solvent	$\lambda_{max}(\log \epsilon)$	Ref.
$C_{20}H_{16}N_2O_3$			
2,4,6-Cycloheptatrien-1-one, 2-hydroxy-5-[(p-methoxyphenyl)azo]-4-phenyl-	n.s.g.	224(4.30),256(4.32), 304(4.01),412(4.46)	40-0085-69
$C_{20}H_{16}N_2O_4$			
2,3-Naphthalenedicarboximide, N-amino-7-methoxy-1-(p-methoxyphenyl)-	BuOH	270(4.67)	39-0838-69C
1,4-Naphthalenediol, 2-(phenylazo)-, diacetate	MeOH	285(4.21),295(4.23), 331(4.31),365s(3.96)	44-2750-69
$C_{20}H_{16}N_2O_5$			
Camptothecin, 10-hydroxy-	MeOH	222(4.70),267(4.44), 330(4.08),382(4.45)	44-1364-69
$C_{20}H_{16}N_2O_8$			
4-Cyclopentene-1,3-diol, 4-methyl-, bis(p-nitrobenzoate)	EtOH	260(4.42)	95-0750-69
$C_{20}H_{16}N_2S$			
Lepidine, 2-(4-methyl-2-phenyl-5-thiazolyl)-	n.s.g.	352(4.56)	97-0186-69
Quinoline, 2-(2,4-dimethyl-5-thiazolyl)-4-phenyl-	n.s.g.	348(4.51)	97-0186-69
$C_{20}H_{16}N_4O_4S$			
Acetophenone, 2'-(phenylthio)-, 2,4-dinitrophenylhydrazone	EtOH	368(4.40)	7-0787-69
Acetophenone, 3'-(phenylthio)-, 2,4-dinitrophenylhydrazone	EtOH	378(4.45)	7-0787-69
$C_{20}H_{16}N_4O_5$			
Acetophenone, 2'-phenoxy-, 2,4-dinitrophenylhydrazone	EtOH	374(4.44)	7-0787-69
Acetophenone, 3'-phenoxy-, 2,4-dinitrophenylhydrazone	EtOH	380(4.45)	7-0787-69
$C_{20}H_{16}N_4O_6$			
Barbituric acid, 5,5'-[1-(2-naphthyl)ethylidene]di-	EtOH	257(4.28)	103-0827-69
$\Delta^{3,3'}$-Biindoline]-2,2'-dione, 1,1'-bis-[(methylcarbamoyl)oxy]-	dioxan	459(3.62)	24-3691-69
$C_{20}H_{16}N_6O_4$			
Cinnamic acid, p-azido-, ethylene ester	dioxan	312(4.72)	73-3811-69
$C_{20}H_{16}O_2$			
Benzophenone, 4-methoxy-2-phenyl-	MeCN	245(4.39),285(3.89)	44-2736-69
9H-Cyclopropa[e]pyrene-9-carboxylic acid, 8b,9a-dihydro-, ethyl ester, endo	EtOH	221(4.69),260(4.55), 292(4.13),305(4.17)	39-1427-69C
exo	EtOH	221(4.71),261(4.55), 292(4.13),305(4.17)	39-1427-69C
Dicyclopent[a,c]anthracene-7,12-dione, 1,2,3,4,5,6-hexahydro-	C_6H_{12}	201f(4.61),260f(4.60), 277s(4.19),329(3.65), 360(3.78),420s(2.48)	39-0061-69B
2-Propanone, 1-(2,6-diphenyl-4H-pyran-4-ylidene)-	MeCN	230(4.09),270(4.11), 297(4.26),375(4.39)	44-2736-69
2H-Pyran-$\Delta^{2,\alpha}$-acetaldehyde, α-methyl-4,6-diphenyl-	MeCN	228(4.18),283(4.48), 340(4.13)	44-2736-69

Compound	Solvent	λ_{max}(log ε)	Ref.
Spirobi[2H-1-benzopyran], 3,3'-trimethylene-	dioxan	260(4.35),268(4.27), 297(3.75),307(3.68)	5-0162-69B
$C_{20}H_{16}O_3$			
Cyclobuta[b]naphthalene-3,8-dione, 2a,8a-dihydro-2a-methoxy-1-methyl-	EtOH	230(4.57),250(4.47), 350(2.84)	44-0520-69
17H-Cyclopenta[a]phenanthren-17-one, 16-acetoxy-15,16-dihydro-11-methyl-	EtOH	263(4.87),288(4.52), 301(4.35),342(3.12), 357(3.37),375(3.42)	39-2484-69C
1-Naphthoic acid, 8-p-toluyl-, methyl ester	M H_2SO_4	267(4.5),290(4.4)	35-0477-69
	dioxan	290(4.9)	35-0477-69
1H,3H-Naphtho[1,8-cd]pyran-1-one, 3-methoxy-3-p-tolyl-	M H_2SO_4	305(4.2),327(4.1)	35-0477-69
	dioxan	297(4.3),308(4.3), 326(4.3)	
Phenanthro[1,2-b]furan-1-carboxylic acid, 2-methyl-, ethyl ester	C_6H_{12}	213(4.42),226(4.42), 258(4.85),266(5.02), 296(4.41),302(3.94), 322(3.07),331(2.90), 338(3.30),346(2.84), 354(3.33)	33-1461-69
Phenanthro[1,2-b]furan-2-carboxylic acid, 1-methyl-, ethyl ester	C_6H_{12}	204(4.49),231(4.01), 271(4.80),291(4.66), 298(4.59),313(4.54), 329(3.52),337(3.20), 345(3.76),353(3.16), 361(3.83)	33-1461-69
Psoralen, 4'-ethyl-8-methyl-4-phenyl-	MeOH	225(3.80),255(4.45), 310(3.72)	42-1014-69
$C_{20}H_{16}O_4$			
Otobain, tetradehydro-	EtOH	221(4.60),241(4.67), 297(3.89),312(3.81), 352(3.52)	39-0693-69C
$C_{20}H_{16}O_4S$			
3,4-Thiophenedicarboxylic acid, 2,5-diphenyl-, dimethyl ester	MeOH	206(4.49),240(4.34), 295(4.12)	77-1129-69
$C_{20}H_{16}O_5$			
3-Benzofurancarboxaldehyde, 5-allyl-7-methoxy-2-[3,4-(methylenedioxy)-phenyl]-	EtOH	256(4.32),335(4.22)	12-1011-69
$C_{20}H_{16}O_6$			
3-Benzofurancarboxylic acid, 5-allyl-7-methoxy-2-[3,4-(methylenedioxy)-phenyl]-	EtOH	322(4.39)	12-1329-69
Cycloartocarpesin	EtOH	278(4.15),334(4.18)	2-0101-69
Hibalactone, (±)-	EtOH	236(4.17),292(4.10), 332(4.23)	39-2470-69C
(-)-	EtOH	237(4.16),293(4.12), 332(4.24)	39-2470-69C
Isoelliptone, (±)-	EtOH	236(4.59),255(4.09), 276(4.01),300s(3.80), 335(3.55)	31-0789-69
Isohibalactone, (-)-	EtOH	238(4.09),295(4.04), 338(4.21)	39-0693-69C
Naphtho[1,2-d]-1,3-dioxole-7,8-dimethanol, 9-[3,4-(methylenedioxy)phenyl]-	EtOH	223(4.51),248(4.64), 296(3.89),310(3.80), 350(3.54)	39-0693-69C

Compound	Solvent	$\lambda_{max}(\log \epsilon)$	Ref.
Viridin	n.s.g.	300(4.22)	77-0839-69
$C_{20}H_{16}O_7$			
Anthraquinone, 1,7-diacetoxy-3-methoxy-6-methyl-	EtOH	270(4.67),331(3.71)	23-0767-69
2-Butyne-1,4-dione, 1-[3,4-(methylenedioxy)phenyl]-4-(3,4,5-trimethoxyphenyl)-	EtOH	340(4.19)	36-0176-69
Flavone, 5,7-diacetoxy-4'-methoxy-	EtOH	230(4.18),259(4.18), 324(4.08)	105-0397-69
Naphtho[2,3-c]furan-1(3H)-one, 9-(3,4-dihydroxy-5-methoxyphenyl)-7-hydroxy-6-methoxy-	EtOH	259(4.36),315(3.88), 356(3.64)	23-4495-69
	EtOH-base	278(--)	23-4495-69
$C_{20}H_{16}O_8$			
2,3,6,7-Biphenylenetetrol, tetraacetate	EtOH	251(4.75),344(3.79), 363(3.91)	39-2579-69C
Plicatinaphthol	EtOH	264(4.59),312(3.89), 323(3.88),365(3.68)	23-0457-69
	EtOH-base	275(--)	23-0457-69
Succinic acid, piperonylpiperonylidene-, hydrate	EtOH	232(4.24),287(4.15), 308(3.95)	39-2470-69C
(-)-acid	EtOH	229(4.20),285(4.14), 306(3.99)	39-2470-69C
(-)-acid, quinine salt	EtOH	232(4.68),284(4.25), 311(4.12)	39-2470-69C
(+)-acid, diquinine salt	EtOH	231(4.92),281(4.39), 316(4.25)	39-2470-69C
$C_{20}H_{16}O_{10}$			
Dibenzo-p-dioxin-1,2,6,7-tetrol, tetraacetate	EtOH	229(4.53),288(3.41)	23-0733-69
$C_{20}H_{16}S$			
Thiophene, 2-(p-styrylstyryl)-	DMF	368(4.78)	33-2521-69
$C_{20}H_{17}BrCl_2O_4$			
Anthracene, 9-(bromomethylene)-10-(dichloromethylene)-9,10-dihydro-2,3,6,7-tetramethoxy-	EtOH	263(4.46),306(4.34)	7-0671-69
$C_{20}H_{17}ClN_2O_2$			
2-Pyrazoline, 3-(chloroacetyl)-1-cinnamoyl-4-phenyl-	$C_2H_4Cl_2$	280(4.24),328(4.45)	94-2105-69
$C_{20}H_{17}ClN_2O_3$			
Furo[2,3-d]pyrimidine-2,4(1H,3H)-dione, 3-(p-chlorophenyl)-5,6-dihydro-6,6-dimethyl-5-phenyl-	EtOH	210(4.28),220(4.28), 268(4.19)	87-0339-69
Furo[2,3-d]pyrimidine-2,4(1H,3H)-dione, 5-(p-chlorophenyl)-5,6-dihydro-6,6-dimethyl-3-phenyl-	EtOH	212s(4.21),222(4.24), 265(3.96)	87-0339-69
$C_{20}H_{17}ClO_7$			
2-(o-Acetoxy-α-methylstyryl)-1-benzopyrylium perchlorate	$MeOH-HClO_4$	261(4.03),298(3.81), 460(4.21)	5-0226-69E
$C_{20}H_{17}Cl_2NO_5$			
2-Chloro-1-methyl-3-(2-phenyl-4H-1-benzopyran-4-ylidene)-1-pyrrolinium perchlorate	MeCN	235(4.03),243(4.30), 455(4.28)	4-0803-69

Compound	Solvent	$\lambda_{max}(\log \epsilon)$	Ref.
$C_{20}H_{17}IN_4O$			
1,4-Dihydro-1,3-dimethyl-4-oxo-6,7-diphenylpteridinium iodide	EtOH	224(4.52),286(4.25), 393(4.12)	39-2415-69C
$C_{20}H_{17}Li$			
Ethane, 1,1,2-triphenyl-, 1-lithium derivative	THF	482(4.48)	35-2456-69
$C_{20}H_{17}N$			
Acridan, 9-benzyl-	hexane	277(4.04)	78-1125-69
Benzo[f]quinoline, 5,6-dihydro-3-p-tolyl-	dioxan	232(4.02),296(4.31), 304(4.33),327(4.51)	24-1202-69
$C_{20}H_{17}NO$			
Benzophenone, 2-o-toluidino-	EtOH	233s(4.19),255(4.29), 300s(3.67),360s(3.29), 405(3.86)	22-3523-69
Benzophenone, 2-p-toluidino-	EtOH	234s(4.15),257(4.32), 280s(4.08),415(3.85)	22-3523-69
Ketone, 2,4-dihydro-4-methyl-1H-cyclopenta[b]quinolin-5-yl phenyl	EtOH	225(4.3),275(4.0), 305(3.5),475(4.3)	122-0431-69
$C_{20}H_{17}NO_2$			
Benzophenone, 2-o-anisidino-	EtOH	229s(4.21),255(4.34), 305(3.88),415(3.85)	22-3523-69
Benzophenone, 2-p-anisidino-	EtOH	203(4.54),254(4.33), 276s(4.12),404(3.91)	22-3523-69
2-Pyrrolidinone, 1-methyl-3-(2-phenyl-4H-1-benzopyran-4-ylidene)-	MeCN	242(4.50),273(4.55), 355(4.34)	4-0803-69
3-Pyrroline, 1,2-dibenzoyl-2-methyl-5-methylene-	EtOH	244(4.15),278s(3.78), 316s(3.18)	5-0137-69D
9H-Pyrrolo[1,2-a]indole, 2-acetyl-7-(benzyloxy)-	EtOH	202(4.58),243(3.97), 292(4.34),312(4.33)	88-0101-69
$C_{20}H_{17}NO_3$			
1-Butene-1-carboxylic acid, 3-ethoxy-3-hydroxy-2,4-diphenyl-1-cyano-, lactone	CH_2Cl_2	231(3.82),308(4.22)	24-0319-69
2-(α-Carboxy-β-hydroxystyryl)isoquinolinium hydroxide, inner salt, ethyl ester	EtOH	226(4.75),267(4.26), 326(3.62),349(3.63), 407(3.42)	24-0915-69
	$EtOH-H_2SO_4$	233(4.58),342(3.69)	24-0915-69
$C_{20}H_{17}NO_4$			
3-Buten-1-ol, 3-nitro-4-(9-phenanthryl)-, acetate	EtOH	340(3.86)	87-0157-69
Pyrrole-2-carboxylic acid, 5-benzoyl-4-hydroxy-1-o-tolyl-, methyl ester	MeOH	259(4.16),299(4.02), 332(3.98)	78-0527-69
$C_{20}H_{17}NO_5$			
Chelidonine, didehydro-	EtOH	235s(3.94),293(3.78)	23-3701-69
	EtOH-HCl	241(--),302(--), 394(--)	23-3701-69
hydrobromide	EtOH	233(4.04),298(3.99), 397(3.23)	23-3701-69
Fumariline	EtOH	203(4.60),237(4.31), 263(4.05),294(3.66), 355(3.51)	23-3593-69
Imenine	EtOH	240(4.15),275(4.38), 345(3.58),438(3.42)	77-1217-69

Compound	Solvent	$\lambda_{max}(\log \epsilon)$	Ref.
2,3-Indolizinedicarboxylic acid, 1-phenacyl-, dimethyl ester	MeOH	238(4.34),250(4.32), 345(3.85)	39-1143-69C
$C_{20}H_{17}NO_6$			
Adlumidine	EtOH	200(4.69),220(4.47), 236s(4.12),294(3.79), 323(3.73)	78-5059-69
Bicuculline	EtOH	202(4.74),221(4.47), 238s(4.09),298(3.76), 325(3.72)	78-5059-69
Capnoidine	EtOH	200(4.68),220(4.47), 235s(4.13),294(3.85), 323(3.78)	78-5059-69
Sibiricine	n.s.g.	205(4.80),240(3.94), 291(3.91),313s(3.99)	23-3585-69
$C_{20}H_{17}N_3$			
4H-Isoquino[8,1-ab]phenazine, 5,6-dihydro-6,11-dimethyl-	C_6H_{12}	467(--)	83-0487-69
	MeOH	480(3.88)	83-0487-69
$C_{20}H_{17}N_3O_2$			
1(2H)-Phenanthrone, 3,4-dihydro-, (p-nitrophenyl)hydrazone	EtOH	229(4.55),274s(4.12), 304(4.04),328(4.03), 412(4.59)	24-2384-69
$C_{20}H_{17}N_3O_3$			
4H-Quinolizine-1-carboxylic acid, 2-(benzylamino)-3-cyano-4-oxo-	EtOH	265(4.69),302(4.22), 391(4.06)	95-0203-69
$C_{20}H_{17}N_3O_4$			
Acetic acid, 2-[1,4-dihydro-1,4-dioxo-3-(N-phenylacetamido)-2-naphthyl]-hydrazide	alkali	575(3.73)	95-0007-69
$C_{20}H_{17}N_3O_5$			
Furo[2,3-d]pyrimidine-2,4(1H,3H)-dione, 5,6-dihydro-6,6-dimethyl-3-(p-nitrophenyl)-5-phenyl-	EtOH	208(4.39),211(4.40), 267(4.26)	87-0339-69
$C_{20}H_{17}O_3PS$			
Phenothiaphosphine, 2,8-dimethyl-10-phenyl-, 5,5,10-trioxide	EtOH	224(4.71),269(3.61), 277(3.64),287(3.63)	78-3919-69
$C_{20}H_{17}PS$			
Phenothiaphosphine, 2,8-dimethyl-10-phenyl-	EtOH	226(4.54),269(4.09), 304(3.76)	78-3919-69
$C_{20}H_{18}$			
11H-Benzo[a]fluorene, 11-isopropyl-	EtOH	256(4.74),266(4.94), 286s(4.05),296(4.18), 307(4.17),318(4.05), 332(3.36),346(3.05)	22-2115-69
11H-9,10-endo-Cyclopropanthracene, 13-cyclopropyl-9,10,12,13-tetrahydro-	C_6H_{12}	266(3.14),273(3.23)	88-0085-69
o-Terphenyl, 4',5'-dimethyl-	hexane	236(4.49)	24-1928-69
p-Terphenyl, 2'-ethyl-	hexane	264(4.32)	89-0753-69
$C_{20}H_{18}BrNO_5$			
Chelidonine, didehydro-, hydrobromide	EtOH	233(4.04),298(3.99), 397(3.23)	23-3701-69

Compound	Solvent	λ max(log ε)	Ref.
Maleic acid, (p-bromo-N-phenacylanil-ino)-, dimethyl ester	MeOH	243(4.30),288(4.26)	78-0527-69
$C_{20}H_{18}Br_2N_2$			
3-Buten-2-one, 4-(p-bromophenyl)-, azine	$CHCl_3$	335(4.71),345(4.70)	22-1367-69
Cinnamaldehyde, p-bromo-α-methyl-, azine	$CHCl_3$	341(4.80)	22-1367-69
$C_{20}H_{18}Br_2N_2O_4S_2$			
p-Toluenesulfonamide, N,N'-bis(2,5-di-bromo-p-phenylene)bis-	EtOH	225(4.62),257s(4.12)	117-0287-69
$C_{20}H_{18}FeO$			
Acrylophenone, 3-ferrocenyl-4'-methyl-	MeOH	325(4.27)	73-2771-69
	n.s.g.	269(4.2),315(4.3), 385(3.5),475(3.3)	65-1008-69
2-Propen-1-one, 1-ferrocenyl-3-p-tolyl-	MeOH	316(4.29)	73-2771-69
$C_{20}H_{18}FeO_2$			
Acrylophenone, 3-ferrocenyl-4'-methoxy-	MeOH	328(4.40)	73-2771-69
2-Propen-1-one, 1-ferrocenyl-3-(p-meth-oxyphenyl)-	MeOH	338(4.31)	73-2771-69
$C_{20}H_{18}GeO$			
Germane, acetyltriphenyl-	heptane	191(4.96),366(2.59)	59-0765-69
$C_{20}H_{18}HgO_4$			
Mercuty, bis[2-oxo-6-(1-pentynyl)-2H-pyran-5-yl)-	MeOH	233(4.20),250s(4.07), 325(4.32)	1-1059-69
Mercury, [1-(5-oxo-2(5H)-furylidene)-2-hexynyl][2-oxo-6-(1-pentynyl)-2H-pyran-5-yl]-	MeOH	231(4.12),328(4.51)	1-1059-69
$C_{20}H_{18}INO_5$			
Maleic acid, (p-iodo-N-phenacylanil-ino)-, dimethyl ester	MeOH	242(4.40),289(4.15)	78-0527-69
$C_{20}H_{18}N_2$			
Acetophenone, diphenylhydrazone	EtOH	243(4.2),292(4.0), 350(3.8)	28-0137-69B
Benzophenone, methylphenylhydrazone	EtOH	254(4.35),290s(3.69), 364(3.87)	39-1703-69C
	EtOH-HCl	231(4.10),285(4.25), 429(3.48)	39-1703-69C
Indole, 1,3-dimethyl-2-[(2-methyl-3H-in-dol-3-ylidene)methyl]-, perchlorate	MeOH	510(3.34)	24-1347-69
Indolo[2,3-a]carbazole, 5,6,11,12-tetrahydro-11,12-dimethyl-	CH_2Cl_2	258(4.33),354s(4.33), 367(4.47),385(4.40)	24-1198-69
m-Tolualdehyde, diphenylhydrazone	EtOH	238(4.3),308s(4.1), 341(4.3)	28-0137-69B
o-Tolualdehyde, diphenylhydrazone	EtOH	238(4.2),308s(4.0), 342(4.3)	28-0137-69B
p-Tolualdehyde, diphenylhydrazone	EtOH	239(4.3),306s(4.1), 341(4.4)	28-0137-69B
$C_{20}H_{18}N_2O$			
m-Anisaldehyde, diphenylhydrazone	EtOH	210s(4.5),238(4.2), 303s(4.0),342(4.3)	28-0137-69B

Compound	Solvent	λ_{max}(log ε)	Ref.
o-Anisaldehyde, diphenylhydrazone	EtOH	203(4.6),239(4.2), 307(4.0),348(4.3)	28-0137-69B
p-Anisaldehyde, diphenylhydrazone	EtOH	244(4.2),309(4.2), 343(4.4)	28-0137-69B
Benz[c]acridine-7-carboxamide, 5,6-dihydro-5,5-dimethyl-	MeOH	214(4.53),225(4.19), 259s(4.31),267(4.51), 302(3.74),316(3.76), 332(3.90),346(3.96)	4-0361-69
2-Indolinone, 3-[(1,3-dimethylindol-2-yl)methylene]-1-methyl-	MeOH	415(4.18)	24-1347-69
$C_{20}H_{18}N_2O_2$			
5H-Benz[a]phenoxazin-5-one, 9-(diethylamino)-	EtOH	555(4.66)	73-0221-69
$C_{20}H_{18}N_2O_2S$			
2H-1,2,4-Benzothiadiazine, 4-benzyl-3,4-dihydro-2-phenyl-, 1,1-dioxide	EtOH	209(4.56),258(4.09), 320(3.56)	7-0590-69
$C_{20}H_{18}N_2O_3$			
Furo[2,3-d]pyrimidine-2,4(1H,3H)-dione, 5,6-dihydro-6,6-dimethyl-3,5-diphenyl-	EtOH	213(4.26),266(4.22)	87-0339-69
$C_{20}H_{18}N_2O_4$			
Acetanilide, 2'-[(1,2-dihydro-1-methoxy-4-methyl-2-oxo-2-quinolyl)carbonyl]-	EtOH	232(4.74),275(4.27), 340(4.13)	18-2952-69
$C_{20}H_{18}N_2O_5$			
Isoquinoline, 4-(5,6-dimethoxy-2,1-benzisoxazol-3-yl)-6,7-dimethoxy-	EtOH	225(4.47),350(4.05)	78-5365-69
hydrochloride	EtOH	253(4.52),320(4.23)	78-5365-69
$C_{20}H_{18}N_2S_2$			
2H-1,3,5-Thiadiazine-2-thione, 5-benzyltetrahydro-3-(1-naphthyl)-	MeOH	242(3.90),295(4.33)	73-2952-69
2H-1,3,5-Thiadiazine-2-thione, 5-benzyltetrahydro-3-(2-naphthyl)-	MeOH	256(4.64),290(4.44)	73-2952-69
$C_{20}H_{18}N_2Si$			
Silane, (1-diazoethyl)triphenyl-	C_6H_{12}	207(4.61),220(4.48), 224s(4.44),253(3.68), 426(1.75)	23-4353-69
$C_{20}H_{18}N_4O_3S$			
2-Propanone, 1-[1,2-dihydro-4-(methylsulfonyl)-2-(1-phthalazinyl)-1-phthalazinyl]-	EtOH	245s(4.06),362(4.27)	95-0959-69
$C_{20}H_{18}N_4O_4$			
5,5'-Bihydantoin, 3,3'-dimethyl-5,5'-diphenyl-	MeOH	215s(4.22)	44-1133-69
$C_{20}H_{18}N_8O_2$			
Hydrazine, 1,1'-isophthaloylbis[2-picolinimidoyl-	H_2SO_4	262(4.50)	4-0965-69
$C_{20}H_{18}OSi$			
Silane, acetyltriphenyl-	heptane	194(5.04),375(2.61)	59-0765-69

Compound	Solvent	λ_{max}(log ϵ)	Ref.
$C_{20}H_{18}O_2$			
Bicyclo[3.1.0]hex-2-en-3-ol, acetate	C_6H_{12}	228(4.11),260s(3.19), 276(3.00)	35-0434-69
Naphthalene, 1-(3,4-dimethoxystyryl)-	DMF	345(4.39)	33-2521-69
Naphthalene, 2-(3,4-dimethoxystyryl)-	DMF	286(4.28),338(4.59)	33-2521-69
$C_{20}H_{18}O_3$			
α,2'-Stilbenedicarboxylic acid, α'-tert-butyl-, cyclic anhydride	EtOH	238(4.08),273(3.85), 315(3.64)	35-0388-69
$C_{20}H_{18}O_4$			
Benzyl alcohol, 2,2'-ethynylenedi-, dimethyl ester	EtOH	285(4.44),293s(4.31), 304(4.37)	44-0874-69
Bicyclo[1.1.0]butane-2,4-dicarboxylic acid, 1,3-diphenyl-, dimethyl ester, endo-endo	MeOH	219(4.22),269(4.13)	35-4612-69
exo-endo	MeOH	220(4.10)	35-4612-69
exo-exo	MeOH	221(4.12)	35-4612-69
Cyclohepta[def]fluorene-4,8-diol, 4,8,9,10-tetrahydro-, diacetate	EtOH	214(4.43),227s(4.23), 241(4.10),255s(4.29), 263(4.41),279(4.04)	39-1427-69C
Eupomatene	EtOH	235(4.42),266(4.44), 312(4.38)	12-1011-69
Isoeupomatene	EtOH	219(4.50),300(4.38), 314(4.39)	12-1011-69
$C_{20}H_{18}O_5$			
3-Benzofuranmethanol, 5-allyl-7-methoxy-2-[3,4-(methylenedioxy)phenyl]-	EtOH	219(4.50),312(4.37)	12-1011-69
Cubebinol	EtOH	228(4.28),292(4.12), 310(4.00),321s(3.91)	39-2470-69C
Furan, 2,3-dihydro-3,4-dipiperonyl-, (+)-	EtOH	236(4.07),287(3.94)	39-2470-69C
Isocubebinic ether	EtOH	236(4.88),289(3.95)	39-2470-69C
	n.s.g.	237(3.86),292(3.92)	77-0341-69
$C_{20}H_{18}O_6$			
Carpanone	MeOH	243(3.93),262(4.04), 298(3.71)	88-5159-69
Hinokinin, (±)-	EtOH	236(3.86),287(3.91)	39-2470-69C
Isohinokinin, (±)-	EtOH	236(3.92),287(3.92)	39-2470-69C
β-Lumicolchicone, 5,6-dehydro-	MeOH	245(4.30),293(4.40), 360(3.70)	32-1059-69
Sesamin, (+)-	n.s.g.	236(3.91),286(3.89)	39-2477-69C
Viridiol	n.s.g.	250(4.47),317(4.07)	77-0839-69
	EtOH-base	290(--),500(--)	77-0839-69
$C_{20}H_{18}O_7$			
Anthraquinone, 3-acetoxy-1,6,8-trimethoxy-2-methyl-	EtOH	224s(4.40),242(3.28), 280(4.42),348(3.70), 391(3.70)	39-2763-69C
Anthraquinone, 1,3,6,8-tetrahydroxy-2-(tetrahydro-6-methyl-2H-pyran-2-yl)-	MeOH	223(4.30),263(4.16), 292(4.25),312(4.00), 452(3.94)	78-1497-69
$C_{20}H_{18}O_8$			
Isoelliptic acid	EtOH	233(4.56),278(4.02), 340(3.81)	31-0789-69

Compound	Solvent	$\lambda_{max}(\log \epsilon)$	Ref.
$C_{20}H_{18}O_9$			
3(1H)-Isobenzofuranone, 1-hydroxy-1-(2,3-dicarbomethoxy-4-methoxyphenyl)-4-methoxy-	n.s.g.	305(3.78)	39-2059-69C
$C_{20}H_{18}O_{11}$			
Dibenzo-p-dioxin-2(1H)-one, 1,4a,6,7-tetraacetoxy-4a,10a-dihydro-, trans	EtOH	217(4.25),275(3.30)	23-0733-69
$C_{20}H_{19}AsIN$			
5-Methyl-5-phenyldibenz[b,e][1,4]aza-arsepinium iodide	EtOH	260(4.23),330(3.49)	101-0117-69E
$C_{20}H_{19}Br_2Cl_3O_4$			
Anthracene, 9-(dibromomethyl)-10-(trichloromethyl)-9,10-dihydro-2,3,6,7-tetramethoxy-	EtOH	293(3.97)	7-0671-69
$C_{20}H_{19}Cl$			
Anthracene, 9-chloro-10-cyclohexyl-	EtOH	253s(4.91),259(5.17), 342(3.52),359(3.84), 378(4.04),399(4.02)	78-3485-69
$C_{20}H_{19}ClN_2O$			
4-Quinolinol, 5-chloro-3-[3,4-dihydro-2(1H)-isoquinolyl)methyl]-2-methyl-	EtOH	235(4.33),248(4.47), 253(4.30),317(3.99), 328(3.97)	2-1186-69
4-Quinolinol, 7-chloro-3-[3,4-dihydro-2(1H)-isoquinolyl)methyl]-2-methyl-	EtOH	237s(4.36),245(4.53), 255(4.42),315(3.96), 328(3.90)	2-1186-69
$C_{20}H_{19}ClN_2O_5$			
Isoquinoline, 4-(5,6-dimethoxy-2,1-benzisoxazol-3-yl)-6,7-dimethoxy-, hydrochloride	EtOH	253(4.52),320(4.23)	78-5365-69
$C_{20}H_{19}ClO_7$			
Flavone, 3'-chloro-2',3,5,7,8-pentamethoxy- (unchanged by $AlCl_3$)	EtOH	258(4.19),333(3.94)	39-2418-69C
$C_{20}H_{19}Cl_2NO$			
2-Propen-1-one, 3-[bis(2-chloroethyl)-amino]-1-fluoren-2-yl-	EtOH	210(4.48),317(4.25), 349(4.48)	24-3139-69
$C_{20}H_{19}IN_2O_4$			
1-(4,6-Dicarboxy-2-biphenylyl)-2-methylpyrazolium iodide, dimethyl ester	EtOH	263(3.80)	44-4134-69
$C_{20}H_{19}IN_2S$			
2-(6-Anilino-1,3,5-hexatrienyl)-3-methylbenzothiazolium iodide	MeOH-NaOMe	487(4.40)	22-1284-69
$C_{20}H_{19}IN_2Se$			
2-(6-Anilino-1,3,5-hexatrienyl)-3-methylbenzoselenazolium iodide	MeOH MeOH-NaOMe	622(4.95) 480(4.57)	22-1284-69 22-1284-69
$C_{20}H_{19}N$			
1,3-Pentadienylamine, N,N-dimethyl-5-phenalen-1-ylidene-	hexane	277(4.36),327(4.37), 381(3.85),504(4.88)	24-2301-69

Compound	Solvent	λ_{max}(log ε)	Ref.
1,3-Pentadienylamine, N,N-dimethyl-5-phenalen-1-ylidene- (cont.)	MeCN	280(4.32),331(4.23), 382(3.86),517(4.76)	24-2301-69
$C_{20}H_{19}NO_2$			
2-Cyclobuten-1-one, 2-acetyl-3-(dimethylamino)-4,4-diphenyl-	dioxan	250(4.23),285(4.30)	33-2641-69
Pyrrole-3-carboxylic acid, 1-methyl-2,4-diphenyl-, ethyl ester	MeOH	203(5.57),222(5.34), 280(3.98)	77-0066-69
2-Pyrroline, 1,5-dibenzoyl-2,5-dimethyl-	EtOH	242(4.23)	5-0137-69D
$C_{20}H_{19}NO_3$			
1-Cyclobutene-1-carboxylic acid, 2-(dimethylamino)-4-oxo-3,3-diphenyl-, methyl ester	dioxan	242(4.28),270(4.32)	33-2641-69
5H-Dibenzo[a,d]cycloheptene-5-carboxamide, N,N'-diethyl-10,11-dihydro-10,11-dioxo-	EtOH	259(4.06)	23-2827-69
	MeCN	256(4.05)	23-2827-69
2H-Isoxazolo[3,2-a]isoquinoline-1-carboxylic acid, 5,6-dihydro-2-phenyl-, ethyl ester	$CHCl_3$	245(4.09),316(3.65)	24-0904-69
$C_{20}H_{19}NO_4$			
Diacetamide, N-(1,2-dihydro-3-hydroxy-4-phenanthryl)-, acetate	EtOH	217(4.40),238(4.74), 304s(3.96),317(4.05), 336(3.50)	24-2384-69
$C_{20}H_{19}NO_5$			
Berberine	pH 7.5	345(4.3),420(3.9)	57-0755-69D
Ketone, 6,7-dimethoxy-1-isoquinolyl 3,4-dimethoxyphenyl	HCl	280(4.47),349(3.97)	83-0572-69
	MeOH	233(4.56),266(4.57), 322(4.31)	83-0572-69
Parfumine	EtOH	235(4.42),260(4.10), 290s(--),358(3.42)	30-1255-69F
$C_{20}H_{19}NO_6$			
Corlumidine	EtOH	235s(4.11),290(3.70), 323(3.73)	78-5059-69
Ochrobirine	MeOH	205(4.80),240(3.94), 291(3.91)	23-3589-69
$C_{20}H_{19}NO_8$			
2-Naphthoic acid, 3-(4-carbamyl-3,6-dihydroxy-5-oxo-1,2,5,6-tetrahydrobenzyl)-4,8-dihydroxy-4-methyl-1-oxo-1,2,3,4-tetrahydro-, γ-lactone	EtOH	265(4.54),335(3-95)	30-1255-69F
ε-lactone	EtOH	229s(4.38),267(4.37), 320(3.51),362(3.64)	30-1255-69F
$C_{20}H_{19}N_9O_2$			
Isoalloxazine, 10-[3-(6-amino-9H-purin-9-yl)propyl]-7,8-dimethyl-	pH 7	265(4.54),368(3.94), 450(3.97)	44-3240-69
	MeOH	267(4.63),350(3.87), 443(4.03)	44-3240-69
$C_{20}H_{19}O_5P$			
2(5H)-Furanone, 5-acetyl-3-(diphenylphosphinyl)-4-methoxy-5-methyl-	MeCN	225(4.39),246(4.12)	35-2293-69

Compound	Solvent	$\lambda_{max}(\log \epsilon)$	Ref.
$C_{20}H_{20}$			
Anthracene, 9-cyclohexyl-	EtOH	251s(5.09),255(5.20), 333(3.54),349(3.81), 366(3.95),386(3.94)	78-3485-69
Benzo[j]fluoranthene, 4,5,6,6a,6b,7,-8,12b-octahydro-, anti-cis	EtOH	268(3.04),276(3.02)	25-0877-69
syn-cis	EtOH	267(3.07),274(3.03)	25-0877-69
$C_{20}H_{20}AsI$			
Ethyltriphenylarsonium iodide	n.s.g.	226(4.4),252(3.4), 258(3.5),264(3.6), 270(3.5)	120-0195-69
$C_{20}H_{20}AuP$			
Gold, ethyl(triphenylphosphine)-	dioxan	275(3.62)	39-0276-69B
$C_{20}H_{20}Br_2Cl_2O_4$			
Anthracene, 9-(dibromomethyl)-10-(dichloromethyl)-9,10-dihydro-2,3,6,7-tetramethoxy-	EtOH	292(3.94)	7-0671-69
$C_{20}H_{20}Br_4O_4$			
Anthracene, 9,10-bis(dibromomethyl)-9,10-dihydro-2,3,6,7-tetramethoxy-	EtOH	292(3.96)	7-0671-69
$C_{20}H_{20}ClNO_4S_2$			
N-(4,5-Diphenyl-1,3-dithiol-2-ylidene)-piperidinium perchlorate	EtOH	234(4.39),320(4.13)	94-1924-69
$C_{20}H_{20}ClNO_5$			
[(10,11-Dihydro-9H-benzo[a]xanthen-8-yl)methylene]dimethylammonium perchlorate	MeCN	217(4.63),254(4.37), 288(4.17),327(4.05), 350(3.86),368(3.83), 462(4.43),468(4.45), 518(4.17)	44-2736-69
Dimethyl[1-methyl-2-(2-phenyl-4H-1-benzopyran-4-ylidene)ethylidene]-ammonium perchlorate	MeCN	237(4.28),274(3.79), 310(4.14),522(4.43)	44-2736-69
$C_{20}H_{20}ClNO_5S$			
4-[p-(Dimethylamino)styryl]-7-methoxy-1-benzothiopyrylium perchlorate	CH_2Cl_2	265(4.12),268(4.12), 273(4.13),304(3.77), 339(3.80),390(3.88), 428(3.61),685(4.91)	2-0017-69
$C_{20}H_{20}FeO$			
Ferrocene, (α-methoxycinnamyl)-, (E)-	EtOH	253(4.29)	101-0361-69C
Ferrocene, (3-methoxy-3-phenylpropen-yl)-, (E)-	EtOH	281(4.01)	101-0361-69C
$C_{20}H_{20}INO_4$			
7,8-Dimethoxy-2-[3,4-(methylenedioxy)-phenethyl]isoquinolinium iodide	EtOH	258(4.56),293(3.83)	78-1881-69
$C_{20}H_{20}N_2$			
9-Anthronitrile, 10-[(diethylamino)-methyl]-	EtOH	259(5.14),369(3.90), 388(4.01)	23-2827-69
3-Buten-2-one, 4-phenyl-, azine	$CHCl_3$	327(4.71),342(4.70)	22-1367-69
Cinnamaldehyde, α-methyl-, azine	$CHCl_3$	335(4.76)	22-1367-69

Compound	Solvent	$\lambda_{max}(\log \epsilon)$	Ref.
Cyclohepta[c]pyrrole-4-carbonitrile, 2,4-dihydro-1,3,5,7-tetramethyl-2-phenyl-	EtOH	320(3.95)	104-2014-69
Cyclohepta[c]pyrrole-4-carbonitrile, 2,6-dihydro-1,3,5,7-tetramethyl-2-phenyl-	EtOH	252(4.42)	104-2014-69
Cyclohepta[c]pyrrole-6-carbonitrile, 2,6-dihydro-1,3,5,7-tetramethyl-2-phenyl-	EtOH	248(4.49)	104-2014-69
$C_{20}H_{20}N_2O$			
Pyrrolo[1,2-a:5,4-b']diindol-12(5aH)-one, 6,10b,11,11a-tetrahydro-5a,10b,11a-trimethyl-	EtOH	241(4.52),265s(3.89), 301(3.62),406(3.46)	39-2703-69C
$C_{20}H_{20}N_2O_2$			
2H-Pyrido[4,3-b]indole-2-carboxylic acid, 1,3,4,5-tetrahydro-4-phenyl-, ethyl ester	MeOH	221(4.6),280(3.9), 291(3.8)	33-0629-69
$C_{20}H_{20}N_2O_3$			
23-Oxastrychnine	EtOH	242(4.07),280(3.40)	33-1564-69
$C_{20}H_{20}N_2O_4$			
o-Acetotoluidide, α-(1,2-dihydro-1-methoxy-4-methyl-2-oxo-3-quinolyl)-α-hydroxy-	EtOH	230(4.77),276(3.86), 331(3.76)	18-2952-69
$C_{20}H_{20}N_2O_4S$			
Spiro[2H-1-benzopyran-2,2'-benzothiazoline], 3-isopropyl-8-methoxy-3'-methyl-6-nitro-	EtOH	217(4.51),255(4.28), 285s(4.03),344(3.86)	22-3329-69
$C_{20}H_{20}N_2O_4S_2$			
Dibenzenesulfonamide, N-[p-(dimethylamino)phenyl]-	EtOH-1% $CHCl_3$	275(4.23)	44-2083-69
$C_{20}H_{20}N_2O_6S$			
4H-Furo[3,4-d][1,3]thiazine-2-acetic acid, 1,2,5,7-tetrahydro-7-oxo-α-phthalimido-, tert-butyl ester	EtOH	216(4.62),268(3.67)	44-1582-69
$C_{20}H_{20}N_2S$			
1,3,5,7-Tetramethyl-2-phenylcyclohepta-[c]pyrrolium thiocyanate	$C_2H_4Cl_2$	281(4.54),365(3.93), 710(3.30)	104-2014-69
$C_{20}H_{20}N_2S_2$			
3,3'-Biindole, 1,1'-dimethyl-2,2'-bis-(methylthio)-	EtOH	225(4.72),295(4.31)	94-0550-69
$C_{20}H_{20}N_4$			
Benzimidazole, 2,2'-vinylenebis[5,6-dimethyl-, cis	EtOH	255s(3.99),270(3.99), 279(3.99),283(3.99), 289(4.02),335(3.78), 355(3.81),375(3.68), 398(3.55),428(2.88)	87-0818-69
trans	EtOH	225s(4.07),358(4.35), 375(4.48),395(4.34)	87-0818-69
Malononitrile, [bis[p-(dimethylamino)-phenyl]methylene]-	MeCN	265(4.36),310(3.57), 320(3.60),430(4.69)	44-2146-69

Compound	Solvent	$\lambda_{max}(\log \epsilon)$	Ref.
$C_{20}H_{20}N_4O$			
Benzamide, N-[3-(3-amino-6-phenylpyrazinyl)propyl]-	MeOH	278(4.23),341(3.98)	95-1652-69
	MeOH-HCl	272(4.27),355(3.89)	95-1652-69
$C_{20}H_{20}N_4O_4$			
3,6-Heptadien-2-one, 5-methyl-7-phenyl-, 2,4-dinitrophenylhydrazone	$CHCl_3$	378(4.60)	22-3719-69
Mesoxalic acid, diethyl ester, 2-[(o-2-benzimidazolylphenyl)hydrazone]	MeOH	220(4.47),246(4.21), 302(4.44),362(4.32)	4-0605-69
Pyrazole-4-carbamic acid, 3-[(benzyloxy)methyl]-5-carbamoyl-, benzyl ester	n.s.g.	210(4.37),253(3.49)	88-0289-69
Spiro[2H-furo[3,2-b]indole-2,2'-indolin]-3'-ol, 3,3a,4,8b-tetrahydro-3',3a,8b-trimethyl-1,4-dinitroso-	EtOH	222s(4.18),291(4.19), 296(4.19)	39-2703-69C
$C_{20}H_{20}N_6O_2$			
Propionitrile, 3,3'-[ethylenebis(p-phenylenenitrosimino)]di-	EtOH	215(4.23),279(4.06)	103-0623-69
$C_{20}H_{20}O$			
Ketone, phenyl 3-phenyl-2-norbornyl, endo-2 exo-3	EtOH	245(4.09),280s(3.25)	101-0361-69B
$C_{20}H_{20}O_2$			
3,5-Cyclohexadiene-1,2-dimethanol, 3,6-diphenyl-, cis	EtOH	328(4.30)	22-3662-69
trans	EtOH	230s(3.94),345(4.38)	22-3662-69
$C_{20}H_{20}O_4$			
Benzofuran, 2,3-dihydro-7-methoxy-3-methyl-2-[3,4-(methylenedioxy)-phenyl]-5-propenyl-	EtOH	279(4)	12-1011-69
Chalcone, 2',4,4'-trihydroxy-3'-(3-methyl-2-butenyl)-	MeOH	374(4.34)	2-1072-69
Dibenzo[a,e]cyclooctene, 5,6-didehydro-11,12-dihydro-2,3,8,9-tetramethoxy-	n.s.g.	332(4.45)	89-0447-69
Eupomatene, dihydro-	EtOH	217(4.52),315(4.28)	12-1011-69
Flavanone, 4',7-dihydroxy-8-(3-methyl-2-butenyl)- (isobavachin)	MeOH	284(4.02)	2-1072-69
α,2-Stilbenedicarboxylic acid, α'-tert-butyl-	EtOH	207(4.40),225(3.92), 282(3.03)	35-0388-69
Succinic acid, 2-(diphenylmethylene)-3-methyl-, 1-ethyl ester	EtOH	261(4.00)	22-1344-69
$C_{20}H_{20}O_5$			
3-Benzofuranmethanol, 7-methoxy-2-[3,4-(methylenedioxy)phenyl]-5-propyl-	EtOH	310(4.38)	12-1011-69
Cubebinol, dihydro-	EtOH	235(3.63),290(3.86)	39-2470-69C
Isoflavone, 3',7-diethoxy-4'-methoxy-	EtOH	262(4.28),288(4.14)	2-0118-69
Xanthen-9-one, 1-hydroxy-3,7-dimethoxy-2-(3-methyl-2-butenyl)-	MeOH	235(4.47),264(4.51), 305(4.16),372(3.74)	78-2787-69
$C_{20}H_{20}O_6$			
Averythrin, dihydro-	EtOH	224(4.51),294(4.55), 325(4.03),457(3.98)	39-2763-69C
3,6(2H,5H)-Benzofurandione, 2-(3,4-dimethoxybenzylidene)-4-methoxy-5,5-dimethyl-	MeOH	272(4.4),308(4.2), 385(4.4)	2-0540-69

Compound	Solvent	λ_{max}(log ϵ)	Ref.
3(2H)-Benzofuranone, 2-(3,4-dimethoxy-benzylidene)-4,6-dimethoxy-5-methyl-	MeOH	256(4.2),395(4.5)	2-0540-69
2,3-Biphenylenediol, 6,7-dimethoxy-1,8-dimethyl-, 2,3-diacetate	EtOH	257s(4.75),264(4.86), 354(4.00),372(4.10)	39-2579-69C
1,4-Butanediol, 2-piperonyl-3-piperonylidene-	EtOH	266(3.92),289(3.94)	39-2470-69C
Carpanone, dihydro-	n.s.g.	243(4.25),300(3.79)	88-5159-69
Chalcone, 2-acetoxy-4,4',6-trimethoxy-	EtOH	355(4.50)	78-2367-69
Colchicine, 7-deacetamino-7-oxo-	pH 1	244(4.53),347(4.25)	63-0366-69
	pH 13	246(4.46),349(4.21)	63-0366-69
	EtOH	242(4.53),346(4.24)	63-0366-69
Coleon B quinone, monomethoxy-	EtOH	254(4.26),320(3.10), 415(3.49)	33-1685-69
Cubebin, (-)	EtOH	235(3.90),287(3.89)	39-2470-69C
Fibleucin	n.s.g.	208(4.05),230(3.90)	77-0653-69
β-Lumicolchicone	MeOH	230(4.11),254(4.20), 285(3.79),332(4.01), 362(4.10)	32-1059-69
Piperonylic acid, ester with 2'-hydroxy-3'-methoxy-5'-propylacetophenone	EtOH	257(3.05),305(3.06)	12-1011-69
α,2-Stilbenedicarboxylic acid, 3,5-dimethoxy-, dimethyl ester	MeOH	280(4.31)	44-3192-69
$C_{20}H_{20}O_7$			
Artocarpesin, dihydrooxy-	EtOH	265(4.26),348(4.14)	2-0101-69
2-Butyne-1,4-diol, 1-[3,4-(methylenedioxy)phenyl]-4-(3,4,5-trimethoxyphenyl)-	MeOH	282(3.57)	36-0176-69
Flavone, 2',3',5,6,7-pentamethoxy-	EtOH	259(4.31),300(4.32)	2-0110-69
Flavone, 3',4',5,6,7-pentamethoxy-	n.s.g.	265(4.41),320(4.20)	2-0940-69
Flavone, 3',4',5,6,7-pentamethoxy-	EtOH	240(4.41),265(4.22), 328(4.46)	78-1603-69
Isoflavone, 2',3',4',6',7-pentamethoxy-	EtOH	238(4.45),247(4.41), 288(4.13),303s(4.05)	39-0887-69C
Isoflavone, 2',4',5',6,7-pentamethoxy-	$CHCl_3$	258(4.26),304(4.21)	39-1787-69C
Isoflavone, 2',4',5',6',7-pentamethoxy-	EtOH	227(4.32),272(4.21), 310(3.89)	39-0887-69C
Philenopteran, 9-O-methyl-, acetate	EtOH	278(3.58)	39-0887-69C
$C_{20}H_{20}O_8$			
Altersolanol B, diacetate	EtOH	217(4.54),267(4.15), 282(3.96),423(3.62)	23-0767-69
Flavone, 5-hydroxy-3,3',4',6,7-pentamethoxy-	EtOH	239(4.12),256(4.27), 273(4.23),345(4.35)	18-1398-69
	EtOH	255(4.25),274(4.21), 346(4.31)	44-1460-69
	MeOH	254(4.27),273s(4.25), 348(4.35)	102-0511-69
$C_{20}H_{20}O_{12}$			
2-Cyclohexene-1,4-dione, 5-hydroxy-6-[(1,5,6-trihydroxy-4-oxo-2-cyclohexen-1-yl)oxy]-, tetraacetate	EtOH	218(4.24)	23-0733-69
$C_{20}H_{21}ClN_2$			
Isoindole, 2-(2-aminocyclohexyl)-1-(p-chlorophenyl)-, trans	EtOH	224(4.61),275(3.69), 286(3.70),348(4.10)	44-1720-69

Compound	Solvent	$\lambda_{max}(\log \epsilon)$	Ref.
$C_{20}H_{21}ClN_2O$			
1(2H)-Quinolinecarboxamide, N-allyl-6-chloro-3,4-dihydro-2-methyl-4-phenyl-	n.s.g.	258(4.20)	40-1047-69
$C_{20}H_{21}ClN_2OS$			
Thiourea, N-(2-benzoyl-4-chlorophenyl)-N'-cyclohexyl-	n.s.g.	244(4.47),290s(--)	40-0917-69
$C_{20}H_{21}ClN_2O_2$			
Urea, N-(2-benzoyl-4-chlorophenyl)-N'-cyclohexyl-	n.s.g.	243(4.59),276s(--), 360(3.91)	40-0917-69
$C_{20}H_{21}ClN_2O_4$			
Malonic acid, [(2-amino-5-chloro-α-phenylbenzylidene)amino]-, diethyl ester	EtOH	237(4.45),370(3.76)	111-0239-69
$C_{20}H_{21}N$			
Cyclohexylamine, N-(diphenylvinylidene)-	C_6H_{12}	277(4.28)	88-5093-69
6H-Dibenz[cd,h]azulene, 2-[(dimethylamino)methyl]-7,11b-dihydro-	EtOH	237s(--),243(4.24), 249s(--),274(3.97), 283s(--),295s(--), 305s(--)	87-0444-69
$C_{20}H_{21}NO$			
1-Indanone, 2-benzylidene-3-(sec-butylamino)-	hexane	231(4.09),237(4.09), 262(4.06),315(4.32), 325s(4.22)	44-0596-69
1-Indanone, 2-benzylidene-3-(isopropylmethylamino)-	n.s.g.	228(4.15),232s(4.11), 259(3.96),300s(4.08), 312(4.39),322(4.30)	44-0596-69
1-Indanone, 3-(isopropylamino)-2-(p-methylbenzylidene)-	hexane	237(4.15),244(4.14), 263s(3.97),326(4.40), 340s(4.35)	44-0596-69
Spiro[cyclohexane-1,2'(1'H)-quinolin]-2-one, 3',4'-dihydro-1'-phenyl-	MeOH	252(4.25),280s(3.9)	24-3486-69
$C_{20}H_{21}NOS_2$			
1-Piperidinecarbodithioic acid, ester with 2-mercapto-2-phenylacetophenone	EtOH	251(4.39),282(4.10)	94-1924-69
$C_{20}H_{21}NO_2$			
1-Azaspiro[3.5]nonan-2-one, 3-phenoxy-1-phenyl-	EtOH	255(4.14)	78-4421-69
Koenimbin, N-methyl-	EtOH	242(4.55),301(4.26), 339(3.83),357(3.75), 371(3.68)	31-0790-69
Spiro[cyclopentane-1,3'(2'H)-quinolin]-2'-one, 1',4'-dihydro-4'-methoxy-4'-phenyl-	EtOH	256(4.03),287(3.49), 296(3.39)	94-1290-69
$C_{20}H_{21}NO_2S$			
Thiocinnamamide, α-ethyl-β-hydroxy-N,N-dimethyl-, benzoate	MeOH	231(4.39),268(4.24)	5-0073-69E
$C_{20}H_{21}NO_3$			
Koenigicine	EtOH	239(4.56),300(4.52), 361(3.97)	31-0790-69
$C_{20}H_{21}NO_4$			
Canadine, (+)-	EtOH	230s(4.02),292(3.88)	78-1881-69

Compound	Solvent	$\lambda_{max}(\log \epsilon)$	Ref.
Isoquinoline, 7,8-dimethoxy-1-veratryl-	EtOH	215(3.71),245(3.68), 295(2.85),370(2.71)	44-2665-69
$C_{20}H_{21}NO_5$			
Estra-1,3,5(10),8(9)-tetraen-17-one, 3-acetoxy-11β-nitro-	EtOH	271(4.11)	31-1018-69
$C_{20}H_{21}NO_6$			
Adluminediol	EtOH	237(3.92),292(3.91)	78-5059-69
Aknadilactam	EtOH	232s(3.84),267(3.98)	88-3287-69
9-Anthroic acid, 10-(aminomethyl)-9,10-dihydro-, ethyl ester, oxalate	EtOH	257(2.66),262(2.73)	33-2197-69
Bicucullinediol	EtOH	205(4.76),233(3.98), 291(3.92)	78-5059-69
Capnoidinediol	EtOH	237(3.92),292(3.91)	78-5059-69
$C_{20}H_{21}NS_2$			
Phenethylamine, N-(3,3-di-2-thienylallyl)-α-methyl-, hydrochloride	MeOH	266(4.22),290(4.22)	111-0228-69
Piperidine, 1-(4,5-diphenyl-1,3-dithiol-2-yl)-	EtOH	229(4.20),305(3.72), 336(3.77)	94-1924-69
$C_{20}H_{21}N_3$			
Benzimidazo[1,2-c]quinazoline, 5-cyclohexyl-5,6-dihydro-	MeOH	243(4.19),248s(4.11), 275(4.17),281(4.25), 290s(4.23),302(4.25), 315s(4.00)	4-0605-69
$C_{20}H_{21}N_3O$			
Anthranilonitrile, 5-ethoxy-N-(α-1-pyrrolidinylbenzylidene)-	EtOH	217(4.50),250(4.01), 329(3.72)	22-2008-69
Yohimban-16-carbonitrile, 17-oxo-	pH 13	280(4.12)	24-3248-69
	EtOH	276(3.92),282(3.92), 290(3.86)	24-3248-69
$C_{20}H_{21}N_3O_2$			
Methanol, (4-amino-3-methoxyphenyl)-bis(p-aminophenyl)-	pH 4.62	552(4.84)	64-0542-69
2-Propanone, 1-(2,4-dimethyl-5-phenylimidazol-1-yl)-1-phenyl-, oxime, 3-oxide	MeOH	258(4.27)	48-0746-69
isomer	MeOH	260(4.16)	48-0746-69
3,4-Secoyohimbane-3,17-dione, 4-cyano-	MeOH	208(4.37),238(4.11), 312(4.26)	35-4317-69
$C_{20}H_{21}N_3O_2S$			
Benzamide, N-[4-(1-adamantylcarbonyl)-1,2,3-thiadiazol-5-yl]-	EtOH	238(4.26),299(4.20)	18-1617-69
$C_{20}H_{21}N_3O_3S$			
Hydrocinnamic acid, α-(dicyanomethylene)-β-[(methylthio)morpholinomethylene]-, ethyl ester	C_6H_{12}	477(3.81)	5-0073-69E
$C_{20}H_{21}N_3O_5$			
Pyrrolo[2,3-b]indole-2-carboxylic acid, 1,2,3,3a,8,8a-hexahydro-3a-(5-nitrosalicyl)-, ethyl ester, hydrochloride	EtOH	240(4.07),310(3.95)	35-2792-69
	2N NaOH	422(4.30)	35-2792-69

Compound	Solvent	$\lambda_{max}(\log \epsilon)$	Ref.
$C_{20}H_{21}O_6P$			
2-Hexenoic acid, 2-(diphenylphosphinyl)-4-hydroxy-3-methoxy-4-methyl-5-oxo-	MeCN	223(4.29),265(3.87)	35-2293-69
$C_{20}H_{22}BF_4NO$			
6-Methoxy-1,3,5,7-tetramethyl-N-phenyl-cyclohepta[c]pyrrolium tetrafluoro-borate	$C_2H_4Cl_2$	265(4.23),315(4.56), 354(3.92),632(3.18)	104-0947-69
$C_{20}H_{22}BrNO$			
1-Indanone, 3-bromo-2-[α-(tert-butyl-amino)benzyl]-, hydrobromide	$CHCl_3$	251(4.15),296(3.59)	44-0596-69
$C_{20}H_{22}ClNO_3$			
Cyclohexanebutyramide, N-(3-chloro-1,4-dihydro-1,4-dioxo-2-naphthyl)-	MeOH	247(4.29),253(4.32), 286(3.97),336(3.52)	4-0909-69
$C_{20}H_{22}Cl_2O_9$			
1,2,3,4-Cyclohexanetetrol, 5,6-dichloro-1-(hydroxymethyl)-, 1,2,3-triacetate α-benzoate	MeOH	230(4.02),274(3.98), 281(2.90)	44-3898-69
$C_{20}H_{22}N_2$			
1,3-Azulenedivaleronitrile	EtOH	231(4.47),282(4.74), 349(3.70),366(3.58), 625(2.50),682(2.41), 765(1.94)	44-2375-69
$C_{20}H_{22}N_2O$			
Indole-2-carboxamide, N-tert-butyl-3-methyl-6-phenyl-	EtOH	255(4.6),313(4.5)	24-0678-69
Vincanicine	EtOH	248(3.90),293(3.29), 376(4.04)	105-0386-69
$C_{20}H_{22}N_2OS$			
Atropamide, N-benzoyl-β-(diethylamino)-thio-	$CHCl_3$	300(4.26),405(4.23)	4-0037-69
$C_{20}H_{22}N_2O_2$			
3H-1-Benzazepine-4-carboxylic acid, 4,5-dihydro-2-(methylamino)-5-phenyl-, ethyl ester	EtOH	263(4.1)	33-1929-69
1,4,2,5-Dioxadiazin, 3,6-dimesityl-	EtOH	250(4.26)	32-0165-69
Gardnutine	MeOH	224(4.63),260(3.64), 295(3.73),302s(3.63)	88-1485-69
3-Isoxazoline, 2-methyl-5-morpholino-	n.s.g.	250s(4.0),315(3.9)	94-2201-69
1H-Naphth[2,3-d]imidazole-4,9-dione, 2-(3-cyclohexylpropyl)-	pH 1	246(4.61),250(4.72), 272(4.43),277s(--)	4-0909-69
	pH 13	263(4.61)	4-0909-69
	MeOH	247(4.69),273(4.41), 277(4.40),333(3.46)	4-0909-69
1,2,4-Oxadiazole, 3,5-dimesityl-, 4-oxide	EtOH	283(3.94)	32-0165-69
boron trifluoride adduct	EtOH	280(3.66)	32-0165-69
Pyrrolo[1,2-a:5,4-b']diindole-11a,12(12H)-diol, 5a,6,10b,11-tetrahydro-5a,10b,12-trimethyl-	EtOH	254(4.17),294(3.72)	39-2703-69C
3-Quinolineacetamide, 1,2,3,4-tetrahy-dro-N,N,1-trimethyl-2-oxo-4-phenyl-	EtOH	256(4.0)	33-1929-69

Compound	Solvent	$\lambda_{max}(\log \epsilon)$	Ref.
$C_{20}H_{22}N_2O_3$			
Gardnutine, hydroxy-	MeOH	223(4.61),261(3.69), 295(3.79),301s(3.70)	88-1485-69
Wieland-Gumlich aldehyde derivative by-product m. 242-3°	EtOH-HCl	257(4.06),283s(3.55)	33-1564-69
	aq. KOH	240(3.84),292(3.49)	33-1564-69
$C_{20}H_{22}N_2O_5$			
1,3-Dioxolane, 2,2'-(azoxydi-p-phenylene)bis[2-methyl-	$CHCl_3$	267(4.00),330(4.33)	78-4241-69
$C_{20}H_{22}N_4$			
5,5'-Bibenzimidazole, 2,2'-diethyl-1,1'-dimethyl-	MeOH	231(4.75),297(4.31)	22-1926-69
Propionitrile, 3,3'-[ethylenebis(p-phenyleneimino)]di-	EtOH	253(3.99),299(3.16)	103-0623-69
$C_{20}H_{22}N_4O_2$			
3-Pyrazolidinone, 1,1'-(ethylenedi-p-phenylene)di-	EtOH	252(4.36)	103-0623-69
$C_{20}H_{22}N_4O_3S$			
Thiamine, 2-phenyloxalyl-	EtOH	233(4.09),274(3.83), 420(3.90+)	94-0128-69
$C_{20}H_{22}N_4O_4S_2$			
Ketone, [3-[(4-amino-2-methyl-5-pyrimidinyl)methyl]-5-(2-hydroxyethyl)-4-methyl-4-thiazolin-2-ylidene]hydroxymethyl 2-thienyl, acetate	EtOH	232(4.26),274(4.10), 415(4.43)	94-0128-69
$C_{20}H_{22}N_4O_5S$			
Ketone, [3-[(4-amino-2-methyl-5-pyrimidinyl)methyl]-5-(2-hydroxyethyl)-4-methyl-4-thiazolin-2-ylidene]hydroxymethyl 2-furyl, acetate	EtOH	232(4.18),277(4.09), 410(4.41)	94-0128-69
$C_{20}H_{22}N_4O_6$			
β-Alanine, N,N'-(ethylenedi-p-phenylene)bis[N-nitroso-	EtOH	217(4.33),278(4.18)	103-0623-69
$C_{20}H_{22}N_4O_7$			
β-Alanine, N,N'-(oxydi-p-phenylene)bis-[N-nitroso-, dimethyl ester	EtOH	222(4.34),283(4.24)	103-0623-69
$C_{20}H_{22}N_8O_6S_2$			
9H-Purine, 6,6"-dithiobis[9-(2-deoxy-α-D-erythro-pentofuranosyl-	EtOH-1% DMSO	290(4.47)	87-1117-69
9H-Purine, 6,6"-dithiobis[9-(2-deoxy-β-D-erythro-pentofuranosyl)-	EtOH-1%	289(4.47)	87-1117-69
$C_{20}H_{22}N_8O_8S_2$			
9H-Purine, 6,6'-dithiobis[9-β-D-arabinofuranosyl-	EtOH-1% DMSO	288(4.45)	87-1117-69
$C_{20}H_{22}O$			
4,6,8,10,12-Tridecapentaen-3-one, 6-methyl-13-phenyl-, all trans	pet ether	371(4.87),392(5.02), 415(3.97)	88-4049-69
4,6,8,10,12-Tridecapentaen-3-one, 7-methyl-13-phenyl-, all trans	pet ether	285(3.90),295(3.94), 372(4.86),390(4.99), 412(4.90)	88-4049-69

Compound	Solvent	λ_{max}(log ε)	Ref.
4,6,8,10,12-Tridecapentaen-3-one, 8-methyl-13-phenyl-, all trans (asperenone)	pet ether	372(4.85),392(4.98), 415(4.91)	88-4049-69
$C_{20}H_{22}O_2$			
Anthrone, 10-hydroxy-3,10-diisopropyl-	EtOH	211(4.37),238s(3.89), 283(4.23)	39-2266-69C
B-Homoestra-1,3,5(10),8,14-pentaen-17-one, 3-methoxy-	EtOH	287(4.38)	31-0571-69
$C_{20}H_{22}O_2S$			
2H-Thiopyran, 3-acetyltetrahydro-4-hydroxy-4-methyl-2,6-diphenyl-	EtOH	220(4.26)	39-1647-69C
$C_{20}H_{22}O_3$			
Ketone, 9-ethyl-9,10-dihydro-1,3-dihydroxy-6,9-dimethyl-4-phenanthryl methyl	EtOH	323(4.47),370s(3.76)	2-0873-69
$C_{20}H_{22}O_4$			
2H,8H-Benzo[1,2-b:5,4-b']dipyran-2-one, 10-(1,1-dimethylallyl)-5-methoxy-8,8-dimethyl-	EtOH	225(4.33),264s(4.36), 271(4.40),330(4.02), 342s(4.01)	12-2175-69
Benzofuran, 7-methoxy-3-methyl-2-(3-hydroxy-4-methoxyphenyl)-5-propyl-	EtOH	308(4.45)	12-1011-69
[Bi-1,4-cyclohexadien-1-yl]-3,3',6,6'-tetrone, 5,5'-di-tert-butyl-	n.s.g.	252(4.09)	64-0547-69
Phenol, 4-(2,3-dihydro-7-methoxy-3-methyl-5-propenyl-2-benzofuranyl)-2-methoxy-	EtOH	274(4.34)	12-1011-69
	KOH	268(4.54)	12-1011-69
$C_{20}H_{22}O_5$			
2H-1-Benzopyran-3-carboxylic acid, 6-(3,7-dimethyl-2,6-octadienyl)-7-hydroxy-2-oxo-	MeOH	253(3.83),295(3.82), 348(4.01),440(3.14)	32-0308-69
Estra-1,3,5(10)-triene-4-carboxylic acid, 1-methoxy-6,17-dioxo-	EtOH	217(4.30),262(3.82)	39-1234-69C
Estra-1,3,5(10)-triene-6,17-dione, 3-acetoxy-9α-hydroxy-	MeOH	251(4.00)	44-0116-69
$C_{20}H_{22}O_6$			
p-Benzenediacrylic acid, α,α'-diacetyl-, diethyl ester	EtOH	205(4.31),225(4.08), 325(4.53)	23-4076-69
1,4-Butanediol, 2,3-dipiperonyl-, meso-	EtOH	234(3.91),287(3.91)	39-2470-69C
Cubebin, dihydro-, (-)	EtOH	235(3.91),287(3.90)	39-2470-69C
Oroselol, 9-methoxy-O-senecionyl-dihydro-	ether	318(4.16)	24-1673-69
Propiophenone, 4'-hydroxy-3''',5'-dimethoxy-3',4'''-oxydi-	EtOH	276(4.27),294(4.24)	44-0585-69
$C_{20}H_{22}O_7$			
Elephantin	MeOH	215(4.40)	44-3867-69
Eupatundin, dehydro-	EtOH	213(4.27),296(1.72)	44-3876-69
$C_{20}H_{22}O_8$			
Acetic acid, [[α-(2-hydroxy-p-anisoyl)-4,5-dimethoxy-o-tolyl]oxy]-, methyl ester	EtOH	229(4.25),278(4.20), 316(3.92)	18-0199-69
Capenicinone	MeOH	253(3.99)	78-4835-69

Compound	Solvent	λmax(log ε)	Ref.
$C_{20}H_{22}O_9$			
5-Cyclohexene-1,2,3,4-tetrol, 1-(hydroxymethyl)-, 2,3,4-triacetate α-benzoate	MeOH	274(2.98),281(2.89)	44-3898-69
Olivin	EtOH	230(4.27),277(4.56), 355s(3.64),408(3.59)	105-0257-69
	EtOH-EtONa	288(4.53),314(4.03), 412(4.21)	105-0257-69
	EtOH-NaOAc	223(4.35),320(4.61), 410(3.75)	105-0257-69
	EtOH-borax	227(4.27),277(4.59), 325(3.63),408(4.08)	105-0257-69
	EtOH-$AlCl_3$	233(4.29),282(4.47), 327(3.83),343(3.85), 426(4.01)	105-0257-69
$C_{20}H_{23}BrN_4O_3$			
2-Propanol, 1-[(2-bromo-6-nitro-9-acridinyl)amino]-3-(diethylamino)-, dihydrochloride	EtOH	257(4.38),283(4.45), 322s(4.02),372s(3.66), 443(3.74)	103-0707-69
$C_{20}H_{23}BrO_8$			
Propiophenone, 2-bromo-2'-hydroxy-3-(3-hydroxy-4-methoxyphenyl)-3,3',4',6'-tetramethoxy-	CH_2Cl_2	311(4.39)	24-0112-69
$C_{20}H_{23}ClN_2O$			
1(2H)-Quinolinecarboxamide, 6-chloro-N-ethyl-3,4-dihydro-2,3-dimethyl-4-phenyl-	n.s.g.	258(4.03)	40-1047-69
$C_{20}H_{23}IN_2O$			
4-[p-(Dimethylamino)styryl]-2,3-dihydro-5-methyl-1,5-benzoxazepinium iodide	MeOH	528(5.20)	124-1178-69
$C_{20}H_{23}N$			
1-Indanamine, 2-benzylidene-N-tert-butyl-	hexane	251(4.42)	88-1717-69
Indene-2-methylamine, N-tert-butyl-α-phenyl-	hexane	260(4.20),265(4.18)	88-1717-69
$C_{20}H_{23}NO$			
Azetidine, 3-benzoyl-1-tert-butyl-2-phenyl-, cis	isooctane	240(4.04)	44-0310-69
trans	isooctane	242(4.19)	44-0310-69
Pyridine, 1,2,5,6-tetrahydro-4-(p-methoxybenzyl)-1-methyl-2-phenyl-	EtOH	235(3.45),277(3.37), 284(3.32)	44-4158-69
$C_{20}H_{23}NO_2$			
Koenimbin, dihydro-N-methyl-	EtOH	247(4.49),266(4.33), 314(4.29),337(3.73), 352(3.51)	31-0790-69
$C_{20}H_{23}NO_2S_2$			
1-Propanol, 3-[(β-hydroxy-α-methylphenethyl)amino]-1,1-di-2-thienyl-	MeOH	233(4.34)	111-0228-69
$C_{20}H_{23}NO_3$			
Koenigicine, dihydro-	EtOH	238(4.58),267(4.25), 325(4.21)	31-0790-69

Compound	Solvent	$\lambda_{max}(\log \epsilon)$	Ref.
$C_{20}H_{23}NO_3S$			
Indole-3-pentanol, p-toluenesulfonate	EtOH	225(4.55),274(3.77), 283(3.78),291(3.72)	78-4843-69
$C_{20}H_{23}NO_4$			
Morphinan-7-one, 5,6,8,14-tetradehydro-2,3,6-trimethoxy-17-methyl-	EtOH	205(4.64),239(4.25), 285(3.92)	88-1771-69
Pyrrole-2,3-dicarboxylic acid, 1-cyclohexyl-4-phenyl-, dimethyl ester	MeOH	246(4.15),298(3.48)	78-0527-69
Spiro[7H-benzo[de]quinoline-7,1'-[2,5]-cyclohexadien]-4'-one, 1,2,3,8,9,9a-hexahydro-6-hydroxy-2',5-dimethoxy-1-methyl-	EtOH	235(4.19),285(3.87)	94-0814-69
Spiro[7H-benzo[de]quinoline-7,1'-[3,5]-cyclohexadien]-2'-one, 1,2,3,8,9,9a-hexahydro-6-hydroxy-4',5-dimethoxy-1-methyl-	MeOH	234s(4.18),243s(4.03), 293(3.96),313s(3.89)	39-0004-69C
Thalisopavine	EtOH	289(4.06)	44-1062-69
$C_{20}H_{23}NO_6$			
Corlumidinediol	EtOH	236(3.93),287(3.86)	78-5059-69
Estra-1,3,5(10)-trien-17-one, 3,9α-dihydroxy-11β-nitro-, 3-acetate	EtOH	268(2.90),273(2.85)	31-1018-69
$C_{20}H_{23}NO_{12}$			
α-D-Galactopyranoside, o-nitrophenyl, tetraacetate	$C_2H_4Cl_2$	258(3.54),314(3.26)	18-1052-69
β-isomer	$C_2H_4Cl_2$	255(3.54),306(3.23)	18-1052-69
β-D-Glucopyranoside, o-nitrophenyl, tetraacetate	$C_2H_4Cl_2$	256(3.56),305(3.23)	18-1052-69
α-D-Glucopyranoside, p-nitrophenyl, tetraacetate	$C_2H_4Cl_2$	294(4.04)	18-1052-69
β-isomer	$C_2H_4Cl_2$	295(4.04)	18-1052-69
$C_{20}H_{23}NS$			
Dibenzo[b,e]thiepin, 6,11-dihydro-2-methyl-11-[3-(dimethylamino)propylidene]-, cis	heptane	230(4.31),262(3.99), 310(3.29)	73-1015-69
trans	heptane	231(4.39),263(3.98), 310(3.39)	73-1015-69
$C_{20}H_{23}N_3$			
Phthalazine, 1-benzyl-4-[3-(dimethylamino)propyl]-	EtOH	217(4.9),249(4.2), 266(4.1)	33-1376-69
$C_{20}H_{23}N_3NaO_8PS$			
Benzamide, N-[1-(2-deoxy-β-D-erythro-pentofuranosyl)-1,2-dihydro-2-oxo-4-pyrimidinyl]-, monoacetate O-ester with S-ethyl phosphorothioate, sodium salt	pH 1	257(4.11),315(4.30)	35-1522-69
	H_2O	259(4.32),302(4.06)	35-1522-69
$C_{20}H_{23}N_3O$			
Yohimban-16-carbonitrile, 17-hydroxy-	EtOH	276(3.77),282(3.80), 289(3.74)	24-3248-69
$C_{20}H_{23}N_3O_2$			
Spiro[indoline-3,1'(5'H)-indolizine]-$\Delta^{2,\alpha}$-acetic acid, α-cyano-2',3',6',-7',8,8'a-hexahydro-, ethyl ester	EtOH	335(4.30)	94-2306-69

Compound	Solvent	λ_{max}(log ε)	Ref.
$C_{20}H_{23}N_3O_4$			
Pyrrole-3-carboxylic acid, 5-[(4-carboxy-3,5-dimethyl-2H-pyrrol-2-ylidene)cyanomethyl]-2,4-dimethyl-, diethyl ester	EtOH	495(4.57)	5-0116-69A
	EtOH-HCl	564(4.58)	5-0116-69A
$C_{20}H_{23}N_3O_8$			
1H-1,2,3-Triazole, 1-α-D-glucopyranosyl-4-phenyl-, 3',4',6'-triacetate	EtOH	243(4.20),275s(2.90)	4-0639-69
$C_{20}H_{24}$			
[4,4]Paracyclophane, tetracyanoethylene complex	CH_2Cl_2	476(3.22)	35-3553-69
$C_{20}H_{24}BrP$			
Benzyl-2-cyclohexen-1-ylmethylphenylphosphonium bromide	MeOH	216s(4.16),255s(2.78), 260(2.93),266(3.00), 273(2.88)	24-3546-69
$C_{20}H_{24}ClNO_4$			
1H-Cyclohepta[c]quinolizinium perchlorate, 2,3,4,7,8,9,10,11-octahydro-6-phenyl-	$CHCl_3$	288(4.11)	78-4161-69
$C_{20}H_{24}ClNO_5$			
Estra-1,3,5(10)-trien-17-one, 3-(2-chloroethoxy)-9α-hydroxy-11β-nitro-	EtOH	277(3.20),285(3.13)	31-1018-69
$C_{20}H_{24}ClNO_6S$			
Alanine, 3-[[α-(2,4-dimethoxyphenacyl)-p-hydroxybenzyl]thio]-, hydrochloride	n.s.g.	213(4.13),229(4.14), 271(3.93),306(3.79)	7-0624-69
$C_{20}H_{24}ClNO_8$			
Mohinine, perchlorate	EtOH	231(4.56),272(4.23), 308(3.99)	102-1559-69
$C_{20}H_{24}F_2O_3$			
Androst-4-ene-3,17-dione, 19-hydroxy-6α,7α-(difluoromethylene)-	EtOH	248(4.11)	78-1219-69
$C_{20}H_{24}INO_2$			
4-Benzylidene-1,2,3,4-tetrahydro-2,2-dimethyl-7,8-dimethoxyisoquinolinium iodide	EtOH	223(4.36),312(4.29)	78-0101-69
$C_{20}H_{24}INO_4$			
3,4-Dihydro-6,7-dimethoxy-1-(3,4-dimethoxyphenyl)-2-methylisoquinolinium iodide	EtOH	250(4.32),312(4.07), 365(4.17)	1-0244-69
Mohinine, iodide	EtOH	228(4.78),270(4.07), 330(4.10)	102-1559-69
$C_{20}H_{24}IN_3O$			
1,3-Dimethyl-2-[2-(5-piperidino-2-furyl)vinyl]benzimidazolium iodide	MeOH	490(4.32)	103-0246-69
$C_{20}H_{24}I_2N_4$			
1,1',2,2',3,3'-Hexamethyl-5,5'-bibenzimidazolium diiodide	MeOH	228(4.79),296(4.27)	22-1926-69
	MeOH-H_2SO_4	228(4.81),296(4.27)	22-1926-69
	MeOH-NaOH	228(4.79),295(4.26)	22-1926-69

Compound	Solvent	λ_{max}(log ϵ)	Ref.
$C_{20}H_{24}N_2$			
Pyrrolidine, 1,1'-(3,5-biphenylylene)di-	EtOH	237s(4.61),249(4.63), 334(3.60)	94-2126-69
$C_{20}H_{24}N_2O$			
2ξ-Ajmaline, 19,20-didehydro-1-demethyl-17,21-dideoxy-11-methoxy-	MeOH	295(3.66)	88-1485-69
Atropaldehyde, β-(benzylamino)-β-(tert-butylamino)-	EtOH	280s(--),314(4.01)	44-3451-69
7,13b-Methano-13bH-indolo[3,2-e][2]benzazocin-2(1H)-one, 3,4,4a,5,6,7,8,13-octahydro-6,13-dimethyl-, cis	EtOH	229(4.57),286(3.92), 294(3.89)	44-3165-69
trans	EtOH	228(4.56),285(3.89), 294(3.89)	44-3165-69
Propionamidine, 2-benzoyl-N-tert-butyl-N'-phenyl-	EtOH	242(4.32)	44-3451-69
Propionamidine, 2-benzoyl-N,N-diethyl-N'-phenyl-	EtOH	245(4.36)	44-3451-69
Vobasinediol, anhydro-	MeOH	224(4.85),285(3.98), 293(3.88)	33-0701-69
$C_{20}H_{24}N_2O_2$			
Akuammilinol	EtOH	266(3.72)	33-0701-69
Aniline, N,N'-ethanediylidenebis[2,4,6-trimethyl-, N,N'-dioxide	EtOH	245(3.71),349(4.20)	39-1073-69C
Benzylidenimine, 4-methyl-N-2-salsolidyl-	n.s.g.	270(4.55),313(4.61)	65-2723-69
Benzylidenimine, α-methyl-N-2-salsolidyl-	n.s.g.	236(4.50),287(3.70), 320(3.40)	65-2723-69
Diaboline, 1-deacetyl-1-methyl-	MeOH	260(4.10),309(3.56)	33-1564-69
Diaboline, 1-deacetyl-O-methyl-	EtOH	243(3.94),298(3.55)	33-1564-69
17βH-Diaboline, 1-deacetyl-O-methyl-	EtOH	246(3.88),300(3.47)	33-1564-69
1,2-Diazocine, 3,4,5,6,7,8-hexahydro-3,8-bis(p-methoxyphenyl)-, cis	$CHCl_3$	361(1.56)	35-3226-69
trans	$CHCl_3$	379(2.03)	35-3226-69
Gardnerine	MeOH	229(4.56),269(3.70), 298(3.77)	88-1485-69
3-Indolinol, 2,3,3'-trimethyl-2,2'-methylenedi-	EtOH	244(4.20),294(3.65)	39-2703-69C
Isodihydropleiocarpamine	EtOH	251(4.07),303(3.49)	33-0033-69
Morpholine, 4,4'-(3,5-biphenylylene)di-	EtOH	246(4.60),314(3.38)	94-2126-69
1,2,4-Oxadiazolidin-5-one, 2-(1-ethylcyclohexyl)-4-(1-naphthyl)-	EtOH	241(3.81)	39-1073-69C
Spiro[isoquinoline-1(2H),4'-piperidine]-4,6-diol, 1'-benzyl-3,4-dihydro-, dihydrochloride	EtOH	282(3.26)	94-2353-69
$C_{20}H_{24}N_2O_3$			
Carbazole-1,4-dione, 9-butyl-3-(butylamino)-7-hydroxy-	$CHCl_3$	354(4.23)	18-2043-69
Cyclohexanebutyramide, N-(3-amino-1,4-dihydro-1,4-dioxo-2-naphthyl)-	MeOH	268(4.34),327(3.34), 452(3.38)	4-0909-69
Voacangine, 4-deethyl-	MeOH	224(4.42),283(3.98), 291s(3.95),297s(3.92)	27-0400-69
$C_{20}H_{24}N_2O_4$			
Acetophenone, 3',4'-dimethoxy-, azine	C_6H_{12}	280s(4.27),322(4.51)	39-0836-69C
β-Alanine, N,N'-(ethylenedi-p-phenylene)di-	EtOH	252(4.41),300(3.64)	103-0623-69

Compound	Solvent	$\lambda_{max}(\log \epsilon)$	Ref.
$C_{20}H_{24}N_2O_5$			
Thebaine, 6,7-dihydro-6-methoxy-7-oxo-, oxime	EtOH	234(4.34)	77-0057-69
	EtOH-NaOEt	273(4.23)	77-0057-69
$C_{20}H_{24}N_2O_8$			
4-Penten-1-ol, 5,5'-p-phenylenebis[4-nitro-, diacetate	EtOH	330(4.31)	87-0157-69
$C_{20}H_{24}N_4$			
1,2-Cyclohexanedione, bis(methylphenylhydrazone)	MeOH	304(4.25),345(4.05)	24-1198-69
$C_{20}H_{24}N_4O_2S_3$			
Thiamine, benzylxanthogenate	EtOH	230(4.37),268(3.37), 306(4.17)	94-2299-69
Thiamine, S-dithiobenzyloxycarbonyl-	EtOH	238s(4.13),268(3.97), 314(4.14)	94-2299-69
$C_{20}H_{24}N_4O_4$			
4H-Cyclopropa[b]naphthalen-4-one, 1,1a,2,5,6,6a,7,7a-octahydro-1,1,6a-trimethyl-, 2,4-dinitrophenylhydrazone	$CHCl_3$	388(4.50)	32-0219-69
15-Nor-α-cyperone, 2,4-dinitrophenylhydrazone, (-)-	$CHCl_3$	400(4.45)	32-0231-69
β-isomer	$CHCl_3$	406(4.58)	32-0231-69
2-Tetrazene-1,4-dicarboxylic acid, 1,4-dibenzyl-, diethyl ester	$(MeOCH_2)_2$	230(4.20),268(3.89), 297(3.81),377(4.65)	35-6452-69
$C_{20}H_{24}N_6$			
2-Pyrazoline, 1,1'-(ethylenedi-p-phenylene)bis[3-amino-	EtOH	278(4.44)	103-0623-69
$C_{20}H_{24}N_{10}O_6S_2$			
9H-Purine, 6,6"-dithiobis[2-amino-9-(2-deoxy-α-D-erythro-pentofuranosyl)-	EtOH-1% DMSO	322(4.33)	87-1117-69
β-isomer	EtOH-1%	322(4.32)	87-1117-69
9H-Purine, 6,6"-dithiobis[3-amino-9-(2-deoxy-β-D-erythro-pentofuranosyl)-	EtOH-1% DMSO	322(4.32)	87-1117-69
$C_{20}H_{24}O$			
6(5H)-Chrysenone, 6a,7,8,9,10,10a,11,12-octahydro-5,5-dimethyl-	EtOH	264(4.13)	22-2110-69
$C_{20}H_{24}O_2$			
9,10-Anthracenediol, 9,10-dihydro-2,9-diisopropyl-	EtOH	217(4.18),265(2.77), 273(2.63)	39-2266-69C
9,10-Anthracenediol, 9,10-dihydro-9,10-dipropyl-, cis	EtOH	213(4.28),262(2.78), 271s(2.62)	39-2266-69C
Crocetindialdehyde	benzene	422(3.86),445(5.07), 473(5.06)	33-0806-69
D-Homoestra-1,3,5(10),14-tetraen-17a-one, 3-methoxy-	EtOH	278(3.3)	70-1655-69
3,5,7,9,11,13,15-Octadecaheptaene-2,17-dione, 7,12-dimethyl-	pet ether	405(--),427(--), 455(--)	39-0429-69C
	$CHCl_3$	441(5.04),467(5.02)	39-0429-69C
$C_{20}H_{24}O_3$			
Estra-1,3,5(10)-trien-6-one, 3-acetoxy-	EtOH	209(4.44),248(4.04), 297(3.38)	39-1234-69C

Compound	Solvent	λ_{max}(log ε)	Ref.
D-Homoestra-1,3,5(10),16-tetraen-17a-one, 17-hydroxy-3-methoxy-	EtOH	219(3.96),269(3.89), 286s(3.62)	44-3829-69
$C_{20}H_{24}O_4$			
1,3-Azulenedipropionic acid, diethyl ester	EtOH	244(3.98),282(4.63), 350(3.66),367(2.56)	44-2375-69
Chromone, 5,7-dihydroxy-2-methyl-6,8-bis(3-methyl-2-butenyl)-	MeOH	263(4.34),302(3.88)	2-1072-69
1-Cyclopentene-1-heptanoic acid, 3-hydroxy-5-oxo-2-styryl-	MeOH	325(4.56)	88-1615-69
Dihydrodehydrodiisoeugenol	EtOH	250(4.25),290(3.90)	12-1011-69
Neoadenostylone	EtOH	222(3.99),302(4.17)	73-1739-69
$C_{20}H_{24}O_4S_8$			
2,3,4,5,10,11,12,13-Octathiatricyclo-[12.2.2.2^{6,9}]eicosa-6,8,14,16,17,19-hexaene, 7,15,17,19-tetraethoxy-	$CHCl_3$	371(4.01)	77-0847-69
$C_{20}H_{24}O_5$			
Estra-1,3,5(10)-trien-6-one, 17β-acetoxy-3,9α-dihydroxy-	MeOH	256(4.26)	44-0116-69
Isosequirin, trimethyl-	EtOH	224(4.24),285(3.75)	39-1921-69C
4-Pentene-1,2-diol, 3-(3,4-dimethoxy-phenyl)-5-(p-methoxyphenyl)- (tri-methylsequirin C)	EtOH	265(4.46)	39-1921-69C
2H-Pyran-4-ol, 5-(3,4-dimethoxyphenyl)-tetrahydro-2-(p-methoxyphenyl)-	EtOH	227(4.30),276(3.66), 280(3.65)	39-1921-69C
Sequirin B, trimethyl-	EtOH	227(4.39),275(3.68), 280(3.68)	39-1921-69C
$C_{20}H_{24}O_6$			
Isoflavan, 2',3',4',6',7-pentamethoxy-	EtOH	283(3.80),288(3.72)	39-0887-69C
Isotaxiresinol, 6-methyl ether	n.s.g.	214(4.5),384(3.9)	1-2021-69
Longocarpan, dimethyl ether	EtOH	283(3.8),288(3.71)	39-0887-69C
Phorbolactone, hemi-acetal	MeOH	231(3.86),325(1.77)	64-0080-69
$C_{20}H_{24}O_7$			
Bahia II	EtOH	212(4.29)	23-2849-69
Euparotin	EtOH	213(4.25)(end abs.)	44-3876-69
Eupatundin	MeOH	209(4.21)	44-3876-69
Gibberellin A_{26}, methyl ester	n.s.g.	288(2.45)	88-2077-69
$C_{20}H_{24}O_8$			
Capenicinone, dihydro-	MeOH	253(4.01)	78-4835-69
10-Epieupatoroxin	MeOH	214(4.18)	44-3876-69
Eupatoroxin	MeOH	213(4.11)	44-3876-69
6-Isochromanacetic acid, 4a,5,6,7,8,8a-hexahydro-5,7-dihydroxy- ,4-dimethyl-ene-3-oxo-8-vinyl-, methyl ester, diacetate	MeOH	208(4.24)(end abs.)	44-3903-69
$C_{20}H_{25}BrN_2O_{10}$			
3-Carbamoyl-1-β-D-glucopyranosylpyridin-ium bromide, 2',3',4',6'-tetraacetate	EtOH	266(3.70)	39-0192-69B
$C_{20}H_{25}BrO_3$			
Estra-4,6-dien-3-one, 4-bromo-17β-hydroxy-, acetate	EtOH	300(4.31)	94-1212-69

Compound	Solvent	$\lambda_{max}(\log \epsilon)$	Ref.
$C_{20}H_{25}Br_4NO_7$			
1,2-Dihydro-9-hydroxy-3,4-dimethoxy-10-methyl-1,2-dioxoacridinium tribromide, bis(dimethyl acetal), hypobromite	EtOH	218(4.57),248(4.58), 291(3.99),364(4.29)	12-1477-69
$C_{20}H_{25}ClN_2O$			
Macusine B, chloride	MeOH	218(4.66),270(3.88), 280(3.85),288(3.77)	33-0033-69
$C_{20}H_{25}ClN_2O_2$			
Hemitoxiferin	EtOH	246(3.93),301(3.50)	33-1564-69
$C_{20}H_{25}ClO_3$			
3,5-Estradien-7-one, 3-chloro-17β-hydroxy-, acetate	MeOH	284(4.45)	97-0421-69
$C_{20}H_{25}ClO_4$			
Crotophorbolone, 20-chloro-20-deoxy-	MeOH	240(3.69)	5-0158-69H
$C_{20}H_{25}ClO_7$			
Eupachlorin	MeOH	212(4.20)	44-3876-69
Eupatundin chlorohydrin	MeOH	212(4.18)	44-3876-69
$C_{20}H_{25}FO_3$			
B-Homoandrost-4-ene-3,17-dione, 7β,19-epoxy-7α-fluoro-	EtOH	240(4.04)	78-1219-69
$C_{20}H_{25}NO$			
2,5-Cyclohexadien-1-one, 2,6-di-tert-butyl-4-(phenylimino)-	n.s.g.	285(4.33)	70-2460-69
$C_{20}H_{25}NO_2$			
Benzyl alcohol, α-[2-(diethylamino)-ethyl]-, benzoate	dioxan	273(3.07),281(2.97)	65-1861-69
Benzyl alcohol, α-[2-(dimethylamino)-ethyl]-α-ethyl-, benzoate	dioxan	273(3.15),280(2.99)	65-1861-69
Isoquinoline, 1,2,3,4-tetrahydro-6,7-dimethoxy-2-methyl-1-phenethyl-, (R)-(-)-	MeOH	230s(3.96),283(3.62), 291s(3.53)	33-0678-69
hydrobromide	MeOH	231(3.93),285(3.54), 291s(3.49)	33-0678-69
1-Penten-3-ol, 5-[(p-methoxybenzyl)-methylamino]-3-phenyl-	EtOH	219(4.11),275(3.42), 281(3.39)	44-4158-69
$C_{20}H_{25}NO_3$			
Delnudine	EtOH	300(1.76)	88-5335-69
2,4-Pentadienoic acid, 2-acetyl-5-phenyl-5-piperidino-, ethyl ester	EtOH	253(4.16),414(4.16)	70-0363-69
2-Stilbenol, 6-[2-(dimethylamino)ethyl]-3,4'-dimethoxy-, trans	EtOH	216(4.45),299(4.32)	78-5475-69
	EtOH-NaOH	218(4.43),256(4.21), 296(4.18),360(3.92)	78-5475-69
3-Stilbenol, 2-[2-(dimethylamino)ethyl]-4,4'-dimethoxy-, trans	EtOH	223(4.46),257(4.37), 309(4.29)	78-5475-69
	EtOH-NaOH	218(4.20),257(3.98), 317(4.37)	78-5475-69
$C_{20}H_{25}NO_4$			
Erythroculine	EtOH	304(3.62)	88-0153-69

Compound	Solvent	$\lambda_{max}(\log \epsilon)$	Ref.
Isoquinoline, 1-(3,4-dimethoxyphenyl)-1,2,3,4-tetrahydro-6,7-dimethoxy-2-methyl-	EtOH	228s(4.20),281(3.78)	1-0244-69
Isoquinoline, 1,2,3,4-tetrahydro-5,6-dimethoxy-1-veratryl-, hydrochloride	H_2O	230(4.22),278(3.62)	44-2665-69
Laudanine	EtOH	283(3.92)	39-2030-69C
19-Nor-17α-pregn-4-en-20-yn-3-one, 17-hydroxy-, nitrate	MeOH	239(4.25)	78-0761-69
$C_{20}H_{25}NO_5$			
Estra-1,3,5(10)-triene-3,17β-diol, 2-nitro-, acetate	EtOH	294(3.90)	44-3699-69
7-Isoquinolinol, 1,2,3,4-tetrahydro-1-(4-hydroxy-3,5-dimethoxyphenethyl)-6-methoxy-, (R)-(-)-	EtOH	230s(4.17),283(3.70)	33-1228-69
$C_{20}H_{25}NO_7$			
1,2,9-Acridantrione, 3,4-dimethoxy-10-methyl-, 1,2-bis(dimethyl acetal)	EtOH	213(4.25),227(4.18), 280s(4.23),287(4.30), 325(4.09)	12-1477-69
$C_{20}H_{25}N_3O_4S$			
Benzimidazole, 2-[p-(dimethylamino)-styryl]-1-methyl-, methosulfate	MeOH	401(4.24)	22-1926-69
$C_{20}H_{26}$			
Chrysene, 1,2,3,4,4a,5,6,11,12,12a-decahydro-11,11-dimethyl-	EtOH	274(3.64)	22-2115-69
1,3,5-Cycloheptatriene, 7-(2,4-di-tert-butyl-2,4-cyclopentadien-1-ylidene)-	CH_2Cl_2	376(4.02)	89-0881-69
$C_{20}H_{26}BrNO_2$			
Isoquinoline, 1,2,3,4-tetrahydro-6,7-dimethoxy-2-methyl-1-phenethyl-, (R)-(-)-, hydrobromide	MeOH	231(3.93),285(3.54), 291s(3.49)	33-0678-69
$C_{20}H_{26}BrP$			
Benzylcyclohexylmethylphenylphosphonium bromide	MeOH	217s(4.15),255s(2.79), 260(2.95),266(3.03), 273(2.90)	24-3546-69
$C_{20}H_{26}ClNO_4$			
2,3,4,4a,5,7,8,9,10,11-Decahydro-6-phenyl-1H-cyclohepta[c]quinolizinium perchlorate	$CHCl_3$	330(4.00)	78-4161-69
Isoquinoline, 1,2,3,4-tetrahydro-5,6-dimethoxy-1-veratryl-, hydrochloride	H_2O	230(4.22),278(3.62)	44-2665-69
Spiro[cyclohexane-1,2'-[2H]indoline], 4',5',6',7'-tetrahydro-1'-methyl-3'-phenyl-, perchlorate	EtOH	285(3.81)	78-4161-69
$C_{20}H_{26}ClN_3O_6S_2$			
Sporidesmin D (ethanolate)	MeOH	216(4.45),252(4.00), 300(3.28)	39-1564-69C
$C_{20}H_{26}Cl_2N_2O_2$			
Spiro[isoquinoline-1(2H),4'-piperidine]-4,6-diol, 1'-benzyl-3,4-dihydro-, dihydrochloride	EtOH	282(3.26)	94-2353-69

Compound	Solvent	λmax(log ε)	Ref.
$C_{20}H_{26}F_2O_2$			
Androst-4-en-3-one, 6α,7α-(difluoro-methylene)-17β-hydroxy-	EtOH	246(4.14)	78-1219-69
$C_{20}H_{26}F_2O_3$			
Androst-4-ene-3,17-dione, 7α-(difluoro-methyl)-19-hydroxy-	EtOH	244(4.08)	78-1219-69
$C_{20}H_{26}N_2$			
1,1'-Azopropane, 1,1'-dimethyl-1,1'-diphenyl-, trans	isooctane	376(1.56)	35-1710-69
	MeOH	375(1.57)	35-1710-69
	MeCN	375(1.54)	35-1710-69
1H-1-Benzazepine, 4-[(dimethylamino)-methyl]-2,3,4,5-tetrahydro-1-methyl-5-phenyl-	EtOH	260(3.6)	33-1929-69
7,13b-Methano-13bH-indolo[3,2-e][2]-benzazocine, 1,2,3,4,4a,5,6,7,8,13-decahydro-6,13-dimethyl-, cis	EtOH	231(4.56),286(3.88), 293(3.86)	44-3165-69
trans	EtOH	231(4.54),285(3.88), 293(3.85)	44-3165-69
Quinoline, 4-anilino-3-ethyl-1,2,3,4-tetrahydro-2-propyl-	MeOH	251(4.43),302(3.87)	18-2885-69
Quinoline, 3-[2-(dimethylamino)ethyl]-1,2,3,4-tetrahydro-1-methyl-4-phenyl-	EtOH	258(4.0),305(3.5)	33-1929-69
$C_{20}H_{26}N_2O$			
14-Epivincaminol, deoxy-	n.s.g.	228(4.54),282(3.86)	28-1442-69A
Pseudoyohimban, 10-methoxy-, (±)-, hydrobromide	EtOH	224(4.42),281(3.95), 295s(3.90)	4-0577-69
Vincaminol, deoxy-	n.s.g.	228(4.54),282(3.86)	28-1442-69A
Yohimban, 10-methoxy-, (±)-	EtOH	225(4.47),280(3.97), 296s(3.90)	4-0577-69
$C_{20}H_{26}N_2O_2$			
Picralinol	EtOH	243(3.80),293(3.46)	33-0701-69
$C_{20}H_{26}N_2O_3$			
Iboxygaine hydroxyindolenine	EtOH	227(4.12),265(3.73), 285(3.79),313(3.68)	44-0412-69
$C_{20}H_{26}N_2O_{10}$			
Nicotinamide, 1-β-D-glucopyranosyl-1,4-dihydro-, 2',3',4',6'-tetraacetate	EtOH	331(3.79)	39-0192-69B
Nicotinamide, 1-β-D-glucopyranosyl-1,6-dihydro-, 2',3',4',6'-tetraacetate	EtOH	262(3.39),338(3.85)	39-0192-69B
$C_{20}H_{26}N_4O$			
Benzo[c]cinnoline, 3,8-bis(diethyl-amino)-, 5-oxide	C_6H_{12}	204(4.31),247s(4.32), 271(4.43),311(4.70), 320s(4.63),349s(4.35), 474(3.60),497(3.52)	4-0523-69
	EtOH	303(4.16),315s(4.13), 338(4.75),355s(4.48), 519(3.50)	4-0523-69
$C_{20}H_{26}N_4O_2$			
Benzo[c]cinnoline, 3,8-bis(diethyl-amino)-, 5,6-dioxide	C_6H_{12}	205(4.32),227(4.08), 260(4.20),311(4.53), 322s(4.53),336(4.59), 354s(4.37),388s(3.34),	4-0523-69

(continued on next page)

Compound	Solvent	λ_{max}(log ε)	Ref.
Benzo[c]cinnoline, 3,8-bis(diethyl-amino)-, 5,6-dioxide (cont.)		470(3.54),495(3.38)	4-0523-69
	EtOH	230(4.21),257(4.21), 319(4.77),335(4.74), 350s(4.57),533(3.56)	4-0523-69
Biurea, 1,6-dimesityl-	EtOH	256(2.7)	28-0730-69A
$C_{20}H_{26}N_4O_4$			
3H-Cyclopropa[b]naphthalen-3-one, 1,1a,2,2a,4,5,6,6a,7,7a-decahydro-1,1,6a-trimethyl-, 2,4-dinitrophenyl-hydrazone	$CHCl_3$	368(4.43)	32-0247-69
2(3H)-Naphthalenone, 4,4a,5,6,7,8-hexa-hydro-7-isopropyl-4a-methyl-, 2,4-dinitrophenylhydrazone	$CHCl_3$	385(4.48)	32-0247-69
$C_{20}H_{26}N_4O_5$			
1(2H)-Purinebutyric acid, 4-amino-α-carboxyamino)-2-oxo-, α-benzyl tert-butyl ester	EtOH	274(4.00)	78-5989-69
$C_{20}H_{26}N_4O_6$			
[2,2'-Bi-2H-indazole]-5,5'-dicarboxylic acid, 3,3',3a,3'a,4,4',5,5',6,6',7,7'-dodecahydro-3,3'-dioxo-, diethyl ester	EtOH	206(3.61),238(3.84), 255s(3.68)	39-2783-69C
[2,2'-Bi-2H-indazole]-7,7'-dicarboxylic acid, 3,3',3a,3'a,4,4',5,5',6,6',7,7'-dodecahydro-3,3'-dioxo-, diethyl ester	EtOH	205(3.41),238(3.65), 257s(3.42)	39-2783-69C
$C_{20}H_{26}N_6OS_3$			
Spiro[furo[2,3-d]thiazole-2(3H),4'-imid-azolidine]-2',5'-dithione, 1',3'-di-allyl-3-[(4-amino-2-methyl-5-pyrimi-dinyl)methyl]tetrahydro-3a-methyl-	EtOH	325(4.19)	94-0910-69
$C_{20}H_{26}O$			
Pulchellin F	n.s.g.	214(4.11)	102-0661-69
$C_{20}H_{26}O_2$			
9β,10α-Androsta-4,6-diene-3,17-dione, 18-methyl-	n.s.g.	284(4.40)	54-0752-69
2,5-Cyclohexadien-1-one, 2,6-di-tert-butyl-4-hydroxy-4-phenyl-	MeOH	235s(4.07)	64-0547-69
Estra-1,3,5(10)-trien-3-ol, acetate	C_6H_{12}	214(3.97),270(2.92), 276(2.89)	39-1234-69C
18-Homo-5α-androsta-1,4-dien-3-one, 17β,18a-epoxy-	EtOH	243(4.09)	73-3479-69
1(2H)-Naphthalenone, 3,4-dihydro-2-(2-isobutyrylcyclohexyl)-	EtOH	206(4.30),248(4.10), 290(3.21)	22-2110-69
19-Nor-17α-pregn-4-en-20-yn-3-one, 17-hydroxy-	MeOH	240(4.2)	48-0671-69
Vitamin A_2 acid, all-trans	EtOH	305(4.14),373(4.61)	22-3252-69
9-cis	EtOH	303(4.26),368(4.53)	22-3252-69
$C_{20}H_{26}O_3$			
Estra-1,3,5(10)-triene-3,17β-diol, 17-acetate	EtOH	281(3.30),288(3.27)	94-1206-69
Estra-1,3,5(10)-trien-17-one, 11-hydr-oxy-3-methoxy-9-methyl-	n.s.g.	278(3.33)	23-1989-69

Compound	Solvent	$\lambda_{max}(\log \epsilon)$	Ref.
Fukujusone	EtOH	250(3.95)	31-1129-69
D-Homoestra-1,3,5(10)-trien-17a-one, 3-methoxy-	EtOH	277(3.2)	70-1655-69
19-Norandrosta-4,6-dien-3-one, 17β-acetoxy-	EtOH	282(4.43)	94-1212-69
Taxodione	MeOH	320(4.40),332(4.41), 400(3.30)	44-3912-69
$C_{20}H_{26}O_4$			
4-Cyclopentene-1,3-dione, 4-[1-(hexahydro-8-methyl-2-oxo-4H-3a,7-ethanobenzofuran-4-yl)ethyl]-2,2-dimethyl-	EtOH	235(4.11)	94-2036-69
Estra-1,3,5(10)-triene-2,3,17β-triol, 17-acetate	EtOH	253(2.61),288(3.63)	94-1206-69
Oopodin	EtOH	233(4.35)	105-0314-69
$C_{20}H_{26}O_5$			
Crotophorbolone	MeOH	196(4.18),241(3.68), 335(1.83)	5-0158-69B
1,3-Cyclohexanedicarboxylic acid, 6-oxo-1-phenethyl-, diethyl ester	EtOH	255(3.27)	77-1253-69
2-Oxaandrosta-4,6-diene-1-carboxylic acid, 17β-hydroxy-17-methyl-3-oxo-	EtOH	274(4.32)	19-0275-69
	EtOH	274(4.28)	78-4257-69
1,2-Pentanediol, 3-(3,4-dimethoxyphenyl)-5-(p-methoxyphenyl)-	EtOH	278(3.69),285(3.62)	39-1921-69C
Phorbobutanone	MeOH	227s(3.75),310s(1.89), 330s(1.78)	64-0099-69
13-Tetradecenoic acid, 11-hydroxy-9,12-dioxo-14-phenyl-	MeOH	295(4.35)	88-1615-69
$C_{20}H_{26}O_6$			
1,3-Cyclohexanedicarboxylic acid, 1-(m-methoxybenzyl)-6-oxo-, diethyl ester	EtOH	275(3.41)	77-1253-69
Phorbobutanone, hydroxy-	MeOH	227(3.76),326(2.15)	5-0142-69E
$C_{20}H_{26}O_7$			
Elephantin, tetrahydro-	MeOH	211(4.03)	44-3867-69
D-arabino-Tetritol, 1-C-[5-(p-methoxyphenyl)-2-furyl]-1,3,4-tri-O-methyl-, acetate	MeOH	287(4.34),306(4.08)	65-0119-69
$C_{20}H_{27}$			
(3,5-Di-tert-butyl-1,3-cyclopentadien-1-yl)cycloheptatrienylium	CH_2Cl_2	234(4.33),284(3.82), 308(3.38),497(4.34)	89-0881-69
$C_{20}H_{27}ClN_2O_4S$			
2-[[3-(Dimethylamino)-5,5-dimethyl-2-cyclohexen-1-ylidene]methyl]-3-ethylbenzothiazolium perchlorate	EtOH	486(5.23)	65-1732-69
$C_{20}H_{27}Cl_2N_4O_8P$			
Uridine, 5'-[hydrogen[p-[bis(2-chloroethyl)amino]benzyl]phosphoramidate]	pH 2	261(4.26)	65-0668-69
	pH 12	258(4.21)	65-0668-69
$C_{20}H_{27}NO_3$			
Taxodione, 7-amino-	MeOH	265(3.62),295(3.58), 485(4.18)	44-3912-69

Compound	Solvent	$\lambda_{max}(\log \epsilon)$	Ref.
$C_{20}H_{27}NO_4$			
10-Azaestra-2,4-dien-1-one, 3-ethoxy-17β-hydroxy-, formate	EtOH	215(4.52),234s(--), 293(3.77)	22-0117-69
$C_{20}H_{27}NO_5$			
Estr-4-en-3-one, 17β-hydroxy-2α-nitro-, acetate	EtOH	246(4.18)	44-3699-69
19-Norpregn-4-ene-3,20-dione, 17-hydroxy-, nitrate	MeOH	239(4.25)	78-0761-69
Phalenopsine T	EtOH	210(3.93),247(2.04), 252(2.18),258(2.28), 264(2.15)	1-1151-69
$C_{20}H_{27}N_3O$			
Phenol, 2,6-di-tert-butyl-4-(3-phenyl-1-triazeno)-	n.s.g.	242(4.19),292(3.95), 357(4.35)	70-2460-69
$C_{20}H_{28}$			
Azulene, 1,3-dipentyl-	C_6H_{12}	241(4.10),282(4.73), 351(3.70),368(3.66), 588s(2.36),608s(2.42), 632(2.49),663(2.41), 695(2.42),735s(2.04), 776(--)	44-2375-69
$C_{20}H_{28}Br_2NiP_2$			
Nickel, [2,2'-biphenylylenebis[diethylphosphine]]dibromo-	EtOH	315s(2.95),390s(2.60), 535(1.76),810(1.69)	39-1097-69A
	$CHCl_3$	589(2.45),800(2.03)	39-1097-69A
$C_{20}H_{28}ClNO_7$			
[2-(1-Acetonyl-7,8-dimethoxy-2-naphthyl)ethyl]trimethylammonium perchlorate	H_2O	289(3.76),300(3.72), 337(3.36)	49-0163-69
$C_{20}H_{28}Cl_2NiP_2$			
Nickel, [2,2'-biphenylylenebis[diethylphosphine]]dichloro-	benzene	325s(3.44),352s(3.38), 505(2.39),830(2.09), 1000s(1.70)	39-1097-69A
	$CHCl_3$	503(2.74),780(1.58)	39-1097-69A
$C_{20}H_{28}F_2O_2$			
Androst-4-en-3-one, 7α-(difluoromethyl)-17β-hydroxy-	EtOH	242(4.16)	78-1219-69
$C_{20}H_{28}IP$			
Dimesityldimethylphosphonium iodide	C_6H_{12}	237(4.36),278(3.43), 286(3.45)	65-1544-69
$C_{20}H_{28}I_2NiP_2$			
Nickel, [2,2'-biphenylylenebis[diethylphosphine]]diiodo-	benzene	304(3.80),350(3.70), 433(3.58),865(2.77), 1000s(2.12)	39-1097-69A
	$CHCl_3$	425(3.63),843(2.85)	39-1097-69A
$C_{20}H_{28}N_2O$			
Spiro[cyclohexane-1,5'[3H]indole]-2'-carboxamide, N-tert-butyl-4',6'-dimethyl-	EtOH	277(3.8)	24-1876-69
Yohimban, 2,7-dihydro-10-methoxy-, (±)-	pH 1	239(4.10),305(3.42)	4-0577-69

Compound	Solvent	λ max (log ε)	Ref.
Yohimban, 2,7-dihydro-10-methoxy-, (±)- (cont.)	N HCl	238(3.61),278(3.18), 305(2.95)	4-0577-69
	EtOH	246(4.07),315(3.50)	4-0577-69
$C_{20}H_{28}N_2O_2$			
Corynantheol, dihydro-10-methoxy-	EtOH	227(4.46),282(3.95), 294s(3.90)	78-5329-69
3-Epihunterburnine, dihydro-O-methyl-	EtOH	226(4.30),281(3.83), 294s(3.77)	78-5319-69
Isocorynantheidic acid, methyl ester	EtOH	225(4.37),282(3.82), 290(3.72)	24-3963-69
$C_{20}H_{28}N_4O_4$			
1(2H)-Naphthalenone, 3,4,4a,5,6,7,8,8a-octahydro-7-isopropyl-4a-methyl-, 2,4-dinitrophenylhydrazone	$CHCl_3$	366(4.50)	32-0247-69
isomer	$CHCl_3$	364(4.47)	32-0247-69
$C_{20}H_{28}O$			
3-Dehydroretinol	EtOH	318(4.39)	22-3247-69
A-Homoestra-2,4,5(10)-trien-17β-ol, 17α-methyl-	EtOH	268(3.57)	33-0121-69
1,6,8-Nonatrien-4-yn-3-ol, 3,7-dimethyl-9-(2,6,6-trimethyl-1-cyclohexen-1-yl)-	EtOH	294(4.50)	22-3247-69
isomer	EtOH	299(4.37)	22-3247-69
19-Norpregna-3,5-dien-20-one	MeOH	228(4.20),235(4.21)	73-0458-69
Vitamin A aldehyde	benzene	380(4.64)	78-5383-69
	C_6H_{12}	370(4.68)	78-5383-69
	hexane	366(4.70)	78-5383-69
	MeOH	377(4.48)	78-5383-69
	acetone	371(4.68)	78-5383-69
	dioxan	374(4.66)	78-5383-69
	CCl_4	378(4.69)	78-5383-69
	CS_2	390(4.58)	78-5383-69
	DMF	380(4.66)	78-5383-69
$C_{20}H_{28}O_2$			
9β,10α-Androsta-4,6-dien-3-one, 17β-hydroxy-18-methyl-	n.s.g.	287(4.40)	54-0752-69
9β,10α-Androst-4-ene-3,17-dione, 18-methyl-	EtOH	240(4.21)	44-0107-69
	EtOH	241(4.21)	54-0752-69
3β,5-Cycloestr-1(10)-en-17β-ol, acetate	EtOH	207(3.78)	22-0613-69
3β,5-Cycloestr-9-en-17β-ol, acetate	EtOH	208(4.04)	22-0613-69
18-Homo-5α-androst-1-en-3-one, 17β,18a-epoxy-	EtOH	231(3.98)	73-3479-69
B-Homoestra-1,3,5(10)-trien-17β-ol, 3-methoxy-	EtOH	276(3.28),285(3.23)	31-0571-69
Kaur-16-en-19-al, 12-oxo-	EtOH	276(1.71),295(1.86)	88-0599-69
1(2H)-Naphthalenone, 4a,5,6,7,8,8a-hexahydro-4-[2-(3-furyl)ethyl]-4a,8,8-trimethyl-	n.s.g.	204(4.13)	78-2233-69
2,4,7-Nonatrienal, 3,7-dimethyl-6-oxo-9-(2,6,6-trimethyl-1-cyclohexen-1-yl)-	EtOH	290(3.01)	44-3039-69
Solidagenone, 9-deoxy-	n.s.g.	215(3.80)	78-2233-69
$C_{20}H_{28}O_2S$			
Benzo[c]thiophen-4(5H)-one, 6,7-dihydro-5-(2-isobutyrylcyclohexyl)-1,3-dimethyl-	EtOH	220(4.05),265(4.00), 295s(3.35)	22-2110-69

Compound	Solvent	λ_{max}(log ϵ)	Ref.
$C_{20}H_{28}O_3$			
10α-Estr-4-en-3-one, 17β-acetoxy-	EtOH	303(4.31)	44-1447-69
D-Homoestra-1,3,5(10)-triene-14α,17aα-diol, 3-methoxy-	EtOH	278(2.9)	70-1655-69
17aβ-isomer	EtOH	279(2.9)	70-1655-69
A-Norestr-3(5)-en-2-one, 17β-acetoxy-1β-methyl-	EtOH	239(4.17)	78-1367-69
A-Nor-5α-estr-1(10)-en-2-one, 17β-acetoxy-1-methyl-	EtOH	242(4.22)	78-1367-69
17a-Oxa-9β,10α-androst-4-ene-3,17-dione, 18-methyl-	n.s.g.	240(4.23)	54-0752-69
Solidagenone II	n.s.g.	223(4.03)	78-2233-69
Taxodone	MeOH	316(4.30)	44-3912-69
$C_{20}H_{28}O_3S$			
Bicyclo[5.2.0]non-8-en-3-ol, 1,7,8,9-tetramethyl-, p-toluenesulfonate	MeOH	225(4.09),256(2.64), 262(2.75),267(2.70), 273(2.66)	24-0275-69
$C_{20}H_{28}O_4$			
Estr-4-en-3-one, 14α,17β-dihydroxy-, 17-acetate, d-	EtOH	241(4.18)	44-3530-69
19-Norandrost-4-en-3-one, 4,17β-dihydroxy-, 17-acetate	EtOH	276(4.12)	94-1206-69
19-Nor-10ξ,14β,17α-cortexone, 8-hydroxy-	EtOH	243(4.18)	33-2156-69
Taxoquinone	MeOH	276(4.08),408(2.90)	44-3912-69
$C_{20}H_{28}O_5$			
3,9-Dioxapentacyclo[6.4.0.0^{2,7}.0^{4,11}.-0^{5,10}]dodecan-6-one, 2,6,8,10-tetraethyl-12-hydroxy-, acetate	EtOH	225(3.54)	44-4056-69
Estr-4-en-3-one, 10β,11β,17β-trihydroxy-, 17-acetate, dl-	EtOH	237(4.13)	44-3530-69
$C_{20}H_{28}O_7$			
Hydroxybisdehydrophorbol hemiketal	MeOH	196(4.09),238(3.72), 336(1.85)	64-0080-69
$C_{20}H_{29}BrO_3$			
A-Nor-5α,10α-estran-2-one, 17-acetoxy-3α-bromo-1α-methyl-	EtOH	319(2.16)	78-1367-69
$C_{20}H_{29}ClN_2O_2$			
Hunterburnine, dihydro-, α-methochloride	MeOH	274(3.96),301(3.66), 309s(3.60)	78-5319-69
$C_{20}H_{29}N$			
Podocarpa-6,8,11,13-tetraen-15-amine, 13-isopropyl-, hydrochloride	MeOH	220(4.47),265(3.97)	5-0193-69H
$C_{20}H_{29}NO$			
Dehydroabietylamine, 7-oxo-, hydrochloride	MeOH	255(4.04)	5-0193-69H
$C_{20}H_{29}N_3O_2$			
1,4-Phthalazinedione, 5-[2-(dipentylamino)vinyl]-2,3-dihydro-	THF	320(3.8),370(4.1)	24-3241-69

Compound	Solvent	$\lambda_{max}(\log \epsilon)$	Ref.
$C_{20}H_{30}ClNO_7$			
[2-[1-(2-Hydroxypropyl)-7,8-dimethoxy-2-naphthyl]ethyl]trimethylammonium perchlorate	H_2O	290(3.78),301(3.75), 338(3.37)	49-0163-69
$C_{20}H_{30}Cl_4Rh_2$			
Rhodium, di-μ-chlorodichlorobis(pentamethyl-π-cyclopentadienyl)di-	H_2O	205(3.99),224(4.29), 382(3.62)	39-1299-69A
	EtOH	210(4.11),248(4.53), 413(3.62)	39-1299-69A
$C_{20}H_{30}N_2O_2$			
p-Benzoquinone, 2,5-bis(cyclohexylmethylamino)-	EtOH	227(4.39),378(4.32), 522(2.72)	39-1245-69C
$C_{20}H_{30}N_2O_4$			
4H-Quinolizin-4-one, 2-(2-amino-5-methoxyphenyl)-1,2,3,6,7,8,9,9a-octahydro-1-hydroxy-8-[1-(hydroxymethyl)propyl]-	EtOH	236(3.93),302(3.51)	78-5319-69
$C_{20}H_{30}O$			
Androst-4-en-3-one, 6α-methyl-	EtOH	241(4.16)	44-2288-69
$C_{20}H_{30}O_2$			
Agbaninol	MeOH	214(3.62)	39-2153-69C
Androst-4-en-3-one, 6β-hydroxy-6α-methyl-	EtOH	239(3.95)	44-2288-69
9β,10α-Androst-4-en-3-one, 17β-hydroxy-18-methyl-	EtOH	242(4.21)	44-0107-69
	n.s.g.	242(4.22)	54-0752-69
13-Oxadispiro[5.0.5.1]trideca-1,4-dien-3-one, 2,4-di-tert-butyl-	n.s.g.	263(4.11)	70-1732-69
$C_{20}H_{30}O_2S$			
Androst-4-en-3-one, 17β-hydroxy-7α-mercapto-17α-methyl-	EtOH	239(4.13)	94-0011-69
$C_{20}H_{30}O_3$			
Agbanindiol B	MeOH	211(3.67)	39-2153-69C
8-Episolidaganone	n.s.g.	207(3.79)	78-2233-69
A-Nor-5α,10α-estran-2-one, 17β-acetoxy-1α-methyl-	EtOH	294(1.57)	78-1367-69
A-Norestr-5(10)-en-2α-ol, 17β-acetoxy-1β-methyl-	MeCN	194s(3.83)	78-1367-69
Sesquiterpene from Brickellia Guatemaliensis	n.s.g.	238(4.47)	88-5109-69
$C_{20}H_{30}O_4$			
5-Heptenoic acid, 7-[2-(3-hydroxy-1-octenyl)-5-oxo-3-cyclopenten-1-yl]-(15-hydroxy-9-oxoprost-5,10,13-trienoic acid, 5-cis-10,13-trans)	EtOH	217(4.00)	44-3552-69
Labda-8,13-diene-15,19-dioic acid, trans	EtOH	215(4.20)	12-0491-69
9,10-Phenanthrenedicarboxylic acid, 1,2,3,4,5,6,7,8,8a,9,10,10a-dodecahydro-, diethyl ester	n.s.g.	208(4.16)	22-0962-69

Compound	Solvent	λ_{max}(log ε)	Ref.
$C_{20}H_{30}O_5$			
4,6-Benzo[b]furandiol, 2-(1-hydroxy-1-methylethyl)-7-isobutyryl-5-isopentyl-2,3-dihydro-	EtOH	292(4.18),350(3.24)	39-0938-69C
$C_{20}H_{31}NO$			
Azacyclotridecan-2-one, 1-(2,6-xylyl)-, biradical	MeOH	242(3.87),328(4.13)	88-2281-69
2,5-Cyclohexadien-1-one, 2,6-di-tert-butyl-4-(piperidinomethylene)-	hexane	380(4.53)	70-0580-69
$C_{20}H_{31}NO_6$			
Echiumine, trans	EtOH	217(4.01)	73-1832-69
$C_{20}H_{32}$			
1,3,6,10-Cyclotetradecatetraene, 14-isopropyl-3,7,11-trimethyl- (cembrene)	n.s.g.	245(3.86)	105-0007-69
1,5,10-Cyclotetradecatriene, 12-isopropyl-1,5-dimethyl-9-methylene-(isocembrene)	n.s.g.	238(4.3)	105-0007-69
$C_{20}H_{32}Br_2N_4O_2$			
N,N,N,N',N'-Pentamethyl-N'-[2-[(4-nitro-1-naphthyl)amino]ethyl]-N,N'-trimethylenediammonium dibromide (also spectra in biological media)	pH 6.5 EtOH	322(3.40),428(4.16) 333s(3.45),414(4.19)	35-5136-69 35-5136-69
$C_{20}H_{32}O_2$			
7-Abieten-18-oic acid, methyl ester	n.s.g.	204(3.83)	44-1550-69
8-Abieten-18-oic acid, methyl ester	n.s.g.	197(3.89)	44-1550-69
13-Abieten-18-oic acid, methyl ester	n.s.g.	191(3.98)	44-1550-69
Copaiferic acid	EtOH	218(4.02)	7-0539-69
$C_{20}H_{32}O_3$			
Agbanindiol B, dihydro-	MeOH	212(3.72)	39-2153-69C
Labda-8,13-dien-15-oic acid, 19-hydroxy-, trans	EtOH	210(4.10)	12-0491-69
Stachysolon	EtOH	239(4.05)	105-0005-69
$C_{20}H_{32}O_4$			
Portulal	EtOH	285(1.43)	88-0359-69
Prosta-10,13-dienoic acid, 15-hydroxy-9-oxo-, 15S-	EtOH	217(4.07)	44-3552-69
Prostaglandin B_1, dl-	EtOH	278(4.42)	94-0408-69
$C_{20}H_{32}O_5$			
Taxa-4(20),11-diene-5α,7β,9α,10β,13α-pentol	n.s.g.	224(3.94)	77-1282-69
$C_{20}H_{33}NO_2$			
Bicyclo[9.3.1]pentadec-11(15)-en-12-one, 13-methyl-15-morpholino-	EtOH	339(4.28)	39-0592-69C
Myristohydrxamic acid, N-phenyl-	EtOH	247(4.02)	42-0831-69
$C_{20}H_{34}O$			
Isoabienol	EtOH	226(4.33)	65-0451-69
Neoabienol	n.s.g.	238(4.46)	105-0210-69

Compound	Solvent	$\lambda_{max}(\log \epsilon)$	Ref.
$C_{20}H_{34}O_5$			
α-L-Arabinopyranoside, eudesm-4(14)-en-11-yl	EtOH	210(2.33)	44-3697-69
Prost-13-enoic acid, 9α,11α-dihydroxy-15-oxo-, trans	EtOH	233(4.04)	44-3552-69
$C_{20}H_{35}N_3O$			
Phenol, 2,6-di-tert-butyl-4-(3,3-diisopropyl-1-triazeno)-	n.s.g.	234(4.05),291(4.12), 329(4.09)	70-2460-69
Phenol, 2,6-di-tert-butyl-4-(3,3-dipropyl-1-triazeno)-	n.s.g.	234(4.06),291(4.12), 328(4.12)	70-2460-69
$C_{20}H_{38}N_2$			
1,1'-Azopropane, 1,1'-dicyclohexyl-1,1'-dimethyl-	hexane	381(1.34)	35-1710-69
	isooctane	381(1.34)	35-1710-69
	MeCN	380(1.36)	35-1710-69
$C_{20}H_{38}O_2$			
Eicosanal, 3-oxo-	isooctane	268(3.78)	88-5205-69
Eicosanal, 3-oxo-, copper chelate	isooctane	245(3.86),302(3.95)	88-5205-69

Compound	Solvent	λ_{max}(log ε)	Ref.
$C_{21}H_{11}Cl_5O_5$			
Acetic acid, (4-benzoyl-3-hydroxyphen-oxy)-, pentachlorophenyl ester	$CHCl_3$	283(4.21),325(3.91)	39-0037-69C
$C_{21}H_{12}ClNO_3$			
Spiro[1,4,2-dioxazole-5,9'(10'H)-phenan-thren]-10'-one, 3-(o-chlorophenyl)-	EtOH	250(4.59),275(4.13)	23-1473-69
$C_{21}H_{12}N_4$			
1,2-Diazaspiro[4.4]nona-1,3,6,8-tetra-ene-3,4-dicarbonitrile, 6,9-diphenyl-	$CHCl_3$	242(4.34),332(4.20)	64-0536-69
$C_{21}H_{12}N_4O_3$			
Furo[3,2-e]-s-triazolo[1,5-c]pyrimidine, 8,9-di-2-furyl-2-phenyl-	EtOH	239(4.34),259(4.34), 271s(4.31),352(4.28)	95-1434-69
$C_{21}H_{12}O_7$			
[1,1'-Binaphthalene]-3,4,5',8'-tetrone, 6',7'-epoxy-6',7'-dihydro-5-hydroxy-4'-methoxy-	EtOH	361(3.59)	39-2059-69C
$C_{21}H_{13}BrN_4O_5$			
Benzofuran, 2-(p-bromobenzoyl)-, 2,4-dinitrophenylhydrazone	dioxan	393(4.50)	32-1273-69
isomer	dioxan	393(4.53)	32-1273-69
$C_{21}H_{13}BrO_2$			
Spiro[2H-1-benzopyran-2,3'-[3H]naphtho-[2,1-b]pyran], 6-bromo-	dioxan	244(4.91),300(3.99), 312(4.05),334(3.75), 349(3.76)	5-0162-69B
$C_{21}H_{13}Br_3O_5$			
Acetic acid, (4-benzoyl-3-hydroxyphen-oxy)-, 2,4,6-tribromophenyl ester	$CHCl_3$	285(4.25),325(3.97)	39-0037-69C
$C_{21}H_{13}ClN_2O_2$			
Spiro[Δ^2-1,2,4-oxadiazoline-5,9'(10'H)-phenanthren]-10'-one, 3-(o-chloro-phenyl)-	EtOH	265(4.46),290(4.44)	23-1473-69
$C_{21}H_{13}ClN_4O_5$			
Benzofuran, 2-(p-chlorobenzoyl)-, 2,4-dinitrophenylhydrazone	dioxan	393(4.51)	32-1273-69
isomer	dioxan	393(4.48)	32-1273-69
$C_{21}H_{13}Cl_3O_5$			
Acetic acid, (4-benzoyl-3-hydroxyphen-oxy)-, 2,4,6-trichlorophenyl ester	$CHCl_3$	285(4.23),322(3.96)	39-0037-69C
$C_{21}H_{13}FN_4O_5$			
Benzofuran, 2-(p-fluorobenzoyl)-, 2,4-dinitrophenylhydrazone	dioxan	393(4.50)	32-1273-69
isomer	dioxan	393(4.54)	32-1273-69
$C_{21}H_{13}IN_4O_5$			
Benzofuran, 2-(p-iodobenzoyl)-, 2,4-dinitrophenylhydrazone	dioxan	393(4.50)	32-1273-69
isomer	dioxan	393(4.45)	32-1273-69

Compound	Solvent	λ_{max}(log ε)	Ref.
$C_{21}H_{13}N$			
Dibenz[a,h]acridine	dioxan	250(4.29),262(4.28), 270(4.41),289(4.93), 297(4.99),318(4.16), 333(4.06),346(3.76), 354(3.76),370(4.10), 383(3.63),395(4.29)	24-1202-69
$C_{21}H_{13}NO$			
Phenanthro[9,10-d]oxazole, 2-phenyl-	EtOH	260(4.69),302(4.12)	77-0200-69
$C_{21}H_{13}N_5O_2$			
Pyridine, 2,6-bis(5-phenyl-1,3,4-oxadia-zol-2-yl)-	H_2SO_4	278(4.45),349(4.49)	4-0965-69
$C_{21}H_{14}$			
Azuleno[5,6,7-cd]phenalene, 4-methyl-	EtOH	247(4.35),318(5.11), 361(3.79),406(3.92), 427(4.47),437(4.25), 453(4.98),617(2.81), 635(2.92),669(3.14), 704(3.09),744(3.27)	24-2301-69
$C_{21}H_{14}BrClO_6$			
3-(5-Bromo-2-hydroxystyryl)naphtho-[2,1-b]pyrylium perchlorate	MeOH-$HClO_4$	235(4.53),306(3.91), 333(3.97),519(4.42)	5-0162-69B
$C_{21}H_{14}N_2$			
1H-Phenanthro[9,10-d]imidazole, 2-phenyl-	EtOH	261(4.75),310(4.27)	77-0200-69
$C_{21}H_{14}N_2O$			
Indolo[1,2-a]quinazolin-5(6H)-one, 7-phenyl-	EtOH	378(3.00)	44-0887-69
$C_{21}H_{14}N_2O_2S$			
2-Benzimidazolinethione, 1,3-dibenzoyl-	MeOH	249(4.49),298(4.30)	97-0152-69
	dioxan	302(4.22)	4-0023-69
$C_{21}H_{14}N_2O_5$			
Ether, benzyl p-[(2,4-dinitrophenyl)-ethynyl]phenyl]-	dioxan	380(4.66)	56-1469-69
$C_{21}H_{14}N_4O_5$			
Benzofuran, 2-benzoyl-, 2,4-dinitrophen-ylhydrazone	dioxan	394(4.55)	32-1273-69
isomer	dioxan	394(4.50)	32-1273-69
$C_{21}H_{14}N_4O_6$			
Phthalide, 3-benzoyl-, 2,4-dinitrophen-ylhydrazone	EtOH	220s(4.44),256(4.17), 350(4.41)	12-0577-69
$C_{21}H_{14}O_2$			
Spiro[2H-1-benzopyran-2,3'-[3H]naphtho-[2,1-b]pyran	dioxan	243(4.83),300(3.98), 313(4.02),335(3.66), 349(3.73)	5-0162-69B
$C_{21}H_{14}O_4$			
Anthraquinone, 1-(benzyloxy)-8-hydroxy-	$CHCl_3$	256(4.45),278(4.14), 420(4.02),430s(4.02)	114-0109-69A

Compound	Solvent	λ_{max}(log ϵ)	Ref.
Anthraquinone, 1-(benzyloxy)-8-hydroxy- (cont.)	EtOH	224(4.69),255(4.38), 279s(4.04),420(4.00), 427s(3.99)	114-0109-69A
	dioxan	224(4.65),256(4.40), 278s(4.04),396s(3.93), 415(4.00),425s(3.98)	114-0109-69A
	CCl_4	256(4.37),278(4.14), 279s(4.00),404s(3.97), 416(4.01),430s(3.99)	114-0109-69A
	pyridine	420(4.01),430s(3.99)	114-0109-69A
$C_{21}H_{14}O_7$			
[1,1'-Binaphthalene]-3,4,8'(5'H)-trione, 6',7'-epoxy-6',7'-dihydro-5,5'-dihydroxy-4'-methoxy-	EtOH	235(4.50),301(3.78), 443(3.80)	39-2059-69C
$C_{21}H_{15}$			
5-Phenyldibenzo[a,d]cycloheptenylium perchlorate	H_2SO_4	240(4.26),268(4.10), 312(5.01),384(3.76), 414(3.70),528(3.65), 564(3.69)	5-0048-69A
$C_{21}H_{15}Cl$			
Anthracene, 9-benzyl-10-chloro-	EtOH	251s(4.85),258(5.07), 341(3.52),358(3.84), 377(4.06),398(4.05)	78-3485-69
Anthracene, 9-chloro-10-o-tolyl-	EtOH	253s(4.76),260(4.93), 340(3.45),357(3.77), 376(3.98),396(3.96)	78-3485-69
Anthracene, 9-chloro-10-p-tolyl-	EtOH	250s(4.84),259(5.08), 342(3.60),357(3.86), 376(4.05),397(4.01)	78-3485-69
$C_{21}H_{15}ClN_2$			
Imidazole, 2-(p-chlorophenyl)-4,5-diphenyl-	MeOH	310(4.43)	44-3981-69
Imidazole, 4-(p-chlorophenyl)-2,5-diphenyl-	MeOH	302(4.45)	44-3981-69
$C_{21}H_{15}ClO$			
Anthracene, 9-chloro-10-(p-methoxyphenyl)-	EtOH	253s(4.94),261(5.23), 342(3.53),358(3.87), 376(4.08),399(4.06)	78-3485-69
$C_{21}H_{15}ClO_4S$			
2,4-Diphenyl-1-benzothiopyrylium perchlorate	HOAc-$HClO_4$	270(3.37),300(3.99), 402(3.40)	2-0191-69
$C_{21}H_{15}ClO_4S_2$			
2,4,5-Triphenyl-1,3-dithiol-1-ium perchlorate	50% H_2SO_4	231s(4.35),253s(4.14), 310(4.00),403(4.19)	94-1931-69
$C_{21}H_{15}N$			
10H-Indolo[1,2-a]indole, 11-phenyl-	EtOH	222(4.46),257(4.57), 275s(4.27),315(4.26)	44-2988-69
$C_{21}H_{15}NO$			
3-Isoquinolinol, 1,4-diphenyl-	EtOH	360(3.92),435(3.50)	39-1729-69C
	ether	360(4.10)	39-1729-69C
	EtOH-ether	362(4.05),435(3.00)	39-1729-69C

Compound	Solvent	$\lambda_{max}(\log \epsilon)$	Ref.
Oxazole, 2,4,5-triphenyl-	EtOH	307(4.30)	77-0200-69
$C_{21}H_{15}NOS_4$			
2,4-Thiazolidinedione, 3-methyl-5-[α-[(4-phenyl-1,3-dithiol-2-ylidene)-methyl]benzylidene]-2-thio-	$CHCl_3$	338(3.90),542(4.74)	24-1580-69
2,4-Thiazolidinedione, 3-methyl-5-[α-[(5-phenyl-3H-1,2-dithiol-3-ylidene)-methyl]benzylidene]-2-thio-	$CHCl_3$	242(4.35),321(4.18), 572(4.62)	24-1580-69
$C_{21}H_{15}NO_2$			
Spiro[2H-benz[f]indene]-2,4'(1'H)-iso-quinoline]-1',3'(2'H)-dione, 1,3-dihydro-	EtOH	230(6.00),272(4.89), 282(4.93),290(4.81), 308(4.15),322(4.20)	44-3664-69
$C_{21}H_{15}NO_2S$			
Thiobenzamide, N,N-dibenzoyl-	C_6H_{12}	224(4.46),233(4.40), 500(2.18)	18-2973-69
$C_{21}H_{15}NO_3$			
Anthrone, 4-(benzyloxy)-5-hydroxy-10-imino-	EtOH	218(4.73),249(4.35), 283(3.96),372(3.61), 519(3.77)	114-0109-69A
	dioxan	216(4.55),251s(4.29), 278s(4.04),368(3.68), 401s(3.64),514s(3.12), 544(3.18),571s(3.08)	114-0109-69A
	$CHCl_3$	250(4.37),283(4.03), 368(3.65),406(3.53), 534(3.58)	114-0109-69A
	CCl_4	274(4.09),353(3.65), 361(3.66),406(3.67), 519(2.89),556(2.91), 589(2.77)	114-0109-69A
	pyridine	373(3.68),398s(3.64), 539(3.34),565s(3.26)	114-0109-69A
Ether, benzyl p-[(p-nitrophenyl)ethyn-yl]phenyl	dioxan	354(4.98)	56-1469-69
$C_{21}H_{15}N_3$			
as-Triazine, 3,5,6-triphenyl-	MeOH	227(4.55),261(4.56), 386(2.65)	88-3147-69
$C_{21}H_{15}N_3O_3$			
4(3H)-Quinazolinone, 2-benzyl-6-nitro-3-phenyl-	EtOH	260s(4.06),320(4.21)	18-3198-69
4(3H)-Quinazolinone, 2-benzyl-7-nitro-3-phenyl-	EtOH	250(4.55),320(3.53), 345(3.59)	18-3198-69
$C_{21}H_{15}N_3O_3S$			
Benzimidazole, 1-benzoyl-2-[(p-nitro-benzyl)thio]-	dioxan	263(4.37)	4-0023-69
$C_{21}H_{15}N_5O_2$			
Benzophenone, 4-nitro-, 4-quinazolinyl-hydrazone	EtOH	282(4.59),380(4.39)	2-0412-69
Benzophenone, 4-nitro-, 2-quinoxalinyl-hydrazone	EtOH	242(4.35),305(4.23), 388(4.21)	2-0412-69

Compound	Solvent	λ_{max}(log ε)	Ref.
$C_{21}H_{15}N_7$			
Pyridine, 2,6-bis(5-phenyl-s-triazol-3-yl)-	H_2SO_4	265(4.54),338(4.35)	4-0965-69
$C_{21}H_{15}O$			
5-(p-Hydroxyphenyl)dibenzo[a,d]tropylium (perchlorate)	H_2SO_4	241(4.31),268(4.15), 312(5.07),387(3.90), 409(3.90),525(3.68), 563(3.73)	5-0048-69A
$C_{21}H_{16}$			
Anthracene, 9-benzyl-	EtOH	250s(4.73),256(4.85), 333(3.54),343(3.83), 367(4.03),386(4.02)	78-3485-69
Anthracene, 9-o-tolyl-	EtOH	245s(4.86),254(5.11), 331(3.52),346(3.84), 364(4.03),383(4.02)	78-3485-69
Anthracene, 9-p-tolyl-	EtOH	247s(4.90),255(5.12), 331(3.54),346(3.84), 364(4.00),383(3.97)	78-3485-69
$C_{21}H_{16}BrN_3$			
1H-1,2,3-Triazole, 1-(p-bromophenyl)-5-(diphenylmethyl)-	EtOH	210(5.60)	4-0251-69
$C_{21}H_{16}Cl_2S$			
Ethylene, 1,1-bis(p-chlorophenyl)-2-(p-tolylthio)-	C_6H_{12}	235s(4.36),266(4.18), 317(4.31)	39-0932-69B
	DMF	319(4.28)	39-0932-69B
$C_{21}H_{16}NNaO_2$			
Propene, 3-aci-nitro-1,2,3-triphenyl-, sodium salt	EtOH	236(4.35),306(4.29)	44-0984-69
$C_{21}H_{16}N_2$			
Imidazole, 2,4,5-triphenyl-	EtOH	308(4.34)	77-0200-69
	EtOH	230(4.5),298(4.4)	95-0783-69
Pyrazole, 3,4,5-triphenyl-	EtOH	250(4.4)	95-0783-69
	ether	250(4.46)	22-1667-69
3H-Pyrazole, 3,3,5-triphenyl-	ether	236(4.51),300s(3.66)	22-1667-69
$C_{21}H_{16}N_2O_2$			
1-Pyrazolol, 3,4,5-triphenyl-, 2-oxide	EtOH	260(4.4)	44-0194-69
$C_{21}H_{16}N_2O_3$			
Naphtho[2,3-h]quinoline-7,12-dione, 5-morpholino-	EtOH	290(4.36),450(3.62)	104-0742-69
$C_{21}H_{16}N_2O_4$			
Pyrrole-3,4-dicarboxylic acid, 2-cyano-1,5-diphenyl-, dimethyl ester	MeCN	235(4.34),285(4.00)	44-2146-69
$C_{21}H_{16}N_6O_2$			
Malononitrile, α-[2-(2-pyridyl)hydrazino)benzyl]-	EtOH	292(4.32),320(4.29)	2-0412-69
1,4-Phthalazinedione, 5-(3,5-diphenyl-1-formazano)-2,3-dihydro-	DMF	496(4.18)	104-1274-69

Compound	Solvent	$\lambda_{max}(\log \epsilon)$	Ref.
$C_{21}H_{16}O$			
Anthracene, 9-(p-methoxyphenyl)-	EtOH	247s(4.98),255(5.21), 332(3.54),346(3.83), 364(4.01),383(3.99)	78-3485-69
Ether, benzyl p-(phenylethynyl)phenyl	dioxan	309f(4.88)	56-1469-69
$C_{21}H_{16}OS$			
1,3-Oxathiole, 2,4,5-triphenyl-	EtOH	225(4.27),342(3.83)	44-2006-69
	MeCN	226(--),342(--)	44-2006-69
Thiophene, 2-[p-[(p-methoxyphenyl)-ethynyl]styryl]-	DMF	358(4.75)	35-2521-69
$C_{21}H_{16}O_2$			
Benzofuran, 6-(benzyloxy)-2-phenyl-	EtOH	282s(4.12),294s(4.23), 314(4.51),327(4.43)	12-2395-69
Fluorene-9-acetic acid, 9-phenyl-	EtOH	236(4.11),267(4.16), 270(4.16),278s(--), 294(3.68),305(3.85)	44-2401-69
4H-Pyran-$\Delta^{4,\gamma}$-crotonaldehyde, 2,6-diphenyl-	MeCN	245(4.14),302(4.14), 421(4.59)	44-2736-69
Stilbene, 3,4-(methylenedioxy)-4'-phenyl-	DMF	345(4.64)	33-2521-69
$C_{21}H_{16}O_4$			
Benzophenone, 2-hydroxy-4-methoxy-, benzoate	$CHCl_3$	285(4.02)	39-0034-69C
Benzophenone, 4-hydroxy-2-methoxy-, benzoate	$CHCl_3$	none	39-0034-69C
Carbonic acid, methyl ester, ester with 5-(α-hydroxybenzylidene)-1,3-cyclopentadien-1-yl phenyl ketone	EtOH	237(4.17),323(4.25)	39-2464-69C
$C_{21}H_{16}O_5$			
Acetic acid, (4-benzoyl-3-hydroxyphenoxy)phenyl ester	$CHCl_3$	285(4.23),325(3.96)	39-0037-69C
$C_{21}H_{16}O_6$			
Collinusin, dehydro-	$CHCl_3$	261(4.78),296(4.03), 312(4.03),352(3.74)	78-2815-69
Robustone	EtOH	230(4.37),285(4.61)	39-0365-69C
$C_{21}H_{16}O_7$			
Diphyllin	EtOH	230(4.23),268(4.60), 294(3.81),312(3.78), 325(3.77),360(3.54)	78-2815-69
Furano[2",3":3,4]isoflavone, 4',5'-dimethoxy-2'-acetoxy-	EtOH	230(4.53),247s(4.48), 312s(3.81)	18-1693-69
Furano[3",2":6,7]isoflavone, 3',4'-dimethoxy-2'-acetoxy-	EtOH	239(4.53),306(4.02)	18-1693-69
$C_{21}H_{16}O_9$			
Benzofuro[3,2-c]benzopyran-6-one, 8,9-diacetoxy-3,4-dimethoxy-	EtOH	242(4.35),266(4.00), 332(4.49),348(4.43)	31-0122-69
$C_{21}H_{16}S_2$			
2H-1-Benzothiopyran, 4-phenyl-2-(phenylthio)-	EtOH	235(4.35),255s(4.22)	2-0315-69
$C_{21}H_{17}BrO$			
Anisole, p-(1-bromo-2,2-diphenylvinyl)-	C_6H_{12}	240(4.34),302(4.06)	35-6734-69

Compound	Solvent	$\lambda_{max}(\log \epsilon)$	Ref.
Anisole, p-(2-bromo-1,2-diphenylvinyl)-	C_6H_{12}	239(4.30),304(4.03)	35-6734-69
$C_{21}H_{17}ClO_4S$			
1,2-Diphenyl-1H-2-benzothiopyranium perchlorate	MeOH	202(4.67),285(4.13)	44-3202-69
$C_{21}H_{17}ClO_7$			
2,3-Dihydro-3-salicylidene-1H-cyclopenta[b]benzopyrylium perchlorate, acetate	$MeOH-HClO_4$	255(3.93),307(4.07), 532(4.58)	5-0226-69E
$C_{21}H_{17}Cl_2N_5O_2$			
3-Formazancarboxanilide, 1,5-bis(p-chlorophenyl)-2'-hydroxy-N-methyl-	EtOH	432(4.35)	22-4390-69
	EtOH-KOH	512(4.76)	22-4390-69
$C_{21}H_{17}N$			
Dibenz[a,h]acridine, 5,6,12,13-tetrahydro-	dioxan	231(4.19),289(4.14), 300(4.17),336(4.49), 345(4.51)	24-1202-69
Indole, 1-benzyl-3-phenyl-	EtOH	225(4.51),269(4.17), 278s(--),295s(--)	39-1537-69C
Indole, 2-(diphenylmethyl)-	EtOH	291(3.85)	44-2988-69
$C_{21}H_{17}NO$			
Indole-2-methanol, α,α-diphenyl-	EtOH	292(3.84)	44-2988-69
$C_{21}H_{17}NOS$			
Spiro[benzothiazoline-2,5'-[3H]naphtho-[2,1-b]pyran], 2,3-dimethyl-	$(MeOCH_2)_2$	290s(--),303(4.11), 316(4.02),342(3.80), 357(3.79)	78-3251-69
$C_{21}H_{17}NO_2$			
Acetamide, 2-benzoyl-2,2-diphenyl-	EtOH	240s(4.06)	22-2756-69
Benz[c]acridin-7-ol, 5,6-dimethyl-, acetate	MeOH	223(4.57),232s(4.47), 273s(4.65),292(4.69), 295(4.69),326(3.70), 342(3.81),356(3.77), 373(3.74),392(3.68)	4-0361-69
2-Isoxazoline, 3,4,5-triphenyl-, 2-oxide, trans	EtOH	285(4.25)	44-0984-69
Propene, 1-nitro-1,2,3-triphenyl-	EtOH	223s(4.02),243(4.14)	44-0984-69
Propene, 3-nitro-1,2,3-triphenyl-	EtOH	233(3.81),262(4.08)	44-0984-69
Propene, 3-aci-nitro-1,2,3-triphenyl-	EtOH	233(4.36),285(4.34)	44-0984-69
sodium salt	EtOH	236(4.35),306(4.29)	44-0984-69
$C_{21}H_{17}NO_3$			
Acrylophenone, 4'-(benzyloxy)-2'-hydroxy-3-(2-pyridyl)-	EtOH	264(4.03),276(4.02), 315(4.34),345s(4.27)	95-0375-69
	EtOH-HCl	250s(3.98),265s(4.01), 306(4.27),357(4.14)	95-0375-69
	EtOH-NaOH	251(4.26),305(4.34), 429(3.93)	95-0375-69
Acrylophenone, 4'-(benzyloxy)-2'-hydroxy-3-(3-pyridyl)-	EtOH	270s(4.03),311(4.34), 345s(4.25)	95-0375-69
	EtOH-HCl	270s(4.20),287(4.28), 349(4.14)	95-0375-69
	EtOH-NaOH	251(4.20),305(4.31), 427(3.92)	95-0375-69
Acrylophenone, 4'-(benzyloxy)-2'-hydroxy-3-(4-pyridyl)-	EtOH	270s(4.20),288(4.30), 347(4.15)	95-0375-69

Compound	Solvent	λ_{max}(log ε)	Ref.
Acrylophenone, 4'-(benzyloxy)-2'-hydroxy-3-(4-pyridyl)- (cont.)	EtOH-HCl	272s(4.30),287(4.39), 363(4.08)	95-0375-69
	EtOH-NaOH	254(4.22),283(4.27), 431(3.87)	95-0375-69
Ether, benzyl p-(p-nitrostyryl)phenyl, (E)-	dioxan	395(4.61)	56-1469-69
$C_{21}H_{17}NO_4$			
Pyrrolo[1,2-f]phenanthridine-1-acetic acid, 2-carboxy-, dimethyl ester	MeOH	209(4.29),214(4.29), 243(4.52),259s(4.74), 265(4.83),276s(4.39), 285(4.22),312(3.90), 331s(3.73)	39-2311-69C
$C_{21}H_{17}NO_5$			
3,4-Pyridinedicarboxylic acid, 1,6-dihydro-6-oxo-1,2-diphenyl-, dimethyl ester	MeOH	205(4.65),250(3.99), 355(3.82)	77-1129-69
$C_{21}H_{17}N_5O_8$			
Pyridine, 2,6-bis(2,4-dinitrobenzyl)-4-ethyl-	EtOH	239(4.41)	22-4425-69
$C_{21}H_{17}N_7O_6$			
3-Formazancarboxanilide, 2'-hydroxy-N-methyl-1,5-bis(p-nitrophenyl)-	EtOH	445(4.40)	22-4390-69
	EtOH-KOH	620(4.76)	22-4390-69
$C_{21}H_{18}BF_2NO_2$			
Naphtho[1,2-e]-1,3,2-dioxaborin, 2,2-difluoro-4-[p-(dimethylamino)styryl]-	MeCN	252(4.03),275(4.00), 342(3.92),570(4.86)	4-0029-69
Naphtho[2,1-e]-1,3,2-dioxaborin, 2,2-difluoro-4-[p-(dimethylamino)styryl]-	MeCN	227(4.23),265(4.11), 329(3.97),569(4.95)	4-0029-69
$C_{21}H_{18}ClNO_4S_2$			
4'-(Dimethylamino)-4-(2-thienyl)thioflavylium perchlorate	HOAc-$HClO_4$	262(3.81),370(3.36), 635(3.65)	2-0191-69
$C_{21}H_{18}FeO_2$			
Acrylophenone, 3-(1'-acetylferrocenyl)-	EtOH	270(4.05),334(4.19)	73-2235-69
Acrylophenone, 3-(2-acetylferrocenyl)-	EtOH	249(4.33),311(4.24)	73-2235-69
2-Propen-1-one, 1-(1'-acetylferrocenyl)-3-phenyl-	EtOH	226(3.19),320(4.05)	73-2235-69
$C_{21}H_{18}FeO_3S$			
2-Propen-1-one, 3-(1',2-diacetylferrocenyl)-1-(2-thienyl)-	EtOH	262(4.13),302(4.22)	73-2235-69
$C_{21}H_{18}Fe_2O$			
Ferrocenyl ketone	$CHCl_3$	365(3.5),475(3.2)	101-0283-69A
	$CHCl_3$-CF_3COOH	385s(3.7),625(3.7)	101-0283-69A
$C_{21}H_{18}N_2$			
Acridine, 9-[p-(dimethylamino)phenyl]-	n.s.g.	250(4.46),340(3.50), 320(3.33),400(3.58)	103-0420-69
Cinnamaldehyde, diphenylhydrazone	EtOH	255(4.2),276(4.1), 359(4.6)	28-0137-69B
5,12-Methanodibenzo[b,f][1,4]diazocine, 6,11-dihydro-13-phenyl-	EtOH	286(3.31)	39-0882-69C

Compound	Solvent	λ_{max}(log ε)	Ref.
$C_{21}H_{18}N_2O$			
Benzo[g]quinoline, 4-p-phenetidino-	EtOH	247(4.84),403(4.09)	103-0223-69
$C_{21}H_{18}N_2O_2$			
Benzamide, o-[(α-phenylphenacyl)amino]-	EtOH	255(4.34),344(3.76)	44-0887-69
Naphtho[2,3-h]quinoline-7,12-dione, 5-(butylamino)-	EtOH	310(4.41),520(3.83)	104-0742-69
$C_{21}H_{18}N_2O_2S$			
2(1H)-Pyrimidinethione, 5,6-dihydro-4-(1-hydroxy-2-naphthyl)-6-(p-methoxyphenyl)-	$CHCl_3$	261(3.58),268(4.62), 287(3.74),298(3.71), 376(3.83)	115-0017-69
$C_{21}H_{18}N_2O_4$			
Diacetanilide, 2"-[(1,2-dihydro-4-methyl-2-oxo-3-quinolyl)carbonyl]-	EtOH	328(3.82)	18-2952-69
$C_{21}H_{18}N_2O_5$			
Camptothecine, 10-methoxy-	MeOH	220(4.70),264(4.47), 293(3.76),312(3.93), 328(4.08),365s(4.44), 379(4.50)	44-1364-69
$C_{21}H_{18}N_4O_2$			
1H-Pyrazolo[3,4-b]pyrazin-6-ol, 5-methyl-1,3-diphenyl-	EtOH	252(4.40),270(4.45)	32-0463-69
$C_{21}H_{18}N_6$			
Melamine, N^1,N^3,N^5-trimethyl-	n.s.g.	271(3.12)	42-0595-69
$C_{21}H_{18}N_6O_9S_3$			
s-Triazine, 2,4,6-tris(m-sulfoanilino)-	n.s.g.	275(3.77)	42-0595-69
s-Triazine, 2,4,6-tris(p-sulfoanilino)-	n.s.g.	288(4.20)	42-0595-69
$C_{21}H_{18}N_8O_{10}$			
2,4-Hexadienoic acid, 6-formyl-2-methyl-4-oxo-, methyl ester, 4,6-bis-(2,4-dinitrophenylhydrazone)	$CHCl_3$	377(4.74)	40-0096-69
$C_{21}H_{18}O$			
Anisole, p-(p-phenylstyryl)-	DMF	339(4.67)	33-2521-69
Ether, benzyl p-styrylphenyl, (E)-	dioxan	318f(4.64)	56-1469-69
2,4,6,8-Nonatetraenal, 2,9-diphenyl-, m. 107°	CH_2Cl_2	242(4.29),280(4.13), 395(5.17)	24-0623-69
m. 122°	CH_2Cl_2	242(4.18),280(3.98), 395(4.99)	24-0623-69
2-Propanone, 1,1,1-triphenyl-	heptane	189(4.97),302(2.50)	59-0765-69
$C_{21}H_{18}O_2$			
Cyclopenta[1,2-b:1,5-b']bis[1]benzopyran, 6,7-dihydro-3,10-dimethyl-	dioxan	255(4.19),290(3.81), 370(3.56)	5-0162-69B
$C_{21}H_{18}O_2S$			
Sulfone, benzyl 1,2-diphenylvinyl	EtOH	271(4.14)	78-3193-69
$C_{21}H_{18}O_4$			
3,5-Cyclohexadiene-1,2-dicarboxylic acid, 3,6-diphenyl-, monomethyl ester, cis	EtOH	232s(3.95),340(4.16)	22-3662-69

Compound	Solvent	$\lambda_{max}(\log \epsilon)$	Ref.
$C_{21}H_{18}O_6$			
3-Benzofurancarboxylic acid, 5-allyl-7-methoxy-2-[3,4-(methylenedioxy)-phenyl]-	EtOH	325(4.30)	12-1011-69
Collinusin	EtOH	247(4.19),347(4.02)	78-2815-69
Derrubone	EtOH	269(4.48),292(4.25)	39-0365-69C
2(3H)-Furanone, 3-[bis(p-hydroxyphenyl)-methylene]dihydro-, diacetate	EtOH	225(4.26),286(4.14)	44-3792-69
Glycyrol	EtOH	227(4.49),244(4.36), 256s(4.26),347(4.42), 356s(4.38)	94-0729-69
α-Isoderrubone	EtOH	263(4.44),291(4.18)	39-0365-69C
β-Isoderrubone	EtOH	268(4.49),295(4.22)	39-0365-69C
Isoglycyrol	EtOH	224s(4.38),248(4.34), 256s(4.27),348(4.40), 356s(4.36)	94-0729-69
$C_{21}H_{18}O_7$			
5H-Furo[3,2-][1]benzopyran-5-one, 2,3-dihydro-6-(2-hydroxy-4,5-dimethoxy-phenyl)-, acetate	EtOH	276(4.17),307(4.17)	18-1693-69
$C_{21}H_{18}O_8$			
Anthraquinone, 1,3-diacetoxy-6,8-dimeth-oxy-2-methyl-	EtOH	229(4.35),245(4.35), 280(4.44),336(3.60), 339(3.71)	39-2763-69C
Benzo[b]furan-3-methanol, 2-(2-hydroxy-4-methoxyphenyl)-5,6-(methylene-dioxy)-, diacetate	EtOH	274(4.15),320(4.34)	18-1408-69
Flavone, 3,5-diacetoxy-4',7-dimethoxy-	EtOH	230(4.38),255(4.31), 322(4.44)	105-0397-69
Flavone, 5,7-diacetoxy-3,4'-dimethoxy-	EtOH	255(4.33),339(4.43)	105-0397-69
$C_{21}H_{18}O_9$			
2(4aH)-Dibenzofuranone, 3,7,9-triacet-oxy-6-acetyl-4a-methyl-	EtOH	210(4.36),243(4.25), 265s(4.04),320s(3.73), 363(4.04)	54-0851-69
$C_{21}H_{19}BrO_7S$			
Vernolepin, p-bromobenzenesulfonate	MeOH	235(4.14),265(2.78)	44-3903-69
$C_{21}H_{19}ClN_2O_3$			
Furo[2,3-d]pyrimidine-2,4(1H,3H)-dione, 5-(p-chlorophenyl)-5,6-dihydro-6,6-dimethyl-3-p-tolyl-	EtOH	209(4.53),222(4.42), 267(4.04)	87-0339-69
$C_{21}H_{19}ClN_2O_4$			
Furo[2,3-d]pyrimidine-2,4(1H,3H)-dione, 5-(p-chlorophenyl)-5,6-dihydro-3-(p-methoxyphenyl)-6,6-dimethyl-	EtOH	208(4.54),225(4.51), 267(4.15)	87-0339-69
$C_{21}H_{19}Cl_3N_2O_2$			
Strychnine, 2,14,22-trichloro-	EtOH	258(4.21),288(3.59), 297s(3.51)	33-1564-69
$C_{21}H_{19}Li$			
Propane, 1,1,3-triphenyl-, 1-lithium derivative	THF	492(4.48)	35-2456-69

Compound	Solvent	λ max(log ε)	Ref.
$C_{21}H_{19}N$			
Acridan, 9-benzyl-9-methyl-	MeOH	287(4.20)	36-0335-69
Acridan, 10-benzyl-9-methyl-	MeOH	288(4.20)	36-0335-69
Acridan, 9-(α-methylbenzyl)-	hexane	278(4.21)	78-1125-69
$C_{21}H_{19}NO$			
1'-Acrylonaphthone, 3-[p-(dimethyl-amino)phenyl]-	EtOH	244(4.1),335(4.1), 404(4.35)	2-0311-69
Benzophenone, 4-(dimethylamino)-2-phenyl-	MeCN	241(4.41),346(4.05)	44-2736-69
Benzyl alcohol, o-(1-phenyl-2-isoindo-linyl)-	EtOH	258(3.83),263s(3.78), 272s(3.65)	44-1720-69
$C_{21}H_{19}NO_2$			
Benz[c]acridin-7-ol, 5,6-dihydro-5,5-dimethyl-, acetate	MeOH	225(4.44),258s(4.50), 265(4.60),299(4.02), 312(4.02),327(4.08), 343(4.13)	4-0361-69
1,3-Cyclopentadiene-1-carboxaldehyde, 3-(dimethylamino)-5-(hydroxymethyl-ene)-2,4-diphenyl-	CH_2Cl_2	245(4.46),296(4.46), 335(4.18),485(3.87)	24-0623-69
9H-Pyrrolo[1,2-a]indole, 2-acetyl-7-(benzyloxy)-6-methyl-	EtOH	207(4.58),244(3.92), 293(4.31)	88-0101-69
$C_{21}H_{19}NO_2S$			
2-Thiophenecarboxaldehyde, 5-morpholino-3,4-diphenyl-	EtOH	270(4.0),360(4.3)	48-0827-69
$C_{21}H_{19}NO_3$			
1,3-Butadiene-1-carboxylic acid, 1-cyano-3-ethoxy-2,4-diphenyl-	CH_2Cl_2	258(4.12),310(3.99), 380s(--)	24-0319-69
1,4-Isoquinolinedione, 2,3-dihydro-3-(2-oxocyclohexyl)-3-phenyl-	EtOH	218(4.60),295(3.38), 350(2.78)	33-1810-69
$C_{21}H_{19}NO_4$			
4-Penten-1-ol, 5-(9-anthryl)-4-nitro-, acetate	EtOH	348(3.58)	87-0157-69
4-Penten-1-ol, 4-nitro-5-(9-phenan-thryl)-, acetate	EtOH	340(3.87)	87-0157-69
Pyrrole-2-carboxylic acid, 5-benzoyl-1-(o-ethylphenyl)-4-hydroxy-, methyl ester	MeOH	259(4.15),302(4.00), 334(3.98)	78-0527-69
Pyrrole-2,3-dicarboxylic acid, 4-phenyl-1-p-tolyl-, dimethyl ester	MeOH	233(4.40),273(3.91)	78-0527-69
Spiro[2,5-cyclohexadiene-1,7'(6'H)-indeno[1,2-b]quinolizin]-4,6'-dione, 1',2',3',4',12',12'a-hexahydro-9',10'-dihydroxy-	EtOH	220(4.33),250(4.37), 296(3.63),350(3.93)	39-1309-69C
$C_{21}H_{19}NO_5$			
Pyrrole-2,3-dicarboxylic acid, 1-(p-methoxyphenyl)-4-phenyl-, dimethyl ester	MeOH	236(4.38),287s(3.93)	78-0527-69
$C_{21}H_{19}N_3O_3$			
Benz[c]acridine-7-carbamic acid, 5,6-dihydro-5,5-dimethyl-N-nitroso-, methyl ester	MeOH	215(4.66),225s(4.47), 262s(3.58),269(4.57), 304(3.92),320(3.93), 335(4.09),349(4.14)	4-0361-69

Compound	Solvent	λ_{max}(log ϵ)	Ref.
$C_{21}H_{19}N_7O_2$			
Hydrazine, 1,1'-(2,6-pyridinedicarbox-imidoyl)bis[2-benzoyl-, (hydrate)	H_2SO_4	270(4.52)	4-0965-69
$C_{21}H_{19}O_2P$			
Acetic acid, (triphenylphosphoranyli-dene)-, methyl ester	EtOH	262(3.66),273(3.64), 278(3.48)	39-1100-69C
$C_{21}H_{19}O_3PS$			
Phenothiaphosphine, 2,8-dimethyl-10-p-tolyl-, 5,5,10-trioxide	EtOH	225(4.63),269s(3.67), 279(3.67),288(3.63)	78-3919-69
$C_{21}H_{19}PS$			
Phenothiaphosphine, 2,8-dimethyl-10-p-tolyl-	EtOH	227(4.51),269(4.10), 304(3.74)	78-3919-69
$C_{21}H_{20}$			
11H-Benzo[a]fluorene, 11-isopropyl-4-methyl-	EtOH	259(4.71),269(4.91), 286(4.09),296(4.19), 310(4.10),320(3.97), 332(3.39),340s(3.10), 349(3.32)	22-2115-69
Propane, 1,1,3-triphenyl-, 1-lithium derivative	THF	492(4.48)	35-2456-69
$C_{21}H_{20}BrNO_5$			
11-Hydroxy-2,3,9,10-tetramethoxybenz-[a]acridizinium bromide	EtOH	220(4.05),240s(4.12), 244(4.13),275s(4.12), 305(4.34),312s(4.33), 337(4.08),367s(3.71), 415s(3.69),438(3.78)	44-1349-69
$C_{21}H_{20}ClNO_2$			
Indole-2-carboxylic acid, 5-chloro-1-(cyclopropylmethyl)-3-phenyl-	MeOH	242(4.53),303(4.16)	94-1263-69
$C_{21}H_{20}ClNO_4$			
6,7-Dihydro-1-methyl-2,3-diphenyl-5H-1-pyrindinium perchlorate	EtOH	245(4.01),305(3.95)	94-2209-69
Dimethyl[4-[3-(1-naphthyl)allylidene]-2,5-cyclohexadien-1-ylidene]ammonium perchlorate	HOAc-1% $HClO_4$	288(4.05),340(3.91), 565(4.41)	2-0311-69
$C_{21}H_{20}ClNO_5$			
[2-(2,6-Diphenyl-4H-pyran-4-ylidene)-ethylidene]dimethylammonium perchlorate	MeCN	226(4.20),252s(4.10), 264s(4.09),336(4.23), 422(4.62),443(4.60), 544(4.14)	44-2736-69
[2-(4,6-Diphenyl-2H-pyran-2-ylidene)-ethylidene]dimethylammonium perchlorate	MeCN	239(4.25),316(4.61), 476(4.33)	44-2736-69
1-Methyl-2-[(2-phenyl-4H-1-benzopyran-4-ylidene)methyl]-1-pyrrolinium perchlorate	MeCN	238(4.28),317(4.22), 430(4.56)	44-2736-69
$C_{21}H_{20}ClN_3O_3$			
Strychnine, 2-chloro-11-oxo-, 11-oxime	EtOH	238(4.03),297(3.72), 314(3.74)	33-1564-69
	EtOH-base	302(4.31),312(4.30)	33-1564-69

Compound	Solvent	λ_{max}(log ϵ)	Ref.
Strychnine, 2-chloro-11-oxo-, 11-oxime, hydrochloride	EtOH	237(4.22),297(3.98), 310(3.96)	33-1564-69
$C_{21}H_{20}ClN_3O_6$			
3-Furoic acid, 4-(p-chlorophenyl)-4,5-dihydro-5,5-dimethyl-2-[3-(p-nitrophenyl)ureido]-, methyl ester	EtOH	208(4.42),225(4.33), 274(4.04),326(4.36)	87-0339-69
$C_{21}H_{20}Cl_2N_2O_2$			
Strychnine, 2,14-dichloro-	EtOH	258(4.20),288(3.57), 297s(3.49)	33-1564-69
$C_{21}H_{20}Cl_2N_2O_4$			
3-Furoic acid, 4-(p-chlorophenyl)-2-[3-(p-chlorophenyl)ureido]-4,5-dihydro-5,5-dimethyl-, methyl ester	EtOH	208(4.45),224(4.18), 250(4.15),285(4.43)	87-0339-69
$C_{21}H_{20}FeN_2O_4S$			
Tryptophan, N-(ferrocenesulfonyl)-	EtOH	426(2.18)	49-1552-69
$C_{21}H_{20}NOPS$			
Formanilide, 1-(di-p-tolylphosphinyl)-thio-	EtOH	329(4.08),441(2.18)	18-2975-69
$C_{21}H_{20}NO_3PS$			
Formanilide, 1-[bis(p-methoxyphenyl)-phosphinyl]thio-	EtOH	328(4.11),442(2.06)	18-2975-69
$C_{21}H_{20}N_2O$			
o-Acetotoluidide, α-(p-anilinophenyl)-	EtOH	239s(--),289(4.38)	78-0847-69
Benzoic acid, 2,4,6-trimethyl-, (1-naphthylmethylene)hydrazide	EtOH	230(4.5),261s(4.0), 325(4.3),341(4.2)	28-0248-69B
Benzoic acid, 2,4,6-trimethyl-, (2-naphthylmethylene)hydrazide	EtOH	212(4.4),235(4.3), 244(4.3),264(4.5), 274(4.5),303(4.5), 316s(4.4),330s(3.8), 348(3.4)	28-0248-69B
$C_{21}H_{20}N_2O_2$			
Benz[c]acridine-7-carbamic acid, 5,6-dihydro-5,5-dimethyl-, methyl ester	MeOH	213(4.55),228(4.40), 260s(4.49),268(4.61), 302(3.98),316(4.01), 330(4.09),345(4.14)	4-0361-69
5,10-o-Benzene-1H-pyrazolo[1,2-b]-phthalazine-1,3(2H)-dione, 2,2-diethyl-5,10-dihydro-	MeOH	262(3.76)	44-3181-69
1,5-Benzoxazepine, 4-[3-(2,3-dihydro-1,5-benzoxazepin-4-yl)allylidene]-2,3,4,5-tetrahydro-, perchlorate	EtOH	498(5.20)	124-1178-69
	$CHCl_3$	540(5.20)	124-1178-69
$C_{21}H_{20}N_2O_3$			
4-Isoquinolinecarbonitrile, 6,8-dimethoxy-3-(p-methoxyphenethyl)-	EtOH	256(4.63),340(3.72)	95-1492-69
$C_{21}H_{20}N_2O_4$			
Furo[2,3-d]pyrimidine-2,4(1H,3H)-dione, 5,6-dihydro-3-(p-methoxyphenyl)-6,6-dimethyl-5-phenyl-	EtOH	208(4.42),223(4.32), 267(4.20)	87-0339-69
Pyrrole-3-carboxylic acid, 1-benzyl-4-[(p-methoxyphenyl)carbamoyl]-	EtOH	244(4.30),288(4.22)	94-2461-69

Compound	Solvent	$\lambda_{max}(\log \epsilon)$	Ref.
$C_{21}H_{20}N_2S_2$			
2H-1,3,5-Thiadiazine-2-thione, 5-benzyl-tetrahydro-3-(1-naphthylmethyl)-	MeOH	255(3.82),286(4.15)	73-2952-69
$C_{21}H_{20}N_2Si$			
Silane, (1-diazopropyl)triphenyl-	C_6H_{12}	221s(4.58),255(3.90), 426(1.77)	23-4353-69
$C_{21}H_{20}N_4O_5$			
L-Isoleucine, N-(2-quinoxalinecarbonyl)-, p-nitrophenyl ester	EtOH	207(4.38),244(4.55), 278(3.97),318(3.88)	87-0141-69
$C_{21}H_{20}O$			
[m-Terphenyl]-5'-ol, 4'-isopropyl-	EtOH	206(4.58),235(4.43), 264(4.22),300(3.65)	44-0444-69
$C_{21}H_{20}O_3$			
3(4H)-Dibenzofuranone, 4a,9b-dihydro-6-(6-hydroxy-m-tolyl)-8,9b-dimethyl-	EtOH	299(3.81)	44-2966-69
	EtOH-EtONa	321(3.90)	44-2966-69
Naphthalene, 2-(3,4,5-trimethoxystyryl)-	DMF	289(4.30),333(4.60)	33-2521-69
$C_{21}H_{20}O_4$			
Benzo[1,2-b:3,4-b']bisbenzofuran-7(6H)-one, 5a,7a,12a,12b-tetrahydro-12a-hydroxy-2,9,12b-trimethyl-	EtOH	287(3.78),293(3.75)	44-2966-69
	EtOH-EtONa	242(4.15),304(3.84)	44-2966-69
Benzo[1,2-b:5,4-b']bisbenzofuran-12(6H)-one, 5a,6a,11b,12a-tetrahydro-5a-hydroxy-2,10,11b-trimethyl-	EtOH	288(3.73),294(3.71)	44-2966-69
	EtOH-EtONa	300(3.85)	44-2966-69
$C_{21}H_{20}O_6$			
3-Benzofurancarboxylic acid, 7-methoxy-2-[3,4-(methylenedioxy)phenyl]-5-propyl-, methyl ester	EtOH	325(4.29)	12-1011-69
2(3H)-Furanone, dihydro-3-piperonylidene-4-veratryl-	EtOH	233(4.22),287(4.02), 295(4.01),332(4.21)	39-0693-69C
Ketone, 7-hydroxy-5-methoxy-2,2-dimethyl-2H-1-benzopyran-6-yl piperonyl	EtOH	230s(4.13),235(4.18), 265(4.51),289s(4.11), 355(3.68)	39-0365-69C
Phenanthrenequinone, 1,4,8-trihydroxy-2-methyl-3-(4-methylvaleryl)-	EtOH	235(4.38),250(4.22), 294(3.82),500(3.89)	22-3100-69
Robustic acid, dihydrodemethyl-	EtOH	248s(4.13),325(4.18)	12-1923-69
	EtOH-NaOH	260(4.24),340(4.28)	12-1923-69
$(C_{21}H_{20}O_6)_n$			
Poly[oxy[2-[(3,4,5-trimethyl-p-benzoquinon-2-yl)methyl]trimethylene]-oxyisophthaloyl]	95% THF	262(4.17),292(3.52)	47-1269-69
Poly[oxy[2-[(3,4,5-trimethyl-p-benzoquinon-2-yl)methyl]trimethylene]-oxyterephthaloyl]	95% THF	253(4.41),292(3.68)	47-1269-69
$C_{21}H_{20}O_7$			
2(3H)-Furanone, dihydro-3-[hydroxybis-(4-acetoxyphenyl)methyl]-	EtOH	222(4.15)	44-3792-69
Piperonylic acid, ester with methyl (5-allyl-2-hydroxy-3-methoxyphenyl)acetate	EtOH	269(3.99),299(3.97)	12-1011-69

Compound	Solvent	λ_{max}(log ε)	Ref.
$C_{21}H_{20}O_8$			
Acetic acid, [2-(7-hydroxy-4-oxo-4H-1-benzopyran-3-yl)-4,5-dimethoxyphenoxy]-, ethyl ester	EtOH	255(4.53),297(4.28), 336(4.01)	18-0199-69
Flavone, 6-hydroxy-3',4',5,6-tetramethoxy-, acetate	EtOH	239(4.40),261s(4.18), 330(4.40)	40-1270-69
Philenopteran, diacetate	EtOH	283(3.71)	39-0887-69C
$C_{21}H_{20}O_9$			
Frangulin A (4 λ,5 ε)	n.s.g.	225(4.52),264(4.28), 300(4.15),430(3.97), ?(4.05)	64-1408-69
$C_{21}H_{20}O_{10}$			
Kaempferol 7-rhamnoside	EtOH and EtOH-NaOAc	257(4.25),268(4.25), 325(4.04),370(4.37)	95-0872-69
	EtOH-AlCl$_3$	269(4.38),354(3.92), 429(4.42)	95-0872-69
$C_{21}H_{20}O_{11}$			
Quercetin 7-rhamnoside	EtOH	257(4.31),377(4.26)	95-0872-69
	EtOH-AlCl$_3$	269(4.37),437(4.35)	95-0872-69
$C_{21}H_{21}ClN_2$			
1,2-Diazaspiro[5.5]undeca-2,4-diene, 3-chloro-4,5-diphenyl-, hydrochloride, (hydrate)	n.s.g.	372(3.68)	39-1338-69C
$C_{21}H_{21}ClN_2O_2$			
Strychnine, 2-chloro-	EtOH	259(4.20),288(3.61), 297s(3.52)	33-1564-69
$C_{21}H_{21}ClN_2O_4$			
3-[(1,3-Dimethylindol-2-yl)methylene]-1,2-dimethyl-3H-indolium perchlorate	MeOH	505(3.58)	24-1347-69
3-Furoic acid, 4-(p-chlorophenyl)-4,5-dihydro-5,5-dimethyl-2-(3-phenylureido)-, methyl ester	EtOH	207(4.53),224(4.28), 241(4.18),288(4.51)	87-0339-69
3-Furoic acid, 2-[3-(p-chlorophenyl)-ureido]-4,5-dihydro-5,5-dimethyl-4-phenyl-, methyl ester	EtOH	207(4.60),247(4.28), 287(4.53)	87-0339-69
$C_{21}H_{21}ClO_7S$			
Benzophenone, 3-chloro-4-hydroxy-2'-mercapto-2,4',6'-trimethoxy-6-methyl-, diacetate	MeOH	209(4.70),242s(4.23), 311(3.89)	44-1463-69
$C_{21}H_{21}IN_2O$			
2-(6-Anilino-1,3,5-hexatrienyl)-3-ethylbenzoxazolium iodide	MeOH	575(4.99)	22-1284-69
	MeOH-NaOMe	466(4.76)	22-1284-69
$C_{21}H_{21}NO$			
Pyridine, 4-(ethoxymethyl)-3-methyl-2,5-diphenyl-	n.s.g.	212(4.32),246(4.10), 286(3.94)	39-2255-69C
2(1H)-Pyridone, 1-butyl-4,6-diphenyl-	EtOH	243(4.36),267s(4.11), 326(3.82)	48-0061-69
$C_{21}H_{21}NO_2$			
Cyclopent[a]inden-8(3H)-one, 2-[(dimethylamino)methyl]-3a,8a-dihydro-3a-hydroxy-8a-phenyl-	EtOH	248(4.06),294(3.23)	104-2090-69

Compound	Solvent	λ max(log ε)	Ref.
$C_{21}H_{21}NO_4$			
Isonicotinic acid, 1,2-dibenzyl-1,2,5,6-tetrahydro-3-hydroxy-6-oxo-, methyl ester	90% EtOH	252(3.90)	94-0467-69
Ochotensine	EtOH	226(4.41),287(4.20)	23-2501-69
$C_{21}H_{21}NO_5$			
Acetophenone, 2-(7,8-dimethoxy-4-isoquinolyl)-3',4'-dimethoxy-, hydrochloride	EtOH	233(4.61),275(4.16), 360(3.52)	78-0101-69
Maleic acid, (N-phenacyl-p-toluidino)-, dimethyl ester	MeOH	246(4.34),290(4.36)	78-0527-69
$C_{21}H_{21}NO_6$			
Adlumine	EtOH	201(4.45),220(4.24), 236s(3.93),286(3.38), 324(3.45)	78-5059-69
Corlumine	EtOH	202(4.75),220(4.51), 236s(4.16),286(3.66), 320(3.74)	78-5059-69
β-Hydrastine	EtOH	202(4.79),218(4.53), 238s(4.15),298(3.86), 316(3.63)	78-5059-69
Maleic acid, (N-phenacyl-p-anisidino)-, dimethyl ester	MeOH	244(4.33),288(4.35)	78-0527-69
$C_{21}H_{21}NS$			
Butylamine, N-(4,6-diphenyl-2H-thiopyran-2-ylidene)-	EtOH	221(4.25),270(4.48), 376(3.80)	48-0061-69
$C_{21}H_{21}N_3O_3$			
Diaboline, 1-(cyanoformyl)-1-deacetyl-, hydrochloride	EtOH	278(3.85),290(3.83)	33-1564-69
Strychnine, 11-oxo-, 11-oxime	EtOH	234(4.24),290(3.78), 314(3.86)	33-1564-69
	EtOH-base	292(4.23),310(4.24)	33-1564-69
hydrochloride	EtOH	234(4.20),290(3.79), 310(3.84)	33-1564-69
$C_{21}H_{21}N_3O_4S$			
Propionamide, N-(3,3'-dimethyl-6-nitrospiro[2H-1-benzopyran-2,2'-benzothiazolin]-5'-yl)-2-methyl-	$(MeOCH_2)_2$	232(4.65),268(4.55), 311(4.22),341s(--)	78-3251-69
$C_{21}H_{21}N_3O_6$			
Dehydrocubebin, semicarbazone, (±)-	EtOH	232(4.06),292(4.10), 314(4.06)	39-2470-69C
3-Furoic acid, 4,5-dihydro-5,5-dimethyl-2-[3-(p-nitrophenyl)ureido]-4-phenyl-, methyl ester	EtOH	207(4.51),220s(4.40), 278(4.16),327(4.42)	87-0339-69
$C_{21}H_{21}N_3S$			
1-Naphthaldehyde, 4-mesityl-3-thiosemicarbazone	EtOH	210(4.7),235(4.5), 266(4.0),272(4.0), 345(4.4),366s(4.3)	28-0248-69B
2-Naphthaldehyde, 4-mesityl-3-thiosemicarbazone	EtOH	240(4.3),250s(4.2), 274(4.2),285(4.3), 325(4.3),341(4.2)	28-0248-69B

Compound	Solvent	λ_{max}(log ε)	Ref.
$C_{21}H_{21}P$			
Phosphine, tri-m-tolyl-	Me methacrylate	292(3.69)	47-0265-69
Phosphine, tri-o-tolyl-	Me methacrylate	292(3.91)	47-0265-69
Phosphine, tri-p-tolyl-	Me methacrylate	292(3.71)	47-0265-69
$C_{21}H_{22}$			
Spiro[cyclopropane-1,7'-norcarane], 2,3-diphenyl-, trans, (±)-	EtOH	238(4.2),268s(3.2), 275s(3.0)	44-2217-69
$C_{21}H_{22}AsI$			
Triphenylpropylarsonium iodide	n.s.g.	229(4.4),252(3.2), 258(3.4),264(3.5), 270(3.5),290(3.5)	120-0195-69
$C_{21}H_{22}ClN_3$			
Quinaldine, 4-(N-benzylpiperazinyl)-7-chloro-	EtOH	227(4.52),248s(3.19), 315(3.80)	2-1186-69
$C_{21}H_{22}Cl_2N_2$			
1,2-Diazaspiro[5.5]undeca-2,4-diene, 3-chloro-4,5-diphenyl-, hydrochloride, (hydrate)	n.s.g.	372(3.68)	39-1338-69C
$C_{21}H_{22}INO_4$			
Palmatine iodide	MeOH	225(4.65),265(4.53), 272s(4.50)	120-0309-69
$C_{21}H_{22}N_2O$			
Acrylic acid, 2,3-diphenyl-, cyclohexylidenehydrazide, (E)-	MeCN	286(4.39)	39-1338-69C
$C_{21}H_{22}N_2O_2$			
Kopsanone, N_α-formyl-	MeOH	207(4.33),252(4.07), 282(3.62),289(3.63)	20-0271-69
$C_{21}H_{22}N_2O_3$			
7H-Dibenzo[d,f]azonine-7-carbonitrile, 5,6,8,9-tetrahydro-2-(hydroxymethyl)-3-methoxy-, acetate	EtOH	284(3.45)	88-0153-69
Scandine	EtOH	212(4.47),258(3.94), 267s(3.87),284s(3.50)	33-1886-69
$C_{21}H_{22}N_2O_3S$			
Spiro[2H-1-benzopyran-2,2'-benzothiazoline], 8-tert-butyl-3,3'-dimethyl-6-nitro-	$(MeOCH_2)_2$	227s(--),246s(--), 272(4.31),302s(--), 343s(--)	78-3251-69
Spiro[2H-1-benzopyran-2,2'-benzothiazoline], 8-isopropyl-3,3',5-trimethyl-6-nitro-	$(MeOCH_2)_2$	260(4.32),268s(--), 303s(--),360s(--)	78-3251-69
$C_{21}H_{22}N_2O_4$			
3-Furoic acid, 4,5-dihydro-5,5-dimethyl-4-phenyl-2-(3-phenylureido)-, methyl ester	EtOH	208(4.59),240(4.25), 286(4.53)	87-0339-69

Compound	Solvent	$\lambda_{max}(\log \epsilon)$	Ref.
$C_{21}H_{22}N_2O_4S$			
Spiro[2H-1-benzopyran-2,2'-benzothiazoline], 3-butyl-8-methoxy-3'-methyl-6-nitro-	EtOH	217(4.54),253(4.33), 285s(4.06),344(3.92)	22-3329-69
$C_{21}H_{22}N_4O_3S$			
1-Benzimidazolecarboxamide, N-cyclohexyl-2-[(p-nitrobenzyl)thio]-	dioxan	284(4.29)	4-0023-69
$C_{21}H_{22}N_4O_4$			
Hydrocinnamaldehyde, α-cyclohexylidene-, 2,4-dinitrophenylhydrazone	EtOH	259(4.24),295(4.01), 388(4.51)	39-0460-69C
$C_{21}H_{22}O$			
2-Cyclohexen-1-one, 2-isopropyl-3,5-diphenyl-	EtOH	204(4.25),260(3.99)	44-0444-69
$C_{21}H_{22}O_5$			
Trachyphyllin, acetate	EtOH	225(4.36),260(4.39), 269(4.45),324(3.98), 347(4.08)	12-2175-69
$C_{21}H_{22}O_6$			
2(3H)-Furanone, dihydro-3-piperonyl-4-veratryl-	EtOH	233(4.05),284(3.89)	39-0693-69C
$C_{21}H_{22}O_7$			
2-Anthraceneacetic acid, 9,10-dihydro-α,4,5,7-tetramethoxy-10-oxo-, methyl ester	EtOH	230(3.99),280s(4.11), 283(4.12),306s(3.80), 375(3.40)	105-0257-69
Crotonic acid, 3-methyl-, 8-ester with 8,9-dihydro-9-hydroxy-8-(1-hydroxy-1-methylethyl)-2H-furo[2,3-h]-1-benzopyran-2-one, acetate	ether	244(3.61),256(3.56), 297s(4.01),319(4.15)	24-1673-69
Flavone, 4',7-diethoxy-5-hydroxy-3',8-dimethoxy-	EtOH	254(4.26),275(4.32), 339(4.31)	18-2327-69
Melanoxin, O-diacetyl-	MeOH	210(4.24),225(4.14), 293(3.71)	78-4409-69
Tomasin	n.s.g.	218(4.50),250(4.36), 301(4.07)	105-0299-69
$C_{21}H_{22}O_8$			
Flavone, 3,3',4,5,6,7-hexamethoxy-	EtOH	242(4.32),251s(4.32), 266s(4.21),333(4.39)	44-1460-69
$C_{21}H_{22}O_9$			
1H-Naphtho[2,3-c]pyran-1-one, 3,4-dihydro-6,9-dihydroxy-7,8,10-trimethoxy-3-methyl-, diacetate	EtOH	254(4.55),300(3.57), 350(3.32)	23-1561-69
$C_{21}H_{23}ClN_2O$			
1(2H)-Quinolinecarboxamide, N-allyl-6-chloro-3,4-dihydro-2,3-dimethyl-4-phenyl-	n.s.g.	259(4.19)	40-1047-69
$C_{21}H_{23}FeNO_4S$			
Alanine, N-(ferrocenesulfonyl)-3-phenyl-, ethyl ester	EtOH	431(2.21)	49-1552-69

Compound	Solvent	λ$_{max}$(log ε)	Ref.
$C_{21}H_{23}NO_2$			
Spiro[cyclopentane-1,3'(2'H)-quinolin]-2'-one, 4'-ethoxy-1',4'-dihydro-4'-phenyl-	EtOH	255(4.05),286(3.40), 294(3.32)	94-1290-69
$C_{21}H_{23}NO_2S$			
Thiocinnamamide, N,N-diethyl-β-hydroxy-α-methyl-, benzoate	MeOH	232(4.38),264(4.26)	5-0073-69E
Thiocinnamamide, β-hydroxy-N,N-dimethyl-α-propyl-, benzoate	MeOH	231(4.39),264(4.21)	5-0073-69E
$C_{21}H_{23}NO_3$			
Koenigicine, N-methyl-	EtOH	241(4.55),303(4.56), 346(3.92),361(3.82)	31-0790-69
$C_{21}H_{23}NO_4$			
Isoquinoline, 1-(2-methyl-3,4-dimethoxy-benzylidene)-2-methyl-6,7-(methylene-dioxy)-1,2,3,4-tetrahydro-	EtOH	261(4.06),324(4.13)	95-0869-69
4H-Quinolizin-4-one, 2-(3,4-dihydroxy-phenyl)octahydro-3-(p-hydroxyphen-yl)-, cis	EtOH	214(4.23),225(4.17), 284(3.67)	39-1309-69C
$C_{21}H_{23}NO_5$			
Alkaloid CC-20 from Colchicum cornigerum	EtOH	243(4.27),280(3.78)	73-3540-69
Fumaricine	EtOH	207(4.74),235(3.94), 288(3.74)	23-3593-69
$C_{21}H_{23}N_3O_2$			
2-Propanone, 1-(2-ethyl-4-methyl-5-phenylimidazol-1-yl)-1-phenyl-, oxime, 3-oxide	MeOH	261(4.16)	48-0746-69
isomer	MeOH	259(4.18)	48-0746-69
$C_{21}H_{23}N_3O_3$			
Methanol, (4-amino-3,5-dimethoxyphenyl)-bis(p-aminophenyl)-	pH 4.62	559(4.77)	64-0542-69
Methanol, bis(4-amino-3-methoxyphenyl)-(p-aminophenyl)-	pH 4.62	566(4.80)	64-0542-69
$C_{21}H_{23}N_3O_4$			
Diaboline, 1-carbamoyl-17β-carboxy-1-deacetyl-17-deoxy-, hydrochloride	EtOH	244(3.99),279(3.22)	33-1564-69
$C_{21}H_{23}N_3O_6$			
Cubebin, semicarbazone	EtOH	234(4.15),287(3.93)	39-2470-69C
$C_{21}H_{24}BrNO_4$			
13,13a-Didehydro-10,11-dihydroxy-2,3-dimethoxy-7,13-dimethylberbinium bromide	EtOH	232(4.18),330(4.20)	35-4009-69
$C_{21}H_{24}ClN_5O_4$			
Purine-7-butyric acid, α-(carboxyamino)-6-chloro-, N-benzyl tert-butyl ester	EtOH	270(3.87)	78-5971-69
9H-Purine-9-butyric acid, α-(carboxy-amino)-6-chloro-, N-benzyl tert-butyl ester	EtOH	266(3.97)	78-5971-69

Compound	Solvent	λ_{max}(log ε)	Ref.
$C_{21}H_{24}N_2O$			
1,5-Benzoxazepine, 4-[p-(dimethylamino)-styryl]-2,3-dihydro-2,2-dimethyl-, perchlorate	MeOH	521(5.18)	124-1178-69
1(2H)-Naphthalenone, 3,4,4a,5,8,8a-hexa-hydro-4β-[N-(2-indol-3-ylethyl)form-imidoyl]-	n.s.g.	275(3.74),282(3.76), 291(3.71)	35-7315-69
1,4-Pentadien-3-one, 1,1-bis(dimethyl-amino)-2,4-diphenyl-	dioxan	225(4.3),280(4.1)	88-4279-69
$C_{21}H_{24}N_2O_2$			
Alstonerine	EtOH	231(4.58),261(4.01), 284(3.92),293(3.85)	77-1306-69
Aspidofractinine, 1-acetyl-3-oxo-	EtOH	251(4.09),277s(3.54), 286s(3.47)	33-0076-69
22-Epikopsanol, 1-formyl-	MeOH	208(4.32),256(4.11), 283(3.67),291(3.66)	20-0271-69
Kopsanol, 1-formyl-	MeOH	208(4.28),254(4.04), 283(3.60),290(3.61)	20-0063-69
$C_{21}H_{24}N_2O_3$			
Aspidofractinine, 1-carboxy-3-oxo-, methyl ester	EtOH	240(4.12),280(3.37), 287(3.33)	33-0076-69
17βH-Diaboline, 1-deacetyl-1-formyl-O-methyl-	EtOH	249(4.05),280(3.56)	33-1564-69
19-Epiajmalicine	MeOH	224(4.62),240(4.18), 274(3.85),282(3.87), 290(3.79)	24-3558-69
19-Epi-3-isoajmalicine	MeOH	224(4.62),240(4.17), 274(3.90),282(3.92), 290(3.85)	24-3558-69
Ervinidinine	EtOH	228(3.86),298(3.84), 332(3.98)	105-0058-69
Geissoschizine, dibenzoyltartrate	MeOH	223(4.78),268s(4.02), 273(4.03),282s(3.99), 290(3.80)	35-7349-69
	MeOH-NaOMe	227(4.80),276(4.39)	35-7349-69
Preakuammicine	EtOH	262(3.78)(changing)	35-5874-69
Rhazine	EtOH	228(4.52),279(3.85), 293(3.71)	42-0635-69
Yohimban, 18-carboxy-17-oxo-, methyl ester, (±)-	pH 13	224(4.51),280(4.31), 287(4.30)	24-3248-69
	EtOH	225(4.59),273(3.92), 281(3.90),289(3.83)	24-3248-69
Yohimbinone	EtOH	224(4.50),272(3.92), 281(3.95),287(3.89)	24-3248-69
	n.s.g.	227(4.55),283(3.84), 290(3.77)	35-7315-69
$C_{21}H_{24}N_2O_4$			
19-Epi-3-isoajmalicine, 16,17-dihydro-17-oxo-	MeOH	224(4.55),281(3.86), 289(3.72)	24-3558-69
Formosanine, dl-	MeOH	208(4.49),243(4.25)	24-3558-69
3-Isoajmalicine, 16,17-dihydro-17-oxo-	MeOH	223(4.56),282(3.85), 289(3.79)	24-3558-69
7-Isoformosanine	MeOH	208(4.48),243(4.25)	24-3558-69
3-Pyridinepropionic acid, 1-[2-(2-form-ylindol-3-yl)ethyl]-1,4,5,6-tetrahy-dro-2-methyl-4-oxo-, methyl ester	MeOH	224(4.13),316(4.40)	24-0310-69
Vincaricine	EtOH	236(3.85),308(3.48)	105-0440-69

Compound	Solvent	λ max (log ε)	Ref.
Vincaricine (cont.)	70% $HClO_4$	255(3.97),343(4.02)	105-0440-69
$C_{21}H_{24}N_2O_9S$			
Benzimidazole, 2-(β-D-glucopyranosyl-thio)-, 2',3',4',6'-tetraacetate	MeOH	250s(3.77),255(3.78), 282(4.10),288(4.06)	97-0152-69
	EtOH	253(3.83),283(4.13), 289(4.10)	48-0997-69
2-Benzimidazolethione, 1-β-D-glucopyran-osyl-, 2',3',4',6'-tetraacetate	MeOH	250(3.95),298s(4.34), 308(4.45)	97-0152-69
	EtOH	220(4.30),248(4.01), 308(4.48)	48-0997-69
$C_{21}H_{24}N_4O$			
Anthranilonitrile, 5-ethoxy-N-[α-(4-methylpiperazino)benzylidene]-	EtOH	218(4.49),250(4.01), 330(3.74)	22-2008-69
$C_{21}H_{24}N_4O_2$			
2-Cyclohexen-1-one, 3-[(p-methoxyphen-yl)azo]-2-methyl-, (p-methoxyphenyl)-hydrazone	$CHCl_3$	332(4.10),451(4.55)	24-1379-69
$C_{21}H_{24}N_4O_8$			
Isohydrocoriamyrtin, 2,4-dinitrophenyl-hydrazone	EtOH	260(4.15),290(3.90), 375(4.49)	78-1109-69
Jalaric acid, 2,4-dinitrophenylhydrazone	EtOH	335(4.35)	78-3841-69
$C_{21}H_{24}O_3$			
Pregna-1,4,6,15-tetraene-3,20-dione, 17α-hydroxy-	MeOH	222(4.07),257(3.95), 298(4.08)	87-0393-69
$C_{21}H_{24}O_4$			
Benzofuran, 2-(3,4-dimethoxyphenyl)-2,3-dihydro-7-methoxy-3-methyl-5-propenyl-	EtOH	276(4.21)	12-1011-69
$C_{21}H_{24}O_5$			
Melanoxin, O-diethyldehydro-	MeOH	221(4.30),285(4.17), 324(4.50)	78-4409-69
$C_{21}H_{24}O_6$			
7α,10aβ-Gibba-3,4a(4b)-diene-1α,10β-dicarboxylic acid, 1,7-dimethyl-2,8-dioxo-, dimethyl ester	EtOH	309(4.14)	78-1293-69
$C_{21}H_{24}O_7$			
Isovaleric acid, 8-ester with 8,9-dihy-dro-9-hydroxy-8-(1-hydroxy-1-methyl-ethyl)-2H-furo[2,3-h]-1-benzopyran-2-one, acetate	ether	244(3.62),256(3.56), 319(4.14)	24-1673-69
$C_{21}H_{24}O_8$			
Altersolanol A, acetonide, dimethyl ether	EtOH	217(4.58),267(4.17), 413(3.61)	23-0777-69
Hydrocinnamic acid, α-(3,4-dimethoxy-α-methylphenethyl)-α-hydroxy-3,4-(methylenedioxy)-	EtOH	231(4.05),282(3.81)	39-0693-69C
$C_{21}H_{24}O_{10}$			
Neriaphin	EtOH	223(4.31),233s(4.27), 259(4.34),274(4.27), 306(3.78),412(3.47)	88-0471-69

Compound	Solvent	λ_{max}(log ϵ)	Ref.
$C_{21}H_{25}BrO_3$			
Androsta-1,4,6-trien-3-one, 17β-acetoxy-2-bromo-	EtOH	223(4.18),270(4.13), 306(4.00)	94-1212-69
$C_{21}H_{25}BrO_4$			
Androsta-4,6-dien-3-one, 17β-acetoxy-4-bromo-2β,19-epoxy-	EtOH	303(4.32)	94-1212-69
$C_{21}H_{25}ClN_2O_3$			
23,24-Seco-C-nor-13β-strychninium, 23-hydroxy-19-methyl-, chloride	EtOH	256(4.07),288s(3.45)	33-1564-69
$C_{21}H_{25}ClN_6O_4$			
9H-Purine-9-butyric acid, 2-amino-α-(carboxyamino)-6-chloro-, N-benzyl tert-butyl ester	EtOH	223(4.41),247(3.78), 311(3.86)	78-5971-69
$C_{21}H_{25}NO$			
Benzoxazole, 4,6-di-tert-butyl-2-phenyl-	EtOH	232(4.20),299(4.49), 301(4.50),321(4.21)	18-3559-69
$C_{21}H_{25}NO_2$			
A-Homo-19-nor-17α-pregna-1(10),2,4a-trien-20-yn-3-one, 17-hydroxy-, oxime	EtOH	233(4.23),303(4.03)	33-0121-69
$C_{21}H_{25}NO_3$			
Acetaldehyde, [[(1,4-dimethyl-2-dibenzofuranyl)methylene]amino]-, diethyl acetal	EtOH	225(4.32),251s(--), 257(4.66)	4-0379-69
Koenigicine, dihydro-N-methyl-	EtOH	242(4.58),273(4.28), 318(4.22)	31-0790-69
Phenol, 3,5-di-tert-butyl-4-nitroso-, benzoate	hexane	761(1.71)	18-3611-69
$C_{21}H_{25}NO_4$			
Dibenzo[a,g]quinolizine, 5,6,7,8,13,13a-hexahydro-1,2,10,11-tetramethoxy-	EtOH	215(5.00),235(4.20), 288(3.79)	44-2665-69
Dibenzo[a,g]quinolizine, 5,6,7,8,13,13a-hexahydro-3,4,10,11-tetramethoxy-, hydrochloride	H_2O	225s(4.23),282(3.72)	44-2665-69
Glaucine, d-	EtOH	219(4.58),282(4.18), 302(4.17)	44-2803-69
Homoprotoberberine	MeOH	227s(4.27),285(3.85)	39-0874-69C
Oxazolidine-4,5-dione, 2-(1-adamantyl)-2-ethoxy-N-phenyl-	EtOH	267(3.72)	78-2909-69
$C_{21}H_{25}NO_5$			
Alkaloid CC-10 from Cochicum cornigerum	EtOH	240(4.26),280s(3.65), 305s(3.42)	73-3540-69
8H-Dibenzo[a,g]quinolizin-11-ol, 5,6,13,13a-tetrahydro-2,3,9,10-tetramethoxy-	EtOH	224(3.96),273s(3.36), 283(3.55),292s(3.38)	44-1349-69
hydrobromide	EtOH	224s(3.91),273s(3.34), 282(3.53),288s(3.51), 290(3.42)	44-1349-69
Isoquino[2,1-b][2]benzazepine-2,10-diol, 5,6,8,13,14,14a-hexahydro-3,9,11-trimethoxy-	EtOH	230s(4.16),282(3.82)	33-1228-69

Compound	Solvent	λ max (log ε)	Ref.
Isoquinoline, 1,2,3,4-tetrahydro-6,7-(methylenedioxy)-1-(α-hydroxy-2-methyl-3,4-dimethoxybenzyl)-2-methyl-, levo-α-	MeOH	290(3.46)	95-0869-69
Isoquinoline, 1,2,3,4-tetrahydro-6,7-(methylenedioxy)-1-(3,4,5-trimethoxyphenethyl)-, hydrochloride	EtOH	230s(4.03),292s(3.66)	33-1228-69
19-Nor-17α-pregna-1,3,5(10)-trien-20-yne-9α,17β-diol, 3-methoxy-11β-nitro-	EtOH	278(3.39),285(3.34)	31-1018-69
$C_{21}H_{25}NO_6$			
Adluminediol	EtOH	230(4.03),288(3.78)	78-5059-69
Corluminediol	EtOH	205(4.80),234(3.98), 288(3.80)	78-5059-69
β-Hydrastinediol	EtOH	205(4.92),236(4.18), 287(3.88)	78-5059-69
$C_{21}H_{25}NO_6S$			
Alanine, 3-[[α-(2,4-dimethoxyphenacyl)-p-meth xybenzyl]thio]-, hydrochloride	n.s.g.	213(4.11),228(4.16), 271(3.92),306(3.75)	7-0624-69
$C_{21}H_{25}N_3$			
1H-1,5-Benzodiazepine, 4-[p-(dimethylamino)styryl]-2,3-dihydro-2,3-dimethyl-, hydrobromide	EtOH	532(3.74)	124-0740-69
$C_{21}H_{26}$			
Tricyclo[14.2.2.1^{2,6}]heneicosa-2,4,6(21),16,18,19-hexaene	hexane	250(4.16)	88-4551-69
$C_{21}H_{26}Br_2O_3$			
Androsta-4,6-dien-3-one, 17β-acetoxy-2,2-dibromo-	EtOH	300(4.36)	94-1212-69
$C_{21}H_{26}ClN_3OS$			
1-Piperazineethanol, 4-[3-(2-chlorophenothiazin-10-yl)propyl]-	EtOH	208(4.66),229(4.46), 244(4.48),255(4.32), 314(3.43)	9-0249-69
$C_{21}H_{26}INO_5$			
3,4-Dihydro-6,7-dimethoxy-2-methyl-1-(3,4,5-trimethoxyphenyl)isoquinolinium iodide	EtOH	248(4.31),310(4.05), 364(4.13)	1-0244-69
$C_{21}H_{26}N_2$			
Isoindole, 1,1'-methylenebis[4,7-dihydro-2,3-dimethyl-	EtOH	210(4.18)	77-0795-69
Pyrrolidine, 1,1'-(2-methyl-3,5-biphenylylene)di-	EtOH	239(4.53),320(3.59)	94-2126-69
$C_{21}H_{26}N_2O_2$			
Aspidofractinine, 1-carboxy-, methyl ester	EtOH	244(4.15),280(3.41), 288(3.38)	33-0076-69
Benzylidenimine, α,4-dimethyl-N-2-salsolidyl-	n.s.g.	235(4.26),286(3.78), 317(3.48)	65-2723-69
17βH-Diaboline, 1-deacetyl-O,1-dimethyl-	EtOH	261(4.01),311(3.48)	33-1564-69
Pyrrole, 2,2'-benzylidenebis[3-methoxy-4,5-dimethyl-	MeOH	225(4.12)	88-0409-69

Compound	Solvent	λ_{max}(log ε)	Ref.
$C_{21}H_{26}N_2O_3$			
Ervinidinine, dihydro-	EtOH	248(3.40),308(3.62)	105-0058-69
4-Pyridineacetic acid, 5-ethyl-1,2,3,4-tetrahydro-1-(2-indol-3-ylethyl)-2-oxo-, ethyl ester	n.s.g.	221(4.58),266(3.96), 289(3.78)	35-7342-69
Yohimbine, (±)-	EtOH	226(4.55),281(3.84), 290(3.76)	24-3248-69
β-Yohimbine, (±)-	EtOH	226(4.54),282(3.86), 289(3.79)	24-3248-69
$C_{21}H_{26}N_2O_4$			
Isohydrocoriamyrtin, phenylhydrazone	EtOH	248(3.88),303(4.07), 336(4.35)	78-1109-69
$C_{21}H_{26}N_2O_5$			
9,17-Secomorphinan-6,7-dione, 8,9,10,14-tetradehydro-4,5α-epoxy-3-methoxy-17,17-dimethyl-, 6-(dimethyl acetal), oxime	EtOH	257(4.26),350(4.10)	77-0057-69
$C_{21}H_{26}N_2S_2$			
Phenothiazine, 10-[2-(1-methyl-2-piperidyl)ethyl]-2-(methylthio)-	EtOH	227(4.09),264(4.45), 318(3.6)	9-0249-69
$C_{21}H_{26}N_4O_3$			
2-Propanol, 1-(diethylamino)-3-[(2-methyl-6-nitro-9-acridinyl)amino]-, dihydrochloride	EtOH	249(4.36),281(4.30), 315(4.08),357s(3.56), 445(3.74)	103-0707-69
2-Propanol, 1-(diethylamino)-3-[(4-methyl-6-nitro-9-acridinyl)amino]-, dihydrochloride	EtOH	246(4.40),280(4.41), 313(4.07),355s(3.38), 440(3.79)	103-0707-69
$C_{21}H_{26}N_4O_4$			
4H-Cyclopropa[b]naphthalen-4-one, 1,1a,2,5,6,6a,7,7a-octahydro-1,1,3,6a-tetramethyl-, 2,4-dinitrophenylhydrazone	$CHCl_3$	393(4.48)	32-0219-69
Ornithine, N^2-benzoyl-N^5-(4,5,6,7-tetrahydro-4-oxo-3H-cyclopentapyrimidin-2-yl)-, ethyl ester	H_2O	229(4.26),293(3.94)	18-3314-69
$C_{21}H_{26}N_6O_4$			
9H-Purine-9-butyric acid, 6-amino-α-(carboxyamino)-, N-benzyl tert-butyl ester	EtOH	262(4.14)	78-5971-69
$C_{21}H_{26}O$			
2,5-Cyclohexadien-1-one, 4-benzylidene-2,6-di-tert-butyl-	hexane	344(4.48)	70-0580-69
$C_{21}H_{26}O_2$			
Cannabinol	EtOH	223(4.53),285(4.22), 299s(4.06)	33-1102-69
6H-Dibenzo[b,d]pyran-1-ol, 6,6,9-trimethyl-3-pentyl-	EtOH	225s(4.38),282(4.09), 311(3.99)	33-1102-69
Pregna-3,5,16-triene-7,20-dione	EtOH	242(4.07),281(4.30)	2-1084-69
9β,10α-Pregna-4,6,8(14)-triene-3,20-dione	EtOH	271(3.89),283(3.92), 348(4.41)	33-1157-69
Pregn-4-en-20-yne-3,6-dione	EtOH	250(4.03)	44-3502-69

Compound	Solvent	λ max (log ε)	Ref.
$C_{21}H_{26}O_3$			
Estra-1,3,5(10)-trien-17-one, 3-hydroxy-6β-methyl-, acetate	EtOH	268(2.93),275(2.90)	33-0453-69
Pregna-4,8-diene-3,11,20-trione	n.s.g.	241(4.21)	23-1989-69
Pregna-1,4,15-triene-3,20-dione, 17α-hydroxy-	MeOH	243(4.21)	87-0393-69
Pregna-4,6,15-triene-3,20-dione, 17α-hydroxy-	MeOH	284(4.40)	87-0393-69
$C_{21}H_{26}O_4$			
Androsta-4,6-diene-3,17-dione, 19-acetoxy-	EtOH	277(4.27)	78-1219-69
$C_{21}H_{26}O_5$			
7a-Gibba-3,4a(4b)-diene-10β-carboxylic acid, 1,7-dimethyl-2-oxo-8-(ethylenedioxy)-, methyl ester	EtOH	308(4.24)	78-1293-69
$C_{21}H_{26}O_6$			
Quassinol, methyl ether	EtOH	314(4.28)	7-0230-69
	EtOH-NaOH	389(4.69)	7-0230-69
$C_{21}H_{26}O_8$			
Eriophyllin	EtOH	210(4.62)(end abs.)	102-2381-69
$C_{21}H_{26}O_{10}$			
Benzenepentol, pentapropionate	MeOH	266(2.86)	34-0118-69
Dehydroginkgolide A (spectra after equilibrium)	MeOH	264(3.0),333(1.6), 388s(0.9)	77-1467-69
	THF	391(1.6),411(1.5)	77-1467-69
Neriaphin reduction product	EtOH	238(4.68),246(4.62), 284(3.68),296(3.80), 309(3.80),338(3.58), 351(3.60)	88-0471-69
$C_{21}H_{27}BrN_2O_{10}$			
3-Carbamoyl-1-β-D-glucopyranosyl-4-picolinium bromide, 2',3',4',6'-tetraacetate	EtOH	261(3.69)	39-0192-69B
$C_{21}H_{27}BrO_3$			
Androsta-4,6-dien-3-one, 17β-acetoxy-2α-bromo-	EtOH	287(4.34)	94-1212-69
Androsta-4,6-dien-3-one, 17β-acetoxy-4-bromo-	EtOH	300(4.35)	94-1212-69
$C_{21}H_{27}ClN_2O_2$			
1-Deacetyl-O,4-dimethyldiabolinium chloride	EtOH	244(3.91),298(3.50)	33-1564-69
1-Deacetyl-O,4-dimethyl-17βH-diabolinium chloride	EtOH	246(3.90),300(3.46)	33-1564-69
$C_{21}H_{27}IN_2O_2$			
1-Deacetyl-4-ethyldiabolinium iodide	MeOH	298(3.49)	33-1564-69
$C_{21}H_{27}NO_3$			
A-Homoestra-1(10),2,4a-triene-4,17-dione, cyclic 17-(ethylene acetal)-, oxime	EtOH	233(4.19),303(4.00)	33-0121-69

Compound	Solvent	$\lambda_{max}(\log \epsilon)$	Ref.
$C_{21}H_{27}NO_5$			
Alkaloid CC-2 from Colchicum cornigerum	EtOH	210(4.60),236s(3.80), 289(3.30)	73-3540-69
10-Azaestra-2,4-dien-1-one, 3,17β-diacetoxy-	EtOH	237(3.72),310(3.83)	22-0117-69
Isoquinoline, 1,2,3,4-tetrahydro-6,7-dimethoxy-2-methyl-1-(3,4,5-trimethoxyphenyl)-	EtOH	281(3.57)	1-0244-69
4-Isoquinolinol, 1,2,3,4-tetrahydro-2-(3,4-dimethoxyphenethyl)-6,7-dimethoxy-	EtOH	230s(3.96),284(3.49)	78-1881-69
7-Isoquinolinol, 1,2,3,4-tetrahydro-1-(4-hydroxy-3,5-dimethoxyphenethyl)-6-methoxy-2-methyl-, (R)-(-)-	MeOH	230s(4.17),283(3.71), 294s(3.54)	33-0678-69
hydrochloride	MeOH	239s(4.18),283(3.71), 294s(3.54)	33-0678-69
(S)-(+)-	MeOH	230s(4.16),283(3.71), 294s(3.53)	33-0678-69
hydrochloride	MeOH	230s(4.16),283(3.68), 293s(3.52)	33-0678-69
Kreysiginine, (-)-	EtOH	212(4.60),231s(4.02), 275(3.07),282(3.07)	73-3540-69
$C_{21}H_{27}NO_6$			
10-Azaestra-2,4-diene-4-carboxylic acid, 3,17β-dihydroxy-1-oxo-, ethyl ester, 17-formate	EtOH	277s(3.95),284(3.98)	22-0117-69
$C_{21}H_{27}N_3O_3$			
2,2'-Bipyrrole, 3,3'-dimethoxy-5-[(3-methoxy-4,5-dimethyl-2H-pyrrol-2-ylidene)methyl]-4,4',5'-trimethyl-	acetone	505(4.35)	88-0409-69
	acetone-HCl	558s(4.58),589(4.77)	88-0409-69
$C_{21}H_{27}N_5O_2$			
Piperidine, 1-[4-[[p-(diethylamino)-phenyl]azo]-2-nitrophenyl]-	benzene	290(3.58),430(4.47)	73-2092-69
$C_{21}H_{28}$			
Chrysene, 1,2,3,4,4a,5,6,11,12,12a-decahydro-7,11,11-trimethyl-	EtOH	270(3.67)	22-2115-69
$C_{21}H_{28}ClNO_4$			
1,2,3,4,4a,5,6,8,9,10,11,12-Dodecahydro-6-phenylcycloocta[c]quinolizinium perchlorate	$CHCl_3$	328(3.87)	78-4161-69
$C_{21}H_{28}N_2O$			
2'H-Androsta-4,16-dieno[16,17-c]pyrazol-3-one, 5'-methyl-	EtOH	235(4.06)	22-1256-69
$C_{21}H_{28}N_2O_2$			
2β,19α-Vallesine	EtOH	259(4.12),289(3.35)	88-2523-69
$C_{21}H_{28}N_2O_3$			
Cylindrocarpinol, N-formyl-	MeOH	220(4.41),260(3.93), 292s(3.56)	39-0417-69C
$C_{21}H_{28}N_2O_4$			
1-Naphthamide, 1,2,3,4,4a,5,6,7,8,8a-decahydro-5,6,7-trihydroxy-N-(2-indol-3-ylethyl)-	EtOH	181(3.59),189(3.49)	35-7315-69

Compound	Solvent	λ max(log ε)	Ref.
$C_{21}H_{28}N_2O_{10}$			
Nicotinamide, 1-β-D-glucopyranosyl-1,4-dihydro-4-methyl-, 2',3',4',6'-tetraacetate	EtOH	317(3.66)	39-0192-69B
Nicotinamide, 1-β-D-glucopyranosyl-1,6-dihydro-4-methyl-, 2',3',4',6'-tetraacetate	EtOH	253(3.73),332(3.69)	39-0192-69B
$C_{21}H_{28}N_4O_5S$			
2-Butanone, 1-[3-[(4-amino-2-methyl-5-pyrimidinyl)methyl]-5-(2-hydroxyethyl)-4-methyl-4-thiazolin-2-ylidene]-1-hydroxy-3-methyl-, diacetate	EtOH	233(4.31),273(3.92), 362(4.41)	94-0128-69
$C_{21}H_{28}O$			
Pregn-4-en-20-yn-3-one	EtOH	240(4.21)	44-3502-69
17α-Pregn-4-en-20-yn-3-one	EtOH	240(4.21)	44-3502-69
$C_{21}H_{28}O_2$			
A-Homoestra-2,4,5(10)-trien-17-one, cyclic ethylene acetal	EtOH	270(3.61)	33-0121-69
1(2H)-Naphthalenone, 3,4-dihydro-2-(2-isobutyrylcyclohexyl)-8-methyl-	EtOH	211(4.31),255(4.08), 296(3.37)	22-2110-69
19-Norpregna-4,16-diene-3,20-dione, 16-methyl-	EtOH	243(4.38)	24-0643-69
19-Norpregna-1,3,5(10)-trien-11-one, 20β-hydroxy-9β-methyl-	EtOH	260(2.63),267(2.72), 274(2.69)	23-3489-69
Vitamin A_2 acid, methyl ester, 9-cis	EtOH	305(4.26),371(4.54)	22-3252-69
all trans	EtOH	307(4.11),380(4.59)	22-3252-69
$C_{21}H_{28}O_3$			
Androsta-3,5-diene-7,17-dione, cyclic 17-(ethylene acetal)	n.s.g.	280(4.42)	22-0617-69
Androsta-4,6-dien-3-one, 17β-acetoxy-	EtOH	282(4.43)	94-1212-69
5α-Androsta-8,14-dien-16-one, 3β-acetoxy-	EtOH	291(4.17)	39-0554-69C
Androst-5-eno[6,5,4-bc]furan-3-one, 17β-hydroxy-5'-methyl-	EtOH	209(4.18),230s(3.87), 296(3.50)	94-2586-69
Cyclohexanone, 2-[2-(5-hydroxy-2,7-dimethyl-2H-1-benzopyran-2-yl)-ethyl]-3,3-dimethyl-	n.s.g.	229(4.34),278(3.91), 285s(--)	88-3017-69
Etiojerv-4-ene-3,11-dione, 17α-hydroxy-17β-vinyl-	MeOH	235(4.17)	78-3145-69
18-Nor-D-homoandrosta-5,13(17a)-dien-17-one, 3β-acetoxy-	EtOH	241(4.15)	22-1673-69
9β,10α-Pregna-4,6-diene-3,20-dione, 8-hydroxy-	EtOH	280(4.43)	33-1157-69
9β,10α-Pregna-4,6-diene-3,20-dione, 9-hydroxy-	EtOH	283(4.41)	33-1157-69
Pregn-4-ene-3,20-dione, 9α,11α-epoxy-	n.s.g.	240(4.17)	23-1989-69
5α-Pregn-9(11)-ene-3,12,20-trione	EtOH	236(4.11)	44-0409-69
5β-Pregn-9(11)-ene-3,12,20-trione	EtOH	237(4.11)	44-0409-69
$C_{21}H_{28}O_4$			
Androsta-4,6-dien-3-one, 17β,19-dihydroxy-, 17-acetate	EtOH	282(4.41)	94-1212-69
Androst-4-en-3-one, 17β-acetoxy-2β,19-epoxy-	EtOH	242(4.16)	94-1206-69
Androst-5-en-3-one, 17β-acetoxy-4β,19-epoxy-	EtOH	313(2.42)	88-1925-69

Compound	Solvent	λ max(log ε)	Ref.
Cyclopropa[5,6]-A-nor-5β-androstan-3-one, 3',19-epoxy-3',6β-dihydro-17β-hydroxy-, acetate	EtOH	204(3.63)	88-1925-69
Hirundigenin, anhydro-	EtOH	289(1.95)	33-1175-69
2-Oxaandrosta-4,6-dien-3-one, 17β-hydroxy-17-methyl-, acetate	EtOH	272(4.32)	78-4257-69
5β-Pregna-7,14-diene-6,20-dione, 2β,3β-dihydroxy-	EtOH	296(4.11)	77-0402-69
5β-Pregna-8,14-diene-6,20-dione, 2,3-dihydroxy-	EtOH	246(4.08)	77-0402-69
Retinoic acid, 4,5-didehydro-5,6-dihydro-6-hydroxy-3-oxo-, methyl ester	EtOH	230(4.15),343(4.65)	22-3252-69
isomer	EtOH	230(4.18),342(4.52)	22-3252-69
Vitamin A2 acid, methyl ester, endocyclic peroxide, 9-cis	EtOH	326s(4.45),338(4.48), 347s(4.45)	22-3252-69
exocyclic peroxide	EtOH	295s(4.33),309(4.45), 320(4.42)	22-3252-69
Vitamin A_2 acid, methyl ester, endocyclic peroxide, trans	EtOH	322(4.47),335(4.53), 350(4.43)	22-3252-69
Vitamin A_2 acid, methyl ester, exocyclic peroxide	EtOH	294(4.39),307(4.50), 320(4.37)	22-3252-69
$C_{21}H_{28}O_5$			
14β,17α-Cortexone, 8,19-epoxy-19-hydroxy-	EtOH	244(4.16)	33-2156-69
Crotophorbolone, 20-methyl ether	MeOH	195(4.20),238(3.70)	5-0158-69H
2-Oxaandrosta-4,6-diene-1-carboxylic acid, 17β-hydroxy-17-methyl-3-oxo-, methyl ester	EtOH	274(4.33)	78-4257-69
2-Oxaandrosta-4,6-dien-3-one, 1ξ,17β-dihydroxy-17α-methyl-, 1-acetate	EtOH	274(4.34)	78-4257-69
$C_{21}H_{28}O_6$			
1-Oxaspiro[5.5]undec-3-en-2-one, 3,3'-methylenebis[4-hydroxy-	MeOH	253(4.32)	83-0075-69
Vincetogenin	EtOH	216s(3.79)	33-1175-69
$C_{21}H_{28}O_7$			
D-glycero-D-galacto-Octose, 3,6-anhydro-2-deoxy-4,5:7,8-di-O-isopropylidene-1-C-(p-methoxyphenyl)-	MeOH	275(4.23)	65-0119-69
D-glycero-D-talo-Octose, 3,6-anhydro-2-deoxy-4,5:7,8-di-O-isopropylidene-1-C-(p-methoxyphenyl)-	MeOH	275(4.23)	65-0119-69
$C_{21}H_{28}O_8$			
Euparotin, dihydro-13-methoxy-	MeOH	218(3.80)	44-3876-69
Eupatundin, dihydro-13-methoxy-	MeOH	217(4.05)	44-3876-69
$C_{21}H_{29}BrO_4$			
Androst-4-en-3-one, 4-bromo-17β,19-dihydroxy-, 17-acetate	EtOH	262(4.06)	94-1212-69
$C_{21}H_{29}ClO_4$			
Androst-4-en-3-one, 4-chloro-17β,19-dihydroxy-, 17-acetate	EtOH	256(4.23)	94-1212-69
$C_{21}H_{29}N$			
13-Azabicyclo[8.2.1]trideca-10,12-diene, 13-(3,4,5-trimethylphenyl)-	hexane	241(4.12)	78-5357-69

Compound	Solvent	$\lambda_{max}(\log \epsilon)$	Ref.
$C_{21}H_{29}NO_2$			
Etiojerva-1,4-dien-3-one, 17α-acetyl-amino-	EtOH	244(4.23)	78-4551-69
$C_{21}H_{29}NO_5$			
Alkaloid CC-3b from Colchicum corni-gerum	EtOH	208(4.60),226s(3.95), 284(2.98)	73-3540-69
5α-Androst-9(11)-en-12-one, 3β-hydroxy-17-nitro-, acetate	EtOH	237(4.16)	44-0409-69
Pregn-4-ene-3,20-dione, 17-hydroxy-, nitrate	MeOH	240(4.24)	78-0761-69
$C_{21}H_{29}NO_6P_2$			
Phosphonic acid, 9-acridanylidenedi-, tetraethyl ester	MeOH	209(4.61),282(4.31), 292(4.11),332(3.94)	44-1420-69
$C_{21}H_{29}N_3O$			
Phenol, 2,6-di-tert-butyl-4-(3-methyl-3-phenyl-1-triazeno)-	n.s.g.	242(4.20),297(3.99), 357(4.35)	70-2460-69
$C_{21}H_{29}N_3OS$			
Spiro[1]benzothieno[2,3-d]pyrimidine-2(1H),1'-cyclohexan]-4(3H)-one, 3-(cyclohexylideneamino)-5,6,7,8-tetrahydro-	MeOH	230(4.27),333(3.51)	48-0402-69
$C_{21}H_{29}N_3O_3$			
Androst-4-en-3-one, 6β-azido-17β-hydroxy-, acetate	MeOH	235(4.10)	48-0445-69
$C_{21}H_{30}N_2O$			
2'H-Androsta-5,16-dieno[16,17-c]pyrazol-3β-ol, 5'-methyl-	EtOH	225(3.74)	22-1256-69
$(C_{21}H_{30}N_2O_6)_n$			
Poly[oxy[2-[(3,5,6-trimethyl-p-benzo-quinon-2-yl)methyl]trimethylene]-oxycarbonyliminohexamethylene-iminocarbonyl]	95% THF	264(4.39),291(3.62)	47-1269-69
$C_{21}H_{30}N_2O_8$			
3,4-Furandicarboxylic acid, 5-(tert-butylimino)-2-[2-carboxy-1-(1-pyrrolidinyl)vinyl]-2,5-dihydro-2-methoxy-, trimethyl ester	ether	255(4.03),297(4.18)	24-1656-69
$C_{21}H_{30}O$			
17α-Pregna-4,20-dien-3-one	EtOH	240(4.21)	44-3502-69
$C_{21}H_{30}O_2$			
Androsta-4,16-dieno[17,16-b]furan-3-one, 4',5',16α,17α-tetrahydro-	EtOH	241(4.22)	73-3479-69
Cannabidiol, (-)	EtOH	232s(4.04),274(3.12), 282(3.10)	33-1102-69
Δ^8-Cannabinol, tetrahydro-, 6a,10a-trans	EtOH	230s(4.07),275(3.22), 282(3.22)	33-1102-69
Δ^9-Cannabinol, tetrahydro-, 6a,10a-trans	EtOH	230s(4.03),276(3.20), 283(3.21)	33-1102-69
Δ^8-Dibenzo[b,d]pyran-3-ol, 6,6,9-tri-methyl-1-pentyltetrahydro-, 6a,10a-trans	EtOH	226s(4.03),287(3.47)	33-1102-69

Compound	Solvent	λ_{max}(log ε)	Ref.
18,19-Dinor-10β-pregn-13-ene-3,20-dione, 5β,14β-dimethyl-	EtOH	257(4.14)	22-3265-69
D-Homo-18-nor-9β,10α-androst-4-ene-3,17a-dione, 13-ethyl-	EtOH	240(4.23)	44-0107-69
Olivetol, 4-(3-3,4-trans-p-mentha-1,8-dienyl)-	EtOH	221(3.92),286(3.30)	33-1102-69
Retinoic acid, methyl ester, 11-cis	EtOH	246(4.12),349(4.47)	22-3242-69
$C_{21}H_{30}O_3$			
5α,8β,9β-Androst-14-en-16-one, 3β-acetoxy-	EtOH	234(4.19)	39-0554-69C
10α-Androst-5-en-3-one, 17β-acetoxy-	benzene	298(2.09)	33-0173-69
	EtOH	223(2.97),295(2.08)	33-0173-69
3[4→5β]-abeo-4,6α-Cycloandrostane-3,7-dione, 17β-hydroxy-4,4-dimethyl-	EtOH	230(3.34)(end abs.)	33-2436-69
Gon-4-en-3-one, 17β-acetoxy-13β-ethyl-	EtOH	241(4.22)	44-3530-69
A-Nor-5ξ-androstane-3,7-dione, 17β-hydroxy-6-isopropylidene-	EtOH	252(3.76)	33-2436-69
Pregn-4-ene-3,20-dione, 11-hydroxy-	n.s.g.	242(4.20)	23-1989-69
9β,10α-Pregn-4-ene-3,20-dione, 8-hydroxy-	EtOH	239(4.16)	33-1157-69
9β,10α-Pregn-4-ene-3,20-dione, 9-hydroxy-	EtOH	241(4.21)	33-1157-69
1-Propanone, 1,1',1"-s-phenenyltris-[2,2-dimethyl-	EtOH	222(4.48),247s(3.88), 290(2.79),316s(2.60)	1-0751-69
$C_{21}H_{30}O_4$			
5α-Androstane-Δ$^{2,\alpha}$-acetic acid, 17β-hydroxy-3-oxo-	EtOH	230s(--),240(4.00)	44-3032-69
5α-Androstane-Δ$^{16,\alpha}$-acetic acid, 3β-hydroxy-17-oxo-	MeOH	241(4.07)	44-0237-69
5α-Androstane-Δ$^{2,\alpha}$-acetic acid, 3,3,17β-trihydroxy-, γ-lactone	EtOH	217(4.13)	44-3032-69
6H-Benzo[b][1,2]dioxeto[3',4':4,5]benzo-[1,2-d]pyran-1-ol, 6a,7,7a,9a,10,10a-hexahydro-6,6,9a-trimethyl-3-pentyl-, cis	EtOH	222s(3.94),280(3.34)	35-5190-69
Gon-4-en-3-one, 13β-ethyl-14α,17β-dihydroxy-, 17-acetate, d-	EtOH	241(4.21)	44-3530-69
Hirundigenin, anhydrodihydro-	EtOH	202(4.06)	33-1175-69
Isohirundigenin, anhydrodihydro-	EtOH	214(3.76),292(1.88)	33-1175-69
9β,10α-Pregn-4-ene-3,20-dione, 9,21-dihydroxy-	EtOH	240(4.20)	33-1157-69
$C_{21}H_{30}O_5$			
Androst-4-en-3-one, 4,17β,19-trihydroxy-, 17-acetate	EtOH	278(4.14)	94-1206-69
Androst-15-en-17-one, 14β-hydroperoxy-3β-hydroxy-, 3-acetate	MeOH	212(3.94)	23-3693-69
Gon-4-en-3-one, 13β-ethyl-6β,10β,17β-trihydroxy-, 17-acetate	EtOH	233(4.14)	44-3530-69
Hirundigenin	EtOH	196(3.92)(end abs.)	33-1175-69
$C_{21}H_{30}O_6$			
Nigakilactone A	EtOH	271(3.68)	88-3013-69
$C_{21}H_{30}O_{12}$			
D-gluco-Octose, 3,7-anhydro-2-deoxy-6-O-β-D-galactopyranosyl-1-C-(p-methoxyphenyl)-	MeOH	279(4.21)	65-0119-69

Compound	Solvent	λ_{max}(log ϵ)	Ref.
D-gluco-Octose, 3,7-anhydro-2-deoxy-6-O-α-D-glucopyranosyl-1-C-(p-methoxyphenyl)-	MeOH	276(4.20)	65-0119-69
$C_{21}H_{31}ClO_2$			
6H-Dibenzo[b,d]pyran-3-ol, 9-chloro-hexahydro-6,6,9-trimethyl-1-pentyl-, 6a,10a-trans	EtOH	228s(4.08),284(3.51)	33-1102-69
$C_{21}H_{31}IN_2O_2$			
Corynantheol, dihydro-10-methoxy-, methiodide	EtOH	275(3.96),296s(3.75), 309s(3.63)	78-5329-69
3-Epihunterburnine, dihydro-O-methyl-, methiodide	EtOH	218(4.63),276(3.90), 296s(3.68),308s(3.55)	78-5319-69
$C_{21}H_{31}NO_2$			
Etiojerv-4-en-3-one, 17α-acetamido-	EtOH	240(4.21)	78-4551-69
$C_{21}H_{31}NO_3$			
Benzoic acid, p-butoxy-, 3-[(dimethyl-amino)methyl]-2-norbornyl ester, hydrochloride	H_2O	262(4.24)	36-0553-69
$C_{21}H_{31}NO_4$			
Androst-5-en-3β-ol, 6-nitro-, acetate	EtOH	262(3.18)	22-1632-69
$C_{21}H_{31}NO_7$			
Lasiocarpine, didehydro-	n.s.g.	219(4.02)	39-1155-69C
$C_{21}H_{31}N_3O$			
1(4H)-Naphthalenone, 2-[[3-(dimethyl-amino)propyl]amino]-4-(hexylimino)-	EtOH	238(4.41),271(4.14), 280s(4.11),337(3.68), 423(3.68)	39-1799-69C
$C_{21}H_{32}ClNO_8$			
[2-[1-(2,4-Dihydroxybutyl)-7,8-dimethoxy-2-naphthyl]ethyl]trimethylammonium perchlorate	H_2O	290(3.82),301(3.79), 338(3.41)	49-0163-69
$C_{21}H_{32}N_2OS$			
2'H-5α-Androst-2-eno[3,2-c]pyrazol-17β-ol, 7-mercapto-17α-methyl-	EtOH	224(3.72)	94-0011-69
$C_{21}H_{32}O$			
5α-Pregn-17-en-21-al, trans	EtOH	245(4.22)	39-0460-69C
5α-Pregn-1-en-3-one	n.s.g.	227(3.98)	78-3075-69
17α-Pregn-4-en-3-one	EtOH	240(4.21)	44-3502-69
Tricyclo[10.3.3.1^{1,12}]nonadecan-19-one, 14,17-dimethylene-	hexane	305(1.50)	88-3653-69
$C_{21}H_{32}O_2$			
Androst-4-en-3-one, 2α-ethyl-17β-hydroxy-	EtOH	241(4.17)	44-2288-69
Androst-4-en-3-one, 17β-hydroxy-2α,6β-dimethyl-	EtOH	241(4.19)	44-2288-69
9β,10α-Androst-4-en-3-one, 17-hydroxy-17α,18-dimethyl-	n.s.g.	242(4.29)	54-0752-69
6aβ,12aβ,12bα-Benz[a]anthracen-9(1H)-one, 4aβ-ethyl-2,3,4,4a,5,6,6a,7,10,-11,11a,12,12a,12b-tetradecahydro-4-hydroxy-7-methyl-	EtOH	243(4.25)	44-0107-69

Compound	Solvent	$\lambda_{max}(\log \epsilon)$	Ref.
Cannabinerol, 6-cis	EtOH	272(3.03),282(3.02)	88-5349-69
Cyclohexanecarboxylic acid, 2-(m-isopropylphenethyl)-1β,3α-dimethyl-, methyl ester	EtOH	264(2.65),271(2.40)	44-1459-69
6H-Dibenzo[b,d]pyran-1-ol, 6a,7,8,9,10,10a-hexahydro-6,6,9-trimethyl-3-pentyl-, cis	EtOH	273(3.03),282(2.95)	88-5349-69
trans	EtOH	274(3.03),281(3.02)	88-5349-69
D-Homo-18-nor-9β,10α-androst-4-en-3-one, 13-ethyl-17aβ-hydroxy-	EtOH	243(4.21)	44-0107-69
5α-Pregn-17-en-21-al, 3β-hydroxy-, trans	EtOH	245(4.16)	39-0460-69C
5α,17α-Pregn-14-en-20-one, 3β-hydroxy-	EtOH	278(1.72)	33-2583-69
$C_{21}H_{32}O_2S$			
Androst-4-en-3-one, 17β-hydroxy-17α-methyl-7α-(methylthio)-	EtOH	242(4.22)	94-0011-69
$C_{21}H_{32}O_3$			
Podocarpan-15-oic acid, 13-methyl-13-vinyl-7-oxo-, methyl ester	EtOH	249(3.91)	88-1683-69
Podocarp-8-en-15-oic acid, 13β-isopropyl-11-oxo-, methyl ester	EtOH	246(4.06)	44-3464-69
Podocarp-9(11)-en-15-oic acid, 13-isopropyl-12-oxo-, methyl ester	n.s.g.	238(4.10)	44-2775-69
isomer	n.s.g.	241(4.18)	44-2775-69
Pregn-4-en-3-one, 17α,20α-dihydroxy-	EtOH	241(4.18)	105-0128-69
Pregn-4-en-3-one, 17α,20β-dihydroxy-	EtOH	242(4.21)	105-0128-69
$C_{21}H_{32}O_3S$			
Androstan-3-one, 17β-hydroxy-2-(hydroxymethylene)-7-mercapto-17α-methyl-	EtOH	285(3.95)	94-0011-69
$C_{21}H_{32}O_4$			
5α-Androstane-$\Delta^{2,\alpha}$-acetic acid, 3,17β-dihydroxy-	EtOH	220(4.30)	44-3032-69
Benzo[1,2:3,4]dicycloheptene-6,6(1H)-dicarboxylic acid, 2,3,4,5,5a,7,7a,-8,9,10,11,12-dodecahydro-, ethyl methyl ester	EtOH	209(3.96)	22-4493-69
Bicyclo[3.2.0]hept-2-ene-1-heptanoic acid, 2-hexanoyl-7-oxo-, methyl ester	EtOH	225(3.70)	88-1639-69
1-Cyclopentene-1-heptanoic acid, 5-oxo-2-(3-oxo-1-octenyl)-, methyl ester	MeOH	296(4.36)	88-1615-69
Kolavic acid, methyl ester	MeOH	220(4.26)	39-2153-69C
Labda-8,13(15)-diene-15,19-dioic acid, 15-methyl ester, trans	EtOH	215(4.17)	12-0491-69
Labda-8,13(15)-diene-15,19-dioic acid, 19-methyl ester, trans	EtOH	215(4.14)	12-0491-69
Podocarp-8-en-15-oic acid, 11α-hydroxy-13β-isopropyl-7-oxo-, methyl ester	EtOH	247(4.08)	44-3464-69
Pregn-17(20)-en-21-oic acid, 3α,14-dihydroxy-	EtOH	224(4.11)	35-1228-69
Prostaglandin PGB_1, 15-dehydro-, methyl ester	MeOH	296(4.36)	88-2771-69
$C_{21}H_{33}NO$			
18,19-Dinor-10α-pregn-13-en-20-one, 3α-amino-5β,14β-dimethyl-	EtOH	257(4.05)	22-3265-69
Pregn-1-en-3-one, oxime	n.s.g.	235(4.21)	78-3075-69

Compound	Solvent	$\lambda_{max}(\log \epsilon)$	Ref.
$C_{21}H_{33}NOS$			
Androst-4-en-3-one, 17β-[(2-mercapto-ethyl)amino]-	n.s.g.	241(4.02)	23-0160-69
$C_{21}H_{33}NO_3$			
7a-Aza-B-homopregn-5-en-7-one, 3β,20β-dihydroxy-	EtOH	222(4.16)	2-1084-69
$C_{21}H_{33}NO_5$			
5α-Androstane-3α,17β-diol, 17-acetate, 3-nitrate	C_6H_{12}	186(3.76)(end abs.)	78-0761-69
5α-Androstane-3β,17β-diol, 3-acetate, 17-nitrate	C_6H_{12}	188(3.76)(end abs.)	78-0761-69
5α-Androstane-3β,17β-diol, 17-acetate, 3-nitrate	C_6H_{12}	186(3.80)(end abs.)	78-0761-69
5β-Androstane-3α,17β-diol, 17-acetate, 3-nitrate	C_6H_{12}	186(3.80)(end abs.)	78-0761-69
$C_{21}H_{33}NO_7$			
Lasiocarpine, trans	EtOH	219(4.07)	73-1832-69
$C_{21}H_{33}N_5O_3S$			
Formamide, N-[(4-amino-2-methyl-5-pyr-imidyl)methyl]-N-[4-hydroxy-1-methyl-2-[(2,2,6,6-tetramethyl-4-oxopiperi-dino)thio]-1-butenyl]-	EtOH	235(4.17),276s(3.77)	18-1942-69
$C_{21}H_{34}Br_2N_4O_2$			
N,N,N,N',N'-Pentamethyl-N'-[3-[(4-nitro-1-naphthyl)amino]propyl]-N,N'-trimeth-ylenediammonium dibromide (also spectra in biological media)	pH 6.5	322(3.40),443(4.19)	35-5136-69
	EtOH	333s(3.45),428(4.21)	35-5136-69
$C_{21}H_{34}N_2O_4$			
1-Cyclopentene-1-heptanoic acid, 5-oxo-2-(3-oxo-1-octenyl)-, methyl ester, 3,5-dioxime	MeOH	308(4.58),317(4.58)	88-2771-69
$C_{21}H_{34}O_2$			
Copaiferic acid, methyl ester	EtOH	219(4.14)	7-0539-69
1-Oxaspiro[2.5]octa-4,7-dien-6-one, 5,7-di-tert-butyl-2,2-dipropyl-	n.s.g.	262(4.12)	70-1732-69
$C_{21}H_{34}O_3$			
5α,17α-Pregnan-20-one, 3β,14β-dihydroxy-	EtOH	277(1.91)	33-2583-69
$C_{21}H_{34}O_3S$			
Podocarpane-6α,7α,17-triol, 13β-isopro-pyl-, cyclic 6,7-thiocarbonate	MeOH	238(4.49)	78-1335-69
$C_{21}H_{34}O_4$			
1-Cyclopentene-1-heptanoic acid, 5-oxo-2-(3-oxooctyl)-, methyl ester	MeOH	238(4.13)	88-2771-69
5α,17α-Pregnan-20-one, 3β,8β,14β-tri-hydroxy-	EtOH	275(2.13)	33-2583-69
$C_{21}H_{34}O_7$			
Stephanol	EtOH	210(3.70)(end abs.)	94-2448-69

Compound	Solvent	$\lambda_{max}(\log \epsilon)$	Ref.
$C_{21}H_{35}NO_2$			
Tetradecanohydroxamic acid, N-p-tolyl-	EtOH	248(4.04)	42-0831-69
$C_{21}H_{36}$			
Benzene, 1,3,5-trineopentyl-	hexane	199(4.74),203(4.76), 217s(4.04),259s(3.96), 265(2.40),271(2.32)	1-0751-69
$C_{21}H_{37}NO_3S_2$			
2-Pyridinol, 1-acetyl-4-tert-butyl-3,6-bis(tert-butylthio)-1,2,3,6-tetrahydro-, acetate	MeOH	200(4.19)(end abs.)	44-0660-69

Compound	Solvent	$\lambda_{max}(\log \epsilon)$	Ref.
$C_{22}H_9ClO_4$			
5,7,12,14-Pentacenetetrone, 1-chloro-	dioxan	265(4.79),340(3.90)	104-2154-69
5,7,12,14-Pentacenetetrone, 2-chloro-	dioxan	270(4.86),340(3.98)	104-2154-69
$C_{22}H_{10}O_4$			
5,7,12,14-Pentacenetetrone	dioxan	270(4.90),335(3.91)	104-2154-69
	$CHCl_3$	270(4.06),343(4.14)	77-0563-69
$C_{22}H_{10}O_7S$			
1-Pentacenesulfonic acid, 5,7,12,14-tetraoxo-	H_2O	270(4.85),345(3.88)	104-2154-69
2-Pentacenesulfonic acid, 5,7,12,14-	H_2O	270(4.84),345(3.96)	104-2154-69
$C_{22}H_{12}ClNO_5$			
Spiro[1,4,2-dioxazole-5,9'(10'H)-phenanthren]-10'-one, 3-[3-chloro-4,5-(methylenedioxy)phenyl]-	EtOH	252(4.64),277(4.23)	23-1473-69
$C_{22}H_{12}Cl_4O$			
4H-Pyran, 2,6-diphenyl-4-(2,3,4,5-tetrachloro-2,4-cyclopentadien-1-ylidene)-	benzene	466(4.57)	83-0886-69
	C_6H_{12}	450(4.56),474s(4.51)	83-0886-69
	MeOH	460(4.58)	83-0886-69
	CH_2Cl_2	253(4.19),340(4.28), 464(4.57)	83-0886-69
$C_{22}H_{12}Cl_4S$			
4H-Thiopyran, 2,6-diphenyl-4-(2,3,4,5-tetrachloro-2,4-cyclopentadien-1-ylidene)-	benzene	510(4.52)	83-0886-69
	C_6H_{12}	494(--)	83-0886-69
	CH_2Cl_2	245(4.34),345(4.14), 505(4.55)	83-0886-69
	MeOH	502(--)	83-0886-69
$C_{22}H_{12}N_2$			
Cyclohepta[4,5]naphtho[6,7,8-bcd:-9,10,1-dcb], 4,9-dihydro-	dioxan	271(4.54),284(4.61), 295(4.81),320(3.85), 332(3.83),346(3.93), 363(3.92),396(3.01), 410(3.07),435(3.05), 465(2.86)	24-2728-69
1,3-Cyclopentadiene-1,2-dicarbonitrile, 5-(2,3-diphenyl-2-cyclopropen-1-ylidene)-	MeCN	234(4.16),258(4.11), 268(4.10),368(4.47)	33-2472-69
	H_2SO_4	255s(4.24),270s(4.28), 315(4.49),364(4.30), 441s(3.54)	33-2472-69
1,3-Cyclopentadiene-1,3-dicarbonitrile, 5-(2,3-diphenyl-2-cyclopropen-1-ylidene)-	MeCN	248(4.39),256s(4.30), 269(4.13),358(4.50)	33-2472-69
	H_2SO_4	270s(4.44),307(4.65), 361(4.36),416s(3.54)	33-2472-69
2,5-Cyclopentadiene-1,2-dicarbonitrile, 4-(2,3-diphenyl-2-cyclopropen-1-ylidene)-	MeCN	235s(4.39),256s(4.66), 261(4.70),270s(4.60), 347(4.72)	33-2472-69
	H_2SO_4	245(3.95),313(4.50), 324(4.47),347(4.34), 369s(4.19)	33-2472-69
	C_6H_{12}	367(--)	33-2472-69
(and other solvents without ε)	CCl_4	360(--)	33-2472-69
$C_{22}H_{12}N_2S_2$			
Phenothiazino[4,3-c]phenothiazine	$C_6H_3Cl_3$	332(4.24),440(3.85), 700(4.68)	24-2728-69

Compound	Solvent	$\lambda_{max}(\log \epsilon)$	Ref.
Phenothiazino[4,3-c]phenothiazine (cont.)	dioxan	264(4.32),298(4.56)	24-2728-69
$C_{22}H_{13}Br$			
Pyrene, 2-(p-bromophenyl)-	EtOH	236(4.62),279(5.38), 308(4.68),321(4.95), 338(5.22)	24-2301-69
$C_{22}H_{13}BrN_2$			
Benzo[b]phenazine, 6-bromo-11-phenyl-	EtOH	254(4.60),288(4.94)	44-1651-69
$C_{22}H_{13}BrN_2S_2$			
2H-1,4-Benzothiazine, 2-(2-benzothiazolylmethylene)-3-(p-bromophenyl)-	EtOH	254(4.50),327(4.32), 420(4.03)	32-0323-69
	EtOH-12N HCl	239s(4.49),340(4.37), 471(4.17)	32-0323-69
$C_{22}H_{13}Cl$			
Pyrene, 2-(p-chlorophenyl)-	EtOH	223(4.64),236(4.35), 277(5.19),308(4.44), 322(4.73),338(5.02)	24-2301-69
$C_{22}H_{13}N_3O_6$			
1-Cyclopropene, 1,2-diphenyl-3-(2,4,6-trinitrobenzylidene)-	CH_2Cl_2	253(4.43),410s(--), 495(4.25)	24-0319-69
$C_{22}H_{14}$			
Indeno[2,1-a]indene, 5-phenyl-	EtOH	260(4.7),285(4.8), 402(4.1),430(4.1)	5-0076-69C
Pentacene	benzene	576(3.72)	44-1734-69
Pyrene, 1-phenyl-	EtOH	236(4.82),244(4.85), 267(4.58),277(4.80), 312(4.17),328(4.53), 342(4.66)	24-2301-69
Pyrene, 2-phenyl-	EtOH	273(4.86),308(4.07), 320(4.46),336(4.73)	24-2301-69
Pyrene, 4-phenyl-	EtOH	234(4.76),242(5.00), 256(4.22),266(4.54), 276(4.81),311(4.19), 324(4.52),339(4.67)	24-2301-69
$C_{22}H_{14}Br_2$			
Indene, 3-bromo-1-(α-bromobenzylidene)-2-phenyl-, cis	EtOH	247(4.35),277(4.28), 340(4.06)	44-0879-69
trans	EtOH	248(4.25),274(4.26), 338(4.01)	44-0879-69
$C_{22}H_{14}Br_2O$			
Ketone, 1,1-dibromo-2-phenylinden-3-yl phenyl	ether	257(4.47),337s(3.55)	44-0879-69
$C_{22}H_{14}ClNO_4$			
Spiro[1,4,2-dioxazole-5,9'(10'H)-phenanthren]-10'-one, 3-(2-chloro-4-methoxyphenyl)-	EtOH	255(4.65),275(4.32)	23-1473-69
$C_{22}H_{14}Cl_2S$			
Benzo[b]thiophene, 5-chloro-2-[p-(p-chlorostyryl)phenyl]-	DMF	358(4.81)	33-1282-69

Compound	Solvent	λ_{max}(log ε)	Ref.
$C_{22}H_{14}I_2$			
Indene, 3-iodo-1-(α-iodobenzylidene)-2-phenyl-, cis	EtOH	247(4.34),288(4.15),348(4.04)	44-0879-69
trans	ether	248(4.30),289(4.16),297s(4.15),339s(4.03),348(4.07)	44-0879-69
$C_{22}H_{14}N_2$			
Carbazolo[2,1-a]carbazole, 5,12-dihydro-	dioxan	260(4.64),290(4.76),308(4.66),322(4.52),335(4.70),361(4.27),380(4.53)	24-2728-69
Carbazolo[3,4-a]carbazole, 5,8-dihydro-	dioxan	260(4.60),306(4.78),346(4.22),365(3.95),384(4.07)	24-2728-69
Carbazolo[3,4-c]carbazole, 5,10-dihydro-	EtOH	232(4.74),275(4.48),285(4.56),303(4.54),312(4.68),331(4.14),345(4.08),359(4.02),376(3.98)	24-2728-69
$C_{22}H_{14}N_2O$			
Benzo[b]phenazine, 6-phenyl-, 12-oxide	EtOH	242(4.56),244(4.20),286(4.89)	44-1651-69
$C_{22}H_{14}N_2O_2$			
Quinoxaline, 2,3-dibenzoyl-	EtOH	257(4.68),324(3.90)	44-1651-69
$C_{22}H_{14}N_2O_3$			
Xanthene-2,2(1H)-dicarbonitrile, 1-hydroxy-4-methyl-9-oxo-1-phenyl-	MeOH	309(3.99),461(4.47)	44-2407-69
	CH_2Cl_2	259(4.17),265(4.20),306(3.98)	44-2407-69
$C_{22}H_{14}N_2O_6$			
Phenanthro[9,10-d]-1,3-dioxole, 5,10-dinitro-2-p-tolyl-	dioxan	306(4.63),334s(4.26),476(3.29)	24-2384-69
$C_{22}H_{14}N_2S$			
5H-Benzo[a]phenothiazine, 5-(phenylimino)-	MeCN	230(4.44),268(4.37),318(4.29),360(4.05),470(4.05)	44-1691-69
$C_{22}H_{14}N_2S_2$			
2H-1,4-Benzothiazine, 2-(2-benzothiazolylmethylene)-3-phenyl-	EtOH	253(4.42),326(4.31),416(4.11)	32-0323-69
	EtOH-12N HCl	236s(4.47),345(4.41),490(4.13)	32-0323-69
Phenothiazino[2,1-a]phenthiazine, 5,13-dihydro-	dioxan	280(4.80),295(4.64),330(3.78),382(3.78),430s(3.50)	24-2728-69
Phenothiazino[4,3-c]phenothiazine, 8,16-dihydro-	dioxan	264(4.40),280(4.80),295(4.62),340(3.89),430(3.31)	24-2728-69
$C_{22}H_{14}N_4$			
2,3-Diazaphenazine, 1,4-diphenyl-	EtOH	245(4.79),274s(4.30),345(3.85)	44-1651-69

Compound	Solvent	λ_{max}(log ε)	Ref.
$C_{22}H_{14}N_6O_{12}S_2$			
2,7-Naphthalenedisulfonic acid, 4,5-dihydroxy-3,6-bis[(p-nitrophenyl)-azo]-	pH 2.5 and 6.5	540(4.38),630(4.18)	70-0247-69
	pH 11.5	590(4.28)	70-0247-69
$C_{22}H_{14}O_2$			
Benzo[1,2-b:5,4-b']difuran, 2,6-diphenyl-	MeCN	284s(4.59),290(4.60), 321s(4.51),338(4.67), 351(4.61)	12-2395-69
Lactone, m. 167°	EtOH	247(4.34),287(4.11)	44-3076-69
$C_{22}H_{14}O_7$			
[1,1'-Binaphthalene]-3,4,5',8'-tetrone, 6',7'-epoxy-6',7'-dihydro-4,5'-dimethoxy-	EtOH	244(4.31),348(3.54), 418(3.78)	39-2059-69C
$C_{22}H_{14}S$			
Benzo[b]naphtho[1,2-d]thiophene, 5-phenyl-	$CHCl_3$	249(4.71),258(4.76), 281(4.67),305(4.34), 321s(3.93),353(3.61)	44-1310-69
$C_{22}H_{15}Br$			
Indene, 1-(α-bromobenzylidene)-2-phenyl-, cis	EtOH	252s(4.34),274(4.38), 339(4.00)	44-0879-69
$C_{22}H_{15}BrN_2O$			
Ketone, 3-(α-bromobenzyl)-2-quinoxalinyl phenyl	EtOH	254(4.51),327(3.81)	44-1651-69
$C_{22}H_{15}BrN_2O_2S$			
Thieno[3,4-b]quinoxaline, 1-bromo-1,3-dihydro-1,3-diphenyl-, 2,2-dioxide	EtOH	248(4.45),279s(4.00), 326(3.86)	44-1651-69
$C_{22}H_{15}BrO$			
Benzofuran, 5-bromo-2-(p-styrylphenyl)-	DMF	356(4.83)	33-1282-69
$C_{22}H_{15}Cl$			
Anthracene, 2-(p-chlorostyryl)-	DMF	313(4.90),328(4.95), 357(4.22),374(4.34), 393(4.26)	33-2521-69
$C_{22}H_{15}ClNS$			
2-(p-Chlorophenyl)-4,6-diphenyl-1,3-thiazin-1-ium (perchlorate)	MeCN	421(4.38)	24-0269-69
4-(p-Chlorophenyl)-2,6-diphenyl-1,3-thiazin-1-ium (perchlorate)	MeCN	422(4.46)	24-0269-69
6-(p-Chlorophenyl)-2,4-diphenyl-1,3-thiazin-1-ium (perchlorate)	MeCN	424(4.41)	24-0269-69
$C_{22}H_{15}ClN_2O_3$			
Spiro[Δ^2-1,2,4-oxadiazoline-3,9'(10'H)-phenanthren]-10'-one, 3-(2-chloro-4-methoxyphenyl)-	EtOH	277(4.31)	23-1473-69
$C_{22}H_{15}ClO$			
Benzofuran, 2-[p-(m-chlorostyryl)-phenyl]-	DMF	356(4.77)	33-1282-69
Benzofuran, 2-[p-(p-chlorostyryl)-phenyl]-	DMF	358(4.82)	33-1282-69
Benzofuran, 5-chloro-2-(p-styrylphenyl)-	DMF	357(4.81)	33-1282-69

Compound	Solvent	$\lambda_{max}(\log \epsilon)$	Ref.
$C_{22}H_{15}ClS$			
Benzo[b]thiophene, 2-[p-(m-chlorostyryl)phenyl]-	DMF	357(4.76)	33-1282-69
Benzo[b]thiophene, 2-[p-(p-chlorostyryl)phenyl]-	DMF	358(4.77)	33-1282-69
Benzo[b]thiophene, 5-chloro-2-(p-styrylphenyl)-	DMF	356(4.78)	33-1282-69
Benzo[b]thiophene, 6-(p-chlorostyryl)-2-phenyl-	DMF	358(4.74)	33-1282-69
$C_{22}H_{15}Cl_5O_5$			
Benzophenone, 2-hydroxy-4-[[2-(pentachlorophenoxy)ethoxy]methoxy]-	EtOH	287(4.24),325(3.99)	39-0034-69C
Benzophenone, 2-hydroxy-4-[[2-(pentachlorophenoxy)methoxy]ethoxy]-	$CHCl_3$	290(4.18),325(3.93)	39-0034-69C
$C_{22}H_{15}FO$			
Benzofuran, 2-[p-(p-fluorostyryl)phenyl]-	DMF	354(4.77)	33-1282-69
$C_{22}H_{15}FS$			
Benzo[b]thiophene, 2-[p-(p-fluorostyryl)phenyl]-	DMF	354(4.78)	33-1282-69
$C_{22}H_{15}NO_2$			
3-Indolinone, 2-(α-phenylphenacylidene)-	EtOH	482(3.82)	44-0887-69
$C_{22}H_{15}NO_2S$			
9-Phenanthrenecarbonitrile, 10-(p-tolylsulfonyl)-	n.s.g.	208(4.42),234(4.46), 261(4.59),330(4.09), 358(3.42),375(3.38)	88-3917-69
$C_{22}H_{15}N_3$			
11H-Dibenzo[a,c]pyrrolo[2,3-i]phenazine, 12,13-dihydro-	DMF	270(4.57),320(4.59), 412(3.98),465(4.21)	103-0199-69
6H-Pyrrolo[2,3-g]quinoxaline, 2,3-diphenyl-	MeOH	245(4.39),295(4.64), 376(4.22)	103-0199-69
$C_{22}H_{15}N_3O_2S$			
4H-1-Benzothiopyran, 4-[[(p-nitrophenyl)azo]methylene]-2-phenyl-	EtOH	400(3.42),525(3.05)	2-0311-69
$C_{22}H_{15}N_3O_3$			
4(3H)-Quinazolinone, 2-(1,2-diphenylvinyl)-6-nitro-	EtOH	260(4.30),305s(4.20), 375(4.17)	18-3198-69
4(3H)-Quinazolinone, 2-(1,2-diphenylvinyl)-7-nitro-	EtOH	216(4.59),256(4.16), 323(4.23)	18-3198-69
$C_{22}H_{15}N_5O_{13}S_3$			
2,7-Naphthalenedisulfonic acid, 4,5-dihydroxy-3-[(p-nitrophenyl)azo]-6-[(p-sulfophenyl)azo]-	pH 1.8 and 6.8	540(4.55),630(4.26)	70-0247-69
	pH 11.8	570(4.45)	70-0247-69
$C_{22}H_{16}$			
Anthracene, 2-styryl-	DMF	310(4.91),326(4.95), 355(4.15),373(4.28), 393(4.20)	33-2521-69
Benzocyclobutadiene-anthracene adduct	EtOH	262s(3.40),268(3.57), 274(3.62)	78-1661-69

Compound	Solvent	$\lambda_{max}(\log \epsilon)$	Ref.
Ethylene, 1,2-di-1-naphthyl-	DMF	346(4.35)	33-2521-69
Ethylene, 1,2-di-2-naphthyl-, trans	DMF	277(4.64),288(4.52), 333(4.72)	33-2521-69
Ethylene, 1-(1-naphthyl)-2-(2-naphthyl)-	DMF	340(4.46)	33-2521-69
Indene, 1-benzylidene-2-phenyl-, cis	ether	274(4.49),345(4.16)	44-0879-69
trans	ether	265(4.39),311(3.97)	44-0879- 9
Phenanthrene, 9-styryl-	C_6H_{12}	255(4.81),272s(4.19), 286(4.03),298(4.10), 324(2.57),332(3.56), 340(2.48),348(2.43)	44-1923-69
Stilbene, 4-(phenylethynyl)-	DMF	343(4.77)	33-2521-69
$(C_{22}H_{16})_n$			
Diphenylbenzofulvene, dimer A	ether	218(4.54),286(4.49)	44-0879-69
dimer B	ether	272(4.36)	44-0879-69
dimer C	ether	288(4.53),345(4.81), 368(3.77),386(3.57)	44-0879-69
$C_{22}H_{16}Br_2N_2$			
Quinoxaline, 2,3-bis(α-bromobenzyl)-	EtOH	247(4.46),329(3.81)	44-1651-69
$C_{22}H_{16}ClNOS$			
Ketone, p-chlorophenyl (β-thiobenzoyl-aminostyryl)	MeCN	567(2.48)	24-0269-69
Thione, p-chlorophenyl (β-benzoylamino-styryl)	MeCN	520(2.31)	24-0269-69
Thione, (p-chloro-β-benzoylaminostyryl) phenyl	MeCN	520(2.31)	24-0269-69
$C_{22}H_{16}ClNO_4S$			
2,4,6-Triphenyl-1,3-thiazin-1-ium perchlorate	MeCN	418(4.41)	24-0269-69
$C_{22}H_{16}Cl_2$			
Benzene, 1-(2,4-dichlorostyryl)-4-styryl-	DMF	361(4.81)	33-2521-69
$C_{22}H_{16}N_2$			
1-Naphthaldehyde, azine	EtOH	209(4.8),222(4.8), 250s(4.4),355(4.5)	28-0248-69B
2-Naphthaldehyde, azine	EtOH	218(4.6),253s(4.4), 264(4.5),273(4.6), 284(4.5),326(4.7), 360s(4.4)	28-0248-69B
Pyridazine, 3,4,6-triphenyl-	EtOH	272(4.45)	35-0676-69
$C_{22}H_{16}N_2O$			
4-Isoxazoline-4-carbonitrile, 2,3,5-triphenyl-	EtOH	280(4.47),382(4.38)	88-0823-69
Quinoxaline, 3-benzoyl-2-benzyl-	EtOH	230(4.23),259(4.34), 326(4.05)	44-1651-69
$C_{22}H_{16}N_2O_2$			
Cinnoline, 3,4-dibenzoyl-1,4-dihydro-	n.s.g.	255(4.15),350s(3.72), 390(3.77)	39-1795-69C
Quinoxaline, 3-benzoyl-2-benzyl-, 1-oxide	EtOH	243(4.51),248(4.41), 276(4.16),326(3.86)	44-1651-69

Compound	Solvent	λ_{max}(log ϵ)	Ref.
$C_{22}H_{16}N_2O_2S$			
Thieno[3,4-b]quinoxaline, 1,3-dihydro-1,3-diphenyl-, 2,2-dioxide	EtOH	222(4.57),240(4.56), 297(4.03),327(3.92)	44-1651-69
$C_{22}H_{16}N_2O_3$			
Carbazole-1,2-dicarboximide, N-(2-hydroxyethyl)-3-phenyl-	EtOH	222(4.72),242s(4.43), 280(4.66),360(3.98), 420(3.71)	95-0058-69
$C_{22}H_{16}N_2S$			
4H-1-Benzothiopyran, 2-phenyl-4-[(phenylazo)methylene]-	EtOH	315(3.89),472(4.35)	2-0311-69
$C_{22}H_{16}N_4O_2$			
Phenol, 4,4'-[1,7-naphthylenebis(azo)]-di-	EtOH	357(4.58),400s(--)	124-0179-69
	EtOH-base	417(4.61),490s(4.49)	124-0179-69
$C_{22}H_{16}N_4O_8$			
Toluene-α,α-diol, α-[(p-nitrophenyl)-azo]-, acetate, p-nitrobenzoate	dioxan	270(4.40),405(2.66)	39-1571-69C
$C_{22}H_{16}N_4O_8S_2$			
2,7-Naphthalenedisulfonic acid, 4,5-dihydroxy-3,6-bis(phenylazo)-	pH 1.5	540(4.53),630(4.26)	70-0247-69
	pH 7.0	540(4.53),630(4.26)	70-0247-69
	pH 11.5	535(4.36)	70-0247-69
	30% NaOH	610(4.62)	70-0247-69
$C_{22}H_{16}O$			
Benzofuran, 2-phenyl-6-styryl-	DMF	351(4.73)	33-1282-69
Benzofuran, 2-(p-styrylphenyl)-	DMF	355(4.80)	33-1282-69
Ketone, phenyl 2-phenylinden-3-yl	ether	249(4.28),291(4.26)	44-0879-69
2-Naphthol, 1,3-diphenyl-	EtOH	255(4.61),310(3.98)	24-3405-69
$C_{22}H_{16}O_2$			
Cyclobut[a]inden-7-one, 2a,7a-dihydro-, dimer	EtOH	234(4.62),259(4.14), 305(3.62)	39-2656-69C
Spiro[2H-1-benzopyran-2,3'-[3H]naphtho-[2,1-b]pyran], 2'-methyl-	dioxan	246(4.89),300(4.08), 313(4.09),335(3.82), 349(3.90)	5-0162-69B
Spiro[2H-1-benzopyran-2,3'-[3H]naphtho-[2,1-b]pyran], 3-methyl-	dioxan	285(4.86),301(4.02), 313(4.02),335(3.73), 349(3.81)	5-0162-69B
Spiro[2H-1-benzopyran-2,3'-[3H]naphtho-[2,1-b]pyran], 6-methyl-	dioxan	243(4.86),300(4.00), 312(4.06),334(3.72), 349(3.76)	5-0162-69B
$C_{22}H_{16}O_3$			
1,4-Butanedione, 2,3-epoxy-1,2,4-triphenyl-, cis	EtOH	256(4.42)	77-0199-69
$C_{22}H_{16}O_4$			
Benz[a]anthracene-5,7-diol, diacetate	EtOH	256(4.60),272(4.62), 281(4.87),292(4.91), 318(3.72),334(3.89), 348(3.93),362(3.78), 385(2.84)	44-0874-69
Benz[a]anthracene-5,12-diol, diacetate	EtOH	260(4.64),268(4.70), 278(4.88),288(4.87), 302(3.97),336(3.89), 350(3.95),362(3.83), 386(3.20)	44-0874-69

Compound	Solvent	λmax(log ε)	Ref.
Dibenzo[a,e]cyclodecene-5,8-diol, 6,7,13,14-tetradehydro-5,8-dihydro-, diacetate	EtOH	221(4.45),284(4.22), 307(4.18)	44-0874-69
Glycidic acid, 3-benzoyl-2,3-diphenyl-, cis	C_6H_{12}	285s(2.89),324(2.11), 337(2.11)	28-1157-69A
	EtOH	252(4.15),325(2.58)	28-1157-69A
$C_{22}H_{16}O_5$			
Anthra[2,3-b]benzofuran-1,7,12(2H)-trione, 3,4-dihydro-6-hydroxy-3,3-dimethyl-	MeOH	495(3.74)	64-0137-69
1,3-Propanedione, 2-hydroxy-1-(o-hydroxyphenyl)-3-phenyl-, 2-benzoate	EtOH	248s(4.14),283(3.18), 316(3.50)	18-3345-69
$C_{22}H_{16}O_7$			
[1,1'-Binaphthalene]-3,4,8'(5'H)-trione, 6',7'-epoxy-6',7'-dihydro-5'-hydroxy-4',5-dimethoxy-	EtOH-4% dioxan	234(4.58),330(3.70), 419(3.85)	39-2059-69C
$C_{22}H_{16}O_8$			
6H,12H-Cyclobuta[1,2-c:3,4-c']bis[1]-benzopyran-6,12-dione, 6a,6b,12a,12b-tetrahydro-6b,12b-diacetoxy-	MeOH	211(4.58),273(3.65), 279(3.63)	24-3122-69
$C_{22}H_{16}S$			
Benzo[b]cyclobuta[d]thiophene, 2a,7b-dihydro-1,2-diphenyl-	EtOH	253(4.26),260s(4.21), 298(4.18)	88-4791-69
Benzo[b]cyclobuta[d]thiophene, 2a,7b-dihydro-2,2a-diphenyl-	EtOH	250(4.34),305(3.30)	88-4791-69
Benzo[b]thiophene, 2-phenyl-5-styryl-	DMF	314(4.79)	33-1282-69
Benzo[b]thiophene, 2-phenyl-6-styryl-	DMF	354(4.74)	33-1282-69
Benzo[b]thiophene, 2-(p-styrylphenyl)-	DMF	355(4.77)	33-1282-69
$C_{22}H_{16}S_2$			
Thiophene, 2,2'-(2,6-naphthylenedivinylene)di-	DMF	300(4.37),378(4.86), 400(4.80)	33-2521-69
$C_{22}H_{17}BrO$			
Indene, 2-bromo-1-methoxy-1,3-diphenyl-	MeOH	234(4.50),293(3.75), 309s(3.60)	44-1124-69
$C_{22}H_{17}Br_3O_5$			
Benzophenone, 2-hydroxy-4-[2-(2,4,6-tribromophenoxy)methoxy]ethoxy]-	$CHCl_3$	290(4.19),325(3.93)	39-0034-69C
$C_{22}H_{17}Cl$			
Benzene, 1-(p-chlorostyryl)-4-styryl-	DMF	360(4.81)	33-2521-69
$C_{22}H_{17}ClIN$			
8-(p-Chlorophenyl)-5,6-dihydro-4H-pyrido[3,2,1-de]phenanthridinium iodide	EtOH	250(4.60),270s(4.35), 335(4.09),370s(3.87)	2-0848-69
$C_{22}H_{17}ClO_6$			
3-(2-Hydroxy-5-methylstyryl)naphtho-[2,1-b]pyrylium perchlorate	MeOH-$HClO_4$	233(4.37),307(3.81), 338(3.86),535(4.40)	5-0162-69B
$C_{22}H_{17}Cl_2N_3O_4$			
2-Formyl-1-phenylquinolinium perchlorate, (p-chlorophenyl)hydrazone	EtOH	271(3.35),317(3.0), 473(3.40)	124-0370-69

Compound	Solvent	$\lambda_{max}(\log \epsilon)$	Ref.
$C_{22}H_{17}Cl_3O_5$			
Benzophenone, 2-hydroxy-4-[2-[(2,4,5-trichlorophenoxy)methoxy]ethoxy]-, α-form	$CHCl_3$	290(4.24),329(4.00)	39-0034-69C
β-form	$CHCl_3$	288(4.24),327(4.09)	39-0034-69C
Benzophenone, 2-hydroxy-4-[2-[(2,4,6-trichlorophenoxy)methoxy]ethoxy]-	$CHCl_3$	290(4.27),330(3.99)	39-0034-69C
$C_{22}H_{17}IN_2O_2$			
5,6-Dihydro-8-(p-nitrophenyl)-4H-pyrido-[3,2,1-de]phenanthridinium iodide	EtOH	214(4.59),252(4.55), 272s(4.47),340(3.92), 380s(3.65)	2-0848-69
$C_{22}H_{17}N$			
12H-Benzo[f]benzo[6,7]cyclohepta[1,2-b]-quinoline, 13,14-dihydro-	dioxan	259(4.64),282(4.62), 293(4.49),324(3.61), 339(3.83),355(3.91)	24-1202-69
1-Naphthylamine, N-[1-(2-naphthyl)ethyl-idene]-	EtOH	232(4.67),248s(4.60), 296(4.26),336s(3.65)	56-0749-69
2-Naphthylamine, N-[1-(2-naphthyl)ethyl-idene]-	EtOH	252(4.68),282(4.20), 325(3.68)	56-0749-69
$C_{22}H_{17}NOS$			
Benzamide, N-(2-benzoyl-1-phenylvinyl)-thio-	MeCN	567(2.43)	24-0269-69
$C_{22}H_{17}NO_2$			
Benzyl alcohol, α-[(benzylideneamino)-methylene]-, benzoate	EtOH	230(4.39),338(4.46)	77-1215-69
9H-Dibenzo[a,c]carbazole, 3,6-dimethoxy-	n.s.g.	208(4.53),234s(4.43), 268(3.79),307(4.38), 324s(4.26),376(3.61), 396(3.66)	117-0047-69
$C_{22}H_{17}NO_2S$			
Acrylonitrile, 2,3-diphenyl-3-(p-tolyl-sulfonyl)-, (Z)-	n.s.g.	287(4.01)	88-3917-69
$C_{22}H_{17}NO_4S$			
6,8a-Epithio-8aH-carbazole-7,8-dicarb-oxylic acid, 6,9-dihydro-6-phenyl-, dimethyl ester	EtOH	232(4.56),247(4.48), 316(4.34),360(4.06)	95-0058-69
$C_{22}H_{17}N_3$			
6H-Pyrrolo[2,3-g]quinoxaline, 7,8-dihydro-2,3-diphenyl-	MeOH	255(4.42),295(4.36), 425(4.17)	103-0199-69
as-Triazine, 5,6-diphenyl-3-p-tolyl-	MeOH	220s(--),287(4.51), 390s(--)	88-3147-69
s-Triazolo[3,4-a]isoquinoline, 2,3-dihydro-2,3-diphenyl-	MeCN	275s(--),292(4.38), 398(3.99)	24-3176-69
s-Triazolo[4,3-a]quinoline, 1,2-dihydro-1,2-diphenyl-	MeCN	214(4.50),258(5.12), 295(4.47),447(3.81)	24-3176-69
$C_{22}H_{17}N_3O$			
Malononitrile, [1-(dimethylamino)-2-(2-phenyl-4H-1-benzopyran-4-ylidene]-ethylidene]-	MeCN	290(3.95),310(3.97), 358(3.79),523(4.84)	4-0803-69

Compound	Solvent	$\lambda_{max}(\log \epsilon)$	Ref.
$C_{22}H_{17}N_3O_4$			
3-Pyrroline-3,4-dicarboxylic acid, 2,2-dicyano-1,5-diphenyl-, dimethyl ester	MeCN	230(4.20),265s(3.45), 272s(3.40)	44-2146-69
$C_{22}H_{17}N_3O_6$			
1,1-Ethanediol, 1-[(p-nitrophenyl)azo]-, dibenzoate	dioxan	277(4.30),281s(4.29), 410(2.58)	39-1571-69C
Toluene-α,α-diol, α-[(p-nitrophenyl)-azo]-, acetate benzoate	dioxan	279s(4.31),283(4.31), 410(2.68)	39-1571-69C
$C_{22}H_{18}$			
Benzene, o-distyryl-	DMF	287(4.60),323(4.48)	33-2521-69
Benzene, p-distyryl-	DMF	357(4.79)	33-2521-69
9,10-Ethanoanthracene, 9,10-dihydro-11-phenyl-	EtOH	253(2.94),259(3.04), 265(3.19),272(3.24)	78-1661-69
Indene, 3-benzyl-2-phenyl-	EtOH	293(4.34)	44-0879-69
Naphthalene, 1,2-dihydro-2,6-diphenyl-	MeOH	248(4.75)	88-4509-69
Naphthalene, 1,2-dihydro-3,7-diphenyl-	MeOH	260(4.49),317(4.53)	88-4509-69
Phenanthrene, 9-(α-methylbenzyl)-	C_6H_{12}	248s(4.69),254(4.80), 270(4.32),277(4.15), 285(4.03),297(4.08), 317(2.36),324(2.38), 331(2.45),339(2.38), 347(2.36)	44-1923-69
Pyrene, 2-(1-cyclohexen-1-yl)-	EtOH	233(4.26),268(4.94), 308(4.18),321(4.53), 337(4.81)	24-2301-69
$C_{22}H_{18}ClNO_5$			
2-(Diphenylmethyl)-4-phenylisoxazolium perchlorate	CH_2Cl_2	230(4.17)	44-3451-69
$C_{22}H_{18}Cl_2N_2S_2Se$			
Benzothiazolium, 2,2'-(2,5-selenophene-diylidenedimethylidyne)bis[3-methyl-, perchlorate	H_2O	342(4.47),496(4.65), 530(4.65)	124-0824-69
$C_{22}H_{18}F_2N_2$			
1,2-Naphthalenedicarbonitrile, 7-fluoro-4-(α,α-dimethyl-p-fluorobenzyl)-3,4-dihydro-4-methyl-	MeCN	244(4.13),295(3.89), 335(3.58)	44-1923-69
$C_{22}H_{18}FeN_2O_6$			
Iron, tricarbonyl(6,7,8,14-tetradehydro-4,5α-epoxy-3,6-dimethoxymorphinan-17-carbonitrile)-	n.s.g.	252(4.11),287(3.70)	12-0971-69
$C_{22}H_{18}Fe_2O_2$			
[1,1"-Biferrocene]-2,2"-dicarboxalde-hyde, (1R,1"R)-(-)-	EtOH	456(3.01)	49-1515-69
$C_{22}H_{18}IN$			
5,6-Dihydro-8-phenyl-4H-pyrido[3,2,1-de]phenanthridinium iodide	EtOH	250(4.57),271s(4.30), 336(4.05),370s(3.78)	2-0848-69
$C_{22}H_{18}N_2$			
Acridine, 9-(1-methyl-5-indolinyl)-	n.s.g.	355(3.88),400(3.80), 473(--)	103-0420-69

Compound	Solvent	$\lambda_{max}(\log \epsilon)$	Ref.
Naphthalene, 1,7-dianilino-	dioxan	257(4.53),305(4.66), 338(4.20),360s(3.63), 375s(3.20)	24-2728-69
Quinoxaline, 2,3-dibenzyl-	EtOH	239(4.57),294s(3.98), 320(4.23)	44-1651-69
$C_{22}H_{18}N_2O$			
Imidazole, 2-(p-methoxyphenyl)-4,5-diphenyl-	MeOH	300(4.43)	44-3981-69
Imidazole, 5-(p-methoxyphenyl)-2,4-diphenyl-	MeOH	299(4.38)	44-3981-69
2-Pyrazoline, 4-(phenoxymethylene)-1,3-diphenyl-	n.s.g.	230(4.22),263(4.13)	28-1063-69B
2-Quinoxalinemethanol, 3-benzyl-α-phenyl-	EtOH	240(4.47),311s(3.89), 321(4.00)	44-1651-69
$C_{22}H_{18}N_2O_2$			
Naphtho[2,3-h]quinoline-7,12-dione, 5-piperidino-	EtOH	300(4.36),480(3.73)	104-0742-69
$C_{22}H_{18}N_2O_3$			
Acetamide, N-(7-methyl-1-p-tolyl-2,3-naphthalenedicarboximido)-	BuOH	265(4.63)	39-0838-69C
1H-Benz[g]indazol-5-ol, 1-acetyl-3-methyl-4-phenyl-, acetate	EtOH	240(4.40),261(4.55), 329(3.60),344(3.64)	103-0254-69
$C_{22}H_{18}N_2O_4$			
1,1-Ethanediol, 1-(phenylazo)-, dibenzoate	dioxan	275(4.13),282s(4.09), 403(2.35)	39-1571-69C
Toluene-α,α-diol, α-(phenylazo)-, acetate benzoate	dioxan	276(4.13),281s(4.12), 401(2.50)	39-1571-69C
$C_{22}H_{18}N_2O_5$			
Acetamide, N-[7-methoxy-1-(p-methoxyphenyl)-2,3-naphthalenedicarboximido]-	EtOH	270(4.64),318s(3.92)	39-0838-69C
$C_{22}H_{18}N_2O_6$			
Camptothecine, 10-hydroxy-, acetate	MeOH	218(4.39),253(4.24), 290(3.65),365(4.13)	44-1364-69
$C_{22}H_{18}N_2S_2Se$			
Benzothiazoline, 2,2'-(2,5-selenophenediyldimethylidyne)bis[3-methyl-	benzene	327(4.3),440(4.76), 466(4.75)	124-0824-69
$C_{22}H_{18}N_2S_3$			
2,4-Thiazolidinedithione, 3-ethyl-5-[2-(1-phenyl-2(1H)-quinolylidene)-ethylidene]-	BuOH	253(4.35),296(4.04), 341(3.93)	65-2116-69
2,4-Thiazolidinedithione, 3-ethyl-5-[2-(1-phenyl-4(1H)-quinolylidene)-ethylidene]-	BuOH	244(4.10),255(4.07)	65-2116-69
$C_{22}H_{18}N_4$			
Imidazole, 4,5-diphenyl-2-(p-tolylazo)-	n.s.g.	290(4.17),415(4.40)	104-0763-69
$C_{22}H_{18}N_4O_2S$			
Imidazole, 5-amino-4-(benzylthio)-2-(p-nitrophenyl)-1-phenyl-	EtOH	420(4.67)	44-3661-69

Compound	Solvent	$\lambda_{max}(\log \epsilon)$	Ref.
$C_{22}H_{18}O$			
4H-Cyclopenta[c]furan, 4-(diphenylmethylene)-1,3-dimethyl-	EtOH	258(4.16),375(4.04)	88-1299-69
2-Indanone, 1-methyl-1,3-diphenyl-	CH_2Cl_2	261(3.11),268(3.15), 276(3.10)	24-1597-69
$C_{22}H_{18}OS$			
Acrylophenone, 2-(methylthio)-3,3-diphenyl-	EtOH	251(4.38)	78-3675-69
$C_{22}H_{18}O_2$			
Fluorene-1-acetic acid, 9-phenyl-, methyl ester	EtOH	236s(--),259s(--), 269(4.23),275s(--), 282(4.00),293(3.71), 304(3.86),365s(--)	44-2401-69
2,5-Norbornadiene-2-carboxylic acid, 7-(diphenylmethylene)-, methyl ester	EtOH	218s(4.31),225s(4.27), 250(4.29),305s(3.20)	89-0209-69
$C_{22}H_{18}O_3$			
Chalcone, 4'-(benzyloxy)-2'-hydroxy-	EtOH	256(3.79),324(4.37), 345(4.35)	95-0375-69
	EtOH-HCl	256(3.79),324(4.37), 345(4.35)	95-0375-69
	EtOH-NaOH	250(4.11),312(4.33), 418(3.95)	95-0375-69
$C_{22}H_{18}O_4$			
Isophthalic acid, 4,5-diphenyl-, dimethyl ester	EtOH	239(4.31),300(3.47)	44-4134-69
$C_{22}H_{18}O_5$			
1,3-Cyclopentadiene-1,2-dicarboxylic acid, 5-(hydroxymethylene)-3,4-diphenyl-, dimethyl ester	EtOH	287(4.38),335(4.16), 375s(3.74)	33-0396-69
7-Oxabicyclo[2.2.1]hepta-2,5-diene-2,3-dicarboxylic acid, 5,6-diphenyl-, dimethyl ester	EtOH	228(4.32),240s(4.19), 262s(4.08),348(3.28)	33-0396-69
7-Oxabicyclo[4.1.0]hepta-2,4-diene-3,4-dicarboxylic acid, 2,5-diphenyl-, dimethyl ester	EtOH	221s(4.13),235s(4.08), 328(4.23)	33-0396-69
3-Oxatetracyclo[3.2.0.0^{2,7}.0^{4,6}]heptane-1,5-dicarboxylic acid, 6,7-diphenyl-, dimethyl ester	EtOH	220(4.23)	33-0396-69
[p-Ter phenyl]-2',3'-dicarboxylic acid, 5'-hydroxy-, dimethyl ester	EtOH	239(4.41),263s(4.00), 298(3.57)	33-0396-69
$C_{22}H_{18}O_6$			
Robustone, methyl ether	EtOH	230(4.41),271(4.61)	39-0365-69C
$C_{22}H_{18}O_7$			
Justicidin A	EtOH	259(4.37),296(3.72), 309(3.72),345(3.36)	102-1499-69
Robustin	EtOH	230(4.38),266(4.45), 348(4.23)	39-0365-69C
$C_{22}H_{18}O_8$			
Furan, 3-piperonyloyl-4-(3,4,5-trimethoxybenzoyl)-	EtOH	282(4.20),310(4.23)	36-0176-69

Compound	Solvent	λ_{max}(log ε)	Ref.
$C_{22}H_{18}O_9$			
Anthraquinone, 1,3,8-trihydroxy-6-methoxy-2-methyl-, triacetate	EtOH	213(4.51),274(4.59), 337(3.75),373s(3.58)	39-2763-69C
3(2H)-Benzofuranone, 4,6,7-trihydroxy-2-(p-methoxybenzylidene)-, triacetate	MeOH	255(4.2),340(4.0), 405(4.0)	2-0543-69
Flavone, 3,4',5-triacetoxy-7-methoxy-	EtOH	255(4.38),303(4.44)	105-0397-69
Flavone, 3,5,7-triacetoxy-4'-methoxy-	EtOH	253(4.39),305(4.38)	105-0397-69
$C_{22}H_{19}BrO_2$			
Ethylene, 1-bromo-1,2-bis(p-methoxyphenyl)-2-phenyl-, cis	C_6H_{12}	248(4.42),308(4.13)	35-6734-69
trans	C_6H_{12}	241(4.40),308(4.11)	35-6734-69
Ethylene, 1-bromo-2,2-bis(p-methoxyphenyl)-1-phenyl-	C_6H_{12}	242(4.37),310(4.04)	35-6734-69
$C_{22}H_{19}BrO_7$			
Elephantol, p-bromobenzoate	MeOH	246(4.40)	44-3867-69
$C_{22}H_{19}ClN_2$			
Isoindole, 1-(p-chlorophenyl)-2-[o-(methylamino)benzyl]-	EtOH	245s(4.29),276(3.81), 287(3.81),351(3.83), 367s(3.07)	44-1720-69
$C_{22}H_{19}ClO_4$			
3,5-Cyclohexadiene-1,2-dicarboxylic acid, 3-(p-chlorophenyl)-6-phenyl-, dimethyl ester, trans	EtOH	233s(3.95),338(4.39)	22-3662-69
2,4,6-Octatrienedioic acid, 3-(p-chlorophenyl)-6-phenyl-, cis-trans-cis	THF	319(4.45)	22-3665-69
$C_{22}H_{19}ClO_4S_2$			
3,4-Dimethyl-2-[(3-methyl-4H-1-benzothiopyran-4-ylidene)methyl]-1-benzothiopyrylium perchlorate	CH_2Cl_2	292(4.46),332(3.97), 390(3.61),540s(3.48), 578(3.74),645(3.38), 738s(3.15)	2-0017-69
4,6-Dimethyl-2-[(6-methyl-4H-1-benzothiopyran-4-ylidene)methyl]-1-benzothiopyrylium perchlorate	CH_2Cl_2	249(4.45),253(4.47), 262(4.45),297s(3.94), 328(3.87),395(3.82), 585s(4.41),628(4.73), 742s(3.23)	2-0017-69
4,8-Dimethyl-2-[(8-methyl-4H-1-benzothiopyran-4-ylidene)methyl]-1-benzothiopyrylium perchlorate	CH_2Cl_2	235s(4.42),260s(4.29), 286(4.18),320(3.92), 414(3.77),575s(4.46), 613(4.92),726(3.23), 760s(3.20)	2-0017-69
$C_{22}H_{19}ClO_6S_2$			
6-Methoxy-2-[(6-methoxy-4H-1-benzothiopyran-4-ylidene)methyl]-4-methyl-1-benzothiopyrylium perchlorate	CH_2Cl_2	285(4.15),335(3.67), 360(3.63),420(3.49), 602s(4.09),636(4.30), 740s(3.16)	2-0017-69
8-Methoxy-2-[(8-methoxy-4H-1-benzothiopyran-4-ylidene)methyl]-4-methyl-1-benzothiopyrylium perchlorate	CH_2Cl_2	292(4.46),332(3.97), 381(3.61),455(3.78), 589s(4.42),623(4.85), 740s(3.18)	2-0017-69
$C_{22}H_{19}ClO_7$			
1,2,3,4-Tetrahydro-4-salicylidenexanthylium perchlorate, acetate	MeOH-$HClO_4$	241(4.25),249(4.27), 258(4.22),352(3.76), 368(3.76),450(4.01)	5-0226-69E

Compound	Solvent	λ max (log ε)	Ref.
$C_{22}H_{19}ClO_9$			
Flavone, 3'-chloro-2',5-dihydroxy-3,7,8-trimethoxy-, diacetate	EtOH	256(4.40),305(4.02), 318s(4.00)	39-2418-69C
$C_{22}H_{19}Cl_2NO_5$			
2-Chloro-3-(2,6-diphenyl-4H-pyran-4-ylidene)-1-methyl-1-pyrrolinium perchlorate	MeCN	247(4.18),273(4.21), 298(4.20),385(4.42), 525(3.60)	4-0803-69
$C_{22}H_{19}Fe_2N$			
[1,1"-Biferrocene]-2-carbonitrile, 2'-methyl-, (1R,1"S)-(-)-	EtOH	447(2.86)	49-1515-69
$C_{22}H_{19}N$			
5H-Benzo[f]benzo[6,7]cyclohepta[1,2-b]-quinoline, 6,12,13,14-tetrahydro-	dioxan	232(4.19),287(4.19), 320(4.42)	24-1202-69
$C_{22}H_{19}NOS_2$			
Imidocarbonic acid, benzoyldithio-, dibenzyl ester	EtOH	247(4.17),286(4.06)	39-2794-69C
$C_{22}H_{19}NO_2$			
2-Pyrrolidinone, 3-(2,6-diphenyl-4H-pyran-4-ylidene)-1-methyl-	MeCN	270(4.36),356(4.42)	4-0803-69
$C_{22}H_{19}NO_3$			
2H-Pyran-2,4(3H)-dione, 6-phenyl-3-[α-(1-pyrrolidinyl)benzylidene]-	MeOH	275(3.93),358(3.89)	44-2527-69
$C_{22}H_{19}NO_4$			
9aH-Pyrido[1,2-f]phenanthridine-7,9-dicarboxylic acid, 9a-methyl-, dimethyl ester	MeOH	211(4.52),248s(4.29), 253(4.30),261s(4.28), 300(4.29),404(4.25)	39-2311-69C
$C_{22}H_{19}NO_5$			
Acetophenone, 2',4'-bis(benzyloxy)-2-nitro-	EtOH	277(4.15),305(3.92)	104-1803-69
$C_{22}H_{19}N_3O$			
Benzamide, N-(α-2-benzimidazolylphenethyl)-	EtOH	275(4.11),282(4.15)	94-2381-69
Δ^3-1,3,4-Oxadiazoline, 5-(phenylimino)-2,2-di-p-tolyl-	hexane	224(4.30),325(3.82), 360s(--)	44-3230-69
$C_{22}H_{19}N_3O_3$			
Δ^3-1,3,4-Oxadiazoline, 2,2-bis(p-methoxyphenyl)-5-(phenylimino)-	hexane	230(4.38),324(3.81), 357s(--)	44-3230-69
$C_{22}H_{19}N_5O$			
[1,2'(1'H)-Biphthalazine]-4'-carbonitrile, 1'-(1-methyl-2-oxobutyl)-	EtOH	250s(4.00),378(4.28)	95-0959-69
$C_{22}H_{19}N_5O_2$			
Isoalloxazine, 7,8-dimethyl-10-(2-indol-3-ylethyl)-	propylene glycol	271(4.52),360(3.88), 445(4.05)	44-3240-69
$C_{22}H_{19}N_5O_8$			
Pyridine, 2,6-bis(2,4-dinitrobenzyl)-4-isopropyl-	EtOH	242(4.43)	22-4425-69

Compound	Solvent	$\lambda_{max}(\log \epsilon)$	Ref.
Pyridine, 2,6-bis(2,4-dinitrobenzyl)-4-propyl-	EtOH	242(4.42)	22-4425-69
$C_{22}H_{20}$			
1,3,5,7,9,11,13,15-Cyclohexadecaoctaene, 1-phenyl-	hexane	295(4.36),440(3.08)	88-4575-69
Propene, 2-benzyl-1,3-diphenyl-	EtOH	246(3.89),266s(3.54), 270s(3.39)	1-2409-69
Pyrene, 2-cyclohexyl-	EtOH	236(4.85),246(5.14), 253(4.38),264(4.56), 276(4.76),296(3.80), 307(4.20),320(4.62), 337(4.85)	24-2301-69
o-Styryldiphenylethane	EtOH	290(4.27)	44-0879-69
$C_{22}H_{20}Br_2O_2$			
2,2'-Spirobis[2H-1-benzopyran], 6,6'-dibromo-3-pentyl-	dioxan	235(4.68),260(4.29), 311(3.62)	5-0162-69B
$C_{22}H_{20}ClNO_4$			
6,11-Dihydro-6-methyl-12-phenyl-6,11-ethanobenzo[b]quinolizinium perchlorate	n.s.g.	265s(3.74),272(3.76)	44-1700-69
6,11-Dihydro-11-methyl-12-phenyl-6,11-ethanobenzo[b]quinolizinium perchlorate	n.s.g.	266s(3.71),270(3.73)	44-1700-69
$C_{22}H_{20}ClNO_6$			
Acetyl[2-(4,6-diphenyl-2H-pyran-2-ylidene)ethylidene]methylammonium perchlorate	MeCN	255(4.41),280(4.37), 310s(4.08)	44-2736-69
$C_{22}H_{20}F_6$			
Naphthalene, 1-(α,α-dimethylbenzyl)-1,2-dihydro-3,4-bis(trifluoromethyl)-1-methyl-	MeCN	223s(4.24),276(3.83)	44-1923-69
$C_{22}H_{20}FeN_2O_6$			
Iron, tricarbonyl(6,7,8,14-tetradehydro-4-hydroxy-3,6-dimethoxymorphinan-17-carbonitrile)-	$CHCl_3$	255(4.01),277(3.81)	12-0971-69
$C_{22}H_{20}FeN_2O_7$			
Iron, tricarbonyl(N-cyanonorthebaine)-	$CHCl_3$	252(4.07),287(3.70)	12-0971-69
$C_{22}H_{20}Fe_2O$			
1,1"-Biferrocene, 2,2"-(oxydimethylene)-	C_6H_{12}	206(4.54),227(4.67), 270s(3.98),300(3.95)	78-3477-69
[1,1"-Biferrocene]-2-carboxaldehyde, 2"-methyl-, (1R,1"S)-(+)-	EtOH	461(2.98)	49-1515-69
Biferrocenyl, 5,5'-(β-oxapropylene)-, (1R,1'R)-(-)-	EtOH	448(2.80)	49-1515-69
$C_{22}H_{20}Fe_2O_2$			
[1,1"-Biferrocene]-2-carboxylic acid, 2"-methyl-, (1S,1"R)-(-)-	EtOH	448(2.77)	49-1515-69
$C_{22}H_{20}Fe_2S$			
1,1"-Biferrocene, 2,2'-(thiodimethylene)-	C_6H_{12}	206(4.57),226(4.64), 266s(3.97),301(3.86)	78-3477-69

Compound	Solvent	λmax(log ε)	Ref.
$C_{22}H_{20}N_2$			
4,4'-Bipyridine, tetrahydro-1,1'-diphenyl-	EtOH	291(4.59)	44-3672-69
1,2-Naphthalenedicarbonitrile, 4-(α,α-dimethylbenzyl)-3,4-dihydro-4-methyl-	MeCN	240(4.05),245(4.10), 313(3.91)	44-1923-69
$C_{22}H_{20}N_2O$			
1,3-Butadiene-1,1-dicarbonitrile, 3-ethoxy-2,4-di-p-tolyl-	CH_2Cl_2	258(4.19),333(4.18)	24-0319-69
$C_{22}H_{20}N_2O_2$			
2,2'-Biquinoline, 4,4'-diethoxy-, cuprous chelate	iso-AmOH	538(3.86)	3-0344-69
$C_{22}H_{20}N_2O_6$			
Pyrrole-2,4-dicarboxylic acid, 1-benzyl-3-[(p-methoxyphenyl)carbamoyl]-, 4-methyl ester	EtOH	220(4.53)	94-2461-69
$C_{22}H_{20}N_2O_{12}$			
[3,3'-Bipyridine]-2,2',5,5',6,6'-hexol, hexaacetate	$CHCl_3$	277(4.23)	83-0264-69
$C_{22}H_{20}N_2S_2$			
2H-1,3,5-Thiadiazine-2-thione, 5-benzyl-3-(4-biphenylyl)tetrahydro-	MeOH	250(4.29),275(4.38)	73-2952-69
$C_{22}H_{20}N_4O_2$			
Benzo[1mn]diimidazo[1,2-c:1',2'-j][3,8]-phenanthroline-6,13-dione, 1,2,8,9-tetrahydro-2,2,9,9-tetramethyl-	HCOOH	<u>382</u>(--),406(4.57)	121-1067-69
Nipecotamide, 3,5-dicyano-N,1-dimethyl-6-oxo-2,4-diphenyl-	base	261(4.21)	24-4147-69
$C_{22}H_{20}N_4O_6S_2$			
[$\Delta^{2,2'}$-Bi-2H-1,4-benzothiazine]-7,7'-dipropionic acid, α,α'-diamino-5,5'-dihydroxy-	2N HCl	246(4.25),266s(4.15), 330(4.16),551s(4.22), 587(4.42)	32-1193-69
	pH 13	246(4.34),313(4.11), 482(4.33)	32-1193-69
$C_{22}H_{20}N_4O_7S_2$			
[$\Delta^{2,2'}$-Bi-2H-1,4-benzothiazine]-7,7'-dipropionic acid, α,α'-diamino-3,4-dihydro-5,5'-dihydroxy-3-oxo-	pH 13	240(4.56),327(4.04), 452(4.14)	32-0323-69
$C_{22}H_{20}N_6O_4$			
Cinnamic acid, p-azido-, tetramethylene ester	dioxan	312(4.73)	73-3811-69
$C_{22}H_{20}O$			
Propiophenone, 2-benzyl-3-phenyl-	EtOH	245(4.20),285s(3.10), 320(2.09)	1-2409-69
$C_{22}H_{20}O_2$			
p-Dioxane, 2,2,6-triphenyl-	EtOH	257(2.63)	22-2048-69
Stilbene, 3,4-dimethoxy-4'-phenyl-	DMF	346(4.64)	33-2521-69

Compound	Solvent	λ_{max}(log ε)	Ref.
$C_{22}H_{20}O_4$			
3,5-Cyclohexadiene-1,2-dicarboxylic acid, 3,6-diphenyl-, dimethyl ester, cis	EtOH	234s(4.11),338(4.30)	22-3662-69
2,4,6-Octatrienedioic acid, 3,6-diphenyl-, dimethyl ester, cis-trans-cis	THF	318(4.45)	22-3665-69
cis-trans-trans	THF	312(4.52)	22-3665-69
trans-trans-trans	THF	307(4.61)	22-3665-69
p-Toluic acid, 2,6-bis(benzyloxy)-	EtOH	215(3.53),281(3.41)	39-1721-69C
$C_{22}H_{20}O_4S_2$			
Benzofuran, 5-allyl-3-(1,2-dithiolan-2-yl)-7-methoxy-2-[3,4-(methylenedioxy)phenyl]-	EtOH	308(4.32)	12-1011-69
$C_{22}H_{20}O_6$			
Cubebinic ether	EtOH	226(4.37),292(4.12), 310(4.98),321s(3.88)	39-2470-69C
Glycyrol, 5-O-methyl-	EtOH	229(4.50),247s(4.35), 264s(4.14),346(4.38), 362s(4.32)	94-0729-69
Isoglycyrol, monomethyl ether	EtOH	208(4.49),225(4.33), 247(4.29),256(4.19), 264(4.12),347(4.35), 357(4.29)	94-0729-69
Robustic acid	EtOH	233(4.42),258(4.48), 345(4.27)	39-0365-69C
Robustone, dihydro-, methyl ether	EtOH	260(4.41),294(4.27)	39-0365-69C
$C_{22}H_{20}O_8$			
2,3,6,7-Biphenylenetetrol, 1,8-dimethyl-, tetraacetate	EtOH	252(4.82),259(4.99), 342(3.94),363(4.02)	39-2579-69C
$C_{22}H_{20}O_9$			
2(4aH)-Dibenzofuranone, 6-acetyl-3,7,9-trihydroxy-4,4a-dimethyl-, triacetate	EtOH	210(4.24),243(4.14), 267(4.05),320s(3.76), 363(3.97)	54-0851-69
Eupatorin, diacetate	EtOH	233(4.44),265(4.26), 319(4.57)	78-1603-69
Flavone, 3',6-diacetoxy-4',5,7-trimethoxy-	EtOH	230s(4.34),264(4.04), 317(4.44)	40-1270-69
Flavone, 4',6-diacetoxy-3',5,7-trimethoxy-	EtOH	238(4.34),263(4.25), 313(4.36)	40-1270-69
Flavone, 5,6-diacetoxy-3',4',7-trimethoxy-	EtOH	238(4.39),260s(4.47), 333(4.39)	40-1270-69
$C_{22}H_{21}BN_2O_8S$			
Uridine, cyclic 2',3'-benzeneboronate 5'-p-toluenesulfonate	EtOH	260(4.33)	65-0434-69
$C_{22}H_{21}BrO_2$			
2,2'-Spirobi[2H-1-benzopyran], 6'-bromo-3-pentyl-	dioxan	230(4.61),255(4.31), 308(3.68)	5-0162-69B
$C_{22}H_{21}ClN_2O$			
4-Quinolinol, 5-chloro-3-(N-pyrrolidinomethyl)-2-styryl-	EtOH	250(4.29),285(3.98), 330(3.94)	2-1186-69
$C_{22}H_{21}ClN_2O_2$			
4-Quinolinol, 5-chloro-3-(morpholinomethyl)-2-styryl-	EtOH	255(4.40),285(4.03), 330(4.35)	2-1186-69

Compound	Solvent	λ_{max}(log ϵ)	Ref.
$C_{22}H_{21}F_7O_3$			
Estrone, heptafluorobutyrate	heptane	266(4.29),274(4.20)	13-0739-69A
$C_{22}H_{21}FeNO_6$			
Iron, tricarbonyl(6,7,8,14-tetradehydro-4,5α-epoxy-3,6-dimethoxy-17-methyl-morphinan)-	EtOH	242(4.35),287(3.74)	12-0971-69
	$CHCl_3$	252(4.01),287(3.72)	12-0971-69
$C_{22}H_{21}Fe_2NO$			
[1,1"-Biferrocene]-2-carboxamide, 2"-methyl-, (1S,1"R)-(+)-	EtOH	447(2.86)	49-1515-69
$C_{22}H_{21}N$			
Aniline, p-(9,10-dihydro-9-anthryl)-N,N-dimethyl-	MeCN	306(3.42)	88-3829-69
1,3,5-Heptatrienylamine, N,N-dimethyl-7-phenalen-1-ylidene-	hexane	277(--),324(--), 411(--),531(--)	24-2301-69
	MeCN	277(4.50),324(4.40), 415(4.11),541(4.77)	24-2301-69
$C_{22}H_{21}NO_2$			
3-Pyrroline, 1,5-dibenzoyl-3,4,5-tri-methyl-2-methylene-	EtOH	232(4.32),276s(3.85), 320s(3.18)	5-0137-69D
$C_{22}H_{21}NO_3$			
1,3-Cyclopentadiene-1-carboxaldehyde, 3-(dimethylamino)-5-(hydroxymethyl-ene)-4-(p-methoxyphenyl)-2-phenyl-	CH_2Cl_2	247(4.50),290(4.45), 365(4.20),485(3.88)	24-0623-69
$C_{22}H_{21}NO_4$			
5-Hexen-1-ol, 5-nitro-6-(9-phenan-thryl)-, acetate	EtOH	340(3.89)	87-0157-69
$C_{22}H_{21}NO_5$			
Benzyl alcohol, 2,4-bis(benzyloxy)-α-(nitromethyl)-	EtOH	276(3.65)	104-1803-69
$C_{22}H_{21}NO_8$			
11bH-Benzo[c]quinolizine-1,2,3,4-tetra-carboxylic acid, 6-methyl-, tetra-methyl ester	MeOH	208(4.30),225s(4.04), 268s(4.39),280(4.48), 290s(4.40),400(3.83)	39-2316-69C
	MeOH-$HClO_4$	213(4.61),257(4.62), 341(4.15)	39-2316-69C
$C_{22}H_{21}N_3$			
Quinoline, 3-(3,4-dihydro-1-isoquinol-yl)-4-(1-pyrrolidinyl)-, hydrochlor-ide	EtOH	368(3.59)	2-1010-69
$C_{22}H_{21}N_5O$			
Benzamide, N-[3-(3-amino-6-indol-3-yl-pyrazinyl)propyl]-	MeOH	230(4.45),274(4.26), 363(3.88)	95-1652-69
	MeOH-HCl	226(4.56),308(4.30), 410(3.71)	95-1652-69
$C_{22}H_{22}$			
11H-Benzo[a]fluorene, 11-isopropyl-1,4-dimethyl-	EtOH	262(4.69),271(4.85), 287(4.06),296(4.13), 313(4.03),320(3.95), 326(3.92),334(3.64), 352(3.53)	22-2115-69

Compound	Solvent	λ_{max}(log ε)	Ref.
o-Terphenyl, 3',4',5',6'-tetramethyl-	hexane	226(4.38)	24-1928-69
$C_{22}H_{22}ClNO_4$			
5,6,7,8-Tetrahydro-1-methyl-2,3-diphenylquinolizinium perchlorate	EtOH	244(4.00),304(3.92)	94-2209-69
$C_{22}H_{22}ClNO_5$			
[2-(2,6-Diphenyl-4H-pyran-4-ylidene)-1-methylethylidene]dimethylammonium perchlorate	MeCN	247(4.12),273(4.01), 328(4.03),420(4.38)	44-2736-69
[2-(4,6-Diphenyl-2H-pyran-2-ylidene)-1-methylethylidene]dimethylammonium perchlorate	MeCN	241(4.18),313(4.34), 356(3.98),366s(4.00), 380(4.21)	44-2736-69
[2-(4,6-Diphenyl-2H-pyran-2-ylidene)-propylidene]dimethylammonium perchlorate	MeCN	237(4.18),310(4.51), 466(4.31),484(4.29)	44-2736-69
$C_{22}H_{22}ClNS$			
2-Thiopheneallylamine, N-(β-chloro-α-methylphenethyl)-γ-phenyl-	MeOH	244(4.00),285(4.02)	111-0228-69
$C_{22}H_{22}ClN_3$			
Quinoline, 3-(3,4-dihydro-1-isoquinolyl)-4-(1-pyrrolidinyl)-, hydrochloride	EtOH	368(3.59)	2-1010-69
$C_{22}H_{22}FeO_6$			
Iron, (androsta-1,4-diene-3,11,17-trione)tricarbonyl-	EtOH	215(4.26)	101-0342-69A
$C_{22}H_{22}Fe_2$			
1,1"-Biferrocene, 2,2"-dimethyl-, (1R,1"R)-(-)-	EtOH	443(2.38)	49-1515-69
Biferrocene, 2,7-dimethyl-	C_6H_{12}	202(4.66),215s(4.55), 250s(3.88),300s(3.30)	78-3477-69
Biferrocene, 2,10-dimethyl-	C_6H_{12}	197(4.56),222(4.68), 255s(4.01),294(3.85)	78-3477-69
$C_{22}H_{22}Fe_2O$			
[1,1"-Biferrocene]-2-methanol, 2"-methyl-, (1R,1"S)-(-)-	EtOH	451(2.70)	49-1515-69
Biferrocene-10-methanol, 2-methyl-	C_6H_{12}	197(4.56),222(4.64), 255s(4.01),296(3.85)	78-3477-69
$C_{22}H_{22}IPS$			
2,8,10-Trimethyl-10-p-tolylphenothiaphosphinium iodide	EtOH	220(4.65),267(4.07), 304s(3.77),318(3.81)	78-3919-69
$C_{22}H_{22}N_2$			
Benzophenone, isopropylphenylhydrazone	EtOH	245(3.87),300s(3.30), 350(3.34)	39-1703-69C
$C_{22}H_{22}N_2O_2$			
1,5-Diazocine-2,4(3H,7H)-dione, 3,3,7,7-tetramethyl-6,8-diphenyl-	MeCN	233(3.99)	88-4497-69
$C_{22}H_{22}N_2O_3$			
Benzaldehyde, 2,4,6-trimethoxy-, diphenylhydrazone	EtOH	203(4.6),247(4.2), 340(4.3)	28-0137-69B

Compound	Solvent	$\lambda_{max}(\log \epsilon)$	Ref.
$C_{22}H_{22}N_2O_4$			
Benzo[a]phenazine-8,11-dicarboxylic acid, 7,9,10,12-tetrahydro-, diethyl ester	EtOH	229(4.69),277(4.39), 349(3.88),365(3.94)	33-0322-69
$C_{22}H_{22}N_2S$			
4(1H)-Pyrimidinethione, 1-cyclohexyl-2,5-diphenyl-	EtOH	237(4.23),348(4.30)	4-0037-69
$C_{22}H_{22}N_3$			
Azapseudoisocyanine anion	EtOH at 100°K	286(4.4),380(4.3), 400(4.9),422(5.1)	62-0097-69B
$C_{22}H_{22}N_4$			
Pyrazole, 1,1'-ethylenebis[5-methyl-4-phenyl-	MeOH	242(4.45)	44-3639-69
$C_{22}H_{22}N_4O_3S$			
3-Pentanone, 2-[1,2-dihydro-4-(methyl-sulfonyl)-2-(1-phthalazinyl)-1-phthalazinyl]-	EtOH	250s(3.92),368(4.11)	95-0959-69
$C_{22}H_{22}N_4O_4$			
5,5'-Bihydantoin, 3,3'-diethyl-5,5'-diphenyl-	MeOH	215s(4.22)	44-1133-69
$C_{22}H_{22}N_8O_4S_4$			
2H-1,3-Thiazine-2,4(3H)-dione, 3,3'-ethylenebis[dihydro-2-thio-, 4,4'-bis[(p-nitrophenyl)hydrazone]	EtOH	239(4.01),304(3.99), 404(4.36)	103-0049-69
$C_{22}H_{22}O_2$			
Cyclobutene, 3,4-dibenzoyl-1,2,3,4-tetramethyl-	hexane	244(4.17)	24-1928-69
2,2'-Spirobi[2H-1-benzopyran], 3-pentyl-	dioxan	259(4.33),301(3.80)	5-0162-69B
$C_{22}H_{22}O_3$			
Benzhydrol, α-[[(β-hydroxyphenethyl)-oxy]methyl]-	EtOH	256(2.74)	22-2048-69
$C_{22}H_{22}O_4$			
Phthalic acid, 3,4,5,6-tetrahydro-3,6-diphenyl-, dimethyl ester	n.s.g.	264(3.49),268(3.28)	22-3662-69
$C_{22}H_{22}O_5$			
Phenol, 2-methoxy-4-(7-methoxy-3-methyl-5-propenyl-2-benzofuranyl)-, acetate	EtOH	248(4.21),305(4.44)	12-1011-69
$C_{22}H_{22}O_6$			
2H-1-Benzopyran, 6-allyl-2,3,8-trimeth-oxy-2-[3,4-(methylenedioxy)phenyl]-	EtOH	231(4.57),270(4.17), 280(4.20)	12-1011-69
Dehydrocubebin ethyl acetal	EtOH	276(4.16),287s(4.05), 305(3.91)	39-2470-69C
Dihydrocubebinic ether	EtOH	234(3.90),290(3.98)	39-2470-69C
Robustic acid, dihydro-	EtOH	252s(3.98),332(4.20)	12-1923-69
	EtOH	240s(4.17),334(4.36)	39-0365-69C
	EtOH-NaOH	280(4.23),349(4.27)	12-1923-69

Compound	Solvent	λmax(log ε)	Ref.
$C_{22}H_{22}O_7$			
6H-Benzofuro[3,2-c][1]benzopyran-6-one, 9-hydroxy-2-(3-hydroxy-3-methylbutyl)-1,3-dimethoxy-	EtOH	208(4.45),227(4.37), 246(4.24),262s(4.01), 345(4.30),357s(4.24)	94-0729-69
$C_{22}H_{22}O_8$			
Benzophenone, 2,2',4-trihydroxy-4'-methoxy-6,6'-dimethyl-, triacetate	EtOH	213(4.3),231(4.16), 283(3.52)	39-1721-69C
$C_{22}H_{22}O_9$			
Altersolanol B, triacetate	EtOH	265(4.38),348(3.36), 379(3.39)	23-0767-69
$C_{22}H_{22}O_{10}$			
4,9-Ethenonaphtho[2,3-d]-1,3-dioxole-5,6,10,11-tetracarboxylic acid, 4,4a,7,9-tetrahydro-7-methyl-, tetramethyl ester	EtOH	249(3.34),342(2.01)	12-1721-69
$C_{22}H_{22}O_{11}$			
Altersolanol A, triacetate	EtOH	267(4.31),384(3.43)	23-0777-69
$C_{22}H_{22}O_{12}$			
Flavone, 3,3',4',7-tetrahydroxy-5-methoxy-, 3-β-D-glucopyranoside	MeOH	253(4.39),349(4.31)	24-0792-69
$C_{22}H_{23}BrO_4$			
Benzoic acid, p-bromo-, ester with 5,9-epoxy-2-(2-hydroxy-1-methylethyl)-1-vinyl-p-mentha-4,8-dien-3-one	EtOH	245(4.31),275s(3.60)	39-1491-69C
$C_{22}H_{23}ClN_2O_4$			
3-Furoic acid, 4-(p-chlorophenyl)-4,5-dihydro-5,5-dimethyl-2-(3-p-tolylureido)-, methyl ester	EtOH	207(4.49),223(4.28), 254(4.19),289(4.40)	87-0339-69
$C_{22}H_{23}ClN_2O_5$			
3-Furoic acid, 4-(p-chlorophenyl)-4,5-dihydro-5,5-dimethyl-2-[3-(p-methoxyphenyl)ureido]-, methyl ester	EtOH	208(4.32),224(4.28), 230s(4.20),290(4.30)	87-0339-69
$C_{22}H_{23}FeNO_6$			
Iron, tricarbonyl(6,7,8,14-tetradehydro-3,6-dimethoxy-17-methylmorphinan-4-ol)-	$CHCl_3$	252(4.07),272(3.75)	12-0971-69
$C_{22}H_{23}NOS$			
Benzyl alcohol, α-[1-[(γ-2-thienylcinnamyl)amino]ethyl]-	MeOH	244(3.87),285(3.89)	111-0228-69
$C_{22}H_{23}NO_2$			
3-Pyrroline, 1,5-dibenzoyl-3,4,5-trimethyl-2-methylene-, dihydro deriv.	EtOH	230s(4.15)	5-0137-69D
$C_{22}H_{23}NO_3$			
Dibenzo[f,h]pyrrolo[1,2-b]isoquinolin-6-ol, 9,11,12,13,13a,14-hexahydro-2,3-dimethoxy-	MeOH	260(4.74),287(4.47), 344(3.05),362(2.75)	5-0154-69A

Compound	Solvent	λmax(log ε)	Ref.
$C_{22}H_{23}NO_6$			
Bractavine	MeOH	230s(4.27),284(3.80)	106-0635-69
Orientalidine	EtOH	231s(4.25),286(3.78)	73-0875-69
$C_{22}H_{23}NO_7$			
α-Narcotine	EtOH	209(4.83),251s(4.19), 289(3.60),308(3.69)	78-5059-69
β-Narcotine	EtOH	211(4.77),237s(4.20), 291(3.56),312(3.66)	78-5059-69
$C_{22}H_{23}NS$			
2-Thiopheneallylamine, N-(α-methyl-phenethyl)-γ-phenyl-	MeOH	244(3.95),286(4.08)	111-0228-69
$C_{22}H_{23}N_2O_8P$			
Thymine, 1-(2-deoxy-β-D-threo-pentfur-anosyl)-, 5-(diphenyl phosphate)	dioxan	263(3.97)	69-4889-69
$C_{22}H_{23}N_3O_3$			
17βH-Diaboline, 1-(cyanoformyl)-1-de-acetyl-O-methyl-	EtOH	279(3.77),294(3.79)	33-1564-69
$C_{22}H_{23}N_3O_5$			
Tryptophan, N-(N-carboxy-DL-alanyl)-, N-benzyl ester, DL-	EtOH	221(4.53),282(3.78), 290(3.74)	39-1003-69C
$C_{22}H_{23}N_3O_6$			
2-Azabicyclo[2.2.2]octane, 2-(3,5-di-nitrobenzoyl)-6-(p-methoxybenzyl)-	EtOH	223(4.14),283(3.32)	22-0781-69
$C_{22}H_{23}N_5O_6S$			
7H-Pyrrolo[2,3-d]pyrimidine-5-carboni-trile, 4-amino-7-(2,3-O-isopropyli-dene-β-D-ribofuranosyl)-, 5'-p-toluenesulfonate	pH 1	213(4.38),273(4.03)	35-2102-69
	pH 11	210(4.39),272(3.95)	35-2102-69
	EtOH	221(4.39),273(4.05)	35-2102-69
$C_{22}H_{24}AsI$			
Butyltriphenylarsonium iodide	n.s.g.	230(4.3),252(3.2), 258(3.3),264(3.4), 271(3.4),290(3.6)	120-0195-69
$C_{22}H_{24}ClN_3O_3$			
1-(Cyanoformyl)-1-deacetyl-4-methyl-diabolinium chloride	EtOH	241(4.00),280(3.26)	33-1564-69
19-Methyl-11-oxostrychninium chloride, 11-oxime	EtOH	232(4.21),290(3.83), 310(3.87)	33-1564-69
	EtOH-base	310(4.28)	33-1564-69
$C_{22}H_{24}ClN_3O_8S_3$			
Sporidesmin E, diacetate	MeOH	218(4.65),252(4.18), 302(3.60)	39-1665-69C
$C_{22}H_{24}ClO_4P$			
Benzhydrylmethylphenylpropylphosphonium perchlorate	MeOH	220s(4.29),260s(3.14), 267(3.16),273s(3.96)	24-3546-69
$C_{22}H_{24}Cl_2N_2O_8S$			
Ammonium, [2-(2-phenyl-4H-1-benzothio-pyran-4-ylidene)propanediylidene]-bis[dimethyl-, diperchlorate	MeCN	267(4.24),303(4.44), 400(4.24),520(4.08)	44-2736-69

Compound	Solvent	λ_{max}(log ε)	Ref.
$C_{22}H_{24}Cl_2N_2O_9$			
Ammonium, [2-(2-phenyl-4H-1-benzopyran-4-ylidene)propanediylidene]bis-[dimethyl-, diperchlorate	MeCN	243s(4.31),266s(4.24), 313(4.49),428(4.20), 453(4.19)	44-2736-69
$C_{22}H_{24}Cl_2O_{10}$			
1,2,3,4-Cyclohexanetetrol, 5,6-dichloro-1-(hydroxymethyl)-, 1,2,3,4-tetra-acetate benzoate	MeOH	230(4.13),274(3.01), 281(2.92)	44-3898-69
$C_{22}H_{24}F_3NO_5$			
Thebaine methotrifluoroacetate	CF_3COOH	322(3.85)(changing)	39-1215-69C
$C_{22}H_{24}INS$			
2-(Butylmethylamino)-4,6-diphenylthio-pyrylium iodide	EtOH	281(4.30),326(4.16), 401(4.06)	48-0061-69
$C_{22}H_{24}N_2$			
3,4-Diazabicyclo[4.2.0]octa-2,4-diene, 1,6,7,8-tetramethyl-2,5-diphenyl-, trans	MeOH	269(4.05)	24-1928-69
Indolo[2,3-a]quinolizine, 12-benzyl-1,2,3,4,6,7,12,12b-octahydro-	3% HCl	281(3.95)	24-3959-69
	MeOH	286(3.90)	24-3959-69
$C_{22}H_{24}N_2OS$			
Atropamide, N-benzoyl-β-(cyclohexyl-amino)thio-	$CHCl_3$	282s(4.09),300(4.10), 400(4.26)	4-0037-69
$C_{22}H_{24}N_2O_3$			
2H-Pyrrole-2-propionic acid, 3-hydroxy-5-(dimethylamino)-4-phenyl-, benzyl ester	MeOH	225(4.08),287(4.23)	78-5721-69
$C_{22}H_{24}N_2O_4$			
Benzo[c]cinnoline, 3,8-bis[(tetrahydro-2H-pyran-2-yl)oxy]-	EtOH	202(4.45),218s(4.10), 257(3.85),287s(4.13), 309(3.86),392(3.45)	4-0523-69
3-Furoic acid, 4,5-dihydro-5,5-dimethyl-4-phenyl-2-(3-p-tolylureido)-, methyl ester	EtOH	207(4.63),233(4.29), 288(4.50)	87-0339-69
Kopsine, lactone	EtOH	243(4.11),280(3.37), 287(3.33)	33-0076-69
$C_{22}H_{24}N_2O_4S_2$			
Dibenzenesulfonamide, N-[p-(diethyl-amino)phenyl]-	EtOH-1% $CHCl_3$	275(4.21)	44-2083-69
Di-p-toluenesulfonamide, N-[p-(dimeth-ylamino)phenyl]-	EtOH-1% $CHCl_3$	236(4.41),275(4.19)	44-2083-69
$C_{22}H_{24}N_2O_5$			
Benzo[c]cinnoline, 3,8-bis[(tetrahydro-2H-pyran-2-yl)oxy]-, 5-oxide	EtOH	208(4.31),250s(4.66), 261(4.70),298s(4.23), 335(4.27),402(3.65), 420(3.62)	4-0523-69
3-Furoic acid, 4,5-dihydro-5,5-dimethyl-4-phenyl-2-[3-(p-methoxyphenyl)-ureido]-, methyl ester	EtOH	207(4.53),230(4.20), 252(4.19),288(4.50)	87-0339-69
Kopsine pseudoacid	EtOH	240(4.08),278(3.33), 286(3.32)	33-0076-69

Compound	Solvent	$\lambda_{max}(\log \epsilon)$	Ref.
$C_{22}H_{24}N_2O_6$			
Benzo[c]cinnoline, 3,8-bis[(tetrahydro-2H-pyran-2-yl)oxy]-, 5,6-dioxide	EtOH	224s(4.06),242s(4.29), 266(4.68),283(4.59), 350(4.00),404(3.50), 425(3.49)	4-0523-69
$C_{22}H_{24}N_2O_8$			
2-Naphthalenecarboxylic acid, 3-[4-carbamyl)-3,6-dihydroxy-2-(dimethylamino)-5-oxo-1,2,5,6-tetrahydrobenzyl]-4,8-dihydroxy-4-methyl-1-oxo-1,2,3,4-tetrahydro-, γ-lactone	EtOH	265(4.09),340(3.39)	30-1255-69F
2H-Pyran, 2,2'-[(2,2'-dimethyl-4,4'-biphenylylene)dioxy]bis[tetrahydro-	EtOH	225(4.48),268s(4.09), 325s(3.60)	4-0523-69
$C_{22}H_{24}N_2O_{10}$			
Quinazoline, 4-(β-D-glucopyranosyloxy)-, 2',3',4',6'-tetraacetate	n.s.g.	263(3.6),298(3.5), 309(3.5)	106-0035-69
4(3H)-Quinazolinone, 3-β-D-glucopyranosyl-, 2',3',4',6'-tetraacetate	n.s.g.	225(4.5),266f(3.8), 302f(3.5)	106-0035-69
$C_{22}H_{24}N_4O_4S$			
Acetophenone, 2-[3-[(4-amino-2-methyl-5-pyrimidinyl)methyl]-5-(2-hydroxyethyl)-4-methyl-4-thiazolin-2-ylidene]-2-hydroxy-, 2-acetate	EtOH	229(4.31),271(3.92), 392(4.41)	94-0128-69
$C_{22}H_{24}N_4O_5S_2$			
Ketone, [3-[(4-amino-2-methyl-5-pyrimidinyl)methyl]-5-(2-hydroxyethyl)-4-methyl-4-thiazolin-2-ylidene]hydroxymethyl 2-thienyl, diacetate	EtOH	233(4.23),274(4.0), 414(4.36)	94-0128-69
$C_{22}H_{24}N_4O_6S$			
Ketone, [3-[(4-amino-2-methyl-5-pyrimidinyl)methyl]-5-(2-hydroxyethyl)-4-methyl-4-thiazolin-2-ylidene]hydroxymethyl 2-furyl, diacetate	EtOH	231(4.16),277(4.14), 409(4.46)	94-0128-69
$C_{22}H_{24}O_4$			
Dibenzo[a,e]cyclodecene-5,8-diol, 5,6,7,8,13,14-hexahydro-, diacetate	EtOH	266(4.36),273(3.36)	44-0874-69
Tricyclo[10.2.2.2^{4,7}]octadeca-4,6,12,14,15,17-hexaene-9,10-dicarboxylic acid, dimethyl ester, cis	EtOH	215(4.16),271(2.53), 282(2.33)	35-3517-69
trans	EtOH	214(4.15),271(2.60), 283(2.42)	35-3517-69
$C_{22}H_{24}O_5$			
Estra-1,3,5(10),9(11)-tetraen-6-one, 3,17β-diacetoxy-	MeOH	240(4.38),264(4.17)	44-0116-69
$C_{22}H_{24}O_6$			
9-Anthraldehyde, 10-(ethoxymethyl)-2,3,6,7-tetramethoxy-	EtOH	262(4.88),272s(4.68), 407(4.20)	7-0671-69
2H-1-Benzopyran, 2,3,8-trimethoxy-2-[3,4-(methylenedioxy)phenyl]-6-propyl-	EtOH	229(4.57),268(4.14), 278(4.18)	12-1011-69

Compound	Solvent	λ_{max}(log ε)	Ref.
Dispiro[oxetane-3,7'(8'H)-[6H,15H]dibenzo[b,i][1,4,8,11]tetraoxatetradecin-16'(17'H),3"-oxetane]	MeOH	272(3.63)	25-1271-69
$C_{22}H_{24}O_7$			
Eupatilin, diethyl ether	MeOH	240(4.33),266(4.07), 338(4.41)	78-1603-69
Eupatorin, diethyl ether	MeOH	240(4.34),265(4.10), 338(4.43)	78-1603-69
Luteolin, 6-methoxy-, triethyl deriv.	EtOH	244(4.26),277(4.15), 342(4.42)	31-0355-69
	EtOH-$AlCl_3$	262(4.15),293(4.26), 368(4.42)	31-0355-69
$C_{22}H_{24}O_8$			
Longocarpan, diacetate	EtOH	283(3.70)	39-0887-69C
$C_{22}H_{24}O_{10}$			
Flavanone, 5,7-dihydroxy-8-methoxy-, 7-β-D-glucopyranoside	n.s.g.	287(4.19),340s(3.61)	88-1471-69
Poriolin	EtOH	292(4.17),347(3.41)	12-0483-69
	EtOH-NaOEt	250s(4.09),292(4.20), 348(3.51),428(3.40)	12-0483-69
	EtOH-NaOAc	292(4.17),347(3.41)	12-0483-69
	EtOH-$AlCl_3$	313(4.14),400(3.25)	12-0483-69
$C_{22}H_{25}BrO_5$			
Gaillardin, deacetyldihydro-, p-bromobenzoate	MeOH	245(4.24)	88-0973-69
$C_{22}H_{25}BrO_8$			
Euparotin, bromoacetate	MeOH	212(4.08)(end abs.)	44-3786-69
$C_{22}H_{25}NO$			
Azetidine, 1-cyclohexyl-2-phenyl-3-benzoyl-, cis	isooctane	240(4.02)	44-0310-69
trans	isooctane	242(4.18)	44-0310-69
7H-6,18-Methenocycloheptacyclopentadecene-11-carbonitrile, 8,9,10,11,12,13,-14,15,16,17-decahydro-12-oxo-	C_6H_{12}	233(4.39),282(4.65), 350(3.71),368(3.65), 580s(2.33),605(2.39), 629(2.48),660(2.40), 689(2.42),732(2.04), 770(2.00)	44-2375-69
$C_{22}H_{25}NOS$			
2-Thiophenemethanol, α-[2-[(α-methylphenethyl)amino]ethyl]-α-phenyl-	MeOH	234(3.89)	111-0228-69
$C_{22}H_{25}NO_2S$			
2-Thiophenemethanol, α-[2-[(β-hydroxy-α-methylphenethyl)amino]ethyl]-α-phenyl-	MeOH	235(3.95)	111-0228-69
$C_{22}H_{25}NO_5$			
C-Homoberbine, 9,10,11-trimethoxy-2,3-(methylenedioxy)-	EtOH	230s(4.13),289(3.75)	33-1228-69
Isoquinoline, 1-(2-methylveratrylidene)-2-methyl-6,7-(methylenedioxy)-8-methoxy-1,2,3,4-tetrahydro-	EtOH	257(4.11),305(4.04)	95-0869-69

Compound	Solvent	λ max (log ε)	Ref.
$C_{22}H_{25}N_3O_2$			
2-Propanone, 1-(2,4-dimethyl-5-p-tolyl-imidazol-1-yl)-1-p-tolyl-, oxime, 3-oxide	MeOH	261(4.22)	48-0746-69
Yohimban-16-carbonitrile, 17-acetoxy-	EtOH	276(3.73),282(3.75), 289(3.69)	24-3248-69
$C_{22}H_{25}N_3O_4$			
Methanol, tris(4-amino-3-methoxyphenyl)-	pH 4.62	580(4.83)	64-0542-69
$C_{22}H_{25}N_3O_9$			
1H-1,2,3-Triazole, 1-β-D-galactopyranosyl-4-phenyl-, 2',3',4',6'-tetraacetate	EtOH	243(4.24),247s(2.88)	4-0639-69
1H-1,2,3-Triazole, 1-β-D-galactopyranosyl-5-phenyl-, 2',3',4',6'-tetraacetate	EtOH	241(4.00)	4-0639-69
1H-1,2,3-Triazole, 1-β-D-glucopyranosyl-4-phenyl-, 2',3',4',6'-tetraacetate	EtOH	243(4.26),260s(3.92), 277s(2.94)	4-0639-69
1H-1,2,3-Triazole, 1-β-D-glucopyranosyl-5-phenyl-, 2',3',4',6'-tetraacetate	EtOH	242(4.04)	4-0639-69
$C_{22}H_{26}ClN_3O_4$			
1-Carbamoyl-17β-carboxy-1-deacetyl-17-deoxy-4-methyldiabolinium chloride	EtOH	244(4.02),280(3.18)	33-1564-69
	EtOH-base	224(3.97),281(3.21)	33-1564-69
$C_{22}H_{26}F_2O_4$			
Androst-4-ene-3,17-dione, 19-acetoxy-6α,7α-(difluoromethylene)-	EtOH	243(4.13)	78-1219-69
$C_{22}H_{26}N_2$			
3H-Indole, 3,3-dimethyl-2-[2-(2,3,3-trimethylindolinomethyl)]-	C_6H_{12}	251(4.22),275s(3.74), 286(3.70)	32-1236-69
	EtOH	248(4.13),275s(3.80)	32-1236-69
$C_{22}H_{26}N_2O_3$			
Diaboline, O-methyl-	EtOH	250(4.09),277s(3.43), 285s(3.34)	33-1564-69
17βH-Diaboline, O-methyl-	EtOH	253(4.09),277s(3.44), 286s(3.32)	33-1564-69
5'H-Pregna-1,4,6,15-tetraeno[2,1-c]-pyrazole-3,20-dione, 1β,2β-dihydro-17-hydroxy-	MeOH	244(3.83),290(4.35)	87-0393-69
Voacangine, 4,20-dehydro-	MeOH	222(4.28),285(3.81), 289s(3.79),296s(3.74)	27-0400-69
$C_{22}H_{26}N_2O_4$			
17βH-Diaboline, 1-carboxy-1-deacetyl-O-methyl-, methyl ester	EtOH	242(3.71),280(2.94)	33-1564-69
Voachalotine oxindole	MeOH	210(4.59),257(3.93), 290s(--)	20-0523-69
$C_{22}H_{26}N_2O_5$			
2,6-Methano-2,5,6,7-tetrahydro-1(1H)-benzazonine, 3-acetyl-4-amino-7-hydroxy-7-methyl-, triacetyl deriv.	$CHCl_3$	309(4.02)	1-3567-69
$C_{22}H_{26}N_2O_6$			
4-Piperidineacetic acid, 5-acetyl-5-carboxy-1-(2-indol-3-ylethyl)-2-oxo-, dimethyl ester	EtOH	221(4.53),276(3.71), 284(3.75),292(3.69)	35-7359-69

Compound	Solvent	$\lambda_{max}(\log \epsilon)$	Ref.
$C_{22}H_{26}N_2O_9S$			
Benzimidazole, 2-(β-D-glucopyranosylthio)-1-methyl-, 2',3',4',6'-tetraacetate	MeOH	260s(3.88),283(4.11), 288(4.09)	97-0152-69
2-Benzimidazolinethione, 1-β-D-glucopyranosyl-3-methyl-, 2',3',4',6'-tetraacetate	MeOH	247(4.05),300s(4.30), 308(4.42)	97-0152-69
$C_{22}H_{26}N_4$			
3,3'-Bi-2-pyrazoline, 5,5'-dimethyl-1,1'-di-p-tolyl-	EtOH	231(3.85),260(3.85), 318(3.69),400(4.48)	22-1683-69
$C_{22}H_{26}N_4O_2$			
Indoline, 2,3,3,3',3'-pentamethyl-2,2'-methylenebis[1-nitroso-	EtOH	224(4.17),292s(4.20), 296(4.21)	32-1236-69
$C_{22}H_{26}N_4O_6$			
β-Alanine, N,N'-(ethylenedi-p-phenylene)bis[N-nitroso-, dimethyl ester	EtOH	215(4.31),278(4.17)	103-0623-69
$C_{22}H_{26}N_4O_8$			
Isohydrocoriamyrtin, methyl-2,4-dinitrophenylhydrazone	EtOH	267(3.76),283(3.69), 396(4.03)	78-1117-69
Isohydrocoriamyrtin, 11-O-methyl-, 2,4-dinitrophenylhydrazone	EtOH	260(4.08),290(3.81), 375(4.44)	78-1109-69
1H-1,2,3-Triazole, 1-(2-acetamido-2-deoxy-β-D-glucopyranosyl)-4-phenyl-, 3',4',6'-triacetate	EtOH	242(4.30),247s(3.04), 285s(2.34)	4-0639-69
1H-1,2,3-Triazole, 1-(2-acetamido-2-deoxy-β-D-glucopyranosyl)-5-phenyl-, 3',4',6'-triacetate	EtOH	242(4.02)	4-0639-69
$C_{22}H_{26}O_3$			
Pregna-1,4,6-triene-3,20-dione, 17α-hydroxy-16-methylene-	MeOH	221(4.10),256(3.98), 299(4.12)	87-0631-69
$C_{22}H_{26}O_4$			
Octanedioic acid, 3,6-diphenyl-, dimethyl ester	EtOH	243(2.30),247(2.44), 254(2.57),258(2.66), 263(2.55),267(2.40)	22-3665-69
$C_{22}H_{26}O_5$			
2H-1-Benzopyran-3-carboxylic acid, 7-hydroxy-6-(3,7-dimethyl-2,6-octadienyl)-2-oxo-, ethyl ester	MeOH	251(3.82),260(3.75), 365(4.28),430(3.62)	32-0308-69
2H-1-Benzopyran-3-carboxylic acid, 7-hydroxy-8-(3,7-dimethyl-2,6-octadienyl)-2-oxo-, ethyl ester	MeOH	262(3.88),358(4.46), 425(3.55)	32-0308-69
$C_{22}H_{26}O_6$			
Arctigenin, methyl ether	MeOH	207(4.76),230(4.39), 279(3.93)	88-3803-69
Estra-1,3,5(10)-trien-6-one, 3,9α,17β-trihydroxy-, 3,17-diacetate	MeOH	210(4.27),243(3.89)	44-0116-69
3,9-Phenanthrenedione, 4,4a-dihydro-8-hydroxy-7-isopropyl-5,6,10-trimethoxy-1,4aβ-dimethyl-	EtOH	208(4.46),290(4.27), 330s(3.81),390(3.64)	33-1685-69
2H-Pyran-4-ol, 5-(3,4-dimethoxyphenyl)-tetrahydro-2-(p-methoxyphenyl)-	EtOH	227(4.22),275(3.68), 280(3.67)	39-1921-69C

Compound	Solvent	λ_{max}(log ε)	Ref.
Sequirin B, trimethyl-, acetate	EtOH	227(4.31),275(3.63), 280(3.63)	39-1921-69C
$C_{22}H_{26}O_7$			
D-gluco-Octose, 3,6-anhydro-5-O-benzyl-2-deoxy-1-C-(p-methoxyphenyl)-	MeOH	278(4.26)	65-0119-69
$C_{22}H_{26}O_8$			
Benzoic acid, 4-ethoxy-3-methoxy-, ester with 4'-ethoxy-2'-hydroxy-3',6'-dimethoxyacetophenone	EtOH	265(4.17),300(3.99)	18-2327-69
Euparotin, acetate	EtOH	210(4.26)(end abs.)	44-3876-69
Malonic acid, (m-phenylenedimethylidyne)di-, tetraethyl ester	EtOH	277(4.58)	23-4076-69
Malonic acid, (p-phenylenedimethylidyne)di-, tetraethyl ester	EtOH	203(4.34),221(4.11), 320(4.55)	23-4076-69
1,3-Propanedione, 1-(4-ethoxy-2-hydroxy-3,6-dimethoxyphenyl)-3-(4-ethoxy-3-methoxyphenyl)-	EtOH	298(4.31),387(4.17), 405s(4.10)	18-2327-69
$C_{22}H_{26}O_9$			
Olivin, 6,9-di-O-methyl-	EtOH	225(4.38),275(4.65), 325(3.68),340(3.68), 379(3.89)	105-0257-69
$C_{22}H_{27}BrN_2O_4$			
Isohydrocoriamyrtin, 11-O-methyl-, (p-bromophenyl)hydrazone	EtOH	255(3.92),313(4.28), 343(4.41)	78-1109-69
$C_{22}H_{27}Br_2P$			
Phosphorane, (2,7-dibromofluoren-9-ylidene)tripropyl-	$CHCl_3$	261(4.64),269(4.63), 302(4.40),401(3.37)	5-0001-69D
$C_{22}H_{27}ClN_2O_2$			
4-Allyl-1-deacetyldiabolinium chloride	EtOH	245(3.92),299(3.49)	33-1564-69
$C_{22}H_{27}ClO_2$			
18-Nor-9β,10α,17α-pregna-4,6-dien-20-yn-3-one, 21-chloro-13-ethyl-17-hydroxy-	n.s.g.	287(4.35)	54-0752-69
18-Nor-9β,10α-pregna-1,4,6-triene-3,20-dione, 6-chloro-13-ethyl-	n.s.g.	227(4.07),255(4.02), 302(4.02)	54-0737-69
$C_{22}H_{27}ClO_3$			
18-Nor-9β,10α-androsta-1,4,6-trien-3-one, 6-chloro-13-ethyl-17β-hydroxy-, acetate	n.s.g.	255(4.02),302(4.02)	54-0752-69
$C_{22}H_{27}ClO_8$			
Eupachlorin, acetate	MeOH	212(4.20)	44-3876-69
$C_{22}H_{27}FO_3$			
18-Nor-9β,10α-androsta-1,4,6-trien-3-one, 13-ethyl-6-fluoro-17β-hydroxy-, acetate	n.s.g.	255(4.02),301(4.05)	54-0752-69
$C_{22}H_{27}IN_2O$			
4-[p-(Dimethylamino)styryl]-2,3-dihydro-2,2,5-trimethyl-1,5-benzoxazepinium iodide	MeOH	528(5.23)	124-1178-69

Compound	Solvent	λ_{max}(log ε)	Ref.
$C_{22}H_{27}IN_2O_2$			
4-Allyl-1-deacetyldiabolinium iodide	MeOH	298(3.49)	33-1564-69
$C_{22}H_{27}NO_3$			
Hydrocinnamic acid, α,α-dimethyl-β-(2-methyl-N-phenylpropionamido)-, methyl ester	EtOH	259(2.66),264(2.54)	77-0936-69
$C_{22}H_{27}NO_3S$			
Tetrahydro-3-phenylquinolizinium p-toluenesulfonate	H_2O	217(4.29),224(4.26), 244(4.13)	87-0563-69
$C_{22}H_{27}NO_5$			
8H-Dibenzo[a,g]quinolizine, 5,6,13,13a-tetrahydro-2,3,9,10,11-pentamethoxy-	75% EtOH	224s(3.97),275s(3.34), 283(3.48),286s(3.44), 293s(3.30)	44-1349-69
C-Homoberbin-10-ol, 2,3,9,11-tetramethoxy-	EtOH	230s(4.20),282(3.80)	33-1228-69
Isoquinoline, 1,2,3,4-tetrahydro-8-methoxy-2-methyl-1-(2-methylveratryl)-6,7-(methylenedioxy)-	EtOH	278(3.47)	95-0869-69
Kreysigine, (-)-	EtOH	218(4.62),257(4.10), 293(3.67)	73-3540-69
$C_{22}H_{27}NO_7$			
α-Narcotinediol	EtOH	205(4.82),235(4.24), 281(3.59)	78-5059-69
$C_{22}H_{27}N_3O_6$			
Isohydrocoriamyrtin, 11-O-methyl-, (p-nitrophenyl)hydrazone	EtOH	251(3.71),296(3.65), 326(3.46),397(4.44)	78-1117-69
$C_{22}H_{28}$			
Biphenyl, 3,4'-decamethylene-	hexane	248(4.22)	88-4551-69
[1,9]Paracyclophane, tetracyanoethylene complex	CH_2Cl_2	509(3.24)	35-3553-69
$C_{22}H_{28}ClN_3O_7S_2$			
Sporidesmin D, monoacetate	MeOH	217(4.47),252(4.07), 302(3.34)	39-1564-69C
$C_{22}H_{28}Cl_2O_3$			
Androst-4-en-3-one, 6-(dichloromethylene)-17β-hydroxy-, acetate	EtOH	242(3.99),268(3.88)	35-2062-69
Androst-5-en-3-one, 4-(dichloromethylene)-17β-hydroxy-, acetate	EtOH	245(3.69),268(3.72)	35-2062-69
$C_{22}H_{28}INO_4$			
Glaucine, methiodide	MeOH	283(4.20),303(4.22)	78-3667-69
$C_{22}H_{28}N_2$			
Indoline, 2,3,3,3',3'-pentamethyl-2,2'-methylenedi-	EtOH	244(4.19),292(3.72)	32-1236-69
$C_{22}H_{28}N_2NiP_2S_2$			
2,2'-Biphenylylenebisdiethylphosphine, dithiocyanatonickel complex	$CHCl_3$	455(3.34)	39-1097-69A

Compound	Solvent	$\lambda_{max}(\log \epsilon)$	Ref.
$C_{22}H_{28}N_2O$			
1(4H)-Naphthalenone, 2-(cyclohexyl-amino)-4-(cyclohexylimino)-	EtOH	338(3.69),428(3.72)	39-1799-69C
$C_{22}H_{28}N_2O_2$			
19,20-Secostrychnine, 21,22-dihydro-19-methyl-	EtOH	305(3.84)	56-0699-69
$C_{22}H_{28}N_2O_3$			
Benz[a]azulene-10-carboxylic acid, 10-cyano-1,2,3,4,4a,4b,10,10a-octahydro-10a-morpholino-, ethyl ester	EtOH	282(3.53)	77-0739-69
Ervinceine	EtOH	248(4.12),328(4.26)	105-0280-69
17,18-Secoyohimban-17-oic acid, 1-form-yl-, ethyl ester	n.s.g.	247(4.23),291(3.63), 299(3.58)	35-7342-69
$C_{22}H_{28}N_2O_4$			
Isohydrocoriamyrtin, methylphenylhydra-zone	EtOH	248(3.89),305s(4.24), 328(4.43)	78-1109-69
Isohydrocoriamyrtin, 11-O-methyl-, phenylhydrazone	EtOH	248(3.89),303(4.07), 337(4.36)	78-1117-69
$C_{22}H_{28}N_2O_5$			
Cylindrocarine, 1-formyl-20-hydroxy-	MeOH	219(4.55),259(4.18), 291s(3.54)	39-0417-69C
Pyrrolo[3,4-b]pyrrole-4,6-dione, 1-benz-yl-3-carbomethoxy-5-(p-methoxyphenyl-carbamoyl)-1,4,5,6-tetrahydro-	EtOH	220(4.41)	94-2461-69
Thebaine, 6,7-dihydro-6,6-diethoxy-7-oxo-, oxime	EtOH	234(4.36)	77-0057-69
Voacristine pseudoindoxyl, hydrochlor-ide	EtOH	228(4.39),253s(--)	44-0412-69
$C_{22}H_{28}N_2O_6$			
8H-Oxazolo[3,2-a]pyrrolo[2,1-c]pyrazine-2-carboxylic acid, 5β-benzyl-2,3,5,6,-9,10,10a,10b-octahydro-10b-hydroxy-2-isopropyl-3,6-dioxo-, ethyl ester	MeOH	252(2.29),258(2.32), 264(2.21)	33-1549-69
$C_{22}H_{28}N_4O_6$			
2(3H)-Naphthalenone, 4,4a,5,6,7,8-tetra-hydro-7β-(1-hydroxy-1-methylethyl)-4aβ-methyl-, 2,4-dinitrophenyl-hydrazone, acetate	$CHCl_3$	382(4.53)	32-0231-69
3-Quinolinecarboxylic acid, 1-(2,4-di-nitroanilino)-1,4,5,6,7,8-hexahydro-2,4,4,7-tetramethyl-, ethyl ester	MeOH	215(4.13),256(3.92), 326(4.17)	44-0380-69
	EtOH	326(4.25)	44-1465-69
	pH 13	222(3.96),404(4.05), 460(3.87)	44-0380-69
$C_{22}H_{28}N_4O_8$			
Isohydrocoriamyrtin, 11-O-methyl-, methyl(2,4-dinitrophenyl)hydrazone	EtOH	267(3.96),283(3.94), 396(4.26)	78-1117-69
$C_{22}H_{28}O$			
Benzophenone, 2,2',3,3',4,4',5,5',6-nonamethyl-	EtOH	268(4.22),300s(3.54)	39-0536-69C
Estra-1,3,5(10)-triene, 3-methoxy-17-propenylidene-	EtOH	278(3.32),287(3.28)	35-3289-69

Compound	Solvent	$\lambda_{max}(\log \epsilon)$	Ref.
$C_{22}H_{28}O_2$			
13,24-Cyclo-18,21-dinor-9β,10α-chola-4,22-diene-3,20-dione	n.s.g.	238(4.36)	54-0737-69
18-Nor-9β,10α-pregna-4,6-dien-20-yn-3-one, 13-ethyl-17-hydroxy-	n.s.g.	287(4.40)	54-0752-69
18-Nor-9β,10α-pregna-1,4,6-triene-3,20-dione, 13-ethyl-	n.s.g.	225(4.08),252(3.97), 302(4.09)	54-0737-69
$C_{22}H_{28}O_3$			
Estr-4-en-3-one, 17β-acetoxy-17α-ethynyl-	MeOH	240(4.2)	48-0671-69
B-Homoestra-1,3,5(10),8-tetraen-17β-ol, 3-methoxy-, acetate	EtOH	254(4.08)	31-0571-69
D-Homo-B-nor-9β-estra-1,3,5(10),14-tetraen-17β-ol, 3-methoxy-8β-methyl-, acetate	MeOH	280(3.44)	78-4011-69
$C_{22}H_{28}O_4$			
Androsta-4,6-dien-3-one, 1α,2α-epoxy-17β-hydroxy-, propionate	EtOH	291(4.32)	19-0275-69
Benzoic acid, p-(3,5-di-tert-butyl-2-hydroxyphenoxy)-, methyl ester	EtOH	255(4.29)	44-0550-69
$C_{22}H_{28}O_5$			
Benz[e]azulene-3,8-dione, 3a,4,6a,7,9,-10,10a,10b-octahydro-10a-hydroxy-5-(hydroxymethyl)-7-isopropylidene-2,10-dimethyl-, 5-acetate	EtOH	205(4.15),235(4.09)	88-2399-69
13,15-Seco-4-tiglia-1,6,14-triene-3,13-dione, 9,20-dihydroxy-, 20-acetate	MeOH	196(4.15),237(4.08), 313(2.19)	5-0158-69H
$C_{22}H_{28}O_6$			
Crotophorbolone, 20-acetate	MeOH	202(4.11),245(3.70), 334(1.85)	5-0158-69H
2-Oxaandrosta-4,6-diene-1β-carboxylic acid, 17β-hydroxy-3-oxo-, methyl ester, acetate	n.s.g.	274(4.32)	19-0275-69
3,9-Phenanthrenedione, 1,2,4,4a-tetrahydro-8-hydroxy-7-isopropyl-5,6,10-trimethoxy-1,4a-dimethyl-	EtOH	263(4.07),302(3.80), 375(3.76)	33-1685-69
Phorbobutanone, 20-monoacetate	MeOH	230s(3.48),306(1.93), 330s(1.81)	64-0099-69
	CH_2Cl_2	311(1.85),330(1.73)	64-0099-69
Quassinol, dimethyl ether	EtOH	313(4.29)	7-0230-69
$C_{22}H_{28}O_7$			
14(13→12)-abeo-Tiglia-1,6-diene-3,13-dione, 4,9,12α-trihydroxy-20-acetoxy-	MeOH	235(3.73),327(2.07)	5-0142-69E
$C_{22}H_{29}ClO_2$			
18-Nor-9β,10α-pregna-4,6-diene-3,20-dione, 6-chloro-13-ethyl-	n.s.g.	287(4.32)	54-0737-69
18-Nor-9β,10α-pregn-4-en-20-yn-3-one, 21-chloro-13-ethyl-17-hydroxy-	n.s.g.	242(4.19)	54-0752-69
$C_{22}H_{29}ClO_3$			
18-Nor-9β,10α-androsta-4,6-dien-3-one, 17β-acetoxy-6-chloro-13-ethyl-	n.s.g.	288(4.32)	54-0752-69
19-Nor-5α,17α-carda-14,20(22)-dienolide, 5-chloro-3β-hydroxy-	EtOH	208(4.25)	33-0482-69

Compound	Solvent	λ max(log ε)	Ref.
$C_{22}H_{29}Cl_2NO_2$			
Acetamide, 2,2-dichloro-N-(13-isopropyl-7-oxopodocarpa-8,11,13-trien-15-yl)-	MeOH	254(4.00)	5-0193-69H
$C_{22}H_{29}Cl_3O_3$			
Androst-5-en-3-one, 17β-hydroxy-4α-(trichloromethyl)-, acetate	EtOH	270(1.72)	35-2062-69
$C_{22}H_{29}FO_3$			
18-Nor-9β,10α-androsta-4,6-dien-3-one, 17β-acetoxy-13-ethyl-6-fluoro-	n.s.g.	285(4.35)	54-0752-69
$C_{22}H_{29}NO_2$			
9β,10α-Pregn-4-ene-18-carbonitrile, 3,20-dioxo-	n.s.g.	241(4.22)	54-0737-69
$C_{22}H_{29}NO_3$			
Denudatine, oxo-	n.s.g.	230(3.54)	88-4369-69
Etiojerva-1,3,5(10)-trien-3-ol, 17α-acetamido-, 3-acetate	EtOH	271(3.04),278(3.04)	78-4551-69
19-Nor-9β,17α-pregna-4,20-diene-20-carbamic acid, 17-hydroxy-N-methyl-3-oxo-, γ-lactone	EtOH	240s(4.13)	44-3022-69
9β,10α-Pregn-4-ene-18-carbonitrile, 18,20β-epoxy-20-hydroxy-3-oxo-	n.s.g.	241(4.22)	54-0737-69
9β,10α-Pregn-4-ene-18-carbonitrile, 2β-hydroxy-3,20-dioxo-	n.s.g.	240(4.12)	54-0737-69
$C_{22}H_{29}NO_5$			
Isoquinoline, 1,2,3,4-tetrahydro-6,7-dimethoxy-1-(3,4,5-trimethoxyphenethyl)-, hydrobromide	EtOH	230s(4.18),280(3.57)	33-1228-69
$C_{22}H_{29}N_3$			
Benzo[g]quinoline, 4-[[4-(diethylamino)-1-methylbutyl]amino]-	EtOH	243(4.72),254(4.69), 304(3.50),390(3.92)	103-0223-69
Quinazoline, 2-[3-(dimethylamino)propyl]-4-mesityl-1,2-dihydro-, hydrochloride	EtOH	227(4.7),385(3.4)	33-2351-69
$C_{22}H_{29}N_3O_2$			
1,4-Phthalazinedione, 5-[2-(dicyclohexylamino)vinyl]-2,3-dihydro-	THF	270(3.4),310(3.5), 380(4.0)	24-3241-69
$C_{22}H_{29}N_3O_3$			
A-Homoestra-1(10),2,4a-triene-4,17-dione, cyclic 17-(ethylene acetal)-, semicarbazone	EtOH	248(4.21),322(4.30)	33-0121-69
$C_{22}H_{30}$			
Biphenyl, 3-butyl-4'-hexyl-	hexane	252(4.29)	88-4551-69
$C_{22}H_{30}Cl_2O_3$			
5α-Androstan-3-one, 17β-acetoxy-2-(dichloromethylene)-	EtOH	252(3.86)	35-2062-69
$C_{22}H_{30}N_2O$			
Androst-4-en-3-one, 17β-pyrazol-3-yl-	MeOH	241(4.22)	73-0442-69
1,3,4-Oxadiazole, 2,5-di-1-adamantyl-	EtOH	250(1.36)	18-1617-69

Compound	Solvent	λ_{max}(log ε)	Ref.
18,19-Secostrychnidine, 21,22-dihydro-19-methyl-	EtOH	258(3.96),304(3.45)	56-0699-69
$C_{22}H_{30}N_2O_2$			
2β,19α-Aspidospermine, (±)-	EtOH	256(4.05),285(3.55)	88-2523-69
23-Hydroxydihydrochano-N_b-methyldihydrochanodihydrostrychnine	EtOH	255(4.12)	56-0699-69
$C_{22}H_{30}N_2O_3$			
Cylindrocarine, N-acetyl-17-demethyl-	MeOH	212(4.47),250(4.01), 278(3.58),288(3.50)	39-0417-69C
Cylindrocarine,1-methyl-	MeOH	218(4.42),265(3.82), 303(3.34)	39-0417-69C
Ervinceine, dihydro-	EtOH	249(3.78),305(3.73)	105-0280-69
$C_{22}H_{30}N_2O_4$			
13,14-Diazatricyclo[6.4.1.1^{2,7}]tetradeca-3,5,9,11-tetraene-13,14-dicarboxylic acid, di-tert-butyl ester	EtOH	233(4.14),238(4.16)	35-3616-69
$C_{22}H_{30}N_2O_4S_2$			
Dithieno[2,3-c:2',3'-h][1,6]diazecine-2,8-dicarboxylic acid, 5,11-diethyl-4,5,6,10,11,12-hexahydro-3,9-dimethyl-, dimethyl ester	EtOH	269(4.35)	44-0333-69
Dithieno[2,3-c:3',2'-h][1,6]diazecine-2,8-dicarboxylic acid, 5,11-diethyl-4,5,6,10,11,12-hexahydro-3,7-dimethyl-, dimethyl ester	EtOH	261(4.35),270s(4.33)	44-0333-69
$C_{22}H_{30}N_2O_6$			
Malonic acid, acetamido[[2-(2-hydroxy-1,1-dimethylethyl)indol-3-yl]methyl]-, diethyl ester	EtOH	223(4.46),283(3.78), 291(3.72)	39-1003-69C
$C_{22}H_{30}N_4O_4$			
1,6-Hexanediamine, N,N'-bis(6-nitro-3,4-xylyl)-	EtOH	237(4.69),294(4.17), 442(4.13)	44-3240-69
	EtOH-HCl	237(4.70),293(4.15), 442(4.13)	44-3240-69
	EtOH-NaOH	235s(4.38),295(4.16), 445(4.12)	44-3240-69
4-Octen-1-one, 1-(1-cyclopenten-1-yl)-5-propyl-, 2,4-dinitrophenylhydrazone	EtOH	379(4.19)	33-1996-69
9-Tridecen-5-ynal, 10-propyl-, 2,4-dinitrophenylhydrazone	EtOH	360(4.28)	33-1996-69
$C_{22}H_{30}N_4O_5S$			
2-Butanone, 1-[3-[(4-amino-2-methyl-5-pyrimidinyl)methyl]-5-(2-hydroxyethyl)-4-methyl-4-thiazolin-2-ylidene]-1-hydroxy-3,3-dimethyl-, diacetate	EtOH	233(4.30),274(4.07), 363(4.42)	94-0128-69
2-Pentanone, 1-[3-[(4-amino-2-methyl-5-pyrimidinyl)methyl]-5-(2-hydroxyethyl)-4-methyl-4-thiazolin-2-ylidene]-1-hydroxy-3-methyl-, diacetate	EtOH	373(4.42)	94-0128-69
2-Pentanone, 1-[3-[(4-amino-2-methyl-5-pyrimidinyl)methyl]-5-(2-hydroxyethyl)-4-methyl-4-thiazolin-2-ylidene]-1-hydroxy-4-methyl-, diacetate	EtOH	232(4.24),273(3.90), 373(4.42)	94-0128-69

Compound	Solvent	λ_{max}(log ε)	Ref.
$C_{22}H_{30}N_6O_{13}$			
2-Azetidinecarboxylic acid, 1-[5-(2-amino-2-deoxy-L-xylonamido)-1,5-dideoxy-1-(3,4-dihydro-2,4-dioxo-1(2H)-pyrimidinyl)-β-D-allofuranuronoyl]-3-ethylidene-, monocarbamate, (E)-L-, (polyoxin K)	0.05N HCl	259(3.95)	35-7490-69
	0.05N NaOH	262(3.86)	35-7490-69
$C_{22}H_{30}O$			
Estra-1,3,5(10)-triene, 3-methoxy-17-propylidene-	EtOH	278(3.28),287(3.26)	77-0043-69
$C_{22}H_{30}O_2$			
4'H-Androsta-1,4,16-trieno[17,16-b]pyran-3-one, 5',6',16α,17α-tetrahydro-	EtOH	246(4.18)	73-3479-69
2H-1-Benzopyran-5-ol, 2-[2-(4,8-dimethyl-3,7-nonadienyl)-2,7-dimethyl-	MeOH	230(4.24),280(3.77), 289(3.77)	32-0308-69
Dinorchola-1,4-dien-3-one, 16β,23-epoxy-	EtOH	244(4.26)	73-3479-69
1(2H)-Naphthalenone, 3,4-dihydro-2-(2-isobutyrylcyclohexyl)-5,8-dimethyl-, m. 73°	EtOH	212(4.33),254(4.03), 303(3.34)	22-2110-69
m. 131°	EtOH	212(4.32),254(4.03) 303(3.34)	22-2110-69
18-Nor-9β,10α-pregna-4,6-diene-3,20-dione, 13-ethyl-	n.s.g.	285(4.41)	54-0737-69
18-Nor-9β,10α,17α-pregna-4,6-diene-3,20-dione, 13-ethyl-	n.s.g.	286(4.35)	54-0737-69
18-Nor-9β,10α-pregn-4-en-20-yn-3-one, 13-ethyl-17-hydroxy-	EtOH	241(4.21)	44-0107-69
	n.s.g.	243(4.21)	54-0752-69
$C_{22}H_{30}O_3$			
Androsta-2,4-diene-3-carboxaldehyde, 17β-acetoxy-	EtOH	280(4.39),295(4.32)	78-4535-69
Androsta-3,5-diene-3-carboxaldehyde, 17β-acetoxy-	EtOH	288(4.44)	78-4535-69
Androsta-1,4-dien-3-one, 17β-acetoxy-2-methyl-	EtOH	247(4.21)	44-2288-69
2H-1-Benzopyran-5-ol, 2-[2-(5,5-dimethyl-1-oxaspiro[2.5]oct-4-yl)ethyl]-2,7-dimethyl-	n.s.g.	231(4.41),281(3.83)	88-3017-69
B-Homoestra-1,3,5(10)-trien-17β-ol, 3-methoxy-, acetate	EtOH	276(3.26),283(3.22)	31-0571-69
19-Norpregna-1(10),5-dien-20-one, 3β-acetoxy-	MeOH	237(3.98)	73-0458-69
19-Norpregna-3,5-dien-20-one, 3-acetoxy-	MeOH	234(4.27)	73-0458-69
14β,17α-Pregn-4-ene-3,20-dione, 16β,17β-epoxy-16 -methyl-	MeOH	241(4.20)	73-0451-69
9β,10α-Pregn-4-ene-3,20-dione, 17α-hydroxy-18-methyl-	n.s.g.	242(4.22)	54-0737-69
Siccanochromene E	EtOH	234(4.42),287(4.14), 293(4.15)	88-2457-69
$C_{22}H_{30}O_4$			
6H-Dibenzo[b,d]pyran-4-carboxylic acid, 6a,7,8,10a-tetrahydro-1-hydroxy-6,6,9-trimethyl-3-pentyl-	EtOH	216(4.44),250s(3.51), 285s(3.23)	88-2339-69
Estra-1(10),5-diene-3β,17β-diol, diacetate	MeOH	236(4.21)	73-0458-69

Compound	Solvent	λ_{max}(log ϵ)	Ref.
$C_{22}H_{30}O_5$			
14β,17α-Cortexone, 8,19-epoxy-19-methoxy-	EtOH	241(4.23)	33-2156-69
Estr-4-en-3-one, 11β,17β-diacetoxy-, d-	MeOH	239(4.22)	44-3530-69
Estr-4-en-3-one, 12α,17β-diacetoxy-, l-	EtOH	240(4.21)	44-3530-69
4-Oxaandrost-5-en-3-one, 6-acetyl-17β-hydroxy-, acetate	MeOH	252(3.99)	44-2767-69
$C_{22}H_{30}O_6$			
1-Oxaspiro[5.5]undec-3-en-2-one, 3,3'-ethylidenebis[4-hydroxy-	MeOH	253(4.22)	83-0075-69
$C_{22}H_{30}O_7$			
Tolypolide F_1	EtOH	207(3.95)	88-0091-69
$C_{22}H_{31}FO_4$			
Pregn-4-ene-3,20-dione, 6α-fluoro-9α,21-dihydroxy-16α-methyl-	MeOH	236(4.15)	5-0168-69F
Pregn-4-ene-3,20-dione, 6α-fluoro-11β,21-dihydroxy-16α-methyl-	MeOH	237(4.18)	5-0168-69F
Pregn-4-ene-3,20-dione, 6β-fluoro-11α,21-dihydroxy-16α-methyl-	MeOH	232(4.06)	5-0168-69F
11β-form	MeOH	232(4.05)	5-0168-69F
$C_{22}H_{31}NO$			
2-Aziridinone, 1,3-di-1-adamantyl-	hexane	252(2.17)	77-0049-69
	EtOH	241(2.34)	77-0049-69
$C_{22}H_{31}NO_2$			
Acetamide, N-(13-isopropyl-7-oxopodocarpa-8,11,13-trien-15-yl)-	MeOH	253(4.02)	5-0193-69H
9β,10α-Pregn-4-ene-20-carbonitrile, 20-hydroxy-3-oxo-	n.s.g.	241(4.21)	54-0737-69
$C_{22}H_{31}NO_3$			
Androsta-1,4-dien-3-one, 17β-acetoxy-, O-methyloxime	EtOH	245s(4.03),269(4.18)	78-1219-69
$C_{22}H_{31}NO_5$			
Estra-3,5-dien-17β-ol, 3-ethoxy-2β-nitro-, acetate	EtOH	246(4.31)	44-3699-69
$C_{22}H_{31}N_3O_3$			
Androst-4-en-3-one, 6β-azido-17β-hydroxy-17α-methyl-, acetate	MeOH	235(4.16)	48-0445-69
$C_{22}H_{32}BF_4NO_4$			
10-Azoniaestra-1,3,5(10)-triene, 1,3-diethoxy-17β-hydroxy-, tetrafluoroborate, 17-formate	EtOH	241(4.09),276(3.90)	22-0117-69
$C_{22}H_{32}ClNO_8$			
[2-[1-(2-Hydroxypropyl)-7,8-dimethoxy-2-naphthyl]ethyl]trimethylammonium perchlorate, acetate	H_2O	290(3.76),301(3.72), 338(3.35)	49-0163-69
$C_{22}H_{32}N_2$			
Ethylenediamine, N,N'-di-tert-butyl-1,2-diphenyl-	EtOH	249(2.78)	35-2653-69

Compound	Solvent	λ max (log ε)	Ref.
Propionitrile, 3-(androsta-3,5-dien-17β-ylamino)-	EtOH	228(4.30),235(4.34), 243s(4.08)	94-1749-69
$C_{22}H_{32}N_2O$			
5α-Androstan-3-one, 17β-pyrazol-3-yl-	MeOH-HCl	221(3.57)	73-0442-69
Androst-4-en-3β-ol, 17β-pyrazol-3-yl-	MeOH-HCl	220(3.42)	73-0442-69
$C_{22}H_{32}N_4O_8S_2$			
Hexamethylenediamine, 1,6-dimethoxy-3,4-dinitroso-N,N'-bis(p-tolylsulfonyl)-	dioxan	288(3.89)	88-2651-69
$C_{22}H_{32}O_2$			
4'H-5α-Androsta-1,16-dieno[17,16-b]pyran-3-one, 5',6',16α,17α-tetrahydro-	EtOH	229(3.99)	73-3479-69
4'H-5α-Androsta-1,16-dieno[17,16-b]pyran-3-one, 5',6',16β,17α-tetrahydro-	EtOH	229(3.98)	73-3479-69
Androsta-3,5-dien-3-ol, 6-methyl-, acetate	EtOH	245(4.18)	44-2288-69
9β,10α-Androsta-4,6-dien-3-one, 17α-ethyl-17-hydroxy-18-methyl-	n.s.g.	286(4.39)	54-0752-69
5α-Bisnorchol-1-en-3-one, 16β,23-oxido-	EtOH	229(4.20)	73-3479-69
Δ^8-Cannabinol, tetrahydro-1'-methyl-, 6a,10a-trans	EtOH	230s(4.02),276(3.19), 282(3.20)	33-1102-69
Δ^9-Cannabinol, tetrahydro-1'-methyl-, 6a,10a-trans	EtOH	230s(4.05),275(3.43), 281(3.44)	33-1102-69
18-Nor-9β,10α-pregna-4,6-dien-3-one, 13-ethyl-20β-hydroxy-	n.s.g.	288(4.41)	54-0737-69
18-Nor-9β,10α-pregna-4,20-dien-3-one, 13-ethyl-17-hydroxy-	n.s.g.	243(4.20)	54-0752-69
18-Nor-9β,10α-pregn-4-ene-3,20-dione, 13-ethyl-	n.s.g.	242(4.23)	54-0737-69
Resorcinol, 5-methyl-2-(3,7,11-trimethyl-2,6,10-dodecatrienyl)- (grifolin)	MeOH	275(2.88),282(2.88)	32-0308-69
Resorcinol, 5-methyl-4-(3,7,11-trimethyl-2,6,10-dodecatrienyl)-	MeOH	283(3.55)	32-0308-69
Retinol, acetate	benzene	333(4.72)	78-5383-69
	hexane	324(4.72)	78-5383-69
	C_6H_{12}	325(4.72)	78-5383-69
	CCl_4	333(4.71)	78-5383-69
	dioxan	328(4.70)	78-5383-69
	DMF	331(4.72)	78-5383-69
$C_{22}H_{32}O_3$			
Androst-4-en-3-one, 17β-acetoxy-2α-methyl-	EtOH	240(4.16)	44-2288-69
Cyclopropa[5,6]-A-nor-5β-estran-3-one, 3',6α-dihydro-17β-hydroxy-3',3'-dimethyl-, acetate	EtOH	216(3.75)	88-1925-69
19-Norandrost-5-en-3-one, 17β-acetoxy-4,4-dimethyl-	EtOH	296(2.00)	88-1925-69
18-Nor-9β,10α-androst-4-en-3-one, 17β-acetoxy-13-ethyl-	n.s.g.	242(4.20)	54-0752-69
Pregn-5-ene-21-carboxaldehyde, 3β-hydroxy-20-oxo-	MeOH	272(3.40)	73-0442-69
$C_{22}H_{32}O_3S$			
Androst-4-en-3-one, 17β-hydroxy-7α-mercapto-17α-methyl-, 7-acetate	EtOH	239(4.29)	94-0011-69

Compound	Solvent	λ_{max}(log ε)	Ref.
$C_{22}H_{32}O_4$			
Agbanindiol B, monoacetate	MeOH	212(3.72)	39-2153-69C
$C_{22}H_{32}O_5$			
18-Norpregn-4-ene-3,20-dione, 13-ethyl-14α,17β,21-trihydroxy-	MeOH	241(4.17)	5-0161-69F
18-Norpregn-4-ene-3,20-dione, 13-ethyl-15α,17α,21-trihydroxy-	MeOH	242(4.12)	5-0161-69F
Torilin	EtOH	236(4.29)	78-4751-69
$C_{22}H_{32}O_6$			
Nigakilactone B	EtOH	272(3.83)	88-3013-69
$C_{22}H_{33}NO$			
4'H-Androsta-2,4-dieno[3,2-b]pyrrol-17β-ol, 2β,5'-dihydro-17-methyl-	EtOH	237(4.34)	19-0269-69
$C_{22}H_{33}NO_4$			
Pregn-5-en-16-one, 3β-hydroxy-20-methyl-21-nitro-	EtOH	285(3.1)	88-0583-69
	EtOH-NaOH	235(4.0)	88-0583-69
$C_{22}H_{34}N_2OS$			
2'H-5α-Androst-2-eno[3,2-c]pyrazol-17β-ol, 17α-methyl-7-(methylthio)-	EtOH	224(3.67)	94-0011-69
$C_{22}H_{34}N_2O_2$			
Tryptophan, 6-isopentyl-2-tert-pentyl-, methyl ester	n.s.g.	228(4.54),286(3.83)	88-4457-69
$C_{22}H_{34}O_2$			
Androst-4-en-3-one, 17α-ethyl-17β-hydroxy-18-methyl-, dl-	EtOH	242(4.21)	44-0107-69
	n.s.g.	242(4.20)	54-0752-69
12,14-(2-Oxapropano)abiet-7,9(11)-dien-ol	EtOH	238(--),245(4.13), 255(--)	44-1940-69
5α-Pregnan-21-al, 3β-hydroxy-20-methylene-	EtOH	222(3.67)	13-0637-69B
$C_{22}H_{34}O_2S$			
Androst-4-en-3-one, 7α-(ethylthio)-17β-hydroxy-17α-methyl-	EtOH	242(4.19)	94-0011-69
$C_{22}H_{34}O_2S_2$			
Androst-4-en-3-one, 17β-hydroxy-7α-[(2-mercaptoethyl)thio]-17α-methyl-	EtOH	241(4.20)	94-0011-69
$C_{22}H_{34}O_3$			
5α-Androstane-$\Delta^{3,\alpha}$-acetic acid, 17β-hydroxy-, methyl ester	MeOH	225(4.18)	39-2728-69C
$C_{22}H_{34}O_3S$			
Androstan-3-one, 17β-hydroxy-2-(hydroxymethylene)-7-(methylthio)-17α-methyl-	EtOH	283(3.97)	94-0011-69
Androst-4-en-3-one, 7α-(ethylsulfinyl)-17β-hydroxy-17α-methyl-	EtOH	235(4.06)	94-0011-69
$C_{22}H_{34}O_4$			
1-Cyclopentene-1-heptanoic acid, 2-(3-hydroxy-1-octynyl)-5-oxo-, ethyl ester	EtOH	272(4.27)	94-0408-69

Compound	Solvent	λ_{max}(log ε)	Ref.
Labda-8,13-diene-15,19-dioic acid, dimethyl ester, trans	EtOH	215(4.22)	12-0491-69
Labda-8(20),13-dien-15-oic acid, 19-acetoxy-	EtOH	219(4.10)	12-1681-69
Solidagonic acid	EtOH	218(3.96)	18-0812-69
$C_{22}H_{34}O_4S$			
Androst-4-en-3-one, 7α-(ethylsulfonyl)-17β-hydroxy-17α-methyl-	EtOH	242(4.20)	94-0011-69
$C_{22}H_{35}NO$			
18,19-Dinor-10α-pregn-13-en-20-one, 3β-(methylamino)-5β,14β-dimethyl-	EtOH	253(4.03)	22-3265-69
10β-	EtOH	258(4.01)	22-3265-69
$C_{22}H_{35}NO_2$			
Androst-5-en-17-one, 3α-[(2-hydroxyethyl)methylamino]-	EtOH	197(3.95)	39-2504-69C
$C_{22}H_{36}N_2$			
1,3-Propanediamine, N-androsta-3,5-dien-17β-yl-, dihydrochloride	EtOH	228s(4.00),235(4.08), 243s(3.90)	94-1749-69
$C_{22}H_{36}N_2O$			
5α-Androstan-3β-ol, 17β-(2-pyrazolin-3-yl)- (after ten months)	MeOH-HCl	221(3.78)	73-0442-69
$C_{22}H_{36}N_2O_2$			
Hydroquinone, 2,5-bis[(cyclohexylmethylamino)methyl]-	EtOH	302(3.67)	39-1245-69C
Hydroquinone, 2,5-bis(hexahydro-1(2H)-azocinyl)methyl]-	EtOH	303(3.70)	39-1245-69C
$C_{22}H_{36}N_4O_2$			
p-Benzoquinone, 2,5-dimethyl-3,6-bis-[(2-piperidinoethyl)amino]-	EtOH	221(4.41),353(4.39), 535(2.12)	39-1325-69C
$C_{22}H_{36}O_2$			
Abienol, acetate, trans	EtOH	232(4.40)	65-0451-69
Isoabienol, acetate	EtOH	225(4.24)	65-0451-69
$C_{22}H_{36}O_4$			
1-Cyclopentene-1-heptanoic acid, 2-(3-hydroxy-1-octenyl)-5-oxo-, ethyl ester	EtOH	278(4.32)	94-0408-69
5(6H)-Indanone, 1β-tert-butoxy-7aβ-ethyl-6-[2-(2-ethyl-1,3-dioxolan-2-yl)ethyl]-7,7a-dihydro-	EtOH	250(4.07)	44-0107-69
$C_{22}H_{36}O_5$			
Ancepsenolide, hydroxy-	EtOH	209(4.20)	44-1989-69
$C_{22}H_{39}NO_4$			
4-Octadecenoic acid, 2-acetamido-3-oxo-, ethyl ester, trans, m. 72-4°	EtOH	235(3.90),293(4.01)	104-1973-69
isomer m 64°	EtOH	235(4.15)	104-1973-69
$C_{22}H_{39}N_3O$			
Phenol, 2,6-di-tert-butyl-4-(3,3-dibutyl-1-triazeno)-	n.s.g.	234(4.07),291(4.13), 327(4.14)	70-2460-69

Compound	Solvent	$\lambda_{max}(\log \epsilon)$	Ref.
$C_{22}H_{41}BrN_4O_2$			
[3-(Diethylamino)-2,4-bis(diethylcarbamoyl)-2-cyclobuten-1-ylidene]diethylammonium bromide	n.s.g.	295(4.48)	78-3447-69
$C_{22}H_{41}NO_4$			
Octadecanoic acid, 2-acetamido-3-oxo-, ethyl ester	EtOH	259(2.86)	104-1973-69
$C_{22}H_{41}O_4P$			
Tributyl[1-(dicarboxymethyl)-2-methylpropyl]phosphonium hydroxide, cyclic isopropylidene ester, inner salt	MeOH	260(4.1)	49-0576-69
$C_{22}H_{43}N_9O_{11}$			
L-Alanine, L-alanyl-L-isoleucyl-N(ω)-nitro-L-arginyl-L-seryl-, acetate	HOAc	269(3.72)	39-0401-69C

Compound	Solvent	λ_{max}(log ε)	Ref.
$C_{23}H_{12}OS$			
14H-Benzo[b]benzo[3,4]fluoreno[2,1-d]-thiophen-14-one	$CHCl_3$	253s(4.57),271(4.81), 305s(4.54),314(4.62), 333(4.15),347(3.98)	44-1310-69
$C_{23}H_{13}NO_2S$			
Naphtho[2,3-h]quinoline-7,12-dione, 5-(phenylthio)-	EtOH	283(4.42),430(3.81)	104-0742-69
$C_{23}H_{13}N_3OS$			
5H-Benzo[a]phenothiazine, 5-(2-benzoxa-zolylimino)-	DMF	280(4.02),365(3.78), 560(4.00)	44-1691-69
$C_{23}H_{13}N_3S_2$			
5H-Benzo[a]phenothiazine, 5-(2-benzoxa-zolylimino)-	DMF	218(4.37),263(4.08), 340(3.96),370(3.93), 519(3.97)	44-1691-69
$C_{23}H_{14}N_4S$			
5H-Benzo[a]phenothiazine, 5-(2-benzimid-azolylimino)-	DMF	283(4.31),348(4.18), 374(4.22),535(4.29)	44-1691-69
$C_{23}H_{14}O_2$			
14H-Indeno[1,2-b]phenanthro[9,10-e]-p-dioxin	dioxan	367(3.47)	24-3747-69
$C_{23}H_{14}O_6$			
Sphagnorubin	MeOH-HCl	240(4.77),291(4.75), 546(4.90)	64-1211-69
$C_{23}H_{14}S$			
14H-Benzo[b]benzo[3,4]fluoreno[2,1-d]-thiophene	$CHCl_3$	251(4.62),259(4.61), 277(4.43),298(4.00), 311(4.15),334s(4.27), 348(4.49),367(4.50)	44-1310-69
$C_{23}H_{15}BrOS$			
2H-Pyran-2-thione, 3-bromo-4,5,6-tri-phenyl-	EtOH	292(4.26),420(4.01)	39-1950-69C
$C_{23}H_{15}Br_2N_3O_2$			
4H-1,2-Diazepine, 3,7-bis(p-bromophen-yl)-5-(p-nitrophenyl)-	EtOH	206(4.59),283(4.61)	1-3125-69
$C_{23}H_{15}NO$			
5H-Pyrrolo[2,1-a]isoindol-5-one, 1,2-diphenyl-	EtOH	240(4.47),345(3.86), 271(4.45)	88-0887-69
$C_{23}H_{15}NO_4$			
2H-Pyran-2-one, 3-nitro-4,5,6-triphenyl-	EtOH	253(4.29),355(4.09)	39-1950-69C
$C_{23}H_{16}$			
7H-Benzo[c]fluorene, 5-phenyl-	$CHCl_3$	246(4.51),330(4.31), 344(4.32)	44-1310-69
$C_{23}H_{16}Br_2N_2$			
4H-1,2-Diazepine, 3,7-bis(p-bromophen-yl)-5-phenyl-	EtOH	215(4.14),267(4.30), 300(4.20)	1-3125-69

Compound	Solvent	λ_{max}(log ε)	Ref.
$C_{23}H_{16}Cl_2NOPS$			
Formamide, 1-[bis(p-chlorophenyl)phosphinyl]-N-1-naphthylthio-	EtOH	335(3.76),388s(1.83)	18-2975-69
$C_{23}H_{16}N_2$			
Benzo[b]phenazine, 6-methyl-11-phenyl-	EtOH	250(4.71),288(4.98), 322(3.93)	44-1651-69
$C_{23}H_{16}N_2O$			
1H-Benz[g]indazol-5-ol, 3,4-diphenyl-	EtOH	223(4.63),256(4.56), 278s(--),318(4.07)	103-0254-69
Δ^2-Cyclopenta[b]quinoxalin-2-ol, 1,3-diphenyl-	EtOH	224(4.34),266(4.13), 310(4.06),325(4.00)	44-1651-69
$C_{23}H_{16}N_2O_2$			
2,4,6-Cycloheptatrien-1-one, 2-hydroxy-5-(1-naphthylazo)-4-phenyl-	n.s.g.	215(4.51),258(4.21), 312s(3.84),427(4.22), 456(4.23)	40-0085-69
2,4,6-Cycloheptatrien-1-one, 2-hydroxy-5-(2-naphthylazo)-4-phenyl-	n.s.g.	265(4.41),305s(4.22), 414(4.48)	40-0085-69
$C_{23}H_{16}N_2O_3$			
1,3-Butadiene-1-carboxylic acid, 3-amino-2,4-diphenyl-1-(p-nitrophenyl)-, lactam	CH_2Cl_2	365(4.42)	24-0319-69
1-Buten-3-one, 1-cyano-2,4-diphenyl-1-(p-nitrophenyl)-	CH_2Cl_2	304(4.15)	24-0319-69
$C_{23}H_{16}N_2O_4$			
Xanthene-2,2(1H)-dicarbonitrile, 1-hydroxy-1-(p-methoxyphenyl)-4-methyl-9-oxo-	MeOH	296(4.20),461(4.41)	44-2407-69
	CH_2Cl_2	256(4.22),264(4.22), 300(4.09)	44-2407-69
$C_{23}H_{16}N_2S_2$			
2H-1,4-Benzothiazine, 2-(2-benzothiazolylmethylene)-3-p-tolyl-	EtOH	257(4.42),327(4.21), 418(4.03)	32-0323-69
	EtOH-HCl	252(4.30),350(4.23), 493(3.89)	32-0323-69
$C_{23}H_{16}O$			
Pyrene, 2-(p-methoxyphenyl)-	EtOH	231(4.67),283(5.19), 307(4.65),321(4.77), 337(5.05)	24-2301-69
$C_{23}H_{16}OS$			
2H-Thiopyran-2-one, 4,5,6-triphenyl-	EtOH	254(4.22),358(3.77)	39-1950-69C
$C_{23}H_{16}O_2$			
Anthracene, 2-[3,4-(methylenedioxy)-styryl]-	DMF	307(4.63),327(4.68), 340(4.72),377(4.42), 396(4.37)	35-2521-69
Benzo[1,2-b:5,4-b']difuran, 8-methyl-2,6-diphenyl-	MeCN	284s(4.51),293(4.53), 325s(4.48),337(4.58), 351(4.52)	12-2395-69
2,2'-Spirobi[2H-1-benzopyran], 3-phenyl-	dioxan	266(4.30),292(4.31)	5-0162-69B
Stilbene, 3,4-(methylenedioxy)-4'-(phenylethynyl)-	DMF	355(4.74)	33-2521-69

Compound	Solvent	$\lambda_{max}(\log \epsilon)$	Ref.
$C_{23}H_{16}O_2S$			
Benzo[b]thiophene, 2-[p-[3,4-(methylene-dioxy)styryl]phenyl]-	DMF	366(4.77)	33-1282-69
Benzo[b]thiophene, 5-[3,4-(methylenedi-oxy)styryl]-2-phenyl-	DMF	308(4.56),328(4.57)	33-1282-69
3H-Naphtho[2,3-c][1,2]oxathiole, 4,9-di-phenyl-, 1-oxide	EtOH	239(4.76),301(4.15), 333(3.78)	44-1310-69
$C_{23}H_{16}O_3$			
Benzofuran, 2-[p-[3,4-(methylenedioxy)-styryl]phenyl]-	DMF	366(4.81)	33-1282-69
Benzofuran, 6-[3,4-(methylenedioxy)-styryl]-2-phenyl-	DMF	284(4.03),361(4.74)	33-1282-69
1,3-Indandione, 2-(benzoylmethyl)-2-phenyl-	EtOH	227(4.66),245(4.51)	103-0251-69
$C_{23}H_{16}O_4$			
8H-Dibenzo[b,h]xanthene-8,13(13aH)-dione, 5-hydroxy-6,13a-dimethyl-	EtOH	237(4.60),264(4.20), 313(4.14),366(3.99), 486(3.88)	88-3579-69
$C_{23}H_{16}O_8$			
Coumarin, 4,4'-methylenebis[3-acetoxy-	EtOH	277(4.31),315s(--)	87-0531-69
$C_{23}H_{17}BrN_2$			
4H-1,2-Diazepine, 5-(p-bromophenyl)-3,7-diphenyl-	EtOH	225(4.39),263(4.59), 294(4.48)	1-3125-69
$C_{23}H_{17}BrO_2$			
Benzofuran, 5-bromo-2-[p-(p-methoxy-styryl)phenyl]-	DMF	365(4.81)	33-1282-69
$C_{23}H_{17}ClN_2$			
4H-1,2-Diazepine, 5-(p-chlorophenyl)-3,7-diphenyl-	EtOH	209(4.48),261(4.52), 292(4.38)	1-3125-69
$C_{23}H_{17}ClN_2O_4S$			
Spiro[2H-1-benzopyran-2,2'-benzothiazo-line], 3-(p-chlorophenyl)-8-methoxy-3'-methyl-6-nitro-	EtOH	215(4.48),293(4.33), 332s(4.08)	22-3329-69
$C_{23}H_{17}ClN_2S$			
4H-Thiopyran-4-one, 3-chloro-2,6-di-phenyl-, phenylhydrazone	EtOH	254(4.34),381(4.18)	39-0315-69C
$C_{23}H_{17}ClO$			
Anisole, p-[[p-(p-chlorostyryl)phenyl]-ethynyl]-	DMF	350(4.79)	33-2521-69
Benzofuran, 2-[p-(p-chlorostyryl)phen-yl]-6-methyl-	DMF	362(4.81)	33-1282-69
$C_{23}H_{17}ClOS$			
Benzo[b]thiophene, 5-chloro-2-[p-(p-methoxystyryl)phenyl]-	DMF	364(4.78)	33-1282-69
$C_{23}H_{17}ClO_2$			
Benzofuran, 5-chloro-2-[p-(p-methoxy-styryl)phenyl]-	DMF	365(4.81)	33-1282-69

Compound	Solvent	$\lambda_{max}(\log \epsilon)$	Ref.
$C_{23}H_{17}ClO_5S$			
4-(p-Hydroxystyryl)-2-phenyl-1-benzothiopyrylium perchlorate	CH_2Cl_2	262(4.13),555(3.42), 695s(3.06)	2-0311-69
$C_{23}H_{17}ClS$			
Benzo[b]thiophene, 2-[p-(p-chlorostyryl)phenyl]-6-methyl-	DMF	361(4.79)	33-1282-69
$C_{23}H_{17}N$			
Pyridine, 2,4,6-triphenyl-	MeOH	309(3.94)	5-0093-69G
$C_{23}H_{17}NO$			
5H-Isoindolo[1,2-b][3]benzazepin-5-one, 7,8-dihydro-1,3-diphenyl-	EtOH	292(3.87),356(4.38)	88-0887-69
$C_{23}H_{17}NO_2$			
Benzoic acid, o-(3,4-diphenylpyrrol-2-yl)-	EtOH	267(4.08)	88-0887-69
Cyclopenta[ij]pyrido[2,1,6-de]quinolizine-4-carboxylic acid, 9-phenyl-, ethyl ester	EtOH	279(4.69),308s(4.30), 349(4.29),369s(4.23), 393(4.17),425(4.01), 447(4.41),540s(2.38), 572(2.49),622(2.46)	39-0239-69C
$C_{23}H_{17}NO_4$			
Pyrrole-2-carboxylic acid, 5-benzoyl-4-hydroxy-1-(1-naphthyl)-, methyl ester	MeOH	224(4.85),262(4.20), 287(4.16),296(4.16), 332(3.99)	78-0527-69
$C_{23}H_{17}NS$			
Aniline, N-(4,6-diphenyl-2H-thiopyran-2-ylidene)-	EtOH	227(4.37),281(4.33), 400(3.71)	48-0061-69
$C_{23}H_{17}NS_2$			
Aniline, N-[α-[(4-phenyl-1,3-dithiol-2-ylidene)methyl]benzylidene]-	$CHCl_3$	245(4.49),408(3.34)	24-1580-69
Aniline, N-[α-[(5-phenyl-3H-1,2-dithiol-3-ylidene)methyl]benzylidene]-	$CHCl_3$	246(4.64),459(4.29)	24-1580-69
$C_{23}H_{17}N_3O_2$			
4H-1,2-Diazepine, 5-(m-nitrophenyl)-3,7-diphenyl-	EtOH	211(4.48),256(4.60), 291s(4.32)	1-3125-69
4H-1,2-Diazepine, 5-(p-nitrophenyl)-3,7-diphenyl-	EtOH	213(4.34),245(4.22),	1-3125-69
$C_{23}H_{18}$			
Indene-anthracene adduct	EtOH	244(3.36),249(3.49), 255(3.65),261(3.71), 269(3.74),276(3.75)	78-1661-69
Indene, 1-ethylidene-2,3-diphenyl-	pentane	242(4.50),262(4.47), 400(3.42)	88-3867-69
Spiro[cyclopropane-1,1'-inden], 2',3'-diphenyl-	pentane	240(4.15),285(3.76)	88-3867-69
$C_{23}H_{18}BrNO$			
Phthalimidine, 3-(α-bromobenzylidene)-2-phenethyl-	EtOH	220(4.54),268(4.04), 328(4.12)	88-0887-69
stereoisomer	EtOH	224(4.50),268(4.00), 332(4.13)	88-0887-69

Compound	Solvent	$\lambda_{max}(\log \epsilon)$	Ref.
$C_{23}H_{18}ClNO_4S$			
4,6-Diphenyl-2-p-tolyl-1,3-thiazin-1-ium perchlorate	MeCN	430(4.34)	24-0269-69
$C_{23}H_{18}Cl_2O$			
Anisole, p-[p-(2,4-dichlorostyryl)-styryl]-	DMF	369(4.77)	33-2521-69
$C_{23}H_{18}NOPS$			
Formamide, 1-(diphenylphosphinyl)-N-1-naphthylthio-	EtOH	343(3.84),436s(2.19)	18-2975-69
$C_{23}H_{18}NPS_2$			
Formamide, 1-(diphenylphosphinothiol)-N-1-naphthylthio-	EtOH	352(3.87),446s(2.10)	18-2975-69
$C_{23}H_{18}N_2$			
4H-1,2-Diazepine, 3,5,7-triphenyl-	EtOH	209(4.47),258(4.50), 293(4.35)	1-3125-69
1-Naphthaldehyde, diphenylhydrazone	EtOH	210(4.9),235(4.4), 255s(4.2),366(4.3)	28-0248-69B
2-Naphthaldehyde, diphenylhydrazone	EtOH	203(4.6),236(4.6), 271(4.1),293(4.0)	28-0248-69B
1-Naphthylamine, N,N'-3-propen-1-yl-3-ylidenebis-	EtOH	376(4.37)	24-2609-69
hydrochloride	MeOH	240s(--),255s(--), 280(4.07),378(4.23)	59-1075-69
perchlorate	EtOH	377(4.36)	24-2609-69
2-Naphthylamine, N,N'-3-propen-1-yl-3-ylidenebis-	EtOH	389(4.67)	24-2609-69
perchlorate	EtOH	397(4.63)	24-2609-69
$C_{23}H_{18}N_2O$			
Quinoxaline, 3-benzoyl-2-(α-methyl-benzyl)-	EtOH	246(4.58),323(3.95)	44-1651-69
$C_{23}H_{18}N_2OS$			
4-Thiazoline, 4-benzoyl-3-methyl-5-phenyl-2-(phenylimino)-	MeCN	250(4.54),336(3.76), 418(3.85)	23-3557-69
$C_{23}H_{18}N_2O_2S$			
Thieno[3,4-b]quinoxaline, 1,3-dihydro-1-methyl-1,3-diphenyl-, 2,2-dioxide	EtOH	241(4.53),324(3.86)	44-1651-69
$C_{23}H_{18}N_2O_3S$			
Spiro[2H-1-benzopyran-2,2'-benzothiazoline], 3'-benzyl-3-methyl-6-nitro-	$(MeOCH_2)_2$	254s(--),269(4.32), 299s(--),332s(--)	78-3251-69
$C_{23}H_{18}N_2O_4$			
2H-Cyclopenta[c]pyridazine-3,4-dicarboxylic acid, 5,7-diphenyl-, dimethyl ester	$CHCl_3$	261(4.40),305(3.75)	64-0536-69
1,2-Diazaspiro[4.4]nona-1,3,6,8-tetraene-3,4-dicarboxylic acid, 6,9-diphenyl-, dimethyl ester	ether	264(4.58),335(3.63), 360(3.36)	64-0536-69
$C_{23}H_{18}N_2O_4S$			
Spiro[2H-1-benzopyran-2,2'-benzothiazoline], 8-methoxy-3'-methyl-6-nitro-3-phenyl-	EtOH	218(4.03),301(3.83), 340s(3.49)	22-3329-69

Compound	Solvent	$\lambda_{max}(\log \epsilon)$	Ref.
$C_{23}H_{18}N_2S$			
4(1H)-Pyrimidinethione, 1-benzyl-2,5-diphenyl-	$CHCl_3$	352(4.28)	4-0037-69
2H-Thiopyran-2-one, 4,6-diphenyl-, phenylhydrazone	EtOH	276(4.38),403(3.69)	48-0061-69
$C_{23}H_{18}O$			
Anisole, p-[(p-styrylphenyl)ethynyl]-	DMF	347(4.78)	33-2521-69
Anisole, p-[p-(phenylethynyl)styryl]-	DMF	350(4.77)	33-2521-69
Anthracene, 2-(p-methoxystyryl)-	DMF	320(4.79),335(4.85), 359(4.23),377(4.35), 395(4.29)	33-2521-69
Benzofuran, 5-methyl-2-(p-styrylphenyl)-	DMF	357(4.80)	33-1282-69
Benzofuran, 6-methyl-2-(p-styrylphenyl)-	DMF	359(4.76)	33-1282-69
2-Cyclopenten-1-one, 3,4,4-triphenyl-	EtOH	290(4.23)	35-6868-69
2-Cyclopenten-1-one, 3,5,5-triphenyl-	EtOH	289(4.36)	35-6868-69
Furan, 3,4-diphenyl-2-p-tolyl-	EtOH	244(4.28),289(4.22)	39-1950-69C
2H-Pyran, 2,4,6-triphenyl-	EtOH	265(4.32),340(4.29)	88-2195-69
$C_{23}H_{18}OS$			
Benzo[b]thiophene, 2-[p-(p-methoxystyryl)phenyl]-	DMF	362(4.77)	33-1282-69
Benzo[b]thiophene, 5-(p-methoxystyryl)-2-phenyl-	DMF	312(4.77),326(4.77)	33-1282-69
Benzo[b]thiophene, 6-(p-methoxystyryl)-2-phenyl-	DMF	362(4.77)	33-1282-69
2-Naphthalenemethanol, 3-mercapto-1,4-diphenyl-	EtOH	231(4.44),253(4.51), 278s(3.95),293(3.91), 304(3.89),346s(3.25)	44-1310-69
$C_{23}H_{18}O_2$			
Benzofuran, 2-[p-(p-methoxystyryl)-phenyl]-	DMF	363(4.81)	33-1282-69
Benzofuran, 5-methoxy-2-(p-styryl-phenyl)-	DMF	360(4.79)	33-1282-69
Benzofuran, 6-(p-methoxystyryl)-2-phenyl-	DMF	290(3.98),358(4.75)	33-1282-69
2-Cyclopropene-1-carboxylic acid, 1,2,3-triphenyl-, methyl ester	EtOH	227(4.63),233(4.58), 296(4.48),310(4.64), 326(4.57)	44-0670-69
2H-Pyran-2-ol, 2,4,6-triphenyl-	EtOH	233(4.15),274(4.30), 345(4.29)	88-2195-69
Stilbene, 4-[3,4-(methylenedioxy)styryl]-	DMF	367(4.79)	33-2521-69
$C_{23}H_{18}O_3$			
5H-Dinaphtho[1,2-b:2',1'-d]pyran-5,14(6H)-dione, 6a,14a-dihydro-8,14a-dimethyl-	MeOH	230(4.85),285s(--), 294(4.07),309s(--), 323s(--)	44-3710-69
	MeOH-NaOH	254(4.78),342(4.03)	44-3710-69
$C_{23}H_{18}O_7$			
2H,6H-Benzo[1,2-b:5,4-b']dipyran-6-one, 5-hydroxy-2,2-dimethyl-7-[3,4-(methylenedioxy)phenyl]-, acetate	EtOH	229(4.45),269(4.58), 290s(4.35),326(3.97)	39-0365-69C
$C_{23}H_{18}O_8$			
Diphyllin, O-acetyl- (4 λ,3 ε)	EtOH	260(4.74),290s(4.04), 315s(3.80),352(?)	78-2815-69

Compound	Solvent	$\lambda_{max}(\log \epsilon)$	Ref.
$C_{23}H_{18}S$			
Benzo[b]thiophene, 5-methyl-2-(p-styryl-phenyl)-	DMF	357(4.76)	33-1282-69
Benzo[b]thiophene, 6-methyl-2-(stilben-4-yl)-	DMF	359(4.77)	33-1282-69
$C_{23}H_{19}ClN_2O_6$			
4-Acetamido-2,3,5-triphenyloxazolium perchlorate	MeCN	251(4.34),312(4.25)	18-2310-69
2-Benzyl-4-(p-nitrobenzyl)isoquinolin-ium perchlorate	EtOH	237(4.50),340(3.70)	78-0101-69
$C_{23}H_{19}Cl_2NO_5$			
[1-Chloro-2-(3-phenyl-1H-naphtho[2,1-b]-pyran-1-ylidene]ethylidene]dimethyl-ammonium perchlorate	MeCN	235(4.60),300(4.18), 385(4.16),470(3.81)	4-0803-69
[1-Chloro-2-(2-phenyl-4H-naphtho[1,2-b]-pyran-4-ylidene)ethylidene]dimethyl-ammonium perchlorate	MeCN	215(4.59),279(4.32), 303(4.03),335(4.05), 465(4.68)	4-0803-69
$C_{23}H_{19}Cl_2N_3O_5$			
2-Formyl-1-(p-methoxyphenyl)quinolinium perchlorate, (p-chlorophenyl)hydra-zone	EtOH	269(3.12),316(2.56), 343(2.56),474(2.95)	124-0370-69
$C_{23}H_{19}Fe$			
Ferrocenyldiphenylmethylium cation	58% H_2SO_4	258(4.06),365(4.08), 560(3.37)	35-0509-69
$C_{23}H_{19}N$			
Phenalene, 1-[3-(N-methylanilino)-2-propenylidene]-	hexane	275(4.36),390(3.94), 486(4.75)	24-2301-69
	MeCN	276(4.30),391(3.88), 495(4.68)	24-2301-69
$C_{23}H_{19}NO$			
1-Indanone, 3-(benzylamino)-2-benzyli-dene-	hexane	228(4.16),265s(4.00), 310(4.35),326s(4.18)	44-0596-69
Phthalimidine, 3-benzylidene-2-phen-ethyl-	EtOH	221(4.52),268(4.00), 326(4.15)	88-0887-69
geometric isomer	EtOH	222(4.53),269(4.05), 324(4.15)	88-0887-69
4(1H)-Pyridone, 2,3-dihydro-1,2,6-tri-phenyl-	EtOH	249(4.21),353(4.33)	18-1357-69
$C_{23}H_{19}NO_2$			
Indole, 3-benzyl-2-(p-methoxybenzoyl)-	EtOH	220(4.51),327(4.30)	78-0227-69
1,4-Isoquinolinedione, 2,3-dihydro-3-(m-methylbenzyl)-3-phenyl-	EtOH	216(4.67),295(3.38), 350(2.90)	33-1810-69
1H-Naphtho[2,1-b]pyran-$\Delta^{1,\alpha}$-acetamide, N,N-dimethyl-3-phenyl-	MeCN	256(4.56),298(4.31), 363(4.13)	4-0803-69
4H-Naphtho[1,2-b]pyran-$\Delta^{4,\alpha}$-acetamide, N,N-dimethyl-2-phenyl-	MeCN	272(4.57),328(4.25), 360(4.30)	4-0803-69
$C_{23}H_{19}NO_6$			
Benz[c]acridan-3,4-dicarboxylic acid, 6-methoxy-12-methyl-7-oxo-, dimethyl ester	EtOH	283(4.66),358(4.27), 422(3.81),442(3.81)	12-1721-69

Compound	Solvent	λ max(log ε)	Ref.
$C_{23}H_{19}N_3O_7$			
p-Anisic acid, α-hydroxy-α-[(p-nitrophenyl)azo]-, benzyl ester, acetate	dioxan	267(4.46),273s(4.45), 410(2.68)	39-1571-69C
$C_{23}H_{19}N_4O_2$			
2-Formyl-1-(p-methoxyphenyl)quinolinium (p-nitrophenyl)hydrazone	EtOH-acid	471(2.86)	124-0370-69
$C_{23}H_{19}N_5O$			
[1,2'-(1'H)-Biphthalazine]-4'-carbonitrile, 1'-(2-oxocyclohexyl)-	EtOH	255s(4.01),379(4.31), 393s(4.25)	95-0959-69
$C_{23}H_{19}N_5O_5$			
2-[1-(2-Hydroxy-5-nitrophenyl)azo]-4,5-diphenylimidazole, acetic acid adduct	50% EtOH	471(4.43)	3-1690-69
	+ base	511(4.43)	3-1690-69
$C_{23}H_{19}N_5O_7$			
Uracil, 1-(3-azido-3-deoxy-β-D-arabinofuranosyl)-, 2',5'-dibenzoate	MeOH	232(4.37),261(3.91)	94-1188-69
Uracil, 1-(5-azido-5-deoxy-β-D-xylofuranosyl)-, 2',3'-dibenzoate	EtOH	230(4.52),260(4.11)	94-0798-69
$C_{23}H_{19}O$			
5-(4-Hydroxy-3,5-xylyl)-5H-dibenzo[a,d]-cycloheptenylium (perchlorate)	H_2SO_4	241(4.33),267(4.16), 311(5.08),379(3.83), 413(3.64),524(3.65), 561(3.75)	5-0048-69A
$C_{23}H_{20}BF_2NO_2$			
Naphtho[1,2-e]-1,3,2-dioxaborin, 2,2-difluoro-4-[4-[p-(dimethylamino)-phenyl]-1,3-butadienyl]-	MeCN	224(4.48),305(3.86), 360(3.99),425(3.96), 620(4.74)	4-0029-69
$C_{23}H_{20}ClNO_4$			
6-[Dibenzo[a,d]cyclohepten-5-ylidene]-1,4-cyclohexadien-3-ylidene-N,N-dimethylimmonium perchlorate	MeCN	270(4.25),284s(4.15), 385(4.39),457(4.12)	5-0048-69A
$C_{23}H_{20}ClN_3O_4$			
2-Formyl-1-phenylquinolinium perchlorate, (o-tolylhydrazone)	EtOH	272(3.35),326(2.89), 478(3.28)	124-0370-69
2-Formyl-1-phenylquinolinium perchlorate, (p-tolylhydrazone)	EtOH	276(3.61),323(3.12), 488(3.64)	124-0370-69
$C_{23}H_{20}ClN_3O_5$			
2-Formyl-1-(p-methoxyphenyl)quinolinium perchlorate, phenylhydrazone	EtOH	271(3.41),340(3.0), 478(2.95)	124-0370-69
2-Formyl-1-phenylquinolinium perchlorate, (o-methoxyphenyl)hydrazone	EtOH	258(3.28),328(2.92), 492(3.55)	124-0370-69
2-Formyl-1-phenylquinolinium perchlorate, (p-methoxyphenyl)hydrazone	EtOH	276(3.24),326(2.84), 502(3.34)	124-0370-69
$C_{23}H_{20}Fe_2O$			
2-Propen-1-one, 1,3-diferrocenyl-	EtOH	320(4.25),385(3.64)	73-2235-69
$C_{23}H_{20}N_2$			
Acridine, 9-(1,2,3,4-tetrahydro-1-methyl-6-quinolyl)-	n.s.g.	250(4.72),324(3.62), 355(3.88),425(3.72)	103-0420-69
Crotononitrile, 3-(benzylamino)-2,4-diphenyl-	$CHCl_3$	298(4.24)	39-2132-69C

Compound	Solvent	$\lambda_{max}(\log \epsilon)$	Ref.
Quinoxaline, 2-benzyl-3-(α-methylbenzyl)-	EtOH	240(4.48),311s(3.98), 321(4.04)	44-1651-69
$C_{23}H_{20}N_2O$			
4H-Naphtho[1,2-b]pyran-$\Delta^{4,\alpha}$-acetamidine, N,N-dimethyl-2-phenyl-, perchlorate	MeCN	270(4.50),365(4.25)	4-0803-69
$C_{23}H_{20}N_2OS$			
Atropamide, N-benzoyl-β-(benzylamino)-thio-	$CHCl_3$	302(4.21),398(4.30)	4-0037-69
$C_{23}H_{20}N_2O_2$			
Naphtho[2,3-h]quinoline-7,12-dione, 5-(cyclohexylamino)-	EtOH	310(4.42),518(3.93)	104-0742-69
$C_{23}H_{20}N_2O_3$			
Benzo[3,4]cyclobuta[1,2-b]phenazine, 6,7,8-trimethoxy-10,11-dimethyl-	EtOH	283(4.57),314(4.68), 326(4.70),400s(3.99), 423(4.18),450(4.18)	39-0646-69C
$C_{23}H_{20}N_2O_4S$			
3H-Indol-3-one, 5-ethoxy-2-phenyl-, O-(p-tolylsulfonyl)oxime	EtOH	279(4.58),481(3.35)	22-2008-69
$C_{23}H_{20}N_2S_2$			
Imidazole, 2,4-bis(benzylthio)-5-phenyl-	EtOH	277(4.16)	23-1123-69
anion	H_2O	282(4.13)	23-1123-69
cation	H_2O	277(4.18)	23-1123-69
$C_{23}H_{20}N_4$			
Imidazole, 4,5-diphenyl-2-(2,4-xylylazo)-	n.s.g.	292(4.08),407(4.31)	104-0763-69
$C_{23}H_{20}O$			
Anisole, p-(p-styrylstyryl)-	DMF	364(4.79)	33-2521-69
Ether, ethyl 1,2,3-triphenyl-2-cyclopropen-1-yl	EtOH	223(4.44),228(4.42), 302(4.34),317(4.34)	44-2728-69
2-Indanone, 1,3-dimethyl-1,3-diphenyl-, cis	C_6H_{12}	260(3.09),267(3.23), 274(3.24)	24-1597-69
trans	C_6H_{12}	261(3.10),267(3.17), 274(3.10)	24-1597-69
Indene, 2-methoxy-1-methyl-1,3-diphenyl-	$C_6H_{11}Me$	268s(4.00),273(4.01)	24-1597-69
$C_{23}H_{20}O_2$			
3-Buten-2-one, 4-methoxy-1,1,3-triphenyl-	EtOH	275(4.07)	24-3405-69
1,4-Pentanedione, 1,2,2-triphenyl-	EtOH	239(3.99),316(2.49)	35-6868-69
1,4-Pentanedione, 1,3,3-triphenyl-	EtOH	243(4.14),277(3.20)	35-6868-69
$C_{23}H_{20}O_5$			
Benzophenone, 2'-(hydroxymethyl)-2,4-dimethoxy-, benzoate	EtOH	215(4.29),231(4.35), 278(4.00),312(3.88)	39-1873-69C
$C_{23}H_{20}O_6$			
2,3-Furandicarboxylic acid, 4-benzoyl-5-phenyl-, diethyl ester	EtOH	266(4.23),316(4.22)	88-2053-69
$C_{23}H_{20}O_7$			
α-Isoderrubone, monoacetate	EtOH	225(4.35),259(4.39), 295(3.95),324s(3.82)	39-0365-69C

Compound	Solvent	λ_{max}(log ε)	Ref.
β-Isoderrubone, monoacetate	EtOH	225s(4.33),255(4.34), 294(4.24)	39-0365-69C
Isoglycyrol, monoacetate	EtOH	207(4.50),221(4.44), 226s(4.42),242(4.31), 254s(4.15),262s(4.06), 274s(3.86),288(3.66), 312s(4.05),339(4.47), 353(4.42)	94-0729-69
Robustin, methyl ether	EtOH	232(4.39),269(4.44), 354(4.23)	39-0365-69C
$C_{23}H_{20}O_8$			
Acetic acid, [4,5-dimethoxy-2-(5-oxo-5H-furo[3,2-g][1]benzopyran-6-yl)-phenoxy]-, ethyl ester	EtOH	237(4.54),297(4.10)	31-0789-69
$C_{23}H_{20}O_{10}$			
Flavone, 3',4',6-trihydroxy-5,7-dimethoxy-, triacetate	EtOH	240(4.07),262(4.14), 307(4.18)	40-1270-69
Flavone, 3',5,6-trihydroxy-4',7-dimethoxy-, triacetate	EtOH	230s(4.38),262(4.13), 319(4.48)	40-1270-69
Flavone, 4',5,6-trihydroxy-3',7-dimethoxy-, triacetate	EtOH	239(4.35),261(4.18), 311(4.37)	40-1270-69
$C_{23}H_{21}BrO_3$			
Ethylene, bromotris(p-methoxyphenyl)-	C_6H_{12}	249(4.44),311(4.11)	35-6734-69
$C_{23}H_{21}ClN_2O$			
1(2H)-Quinolinecarboxamide, 6-chloro-3,4-dihydro-N-methyl-2,4-diphenyl-	n.s.g.	258(4.08)	40-1047-69
1(2H)-Quinolinecarboxanilide, 6-chloro-3,4-dihydro-2-methyl-4-phenyl-	n.s.g.	238(4.36),268(4.35)	40-1047-69
$C_{23}H_{21}Fe_2N$			
[1,1"-Biferrocene]-2-acetonitrile, 2"-methyl-	C_6H_{12}	197(4.56),221(4.64), 255s(4.00),294(3.80)	78-3477-69
$C_{23}H_{21}N$			
5-[p-(Dimethylamino)phenyl]-5H-dibenzo-[a,d]cycloheptenylium (dication) perchlorate	H_2SO_4	243(4.34),267(4.13), 312(5.11),388(3.88), 411(4.00),527(3.62), 562(3.64)	5-0048-69A
$C_{23}H_{21}NO$			
Ketone, 4-biphenylyl 1-methyl-2-phenyl-3-azetidinyl, cis	isooctane	282(4.36)	44-0310-69
trans	isooctane	282(4.42)	44-0310-69
$C_{23}H_{21}NO_2$			
1,3-Pentanedione, 5-anilino-1,5-diphenyl-	EtOH	249(4.28),311(4.24)	18-1357-69
	EtOH-HCl	319(4.22)	18-1357-69
	EtOH-NaOH	247(4.34),332(4.25)	18-1357-69
$C_{23}H_{21}N_3$			
Imidazole, 4-[p-(dimethylamino)phenyl]-2,5-diphenyl-	MeOH	312(4.36)	44-3981-69
$C_{23}H_{21}N_3O_6S$			
4-Thiouracil, 1-(3-amino-3-deoxy-β-D-arabinofuranosyl)-, 2',5'-dibenzoate	EtOH	233(4.52),341(4.34)	94-1188-69

Compound	Solvent	λmax(log ε)	Ref.
$C_{23}H_{21}N_3O_7$			
Uracil, 1-(3-amino-3-deoxy-β-D-arabinofuranosyl)-, 2',5'-dibenzoate	MeOH	232(4.48),263(4.01)	94-1188-69
$C_{23}H_{21}N_5O_8$			
Pyridine, 4-tert-butyl-2,6-bis(2,4-dinitrobenzyl)-	EtOH	241(4.44)	22-4425-69
$C_{23}H_{22}$			
1-Butene, 3-benzyl-2,4-diphenyl-	EtOH	237(3.74),269(3.21)	1-2409-69
$C_{23}H_{22}BN_5O_6S$			
Adenosine, cyclic 2',3'-benzeneboronate, 5'-p-toluenesulfonate	MeOH	260(4.16)	78-0477-69
$C_{23}H_{22}ClNO_4$			
6,11-Dihydro-6,11-dimethyl-12-phenyl-6,11-ethanoacridizinium perchlorate	n.s.g.	266s(3.71),273(3.74)	44-1700-69
$C_{23}H_{22}ClNO_5$			
2-[(2,6-Diphenyl-4H-pyran-4-ylidene)-methyl]-1-methyl-1-pyrrolinium perchlorate	MeCN	253(4.01),280(3.95), 335(4.19),423(4.63), 445(4.52)	44-2736-69
2-[(4,6-Diphenyl-2H-pyran-2-ylidene)-methyl]-1-methyl-1-pyrrolinium perchlorate	MeCN	275(5.00),465(4.30)	44-2736-69
$C_{23}H_{22}Fe_2$			
1,1"-Biferrocene, 2-methyl-2"-vinyl-, (1S,1"R)-(+)-	EtOH	451(2.79)	49-1515-69
$C_{23}H_{22}Fe_2O$			
Ketone, methyl 2-(2-methylferrocenyl)-ferrocenyl, (1"S,2R)-(-)-	EtOH	458(2.92)	49-1515-69
$C_{23}H_{22}Fe_2O_2$			
[1,1"-Biferrocene]-2-carboxylic acid, 2"-methyl-, methyl ester, (1S,1"R)-(-)-	EtOH	450(2.70)	49-1515-69
$C_{23}H_{22}N_2$			
Acridine, 9-[p-(diethylamino)phenyl]-	n.s.g.	250(4.46),340(3.50), 312(3.33),400(3.58)	103-0420-69
Indoline, 1,3,3-trimethyl-2-[N-(1-naphthyl)formimidoyl]methylene]-, perchlorate	EtOH	386(4.65)	124-0179-69
Indoline, 1,3,3-trimethyl-2-[N-(2-naphthyl)formimidoyl]methylene]-, perchlorate	EtOH	377(4.54)	124-0179-69
$C_{23}H_{22}N_2O$			
Benz[c]acridine, 5,6-dimethyl-7-morpholino-	MeOH	226(4.55),266s(4.47), 274(4.52),290s(4.54), 300(4.65),312s(4.39), 346(3.63),362(3.74), 381(3.85),398(3.86)	4-0361-69
Butyronitrile, 3-(benzylamino)-3-hydroxy-2,4-diphenyl-	EtOH	260(4.04),303(4.14)	39-2132-69C
	EtOH-HCl	268(4.22),272(4.22)	39-2132-69C
	EtOH-NaOH	252(4.07),306(4.31)	39-2132-69C

Compound	Solvent	λ_{max}(log ε)	Ref.
$C_{23}H_{22}N_2O_6$			
Pyrrole-2,4-dicarboxylic acid, 1-benzyl-3-[(p-methoxyphenyl)carbamoyl]-, dimethyl ester	EtOH	248(4.37)	94-2461-69
Pyrrole-3,4-dicarboxylic acid, 1-benzyl-2-[(p-methoxyphenyl)carbamoyl]-, dimethyl ester	EtOH	281(4.14)	94-2461-69
$C_{23}H_{22}N_2O_9S$			
Uracil, 1-β-D-xylofuranosyl-, 2'-benzoate 5'-p-toluenesulfonate	EtOH	226(4.46),259(4.11)	94-0798-69
$C_{23}H_{22}N_4O_2$			
Butyric acid, 5-ethyl-1,3-diphenyl-1H-pyrazolo[3,4-b]pyrazin-6-yl ester	EtOH	253(4.34),270(4.39)	32-0463-69
$C_{23}H_{22}N_4O_3S$			
Cyclohexanone, 2-[1,2-dihydro-4-(methylsulfonyl)-2-(1-phthalazinyl)-1-phthalazinyl]-	EtOH	250s(3.99),370(4.23)	95-0959-69
$C_{23}H_{22}N_4O_5$			
Hypoxanthine, 9-(diphenylmethyl)-1-β-D-ribofuranosyl-	pH 1	253(4.09)	44-2646-69
	pH 7	253(4.09)	44-2646-69
	pH 13	253(4.08)	44-2646-69
$C_{23}H_{22}O$			
2c-Indanol, 1t,3c-dimethyl-1r,3t-diphenyl-	C_6H_{12}	251(2.80),259(2.94), 265(2.98),272(2.87)	24-1597-69
isomer	C_6H_{11}Me-isopentane	258(2.98),264(3.10), 271(3.08)	24-1597-69
$C_{23}H_{22}O_3$			
Stilbene, 3,4,5-trimethoxy-4'-phenyl-	DMF	340(4.66)	33-2521-69
$C_{23}H_{22}O_4$			
3,5-Cyclohexadiene-1,2-dicarboxylic acid, 3-phenyl-6-p-tolyl-, dimethyl ester, trans	EtOH	234s(3.97),337(4.36)	22-3662-69
2,4,6-Octatrienedioic acid, 3-phenyl-6-p-tolyl-, dimethyl ester, cis-trans-cis	THF	320(4.43)	22-3665-69
p-Toluic acid, 2,6-bis(benzyloxy)-, methyl ester	EtOH	218(4.56),277(3.43)	39-1721-69C
$C_{23}H_{22}O_5$			
Acetophenone, 4',6'-bis(benzyloxy)-2'-hydroxy-3'-methoxy-	EtOH	290(4.24),322(3.53)	18-2327-69
o-Xylene-α,α'-diol, α-(2,4-dimethoxyphenyl)-, α'-benzoate	EtOH	215(4.26),231(4.31), 279(3.53)	39-1873-69C
$C_{23}H_{22}O_6$			
Derrubone, dimethyl ether	EtOH	264(4.41),290(4.23)	39-0365-69C
Glycyrol, dimethyl ether	EtOH	209(4.55),230(4.53), 243(4.39),254(4.25), 263(4.13),344(4.43), 357(4.36)	94-0729-69
Robustic acid, methyl ether	EtOH	232(4.40),261s(4.40), 268(4.41),354(4.25)	39-0365-69C

Compound	Solvent	$\lambda_{max}(\log \epsilon)$	Ref.
$C_{23}H_{22}O_7$			
Naphtho[2,3-c]furan-1(3H)-one, 6,7-dimethoxy-9-(3,4,5-trimethoxyphenyl)-	EtOH	259(4.60),315(3.85), 350(3.57)	23-4495-69
$C_{23}H_{22}O_9$			
Acetic acid, [2-(7-hydroxy-4-oxo-4H-1-benzopyran-3-yl)-4,5-dimethoxyphenoxy]-, ethyl ester, acetate	EtOH	294(4.12)	18-0199-69
$C_{23}H_{22}O_{10}$			
Flavone, 3,3'-dihydroxy-4',5,6,7-tetramethoxy-, diacetate	EtOH	235(4.45),263(4.40), 316(4.49)	44-1460-69
Flavone, 4',5-dihydroxy-3,3',6,7-tetramethoxy-, diacetate	EtOH	233s(4.57),246s(4.50), 322(4.56)	102-1515-69
$C_{23}H_{23}BrO_2$			
2,2'-Spirobi[2H-1-benzopyran], 6-bromo-6'-methyl-3-pentyl-	dioxan	230(4.46),360(3.53)	5-0162-69B
$C_{23}H_{23}Cl_2NO_5$			
[4-[2-(p-Chlorophenyl)-4H-1-benzopyran-4-ylidene]-2-butenylidene]diethylammonium perchlorate	MeCN	248(4.33),320(4.24), 333s(4.16),513(4.70), 543s(4.52),715(3.32)	44-2736-69
$C_{23}H_{23}IN_2$			
2-(6-Anilino-1,3,5-hexatrienyl)-1-ethylquinolinium iodide	MeOH	620(4.84)	22-1284-69
	MeOH-NaOMe	530(4.57)	22-1284-69
$C_{23}H_{23}NO_2$			
Ethanol, 2-(dimethylamino)-1,1-diphenyl-, benzoate	dioxan	260(3.21),281(2.96)	65-1861-69
$C_{23}H_{23}NO_3$			
Cyclopent[a]inden-8(3H)-one, 3,3a-dihydro-3a-hydroxy-2-(morpholinomethyl)-8a-phenyl-	EtOH	248(4.06),292(3.04)	104-2090-69
$C_{23}H_{23}NO_4$			
1,3-Cyclopentadiene-1-carboxaldehyde, 3-(dimethylamino)-5-(hydroxymethylene)-2,4-bis(p-methoxyphenyl)-	CH_2Cl_2	247(4.38),274(4.34), 310(4.22),375(4.13), 485(3.77)	24-0623-69
$C_{23}H_{23}NO_9$			
1H-Benzo[c]quinolizine-1,2,3,4-tetracarboxylic acid, 1-(methoxymethyl)-, tetramethyl ester	MeOH	209(4.36),223(4.46), 290(4.36),413(3.73)	39-2316-69C
	$MeOH-HClO_4$	240(4.23),342(3.94)	39-2316-69C
4H-Pyrrolo[1,2-a][1]benzazepine-1,2,3,4-tetracarboxylic acid, 4-(methoxymethyl)-, tetramethyl ester	MeOH	209(4.60),220(4.53), 240(4.53),295(4.30), 307(4.30),358(3.85), 480(4.08)	39-2316-69C
	$MeOH-HClO_4$	213(4.52),245(4.53), 315s(3.90)	39-2316-69C
$C_{23}H_{23}N_3$			
Quinoline, 3-(3,4-dihydro-1-isoquinolyl)-4-piperidino-, dihydrochloride	EtOH	223(4.40),372(3.80)	2-1010-69
$C_{23}H_{23}N_5O_2$			
3-Formazancarboxanilide, 2'-hydroxy-N-methyl-1,5-di-p-tolyl-	EtOH	440(4.11)	22-4390-69
	EtOH-KOH	505(4.42)	22-4390-69

Compound	Solvent	λ_{max}(log ε)	Ref.
Piperidine, 1-[4-[(p-anilinophenyl)azo]-2-nitrophenyl]-	benzene	350(3.79),430(4.49)	73-2092-69
$C_{23}H_{23}N_5O_4$			
3-Formazancarboxanilide, 2'-hydroxy-1,5-bis(p-methoxyphenyl)-N-methyl-	EtOH	460(4.27)	22-4390-69
	EtOH-KOH	510(4.56)	22-4390-69
$C_{23}H_{24}BrP$			
Fluoren-9-ylmethylphenylpropylphosphonium bromide	MeOH	220s(4.47),266(4.20), 298s(3.11)	24-3546-69
$C_{23}H_{24}ClNO_4$			
Dimethyl[β-(7-phenyl-2,4,6-heptatrienylidene)phenethylidene]ammonium perchlorate	CH_2Cl_2	260(4.36),303(4.28), 475(5.02)	24-0623-69
$C_{23}H_{24}FeN_2O_4S$			
Tryptophan, N-(ferrocenylsulfonyl)-, ethyl ester	EtOH	426(2.24)	49-1552-69
$C_{23}H_{24}Fe_2$			
1,1"-Biferrocene, 2-ethyl-2"-methyl-, (1R,1"S)-(-)-	EtOH	452(2.74)	49-1515-69
Ferrocene, 1,1'-trimethylenedi-	n.s.g.	325s(2.36),445(2.38)	78-6001-69
$C_{23}H_{24}Fe_2O$			
1,1"-Biferrocene, 2-(methoxymethyl)-2"-methyl-	C_6H_{12}	197(4.58),222(4.66), 255s(4.03),296(3.85)	78-3477-69
$C_{23}H_{24}N_2$			
Benzylidenimine, α-methyl-N-(N-benzyl-1-phenylethylamino)-	n.s.g.	233(3.95),313(2.90)	65-2723-69
$C_{23}H_{24}N_2O$			
Benzylidenimine, 4-methoxy-N-(N-benzyl-α-phenylethylamino)-	n.s.g.	295(4.40)	65-2723-69
$C_{23}H_{24}N_2O_2$			
4-Oxa-1,2-diazaspiro[4.5]dec-2-ene, 1-acetyl-3-(1,2-diphenylvinyl)-	n.s.g.	324(4.35)	39-1338-69C
1H-Pyrazolo[1,2-a]pyridazine-1,3(2H)-dione, 2,2-diethyl-5,8-dihydro-5,8-diphenyl-	C_6H_{12}	259(3.34)	44-3181-69
	EtOH	259(3.36)	44-3181-69
$C_{23}H_{24}N_2O_5$			
Aspidofractinine, 1,11-dicarboxy-10,11-dehydro-3-oxo-, dimethyl ester	EtOH	239(4.13),290s(3.9), 316(4.16)	33-0076-69
$C_{23}H_{24}N_8O_8$			
7H-1,4a-Methanonaphthalen-7-one, 8a-[2-(2,4-dinitrophenyl)hydrazino]-octahydro-, 2,4-dinitrophenylhydrazone	$CHCl_3$	352(4.36)	78-5281-69
$C_{23}H_{24}O$			
Phenanthrene, 2-benzylidene-2,3,4,4a,9,10-hexahydro-6-methoxy-4aβ-methyl-	n.s.g.	288(4.30)	12-1711-69

Compound	Solvent	λ_{max}(log ε)	Ref.
$C_{23}H_{24}O_2$			
1(2H)-Phenanthrone, 2-benzylidene-3,4,4a,9,10,10aβ-hexahydro-6-methoxy-4aβ-methyl-	n.s.g.	283(4.37)	12-1711-69
2,2'-Spirobi[2H-1-benzopyran], 6'-methyl-3-pentyl-	dioxan	259(4.39),308(3.70)	5-0162-69B
$C_{23}H_{24}O_3$			
1-Phenanthrol, 2,3,4,4a,9,10-hexahydro-7-methoxy-4a-methyl-, benzoate	EtOH	228(4.36),275(3.54), 285(3.46)	44-1962-69
$C_{23}H_{24}O_7$			
6H-Benzofuro[3,2-c][1]benzopyran-6-one, 9-hydroxy-1,3-dimethoxy-2-(3-methoxy-3-methylbutyl)-	EtOH	209(4.39),228(4.33), 246(4.22),262s(4.02), 346(4.27),356(4.21)	94-0729-69
β-Toxicarol, dehydrodihydro-, methyl ether	EtOH	237(4.40),279(4.46), 311(4.30),320s(4.16)	39-0365-69C
$C_{23}H_{24}O_8$			
Flavone, 4',7-diethoxy-5-hydroxy-3',8-dimethoxy-, acetate	EtOH	246(4.39),264(4.27), 340(4.41)	18-2327-69
$C_{23}H_{25}ClN_2O_6$			
Dimethyl[1-morpholino-2-(2-phenyl-4H-1-benzopyran-4-ylidene)ethylidene]-ammonium perchlorate	MeCN	233(4.40),295(4.23), 385(4.32)	4-0803-69
$C_{23}H_{25}F_7O_3$			
Androsta-3,5-dien-17-one, 3-hydroxy-, heptafluorobutyrate	heptane	227(4.16),232(4.15)	13-0739-69A
$C_{23}H_{25}IN_2O_2$			
4-[3-(2,3-Dihydro-5-methyl-1,5-benzoxazepin-4(5H)-ylidene)propenyl]-2,3-dihydro-5-methyl-1,5-benzoxazepinium iodide	EtOH	496(5.30)	124-1178-69
$C_{23}H_{25}NO$			
Bicyclomahanimbine	EtOH	242(4.61),255(4.43), 260(4.40),305(4.21), 331(3.62)	88-3857-69
Curryangine	n.s.g.	228(4.37),239(4.52), 266(4.08),308(4.12)	2-1060-69
Cyclomahanimbine	EtOH	246(4.51),251(4.28), 257(4.06),307(3.96), 341(3.35)	88-3857-69
Mahanimbidine	EtOH	241(4.62),257(4.37), 307(4.20),335(3.51)	88-3857-69
$C_{23}H_{25}NO_3$			
Dibenzo[f,h]pyrrolo[1,2-b]isoquinoline, 9,11,12,13,13a,14-hexahydro-2,3,6-trimethoxy-	MeOH	259(4.71),286(4.46), 342(2.99),360(2.61)	5-0154-69A
$C_{23}H_{25}NO_5$			
2-Stilbenecarboxylic acid, 3,5-dimethoxy-α-(piperidinocarbonyl)-	MeOH	278(4.27)	44-3192-69
$C_{23}H_{25}NO_6$			
Orientalidine methine	EtOH	230s(4.80),277(4.04), 305s(3.45)	73-0875-69

Compound	Solvent	$\lambda_{max}(\log \epsilon)$	Ref.
$C_{23}H_{25}N_2$			
Malachite Green cation	98% HOAc	428(4.30),621(5.02)	39-1068-69C
$C_{23}H_{25}N_3O_3$			
Neoechinuline	n.s.g.	231(4.51),287(4.12), 420(3.99)	88-4457-69
$C_{23}H_{25}N_3O_5$			
Tryptophan, N-(N-carboxy-DL-alanyl)-, N-benzyl methyl ester, DL-	EtOH	220(4.55),282(3.80), 290(3.73)	39-1003-69C
$C_{23}H_{25}N_9O_2$			
Isoalloxazine, 10-[6-(6-amino-9H-purin-9-yl)hexyl]-7,8-dimethyl-	propylene glycol	268(4.65),355(3.89), 445(4.05)	44-3240-69
$C_{23}H_{26}ClN_3O_3$			
1-(Cyanoformyl)-1-deacetyl-O,4-dimethyl-17ξ-diabolinium chloride	EtOH	278(3.94)	33-1564-69
$C_{23}H_{26}Cl_6O_4$			
Androsta-3,5-diene-3,17β-diol, bis-(trichloroacetate)	10:1 MeOH: dioxan	237(4.3)	78-1717-69
$C_{23}H_{26}N_2O_4$			
Akuammiline	EtOH	266(3.73)	33-0701-69
Aspidofractinine, 1,11α-dicarboxy-3,4-dehydro-, dimethyl ester	EtOH	245(4.09),279(3.34), 287(3.29)	33-0076-69
$C_{23}H_{26}N_2O_5$			
Aspidofractinine, 1,11α-dicarboxy-3-oxo-, dimethyl ester	EtOH	243(4.08),281(3.30), 289(3.26)	33-0076-69
Aspidofractinine, 1,11β-dicarboxy-3-oxo-, dimethyl ester	EtOH	242(4.11),283(3.38), 290s(3.35)	33-0076-69
$C_{23}H_{26}N_2O_{10}$			
Uridine, 5-[(benzyloxy)methyl]-, 2',3',5'-triacetate	EtOH	210(4.17),262(3.98)	73-1696-69
$C_{23}H_{26}N_2O_{10}S$			
Benzimidazole, 1-acetyl-2-(β-D-glucopyranosylthio)-	EtOH	243(4.26),248(4.23), 284(4.04)	48-0997-69
$C_{23}H_{26}O_4$			
Pregna-1,4,6,15-tetraene-3,20-dione, 17α-acetoxy-	MeOH	220(4.02),256(3.92), 297(4.04)	87-0393-69
$C_{23}H_{26}O_8$			
Flavone, 3,3'-diethoxy-4',5,6,7-tetramethoxy-	EtOH	233(4.27),242(4.23), 249s(4.11),264s(4.11)	44-1460-69
$C_{23}H_{26}O_9$			
Olivin, 3',4'-O-isopropylidene-	EtOH	231(4.25),276(4.58), 326(3.60),406(4.09)	105-0257-69
$C_{23}H_{26}O_{10}$			
Flavone, 3,3',4',5,5',6,7,8-octamethoxy-(exoticin)	EtOH	209(4.72),255(4.26), 274s(4.24),334(4.30)	2-0636-69

Compound	Solvent	$\lambda_{max}(\log \epsilon)$	Ref.
$C_{23}H_{27}ClO_4$			
Benzaldehyde, 5-chloro-2,4-dihydroxy-6-methyl-3-[5-(1,2,6-trimethyl-3-oxo-Δ^4-cyclohexyl)-3-methyl-2,4-pentadienyl]-	MeOH	234(4.66),292(4.11), 347(3.98)	78-1323-69
$C_{23}H_{27}Cl_3O_4$			
Androsta-3,5,7-triene-3,17β-diol, 17-acetate trichloroacetate	MeOH	325(4.2)	78-1717-69
$C_{23}H_{27}NO_4$			
Cryptopleurine	EtOH	232(4.01),280(3.75)	12-1805-69
$C_{23}H_{27}NO_6$			
Orientalidine methine, dihydro-	EtOH	240s(4.22),287(3.74)	73-0875-69
$C_{23}H_{27}NO_8$			
Pregna-1,4-diene-3,11,20-trione, 17,21-dihydroxy-, 21-acetate nitrate	MeOH	238(4.21)	78-0761-69
$C_{23}H_{27}N_3O_5$			
Methanol, (4-amino-3,5-dimethoxyphenyl)-bis(4-amino-3-methoxyphenyl)-	pH 4.62	593(4.79)	64-0542-69
Methanol, bis(4-amino-3,5-dimethoxyphenyl)(p-aminophenyl)-	pH 4.62	592(4.77)	64-0542-69
$C_{23}H_{27}N_5O_4$			
Azetidine, 2,3-bis(tert-butylimino)-1,4-bis(p-nitrophenyl)-	EtOH	240(4.15),288(4.05), 376(4.37)	78-1467-69
$C_{23}H_{28}N_2O$			
Indole-2-carboxamide, N,1-di-tert-butyl-3-phenyl-	EtOH	225(4.5),283(4.0)	24-0678-69
Morpholine, 4-[6-(α-anilinobenzyl)-1-cyclohexen-1-yl]-	MeOH	249(4.26),299(3.41)	88-3549-69
3-Penteno-2',6'-xylidide, 3,4-dimethyl-2-(2,6-xylylimino)-	EtOH	264(4.2)	24-1876-69
$C_{23}H_{28}N_2O_2$			
23-Hydroxychano-N_b-methyldihydrochano-dihydrostrychnine	EtOH	305(3.83)	56-0699-69
$C_{23}H_{28}N_2O_4$			
Aspidofractinine, 1,11α-dicarboxy-, dimethyl ester	EtOH	247(4.09),281(3.26), 289(3.22)	33-0076-69
$C_{23}H_{28}N_2O_5$			
Aspidofractinine, 1,11α-dicarboxy-3β-hydroxy-, dimethyl ester	EtOH	246(4.13),280(3.34), 287s(3.29)	33-0076-69
Phthalimidine, 3-[6-[2-(dimethylamino)-ethyl]veratrylidene]-6,7-dimethoxy-	EtOH	222s(4.41),296(4.09), 365(4.32)	88-5121-69
3-Pyridinepropionic acid, 1,4,5,6-tetrahydro-1-[2-[2-(hydroxymethyl)indol-3-yl]ethyl]-2-methyl-4-oxo-, methyl ester, acetate	MeOH	218(4.58),281(3.96), 336(4.20)	24-0310-69
$C_{23}H_{28}N_4O_2$			
Azetidine, 2,3-bis(tert-butylimino)-1-(p-nitrophenyl)-4-phenyl-	EtOH	227s(4.51),292(3.70), 380(4.38)	78-1467-69
Azetidine, 2,3-bis(tert-butylimino)-4-(p-nitrophenyl)-1-phenyl-	EtOH	224(4.33),260(3.93), 296(4.11),305(4.10)	78-1467-69

Compound	Solvent	λmax(log ε)	Ref.
$C_{23}H_{28}N_4O_8$			
Isohydrocoriamyrtin, 11-O-ethyl-, 2,4-dinitrophenylhydrazone	EtOH	259(4.04),290(3.75), 372(4.39)	78-1109-69
$C_{23}H_{28}O_2S$			
21-Norchola-4,17(20),22-trien-24-oic acid, 20-mercapto-3-oxo-, γ-(thiolactone)	EtOH	239(4.29),328(4.15)	87-0001-69
$C_{23}H_{28}O_3$			
Carda-4,14,20(22)-trienolide, 3-oxo-	n.s.g.	232(4.35)	5-0168-69A
3'H-Cyclopropa[1,2]pregna-1,4,6-triene-3,20-dione, 1,2-dihydro-17-hydroxy-16-methylene-	MeOH	282(4.31)	87-0631-69
Pregna-1,4,6-triene-3,20-dione, 16,17α-epoxy-6,16β-dimethyl-	MeOH	227(4.17),255(3.96), 304(4.09)	87-0631-69
$C_{23}H_{28}O_4$			
Carda-4,20(22)-dienolide, 14,15β-epoxy-3-oxo-	n.s.g.	225(4.35)	5-0110-69G
Pregna-4,6,15-triene-3,20-dione, 17α-acetoxy-	MeOH	283(4.43)	87-0393-69
$C_{23}H_{28}O_5$			
Estra-1,3,5(10)-trien-6-one, 1,17β-diacetoxy-4-methyl-	EtOH	250(4.05),303(3.55)	39-1234-69C
$C_{23}H_{28}O_6$			
Estra-1,3,5(10)-triene-1,2-dicarboxylic acid, 3-methoxy-17-oxo-, dimethyl ester	n.s.g.	215(4.31),301(3.51)	12-2255-69
Estra-1,3,5(10)-trien-6-one, 3,17β-diacetoxy-9α-hydroxy-1-methyl-	MeOH	252(3.88)	44-0116-69
$C_{23}H_{28}O_6S$			
5α,17α-Carda-14,20(22)-dienolide, 3β,5-dihydroxy-19-oxo-, cyclic sulfite	EtOH	211(4.20)	33-0482-69
$C_{23}H_{28}O_7$			
3,9-Phenanthrenedione, 1,2,4,4a,10,10a-hexahydro-5,6,8-trihydroxy-7-isopropyl-1,4a-dimethyl-, 5,6-diacetate	EtOH	220(4.31),267(4.00), 341(3.73)	33-1685-69
Quassinol, acetate, methyl ether	EtOH	278(4.19)	7-0230-69
$C_{23}H_{28}O_8$			
15,16,17-Trinortiglia-1,6-diene-3,13-dione, 4,9,20-triacetoxy-	MeOH	241(3.73),313(2.00)	5-0158-69H
$C_{23}H_{28}O_{11}$			
Nivalenol, tetraacetyl-	MeOH	227(3.90)	88-2823-69
$C_{23}H_{29}ClN_2O$			
Acridine, 3-chloro-9-[4-(diethylamino)-1-methylbutylamino]-7-methoxy-, hydrochloride	H_2O	226(5.39),269(5.74), 281(5.61),328(3.41), 344(3.59),358(3.24), 426(3.91),447(3.84)	80-0893-69
$C_{23}H_{29}ClO_3$			
3'H-Cyclopropa[1,2]androsta-1,4,6-trien-3-one, 4-chloro-1β,2β-dihydro-17β-hydroxy-7-methyl-, acetate	MeOH	217(3.78),307(4.32)	24-2565-69

Compound	Solvent	$\lambda_{max}(\log \epsilon)$	Ref.
$C_{23}H_{29}ClO_4$			
Benzaldehyde, 5-chloro-2,4-dihydroxy-6-methyl-3-[3-methyl-5-(1,2,6-trimethyl-3-oxocyclohexyl)-2,4-pentadienyl]-	MeOH	230(4.55),293(4.05), 347(4.00)	78-1323-69
$C_{23}H_{29}ClO_6$			
Pregn-4-ene-3,11-dione, 4-chloro-17,20:-20,21-bis(methylenedioxy)-	EtOH	254(4.20)	44-1455-69
$C_{23}H_{29}NO$			
Cyclohexanone, 2-(α-anilinobenzyl)-4-tert-butyl-	MeOH	248(4.10),296(3.28)	88-3549-69
$C_{23}H_{29}NO_3$			
Estra-1,3,5(10)-trien-17β-ol, 3-methoxy-17α-(3-methyl-5-isoxazolyl)-	MeOH	217(4.22),278(3.30), 287(3.27)	24-3324-69
$C_{23}H_{29}NO_5$			
Estra-1,3,5(10)-trien-17-one, 3-(cyclopentyloxy)-9α-hydroxy-11β-nitro-	EtOH	277(3.19)	31-1018-69
C-Homoberbine, 2,3,9,10,11-pentamethoxy-	EtOH	230s(4.15),282(3.62)	33-1228-69
$C_{23}H_{29}N_3$			
Azetidine, 2,3-bis(tert-butylimino)-1,4-diphenyl-	EtOH	228(4.36),298s(3.70), 329(3.98)	78-1467-69
$C_{23}H_{29}N_3O$			
Dibenzo[b,f][1,4]diazocine, 5,6,11,12-tetrahydro-5-methyl-12-(3-piperidinopropionyl)-	EtOH	260(4.04),306(3.53)	39-0882-69C
$C_{23}H_{29}N_3O_4$			
Pregna-4,6-diene-3,20-dione, 6-azido-17α-hydroxy-, acetate	MeOH	240(3.91),282(3.90)	48-0919-69
$C_{23}H_{29}N_3S_2$			
Pyrrole, 2,3,4-trimethyl-1,5-bis-[(3,4,5-trimethylpyrrol-2-yl)-thiocarbonyl]-	$CHCl_3$	316(4.13),399(4.21), 512(3.74)	12-0239-69
	10N HCl	268(4.11),403(4.38), 466(4.31),510(4.33)	12-0239-69
$C_{23}H_{30}$			
Tricyclo[6.2.2.1^{2,6}]tricosa-2,4,6(23),18,20,21-hexaene	hexane	252(4.27)	88-4551-69
$C_{23}H_{30}ClIO_3$			
3'H-Cycloprop[1,2]androsta-1,4-dien-3-one, 4-chloro-1β,2β-dihydro-17β-hydroxy-7β-(iodomethyl)-, acetate	MeOH	256(4.08)	24-2565-69
$C_{23}H_{30}F_2O_4$			
3'H-Cyclopropa[6,7]pregna-4,6-diene-3,20-dione, 3',3'-difluoro-6α,7α-dihydro-11β,21-dihydroxy-16α-methyl-	EtOH	248(4.17)	78-1219-69
6β,7β-	EtOH	257(4.21)	78-1219-69
$C_{23}H_{30}N_2$			
Aniline, N,N'-1-propen-1-yl-3-ylidenebis[4-butyl-, hydrochloride	MeOH	248(3.85),282s(--), 291(3.34),300s(--), 387(4.43)	59-1075-69

Compound	Solvent	λ_{max}(log ε)	Ref.
Indene, 1,3-bis[2-(dimethylamino)ethyl]-2-phenyl-, dihydrochloride	EtOH	287(4.21)	87-0513-69
$C_{23}H_{30}N_2O_4$			
Isohydrocoriamyrtin, 11-O-methyl-, methylphenylhydrazone	EtOH	248(4.11),305s(4.24), 328(4.44)	78-1109-69
$C_{23}H_{30}N_2O_5$			
Malonic acid, acetamido[[2-(1,1-dimethylallyl)indol-3-yl]methyl]-, diethyl ester	EtOH	225(4.51),283(3.86), 292(3.82)	39-1003-69C
$C_{23}H_{30}N_6O_{15}$			
Polyoxin F	0.05N HCl	276(4.06)	35-7490-69
	0.05N NaOH	271(3.87)	35-7490-69
$C_{23}H_{30}O_2$			
18-Nor-9β,10α-pregna-4,6-diene-3,20-dione, 13-allyl-	n.s.g.	287(4.42)	54-0737-69
$C_{23}H_{30}O_2S$			
21-Norchola-5,17(20),22-trien-24-oic acid, 3β-hydroxy-20-mercapto-, γ-(thiolactone)	EtOH	325(4.20)	87-0001-69
$C_{23}H_{30}O_3$			
Anhydrocanarigenin	n.s.g.	220s(4.46),226(4.50), 233(4.46),242(4.23)	5-0110-69G
5α-Carda-14,20(22)-dienolide, 3-oxo-	MeOH	215(4.16)	5-0136-69F
5β-Carda-14,20(22)-dienolide, 3-oxo-	n.s.g.	208(4.26)	5-0168-69A
Pregna-4,6-diene-3,20-dione, 16,17α-epoxy-6,16β-dimethyl-	MeOH	289(4.39)	87-0631-69
Pregna-4,6-diene-3,20-dione, 17-hydroxy-6-methyl-16-methylene-	MeOH	289(4.39)	87-0631-69
Pregna-4,6,15-triene-3,20-dione, 17-hydroxy-6,16-dimethyl-	MeOH	287(4.39)	87-0631-69
Pregn-5-eno[6,5,4-bc]furan-3,20-dione, 5'-methyl-	EtOH	209(4.20),299(3.51)	94-2586-69
$C_{23}H_{30}O_4$			
Androst-5-eno[6,5,4-bc]furan-3-one, 17β-hydroxy-5'-methyl-, acetate	EtOH	211(4.18),298(3.52)	94-2586-69
Carda-4,20(22)-dienolide, 14,15β-epoxy-3β-hydroxy-	n.s.g.	208(4.26)	5-0110-69G
5ξ-Card-20(22)-enolide, 14,15β-epoxy-3-oxo-	n.s.g.	216(4.20)	5-0110-69G
19-Norpregn-4-ene-3,20-dione, 17α-acetoxy-16-methylene-	EtOH	240(4.26)	24-0643-69
Pregna-5,17(20)-dien-21-al, 3,11-dioxo-, cyclic 3-(ethylene acetal)	EtOH	242(4.14)	39-0460-69C
Pregna-4,14-diene-3,20-dione, 21-acetoxy-	n.s.g.	242(4.23)	5-0168-69A
Pregna-5,16-diene-7,20-dione, 3β-acetoxy-	EtOH	235(4.36)	2-1084-69
5α-Pregna-9(11),16-diene-16,20-dione, 3β-acetoxy-	EtOH	229(4.23)	44-0409-69
5α-Pregna-14,16-diene-11,20-dione, 3β-acetoxy-	EtOH	304(4.02)	94-1401-69

Compound	Solvent	λ_{max}(log ϵ)	Ref.
$C_{23}H_{30}O_4S$			
2,5-Cyclohexadien-1-one, 2,6-di-tert-butyl-4-(2-hydroxyethylidene)-, p-toluenesulfonate	hexane	298(4.48)	70-0580-69
$C_{23}H_{30}O_5$			
Androsta-1,4-dien-3-one, 2,17β-diacet-oxy-	EtOH	247(4.22),315(3.08)	33-0459-69
Hirundigenin, O-acetylanhydro-	EtOH	198(4.21),290(1.93)	33-1175-69
5α-Pregn-16-ene-11,20-dione, 14β,15β-epoxy-3β-hydroxy-, acetate	EtOH	240(3.88)	94-1401-69
Pregn-5-eno[6,5,4-bc]furan-3,20-dione, 17,21-dihydroxy-5'-methyl-	EtOH	210(4.18),230s(3.86), 297(3.53)	94-2586-69
17α-Strophanthidin, 14-anhydro-	EtOH	210(4.27)	33-0482-69
$C_{23}H_{30}O_6$			
Coleon B, dihydro-, tetramethyl ether	EtOH	250(3.97),268(3.90), 296(3.88),319s(3.76)	33-1685-69
14β,17α-Cortexone, 21-O-acetyl-8,19-epoxy-19-hydroxy-	EtOH	244(4.08)	33-2156-69
12α-Etiojerv-4-ene-3,11-dione, 17β-acetoxyacetyl-17α-hydroxy-	MeOH	235(4.27)	78-3145-69
C-Nor-D-homocortisone, 21-acetate	MeOH	234(4.23)	78-3145-69
2-Oxaandrosta-4,6-dien-3-one, 1,17-di-acetoxy-17α-methyl-	EtOH	274(4.33)	78-4257-69
9(1H)-Phenanthrone, 2,3,4,4a,10,10c-hexahydro-5,6,8-trihydroxy-7-iso-propyl-1,4a-dimethyl-, 5,6-diacetate	EtOH	220(4.39),266(4.01), 341(3.74)	33-1685-69
5α,14β-Pregn-16-ene-11,15,20-trione, 3β-acetoxy-14-hydroxy-	EtOH	241(4.04)	94-1401-69
17α-Strophanthidinic acid, 14-anhydro-	EtOH	214(4.22)	33-0482-69
$C_{23}H_{30}Si_2$			
2,6-Disilahepta-3,4-diene, 2,2,6,6-tetramethyl-3-phenyl-5-styryl-	n.s.g.	290(4.45)	101-0291-69C
$C_{23}H_{31}ClO_2$			
3'H-Cyclopropa[6,7]-18-nor-9β,10α-pregna-4,6-diene-3,20-dione, 6α-chloro-13-ethyl-6,7α-dihydro-	n.s.g.	255(4.03)	54-0737-69
$C_{23}H_{31}ClO_3$			
Benzaldehyde, 5-chloro-2,4-dihydroxy-6-methyl-3-(3,7,11-trimethyl-2,6,10-dodecatrienyl)-	MeOH	228(4.13),293(4.02), 345(3.89)	78-1323-69
3'H-Cycloprop[1,2]androsta-1,4-dien-3-one, 4-chloro-1β,2β-dihydro-17β-hydroxy-7β-methyl-, acetate	MeOH	256(4.05)	24-2565-69
$C_{23}H_{31}ClO_4$			
Benzaldehyde, 5-chloro-2,4-dihydroxy-6-methyl-3-[3-methyl-5-(1,2,6-trimethyl-3-oxocyclohexyl)-2-pentenyl]-	MeOH	231(4.36),293(4.08), 346(3.96)	78-1323-69
$C_{23}H_{31}ClO_5$			
17α-Strophanthidin, 14ξ-chloro-14-deoxy-	EtOH	215(4.20)	33-0482-69
$C_{23}H_{31}ClO_6$			
Pregn-4-en-3-one, 4-chloro-11β-hydroxy-17,20:20,21-bis(methylenedioxy)-	MeOH	255(4.16)	44-1455-69

Compound	Solvent	λ_{max}(log ε)	Ref.
$C_{23}H_{31}Cl_3O_4$			
5α-Androst-2-ene-3,17β-diol, 17-acetate trichloroacetate	C_6H_{12}	220s(3.08)	78-1679-69
5β-Androst-2-ene-3,17β-diol, 17-acetate trichloroacetate	C_6H_{12}	225(3.08)	78-1679-69
$C_{23}H_{31}NO_2$			
1'H-Pregn-5-eno[6,5,4-bc]pyrrole-3,20-dione, 5'-methyl-	EtOH	210(4.07),256(4.06), 310(3.61)	94-2586-69
$C_{23}H_{31}NO_3$			
Androst-4-en-3-one, 17β-hydroxy-17α-(3-methylisoxazol-5-yl)-	MeOH	239(4.22)	24-3324-69
1'H-Androst-5-eno[6,5,4-bc]pyrrol-3-one, 17β-hydroxy-5'-methyl-, acetate	EtOH	209(4.07),257(4.06), 310(3.61)	94-2586-69
$C_{23}H_{31}NO_5$			
Isoquinoline, 1,2,3,4-tetrahydro-6,7-dimethoxy-2-methyl-1-(3,4,5-trimethoxyphenethyl)-, S-(+)-	MeOH	230s(4.18),281(3.58), 292(3.47)	33-0678-69
hydrobromide	MeOH	230s(4.22),282(3.60), 292s(3.48)	33-0678-69
$C_{23}H_{31}NO_6$			
5α-Pregn-9(11)-ene-12,20-dione, 3β-hydroxy-17ξ-nitro-, acetate	EtOH	237(4.06)	44-0409-69
$C_{23}H_{31}NO_7$			
Pregn-4-ene-3,20-dione, 17,21-dihydroxy-, 21-acetate, nitrate	C_6H_{12}	231(4.28)	78-0761-69
$C_{23}H_{31}N_3O_4$			
Pregn-4-ene-3,20-dione, 6α-azido-17α-hydroxy-, acetate	MeOH	238(4.14)	48-0445-69
Pregn-4-ene-3,20-dione, 6β-azido-17α-hydroxy-, acetate	MeOH	235(4.13)	48-0445-69
$C_{23}H_{31}N_3O_5$			
Pregn-4-ene-3,20-dione, 6β-azido-7α,17α-dihydroxy-, 17-acetate	MeOH	235(4.03)	48-0919-69
$C_{23}H_{31}N_3O_6$			
1(2H)-Pyrimidinebutyric acid, α-(carboxyamino)-4-ethoxy-5-methyl-2-oxo-, α-benzyl tert-butyl ester	EtOH	284(3.78)	78-5989-69
$C_{23}H_{32}Cl_2O_4$			
5α-Androst-2-ene-3,17β-diol, 17-acetate dichloroacetate	EtOH	215s(3.00)	35-2062-69
$C_{23}H_{32}N_2O_3$			
5α-Androsteno[3,2-c]pyridazin-6'(1'H)-one, 17β-acetoxy-	EtOH	295(3.31)	44-3032-69
$C_{23}H_{32}N_2O_4$			
Pregna-5,16-diene-7,20-dione, 3β-acetoxy-, dioxime	EtOH	237(4.43)	2-1084-69
$C_{23}H_{32}N_6O_{13}$			
Polyoxin H	0.05N HCl	265(3.88)	35-7490-69

Compound	Solvent	$\lambda_{max}(\log \epsilon)$	Ref.
Polyoxin H (cont.)	0.05N NaOH	266(3.79)	35-7490-69
$C_{23}H_{32}N_6O_{14}$			
Polyoxin A	0.05N HCl	262(3.94)	35-7490-69
	0.05N NaOH	264(3.80)	35-7490-69
$C_{23}H_{32}O_2$			
Estra-1,3,5(10)-trien-17β-ol, 1,3,4-trimethyl-, acetate	C_6H_{12}	270(2.58)	78-1679-69
D-Homo-18-nor-9β,10α,17α-pregn-4-en-20-yn-3-one, 13-ethyl-17a-hydroxy-	EtOH	242(4.23)	44-0107-69
18-Nor-9β,10α-pregna-4,6-diene-3,20-dione, 13-propyl-	n.s.g.	287(4.42)	54-0737-69
18-Nor-9β,10α,17α-pregna-4,6-diene-3,20-dione, 13-propyl-	n.s.g.	287(4.40)	54-0737-69
18-Nor-9β,10α-pregn-4-ene-3,20-dione, 13-allyl-	n.s.g.	241(4.23)	54-0737-69
$C_{23}H_{32}O_2S$			
21-Nor-5α-chola-17(20),22-dien-24-oic acid, 3-hydroxy-20-mercapto-, γ-(thiolactone)	EtOH	325(4.28)	87-0001-69
$C_{23}H_{32}O_3$			
Androsta-1,4-dien-3-one, 2-ethyl-17β-hydroxy-, acetate	EtOH	248(4.18)	44-2288-69
Androst-4-ene-$\Delta^{3,\alpha}$-acetaldehyde, 17β-hydroxy-, acetate	EtOH	301(4.46)	78-4535-69
Benzaldehyde, 2,4-dihydroxy-6-methyl-3-(3,7,11-trimethyl-2,6,10-dodecatrienyl)-	MeOH	223(4.18),233s(4.06), 297(4.20),340s(3.59)	78-1323-69
5α-Carda-14,20(22)-dienolide, 3β-hydroxy-	MeOH	214(4.17)	5-0136-69F
5β-Carda-14,20(22)-dienolide, 3α-hydroxy-	n.s.g.	209(4.23)	5-0168-69A
14α-Carda-4,20(22)-dienolide, 3β-hydroxy-	n.s.g.	208(4.24)	5-0110-69G
3'H-Cycloprop[1,2]androsta-1,4-dien-3-one, 1,2-dihydro-17β-hydroxy-7β-methyl-, acetate	MeOH	241(4.17)	24-2565-69
Pregna-4,15-diene-3,20-dione, 17α-hydroxy-6α,16-dimethyl-	MeOH	241(4.21)	87-0393-69
β-Uzarigenin, anhydro-	EtOH	201(4.23),206(4.23), 211(4.22)	33-2583-69
$C_{23}H_{32}O_4$			
Androst-5-ene-3,7-dione, 17β-acetoxy-4,4-dimethyl-	C_6H_{12}	230(4.08),299(1.67), 339(1.48)	33-2436-69
	EtOH	239(4.05),290(1.89), 332(1.61)	33-2436-69
	dioxan	290(1.75),340(1.45)	33-2436-69
Benzaldehyde, 2,4-dihydroxy-6-methyl-3-[3-methyl-5-(1,2,6-trimethyl-3-oxo-cyclohexyl)-3-pentenyl]-	MeOH	223(4.19),233s(4.06), 295(4.22),340s(3.57)	78-1323-69
Carda-4,20(22)-dienolide, 3β,14β-di-hydroxy- (canarigenin)	n.s.g.	212s(4.23)	5-0110-69G
5β-Card-20(22)-enolide, 14β-hydroxy-3-oxo-	n.s.g.	218(4.18)	5-0168-69A
8-Chromancarboxaldehyde, 5-hydroxy-2,7-dimethyl-2-[2-(1,2,6-trimethyl-3-oxo-cyclohexyl)ethyl]-	MeOH	224(4.11),233(4.13), 286(4.13),315s(3.80)	78-1323-69

Compound	Solvent	λ_{max}(log ϵ)	Ref.
β-Coroglaucigenin, anhydro-	EtOH	201(4.23),212(4.21)	33-2583-69
Cyclopropa[5,6]-A-nor-5α-androstane-3,7-dione, 3',6β-dihydro-17β-hydroxy-3',3'-dimethyl-	EtOH	230(3.19)(end abs.)	33-2436-69
Cyclopropa[5,6]-A-nor-5β-androstane-3,7-dione, 3',6α-dihydro-17β-hydroxy-3',3'-dimethyl-	EtOH	230(3.53)(end abs.)	33-2436-69
A-Nor-5ξ-androstane-3,7-dione, 17β-hydroxy-6-isopropylidene-, acetate	EtOH	254(3.77)	33-2436-69
Pregna-5,17(20)-dien-21-al, 3-(ethylenedioxy)-11β-hydroxy-, trans	EtOH	245(4.25)	39-0460-69C
Pregn-4-ene-16α-acetic acid, 3,20-dioxo-	EtOH	237(4.18)	44-3779-69
Pregn-4-ene-6,20-dione, 21-acetoxy-	EtOH	249(3.86)	73-0344-69
5α-Pregn-9(11)-ene-12,20-dione, 3β-acetoxy-	EtOH	238(4.10)	44-0409-69
Uzarigenone	MeOH	217(4.26)	5-0136-69F
$C_{23}H_{32}O_5$			
5α-Androst-14-en-16-one, 3β,17β-diacetoxy-	EtOH	236(4.22)	39-0554-69C
5α,14β-Androst-15-en-17-one, 3β,16-di-	EtOH	238(3.81)	39-0554-69C
Digitoxigenin, 15-oxo-	EtOH	215(4.19)	94-0515-69
Gon-4-en-3-one, 12α,17β-diacetoxy-13β-ethoxy-, 1-	EtOH	239(4.25)	44-3530-69
Hirundigenin, acetyldihydroanhydro-	EtOH	203(4.01)	33-1175-69
Securigenol	n.s.g.	216(4.20),277s(1.5)	105-0110-69
$C_{23}H_{32}O_6$			
Androst-15-en-17-one, 3β,14-dihydroxy-, 3-acetate 14-peroxyacetate	EtOH	212(3.92)	23-3693-69
Pregna-3,5-diene-6-carboxaldehyde, 11β,17α,21-trihydroxy-3-methoxy-20-oxo-	EtOH	219(4.04),323(4.19)	78-1155-69
5α,14β-Pregn-16-ene-11,20-dione, 3β-acetoxy-14,15β-dihydroxy-	EtOH	223(3.84)	94-1401-69
17α-Strophanthidin	EtOH	218(4.20)	33-0482-69
$C_{23}H_{32}O_7$			
Nigakilactone A, acetate	MeOH	273(3.60)	88-3013-69
17α-Strophanthidinic acid	EtOH	220(4.16)	33-0482-69
$C_{23}H_{32}O_8$			
Kurameric acid	MeOH	243(4.11),278(3.36), 288(3.30)	78-2723-69
	MeOH-base	235s(4.10),278(3.36), 288(3.30)	78-2723-69
$C_{23}H_{32}O_{10}$			
Paucin	n.s.g.	207(3.98)	88-0515-69
$C_{23}H_{33}ClO_3$			
Androst-4-en-3-one, 4-chloro-17β-hydroxy-1α,7β-dimethyl-, acetate	MeOH	259(4.11)	24-2565-69
$C_{23}H_{33}NO_3$			
Androst-5-ene-3β,17β-diol, 17α-(3-methyl-5-isoxazolyl)-	MeOH	193(4.08),218(3.91)	24-3324-69
19-Norpregna-3,5-dien-20-one, 3-acetoxy-, O-methyloxime	MeOH	233(4.25)	73-0458-69

Compound	Solvent	λ_{max}(log ε)	Ref.
$C_{23}H_{33}NO_3S_2$			
2-Pyridinol, 1-acetyl-3,6-bis(tert-butylthio)-1,2,3,6-tetrahydro-4-phenyl-, acetate	hexane MeOH	200(4.63),248(4.24) 204(4.63),249(4.32)	44-0660-69 44-0660-69
$C_{23}H_{33}NO_4$			
Acetamide, N-(17β-hydroxy-3-oxoandrost-4-en-6β-yl)-, acetate	MeOH	241(4.13)	48-0445-69
$C_{23}H_{33}NO_4S$			
Cysteine, N-[(3-oxoandrost-4-en-17β-yl)-carbonyl]-	n.s.g.	242(4.10)	23-0160-69
$C_{23}H_{33}NO_5$			
Androsta-3,5-dien-17β-ol, 3-ethoxy-2β-nitro-, acetate	EtOH	248(4.34)	44-3699-69
Androst-5-en-7-one, 3β,17β-diacetoxy-, oxime	EtOH	237(4.14)	2-1084-69
7a-Aza-B-homoandrost-5-en-7-one, 3β,17β-diacetoxy-	EtOH	222(4.16)	2-1084-69
$C_{23}H_{33}N_3O_3$			
Hydroneoechinuline	EtOH	228(4.57),285(3.86), 295(3.82)	88-4457-69
$C_{23}H_{34}N_2O_2$			
Androst-4-en-3β-ol, 17β-[5-(hydroxymethyl)pyrazol-3-yl]-	MeOH-HCl	215(4.04)	73-0442-69
Androst-5-en-3β-ol, 17β-[5-(hydroxymethyl)pyrazol-3-yl]-	MeOH	215(3.98)	73-0442-69
2-Bornanone, 3,3'-(trimethylenedinitrilo)di-, (+)-	MeOH	217(4.16),283(2.41), 376(1.88)	39-2044-69C
$C_{23}H_{34}O_2$			
Δ^8-Cannabinol, tetrahydro-1',1'-dimethyl-, 6a,10a-trans	EtOH	230s(4.09),276(3.16), 282(3.17)	33-1102-69
Δ^9-Cannabinol, tetrahydro-1',1'-dimethyl-, 6a,10a-trans	EtOH	230s(4.00),276(3.25), 283 (3.26)	33-1102-69
18-Nor-9β,10α-androst-4-en-3-one, 17α-allyl-13-ethyl-17-hydroxy-	n.s.g.	243(4.22)	54-0752-69
18-Nor-9β,10α-pregn-4-ene-3,20-dione, 13-propyl-	n.s.g.	242(4.22)	54-0737-69
$C_{23}H_{34}O_2S$			
Androst-4-en-3-one, 7α-(allylthio)-17β-hydroxy-17α-methyl-	EtOH	242(4.19)	94-0011-69
$C_{23}H_{34}O_3$			
Androst-4-en-3-one, 17β-acetoxy-2α-ethyl-	EtOH	240(4.16)	44-2288-69
12,14-(2-Oxapropano)abieta-7,9(11)-dienoic acid, methyl ester	EtOH	237(--),244(4.18), 253(--)	44-1940-69
5α-Pregn-17(20)-en-21-al, 3β-hydroxy-, acetate	EtOH	245(4.19)	78-4535-69
Pregn-16-en-20-one, 3β-acetoxy-	MeOH	239(3.98)	24-1253-69
$C_{23}H_{34}O_4$			
5α-Androstane-$\Delta^{2,\alpha}$-acetic acid, 3-ethoxy-3,17β-dihydroxy-, γ-lactone	EtOH	215(4.14)	44-3032-69

Compound	Solvent	λ_{max}(log ε)	Ref.
5α-Card-20(22)-enolide, 3β,14β-dihydroxy- (uzarigenin)	MeOH	217(4.21)	5-0136-69F
Digitoxigenin	n.s.g.	217(4.21)	5-0168-69A
12,14-(2-Oxapropano)abiet-8(9)-enoic acid, 7-oxo-, methyl ester	EtOH	249(4.11)	44-1940-69
5α-Pregn-16-en-20-one, 3β,12α-dihydroxy-, 3-acetate	MeOH	238(3.95)	44-1606-69
$C_{23}H_{34}O_5$			
Digitoxigenin, 15β-hydroxy-	EtOH	217(4.19)	94-0515-69
14α-Digitoxigenin, 15α-hydroxy-	EtOH	218(4.20)	94-0515-69
17α-Digitoxigenin, 15β-hydroxy-	EtOH	217(4.17)	94-0515-69
$C_{23}H_{34}O_6$			
Bipindogenin	EtOH	217(4.20)	105-0164-69
Card-20(22)-enolide, 1β,3β,5β,14β-tetrahydroxy- (evonogenin)	n.s.g.	218(4.16)	105-0321-69
$C_{23}H_{35}Cl_2N$			
Androsta-3,5-dien-17β-amine, N,N-bis-(2-chloroethyl)-	EtOH	229(4.26),235(4.28), 243s(4.08)	94-1749-69
$C_{23}H_{35}NO$			
Estra-1,3,5(10)-trien-3-ol, 2-[(diethylamino)methyl]-	EtOH	229(3.71),288(3.48)	39-1234-69C
$C_{23}H_{35}NO_2S$			
Androstane-2α-carbonitrile, 7α-(ethylthio)-17β-hydroxy-17-methyl-3-oxo-	MeOH	233(4.08)	94-0011-69
Androstano[3,2-c]isoxazol-17β-ol, 7α-(ethylthio)-17α-methyl-	EtOH	224(4.00)	94-0011-69
Androst-2-eno[2,3-d]isoxazol-17β-ol, 7α-(ethylthio)-17α-methyl-	EtOH	225(3.98)	94-0011-69
$C_{23}H_{35}NO_7$			
5β-Androstane-3α,11β,17β-triol, 3,17-diacetate, nitrate	C_6H_{12}	186(3.77)(end abs.)	78-0761-69
$C_{23}H_{35}N_5O$			
Acetamide, N-(13-isopropyl-7-oxopodocarpa-8,11,13-trien-15-yl)-, 7-(amidinohydrazone), hydrochloride	MeOH	281(4.28)	5-0193-69H
$C_{23}H_{35}O_4P$			
Estra-1,3,5(10)-trien-3-ol, 2-methyl-, diethyl phosphate	EtOH	217(3.90),273(3.10), 281(3.12)	39-1234-69C
$C_{23}H_{36}N_2OS$			
2'H-5α-Androst-2-eno[3,2-c]pyrazol-17β-ol, 7-(ethylthio)-17α-methyl-	EtOH	224(3.74)	94-0011-69
$C_{23}H_{36}N_2O_2S$			
2'H-5α-Androst-2-eno[3,2-c]pyrazol-17β-ol, 7α-(ethylsulfinyl)-17α-methyl-	EtOH	224(3.73)	94-0011-69
$C_{23}H_{36}N_2O_3S$			
2'H-5α-Androst-2-eno[3,2-c]pyrazol-17β-ol, 7α-(ethylsulfonyl)-17α-methyl-	EtOH	224(3.67)	94-0011-69

Compound	Solvent	λ max(log ε)	Ref.
$C_{23}H_{36}N_4O_4$			
Aspochracin	MeOH	297(4.47)	88-0695-69
$C_{23}H_{36}O_2S$			
Androst-4-en-3-one, 17β-hydroxy-7α-(isopropylthio)-17α-methyl-	EtOH	242(4.23)	94-0011-69
Androst-4-en-3-one, 17β-hydroxy-17α-methyl-7α-(propylthio)-	EtOH	242(4.19)	94-0011-69
$C_{23}H_{36}O_3S$			
5α-Androstan-3-one, 7-(ethylthio)-17β-hydroxy-2-(hydroxymethylene)-17α-methyl-	EtOH	278(4.08)	94-0011-69
$C_{23}H_{36}O_4$			
Labda-8(14),13(15)-dien-16-oic acid, 19-acetoxy-, methyl ester	EtOH	221(4.13)	12-1681-69
5α,17α-Pregnan-20-one, 3β,14β-dihydroxy-, 3-acetate	C_6H_{12}	219(2.06),281(1.71)	33-2583-69
Solidagonic acid, methyl ester	EtOH	218(4.19)	18-0812-69
$C_{23}H_{37}NO$			
18,19-Dinor-10α-pregn-13-en-20-one, 3β-(dimethylamino)-5β,14β-dimethyl-	EtOH	258(4.14)	22-3265-69
10β-	EtOH	258(4.09)	22-3265-69
$C_{23}H_{37}NOS$			
Pregn-4-en-3-one, 20β-[(2-mercaptoethyl)amino]-	n.s.g.	242(4.03)	23-0160-69
$C_{23}H_{37}NO_2$			
Ethanol, 2,2'-(androsta-3,5-dien-17β-ylimino)di-	EtOH	228(4.29),235(4.32), 243s(4.11)	94-1749-69
$C_{23}H_{38}N_2O$			
18,19-Dinor-10β-pregn-13-en-20-one, 3β-(dimethylamino)-5β,14β-dimethyl-, oxime	EtOH	242(4.03)	22-3265-69
$C_{23}H_{38}O_4$			
2(3H)-Naphthalenone, 5β-tert-butoxy-4aβ-ethyl-1-[2-(2-ethyl-1,3-dioxolan-2-yl)ethyl]-4,4a,5,6,7,8-hexahydro-, dl-	EtOH	256(4.09)	44-0107-69
$C_{23}H_{42}N_4O_4$			
Aspochracin, hexahydro-	n.s.g.	none	88-0695-69
$C_{23}H_{43}NOS_3$			
Pyridine, 1-acetyl-4-tert-butyl-2,3,6-tris(tert-butylthio)-1,2,3,6-tetrahydro-	hexane	194(4.34)	44-0660-69
	MeOH	205(4.35)	44-0660-69
$C_{23}H_{51}B_{10}N$			
Tetrabutylammonium [7,10^2]hemiousenide	MeCN	264(3.92),512(4.16)	35-0323-69

Compound	Solvent	λ_{max}(log ϵ)	Ref.
$C_{24}H_{12}N_2$			
Diacenaphtho[1,2-b:1',2'-e]pyrazine	H_2SO_4	233(4.59),296(4.24), 348(4.26),410(4.52)	95-0789-69
$C_{24}H_{15}Cl_2NO_4$			
1-Cyclopropene, 1,2-bis(p-chlorophenyl)-3-(4-nitro-α-carbomethoxybenzylidene)-	CH_2Cl_2	268s(--),278(4.51), 312s(--),393(4.23)	24-0319-69
$C_{24}H_{15}NO_2$			
2-Penten-4-ynenitrile, 3-hydroxy-2,5-diphenyl-, benzoate	EtOH	209(4.01),237(4.28), 331(4.37)	39-0915-69C
$C_{24}H_{15}NO_2S$			
Naphtho[2,3-h]quinoline-7,12-dione, 5-(p-tolylthio)-	EtOH	287(4.39),430(3.81)	104-0742-69
$C_{24}H_{16}$			
$\Delta^{1,1'}$-Biacenaphthene	benzene	298(3.54),323(3.70), 340(3.99),360(4.38), 380(4.70),403(4.77)	24-3599-69
	C_6H_{12}	216(3.85),254(4.54), 286(3.50)	24-3599-69
Triphenylene, 2-phenyl-	n.s.g.	260s(4.78),268(4.86), 288s(4.39),301s(4.29)	18-0766-69
$C_{24}H_{16}Cl_2O_{10}$			
2,2'-p-Phenylenebis(1-benzo[b]pyrylium) diperchlorate	MeCN	255(4.27),280s(3.90), 395(4.30)	4-0623-69
$C_{24}H_{16}N_2O$			
2-Naphthol, 1-(9-anthrylazo)-	$CHCl_3$	247(4.80),260(4.79), 300s(4.14),555(4.18)	22-3538-69
$C_{24}H_{16}N_2O_4$			
[$\Delta^{3,3'}$-Biindoline]-2,2'-dione, 1-(benzoyloxy)-1'-methyl-	dioxan	465(3.60)	24-3691-69
Malononitrile, [2-[3-(α-hydroxybenzylidene)-4-oxo-2-chromanylidene]propylidene]-, acetate	CH_2Cl_2	251(4.34),295s(4.04), 372s(4.04),463(4.34)	44-2407-69
Xanthene-2,2(1H)-dicarbonitrile, 1-hydroxy-4-methyl-9-oxo-1-phenyl-, acetate	CH_2Cl_2	256(4.27),262(4.28), 296(4.00),320s(3.81)	44-2407-69
$C_{24}H_{16}N_2S$			
Quinoline, 4-phenyl-2-(2-phenyl-4-thiazolyl)-	n.s.g.	345(3.96)	97-0186-69
$C_{24}H_{16}N_2S_2$			
Benzothiazole, 2-[(3,5-diphenyl-2H-1,4-thiazin-2-ylidene)methyl]-	EtOH	264(4.35),328(4.20), 390(4.18),409s(4.15), 440s(4.00)	32-0323-69
	EtOH-HCl	235(4.42),260(4.46), 356(4.20),508(4.22)	32-0323-69
3H-1,2-Dithiole-$\Delta^{3,\alpha}$-acetonitrile, 5-phenyl-α-(N-phenylbenzimidoyl)-	$CHCl_3$	242(4.68),302(4.33), 450(4.27)	24-1580-69
$C_{24}H_{16}N_4O_2$			
Phenoxazine, 4,4'-azodi-	n.s.g.	240(4.85),317(4.77), 460(3.07)	25-0953-69

Compound	Solvent	λ_{max}(log ε)	Ref.
$C_{24}H_{16}N_6O_4$			
Cinnamic acid, p-azido-, m-phenylene ester	dioxan	319(4.79)	73-3811-69
Cinnamic acid, p-azido-, p-phenylene ester	dioxan	316(4.73)	73-3811-69
$C_{24}H_{16}O$			
Naphtho[2,3-c]furan, 1,3-diphenyl-	benzene	367(3.75),383(3.75), 524(3.90),546(3.90)	44-0538-69
$C_{24}H_{16}O_3$			
Cyclopent[a]indene-3,8-dione, 3a,8a-dihydro-3a-hydroxy-1,2-diphenyl-, acetate	EtOH	231(4.47),300(3.96), 320(3.96)	44-0520-69
7H-Furo[3,2-g][1]benzopyran-7-one, 9-methyl-3,5-diphenyl-	MeOH	222(3.82),254(4.28), 307(4.01)	42-1014-69
$C_{24}H_{16}O_4$			
25,26,27,28-Tetraoxapentacyclo[20.2.1.-$1^{4,7}.1^{10,13}.1^{16,19}$]octacosa-2,4,6,8,-10,12,14,16,18,20,22,24-dodecaene	ether	256(4.08),328s(4.69), 343(5.02),360(5.16), 480s(2.90)	12-1951-69
geometric isomer	ether	260(4.16),335s(4.72), 349(5.03),362(5.19), 370s(5.11),406s(3.10), 464s(2.56),519(2.43), 574s(2.36)	12-1951-69
$C_{24}H_{16}O_9$			
[1,1'-Binaphthalene]-3,4,8'(5'H)-trione, 6',7'-epoxy-6',7'-dihydro-4',5,5'-trihydroxy-, 4',5'-diacetate	EtOH	245(4.39),292(3.79), 443(3.75)	39-2059-69C
$C_{24}H_{16}S$			
Naphtho[2,3-c]thiophene, 1,3-diphenyl-	benzene	513(3.94)	44-0538-69
$C_{24}H_{17}ClN_2$			
Pyrimidine, 2-chloro-4,5-diphenyl-6-styryl-	EtOH	238(4.25),266(4.20), 346(4.39)	103-0825-69
$C_{24}H_{17}ClN_2O_4$			
2-Isoindolinepropionanilide, 2'-benzoyl-4'-chloro-1,3-dioxo-	MeOH	232(4.57)	36-0830-69
$C_{24}H_{17}NO_2$			
Isoxazole, 4-benzoyl-3-phenyl-5-styryl-	MeOH	229(4.32),235(4.31), 259(4.28),324(4.46)	32-0753-69
$C_{24}H_{17}NO_4$			
1-Cyclopropene, 1,2-diphenyl-3-(p-nitro-α-carbomethoxybenzylidene)-	CH_2Cl_2	246(4.42),268(4.37), 293(4.27),305s(--), 392(4.12)	24-0319-69
$C_{24}H_{17}N_3O$			
11H-Dibenzo[a,c]pyrrolo[2,3-i]phenazine, 11-acetyl-12,13-dihydro-	DMF	400(4.12),425(4.23)	103-0199-69
$C_{24}H_{17}N_3O_3$			
Isoquinaldonitrile, 2-benzoyl-1,2-dihydro-1-(o-nitrobenzyl)-	MeOH	228(4.41),281(4.00), 312(3.97)	44-3786-69

Compound	Solvent	λ_{max}(log ϵ)	Ref.
$C_{24}H_{17}N_3O_6$			
1-Cyclopropene, 1,2-di-p-tolyl-3-(2,4,6-trinitrobenzylidene)-	CH_2Cl_2	259(4.38),278(4.40), 398(4.00),508(4.30)	24-0319-69
$C_{24}H_{17}N_3O_8$			
1-Cyclopropene, 1,2-bis(p-methoxyphenyl)-3-(2,4,6-trinitrobenzylidene)-	CH_2Cl_2	278s(--),295(4.37), 398(4.10),524(4.33)	24-0319-69
$C_{24}H_{18}$			
Naphthalene, 1-(p-phenylstyryl)-	DMF	285(4.18),342(4.51)	33-2521-69
Naphthalene, 2-(p-phenylstyryl)-	DMF	293(4.41),340(4.73)	33-2521-69
$C_{24}H_{18}N_2O$			
1H-Benz[g]indazol-5-ol, 3-methyl-1,4-diphenyl-	EtOH	221(4.64),266(4.50), 320(3.95),342(3.90)	103-0251-69
2(1H)-Pyrimidinone, 4,5-diphenyl-6-styryl-	EtOH	238(4.36),350(4.23)	103-0825-69
$C_{24}H_{18}N_2OS$			
Thiourea, 1-(p-benzoylphenyl)-3-(1-naphthyl)-	EtOH	293(4.36),307(4.40), 323(4.40)	34-0506-69
Thiourea, 1-(p-benzoylphenyl)-3-(2-naphthyl)-	EtOH	319(4.47)	34-0506-69
$C_{24}H_{18}N_2OS_2$			
2H-1,4-Benzothiazine, 2-(2-benzothiazolylmethylene)-7-methoxy-3-p-tolyl-	EtOH	258(4.48),322(4.27), 434(4.26)	32-0323-69
	EtOH-HCl	264(4.51),346(4.35), 510(4.23)	32-0323-69
$C_{24}H_{18}N_2O_4$			
Benzo[3,4]cyclobuta[1,2-b]phenazine-8,9-diol, 6,7-dimethyl-, diacetate	EtOH	230(4.31),243s(4.36), 247(4.40),292(4.86), 299(4.86),390(4.17), 411(4.44),436(4.50)	39-2579-69C
$C_{24}H_{18}N_4$			
5H,11H-Benzotriazolo[2,1-a]benzotriazole, 5,11-diphenyl-	n.s.g.	212(4.39),223s(--), 288(4.47),355(3.39)	39-0752-69C
$C_{24}H_{18}N_6O_3$			
Benzoic acid, [[1-[(α-hydroxybenzylidene)amino]-1H-1,2,3-triazol-5-yl]-methylene]hydrazide, benzoate	EtOH	208(4.03),245(3.99), 281(4.16)	39-1416-69C
$C_{24}H_{18}N_6S$			
Sulfur diimide, bis[p-(phenylazo)-phenyl]-	hexane	330(--),475(--)	48-0621-69
$C_{24}H_{18}O$			
Furan, 2-phenyl-5-(p-styrylphenyl)-	DMF	288(4.19),368(4.74)	33-1282-69
Naphthalene, 1-(p-phenoxystyryl)-	DMF	335(4.41)	33-2521-69
Naphthalene, 2-(p-phenoxystyryl)-	DMF	287(4.45),330(4.63)	33-2521-69
$C_{24}H_{18}OS$			
Thione, 8-benzyl-1-naphthyl phenyl, S-oxide	$CHCl_3$	293(4.00),313(4.00), 330(3.93)	77-1483-69
isomer	$CHCl_3$	320(4.15)	77-1483-69

Compound	Solvent	λ_{max}(log ε)	Ref.
$C_{24}H_{18}O_2$			
Acrylophenone, 3,3"-m-phenylenedi-	EtOH	307(4.66)	23-4076-69
Acrylophenone, 3,3"-p-phenylenedi-	EtOH	233(4.13),275(4.09), 355(4.68)	23-4076-69
6H-Benzo[a][1]benzopyrano[2,3-g]xanthene, 7,8-dihydro-	dioxan	245(4.89),300(4.11), 313(4.08),335(3.79), 349(3.86)	5-0162-69B
Benzo[1,2-b:3,4-b']difuran, 5-ethyl-2,7-diphenyl-	MeCN	305(4.62)	12-2395-69
$C_{24}H_{18}O_2S$			
Benzo[b]thiophene, 6-methyl-2-[p-[3,4-(methylenedioxy)styryl]phenyl]-	DMF	367(4.77)	33-1282-69
$C_{24}H_{18}O_3$			
Benzofuran, 5-methyl-2-[p-[3,4-(methylenedioxy)styryl]phenyl]-	DMF	366(3.78)	33-1282-69
Benzofuran, 6-methyl-2-[p-[3,4-(methylenedioxy)styryl]phenyl]-	DMF	368(4.79)	33-1282-69
Stilbene, 4'-[(p-methoxyphenyl)ethynyl]-3,4-(methylenedioxy)-	DMF	359(4.78)	33-2521-69
$C_{24}H_{19}ClO_5S$			
4-(p-Methoxystyryl)-2-phenyl-1-benzothiopyrylium perchlorate	CH_2Cl_2	270(4.29),400(3.78), 570(4.31)	2-0311-69
$C_{24}H_{19}NO$			
2(1H)-Pyridone, 1-benzyl-4,6-diphenyl-	EtOH	243(4.33),268(4.11), 327(3.80)	48-0061-69
$C_{24}H_{19}NO_2$			
[2.2](2,7)-Naphthalenophane, 10-nitro-	EtOH	217(5.1),286(3.8)	88-4567-69
2-Pyrrolidinone, 1-methyl-3-(3-phenyl-1H-naphtho[2,1-b]pyran-1-ylidene)-	MeCN	225(4.55),258(4.53), 297(4.30),355(4.16)	4-0803-69
$C_{24}H_{19}NS$			
Benzylamine, N-(4,6-diphenyl-2H-thiopyran-2-ylidene)-	EtOH	271(4.46),376(3.86)	48-0061-69
2(1H)-Pyridinethione, 1-methyl-4,5,6-triphenyl-	EtOH	292(4.34),378(3.00)	39-1950-69C
$C_{24}H_{19}N_3$			
4H-6-Azaphenaleno[5,6-a]phenazine, 5,6-dihydro-6,12-dimethyl-	C_6H_{12}	498(--)	83-0487-69
	MeOH	470(3.93)	83-0487-69
$C_{24}H_{19}N_3O$			
6H-Pyrrolo[2,3-g]quinoxaline, 6-acetyl-7,8-dihydro-2,3-diphenyl-	DMF	270(4.62),380(4.28)	103-0199-69
$C_{24}H_{19}N_3O_7$			
2-Cyclopropene, 3-ethoxy-1,2-diphenyl-1-(2,4,6-trinitrobenzyl)-	CH_2Cl_2	260(4.38)	24-0319-69
$C_{24}H_{20}$			
1,3-Butadiene, 1-phenyl-4-(p-styrylphenyl)-	DMF	375(4.92),395(4.79)	33-2521-69
5,16:8,13-Diethenodibenzo[a,g]cyclododecene, 6,7,14,15-tetrahydro-, anti	EtOH	238(4.5),280(3.8), 315(3.8)	35-2374-69
syn	EtOH	220(4.9),298f(4.0)	35-2374-69

Compound	Solvent	$\lambda_{max}(\log \epsilon)$	Ref.
$C_{24}H_{20}AsI$			
Tetraphenylarsonium iodide	n.s.g.	230(4.4),253(3.2), 259(3.3),264(3.4), 271(3.4)	120-0195-69
$C_{24}H_{20}AsN$			
Diphenylamine, 4-(diphenylarsino)-	EtOH	211(4.53),240s(4.09), 300(4.40)	101-0117-69E
$C_{24}H_{20}Br_2O_2$			
2-Propen-1-ol, 3,3'-p-phenylenebis[1-(p-bromophenyl)-	EtOH	292(4.55)	124-1189-69
$C_{24}H_{20}Cl_2N_2O_7$			
Thymine, 1-(2-deoxy-α-D-erythro-pentofuranosyl)-, 3',5'-bis(p-chlorobenzoate)	EtOH	242(4.54)	44-3806-69
anomer	EtOH	242(4.54)	44-3806-69
$C_{24}H_{20}Cl_2N_2O_7S$			
Uracil, 5-(methylthio)-1-(2-deoxy-α-D-erythro-pentofuranosyl)-, 3',5'-bis-(p-chlorobenzoate)	EtOH	242(4.61)	44-3806-69
$C_{24}H_{20}Cl_2O_2$			
2-Propen-1-ol, 3,3'-p-phenylenebis[1-(p-chlorophenyl)-	EtOH	292(4.54)	124-1189-69
$C_{24}H_{20}FeO$			
Ferrocene, 1,1'-[(diphenylmethylene)-oxymethylene]-	$CHCl_3$	444(2.34)	39-2260-69C
Iron, π-cyclopentadienyl(dihydro-1,1-diphenyl-π-1H-cyclopenta[c]furanyl)-	$CHCl_3$	438(2.06)	39-2260-69C
$C_{24}H_{20}IN_3O_2S_3$			
3-Methyl-2-[[6-methyl-4-[(3-methyl-2-benzothiazolinylidene)methyl]-2H-thiopyran-2-ylidene]methyl]-6-nitrobenzothiazolium iodide	EtOH-2% $MeNO_2$	609(5.07)	124-0512-69
3-Methyl-2-[[6-methyl-4-[(3-methyl-6-nitro-2-benzothiazolinylidene)methyl]-2H-thiopyran-2-ylidene]methyl]-benzothiazolium iodide	EtOH-2% $MeNO_2$	621(4.98)	124-0512-69
$C_{24}H_{20}N_2$			
4H-1,2-Diazepine, 3,7-diphenyl-5-p-tolyl-	EtOH	216(4.32),264(4.42), 295(4.34)	1-3125-69
$C_{24}H_{20}N_2O$			
4H-1,2-Diazepine, 5-(p-methoxyphenyl)-3,7-diphenyl-	EtOH	209(4.56),270(4.50), 293s(4.46)	1-3125-69
2(1H)-Pyrimidinone, 3,4-dihydro-4,5-diphenyl-6-styryl-	EtOH	232(4.17),262(4.34), 324(4.32)	103-0825-69
$C_{24}H_{20}N_2OS$			
4-Thiazoline, 2-(phenylimino)-3-methyl-4-p-toluoyl-5-phenyl-	MeCN	255(4.45),340(3.75), 420(3.82)	23-3557-69

Compound	Solvent	λ max(log ε)	Ref.
$C_{24}H_{20}N_2O_2$			
1,3-Indandione, 2-acetonyl-2-phenyl-, 2-(phenylhydrazone)	EtOH	226(4.67),268(4.35), 300s(--)	103-0251-69
$C_{24}H_{20}N_2O_2S$			
Thieno[3,4-b]quinoxaline, 1,3-dihydro-1,3-dimethyl-1,3-diphenyl-, 2,2-dioxide	EtOH	241(4.49),325(3.88)	44-1651-69
Thieno[3,4-b]quinoxaline, 1,4-dihydro-1,4-dimethyl-1,3-diphenyl-, 2,2-dioxide	EtOH	219(4.57),245(4.72), 327(4.11)	44-1651-69
$C_{24}H_{20}N_2O_3$			
4-Cinnolinol, 3,4-dibenzoyl-1-ethyl-1,4-dihydro-	n.s.g.	255(4.16),348(3.61), 400s(3.23)	39-1795-69C
$C_{24}H_{20}N_2O_3S$			
Spiro[2H-1-benzopyran-2,2'-benzothiazoline], 3-methyl-6-nitro-3'-phenethyl-	$(MeOCH_2)_2$	254(4.35),267(4.35), 301(4.02),336s(--)	78-3251-69
$C_{24}H_{20}N_2O_5$			
1-Cyclopropene, 1,2-diphenyl-3-(2,4-dinitrobenzylidene)-	CH_2Cl_2	273(4.40)	24-0319-69
Isoquinoline, 3,4-dihydro-6,7-(methylenedioxy)-1-[2-nitro-4-(benzyloxy)-benzyl]-	EtOH	234(4.35),260(3.91), 312(3.92)	2-0755-69
hydrochloride	EtOH	245(4.46),272(4.22)	2-0755-69
$C_{24}H_{20}N_4O_8S_2$			
2,7-Naphthalenedisulfonic acid, 4,5-dihydroxy-3,6-bis(p-tolylazo)-	pH 2.5	560(4.58),640(4.15)	70-0247-69
	pH 7.0	580(4.58),640(4.15)	70-0247-69
	pH 11.5	540(4.40)	70-0247-69
	30% NaOH	600(4.61)	70-0247-69
$C_{24}H_{20}NiO_4P_2S_2Se_2$			
Nickel, bis(dihydrogen phosphoroselenothioato)-, O,O,O,O-tetraphenyl ester	$CHCl_3$	248(4.00),339(4.26), 430(2.96),546(2.02), 736(1.98)	70-0199-69
$C_{24}H_{20}O$			
Benzofuran, 4,6-dimethyl-2-(p-styrylphenyl)-	DMF	363(4.72)	33-1282-69
Benzofuran, 5,6-dimethyl-2-(p-styrylphenyl)-	DMF	361(4.78)	33-1282-69
Benzofuran, 5,7-dimethyl-2-(p-styrylphenyl)-	DMF	358(4.78)	33-1282-69
$C_{24}H_{20}OS$			
Benzo[b]thiophene, 2-[p-(p-methoxystyryl)phenyl]-5-methyl-	DMF	364(4.78)	33-1282-69
Benzo[b]thiophene, 2-[p-(p-methoxystyryl)phenyl]-6-methyl-	DMF	365(4.79)	33-1282-69
$C_{24}H_{20}O_2$			
Anthracene, 2-(3,4-dimethoxystyryl)-	DMF	310(4.59),325(4.69), 340(4.72),377(4.39), 397(4.36)	33-2521-69
Benzene, 1,2-dimethoxy-4-[(p-styrylphenyl)ethynyl]-	DMF	348(4.75)	33-2521-69

Compound	Solvent	$\lambda_{max}(\log \epsilon)$	Ref.
Benzofuran, 2-[p-(p-methoxystyryl)-phenyl]-5-methyl-	DMF	364(4.83)	33-1282-69
Benzofuran, 2-[p-(p-methoxystyryl)-phenyl]-6-methyl-	DMF	366(4.82)	33-1282-69
Benzoic acid, p-(9,10-dihydro-9,10-ethanoanthracen-11-yl)-, methyl ester	EtOH	250(4.25),273(3.65), 284(3.05)	78-1661-69
5H-1-Benzopyran-8-acrolein, 6,7-dihydro-2,4-diphenyl-	MeCN	242(4.25),255(4.26), 270(4.30),345s(4.11), 365(4.18),384(4.20), 470(4.13),510s(4.06)	44-2736-69
Stilbene, 3,4-dimethoxy-4'-(phenylethynyl)-	DMF	357(4.72)	33-2521-69
Stilbene, 4-methoxy-4'-[(p-methoxyphenyl)ethynyl]-	DMF	354(4.79)	33-2521-69
$C_{24}H_{20}O_2S$			
Benzo[b]thiophene, 2-[p-(3,4-dimethoxystyryl)phenyl]-	DMF	366(4.76)	33-1282-69
$C_{24}H_{20}O_3$			
Benzofuran, 2-[p-(3,4-dimethoxystyryl)-phenyl]-	DMF	366(4.79)	33-1282-69
Benzofuran, 6-(3,4-dimethoxystyryl)-2-phenyl-	DMF	361(4.75)	33-1282-69
Benzofuran, 5-methoxy-2-[p-(p-methoxystyryl)phenyl]-	DMF	367(4.83)	33-1282-69
Stilbene, 4'-(p-methoxystyryl)-3,4-(methylenedioxy)-	DMF	372(4.78)	33-2521-69
$C_{24}H_{20}O_4$			
2,5-Norbornadiene-2,3-dicarboxylic acid, 7-(diphenylmethylene)-, dimethyl ester	EtOH	215s(4.29),223s(4.25), 249(4.27),307s(2.88)	89-0209-69
$C_{24}H_{20}O_4P_2PdS_2Se_2$			
Palladium, bis(dihydrogen phosphoroselenothioato)-, O,O,O,O-tetraphenyl ester	$CHCl_3$	242(4.00),313(4.46), 480(2.49)	70-0199-69
$C_{24}H_{20}O_6$			
Isoflavone, 7-(benzyloxy)-2'-hydroxy-4',5'-dimethoxy-	EtOH	245s(4.35),267(4.27), 299(4.32)	18-0233-69
$C_{24}H_{20}O_7$			
Flavone, 4'-(benzyloxy)-5,7-dihydroxy-3,8-dimethoxy-	EtOH	276(4.39),320(4.25)	31-0349-69
$C_{24}H_{20}O_8$			
Naphtho[1,2-d]-1,3-dioxole-7,8-dimethanol, 9-[3,4-(methylenedioxy)phenyl]-, diacetate	EtOH	223(4.46),250(4.62), 295(3.87),312(3.85), 352(3.56)	39-0693-69C
$C_{24}H_{20}O_{11}$			
Eupafolin, tetraacetate	EtOH	265(4.25),301(4.26)	78-1603-69
Flavone, 3',4',5,7-tetrahydroxy-6-methoxy-, tetraacetate	EtOH	265(4.37),302(4.39)	31-0355-69
$C_{24}H_{20}Si$			
Silane, tetraphenyl-	C_6H_{12}	265(3.18)	101-0017-69B

Compound	Solvent	λ_{max}(log ε)	Ref.
$C_{24}H_{21}ClN_2O_4S$			
4-[N-[p-(Dimethylamino)phenyl]formimidoyl]-2-phenyl-1-benzothiopyrylium perchlorate	CH_2Cl_2	262(4.28),305(3.94), 400(3.53),792(4.1)	2-0311-69
$C_{24}H_{21}IN_2S_3$			
3-Methyl-2-[[6-methyl-4-[(3-methyl-2-benzothiazolinylidene)methyl]-2H-thiopyran-2-ylidene]methyl]benzothiazolium iodide	EtOH-2% $MeNO_2$	610(5.06)	124-0512-69
$C_{24}H_{21}NO_5$			
Maleic acid, (1-naphthylphenacylamino)-, dimethyl ester	MeOH	239(4.26),274(4.26), 287(4.20)	78-0527-69
$C_{24}H_{21}NO_6$			
Compound A from phenanthridine and methyl propiolate reaction	MeOH	211(4.56),253(4.48), 269s(4.39),286s(4.25), 414(3.89)	39-2311-69C
Compound B from phenanthridine and methyl propiolate reaction	MeOH	211(4.43),260(4.47), 287s(4.23),323s(3.90), 406(3.84),441s(3.73)	39-2311-69C
4-Isoindolinecarboxylic acid, 3a,4,7,7a-tetrahydro-7-hydroxy-1,3-dioxo-2-phenyl-, ethyl ester, benzoate, cis	EtOH	227(4.31),275(3.15), 282(3.08)	78-4315-69
$C_{24}H_{21}N_3O_2S_2$			
2-Benzothiazolinone, 3,3'-[(phenylimino)diethylene]bis-	MeOH	213(4.79),250(4.52), 288(3.64)	4-0163-69
$C_{24}H_{21}N_3S$			
Imidazole, 5-(benzylideneamino)-4-(benzylthio)-1-methyl-2-phenyl-	EtOH	252(4.26),374(4.08)	44-3661-69
	EtOH-acid	224(4.24),273(4.07)	44-3661-69
$C_{24}H_{22}BrFN_2O_2$			
1-Benzyl-3-[3-(carboxymethyl)-5-fluoroindol-2-yl]pyridinium bromide, ethyl ester	EtOH	221(4.55),247(4.14), 322(4.21)	22-4154-69
$C_{24}H_{22}ClN_3O_5$			
2-Formyl-1-(p-methoxyphenyl)quinolinium perchlorate, o-tolylhydrazone	EtOH	265(3.32),335(2.83), 478(3.18)	124-0370-69
2-Formyl-1-(p-methoxyphenyl)quinolinium perchlorate, p-tolylhydrazone	EtOH	270(3.22),340(2.77), 488(2.90)	124-0370-69
$C_{24}H_{22}ClN_3O_6$			
2-Formyl-1-(p-methoxyphenyl)quinolinium perchlorate, (o-methoxyphenyl)hydrazone	EtOH	261(3.24),334(2.81), 494(2.93)	124-0370-69
2-Formyl-1-(p-methoxyphenyl)quinolinium perchlorate, (p-methoxyphenyl)hydrazone	EtOH	270(3.14),321(2.81), 505(2.50)	124-0370-69
$C_{24}H_{22}ClN_5O_9$			
Benzophenone, 5-chloro-2-(2-aminopivalylamido)-, picrate	MeOH	238(4.58),350(4.26)	36-0830-69
$C_{24}H_{22}Fe$			
Ferrocene, 1,1'-[1-(2,4-cyclopentadien-1-yl)-3-(2,4-cyclopentadien-1-ylidene)-1-methyltrimethylene]-	n.s.g.	380(3.16),450(2.98)	78-6001-69

Compound	Solvent	λ max (log ε)	Ref.
$C_{24}H_{22}N_2$			
Acridine, 9-(1-ethyl-1,2,3,4-tetrahydro-6-quinolyl)-	n.s.g.	250(4.74),305(3.38), 355(3.77),455(3.69)	103-0420-69
$\Delta^{1(2H),\alpha}$-Naphthaleneacetonitrile, α-[1-amino-2-(3,4-dihydro-1-naphthyl)-vinyl]-3,4-dihydro-	EtOH	273(4.03),285(4.07), 294(4.05),382(3.96)	39-0217-69C
Quinoxaline, 2,3-bis(α-methylbenzyl)-	EtOH	238(4.58),310(4.00), 321(4.08)	44-1651-69
$C_{24}H_{22}N_2O$			
4(3H)-Quinazolinone, 2-(2,2-diphenyl-ethyl)-3-ethyl-	EtOH	223(4.49),272(3.91)	115-0057-69
$C_{24}H_{22}N_2O_2$			
4(3H)-Quinazolinone, 2-(p-methoxy-β-phenylphenethyl)-3-methyl-	EtOH	228(4.57),267(4.16)	115-0057-69
$C_{24}H_{22}N_2O_2S_3$			
2,4-Thiazolidinedithione, 3-ethyl-5-[2-[6-methoxy-1-(p-methoxyphenyl)-2(1H)-quinolylidene]ethylidene]-	BuOH	256(4.35),292(4.07), 350(3.92)	65-2116-69
2,4-Thiazoldinedithione, 3-ethyl-5-[2-[6-methoxy-1-(p-methoxyphenyl)-4(1H)-quinolylidene]ethylidene]-	BuOH	261(4.29),315(4.03)	65-2116-69
$C_{24}H_{22}N_2O_7$			
Acetamide, N-[1-(3,4-dimethoxyphenyl)-6,7-dimethoxy-2,3-naphthalenedicarb-oximido]-	HOAc	282(4.87)	39-0838-69C
3,8-Diazabicyclo[3.2.1]oct-6-ene-6,7-dicarboxylic acid, 8-benzyl-3-(p-methoxyphenyl)-2,4-dioxo-, dimethyl ester	EtOH	265(3.61)	94-2461-69
6H-4,6a,11-Metheno-2H-pyrazino[1,2-b]-isoquinoline-11,12(10aH)-dicarboxylic acid, 1,3,4,11a-tetrahydro-2-(p-meth-oxyphenyl)-1,3-dioxo-, dimethyl ester	EtOH	226(4.15),273(3.56), 279(3.53)	94-2461-69
Pyrrolo[3,4-b]pyrrole-3,3a(1H)-dicarb-oxylic acid, 1-benzyl-4,5,6,6a-tetra-hydro-5-(p-methoxyphenyl)-4,6-dioxo-, dimethyl ester	EtOH	231(4.24),302(4.20)	94-2461-69
$C_{24}H_{22}N_2S_3$			
2,4-Thiazolidinedithione, 3-ethyl-5-[2-(6-methyl-1-p-tolyl-2(1H)-quinolyli-dene)ethylidene]-	BuOH	255(4.46),297(4.17), 343(4.08)	65-2116-69
$C_{24}H_{22}N_4$			
Imidazole, 1-methyl-4,5-diphenyl-2-(2,4-xylylazo)-	n.s.g.	397(4.25)	104-0763-69
Imidazole, 2-(2,4,6-trimethylphenylazo)-4,5-diphenyl-	n.s.g.	292(4.12),397(4.34)	104-0763-69
$C_{24}H_{22}O_2$			
Benzene, m-bis(p-methoxystyryl)-	DMF	327(4.77)	33-2521-69
Benzene, o-bis(p-methoxystyryl)-	DMF	300(4.62),337(4.50)	33-2521-69
Benzene, p-bis(p-methoxystyryl)-	DMF	369(4.83)	33-2521-69
2-Propen-1-ol, 3,3'-p-phenylenebis-[1-phenyl-	EtOH	292(4.60)	124-1189-69
Stilbene, 3,4-dimethoxy-4'-styryl-	DMF	368(4.77)	33-2521-69

Compound	Solvent	λmax(log ε)	Ref.
$C_{24}H_{22}O_4$			
[p-Terphenyl]-2',5'-dicarboxylic acid, diethyl ester	EtOH	233(4.47),303(3.72)	94-1591-69
$C_{24}H_{22}O_5$			
6H-Benzofuro[3,2-c][1]benzopyran, 3-(benzyloxy)-6a,11a-dihydro-8,9-dimethoxy-	EtOH	287(3.88),298(3.85)	18-0233-69
Mesuagin	EtOH	235(4.31),286(4.40), 362(3.79)	44-3784-69
$C_{24}H_{22}O_7$			
Glycyrol, 5-O-methyl-, monoacetate	EtOH	232(4.44),239s(4.36), 252s(4.08),260s(3.90), 274s(3.53),289s(3.35), 301s(3.79),310s(4.08), 326s(4.38),336(4.52), 351(4.44)	94-0729-69
$C_{24}H_{22}O_8$			
7-Oxabicyclo[2.2.1]hept-2-ene, 2-piperonyloyl-3-(3,4,5-trimethoxybenzoyl)-	EtOH	277(4.11),320(4.12)	36-0176-69
$C_{24}H_{22}O_{11}$			
Centaureidin, triacetate	EtOH	328(4.00),344(4.06)	18-2701-69
Flavone, 3,4',5-triacetoxy-3',6,7-trimethoxy-	EtOH	241(4.29),308(4.13)	18-1398-69
Flavone, 3',5,7-triacetoxy-3,4',8-trimethoxy-	EtOH	247s(4.24),351(4.03), 360s(4.02)	18-2701-69
$C_{24}H_{23}BrN_2O_2$			
1-Benzyl-3-[3-(carboxymethyl)indol-2-yl]pyridinium bromide, ethyl ester	EtOH	214(4.63),249(4.26), 329(4.23)	22-4154-69
$C_{24}H_{23}ClN_2O$			
1(2H)-Quinolinecarboxamide, 6-chloro-3,4-dihydro-N,2-dimethyl-3,4-diphenyl-	n.s.g.	259(4.17)	40-1047-69
1(2H)-Quinolinecarboxamide, 6-chloro-N-ethyl-3,4-dihydro-3,4-diphenyl-	n.s.g.	258(4.13)	40-1047-69
$C_{24}H_{23}NO$			
Azetidine, 1-ethyl-2-phenyl-3-(p-phenylbenzoyl)-, cis	isooctane	282(4.35)	44-0310-69
trans	isooctane	282(4.42)	44-0310-69
$C_{24}H_{24}Ca_3O_{28}P_2$			
Bis(L-ascorbic acid-3,3')phosphate, calcium salt	pH 1	235(4.26)	94-0387-69
	pH 13	260(4.45)	94-0387-69
$C_{24}H_{24}ClNO_5$			
[(6,7-Dihydro-2,4-diphenyl-5H-1-benzopyran-8-yl)methylene]dimethylammonium perchlorate	MeCN	238(4.22),305(4.48), 353(3.98),493(4.27)	44-2736-69
$C_{24}H_{24}Fe_2O_2$			
[1,1"-Biferrocene]-2-methanol, 2"-methyl-, acetate	C_6H_{12}	198(4.63),222(4.67), 255s(4.01),295(3.82)	78-3477-69

Compound	Solvent	λ_{max}(log ε)	Ref.
$C_{24}H_{24}N_2$			
4,4'-Bipyridine, 1,1'-dibenzyl-1,1',4,4'-tetrahydro-	EtOH	228(4.10),285(3.56)	44-3672-69
1,2-Naphthalenedicarbonitrile, 3,4-dihydro-4,7-dimethyl-4-(p,α,α-trimethylbenzyl)-	MeCN	250(4.18),307(3.90), 345s(3.58)	44-1923-69
$C_{24}H_{24}N_2O_2$			
Benzo[1,2-b:4,5-b']bis[1,4]oxazine, 4,9-dibenzyl-2,3,4,7,8,9-hexahydro-	EtOH	260(3.89),333(3.84)	39-1325-69C
$C_{24}H_{24}N_2O_3$			
Quinoxalone, 1,3-bis(p-methoxybenzyl)-3,4-dihydro-	EtOH	228(4.58),276s(3.74), 285s(3.68),308(3.61)	106-0308-69
$C_{24}H_{24}N_2O_4$			
1H,3H-4,9b-(Iminoethano)-3a,1-propenonaphth[2,1-c]isoxazol-15-one, 4β,5-dihydro-9-hydroxy-8-methoxy-12-methyl-3-phenyl-	EtOH	235(4.02),317(3.09), 361(2.94)	77-1411-69
	EtOH-NaOEt	254(4.09),298(3.64), 361(3.17)	77-1411-69
Morphinan-6-one, 7,8-didehydro-4,5α-epoxy-14-(N-hydroxyanilino)-3-methoxy-17-methyl-	EtOH	240(3.96),280(3.40), 350s(2.80)	77-1411-69
$C_{24}H_{24}N_2O_8$			
Fumaric acid, (2,2'-biphenylylenediimino)di-, tetramethyl ester	MeOH	238(4.37),331(4.43)	23-3545-69
$C_{24}H_{24}N_4O_2$			
3-Pyrazolin-5-one, 3,3'-vinylenebis-[2,4-dimethyl-1-phenyl-, (E)-	EtOH	225(4.30),318(4.35)	94-1309-69
$C_{24}H_{24}N_4O_6S_2$			
[$\Delta^{2,2'}$-Bi-2H-1,4-benzothiazine]-7,7'-dipropionic acid, α,α'-diamino-5,5'-dihydroxy-, dimethyl ester	MeOH	232(4.27),281(4.26), 303(4.24),465(4.31), 481s(4.20)	32-1193-69
	MeOH-HCl	246(4.26),267s(4.12), 327(4.21),550s(4.34), 586(4.48)	32-1193-69
$C_{24}H_{24}O_8$			
Carpanone, dihydro-, diacetate	MeOH	236s(3.96),291(3.88)	88-5159-69
Plicatinaphthol, tetramethyl ether	EtOH	261(4.66),310s(3.90), 320(3.93),351(3.58)	23-0457-69
Succinic acid, piperonylpiperonylidene-, diethyl ester	EtOH	232(4.19),288(4.07), 312(3.99)	39-2470-69C
$C_{24}H_{24}O_{11}$			
2-Anthraceneacetic acid, 1,2,3,4-tetrahydro-3,5,7,10-tetrahydroxy-α-methoxy-4-oxo-, methyl ester, 2,5,7-triacetate	EtOH	225(4.42),260s(4.62), 268(4.70),293s(3.80), 304(3.82),385(3.86)	105-0257-69
$C_{24}H_{24}O_{12}$			
Altersolanol A, tetraacetate	EtOH	210(4.46),266(4.34), 380(3.44)	23-0767-69
$C_{24}H_{25}ClN_2O_6$			
1-Methyl-2-morpholino-3-(2-phenyl-4H-1-benzopyran-4-ylidene)-1-pyrrolinium perchlorate	MeCN	239(4.36),300(4.19), 410(4.31)	4-0803-69

Compound	Solvent	λ_{max}(log ε)	Ref.
$C_{24}H_{25}ClO_7$			
2-(o-Acetoxy-β-pentylstyryl)-1-benzopyrylium perchlorate	MeOH-$HClO_4$	261(4.05),298(3.88), 460(4.33)	5-0226-69E
$C_{24}H_{25}FN_2O_2$			
Indole-3-acetic acid, 2-(N-benzyl-1,2,5,6-tetrahydro-3-pyridyl)-5-fluoro-, ethyl ester	EtOH	225(4.40),299(4.23)	22-4154-69
$C_{24}H_{25}N$			
4-Stilbenamine, N,N-diethyl-4'-phenyl-	DMF	380(4.65)	33-2521-69
$C_{24}H_{25}NO_2$			
Cyclopent[a]inden-8(3H)-one, 3a,8a-dihydro-3a-hydroxy-8a-phenyl-2-(piperidinomethyl)-	EtOH	248(4.07),294(3.23)	104-2090-69
$C_{24}H_{25}NO_4$			
9H-Phenanthro[9,10-b]quinolizin-9-one, 11,12,13,14,14a,15-hexahydro-2,3,6-trimethoxy-	EtOH	254(4.57),262(4.56), 284(4.35),338(3.97)	39-1309-69C
$C_{24}H_{25}NO_9$			
Carbonic acid, ethyl ester, ester with 4-hydroxy-3,5-dimethoxybenzoic acid ester with N-(4-hydroxybutyl)phthalimide	MeOH	220(4.56),233s(4.18), 242(4.14),258(3.98), 298(3.57)	78-5155-69
$C_{24}H_{25}N_3O$			
o-Cresol, 5-(benzo[g]quinolin-4-ylamino)-α-(diethylamino)-	EtOH	248(4.76),395(4.01)	103-0223-69
$C_{24}H_{25}N_5O_8$			
Benzamide, N-[9-(6-deoxy-β-L-galactopyranosyl)-9H-purin-6-yl]-, triacetate	pH 1	288(4.41),253s(4.05)	4-0949-69
	pH 7	255s(4.10),279(4.35)	4-0949-69
	pH 13	302(4.13)	4-0949-69
$C_{24}H_{26}Cl_2N_2O_9$			
Ammonium, [2-(2,6-diphenyl-4H-pyran-4-ylidene)propanediylidene]bis[dimethyl-, diperchlorate	MeCN	268(4.14),355(4.41), 430(4.66)	44-2736-69
Ammonium, [2-(4,6-diphenyl-2H-pyran-2-ylidene)propanediylidene]bis[dimethyl-, diperchlorate	MeCN	244(4.13),315(4.51), 343(4.50),472(4.38)	44-2736-69
$C_{24}H_{26}NO_3P$			
Phosphine oxide, diphenyl[(1,2,3,4-tetrahydro-6,7-dimethoxy-1-isoquinolyl)methyl]-	EtOH	260(3.36),265(3.45), 272(3.52),283(3.56), 286(3.56)	39-0094-69C
$C_{24}H_{26}N_2$			
2,4-Xylidine, N,N'-(2,5-dimethyl-2,5-cyclohexadiene-1,4-diylidene)di-	EtOH	227s(4.10),302(4.39), 465(3.77)	78-2291-69
$C_{24}H_{26}N_2O_2$			
Indole-3-acetic acid, 2-(1-benzyl-1,2,5,6-tetrahydro-3-pyridyl)-, ethyl ester	EtOH	227(4.52),296(4.25)	22-4154-69
Pyridazine, 3,6-diphenyl-4-[3-[(tetrahydro-2H-pyran-2-yloxy)propyl]-	MeOH	262(4.38)	4-0497-69

Compound	Solvent	λ_{max}(log ε)	Ref.
$C_{24}H_{26}N_2O_4$			
1,4-Cyclohexanedicarboxylic acid, 2,3-bis(phenylimino)-, diethyl ester	EtOH	268(4.11),287(4.36)	33-0322-69
$C_{24}H_{26}N_2O_5$			
23,24-Seco-C-nor-13β-strychnine, 23-hydroxy-, diacetate, hydrochloride	EtOH	258(4.03),286s(3.6)	33-1564-69
$C_{24}H_{26}N_4O_2$			
3-Pyrazolin-5-one, 3,3'-ethylenebis-[2,4-dimethyl-1-phenyl-	EtOH	249(4.29),278(4.27)	94-1309-69
$C_{24}H_{26}N_4O_5S$			
Acetophenone, 2-[3-[(4-amino-2-methyl-5-pyrimidinyl)methyl]-5-(2-hydroxy-ethyl)-4-methyl-4-thiazolin-2-ylidene]-2-hydroxy-, diacetate	EtOH	234(4.32),272(3.90), 391(4.40)	94-0128-69
Thiamine, 2-phenyloxalyl-, diacetate	EtOH	234(4.32),272(3.90), 391(4.40)	94-0343-69
$C_{24}H_{26}O$			
Ether, bis(3-inden-3-ylpropyl)	EtOH	223(4.21),252(4.27), 279(3.16),289(2.83)	39-0345-69C
$C_{24}H_{26}O_2$			
4-Phenanthrone, 1,2,3,4-tetrahydro-3-(2-isobutyryl-1-cyclohexyl)-	EtOH	248s(4.68),252(4.77), 275(3.94),286(4.06), 297(3.94),335(3.35), 346(3.37)	22-2110-69
2,2'-Spirobi[2H-1-benzopyran], 6,6'-dimethyl-3-pentyl-	dioxan	260(4.30),310(3.75)	5-0162-69B
$C_{24}H_{26}O_3$			
11H-Benzo[a]fluorene, 7-isopropyl-1,2,9-trimethoxy-11-methyl-	n.s.g.	272(4.68),279(4.79), 302(4.23),326(4.04)	39-2225-69C
$C_{24}H_{26}O_4$			
Cinnamic acid, 4,4'-ethylenedi-, diethyl ester	EtOH	221(4.49),288(4.72), 294(4.72)	35-3517-69
Naphthalene-1,2-dicarboxylic acid, 3,4-dihydro-4-(α,α-dimethylbenzyl)-4-methyl-, dimethyl ester	MeCN	232(4.22),299(4.02)	44-1923-69
$C_{24}H_{26}O_5$			
2H,6H,14H-Dipyrano[2,3-a:2',3'-c]xanthen-14-one, 3,4,7,8-tetrahydro-12-methoxy-2,2,6,6-tetramethyl-	MeOH	230(4.44),265(4.53), 313(4.19),367(3.83)	78-2787-69
Xanthen-9-one, 1,3-dihydroxy-7-methoxy-2,4-bis(3-methyl-2-butenyl)-	MeOH	235(4.49),269(4.49), 317(4.13),380(3.66)	78-2787-69
$C_{24}H_{26}O_7$			
Decursidin	EtOH	222(4.70),300(4.04), 324(4.24)	95-0549-69
$C_{24}H_{26}O_{13}$			
Chrysosplenoside D	EtOH	255(4.46),273(4.46), 342(4.48)	95-0702-69

Compound	Solvent	$\lambda_{max}(\log \epsilon)$	Ref.
$C_{24}H_{27}ClF_2O_4$			
Pregna-1,4-diene-3,20-dione, 17α-acetoxy-6β-chloro-6α,7α-difluoromethylene-	MeOH	245(4.24)	78-1219-69
$C_{24}H_{27}Fe_2N$			
[1,1"-Biferrocene]-2-methylamine, N,N,2"-trimethyl-	C_6H_{12}	197(4.62),222(4.69), 255s(4.03),296(3.86)	78-3477-69
	EtOH	450(2.08)	49-1515-69
$C_{24}H_{27}NO_4$			
4H-Quinolizin-4-one, 1,6,7,8,9,9a-hexahydro-2-(3,4-dimethoxyphenyl)-3-(p-methoxyphenyl)-	EtOH	228(4.35),285(3.98), 320(4.04)	39-1903-69C
Tylophorine	MeOH	258(4.77),288(4.47), 340(3.52),357(3.40)	5-0154-69A
$C_{24}H_{27}NO_5$			
2-Stilbenecarboxylic acid, 3,5-dimethoxy-α-(piperidinocarbonyl)-, methyl ester	MeOH	280(4.31)	44-3192-69
$C_{24}H_{27}NO_6$			
Orientalidine bismethine	EtOH	230(3.41),255(4.34), 296(4.20)	73-0875-69
$C_{24}H_{27}N_3O_5$			
Tryptophan, N-(N-carboxy-DL-alanyl)-, N-benzyl ethyl ester, DL-	EtOH	220(4.55),282(3.78), 290(3.71)	39-1003-69C
$C_{24}H_{27}N_3O_6$			
Lysine, N^6-carboxy-N^2-[(2-phenyl-2-oxazolin-4-yl)carbonyl]-, N^6-benzyl ester, L-	MeOH	245(4.06)	39-1610-69C
$C_{24}H_{28}Cl_2N_2O_4S_3$			
p-Toluenesulfonamide, N,N'-[(2,5-dichloro-3,4-thiophenediyl)dimethylene]-bis[N-ethyl-	EtOH	233(4.43)	44-0333-69
$C_{24}H_{28}F_2O_4$			
Pregna-1,4-diene-3,20-dione, 17α-acetoxy-6α,7α-(difluoromethylene)-	EtOH	243(4.20)	78-1219-69
$C_{24}H_{28}N_2O$			
2',6'-Formoxylidide, 1-(N-2,6-xylyl-1-cyclohexene-1-carboximidoyl)-	EtOH	238(4.44),334(2.7)	24-1876-69
Spiro[cyclohexane-1,3'-[3H]indole]-2'-carboxy-3",5"-xylidide, 4',6'-dimethyl-	EtOH	280(4.1)	24-1876-69
$C_{24}H_{28}N_2O_2$			
Indole-3-acetic acid, 2-(N-benzyl-3-piperidyl)-, ethyl ester	EtOH	222(4.62),281(4.04), 290(3.95)	22-4154-69
Indole-2-carboxamide, 3-benzoyl-N,1-di-tert-butyl-	EtOH	216(4.6),252(4.2), 321(3.8)	24-0678-69
$C_{24}H_{28}N_2O_3S$			
Spiro[2H-1-benzopyran-2,2'-benzothiazoline], 3-methyl-6-nitro-3'-octyl-	$(MeOCH_2)_2$	251s(--),268(4.40), 303(4.08),334s(--)	78-3251-69

Compound	Solvent	$\lambda_{max}(\log \epsilon)$	Ref.
$C_{24}H_{28}N_2O_4$			
Akuammilinol, diacetate	EtOH	268(3.74)	33-0701-69
$C_{24}H_{28}N_2O_4S_2$			
Di-p-toluenesulfonamide, N-[p-(diethyl-amino)phenyl]-	EtOH-1% $CHCl_3$	236(4.46),275(4.39)	44-2083-69
Di-3,4-xylenesulfonamide, N-[p-(dimeth-ylamino)phenyl]-	EtOH-1% $CHCl_3$	238(4.41),275(4.26)	44-2083-69
$C_{24}H_{28}N_2O_5$			
Voachalotine oxindole, acetate	MeOH	210(4.50),258(3.82), 290s(--)	20-0523-69
$C_{24}H_{28}N_2O_6$			
o-Veratronitrile, 6-[2,3,4,5-tetrahydro-1-hydroxy-7,8-dimethoxy-3-methyl-1H-3-benzazepin-2-yl)-, acetate	MeOH	233s(4.24),290(3.79), 307(3.74)	88-5121-69
$C_{24}H_{28}N_4O_6$			
2-Pentanone, 5,5-bis(3,5-dimethyl-2-furyl)-4-methyl-, 2,4-dinitro-phenylhydrazone	EtOH	228(4.50),362(4.36)	35-4739-69
$C_{24}H_{28}O_2$			
2-Cyclohexen-1-one, 2-(2,5-dimethylben-zyl)-3-[(2,5-dimethylbenzyl)oxy]-	EtOH	204(4.43),269(4.28)	44-2192-69
$C_{24}H_{28}O_3$			
14-Anhydroscillarenone	MeOH	236(4.26),299(3.77)	88-3033-69
$C_{24}H_{28}O_4$			
9,19-Cyclo-9β-pregna-5,16-diene-2,7,11,20-tetrone, 4,4,14-trimethyl-	EtOH	236(4.19),266s(3.81)	39-1050-69C
3'H-Cyclopropa[1,2]-24-nor-20ξ-chola-1,4,6,15-tetraen-23-oic acid, 1β,2β-dihydro-17,20-dihydroxy-3-oxo-, γ-lactone	MeOH	282(4.32)	87-0393-69
$C_{24}H_{28}O_5$			
19-Nor-9β-pregna-1(10),5,16-triene-2,7,11,20-tetrone, 9-(hydroxy-methyl)-4,4,14-trimethyl-	EtOH-HCl	238(4.12),288(4.26)	39-1050-69C
Scilliglaucosidin-3-one	MeOH	245(4.08),299(3.71)	44-3894-69
$C_{24}H_{28}O_6$			
1,3-Azulenedipropionic acid, α,α'-di-acetyl-, diethyl ester	EtOH	234(4.00),282(4.61), 285s(4.58),348(3.65), 365(3.57),610(2.47), 650(2.39),830(1.94)	44-2375-69
$C_{24}H_{28}O_7$			
Anthraquinone, 2-(1-hydroxyhexyl)-1,3,6,8-tetramethoxy-	EtOH	224(4.40),283(4.56), 335(3.77),400s(3.48)	39-2763-69C
Eupafolin, tetraethyl ether	EtOH	238(4.24),267(3.90)	78-1603-69
Pseudorhodomyrtoxin	EtOH	278(4.45),312(4.38)	39-2403-69C
Xanthyletin, 3',4'-dihydroxy-3',4'-dihydro-, angeloyl isovaleroyl diester	MeOH	222(4.52),300(4.04), 324(4.23)	95-0549-69

Compound	Solvent	$\lambda_{max}(\log \epsilon)$	Ref.
$C_{24}H_{28}O_8$			
Flavone, 3,3',5-triethoxy-4',6,7-tri-methoxy-	EtOH	233(4.20),242(4.17), 250s(4.17),265(4.00)	44-1460-69
Flavone, 3',5,7-triethoxy-3,4',6-tri-methoxy-	EtOH	258s(4.18),358(4.10)	18-2701-69
Flavone, 3',5,7-triethoxy-3,4',8-tri-methoxy-	EtOH	253(4.35),272(4.33), 353(4.31)	18-2701-69
[$\Delta^{14,15}$-Iso]phorbolactone, hemiacetal, 12,20-diacetate	MeOH	230(4.27),340(1.75)	64-0080-69
Phorbolactone, hemiacetal, 12,20-di-acetate	MeOH	233(3.83),325(1.73)	64-0080-69
$C_{24}H_{29}ClF_2O_4$			
Pregn-4-ene-3,20-dione, 17α-acetoxy-6β-chloro-6α,7α-(difluoromethylene)-	EtOH	247(4.02)	78-1219-69
$C_{24}H_{29}ClN_2O$			
1(2H)-Quinolinecarboxamide, 6-chloro-N-cyclohexyl-3,4-dihydro-2,3-dimethyl-4-phenyl-	n.s.g.	258(4.12)	40-1047-69
$C_{24}H_{29}ClO_4$			
18-Nor-9β,10α-pregna-1,4,6-triene-3,20-dione, 17α-acetoxy-6-chloro-13-ethyl-	n.s.g.	301(4.03)	54-0737-69
Pregna-4,6-diene-3,20-dione, 17-acetoxy-4-chloro-1α,2α-methylene-	MeOH	215(3.76),295(4.27)	24-2570-69
$C_{24}H_{29}Cl_5OS$			
2,5-Cyclohexadien-1-one, 2,4,6-tri-tert-butyl-4-[(pentachlorophenyl)-thio]-	MeOH	242s(4.35)	64-0547-69
$C_{24}H_{29}Cl_5O_2$			
2,5-Cyclohexadien-1-one, 2,4,6-tri-tert-butyl-4-(pentachlorophenoxy)-	C_6H_{12}	231s(4.31)	64-0547-69
$C_{24}H_{29}FO_4$			
Pregna-4,6-diene-3,20-dione, 17-acetoxy-6-fluoro-1α,2α-methylene-	MeOH	281(4.28)	24-2570-69
$C_{24}H_{29}NO_3$			
2H-Quinolizine, 1,3,4,6,9,9a-hexahydro-8-(3,4-dimethoxyphenyl)-7-(p-methoxy-phenyl)-	EtOH	235(4.30),284(3.97)	39-1309-69C
$C_{24}H_{29}NO_5$			
3-Isoxazolecarboxylic acid, 5-(3,17-dihydroxyestra-1,3,5(10)-trien-17α-yl)-, ethyl ester	n.s.g.	223(3.93),281(3.29)	24-3324-69
$C_{24}H_{29}N_3O_6$			
Methanol, bis(4-amino-3,5-dimethoxy-phenyl)(4-amino-3-methoxyphenyl)-	pH 4.62	604(4.75)	64-0542-69
3,4-Seco-20α-yohimban-16β-carboxylic acid, 4-cyano-18β-hydroxy-11,17α-dimethoxy-3-oxo-, methyl ester	MeOH	215(4.41),232s(4.11), 260(3.81),336(4.34)	35-4317-69
$C_{24}H_{30}BrClO_4$			
Pregn-4-ene-3,20-dione, 7α-bromo-6β-chloro-17-hydroxy-1α,2α-methylene-, acetate	MeOH	235(4.10)	24-2570-69

Compound	Solvent	$\lambda_{max}(\log \epsilon)$	Ref.
$C_{24}H_{30}BrFO_4$			
Pregn-4-ene-3,20-dione, 7α-bromo-6β-fluoro-17-hydroxy-1α,2α-methylene-, acetate	MeOH	228(4.08)	24-2570-69
$C_{24}H_{30}Br_2O_4$			
Pregn-4-ene-3,20-dione, 6β,7α-dibromo-17-hydroxy-1α,2α-methylene-, acetate	MeOH	240(4.08)	24-2570-69
$C_{24}H_{30}ClNO_6$			
2-(1-Allyl-3-butenyl)-3-ethyl-6,7-dihydro-9,10-dimethoxybenzo[a]quinolizinium perchlorate	MeOH	225s(4.06),233s(4.02), 255s(3.77),282(4.06), 356(3.98)	24-1779-69
$C_{24}H_{30}ClN_3O_8S_2$			
Sporidesmin D, diacetate	MeOH	218(4.50),253(4.04), 307(3.30)	39-1564-69C
$C_{24}H_{30}Cl_2O_4$			
Pregn-4-ene-3,20-dione, 6β,7α-dichloro-17-hydroxy-1α,2α-methylene-, acetate	MeOH	234(4.10)	24-2570-69
$C_{24}H_{30}FNO_7$			
Pregna-1,4-diene-3,20-dione, 6α-fluoro-11α,21-dihydroxy-16α-methyl-, 21-acetate nitrate	MeOH	240(4.25)	78-0761-69
11β-	MeOH	237(4.24)	78-0761-69
$C_{24}H_{30}F_2O_4$			
Pregn-4-ene-3,20-dione, 17α-acetoxy-6α,7α-(difluoromethylene)-	EtOH	246(4.18)	78-1219-69
$C_{24}H_{30}N_2O$			
Indole-2-carboxamide, N,1-di-tert-butyl-3-methyl-6-phenyl-	EtOH	257(4.6),302(4.3)	24-0678-69
$C_{24}H_{30}N_4NaO_{14}P$			
Thymidine, thymidylyl(5→5')-diacetate	H_2O	267(4.29)	35-1522-69
$C_{24}H_{30}O$			
Androsta-3,5,16-triene-3-carboxaldehyde, 6-methyl-17-(1-propynyl)-	EtOH	230(4.14),236(4.10), 302(4.40)	78-4535-69
$C_{24}H_{30}O_2$			
Anthrone, 10-hydroxy-3,10-dineopentyl-	EtOH	213(4.35),240s(3.91), 287(4.26)	39-2266-69C
Benzo[4,5,6]pregn-4-ene-3,20-dione	EtOH	211(4.35),260(4.06), 303(3.40)	94-2586-69
$C_{24}H_{30}O_3$			
Δ^{14}-Anhydrobufalin, 3-dehydro-	MeOH	302(2.32)	94-1711-69
3'H-Cyclopropa[1,2]pregna-1,4,6,15-tetraene-3,20-dione, 1β,2β-dihydro-17-hydroxy-6,16-dimethyl-	MeOH	287(4.28)	87-0631-69
3'H-Cyclopropa[1,2]pregna-1,4,6-triene-3,20-dione, 1β,2β-dihydro-17-hydroxy-6-methyl-16-methylene-	MeOH	287(4.28)	87-0631-69
3'H-Cyclopropa[1,2]pregna-1,4,6-triene-3,20-dione, 16α,17-epoxy-1β,2β-dihydro-6,16-dimethyl-	MeOH	287(4.27)	87-0631-69

Compound	Solvent	$\lambda_{max}(\log \epsilon)$	Ref.
Scillarenin	MeOH	300(3.77)	88-3033-69
$C_{24}H_{30}O_4$			
9,19-Cyclo-9β-pregna-5,16-diene-7,11,20-trione, 2-hydroxy-4,4,14-trimethyl-	EtOH	237(4.17),270s(3.81)	39-1050-69C
isomer	EtOH	237(3.18),267s(3.74)	39-1050-69C
9,19-Cyclo-5α,8β-pregn-16-ene-2,7,11,20-tetrone, 4,4,14-trimethyl-	EtOH	235(4.00)	39-1050-69C
3'H-Cyclopropa[1,2]pregna-1,4,6-triene-3,20-dione, 1β,2β-dihydro-17-hydroxy-, acetate	MeOH	282(4.33)	24-2570-69
B(9a)-Homo-19-nor-8ξ,9ξ-pregna-5(10),16-diene-2,7,11,20-tetrone, 4,4,14-trimethyl-	EtOH	237(3.96)	39-1050-69C
$C_{24}H_{30}O_5$			
Chola-1,4-dien-24-oic acid, 3,7,12-trioxo-	EtOH	233(3.96)	32-1252-69
Chola-1,4-dien-24-oic acid, 3,7,12-trioxo-	EtOH	237(4.11)	32-1252-69
19-Nor-9β-pregna-5(10),16-diene-2,7,11,20-tetrone, 9-(hydroxy-methyl)-4,4,14-trimethyl-	EtOH	238(4.01)	39-1050-69C
Pregn-4-ene-3,20-dione, 17-acetoxy-6β,7β-epoxy-1α,2α-methylene-	MeOH	240(4.19)	24-2570-69
$C_{24}H_{30}O_7$			
Crotophorbolone enol, 13,20-diacetate	MeOH	197(4.38),240s(3.74), 336(1.82)	64-0091-69
$C_{24}H_{31}BrO_5$			
Pregn-4-ene-3,20-dione, 7α-bromo-6β,17-dihydroxy-1α,2α-methylene-, 17-acetate	MeOH	232(4.10)	24-2570-69
$C_{24}H_{31}ClO_4$			
18-Nor-9β,10α-pregna-4,6-diene-3,20-dione, 17α-acetoxy-6-chloro-13-ethyl-	n.s.g.	288(4.32)	54-0737-69
Pregn-4-ene-3,20-dione, 17α-acetoxy-6β-chloro-16-methylene-	MeOH	240(4.18)	87-0631-69
$C_{24}H_{31}ClO_5$			
Pregn-4-ene-3,20-dione, 6α-chloro-7β,17-dihydroxy-1α,2α-methylene-, 17-acetate	MeOH	235(4.12)	24-2570-69
Pregn-4-ene-3,20-dione, 6β-chloro-7α,17-dihydroxy-1α,2α-methylene-, 17-acetate	MeOH	236(4.03)	24-2570-69
Pregn-4-ene-3,20-dione, 7-chloro-6β,17-dihydroxy-1α,2α-methylene-, 17-acetate	MeOH	231(4.11)	24-2570-69
$C_{24}H_{31}Cl_3O_2$			
2,5-Cyclohexadien-1-one, 2,4,6-tri-tert-butyl-4-(2,4,6-trichlorophenoxy)-	MeOH	230s(4.26)	64-0547-69
$C_{24}H_{31}FO_6$			
Pregna-4,6-dien-3-one, 9α-fluoro-11-hydroxy-16α-methyl-17α,20:20,21-bis(methylenedioxy)-	EtOH	279(4.26)	78-1219-69

Compound	Solvent	$\lambda_{max}(\log \epsilon)$	Ref.
$C_{24}H_{31}NO_5$			
3-Isoxazolecarboxylic acid, 5-(17-hydroxy-3-oxoestr-4-en-17α-yl)-, ethyl ester	MeOH	240(4.28)	24-3324-69
$C_{24}H_{31}NO_6$			
Estra-1,3,5(10)-trien-17-one, 3,9α-dihydroxy-11β-nitro-, 3-hexanoate	EtOH	268(2.94),276(2.88)	31-1018-69
$C_{24}H_{32}$			
Biphenyl, 3,4'-dodecamethylene-	hexane	252(4.29)	88-4551-69
[6,6]Paracyclophane, tetracyanoethylene complex	CH_2Cl_2	490(3.43)	35-3553-69
$C_{24}H_{32}Cl_2N_4O_{10}P$			
Uridine-5'-phosphoramidic acid, N-[(4-dichloroethylaminophenyl)-2-carbethoxyethyl]-	pH 7	260(4.27)	65-0668-69
$C_{24}H_{32}N_2O_{10}S$			
Alanine, 3-[2-[(2-acetamido-2-carboxyethyl)thio]-3,4-dihydroxyphenyl]-N-acetyl-, diethyl ester, diacetate	MeOH	268(4.39)	32-0969-69
$C_{24}H_{32}O_2$			
4'H-Benzo[4,5,6]pregna-4,6-diene-3,20-dione, 5',6'-dihydro-	EtOH	295(4.26)	94-2586-69
4,6,8,10,12,14,16-Eicosaheptaene-3,18-dione, 4,8,13,17-tetramethyl-	$CHCl_3$	421(4.87),433(5.05), 471(5.03)	39-0429-69C
$C_{24}H_{32}O_2S$			
Androsta-5,16-dien-3β-ol, 17-(5-methoxy-2-thienyl)-	EtOH	297(4.07)	87-0001-69
$C_{24}H_{32}O_3$			
5β-Bufa-8(14),20,22-trienolide, 3β-hydroxy-	MeOH	301(2.36)	94-1711-69
5β-Bufa-14,20,22-trienolide, 3β-hydroxy-	MeOH	302(2.40)	94-1711-69
14α-Bufa-5,20,22-trienolide, 3β-hydroxy-	MeOH	300(3.77)	94-1711-69
$C_{24}H_{32}O_3S$			
Androst-4-en-3-one, 17ξ-hydroxy-17-(5-methoxy-2-thienyl)-	EtOH	251(4.30)	87-0001-69
$C_{24}H_{32}O_4$			
14α-Artebufogenin	MeOH	300(2.23)	94-1711-69
14β-Artebufogenin	MeOH	301(2.10)	94-1711-69
5β-Bufa-20,22-dienolide, 14α,15α-epoxy-3β-hydroxy-	MeOH	301(2.21)	94-1711-69
5β-Carda-14,20(22)-dienolide, 3β-hydroxy-, formate	n.s.g.	215(4.24)	5-0168-69A
Chola-1,4-dien-24-oic acid, 3,12-dioxo-	EtOH	243(3.97)	32-1252-69
Estra-1,3,5(10)-triene-1,17β-diol, 2,4-dimethyl-, diacetate	EtOH	272(3.66)	44-2288-69
B(9a)-Homo-19-nor-5ξ,8ξ,9ξ-pregn-1(10)-ene-2,7,11,20-tetrone, 4,4,14-trimethyl-	EtOH	238(4.15)	39-1050-69C
18-Nor-9β,10α-pregna-4,6-diene-3,20-dione, 17α-acetoxy-13-ethyl-	n.s.g.	285(4.43)	54-0737-69

Compound	Solvent	λ_{max}(log ϵ)	Ref.
Pregn-8-ene-2,7,11,20-tetrone, 4,4,14α-trimethyl-	EtOH	270(3.90)	39-1047-69C
Siccanochromenic acid, methyl ester	EtOH	256(4.34),262(4.36), 293(3.50),330(3.31)	88-2457-69
$C_{24}H_{32}O_5$			
Chol-1-en-24-oic acid, 3,7,12-trioxo-	EtOH	228(3.79)	32-1252-69
Chol-4-en-24-oic acid, 3,7,12-trioxo-	EtOH	278(3.90)	32-1252-69
Chol-5-en-24-oic acid, 3,7,12-trioxo-	EtOH	277(3.97)	32-1252-69
Coumarin, 7-[(10,11-dihydroxy-3,7,11-trimethyl-2,6-dodecadienyl)oxy]-	n.s.g.	324(4.20)	105-0191-69
Pregna-4,6-diene-3,20-dione, 11β,21-dihydroxy-16α-methyl-, 21-acetate	EtOH	284(4.40)	78-1219-69
Pregn-4-en-21-al, 17,20α-dihydroxy-3,11-dioxo-, cyclic acetal with acetone	MeOH	238(4.20)	44-3513-69
20β-	MeOH	238(4.19)	44-3513-69
$C_{24}H_{32}O_6$			
Anthracene, 9,10-diethyl-9,10-dihydro-1,2,3,5,6,7-hexamethoxy-, 9,10-trans	hexane	255s(2.86),261(2.91), 271s(2.81)	77-1032-69
14β,17α-Cortexone, 21-O-acetyl-8,19-epoxy-19-methoxy-	EtOH	241(4.09)	33-2156-69
Pregna-4,6-diene-3,20-dione, 21-acetoxy-11β,17α-dihydroxy-6-methyl-	EtOH	289(4.38)	78-1155-69
Pregn-4-ene-3,20-dione, 21-acetoxy-11β,17α-dihydroxy-6-methylene-	EtOH	260(4.07)	78-1155-69
Pregn-4-en-21-oic acid, 17,20α-dihydroxy-3,11-dioxo-, cyclic acetal with acetone	MeOH	238(4.17)	44-3513-69
20β-	MeOH	238(4.18)	44-3513-69
17α-Strophanthidinic acid, 14-anhydro-, methyl ester	EtOH	214(4.20)	33-0482-69
$C_{24}H_{32}O_7$			
Bersaldegenin	MeOH	297(3.72)	88-1709-69
Labda-8(20),13-diene-15,16-dioic anhydride, 3α,6β-dihydroxy-, diacetate	MeOH	206(3.92)	32-0276-69
$C_{24}H_{32}O_8$			
Shikokianin	EtOH	237(3.98)	18-1778-69
Tiglia-1,6-diene-3,13-dione, 15,20-diacetoxy-4,9-dihydroxy-	MeOH	195(4.10),244(3.66), 335(3.85)	64-0091-69
$C_{24}H_{32}O_{10}$			
8β-Podocarp-13-en-15-oic acid, 8,12β,13,14-tetracarboxy-, 13,14,15-trimethyl ester	EtOH	238(3.67)	44-1257-69
$C_{24}H_{33}NO_3$			
9β,10α-Pregn-4-ene-18-carbonitrile, 20,20-(ethylenedioxy)-3-oxo-	n.s.g.	241(4.23)	54-0737-69
$C_{24}H_{33}NO_4$			
Podocarpa-8,11,13-triene-15-carbamic acid, 13-acetyl-7-oxo-, tert-butyl ester	MeOH	251(4.19)	5-0193-69H
$C_{24}H_{33}N_3O_7S$			
Pregn-4-ene-3,20-dione, 6β-azido-7α,17-dihydroxy-, 17-acetate 7-methanesulfonate	MeOH	235(4.11)	48-0919-69

Compound	Solvent	λ_{max}(log ε)	Ref.
$C_{24}H_{34}$			
Chrysene, 1,2,3,4,4a,5,6,11,12,12a-decahydro-10-isopropyl-7,11,11-trimethyl-	EtOH	260(3.70)	22-2115-69
$C_{24}H_{34}O_2$			
1(2H)-Naphthalenone, 3,4-dihydro-2-(2-isobutyrylcyclohexyl)-8-isopropyl-5-methyl-	EtOH	213(4.31),255(3.97), 303(3.31)	22-2110-69
18-Nor-9β,10α-androsta-4,6-dien-3-one, 13-ethyl-17β-hydroxy-17-(2-methylallyl)-	n.s.g.	288(4.41)	54-0752-69
$C_{24}H_{34}O_2S$			
5α-Androst-16-en-3β-ol, 17-(5-methoxy-2-thienyl)-	EtOH	297(4.08)	87-0001-69
$C_{24}H_{34}O_3$			
Pregna-1,8-diene-3,20-dione, 2-hydroxy-4,4,14α-trimethyl-	EtOH	268(4.06)	39-1047-69C
Pregna-5,16-dien-20-one, 3β-acetoxy-21-methyl-	MeOH	240(4.11)	44-3754-69
5α-Pregnane-5-acrolein, 3,20-dioxo-, (E)-	dioxan	221(4.14)	39-0460-69C
5α-Pregn-16-en-20-one, 3β-acetoxy-12-methylene-	MeOH	237(3.89)	44-1606-69
Spiro[cyclopropane-1,6'-[18]nor-9β,10α-androst-4-en]-3'-one, 13'-ethyl-17'β-hydroxy-, acetate	n.s.g.	253(4.10)	54-0752-69
$C_{24}H_{34}O_3S$			
Androst-5-ene-3β,17ξ-diol, 17-(5-methoxy-2-thienyl)-	EtOH	254(4.88)	87-0001-69
$C_{24}H_{34}O_4$			
Androsta-3,5-diene-3,17β-diol, 2α-methyl-, diacetate	EtOH	237(4.23)	44-2288-69
Androst-4-en-3-one, 6β-acetyl-17β-hydroxy-2α-methyl-, acetate	EtOH	246(4.09)	44-2288-69
	EtOH-KOH	428(4.03)	44-2288-69
Cannabidiolcarboxylic acid, ethyl ester	EtOH	224(4.44),273(4.16), 308(3.73)	33-1102-69
5β-Chola-20(21),22-dien-24-oic acid, 14,21-epoxy-3β-hydroxy-	EtOH	298(4.04)	94-1698-69
Chol-4-en-24-oic acid, 3,12-dioxo-	EtOH	238(3.81)	32-1252-69
Chol-5-en-24-oic acid, 7,12-dioxo-	EtOH	237(3.96)	32-1252-69
9,19-Cyclo-5α,9β-pregnane-7,11,20-trione, 2α-hydroxy-4,4,14-trimethyl-	EtOH	209(3.67)	39-1050-69C
2β-	EtOH	215(3.54)	39-1050-69C
18-Nor-9β,10α-pregn-4-ene-3,20-dione, 17α-acetoxy-13-ethyl-	n.s.g.	243(4.19)	54-0737-69
5α-Pregn-8-ene-7,11,20-trione, 2β-hydroxy-4,4,14-trimethyl-	EtOH	272(3.95)	39-1050-69C
Presiccanochromenic acid, methyl ester	EtOH	223(4.29),272(4.04), 306(3.52)	88-2457-69
$C_{24}H_{34}O_5$			
Agbanindiol B	MeOH	212(3.83)	39-2153-69C
Androst-4-ene-19-carboxaldehyde, 3,17-dioxo-, cyclic 17,19-bis(ethylene acetal)	EtOH	244(4.08)	88-4543-69

Compound	Solvent	λ_{max}(log ε)	Ref.
Compound, m. 216-218°	EtOH	285(2.31)	100-0115-69
Digitoxigenin, formate	n.s.g.	217(4.20)	5-0168-69A
Pregn-4-ene-3,11-dione, 17,20α,21-trihydroxy-, cyclic 17,20α-acetal with acetone	MeOH	238(4.18)	44-3513-69
Pregn-4-ene-3,11-dione, 17,20β,21-trihydroxy-, cyclic 17,20β-acetal with acetone	MeOH	238(4.19)	44-3513-69
Pregn-4-ene-3,11-dione, 17,20α,21-trihydroxy-, cyclic 20,21-acetal with acetone	MeOH	238(4.19)	44-3513-69
20β-	MeOH	238(4.19)	44-3513-69
$C_{24}H_{34}O_6$			
1-Oxaspiro[5.5]undec-3-en-2-one, 3,3'-butylidenebis[4-hydroxy-	MeOH	253(4.30)	83-0075-69
1-Oxaspiro[5.5]undec-3-en-2-one, 3,3'-isobutylidenebis[4-hydroxy-	MeOH	253(4.28)	83-0075-69
Pregna-3,5-dien-20-one, 11β,17α,21-trihydroxy-3-methoxy-, 21-acetate	EtOH	240(4.29)	78-1155-69
$C_{24}H_{34}O_7$			
Nigakilactone C	MeOH	265(3.63)	88-3013-69
$C_{24}H_{34}O_{10}S_2$			
D-Mannitol, 1,2,3,4-tetra-O-methyl-, di-p-toluenesulfonate	EtOH	262(3.08),273(3.00)	44-3845-69
$C_{24}H_{35}NO_5$			
Androst-4-en-3-one, 17β-hydroxy-17α-methyl-2α-(2-nitroethyl)-, acetate	EtOH	241(4.11)	19-0269-69
$C_{24}H_{35}N_3OS$			
Spiro[cycloheptane-1,2'-[2H]cyclohepta-[4,5]thieno[2,3-d]pyrimidin]-4'(3'H)-one, 3'-(cycloheptylideneamino)-1',5',6',7',8',9'-hexahydro-	MeOH	229(4.19),249s(4.08), 332(3.35)	48-0402-69
$C_{24}H_{35}N_3O_3$			
Androsta-3,5-diene-6-carboxaldehyde, 17β-hydroxy-3-methyl-, semicarbazone, acetate	EtOH	229(3.90),295(4.42), 306(4.52),317(4.42)	78-4535-69
$C_{24}H_{35}N_7O$			
Dehydroabietylamine, 7-[2-(4-amino-1,2,4-triazol-3-yl)hydrazono]-N-acetyl-, hydrochloride	MeOH	293(4.25)	5-0193-69H
$C_{24}H_{36}ClNO_4$			
1-[3,7-Dimethyl-9-(2,6,6-trimethyl-1-cyclohexen-1-yl)-2,4,6,8-nonatetraenylidene]pyrrolidinium perchlorate, all trans	EtOH	458(4.54)	35-2141-69
	MeCN	452(4.55)	35-2141-69
	HOAc	454(4.57)	35-2141-69
	acetone	450(4.58)	35-2141-69
	dioxan	450(4.58)	35-2141-69
	DMSO	452(4.54)	35-2141-69
	MeI	480(4.55)	35-2141-69
	EtBr	477(4.56)	35-2141-69
	CH_2Cl_2	496(4.57)	35-2141-69
	$CHCl_3$	481(4.59)	35-2141-69
	$CHBr_3$	487(4.55)	35-2141-69

Compound	Solvent	λ_{max}(log ϵ)	Ref.
1-[3,7-Dimethyl-9-(2,6,6-trimethyl-1-cyclohexen-1-yl)-2,4,6,8-nonatetraenylidene]pyrrolidinium perchlorate, all trans (cont.)	$C_2H_4Cl_2$	497(4.59)	35-2141-69
	PhCl	477(4.55)	35-2141-69
	PhBr	480(4.54)	35-2141-69
	PhSMe	474(4.57)	35-2141-69
	$C_2H_2Cl_2$cis	508(4.57)	35-2141-69
	trans	474(4.60)	35-2141-69
$C_{24}H_{36}N_2O_2$			
Aniline, N-tert-butyl-N-[[p-(tert-butylhydroxyamino)-α-methylbenzyl]oxy]-p-ethyl-	EtOH	250(4.05)	39-1459-69C
$C_{24}H_{36}O_2$			
Chol-22-en-24-oic acid, 20α-hydroxy-, γ-lactone	EtOH	213(4.03)	88-0961-69
20β-	EtOH	216(3.96)	88-0961-69
18-Nor-9β,10α-androst-4-en-3-one, 13-ethyl-17β-hydroxy-17-(2-methylallyl)-	n.s.g.	240(4.22)	54-0752-69
$C_{24}H_{36}O_3$			
Pregn-5-en-20-one, 17α-acetoxy-6-methyl-	C_6H_{12}	198(3.95)	87-0548-69
$C_{24}H_{36}O_3S$			
5α-Androstane-3β,17ξ-diol, 17-(5-methoxy-2-thienyl)-	EtOH	258(3.80)	87-0001-69
$C_{24}H_{36}O_4$			
5α-Pregn-16-en-20-one, 3β,12β-dihydroxy-12α-methyl-, 3-acetate	MeOH	242(3.92)	44-1606-69
$C_{24}H_{36}O_4S$			
Acetic acid, [(17β-hydroxy-17-methyl-3-oxoandrost-4-en-7α-yl)thio]-, ethyl ester	EtOH	241(4.22)	94-0011-69
$C_{24}H_{36}O_5$			
Pregn-4-en-3-one, 17,20α-(isopropylidenedioxy)-11β,21-dihydroxy-	MeOH	242(4.19)	44-3513-69
20β-	MeOH	242(4.19)	44-3513-69
Pregn-4-en-3-one, 20α,21-(isopropylidenedioxy)-11β,17-dihydroxy-	MeOH	242(4.19)	44-3513-69
20β-	MeOH	242(4.18)	44-3513-69
$C_{24}H_{36}O_6$			
Taxa-4(20),11-diene-5α,9α,10β,13α-tetrol, 9α,10β-diacetate	n.s.g.	223(3.68)	77-1282-69
$C_{24}H_{37}ClN_2O_6$			
Heliotrine, 4-[(2,3-dihydro-1-hydroxy-1H-pyrrolizin-7-yl)methyl]-, chloride	n.s.g.	215(4.13)	88-3603-69
$C_{24}H_{37}NO_3$			
18,19-Dinor-10α-pregn-13-ene-3α-carbamic acid, 5,14-dimethyl-20-oxo-, ethyl ester	EtOH	258(4.22)	22-3265-69
$C_{24}H_{38}N_2OS$			
2'H-5α-Androst-2-eno[3,2-c]pyrazol-17β-ol, 7-(isopropylthio)-17α-methyl-	EtOH	225(3.69)	94-0011-69

Compound	Solvent	λ_{max}(log ϵ)	Ref.
2'H-5α-Androst-2-eno[3,2-c]pyrazol-17β-ol, 17α-methyl-7-(propylthio)-	EtOH	225(3.70)	94-0011-69
$C_{24}H_{38}N_2O_3$			
18,19-Dinor-10α-pregn-13-ene-3α-carbamic acid, 5β,14β-dimethyl-20-oxo-, ethyl ester, 20-oxime	EtOH	243(4.21)	22-3265-69
$C_{24}H_{38}O_2S$			
Androst-4-en-3-one, 7α-(butylthio)-17β-hydroxy-17α-methyl-	EtOH	242(4.13)	94-0011-69
Androst-4-en-3-one, 7α-(tert-butylthio)-17β-hydroxy-17α-methyl-	EtOH	243(4.20)	94-0011-69
$C_{24}H_{38}O_3$			
5α-Pregnan-21-al, 3β-acetoxy-20-methyl-ene-	EtOH	223(3.84)	13-0637-69B
$C_{24}H_{38}O_3S$			
Androstan-3-one, 17β-hydroxy-2-(hydroxy-methylene)-7-(isopropylthio)-17α-methyl-	EtOH	280(4.06)	94-0011-69
Androstan-3-one, 17β-hydroxy-2-(hydroxy-methylene)-17α-methyl-7-(propylthio)-	EtOH	285(3.99)	94-0011-69
$C_{24}H_{40}N_2O_2$			
Hydroquinone, 2,3-bis[(N-methylcyclohex-ylamino)methyl]-5,6-dimethyl-	EtOH	302(3.62)	39-1245-69C
Hydroquinone, 2,5-bis[(N-methylcyclohex-ylamino)methyl]-3,6-dimethyl-	EtOH	265(3.34),305(3.83)	39-1245-69C
$C_{24}H_{40}O_3$			
Hexadecanophenone, 4'-ethoxy-2'-hydroxy-	n.s.g.	270(3.98)	104-1609-69

Compound	Solvent	$\lambda_{max}(\log \epsilon)$	Ref.
$C_{25}H_{14}ClNO_3$			
4H-Pyrano[3,2-h]quinolin-4-one, 3-benzoyl-6-chloro-2-phenyl-	EtOH	272(4.58)	23-0105-69
Spiro[chrysene-6(5H),5'-[1,4,2]dioxazol]-5-one, 3'-(o-chlorophenyl)-	EtOH	270(4.38),278(4.44), 288(4.45),355(3.76)	23-1473-69
$C_{25}H_{14}N_4$			
Anthracene, 9-(p-tricyanovinylanilino)-	$CHCl_3$	258(5.10),365(3.76), 385(3.84),485(4.55)	22-1659-69
Spiro[indene-1,3'-[3H]pyrazole]-4',5'-dicarbonitrile, 2,3-diphenyl-	$CHCl_3$	268(4.28),298(3.08), 310(3.09),325(2.98), 342(2.82)	64-0536-69
$C_{25}H_{15}ClN_2O_2$			
Spiro[chrysene-6(5H),5'-[Δ^2-1,2,4]oxadiazolin]-5-one, 3'-(o-chlorophenyl)-	EtOH	265(4.53),360(3.98)	23-1473-69
$C_{25}H_{15}N$			
Benzo[5,6]phenanthro[4,3-f]quinoline	dioxan	235(4.7),260(4.8), 315(4.3),350s(4.0), 392(2.9),408(2.8)	88-3597-69
$C_{25}H_{16}O_2$			
3,3'-Spirobi[3H-naphtho[2,1-b]pyran]	dioxan	247(5.06),300(4.22), 313(4.27),336(3.99), 350(4.05)	5-0162-69B
$C_{25}H_{17}ClO_4S$			
1,3-Diphenyl-4-thiaphenanthrenium perchlorate	HOAc-1% $HClO_4$	300(4.00),420(4.25)	2-0191-69
2,4-Diphenyl-1-thiaphenanthrenium perchlorate	HOAc-1% $HClO_4$	260(4.43),320(4.24), 385(4.0),470(4.04)	2-0191-69
$C_{25}H_{17}NOS$			
Benzo[b]thiophen-3(2H)-one, 2-[2-(1-phenyl-2(1H)-quinolylidene)-ethylidene]-	BuOH	573(5.11),609(5.07)	65-1829-69
Benzo[b]thiophen-3(2H)-one, 2-[2-(1-phenyl-4(1H)-quinolylidene)-ethylidene]-	BuOH	605(5.06),649(5.11)	65-1829-69
$C_{25}H_{17}NO_3$			
2-Penten-4-ynenitrile, 3-hydroxy-2-(p-methoxyphenyl)-5-phenyl-, benzoate	EtOH	211(4.26),238(4.57), 348(4.64)	39-0915-69C
$C_{25}H_{17}N_5O$			
[1,2'(1'H)-Biphthalazine]-4'-carbonitrile, 1'-phenacyl-	EtOH	245(4.46),285(3.81), 372(4.33)	95-0959-69
$C_{25}H_{18}N_2O_2$			
2-Buten-1,4-dione, 1,4-diphenyl-2-(4-phenylpyrazol-3-yl)-	EtOH	244(4.26),266(4.24), 450(4.45)	39-2464-69C
$C_{25}H_{18}N_2O_5$			
Bicyclo[2.2.2]oct-7-ene-2,3,5,6-tetracarboxylic 2,3:5,6-diimide, 1-formyl-N,N'-diphenyl-	dioxan	219(4.45)	78-4315-69

Compound	Solvent	$\lambda_{max}(\log \epsilon)$	Ref.
$C_{25}H_{18}N_2S$			
Quinoline, 2-(4-methyl-2-phenyl-5-thiazolyl)-4-phenyl-	n.s.g.	356(4.59)	97-0186-69
$C_{25}H_{18}N_6O_2$			
Propionitrile, α-cyano-β-(2-quinolylhydrazino)-β-(p-nitrophenyl)-β-phenyl-	EtOH	280(4.47),310(4.26), 364(4.23)	2-0412-69
$C_{25}H_{18}O_3$			
Cyclobuta[b]naphthalene-3,8-dione, 2a,8a-dihydro-2a-methoxy-1,2-diphenyl-	EtOH	225(4.55),255(4.25), 288s(4.10),355s(3.37)	44-0520-69
$C_{25}H_{19}BrN_2Si$			
Silane, (p-bromo-α-diazobenzyl)triphenyl-	C_6H_{12}	222(4.57),289(4.29), 437(1.98)	23-4353-69
$C_{25}H_{19}ClN_2O_4$			
2-Isoindolinepropionanilide, 2'-benzoyl-4'-chloro-N-methyl-1,3-dioxo-	MeOH	218(4.83)	36-0830-69
$C_{25}H_{19}ClN_2Si$			
Silane, (p-chloro-α-diazobenzyl)-triphenyl-	C_6H_{12}	222(4.55),287(4.37), 320s(3.36),441(1.98)	23-4353-69
$C_{25}H_{19}ClN_4O_7$			
Pyridine, 3-chloro-4,5-dimethyl-2,6-diphenyl-, picrate	EtOH	214(4.29),232(4.20), 287(3.82)	39-2249-69C
$C_{25}H_{19}FN_2Si$			
Silane, (α-diazo-p-fluorobenzyl)-triphenyl-	C_6H_{12}	193s(4.31),196s(4.45), 201s(4.59),209(4.65), 221s(4.54),276(4.23), 309s(3.41),438(1.91)	23-4353-69
$C_{25}H_{19}NO$			
Benz[c]acridine, 5,6-dimethyl-7-phenoxy-	MeOH	222(4.70),234s(4.50), 276s(4.67),284(4.71), 297(4.74),328(3.71), 342(3.82),357(3.82), 373(3.77),393(3.71)	4-0361-69
Ketone, 5-(α-anilinobenzylidene)-1,3-cyclopentadien-1-yl phenyl	EtOH	237(4.19),289(4.21), 358(4.04),434(4.27)	39-2464-69C
$C_{25}H_{19}N_3O$			
2-Naphthol, 1-[(3-methyl-2-phenyl-1-indolizinyl)azo]-	EtOH	222(4.65),250(4.37), 510(4.41),537(4.37)	39-1279-69C
2-Naphthol, 1-[(5-methyl-2-phenyl-1-indolizinyl)azo]-	EtOH	219(4.66),250(4.43), 502(4.44),531(4.44)	39-1279-69C
$C_{25}H_{19}N_3O_4$			
Benzoic acid, 2-[1,4-dihydro-1,4-dioxo-3-(N-phenylacetamido)-2-naphthyl]-hydrazide	alkali	615(3.99)	95-0007-69
$C_{25}H_{20}$			
1H-Cycloheptatriene, 1,4,7-triphenyl-	MeOH	244(4.30),325(4.18)	88-4401-69
7H-Cycloheptatriene, 1,4,5-triphenyl-	MeOH	251(4.06),317(4.04)	88-4401-69

Compound	Solvent	λ_{max}(log ε)	Ref.
$C_{25}H_{20}Br_2N_2O$			
Pyrimidine, 4-(α,β-dibromophenethyl)-2-methoxy-5,6-diphenyl-	EtOH	302(4.04)	103-0825-69
$C_{25}H_{20}ClNO_4S_2$			
3-[p-(Dimethylamino)phenyl]-1-(2-thienyl)naphtho[2,1-b]thiopyrylium perchlorate	HOAc-1% $HClO_4$	250(4.5),400(3.12), 650(3.56)	2-0191-69
$C_{25}H_{20}Cl_2N_2O_2S$			
4H-Thiopyran-4-one, 3,5-dichloro-2,6-bis(p-methoxyphenyl)-, phenylhydrazone	EtOH	250(4.46),381(4.43)	39-0315-69C
$C_{25}H_{20}Cl_2N_2O_8S$			
Isobarbituric acid, 1-(2-deoxy-α-D-erythro-pentofuranosyl)-5-thio-, 5-acetate, 3',5'-(p-chlorobenzoate)	EtOH	241(4.56),272(4.03)	44-3806-69
$C_{25}H_{20}FeO$			
2-Propen-1-one, 1-(m-ferrocenylphenyl)-3-phenyl-	MeOH	307(4.47)	73-2771-69
2-Propen-1-one, 1-(p-ferrocenylphenyl)-3-phenyl-	MeOH	325(4.34)	73-2771-69
2-Propen-1-one, 3-(m-ferrocenylphenyl)-1-phenyl-	MeOH	286(4.34)	73-2771-69
2-Propen-1-one, 3-(p-ferrocenylphenyl)-1-phenyl-	MeOH	347(4.39)	73-2771-69
$C_{25}H_{20}GeN_2$			
Germane, (α-diazobenzyl)triphenyl-	C_6H_{12}	217s(4.51),284(4.11), 451(1.85)	23-4353-69
$C_{25}H_{20}GeO$			
Germane, benzoyltriphenyl-	heptane	253(4.21),418(2.43)	59-0765-69
$C_{25}H_{20}N_2$			
Benzophenone, diphenylhydrazone	EtOH	250(4.34),300(3.93), 364(3.84)	39-1703-69C
$C_{25}H_{20}N_2O$			
2H-Cyclopenta[b]quinoxalin-2-one, 1,4-dihydro-1,4-dimethyl-1,3-diphenyl-	EtOH	222(4.33),226(4.34), 263(4.01),306(4.00), 323(3.94)	44-1651-69
Pyrimidine, 2-methoxy-4,5-diphenyl-6-styryl-	EtOH	232(4.35),256(4.20), 346(4.45)	103-0825-69
$C_{25}H_{20}N_2O_3$			
1,3-Butadiene-1-carbonitrile, 3-ethoxy-1-(p-nitrophenyl)-2,4-diphenyl-	CH_2Cl_2	260(4.26),315(4.19), 380s(--)	24-0319-69
Propionic acid, 2-[o-(2-hydroxy-1-naphthylazo)phenyl]-2-phenyl-	EtOH	228(4.59),315(3.98), 485(4.08)	44-2799-69
$C_{25}H_{20}N_2Si$			
Silane, (α-diazobenzyl)triphenyl-	C_6H_{12}	222(4.49),280s(4.12), 308s(3.30),440(1.83)	23-4353-69
$C_{25}H_{20}N_4O_3S$			
Acetophenone, 2-[1,2-dihydro-4-(methylsulfonyl)-2-(1-phthalazinyl)-1-phthalazinyl]-	EtOH	245(4.40),280s(3.80), 363(4.24)	95-0959-69

Compound	Solvent	$\lambda_{max}(\log \epsilon)$	Ref.
$C_{25}H_{20}O$			
Furan, 2-(p-styrylphenyl)-5-p-tolyl-	DMF	297(4.09),374(4.75)	33-1282-69
$C_{25}H_{20}OSi$			
Silane, benzoyltriphenyl-	heptane	257(4.21),425(2.47)	59-0765-69
$C_{25}H_{20}O_2$			
Furan, 2-[p-(p-methoxystyryl)phenyl]-5-phenyl-	DMF	298(4.04),374(4.81)	33-1282-69
$C_{25}H_{20}O_3$			
2-Furoic acid, 3,4-diphenyl-5-p-tolyl-, methyl ester	EtOH	233(4.35),308(4.47)	39-1950-69C
$C_{25}H_{20}O_4$			
3,3'(4H,4'H)-Spirobi[2H-naphtho[2,3-b]-[1,4]dioxepin]	$CHCl_3$	244(4.92),275(4.14), 285(4.13),314(3.64), 328(3.79)	78-5427-69
$C_{25}H_{20}O_5$			
2-Naphthoic acid, 3-[p-(benzyloxy)benzylidene]-1,2,3,4-tetrahydro-6-hydroxy-4-oxo-	EtOH	230(4.34),339(4.15)	2-0561-69
$C_{25}H_{20}S$			
Thiophene, 2-(p-styrylphenyl)-5-p-tolyl-	DMF	374(4.74)	33-1282-69
$C_{25}H_{21}ClN_2O_6$			
1,3-Dimethyl-6-(m-nitrostyryl)-2-phenylcyclohepta[c]pyrrolium perchlorate (4 λ,3 ε)	$C_2H_4Cl_2$	300(4.32),340(4.62), 470(4.37),660(?)	104-2179-69
$C_{25}H_{21}ClN_2O_7$			
6-(2-Hydroxy-5-nitrostyryl)-1,3-dimethyl-2-phenylcyclohepta[c]pyrrolium perchlorate	$C_2H_4Cl_2$	330(4.23),510(4.40)	104-2179-69
$C_{25}H_{21}ClO$			
Benzofuran, 5-chloro-2-[p-(p-isopropylstyryl)phenyl]-	DMF	360(4.84)	33-1282-69
$C_{25}H_{21}ClS$			
Benzo[b]thiophene, 5-chloro-2-[p-(p-isopropylstyryl)phenyl]-	DMF	359(4.80)	33-1282-69
$C_{25}H_{21}N$			
Aniline, N-methyl-N-(5-phenalen-1-ylidene-1,3-pentadienyl)-	hexane	278(4.23),340(4.31), 511(4.84)	24-2301-69
	MeCN	281(4.36),342(4.37), 515(4.91)	24-2301-69
$C_{25}H_{21}NO_3$			
Morpholine, 4-[(2-phenyl-4H-naphtho[1,2-b]pyran-4-ylidene]acetyl]-	MeCN	258(4.55),350(4.30)	4-0803-69
$C_{25}H_{21}NO_8$			
3-Carboxy-1-D-ribofuranosylpyridinium hydroxide, inner salt, 3',5'-dibenzoate	MeOH	273(3.68)	39-0918-69C

Compound	Solvent	λ_{max}(log ϵ)	Ref.
$C_{25}H_{21}N_3O_2$			
4H-1,2-Diazepine, 5-(m-nitrophenyl)-3,7-di-p-tolyl-	EtOH	220(4.43),264(4.60), 305s(4.32)	1-3125-69
$C_{25}H_{22}$			
Anthracene, 2-(p-isopropylstyryl)-	DMF	314(4.88),329(4.94), 356(4.20),374(4.34), 394(4.25)	33-2521-69
Phenanthrene, 7-isopropyl-1-styryl-	DMF	328(4.44)	33-2521-69
$C_{25}H_{22}BF_2NO_2$			
Naphth[1,2-e]-1,3,2-dioxaborin, 2,2-difluoro-4-[β-(5-julolidinyl)vinyl]-	MeCN	350(3.84),610(4.80)	4-0029-69
$C_{25}H_{22}ClNO_4$			
1,3-Dimethyl-2-phenyl-6-styrylcyclohepta[c]pyrrolium perchlorate	$C_2H_4Cl_2$	270(4.2),340(4.39), 440(4.56),670(3.47)	104-2179-69
$C_{25}H_{22}ClNO_4S$			
4-[p-(Dimethylamino)phenyl]-2,6-diphenylthiopyrylium perchlorate	CH_2Cl_2	262(4.28),294(4.95), 372(4.12),712(5.0)	2-0311-69
$C_{25}H_{22}ClNO_5$			
Dimethyl[α-[(2-phenyl-4H-1-benzopyran-4-ylidene]methyl]benzylidene]ammonium perchlorate	MeCN	238(4.34),320(4.15), 445(4.55)	44-2736-69
6-(o-Hydroxystyryl)-1,3-dimethyl-2-phenylcyclohepta[c]pyrrolium perchlorate	$C_2H_4Cl_2$	295(4.37),330(4.36), 470(4.30),600(3.47)	104-2179-69
$C_{25}H_{22}Cl_2N_2O_7S$			
Uridine, 2'-deoxy-5-(ethylthio)-, 3',5'-bis(p-chlorobenzoate)	EtOH	242(4.57)	44-3806-69
$C_{25}H_{22}Cl_2O_8$			
2,6-Cresotic acid, 4-methoxy-, 4,6-dichloro-5-methyl-m-phenylene ester	EtOH	231(4.18),270(4.49), 308(4.10)	39-1721-69C
$C_{25}H_{22}Fe_2O$			
1,4-Pentadien-3-one, 1,5-diferrocenyl-	EtOH	207(4.56),340(4.25), 402(3.62),527(3.71)	18-3273-69
$C_{25}H_{22}Fe_2O_2$			
2-Propen-1-one, 1-(1'-acetylferrocenyl)-3-ferrocenyl-	EtOH	270(4.11),329(4.19)	73-2235-69
2-Propen-1-one, 3-(1'-acetylferrocenyl)-1-ferrocenyl-	EtOH	256(4.10),319(4.15)	73-2235-69
$C_{25}H_{22}INS$			
2-(Benzylmethylamino)-4,6-diphenylthiopyrylium iodide	EtOH	280(4.21),326(4.06), 401(4.06)	48-0061-69
$C_{25}H_{22}IN_3O_3S_3$			
2-[[4-[(6-Methoxy-3-methyl-2-benzothiazolinylidene)methyl]-6-methyl-2H-thiopyran-2-ylidene]methyl]-3-methyl-6-nitrobenzothiazolium iodide	EtOH-2% MeOH	618(4.95)	124-0512-69
6-Methoxy-3-methyl-2-[[6-methyl-4-[(3-methyl-6-nitro-2-benzothiazolinylidene)methyl]-2H-thiopyran-2-ylidene]methyl]benzothiazolium iodide	EtOH-2%	637(4.86)	124-0512-69

Compound	Solvent	$\lambda_{max}(\log \epsilon)$	Ref.
$C_{25}H_{22}N_2$			
Naphtho[2,1-b]pyrazole, 5,5-dimethyl-1,3-diphenyl-	EtOH	262(4.35),282s(--), 292s(--),302s(--)	103-0251-69
$C_{25}H_{22}N_2O$			
Pyrimidine, 2-methoxy-4-phenethyl-	EtOH	288(4.01)	103-0825-69
1H-Pyrrolo[3,4-b]quinoline-3-ethanol, 2,3-dihydro-α,α-diphenyl-	EtOH	232(4.51),298(3.62), 300(3.64),306(3.75), 313(3.72),319(3.87)	44-3853-69
$C_{25}H_{22}N_2S_2$			
2H-1,3,5-Thiadiazine-2-thione, tetrahydro-3,5-bis(1-naphthylmethyl)-	MeOH	250(4.00),285(4.37)	73-2952-69
$C_{25}H_{22}O$			
Anisole, p-[p-(4-phenyl-1,3-butadienyl)-styryl]-	DMF	381(4.91),400(4.78)	33-2521-69
Benzofuran, 2-[p-(p-isopropylstyryl)-phenyl]-	DMF	358(4.81)	33-1282-69
Benzofuran, 4,6,7-trimethyl-2-(p-styryl-phenyl)-	DMF	363(4.75)	33-1282-69
$C_{25}H_{22}O_2$			
Benzofuran, 2-[p-(p-methoxystyryl)-phenyl]-4,6-dimethyl-	DMF	369(4.77)	33-1282-69
Benzofuran, 2-[p-(p-methoxystyryl)-phenyl]-5,6-dimethyl-	DMF	367(4.82)	33-1282-69
Benzofuran, 2-[p-(p-methoxystyryl)-phenyl]-5,7-dimethyl-	DMF	365(4.81)	33-1282-69
$C_{25}H_{22}O_2S$			
Benzo[b]thiophene, 2-[p-(3,4-dimethoxy-styryl)phenyl]-6-methyl-	DMF	368(4.78)	33-1282-69
$C_{25}H_{22}O_3$			
Benzofuran, 2-[p-(3,4-dimethoxystyryl)-phenyl]-5-methyl-	DMF	367(4.78)	33-1282-69
Benzofuran, 2-[p-(3,4-dimethoxystyryl)-phenyl]-6-methyl-	DMF	368(4.79)	33-1282-69
Stilbene, 4-[(3,4-dimethoxyphenyl)-ethynyl]-4'-methoxy-	DMF	357(4.77)	33-2521-69
Stilbene, 3,4,5-trimethoxy-4'-(phenyl-ethynyl)-	DMF	353(4.73)	33-2521-69
Styrene, β-2-anthryl-3,4,5-trimethoxy-	DMF	323(4.79),335(4.77), 357(4.28),376(4.38), 396(4.34)	33-2521-69
$C_{25}H_{22}O_3S$			
Benzo[b]thiophene, 2-[p-(3,4,5-trimeth-oxystyryl)phenyl]-	DMF	363(4.78)	33-1282-69
$C_{25}H_{22}O_4$			
Benzofuran, 2-phenyl-6-(3,4,5-trimeth-oxystyryl)-	DMF	286(4.05),357(4.75)	33-1282-69
Benzofuran, 2-[p-(3,4,5-trimethoxy-styryl)phenyl]-	DMF	362(4.81),380(4.67)	33-1282-69
$C_{25}H_{22}O_6$			
Isoflavone, 7-(benzyloxy)-2',4',5'-tri-methoxy-	EtOH	248(4.40),297(4.32)	18-0233-69

Compound	Solvent	λ_{max}(log ε)	Ref.
$C_{25}H_{22}O_8$			
Derrubone, diacetate	EtOH	255(4.35),290(4.07)	39-0365-69C
Flavone, 3'-(benzyloxy)-5,7-dihydroxy-3,4',6-trimethoxy-	EtOH	255(4.28),272(4.21),348(4.34)	18-2701-69
	EtOH-NaOAc	275(4.34),298(4.05),364(4.22)	18-2701-69
	EtOH-$AlCl_3$	261(4.24),366(4.31)	18-2701-69
Flavone, 3'-(benzyloxy)-5,7-dihydroxy-3,4',8-trimethoxy-	EtOH	258(4.30),276(4.27),340(4.20),362s(4.20)	18-2701-69
Glycyrol, diacetate	EtOH	204(4.63),228(4.56),236s(4.49),259s(3.99),290s(3.90),317s(4.32),329(4.42),344(4.32)	94-0729-69
$C_{25}H_{22}O_{12}$			
Axillarin, tetraacetate	EtOH	256(4.16),320(4.06),340s(4.03)	18-1649-69
Flavone, 3',4',5,7-tetrahydroxy-3,8-dimethoxy-, tetraacetate	EtOH	249s(4.23),255(4.24),311(3.93),354(4.05)	18-1649-69
$C_{25}H_{22}S$			
Benzo[b]thiophene, 2-[p-(p-isopropylstyryl)phenyl]-	DMF	358(4.79)	33-1282-69
2H-1-Benzothiopyran, 5,6,7,8-tetrahydro-2-(2-naphthyl)-4-phenyl-	n.s.g.	224(4.4),243(4.5),272(3.7),385(3.2)	104-1660-69
$C_{25}H_{23}AsIN$			
(p-Anilinophenyl)methyldiphenylarsonium iodide	EtOH	215(4.67),315(4.44)	101-0117-69E
$C_{25}H_{23}BrN_2O$			
1,2-Dibenzylimidazolium 3-(p-bromophenacylide)	n.s.g.	245s(4.13),310(3.67)	103-0638-69
$C_{25}H_{23}ClN_2O$			
1(2H)-Quinolinecarboxamide, N-allyl-6-chloro-3,4-dihydro-2,4-diphenyl-	n.s.g.	256(4.10)	40-1047-69
$C_{25}H_{23}ClN_2O_7$			
3-Carbamoyl-1-D-ribofuranosylpyridinium chloride, 3',5'-dibenzoate	MeOH	268(3.74)	39-0199-69C
$C_{25}H_{23}IN_2OS_3$			
2-[[4-[(6-Methoxy-3-methyl-2-benzothiazolinylidene)methyl]-6-methyl-2H-thiopyran-2-ylidene]methyl]-3-methylbenzothiazolium iodide	EtOH-2% $MeNO_2$	617(4.88)	124-0512-69
6-Methoxy-3-methyl-2-[[6-methyl-4-[(3-methyl-2-benzothiazolinylidene)methyl]-2H-thiopyran-2-ylidene]methyl]-benzothiazolium iodide	EtOH-2% $MeNO_2$	623(4.98)	124-0512-69
$C_{25}H_{23}IN_2O_5$			
3,4-Dihydro-2-methyl-6,7-(methylenedioxy)-1-[4-(benzyloxy)-2-nitrobenzyl]-isoquinolinium iodide	EtOH	255(4.49),272(4.33)	2-0755-69
$C_{25}H_{23}NO$			
1-Indanone, 2-benzylidene-3-(N-isopropylanilino)-	hexane	231(4.17),238(4.15),259(4.31),307(4.39),317(4.42),331(4.34)	44-0596-69

Compound	Solvent	λ max(log ε)	Ref.
4(1H)-Pyridone, 3-benzyl-2,3-dihydro-6-methyl-1,2-diphenyl-	EtOH	240(3.66),333(4.23)	18-1357-69
	EtOH-HCl	245(3.56),331(4.13)	18-1357-69
4(1H)-Pyridone, 2,3-dihydro-6-phenethyl-1,2-diphenyl-	EtOH	238(3.65),330(4.24)	18-1357-69
	EtOH-HCl	236(3.72),326(4.12)	18-1357-69
$C_{25}H_{23}NOS$			
Crotonamide, 4-benzoyl-N,N-dimethyl-3,4-diphenyl-thio-	EtOH	255(4.00)	39-1950-69C
$C_{25}H_{23}NO_2$			
Inden-1-ol, 2-morpholino-1,3-diphenyl-	MeCN	242(4.31),312(4.16), 358(3.71)	44-1124-69
$C_{25}H_{23}NO_8$			
5H-Pyrrolo[1,2-a]azepine-5,6,7,8-tetracarboxylic acid, 3-styryl-, tetramethyl ester, trans	MeOH	207(4.41),220s(4.15), 310(4.36),473(4.32)	39-2316-69C
$C_{25}H_{23}N_3O_3S$			
4-Thiazoline, 4-benzoyl-3-isopropyl-5-(m-nitroanilino)-2-phenyl-	MeCN	252(4.36),355(3.98), 424(4.10)	23-3557-69
$C_{25}H_{24}$			
Benzene, 1-(p-isopropylstyryl)-4-styryl-	DMF	360(4.82)	33-2521-69
$C_{25}H_{24}BF_4NO$			
6-Methoxy-1,3,5-trimethyl-2,7-diphenyl-cyclohepta[c]pyrrolium tetrafluoroborate	$C_2H_4Cl_2$	319(4.63),628(3.27)	104-0947-69
$C_{25}H_{24}Br_2ClP$			
Phosphorane, (p-chlorophenyl)(2,7-dibromofluoren-9-ylidene)dipropyl-	$CHCl_3$	260(4.78),298(4.59), 394(3.43)	5-0001-69D
$C_{25}H_{24}Br_3P$			
Phosphorane, (p-bromophenyl)(2,7-dibromofluoren-9-ylidene)dipropyl-	$CHCl_3$	260(4.69),298(4.59), 395(3.44)	5-0001-69D
$C_{25}H_{24}N_2OS$			
4-Thiazoline, 5-anilino-4-benzoyl-3-isopropyl-2-phenyl-	MeCN	250(4.22),350(3.10), 415(4.09)	23-3557-69
$C_{25}H_{24}N_2O_5$			
Isoquinoline, 1,2,3,4-tetrahydro-2-methyl-6,7-(methylenedioxy)-1-[2-nitro-4-(benzyloxy)benzyl]-	EtOH	290(3.87)	2-0755-69
$C_{25}H_{24}N_4O_4S_2$			
Ketone, [3-[(4-amino-2-methyl-5-pyrimidinyl)methyl]-5-(2-hydroxyethyl)-4-methyl-4-thiazolin-2-ylidene]hydroxymethyl 2-thienyl, benzoate	EtOH	230(4.47),276(3.80), 413(4.47)	94-0128-69
$C_{25}H_{24}N_4O_5S$			
Ketone, [3-[(4-amino-2-methyl-5-pyrimidinyl)methyl]-5-(2-hydroxyethyl)-4-methyl-4-thiazolin-2-ylidene]hydroxymethyl 2-furyl, benzoate	EtOH	230(4.48),276(4.19), 412(4.45)	94-0128-69

Compound	Solvent	$\lambda_{max}(\log \epsilon)$	Ref.
$C_{25}H_{24}O_2$			
2-Indanol, 1,3-dimethyl-1,3-diphenyl-, acetate	ether	253(2.79),258(2.96), 265(3.01),272(2.92)	24-1597-69
stereoisomer	ether	253(2.79),258(2.96), 265(3.01),272(2.92)	24-1597-69
$C_{25}H_{24}O_3$			
Stilbene, 3,4-dimethoxy-4'-(p-methoxy-styryl)-	DMF	373(4.83)	33-2521-69
Stilbene, 3,4,5-trimethoxy-4'-styryl-	DMF	366(4.81)	33-2521-69
$C_{25}H_{24}O_5$			
2H,6H-Benzo[1,2-b:5,4-b']dipyran-2,6-dione, 7,8-dihydro-5-hydroxy-7,8-dimethyl-10-(3-methyl-2-butenyl)-4-phenyl-	EtOH	286(4.42),337(3.94)	44-4203-69
	EtOH-base	286(--),315(--), 425(--)	44-4203-69
Chandalone	EtOH	226(4.50),285(4.72)	39-0374-69C
Mesuagin, O-methyl-	EtOH	232(4.64),285(4.37)	44-3784-69
Warangalone	EtOH	227(4.56),287(4.73), 340s(4.00)	39-0374-69C
$C_{25}H_{24}O_8$			
2,6-Cresotic acid, 4-methoxy-5-methyl-m-phenylene ester	EtOH	220(4.3),268(4.29), 300(3.88)	39-1721-69C
$C_{25}H_{24}O_{12}$			
2-Anthraceneacetic acid, 1,2,3,4-tetra-hydro-3,5,7,10-tetrahydroxy-α-methoxy-4-oxo-, tetraacetate	EtOH	250s(4.75),258(4.84), 305(3.90),315(3.83), 355(3.54)	105-0257-69
$C_{25}H_{25}BrN_2O_2$			
1-Benzyl-3-[3-(carboxymethyl)-5-methyl-indol-2-yl)pyridinium bromide, ethyl ester	EtOH	215(4.63),250(4.23), 330(4.20)	22-4154-69
$C_{25}H_{25}Br_2P$			
Phosphorane, (2,7-dibromofluoren-9-ylidene)phenyldipropyl-	$CHCl_3$	260(4.69),268(4.59), 297(4.54),394(3.46)	5-0001-69D
$C_{25}H_{25}ClN_2O$			
1(2H)-Quinolinecarboxamide, 6-chloro-N-ethyl-3,4-dihydro-2-methyl-3,4-di-phenyl-	n.s.g.	259(4.19)	40-1047-69
$C_{25}H_{25}ClN_2O_2$			
1(2H)-Quinolinecarboxamide, 6-chloro-3,4-dihydro-3-(p-methoxyphenyl)-2-methyl-4-phenyl-	n.s.g.	260(4.19)	40-1047-69
$C_{25}H_{25}NO$			
Azetidine, 1-isopropyl-2-phenyl-3-(p-phenylbenzoyl)-, cis	isooctane	282(4.36)	44-0310-69
trans	isooctane	282(4.42)	44-0310-69
$C_{25}H_{26}ClNO_4S$			
[4-(2,6-Diphenyl-4H-thiopyran-4-ylid-ene)-2-butenylidene]diethylammonium perchlorate	MeCN	243(4.30),286(3.81), 302(3.79),339(4.02), 540(4.83),576(4.74)	44-2736-69

Compound	Solvent	λ max(log ε)	Ref.
$C_{25}H_{26}ClNO_5$			
[1-(6,7-Dihydro-2,4-diphenyl-5H-1-benzo-pyran-8-yl)ethylidene]dimethylammonium perchlorate	MeCN	239(4.21),278(4.26), 360(3.80),374(3.77), 465(3.95)	44-2736-69
[4-(2,6-Diphenyl-4H-pyran-4-ylidene)-2-butenylidene]diethylammonium perchlorate	MeCN	234(4.17),248(4.15), 274(3.71),285(3.69), 340(4.17),516(4.88), 549(4.86)	44-2736-69
[4-(4,6-Diphenyl-2H-pyran-2-ylidene)-2-butenylidene]diethylammonium perchlorate	MeCN	263(4.26),300s(4.16), 325(4.43),397s(4.18), 558(4.56),595(4.55)	44-2736-69
$C_{25}H_{26}N_2$			
Indole, 1-[α-[2-(dimethylamino)ethyl]-benzyl]-3-phenyl-, oxalate	EtOH	225(4.47),269(4.13), 282s(--),296s(--)	39-1537-69
$C_{25}H_{26}N_2O_3$			
Isoquinoline, 1-[2-amino-4-(benzyloxy)-benzyl]-1,2,3,4-tetrahydro-2-methyl-6,7-(methylenedioxy)-	EtOH	239(4.05),292(3.95)	2-0746-69
1,4-Isoquinolinedione, 2,3-dihydro-3-(2-morpholino-2-cyclohexen-1-yl)-3-phenyl-	EtOH	220(4.60),295(3.42), 350(3.00)	33-1810-69
$C_{25}H_{26}N_4O_2$			
1-Piperazinecarboxylic acid, 4-[3-(3,4-dihydro-1-isoquinolyl)-4-quinolyl]-, ethyl ester	EtOH	232(4.51),322(3.76), 328(3.76),338(4.78)	2-1010-69
$C_{25}H_{26}N_6O_6$			
Barbituric acid, 5,5'-[bis[p-(dimethyl-amino)phenyl]methylene]di-	EtOH	243(4.29),360(4.51)	103-0827-69
$C_{25}H_{26}O_5$			
1,4-Pentadien-3-one, 1,5-bis[2-(allyl-oxy)-3-methoxyphenyl]-	EtOH	312(4.27)	12-1011-69
$C_{25}H_{26}O_9$			
Sequirin B, tetraacetate	EtOH	219(4.18),269(2.97)	39-1921-69C
$C_{25}H_{26}O_{11}$			
Altersolanol A, acetonide, triacetate	EtOH	210(4.49),265(4.31), 384(3.48)	23-0777-69
$C_{25}H_{27}IN_2O$			
1,2,3-Trimethyl-2-[6-(N-phenylacetamido)-1,3,5-hexatrienyl-3H-indolium iodide	MeOH	485(4.76)	22-1284-69
$C_{25}H_{27}N_5O_4$			
3-Formazancarboxanilide, 1,5-bis(o-eth-oxyphenyl)-2'-hydroxy-N-methyl-	EtOH	455(4.29)	22-4390-69
	EtOH-KOH	517(4.66)	22-4390-69
$C_{25}H_{28}$			
Phenanthrene, 2-benzylidene-2,3,4,4a,-9,10-hexahydro-7-isopropyl-4aβ-methyl-	n.s.g.	216(4.75),285(4.52)	12-1711-69
$C_{25}H_{28}N_2O_2$			
1,5-Benzoxazepine, 4-[3-(2,3-dihydro- (name completed on next page)	EtOH	498(5.08)	124-1178-69
	$CHCl_3$	555(5.00)	124-1178-69

Compound	Solvent	λ_{max}(log ϵ)	Ref.
2,2-dimethyl-1,5-benzoxazepin-4-yl)-allylidene]-2,3,4,5-tetrahydro-2,2-dimethyl-, perchlorate			
Indole-3-acetic acid, 2-(1-benzyl-1,2,5,6-tetrahydro-3-pyridyl)-5-methyl-, ethyl ester	EtOH	227(4.43),298(4.18)	22-4154-69
$C_{25}H_{28}N_2O_4$			
Cumalinic acid, 6-[2,2-bis(p-dimethylaminophenyl)vinyl]-5,6-dihydro-, methyl ester	$CHCl_3$	270(4.4)	24-2835-69
$C_{25}H_{28}N_2O_7$			
Malonic acid, [[5-(benzyloxy)-2-carboxy-1H-pyrrolo[2,3-c]pyridin-3-yl]methyl]-, triethyl ester, hydrochloride	EtOH	284(4.26),292(4.33), 350(3.65)	35-2338-69
$C_{25}H_{28}N_4O_5$			
Aspercolorin	n.s.g.	210(4.45),226(4.28), 260(4.13),315(3.62)	119-0035S-69
$C_{25}H_{28}N_8O_8$			
1,6(2H,5H)-Naphthalenedione, hexahydro-3,3,8a-trimethyl-, bis(2,4-dinitrophenylhydrazone), cis	EtOH	368(4.24)	78-5275-69
$C_{25}H_{28}OS_2$			
2-Bornanone, 3-carbonyl-, 3-(dibenzyl mercaptole)	dioxan	330(4.08),375(2.66)	28-0186-69A
$C_{25}H_{28}O_3$			
2(3H)-Phenanthrone, 4,4a,9,10-tetrahydro-7-methoxy-1-(m-methoxyphenethyl)-4aβ-methyl-	EtOH	224(4.32),243(4.19)	44-3717-69
$C_{25}H_{28}O_5$			
Xanthen-9-one, 1-hydroxy-3,7-dimethoxy-2,4-bis(3-methyl-2-butenyl)-	MeOH	238(4.49),269(4.58), 301(4.02),386(3.68)	78-2787-69
$C_{25}H_{28}O_9$			
Eurycomalactone, enol triacetate	EtOH	285(4.37)	77-0821-69
3,9-Phenanthrenedione, 1,2,4,4a-tetrahydro-5,6,8,10-tetrahydroxy-7-isopropyl-1β,4aβ-dimethyl-, 5,6,10-triacetate	EtOH	254(4.03),291(3.79), 365(3.69)	33-1685-69
$C_{25}H_{28}O_{13}$			
Chrysosplenoside B	EtOH	254(4.32),274(4.31), 341(4.34)	95-0702-69
$C_{25}H_{29}ClO_4$			
3'H-Cyclopropa[1,2]pregna-1,4,6-triene-3,20-dione, 6-chloro-1β,2β-dihydro-17α-hydroxy-16-methylene-, acetate	MeOH	282(4.23)	87-0631-69
$C_{25}H_{29}F_7O_3$			
Progesterone, heptafluorobutyrate	heptane	227(4.16),232(4.15)	13-0739-69A

Compound	Solvent	λ max(log ε)	Ref.
$C_{25}H_{29}NO$			
Indone, 2-[α-(cyclohexylisopropylamino)-benzyl]-	n.s.g.	237(4.60),242(4.65), 257s(3.72),318(3.00), 330(2.93),391(2.93)	44-0596-69
$C_{25}H_{30}$			
Bicyclo[5.4.1]dodeca-2,5,7,9,11-pentaene, 4-(2,4-di-tert-butyl-2,4-cyclopentadien-1-ylidene)-	C_6H_{12}	238s(4.38),243(4.40), 249s(4.34),327(4.36), 355s(4.34),393(4.00)	88-2093-69
cation	EtOH-$HClO_4$	274(4.37),316(4.61), 391(3.99),564(4.06)	88-2093-69
$C_{25}H_{30}Fe_2IN$			
Trimethyl[[2-(2-methylferrocenyl)ferrocenyl]methyl]ammonium iodide	EtOH	442(2.43)	49-1515-69
$C_{25}H_{30}N_2$			
1H,5H-Benzo[ij]quinolizine, 9,9'-methylenebis[2,3,6,7-tetrahydro-	98% HOAc	638(5.23)	39-1068-69B
$C_{25}H_{30}N_3$			
Crystal Violet (cation)	98% HOAc	589(5.07)	39-1068-69B
$C_{25}H_{30}O_3$			
Estra-1,3,5(10)-triene-16α,17β-diol, 3-(benzyloxy)-	MeOH	278(3.27),287(3.23)	13-0591-69B
9β-Estr-5(10)-en-3-one, 17β-hydroxy-, 17-benzoate	EtOH	228(4.27),274(2.96), 281(2.89)	73-2459-69
$C_{25}H_{30}O_4$			
3'H-Cyclopropa[1,2]pregna-1,4,6-triene-3,20-dione, 1β,2β-dihydro-17-hydroxy-16-methylene-, acetate	MeOH	282(4.31)	87-0631-69
Pregna-1,4,6-triene-3,20-dione, 17α-hydroxy-6-methyl-16-methylene-, acetate	MeOH	227(4.16),253(3.98), 303(4.10)	87-0631-69
$C_{25}H_{30}O_8$			
Flavone, 3',4',5,7-tetraethoxy-3,6-dimethoxy-	EtOH	240(4.31),335(4.36)	18-1649-69
9(1H)-Phenanthrone, 2,3,4,4a-tetrahydro-5,6,8,10-tetrahydroxy-7-isopropyl-1β,4aβ-dimethyl-, 5,6,10-triacetate	EtOH	228(4.09),257(4.12), 288(3.90),361(3.77)	33-1685-69
$C_{25}H_{31}BrO_6$			
Pregn-4-ene-3,20-dione, 7α-bromo-6β,17-dihydroxy-1α,2α-methylene-, 6-formate 17-acetate	MeOH	230(4.11)	24-2570-69
$C_{25}H_{31}ClO_4$			
Pregna-4,6-diene-3,20-dione, 4-chloro-17-hydroxy-7-methyl-1α,2α-methylene-, acetate	MeOH	218(3.80),308(4.30)	24-2565-69
$C_{25}H_{31}ClO_6$			
Benzaldehyde, 5-chloro-2,4-dihydroxy-3-[5-(3-hydroxy-1,2,6-trimethyl-5-oxocyclohexyl)-3-methyl-2,4-pentadienyl]-6-methyl-, 3-acetate	MeOH	239(4.60),293(4.07), 347(3.98)	78-1323-69

Compound	Solvent	λ_{max}(log ϵ)	Ref.
Pregn-4-ene-3,20-dione, 7α-chloro-6β,17-dihydroxy-1α,2α-methylene-, 6-formate 17-acetate	MeOH	230(4.10)	24-2570-69
$C_{25}H_{31}F_3O_6$			
Pregn-4-en-3-one, 9α-fluoro-11β-hydroxy-16α-methyl-17α,20:20,21-bis(methylenedioxy)-6α,7α-(difluoromethylene)-	MeOH	243(4.09)	78-1219-69
6β,7β-	MeOH	251(4.06)	78-1219-69
$C_{25}H_{31}NOS$			
5α-Androst-1-eno[2,3-b]-2'H-[1',4']-benzothiazine, 17β-hydroxy-	EtOH	253(4.38)	7-0163-69
$C_{25}H_{31}NO_7$			
Oxobundlin A	MeOH	233(4.61),292(4.28)	88-2249-69
$C_{25}H_{31}N_3O_7$			
Methanol, tris(4-amino-3,5-dimethoxyphenyl)-	pH 4.62	616(4.75)	64-0542-69
$C_{25}H_{32}ClIO_4$			
Pregn-4-ene-3,20-dione, 4-chloro-17-hydroxy-7β-(iodomethyl)-1α,2α-methylene-, acetate	MeOH	255(4.10)	24-2565-69
$C_{25}H_{32}Cl_6O_6$			
5α-Androstane-2,3,17β-triol, 17-acetate bis(trichloroacetate)	EtOH	220(2.75)(end abs.)	78-1679-69
5β-	EtOH	220(2.83)(end abs.)	78-1679-69
$C_{25}H_{32}F_2O_5$			
Pregn-4-ene-3,20-dione, 11β,21-dihydroxy-16α-methyl-6α,7α-(difluoromethylene)-, 21-acetate	EtOH	246(4.15)	78-1219-69
6β,7β-	EtOH	255(4.22)	78-1219-69
$C_{25}H_{32}N_2O_5S_2$			
Cylindrocarine, 1-acetyl-20-hydroxy-, S-methyl dithiocarbonate	MeOH	218(4.75),258(4.27), 279(4.27),360s(2.75)	39-0417-69C
$C_{25}H_{32}O_3$			
3'H-Cyclopropa[1,2]pregna-1,4,6-trien-3-one, 16α,17:20,21-diepoxy-1β,2β-dihydro-6,16,20-trimethyl-	MeOH	287(4.26)	87-0631-69
$C_{25}H_{32}O_4$			
Chola-4,14,22-trien-24-oic acid, 3,21-dioxo-, methyl ester	MeOH	236(4.31)	88-3033-69
Chola-4,14,22-trien-24-oic acid, 20,21-dioxo-, methyl ester	MeOH	233(4.36)	88-3033-69
Pregna-2,4,6-triene-2-carboxaldehyde, 17α-acetoxy-6-methyl-20-oxo-	EtOH	358(4.20)	78-4535-69
$C_{25}H_{32}O_5$			
19-Norpregna-1,3,5(10)-trien-6-one, 1,20β-diacetoxy-4-methyl-	EtOH	223(3.71),251(3.86), 304(3.29)	39-1234-69C
$C_{25}H_{32}O_6$			
Estra-1,3,5(10)-triene-1,2,17β-triol, 4-methyl-, triacetate	EtOH	205(4.48+),267(2.60)	33-0459-69

Compound	Solvent	λ_{max}(log ε)	Ref.
Pregna-3,5,9(11)-triene-6-carboxaldehyde, 17α,21-dihydroxy-3-methoxy-20-oxo-, 21-acetate	EtOH	219(4.09),321(4.19)	78-1155-69
Pregn-4-ene-$\Delta^{3,\alpha}$-acetic acid, 17α,21-dihydroxy-α-(2-hydroxyethyl)-11,20-dioxo-, γ-lactone	EtOH	292(4.40)	44-3796-69
Pregn-5-eno[6,5,4-bc]furan-3-one, 5'-methyl-17,20:20,21-bis(methylenedioxy)-	EtOH	210(4.15),230s(3.88), 297(3.50)	94-2586-69
$C_{25}H_{32}O_7$			
Estra-1,3,5(10)-triene-1,2-dicarboxylic acid, 3-methoxy-17-oxo-, dimethyl ester, cyclic 17-(ethylene acetal)	n.s.g.	215(4.39),302(3.57)	12-2255-69
$C_{25}H_{32}O_{14}$			
Secologanin, tetraacetate	EtOH	234(4.02)	39-0721-69C
	n.s.g.	234(3.98)	39-1187-69C
$C_{25}H_{32}O_{15}$			
Kingiside, tetraacetate	EtOH	233(4.06)	88-2725-69
$C_{25}H_{33}BrO_5$			
5β-Pregnan-11-one, 17α-bromo-3α,20β-dihydroxy-, diacetate	EtOH	317(2.16)	23-3489-69
Pregn-4-ene-3,20-dione, 7α-bromo-17-hydroxy-6β-methoxy-1α,2α-methylene-, acetate	MeOH	230(4.08)	24-2570-69
$C_{25}H_{33}ClNO_2PS$			
Phenothiaphosphinic acid, 2-chloro-8-methyl-, dicyclohexylamine salt	EtOH	219(4.40),258s(3.86), 272(4.14),288s(3.86), 312(3.44)	78-3919-69
$C_{25}H_{33}ClO_4$			
Pregn-4-ene-3,20-dione, 4-chloro-17-hydroxy-7β-methyl-1α,2α-methylene-, acetate	MeOH	256(4.05)	24-2565-69
$C_{25}H_{33}ClO_5$			
Pregn-4-ene-3,20-dione, 7α-chloro-6β-methoxy-1α,2α-methylene-, acetate	MeOH	228(4.06)	24-2570-69
$C_{25}H_{33}FN_2O_5$			
2'H-Pregna-2,4-dieno[3,2-c]pyrazol-20-one, 9-fluoro-11β,16α,17,21-tetrahydroxy-, cyclic 16,17-acetal with acetone	EtOH	260(3.97)	22-1256-69
$C_{25}H_{33}NO_5$			
Acetamide, N-(17-hydroxy-3,20-dioxopregna-4,6-dien-6-yl)-, acetate	MeOH	288(4.08)	48-0919-69
$C_{25}H_{33}NO_7$			
Bundlin A	MeOH	227(4.90)	88-2249-69
Pyruvamide, N-(7,13-dihydroxy-1,4,10,19-tetramethyl-17,18-dioxo-16-oxabicyclo[13.2.2]nonadeca-3,5,9,11-tetraen-2-yl)-, all-E (T-2636-C antibiotic)	EtOH	227(4.67)	88-2239-69

Compound	Solvent	λ_{max}(log ε)	Ref.
$C_{25}H_{33}NO_{11}$			
Ipecoside, deacetyl-, hydrochloride	EtOH	234(4.16),285(3.61)	39-1187-69C
Isoipecoside, deacetyl-, hydrochloride	EtOH	235(4.14),285(3.60)	39-1187-69C
$C_{25}H_{33}N_3O$			
12-Azabicyclo[9.2.1]tetradeca-11(14),13-diene, 2-ethyl-13-[(3-methoxy-5-pyrrol-2-yl-2H-pyrrol-2-ylidene)methyl]-, hydrochloride	MeOH	500s(4.50),530(4.88)	35-1263-69
	MeOH-KOH	467(4.49)	35-1263-69
$C_{25}H_{33}N_3O_6$			
Pregn-4-ene-3,20-dione, 6β-azido-7α,17α-dihydroxy-, diacetate	MeOH	235(4.01)	48-0919-69
$C_{25}H_{34}N_2O_3$			
2'H-Androsta-5,16-dieno[16,17-c]pyrazol-3β-ol, 2'-acetyl-5'-methyl-, acetate	EtOH	235(3.83)	22-1256-69
Pyrazole-3-carboxylic acid, 5-(3-oxa-androst-4-en-17β-yl)-, ethyl ester	MeOH	240(4.34)	73-0442-69
$C_{25}H_{34}O$			
Apocarotenal	hexane	411(4.88)	78-5383-69
	C_6H_{12}	416(4.86)	
	benzene	427(4.37)	
	dioxan	425(4.86)	
	acetone	418(4.85)	
	CCl_4	423(4.85)	
	CS_2	448(4.78)	
	DMF	429(4.85)	
	1:1 CS_2:-hexane	427(4.83)	
	1:1 CS_2:-C_6H_{12}	432(4.81)	
$C_{25}H_{34}O_3$			
Pregna-3,5-diene-3-acetic acid, α-(2-hydroxyethyl)-20-oxo-, γ-lactone	EtOH	239s(4.40)	44-3796-69
$C_{25}H_{34}O_4$			
Androsta-3,5-diene-3-acetic acid, 17β-hydroxy-α-(2-hydroxyethyl)-, γ-lactone, acetate	EtOH	293(4.43)	44-3796-69
12'-Apofucoxanthinal	$CHCl_3$	379(4.68),400(4.88), 422(4.85)	39-0429-69C
5α-Carda-14,20(22)-dienolide, 3β-acetoxy-	MeOH	214(4.25)	5-0136-69F
5β-Carda-14,20(22)-dienolide, 3α-acetoxy-	n.s.g.	214(4.24)	5-0168-69A
Pregn-4-ene-3,20-dione, 17-acetoxy-7β-methyl-1α,2α-methylene-	MeOH	240(4.17)	24-2565-69
$C_{25}H_{34}O_5$			
Canarigenin, 3-acetate	n.s.g.	216s(4.28)	5-0110-69G
$C_{25}H_{34}O_6$			
5β-Androst-1-ene-2,19-dicarboxaldehyde, 3,17-dioxo-, cyclic 17,19-bis(ethylene acetal)	EtOH	251(4.07)	88-4547-69
Isohirundigenin, O-acetyldihydroanhydro-	EtOH	215(3.93),289f(2.08), 298(2.09),308(2.00)	33-1175-69

Compound	Solvent	λ max(log ε)	Ref.
Pregn-4-en-21-oic acid, 17,20α-dihydroxy-3,11-dioxo-, methyl ester, cyclic acetal with acetone	MeOH	238(4.20)	44-3513-69
17,20β-	MeOH	238(4.15)	44-3513-69
$C_{25}H_{34}O_7$			
Isohellebrigeninic acid, methyl ester	MeOH	300(4.32)	44-3894-69
Pregna-3,5-diene-6-carboxaldehyde, 11β,17α,21-trihydroxy-3-methoxy-20-oxo-, 21-acetate	EtOH	218(4.06),324(4.22)	78-1155-69
$C_{25}H_{35}ClO_7$			
Pregn-4-en-3-one, 4-chloro-11-(methoxymethyl)-17,20:20,21-bis(methylenedioxy)-	EtOH	255(4.17)	44-1455-69
$C_{25}H_{35}NO_4$			
Androst-5-ene-3β,17β-diol, 17α-(3-methyl-5-isoxazolyl)-, 3-acetate	MeOH	217(3.89)	24-3324-69
$C_{25}H_{35}NO_5$			
Acetamide, N-(17-hydroxy-3,20-dioxopregn-4-en-6β-yl)-, acetate	MeOH	239(4.11)	48-0445-69
$C_{25}H_{35}NO_6$			
Acetamide, N-(7α,17-dihydroxy-3,20-dioxopregn-4-en-6β-yl)-, 17-acetate	MeOH	238(4.11)	48-0919-69
$C_{25}H_{35}NO_7$			
Lactamide, N-(7,13-dihydroxy-1,4,10,19-tetramethyl-17,18-dioxo-16-oxabicyclo[13.2.2]nonadeca-3,5,9,11-tetraen-2-yl)-	EtOH	234(4.33)	88-2239-69
Pyruvamide, N-(7,13-dihydroxy-1,4,10,19-tetramethyl-17,18-dioxo-16-oxabicyclo[13.2.2]nonadeca-3,9,11-trien-2-yl)- (T-2636-F)	EtOH	228(4.68)	88-2239-69
$C_{25}H_{35}N_3O_3$			
Pyrazole-3-carboxylic acid, 5-(3-oxoandrost-4-en-17β-yl)-, ethyl ester, 5-oxime	MeOH	239(4.42)	73-0442-69
$C_{25}H_{36}ClNO_{10}$			
[2-[1-(2,4-Dihydroxybutyl)-7,8-dimethoxy-2-naphthyl]ethyl]trimethylammonium perchlorate, diacetate	H_2O	290(3.75),301(3.72), 338(3.35)	49-0163-69
$C_{25}H_{36}N_2O_2$			
1,4-Naphthalenediol, 2-(N-methylcyclohexylamino)-3-[(N-methylcyclohexylamino)methyl]-	EtOH	240(4.15),250s(4.09), 264(3.98),325s(3.27)	39-1245-69C
$C_{25}H_{36}N_2O_3$			
Pyrazole-3-carboxylic acid, 5-(3β-hydroxyandrost-5-en-17β-yl)-, ethyl ester	MeOH	222(3.66),245(3.21), 250(3.11)	73-0442-69
Pyrazole-3-carboxylic acid, 5-(3-oxo-5α-androstan-17β-yl)-, ethyl ester	MeOH	220(4.02)	73-0442-69

Compound	Solvent	$\lambda_{max}(\log \epsilon)$	Ref.
$C_{25}H_{36}N_2O_4S$			
Indolo[2,3-a]quinolizine, 1,2,3,4,6,7,-12,12b-octahydro-, d-camphorsulfonate	MeOH	270(3.80),278(3.80), 287(3.72)	24-3959-69
$C_{25}H_{36}O$			
19,21,27-Trinorcholesta-1,3,5(10),-17(20)-tetraene, 3-methoxy-	EtOH	278(3.28),287(3.26)	77-0043-69
$C_{25}H_{36}O_4$			
Androsta-2,4-diene-3,17β-diol, 2,6β-dimethyl-, diacetate	EtOH	269(4.03)	44-2288-69
Androsta-3,5-diene-3,17β-diol, 2α,6-dimethyl-, diacetate	EtOH	247(4.33)	44-2288-69
Androsta-3,5-diene-3,17β-diol, 2α-ethyl-, diacetate	EtOH	238(4.31)	44-2288-69
Chola-5,22-dien-24-oic acid, 20,21-epoxy-3β-hydroxy-, methyl ester	MeOH	222(4.03)	88-3029-69
5β-Chola-20(21),22-dien-24-oic acid, 14,21-epoxy-3β-hydroxy-, methyl ester	EtOH	304(4.50)	94-1698-69
$C_{25}H_{36}O_5$			
Androst-5-en-7-one, 17β-acetoxy-3-(ethylenedioxy)-4,4-dimethyl-	C_6H_{12}	348(1.56)	33-2436-69
	EtOH	244(4.08),331(1.75)	33-2436-69
18-Nor-5β-pregn-12-en-11-one, 3α,20β-diacetoxy-12-methyl-	EtOH	250(4.05)	23-3489-69
5α-Pregn-16-ene-12,20-dione, 3β-acetoxy-, cyclic 12-(ethylene acetal)	MeOH	236(3.88)	44-1606-69
13,14-Seco-12,14-cyclo-5β-pregn-12-en-11-one, 3α,20β-diacetoxy-	EtOH	252(4.16)	23-3489-69
Uzarigenin, 3-acetate	MeOH	217(4.18)	5-0136-69F
$C_{25}H_{36}O_6$			
5α-Androstane-$\Delta^{2,\alpha}$-acetic acid, 3β,17β-dihydroxy-, diacetate	EtOH	222(4.30)	44-3032-69
5α-Androstane-$\Delta^{16,\alpha}$-acetic acid, 3β,17β-dihydroxy-, diacetate	MeOH	218(4.11)	44-0237-69
5β-Androstane-19-carboxaldehyde, 2-(hydroxymethylene)-3,17-dioxo-, cyclic 17,19-bis(ethylene acetal)	EtOH	286(3.66)	88-4547-69
Digitoxigenin, 15β-hydroxy-, 3-monoacetate	EtOH	216(4.22)	94-0515-69
14α-Digitoxigenin, 15α-hydroxy-, monoacetate	EtOH	218(4.20)	94-0515-69
17α-Digitoxigenin, 15β-hydroxy-, monoacetate	EtOH	218(4.14)	94-0515-69
Spiro[cyclopropane-1,8'-tricyclo-[3.2.1.0$^{2',7'}$]oct-3'-ene-3',6',7'-tricarboxylic acid, tri-tert-butyl-ester, endo	EtOH	230(3.66)	24-4164-69
exo	EtOH	230(3.75)	24-4164-69
$C_{25}H_{37}NO$			
9β,10α-D-Homoandrosta-3,5-dien-17a-one, 18-methyl-3-pyrrolidino-, dl-	EtOH	277(4.40)	44-0107-69
$C_{25}H_{37}NO_3$			
Podocarpa-8,11,13-triene-15-carbamic acid, 13-isopropyl-7-oxo-, tert-butyl ester	MeOH	253(4.05)	5-0193-69H

Compound	Solvent	$\lambda_{max}(\log \epsilon)$	Ref.
$C_{25}H_{37}NO_4$			
Podocarpa-8,11,13-triene-15-carbamic acid, 13-(1-hydroxy-1-methylethyl)-7-oxo-, tert-butyl ester	MeOH	253(4.04)	5-0193-69H
$C_{25}H_{37}NO_5$			
7a-Aza-B-homopregn-5-en-7-one, 3β,20β-diacetoxy-	EtOH	219(4.29)	2-1084-69
Pregn-5-en-7-one, 3β,20β-diacetoxy-, oxime	EtOH	237(4.08)	2-1084-69
$C_{25}H_{37}N_3O_5$			
5β-Pregnan-11-one, 17α-azido-3α,20β-dihydroxy-, diacetate	EtOH	321(2.16)	23-3489-69
$C_{25}H_{38}N_2O_5$			
4H-Quinolizin-4-one, 2-(2-amino-5-methoxyphenyl)-7α-ethyl-1,2,3,6,7,8,9,9a-octahydro-1α-hydroxy-8β-[2-(tetrahydro-2H-pyran-2-yl)oxy]ethyl]-, allo-	EtOH	235(3.92),302(3.52)	78-5319-69
	EtOH-HCl	276(3.29),283(3.26)	78-5319-69
4H-Quinolizin-4-one, 2-(2-amino-5-methoxyphenyl)-1,2,3,6,7,8,9,9a-octahydro-1α-hydroxy-8α-[1-[[(tetrahydro-2H-pyran-2-yl)oxy]methyl]propyl]-, allo-	EtOH	237(3.92),303(3.52)	78-5319-69
	EtOH-HCl	259(4.13),298(3.49)	78-5319-69
1,2-Secohydroquinine, 1',2',3',4'-tetrahydro-2'-oxo-2-[(tetrahydro-2H-pyran-2-yl)oxy]-, allo-	EtOH	235(3.86),302(3.51)	78-5329-69
$C_{25}H_{38}O_2$			
Δ^8-Cannabinol, 5'-ethyltetrahydro-1',2'-dimethyl-, 6a,10a-trans	EtOH	230s(4.10),276(3.29), 282(3.30)	33-1102-69
Δ^9-Cannabinol, 5'-ethyltetrahydro-1',2'-dimethyl-, 6a,10a-trans	EtOH	230s(4.02),276(3.19), 282(3.20)	33-1102-69
$C_{25}H_{39}NO_5$			
5β-Pregnan-11-one, 12α-amino-3α,20β-diacetoxy-	EtOH	310(2.28)	23-3489-69
$C_{25}H_{39}NO_7$			
1,4-(Iminomethano)inden-8-one, 9-ethylhexahydro-2-hydroxy-3,7-dimethoxy-4-(methoxymethyl)-7a-(8-methoxy-4-oxobicyclo[3.2.1]oct-6-yl)-	EtOH	245(3.50)	88-2239-69
$C_{25}H_{39}O_4P$			
Tributyl[α-(dicarboxymethyl)benzyl]-phosphonium hydroxide, cyclic isopropylidene ester, inner salt	MeOH	260(4.2)	49-0576-69
$C_{25}H_{40}$			
A,B-Dinorcholesta-2,5-diene	C_6H_{12}	238(3.97),245(3.99), 253(3.82)	44-2297-69
$C_{25}H_{40}N_2OS$			
2'H-5α-Androst-2-eno[3,2-c]pyrazol-17β-ol, 7-(butylthio)-17α-methyl-	EtOH	225(3.68)	94-0011-69
2'H-5α-Androst-2-eno[3,2-c]pyrazol-17β-ol, 7-(tert-butylthio)-17α-methyl-	EtOH	223(3.69)	94-0011-69

Compound	Solvent	$\lambda_{max}(\log \epsilon)$	Ref.
$C_{25}H_{40}N_2O_3$			
2-Pyrazoline-5-carboxylic acid, 3-(3β-hydroxy-5α-androstan-17β-yl)-, ethyl ester (after ten months)	MeOH	220(3.82)	73-0442-69
$C_{25}H_{40}N_2O_4$			
2H-Quinolizin-1-ol, 2-(2-amino-5-methoxy henyl)-7-ethyl-1,3,4,6,7,8,9,9a-octahydro-8β-[2-[(tetrahydro-2H-pyran-2-yl)oxy]ethyl]-, allo-	EtOH	235(3.85),300(3.46)	78-5319-69
$C_{25}H_{40}O$			
A,B-Dinorcholest-3(5)-en-2-one	EtOH	235(4.13)	44-2297-69
$C_{25}H_{40}O_3S$			
5α-Androstan-3-one, 7-(butylthio)-17β-hydroxy-2-(hydroxymethylene)-17α-methyl-	EtOH	281(4.01)	94-0011-69
5α-Androstan-3-one, 7-(tert-butylthio)-17β-hydroxy-2-(hydroxymethylene)-17α-methyl-	EtOH	285(3.95)	94-0011-69
$(C_{25}H_{40}O_4Sn)_n$			
Tin, tributyl(dihydrogen benzylmalonato)-, cyclic isopropylidene ester polymers	isobutyronitrile	267(4.31)(anom.)	5-0001-69A
$C_{25}H_{42}O$			
A,B-Dinor-5β-cholestan-2-one	EtOH	289(1.15)	44-2297-69
$C_{25}H_{45}ClN_2O$			
4(3H)-Pyrimidinone, 6-chloro-5-decyl-2-undecyl-	MeOH	233(3.70),281(3.82)	44-2972-69
$C_{25}H_{54}ClIrOP_2$			
Iridium, carbonylchlorobis(tributylphosphine)-	toluene	332(2.94),376(3.01), 432(2.20)	64-0770-69

Compound	Solvent	λ_{max}(log ε)	Ref.
$C_{26}H_{13}N_3O_3$			
Cyclopent[5,6]isoindolo[2,1-a]benzimidazole-1,3,11(2H)-trione, 2-(2-quinolyl)-	$C_{10}H_7Cl$	426(4.56),446(4.58)	33-1259-69
$C_{26}H_{14}ClNO_5$			
Spiro[chrysene-6(5H),5'-[1,4,2]dioxazol]-5-one, 3'-[3-chloro-4,5-(methylenedioxy)phenyl]-	EtOH	265(4.54),290(4.52), 355(3.77)	23-1473-69
$C_{26}H_{14}N_2O_5$			
1,4-Naphthoquinone, 2,3-epoxy-2,3-dihydro-5-hydroxy-8-(1-hydroxybenz[a]phenazin-5-yl)-	EtOH	239(4.60),267(4.72), 363(3.92),441(4.16)	39-2059-69C
$C_{26}H_{15}Br$			
Benzo[k]fluoranthene, 7-bromo-12-phenyl-	n.s.g.	216(4.50),248(4.64), 271(4.11),288(4.06), 299(4.46),331(4.57), 387(3.81)	44-0371-69
$C_{26}H_{15}ClN_2O_4$			
Spiro[chrysene-6(5H),5'-[Δ^2-1,2,4]oxadiazolin]-5-one, 3'-[3-chloro-4,5-(methylenedioxy)phenyl]-	EtOH	264(4.53),290(3.48), 360(3.78)	23-1473-69
$C_{26}H_{16}$			
Azuleno[5,6,7-cd]phenalene, phenyl-	EtOH	248(4.40),321(5.14), 363(3.83),381(3.71), 407(3.96),429(4.50), 454(4.92),618(2.83), 637(2.93),670(3.15), 705(3.10),748(3.28)	24-2301-69
Benzo[k]fluoranthene, 7-phenyl-	EtOH	216(4.63),246(4.63), 270(4.25),286(4.26), 298(4.49),309(4.58), 364(3.72),384(3.91)	35-0371-69
9,9'-Bifluorenylidene	heptane	245(4.9),265(4.6), 270s(4.5),440(4.3)	54-0801-69
Naphthalene, 1,5-bis(phenylethynyl)-	EtOH	208(4.67),341(4.52), 355(4.43)	35-0371-69
Naphthalene, 1,8-bis(phenylethynyl)-	EtOH	209(4.68),243(4.82), 265(4.41),344(4.47), 365(4.43)	35-0371-69
Pyrene, 2-(1-naphthyl)-	EtOH	246(4.79),253(4.81), 275(4.65),294(4.60), 308(4.55),321(4.74), 338(4.95)	24-2301-69
Pyrene, 2-(2-naphthyl)-	EtOH	236(4.70),267(5.37), 295(5.14),308(4.98), 321(4.93),338(5.14)	24-2301-69
$C_{26}H_{16}ClNOS_2$			
Cinnamonitrile, α-(p-chlorobenzoyl)-β-[(4-phenyl-1,3-dithiol-2-ylidene)-methyl]-	$CHCl_3$	295(4.00),534(4.61)	24-1580-69
$C_{26}H_{16}ClNO_4$			
Spiro[chrysene-6(5H),5'-[1,4,2]dioxazol]-5-one, 3'-(2-chloro-4-methoxyphenyl)-	EtOH	264(4.43),282(4.56), 360(3.97)	23-1473-69

Compound	Solvent	λ_{max}(log ε)	Ref.
$C_{26}H_{16}Cl_2N_2O_2$			
Benzophenone, 2,2"-azobis[5-chloro-	MeOH	252(4.50),340(4.12)	78-5205-69
$C_{26}H_{16}N_2O_5$			
1(2H)-Naphthalenone, 2,3-epoxy-3,4-dihydro-4,5-dihydroxy-8-(1-hydroxybenzo[a]phenazin-6-yl)-	EtOH	240(4.64),266(4.70), 330s(3.95),443(4.17)	39-2059-69C
$C_{26}H_{16}N_4O$			
Anthracene, 9-p-anisidino-10-(tricyanovinyl)-	$CHCl_3$	256(5.24),438(3.74), 655(3.70)	22-1651-69
$C_{26}H_{16}O_2$			
Biphenylene, 2,6-dibenzoyl-	EtOH	264(4.89),295(4.71), 353(4.25),382(4.34)	39-2789-69C
$\Delta^{9,9'}$-Bixanthene (at -45°)	isopentane-$C_6H_{11}Me$	280s(4.2),370s(4.3)	24-3033-69
$C_{26}H_{16}O_3$			
Biphenylene, 2-benzoyl-6-[3-(2-furyl)acryloyl]-	EtOH	264(4.34),317s(4.09), 368(4.17),386(4.20)	39-2789-69C
$C_{26}H_{17}BrO$			
Benzofuran, 5-bromo-2-[p-[2-(1-naphthyl)vinyl]phenyl]-	DMF	365(4.73)	33-1282-69
Benzofuran, 5-bromo-2-[p-[2-(2-naphthyl)vinyl]phenyl]-	DMF	300(4.13),367(4.88), 387(4.74)	33-1282-69
$C_{26}H_{17}BrO_{12}$			
γ-Rubromycin, bromo-	$CHCl_3$	323(4.30),361(4.04), 376(4.02),484(3.85), 513(3.89),550(3.66)	24-0126-69
$C_{26}H_{17}ClN_2O_3$			
Spiro[chrysene-6(5H),5'-[Δ^2-1,2,4]oxadiazolin]-5-one, 3'-(2-chloro-4-methoxyphenyl)-	EtOH	268(4.64),288(4.67), 360(3.93)	23-1473-69
$C_{26}H_{17}ClO$			
Benzofuran, 5-chloro-2-[p-[2-(1-naphthyl)vinyl]phenyl]-	DMF	365(4.73)	33-1282-69
Benzofuran, 5-chloro-2-[p-[2-(2-naphthyl)vinyl]phenyl]-	DMF	301(4.16),367(4.88), 387(4.74)	33-1282-69
Naphtho[2,1-b]furan, 2-[p-(p-chlorostyryl)phenyl]-	DMF	326(4.36),374(4.84)	33-1282-69
$C_{26}H_{17}ClO_2$			
Benzofuran, 2-[p-[2-[5-(p-chlorophenyl)-2-furyl]vinyl]phenyl]-	DMF	313(4.20),394(4.81), 418(4.68)	33-1282-69
$C_{26}H_{17}ClS$			
Benzo[b]thiophene, 5-chloro-2-[p-[2-(1-naphthyl)vinyl]phenyl]-	DMF	364(4.70)	33-1282-69
Benzo[b]thiophene, 5-chloro-2-[p-[2-(2-naphthyl)vinyl]phenyl]-	DMF	302(4.20),366(4.84)	33-1282-69
$C_{26}H_{17}NO_4$			
4H-Naphtho[1,2-b]pyran-4-one, 2,3-dihydro-3-(p-methylbenzylidene)-2-phenyl-	$CHCl_3$	260(4.08),266(4.12), 280(3.57),366(3.33)	115-0017-69

Compound	Solvent	λ max(log ε)	Ref.
$C_{26}H_{18}$			
Anthracene, 9,10-diphenyl-	benzene	395(4.13)	44-1734-69
Anthracene, 2-[2-(1-naphthyl)vinyl]-	DMF	284(4.54),333(4.66), 377(4.41),393(4.36)	33-2521-69
Anthracene, 2-[2-(2-naphthyl)vinyl]-	DMF	282(4.51),293(4.70), 319(4.83),334(4.97), 359(4.34),377(4.48), 397(4.45)	33-2521-69
Naphthalene, 1-[p-(phenylethynyl)-styryl]-	DMF	303(4.26),353(4.67)	33-2521-69
Naphthalene, 2-[p-(phenylethynyl)-styryl]-	DMF	297(4.38),353(4.81)	33-2521-69
$C_{26}H_{18}Br_2N_4$			
s-Tetrazine, 1,4-bis(p-bromophenyl)-1,4-dihydro-3,6-diphenyl-	EtOH	275(4.28),337(4.27)	35-2443-69
$C_{26}H_{18}N_2O$			
1,1'-Azoxyfluorene	EtOH	264(4.50),268(4.51), 272s(4.48),290(4.10), 326(4.21)	39-0345-69C
$C_{26}H_{18}N_4O$			
Ether, bis(α-diazo-α-phenyl-p-tolyl)	dioxan	285(4.67),525(2.37)	47-3313-69
$C_{26}H_{18}O$			
Benzofuran, 2-[p-[2-(1-naphthyl)vinyl]-phenyl]-	DMF	366(4.73)	33-1282-69
Benzofuran, 2-[p-[2-(2-naphthyl)vinyl]-phenyl]-	DMF	301(4.15),366(4.85), 386(4.71)	33-1282-69
Benzofuran, 6-[2-(1-naphthyl)vinyl]-2-phenyl-	DMF	364(4.66)	33-1282-69
Benzofuran, 6-[2-(2-naphthyl)vinyl]-2-phenyl-	DMF	290(4.20),301(4.23), 361(4.80)	33-1282-69
Naphtho[1,2-b]furan, 2-(p-styrylphenyl)-	DMF	303(4.32),366(4.78), 385(4.66)	33-1282-69
Naphtho[2,1-b]furan, 2-(p-styrylphenyl)-	DMF	325(4.35),370(4.83)	33-1282-69
$C_{26}H_{18}O_2$			
4H-Naphtho[1,2-b]pyran-4-one, 3-benzylidene-2,3-dihydro-2-phenyl-	$CHCl_3$	248(3.67),273(3.75), 307(3.63),340(3.74), 388(3.51)	115-0017-69
$C_{26}H_{18}O_{12}$			
γ-Isorubromycin	$CHCl_3$	324(4.29),362s(4.08), 438(3.77),470(3.89), 495(4.02),532(3.93)	24-0126-69
γ-Rubromycin	$CHCl_3$	317(4.35),349(4.12), 364(4.06),484(3.87), 513(3.91),551(3.68)	24-0126-69
$C_{26}H_{18}S$			
Benzo[b]thiophene, 2-[p-[2-(1-naphthyl)-vinyl]phenyl]-	DMF	363(4.70)	33-1282-69
Benzo[b]thiophene, 2-[p-[2-(2-naphthyl)-vinyl]phenyl]-	DMF	301(4.24),365(4.84)	33-1282-69
Benzo[b]thiophene, 6-[2-(1-naphthyl)-vinyl]-2-phenyl-	DMF	365(4.67)	33-1282-69
Benzo[b]thiophene, 6-[2-(2-naphthyl)-vinyl]-2-phenyl-	DMF	303(4.19),365(4.80)	33-1282-69

Compound	Solvent	λ_{max}(log ε)	Ref.
Naphtho[2,1-b]thiophene, 2-(p-styryl-phenyl)-	DMF	370(4.79)	33-1282-69
$C_{26}H_{18}S_2$			
Thiophene, 2,2'-[ethynylenebis(p-phen-ylenevinylene)]di-	DMF	378(4.94)	33-2521-69
$C_{26}H_{19}ClN_4O_3$			
Carbazole-1-carboxanilide, 4'-chloro-2-hydroxy-3-[(p-methoxyphenyl)azo]-	dioxan	239(4.57),250(4.43), 291(4.75),316(4.68), 457(4.52)	2-1065-69
$C_{26}H_{19}NO_2S$			
Benzo[b]thiophen-3(2H)-one, 2-[2-[1-(p-methoxyphenyl)-2(1H)-quinolylidene]-ethylidene]-	BuOH	574(5.07),609(5.03)	65-1829-69
$C_{26}H_{19}N_3$			
Pyridine, 2,2'-(2,5-diphenylpyrrole-3,4-diyl)di-	EtOH	235s(4.2),265(4.3), 302(4.3)	95-0783-69
Pyridine, 3,3'-(2,5-diphenylpyrrole-3,4-diyl)di-	EtOH	267(4.4),300(4.3)	95-0783-69
Pyridine, 4,4'-(2,5-diphenylpyrrole-3,4-diyl)di-	EtOH	267(4.4),310s(4.1)	95-0783-69
$C_{26}H_{19}N_3O$			
Malononitrile, [1-(dimethylamino)-2-(2-phenyl-4H-naphtho[1,2-b]pyran-4-ylidene)ethylidene]-	MeCN	270(4.35),368(4.09), 385(4.10),490(4.17), 522(4.38)	4-0803-69
$C_{26}H_{19}N_3O_2$			
Ketone, 1-[(2-hydroxy-1-naphthyl)azo]-2-phenyl-3-indolizinyl methyl	EtOH	219(4.47),478(4.27), 508(4.29)	39-1279-69C
$C_{26}H_{19}N_5O$			
[1,2'(1'H)-Biphthalazine]-4'-carbo-nitrile, 1'-(α-methylphenacyl)-	EtOH	245(4.37),280(3.76), 377(4.26),392s(4.18)	95-0959-69
$C_{26}H_{20}$			
Bicyclo[4.2.0]octa-1,3,5-triene, 7,7,8-triphenyl-	EtOH	255(3.18),261(3.32), 266(3.40),273(3.33)	44-0461-69
Biphenylene, 2,6-dibenzyl-	EtOH	250(4.82),258(4.99), 269s(3.93),334s(3.63), 338s(3.69),347(3.93), 353s(3.78),366(4.11)	39-2789-69C
Ethylene, tetraphenyl-	dioxan	240(4.39),310(4.15)	47-3313-69
Naphthalene, 1,3-distyryl-	DMF	321(4.75)	33-2521-69
Naphthalene, 1,4-distyryl-	DMF	375(4.58)	33-2521-69
Naphthalene, 1,5-distyryl-	DMF	355(4.52)	33-2521-69
Naphthalene, 1,6-distyryl-	DMF	323(4.85)	33-2521-69
Naphthalene, 1,7-distyryl-	DMF	295(4.61),325(4.85), 340(4.88),382(4.36)	33-2521-69
Naphthalene, 2,3-distyryl-	DMF	304(4.81)	33-2521-69
Naphthalene, 2,6-distyryl-	DMF	283(4.48),293(4.52), 360(4.85),381(4.75)	33-2521-69
Stilbene, p,p'-diphenyl-	DMF	345(4.78)	33-2521-69
Stilbene, 4-[2-(1-naphthyl)vinyl]-	DMF	365(4.74)	33-2521-69
Stilbene, 4-[2-(2-naphthyl)vinyl]-	DMF	302(4.24),367(4.87)	33-2521-69
p-Terphenyl-, 4-styryl-	DMF	337(4.77)	33-2521-69

Compound	Solvent	λ max (log ε)	Ref.
$C_{26}H_{20}Br_2O$			
Ether, bis(2-bromo-5-phenylbenzyl)	n.s.g.	256(4.54)	40-0085-69
$C_{26}H_{20}Cl_2O_{10}$			
1-Benzopyrylium, 2,2'-p-phenylenebis-[4-methyl-, diperchlorate	MeCN	248(4.40),292(4.43), 325(4.15),365(4.18)	4-0623-69
$C_{26}H_{20}N_2$			
Benzophenone, azine	n.s.g.	280(4.27),315(4.20)	47-3313-69
1,4-Diazabutadiene, 1,2,3,4,-tetra-phenyl-	EtOH	264(4.44)	78-1467-69
$C_{26}H_{20}N_2O$			
1-Cyclopropene, 3-ethoxy-1,2-diphenyl-3-(α,α-dicyanobenzyl)-	CH_2Cl_2	224s(--),229(4.35), 232(--),291s(--), 303(4.45),316(--)	24-0319-69
$C_{26}H_{20}N_2O_2$			
1H-Benz[g]indazol-5-ol, 3-methyl-1,4-diphenyl-, acetate	EtOH	224(4.61),257(4.57), 301(4.01),340(3.65)	103-0251-69
1H-[1,3]Oxazino[3',4':1,2]pyrrolo-[3,4-b]quinolin-1-one, 3,4,4a,11-tetrahydro-3,3-diphenyl-	EtOH	230(4.55),306(3.67), 313(3.72),319(3.87)	44-3853-69
$C_{26}H_{20}N_2S_2$			
Benzothiazole, 2-[(3,5-di-p-tolyl-2H-1,4-thiazin-2-ylidene)methyl]-	EtOH	269(4.56),333(4.29), 392(4.22),410s(4.18), 444s(4.10)	32-0323-69
	EtOH-HCl	235(4.35),270(4.45), 357(4.22),518(4.15)	32-0323-69
$C_{26}H_{20}N_2S_3$			
2,4-Thiazolidinedithione, 3-ethyl-5-[2-(4-phenylbenzo[f]quinolin-3(4H)-ylidene)ethylidene]-	BuOH	236(4.53),278(4.35)	65-2116-69
$C_{26}H_{20}N_4$			
s-Tetrazine, 1,4-dihydro-1,3,4,6-tetra-phenyl-	EtOH	271(4.25),332(4.11)	35-2443-69
$C_{26}H_{20}O$			
Acetophenone, 2,2,2-triphenyl-	heptane	251(4.01),333(2.29)	59-0765-69
Ether, phenyl p-(p-phenylstyryl)phenyl	DMF	335(4.69)	33-2521-69
$C_{26}H_{20}O_5$			
10,13-Epoxybenzo[b]triphenylene-11,12-dicarboxylic acid, 10,11,12,13-tetra-hydro-, dimethyl ester, trans	EtOH	253(5.08),262(5.27), 278s(4.31),288(4.25), 297s(4.01),310(3.70), 317s(3.26),324s(3.08), 332(3.04),339(2.93), 348(2.86)	12-1449-69
$C_{26}H_{21}Cl_2NO_5$			
2-Cyclopropene, 3-ethoxy-1,2-bis(p-chlorophenyl)-1-(4-nitro-α-carbo-methoxybenzyl)-	CH_2Cl_2	279(4.46)	24-0319-69

Compound	Solvent	$\lambda_{max}(\log \epsilon)$	Ref.
$C_{26}H_{21}Cl_4N$			
Pyridine, 1-butyl-1,4-dihydro-2,6-diphenyl-4-(2,3,4,5-tetrachloro-2,4-cyclopentadien-1-ylidene)-	benzene	460(4.62)	83-0886-69
	C_6H_{12}	462(4.62)	83-0886-69
	MeOH	446(4.62)	83-0886-69
	CH_2Cl_2	278(4.02),451(4.62)	83-0886-69
$C_{26}H_{21}NO$			
Benzaldehyde, O-trityloxime, anti	EtOH	260(4.26)	88-1841-69
$C_{26}H_{21}NO_4$			
p-Benzotoluidide, N-(4-formyl-2-hydroxy-1,3-butadienyl)-, benzoate	EtOH	229(4.23),332(4.40)	78-2035-69
1-Cyclopropene, 1,2-di-p-tolyl-3-(4-nitro-α-carbomethoxybenzylidene)-	CH_2Cl_2	252s(--),277(4.46), 309s(--),393(4.22)	24-0319-69
4-Penten-1-ol, 4-nitro-5-(9-phenanthryl)-, benzoate	EtOH	342(3.89)	87-0157-69
$C_{26}H_{21}N_3O_4$			
Acetic acid, phenyl-, 2-[1,4-dihydro-1,4-dioxo-3-(N-phenylacetamido)-2-naphthyl]hydrazide	alkali	580(3.76)	95-0007-69
$C_{26}H_{22}$			
Ethane, 1,1,2,2-tetraphenyl-	EtOH	225(4.42),263(3.08), 271(2.93)	108-0165-69
o-Xylene, α,α,α'-triphenyl-	EtOH	244(2.69),250(2.86), 257(2.99),260(3.02), 263(3.08),265s(2.97), 270(2.95)	44-0461-69
$C_{26}H_{22}Cl_2NO_2PS$			
Phenothiaphosphinic acid, 2,8-dichloro-, dibenzylammonium salt	EtOH	216(4.51),274(4.32), 289s(4.09),312s(3.68)	78-3919-69
$C_{26}H_{22}F_2NO_2PS$			
Dibenzo[b,e]thiaphospho(V)rin, 3,7-difluoro-5-hydroxy-5-oxo-, dibenzylammonium salt	EtOH	223(4.36),242s(3.85), 264(4.18),285s(3.44), 315(3.60)	39-2424-69C
$C_{26}H_{22}N_2$			
Spiro[2,4-cyclohexadiene-1,2'(1'H)-quinoline], 3',4'-dihydro-1'-phenyl-6-(phenylimino)-	MeOH	244(4.19),294(4.04), 371(3.36)	24-3486-69
$C_{26}H_{22}N_2OP_2S_2$			
Formimidic acid, 1-(diphenylphosphinothioyl)thio-, anhydrosulfide with 1-(diphenylphosphinyl)thioformimidic acid	EtOH	253(4.29),260-275s(--) (two shoulders)	18-2975-69
$C_{26}H_{22}N_2O_2$			
2,3-Naphthalenedicarboxamide, N,N'-dibenzyl-	EtOH	233(4.89),268s(3.94), 278s(3.85),316(2.86), 329(2.91)	44-2418-69
$C_{26}H_{22}N_2O_2P_2S$			
Formimidic acid, 1-(diphenylphosphinyl)-thio-, anhydrosulfide	EtOH	253s(4.26),265s(4.17), 272s(4.07)	18-2975-69

Compound	Solvent	λ_{max}(log ε)	Ref.
$C_{26}H_{22}N_2O_3$			
2-Cyclopropene, 3-ethoxy-1-[1-(p-nitrophenyl)-1-cyanoethyl]-1,2-diphenyl-	CH_2Cl_2	275(4.40)	24-0319-69
$C_{26}H_{22}N_2P_2S_3$			
Formimidic acid, 1-(diphenylphosphinothioyl)thio-, anhydrosulfide	MeCN	257s(4.20)	18-2975-69
$C_{26}H_{22}N_2Si$			
Silane, (α-diazophenethyl)triphenyl-	C_6H_{12}	220s(4.54),258(3.85), 424(1.67)	23-4353-69
$C_{26}H_{22}N_4$			
Benzil, bis(phenylhydrazone), amphi	EtOH	242(4.31),322(4.28), 399(4.08)	88-2697-69
anti	EtOH	240(4.53),298(4.26), 341(4.54)	88-2697-69
syn	EtOH	298(4.15),368(4.57)	88-2697-69
$C_{26}H_{22}N_4O$			
Pyrazino[2,3-f]quinazolin-10(9H)-one, 9-butyl-2,3-diphenyl-	EtOH	212(4.44),226(4.43), 243(3.47),270(4.41), 308(3.92),375(4.18)	2-1166-69
$C_{26}H_{22}N_4O_3S$			
Propiophenone, 2-[1,2-dihydro-4-(methylsulfonyl)-2-(1-phthalazinyl)-1-phthalazinyl]-	EtOH	245(4.33),280(3.79), 368(4.21)	95-0959-69
$C_{26}H_{22}OS$			
Thiophene, 2-[(p-methoxystyryl)phenyl]-5-p-tolyl-	DMF	378(4.77)	33-1282-69
$C_{26}H_{22}O_2$			
Furan, 2-[p-(p-methoxystyryl)phenyl]-5-p-tolyl-	DMF	299(4.15),376(4.81)	33-1282-69
$C_{26}H_{22}O_3$			
2,5-Cyclohexadien-1-one, 4-methoxy-4-(p-methoxyphenyl)-2,6-diphenyl-	n.s.g.	230(4.54),275(3.95), 282(3.91)	24-3795-69
$C_{26}H_{22}O_5$			
11,14-Epoxytribenzo[a,c,g]cyclodecene-12,13-dicarboxylic acid, 11,12,13,14-tetrahydro-, dimethyl ester, trans-trans	EtOH	256s(4.05),262(4.17), 268s(4.14),278s(4.08), 298(3.71)	12-1449-69
$C_{26}H_{22}O_8$			
Anthraquinone, 1,2-dihydroxy-3-(2-hydroxy-4,4-dimethyl-6-oxo-1-cyclohexen-1-yl)-, 1,3-diacetate	MeOH	565(3.66)	64-0137-69
[o-Terphenyl]-3',4',5',6'-tetracarboxylic acid, tetramethyl ester	$CHCl_3$	244(4.29),297(3.56)	44-4134-69
$C_{26}H_{23}BrO_2$			
Spiro[2H-1-benzopyran-2,2'-[2H]naphtho[1,2-b]pyran], 6-bromo-3-pentyl-	dioxan	258(4.69),267(4.78), 311(3.92),320(3.92), 334(3.75),350(3.58)	5-0162-69B

Compound	Solvent	$\lambda_{max}(\log \epsilon)$	Ref.
Spiro[2H-1-benzopyran-2,3'-[3H]naphtho-[1,2-b]pyran], 6-bromo-3-pentyl-	dioxan	245(4.86),301(4.02), 314(4.05),335(3.79), 349(3.83)	5-0162-69B
$C_{26}H_{23}ClN_2O_6$			
1,3,5-Trimethyl-6-(m-nitrostyryl)-2-phenylcyclohepta[c]pyrrolium perchlorate	$C_2H_4Cl_2$	274(4.38),345(4.38), 430(4.40),608(3.30)	104-2179-69
$C_{26}H_{23}ClN_2O_7$			
6-(2-Hydroxy-5-nitrostyryl)-1,3,5-trimethyl-2-phenylcyclohepta[c]pyrrolium perchlorate	$C_2H_4Cl_2$	330(4.30),460(4.15), 660(3.53)	104-2179-69
$C_{26}H_{23}ClO_2$			
Spiro[2H-1-benzopyran-2,3'-[3H]naphtho-[2,1-b]pyran], 7-chloro-3-pentyl-	dioxan	245(4.88),301(4.11), 313(4.12),334(3.79), 348(3.89)	5-0162-69B
$C_{26}H_{23}NO_5$			
2-Cyclopropene, 3-ethoxy-1,2-diphenyl-1-(p-nitro-α-carbomethoxybenzyl)-	CH_2Cl_2	276(4.41)	24-0319-69
$C_{26}H_{23}NO_6$			
α-D-xylo-Tetrofuranoside, methyl 3-O-benzyl-4-C-(1-carboxybenzo[f]quinolin-3-yl)-	n.s.g.	243s(4.49),256(4.51)	65-1413-69
$C_{26}H_{23}NO_8$			
3-Carboxy-1-D-ribofuranosyl-4-picolinium hydroxide, inner salt, 3',5'-dibenzoate	MeOH	274(3.68)	39-0918-69C
$C_{26}H_{24}$			
1,2,3-Hexatriene, 5,5-dimethyl-1,1,4-triphenyl-	C_6H_{12}	258(4.36),369(4.41)	56-1843-69
$C_{26}H_{24}Br_2$			
1,3-Hexadiene, 2,3-dibromo-5,5-dimethyl-1,1,4-triphenyl-	C_6H_{12}	283(4.05)	56-1843-69
geometric isomer	C_6H_{12}	292(4.20)	56-1843-69
$C_{26}H_{24}Br_2NP$			
Phosphorane, (2,7-dibromofluoren-9-ylidene)(p-cyanophenyl)dipropyl-	$CHCl_3$	260(4.70),300(4.63), 390(3.41)	5-0001-69D
$C_{26}H_{24}Cl_2O_6$			
3-(4-Chloro-2-hydroxy-α-pentylstyryl)-naphtho[2,1-b]pyrylium perchlorate	MeOH-$HClO_4$	243(4.63),306(4.11), 333(4.03),530(4.45)	5-0162-69B
$C_{26}H_{24}N_2O$			
4H-1,2-Diazepine, 5-(p-methoxyphenyl)-3,7-di-p-tolyl-	EtOH	209(4.54),271(4.46), 302(4.44)	1-3125-69
$C_{26}H_{24}N_2OS$			
21-Oxa-23-thia-2,24-diazapentacyclo-[$16.2.1.1^{3,6}.1^{8,11}.1^{13,16}$]tetracosa-2,4,6(24),7,9,11,13(22),14,16,18,20-undecaene, 5,14-diethyl-4,15-dimethyl- (Soret band)	EtOH	391(5.00)	77-1480-69

Compound	Solvent	$\lambda_{max}(\log \epsilon)$	Ref.
$C_{26}H_{24}N_2O_2$			
21,23-Dioxa-22,24-diazapentacyclo-[16.2.1.1^{3,6}.1^{8,11}1^{13,16}]tetracosa-2,4,6(24),7,9,11,13(22),14,16,18,20-undecaene, 4,15-diethyl-5,14-dimethyl- (Soret band)	EtOH	380(5.27)	77-1480-69
$C_{26}H_{24}N_2O_4$			
Carbazole-1,2-dicarboximide, N-(formylmethyl)-3-phenyl-, N-(diethyl acetal)	EtOH	222(4.45),243s(4.14), 282(4.37),360(3.96), 420(3.62)	95-0058-69
$C_{26}H_{24}N_4O_9$			
β-Lumicolcicone, mono[(2,4-dinitrophenyl)hydrazone]	MeOH	227(3.93),257(3.94), 325(3.65),410(4.08)	32-1059-69
$C_{26}H_{24}O$			
Benzofuran, 2-[p-(p-isopropylstyryl)-phenyl]-5-methyl-	DMF	360(4.82)	33-1282-69
Benzofuran, 2-[p-(p-isopropylstyryl)-phenyl]-6-methyl-	DMF	361(4.81)	33-1282-69
Phenanthrene, 7-isopropyl-1-(p-methoxystyryl)-	DMF	276(4.53),335(4.49)	33-2521-69
$C_{26}H_{24}O_2$			
Benzofuran, 2-[p-(p-methoxystyryl)-phenyl]-4,6,7-trimethyl-	DMF	368(4.78)	33-1282-69
Spiro[2H-1-benzopyran-2,2'-[2H]naphtho-[1,2-b]pyran], 3-pentyl-	dioxan	258(4.71),267(4.79), 310(3.91),334(3.75), 350(3.60)	5-0162-69B
Spiro[2H-1-benzopyran-2,2'-[2H]naphtho-[1,2-b]pyran], 3'-pentyl-	dioxan	257(4.70),265(4.73), 299(3.89),308(3.91), 336(3.61),351(3.54)	5-0162-69B
Spiro[2H-1-benzopyran-2,3'-[3H]naphtho-[2,1-b]pyran], 2'-pentyl-	dioxan	243(4.90),300(4.08), 312(4.08),336(3.72), 350(3.75)	5-0162-69B
Spiro[2H-1-benzopyran-2,3'-[3H]naphtho-[2,1-b]pyran], 3-pentyl-	dioxan	245(4.88),301(4.08), 314(4.08),336(3.81), 350(3.89)	5-0162-69B
Tetracyclopent[a,c,h,j]anthracene-7,14-dione, 1,2,3,4,5,6,8,9,10,11,12,13-dodecahydro-	C_6H_{12}	220(4.44),226(4.46), 265f(4.60),290(4.20), 362(3.97)	39-0061-69B
$C_{26}H_{24}O_3$			
Spiro[2H-1-benzopyran-2,3'-[3H]naphtho-[2,1-b]pyran]-6-ol, 3-pentyl-	dioxan	246(4.91),300(4.01), 313(4.08),334(3.95), 349(3.91)	5-0162-69B
$C_{26}H_{24}O_3S$			
Benzo[b]thiophene, 6-methyl-2-[p-(3,4,5-trimethoxystyryl)phenyl]-	DMF	365(4.77)	33-1282-69
$C_{26}H_{24}O_4$			
Benzofuran, 5-methyl-2-[p-(3,4,5-trimethoxystyryl)phenyl]-	DMF	364(4.81),382(4.67)	33-1282-69
Benzofuran, 6-methyl-2-[p-(3,4,5-trimethoxystyryl)phenyl]-	DMF	365(4.81)	33-1282-69
Stilbene, 4'-[(3,4-dimethoxyphenyl)-ethynyl]-3,4-dimethoxy-	DMF	361(4.77)	33-2521-69

Compound	Solvent	$\lambda_{max}(\log \epsilon)$	Ref.
$C_{26}H_{24}O_6$			
Mesuagin, acetate	EtOH	235(4.60),285(4.20), 335(3.98)	44-3784-69
$C_{26}H_{24}O_7$			
Flavone, 2'-(benzyloxy)-3',5,7,8-tetra-methoxy-	MeOH	263(4.39),323s(3.61)	2-0110-69
Flavone, 3'-(benzyloxy)-4',5,6,7-tetra-methoxy-	EtOH	240(4.37),258s(4.20), 331(4.37)	24-0112-69
Flavone, 6-(benzyloxy)-3',4',5,7-tetra-methoxy-	EtOH	241(4.04),265(4.21), 329(4.46)	40-1270-69
$C_{26}H_{24}O_8$			
Flavone, 4'-(benzyloxy)-3-hydroxy-3',5,6,7-tetramethoxy-	EtOH	237s(4.29),255(4.37), 360(4.39)	18-1398-69
$C_{26}H_{24}O_{11}$			
2-Anthraceneacetic acid, 4,5,7,10-tetra-hydro-α-methoxy-, methyl ester, tetra-acetate	EtOH	227(4.14),254s(4.79), 262(5.13),355(3.71), 374(3.71),394(3.69)	105-0257-69
$C_{26}H_{24}S$			
Benzo[b]thiophene, 2-[p-(p-isopropyl-styryl)phenyl]-5-methyl-	DMF	359(4.78)	33-1282-69
Benzo[b]thiophene, 2-[p-(p-isopropyl-styryl)phenyl]-6-methyl-	DMF	361(4.77)	33-1282-69
$C_{26}H_{25}BrN_3O$			
α-Antipyrinyl-α-(m-bromophenyl)-p-(di-methylamino)benzylium (picrate)	pH 4	540(4.26)	122-0138-69
α-Antipyrinyl-α-(p-bromophenyl)-p-(di-methylamino)benzylium (picrate)	pH 4	545(4.22)	122-0138-69
$C_{26}H_{25}ClN_2O$			
1(2H)-Quinolinecarboxamide, N-allyl-6-chloro-3,4-dihydro-2-methyl-3,4-diphenyl-	n.s.g.	259(4.19)	40-1047-69
$C_{26}H_{25}ClN_2O_7$			
3-Carbamoyl-1-D-ribofuranosyl-4-picolin-ium chloride, 3',5'-dibenzoate	MeOH	265s(3.70)	39-0199-69C
$C_{26}H_{25}ClN_3O$			
α-Antipyrinyl-α-(m-chlorophenyl)-p-(di-methylamino)benzylium (picrate)	pH 4	540(4.13)	122-0138-69
α-Antipyrinyl-α-(p-chlorophenyl)-p-(di-methylamino)benzylium (picrate)	pH 4	545(4.28)	122-0138-69
$C_{26}H_{25}IN_2O_2S_3$			
6-Methoxy-2-[[4-[(6-methoxy-3-methyl-2-benzothiazolinylidene)methyl]-6-meth-yl-2H-thiopyran-2-ylidene]methyl]-3-methylbenzothiazolium iodide	EtOH-2% $MeNO_2$	629(5.06)	124-0512-69
$C_{26}H_{25}NO$			
4-Stilbenamine, 4'-(2-benzofuranyl)-	DMF	325(4.18),400(4.70)	33-1282-69
$C_{26}H_{25}NO_2$			
3-Buten-2-one, 4-morpholino-1,1,4-tri-phenyl-	$CHCl_3$	232(4.12),317(4.27)	24-3428-69

Compound	Solvent	$\lambda_{max}(\log \epsilon)$	Ref.
$C_{26}H_{25}NS$			
Benzo[b]thiophene, 2-[p-(diethylamino)-stilben-4-yl]-	DMF	320(4.26),398(4.67)	33-1282-69
Benzo[b]thiophene, 5-[p-(diethylamino)-styryl]-2-phenyl-	DMF	367(4.66)	33-1282-69
$C_{26}H_{25}N_4O_3$			
α-Antipyrinyl-p-(dimethylamino)-α-(m-nitrophenyl)benzylium (picrate)	pH 4	532(4.13)	122-0138-69
α-Antipyrinyl-p-(dimethylamino)-α-(p-nitrophenyl)benzylium (picrate)	pH 4	540(4.46)	122-0138-69
$C_{26}H_{26}N_2$			
Spiro[cyclohexane-1,2'(1'H)-quinoline], 3',4'-dihydro-1'-phenyl-2-(phenylimino)-	MeOH	250(4.02),285(4.01)	24-3486-69
$C_{26}H_{26}N_2O_3$			
1H,10H-3a,9b-Diazabenzo[a]naphth[2,1,8-cde]azulene-10-carboxylic acid, 2,3,4,5,11,12-hexahydro-10-hydroxy-12-phenyl-, methyl ester	MeOH	213(4.54),227s(4.44), 305(4.48),316(4.41)	44-0330-69
4(3H)-Quinazolinone, 2-[2,2-bis(p-methoxyphenyl)ethyl]-3-ethyl-	EtOH	229(4.68),268(4.31)	115-0057-69
$C_{26}H_{26}N_2O_{12}$			
Helianthoidin, dinitro-	EtOH	223(4.47),244s(4.35), 298(3.87),344(3.95)	39-0693-69C
$C_{26}H_{26}N_3O$			
α-Antipyrinyl-p-(dimethylamino)-α-phenylbenzylium (picrate)	pH 4	540(4.42)	122-0138-69
$C_{26}H_{26}N_4$			
2,2'-Biquinoline, 4,4'-dipyrrolidino-, cuprous chelate	iso-AmOH	536(3.98)	3-0344-69
$C_{26}H_{26}N_6$			
Naphthalene, 1,5-bis(p-dimethylaminophenylazo)-	EtOH	440(4.63)	124-0179-69
$C_{26}H_{26}O$			
Anisole, p-[p-(p-isopropylstyryl)styryl]-	DMF	367(4.84)	33-2521-69
Cyclopentanone, 3,3-dimethyl-4-trityl- (uncertain if all are maxima)	C_6H_{12}	230(3.99),240(3.86), 250(3.38),260s(3.01), 265s(2.96),272s(2.76)	39-0449-69B
$C_{26}H_{26}O_2$			
2-Hexyne-1,4-diol, 5,5-dimethyl-1,1,4-triphenyl-	EtOH	252(2.69),258(2.78), 264(2.65)	56-1843-69
2-Propen-1-ol, 3,3'-p-phenylenebis[1-p-tolyl-	EtOH	292(4.50)	124-1189-69
$C_{26}H_{26}O_2S$			
Androsta-5,9(11)-dieno[6,5,4-bc]thiophene-3,17-dione, 5'-phenyl-	dioxan	232(4.19),278(3.94), 331(3.82)	94-2586-69

Compound	Solvent	$\lambda_{max}(\log \epsilon)$	Ref.
$C_{26}H_{26}O_3$			
1,3-Cyclohexanedione, 2-[4-(diphenylmethylene)-3,4-dihydro-2H-pyran-2-yl]-5,5-dimethyl-	dioxan	239(4.2),286(4.37)	24-3950-69
$C_{26}H_{26}O_4$			
Benzene, m-bis(3,4-dimethoxystyryl)-	DMF	333(4.76)	33-2521-69
Benzene, o-bis(3,4-dimethoxystyryl)-	DMF	322(4.54),344(4.52)	33-2521-69
Benzene, p-bis(3,4-dimethoxystyryl)-	DMF	377(4.85)	33-2521-69
2-Propen-1-ol, 3,3'-p-phenylenebis-[1-(p-methoxyphenyl)-	EtOH	292(4.53)	124-1189-69
Stilbene, 3,4,5-trimethoxy-4'-(p-methoxystyryl)-	DMF	371(4.83)	33-2521-69
$C_{26}H_{26}O_5$			
2H,6H-Benzo[1,2-b:5,4-b']dipyran-2,6-dione, 7,8-dihydro-5-methoxy-7a,8-dimethyl-10-(3-methyl-2-butenyl)-4-phenyl-	EtOH	270(4.38),330(4.04)	44-4203-69
$C_{26}H_{26}O_6$			
Lonchocarpic acid	EtOH	235s(4.25),268(4.35), 344(3.98)	39-0374-69C
Scandenin	EtOH	234(4.62),286(4.18), 343(4.22)	39-0374-69C
$C_{26}H_{26}O_{12}$			
2-Anthraceneacetic acid, 1,2,3,4-tetrahydro-3,5,6,8-tetrahydroxy- -methoxy-4-oxo-, ethyl ester, 3-formate 5,6,8-triacetate	EtOH	259(4.76),304(3.72), 358(3.50)	105-0494-69
2-Anthraceneacetic acid, 1,2,3,4-tetrahydro-3,5,7,10-tetrahydroxy- -methoxy-4-oxo-, methyl ester, tetraacetate	EtOH	250s(4.81),258(4.93), 304(3.98),315(3.91)	105-0257-69
$C_{26}H_{26}O_{13}$			
Altersolanol A, pentaacetate	EtOH	210(4.51),268(4.34), 383(3.41)	23-0767-69
$C_{26}H_{27}Br_2OP$			
Phosphorane, (2,7-dibromofluoren-9-ylidene)(p-methoxyphenyl)dipropyl-	$CHCl_3$	260(4.66),270(4.70), 300(4.44),397(3.25)	5-0001-69D
$C_{26}H_{27}ClN_2O_2$			
1(2H)-Quinolinecarboxamide, 6-chloro-N-ethyl-3,4-dihydro-3-(p-methoxyphenyl)-2-methyl-4-phenyl-	n.s.g.	259(4.21)	40-1047-69
$C_{26}H_{27}NO$			
Azetidine, 1-tert-butyl-2-phenyl-3-(p-phenylbenzoyl)-, cis	isooctane	282(4.36)	44-0310-69
trans	isooctane	282(4.42)	44-0310-69
$C_{26}H_{27}NO_4$			
Flavinantine, O-benzyl-	MeOH	235(4.15),282(3.83)	39-1063-69C
$C_{26}H_{27}N_4O$			
α-(p-Aminophenyl)-α-antipyrinyl-p-(dimethylamino)benzylium (picrate)	pH 4	567(4.04)	122-0138-69

Compound	Solvent	λ_{max}(log ε)	Ref.
$C_{26}H_{28}F_2O_3$			
3'H-Cycloprop[6,7]estra-4,6-dien-3-one, 3',3'-difluoro-6α,7α-dihydro-17β-hydroxy-, benzoate	EtOH	234(4.34)	78-1219-69
6β,7β-	EtOH	237(4.36)	78-1219-69
$C_{26}H_{28}Fe_2$			
[1,1]Ferrocenophane, 1,1,12,12-tetramethyl-	n.s.g.	325s(2.33),464(2.45)	78-6001-69
$C_{26}H_{28}N_2O_{11}$			
3,4,5-Isoxazolidinetricarboxylic acid, 5-(N-acetoxyanilino)-3-(carboxymethyl)-2-phenyl-, tetramethyl ester	MeOH	231(3.89)	24-2346-69
$C_{26}H_{28}N_4O_5S$			
3-Buten-2-one, 1-[3-[(4-amino-2-methyl-5-pyrimidinyl)methyl]-5-(2-hydroxyethyl)-4-methyl-4-thiazolin-2-ylidene]-1-hydroxy-4-phenyl-, diacetate	EtOH	229(4.31),290(4.18), 307(4.08),438(4.40)	94-0128-69
$C_{26}H_{28}N_4O_9$			
Phorbolactone, hemiacetal, 2,4-dinitrophenylhydrazone	EtOH	256(4.13),290s(3.71), 380(4.42)	64-0080-69
$C_{26}H_{28}N_6$			
Quinoxaline, 2-[5-methyl-4-[3-(4-o-tolyl-1-piperazinyl)propenyl]-pyrazol-1-yl]-	EtOH	252(4.49),342(4.15), 351(4.14)	36-0432-69
$C_{26}H_{28}N_6O$			
1-Propanone, 1-[5-methyl-1-(2-quinoxalinyl)-4-pyrazolyl]-3-[4-(o-tolyl)-1-piperazinyl]-	EtOH	238(4.37),267(4.43), 333(4.11)	36-0432-69
hydrochloride	EtOH	238(4.35),267(4.44), 331(4.06)	36-0432-69
$C_{26}H_{28}O_4$			
Estra-1,3,5(10)-trien-6-one, 1,17β-dihydroxy-4-methyl-, 17-benzoate	EtOH	218(4.20),232(4.38), 262(3.81),332(3.36)	39-1234-69C
$C_{26}H_{28}O_6$			
Scandenin, dihydro-	EtOH	252s(4.07),332(4.26)	12-1923-69
	EtOH-NaOH	263(4.30),335(4.34)	12-1923-69
$C_{26}H_{28}O_8$			
Helianthoidin	EtOH	227(4.29),283(3.80)	39-0693-69C
$C_{26}H_{29}ClN_2O_2$			
4-Benzyl-1-deacetyldiabolinium chloride	MeOH	299(3.49)	33-1564-69
$C_{26}H_{29}N_3O_{11}$			
L-Threonine, N,N'-[(2-hydroxy-4,9-dimethyl-3-oxo-3H-phenoxazine-1,6-diyl)dicarbonyl]di-, dimethyl ester	$CHCl_3$-7% MeOH	417(4.22)	24-3205-69
$C_{26}H_{30}Cl_2N_2O_{11}$			
Ammonium, [2-[4,6-bis(p-methoxyphenyl)-2H-pyran-2-ylidene]propanediylidene]-bis[dimethyl-, diperchlorate	MeCN	266(4.16),310(4.28), 407(4.56),479(4.45)	44-2736-69

Compound	Solvent	λ_{max}(log ε)	Ref.
$C_{26}H_{30}INO_2$			
1-(8a-Benzyl-3,3a,8,8a-tetrahydro-3a-hydroxy-8-oxocyclopent[a]inden-2-yl)-methyl]-1-methylpiperidinium iodide	EtOH	248(4.03),290(3.15)	104-2090-69
$C_{26}H_{30}N_2O$			
Benzamide, N-[2-(3-cyclopenten-1-yl)-butyl]-N-(2-indol-3-ylethyl)-	n.s.g.	221(4.58),274(3.74), 282(3.76),291(3.69)	35-7333-69
1(4H)-Naphthalenone, 2-anilino-4-[(4-cyclohexylbutyl)imino]-	EtOH	252(4.27),280(4.40), 340s(3.74),440(3.70)	39-1799-69C
1(4H)-Naphthalenone, 2-[(4-cyclohexyl-butyl)amino]-4-(phenylimino)-	EtOH	240(4.50),270s(4.20), 280s(4.18),335(3.80), 453(3.85)	39-1799-69C
$C_{26}H_{30}N_2O_4$			
1,4,5,8-Naphthalenetetracarboxylic 1,8:4,5-diimide, N,N'-dihexyl-	HCOOH	360(--),382(4.37)	121-1067-69
$C_{26}H_{30}N_2O_6$			
3β,20α-Yohimban-19α-acetic acid, α-acet-yl-16-carboxy-16,17-didehydro-17-hydroxy-, dimethyl ester	pH 13	220(4.72),284(4.35)	44-3545-69
	EtOH	225(4.60),261(4.11), 283(3.90),290(3.84)	44-3545-69
19β-	EtOH	226(4.62),260(4.11), 282(3.95),290(3.84)	44-3545-69
$C_{26}H_{30}N_4O_{10}$			
L-Threonine, N,N'-[(2-amino-1,9-dimeth-yl-3-oxo-3H-phenoxazine-4,6-diyl)di-carbonyl]di-, dimethyl ester	$CHCl_3$	433(4.37),450(4.34)	24-3205-69
$C_{26}H_{30}N_6O_2$			
17,18-Secoyohimban-17-ol, 10-[(1-phenyl-1H-tetrazol-5-yl)oxy]-	EtOH	231(4.63),288(3.94), 293s(3.93)	78-5319-69
$C_{26}H_{30}N_8O_4S_4$			
2H-1,3-Thiazine-2,4(3H)-dione, 3,3'-hexamethylenebis[dihydro-2-thio-, 4,4'-bis[(p-nitrophenyl)hydrazone]	EtOH	238(4.43),305(4.36), 400(4.66)	103-0049-69
$C_{26}H_{30}O_2$			
Androsta-4,6-dien-3-one, 2-benzylidene-17β-hydroxy-	MeOH	231(4.18),318(4.38)	94-2319-69
1-Norbornanol, 7-(2,2-diphenylvinyl)-3,3,7-trimethyl-, acetate	EtOH	251(4.24)	39-2634-69C
Picene, 5,6,6a,6b,7,8,12b,13-octahydro-3,10-dimethoxy-6aα,12bβ-dimethyl-	CH_2Cl_2	250(4.31)	44-3717-69
6aβ-	CH_2Cl_2	250(4.28)	44-3717-69
$C_{26}H_{30}O_2S$			
Androst-5-eno[6.5.4-bc]thiophen-3-one, 17β-hydroxy-5'-phenyl-	EtOH	232(4.18),278(3.96), 333(3.81)	94-2586-69
$C_{26}H_{30}O_3$			
Androst-5-eno[6,5,4-bc]furan-3-one, 17β-hydroxy-5'-phenyl-	EtOH	226(4.26),249s(3.97), 348(4.10)	94-2586-69
Estra-1,3,5(10)-triene-1,17β-diol, 4-methyl-, 17-benzoate	EtOH	216(3.98),231(4.06), 286(3.15)	39-1234-69C

Compound	Solvent	$\lambda_{max}(\log \epsilon)$	Ref.
$C_{26}H_{30}O_4$ 1,5-Cyclohexadiene-1-carboxylic acid, 3,5-di-tert-butyl-3-(2-naphthyloxy)-4-oxo-, methyl ester	C_6H_{12}	312(3.67)	64-0547-69
$C_{26}H_{30}O_6$ Benzo[4,5,6]pregn-4-ene-5',6'-dicarboxylic acid, 3,20-dioxo-	n.s.g.	229(4.45),308(3.22)	94-2586-69
Gedunin, 7-deacetyl-7-oxo-	EtOH	221(4.11)	42-0682-69
$C_{26}H_{30}O_7$ Bufa-1,4,20,22-tetraenolide, 8,12β,14-trihydroxy-3-oxo-,12-acetate	EtOH	241(4.18),300(4.21)	119-0186-69
$C_{26}H_{30}O_8$ Helianthoidin, dihydro-	EtOH	232(4.03),283(3.80)	39-0693-69C
$C_{26}H_{30}O_9$ Bufa-4,20,22-trienolide, 1α,2α-epoxy-6,8,12,14-tetrahydroxy-3-oxo-, 6-acetate	EtOH	223(4.18),300(3.77)	119-0186-69
$C_{26}H_{30}O_{11}$ Rubratoxin B	n.s.g.	250(3.99)	88-0367-69
$C_{26}H_{31}Cl_3O_5$ Pregna-3,5,7-trien-20-one, 3,17α-dihydroxy-6-methyl-, 17-acetate trichloroacetate	MeOH	320(4.15)	78-1717-69
$C_{26}H_{31}NO$ 5α-Androst-2-eno[3,2-b]quinolin-17-one	EtOH	210(5.11),233(4.99), 238(5.05),288(3.91), 294(3.96),301(3.96), 307(4.16),313(4.04), 321(4.40)	39-1419-69C
$C_{26}H_{32}Cl_2O_5$ 5β-Chola-3,6-dien-24-oic acid, 3,7-dichloro-4,6-diformyl-12-oxo-	EtOH	252(3.97)	32-1243-69
$C_{26}H_{32}Fe_2N_2$ [1,1"-Biferrocene]-2,2"-bis(methylamine), N,N,N',N'-tetramethyl-	C_6H_{12}	201(4.77),215s(4.63), 250s(3.96),300s(3.31)	78-3477-69
(1R,1"R)-(-)-	EtOH	440(2.49)	49-1515-69
symmetric	C_6H_{12}	198(4.65),222(4.68), 255s(4.03),298(3.88)	78-3477-69
$C_{26}H_{32}N_2O$ 2'H-Androsta-2,4-dieno[3,2-c]pyrazol-17β-ol, 5'-phenyl-	EtOH	258(4.37)	94-2319-69
$C_{26}H_{32}N_2O_4S_2$ Di-3,4-xylenesulfonamide, N-[p-(diethylamino)phenyl]-	EtOH-1% $CHCl_3$	238(4.38),280(4.30)	44-2083-69
$C_{26}H_{32}N_2O_5$ 5β-Cholan-24-oic acid, 2,6-dicyano-3,7,12-trioxo-	EtOH	233(3.52)	32-1243-69

Compound	Solvent	λmax(log ε)	Ref.
5β-Cholan-24-oic acid, 4,6-dicyano-3,7,12-trioxo-	EtOH	231(3.71)	32-1243-69
Ketone, 4b,5,9,13b,15,16-hexahydro-2,3,11,12-tetramethoxy-6H,8H-pyrimido[2,1-a:4,3-a']diisoquinolin-5-yl methyl	EtOH-HCl	249(4.32),296(3.88), 316(408),376(4.08)	39-0085-69C
	EtOH-NaOH	224(4.30),282s(3.87), 286(3.89),291s(3.86)	39-0085-69C
Picralinol, N-acetyl-, diacetate	EtOH	253(4.07),279(3.57)	33-0701-69
$C_{26}H_{32}N_4O_6$			
Torilin, decaetyl-, 2,4-dinitrophenylhydrazone	$CHCl_3$	388(4.51)	78-4751-69
$C_{26}H_{32}OS$			
Androst-5-eno[6,5,4-bc]thiophene-17β-ol, 5'-phenyl-	EtOH	294(4.16)	94-2586-69
$C_{26}H_{32}O_2$			
Androst-4-en-6-one, 4-benzoyl-	EtOH	249(4.17)	94-2586-69
Cyclohexanone, 2-isopropyl-4-isovaleryl-3,5-diphenyl-	EtOH	248(2.26),253(2.41), 259(2.57),266(2.50), 268(2.28),295(1.98)	44-0444-69
$C_{26}H_{32}O_3$			
Androst-4-en-6-one, 4-hydroxy-, benzoate	EtOH	234(4.24)	94-2586-69
5β,19 -Cyclo-9β-androstan-3-one, 17β-hydroxy-, benzoate	EtOH	230(4.20),267(2.92), 274(2.98),281(2.98)	73-2459-69
$C_{26}H_{32}O_4$			
3'H-Cyclopropa[1,2]pregna-1,4,6-triene-3,20-dione, 1β,2β-dihydro-17-hydroxy-6-methyl-16-methylene-, acetate	MeOH	285(4.27)	87-0631-69
Spiro[2,4,5-cycloheptatriene-1,1'(3'aH)-pentalene]-2',3'-dicarboxylic acid, 4',6'-di-tert-butyl-, dimethyl ester	EtOH	232(4.34),280(3.90), 334s(3.11)	89-0881-69
$C_{26}H_{32}O_7$			
Bersaldegenin, 1,3,5-orthoacetate	MeOH	298(3.70)	88-1709-69
Melianthugenin	MeOH	299(3.77)	119-0191-69
$C_{26}H_{32}O_8$			
Crotophorbolone, 4,9,20-triacetate	MeOH	199(4.18),242(3.76)	5-0158-69H
$C_{26}H_{32}O_{11}$			
Rubratoxin A	n.s.g.	204(4.50),225s(--), 252s(3.65)	88-0367-69
$C_{26}H_{32}O_{15}$			
2-Norpinene-6-carboxylic acid, 1-(β-D-glucopyranosyloxy)-6-(hydroxymethyl)-2-methyl-4-oxo-, pentaacetate	EtOH	242(3.86)	78-1825-69
$C_{26}H_{33}BrO_6$			
Pregn-4-ene-3,20-dione, 7α-bromo-6β,17-dihydroxy-1 ,2 -methylene-, diacetate	MeOH	231(4.12)	24-2570-69
$C_{26}H_{33}ClO_6$			
Pregn-4-ene-3,20-dione, 7α-chloro-6β,17-dihydroxy-1α,2α-methylene-, diacetate	MeOH	230(4.10)	24-2570-69

Compound	Solvent	$\lambda_{max}(\log \epsilon)$	Ref.
$C_{26}H_{33}NO_4$			
5α-Androstan-3-one, 17β-hydroxy-2-(p-nitrobenzylidene)-	EtOH	213s(3.59),311(4.21)	94-2319-69
$C_{26}H_{33}NO_6$			
Estra-1,3,5(10)-trien-17-one, 3,9α-dihydroxy-11β-nitro-, 3-cyclopentanepropionate	EtOH	267(2.96),276(2.88)	31-1018-69
$C_{26}H_{33}N_3O_3$			
2'H-5α-Androst-2-eno[3,2-c]pyrazol-17β-ol, 5'-(p-nitrophenyl)-	EtOH	228(4.06),326(4.14)	94-2319-69
$C_{26}H_{34}N_2O$			
2'H-5α-Androst-2-eno[3,2-c]pyrazol-17-ol, 5'-phenyl-	EtOH	256(4.16)	94-2319-69
5β-	EtOH	256(4.17)	94-2319-69
$C_{26}H_{34}N_2O_2$			
2-Piperidone, 1-phenethyl-3-[5-(phenethylamino)valeryl]-	dioxan	268(3.87)	95-1029-69
$C_{26}H_{34}N_4O_{10}$			
[2,2'-Bi-2H-indazole]-5,5,5',5'(3H,3'H)-tetracarboxylic acid, 3a,3'a,4,4',6,-6',7,7'-octahydro-3,3'-dioxo-, tetraethyl ester	EtOH	203(4.09),238(4.16), 256s(3.99)	39-2783-69C
$C_{26}H_{34}O_2$			
Resorcinol, 2,4-di-p-mentha-1,8-dien-3-yl-, trans	EtOH	232s(4.02),284(3.3)	33-1102-69
$C_{26}H_{34}O_2S$			
Androst-4-en-3-one, 17β-hydroxy-17α-methyl-7α-(phenylthio)-	EtOH	246(4.22)	94-0011-69
$C_{26}H_{34}O_4$			
5β-Bufa-14,20,22-trienolide, 3-hydroxy-, acetate	MeOH	302(2.39)	94-1711-69
Pregn-4-ene-3,20-dione, 6β-(3-acetoxy-1-propynyl)-	EtOH	239(4.19)	94-2586-69
$C_{26}H_{34}O_5$			
5β-Bufa-20,22-dienolide, 14α,15α-epoxy-3β-hydroxy-, acetate	MeOH	301(2.07)	94-1711-69
$C_{26}H_{34}O_7$			
Acetic acid, (p-menth-3-yloxy)-, ester with 2,3-dihydro-2-(1-hydroxy-1-methylethyl)-7H-pyrano[2,3-g]-1,4-benzodioxin-7-one	EtOH	230(4.26),253(3.78), 260(3.72),297(3.95), 343(4.10)	39-0526-69C
Cholanic acid, 2,6-diformyl-3,7,12-trioxo-	EtOH	283(3.80),364(3.72)	32-1243-69
Cholanic acid, 4,6-diformyl-3,7,12-trioxo-	EtOH	298(3.66)	32-1243-69
$C_{26}H_{34}O_7S$			
7,9a-Methano-9aH-cyclopenta[b]heptalen-4(1H)-one, 2,3,4a,5,6,7,8,9,10,11-decahydro-2,8,11,12-tetrahydroxy-1,1,8-trimethyl-, 2-p-toluenesulfonate	EtOH	223(4.22),253(3.95)	94-2036-69

Compound	Solvent	λ_{max}(log ε)	Ref.
$C_{26}H_{34}O_8$			
Bersaldegenin, 3-acetate	MeOH	297(3.71)	88-1709-69
Pregna-3,5-diene-6-carboxaldehyde, 11β,17α,21-trihydroxy-3-methoxy-, 21-acetate 11-formate	EtOH	218(4.05),322(4.22)	78-1155-69
$C_{26}H_{34}O_9$			
3,10a-Ethanophenanthrene-1,2,8-tricarboxylic acid, 1'-hydroxy-4b,8-dimethyl-2'-oxo-, trimethyl ester	EtOH	236(3.66)	44-1257-69
$C_{26}H_{34}O_{15}$			
1H,3H-Pyrano[3,4-c]pyran-1-one, 6-(β-D-glucopyranosyloxy)-4,4a,5,6-tetrahydro-5-(1-hydroxyethyl)-, pentaacetate	EtOH	243(3.92)	88-2725-69
$C_{26}H_{35}ClO_4$			
Pregna-3,5-dien-20-one, 6-chloro-3-ethoxy-17α-hydroxy-16-methylene-, acetate	MeOH	252(4.34)	87-0631-69
$C_{26}H_{35}Cl_3O_2$			
2,5-Cyclohexadien-1-one, 2,4,6-tri-tert-butyl-4-[(2,4,6-trichloro-3,5-xylyl)-oxy]-	MeOH	231s(4.23)	64-0547-69
$C_{26}H_{35}NO_6$			
5α-Androst-15-ene-14β-carbonitrile, 3β,16,17β-triacetoxy-	EtOH	213(3.72)	39-0554-69C
$C_{26}H_{35}NO_8$			
Veronamine, (-)-	EtOH	275(3.44),282(3.43)	88-4951-69
$C_{26}H_{35}N_3O_2$			
1H-Naphth[2,3-d]imidazole-4,9-dione, 1-[2-(diethylamino)ethyl]-2-(3-cyclohexylpropyl)-	pH 1	248(4.62),280(4.19)	4-0909-69
	pH 7	248(4.63),283(4.15)	4-0909-69
	pH 13	249(4.50),283(4.16)	4-0909-69
	MeOH	248(4.65),275(4.16), 281(4.17),332(3.53)	4-0909-69
$C_{26}H_{36}BrNO_7$			
5α-Pregnan-20-one, 5α-bromo-3β,6β,17α-trihydroxy-16-methylene-, 3,17-diacetate 6-nitrite	EtOH	234(3.30)	24-0643-69
$C_{26}H_{36}NO_2PS$			
Phenothiaphosphinic acid, 2,8-dimethyl-, dicyclohexylamine salt	EtOH	224(4.69),269(4.91), 286s(4.73),314s(3.92)	78-3919-69
$C_{26}H_{36}N_2$			
Ethylenediamine, N,N'-dicyclohexyl-1,2-diphenyl-, meso-	EtOH	242(2.79)	35-2653-69
$C_{26}H_{36}N_2O$			
1(4H)-Naphthalenone, 2-[(4-cyclohexylbutyl)amino]-4-(cyclohexylimino)-	EtOH	239(4.41),272(4.11), 280s(4.09),338(3.70), 425(3.70)	39-1799-69C
$C_{26}H_{36}O$			
3'H-Cycloprop[2,3]-5α-androst-2-en-17β-ol, 2β,3β-dihydro-3'-phenyl-	EtOH	226(4.11),256s(2.59), 263(2.73),269(2.81), 276(2.71)	94-2319-69

Compound	Solvent	λmax(log ε)	Ref.
$C_{26}H_{36}O_3$			
9,19-Cyclo-24-nor-5α,9β-chol-11-en-23-oic acid, 16-hydroxy-4,4,14-trimethyl-3-oxo-, δ-lactone	EtOH	210(4.15)	32-0915-69
8,12-Etheno-8βH-cyclopenta[a]phenanthrene-4-carboxylic acid, 1,2,3,4,5,-6,7,9,10,11,12,13,14,15-tetradecahydro-1'-isopropyl-4β,10β-dimethyl-15-oxo-, methyl ester	n.s.g.	273(3.96),340(1.92)	44-4016-69
1H-2,3,10aα-Methenocyclobut[1,7]indeno-[5,4-a]naphthalene-8-carboxylic acid, 2,2a,3,3a,4,4a,4b,5,6,7,8,8a,9,10,-10b,10c-hexadecahydro-3-isopropyl-4b,8-dimethyl-1-oxo-, methyl ester	EtOH	298(1.81)	44-4016-69
19-Nor-$\Delta^{1,3,5(10)}$-tigogenin-3-ol	EtOH	280(3.40)	88-2113-69
$C_{26}H_{36}O_4$			
Androst-4-en-17β-ol, 3-(4,5-dihydro-2-oxo-3-furylidene)-17α-methyl-, acetate	EtOH	295(4.41)	44-3796-69
Benzoic acid, p-[(1,3,5-tri-tert-butyl-4-oxo-2,5-cyclohexadien-1-yl)oxy]-, methyl ester	EtOH	250(4.30)	44-0550-69
Benzoic acid, p-[(1,3,5-tri-tert-butyl-6-oxo-2,4-cyclohexadien-1-yl)oxy]-, methyl ester	EtOH	255(3.30),308(3.12)	44-0550-69
Bovinone	EtOH	287(4.31)	39-2398-69C
5α,14α-Bufa-20,22-dienolide, 3β-acetoxy-	EtOH	298(3.75)	13-0637-69B
$C_{26}H_{36}O_5$			
18-Nor-9β,10α-pregna-3,5-dien-20-one, 3,17α-diacetoxy-13-ethyl-	n.s.g.	236(4.24)	54-0737-69
$C_{26}H_{36}O_6$			
19-Norpregna-1(10),9(11)-dien-20-one, 3β,6α,21-trihydroxy-5β,16α-dimethyl-, 3,21-diacetate	MeOH	226(4.04)	5-0152-69F
Pregn-4-ene-3,11-dione, 17,20α,21-trihydroxy-, 21-acetate 17,20-cyclic acetal with acetone	MeOH	238(4.19)	44-3513-69
20β-	MeOH	239(4.20)	44-3513-69
$C_{26}H_{36}O_{10}$			
Butyric acid, benzenepentayl ester	MeOH	266(2.84)	34-0118-69
Isobutyric acid, benzenepentayl ester	MeOH	264(2.85)	34-0118-69
$C_{26}H_{37}NO_8$			
Malaxin	MeOH	257(4.2)	78-2723-69
$C_{26}H_{38}B_2Cl_2N_2$			
1,3,2,4-Diazadiborilidine, 1,3-bis-(p-chlorophenyl)-2-(1,1-diethylpropyl)-4,5,5-triethyl-	EtOH	266(4.7),272(3.04), 280(2.87)	44-2579-69
$C_{26}H_{38}B_2N_4O_4$			
1,3,2,4-Diazadiborilidine, 2-(1,1-diethylpropyl)-4,5,5-triethyl-1,3-bis(p-nitrophenyl)-	EtOH	216(4.2),283(4.34)	44-2579-69

Compound	Solvent	$\lambda_{max}(\log \epsilon)$	Ref.
$C_{26}H_{38}N_2O$			
1(4H)-Naphthalenone, 4-[(4-cyclohexylbutyl)imino]-2-(hexylamino)-	EtOH	239(4.38),272(4.0), 280s(4.08),338(3.66), 425(3.66)	39-1799-69C
$C_{26}H_{38}N_2O_2$			
1,4-Naphthalenediol, 2,3-bis[(N-methylcyclohexylamino)methyl]-	EtOH	240(4.24),254(4.06), 310s(3.31),330(3.28)	39-1245-69C
$C_{26}H_{38}N_8O_6S_2$			
3-Morpholinone, 6,6'-dithiobis[4-[(4-amino-2-methyl-5-pyrimidinyl)methyl]-6-(2-hydroxyethyl)-5-methyl-	EtOH	239(4.21),278(3.81)	94-0343-69
$C_{26}H_{38}O_3$			
5α-Cholesta-7,22-dien-24-al, 3β-acetoxy-	ether	217(4.37)	24-2629-69
$C_{26}H_{38}O_4$			
5β-Chola-20(21),22-dienoic acid, 14,21-epoxy-3β-hydroxy-, ethyl ester	MeOH	305(4.53)	94-1698-69
$C_{26}H_{38}O_5$			
5β-Chola-20(21),22-dienoic acid, 14β,15β-epoxy-3-hydroxy-21-methoxy-, methyl ester	EtOH	290(4.3)	94-1706-69
5β,14α-Chola-20(21),22-dienoic acid, 3-hydroxy-21-methoxy-15-oxo-, methyl ester	EtOH	290(4.2)	94-1706-69
$C_{26}H_{38}O_6$			
Pregn-4-en-3-one, 11β,17,20α,21-tetrahydroxy-, cyclic 17,20-acetal with acetone, 21-acetate	MeOH	242(4.19)	44-3513-69
20β-	MeOH	242(4.19)	44-3513-69
$C_{26}H_{38}O_8$			
Taxa-4(20),11-diene-5α,7β,9α,10β,13α-pentol, 9α,10β,13α-triacetate	n.s.g.	225(3.80)	77-1282-69
$C_{26}H_{40}B_2N_2$			
1,3,2,4-Diazadiborilidine, 2-(1,1-diethylpropyl)-4,5,5-triethyl-1,3-diphenyl-	EtOH	216(4.2),233(3.7), 265(2.92),272(2.72)	44-2579-69
$C_{26}H_{40}IN_3$			
2a,10-Di-tert-butyl-2,9-bis(dimethylamino)-1-methyl-2a,3,8,8a-tetrahydro-3,8-ethenonaphtho[2,3-b]azetium iodide	MeOH	212(4.49),263s(3.18)	89-0675-69
$C_{26}H_{40}N_2O_2$			
Cumidine, N-tert-butyl-N-[[p-tert-butylhydroxyamino-α,α-dimethylbenzyl]oxy]-	EtOH	250(4.06)	39-1459-69C
$C_{26}H_{40}O$			
B-Norcholesta-1,4-dien-3-one	n.s.g.	244(4.15)	73-0681-69
19-Norcholesta-1,3,5(10)-trien-3-ol	EtOH	210(3.88),221s(3.82), 282(3.23)	39-1240-69C

Compound	Solvent	λ_{max}(log ε)	Ref.
$C_{26}H_{40}O_2$			
4,7-Methanoindene-1,8-dione, 2,3a,5,7-tetra-tert-butyl-3a,4,7,7a-tetrahydro-	isooctane	193(4.12),226(3.80), 324(1.79),337(1.91), 351(1.86)	35-6785-69
	MeOH	203(3.88),231(3.80), 334(1.96)	35-6785-69
21-Norcholesta-5,16-dien-20-one, 3β-acetoxy-	MeOH	240(4.11)	44-3759-69
Pseudoionone, photodimer II	EtOH	283s(2.28)	18-1153-69
Pseudoionone, photodimer III	EtOH	283s(2.30)	18-1153-69
$C_{26}H_{40}O_3$			
Coumarin, 4-heptadecyl-7-hydroxy-	MeOH	323(4.18)	61-0845-69
$C_{26}H_{41}NO$			
Alkaloid C from Buxus Balearica	MeOH	239(4.52)	105-0020-69
$C_{26}H_{41}N_3$			
3,8-Ethenonaphtho[2,3-b]azete, 2a,10-di-tert-butyl-2,9-bis(dimethylamino)-1-methyl-1,2,2a,3,8,8a-hexahydro-	MeOH	264s(3.20)	89-0675-69
$C_{26}H_{42}O_4$			
2,3-Seco-B-norcholest-4-ene-2,3-dioic acid	EtOH	225(4.10)	44-2297-69
$C_{26}H_{44}O_3$			
Hexadecanophenone, 4'-butoxy-2'-hydroxy-	n.s.g.	270(3.95)	104-1609-69
Octadecanophenone, 4'-ethoxy-2'-hydroxy-	n.s.g.	270(4.08)	104-1609-69
$C_{26}H_{48}O_2$			
Pseudoionone, photodimer III reduction product	EtOH	283s(1.89)	18-1153-69
$C_{26}H_{50}O_2$			
Pseudoionone, photodimer II reduction product	EtOH	282(2.68)	18-1153-69

Compound	Solvent	$\lambda_{max}(\log \epsilon)$	Ref.
$C_{27}H_{15}NO_2S$			
Naphtho[2,3-h]quinoline-7,12-dione, 5-(2-naphthylthio)-	EtOH	285(4.46),425(3.84)	104-0742-69
$C_{27}H_{15}N_3O_2$			
2,2,3(1H)-Dibenzofurantricarbonitrile, 1-hydroxy-1,4-diphenyl-	MeOH	329(4.26),369s(4.10), 485(3.62)	44-2407-69
	CH_2Cl_2	251(4.40),295(3.88), 344(3.74)	44-2407-69
2,3,3(4H)-Dibenzofurantricarbonitrile, 4-hydroxy-1,4-diphenyl-	MeOH	255s(4.30),298(4.11), 316s(4.06),372(4.00), 482(4.06)	44-2407-69
	CH_2Cl_2	244(4.20),367(4.30)	44-2407-69
$C_{27}H_{16}N_2O_3$			
Xanthene-2,2(1H)-dicarbonitrile, 1-hydroxy-9-oxo-1,4-diphenyl-	MeOH	309(4.10),453(4.59)	44-2407-69
	CH_2Cl_2	260(4.27),308(3.98)	44-2407-69
$C_{27}H_{16}N_4$			
2H-Cyclopenta[c]pyridazine-3,4-dicarbonitrile, 5,6,7-triphenyl-	$CHCl_3$	262(4.46),308(3.64)	64-0536-69
$C_{27}H_{16}O_4$			
9,9'-Spirobi[fluorene]-2,2'-dicarboxylic acid	EtOH	222(4.75),240s(4.45), 282s(4.59),288(4.61), 301s(4.32),314(4.39)	33-1202-69
$C_{27}H_{17}BF_2O_3$			
Naphtho[1,2-e]-1,3,2-dioxaborin, 2,2-difluoro-4-[(2-phenylbenzopyran-4-ylidene)methyl]-	MeCN	235(4.51),339(4.20), 538(4.74)	4-0029-69
$C_{27}H_{17}N$			
9-Fluorenylidenemethylamine, N-(9-fluorenylidene)-	dioxan	237(4.72),263(4.71), 304(4.06),332(4.18), 439(4.27),453(4.28)	89-0772-69
$C_{27}H_{18}$			
11H-Diindeno[2,1-b:1',2'-h]fluorene, 13,15-dihydro-	C_6H_{12}	256s(4.90),262(4.95), 280s(4.53),311(4.06), 318(4.07),327(4.33), 334(4.35),342(4.65)	33-1023-69
$C_{27}H_{18}N_2$			
1H-Phenanthro[9,10-d]imidazole, 1,2-diphenyl-	EtOH	260(4.83),306(4.20)	77-0200-69
$C_{27}H_{18}N_2O_2$			
Carbazole-1,2-dicarboximide, 9-methyl-N,3-diphenyl-	EtOH	228(4.83),254(4.52), 295(4.75),362(4.02), 425(3.79)	95-0058-69
$C_{27}H_{18}O$			
Ketone, methyl 9,9'-spirobifluoren-2-yl	EtOH	212(4.78),225(4.74), 249s(4.33),273(4.37), 279(4.39),283(4.39), 295(4.41),307(4.41), 326(4.14)	33-1202-69

Compound	Solvent	$\lambda_{max}(\log \epsilon)$	Ref.
$C_{27}H_{18}O_2$			
Dinaphtho[1,2-c:1',2'-c']cyclopenta-[1,2-b:1,5-b']dipyran, 6,7-dihydro-	dioxan	248(5.10),302(4.29), 314(4.32),337(4.11), 349(4.11)	5-0162-69B
Spiro[2H-1-benzopyran-2,2'-[2H]naphtho-[1,2-b]pyran], 3-phenyl-	dioxan	281(4.59),338(4.10)	5-0162-69B
Spiro[2H-1-benzopyran-2,2'-[2H]naphtho-[1,2-b]pyran], 3'-phenyl-	dioxan	258(4.66),266(4.74), 290(4.31),351(3.63)	5-0162-69B
Spiro[2H-1-benzopyran-2,3'-[3H]naphtho-[2,1-b]pyran], 2'-phenyl-	dioxan	245(4.85),289(4.36), 299(4.36),314(4.33), 349(3.87)	5-0162-69B
Spiro[2H-1-benzopyran-2,3'-[3H]naphtho-[2,1-b]pyran], 3-phenyl-	dioxan	243(4.70),257(4.59), 314(4.15),326(4.19), 355(4.18)	5-0162-69B
$C_{27}H_{18}O_5$			
Xylerythrin, 5-O-methyl-	dioxan	246(4.37),357(4.03), 448(4.24)	1-2583-69
$C_{27}H_{19}BrN_2O_2$			
Benzil, mono[N-phenyl-N-(p-bromobenz-oyl)hydrazone]	n.s.g.	253(4.43)	44-0231-69
Benzoic acid, p-bromo-, 1,2-diphenyl-2-(phenylazo)vinyl ester	n.s.g.	251(4.62),348(4.36), 454(2.68)	44-0231-69
$C_{27}H_{19}ClO_2$			
Benzofuran, 2-[p-[2-[5-(p-chlorophenyl)-2-furyl]vinyl]phenyl]-5-methyl-	DMF	317(4.13),395(4.85), 418(4.72)	33-1282-69
$C_{27}H_{19}ClO_7$			
3-[2-(2-Acetoxy-1-naphthyl)vinyl]-naphtho[2,1-b]pyrylium perchlorate	$MeOH-HClO_4$	310(4.23),512(4.47)	5-0226-69E
$C_{27}H_{19}NO_2$			
9-Phenanthrenecarboxanilide, 9,10-di-hydro-10-oxo-9-phenyl-	$CHCl_3$	335(3.50)	22-2756-69
$C_{27}H_{19}N_3O_4$			
Benzil, mono[N-phenyl-N-(p-nitrobenz-oyl)hydrazone]	n.s.g.	259(4.48)	44-0231-69
α-Stilbenol, α'-(phenylazo)-, p-nitro-benzoate	n.s.g.	255(4.49),347(4.30), 450(2.68)	44-0231-69
$C_{27}H_{19}N_3O_6$			
Toluene-α,α-diol, α-[(p-nitrophenyl)-azo]-, dibenzoate	dioxan	278(4.29),283(4.29), 410(2.70)	39-1571-69C
$C_{27}H_{20}N_2$			
Imidazole, 1,2,4,5-tetraphenyl-	EtOH	285(4.20)	77-0200-69
$C_{27}H_{20}N_2OS$			
4-Thiazoline, 2-(1-naphthylimino)-3-methyl-5-phenyl-4-benzoyl-	MeCN	250(4.47),320(3.81), 420(4.86)	23-3557-69
$C_{27}H_{20}N_2O_3$			
1H-Benz[g]indazol-5-ol, 1-acetyl-3,4-diphenyl-, acetate	EtOH	225(4.68),243(4.58), 262(3.72),332(3.81), 348(3.85)	103-0254-69

Compound	Solvent	λ_{max}(log ε)	Ref.
$C_{27}H_{20}N_2O_4$			
Spiro[indene-1,3'-[3H]pyrazole]-4',5'-dicarboxylic acid, 2,3-diphenyl-, dimethyl ester	$CHCl_3$	241(4.46),265(4.36), 342(3.83),365(3.81)	64-0536-69
Toluene-α,α-diol, α-(phenylazo)-, dibenzoate	dioxan	276(4.16),281s(4.15), 401(2.52)	39-1571-69C
$C_{27}H_{20}O$			
Benzofuran, 5-methyl-2-[p-[2-(1-naphthyl)vinyl]phenyl]-	DMF	366(4.73)	33-1282-69
Benzofuran, 5-methyl-2-[p-[2-(2-naphthyl)vinyl]phenyl]-	DMF	301(4.23),367(4.88), 387(4.74)	33-1282-69
Benzofuran, 6-methyl-2-[p-[2-(1-naphthyl)vinyl]phenyl]-	DMF	370(4.72)	33-1282-69
Benzofuran, 6-methyl-2-[p-[2-(2-naphthyl)vinyl]phenyl]-	DMF	302(4.20),369(4.85), 388(4.73)	33-1282-69
Naphthalene, 1-[p-[(p-methoxyphenyl)-ethynyl]styryl]-	DMF	355(4.70)	33-2521-69
Naphthalene, 2-[p-[(p-methoxyphenyl)-ethynyl]styryl]-	DMF	297(4.39),357(4.84)	33-2521-69
$C_{27}H_{20}OS$			
Acrylophenone, 3,3-diphenyl-2-(phenylthio)-	EtOH	254(4.42)	78-3675-69
Naphtho[2,1-b]thiophene, 2-[p-(p-methoxystyryl)phenyl]-	DMF	376(4.81)	33-1282-69
$C_{27}H_{20}O_2$			
Benzofuran, 5-methoxy-2-[p-[2-(1-naphthyl)vinyl]phenyl]-	DMF	369(4.74)	33-1282-69
Benzofuran, 5-methoxy-2-[p-[2-(2-naphthyl)vinyl]phenyl]-	DMF	300(4.22),369(4.86), 388(4.73)	33-1282-69
Naphtho[1,2-b]furan, 2-[p-(p-methoxystyryl)phenyl]-	DMF	305(4.23),372(4.83), 390(4.70)	33-1282-69
Naphtho[2,1-b]furan, 2-[p-(p-methoxystyryl)phenyl]-	DMF	327(4.29),376(4.85), 395(4.72)	33-1282-69
p-Terphenyl, 4-[3,4-(methylenedioxy)-styryl]-	DMF	351(4.72)	33-2521-69
$C_{27}H_{20}O_{12}$			
α-Rubromycin	$CHCl_3$	319(4.32),352(4.06), 365(4.05),415(3.75), 484(3.90)	24-0126-69
β-Rubromycin	$CHCl_3$	316(4.36),350(4.10), 364(4.08),504(3.81)	24-0126-69
$C_{27}H_{20}S$			
Benzo[b]thiophene, 5-methyl-2-[p-[2-(1-naphthyl)vinyl]phenyl]-	DMF	364(4.70)	33-1282-69
Benzo[b]thiophene, 5-methyl-2-[p-[2-(2-naphthyl)vinyl]phenyl]-	DMF	301(4.20),367(4.84)	33-1282-69
Benzo[b]thiophene, 6-methyl-2-[p-[2-(1-naphthyl)vinyl]phenyl]-	DMF	368(4.70)	33-1282-69
$C_{27}H_{21}NOS$			
Benzo[b]thiophen-3(2H)-one, 2-[2-(6-methyl-1-p-tolyl-2(1H)-quinolylidene)ethylidene]-	BuOH	577(4.98),612(4.99)	65-1829-69

Compound	Solvent	$\lambda_{max}(\log \epsilon)$	Ref.
Benzo[b]thiophen-3(2H)-one, 2-[2-(6-methyl-1-p-tolyl-4(1H)-quinolylidene)ethylidene]-	BuOH	610(5.08),656(5.19)	65-1829-69
$C_{27}H_{21}NO_3S$			
Benzo[b]thiophen 3(2H)-one, 2-[2-[6-methoxy-1-(p-methoxyphenyl)-2(1H)-quinolylidene]ethylidene]-	BuOH	586(5.10),618(5.11)	65-1829-69
Benzo[b]thiophen-3(2H)-one, 2-[2-[6-methoxy-1-(p-methoxyphenyl)-4(1H)-quinolylidene]ethylidene]-	BuOH	617(5.07),661(5.22)	65-1829-69
$C_{27}H_{22}BrN_3O$			
1-(3-Benzyl-2-oxo-1,5-diphenyl-4-imidazolin-4-yl)pyridinium bromide	EtOH	257(4.15),398(4.06)	78-3267-69
$C_{27}H_{22}Cl_2N_4O_7S$			
Uracil, 1-(2-deoxy-α-D-erythro-pentofuranosyl)-5-[(imidazol-2-ylthio)-methyl]-, 3',5'-bis(p-chlorobenzoate)	$CHCl_3$	248(4.73),270s(4.35),282s(4.20)	63-0809-69
β-	$CHCl_3$	251(4.64),270s(4.43),282s(4.25)	63-0809-69
$C_{27}H_{22}N_2$			
2-Biphenylamine, N,N'-1-propen-1-yl-3-ylidenebis-, hydrochloride	MeOH	223s(--),229(4.21),265s(--),381(4.25)	59-1075-69
4-Biphenylamine, N,N'-1-propen-1-yl-3-ylidenebis-, hydrochloride	MeOH	233(4.16),244s(--),256(4.28),275s(--),306s(--),407(4.77)	59-1075-69
$C_{27}H_{22}N_2O$			
1-Cyclopropene, 3-ethoxy-3-(2-phenyl-1,1-dicyanoethyl)-1,2-diphenyl-	CH_2Cl_2	224(4.39),231(4.38),289(4.35),303(4.50),319(4.41)	24-0319-69
4(3H)-Quinazolinone, 3-methyl-2-[β-naphthyl)phenethyl]-	EtOH	223(4.94),283(4.02)	115-0057-69
$C_{27}H_{22}N_2OS_3$			
2,4-Thiazolidinedithione, 3-ethyl-5-[2-[4-(p-methoxyphenyl)benzo[f]quinolin-3(4H)-ylidene]ethylidene]-	BuOH	237(4.59),283(4.34)	65-2116-69
$C_{27}H_{22}N_2O_6$			
4,8-Ethenobenzo[1,2-c:4,5-c']dipyrrole-4(1H)-carboxylic acid, 2,3,3a,4a,5,-6,7,7a,8,8a-decahydro-1,3,5,7-tetraoxo-2,6-diphenyl-, ethyl ester	EtOH	217(4.25)	78-4315-69
$C_{27}H_{22}N_6$			
Aniline, N,N'-1-propen-1-yl-3-ylidenebis[4-(phenylazo)-, hydrochloride	MeOH	249(4.24),306s(--),324s(--),435(4.78)	59-1075-69
$C_{27}H_{22}O$			
Anisole, p-(p-4-biphenylylstyryl)-	DMF	347(4.75)	33-2521-69
Anisole, p-[p-[2-(1-naphthyl)vinyl]-styryl]-	DMF	373(4.76)	33-2521-69
Anisole, p-[p-[2-(2-naphthyl)vinyl]-styryl]-	DMF	373(4.86)	33-2521-69

Compound	Solvent	λ max (log ε)	Ref.
$C_{27}H_{23}Cl_2NO_5$			
[4-[1-Chloro-2-(2,5-diphenyl-4H-pyran-4-ylidene)ethylidene]-2,5-cyclohexadien-1-ylidene]dimethylammonium perchlorate	MeCN	236(4.19),272(4.22), 406(4.30),635(4.80)	4-0803-69
$C_{27}H_{23}N$			
Acridan, 9-(2,2-diphenylethyl)-	hexane	276(4.13)	78-1125-69
$C_{27}H_{23}NO_2$			
Acetophenone, 4'-(dimethylamino)-2-(2,6-diphenyl-4H-pyran-4-ylidene)-	MeCN	248(4.32),300(4.15), 420(4.06)	4-0803-69
$C_{27}H_{23}N_5O$			
Semicarbazide, 2-methyl-4,4-diphenyl-1-[α-(phenylazo)benzylidene]-	dioxan	250(4.34),364(4.1)	24-3082-69
$C_{27}H_{24}ClNO_5$			
[α-[(2,6-Diphenyl-4H-pyran-4-ylidene)-methyl]benzylidene]dimethylammonium perchlorate	MeCN	255(4.16),340(4.16), 432(4.60),453(4.63)	44-2736-69
[α-[(4,6-Diphenyl-2H-pyran-2-ylidene)-methyl]benzylidene]dimethylammonium perchlorate	MeCN	245(4.27),318(4.42), 500(4.21)	44-2736-69
$C_{27}H_{24}Cl_2N_4$			
1,4-Cyclohexadiene-1,4-diamine, 3,6-bis-[(p-chlorophenyl)imino]-N-mesityl-	EtOH	272(4.25),352s(4.07), 402(4.18)	78-4153-69
$C_{27}H_{24}Fe_2O_3$			
2-Propen-1-one, 1,3-bis(1'-acetylferrocenyl)-	EtOH	263(4.23),325(4.22)	73-2235-69
$C_{27}H_{24}N_8O_4$			
Isoalloxazine, 10,10'-trimethylenebis-[7,8-dimethyl-	propylene glycol	268(4.75),355(4.18), 443(4.31)	44-3240-69
$C_{27}H_{24}O_4$			
2,5-Cyclohexadien-1-one, 4-methoxy-2,6-bis(p-methoxyphenyl)-4-phenyl-	n.s.g.	229(4.50),315(3.61)	24-3795-69
$C_{27}H_{24}O_9$			
Flavone, 3'-(benzyloxy)-5,7-dihydroxy-3,4',6-trimethoxy-, 7-acetate	EtOH	257(4.64),274s(4.50), 352(4.60)	18-2701-69
	EtOH-$AlCl_3$	259(4.56),360(4.57)	18-2701-69
$C_{27}H_{25}BrN_4O_4S$			
Acetophenone, 2-[3-[(4-amino-2-methyl-5-pyrimidinyl)methyl]-5-(2-hydroxy-ethyl)-4-methyl-4-thiazolin-2-ylidene]-4'-bromo-2-hydroxy-, 2-benzoate	EtOH	232(4.45),270(4.01), 398(4.38)	94-0128-69
$C_{27}H_{25}Cl_5O_2$			
2,4-Cyclohexadien-1-one, 4-benzoyl-2,6-di-tert-butyl-6-(pentachlorophenoxy)-	C_6H_{12}	305(3.71)	64-0547-69
$C_{27}H_{25}IN_2OS$			
1-Ethyl-2-[6-(N-phenylacetamido)-1,3,5-hexatrienyl]naphtho[1,2-d]thiazolium iodide	MeOH	470(4.68)	22-1284-69

Compound	Solvent	$\lambda_{max}(\log \epsilon)$	Ref.
$C_{27}H_{25}NO_7$			
9H-Phenanthro[9,10-b]quinolizin-9-one, 11,12,13,14,14a,15-hexahydro-2,3,6-trihydroxy-, triacetate	EtOH	243s(4.67),255(4.68), 285s(4.37),320(4.07)	39-1309-69C
$C_{27}H_{26}BrNO_6$			
α-D-xylo-Tetrofuranose, 3-O-benzyl-4-C-(5-bromo-4-carboxy-2-quinolyl)-1,2-O-cyclohexylidene-	n.s.g.	252(3.98),304(3.59)	65-1413-69
$C_{27}H_{26}ClNO_5$			
Diethyl[4-(3-phenyl-1H-naphtho[2,1-b]-pyran-1-ylidene)-2-butenylidene]-ammonium perchlorate	MeCN	244(4.52),268(4.37), 301(3.89),367(4.11), 518(4.70)	44-2736-69
$C_{27}H_{26}Cl_2N_2O_7S$			
Uridine, 5-(butylthio)-2'-deoxy-, 3',5'-bis(p-chlorobenzoate)	EtOH	242(4.60)	44-3806-69
$C_{27}H_{26}F_{14}O_4$			
Androsta-3,5-diene-3,17β-diol, bis-(heptafluorobutyrate)	heptane	228(4.21),231(4.20)	13-0739-69A
$C_{27}H_{26}N_4O_4S$			
Acetophenone, 2-[3-[(4-amino-2-methyl-5-pyrimidinyl)methyl]-5-(2-hydroxy-ethyl)-4-methyl-4-thiazolin-2-ylidene]-2-hydroxy-, 2-benzoate	EtOH	232(4.47),273(3.92), 394(4.40)	94-0128-69
$C_{27}H_{26}O_2$			
4,7-Methanocyclobuta[b]naphthalene-3,8-dione, 2a,3a,4,5,6,7,7a,8a-octahydro-2a,8a-dimethyl-1,2-diphenyl-	EtOH	222(4.30),295(4.07)	39-0105-69C
Phenanthrene, 7-isopropyl-1-(3,4-dimethoxystyryl)-	DMF	340(4.48)	33-2521-69
Spiro[2H-1-benzopyran-2,2'-[2H]naphtho-[1,2-b]pyran], 6-methyl-3-pentyl-	dioxan	258(4.70),267(4.78), 310(3.93),320(3.93), 334(3.76),350(3.60)	5-0162-69B
Spiro[2H-1-benzopyran-2,3'-[3H]naphtho-[2,1-b]pyran], 6-methyl-3-pentyl-	dioxan	245(4.90),301(4.06), 314(4.11),336(3.81), 350(3.89)	5-0162-69B
Spiro[2H-1-benzopyran-2,3'-[3H]naphtho-[2,1-b]pyran], 7-methyl-3-pentyl-	dioxan	245(4.87),300(4.08), 313(4.09),335(3.79), 349(3.86)	5-0162-69B
$C_{27}H_{26}O_3$			
Spiro[2H-1-benzopyran-2,2'-[2H]naphtho-[1,2-b]pyran], 6-methoxy-3-pentyl-	dioxan	258(4.69),267(4.79), 320(3.96),333(3.90), 350(3.64)	5-0162-69B
$C_{27}H_{26}O_9$			
Phenanthrenequinone, 1,4,8-trihydroxy-2-methyl-3-(4-methylvaleryl)-, triacetate	EtOH	257(4.50),349(3.75)	22-3100-69
$C_{27}H_{26}O_{11}$			
2-Anthraceneacetic acid, 4,5,6,8-tetrahydroxy-α-methoxy-, ethyl ester, tetraacetate	EtOH	227(4.26),263(5.20), 322(3.50),340(3.60), 356(3.72),376(3.70), 396(3.67)	105-0494-69

Compound	Solvent	λ_{max} (log ε)	Ref.
$C_{27}H_{27}Br_2OP$			
Acetophenone, 4-[(2,7-dibromofluoren-9-ylidene)dipropylphosphoranyl]-	$CHCl_3$	258(4.72),297(4.63), 394(3.38)	5-0001-69D
$C_{27}H_{27}ClN_2O_2$			
1(2H)-Quinolinecarboxamide, N-allyl-6-chloro-3,4-dihydro-3-(p-methoxyphenyl)-2-methyl-4-phenyl-	n.s.g.	258(4.22)	40-1047-69
$C_{27}H_{27}ClN_2O_6$			
Dimethyl[1-morpholino-2-(2-phenyl-4H-naphtho[1,2-b]pyran-4-ylidene)-ethylidene]ammonium perchlorate	MeCN	277(4.33),328(4.15), 385(4.19)	4-0803-69
$C_{27}H_{27}ClO_6$			
3-(2-Hydroxy-4-methyl-α-pentylstyryl)-naphtho[2,1-b]pyrylium perchlorate	MeOH-$HClO_4$	235(4.49),298(4.00), 306(4.04),340(4.07), 549(4.65)	5-0162-69B
$C_{27}H_{27}ClO_7$			
3-(2-Hydroxy-4-methoxy-α-pentylstyryl)-naphtho[2,1-b]pyrylium perchlorate	MeOH-$HClO_4$	307(3.85),346(4.03), 575(4.71)	5-0162-69B
$C_{27}H_{27}NO_6$			
α-D-xylo-Tetrofuranose, 3-O-benzyl-4-C-(4-carboxy-2-quinolyl)-1,2-O-cyclohexylidene-	n.s.g.	235(4.38),308(3.58)	65-1413-69
$C_{27}H_{28}N_2O_2$			
1H-Naphth[2,3-d]imidazole-4,9-dione, 1-benzyl-2-(3-cyclohexylpropyl)-	MeOH	248(4.67),280(4.18), 332(3.52)	4-0909-69
$C_{27}H_{28}N_2O_{13}$			
3a,9a-Butano-3H-benz[f]indazole-4,9-dione, 8,10,11,12,13-pentahydroxy-6-methoxy-12-methyl-, pentaacetate	EtOH	250(4.38),275s(4.02), 322(3.59)	23-0767-69
$C_{27}H_{28}N_3O$			
α-Antipyrinyl-p-(dimethylamino)-α-m-tolylbenzylium (picrate)	pH 4	540(4.30)	122-0138-69
α-Antipyrinyl-p-(dimethylamino)-α-p-tolylbenzylium (picrate)	pH 4	540(4.30)	122-0138-69
$C_{27}H_{28}N_3O_2$			
α-Antipyrinyl-p-(dimethylamino)-α-(m-methoxyphenyl)benzylium (picrate)	pH 4	540(4.33)	122-0138-69
α-Antipyrinyl-p-(dimethylamino)-α-(p-methoxyphenyl)benzylium (picrate)	pH 4	545(4.57)	122-0138-69
$C_{27}H_{28}O_2$			
Stilbene, 4'-(p-isopropylstyryl)-3,4-dimethoxy-	DMF	369(4.81)	33-2521-69
$C_{27}H_{28}O_5$			
Stilbene, 4'-(3,4-dimethoxystyryl)-3,4,5-trimethoxy-	DMF	374(4.83)	33-2521-69
Warangalone, dimethyl ether	EtOH	225(4.41),276(4.72), 325(3.94)	39-0374-69C

Compound	Solvent	λ_{max}(log ε)	Ref.
$C_{27}H_{28}O_6$			
Lonchocarpenin	EtOH	234(4.37),261(4.49), 348(4.21)	39-0374-69C
$C_{27}H_{28}O_{10}$			
Isosequirin, pentaacetate	MeOH	220(4.48),272(3.59)	39-1921-69C
$C_{27}H_{29}NO_3$			
Estra-1,3,5(10)-triene-3,17β-diol, 17α-(3-phenyl-5-isoxazolyl)-	MeOH	235(4.23),270(3.33), 277(3.39)	24-3324-69
$C_{27}H_{30}Br_2NP$			
Aniline, p-[(2,7-dibromofluoren-9-ylidene)dipropylphosphoranyl]-N,N-dimethyl-	$CHCl_3$	260(4.82),270(5.04), 297(4.37),401(3.17)	5-0001-69D
$C_{27}H_{30}O_4$			
Estra-1,3,5(10)-trien-6-one, 17β-hydroxy-1-methoxy-4-methyl-, benzoate	EtOH	220s(4.28),231(4.40), 258(3.80),327(3.41)	39-1234-69C
$C_{27}H_{30}O_6$			
Crotophorbolone, 20-benzoate	MeOH	228(4.32)	5-0158-69H
Phorbobutanone, 20-monobenzoate	MeOH	226(4.32),278s(3.04), 302s(1.86),332s(1.79)	64-0099-69
Spiro[cyclohexane-1,3'-[1H,3H]pyrano-[4,3-b][1]benzopyran]-1'-one, 4',10'-dihydro-10'-(4-hydroxy-2-oxo-1-oxaspiro[5.5]undec-3-en-3-yl)-	MeOH	250(3.79),286(3.40)	83-0075-69
$C_{27}H_{30}O_{14}$			
Isorhoifolin	MeOH	268(4.59),338(4.66)	24-2083-69
$C_{27}H_{30}O_{15}$			
Aloe-emodin, 1,8-di-β-glucoside	60% EtOH	222(4.46),260(4.57), 388(3.92)	88-3751-69
Apigenin, 6,8-di-β-D-glucopyranosyl-	EtOH	273(4.19),337(4.20)	28-0980-69A
Luteolin, 7β-neohesperidoside	MeOH	254(4.30),265s(4.26), 349(4.34)	24-3009-69
$C_{27}H_{31}ClF_4O_6$			
3'H-Cyclopropa[6,7]pregna-4,6-diene-3,20-dione, 3',3'-difluoro-6α,7α-dihydro-11β,21-dihydroxy-16α-methyl-, 21-acetate chlorodifluoroacetate	EtOH	244(4.14)	78-1219-69
6β,7β-	EtOH	252(4.19)	78-1219-69
$C_{27}H_{31}Fe_2N$			
Piperidine, 1-[[2-(2-methylferrocenyl)-ferrocenyl]methyl]-	C_6H_{12}	197(4.65),222(4.71), 255s(4.05),296(3.88)	78-3477-69
$C_{27}H_{31}NO_3$			
Estr-4-en-3-one, 17β-hydroxy-17α-(3-phenyl-5-isoxazolyl)-	MeOH	219(4.16),241(4.51)	24-3324-69
$C_{27}H_{31}N_3O_5$			
Tryptophan, N-(n-carboxy-L-alanyl)-2-(1,1-dimethylallyl)-, N-benzyl ester	EtOH	223(4.51),282(3.90), 291(3.85)	39-1003-69C

Compound	Solvent	λ_{max}(log ϵ)	Ref.
$C_{27}H_{31}N_3O_7$			
Malonic acid, [2-(carboxyamino)propionamido](indol-3-ylmethyl)-, N-benzyl diethyl ester, DL-	EtOH	218(4.56),282(3.82), 290(3.74)	39-1003-69C
$C_{27}H_{32}N_2O_2$			
Carbamic acid, [2-(3-cyclopenten-1-yl)-butyl]-(2-indol-3-ylethyl)-, benzyl ester	n.s.g.	222(4.45),275(3.60), 282(3.69),291(3.62)	35-7333-69
$C_{27}H_{32}N_2O_5$			
9,18-Methano-6H,8H-[1,5]diazocino[2,1-a:4,5-a']diisoquinolin-19-one,	EtOH-HCl	238(4.17),289(3.82), 314(3.77),370(3.80)	39-0085-69C
5,9,10,12,13,17b,18,18c-octahydro-2,3,15,16-tetramethoxy-	EtOH-NaOH	222(4.29),284(3.87), 288s(3.85)	39-0085-69C
$C_{27}H_{32}O_3$			
Estra-1,3,5(10)-trien-17β-ol, 1-methoxy-4-methyl-, benzoate	EtOH	218(4.20),230(4.36), 284(3.43)	39-1234-69C
$C_{27}H_{32}O_6$			
Benzo[4,5,6]pregn-4-ene-4'α,5'α-dicarb-	EtOH	251(4.04)	94-2586-69
oxylic anhydride, 6α,6'α-epoxy-3',4',5',6'-tetrahydro-6'-methyl-3,20-dioxo-	EtOH	251(4.04)	94-2604-69
$C_{27}H_{32}O_9$			
Atalantin	n.s.g.	210(4.20)	2-0870-69
$C_{27}H_{32}O_{10}$			
9(1H)-Phenanthrone, 2,3,4,4a-tetrahydro-3,5,6,8,10-pentahydroxy-7-isopropyl-1,4a-dimethyl-, 3,5,6,10-tetraacetate	EtOH	229(4.08),256(4.11), 290(3.89),360(3.76)	33-1685-69
$C_{27}H_{32}O_{15}$			
Rubrofusarin, gentiobioside	EtOH	223(4.34),255s(4.32), 277(4.61),323(3.31), 399(3.73)	94-0458-69
$C_{27}H_{33}NO_3$			
Estr-4-ene-3β,17β-diol, 17α-(3-phenyl-5-isoxazolyl)-	MeOH	241(4.18),275(3.00)	24-3324-69
$C_{27}H_{33}N_3O_7$			
3H-Pyrrolo[3,4-b]quinolin-3-one, 1,2-dihydro-2-[2-(hydroxymethyl)-4,5-dioxo-5-(1-pyrrolidinyl)valeryl]-, acetate, 4-(diethyl acetal)	EtOH	246(4.61),303(3.95), 319s(3.77)	44-3853-69
$C_{27}H_{33}OP$			
Phosphine oxide, trimesityl-	C_6H_{12}	279(3.48),288(3.50)	65-1544-69
$C_{27}H_{33}P$			
Phosphine, trimesityl-	C_6H_{12}	313(4.20)	65-1544-69
$C_{27}H_{34}N_2O_4$			
Carbamic acid, [2-(3,4-dihydroxycyclopentyl)butyl](2-indol-3-ylethyl)-, benzyl ester, compound with 1,3,5-trinitrobenzene	n.s.g.	220(4.51),274(3.69), 281(3.72),290(3.65)	35-7333-69

Compound	Solvent	$\lambda_{max}(\log \epsilon)$	Ref.
$C_{27}H_{34}N_2O_9$			
Isovincoside, hydrochloride	EtOH	227(4.48),273(3.86), 281(3.85),289(3.80)	39-1193-69C
Vincoside, hydrochloride	EtOH	228(4.49),272(3.87), 281(3.85),290(3.79)	39-1193-69C
$C_{27}H_{34}O_5$			
14,21-Cyclo-14β,17α-pregna-3,15,20-triene-21-carboxylic acid, 17-acetyl-3β-hydroxy-, methyl ester, acetate	MeOH	234(3.60)	78-4579-69
$C_{27}H_{34}O_7$			
Pregna-3,5-diene-3-acetic acid, α-(2-hydroxyethyl)-17,20:20,21-bis(methylenedioxy)-11-oxo-, γ-lactone	EtOH	233s(4.33),240s(4.34)	44-3796-69
Pregn-4-ene-$\Delta^{3,\alpha}$-acetic acid, α-(2-hydroxyethyl)-17,20:20,21-bis-(methylenedioxy)-11-oxo-, γ-lactone	EtOH	292(4.43)	44-3796-69
$C_{27}H_{35}NO_4S$			
Podocarpa-8,11,13-trien-15-ylamine, 13-isopropenyl-7-oxo-, p-toluenesulfonic acid derivative	MeOH	224(4.38),229(4.38), 239(4.37)	5-0193-69H
$C_{27}H_{35}NO_6$			
14β-Pregn-5-en-20-one, 3β,8,12β,14-tetrahydroxy-, 12-nicotinate	EtOH	264(3.60)	94-2391-69
$C_{27}H_{35}NO_8$			
Pyruvamide, N-(7,13-dihydroxy-1,4,10,19-tetramethyl-17,18-dioxo-16-oxabicyclo-[13.2.2]nonadeca-3,5,9,11-tetraen-2-yl)-, 13-acetate	EtOH	227(4.67)	88-2239-69
$C_{27}H_{35}NO_{12}$			
Isoipecoside	n.s.g.	234(4.14),285(3.59)	39-1187-69C
$C_{27}H_{35}N_2O_5$			
5,11b-Didehydro-1'-oxo-1',8'a-seco-12,13-dinoremetinium	EtOH-HCl	244(4.24),288(3.85), 303(3.96),355(4.00)	39-0101-69C
$C_{27}H_{36}N_2O$			
Morpholine, 4-[6-(α-anilinobenzyl)-4-tert-butyl-1-cyclohexen-1-yl]-	MeOH	249(4.17),299(3.35)	88-3549-69
$C_{27}H_{36}N_2O_9$			
Isovincoside, dihydro-, hydrochloride	EtOH	227(4.44),273(3.90), 281(3.90),290(3.81)	39-1193-69C
Vincoside, dihydro-, hydrochloride	EtOH	227(4.45),272(3.86), 281(3.85),290(3.78)	39-1193-69C
$C_{27}H_{36}N_4O_6$			
Podocarp-8-en-15-oic acid, 13β-isopropyl-7-oxo-, methyl ester, 7-(2,4-dinitrophenylhydrazone)	EtOH	248(4.02)	44-3464-69
$C_{27}H_{36}O_2$			
Naphtho[1',2':16,17]androsta-4,16-diene-3,3'(2'H)-dione, 5',6',7',8',8'a,16-hexahydro-	MeOH	240(4.46)	44-1606-69

Compound	Solvent	$\lambda_{max}(\log \epsilon)$	Ref.
$C_{27}H_{36}O_2S$			
Androst-4-en-3-one, 7α-(benzylthio)-17β-hydroxy-17α-methyl-	EtOH	243(4.25)	94-0011-69
$C_{27}H_{36}O_4$			
10'-Apofucoxanthinal	$CHCl_3$	405(4.80),426(5.00), 453(4.99)	39-0429-69C
$C_{27}H_{36}O_5$			
Surangin A	EtOH	222(4.44),296(4.32), 325s(4.21)	78-1453-69
	EtOH-HCl	222(4.46),295(4.35), 325(4.21)	78-1453-69
	EtOH-KOH	225(4.25),257(4.01), 332(4.45)	78-1453-69
$C_{27}H_{36}O_7$			
5α-Pregna-14,16-dien-20-one, 3β,11β,12β-trihydroxy-, triacetate	EtOH	301(4.13)	94-0324-69
$C_{27}H_{36}O_8$			
5α-Pregn-16-en-20-one, 3β,11β,12β-triacetoxy-14β,15β-epoxy-	EtOH	242(3.71)	94-0324-69
$C_{27}H_{36}O_9$			
5α-Pregn-16-ene-15,20-dione, 3β,11β,12β,14β-tetrahydroxy-, 3,11,12-triacetate	EtOH	247(3.93)	94-0324-69
$C_{27}H_{36}O_{10}$			
β-D-Glucopyranosiduronic acid, 3,11,20-trioxopregn-4-en-21-yl	MeOH	238(4.19)	69-1188-69
$C_{27}H_{36}O_{11}$			
β-D-Glucopyranosiduronic acid, 17-hydroxy-3,11,20-trioxopregn-4-en-21-yl, barium salt	MeOH	238(4.20)	69-1188-69
$C_{27}H_{36}O_{15}$			
5-Epiloganin, pentaacetate	EtOH	234(4.05)	39-0721-69C
Loganin, pentaacetate	EtOH	232(4.01)	39-0721-69C
$C_{27}H_{37}ClO_6$			
Pregn-4-ene-3,11-dione, 4-chloro-17,20,20,21-tetrahydroxy-, cyclic 17,20:20,21-diacetal with acetone	EtOH	254(4.21)	44-1455-69
$C_{27}H_{37}NO_8$			
Lactamide, N-(7,13-dihydroxy-1,4,10,19-tetramethyl-17,18-dioxo-16-oxabicyclo-[13.2.2]nonadeca-3,5,9,11-tetraen-2-yl)-, 13-acetate (T-2636D)	EtOH	229(4.68)	88-2239-69
$C_{27}H_{37}NO_{10}$			
β-D-Glucopyranosiduronamide, 17-hydroxy-3,11,20-trioxopregn-4-en-21-yl	MeOH	238(4.20)	69-1188-69
$C_{27}H_{37}N_3O_3$			
Androsta-3,5-diene-3-carboxaldehyde, 17β-hydroxy-6-methyl-17α-1-propynyl-, semicarbazone, acetate	EtOH	294(4.46),306(4.60), 317(4.54)	78-4535-69

Compound	Solvent	$\lambda_{max}(\log \epsilon)$	Ref.
$C_{27}H_{38}O_2$			
Albolic acid	hexane	211(4.30)	88-2929-69
Naphth[1',2':16,17]androsta-5,16-dien-3'(2'αH)-one, 5',6',7',8',8'a,16-hexahydro-3β-hydroxy-	MeOH	239(4.15)	44-1606-69
$C_{27}H_{38}O_3$			
Benzaldehyde, 2,4-dihydroxy-3,5-bis-(3,7-dimethyl-2,6-octadienyl)-	MeOH	234(4.10),290(4.11), 332(3.67)	32-0308-69
$C_{27}H_{38}O_4$			
8,12-Etheno-8βH-cyclopenta[a]phenanthrene-4-carboxylic acid, 1,2,3,4,5,6-7,9,10,11,12,13,14,15-tetradecahydro-1'-isopropyl-17-methoxy-4β,10β-dimethyl-15-oxo-, methyl ester	n.s.g.	238(4.05),295(1.05)	44-4016-69
8,12-Etheno-8βH-cyclopenta[a]phenanthrene-4-carboxylic acid, 1,2,3,4,5,6-7,9,10,11,12,13,14,17-tetradecahydro-1'-isopropyl-15-methoxy-4β,10β-dimethyl-17-oxo-, methyl ester	n.s.g.	242(3.99)	44-4016-69
1H-2,3,10a-Methenocyclobut[1,7]indeno-[5,4-a]naphthalene-8-carboxylic acid, hexadecahydro-3-isopropyl-2a-methoxy-4b,8-dimethyl-1-oxo-, methyl ester	n.s.g.	247(4.10),290(2.23)	44-4016-69
19-Norcholesta-1,3,5(10)-triene-6,11-dione, 9α-hydroxy-3-methoxy-	EtOH	211(4.01),228(3.95), 283(4.09)	39-1240-69C
19-Norpregn-4-ene-3,20-dione, 17-hydroxy-16-methylene-, hexanoate	EtOH	239(4.21)	24-0643-69
$C_{27}H_{38}O_5$			
5β-Chola-20(21),22-dien-24-oic acid, 3β-acetoxy-14,21-epoxy-, methyl ester	EtOH	304(4.40)	94-1698-69
$C_{27}H_{38}O_6$			
5β-Chola-20(21),22-dien-24-oic acid, 16β-acetoxy-14,21-epoxy-3β-hydroxy-, methyl ester	EtOH	300(4.38)	94-1698-69
Pregn-4-ene-3,11-dione, 17,20,20,21-tetrahydroxy-, cyclic 17,20:20,21-diacetal with acetone	EtOH	241(4.22)	44-1455-69
$C_{27}H_{38}O_7$			
Atis-15-ene-17,18-dioic acid, 15-carboxy-13,14-epoxy-13-isopropyl-, trimethyl ester	EtOH	237(3.66)	44-1257-69
Atis-15-ene-17,18-dioic acid, 15-carboxy-13-isopropyl-14-oxo-, trimethyl ester	EtOH	239(4.18),312(2.72)	44-1257-69
14α-Digitoxigenin, 15α-acetoxy-	EtOH	218(4.20)	94-0515-69
$C_{27}H_{38}O_9$			
β-D-Glucopyranosiduronic acid, 3,20-dioxopregn-4-en-21-yl	MeOH	240(4.21)	69-1188-69
5α-Pregn-16-en-20-one, 3β,11β,12β,14β,-15α-pentahydroxy-, 3,11,12-triacetate	EtOH	230(3.90)	94-0324-69
$C_{27}H_{38}O_{10}$			
β-D-Glucopyranosiduronic acid, 11β-hydroxy-3,20-dioxopregn-4-en-21-yl	MeOH	241(4.20)	69-1188-69

Compound	Solvent	λ_{max}(log ε)	Ref.
β-D-Glucopyranosiduronic acid, 17-hydroxy-3,20-dioxopregn-4-en-21-yl	MeOH	241(4.23)	69-1188-69
$C_{27}H_{38}O_{11}$			
β-D-Glucopyranosiduronic acid, 11β,17-dihydroxy-3,20-dioxopregn-4-en-21-yl	MeOH	242(4.21)	69-1188-69
barium salt	H_2O	248(4.20)	69-1188-69
$C_{27}H_{39}ClO_6$			
Pregn-4-en-3-one, 4-chloro-11β,17,20,-20,21-pentahydroxy-, cyclic 17,20:-20,21-diacetal with acetone	EtOH	254(4.18)	44-1455-69
$C_{27}H_{39}NO_2$			
Korsininedione	n.s.g.	252(2.6),303(2.19)	105-0054-69
$C_{27}H_{39}NO_5$			
Podocarpa-8,11,13-triene-15-carbamic acid, 13-(1-hydroxy-1-methylethyl)-7-oxo-, tert-butyl ester, acetate	MeOH	251(4.04)	5-0193-69H
$C_{27}H_{39}NO_{11}$			
β-D-Glucopyranosiduronic acid, 17-hydroxy-3,11,20-trioxopregn-4-en-21-yl, ammonium salt	MeOH	238(4.19)	69-1188-69
$C_{27}H_{40}N_4O_3S$			
2-Pentene-3-sulfenic acid, 2-[N-[(4-amino-2-methyl-5-pyrimidinyl)methyl]formamido]-5-hydroxy-, 2,6-di-tert-butyl-p-tolyl ester	EtOH	235(4.22),277(4.03)	18-1942-69
$C_{27}H_{40}O$			
19-Norcholesta-1,3,5(10)-trien-6-one, 4-methyl-	C_6H_{12}	216(4.13),250(3.98), 300(3.24)	39-1240-69C
$C_{27}H_{40}O_2$			
19-Norcholesta-1,3,5(10)-trien-6-one, 1-hydroxy-4-methyl-	EtOH	228(4.23),259(3.93), 330(3.60)	39-1240-69C
19-Norcholesta-1,3,5(10)-trien-6-one, 3-methoxy-	C_6H_{12}	220(4.38),250(3.96), 257s(3.90),320(3.49)	39-1240-69C
$C_{27}H_{40}O_3$			
19-Norcholesta-1,3,5(10)-trien-11-one, 9β-hydroxy-3-methoxy-	EtOH	218(4.00),277(3.31), 283(3.29)	39-1240-69C
$C_{27}H_{40}O_5$			
Chola-7,9(11)-dien-24-oic acid, 3α,12α-dihydroxy-, methyl ester, 3-acetate	EtOH	238s(4.14),247(4.21), 255s(4.09)	39-2723-69C
3α,12β-	EtOH	241s(4.19),247(4.23), 256s(4.07)	39-2723-69C
$C_{27}H_{40}O_6$			
Pregn-4-en-3-one, 11β,17,20,20,21-pentahydroxy-, cyclic 17,20:20,21-diacetal with acetone	EtOH	241(4.20)	44-1455-69
$C_{27}H_{41}N$			
Solanthrene	EtOH	228(4.27),235(4.29), 243(4.09)	94-2370-69

Compound	Solvent	λ_{max}(log ε)	Ref.
$C_{27}H_{41}NO_3$			
Buxene	MeOH	243(3.94)	88-4423-69
$C_{27}H_{41}NO_4$			
7a-Aza-B-homo-22β-spirost-5-en-7-one, 3β-hydroxy-	EtOH	222(4.20)	2-1084-69
$C_{27}H_{41}NO_6$			
Acetamide, N-(3α,20β-dihydroxy-11-oxo-5β-pregnan-12α-yl)-, diacetate	EtOH	312(2.17)	23-3489-69
$C_{27}H_{41}NO_{11}$			
β-D-Glucopyranosiduronic acid, 11β,17-dihydroxy-3,20-dioxopregn-4-en-21-yl, ammonium salt	MeOH	242(4.20)	69-1188-69
$C_{27}H_{42}$			
3α,5α-Cyclocholesta-6,8(14)-diene	hexane	262(4.39)	39-2098-69C
$C_{27}H_{42}O$			
Cholesta-5,16,20(22)-trien-3β-ol	isooctane	243(4.22)	44-3767-69
19-Norcholesta-1,3,5(10)-triene, 3-methoxy-	C_6H_{12}	210(4.05),222(3.94), 280(3.94),288(3.91)	39-1234-69C
19-Norcholesta-1,3,5(10)-trien-1-ol, 4-methyl-	EtOH	226(3.72),284(3.38)	44-1601-69
19-Norcholesta-1,3,5(10)-trien-2-ol, 4-methyl-	EtOH	223(3.79),282(3.36)	44-1610-69
19-Norcholesta-1,3,5(10)-trien-4-ol, 2-methyl-	EtOH	227(3.73),277(3.23), 283(3.26)	44-1601-69
$C_{27}H_{42}O_2$			
Cholesta-1,4-dien-3-one, 2-hydroxy-	EtOH	204(4.16),254(4.18)	33-0459-69
	NaOH	231(4.35),350(3.38)	33-0459-69
Cholesta-5,17(20)-dien-22-one, 3β-hydroxy-, cis	MeOH	253(3.89)	44-3767-69
5β-Cholest-1-en-3-one, 4β,5-epoxy-	EtOH	231(3.89)	33-0459-69
B-Norcholest-4-en-3-one, 2-(hydroxymethylene)-	EtOH	250(4.05),307(3.83)	44-2297-69
$C_{27}H_{42}O_6$			
Podecdysone B	EtOH	244(4.12)	77-0402-69
$C_{27}H_{43}NO$			
Cyclopenta[5,6]naphth[1,2-d]azepin-2(3H)-one, 8β-(1,5-dimethylhexyl)-4,5,5a,5b,6,7,7a,8,9,10,10a,10b-tetradecahydro-5a,7a-dimethyl-	EtOH	269(4.26)	12-0271-69
$C_{27}H_{43}NO_2$			
Solacongestidine, 23-oxo-	EtOH	267(2.52),277(2.45), 405(1.83)	44-1577-69
Solacongestidine, 24-oxo-	EtOH	270(2.17),345(1.57)	44-1577-69
$C_{27}H_{43}N_3O$			
Cholest-4-en-3-one, 6α-azido-	MeOH	236(4.08)	48-0445-69
$C_{27}H_{44}F_2$			
5α-Cholest-9-ene, 3β,6β-difluoro-	C_6H_{12}	199(4.08)	13-0051-69A

Compound	Solvent	$\lambda_{max}(\log \epsilon)$	Ref.
$C_{27}H_{44}N_2O$			
5α-Cholestan-3-one, 2-diazo-	n.s.g.	258(3.91)	22-3166-69
$C_{27}H_{44}O$			
Cholesta-2,4-dien-6β-ol	C_6H_{12}	265(3.81)	78-3925-69
Cholest-4-en-3-one	EtOH	242(4.23)	94-1255-69
5α-Cholest-1-en-3-one	EtOH	231(4.00)	94-1255-69
	n.s.g.	229(3.95)	22-3166-69
A-Nor-5α-cholestan-2-one, 1-methylene-	C_6H_{12}	225(4.22)	22-3166-69
1(10)-Seco-5α-cholesta-1,10(19)-dien-3-one	C_6H_{12}	215(3.97)	22-3166-69
$C_{27}H_{44}O_2$			
Cholecalciferol, 25-hydroxy-	EtOH	264(4.24)	13-0567-69A
	n.s.g.	265(4.25)	54-1080-69
Cholesta-5,7-diene-3β,25-diol	EtOH	271(3.99),281(4.02), 293(3.79)	13-0567-69A
	n.s.g.	263s(3.91),271(4.06), 282(4.08),293(3.85)	54-1080-69
Cholest-5-en-4-one, 6-hydroxy-	n.s.g.	294(4.00)	88-3753-69
ferric chloride complex	n.s.g.	535(3.23)	88-3753-69
5α-Cholest-17(20)-en-22-one, 3β-hydroxy-, cis	MeOH	253(3.90)	44-3767-69
9,10-Secocholesta-5(10),6,8-triene-3β,25-diol, 6-cis	EtOH	256(3.92)	13-0567-69A
$C_{27}H_{44}O_6$			
3-Epicrustecdysone, 2-deoxy-	EtOH	243(4.09)	12-1059-69
Ponasterone A	MeOH	244(4.09),326(2.11)	22-3475-69
$C_{27}H_{44}O_7$			
5β-Cholest-7-en-6-one, 2β,3β,11α,14α,-20,22-hexahydroxy-	MeOH	243(4.01)	77-0546-69
$C_{27}H_{45}BrO_2$			
5β-Cholestan-3-one, 6β-bromo-5-hydroxy-	dioxan	280(1.36)	12-0807-69
$C_{27}H_{45}FO$			
5α-Cholestan-3-one, 2α-fluoro-	n.s.g.	293(1.28)	22-3166-69
19-Nor-6β-cholest-9-en-5-ol, 3β-fluoro-5-methyl-, 4-abeo(5→6)-	n.s.g.	207(4.08)	22-1758-69
$C_{27}H_{45}NO$			
4-Aza-A-homocholest-4a-en-3-one	EtOH	244(4.19)	94-1255-69
3-Aza-A-homo-5α-cholest-1-en-4-one	EtOH	237(4.03)	94-1255-69
4-Aza-A-homo-5α-cholest-1-en-3-one	EtOH	218(4.14)	94-1255-69
Solacongestidine	EtOH	239(2.56)	44-1577-69
	EtOH-HCl	222(3.19)	44-1577-69
$C_{27}H_{45}NO_2$			
5α-Cholestane-2,3-dione, 2-oxime	dioxan	235(3.91)	22-3166-69
	dioxan-NaOH	310(4.20)	22-3166-69
Solafloridine	EtOH	240(2.41)	44-1577-69
$C_{27}H_{45}O_9P_3Ru_3$			
Ruthenium, nonacarbonyltris(triethylphosphine)-, triangulo-	$C_6H_{11}Me$	350s(4.0),460(4.0)	101-0289-69A

Compound	Solvent	$\lambda_{max}(\log \epsilon)$	Ref.
$C_{27}H_{46}O_2$			
18,19-Dinorcholest-9-ene-3,6-diol, 5,14-dimethyl-	EtOH	205(3.91),207(3.88), 210(3.79),215(3.57)	78-3925-69
18,19-Dinorcholest-13(17)-ene-3,6-diol, 5,14-dimethyl-	C_6H_{12}	195(4.07),200(4.01), 205(3.95),210(3.87)	78-3925-69
$C_{27}H_{47}NO_3$			
5α-Cholestan-3β-ol, nitrate	C_6H_{12}	190(3.78)	78-0761-69

Compound	Solvent	$\lambda_{max}(\log \epsilon)$	Ref.
$C_{28}H_{16}Cl_2N_2$			
$\Delta^{1,1'}$-Bi-1H-isoindole, 5,5'-dichloro-3,3'-diphenyl-	iso-PrOH	250(4.62),333(4.14), 455(4.69)	44-0649-69
$C_{28}H_{16}N_2O_5$			
1H-Benzo[b]cyclopropa[3,4]cyclopenta-[1,2-e]pyran-1,8a(7H)-dicarboximide, 1-cyano-1a,8-dihydro-8-hydroxy-7-oxo-1a,8-diphenyl-	CH_2Cl_2	242(4.40),272s(3.93), 300(3.93)	44-2407-69
Xanthene-2,3-dicarboximide, 2-cyano-1,2-dihydro-1-hydroxy-9-oxo-1,4-diphenyl-	MeOH	308(4.05),492(3.88)	44-2407-69
	CH_2Cl_2	273(4.17),321(4.06)	44-2407-69
$C_{28}H_{16}N_6$			
Bi-6H-indolo[2,3-b]quinoxaline	H_2SO_4	287(5.00),478(4.99)	47-1803-69
9,9'-Bi-6H-indolo[2,3-b]quinoxaline	H_2SO_4	287(4.95),425(4.98), 451(5.07)	47-1803-69
$C_{28}H_{16}N_6O$			
6H-Indolo[2,3-b]quinoxaline, oxybis-	H_2SO_4	285(4.99),480(4.90)	47-1803-69
6H-Indolo[2,3-b]quinoxaline, 9,9'-oxybis-	H_2SO_4	282(4.90),423(4.93), 442(4.96)	47-1803-69
$C_{28}H_{16}O_2$			
Diphenanthro[9,10-b:9',10'-e]-p-dioxin	dioxan	294(4.17),306(4.08), 359(3.32),373(3.29)	24-3747-69
$C_{28}H_{16}O_3$			
Acenaphtho[1,2-j]phenanthrene-4,5-dicarboxylic anhydride, 3b,4,5,5a-tetrahydro-	dioxan	255(4.32),266(4.36), 290(3.64),300(3.71), 314(3.78),339(3.94), 358(4.30),378(4.60), 400(4.66)	24-3599-69
isomer	dioxan	288(3.58),300(3.70), 314(3.77),338(3.96), 357(4.25),378(4.55), 400(4.58)	24-3599-69
$C_{28}H_{17}Cl_2N_3$			
Indole, 5-chloro-3-[(5-chloro-2-phenyl-3H-indol-3-ylidene)amino]-2-phenyl-	EtOH	270(4.54),571(4.04)	22-1234-69
$C_{28}H_{17}NO$			
Anthrone, 10-(9-anthrylimino)-	ether	263(5.12),315(3.83), 375(3.87),391(3.99), 413(3.91),550(3.00)	22-3538-69
$C_{28}H_{18}$			
Benzo[b]biphenylene, 5,10-diphenyl-	EtOH	230(4.43),250(4.41), 278(4.59),304(4.53), 355s(4.43),380(3.49), 400(3.50)	44-0538-69
1,1'-Bicyclopent[fg]acenaphthylene, 1,1',2,2'-tetrahydro-	EtOH	240(4.89),309(3.98), 316s(4.08),322(4.25), 329s(4.16),345(4.24), 352s(4.11),361(4.14)	35-3689-69
Heptazethrene	C_6H_{12}	227(4.76),238(4.58), 253(4.42),264(4.60), 274(4.90),292(4.30), 296(4.27),310(4.22),	5-0043-69A
(continued on next page)			

Compound	Solvent	λ_{max}(log ϵ)	Ref.
Heptazethrene (cont.)		323(4.42),338(4.65), 348(4.50),357(4.82), 374(3.76)	
Pyrene, 1-[2-(2-naphthyl)vinyl]-	DMF	278(4.51),313(4.28), 387(4.65)	33-2521-69
$C_{28}H_{18}Cl_2N_4$			
2,5-Bis(p-chlorophenyl)-1,2-dihydro-3,6-diphenylpyrazolo[4,3-c]pyrazolium hydroxide, inner salt	benzene	297(4.17),408(4.05)	77-1393-69
$C_{28}H_{18}N_2$			
9-Anthrylamine, N-(10-imino-9(10H)-anthrylidene)-	ether	264(5.10),385(3.89), 395(3.98),414(3.91), 510(3.14)	22-3538-69
9,9'-Azoanthracene	$CHCl_3$	248(5.08),300s(4.11), 380(3.80),477(3.97), 600s(3.72)	22-3538-69
$C_{28}H_{18}N_2O$			
Anthrone, 10-(9-anthrylimino)-, oxime	ether	264(4.48),280(4.27), 378(3.75),394(3.89), 415(3.82)	22-3538-69
$C_{28}H_{18}N_2O_4$			
Xanthene-2,2(1H)-dicarbonitrile, 1-hydroxy-1-(p-methoxyphenyl)-9-oxo-4-phenyl-	MeOH	301(4.24),464(4.54)	44-2407-69
	CH_2Cl_2	260(4.32),307(4.00)	44-2407-69
Xanthene-2,2(1H)-dicarbonitrile, 1-hydroxy-4-(p-methoxyphenyl)-9-oxo-1-phenyl-	MeOH	312(4.16),465(4.60)	44-2407-69
	CH_2Cl_2	264(4.40),309(4.09), 323s(4.06)	44-2407-69
$C_{28}H_{18}N_2O_6$			
Uracil, 5'-O-trityl-2,2'-anhydro-1-β-D-arabinofuranosyl-	EtOH	224s(--),249(3.82)	23-0495-69
$C_{28}H_{18}N_2S_2$			
$\Delta^{2,2'}$-Bi-2H-1,4-benzothiazine, 3,3'-diphenyl-	EtOH	263(4.59),297s(4.15), 344(4.03),466(3.78)	32-0323-69
	EtOH-HCl	256(4.48),312(4.10), 360(4.00),594(3.78)	32-0323-69
$C_{28}H_{18}N_4$			
2,2'-Biquinoline, 4,4'-dipyridino-, cuprous chelate	iso-AmOH	552(4.00)	3-0344-69
$C_{28}H_{18}O_2$			
10,10'-Bianthrone	EtOH	267(4.43),297(3.94), 315(3.76)	18-1377-69
Phenanthro[9,10-b]-p-dioxin, 2,3-diphenyl-	benzene	364(3.25)	24-3747-69
$C_{28}H_{19}BF_2O_3$			
Naphtho[1,2-e]-1,3,2-dioxaborin, 2,2-dilfuoro-4-[-(2-phenyl-3-benzopyranyl)vinyl]-	MeCN	238(4.46),340(4.15), 380s(3.86),520s(4.53), 540(4.86)	4-0029-69
$C_{28}H_{19}BrO$			
Benzofuran, 5-bromo-2-[p-(p-phenylstyryl)phenyl]-	DMF	369(4.90)	33-1282-69

Compound	Solvent	$\lambda_{max}(\log \epsilon)$	Ref.
$C_{28}H_{19}ClO$			
Benzofuran, 5-chloro-2-[p-(p-phenylstyryl)phenyl]-	DMF	369(4.90)	33-1282-69
$C_{28}H_{19}ClO_4S_2$			
1-Methyl-3-[1H-naphtho[2,1-b]thiopyran-1-ylidenemethyl)naphtho[2,1-b]thiopyrylium perchlorate	CH_2Cl_2	245(4.81),276(4.45), 325(4.23),396(4.14), 608s(4.47),652(4.98), 781(3.55)	2-0017-69
4-Methyl-2-[4H-naphtho[1,2-b]thiopyran-4-ylidenemethyl)naphtho[1,2-b]thiopyrylium perchlorate	CH_2Cl_2	254(4.64),312s(4.15), 352s(3.74),450(3.69), 650(3.92)	2-0017-69
$C_{28}H_{19}ClS$			
Benzo[b]thiophene, 5-chloro-2-[p-(p-phenylstyryl)phenyl]-	DMF	368(4.86)	33-1282-69
$C_{28}H_{19}Cl_2N$			
Pyrrole, 3,4-bis(p-chlorophenyl)-2,5-diphenyl-	EtOH	266(4.3),315(4.2)	95-0783-69
$C_{28}H_{19}NO_2$			
Bianthrone, monoxime	EtOH	260(4.35)	22-3538-69
$C_{28}H_{19}N_3$			
Indole, 2-phenyl-3-[(2-phenyl-3H-indol-3-ylidene)amino]-	EtOH	291(4.38),568(3.91)	22-1234-69
$C_{28}H_{20}$			
Anthracene, 2-(p-phenylstyryl)-	DMF	327(4.86),340(4.93), 360(4.39),379(4.52), 397(4.48)	33-2521-69
Butatriene, tetraphenyl-	C_6H_{12}	274(4.56),318(3.58), 420(4.63)	56-1843-69
Cyclobut[b]anthracene, 1,2-dihydro-3,10-diphenyl-	C_6H_{12}	214(4.58),261(5.01), 373(4.02)	44-0538-69
Cyclobuta[l]phenanthrene, 2a,10b-dihydro-1,2-diphenyl-	EtOH	264(4.58),296s(4.20), 311(4.01)	88-3843-69
Stilbene, 4-[(4-biphenylyl)ethynyl]-	DMF	349(4.88)	33-2521-69
Stilbene, 4-phenyl-4'-(phenylethynyl)-	DMF	355(4.85)	33-2521-69
$C_{28}H_{20}BrO_2P$			
Ketone, 2-benzofuranyl bromo(triphenylphosphoranylidene)methyl	EtOH	225(4.41),275(4.21), 345(4.03)	65-0860-69
$C_{28}H_{20}Br_3O_2P$			
[(2-Benzofuranylcarbonyl)dibromomethyl]-triphenylphosphonium bromide	EtOH	235(4.33),245(4.33)	65-0860-69
$C_{28}H_{20}ClO_2P$			
Ketone, 2-benzofuranyl chloro(triphenylphosphoranylidene)methyl	EtOH	225(4.36),274(4.11), 345(4.07)	65-0860-69
$C_{28}H_{20}Cl_2$			
Benzene, 1-(2,4-dichlorostyryl)-4-(p-phenylstyryl)-	DMF	373(4.86)	33-2521-69
$C_{28}H_{20}IO_2P$			
Ketone, 2-benzofuranyl iodo(triphenylphosphoranylidene)methyl	EtOH	225(4.39),292(4.23)	65-0860-69

Compound	Solvent	λ_{max}(log ϵ)	Ref.
$C_{28}H_{20}N_2$			
Bianthrone, diimine	$CHCl_3$	252(4.47)	22-3538-69
$C_{28}H_{20}N_2O$			
Bianthrone imine, oxime	EtOH	245(4.32)	22-3538-69
$C_{28}H_{20}N_2O_2$			
Bianthrone, dioxime	$CHCl_3$	237(4.43)	22-3538-69
$C_{28}H_{20}O$			
Anthracene, 2-(p-phenoxystyryl)-	DMF	318(4.83),331(4.86), 357(4.26),375(4.37), 395(4.30)	33-2521-69
Benzofuran, 2,3-diphenyl-6-styryl-	DMF	286(4.33),352(4.70)	33-1282-69
Benzofuran, 2-phenyl-6-(p-phenylstyryl)-	DMF	295(4.08),365(4.82)	33-1282-69
Benzofuran, 2-[p-(p-phenylstyryl)-phenyl]-	DMF	368(4.89)	33-1282-69
Benzofuran, 5-phenyl-2-(p-styrylphenyl-	DMF	359(4.84)	33-1282-69
Dibenzofuran, 3,7-distyryl-	DMF	369(4.91)	33-1282-69
Ether, phenyl p-[p-(phenylethynyl)-styryl]phenyl	DMF	345(4.70)	33-2521-69
Furan, 2-[p-[2-(2-naphthyl)vinyl]-phenyl]-5-phenyl-	DMF	282(4.28),297(4.13), 308(4.22),377(4.82)	33-1282-69
$C_{28}H_{20}OS$			
Benzo[b]thiophene, 2-[p-(p-phenoxy-styryl)phenyl]-	DMF	360(4.79)	33-1282-69
Phenoxathiin, 2,8-distyryl-	DMF	298(4.86)	33-1282-69
$C_{28}H_{20}O_2$			
Benzofuran, 2-[p-(p-phenoxystyryl)phen-yl]-	DMF	360(4.84)	33-1282-69
Benzofuran, 6-(p-phenoxystyryl)-2-phen-yl-	DMF	355(4.78)	33-1282-69
3,3'-Spirobi[3H-naphtho[2,1-b]pyran], 2,2'-trimethylene-	dioxan	249(5.08),301(4.29), 314(4.33),335(4.08), 349(4.15)	5-0162-69B
Spiro[2H-naphtho[1,2-b]pyran-2,3'-[3H]-naphtho[2,1-b]pyran, 3,2'-trimethyl-ene-	dioxan	241(4.90),258(4.72), 267(4.79),300(4.14), 313(4.18),334(4.02), 369(4.02)	5-0162-69B
$C_{28}H_{20}O_4S$			
Thiiran, 2,3-dibenzoyl-2,3-diphenyl-, 1,1-dioxide, cis	EtOH	233s(4.33),254(4.29)	35-2097-69
	MeCN	232s(4.33),254(4.30)	35-2097-69
$C_{28}H_{20}S$			
Benzo[b]thiophene, 2-(p-biphenylyl)-6-styryl-	DMF	281(4.16),366(4.83)	33-1282-69
Benzo[b]thiophene, 2-phenyl-5-(p-phenyl-styryl)-	DMF	329(4.83)	33-1282-69
Benzo[b]thiophene, 2-phenyl-6-(p-phenyl-styryl)-	DMF	367(4.85)	33-1282-69
Benzo[b]thiophene, 2-[p-(p-phenylsty-ryl)phenyl]-	DMF	367(4.85)	33-1282-69
Dibenzothiophene, 2,8-distyryl-	DMF	319(4.75)	33-1282-69
Dibenzothiophene, 3,7-distyryl-	DMF	366(4.93),384(4.85)	33-1282-69
Unknown compound	ether	267(4.49),327(4.18)	44-0879-69

Compound	Solvent	$\lambda_{max}(\log \epsilon)$	Ref.
$C_{28}H_{20}S_2$			
Thianthrene, 2,7-distyryl-	DMF	329(4.88)	33-1282-69
$C_{28}H_{21}Br_2O_2P$			
[(2-Benzofuranylcarbonyl)bromomethyl]-triphenylphosphonium bromide	EtOH	272(4.28),325(3.87)	65-0860-69
$C_{28}H_{21}ClO_7$			
3-[2-(2-Acetoxy-1-naphthyl)-1-methyl-vinyl]naphtho[2,1-b]pyrylium perchlorate	MeOH-$HClO_4$	306(4.12),521(4.40)	5-0226-69E
$C_{28}H_{21}Cl_2NO_2$			
o-Cinnamotoluidide, α',α'-bis(p-chlorophenyl)-α'-hydroxy-	EtOH	296(4.31)	115-0057-69
$C_{28}H_{21}N$			
Pyrrole, 2,3,4,5-tetraphenyl-	H_2SO_4	314(4.08),367(4.09)	22-1667-69
$C_{28}H_{21}NO$			
2-Pyrrolin-5-one, 2,3,4,4-tetraphenyl-	EtOH	227(4.38),309(3.75)	22-1667-69
2H-Pyrrol-2-ol, 2,3,4,5-tetraphenyl-	C_6H_{12}	235s(4.33),325s(3.72)	22-1667-69
	EtOH	235s(4.33),325s(3.72)	22-1667-69
	ether	242(4.42),310s(3.78)	22-1667-69
$C_{28}H_{21}NO_2$			
Hydroperoxide, 2,3,4,5-tetraphenyl-2H-pyrrol-2-yl	ether	242(4.34),310s(3.78)	22-1667-69
$C_{28}H_{21}NO_3$			
Benzamide, N-(benzoyldiphenylacetyl)-	EtOH	244(4.40)	22-2756-69
$C_{28}H_{21}N_3$			
Indole, 3,3'-iminobis[2-phenyl-	EtOH	209(4.73),250(4.54), 309(4.45),360(4.11)	22-1234-69
$C_{28}H_{21}O_2P$			
Ketone, 2-benzofuranyl (triphenylphosphoranylidene)methyl	EtOH	225(4.35),274(4.15), 329(4.31)	65-0860-69
$C_{28}H_{22}$			
Benzene, p-(p-phenylstyryl)-4-styryl-	DMF	368(4.88)	33-2521-69
Biphenyl-, 4,4'-distyryl-	DMF	354(4.89)	33-2521-69
Naphthalene, 1-[p-(4-phenyl-1,3-butadienyl)styryl]-	DMF	384(4.85)	33-2521-69
Naphthalene, 2-[p-(4-phenyl-1,3-butadienyl)styryl]-	DMF	312(4.20),384(4.96), 404(4.82)	33-2521-69
Stilbene, 4-(2,2-diphenylvinyl)-	DMF	357(4.69)	33-2521-69
$C_{28}H_{22}BrO_2P$			
[(2-Benzofuranylcarbonyl)methyl]triphenylphosphonium bromide	EtOH	277(4.18),291(4.16), 320(4.31)	65-0860-69
$C_{28}H_{22}N_2$			
5,12:6,11-Di-o-benzenodibenzo[a,e]cyclooctene-5,11(6H,12H)-diamine	$CHCl_3$	241(3.70),271(3.18), 281(2.95)	44-4166-69
$C_{28}H_{22}N_2O_3$			
p-Anisic acid, 1,2-diphenyl-2-(phenylazo)vinyl ester	n.s.g.	260(4.43),350(4.30), 452(2.70)	44-0231-69

Compound	Solvent	λ_{max}(log ϵ)	Ref.
Benzil, mono[N-phenyl-N-(p-methoxybenzoyl)hydrazone]	n.s.g.	256(4.45)	44-0231-69
$C_{28}H_{22}O$			
Benzofuran, 4,6-dimethyl-2-[p-[2-(2-naphthyl)vinyl]phenyl]-	DMF	302(4.18),371(4.84)	33-1282-69
Benzofuran, 5,6-dimethyl-2-[p-[2-(1-naphthyl)vinyl]phenyl]-	DMF	373(4.70)	33-1282-69
Benzofuran, 5,6-dimethyl-2-[p-[2-(2-naphthyl)vinyl]phenyl]-	DMF	302(4.08),371(4.85), 389(4.72)	33-1282-69
Benzofuran, 5,7-dimethyl-2-[p-[2-(1-naphthyl)vinyl]phenyl]-	DMF	367(4.72)	33-1282-69
Benzofuran, 5,7-dimethyl-2-[p-[2-(2-naphthyl)vinyl]phenyl]-	DMF	302(4.23),368(4.83), 387(4.69)	33-1282-69
Ether, phenyl p-(p-styrylstyryl)phenyl	DMF	362(4.83)	33-2521-69
$C_{28}H_{22}O_2$			
Acetylene, bis[p-(benzyloxy)phenyl]-	n.s.g.	313f(4.92)	56-1469-69
3,3'-Spirobi[3H-naphtho[2,1-b]pyran], 2-isopropyl-	dioxan	248(5.03),287(4.05), 300(4.16),312(4.19), 335(3.98),349(4.03)	5-0162-69B
3,3'-Spirobi[3H-naphtho[2,1-b]pyran], 2-propyl-	dioxan	248(5.07),287(4.03), 300(4.16),312(4.20), 335(3.94),349(4.00)	5-0162-69B
$C_{28}H_{22}O_6$			
Benzophenone, 4,4"-(ethylenedioxy)bis-[2-hydroxy-	DMF	290(4.46),320(4.29)	39-0034-69C
$C_{28}H_{23}ClN_2O$			
1(2H)-Quinolinecarboxanilide, 6-chloro-3,4-dihydro-2,4-diphenyl-	n.s.g.	240(4.26),267(4.28)	40-1047-69
$C_{28}H_{23}ClN_2O_4S$			
4-[N-[p-(Dimethylamino)phenyl]formimidoyl]-2-phenylnaphtho[1,2-b]thiopyrylium perchlorate	CH_2Cl_2	242(4.48),305(4.14), 394(5.04),445(4.6), 810(5.12)	2-0311-69
$C_{28}H_{23}NO$			
Pyrrolidine, 3,4-epoxy-2,3,4,5-tetraphenyl-	EtOH	258(4.30),265(4.26), 269s(4.15),315s(2.95)	22-1667-69
$C_{28}H_{23}NO_2$			
o-Cinnamotoluidide, α'-hydroxy-α',α'-diphenyl-	EtOH	298(4.26)	115-0057-69
$C_{28}H_{23}NO_3$			
6,11-Epoxy-3H-naphth[2,3-d]azepine-3-carboxylic acid, 5a,6,11,11a-tetrahydro-6,11-diphenyl-, dimethyl ester, endo	EtOH	234(4.18)	44-2888-69
$C_{28}H_{24}N_2$			
4-Phenanthrone, 1,2,3,4-tetrahydro-, azine	$CHCl_3$	328s(4.27),342(4.34)	24-2384-69
$C_{28}H_{24}N_2O_2$			
1,2-Acenaphthenedicarboxamide, N,N'-dibenzyl-, trans (continued on next page)	EtOH	227(4.88),269s(3.73), 279(3.92),289(4.01), 293s(3.90),300(3.79),	44-2418-69

Compound	Solvent	λ_{max}(log ϵ)	Ref.
1,2-Acenaphthenedicarboxamide, N,N'-dibenzyl-, trans (cont.)		306(3.49),317(2.99), 320(2.92)	
2H-Pyrrolo[3,4-b]quinoline-2-carboxylic acid, 3-(2,2-diphenylvinyl)-1,3-dihydro-, ethyl ester	EtOH	227(4.68),252(4.34), 292(3.76),300(3.70), 307(3.81),313(3.76), 321(3.94)	44-3853-69
$C_{28}H_{24}O_2$			
Butyric acid, 2,3,4,4-tetraphenyl-	EtOH	260(2.92)	88-1883-69
p-Dioxane, 2,2,6,6-tetraphenyl-	EtOH	204(4.51),256(2.81)	22-2048-69
Naphthalene, 1,3-bis(p-methoxystyryl)-	DMF	278(4.37),333(4.79)	33-2521-69
Naphthalene, 1,4-bis(p-methoxystyryl)-	DMF	388(4.62)	33-2521-69
Naphthalene, 1,5-bis(p-methoxystyryl)-	DMF	367(4.59)	33-2521-69
Naphthalene, 1,6-bis(p-methoxystyryl)-	DMF	335(4.86)	33-2521-69
Naphthalene, 1,7-bis(p-methoxystyryl)-	DMF	300(4.57),338(4.84), 352(4.88),390(4.44)	33-2521-69
Naphthalene, 2,3-bis(p-methoxystyryl)-	DMF	316(4.78)	33-2521-69
Naphthalene, 2,6-bis(p-methoxystyryl)-	DMF	297(4.48),370(4.89), 389(4.81)	33-2521-69
Stilbene, 3,4'-bis(benzyloxy)-	EtOH	321(4.37)	39-1309-69C
Stilbene, 4,4'-bis(benzyloxy)-, trans	dioxan	326f(4.86)	56-1469-69
$C_{28}H_{24}O_6$			
5H-Cycloprop[cd]inden-2,2a,2b-tricarboxylic acid, 5-(diphenylmethylene)-5a,5b-dihydro-, trimethyl ester	EtOH	230s(4.21),295(4.26)	89-0209-69
1,2,4-Methenopentalene-1,5,6-tricarboxylic acid, 3-(diphenylmethylene)-1,2,3,3a,4,6a-hexahydro-, trimethyl ester	EtOH	231(4.29),276(4.18)	89-0209-69
$C_{28}H_{25}N_3O_7S$			
2H-Furo[2',3':4,5]oxazolo[3,2-b]-as-triazin-2-one, 5a,7,8,8a-tetrahydro-8-hydroxy-7-[(trityloxy)methyl]-, methanesulfonate	EtOH	211(4.39),260(3.82)	73-0618-69
$C_{28}H_{26}N_2O_3$			
2H-Pyrrolo[3,4-b]quinoline-2-carboxylic acid, 1,3-dihydro-3-(2,2-diphenyl-2-hydroxyethyl)-, ethyl ester	EtOH	227s(4.57),231(4.60), 234s(4.56),295(3.64), 300(3.64),308(3.78), 314(3.75),322(3.91)	44-3853-69
$C_{28}H_{26}N_4O$			
Benzamide, N-benzyl-N-(3,3-dibenzyl-1-triazeno)-	EtOH	294(4.18)	44-2997-69
$C_{28}H_{26}O_3$			
Benzhydrol, α,α'-(oxydimethylene)di-	EtOH	205(4.57),256(2.88)	22-2048-69
1,3-Dioxolane, 2-benzhydryl-4,4-diphenyl-	EtOH	202(4.57),258(2.97)	22-2048-69
$C_{28}H_{26}O_5$			
2,5-Cyclohexadien-1-one, 4-methoxy-2,4,6-tris(p-methoxyphenyl)-	n.s.g.	235(4.53),316(3.68)	24-3795-69
$C_{28}H_{26}O_8$			
Acetic acid, [2-[7-(benzyloxy)-4-oxo-4H-1-benzopyran-3-yl]-4,5-dimethoxyphenoxy]-, ethyl ester	EtOH	248(4.39)	18-0199-69

Compound	Solvent	$\lambda_{max}(\log \epsilon)$	Ref.
$C_{28}H_{26}O_{10}$			
Benzo[1,2:4,5]dicycloheptene-2,4,8,10-tetracarboxylic acid, 3,9-dihydro-3,9-dioxo-, tetraethyl ester	MeOH	244(4.19),303(4.96), 340s(4.26),400(3.18)	64-0464-69
$C_{28}H_{27}Fe_2N$			
1,1"-Biferrocene-2-methylamine, 2"-methyl-N-phenyl-	C_6H_{12}	199(4.80),222(4.67), 295(3.96)	78-3477-69
$C_{28}H_{27}NO_3$			
1,5-Ethenocyclopent[c]azepine-2(1H)-carboxylic acid, 5,5a,8,8a-tetrahydro-7,8a-dimethyl-8-oxo-5a,6-diphenyl-, ethyl ester	EtOH	227(4.44),270s(3.83)	44-2888-69
6,9-Methano-3H-3-benzazepine-3-carboxylic acid, 5a,6,9,9a-tetrahydro-6,9-dimethyl-10-oxo-7,8-diphenyl-, ethyl ester	EtOH	251(4.07),281(4.20)	44-2888-69
$C_{28}H_{27}NO_5$			
2-Cyclopropene, 3-ethoxy-1,2-di-p-tolyl-1-(4-nitro-α-carbomethoxybenzyl)-	CH_2Cl_2	277(4.21)	24-0319-69
$C_{28}H_{28}N_2$			
1,2-Naphthalenedicarbonitrile, 3,4-dihydro-7-isopropenyl-4-methyl-4-(α,α-dimethyl-4-isopropenyl)-	MeCN	206(4.54),260(4.54), 310(3.92),360s(2.30)	44-1923-69
$C_{28}H_{28}N_2OS$			
4-Thiazoline, 5-anilino-4-cyclohexanecarbonyl-2,3-diphenyl-	MeCN	250(4.20),320(3.98), 416(4.03)	23-3557-69
$(C_{28}H_{28}N_2O_6)_n$			
Poly[oxy[2-[(3,5,6-trimethyl-p-benzoquinon-2-yl)methyl]trimethylene]oxycarbonylimino-p-phenylenemethylene-p-phenyleneiminocarbonyl]	95% THF	251(4.75)	47-1269-69
$C_{28}H_{28}N_4Ni$			
Nickel, [2,3,7,8,12,13,17,18-octamethylporphinato(2-)]-	$CHCl_3$	290(4.10),332(4.14), 390(5.23),518(4.03), 555(4.50)	39-0655-69C
$C_{28}H_{28}N_4O_5S$			
Acetophenone, 2-[3-[(4-amino-2-methyl-5-pyrimidinyl)methyl]-5-(2-hydroxyethyl)-4-methyl-4-thiazolin-2-ylidene]-2-hydroxy-4'-methoxy-, 2-benzoate	EtOH	227(4.45),272(4.15), 397(4.44)	94-0128-69
4-Thiazoline, 5-(m-nitroanilino)-4-(p-nitrobenzoyl)-3-cyclohexyl-2-phenyl-	MeCN	265(4.40),455(4.14)	23-3557-69
4-Thiazoline, 5-(p-nitroanilino)-4-(p-nitrobenzoyl)-3-cyclohexyl-2-phenyl-	MeCN	250(4.32),290(4.26), 375(4.09),460(4.17)	23-3557-69
$C_{28}H_{28}O_9$			
16α,24-Cyclo-13,14-secoergosta-2,4,6,-25(27)-tetraene-18,26-dioic acid, 14α,17-epoxy-13,14,20,22-tetrahydroxy-1,15-dioxo-, γ-lactone δ-lactone	n.s.g.	320(3.81)	88-1765-69

Compound	Solvent	λ_{max}(log ε)	Ref.
16α,24-Cyclo-13,14-secoergosta-2,4,6-triene-18,26-dioic acid, 14α,17:14,27-diepoxy-13,20,22-trihydroxy-1,15-dioxo-, γ-lactone δ-lactone	n.s.g.	328(3.70)	88-1765-69
$C_{28}H_{28}O_{10}$			
Anthraquinone, 2-hexyl-1,3,6,8-tetrahydroxy-, tetraacetate	EtOH	212(4.50),263(4.68), 339(3.81)	39-2763-69C
$C_{28}H_{28}O_{11}$			
Cleistanthin	EtOH	262(4.80),294(3.99), 315(4.01),335(3.68)	78-2815-69
$C_{28}H_{29}ClN_2O$			
1(2H)-Quinolinecarboxamide, 6-chloro-N-cyclohexyl-3,4-dihydro-2,4-diphenyl-	n.s.g.	258(4.18)	40-1047-69
$C_{28}H_{29}NO_6$			
α-D-xylo-Tetrofuranose, 3-O-benzyl-4-C-(4-carboxy-5-methyl-2-quinolyl)-1,2-O-cyclohexylidene-	n.s.g.	236(4.50),322(3.72)	65-1413-69
$C_{28}H_{30}ClNO_5$			
[3-(6,7-Dihydro-2,4-diphenyl-5H-1-benzopyran-8-yl)allylidene]diethylammonium perchlorate	MeCN	262(4.16),326(4.40), 340s(4.30),408(4.09), 424(4.20),570(4.54), 605(4.42),652s(4.28)	44-2736-69
$C_{28}H_{30}N_2O_5$			
Cylindrocarine, 1-benzoyl-20-oxo-	MeOH	225(4.55),250(3.13), 294(3.11)	39-0417-69C
$C_{28}H_{30}O_2$			
4-Hexyn-3-ol, 6-ethoxy-2,2-dimethyl-3,6,6-triphenyl-	EtOH	252(2.72),258(2.80), 264(2.67)	56-1843-69
$C_{28}H_{30}O_5$			
Estra-1,3,5(10)-trien-6-one, 1,17β-dihydroxy-4-methyl-, 1-acetate benzoate	EtOH	221(4.03),229(3.95), 250(3.63),304(3.04)	39-1234-69C
$C_{28}H_{30}O_6$			
Benzene, m-bis(3,4,5-trimethoxystyryl)-	DMF	325(4.79)	33-2521-69
Benzene, o-bis(3,4,5-trimethoxystyryl)-	DMF	304(4.57),336(4.49)	33-2521-69
Benzene, p-bis(3,4,5-trimethoxystyryl)-	DMF	372(4.82)	33-2521-69
Lonchocarpic acid, dimethyl ether	EtOH	234(4.47),270(4.54), 353(4.32)	39-0374-69C
2-Propen-1-ol, 3,3'-p-phenylenebis[1-(2,4-dimethoxyphenyl)-	EtOH	292(4.52)	124-1189-69
2H,6H-Pyrano[3,2-b]xanthen-6-one, 5,9,10-trihydroxy-2,2-dimethyl-12-(5-methyl-2-isopropenylhex-4-enyl)-	MeOH	240(4.36),283(4.66), 338(4.32)	88-4893-69
$C_{28}H_{30}O_9$			
Physalin B	n.s.g.	222(4.00)	88-1765-69
8,14-Secobufa-4,6,20,22-tetraenolide, 1α,2α-epoxy-3β,12β-dihydroxy-8,14-dioxo-, diacetate	EtOH	279(4.22)	119-0186-69
$C_{28}H_{30}O_{10}$			
Physalin A	n.s.g.	218s(4.00)	88-1083-69

Compound	Solvent	λ_{max}(log ϵ)	Ref.
$C_{28}H_{30}O_{13}$			
Olivin, 2,6,8,9-tetraacetate	EtOH	252s(4.69),259(4.79), 305(3.81),314(3.81)	105-0257-69
$C_{28}H_{30}O_{14}$			
Pseudobaptisin	MeOH	220(4.48),248s(4.33), 262(4.35),292(4.22)	24-3006-69
$C_{28}H_{31}BrO_5$			
Hirundigenin, bromobenzoylanhydro-	EtOH	200(4.75),244(4.44), 270(3.13),282(2.84)	33-1175-69
$C_{28}H_{31}N_4O$			
Antipyrinylbis[p-(dimethylamino)phenyl]-methylium (picrate)	pH 4	607(4.70)	122-0138-69
$C_{28}H_{32}N_2O_2$			
2-Propen-1-ol, 3,3'-p-phenylenebis[1-p-(dimethylamino)phenyl]-	EtOH	260(4.54),287(4.49)	124-1189-69
$C_{28}H_{32}N_2O_4$			
Cylindrocarine, 1-benzoyl-	MeOH	213(4.32),226s(4.12), 275(3.63),290(3.40)	39-0417-69C
2'H-Pregna-2,4-dieno[3,2-c]pyrazole-11,20-dione, 17,21-dihydroxy-2'-phenyl-	dioxan	269(4.26)	22-1256-69
$C_{28}H_{32}N_2O_5$			
Cylindrocarine, 1-benzoyl-20-hydroxy-	MeOH	211(4.69),226(3.48), 275(3.13)	39-0417-69C
$C_{28}H_{32}O_3$			
Pregn-5-eno[6,5,4-bc]furan-3,20-dione, 5'-phenyl-	EtOH	226(4.30),245s(3.99), 348(4.13)	94-2586-69
17-iso-	EtOH	226(4.31),245s(4.00), 349(4.13)	94-2586-69
$C_{28}H_{32}O_3S$			
Androst-5-eno[6,5,4-bc]thiophen-3-one, 17β-hydroxy-5'-phenyl-, acetate	EtOH	232(4.18),238s(4.16), 278(3.95),332(3.81)	94-2586-69
$C_{28}H_{32}O_4$			
Estra-1,3,5(10)-triene-1,17β-diol, 4-methyl-, 1-acetate benzoate	EtOH	230(4.12),270(3.04)	39-1234-69C
$C_{28}H_{32}O_9$			
1-Hexanone, 1-(9,10-diacetoxy-1,3,6,8-tetramethoxy-2-anthryl)-	EtOH	249s(4.48),273(4.90), 360s(3.56),376(3.66), 395(3.72),415s(3.54)	39-2763-69C
$C_{28}H_{32}O_{10}$			
Isotaxiresinol, 6-methyl ether, tetra-acetate	n.s.g.	212(4.6),268s(3.5), 275s(3.55),283(3.6)	1-2021-69
$C_{28}H_{33}NO_3$			
Androst-4-en-3-one, 17β-hydroxy-17α-(3-phenyl-5-isoxazolyl)-	MeOH	241(4.48)	24-3324-69
$C_{28}H_{33}NO_5$			
Isoquinoline, 1-[4-(benzyloxy)-3,5-di-methoxyphenethyl]-1,2,3,4-tetrahydro-6,7-dimethoxy-, hydrochloride	EtOH	230s(4.28),280(3.61)	33-1228-69

Compound	Solvent	λ_{max}(log ε)	Ref.
$C_{28}H_{33}N_3O_6$			
2'H-Pregna-2,4-dieno[3,2-c]pyrazol-20-one, 11β,17,21-trihydroxy-2'-(p-nitrophenyl)-	EtOH	224(4.06),299(4.26)	22-1256-69
$C_{28}H_{34}O_2$			
3'H-Cycloprop[2,3]androsta-2,4,6-trien-17β-ol, 2β,3β-dihydro-3'-phenyl-, acetate	EtOH	255(4.54)	94-2319-69
$C_{28}H_{34}O_3$			
Androst-4-en-3-one, 2-benzylidene-17β-hydroxy-, acetate	EtOH	234(4.05),266(4.08), 306(4.14)	94-2319-69
$C_{28}H_{34}O_6$			
1-Oxaspiro[5.5]undec-3-en-2-one, 3,3'-phenethylidenebis[4-hydroxy-	MeOH	254(4.22)	83-0075-69
$C_{28}H_{34}O_8$			
16,17-Seco-24-norchola-1,20(22)-diene-16,21-dioic acid, 14,15-epoxy-7,17,23-trihydroxy-4,4,8-trimethyl-3-oxo-, 16,17:21,23-dilactone, acetate	MeOH	217(4.11)	78-5007-69
$C_{28}H_{34}O_9$			
Photogedunin	MeOH	224(4.14)	78-5007-69
Unnamed diacid	MeOH	228(4.11)	78-5007-69
$C_{28}H_{34}O_{13}$			
Olivin, (3-acetyloliosyl)-	EtOH	227(4.39),275(4.72), 307(3.84),318(3.86), 332s(3.67),405(3.99)	105-0483-69
	EtOH-KOH	227(4.37),275(4.70), 307(3.81),318(3.82), 332s(3.64),410(3.90)	105-0483-69
$C_{28}H_{34}O_{14}$			
Chalcone, 2',4',6'-trihydroxy-4-methoxy-, 4'-(2-O-α-L-rhamnopyranosyl-β-D-glucopyranoside)	MeOH	363(4.56)	24-0785-69
$C_{28}H_{34}O_{15}$			
Flavanone, 5,7-dihydroxy-4'-methoxy-, 7-(2-O-α-L-rhamnopyranosyl-β-D-glucopyranoside)	MeOH	283(4.26),330s(3.53)	24-0785-69
$C_{28}H_{34}O_{17}$			
Scandoside, hexaacetate	EtOH	230(4.01)	94-1942-69
$C_{28}H_{35}NO_3$			
Androst-5-ene-3β,17β-diol, 17α-(3-phenyl-5-isoxazolyl)-	MeOH	242(4.22)	24-3324-69
$C_{28}H_{35}N_3O_6$			
1,2-Secohydroquinine, 1-carbamoyl-1',2',3',4'-tetrahydro-2-hydroxy-2'-oxo-, 9-benzoate	EtOH	235(4.25),258(4.16), 300(3.48)	78-5319-69
allo-	EtOH	234(4.21),258(4.15), 298(3.51)	78-5319-69

Compound	Solvent	λ_{max}(log ε)	Ref.
$C_{28}H_{36}IP$			
Trimesitylmethylphosphonium iodide	C_6H_{12}	246(4.56),284(3.70), 292(3.70)	65-1544-69
$C_{28}H_{36}N_2O$			
Formanilide, 1-[N-(2,6-diethylphenyl)-1-cyclohexene-1-carboximidoyl]-2',6'-diethyl-	EtOH	242(4.3),337(2.7)	24-1876-69
$C_{28}H_{36}O$			
Pregn-5-eno[6,5,4-bc]furan, 5'-phenyl-	EtOH	226(3.90),233s(3.82), 300(4.42),312s(4.35)	94-2586-69
$C_{28}H_{36}O_2$			
3'H-Cyclopropa[2,3]androsta-2,4-dien-17β-ol, 2β,3β-dihydro-3'-phenyl-, acetate	EtOH	235(4.11),271(3.04), 278(2.85)	94-2319-69
$C_{28}H_{36}O_3$			
5α-Androstane-$\Delta^{3,\alpha}$-acetaldehyde, 17β-hydroxy-, benzoate	EtOH	236(4.34)	39-0460-69C
5α-Androstan-3-one, 2-benzylidene-17-hydroxy-, acetate	EtOH	224(3.85),231s(3.76),	94-2319-69
Photogedunin, 1,2-dihydro-23-deoxy-	MeOH	210(3.97)	78-5007-69
$C_{28}H_{36}O_6$			
Lineolonone, 12-benzoyl-	EtOH	233(4.10),278(3.13)	94-2391-69
Withaferin A, 4-dehydro-	EtOH	223(4.17)	44-3858-69
$C_{28}H_{36}O_8$			
Hellebrigenin, 3,5-diacetate	MeOH	298(3.75)	44-3894-69
$C_{28}H_{36}O_9$			
Bersaldegenin, 1,3-diacetate	MeOH	298(3.69)	88-1709-69
Photogedunin, dihydro-	MeOH	205(3.85)	78-5007-69
	MeOH-base	227(4.00)	78-5007-69
$C_{28}H_{37}BrO_2$			
3'H-Cycloprop[2,3]-5α-androst-2-en-17β-ol, 3'-(p-bromophenyl)-2β,3β-dihydro-, acetate	EtOH	234(4.28),270(2.91), 278(2.92),287(2.72)	94-2319-69
$C_{28}H_{37}NO_4$			
3'H-Cycloprop[2,3]-5α-androst-2-en-17β-ol, 2β,3β-dihydro-3'-(o-nitrophenyl)-, acetate	EtOH	220(4.12),258s(3.51), 311(3.27)	94-2319-69
3'H-Cyclopropa[1,2]pregna-1,4,6-triene-3,20-dione, 1β,2β-dihydro-17-hydroxy-4-(1-pyrrolidinyl)-, acetate	MeCN	283(4.18),389(3.33)	24-2570-69
$C_{28}H_{37}NO_5$			
3'H-Cyclopropa[1,2]pregna-1,4,6-triene-3,20-dione, 1β,2β-dihydro-17-hydroxy-4-morpholino-, acetate	MeCN	282(4.18),389(3.18)	24-2570-69
$C_{28}H_{38}Fe_2I_2N_2$			
Ammonium ([1,1"-biferrocene]-2,2"-diyldimethylene)bis[trimethyl-, diiodide	EtOH	435(2.46)	49-1515-69

Compound	Solvent	λ_{max}(log ε)	Ref.
$C_{28}H_{38}N_2O_4$			
Pregna-4,6-diene-3,20-dione, 17-acetoxy-1α,2α-methylene-4-piperazino-	MeCN	282(4.17),392(3.20)	24-2570-69
$C_{28}H_{38}O_2$			
3'H-Cycloprop[2,3]-5α-androst-2-en-17β-ol, 2β,3β-dihydro-3'-phenyl-	EtOH	225(4.13),258s(2.68), 264(2.79),271(2.84), 278(2.73)	94-2319-69
$C_{28}H_{38}O_4$			
6'H-Indeno[4',5':16,17]-5α-androst-16-ene-6',12-dione, 1',2',3',3'a,16,17-hexahydro-3β-hydroxy-, acetate	MeOH	240(4.14)	44-1606-69
$C_{28}H_{38}O_6$			
Androsta-2,4-diene-3,17β-diol, 6-(1-hydroxyethylidene)-, triacetate	EtOH	288(3.97)	44-2288-69
Withaferin A	EtOH	214(4.24),335(2.22)	44-3858-69
$C_{28}H_{38}O_8$			
16,17-Seco-24-norchol-20(22)-ene-16,21-dioic acid, 3,7,17,23-tetrahydroxy-4,4,8-trimethyl-, 16,17:21,23-dilactone 7-acetate	MeOH	208(3.97)	78-5007-69
$C_{28}H_{38}O_{10}$			
β-D-Glucopyranosiduronic acid, 3,11,20-trioxopregn-4-en-21-yl, methyl ester	MeOH	238(4.21)	69-1188-69
$C_{28}H_{38}O_{11}$			
β-D-Glucopyranosiduronic acid, 17-hydroxy-3,11,20-trioxopregn-4-en-21-yl, methyl ester	MeOH	238(4.20)	69-1188-69
$C_{28}H_{39}NO_2$			
5α-Androstan-3-one, 2-[p-(dimethylamino)benzylidene]-17-hydroxy-	EtOH	253(3.98),328s(3.86), 388(4.33)	94-2319-69
$C_{28}H_{39}N_3O_9S$			
Glutathione, S-(2,3,17β-trihydroxyestra-1,3,5(10)-trien-4(1)-yl-	50% EtOH	261(3.59),302(3.59)	13-0711-69A
$C_{28}H_{40}O$			
Ergostatetraenone	MeOH	214(4.06),298(4.27)	5-0194-69D
Tachysterol	iso-PrOH	272s(4.32),281(4.39), 292s(4.31)	95-0919-69
$C_{28}H_{40}O_2$			
Albolic acid, methyl ester	EtOH	210(4.30)	88-2929-69
$C_{28}H_{40}O_2S_4$			
9,19-Cyclo-5α,9β-pregnane-2,7,11,20-tetrone, 4,4,14-trimethyl-, cyclic 2,20-bis(ethylene mercaptole)	EtOH	229s(3.34)	39-1050-69C
$C_{28}H_{40}O_3$			
Oxepino[4,3-b]benzofuran-1(10bH)-one, 3,5,7,9-tetra-tert-butyl-	MeOH	272(3.3),280(3.3), 350(3.9)	44-1198-69
free radical	$CHCl_3$	700(4.00)	44-1198-69

Compound	Solvent	λ_{max}(log ε)	Ref.
$C_{28}H_{40}O_4$			
Acetic acid, [p-[(1,3,5-tri-tert-butyl-4-oxo-2,5-cyclohexadien-1-yl)oxy]-phenyl]-, ethyl ester	MeOH	242s(4.04)	64-0547-69
Bovinone, dimethyl ether	EtOH	283(4.17)	39-2398-69C
5α-Carda-14,20(22)-dienolide, 3-(2,2-dimethyl-1,3-propanedioxy)-	MeOH	214(4.19)	5-0136-69F
$C_{28}H_{40}O_5$			
5β-Chola-20(21),22-dienoic acid, 3β-acetoxy-14,21-epoxy-, ethyl ester	EtOH	304(4.45)	94-1698-69
Δ^4-Tigogenin-3-one, 10-acetoxy-	EtOH	242(4.18)	88-2113-69
Withaferin A, 27-deoxy-2,3-dihydro-	EtOH	227(3.85)	44-3858-69
$C_{28}H_{40}O_6$			
Pregn-8-ene-7,11-dione, 2α,20α-dihydr-oxy-4,4,14α-trimethyl-, diacetate	EtOH	269(3.82)	39-1047-69C
Withaferin A, 2,3-dihydro-	EtOH	217(3.94)	44-3858-69
$C_{28}H_{40}O_7$			
Hirundoside A	EtOH	199(4.25),286(1.58)	33-1175-69
$C_{28}H_{40}O_8$			
Taxa-4(20),11-diene-5α,9α,10β,13α-tetrol, tetraacetate	n.s.g.	221(3.84)	77-1282-69
$C_{28}H_{40}O_9$			
β-D-Glucopyranosiduronic acid, 3,20-dioxopregn-4-en-21-yl, methyl ester	MeOH	241(4.22)	69-1188-69
$C_{28}H_{40}O_{10}$			
β-D-Glucopyranosiduronic acid, 11β-hy-droxy-3,20-dioxopregn-4-en-21-yl, methyl ester	MeOH	239(4.21)	69-1188-69
β-D-Glucopyranosiduronic acid, 17-hy-droxy-3,20-dioxopregn-4-en-21-yl, methyl ester	MeOH	240(4.21)	69-1188-69
$C_{28}H_{40}O_{11}$			
β-D-Glucopyranosiduronic acid, 11β,17-dihydroxy-3,20-dioxopregn-4-en-21-yl, methyl ester	MeOH	242(4.21)	69-1188-69
$C_{28}H_{40}O_{12}$			
Nervosinic acid	MeOH	244(4.1),281(3.3), 292(3.3)	78-2723-69
	MeOH-base	240s(4.1)	78-2723-69
$C_{28}H_{41}BrO_5$			
5α-Card-20(22)-enolide, 15α-bromo-14β-hydroxy-3-(2,2-dimethyl-1,3-propane-dioxy)-	MeOH	214(4.17)	5-0136-69F
15β-	MeOH	214(4.17)	5-0136-69F
$C_{28}H_{41}NO$			
3'H-Cycloprop[2,3]-5α-androst-2-en-17β-ol, 3-[p-(dimethylamino)phenyl]-2β,3β-dihydro-	MeOH	254(4.31),302(3.27)	94-2319-69

Compound	Solvent	λ_{max}(log ε)	Ref.
$C_{28}H_{42}O_2$			
[Bi-2,5-cyclohexadien-1-yl]-4,4'-dione, 3,3',5,5'-tetra-tert-butyl-	C_6H_{12}	242(3.90)	64-0547-69
2,5-Cyclohexadien-1-one, 2,4,6-tri-tert-butyl-4-(p-tert-butylphenoxy)-	C_6H_{12}	241(4.03)	64-0547-69
	MeOH	242(3.99)	64-0547-69
19-Norcholesta-1,3,5(10)-trien-3-ol, acetate	C_6H_{12}	210(4.00),217(3.95), 270(2.79),277(2.75)	39-1240-69C
19-Norcholesta-1,3,5(10)-trien-6-one, 1-methoxy-4-methyl-	C_6H_{12}	219(4.35),248(3.84), 254(3.83),320(3.54)	39-1240-69C
19-Norcholesta-1,3,5(10)-trien-6-one, 2-methoxy-4-methyl-	EtOH	207(4.25),223(4.26), 229(4.25),270(4.19)	39-1240-69C
$C_{28}H_{42}O_3$			
27-Norcholesta-5,7-dien-25-one, 3β-acetoxy-	n.s.g.	263s(3.91),271(4.06), 282(4.08),293(3.85)	54-1080-69
$C_{28}H_{42}O_4$			
25-Homochola-5,7-dienoic acid, 3β-hydroxy-, methyl ester, 3-acetate	EtOH	271(4.02),282(4.04), 293(3.80)	13-0567-69A
$C_{28}H_{42}O_5$			
5α-Card-20(22)-enolide, 3-(2,2-dimethyl-1,3-propanedioxy)-14β-hydroxy-	MeOH	216(4.21)	5-0136-69F
5β-	n.s.g.	217(4.28)	5-0168-69A
Chola-7,9(11)-dien-24-oic acid, 3α-acetoxy-12α-methyl-, methyl ester	EtOH	240s(4.16),248(4.22), 257s(4.09)	39-2723-69C
12β-	EtOH	238s(4.00),246(4.06), 255s(3.93)	39-2723-69C
$C_{28}H_{43}NO$			
Cholesta-3,5-dieno[4,3-d]isoxazole	EtOH	230(4.18)	88-5271-69
Cholest-4-ene-4-carbonitrile, 3-oxo-	EtOH	250(4.23)	88-5271-69
$C_{28}H_{44}O$			
Calciferol, 5,6-trans	iso-PrOH	273(4.39)	95-0919-69
Isocalciferol	iso-PrOH	277(4.55),288(4.64), 300(4.52)	95-0919-69
Isocalciferol, 5,6-cis	iso-PrOH	276s(4.51),287(4.58), 298s(4.42)	95-0919-69
Isotachysterol	iso-PrOH	279(4.55),290(4.66), 302(4.52)	95-0919-69
19-Norcholesta-1,3,5(10)-triene, 3-methoxy-1-methyl-	C_6H_{12}	212(4.14),225s(3.97), 281(3.18),288(3.20)	39-1240-69C
Precalciferol	iso-PrOH	261(3.95)	95-0919-69
$C_{28}H_{44}O_2$			
Cholest-5-en-3-one, 4-(hydroxymethylene)-	EtOH	232(4.18),337(3.53)	88-5271-69
5α-Ergosta-7,22-dien-6-one, 3β-hydroxy-	EtOH	244(4.16)	33-2428-69
$C_{28}H_{44}O_{12}$			
Hydronervosinic acid	MeOH	244(4.1),281(3.3), 292(3.3)	78-2723-69
$C_{28}H_{45}BrO$			
A-Homo-5α-cholest-2-en-4-one, 3-bromo-	C_6H_{12}	263(3.59)	22-3194-69
$C_{28}H_{45}ClO$			
A-Homo-5α-cholest-2-en-4-one, 3-chloro-	C_6H_{12}	252(4.00)	22-3194-69

Compound	Solvent	$\lambda_{max}(\log \epsilon)$	Ref.
$C_{28}H_{46}O$			
Cholest-3-en-2-one, 3-methyl-	EtOH	242(3.98)	32-0176-69
Cholest-4-en-6-one, 4-methyl-	EtOH	252(3.72)	39-1128-69C
Cholest-5-en-4-one, 6-methyl-	EtOH	253(3.73)	39-1128-69C
5α-Cholest-2-en-4-one, 3-methyl-	EtOH	237(3.91)	32-0206-69
A-Homo-5α-cholest-2-en-4-one	C_6H_{12}	228(3.85)	22-3194-69
A-Norcholest-5-en-2-one, 3,3-dimethyl-	C_6H_{12}	300(1.97)	22-3185-69
$C_{28}H_{46}O_2$			
5α-Cholestan-3-one, 2-(hydroxymethyl-ene)-	MeOH	202(4.26),286(4.17)	22-3166-69
$C_{28}H_{47}FO$			
5α-Cholest-9-ene, 3β-fluoro-6β-methoxy-	C_6H_{12}	199(4.05)	13-0051-69A
$C_{28}H_{48}O$			
A-Homo-5α-cholestan-4-one	C_6H_{12}	296(1.52)	22-3194-69
A-Nor-5β-cholestan-2-one, 3,3-dimethyl-	EtOH	291(1.45)	22-3189-69
$C_{28}H_{48}O_3$			
Octadecanophenone, 4'-butoxy-2'-hydroxy-	n.s.g.	270(3.9)	104-1609-69
$C_{28}H_{50}OS$			
5α-Cholestane, 6α-(R)-(methylsulfinyl)-	hexane	214(3.45)	39-1166-69C
	EtOH	216s(3.15)	39-1166-69C
5α-Cholestane, 6β-(R)-(methylsulfinyl)-	hexane	211(3.58),236s(2.78)	39-1166-69C
	EtOH	220s(2.85)	39-1166-69C
5α-Cholestane, 6β-(S)-(methylsulfinyl)-	hexane	210(3.58),232s(2.90)	39-1166-69C
	EtOH	217(3.00)	39-1166-69C
$C_{28}H_{52}O_8$			
Acetone, 1-undecanoyloxy-3-hydroxy-, dimer	dioxan	224(1.87)	23-1249-69

Compound	Solvent	$\lambda_{max}(\log \epsilon)$	Ref.
$C_{29}H_{17}Br_3O$			
2,4-Cyclopentadien-1-one, 2,3,4-tris-(p-bromophenyl)-5-phenyl-	benzene	345(3.95),514(3.19)	64-0685-69
2,4-Cyclopentadien-1-one, 2,3,5-tris-(p-bromophenyl)-4-phenyl-	benzene	341(3.90),517(3.29)	64-0685-69
$C_{29}H_{17}Cl_3O$			
2,4-Cyclopentadien-1-one, 2,3,4-tris-(p-chlorophenyl)-5-phenyl-	benzene	244(3.93),515(3.16)	64-0685-69
2,4-Cyclopentadien-1-one, 2,3,5-tris-(p-chlorophenyl)-4-phenyl-	benzene	340(3.88),519(3.25)	64-0685-69
$C_{29}H_{17}N_3O_3$			
2,2,3(1H)-Dibenzofurantricarbonitrile, 1-hydroxy-1,4-diphenyl-, acetate	C_6H_{12}	234(4.17),254s(4.04), 324s(4.06),366(4.27), 380s(4.19)	44-2407-69
2,2,3(4H)-Dibenzofurantricarbonitrile, 4-hydroxy-1,4-diphenyl-, acetate	CH_2Cl_2	255(4.56),296(4.06), 340s(3.87)	44-2407-69
$C_{29}H_{18}Br_2O$			
2,4-Cyclopentadien-1-one, 2,3-bis(p-bromophenyl)-4,5-diphenyl-	benzene	342(3.88),512(3.15)	64-0685-69
$C_{29}H_{18}Cl_2O$			
2,4-Cyclopentadien-1-one, 2,3-bis(p-chlorophenyl)-4,5-diphenyl-	benzene	342(3.85),514(3.16)	64-0685-69
$C_{29}H_{18}N_2O_4$			
Xanthene-2,2(1H)-dicarbonitrile, 1-hydroxy-9-oxo-1,4-diphenyl-, acetate	CH_2Cl_2	257(4.29),304(4.00)	44-2407-69
$C_{29}H_{18}N_2O_5$			
Xanthene-2,3-dicarboximide, 2-cyano-1,2-dihydro-1-hydroxy-N-methyl-9-oxo-1,4-diphenyl-	MeOH	304s(4.06),492(3.97)	44-2407-69
	CH_2Cl_2	269(4.22),320(4.11)	44-2407-69
$C_{29}H_{18}N_4O_2$			
2,2,3(1H)-Acridantricarbonitrile, 1-hydroxy-10-methyl-9-oxo-1,4-diphenyl-	MeOH	245(4.60),345(4.24), 424(3.96)	44-2407-69
	CH_2Cl_2	250s(4.27),276(4.30), 288(4.32),336(4.08), 420(3.63)	44-2407-69
Cyclopropa[4,5]cyclopenta[1,2-b]quinoline-1,1,8a-tricarbonitrile, 1a,2,7,8-tetrahydro-8-hydroxy-2-methyl-7-oxo-1a,8-diphenyl-	CH_2Cl_2	249(4.35),283(3.57), 298(3.52),330(3.99), 343(4.02)	44-2407-69
8,8,9(6H)-Phenanthridinetricarbonitrile, 5,7-dihydro-7-hydroxy-5-methyl-6-oxo-7,10-diphenyl-	MeOH	230s(4.64),282s(4.08), 359(4.24),445s(3.65)	44-2407-69
	CH_2Cl_2	237(4.53),263(4.25), 296(3.84),345(3.93), 410s(3.59)	44-2407-69
$C_{29}H_{19}BF_2O_3$			
Naphtho[1,2-e]-1,3,2-dioxaborin, 2,2-difluoro-4-[(2,6-diphenyl-4-pyranylidene)methyl]-	MeCN	247(4.49),266(4.30), 526(4.82)	4-0029-69
Naphtho[2,1-e]-1,3,2-dioxaborin, 2,2-difluoro-4-[(2,6-diphenyl-4-pyranylidene)methyl]-	MeCN	278(4.67),370(4.13), 410(4.07),520(4.82)	4-0029-69

Compound	Solvent	λ_{max}(log ϵ)	Ref.
$C_{29}H_{19}ClO_4S_3$			
2-Phenyl-4-[[4-(2-thienyl)-2H-1-benzothiopyran-2-ylidene]methyl]-1-benzothiopyrylium perchlorate	CH_2Cl_2	290(3.92),390(3.7), 670(4.25)	2-0311-69
$C_{29}H_{20}$			
4H-Cyclobuta[l]cyclopenta[def]phenanthrene, 7b,9a-dihydro-	EtOH	266(4.59),296s(4.28), 312s(4.02)	88-3843-69
$C_{29}H_{20}N_2O$			
1H-Benz[g]indazole, 5-hydroxy-1,3,4-triphenyl-	EtOH	220(4.64),258(4.44), 270s(--),320(3.93), 340(3.88)	103-0251-69
$C_{29}H_{20}O$			
2,4-Cyclopentadien-1-one, 2,3,4,5-tetraphenyl-	benzene	342(3.83),512(3.12)	64-0685-69
$C_{29}H_{20}O_2$			
Indene-$\Delta^{1,\alpha}$-methanol, α,2-diphenyl-, benzoate	ether	236(4.50),273(4.52), 336(4.10)	44-0879-69
9,9'-Spirobifluorene, 2,2'-diacetyl-	EtOH	227(4.68),245s(4.34), 294s(4.61),298(4.63), 324(4.50)	33-1202-69
$C_{29}H_{20}O_3$			
Benzofuran, 2-[p-[3,4-(methylenedioxy)-styryl]phenyl]-5-phenyl-	DMF	368(4.86),387(4.73)	33-1282-69
3,7-Dioxatricyclo[4.1.0.0^{2,4}]heptan-5-one, 1,2,4,6-tetraphenyl-, trans	EtOH	258s(3.40),330(2.56)	28-1157-69A
9,9'-Spirobi[fluorene]-2-butyric acid, γ-oxo-	EtOH	214s(4.71),219s(4.71), 224s(4.70),248s(4.39), 262s(4.37),265s(4.39), 268s(4.42),272(4.44), 279s(4.42),282s(4.41), 295(4.41),306(4.37), 325(4.14)	33-1202-69
$C_{29}H_{20}O_4$			
3,6-Dioxabicyclo[3.1.0]hexan-2-one, 4-benzoyl-1,4,5-triphenyl-	EtOH	254(4.23),329(2.48)	28-1157-69A
$C_{29}H_{20}O_6$			
Xylerythrin, 5-O-methyl-, acetate	EtOH	246(4.33),384(4.18)	1-2583-69
$C_{29}H_{21}ClN$			
1-(p-Chlorophenyl)-2,4,6-triphenylpyridinium(perchlorate)	MeOH	311(4.53)	5-0093-69G
$C_{29}H_{21}ClO_7$			
9,10-Dihydro-8-[(2-hydroxy-1-naphthyl)-methylene]-8H-cyclopenta[b]naphtho-[1,2-e]pyrylium perchlorate	MeOH-$HClO_4$	304(4.12),504(4.47)	5-0226-69E
$C_{29}H_{22}$			
Cyclobuta[l]phenanthrene, 2a,10b-dihydro-4-methyl-1,2-diphenyl-, cis	EtOH	267(4.60),299s(4.22), 313(4.04)	88-3843-69
$C_{29}H_{22}ClNO_4$			
1,2,4,6-Tetraphenylpyridinium perchlorate	MeOH	310(4.50)	5-0093-69G

Compound	Solvent	$\lambda_{max}(\log \epsilon)$	Ref.
$C_{29}H_{22}N_2O_2$			
1,3-Indandione, 2-phenacyl-2-phenyl-, 2-(phenylhydrazone)	EtOH	227(4.65),251(4.30), 284s(--)	103-0251-69
$C_{29}H_{22}N_2O_4$			
1,2-Diazaspiro[4.4]nona-1,3,6,8-tetraene-3,4-dicarboxylic acid, 6,7,8-triphenyl-, dimethyl ester	ether	250(4.37),315(3.99)	64-0536-69
$C_{29}H_{22}N_4$			
4,4-Imidazolidinedicarbonitrile, 1,2,3,5-tetraphenyl-	MeCN	237(4.29),287s(3.28)	44-2146-69
$C_{29}H_{22}O$			
Anisole, p-[p-[(4-biphenylyl)ethynyl]-styryl]-	DMF	358(4.84)	33-2521-69
Anisole, p-[[p-(p-phenylstyryl)phenyl]-ethynyl]-	DMF	360(4.87)	33-2521-69
Benzofuran, 5-benzyl-2-(p-styrylphenyl)-	DMF	358(4.81)	33-1282-69
Benzofuran, 5-methyl-2-[p-(p-phenyl-styryl)phenyl]-	DMF	369(4.88)	33-1282-69
Benzofuran, 6-methyl-2-[p-(p-phenyl-styryl)phenyl]-	DMF	371(4.88)	33-1282-69
1,4-Pentadien-3-one, 1,1,5,5-tetra-phenyl-	EtOH	230s(4.38),311(4.05), 336s(4.00)	39-1755-69C
2H-Pyran, 2,2,4,6-tetraphenyl-	EtOH	254(4.36),353(4.13)	22-3638-69
$C_{29}H_{22}OS$			
Benzo[b]thiophene, 2-(p-biphenylyl)-6-(p-methoxystyryl)-	DMF	283(4.24),372(4.83)	33-1282-69
Benzo[b]thiophene, 5-methyl-2-[p-(p-phenoxystyryl)phenyl]-	DMF	360(4.80)	33-1282-69
Benzo[b]thiophene, 6-methyl-2-[p-(p-phenoxystyryl)phenyl]-	DMF	362(4.80)	33-1282-69
$C_{29}H_{22}O_2$			
Benzofuran, 5-methoxy-2-[p-(p-phenyl-styryl)phenyl]-	DMF	372(4.88)	33-1282-69
Benzofuran, 6-(p-methoxystyryl)-2,3-diphenyl-	DMF	297(4.32),358(4.71)	33-1282-69
Benzofuran, 2-[p-(p-methoxystyryl)phen-yl]-5-phenyl-	DMF	366(4.86)	33-1282-69
Benzofuran, 5-methyl-2-[p-(p-phenoxy-styryl)phenyl]-	DMF	362(4.83)	33-1282-69
Benzofuran, 6-methyl-2-[p-(p-phenoxy-styryl)phenyl]-	DMF	364(4.82)	33-1282-69
9,9'-Spirobi[fluorene]-2-butyric acid	EtOH	208(4.73),220(4.76), 226(4.74),241(4.44), 266(4.41),275(4.39), 298(4.05),310(4.17)	33-1202-69
$C_{29}H_{22}O_2S$			
Naphthalene, 1-(tritylsulfonyl)-	EtOH	287(3.82)	78-2987-69
Naphthalene, 2-(tritylsulfonyl)-	EtOH	260(--),267(3.72)	78-2987-69
$C_{29}H_{22}O_5$			
Flavone, 2',7-bis(benzyloxy)-3-hydroxy-	EtOH	254(4.31),318(4.30)	104-1803-69

Compound	Solvent	$\lambda_{max}(\log \epsilon)$	Ref.
$C_{29}H_{22}S$			
Benzo[b]thiophene, 5-methyl-2-[p-(p-phenylstyryl)phenyl]-	DMF	368(4.87)	33-1282-69
Benzo[b]thiophene, 6-methyl-2-[p-(p-phenylstyryl)phenyl]-	DMF	370(4.87)	33-1282-69
$C_{29}H_{23}F_3N_2O_4$			
Uridine, 2',3'-didehydro-2',3'-dideoxy-5-(trifluoromethyl)-5'-O-trityl-	MeOH	260(3.95)	87-0543-69
$C_{29}H_{23}NO$			
2H-Pyrrole, 2-methoxy-2,3,4,5-tetraphenyl-	EtOH	236s(4.40),250s(4.32), 275s(4.13),325s(3.86)	22-1667-69
$C_{29}H_{23}NO_2$			
1-Pyrroline, 3,4-epoxy-5-methoxy-2,3,4,5-tetraphenyl-	EtOH	252(4.20)	22-1667-69
$C_{29}H_{23}NO_4$			
3-Pyrroline, 4-acetyl-1,2,2-tribenzoyl-3-methyl-5-methylene-	EtOH	234(4.53),316(4.00)	5-0137-69D
$C_{29}H_{23}NO_5$			
Chalcone, 2',4'-bis(benzyloxy)-α-nitro-	$CHCl_3$	292(4.35),314(4.35)	104-1803-69
$C_{29}H_{23}O_2P$			
1-Propanone, 1-(2-benzofuranyl)-2-(triphenylphosphoranylidene)-	EtOH	275(4.18),340(4.08)	65-0860-69
$C_{29}H_{24}$			
Phenanthrene, 7-isopropyl-1-[2-(1-naphthyl)vinyl]-	DMF	344(4.44)	33-2521-69
$C_{29}H_{24}BrO_2P$			
[1-(2-Benzofuranylcarbonyl)ethyl]triphenylphosphonium bromide	EtOH	270(3.91),317(4.33)	65-0860-69
$C_{29}H_{24}ClNO_4S$			
4-[p-(Dimethylamino)styryl]-2-phenylnaphtho[1,2-b]thiopyrylium perchlorate	CH_2Cl_2	242(4.55),305(4.26), 375(4.23),420(4.0), 712(4.6)	2-0311-69
$C_{29}H_{24}O$			
Anisole, p-[p-(2,2-diphenylvinyl)styryl]-	DMF	364(4.73)	33-2521-69
Anisole, p-[p-(p-phenylstyryl)styryl]-	DMF	375(4.89)	33-2521-69
Anisole, p-(α-phenyl-p-styrylstyryl)-	DMF	360(4.70)	33-2521-69
Benzofuran, 4,6,7-trimethyl-2-[p-[2-(1-naphthyl)vinyl]phenyl]-	DMF	374(4.70)	33-1282-69
Benzofuran, 4,6,7-trimethyl-2-[p-[2-(2-naphthyl)vinyl]phenyl]-	DMF	302(4.23),372(4.81)	33-1282-69
Naphtho[2,1-b]furan, 2-[p-(p-isopropylstyryl)phenyl]-	DMF	325(4.36),373(4.85), 393(4.71)	33-1282-69
$C_{29}H_{24}O_2$			
Benzene, 1-(p-methoxystyryl)-4-(p-phenoxystyryl)-	DMF	368(4.85)	33-2521-69
3,3'-Spirobichroman, 6,6'-diphenyl-	dioxan	256(4.69),262(4.71)	78-2155-69
3,3'-Spirobi[3H-naphtho[2,1-b]pyran], 2-butyl- (continued on next page)	dioxan	248(5.04),287(4.04), 300(4.15),312(4.19),	5-0162-69B

Compound	Solvent	λ_{max}(log ϵ)	Ref.
		335(3.98),349(4.03)	
3,3'-Spirobi[3H-naphtho[2,1-b]pyran], 2-isobutyl-	dioxan	248(5.04),287(4.09), 300(4.20),312(4.22), 335(4.02),349(4.06)	5-0162-69B
$C_{29}H_{24}O_4$			
7,7'(8H,8'H)-Spirobi[6H-dibenzo[f,h]-[1,5]dioxonin]	CH_2Cl_2	249(4.23),282(3.97)	78-5431-69
$C_{29}H_{25}ClN_2O$			
1(2H)-Quinolinecarboxanilide, 6-chloro-3,4-dihydro-2-methyl-3,4-diphenyl-	n.s.g.	235s(--),268(4.29)	40-1047-69
$C_{29}H_{25}F_3N_2O_5$			
Uridine, 2'-deoxy-5-(trifluoromethyl)-5'-O-trityl-	MeOH	262(3.98)	87-0543-69
$C_{29}H_{25}NO_3$			
o-Cinnamotoluidide, α'-hydroxy-4-methoxy-α',α'-diphenyl-	EtOH	319(4.47)	115-0057-69
6,11-Epoxy-3H-naphth[2,3-d]azepine-3-carboxylic acid, 5a,6,11,11a-tetrahydro-6,11-diphenyl-, ethyl ester, endo	EtOH	235(4.10)	44-2888-69
$C_{29}H_{26}N_4O_4S$			
9H-Purine-6(1H)-thione, 9-(5-O-trityl-β-D-arabinofuranosyl)-	EtOH	323(4.29)	23-1095-69
$C_{29}H_{26}O_5$			
Perezinone, dibenzoate	MeOH	204(4.34),234(4.60), 268(4.06),315(3.89), 327(3.93)	23-2465-69
$C_{29}H_{26}O_{10}$			
Flavone, 3'-(benzyloxy)-5,7-dihydroxy-3,4',8-trimethoxy-, diacetate	EtOH	250(4.04),369(4.22)	18-2701-69
$C_{29}H_{27}Fe_2N$			
Benzylamine, N,N-([1,1"-biferrocene]-2,2"-diyldimethylene)-	C_6H_{12}	202(4.67),225(4.70), 257s(4.05),300(3.92)	78-3477-69
$C_{29}H_{28}F_{14}O_5$			
Pregna-3,5-dien-20-one, 3,21-dihydroxy-, bis(heptafluorobutyrate)	heptane	226(4.20),231(4.18)	13-0739-69A
$C_{29}H_{28}N_4$			
Quinoline, 4-(4-benzyl-1-piperazinyl)-3-(3,4-dihydro-1-isoquinolyl)-	EtOH	342(4.63)	2-1010-69
$C_{29}H_{28}O_3$			
Cyclopenta[c]pyran-6(1H)-one, 3,4,4a,5-tetrahydro-3-hydroxy-3,4,4-trimethyl-1,1,7-triphenyl-	n.s.g.	240(4.18)	88-0909-69
$C_{29}H_{28}O_4$			
Taxoquinone	MeOH	276(4.08),408(2.90)	44-3912-69

Compound	Solvent	λ_{max}(log ε)	Ref.
$C_{29}H_{28}O_5$			
2,5-Cyclohexadien-1-one, 4-ethoxy-2,4,6-tris(p-methoxyphenyl)-	n.s.g.	235(4.37),315(3.46)	24-3795-69
$C_{29}H_{29}BrN_4O_8$			
3-(Diphenylmethyl)-3,6-dihydro-6-oxo-1-β-D-ribofuranosylpurinium bromide, 2',3',5'-triacetate	pH 1	253(3.99),280s(3.55)	44-2646-69
	pH 7	255(3.86),300(3.50)	44-2646-69
$C_{29}H_{29}N_3O_3S$			
4-Thiazoline, 5-(m-nitroanilino)-4-p-toluoyl-3-cyclohexyl-2-phenyl-	MeCN	233(3.93),350(4.05), 430(4.15)	23-3557-69
4-Thiazoline, 5-(p-nitroanilino)-4-p-toluoyl-3-cyclohexyl-2-phenyl-	MeCN	255(4.16),280(4.16), 365(4.09),440(4.11)	23-3557-69
$C_{29}H_{29}N_3O_{10}S_2$			
as-Triazine-3,5(2H,4H)-dione, 2-(5-O-trityl-β-D-ribofuranosyl)-, 2',3'-dimethanesulfonate	EtOH	212(4.39),259(3.78)	73-0618-69
$C_{29}H_{30}ClNO_6$			
3-Ethyl-6,7-dihydro-9,10-dimethoxy-2-[2-(1-naphthyl)ethyl]benzo[a]quinolizinium perchlorate	MeOH	216(4.62),282(4.12), 312s(4.09),362(4.00)	24-1779-69
$C_{29}H_{30}N_2OS$			
4-Thiazoline, 5-anilino-3-cyclohexyl-2-phenyl-4-p-toluoyl-	MeCN	260(4.16),350(3.97), 420(4.09)	23-3557-69
$C_{29}H_{30}N_2O_4$			
Vincamajine, benzoate	EtOH	231(4.26),283(3.45), 295(3.47)	25-1387-69
$C_{29}H_{30}N_4Ni$			
Nickel, [3,7-diethyl-2,8,12,15,18-pentamethylporphinato(2-)]-	$CHCl_3$	293(4.00),330(3.99), 400(5.19),522(3.99), 558(4.18)	39-0655-69C
$C_{29}H_{30}N_4NiO$			
Nickel, [1a,19a-dihydro-1a,3,5,6,10,11,-15,16,19a-nonamethyloxireno[b]porphinato(2-)]-	$CHCl_3$	260(4.23),283s(4.14), 311s(4.07),381(4.29), 421(4.53),441s(4.49), 524(3.59),580(3.61), 694(4.18)	39-0655-69C
Nickel, [3,3,5,7,8,12,13,17,18-nonamethyl-2(3H)-porphinato(2-)]-	$CHCl_3$	415(4.52),562s(3.69), 617(3.70),695(4.03)	39-0655-69C
$C_{29}H_{31}ClN_2O$			
1(2H)-Quinolinecarboxamide, 6-chloro-N-cyclohexyl-3,4-dihydro-2-methyl-3,4-diphenyl-	n.s.g.	260(4.10)	40-1047-69
$C_{29}H_{31}N_3O$			
21-Oxa-22,23,24-triazapentacyclo[16.2.-1.1^{3,6}.1^{8,11}.1^{13,16}]tetracosa-2,4,6(24),7,9,11,13(22),14,16,18,20-undecaene, 5,9,14-triethyl-4,10,15-trimethyl-	EtOH	374(4.97),398(4.97)	77-1480-69

Compound	Solvent	λ_{max}(log ε)	Ref.
$C_{29}H_{31}N_3O_4$			
20,21-Seco-2ξ-ajmaline, 4-benzoyl-20-cyano-4-de(hydroxymethyl)-2-hydroxy-, 17-acetate	6N HCl	234(4.13),243(4.03), 294(3.90)	88-0901-69
	EtOH	244(4.14),292(3.40)	88-0901-69
$C_{29}H_{31}N_3S$			
21-Thia-22,23,24-triazapentacyclo-[16.2.1.1^{3,6}.1^{8,11}.1^{13,16}]tetracosa-2,4,6(24),7,9,11,13(22),14,16,18,20-undecaene, 5,9,14-triethyl-4,10,15-trimethyl-	EtOH	403(5.07)	77-1480-69
$C_{29}H_{32}N_4$			
Porphine, 3,7-diethyl-2,8,12,15,18-pentamethyl-	$CHCl_3$	407(5.26),504(4.02), 538(3.78),574(3.77), 602(3.50),630(2.99)	39-0655-69C
$C_{29}H_{32}N_4S$			
21H-5-Thiaporphine, 13,17-diethyl-2,3,7,8,12,18-hexamethyl-	$CHCl_3$	304(4.33),384(4.67), 558(3.91),578(4.02), 599(4.15),629(4.35)	88-3689-69
hydrobromide	$CHCl_3$	309(4.30),390(4.80), 529(3.55),566(3.71), 610s(3.85),666(4.49)	88-3689-69
$C_{29}H_{33}ClN_4Ni$			
Nickel(1+), octadehydro-1,2,3,7,8,12,-13,17,18,19-decamethylcorrinato)-, chloride	$CHCl_3$	275(4.35),353(4.37), 390s(3.97),556(4.05)	39-0655-69C
$C_{29}H_{33}NO$			
Azetidine, 1-(1,1-diethylpropyl)-2-phenyl-3-(p-phenylbenzoyl)-, cis	isooctane	282(4.36)	44-0310-69
trans	isooctane	282(4.43)	44-0310-69
$C_{29}H_{33}NO_5$			
C-Homoberbine, 10-(benzyloxy)-2,3,9,11-tetramethoxy-	EtOH	230s(4.32),283(3.76)	33-1228-69
$C_{29}H_{34}O_8$			
Meliac-14(15)-enoic acid, 3β-acetoxy-1-oxo-, methyl ester	n.s.g.	226(3.97)	39-2439-69C
$C_{29}H_{34}O_{10}$			
Atalantin, acetate	n.s.g.	210(4.15)	2-0870-69
$C_{29}H_{35}N_3O_2$			
16,17-Seco-E-nor-20ξ-yohimban, 16-(3,4-dihydro-6,7-dimethoxy-1-isoquinolyl)-, dihydrochloride	EtOH	220(4.65),248(4.18), 273(3.98),282(4.01), 290(4.01),306(3.93), 360(3.87)	44-1572-69
$C_{29}H_{35}N_3O_5$			
Tryptophan, N-(N-carboxy-L-alanyl)-2-(1,1-dimethylallyl)-, N-benzyl ethyl ester, DL-	EtOH	225(4.51),283(3.89), 291(3.85)	39-1003-69C

Compound	Solvent	λ_{max}(log ε)	Ref.
$C_{29}H_{35}N_3O_7$			
Pyrrole-3-carboxylic acid, 5,5'-[(4-carboxy-3,5'-dimethyl-2H-pyrrol-2-ylidene)oxoethylene]bis[2,4-dimethyl-, triethyl ester	EtOH	484(4.60)	5-0105-69A
	EtOH-HCl	524(4.72)	5-0105-69A
$C_{29}H_{36}N_2O_4$			
2H-Benzo[a]quinolizine, 2-[(3,4-dihydro-6,7-dimethoxy-1-isoquinolyl)methylene]-3-ethyl-1,3,4,6,7,11b-hexahydro-9,10-dimethoxy-, dihydrochloride	pH 1	238(4.20),289(3.80), 310(3.84),359(3.88)	39-0094-69C
$C_{29}H_{36}N_2O_{10}$			
Isovincoside, N-acetyl-	EtOH	231(4.46),273(3.86), 280(3.84),289(3.78)	39-1193-69C
Vincoside, N-acetyl-	EtOH	230(4.49),273(3.85), 281(3.84),289(3.75)	39-1193-69C
$C_{29}H_{36}N_4O_3$			
Isohydrocoriamyrtin, 11-methylphenylhydrazino-, methylphenylhydrazone	EtOH	248(4.27),305s(4.30), 328(4.49)	78-1109-69
$C_{29}H_{36}O_7$			
Benz[4,5,6]androst-4-ene-5',6'-dicarboxylic acid, 17β-hydroxy-3-oxo-, dimethyl ester, propionate	EtOH	231(4.55),303s(3.18), 310(3.19)	94-2586-69
Benzo[4,5,6]pregn-4-ene-5',6'-dicarboxylic acid, 4'-methyl-4',6-oxido-3,20-dioxo-, dimethyl ester	EtOH	245(4.17)	94-2604-69
$C_{29}H_{36}O_{17}$			
Scandoside, methyl ester, hexaacetate	EtOH	233(4.00)	94-1942-69
$C_{29}H_{37}N_3O_2$			
16,17-Seco-E-nor-20ξ-yohimban, 16-(1,2,3,4-tetrahydro-6,7-dimethoxy-1-isoquinolyl)-, dihydrochloride	EtOH	223(4.65),281(4.06), 288s(3.97)	44-1572-69
$C_{29}H_{37}N_3O_3$			
17,18-Seco-20ξ-yohimban-17-amide, N-(3,4-dimethoxyphenethyl)-	EtOH	227(4.65),279(4.02)	44-1572-69
$C_{29}H_{38}N_2O_{10}$			
Isovincoside, N-acetyldihydro-	EtOH	229(4.48),273(3.87), 281(3.86),290(3.76)	39-1193-69C
Vincoside, N-acetyldihydro-	EtOH	228(4.51),273(3.83), 281(3.82),289(3.76)	39-1193-69C
$C_{29}H_{38}N_4O_7$			
2,6-Methano-1H-naphth[2,1-d]oxocin-9-carboxylic acid, 2,3,5,6,7,8,8a,9,-10,11,12,12a-dodecahydro-13-isopropyl-9,12a-dimethyl-7-oxo-, methyl ester, 7-(2,4-dinitrophenylhydrazone)	EtOH	384(4.34)	44-1940-69
$C_{29}H_{38}O_7$			
Benz[4,5,6]pregn-4-ene-5',6'-dicarboxylic acid, 5',6'-dihydro-4'-methyl-4',6-oxido-3,20-dioxo-, dimethyl ester, cis	EtOH	255(3.97)	94-2604-69

Compound	Solvent	λ_{max}(log ε)	Ref.
Benz[4,5,6]pregn-4-ene-5',6'-dicarboxylic acid, 5',6'-dihydro-4'-methyl-4',6-oxido-3,20-dioxo-, dimethyl ester, trans	EtOH	252(3.96)	94-2604-69
Surangin B	EtOH	222(4.43),295(4.29), 329(4.24)	78-1453-69
	EtOH-HCl	222(4.53),295(4.33), 325(4.16)	78-1453-69
	EtOH-KOH	225(4.25),258(4.06), 333(4.52)	78-1453-69
$C_{29}H_{38}O_8$			
Isohellebrigeninic acid, dimethyl ester, diacetate	MeOH	300(4.39)	44-3894-69
$C_{29}H_{39}NO_4$			
Pregna-4,6-diene-3,20-dione, 17-acetoxy-1α,2α-methylene-4-piperidino-	MeCN	282(4.16),395(3.22)	24-2570-69
$C_{29}H_{40}N_2O_5$			
Benzo[i]phenanthridin-2(3H)-one, 8-[1-(1-acetyl-5-methyl-3-oxo-2-piperidyl)ethyl]-4,4a,4b,5,9,10,10a,10b,-11,12-decahydro-5β-hydroxy-4aβ,7-dimethyl-, 6-oxide	n.s.g.	238(4.22),289(4.03)	88-3353-69
$C_{29}H_{40}O_4$			
Naphth[1',2':16,17]-5α-androst-16-ene-3'(2'αH),12-dione, 5',6',7',8',8'a,16-hexahydro-3β-hydroxy-, acetate	MeOH	239(4.15)	44-1606-69
$C_{29}H_{40}O_6$			
Androsta-2,4-diene-3,17β-diol, 2-ethyl-6-(1-hydroxyethylidene)-, triacetate	EtOH	293(4.04)	44-2288-69
$C_{29}H_{40}O_7$			
5β-Chola-20(21),22-dien-24-oic acid, 14,21-epoxy-3β,11α-dihydroxy-, methyl ester, diacetate	MeOH	304(4.19)	94-1698-69
5β-Chola-20(21),22-dien-24-oic acid, 14,21-epoxy-3β,16-dihydroxy-, methyl ester, diacetate	EtOH	295(4.27)	94-1698-69
$C_{29}H_{41}FO_7$			
Pregn-4-ene-3,20-dione, 9-fluoro-11β,16α,17-trihydroxy-21-[(tetrahydro-2H-pyran-2-yl)oxy]-, cyclic 16,17-acetal with acetone	EtOH	238(4.26)	22-1256-69
$C_{29}H_{42}O_3$			
19-Norcholesta-1,3,5(10)-trien-6-one, 1-hydroxy-4-methyl-, acetate	C_6H_{12}	210(4.40),245(3.90), 299(3.30)	39-1240-69C
28-Norolean-12-ene-3,16,22-trione	EtOH	291(3.77),310(4.12)	94-0474-69
$C_{29}H_{42}O_5$			
Diosgenin, 7-oxo-, acetate	EtOH	237(4.23)	94-2031-69
22β-Spirost-5-en-7-one, 3β-acetoxy-	EtOH	236(4.07)	2-1084-69

Compound	Solvent	λ max(log ε)	Ref.
$C_{29}H_{42}O_6$			
Chola-7,9(11)-dien-24-oic acid, 3α,12α-diacetoxy-, methyl ester	EtOH	240s(4.10),249(4.16), 258s(4.05)	39-2723-69C
3α,12β-	EtOH	238s(4.15),247(4.20), 255s(4.06)	39-2723-69C
Withaferin A, 2,7-deoxy-2,3-dihydro-3-methoxy-	EtOH	227(3.91)	44-3858-69
$C_{29}H_{42}O_7$			
Withacnistin, 18-deacetyl-2,3-dihydro-3-methoxy-	EtOH	228(3.89)	44-3858-69
Withaferin A, 2,3-dihydro-3-methoxy-	EtOH	217(3.98)	44-3858-69
$C_{29}H_{42}O_9$			
β-Uscharidin, tetrahydro-	EtOH	217(4.21)	33-2276-69
$C_{29}H_{43}NO_5$			
7a-Aza-B-homo-22β-spirost-5-en-7-one, 3β-acetoxy-	EtOH	220(4.22)	2-1084-69
22β-Spirost-5-en-7-one, 3β-acetoxy-, oxime	EtOH	237(4.13)	2-1084-69
$C_{29}H_{44}O_2$			
19-Norcholesta-1,3,5(10)-trien-1-ol, 4-methyl-, acetate	C_6H_{12}	229(3.66),265(3.20)	39-1240-69C
19-Norcholesta-1,3,5(10)-trien-3-ol, 1-methyl-, acetate	EtOH	216(3.93),273(2.64)	39-1240-69C
$C_{29}H_{44}O_3$			
Cholesta-1,4-dien-3-one, 2-acetoxy-	EtOH	250(4.22)	33-0459-69
Cholesta-5,17(20)-dien-22-one, 3β-acetoxy-, 5-cis	MeOH	253(3.90)	44-3767-69
5α-Cholesta-7,22-dien-24-one, 3β-acetoxy-	ether	222(4.29)	24-2629-69
19-Norcholesta-1,3,5(10)-triene-4-carboxylic acid, 1-methoxy-, methyl ester	C H	215(4.42),223(4.38), 258(4.05)	39-1240-69C
Stigmasta-5,22,24(28)-trien-29-oic acid, 3β-hydroxy-	EtOH	262(4.28)	35-1248-69
$C_{29}H_{44}O_4$			
Stigmasta-5,24(28)-dien-29-oic acid, 3β,22,23-trihydroxy-, γ-lactone	EtOH	216(4.18)	35-1248-69
$C_{29}H_{44}O_5$			
5α,25D-Spirost-8(9)-ene-3β,12α-diol, 3-acetate	MeOH	196(4.08)	78-2603-69
5α-Stigmast-24(28)-en-29-oic acid, 5,6-epoxy-3β,22,23-trihydroxy-, γ-lactone	EtOH	214(4.08)	35-1248-69
$C_{29}H_{44}O_9$			
Alliotoxin	EtOH	217(4.18)	105-0163-69
$C_{29}H_{44}O_{11}$			
Glucoevonogenin	EtOH	218(4.16)	105-0167-69
$C_{29}H_{46}N_2O$			
Cholest-5-en-3-one, 2-diazo-4,4-dimethyl-	EtOH	290(3.85)	22-3180-69

Compound	Solvent	λ_{max}(log ε)	Ref.
$C_{29}H_{46}O$			
Cholesta-1,4-dien-3-one, 1,4-dimethyl-	EtOH	249(4.00)	22-3180-69
Cholesta-1,5-dien-3-one, 4,4-dimethyl-	C_6H_{12}	203(4.06),221(3.99)	22-3180-69
Cholesta-2,5-dien-7-one, 4,4-dimethyl-	C_6H_{12}	235(4.09)	22-1236-69
Cholesta-3,5-dien-7-one, 3,4-dimethyl-	C_6H_{12}	282(4.38)	22-1236-69
Cholest-5-en-7-one, 3-methyl-4-methylene-	C_6H_{12}	253(4.07)	22-1236-69
19-Norcholesta-1,5(10)-dien-3-one, 1,4,4-trimethyl-	EtOH	323(3.45)	22-3180-69
19-Norcholesta-1(10),5-dien-3-one, 1,4,4-trimethyl-	EtOH	238(4.22),247(4.22), 258(4.00)	22-3180-69
A-Norcholest-5-en-2-one, 3,3-dimethyl-1-methylene-	EtOH	228(3.90)	22-3180-69
1(10)-Secocholesta-1,5,10(19)-trien-3-one, 4,4-dimethyl-	EtOH	219(4.56)	22-3180-69
Unnamed compound	n.s.g.	272(3.95)	2-1279-69
$C_{29}H_{46}O_2$			
5β,9β,10α-Ergost-7-en-3-one, 2-(hydroxymethylene)-	n.s.g.	282(3.91)	39-0674-69C
$C_{29}H_{46}O_3$			
Cholesta-5,7-diene-3β,25-diol, 3-acetate	n.s.g.	263s(3.91),271(4.06), 282(4.08),293(3.84)	54-1080-69
5α-Cholest-17(20)-en-22-one, 3β-acetoxy-	MeOH	253(3.90)	44-3767-69
$C_{29}H_{46}O_4$			
9,11-Octadecadienoic acid, 13-(4-tert-butyl-2-hydroxyphenoxy)-, methyl ester	EtOH	234(4.45),283(3.56)	23-2106-69
$C_{29}H_{46}O_4S$			
5α-Ergosta-7,22-dien-6-one, 3β-hydroxy-, methanesulfonate	EtOH	243(4.15)	33-2428-69
$C_{29}H_{46}O_7$			
Ajugasterone B	MeOH	244(4.03)	77-0082-69
$C_{29}H_{46}O_8$			
Crustecdysone, 2-acetate	EtOH	242(4.12)	12-1045-69
Crustecdysone, 3-acetate	EtOH	243(4.09)	12-1045-69
Crustecdysone, 22-acetate	EtOH	243(4.03)	12-1045-69
$C_{29}H_{47}FO_2$			
5α-Cholest-9-en-6β-ol, 3β-fluoro-, acetate	C_6H_{12}	199(4.19)	13-0051-69A
4,6β-Cyclo-4,5-seco-18-norcholest-9-en-5α-ol, 3β-fluoro-5-methyl-, acetate	n.s.g.	207(4.06)	22-1758-69
$C_{29}H_{47}NO_2$			
Cholest-5-ene-2,3-dione, 4,4-dimethyl-, 2-oxime	EtOH	236(3.89)	22-3180-69
$C_{29}H_{47}NO_3$			
Benz[c]indeno[5,4-e]azepin-6(1H)-one, 1β-(1,5-dimethylhexyl)-2,3,3a,3b,4,5,8,9,10,10a,10b,11,12,12a-tetradecahydro-8β-hydroxy-10aβ,12aβ-dimethyl-, acetate	n.s.g.	218(4.01)	2-1167-69
5α-Cholestane-2,3-dione, 2-(O-acetyloxime)	n.s.g.	233(3.82)	22-3166-69

Compound	Solvent	λmax(log ε)	Ref.
$C_{29}H_{47}NO_6$			
5α-Cholestan-6-one, 3β,5-dihydroxy-, 3-acetate nitrate	C_6H_{12}	186(3.95)(end abs.)	78-0761-69
$C_{29}H_{48}N_2O$			
5α-Cholestan-3-one, 2-diazo-4,4-dimethyl-	n.s.g.	232(4.08)	22-3173-69
$C_{29}H_{48}O$			
Cholest-3-en-2-one, 3,4-dimethyl-	C_6H_{12}	244(4.04)	22-3953-69
Cholest-5-en-7-one, 3α,4α-dimethyl-	EtOH	227(3.95)	22-1236-69
5α-Cholest-1-en-3-one, 4,4-dimethyl-	C_6H_{12}	221(3.85)	22-3173-69
Stigmast-4-en-3-one	EtOH	241(4.19)	94-0163-69
$C_{29}H_{48}OS$			
Cholest-3-eno[3,4-b][1,4]oxathiin, 5',6'-dihydro-	hexane	227(3.69)	94-0355-69
$C_{29}H_{49}BrO$			
5α-Cholestan-3-one, 2α-bromo-4,4-dimethyl-	C_6H_{12}	294(1.49)	22-3953-69
5α-Cholestan-3-one, 2β-bromo-4,4-dimethyl-	C_6H_{12}	294(1.66)	22-3953-69
$C_{29}H_{49}FO$			
Cholestan-2-one, 5β-fluoro-3,4-dimethyl-	C_6H_{12}	294(1.62)	22-3953-69
5α-Cholestan-3-one, 2α-fluoro-4,4-dimethyl-	C_6H_{12}	296(1.42)	22-3953-69
5α-Cholestan-3-one, 2β-fluoro-4,4-dimethyl-	C_6H_{12}	294(1.70)	22-3173-69
	C_6H_{12}	294(1.32)	22-3953-69
$C_{29}H_{49}NO$			
Cholestane-$\Delta^{3,\alpha}$-acetamide	MeOH	220(4.33)	39-2728-69C
$C_{29}H_{49}NO_2$			
5α-Cholestane-2,3-dione, 4,4-dimethyl-, 2-oxime	EtOH	238(3.82)	22-3173-69
$C_{29}H_{49}NO_5$			
5α-Cholestane-3β,19-diol, 3-acetate 19-nitrate	C_6H_{12}	186(3.78)(end abs.)	78-0761-69
5β-Cholestane-3β,19-diol, 3-acetate 19-nitrate	C_6H_{12}	186(3.77)(end abs.)	78-0761-69
$C_{29}H_{50}N_2O_2$			
5α-Cholestane-2,3-dione, 4,4-dimethyl-, dioxime	EtOH	225(3.62)	22-3173-69
$C_{29}H_{50}O$			
5α-Cholestan-3-one, 4,4-dimethyl-	C_6H_{12}	282(1.52)	22-3953-69
$C_{29}H_{50}OS$			
Ethanol, 2-(cholest-5-en-4α-ylthio)-	EtOH	212(3.23)	94-0355-69
$C_{29}H_{51}NO_2$			
Docosanohydroxamic acid, N-p-tolyl-	EtOH	249(4.10)	42-0831-69
$C_{29}H_{52}OS$			
5α-Cholestane, 6α-(R)-(ethylsulfinyl)-	hexane	214(3.43)	39-1668-69C
	EtOH	219s(3.04)	39-1668-69C

Compound	Solvent	$\lambda_{max}(\log \epsilon)$	Ref.
5α-Cholestane, 6α-(S)-(ethylsulfinyl)-	hexane	211(3.62),233s(3.00)	39-1668-69C
	EtOH	221(3.08)	39-1668-69C
5α-Cholestane, 6β-(R)-(ethylsulfinyl)-	hexane	212(3.62),236s(2.90)	39-1668-69C
	EtOH	222(2.95)	39-1668-69C
5α-Cholestane, 6β-(S)-(ethylsulfinyl)-	hexane	211(3.58),232s(3.00)	39-1668-69C
	EtOH	220(3.00)	39-1668-69C

Compound	Solvent	λ_{max}(log ϵ)	Ref.
$C_{30}H_{14}N_2$			
Anthraceno[2',3':2,3]fluoranthene, 1',4'-dicyano-	benzene	285(5.69),354(5.87), 362(5.84),435(4.73), 458(4.84),486(4.93), 522(4.92)	78-5639-69
$C_{30}H_{14}O_4$			
6,8,15,17-Heptacenetetrone	H_2SO_4	231(4.43),271(4.41), 323(4.76),427s(4.00), 470s(4.04),590(4.18)	77-0563-69
$C_{30}H_{15}N_3O_3$			
Cyclopent[5,6]isoindolo[2,1-a]perimidine-9,11,13(10H)-trione, 10-(2-quinolyl)-	$C_{10}H_7Cl$	429(4.57),448(4.62)	33-1259-69
$C_{30}H_{16}N_2O_4$			
s-Indacene-1,3,5,7(2H,6H)-tetrone, 2,6-di-2-quinolyl-	$C_{10}H_7Cl$	450s(4.74),475(4.99)	33-1259-69
$C_{30}H_{16}N_2O_5$			
s-Indacene-1,3,5,7(2H,6H)-tetrone, 2-(3-hydroxy-2-quinolyl)-6-(2-quinolyl)-	$C_{10}H_7Cl$	452(4.73),477(5.02)	33-1259-69
$C_{30}H_{16}N_2O_6$			
s-Indacene-1,3,5,7(2H,6H)-tetrone, 2,6-bis(3-hydroxy-2-quinolyl)-	$C_{10}H_7Cl$	453(4.68),481(5.08)	33-1259-69
$C_{30}H_{17}ClO_4$			
Spiro[phenanthro[9,10-b]-p-dioxin-2(3H),2'-phenanthro[9,10-d][1,3]-dioxole, 3-chloro-	benzene	292(4.23),305(4.36), 317(4.14),340(3.27), 357(3.38),375(3.19)	24-3747-69
$C_{30}H_{18}$			
Anthracene, 9,10-bis(phenylethynyl)-	benzene	455(4.52)	44-1734-69
1,2-Butadiene, 1,4-difluoren-9-ylidene-	CH_2Cl_2	254(4.74),278(4.54), 475(4.82),505(4.99)	78-0955-69
$C_{30}H_{18}Br_2O_2$			
Acetophenone, 2,2"-(9,10-phenanthrenediylidene)bis[4'-bromo-	n.s.g.	253(4.84),261(4.86), 284(4.61),308(4.49), 360(4.00)	88-0457-69
$C_{30}H_{18}Cl_4O_3$			
1H-Cyclopenta[b][1,4]benzodioxin-1-one, 5,6,7,8-tetrachloro-3a,9a-dihydro-9a-methyl-2,3,3a-triphenyl-	MeOH	309(4.06)	77-0467-69
$C_{30}H_{18}N_2O_6$			
[$\Delta^{3,3'}$-Biindoline]-2,2'-dione, 1,1'-bis(benzyloxy)-	dioxan	453(3.71)	24-3691-69
Xanthene-2,3-dicarboximide, 2-cyano-1,2-dihydro-1-hydroxy-9-oxo-1,4-diphenyl-, acetate	CH_2Cl_2	275(4.08),323(4.03)	44-2407-69
$C_{30}H_{18}O_6$			
3,3'-Biphthalide, 3,3'-dibenzoyl-, m. 240-242°	EtOH	252(4.35),283s(3.81)	12-0577-69
isomer m. 223-224°	EtOH	252(4.28),284s(3.81)	12-0577-69

Compound	Solvent	λ_{max}(log ε)	Ref.
$C_{30}H_{19}NO_2$			
2-Butene, 1,4-difluoren-9-ylidene-2-nitro-	$CHCl_3$	420(4.26)	24-1707-69
$C_{30}H_{20}$			
$\Delta^{5,5'}$-Bi-5H-dibenzo[a,d]cycloheptene, syn	dioxan	282(4.46)	24-1453-69
2-Butene, 1,4-difluoren-9-ylidene-	$CHCl_3$	470(4.81)	24-1707-69
4H-Cyclopenta[def]phenanthrene, dimer	CH_2Cl_2	274(4.56),285s(4.40), 303s(3.60),313s(3.85)	88-3843-69
$C_{30}H_{20}Br_2$			
Cyclobutene, 1,2-dibromo-3,4-bis(diphenylmethylene)-	CH_2Cl_2	253(4.54),287(4.63)	88-2531-69
1,2,4,5-Hexatetraene, 3,4-dibromo-1,1,6,6-tetraphenyl-	dioxan	303(4.54)	25-0491-69
$C_{30}H_{20}Br_2O_2$			
Methylium, (3,4-dioxo-1-cyclobuten-1,2-ylene)bis[diphenyl-, dibromide	MeCN	260(4.32),270(4.32), 346(3.96),460(4.14)	88-2531-69
$C_{30}H_{20}F_2$			
Benzene, 1,4-difluoro-2,3,5,6-tetraphenyl-	THF	242(4.65)	39-0489-69C
$C_{30}H_{20}O$			
Naphtho[1,2-b]furan, 2-[p-[2-(1-naphthyl)vinyl]phenyl]-	DMF	377(4.74)	33-1282-69
Naphtho[1,2-b]furan, 2-[p-[2-(2-naphthyl)vinyl]phenyl]-	DMF	289(4.36),375(4.86), 395(4.73)	33-1282-69
Naphtho[2,1-b]furan, 2-[p-[2-(1-naphthyl)vinyl]phenyl]-	DMF	380(4.78)	33-1282-69
Naphtho[2,1-b]furan, 2-[p-[2-(2-naphthyl)vinyl]phenyl]-	DMF	326(4.51),379(4.88), 399(4.75)	33-1282-69
$C_{30}H_{20}O_2$			
1,2-Cyclobutanedione, 3,4-bis(diphenylmethylene)-	MeCN	252(4.05),274(4.11), 315s(3.80),465s(3.79), 499(4.08),620(3.80)	88-2531-69
1,3-Cyclobutanedione, 2,4-bis(diphenylmethylene)-	MeOH	311(4.02),347(3.91), 437s(4.55),462(4.73)	39-1755-69C
	MeOH-KOH	256(4.21),324(4.39)	39-1755-69C
$C_{30}H_{20}O_3$			
Unidentified compound	MeCN	308(4.02),415(3.97)	88-2531-69
$C_{30}H_{20}O_4S$			
Thianthrene, 2,7-bis[3,4-(methylenedioxy)styryl]-	DMF	350(4.88)	33-1282-69
$C_{30}H_{20}O_5$			
[30]Annulene pentoxide	EtOH	234(4.54),299(4.24), 319(4.30),410(4.87), 480s(3.82)	12-1951-69
geometric isomer	ether	298(4.36),313(4.36), 400s(4.88),412(4.92), 465s(3.91),500s(3.83)	12-1951-69
Dibenzofuran, 3,7-bis[3,4-(methylenedioxy)styryl]-	DMF	382(4.93)	33-1282-69

Compound	Solvent	λ_{max}(log ϵ)	Ref.
$C_{30}H_{20}O_7$			
Xylerythrin, diacetate	EtOH	242(4.31),366(4.14), 400s(4.09)	1-2583-69
$C_{30}H_{20}O_{11}$			
Fukugetin, (±)-	MeOH	224s(4.57),275(4.33), 288(4.35),345(4.13)	88-0121-69
$C_{30}H_{20}O_{12}$			
Lumiluteoskyrin	n.s.g.	285(4.2),525(4.1)	88-0767-69
$C_{30}H_{20}S$			
Naphtho[2,1-b]thiophene, 2-[p-[2-(1-naphthyl)vinyl]phenyl]-	DMF	380(4.75)	33-1282-69
Naphtho[2,1-b]thiophene, 2-[p-[2-(2-naphthyl)vinyl]phenyl]-	DMF	288(4.23),379(4.86)	33-1282-69
$C_{30}H_{20}S_2$			
Thieno[3,4-c]thiophene-5-S^{IV}, 1,3,4,6-tetraphenyl-	$C_2H_4Cl_2$	255(4.23),296(4.30), 553(4.11)	35-3952-69
$C_{30}H_{21}ClO$			
Furan, 5-[p-(p-chlorostyryl)phenyl]-2,3-diphenyl-	DMF	302(4.33),373(4.75)	33-1282-69
$C_{30}H_{21}NO_2$			
Butane, 1,4-difluoren-9-ylidene-2-nitro-	MeOH	230(4.58),247(4.57), 256(4.61),400(4.02)	24-1707-69
	$CHCl_3$	252(4.73),260(4.89), 300(4.29),315(4.29)	24-1707-69
$C_{30}H_{22}$			
Acetylene, bis(p-styrylphenyl)-	DMF	362(4.92)	33-2521-69
Anthracene, 9-(p-styrylstyryl)-	DMF	345(4.48),392(4.36)	33-2521-69
5,5'-Bi-5H-dibenzo[a,d]cycloheptene	dioxan	231s(4.54),246s(4.32), 296(4.32)	24-1453-69
Cyclobutene, 1,2-bis(diphenylmethylene)-	$CHCl_3$	280(4.49),305s(4.39)	77-1219-69
p-Terphenyl, 4-[2-(2-naphthyl)vinyl]-	DMF	297(4.40),350(4.81)	33-2521-69
$C_{30}H_{22}Br_2$			
Cyclobutane, 1,2-dibromo-3,4-bis(diphenylmethylene)-	$CHCl_3$	262(4.39),378(4.16)	77-1219-69
$C_{30}H_{22}Br_2O_{10}$			
Benzo[ghi]perylene-4,11-dione, 1,2-diacetyl-6,9-dibromo-1,2-dihydro-5,10-dihydroxy-3,7,8,12-tetramethoxy-	EtOH	220(4.41),268(4.30), 353(3.26),495(4.17), 579(3.75),643(3.04)	39-1219-69C
$C_{30}H_{22}Cl_2N_2O_2S_2$			
Acetoacetic acid, 2-mercapto-1-thio-, anhydrosulfide with N-(p-chlorophenyl)thiobenzimidic acid N-(p-chlorophenyl)benzimidate	MeOH	202(4.84),246s(4.22)	77-1128-69
$C_{30}H_{22}N_2$			
Indolo[2,3-a]carbazole, 5,6,11,12-tetrahydro-11,12-diphenyl-	CH_2Cl_2	255(4.46),353(4.29), 369(4.46),388(4.40)	24-1198-69

Compound	Solvent	λ_{max}(log ϵ)	Ref.
$C_{30}H_{22}N_6O_2$			
2H-1,4-Benzodiazepin-2-one, 7,7'-azobis-[1,3-dihydro-5-phenyl-	MeOH	224(4.45),256(4.41), 364(4.18)	88-3201-69
$C_{30}H_{22}O$			
Benzofuran, 2-phenyl-4,6-distyryl-	DMF	330(4.81),360(4.67)	33-1282-69
Benzofuran, 2-phenyl-5,6-distyryl-	DMF	317(4.71),350(4.60)	33-1282-69
Benzofuran, 6-styryl-2-(p-styrylphenyl)-	DMF	324(4.30),382(4.91)	33-1282-69
Furan, 2,3-diphenyl-5-(p-styrylphenyl)-	DMF	300(4.33),370(4.75)	33-1282-69
Furan, 2-phenyl-5-[p-(p-phenylstyryl)-phenyl]-	DMF	310(4.20),380(4.86)	33-1282-69
$C_{30}H_{22}O_2$			
Furan, 2-[p-(p-phenoxystyryl)phenyl]-5-phenyl-	DMF	297(4.10),373(4.82)	33-1282-69
Succinaldehyde, bis(diphenylmethylene)-	$CHCl_3$	300(4.21),335s(4.06)	77-1219-69
$C_{30}H_{22}O_3$			
4-Pentenoic acid, 2-(diphenylmethylene)-3-oxo-5,5-diphenyl-	EtOH	228s(4.41),265(4.20), 304s(4.09)	39-1755-69C
$C_{30}H_{22}O_4$			
Biphenyl, 4,4'-bis[3,4-(methylenedioxy)-styryl]-	DMF	370(4.91)	33-2521-69
$C_{30}H_{22}S$			
Benzo[b]thiophene, 6-styryl-2-(p-styryl-phenyl)-	DMF	383(4.92)	33-1282-69
$C_{30}H_{23}ClO_7$			
8,9,10,11-Tetrahydro-8-[(2-hydroxy-1-naphthyl)methylene]benzo[a]xanthyl-ium perchlorate, acetate	MeOH-$HClO_4$	246(4.62),302(4.20), 480(4.26)	5-0226-69E
$C_{30}H_{23}NO_3$			
Benzamide, N-(2,3-dihydro-3-phenylphen-anthro[9,10-b]-p-dioxin-2-yl)-N-methyl-	MeOH	252(4.56),308(3.98), 356(3.08)	5-0066-69D
$C_{30}H_{24}$			
Benzene, 1,2,4-tristyryl-	DMF	324(4.78),353(4.73)	33-2521-69
Benzene, 1,3,5-tristyryl-	DMF	317(4.96)	33-2521-69
5,16-o-Benzeno-6,15:9,12-diethenocyclo-dodeca[b]naphthalene, 5,7,8,13,14,16-hexahydro-	isooctane	228(4.46),264(3.70), 271s(3.56),278(3.57), 293s(2.77),315s(2.40)	77-1358-69
$\Delta^{5,5'}$-Bi-5H-dibenzo[a,d]cycloheptene, 10,10',11,11'-tetrahydro-	dioxan	274(4.20)	24-1453-69
Butadiene, 1-phenyl-4-[p-(p-phenylsty-ryl)phenyl]-	DMF	385(4.97),405(4.82)	33-2521-69
Stilbene, 4,4'-distyryl-	DMF	385(4.96)	33-2521-69
$C_{30}H_{24}Br_2N_4$			
3,3'-Bi-2-pyrazoline, 1,1'-bis(p-bromo-phenyl)-5,5'-diphenyl-	EtOH	255(3.99),308(3.83), 401(4.64)	22-1683-69
$C_{30}H_{24}Br_3CoO_6$			
Cobalt, tris(2-bromo-1-phenyl-1,3-but-anedionato)-	$CHCl_3$	268(4.42),285(4.46), 368(4.12)	2-0628-69

Compound	Solvent	λ max (log ε)	Ref.
$C_{30}H_{24}Br_3CrO_6$			
Chromium, tris(2-bromo-1-phenyl-1,3-butanedionato)-	$CHCl_3$	267(4.40),275(4.42), 377(4.23)	2-0628-69
$C_{30}H_{24}ClNO_4$			
2,4,6-Triphenyl-1-p-tolylpyridinium perchlorate	MeOH	309(4.50)	5-0093-69G
$C_{30}H_{24}ClNO_5$			
1-(p-Methoxyphenyl)-2,4,6-triphenylpyridinium perchlorate	MeOH	308(4.50)	5-0093-69G
$C_{30}H_{24}Cl_3CoO_6$			
Cobalt, tris(2-chloro-1-phenyl-1,3-butanedionato)-	$CHCl_3$	266(4.43),285(4.48), 365(4.01)	2-0628-69
$C_{30}H_{24}Cl_3CrO_6$			
Chromium, tris(2-chloro-1-phenyl-1,3-butanedionato)-	$CHCl_3$	264(4.45),275(4.42), 374(4.39)	2-0628-69
$C_{30}H_{24}CoI_3O_6$			
Cobalt, tris(2-iodo-1-phenyl-1,3-butanedionato)-	$CHCl_3$	270(4.42),285(4.41), 370(4.20)	2-0628-69
$C_{30}H_{24}CrI_3O_6$			
Chromium, tris(2-iodo-1-phenyl-1,3-butanedionato)-	$CHCl_3$	270(4.38),280(4.23), 365(4.21)	2-0628-69
$C_{30}H_{24}N_2O_2S_2$			
Acetoacetic acid, 2-mercapto-1-thio-, anhydrosulfide with N-phenylthiobenzimidic acid N-phenylbenzimidate	MeOH	201(4.83),240s(4.10), 267s(3.58)	77-1128-69
$C_{30}H_{24}N_2O_9S$			
Uracil, 6-(β-D-ribofuranosylthio)-, 2',3',5'-tribenzoate	pH 1	231(4.42),274(4.11)	87-0653-69
	pH 11	232(4.55),284(3.96)	87-0653-69
$C_{30}H_{24}O$			
Benzofuran, 4,6-dimethyl-2-[p-(p-phenylstyryl)phenyl]-	DMF	375(4.81)	33-1282-69
Benzofuran, 5,6-dimethyl-2-[p-(p-phenylstyryl)phenyl]-	DMF	374(4.88)	33-1282-69
Benzofuran, 5,7-dimethyl-2-[p-(p-phenylstyryl)phenyl]-	DMF	371(4.88)	33-1282-69
$C_{30}H_{24}O_2$			
Benzene, 1,2-dimethoxy-4-[[p-(p-phenylstyryl)phenyl]ethynyl]-	DMF	362(4.85)	33-2521-69
Benzofuran, 5-benzyl-2-[p-(p-methoxystyryl)phenyl]-	DMF	365(4.82)	33-1282-69
1,2-Cyclobutanediol, 3,4-bis(diphenylmethylene)-	EtOH	262(4.33),369(4.24)	88-2531-69
3,3'-Spirobi[3H-naphtho[2,1-b]pyran], 2-(3-methyl-2-butenyl)-	dioxan	248(5.04),287(4.00), 300(4.16),312(4.20), 334(3.92),349(3.98)	5-0162-69B
Stilbene, 4'-[(4-biphenylyl)ethynyl]-3,4-dimethoxy-	DMF	364(4.82)	33-2521-69

Compound	Solvent	λ_{max}(log ϵ)	Ref.
$C_{30}H_{24}O_2S$			
Dibenzothiophene, 2,8-bis(p-methoxystyryl)-, (E,E)-	DMF	323(4.74)	33-1282-69
Dibenzothiophene, 3,7-bis(p-methoxystyryl)-, (E,E)-	DMF	376(4.95),393(4.88)	33-1282-69
$C_{30}H_{24}O_2S_2$			
Thianthrene, 2,7-bis(p-methoxystyryl)-, (E,E)-	DMF	343(4.90)	33-1282-69
$C_{30}H_{24}O_3$			
Dibenzofuran, 3,7-bis(p-methoxystyryl)-	DMF	377(4.92)	33-1282-69
$C_{30}H_{24}O_3S$			
Phenoxathiin, 2,8-bis(p-methoxystyryl)-	DMF	304(4.79),351(4.59)	33-1282-69
$C_{30}H_{24}O_7$			
Anhydroolivinic acid, benzhydryl ester	EtOH	257(4.05),272(3.98), 314(4.08),370(3.98)	105-0494-69
$C_{30}H_{24}O_9$			
Chrysoaphin-sl-3	$CHCl_3$	269(4.58),287s(4.24), 324(3.74),385s(4.29), 400(4.40),459(4.05), 489(4.09)	39-0627-69C
$C_{30}H_{24}O_{10}$			
Elsinochrome A	EtOH	219(4.76),262(4.59), 280(4.55),330(3.74), 430(4.40),455(4.44), 525(4.1),563(4.21)	39-1219-69C
$C_{30}H_{25}AsBiBr_3$			
Tetraphenylarsonium tribromophenylbismuthate	CH_2Cl_2	362(3.60)	101-0099-69E
$C_{30}H_{25}AsBiCl_3$			
Tetraphenylarsonium trichlorophenylbismuthate	CH_2Cl_2	326(3.48)	101-0099-69E
$C_{30}H_{25}AsBiI_3$			
Tetraphenylarsonium triiodophenylbismuthate	CH_2Cl_2	308(4.39),425(3.72)	101-0099-69E
$C_{30}H_{25}BrIO_2P$			
[1-(2-Benzofuranylcarbonyl)-1-bromopropyl]triphenylphosphonium iodide	EtOH	226(4.19),280(4.13), 340(4.19)	65-0860-69
$C_{30}H_{25}I_2O_2P$			
[1-(2-Benzofuranylcarbonyl)-1-iodopropyl]triphenylphosphonium iodide	EtOH	225(4.40),270(3.81), 325(4.05)	65-0860-69
$C_{30}H_{25}N_3$			
Carbazol-1(2H)-one, 3,4-dihydro-9-phenyl-, diphenylhydrazone	MeOH	301(4.29),371(4.13)	24-1198-69
$C_{30}H_{25}N_5O_3S$			
Glycine, N-[[2-(5H-benzo[a]phenothiazin-5-ylideneamino)-1-benzimidazolyl]carbonyl]-, butyl ester	$CHCl_3$	240(4.45),277(4.27), 355(4.15),388(4.16), 560(4.27)	44-1691-69

Compound	Solvent	$\lambda_{max}(\log \epsilon)$	Ref.
$C_{30}H_{26}N_4$			
Imidazole, 1-benzyl-4,5-diphenyl-2-(2,4-xylylazo)-	n.s.g.	397(4.25)	104-0763-69
$C_{30}H_{26}O_2$			
Benzene, 1-(p-methoxy-β-phenylstyryl)-4-(p-methoxystyryl)-	DMF	366(4.73)	33-2521-69
Biphenyl, 4,4'-bis(p-methoxystyryl)-	DMF	364(4.91)	33-2521-69
2,2'-Spirobi[2H-naphtho[1,2-b]pyran], 3-pentyl-	dioxan	256(4.85),268(4.91), 310(4.01),320(4.01), 335(3.94),350(3.85)	5-0162-69B
3,3'-Spirobi[3H-naphtho[2,1-b]pyran], 2-pentyl-	dioxan	249(5.07),301(4.24), 313(4.28),336(4.04), 350(4.11)	5-0162-69B
Spiro[2H-naphtho[1,2-b]pyran-2,3'-[3H]-naphtho[2,1-b]pyran], 2'-pentyl-	dioxan	240(4.88),259(4.72), 268(4.81),300(4.11), 312(4.18),335(4.03), 350(3.98)	5-0162-69B
Spiro[2H-naphtho[1,2-b]pyran-2,3'-[3H]-naphtho[2,1-b]pyran], 3-pentyl-	dioxan	241(4.84),257(4.69), 266(4.74),301(4.09), 313(4.11),336(3.98), 350(3.99)	5-0162-69B
Stilbene, 4-[2,2-bis(p-methoxyphenyl)-vinyl]-	DMF	364(4.64)	33-2521-69
Stilbene, 3,4-dimethoxy-4'-(p-phenyl-styryl)-	DMF	379(4.88)	33-2521-69
Stilbene, 4'-(2,2-diphenylvinyl)-3,4-dimethoxy-	DMF	368(4.72)	33-2521-69
$C_{30}H_{26}O_3$			
Stilbene, 3,4-dimethoxy-4'-(p-phenoxy-styryl)-	DMF	371(4.83)	33-2521-69
$C_{30}H_{26}O_{10}$			
Elsinochrome B	EtOH	262(4.54),339(3.70), 455(4.33),525(4.12), 563(4.17)	39-1219-69C
Xanthoaphin-sl-2	ether	257(4.47),282(4.58), 306s(3.63),357(3.86), 340s(3.48),376(4.18), 403(3.71),430(3.86), 458(3.90)	39-0627-69C
$C_{30}H_{26}O_{12}$			
α-Rubromycin, O,O',O"-trimethyl-	$CHCl_3$	304(4.32),347(4.04), 376(3.86),427(3.60)	24-0126-69
$C_{30}H_{26}O_{13}$			
Tribuloside (also in other solvents)	EtOH	270(4.35),305s(--), 315(4.48),355s(--)	102-0299-69
	EtOH-NaOAc	278(--),315(--), 360s(--)	102-0299-69
$C_{30}H_{26}O_{14}$			
Floccosin	EtOH	205(4.67),224s(4.46), 272(4.34),288(4.29), 335(3.93)	23-1561-69

Compound	Solvent	$\lambda_{max}(\log \epsilon)$	Ref.
$C_{30}H_{27}ClN_2O_2$			
1(2H)-Quinolinecarboxanilide, 6-chloro-3,4-dihydro-3-(p-methoxyphenyl)-2-methyl-4-phenyl-	n.s.g.	235(4.55),268(4.34)	40-1047-69
$C_{30}H_{27}CoO_6$			
Cobalt, tris(1-phenyl-1,3-butanedionato)-	$CHCl_3$	265(4.74),278(4.70), 355(4.05)	2-0628-69
$C_{30}H_{27}CrO_6$			
Chromium, tris(1-phenyl-1,3-butanedionato)-	$CHCl_3$	259(4.43),274(4.28), 359(4.36)	2-0628-69
$C_{30}H_{27}NO_4$			
o-Cinnamotoluidide, α'-hydroxy-α',α'-bis(p-methoxyphenyl)-	EtOH	291(4.55)	115-0057-69
$C_{30}H_{27}N_3O_4$			
3-Isoxazolidinone, 4-[[[3-hydroxy-5-(hydroxymethyl)-2-methyl-4-pyridyl]methylene]amino]-2-trityl-, DL-	MeOH	218(4.53),258(4.11), 340(3.78)	78-0163-69
$C_{30}H_{27}O_8P$			
2H-Phosphole-2,3,4,5-tetracarboxylic acid, 1,1-dihydro-1,1,2-triphenyl-, tetramethyl ester	EtOH	228(4.45),268s(3.91), 275s(3.83),344(4.28)	39-1100-69C
$C_{30}H_{28}ClN_2O_4PS$			
3-Ethyl-2-[2-[(triphenylphosphoranylidene)amino]propenyl]benzothiazolium perchlorate	EtOH	398(5.76)	65-1732-69
$C_{30}H_{28}O_4$			
Naphthalene, 1,3-bis(3,4-dimethoxystyryl)-	DMF	340(4.79)	33-2521-69
Naphthalene, 1,5-bis(3,4-dimethoxystyryl)-	DMF	369(4.62)	33-2521-69
Naphthalene, 1,6-bis(3,4-dimethoxystyryl)-	DMF	343(4.82)	33-2521-69
Naphthalene, 2,3-bis(3,4-dimethoxystyryl)-	DMF	327(4.76)	33-2521-69
Naphthalene, 2,6-bis(3,4-dimethoxystyryl)-	DMF	290(4.34),377(4.89), 396(4.83)	33-2521-69
$C_{30}H_{28}O_5$			
2,5-Cyclohexadien-1-one, 4-(allyloxy)-2,4,6-tris(p-methoxyphenyl)-	n.s.g.	235(4.49),315(3.59)	24-3795-69
p-Toluic acid, 2,6-bis(benzyloxy)-, 5-methoxy-m-tolyl ester	EtOH	218(4.1),279(3.55)	39-1721-69C
$C_{30}H_{28}O_{10}$			
Elsinochrome C	EtOH	272(4.49),339(3.67), 355(3.62),455(4.26), 523(4.05),563(4.13)	39-1219-69C
$C_{30}H_{29}BrNO_2P$			
[(3,4-Dihydro-6,7-dimethoxy-1(2H)-isoquinolylidene)methyl]triphenylphosphonium bromide	EtOH	227(4.61),276(4.01), 331(4.35)	39-0094-69C
	EtOH-HCl	207(4.66),223(4.54), 256(4.05),327(4.10), 389(3.78)	39-0094-69C

Compound	Solvent	λ max (log ε)	Ref.
$C_{30}H_{29}BrN_4O_3$			
2-Porphinecarboxylic acid, 18-acetyl-13-bromo-3,7,8,12,17-pentamethyl-, ethyl ester	$CHCl_3$	421(5.32),515(4.14), 551(4.06),584(3.87), 639(3.58)	77-0767-69
$C_{30}H_{30}Cl_2O_8$			
6,23:11,18-Dimethanodibenzo[a,k]cycloeicosenediylium, 25,26-didehydro-7,8,9,10,19,20,21,22-octahydro-, diperchlorate	98% H_2SO_4	294(5.10)	77-1170-69
$C_{30}H_{30}CuO_8$			
Copper, bis(hydrogen 4-benzoyl-2-methyl-3-oxocyclopentacarboxylato)-, dimethyl ester	EtOH	245(3.82),320(3.98)	39-1845-69C
$C_{30}H_{30}N_4O_2$			
Porphine, 3,7-diacetyl-2,8,12,13,17,18-hexamethyl-	$CHCl_3$	423(5.18),516(4.06), 553(3.96),587(3.83), 640(3.70)	77-0767-69
$C_{30}H_{30}O_8$			
Gossypol	EtOH	237(4.78),277s(4.42), 285s(4.42),375(4.20)	25-1738-69
	MeCN	234(4.95),278s(4.52), 287(4.56),365(4.22)	25-1738-69
	dioxan	234(4.95),279s(4.54), 288(4.60),365(4.34)	25-1738-69
	$CHCl_3$	236(4.89),286(4.59), 361(4.33)	25-1738-69
$C_{30}H_{31}BrNO_2P$			
Triphenyl[(1,2,3,4-tetrahydro-6,7-dimethoxy-1-isoquinolyl)methyl]phosphonium bromide, hydrobromide	EtOH-HCl	206(4.80),225(4.48), 270(3.68),276(3.70)	39-0094-69C
$C_{30}H_{31}IN_4O_2$			
2,9-Bis(dimethylamino)-2a,3,8,8a-tetrahydro-1-methyl-7-nitro-2a,10-diphenyl-3,8-ethenonaphtho[2,3-b]azetium iodide	MeOH	300s(3.87),390(3.11), 414(3.08)	89-0675-69
$C_{30}H_{32}IN_3$			
2,9-Bis(dimethylamino)-2a,3,8,8a-tetrahydro-1-methyl-2a,10-diphenyl-3,8-ethenonaphtho[2,3-b]azetium iodide	MeOH	219(4.69),325(3.51)	89-0675-69
$C_{30}H_{32}N_2$			
Ethylenediamine, N,N'-dibenzyl-1,2-dimethyl-1,2-diphenyl-	EtOH	242(2.88)	35-2653-69
Ethylenediamine, N,N'-dibenzyl-1,2-di-o-tolyl-	EtOH	255(2.08)	35-2653-69
Ethylenediamine, N,N'-dibenzyl-1,2-di-p-tolyl-	EtOH	248(2.18)	35-2653-69
$C_{30}H_{32}N_2O_2$			
Ethylenediamine, N,N'-dibenzyl-1,2-bis-(p-methoxyphenyl)-	n.s.g.	230(4.26),274(3.38), 282(3.26)	35-2653-69

Compound	Solvent	$\lambda_{max}(\log \epsilon)$	Ref.
$C_{30}H_{32}N_4Ni$			
Nickel, [2,7-diethyl-3,8,12,13,17,18-hexamethylporphinato(2-)]-	$CHCl_3$	292(4.08),333(4.11), 392(5.33),518(4.06), 553(4.54)	39-0655-69C
Nickel, [2,18-diethyl-3,7,8,12,13,17-hexamethylporphinato(2-)]-	$CHCl_3$	291(4.03),333(4.08), 392(5.27),519(4.01), 554(4.49)	39-0655-69C
$C_{30}H_{32}O_9$			
3',3'''-Bipropiophenone, 4'-hydroxy-4'''-(2-hydroxy-3-methoxy-5-propionylphenoxy)-5',5'''-dimethoxy-	EtOH	282(4.41)	44-0585-69
[m-Terphenyl]-2,2",4'-triol, 3,3",5'-trimethoxy-5,5"-dipropionyl-, 2-propionate	EtOH	252(4.55)	44-0585-69
$C_{30}H_{33}N_3O_{13}$			
L-Threonine, N,N'-[(2-hydroxy-4,9-dimethyl-3-oxo-3H-phenoxazine-1,6-diyl)dicarbonyl]di-, dimethyl ester, 3,3'-diacetate	$CHCl_3$	416(4.19)	24-3205-69
$C_{30}H_{34}N_2O_4$			
Estra-1,3,5(10)-trien-17β-ol, 3-methoxy-17-(3-methyl-5-isoxazolyl)-, carbanilate	MeOH	232(4.43),278(3.40), 288(3.29)	24-3324-69
$C_{30}H_{34}N_2O_5$			
Cylindrocarine, 1-cinnamoyl-20-hydroxy-	MeOH	215(4.61),249s(4.01), 286(4.35),293s(4.28), 317s(3.17)	39-0417-69C
2'H-Pregna-2,4-dieno[3,2-c]pyrazol-11-one, 17,20:20,21-bis(methylenedioxy)-2'-phenyl-	EtOH	269(4.26)	22-1256-69
$C_{30}H_{34}N_2O_6$			
Cylindrocarine, 1-benzoyl-20-hydroxy-, acetate	MeOH	213(4.35),275(3.79), 292(3.76)	39-0417-69C
$C_{30}H_{34}N_4$			
Porphine, 2,7-diethyl-3,8,12,13,17,18-hexamethyl-	$CHCl_3$	272(3.90),332s(4.20), 378s(4.88),403(5.17), 503(4.08),534(3.86), 571(3.71),622(3.47), 652(3.04)	39-0655-69C
$C_{30}H_{34}N_4O_{12}$			
L-Threonine, N,N'-[(2-amino-1,9-dimethyl-3-oxo-3H-phenoxazine-4,6-diyl)-dicarbonyl]di-, dimethyl ester, diacetate	$CHCl_3$	433(4.39),450(4.36)	24-3205-69
$C_{30}H_{34}O_2$			
2-Hexyne, 1,4-diethoxy-5,5-dimethyl-1,1,4-triphenyl-	EtOH	253(2.71),259(2.77), 264(2.66)	56-1843-69
$C_{30}H_{34}O_4$			
Sophoradochromene	EtOH	380(4.50)	94-1302-69
	EtOH-NaOEt	420(4.50)	94-1302-69
	EtOH-$AlCl_3$	435(4.57)	94-1302-69

Compound	Solvent	λmax(log ε)	Ref.
Sophoranochromene	EtOH	286(4.20)	94-1302-69
	EtOH-NaOH	262(4.22),348(4.49)	94-1302-69
$C_{30}H_{34}O_6$			
Androsta-4,6-dien-3-one, 6,17β-diacetoxy-4-benzoyl-	EtOH	250(4.20),292(4.34)	94-2586-69
$C_{30}H_{34}O_8$			
Benzene, 1,5-bis[3-oxo-3-(2,4-dimethoxyphenyl)propyl]-2,4-dimethoxy-	EtOH	227(4.61),268(4.44), 304(4.25)	23-1529-69
Benzene, 2,4-dimethoxy-1-[1-oxo-3-(2,4-dimethoxyphenyl)propyl]-5-[3-oxo-3-(2,4-dimethoxyphenyl)propyl]-	EtOH	227(4.59),267(4.42), 304(4.23)	23-1529-69
2-Propen-1-ol, 3,3'-p-phenylenebis[1-(2,4,6-trimethoxyphenyl)-, (E,E)-	EtOH	267(4.05)	124-1189-69
$C_{30}H_{34}O_{13}$			
1(2H)-Anthracenone, 3-(3,4-dihydroxy-1-methoxy-2-oxopentyl)-3,4-dihydro-2,8-dihydroxy-6,9-dimethoxy-, tetraacetate	EtOH	225(4.24),268(4.60), 328(3.88),354(3.65)	105-0257-69
$C_{30}H_{35}NO_5$			
Pregna-4,6-dieno[6,7-d]oxazole-3,20-dione, 6α,7α-dihydro-17-hydroxy-2'-phenyl-, acetate	MeOH	240(4.31)	48-0919-69
$C_{30}H_{35}N_3O_6$			
1,2-Secohydroquinine, 1-cyano-1',2',3',4'-tetrahydro-2-hydroxy-2'-oxo-, 2-acetate 9-benzoate, allo	EtOH	234(4.21),258(4.16), 299(3.48)	78-5329-69
1,6-Secohydroquinine, 1-cyano-1',2',3',4'-tetrahydro-6-hydroxy-2'-oxo-, 6-acetate 9-benzoate	EtOH	234(4.21),257(4.12), 298(3.44)	78-5329-69
allo	EtOH	235(4.22),259(4.16), 300(3.49)	78-5319-69
$C_{30}H_{35}N_3O_7$			
2'H-Pregna-2,4-dieno[3,2-c]pyrazol-11β-ol, 17,20:20,21-bis(methylenedioxy)-2'-(p-nitrophenyl)-	EtOH	226(4.10),299(4.20), 313(4.20)	22-1256-69
$C_{30}H_{36}N_2O_3$			
1'H-Androst-4-eno[3,2-c]pyrazol-17β-ol, 1'-acetyl-5'-phenyl-, acetate	EtOH	222(4.23),262s(4.09), 270(4.10),282s(4.04), 315(3.94),330s(3.83)	94-2319-69
$C_{30}H_{36}N_2O_5$			
Cylindrocarine, 1-hydrocinnamoyl-20-hydroxy-	MeOH	213(4.66),260(3.95), 292(3.69)	39-0417-69C
$C_{30}H_{36}N_4OS$			
Biline-1(24H)-thione, 8,12-diethyl-19-methoxy-2,3,7,13,17,18-hexamethyl-	$CHCl_3$	323(4.60),337s(4.59), 412(4.68),732(3.95)	88-3689-69
$C_{30}H_{36}O_4$			
Sophoradin	EtOH	380(4.60)	94-1299-69
	EtOH-NaOEt	480(4.70)	94-1299-69
Sophoranone	EtOH	286(4.14)	94-1299-69

Compound	Solvent	λ_{max}(log ε)	Ref.
$C_{30}H_{36}O_{10}$			
Photogedunin, acetate	n.s.g.	220(4.12)	78-5007-69
$C_{30}H_{37}Br_{2}NO_{2}$			
2'H-1'-Benzopyrano[3',2':2,3]-5α-androstan-17-one, 6',8'-dibromo-3-pyrrolidino-	C_6H_{12}	238(4.59),244(4.54), 327(3.71),340(3.69)	39-1419-69C
$C_{30}H_{37}NO_{4}$			
Androst-5-ene-3β,17β-diol, 17α-(3-phenyl-5-isoxazolyl)-, 3-acetate	MeOH	241(4.22)	24-3324-69
$C_{30}H_{37}NO_{6}$			
Benzamide, N-(7α,17-dihydroxy-3,20-dioxopregn-4-en-6β-yl)-, 17-acetate	MeOH	235(4.26)	48-0919-69
$C_{30}H_{38}N_{2}O_{3}$			
1'H-5α-Androstano[3,2-c]pyrazol-17β-ol, 1-acetyl-5'-phenyl-, acetate	EtOH	275(4.27)	94-2319-69
5β-	EtOH	275(4.27)	94-2319-69
$C_{30}H_{38}O_{5}$			
Uzarigenin, O-benzoyl-	EtOH	226(4.35),272(2.90), 279(2.80)	33-2583-69
$C_{30}H_{38}O_{6}$			
Benzene, 1,5-bis[3-(2,4-dimethoxyphenyl)propyl]-2,4-dimethoxy-	EtOH	227(4.48),280(4.06)	23-1529-69
$C_{30}H_{38}O_{8}$			
Chlorothrycin, methanolysis product	EtOH	225(4.25),257s(4.00)	33-0127-69
	EtOH-NaOH	216(4.22),258(4.23)	33-0127-69
$C_{30}H_{39}Br_{2}NO_{2}$			
2'H-1'-Benzopyrano[3',2':2,3]-5α-androstan-17β-ol, 6',8'-dibromo-3-pyrrolidino-	C_6H_{12}	238(4.51),245(4.46), 328(3.65),340(3.62)	39-1419-69C
$C_{30}H_{39}N_{3}O_{2}$			
16,17-Seco-E-nor-20ξ-yohimban, 16-(1,2,3,4-tetrahydro-6,7-dimethoxy-2-methyl-1-isoquinolyl)-, dihydrochloride	EtOH	220(4.73),273s(4.10), 279(4.11),288(4.08)	44-1572-69
$C_{30}H_{40}N_{2}$			
13-Azabicyclo[8.2.1]trideca-10,12-diene, 13,13'-p-phenylenebis-	hexane	248(4.34)	78-5357-69
$C_{30}H_{40}N_{2}O_{4}$			
1,5-Cyclohexadiene-1-carbonitrile, 3,3'-dioxybis[3,5-di-tert-butyl-4-oxo-	C_6H_{12}	312(3.51)	64-0547-69
$C_{30}H_{40}O$			
8'-Apo-β-carotenal (also other solvents)	hexane	478(5.00)	78-5383-69
	C_6H_{12}	456(5.07),483(4.97)	78-5383-69
$C_{30}H_{40}O_{6}$			
Bovinone, diacetate	EtOH	260(4.21),348(3.35)	39-2398-69C

Compound	Solvent	λmax(log ε)	Ref.
$C_{30}H_{40}O_7$			
Withacnistin	EtOH	215(4.21)	44-3858-69
$C_{30}H_{41}FN_2O_6$			
2'H-Pregna-2,4-dieno[3,2-c]pyrazol-20-one, 9-fluoro-11β,16α,17-trihydroxy-21-[(tetrahydro-2H-pyran-2-yl)oxy]-, cyclic 16,17-acetal with acetone	EtOH	260(3.97)	22-1256-69
$C_{30}H_{41}N_3O_5$			
Alangamide	EtOH and EtOH-HCl	211(4.66),227s(4.13), 286(3.85)	2-0635-69
	EtOH-NaOH	222(4.74),230s(4.19), 289(3.93),302s(3.82)	2-0635-69
$C_{30}H_{42}O_4$			
Sophoradochromene, octahydro-	EtOH	289(4.20)	94-1302-69
$C_{30}H_{42}O_5$			
5H-3,5aβ-(Epoxymethano)chryseno[2,1-c]-oxepin-5,6,11(1H,7H)-trione, tetradecahydro-2,2,7a,7b,10,10,13-heptamethyl-	EtOH	203(4.03)	44-3135-69
$C_{30}H_{42}O_6$			
Withaferin A, 2,7-deoxy-2,3-dihydro-, acetate	EtOH	226(3.96)	44-3858-69
$C_{30}H_{42}O_{10}$			
Taxa-4(20),11-diene-2α,5α,9α,10β,13α-pentol, pentaacetate	n.s.g.	217(4.06)	77-1282-69
Taxa-4(20),11-diene-5α,7β,9α,10β,13α-pentol, pentaacetate	n.s.g.	211(3.91)	77-1282-69
$C_{30}H_{44}$			
Biphenyl, 4,4'-(octadecamethylene)-	MeOH	260(4.40)	97-0342-69
1,3,5,7,9-Decapentaene, 3,8-dimethyl-1,10-bis(2,6,6-trimethyl-1-cyclohexen-1-yl)-, all trans	hexane	373(4.89)	78-5383-69
	C_6H_{12}	375(4.89)	78-5383-69
	benzene	383(4.87)	78-5383-69
	CCl_4	383(4.85)	78-5383-69
	dioxan	377(4.85)	78-5383-69
	acetone	377(4.97)	78-5383-69
	CS_2	398(4.76)	78-5383-69
	DMF	381(4.93)	78-5383-69
(also other mixtures)	80% hexane-20% CS_2	378(4.87)	78-5383-69
$C_{30}H_{44}N_6O_{11}$			
β-D-Glucopyranosiduronic acid, 17-hydroxy-3,11,20-trioxopregn-4-en-21-yl, methyl ester, 3,20-disemicarbazone	MeOH	269(4.51)	69-1188-69
$C_{30}H_{44}O_9$			
12β,14-Cyclo-13,14-seco-5α-spirosta-9(11),13(18)-dien-3β-ol, 12-methyl-, acetate	MeOH	205(3.98),210(3.65), 215(3.23)(end abs.)	78-2603-69
$C_{30}H_{46}N_6O_9$			
β-D-Glucopyranosiduronic acid, 3,20-dioxopregn-4-en-21-yl, methyl ester, 3,20-disemicarbazone	MeOH	269(4.51)	69-1188-69

Compound	Solvent	$\lambda_{max}(\log \epsilon)$	Ref.
$C_{30}H_{46}O$			
Glochidone	EtOH	229(3.99),334(1.9)	39-1710-69C
$C_{30}H_{46}O_2$			
Mollugogenol C	EtOH	243(4.55),252(4.61), 261(4.45)	42-1061-69
$C_{30}H_{46}O_3$			
5α-Ergosta-7,22-dien-6-one, 3α-acetoxy-	EtOH	243(4.18)	33-2428-69
5α-Ergosta-7,22-dien-6-one, 3β-acetoxy-	EtOH	243(4.15),311(2.06)	33-2428-69
	dioxan	241(4.20),318(2.13), 327(2.17),340(2.06)	33-2428-69
5β-Ergosta-7,22-dien-6-one, 3β-acetoxy-	EtOH	245(4.17),314(2.18)	33-2428-69
	dioxan	242(4.23),323(2.16), 335(2.22),350(2.11)	33-2428-69
8α,9β-Lanost-2-ene-1,7,11-trione	EtOH	223(3.95),305(2.09)	39-1079-69C
8β,9α-Lanost-2-ene-1,7,11-trione	EtOH	222(4.06),295(2.15)	39-1079-69C
$C_{30}H_{46}O_4$			
Bovinone, leuco-tetramethyl ether	EtOH	284(3.61)	39-2398-69C
$C_{30}H_{46}O_5$			
12β,14-Cyclo-13,14-seco-5α-spirost-13(18)-ene-3β,11α-diol, 12-methyl-, 3-acetate	MeOH	205(3.47)(end abs.)	78-2603-69
3β,11β-diol (end absorptions)	MeOH	200(3.81),205(3.62), 210(3.19),220(2.26)	78-2603-69
$C_{30}H_{47}NO_4$			
18α,19βH-Ursane-21,22-dione, 20,28-epoxy-3β-hydroxy-, 22-oxime	EtOH	254(3.70)	73-0240-69
	EtOH-KOH	312(4.00)	73-0240-69
$C_{30}H_{48}O$			
Leucotylidiene	EtOH	244(4.59),252(4.66), 261(4.48)	94-0279-69
21-Noreburica-8,20(22),23-trien-3β-ol	EtOH	241(4.45)	39-1047-69C
$C_{30}H_{48}O_2$			
Mollugogenol B	n.s.g.	243(4.17),251(4.23), 261(4.07)	42-0096-69
$C_{30}H_{48}O_3$			
5α-Ergost-7-en-6-one, 3α-acetoxy-	EtOH	243(4.15)	33-2428-69
5α-Ergost-7-en-6-one, 3β-acetoxy-	EtOH	245(4.15)	33-2428-69
19-Nor-5β-cholesta-9,11-dien-6β-ol, 3β-methoxy-5-methyl-, acetate	n.s.g.	249(4.40)	88-3575-69
$C_{30}H_{48}O_7$			
Crustecdysone, 20,22-acetonide	EtOH	242(4.09)	12-1045-69
$C_{30}H_{49}NO$			
Cholest-5-en-3-one, 4-[(dimethylamino)-methylene]-	EtOH	240(4.02),335(3.98)	88-5271-69
Homoazaoleana-11,13(18)-dien-30-ol	EtOH	242(4.51),250(4.55), 260(4.35)	104-1581-69
$C_{30}H_{50}O_2$			
5α-Cholestane-$\Delta^{3,\alpha}$-acetic acid, methyl ester	MeOH	223(3.71)	39-2728-69C
5β-	MeOH	223(3.70)	39-2728-69C

Compound	Solvent	λmax(log ε)	Ref.
Cholest-9-en-4β-ol, 5β-methyl-, acetate	n.s.g.	202(4.05)(end abs.)	88-0105-69
Cholest-5-en-2-one, 3β-hydroxy-3α,4,4-trimethyl-	C_6H_{12}	282(1.86)	22-3189-69
	EtOH	277(1.83)	22-3189-69
Cholest-5-en-3-one, 2β-hydroxy-2α,4,4-trimethyl-	EtOH	308(1.54)	22-3189-69
Cycloartenol, 11-dehydro-	MeOH	213(4.18)	39-0332-69C
A-Norcholest-5-en-2β-ol, 2α-acetyl-3,3-dimethyl-	EtOH	292(1.73)	22-3189-69
$C_{30}H_{54}OS$			
5α-Cholestane, 6α-(R)-(isopropylsulfinyl)-	hexane	210(3.57),230s(3.15)	39-1668-69C
	EtOH	223(3.00)	39-1668-69C
5α-Cholestane, 6α-(S)-(isopropylsulfinyl)-	hexane	208(3.61),227s(3.15)	39-1668-69C
	EtOH	220(3.00)	39-1668-69C
5α-Cholestane, 6β-(R)-(isopropylsulfinyl)-	hexane	212(3.60),230s(2.95)	39-1668-69C
	EtOH	223(2.90)	39-1668-69C
5α-Cholestane, 6β-(S)-(isopropylsulfinyl)-	hexane	211(3.53),229s(3.00)	39-1668-69C
	EtOH	223(2.90)	39-1668-69C
5α-Cholestane, 6α-(R)-(propylsulfinyl)-	hexane	215(3.48),230s(3.15)	39-1668-69C
	EtOH	217(3.11)	39-1668-69C
5α-Cholestane, 6α-(S)-(propylsulfinyl)-	hexane	211(3.49),227s(3.18)	39-1668-69C
	EtOH	219(3.04)	39-1668-69C
5α-Cholestane, 6β-(R)-(propylsulfinyl)-	hexane	210(3.66),232s(3.08)	39-1668-69C
	EtOH	222(3.00)	39-1668-69C
5α-Cholestane, 6β-(S)-(propylsulfinyl)-	hexane	209(3.54),230s(3.04)	39-1668-69C
	EtOH	218s(3.04)	39-1668-69C

Compound	Solvent	$\lambda_{max}(\log \epsilon)$	Ref.
$C_{31}H_{17}N$			
Diacenaphtho[1,2-b:1',2'-d]pyridine, 5-phenyl-	dioxan	348(4.65),410(4.03)	95-0789-69
Diacenaphtho[1,2-b:1',2'-e]pyridine, 4-phenyl-	dioxan	328(4.94),380(4.05), 400(4.22)	95-0789-69
$C_{31}H_{18}Br_2Cl_3P$			
Phosphorane, tris(p-chlorophenyl)(2,7-dibromofluoren-9-ylidene)-	$CHCl_3$	289(4.46),379(3.35)	5-0001-69D
$C_{31}H_{18}N_4$			
Anthracene, 9-(p-tricyanovinylanilino)-10-phenyl-	$CHCl_3$	262(4.95),370(3.70), 382(3.92),487(4.52)	22-1659-69
$C_{31}H_{20}Br_2ClP$			
Phosphorane, (p-chlorophenyl)(2,7-dibromofluoren-9-ylidene)diphenyl-	$CHCl_3$	258(4.64),292(4.52), 384(3.57)	5-0001-69D
$C_{31}H_{20}Br_3P$			
Phosphorane, (p-bromophenyl)(2,7-dibromofluoren-9-ylidene)diphenyl-	$CHCl_3$	292(4.51),384(3.55)	5-0001-69D
$C_{31}H_{20}O_2$			
2,2'-Spirobi[2H-naphtho[1,2-b]pyran], 3-phenyl-	dioxan	267(4.79),281(4.62), 327(4.20),337(4.21), 352(4.16)	5-0162-69B
3,3'-Spirobi[3H-naphtho[2,1-b]pyran], 2-phenyl-	dioxan	247(4.99),314(4.35), 327(4.30),350(4.31)	5-0162-69B
Spiro[2H-naphtho[1,2-b]pyran-2,3'-[3H]-naphtho[2,1-b]pyran], 2'-phenyl-	dioxan	237(4.77),244(4.70), 283(4.66),315(4.26), 336(4.27),350(4.26)	5-0162-69B
$C_{31}H_{21}Br_2P$			
Phosphorane, (2,7-dibromofluoren-9-ylidene)triphenyl-	$CHCl_3$	264(4.63),272(4.53), 290(4.53),384(3.52)	5-0001-69D
$C_{31}H_{21}ClO_4$			
9-(5-Fluoren-9-ylidene-1,3-pentadienyl)-fluoren-9-ylium perchlorate (relative absorbancies given)	CH_2Cl_2	260(4.89),360(4.45), 460(4.34),700(4.65), 770(5.00)	78-0955-69
$C_{31}H_{21}ClO_4S_2$			
2-Phenyl-4-[(2-phenyl-4H-1-benzothiopyran-4-ylidene)methyl]-1-benzothiopyrylium perchlorate	CH_2Cl_2	255(4.62),390(4.15), 707(4.52)	2-0311-69
2-Phenyl-4-[(4-phenyl-2H-1-benzothiopyran-2-ylidene)methyl]-1-benzothiopyrylium perchlorate	CH_2Cl_2	280(3.86),388(3.66), 650(4.2)	2-0311-69
$C_{31}H_{22}N_2$			
1-Anthramine, N,N'-1-propen-1-yl-3-ylidenebis-, hydrochloride	MeOH	252(5.26),282s(--), 310(4.37),343s(--), 355s(--),417(4.27)	59-1075-69
$C_{31}H_{22}N_2O_2$			
1H-Benz[g]indazol-5-ol, 1,3,4-triphenyl-, acetate	EtOH	222(4.76),252(4.68), 310(4.17),342(3.83)	103-0251-69

Compound	Solvent	$\lambda_{max}(\log \epsilon)$	Ref.
$C_{31}H_{23}IN_4$			
1-(2-Quinolyl)-4-[3-[1-(2-quinolyl)-4-(1H)-pyridylidene]propenyl]pyridinium iodide	MeOH	672(5.20)	22-1275-69
$C_{31}H_{24}N_4O_7S$			
Purine-6(1H)-thione, 7-β-D-ribofuranosyl-, 2',3',5'-tribenzoate	pH 1	243(4.54),345(4.24)	44-2646-69
	pH 7	235(4.59),342(4.18)	44-2646-69
	pH 13	226(4.54),319(4.18)	44-2646-69
$C_{31}H_{24}O$			
Anthracene, 9-[p-(p-methoxystyryl)-styryl]-	DMF	352(4.57),393(4.45)	33-2521-69
$C_{31}H_{24}O_2$			
Furan, 5-[p-(p-methoxystyryl)phenyl]-2,3-diphenyl-	DMF	310(4.28),375(4.80)	33-1282-69
$C_{31}H_{24}O_3$			
2,4-Cyclopentadien-1-one, 2,3-bis(p-methoxyphenyl)-4,5-diphenyl-	benzene	370(3.96),512(3.08)	64-0685-69
4-Pentenoic acid, 2-(diphenylmethylene)-3-oxo-5,5-diphenyl-, methyl ester	EtOH	271(4.21),337s(3.96)	39-1755-69C
$C_{31}H_{25}NO$			
2-Cyclopropene, 3-ethoxy-1,2-diphenyl-1-(α-cyanobenzhydryl)-	CH_2Cl_2	216(4.51),267s(--), 279(4.14),290(--)	24-0319-69
$C_{31}H_{26}$			
Phenanthrene, 7-isopropyl-1-(p-phenylstyryl)-	DMF	285(4.52),341(4.62)	33-2521-69
Stilbene, 4-[(4-biphenylyl)ethynyl]-4'-isopropyl-	DMF	353(4.86)	33-2521-69
$C_{31}H_{26}Cl_2N_2O_8$			
Uracil, 5-[(benzyloxy)methyl]-1-(2-deoxy-α-D-ribofuranosyl)-, 3',5'-bis(p-chlorobenzoate)	$CHCl_3$	245(4.77),270s(4.30), 282s(4.04)	63-0809-69
β-	$CHCl_3$	247(4.78),270s(4.36), 282s(4.05)	63-0809-69
$C_{31}H_{26}N_2O_2$			
Acetophenone, 2,2'-(trimethylenedinitrilo)bis[2-phenyl-	MeOH	251(3.84)	39-2044-69C
$C_{31}H_{26}N_4O$			
Ketone, methyl 1,3,7,9-tetraphenyl-1,2,7,8-tetraazaspiro[4.4]nona-2,8-dien-6-yl	n.s.g.	242(4.54),339(4.47)	28-1063-69B
$C_{31}H_{26}N_4O_2$			
1,2,7,8-Tetraazaspiro[4.4]nona-2,8-diene-6-carboxylic acid, 1,3,7,9-tetraphenyl-, methyl ester	n.s.g.	242(4.51),343(4.48)	28-1063-69B
$C_{31}H_{26}O$			
Benzofuran, 2-[p-(p-isopropylstyryl)-phenyl]-5-phenyl-	DMF	362(4.86),380(4.70)	33-1282-69
Benzofuran, 4,6,7-trimethyl-2-[p-(p-phenylstyryl)phenyl]-	DMF	375(4.85)	33-1282-69

Compound	Solvent	λ_{max}(log ε)	Ref.
$C_{31}H_{26}O_4$			
Benzofuran, 5-phenyl-2-[p-(3,4,5-trimethoxystyryl)phenyl]-	DMF	365(4.85)	33-1282-69
$C_{31}H_{26}O_7$			
Flavone, 3',7-bis(benzyloxy)-5-hydroxy-4',6-dimethoxy-	EtOH	244(4.29),277(4.24), 339(4.39)	31-0355-69
	EtOH-$AlCl_3$	261(4.16),293(4.25), 364(4.38)	31-0355-69
$C_{31}H_{26}O_8$			
Flavone, 3',4'-bis(benzyloxy)-5,7-dihydroxy-3,6-dimethoxy-	EtOH	255(4.27),273(4.23), 347(4.35)	18-1649-69
	EtOH-NaOAc	276(4.36),322(4.38), 364(4.23)	18-1649-69
	EtOH-NaOEt	277(4.47),298(4.18), 312(4.18),384(4.27)	18-1649-69
	EtOH-$AlCl_3$	262(--),285(--), 368(--)	18-1649-69
Flavone, 3',4'-bis(benzyloxy)-5,7-dihydroxy-3,8-dimethoxy-	EtOH	258(4.28),277(4.32), 339(4.21),355(4.20)	18-1649-69
	EtOH-NaOAc	258(4.04),284(4.38), 320(4.18),366(4.08)	18-1649-69
	EtOH-NaOEt	233(4.21),287(4.40), 316(4.11),390(3.98)	18-1649-69
	EtOH-$AlCl_3$	263(--),287(--), 306s(--),360(--), 420(--)	18-1649-69
$C_{31}H_{26}O_9$			
Chrysoaphin-sl-1	$CHCl_3$	266(4.57),312(3.62), 327(3.67),382(4.31), 403(4.45),430(3.80), 456(4.03),486(4.09)	39-0627-69C
$C_{31}H_{26}S$			
Benzo[b]thiophene, 2-(p-biphenylyl)-6-(p-isopropylstyryl)-	DMF	284(4.19),368(4.83)	33-1282-69
$C_{31}H_{27}ClN_2O_4$			
1-[p-(Dimethylamino)phenyl]-2,4,6-triphenylpyridinium perchlorate	MeOH	307(4.63),438(3.44)	5-0093-69G
$C_{31}H_{27}NO_{10}$			
2-Anthraceneacetanilide, 4,5,6,8-tetrahydroxy-α-methoxy-, tetraacetate	EtOH	264(5.20),320(3.36), 340(3.59),356(3.71), 374(3.70),396(3.69)	105-0494-69
$C_{31}H_{27}N_3O_3$			
9-Acridanone, 2,7-dimethyl-4,5-bis[(salicylideneamino)methyl]-	EtOH	316(3.83),390(3.74), 405(3.79)	18-1934-69
cobalt complex	EtOH	500(3.32),571(3.04), 1370(2.68)	18-1934-69
$C_{31}H_{28}$			
Benzene, 1-(p-isopropylstyryl)-4-(p-phenylstyryl)-	DMF	372(4.90)	33-2521-69
Cyclopropane, 3-(2,2-diphenylvinyl)-1,1-dimethyl-2,2-diphenyl-	tert-BuOH	271(4.39)	35-1718-69

Compound	Solvent	λ_{max}(log ε)	Ref.
$C_{31}H_{28}ClN_2O_4PS$			
3-Ethyl-2-[4-(triphenylphosphoranylidene)amino]-1,3-butadienyl]benzothiazolium perchlorate	EtOH	500(4.96)	65-1732-69
$C_{31}H_{28}IO_2P$			
[1-(2-Benzofuranylcarbonyl)-1-methylphenyl]triphenylphosphonium iodide	EtOH	274(4.05),316(4.33)	65-0860-69
$C_{31}H_{28}N_4O_4S$			
2'-Acetonaphthone, 2-[3-[(4-amino-2-methyl-5-pyrimidinyl)methyl]-5-(2-hydroxyethyl)-4-methyl-4-thiazolin-2-ylidene]-2-hydroxy-, monobenzoate	EtOH	216(4.62),231(4.31), 264(3.92),402(4.39)	94-0128-69
$C_{31}H_{28}O_3$			
Stilbene, 4-[2,2-bis(p-methoxyphenyl)-vinyl]-4'-methoxy-	DMF	370(4.73)	33-2521-69
Stilbene, 3,4-dimethoxy-4'-(p-methoxy-β-phenylstyryl)-	DMF	369(4.72)	33-2521-69
$C_{31}H_{29}NO_6$			
α-D-xylo-Tetrofuranose, 3-O-benzyl-4-C-(1-carboxybenzo[f]quinolin-3-yl)-1,2-O-cyclohexylidene-	n.s.g.	243s(4.52),256(4.56)	65-1413-69
$C_{31}H_{30}N_4O_4S$			
9H-Purine-6-thiol, 9-(2,3,4-tri-O-benzyl-α-D-arabinopyranosyl)-	pH 1	323(4.31)	87-1125-69
	pH 7	321(4.29)	87-1125-69
	pH 13	312(4.34)	87-1125-69
β-anomer	pH 13	313(4.37)	87-1125-69
	MeOH	322(4.36)	87-1125-69
$C_{31}H_{30}N_4O_5$			
Hypoxanthine, 9-(2,3,4-tri-O-benzyl-α-D-arabinopyranosyl)-	pH 1	250(4.05)	87-1125-69
	pH 7	249(4.07)	87-1125-69
	pH 13	253(4.13)	87-1125-69
β-anomer	pH 1	251(4.05)	87-1125-69
	pH 7	249(4.08)	87-1125-69
	pH 13	253(4.14)	87-1125-69
$C_{31}H_{30}O_2$			
Hardwick acid, methyl ester	EtOH	215(4.06)	7-0539-69
$C_{31}H_{31}N_5O_4$			
Adenine, 9-(2,3,4-tri-O-benzyl-α-D-arabinopyranosyl)-	pH 1	257(4.10)	44-0092-69
	pH 7	260(4.12)	44-0092-69
	pH 13	260(4.12)	44-0092-69
β-anomer	pH 1	257(4.12)	44-0092-69
	pH 7	260(4.13)	44-0092-69
	pH 13	260(4.13)	44-0092-69
$C_{31}H_{32}N_2O_5S$			
Dimethylsulfoxonium (1,4-dihydro-4-oxo-5-O-trityl-β-D-arabinofuranosyl)-2-pyrimidinyl]methylide	MeOH	230(4.38),279(4.37)	35-7752-69
$C_{31}H_{32}N_2O_6$			
D-Allitol, 2,5-anhydro-3,4,6-tri-O-benzyl-1-deoxy-1-(3,4-dihydro-2,4-dioxo-1(2H)-pyrimidinyl)-	EtOH	216(4.16),264(4.01)	73-1684-69

Compound	Solvent	λ max (log ε)	Ref.
$C_{31}H_{32}N_2O_6S$			
Dimethylsulfoxonium [1,4-dihydro-4-oxo-1-(5-O-trityl-β-D-arabinofuranosyl)-2-pyrimidinyl]methylide	MeOH	231(4.34),280(4.33)	35-7752-69
$C_{31}H_{32}N_4O_2$			
Porphine, 3,7-diacetyl-12-ethyl-2,8,13,17,18-pentamethyl-	$CHCl_3$	422(5.21),515(4.06), 553(3.96),586(3.79), 640(3.68)	77-0767-69
$C_{31}H_{32}N_4O_3$			
2-Porphinecarboxylic acid, 18-acetyl-3,7,8,12,13,17-hexamethyl-, ethyl ester	$CHCl_3$	419(5.27),513(4.11), 550(3.98),585(3.85), 638(3.72)	77-0767-69
$C_{31}H_{32}N_6O_4$			
9H-Purine, 2,6-diamino-9-(2,3,5-tri-O-benzyl-β-D-arabinofuranosyl)-	pH 1	254(4.04),292(3.95)	87-0498-69
	pH 7, 13	256(3.95),279(3.96)	87-0498-69
	EtOH	256(4.03),283(3.95)	87-0498-69
$C_{31}H_{32}O_{12}$			
1,3,4,6,9,10-Phenanthrenehexol, 2-isopropyl-5,8-dimethyl-, hexaacetate	EtOH	237(4.16),264(4.58), 285(4.25),317(4.02), 329(4.00),360(2.88), 379(2.88)	33-1685-69
$C_{31}H_{34}D_3N_4Ni$			
Nickel(1+), (octadehydro-3,8-diethyl-2,7,12,13,17,18,19-heptamethyl-1-methyl-1-d_3-corrinato)-, iodide	$CHCl_3$	267(4.49),271s(4.49), 352(4.49),397s(3.80), 457(3.67),562(3.52)	39-0655-69C
perchlorate	$CHCl_3$	274(4.48),352(4.50), 454(3.69),559(4.14)	39-0655-69C
$C_{31}H_{34}IN_3O$			
2,9-Bis(dimethylamino)-2a,3,8,8a-tetrahydro-5-methoxy-1-methyl-2a,10-diphenyl-3,8-ethenonaphtho[2,3-b]azetium iodide	MeOH	210(4.66),293(3.74), 325(3.75)	89-0675-69
$C_{31}H_{34}N_2$			
Toluene, α,α-bis(2,3,6,7-tetrahydro-1H,5H-benzo[ij]quinolizin-9-yl)-	98% HOAc	440(4.24),657(4.04)	39-1068-69B
$C_{31}H_{34}N_4Ni$			
Nickel, [2,17-diethyl-3,5,7,8,12,13,18-heptamethylporphinato(2-)]-	$CHCl_3$	408(5.23),537(3.99), 569(4.13)	39-0655-69C
Nickel, [13,17-diethyl-2,3,5,7,8,12,18-heptamethylporphinato(2-)]-	$CHCl_3$	292(4.06),330(4.06), 401(5.29),521(4.06), 556(4.25)	39-0655-69C
Nickel, [2,3,7,8,12,13,17,18-octamethyl-5-propylporphinato(2-)]-	$CHCl_3$	299(4.03),345(4.11), 408(5.22),534(3.96), 571(4.08)	39-0655-69C
Nickel, [2,7,12-triethyl-3,8,13,17,18-pentamethylporphinato(2-)]-	$CHCl_3$	291(4.00),333(4.04), 392(5.25),476(3.29), 518(3.98),553(4.47)	39-0655-69C
$C_{31}H_{34}O_7S$			
19-Nor-9β-pregna-1(10),5,16-triene-2,7,11,20-tetrone, 9-(hydroxymethyl)-4,4,14-trimethyl-, p-toluenesulfonate	EtOH	227(4.48),288(4.41)	39-1050-69C

Compound	Solvent	λ_{max}(log ε)	Ref.
$C_{31}H_{34}O_{13}$			
1(2H)-Anthracenone, 3,4-dihydro-2,6,8,9-tetrahydroxy-3-[methoxy(2,2,5-trimethyl-1,3-dioxolan-4-yl)carbonyl]methyl]-, tetraacetate	EtOH	252s(4.69),259(4.79), 305(3.81),360(3.44)	105-0257-69
$C_{31}H_{35}BrO_9$			
Neophorbol, 3-O-(p-bromobenzoyl)-, 13,20-diacetate	MeOH	245(4.32),282(2.94)	5-0130-69E
$C_{31}H_{35}FN_2O_5$			
2'H-Pregna-2,4,6-trieno[3,2-c]pyrazol-20-one, 9-fluoro-11β,16α,17,21-tetrahydroxy-2'-phenyl-, cyclic 16,17-acetal with acetone	EtOH	228(4.11),311(4.30)	22-1256-69
$C_{31}H_{36}FN_3O_7$			
2'H-Pregna-2,4-dieno[3,2-c]pyrazol-20-one, 9-fluoro-11β,16α,17,21-tetrahydroxy-2'-(p-nitrophenyl)-, cyclic 16,17-acetal with acetone	EtOH	226(4.09),317(4.22)	22-1256-69
$C_{31}H_{36}F_2N_2O_5$			
2'H-Pregna-2,4-dieno[3,2-c]pyrazol-20-one, 9-fluoro-2'-(p-fluorophenyl)-11β,16α,17,21-tetrahydroxy-, cyclic 16,17-acetal with acetone	EtOH	262(4.20)	22-1256-69
$C_{31}H_{36}N_2O_4$			
2'H-Pregna-2,4,6-trieno[3,2-c]pyrazol-20-one, 16α,17,21-trihydroxy-2'-phenyl-, cyclic 16,17-acetal with acetone	EtOH	223(4.08),314(4.27)	22-1256-69
$C_{31}H_{36}O_4$			
Spiro[bicyclo[5.4.1]dodeca-2,5,7,9,11-pentaene-4,1'(3'aH)-pentalene]-2',3'-dicarboxylic acid, 4',6'-di-tert-butyl-, dimethyl ester	C_6H_{12}	236(4.44),277(4.62), 313s(3.60)	88-2093-69
isomer	C_6H_{12}	237(4.48),276(4.60), 315s(3.63)	88-2093-69
$C_{31}H_{36}O_7S$			
19-Nor-9β-pregna-1(10),5,16-triene-2,11,20-trione, 7β-hydroxy-9-(hydroxymethyl)-4,4,14-trimethyl-, 9-p-toluenesulfonate	EtOH	227(4.40),243s(4.13), 296(4.03)	39-1050-69C
19-Nor-9β-pregna-1(10),5,16-triene-7,11,20-trione, 2α-hydroxy-9-(hydroxymethyl)-4,4,14-trimethyl-, 9-p-toluenesulfonate	EtOH	226(4.44),243s(4.16), 285(4.20)	39-1050-69C
2β-	EtOH	226(4.41),243s(4.13), 285(4.19)	39-1050-69C
$C_{31}H_{37}ClN_4NiO_4$			
Nickel(1+), (octadehydro-3,8-diethyl-1,2,7,12,13,17,18,19-octamethylcorrinato)-, perchlorate	$CHCl_3$	276(4.49),353(4.53), 559(4.16)	39-0655-69C

Compound	Solvent	λ_{max}(log ε)	Ref.
$C_{31}H_{37}FN_2O_5$			
2'H-Pregna-2,4-dieno[3,2-c]pyrazol-20-one, 9-fluoro-11β,16α,17,21-tetrahydroxy-2'-phenyl-, cyclic 16,17-acetal with acetone	EtOH	262(4.25)	22-1256-69
1'H-Pregn-4-eno[3,2-c]pyrazol-20-one, 9-fluoro-11β,16α,17,21-tetrahydroxy-1'-phenyl-, cyclic 16,17-acetal with acetone	EtOH	299(4.46)	22-1256-69
$C_{31}H_{37}FN_4Ni$			
Nickel, (octadehydro-8,12-diethyl-1,2,3,7,13,17,18,19-octamethylcorrinato)-, fluoride	$CHCl_3$	275(4.35),352(4.38), 402s(3.64),456(3.54), 559(4.01)	39-0655-69C
$C_{31}H_{37}IN_4Ni$			
Nickel, (octadehydro-3,8-diethyl-1,2,7,12,13,17,18,19-octamethylcorrinato)-, iodide	$CHCl_3$	269(4.47),352(4.45), 402s(3.84),453s(3.70), 561(4.09)	39-0655-69C
$C_{31}H_{38}O_7$			
14,22:15α,24:16β,20-Tricyclo-26,27-dinor-14β,17α-cholesta-5,23-diene-21,25-dioic acid, 17-acetyl-3β-hydroxy-, dimethyl ester, acetate	MeOH	245(3.80)	78-4579-69
$C_{31}H_{39}BrN_4$			
Biline, 1-bromo-3,8,19-triethyl-10,23-dihydro-2,7,12,13,17,18-hexamethyl-, dihydrobromide	$CHCl_3$	376(4.16),442s(4.23), 461(4.50),504s(4.66), 533(5.19)	39-0655-69C
$C_{31}H_{39}FN_2O_5$			
2'H-5α-Pregn-2-eno[3,2-c]pyrazol-20-one, 9-fluoro-11β,16α,17,21-tetrahydroxy-2'-phenyl-, cyclic 16,17-acetal with acetone	dioxan	258(4.12)	22-1256-69
5β-	dioxan	256(4.14)	22-1256-69
$C_{31}H_{39}F_3O_9$			
Pregna-4,6-dien-20-one, 21-acetoxy-9α-fluoro-11β,16α,17α-trihydroxy-2,3-(difluoromethylene)-3-(2-acetoxyethoxy)-, 16,17-acetonide	EtOH	245(4.30)	78-1219-69
$C_{31}H_{39}N_5O_5$			
Ergocornine	MeOH	311(3.91)	33-1549-69
Ergocorninine	MeOH	241(4.31),313(3.92)	33-1549-69
$C_{31}H_{42}N_2O_6$			
Batrachotoxin	MeOH-HCl	234(3.99),262(3.70)	35-3931-69
$C_{31}H_{42}O_7$			
Surangin B, methyl ether	EtOH	210(4.51),224(4.50), 300(4.17)	78-1453-69
$C_{31}H_{44}O_2$			
1H,8H-Benzo[1,2-c:3,4-c']bis[2]benzopyran, 4,4a,8a,9,12,12a,14,14a-octahydro-3,8,8,11,14,14-hexamethyl-5-pentyl-	EtOH	236s(4.03),279(3.35), 288(3.34)	33-1102-69

Compound	Solvent	λ_{max}(log ϵ)	Ref.
Resorcinol, 2,4-di-p-mentha-1,8-dien-3-yl)-5-pentyl-, trans	EtOH	232s(4.31),284(3.61)	33-1102-69
$C_{31}H_{45}NO_4$			
Glycyrrhetic acid Δ^{18}-lactam	EtOH	282(4.12)	104-1581-69
$C_{31}H_{45}NO_5$			
3(4H)-Pyridone, 2-(3β,16β-diacetoxy-pregn-5-en-20α-yl)-5,6-dihydro-5-methyl-	MeOH	270s(2.24),400(2.01)	5-0159-69C
$C_{31}H_{46}N_4O_3$			
Glycyrrhetic acid derivative (tetrazole)	EtOH	248(4.08)	104-1581-69
$C_{31}H_{46}O_4$			
Glycyrrhetic acid, 18-dehydro-, methyl ester	EtOH	283(4.14)	104-1581-69
19-Norcholesta-1,3,5(10)-triene-1,2-diol, 4-methyl-, diacetate	EtOH	205(4.51),270(2.74)	33-0459-69
28-Noroleana-12,17-dien-16-one, 3β,21β-dihydroxy-, 3-acetate	EtOH	300(4.09)	44-3135-69
$C_{31}H_{46}O_5$			
Chola-5,20,22-trien-24-oic acid, 21-methoxy-3β-[(tetrahydro-2H-pyran-2-yl)oxy]-, methyl ester	MeOH	295(4.32)	88-3029-69
$C_{31}H_{46}O_8$			
Taxa-4(20),11-diene-2α,5α,7β,10β-tetrol, 5α,7β,10β-triacetate (2-methylbutyrate)	n.s.g.	209(3.96)	77-1282-69
$C_{31}H_{46}O_9$			
Taxa-4(20),11-diene-2α,5α,7β,9α,10β-pentol, 7β,9α,10β-triacetate 2-(2-methylbutyrate)	n.s.g.	215(4.02)	77-1282-69
$C_{31}H_{47}NO_4$			
1H-Chryseno[2,1-c]azepine-12-carboxylic acid, eicosahydro-5,5,7a,7b,9a,12,15b-heptamethyl-3,15-dioxo-, methyl ester	EtOH	244(4.08)	104-1581-69
4(1H)-Pyridone, 1-acetyl-2,3-dihydro-6-(3β-hydroxy-5α-pregnan-20α-yl)-3-methyl-, acetate	MeOH	246(3.98)	5-0159-69C
Solacongestidine, 24-oxo-, N,O-diacetate	EtOH	275(3.60)	44-1577-69
$C_{31}H_{48}O_4$			
Lantic acid, methyl ester	n.s.g.	207(3.8)	42-0100-69
$C_{31}H_{48}O_7$			
Cephalosporin P_1, monodeacetyl-	MeOH	220(3.90)	78-3341-69
$C_{31}H_{48}O_9$			
Crustecdysone, 2,3-diacetate	EtOH	242(4.06)	12-1045-69
Crustecdysone, 2,22-diacetate	EtOH	243(4.07)	12-1045-69
Crustecdysone, 3,22-diacetate	EtOH	243(4.09)	12-1045-69
$C_{31}H_{49}NO_3$			
Cholest-5-ene-2,3-dione, 4,4-dimethyl-, 2-(O-acetyloxime)	EtOH	218(3.98)	22-3180-69

Compound	Solvent	λ_{max}(log ε)	Ref.
Solacongestidine, diacetyl derivative	EtOH	235(3.90)	44-1577-69
$C_{31}H_{50}ClNO_5$			
5α-Pregnane-3β,16β-diol, 20α-[(2R,3R,5S)-1-chloro-3-hydroxy-5-methyl-2-piperidyl]-, 3,16-diacetate	MeOH	280(2.53)	5-0159-69C
	dioxan	284(2.51)	24-4080-69
(2R,3S,5R)-	MeOH	287(2.66)	5-0159-69C
(2S,3R,5S)-	dioxan	277(2.60)	24-4080-69
(2S,3S,5S)-	dioxan	273(2.52)	24-4080-69
$C_{31}H_{50}N_2O_6$			
5α-Pregnane-3β,16β-diol, 20α-[(2R,3R,5R)-3-hydroxy-5-methyl-1-nitroso-2-piperidyl]-, 3,16-diacetate	MeOH	237(3.82),362(2.28)	5-0159-69C
(2R,3S,5R)-	MeOH	242(3.84),365(2.00)	5-0159-69C
$C_{31}H_{50}O_3$			
5β-Cholest-1-en-3-one, 5,6β-dihydroxy-4,4-dimethyl-, cyclic acetal with acetaldehyde	EtOH	228(4.03)	88-4077-69
Olean-13(18)-en-24-oic acid, 3α-hydroxy-, methyl ester	EtOH	206(3.73)	102-2083-69
$C_{31}H_{50}O_5$			
5β-Cholestan-6-one, 3α,5-diacetoxy-	EtOH	292(1.72)	44-2768-69
$C_{31}H_{51}NO_3$			
5α-Cholestane-2,3-dione, 4,4-dimethyl-, 2-(O-acetyloxime)	n.s.g.	218(3.93)	22-3173-69
$C_{31}H_{52}O_7$			
Cephalosporin P_1, monodeacetyl-, hydrogenation product	MeOH	222(3.90)	78-3341-69
$C_{31}H_{56}OS$			
5α-Cholestane, 6α-(R)-(butylsulfinyl)-	hexane	212(3.54),231s(3.08)	39-1668-69C
	EtOH	217(3.04)	39-1668-69C
5α-Cholestane, 6α-(S)-(butylsulfinyl)-	hexane	205(3.73),223s(3.11)	39-1668-69C
	EtOH	220(3.00)	39-1668-69C
5α-Cholestane, 6α-(S)-(tert-butylsulfinyl)-	hexane	207s(3.54),230s(3.15)	39-1668-69C
	EtOH	225s(3.11)	39-1668-69C
5α-Cholestane, 6β-(R)-(butylsulfinyl)-	hexane	209(3.66),234s(2.95)	39-1668-69C
	EtOH	222(2.95)	39-1668-69C
5α-Cholestane, 6β-(S)-(butylsulfinyl)-	hexane	209(3.60),233s(2.95)	39-1668-69C
	EtOH	218(3.00)	39-1668-69C
5α-Cholestane, 6β-(S)-(tert-butylsulfinyl)-	hexane	207(3.62),233s(2.90)	39-1668-69C
	EtOH	226(2.85)	39-1668-69C

Compound	Solvent	λmax(log ε)	Ref.
$C_{32}H_{12}Cl_6$			
Dibenzo[de,qr]hexacene, hexachloro-	$C_6H_3Cl_3$	333(4.20),352(4.09), 372(4.09),390(4.21), 460(3.42),494(3.54), 551(3.75),604(4.32), 660(4.82)	5-0043-69A
$C_{32}H_{18}$			
Dibenzo[de,qr]hexacene	$C_6H_3Cl_3$	383(4.08),403(4.24), 427(3.64),455(3.48), 480(3.60),540(3.82), 575(4.30),623(4.80)	5-0043-69A
2,4-Hexadiyne, 1,6-fluoren-9-ylidene-	CH_2Cl_2	263(4.76),272(4.82), 357(4.32),375(4.48), 393(4.50),417(4.65), 457(4.66)	78-0955-69
$C_{32}H_{18}N_6Na_4O_{18}S_6$			
5H,19H-Dibenzo[c,o][1,13,2,7,8,14,19,-20]dithiahexaazacyclotetracosine-3,17,28,33-tetrasulfonic acid, 14,28-dihydroxy-, 6,6,20,20-tetraoxide, tetrasodium salt	H_2O	316(4.15),382(3.89), 495(4.39)	73-35 9-69
$C_{32}H_{19}N_5O_2$			
1,3-Indandione, 5,6-bis(2-benzimidazol-yl)-2-(2-quinolyl)-	$C_{10}H_7Cl$	433(4.39),456(4.43)	33-1259-69
$C_{32}H_{20}$			
Benzo[3,4]cyclobut[1,2-b]anthracene, 5,12-diphenyl-	EtOH	240(4.49),293s(4.69), 303(4.81),325(4.66), 347(4.13),364(4.06), 386(3.92),410(3.55), 434(3.41)	44-0538-69
Benzo[k]fluoranthene, 7,12-diphenyl-	EtOH	252(4.80),270(4.48), 298(4.63),310(4.79), 366(3.90),386(4.18)	35-0371-69
Benzo[4,5]pentaleno[1,2-b]naphthalene, 5,12-diphenyl-	C_6H_{12}	275(4.74),320(4.69), 441(4.07),470(4.10)	44-0538-69
Dibenzo[de,qr]hexacene, 7,16-dihydro-	benzene	298(4.02),309(4.18), 340(4.15),355(4.42), 374(4.60),400(4.64)	5-0043-69A
7H,10H-Dibenzo[de,lm]hexacene	benzene	296(4.49),308(4.50), 324(4.68),340(4.80), 356(4.55),380(4.23), 402(4.24),420(3.90)	5-0208-69D
2,3,4-Hexatriene, 1,6-difluoren-9-ylidene-	CH_2Cl_2	256(4.78),281(4.58), 455(4.61),488(4.95), 522(5.13)	78-0955-69
2-Hexen-4-yne, 1,6-difluoren-9-ylidene-, (E)-	CH_2Cl_2	240(4.73),268(4.67), 332(4.09),435(4.71), 465(4.68)	78-0955-69
$C_{32}H_{20}N_2O_2$			
11,12'-Bi-12H-benzo[b]phenoxazine	MeCN	236(4.90),260(4.74), 310(4.30),368(4.29)	44-1691-6-
$C_{32}H_{20}N_4O$			
Ethenetricarbonitrile, [N-(10-phenyl-9-anthryl)-p-anisidino]-	$CHCl_3$	263(5.02),377(4.25)	22-1651-69

Compound	Solvent	$\lambda_{max}(\log \epsilon)$	Ref.
$C_{32}H_{21}N_3$			
Benzo[a]phenazine, 5-(N-2-naphthylanil-ino)-	EtOH	253(4.57),266s(4.44), 297s(4.09),324(3.82), 396(3.53),518(3.81)	44-3456-69
$C_{32}H_{21}N_5$			
Ethenetricarbonitrile, [N-methyl-p-[(10-phenyl-9-anthryl)amino]anilino]-	$CHCl_3$	263(4.98),314(4.19), 358(3.99),380(4.09), 400(4.15)	22-1659-69
$C_{32}H_{22}$			
Butadiyne, bis(p-styrylphenyl)-	DMF	367(4.99),393(4.81)	33-2521-69
2,4-Hexadiene, 1,6-difluoren-9-ylidene-, cis-cis	CH_2Cl_2	247(4.88),269(4.71), 400(4.52),424(4.75), 448(4.95),477(4.84)	78-0955-69
cis-trans	CH_2Cl_2	247(4.92),270(4.71), 425(4.79),451(4.97), 482(4.91)	78-0955-69
trans-trans	CH_2Cl_2	248(4.85),273(4.66), 282(4.49),405(4.49), 427(4.80),456(5.03), 488(5.02)	78-0955-69
Naphthalene, 1-[p-[(4-biphenylyl)ethyn-yl]styryl]-	DMF	359(4.78)	33-2521-69
Naphthalene, 2-[p-[(4-biphenylyl)ethyn-yl]styryl]-	DMF	303(4.46),360(4.91)	33-2521-69
$C_{32}H_{22}O$			
Benzofuran, 6-[2-(2-naphthyl)vinyl]-2,3-diphenyl-	DMF	280(4.39),361(4.77)	33-1282-69
Benzofuran, 2-[p-[2-(1-naphthyl)vinyl]-phenyl]-5-phenyl-	DMF	368(4.77)	33-1282-69
Benzofuran, 2-[p-[2-(2-naphthyl)vinyl]-phenyl]-5-phenyl-	DMF	368(4.90),388(4.76)	33-1282-69
Naphtho[1,2-b]furan, 2-[p-(p-phenylsty-ryl)phenyl]-	DMF	308(4.29),377(4.89)	33-1282-69
Naphtho[2,1-b]furan, 2-[p-(p-phenylsty-ryl)phenyl]-	DMF	328(4.38),381(4.91)	33-1282-69
$C_{32}H_{22}O_4$			
Acetylene, bis[p-[3,4-(methylenedioxy)-styryl]phenyl]-	DMF	376(4.96)	33-2521-69
$C_{32}H_{22}S$			
Benzo[b]thiophene, 2-(p-biphenylyl)-6-[2-(1-naphthyl)vinyl]-	DMF	373(4.76)	33-1282-69
Naphtho[2,1-b]thiophene, 2-[p-(p-phenyl-styryl)phenyl]-	DMF	381(4.86)	33-1282-69
$C_{32}H_{23}Br_2OP$			
Phosphorane, (2,7-dibromofluoren-9-ylidene)(p-methoxyphenyl)diphenyl-	$CHCl_3$	255(4.76),294(5.58), 384(3.60)	5-0001-69D
$C_{32}H_{24}$			
Naphthalene, 2-[p-(2,2-diphenylvinyl)-styryl]-	DMF	302(4.17),367(4.78)	33-2521-69
Stilbene, 4-[2-(1-naphthyl)vinyl]-4'-phenyl-	DMF	378(4.84)	33-2521-69
Stilbene, 4-[2-(2-naphthyl)vinyl]-4'-phenyl-	DMF	310(4.22),378(4.95)	33-2521-69

Compound	Solvent	$\lambda_{max}(\log \epsilon)$	Ref.
p-Terphenyl, 4-(p-phenylstyryl)-	DMF	349(4.82)	33-2521-69
$C_{32}H_{24}N_2O_3$			
Acetamide, N-[10-(9,10-dihydro-10-oxo-9-anthryl)-9(10H)-anthrylidene]-, 10-(O-acetyloxime)	$CHCl_3$	265(4.46)	22-3538-69
$C_{32}H_{24}O$			
Furan, 2,5-bis(p-styrylphenyl)-	DMF	288(4.28),333(4.42), 396(4.90)	33-1282-69
$C_{32}H_{24}O_8$			
2(3H)-Benzofuranone, 6,6-dihydroxy-3-(p-hydroxyphenyl)-4,7-diphenyl-, triacetate	EtOH	220s(4.64),250s(4.30), 290s(3.75)	1-2583-69
$C_{32}H_{24}S$			
Thiophene, 2,5-bis(p-styrylphenyl)-	DMF	286(4.36),397(4.89)	33-1282-69
$C_{32}H_{26}$			
Butadiene, 1,4-bis(p-styrylphenyl)-	DMF	282(4.20),403(5.02), 425(4.88)	33-2521-69
$C_{32}H_{26}N_2$			
Indolo[2,3-a]carbazole, 5,6,11,12-tetrahydro-11,12-dibenzyl-	CH_2Cl_2	264(4.28),370(4.38), 388(4.31)	24-1198-69
$C_{32}H_{26}N_2O_6S_2$			
[$\Delta^{2,2'}$-Bi-2H-1,4-benzothiazine]-5,5'-diol, 8,8'-bis[(2-hydroxy-p-tolyl)-oxy]-7,7'-dimethyl-	EtOH	226s(4.57),281(4.27), 287(4.26),308(4.24), 465(4.30),484(4.29)	32-0323-69
	EtOH-HCl	255s(4.26),333(4.17), 596(4.38)	32-0323-69
	pH 13	240(4.57),296(4.24), 487(4.34)	32-0323-69
$C_{32}H_{26}O_2$			
Acetylene, bis[p-(p-methoxystyryl)-phenyl]-	DMF	373(4.96)	33-2521-69
Anthracene, 9-[p-(3,4-dimethoxystyryl)-styryl]-	DMF	357(4.54),392(4.46)	33-2521-69
2-Naphthalenemethanol, α,α'-(p-phenyl-enedivinylene)di-	EtOH	292(4.48)	124-1189-69
$C_{32}H_{26}O_2S$			
Benzo[b]thiophene, 6-(p-methoxystyryl)-2-[p-(p-methoxystyryl)phenyl]-	DMF	389(4.93)	33-1282-69
$C_{32}H_{26}O_3$			
Benzofuran, 4,6-bis(p-methoxystyryl)-2-phenyl-	DMF	340(4.86)	33-1282-69
Benzofuran, 5,6-bis(p-methoxystyryl)-2-phenyl-	DMF	325(4.81)	33-1282-69
Benzofuran, 6-(p-methoxystyryl)-2-[p-(p-methoxystyryl)phenyl]-	DMF	276(4.39),390(4.93)	33-1282-69
$C_{32}H_{26}O_4$			
2,4-Cyclopentadien-1-one, 2,3,4-tris-(p-methoxyphenyl)-5-phenyl-	benzene	370(3.97),535(3.10)	64-0685-69

Compound	Solvent	λ_{max}(log ε)	Ref.
2,4-Cyclopentadien-1-one, 2,3,5-tris-(p-methoxyphenyl)-4-phenyl-	benzene	386(3.87),550(3.24)	64-0685-69
$C_{32}H_{26}O_8$			
6H,14H-2,4,9,11-Tetraoxadibenzo[bc,kl]-coronene-6,14-dione, 1,3,3a,8,10,10a-hexahydro-7,13-dimethoxy-1β,3α,8β,10α-tetramethyl-	EtOH	335(4.00),416s(3.93), 434(4.45),498(3.93), 531(3.92)	39-0631-69C
	H_2SO_4	277(4.33),295(4.22), 472(4.46),533(4.19), 576(4.35)	39-0631-69C
	Ac_2O-H_2SO_4	300(4.12),332(4.30), 366(4.11),432(3.83), 599(4.29)	39-0631-69C
after one hour	CF_3COOH	271(4.07),338(4.29), 383(4.08),461(3.80), 661(4.26),845(2.86), 973(3.06)	39-0631-69C
7H,14H-2,4,9,11-Tetraoxadibenzo[bc,kl]-coronene-7,14-dione, 1,3,3a,8,10,10a-hexahydro-6,11-dimethoxy-1β,3α,8β,10α-tetramethyl-	EtOH	254(4.29),275(4.37), 297(4.59),307(4.59), 425(4.29),523(3.81), 563(4.35),611(4.64)	39-0631-69C
	H_2SO_4	275(4.33),290(4.26), 345(3.32),466(4.42), 537(4.26),554(3.97), 582(4.60)	39-0631-69C
$C_{32}H_{26}O_{11}$			
Viopurpurin, trimethyl-	EtOH	218(4.15),270(4.38), 278(4.39),355(3.09)	23-1223-69
	$CHCl_3$	455(3.01)	23-1223-69
$C_{32}H_{27}BrN_2O_8$			
3-Carbamoyl-1-D-ribofuranosylpyridinium bromide, 2',3',5'-tribenzoate	MeOH	270(3.85)	39-0199-69C
$C_{32}H_{27}N_3O_2$			
Indole, 5-ethoxy-3-[(5-ethoxy-2-phenyl-3H-indol-3-ylidene)amino]-2-phenyl-	EtOH	291(4.52),568(4.03)	22-1234-69
$C_{32}H_{27}O_5P$			
Fumaric acid, [benzoyl(triphenylphosphoranylidene)methyl]-, dimethyl ester	EtOH	225(4.55),267(3.88), 274(3.88),307(3.75), 367(3.77)	39-1100-69C
Maleic acid, [benzoyl(triphenylphosphoranylidene)methyl]-, dimethyl ester	EtOH	228(4.53),266(3.90), 274(3.86),305(3.74), 403(3.51)	39-1100-69C
$C_{32}H_{28}$			
Benzocyclobutene, 2a,3,4,5,6,6a-hexahydro-1,2,3,6-tetraphenyl-	hexane	228s(4.36),288(4.13)	35-0523-69
$C_{32}H_{28}O_4S$			
Dibenzothiophene, 2,8-bis(3,4-dimethoxystyryl)-, (E,E)-	DMF	330(4.82)	33-1282-69
$C_{32}H_{28}O_4S_2$			
Thianthrene, 2,7-bis(3,4-dimethoxystyryl)-, (E,E)-	DMF	349(3.91)	33-1282-69

Compound	Solvent	λ_{max}(log ϵ)	Ref.
$C_{32}H_{28}O_7$			
Flavone, 3',5-bis(benzyloxy)-4',7,8-trimethoxy-	EtOH	244s(4.39),272(4.32), 343(4.37)	24-0112-69
Flavone, 3',6-bis(benzyloxy)-4',5,7-trimethoxy-	EtOH	241(4.40),266(4.18), 329(4.41)	40-1270-69
Flavone, 3',7-bis(benzyloxy)-3',5,7-trimethoxy-	EtOH	242(4.37),267(4.20), 331(4.45)	40-1270-69
$C_{32}H_{28}O_8$			
Flavone, 3,4'-bis(benzyloxy)-5-hydroxy-3',6,7-trimethoxy-	EtOH	259(4.33),273(4.29), 347(4.34)	18-1398-69
	EtOH-NaOH	294(4.32),380s(3.69)	18-1398-69
	EtOH-NaOAc	259(4.32),274(4.28), 346(4.32)	18-1398-69
	EtOH-$AlCl_3$	269(4.32),284(4.31), 376(4.33)	18-1398-69
Flavone, 3',4'-bis(benzyloxy)-5-hydroxy-3,6,7-trimethoxy-	EtOH	239(4.28),256(4.31), 274(4.26),345(4.39)	18-1649-69
	EtOH-$AlCl_3$	265(--),286(--), 366(--)	18-1649-69
$C_{32}H_{29}ClO_7$			
3-[2-(2-Acetoxy-1-naphthyl)-1-pentylvinyl]naphtho[2,1-b]pyrylium perchlorate	MeOH-$HClO_4$	306(4.19),518(4.50)	5-0226-69E
$C_{32}H_{29}N_3O_3S$			
4-Thiazoline, 3-cyclohexyl-5-(1-naphthylamino)-4-(p-nitrobenzoyl)-2-phenyl-	MeCN	272(4.36),457(4.16)	23-3557-69
$C_{32}H_{29}O_4P$			
Phosphoric acid, dimethyl-5-methyl-2,3,4,5-tetraphenyl-1,3-cyclopentadien-1-yl ester	EtOH	244(4.2),316(4.0)	39-2605-69C
$C_{32}H_{30}N_2O_7S$			
4-Thiouridine, 5'-O-trityl-, 2',3'-diacetate	EtOH	242(3.38),329(4.28)	94-0181-69
$C_{32}H_{30}N_4O_8S$			
Acetamide, N,N'-[dithiobis[1-(4-isopropylidene-5-oxo-2-oxazolin-2-yl)vinylene]]bis[2-phenoxy-	EtOH	366(4.49)	88-3381-69
$C_{32}H_{30}O_2$			
3,3'-Spirobi[3H-naphtho[2,1-b]pyran], 2-heptyl-	dioxan	248(5.07),287(4.05), 300(4.19),312(4.22), 335(3.96),349(4.02)	5-0162-69B
$C_{32}H_{30}O_4$			
Biphenyl, 4,4'-bis(3,4-dimethoxystyryl)-	DMF	372(4.91)	33-2521-69
Stilbene, 4'-[2,2-bis(p-methoxyphenyl)-vinyl]-3,4-dimethoxy-	DMF	372(4.74)	33-2521-69
$C_{32}H_{30}O_{14}$			
[4,4'-Bixanthene]-10a,10'a(5H,5'H)-dicarboxylic acid, 6,6',7,7'-tetrahydro-1,1',5,5',8,8'-hexahydroxy-6β,6'β-dimethyl-9,9'-dioxo-, dimethyl ester	EtOH	248(4.26),337(4.50)	119-0020S-69

Compound	Solvent	$\lambda_{max}(\log \epsilon)$	Ref.
$C_{32}H_{31}NO_3$			
Tuuranine, N,O-dibenzyl-	EtOH	267(3.93),276(3.96), 301(3.61)	2-0945-69
$C_{32}H_{32}$			
Naphthalene, 2,6-bis(p-isopropylstyryl)-	DMF	284(4.46),295(4.51), 365(4.89),386(4.81)	33-2521-69
$C_{32}H_{32}NO_4P$			
Acetamide, N-[4,5-dimethoxy-2-[triphenylphosphoranylidene)acetyl]phenethyl]-	EtOH-NaOH	214(4.69),267(4.05), 275(4.04),292(4.07)	39-0094-69C
$C_{32}H_{32}N_2O_5$			
Isoquinoline, 2-benzyl-1-[4-(benzyloxy)-2-nitrobenzyl]-1,2,3,4-tetrahydro-6,7-dimethoxy-	EtOH	284(3.90)	2-0945-69
$C_{32}H_{32}N_4$			
1,2-Cyclohexanedione, bis(benzylphenylhydrazone)	MeOH	303(4.21),343(4.01)	24-1198-69
$C_{32}H_{32}O_6$			
Naphthalene, 2,6-bis(3,4,5-trimethoxystyryl)-	DMF	297(4.38),372(4.89), 391(4.82)	33-2521-69
$C_{32}H_{33}FN_4O_4$			
2,18-Porphinedipropionic acid, 10-fluoro-3,8,13,17-tetramethyl-, dimethyl ester, meso-	n.s.g.	399(5.32),496(4.20), 528(3.68),567(3.75), 621(3.20)	25-0654-69
$C_{32}H_{33}NO_4$			
Isoquinoline, 6-(benzyloxy)-1-[p-(benzyloxy)benzyl]-1,2,3,4-tetrahydro-5,7-dimethoxy-, dl-	MeOH	276(3.58),282(3.57)	44-3884-69
$C_{32}H_{34}ClNO_6$			
2-(α-Benzylphenethyl)-3-ethyl-6,7-dihydro-9,10-dimethoxybenzo[a]quinolizinium perchlorate	MeOH	209(4.67),227s(4.21), 270s(4.14),285(4.26), 357(4.17)	24-1779-69
$C_{32}H_{34}N_2O_3$			
Isoquinoline, 1-[2-amino-4-(benzyloxy)-benzyl]-2-benzyl-1,2,3,4-tetrahydro-6,7-dimethoxy-	EtOH	288(3.82)	2-0945-69
$C_{32}H_{34}N_2O_5S$			
Dimethylsulfoxonium [1-(2-deoxy-5-O-trityl-β-D-threo-pentofuranosyl)-1,4-dihydro-5-methyl-4-oxo-2-pyrimidinyl]-methylide	MeOH	231(4.31),280(4.33)	35-7752-69
$C_{32}H_{34}O_{15}$			
Olivin, hexaacetate	EtOH	252s(4.70),258(4.80), 303(3.88),312s(3.80), 359(3.58)	105-0257-69
$C_{32}H_{35}NO_4$			
2-Oxazolin-5-one, 4-[3,5-diisopropyl-4-(3-isopropyl-4-methoxyphenoxy)benzylidene]-2-phenyl-	EtOH	362(4.58),373(4.68), 390(4.59)	94-2176-69

Compound	Solvent	λ_{max}(log ε)	Ref.
$C_{32}H_{36}CuN_4$			
Etioporphyrin II, copper complex	$CHCl_3$	328(4.22),399(5.51), 525(4.09),562(4.39)	39-0176-69C
$C_{32}H_{36}N_2O_9$			
Veneserpine	EtOH	225(4.65),274(4.18), 292(4.08)	25-1388-69
$C_{32}H_{36}N_4Ni$			
Nickel, (12,17,20-triethyl-2,3,7,8,13,18-hexamethylporphinato)-	$CHCl_3$	299(4.02),347(4.11), 408(5.21),493s(3.46), 535(3.95),569(4.06)	39-0655-69C
$C_{32}H_{36}O_{14}$			
Olivin, 1-deoxo-, hexaacetate	EtOH	220(4.38),237(4.26), 275(3.87),340(3.95)	105-0257-69
$C_{32}H_{37}NO_5$			
Cinnamic acid, α-benzamido-3,5-diisopropyl-4-(3-isopropyl-4-methoxyphenoxy)-	EtOH	227(4.49),293(4.37)	94-2176-69
$C_{32}H_{38}Hg$			
Mercury, bis(2-phenyl-2-bornen-3-yl)-	n.s.g.	225(3.62),246(3.62), 280(3.62)	88-3521-69
$C_{32}H_{38}N_4$			
Etioporphyrin II	$CHCl_3$	400(5.30),499(4.20), 534(4.07),567(3.88), 621(3.75)	39-0176-69C
$C_{32}H_{39}ClN_4Ni$			
Nickel(1+), (octadehydro-1,3,8-triethyl-2,7,12,13,17,18,19-heptamethylcorrinato)-, chloride	$CHCl_3$	272(4.46),353(4.48), 456(3.67),557(4.11)	39-0655-69C
$C_{32}H_{39}ClN_4NiO_4$			
Nickel(1+), (octadehydro-1,3,8-triethyl-2,7,12,13,17,18,19-heptamethylcorrinato)-, perchlorate	$CHCl_3$	275(4.44),353(4.47), 456(3.65),557(4.10)	39-0655-69C
$C_{32}H_{39}IN_4Ni$			
Nickel(1+), (octadehydro-1,3,8-triethyl-2,7,12,13,17,18,19-heptamethylcorrinato)-, iodide	$CHCl_3$	273(4.38),354(4.40), 449s(3.62),558(4.01)	39-0655-69C
$C_{32}H_{39}N_3O_7$			
Malonic acid, [2-(carboxyamino)propionamido][2-(1,1-dimethylallyl)indol-3-yl]methyl]-, N-benzyl diethyl ester, L-	EtOH	223(4.51),283(3.91), 291(3.85)	39-1003-69C
$C_{32}H_{40}N_2O_{20}$			
Pyrimidine, 2,4-bis(β-D-glucopyranosyloxy)-, 2',2",3',3",4',4",6',6"-octaacetate	50% EtOH	260(3.89)	39-0203-69C
2(1H)-Pyrimidinone, 1-β-D-glucopyranosyl-4-(β-D-glucopyranosyloxy)-, octaacetate	50% EtOH	280(3.82)	39-0203-69C
Uracil, 1,3-di-β-D-glucopyranosyl-, octaacetate	50% EtOH	260(3.98)	39-0203-69C

Compound	Solvent	λmax(log ε)	Ref.
$C_{32}H_{40}O_4$			
1'βH-Cyclobuta[16,17]-18,20α-cyclopregn-5-ene-3β,20-diol, 3',4'-dihydro-, 3-acetate benzoate	EtOH	229(4.21),274(3.13), 281(3.05)	44-3779-69
$C_{32}H_{41}N_5O_5$			
α-Ergocryptine	MeOH	241(4.31),313(3.95)	33-1549-69
β-Ergocryptine	MeOH	241(4.31),312(3.94)	33-1549-69
α-Ergocryptinine	MeOH	242(4.30),313(3.94)	33-1549-69
β-Ergocryptinine	MeOH	312(3.93)	33-1549-69
$C_{32}H_{42}N_2O_6$			
4H-Quinolizin-4-one, 2-[2-amino-5-methoxyphenyl)-7α-ethyl-1,2,3,6,7,8,9,9a-octahydro-12-hydroxy-8β-[2-(tetrahydro-2H-pyran-2-yl)oxy]ethyl]-, benzoate	EtOH	233(4.33),304(3.53)	78-5329-69
allo-	EtOH	233(4.32),285(3.29), 307(3.50)	78-5329-69
1,6-Secohydroquinine, 1',2',3',4'-tetrahydro-2'-oxo-6-[(tetrahydro-2H-pyran-2-yl)oxy]-, benzoate, allo-	EtOH	233(4.20),258(4.15), 296(3.49)	78-5329-69
$C_{32}H_{42}O_8$			
Acrovestone	EtOH	230(4.47),297(4.36), 342(4.09)	2-0873-69
	EtOH-NaOH	336(4.45)	2-0873-69
Withacnistin, acetate	EtOH	217(4.19)	44-3858-69
Withaferin A, diacetate	EtOH	214(4.26)	44-3858-69
$C_{32}H_{43}N_3O_3$			
1'H-5α-Androstano[3,2-c]pyrazol-17β-ol, 1'-acetyl-5'-[p-(dimethylamino)phenyl]-, 17-acetate	MeOH	230(4.13),255(4.13), 310(4.16)	94-2319-69
$C_{32}H_{44}N_2$			
13-Azabicyclo[8.2.1]trideca-10,12-diene, N,N'-(2,3-dimethyl-p-phenylene)bis-	C_6H_{12}	239(4.32)	78-5357-69
13-Azabicyclo[8.2.1]trideca-10,12-diene, N,N'-(2,5-dimethyl-p-phenylene)bis-	C_6H_{12}	238(4.22)	78-5357-69
$C_{32}H_{44}N_2O_6$			
Homobatrachotoxin	MeOH-HCl	233(3.95),264(3.70)	35-3931-69
$C_{32}H_{44}O_6$			
9,19-Cyclo-9β-lanost-24-en-26-oic acid, 16,23-epoxy-12,23-dihydroxy-3-oxo-, γ-lactone, acetate	EtOH	205(4.10)	32-0915-69
$C_{32}H_{44}O_{12}$			
Taxa-4(20),11-diene-2α,5α,7β,9α,10β,13α-hexol, hexaacetate	n.s.g.	213(4.04)	77-1282-69
$C_{32}H_{46}O_7$			
29-Nordammar-17(20)-en-21-oic acid, 16β-hydroxy-3,6,7-trioxo-, methyl ester, acetate	MeOH-acid	219(3.92),282(3.65)	78-3341-69
$C_{32}H_{46}O_8$			
Withacnistin, 3-ethoxy-2,3-dihydro-	EtOH	227(3.93)	44-3858-69

Compound	Solvent	$\lambda_{max}(\log \epsilon)$	Ref.
$C_{32}H_{48}$			
Biphenyl, 4,4'-(eicosamethylene)-	MeOH	260(4.34)	97-0342-69
$C_{32}H_{48}O_2$			
Spiro[chroman-2,1'-[3,5]cyclohexadien]-2'-one, 3',5',6,8-tetra-tert-butyl-3,4-dimethyl-	C_6H_{12}	270(3.79),283(3.81), 306(3.65),335(2.77), 349(2.84),361(2.77), 375(2.44)	44-0227-69
Spiro[chroman-3,1'-[3,5]cyclohexadien]-2'-one, 3',5',6,8-tetra-tert-butyl-2,4-dimethyl-	C_6H_{12}	227s(3.66),284(3.78), 308(3.76),375s(2.52)	44-0227-69
$C_{32}H_{48}O_4$			
Isobauren-7,11-dione, 3β-acetate	n.s.g.	269(3.90)	102-0781-69
$C_{32}H_{48}O_5$			
5β-Ergosta-7,22-dien-6-one, 2β,3β-diacetoxy-	EtOH	247(4.19)	33-2428-69
$C_{32}H_{50}O_2$			
Leucotylidiene, 6-O-acetyl-	EtOH	244(4.73),252(4.79), 261(4.64)	94-0279-69
21-Noreburica-8,20(22),24(28)-trien-3β-ol, acetate	EtOH	208(4.09)	39-1047-69C
Pachysandiol A, monoacetate	EtOH	238s(--),245(4.39), 252s(--)	95-1358-69
Phenol, 2,2'-(1,2-dimethylethylene)bis-[4,6-di-tert-butyl-	C_6H_{12}	277(3.88),282(3.88)	44-0227-69
$C_{32}H_{50}O_7$			
Cephalosporin P_1, monodeacetyl-, methyl ester	EtOH	220(3.93)	78-3341-69
$C_{32}H_{52}O_3$			
Cholest-5-en-2-one, 3β-hydroxy-3α,4,4-trimethyl-, acetate	EtOH	292(2.04)	22-3189-69
Eburicoic acid, methyl ester	EtOH	243(2.88),252(2.73)	95-1149-69
Urs-12-ene-3β,11β-diol, 3-acetate	EtOH	210(3.75)	102-2083-69
$C_{32}H_{52}O_7$			
Cephalosporin P_1, monodeacetyl-, methyl ester hydrogenation product	MeOH	230(3.93)	78-3341-69
$C_{32}H_{54}O_3$			
A-Nor-5β-cholestan-6-one, 2β-hydroxy-2-(hydroxymethyl)-3,3-dimethyl-, cyclic acetal with acetone	EtOH	293(1.90)	22-3189-69

Compound	Solvent	λ_{max}(log ε)	Ref.
$C_{33}H_{20}N_4$			
1,2-Diazaspiro[4.4]nona-1,3,6,8-tetraene-3,4-dicarbonitrile, 6,7,8,9-tetraphenyl-	$CHCl_3$	244(4.45),330(4.00), 365(3.75)	64-0536-69
$C_{33}H_{23}ClO_4$			
9-(7-Fluoren-9-ylidene-1,3,5-heptatrienyl)fluoren-9-ylium perchlorate (relative absorbances given)	CH_2Cl_2	260(4.99),360(4.50), 500(4.30),755(4.57), 838(5.00)	78-0955-69
$C_{33}H_{23}ClO_4S_2$			
2-Phenyl-4-[3-(2-phenyl-4H-1-benzothiopyran-4-ylidene)propyl]-1-benzothiopyrylium perchlorate	CH_2Cl_2	262(3.76),400(3.58), 780(4.18)	2-0311-69
$C_{33}H_{23}N$			
Isoquinoline, 5,6,7,8-tetraphenyl-	EtOH	289(4.16),333(3.76)	39-1758-69C
Quinoline, 5,6,7,8-tetraphenyl-	EtOH	244(4.37),312(3.61)	39-1758-69C
$C_{33}H_{23}N_5$			
Ethenetricarbonitrile, [p-(dimethylamino)-N-(10-phenyl-9-anthryl)-anilino]-	$CHCl_3$	259(5.03),320(4.13), 407(4.15)	22-1659-69
$C_{33}H_{24}O$			
Benzofuran, 5-benzyl-2-[p-[2-(1-naphthyl)vinyl]phenyl]-	DMF	367(4.73)	33-1282-69
Benzofuran, 5-benzyl-2-[p-[2-(2-naphthyl)vinyl]phenyl]-	DMF	302(4.22),368(4.89), 387(4.75)	33-1282-69
$C_{33}H_{24}O_2$			
11,11'-Spirobi[11H-benzo[b]fluorene]-6,6'-(7H,7'H)-dione, 8,8',9,9'-tetrahydro-	EtOH	212s(4.65),217(4.66), 233s(4.63),255(4.89), 272s(4.66),282s(4.42), 292(4.25),333(3.79)	33-1202-69
$C_{33}H_{24}O_6$			
9,9'-Spirobi[fluorene]-2,2'-dibutyric acid, γ,γ'-dioxo-	EtOH	227(4.57),243s(4.23), 292s(4.47),298(4.50), 324(4.34)	33-1202-69
$C_{33}H_{25}N_3O_5$			
Cyclopenta[c]quinolizine-1,2-dicarboxylic acid, 3-[(1-hydroxy-2-naphthyl)-azo]-5-methyl-4-phenyl-, dimethyl ester	EtOH	218(4.57),254(4.33)	39-1279-69C
$C_{33}H_{26}Br_2NP$			
Aniline, p-[(2,7-dibromofluoren-9-ylidene)diphenylphosphoranyl]-N,N-dimethyl-	$CHCl_3$	260(4.64),269(4.64), 295(4.68),388(3.45)	5-0001-69D
$C_{33}H_{26}N_2O_3$			
3,5-Pyrazolidinedione, 4-[4-(diphenylmethylene)-3,4-dihydro-2H-pyran-2-yl]-1,2-diphenyl-	dioxan	237(4.42),290(4.39)	24-3950-69
$C_{33}H_{27}NO$			
1-Isoindolinol, 2,3,3-triphenyl-1-o-tolyl-	C_6H_{12}	260s(4.0)	104-1110-69
	0.1N H_2SO_4	305(4.15)	104-1110-69

Compound	Solvent	λmax(log ε)	Ref.
$C_{33}H_{27}NO_2$			
1-Isoindolinol, 1-(o-methoxyphenyl)-2,3,3-triphenyl-	C_6H_{12}	270s(3.5)	104-1110-69
	0.1N H_2SO_4	265(3.77),302(3.94)	104-1110-69
$C_{33}H_{28}$			
11,11'-Spirobi[11H-benzo[b]fluorene], 6,6',7,7',8,8',9,9'-octahydro-	EtOH	211(4.79),224(4.80), 231(4.83),243(4.51), 268(4.44),273s(4.42), 278s(4.39),285s(4.24), 292s(4.01),298s(3.97), 305(4.20),311s(4.12), 317(4.45)	33-1202-69
$C_{33}H_{28}N_2O$			
2-Cyclopenten-1-one, 2,3-diphenyl-4,5-bis(2,6-xylylimino)-	EtOH	240(4.40),330(4.19)	88-3403-69
$C_{33}H_{28}O_4$			
9,9'-Spirobi[fluorene]-2,2'-dibutyric acid	EtOH	213(4.76),221(4.78), 226s(4.75),245(4.43), 258s(4.44),267(4.49), 277(4.48),283s(4.37), 288s(4.27),299(4.13), 304s(3.99),312(4.33)	33-1202-69
$C_{33}H_{30}Cl_2FeN_{12}O_8$			
Iron(2+), tris(picolinaldehyde 2-pyridylhydrazone)-, diperchlorate	H_2O	463(4.05)	39-0819-69A
$C_{33}H_{30}N_4O$			
1-Butanone, 1-(1,3,7,9-tetraphenyl-1,2,7,8-tetraazaspiro[4.4]nona-2,8-dien-6-yl)-	n.s.g.	242(4.32),342(4.27)	28-1063-69B
$C_{33}H_{30}N_4O_6S$			
9H-Purine-6(1H)-thione, 9-(5-O-trityl-β-D-arabinofuranosyl)-, 2',3'-diacetate	EtOH	323(4.25)	23-1095-69
$C_{33}H_{30}O_3$			
Benzene, 1,2,4-tris(p-methoxystyryl)-	DMF	340(4.78)	33-2521-69
Benzene, 1,3,5-tris(p-methoxystyryl)-	DMF	330(4.99)	33-2521-69
$C_{33}H_{32}Br_3N_3O_5$			
1-Cyclopentene-1,2-dicarboxylic acid, 4-(4-bromo-2,6-xylidino)-3,5-bis-[(4-bromo-2,6-xylyl)imino]-4-hydroxy-, dimethyl ester	EtOH	253(4.49),370(4.08)	88-3407-69
$C_{33}H_{32}N_2OS$			
4-Thiazoline, 3-cyclohexyl-5-(1-naphthylamino)-2-phenyl-4-p-toluoyl-	MeCN	260(4.31),350(3.94), 420(4.17)	23-3557-69
$C_{33}H_{32}O_2$			
9,9'-Spirobi[fluorene]-2,2'-dibutanol	EtOH	212(4.78),221(4.80), 246s(4.47),267(4.52), 277(4.51),283s(4.38), 288s(4.27),299(4.13), 304s(3.99),311(4.33)	33-1202-69
3,3'-Spirobi[3H-naphtho[2,1-b]pyran], 2-octyl-	dioxan	248(5.09),287(4.08), 300(4.20),312(4.24), 334(3.98),349(4.06)	5-0162-69B

Compound	Solvent	λ_{max}(log ε)	Ref.
$C_{33}H_{32}O_5$			
Stilbene, 4'-[2,2-bis(p-methoxyphenyl)-vinyl]-3,4,5-trimethoxy-	DMF	370(4.73)	33-2521-69
$C_{33}H_{33}AlO_6S_3$			
Aluminum, tris(3-phenyl-2,4-pentanedionato)-	CH_2Cl_2	252(4.53),291(4.43)	88-2255-69
$C_{33}H_{33}O_8P$			
2H-Phosphole-2,3,4,5-tetracarboxylic acid, 1,1-dihydro-1,1,2-tri-p-tolyl-, tetramethyl ester	EtOH	236(4.43),280s(3.70), 343(4.28)	39-1100-69C
	HCl	239(4.38),272s(4.01), 279s(3.89)	39-1100-69C
$C_{33}H_{34}O$			
2,5-Cyclohexadien-1-one, 2,6-di-tert-butyl-4-(p,α-diphenylbenzylidene)-	MeOH	283(4.28),387(4.50)	44-2472-69
$C_{33}H_{34}O_3S$			
Androst-5-eno[6,5,4-bc]thiophen-3-one, 17β-hydroxy-5'-phenyl-, benzoate	EtOH	231(4.48),277(3.99), 333(3.82)	94-2586-69
$C_{33}H_{35}AsO_2$			
Arsenic, (3,4-dipivaloyl-2,4-cyclopentadien-1-ylidene)triphenyl-	EtOH	220(4.59),264(4.06), 270(4.09),304(4.21)	39-2464-69C
$C_{33}H_{36}N_4NiO_2$			
Nickel, [hydrogen 2,17-diethyl-3,7,8,12,13,18-hexamethyl-5-porphinecarboxylato(2-)]-, ethyl ester	$CHCl_3$	401(5.17),530(4.21), 568(4.38)	39-0655-69C
$C_{33}H_{36}N_4O_4$			
Canthiumine	n.s.g.	279(4.39)	20-0583-69
$C_{33}H_{36}N_4O_5$			
Cinnamamide, α-[2-[α-(dimethylamino)-β-hydroxyhydrocinnamamido]-3-methylcrotonamido]-N-(p-hydroxystyryl)-	EtOH	270-294(4.43)(end abs.)	78-0937-69
$C_{33}H_{36}O_{16}$			
Neriaphin, hexaacetate	EtOH	220(4.25),260(4.50), 287(3.64),297(3.70), 367(3.33)	88-0471-69
$C_{33}H_{37}N_3O_6$			
Tryptophan, N-[3-(p-tert-butoxyphenyl)-N-carboxy-L-alanyl]-, N-benzyl methyl ester, L-	EtOH	274(3.80),281(3.81), 290(3.74)	33-1058-69
$C_{33}H_{38}ClIrOP_2$			
Iridium, bis(butyldiphenylphosphine)-carbonylchloro-	toluene	336(3.44),383(3.56), 436(2.77)	64-0770-69
$C_{33}H_{38}N_4Ni$			
Nickel, [2,3,7,8,12,13,17,18-octamethyl-5-pentylporphinato(2-)]-	$CHCl_3$	299(4.01),345(4.09), 408(5.21),497s(3.44), 534(3.94),569(4.06)	39-0655-69C
Nickel, [2,5,7,17-tetraethyl-3,8,12,13,18-pentamethylporphinato(2-)]-	$CHCl_3$	302(4.02),350(4.11), 410(5.18),503s(3.49), 538(5.93),574(3.04), 624(3.24)	39-0655-69C

Compound	Solvent	$\lambda_{max}(\log \epsilon)$	Ref.
$C_{33}H_{38}N_5O_{14}P$			
Thymidine, thymidyl-(3'→5')-, ester with hydracrylonitrile, 3-(3-benzoylpropionate)	EtOH	250(4.35),263s(4.31)	35-3360-69
$C_{33}H_{38}O_5S$			
Androstan-3-one, 5α,17β-dihydroxy-6β-mercapto-, 6,17-dibenzoate	EtOH	234(3.39),272(3.03)	94-2586-69
$C_{33}H_{39}BrO_{10}$			
Phorbol, 3-O-(p-bromobenzoyl)-3-deoxo-3-hydroxy-, 12,13,20-triacetate	MeOH	245(4.31),282(2.96)	5-0130-69E
$C_{33}H_{41}ClCoN_4O_4$			
Cobalt(1+), (octadehydro-2,8,12,18-tetraethyl-1,3,7,13,17,19-hexamethylcorrinato)-, perchlorate	$CHCl_3$	282(4.43),342(4.35), 498(4.11),537s(3.91), 566(3.89)	39-0176-69C
$C_{33}H_{41}ClN_4NiO_4$			
Nickel(1+), (octadehydro-2,8,12,18-tetraethyl-1,3,7,13,17,19-hexamethylcorrinato)-, perchlorate	EtOH	276(4.53),353(4.51), 401s(3.80),454(3.68), 559(4.14)	39-0176-69C
$C_{33}H_{42}N_2O_{10}$			
2(1H)-Pyrimidinone, 1-β-D-glucopyranosyl-4-(β-D-glucopyranosyloxy)-5-methyl-, octaacetate	50% EtOH	290(3.81)	39-0203-69C
$C_{33}H_{42}N_4O_5$			
Tyramine, N,N-dimethyl-β-phenylseryl-valylphenylalanyl-	EtOH	225s(4),278(3.20), 286(3.11)	78-0937-69
$C_{33}H_{42}O_{13}$			
β-D-Glucopyranosiduronic acid, 3,17β-dihydroxyestra-1,3,5(10)-trien-16β-yl, methyl ester, 2,3,4,17-tetraacetate	MeOH	280(3.32),287(3.26)	13-0591-69B
$C_{33}H_{44}F_4$			
7,11b-Etheno-11bαH-cyclopenta[a]chrysene, 3α-(1,5-dimethylhexyl)-8,9,10,11-tetrafluorotetradecahydro-3aα,5bα-dimethyl-	hexane	217(3.83),265s(2.89)	39-2098-69C
Naphthalene, 5-[2-[1β-(1,5-dimethylhexyl)-3a,4,5,6,7,7a-hexahydro-5β-isopropenyl-7aβ-methyl-4α-indanyl]-ethyl]-1,2,3,4-tetrafluoro-	hexane	220(4.90),262(3.51), 274(3.78),283(3.85), 293(3.48)	39-2098-69C
$C_{33}H_{45}NO_5$			
2-Biphenylol, 2'-(2-amino-4-tert-butyl-5-methoxyphenoxy)-4,4'-di-tert-butyl-5,5'-dimethoxy-	EtOH	208(4.72),303(4.08)	39-0258-69C
	EtOH-HCl	295(3.95)	39-0258-69C
	EtOH-NaOH	307(3.95)	39-0258-69C
$C_{33}H_{46}N_2$			
Cholesta-2,4-dieno[2,3-b]quinoxaline	EtOH	219(4.53),269(4.41), 347(4.21),362(4.19)	33-0459-69
$C_{33}H_{46}O_9$			
Withacnistin, 2,3-dihydro-3-methoxy-, acetate	EtOH	229(3.87)	44-3858-69

Compound	Solvent	λ_{max}(log ε)	Ref.
$C_{33}H_{47}NO_{11}$			
12,14-Secocevane-12,14-dione, 4,9-epoxy-3α,4β,16β,17,20-pentahydroxy-, 3,4,16-triacetate	EtOH	290(1.61)	78-3975-69
$C_{33}H_{47}NS$			
Cholest-1-eno[2,3-b]-2'H-[1',4']benzothiazine	EtOH	254(4.39)	7-0182-69
$C_{33}H_{48}$			
7,11b-Etheno-11bH-cyclopenta[a]chrysene, 3α-(1,5-dimethylhexyl)-1,2,3,3a,4,5,-5a,5b,6,7,12,13,13a,13b-tetradecahydro-3a,5b-dimethyl-	hexane	216(3.70)	39-2098-69C
$C_{33}H_{48}N_4O_4$			
18,19-Dinorcholest-13(17)-en-16-one, 5,14-dimethyl-, 2,4-dinitrophenylhydrazone	EtOH	394(4.31)	78-0149-69
$C_{33}H_{48}O_{10}$			
Taxa-4(20),11-diene-2α,5α,7β,9α,10β-pentol, 5,7,9,10-tetraacetate (2-methylbutyrate)	n.s.g.	214(3.96)	77-1282-69
$C_{33}H_{50}O_6$			
Holothurinogenin, 17-deoxy-7,8-dihydro-12β-methoxy-22,25-oxido-, 3-acetate	EtOH	none	78-1897-69
$C_{33}H_{50}O_{10}$			
Crustecdysone, triacetate	EtOH	242(4.12)	12-1045-69
$C_{33}H_{51}NO_5$			
Solafloridine, triacetate	EtOH	238(3.60)	44-1577-69
$C_{33}H_{52}O_4$			
Olean-13(18)-en-24-oic acid, 3α-acetoxy-, methyl ester	EtOH	207(3.95)	102-2083-69
$C_{33}H_{52}O_5$			
A-Norcholest-5-en-7-one, 2β-acetoxy-2α-(acetoxymethyl)-3,3-dimethyl-	EtOH	241(4.20)	22-3189-69
$C_{33}H_{52}O_7$			
Crustecdysone, diacetonide	EtOH	242(4.07)	12-1045-69
$C_{33}H_{54}O_5$			
A-Nor-5β-cholestan-6-one, 2β-acetoxy-2α-(acetoxymethyl)-3,3-dimethyl-	EtOH	295(1.90)	22-3189-69
A-Nor-5β-cholestan-7-one, 2β-acetoxy-2α-(acetoxymethyl)-3,3-dimethyl-	EtOH	202(1.85)	22-3189-69
$C_{33}H_{54}O_{11}$			
Ponasteroside A	EtOH	245(4.14)	78-3909-69

Compound	Solvent	λ max (log ε)	Ref.
$C_{34}H_{20}$			
Naphthacene, 5,12-bis(phenylethynyl)-	benzene	548(4.37)	44-1734-69
$C_{34}H_{23}BrO_2S_2$			
4H,8H-Cyclobuta[1,2-b:3,4-b']bisthiopyran-4,8-dione, 3-bromo-4a,4b,8a,8b-tetrahydro-2,4b,6,8b-tetraphenyl-	$CHCl_3$	281(4.34),339(4.05)	18-3005-69
$C_{34}H_{23}NO$			
1H-Dibenz[e,g]isoindol-1-one, 2,11b-dihydro-2,3,11b-triphenyl-	$CHCl_3$	278(4.30),342(3.92)	22-2756-69
Phenanthrene, 9-benzoyl-10-(N-phenylbenzimidoyl)-	$CHCl_3$	256(4.82),400s(3.05)	22-2756-69
$C_{34}H_{23}NO_3$			
9-Phenanthrenecarboxanilide, N-benzoyl-9,10-dihydro-10-oxo-9-phenyl-	$CHCl_3$	280s(4.15),326(3.51)	22-2756-69
$C_{34}H_{24}$			
Naphthalene, 1,5-bis[2-(1-naphthyl)-vinyl]-	DMF	306(4.18),368(4.58)	33-2521-69
Naphthalene, 1,6-bis[2-(1-naphthyl)-vinyl]-	DMF	346(4.73)	33-2521-69
Naphthalene, 1,7-bis[2-(1-naphthyl)-vinyl]-	DMF	302(4.26),379(4.79)	33-2521-69
Naphthalene, 2,3-bis[2-(1-naphthyl)-vinyl]-	DMF	330(4.65)	33-2521-69
Naphthalene, 2,6-bis[2-(1-naphthyl)-vinyl]-	DMF	303(4.23),380(4.81)	33-2521-69
2,4,6-Octatriene, 1,8-difluoren-9-ylidene-	CH_2Cl_2	256(4.75),278(4.53), 421(4.53),445(4.88), 473(5.12),507(5.14)	78-0955-69
Stilbene, 4-[(4-biphenylyl)ethynyl]-4'-phenyl-	DMF	363(4.93)	33-2521-69
$C_{34}H_{24}O$			
Benzofuran, 2,3-diphenyl-6-(p-phenylstyryl)-, (E)-	DMF	304(4.38),365(4.81)	33-1282-69
Benzofuran, 5-phenyl-2-[p-(p-phenylstyryl)phenyl]-, (E)-	DMF	371(4.91)	33-1282-69
$C_{34}H_{24}O_2$			
Benzofuran, 2-[p-(p-phenoxystyryl)-phenyl]-5-phenyl-	DMF	363(4.88)	33-1282-69
$C_{34}H_{24}O_2S_2$			
4H,8H-Cyclobuta[1,2-b:3,4-b']bisthiopyran-4,8-dione, 4a,4b,8a,8b-tetrahydro-2,4b,6,8b-tetraphenyl-	$CHCl_3$	283(4.38),335(4.12)	18-3005-69
$C_{34}H_{25}BF_2O_3$			
Naphtho[1,2-e]-1,3,2-dioxaborin, 2,2-difluoro-4-[2-(2,4-diphenyl-5,6,7,8-tetrahydrobenzopyran-8-yl)vinyl]-	MeCN	265(4.21),296s(3.94), 341s(4.04),415(3.74), 660(4.46),704(4.45)	4-0029-69
$C_{34}H_{25}NO$			
2-Pyrrolin-5-one, 1,2,3,4,4-pentaphenyl-	EtOH	300s(3.80)	22-2756-69
$C_{34}H_{26}$			
Benzene, p-bis(p-phenylstyryl)-	DMF	380(4.95)	33-2521-69

Compound	Solvent	$\lambda_{max}(\log \epsilon)$	Ref.
Benzene, 1-[2-(1-naphthyl)vinyl]-3,5-distyryl-	toluene	315(4.84)	104-0903-69
Benzene, 1-[2-(2-naphthyl)vinyl]-3,5-distyryl-	toluene	315(5.02)	104-0903-69
Naphthalene, 2,3,6-tristyryl-	DMF	333(4.85)	33-2521-69
Stilbene, 4-(2,2-diphenylvinyl)-4'-phenyl-	DMF	368(4.80)	33-2521-69
$C_{34}H_{26}N_2O_8$			
4(3H)-Quinazolinone, 3-β-D-ribofuranosyl-, 2',3',5'-tribenzoate	MeOH	229(4.85),263(4.02), 272(3.99),301(3.53), 312(3.43)	4-0089-69
$C_{34}H_{26}N_8O_{14}$			
Ethane, 1,2-bis(4-methyl-2-quinolyl)-, dipicrate	EtOH	308(3.49),317(3.56)	39-2246-69C
$C_{34}H_{26}O_2$			
Butadiyne, bis[p-(p-methoxystyryl)-phenyl]-	DMF	381(5.01)	33-2521-69
$C_{34}H_{27}Br_2O_3P$			
Phosphorane, (2,7-dibromofluoren-9-ylidene)tris(p-methoxyphenyl)-	$CHCl_3$	252(4.90),266(4.70), 278(4.55),295(4.54), 384(3.50)	5-0001-69D
$C_{34}H_{27}N_3O_7$			
1H-1,2,3-Triazole, 4-phenyl-1-β-D-ribofuranosyl-, 2',3',5'-tribenzoate	EtOH	232(4.67),275s(3.59), 281(3.46)	4-0639-69
1H-1,2,3-Triazole, 5-phenyl-1-β-D-ribofuranosyl-, 2',3',5'-tribenzoate	EtOH	231(4.65)	4-0639-69
$C_{34}H_{28}N_2$			
2-Anilino-1,1-dibenzyl-3-phenyl-1H-isoindolium hydroxide, inner salt	MeCN	265(4.15),346(4.00), 460(4.02)	35-3670-69
$C_{34}H_{28}N_2O$			
Acetamide, N-(1,1,3,3-tetraphenyl-2-isoindolinyl)-	CH_2Cl_2	252(3.58),259(3.54), 266(3.49),272(3.35)	44-0461-69
$C_{34}H_{28}O_2$			
Naphthalene, 2-[p-[2,2-bis(p-methoxyphenyl)vinyl]styryl]-	DMF	373(4.77)	33-2521-69
$C_{34}H_{29}ClN_2$			
2-Anilino-1,1-dibenzyl-3-phenyl-1H-isoindolium chloride	MeCN	284(3.60),415(4.13)	35-3670-69
$C_{34}H_{29}N_3O_3S$			
4-Thiazoline, 4-benzoyl-5-biphenylyl-3-cyclohexyl-2-(p-nitrophenylimino)-	MeCN	260(4.71),340(3.74), 420(3.82)	23-3557-69
$C_{34}H_{30}N_4O_5S$			
Acetophenone, 2-[3-[(4-amino-2-methyl-5-pyrimidinyl)methyl]-5-(2-hydroxyethyl)-4-methyl-4-thiazolin-2-ylidene]-2-hydroxy-, dibenzoate	EtOH	232(4.30),270(3.89), 394(4.41)	94-0128-69
$C_{34}H_{30}O_4$			
Acetylene, bis[p-(3,4-dimethoxystyryl)-phenyl]-	DMF	378(4.95)	33-2521-69

Compound	Solvent	λ max (log ε)	Ref.
$C_{34}H_{30}O_{16}$ Floccosin, diacetate	EtOH	205(4.69),222s(4.47), 270(4.32),278(4.27), 352(3.94)	23-1561-69
$C_{34}H_{31}N_3O_3S$ 4-Thiazoline, 5-(4-biphenylylamino)-3-cyclohexyl-4-(p-nitrobenzoyl)-2-phenyl-	MeCN	260(4.71),340(3.74), 420(3.82)	23-3557-69
$C_{34}H_{31}N_3O_{12}S$ Acetamide, N-(1-β-D-galactopyranosyl-1,2-dihydro-2-oxo-4-pyrimidinyl)-, tribenzoate, methanesulfonate	EtOH	232(4.65),282(3.82), 300(3.78)	94-0416-69
$C_{34}H_{32}INO_4$ 10,11-Bis(benzyloxy)protoberbinium iodide	EtOH	218(4.57),265(4.28), 288(4.41),308(4.34), 340s(4.08),370(3.67)	35-4009-69
$C_{34}H_{32}N_4O_6S$ Thiamine, (hydroxymethyl)-, O,O',O"-tribenzoate	EtOH	233(4.19),275(3.68)	94-0343-69
$C_{34}H_{32}N_6$ Succinonitrile, 2,3-dimethyl-2,3-bis-[m-(phenylazo)phenethyl]-	toluene	322(4.52),440(3.12)	49-1451-69
$C_{34}H_{32}N_8$ Butyronitrile, 2,2'-azobis[2-methyl-4-[m-(phenylazo)phenyl]-	$CHCl_3$	321(4.51),440(3.20)	49-1451-69
$C_{34}H_{32}O$ Dibenzofuran, 3,7-bis(p-isopropyl-styryl)-	DMF	373(4.93)	33-1282-69
$C_{34}H_{32}O_6S_2$ Thianthrene, 2,7-bis(3,4,5-trimethoxy-styryl)-	DMF	342(4.89)	33-1282-69
$C_{34}H_{32}O_7$ Dibenzofuran, 3,7-bis(3,4,5-trimethoxy-styryl)-, (E,E)-	DMF	379(4.90)	33-1282-69
$C_{34}H_{32}S$ Dibenzothiophene, 3,7-bis(p-isopropyl-styryl)-, (E,E)-	DMF	370(4.95),388(4.88)	33-1282-69
$C_{34}H_{32}S_2$ Thianthrene, 2,7-bis(p-isopropylstyr-yl)-, (E,E)-	DMF	335(4.92)	33-1282-69
$C_{34}H_{34}$ Biphenyl, 4,4'-bis(p-isopropylstyryl)-	DMF	359(4.91)	33-2521-69
$C_{34}H_{34}CuN_6$ Copper, [13,17-diethyl-2,8,12,18-tetra-methyl-3,7-porphinedipropionitril-ato(2-)]-	$CHCl_3$	401(5.19),528(3.83), 564(4.05)	39-0176-69C

Compound	Solvent	$\lambda_{max}(\log \epsilon)$	Ref.
$C_{34}H_{34}HgN_4O_8S_2$			
Mercury, bis[N-(2-isopropylidene)-3,7-dioxo-4-thia-1-azabicyclo[3.2.0]hept-6-yl)-2-phenoxypropionimidato]-	CH_2Cl_2	272(4.32)	88-3385-69
$C_{34}H_{34}N_4O_4$			
Pyromethylphaeophorbide b	dioxan	282(4.31),326(4.48), 370(4.39),415(4.82), 438(5.24),532(4.09), 555(3.89),602(3.92), 656(4.50)	5-0177-69E
$C_{34}H_{34}O_6$			
Biphenyl, 4,4'-bis(3,4,5-trimethoxystyryl)-	DMF	367(4.91)	35-2521-69
$C_{34}H_{34}O_7$			
D-1-gluco-Octose, 3,7-anhydro-2-deoxy-1-C-(p-methoxyphenyl)-8-O-trityl-	MeOH	275(4.18)	65-0119-69
$C_{34}H_{34}O_{10}$			
Vioxanthin, tetramethyl-	EtOH	264(4.72),307(3.97), 320(3.95),350(3.73)	23-1223-69
$C_{34}H_{34}O_{14}$			
Floccosin, tetra-O-methyl-	EtOH	208(4.71),229s(4.44), 270(4.36),335(3.78)	23-1561-69
$C_{34}H_{36}N_2O_6$			
7-Isoquinolinol, 1,2,3,4-tetrahydro-6-methoxy-1-salicyl-8-[(1,2,3,4-tetrahydro-6-methoxy-1-salicyl-7-isoquinolyl)oxy]-	MeOH	283(3.98),295s(3.76)	94-1977-69
7-Isoquinolinol, 1,2,3,4-tetrahydro-6-methoxy-1-[5-[(1,2,3,4-tetrahydro-6-methoxy-1-salicyl-7-isoquinolyl)oxy]-salicyl]-	MeOH	283(4.04),295s(3.76)	94-1977-69
$C_{34}H_{37}NO_5$			
Isoquinoline, 7-(benzyloxy)-1-[4-(benzyloxy)-3,5-dimethoxyphenethyl]-1,2,3,4-tetrahydro-6-methoxy-, (R)-(+)-	MeOH	230s(4.30),280(3.63), 290s(3.55)	33-0678-69
hydrochloride	MeOH	230s(4.31),282(3.60), 292s(3.46)	33-0678-69
(S)-(-)-	MeOH	230s(4.34),280(3.68), 290s(3.57)	33-0678-69
hydrochloride	MeOH	230s(4.31),282(3.59), 291s(3.48)	33-0678-69
$C_{34}H_{38}N_2O_{12}$			
Isovincoside, lactam, tetraacetate	EtOH	228(4.46),273(3.93), 280(3.91),289(3.78)	39-1193-69C
Vincoside, lactam, tetraacetate (4 λ,3 ε)	EtOH	228(4.46),273(3.93), 290(3.89),?(3.78)	39-1193-69C
$C_{34}H_{38}N_4Ni$			
Nickel, [3,8,12,13,18-pentamethyl-5-propenyl-2,7,17-triethylporphinato(2-)]-	$CHCl_3$	296(4.03),345(4.10), 405(5.19),528(3.96), 564(4.16)	39-0655-69C

Compound	Solvent	λ_{max}(log ϵ)	Ref.
$C_{34}H_{40}N_2O_2$			
p-Toluidine, N-tert-butyl-N-[[p-tert-butylhydroxyamino)-α-phenylbenzyl]-oxy]-α-phenyl-	EtOH	252(4.19)	39-1459-69C
$C_{34}H_{40}N_4Ni$			
Nickel, [5-butyl-2,17-diethyl-3,7,8,-12,13,18-hexamethylporphinato(2-)]-	$CHCl_3$	409(5.22),498s(3.49), 535(3.97),570(4.08)	39-0655-69C
Nickel, [2,7,17-triethyl-3,8,12,13,18-pentamethyl-5-propylporphinato(2-)]-	$CHCl_3$	301(3.99),348(4.07), 410(5.17),502s(3.43), 536(3.91),574(4.03)	39-0655-69C
$C_{34}H_{41}ClN_4NiO_6$			
Nickel(1+), (hydrogen octadehydro-3,8-diethyl-2,7,12,13,17,18,19-hepta-methyl-1-corrinacetato)-, perchlorate, ethyl ester	$CHCl_3$	271s(4.41),278(4.41), 355(4.44),456(3.62), 563(4.08)	39-0655-69C
$C_{34}H_{41}IN_4NiO_2$			
Nickel(1+), (hydrogen octadehydro-3,8-diethyl-2,7,12,13,17,18,19-hepta-methyl-1-corrinacetato)-, iodide, ethyl ester	$CHCl_3$	267(4.45),273s(4.44), 354(4.44),399s(3.93), 449s(3.78),565(4.09)	39-0655-69C
$C_{34}H_{43}ClN_4NiO_4$			
Nickel(1+), (1-butyloctadehydro-3,8-diethyl-2,7,12,13,17,18,19-hepta-methylcorrinato)-, perchlorate	$CHCl_3$	276(4.45),354(4.50), 385s(4.01),438s(3.78), 556(4.10)	39-0655-69C
$C_{34}H_{43}IN_4Ni$			
Nickel(1+), 1-butyloctadehydro-3,8-diethyl-2,7,12,13,17,18,19-hepta-methylcorrinato)-, iodide	$CHCl_3$	274(4.42),354(4.45), 452(3.65),558(4.06)	39-0655-69C
$C_{34}H_{44}N_4O_5$			
Tyramine, N,N-dimethyl-β-phenylseryl-valylphenylalanyl-O-methyl-	EtOH	225s(4.60),279(3.54), 286(3.48)	78-0937-69
$C_{34}H_{44}O_3$			
19-Norcholesta-1,3,5(10)-trien-6-one, 1-hydroxy-4-methyl-, benzoate	C_6H_{12}	275(4.30),283(4.26), 300(4.12)	39-1240-69C
$C_{34}H_{44}O_9$			
Salannin	n.s.g.	208(4.32)	102-1817-69
$C_{34}H_{44}O_{13}$			
β-D-Glucopyranosiduronic acid, 3,11,20-trioxopregn-4-en-21-yl, methyl ester, triacetate	MeOH	238(4.21)	69-1188-69
$C_{34}H_{44}O_{14}$			
β-D-Glucopyranosiduronic acid, 17-hy-droxy-3,11,20-trioxopregn-4-en-21-yl, methyl ester, 2,3,4-triacetate	MeOH	238(4.20)	69-1188-69
$C_{34}H_{44}O_{16}$			
Olivin, (3-acetyloliosyl)olivosyl-	EtOH	228(4.46),274(4.73), 318(3.87),405(4.10)	105-0483-69

Compound	Solvent	$\lambda_{max}(\log \epsilon)$	Ref.
$C_{34}H_{46}F_4O$			
4b,13b-Etheno-5H-cyclopenta[a]triphenyl-ene, 11β-(1,5-dimethylhexyl)-1,2,3,4-tetrafluoro-6,7,8,8a,8b,9,10,10a,11,-12,13,13a-dodecahydro-6β-methoxy-8aβ,10aβ-dimethyl-	hexane	217(3.83),270(2.74)	39-2098-69C
$C_{34}H_{46}O_8$			
Bovinone, leuco-tetraacetate	EtOH	270(3.37)	39-2398-69C
$C_{34}H_{46}O_{13}$			
β-D-Glucopyranosiduronic acid, 11β-hydroxy-3,20-dioxopregn-4-en-21-yl, methyl ester, 2,3,4-triacetate	MeOH	240(4.20)	69-1188-69
β-D-Glucopyranosiduronic acid, 17-hydroxy-3,20-dioxopregn-4-en-21-yl, methyl ester, 2,3,4-triacetate	MeOH	241(4.22)	69-1188-69
$C_{34}H_{46}O_{14}$			
β-D-Glucopyranosiduronic acid, 11β,17-dihydroxy-3,20-dioxopregn-4-en-21-yl, methyl ester, 2,3,4-triacetate	MeOH	242(4.21)	69-1188-69
$C_{34}H_{48}O$			
2'H-1'-Benzopyrano[3',2':2,3]cholest-3-ene	EtOH	248(4.23),256(4.11), 271(3.74),350(3.74)	39-1419-69C
$C_{34}H_{48}O_9$			
Withacnistin, 3-ethoxy-2,3-dihydro-, acetate	EtOH	227(3.95)	44-3858-69
$C_{34}H_{49}ClO_8$			
4-Heptadecyl-7-hydroxy-2',4'-dimethoxy-flavylium perchlorate	MeOH	466(4.74)	61-0845-69
$C_{34}H_{49}N$			
5α-Cholest-2-eno[3,2-b]quinoline	C_6H_{12}	212(4.66),232(4.53), 237(4.56),275(3.48), 294(3.45),300(3.45), 307(3.61),313(3.50), 321(3.78)	39-1419-69C
$C_{34}H_{50}O_2$			
4'H-1'-Benzopyrano[3',2':2,3]cholestan-4'-ol	EtOH	260(4.17),270(4.15), 297(3.90)	39-1419-69C
$C_{34}H_{50}O_3$			
5α-Cholestan-3-one, 19-hydroxy-, benzoate	EtOH	231(4.13),274(2.96)	23-1495-69
$C_{34}H_{50}O_9S$			
Crustecdysone, 2-p-toluenesulfonate	EtOH	226(4.37),242s(4.16)	12-1059-69
$C_{34}H_{52}N_2O_2S$			
p-Toluenesulfonic acid, 5α-cholest-1-en-3-ylidenehydrazide	C_6H_{12}	275(4.28)	25-1093-69
	$CHCl_3$	415(2.11)	25-1093-69
$C_{34}H_{52}O_3$			
5α-Cholestane-3β,19-diol, 19-benzoate	EtOH	231(4.12),274(2.98)	23-1495-69

Compound	Solvent	$\lambda_{max}(\log \epsilon)$	Ref.
$C_{34}H_{52}O_8$			
Cephalosporin P_1, methyl ester	MeOH	220(4.04)	78-3341-69
$C_{34}H_{54}OS$			
5α-Cholestane, 3α-(R)-benzylsulfinyl-	hexane	219(3.90),243(3.50)	39-2159-69C
	EtOH	218(3.91)	39-2159-69C
5α-Cholestane, 3α-(S)-benzylsulfinyl-	hexane	219(4.00),242(3.62)	39-2159-69C
	EtOH	218(4.00)	39-2159-69C
5α-Cholestane, 6α-(R)-benzylsulfinyl-	hexane	221(4.03),240s(3.52)	39-2159-69C
	EtOH	218s(4.00)	39-2159-69C
5α-Cholestane, 6α-(S)-benzylsulfinyl-	hexane	220(4.03),240s(3.59)	39-2159-69C
	EtOH	218s(4.02)	39-2159-69C
5α-Cholestane, 6β-(R)-benzylsulfinyl-	hexane	219(4.02),246(3.40)	39-2159-69C
	EtOH	220(3.98),240s(3.45)	39-2159-69C
5α-Cholestane, 6β-(S)-benzylsulfinyl-	hexane	219(3.99),245(3.56)	39-2159-69C
	EtOH	218s(3.97)	39-2159-69C

Compound	Solvent	$\lambda_{max}(\log \epsilon)$	Ref.
$C_{35}H_{20}Cl_4O_3$			
1H-Cyclopenta[b][1,4]benzodioxin-1-one, 5,6,7,8-tetrachloro-3a,9a-dihydro-2,3,3a,9a-tetraphenyl-	MeOH	309(4.04)	77-0467-69
$C_{35}H_{25}Br_2P$			
Phosphorin, 1,1-bis(p-bromophenyl)-1,1-dihydro-2,4,6-triphenyl-	benzene	337(4.24),383(3.86), 519(3.99)	88-1231-69
$C_{35}H_{25}P$			
Spiro[5H-dibenzophosphole-5,1'-phosphorin], 2',4',6'-triphenyl-	benzene	336(4.05),474(3.80)	88-1231-69
$C_{35}H_{26}$			
1,3-Cyclopentadiene, 1,2,3,4,5-pentaphenyl-	C_6H_{12}	246(4.43),267(4.36), 340(4.07)	39-2605-69C
Naphthalene, 5-methyl-1,2,3,4-tetraphenyl-	n.s.g.	225(4.76),244(4.81), 301(4.21)	39-0748-69C
Naphthalene, 6-methyl-1,2,3,4-tetraphenyl-	n.s.g.	223(4.77),245(4.86), 295(4.14)	39-0748-69C
$C_{35}H_{26}ClNO_5$			
1-(p-Phenoxyphenyl)-2,4,6-triphenylpyridinium perchlorate	MeOH	310(4.52)	5-0093-69G
$C_{35}H_{26}N_2O_4$			
1,2-Diazaspiro[4.4]nona-1,3,6,8-tetraene-3,4-dicarboxylic acid, 6,7,8,9-tetraphenyl-, dimethyl ester	ether	256(4.42),315(3.88)	64-0536-69
$C_{35}H_{26}O$			
Benzofuran, 5-benzyl-2-[p-(p-phenylstyryl)phenyl]-	DMF	370(4.89)	33-1282-69
$C_{35}H_{26}O_2$			
Hydroperoxide, 1,2,3,4,5-pentaphenyl-2,4-cyclopentadien-1-yl	ether	356(3.86)	28-0622-69A
$C_{35}H_{27}N$			
Pyrrole, 2,3,4,5-tetraphenyl-1-p-tolyl-	$CHCl_3$	262(4.50),300s(4.25)	22-2765-69
$C_{35}H_{27}NO$			
Pyrrole, N-(p-methoxyphenyl)-2,3,4,5-tetraphenyl-	$CHCl_3$	262(4.50),300s(4.30)	22-2765-69
2-Pyrrolin-5-one, 2,3,4,4-tetraphenyl-1-p-tolyl-	EtOH	300s(3.80)	22-2756-69
$C_{35}H_{27}NO_2$			
2-Pyrrolin-5-one, 1-(p-methoxyphenyl)-2,3,4,4-tetraphenyl-	EtOH	300s(3.80)	22-2756-69
$C_{35}H_{27}P$			
Phosphorin, 1,1-dihydro-1,1,2,4,6-pentaphenyl-	benzene	342(4.23),381(3.80), 515(3.95)	88-1231-69
$C_{35}H_{28}O_2$			
12,12'-(6H,6'H)-Spirobi[cyclohepta[b]fluorene]-6,6'-dione, 7,7',8,8',9,-9',10,10'-octahydro-	EtOH	209(4.64),235s(4.60), 253(4.82),272s(4.52), 293(4.28),310s(3.75)	33-1202-69

Compound	Solvent	$\lambda_{max}(\log \epsilon)$	Ref.
$C_{35}H_{28}Si$			
Silacyclopenta-2,4-diene, 1-methyl-1,2,3,4,5-pentaphenyl-	C_6H_{12}	219(4.4),249(4.4), 357(3.9)	6-0203-69
$C_{35}H_{30}O_{10}$			
Flavone, 5,7-diacetoxy-3',4'-bis(benzyloxy)-3,6-dimethoxy-	EtOH	252(4.32),350(4.21)	18-1649-69
$C_{35}H_{32}$			
12,12'(6H,6'H)-Spirobi[cyclohepta[b]-fluorene], 7,7',8,8',9,9',10,10'-octahydro-	EtOH	210(4.79),225(4.81), 231(4.85),245(4.85), 270(4.43),280(4.39), 285s(4.30),290s(4.20), 302(4.20),308(4.08), 315(4.43)	33-1202-69
$C_{35}H_{32}O_4$			
9,9'-Spirobi[fluorene]-2,2'-divaleric acid	EtOH	212(4.77),221(4.78), 226s(4.73),246(4.43), 258s(4.43),267(4.47), 277(4.46),283s(4.36), 288s(4.25),300(4.12), 304s(3.97),312(4.31)	33-1202-69
$C_{35}H_{34}N_4O_{12}$			
Resorcinol, 2-(p-mentha-1,8-dien-3-yl)-5-pentyl-, bis(3,5-dinitrobenzoate)	EtOH	251(4.24)	33-1102-69
Resorcinol, 4-(p-mentha-1,8-dien-3-yl)-5-pentyl-, bis(3,5-dinitrobenzoate)	EtOH	255(4.32)	33-1102-69
$C_{35}H_{36}N_4Ni$			
Nickel, [1-benzylidene-1,22-dihydro-2,3,7,8,12,13,17,18,19-nonamethyl-bilinato(2-)]-	$CHCl_3$	276(4.22),318(4.22), 344s(4.26),355(4.27), 401s(4.43),418(4.70), 500s(3.59),737s(3.86), 802(4.16)	39-0655-69C
$C_{35}H_{37}NO_5$			
C-Homo-13a-berbine, 2,10-bis(benzyloxy)-3,9,11-trimethoxy-, 14aR-(+)-	EtOH	230s(4.33),282(3.74)	33-1228-69
14aS-(-)-	EtOH	230s(4.34),282(3.74)	33-1228-69
$C_{35}H_{37}NO_8$			
Isoquinoline, 1-[p-(benzyloxy)benzyl]-6-(benzyloxy)-1,2,3,4-tetrahydro-5,7-dimethoxy-2-methyl-, oxalate, dl-	MeOH	277(3.53),283(3.52)	44-3884-69
$C_{35}H_{39}ClN_6NiO_4$			
Nickel(1+), (octadehydro-8,12-diethyl-1,3,7,13,17,19-hexamethyl-2,18-corr-indipropionitrilato)-, perchlorate	EtOH	277(4.47),354(4.44), 375s(3.88),456(3.74), 562(4.09)	39-0176-69C
$C_{35}H_{39}N_5O_5$			
Ergocristine	MeOH	312(3.91)	33-1549-69
Ergocristinine	MeOH	313(3.96)	33-1549-69
$C_{35}H_{40}N_4NiO_2$			
Nickel, [hydrogen 2,17-diethyl-3,7,8,12,13,18-hexamethylporphine-propionato(2-)]-, ethyl ester	$CHCl_3$	299(3.91),349(4.08), 409(5.16),536(3.98), 572(4.06)	39-0655-69C

Compound	Solvent	λ_{max}(log ϵ)	Ref.
$C_{35}H_{42}N_2O_{18}S$			
2-Benzimidazolinethione, 1,3-di-β-D-glucopyranosyl-, octaacetate	MeOH	252(3.91),302s(4.32), 311(4.42)	97-0152-69
	EtOH	222(4.29),252(3.95), 312(4.41)	48-0997-69
$C_{35}H_{42}N_6$			
Biline-2,18-dipropionitrile, 8,12-diethyl-10,23-dihydro-1,3,7,13,17,19-hexamethyl-, dihydrobromide	$CHCl_3$	374(4.17),435s(4.15), 454(4.55),500s(4.58), 529(5.20)	39-0176-69C
$C_{35}H_{43}ClN_4NiO_6$			
Nickel(1+), [hydrogen octadehydro-3,8-diethyl-2,7,12,13,17,18,19-heptamethyl-1-corrinpropionato)-, perchlorate, ethyl ester	$CHCl_3$	275(4.46),355(4.49), 458s(3.73),559(4.11)	39-0655-69C
$C_{35}H_{43}IN_4NiO_2$			
Nickel(1+), [hydrogen octadehydro-3,8-diethyl-2,7,12,13,17,18,19-heptamethyl-1-corrinpropionato)-, iodide, ethyl ester	$CHCl_3$	272s(4.48),280(4.49), 354(4.51),450s(3.73), 563(4.09)	39-0655-69C
$C_{35}H_{44}F_4O_2$			
4b,13b-Etheno-5H-cyclopenta[a]triphenylen-6-ol, 11β-(1,5-dimethylhexyl)-1,2,3,4-tetrafluoro-6,7,8,8a,10,-10a,11,12,13,13-decahydro-8aβ,10aβ-dimethyl-, acetate	hexane	220(3.98),268(2.70)	39-2098-69C
$C_{35}H_{44}N_2O_{18}$			
Benzimidazoline, 1,3-di-β-D-glucopyranosyl-, octaacetate	EtOH	215(4.58),261(3.88), 309(3.81)	48-0997-69
$C_{35}H_{44}O_{15}$			
Gibberellenic acid, 2-O-β-glucosyl-, tetraacetate, dimethyl ester	n.s.g.	253(4.23)	88-2081-69
$C_{35}H_{46}N_4O_2$			
Tetrapyrrotriene-1,10-dione, 2,3,4,5,6,7,8,9-octaethyl-	$CHCl_3$	369(4.69),650(4.15)	5-0167-69E
$C_{35}H_{47}N_5O_5$			
Lasiodine B	EtOH	250(3.90)(end abs.)	78-0937-69
$C_{35}H_{48}O$			
5β-Lumista-7,22-dien-3-one, 2-benzylidene-	n.s.g.	289(4.18)	39-0674-69C
$C_{35}H_{50}$			
1,3,5,7,9,11,13-Tetradecaheptaene, 3,7,12-trimethyl-1,14-bis(2,6,6-trimethyl-1-cyclohexen-1-yl)-, all trans	hexane	413(5.02)	78-5383-69
	C_6H_{12}	418(5.05)	
	benzene	429(5.05)	
	CCl_4	427(5.04)	
	dioxan	425(5.00)	
	acetone	420(5.03)	
	DMF	427(5.05)	
	CS_2	448(4.98)	
(also other solvent mixtures)	80% hexane-20% CS_2	423(5.01)	

Compound	Solvent	λ_{max}(log ε)	Ref.
$C_{35}H_{52}N_4O_4$			
Stigmast-4-en-3-one, 2,4-dinitrophenyl-hydrazone	EtOH	389(4.8)	94-0163-69
$C_{35}H_{52}N_4O_5$			
Stigmastane-3,6-dione, mono(2,4-dinitro-phenylhydrazone)	EtOH	368(4.18)	94-0163-69
$C_{35}H_{52}O_{11}$			
Crustecdysone, tetraacetate	EtOH	242(4.04)	12-1045-69
$C_{35}H_{54}O_7$			
Cephalosporin P_1, deacetyl-, methyl ester, acetonide	EtOH	220(3.75)	78-3341-69

Compound	Solvent	λ_{max}(log ε)	Ref.
$C_{36}H_{18}S_3$			
1,20:3,5:6,8:10,12:13,15:17,19-Hexaethenocyclooctadeca[1,2-c:7,8-c':13,14-c"]trithiophene	THF	260(5.03),290s(4.59)	5-0024-69D
$C_{36}H_{24}O$			
Dibenzofuran, 3,7-bis[2-(1-naphthyl)-vinyl]-, (E,E)-	DMF	385(4.87)	33-1282-69
$C_{36}H_{24}O_6$			
[36]Annulene hexaoxide	$CHCl_3$	315(4.52),325s(4.41), 412s(4.69),425(4.78)	12-1951-69
$C_{36}H_{24}S_2$			
Thianthrene, 2,7-bis[2-(1-naphthyl)-vinyl]-, (E,E)-	DMF	350(4.78)	33-1282-69
Thianthrene, 2,7-bis[2-(2-naphthyl)-vinyl]-, (E,E)-	DMF	290(4.62),345(4.97)	33-1282-69
$C_{36}H_{26}$			
Biphenyl, 4,4'-bis[2-(1-naphthyl)vinyl]-	DMF	370(4.84)	33-2521-69
$C_{36}H_{26}O$			
Furan, 2,3-diphenyl-5-[p-(p-phenylstyryl)phenyl]-, (E)-	DMF	315(4.36),380(4.83)	33-1282-69
$C_{36}H_{26}O_2$			
Furan, 5-[p-(p-phenoxystyryl)phenyl]-2,3-diphenyl-	DMF	306(4.27),373(4.76)	33-1282-69
$C_{36}H_{28}$			
1,3-Cyclopentadiene, 5-methyl-1,2,3,4,5-pentaphenyl-	n.s.g.	245(4.44),330(3.9)	28-1412-69B
Ethylene, 1-phenyl-1,2-bis(p-styryl-phenyl)-	DMF	375(4.87)	33-2521-69
Stilbene, 4'-phenyl-3,5-distyryl-	toluene	320(4.99)	104-0903-69
$C_{36}H_{28}N_2$			
2H-Dibenzo[e,g]isoindole, 2-[p-(dimethylamino)phenyl]-1,3-diphenyl-	$CHCl_3$	260(4.66),300s(4.44), 330s(4.02)	22-2765-69
$C_{36}H_{28}N_2O$			
1H-Dibenz[e,g]isoindol-1-one, 2-[p-(dimethylamino)phenyl]-2,11b-dihydro-3,11b-diphenyl-	$CHCl_3$	274(4.48),350(3.85)	22-2756-69
Phenanthrene, 9-[N-(p-dimethylaminophenyl)benzimidoyl]-10-benzoyl-	$CHCl_3$	258(4.84),390(3.99)	22-2756-69
$C_{36}H_{28}N_4O_8S_2$			
[$\Delta^{2,2'}$-Bi-2H-1,4-benzothiazine]-7,7'-dipropionic acid, α,α'-dibenzamido-5,5'-dihydroxy-	pH 13	239s(4.56),312(4.15), 483(4.37)	32-1193-69
	MeOH-HCl	267s(4.20),327(4.17), 553s(4.34),587(4.47)	32-1193-69
$C_{36}H_{28}O_2$			
Spiro[bicyclo[3.2.0]heptane-2,1'-cyclobutane]-2',6-dione, 3',7-dibenzylidene-3,4-diphenyl-	EtOH	220s(4.41),227(4.31), 233(4.33),318(4.70)	77-1281-69

Compound	Solvent	λ max(log ε)	Ref.
$C_{36}H_{28}O_4$			
Spiro[cyclopentane-1,3(2'H)-cyclopenta-[b]pyran]-2,2',4'(5'H)-trione, 6',7'-dihydro-4,5,5',6'-tetraphenyl-	EtOH	292(4.04)	23-2853-69
$C_{36}H_{29}NO_6$			
Chalcone, 2,2',4'-tris(benzyloxy)-α-nitro-	$CHCl_3$	288(4.18),320(4.17)	104-1803-69
$C_{36}H_{29}O_2P$			
2-Cyclopenten-1-one, 5-methyl-2,3,4-triphenyl-5-(diphenylphosphinyl)-, P-oxide	EtOH	302(4.04)	88-4335-69
$C_{36}H_{30}AsBiBr_2$			
Tetraphenylarsonium dibromodiphenyl-bismuthate	CH_2Cl_2	309(3.83)	101-0099-69E
$C_{36}H_{30}AsBiCl_2$			
Tetraphenylarsonium dichlorodiphenyl-bismuthate	CH_2Cl_2	295(3.68)	101-0099-69E
$C_{36}H_{30}AsBiI_2$			
Tetraphenylarsonium diiodophenyl-bismuthate	CH_2Cl_2	294(4.10),342(4.00)	101-0099-69E
$C_{36}H_{30}Cl_2P_2Pt$			
Platinum, dichlorobis(triphenylphos-phine)-, cis	n.s.g.	267s(4.05),275s(3.94), 328(2.73)	101-0457-69B
trans	n.s.g.	260(4.22),283(4.44), 320s(3.20)	101-0457-69B
$C_{36}H_{30}N_2$			
Pyrrole, N-[p-(dimethylamino)phenyl]-2,3,4,5-tetraphenyl-	$CHCl_3$	276(4.68),300s(4.41)	22-2765-69
$C_{36}H_{30}O_2$			
Stilbene, 4-[2,2-bis(p-methoxyphenyl)-vinyl]-4'-phenyl-	DMF	375(4.80)	33-2521-69
$C_{36}H_{30}O_3$			
Stilbene, 4-[2,2-bis(p-methoxyphenyl)-vinyl]-4'-phenoxy-	DMF	368(4.75)	33-2521-69
$C_{36}H_{30}O_4$			
Butadiyne, bis[p-(3,4-dimethoxystyryl)-phenyl]-	DMF	389(5.01)	33-2521-69
$C_{36}H_{30}Si$			
Silacyclopenta-2,4-diene, 1-ethyl-1,2,3,4,5-pentaphenyl-	C_6H_{12}	220(4.4),249(4.4), 360(3.9)	6-0203-69
$C_{36}H_{34}$			
Acetylene, bis[p-(p-isopropylstyryl)-phenyl]-	DMF	367(4.96)	33-2521-69
$C_{36}H_{34}N_2O_6S_2$			
$\Delta^{2,2'}$-Bi-2H-benzothiazine, 5,5'-dimeth-oxy-8,8'-bis[(2-methoxy-p-tolyl)oxy]-7,7'-dimethyl-	EtOH	230s(4.60),275(4.40), 301(4.32),472(4.36), 487(4.34)	32-0323-69

Compound	Solvent	$\lambda_{max}(\log \epsilon)$	Ref.
$\Delta^{2,2'}$-Bi-2H-1,4-benzothiazine, 5,5'-dimethoxy-8,8'-bis[(2-methoxy-p-tolyl)-oxy]-7,7'-dimethyl- (cont.)	EtOH-HCl	266s(4.33),326(4.20), 574s(4.40),596(4.45)	32-0323-69
$C_{36}H_{34}O$			
Benzofuran, 6-(p-isopropylstyryl)-2-[p-(p-isopropylstyryl)phenyl]-	DMF	330(4.37),385(4.94)	33-1282-69
$C_{36}H_{34}O_6$			
Acetylene, bis[p-(3,4,5-trimethoxystyryl)phenyl]-	DMF	373(4.94)	33-2521-69
$C_{36}H_{34}O_{12}$			
Flavone, 4',7-bis(benzyloxy)-3,3'-dihydroxy-5-methoxy-, 3-β-D-glucopyranoside	MeOH	253(4.35),343(4.21)	24-0792-69
$C_{36}H_{34}O_{16}$			
Floccosin, di-O-methyl-, diacetate	EtOH	212(4.69),229s(4.44), 270(4.36),338(3.78)	23-1561-69
$C_{36}H_{34}S$			
Benzo[b]thiophene, 6-(p-isopropylstyryl)-2-[p-(p-isopropylstyryl)phenyl]-	DMF	386(4.94)	33-1282-69
$C_{36}H_{36}$			
Stilbene, 4,4'-bis(p-isopropylstyryl)-	DMF	389(4.98)	33-2521-69
$C_{36}H_{36}IN_2PS$			
2-[[5,5-Dimethyl-3-[(triphenylphosphoranylidene)amino]-2-cyclohexen-1-ylidene]methyl]-3-ethylbenzothiazolium iodide	EtOH	506(2.45)	65-1732-69
$C_{36}H_{36}N_2O_5$			
Tiliacorine	EtOH	295(3.91)	78-3091-69
Tiliacorinine	EtOH	290(3.95)	78-3091-69
$C_{36}H_{36}N_2O_7$			
Mosine	MeOH	290(4.09)	102-1559-69
dimethiodide	MeOH	290(4.01)	102-1559-69
$C_{36}H_{36}N_4$			
Indoline, 2,2'-[1,4-naphthylenebis(nitriloethanediylidene)]bis[1,3,3-trimethyl-, dihydrochloride	EtOH	364(4.66),422(4.70)	124-0179-69
Indoline, 2,2'-[1,7-naphthylenebis(nitriloethanediylidene)]bis[1,3,3-trimethyl-, dihydrochloride	EtOH	378(4.73),405s(4.61)	124-0179-69
	EtOH-HCl	407(4.83),426(4.72)	124-0179-69
$C_{36}H_{36}O_6$			
Benzene, 1,2,4-tris(3,4-dimethoxystyryl)-	DMF	350(4.78)	33-2521-69
Benzene, 1,3,5-tris(3,4-dimethoxystyryl)-	DMF	336(4.96)	33-2521-69
$C_{36}H_{36}O_{19}$			
Flavone, 3,3',4',7-tetrahydroxy-5-methoxy-, 3-β-D-galactopyranoside, heptaacetate	MeOH	263(4.37),296(4.10), 330(4.13)	24-0792-69

Compound	Solvent	λ_{max}(log ε)	Ref.
Flavone, 3,3',4',7-tetrahydroxy-5-methoxy-, 3-β-D-glucopyranoside, heptaacetate	MeOH	264(4.55),330(4.30)	24-0792-69
$C_{36}H_{38}N_2O_6$			
Dryadodaphnine	EtOH	285(3.90)	39-1627-69C
	EtOH-KOH	288(3.91),305(3.91)	39-1627-69C
$C_{36}H_{38}N_4O_4$			
Protoporphyrin IX	$CHCl_3$	507(4.13),542(4.08), 577(3.83),632(3.68)	65-2558-69
$C_{36}H_{38}N_4O_6$			
Photoprotoporphyrin, dimethyl ester, isomer 2	dioxan	422(5.07),500(3.92), 565)4.21),608(3.89), 668(4.76)	5-0173-69J
isomer 3	dioxan	436(4.98),500(3.91), 565(4.18),613(3.85), 671(4.75)	5-0173-69J
$C_{36}H_{38}O_{10}$			
Globiflorin $3B_1$, hexamethyl ether	EtOH	275s(3.86),280(3.88)	39-2572-69C
Globiflorin $3B_2$, methyl ether	EtOH	275s(3.87),280(3.89)	39-2572-69C
$C_{36}H_{40}N_2O_6$			
7-Isoquinolinol, 1,2,3,4-tetrahydro-1-[5-hydroxy-2-[[1,2,3,4-tetrahydro-1-(m-hydroxybenzyl)-6-methoxy-2-methyl-7-isoquinolyl]oxy]benzyl]-6-methoxy-2-methyl-	MeOH	283(3.87)	94-1977-69
7-Isoquinolinol, 1,2,3,4-tetrahydro-6-methoxy-2-methyl-1-salicyl-8-[(1,2,3,4-tetrahydro-6-methoxy-2-methyl-1-salicyl-7-isoquinolyl)oxy]-	MeOH	283(4.01),295s(3.61)	39-0502-69C
$C_{36}H_{40}N_3P$			
Phosphorus(1+), tris(dimethylaminato)-[2,3,4,5-tetraphenyl-2,4-cyclopentadien-1-yl)methyl]-, hydroxide, inner salt	EtOH-2% Et_3N	325(3.93)	39-2605-69C
$C_{36}H_{40}N_4O_5$			
2,18-Porphinedipropionic acid, 7,12-diethyl-3,8,13,17-tetramethyl-β^2-oxo-, dimethyl ester	NaOMe	396(5.24),497(4.11), 533(3.98),568(3.80), 620(3.58)	35-1232-69
	$CHCl_3$	410(5.35),509(4.08), 544(4.20),573(4.03), 630(3.32)	35-1232-69
$C_{36}H_{40}O_{19}$			
Castalin, nonamethyl-	dioxan	233(4.61)	5-0186-69A
$C_{36}H_{42}N_4O_4$			
Mesoporphyrin IX, dimethyl ester, AlOH complex	$CHCl_3$	335(4.39),409(4.37), 536(4.18),573(4.23)	5-0115-69H
Bi(OH)NO_2 complex	$CHCl_3$	352(4.70),460(4.92), 574(4.09)	5-0115-69H
Co(II) complex	$CHCl_3$	325(4.21),392(5.25), 521(4.02),553(4.33)	5-0115-69H

Compound	Solvent	λ_{max} (log ε)	Ref.
Mesoporphyrin IX, dimethyl ester, FeCl complex	$CHCl_3$	379(4.98),510(3.95), 533(3.97),639(3.63)	5-0115-69H
GaOH complex	$CHCl_3$	335(4.31),402(5.63), 534(4.17),572(4.32)	5-0115-69H
$GeCl_2$ complex	$CHCl_3$	345(4.43),410(5.47), 538(4.21),574(4.18)	5-0115-69H
Pd(II) complex	$CHCl_3$	329(4.11),393(5.24), 512(4.13),546(4.62)	5-0115-69H
Zn complex (also other metal complex compounds)	$CHCl_3$	333(4.32),404(5.33), 536(4.18),572(4.28)	5-0115-69H
$C_{36}H_{42}N_4O_7$			
Tryptophan, N-[3-(p-tert-butoxyphenyl)-N-(N-carboxy-L-alanyl)-L-alanyl]-, N-benzyl dimethyl ester, L-	EtOH	275(3.78),283(3.79), 291(3.72)	33-1058-69
$C_{36}H_{43}FN_2O_5$			
1'H-Pregna-4,6-dieno[3,2-c]pyrazol-20-one, 6-fluoro-16α,17-dihydroxy-1'-phenyl-21-[(tetrahydro-2H-pyran-2-yl)oxy]-, cyclic acetal with acetone	EtOH	250(4.09),320(4.55)	22-1256-69
2'H-Pregna-2,4,6-trieno[3,2-c]pyrazol-20-one, 6-fluoro-16α,17-dihydroxy-2'-phenyl-21-[(tetrahydro-2H-pyran-2-yl)-oxy]-, cyclic acetal with acetone	EtOH	224(4.09),311(4.30)	22-1256-69
$C_{36}H_{44}ClFeN_4$			
Iron(1+), [2,3,7,8,12,13,17,18-octa-ethylporphinato(2-)]-, chloride	benzene	391(5.28),487(4.11), 496(4.13),520(3.61), 544(3.21),593(3.61), 617(3.65),647(4.86)	35-7485-69
$C_{36}H_{44}CuN_4O_2$			
Copper, [2,3,7,7,12,12,17,18-octaethyl-dihydro-8,13-porphinedionato(2-)]-	$CHCl_3$	391(4.76),436(4.94), 580(4.20),623(4.67)	5-0167-69E
Copper, [2,3,7,7,12,13,17,17-octaethyl-dihydro-8,18-porphinedionato(2-)]-	$CHCl_3$	383(4.69),426(5.98), 483(3.26),521(3.50), 549(3.34),646(4.12), 660(4.06),706(4.79)	5-0167-69E
Copper, [2,3,7,7,12,13,18,18-octaethyl-dihydro-8,17-porphinedionato(2-)]-	$CHCl_3$	440(4.97),540(3.63), 583(4.08),667(4.60)	5-0167-69E
Copper, [2,3,7,7,13,13,17,18-octaethyl-dihydro-8,12-porphinedionato(2-)]-	$CHCl_3$	398(4.72),434(4.98), 544(3.61),588(3.76), 704(4.70)	5-0167-69E
Copper, [2,3,8,8,12,12,17,18-octaethyl-dihydro-7,13-porphinedionato(2-)]-	$CHCl_3$	372(4.54),418(4.95), 470(3.28),507(3.51), 542(3.53),595(3.99), 648(4.69)	5-0167-69E
Copper, [3,3,7,8,12,13,17,18-octaethyl-20-hydroxy-2(3H)-porphinato(2-)]-	dioxan	304(4.15),340(4.39), 410s(4.73),430(5.13), 451(4.37),558s(3.67), 600(4.17),639(4.30), 696(3.58)	5-0135-69C
$C_{36}H_{44}CuN_4O_3$			
Copper, [3,3,8,8,13,13,17,18-octaethyl-8,13-dihydro-2,7,12(3H)-porphinetri-onato(2-)]-	$CHCl_3$	434(4.88),671(4.24), 721(4.73)	5-0167-69E

Compound	Solvent	λ_{max}(log ε)	Ref.
$C_{36}H_{44}FN_3O_8$ 2'H-Pregna-2,4-dieno[3,2-c]pyrazol-20-one, 9-fluoro-11β,16α,17-trihydroxy-2'-(p-nitrophenyl)-21-[(tetrahydro-2H-pyran-2-yl)oxy]-, cyclic 16,17-acetal with acetone	EtOH	224(4.14),312(4.22)	22-1256-69
$C_{36}H_{44}F_2N_2O_6$ 2'H-Pregna-2,4-dieno[3,2-c]pyrazol-20-one, 9-fluoro-2'-(p-fluorophenyl)-11β,16α,17-trihydroxy-21-[(tetrahydro-2H-pyran-2-yl)oxy]-, cyclic 16,17-acetal with acetone	EtOH	260(4.20)	22-1256-69
$C_{36}H_{44}N_4NiO$ Nickel, [3,3,7,8,12,13,17,18-octaethyl-2(3H)-porphinato(2-)]-	dioxan	295(4.22),315(4.23), 371(4.55),411(5.01), 506(3.56),531s(3.60), 544(3.77),570(3.91), 588(3.85),615(4.74)	5-0135-69C
$C_{36}H_{44}N_4NiO_2$ Nickel, [3,3,7,8,12,13,17,18-octaethyl-20-hydroxy-2(3H)-porphinato(2-)]-	dioxan	296(4.20),337(4.31), 390s(4.44),419(4.83), 428(4.83),548s(3.53), 590(4.02),631(4.41)	5-0135-69C
$C_{36}H_{44}N_4OZn$ Zinc, [3,3,7,8,12,13,17,18-octaethyl-2(3H)-porphinato(2-)]-	dioxan	267(3.94),316s(4.21), 331(4.30),377s(4.54), 398(4.80),419(5.36), 517s(3.49),527(3.56), 543(3.62),569(4.01), 589(3.78),617(4.89)	5-0135-69C
$C_{36}H_{44}N_4O_2$ 1,1,2,2-Ethanetetracarbonitrile, 1-(3β-hydroxyergosta-5,8(14),9(11),22-tetraen-7α-yl)-, acetate	C_6H_{12}	280(3.82)	88-0207-69
$C_{36}H_{44}OP_2$ Diphosphine, tetramesityl-, oxide	C_6H_{12}	276(4.10)	65-1544-69
$C_{36}H_{44}O_2P_2$ Diphosphine, tetramesityl-, dioxide	C_6H_{12}	278(3.94)	65-1544-69
$C_{36}H_{44}P_2$ Diphosphine, tetramesityl-	C_6H_{12}	288(4.10),298(4.10)	65-1544-69
$C_{36}H_{45}FN_2O_6$ 2'H-Pregna-2,4-dieno[3,2-c]pyrazol-20-one, 9-fluoro-11β,16α,17-trihydroxy-2'-phenyl-21-[(tetrahydro-2H-pyran-2-yl)oxy]-, cyclic 16,17-acetal with acetone	EtOH	262(4.25)	22-1256-69
1'H-Pregn-4-eno[3,2-c]pyrazol-20-one, 9-fluoro-11β,16α,17-trihydroxy-1'-phenyl-21-[(tetrahydro-2H-pyran-2-yl)oxy]-, cyclic 16,17-acetal with acetone	EtOH	299(4.48)	22-1256-69

Compound	Solvent	λ_{max}(log ϵ)	Ref.
$C_{36}H_{46}ClFeN_4$			
Iron, chloro[2,3,7,8,12,13,17,18-octaethyl-2,3-dihydroporphinato(2-)]-, trans	benzene	376(4.95),471(3.91), 510s(3.75),559(3.78), 603(4.38),751(3.46)	35-7485-69
$C_{36}H_{46}N_2O_{12}$			
Neoantimycin	EtOH	227(4.48),319(3.80)	78-2193-69
	EtOH-HCl	227(--),311(--)	78-2193-69
	EtOH-NaOH	222(--),342(--)	78-2193-69
$C_{36}H_{46}N_4O$			
2(3H)-Porphinone, 3,3,7,8,12,13,17,18-octaethyl-	$CHCl_3$	394s(5.04),408(5.19), 493s(3.72),512(3.90), 550(4.04),588(3.73), 615(3.26),642(4.52)	39-0564-69C
5(2H)-Porphinone, 2,3,7,8,12,13,17,18-octaethyl-	$CHCl_3$	405(5.16),547s(3.68), 587(3.96),633(4.21)	39-0564-69C
cation	$CHCl_3$-HOAc	414(5.25),535(3.79), 576(3.73),630s(4.13), 682(4.54)	39-0564-69C
dication	$CHCl_3$-CF_3COOH	415(5.39),559(4.02), 612(4.04)	39-0564-69C
anion	$CHCl_3$-EtOH-KOH	405s(4.86),419(5.12), 495(3.42),531(3.66), 573(3.67),620(3.79), 673(4.25)	39-0564-69C
$C_{36}H_{46}N_4O_2$			
1,1,2,2-Ethanetetracarbonitrile, 1-(3β-hydroxyergosta-5,8(14),22-trien-7α-yl)-, acetate	C_6H_{12}	213(3.91)	88-0207-69
2,10-Porphinedione, 3,3,7,8,12,13,17,18-octaethyl-3,24-dihydro-	dioxan	281(4.24),296s(4.16), 328s(4.27),376(4.62), 399s(4.68),411(4.72), 554s(3.71),600(3.89), 635(4.02),687(4.05)	5-0135-69C
2,12-Porphinedione, 3,3,7,8,13,13,17,18-octaethyl-3,13-dihydro-	benzene	398(5.13),409(5.26), 481(3.51),508(3.76), 548(3.90),626(3.56), 652(3.76),687(4.79)	39-0564-69C
2,15-Porphinedione, 3,3,7,8,12,13,17,18-octaethyl-3,24-dihydro-	dioxan	308(4.46),396(4.91), 404(4.91),418s(4.72), 513s(3.64),546(3.91), 588(4.09),634(3.92)	5-0135-69C
Porphinedione, octaethyldihydro- (compound I)	$CHCl_3$	413(5.17),436(5.04), 537(3.53),591(4.00), 627(4.37)	5-0167-69E
Porphinedione, octaethyldihydro- (compound II)	$CHCl_3$	421(4.94),442(4.96), 543(3.96),586(4.22), 592(4.21),635(4.28)	5-0167-69E
Porphinedione, octaethyldihydro- (compound III)	$CHCl_3$	408(4.96),426s(--), 485s(--),523(3.70), 560(4.01),603(4.10), 654(4.52)	5-0167-69E
Porphinedione, octaethyldihydro- (compound IV)	$CHCl_3$	402(4.93),425(5.10), 486(3.64),516(4.01), 552(4.08),613(3.81), 640(3.69),673(4.65)	5-0167-69E

Compound	Solvent	λmax(log ε)	Ref.
Porphinedione, octaethyldihydro- (compound V)	$CHCl_3$	402(5.22),413(5.28), 488(3.61),514(3.80), 556(4.00),623(3.78), 628s(--),654(3.89), 691(4.74)	5-0167-69E
2(3H)-Porphinone, 3,3,7,8,12,13,17,18-octaethyl-20-hydroxy-	dioxan	308(4.33),336(4.45), 387s(4.80),402(5.02), 421(5.17),482s(3.41), 509(3.67),549(3.84), 581(4.11),591(4.08), 641(4.24)	5-0135-69C
$C_{36}H_{46}N_4O_3$			
2,7,12(3H)-Porphinetrione, 3,3,8,8,13,13,17,18-octaethyl-8,13-dihydro-	$CHCl_3$	401(4.83),418(4.87), 599(4.22),621s(--), 690(4.23)	5-0167-69E
2,13,17(3H)-Porphinetrione, 3,3,7,8,12,12,18,18-octaethyl-12,18-dihydro-	$CHCl_3$	410s(--),426(4.91), 461(4.69),585s(--), 630(4.34),669(4.14), 695(3.85),736(4.74)	5-0167-69E
$C_{36}H_{46}N_4O_6$			
Tyramine, N,N-dimethyl-O-acetyl-β-phenylserylvalylphenylalanyl-O-methyl-	EtOH	225s(4.73),278(3.56), 286(3.50)	78-0937-69
$C_{36}H_{48}Cl_2N_4Sn$			
Tin, dichloro[2,3,7,8,12,13,17,18-octaethyl-2,3,7,8-tetrahydroporphinato-(2-)]-	pyridine	372(4.7),396(5.0), 405(5.2),560(4.3), 610(4.8)	88-3077-69
$C_{36}H_{48}N_4$			
Porphine, 2,3,7,8,12,13,17,18-octaethyl-2,3-dihydro-, cis	C_6H_{12}	380(4.98),393(5.24), 497(4.13),651(4.94)	88-1145-69
trans	C_6H_{12}	646(5.02)	88-1141-69
	C_6H_{12}	391(5.29),497(4.16), 647(5.02)	88-1145-69
$C_{36}H_{48}O_4$			
28-Norolean-12,17(22)-dien-16-one, 3β,12β-dihydroxy-, 3-benzoate	EtOH	231(4.32)	44-3135-69
$C_{36}H_{49}NO$			
1H-Chryseno[2,1-b]carbazole-2-methanol, 2,3,4,4a,5,6,6a,6b,7,8,8a,9,10,15,-15a,15b-hexadecahydro-2,4a,6a,6b,9,9,15a-heptamethyl-	EtOH	232(4.57),242(4.41), 250(3.93),260(4.70), 282(4.55)	104-1931-69
$C_{36}H_{50}N_4$			
Porphine, 2,3,7,8,12,13,17,18-octaethyl-2,3,7,8-tetrahydro-	C_6H_{12}	582(4.50)	88-1141-69
Porphine, 2,3,7,8,12,13,17,18-octaethyl-2,3,12,13-tetrahydro-	C_6H_{12}	722(5.32)	88-1141-69
$C_{36}H_{50}N_6O_{13}$			
β-D-Glucopyranosiduronic acid, 3,11,20-trioxopregn-4-en-21-yl methyl ester, 3,20-disemicarbazone, 2,3,4-triacetate	MeOH	269(4.48)	69-1188-69

Compound	Solvent	λ_{max}(log ε)	Ref.
$C_{36}H_{50}N_6O_{14}$			
β-D-Glucopyranosiduronic acid, 17-hydroxy-3,11,20-trioxopregn-4-en-21-yl methyl ester, 3,20-disemicarbazone, 2,3,4-triacetate	MeOH	269(4.53)	69-1188-69
$C_{36}H_{51}NO$			
1,4a-(Epoxymethano)-4aH-chryseno[2,1-b]carbazole, 1,2,3,4,5,6,6a,6b,7,8,-8a,9,10,15,15a,15b,16,17,17a,17b-eicosahydro-2,2,6a,6b,9,9,15a-heptamethyl-	EtOH	230(4.56),284(3.87)	104-1931-69
$C_{36}H_{51}NO_5$			
16-Isosolanocapsine, N-carboxy-3-deamino-O-methyl-3-oxo-, benzyl ester	dioxan	258(2.74)	5-0159-69C
$C_{36}H_{52}N_6O_{12}$			
β-D-Glucopyranosiduronic acid, 3,20-dioxopregn-4-en-21-yl methyl ester, 3,20-disemicarbazone, 2,3,4-triacetate	MeOH	269(4.51)	69-1188-69
$C_{36}H_{52}N_6O_{13}$			
β-D-Glucopyranosiduronic acid, 11β-hydroxy-3,20-dioxopregn-4-en-21-yl methyl ester, 3,20-disemicarbazone, 2,3,4-triacetate	MeOH	269(4.51)	69-1188-69
β-D-Glucopyranosiduronic acid, 17-hydroxy-3,20-dioxopregn-4-en-21-yl methyl ester, 3,20-disemicarbazone, 2,3,4-triacetate	MeOH	268(4.56)	69-1188-69
$C_{36}H_{52}N_6O_{14}$			
β-D-Glucopyranosiduronic acid, 11β,17-dihydroxy-3,20-dioxopregn-4-en-21-yl methyl ester, 3,20-disemicarbazone, 2,3,4-triacetate	MeOH	268(4.55)	69-1188-69
$C_{36}H_{53}NO_5$			
16-Isosolanocapsine, N-carboxy-3-deamino-3β-hydroxy-O-methyl-, benzyl ester	dioxan	258(2.39)	5-0159-69C
$C_{36}H_{54}O_4$			
5α-Cholestane-3β,19-diol, 3-acetate benzoate	EtOH	231(4.14),274(3.04)	23-1495-69
5α-Cholestan-3-one, 19-hydroxy-, cyclic ethylene acetal, benzoate	EtOH	231(4.13),274(2.96)	23-1495-69
$C_{36}H_{56}O_4$			
A-Nor-5α-androstane-1α,17β-diol, 2β-[(17β-hydroxy-1β-methyl-A-nor-5α-estr-9-en-3α-yl)oxy]-	MeCN	200s(3.81)	78-1367-69
A-Nor-5α-androstane-1α,17β-diol, 2β-[(17β-hydroxy-1β-methyl-A-norestr-5(10)-en-2α-yl)oxy]-	MeCN	220(3.81)	78-1367-69
$C_{36}H_{58}O_4$			
2,5-Cyclohexadien-1-one, 4,4'-dioxybis-[2,4,6-tri-tert-butyl-	MeOH	241(4.11)	64-0547-69

Compound	Solvent	λ_{max}(log ϵ)	Ref.
$C_{37}H_{22}O_4$			
8c,22a:16b,17a-Diepoxy-22H-indeno[1,2-b]phenanthro[9,10-e]phenanthro[9',-10':3,4]cyclobuta[1,2-e][1,4]dioxin	dioxan	273(4.43)	24-3747-69
$C_{37}H_{25}N$			
Phenanthridine, 7,8,9,10-tetraphenyl-	EtOH	222(4.67),264(4.71)	39-1758-69C
$C_{37}H_{27}AsO_2$			
Arsenic, (3,4-dibenzoyl-2,4-cyclopenta-dien-1-ylidene)triphenyl-	EtOH	220(4.63),241s(4.38), 270(4.08),299(4.29), 348s(4.08)	39-2464-69C
$C_{37}H_{30}$			
1,3-Cyclopentadiene, 5-ethyl-1,2,3,4,5-pentaphenyl-	n.s.g.	245(4.44),330(3.90)	28-1412-69B
$C_{37}H_{30}BrIrOP_2$			
Iridium, bromocarbonylbis(triphenyl-phosphine)-	toluene	336(3.27),390(3.58), 444(2.43)	64-0770-69
$C_{37}H_{30}BrIrO_7P_2$			
Iridium, bromocarbonylbis(phosphorous acid)-, hexaphenyl ester	toluene	327(3.32),349(3.23), 387(3.54),440(2.59)	64-0770-69
$C_{37}H_{30}ClIrOP_2$			
Iridium, carbonylchlorobis(triphenyl-phosphine)-	toluene	338(3.34),288(3.55), 440(2.64)	64-0770-69
$C_{37}H_{30}ClIrO_7P_2$			
Iridium, carbonylchlorobis(phosphorous acid)-, hexaphenyl ester	toluene	321(3.43),344(3.40), 383(2.99),436(2.76)	64-0770-69
$C_{37}H_{30}FeO_2$			
Ferrocenemethanol, 1',2-[(diphenylmeth-ylene)oxymethylene]-α,α-diphenyl-	$CHCl_3$	442(2.34)	39-2260-69C
$C_{37}H_{30}N_2O_{11}S$			
Uracil, 3-benzoyl-β-D-xylofuranosyl-, 2',3'-dibenzoate, 5'-p-toluene-sulfonate	EtOH	230(4.68),260s(4.37)	94-0798-69
$C_{37}H_{30}O$			
1-Cyclobutene, 4-ethoxy-3-(diphenyl-methylene)-1,2,4-triphenyl-	$CHCl_3$	349(4.42),358(4.41)	88-3009-69
$C_{37}H_{30}O_7$			
Flavone, 3',4',7-tris(benzyloxy)-5-hydroxy-6-methoxy-	EtOH	243(4.31),278(4.25), 337(4.43)	31-0355-69
	EtOH-$AlCl_3$	259(4.21),295(4.28), 360(4.38)	31-0355-69
$C_{37}H_{31}P$			
Phosphorin, 1,1-dihydro-2,4,6-triphenyl-1,1-di-p-tolyl-	benzene	340(4.27),516(4.02)	88-1231-69
$C_{37}H_{32}Br_3N_3O_2$			
1-Cyclopentene-1-carboxylic acid, 3,4,5-tris[(4-bromo-2,6-xylyl)imino]-2-phenyl-, methyl ester	EtOH	250(4.56),252(4.58), 310(4.26)	88-3407-69

Compound	Solvent	λ_{max}(log ε)	Ref.
$C_{37}H_{32}O_2$			
5,7:14,16-Dietheno-2,19-hexanocyclotrideca[1,2-a 1,13-a']diindene-8,21(9H)-dione, 10,11,12,13-tetrahydro-	EtOH	227(4.57),243s(4.23), 292s(4.47),298(4.50), 324s(4.34)	33-1202-69
$C_{37}H_{32}O_3$			
Naphthalene, 2,3,6-tris(p-methoxystyryl)-	DMF	350(4.89)	33-2521-69
$C_{37}H_{32}O_5$			
Chalcone, 2',3,4'-tris(benzyloxy)-4-methoxy-	MeOH	354(4.27)	2-0305-69
$C_{37}H_{32}O_{17}$			
α-Rubromycin, dihydro-, pentaacetate	$CHCl_3$	261(4.89),286(4.23), 298(4.24),334(4.27), 362s(--)	24-0126-69
$C_{37}H_{32}Si$			
Silacyclopenta-2,4-diene, 1,2,3,4,5-pentaphenyl-1-propyl-	C_6H_{12}	220(4.4),249(4.4), 360(3.9)	6-0203-69
$C_{37}H_{34}BrNO_3$			
3-Acetyl-9,10-bis(benzyloxy)-6,7-dihydro-2-phenethylbenzo[a]quinolizinium bromide	MeOH	235s(3.10),265s(2.94), 297(3.06),378(3.19), 470s(2.90)	24-1779-69
$C_{37}H_{36}$			
5,7:14,16-Dietheno-2,19-hexanocyclotrideca[1,2-a:1,13-a']diindene, 8,9,10,11,12,13-hexahydro-	EtOH	214(4.75),229(4.75), 263s(4.49),272(4.57), 281s(4.50),292s(4.31), 305(4.04),318(4.02)	33-1202-69
$C_{37}H_{36}Br_2N_3P$			
Aniline, 4,4',4"-[(2,7-dibromofluoren-9-ylidene)phosphoranylidyne]tris-N,N-dimethyl-	$CHCl_3$	261(4.78),270(4.98), 289(4.84),394(3.12)	5-0001-69D
$C_{37}H_{36}O_2$			
9,9'-Spirobifluorene, 2,2'-diacetyl-7,7'-dibutyl-	EtOH	215(4.64),232(4.63), 247s(4.32),295s(4.55), 305(4.62),329(4.58)	33-1202-69
$C_{37}H_{36}O_4$			
9,9'-Spirobi[fluorene]-2,2'-dihexanoic acid	EtOH	213(4.51),226s(4.51), 249(4.42),267(4.48), 277(4.47),283s(4.37), 288s(4.26),299(4.12), 303s(3.98),312(4.33)	33-1202-69
$C_{37}H_{37}N_5O_4$			
2,18-Porphinedipropionic acid, 10-cyano-3,7,12,17-tetramethyl-8,13-divinyl-, dimethyl ester	$CHCl_3$	409(5.17),517(4.00), 555(4.08),589(3.76), 642(4.04)	39-0517-69C
	HCOOH	417(5.42),530s(3.46), 573(3.99),624(3.98)	39-0517-69C
	hemin in ether	383(4.69),505(3.70), 552(3.71),616s(3.52)	39-0517-69C
copper complex	$CHCl_3$	413(--),553(--), 595(--)	39-0517-69C

Compound	Solvent	λ_{max}(log ε)	Ref.
$C_{37}H_{38}BrN_3O_{11}$			
Phorbol, 12,13-diacetate, 20-O-[2'-bromo-4'-nitro-azobenzene-4-carbonyl]-	EtOH	333(4.42)	5-0130-69E
$C_{37}H_{38}N_2O_6$			
Coclobine	EtOH	230(5.00),274(4.15), 300s(3.84)	95-1163-69
$C_{37}H_{38}N_4O_7$			
Phaeophorbide b, 10(R)-methoxy-	dioxan	222(4.47),294(4.41), 332(4.41),368(4.48), 415(4.85),437(5.25), 525(4.10),556(3.89), 600(3.89),656(4.56)	5-0177-69E
10(S)-	dioxan	293(4.42),333(4.41), 369(4.45),415(4.86), 436(5.27),524(4.11), 556(3.90),600(3.89), 655(4.55)	5-0177-69E
$C_{37}H_{40}$			
9,9'-Spirobifluorene, 2,2'-dibutyl-7,7'-diethyl-	EtOH	215(4.78),223(4.77), 258s(4.45),270(4.53), 280(4.52),290s(4.32), 303(4.19),307s(4.06), 315(4.31)	33-1202-69
$C_{37}H_{40}N_2O_6$			
Berbamine	EtOH	283(3.91)	120-0159-69
$C_{37}H_{40}O_2$			
9,9'-Spirobi[fluorene]-2,2'-dihexanol	EtOH	213(4.82),221(4.83), 226(4.81),246(4.43), 267(4.48),277(4.47), 283s(4.37),288s(4.27), 300(4.12),304s(3.97), 312(4.32)	33-1202-69
$C_{37}H_{42}N_2O_6$			
Isoliensinine	EtOH	286(4.01)	39-0298-69C
$C_{37}H_{42}O_{11}$			
Spiro[2,5-cyclohexadiene-1,6'-dibenzo-[d,f][1,3]dioxepin]-4-one, 5-[2-hydroxy-5-(1-hydroxypropyl)-3-methoxyphenyl]-2',10'-bis(1-hydroxypropyl)-3,4',8'-trimethoxy-	EtOH	255(4.26),285(4.01)	44-0580-69
$C_{37}H_{43}NO_{13}$			
Tolypomycinone	EtOH	228(4.51),278(4.32), 307(4.19),396(3.71)	88-0091-69
$C_{37}H_{44}N_2O_{14}$			
Isovincoside, N-acetyl-, tetraacetate	EtOH	230(4.43),272(3.85), 281(3.84),289(3.80)	39-1193-69C
Vincoside, N-acetyl-, tetraacetate	EtOH	230(4.43),273(3.86), 281(4.84),289(3.80)	39-1193-69C

Compound	Solvent	$\lambda_{max}(\log \epsilon)$	Ref.
$C_{37}H_{45}ClCoN_4O_8$			
Cobalt(1+), (dihydrogen octadehydro-1,2,3,7,13,17,18,19-octamethyl-8,12-corrindipropionato)-, perchlorate, ethyl ester	$CHCl_3$	280(4.46),352(4.35), 499(4.13),542(3.90), 568s(3.87)	39-0176-69C
$C_{37}H_{45}ClN_4NiO_8$			
Nickel(1+), (dihydrogen octadehydro-8,12-diethyl-1,3,7,13,17,19-hexamethyl-2,8-corrindipropionato)-, perchlorate, dimethyl ester	EtOH	276(4.48),354(4.47), 457(3.66),563(4.13)	39-0176-69C
Nickel(1+), (dihydrogen octadehydro-1,2,3,7,13,17,18,19-octamethyl-8,12-corrindipropionato)-, perchlorate, ethyl ester	EtOH	273(4.51),351(4.54), 457(3.71),562(4.16)	39-0176-69C
$C_{37}H_{45}NO_{13}$			
Tolypomycinone, dihydro-	EtOH	229(4.60),303(4.32), 444(3.97)	88-0091-69
isomer	EtOH	228(4.34),277(4.27), 302s(--),332(3.92), 407(3.59)	88-0091-69
$C_{37}H_{46}N_2O_{12}$			
Neoantimycin, oxo-, methyl ether	EtOH	293(3.36)	78-2193-69
$C_{37}H_{46}N_2O_{14}$			
Isovincoside, N-acetyldihydro-, tetraacetate	EtOH	231(4.42),273(3.89), 282(3.89),292(3.80)	39-1193-69C
Vincoside, N-acetyldihydro-, tetraacetate	EtOH	228(4.43),273(3.87), 281(3.86),289(3.79)	39-1193-69C
$C_{37}H_{47}NO_3$			
2H-Chryseno[2,1-b]carbazole-2-carboxylic acid, 3,4,4a,5,6,6a,6b,7,8,8a,-9,10,15,15a,15b,16-hexadecahydro-2,4a,6a,6b,9,9,15a-heptamethyl-16-oxo-, methyl ester	EtOH	230(4.66),292(3.91)	104-1931-69
$C_{37}H_{47}NO_{13}$			
Tolypomycinone, tetrahydro-	EtOH	233(4.42),302(4.29), 441(3.96)	88-0091-69
$C_{37}H_{48}N_2O_5$			
18αH-Indolo[3,2-b]glycyrrhetinic acid, 5'-nitro-, methyl ester	EtOH	262(4.46),338(3.43)	104-1931-69
$C_{37}H_{48}N_2O_{12}$			
Neoantimycin, methyl ether	EtOH	293(3.32)	78-2193-69
$C_{37}H_{48}N_4$			
Porphine, 2,2,7,8,12,13,17,18-octaethyl-2,3-dihydro-3-methylene-	$CHCl_3$	405(5.24),501(4.08), 507(4.09),534(4.26), 601(3.69),627(3.53), 659(4.68)	5-0167-69E
$C_{37}H_{48}N_4O_4$			
Biline-2,18-dipropionic acid, 8,12-diethyl-10,23-dihydro-1,3,7,13,17,19-hexamethyl-, dimethyl ester, di-HBr	$CHCl_3$	372(4.14),433s(4.17), 457(4.66),498s(4.58), 529(5.17)	39-0176-69C

Compound	Solvent	λmax(log ε)	Ref.
Biline-2,18-dipropionic acid, 10,23-dihydro-1,3,7,8,12,13,17,19-octamethyl-, diethyl ester, dihydrobromide	$CHCl_3$	369(4.16),434(4.30), 456(4.81),490(4.80), 524(5.05)	39-0176-69C
$C_{37}H_{49}NO_3$			
18αH-Indolo[3,2-b]olean-12-en-30-oic acid, 11-oxo-, methyl ester	EtOH	232(4.57),284(3.89)	104-1931-69
18βH-Indolo[3,2-b]olean-12-en-30-oic acid, 11-oxo-, methyl ester	EtOH	232(4.62),284(3.95)	104-1931-69
$C_{37}H_{50}N_4O$			
2-Porphinol, 3,3,7,8,12,13,17,18-octaethyl-2,3-dihydro-2-methyl-	$CHCl_3$	395(5.27),498(4.13), 524(3.54),545s(--), 593(3.62),645(4.66)	5-0167-69E
$C_{37}H_{51}NO_2$			
4aH-Chryseno[2,1-b]carbazole-4aβ-carboxylic acid, octadecahydro-2,2,6aα,6bβ,-9,9,15aβ-heptamethyl-, methyl ester	EtOH	228(4.61),290(3.87)	104-1931-69
$C_{37}H_{53}NO$			
1,4a-(Epoxymethano)-4aH-chryseno[2,1-b]-carbazole, eicosahydro-2,2,6a,6b,9,9,-10,15a-octamethyl-	EtOH	232(4.51),284(3.82)	104-1931-69
$C_{37}H_{55}NO$			
Phenol, o-(N-9,19-cyclo-9β-lanost-24-en-3β-ylformimidoyl)-	dioxan	218(4.44),255(4.21), 316(3.76),417(1.98)	78-3783-69
Phenol, o-(N-lanosta-8,24-dien-3β-yl formimidoyl)-	dioxan	216(4.43),255(4.18), 316(3.69),417(2.63)	78-3783-69
$C_{37}H_{57}NO$			
Phenol, o-(N-9,19-cyclo-9β-lanost-3β-ylformimidoyl)-	dioxan	224(4.42),256(4.18), 317(3.90),417(2.12)	78-3783-69
Phenol, o-(N-lanost-8-en-3β-ylformimidoyl)-	dioxan	220(4.49),255(4.21), 316(3.74),417(1.96)	78-3783-69
$C_{37}H_{66}BrIrOP_2$			
Iridium, bromocarbonylbis(tricyclohexylphosphine)-	toluene	340(3.50),384(3.49), 436(2.86)	64-0770-69
$C_{37}H_{66}ClIrOP_2$			
Iridium, carbonylchlorobis(tricyclohexylphosphine)-	toluene	337(3.38),380(3.53), 432(2.80)	64-0770-69
$C_{37}H_{66}IIrOP_2$			
Iridium, carbonyliodobis(tricyclohexylphosphine)-	toluene	350(3.37),391(3.44), 442(2.77)	64-0770-69

Compound	Solvent	$\lambda_{max}(\log \epsilon)$	Ref.
$C_{38}H_{18}O_4$			
7,9,18,20-Nonacenetetrone	H_2SO_4	300(4.54),390(4.80), 745(4.47),829(4.46)	77-0563-69
$C_{38}H_{22}$			
Benzene, 1,2,4,5-tetrakis(phenylethyn-yl)-	$CHCl_3$	320(5.0),360s(4.7)	88-5167-69
Diindeno[2,1-a:2',1'-g]-s-indacene, 7,14-diphenyl-	n.s.g.	275(4.5),310(4.6), 335s(4.8),350(4.9), 495(4.3),510(4.4)	88-5167-69
Indeno[2,1-a]indene, 5-phenyl-2,3-bis-(phenylethynyl)-	n.s.g.	330(4.9),337(4.9), 450(4.2),470(4.2)	88-5167-69
Pentacene, 6,13-bis(phenylethynyl)-	benzene	655(4.41)	44-1734-69
$C_{38}H_{24}S$			
Acenaphtho[5,6-cd]thio(S^{IV})pyran, 1,3,6,7-tetraphenyl-	CH_2Cl_2	256(4.42),273(4.40), 301(4.31),415(4.86), 565(3.59),610(3.54), 665(3.37)	35-3953-69
$C_{38}H_{26}$			
Acetylene, bis[p-[2-(1-naphthyl)vinyl]-phenyl]-	DMF	379(4.93)	33-2521-69
Anthracene, 1,2,3,4-tetraphenyl-	n.s.g.	269(4.88),360(3.73), 379(3.86),400(3.78)	39-0765-69C
Phenanthrene, 1,2,3,4-tetraphenyl-	n.s.g.	269(4.68),307(4.16)	39-0765-69C
$C_{38}H_{28}$			
Anthracene, 9-(3,5-distyrylstyryl)-	toluene	315(4.88)	104-0903-69
Naphthalene, 1,5-bis(p-phenylstyryl)-	DMF	291(4.69),369(4.74)	33-2521-69
Naphthalene, 1,6-bis(p-phenylstyryl)-	DMF	344(4.96)	33-2521-69
Naphthalene, 2,6-bis(p-phenylstyryl)-	DMF	330(4.56),380(4.96)	33-2521-69
Stilbene, 3,5-bis[2-(1-naphthyl)vinyl]-	toluene	335(4.83)	104-0903-69
Stilbene, 3,5-bis[2-(2-naphthyl)vinyl]-	toluene	330(5.00)	104-0903-69
Stilbene, 4,4'-bis[2-(1-naphthyl)vinyl]-	DMF	397(4.95)	33-2521-69
Stilbene, 3-[2-(1-naphthyl)vinyl]-5-[2-(2-naphthyl)vinyl]-	toluene	330(4.93)	104-0903-69
$C_{38}H_{28}N_2O_{10}$			
2-Oxazolin-5-one, 4,4'-[(6,6'-dihydroxy-5,5'-dimethoxy-3,3'-biphenylylene)di-methylidyne]bis[2-phenyl-, acetate	$CHCl_3$	380(4.88),402(4.88)	18-1752-69
$C_{38}H_{28}O_2$			
Naphthalene, 1,4-bis(p-phenoxystyryl)-	DMF	285(4.36),388(4.66)	33-2521-69
Naphthalene, 1,5-bis(p-phenoxystyryl)-	DMF	277(4.62),363(4.61)	33-2521-69
Naphthalene, 1,6-bis(p-phenoxystyryl)-	DMF	332(4.89)	33-2521-69
Naphthalene, 2,3-bis(p-phenoxystyryl)-	DMF	314(4.81)	33-2521-69
$C_{38}H_{30}Cl_3NO_{10}$			
Ammonium, [1-chloro-2-(4,6-diphenyl-2H-pyran-2-ylidene)-2-(4,6-diphenylpyr-ylium-2-yl)ethylidene]dimethyl-, diperchlorate	MeCN	280(4.51),338(4.67), 558(4.52),638(4.48)	4-0803-69
Ammonium, [1-chloro-2-(2,6-diphenyl-4H-pyran-4-ylidene)-2-(2,6-diphenylpyr-ylium-4-yl)ethylidene]dimethyl-, diperchlorate	MeCN	232(4.42),275(4.11), 409(4.36),443(4.26), 532(4.02)	4-0803-69

Compound	Solvent	λ_{max}(log ε)	Ref.
$C_{38}H_{30}O_8$			
[5,5'-Bi-5H-dibenzo[a,d]cycloheptene]-10,10',11,11'-tetrol, tetraacetate	MeCN	248(4.37)	23-2827-69
$C_{38}H_{32}N_2O_9$			
2,3-Indolizinedicarboxylic acid, 1,1'-(oxydibenzylidene)di-, tetramethyl ester	MeOH	226(4.78),251s(4.73), 257(4.77),333(4.64), 345s(4.61)	39-1143-69C
$C_{38}H_{32}O_2$			
Dispiro[4.1.4.1]dodec-2-ene-1,8-dione, 2-benzyl-9-benzylidene-6,12-diphenyl-	EtOH	231s(3.97),308(4.24)	39-1868-69C
Ethylene, 1,2-bis[p-(p-methoxystyryl)-phenyl]-1-phenyl-	DMF	283(4.32),384(4.89)	33-2521-69
$C_{38}H_{32}O_7$			
Flavone, 3',4',6-tris(benzyloxy)-5,7-dimethoxy-	EtOH	241s(4.43),265(4.23), 329(4.47)	40-1270-69
Flavone, 4',5,7-tris(benzyloxy)-3',8-dimethoxy-	EtOH	247s(4.36),272(4.37), 338(4.40)	18-2327-69
$C_{38}H_{32}O_{11}$			
Anthracene-2-acetic acid, 4,5,6,8-tetra-hydroxy-α-methoxy-, (diphenylmethyl) ester, tetraacetate	EtOH	264(5.15),340(3.58), 358(3.68),376(3.67), 396(3.63)	105-0494-69
$C_{38}H_{34}O_8$			
Acetophenone, 4',6'-bis(benzyloxy)-2'-hydroxy-3'-methyl-α-[4-(benzyloxy)-3-methoxybenzoyl]-	EtOH	301(4.17),390(4.44), 408s(4.39)	18-2327-69
$C_{38}H_{34}O_{10}$			
Elsinochrome A, tetraacetyldihydro-	EtOH	213(4.55),277(4.19), 286(4.15),363(3.46), 380(3.49),467(4.00), 496(4.06)	39-1219-69C
$C_{38}H_{34}O_{14}$			
Julimycin B-II	MeOH	245(4.2),287(4.3), 458(4.0)	78-3007-69
$C_{38}H_{34}O_{15}$			
Julichrome Q1.4	MeOH	249(4.5),398(4.0), 450(3.9)	78-3007-69
$C_{38}H_{34}O_{18}$			
Floccosin, tetraacetate A	EtOH	208(4.65),259(4.35), 315(3.47)	23-1561-69
Floccosin, tetraacetate B	EtOH	208(4.72),259(4.33), 315(3.48)	23-1561-69
$C_{38}H_{34}Si$			
Silacyclopenta-2,4-diene, 1-butyl-1,2,3,4,5-pentaphenyl-	C_6H_{12}	220(4.4),249(4.4), 365(3.9)	6-0203-69
$C_{38}H_{36}O$			
Furan, 2,5-bis[p-(p-isopropylstyryl)-phenyl]-, (E,E)-	DMF	284(4.27),337(4.49), 398(4.93)	33-1282-69

Compound	Solvent	λ_{max}(log ε)	Ref.
$C_{38}H_{36}O_4$			
Bicyclo[2.2.0]hex-5-ene-2,3-dipropionic acid, 2,3,5,6-tetraphenyl-, dimethyl ester	EtOH	207(4.56),230s(4.14), 314(4.46)	44-0670-69
$C_{38}H_{36}O_{15}$			
Julichrome Q1.3	MeOH	220s(4.5),287(4.2), 450(3.7)	78-3007-69
$C_{38}H_{36}O_{16}$			
Julichrome Q3.4	MeOH	209(4.48),267(4.25), 370(3.94)	78-3007-69
$C_{38}H_{36}S$			
Thiophene, 2,5-bis[p-(p-isopropylstyryl)phenyl]-, (E)-	DMF	287(4.44),398(4.91)	33-1282-69
$C_{38}H_{38}$			
1,3-Butadiene, 1,4-bis[p-(p-isopropylstyryl)phenyl]-	DMF	285(4.26),406(5.04), 426(4.90)	33-2521-69
$C_{38}H_{38}N_2O$			
17,18-Secoyohimban-17-one, 17-trityl-, hydrochloride	n.s.g.	224(4.64),283(3.84), 290(3.76)	35-7342-69
$C_{38}H_{38}N_4O_2S_2$			
Hydantoin, 5,5'-(2-ethyl-1-propyltrimethylene)bis[1,3-diphenyl-2-thio-	EtOH	242(4.17),281(4.09)	94-2436-69
$C_{38}H_{39}ClN_2O$			
17,18-Secoyohimban-17-one, 17-trityl-, hydrochloride	n.s.g.	224(4.64),283(3.84), 290(3.76)	35-7342-69
$C_{38}H_{40}Br_2O_2$			
Dispiro[4.1.4.1]dodecane-1α,8β-diol, 2α,9β-bis(p-bromobenzyl)-6,12-diphenyl-	EtOH	236s(4.12)	39-1868-69C
$C_{38}H_{40}N_2$			
Aniline, 4,4'-[ethynylenebis(p-phenylenevinylene)]bis[N,N-diethyl-	DMF	300(4.24),417(5.00)	33-2521-69
$C_{38}H_{40}N_6O_4$			
Protoporphyrin IX, cyano-4(2)-(2-cyanoethyl)-4(2)-devinyl-3,4(1,2)-dihydro-, dimethyl ester, meso-	$CHCl_3$	408(5.13),527(3.81), 569(4.09),605(3.60), 659(4.08)	39-0517-69C
	HCOOH	421(5.48),543s(3.44), 590(4.01),639(4.32)	39-0517-69C
	hemin in ether	384(4.99),593(3.99), 628(3.90)	39-0517-69C
copper complex	$CHCl_3$	417(--),571(--), 613(--)	39-0517-69C
$C_{38}H_{40}O_2$			
Dispiro[4.1.4.1]dodecane-1α,8β-diol, 2α,9β-dibenzyl-6,12-diphenyl-, trans	EtOH	311(4.72)	39-1868-69C

Compound	Solvent	$\lambda_{max}(\log \epsilon)$	Ref.
$C_{38}H_{42}N_4Ni$			
Nickel, [1-benzylidene-3,8,8-triethyl-1,22-dihydro-2,7,12,13,17,19-hexamethylbilinato(2-)]-	$CHCl_3$	277(4.24),319s(4.24),345s(4.28),355(4.29),402s(4.45),419(4.72),436s(3.63),736s(3.87),801(4.17)	39-0655-69C
$C_{38}H_{44}CuN_4O_4$			
Copper, [dihydrogen 8,12-diethyl-3,7,13,17-tetramethyl-2,18-porphinedipropionato(2-)]-, diethyl ester	$CHCl_3$	328(4.21),398(5.54),526(4.09),563(4.37)	39-0176-69C
$C_{38}H_{44}O_{12}$			
β-D-Glucopyranosiduronic acid, 3-(benzyloxy)-17-oxoestra-1,3,5(10)-trien-16α-yl methyl ester, triacetate	dioxan	278(3.29),287(3.26)	13-0591-69B
$C_{38}H_{46}O_{12}$			
β-D-Glucopyranosiduronic acid, 3-(benzyloxy)-17β-hydroxyestra-1,3,5(10)-trien-16α-yl methyl ester, 2,3,4-acetate	MeOH	278(3.38),287(3.35)	13-0591-69B
$C_{38}H_{48}N_4O_3$			
2(3H)-Porphinone, 3,3,7,8,12,13,17,18-octaethyl-10-hydroxy-, acetate	dioxan	266(4.00),315s(4.21),324(4.23),367s(4.56),390s(4.97),407(5.27),494s(3.81),510(4.06),546(4.02),587(3.76),595s(3.63),644(4.49)	5-0135-69C
2(3H)-Porphinone, 3,3,7,8,12,13,17,18-octaethyl-15-hydroxy-, acetate	dioxan	268(4.02),313s(4.20),323(4.23),365s(4.60),389s(5.04),403(5.25),494s(3.83),508(4.01),546(4.03),584(3.81),614(3.17),641(4.51)	5-0135-69C
2(3H)-Porphinone, 3,3,7,8,12,13,17,18-octaethyl-20-hydroxy-, acetate	dioxan	267(3.96),315s(4.19),327(4.25),369s(4.63),393s(5.04),408(5.26),493s(3.80),510(4.00),548(4.04),586(3.79),614(3.19),641(4.47)	5-0135-69C
$C_{38}H_{51}NO_3$			
1H-Chryseno[2,1-b]carbazole-2-carboxylic acid, 2,3,4a,5,6,6a,6b,7,8,8a,9,-10,15,15a,15b,16,17b-octadecahydro-2α,4aβ,6aα,6bβ,9,9,13,15aβ-octamethyl-16-oxo-, methyl ester	EtOH	238(4.63),288(3.91)	104-1931-69
isomer	EtOH	232(4.59),288(3.92)	104-1931-69
$C_{38}H_{51}NO_4$			
1H-Chryseno[2,1-b]carbazole-2-carboxylic acid, octadecahydro-13-methoxy-2α,4aβ,6aα,6bβ,9,9,13aβ-heptamethyl-16-oxo-, methyl ester	EtOH	236(4.83),282(4.16)	104-1931-69
$C_{38}H_{55}Br_2NO$			
2'H-1'-Benzopyrano[3',2':2,3]cholestane, 6',8'-dibromo-3-pyrrolidino-	C_6H_{12}	238(4.33),245(4.27),328(3.44),339(3.42)	39-1419-69C

Compound	Solvent	λ_{max}(log ϵ)	Ref.
$C_{38}H_{56}N_4O_6$			
Compound from Cetraria nivalis, acetate, 2,4-dinitrophenylhydrazone	n.s.g.	372(4.61)	1-3038-69
$C_{38}H_{57}NO$			
2'H-1'-Benzopyrano[3',2':2,3]cholestane, 3-pyrrolidino-	C_6H_{12}	224(4.51),312(3.76), 323(3.69)	39-1419-69C

Compound	Solvent	λ max(log ε)	Ref.
$C_{39}H_{36}O_2$			
5,7:15,17-Dietheno-2,20-heptano-8H-cyclotetradeca[1,2-a:1,14-a']diindene-8,22-dione, 9,10,11,12,13,14-hexahydro-	EtOH	212(4.59),235(4.64), 247s(4.43),306(4.62), 323s(4.63)	33-1202-69
$C_{39}H_{38}N_4O_3$			
Isocoriamyrtin, 11-diphenylhydrazino-, diphenylhydrazone	EtOH	246(4.19),306(4.31), 334(4.34)	78-1109-69
$C_{39}H_{40}$			
5,7:15,17-Dietheno-2,20-heptano-8H-cyclotetradeca[1,2-a:1,14-a']di-indene, 9,10,11,12,13,14-hexahydro-	EtOH	215(4.73),224(4.72), 229s(4.70),254s(4.44), 262s(4.49),270(4.57), 279s(4.53),287s(4.40), 292s(4.28),304(4.15), 310s(4.05),317(4.26)	33-1202-69
$C_{39}H_{40}O_4$			
9,9'-Spirobi[fluorene]-2,2'-diheptanoic acid	EtOH	213(4.77),221(4.79), 226s(4.76),246(4.43), 257s(4.44),267(4.49), 277(4.48),283s(4.37), 288s(4.27),299(4.13), 304s(3.99),312(4.33)	33-1202-69
$C_{39}H_{40}O_5$			
Phorbobutanone, 20-trityl ether	MeOH	260s(3.48),310s(1.90), 330s(1.78)	64-0099-69
$C_{39}H_{44}N_2O_7$			
Thalidasine	EtOH	275(3.66),282(3.66)	44-3884-69
$C_{39}H_{44}N_4O$			
Pycnanthinol	EtOH	255(4.11),315(3.52), 327(3.53)	33-0033-69
$C_{39}H_{45}O_9P_3Ru_3$			
Ruthenium, nonacarbonyltris(diethyl-phenylphosphine)tri-, triangulo-	$C_6H_{11}Me$	375(4.1),475(4.0)	101-0289-69A
$C_{39}H_{47}NO_{18}$			
Ipecoside, hexaacetate	EtOH	222(4.44),264(3.42), 273(3.30)	39-1187-69C
Isoipecoside, hexaacetate	EtOH	223(4.44),263(3.41), 272(3.31)	39-1187-69C
$C_{39}H_{48}O_3$			
1,5-Octanedione, 2-isopropyl-1-(3-isopropyl-4-oxo-2,6-diphenylcyclohexyl)-7-methyl-3-phenyl-	EtOH	248(2.48),253(2.64), 259(2.80),266(2.75), 268(2.59),299(2.32)	44-0444-69
$C_{39}H_{49}ClCoN_4O_8$			
Cobalt(1+), (dihydrogen octadehydro-8,12-diethyl-1,3,7,13,17,19-hexamethyl-2,18-corrindipropionato)-, perchlorate, diethyl ester	$CHCl_3$	282(4.48),350(4.36), 499(4.13),546s(3.90), 568s(3.87)	39-0176-69C

Compound	Solvent	λ_{max}(log ε)	Ref.
$C_{39}H_{49}ClN_4NiO_8$			
Nickel(1+), (dihydrogen octadehydro-2,18-diethyl-1,3,7,13,17,19-hexamethyl-8,12-corrindipropionato)-, perchlorate, diethyl ester	EtOH	276(4.49),353(4.51),405(3.76),452(3.68),561(4.12)	39-0176-69C
Nickel(1+), (dihydrogen octadehydro-8,12-diethyl-1,3,7,13,17,19-hexamethyl-2,18-corrindipropionato)-, perchlorate, diethyl ester	EtOH	276(4.50),353(4.50),449(3.73),561(4.13)	39-0176-69C
$C_{39}H_{49}N_5O_7$			
Lasiodine A	EtOH	281(4.49)	78-0937-69
$C_{39}H_{52}N_4O_4$			
Biline-2,18-dipropionic acid, 8,12-diethyl-10,23-dihydro-1,3,7,13,17,19-hexamethyl-, diethyl ester, dihydrobromide	$CHCl_3$	375(4.21),436(4.06),458(4.41),532(5.35)	39-0176-69C
Biline-8,12-dipropionic acid, 2,18-diethyl-10,23-dihydro-1,3,7,13,17,19-hexamethyl-, diethyl ester, dihydrobromide	$CHCl_3$	373(4.12),458(4.45),526(5.27)	39-0176-69C
$C_{39}H_{60}O_4S$			
5α-Cholestan-3-one, 2α-hydroxy-, p-cyclohexylbenzenesulfonate	C_6H_{12}	210(4.33),226(4.16)	22-3166-69

Compound	Solvent	$\lambda_{max}(\log \epsilon)$	Ref.
$C_{40}H_{24}N_4$			
$\Delta^{14,14'}$-Bi-14H-dibenzo[4,5:6,7][1,3]-diazepino[2,1-a]isoindole, cis	90% EtOH	212(4.88),251(4.86), 320(4.28),454(4.12)	78-5465-69
trans	90% EtOH	214(4.89),253(4.81), 258s(4.80),304s(4.49), 333s(4.19),383(3.95), 476(4.10)	78-5465-69
$C_{40}H_{26}N_4O$			
Phthalimidine, 2-(2'-amino-2-biphenylyl)-3(14H)-dibenzo[4,5:6,7][1,3]diazepino[2,1-a]isoindol-14-ylidene-, cis	90% EtOH	212(4.83),360s(3.90), 482(4.13)	78-5465-69
trans	90% EtOH	225(4.83),300s(4.17), 340s(3.89),504(3.62)	78-5465-69
$C_{40}H_{28}O$			
Dibenzofuran, 3,7-bis(p-biphenylvinyl)-	DMF	384(4.99)	33-1282-69
Furan, 2,5-bis[p-[2-(1-naphthyl)-vinyl]phenyl]-, (E,E)-	DMF	405(4.89)	33-1282-69
$C_{40}H_{28}O_2S$			
Dibenzothiophene, 2,8-bis(p-phenoxystyryl)-, (E,E)-	DMF	325(4.94)	33-1282-69
Dibenzothiophene, 3,7-bis(p-phenoxystyryl)-, (E,E)-	DMF	374(4.97),390(4.89)	33-1282-69
$C_{40}H_{28}O_2S_2$			
Thianthrene, 2,7-bis(p-phenoxystyryl)-, (E,E)-	DMF	339(4.94)	33-1282-69
$C_{40}H_{28}O_3$			
Dibenzofuran, 3,7-bis(p-phenoxystyryl)-, (E,E)-	DMF	377(4.94)	33-1282-69
$C_{40}H_{28}S$			
Thiophene, 2,5-bis[p-[2-(1-naphthyl)-vinyl]phenyl]-, (E,E)-	DMF	403(4.91)	33-1282-69
$C_{40}H_{28}S_2$			
Thianthrene, 2,7-bis(p-phenylstyryl)-, (E,E)-	DMF	350(5.04)	33-1282-69
$C_{40}H_{30}$			
1,3-Butadiene, 1,4-bis[p-[2-(1-naphthyl)vinyl]phenyl]-	DMF	410(5.02)	33-2521-69
Stilbene, 3-[2-(1-naphthyl)vinyl]-4'-phenyl-5-styryl-	toluene	330(4.96)	104-0903-69
Stilbene, 3-[2-(2-naphthyl)vinyl]-4'-phenyl-5-styryl-	toluene	330(5.01)	104-0903-69
$C_{40}H_{34}N_2O_4$			
1H-Pyrazolo[5,1-a]isoindole-2,3-dicarboxylic acid, 8,8-dibenzyl-3a,8-dihydro-1,3a-diphenyl-, dimethyl ester	MeCN	250(3.92),332(3.97)	35-3670-69
$C_{40}H_{34}N_6$			
Quinoline, 4,4'-(1,4-piperazinediyl)bis-[3-(3,4-dihydro-1-isoquinolyl)-	EtOH	225(4.78),282(4.42), 329(4.22)	2-1010-69

Compound	Solvent	λ_{max}(log ϵ)	Ref.
$C_{40}H_{35}NO_3$			
Acetic acid, diphenyl-, α-(diphenylmethylene)-γ-morpholinocinnamyl ester	$CHCl_3$	234(4.31),345(4.11)	24-3428-69
$C_{40}H_{36}N_2O_{12}$			
3,3'-Biphenylenediacrylic acid, α,α'-dibenzamido-6,6'-dihydroxy-5,5'-dimethoxy-, dimethyl ester, diacetate	EtOH	225(4.67),295(4.49)	18-1752-69
$C_{40}H_{38}$			
p-Terphenyl, 4,4"-bis(p-isopropylstyryl)-	DMF	357(4.94)	33-2521-69
$C_{40}H_{38}O_6$			
Naphthalene, 2,3,6-tris(3,4-dimethoxystyryl)-	DMF	361(4.88)	33-2521-69
$C_{40}H_{40}N_2O_{12}$			
3,3'-Biphenyldipropionic acid, α,α'-dibenzamido-6,6'-dihydroxy-5,5'-dimethoxy-	EtOH	281(3.65)	18-1752-69
$C_{40}H_{40}O_{15}$			
Julichrome Q1.3, 8,8'-dimethyl ether	dioxan	229(4.54),375(3.75)	78-3007-69
$C_{40}H_{42}O_{12}$			
Globiflorin $3B_1$, hexamethyl ether, diacetate	EtOH	274s(3.84),279(3.86)	39-2572-69C
Globiflorin $3B_2$, hexamethyl ether, diacetate	EtOH	274s(3.83),279(3.85)	39-2572-69C
$C_{40}H_{44}Cl_2N_4$			
5H-3,5-Ethanodipyrrolo[2,3-d:1',2',3'-lm]carbazolium, 13-ethylidene-7-(12-ethylidene-1,2,3a,4,5,7-hexahydro-3-methyl-3,5-ethano-3H-pyrrolo[2,3-d]-carbazolium-6-yl)-1,2,3a,4-tetrahydro-3-methyl-, dichloride	pH 1	260(3.96),334(3.90), 468(3.94)	33-0689-69
	H_2O	274(4.05),340(4.07)	33-0689-69
	MeOH	278(4.14),348(4.15)	33-0689-69
$C_{40}H_{44}N_4O_2$			
Pycnanthine	EtOH	255(4.16),311(3.64), 326(3.62)	33-0033-69
$C_{40}H_{46}N_4O_2$			
Pycnanthine, 6',7'-dihydro- (pleiomutinin)	EtOH	255(4.19),311(3.66), 326(3.64)	33-0033-69
$C_{40}H_{46}N_4O_{10}$			
2,17-Porphinedihydracrylic acid, 8,12-bis(2-carboxyethyl)-3,7,13,18-tetramethyl-, tetramethyl ester	$CHCl_3$	500(4.17),535(4.08), 570(4.04),623(3.63)	65-2558-69
$C_{40}H_{47}I_3N_4$			
Anhydrocalebassine diiodide hydroiodide	pH 1	252(4.25),318(3.87), 450(4.4)	33-0689-69
after 18 hours	pH 7	226(4.73),297s(3.80), 412(3.82)	33-0689-69

Compound	Solvent	λ_{max}(log ε)	Ref.
$C_{40}H_{48}N_4O_2$			
8',9'-Chanopycnanthine, 6',7',8',9'-tetrahydro-	EtOH	255(4.10),311(3.59),326(3.56)	33-0033-69
$C_{40}H_{48}O_{13}$			
β-D-Glucopyranosiduronic acid, 17β-hydroxy-3-(benzyloxy)estra-1,3,5(10)-trien-16α-yl methyl ester, tetraacetate	MeOH	278(3.40),287(3.32)	13-0591-69B
$C_{40}H_{51}N_5O_7$			
Lasiodine A, O-methyl-	EtOH	283(4.45)(end abs.)	78-0937-69
$C_{40}H_{53}N_5O_7$			
Lasiodine A, dihydro-N-methyl-	EtOH	227s(4.62),278(3.64),286(3.55)	78-0937-69
$C_{40}H_{54}O_4$			
Fucoxanthinol	EtOH	452(4.94)	39-0429-69C
$C_{40}H_{56}$			
β-Carotene (also in other solvents)	hexane	450(5.15),478(5.10)	78-5383-69
	C_6H_{12}	454(5.16),483(5.10)	78-5383-69
trans	C_6H_{12}	456(--)	44-3039-69
Lycopene	hexane	442(5.07),467(5.26),497(5.22)	78-5383-69
	C_6H_{12}	446(5.05),473(5.24),505(5.17)	78-5383-69
	benzene	454(5.03),483(5.21),518(5.14)	78-5383-69
	CCl_4	452(5.06),480(5.24),512(5.18)	78-5383-69
	dioxan	450(5.05),476(5.24),507(5.17)	78-5383-69
	acetone	444(5.05),471(5.23),512(5.18)	78-5383-69
	CS_2	473(4.97),502(5.14),540(5.09)	78-5383-69
(and many mixed solvents)	DMF	452(5.06),478(5.24),512(5.17)	78-5383-69
$C_{40}H_{56}O$			
2,5,7,9,11,13,15,17-Octadecaoctaen-4-one, 3,7,12,16-tetramethyl-1,18-bis-(2,6,6-trimethyl-1-cyclohexen-1-yl)-	C_6H_{12}	438(--)	44-3039-69
$C_{40}H_{56}O_3$			
Pyrenoxanthin	pet ether	420(--),448(--),472(--)	44-4207-69
	$CHCl_3$	432s(4.87),454(5.01),482(4.95)	44-4207-69

Compound	Solvent	λ_{max}(log ϵ)	Ref.
$C_{41}H_{26}O_{26}$			
Castalagin	dioxan	229(4.89)	5-0186-69A
$C_{41}H_{27}ClO_4S_2$			
2-Phenyl-4-[3-(2-phenyl-4H-1-naphtho-[1,2-b]thiopyran-4-ylidene)propenyl]-naphtho[1,2-b]thiopyrylium perchlorate	CH_2Cl_2	250(4.71),302(4.48), 405(4.36),792(5.31)	2-0311-69
$C_{41}H_{28}O$			
2,4-Cyclopentadien-1-one, 2,3-bis-(p-biphenylyl)-4,5-diphenyl-	benzene	361(4.05),521(3.32)	64-0685-69
$C_{41}H_{31}O_2P$			
2-Cyclopenten-1-one, 5-(diphenylphos-phinyl)-2,3,4,5-tetraphenyl-, cis	EtOH	311(3.97)	88-4335-69
trans	EtOH	311(3.98)	88-4335-69
3-Cyclopenten-1-one, 2-(diphenylphos-phinyl)-2,3,4,5-tetraphenyl-	EtOH	270(4.01)	88-4335-69
$C_{41}H_{40}O_2$			
5,7:16,18-Dietheno-2,21-octanocyclopen-tadeca[1,2-a:1,15-a']diinden-8,23(9H)-dione, 10,11,12,13,14,15-hexahydro-	EtOH	212(4.61),233(4.60), 246s(4.39),306(4.58), 323s(4.51)	33-1202-69
$C_{41}H_{40}O_{14}$			
[2,2'-Bianthracene]-8,9',10,10'-tetrone, 8aα,10aβ-epoxy-5,6,7,8a,9,10a-hexa-hydro-6β,9-dihydroxy-5β,5'-bis(1-hydroxyethyl)-1,1',8'-trimethoxy-6,6'-dimethyl-, 5,5'-diacetate	dioxan	228(4.54),268(4.54), 370(3.91)	78-3007-69
$C_{41}H_{43}N_3O_7$			
1H-Pyrrolo[3,4-b]quinolin-1-one, 3-(2,2-diphenylvinyl)-2,3-dihydro-2-[2-(hy-droxymethyl)-4,5-dioxo-5-(1-pyrroli-dinyl)valeryl]-, acetate 4-(diethyl acetal)	EtOH	247(4.76),260s(3.34), 295(4.11)	44-3853-69
$C_{41}H_{44}$			
5,7:16,18-Dietheno-2,21-octanocyclo-pentadeca[1,2-a:1,15-a']diindene, 8,9,10,11,12,13,14,15-octahydro-	EtOH	215(4.80),234(4.79), 228s(4.76),248s(4.47), 271(4.56),281(4.55), 287s(4.43),292s(4.33), 303(4.19),308s(4.05), 316(4.31)	33-1202-69
$C_{41}H_{45}N_3O_6$			
1H-Pyrrolo[3,4-b]quinoline, 3-(2,2-di-phenylvinyl)-2,3-dihydro-2-[2-(acet-oxymethyl)-4,5-dioxo-5-(1-pyrrolidi-nyl)valeryl]-, 4-(diethyl acetal)	EtOH	229(4.71),256(4.30), 295(3.74),302(3.69), 308(3.79),314(3.74), 323(3.92)	44-3853-69
$C_{41}H_{46}O_{22}$			
Rubrofusarin, gentiobioside, acetate	EtOH	223(4.46),255s(4.46), 276(4.71),322(3.49), 395(3.79)	94-0458-69

Compound	Solvent	$\lambda_{max}(\log \epsilon)$	Ref.
$C_{41}H_{47}N_3O_8$			
1H-Pyrrolo[3,4-b]quinoline, 2-[2-(acetoxymethyl)-4,5-dioxo-5-(1-pyrrolidinyl)valeryl]-2,3-dihydro-3-(1,2-dihydroxy-2,2-diphenylethyl)-, 4-(diethyl acetal)	EtOH	228s(4.72),232(4.74), 235s(4.71),295(3.72), 302(3.74),308(3.85), 315(3.80),322(3.98)	44-3853-69
$C_{41}H_{48}I_2N_4O$			
Anhydro-O-methylisocalebassine diiodide	pH 1	226(4.53),252(4.18), 318(3.73),445(4.38)	33-0689-69
	H_2O	226(4.61),268s(4.10), 296s(3.85),392(4.10)	33-0689-69
	NaOH	226(4.63),268s(4.10), 296s(3.85),394(4.09)	33-0689-69
$C_{41}H_{48}N_4O_4$			
Umbellamine	EtOH	232(4.59),248s(4.04), 276s(3.86),288s(4.01), 295(4.06),299s(4.03)	33-0089-69
	EtOH-HCl	224(4.67),248(4.00), 286s(4.03),295(4.06)	33-0089-69
	EtOH-KOH	234(4.61),298(3.96), 309(3.98)	33-0089-69
	12N HCl	222(4.66),262(4.09), 283(3.88),294(3.90), 322s(3.63)	33-0089-69
$C_{41}H_{49}BrO_2$			
Dispiro[5.0.5.1]trideca-1,4,8,11-tetraene-3,10-dione, 13-(p-bromophenyl)-2,4,9,11-tetra-tert-butyl-13-phenyl-	MeOH	268(4.30),332(4.21)	44-1203-69
$C_{41}H_{49}ClO_2$			
Dispiro[5.0.5.1]trideca-1,4,8,11-tetraene-3,10-dione, 2,4,9,11-tetra-tert-butyl-13-(p-chlorophenyl)-13-phenyl-	MeOH	268(4.29),332(4.22)	44-1203-69
$C_{41}H_{50}O_2$			
Dispiro[5.0.5.1]trideca-1,4,8,11-tetraene-3,10-dione, 2,4,9,11-tetra-tert-butyl-13,13-diphenyl-	MeOH	268(4.27),333(4.22)	44-1203-69
$C_{41}H_{52}N_2O_2$			
Isocalciferol, p-(phenylazo)benzoate, 5,6-cis	iso-PrOH	230(4.18),278s(4.57), 288(4.67),300(4.61), 325(4.41)	95-0919-69
$C_{41}H_{59}NO_6S_2$			
1,4-Dithia-7-azaspiro[4.5]decane-7-carboxylic acid, 6-(3β,16α-dihydroxy-5α-pregnan-20α-yl)-10-methyl-, benzyl ester, diacetate	MeOH	237s(2.96)	5-0159-69C
$C_{41}H_{64}O_4S$			
5α-Cholestan-3-one, 2α-hydroxy-4,4-dimethyl-, p-cyclohexylbenzenesulfonate	C_6H_{12}	226(3.56)	22-3173-69
$C_{41}H_{71}NO_{15}$			
Erythromycin A, 2",4'-diacetate	$CHCl_3$	292(1.46)	56-0763-69

Compound	Solvent	λ max (log ε)	Ref.
$C_{42}H_{26}N_2$			
Δ10,10'-Bi-10H-indolo[1,2-a]indole, 11,11'-diphenyl-	EtOH	251(4.66),271(4.59), 285s(4.54),301(4.49), 319(4.44),414(3.76), 480(3.89),624(4.43)	44-2988-69
$C_{42}H_{28}N_2$			
10,10'-Bi-10H-indolo[1,2-a]indole, 11,11'-diphenyl-	EtOH	222(4.62),258(4.67), 322(4.30)	44-2988-69
$C_{42}H_{30}$			
Anthracene, 9-[3-[2-(2-naphthyl)vinyl]-5-styrylstyryl]-	toluene	330(4.96)	104-0903-69
Benzene, 1-[2-(1-naphthyl)vinyl]-3,5-bis[2-(2-naphthyl)vinyl]-	toluene	330(4.93)	104-0903-69
Benzene, 1-[2-(2-naphthyl)vinyl]-3,5-bis[2-(1-naphthyl)vinyl]-	toluene	330(5.03)	104-0903-69
Bibenzyl, α,α'-bis(fluoren-9-ylidene-methyl)-	EtOH-10% THF	230(4.88),248(4.73), 258(4.86),290(4.42), 303(4.41),318(4.47)	24-1707-69
p-Terphenyl, 4,4"-bis[2-(1-naphthyl)-vinyl]-	DMF	368(4.94)	33-2521-69
$C_{42}H_{30}N_2$			
10H-Indolo[1,2-a]indole, 10-[2-(diphen-ylmethyl)indol-3-yl]-11-phenyl-	EtOH	223(4.82),257(4.58), 285s(4.40),292(4.31), 317(4.20)	44-2988-69
$C_{42}H_{30}O$			
Benzofuran, 2-phenyl-4,6-bis(p-phenyl-styryl)-, (E,E)-	DMF	351(4.96)	33-1282-69
$C_{42}H_{30}O_2$			
Acetylene, bis[p-(p-phenoxystyryl)-phenyl]-	DMF	369(4.97)	33-2521-69
$C_{42}H_{32}$			
Stilbene, 3,5-bis(p-phenylstyryl)-	toluene	335(5.08)	104-0903-69
$C_{42}H_{32}O$			
4-Pentenophenone, 3-(diphenylmethylene)-2,5,5-triphenyl-	C_6H_{12}	259(4.24),284(4.18)	104-0082-69
$C_{42}H_{34}O_2$			
3-Oxabicyclo[3.1.0]hexane-6-methanol, α,α,2,2,4,4-hexaphenyl-	C_6H_{12}	260(3.28)	104-0082-69
$C_{42}H_{34}O_{14}$			
Globiflorin $3B_1$, octaacetate	EtOH	273(3.68),280s(3.66)	39-2572-69C
Globiflorin $3B_2$, octaacetate	EtOH	273(3.60),280s(3.58)	39-2572-69C
$C_{42}H_{35}N_3O_2$			
1H-s-Triazolo[5,1-a]isoindol-2(3H)-one, 5,5-dibenzyl-5,9b-dihydro-1-(p-meth-oxyphenyl)-3,9b-diphenyl-	MeCN	255(4.26)	35-3670-69
$C_{42}H_{36}$			
Hexa-m-phenylene, 4,6',4",6‴,4⁗,6""'-hexamethyl-	THF	241(5.02)	5-0024-69D

Compound	Solvent	$\lambda_{max}(\log \epsilon)$	Ref.
$C_{42}H_{36}N_2O_{16}$			
11b,11'b-Bi-11bH-benzo[a]quinolizine]-1,1',2,2',3,3',4,4'-octacarboxylic acid, octamethyl ester	n.s.g.	230(3.35),270(3.47), 450(3.03)	77-1044-69
$C_{42}H_{36}O_3$			
1,2,3-Cyclopropanetrimethanol, α,α,α',α',α",α"-hexaphenyl-, trans	C_6H_{12}	260(3.18)	104-0082-69
$C_{42}H_{38}O_4$			
Benzene, 1,2,4,5-tetrakis(p-methoxystyryl)-	DMF	354(4.96)	33-2521-69
$C_{42}H_{48}O_{22}$			
Rubrofusarin, gentiobioside, acetate, methyl ether	EtOH	224(4.41),252s(4.46), 270(4.66),325(3.65), 341(3.72),368(3.77)	94-0458-69
$C_{42}H_{50}N_2O_9$			
Adiantifoline	EtOH	283(4.51),292s(--), 302(4.39),312(4.34)	100-0029-69
$C_{42}H_{50}N_4O_3$			
8',9'-Chanopycnanthine, N'-acetyl-6',7',8',9'-tetrahydro-	EtOH	255(4.12),311(3.61), 326(3.58)	33-0033-69
$C_{42}H_{50}N_4O_4$			
Umbellamine, O-methyl-	EtOH	230(4.59),250s(4.09), 294(4.09)	33-0089-69
$C_{42}H_{54}N_4O_6$			
15,15'-Biaspidospermidine, 1,1'-diacetyl-16,16',17,17'-tetrahydroxy-	EtOH	236(4.29),297(4.06)	31-0575-69
	EtOH-NaOH	243(4.23),302(3.93), 335s(3.83)	31-0575-69
$C_{42}H_{58}O_6$			
Fucoxanthin (also in other solvents)	hexane	427(--),450(--), 476(--)	39-0429-69C
	CS_2	450s(--),478(5.13), 508(5.03)	39-0429-69C
$C_{42}H_{62}O_5$			
5α-Androstan-3-one, 17β-hydroxy-2α-(17β-hydroxy-5α-androst-2-en-3-yl)-, diacetate	C_6H_{12}	288(1.70)	78-1707-69
$C_{42}H_{62}O_6$			
5α-Androstan-3-one, 17β-hydroxy-2α-(17β-hydroxy-A-nor-5α-androstan-2α-yl)carbonyl-, diacetate	EtOH	290(4.04),315(4.31)	78-1707-69
$C_{42}H_{62}O_7$			
A-Nor-5α-androstane-1α,17β-diol, 2β-[(17β-hydroxy-1β-methyl-A-norestr-5(10)-en-2α-yl)oxy]-, triacetate	MeCN	201s(3.84)	78-1367-69
isomer	MeCN	218(3.83)	78-1367-69
$C_{42}H_{62}O_{15}$			
Gitoxin, cyclic 3''',4'''-carbonate	EtOH	218(4.16)	94-0682-69

Compound	Solvent	λ_{max}(log ϵ)	Ref.
$C_{43}H_{31}P$			
Phosphorin, 1,1-dihydro-1,1-di-1-naphthyl-2,4,6-triphenyl-	benzene	346(4.63),441(4.25), 525(4.02)	88-1231-69
$C_{43}H_{42}ClIrOP_2$			
Iridium, carbonylchlorobis(tribenzylphosphine)-	toluene	330(3.36),379(3.56), 427(2.67)	64-0770-69
Iridium, carbonylchlorobis(tri-p-tolylphosphine)-	toluene	337(3.46),387(3.58), 440(2.72)	64-0770-69
$C_{43}H_{46}O_2$			
Dispiro[5.0.5.1]trideca-1,4,8,11-tetraene-3,10-dione, 2,4,9-tri-tert-butyl-11,13,13-triphenyl-	MeOH	260(4.27),341(4.20)	44-1203-69
$C_{43}H_{46}O_{22}$			
Glucofrangulin A, acetate	n.s.g.	212(4.57),264(4.56), 360(4.26)	64-1408-69
$C_{43}H_{50}N_4O_2$			
5-Porphinol, 2,3,7,8,12,13,17,18-octaethyl-, benzoate	$CHCl_3$	402(5.24),501(4.21), 533(3.82),571(3.79), 624(3.28)	39-0564-69C
$C_{43}H_{50}N_4O_5$			
Umbellamine, acetate	EtOH	230(4.52),251s(4.11), 285(3.98),289(3.99), 293(4.00)	33-0089-69
$C_{43}H_{52}O_4$			
Benzoic acid, p-(2,4,9,11-tetra-tert-butyl-3,10-dioxo-13-phenyldispiro-[5.0.5.1]trideca-1,4,8,11-tetraen-13-yl)-, methyl ester	MeOH	258(4.42),330(4.21)	44-1203-69
$C_{43}H_{54}N_2O_{14}$			
Tolypomycin Y	EtOH	230(4.46),290(4.37), 337(4.10)	88-0097-69
$C_{43}H_{58}O_6$			
Bronianone	n.s.g.	250(4.02),278(3.85), 365(3.81)	77-0879-69
$C_{43}H_{66}O_{16}$			
Gitoxin, 3'''-(methyl carbonate)	EtOH	218(4.18)	94-0682-69
Gitoxin, 4'''-(methyl carbonate)	EtOH	219(4.18)	94-0682-69
Gitoxin, 16-(methyl carbonate)	EtOH	215(4.17)	94-0682-69
$C_{43}H_{72}N_4O_{14}S$			
Leucomycin A_3, 7-(thiosemicarbazone)	EtOH	232(4.54),271(4.37)	94-0844-69

Compound	Solvent	$\lambda_{max}(\log \epsilon)$	Ref.
$C_{44}H_{10}Cl_{20}N_4$			
Porphine, 5,10,15,20-tetrakis(pentachlorophenyl)-	benzene	422(5.57),515(4.37), 591(3.85),663(3.56)	4-0927-69
copper chelate	benzene	419(5.36),500s(--), 543(4.00),577(3.54)	4-0927-69
$C_{44}H_{10}F_{20}N_4$			
Porphine, 5,10,15,20-tetrakis(pentafluorophenyl)-	benzene	417(5.37),508(4.28), 586(3.76),659(3.54)	4-0927-69
	DMF	410(5.42),504(4.20), 579(3.71),654(3.57)	4-0927-69
copper chelate	DMF	412(5.34),503(3.61), 538(4.11),572(3.66)	4-0927-69
$C_{44}H_{26}$			
5,5'-Biindeno[2,1-a]indene, 10,10'-diphenyl-	EtOH	278(4.56),475(4.32)	5-0076-69C
$C_{44}H_{28}ClFeN_4$			
Iron, chloro[5,10,15,20-tetraphenylporphinato(2-)]-	$CHCl_3$	378(4.76),415(5.03), 510(4.11),578(3.52), 657(3.45),691(3.49)	35-2403-69
	pyridine	442(5.14),509(4.00), 530(4.00),655(3.28), 698(3.38)	35-2403-69
$C_{44}H_{30}$			
Butadiyne, bis[p-(p-phenylstyryl)-phenyl]-	DMF	303(4.37),385(5.10)	33-2521-69
$C_{44}H_{30}Cl_2$			
Propene, 1,1-bis(p-chlorophenyl)-3-(2,3,4,5-tetraphenyl-2,4-cyclopentadien-1-ylidene)-	C_6H_{12}	264(4.47),402(4.53), 545(3.48)	117-0047-69
$C_{44}H_{32}$			
Anthracene, 9-[3-(p-phenylstyryl)-5-styrylstyryl]-	toluene	335(4.95)	104-0903-69
$\Delta^{5,5'}$-Biindeno[2,1-a]indene, 5a,5'a,-10,10',10a,10'a-hexahydro-10,10'-diphenyl-	EtOH	245(4.2),320(4.2)	5-0076-69C
Ethylene, 1,2-bis[p-[2-(1-naphthyl)-vinyl]phenyl]-1-phenyl-	DMF	386(4.85)	33-2521-69
Propene, 1,1-diphenyl-3-(2,3,4,5-tetraphenyl-2,4-cyclopentadien-1-ylidene)-	C_6H_{12}	260(4.61),390(4.56), 510(3.45)	117-0047-69
Stilbene, 3,5-bis[2-(1-naphthyl)vinyl]-4'-phenyl-	toluene	330(4.97)	104-0903-69
Stilbene, 3,5-bis[2-(2-naphthyl)vinyl]-4'-phenyl-	toluene	330(5.07)	104-0903-69
Stilbene, 3-[2-(1-naphthyl)vinyl]-5-[2-(2-naphthyl)vinyl]-	toluene	330(5.04)	104-0903-69
$C_{44}H_{32}Cl_4N_6O_8$			
Quinolinium, 2,2'-[[bis(p-chlorophenyl)-tetrazanediylidene]methylidyne]bis-[1-phenyl-, diperchlorate	EtOH	220(3.31),318(2.90)	124-0370-69

Compound	Solvent	λ_{max}(log ϵ)	Ref.
$C_{44}H_{32}N_4$			
Cycloocta[1,2-b:5,6-b']diquinoxaline, 6,7,14,15-tetrahydro-6,7,14,15-tetraphenyl-	EtOH	254(4.15),288(4.69)	44-1651-69
$C_{44}H_{34}Cl_2N_6O_8$			
Quinolinium, 2,2'-[(diphenyltetrazane-diylidene)dimethylidyne]bis[1-phenyl-, diperchlorate	EtOH	235(3.35),340(2.97)	124-0370-69
$C_{44}H_{35}ClP_2Pt$			
Platinum, chloro(phenylethynyl)bis-(triphenylphosphine)-, trans	n.s.g.	264(4.60),314(4.12), 335s(3.60)	101-0457-69B
$C_{44}H_{38}ClNO_{14}$			
3-Carboxy-1-D-ribofuranosylpyridinium chloride, D-ribofuranosyl ester, tetrabenzoate	MeOH	265(3.76)	39-0918-69C
$C_{44}H_{38}N_4$			
Tetrabenzo[b,g,l,q]porphine, 1,4,8,11,15,18,22,25-octamethyl-	$C_{10}H_7Cl$	375(4.15),400(4.55), 428(5.24),445(5.37), 540(3.53),574s(4.10), 582(4.17),602s(4.13), 614(4.51),620s(4.72), 627(4.92),674(4.68)	77-0345-69
$C_{44}H_{52}CuN_4O_8$			
Copper, [tetrahydrogen 3,7,13,17-tetramethyl-2,8,12,18-porphinetetrapropionato(2-)], tetraethyl ester	$CHCl_3$	327(4.23),399(5.55), 528(4.10),564(4.37)	39-0176-69C
$C_{44}H_{54}O_2$			
2,5-Cyclohexadien-1-one, 4,4'-(2,3-diphenylbutanediylidene)bis[2,6-di-tert-butyl-	$CHCl_3$	327(4.7)	44-1211-69
$C_{44}H_{55}N_5O_9$			
Lasiodine A, N-acetyl-O-methyl-, acetate	EtOH	284(4.27)	78-0937-69
$C_{44}H_{58}O_6$			
Anhydrofucoxanthin, acetate	benzene	444s(--),461(5.02), 483(--)	39-0429-69C
	EtOH	235(3.35),267(--), 320(--),334(--)	39-0429-69C
$C_{44}H_{59}ClO_7$			
Fucoxanthin, chloroacetate	pentane	425s(--),448(5.05), 480(--)	39-0429-69C
$C_{44}H_{60}O_7$			
Fucoxanthin, acetate	pentane	425s(--),449(5.04), 478(--)	39-0429-69C
$C_{44}H_{61}Cl_3O_6$			
5α-Androst-2-ene-3,17β-diol, 2-(17β-hydroxy-5α-androst-2-en-3-yl)-, 17,17-diacetate trichloroacetate	EtOH	224(4.19)	78-1707-69

Compound	Solvent	$\lambda_{max}(\log \epsilon)$	Ref.
$C_{44}H_{64}O_{17}$			
Gitoxin, 3''',4'''-carbonate 16-(methyl carbonate)	EtOH	217(4.17)	94-0682-69
$C_{44}H_{68}O_{16}$			
Gitoxin, 4'''-(ethyl carbonate)	EtOH	219(4.15)	94-0682-69
Gitoxin, 16-(ethyl carbonate)	EtOH	216(4.18)	94-0682-69
$C_{45}H_{30}Br_3CoO_6$			
Cobalt, tris(2-bromo-1,3-diphenyl-1,3-propanedionato)-	$CHCl_3$	280s(4.53),290(4.44), 385(4.05)	2-0628-69
$C_{45}H_{30}Br_3CrO_6$			
Chromium, tris(2-bromo-1,3-diphenyl-1,3-propanedionato)-	$CHCl_3$	375(4.38),393(4.32)	2-0628-69
$C_{45}H_{30}Cl_3CrO_6$			
Chromium, tris(2-chloro-1,3-diphenyl-1,3-propanedionato)-	$CHCl_3$	370(4.40),390(4.44)	2-0628-69
$C_{45}H_{30}CoI_3O_6$			
Cobalt, tris(2-iodo-1,3-diphenyl-1,3-propanedionato)-	$CHCl_3$	280s(4.30),295(4.32), 390(3.98)	2-0628-69
$C_{45}H_{30}CrI_3O_6$			
Chromium, tris(2-iodo-1,3-diphenyl-1,3-propanedionato)-	$CHCl_3$	375(4.38),394(4.45)	2-0628-69
$C_{45}H_{33}CoO_6$			
Cobalt, tris(1,3-diphenyl-1,3-propanedionato)-	$CHCl_3$	275s(4.58),288(4.60), 380(4.10)	2-0628-69
$C_{45}H_{33}CrO_6$			
Chromium, tris(1,3-diphenyl-1,3-propanedionato)-	$CHCl_3$	360(4.45),385(4.53)	2-0628-69
$C_{45}H_{40}ClNO_{14}$			
3-Carboxy-1-D-ribofuranosyl-4-picolinium chloride, D-ribofuranosyl ester, tetrabenzoate	MeOH	265(3.80)	39-0918-69C
$C_{45}H_{42}O_2$			
Dispiro[5.0.5.1]trideca-1,4,8,11-tetraene-3,10-dione, 2,4-di-tert-butyl-9,11,13,13-tetraphenyl-	MeOH	255(4.21),350(4.18)	44-1203-69
$C_{45}H_{47}N_8O_{13}P$			
Guanosine, 5'-(O-p-methoxy-α,α-diphenylbenzyl)thymidyl-(3'→5')-N-acetyl-2'-deoxy-, ester with hydracrylonitrile	EtOH	261(4.34)	35-3360-69
$C_{45}H_{57}ClCoN_4O_{12}$			
Cobalt(1+), (tetrahydrogen octadehydro-1,3,7,13,17,19-hexamethyl-2,8,12,18-corrintetrapropionato)-, perchlorate, tetraethyl ester	$CHCl_3$	283(4.47),352(4.36), 500(4.09),547(3.90), 571(3.89)	39-0176-69C

Compound	Solvent	λ_{max}(log ε)	Ref.
$C_{45}H_{57}ClN_4NiO_{12}$			
Nickel(1+), (tetrahydrogen octadehydro-1,3,7,13,17,19-hexamethyl-2,8,12,18-corrintetrapropionato)-, perchlorate, tetraethyl ester	EtOH	276(4.52),354(4.52), 446(3.67),566(4.13)	39-0176-69C
$C_{45}H_{57}N_3O_9$			
Beauvericin	EtOH	204(4.13),248(3.12), 256(3.06),263(2.94)	88-4255-69
$C_{45}H_{58}N_2O_{14}$			
Tolypocyclonide	EtOH	248(4.29),290(4.28), 330s(--)	88-0097-69
$C_{45}H_{60}N_4O_8$			
Biline-2,8,12,18-tetrapropionic acid, 10,23-dihydro-1,3,7,13,17,19-hexamethyl-, diethyl ester, dihydrobromide	$CHCl_3$	373(4.16),454(4.62), 525(5.29)	39-0176-69C
$C_{45}H_{62}$			
1,3,5,7,9,11,13,15,17,19,21-Docosaundecaene, 3,7,11,16,20-pentamethyl-1,22-bis(2,6,6-trimethyl-1-cyclohexen-1-yl)-, all-E-	hexane	476(5.22),505(5.15)	78-5383-69
	C_6H_{12}	483(5.22),515(5.15)	
	benzene	492(5.21),526(5.13)	
	CCl_4	490(5.18),523(5.10)	
	dioxan	487(5.23),520(5.15)	
	acetone	480(5.23),512(5.17)	
	DMF	490(5.18),520(5.10)	
	CS_2	512(5.13),549(5.05)	
	1:1 CS_2-hexane	497(5.18),529(5.11)	
	1:1 CS_2-C_6H_{12}	497(5.17),531(4.10)	
$C_{45}H_{68}O_{18}$			
Gitoxin, 4''',16-bis(methyl carbonate)	EtOH	216(4.19)	94-0682-69
$C_{46}H_{26}N_8O_4$			
1,1,2,2-Ethanetetracarbonitrile, 1,2-bis[10-(p-nitroanilino)-9-anthryl]-	$CHCl_3$	260(4.92),322(4.35), 404(4.32)	22-1651-69
$C_{46}H_{32}$			
Anthracene, 9-[3,5-bis[2-(1-naphthyl)-vinyl]styryl]-	toluene	330(4.80)	104-0903-69
Anthracene, 9-[3,5-bis[2-(2-naphthyl)-vinyl]styryl]-	toluene	330(5.00)	104-0903-69
Anthracene, 9-[3-[2-(1-naphthyl)vinyl]-5-[2-(2-naphthyl)vinyl]styryl]-	toluene	330(4.94)	104-0903-69
$C_{46}H_{34}$			
Naphthalene, 1-[3,5-bis(p-phenylstyryl)-styryl]-	toluene	335(5.07)	104-0903-69
Naphthalene, 2-[3,5-bis(p-phenylstyryl)-styryl]-	toluene	335(5.10)	104-0903-69
$C_{46}H_{34}N_6O_5$			
Glycine, N-[[2-[11-(12H-benzo[b]phenoxazin-12-yl)-3H-benzo[b]phenoxazin-3-ylidene]amino]-1-benzimidazolylcarbonyl]-, butyl ester	$CHCl_3$	460(4.31),548(4.45), 597(4.44),650(4.20)	44-1691-69

Compound	Solvent	λmax(log ε)	Ref.
$C_{46}H_{34}O$			
Benzofuran, 4,6,7-tristyryl-2-(p-styrylphenyl)-, all-E-	DMF	365(4.88),404(4.86)	33-1282-69
$C_{46}H_{36}O_2$			
Propene, 1,1-bis(p-methoxyphenyl)-3-(2,3,4,5-tetraphenyl-2,4-cyclopentadien-1-ylidene)-	C_6H_{12}	264(4.40),420(4.59)	117-0047-69
$C_{46}H_{38}Cl_2N_6O_8$			
Quinolinium, 2,2'-[(di-o-tolyltetrazanediylidene)dimethylidyne]bis[1-phenyl-, diperchlorate	EtOH	226(3.4),238(3.38), 333(3.2)	124-0370-69
$C_{46}H_{42}N_6O_8$			
Guanosine, 5'-O-trityl-, 2'-ester with N-carboxy-3-phenyl-L-alanine N-benzyl ester	EtOH	212(4.50),257(4.13), 285s(3.78)	73-3755-69
$C_{46}H_{44}O_{17}$			
Flavone, 4',7-bis(benzyloxy)-3,3'-dihydroxy-5-methoxy-, 3-β-D-glucopyranoside, pentaacetate	MeOH	265(4.41),333(4.35)	24-0792-69
$C_{46}H_{46}O_8$			
Benzene, 1,2,4,5-tetrakis(3,4-dimethoxystyryl)-	DMF	358(4.88)	33-2521-69
$C_{46}H_{62}Cl_4O_8$			
5α-Androst-2-ene-3,17β-diol, 3,3'-(tetrachlorosuccinate), diacetate	dioxan	225s(3.41)	35-2062-69
$C_{46}H_{62}O_3$			
Chlorobiumquinone, trans	isooctane	245s(4.49),250(4.51), 255s(4.49),265s(4.34), 325(3.48)	35-6889-69
$C_{47}H_{37}N_2P$			
Phosphorin, 1,1-bis(diphenylamino)-1,1-dihydro-2,4,6-triphenyl-	C_6H_{12}	330(4.28),439(3.93)	89-0770-69
$C_{47}H_{66}O_{22}$			
Olivomycin D	n.s.g.	227(4.34),278(4.68), 307(3.77),318(3.80), 332s(3.58),410s(4.03)	105-0483-69
$C_{47}H_{70}O_{20}$			
Gitoxin, 3''',4''',16-tris(methyl carbonate)	EtOH	215(4.12)	94-0682-69
$C_{47}H_{72}O_{18}$			
Gitoxin, 4''',16-bis(ethyl carbonate)	EtOH	216(4.16)	94-0682-69
$C_{48}H_{28}$			
Diacenaphtho[1,2-j:1',2'-l]fluoranthene, 3b-(1-acenaphthylenyl)-3b,3c,15b,15c-tetrahydro-	C_6H_{12}	256(4.60),268(4.58)	24-3599-69
	benzene	292(4.18),320(4.32), 342(4.16),363(4.32), 384(4.64),409(4.72)	24-3599-69

Compound	Solvent	$\lambda_{max}(\log \epsilon)$	Ref.
$C_{48}H_{34}$			
Anthracene, 9-[3-[2-(2-naphthyl)vinyl]-5-(p-phenylstyryl)styryl]-	toluene	335(5.10)	104-0903-69
$C_{48}H_{36}$			
Benzene, 1,3,5-tris(p-phenylstyryl)-	DMF	337(5.15)	33-2521-69
$C_{48}H_{36}O$			
Furan, 2,5-bis(2,4-distyrylphenyl)-, all-E-	DMF	319(4.91),399(4.73)	33-1282-69
$C_{48}H_{36}O_2$			
Ethylene, 1,2-bis[p-(p-phenoxystyryl)-phenyl]-1-phenyl-	DMF	335(4.66),376(4.83)	33-2521-69
$C_{48}H_{42}N_2$			
Aniline, 4,4'-[3-(2,3,4,5-tetraphenyl-2,4-cyclopentadien-1-ylidene)propen-ylidene]bis[N,N-dimethyl-	C_6H_{12}	243(4.50),343(4.24), 450(4.22)	117-0047-69
$C_{48}H_{72}O_3$			
p-Benzoquinone, 2-methoxy-5-methyl-5,6-bis(3,7,11,15-tetramethyl-2,6,10,14-hexadecatetraenyl)-	C_6H_{12}	288(3.86),310(3.93), 375s(2.71)	44-0227-69
$C_{49}H_{49}N_6O_{13}P$			
Thymidine, N-benzoyl-2'-deoxycytidylyl-(5'→3')-5'-O-(p-methoxy-α,α-diphenyl-benzyl)-, ester with hydracryloni-trile	EtOH	262(4.50)	35-3360-69
$C_{49}H_{68}O_7S$			
5α-Androst-2-ene-3,17β-diol, 2-(17β-hy-droxy-5α-androst-2-en-3-yl)-, 2,17-diacetate, p-toluenesulfonate	EtOH	224(4.19)	78-1707-69
$C_{50}H_{36}$			
Anthracene, 9-[3,5-bis(p-phenylstyryl)-styryl]-	toluene	330(5.01)	104-0903-69
$C_{50}H_{42}O_5$			
Benzofuran, 4,6,7-tris(p-methoxysty-ryl)-2-[p-(p-methoxystyryl)phenyl]-	DMF	377(4.93),414(4.93)	33-1282-69
$C_{50}H_{49}N_8O_{12}P$			
Adenosine, 5'-O-(p-methoxy-α,α-diphenyl-benzyl)thymidyl-(3'→5')-N-benzoyl-2'-deoxy-, ester with hydracrylonitrile	EtOH	275(4.43)	35-3360-69
$C_{50}H_{50}N_4O_8S_2$			
5,5'-Bibenzimidazole, 2,2'-bis(p-meth-oxystyryl)-1,1'-dimethyl-, bis(meth-yl p-toluenesulfonate)	MeOH	372(4.44)	22-1926-69
$C_{50}H_{68}$			
1,3,5,7,9,11,13,15,17,19,21,23,25-Hexa-cosatridecaene, 3,7,11,15,20,24-hexa-methyl-1-cyclohexen-1-yl)-, all-E-	hexane	500(5.33),534(5.27)	78-5383-69
	C_6H_{12}	505(5.31),537(5.23)	78-5383-69
	benzene	518(5.26),552(5.17)	78-5383-69
	dioxan	512(5.31),546(5.23)	78-5383-69
	acetone	505(5.34),537(5.27)	78-5383-69

Compound	Solvent	λmax(log ε)	Ref.
1,3,5,7,9,11,13,15,17,19,21,23,25-Hexacosatridecaene, 3,7,11,15,20,24-hexamethyl-1-cyclohexen-1-yl)- (cont.)	CCl_4	515(5.24),549(5.16)	78-5383-69
	DMF	512(5.25),549(5.16)	78-5383-69
	CS_2	537(5.26),578(5.17)	78-5383-69
	1:1 CS_2-hexane	520(5.28),555(5.19)	78-5383-69
	1:1 CS_2-C_6H_{12}	523(5.27),558(5.17)	78-5383-69
$C_{50}H_{76}O_{20}$			
Gitoxin, 3''',4''',16-tris(ethyl carbonate)	EtOH	216(4.15)	94-0682-69
$C_{50}H_{78}O_8$			
3,3',4,4'-Biphenyltetrol, 5,5'-dipentadecyl-, tetraacetate	EtOH	256(4.33)	18-2375-69
$C_{51}H_{42}O_3$			
m-Terphenyl, 4,4"-bis(p-methoxystyryl)-5'-[p-(p-methoxystyryl)phenyl]-	DMF	347(5.18)	33-2521-69
$C_{51}H_{45}O_9P_3Ru_3$			
Ruthenium, nonacarbonyltris(ethyldiphenylphosphine)tri-, triangulo-	$C_6H_{11}Me$	365s(4.1),495(4.1)	101-0289-69A
$C_{51}H_{56}O_9$			
Olean-12-ene-11,16,21-trione, 3β,22α,28-trihydroxy-, tribenzoate	EtOH	232(4.83)	44-3135-69
$C_{51}H_{63}N_5O_9$			
Tryptophan, N-[N-[N-(3-tert-butoxy-N-carboxy-L-alanyl)-L-alanyl]-3-(p-tert-butoxyphenyl)-L-alanyl]-, N-(α,α-dimethyl-p-phenylbenzyl)methyl ester, L-	EtOH	255(4.32)	33-1058-69
$C_{52}H_{36}$			
Stilbene, 3,5-bis[2-(9-anthryl)vinyl]-4'-phenyl-	toluene	335(4.87),390(--)	104-0903-69
$C_{52}H_{36}Cl_2O_{11}$			
Pyrylium, 2-[(4,6-diphenyl-2H-pyran-2-ylidene)(4,6-diphenylpyrylium-2-yl)-methyl]-4,6-diphenyl-, diperchlorate	CH_2Cl_2	250(4.42),385(4.43), 351(4.82),600(4.62)	4-0623-69
Pyrylium, 4-[(4,6-diphenyl-2H-pyran-2-ylidene)(2,6-diphenylpyrylium-4-yl)-methyl]-2,6-diphenyl-, diperchlorate	MeCN	239(4.55),268(4.53), 361(4.52),585(4.49)	4-0623-69
$C_{52}H_{40}P_2Pt$			
Platinum, bis(phenylethynyl)bis(triphenylphosphine)-, cis	n.s.g.	254(4.64),314(4.41)	101-0457-69B
trans	n.s.g.	266(4.69),290s(4.51), 348(4.39)	101-0457-69B
$C_{52}H_{44}O_5$			
Furan, 2,5-bis[2,4-bis(p-methoxystyryl)-phenyl]-, all-E-	DMF	337(4.96),403(4.77)	33-1282-69
$C_{52}H_{48}O_4$			
Dispiro[3H-naphtho[2,1-b]pyran-3,6'-[6H,14H]dibenzo[b,h][1,7]dioxacyclododecin-14',3"-[3H]naphtho[2,1-b]pyran], 7',15'-dipentyl-	dioxan	262(4.55),272(4.54), 315(4.20),339(4.00), 356(3.91),408(3.67)	64-0652-69

Compound	Solvent	λ_{max}(log ε)	Ref.
$C_{53}H_{54}N_5O_{15}P$			
Thymidine, thymidyl-(5'→3')-5'-O-(p-methoxy-α,α-diphenylbenzyl)-, ester with hydracrylonitrile, mono-(3-benzoylpropionate)	EtOH	245(4.37)	35-3360-69
$C_{53}H_{72}O_8$			
Amitenone	EtOH	286(4.47)	39-2398-69C
	EtOH-NaOH	325(--)	39-2398-69C
$C_{53}H_{81}NO_3$			
Rhodoquinone-9	EtOH	285(4.04),515(3.09)	88-1969-69
$C_{54}H_{30}$			
[13]Helicene	dioxan	225(5.0),250(5.1), 310f(4.6),402(3.9), 430s(3.8)	88-3683-69
Pentacene, 5,7,12,14-tetrakis(phenyl-ethynyl)-	o-$C_6H_4Cl_2$	705(4.43)	44-1734-69
$C_{54}H_{32}$			
Ethylene, 1-(2-benzo[c]phenanthryl)-2-(2-octaheliceny1)-	dioxan	270(5.0),310s(4.9), 350s(4.8),445s(3.0)	88-3683-69
$C_{54}H_{36}$			
o,p,p,o,p,p,o,p,p-Nonaphenylene	$CHCl_3$	283(4.92)	5-0030-69D
$C_{54}H_{36}O_3$			
Spiro[3,5-cyclohexadiene-1,6'-dibenzo-[d,f][1,3]dioxepin]-2-one, 2',3,4',5,8',10'-hexaphenyl-	$CHCl_3$	370(3.7)	44-2027-69
$C_{54}H_{48}O_6$			
m-Terphenyl, 4,4"-bis(3,4-dimethoxy-styryl)-5'-[p-(3,4-dimethoxystyryl)-phenyl]-	DMF	351(5.13)	33-2521-69
$C_{54}H_{86}O_2$			
Cholest-5-en-7-one, dimer 4	dioxan	289(2.66)	88-4589-69
$C_{54}H_{90}S_2$			
Cholest-4-en-4-yl disulfide	hexane	224(4.30)	94-0355-69
$C_{56}H_{34}N_8$			
Phthalocyanine, tetraphenyl-	PhCl	355(4.85),615(4.52), 649(4.68),679(5.09), 713(5.13)	65-2129-69
	pyridine	360(4.81),623(4.52), 693(5.16)	65-2129-69
cobalt complex	$C_{10}H_7Cl$	350(4.51),620(4.34), 690(5.02)	65-2129-69
manganese complex	$C_6H_3Cl_3$	375(4.28),530(3.86), 670(3.92),745(4.66)	65-2129-69
lead complex	PhCl	370(4.87),655(4.31), 725(5.10)	65-2129-69
palladium complex	PhCl	340(4.87),616(4.93), 676(4.89)	65-2129-69
$SnCl_2$ complex	PhCl	375(4.72),645(4.66), 720(5.42)	65-2129-69

Compound	Solvent	λmax(log ε)	Ref.
Phthalocyanine, tetraphenyl-, VO complex	PhCl	355(4.82),643(4.53), 715(5.27)	65-2129-69
	$C_6H_3Cl_3$	355(4.85),645(4.61), 715(5.37)	65-2129-69
zinc complex (also other metal complexes)	$C_{10}H_7Cl$	365(4.99),625(4.59), 698(5.28)	65-2129-69
$C_{56}H_{56}O_{26}$			
Castalagin, pentadeca-O-methyl-	dioxan	225(4.87)	5-0186-69A
$C_{57}H_{54}O_9$			
m-Terphenyl, 4,4"-bis(3,4,5-trimethoxy-styryl)-5'-[p-(3,4,5-trimethoxystyryl)phenyl]-	DMF	347(5.16)	33-2521-69
$C_{58}H_{39}Br$			
Benz[e]-as-indacene, 3-bromo-3,6a-dihydro-1,2,3,4,5,6,6a-heptaphenyl-	C_6H_{12}	235(4.71),270(4.47), 310(4.15),365(3.87), 390(3.79)	39-2605-69C
$C_{58}H_{39}Cl$			
Benz[e]-as-indacene, 3-chloro-3,6a-dihydro-1,2,3,4,5,6,6a-heptaphenyl-	C_6H_{12}	265(4.56),308(4.13), 353s(3.90),365(3.93), 395s(3.66)	39-2605-69C
$C_{58}H_{40}$			
Benz[e]-as-indacene, 3,6a-dihydro-1,2,3,4,5,6,6a-heptaphenyl-	C_6H_{12}	235(4.7),258s(4.58), 310(4.39),362(4.13)	39-2605-69C
	$CHCl_3$	255(4.57),314(4.34), 366(4.11)	39-2605-69C
Pentatetraene, tetraphenyl-, dimer	n.s.g.	296(4.56),370(3.77)	35-6112-69
$C_{58}H_{40}O$			
Benz[e]-as-indacen-3-ol, 3,6a-dihydro-1,2,3,4,5,6,6a-heptaphenyl-	C_6H_{12}	260(4.55),298s(4.16), 310(4.23),365(4.01)	39-2605-69C
$C_{58}H_{40}O_2$			
Substance A, m. 242-244°	n.s.g.	245(4.59),282(4.53), 348(4.05)	39-2605-69C
Substance B, m. 286-288°	n.s.g.	243(4.61),260s(4.47), 300(4.47),312(4.49)	39-2605-69C
$C_{58}H_{42}N_6O_5$			
Glycine, N-[[2-[[2-phenyl-11-(2-phenyl-12H-benzo[b]phenoxazin-12-yl)-3H-benzo[b]phenoxazin-3-ylidene]amino]-1-benzimidazolyl]carbonyl]-, butyl ester	$CHCl_3$	279(4.95),310(4.40), 478(4.36),540(4.51), 580(4.42),630(4.18)	44-1691-69
$C_{58}H_{42}O_2$			
Substance A, dihydro derivative	C_6H_{12}	255(4.45),275(4.42), 355(3.59)	39-2605-69C
Substance B, dihydro derivative	C_6H_{12}	253(4.50),292(4.4), 305s(4.36),322s(4.20)	39-2605-69C
$C_{59}H_{42}O$			
Benz[e]-as-indacene, 3,6a-dihydro-3-methoxy-1,2,3,4,5,6,6a-heptaphenyl-	C_6H_{12}	255(4.61),315(4.45)	39-2605-69C

Compound	Solvent	λ max (log ε)	Ref.
$C_{60}H_{42}$			
m-Terphenyl, 4,4"-bis[2-(1-naphthyl)-vinyl]-5'-[p-[2-(1-naphthyl)vinyl]-phenyl]-	DMF	298(4.56),353(5.05)	33-2521-69
m-Terphenyl, 4,4"-bis[2-(2-naphthyl)-vinyl]-5'-[p-[2-(2-naphthyl)vinyl]-phenyl]-	DMF	294(4.80),349(5.22)	33-2521-69
$C_{60}H_{52}N_{10}$			
Succinonitrile, tetrakis[m-(phenylazo)-phenethyl]-	toluene	322(4.82),440(3.42)	49-1451-69
$C_{60}H_{52}N_{12}$			
Butyronitrile, 2,2'-azobis[2-[m-(phenyl-azo)phenethyl]-4-[m-(phenylazo)-phenyl]-	$CHCl_3$	321(4.81),440(3.48)	49-1451-69
$C_{62}H_{86}N_{12}O_{16}$			
Actinomycin D	H_2O	420(4.4)	10-0405-69A
	EtOH	420s(4.4),445(4.4)	10-0405-69A
$C_{63}H_{45}AsO_9Ru_3$			
Ruthenium, nonacarbonyltris(triphenyl-arsine)tri-, triangulo-	benzene	425(4.0),465s(3.9)	101-0289-69A
$C_{63}H_{45}O_9P_3Ru_3$			
Ruthenium, nonacarbonyltris(triphenyl-phosphine)tri-, triangulo-	$C_6H_{11}Me$	395(4.1),508(4.1)	101-0289-69A
$C_{63}H_{45}O_9Ru_3Sb$			
Ruthenium, nonacarbonyltris(triphenyl-stibine)tri-, triangulo-	benzene	440s(3.8)	101-0289-69A
$C_{66}H_{48}$			
m-Terphenyl, 4,4"-bis(p-phenylstyryl)-5'-[p-(p-phenylstyryl)phenyl]-	DMF	353(5.21)	33-2521-69
$C_{66}H_{70}N_8O_{22}P_2$			
Thymidine, thymidyl-(5'→3')-thymidylyl-(5'→3')-5'-O-(p-methoxy-α,α-diphenyl-benzyl)-, diester with hydracryloni-trile, mono(3-benzoylpropionate)	EtOH	238(4.48),250(4.47), 263(4.46)	35-3360-69
$C_{68}H_{56}N_4$			
5H,7H-s-Tetrazino[6,1-a:3,4-a']diiso-indole, 7,7,14,14-tetrabenzyl-4b,11b,12,14a-tetrahydro-4b,5,11b,12-tetraphenyl-	MeCN	245(4.54),290(4.35), 400(3.76)	35-3670-69
$C_{71}H_{51}O_2P$			
Phosphorin, 1,1-dihydro-2,4,6-triphen-yl-1,1-bis[(5'-phenyl-m-terphenyl-2'-yl)oxy]-	C_6H_{12}	251(4.96),323(4.23), 425(4.12)	89-0770-69
$C_{76}H_{52}N_2O_2$			
Pyridinium, 4,4'-p-phenylenebis[1-(2'-hydroxy-m-terphenyl-5'-yl)-2,6-di-phenyl-, dihydroxide, bis inner salt (also in other solvents)	MeOH	330(4.89),560(3.88)	5-0093-69G
	EtOH	332(4.87),603(3.95)	5-0093-69G
	PrOH	334(--),623(--)	5-0093-69G

Compound	Solvent	$\lambda_{max}(\log \epsilon)$	Ref.
$C_{96}H_{66}Fe_2N_8O$			
μ-Oxobis(tetraphenylporphine iron(III)), xylene solvate	benzene	408(5.08),572(4.03), 612(3.68)	35-2403-69
$C_{158}H_{224}N_{36}O_{34}S$			
Thyrocalcitonin (pig reduced), 1-de-L-cysteine-2-de-L-serine-3-de-L-asparagine-4-de-L-leucine-5-de-L-serine-6-de-L-threonine-7-de-L-cysteine-8-de-L-valine-9-de-L-leucine-10-(3-tert-butoxy-N-carboxy-L-alanine)-12-[3-(p-tert-butoxyphenyl)-L-alanine]-23-(3-tert-butoxy-L-alanine)-31-(L-2-amino-3-tert-butoxybutyric acid)-, 30-tert-butyl 10-(α,α-dimethyl-p-phenylbenzyl) ester, diacetate	EtOH	259(4.32)	33-1058-69

1- -69, Acta Chem. Scand., 23 (1969)
0144 K.E. Stensiö and C.A. Wachtmeister
0159 O. Buchardt, P.L. Kumler and C. Lohse
0244 K. Leander et al.
0286 P.M. Boll et al.
0294 K. Undheim and M. Gacek
0371 K. Undheim and V. Nordal
0597 E. Berner and P. Kolsaker
0703 J. Skramstad
0751 P. Martinson
1059 K. Hauge
1151 S. Brandange and B. Lüning
1461 K. Hauge
1704 K. Undheim et al.
2021 H. Erdtman and K. Tsuno
2065 K. Undheim et al.
2149 O. Buchardt et al.
2177 M. Elander et al.
2403 T. Hase
2409 T. Hase
2437 K. Undheim and J. Røe
2475 K. Undheim and T. Greibrokk
2488 K. Undheim and M. Gacek
2578 J. Bergman and H. Erdtman
2583 J. Gripenberg and J. Martikkala
2888 G. Kjellin and J. Sandström
2989 O.H. Mattson and I. Nyberg
3038 T. Bruun
3125 O. Buchardt et al.
3567 H. Lund and A.D. Thomsen

2- -69, Indian J. Chem., 7 (1969)
0001 P.C. Srivastava
0009 B.D. Tilak and S.S. Gupte
0017 B.D. Tilak and S.K. Jain
0040 V.M. Chari et al.
0049 N.R. Krishnaswamy et al.
0061 A.C. Jain et al.
0101 P.C. Parsatharathy et al.
0110 S.C. Datta et al.
0115 V.K. Ahluwalia et al.
0118 M.R. Parthasarathy et al.
0129 P.B. Talukdar and S.K. Sengupta
0135 S. Ghosal and S.K. Dutta
0191 B.D. Tilak and G.T. Panse
0196 P.K. Grover and N. Anand
0215 J. Chakravarty and U.R. Ghatak
0266 C.P. Prabhakaran and C.C.C. Patel
0305 A.C. Jain et al.
0307 N.L. Dutta and C. Anasim
0310 R. Raghavan et al.
0311 B.D. Tilak and G.T. Panse
0315 B.D. Tilak and G.T. Panse
0412 T. George and D.V. Mehta
0419 B.D. Tilak and K.N. Subbaswami
0422 B.D. Tilak et al.
0450 O.P. Vig et al.
0452 B.S. Joshi and D.F. Rane
0463 B.R. Samant and A.B. Kulkarni
0536 N.S. Narasimhan and M.V. Paradkar
0540 A.C. Jain et al.
0543 A.C. Jain et al.
0557 V.A. Pol et al.
0561 B.D. Hosangadi et al.
0583 T.R. Vakula and V.R. Srinavasan
0628 P.R. Singh and R. Sahai
0635 S.C. Pakrashi and E. Ali
0636 B.S. Joshi and V.N. Kamat
0637 S.L. Jindal and B.D. Tilak
0643 J. Alexander et al.
0662 S. Seshadri et al.
0667 S. Seshadri et al.
0678 B.D. Paul and P.K. Bose
0737 B.D. Tilak and S.L. Jindal
0746 S.C. Datta et al.
0755 S. Narayanaswami et al.
0841 T.R. Govindachari et al.
0848 K. Nagurajan et al.
0870 M.R. Thakar and B.K. Sabata
0873 T.R. Govindachari et al.
0876 S.N. Balasubramanyam and M. Sivarajan
0940 G. Srimannayana and N.V. SubbaRao
0945 S. Narayanaswami et al.
0948 B.D. Tilak and S.L. Jindal
0952 K.N. Arjungi et al.
0959 T. George et al.
0964 K. Nagarajan and V. Ranga Rao
1010 S.S. Chakravorti et al.
1051 Y.H. Deshpande and V.R. Rao
1060 N.L. Dutta et al.
1065 M.R.B. Bhagwanth et al.
1072 A.C. Jain et al.
1084 H. Singh and S. Padmanabhan
1111 O.P. Vig et al.
1114 R.B. Tirodkar and R.N. Usgaonkar
1166 C.M. Gupta et al.
1167 M.S. Ahmad et al.
1169 Z.G. Hajos and M.W. Goldberg
1182 A.C. Jain et al.
1186 A.C. DasGupta et al.
1279 A.K. Devi et al.

3- -69, Anal. Chem., 41 (1969)
0344 A.A. Schilt and W.C. Hoyle
0529 D.T. Haworth and J.H. Munroe
1652 T.M. Florence et al.
1690 L.E. Mattison et al.
1750 H.O. Friestad et al.

4- -69, J. Heterocyclic Chem., 6 (1969)
0005 G. Garcia-Muñoz et al.
0023 E. Dyer and C.E. Minnier
0029 J.A. Van Allan and G.A. Reynolds
0037 R.W. Lamon
0053 V. Sunjic et al.
0089 M.G. Stout and R.K. Robins
0093 S.F. Martin and R.N. Castle
0131 W.L. Nelson et al.
0163 P. Sohar et al.
0191 L.R. Worden et al.
0207 J.F. Gerster et al.
0215 B.C. Hinshaw et al.
0247 C.R. Johnson and C.B. Thanawalla
0251 M.W. Barker and J.H. Gardner
0253 A.C. Ranade and H. Gilman
0279 G. Casini et al.
0317 R.C. Bertelson et al.
0361 N.H. Cromwell and L.A. Nielsen

0375 G.A. Reynolds and J.A. Van Allan
0379 A.N. Fujiwara et al.
0389 A.N. Fujiwara et al.
0397 W.J. Humphlett
0405 A. Giner-Sorolla
0407 R.F. Meyer
0415 W.W. Paudler and H.G. Shin
0465 G. Galiazzo et al.
0491 E.F. Elslager
0497 P. Raffey and J.P. Verge
0507 P.L. Southwick et al.
0523 F.E. Kempter and R.N. Castle
0533 R.N. Castle et al.
0545 C. Wijnberger and C.L. Habraken
0577 G.L. Morrison et al.
0593 I. Wempen, H.U. Blank and J.J. Fox
0599 S.C. Bell and P.H.L. Wei
0605 R.H. Spector and M.M. Joullie
0613 A. Rosowsky et al.
0623 G.A. Reynolds and J.A. Van Allan
0635 M. Wilhelm and P. Schmidt
0639 M.T. Garcia-Lopez et al.
0681 A. Warshawsky and D. Ben-Ishai
0697 M.M. Joullie and J.K. Puthenpurayil
0707 F.W. Short and L.M. Long
0735 M. Israel and L.L. Jones
0769 A. Gasco et al.
0771 E. Campaigne, D. McClure and J. Ashby
0783 G. Adembri et al.
0797 H. Ogura, T. Itoh and K. Kikuchi
0803 J.A. Van Allan et al.
0809 W.F. Gilmore and R.N. Clark
0819 D.H. Kim and A.A. Santilli
0835 W.A. Remers et al.
0841 J.H. Finley and G.P. Volpp
0859 S. Ghersetti et al.
0875 E. Campaigne and J. Ashby
0885 E. Campaigne, J. Ashby and S.W. Osborne
0909 F.I. Carroll and J.T. Blackwell
0917 B.M. Monroe
0927 F.R. Longo et al.
0937 T.J. McCord et al.
0947 J.B. Wright
0949 L.V. Fisher et al.
0955 E.P. Lira
0965 P.M. Hergenrother
0987 A.G. Anderson, Jr. and D.R. Fagerburg
0995 D. Lipkin et al.

5- -69A, Ann. Chem. Liebigs, 721 (1969)
0001 R. Sommer et al.
0043 R.K. Erünlü
0048 B. Föhlisch
0105 A. Treibs and F.-H. Kreuzer
0116 A. Treibs and F.-H. Kreuzer
0121 E. Zbiral and H. Hengstberger
0154 W. Wiegrebe et al.
0168 W. Fritsch et al.
0186 W. Mayer et al.

5- -69B, Ann. Chem. Liebigs, 722 (1969)
0162 C. Schiele et al.

5- -69C, Ann. Chem. Liebigs, 723 (1969)
0061 E. Müller et al.
0076 E. Müller et al.
0135 H.H. Inhoffen and A. Gossauer
0159 H. Ripperger and K. Schreiber

5- -69D, Ann. Chem. Liebigs, 724 (1969)
0001 H. Goetz and B. Klabuhn
0024 F. Binnig et al.
0030 H. Meyer and H.A. Staab
0066 K.R. Eicken
0091 R. Lemke
0094 H.-G. Franck et al.
0137 K. Jacob et al.
0194 K. Petzold and K. Kieslich
0208 R.K. Erünlü
0217 H. Wamhoff and F. Korte
0226 H. Schildknecht and G. Hatzmann

5- -69E, Ann. Chem. Liebigs, 725 (1969)
0001 E. Schmitz et al.
0052 G. Kaupp and H. Prinzbach
0073 R. Gompper and W. Elser
0099 F. Dallacker and A. Weiner
0130 E. Hecker et al.
0142 H. Bartsch and E. Hecker
0167 H.H. Inhoffen and W. Nolte
0177 H. Wolf et al.
0226 C. Schiele and M. Stepec

5- -69F, Ann. Chem. Liebigs, 726 (1969)
0100 E.C. Taylor and K. Lenard
0136 U. Stache et al.
0152 K. Kieslich and G. Schulz
0161 K. Kieslich et al.
0168 K. Kieslich et al.
0201 W. Pfleiderer and M. Shanshal
0216 D. Hausigk

5- -69G, Ann. Chem. Liebigs, 727 (1969)
0035 W. Walter and P.-M. Pell
0093 K. Dimroth and C. Reichardt
0106 H.-W. Wanzlick et al.
0110 W. Fritsch et al.

5- -69H, Ann. Chem. Liebigs, 728 (1969)
0001 A. Roedig and W. Wenzel
0115 A. Treibs
0158 H.W. THielmann and E. Hecker
0193 C. Rufer et al.

5- -69I, Ann. Chem. Liebigs, 729 (1969)
0064 E.V. Dehmlow
0152 G. Werner and R. Schickfluss
0231 K. Reinhold and P. Renz
0237 H. Hartmann et al.

5- -69J, Ann. Chem. Liebigs, 730 (1969)
0173 H.H. Inhoffen et al.
0191 T. Suhadole and D. Hadzi

6- -69, Ann. Chim.(Paris), 4 (1969)
0203 B. Résibois et al.
0277 N. Ronzani and J. Wiemann

7- -69, Ann. chim.(Rome), 59 (1969)
0163 V. Carelli et al.
0182 V. Carelli et al.
0230 G.C. Casinovi
0315 L. Marchetti and G. Tosi
0335 G. Minardi, P. Schenone and G. Bignardi
0451 E. Bellasio and E. Testa
0539 F. Delle Monache et al.
0552 T. LaNoce et al.
0590 M. Ghelardoni, V. Pestellini and F. Russo
0624 R. Calcinari
0658 G. Pifferi et al.
0671 A. Arcoleo, T. Garofano and M.C. Aversa
0712 L. Marchetti and G. Tosi
0787 P. Finocchiaro
0922 A. Arcoleo et al.

9- -69, Appl. Spectroscopy, 23 (1969)
0249 A. Mustafa et al.
0254 A. Mustafa et al.

10- -69A, Arch. Biochem. Biophys., 129 (1969)
0405 R.B. Homer

10- -69B, Arch. Biochem. Biophys., 130 (1969)
0312 S. Schwimmer

10- -69C, Arch. Biochem. Biophys., 131 (1969)
0655 M. Rieber and G. Bemski

10- -69D, Arch. Biochem. Biophys., 132 (1969)
0001 S. Fukui et al.
0205 L.L. Ingraham and H. Johansen

10- -69E, Arch. Biochem. Biophys., 133 (1969)
0436 J.M. Whiteley, J.H. Drais and F.M. Huennekens

10- 69F, Arch. Biochem. Biophys., 134 (1969)
0214 H.C. Sorensen and L.L. Ingraham
0414 F. Schlenk and C.R. Zydek-Cwick

11- -69A, Arkiv Kemi 30 (1969)
0277 G. Claeson
0445 J. Santesson and C.A. Wachtmeister
0455 J. Santesson
0511 G. Claeson

11- -69B, Arkiv Kemi 31 (1969)
0031 L.B. Agenäs
0159 L.B. Agenäs

12- -69, Australian J. Chem., 22 (1969)
0161 M.T.W. Hearn and A.D. Ward
0185 L.K. Dalton et al.
0221 J.R. Cannon et al.
0239 P.S. Clezy and G.A. Smythe
0271 M.S. Ahmad et al.
0447 F.N. Lahey et al.
0483 W.E. Hillis and N. Ishikura
0491 R.A. Marty and R.M. Carman
0495 R.J. Park and M.D. Sutherland
0577 D.G. Buckley et al.
0721 R.W. Green
0761 A.J. Birch and G.S.R. SubbaRao
0765 W.D. Crow and I. Gosney
0775 S.R. Johns et al.
0793 S.R. Johns et al.
0801 S.R. Johns et al.
0807 D.J. Collins et al.
0935 R.K. Norris and S. Sternhell
0971 A.J. Birch and A. Fitton
0977 D.P. Kelly et al.
0997 T. Teitei and L.K. Dalton
1011 R.S. McCredie et al.
1033 D.W. Connell and M.D. Sutherland
1045 M.N. Galbraith and D.H.S. Horn
1059 M.N. Galbraith and D.H.S. Horn
1283 N.K. Hart, S.R. Johns and J.A. Lamberton
1329 E. Ritchie and W.C. Taylor
1449 A.P. Bindra et al.
1457 R.F.C. Brown and M. Butcher
1477 R.H. Prager and H.M. Thredgold
1493 R.H. Prager and H.M. Thredgold
1503 R.H. Prager and H.M. Thredgold
1525 L.K. Dalton and T. Teitei
1531 J.W. L der et al.
1571 L. Dubicki and R.L. Martin
1681 R.M. Carman et al.
1711 C.R. Bennett et al.
1721 J.A. Diment et al.
1737 G.W.K. Cavill and P.J. Williams
1745 D.B. Paul and H.J. Rodda
1759 D.B. Paul and H.J. Rodda
1803 J. Mohandas et al.
1805 N.R. Farnsworth et al.
1915 D.J. Brecknell et al.
1923 A.J. Birch et al.
1933 A.J. Birch et al.
1951 J.A. Elix
2025 S.D. Sarat
2037 A.J. Birch and G. SubbaRao
2175 E.V. Lassak and J.T. Pinhey
2219 S.R. Johns et al.
2251 E.J. Browne
2255 A.J. Birch and B. McKague
2351 E.W. Della and M. Kendall
2395 R.G. Cooke and R.M. McQuilkin
2403 E.F.L.J. Anet
2489 F. Balkan et al.
2493 A.J. Birch et al.
2497 A.W.K. Chan and W.D. Crow
2581 B.D. Batts and E. Spinner
2595 B.D. Batts and E. Spinner
2611 B.D. Batts and E. Spinner

13- -69A, Steroids, 13 (1969)
0051 J.M. Coxon et al.
0567 J.A. Campbell et al.
0711 P.H. Jellinck and J.S. Elce

0739 L.A. Dehennin and R. Scholler

13- -69B, Steroids, 14 (1969)
0067 R.O. Mummu et al.
0591 J.P. Joseph et al.
0637 C.R. Engel et al.
0729 J.P. Gratz and D. Rosenthal

16- -69, Bol. inst. quim. univ. na. auton. Mex., 21 (1969)
0007 M. Salmon, E. Diaz and F. Watts
0066 J. Romo et al.
0082 J. Romo and C. Lopez V.
0092 J. Romo
0226 C. Aguilar et al.
0241 F. Walls et al.

17- -69, Boll. sci. fac. chim. ind. Bologna, 27 (1969)
0095 G. DiModica and L. Falletti
0145 C. Dell'Erba, G. Guanti and G. Garbarino

18- -69, Bull. Chem. Soc. Japan, 42 (1969)
0010 A. Kuboyama et al.
0168 T. Komorita et al.
0181 O. Tsuge et al.
0191 C. Shin et al.
0199 K. Fukui et al.
0210 R. Tanikaga
0233 K. Fukui et al.
0258 T. Sasaki and T. Yoshioka
0443 Y. Tezuka et al.
0556 T. Sasaki and T. Yoshioka
0560 H. Obara et al.
0569 Y. Omote et al.
0750 H. Kawashima and I. Kumashiro
0766 T. Sato, S. Shimada and K. Hata
0812 S. Kusumoto et al.
0821 K. Terauchi and H. Sakurai
0826 T. Sasaki and T. Yoshioka
0843 T. Irie et al.
0854 M. Kinoshita and S. Umezawa
0960 S. Matusmoto et al.
1052 Y. Tsuzuki et al.
1098 N. Sugiyama et al.
1153 N. Sugiyama et al.
1353 N. Sugiyama et al.
1357 N. Sugiyama et al.
1377 N. Sugiyama et al.
1398 K. Fukui et al.
1404 S. Hayashi et al.
1408 K. Fukui and M. Nakayama
1617 T. Sasaki, S. Eguchi and T. Toru
1649 K. Fukui et al.
1693 K. Fukui et al.
1752 Y. Omote et al.
1776 S. Yamamura et al.
1778 T. Kubota and I. Kubo
1831 H. Shizuka et al.
1934 H. Okawa and T. Yoshino
1942 K. Murayama and T. Yoshioka
1964 S.Oae, K. Sakai and N. Kunieda
1971 K. Fukui et al.
2013 T. Ando et al.
2033 T. Tsuji et al.
2043 K. Sugita and J. Kumonotani
2044 Y. Noda
2090 Y. Omote et al.
2264 Y. Torii et al.
2282 Y. Ogata and Y. Kosugi
2310 A. Chinone et al.
2319 K. Sato et al.
2323 T. Yonezawa et al.
2327 K. Fukui et al.
2375 T. Kato and J. Kumonotani
2380 K. Fukui et al.
2386 M. Oda et al.
2389 C. Kashima et al.
2395 K. Fukui et al.
2405 Y. Naya and M. Kotake
2453 H. Ito et al.
2589 K. Sato and M. Hirayama
2593 J. Katsuhara
2614 M. Hirota and F. Shinozaki
2662 K. Sugiura and M. Goto
2690 N. Sugiyama et al.
2695 K. Sugita and J. Kumonotani
2701 T. Horie
2732 K. Morita and T. Kobayashi
2736 T. Sasaki and S. Eguchi
2885 M. Funabashi et al.
2899 S. Oae et al.
2924 K. Wakabayashi et al.
2931 K. Wakabayashi et al.
2952 T. Sakan et al.
2973 T. Matsuura and I. Saito
2975 I. Ojima et al.
3005 N. Sugiyama et al.
3008 T. Sasaki and T. Yoshioka
3011 K. Yamada et al.
3016 Y. Omote et al.
3198 R.K. Thakkar and S.R. Patel
3233 A. Omori et al.
3273 H. Kono et al.
3277 T. Nozoe et al.
3306 K. Koyano and H. Suzuki
3314 K. Anzai
3318 Y. Inouye and H. Kakisawa
3335 T. Sasaki et al.
3342 M. Suzuki et al.
3345 H. Obara and J. Onodera
3539 M. Kawana and S. Emoto
3556 A. Ohno et al.
3559 R. Okazaki et al.
3565 T. Yamada
3596 C. Kashima et al.
3611 R. Okazaki et al.
3615 T. Sakai and Y. Hirose

19- -69, Bull. Acad. Polon. Sci., 17 (1969)
0013 R.A. Kolinski and B. Korybut-Daszkiewicz
0079 W. Sobotka and W. Gruszecki
0145 S. Mejer and K. Kalinowska
0253 L. Skulski
0269 M. Kocor and W. Kroszczynski
0275 M. Kocor et al.
0469 A. Siewinski
0475 A. Siewinski

20- -69, Bull. soc. chim. Belges, 78 (1969)
0017 H. De Pooter and N. Schamp
0063 J.C. Braekman et al.
0271 C. Hootele and J.C. Braekman
0277 M. Verzele et al.
0523 J.C. Braekman et al.
0553 A.J. Hubert and G. Anthoine
0583 G. Boulvin et al.

21- -69, Bull. soc. chim. biol., 51 (1969)
1511 W. Guschlbauer and M. Privat de Garilhe

22- -69, Bull. soc. chim. France, (1969)
0117 D. Bertin and J. Perronnet
0189 J. Bascoul and A.C. de Paulet
0198 M. Cariou
0205 M. Cariou
0210 M. Cariou
0217 M. Cariou
0231 S, Gelin and R. Gelin
0232 J.-P. Dalle et al.
0239 M. Mousseron-Canet and J.-P. Chabaud
0245 M. Mousseron-Canet and J.-P. Chabaud
0313 R. Danion-Bougot and R. Carrie
0534 S. Deswarte
0545 S. Deswarte
0601 D. Gagniant et al.
0607 P. Cagniant et al.
0613 M. Mousseron-Canet and J.-L. Borgna
0617 M. Mailloux et al.
0628 D. Ricard and J. Cantacuzene
0781 J.P. Béngué and M. Fétizon
0817 G. Dumenil et al.
0903 E. Elkik
0948 J. Streith et al.
0962 H. Christol et al.
0985 P. Cagniant et al.
0991 P. Cagniant et al.
1234 J. Schmitt et al.
1236 A. Abad et al.
1251 J.P. Bégué and M. Fétizon
1256 H. Carpio et al.
1275 J. Metzger et al.
1284 J. Metzger et al.
1299 J.F. Biellmann and H.J. Callot
1344 F. Bourelle-Wargnier and B. Gastambide
1349 G. Descotes and P. Robbe
1367 J. Elguero et al.
1383 S. Gelin and R. Gelin
1632 G. Defaye and M. Fétizon
1651 G. Cauquis and Y. Thibaud
1659 G. Cauquis and Y. Thibaud
1664 G. Rio and J. Berthelot
1667 G. Rio et al.
1673 M. Fétizon et al.
1683 J. Elguero et al.
1724 A. Robert and A. Foucaud
1758 J. Brial and M. Mousseron-Canet
1920 R. Bucourt et al.
1926 J. Schoenleber et al.
1981 E. Marechal et al.
2008 J. Schmitt et al.
2013 P. Moreau et al.
2021 P. Moreau et al.
2045 G. Thuillier et al.
2048 R. Soulier and J. Soulier
2076 P. Ribereau and P. Pastour
2079 R. Martin and J.-M. Betoux
2095 R. Fraisse-Jullien and C. Frejaville
2110 A. Reisse et al.
2115 A. Reisse et al.
2175 J. Streith and J.M. Cassal
2355 J. Itier and A. Casadevall
2365 C. Gourmelon and Y. Graff
2370 C. Gourmelon and Y. Graff
2372 M. Perry and Y. Maroni-Barnaud
2415 M. Julia and M. Maumy
2492 J. Daunis et al.
2508 G.J. Martin et al.
2692 F. Millot and F. Tervier
2756 J. Rigaudy et al.
2765 J. Rigaudy and J. Baranne-Lafont
2820 R. Jacquier et al.
2894 J. Armand and J.P. Guetté
3002 F. Tervier et al.
3100 M. Lounasmaa and J. Zylber
3133 A. LeBerre and C. Renault
3139 A. LeBerre and C. Renault
3166 M. Avaro and J. Levisalles
3173 M. Avaro and J. Levisalles
3180 M. Avaro and J. Levisalles
3185 J. Alais and J. Levisalles
3189 J. Alais et al.
3194 J. Levisalles et al.
3199 L. Mion et al.
3232 Y. Bahurel et al.
3242 M. Mousseron-Canet and J.-L. Olivé
3247 J.-L. Olivé et al.
3252 J.-L. Olivé and M. Mousseron-Canet
3265 F. Frappier et al.
3281 P. Courtot et al.
3306 J.L. Aubagnac et al.
3316 J.L. Aubagnac et al.
3329 R. Guglielmetti and J. Metzger
3475 K. Nakanishi
3523 J. Itier and A. Casadevall
3538 J. Rigaudy et al.
3609 G. Rio and J. Berthelot
3612 R.M. Ruiz et al.
3638 J.P. Montillier and J. Dreux
3662 P. Courtot and R. Rumin
3665 P. Courtot and R. Rumin
3670 J. Daunis et al.
3675 J. Daunis et al.
3694 R. Jacquier et al.
3719 M. de Botton
3883 P. Duhamel et al.
3927 K. Antonakis and F. Leclercq
3953 J. Levisalles and M. Rudler-Chauvin
3981 J.-C. Limasset et al.
4004 M. Robba et al.
4068 H. Christol et al.
4091 S. Gelin and R. Gelin
4108 J.-P. Fleury et al.
4151 G. Descotes et al.
4154 D.K.M. Duc and M. Fétizon

4159 C. Sabate-Alduy and J. Lematre
4321 J.C. Jallageas et al.
4390 L. Oliveros and H. Wahl
4425 C. Mercier and J.-P. Dubose
4447 D. Molho and M. Giraud
4493 H. Christol et al.
4501 H. Christol et al.
4569 J.C. Halle et al.
4590 F. Gracian et al.

23- -69, Can. J. Chem., 47 (1969)
0051 H. Izawa et al.
0075 G.B. Howarth et al.
0081 G.B. Howarth et al.
0105 D.R. Patel, C.S. Choxi and S.R. Patel
0137 E. Piers and R.J. Keziere
0160 O.H. Wheeler and C. Reyes-Zamora
0433 K. Wiesner et al.
0449 W.A. Ayer et al.
0457 H. MacLean and B.F. MacDonald
0495 K.G. Ogilvie and D. Iwacha
0497 W.B. Watkins and R.N. Seelye
0511 G.M. Cree et al.
0515 D. Berncy and P. Deslongchamps
0687 B.D. Challand
0733 J.Y. Savoie and P. Brassard
0767 A. Stoessl
0777 A. Stoessl
0785 G.S. Bajwa and R.K. Brown
0831 E. Piers et al.
1095 J.P. Bell
1117 J.T. Edward and J.K. Liu
1123 J.T. Edward and J.K. Liu
1129 B.M. Lynch et al.
1139 R.W. Doskotch and F. El-Feraly
1169 S.D. Saraf
1223 A.S. Ng, G. Just and F. Blank
1249 L.R. Garson et al.
1473 W.I. Awad and M. Sobhy
1495 Y. Watanabe et al.
1529 T.R. Kasturi and K.M. Damodaran
1561 F. Blank et al.
1943 E. Morita and M.W. Dietrich
1981 H. Veschambre and D. Vochelle
1989 J.W. ApSimon
2029 E. Kiehlmann and P.-W. Loo
2061 P.A. Crooks and B. Robinson
2087 G.M. Strunz et al.
2106 T. Kato et al.
2263 R.I. Zalewski and G.E. Dunn
2355 J.C. Doty et al.
2391 R.C. Bansal et al.
2403 J. Allard and N. Dufort
2457 W.A. Ayer and B. Altenkirk
2465 P. Joseph-Nathan and M.P. Gonzalez
2501 R.B. Kelly and B.A. Beckett
2751 I. Brown et al.
2781 E. Cavalieri and S. Horoupian
2803 S.D. Saraf
2827 M.A. Davis et al.
2849 A. Romo de Vivar and A. Ortega
2853 W.D. Barker et al.
2898 G.D. Thorn
3266 T. Mori et al.
3489 O.E. Edwards and T. Sano
3545 S.K. Khetan and M.V. George
3557 J.W. Lown, G. Dallas and T.W. Maloney
3585 R.H.F. Manske et al.
3589 R.H.F. Manske et al.
3593 D.B. MacLean et al.
3631 W.M.J. Strachan et al.
3647 S. McLean et al.
3661 R.A. Bell and M.B. Gravestock
3693 A. Afonso
3700 G.M. Strunz et al.
3701 M.H. Benn and R.E. Mitchell
3997 P. Yates et al.
4011 W.M.J. Strachan and E. Buncel
4041 K.R. Kopecky and S. Evani
4076 H.L. Holmes and D.J. Currie
4129 E. Buncel et al.
4299 E. Piers et al.
4307 E. Piers et al.
4313 L.R.C. Barclay et al.
4353 A.G. Brook and P.F. Jones
4393 G. Fodor et al.
4483 E. Costakis et al.
4489 F.G. Jiminez et al.
4495 H. MacLean and B.F. MacDonald
4503 R.M. Bowman et al.
4655 A. Corsini and E.J. Billo

24- -69, Chem. Ber., 102 (1 69)
0112 L. Farkas et al.
0126 H. Brockmann et al.
0269 R.R. Schmidt and D. Schwille
0275 R. Crigee, G. Bolz and R. Askani
0310 E. Winterfeldt et al.
0319 T. Eicher and A.-M. Hansen
0342 M. Fischer
0388 K. Schank et al.
0494 F. Cramer et al.
0522 G. Dietz, W. Fiedler and G. Faust
0568 H. Quast and E. Schmitt
0603 E. Hoyer et al.
0623 C. Jutz and R. Heinicke
0643 W. Mehrhof et al.
0678 B. Zeeh
0728 E. Bayer and E. Breitmaier
0785 H. Wagner et al.
0792 L. Hörhammer et al.
0864 F. Bohlmann et al.
0904 H. Seidl, R. Huisgen and R. Knorr
0915 R. Huisgen, H. Seidl and J. Wolff
1037 F. Bohlmann and C. Zdero
1071 W. Meyer zu Reckendorf
1129 W. Steglich and G. Hofle
1198 W. Moldenhauer and H. Simon
1202 W. Schroth et al.
1253 R. Tschesche et al.
1309 W. Flitsch and E. Gerstmann
1347 H. von Dobeneck et al.
1379 B. Eistert et al.
1453 A. Schönberg et al.
1529 L. Szabo and C. Szanay
1580 H. Behringer and J. Falkenberg
1597 G. Quinkert et al.
1656 E. Winterfeldt et al.

1673 F. Bohlmann and M. Grenz
1679 F. Bohlmann and C. Zdero
1682 F. Bohlmann et al.
1691 F. Bohlmann and C. Zdero
1707 T. Severin and I. Schnabel
1739 K. Fickentscher
1774 F. Bohlmann and V.S. Bhaskar Rao
1779 H.-J. Teuber and K.D. Schröder
1789 D. Schönleber
1865 G. Snatzke and H. Langen
1876 B. Zeeh
1928 G. Maier et al.
1961 W. Geiger et al.
2057 R. Tschesche et al.
2083 H. Wagner et al.
2093 S. Hunig et al.
2117 W. Walter and K.J. Reubke
2153 P. Pachaly
2163 T. Severin, H. Lerche and D. Batz
2211 F. Bohlmann and C. Zdero
2216 M. Regitz and W. Anschutz
2301 C. Jutz et al.
2336 E. Winterfeldt and W. Krohn
2346 E. Winterfeldt et al.
2378 K. Fickentscher
2384 H. Dannenberg and E. Meyer
2414 F. Dallacker and Z. Damo
2471 M. Verbeek et al.
2493 D. Meuche and S. Huneck
2502 D. Meuche and S. Huneck
2565 H. Hofmeister, G. Schulz and R. Wiechert
2570 L. Laurent et al.
2609 G. Fischer
2629 W. Sucrow and B. Radüchel
2685 W. Ried and P. Weidemann
2691 G. Rücker
2697 G. Rücker
2728 M. Zander and W.H. Franke
2739 H. Wamhoff
2835 H. Behringer and P. Heckmaier
2864 H. Ripperger and K. Schreiber
2877 R. Brossmer and D. Ziegler
2987 R.W. Hoffmann and K.R. Eicken
3000 R. Walentowski and H.W. Wanzlick
3006 H. Wagner et al.
3009 H. Inouye et al.
3033 G. Kortum and P. Krieg
3082 F.A. Neugebauer and M. Jenne
3122 N. Matzat, H. Wamhoff and F. Korte
3127 A. Roedig and W. Wenzel
3139 G. Fischer and K. Lohs
3159 T. Eicher et al.
3176 T. Eicher et al.
3205 H. Brockmann and E. Schulze
3241 K.D. Gundermann and D. Schedlitzki
3248 L. Töke et al.
3268 W. Flitsch et al.
3293 F. Bohlmann and P. Hänel
3298 F. Bohlmann and H. Mönch
3304 R. Askani
3324 H. Laurent and G. Schulz
3405 R. Huisgen et al.
3428 L.A. Feiler and R. Huisgen
3486 M. Fischer and F. Wagner
3546 W.D. Balzer
3558 E. Winterfeldt et al.
3599 M. Zander
3647 S. Brodka and H. Simon
3666 G. Simchen and W. Kramer
3691 L. Capuano and W. Ebner
3735 G. Bonola and E. Sianesi
3747 S. Farid and D. Hess
3765 F. Bohlmann and C. Zdero
3775 H. Boshagen and W. Geiger
3795 K. Dimroth and H. Thomas
3818 T. Sasaki and M. Murata
3877 M. Regitz and J. Ruter
3950 F. Eiden and M. Peglow
3959 L. Novak and C. Szanty
3963 C. Szanty and M. Barczai-Beke
3996 M. Avram et al.
4017 F. Bohlmann and G. Haffer
4032 G. Barlin and W. Pfleiderer
4080 H. Ripperger and K. Schreiber
4104 W. Steglich et al.
4147 G. Dietz et al.
4164 W. Eberbach and H. Prinzbach
4199 W. Meyer zu Reckendorf and J.C. Jochims
4209 W. Bohlmann and J. Schuber

25- -69, <u>Chem. and Ind.(London)</u>, (1969)
0047 K. Bhatti et al.
0075 M. Oda et al.
0107 R. Bonnett and A.F. McDonagh
0135 J.S. Walia and P.S. Walia
0171 A.W. Archer and P.A. Claret
0269 T.C. Shields et al.
0328 A. Chatterjee and S.C. Basa
0381 A. Modro et al.
0458 D.H. Kim and A.A. Santilli
0491 M. Higashi et al.
0549 B.K. Chowdhury and D.P. Chakraborty
0619 T.C. Shields and W.E. Billups
0654 M.J. Billig and E.W. Baker
0695 D.E. Fenton
0877 D.V. Hertzler et al.
0953 K.S. Balachandran and I. Bhatnagar
0983 M.F. Dutton and J.G. Heathcote
1018 F.L.C. Baranyovits and R. Ghosh
1056 S.K. Talapatra et al.
1093 L. Caglioti and G. Rosini
1271 A.W. Archer and P.A. Claret
1387 B. Mukherjee et al.
1388 A. Chatterjee et al.
1662 D.P. Chakraborty et al.
1738 A.B. Wood et al.
1779 N. Hayashi et al.

27- -69, <u>Chimia</u>, <u>23</u> (1969)
0036 W. Sucrow
0037 H. Prinzbach and H.D. Martin
0155 E. Fischer et al.
0400 A. Goldblatt et al.
0405 L. Hoesch and A.S. Dreiding
0411 A. Yogev

28- -69A, Compt. rend., 268 (1969) Series C
0075 D. Plouin and R. Glénat
0078 P. Grammaticakis
0082 R. Toubiana
0186 A.M. Lamazouère et al.
0194 G. Jacob and P. Cagniant
0279 C. Laurence and R. Chiron
0340 N. Ardoin et al.
0536 M. Lamant and G. LeGuillanton
0557 P. Dupuis and J. Néel
0622 J.J. Basselier and J.P. LeRoux
0730 P. Grammaticakis
0864 G. LeGuillanton and M. Lamant
0870 J. Roggero and C. Divorne
0980 J. Chopin et al.
0986 L. David and A. Kergomard
1157 G. Rio et al.
1160 J.P. Coat and S. David
1170 E. Touboul et al.
1442 L. Olivier, J. Levy and J. LeMen
1535 S. Bancel and P. Cresson
1549 V. Bertin et al.
1703 P. Grammaticakis
1808 S. Bancel and P. Cresson
1844 J. Tohier and M.B. Fleury
1878 G. Doucet-Baudry
2028 C. Daremon and R. Rambaud
2034 M.F. Chasle and A. Foucaud
2040 M. Bruni, M.M. Geistel and A. Pousse
2217 D.Q. Quan and L. Cobian

28- -69B, Compt. rend., 269 (1969) Series C
0137 P. Grammaticakis
0248 P. Grammaticakis
0346 R. Maurin, E. Senft and M. Bertrand
0615 C. Divorne and J. Roggero
0639 N. Langlois, B.C. Dao and P. Potier
0642 F. Venien
0781 A. Resplandy et al.
0865 J.P. Chabaud et al.
1063 A. Aspect et al.
1315 Y. Bessière-Chrétien and B. Meklate
1343 D. Bouin and A. Friedmann
1550 P. Bichaut et al.
1562 Dang Quoc Quan
1654 M. Julia, R. Ketoh and R. Labia

30- -69A, Doklady Akad. Nauk S.S.S.R., 184 (1969)
0108 N.I. Shergina et al.
0355 F.I. Luknitskii et al.
0605 V.I. Minkin et al.
0867 T.N. Pliev et al.
1113 T.N. Pliev

30- -69C, Doklady Akad. Nauk S.S.S.R., 186 (1969)
0620 M.F. Shostakovskii
1079 K.K. Babievskii et al.

30- -69F, Doklady Akad. Nauk S.S.S.R., 189 (1969)
0338 V.M. Potapov and G.V. Kiryushkina
0559 V.M. Potapov et al.
1255 A.I. Gurevich and M.N. Kolosov
1262 I.A. Izrailov et al.

31- -69, Experientia, 25 (1969)
0122 K. Fukui et al.
0349 K. Fukui et al.
0354 D.S. Bhakhani and M.M. Dhar
0354B K. Fukui et al.
0355 K. Fukui et al.
0571 E. Galantay and H.P. Weber
0575 E.C. Miranda and B. Gilbert
0789 K. Fukui et al.
0790 S.P. Kureel et al.
0907 Y. Asekawa et al.
1018 G. Baldratti et al.
1129 Y. Shimizu et al.
1139 S. Hayashi et al.
1237 I. Chibata et al.

32- -69, Gazz. chim. ital., 99 (1969)
0029 E. Fattorusso et al.
0165 S. Morrocchi et al.
0176 L. Mangoni and V. Dovinola
0206 V. Dovinola and L. Mangoni
0219 F. Fringuelli et al.
0231 F. Fringuelli et al.
0247 A. Taticchi et al.
0260 L. Canonica et al.
0276 L. Canonica et al.
0308 G. Cardillo et al.
0323 R.A. Nicolaus et al.
0431 L. Minale et al.
0463 M. Guarneri and P. Giori
0535 C. Dell'Erba et al.
0565 S. Morrocchi et al.
0588 M. Ruccia et al.
0612 G. Cardillo, R. Cricchio and L. Merlini
0677 S. Carboni et al.
0711 F. Piozzi et al.
0753 G. Renzi et al.
0848 M. Ferrari et al.
0915 S. Corsano et al.
0969 E. Fattorusso et al.
1000 G. Rigatti et al.
1059 L. Canonica et al.
1115 G. Cignarella and G.G. Gallo
1150 M.L. Scarpati and M. Guiso
1177 F. Bordin et al.
1193 G. Prota et al.
1236 G. Berti, A. DaSettimo and E. Nannipieri
1243 A.M. Bellini et al.
1252 R. Rocchi et al.
1273 M. Ghelardoni, V. Pestellini and C. Musante

33- -69, Helv. Chim. Acta, 52 (1969)
0033 A.A. Gorman et al.
0076 A. Guggisberg et al.
0089 Y. Morita, M. Hesse and H. Schmid
0121 P. Wieland and G. Anner
0127 W. Keller-Schierlein et al.
0173 S. Kuwata amd K. Schaffner

0322 H.R. Schweizer
0388 B. Maurer and W. Keller-Schierlein
0396 H. Prinzbach and P. Vogel
0453 P. Wieland and G. Anner
0459 M.L. Mihailovic et al.
0478 O. Sticher and H. Schmid
0482 A. Manzetti and M. Ehrenstein
0584 P. Vogel et al.
0629 A. Ebnöther et al.
0678 A. Brossi et al.
0689 K.W. Gemmell et al.
0701 J.J. Dugan et al.
0720 M. Fräter-Schröder, R. Good and C.H. Eugster
0725 G. Ganter and J.-F. Moser
0789 K. Seibold et al.
0806 C.H. Eugster et al.
0808 T. Matsumoto et al.
0956 G. Kaupp and H. Prinzbach
0967 C. Ganter and J.-F. Moser
0971 D. Bellus et al.
1010 D. Bellus and K. Schaffner
1023 L. Chardonnens and T. Stauner
1030 D. Felix et al.
1058 B. Riniker et al.
1091 L. Chardonnens and L. Avar
1102 T. Petrzilka et al.
1157 H. Els et al.
1165 H. Bernotat-Wulf et al.
1175 K. Stockel et al.
1202 G. Haas and V. Prelog
1228 A. Brossi and S. Teitel
1237 K. Schaffner-Sabba
1249 A.F. Thomas et al.
1259 B.K. Manukian et al.
1282 A.E. Siegrist and H.R. Meyer
1354 W. Regel and W. von Philipsborn
1376 A. Marxer, F. Hofer and U. Salzmann
1461 K. Stockel et al.
1495 K. Heusler and H. Loeliger
1549 P.A. Stadler et al.
1564 J.R. Hymon et al.
1624 J. Reiner and W. Jenny
1685 M. Ribi et al.
1810 I. Felner and K. Schenker
1886 K. Bernauer et al.
1911 P. Pfaffli and C. Tamm
1929 A. Vogel et al.
1996 M. Stoll and I. Flament
2004 H.J. Veith and M. Hesse
2156 W. Merkel and M. Ehrenstein
2197 R. Blaser et al.
2221 H.R. Schweizer
2236 H.R. Schweizer
2276 F. Brugschweiler et al.
2351 A. Marxer et al.
2417 H. Dahn and M. Ballenegger
2428 H. Scherrer
2436 S. Domb et al.
2465 U. Joss and H. Schaltegger
2472 H. Prinzbach and E. Woischnik
2521 A.E. Siegrist et al.
2555 E. Hardegger et al.
2583 R. Elber, E. Weiss and T. Reichstein
2641 H.-J. Gais et al.

34- -69, J. Chem. Eng. Data, 14 (1969)
0112 D.C. Berndt
0116 M.J. Kamlet
0118 A.J. Fatiadi
0119 W.F. Gilmore and R.N. Clark
0125 T.L. Jacobs and N. Juster
0278 D.C. Bhura and S.G. Tandon
0506 D.M. Wiles and T. Suprunchuk

35- -69, J. Am. Chem. Soc., 91 (1969)
0150 W.C. Wildman and D.T. Bailey
0157 W.J. McGahren et al.
0215 O.L. Chapman and R.A. Fugiel
0323 K.M. Harmon et al.
0355 H. Bock, H. Alt and H. Seidl
0366 R.C. Kerber and A. Porter
0371 B. Bossenbroek et al.
0388 J.E. Mulvaney et al.
0434 H.E. Zimmerman et al.
0456 A. Padwa et al.
0462 A. Padwa and D. Eastman
0477 D.P. Weeks and G.W. Zuorick
0509 E.A. Hill and R. Wiesner
0517 T.T. Howarth, G.P. Murphy and T.M. Harris
0523 E.H. White et al.
0645 D.J. Sandman and K. Mislow
0676 F.E. Henoch et al.
0711 J.L. Brewbaker and H. Hart
0731 C.D. Snyder and H. Rapoport
0737 R.K. Chaturvedi and G.L. Schmir
0758 R.J. Wilson et al.
0760 G.W. Brown and F. Sondheimer
0779 W.F. Erman
0879 H.E. Zimmerman and N. Lewin
0902 V. Boekelheide and E. Sturn
0918 B.M. Trost
0924 A.G. Anderson, Jr., and D.M. Forkey
1028 N. Furutachi et al.
1036 K. Wiesner and T. Inaba
1085 B.E. Ludwig and G.R. McMillan
1176 J.C. Sheehan and M. Mehdi Nafissi-V
1179 W.H. Pirkle and L.H. McKendry
1206 M. Polk, M. Siskin and C.C. Price
1228 F. Sondheimer et al.
1232 M.T. Cox et al.
1239 S.W. Staley and T.J. Henry
1248 J.A. Edwards et al.
1256 C.G. Overberger et al.
1258 F.L. Hedberg and H. Rosenberg
1263 H.H. Wasserman et al.
1264 H.H. Wasserman et al.
1401 R.B. Morin et al.
1522 A.F. Cook et al.
1580 N.H. Pirkle et al.
1582 S. Irie, F. Egami and Y. Inouye
1665 B.A. Hess, Jr., et al.
1672 B.A. Hess, Jr. and V. Boekelheide
1710 D.J. Severn and E.M. Kosower
1718 H.E. Zimmerman and P.S. Mariano
1822 J. Steigman et al.
1848 K.B. Sharpless and E.E. van Tamelen
1862 R. Filler and E.W. Choe
2062 J. Libman et al.
2097 D.C. Dittmer et al.

2102 R.L. Tolman, R.K. Robins and L.B. Townsend
2134 G.A. Ellestad et al.
2141 C.S. Irving et al.
2154 G.J.D. Peddle and R.W. Walsingham
2183 H. Hart and R.K. Murray, Jr.
2279 Z. Yoshida et al.
2293 F. Ramirez and G.V. Loewengart
2299 E. Wenkert, F. Haviv and A. Zeitlin
2325 E.M. Kosower et al.
2338 B. Frydman et al.
2342 F. Ziegler et al.
2358 R.E. Barnett and W.P. Jencks
2371 G. Stork and M. Marx
2374 H.H. Wasserman and P.M. Keehn
2375 H.H. Wasserman et al.
2403 E.B. Fleischer and T.S. Srivastava
2443 W.M. Tolles et al.
2456 R. Waack and M.A. Doran
2467 J.D. Margerum and C.T. Petrusis
2653 A. Padwa et al.
2792 G.M. Loudon et al.
2800 M.A. Schwartz et al.
2806 S. Danishefsly and B.H. Migdalof
2808 W.J. McGahren and M.P. Kunstmann
2818 W.R. Dolbier and W.M. Williams
3020 J.S. Haywood-Farmer and R.E. Pincock
3075 Y.F. Shealy and J.D. Clayton
3226 C.G. Overberger et al.
3281 A.F. Thomas
3289 P. Rona and P. Crabbe
3299 R.S. Bly and S.U. Koock
3308 L.D. Quin, J.G. Bryson and C.G. Moreland
3316 H.E. Zimmerman et al.
3360 R.L. Letsinger et al.
3373 D.M. Lemal, J.V. Staros and V. Austel
3375 W. Jetz and W.A.G. Graham
3383 J.T. D'Agostino and H.H. Jaffe
3391 D.A. Shuman, R.K. Robins and M.J. Robins
3517 H.J. Reich and D.J. Cram
3553 M. Sheehan and D.J. Cram
3558 R.C. Hahn et al.
3616 L.A. Paquette et al.
3634 B.C. Pal et al.
3664 L.L. Barber, O.L. Chapman and Jean D. Lassila
3670 B. Singh
3676 J.E. McMurry
3689 B.M. Trost and G.M. Bright
3724 A. Calder et al.
3931 T. Tokuyama, J. Daly and B. Witkop
3952 M.P. Cava and G.E.M. Husbands
3953 J.M. Hoffman, Jr., and R.H. Schlessinger
3970 L.A. Paquette et al.
3973 L.A. Paquette and J.C. Philips
3974 M.J. Strauss and H. Schran
3982 G.B. Porter
3998 S.W. Staley and D.W. Reichard
4000 A. Padwa et al.
4009 M. Shamma and C.D. Jones
4155 R.R. Minesinger and M.J. Kamlet
4317 J.D. Albright and L. Goldman
4490 W.H. Glaze and T.L. Brewer
4494 J.W. Hanifin and E. Cohen
4532 T. Hata et al.
4598 T. Hino and M. Nakagawa
4612 R.B. Woodward and D.L. Dalrymple
4714 L.A. Paquette et al.
4734 J.A. Elix and M.V. Sargent
4739 P. Yates et al.
4749 P. Yates and E.S. Hand
4766 J. Ciabattoni and E.C. Nathan, III
4824 T.L. Jacobs and R.S. Macomber
4912 E.T. Kaiser and K.W. Lo
4934 M. Bodanszky et al.
5136 E.J. Gabbay
5181 E.C. Taylor et al.
5190 R.K. Razdan and V.V. Kane
5192 J.E. Shields and J. Bornstein
5246 Z. Rappoport and A. Gal
5264 Y. Kondo and B. Witkop
5270 S.J. Benkovic et al.
5296 L.A. Paquette et al.
5307 H.E. Zimmerman and G.E. Samuelson
5392 K. Klier
5409 J. Nagyvary
5415 T.C. Shields and A.N. Kurtz
5510 J.L. Kice and G.J. Kasperek
5694 M.B. D'Amore and R.G. Bergman
5792 E.F. Ullman et al.
5862 D.J. Goldsmith and C.F. Phillips
5870 W.S. Trahanovsky and D.K. Wells
5871 D.K. Wells and W.S. Trahanovsky
5872 A.I. Scott and P.L. Cherry
5874 A.I. Scott and A.I. Qureshi
5937 L. Edwards and J.W. Raymonda
6038 T.L. Jacobs et al.
6049 W.A. Henderson, Jr., et al.
6083 H.W. Smith and H. Rapoport
6102 D.Y. Curtin and S.R. Byrn
6107 L.A. Paquette and G.R. Krow
6112 K.W. Ratts and R.D. Partos
6122 J.H. Boyer and R. Selvarajan
6199 T.F. Spande, A. Fontana and B. Witkop
6362 P.D. Bartlett et al.
6391 W.M. Jones and C.L. Ennis
6404 P.H. Campbell et al.
6432 M.J. Jorgenson
6444 J.E. Baldwin and S.M. Krueger
6452 S.F. Nelson and D.H. Heath
6470 G. Büchi and J.A. Carlson
6473 G. Büchi et al.
6613 C.G. Pitt
6683 Y. Shvo and H. Shanan-Atidi
6689 Y. Shvo and H. Shanan-Atidi
6703 D. Landini et al.
6707 R.F. Heck
6734 Z. Rappoport and Y. Apeloig
6742 P. Bemporad et al.
6766 E.L. Allred and R.L. Smith
6785 E.W. Garbisch and R.F. Sprecher
6868 P. Yates et al.
6889 W.E. Bondinell, C.D. Snyder and H. Rapoport
6891 J.D. Bower and R.H. Schlessinger

7166 Y. Fujiwara
7187 F.E. Brot and M.L. Bender
7315 E.E. van Tamelen et al.
7333 E.E. van Tamelen et al.
7342 E.E. van Tamelen and J.B. Hester
7349 E.E. van Tamelen and I.G. Wright
7359 E.E. van Tamelen et al.
7485 H.W. Whitlock, Jr., et al.
7490 K. Isono et al.
7518 J. Griffiths and F. Sondheimer
7520 G.H. Mitchell and F. Sondheimer
7534 P.J. Whitman and B.M. Trost
7752 T. Kunieda and B. Witkop
7763 M.G. Waite et al.
7769 S. Masamune, S. Takada and R.T. Seidner
7780 G. Stork and P.L. Stetter

36- -69, J. Pharm. Sci., 58 (1969)
0047 T. George et al.
0176 E.R. Trumbull and J.R. Cole
0335 G.A. Digenis
0340 A.R. Martin et al.
0432 V.P. Arya et al.
0460 M. Grifantini et al.
0490 C.F. Barfknecht and T.R. Westby
0553 M.R. Boots and S.G. Boots
0675 K.H. Lee and T.O. Soine
0830 M. Steinman and J.G. Topliss
0857 P. Franchetti et al.
0860 N. Castagnoli, Jr., et al.
0877 R.W. Doskotch and F.S. El-Feraly
0894 W.L. Nelson and K.F. Nelson
1038 H.C. Wormser

38- -69B, J. Chem. Phys., 51 (1969)
0033 J.R. Lombardi et al.
0045 M.B. Robin et al.
0052 H. Basch et al.
4881 E. Hayon

39- -69A, J. Chem. Soc., Sect. A (1969)
0133 E.W. Abel et al.
0173 J.R. Horder and M.F. Lappert
0392 L. Santucci and C. Triboulet
0526 N.M.D. Brown and P. Bladon
0819 C.F. Bell and D.R. Rose
0893 J.R. Majer et al.
0987 M.I. Bruce et al.
1097 D.W. Allen et al.
1152 D.C. Bradley and M.H. Gitlitz
1299 B.L. Booth et al.
1305 G. DiLonardo and C. Zauli
1587 J. Mason

39- -69B, J. Chem. Soc., Sect. B (1969)
0054 P.F. Holt and H. Lindsay
0061 R.R. Hill and G.H. Mitchell
0096 E. Kalatzis
0111 A.P. Downing et al.
0192 A.C. Lovesey and W.C.J. Ross
0201 J.R. Majer and D. Phillips
0207 J.F. Corbett
0270 D.J. Brown and P.B. Ghosh
0276 B.J. Gregory and C.K. Ingold
0299 J.V. Greenhill
0333 G.B. Barlin and W.V. Brown
0446 M.H. Palmer and P.S. McIntyre
0449 A.J. Bellamy
0481 R. Baker and M.J. Spillett
0539 M.H. Palmer and P.S. McIntyre
0544 A. Fischer et al.
0554 E. Baciocchi and A. Schiroli
0581 R. Baker and M.J. Spillett
0616 N. Filipescu and J.M. Menter
0710 I.R. Bellobono et al.
0725 S. Chatterjee
0742 M. Bossa et al.
0802 A.G. Briggs et al.
0823 J.F. Corbett
0827 J.F. Corbett
0873 A. Davies and K.D. Warren
0922 J.W. Eastes et al.
0932 P. Beltrame et al.
1068 C.C. Barker and G. Hallas
1100 P. Beltrame et al.
1135 A. Fischer and I.J. Miller
1143 B. Halpern et al.
1161 P.H. Emslie et al.
1178 A.A. Abdallah et al.
1182 M. Bossa
1185 A.R. McCarthy et al.

39- -69C, J. Chem. Soc., Sect. C (1969)
0004 T. Kametani et al.
0016 D.H.R. Barton et al.
0029 A.K. Das Gupta et al.
0033 A.K. Das Gupta and K.R. Das
0034 F.S.H. Head
0037 F.S.H. Head and G. Lund
0056 J.A. Ballantine et al.
0068 S.H. Graham et al.
0070 R.K. Grantham and D. Meth-Cohn
0085 N. Whittaker
0094 N. Whittaker
0101 H.T. Openshaw et al.
0105 J.A. Barltrop and D. Giles
0114 Z. Neiman et al.
0120 R.J.W. Cremlyn and R. Hornby
0124 D.D. Chapman, W.J. Musliner and J.W. Gates
0133 S. David and H. Hirshfeld
0152 A. Albert
0160 N. Heap et al.
0176 R. Grigg et al.
0199 M. Jarman and W.C.J. Ross
0203 G.T. Rogers et al.
0211 L.P. Anderson et al.
0217 P.J. Hattersley, I.M. Lockhart and M. Wright
0227 E.W. Duck et al.
0239 R.P. Cunningham et al.
0245 T. Nishiwaki
0252 D.W. Allen and I.T. Millar
0258 F.R. Hewgill, W.L. Spencer and S.H. Tay
0268 F.H. Jackson and A.T. Peters
0281 J.E. Atkinson and J.R. Lewis
0298 T. Kametani et al.
0315 I. El-Sayed El-Kholy and F.K. Rafla

0332 D.H.R. Barton et al.
0345 Y. Yost and H.R. Gutman
0365 A.J. East, W.D. Ollis and R.E. Wheeler
0374 C.P. Falshaw et al.
0397 A.H. Lamberton and H.M. Yusuf
0401 B. Green and L.R. Garson
0417 B.V. Milborrow and C. Djerassi
0429 R. Bonnett et al.
0460 W. Nagata and Y. Hayase
0474 G. Eglinton et al.
0486 I. Carpenter et al.
0489 G. Camaggi and F. Gozzo
0502 T. Kametani and I. Noguchi
0517 D.B. Morell and A.W. Nichol
0520 T. Kametani et al.
0526 F.M. Dean and B. Parton
0536 N.E. Alexandrou
0554 A.C. Campbell, W. Lawrie and J. McLean
0559 P. Crooij and J. Eliaers
0564 R. Bonnett et al.
0592 J.R. Hargreaves et al.
0595 E. Houghton and J.E. Saxton
0600 C.N. O'Callaghan and D. Twomey
0603 J.A. Baker and P.V. Chatfield
0608 J.K. Groves and N. Jones
0627 H.J. Banks, D.W. Cameron and J.C.A. Craik
0631 D.W. Cameron et al.
0646 J.F.W. McOmie and M.L. Watts
0652 A. Lawson and R.B. Tinkler
0655 R. Grigg et al.
0674 K.D. Bingham et al.
0683 R.K. Bentley et al.
0685 R.K. Bentley et al.
0693 R.S. Burden, L. Crombie and D.A. Whiting
0704 D.J. Aberhart et al.
0721 A.R. Battersby et al.
0728 E. Beska, P. Rapos and P. Winternitz
0742 C.D. Campbell and C.W. Rees
0748 C.D. Campbell and C.W. Rees
0752 C.D. Campbell and C.W. Rees
0756 C.W. Rees and R.C. Storr
0760 C.W. Rees and R.C. Storr
0765 C.W. Rees and R.C. Storr
0769 R.W. Hoffman et al.
0772 R.S. Atkinson and C.W. Rees
0778 R.S. Atkinson and C.W. Rees
0787 S.D. Andrews, A.C. Day and A.N. McDonald
0791 M.W. Winkley and R.K. Robins
0796 D.E. Ames et al.
0801 T. Kametani et al.
0830 R.K. Bentley et al.
0836 F.G. Baddar et al.
0838 F.G. Baddar et al.
0861 L. Stephenson et al.
0874 T. Kametani et al.
0879 W.A. Kennedy and T.B.H. McMurry
0882 N.J. Harper and J.M. Sprake
0887 A. Pelter and I. Amenechi
0915 H.N. Al-Jallo et al.
0918 M. Jarman
0928 K.J.M. Andrews, W.E. Barber and B.P. Tong
0938 D.M. Cahill and P.V.R. Shannon
0944 A.J. Hubert and H. Reimlinger
0978 R.P. Houghton and E.S. waight
1003 E. Houghton and J.E. Saxton
1016 L. Crombie et al.
1024 L. Crombie et al.
1028 E.W. Collington and G. Jones
1047 D.H.R. Barton et al.
1050 D.H.R. Barton et al.
1063 T. Kametani et al.
1065 A.C. Day and R.N. Inwood
1073 J.E. Baldwin et al.
1079 J. Scotney and E.V. Truter
1081 A.M. Gorringe et al.
1086 T. Sasaki et al.
1088 L. Bartlett et al.
1096 R.K. Bentley, E.R.H. Jones and V. Thaller
1100 N.E. Waite et al.
1109 S.H. Harper et al.
1120 G.C. Barrett and A.R. Khokhar
1128 J.R. Bull
1143 R.M. Acheson and J.N. Bridsen
1149 H.J. Sare and E.F.M. Stephenson
1155 A.R. Mattocks
1166 D.N. Jones et al.
1184 M. Brown and R.H. Thomson
1187 A.R. Battersby et al.
1193 A.R. Battersby et al.
1215 R.T. Channon et al.
1219 R.J.J. Ch. Lousberg et al.
1234 R.C. Cambie et al.
1240 R.C. Cambie et al.
1245 D.W. Cameron et al.
1279 J.M. Tedder, K.H. Todd and W.K. Gibson
1282 A. Kreutzberger and M.F.G. Stevens
1309 J.M. Paton et al.
1325 I. Baxter et al.
1334 A.J. Hubert
1338 J.W. Lown
1346 C.W. Shoppee and R.E. Lack
1353 J.M. Blatchly et al.
1408 J. Clark et al.
1416 H. El Khadem et al.
1419 M.S. Manhas and J.R. McCoy
1427 R. Munday and I.D. Sutherland
1449 A.M. Gorringe et al.
1459 A. Calder and A.R. Forrester
1474 C.W. Rees and R.C. Storr
1491 K. Takeda et al.
1499 J.D. Hobson and J.R. Malpass
1518 B.S. Joshi, V.N. Kamat and D.F. Rane
1537 C.R. Ganellin and H.F. Ridley
1540 E. Byrne and P.V.R. Shannon
1542 E.M. Richards et al.
1564 W.D. Jamieson and R. Rahman
1567 H.D. Locksley and I.G. Murray
1571 W.A.F. Gladstone
1593 J. Maguire, D. Paton and F.L. Rose
1610 J.G.D. Carpenter and J.W. Moore

1619 T.B.H. McMurry and R.C. Mollan
1625 S. Seto et al.
1627 I.R.C. Bick et al.
1632 T.L. Eggerichs et al.
1635 W.L.F. Armarego and T. Kobayashi
1643 J.E. Dickeson and L.A. Summers
1647 G.C. Forward and D.A. Whiting
1660 R.E. Banks et al.
1665 R. Rahman, S. Safe and A. Taylor
1668 D.N. Jones, D. Mundy and R.D. Whitehouse
1678 D.W. Jones
1681 K.L. Stuart, C. Chambers and D. Byfield
1684 R. Harrison et al.
1703 D.C. Iffland et al.
1710 W.H. Hui and M.L. Fung
1721 M. Afzal, J.S. Davies and C.H. Hassall
1729 D.W. Jones
1751 J. Clark and W. Pendergast
1755 G.A. Taylor
1758 G.W.J. Fleet and I. Fleming
1787 R.V.M. Campbell et al.
1795 D.E. Ames, H.R. Ansari and A.W. Ellis
1799 F.J. Bullock et al.
1803 M.S. Baird and C.B. Reese
1808 M.S. Baird and C.B. Reese
1813 R.F. Curtis and J.A. Taylor
1828 A. Chaudhuri and P.K. Sarina
1845 M. Elliott, N.F. Janes and K.A. Jeffs
1860 R. Nery
1866 R.E. Banks et al.
1868 G.C. Forward and D.A. Whiting
1873 J.S. Davies et al.
1883 J. Clark and P.N.T. Murdock
1901 A.J. Boulton et al.
1921 N.A.R. Hatam and D.A. Whiting
1935 V. Askam and R.H.L. Deeks
1950 I. El-Sayed El-Kholy et al.
1982 I.G.C. Coutts et al.
1987 E.C. Campbell et al.
2030 T. Kametani et al.
2034 T. Kametani et al.
2044 R.H. McDougall and S.H. Malik
2053 A. Rashid and G. Read
2059 G. Read, A. Rashid and L.C. Vining
2098 I.F. Eckhard et al.
2123 J.P. Clayton
2127 D.J. Neadle and R.J. Pollitt
2132 C.R. Ganellin and C.J.S. Stolz
2134 K.B. Prasad et al.
2153 D.E.U. Ekong and J.I. Okegun
2159 D.N. Jones and W. Higgins
2164 D.F. Bowman and F.R. Hewgill
2173 R.E. Atkinson et al.
2186 M. Atkinson and A.M. Horsington
2187 D. Giles and W.B. Turner
2198 M. Greenhalgh et al.
2201 B. Jackson et al.
2225 K.W. Bentley et al.
2246 A.M. Jones and C.A. Russell
2249 R.L. Jones and C.W. Rees
2251 R.L. Jones and C.W. Rees
2255 R.L. Jones and C.W. Rees
2260 E.S. Bolton et al.
2266 D. Cohen, L. Hewitt and I.T. Millar
2270 J. Adamson et al.
2311 R.M. Acheson and M.S. Verlander
2316 R.M. Acheson and J.K. Stubbs
2319 N.E. Alexandrou and D.N. Nicolaides
2327 K.H. Pegel and W.G. Wright
2357 H.S. Blair and G.A.F. Roberts
2379 A. Albert
2398 P.C. Beaumont and R.L. Edwards
2403 N.H. Anderson et al.
2415 Z. Neiman, F. Bergmann and A.K. Meyer
2418 A.E. Bird and A.C. Marshall
2421 I. Carpenter et al.
2424 I. Granoth, A. Kalir and Z. Pelah
2439 D.A.H. Taylor
2443 S.D. Andrews, A.C. Day and R.N. Inwood
2464 D. Lloyd and N.W. Preston
2470 J.E. Battlerbee et al.
2477 R.S. Burden and L. Crombie
2484 M.M. Coombs
2489 D.A. Crombie and S. Shaw
2504 D.D. Evans and J. Hussey
2509 I. Fleming and J.B. Mason
2514 F.G. Holliman
2517 R.B. Herbert and F.G. Holliman
2527 L. Bartlett et al.
2559 L.P. Anderson et al.
2572 A. Pelter et al.
2579 J.F.W. McOmie and D.E. West
2587 W.A.F. Gladstone et al.
2602 T. Kametani et al.
2605 M.J. Gallagher and I.D. Jenkins
2618 A.K. Das Gupta et al.
2620 D.J. Brown and P.W. Ford
2631 V. Skaric and B. Gaspert
2634 W. Parker, R.A. Raphael and J.S. Roberts
2656 E.W. Collington and G. Jones
2703 G. Berti et al.
2710 G.I. Fray and D.P.S. Smith
2720 D.J. Brown and P.W. Ford
2723 T. Dahl et al.
2728 A.K. Bose et al.
2738 A. Jackson et al.
2747 P.J.N. Brown et al.
2750 D.T. Drewry and R.M. Scrowston
2763 M.V. Sargent et al.
2767 C.W. Shoppee and B.C. Newman
2783 V. Skaric et al.
2786 K. Takeda et al.
2789 J.M. Blatchly et al.
2794 B.W. Nash et al.
2799 D.A. Cornforth et al.
2805 C.H. Hassall and G. Wooton

40- -69, <u>Nippon Kagaku Zasshi</u>, <u>90</u> (1969)
0073 K. Isagawa et al.
0085 H. Horino
0096 T. Simura et al.
0207 T. Yamada et al.

0207 T. Yamada et al.
0507 T. Sakan et al.
0713 Y. Kitamura et al.
0716 T. Simura et al.
0917 T. Ishiwaka et al.
1047 Y. Toi et al.
1048 H. Imai et al.
1263 O. Tsuge et al.
1270 K. Fukui et al.

42- -69, J. Indian Chem. Soc., 46 (1969)
0096 P. Chakrabarti et al.
0100 A.K. Barua et al.
0103 A. Chatterjee and R. Raychaudhuri
0148 S.H. Dandegaonker
0182 P.B. Talukdar and S.K. Sengupta
0275 D.K. Chatterjee and K. Sen
0388 P.B. Talukdar and S.K. Sengupta
0415 K.D. Gupta
0595 R.D. Dasai
0635 A. Chatterjee et al.
0682 B.K. Chowdhury and D.P. Chakraborty
0831 V.K. Gupta and S.G. Tandon
0855 M.K. Saxena and M.M. Bokadia
0919 P.S. Jogdeo et al.
0935 R.B. Tirodkar and R.N. Usgaonkar
0945 P.R. Singh and R. Sahai
1014 K.N.H. Pardanani et al.
1061 P. Chakrabarti and A.K. Sanyal

44- -69, J. Org. Chem., 34 (1969)
0036 L. Field and R.B. Barbee
0056 W.E. Parham and D.G. Weetman
0073 A.A. Zimmerman et al.
0077 J.H. Smith and F.M. Menger
0092 A.P. Martinez et al.
0107 J.N. Gardner, B.A. Anderson and E.P. Oliveto
0112 D.M. Piatak
0116 D.M. Piatak et al.
0120 K.H. Dudley and R.W. Chiang
0126 M.D. Bachi et al.
0136 H.E. Smith et al.
0155 W.M. Williams and W.R. Dolbier, Jr.
0165 P. Aeberli and W.J. Houlihan
0183 M. Denzer and H. Ott
0187 J.P. Freeman, J.J. Gannon and D. Surbey
0194 J.P. Freeman and J.J. Gannon
0212 E.E. Schweizer et al.
0224 V.P. Vitullo
0227 C.D. Cook and L.C. Butler
0231 C.S. Russell et al.
0237 M.A. Bielefeld and P. Kurath
0240 M.S.R. Nair et al.
0244 W.J. Wechter
0249 M. Winn and H.E. Zaugg
0310 E. Doomes and N.H. Cromwell
0330 R.N. Schut et al.
0333 D.J. Zwanenburg and H. Wynberg
0340 D.J. Zwanenburg and H. Wynberg
0347 L.H. Klemm et al.
0365 P.C. Thomas et al.
0372 J. Rabinowitz et al.
0380 J. Wolinsky and H.S. Hauer
0384 K.L. Kirk and L.A. Cohen
0395 K.L. Kirk and L.A. Cohen
0409 D. Rosenthal and J.P. Gratz
0412 B. Hwang et al.
0416 A.P. Martinez and W.W. Lee
0431 M.W. Winkley and R.K. Robins
0444 A.T. Nielsen and D.W. Moore
0450 L.A. Paquette and G.V. Meehan
0457 H.L. Slates et al.
0461 L.A. Carpino
0496 J.J. Sims and V.K. Honwad
0505 A.B. Thigpen, Jr., and R. Fuchs
0520 S.P. Pappas et al.
0528 E.C. Friedrich
0534 R. Fuller et al.
0538 M.P. Cava and J.P. Van Meter
0550 T. Matsuura et al.
0580 J.C. Pew and W.J. Connors
0585 J.C. Pew and W.J. Connors
0589 P. Beak and E.M. Monroe
0596 G. Maury and N.H. Cromwell
0649 R.I. Fryer, J.V. Earley and L.H. Sternbach
0660 F.M. Hershenson and L. Bauer
0670 I. Moritani et al.
0698 A.I. Meyers, W.H. Reine and R. Gault
0730 T. Takeshima et al.
0732 A.G. Hortmann et al.
0742 C.A. Cupas and W.A. Roach
0756 R.W. Lamon
0758 H. Hart and J.W. Link
0763 A. Ur Rahman and B.M. Vuano
0794 B.D. Challand et al.
0821 B. Roth and J.Z. Strelitz
0836 L.J. Bellyky et al.
0857 G. Büchi and H. Wüest
0874 H.W. Whitlock, Jr., and J.K. Reed
0879 H.W. Whitlock, Jr., et al.
0887 J.A. Moon et al.
0896 E. Block and E.J. Corey
0973 E. Dyer et al.
0978 U. Wocke and G.B. Brown
1020 I. Wempen and J.J. Fox
1025 A.D. Broom and R.K. Robins
1029 A. Rosenthal and L.B. Nguyen
1062 S.M. Kupchan and A. Yoshitake
1070 R.L. Augustine and W.G. Pierson
1122 G. Büchi and H. Wüest
1124 D.K. Wall et al.
1133 K.H. Dudley and D.L. Bius
1136 D.R. Eckroth and G.M. Love
1147 P.E. Sonnet
1151 D. Lednicer and D.E. Emmert
1176 C.I. Lewis, J.Y. Chang and A.W. Spears
1198 H.D. Becker
1203 H.D. Becker
1211 H.D. Becker
1233 E.J. Corey and E. Block
1257 W. Herz and R.C. Blackstone
1271 R.S.H. Liu and C.G. Krespan
1310 D.C. Dittmer et al.
1349 W. Augstein and C.K. Bradsher
1359 P.K. Martin et al.

1364 M.C. Wani and M.E. Wall
1390 B.A. Otter, E.A. Falco and J.J. Fox
1396 J.A. Montgomery and K. Hewson
1420 O. Redmore
1430 H.O. House and F.A. Richey, Jr.
1447 M. Debono et al.
1455 A. Roy, W.D. Slaunwhite and S. Roy
1457 R.A. Mitchell et al.
1459 H. Takeda et al.
1460 S.M. Kupchan et al.
1463 H. Newman and R.B. Angier
1465 Y.L. Chow and R.H. Quen
1468 D.J. Woodman and Z.L. Murphy
1470 A.T. Nielsen and T.G. Archibald
1474 W.E. Parham et al.
1480 R.F. Borch and D.L. Fields
1547 Kin-Ichi et al.
1550 A.W. Burgstahler, J.N. Marx and D.F. Zinkel
1566 W.C. Lumma, Jr., and G.A. Berchtold
1572 K.R. Williams et al.
1577 Y. Sato et al.
1582 J.E. Dolfini et al.
1592 K. Pilgram and H. Ohse
1601 J. Waters and B. Witkop
1606 G.S. Abernethy, Jr., and M.E. Wall
1627 R.F. Dods and J.S. Roth
1651 E.J. Moriconi et al.
1687 D.P. Harnish et al.
1691 J.A. Van Allan et al.
1695 R. Shapiro and S. Nesnow
1700 C.K. Bradsher and J.A. Stone
1720 P. Aeberli and W.J. Houlihan
1734 D.R. Maulding and B.G. Roberts
1746 P.K. Freeman and R.C. Johnson
1751 P.K. Freeman and R.C. Johnson
1780 A.H. Goldkamp
1786 H. Watanabe et al.
1845 N.C. Yang et al.
1899 W.E. Parham and L.J. Czuba
1907 H.E. Hennis and C.-S. Wang
1923 E. Ciganek
1940 D.K. Black and G.W. Hedrick
1942 D.R. Rodig and N.J. Johnston
1949 D.R. Rodig and N.J. Johnston
1956 V. Boekelheide et al.
1960 V. Boekelheide and R.W. Griffin, Jr.
1962 H.W. Whitlock, Jr., and L.E. Overman
1966 N. Kuwahara et al.
1989 F.J. Schmitz et al.
2006 G.E. Kuhlmann and D.C. Dittmer
2011 S. Nesnow and R. Shapiro
2022 L.L. Replogle, G.C. Peters and J.R. Maynard
2027 H.D. Becker
2038 J.F. Bunnett and N.S. Nadelman
2053 S.C. Mutha and R. Ketcham
2058 W.E. Noland et al.
2083 K.T. Finley et al.
2102 C. Temple, Jr., et al.
2123 M. Kurihara
2129 C.K. Bradsher et al.
2146 W.J. Linn and E. Ciganek
2153 I. Scheinfeld et al.
2157 A. Giner-Sorolla et al.
2160 M.J. Robins and R.K. Robins
2183 J.W. Huffman and J.H. Cecil
2192 J. Correa and R.M. Mainere
2209 H.O. House, T.M. Bare and W.E. Hanners
2217 W.M. Jones and J.M. Walbrick
2235 R.L. Augustine and W.G. Pierson
2239 W.H. Pirkle and M. Dines
2248 S.F. Nelsen
2269 F.D. Greene and J.F. Pazos
2285 R.J. Sundberg and G.S. Kotchmar, Jr
2288 A.J. Liston and P. Toft
2297 W.G. Dauben et al.
2301 W.G. Dauben and G.W. Shaffer
2311 L.R. Worden et al.
2324 H.O. House et al.
2346 R.S. Bly and G.B. Konizer
2375 A.G. Anderson, Jr., and R.D. Breazeale
2393 H. Watanable et al.
2401 A.L. Wilds et al.
2407 K.R. Huffman et al.
2418 E. Grovenstein, Jr., et al.
2462 E.H. White et al.
2472 H.D. Becker
2480 A. Omori, N. Sonoda and S. Tsutsumi
2512 W.G. Dauben et al.
2527 J.F. Stephen and E. Marcus
2579 J. Casanova, Jr., and H.R. Kiefer
2632 J.A. Wright et al.
2636 B.A. Otter et al.
2643 J.P.H. Verheyden and J.G. Moffatt
2646 J.A. Montgomery and H.Thomas
2665 M.P. Cava et al.
2667 J.B. Hendrickson et al.
2676 F.I. Carroll and R. Meck
2715 P. Aeberli and W.J. Houlihan
2720 P. Aeberli and W.J. Houlihan
2728 A. Padwa and D. Eastman
2736 G.A. Reynolds and J.A. Van Allan
2750 K.H. Dudley et al.
2767 P. Narasimha Rao et al.
2768 A.T. Rowland et al.
2775 W. Herz and J.J. Schmid
2782 W. Agosta and D.K. Herron
2788 J.E. Baldwin and H.H. Basson
2792 T.R. Potts and R.E. Harom
2799 R.J. Sundberg and D.E. Blackburn
2803 H. Kaneko and S. Naruto
2817 D.J. Sardella et al.
2866 L.A. Paquette et al.
2879 L.A. Paquette et al.
2888 L.A. Paquette et al.
2901 L.A. Paquette et al.
2966 C.L. Chen, W.J. Connors and W.M. Shinker
2972 S. Yanagida et al.
2983 R.K. Howe
2988 L.J. Dolby and P.D. Lord
2997 W.E. Thun and W.R. McBride
3011 F.N. Jones and and S. Andreades
3022 E. Farkas et al.
3032 M. Debono et al.
3035 P.W. Wegfahrt and H. Rapoport
3039 J.D. Surmatis et al.

3076 J. Blum and Z. Lipshes
3085 W.L. Parker and R.B. Woodward
3093 D.Y. Curtin and Z.M. Holubec
3135 T. Nakano et al.
3161 C. Temple, Jr., et al.
3165 H. Zinnes et al.
3169 J. Wolinsky and H.S. Hauer
3175 H. Wynberg et al.
3181 B.T. Gillis and R.A. Izydore
3187 N. Castagnoli, Jr.
3191 R.P. Mariella and K.H. Brown
3192 N.N. Girotra and N.L. Wendler
3201 W.J. Middleton
3202 C.C. Price and D.M. Follweiler
3213 J.G. Michels and G.C. Wright
3221 K.T. Potts and C. Lovelette
3230 A.M. Cameron et al.
3237 W.W. Paudler and A.G. Zeiler
3240 N.J. Leonard and R.F. Lambert
3273 R.J. Sundberg and F.O. Holcombe, Jr.
3279 P.L. Southwick et al,
3285 I.M. Goldman
3392 A.G. Hortmann and R.E. Youngstrom
3438 J.R. Throckmorton
3451 D.J. Woodman and Z.L. Murphy
3456 D.F. Bowman et al.
3464 W. Herz and J.J. Schmid
3477 S.W. Pelletier, C.W.J. Chang and K.N. Iyer
3484 H. Newman and R.B. Angier
3492 G.B. Cheda and R.H. Hall
3498 G.B. Cheda, R.H. Hall and P.M. Tanna
3502 A.M. Krubiner et al.
3513 M.L. Lewbart and J.J. Schneider
3530 Y.Y. Lin, M. Shibahara and L.L. Smith
3545 F.E. Ziegler and J.G. Sweeny
3552 J.E. Pike et al.
3638 M. L. Sinnott
3639 E.J. Volker and J.A. Moore
3645 J.K. Hecht et al.
3661 S.C. Mutha and R. Ketcham
3664 W.J. Gensler et al.
3670 F. Freeman and T.J. Ito
3672 E.J. Moriconi and R.E. Misner
3697 H. Yoshioka et al.
3699 W. Barbieri et al.
3705 J.F.W. Keana et al.
3710 T.G. Miller
3715 K. Kawakami and H.G. Kuivila
3717 R.E. Ireland et al.
3739 U.R. Ghatak et al.
3754 N.K. Chaudhuri and M. Gut
3759 N.K. Chaudhuri et al.
3767 N.K. Chaudhuri et al.
3779 P. Sunder-Plassman et al.
3784 D.P. Chakraborty and D. Chatterji
3786 J.L. Neumeyer et al.
3789 G. Brieger et al.
3792 H. Torabi et al.
3796 H. Torabi et al.
3806 M.P. Kotick, C. Szantay and T.J. Bardos
3814 N.J. Leonard and R.A. Swaringen, Jr.
3829 T.C. Miller
3837 J. Tadanier et al.
3842 S. Fujii and H. Kobatake
3845 H.B. Sinclair
3853 J.A. Kepler et al.
3858 S.M. Kupchan et al.
3867 S.M. Kupchan et al.
3876 S.M. Kupchan et al.
3884 S.M. Kupchan et al.
3894 S.M. Kupchan et al.
3898 S.M. Kupchan et al.
3903 S.M. Kupchan et al.
3908 S.M. Kupchan et al.
3912 S.M. Kupchan et al.
3981 Y. Ogata et al.
4016 W. Herz and M.G. Nair
4046 P. Yates et al.
4052 P. Yates and P. Singh
4118 R.R. Fraser et al.
4131 J.H. Markgraf et al.
4134 V. Boekelheide and J.E. Nottke
4150 L. Skattebøl and B. Boulette
4154 M. Takeda et al.
4158 M. Takeda et al.
4166 O.L. Chapman and K. Lee
4173 W.V. Rochat and G.L. Gard
4182 R.D. Hoffsommer, D. Taub and N. Wendler
4199 F.J.McEvoy and G.R. Allen, Jr.
4203 G.D. Breck and G.H. Stout
4207 H.Y. Yamamoto et al.

45- -69, J. Pharm. Pharmacol., 21 (1969)
0786 J.E. Forrest et al.

46- -69, J. Phys. Chem., 73 (1969)
1642 C.Y.S. Shen and C.A. Swenson

47- -69, J. Polymer Sci., (1969)
0265 R.J. Eldred
1269 N. Nakabayashi et al.
1803 I. Schopov and N. Popov
3313 L. De Noninck and G. Smets

48- -69, J. prakt. Chem., 311 (1969)
0009 H.G.O. Becker and H.-J. Timpe
0045 J. Fabian et al.
0061 J. Faust et al.
0091 G. Heublein and H. Lauterbach
0153 K. Schulze et al.
0162 G. Drefahl et al.
0260 W. Höbald et al.
0388 E. Fanghänel et al.
0402 K. Gewalde and I. Hofmann
0445 K. Ponsold and G. Schubert
0477 H.G.O. Becker et al.
0614 S. Hauptmann et al.
0621 W.H. Horhold and J. Beck
0646 H.G.O. Becker et al.
0671 H.J. Siemann and S. Schwarz
0746 J. Beger
0786 G. Domschke and H. Oelmann
0800 G. Domschke and H. Oelmann
0827 H. Hartmann and S. Scheithauer
0897 H.G.O. Becker et al.
0919 G. Drefahl et al.

0997 H. Zinner and K. Peseke

49- -69, Monatsh. Chem., 100 (1969)
0001 P. Schuster et al.
0136 T. Kappe et al.
0163 W. Fleischhacker and F. Viebock
0576 P. Margaretha and O.E. Polansky
1274 H. Sterk and T. Kappe
1307 J. Sevcik
1368 A.N. Goud and S. Rajagopal
1451 H. Kammerer and G. Sextro
1479 J. Schantl
1515 K. Schlogl and M. Walser
1540 H. Falk and D. Hofer
1552 H. Falk et al.
1599 H. Hemetsberger et al.
1994 H. Junek, H. Sterk and I. Wrtilek

51- -69, Naturwiss., 56 (1969)
0328 W.M. zu Reckendorf
0515 V. Hahnkamm and G. Gattow

54- -69, Rec. trav. chim., 88 (1969)
0005 J.A. Maassen et al.
0119 H.A. Selling et al.
0307 W. Drenth and G.H.E. Nieuwdorp
0313 D.K. Kettenes et al.
0321 D.J. Zwanenburg and H. Wynberg
0386 A.J. Hoefnagel
0465 H.J.T. Bos and J. Boleij
0562 M.A. Hoefnagel et al.
0641 J.B.F.N. Engberts et al.
0737 R. Van Moorselaer and S.J. Halkes
0752 S.J. Halkes and R. Van Moorselaer
0801 G.P. de Gunst
0805 E. Selegny et al.
0851 J.C. Overeem
0905 A.U. Rahman and H.S.E. Gatica
0989 J.P. Ward and D.A. Van Dorp
1080 S.J. Halkes and N.P. Van Vliet
1244 H. Wynberg and H.J.M. Sinnige
1263 K.B. DeRoos and C.A. Salemink

55- -69, Ricerca Sci., 39 (1969)
0074 E. Baciocchi and P.L. Bocca
0859 P. Franchetti et al.

56- -69, Roczniki Chem., 43 (1969)
0499 M. Draminski and B. Fiszer
0519 W. Kirkov et al.
0699 O. Achmatowicz, Jr., and J. Szychowski
0749 J. Moszew and M. Bala
0763 A. Banaszek et al.
0943 H. Kuczynski and K. Marks
1219 K. Okon et al.
1469 E. Wyrzykiewicz
1641 S. Smolinski et al.
1653 K. Okon et al.
1843 W. Jasiobedzki

57- -69B, Science, 164 (1969)
1083 A.R. Mosier, W.D. Guenzi and L.L. Miller

57- -69D, Science, 166 (1969)
0755 A.K. Krey and F.E. Hahn

59- -69, Spectrochim. Acta, 25A (1969)
0275 K.R. Loos et al.
0313 W.M. Byrne and N.H.P. Smith
0393 G. Kresze and W. Amann
0407 J. Barrett and M.J. Hitch
0447 V. Bertoli and P.H. Plesch
0487 R. Van Der Linde et al.
0501 T.M. McKinney
0765 K. Yates et al.
0989 D.J. Cowley and L.H. Sutcliffe
1075 C.L. Honeybourne and G.A. webb
1167 T.M. Ward and J.B. Weber

60- -69, Trans. Faraday Soc., 65 (1969)
0904 J. Stals et al.
2611 R.D. Gillard and P.R. Mitchell

61- -69, Ber. Bunsengesellschaft Phys. Chem., 73 (1969)
0203 J. Sunkel and H. Staude
0845 D. Moebius et al.

62- -69A, Z. phys. Chem.(Frankfurt), 63 (1969)
0029 P. Bortolus et al.

62- -69B, Z. phys. Chem.(Frankfurt), 64 (1969)
0097 G. Scheibe et al.

63- -69, Z. physiol. Chem., 350 (1969)
0085 S. Edwards and F.H. Marquardt
0366 H.J. Zeitler and H. Niemer
0457 R. Brossmer and D. Ziegler
0473 C. Woenckhaus et al.
0710 H. Guglielmi and B. Athen
0809 H. Guglielmi and B. Athen
1291 W. Rudiger

64- -69, Z. Naturforsch., 24b (1969)
0024 R. Mayer et al.
0038 H. Schildknecht et al.
0080 M. Gschwendt and E. Hecker
0091 H. Bartsen and E. Hecker
0099 H. Bartsen and E. Hecker
0137 H.-W. Wanzlick et al.
0225 H.A.B. Linke
0464 B. Fohlisch and E. Widmann
0524 G. Juppe et al.
0536 H. Dürr and L. Schrader
0542 W. Broser and W. Harber
0547 A. Rieker et al.
0652 C. Schiele and A. Wilhelm
0685 W. Broser et al.
0750 S. Huneck and J. Santesson
0756 S. Huneck and J. Santesson
0770 W. Strohmeier and F.J. Müller
0941 W. Steglich et al.
0997 H.A.B. Linke and D. Pramer
1211 H. Rudolph and E. Vowinkel
1336 W. Grunbein et al.

1353 E. Haug and B. Föhlisch
1408 H. Wagner and H.P. Hörhammer, Jr.
1518 H.A. Brune et al.

65- -69, Zhur. Obshchei Khim., 39 (1969)
0059 T.V. Ershova et al.
0070 I.S. Ioffe et al.
0078 I.S. Ioffe et al.
0102 V.I. Koltunova and V.M. Berezovskii
0119 Y.A. Zhdanov and V.A. Polenov
0303 E.G. Rukhadze et al.
0321 A.I. Bokanov et al.
0373 A.I. Bokanov et al.
0392 A.I. Razumov et al.
0401 E.G. Rukhadze et al.
0434 A.M. Yurkevich et al.
0443 A.V. Bogatskii and S.A. Andronati
0451 P.F. Vlad, A.G. Russo and C.K. Fan
0640 I.S. Ioffe, A.B. Tomchin and E.N. Zhukova
0660 M.K. Verzilina and A.V. Belotsvetov
0668 N.I. Grineva et al.
0860 M.I. Shevchuk et al.
0879 A.E. Lutskii et al.
0941 M.Z. Girshovich and A.V. El'tsov
1008 A.V. Dombrovskii et al.
1152 I.S. Ioffe and A.B. Tomchin
1156 I.S. Ioffe and A.B. Tomchin
1347 N.G. Ivanova et al.
1373 E.A. Smirnov and A.E. Podorol'skaya
1377 E.A. Smirnov and A.E. Podorol'skaya
1416 Y.A. Zhdanov et al.
1544 B.I. Stepanov, E.N. Karpova and A.I. Bokanov
1607 V.I. Vysochin et al.
1615 V.P. Petrov and V.A. Barkhash
1619 N.N. Polle and E.G. Polle
1630 Z.I. Aksel'rod and V.M. Berezovskii
1634 S.M. Shein et al.
1732 I.N. Zhmurova et al.
1766 S.A. Anichkina
1774 V.P. Molosnova et al.
1829 S.V. Lepikhova and G.T. Pilyugin
1835 V.V. Zaitsev et al.
1861 S.V. Bogatkov and E.M. Cherkasova
1954 I.P. Gozman
2071 V.M. Vlasov and G.G. Yakobson
2111 I.S. Ioffe et al.
2116 S.V. Lepikhova and G.T. Pilyugin
2125 L.M. Roitshtein et al.
2129 S.A. Mikhalenko and E.A. Luk'yanets
2326 I.N. Vorozhtsov et al.
2332 T.P. Lobanova, E.I. Berus and V.A. Barkhash
2339 I.S. Ioffe, A.B. Tomchin and E.N. Zhukova
2345 I.S. Ioffe, A.B. Tomchin and E.A. Rusakov
2558 R.P. Evstigneeva et al.
2568 A.V. Bogatskii et al.
2723 A.N. Kost and M.A. Yurovskaya

67- -69, Proc. Iowa Acad. Sci., 76 (1969)
0127 K.W. Kraus et al.

69- -69, Biochemistry, 8 (1969)
0238 R. Shapiro et al.
0733 P.H. Stahl
1042 H. Barrett et al.
1188 V.R. Mattox et al.
1344 P.W. Wigler and H.J. Lee
1844 H. Junck et al.
3067 R.G. von Tigerstrom and M. Smith
3071 W.J. Burrows et al.
4172 O.A. Koleoso et al.
4888 A.F. Russell and J.G. Moffatt
5181 A. Pocker and E.H. Fischer

70- -69, Izvest. Akad. Nauk S.S.S.R., (1969)(English translation pagination)
0131 I.N. Vorozhtsov and V.A. Barkhash
0134 V.P. Molosnova et al.
0199 L.A. Il'ina et al.
0247 S.B. Savvin et al.
0363 Z.A. Krasnaya et al.
0580 A.A. Volod'kin et al.
0609 T.D. Petrova et al.
0694 B.A. Arbusov et al.
0812 V.M. Vlasov and G.G. Yakobson
0826 A.A. Akhrem and T.V. Ilyukhina
0835 V.S. Reznik et al.
0873 K. Shakhidayatov et al.
1012 A. Arbasov et al.
1381 R.R. Shagidullin et al.
1655 A.V. Zakharychev et al.
1664 A.A. Akhrem and A.M. Moiseenkov
1732 B.D. Sviridov et al.
2117 A.N. Volkov et al.
2127 M.Y. Karpeiskii
2201 L.B. Volodarskii and A.Y. Tikhonov
2324 A.G. Budnik et al.
2460 A.A. Efremenko et al.
2601 Y.I. Gunar et al.
2627 E.A. Mistryakov and G.T. Katvalyan
2680 A.A. Akhrem et al.

73- 69, Coll. Czech. Chem. Comm., 34 (1969)
0089 H. Pischel and A. Holy
0221 V. Stuzka et al.
0240 E. Rihova and A. Vystreil
0247 M. Bobek and J. Farkas
0336 L. Novotny et al.
0344 J. Hora
0427 J. Palecek et al.
0442 B. Pelc and J. Hodkova
0451 R. Mickova et al.
0458 R. Mickova and K. Syhora
0572 A. Jurasek et al.
0582 V. Benesova et al.
0618 J. Beranek
0681 V. Sanda et al.
0819 J. Benes et al.
0875 V. Preininger et al.
1015 M. Rajsner et al.
1087 P. Vetesnik et al.
1104 M. Prystas and F. Sorm
1673 M. Bobek et al.
1684 M. Bobek and J. Farkas

1690 M. Bobek et al.
1696 J. Farkas and F. Sorm
1739 J. Harmatha et al.
1832 V. Simanek et al.
1963 M. Rajsner et al.
2092 J. Kavalek et al.
2114 R. Kotva et al.
2235 S. Toma
2278 M. Kuchar et al.
2288 K. Vokac et al.
2306 J. Beranek and J. Gut
2316 M. Prystas and F. Sorm
2459 K. Syhora et al.
2771 S. Toma
2792 Z. Samek et al.
2952 P. Kristian and J. Bernat
2959 M. Kucharczyk et al.
3336 J. Palecek and J. Kuthan
3343 M. Kuchar and B. Kakac
3479 A. Kasal and O. Linet
3540 H. Potesilova et al.
3569 J. Jarkovsky et al.
3732 Z. Stransky and J. Gruz
3740 J. Socha and M. Vecera
3755 J. Zemlicka et al.
3792 K. Sindelar et al.
3801 K. Sindelar et al.
3811 A. Mistr et al.
3912 A. Martron et al.
3936 K. Pelz et al.
4005 M. Uher et al.

77- -69, J. Chem. Soc., Sect. D, (1969)
0004 H. Sakura et al.
0043 P. Rona et al.
0049 E.R. Talaty et al.
0057 K.W. Bentley et al.
0066 K.T. Potts and U.P. Singh
0082 S. Imai et al.
0085 S.W. Russell and B.C.L. Weedon
0086 J. Meinwald et al.
0090 P.M. Collins and P. Gupta
0092 R.T. Channon et al.
0125 J.S. Cridland and S.T. Reid
0133 H. Gusten
0140 J.F. Biellmann and H.J. Callot
0162 D.J. Aberhart and K.H. Overton
0167 J.E. Baldwin and J.E. Brown
0174 P.A. Hart and M.P. Tripp
0179 W.A.F. Gladstone
0199 H.H. Wasserman and A.H. Miller
0200 J.L. Cooper and H.H. Wasserman
0202 M.G. Barlow et al.
0203 W.T. Flowers et al.
0228 F. Toda et al.
0327 R.E. Harmon et al.
0341 J.E. Batterbee et al.
0345 C.O. Bender et al.
0347 F. Kohen et al.
0352 M. Oda and Y. Kitahara
0364 D.H. Iles and A. Ledwith
0369 H. Newman and R.B. Angier
0380 B.W. Finucane and J.B. Thomson
0381 J. Kiburis and J.H. Lister
0402 M.N. Galbraith et al.
0432 T. Sasaki et al.
0462 G.J.D. Peddle and R.W. Walsingham
0463 B.W. Bycroft et al.
0467 W.M. Horspool
0546 S. Imai et al.
0549 C. Dell'Erba et al.
0550 D.H.R. Barton et al.
0563 I. Baxter et al.
0569 K.T. Potts and U.P. Singh
0598 E. Campaigne et al.
0630 C.J.W. Brooks and M.M. Campbell
0645 B.J. Gregory et al.
0653 K. Ito and H. Furukawa
0660 S.I. Goldberg and V.M. Balthis
0662 H. Hikino et al.
0674 R.G. C rlson and D.E. Henton
0677 P.C. Arora and D. Mackay
0680 L.A. Paquette and J.C. Philips
0685 C.T. Bedford and T. Money
0709 J.M. Gourley et al.
0726 M.S. Carson et al.
0737 M. Oda et al.
0739 M. Oda et al.
0767 P.S. Clezy and A.J. Liepa
0773 I.H. Sadler and J.A.G. Stewart
0781 S. Nishida et al.
0795 S.E. Baldwin and H.H. Basson
0810 H. Heaney and T.J. Ward
0821 Le-Van-Thoi et al.
0832 W. Carruthers and M.I. Qureshi
0833 J. Schwartz
0839 J.S. Moffatt et al.
0847 Z.S. Ariyan and R.L. Martin
0861 P.-M. Vay
0879 W.D. Ollis et al.
0893 M. Ohashi et al.
0923 F.I. Carroll and J.T. Blackwell
0925 F.I. Carroll and J.T. Blackwell
0936 S.A. Procter et al.
0953 G. Doddi et al.
0991 J. Auerbach and R.W. Franck
0996 M.N. Afzal et al.
1003 D.A. Plank et al.
1032 J. MacMillan and E.R.H. Walker
1044 A.O. Plunkett
1062 J. Castells and A. Colombo
1083 R.W. Doskotch et al.
1096 M.F. Ansell and A.J. Bignold
1103 E. Baggiolini et al.
1110 F.L. Bach et al.
1120 S.P. Kureel et al.
1128 K.T. Potts et al.
1129 K.T. Potts et al.
1149 A.K. Ganguly and O.Z. Sarre
1150 J.J.H. Simes et al.
1159 R.M. Dodson and J.P. Nelson
1170 R.E. Harmon et al.
1204 S. Masamune et al.
1214 I.S. Ponticello and R.H. Schlessinger
1215 A. Padwa and W. Eisenhardt
1217 M.D. Glick et al.
1219 F. Toda et al.
1246 G. Barth et al.
1253 S.K. Dasgupta et al.

1281 A.T. Nielsen et al.
1282 D.P. Della Casa de Marcano and T.G. Halsall
1301 T. Kametani et al.
1306 J.M. Cook et al.
1335 F. Kido et al.
1348 U. Eisner
1358 D.T. Longone and G.R. Chipman
1359 F.W. Fowler
1368 A.R. Dunn and R.J. Stoodley
1393 J.H. Lee et al.
1396 M. Miyashita et al.
1411 K.W. Bentley et al.
1439 R.A. Abramovitch et al.
1440 D.J. Pointer et al.
1467 Y. Nakadaira et al.
1479 K. Yamada et al.
1480 M.J. Broadhurst and R. Grigg
1483 A.G. Schultz and R.H. Schlessinger

78- -69, Tetrahedron, 25 (1969)
0057 I.R. Bellobono and G. Favini
0095 O. Yonemitsu et al.
0101 D.W. Brown, S.F. Dyke and M. Sainsbury
0149 J.W. Blunt et al.
0163 J.D. McKinney and C.H. Stammer
0223 J.K. Groves and N. Jones
0227 K.M. Biswas and A.H. Jackson
0255 W. Werner
0275 A.C. Jain et al.
0283 A.C. Jain, P. Lal and T.R. Seshadri
0295 M. Ishikawa et al.
0335 K. Kitahonoki et al.
0459 T. Irie et al.
0469 R.W. Doskotch et al.
0477 A.M. Yurkevich et al.
0517 M. Mazharuddin and G. Thyagarajan
0527 S.K. Khetan and M.V. George
0541 T. Matsuura and I. Saito
0549 T. Matsuura and I. Saito
0557 T. Matsuura and I. Saito
0619 J.C. Bloch
0641 A.F. Casy and R.R. Ison
0657 J. Wagner et al.
0737 H. Ripperger and K. Schreiber
0761 G. Snatzke et al.
0783 Y. Hagiwara, M. Kurihara and N. Yoda
0805 H. Ruesch and T.J. Mabry
0837 T. Miyadera and R. Tachikawa
0847 T. Miyadera and R. Tachikawa
0861 T.H. Barr and W.E. Watts
0883 G. Köbrich and H. Büttner
0937 J. Marchand et al.
0955 H. Fischer and H. Fischer
1013 L. Farkas et al.
1037 H. Hikino et al.
1047 T. Tokoroyama et al.
1089 H.A. Brune, H.P. Wolff and Huther
1109 T. Okuda et al.
1117 T. Okuda and K. Konishi
1125 R. Noyori et al.
1155 D. Burn, J.P. Yardley and V. Petrow
1219 C. Beard et al.
1241 T. Takemoto et al.
1293 K. Mori et al.
1323 G.A. Ellestad et al.
1335 E. Fujita et al.
1367 K. Yoshida
1407 L. Jurd
1423 A.J. Neale, K.M. Davies and J. Ellis
1441 J. Fabian and G. Laban
1453 B.S. Joshi et al.
1467 J.A. Deyrup et al.
1497 J.G. Heathcote and M.F. Dutton
1507 J.E. Atkinson, P. Gupta and J.R. Lewis
1603 S.M. Kupchan et al.
1617 H. Dabrowska-Urbanska et al.
1661 H. Nozaki, H. Kato and R. Noyori
1679 J. Libman et al.
1699 J. Libman and Y. Mazur
1707 J. Libman et al.
1717 D. Amar, V. Permutti and Y. Mazur
1785 C. Ehret and G. Ourisson
1825 N. Aimi et al.
1881 M. Sainsbury et al.
1897 J.D. Chanley and C. Rossi
1929 A. Ostaszynski et al.
1939 B.D. Tilak et al.
1947 G.H. Stout and W.J. Balkenhol
1961 G.H. Stout et al.
2035 A.P. Dillon and K.G. Lewis
2041 G. Snatzke and E. Otto
2099 E.H. White, S. Eguchi and J.N. Marx
2117 J.N. Marx and E.H. White
2145 T. Sasaki, S. Eguchi and M. Ohno
2155 S. Smolinski et al.
2193 L. Cagliotti
2223 G. Köbrich and H. Büttner
2233 T. Anthonsen et al.
2275 M. Shamma and L. Novak
2291 V.R. Holland et al.
2367 L. Jurd
2393 A.T. Nielsen and T.G. Archibald
2417 K. Kirahonoki and Y. Takano
2603 J.M. Coxon et al.
2661 J.P. John et al.
2687 T.M. Harris and C.M. Harris
2715 Y. Ogata et al.
2723 K. Nishikawa et al.
2757 Y. Kanaoka et al.
2767 R.R. Arndt, S.H. Eggers and A. Jordaan
2787 A.C. Jain et al.
2815 T.R. Govindachari et al.
2823 D.A. Ben-Efraim and F. Sondheimer
2837 D.A. Ben-Efraim and F. Sondheimer
2865 C.J.W. Brooks and G.H. Draffan
2887 C.J.W. Brooks and G.H. Draffan
2909 T. Sasaki, S. Eguchi and T. Toru
2987 H. Takeuchi, T. Nagai and N. Tokura
3007 N. Tsuji and K. Nagashima
3075 P. Longevialle
3091 K.W. Gopinath and B.R. Pai

3139 D.W. Theobald
3145 T. Masamune, A. Murai and S. Numata
3161 R.T. Gray and H.E. Smith
3193 Y. Shirota, T. Nagai and N. Tokura
3217 O.N. Devgan et al.
3223 C.-L. Chen and F.D. Hostettler
3251 P.H. Vandewyer et al.
3267 A. Marsili, V. Nuti and M.F. Saettone
3277 M.V. Mavrov et al.
3287 P. Beak and W. Messer
3341 T.S. Chou et al.
3447 R. Buyle and H.G. Viehe
3477 G. Marr, R.E. Moore and B.W. Rockett
3485 D. Mosnaim et al.
3501 J. Blum, F. Grauer and E.D. Bergmann
3509 K. Yamada et al.
3527 N.D. Doktorova et al.
3667 T. Kametani et al.
3675 H. Nozaki et al.
3701 S. Rangaswami and S. Sarangan
3767 J. Wolinsky and D. Nelson
3783 G. Adam et al.
3789 A.P. Bindra and J.A. Elix
3841 M.S. Wadia et al.
3855 A.N. Singh et al.
3879 C.E. Loader and H.J. Anderson
3887 M. Shamma and L.D. Stiver
3895 L. Canonica et al.
3903 L. Canonica et al.
3909 H. Hikino et al.
3919 I. Granoth et al.
3925 J.M. Coxon et al.
3975 H. Mohrle and H.-H. Scheltdorf
4005 H.O. Larson et al.
4011 D.J. France, J.J. Hand and M. Los
4057 B.E. Griffin and C.B. Reese
4153 V.R. Holland and B.C. Saunders
4161 A.I. Meyers and S. Singh
4187 S. Imamura and J. Tsuji
4197 R. Sosa and L. Paoloni
4207 A.K. Das Gupta et al.
4241 V.M. Clark et al.
4257 M. Kocor et al.
4291 A. Katritzky and E. Lunt
4315 E.J.J. Grabowski and R.L. Autrey
4339 F.M. Beringer et al.
4375 H.A. Brune and W. Schwab
4409 B.J. Donnelly et al.
4415 D.L. Dreyer
4421 M.S. Manhas et al.
4535 M.J. Grimwade and M.G. Lester
4551 T. Masamune and K. Orito
4579 A.J. Solo et al.
4649 Y. Shvo and I. Belsky
4751 H. Chikamatsu et al.
4767 H. Yoshioka and T.J. Mabry
4825 H. Alt and H. Bock
4835 A. Corbelle et al.
4843 A.H. Jackson and B. Naidoo
4895 D. Helmlinzer and G. Ourisson
4919 Y. Ogata et al.
4933 L. Skattebøl
5007 B.A. Burke et al.
5059 G. Snatzke et al.
5115 J. Altman et al.
5155 S. Sugiura et al.
5205 M. Ogata, H. Matsumoto and H. Kano
5227 G.A. Eagle et al.
5245 T.H. Barr et al.
5267 P.C. Mukharji and A.N. Ganguly
5275 P.C. Mukharji and T.K. Das Gupta
5281 P.C. Mukharji and A.N. Ganguly
5287 P.C. Mukharji et al.
5295 G.H. Stout and J.L. Fries
5319 Y.K. Sawa and H. Matusmura
5329 Y.K. Sawa and H. Matsumura
5357 H. Nozaki, T. Koyama and T. Mori
5365 S.F. Dyke et al.
5383 F. Feichtmayr et al.
5415 S.R. Udupa et al.
5427 S. Smolinski and A. Malata
5431 S. Smolinski et al.
5443 S.J. Rhoads and R.W. Holder
5465 A.P. Bindra and J.A. Elix
5475 N.J. McCorkindale et al.
5601 G. Snatzke, B. Ehrig and H. Klein
5639 E. Clar, A. Mullen and U. Sanigök
5703 F. Duus, E.B. Pedersen and S.-O. Lawesson
5721 R. Fuks and H.G. Viehe
5745 H. Möhrle and Kilian
5761 S. Oae et al.
5807 R.E. Harmon et al.
5879 D. Stauffacher et al.
5893 J. Macmillan and R.J. Pryce
5971 A.J.H. Nollet et al.
5983 A.J.H. Nollet and U.K. Pandit
5989 A.J.H. Nollet and U.K. Pandit
6001 T.H. Barr, H.L. Lentzner and W.E. Watts
6025 M.F. Ansell and T.M. Kafka

80- -69, Revue Romaine Chim., 14 (1969)
0225 L. Fey and N. Bodor
0481 L. Fey and N. Bodor
0893 I. Florea
0941 A.T. Balaban et al.
1191 M. Avram et al.
1323 A.T. Balaban
1617 C. Bodea and E. Broser

83- -69, Arch. Pharm., 302 (1969)
0043 H. Oelschlager et al.
0053 K. Fickentscher et al.
0075 G. Langer and H.J. Roth
0100 D. Block et al.
0264 H. Loth and K. Eichner
0387 H.J. Roth and E. Schumann
0401 M. Huke et al.
0487 K. Rehse
0494 J. Schnekenburger
0572 W. Wiegrebe et al.
0628 F. Eiden and J. Pluckhan
0886 G. Seitz
0965 K.E. Schulte et al.

87- -69, J. Med. Chem., 12 (1969)
0001 W.R. Biggerstaff et al.
0058 T.L. Hullar

0141 S. Gerchakov and H.P. Schultz
0146 R.A. Wiley and J.H. Collins
0157 K. Zee-Chang and C.C. Cheng
0175 R. Vince and J. Donovan
0180 E.W. Maynert
0227 B. Roth
0310 C.W. Ryan et al.
0339 E. Campaigne et al.
0357 E.F. Elslager et al.
0381 W.J. Fanshawe et al.
0393 T.L. Popper et al.
0444 E. Galantay et al.
0483 E.S. Wagner et al.
0498 J.A. Montgomery and K. Hewson
0513 P.M.G. Barin et al.
0531 G.M. Cingolani et al.
0533 M. Swierkowski and D. Shugar
0540 D.E. O'Brien et al.
0543 T.A. Khwaja and C. Heidelberger
0545 C.W. Noell and C.C. Cheng
0548 A.P. Shroff
0563 A.H. Beckett et al.
0617 H.F. Herbrandson and R.H. Wood
0620 H.F. Herbrandson and R.H. Wood
0631 E.L. Shapiro et al.
0636 T. Kobayashi et al.
0653 D.A. Shuman et al.
0658 M.G. Stout et al.
0717 A. Giner-Sorolla
0818 K.C. Tsou et al.
0944 V.J. Bauer et al.
1056 M.H. Fleysher et al.
1066 J.W. McFarland et al.
1117 F. Keller and J.E. Bunker
1124 R.P. Williams et al.
1125 A.P. Martinez et al.

88- -69, Tetrahedron Letters, (1969)
0015 M. Franck-Neumann and C. Buchecker
0059 K.B. Wiberg and A. de Meijere
0067 J. Reisch et al.
0073 A. Chatterjee and B.G. Hasra
0085 P. Dowd and A. Gold
0091 T. Kishi et al.
0097 T. Kishi et al.
0101 Y. Yamada, T. Hirata and M. Matsui
0105 J.M. Coxon and M.P. Hartshorn
0121 M. Konoshima et al.
0133 A.K. Ganguly et al.
0141 R. Tschesche et al.
0145 L. Vollner et al.
0153 Y. Inubushi et al.
0207 A.M. Lautzenheiser and P.W. LeQuesne
0227 L. Bang et al.
0239 J. Fabian and H. Hartmann
0287 E. Gegner
0289 M. Sprinzl, J. Farkas and F. Sorm
0297 S. Otsuka et al.
0301 H. Kukisawa et al.
0307 G.L. Chetty et al.
0355 Y. Koyama et al.
0359 S. Yamazaki et al.
0367 M.O. Moss, A.B. Wood and F.V. Robinson
0373 M. Elliott et al.
0375 M.A. Hems
0379 H. Hart and R.K. Murray, Jr.
0409 H. Bauer
0431 R. Hansel and D. Ohlendorf
0435 H. Fischer and H. Fischer
0443 S. Ito et al.
0457 W.N. Sullivan et al.
0471 K.S. Brown, Jr., et al.
0483 A.C. Campbell et al.
0515 T.G. Waddell and T.A. Geissman
0543 Y. Makisumi and T. Sasatani
0557 W.F. Parham et al.
0575 F.D. Greene and S.S. Hecht
0579 P.M. Weintraub and R.E. Banbury
0583 S.V. Kessar et al.
0599 J.L. Breton et al.
0621 T.W. Mattingly, Jr., and A. Zweig
0625 W.A. Henderson, Jr., and A. Zweig
0647 M.H. Rosen
0651 W.E. Barnett and J. McCormack
0663 R.S. Givens
0691 S. Ito, Y. Fujise and M. Sato
0695 M. Myokei et al.
0737 F.A.L. Anet and D.P. Mullis
0767 S. Seo et al.
0775 S. Ito, K. Sakan and Y. Fujise
0785 U. Wocke et al.
0823 F. Texier and R. Carrie
0857 B.G. McFarland et al.
0869 K. Morikawa and Y. Hirose
0887 A. Marsili and V. Scartoni
0901 K. Mashimo and Y. Sato
0909 D.C. Kleinfelter et al.
0927 T. Okada et al.
0961 S. Sarel et al.
0973 T.A. Dullforce et al.
0979 J.L. Ripoll and J.M. Conia
0995 L.A. Paquette et al.
0999 L.A. Paquette et al.
1063 M. Franck-Neumann and G. Leclerc
1071 K. Birnbaum et al.
1083 T. Matsuura et al.
1089 P.G. Gassman and H.P. Benecke
1125 J. Altman et al.
1141 H.H. Inhoffen et al.
1145 H.H. Inhoffen et al.
1169 H.J. Piek
1173 J.L. Pousset and J. Poisson
1185 Y. Sasada et al.
1203 N. Finch and C.W. Gemenden
1207 U.K. Pandit et al.
1231 G. Markl and A. Merz
1235 H. Monti and M. Bertrand
1243 H.W. Moore et al.
1251 E. Piers, R.W. Britton and W. de Waal
1299 T.S. Cantrell and B.L. Harrison
1303 G. Irick, Jr., and J.G. Pacifici
1343 T. Irie et al.
1443 W. Ried and A.H. Schmidt
1451 G.G. DeAngelis and H.J. Hess
1471 H. Wagner et al.
1485 S. Sakai, A. Kubo and J. Haginiwa
1529 A. deGroot et al.
1549 L.S. Davies and G. Jones

1557 H.J. Petersen
1581 R.C. Anderson and R.H. Fleming
1599 S. Hayashi et al.
1601 Y. Ohta and Y. Hirose
1615 M. Miyano and C.R. Dorn
1619 J.R. Wiseman and B.P. Chong
1639 J.F. Bagli and T. Bogri
1661 N. Finch and H.W. Gschwend
1673 T. Matsuura et al.
1683 S. Mihashi et al.
1697 K. Hartke and W. Uhde
1709 S.M. Kupchan and I. Ognyanov
1717 G. Maury and N.H. Cromwell
1761 T. Miwa, M. Kato and T. Tamano
1765 T. Matsuura and M. Kawai
1771 B. Gregson-Allcot and J.M. Osbond
1799 K. Morikawa and Y. Hirose
1803 A. Sato and H. Mishima
1825 T. Nishina et al.
1837 E.J. Corey and K. Achiwa
1841 E.J. Grubbs and J.A. Villarreal
1857 J.E. Robbers and H.G. Floss
1883 J. Ciabattoni et al.
1909 L.N. Yakhontov et al.
1921 I. Murata et al.
1925 K. Kojima et al.
1929 H. Kakisawa and Y. Inouye
1933 Y. Okamoto et al.
1951 N.M. Mollov and H.B. Dutschewska
1961 T. Kishikawa et al.
1969 H. Ogawa et al.
1995 Y. Nakatani and T. Yamanishi
2049 T. Nishiwaki
2053 M. Takaku et al.
2075 M. Yanagita et al.
2077 N. Takahashi et al.
2081 T. Yokota et al.
2093 H. Prinzbach et al.
2113 S.V. Sunthankar and D.V. Telang
2137 A. Roedig et al.
2157 J. Dehler and K. Fritz
2195 J.-P. Griot et al.
2215 J.M. Patterson and L.T. Burks
2231 Y.F. Shealy and C.A. O'Dell
2239 S. Harada et al.
2249 M. Uramoto et al.
2255 Z. Yoshida et al.
2279 E. Klein and V. Rojahn
2281 M. Fischer
2285 M.T. Doel, A.S. Jones and N. Taylor
2323 S.C. Bisarya, U.R. Nayak and S. Dev
2331 S.K. Dasgupta et al.
2339 R. Mechoulam et al.
2359 H. Igeta et al.
2383 K. Goshima and K. Tokuyama
2399 K.L. Stuart and M. Barrett
2417 F. Bohlmann et al.
2457 S. Nozoe and K.T. Suzuki
2523 Y. Ban and I. Iijima
2531 F. Toda, H. Ishihara and K. Akagi
2535 T. Terashima et al.
2539 B.W. Bycroft et al.
2545 M.W. Miller
2651 E.E.J. Dekker et al.
2659 M. Franck-Neumann and C. Buchecker
2697 R.B. Woodward and C. Wintner
2709 J.R. Owen
2717 I. Moretti and G. Torre
2725 I. Souzu and H. Mitsuhashi
2733 S. Hashimoto et al.
2737 S. Takeuchi et al.
2741 H. Hikino et al.
2767 R. Srinavasan and H. Hiraoka
2771 M. Miyano
2775 B. Fraser-Reid et al.
2807 A. Stoessl et al.
2823 T. Tatsuno et al.
2863 L. Jurd
2875 A. Takamizawa and S. Matsumoto
2901 D. Davis and J.A. Elix
2917 T.C. Jain and J.E. McCloskey
2929 T. Rios and F. Gomez G.
2947 H. Izawa et al.
3003 E. Müller et al.
3009 B. Föhlish
3013 T. Murae et al.
3017 S. Nozoe and K. Hirai
3029 K. Rodscheit et al.
3033 U. Stache et al.
3077 D.G. Whitten and J.C.N. Yau
3091 S. Ito et al.
3101 S. Yamada et al.
3117 L.R. Gizzi and M.M. Joullie
3147 H. Neunhoeffer et al.
3169 F. Kido et al.
3179 Y. Hayasi et al.
3185 S. Ito et al.
3201 H.J. Roth and M. Adomeit
3273 I.F. Eckhard et al.
3279 K. Ikawa et al.
3287 J. Kunimoto et al.
3295 M. Oda and Y. Kitahara
3329 S. Morrocchi et al.
3347 V.R. Mironov et al.
3353 H. Suginome et al.
3357 K. Wada et al.
3381 S. Kukolja et al.
3385 S. Wolfe, C. Ferrari and W.S. Lee
3403 N. Obata and T. Takizawa
3407 T. Takizawa et al.
3481 J. Nagyvary and C.M. Tapiero
3509 M. Gschwendt and E. Hecker
3521 J.M. Coxon et al.
3537 R. Lapouyade et al.
3545 J.M. Conia and J.M. Denis
3547 E. Sato and Y. Kanaoka
3549 S. Tomoda et al.
3553 T. Nishino et al.
3575 I.G. Guest and B.A. Marples
3579 F.M. Dean and L.E. Houghton
3589 J.R. Merchant and D.V. Rege
3597 R.H. Martin and M. Deblecker
3603 C.C.J. Culvenor and L.W. Smith
3613 G.M. Strunz et al.
3639 T. Sasaki and A. Kojima
3653 T. Mori et al.
3683 R.H. Martin et al.
3689 R.L.N. Harris
3693 U.K. Pandit et al.
3729 M. Iguchi et al.

3751 F.J. Muhtadi and M.J.R. Moss
3753 J.P. Pete and M.L. Villaume
3761 L. Bang and G. Ourisson
3803 J. Reisch et al.
3829 C. Pac and H. Sakurai
3843 G. Sugowdz et al.
3857 S.P. Kureel et al.
3867 F. Nerdel et al.
3871 G.W. Perold and G. Ourisson
3879 J.H. Ackerman et al.
3897 R.A. Moss and M.J. Landon
3917 N. Obata et al.
3943 T.H. Kinstle and P.D. Carpenter
3957 C.J. Rostek and W.M. Jones
3977 S. Iwasaki et al.
3993 M.L. Sassiver and R.G. Shepherd
4031 R.D.H. Murray and M.M. Ballantyne
4049 G. Pattenden
4065 K. Sasaki and Y. Hirata
4077 T. Okuno and T. Matsumoto
4117 G. Kresze and H. Grill
4139 J.D. Brewer and J.A. Elix
4205 R.L. Letsinger and R.R. Hautala
4255 R.L. Hamill et al.
4265 G.-A. Hoyer
4273 K. Fickentscher
4279 J. Sauer and H. Krapf
4295 M. Iguchi et al.
4313 G.P. Nilles and R.D. Schuetz
4325 U. Weiss and J.M. Edwards
4335 J.A. Miller
4361 R.H. Schlessinger and G.S. Ponticello
4369 M. Gotz and K. Wiesner
4401 T.Toda, M. Nitta and T. Mukai
4423 W. Dopke and R. Hartel
4457 M. Barbetta et al.
4461 B. Zwanenburg et al.
4475 P.H. Nelson and K.G. Untch
4483 F.A. Davis and R.B. Wetzel
4491 A.G. Anastassiou et al.
4495 R.R. Crenshaw and G.M. Luke
4497 A.B. Evnin et al.
4509 L.A. Paquette and J.F. Kelly
4513 R.H. Schlessinger and A.G. Schultz
4543 K. Oka, Y. Ike and S. Hara
4547 K. Oka, Y. Ike and S. Hara
4551 K. Yamamoto et al.
4559 A. Chatterjee and D. Banerjee
4567 R.W. Griffin, Jr., et al.
4575 G. Schroder et al.
4589 J. Hayashi et al.
4639 N. Finch and J.J. Fitt
4669 R.D. Allan et al.
4673 R.D. Allan et al.
4675 S. Naito and Y. Kaneko
4683 N. Suzuki et al.
4689 M. Ando et al.
4703 F. Bohlmann et al.
4729 T. Kamiya et al.
4753 D.T. Carty
4781 R.K. Murray, Jr., and H. Hart
4785 H. Hart et al.
4789 S.K. Talapatra et al.
4791 W.H.F. Sasse et al.
4829 P.M. Atlani and J.F. Biellmann
4841 R.J. Bastiani et al.
4871 A.K. Dhar and S.K. Bose
4875 I. Adachi et al.
4893 P. Arends
4899 J. Streith and P. Martz
4929 T. Tabata and H. Hart
4933 H. Hart et al.
4951 M. Shamma et al.
4957 M. Sprecher and D. Levy
4983 P. Margarethe and O.E. Polansky
4995 R. Schmiechen
5001 J. Hofman et al.
5049 T.L. Burkoth
5055 E. Gellert and R.E. Summone
5093 J.A. Green and L.A. Singer
5105 I. Lalezari et al.
5109 F. Bohlmann and C. Zdero
5121 H. Rönsch
5135 G. Snatzke and H. Seidler
5159 G.C. Brophy et al.
5167 E. Müller et al.
5205 E.S. Rothman et al.
5223 D.K. Chatterjee and K. Sen
5239 A.G. Anastassiou and J.H. Gebrian
5271 C. Huynh and S. Julia
5307 I. Pomerantz et al.
5325 E. Moriconi et al.
5335 M. Götz and K. Wiesner
5337 Y. Kobayashi et al.
5349 R. Mechoulam and B. Yagen

89- -69, <u>Angew. Chem.</u>, <u>8</u> (1969)(International Edition)

0069 G. Schröder et al.
0132 A. Albert
0135 K. Bauer et al.
0209 H. Prinzbach and W. Auge
0210 M. Franck-Neumann
0271 H. Schumann et al.
0273 V. Jäger and H.G. Viehe
0276 H. Prinzbach and M. Klaus
0343 Y. Yamada et al.
0447 G. Seitz et al.
0448 H. Quast and E. Schmitt
0451 S. Hünig and G. Büttner
0456 H. Hofmann and G. Salbeck
0456B H. Schildknecht and G. Hatzmann
0459 E. Grundemann
0593 K. Hafner and H. Tappe
0598 M.J. Janssen and J. Bos
0673 M. Oda and Y. Kitahara
0675 R. Fuks
0753 W. Brenner et al.
0758 G. Ege and W. Planer
0759 R. Criegee and R. Huber
0770 K. Dimroth et al.
0772 W. Bauer and K. Hafner
0880 M. Klaus et al.
0881 H. Prinzbach and H. Knofel
0882 H. Prinzbach and E. Woischnik
0883 H. Prinzbach et al.
0979 H.G. Viehe et al.
0987 D. Hellwinkel and W. Schenk
0989 W. Walter and H. Weiss

94- -69, Chem. Pharm. Bulletin (Japan), 17 (1969)
0011 H. Kaneko et al.
0089 S. Kobayashi et al.
0128 A. Takamizawa et al.
0140 T. Okamoto et al.
0155 M. Komatsu et al.
0163 S. Hayashi et al.
0181 M. Saneyoshi and F. Sawade
0279 I. Yosioka et al.
0306 H. Saikachi and H. Ogawa
0324 H. Mitsuhashi and M. Fukuoka
0343 A. Takamizawa et al.
0355 A. Ishida et al.
0381 H. Nomura et al.
0387 H. Nomura et al.
0408 Y. Yura and J. Ide
0416 K.A. Watanabe et al.
0454 S. Shibata et al.
0458 M. Kaneda et al.
0467 T. Shioiri et al.
0474 H. Itokawa et al.
0510 Y. Kobayashi and I. Kumadaki
0515 M. Okada and Y. Saito
0550 T. Hino et al.
0582 S. Umio et al.
0629 E. Yoshii and S. Kimoto
0639 T. Nishimura et al.
0682 D. Satoh et al.
0729 T. Saitoh and S. Shibata
0763 H. Igeta et al.
0770 A. Nakamura et al.
0775 K. Kikugawa and T. Ukita
0785 K. Kikugawa et al.
0798 K. Kikugawa et al.
0814 T. Kametani and F. Satoh
0844 H. Ogura et al.
0851 A. Ohta and S. Fujii
0910 A. Takamizawa et al.
0966 A. Terada and Y. Kishida
0980 S. Oida and E. Ohki
1128 A. Yamazaki et al.
1158 K. Kawasaki and H. Matsumura
1188 M. Hirata et al.
1206 Y. Morisawa and K. Tanabe
1212 Y. Morisawa and K. Tanabe
1225 S. Takahashi et al.
1255 M. Kobayashi et al.
1263 S. Inaba et al.
1268 A. Yamazaki
1290 C. Kaneko et al.
1294 C. Kaneko et al.
1299 M. Komatsu et al.
1302 M. Komatsu et al.
1309 I. Ito and T. Ueda
1324 K. Endo and P. de Mayo
1401 D. Satoh and S. Nishii
1467 T. Naito et al.
1485 M. Sano et al.
1564 H. Shirai et al.
1572 K. Mitsuhashi et al.
1578 K. Mitsuhashi et al.
1591 Y. Kishida et al.
1598 S. Takahashi and H. Kano
1698 Y. Kamano and M. Komatsu
1706 Y. Kamano et al.
1711 Y. Kamano
1725 T. Nambara et al.
1749 Y. Nagai
1757 H. Saikachi et al.
1782 T. Nambara et al.
1924 A. Takamizawa and K. Hirai
1931 A. Takamizawa and K. Hirai
1942 H. Inouye et al.
1949 H. Inouye et al.
1977 T. Kametani and I. Noguchi
2031 K. Kaneko et al.
2036 M. Yanai et al.
2054 H. Taguchi et al.
2083 T.O. Kamoto et al.
2105 M. Itoh and A. Sugihara
2126 Y. Kishida et al.
2176 T. Matsuura et al.
2181 T. Kosuge et al.
2201 I. Adachi and H. Kano
2209 I. Adachi
2299 A. Takamizawa et al.
2306 T. Oishi et al.
2314 T. Oishi et al.
2319 S. Hayashi and T. Komeno
2353 T. Kametani et al.
2370 M. Mitsuhashi et al.
2373 M. Saneyoshi and K. Terashima
2381 Y. Kanaoka et al.
2389 T. Kosuge et al.
2391 Y. Shimizu et al.
2411 T. Kato et al.
2417 Y. Kishida and A. Terada
2424 Y. Kishida and N. Nakamura
2436 H. Shirai et al.
2442 K. Tomita and M. Nagano
2448 M. Fukuoka and H. Mitsuhashi
2455 S. Ishiwata and Y. Shiokawa
2461 S. Oida and E. Ohki
2548 H. Toda et al.
2571 S. Ishiguro et al.
2581 A. Yamazaki et al.
2586 T. Komeno et al.
2604 T. Komeno et al.

95- -69, J. Pharm. Soc. Japan, 89 (1969)
0007 M. Akatsuka
0058 G. Kobayashi et al.
0074 E. Hayashi and T. Saito
0129 M. Shimizu
0132 T. Itai et al.
0203 G. Kobayashi et al.
0266 S. Senda and H. Izumi
0372 H. Hikino et al.
0375 S. Akaboshi and T. Kutsuma
0410 T. Tomimori and M. Komatsu
0418 T. Kametani et al.
0460 T. Kato et al.
0469 H. Ogura et al.
0506 K. Matoba et al.
0510 T. Okano and H. Matsumoto
0549 K. Hata and K. Sano
0591 T. Takahashi
0689 Y. Usui
0702 M. Shimizu and N. Morita

0737 M. Tomita et al.
0750 K. Matoba et al.
0759 T. Yoshikawa et al.
0767 T. Yoshikawa et al.
0783 O. Tsuge et al.
0789 O. Tsuge et al.
0851 T. Noro et al.
0869 H. Yamaguchi et al.
0872 M. Yasue et al.
0919 T.Takahashi and R. Yamamoto
0959 E. Oishi
1029 S. Akaboshi and T. Kutsuma
1035 S. Akaboshi and T. Kutsuma
1048 T. Kametani et al.
1061 S. Sakai et al.
1149 H. Hikino et al.
1163 K. Ito et al.
1167 T. Watanabe
1260 T. Takahashi et al.
1276 T. Tomimori and M. Komatsu
1305 T. Okano and T. Takahashi
1313 S.T. Lu, S.-J. Wang and F.-S. Lin
1358 T. Kikuchi et al.
1434 H. Saikachi et al.
1446 H. Saikachi and S. Nakomura
1492 T. Koyama et al.
1566 T. Ozawa et al.
1646 S. Sugiura et al.
1652 S. Sugiura et al.
1691 J. Kunimoto et al.

97- -69, Z. Chemie, 9 (1969)
0022 S. Hauptmann et al.
0025 S. Hoffmann
0059 S. Scheithauer and R. Mayer
0063 D. Merkel
0152 P. Nuhn et al.
0184 W. Schroth et al.
0186 G. Kempter et al.
0269 N. Erdmann
0303 M. Muhlstadt et al.
0342 H. Dehne and A. Zschunke
0385 E. Uhlemann and V. Pohl
0421 H.J. Siemann and D. Onken

100- -69, Lloydia, 32 (1969)
0001 J.C. King et al.
0029 R.W. Doskotch et al.
0108 B.M. Sayagurer et al.
0115 R.W. Doskotch et al.
0503 S.A. Gharbo and A.A.M. Habib

101- -69A, J. Organometallic Chem., 16 (1969)
00P5 O.W. Steward and J.E. Dziedzic
0283 R.E. Hester and M. Cais
0289 J.P. Candlin and A.C. Shortland
0309 J.M. Kliegman and A.C. Cope
0335 J. Otera and R. Okawara
0342 H. Alper and J.T. Edward
0491 B.L. Booth et al.

101- -69B, J. Organometallic Chem., 17 (1969)
0017 J. Nagy et al.
0071 H.A. Meinema, E. Rivarola and J.G. Noltes
0361 R.A. Finnegan and W.H. Mueller
0457 I. Collamati and A. Furlani

101- -69C, J. Organometallic Chem., 18 (1969)
0291 J. Klein and S. Brenner
0307 R.M.G. Roberts
0361 M.J.A. Habib and W.E. Watts

101- -69E, J. Organometallic Chem., 20 (1969)
0099 G. Faraglia
0117 R.A. Earley and M.J. Gallagher

102- -69, Phytochemistry, 8 (1969)
0203 J. Corse and D.C. Patterson
0299 S.P. Bhutani et al.
0305 M.A. Irwin and T.A. Geissman
0315 S. Imre
0501 H.H. Lee
0511 A. Banerji et al.
0645 H. Bohrmann et al.
0661 W. Herz and S.K. Roy
0773 D.P. Chakraborty et al.
0781 D.F. Theumann and J. Comin
0789 R.K. Gupta and M.M. Dhar
1297 T.H. Geissman et al.
1499 D.L. Dreyer and A. Lee
1515 K.H. Lee et al.
1559 T.K. Palit and M.P. Khare
1753 A. Yoshitake and T.A. Geissman
1797 L. Fonzes et al.
1817 L.B. deSilva et al.
2083 G.G. Allan
2371 T.G. Waddell and T.A. Geissman
2381 S.J. Torrance et al.

103- -69, Khim. Geterosikl. Soedin., 5 (1969) (English translation edition)
0012 Z.N. Nazarova et al.
0019 S.M. Glozman et al.
0049 N.M. Turkevich and B.S. Zimenskovskii
0055 S.V. Tsukerman et al.
0062 B.A. Arbusov et al.
0090 D.G. Pereyaslova et al.
0093 N.O. Saldabol and I.B. Mazheika
0130 D.Y. Sniker et al.
0172 V.G. Kharchenko et al.
0180 V.G. Pesin
0190 N.N. Suvorov et al.
0196 A.P. Terent'ev et al.
0199 A.P. Terent'ev et al.
0223 A.F. Bekhli and N.P. Kozyreva
0226 G.A. Klimov et al.
0246 F.T. Pozharskii et al.
0251 E.Y. Ozala et al.
0254 E.Y. Ozala et al.
0278 Z.I. Zelikman and V.G. Kul'nevich
0283 S. Hillers et al.
0316 V.A. Zagorevskii et al.
0321 E.K. Orlova et al.
0372 A.Y. Perkone et al.
0383 S.V. Tsukerman et al.
0410 L.N. Yakhontov et al.
0414 N.O. Saldabol and V.V. Krylova
0420 A.K. Sheinkman et al.

0424 V.I. Shvedov et al.
0427 N.O. Saldabol
0434 Z.N. Nazarova and V.S. Pustovarov
0527 V.M. Zolin and R.B. Zhurin
0529 S.N. Kolodyazhnaya and A.M. Simonov
0568 V.I. Shvedov et al.
0574 T.O. Petrova et al.
0602 N.N. Bulatova and N.N. Suvorov
0623 R.B. Zhurin et al.
0638 A.A. Druzhinina and P.M. Kochergin
0683 A.F. Pozharskii et al.
0707 V.P. Maksimets et al.
0762 G.Y. Dubur and Y.R. Uldrikis
0825 V.F. Sedova and V.P. Mamaev
0827 V.M. Vvedenskii
0830 V.M. Vvedenskii et al.

104- -69, Zhur. Organ. Khim., 5 (1969) (English translation edition)
0082 I.A. D'yakonov et al.
0111 N.I. Ganuschchak et al.
0151 B.K. Strelets and L.S. Etros
0158 K.V. Levshina et al.
0225 Y.I. Mushkin
0257 B.V. Ioffe and L.M. Gershtein
0459 E.S. Levchenko and Z.I. Shokol
0468 V.K. Daukshas et al.
0511 M.I. Shevchuk et al.
0543 T.G. Melent'eva et al.
0547 A.V. El'tsov et al.
0559 A.A. Ginesina et al.
0678 N.I. Ganushchak et al.
0730 M.N. Preobrazhenskaya et al.
0736 A.N. Kost et al.
0742 M.V. Gorelik and M.I. Evstratova
0759 O.A. Shavrygina et al.
0763 A.M. Simonov and Y.P. Andriechikov
0826 I.A. Favorskaya and M.M. Plekhotkina
0903 L.Y. Malkes and N.P. Kovalenko
0947 A.V. El'tsov et al.
0998 T.A. Favorskaya et al.
1110 I.V. Samartseva et al.
1115 A.V. El'tsov and A.A. Ginesina
1165 T.A. Favorskaya et al.
1193 K.M. Ermolaev and V.I. Maimind
1196 V.S. Velezheva et al.
1231 I.I. Reznikova
1249 B.A. Porai-Koshits and I.Y. Kvitko
1274 V.M. Ostrovskaya et al.
1298 G.S. Grinenko et al.
1460 I.Y. Kvitko and T.M. Galkina
1566 V.S. Velezheva et al.
1581 G.A. Tolstikov et al.
1609 Y.A. Gurvich and Y.B. Zimin
1628 Y.E. Gerasimenko and I.N. Shevchuk
1634 Y. Kvitko and B.A. Porai-Koshits
1642 M.V. Gorelik and V.V. Pachkova
1660 V.G. Kharchenko et al.
1696 S.M. Makiu and S.D. Yablonovskaya
1713 A.I. Gurevich et al.
1722 B.G. Kovalev et al.
1739 L.A. Ignatova et al.
1790 F.I. Luknitskii
1797 F.I. Luknitskii et al.
1803 M.I. Budagyants et al.
1903 E.Z. Katsnel'son and C.S. Frunkovskii
1931 K. Ok Kim et al.
1945 R.G. Karpenko et al.
1961 L.I. Bagal et al.
1973 S.A. Soldatova et al.
1993 A.L. Voitsekhovskaya et al.
2004 A.V. El'tsov et al.
2014 A.V. El'tsov et al.
2046 L.I. Bagal et al.
2090 A.K. Aren et al.
2154 N.S. Dokunikhin et al.
2164 N.P. Shusherina et al.
2179 A.V. El'tsov et al.

105- -69, Khim. Prirodn. Soedin., 5 (1969) (English translation edition)
0001 A.Z. Abyshev
0005 T.N. Orgiyan and D.P. Popa
0007 N.K. Kashtanova et al.
0020 I.O. Kurakina et al.
0024 S.F. Matkhalikova et al.
0026 T.U. Rakhmatullaev et al.
0054 R.N. Nuriddinov et al.
0056 L.S. Smirnova et al.
0058 V.M. Malikov and S.Y. Yunusov
0089 Y.A. Berlin et al.
0110 V.V. Zatula, I.D. Kovalev and D.G. Kolesnikov
0115 S.I. Skandarov and S.Y. Yunusov
0128 L.M. Kogan et al.
0156 C.A. Salei et al.
0158 M.N. Mukhametzhanov et al.
0163 I.F. Makarevich
0164 I.F. Makarevich and D.G. Kolesnikov
0167 S.G. Kislicheako et al.
0189 N.E. Ermator et al.
0191 N.P. Kir'yalova and V.Y. Bagirov
0210 M.A. Chirkova and V.A. Pentegova
0257 Y.A. Berlin et al.
0280 D.A. Rakhimov, V.M. Malikov and S.Y. Yunusov
0281 D.M. Tsakadze et al.
0299 A.I. Sokolova et al.
0314 S.V. Serkerov
0321 S.G. Kislicheuko et al.
0380 Z. Faizutdinova et al.
0383 A. Abdusamatov et al.
0386 D.A. Rakhimov et al.
0397 S.A. Popravko, A.I. Gurevich and M.S. Kolosov
0411 S.V. Serkerov
0440 D.A. Rakhimov et al.
0478 Y.A. Berlin et al.
0483 Y.A. Berlin et al.
0494 G.P. Bakhaeva et al.
0506 A.A. Savina, G.K. Nikonov and M.E. Perel'son
0530 S.F. Matkhalikova, V.M. Malikov and S.Y. Yunosov

106- -69, Die Pharmazie, 24 (1969)
0035 G. Wagner and F. Süss
0100 G. Wagner and P. Richter

0196 J. Vahldieck and G. Buchmann
0308 P. Pflegel and G. Wagner
0384 P. Pflegel and G. Wagner
0635 K. Delenk-Heydenreich and S. Pfeifer

108- -69, Israel J. Chem., 7 (1969)
0057 F. Bergmann and D. Diller
0099 H. Wieler-Feilchenfeld et al.
0165 E.D. Bergmann and A.Y. Meyer
0435 J. Altman et al.
0479 D.S. Magrill et al.
0555 A. Agranat et al.

111- -69, Chim. ther., 4 (1969)
0228 K. Thiele et al.
0239 J. Schmitt et al.
0298 C. Breysse et al.

112- -69, Spectroscopy Letters, 2 (1969)
0301 J.M. White
0369 R.H. Pottier et al.

114- 69A, Acta Chim. Acad. Sci. Hung., 59 (1969)
0109 A. Muller et al.
0397 I. Alkonyi and A. Sandor

114- -69B, Acta Chim. Acad. Sci. Hung., 60 (1969)
0151 K. Kormendy et al.
0309 B. Lakatos et al.

114- -69C, Acta Chim. Acad. Sci. Hung., 61 (1969)
0093 G. Hornyak et al.
0181 G. Hornyak et al.

115- -69, J. Chem. United Arab Republic, 12 (1969)
0001 A.E. Sammour and M. Elkasaby
0017 A.E. Sammour and M. Elkasaby
0057 M.H. Nosseir and N.N. Messiha

117- -69, Org. Preps. and Procedures, 1 (1969)
0021 J. Strating and E. Molenaar
0043 J. Szmuszkovicz
0047 F.A. Varron and E.I. Becker
0105 J. Szmuszkovicz
0171 A. Brossi and S. Teitel
0187 S.S. Dua, A.E. Jukes and H. Gilman
0209 T.R. Pampalone
0255 M. Carmack et al.
0271 V. Bocchi and G.P. Gardini
0287 T. Doornbos and J. Strating

118- -69, Synthesis, 1 (1969)
0170 E.B. Pedersen and S.O. Lawesson

119- -69, S. African Chem. Inst. J., 22 (1969)
0037 J.H. Snyders and M.J. Pieterse
0186 A.J. Van Wyk and P.R. Enslin
0191 L.A.P. Anderson and J.M. Koekemoer
0020S P.S. Steyn
0035S P.J. Aucamp and C.W. Holzapfel
0107S M.F. Dutton and J.G. Heathcote
0119S L.A.P. Anderson and J.M. Koekemoer

120- -69, Pakistan J. Sci. Ind. Research, 12 (1969-70)
0012 M. Arshad A. Beg and Y.Z. Abbasi
0159 G.A. Miana et al.
0195 M. Arshad A. Beg and Sammiuzzaman
0309 G.A. Miana et al.

121- -69, J. Macromol. Sci., Pt. A, 3 (1969)
0803 G.B. Butler and B. Iachia
1067 J.B. Hodgkin and J. Heller

122- -69, Reakt. Sposobnost' Org. Soedin., 6 (1969)(English translation edition)
0100 I.V. Tselinskii et al.
0138 V.V. Sinev and E.P. Shepel
0161 G.I. Kolesetskaya et al.
0413 V.K. Krylov et al.
0431 L.E. Kholodov et al.
0455 B.A. Ivin et al.
0490 E.R. Soonike and U.Z. Haldna

123- 69A, Moscow Univ. Chem. Bull., 24 (1969)(English translation edition)
0027 A.I. Busev et al.
0041 A.N. Kost et al.

123- 69C, Moscow Univ. Chem. Bull., 24 (1969)(third issue)
0041 N.I. Krikova et al.
0074 L.S. Voronets et al.

124- -69, Ukrain. Khim. Zhur., 35 (1969)
0179 A.I. Kiprianov and V.Y. Buryak
0370 G.T. Pilyugin et al.
0509 I.V. Smolanka and N.P. Mano
0512 A.I. Tolmachev and E.F. Karaban
0740 L.K. Mushkalo and V.A. Chuiguk
0824 M.Y. Kornilov and E.M. Ruban
1178 V.A. Chuiguk and L.S. Borodulya
1189 V.M. Nikitchenko et al.
1278 L.G. Kovalenko et al.